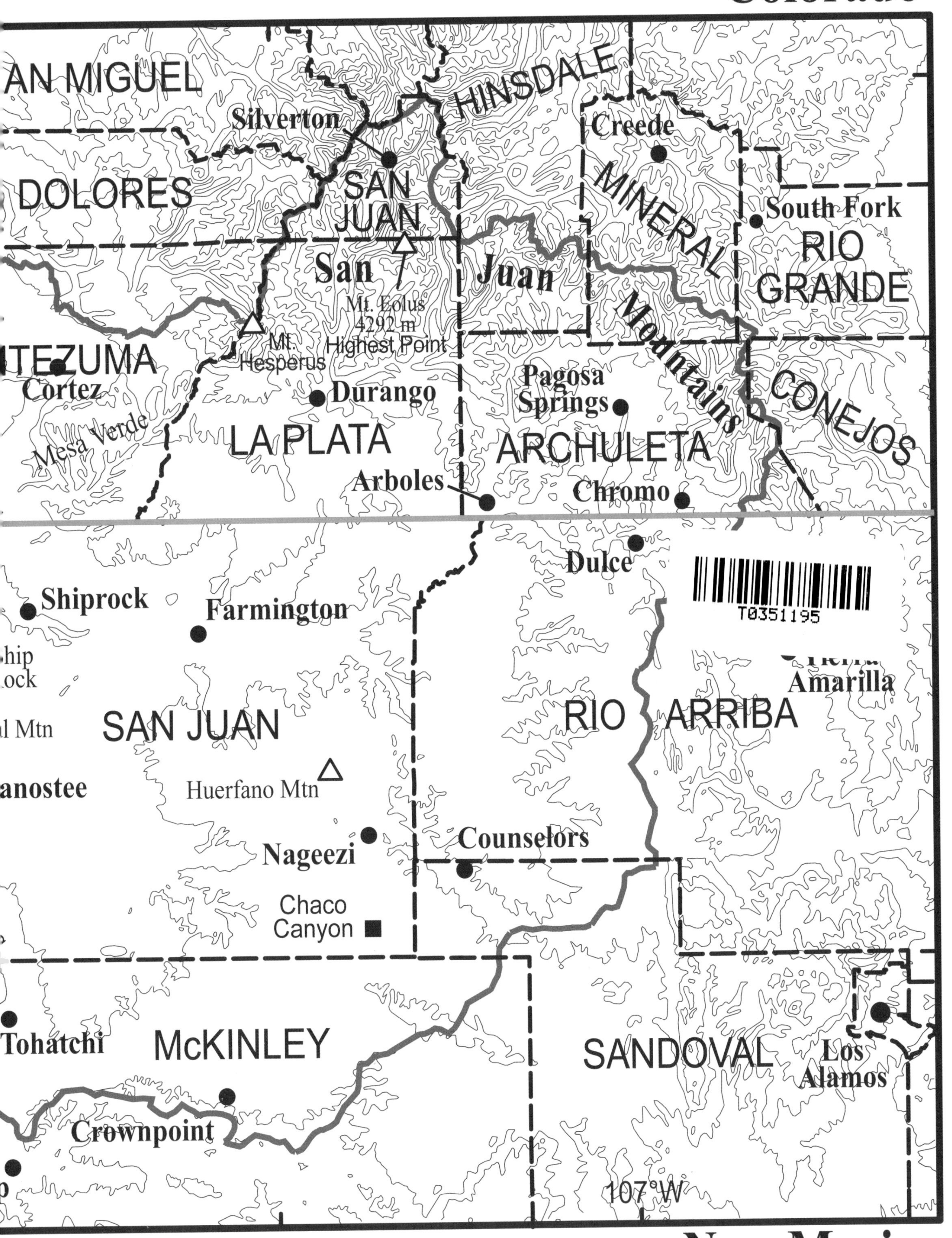
Colorado
AN MIGUEL
HINSDALE
Silverton
Creede
DOLORES
SAN JUAN
MINERAL
South Fork
RIO GRANDE
San
Juan
Mountains
Mt. Eolus
4292 m
Highest Point
Mt. Hesperus
TEZUMA
Cortez
Durango
Pagosa Springs
CONEJOS
Mesa Verde
LA PLATA
ARCHULETA
Arboles
Chromo
Dulce
Shiprock
Farmington
Tierra Amarilla
SAN JUAN
RIO ARRIBA
anostee
Huerfano Mtn
Counselors
Nageezi
Chaco Canyon
Tohatchi
McKINLEY
SANDOVAL
Los Alamos
Crownpoint
107°W
108°W
New Mexico

Flora of the Four Corners Region

Vascular Plants of the San Juan River Drainage

Arizona, Colorado, New Mexico, and Utah

Flora of the Four Corners Region

Vascular Plants of the San Juan River Drainage

Arizona, Colorado, New Mexico, and Utah

Abronia bolackii

Bolack Sand Verbena

Kenneth D. Heil

Steve L. O'Kane, Jr.

Linda Mary Reeves

Arnold Clifford

ISBN 978-1-930723-84-9
Library of Congress Control Number 2012949654
Monographs in Systematic Botany from the Missouri Botanical Garden, Volume 124
ISSN 0161-1542

Scientific Editor and Head: Victoria C. Hollowell
Associate Editor: Allison M. Brock
Associate Editor: Tammy M. Charron
Associate Editor: Laura L. Slown
Press Coordinator: Cirri R. Moran
Proofreader and Production Coordinator: Laurel Anderton

4344 Shaw Blvd.
St. Louis, Missouri 63110, U.S.A.
www.mbgpress.org

Reprinted 2024

Photo credits—All photography by Steve L. O'Kane, Jr. except: *Aliciella subnuda* by Nature's Images; *Astragalus preussii* var. *latus* by Kenneth D. Heil; *Echinocereus engelmannii* var. *variegatus* by Kenneth D. Heil; *Echinocereus triglochidiatus* var. *triglochidiatus* by Nature's Images; *Echinocereus viridiflorus* var. *viridiflorus* by Kenneth D. Heil; *Escobaria missouriensis* by Kenneth D. Heil; *Lewisia nevadensis* by Nature's Images; *Mimulus eastwoodiae* by Kenneth D. Heil; *Mirabilis multiflora* by Nature's Images; *Opuntia phaeacantha* var. *phaeacantha* by Kenneth D. Heil; *Opuntia polyacantha* var. *hystericina* by Kenneth D. Heil; *Pediocactus simpsonii* by Kenneth D. Heil; *Sclerocactus parviflorus* subsp. *parviflorus* by Kenneth D. Heil; *Sophora stenophylla* by Nature's Images; and *Trollius albiflorus* by Nature's Images. Color botanical illustrations by Carolyn Crawford, and color landscape illustrations by Glenn Vandré.

Dedicated in memory of Matthew Bradley Heil

a free spirit with a "good eye"

Asclepias tuberosa L.

Primula parryi A. Gray

FOREWORD

One of my great satisfactions during nearly 60 years of collecting, studying, and enjoying the plants of the American West has been to observe the proliferation of detailed accounts of the plants of particular regions. These are necessary to build up a proper understanding of the complex patterns of variation that characterize this enormously diverse region. In contrast, the realization that plant species throughout the world are seriously endangered by a combination of habitat destruction, alien invasive species, climate change, and selective gathering was not even imagined in the 1940s as a matter worthy of serious attention, whereas now we take these threats for granted and are dealing with them in the best ways that we can devise. Overall, books of this kind build our familiarity with and thus our appreciation of the plants of a given area and thus allow us to understand both the plants and the places they occur more completely.

Certainly, readers of Tony Hillerman's well-appreciated detective stories will appreciate the natural magnificence and geological diversity of the Four Corners region even if they have not had the pleasure of visiting the area personally. The botanical diversity of this region, so well described in this attractive and useful flora, is no less impressive. Clearly, the use of this book will inspire further studies and thus make possible an ever greater depth of knowledge about these plants and their position in the habitats where they occur.

It is clearly of fundamental importance to consider and treat in detail the plants of local areas in order to understand the overall regional trends and patterns. Broad treatments of wider areas are absolutely necessary to tie together and illuminate the overall patterns of variation. More local floras, however, provide the greater focus and precision of detail that only the intensive study of particular areas can provide. In addition, the San Juan River drainage includes parts of four states, and its biota, when included in different state accounts, is often treated there in different ways, or even under different names; for this reason, the local but inclusive approach taken here is all the more important. Only by bringing together information about these plants throughout the region can we understand them properly.

The careful descriptions and fine illustrative material included here will enable people with varying degrees of experience to learn more about and deal with these plants in even more depth—or simply learn to know and enjoy their beauty and interest. As a result of the fieldwork undertaken as part of this project, our knowledge of the flora has been substantially enriched, and gaps in the available knowledge have been identified that will provide years of work and enjoyment for the field-oriented botanists in the future. The pertinent literature has been thoroughly examined during the preparation of this flora, and its findings have been incorporated in the accounts presented here. A serious effort has been made to recruit the most knowledgeable authors, almost 70 of them, for the individual accounts, and also to achieve a degree of consistency across the treatments.

Among the 2355 taxa recorded here are 41 considered to be endemic, or near endemic, to the region, and of course a great deal of the significant genetic variation of all taxa that occur locally is related to the seemingly endless ecological diversity of the area. Not surprisingly, Asteraceae, with 411 taxa in the region; Poaceae, with 260; and Fabaceae, with 159, are the three largest families in the Four Corners region. Less expected, perhaps, is the remarkable diversity of Brassicaceae, with 119 taxa regionally. There are relatively few floras anywhere in the New World in which Brassicaceae are among the several largest families represented.

In summary, this is an outstanding flora that will be of great use not only in the region itself, but in the unending synthesis of the plants of North America and ultimately the world. I congratulate the authors, illustrators, and editors on a job exceedingly well done.

Peter H. Raven
Missouri Botanical Garden, St. Louis, Missouri

Silene laciniata Cavanilles

TABLE OF CONTENTS

PREFACE

As we complete the publication of this flora, botanical studies are undergoing revolutionary revisions. New and comprehensive molecular genetic information is modifying our understanding of phylogeny and classification of vascular plants to an unprecedented degree. At the same time, vast areas of the globe remain botanically unexplored, even in the classical sense. New species of plants and new plant distributions are being discovered, not just in the field, but also in the herbarium and garden. Alterations in plant distributions are quite commonplace and continue to diversify as land use, immigration, and climate change.

The *Flora of the Four Corners Region* is a systematic, multidecade effort involving over sixty contributing botanists, students, artists, and editors coordinated through the San Juan College Herbarium. The San Juan College Herbarium is the largest herbarium in the Four Corners region, with over 50,000 specimens.

We have defined the study area as the drainage basin of the San Juan River, a major tributary of the Colorado River. The study area encompasses a land area approximately the size of the state of Connecticut and is a subregion of the Colorado Plateau. It includes parts of four states in the southwestern United States (Arizona, Colorado, New Mexico, and Utah), one of the regions of North America with the highest rate of plant endemicity. It is a terrain of rugged wild lands containing nearly inaccessible areas, diverse topography, and ever-changing local climates. While rich in natural beauty and resources, the Four Corners region faces rapid development and population growth resulting in extensive habitat fragmentation and loss, erosion, and overuse. The need for a comprehensive flora of the region has never been more urgent.

Floras are available for the Four Corners states: *Arizona Flora* by Thomas H. Kearney and Robert H. Peebles, *Seed Plants of Northern Arizona* by W. B. McDougall, *Colorado Flora: Western Slope* by William A. Weber and Ronald C. Wittmann, *A Flora of New Mexico* by William C. Martin and Charles R. Hutchins, and *A Utah Flora* by Stanley L. Welsh, N. Duane Atwood, Sherel Goodrich, and Larry C. Higgins. The interested reader is encouraged to consult these floras for further study. Although significant effort went into the production of these excellent floras, in general the Four Corners region has received little botanical attention. This is due largely to both geographical and political difficulties of access to important collection sites as well as, until recently, the paucity of botanical science resources and personnel.

With the issuance of this flora, perhaps the last major botanically and endemically rich, but virtually unexplored, region of North America has been catalogued. Continued exploration and discoveries will no doubt result in many welcome additions and revisions for years to come. No flora is ever complete.

Kenneth D. Heil, Supervising Editor

Steve L. O'Kane, Jr., Co-Editor   Linda Mary Reeves, Co-Editor

Arnold Clifford, Consulting Botanist

ACKNOWLEDGMENTS

Many people, students, colleagues, co-workers, friends, and family aided in the production of this book. It would be impossible to thank everyone who played a part in its conception and execution, but to the following people and institutions, in addition to the more than 60 contributors, we are especially grateful.

To the folks and institutions who, despite impossible odds, contributed funding and encouragement: James C. Henderson, Carol Spencer, Michael Tacha, Toni Pendergrass, Glenn Davidson, Marsha Drennon, Gayle Dean, Laurie Gruel, Freida John, Barbara Bollman, John Niebling, Sher Hruska, Terry Peek, Vicky Ramakka, Larry Welsh, Frank Williams, Al Buyok, Lisa Jennings, Kay Brown, and Lisa Wilson of San Juan College; Tommy Bolack, the staff of the B-Square Ranch, and the Bolack Foundation; Kent Applegate and Steve Lynch of BHP Minerals; Annabelle Friddle, whose funds allowed us to purchase reference books; Burlington Resources; and especially Barney Wegener of the Bureau of Land Management, Farmington District, who was the first to believe enough in our project and provide actual cash; and Northern Iowa University for summer salaries.

We couldn't have completed the fieldwork without collecting permits, so we thank Daniela Roth and Jeff Cole (Navajo Nation), Merle Elote (Jicarilla Apache Tribe), Steve Whiteman and Karen Peterson (Southern Ute Tribe), Norm Henderson and John Spence (Glen Canyon NRA), Melissa Memory (Navajo NM), and Jim Ramakka, Wendy Bustard, and Paul Whitefield (Chaco NHP). Private landholders and grazing lease permittees graciously allowed us on their land to collect plants or provided much-needed information; these included Mary and Mauricio Gomez, Salso Gomez, Richard A. Gooding, Gary Kennedy, Buck Navajo, Neil Bedoni, and Calvin Hesuse.

Thanks to all those individuals and institutions who helped with plant information and collection: U.S. Forest Service (Carson and San Juan units), Frank Eldredge, Rich Fleming, Lisa and Dave Hanna, Bill Hatch, Camie Hooley, Callie Vanderbilt, Donna Hobbs, Don Hyder, Joe Mischel, Gerald Williams (interest and encouragement), Glenn Rink, Barbara Ertter, Tom Van Solen and the Van Solen family, Dave Schleser of Nature's Images, Inc., Charles King, Dick Mosley, Al Snyder, Doug Loebig, Mark Heil, David Morin, Phil and Estelle Morin, Sandy Mietty, Joan Michener, Sarah Brinton, Allison Brady, Sterling White, Dale Vandré, William Weber, and Bob Melton. Others allowed us access to herbarium collections: Daniela Roth, Les Landrum, Phil Jenkins, and Tim Hogan. We thank authors who provided assistance, identification, herbarium and/or field work above and beyond what was necessary for their taxa treatments: Ron Hartman, Guy Nesom, Kelly Allred, Stan Welsh, Duane Atwood, Dave Jamieson, Lynne Moore, J. Mark Porter, and C. Barre Hellquist. We also thank Jack and Martha Carter, who helped us to find illustrators at the very last minute.

The talented color illustrators, Carolyn Crawford, who produced the botanical watercolors, and Glenn Vandré, who illustrated the vegetation associations and life zones, provide a fine art quality to this work, which is unusual in modern scientific publications. Garrison Graphics of Farmington, New Mexico, was very helpful with scanning the line drawings.

Students and herbarium assistants tackled the herbarium drudgery of gluing, filing, data entry, and label production with good humor and a positive attitude: Starla White, Cynthia Holmes, Lonnie Herring, Rosanna Flores, Sande Burr, Brean Reed, and especially Aleisha Shevokas, Michelle Szumlinski, Gregory Penn, and Les Lunquist, collections managers for many years of organization. We thank the staff of Missouri Botanical Garden Press, especially Victoria Hollowell, Scientific Editor, who gave us unending positive encouragement and feedback. A very special thanks to Wayne Mietty, plant wrangler, who was always there, ready to go when and where we needed him in the field, and to his wife Sandy for putting up with his lengthy times in the field.

We are particularly grateful to our spouses and family members: Marilyn Heil, Arlene Prather-O'Kane, Timothy Reeves, Casey Clifford III, Casey Clifford, Lena Clifford, and especially the late Sarah Charley, maternal grandmother, herbalist, and weaver, for putting up with this project and providing good counsel so graciously and for so long. No doubt we have omitted the names of many persons we should have mentioned and we thank them as well.

CONTRIBUTORS

Robert P. Adams, Baylor University
Kelly W. Allred, New Mexico State University
Ihsan A. Al-Shebaz, Missouri Botanical Garden
John L. Anderson, Bureau of Land Management
N. D. Atwood, Brigham Young University, Stanley Welsh Herbarium
Daniel F. Austin, Arizona-Sonora Desert Museum
Tina J. Ayers, Northern Arizona University, Deaver Herbarium
Gary I. Baird, Brigham Young University, Boise, Idaho
Susan C. Barber, Oklahoma City College
Theodore M. Barkley, Botanical Research Institute of Texas
Patricia Barlow-Irick
Fred R. Barrie, Missouri Botanical Garden/Field Museum
David L. Bleakley, Bleakley Biological Services
Tommy Bolack, B-Square Ranch, Bolack Museum Foundation
Charlotte M. Christy, Augusta State University
Turner Collins, Evangel University
Carolyn Crawford, Consultant
Robert D. Dorn, Rocky Mountain Environmental Services
J. Mark Egger, WTU Herbarium, Burke Museum of Natural History & Culture
Brian A. Elliott, Rocky Mountain Herbarium
Eve Emschwiller, University of Wisconsin-Madison
Barbara Ertter, University of California, Berkeley
Donald R. Farrar, Iowa State University, Ada Hayden Herbarium
David J. Ferguson, Rio Grande Botanic Garden
Lisa Floyd-Hanna, Prescott College
Sherel Goodrich, U.S. Forest Service
J. Robert Haller, Santa Barbara Botanic Garden
Ronald L. Hartman, University of Wyoming, Rocky Mountain Herbarium
Donald L. Hazlett, University of Northern Colorado
Kenneth D. Heil, San Juan College Herbarium
C. Barre Hellquist, Massachusetts College of Liberal Arts and Science
Peter C. Hoch, Missouri Botanical Garden
Noel H. Holmgren, The New York Botanical Garden
David W. Jamieson, Fort Lewis College
William Jennings, Consultant
Sylvia Kelso, Colorado College
Walter H. Lewis, Washington University in St. Louis
R. John Little, University of California, Berkeley
William J. Litzinger, Prescott College
Timothy K. Lowrey, University of New Mexico
Lynn M. Moore, Consultant
Nancy R. Morin, The Arboretum at Flagstaff
Rodney Myatt, San Jose State University
Jane Mygatt, University of New Mexico Herbarium
Guy L. Nesom, Botanical Research Institute of Texas
Steve L. O'Kane, Jr., Grant Herbarium, University of Northern Iowa
Bruce T. Parfitt, University of Michigan, Flint
Donald J. Pinkava, Arizona State University
J. Mark Porter, Rancho Santa Ana Botanic Garden
Richard K. Rabeler, University of Michigan
Jeffrey S. Redders, U.S. Forest Service
Linda Mary Reeves, San Juan College Herbarium
Timothy Reeves, San Juan College
James L. Reveal, University of Maryland
Glenn Rink, Northern Arizona University
Roland P. Roberts, Towson University
Kenneth R. Robertson, Illinois Natural History Survey
William H. Romme, Colorado State University
Randall W. Scott, Deaver Herbarium, Northern Arizona University
Robert C. Sivinski, State of New Mexico
John R. Spence, National Park Service, Glen Canyon N.R.A.
Roy L. Taylor
Nancy J. Vivrette, Santa Barbara Botanic Garden
Warren L. Wagner, Smithsonian Institution
Gary D. Wallace, U.S. Fish and Wildlife Service
Stanley L. Welsh, Brigham Young University
Thomas L. Wendt, University of Texas, Austin
Alan T. Whittemore, Smithsonian Institution National Arboretum
Michael D. Windham, Duke University

Centaurea maculosa Lam.

About the Bolack Foundation and the B-Square Ranch

The Bolack Foundation, in cooperation with the San Juan College Foundation, supports research activities on the B-Square Ranch and in the San Juan River drainage basin. The Foundation was a major supporter of the Bolack San Juan Basin Flora Project, which has catalogued all of the vascular plants found in the Four Corners region and has now resulted in this publication of a new flora for the region. Many specimens were collected on ranch lands, including the new species Bolack's sand verbena (*Abronia bolackii*).

An archaeological field school is also now in progress on the ranch. This project is uncovering many buildings and artifacts of Ancestral Puebloan (Anasazi) culture tied to those at Aztec Ruins and Chaco Canyon. Mineral studies of agates, garnets, and other regional geological phenomena are under way on the ranch and on Navajo Nation lands. The B-Square Ranch is home to the Bolack Electromechanical Museum and the Bolack Museum of Fish and Wildlife.

The B-Square Ranch, a 12,000-acre lifelong commitment of the late Tom and Tommy Bolack, is a private wildlife refuge, farm, and ranch. It was established in 1957 to demonstrate multiple land use. Because of Tommy Bolack's dedication of over 30 years of implementation and management, the B-Square Ranch continues to function as a working example of conservation. The operation involves land reclamation and production of wildlife habitat. Some 100,000 waterfowl winter at the ranch. Seven artificial lakes cover approximately 75 acres. Many species of native fauna and flora are in abundance on ranch lands.

Agricultural and greenhouse studies have resulted in selection of many plant species and varieties for efficient production under conditions commonly encountered in the Four Corners region. Today the farm's gardens produce over 100 varieties of drought- and disease-resistant nonsurplus crops yearly. Tommy continues to show the fruits of agricultural production which won the B-Square Ranch many ribbons at the New Mexico State Fair; so many, in fact, that the B-Square Ranch exhibits, but no longer formally competes. For the last 40 years the ranch matches the cash award and presents a ribbon to first-place winners.

Over 100,000 visitors come to the B-Square Ranch annually. Visitors include school classes, bird, nature, and 4-H clubs, agricultural tour groups, foreign exchange students, photographers, film crews, and artists. Yearly conservation tours strive to teach humanity's part in improving resource utilization and environmental protection. The B-Square Ranch has been used extensively by San Juan College for biology class field trips, including Field Ornithology, Field Entomology, Field Ecology, and General and Systematic Botany, as well as introductory geological and biological sciences. The ranch is a resource for regional geological formations, resulting in frequent use by geology classes. The B-Square Ranch welcomes visitors and inquiry as well as tours of its museums free of charge. For more information visit the website at www.bolackmuseums.com.

Xanthisma coloradoense (A. Gray) D. R. Morgan & R. L. Hartm.

Asclepias welshii N. H. & P. K. Holmgren

Penstemon rostriflorus Kellogg

Asclepias sanjuanensis K. D. Heil, J. M. Porter & S. L. Welsh

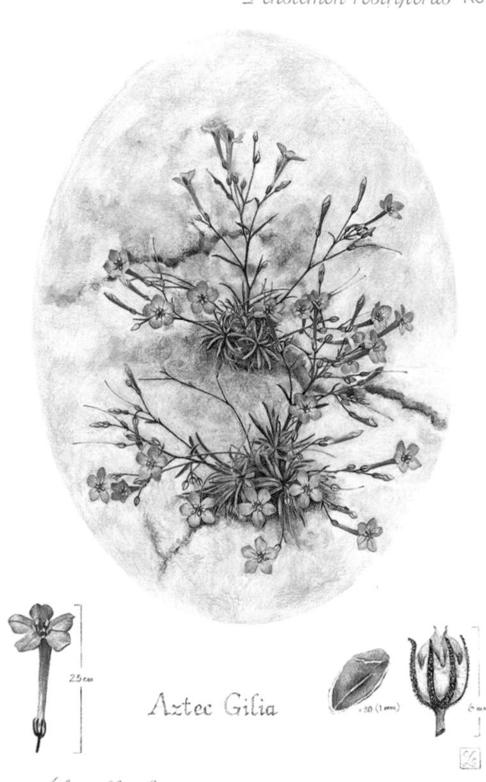

Aliciella formosa (Greene ex Brand) J. M. Porter

INTRODUCTION

The *Flora of the Four Corners Region* is designed to provide a way to identify all of the plants spontaneously growing in the varied environments of the San Juan River drainage. We hope that it will prove useful to beginning and amateur plant enthusiasts, professional botanists, environmental consultants, academic systematists, native nurserymen, ecologists, and government agency personnel.

Expert volunteer authors were solicited for the Four Corners Flora Project and were asked to complete comprehensive treatments following a readable and consistent format. In addition to dichotomous keys and descriptions, line drawings, watercolor paintings, color pencil artwork, and photographs are included as identification guides.

The more than 60 authors contributing to the *Flora* brought a diversity of taxonomic philosophies and levels of experience. Three editors, Kenneth D. Heil (Supervising Editor), Steve L. O'Kane, Jr., and Linda Mary Reeves, were responsible for raising funds, editing manuscripts, writing many treatments, applying for permits, collecting plants, and performing many and varied day-to-day duties. The editors often questioned authors about the validity and distribution of certain taxa and attempted to maintain accuracy and consistency throughout the book. Individual authors, however, had the final say about taxonomic treatments and the recognition of taxa. We have, where possible, let the authors' voices be heard. For this reason, complete uniformity among treatments is not to be expected. Ethnobotanical information is included following a species' description, when known. This information is culled from the literature or from informants. We do not endorse the use of this information except by qualified practitioners.

Numerous constraints, of course, continually plagued the project. These included a tight schedule, limited funds for fieldwork and for herbarium workers, low rainfall, the remoteness, size, and near-inaccessibility of much of the study area, and constraints on the design and content of the book. We have, however, attempted to treat every species, subspecies, and variety of plant known to occur in the study area and to accurately describe their morphology, distributions, and habitat. Identification keys have been constructed to be as simple and user-friendly as possible.

Scope

The *Flora of the Four Corners* includes all of those species, subspecies, and varieties of plants that grow spontaneously in the watershed of the San Juan River, which takes in major portions of Arizona, Colorado, New Mexico, and Utah and is more or less centered on the Four Corners, the only spot in the United States where four states meet at a common point. The inside front and back covers show the major drainage systems of the San Juan River and the major physiographic and political entities in the study area. The entire area encompasses 65,382 square kilometers (25,244 square miles), an area the size of West Virginia and about half the size of Alabama, Arkansas, New York, or North Carolina. The highest point is 4292 m (14083′) at Mt. Eolus in the San Juan Mountains, and the lowest point is 1130 m (3708′) where the San Juan River flows into Lake Powell (the Colorado River). Because of this elevation gradient, vegetation in the study area varies from alpine tundra to coniferous forests, mountain shrublands, lowland sagebrush, blackbrush, and to the sparse communities seen on bare rocks and the scorching sides of low-elevation canyons.

History

Why a *Flora of the Four Corners Region*? After all, there are four state floras that cover this region. The birth of this flora began May 16, 1996, at the Elk Ridge Café in Blanding, Utah. While discussing plants over breakfast, John Anderson, Rich Fleming, Ken Heil, Steve O'Kane, and J. Mark Porter recognized the need for a flora of the region. *A Utah Flora*, besides covering Utah, is also applicable to the desert regions of Colorado and some of New Mexico and Arizona; however, many of the mountain plants of Colorado, New Mexico, and Arizona are missing. *Colorado Flora*: *Western Slope* is also applicable to Utah, New Mexico, and some of Arizona. However, many of the plants found in the Chuska Mountains and those plants of the southern part of the study area are missing and the book lacks full plant descriptions. *Seed Plants of Northern Arizona* is out of print, out of date, and lacks many of the plants found in adjacent New Mexico and Colorado. *A Flora of New Mexico* included little fieldwork in northwestern New Mexico, and the book is out of date and out of print. If a copy can be found at all, it is very expensive. The Four Corners region is unique and, therefore, it was recognized, early on, that extensive and intensive fieldwork was required to adequately represent the flora. The Bureau of

Land Management provided the first seed funding for the project in 1996. In 2000 a joint venture between San Juan College and Tommy Bolack of the Bolack Museum Foundation provided substantial funds and the beginning of the Bolack San Juan Flora Project.

Historical Collectors

Clearly the modern contributors to this flora were not the first collectors to work in the Four Corners area. Listed below are those historical botanical collectors and authors whose work was especially noteworthy within the Four Corners region. Many of these botanists collected throughout the southwest, periodically visiting various localities within the San Juan River drainage. Most of our distinguished botanists are memorialized in plant names from the *Flora* area (Ewan 1950; Ewan and Ewan 1981; Humphrey 1961; Reveal 1972; Williams 2003).

Baker, Charles F. 1872–1927. Entomologist and botanist; collected in Archuleta, La Plata, Mineral, and Montezuma counties, Colorado, and at Aztec, San Juan County, New Mexico.

Barneby, Rupert C. 1911–2000. Specialist in Fabaceae, notably *Astragalus* and *Oxytropis*; collected extensively throughout the American West.

Boissevain, Charles H. 1893–1946. Co-author, with Carol Davidson, of *Colorado Cacti* (1940).

Brandegee, Townshend S. 1843–1925. Botanist for one year with the Hayden Survey in southwestern Colorado; made collections in the Mesa Verde region, Colorado.

Castetter, Edward Franklin. 1896–1978. Ethnobotanist, professor, and vice president at the University of New Mexico; collected at Farmington, Shiprock, and Chaco Canyon, New Mexico.

Clute, Willard Nelson. 1869–1950. Founder of the botanical journal *American Botanist*; collected on the Navajo Nation for two summers; his specimens were sent to Aven Nelson at the University of Wyoming.

Cottam, Walter Pace. 1894–1988. Professor of botany at Brigham Young University and University of Utah; collected in Monument Valley, Arizona, and Utah.

Crandall, Charles Spencer. 1852–1929. Professor of botany and horticulture at Colorado Agricultural College, Fort Collins; collected in the Durango and Silverton, Colorado, regions.

Davidson, Carol. Dates unknown. Co-author, with Charles Boissevain, of *Colorado Cacti* (1940).

Earle, Franklin Sumner. 1856–1929. Botanized at Mancos, Durango, and La Plata Mountains, Colorado, and collected with E. L. Greene and S. M. Tracy.

Eastwood, Alice. 1859–1953. Botanist and curator of botany at California Academy of Sciences; important collections from Silverton, Durango, the La Plata Mountains, Mesa Verde, and McElmo Canyon, Colorado; Monticello, Bluff, Comb Ridge, Mexican Hat, and Willow Creek, Utah.

Gambel, William. 1821–1849. Friend of renowned botanist Thomas Nuttall. Collected a few species of plants in the upper San Juan River drainage on an expedition from Albuquerque to Los Angeles.

Greene, Edward Lee. 1843–1915. First professor of botany at University of California and botanical explorer in New Mexico and Colorado; collected with F. S. Earle and S. M. Mills in the La Plata Mountains, Colorado, and in the vicinity of Durango and Mancos, Colorado.

Harrington, Harold D. 1903–1981. Author of *Manual of the Plants of Colorado* (1954); made collections throughout southwestern Colorado.

Jones, Marcus E. 1852–1934. Field botanist throughout the western United States; collected in Blanding and Monticello, Utah.

Kearney, Thomas Henry. 1874–1956. Botanized extensively throughout Arizona; co-author, with R. H. Peebles, of *Arizona Flora* (1951).

Kelly, George W. 1894–1991. Horticulturist and plant collector of McElmo Canyon, Montezuma County, Colorado; writer of books popularizing botany.

Kittell, M. Teresita. 1892–? Co-author, with I. F. Tidestrom, of *Flora of New Mexico and Arizona* (1941).

McDougall, Walter B. 1883–1980. Author of *Seed Plants of Northern Arizona* (1973); collected throughout the Navajo Indian Reservation.

Nelson, Aven. 1859–1952. Founder of the Rocky Mountain Herbarium, Laramie, Wyoming; identified plants collected by W. N. Clute from the Navajo Nation.

Newberry, John Strong. 1822–1892. Paleontologist, geologist, botanist, and educated in medicine (M.D.); collected in Largo Canyon, New Mexico, and southwestern Colorado.

Owen, Herbert. 1925–2005. Biology professor at Fort Lewis College for 23 years; taught entomology, botany, ecology, plant physiology, and systematic botany and led many students into the field of science; collected plants in La Plata and San Juan counties, Colorado. He planted the seed for the *Flora of the Four Corners*.

Payson, Edwin Blake. 1893–1927. Professor of botany at the University of Wyoming; collected at Mesa Verde, Bayfield, and Archuleta, Colorado.

Peebles, Robert Hibbs. 1900–1955. Collector of Arizona plants and co-author, with T. H. Kearney, of *Arizona Flora* (1951).

Ripley, Harry D. D. 1908–1973. Collecting companion to Rupert C. Barneby.

Rollins, Reed Clark. 1911–1998. Lifelong student of the mustard family (Brassicaceae). Collected extensively in the western U.S. and wrote, among many other publications, *The Cruciferae of Continental North America* (1993).

Rydberg, Per Axel. 1860–1931. Author of *Flora of the Rocky Mountains and Adjacent Plains* (1917) and *Flora of Colorado* (1906); collected throughout southwestern Colorado.

Schmoll, Hazel M. 1891–1990. Curator at the Colorado State Museum; collected in the Pagosa–Piedra region and throughout Mesa Verde National Park, Colorado.

Smith, Charles Piper. 1877–1955. High school teacher in San Jose, California, and later professor of botany at Utah Agricultural College; collected in Hinsdale, La Plata, and Montezuma counties, Colorado, and Rio Arriba and San Juan counties, New Mexico.

Standley, Paul Carpenter. 1884–1963. Co-author, with E. O. Wooton, of *Flora of New Mexico* (1915) and collected in the Chuska and Carrizo Mountains, New Mexico and Arizona.

Tidestrom, Ivar Frederick. 1864–1956. Studied under E. L. Greene. Co-author, with Sister Teresita Kittell, of *Flora of New Mexico and Arizona* (1941).

Tracy, Samuel Mills. 1847–1920. Botanist and agronomist; often collected with F. S. Earle and E. L. Greene; collected in the La Plata Mountains and in the vicinity of Durango and Mancos, Colorado.

Vestal, Paul A. 1908–1978. Noted ethnobotanist, conducted research among the Navajo (Diné) and published *Ethnobotany of the Ramah Navaho* [*sic*] (1952).

Wetherill, Benjamin Alfred. 1861–1950. Guided pack trips for Alice Eastwood in the Mesa Verde, Colorado, region. In 1892 collections were made in Monticello, Montezuma Creek, and McElmo Canyon; in 1895 there were collections from McElmo Canyon, Bluff, Comb Wash, and Willow Creek, Utah.

Whiting, Alfred F. 1912–1978. Published *Ethnobotany of the Hopi* (1939).

Wooton, Elmer Ottis. 1865–1945. Taught chemistry and botany at New Mexico College of Agricultural & Mechanical Arts; co-authored, with P. C. Standley, *Flora of New Mexico* (1915) and made numerous collections throughout northwestern New Mexico and southwestern Colorado.

Methodology

An initial working list of the *Flora* was constructed from the collections at San Juan College Herbarium (SJNM). To this list additional plants and their localities were added from known local herbaria and from available theses and unpublished checklists (e.g., Bleakly 1996; Michener 1960; Michener-Foote and Hogan 1999). This initial list, then, consisted of plants documented by voucher specimens. The list was vastly expanded by the addition of *possible* taxa gleaned from recent floristic treatments as well as from monographic/taxon-specific treatments of major groups and historical floras. Published volumes of *Flora of North America* were also consulted. Based on known voucher specimens and from published sources, a brief habitat description was compiled for each species. Examples of monographic and taxon-specific treatments are Allred (1997) for grasses, Barneby (1964a, b) for *Astragalus*, Carter (1997) for woody plants of New Mexico, Anderson (2001) for cacti, Isely (1998) for Fabaceae, and Rollins (1993) for Brassicaceae. The floras of Wooton and Standley (1915), Tidestrom and Kittell (1941), Rydberg (1906), and McDougall (1973) were especially useful. Of these, Rydberg's and Wooton and Standley's floras included specimen citations and specific locality information that could be directly incorporated in the list.

Once the working list of known and potential taxa was completed, fieldwork began. An attempt was made to survey all habitat types and geological substrates in the study area in every county with an eye to verifying the presence of species predicted to be present. New records were incorporated into the working list at the end of each field season. A master set of voucher specimens is maintained at San Juan College (SJNM). A revised list was then created for the upcoming field season. Most regional Native American tribes graciously allowed us to collect plants on tribal lands.

During the later stages of the project, specimens from the following herbaria were consulted and were added to the improving catalogue: Colorado State University (CS); Deaver Herbarium, Northern Arizona University (ASC); Fort Lewis College (FLD); University of Northern Iowa (ISTC); Museum of Northern Arizona (MNA); Navajo Heritage Herbarium (Window Rock, Arizona); New Mexico State University (NMC); Range Science Herbarium, New Mexico State University (NMCR); Rancho Santa Ana Botanic Garden (RSA); Rocky Mountain Herbarium, University of Wyoming (RM); San Juan College (SJNM); Stanley Welsh Herbarium, Brigham Young University (BRY); University of Colorado Herbarium (COLO); University of New Mexico (UNM); and the herbarium at Mesa Verde National Park. Seven unpublished working versions of the list were created since 1998.

Authors' guidelines based on *Flora of the Rocky Mountains* and the guidelines for the new *Flora of Arizona* project were constructed. A checklist of collected plant taxa of the region, lists of participants, sample treatments, illustrations, and other information for authors was initially sent to authors, illustrators, and benefactors. Later this information was posted on the San Juan College Herbarium website. Authors were recruited based on past publications, expertise in a given plant family, or affiliation with a particular herbarium. Treatments of unassigned families and genera were written by the editors. Most ethnographic information was provided by the editors. An annual report covering yearly activities was produced for benefactors.

Measurements

Plant parts are measured in metric units. Millimeters (mm) accuracy is used up to around 25 mm in length or breadth, whereas centimeters (cm) are used up to 50 cm. Decimeters (dm) are used for plants or plant parts exceeding 50 cm. Meters (m) are used for plants that exceed 1.5 m and for elevational measurements. However, elevation is also given in feet. Listed below are the common plant parts that are measured and used in taxonomic keys.

Calyx – Point of insertion of the pedicel to the tip of the longest calyx teeth; in legumes it is the length of the longest lateral tooth.

Calyx teeth – Base of the sinus to an adjacent lobe or tooth apex.

Corolla lobe – Base of the sinus to an adjacent lobe or tooth apex.

Flower length (total) – Point of insertion of the pedicel (peduncle) to the apex of the longest petal.

Fruit length – Base of the ovary (point of insertion on the receptacle) to the tip excluding style and stigma measurements.

Leaf breadth – Widest portion of the blade.

Leaf measurement – Petiole base to the blade apex.

Peduncle – Uppermost bract to the first flower or a branch of a compound inflorescence.

Petal length – Point of insertion of the petal to its apex.

Petiole – Point of insertion on the stem to the blade base.

Plant height – Length of the plant above ground level including the inflorescence and flowers.

Stipule – Point of insertion on the stem or petiole base to the apex.

Abbreviations

Abbreviations for this text have been kept to a minimum for clarity. The following abbreviations, however, have been used for measurements and distributional information.

m = meters
dm = decimeters
cm = centimeters
mm = millimeters
km = kilometers

Taxonomy

Our principal goal was to document the biodiversity, at the species and infraspecies levels, of the San Juan River watershed. We have attempted to bring the nomenclature up to date. However, species and generic concepts are interpreted by the authors. Names used by the authors follow the rules of the latest *International Code of Botanical Nomenclature* (McNeill et al. 2006). Authority designations follow *Authors of Plant Names* by R. K. Brummit and C. E. Powell (1992).

With such a diverse group of authors, some chose to be conservative, while others followed a more liberal approach. Generally, the species concept involves an organism that is distinct by reproductive isolation. Its distinctness is seen in morphological, cytological, anatomical, chemical, and pollination differences. It is our belief that our authors have researched all sources carefully and their decisions are based on sound review. Ultimately, the taxonomy of a particular group is that of the author. As more collections are made, and more fieldwork is accomplished, taxonomic views of authors will change, and future revisions will be necessary.

Common Names

Because scientific names are not widely known or used by the general public and are thought to be difficult to pronounce, common names are often used. However, many common names are confusing and are not applied according to any logical or consistent system of rules. Many plants do not have a common name and this can be quite distressing to authors of popular guides to plants and to personnel of some government agencies. Often what is done is to conjure up some common name by translating the scientific name into English. For example, the advantage of Osterhout's cryptantha over *Cryptantha osterhoutii* is not immediately apparent (Smith 1977). Most common names have been selected from the USDA Plants Database, USDA Forest Service General Technical Report INT-38, at the discretion of the author, or from local usage. Where possible and useful, local Spanish and Native American names are included, reflecting these important heritages in our area.

Classification

The list of families is set up alphabetically with four major groups of plants: Ferns, Fern Allies, Gymnosperms, and Angiosperms.

FERNS

Family Name	Genera	Species	Infra
1. Aspleniaceae	1	4	
2. Dennstaedtiaceae	1	1	
3. Dryopteridaceae	6	15	
4. Marsileaceae	1	1	
5. Ophioglossaceae	1	10	
6. Polypodiaceae	1	2	
7. Pteridaceae	5	10	

Totals: Families = 7; Genera = 16; Species = 43

FERN ALLIES

Family Name	Genera	Species	Infra
1. Equisetaceae	1	4	1
2. Isoetaceae	1	1	
3. Lycopodiaceae	2	2	
4. Selaginellaceae	1	4	

Totals: Families = 4; Genera = 5; Species = 11; Infraspecific = 1

GYMNOSPERMS

Family Name	Genera	Species	Infra
1. Cupressaceae	1	4	
2. Ephedraceae	1	2	
3. Pinaceae	4	11	2

Totals: Families = 3; Genera = 6; Species = 17; Infraspecific = 2

ANGIOSPERMS

Family Name	Genera	Species	Infra
1. Aceraceae	1	3	
2. Adoxaceae	2	3	1
3. Agavaceae	1	3	
4. Alismataceae	2	5	
5. Amaranthaceae	3	9	
6. Anacardiaceae	2	3	1
7. Apiaceae (Umbelliferae)	19	38	
8. Apocynaceae	2	4	
9. Araceae (Lemnaceae)	2	6	
10. Araliaceae	1	1	
11. Asclepiadaceae	2	17	
12. Asparagaceae	1	1	
13. Asteraceae (Compositae)	121	354	57
14. Berberidaceae	2	4	

Family Name	Genera	Species	Infra
15. Betulaceae	3	5	
16. Bignoniaceae	2	2	
17. Boraginaceae	12	44	9
18. Brassicaceae (Cruciferae)	37	108	11
19. Cactaceae	7	20	5
20. Campanulaceae	2	4	
21. Cannabaceae	2	2	
22. Caprifoliaceae	3	5	
23. Caryophyllaceae	13	41	2
24. Celastraceae	1	1	
25. Ceratophyllaceae	1	1	
26. Chenopodiaceae	15	50	5
27. Cleomaceae	4	6	
28. Clusiaceae (Guttiferae)	1	1	
29. Commelinaceae	2	3	1
30. Convolvulaceae	5	13	
31. Cornaceae	1	2	
32. Crassulaceae	2	4	
33. Crossosomataceae	1	1	
34. Cucurbitaceae	4	7	
35. Cyperaceae	8	98	1
36. Droseraceae	1	1	
37. Elaeagnaceae	2	5	
38. Elatinaceae	1	2	
39. Ericaceae	10	18	
40. Euphorbiaceae	5	14	
41. Fabaceae (Leguminosae)	28	127	32
42. Fagaceae	1	5	3
43. Frankeniaceae	1	1	
44. Fumariaceae	1	2	2
45. Gentianaceae	9	17	
46. Geraniaceae	2	6	1
47. Grossulariaceae	1	9	
48. Haloragaceae	1	2	
49. Hydrangeaceae	2	2	
50. Hydrocharitaceae	2	3	
51. Hydrophyllaceae	4	20	
52. Iridaceae	2	4	
53. Juglandaceae	2	2	
54. Juncaceae	2	20	
55. Juncaginaceae	1	2	
56. Lamiaceae (Labiatae)	19	26	
57. Lentibulariaceae	1	1	
58. Liliaceae	15	29	
59. Linaceae	1	6	
60. Loasaceae	1	9	
61. Malvaceae	7	15	
62. Menyanthaceae	1	1	
63. Moraceae	2	2	

Family Name	Genera	Species	Infra
64. Nyctaginaceae	6	17	1
65. Nymphaeaceae	1	1	
66. Oleaceae	4	6	
67. Onagraceae	7	34	7
68. Orchidaceae	9	17	
69. Orobanchaceae	7	31	
70. Oxalidaceae	1	6	
71. Papaveraceae	2	4	
72. Parnassiaceae	1	3	
73. Pedaliaceae	1	2	
74. Phrymaceae	1	7	1
75. Plantaginaceae	12	47	6
76. Plumbaginaceae	1	1	
77. Poaceae (Gramineae)	85	249	11
78. Polemoniaceae	16	84	9
79. Polygalaceae	1	3	
80. Polygonaceae	9	59	6
81. Portulacaceae	4	11	
82. Potamogetonaceae	2	13	
83. Primulaceae	5	8	
84. Ranunculaceae	13	41	1
85. Rhamnaceae	3	4	
86. Rosaceae	24	63	3
87. Rubiaceae	3	10	
88. Ruppiaceae	1	1	
89. Salicaceae	2	21	2
90. Santalaceae	1	1	
91. Saururaceae	1	1	
92. Saxifragaceae	4	17	
93. Scrophulariaceae	2	2	
94. Simaroubaceae	1	1	
95. Solanaceae	8	25	1
96. Tamaricaceae	1	1	
97. Typhaceae	2	6	
98. Ulmaceae	2	2	
99. Urticaceae	2	2	2
100. Valerianaceae	1	3	
101. Verbenaceae	3	4	
102. Violaceae	1	7	2
103. Viscaceae	2	4	
104. Vitaceae	2	3	
105. Zannichelliaceae	1	1	
106. Zygophyllaceae	3	3	

Totals: Families = 106; Genera = 670; Species = 2046; Infraspecific = 183

Grand Totals: Families = 120; Genera = 697; Species = 2117; Infraspecific = 186

NUMBER OF TAXA PER STATE AND COUNTY

ARIZONA	1323
Apache	1212
Coconino	246
Navajo	729
COLORADO	1751
Archuleta	1048
Conejos	208
Dolores	428
Hinsdale	708
La Plata	1249
Mineral	616
Montezuma	1238
Rio Grande	262
San Miguel	281
San Juan	705
NEW MEXICO	1503
McKinley	818
Rio Arriba	790
Sandoval	324
San Juan	1307
UTAH (San Juan)	1211

Political entity abbreviations used in the *Flora of the Four Corners Region*

State	Abbreviation	County	Abbreviation
Arizona	ARIZ	Apache	Apa
		Coconino	Coc
		Navajo	Nav
Colorado	COLO	Archuleta	Arc
		Conejos	Con
		Dolores	Dol
		Hinsdale	Hin
		La Plata	LPl
		Mineral	Min
		Montezuma	Mon
		Rio Grande	RGr
		San Miguel	SMg
		San Juan	SJn
New Mexico	NMEX	McKinley	McK
		Rio Arriba	RAr
		Sandoval	San
		San Juan	SJn
Utah	UTAH	San Juan	

GEOLOGY

Kenneth D. Heil and Tommy Bolack

The Four Corners region is a land of stark contrasts. As defined by the watershed of the San Juan River, one finds snow-capped volcanic peaks of the San Juan Mountains, lofty volcanic laccoliths with unusual endemic plants, pine forests of the Chuska Mountains, and sedimentary lowlands that support shifting sand and slickrock rims, as well as unusual alcoves with hanging gardens.

No other region matches the Four Corners' diversity in geology. Part of the Colorado Plateau, it is considered to be a region of simple structural geology and a rather stable block of North America that is set off by linear fractures in the earth's crust. However, it has a high number and variety of geologic formations, many of these at different elevations, thus providing habitat for a wide diversity of plants.

The Four Corners region is a geologist's dreamscape with rocks ranging from Precambrian time to the Tertiary Period, a span of 2.5 billion years. The San Juan River and its tributaries are the lifeblood of the Four Corners region. In upper reaches the San Juan, Navajo, Piedra, Animas, and Los Piños rivers form steep, deep, and dark canyons, while Largo, Chaco, and Blanco canyons are wide and tend to meander. Important drainages in the Chuska and Lukachukai Mountains, respectively, are Tsaile and Totsoh creeks. A kaleidoscope of colors ranging from chocolate brown to coffee tan or crimson red are to be seen at the Bisti badlands, Monument Valley, and Valley of the Gods. Scarlet red alcoves with dark-green hanging gardens are found near Bluff, Utah, in Johns and Slickhorn canyons, and numerous other canyons in the San Juan arm of Lake Powell. Golden volcanic tuff and volcanic breccia cliffs are seen throughout the San Juan Mountains.

The single most important event in the sculpting of the San Juan drainage, as it is seen today, was the Laramide orogeny. Mountain-building forces began from the late Cretaceous to the early Tertiary and included folding, faulting, and volcanism. With the folding, faulting, and rising of the San Juan Mountains, basins such as the San Juan, Black Mesa, and Paradox were subsiding. Uplift, weathering, and erosion deposited fluvial sediment in these catchment basins. The result of this deposition can been seen while driving along U.S. 550 between Durango, Colorado, and Cuba, New Mexico. Another result of the Laramide orogeny was the Defiance uplift, a flexure along the Arizona and New Mexico state line that formed the Chuska Mountains. The massive cliffs of Canyon de Chelly were carved by running water from the northwest flank of the Defiance uplift. Strangely, it is in deserts that the erosional power of water is so starkly evident. Because little precipitation falls, rivers and streams cut deeper into their beds more quickly than surrounding areas.

Volcanic activity throughout the San Juan Mountains was going strong by early Tertiary time. One large focal point for volcanic activity can be seen near Silverton, Colorado. A caldera (collapsed volcano) is found north of Silverton and extends for several miles in diameter. Calderas are also found in nearby Lake City and Creede. These volcanoes emitted large amounts of extrusive material that covered the central and eastern San Juan Mountains with several thousand feet of volcanic tuff and flow rock which continued throughout the early Tertiary. Geothermal and hydrothermal processes allowed silica and other minerals to precipitate.

The Tertiary was also the approximate time for the formation of large intrusive igneous rock bodies called laccoliths. Laccoliths are dome-shaped features that form below the earth's surface and are later exposed by uplift, weathering, and erosion. The Four Corners region has a large number of laccoliths that include the following mountain ranges: La Plata, Sleeping Ute, Abajo (Blue), Navajo, and Carrizo. Just outside the San Juan drainage are two other laccoliths, the Henry and La Sal Mountains. Many rare and endemic plants are found on such volcanic laccoliths.

During the last one million years of the Pleistocene Epoch, glaciers migrated southward to cover much of the continent of North America. At the same time, glaciers formed in the San Juan Mountains, added sedimentary material, and carved out the spectacular scenery that is seen today. This beautiful mountain range is sometimes referred to as the American Alps. Many glacial erosional features can be seen in the San Juan Mountains, including cirques, tarns, arêtes, horns, hanging gardens, U-shaped valleys, and glacial striations. Glaciers of this period made it as far south as Durango, Colorado, dumping terminal and lateral moraine deposits. As the streambeds erode and rivers cut deeper, the depositional terraces may be viewed along the entire San Juan River basin. The massive glacial melt also helped to carve many canyons and their tributaries. All of these factors together have produced potentially diverse and numerous plant habitats in the Four Corners region.

GEOLOGIC TIME SCALE AND FORMATIONS PRESENT IN THE FLORA OF THE FOUR CORNERS STUDY AREA (Baars 1995, 2000)

ERA	PERIOD	MILLIONS OF YEARS	FORMATIONS	EXAMPLE LOCATION
Cenozoic	Quaternary	0-1.6	Soil, sand & gravel	San Juan River floodplain
	Tertiary	1.6-65	Terrace gravels	San Juan River floodplain
			Diatremes	Ship Rock, Agathla Peak, Church Rock
			Igneous intrusives	Silverton, Wolf Creek
			San Jose Formation, shale & sandstone	Navajo Lake, Dulce
			Nacimiento Formation, sandstone & mudstone	Aztec, Bloomfield
			Ojo Alamo Sandstone	Farmington
Mesozoic	Cretaceous	65-135	Fruitland-Kirtland Formation, sandstone & shale	Bisti badlands
			Pictured Cliffs Sandstone	Near Fruitland
			Mesa Verde Group, sandstone	Mesa Verde & Black Mesa
			Mancos Shale/Lewis Shale	Black Mesa, Shiprock, Durango, Pagosa Springs
			Dakota Sandstone	Cottonwood, AZ, Four Corners
	Jurassic	135-205	Morrison Formation, sandstone & mudstone	Bluff & Aneth
			Todilto Limestone	East escarpment of Red Valley
			Entrada Sandstone	Baby Rocks & Chilchinbito
			Carmel Formation, siltstone & mudstone	Red Mesa & Bluff
			Navajo Sandstone	Navajo Mountain
			Kayenta Formation, sandstone	Kayenta
			Wingate Sandstone	Beclabito Dome & Lake Powell
	Triassic	205-250	Chinle Formation, conglomerate, sandstone & shale	Chinle & Rock Point
			Moenkopi Formation, siltstone & mudstone	Monument Valley & Lake Powell
Paleozoic	Permian	250-290	De Chelly Sandstone	Canyon de Chelly & Monument Valley
			Cedar Mesa Sandstone	Cedar Mesa
	Pennsylvanian	290-325	Halgaito Shale	Monument Valley
			Hermoso Group, evaporites, limestone, siltstone & sandstone	Monument Valley & San Juan River, Mexican Hat
			Molas Formation, sand & gravel lenses	Molas Lake
	Mississippian	325-355	Redwall/Leadville Formation	Rockwood & Coalbank Pass
	Devonian	355-410	Ouray Limestone	South of Silverton
			Elbert Formation, sandstone, shale & dolomite	Elbert Creek & Rockwood
	Silurian	410-438	Not represented	
	Ordovician	438-510	Not represented	
	Cambrian	510-570	Ignacio Quartzite	Electra Lake
Precambrian		570-4500?	Metamorphics, quartzite & granite	Needle and Grenadier Mountains

CLIMATE

John R. Spence

Climatic patterns in the San Juan River drainage are complex (Keen 1996; Pedersen 1994; Spence 2001). Major air masses and their attendant physical parameters are controlled by the position of the Northern Hemisphere jet stream. From November to late March, fronts move onshore from the Pacific Ocean, bringing cool storms and precipitation in the form of rain or snow to the San Juan River drainage. By spring the Pacific High moves north, blocking storms from entering the Southwest and Rocky Mountains. In late June or July high pressure builds from the Gulf of Mexico (the Bermuda High) and the anticyclonic circulation dominant over the Southwest is displaced northward. The Bermuda High moves north and west causing southern winds circling clockwise around its edges to bring warm moist air from the Gulf of Mexico. This mixes with Pacific moisture from the Gulf of California, bringing monsoonal storms into the Southwest and the Four Corners region. This monsoon season is characterized by thunderstorms, often with heavy precipitation and flooding. The amount and location of precipitation is controlled in part by topography. The average northern extent of this monsoonal precipitation occurs as a frontal boundary from the southwest near the Grand Canyon to the northeast in the southern Rocky Mountains. The monsoon season typically lasts six to eight weeks, dying out by September. Finally, every few years dissipating hurricanes off Baja California move north and east up the Colorado River valley, bringing heavy rains to the Colorado Plateau during September and October.

Decadal and longer-term patterns also affect the climate of the Four Corners region. Recent research has determined that the climate of the American Southwest is affected by sea surface temperature changes in both the Atlantic and Pacific oceans (McCabe et al. 2004). The Pacific Decadal Oscillation (PDO) varies on a time scale of about 20–25 years and fluctuates from a more negative (cooler) to a more positive (warmer) sign. The Atlantic Multidecadal Oscillation (AMO) is similar but occurs on much longer cycles of 50–70 years. The combined sign of these two patterns explains about 50% of the variation in precipitation patterns in the southwestern U.S. For example, when the AMO and PDO are both positive, as occurred in 2004–2005, the Southwest experiences extremely high precipitation. The severe drought in the region that started in 1998–1999 was in part due to a shift in the PDO to a more negative value. Imposed on these complex atmospheric dynamics is a long-term warming trend in the last 100 years that may be increasing the severity and duration of droughts.

Since early in the 20th century there has been a gradual warming of about 0.6°C or 1°F (IPCC 1998). Recent evidence suggests that this warming may be accelerating in the Northern Hemisphere. On the central Colorado Plateau to the west of the Four Corners region, all stations show a gradual increase in temperatures, especially minimum temperatures (Spence 2001). In general, winters have become warmer through much of the low and mid-elevations of the central Colorado Plateau. Another weaker trend has been the gradual but as of yet nonsignificant increase in winter (Pacific) precipitation. This pattern was interrupted with the recent severe drought of the past few years (Hereford, Webb, and Graham 2002). The year 2002 was one of the driest in several centuries in the Southwest. During the 2002 water year, the San Juan River at Bluff, Utah, had the lowest annual streamflow ever recorded (Wilkowske, Allen, and Phillips 2003).

Climate data for 15 weather stations in the Four Corners region are listed in Table 1. The overall climate of the region can be characterized as cold-temperate, with cold winters in most areas except for low-lying basins and relatively warm summers except for the higher mountains. Local climate is profoundly influenced by elevation and topography (orographic effects). A striking pattern is the drop in temperature with increasing elevation. Mean annual temperature varies from a very warm 13.7°C (56.7°F) at Mexican Hat, Utah, to a chilly 0.9°C (33.7°F) at Wolf Creek Pass, Colorado (Table 1; Figure 1). A rise of 100 m generally corresponds to a decrease of about 0.65°C (3.5°F per 1000′) in mean annual temperature. At any given elevation, south- and west-facing slopes are generally warmer than north- and east-facing slopes, especially in winter when low sun angles result in greatly reduced solar radiation on north-facing exposures. Precipitation generally increases with elevation, from a low of around 15.7 cm (6″) at Mexican Hat to a high of 115 cm (45″) at Wolf Creek Pass (Table 1; Figure 2), although the relationship is not as strong as for temperature, and is not linear. Precipitation in the region is bimodal, with a peak in late winter (March) and a second larger peak in late summer (August–September). Above about 2000 m (6700′) much of the precipitation in winter falls as snow, and annual snowfall often exceeds 11.1 m (435″) at Wolf Creek Pass. At any given elevation, soil moisture during the plant growing season usually is greater on north-facing than on south-facing slopes, for two major reasons. First, the north-facing slopes accumulate more snow because of reduced solar radiation in winter, and the melting of this snow in spring recharges the soil moisture. Second, the lower temperatures on north-facing slopes result in lower rates of evaporation and slower depletion of soil moisture reserves.

One way of measuring how much moisture is available for plant growth is to determine the potential evapotranspiration rate (PET; based on Thornthwaite 1948), which is a function primarily of temperature. Potential evapotranspiration rates in cm of rain can be found in Table 1 for 15 climate stations located throughout the San Juan River basin, and range from 81.5 cm (32″) at Mexican Hat to 38.6 cm (15″) at Wolf Creek Pass. The relationship between PET and elevation is not linear, however, and is better described (as is that between elevation and precipitation, Figure 2) by an exponential function. The curves intersect at an elevation of about 2700 m (8860′), which represents a humid-arid boundary: at higher elevations precipitation generally exceeds PET (humid conditions), but at lower elevations PET generally exceeds precipitation (arid conditions). This difference is further illustrated in Figures 3 and 4, showing the annual moisture deficits at Mexican Hat and Silverton, Colorado, respectively. It can be seen that PET greatly exceeds available moisture at Mexican Hat (Figure 3), with only the winter months showing available moisture. At Silverton, on the other hand, precipitation is adequate for much of the year, although there is a short drought period during the late spring and summer growing period (Figure 4). Plant growth in high elevations tends to depend strongly on soil moisture that has been recharged by the abundant winter snowfall. Slope and aspect also strongly control PET at local scales. One consequence resulting from this relationship between precipitation, PET, and elevation in the Four Corners region is that perennial streams are most common at higher elevations. In general, most streams arise in the mountains and flow downslope until they either dry out or reach a larger river or lake. Very few streams originate in the lower basins in the Four Corners region. Streams produced by larger springs at lower elevations tend to be "losing" reaches, as water evaporates at varying distances from the source.

Table 1. Mean annual values for selected climate parameters at stations in the San Juan River drainage. Data are from the Western Regional Climate Center (www.wrcc.dri.edu). The last year of records included is 2004.

STATION	ELEVATION (m)	YEARS (operated)	TEMP. (°C)	PRECIPITATION (cm)	PET[1] (cm)	SNOWFALL (cm)
Wolf Creek Pass, CO	3314.6	44	0.9	115.3	38.6	1107.4
Silverton, CO	2887.9	98	1.8	62.0	41.7	391.2
Lemon Dam, CO	2465.8	23	5.7	77.6	49.1	388.6
Fort Lewis, CO	2367.6	56	6.1	46.0	58.1	200.7
Mesa Verde NP, CO	2215.0	56	9.6	45.2	62.2	205.7
Cedar Point, UT	2105.9	47	8.3	37.6	58.7	170.2
Cortez, CO	1934.6	75	9.4	33.0	61.5	86.4
Chaco Canyon, NM	1925.2	82	9.7	22.4	63.2	38.1
Blanding, UT	1881.6	100	10.2	33.8	64.8	96.5
Navajo Dam, NM	2109.0	41	10.9	32.8	67.8	30.5
Canyon de Chelly, AZ	1747.7	36	11.9	23.4	70.6	15.2
Hovenweep Nat Mn, UT	1632.4	47	10.9	27.9	68.8	48.3
Shiprock, NM	1548.3	78	11.6	17.8	70.6	10.2
Bluff, UT	1345.8	76	12.6	19.6	75.2	20.3
Mexican Hat, UT	1286.6	56	13.7	15.7	81.5	0

[1]Potential evapotranspiration according to Thornthwaite (1948).

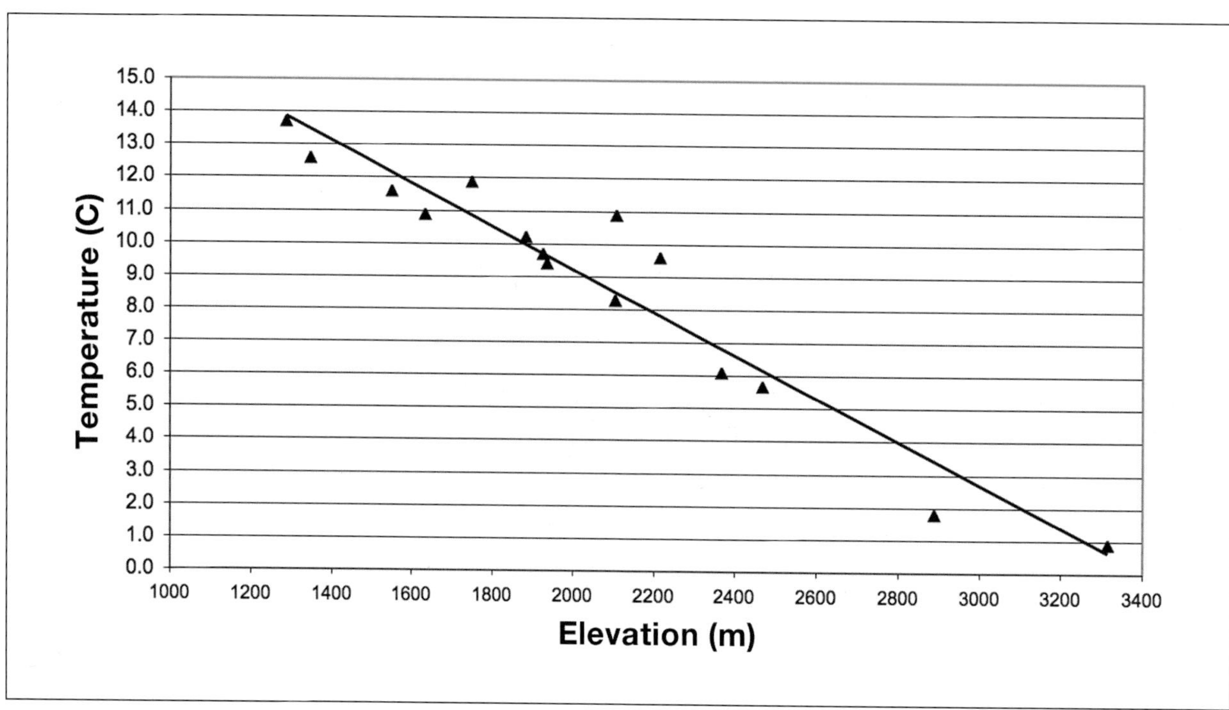

Figure 1. The relationship between mean annual temperature (°C) and elevation (m) for the San Juan River drainage. The data are from 15 weather stations situated throughout the study area.

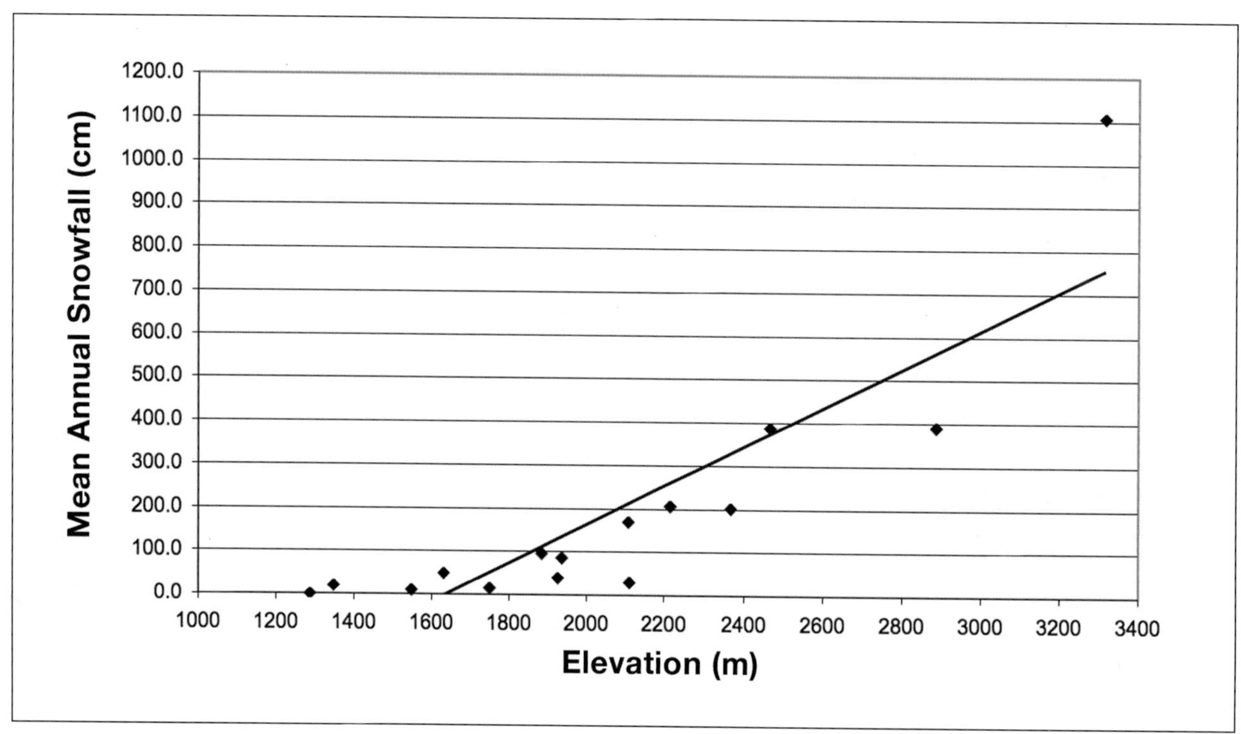

Figure 2. Mean annual snowfall (cm) and elevation (m) for the San Juan River drainage. The data are from 15 weather stations situated throughout the study area. The humid-arid boundary where the line in Figure 1 intersects the line above is at about 2700 m.

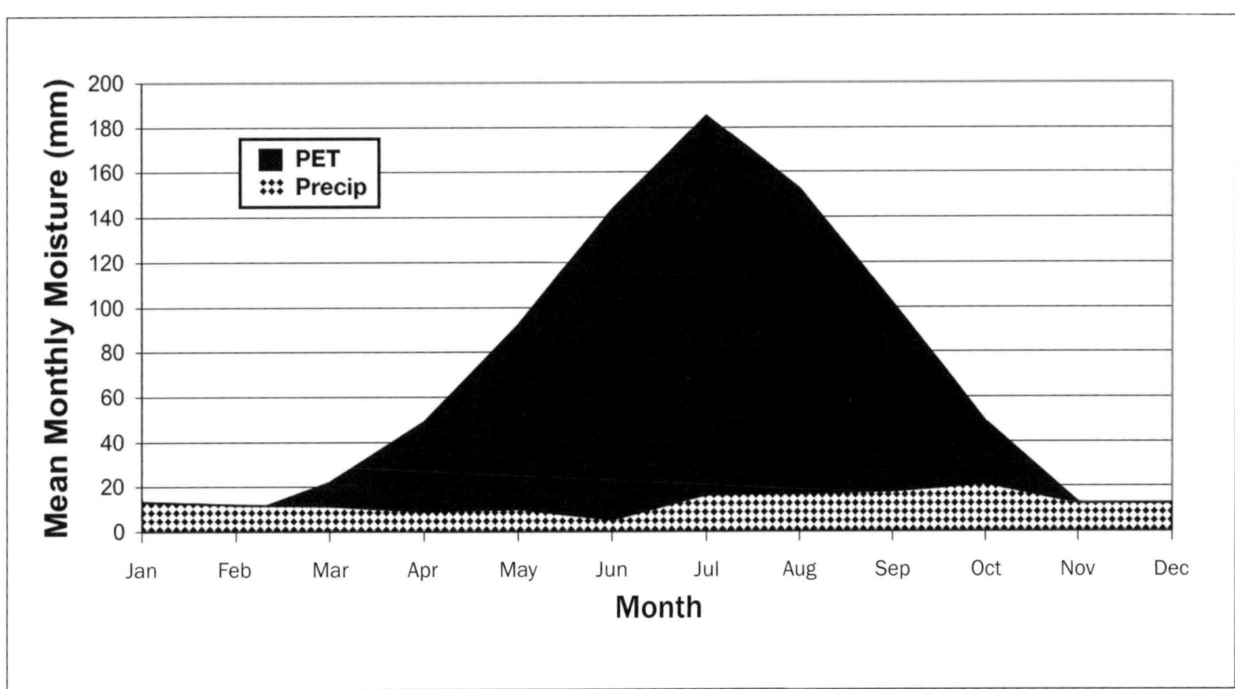

Figure 3. Monthly mean precipitation (cm) and potential evapotranspiration or PET (cm) for Mexican Hat, Utah. Data can be found in Table 1.

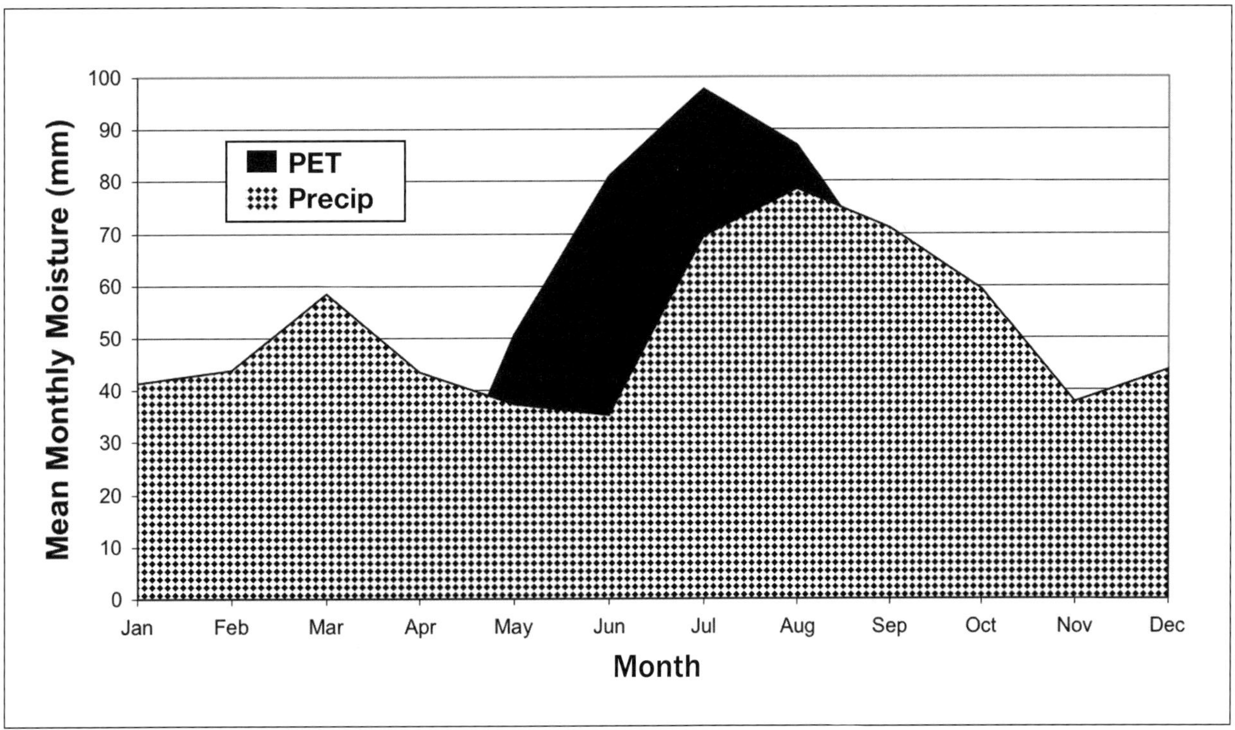

Figure 4. Monthly mean precipitation (cm) and potential evapotranspiration or PET (cm) for Silverton, Colorado. Data can be found in Table 1.

PLANT COMMUNITIES AND VEGETATION PATTERNS

William H. Romme, John R. Spence,
Lisa Floyd-Hanna, and Jeffery S. Redders

Introduction

As moist air masses flow across the southwestern United States, they collide with the massive flanks of the San Juan Mountains in southwestern Colorado and northwestern New Mexico. There the air rises, cools, and condenses the water vapor that has been carried hundreds of miles from the tropical waters of the Pacific Ocean and Gulf of Mexico. The liberated water falls to the ground as rain or snow, and so is born one of the great rivers of the Southwest. The San Juan River plunges steeply from its headwaters near Wolf Creek Pass, flows through the fertile valleys and mesa country of southwestern Colorado and northwestern New Mexico, and finally enters a deep, narrow canyon through which it drops to its confluence with the Colorado River in southeastern Utah.

Along the San Juan River's roughly 400-mile route, it picks up water from numerous tributaries, including the Navajo, Piedra, Los Piños, Animas, La Plata, and Mancos rivers, which arise in other parts of the San Juan Mountains. The flow is augmented during spring snowmelt and after local summer rainstorms by many ephemeral streams that drain the semiarid mesas and basins of the lower terrain lying south and west of the high San Juan Mountains, such as Cañon Largo, the Chaco River, Montezuma and Recapture creeks, Chinle Wash, and Comb Wash.

The immense expanse of country lying within the watershed of the San Juan River encompasses an equally immense variety of soils and local climatic conditions. In the depths of the canyon where the San Juan River enters the Colorado River, precipitation is vanishingly scarce and summer temperatures are almost intolerably high; while at the river's source near 3655 m (12000′) the annual precipitation is greater than almost anywhere else in the Southwest and temperatures are comparable to those in northern Alaska (see the Climate section). Within this remarkably diverse landscape is a similarly diverse flora and vegetation, including numerous endemic species and plant communities found nowhere else on earth. Described here are some of the major vegetation patterns, and an attempt to systematically classify the amazing diversity of vegetation communities in this region.

Major Vegetation Patterns

The distribution of individual species and the diversity of species within plant communities are controlled by variation in local climate, soils, and history. The most striking climatic patterns are those related to elevation, as summarized in the section on climate. At the highest elevations, where precipitation is generally abundant, plant growth and species distributions tend to be limited primarily by low temperatures (Spencer and Romme 1996). For example, the upper timberline in the San Juan Mountains, which occurs at about 3505–3655 m (11500–12000′), appears to be a function primarily of temperature. At elevations higher than this, frosts occur in every month, average daily high temperatures are low, and consequently the effective growing season is so short that large woody plants cannot manufacture enough carbohydrates during the short summer to carry them through the long winter. In contrast, at the lower elevations, moisture availability is usually the primary limiting factor for plants (Floyd-Hanna, Spencer, and Romme 1996). The lower timberline, at around 1830–2135 m (6000–7000′), appears to be a function primarily of drought stress. Precipitation at elevations lower than this is inadequate to supply the water needs of trees except where year-round water occurs, such as at springs or along rivers and streams.

For any given elevation, soil moisture is influenced not only by conditions of precipitation and temperature, but also by local soil conditions. Given the remarkable diversity of geological substrates in San Juan River country (see the Geology section), it is not surprising that soils are equally diverse. In many places, trees grow at lower elevations than would be predicted from elevation alone, because of local variation in moisture availability that is related to variation in soil characteristics or other influences. Trees can be found at the very lowest elevations of the San Juan River watershed wherever supplemental water is provided. The most obvious examples of this are along the few perennial streams in the region. A more subtle example is seen in many arid locales where conifer trees thrive on sandstone outcrops but cannot survive in the silty or sandy soils between outcrops. In this situation, the fractured rock outcrops collect water in deep cracks—water that can be accessed by deep-rooted woody plants—and so the sandstone actually represents a comparatively moist environment.

Deep-rooted woody plants, including trees and large shrubs, generally do best where they can tap deep sources of water, as in fractured bedrock or coarse colluvium. These plants cannot survive in shallow soils atop unfractured bedrock or in soils where the sparse precipitation rapidly drains below their rooting depth. In contrast, grasses and other plants with shallow, fibrous root systems thrive in sandy soils, where the extensive root systems can absorb moisture from the upper soil layer before it drains away. Some of the most severe soil conditions are found in low-precipitation areas where the clayey soils are derived from the Lewis, Mancos, Morrison, and Chinle shale formations. These soils swell when they become wet, preventing water infiltration, and they also typically contain high salt and carbonate concentrations that further reduce soil water potentials and exacerbate the already dry conditions. Many of these sites are almost devoid of plant life, but they support a few plants having various remarkable mechanisms for tolerating these extreme environmental conditions. Many of these plants are endemic to this region, and some are listed as threatened or endangered because of their restricted distribution and low population sizes, e.g., *Sclerocactus mesae-verdae* and *Atriplex pleiantha*, which grow only on the nearly barren Mancos Shales lying to the west and south of the Mesa Verde cuesta. Many additional endemic species in genera like *Astragalus*, *Erigeron*, *Eriogonum*, *Aliciella*, *Gilia* (s.l.), and *Phacelia* also occur on these clay barrens. Some of these, such as *Astragalus cutleri*, *Aliciella formosa*, or *Phacelia splendens*, are also rare and potentially threatened by human activities, although many other endemics are widespread and locally abundant.

Vegetation Classification

The classification and taxonomy of vegetation communities is a less well-developed science than that of plant species. Consequently, there is no single, complete, and universally accepted classification for the vegetation communities of the San Juan River drainage. Several classification systems have been applied in this region, and the literature related to the vegetation and flora of the San Juan River country contains references to all of these disparate—and sometimes inconsistent—vegetation classifications. Because no standard or universally acceptable vegetation classification has yet emerged for this region, and because of the paucity of standard composition data, we cannot present a definitive summary here. Nevertheless, the paragraphs below briefly review eight of the major classification systems that have been used.

First, however, we point out that a major reason for the slow development of a universally satisfactory vegetation classification is the fact that vegetation generally does not exist in discrete units that can be readily differentiated from other types. On the contrary, plant community composition and structure tend to change gradually and continuously, reflecting the "continuum" concept of vegetation which most vegetation ecologists now embrace (e.g., Whittaker 1975; McIntosh 1993). An exception to this pattern of continuous change is seen where the underlying environment changes abruptly, as between an outcrop of sandstone and a sand dune, or between a riparian area and the surrounding arid uplands. In such places vegetation classification may be relatively easy. However, as one ascends the San Juan Mountains, from lower elevation to upper timberline, the changes in forest structure and composition are more subtle, with no obvious boundaries between types. Change in species composition along streams that arise at high elevations and descend long distances exhibit especially subtle patterns. Some classifications focus only on the dominant species, which makes classification somewhat easier, but if one also pays attention to the gradual changes in composition of the less dominant species, the continuum nature of vegetation is very apparent.

Four relatively coarse vegetation classification systems have been commonly used in the southwestern United States. Merriam (1890; 1898) described five *life zones* (Lower Sonoran, Upper Sonoran, Transitional, Canadian, and Hudsonian) that reflect the climate and biotic communities at progressively higher elevations, but these five units are extremely broad and variable. Kuchler (1985) identified *potential natural vegetation* (i.e., the plant community that would exist after a very long period without disturbance);

his map recognizes far more diversity than Merriam's system, but it still is a very coarse classification, and a great deal of actual vegetation diversity is subsumed under relatively few types of potential vegetation. The U.S. Forest Service's *ecological land unit* classification (Bailey 1995) integrates several elements of environment and vegetation to derive units representing a shared ecological potential (Grossman et al. 1998: 5). The nationwide Gap Analysis Program (GAP) has classified and mapped vegetation into a series of *national land cover classes* based on structural and floristic characteristics, and the resulting maps are readily available in digital form (www.natureserve.org). The latter two systems, the ecological land units and the national land cover classes, both incorporate far more vegetational detail than either Merriam's or Kuchler's systems but still represent very broad vegetation units.

Three other vegetation classification systems include far more detail than the systems described above: The U.S. Forest Service has classified and mapped *habitat types* (land units capable of supporting similar vegetation after a long period without disturbance), which synthesize a great deal of information about soils, climate, and vegetation structure (DeVelice et al. 1986), but have been developed primarily for forested plant communities and do not apply to most of the arid portions of the San Juan River drainage. The *Brown-Lowe-Pase* system (Brown, Lowe, and Pase 1980; Brown 1982) integrates climate, topography, and vegetation in a hierarchical classification that has been used for National Park units throughout the southwestern U.S. (Rowlands 1994). In response to some problems with the Brown-Lowe-Pase system when applied to the Colorado Plateau, a similarly structured but refined classification, the *Spence-Romme-Floyd-Hanna-Rowlands (SRFR) system* (Rowlands 1994; Spence et al. 1995; Spence 2002), has been developed more recently and is currently being used in some National Park units.

Another major effort is underway to produce a comprehensive, consistent, and detail-rich classification of the terrestrial vegetation of the United States (Grossman et al. 1998). This effort is being spearheaded by The Nature Conservancy, with substantial input from the Ecological Society of America and various governmental land management agencies. Grossman et al. (1998) summarize the conceptual foundations and structure of the U.S. National Vegetation Classification, as it is called, but the characterization of specific vegetation units is still evolving. The current classification system, which will be updated periodically, can be obtained from The Nature Conservancy's website (www.nature.org).

Descriptions of Terrestrial Communities

It is beyond the scope of this section to provide an exhaustive description of all the many vegetation communities that can be found within the immense drainage of the San Juan River. Moreover, many of the distinctive community types have not yet received enough study to allow us to characterize them in much detail. Rather, in this section we briefly describe some of the major vegetation assemblages along an elevational gradient from the highest peaks to the lowest basins and canyons. Much additional research is needed to complete our understanding of the vegetation in this diverse region, and a plenitude of fascinating ecological stories are waiting to be told. Moreover, given the uncertainty as to which vegetation classification system(s) will eventually be used most widely in the San Juan River drainage (and the possibility that no single system will ever see universal application), our approach is to describe the broad patterns of variation in vegetation, in response to variation in elevation and substrate, rather than describe specific plant communities in detail. Readers can consult the literature cited above for details on any of the commonly used classification systems.

The Alpine Zone

At the highest elevations, above upper tree line, is the alpine zone. The Four Corners region is home to a relatively limited extent of continuous alpine landscape, supported only along the crest of the San Juan Mountains. Small patches of alpine ecosystems may also occur as components in other landscapes, intermixed with spruce-fir forests, aspen forests, mountain grasslands, rock outcrops, or talus slopes. Despite their comparatively small total land area, the treeless alpine ecosystems add immensely to the floristic and vegetational diversity of the Four Corners region. Alpine environments are characterized by cool growing seasons of short duration, cold winters of long duration, high wind conditions, and intense light. Such extreme physical environments generally occur above 3505 m (11500′). Diverse geology and glacial topography add to the complexity and diversity of alpine landscapes of the San Juan country. Pleistocene glaciation and other geologic events have formed substrates in the alpine that vary from steep and rugged to gentle and smooth. Rock outcrops and talus slopes are common. Soils are shallow and rocky on steep slopes and exposed ridges, deeper and more productive on other sites. Mean annual precipitation ranges from about 4–11.7 cm (30–50″). Natural disturbances including nivation (erosion related to late-lying snowbanks), frost action, and solifluction (mass soil movement resulting from freeze-thaw action) commonly occur and shape the distribution of biota in alpine ecosystems.

There is tremendous diversity of species and vegetation communities within the alpine zone. Mosses and lichens constitute a significant contribution to the total flora. Community types often change quickly and abruptly over short distances, due to small-scale topographic changes that exert a significant influence on snow and moisture conditions and the associated vegetation. Alpine plants are uniquely adapted to their environment. High photosynthetic efficiency, resistance to drought, fleshy leaves, evergreen leaves, thick cuticles, epidermal hairs, and dark scales are important adaptations. Alpine vascular plants grow slowly, an adaptation to low temperatures, high desiccation rates, and sudden microclimatic changes. Annual plants are almost absent from our alpine flora. Other adaptations include reduction in growth height, perennial life cycle, herbaceous habit, and accumulation of belowground biomass (Thilenius 1975). Perhaps the best-known symbol of alpine plants is the krummholz form of dwarfed *Picea engelmannii* and other conifers. Growing in association with alpine herbaceous species, krummholz is actually a transition type that occurs between spruce-fir forests of the subalpine climate zone and true alpine.

Four general vegetation types exist within the alpine zone of the San Juan Mountains: fell-field, turf, wetland, and dwarf willow (Baker 1983; Paulsen 1975; Thilenius 1975; Dick-Peddie 1993). A rockland or talus type, dominated by *Senecio atratus*, *Senecio soldanella*, and lichens, is also recognized (Dick-Peddie 1993).

The fell-field type occurs on harsh, windswept sites with shallow, rocky soils. It is dominated by short cushion plants (forbs) and displays a relatively low canopy cover. Common species of this type include *Silene acaulis*, *Paronychia pulvinata*, *Trifolium nanum*, *Trifolium dasyphyllum*, *Stellaria irrigua*, *Phlox caespitosa*, and *Eremogone fendleri*.

The turf type occurs on protected sites away from excessive wind and tends to have deeper, moister, and better-developed soils. It is dominated by forbs and graminoids and usually displays a relatively high canopy cover. Within the general turf type, a number of community types can be recognized including *Carex elynoides*, *Kobresia myosuroides*, *Geum rossii*, and *Festuca thurberi*. A combination turf/fell-field community (*Carex rupestris*/cushion plants) is also recognized (Baker 1983; Dick-Peddie 1993).

The wetland type primarily occurs on poorly drained, low-lying sites where water accumulates. It tends to display high cover and includes the following community types: *Deschampsia cespitosa*, *Potentilla fruticosa*, *Juncus drummondii*, herbaceous wetland, and tall willow. The tall willow type is composed of *Salix planifolia* and *Salix glauca* and occurs along riparian corridors that are often connected to lower-elevation subalpine streams. The herbaceous wetland type is variable and displays a great diversity of species including *Pedicularis groenlandica*, *Caltha leptosepala*, *Rhodiola integrifolia*, *Rhodiola rhodantha*, *Primula parryi*, and species of *Carex*.

The dwarf willow type is composed of *Salix arctica* and *Salix reticulata*. It commonly occurs on sites with a heavy snowpack that extends into the summer.

Although alpine landscapes in the San Juan Mountains are generally remote from roads or settlements, human activities nevertheless have had important impacts on the vegetation. Extensive grazing of cattle and sheep began in the 1870s, when Anglo settlers arrived in increasing numbers (Dishman 1982), and continued into the 20th century. The early sheepmen moved their animals into the alpine zone for summer grazing, where the animals were kept in tightly grouped bands and continuously bedded in the same location for several nights in a row. These practices resulted in large losses of forage through trampling and soil damage from excessive trailing to and from bedding grounds to water (Thilenius 1975). This early livestock grazing was unregulated and quickly became excessive in many areas. Range conditions began deteriorating in the 1890s, and the process accelerated over extensive areas between 1904 and 1920 (DuBois 1903; Dishman 1982). Concern for the condition of alpine ranges resulted in reduction of grazing after establishment of national forests in the San Juan Mountains. Regulation of livestock grazing has improved current rangeland conditions, but the legacies of excessive grazing that occurred a century ago are still with us, especially in fragile environments like the alpine zone (Fleischner 1994).

Mining also was a common activity in alpine ecosystems of the San Juan Mountains. Old supply roads, deserted structures, settling ponds, and mine tailings are common mining features still conspicuous today. When these mines were abandoned, especially in the early days, efforts to limit pollution from them were abandoned also (or never initiated), causing major pollution problems. Abandoned mine tailings often contain soils that are highly acidic or laden with toxic concentrations of heavy metals. Erosion, acid-water runoff, and sedimentation from mines result in adverse impacts to vegetation, streams, and aquatic ecosystems, as can be seen today, for example, in the upper Animas River near Silverton, Colorado.

Another important anthropogenic stress on alpine ecosystems is air pollution from power plants, refineries, and urban areas located upwind from the mountains. Potential changes caused by air pollution include reduced growth and increased mortality of lichens, mosses, and zooplankton, as well as changes in soil pH, soil metal concentrations, water pH, alkalinity, metals, and dissolved oxygen (Haddow et al. 1998). High-elevation lakes appear to be especially sensitive to air pollution (Haddow et al. 1998).

The Subalpine Zone

Lying just below the treeless alpine zone is the subalpine zone. Restricted to the highest mountains, at elevations between about 2745–3505 m (9000–11500′), this zone covers a relatively small portion of the Four Corners region. It is most extensive in the San Juan Mountains but also is found at the tops of the Abajo Mountains, Chuska Mountains, and Navajo Mountain. In the subalpine zone of the San Juan Mountains, annual precipitation averages about 30–40 inches, much of which comes in the form of snow.

The most conspicuous vegetation type in the subalpine zone is evergreen coniferous forest, dominated by Engelmann spruce (*Picea engelmannii*) and subalpine/corkbark fir (*Abies bifolia* var. *bifolia* and var. *arizonica*). Spruce-fir forests cover many of the slopes, ridges, and valleys of the high mountains and are quite variable in composition and structure because of underlying environmental variation throughout this extensive habitat (DeVelice et al. 1986). The understory is typically rich with mesophytic shrubs and herbs including *Vaccinium myrtillus*, *Rubus parviflorus*, *Ribes montigenum*, *Sambucus racemosa*, *Lonicera involucrata*, *Erigeron eximius*, *Geranium richardsonii*, *Ligusticum porteri*, *Mertensia ciliata*, *Arnica cordifolia*, *Aquilegia elegantula*, *Pedicularis racemosa*, *Orthilia secunda*, *Artemisia franserioides*, *Viola canadensis*, *Goodyera oblongifolia*, *Fragaria vesca*, *Oreochrysum parryi*, *Lathyrus leucanthus*, *Pyrola minor*, *Maianthemum stellatum*, *Luzula parviflora*, *Zigadenus elegans*, *Elymus canadensis*, *Carex geyeri*, and *Osmorhiza depauperata* (Redders 2003).

Quaking aspen (*Populus tremuloides*) is also abundant in some stands that have been disturbed by fire, wind-throw, or logging within the last 200 years. Intense disturbances of these kinds have created a coarse-grained mosaic of stands of different age, structure, and composition within the subalpine landscape. Recently disturbed stands may be dominated by plant species that thrive in open environments, e.g., *Chamerion angustifolium*, whereas older stands support a shade-tolerant flora that includes rare and/or restricted species such as the orchids *Calypso bulbosa* and *Goodyera oblongifolia*. Recovery following intense disturbance may be very slow in this cold-stressed environment (Stahelin 1943). One example is where a spruce-fir forest burned in 1879 on Molas Pass. At elevations greater than 10,000 feet near the headwaters of the Animas River, much of the burned area is still dominated today by herbaceous vegetation, with scattered patches of young spruce-fir or aspen forests. At somewhat lower elevations, but still within the subalpine zone, reestablishment of forest vegetation has been more rapid, and areas that burned in 1851 and 1879 now support stands of mature aspen, spruce, and/or fir (Romme, Floyd, and Hanna 2003a).

Although the effects of forest fires remain conspicuous for decades or centuries, fires actually are not very frequent in the subalpine zone, because of the cool, moist conditions that prevail most of the time. Large fires in this zone only occur during summers of prolonged drought, as occurred in 1748, 1851, 1879, and 2002 (e.g., the Missionary Ridge fire in the San Juan Mountains). Many spruce-fir stands in the San Juan Mountains have escaped fire or other intense disturbance for hundreds of years and are among the oldest forests in the Southwest. For example, the forest growing in a moist topographic depression at the head of Martinez Creek in the eastern San Juan Mountains shows no evidence of major disturbance in the last 600+ years (Romme, Floyd, and Hanna 2003a).

Nonforest communities also are important components of the subalpine zone, as well as lower-elevation zones (Redders 2003). Mountain grasslands, also known as parks (larger) or meadows (smaller), occur on upland landforms with well-drained soils, at elevations ranging from about 2285–3505 m (7500–11500′). They often are found on sites that could potentially support forests or woodlands, and the lack of trees may be due to a variety of causes, including competition from herbaceous plants, low temperatures, heavy grazing, soil heaving associated with cold temperatures, and severe fires in the past that locally eliminated the tree seed source or altered soil conditions (Paulsen 1975). The most common mountain grassland type is dominated by *Festuca thurberi* (Redders 2003).

Most of the grasslands in the mountain ranges of the San Juan River drainage have been altered by heavy livestock grazing of the last century and a half. This includes high-elevation grasslands as well as montane and low-elevation meadows and grasslands. Large herds of cattle and sheep were introduced into the San Juan Mountain country in the 1870s and 1880s and foraged with very little regulation until the 1930s (Romme, Floyd, and Hanna 2003a). Sheep grazing began earlier in other parts of the San Juan River drainage, e.g. in the early 1800s in the Chuska Mountains (Savage and Swetnam 1990: 36). Because the composition of southwestern grasslands before the arrival of livestock was undocumented, it will never be known with certainty just what the pre-1900 flora was or how it changed as a result of grazing. However, composition and structure of at least some (and likely many) southwestern grasslands were altered substantially by grazing in the late 19th and early 20th centuries (Fleischner 1994), when livestock removed the more palatable herbaceous species and left the unpalatable, more resistant species. Many grasslands probably have not recovered their original composition, even with regulated grazing or even absence of grazing during the last half-century (Redders 2003).

The Montane Zone

Ponderosa pine, mixed conifer, and aspen forests are common between about 2075 and 2745 m (6800 and 9000′) elevation on the mountains and mesas of the San Juan River country. Ponderosa pine forests are dominated by a canopy of *Pinus ponderosa*, often with a well-developed understory of *Quercus gambelii*. Mixed conifer forests have a complex structure and composition that varies substantially from place to place throughout the region, in response to variation in soils, microclimate, and disturbance history (Redders 2003). Mature stands may be dominated by a variable mix of white fir (*Abies concolor*), Douglas-fir (*Pseudotsuga menziesii*), ponderosa pine, Engelmann spruce, subalpine/corkbark fir, and quaking aspen. Blue spruce (*Picea pungens*) and southwestern white pine (*Pinus strobiformis*) also occur in some stands. Aspen tends to be more abundant where fire, timber harvest, or other disturbances have occurred within the past two centuries.

Understory species composition is as variable as tree species composition. Some of the shrubs and herbs that are commonly associated with mixed conifer forests include *Quercus gambelii*, *Arctostaphylos uva-ursi*, *Vaccinium myrtillus*, *Lonicera involucrata*, *Sambucus racemosa*, *Amelanchier alnifolia*, *Symphoricarpos rotundifolius*, *Mahonia repens*, *Rubus parviflorus*, *Delphinium nelsonii*, *Mertensia fusiformis*, *Arnica cordifolia*, *Erythronium grandiflorum*, *Geranium richardsonii*, *Pedicularis racemosa*, *Erigeron formosissimus*, *Potentilla hippiana*, *Solidago simplex*, *Geranium caespitosum*, *Antennaria rosea*, *Cymopterus lemmonii*, *Lathyrus leucanthus*, *Achillea lanulosa*, *Erigeron eximius*, *Ligusticum porteri*, *Osmorhiza depauperata*, *Mertensia ciliata*, *Aquilegia elegantula*, *Orthilia secunda*, *Artemisia franserioides*, *Viola canadensis*, *Goodyera oblongifolia*, *Fragaria vesca*, *Oreochrysum parryi*, *Pyrola minor*, *Actaea rubra*, *Maianthemum stellatum*, *Luzula parviflora*, *Elymus canadensis*, *Koeleria macranthera*, *Elymus longifolius*, *Poa fendleriana*, *Poa pratensis*, and *Carex geyeri*.

Interspersed among the forest stands are montane grasslands similar to those discussed in the section on the subalpine zone. The most common grassland type in the montane zone is dominated by *Festuca arizonica* (Redders 2003). Extensive areas also are covered by mountain shrubland, also known as Petran chaparral (Keeley and Keeley 1988), which is remarkably diverse and variable in composition. Common dominants include *Quercus gambelii*, *Amelanchier utahensis*, *Symphoricarpos* species, and *Prunus virginiana.* The shrublands are found on many sites where soils and local climate appear suitable for forest, and there is uncertainty as to why these sites do not support forest (Floyd-Hanna, Spencer, and Romme 1996). The most likely explanation is that chronic disturbance over the last several centuries has continually removed the trees and maintained the disturbance-tolerant shrubs. For example, recurrent fire has maintained oak and serviceberry-dominated shrublands on most of the upper portion of Mesa Verde [above about 2285 m (7500′)]. The shrubs resprout readily after fires, which have recurred on average every 100 years during the last several centuries, but the Douglas-fir, piñon (*Pinus edulis*), Utah juniper (*Juniperus osteosperma*), and Rocky Mountain juniper (*J. scopulorum*) are killed and reestablish only very slowly after fire (Floyd et al. 2000). Although fire may be the most common type of disturbance that maintains mountain shrublands where forests could potentially grow, other important disturbances include soil movement and snow-creep on steep slopes (Floyd-Hanna, Spencer, and Romme 1996).

Montane forests, shrublands, and grasslands of the San Juan country have been strongly influenced by land use practices since the late 1800s, such that many stands have very different structure and composition today than before 1880. Formerly frequent fires in ponderosa pine and mixed conifer forests (Wu 1999; Grissino-Mayer et al. 2004) were essentially eliminated as large, unregulated herds of sheep and cattle removed the fine fuels that once carried low-severity fires over extensive areas. Timber harvest in most of the readily accessible areas removed many of the large Douglas-fir, ponderosa pine, and Engelmann spruce (Lynch, Romme, and Floyd 2000), and 20th-century fire exclusion has allowed white fir to increase substantially on many sites.

Arid/Semiarid Basins and Foothills

Moving down and out of the mountains, one enters the extensive arid and semiarid basins of the San Juan River country, a region of mesas, canyons, and broad expanses of rolling terrain. The unifying theme here is low precipitation. The most widespread vegetation types are woodlands, low shrubland-grasslands, and barrens. Less extensive, but of enormous significance to biodiversity, are a variety of riparian and wetland communities.

One of the most widespread forms of vegetation in the Four Corners region is woodland dominated by piñon and juniper (West 1988; Floyd 2003). Though sometimes regarded as a single "type," piñon-juniper woodland exhibits amazing diversity in composition and structure—in large part because of the wide range of environmental conditions in which it occurs. *Pinus edulis* typically grows with *Juniperus osteosperma* and *J. scopulorum* at higher elevations or on mesic sites, and with only *J. osteosperma* at lower elevations. Nearly pure stands of piñon or of juniper may be found at the highest (up to 2745 m or 9000′) or lowest elevations (down to 1220 m or 4000′), respectively. Depending on local moisture availability and fire history, piñon and juniper may form open woodlands or dense stands of closely spaced trees (Romme, Floyd-Hanna, and Hanna 2003b). The open woodland structure generally is associated with very dry sites that cannot support many trees, or sites recovering from past fires. The dense woodland structure is found in places where productivity is relatively high but fires are very infrequent because of limited fuel continuity and natural barriers to fire spread (e.g., cliffs and poorly vegetated slopes). The woodlands in these latter kinds of places may be very old. For instance, some of the piñon-juniper woodlands at the southern end of Mesa Verde had not suffered any major disturbance since the Ancestral Puebloan people abandoned the area 700 years ago (Floyd, Romme, and Hanna 2000; Floyd, Hanna, and Romme 2004) until extensive, drought-induced mortality (either directly or indirectly via *Ips confusus* beetles) occurred from 2002–2004.

Associated with piñon and juniper woodland is a great variety of shrubs and herbs. Local composition varies with geological substrate, local climate, canopy density, and disturbance history. Common shrubs, for example on Mesa Verde, include *Purshia tridentata*, *Quercus gambelii*, *Amelanchier utahensis*, *Artemisia tridentata*, *Artemisia nova*, *Peraphyllum ramosissimum*, *Fendlera rupicola*, and *Yucca baccata*. Common forbs on Mesa Verde include *Penstemon linarioides*, *Pedicularis centranthera*, *Petradoria pumila*, *Cryptantha bakeri*,

Eriogonum racemosum, *Lupinus polyphyllus* var. *ammophilus*, *Astragalus scopulorum*, *Calochortus nuttallii*, *Commandra umbellata*, *Cymopterus bulbosus*, *Cymopterus purpurascens*, *Lomatium triternatum*, *Lomatium grayii*, and *Dieteria bigelovii*. Grasses are abundant in some places, but not in others, depending on local soil conditions and history of grazing and fire. *Poa fendleriana* may attain 80% cover in some areas; other important grasses include *Achnatherum hymenoides*, *Elymus longifolius*, *Elymus smithii*, *Koeleria macranthera*, *Achnatherum lettermanii*, *Hesperostipa comata*, and *Elymus trachycaulus* (Floyd 2003).

Also widespread at middle and low elevations are grasslands or grasslands accompanied by low shrubs such as *Tetradymia canescens*, *Atriplex canescens*, *Artemisia nova*, *Ephedra torreyana*, *Ephedra viridis*, and *Krascheninnikovia lanata*. Dominant grasses include *Bouteloua gracilis*, *Achnatherum hymenoides*, *Pleuraphis jamesii*, *Aristida purpurea*, *Sporobolus airoides*, *Sporobolus cryptandrus*, and, in the spring months, *Vulpia octoflora*. After wet winters these sandy grass-shrub communities support a rich and varied wildflower flora, including for example in Chaco Canyon *Astragalus mollissimus*, *Cryptantha micrantha*, *Cryptantha bakeri*, *Cryptantha flava*, *Eriogonum leptophyllum*, *Eriogonum umbellatum*, *Ipomopsis longiflora*, *Lupinus pusillus*, *Penstemon angustifolius*, *Corydalis aurea*, *Sphaeralcea fendleri*, and *Sphaeralcea coccinea*. Other distinctive perennial species in the region include *Hymenopappus filifolius*, *Rumex hymenosepalus*, and the annual, *Streptanthella longirostris*.

Chronic heavy grazing in many places has reduced or, in some cases, largely eliminated these native forbs and grasses, leading to their replacement by grazing-tolerant native shrubs and succulents (e.g., *Gutierrezia sarothrae*, *Artemisia filifolia*, *Opuntia polyacantha*, and species of *Yucca*), as well as the native annual *Machaeranthera tanacetifolia* and several nonnative annuals (e.g., *Bromus tectorum*, *Salsola tragus*, and *Schismus arabicus*). Today, very few undisturbed stands of grasslands and shrub-steppe survive to provide some indication of pre-European settlement conditions. However, Chaco Culture National Historic Park was fenced in the early 1940s (then Chaco Canyon National Monument), and the resulting vegetation recovery has allowed a glimpse into the biodiversity possible in grazing-free shrublands and grasslands. When protected from grazing, sites with deep sandy soils, such as on the floor of Chaco Canyon, become dense grasslands with *Sporobolus airoides* and *Pleuraphis jamesii* gaining in dominance. *Erigeron* and *Symphyotrichum* species are also common. Upper slopes on Cliffhouse sandstones become more heavily dominated by *Artemisia nova* and *Tetradymia canescens* and other shrubs. In all habitats, release from grazing has been accompanied by an increase in the biotic crust cover and increases in biodiversity. Many of the plant species are either endemic to the Colorado Plateau or are widespread species characteristic of adjacent warm-temperate deserts. Archaeological sites commonly support vigorous stands of wolfberry (*Lycium pallidum*).

Greasewood (*Sarcobatus vermiculatus*) and big sagebrush (*Artemisia tridentata*) often dominate low-lying areas on clay soils, or very deep sandy soils, for example adjacent to Chaco Wash in Chaco Culture National Historic Park. These areas occasionally support standing water for parts of the year or are perched on top of alluvial aquifers. Other shrub species include *Artemisia dracunculus*, *Chrysothamnus viscidiflorus*, and *Ericameria nauseosa*. Important forbs include *Solanum elaeagnifolium*, *Helianthus annuus*, *Abronia fragrans*, and *Mirabilis multiflora*.

In shallower and rockier soils, or in soils with high clay content, one of a variety of low-growing xerophytic shrubs may dominate large areas. The two most widespread of these shrubs are blackbrush (*Coleogyne ramosissima*) on rockier shallow soils and shadscale (*Atriplex confertifolia*) in areas with more abundant clays and carbonate-rich sediments. Blackbrush vegetation tends to be species-poor, with only a few forbs and grasses common among the dominant shrubs, including *Achnatherum hymenoides*, *Pleuraphis jamesii*, *Malacothrix sonchoides*, *Plantago patagonica*, and *Vulpia octoflora*. Shadscale, accompanied by *Atriplex gardneri*, *Atriplex obovata*, *Ephedra torreyana*, and other low shrubs, tends to support many more herbaceous species, including various monocots such as *Androstephium breviflorum*, *Calochortus aureus*, *Calochortus nuttallii*, *Calochortus flexuosus*, and dicot forbs like *Orobanche fasciculata*, *Castilleja chromosa*, *Chamaesyce fendleri*, *Halogeton glomerata*, *Mentzelia albicaulis*, *Solanum elaeagnifolium*, and *Dieteria* (*Machaeranthera*) species.

Much of the low-elevation terrain in the San Juan River country is so dry or has such poorly developed soils that plant cover is extremely low and the areas are best characterized as "barrens." One type of barren is found on large exposed portions of sandstone or igneous bedrock, commonly referred to as "slickrock." Plants can grow only in the cracks and crevices of the rock, where traces of soil have accumulated and where roots can reach deep water sources by growing through the cracks in the bedrock. A variety of shrubs and deep-rooted herbs are commonly found in these sites, e.g., *Cercocarpus intricatus* and *Tetraneuris acaulis*. Although precipitation is very low in most of the places where slickrock barrens occur, nearly all of the rain and snow that does fall runs off of the impermeable rock and accumulates in cracks, crevices, and depressions. In some places, natural depressions in the slickrock accumulate enough water to support localized riparian communities of willows, rushes, and other riparian obligate species. These waterpockets or "tinajas" are very rare, and they provide critical water sources for a variety of wildlife species, from mountain sheep to frogs. Characteristic plant species of these tinajas include *Salix gooddingii*, *Typha latifolia*, and various *Juncus* species (Spence and Henderson 1993).

Another important type of barren is found on extensive outcrops of shale or bentonite of the Mancos or Morrison formations, where precipitation is extremely low (Potter, Reynolds, and Louderbough 1985a, b). The combination of low precipitation, water repellent soils, and high solute concentrations leads to an environment where very few plant species are capable of surviving. Total plant cover in these shale or bentonite barrens may be less than 10%. Some of the species that can tolerate such harsh conditions include *Atriplex corrugata* and *A. gardneri* var. *cuneata*, as well as a number of rare, endemic species such as *Sclerocactus mesae-verdae* and *Atriplex pleiantha*. The presence of *Stanleya pinnata* on Mancos Shales suggests selenium in the substrate. In Mesa Verde, Mancos Shales also support a wide variety of interesting *Eriogonum* species (*E. alatum*, *E. umbellatum*, *E. leptophyllum*), many *Astragalus* species (*A. pattersonii*, *A. praelongus*, *A. wingatanus*), as well as *Aliciella haydenii*, *Gilia sinuata*, and *Symphyotrichum falcatum*. Menefee Shales in Chaco Canyon support very sparse populations of the shrubs *Ephedra torreyana*, *Gutierrezia microcephala*, and *Chrysothamnus viscidiflorus*, plus forbs such as species of *Erigeron*, *Corydalis aurea*, and *Halogeton glomeratus*.

Riparian and Wetland Vegetation

Where water is perennially available, the vegetation is strikingly different from surrounding areas where water is limited during at least a portion of the growing season. Riparian vegetation along rivers and streams exhibits enormous variety from headwaters to confluence with other streams. A few of the more common community types, from higher to lower elevations, are highlighted below, followed by a description of an uncommon but especially pleasing wetland type—the hanging garden.

Evergreen riparian forests are common in the mountains at elevations of about 2440–3050 m (8000–10000′). Subalpine fir, Engelmann spruce, and blue spruce trees occur along the stream channel and throughout the valley floor and commonly display a relatively open and often patchy canopy cover. Common shrubs and herbs include *Alnus incana* subsp. *tenuifolia*, *Salix drummondiana*, *Lonicera involucrata*, *Cornus sericea*, *Cardamine cordifolia*, *Heracleum lanatum*, *Mertensia ciliata*, *Oxypolis fendleri*, *Saxifraga odontoloma*, *Streptopus amplexifolius*, *Equisetum arvense*, *Senecio triangularis*, *Rudbeckia ampla*, *Geranium caespitosum*, and *Calamagrostis canadensis*.

Deciduous riparian forests are common in the mountains at elevations of about 1981–2896 m (6500–9500′). Cottonwood (mostly *Populus angustifolia*) and boxelder (*Acer negundo*) trees occur along the stream channel and throughout the valley floor and commonly display a relatively open and often patchy canopy cover. Common shrubs and herbs include *Alnus incana* subsp. *tenuifolia*, *Lonicera involucrata*, *Cornus sericea*, several species of *Salix*, *Acer glabrum*, *Rosa woodsii*, *Amelanchier alnifolia*, *Crataegus rivularis*, *Cardamine cordifolia*, *Heracleum lanatum*, *Mertensia franciscana*, *Oxypolis fendleri*, *Saxifraga odontoloma*, *Streptopus amplexifolius*, *Equisetum arvense*, *Rudbeckia ampla*, *Geranium caespitosum*, and *Poa pratensis*.

Deciduous riparian forests also extend beyond the mountains, along major rivers and streams flowing through the arid and semiarid basins of the Four Corners region. These forests are dominated by the Rio Grande cottonwood (*Populus deltoides* subsp. *wislizeni*), several species of *Salix*, *Elaeagnus angustifolia*, *Forestiera pubescens*, *Rhus trilobata*, *Tamarix chinensis*, *Juncus arcticus*, *Muhlenbergia asperifolia*, *Equisetum hyemale*, *Euthamia occidentalis*, *Scirpus microcarpus*, *Typha domingensis*, *Typha latifolia*, and *Phragmites australis.*

A wide variety of deciduous riparian shrublands are also found throughout the area. Several species of *Salix* are common dominants above about 2895 m (9500′). *Alnus incana* subsp. *tenuifolia* and *Salix* species occur commonly at middle to high elevations above about 2440 m (8000′). *Cornus sericea*, *Crataegus rivularis*, *Betula occidentalis*, and *Salix* species occur at elevations from about 2135–2745 m (7000–9000′). Other willows and *Tamarix chinensis* are major species in the low-elevation basins. Associated graminoids and forbs are equally diverse. Common herbs associated with the deciduous shrubland at higher elevations include *Caltha leptosepala*, *Cardamine cordifolia*, *Heracleum lanatum*, *Mertensia ciliata*, *Oxypolis fendleri*, *Saxifraga odontoloma*, *Equisetum arvense*, *Senecio triangularis*, *Carex aquatilis*, *Deschampsia cespitosa*, and *Calamagrostis canadensis*. Common herbs at lower elevations include *Juncus arcticus*, *Muhlenbergia asperifolia*, *Equisetum hyemale*, *Euthamia occidentalis*, *Scirpus microcarpus*, *Typha domingensis*, *Typha latifolia*, and *Phragmites australis*.

Some of the most botanically interesting and aesthetically pleasing wetland communities in this region are the hanging gardens of the low-elevation slickrock canyons. After traveling for miles through a generally arid landscape, one suddenly comes upon a small spring or seep supporting a lush growth of ferns, orchids, and other delicate herbs. Hanging gardens and other types of springs are most commonly found in certain geological strata at lower elevations, especially where coarse-grained permeable sandstones such as the Navajo and Cedar Mesa formations overlie and contact more fine-grained fluvial and mudstone deposits like the Kayenta Formation. Springs in general support a diverse regional element of the flora, despite their small areal extent. This is especially true of hanging gardens, which support numerous relictual and disjunct species (Welsh 1989: 36). In addition to the characteristic maidenhair fern, *Adiantum capillus-veneris*, several distinctive Colorado Plateau endemics occur, including *Aquilegia micrantha*, *Cirsium rydbergii*, *Cirsium chellyense*, *Mimulus eastwoodiae*, *Platanthera zothecina*, *Primula specuicola*, *Zigadenus vaginatus*, *Carex curatorum*, and the listed endangered Navajo sedge (*Carex specuicola*). Other species are more representative of prairie vegetation or southern and eastern woodlands, e.g., *Cercis occidentalis*, *Ostrya knowltonii*, *Sorghastrum nutans*, *Panicum virgatum*, and *Andropogon glomeratus*. Even a few boreal-temperate species can be found in hanging gardens, such as *Calamagrostis scopulorum*, *Epipactis gigantea*, *Petrophytum caespitosum*, and *Rosa woodsii*.

Because of the high demand for water in the Southwest, many stream and spring communities have been severely altered or in some cases eliminated by human activities. Dams, such as Navajo Dam, alter river dynamics downstream, primarily by reducing spring flooding and scouring. Unfortunately, many riparian zones in the region are severely degraded by livestock grazing, water pumping, water diversions, and invasion of exotic species. For example, most accessible springs have been impacted by water development and domestic livestock grazing activities, and the exotic saltcedar (*Tamarix* spp.) and Russian olive (*Elaeagnus angustifolia*) have largely replaced the native trees and shrubs in many areas. Nevertheless, intact native riparian communities are critically important for wildlife and contribute greatly to the floristic diversity of the region.

The Four Corners region has several patterns of plant influences with two or more migration pathways. River systems serve as routes for plant movement, for example, from low, hot deserts upward. *Cercis occidentalis* (redbud) is an example, and the pattern can be designated as the "redbud pathway" and includes several other Mojave Desert plants (Welsh et al. 2003) such as *Acamptopappus sphaerocephalus, Brickellia atractyloides, Dyssodia acerosa, Cryptantha nevadensis, Pectocarya heterocarpa, Cladium mariscus, Acleisanthes diffusa, Camissonia parvula, Chylismia parryi, Aliciella latifolia*, and *Chorizanthe brevicornu*. This migration route follows up Glen Canyon and along the San Juan arm of Lake Powell and its tributaries.

Another migration route is the shortgrass prairie community that is found in the Bisti region and extends north to Farmington, New Mexico. Extended distributions of *Corynopuntia clavata, Cylindropuntia imbricata, Dalea cylindriceps*, and *Engelmannia peristenia* occur here. The Bisti region also has an Ice Age relict community of ponderosa pine.

New Mexico's Largo Canyon contains plants from higher elevations. Plants that are typical of higher elevations such as aspen and Douglas-fir find habitat in the upper reaches along steep canyon walls and tributaries where they find ample shade and moisture.

An interesting relict plant community is located north of Durango near Haviland Lake at 2470 m (8150′). Apparently, this is a location of cold air drainage where alpine and subalpine plants are found, such as *Thalictrum alpinum, Spiranthes romanzoffiana, Carex viridula, Pyrola minor*, and disjunct populations of *Salix serissima* and *Carex scirpoidea* var. *scirpoidea*.

Habitat for aquatics is found along rivers, streams, reservoirs, lakes, and canals. Birds, especially ducks, and mammals (including humans), have certainly facilitated the distribution of many aquatics. Bolack Lake is a major stopping site during the migration of ducks and geese. Morgan Lake, as well as Navajo and Williams reservoirs, are favorite boating areas that have experienced the spread of aquatic plants such as *Potamogeton* spp., *Najas marina*, and *Elodea*.

The hanging garden plant *Primula specuicola* (cave primrose) apparently has affinities with boreal species (Welsh et al. 2003). This is also true for *Platanthera zothecina* (alcove bog orchid) and *Zigadenus vaginatus* (sheathed death camas).

The colonization of high-elevation plants from the Rocky Mountain corridor is to be found in the volcanic laccoliths of the Four Corners. These laccoliths include the La Plata, Ute, Abajo, Carrizo, and Navajo mountains. The common corkbark fir of the San Juan Mountains is also found in the Abajo and Chuska mountains. The common white pine of the San Juan Mountains and the limber pine of the Great Basin are both found on Navajo Mountain. At 4033 m (13232′), the La Plata Mountains are the highest of the Four Corners volcanic laccoliths and contain many of the Rocky Mountain affinities. The Carrizo and Ute mountains are the most xeric of the laccoliths. However, they also contain many Rocky Mountain plant species.

NONNATIVE PLANTS

Donald L. Hazlett

Many of the familiar plants in our flora are introductions from distant lands, often from Europe or Asia, that have become established among our native plants. Reasons for their arrival include accidental contaminants in seeds of crop plants, as tended ornamentals that eventually escaped to "grow wild," as medicinal herbs, and as stowaways in animal fur, airplanes, baggage, and international shipments. Most, but not all of these nonnative plant species are from other continents. A few species are from other parts of North America and are plants that did not naturally occur in the vegetation types of our area. Two of these North American species are catalpa (*Catalpa speciosa*), native further east, and biennial wormwood (*Artemisia biennis*), a riparian wormwood that is thought to be native to the Pacific Northwest. Native Americans were among the first to recognize "different" plants in the mix of native plants. They were also correct in associating the presence of these plants with European settlers. For example, an English translation of an indigenous name for tumbleweed (*Salsola*) is "white man's weed." In like fashion, *Plantago major* was referred to by Native Americans by a common name that in English means "white man's foot." As keen observers of nature, Native Americans were well aware that certain plants and insects (i.e., honeybees were called "white man's fly") were not part of the historical mix of plants and animals in their homeland.

Botanists refer to plant species that were not originally a part of the local flora 300 or 400 years ago as exotic, alien, invasive, adventive, naturalized, or simply as nonnative plants. Nonnative plants account for about 12% of the entire Four Corners flora. An inventory of the number of "nonnative" plants that occur among the nearly 2300 plant species occurring in the Four Corners area determined that about 250 species can be labeled as nonnative plants. These species are in 37 plant families—there are 120 plant families in the entire flora—but more than half of these nonnatives are from only three plant families. These high-weed families, and the percentages of nonnatives in them, are Poaceae: 25%, Asteraceae: 16%, and Brassicaceae: 11%. Of the 37 "weed" families that have 10 or more nonnatives, Amaranthaceae has 8% and Fabaceae has 6%.

The geographic distribution, relative abundance, and economic impact of the 250 or so exotic plant species reported in the Four Corners flora are variable. Of the nearly 250 nonnative plants in this flora, 75% occur in Colorado, 75% in New Mexico, 61% in Arizona, and 51% in Utah. These statistics suggest that more appropriate nonnative plant habitat exists (perhaps due to more wetlands) in Colorado and New Mexico than in Arizona and Utah. At least half of the nonnatives occur in each of the four states. Besides the number of species per state, another way to look at nonnative plant distributions is by the number of Four Corners states where each has been collected. By this measure, the proportion of the entire exotic plant inventory that occur in four, three, two, or in only one state, respectively, are 40%, 15%, 15%, and 30%. The 40% of exotic plants that occur in all four states includes ubiquitous and notorious noxious weeds. Among these are a few hardy annuals that flourish even in very dry areas, such as *Halogeton glomeratus* (halogeton), *Salsola tragus* (tumbleweed), and *Erodium cicutarium* (crane's bill) in desert scrubland vegetation. Most of the exotics that occur in all four states, however, are plants of mesic habitats. Among the most notorious of these, all of which have invasive tendencies, are *Acroptilon repens* (Russian thistle), *Bromus inermis* (smooth brome), *Carduus nutans* (musk thistle), *Centaurea diffusa* (diffuse knapweed), *Cirsium arvense* (Canadian thistle), *Elaeagnus angustifolia* (Russian olive), *Linaria dalmatica* (dalmation toadflax), *Malcolmia africana* (African mustard), *Melilotus albus* (white sweet clover), *Melilotus officinalis* (yellow sweet clover), and *Tamarix chinensis* (saltcedar). Among the well-known invasive plant species that do not occur in all four states are *Alhagi maurorum* (camel thorn) [ARIZ, NMEX, and UTAH only], *Centaurea solstitialis* (star thistle) [NMEX only], *Centaurea maculosa* (spotted knapweed) [COLO, NMEX, and UTAH], *Euphorbia esula* (leafy spurge) [COLO, NMEX only], and *Peganum harmala* (African rue) [NMEX only].

A subset of exotic plant species that occur in all four states of the study area are species that have become so well adapted to human disturbances that the best we can do to is to control them from year to year or learn to live with them. These hard-to-eradicate, familiar exotics include *Bromus japonicus* (Japanese brome), *Bromus tectorum* (cheatgrass), *Convolvulus arvensis* (bindweed), *Descurainia sophia* (tansy mustard), *Lactuca serriola* (prickly lettuce), *Portulaca oleracea* (purslane), *Sonchus oleraceus* (sow thistle), *Taraxacum officinale* (dandelion), *Tragopogon dubius* (goatsbeard), and *Tribulus terrestris* (puncture vine).

Noxious Weeds

Each of the Four Corners states has state agencies or departments that identify plants as noxious weeds. The mix of species on noxious weed lists varies both by state and over time. Noxious weed lists may change over time due to the arrival and establishment of additional species or because an established nonnative plant suddenly becomes a more aggressive colonizer. In 2006 the number of plant species officially identified as noxious by the Four Corner states, respectively, were 71 for Colorado, 48 for Arizona, 32 for New Mexico, and 18 for Utah.

The criteria used by each of the four states to designate certain plant species as noxious are variable. Colorado and New Mexico have noxious weed lists that are divided into three categories. Colorado calls these categories A, B, and C lists while New Mexico calls these three categories Class A, Class B, and Class C weeds. In both states the A categories are the most noxious weeds. Otherwise, these lists have differences. For example, a criterion for listing plant species on the Colorado A list is to include the most "feared" plant taxa, plants that are known to be aggressive but that currently are absent or rare in Colorado. The strategy is to detect and extirpate soon after they become established. Plant species on the Colorado B list are required to have state-sanctioned weed management plans. Finally, the Colorado C list includes weeds that local jurisdictions can choose to eradicate, control, or ignore.

As one moves from list A to C on both the Colorado and New Mexico lists, the proportion of the listed plants that actually occurs in a state increases. For Colorado, the number of noxious plants present in the state on lists A, B, and C, respectively, are 17, 40, and 14 or about 30%, 60%, and 90%. In New Mexico the complementary percentages of listed plants present in the state are, for classes A, B, and C, 60%, 75%, and 100%. The only plant on the Colorado A noxious list that occurs in two or more other states is *Alhagi maurorum* (syn. *A. pseudalhagi*; camel thorn), which occurs in all states except Colorado. In contrast, the New Mexico class A category of noxious weeds includes five plants that occur in all four states. These five are *Centaurea diffusa* (diffuse knapweed), *Cirsium arvense* (Canada thistle), *Lepidium latifolium* (perennial pepperweed or tall whitetop), and *Linaria dalmatica* (syn. *L. genistifolia* subsp. *dalmatica*; dalmation toadflax). For Colorado these same five species are on the B, not the A list. This difference illustrates the variable criteria used in Colorado and New Mexico in regard to the designation of a plant to a particular category.

In Utah, only 18 taxa are designated as noxious weeds (Commissioner of Agriculture and Food, Section 4-17-3). With the smallest list of noxious weeds of the four states, it is not surprising that nearly 80% of the noxious plants listed for Utah occur in the Four Corners flora. One notorious nonnative plant species that is on the Utah as well as the Colorado and New Mexico noxious weed lists is *Lythrum salicaria* (purple loosestrife). This nonnative has yet to be collected in the Four Corners area. Another notorious nonnative that is absent from both the Utah and the Arizona noxious weed lists, but of great concern in Colorado and New Mexico, is *Tamarix chinensis* (saltcedar).

About half of the 48 species on the Arizona noxious weed list (Arizona Department of Agriculture) occur in the Four Corners flora. The Arizona list, like the Utah list, is not subdivided into categories as in Colorado and New Mexico. The Arizona noxious weed list illustrates another of the variable criteria used to include a plant species on a noxious weed list. In Arizona the noxious plant list includes several native plant species. Two of these are *Helianthus ciliaris* (Texas blueweed) and *Solanum carolinense* (Carolina horsenettle). Texas blueweed occurs in the Four Corners flora, but because it is native to the area it was not tallied in the inventory of nonnative plants. It was probably included as a noxious weed in Arizona because it has aggressive perennial rhizomes that allow it to become a weed problem in some areas. Likewise, *Solanum carolinense*, an aggressive perennial and poisonous plant, is native to vegetation not too distant to the Four Corners area. Texas blueweed, Carolina horsenettle, and other native plant species are able to thrive in disturbed habitats, such as fallow cropland. This makes these displaced native plants fair game for inclusion on undesirable plant lists. A few other native plants that sometimes find their way onto noxious weed lists are *Iva axillaris* (poverty weed), *Ambrosia tomentosa* (perennial bursage), and *Picradeniopsis oppositifolia* (Plains bahia).

What Is a Weed?

"Weed" is one of those well-known but unscientific terms. Because "weed" is a common term, nearly everyone has tried their hand at a definition. Included here are definitions from folklore, farmers, ranchers, and laymen, as well as definitions associated with different scientific disciplines. An effort was made to focus each definition as narrowly as possible. To better understand personal concepts of "weed," biased or not, consider the following perspectives. The first perspective is a definition of an "undesirable plant species" used in the Undesirable Plant Management Act of 1990. This definition avoids the words "weed" and "native."

National Undesirable Plant Management Act of 1990: "Undesirable Plant Species means plant species that are classified as undesirable, noxious, exotic, injurious, or poisonous, pursuant to State or Federal law." This definition allows for native plants to be undesirable plant species.

Farmer Definition #1. Any plant (native or exotic) that competes with an agricultural plant to the extent that crop production is impacted and economic losses are incurred.

Farmer Definition #2. Any plant (native or exotic) that is "out of place," i.e., growing where it is not wanted. This includes "volunteer" crop plants in a cultivated field of a different crop (i.e., corn in a bean field).

Farmer Definition #3. Plants (native or exotic) that clog irrigation ditches and roadsides to the extent that they must be burnt every fall, winter, and especially on windy days during the spring.

Rancher Definition #1. Any plant (native or exotic) that competes with palatable grass or forage production—one that translates to a decline in "red meat" production.

Rancher Definition #2. Any plant (native or exotic) that can cause death, illness, or injury to livestock, a plant that is in some way detrimental or poisonous to livestock.

Nonbiologist Definition #1. Any plant (native or exotic) that is in the weedy plant family.

Nonbiologist Definition #2. A colorful and/or abundant wildflower (often an exotic) that looks nice in bouquets.

Landowner. Any plant (often an exotic) that reduces the aesthetic or monetary value of real estate.

Weed Specialist. A plant species (native or exotic) that can be controlled with a judicious combination of training workshops, equipment, manual labor, herbicides, and funding.

Hard-line Environmentalist. Exotic plants that occur in native plant associations. Exotic plants in a native plant community are "unnatural," and therefore bad. A native plant is one that grew in this part of North America before 1492 (to pick a convenient date).

Some Scientific Perspectives on Weeds

Entomologist. Weeds are unwanted host plants for certain insect species that damage crops.

Range Scientist. In terms of forage, exotic weeds are often "increaser/poor" species, they increase with overgrazing and provide poor forage.

Ornithologist. Weeds can provide food, habitat, nesting sites, and nesting material for birds. Nonnative thistles provide pappus bristles for bird nests, many nonnative plants provide nesting habitat, and the abundance of seeds produced by nonnative *Amaranthus* and *Chenopodium* species provide food for many birds.

Succession Ecologist. Most weeds are pioneer plant species, usually annuals with a rapid life cycle ("r" strategy), typified by prolific seed production. These annual weeds typically have the ability to grow in full sunlight and to successfully colonize disturbed soils (early successional species).

Ecosystem Ecologist. Weeds contribute biomass to the annual plant functional group, but too many weeds may alter the structure and/or function of an ecosystem. On disturbed sites, especially on slopes, the rapid colonization and growth of weeds can provide the amenities of reducing the severity of wind and soil erosion.

Global Ecologist. Nonnative plants reduce global warming. The premise here is that aggressive, often pioneer nonnative plant species grow in disturbed soils that would otherwise remain bare. Therefore, due to these nonnatives, more atmospheric CO_2 is converted into plant biomass than if these plants were to have remained in Europe. Therefore, the biomass produced by nonnatives helps reduce the levels of atmospheric CO_2 concentrations.

Holistic Ecologist. Humans are a part of nature. Therefore, if humans disperse seeds from one part of the world to another, this is simply part of the dynamics of ecology. Therefore, there is no need to keep track of exotic plants, since once introduced these plants become a "natural" component of an ecosystem.

Plant Taxonomist. A weed is typically viewed as a nonnative, aggressive species. Although some taxonomists avoid collecting weeds, weed collections in herbaria are sometimes the only means available to track weed arrival and migration patterns over time.

Ethnobotanist. A weed is a plant whose benefits to society have not yet been discovered.

Economic Botanist. If a weed becomes too aggressive and common, search for an economic use for this plant. If you find one, the term "weed" can be replaced with the term "crop."

Biodiversity Ecologist #1. In terms of species diversity, exotic plants have increased the plant species diversity of the Four Corners region by about 12%. An axiom of ecological theory suggests that as species diversity increases so does the resilience and resistance of this ecosystem to perturbations. If this axiom is always true, as all good axioms should be, the addition of these nonnative plant species is augmenting ecosystem stability.

Biodiversity Ecologist #2. Genetic or within-species diversity is another component of biodiversity. After a new plant species (a nonnative) arrives in an area it would be surprising if there were no shift in allelic frequencies as these plants adapt to occupy new habitats with different climates and soils. As nonnative plants adapt to a new habitat, they probably select new allele combinations that are manifest as new ecotypes. In some cases hybridizing creates new plant taxa, such as the case of allopolyploid speciation to create *Tragopogon mirus* and *Tragopogon miscellus* (Soltis and Soltis 1989). With so many exotic plants species, there are likely to be many novel alleles selected for within nonnative plant species—a definite increase in plant genetic material (genetic diversity) albeit in the nonnative guild of plants. On the other hand, the native plant species may also alter allelic frequencies as they strive to compete with aggressive nonnative plants.

Weed Medicines

There is an interesting connection between common weedy and common medicinal plants. Throughout the world, it is recognized that the common flora, native or exotic, makes up much of the medicinal flora. Is this correlation due only to availability? After all, what is the practical use of an effective medicinal herb if it is rare and cannot be easily collected and utilized? Availability may well contribute to more frequent uses for common plants, but there may also be other reasons. One of these could be biochemical. A biochemical explanation maintains that the types of biochemicals in common weeds, such as allelopathic and pathogen-detouring compounds, contribute in a significant way to the success of these plants. These same biochemicals, for some reason, may also tend to be effective human medicines (Kay 1966). Why? Perhaps this is a coincidence. Alternatively, the benefits afforded to common weeds by biochemicals are part of the reason they are so abundant and may afford benefits in the form of medicine to us humans, another ruderal species.

ENDEMIC AND NEAR ENDEMIC PLANTS OF THE FOUR CORNERS REGION

Species	Study Area States
Abronia bolackii N. D. Atwood, S. L. Welsh & K. D. Heil	NMEX
Aliciella formosa (Greene ex Brand) J. M. Porter	NMEX
Aliciella haydenii (A. Gray) J. M. Porter subsp. *crandallii* (Rydb.) J. M. Porter	COLO, NMEX
Aliciella cliffordii J. M. Porter	ARIZ, NMEX
Aquilegia micrantha Eastw.	ARIZ, UTAH
Asclepias cutleri Woodson	ARIZ, UTAH
Asclepias sanjuanensis K. D. Heil, J. M. Porter & S. L. Welsh	NMEX
Asclepias welshii N. H. Holmgren & P. K. Holmgren	ARIZ
Astragalus chuskanus Barneby & Spellenb.	ARIZ, NMEX
Astragalus cliffordii S. L. Welsh & N. D. Atwood	NMEX
Astragalus cottamii S. L. Welsh	ARIZ, NMEX, UTAH
Astragalus cronquistii Barneby	COLO, UTAH
Astragalus deterior (Barneby) Barneby	COLO
Astragalus heilii S. L. Welsh & N. D. Atwood	NMEX
Astragalus humillimus A. Gray ex Brandegee	COLO, NMEX
Astragalus micromerius Barneby	NMEX
Astragalus missouriensis Nutt. var. *humistratus* Isely	COLO, NMEX
Astragalus naturitensis Payson	COLO, NMEX, UTAH

Astragalus oocalycis M. E. Jones	COLO, NMEX
Astragalus proximus (Rydb.) Wooton & Standl.	COLO, NMEX
Astragalus cutleri (Barneby) S. L. Welsh	UTAH
Astragalus schmolliae Ced. Porter	COLO
Astragalus tortipes J. L. Anderson & J. M. Porter	COLO
Atriplex pleiantha W. A. Weber	COLO, NMEX, UTAH
Carex curatorum Stacey	UTAH
Carex specuicola J. T. Howell	COLO
Cirsium chellyense R. J. Moore & Frankton	ARIZ, NMEX
Cirsium rydbergii Petr.	UTAH
Cymopterus beckii S. L. Welsh & Goodrich	ARIZ, UTAH
Draba smithii Gilg & O. E. Schulz	COLO
Erigeron abajoensis Cronquist	ARIZ, COLO, NMEX, UTAH
Erigeron canaani S. L. Welsh	UTAH
Erigeron kachinensis S. L. Welsh & Gl. Moore	UTAH
Erigeron religiosus Cronquist	UTAH
Erigeron rhizomatus Cronquist	ARIZ, NMEX
Erigeron sivinskii G. L. Nesom	ARIZ
Eriogonum clavellatum Small	ARIZ, COLO, NMEX, UTAH
Hackelia gracilenta (Eastw.) I. M. Johnst.	COLO
Ipomopsis congesta (Hook.) V. E. Grant subsp. *matthewii* J. M. Porter	ARIZ, NMEX
Ipomopsis polyantha (Rydb.) V. E. Grant	COLO
Pediocactus knowltonii L. D. Benson	NMEX
Penstemon lentus Pennell var. *albiflorus* (D. D. Keck) J. L. Reveal	ARIZ, UTAH
Penstemon lentus Pennell var. *lentus*	ARIZ, COLO, NMEX, UTAH
Penstemon navajoa N. H. Holmgren	UTAH
Penstemon parviflorus Pennell	COLO
Perityle specuicola S. L. Welsh & Neese	UTAH
Phacelia howelliana N. D. Atwood	ARIZ, UTAH
Phacelia indecora J. T. Howell	UTAH
Phacelia splendens Eastw.	NMEX
Phlox caryophylla Wherry	COLO, NMEX
Phlox cluteana A. Nelson	ARIZ, NMEX, UTAH
Physaria navajoensis (O'Kane) O'Kane & Al-Shehbaz	NMEX
Physaria pruinosa (Greene) O'Kane & Al-Shehbaz	COLO, NMEX
Physaria scrotiformis O'Kane	COLO
Psorothamnus thompsoniae (Vail) S. L. Welsh & N. D. Atwood var. *whitingii* (Kearney & Peebles) Barneby	UTAH
Sclerocactus cloverae K. D. Heil & J. M. Porter subsp. *brackii* K. D. Heil & J. M. Porter	NMEX
Sclerocactus mesae-verdae (Boissev. & C. Davidson) L. D. Benson	COLO, NMEX
Townsendia glabella A. Gray	COLO

Primula specuicola Rydberg
Cleome serrulata Pursh
2 cm
5 cm
2 mm

Vegetation associations and life zones

Watercolors by

Glenn Vandré

Alpine north of Silverton, Colorado

Clay badlands east of Navajo Mountain, Arizona

Douglas-fir forest, Haviland Lake, Colorado

Dune community - El Capitan Dunes near Kayenta, Arizona

Hanging gardens near Bluff, Utah

Impounded lake - Tsaile Lake, Arizona

Mancos clay near Mancos, Colorado

Piñon - Utah juniper community - Navajo Mountain

Ponderosa pine at Tsaile

Riparian, Animas River - Farmington, New Mexico

Riparian community near Lake Powell, Utah

Sagebrush (big sage), Navajo Mountain

Slickrock north of Navajo Mountain - Lake Powell, Utah

Subalpine - Little Molas Lake, Colorado

Utah junipers - B-Square Ranch outside Farmington, New Mexico

FLORA OF THE FOUR CORNERS REGION FAMILY KEY

With the kind permission of William A. Weber, this key has been adapted from his excellent family key that appears in *Colorado Flora: Western Slope* (Weber and Wittmann 2001).

1. Plants reproducing by spores; not producing seeds, flowers, or cones .... **Ferns and Fern Allies**, page 37

1' Plants reproducing by seeds; plants producing flowers or cones (the **Seed Plants**) ....(2)

2. Leaves needlelike or scalelike; evergreen trees and shrubs; never with flowers; seeds on the open face of a scale or bract (a "berry" in *Juniperus*) .... **Gymnosperms**, page 38
2' Leaves seldom needlelike or scalelike; plants rarely evergreen; flowers present; seeds contained in a fruit (**Angiosperms**) ....(3)

3. Parasitic or saprophytic; often brightly colored but not saturated green (mistletoe or orchids may be yellowish green) .... **Parasites and Saprophytes**, **Key #1**, page 38
3' Not parasitic; plants producing chlorophyll and therefore with green leaves ....(4)

4. Mature leaves reduced to spines arising from areoles; stems succulent.... **Cactaceae**, Cactus Family, page 387
4' Plants not as above ....(5)

5. Leaves all basal, with circular blades covered with stalked glistening red glands; insectivorous; flowers in a raceme .... **Droseraceae**, Sundew Family, page 491
5' Plants not as above; not insectivorous; inflorescences various ....(6)

6. Submerged aquatic plants, with or without floating leaves .... **Aquatics**, **Key #2**, page 39
6' Terrestrial or semiaquatic, neither submerged nor with floating leaves ....(7)

7. Vines (herbaceous) or lianas (woody), climbing or twining among other plants, often possessing suckers or tendrils .... **Vines**, **Key #3**, page 40
7' Plants not vines, herbaceous or woody ....(8)

8. Flower parts in 3s; leaves mostly parallel-veined; seeds with 1 cotyledon; herbaceous except *Yucca* .... **Monocots**, **Key #4**, page 40
8' Flower parts in (2s) 4s or 5s; leaves mostly net-veined; seeds usually with 2 cotyledons; herbaceous or woody ....(9)

9. Trees or shrubs .... **Woody Dicots**, **Key #5**, page 41
9' Herbaceous, sometimes woody at the base .... **Herbaceous Dicots**, **Key #6**, page 44

FERNS AND FERN ALLIES

Timothy Reeves

1. Plants aquatic, either in water or at the edge of bodies of water, sometimes most conspicuous when washed ashore or at dried or drying edges of ponds ....(2)
1' Plants growing in soil or on rocks, if associated with water, not usually growing in it ....(3)

2. Leaves resembling blades of grass, clustered; submerged in shallow or along the margin of deep mountain lakes and ponds; leaves swollen at base, sporangia embedded in base .... **Isoetaceae,** Quillwort Family
2' Leaves resembling 4-leaf clover, scattered, with a long, slender petiole and a 4-parted blade; usually found at the drying edges of ponds; sporangia contained in hard, beanlike sporocarps .... **Marsileaceae**, Water-clover Family

3. Stems jointed, hollow, green (except fertile stem of *Equisetum arvense*, which is yellowish brown), largely aboveground and erect; leaves tiny, united into sheaths at stem nodes; plants near or on banks of rivers and irrigation ditches (acequias) as well as seasonally dry stream channels ....**Equisetaceae**, Horsetail Family

3′ Stems not jointed, solid, not green, largely belowground; leaves not united into sheaths; plants of various mountainous and canyon terrestrial or rocky habitats ....(4)

4. Leaves very numerous, small (length 1 cm or less), scalelike or bractlike, sessile, borne spirally or opposite in 4 or more ranks on branched stems............(5)

4′ Leaves few, much larger, not scalelike, usually petiolate............(6)

5. Leaves less than 3 mm long; plants rather mosslike with stems mostly prostrate ....**Selaginellaceae**, Spikemoss Family

5′ Leaves 5–10 mm long; plants larger with erect stems, at least in part ....**Lycopodiaceae**, Club-moss Family

6. Leaf divided into 2 parts, a sterile blade, and a stalked, branched fertile portion bearing numerous, exposed, small (0.5–1 mm diameter), yellow or green sporangia; stems erect, without indument ....**Ophioglossaceae**, Adder's-tongue Family

6′ Leaf not divided into 2 parts; sporangia tiny (about 0.4 mm), borne in spots (sori) on surface of leaf blade or associated with revolute leaf margin, appearing brown; stems creeping or erect, hairy or scaly ....(7)

7. Leaves pinnatifid; sori round, on blade surface, without indusia............**Polypodiaceae**, Polypody Family

7′ Leaves pinnate or more divided; sori various, on blade surface or marginal, with or without indusia............(8)

8. Leaves large, coarse, several times compound, forming thicketlike stands; sori usually lacking, if present, marginal; stems with hairs, very long-creeping............**Dennstaedtiaceae**, Bracken Family

8′ Leaves various, not forming thickets; sori on blade surface or marginal; stems with scales, erect or relatively short-creeping............(9)

9. Sporangia marginal, borne at or under revolute leaf margin or on indusial flap............**Pteridaceae**, Maidenhair Fern Family

9′ Sporangia borne on leaf surface, not at margin............(10)

10. Blades once-pinnate, or grasslike and forked, indusium a linear flap along a vein............**Aspleniaceae**, Spleenwort Family

10′ Blades either more divided or the indusium not a linear flap ....**Dryopteridaceae**, Wood Fern Family

GYMNOSPERMS

1. Shrubs with jointed green stems, the leaves represented by small black triangular scales in whorls at the joints; male cones yellow, with stamenlike sporangia; female (ovulate) cones green, of a few loose thin scales............**Ephedraceae**, Ephedra Family

1′ Trees or shrubs with either needlelike or scalelike overlapping leaves, producing berrylike gray cones or cones with woody scales ....(2)

2. Fruit a berrylike cone, the scales fused together and only detectable by their protruding tips; shrubs or small trees with decussately arranged scalelike leaves or flat sharp needles ....**Cupressaceae**, Cypress Family

2′ Fruit a woody cone with spirally arranged scales; small or large trees with needle leaves.....**Pinaceae**, Pine Family

KEY #1 PARASITES AND SAPROPHYTES

1. Plants attached to the bark of trees, or by suckers to the aboveground stems of herbs............(2)

1′ Plants without obvious attachments to the aboveground parts of their host ....(3)

2. Plants attached to the branches of evergreen trees............**Viscaceae**, Mistletoe Family

2′ Threadlike orange or yellow plants attached by suckers to aerial parts of herbs ....**Convolvulaceae** (*Cuscuta*), Dodder or Morning Glory Family

3. Flowers actinomorphic, in a spikelike, erect or nodding raceme ....**Ericaceae** (*Monotropa* and *Pterospora*), Heath Family

3′ Flowers zygomorphic............(4)

4\. Flowers tubular, corolla 5-lobed, the petals united; ovary superior .... **Orobanchaceae** (*Orobanche*), Broomrape Family
4' Flowers with separate petals; ovary inferior .... **Orchidaceae**, Orchid Family

KEY #2 AQUATICS

1\. Plants disclike or thalluslike, without true stems and leaves; free-floating or submerged .... **Araceae**, Arum/Duckweed Family
1' Plants not disclike or thalluslike, with stems and leaves; not free-floating .... (2)

2\. Stems short or lacking, the leaves attached to the bottom, elongate-linear, the tips floating on the surface; flower clusters in dense balls, lower clusters carpellate, and the upper staminate .... **Typhaceae** (*Sparganium*), Cattail Family
2' Plants with definite stems and otherwise not as above .... (3)

3\. Leaves simple, entire or slightly toothed .... (4)
3' Leaves distinctly lobed, compound, or finely dissected .... (15)

4\. Leaves arranged in whorls, linear or oblong .... (5)
4' Leaves not whorled, variously shaped .... (6)

5\. Leaves lax, translucent (only 2 cell layers thick); flowers, when present, sessile (carpellate) or long-pedicelled (staminate) .... **Hydrocharitaceae**, Tapegrass/Frog-bit Family
5' Leaves rather rigid, unless submerged, opaque (more than 2 cell layers thick); flowers sessile in the leaf axils .... **Plantaginaceae** (*Hippuris*), Plantain Family

6\. Leaves nearly orbicular, deeply cordate, very thick and leathery; flowers large, yellow, solitary .... **Nymphaeaceae**, Water-lily Family
6' Leaves narrower, not cordate, flowers not as above .... (7)

7\. Leaves linear or filiform .... (8)
7' Leaves with distinctly broadened blades .... (12)

8\. Flowers in terminal or axillary spikes .... **Potamogetonaceae**, Pondweed Family
8' Flowers in the leaf axils, sessile or on slender, often coiled peduncles .... (9)

9\. Fruit minute, blackish, on an elongate, often coiled peduncle; leaves filiform, over 3 cm long .... **Ruppiaceae**, Ditch-grass Family
9' Flowers and fruits sessile in the leaf axils; leaves shorter than 3 cm .... (10)

10\. Fruit rounded or emarginate, oblong or wider, not beaked .... **Plantaginaceae** (*Callitriche*), Plantain Family
10' Fruit narrowly cylindric, tapered to a beak .... (11)

11\. Fruit flattened, slightly curved, with a stout beak; leaves filiform .... **Zannichelliaceae**, Horned-pondweed Family
11' Fruit terete, straight, the beak whitish, not rigid; leaves linear, flat, the margins finely toothed .... **Hydrocharitaceae** (*Najas*), Tapegrass/Frog-bit Family

12\. Leaves alternate, at least 1 cm long; flowers not sessile in the leaf axils .... (13)
12' Leaves opposite, less than 1 cm long; flowers inconspicuous, sessile in the leaf axils .... (14)

13\. Floating leaves pinnately veined; flowers with showy pink perianth parts .... **Polygonaceae** (*Persicaria*), Knotweed Family
13' Floating leaves with parallel veins, or floating leaves absent; flowers greenish, not showy .... **Potamogetonaceae**, Pondweed Family

14\. Calyx and corolla absent; ovary 4-locular; stipules lacking; floating and submerged leaves often different .... **Plantaginaceae** (*Callitriche*), Plantain Family
14' Calyx and corolla often present; ovary 3- or 5-locular; stipules present; leaves not dimorphic .... **Elatinaceae**, Waterwort Family

15. Leaves 3-foliate, flowers white, petals fringed....................**Menyanthaceae**, Bogbean Family
15′ Leaves not 3-foliate ....(16)

16. Leaves bearing small bladderlike traps; flowers showy, yellow, spurred, on racemes projecting above the water level ....**Lentibulariaceae**, Bladderwort Family
16′ Leaves not bearing bladders; flowers not spurred ....(17)

17. Leaves alternate; flowers white or yellow, conspicuous ....**Ranunculaceae**, Buttercup Family
17′ Leaves whorled; flowers greenish, inconspicuous....................(18)

18. Leaf divisions dichotomous, finely serrate; flowers sessile in the axils of normal leaves ....**Ceratophyllaceae**, Hornwort Family
18′ Leaf divisions pinnate, entire; flowers in an interrupted spike resembling a knotted cord ....**Haloragaceae**, Water-milfoil Family

KEY #3 VINES

1. Leaves simple ....(2)
1′ Leaves compound (Warning: poison ivy keys here!) ....(6)

2. Leaves not lobed....................(3)
2′ Leaves palmately lobed, sometimes only slightly so ....(4)

3. Corolla funnelform; widespread, often weedy....................**Convolvulaceae**, Morning Glory Family
3′ Corolla 2-lipped; rare in our area, native ....**Plantaginaceae** (*Maurandya*), Plantain Family

4. Tendrils absent....................**Cannabaceae** (*Humulus*), Hemp Family
4′ Plants with tendrils....................(5)

5. Herbaceous; fruit a prickly balloon or gourd....................**Cucurbitaceae**, Gourd/Cucumber Family
5′ Woody; fruit a fleshy "grape" ....**Vitaceae**, Grape Family

6. Leaves pinnately compound; flowers sweetpealike (with banner, wings, and keel) or tubular....................(7)
6′ Leaves trifoliolate, palmately 5- to 7-foliolate, or ternately compound; flowers not as above ....(8)

7. Leaflets entire; flowers with banner, wings, and keel (sweetpea type) ....**Fabaceae**, Legume Family
7′ Leaflets serrate; flowers tubular....................**Bignoniaceae**, Catalpa Family

8. Leaves palmately 5- to 7-foliolate ....**Vitaceae**, Grape Family
8′ Leaves not as above ....(9)

9. Leaves with 3 shiny leaflets; flowers greenish; plant short, commonly bearing clusters of greenish white berries....................**Anacardiaceae** (*Toxicodendron*), Sumac Family
9′ Leaves twice ternately compound, or if 3-foliolate, the flowers blue or yellow, with long feathery styles in fruit ....**Ranunculaceae**, Buttercup Family

KEY #4 MONOCOTS

1. Woody plants with stiff, evergreen, daggerlike leaves; large white flowers in racemes or panicles; fruit a large, 3-locular pod....................**Agavaceae**, Agave Family
1′ Herbs ....(2)

2. Tall fernlike plants, the leaves minute, triangular, papery, subtending clusters of filiform green cladodes; flowers small, yellowish; fruit a red berry ....**Asparagaceae**, Asparagus Family
2′ Not as above....................(3)

3. Flowers minute, enclosed in chaffy bracts; typical perianth lacking; flowers arranged in spikes or spikelets (grasses and sedges) ....(4)
3′ Flowers not enclosed in chaffy bracts; perianth of 3 or 6 parts usually present, but may appear papery or chaffy ....(5)

4. Leaves 2-ranked (in 2 rows on the stem), the sheaths usually open, the margins not fused at least at the summit (with a few exceptions); stems cylindric or flattened and almost always hollow; anthers attached to filaments at their middles .... **Poaceae**, Grass Family
4' Leaves 3-ranked, sometimes absent; the sheaths usually closed, the margins fused; stems almost always triangular and solid (a few cylindric and hollow); anthers attached at one end.......... **Cyperaceae**, Sedge Family

5. Flowers with a rudimentary perianth consisting of bristles or scales, or none .... **Typhaceae**, Cattail Family
5' Flowers with sepals and petals (sometimes similar in shape and texture and then called tepals) .... (6)

6. Carpels numerous (over 6), separate and distinct, in a whorl or ball .... **Alismataceae**, Water-plantain Family
6' Carpels 3 or 6, fused into a compound pistil .... (7)

7. Ovary wholly inferior, the floral parts attached to the top of the ovary .... (8)
7' Ovary superior or only partly inferior .... (9)

8. Flowers radially symmetrical; leaves gladiate (sword-shaped) .... **Iridaceae**, Iris Family
8' Flowers bilaterally symmetrical; leaves not gladiate .... **Orchidaceae**, Orchid Family

9. Perianth of 6 chaffy or scalelike similar segments, hardly petal-like, arranged in 2 alternating groups of 3; grasslike plants .... **Juncaceae**, Rush Family
9' Perianth segments petal- or sepal-like, not chaffy or scalelike .... (10)

10. Perianth segments (tepals) minute, greenish; stamens sessile; carpels separating as units at maturity; annual or perennial from a rhizome; grasslike plants of alkaline flats and mountain fens .... **Juncaginaceae**, Arrow-grass Family
10' Not as above .... (11)

11. Flowers not arranged in umbels .... **Liliaceae**, Lily Family
11' Flowers arranged in umbels .... (12)

12. Flowers subtended by a group of papery bracts; stem arising from a bulb .... **Liliaceae** (*Allium*), Lily Family
12' Flowers subtended by a leafy bract (spathe); stem arising from thickened fleshy roots .... **Commelinaceae**, Spiderwort Family

KEY #5 WOODY DICOTS

1. Leaves scalelike, less than 5 mm long, overlapping and appressed to the stem .... **Tamaricaceae**, Tamarisk Family
1' Leaves larger and otherwise not as above .... (2)

2. Leaves and twigs covered by silvery or brownish peltate scales .... **Elaeagnaceae**, Oleaster Family
2' Leaves and twigs not covered by peltate scales .... (3)

3. Leaves opposite or whorled .... (4)
3' Leaves alternate, spirally arranged, or basal .... (18)

4. Leaves evergreen .... (5)
4' Leaves deciduous .... (7)

5. Leaves entire, paler beneath; plants of dense spruce-fir forests and subalpine fens .... **Ericaceae** (*Kalmia*), Heath Family
5' Leaves serrulate or crenate, not pale beneath .... (6)

6. Low, spreading shrubs; leaves elliptic, spreading in a single plane; flowers axillary, reddish .... **Celastraceae**, Staff-tree Family
6' Creeping plants with only slightly woody stems; leaves broadly oval; flowers in pairs, pendent from an erect stalk, white .... **Caprifoliaceae** (*Linnaea*), Honeysuckle Family

7. Leaves linear, often with smaller leaves fascicled in the axils .... (8)
7' Leaves broader, not in axillary fascicles .... (10)

8. Leaves glabrous, appearing terete, the margins revolute; flowers white; restricted to the Mancos and Fruitland Shale formations .... **Frankeniaceae**, Alkali Heath Family
8′ Leaves densely appressed-pubescent, margins not tightly revolute; plants not restricted to the Mancos or Fruitland Shale formations .... (9)

9. Nonaromatic shrub; flowers actinomorphic, lacking petals and with 4 yellow, petaloid sepals; stamens numerous; fruit an achene .... **Rosaceae** (*Coleogyne*), Rose Family
9′ Aromatic shrub; flowers zygomorphic, petals white to pale blue with purplish flecks; 2 fertile stamens; fruit of 4 nutlets .... **Lamiaceae** (*Poliomintha*), Mint Family

10. Leaves compound .... (11)
10′ Leaves simple .... (13)

11. Fruit a berry; pith of older stems more than 1/2 the diameter of the stem .... **Adoxaceae** (*Sambucus*), Moschatel Family
11′ Fruit a samara; pith of older stems less than 1/2 the diameter of the stem .... (12)

12. Fruit a double samaroid schizocarp; pith round in cross section; bundle scars 3 .... **Aceraceae**, Maple Family
12′ Fruit a single samara; pith 4-angled; bundle scars numerous .... **Oleaceae** (*Fraxinus*), Olive Family

13. Some leaves with 3–5 palmate lobes, maplelike .... **Aceraceae**, Maple Family
13′ Leaves not palmately lobed .... (14)

14. Fruit a single samara; pith 4-angled .... **Oleaceae** (*Fraxinus*), Olive Family
14′ Fruit not a samara; pith not 4-angled, usually round .... (15)

15. Lenticels conspicuous; fruit a drupe .... (16)
15′ Lenticels inconspicuous; fruit a capsule or 2–several-seeded berry .... (17)

16. Twigs red; leaves broadly ovate with raised, curved parallel venation; flowers white, in terminal compound cymes; fruit white .... **Cornaceae**, Dogwood Family
16′ Twigs gray-brown; leaves narrowly to broadly oblong; venation inconspicuous; flowers greenish in axillary clusters .... **Oleaceae** (*Forestiera*), Olive Family

17. Flowers yellow, pink, or white; petals united at least at the very base; fruit a 2–several-seeded berry, white or purplish black .... **Caprifoliaceae**, Honeysuckle Family
17′ Flowers white or cream-colored; petals separate; fruit a dry capsule .... **Hydrangeaceae**, Hydrangea Family

18. [from lead 3] Leaves compound .... (19)
18′ Leaves simple .... (25)

19. Leaves spine-margined, evergreen (resembling holly); inner bark yellow .... **Berberidaceae**, Barberry Family
19′ Leaves not spine-margined .... (20)

20. Leaves with 3 leaflets (Warning! Poison ivy is in this family) .... **Anacardiaceae**, Sumac Family
20′ Leaves not trifoliate, usually pinnately compound .... (21)

21. Fruit a legume; leaflets more than 9, entire .... **Fabaceae**, Legume Family
21′ Fruit not a legume; leaflets various, but if numerous, then serrate or with shallow lobes or auricles at the base of the leaflet .... (22)

22. Leaflets 11 or fewer; if more, then the pith not as below .... **Rosaceae**, Rose Family
22′ Leaflets more than 11; branches stout, the pith occupying a major portion of the cross section .... (23)

23. Leaflets entire except for basal auricles; fruit an elongate samara with a central seed .... **Simaroubaceae**, Quassia Family
23′ Leaflets serrate; fruit a drupe or a nut .... (24)

24. Small shrub; fruit a drupe, red, round with a velvety surface; widespread .... **Anacardiaceae**, Sumac Family
24′ Tree to 17 m in height; fruit a stony endocarp (nut) surrounded by a green husk that becomes brittle and papery; Canyon de Chelly .... **Juglandaceae**, Walnut Family

25. Stems with thorns or spines ....(26)
25′ Stems spineless ....(34)

26. Spines (really thorns) often more than 1 cm long, formed by modification of whole branchlet tips ....(27)
26′ Spines shorter, formed at the nodes (modified leaves or stipules) ....(31)

27. Leaves linear, often somewhat thick and succulent ....(28)
27′ Leaves broader, not succulent ....(29)

28. Rigid, short-branched, dense low shrub less than 1 m tall, with many narrowly oblong pale green leaves, the older branches rigid, thornlike; flowers small, greenish, with petals and sepals .... **Crossosomataceae**, Crossosoma Family
28′ Shrubs either over 1 m tall or rigidly branched, with succulent or gray-farinose leaves; flowers lacking petals, subtended by distinctive bracts .... **Chenopodiaceae**, Goosefoot Family

29. Thorns smooth, sharp, reddish, over 2 cm long, derived from branchlets, but scattered along the main branches .... **Rosaceae** (*Crataegus*), Rose Family
29′ Thorns either terminating short branchlets or representing stipules less than 2 cm long ....(30)

30. Leaves with 3 prominent, somewhat parallel veins; low spreading shrub .... **Rhamnaceae** (*Ceanothus*), Buckthorn Family
30′ Leaves without 3 prominent veins; erect shrub or climber .... **Solanaceae** (*Lycium*), Potato Family

31. Leaves linear; young twigs woolly-tomentose .... **Asteraceae** (*Tetradymia*), Sunflower Family
31′ Leaves broader; young twigs not tomentose ....(32)

32. Leaves entire, ovate, acuminate; fruit baseball-sized, green, wrinkled; introduced tree .... **Moraceae** (*Maclura*), Fig/Mulberry Family
32′ Not as above ....(33)

33. Leaves elliptic, entire, toothed, or spine-toothed; fruit red, elliptic .... **Berberidaceae**, Barberry Family
33′ Leaves ovate, lobed and toothed; fruit globose, green, black, or salmon-pink .... **Grossulariaceae**, Gooseberry Family

34. [from lead 25] Leaves pinnately lobed or spine-toothed, leathery; fruit an acorn .... **Fagaceae**, Beech/Oak Family
34′ Leaves not as above; fruit not an acorn ....(35)

35. Leaf blades unequal at the base (one side attached lower than the other) .... **Ulmaceae**, Elm Family
35′ Leaf blades not unequal at the base ....(36)

36. Leaves broadly ovate-cordate or deeply 5–7-lobed, crenate; fruits juicy, blackberrylike (a fleshy catkin); introduced .... **Moraceae**, Fig/Mulberry Family
36′ Not as above ....(37)

37. Leaves palmately lobed (sometimes shallowly so) ....(38)
37′ Leaves not palmately lobed ....(39)

38. Stamens numerous; carpels few to numerous, separate; fruit dry; flowers never tubular .... **Rosaceae**, Rose Family
38′ Stamens 5 or fewer; carpels 2, united, forming a fleshy berry; flowers often tubular .... **Grossulariaceae**, Gooseberry Family

39. Plants only weakly woody; flowers with white, yellow, or pink tepals .... **Polygonaceae** (*Eriogonum*), Knotweed Family
39′ Plants clearly shrubs or trees ....(40)

40. Flowers in catkins, unisexual ....(41)
40′ Flowers in other kinds of inflorescences, mostly bisexual ....(42)

41. Seeds comose (with a tuft of cottony hair); female catkins not conelike.................... **Salicaceae**, Willow Family
41' Seeds without a coma; female catkins conelike.. **Betulaceae**, Birch/Alder Family

42. Flowers in heads, each flower cluster subtended by an involucre of phyllaries; mostly shrubs of low deserts and piñon-juniper woodlands.. **Asteraceae**, Sunflower Family
42' Flowers not as above .... (43)

43. Desert shrubs with farinose pubescence; individual flowers small and inconspicuous .... **Chenopodiaceae**, Goosefoot Family
43' Plants not farinose; flowers conspicuous ....(44)

44. Petals united; flowers usually urn-shaped (with a few exceptions) .... **Ericaceae**, Heath Family
44' Petals distinct; flowers not urn-shaped ....(45)

45. Leaves entire, elliptic or ovate, with 3 (many in *Rhamnus*) prominent veins; stamens opposite the petals.......... .... **Rhamnaceae**, Buckthorn Family
45' Leaves toothed or lobed (rarely entire in *Peraphyllum*, which has inconspicuous lateral veins); stamens numerous or alternate with the petals .... **Rosaceae**, Rose Family

KEY #6 HERBACEOUS DICOTS

1. Cauline leaves in whorls...(2)
1' Cauline leaves not in whorls ....(4)

2. Leaves in a single whorl of broad leaf blades at the top of the stem .... **Cornaceae**, Dogwood Family
2' Leaves in several whorls along the stem ....(3)

3. Stems square; plants with hairs often with recurved hooks....................................... **Rubiaceae**, Madder Family
3' Stems not square; plants glabrous; plants tall, stout .... **Gentianaceae** (*Frasera*), Gentian Family

4. Flowers several to many, sessile in heads that often resemble a single "flower," each flower cluster surrounded or subtended by an involucre ....(5)
4' Flowers not as above ....(6)

5. Involucre papery or umbrella-like, consisting of a single, individual cup; flowers obviously separate and not tightly confined by the involucre .... **Nyctaginaceae**, Four-o'clock Family
5' Involucre not papery or umbrella-like; the flowers in a dense head .... **Asteraceae**, Sunflower Family

6. Perianth none or of a single whorl (tepals or sepals), these all much alike in color and texture.......................... .... **Key 6A**, below
6' Perianth present, of 2 whorls, the outer whorl (sepals) and inner whorl (petals) usually conspicuously different in texture, color, or both ....(7)

7. Petals separate .... **Key 6B**, page 46
7' Petals united (at least at the base) ....(8)

8. Corolla radially symmetrical.. **Key 6C**, page 48
8' Corolla bilaterally symmetrical .... **Key 6D**, page 49

KEY 6A (PERIANTH ABSENT OR OF A SINGLE WHORL)

1. Plants dioecious ....(2)
1' Plants not dioecious ....(4)

2. Leaves simple, not lobed or compound .... **Amaranthaceae**, Amaranth Family
2' Leaves lobed or compound ....(3)

3. Staminate flowers in racemes, carpellate in clusters; fruit nutlike; leaves digitately compound with narrow, serrate leaflets .... **Cannabaceae** (*Cannabis*), Hemp Family

3′ All flowers in open panicles; achenes ribbed, in clusters; leaves ternately compound with small palmately lobed leaflets .... **Ranunculaceae** (*Thalictrum*), Buttercup Family

4. Ovary inferior [Note: Nyctaginaceae is included here although technically the ovary is superior] .... (5)
4′ Ovary superior .... (9)

5. Ovary with 2 locules, 1 ovule in each; fruit 2-seeded .... (6)
5′ Ovary with 1 locule, this with 1–2 ovules (or ovary with 1–3 locules but only 1 locule containing an ovule); fruit 1-seeded .... (7)

6. Petals united at the base; leaves opposite or whorled; flowers often in cymes or paniculate, headlike, clustered, or solitary, never umbels .... **Rubiaceae**, Madder Family
6′ Petals separate; leaves alternate or basal; flowers in umbels .... **Apiaceae**, Parsley Family

7. Leaves alternate, glaucous; flowers greenish white; fruit a drupe .... **Santalaceae**, Sandalwood Family
7′ Leaves opposite; flowers white or pink; fruit an achene .... (8)

8. Leaves simple, entire; flowers pinkish or flesh-colored; fruits hard and bony or with papery wings .... **Nyctaginaceae**, Four-o'clock Family
8′ Leaves pinnately lobed or divided; flowers white; fruits with a delicate parachute of feathery bristles .... **Valerianaceae,** Valerian Family

9. Pistils 2–many in each flower (if 1, the fruit a fleshy, several-seeded berry); stamens usually numerous .... (10)
9′ Pistil 1 in each flower; stamens 1–many (but usually not over 10 in most families) .... (12)

10. Tepals greenish, shorter than the stamens; achenes ribbed; plants with compound leaves, the leaflets ovate, lobed .... **Ranunculaceae** (*Thalictrum*), Buttercup Family
10′ Not as above .... (11)

11. Carpels enclosed in a 4-angled calyx cup; leaves with stipules; flowers in a dense head .... **Rosaceae** (*Sanguisorba*), Rose Family
11′ Carpels not enclosed in the calyx; leaves lacking stipules; flowers not in a dense head, often showy .... **Ranunculaceae**, Buttercup Family

12. Ovary with 2 or more locules .... (13)
12′ Ovary with 1 locule .... (18)

13. Plants with milky juice .... **Euphorbiaceae**, Spurge Family
13′ Plants without milky juice .... (14)

14. Flowers unisexual; ovary on a stalk (in this group individual flowers are reduced to single stamens or a single pistil; the individual, reduced flowers are surrounded by a cuplike involucre that resembles a perianth) .... **Euphorbiaceae**, Spurge Family
14′ Flowers perfect .... (15)

15. Leaves opposite or whorled, entire; stamens 1–many (rarely 2); flowers axillary, solitary, or in small clusters .... **Caryophyllaceae**, Pink Family
15′ Leaves alternate or crowded at the base of the stem, usually toothed; stamens 2; flowers in terminal spikes or racemes .... (16)

16. Inflorescence subtended by a conspicuous white, clawed bract; rare in our area (Canyon de Chelly) .... **Saururaceae**, Lizard-tail Family
16′ Not as above .... (17)

17. Perennials; flowers in spikes; fruit several-seeded .... **Plantaginaceae** (*Besseya*), Plantain Family
17′ Annuals; flowers in racemes; fruit 2-seeded (1 seed in each locule) .. **Brassicaceae** (*Lepidium*), Mustard Family

18. Ovary with several ovules; fruit a several-seeded capsule .... (19)
18′ Ovary with 1 ovule; fruit a 1-seeded achene or utricle .... (20)

19. Perianth of united tepals, pink; leaves oblong, glaucous .... **Primulaceae** (*Glaux*), Primrose Family
19′ Perianth of separate tepals; leaves otherwise .... **Caryophyllaceae**, Pink Family

20. Leaves with stipules either papery or sheathing the stem ....(21)
20′ Stipules none or, if present, herbaceous and not sheathing the stem ....(22)

21. Leaves opposite; stipules not united around the stem .... **Caryophyllaceae** (*Paronychia*), Pink Family
21′ Leaves alternate; stipules united around the stem in a sheath just above the node .... **Polygonaceae**, Knotweed Family

22. Conspicuous, persistent stipules present; leaves opposite ....(23)
22′ Stipules lacking; leaves usually alternate (except in a few Amaranthaceae) ....(24)

23. Stinging hairs not present; plants small and spreading, prostrate, or densely caespitose, stems rarely over 30 cm tall .... **Caryophyllaceae**, Pink Family
23′ Stinging hairs present; plants with erect stems usually over 30 cm tall .... **Urticaceae**, Nettle Family

24. Flowers perfect, the flower clusters subtended by a cuplike involucre; stamens 6–9; fruit an achene .... **Polygonaceae**, Knotweed Family
24′ Flowers perfect or unisexual but not subtended by a cuplike involucre; stamens 1–5; fruit an achene or utricle ....(25)

25. Bracts and perianth papery or membranaceous .... **Amaranthaceae**, Amaranth Family
25′ Bracts and perianth herbaceous to fleshy ....(26)

26. Style and stigma 1; leaves alternate and entire; fruit an achene; annuals .... **Urticaceae** (*Parietaria*), Nettle Family
26′ Styles and stigmas 1–3 (if 1, the leaves toothed); fruit an achene or utricle; annuals or perennials; weedy species, often coarse and scurfy-pubescent .... **Chenopodiaceae**, Goosefoot Family

KEY 6B (PETALS PRESENT, SEPARATE)

1. Stamens alternating with branched staminodia having terminal, yellow, antherlike glands .... **Parnassiaceae**, Grass of Parnassus Family
1′ Stamens lacking alternating branched staminodia ....(2)

2. Ovary inferior, or at least the lower 1/2 fused to the hypanthium or calyx tube ....(3)
2′ Ovary superior (if hypanthium is present, the ovary may seem to be inferior, but it is not embedded in the hypanthium tissues, as in a rose hip) ....(8)

3. Prostrate succulent herbs with rounded, oblanceolate, thick leaves; stamens variable in number; petals yellow .... **Portulacaceae** (*Portulaca*), Purslane Family
3′ Not prostrate succulent herbs ....(4)

4. Two or more styles present .... **Saxifragaceae**, Saxifrage Family
4′ Only 1 style present (stigmas may be lobed) ....(5)

5. Foliage sandpapery, with minutely barbed hairs .... **Loasaceae**, Loasa Family
5′ Plants lacking barbed hairs ....(6)

6. Stamens 2, 4, or 8; petals 2 or 4; style 1, locules usually 4 (2 in *Circaea*) .... **Onagraceae**, Evening Primrose Family
6′ Stamens 5, rarely 4; petals usually 5; styles 2 or more; locules 2–6 ....(7)

7. Locules 4–6; fruit a several-seeded berry; leaves basal, ternately compound .... **Araliaceae**, Ginseng Family
7′ Locules 2; fruit dry, separating into 2 1-seeded mericarps .... **Apiaceae**, Parsley Family

8. Corolla bilaterally symmetrical ....(9)
8′ Corolla radially symmetrical ....(14)

9. Leaves pinnately or palmately compound ....(10)
9′ Leaves simple, entire to deeply lobed, but never truly compound ....(12)

10. Sepals 2, very minute and scalelike; corolla spurred; leaves greatly dissected ... **Fumariaceae**, Fumitory Family
10′ Sepals 4 or 5; corolla not or very inconspicuously spurred; leaves once or twice compound ....(11)

11. Ovary with 1 placenta; petals 5 (a banner, 2 wings, and a keel that consists of 2 partly united petals enclosing the stamens and style); flowers usually shaped like those of a sweetpea .... **Fabaceae**, Legume Family
11′ Ovary with 2 placentae on opposite sides of the ovary; petals 4; stamens exserted .... **Brassicaceae**, Mustard Family

12. Stamens many; carpels more than 1; capsule dehiscing along 1 suture (thus a follicle) .... **Ranunculaceae**, Buttercup Family
12′ Stamens 10 or fewer; ovary of a single or several united carpels; fruit a dehiscent capsule..........................(13)

13. Flowers not spurred, but with a large upper petal (the banner) .... **Polygalaceae**, Milkwort Family
13′ Flowers spurred .... **Violaceae**, Violet Family

14. Stamens of the same number as the petals and opposite them..................(15)
14′ Stamens fewer or more numerous than the petals, or, if the same number, then alternate with them ....(17)

15. Sepals, petals, and stamens each 6 in number, 3 of the sepals petal-like; leaf margins spiny .... **Berberidaceae**, Barberry Family
15′ Sepals, petals, and stamens 2–5 (sepals rarely 6); spineless..................(16)

16. Styles and stigmas 1; sepals usually 5 .... **Primulaceae**, Primrose Family
16′ Styles and stigmas 2 or more; sepals usually 2 .... **Portulacaceae**, Purslane Family

17. Ovary 1 (a single unit), with 1 locule..................(18)
17′ Ovaries more than 1 (several units), or, if 1, then with 2 or more locules..................(29)

18. Stamens 13 or more..................(19)
18′ Stamens 12 or fewer ....(23)

19. Ovary simple (of a single carpel having 1 placenta, 1 style, 1 stigma; many such ovaries may be present in a single flower) .... **Ranunculaceae**, Buttercup Family
19′ Ovary compound (2 or more placentae, styles, or stigmas) .... (20)

20. Placenta free-central or basal .... **Portulacaceae**, Purslane Family
20′ Placenta parietal..................(21)

21. Ovary with 2 parietal placentae; plants usually viscid and ill-smelling .... **Cleomaceae**, Cleome Family
21′ Ovary with 3 or more placentae; plants not viscid or ill-smelling..................(22)

22. Leaves opposite, entire, with minute translucent dots (hold up to light); juice not milky; flowers yellow.................. **Clusiaceae**, St. Johnswort Family
22′ Leaves alternate, toothed or lobed, without translucent dots; juice milky; flowers white or cream .... **Papaveraceae**, Poppy Family

23. Pistil simple (with 1 placenta, style, and stigma) ....(24)
23′ Pistil compound (more than 1 placenta, style, or stigma) ....(25)

24. Stamens and petals attached to the rim of the calyx tube (hypanthium) .... **Rosaceae**, Rose Family
24′ Stamens and petals not attached to the calyx tube.................. **Fabaceae**, Legume Family

25. Ovules attached to base of ovary or to a free-central placenta..................(26)
25′ Ovules attached to 2 or more parietal placentae ....(28)

26. Calyx of united sepals; petals with claws.................. **Caryophyllaceae**, Pink Family
26′ Calyx of separate sepals; petals not stalked ....(27)

27. Sepals 2 or numerous; stamens commonly opposite the petals, sometimes fewer than the petals, sometimes numerous; leaves commonly basal and succulent .... **Portulacaceae**, Purslane Family
27′ Sepals usually 5; stamens not opposite the petals, usually 5 or 10 (rarely 3); leaves usually opposite, not especially succulent.................. **Caryophyllaceae**, Pink Family

28. Ovary with 2 parietal placentae; sepals and petals 4 each .... **Brassicaceae**, Mustard Family
28′ Ovary with 3–5 parietal placentae; sepals and petals 5 .... **Clusiaceae**, St. Johnswort Family

29. Plants with milky juice or stinging hairs; ovary stipitate, exserted from the cyathium .... **Euphorbiaceae**, Spurge Family
29' Plants without milky juice or stinging hairs ....(30)

30. Perianth parts in 2s (rarely) or 4s ....**Brassicaceae**, Mustard Family
30' Perianth parts in 5s or numerous ....(31)

31. Leaves trifoliolate, acrid tasting ....**Oxalidacae**, Woodsorrel Family
31' Leaves not trifoliolate or acrid tasting ....(32)

32. Stamens united in a column around the styles ....**Malvaceae**, Mallow Family
32' Stamens not united in a column around the styles ....(33)

33. Leaves linear or oblong, succulent ....**Crassulaceae**, Stonecrop Family
33' Leaves not as above ....(34)

34. Stamens numerous ....(35)
34' Stamens not more than 10 ....(37)

35. Leaves elliptic, with translucent dots and often minute, black, marginal dots on leaves and petals; stamens tending to be in 5 groups; fruit a capsule ....**Clusiaceae**, St. Johnswort Family
35' Leaves and stamens not as above; fruit an achene ....(36)

36. Stipules lacking; hypanthium not developed ....**Ranunculaceae**, Buttercup Family
36' Stipules present; hypanthium always developed ....**Rosaceae**, Rose Family

37. Petals waxy; anthers opening by terminal pores; leaves often leathery ....**Ericaceae** (*Pyrola*), Heath Family
37' Petals not waxy; anthers opening by slits ....(38)

38. Fruit 5-carpellate, separating at maturity into 5 1-seeded segments (mericarps) ....(39)
38' Fruit not 5-carpellate, or, if so, not separating into mericarps ....(40)

39. Flowers pink or white, mericarps not spiny ....**Geraniaceae**, Geranium Family
39' Flowers yellow; mericarps stoutly spiny (*Tribulus*) or not ....**Zygophyllaceae**, Lignum Vitae Family

40. Petals yellow, copper, or blue, falling within a few hours; capsule 10-locular (1 ovule per locule) .... **Linaceae**, Flax Family
40' Petals white, yellow, or pink-purple, not fugacious; capsule not as above ....**Saxifragaceae**, Saxifrage Family

KEY 6C (PETALS UNITED; FLOWERS RADIALLY SYMMETRICAL)

1. Plants with milky juice ....(2)
1' Plants without milky juice ....(4)

2. Corolla rotate, with a central structure (the gynostegium) consisting of fused stigmas and stamens; corona present, enclosing hornlike structure ....**Asclepiadaceae**, Milkweed Family
2' Corolla bell-shaped, without special hornlike structures ....(3)

3. Leaves opposite ....**Apocynaceae**, Dogbane Family
3' Leaves alternate ....**Campanulaceae**, Bellflower Family

4. Ovary superior ....(5)
4' Ovary inferior or half-inferior ....(18)

5. Stamens more numerous than the corolla lobes, 6–many ....(6)
5' Stamens as many as the corolla lobes or fewer ....(7)

6. Stamens many, united into a tube around the style ....**Malvaceae**, Mallow Family
6' Stamens 6–10, separate and distinct; anthers opening by pores at the basal end; petals waxy .... **Ericaceae** (*Pyrola*), Wintergreen Family

7. Stamens 5, opposite the petals; ovary with 1 locule, and basal or free-central placentation ....(8)

7′ Stamens alternate to the petals or fewer; ovary with more than 1 locule, or, if 1-loculed, then the placentation rarely basal or free-central ....(9)

8. Styles 5; fruit an achene; inflorescence a 1-sided spike........**Plumbaginaceae**, Sea Lavender Family
8′ Style 1; fruit a capsule; inflorescence not a 1-sided spike........**Primulaceae**, Primrose Family

9. Ovary 4-lobed, developing into 4 (or by abortion fewer) 1-seeded nutlets........(10)
9′ Ovary not 4-lobed; fruit a capsule or berry, usually several-seeded........(11)

10. Leaves alternate; stem not square in cross section; not aromatic........**Boraginaceae**, Borage Family
10′ Leaves opposite; stems square in cross section; mostly aromatic........**Lamiaceae**, Mint Family

11. Ovary with 1 locule........(12)
11′ Ovary with 2 or more locules........(14)

12. Leaves basal; flowers solitary, scapose; stoloniferous plants rooted in mud........**Plantaginaceae** (*Limosella*), Plantain Family
12′ Not as above........(13)

13. Leaves opposite or whorled, entire; style 1 or none; plants mostly glabrous; inflorescence not curled in the bud........**Gentianaceae**, Gentian Family
13′ Leaves usually alternate (if opposite, then not entire); styles 2, or single and 2-cleft above; plants mostly hairy; inflorescence commonly curled in the bud........**Hydrophyllaceae**, Waterleaf Family

14. Stigma 3-lobed or style 3-branched; ovary with 3 locules........**Polemoniaceae**, Phlox Family
14′ Stigma entire or 2-lobed, or style 2-cleft; ovary usually with 2 locules........(15)

15. Flowers yellow, in dense terminal spikes or racemes over 20 cm long; filaments hairy........**Scrophulariaceae** (*Verbascum*), Figwort Family
15′ Flowers variously colored, never in spikes or elongate racemes........(16)

16. Styles 2, distinct, each one again 2-cleft; ovules 2 in each locule; flowers axillary, the corolla lavender, with darker pleats; foliage silky-hairy........**Convolvulaceae**, Morning Glory Family
16′ Style 1, or if 2, rarely separate to the base, never again 2-cleft; ovules usually more than 2 per locule; inflorescence various........(17)

17. Style 1, the stigma entire or 2-lobed; fruit a capsule or berry........**Solanaceae**, Nightshade Family
17′ Styles 2 or definitely 2-branched below the stigmas; fruit a capsule........**Hydrophyllaceae**, Waterleaf Family

18. Leaves ternately compound, basal; flowers in a few-flowered, tight, umbel-like cyme; delicate herbs........**Adoxaceae**, Adoxa Family
18′ Not as above........(19)

19. Leaves alternate or basal........**Campanulaceae**, Bellflower Family
19′ Leaves opposite or whorled........(20)

20. Stems creeping, slightly woody; leaves opposite, crenate; flowers 2, pink, pendent from an erect stalk........**Caprifoliaceae** (*Linnaea*), Honeysuckle Family
20′ Stems erect or sprawling, herbaceous; leaves opposite or whorled, entire; flowers white, minute, in cymes........**Rubiaceae**, Madder Family

KEY 6D (PETALS UNITED; FLOWERS BILATERALLY SYMMETRICAL)

1. Ovary superior........(2)
1′ Ovary inferior........(9)

2. Stem 4-angled; leaves opposite........(3)
2′ Stem not 4-angled; leaves opposite, alternate, or basal........(6)

3. Corolla not 2-lipped; corolla with a narrow tube and flaring lobes........**Verbenaceae**, Vervain Family
3′ Corolla 2-lipped, usually strongly so........(4)

4. Calyx tubular, ribbed or pleated; fruit a capsule, loculicidal .... **Phrymaceae**, Monkey-flower Family
4′ Calyx various, not ribbed or pleated; fruit a septicidal capsule or nutlet .... (5)

5. Corolla brownish green; style terminal; gynoecium not lobed or divided; foliage never with a minty odor .... **Scrophulariaceae** (*Scrophularia*), Figwort Family
5′ Corolla various but not brownish green; style arising from the base of the lobed gynoecium; foliage usually with a minty odor .... **Lamiaceae**, Mint Family

6. Corolla papery; leaves basal; inflorescence a spike; fruit a circumscissile capsule .... **Plantaginaceae** (*Plantago*), Plantain Family
6′ Corolla not papery; otherwise not as above .... (7)

7. Calyx split to the base along the lower side; corolla 3–5 cm long and nearly as wide; fruit woody, with 2 long curved claws .... **Pedaliaceae**, Sesame Family
7′ Calyx not split to the base on the lower side; corolla various; fruit not as above .... (8)

8. Corolla not spurred or saclike; fruit loculicidal .... **Orobanchaceae**, Broomrape Family
8′ Corolla base spurred or saclike; fruit dehiscent by terminal slits or pores ...... **Plantaginaceae**, Plantain Family

FERNS AND FERN ALLIES

ASPLENIACEAE Newman SPLEENWORT FAMILY

Timothy Reeves

Terrestrial, lithophytic or epiphytic. Gametophyte green, heart-shaped. **STEMS** erect, sometimes long-creeping, with clathrate scales. **LEAVES** mostly monomorphic, about 1 cm to 3 m long; blades simple to 5-pinnate, rarely subdichotomous, frequently with minute glandular hairs and linear scales. **SORI** borne on veins, each elongate, lunate to linear; indusium usually present, originating on one side of the sorus and shaped like it. **SPORES** monolete; perispore typically winged, reticulate, echinate or perforate. $x = 36$ (rarely 39 or 40). One genus (as treated here) with about 700 diverse species, worldwide. Other treatments recognize about 6–12 genera, here considered part of one very large genus. This family contains many species which hybridize readily, to the delight of both naturalists and horticulturists. Polyploidy is also widespread. Many species are grown as common houseplants. (This treatment follows Wagner, Moran, and Werth 1993.)

Asplenium L. Spleenwort

(Greek *asplenon*, name used by Dioscorides for a fern used for treating spleen diseases; derived from *splen*, spleen) **LEAVES** evergreen in our species. Found predominantly in moist or wet forests, growing in ravines, along streams, on the forest floor, among rocks or on cliffs, or as an epiphyte. Boreal species grow mostly on rock; some associated with particular rock types.

1. Blade subdichotomous or forked, with a few narrow segments, segments 1–2 cm long; not pinnate throughout .... ***A. septentrionale***
1' Blade pinnate throughout or pinnatifid only in the distal 1/3, pinnae undivided, mostly less than 1 cm long.....(2)

2. Pinna oblong-lanceoloate to oblong, mostly 6–10 mm, length 3–5 times width .... ***A. resiliens***
2' Pinna ovate, oblong-ovate, to rhombic, mostly 4–7 mm, length 1–2 times width ....(3)

3. Rachis dark reddish brown throughout .... ***A. trichomanes***
3' Rachis dark only at base, mostly green .... ***A. viride***

Asplenium resiliens Kunze (recoiling) Black-stemmed spleenwort, rock spleenwort, little ebony spleenwort. **STEMS** erect, unbranched; with black scales. **LEAVES** with petioles black, shining, 1.5–3 cm, 1/10–1/4 length of blade; scales blackish brown, filiform; blade linear to narrowly oblanceolate or deltoid, subcoriaceous, pinnate, about 9–20 × 1–2 cm, glabrous; base tapered; apex acute; rachis black, glabrous; pinnae oblong; base usually with lower auricle; margins entire to crenate; apex obtuse. **SORI** 2–5 pairs per pinna, on both basiscopic and acroscopic sides, often confluent in aged portions of fronds. **SPORES** 32 per sporangium. $2n = 108$ (apogamous). In our area known only from Canyon de Chelly National Monument, Arizona, in rock crannies of sandstone; also reported from San Juan County, Utah, on sandstone. ARIZ: Apa; UTAH. 1750 m (5700′). No fertile material seen. Southern United States from southern Nevada and southern Utah to Maryland, south to southern boundary except California; Mexico; Hispaniola, Jamaica; Guatemala; South America; on limestone and other basic rocks (on shell mounds in some Texas populations).

Asplenium septentrionale (L.) Hoffm. (northern) Grassfern, forked spleenwort. **STEMS** erect, branched with dense mats or caespitose, giving a crowded appearance; scales dark reddish brown to black. **LEAVES** with petioles dark reddish brown proximally, green distally, 2–13 cm; glabrous; blade linear, simple or usually pinnate, 0.5 × 0.1–0.4 cm, leathery; base acute; rachis green, lustrous, glabrous; pinnae of pinnate leaves 2 (–4), strongly ascending, linear, 10–30 × 0.75–3 mm; bases acute; margins remotely lacerate; apex acute. **SORI** very long, usually 2–3 per pinna, mostly in pairs parallel to margins, indusium continuous just inside the margin on both sides. **SPORES** 64 per sporangium. $2n = 144$. Our plants were all found along the Piedra River, Colorado, in a riparian community, one found growing in horizontal cracks of large boulders in the river gorge, one in crevices, all on Dakota Sandstone. COLO: Arc, Hin. 2300–2600 m (7600–8500′). Spores: June. Circumboreal in widely scattered locations. Scattered sites from Oregon to District of Columbia with principal distribution in Colorado and New Mexico, West Virginia; Baja California; Europe; Asia. This fern looks very much like a tuft of grass and is probably overlooked in many locations. It is known to hybridize with *A. trichomanes* (*A.* ×*alternifolium* Wulfen) in Europe.

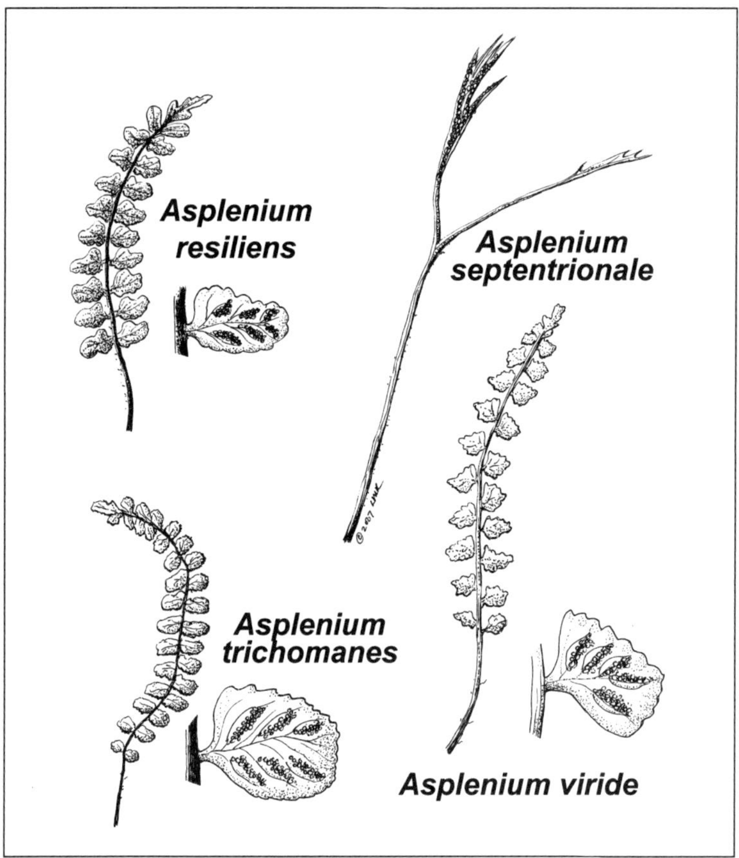

Asplenium trichomanes L. (earlier placed in the genus *Trichomanes*) Maidenhair spleenwort. **STEMS** short-creeping, branched; scales black or with borders brown; scale margins entire to denticulate. **LEAVES** clustered, with petioles reddish brown or blackish brown, shining, 1–4 cm; glabrous or with black, linear-lanceolate or filiform scales at base; blade linear-oblong, pinnate, 3–22 × 0.5–1.5 dm, thin, glabrous or sparsely pubescent; base tapered; apex acute; rachis reddish brown, mostly glabrous; pinna oblong to oval, base broadly cuneate, sometimes with rounded lower auricle, margins entire to shallowly crenate to serrate, apex obtuse. **SORI** 2–4 pairs per pinna, on both basiscopic and acroscopic sides. **SPORES** 27–32 µm diameter; 64 per sporangium. 2*n* = 72 (in our subspecies). On acidic substrates. Most of our specimens are from the Vallecito Lake, Colorado, area, in crevices of cliffs and shaded metaconglomerate rock in ponderosa pine, spruce, fir, and *Juniperus communis* forests. COLO: LPl, RGr. 2400–2600 m (8000–8500′). Spores: May–Sep. The species occurs worldwide. Our plants are **subsp. *trichomanes***, which occurs from the eastern Canadian provinces south to the Carolinas, Georgia, west to the Great Lakes and Oklahoma, southeastern Alaska and British Columbia south to northern California, southwestern Nebraska and southeastern Wyoming and northern Utah south to Texas and Mexico to Guatemala; Europe; Asia; Africa; Australia. Our subspecies, a diploid, is usually restricted to noncalcareous rock substrates. Tetraploid and triploid subspecies also exist. Used by Cherokees as an abortifacient for irregular menses, a breast treatment infusion for "breast diseases" and "acrid humors," a cough medicine infusion for coughs, and a liver aid for "liver complaints."

Asplenium viride Huds. (green) Green spleenwort. **STEMS** short-creeping or ascending, often branched; scales dark reddish brown to black; margins entire to undulate or shallowly dentate. **LEAVES** with petioles reddish brown at base, gradually becoming green, 1–5 cm long; scales dark reddish brown to black, grading into glandular hairs; blade linear, pinnate throughout, about 2–13 × 0.6–1.2 cm, glabrous or with thinly scattered minute hairs; base more or less tapering or truncate; apex acute; rachis green, dull, glabrous or with scattered hairs; pinna deltate to rhombic; base obtuse and often inequilateral; apex rounded to acute. **SORI** 2–4 pairs per pinna, on both basiscopic and acroscopic sides. **SPORES** 64 per sporangium. 2*n* = 72. [*A. trichomanes-ramosum* L.]. Our two collections are from subalpine and alpine sites, one in limestone talus where it is abundant, growing in soil under boulders in a 100-square-meter area. COLO: LPl, SJn. 3450–3650 m (11300–11900′). Spores: Jul–Aug. Alaska, Yukon, Northwest Territories, British Columbia, Alberta, eastern Canadian maritime provinces to Greenland, Washington, northern California, northern Utah, southwestern Colorado, Michigan, and Vermont; Europe; Asia. This species hybridizes with *A. trichomanes*, producing the fertile allotetraploid *A.* ×*adulterinum* Milde on Vancouver Island, British Columbia.

DENNSTAEDTIACEAE Lotsy BRACKEN FAMILY

Timothy Reeves

(For A. W. Dennstaedt, German botanist) Perennial, mostly terrestrial of mesic forests. **STEMS** short- to long-creeping, mostly bearing hairs. **LEAVES** monomorphic, buds with circinate vernation; petioles not articulate, hairy or glabrous; blade mostly 1- or more pinnate, glabrous or hairy; rachis and costae usually grooved adaxially; veins mostly free, pinnate or forking in ultimate segments. **SORI** near or at blade margin, true (inner) indusium present, free or fused with portion of the blade margin to form a cup or pouch or obscured by revolute and modified blade margins. **SPORES** tetrahedral-globose, trilete, very finely granulate. About 17 genera with about 400 species, worldwide, mostly tropical.

Pteridium Gled. ex Scop. Bracken, Brackenfern

(Greek *pteridion*, a small fern) Terrestrial, often forming colonies or thickets, in ours forming conspicuous herbaceous understory in moist to dry forests and in openings. **STEMS** subterranean, slender, up to 60 m or more long; hairs jointed, pale to dark, scales absent. **LEAVES** widely scattered, deciduous; petioles glabrous to pubescent; blade firm, spreading at an angle from petiole, broadly deltate, mostly up to 1 or 2 m tall but in thickets or under small trees becoming scandent and up to 4 m, rarely 7 m long, 2–4-pinnate; segments pinnate, ultimate segments ovate to oblong to linear, base decurrent or surcurrent, margins entire; veins free or joined at margin by commissural vein beneath sori, pinnately 2–3-forked. **SORI** more or less continuous; covered by a recurved, outer false indusium and on the other side by an obscure, membranaceous, inner true indusium. $x = 26$, $n = 52$, $2n = 52, 208$. One species, nearly worldwide; as recognized here consisting of 2 subspecies with 12 varieties. It is likely the most widespread species of vascular plant except perhaps for a few annual weeds. Bracken is aggressive, invading logged and burned forests, as well as pastures, cultivated fields, and roadsides and is a major weed in areas where it takes over agricultural lands (Tryon 1941).

Pteridium aquilinum (L.) Kuhn (of an eagle, from the wing-shaped fronds) Bracken, brackenfern, western brackenfern, fougère des aigles. Plants large, often forming large colonies sometimes covering many square meters. **LEAVES** large, conspicuous, turning yellow or bronze in the fall; petioles erect, pale with a dark, sometimes swollen, base shorter than the blade, 10–100 cm; blade ovate, ovate-triangular, broadly deltate, to nearly pentagonal, 3-pinnate to 3-pinnate-pinnatifid, the lowest pinnae sometimes quite enlarged and with elongate petiolules imparting a tripartite appearance, 30–200 × 15–100 cm; glabrous to sparsely hairy above, blades, rachises, and costae usually densely covered beneath with abundant, contorted, lax, spreading hairs; pinnae (proximal) triangular, distal pinnae oblong; terminal segment of each pinna about 4 times longer than wide, longer ultimate segments less than their width apart, about 1.5–5 mm wide; pinnules at nearly 90° angle to costa; fertile ultimate segments adnate or equally decurrent and surcurrent; outer indusia entire, pilose on margin and surface, hairs like those of the axes. None of the specimens examined is fertile, the species apparently reproducing primarily by rhizomes in our area. Gametophytes and sporelings are rarely reported in nature but should be expected, especially in burned areas. $n = 52$. Dry to mesic forests of ponderosa pine, ponderosa pine/Douglas-fir, ponderosa pine/Gambel's oak, spruce/*Juniperus communis*, spruce/fir, and aspen groves in these forests, in openings, and in full shade to full sun, forming conspicuous colonies, in some areas providing a conspicuous herbaceous layer in montane forests, especially in logged or burned areas, more rarely in drainages, streamlets, stream banks, or seeps; in our area mostly of dry to mesic, noncalcareous soils. ARIZ: Apa; COLO: Arc, Hin, LPl, Min, Mon, SJn; NMEX: RAr, SJn; UTAH. 2000–3050 m (6600–10000′) Sporulating time unknown as no fertile material has been seen. Fertile material should be collected if encountered. Our plants belong to **subsp.** ***aquilinum*** **var**. ***pubescens*** (L.) Underw. Western North America from southeastern Alaska, British Columbia, and Alberta to Washington, Idaho, Montana, south to California, Arizona, and New Mexico, east to South Dakota, western Nebraska south to Texas and Mexico in Baja California, Chihuahua, and Durango. This is our largest fern and is especially common in the Chuska Mountains, growing in logged ponderosa pine forests and in aspen groves. In Colorado bracken is common at a higher elevation in moist forests of aspen, white fir, and Douglas-fir. It is not common in lower drier forests. In Finland, bracken clones have been determined to be about 1500 years old. Bracken is the economically most important fern on earth, having been used to produce potash for making soap and glass, for thatch, swine food, bedding material for animals and man, as packing material for fruits, and as food. It is still used in rural areas of Venezuela for packing and for wrapping heads of curing cheese. Bracken contains the enzyme thiaminase and other mutagenic and carcinogenic compounds. Thiaminase is toxic to livestock when eaten green or dried. The crosiers (fiddleheads) are eaten by people in many cultures and, especially in Japan where large quantities are consumed, may contribute to a high incidence of stomach cancer. Nature enthusiasts have attempted to remove these poisons by boiling in water, but even then the fiddleheads cause cancer of the bladder. Various indigenous peoples (no known local uses) have used var. *pubescens* as a toothache remedy, an anticonvulsive for *Toxicodendron* poisoning, a body deodorant, food (e.g., the roasted starchy centers of rhizomes and young plant tops eaten green), fiber in basketry, camp bedding, and cooking tools to hold, wipe, and serve fish.

DRYOPTERIDACEAE Herter WOOD FERN FAMILY

Timothy Reeves

Sporophyte perennial, lithophytic or terrestrial, rarely epiphytic. Gametophyte terrestrial, aboveground, plano-cordate, glabrous, weakly glandular or pubescent. Sporophyte **STEMS** usually creeping or erect, indument of scales. **LEAVES** with circinate vernation, monomorphic (in ours) or dimorphic; petiole base scaly; blade simple, mostly 1–5 or more pinnate-pinnatifid; ultimate segments with veins pinnate or parallel, simple to forked; blade indument of hairs and/or scales and/or glands, especially on rachis abaxially. **SORI** abaxially on veins (in ours) or at vein tips, or sporangia covering abaxial surface, if in sori then round to elongate; with or without indusium, indusium linear, falcate, or reniform, less often hoodlike, round or cupular. **SPORES** monolete, oblong or reniform, ornamented, often winged perispore; 64 in a sporangium (32 in apogamous species). x = 37, 38, 39, 40, 41, 42. About 60 genera with over 3000 species, worldwide. Some or all of the genera here are occasionally treated in the Aspidiaceae, an illegitimate family name (Flora of North America Editorial Committee 1993; Weber and Wittmann 2001; Werth and Windham 1991).

1. Frond linear, 1-pinnate, leathery; indusium peltate .... ***Polystichum***
1' Frond broader, 1-pinnate-pinnatifid to 3-pinnate-pinnatifid, herbaceous; indusium reniform, linear, hooked, or lacking .... (2)

2. Sorus elongate, indusium linear, hooked, or lacking .... ***Athyrium***
2' Sorus round, indusium reniform, cupular, hood-shaped, or lacking .... (3)

3. Indusium absent, blade deltate, ternate with lowest pinnae matching rest of blade in size and shape and in being petiolate .... ***Gymnocarpium***
3' Indusium present, sometimes obscured in mature sori; blade elliptic to ovate-lanceolate or, if deltate, not appearing ternate, and indusium hoodlike .... (4)

4. Indusium completely underneath the sporangia; cup-shaped, symmetrical, variously divided into narrow to broad segments; blade indument of hairs without prominent cross walls; blade often conspicuously glandular at 10× .... ***Woodsia***
4' Indusium partially to completely covering the sporangia, reniform or hood-shaped; blade indument of scales or hairs with conspicuous brownish cross walls .... (5)

5. Indusium reniform, conspicuous; blade substantial, 40–100 cm long; blade indument of scales, conspicuous and large on lower petiole, narrower above .... ***Dryopteris***
5' Indusium hood-shaped, arching over the sporangia from one side; blade lacking indument or of sparse, inconspicuous hairs with conspicuous brownish cross walls; glandular hairs, if present, inconspicuous even at 30× .... ***Cystopteris***

Athyrium Roth Lady Fern

(Greek *athyros*, without a covering or door, referring to the partially pushed back indusium) Sporophyte terrestrial. **STEMS** ascending or short-creeping, stolons lacking. **LEAVES** monomorphic, generally not evergreen; petioles up to 1/2 the length of the blade, base swollen, dentate; blade lanceolate to oblanceolate, to tripinnate or pinnatifid, herbaceous, apex pinnatifid; pinna segment margin serrulate or crenate; proximal pinnae sessile to short-petiolulate, more or less equilateral; adaxial costal grooves continuous from rachis to costae to costules; vestiture absent or linear to lanceolate scales or abaxially 1-celled glands. **SORI** in a row away from margin, round to elongate, straight U- or J-shaped at distal end; indusium ciliate, shaped like sorus, persistent, attached laterally or absent. **SPORES** brown, rugose. x = 40. About 180 species; tropical, subtropical, and temperate zones worldwide.

1. Sori round, submarginal; indusia absent or occasionally much reduced .... ***A. alpestre***
1' Sori elongate to U- or J-shaped, medial; indusium obvious, with dentate to ciliate margin .... ***A. filix-femina***

Athyrium alpestre (Hoppe) Clairv. (of the mountains) American alpine lady fern, alpine lady fern. **LEAVES** with petioles distally reddish brown to straw-colored, about 10–30 cm long, dark red-brown to blackish swollen base with 2 rows of teeth, basal scales brown, lanceolate to broadly lanceolate, about 13 × 3–4 mm; blade narrowly elliptic to lanceolate, bi- to tripinnate-pinnatifid, 15–60 × 4–25 cm, narrowed proximally, widest below the middle, apex acuminate; pinnae with short stalks, lanceolate to deltate-oblong, apex acute; pinnules pinnatifid, segments oblong, margins crenate to crenulate; rachis with small, light brown scales. **SORI** round or elliptic; indusium absent or tiny, scalelike.

$2n = 80$. [*A. distentifolium* Tausch ex Opiz; *A. americanum* (Butters) Maxon]. Our one collection is from a north slope seep in mixed forest of spruce, aspen, and alder. This represents a range disjunction from central and northwestern Colorado. COLO: Mon. 2725 m (8900′). Our plants are **var. *americanum*** Butters. Southern Alaska, Yukon, British Columbia, Alberta, and western Montana south to eastern California, Nevada, northern Utah, and Colorado; Quebec, Newfoundland, Greenland.

Athyrium filix-femina (L.) Roth (female fern) Lady fern. **LEAVES** with petiole base reddish brown or blackish, straw-colored distally, about 7–60 cm long, scales brown or blackish, linear to ovate, primarily lanceolate, 7–18 × 1–6 mm; blade herbaceous, narrowly elliptic to oblanceolate, up to bipinnate-pinnatifid, 25–130 × 8–25 cm, gradually narrowed to base, broadest at or above the middle, apex acuminate; pinnae sessile or nearly so, linear-oblong, apex acuminate; pinnules linear-oblong to lanceolate, more or less auriculate, apex acute to acuminate; rachis with scales, often with pale glands. **SORI** elliptic to oblong, straight, J-shaped at distal end, or U-shaped; indusium dentate or ciliate. **SPORES** brown. $2n = 80$ (other varieties in the species). [*Polypodium filix-femina* L.]. Although our variety is mapped in *Flora of North America* (1993) throughout our area, specimens have been obtained from only six locations. These are all in mountain canyons of Colorado, occurring in cool, mesic ponderosa pine, spruce, or aspen forests, along rivers, creeks, lakes, springs, or in other wet areas; associated species include alder, bracken, horsetail, and male fern. COLO: Arc, LPl, Min. 2350–2900 m (7700–9500′). Our plants are **var. *californicum*** Butters (of California, referring to the southwestern location of this variety). Southwestern lady fern. Oregon and southeastern South Dakota south to southern California, southern Arizona, and southwestern New Mexico. The species is circumboreal and this or closely related species occur in Mexico, Central America, and South America. The rhizomes and roots of the species have been used by a number of Native Americans for medicine for stoppage of urine, pain, easing labor, bosom pains, sores, and vomiting blood. A few groups used the rhizomes and fiddleheads as food.

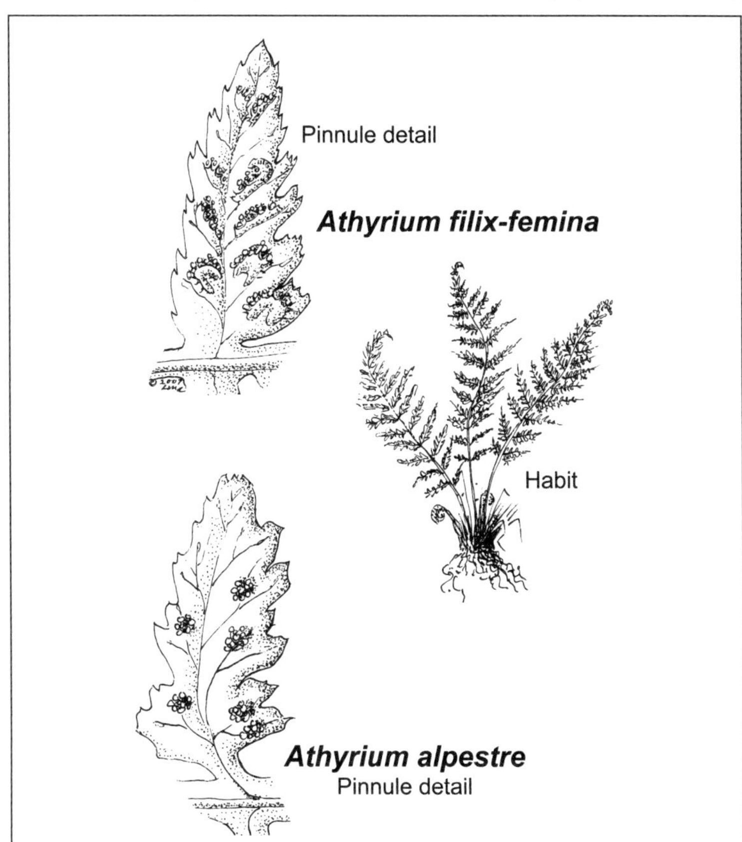

Cystopteris Bernh. Bladder Fern, Brittle Fern

(Greek *kystos*, bladder, + *pteris*, fern, referring to the inflated immature indusium) Sporophyte terrestrial or lithophytic (commonly in ours). **STEMS** short to long-creeping. **LEAVES** monomorphic, not evergreen; petiole 1/3–3 times the length of the blade, with base swollen; blade ovate-lanceolate to elliptic-deltate, to tripinnate-pinnatifid, apex pinnatifid, membranaceous or herbaceous; pinna margin crenulate to serrulate, equilateral to inequilateral; costae adaxially grooved continuously from rachis to costae; glabrous or with multicellular hairs in pinnae axils to unicellular glandular hairs abaxially. **SORI** round, in 1 row on ultimate segments of pinnae, between midrib and margin; indusium ovate to lanceolate, hooded over the sorus toward the segment margin, persistent to obscure when mature. **SPORES** brown, spiny or verrucate. $x = 42$. About 20 species worldwide. A taxonomically problematic and difficult genus due to much variation (especially in *C. fragilis*) and hybrid origin of many species with much apparent polyploidy plus the occasional presence of sterile hybrids. Many of the characters differentiating species are subtle and overlapping. It is vitally important, therefore, for field workers to obtain adequate vegetative material as well as fertile fronds with spores for proper identification. The treatment here of *C. reevesiana* and *C. fragilis* is based upon specimens with spores (observed and measured) and annotations by Christopher Haufler at COLO. Together these two species represent our most common ferns.

1. Blade broadly triangular, shorter than petiole; proximal pinnae with enlarged basiscopic pinnules; stems cordlike, long-creeping, slender, leaf bases more than 1 cm apart; rare .... ***C. montana***
1' Blade elliptic, ovate-lanceolate, or narrowly deltate, equal to or shorter than the petiole; proximal pinnae without enlarged basiscopic pinnules; stems not cordlike, short-creeping, or if long-creeping, leaf bases mostly less than 1 cm apart........ (2)

2. Rachis sometimes with bulblets; rachis, indusia, and midribs of ultimate segments sparsely covered with gland-tipped hairs; blades deltate, usually widest at or near the base; rare .... ***C. utahensis***
2' Rachis without bulblets; rachis, indusia, and midribs of ultimate segments lacking glandular hairs; blades elliptic to lanceolate, widest near the middle........ (3)

3. Lowest pinnae pinnate-pinnatifid to bipinnate; stems mostly long-creeping; spores 33–41 μm diameter; common .... ***C. reevesiana***
3' Lowest pinnae pinnatifid to pinnate-pinnatifid; stems short-creeping; spores 39–60 μm diameter .... (4)

4. Pinnae typically leaving rachis at acute angle, often curving toward blade apex; pinnae along distal 1/3 of blades ovate to narrowly elliptic; pinnae margins crenulate or with rounded teeth; basal basiscopic pinnules of lowest pinnae cuneate to rounded at base........ ***C. tenuis***
4' Pinnae typically perpendicular to rachis, not curving toward blade apex; pinnae along distal 1/3 of blade deltate to ovate; pinnae margins with sharp teeth; basal basiscopic pinnules of lowest pinnae truncate to rounded at base .... ***C. fragilis***

Cystopteris fragilis (L.) Bernh. (easily broken, brittle) Brittle fern, fragile fern. **STEMS** not cordlike, with short internodes and old petiole bases, scales light brownish, mostly lanceolate. **LEAVES** monomorphic, clustered, to 40 cm long, ours much shorter; petiole dark basally, grading to green or straw-colored, shorter or the same length as leaves; blade lanceolate to elliptic, pinnate to bipinnate-pinnatifid, apex acute; pinnae perpendicular to rachis, margins serrate to dentate, base truncate to obtuse. **SORI** present on all leaves; indusium ovate to lanceolate. **SPORES** spiny or verrucate (most of ours), about 39–50 μm diameter. $2n = 168$, 252. [*C. dickieana* R. Sim, *C. fragilis* subsp. *dickieana* (R. Sim) Hyl.]. On ground, in talus, in crevices of rocks or cliffs, limestone, sandstone, granite, rhyolite, in riparian areas, ponderosa pine, spruce, fir, aspen associations to tundra well above timberline. COLO: Arc, Hin, LPl, Min, Mon, RGr, SJn; NMEX: SJn; UTAH. 2340–4285 m (7675–14060′). Spores: late May–Sep. Greenland and northern Alaska south to New Mexico, Nebraska, the southern Great Lakes, New England, and New York; Mexico; circumboreal. This species tends to be confused with the other species of *Cystopteris* but tends to occur at higher elevations and latitudes and reportedly is more likely on cliffs. *C. fragilis* commonly hybridizes with *C. reevesiana* (see that species for discussion). In our area *C. fragilis* forms a hybrid with *C. montana.* Six specimens, covering the range of listed counties, much of the elevation range, and much of the spore date range, have aborted, verrucate spores. This spore type is considered a recessive character and is unknown in *C. reevesiana*. These specimens are interpreted as hybrids between the 4× and 6× types of *C. fragilis* or examples of spore abortion due to gene silencing.

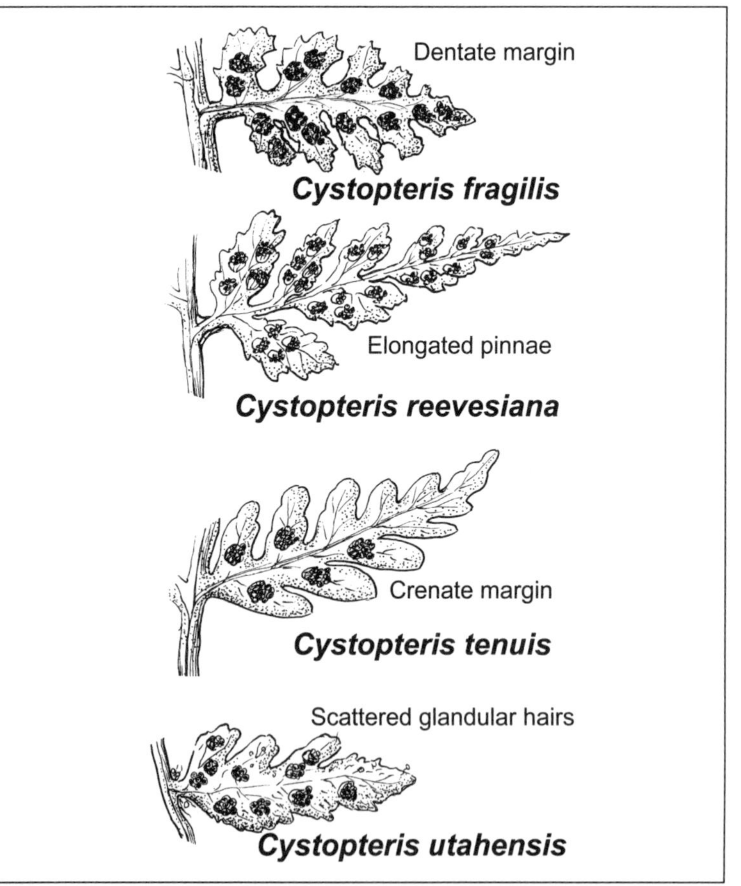

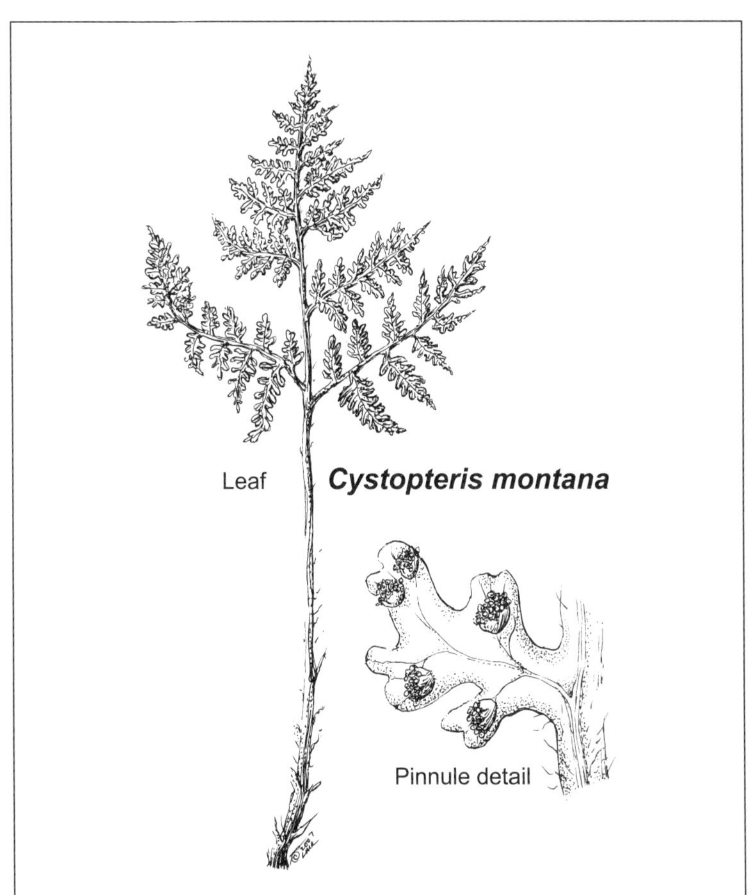

Cystopteris montana (Lam.) Bernh. ex Desv. (mountain) Mountain bladder fern. **STEMS** long-creeping, cordlike, internodes 1–3 cm long, with few old petiole bases; scales light brownish, lanceolate to ovate-lanceolate. **LEAVES** monomorphic, not tightly clustered, to 45 cm long, ours much shorter; petiole brown or black at base, grading to green or straw-colored, about 2–3 times length of blade, somewhat to sparsely scaly; blade elongate-pentagonal, tri- to quadripinnate-pinnatifid; pinnae axils with gland-tipped hairs, pinnae ascending, at acute angle with rachis, margin serrate, base truncate to obtuse. **SORI** equally produced on all leaves and independent of season; indusium cupular, apex truncate, marginal hairs glandular. **SPORES** spiny, 37–42 μm diameter. $2n$ = 168. Rare in moist riparian areas at high elevations on granite under spruce-fir forest, in moss or duff. COLO: LPl, SJn. 3110–3445 m (10200–11300′). Spores: dates unknown, suspected in Aug–Sep. Alaska, Yukon, and southwestern Northwest Territories south to British Columbia and northwest Montana; also west-central Colorado and Saskatchewan; southern Greenland south to Ontario and Quebec as well as Eurasia. This species is a parent, along with *C. fragilis*, of a hybrid found in a remote canyon. The specimen has immature sporangia and it cannot be determined if this is a sterile hybrid or a fertile, undescribed, allopolyploid species. This is the first report of hybridization involving *C. montana* in North America.

Cystopteris reevesiana Lellinger (for Timothy Reeves, born 1947, Southwestern and Mexican fern botanist) Reeves' brittle fern, Reeves' bladder fern, Southwestern brittle fern. **STEMS** not cordlike, with long internodes and persistent petiole bases, scales brownish, lanceolate to ovate. **LEAVES** monomorphic, usually crowded at stem apex, to 40 cm long, but often much shorter in our area; petiole dark purple to straw-colored, shorter than blade length, with sparse scales; blade ovate to elliptic, bi- to tripinnate, apex short-attenuate, glandular hairs occasionally in pinnae axils; pinnae perpendicular to leaf axis, margin dentate to crenate, base truncate to obtuse. **SORI** with indusium cupular to lanceolate. **SPORES** 33–41 μm diameter, echinate. $2n$ = 84. [*C. fragilis* (L.) Bernh. subsp. *tenuifolia* Clute]. A common terrestrial or lithophyte in cracks, talus, boulders, and cliffs; often on sandstone, also on limestone, metaconglomerate, porphyritic andesite, and hornfels; in hanging gardens at lower elevations, riparian communities, ponderosa pine, Douglas-fir, spruce, fir, and aspen, as well as subalpine meadows. ARIZ: Apa, Nav; COLO: Arc, Hin, LPl, Min, Mon, RGr, SJn; NMEX: SJn; UTAH. 1768–3505 m (5800–11500′). Spores: Jul–Oct. Colorado and Utah south to Arizona, New Mexico, western Texas, and northern Mexico. *Cystopteris reevesiana* is one of the diploid ancestors of *C. fragilis* and the two can be difficult to distinguish, especially in smaller specimens. Hybrids between the two are common in the Four Corners region. Ten specimens with aborted, spiny spores are known from our area. They come from COLO: Arc, LPl, RGr, SJn and NMEX: McK and at various elevations, with some occurring at higher elevations (alpine) than *C. reevesiana*. *C. reevesiana* and *C. bulbifera* (L.) Bernh. are the diploid parents of tetraploid *C. utahensis*, which itself can backcross to form a sterile triploid hybrid.

Cystopteris tenuis (Michx.) Desv. (thin, slender) Mackay's brittle fern, upland brittle bladder fern. **STEMS** not cordlike, internodes short, retaining old petiole bases; scales light brownish, lanceolate. **LEAVES** monomorphic, clustered, to 40 cm long, ours generally much shorter; petiole dark at base grading to greenish or straw-colored, shorter than or equaling blade length, sparsely scaly at base; blade lanceolate to elliptic, pinnate to bipinnate-pinnatifid, apex short-attenuate, pinnae axils lacking glandular hairs; pinnae at acute angles to rachis, margins crenulate, base cuneate to obtuse. **SORI** with indusium cupular to lanceolate. **SPORES** 39–50 μm diameter, echinate. $2n = 168$. [*C. fragilis* (L.) Bernh. var. *mackayi* G. Lawson]. Rare, cracks and ledges of sandstone in riparian communities. ARIZ: Apa; COLO: LPl; UTAH. Our three records are from the Weminuche Wilderness, Canyon de Chelly National Monument, and Kane Gulch. The elevation is known only for Kane Gulch: 1935 m (6350′). Spores: dates unknown. In a personal communication, Michael Windham states that records he has seen suggest a broad distribution in sheltered sandstone habitats (in Utah and Arizona). Minnesota east to Nova Scotia, south to Oklahoma and North Carolina; also Arizona Strip, southern Utah, and southern Nevada. This species is probably an allopolyploid from parents *C. protrusa* (Weath.) Blasdell and an unknown diploid *Cystopteris* species.

Cystopteris utahensis Windham & Haufler (of Utah) Utah bladderfern. **STEMS** not cordlike, internodes short, retaining most old petiole bases, scales lanceolate, more or less clathrate. **LEAVES** monomorphic, clustered, to 45 cm long, ours usually shorter; petiole sometimes darker near base or green throughout, shorter than blade length, sparsely scaly; blade deltate to deltate-lanceolate, bipinnate-pinnatifid, apex attenuate, rachis with glandular hairs, misshapen bulblets may be present; pinnae axils with multicellular, gland-tipped hairs; pinnae perpendicular to rachis, margins serrate, base truncate to obtuse. **SORI** with indusium cupular, apex truncate, with scattered, glandular hairs. **SPORES** 39–48 μm diameter, echinate. $2n = 168$. Sandstone canyons. ARIZ: Apa. Our material is from Canyon de Chelly National Monument. Elevations not known. Spores: Aug. Utah and northern Arizona east to northwestern Colorado; also western Texas. *Cystopteris utahensis* is an allopolyploid derived from diploid parents *C. bulbifera* and *C. reevesiana.*

Dryopteris Adans. Wood Fern, Shield Fern

(Greek *drys*, tree, + *pteris*, fern) Terrestrial, rarely lithophytic. **STEMS** short-creeping or erect. **LEAVES** monomorphic; evergreen or nonevergreen in winter; petiole about 1/4–2/3 length of blade, base swollen, sometimes not; blade lanceolate to deltate, pinnate-pinnatifid to tripinnate-pinnatifid, reduced to a pinnatifid apex, herbaceous or slightly leathery, more or less glabrous adaxially; pinnae segment margins entire, crenate, or serrate, more or less spinulose; sessile to petiolulate, often inequilateral; costae adaxially grooved and continuous from rachis to costae to costules; vestiture of linear to ovate scales abaxially, also occasionally with glands. **SORI** round, in 1 row between margin and midrib; indusium round-reniform. **SPORES** brownish, rugose or with folded wings. $x = 41$. Relationships among *Dryopteris* species are complex and confused by the presence of many hybrids. About 150 species; 14 species in Canada and the United States, mostly in temperate Asia.

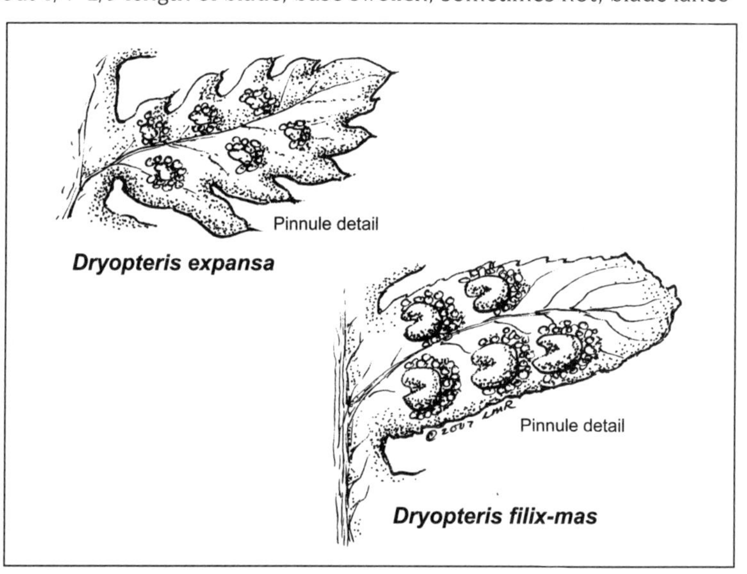

1. Blades tripinnate-pinnatifid at base; sorus small, dotlike .... ***D. expansa***
1′ Blades pinnate-pinnatifid to bipinnate at base; sorus large, prominent .... ***D. filix-mas***

Dryopteris expansa (C. Presl) Fraser-Jenk. & Jermy (expanded out) Northern wood fern, spreading wood fern. **LEAVES** not evergreen, to about 90 × 30 cm, ours generally not as large; petiole generally about 1/3 length of leaf, base scaly, sometimes extending distally; scales diffuse, lighter brown with dark brown central stripe; blade deltoid, tripinnate-pinnatifid, herbaceous, occasionally densely glandular; pinnae lanceolate-oblong; basal pinnae deltate; pinnule margins conspicuously serrate. **SORI** small, not prominent, midway between midvein and margin of segments; indusium non- or sparsely glandular. $2n = 82$. [*Nephrodium expansum* C. Presl; *Dryopteris assimilis* S. Walker; *D. dilatata* (Hoffm.) A. Gray subsp. *americana* (Fischer) Hultén]. Our single collection represents a range extension from the only other Colorado location in Rocky Mountain National Park. The plant was collected on a roadside, south-facing, open talus slope surrounded by spruce-fir forest. COLO: RGr. 3097 m (10160′). Alaska, southern Yukon, and southwestern Northwest Territories, south to California to the southern coast to western Alberta and western Montana and western Colorado; also in northeastern Minnesota to Newfoundland and Greenland. Fiddleheads (with scales removed) and stems used as food by several Native American peoples.

Dryopteris filix-mas (L.) Schott (male fern) Male fern. **LEAVES** not evergreen, about 26–120 × 10–30 cm, ours tending to be smaller; petiole shorter than 1/4 the length of leaf, base scaly, prominently so in ours; scales dispersed, brown, 2 forms: broad and hairlike, with no intermediates; blade ovate-lanceolate, pinnate-pinnatifid to twice pinnate basally, substantial but not coriaceous, not glandular; pinnae lanceolate; basal pinnae ovate-lanceolate, much reduced; pinnule margin lobed to serrate but not conspicuously toothed. **SORI** large and conspicuous, located midway between midvein and margin of segments, indusium without glands. $2n = 164$. [*Polypodium filix-mas* L.]. Scattered locations in shaded, mesic canyons on sandstone, Dakota Sandstone (has calcareous matrix), granite, and metamorphic rock; on cliffs and talus, in shaded cracks, usually along rivers and creeks; woodlands and forests of cottonwood, ponderosa pine, Douglas-fir/ponderosa pine/Gambel's oak, Douglas-fir, and spruce in canyons surrounded by ponderosa pine/aspen/*Juniperus communis*. ARIZ: Apa (Canyon de Chelly National Monument); COLO: Arc, Hin, LPl, Min, Mon, SJn. 1800–3750 m (5900–12300′), here reaching by far the highest known elevation for the species. Throughout the Rocky Mountains from British Columbia, southwestern Alberta, Montana, and Wyoming south to southern Arizona and southern New Mexico and extreme western Texas; also in Wisconsin and Michigan to Quebec, Newfoundland, Nova Scotia, and Greenland; one collection in extreme northern Mexico, near Big Bend; Europe; Asia. According to *Flora of North America* (1993), the taxonomy of *D. filix-mas* is poorly understood. The material (unknown chromosome number) in the southwestern Rocky Mountains (including our area) is morphologically and ecologically different from the tetraploid found in the Northeast and Northwest on limestone. Reportedly, the southwestern Rocky Mountain material closely resembles *D. pseudofilix-mas* (Fée) Rothm. Modern collecting for any food or medicinal use could easily exterminate this species in our area.

Gymnocarpium Newman Oak Fern

(Greek *gymnos*, naked, + *karpos*, fruit, describing the absence of an indusium on the sorus) Terrestrial. **STEMS** long-creeping. **LEAVES** monomorphic, not evergreen; petiole long, about 1.5–3 times length of blade, blade unswollen; blade ovate, ternate, or deltate, bi- or tripinnate-pinnatifid, apex pinnatifid, herbaceous; segment margin entire to crenate-crenulate; proximal pinnae longest, petiolulate, usually inequilateral; costae adaxially grooved, grooves not continuous; indument lacking or of minute glands abaxially and occasionally along costa adaxially. **SORI** round, in 1 row between midrib and margin; indusium absent. **SPORES** brownish, rugose. $x = 40$. Eight species. North-temperate regions in North America; Eurasia.

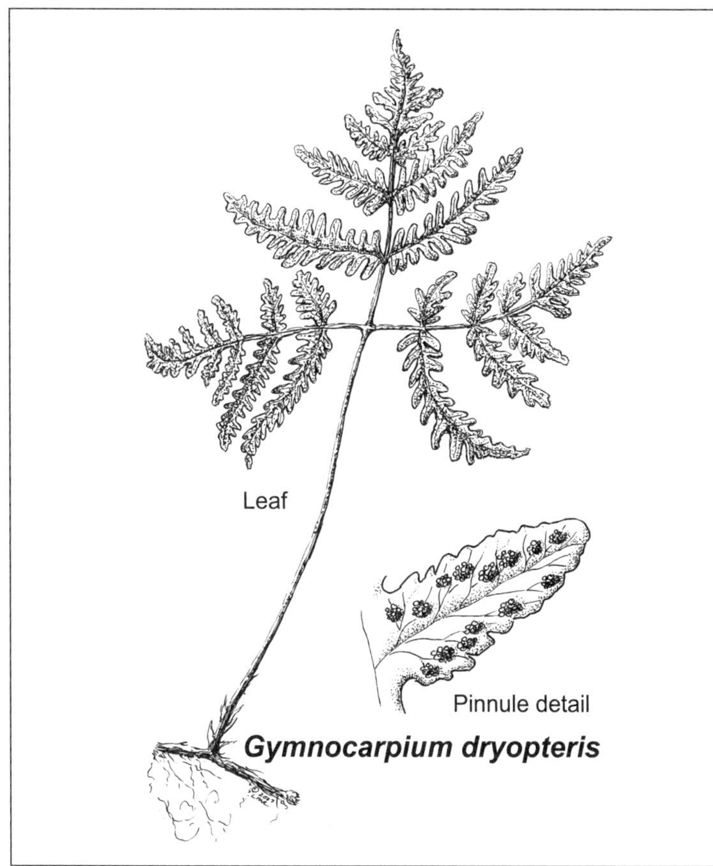

Gymnocarpium dryopteris (L.) Newman (named for the genus *Dryopteris*) Common oak fern, western oak fern. **STEMS** 0.5–1.5 mm diameter; scales 1–4 mm long. **LEAVES** (fertile) mostly 12–42 cm, ours often shorter; petiole 9–28 cm long, with few glandular hairs; scales 2–6 mm long; blade deltate, ternate, bipinnate-pinnatifid, 3–24 cm long, delicate, abaxial surface and rachis glabrous or with few glandular hairs, adaxial surface glabrous; pinna apex entire, rounded; proximal pinnae 2–12 cm long, more or less perpendicular to rachis, with basiscopic pinnules perpendicular to costa; basal basiscopic pinnule usually sessile, pinnatifid or rarely pinnate-pinnatifid; second basal basiscopic pinnule sessile, ultimate segments of proximal pinnae oblong, entire to crenate, apex entire, rounded. **SPORES** 34–39 μm diameter. $2n = 160$. [*Polypodium dryopteris* L.; *Dryopteris linnaeana* C. Chr.; *Lastrea dryopteris* (L.) Bory; *Phegopteris dryopteris* (L.) Fée; *Thelypteris dryopteris* (L.) Sloss.]. Very rare; spruce and spruce-fir forests, deep shaded gorges; mossy and wet, shaded areas; creek banks. COLO: Arc, Hin, LPl.

2633–2896 m (8640–9500′). Alaska, Yukon to the Atlantic, south to Oregon, Arizona, New Mexico, Kansas, and West Virginia; cirumboreal. Known from five collections in our area from three general vicinities.

Polystichum Roth Sword Fern, Christmas Fern, Holly Fern

(Greek *poly*, many, + *stichos*, row, describing the rows of sori) Terrestrials. **STEMS** decumbent or erect. **LEAVES** evergreen and monomorphic; petiole 1/10 to equal the length of blade; blade linear-lanceolate to lanceolate-elliptic, to tripinnate, apex pinnatifid, leathery; segment margins mostly toothed; bases usually inequilateral with acroscopic lobe; costae adaxially grooved and continuous from rachis to costae; scales on costae and sometimes between veins abaxially, glabrous or similarly scaly adaxially. **SORI** round, in 1 to many rows between midrib and margin; indusium peltate, persistent or caducous or, rarely, absent. **SPORES** yellow, brownish, or black, with folds inflated. $x = 41$. About 180 species worldwide. Sterile hybrids tend to be frequent where one or more species grow in close proximity.

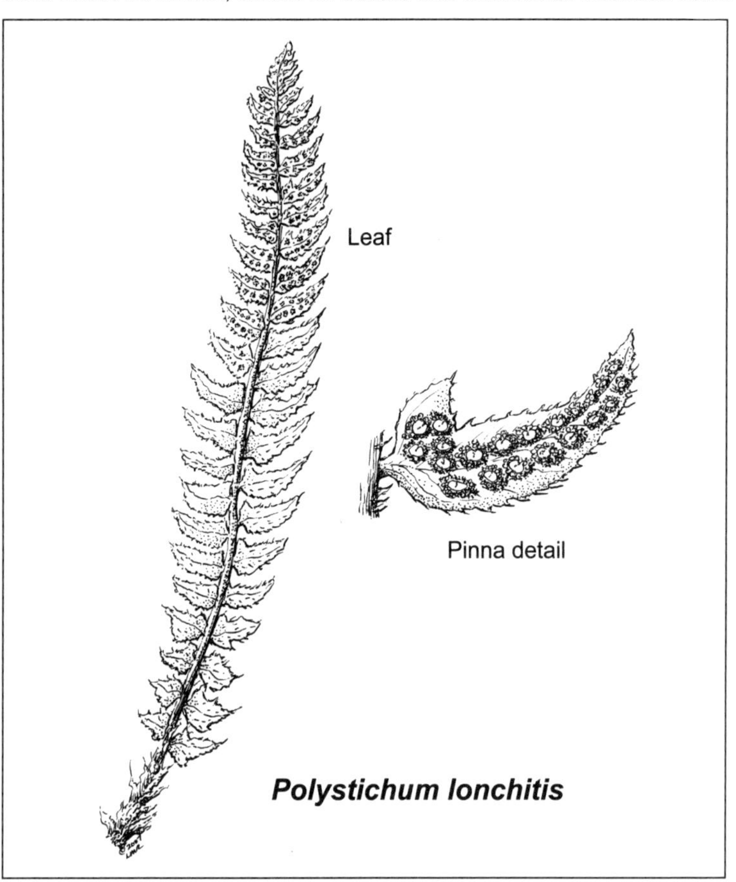

Polystichum lonchitis

Polystichum lonchitis (L.) Roth (a name used by Pliny for a plant with a tongue-shaped leaf) Holly fern. **STEMS** mostly erect. **LEAVES** clustered, sometimes arching, erect, 1–6 dm long; petiole 1/10–1/5 length of blade, dense scales light brown; blade linear, pinnate, base acute; pinnae oblong to lanceolate to falcate, proximal pinnae deltate, 0.4–2.8 cm, base truncate to oblique, auricle well developed; margins spiny; apex acute; scales tiny, dense, restricted to abaxial surface. **SORI** indusium entire or minutely dentate-erose. **SPORES** dark to blackish brown, echinate. $2n = 82$. [*Polypodium lonchitis* L.]. Very rare; cool, mesic sites; spruce and spruce-fir forest communities; on forest floors in shady canyon bottoms, and at the base of cliffs, on rock outcrops and talus. COLO: Arc, LPl. 2438–2682 m (8000–8800′). Spores: Jul–Aug. Southern Alaska and Aleutian Islands, Yukon to Newfoundland, south to California and New Mexico; Europe. Our plants represent a southern range disjunction from northwestern Colorado and northeastern Utah. First fiddleheads appear in early–mid-May.

Woodsia R. Br. Cliff Fern

(for Joseph Woods, English botanist) Lithophyte. **STEMS** somewhat creeping to clustered; usually ascending or erect. **LEAVES** monomorphic, only rarely evergreen; petiole 1/5–3/4 blade length, base not swollen; blade linear to ovate or lanceolate, once or bipinnate-pinnatifid, herbaceous; pinnae with segment margins entire to dentate, base equilateral; costae grooved adaxially and continuous from rachis to costae; glandular hairs often present on both leaf surfaces. **SORI** round, in a row between margin and midrib of ultimate segments; indusium basal, often dissected and persistent. **SPORES** brownish, cristate to rugose. $x = 38, 39, 41$. About 30 species. North-temperate regions and cooler montane tropics. Many species in this genus are often confused with *Cystopteris* species but can be differentiated by the multilobed or fimbriate indusia or persistent petioles. Hybrids among *Woodsia* species are common.

1. Pinna with flattened, multicellular hairs along midrib; petiole brittle .... ***W. scopulina***
1′ Pinna lacking flattened, multicellular hairs along midrib; petiole pliable .... (2)

2. Indusium segments broad for most of length, often divided and filamentous distally; leaves densely glandular, often somewhat viscid .... ***W. plummerae***
2′ Indusium segments narrow, filamentous; leaves glabrescent or moderately glandular .... (3)

3. Pinnule margin smooth to erose; petiole base red-brown to dark purplish .... *W. oregana*
3' Pinnule margin with translucent projections on teeth; petiole base light brown or yellowish .... *W. neomexicana*

Woodsia neomexicana Windham (of New Mexico) New Mexican cliff fern. **STEMS** erect to ascending; with persistent petiole bases, scales lanceolate, brown, some with a darker central stripe. **LEAVES** 5–30 × 1–6 cm; petiole straw-colored to light brownish, brittle; blade linear to lanceolate, pinnate-pinnatifid, glabrous to thinly glandular; pinna ovate-deltate to elliptic, apex rounded or acute; pinnule dentate or lobed, nonshining, thin, with glands and translucent projections on teeth. **SORI** with indusium of filamentous segments. **SPORES** 44–50 μm diameter. $2n = 152$. Crevices of cliffs and outcrops on sandstone, conglomerate, and igneous substrates; sagebrush, piñon, ponderosa pine, Douglas-fir, aspen, and spruce communities. ARIZ: Apa; COLO: Hin, LPl, SJn; NMEX: McK, SJn; UTAH. 1815–3020 m (5950–9900'). Spores: late May–Sep. Arizona, southern Colorado, and New Mexico east to Texas, disjunct to Oklahoma and South Dakota; north-central Mexico. This species, an allotetraploid, is known to hybridize with *W. oregana*, another allotetraploid, to produce sterile tetraploid hybrids. In herbarium collections it is often confused with *Cystopteris* and *W. mexicana*.

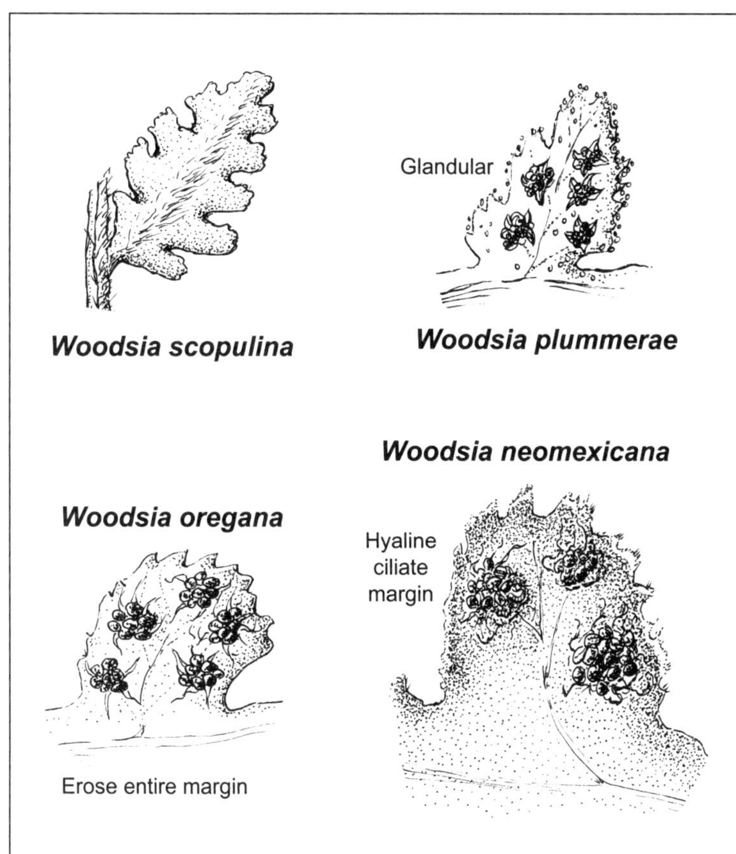

Woodsia oregana D. C. Eaton (of Oregon) Oregon cliff fern. **STEMS** erect to ascending, with persistent petiole bases, scales lanceolate, brownish, sometimes with dark central stripe. **LEAVES** 4–25 × 1–4 cm; petiole red-brown to purplish, pliable; blade linear-lanceolate to ovate, once or bipinnate-pinnatifid, somewhat glandular, occasional hairlike scales on rachis; pinna ovate-deltate to elliptic, apex round to acute, glabrous to glandular; pinnule margin often minutely erose (in ours). **SORI** with indusium of narrow filamentous sections often concealed by developing sporangia. **SPORES** about 45–50 μm diameter (in ours). Crevices of cliffs and outcrops of sandstone and basalt; snakeweed, piñon, juniper, ponderosa pine, spruce, fir, aspen, Douglas-fir, and Gambel's oak communities. ARIZ: Apa; COLO: Hin, Min, Mon, RGr, SJn; NMEX: McK, RAr, SJn; UTAH. 2285–3240 m (7500–10630'). Spores: Jun–Sep. Western Canada to Quebec, south to California, Arizona, New Mexico, and Texas, east to Oklahoma, Kansas, Nebraska, Iowa, Minnesota, Wisconsin, and Michigan. Our plants are **subsp. *cathcartiana*** (B. L. Rob.) Windham. Hybridizes with *W. neomexicana* and *W. plummerae*.

Woodsia plummerae Lemmon (named for Sara Allen Plummer, 1836–1923, a fern and alga botanist) Plummer's cliff fern. **STEMS** erect to ascending, with persistent petiole bases, scales brown or with a dark central stripe, lanceolate. **LEAVES** 5–25 × 1.5–6 cm; petiole red-brownish to purplish, somewhat pliable; blade lanceolate to ovate, bipinnate, densely glandular, often somewhat viscid, glandular hairs abundant on rachis, with few narrow scales; pinna ovate-deltate to elliptic, apex rounded or broadly acute, both surfaces glandular; pinnule dentate, margin densely glandular, occasionally with 1–2-celled translucent projections. **SORI** with indusium of somewhat broad segments, often divided into filamentous tips, often surpassing mature sporangia. **SPORES** 44–50 μm diameter. $2n = 152$. [*W. obtusa* (Spreng.) Torr. var. *glandulosa* D. C. Eaton & M. Faxon; *W. obtusa* var. *plummerae* (Lemmon) Maxon; *W. pusilla* E. Fourn. var. *glandulosa* (D. C. Eaton & Faxon) T. M. C. Taylor]. Known in our area only from Vallecito Creek, where it grows on meta-conglomerate rocks, with Engelmann spruce, Douglas-fir, aspen, ponderosa pine, and *Pinus strobiformis*. COLO: LPl. 2415 m (7920'). Spores: Aug. Southeastern California, Arizona, New Mexico, southern Colorado, western Oklahoma, western Texas. This species may have arisen as an autopolyploid from a Mexican diploid species. *Woodsia plummerae* is difficult to distinguish from *W. oregana* in our area. A number of our specimens have been annotated as hybrids between the two. These occur in ARIZ: Apa; COLO: Hin; NMEX: McK, SJn. Most are sterile hybrids (or spores are too young to determine status), but one specimen is fertile, an allopolyploid. Two other specimens are annotated as *W. plummerae* ×, with the second parent species in question. These are from COLO: Hin and UTAH.

Woodsia scopulina D. C. Eaton (little rock or crag, referring to cliff habitat) Mountain cliff fern, Rocky Mountain cliff fern. **STEMS** erect to ascending, with persistent petiole bases, scales brown or with darker central stripe. **LEAVES** 9–35 × 1–8 cm; petiole red-brown to dark purplish, brittle; blade lanceolate, mostly bipinnate, glandular, especially on rachis; pinna lanceolate-deltate to ovate, with apex rounded to broadly acute or attenuate, flattened hairs along midribs; pinnule margin dentate to lobed, nonshining, thin, with scattered glands and cilia. **SORI** with indusium of narrow filamentous segments, often obscured by developing sporangia. **SPORES** 50–57 μm diameter (in ours). $2n = 152$. [*W. obtusa* Torr. var. *lyallii* Hook.; *W. oregana* D. C. Eaton var. *lyallii* (Hook.) B. Boivin]. Cliffs on a variety of substrates at high elevations in spruce-fir communities. COLO: Hin, Min. 3400–3600 m (11150–11700′). Spores: Jul–Aug. Alaska south to western Canada to Quebec, south to California, New Mexico, Oklahoma, Tennessee, and North Carolina. Our plants are **subsp. *laurentiana*** Windham. A rare species in our study area and certainly a range extension for the subspecies.

EQUISETACEAE Michx. ex DC. HORSETAIL FAMILY

Timothy Reeves

Gametophytes (haploid, minute, rarely seen plants) generally subterrestrial, green, unisexual, with the male smaller than female plants. Sporophytes (diploid visible plants) perennial, with rhizomes. **STEMS** annual or perennial, jointed with distinct nodes, with centers hollow and walls with small canals, often with silica embedded in stems. **LEAVES** minute, in fused whorls subtending the nodes; leaf tips free, toothlike. **CONES** (strobilus) 0.3–10 cm long, with sporangia on the underside of a peltate sporophyll. **SPORES** monomorphic, green (white in hybrids), bearing 4 white, strap-shaped elators. One genus (the two subgenera sometimes treated as separate genera) with about 15 species, many hybrids; worldwide. Many species, due to the embedded silica, were often used by many tribes and other groups as sandpaper or scouring and cleaning aids.

Equisetum L. Horsetail, Scouring Rush, Cola de Caballo, Cañuela, Prêle

(Latin *equis*, horse, + *seta*, bristle) Perennial, rhizomatous. **STEMS** annual or perennial; ridged, center hollow, with a series of large vallecular and small carinal canals; branches present or absent, breaking through base of sheath at nodes. **CONES** 1–4 cm long, terminating green stems or specialized, reproductive, nongreen stems; sporophylls peltate; sporangia 5–10 per sporophyll, pendent, attached to sporophyll inner surface, dehiscing longitudinally. $x = 108$. Fifteen species worldwide. Horsetails and scouring rushes are wetland plants found on riverbanks, along acequias (irrigation ditches), around lakes, at roadsides, in meadows and seeps, and in wet woodlands. The species *E. pratense* Ehrh. is sometimes attributed to our area but no genuine material has been seen. (Hauke 1963, 1978; Cobb, Farnsworth, and Lowe 2005.)

1. Stems branched, with regular whorls of 10 or more branches found at most nodes; cones borne on separate, nongreen stems present briefly in spring .... ***E. arvense***
1′ Stems mostly unbranched, branches, if present, at only a few nodes, in whorls of 1–6; cones borne on green stems, present throughtout growing season or year .... (2)

2. Spore white, malformed; sheath green (especially at upper nodes), often with persistent teeth, and at lower nodes with irregular black band, and/or sheath whitish gray with terminal black band and smudged, irregular basal black band, teeth sometimes persistent; either strongly resembling *E. laevigatum* or *E. hyemale* or intermediate between these two parental species .... ***E. ×ferrissii***
2′ Spore green, spherical; sheath either green or whitish with two crisp, straight black bands; sheath teeth usually promptly deciduous .... (3)

3. Sheath whitish gray with a crisp, straight black band at base, sometimes with a terminal black band; sheath teeth usually shed .... ***E. hyemale***
3′ Sheath green (sometimes some sheaths partially white in fall but without black bands), teeth shed or persistent .... (4)

4. Sheath teeth promptly deciduous on most nodes; teeth on sheath subtending cone numerous, black with thin white margins; cone apex rounded to apiculate with blunt tip; stem ridges flattened or convex; silica ridges (20–45×) undulating and diagonal across ridges; very common .... ***E. laevigatum***

4' Sheath teeth persistent on most nodes; teeth on sheath subtending cone few, white with narrow black centers; cone apex sharply apiculate; stem ridges minutely grooved; silica tubercules (20–45×) in 2 rows on ridges; very rare .... ***E. variegatum***

Equisetum arvense L. (a field or meadow) Field horsetail, common horsetail, bottle brush, prêle des champs. **STEMS** dimorphic; fertile stems nongreen, appearing briefly in spring, to about 20 cm tall, quickly dying after sporulation; sterile stems green, not bearing cones, to 60 (100) cm, regularly branching at most nodes, 10+ long branches per node; sheaths green with dark teeth; branches ascending, usually at least at base, 3–4-angled; first internode of each branch longer than subtending stem sheath; sheath teeth attenuate. **CONES** blunt-tipped, maturing and scales spreading apart. **SPORES** spherical, green, surrounded by 4 elators. $2n$ = about 216. Very common in wet habitats, acequias, river banks, stream banks, edges of ponds, meadows, slopes above wetlands, roadsides, logging roads, marshes, and arroyos. Riparian communities with cottonwood, tamarisk, Russian olive, willow in desert scrub, sagebrush, grassland, piñon-juniper, ponderosa pine, aspen, Douglas-fir, and spruce-fir communities. ARIZ: Apa, Nav; COLO: Arc, Dol, Hin, LPl, Min, Mon, SJn; NMEX: McK, RAr, SJn; UTAH. 1550–3650 m (5000–12000′, the highest elevation recorded for the species). Spores: (Apr–) May. Throughout northern North America, Greenland, Canada, and all the continental United States except for Florida, Louisiana, Mississippi, and South Carolina; Eurasia south to North Africa; Himalayas, central China, Korea, and Japan. See note above under genus description concerning *E. pratense*, a related species.

Equisetum ×ferrissii Clute (pro sp.) (named for Ferris) Intermediate scouring rush. This is the sterile hybrid between *E. hyemale* subsp. *affine* and *E. laevigatum*. **STEMS** monomorphic, 20–180 cm tall; green, tending to be evergreen (primarily in protected sites, i.e., under trees) at least on lowermost internodes; bearing cones; somewhat rough to the touch; ridges 14–32; mostly unbranched; branches, if present, 1–6 per node, short or long, present at a few upper nodes; sheath varying from green with terminal black tip like *E. laevigatum* to whitish gray with terminal and basal black bands like *E. hyemale*; sometimes upper node sheaths green with persistent teeth and middle and/or lower sheaths whitish gray with black bands, the lower band smudged and irregular; our most common and distinctive material resembles *E. hyemale* in having whitish gray sheaths with terminal and basal black bands, the basal band variously smudged and irregular, sheaths sometimes entirely black. **CONES** with pointed apex that is sometimes noticeably slender; scales not spreading apart in age. **SPORES** aborted, white, irregular in shape, not shed from sporangia. Somewhat frequent in wet habitats, especially along irrigation ditches, creek banks, river banks, and disturbed areas including roadsides. Sometimes growing with *E. laevigatum* or *E. arvense*. Riparian communities with cottonwood, Russian olive, tamarisk, and/or willows in sagebrush, piñon-juniper, ponderosa pine, farms, and towns. ARIZ: Apa; COLO: Hin, LPl; NMEX: McK, RAr, SJn (especially frequent here along acequias of the Animas River); UTAH. 1350–2350 m (4400–7700′). Spores: Jun–Nov (and probably throughout winter). Southern British Columbia, southern Alberta, southern Saskatchewan, southeastern Quebec, all United States except Alaska, Kentucky, Mississippi, Tennessee, South Carolina, Georgia, Alabama, and Florida; Mexico in Baja California, Chihuahua, Coahuila, and Durango. This sterile hybrid apparently reproduces vegetatively and occurs at sites where neither parent is present. Multiple collections along 25 miles of irrigation ditches along the Animas River all appear to be derived from the same original source. It may be of interest that this hybrid apparently occurs, in our area, only at lower and middle elevations as compared to both its parental species (but not as low as *E. laevigatum* is found).

Equisetum hyemale L. (winter, alluding to the evergreen nature of the stems) Common scouring rush, rough horsetail, cañuela. **STEMS** monomorphic, 18–220 cm tall; evergreen for more than a year; bearing cones; rough to the touch; ridges 14–50; mostly unbranched; branches, if present, 1–6 per node, short or long, present at a few upper or middle nodes; sheaths whitish gray, with or without a terminal black band, with a basal, crisp, straight black band; sheath teeth promptly deciduous or persistent, the sheath becoming tattered with age and resembling long teeth. **CONES** with pointed apex. **SPORES** spherical, green, surrounded by 4 elators, in mass appearing (at 10×) as "white foam with green specks." $2n = 216$. Somewhat frequent in wet habitats, creek banks, waterfalls, roadsides, river banks, acequia banks, and marshes. Riparian vegetation in sagebrush, piñon-juniper, ponderosa pine, Douglas-fir, spruce-fir, aspen communities. ARIZ: Apa, Nav; COLO: Arc, Hin, LPl, Min, Mon, SJn; NMEX: SJn; UTAH. 1600–3050 m (5250–10000′, the highest known elevation for the species). Spores: mid-May–mid-Jul. Our material is **subsp. *affine*** (Engelm.) Calder & Roy L. Taylor (allied, related). [*Hippochaete hyemalis* (L.) Bruhin subsp. *affinis* (A. Braun) F. Weber]. Alaska, Canada (all provinces), United States (all states), Mexico, Guatemala. *Equisetum hyemale* subsp. *affine* is a parent of *E.* ×*ferrissii* and is collected in our area about as often as that hybrid. The two have not been observed growing together nor have mixed collections of the two been seen.

Equisteum laevigatum A. Braun (smooth) Smooth scouring rush, cola de caballo, cañuela. Annual aboveground, reportedly sometimes evergreen in the southwestern United States, but this not documented by specimens from our area. **STEMS** monomorphic, 20–150 cm tall; smooth to the touch; ridges 10–32, stem ridges flattened or convex; silica ridges (20–45×) undulating and diagonal across ridges; stems mostly unbranched; branches, if present, 1–6 per node, short or long, present at a few upper or middle nodes; sheath green with terminal black band, sometimes partially and irregularly whitish in fall (early stages of stems dying back), without black basal band; teeth usually promptly deciduous, teeth on sheath subtending cone persistent, numerous, black with thin white margins. **CONES** rounded or apiculate with blunt tip. **SPORES** spherical, green, surrounded by 4 elators, in mass appearing (at 10×) as "white foam with green specks." $2n = 216$. [*Hippochaete laevigata* Farw.]. Very common in wet habitats as well as some dry habitats, stream banks, hanging gardens, roadsides, river banks, around boulders in river valleys, along trails, rather dry rocky pastures, below rock outcrops in clay soil, wetlands, banks of acequias, lakeshores, sandy alluvium over limestone, and on high, dry sand dunes. ARIZ: Apa; COLO: Arc, Dol, Hin, LPl, Min, Mon, SMg; NMEX: McK, RAr, SJn; UTAH. 1120–2700 m (3680–8900′). Spores: very late May–mid-Jul. Southeastern British Columbia, southern Alberta, southern Saskatchewan, southern Ontario, southern Quebec, Washington east to Wisconsin, south to California, west and central Texas, Louisiana, Missouri, North Carolina, and Ohio, Mexico in Baja California, Sonora, Durango, and western Chihuahua, also Guatemala. This and *E. arvense* are our most common equisetums, occurring perhaps three times as often as *E. hyemale* and *E.* ×*ferrissii. Equisetum laevigatum* is a parent of *E.* ×*ferrissii* and sometimes grows with it in our area. Small specimens are sometimes misidentified as *E. variegatum*. Resprout growth from damaged underground stems resembles branches without main stems and can be confused with *E. arvense* in a similar growth form. Hopis dried and ground plant; used for ceremonial bread. Kayenta Navajo took an infusion of plant. A cold infusion was used as a lotion for backaches. Ramah Navajo made a compound decoction of plant, used for "lightning infection."

Equisetum variegatum Schleich. ex F. Weber & D. Mohr (smooth, referring to the nonrough texture of the stem ridges) Variegated scouring rush, prêle panachée. **STEMS** monomorphic, unbranched, slender, 6–48 cm tall; evergreen for more than a year; ridges 3–12, stem ridges minutely grooved; silica (20–45×) dots in 2 rows on ridges; sheath green with terminal black band; sheath teeth persistent, with black centers and broad white borders. **CONES** with slender, pointed apex. **SPORES** spherical, green, surrounded by 4 elators, in mass appearing (at 10×) as "white foam with green specks." $2n = 216$. Very rare, known in our area from two collections at Hermosa Park, Colorado. Wet slopes in rocky clay loam soil, abundant, aspen woodland with *Carex, Festuca*, other vegetation closely overgrazed but not the *Equisetum*. COLO: LPl, SJn. 3048 m (10000′). Spores: late Jun. Our material is **subsp. *variegatum***. [*Hippochaete variegata* Bruhin]. Alaska, Canada (all provinces), Washington, Idaho, Montana, south to northern Oregon, northern Utah, western Colorado, Wyoming, and western Nebraska, and Minnesota east to Maine and Connecticut. This northern taxon is not generally referred to our area. Our collections represent a southern range disjunction from northwestern Colorado and northeastern Utah. The stems of this species are small and slender. It differs from *E. laevigatum* in having fewer stem ridges, and these are grooved; in the persistent, prominently white-edged sheath teeth; in the characters of the sheath subtending the cone; and in the silica ornamentation (at 20–45×) of the stem ridges being two rows of tubercules. Small specimens of *E. laevigatum* are sometimes misidentified as *E. variegatum*.

ISOETACEAE Dumort. QUILLWORT FAMILY

Timothy Reeves

Plants perennial, evergreen, caespitose, grasslike, sporophyte heterosporous; gametophytes of 2 sizes, megagametophytes (female) and microgametophytes (male, with usually only a single antheridium); mostly aquatic or semiaquatic, occasionally terrestrial, often grass- or sedgelike in general appearance. Rhizome 2- to 5-lobed, brown, cormlike, grooved, with fine, often branched roots arising from the central groove. **LEAVES** (sporophylls) rushlike, simple, linear, usually spirally inserted, spoonlike at base, apex tapered, small membranous ligule above sporangium, hardened scales and phyllopodia occasionally surrounding; megasporophylls and microsporophylls often produced alternately at bases. **SPORANGIUM** solitary, basal, velum partly or completely covering adaxial surface; megasporangium producing a few to hundreds of megaspores; microsporangium with thousands of microspores. One genus, about 150 species, widespread in mostly temperate and tropical regions worldwide (Tryon and Tryon 1982; Taylor et al. 1993).

Isoetes L. Quillwort

(Greek *isos*, equal, + *etos*, year, referring to evergreen habit) Perennial; rhizome 2- to 3-lobed, globose to spindle-shaped, corky. **LEAVES** several, mainly erect, 1–100 cm long; ligule cordate to deltate, membranous. **SPORANGIUM** ovoid to oblong, 3–15 mm long, unpigmented to brown. **MEGASPORES** white, gray, or black, globose, trilete, smooth, spined, ridged, or tubercled. **MICROSPORES** grayish or brownish, kidney-shaped, monolete, smooth or textured. Size and surface texture of mature, dry megaspores are required for conclusive identification. *Isoetes lacustris* has been reported near our study area in Conejos County, Colorado.

Isoetes bolanderi Engelm. Bolander's quillwort (for Henry Nicholas Bolander, plant collector and botanist, 1832–1897) Plant aquatic, rarely emergent. Rhizome globose, 2-lobed. **LEAVES** deciduous, bright green, but becoming brownish toward base; to 20 cm long, tapering to a thin tip. **SPORANGIUM** brown-streaked on wall; velum covering 1/2 or less the surface of sporangium. **MEGASPORES** white, about 400 μm diameter, somewhat rugulate. **MICROSPORES** brown, about 25 μm diameter, somewhat spiny. $2n = 22$. Alpine, subalpine, or montane lakes and ponds and surrounding muddy shores. ARIZ: Apa; COLO: Arc, Con, Hin, LPl, Min, RGr, SJn; NMEX: McK. 2715–3734 m (8905–12250'). Spores: late summer. British Columbia and Alberta south to Arizona, California, and New Mexico.

LYCOPODIACEAE P. Beauv. ex Mirb. CLUB-MOSS FAMILY

Timothy Reeves

Plants terrestrial, epiphytic or lithophytic; gametophyte prothallium subterranean, fleshy, globose to elliptic, flat or linear, light yellow, monoecious, often with mycorrhizal associations. **STEMS** horizontal (sometimes absent), long-creeping with scattered upright shoots or absent with clustered upright shoots; upright shoots unbranched or dichotomously one–several-branched, or pinnately branched. **LEAVES** scalelike, especially on subterranean stems, needlelike, lanceolate, or ovate, borne in imbricate spirals on upright shoots, mostly ascending or appressed. **SPORANGIUM** kidney-shaped to globose, solitary, borne adaxially near leaf base or in axil of unmodified stem leaf or specialized leaves (sporophylls) clustered into strobili which are terminal on shoots and/or branches. **SPORES** homosporous, trilete, pitted, small-grooved, rugulate, or reticulate. About 10–15 genera with 350–450 species; worldwide, primarily in temperate and tropical montane regions (Wagner and Beitel 1992). Prior to the late 1980s most treatments recognized only two genera worldwide, with *Lycopodium* containing the vast majority of the species (e.g., species here included in *Huperzia*).

1. Horizontal stems absent; upright shoots clustered, short- to long-decumbent; sporangia borne in axils of normal shoot leaves .... ***Huperzia***

1' Horizontal stems present, long-creeping; upright shoots scattered; sporangia borne in axils of modified sporophylls clustered into terminal cones .... ***Lycopodium***

Huperzia Bernh. Gemma Fir-Moss, Fir Club-Moss

(named for Johann Peter Huperz, d. 1816, a German fern horticulturist) Plants lithophytic or terrestrial. **STEMS** absent horizontally; upright shoots evergreen for several or many years, clustered, erect to decumbent, round in cross section, equally dichotomously branched. **LEAVES** mostly not ranked, appressed, mostly not imbricate, ascending to spreading, triangular, lanceolate to oblanceolate, monomorphic or minimally varied in size; basal leaves mostly larger than terminal, margins entire to suberose, roughened by marginal cell papillae. Gemmiferous branchlets and attached gemmae formed among leaves, gemmae abscising at maturity, deltoid, 2.5–6 × 3–6 mm, with 4 leaves flattened into 1 plane, 2 large lateral leaves, and 1 abaxial, 1 adaxial leaf. **SPORANGIUM** kidney-shaped, borne singly at adaxial base of unmodified leaf, fertile leaves in zones, or scattered along shoots. **SPORES** pitted or grooved, concave at equator. $x = 67, 68$. Ten to 15 species; temperate alpine, Arctic regions, and montane tropical in Asia. Since hybrids are common, collectors should take samples of many plants in an area.

Huperzia haleakalae (Brack.) Holub (named for Haleakalae Crater, Maui, Hawaii) Alpine fir-moss. Shoots erect, densely clustered, 8–11 cm, short- to long-decumbent. **LEAVES** spreading-ascending to appressed, lustrous; juvenile growth leaves lanceolate, 4.5–6 (7) mm, apex acute; mature growth leaves ovate, 3–4 mm, apex acute; margins entire. Gemmiferous branchlets on mature portion of shoots; gemmae about 3.5 × 3.5 mm. **SPORES** homosporous, 30–40 μm. Rare; wet, mossy, alpine slopes, or subalpine spruce and spruce-fir forests, shaded north-facing slopes, on rocks or forest floor, near high cliffs, and under willows on lakeshore, granite substrate, San Juan Mountains. COLO: Hin, SJn.

3292–3658 m (10800–12000′, the highest known elevation for the species). Spores: Jul–Aug. Hawaii, Siberia, Alaska, Yukon, British Columbia, Alberta, Washington, northwestern Montana, western Wyoming, and western Colorado. Known from seven collections in our area, from three general vicinities. Colorado material was formerly referred to *Lycopodium* (*Huperzia*) *selago*, a species now recognized as occurring in northeastern North America and north-central and northwestern Canada. Colorado and Wyoming populations are disjunct. The 18+ Colorado collections expand the mapped range in *Flora of North America* (1993). Our material represents a southward range extension.

Lycopodium L. Club-moss

(Greek *lykos*, wolf, + *pous*, *podes*, foot/paw; in reference to the resemblance of the branch tips to a wolf's paw) Plants perennial, terrestrial. **STEMS** horizontal, subterranean or prostrate, occasionally arching; upright shoots scattered, 5–16 mm diameter, round or flat in cross section, unbranched or with up to 4 lateral branchlets. **LEAVES** linear to linear-lanceolate, mostly not imbricate; horizontal stem leaves scattered, appressed, membranous; upright shoot and branch leaves mostly 6-ranked or more, monomorphic, appressed, ascending to spreading, margins entire or weakly dentate. Gemmiferous branchlets and gemmae absent. **STROBILUS** single or multiple, apex blunt to acute; sporophylls reduced and shorter than shoot leaves; sporangia kidney-shaped. $x = 34$. Fifteen to 25 species; mostly temperate and subarctic. In contrast to other genera, hybridization among species in this genus is virtually unknown.

Lycopodium annotinum L. (Latin, *annu*, a year; alluding to the marked separation of annual branches) **STEMS** horizontal, prostrate, to 100 cm long. Upright shoots (2) 10–15 (40) cm, scattered along horizontal stem, with some shoots clustered in groups, mostly simple or few-branched, 1.2–1.6 cm diameter; annual bud constrictions conspicuous on main upright shoots only. **LEAVES** spreading to reflexed, linear-lanceolate, (2.5) 5–8 mm × 0.6–1.2 mm; margins closely and shallowly dentate in distal 1/2; apex sharply pointed. **STROBILUS** solitary, sessile, 15–30 × 3.5–4.5 mm; sporophylls erose, (1.5) 3.5 × 0.7 (–2) mm, abruptly narrowed to pointed tip. $2n = 68$. Very rare, subalpine spruce-fir forest, at one location growing in mosses and with *Huperzia*, creeks and waterfalls, in mossy turf on granite rock, north slopes, San Juan Mountains. COLO: Hin, SJn. 3048–3322 m (10000–10900′). Our elevations are much higher than those listed in *Flora of North America* (1993) and there are higher still collections elsewhere in Colorado. Spores: Jul–early Aug. Alaska across all of northern Canada, south in the west to Washington, Idaho, Utah, Arizona, western Montana, western Wyoming, western Colorado, and north-central New Mexico and in the east to northern Minnesota, Ohio, Tennessee, North Carolina, and New Jersey. Known from four collections in our area, from two general vicinities. Many forms and varieties and some species have been described within *L. annotinum*, which does appear to vary due to environmental conditions.

MARSILEACEAE Mirb. WATER-CLOVER FAMILY

Timothy Reeves

Amphibious or aquatic perennials, often partially submerged or terrestrial in mud or sand. Rhizomes and roots arising at nodes and internodes along length. Gametophytes germinating within spores, with microgametophytes of a few cells, megagametophytes partially breaking out of megaspores and containing 1 archegonium. **STEMS** mainly long-creeping, branching to short or long shoots; hairs present. **LEAVES** alternate, long-petioled, occasionally filiform as in *Pilularia*. **SORI** within hard sporocarps which arise on short stalks near the petiole bases. **SPORES** heterosporous, with the megasporangium producing 1 megaspore and microsporangium giving rise to 20–64 microspores, both occurring within the same sorus. Three genera with about 50 species. Temperate and tropical aquatic habitats worldwide (Flora of North America Editorial Committee 1993).

Marsilea L. Water-clover, Pepperwort

(for Count Luigi Marsigli, 1658–1730, Italian mycologist) Aquatic or semiaquatic, occasionally terrestrial in very moist soil, forming widespread or clumped colonies. Roots mostly at nodes. **LEAVES** appearing similar to a cruciform four-leaf clover; deciduous or evergreen with floating, aquatic leaves larger than leaves from land plants; petiole long; blade palmately compound, divided into 4 pinnae, each cuneate or obdeltate, often with numerous red, brown, or tannish streaks abaxially. **SPOROCARP** attached laterally to stalk apex, tip of stalk often protruding as 2 bumps or teeth; densely to sparsely hairy, dehiscing into 2 valves and releasing a gelatinous receptacle with numerous sori, containing both megasporangia and microsporangia. About 40 species worldwide. Fertile material is generally required for proper species identification.

Marsilea vestita Hook. & Grev. (a little coat, due to the vestiture of hairs) Hairy water-clover. Perennial, forming dense or sometimes diffuse colonies. **LEAVES** with petioles 2–20 cm long, slightly pubescent; pinna pubescent to glabrous, about 5–20 × 4–15 cm. **SPOROCARP** stalk erect, unbranched, mostly arising at petiole base, apex not hooked; sporocarp subglobose to ellipsoid, slightly nodding, about 4–8 mm long × 3–6 mm wide, 2 mm thick, proximal tooth blunt, distal tooth acute, hooked, covered with short reddish hairs. **SORI** 14–22 per sporocarp. [*Marsilea uncinata* A. Braun, *M. mucronata* A. Braun; *M. fournieri* C. Chr.; *M. tenuifolia* Engelm. ex A. Braun]. Rare at shorelines and beaches of lakes, reservoirs, and ponds. Our one collection (sterile) is from a lakeshore at the base of the Abajo Mountains near Monticello, Utah. UTAH. 2362 m (7750′). Sporocarps: collected in Oct. in Colorado just outside our study area. Alberta east to Saskatchewan, south to California, Arizona, New Mexico, Texas, east to Louisiana, Iowa, and Minnesota; Mexico and Peru. Many other segregate species have been named, but characters appear to intergrade with one another, so it is here treated as a single species. Hybrids with *M. macropoda* have been reported in Texas.

OPHIOGLOSSACEAE Martinov ADDER'S-TONGUE FAMILY

Timothy Reeves and Donald R. Farrar

(Greek *ophis*, snake, + *glossa*, tongue, in reference to the fertile tip of the sporophore in the genus *Ophioglossum*) Sporophyte perennial, terrestrial (in ours) or epiphytic. **STEMS** simple, upright, unbranched. **LEAVES** with dilated bases forming a sheath surrounding successive leaf primordia; 1 (–2) per stem, common stalk divided into sterile leaf blade (trophophore) and fertile, spore-bearing part (sporophore); trophophore blades compound (in ours) to simple, rarely absent; veins anastomosing or free, pinnate or flabellate; indument lacking in ours. **SPOROPHORE** pinnately (or ternately) branched (in ours) or simple; sporangium exposed (in ours) or embedded, 0.5–1.5 mm diameter, thick-walled, with thousands of spores. **SPORES** monomorphic, trilete. Five genera, about 70–80 species, nearly worldwide. This family may be only distantly related to other ferns. Recent molecular studies have shown the Ophioglossaceae to be most closely related, among living plants, to *Psilotum*. Chromosome number varies by genus, with *Ophioglossum* having the highest number of chromosomes of any vascular plant, with a high count of $2n = 1262$ (St. John 1929; Tryon and Tryon 1982; Wagner and Wagner 1983, 1986, 1990, 1993).

Botrychium Sw. Grapefern, Moonwort, Botryche

(Latin *botry*, a bunch of grapes, + *oides*, like, referring to the clusters of sporangia) Terrestrial. **STEMS** forming a caudex to 5 mm thick. **TROPHOPHORES** ascending or perpendicular to common stalk, sessile or stalked; blades linear, oblong, or deltate, simple to 5-pinnate, 4–25 × 1–35 cm; pinnae fan-shaped to lanceolate to linear; margins entire to dentate to lacerate, apex rounded or acute; veins free, pinnate or flabellate. **SPOROPHORES** 1 per leaf, 1–3-pinnate, long-stalked, borne at ground level to high on common stalk; sporangia in clusters, sessile to short-stalked, exposed, borne in 2 rows on pinnate branches. About 50–60 species, nearly worldwide. Botrychiums occur mostly at high latitudes or high elevations in disturbed meadows and woods. Our greatest abundance of individuals and species occurs on 10°–45° slopes in grassy sedge subalpine meadows at Molas Pass, Colorado, at approximately 3307 m (10850′). The substrate is arkosic and subarkosic sandstone, andesite porphyry, granite, and mudstone. Our botrychiums occur in mountains, primarily in disturbed sites at high elevations. These include old burns, logged areas, and construction sites, and especially roadsides. Plants even grow in highway gravel. The plants are very difficult to detect in the field, especially in dense grassy subalpine meadows. Plant collectors not specifically seeking botrychiums can easily overlook them altogether. In the following treatment we follow Farrar (2005). Those taxa in bold have voucher specimens from the study area. Other taxa have vouchers from very near the San Juan River drainage.

1. Trophophore (photosynthetic leaf segment) ternate, appearing to be divided into 3 more or less equal segments due to great enlargement of the basal pair of pinnae ....(2)
1′ Trophophore pinnate, without basal pinnae disproportionately enlarged ....(9)

2. Sporophore (spore-bearing leaf segment) absent or tiny and aborted. ....(3)
2′ Sporophore present and conspicuous ....(4)

3. Pinnae apices acute, forming an angle of about 45°, sharply pointed; lamina (blade) thin and membranous ....*B. virginianum*
3′ Pinnae apices obtuse, forming an angle of about 90°, rounded or bluntly pointed; lamina thick and leathery ....***B. multifidum***

4. Trophophore sessile or nearly so, joined to sporophore well above the ground at the top of a common stalk ....(5)
4′ Trophophore long-stalked, joined to the sporophore near or below ground level; common stalk often not visible ....(7)

5. Sporophore stalk as long as or longer than the trophophore, plants often more than 6 inches tall .... *B. virginianum*
5′ Sporophore stalk much shorter than the trophophore, plants usually less than 4 inches tall ....(6)

6. Lobes of basal pinnae elongated and pointed, all but the uppermost pinnae sharply lobed, lustrous .... ***B. lanceolatum***
6′ Lobes of basal pinnae rounded, pinna above the basal pair often undissected, glaucous .... ***B. hesperium***

7. Middle pinnae mostly pinnately compound or lobed .... ***B. multifidum***
7′ Middle pinnae mostly simple and fan-shaped, sometimes palmately cleft into lobes ....(8)

8. Middle pinnae broadly attached to rachis and strongly decurrent .... *B. simplex* var. *simplex*
8′ Middle pinnae narrowly attached to rachis and slightly or not at all decurrent .... ***B. simplex* var. *compositum***

9. Leaf with 2 more or less equal sporophore segments and no trophophore .... *B. paradoxum*
9′ Leaf divided into a trophophore (photosynthetic segment) and a sporophore (spore-bearing segment) ....(10)

10. Basal pinnae pair entire or palmately dissected; pinnae narrowly to broadly fan-shaped in outline, broadest at the outer margin ....(11)
10′ Basal pinnae pair pinnately dissected; pinnae ovate to elliptic in outline, broadest near the base or middle ...(20)

11. Trophophore stalk longer than the distance between the first 2 pairs of pinnae, and sporophore stalk (at the time of spore release) much longer than the total length of the trophophore ....(12)
11′ Trophophore stalk shorter than the distance between the first 2 pairs of pinnae, or sporophore stalk (at the time of spore release) shorter than the total length of the trophophore ....(13)

12. Middle pinnae broadly attached to rachis and strongly decurrent .... *B. simplex* var. *simplex*
12′ Middle pinnae narrowly attached to rachis and slightly or not at all decurrent .... ***B. simplex* var. *compositum***

13. The sides of the basal pinna span less than 60°; pinnae linear to narrowly wedge-shaped ....(14)
13′ The sides of the basal pinna span greater than 60°; pinnae broadly fan-shaped ....(15)

14. Pinnae narrowly wedge-shaped, often shallowly cleft into nonspreading lobes, basal pinnae usually not the largest .... ***B. campestre***
14′ Pinnae linear, often deeply cleft into widely spreading lobes, basal pinnae usually the largest .... *B. lineare*

15. The sides of the basal pinnae span 150° to 180°, pinnae touching to overlapping, especially near the apex ...(16)
15′ The sides of the basal pinnae span 60° to 120°, pinnae well spaced, mostly not touching or overlapping ....(17)

16. Pinna texture delicate; margins finely toothed or crenulate .... *B. crenulatum*
16′ Pinna texture firm; margins entire to undulate, occasionally coarsely toothed or cleft into several segments .... ***B. lunaria***

17. Sporophore tall, its stalk (at time of spore release) equal to or exceeding the total length of the trophophore; sporophore branches spreading, not overlapping; pinnae fan-shaped and entire to shallowly lobed .... ***B. minganense***
17′ Sporophore short, its stalk (at time of spore release) 3/4 or less the entire length of the trophophore; sporophore branches ascending and overlapping; pinnae entire to deeply cleft into 2 (4) lobes ....(18)

18. Pinnae entire or symmetrically cleft, outer margins regularly dentate; basal pinnae often bearing sporangia (all species sometimes have this character) .... *B. ascendens*
18′ Pinnae entire or asymmetrically cleft into a larger upper and smaller lower lobe; basal pinnae seldom bearing sporangia ....(19)

19. Plants in the field with a whitish green appearance (changing to green following collection and storage); outer pinna margins entire to crenulate .... *B. pallidum*
19' Plants dull green; outer pinna margins entire to irregularly toothed .... *B. gallicomontanum*

20. Base of lowermost pinnae angular (90° or less); lobes of basal pinnae divergent (like spread fingers); upper pinnae and lobes of lowermost pinnae narrowly elongate .... (21)
20' Base of lowermost pinnae rounded to cordate (nearly 180°); lobes of basal pinnae parallel to convergent (not spreading); upper pinnae and lobes of lower pinnae ovate .... (22)

21. Trophophore outline broadly triangular (equilateral), as broad as long; lustrous dark green; sporophore ternate; sporangia bright yellow before spore release .... *B. lanceolatum*
21' Trophophore ovate to narrowly triangular in outline, longer than broad; somewhat lustrous to dull green; sporophore pinnate; sporangia dull yellow before spore release .... *B. echo*

22. Trophophore lustrous bright green; all but the uppermost pinnae dissected or lobed on both the upper and lower margins; sporophore pinnately (rarely ternately) divided .... *B. pinnatum*
22' Trophophore glaucous blue-green; pinnae above the basal pair entire or shallowly dissected (often only on the lower margin); sporophore ternately divided .... *B. hesperium*

Botrychium campestre W. H. Wagner & Farrar (of grasslands and plains) Prairie moonwort, dunewort, prairie dunewort, Iowa moonwort. Plant exclusive of its roots 6 (12) cm tall; common stalk usually 5 (10) cm long; underground stems with tiny, nearly spherical bodies 0.4–0.8 mm in diameter, in grapelike clusters. **TROPHOPHORES** dull, whitish green, fleshy, the veins submersed, sessile (usually) to short-stalked, the stalk up to 10% of the total trophophore length; the stalk and midrib fleshy and 2–4 mm broad; blade outline oblong to linear-oblong, commonly widest above the middle, up to 35 × 12 mm; segments linear to oblong to spatulate, the largest strongly asymmetrical and typically bifid, the lower lobe 1/3–2/3 the length of the upper lobe; number of pinnae pairs few, usually less than 6, these mostly remote; segments ascending mostly 30–50°; largest pinnae reaching 7 mm long and 4 mm wide, served by 2 major veins terminating on outer margin with 8–12 veinlets; the lowest pinnae generally narrower and less complex than the 1–3 distal pairs; outer margins mostly shallowly crenate or dentate. **SPOROPHORES** stubby, usually equal to or slightly longer than trophophores but sometimes 1/2 or more longer, all axes fleshy and flattened. **SPORES** finely verrucate, small, (30) 34–38 (44) µm, $n = 45$. Chuska Sandstone; near Whiskey Lake in an open, logged, ponderosa pine community with *Poa fendleriana*, *Packera neomexicana*, *Lithophragma*, and Gambel's oak. NMEX: McK. 2708 m (8885'). Spores: Jun. This typically prairie species occurs from southeastern Alberta and southwestern Saskatchewan, eastern Montana south to north-central and northeastern Colorado and northwestern Kansas to Michigan and extreme southern Ontario. *Botrychium campestre* is one of the smallest moonworts and is a spring–early summer species (in contrast to our other species which mature spores in July–August). Our single collection represents the first for New Mexico and is far disjunct from known locations in northeastern Colorado grasslands and a recent collection from the Rocky Mountains in north-central Colorado.

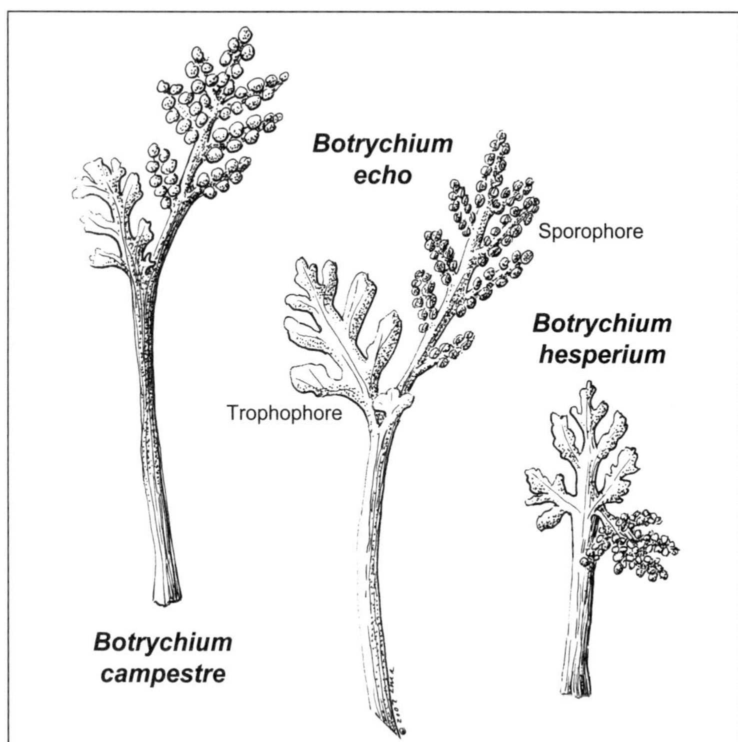

Botrychium echo W. H. Wagner (in reference to the fact that this species seems to repeat the characteristics of other, similar species and for Echo Lake, Colorado, where its distinctions from *B. hesperium* were recognized) Echo moonwort, reflected moonwort. Plant exclusive of its roots 9.5 (3–15) cm tall, the common stalk 6 (2–10) cm tall. **TROPHOPHORES** bright green and shiny in life, nearly sessile to short-stalked, broadly oblong, 2.2 (1–4.5) cm long, pinnate; pinnae narrowly attached to a relatively narrow rachis, remote to approximate, not overlapping, lanceolate to oblanceolate to linear with pointed apices, the pinna bases subsymmetrical, the laminar margins nearly entire; basal pinna pair not exaggerated in length, equal to or somewhat longer than the adjacent pair, spreading or only moderately ascending, not clasping. **SPOROPHORES** relatively short, 3.5 (1.5–8) cm tall, only 1/2 again as long as the sterile segment and only 20% with 1 or more branches that are 1/3 or more as long as main axis of sporangial cluster. **SPORES** 37 (27–53) µm in maximum diameter, irregularly and finely verrucate, the warts small, low, and separated by narrow, shallow channels. $2n = 180$. *Botrychium echo* is our most widespread and abundant species, found at nine sites (see genus description), sometimes in large numbers. In meadows near clumps of Engelmann spruce and in subalpine meadows surrounded by spruce forest in dense vegetation with grasses, *Antennaria*, *Potentilla*, *Achillea*, *Epilobium*, low blueberry, and umbellifers. Also in an area of barren volcanic ash with *B. minganense* and *B. lanceolatum*. COLO: Arc, Hin, LPl, Min, SJn. 3060–3584 m (10040–11758′). Spores: Jul–Aug. Northern Arizona, southeastern Utah, southwestern and central Colorado. *Botrychium echo* is an allotetraploid species with parental species *B. lanceolatum* and *B. lineare*. It is most often confused with *B. hesperium*, which it closely resembles. It can be distinguished from large plants of *B. hesperium* by its pinnately rather than ternately branching sporophore, by its more angular pinnae bases and apices, and by the spreading angle of its pinnae lobes, particularly the inner lower lobes of the basal pinnae. The sterile hybrid between *B. echo* and either *B. lunaria* or *B. minganense* has been collected once in our area. The plant has malformed spores and unusual morphology and does not key to any North American species.

Botrychium hesperium (Maxon & R. T. Clausen) W. H. Wagner & Lellinger (western) Western moonwort. Plant exclusive of its roots 12 (5–20) cm tall, the common stalk 7 (3–13) cm tall. **TROPHOPHORES** gray-green and dull in life, mostly short-stalked, subdeltate, 2.5 (1–5) cm long, pinnate; pinnae broadly attached to a relatively wide rachis, crowded to commonly overlapping, ovate to lanceolate, with rounded apices, the pinnae bases asymmetrical, the laminae margins finely repand; basal pinnae commonly exaggerated, up to twice as long as the adjacent ones, often upright and commonly clasping. **SPOROPHORES** relatively tall, 5 (3–10) cm tall, nearly twice as long as sterile segment, 80% with 1 or more basal branches that are 1/3 or more as long as main axis of sporangial cluster. **SPORES** 37 (29–50) µm in maximum diameter, irregularly and coarsely verrucate, the warts large, prominent, and separated by wide, deep channels. $2n = 180$. *Botrychium hesperium* is rare in our area, being known in low numbers only from Molas Pass (see genus discussion). COLO: SJn. 3303–3309 m (10836–10855′). Spores: Aug. Southeastern Alaska, southwestern Yukon, British Columbia, southwestern Alberta, eastern Washington and Oregon, western Montana, western Wyoming, Idaho, eastern Utah, northern Arizona, and western Colorado. See *B. echo* for differences between these species. Our material is **var. *hesperium***. *B. hesperium* is an allotetraploid with diploid parents *B. lanceolatum* subsp. *lanceolatum* and *B. pallidum*.

Botrychium lanceolatum (S. G. Gmelin) Ångström (broad lance) Triangle moonwort. **TROPHOPHORE** stalk 0–1 mm long; blade dull to shiny green to dark green, deltate, pinnate-bipinnate, to 6 × 7 cm. Pinnae to 5 pairs, ascending, approximate; distance between first and second pinnae not or slightly more than between second and third pairs, linear to broadly lanceolate, entire to divided to tip, margins with distinct lobes or segments, apex acute to rounded, venation pinnate. **SPOROPHORES** pinnate–tripinnate, 1–2.5 times length of trophophores, divided into several equally long branches (all other botrychiums have a single stalk or 1 dominant and 2 smaller). $2n = 90$. *Botrychium lanceolatum* is known in our area from six sites, with several plants at each location including Molas Pass (see genus discussion). Dry, open slopes of tuff with *B. simplex*, *Vaccinium* spp., *Fragaria* spp., *Senecio*, *Shepherdia argentea*, and *Thalictrum*; on bare volcanic ash with

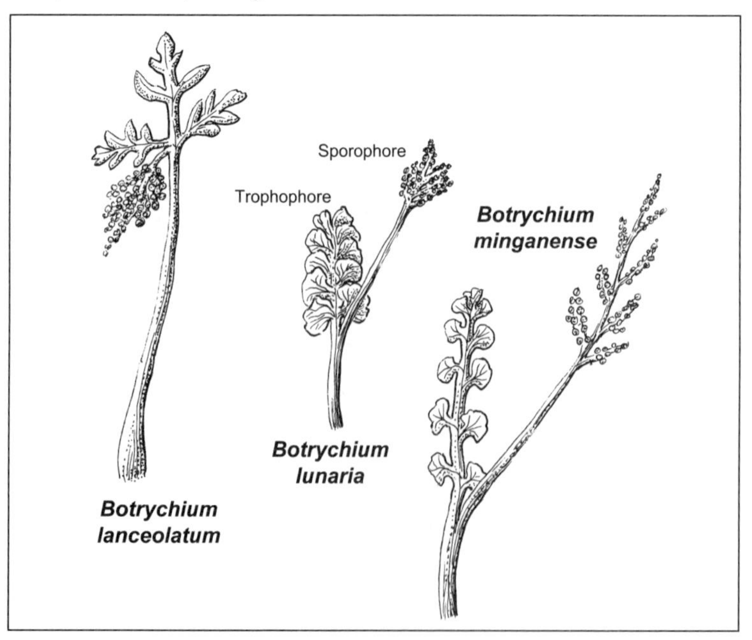

B. echo and *B. minganense*; in the middle of old clearcut skid trail; *Festuca thurberi* grassland; gravelly side of highway; and subalpine grassy (some clumps 2′ tall) meadow opening in spruce. COLO: Hin, Min, SJn. 2743–3584 m (9000–11758′). Spores: Jul–Sep. Western and southern Alaska, southwestern Yukon, southern Alberta, southwestern Saskatchewan, south and east to southern Oregon, western Montana, western Wyoming, western Colorado, eastern Nevada, northern Arizona, and northern New Mexico, and as subsp. *angustifolium* in northeastern Minnesota, the Great Lakes states and provinces, south to eastern Tennessee, western North Carolina, northeastern Georgia, northwestern South Carolina, and northeastward to Maine, eastern Quebec, Newfoundland, and disjunct in western Quebec, northern Labrador, and southern Greenland. Our material is **subsp. *lanceolatum***, which is characterized by the trophophore blade green to pale yellow-green, broad, coarse, succulent; middle and terminal segments usually more than 2 mm wide exclusive of lobes. It is the only subspecies in western North America and is the entity disjunct in western Quebec, northern Labrador, and southern Greenland. Two morphologically and genetically distinct types, provisionally called "red" and "green," occur throughout the subsp. range. Ours are of the "red" type, with red coloration in the common stalk, upwardly curved basal pinnae, and on the basal pinnae, the lower pinnules longer than the upper pinnules. The "green" type is rare in Colorado. These types are in the process of characterization and will receive formal nomenclature. *Botrychium lanceolatum* is a diploid parental species of the allotetraploids *B. echo*, *B. hesperium*, and *B. pinnatum*.

Botrychium lunaria (L.) Sw. (the moon, referring to the half-moon shape of the lower pinnae) Common moonwort. **TROPHOPHORE** stalk 0–1 mm tall; blade dark green, oblong, pinnate, to 10 × 4 cm, thick, fleshy. Pinnae to 9 pairs, spreading, mostly overlapping except in shaded forest forms, distance between first and second pinnae not or slightly more than between second and third pairs, basal pinna pair approximately equal in size and cutting to adjacent pair, broadly fan-shaped, undivided to tip, margins mainly entire or undulate, rarely dentate, apical lobe usually cuneate to spatulate, notched, approximate to adjacent lobes, apex rounded, venation like ribs of fan, midribs absent. **SPOROPHORES** pinnate–bipinnate, 0.8–2 times length of trophophores. $2n = 90$. *Botrychium lunaria* is found at four sites in our area, with modest numbers of plants per site including Molas Pass (see genus description). Subalpine logged areas and spruce-fir forest; in old clearcut skid trails, logged spruce-fir forest with scattered spruce and fir meadows with subarkosic sandstone, small rocky open area on overgrown slopes with *B. minganense* in andesite porphyry and quartzite. COLO: Arc, SJn. 3172–3353 m (10408–11000′). Spores: Aug. Alaska and Yukon east across Canada to Newfoundland and southern Greenland, south and east in the west to eastern California, Nevada, northern Arizona, and northern New Mexico, Montana, western South Dakota, Wyoming, and western Colorado, south in the east to northeastern Minnesota through the Great Lakes states and northern Pennsylvania to northern Vermont, northern New Hampshire, and Maine. Outside North America it occurs in southern South America, Eurasia, New Zealand, and Australia. It is the most widespread of moonworts. *Botrychium lunaria* is a diploid parent of allotetraploids *B. minganense* and *B. pinnatum*. In our area it can only be confused with *B. minganense*. These two are distinguished in the key. Sometimes, the trophophore of *B. lunaria* has a short stalk, but the stalk length seldom equals or exceeds the distance between the first pinna pair as it usually does in *B. minganense*.

Botrychium minganense Vict. (for the Mingan Islands, eastern Quebec) Mingan moonwort. **TROPHOPHORE** stalk 0–2 cm, 0 to 1/5 lenth of trophophore rachis; blade dull green, oblong to linear, pinnate, to 10 × 2.5 cm, firm to herbaceous. Pinnae to 10 pairs, horizontal to slightly ascending, approximate to remote, distance between first and second pinnae not or slightly more than between second and third pairs, basal pinna pair approximately equal in size and cutting to adjacent pair, occasionally basal pinnae and/or some distal pinnae elongate, lobed to tip, nearly circular, fan-shaped or ovate, sides somewhat concave, margins nearly entire, shallowly crenate, occasionally pinnately lobed or divided, apex rounded, venation like ribs of fan with short midrib. **SPOROPHORES** pinnate, bipinnate in very large, robust plants, 1.5–2.5 times length of trophophores. $2n = 180$. *Botrychium minganense* is known from six sites in our area, with moderate numbers of plants including Molas Pass (see genus discussion). Old clearcut skid trails; subalpine logged area and spruce-fir forests; riparian community of *Salix monticola* and mesic forbs; barren volcanic ash with *B. echo* and *B. lanceolatum*; rocky open area of andesite porphyry and quartzite with *B. lunaria*. COLO: Arc, Hin, SJn. 3103–3353 m (10180–11000′). Spores: Aug. Alaska, Yukon, southern Alberta east across Canada to Labrador and Newfoundland, south to California, northern Arizona, northern New Mexico, and northeastern Mexico, east to western Colorado, western South Dakota, northern North Dakota, northern Minnesota, northern Wisconsin, Michigan, northern New York, northern Vermont, northern New Hampshire, and Maine. *Botrychium minganense* is an allotetraploid with parental species *B. lunaria* and *B. pallidum*. It can be confused with both of these (see these two species for details).

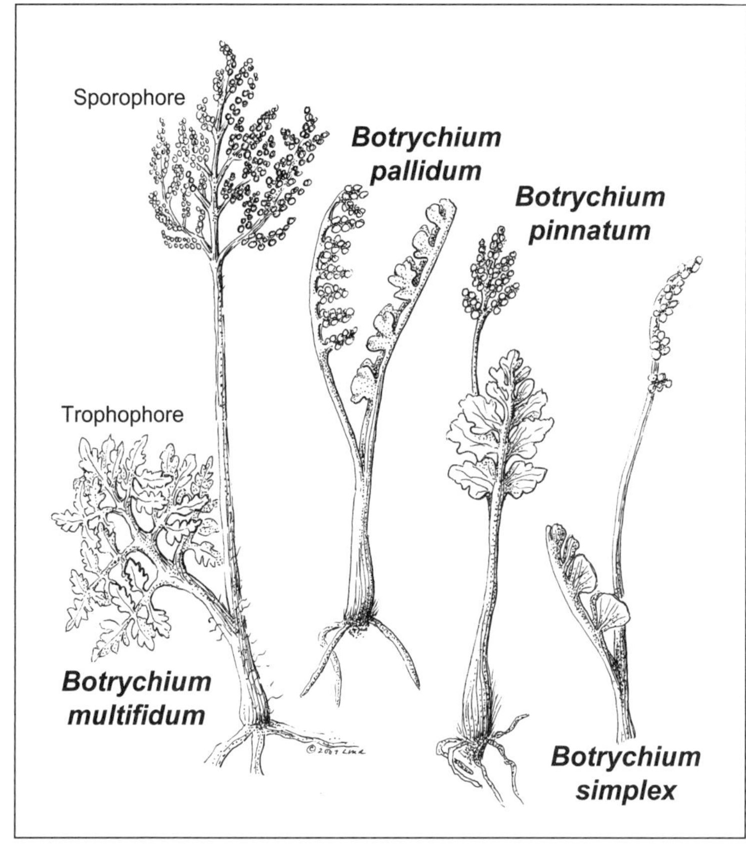

Botrychium multifidum (S. G. Gmel.) Rupr. (much-divided, referring to the trophophore) Leather grapefern. **TROPHOPHORE** stalk 2–15 cm, 0.3–1.2 times length of trophophore rachis; blade shiny green, plane, ternate, bi- to tripinnate, to 25 × 35 cm, leathery. Pinnae to 10 pairs, approximate to remote, horizontal to ascending, distance between first and second pinnae not or slightly more than between second and third pairs, divided to tip; pinnules obliquely ovate, rounded, margins usually ± entire to shallowly crenulate, sometimes inconspicuously and shallowly denticulate, apex rounded, venation pinnate. **SPOROPHORES** bi- to tripinnate, 1.2 times length of trophophores. 2*n* = 90. Our two collections come from Boyce Lake, where 15–20 plants were found in protected areas at the edge of Engelmann spruce and Douglas-fir woods and one plant was found in the center of a grassy meadow with *Salix monticola*, *Agrostis*, *Carex*, and *Mentha arvensis*. COLO: SJn. 2661 m (8730′). Spores: Aug. Western and southern Alaska, southern Yukon across Canada to southern Labrador and Newfoundland and southern Greenland, south in the west to California, Nevada, northern Arizona, and east to western Colorado, Wyoming, northern South Dakota, Minnesota, northeastern Iowa, Michigan, Pennsylvania, West Virginia, Virginia, New Jersey north to Maine; Europe; northwest Asia.

Botrychium pallidum W. H. Wagner (pale, suggesting a very pale, dwarf form of *B. minganense*) Pale moonwort. **TROPHOPHORE** blade longitudinally more or less folded and troughlike when alive, narrowly oblong, up to 4 × 1 cm, pinnate, the pinnae approximate, up to 5 pairs, small, up to 6 mm long, flabellate, the basal or both sides deeply concave, broadly attached, the larger ones ascending and strongly asymmetrical, tending toward 2 lobes, the upper one cleft and larger than the lower; outer margins entire to irregularly crenulate-denticulate; lamina herbaceous, glaucous pale green. **SPOROPHORE** lower pinnae with a large or small branch. **SPORE** diameter 23–28 μm. *n* = 45. *Botrychium pallidum* is known from a single collection from the Molas Pass area, on the shoulder of the highway with bare soil and rock and is found with *B. hesperium* and *B. echo*. COLO: SJn. 3158 m (10360′). Spores: Aug. Southeastern British Columbia east across Canada in southern Alberta, southern Saskatchewan, southern Manitoba, southern Ontario, southern Quebec, and in the U.S. in Montana, Wyoming, central and southwestern Colorado, South Dakota, Minnesota, Wisconsin, Michigan, northern New York, Vermont, New Hampshire, and Maine. All plants west of the Black Hills of South Dakota that have been tested genetically have been shown to be tetraploid, probably allotetraploid with *B. pallidum* as one parent. Plants in our area have not yet been examined genetically. The tetraploid western plants warrant recognition as a distinct species. *Botrychium pallidum* is a diploid parent of *B. minganense* and the two can resemble each other in general appearance in herbarium specimens. *Botrychium pallidum* can most easily be distinguished from *B. minganense*, when alive, by its silvery gray-green color, but this character often does not persist in collected plants, especially if they are stored in a plastic bag. Other characters differentiating the two species are the curvature of the pinna sides and the cutting of their outer margin. The pinna sides of *B. minganense* are concave near the rachis and are more or less straight along the pinna blade. Pinna sides of *B. pallidum* are more concave throughout and most strongly so near their juncture with the outer margin, giving the pinna a mushroom shape. Outer pinnae margins of *B. minganense* are entire, or if lobed, symmetrically so. Outer pinnae margins of *B. pallidum* are often unevenly divided, with the upper lobe longer and broader than the lower lobe. This unequal division can also be seen in the basal branches of the sporophore, those of *B. minganense* being symmetrically branched whereas those of *B. pallidum* are divided into a longer upper and shorter lower branch.

Botrychium pinnatum H. St. John (pinnate) Northwestern moonwort. Plant slender or stout, 3–12 (15.5) cm high; common stalk 1/3–1/2 hypogaean, 3–7 cm tall. **TROPHOPHORES** sessile, or on large plants short-stalked, not glaucous, pinnate, or on large plants bipinnate, the blade oblong or somewhat narrowed to the tip, the simply pinnate blades having 7–11 pinnae which are sessile, broadly winged, and confluent with the rachis, entire or frequently with large rounded pinnate lobings, the pinnae ovate or elliptic, 3–5 cm (up to 2.3 cm in our plants) long, prominently veined with twice-forked veins; trophophore of large plants ovate-deltoid, twice pinnate, 4 (5) cm long, 3 (4) cm wide, the pinnae pinnately cut and closely resembling the whole blade of a smaller plant. **SPOROPHORE** stalk exceeding the trophophore, pinnate to tripinnate (sometimes ternate in large plants), erect, narrow and with ascending branches; spores rounded, tetrahedral, 36–45 μm in diameter, rugose. $2n = 180$. *Botrychium pinnatum* is known in our area from seven sites, some with many plants including Molas Pass (see genus discussion). Subalpine logged area and spruce-fir. COLO: Arc, LPl, Min, SJn. 3048–3353 m (10000–11000′). Spores: Jul–Aug. Alaska, Yukon, western Alberta, Washington to western Montana, Oregon to western Wyoming, south to northern California, northern Nevada, northern Utah, and western and central Colorado. *Botrychium pinnatum* is an allotetraploid with diploid parents *B. lanceolatum* and *B. lunaria*.

Botrychium simplex E. Hitchc. (simple) Least moonwort. **TROPHOPHORE** stalk 0–3 cm tall, 0–1.5 times length of trophophore rachis; blade dull to bright green to whitish green, linear to ovate-oblong to oblong to fully triangular with pinnae arranged ternately, simple to bipinnate (tripinnate), to 7 × 2 cm, fleshy to thin, papery or herbaceous. Pinnae or well-developed lobes to 7 pairs, spreading to ascending, approximate to widely separated, distance between first and second pinnae frequently greater than between second and third pairs, basal pinna pair commonly much larger and more complex than adjacent pair, cuneate to fan-shaped, strongly asymmetric, undivided to divided to tip, basiscopic margins ± perpendicular to rachis, acroscopic margins strongly ascending, basal pinnae often divided into 2 unequal parts, margins usually entire or shallowly sinuate, apex rounded, undivided and boat-shaped to strongly divided and plane, venation pinnate with a midrib in elongated basal pinnae or like ribs of fan and without a midrib in unelongated pinnae. **SPOROPHORES** mainly pinnate, 1–8 times length of trophophores. $2n = 90$. *Botrychium simplex* is known from one location in our area. The plants were found on a dry, open northwest slope of tuff growing with *B. lanceolatum*, *Vaccinium* spp., *Fragaria* spp., *Senecio*, *Shepherdia argentea*, and *Thalictrum*. COLO: Hin. 2743 m (9000′). Southern British Columbia, southern Alberta, extreme southern Saskatchewan and Manitoba, Washington east to North Dakota and Minnesota, Oregon east to Wyoming and several areas throughout South Dakota, south to California, northern Nevada, Utah, and western and central Colorado; Great Lakes states and adjacent provinces, West Virginia, Virginia, eastern Tennessee, western North Carolina north to Maine, southern Quebec, southern Labrador, Newfoundland, and southern Greenland; Europe. Our material is **var. *compositum*** (Lasch) Milde (compounded). The variety was described from Germany and our material may or may not be the same as that in Germany. The variety is in the western part of the species range.

POLYPODIACEAE Bercht. & J. Presl POLYPODY FAMILY

Timothy Reeves and Linda Mary Reeves

Gametophytes cordate or elliptic, glabrous or glandular, antheridia 3-celled. Sporophytes terrestrial, lithophytic or epiphytic; erect, arching, or pendent, with few to numerous roots. **STEMS** variously creeping, variously scaly. **LEAVES** monomorphic to dimorphic, with circinate vernation; petiole variously scaly or not; blade simple to pinnatifid or variously pinnate; rachis grooved or not adaxially; veins free to anastomosing; indument absent to hairy to scaly. **SORI** round or oblong or elongate, rarely marginal or covering much of surface, abaxially on a vein; indusium absent; sporangium with 32 or 64 spores. **SPORES** yellowish, greenish, or translucent, mostly monolete; surface smooth, verrucose, tuberculate, granulate, or spiny. About 40 genera with 500 species. Worldwide, mostly tropical and subtropical. Many tropical species are epiphytic and are of interest horiculturally (Flora of North America Editorial Committee 1993).

Polypodium L. Polypody

(Greek *poly*, many, + *podion*, foot, referring to the knoblike portions of the stem) Plants lithophytic, terrestrial or epiphytic. **STEMS** creeping, branched; often pruinose; scales concolored or bicolored; margins entire to dentate. **LEAVES** monomorphic, not acuminate at tip, to about 60 cm long; petioles straw-colored, flattened, grooved to terete, often somewhat winged; blade ovate to deltate, pinnatifid to pinnate with less than 25 pinnae pairs, usually somewhat scaly or hairy, rachis glabrous to puberulent; scales ovate to linear; segment linear to oblong, margin entire to serrate, apex rounded to attenuate. **SORI** round to oval, often confined to distal region of frond, in 1 to 3 rows lateral to the midrib, indument mostly absent or of sporangiasters. **SPORES** monolete, rugose to tuberculate. $x = 37$. About 100 species. Worldwide distribution, with some reticulate hybridization in North America.

1. Blade rather broad and deltoid, with more or less broad pinnae, sporangiasters absent, immature sori oval .... ***P. hesperium***
1' Blade narrow with remote, narrow pinnae, sporangiasters present, immature sori round .... ***P. saximontanum***

Polypodium hesperium Maxon (western) Western polypody. **STEMS** sometimes whitish pruinose, variable thickness to about 6 mm diameter, bitter to sweet tasting; scales concolored to mottled or grading darker to attachment, lanceolate, margins entire to denticulate. **LEAVES** to about 35 cm; petiole thin; blade oblong to lance-ovate or deltate, pinnatifid, herbaceous to leathery; rachis glabrous to sparsely scaly, scales linear-lanceolate, less than 6 cells wide; segment oblong to linear-lanceolate, less than 1.2 cm wide, margin entire to crenulate, undulate or weakly serrate, apex obtuse to acute. **SORI** oval, less than 3 mm diameter, located midway between margin and midrib; sporangiasters absent. **SPORES** greater than 58 μm diameter, rugose or verrucose to tubercular with shallow surface projections. $2n = 148$. [*P. prolongilobum* Clute; *P. vulgare* L. var. *hesperium* (Maxon) A. Nelson & J. F. MacBr.; *P. vulgare* subsp. *columbianum* (Gilbert) Hultén; *P. vulgare* var. *columbianum* Gilbert]. Cracks in slate, igneous or metaconglomerate rock and cliffs in limited humid, shady canyons. Ponderosa pine, Gambel's oak, aspen, riparian, or semiriparian locations. COLO: LPl, SJn. 2438–2560 m (8000–8400'). Spores: Jun–Oct. Western United States, Baja California, and southern British Columbia east to western Montana, western Colorado, western New Mexico, and Chihuahua, Mexico. Hybridizes to form a sterile hybrid with *P. saximontanum*, which is only known from Vallecito Creek Trail. This species is itself an allopolyploid hybrid product of diploid parents *P. glycyrrhiza* and *P. amorphum*, two western coastal species.

Polypodium saximontanum Windham (rocky mountain) Rocky Mountain polypody. **STEMS** often whitish pruinose, to about 6 mm diameter, bitter to acrid tasting; scales sometimes bicolored, lightish brown with dark central stripe, lanceolate, margin dentate. **LEAVES** to 22 cm long; petiole slender; blade oblong to linear, pinnatifid, somewhat leathery; rachis glabrous to sparsely scaly; scales lance-ovate, more than 6 cells wide; segment oblong, less than 1.2 cm wide, margin entire to crenulate, apex rounded to acute. **SORI** round when immature, less than 3 mm diameter, marginal to midway between margin and midrib; sporangiasters present, less than 40 per sorus, heads often with a few glands. **SPORES** greater than 58 μm diameter, surface tuberculate, projections deep. $2n = 148$. Known in our area only from a few San Juan Mountains collections. Open slopes above creeks, growing in horizontal seams and cracks in Dakota sandstone boulders and metaconglomerate. Ponderosa pine, aspen, Gambel's oak, canyon riparian habitats. COLO: Arc, Hin, LPl. 2350–2600 m (7720–8420'). Spores: Jun–Oct. Eastern Wyoming, adjacent South Dakota south to western Colorado and northern New Mexico. Another allotetraploid species product of diploid parents *P. amorphum* and *P. sibiricum*, two northern species. Hybridizes with *P. hesperium* to form a sterile hybrid, *P. hesperium* ×*saximontanum*, known only from Vallecito Creek Trail, also the type locality of *P. saximontanum.*

PTERIDACEAE Rchb. (ADIANTACEAE) MAIDENHAIR FERN FAMILY

Timothy Reeves and Linda Mary Reeves

(Greek *pteris*, fern, but derived from *pteron*, the word for feather or wing, based on the similarity of a fern frond to a feather) Gametophyte green, obcordate to reniform, glabrous, except farinose in *Notholaena*; antheridia 3-celled. Sporophytes mostly small perennials, lithophytic or terrestrial. **STEMS** compact, creeping, with hairs or scales. **LEAVES** monomorphic or dimorphic; vernation circinate or noncircinate; indument on all surfaces, often hairs and/or glands and/or scales, sometimes farinose; petioles with persistent scales near stem, spines lacking; blades 1- to 6-pinnate; veins often in complex patterns. **SORI** often confluent and forming a continuous band submarginally on abaxial leaf surface, sporangia sometimes covering abaxial surface densely; indusium present or not present, a false indusium when present; spores generally 32 or 64 per sporangium. **SPORES** of 1 kind, brown, black, gray, or yellow, globose to trigonal, occasionally with an equatorial ridge, trilete, cristate to rugose. About 40 genera with about 1000 species. Distribution worldwide, with many genera (the cheilanthoids) in arid and semiarid locales (Flora of North America Editorial Committee 1993).

1. Fertile marginal lobes separate and distinct, not continuous around ultimate segments; sporangia borne on these reflexed indusial flaps; veins prominent, dichotomously branched; leaves deciduous .... ***Adiantum***
1' Fertile margins revolute, mostly continuous around segments; sporangia generally borne on abaxial leaf surface; veins generally obscure, if prominent, pinnately branched; leaves perennial .... (2)

2. Leaves strongly dimorphic, with fertile leaves longer than sterile ones; petioles green to straw-colored distally .... ***Cryptogramma***
2′ Leaves generally monomorphic; petioles brown or black, with or without pubescence, or if light-colored, generally somewhat pubescent .... (3)

3. Blades essentially glabrous viewed at 1×; ultimate segments much longer than wide .... ***Pellaea***
3′ Blades abaxially with white farina or hairs and/or scales; ultimate segments about as long as wide .... (4)

4. Blades with dense white farina on abaxial surface .... ***Argyrochosma***
4′ Blades with hairs and/or scales on abaxial surface .... ***Cheilanthes***

Adiantum L. Maidenhair Fern

(Greek *adiantos*, unwetted, for the rain-shedding leaves) Delicate sporophytes, terrestrial or lithophytic. **STEMS** short to long-creeping, branched; with scales, deep brownish yellow to dark reddish brown or blackish, concolored to bicolored, linear-lanceolate to lanceolate, margin entire to erose, ciliate or denticulate. **LEAVES** suberect to pendent, monomorphic to dimorphic, distant to densely clustered; petiole brownish to dark purplish or blackish, with adaxial groove, glabrous to hispid or strigose; blade lanceolate, ovate, or fan-shaped, membranaceous to papery, primarily glabrous, sometimes with very scattered hairs; ultimate segments short, dark-petioluled, round, fan-shaped, rhomboid or oblong-elliptic, 2.5–30 mm wide, base truncate. **SORI** with sporangia concealed by false indusium formed from inrolled segment margin, gray-green to dark brown, about 1 mm wide; sporangia submarginal. **SPORES** yellow to yellowish brown, tetrahedral to globose, trilete, rugulate to tuberculate, lacking equatorial ridge. x = 29, 30. About 200 species. Almost worldwide, mainly tropical, with greatest diversity in the Andes, although absent from very dry nonriparian areas. Many species are cultivated (Flora of North America Editorial Committee 1993). *Adiantum aleuticum* (Rupr.) C. A. Paris occurs a few miles outside our area in San Juan County, Colorado, where it is common in a canyon.

Adiantum capillus-veneris L. (hair of Venus, apparently referring to the use of the plant as a hair treatment and wash) Common maidenhair, Venus's hair fern, southern maidenhair, culantrillo. **STEMS** short-creeping, slender, scales brown, concolored, luminous, margin entire or with single tooth near base. **LEAVES** pendent or arching, 10–75 cm long; petiole thin, glabrous or glaucous; blade broadly ovate to lanceolate, pinnate to bipinnate, occasionally tripinnate near base, glabrous, rachis glabrous; ultimate segments generally cuneate to fan-shaped or rhomboid or roundish, base cuneate, margins shallowly to deeply lobed to laciniate or denticulate, apex rounded to acute. **SORI** with indusium oblong or crescent-shaped, glabrous. **SPORES** 40–50 μm diameter. $2n$ = 120. At lower elevations in shaded alcoves and ledges, hanging gardens, and canyon wall seepages, especially along rivers and streams and above drip pools, often with *Epipactis gigantea*, *Platanthera zothecina*, *Cirsium rydbergii*, *Mimulus eastwoodii*, and other humidity-loving species. ARIZ: Apa, Nav; COLO: Mon; UTAH. 1120–1829 m (3670–6000′). Spores: Jun–Jul. British Columbia, California, Nevada, Utah, Colorado, south to Arizona, New Mexico, and Texas (especially the limestone Edwards Plateau), east to Oklahoma, Arkansas, and Louisiana; Virginia south to Alabama, Georgia, and Florida. Also disjunct populations in South Dakota, Kentucky, Missouri. Present in Mexico, West Indies, Central America, and South America, as well as tropical and temperate regions in Eurasia and Africa. Although commonly on limestone in many locales, most of our populations grow on sandstone and siltstone.

Argyrochosma (J. Sm.) Windham False Cloak Fern

(Greek *argyros*, silver, + *chosma*, powder, which refers to the white farina on abaxial leaf surfaces) Gametophytes glabrous. Sporophyte plants generally lithophytic. **STEMS** erect to ascending, unbranched; scales concolor, brownish, occasionally black, lanceolate to subulate, margin entire. **LEAVES** monomorphic, 3–30 cm long; petiole dark-colored, with or without adaxial groove, glabrous except for basal scales; blade lanceolate, ovate to deltate, 2- to 6-pinnate, leathery to herbaceous, glabrous or with whitish farina abaxially, glabrous or glandular adaxially; ultimate segments elliptic to deltate, often less than 4 mm wide, base often cordate, stalks dark-colored, shining. **SORI** with false indusia greenish, marginal, narrow; sporangia with 32 or 64 spores, intermixed with farina-producing glands, when present. **SPORES** brown, globose to tetrahedral, rugose or cristate, lacking equatorial ridge. x = 27. About 20 species distributed in the Western Hemisphere. Traditionally placed in *Notholaena* and/or *Pellaea*, two genera of cheilanthoid ferns. *Agyrochosma* is now known to be not closely related to *Notholaena*, but somewhat allied to *Pellaea.* However, the genus' chromosome number is unique among cheilanthoid ferns, among other differences, suggesting that this genus is monophyletic, and therefore, separate.

Argyrochosma limitanea (Maxon) Windham (belonging to the boundary) Southwestern false cloak fern. **STEMS** with brown scales. **LEAVES** 5–30 cm long; petiole reddish brown, brown, or black, the basal scales brown; blade triangular-lanceolate or subpentagonal to deltate, 3- to 5-pinnate (in ours), leathery, white farina covering abaxial surface, glabrous or sparsely glandular adaxially, rachis rounded to slightly flattened adaxially, dark color of stalks continuing into segment bases adaxially. **SORI** with sporangia submarginal on distal 1/2 of secondary veins, each containing 32 spores. $n = 2n = 81$ (apogamous). [*Notholaena limitanea* Maxon; *Cheilanthes limitanea* (Maxon) Mickel; *Pellaea limitanea* (Maxon) C. V. Morton]. South-facing sandstone cliffs and slopes in piñon-juniper woodland communities. In our area it is known only from Comb Ridge, Utah. UTAH. 1525 m (5000′). Spores: Mar. Southeastern California, Arizona, southern Utah, western New Mexico, and west Texas (Chisos Mountains), as well as Sonora and Chihuahua in Mexico. Our plants are **subsp. *limitanea***, which is an asexual triploid. The probable parent species have not been discerned.

Cheilanthes Sw. Lip Fern

(Greek *cheilos*, margin, + *anthus*, flower, referring to the marginal sporangia) Gametophytes glabrous. Sporophyte plants generally lithophytic. **STEMS** long-creeping to compact, usually branched, scales brown to black or bicolored with a dark central stripe, linear-subulate to ovate-lanceolate, margin entire to denticulate. **LEAVES** monomorphic, rigid, clustered to scattered, 4–60 cm long; petiole straw-colored or brownish to blackish, rounded, flattened or with adaxial groove, indument pubescent, scaly, or glabrous; blade linear-oblong, lanceolate, ovate, or elongate-pentagonal, pinnatifid to 4-pinnate, mostly leathery, abaxially pubescent and/or scaly, occasionally glabrous, pubescent to glabrous adaxially, dull; ultimate segments round to elongate or spatulate, less than 4 mm wide, base rounded, truncate or cuneate, stalks shining and dark. **SORI** often borne at thickened vein tips, with false indusia greenish, occasionally whitish, narrow; marginal or nearly marginal, sporangia with 32 or 64 spores, not mixed with farina-producing glands. **SPORES** gray to brown to black or yellowish, rugose, cristate or globose to tetrahedral, lacking equatorial ridge. $x = 30$ (occasionally 29). About 150 species. Western Hemisphere, with a few species in Eurasia, Africa, and Oceania. The largest and most diverse genus of arid-land adapted ferns, lending unexpected great diversity to the fern floras of arid and semiarid Arizona, New Mexico, Texas, and northern Mexico. A genus of "resurrection" ferns, possessing fronds that turn brown in unfavorable dry weather, returning to green with moisture (Flora of North America Editorial Committee 1993; Reeves 1979).

1. Blades pubescent, without scales; vernation circinate .... ***C. feei***
1′ Blades with scales abaxially, pubescent or not; vernation noncircinate .... (2)

2. Blades scaly, without pubescence; ultimate segments glabrous adaxially; stems long-creeping; stem scales concolored .... ***C. fendleri***
2′ Blades scaly and pubescent; ultimate segments densely to sparsely pubescent adaxially, stems compact, stem scales bicolored .... ***C. eatonii***

Cheilanthes eatonii Baker (possibly named for Alvah Augustus Eaton, 1865–1908, self-taught botanist and fern expert) Eaton's lipfern or lip fern. **STEMS** compact; scales bicolored, brown with broad dark central stripe, linear-lanceolate. **LEAVES** tufted, 5–40 cm long, vernation noncircinate; petiole dark brown, rounded adaxially; blade oblong-lanceolate, 3–4-pinnate at base, 1.5–5 cm wide, scales abundant abaxially on main and secondary rachises, linear-lanceolate, mixed with pubescence; ultimate segments round to oval, beadlike, the largest 1–3 mm diameter, abaxially densely tomentose, adaxially pubescent or glabrescent. **SORI** more or less continuous around segment margins; false indusium marginal or nearly so; sporangium with 32 spores. $n = 2n = 90, 120$ (apogamous). [*Cheilanthes castanea* Maxon; *C. eatonii* f. *castanea* (Maxon) Correll]. One collection from a sandstone boulder near rim of deep, narrow canyon; canyon vegetation with riparian plants, *Pinus edulis*, *Artemisia*, *Clematis*, and *Fraxinus*. Known in our area only from Crevasse Canyon, McKinley County, New Mexico. NMEX: McK. 2030 m (6660′). Spores: Aug. Arizona, southern Utah, southern Colorado, New Mexico, through Oklahoma, Texas, south to Zacatecas and Hidalgo, Mexico, east to Arkansas. Also disjunct in Virginia, West Virginia, and Costa Rica. A variable species with triploids and tetraploids of uncertain parentage.

Cheilanthes feei T. Moore (for Antoine Laurent Apollinaire Fée, 1789–1874, French botanist) Slender lip fern or lipfern. **STEMS** compact to short-creeping; scales a mix of concolor brown and striped bicolor, linear-lanceolate, contorted, loosely appressed, persistent. **LEAVES** caespitose, 4–25 cm long, vernation circinate; petiole dark brown or black, rounded adaxially; blade ovate or linear-oblong to lanceolate, pinnatifid to mostly tripinnate, 1–4 cm wide; rachis lacking scales but with dense pubescence; ultimate segments usually round and beadlike, to 3 mm diameter, abaxially densely villous with long hairs, adaxially sparsely hirsute to glabrous. **SORI** more or less continuous around segment margins;

false indusium marginal; sporangia with 32 spores. $n = 2n = 90$ (apogamous). Exposed ridges, rocky crannies, hanging gardens, under ledges, canyon floors with washes, ponds, and permanent streams; commonly on sandstone, also on andesite porphyry and basalt; vegetation communities include blackbrush, shadscale, slickrock, sagebrush, piñon-juniper, ponderosa pine, Douglas-fir. ARIZ: Apa, Coc; COLO: Arc, Hin, LPl, Mon; NMEX: McK, RAr, SJn; UTAH. 1300–2800 m (4265–9200′). Spores: late May–Sep. Southeastern British Columbia southwest to southwestern Alberta, western Montana, Idaho and eastern Washington south to southeastern California, Arizona, New Mexico, Texas, Chihuahua and Coahuila in Mexico; east to Wisconsin, Illinois, and Arkansas, with disjunct populations in Virginia and Kentucky. Another apogamous triploid of unknown parentage.

Cheilanthes fendleri Hook. (for Augustus Fendler, 1813–1883, German pteridologist) Fendler's lipfern or lip fern. **STEMS** long-creeping, slender; scales concolor brown, ovate-lanceolate, mostly straight, loosely appressed. **LEAVES** scattered along stem, 7–30 cm long, ours shorter; vernation noncircinate; petiole dark brown, rounded adaxially; blade lanceolate to ovate-deltate, pinnatifid to 4-pinnate, mostly tripinnate, mostly 1.5–5 cm wide; rachis rounded adaxially, scaly; pinnae glabrous adaxially; ultimate segments round to oblong, somewhat beadlike or mittenish, to 3 mm diameter, abaxially glabrous or a few basal scales, adaxially glabrous. **SORI** more or less continuous around segment margins; false indusium marginal; sporangium with 64 spores. $2n = 60$. Narrow canyons with ponderosa pine, Gambel's oak, and Douglas-fir; on metaconglomerate rock in cracks and seams with a southern exposure. Known from two collections in our area, both from Vallecito Creek, Colorado. COLO: LPl. 2440–2540 m (8000–8330′). Spores: May. Arizona, New Mexico, southern Colorado, western Texas, with a disjunct population in central Texas; northern Mexico.

Cryptogramma R. Br. Parsley Fern, Rock-brake, Cliff-brake

(Greek *cryptos*, hidden, + *gramme*, line, referring to the sori hidden by false indusia) Gametophyte glabrous. Sporophyte plants lithophytic. **STEMS** erect, decumbent or creeping, branched; scales colorless to brownish, concolored to bicolored with a central stripe, ovate to linear, margin entire. **LEAVES** dimorphic, scattered or caespitose; fertile leaves 5–25 cm long; sterile leaves shorter than fertile leaves, 3–20 cm long; petiole dark brown at base, grading to light brown or green near blade, single groove adaxially, scaly; blade deltate to lanceolate, 2- to 4-pinnate, leathery to membranaceous, glabrous abaxially and adaxially to sparsely pubescent, dull to shining; ultimate segments of sterile leaves ovate to obovate or fan-shaped, 4 mm wide or less, margin plane, dentate or laciniate; fertile segments lanceolate to linear, 2 mm wide or less, margins forming false indusia, but later plane. **SORI** with false indusia greenish to brownish, broad, marginal; sporangia abaxially scattered along veins, intermixed with farina glands, each with 64 spores. **SPORES** yellow, tetrahedral, trilete, verrucose, without equatorial ridge. $x = 30$. Up to 11 species. Temperate regions of North America, South America, and Eurasia.

1. Petiole green or straw-colored for entire length (or nearly so); stems 4–20 mm diameter, decumbent to erect, many-branched from base, leaves caespitose, herbaceous to leathery .... ***C. acrostichoides***
1′ Petiole dark in proximal 1/3 (–2/3); stems 1–1.5 mm diameter, creeping, few-branched, leaves scattered along stems, delicate and ephemeral .... ***C. stelleri***

Cryptogramma acrostichoides R. Br. (grasslike) American parsley fern. **STEMS** thick, to 2 cm diameter, branched from base; scales bicolored, dense, lanceolate to linear, to 6 mm long. **LEAVES** caespitose; sterile leaves evergreen, spreading, to 17 cm long; fertile leaves erect, often longer than sterile leaves, to 25 cm; petiole green or straw-colored throughout, rarely (in ours) brown at base, scales bicolored to concolored, becoming sparse above base; blade deltate to cordate, bi- to tripinnate throughout, leathery; segments of sterile leaves oblong to ovate-lanceolate, distal margins crenate to dentate distally; segments of fertile leaves linear, margins revolute, forming false indusia. **SORI** with sporangia coalescing at maturity. $2n = 60$. [*Cryptogramma crispa* (L.) R. Br. ex Hook. subsp. *acrostichoides* (R. Br.) Hultén]. Moist spruce, fir, ponderosa pine, aspen, Douglas-fir, and alpine communities; along creeks, rock slides, and talus slopes of granite or metaconglomerate rock. COLO: Arc, Hin, LPl, Min, Mon, SJn. 2410–3757 m (7920–12325′). Spores: late Jun–mid-Sep. Alaska south through British Columbia and Alberta, south to California and Baja California, Utah, Arizona, Colorado, and north-central New Mexico, as well as Asia. An eastern population extends from north-central Canada east to Ontario and Minnesota.

Cryptogramma stelleri (S. G. Gmel.) Prantl (for Georg Wilhelm Steller, German botanist, zoologist, and explorer, 1709–1746, who worked in Alaska and Siberia as the naturalist on the Bering expedition. Steller's sea cow and Steller's jay are named for him) Slender rock-brake, slender cliff-brake, Steller's rock-brake, fragile rock-brake (also rockbrake or cliffbrake). **STEMS** creeping, slender, few-branched, succulent, scales colorless, sparse, ovate, 0.4 mm long, stems dying back in second year following unfolding of fronds. **LEAVES** scattered, deciduous; sterile leaves 3–15 cm long;

fertile leaves 5–20 cm long; mostly glabrous; petioles dark brown in proximal 1/3 (in ours) to 2/3, green to blade base; blade ovate-lanceolate, pinnate-pinnatifid to bipinnate, herbaceous; segments of sterile leaves ovate-lanceolate to fan-shaped, often lobed; segments of fertile leaves lanceolate to linear, margins reflexed to form continuous false indusia. **SORI** often discrete, with false indusia. $2n = 60$. Rare, with mosses and other ferns in cracks of limestone or hornfels (metamorphic), on cliffs near streams and rivers, and streamside with spruce, narrowleaf cottonwood, *Pinus strobiformis*, and Rocky Mountain maple. COLO: Arc, SJn. 2410–3110 m (7900–10200′). Spores: Aug. Alaska south to Washington and Montana, with scattered populations in northeastern Oregon, northeastern Nevada, Colorado, northern Utah, and northwestern Wyoming. Eastern distribution is from Newfoundland south through Quebec, Ontario to Minnesota, Wisconsin, New York, and New Jersey, with a disjunct population in West Virginia; also northeastern Asia.

Pellaea Link Cliff-brake, Cliffbrake

(Greek *pellos*, dark, referring to the bluish gray leaves) Sporophyte plants rigid, lithophytic. **STEMS** compact to long-creeping, ascending to horizontal, branched; scales brownish concolored or bicolored with dark central stripe, linear-subulate to lanceolate or ovate, margins entire or erose to dentate. **LEAVES** monomorphic to slightly dimorphic, clustered to scattered, to 100 cm long, shorter in ours; petiole black grayish, brownish, or straw-colored, rounded or flattened with a groove adaxially, glabrous to pubescent, often with scales; blade linear to deltate, pinnate to 4-pinnate, leathery, rarely herbaceous, glabrous, pubescent, or scaly abaxially, glabrous adaxially; ultimate segments elliptic, lanceolate to linear, usually 0.5 cm wide or more. **SORI** with false indusia greenish to whitish, narrow, marginal; sporangium with 32 or 64 spores, often intermixed with farina-producing glands. **SPORES** yellow to brown, tetrahedral-globose, rugose or cristate, without equatorial ridge. $x = 29$. About 40 species. Western Hemisphere, with some in Asia, Africa, and Oceania. One of the "resurrection" or cheilanthoid ferns, many species occur in western xeric environments, thus increasing the fern floras of Texas, Arizona, and New Mexico. A poorly defined group, probably polyphyletic, although our species may be closely related to one another.

1. Stem scales bicolored, with dark central region and lighter, brown margin; stipe, rachis, and pinnae rachises glabrous; ultimate segment apex sharp, long, mucronate .... ***P. truncata***

1′ Stem scales uniformly reddish brown or tan; stipe, rachis, and pinnae rachises pubescent; ultimate segment apex not long, sharp, mucronate .... (2)

2. Stipe, rachis, and pinnae rachises densely pubescent with short, curly, appressed hairs on adaxial surface; ultimate segments with midvein on abaxial surface with numerous long hairs .... ***P. atropurpurea***

2′ Stipe, rachis, and pinnae rachises with scattered long hairs; ultimate segments with midvein on abaxial surface glabrous or with a few long hairs .... ***P. glabella***

Pellaea atropurpurea (L.) Link (dark purple) Purple cliff-brake or cliffbrake. **STEMS** ascending, wide, to 1 cm diameter; scales reddish brown or tan, linear-subulate, margins entire to denticulate. **LEAVES** slightly dimorphic, sterile leaves shorter and less divided than fertile leaves, to 50 cm long (ours 28 cm); fiddleheads villous; petiole reddish purple to blackish, shining, rounded adaxially; blade elongate-deltate, pinnate to bipinnate, dull green to grayish green, rachis reddish purple, densely pubescent adaxially with short, curly, appressed hairs; ultimate segments linear-oblong, 10–75 mm long, leathery, with long hairs abaxially along midrib, apex obtuse to slightly mucronate. **SORI** with false hyaline-margined indusium, sometimes white-edged; sporangia long-stalked, with 32 spores, not intermixed with farina-producing glands. $n = 2n = 87$ (apogrmous). One specimen known from Kane Gulch, San Juan County, Utah. Rare in partially shaded cracks and ledges on north-facing outcrops of Cedar Mesa Sandstone in a riparian woodland community. UTAH. 1935 m (6350′). Spores: Jun. Southern Nevada south and east to the coastal prairies of Texas, north and east to Quebec, south to western Florida; Mexico and Guatemala. *P. atropurpurea* is an autopolyploid derivative of an unknown diploid taxon. Hybrids with other pellaeas also occur.

Pellaea glabella Mett. ex Kuhn (smooth) Smooth cliff-brake or cliffbrake. **STEMS** ascending, wide, to 1 cm diameter; scales reddish brown, linear-subulate, margin sinuous, entire to denticulate. **LEAVES** monomorphic, clustered, to 40 cm long (ours 17 cm); fiddleheads sparsely villous; petiole brown or reddish brown, shining, rounded adaxially; blade linear-oblong to ovate-lanceolate, pinnate to bipinnate, to 8 cm wide, bluish green, rachis brownish, glabrous to very sparsely pubescent with long, flaccid hairs; ultimate segments oblong-lanceolate, 5–20 mm long, leathery, occasionally herbaceous, glabrous with occasional thin scales near midrib, apex obtuse. **SORI** with false indusia, borders whitish, erose-denticulate; sporangia long-stalked, with 32 (in ours) spores, not mixed with farina-producing glands. **SPORES** 60–72 μm diameter (in ours). $2n = 116$ (apogamous). Two collections from our area. At base of sandstone cliffs in canyon, sometimes near seeps. COLO: Mon; UTAH. Spores: Jul. Our material belongs to **subsp. *simplex*** (Butters) Á. Löve & D. Löve (simple). Found from British Columbia and Alberta south to Washington, northern Idaho, and northwestern Montana; southern Wyoming, western Colorado, southeastern Utah, northern Arizona, and New Mexico. A very rare species from our area.

Pellaea truncata Goodd. (cut off, possibly referring to the truncated segment base) Spiny cliffbrake or cliff-brake. **STEMS** ascending, wide, to 1 cm diameter; scales bicolored with black central stripe and pale brown edges, linear-subulate, to 0.3 mm wide, margins erose-denticulate. **LEAVES** clustered, slightly dimorphic, with sterile leaves shorter and less divided than fertile, to 45 cm (30 cm in ours); fiddleheads sparsely villous; petiole dark brown or blackish, shining, flattened or slightly grooved adaxially; blade triangular-lanceolate to deltate, mostly bipinnate (occasionally tripinnate on larger fronds), to 18 cm wide (5–6 cm in ours); rachis brownish, glabrous; pinnae mostly perpendicular to rachis; ultimate segments narrowly oblong, to 1 cm long, leathery, glabrous, apex long and sharply mucronate. **SORI** with false indusia, borders whitish, entire; sporangia long-stalked, with 64 spores, intermixed with farina-producing glands. $2n = 58$. [*Pellaea longimucronata* Hook.]. Rare and local at lower elevations on slickrock sandstone, usually on or near the Kayenta Formation. Desert scrub communities including blackbrush, cliffrose, ephedra, and snakeweed. UTAH. 1460–1380 m (4540–4800′). Spores: late Mar–early Oct. Southwestern California east to west Texas (to the western Edwards Plateau), with outlying populations in central Colorado, south to northern Mexico.

SELAGINELLACEAE Willk. SPIKEMOSS FAMILY

Timothy Reeves and Linda Mary Reeves

Sporophytes herbaceous, annual or perennial, occasionally evergreen. **STEMS** leafy, branching dichotomously; rhizophores present or absent at branch forks, base of stems, or throughout. **LEAVES** dimorphic or monomorphic, small or tiny, with adaxial basal ligule, single-veined. **STROBILUS** sometimes cryptic or not well defined, mostly terminal, cylindric, quadrangular, or flattened; sporophylls monomorphic or slightly dimorphic, slightly to somewhat differentiated from sterile leaves; sporangium short-stalked, solitary, in axil of sporophyll, opening with distal slits. **SPORES** heterosporous; megaspores large, generally 4; microspores small, numerous to hundreds. One genus, about 700 species known, probably many more undiscovered. Worldwide, primarily tropical and subtropical, including arid and semiarid environments, where they may add substantially to the lower vascular plant flora. Many species cultivated (*Selaginella kraussiana*, *S. lepidophylla*, *S. braunii*) and introduced, especially in warm, humid climates. Some of the selaginellas are called "resurrection ferns" because they appear to die during unfavorable conditions but regreen when favorable conditions return. Among ferns and allies, selaginellas are distantly related to the Lycopodiaceae and Isoetaceae, although it is often implied that they are more closely related (Flora of North America Editorial Committee 1993; Tryon 1955).

Selaginella P. Beauv. Spikemoss, Selaginella

(Latin *selago*, an old name for *Lycopodium*, but diminutive) Sporophytes terrestrial, lithophytic, hemiepiphytic, or epiphytic (mostly tropical species). Roots branching several times from rhizophore tips. **STEMS** prostrate, creeping, decumbent, caespitose, rosulate, climbing, or erect; minimally to greatly branched; rhizophores stout or filiform. **LEAVES** dimorphic or monomorphic; if monomorphic, linear to lanceolate, greatly overlapping, arranged spirally; if dimorphic, round or oblong to lanceolate, in 4 ranks with 2 ranks of larger, spreading lateral leaves and 2 ranks of smaller, appressed and ascending median leaves, sometimes with axillary leaf at base of each branching dichotomy. **MEGASPORANGIUM** lobed to ovoid. **MICROSPORANGIUM** reniform to ovoid. **MEGASPORES** tetrahedral, sculptured variously, 200–1360 μm diameter. **MICROSPORES** tetrahedral, ovoid or globose, sculptured variously, 20–75 μm diameter. $x = 7–12$. More than 700 species. Worldwide, mostly tropical and subtropical. Eventually this large, diverse genus may be split into more than one genus. Some characters are best seen with a magnification of 20× or higher.

1\. Leaves blunt at tip, lacking terminal setae (hairlike tips); plants very loosely matted, the branches long, distant, and spreading; stems about 1 mm in diameter .... ***S. mutica***

1′ Leaves with terminal whitish or yellowish setae; plants more densely matted, with shorter branches; stems about 2–3 mm in diameter .... (2)

2\. Stems forming looser mats; leaves with terminal setae 0.4–1 mm (in ours) long, setae scabrous, not smooth; leaves bright green; lateral branches spreading, not ascending; strobili inconspicuous .... ***S. underwoodii***

2′ Stems forming dense, cushionlike mats, the branches erect or ascending, strobili erect, elongate .... (3)

3. Leaves with terminal setae 0.5–2 mm long, setae scabrous, not smooth; leaves somewhat dissimilar, gray-green, those on the under and outer sides of the stems curved upward and longer than the upper leaves; branch tips conspicuously white-tipped due to long setae; plants of Colorado in our area .... ***S. densa***

3' Leaves with terminal setae 0.2 (in ours) to 0.7 mm long, setae smooth, not denticulate; leaves all alike, not curved upward, bright green; branch tips not conspicuously white-tipped since the setae are short; plants of Utah in our area .... ***S. watsonii***

Selaginella densa Rydb. (thick, clumpy, referring to the growth habit) Prairie club-moss, Rocky Mountain spikemoss, lesser spikemoss. Sporophyte terrestrial or lithophytic, forming cushionlike mats, occasionally more loosely constructed. **STEMS** somewhat shaggy, short, creeping or decumbent, not readily fragmenting, forked irregularly, ascending, 2–3-forked. **LEAVES** monomorphic in pseudowhorls of 5 or 6, tightly appressed, ascending, green, linear to linear-lanceolate, with abaxial ridges, upper side leaves slightly smaller than underside leaves, 3–5 × 0.4–0.7 mm; underside leaf base long-decurrent, oblique, glabrous; upper side leaves slightly decurrent, oblique, occasionally pubescent; margins long-ciliate; apex slightly keeled to plane, abruptly long-bristled, bristle white to transparent, scabrous. **STROBILUS** solitary, about 1–3 cm long; sporophyll ovate-lanceolate, with well-defined abaxial ridges, ecilate or with the cilia dentiform to piliform and ascending; seta base often strongly broadened and flattened. [*S. scopulorum* Maxon; *S. densa* Rydb. var. *scopulorum* (Maxon) R. M. Tryon; *S. standleyi* Maxon; *S. densa* Rydb. var. *standleyi* (Maxon) R. M. Tryon]. Sunny alpine and subapine areas in rocky/gravelly soil, on talus and scree slopes, in humid canyons by creeks, associated with Dakota Sandstone, metaconglomerate, slickrock, with ponderosa pine, spruce-fir, aspen. COLO: Arc, Con, Hin, LPl, Min, Mon, RGr, SJn. 2350–3925 m (7730–12880'). Spores: Jun–Sep. British Columbia east to southwestern Ontario, in a triangular distribution south to northeastern Arizona, northern New Mexico, and western Texas. Tryon (1955) recognized three varieties based on their distinctive morphology and distinct distributions in the north but states, "At the same time, it must be admitted that identification, particularly of the material from Colorado, is sometimes arbitrary. The intermediates show all transitions between the two extremes in respect to the varietal characters."

Selaginella mutica D. C. Eaton ex Underw. (cut off, referring to the blunt leaves) Bluntleaf spikemoss. Sporophytes terrestrial or lithophytic, forming loose mats. **STEMS** somewhat smooth in appearance, radially symmetrical, long- to short-creeping, not readily fragmenting, forked, lateral branches 1–2-forked. **LEAVES** monomorphic in alternate pseudowhorls of 3, tightly appressed, ascending, green, linear-lanceolate to lanceolate-elliptic, 1–2 × 0.45–0.6 mm, abaxial ridges well defined; base rounded, adnate, pubescent or glabrous; margins transparent-ciliate to denticulate; apex keeled, obtuse or slightly attenuate, blunt to short-setaceous; bristle transparent to greenish or whitish, smooth. **STROBILUS** solitary, about 1–3 cm long; sporophyll ovate-lanceolate to deltate-ovate, abaxial ridges well defined; base glabrous, margins ciliate (ours long-ciliate) to denticulate, apex keeled, blunt to short-setaceous. $2n = 18$. Between seams and layers of sandstone or on basalt in mostly dry sites on slopes or cliffs or ledges, or near creeks or streams with ponderosa pine, piñon, Utah juniper, Gambel's oak, three-leaf sumac, boxelder, Rocky Mountain juniper, Douglas-fir, spruce-fir. ARIZ: Apa, Nav; UTAH. 1720–2970 m (5640–9730'). Spores: Jun–Aug. Southern Wyoming south to eastern Arizona, New Mexico, western Texas, and Chihuahua, Mexico. Our plants are **var. *mutica***.

Selaginella underwoodii Hieron. (for Lucien Marcus Underwood, 1853–1907, pteridologist and founding member of the Board of Scientific Directors of the New York Botanical Garden) Underwood's spikemoss. Sporophyte lithophytic, growing in prostrate, loose mats, occasionally more compact. **STEMS** somewhat shaggy, radially symmetrical, long-creeping, short-creeping, or pendent, not easily fragmenting, irregularly forked, lateral branches spreading, 1–2-forked. **LEAVES** monomorphic, in alternate, weakly imbricate pseudowhorls of 4 on main stem and older lateral branches, in 3 on young lateral branches and secondary branches; loosely appressed, ascending, bright green, linear to narrowly triangular-lanceolate or subulate, about 2.5–3.4 × 0.45–0.5 mm; abaxial ridges prominent; base mostly cuneate and decurrent, rarely rounded and adnate, pubescent or glabrous; margins entire to denticulate or transparent short-ciliate, mostly denticulate toward apex; apex keeled, seta 0.4–1 (in ours) mm, scabrous (in ours); seta transparent to green or greenish yellow, less often white, scabrous. **STROBILI** often paired, 0.5–3.5 cm long; sporophylls lanceolate to ovate-lanceolate or subulate, abaxial ridges prominent, base glabrous with unusual prominent auricles, margins entire, denticulate to short-ciliate, apex keeled, short- to long-setaceous. Moist, rocky canyons on Dakota Sandstone outcrops and boulders or near waterfalls, in narrowleaf cottonwood, ponderosa pine, spruce-fir, and aspen communities. COLO: Hin. 2325–3100 m (7635–10100'). Southern Wyoming south to Arizona, New Mexico, western Texas; Chihuahua, Tamaulipas, and Nuevo León, Mexico.

Selaginella watsonii Underw. (for Serano Watson, 1826–1892, botanist and curator of the Gray Herbarium at Harvard University) Alpine spikemoss, Watson's spikemoss. Sporophytes terrestrial or lithophytic, forming cushionlike mats. **STEMS** slightly shaggy, radially symmetrical, decumbent to long-creeping, not readily fragmenting, irregularly forked, lateral branches strongly ascending, 1–3-forked. **LEAVES** monomorphic, in alternate pseudowhorls of 4, tightly or loosely appressed, ascending, bright green, linear-lanceolate, 3–4 × 0.5–0.7 mm; abaxial ridges prominent; base cuneate, decurrent, glabrous, occasionally puberulent; margins entire to scattered transparent short-ciliate; apex strongly keeled, obtuse, abruptly narrowed to whitish seta, 0.2–0.5 mm long. **STROBILUS** solitary, 0.5–3 cm long; sporophyll lanceolate to ovate-lanceolate, abaxial ridges prominent, base glabrous, margin entire or, rarely, dentate, apex strongly keeled to truncate and short-bristled. Rare, one collection from the dry east slope of Navajo Mountain, Utah; piñon-sagebrush community. UTAH. 2100 m (6900′). Idaho, southwestern Montana, and western Wyoming southwest to California and the Arizona Strip.

GYMNOSPERMS

CUPRESSACEAE Gray CYPRESS FAMILY

Robert P. Adams

Trees or shrubs, evergreen in our flora, roots fibrous to woody. **BARK** fibrous and furrowed or exfoliating in plates. **LEAVES** of adults appressed, scalelike, with a generally visible dorsal oil gland, juvenile leaves decurrent (except both adult and juvenile leaves acicular in *Juniperus communis*), resinous and aromatic, persisting 3–5 years, alternate or in whorls of 3. **POLLEN CONES** maturing and shedding annually, solitary and terminal (except axillary in *J. communis*), oblong, sphorophylls overlapping, bearing 2–10 abaxial microsporangia; pollen spherical, not winged. **SEED CONES** maturing in 1–2 years, borne on a short pedicel, persistent in *Cupressus* and deciduous upon maturity in *Juniperus*, terminal (except axillary in *J. communis*), cone scales overlapping, fused in *Juniperus*, abutting in *Cupressus*, but appearing as spherical to ovoid, cone scales woody to fleshy (resembling a fruit in some *Juniperus*), brown to reddish brown to blue when ripe, opening upon maturity in *Cupressus* but the entire seed cone fused in *Juniperus*. **SEEDS** 1–20 per scale or 1 to numerous per cone, not winged, cotyledons 2–9 (Watson and Eckenwalder 1993). Genera 25–30 (1 in our flora), species 110–130 (5 in our flora); our genus widespread in the Northern Hemisphere (Adams 1993). There are collections of *Cupressus arizonica* from our flora, but these plants are cultivated and not reproducing under natural conditions. Therefore, *C. arizonica* is excluded from this treatment along with cultivars of *Juniperus* (e.g., *J. chinensis* and *J. sabina*). *Juniperus deppeana* Steud. (alligator juniper) is not yet reported for the Four Corners flora; however, it has been found in the Zuni Mountains, just south of the study area, and potential habitat exists in McKinley County, New Mexico.

Juniperus L. Juniper, Cedar, Cedro, Sabino

(Latin *juniperus*, name for juniper) Trees, shrubs; columnar or prostrate; evergreen. Branchlets terete, 2–6-angled, variously oriented, but not in flattened sprays. **LEAVES** opposite in 4 ranks or in whorls of 3; adult leaves closely appressed to divergent, scalelike to subulate, free portion to about 10 mm (to about 15 mm in *J. communis*), abaxial gland visible or not, elongate to oval, sometimes exuding white crystalline deposit. **POLLEN CONES** with 3–7 pairs or trios of sporophylls, each sporophyll with 2–8 pollen sacs. **SEED CONES** maturing in 1 or 2 years, globose to ovoid, berrylike, 2–20 mm, remaining closed, usually glaucous; cone scales persistent, 1–3 pairs, peltate, tightly coalesced, thick and fleshy or fibrous to obscurely woody. **SEEDS** 1–3 per scale, 1–6 per cone, round to faceted, wingless, cotyledons 2–6. $x = 11$. Ranging from Arctic, temperate, and deserts to subtropical and from timberline to sea level. Sixty-six species in the Northern Hemisphere, often a weedy, invasive species in North America (Adams 1993), but generally relictual in isolated populations in central Asia, although weedy (*J. communis*) in Europe. *Juniperus* is the only dioecious (sometimes monoecious) genus of *Cupressaceae* in North America. Numerous cultivars of *Juniperus* species are widely used for landscaping. Mutants or "sports" affecting plant habit and foliage are present in all species and are likely related to single-gene mutations. Many have been given formal names or incorrectly ascribed to hybridization. Gymnocarpy (bare seeds protruding from the cone), caused by insect larvae (Zanoni 1978), is found in most juniper species, particularly in the southwestern United States. Specimens with such aberrations may be almost impossible to identify without chemical data.

1. Leaves of 1 kind, subulate, with basal abscission zone, spreading; cones axillary (*Juniperus* sect. *Juniperus*) ..... ***J. communis***
1' Leaves of 2 kinds, whip (subulate, without basal abscission zone) and scalelike; cones terminal (*Juniperus* sect. *Sabina*) ..... (2)

2. Margins of leaves entire (at 20× and 40×) ..... ***J. scopulorum***
2' Margins of leaves serrate (at 20×) ..... (3)

3. Abaxial glands inconspicuous because embedded in leaf; monoecious, seed cones 8–9 (12) mm, bluish brown, tan beneath glaucous coating, fibrous ..... ***J. osteosperma***
3' Abaxial glands conspicuous; dioecious, seed cones 6–8 mm, reddish blue to brownish blue, resinous ..... ***J. monosperma***

Juniperus communis L. (common) Common juniper. Shrubs prostrate or low with ascending branchlet tips (occasionally spreading shrubs to 3 m). **LEAVES** upturned, to 15 × 1.6 mm, rarely spreading, linear, glaucous stomatal band about as wide as each green marginal band, apex acute and mucronate to acuminate. **SEED CONES** 6–9 mm, shorter than leaves. 2*n* = 22. [*J. canadensis* Lodd. ex Burgsd.; *J. communis* subsp. *depressa* (Pursh) Franco; *J. depressa* Raf. ex M'Murtrie]. Rocky soil, slopes, and summits. ARIZ: Apa; COLO: Arc, Hin, LPl, Min, Mon, RGr, SJn; NMEX: McK, RAr, SJn; UTAH. 2440–3150 m (8000–10335'). Pollen shedding: Apr–May. Canada, high mountains in the western United States, Great Lakes, and New England, and mountain populations as far south as Tennessee and Georgia. *Juniperus communis* varieties were recently reevaluated (Adams and Pandey 2003). Four Corners material belongs to **var. *depressa*** Pursh (depressed or flattened). In Europe, the berries (seed cones) are used for flavoring gin. Used by the Ramah Navajo as an emetic for all ceremonials.

Juniperus monosperma (Engelm.) Sarg. (one seed) One-seed juniper, sabina. Shrubs or small trees; dioecious, to 7 (–12) m, usually branching near the base; crown rounded to flattened-globose. Bark gray to brown, exfoliating in thin strips, that of small branchlets (5–10 mm diameter) smooth, that of larger branchlets exfoliating in either flakes or strips. Branches ascending to erect; branchlets erect, 4–6-sided, about 2/3 as wide as length of scalelike leaf. **LEAVES** green to dark green, abaxial glands elongate, fewer than 1/5 of the glands (on whip leaves) with an evident white crystalline exudate, margins denticulate (at 20×); whip leaves 4–6 mm, glaucous adaxially, scalelike leaves 1–3 mm, not overlapping, or, if so, by less than 1/4 their length, keeled, apex acute to acuminate, spreading. **SEED CONES** maturing in 1 year, 1 size, with straight peduncles, globose to ovoid, 6–8 mm, reddish blue to brownish blue, glaucous, fleshy and resinous, with 1 (–3) seeds. **SEEDS** 4–5 mm. [*J. monosperma* f. *gymnocarpa* (Lemmon) Rehder; *J. occidentalis* var. *gymnocarpa* Lemmon; *J. occidentalis* var. *monosperma* Engelm.; *J. mexicana* var. *monosperma* (Engelm.) Cory; *Sabina monosperma* (Engelm.) Rydb.]. Dry, rocky soils and slopes. NMEX: McK, RAr, San, SJn. 1800–2500 m (5900–8200'). Pollen shedding: Feb–Apr. Arizona, Colorado, New Mexico, and Texas. Reports of hybridization with *J. pinchotii* were refuted (Adams 1975). Ramah Navajo uses include: decoction for postpartum or menstrual pain and a cold infusion for stomachache; ceremonial medicine used in "bath for purification of burial party"; medicine for burns, sweat bath medicine, inner bark given to newborns "to clean out impurities"; cold infusion used for fever; given to sheep for bloating from eating "chamiso"; wood used for fence posts and hogan roofs; boughs used for the sides and roofs of shade houses or special logs for the "Enemy Way Ceremonial"; bark used as platform for sun-drying roasted corn; and wood used to make hunting bows. Navajo uses include: green bark and berries used as a green dye for wool; wood used to make prayer sticks and to make bows for the canopy of the baby's cradleboard; wood used for firewood; and wood made into charcoal and used for smelting silver.

Juniperus osteosperma (Torr.) Little (bone seed) Utah juniper, bone seed juniper, sabina morena. Trees or shrubs, monoecious, to 6 (–12) m, multi- or single-stemmed, crown rounded. Bark exfoliating in thin gray-brown strips, that of smaller and larger branchlets smooth. Branches spreading to ascending; branchlets erect, 3–4-sided in cross section, about as wide as length of scalelike leaves. **LEAVES** light yellow-green, abaxial glands inconspicuous and embedded, exudate absent, margins denticulate (at 20×); whip leaves 3–5 mm, glaucous adaxially; scalelike leaves 1–2 mm, not overlapping, or, if so, by less than 1/10 their length, keeled, apex rounded, acute, or occasionally obtuse, appressed. **SEED CONES** maturing in 1–2 years, of 1–2 sizes, with straight peduncles, globose, (6–) 8–9 (–12) mm, bluish brown, often almost tan beneath glaucous coating, fibrous, with 1 (–2) seeds. **SEEDS** 4–5 mm. [*J. californica* Carrière var. *osteosperma* (Torr.) L. D. Benson; *J. californica* subsp. *osteosperma* (Torr.) A. E. Murray; *J. californica* var. *utahensis* Engelm.; *J. californica* var. *utahense* Vasey; *J. knightii* A. Nelson; *J. megalocarpa* Sudw.; *J. monosperma* var. *knightii* (A. Nelson) Lemmon; *J. occidentalis* Hook. var. *utahensis* (Engelm.) Kent; *J. tetragona* Schltdl. var. *osteosperma* Torr.; *J. utahensis* (Engelm.) Lemmon; J. *utahensis* var. *megalocarpa* (Sudw.) Sarg.; *J. utahensis* var. *cosnino* Lemmon; *Sabina knightii* (A. Nelson) Rydb.; *Sabina megalocarpa* (Sudw.) Cockerell; *Sabina osteosperma* (Torr.) Antoine; *Sabina utahensis* (Engelm.) Rydb.]. Dry, rocky soil and slopes. ARIZ: Apa, Coc, Nav; COLO: Arc, Dol, LPl, Mon, SMg; NMEX: McK, RAr, San, SJn; UTAH. 1680–2380 m (5500–7800'). Pollen shedding: Mar–May. Arizona, California, Colorado, Idaho, Montana, Nevada, New Mexico, Utah, and Wyoming. *J. osteosperma* is the dominant juniper of Utah. Hybridization with *J. occidentalis* reported (Vasek 1966; Terry, Nowak, and Tausch 2000). Navajo uses include: seeds eaten for headaches; used to wash the hair; green timber used to make corrals. The Hopi used the wood for fuel. A major food for wildlife.

Juniperus scopulorum Sarg. (of rocky cliffs) Rocky Mountain juniper, Rocky Mountain redcedar. Trees, dioecious, to 20 m tall, single-stemmed (rarely multistemmed); crown conic to occasionally rounded. Bark brown, exfoliating in thin strips, that of small branchlets (5–10 mm diameter) smooth, that of larger branchlets exfoliating in plates. Branches spreading to ascending; branchlets erect to flaccid, 3–4-sided in cross section, about 2/3 or less as wide as length of scalelike leaves. Leaves light to dark green but often glaucous blue or blue-gray, abaxial gland elliptic, conspicuous, exudate absent, margins entire (at 20× and 40×); whip leaves 3–6 mm, not glaucous adaxially; scalelike leaves 1–3 mm,

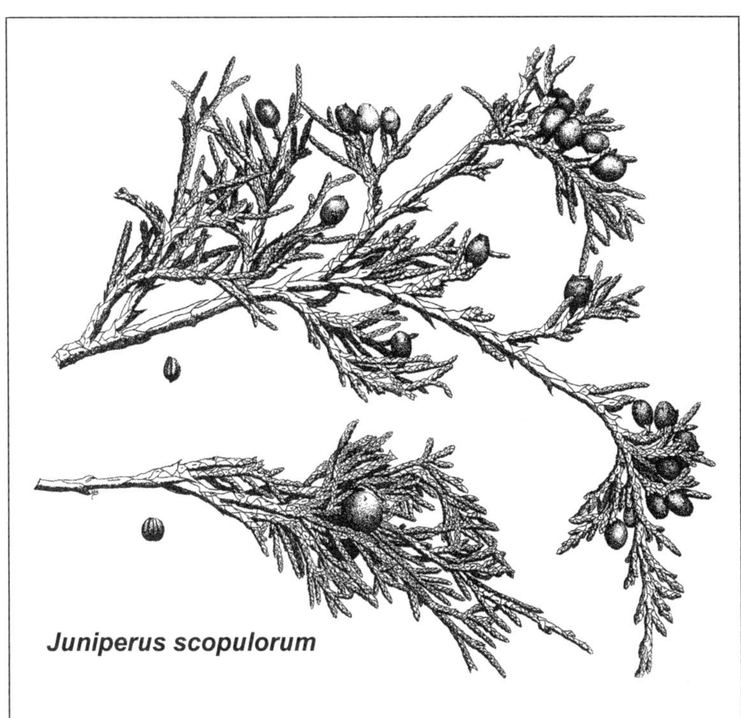
Juniperus scopulorum

not overlapping to overlapping by not more than 1/5 their length, keeled to rounded, apex obtuse to acute, appressed or spreading. **SEED CONES** maturing in 2 years, of 2 distinct sizes, generally with straight peduncles, globose to 2-lobed, 6–9 mm, appearing light blue when heavily glaucous, but dark blue-black beneath glaucous coating when mature (or tan beneath glaucous coating when immature), resinous to fibrous, with (1–) 2 (–3) seeds. **SEEDS** 4–5 mm. 2*n* = 22. [*J. excelsa* sensu Pursh non M. Bieb.; *J. scopulorum* var. *patens* Fassett, ×*fassettii* Boivin (*horizontalis* × *scopulorum*); *J. scopulorum* var. *columnaris* Fassett (environmentally induced by gases from burning coal, see Adams 1982); *J. virginiana* L. var. *montana* Vasey; *J. virginiana* var. *scopulorum* (Sarg.) Lemmon; *J. virginiana* subsp. *scopulorum* (Sarg.) A. E. Murray; *Sabina scopulorum* (Sarg.) Rydb.]. Rocky soils, slopes, and eroded hillsides. ARIZ: Apa, Nav; COLO: Arc, Dol, Hin, LPl, Min, Mon, SJn, SMg; NMEX: McK, RAr, SJn; UTAH. 1950–2750 m (6400–9020′). Pollen shedding: Mar–May. Vancouver Island and Puget Sound, from Alberta to northern Mexico. Hybridizes with *J. virginiana* (Comer, Adams, and Van Haverbecke 1982) in the Missouri River basin. Occasionally used for fence posts. Navajo uses include: plant taken as a "War Dance Medicine"; plant rubbed on the hair for dandruff; pounded mixture of herbs given to patient during the blackening ceremony of the "War Dance." The Kayenta Navajo used the plant for pain. The Ramah Navajo uses include: lotion for headache and stomachache; cold infusion used as a ceremonial medicine to protect from enemies and witches; taken and used as lotion for colds and fever.

EPHEDRACEAE Dumort. EPHEDRA FAMILY

Stanley L. Welsh & N. Duane Atwood

Dioecious shrubs; branches green to olive-green, opposite or whorled, striate. **LEAVES** scalelike, opposite or whorled, more or less connate. **MALE CONES** compound, borne at the nodes or terminal, with 2–8 microsporophylls, these free or with stalks united, with a calyxlike involucre surrounding the stalks. **FEMALE CONES** solitary or whorled, sessile or peduncled, subtended by firm or scarious bracts. **SEEDS** 1–3, hard, somewhat angled to almost terete.

Ephedra L.

The stems simulate those of an *Equisetum*, especially in being green and striate; the differences are obvious, however. **LEAVES** scalelike, either opposite or in whorls of 3. **CONES** laterally produced; stems with a solid, black pith (Cutler 1939; Benson 1943).

1. Leaves and bracts 3 per node; branches whorled; bracts of female cone scales clawed, 6–10 mm wide, scarious ..... ***E. torreyana***

1′ Leaves and bracts 2 per node; branches initially opposite; bracts of female cone scales not clawed, 3–5 mm wide, only the margins scarious ..... ***E. viridis***

Ephedra torreyana S. Watson (for John Torrey, distinguished botanist and colleague of Asa Gray) Torrey's ephedra. Erect shrubs, 2–10 m tall (rarely more). **BRANCHES** blue-green to olive-green, sometimes glaucous, appearing smooth but with many small longitudinal furrows, rigid, terete, to 3.5 mm thick, solitary or whorled at the nodes. **LEAVES** ternate or whorled, 2–5 mm long, dorsimedially thickened, connate for nearly 2/3 their length, at maturity the lobes spreading or recurved, somewhat persistent. **MALE CONES** solitary to several in a whorl, ovate, sessile, 6–8 mm long; bracts ternate in 5–6 whorls, obovate, clawed, scarious except in the center and at the base. **FEMALE CONES**

solitary to several at the nodes, ovoid, 9–13 mm long, sessile, the bracts in 3s in whorls of 5 or 6, obovate, clawed, 6–9 mm long, scarious, the margins minutely toothed and undulate. **SEEDS** solitary or 2, pale brown to yellow-green, scabrous, 7–10 mm long. Dry, sandy or rocky hillsides in blackbrush, salt desert shrub, mountain brush, and piñon-juniper communities. ARIZ: Apa, Nav; COLO: Mon; NMEX: McK, SJn; UTAH. 1060–2330 m (3500–7700′). Texas to Arizona, New Mexico, Utah, and southwest Colorado. This plant serves as food for wildlife and is eaten by livestock, especially by sheep.

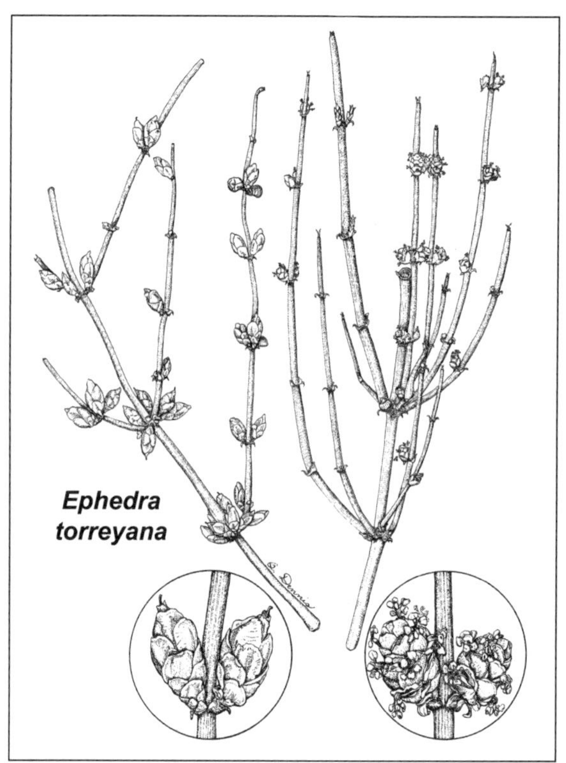

Ephedra viridis Coville (green) Green ephedra, Mormon tea, Brigham's tea. Shrubs 1–15 dm tall, spreading to erect. **BRANCHES** rigid to flexible, bright green to yellow-green or less commonly olive to gray-green, initially opposite, in some finally falsely whorled, typically fastigiate and broomlike. **LEAVES** opposite, 1.5–4 mm long, thickened dorsimedially, deciduous and leaving a thickened, persistent brown base. **MALE CONES** 2 or more, obovoid, sessile, 5–7 mm long, the bracts opposite, 2–4 mm long, membranous, pale yellow, ovate. **FEMALE CONES** obovoid, 6–10 mm long, sessile or pedunculate, with 4–8 pairs of ovate bracts 4–7 mm long. **SEEDS** paired, brown, trigonal, smooth, 5–8 mm long. $2n = 14$.

1. Female cones sessile or nearly so; stems not viscid .... **var. *viridis***
1′ Female cones pedunculate; stems often viscid .... **var. *viscida***

var. *viridis* Green ephedra, Mormon tea. Blackbrush, salt desert shrub, sagebrush, mountain brush, piñon-juniper, rabbitbrush, and mountain brush communities. ARIZ: Apa, Coc, Nav; COLO: Dol, LPl, Mon; NMEX: McK, RAr, SJn; UTAH. 1515–2950 m (5000–9730′). Wyoming to Colorado, Nevada, Oregon, and California. This is the source of Mormon tea (also known by myriad other vernacular names), a yellowish drink made by steeping the branchlets in hot water. It evidently has some physiological activity, but the tea has the flavor of an old stocking soaked in hot water. The plant is not so severely hedged by browsing animals as some of the other species but is still of considerable importance as a browse plant. The drug ephedrine is extracted from certain Old World species.

var. *viscida* (Cutler) L. D. Benson (for viscid or sticky) Cutler's ephedra. Plants mainly buried in sandy soil, with only the branch tips protruding 1–3 dm aboveground; stipes of cones (0) 1–17 mm long, with a pair of bracteate scales above the middle or closely subtending the cones. Mostly in sandy areas with blackbrush, mixed desert shrub, mixed grass, rabbitbrush, and piñon-juniper communities. ARIZ: Apa, Coc, Nav; COLO: Mon; NMEX: McK, SJn; UTAH. 1775–1950 m (5860–6435′). In contemporary floras this plant has been regarded at specific rank as *E. cutleri*. [*E. coryi* E. L. Reed var. *viscida* Cutler; *E. cutleri* Peebles]. However, the length of the stalks of the ovulate cones and the viscid condition of the stems form a continuum with *E. viridis* in a strict sense, especially where the two grow together. Cutler's ephedra is, however, the phase of the species that forms stands in extensive sandy grasslands in southeast Utah and in the Four Corners area of the San Juan Basin, often with only the tips of the stems protruding, almost grasslike, from the sand. Much of the plant is buried within the sandy substrate, which it helps to stabilize. When the sand is blown away, as occasionally, the plants have the shape of a small, treelike shrub.

PINACEAE Spreng. ex Rudolphi PINE FAMILY

J. Robert Haller and Nancy J. Vivrette

Evergreen trees in our flora, usually excurrent (with undivided main trunk); resinous, aromatic; monoecious. **LEAVES** needlelike or narrowly linear, arranged spirally on the branches or clumped (fascicled) on short lateral shoots. **POLLEN CONES** (male) small, soft, with spirally arranged pollen-bearing scales (microsporophylls), drying after shedding pollen (anthesis) and soon deciduous. **SEED CONES** (female) mostly large, becoming woody or papery, maturing and usually releasing seeds in 1 or 2 growing seasons; scales overlapping, spirally arranged. **SEEDS** 2 per scale, borne on upper (adaxial) side, each with attached membranous wing (lacking in *Pinus edulis*). x = 12, 13. (Burns and Honkala 1990; Little 1979; Silba 1986; Thieret 1993). Ten genera, approximately 200 species, including 4 genera and 12 species in our flora. *Pinus*, with about 100 species, is the largest genus. The indigenous distribution of the pine family is limited almost entirely to the Northern Hemisphere, where Pinaceae often dominate large stands of natural vegetation, especially at higher elevations and latitudes. Pinaceae provide most of the world's softwood timber, obtained both from native stands and from plantations of exotic species, the latter especially in the temperate Southern Hemisphere.

1. Leaves (needles) in bundles of 2 or more (commonly solitary in seedlings but very rarely so in more mature trees); seed cones with a thick, woody central axis; terminal portions of scales (apophyses) often distinctly thickened .... ***Pinus***
1′ Leaves solitary, not in bundles; seed cones lacking a thick, woody central axis; terminal portion of scales not distinctly thickened .... (2)

2. Seed cones with conspicuous 3-pronged bracts extending beyond the end of each scale; leaf scars slightly raised at proximal ends; nearly flush with twig surface at distal ends .... ***Pseudotsuga***
2′ Seed cones without 3-pronged bracts between the cone scales; leaf scars not raised at proximal ends .... (3)

3. Leaves attached directly to twig, forming scars flush with twig surface or nearly so; seed cones growing upright on upper branches, disintegrating and releasing seeds in autumn, when each scale separates from the spikelike central axis and falls away .... ***Abies***
3′ Leaves attached to tips of peglike projections from twig which remain after leaves drop; seed cones more or less pendent from branches, falling intact after seeds are shed at the end of summer .... ***Picea***

Abies Mill. Fir, Balsam Fir, He'kwpa (Hopi)

(Classical name of a European fir) Tree with crown usually spirelike, especially when young; branches usually whorled; leaf scars prominent, flush with twig surface or slightly raised or depressed all around. **BARK** thin, smooth, with resin blisters when young, becoming furrowed and/or flaky in age. **LEAVES** (needles) borne singly, arranged spirally, but either twisting near the base to grow upright on both sides of the twig in parallel rows (typically in the sun), or horizontally on both sides of the twig (2-ranked) (typically in the shade). **POLLEN CONES** grouped, leaving gall-like swellings after falling. **SEED CONES** maturing by late summer of the first year, growing upright on upper branches, disintegrating and falling scale by scale when mature, leaving a slender central "spike." **SEEDS** winged.

1. Leaves on lower branches (1.5–) 2–3 cm, dark bluish green, basal portion usually diverging distally (forward) from attachment point by about 45°, then curving upward; seed cones dark brown to purplish; common tree, often dominant in subalpine forests at elevations from 2400–3600 m (8000–11800′), to timberline .... ***A. bifolia***
1′ Leaves on lower branches (2.5–) 3–6 cm, pale grayish green, basal portion usually diverging laterally from twig attachment point by nearly 90°, then curving upward; seed cones grayish green; common tree in montane forests at elevations from 1900–2600 m (6200–8500′), higher on dry, sunny slopes .... ***A. concolor***

Abies bifolia A. Murray (two leaves) Mountain alpine fir, Rocky Mountain fir, alpine fir, white balsam, balsam fir. Tree to about 30 m; trunk 0.7 m wide. **BARK** gray and smooth on young trees, thickening (to 18 cm), darkening, and developing deep, longitudinal furrows. **LEAVES** 1–2.5 cm long, mostly 2-ranked, adaxial (upper) surface light green to bluish green, usually glaucous. **SEED CONES** cylindric, dark purple-blue to grayish purple, 5–10 cm long, 3–3.5 cm wide. This species, along with *Picea engelmannii*, dominates moist subalpine forests up to timberline in our area. We are accepting *Abies bifolia* as a species distinct from *A. lasiocarpa* (Hook.) Nutt. with some reservations. These two taxa seem very close and are distinguished primarily by differences in terpene chemistry and by color differences in the periderm of freshly picked needles. A large portion of their combined geographic ranges consists of intergrading populations, which could be the result of either introgression or of the initial stages of differentiation. If these interpopulation differences have developed to the point where they deserve formal taxonomic recognition, the varietal level would have been preferred rather than full species. However, any taxonomic decision made in a situation such as this will be more or less arbitrary. In a rather different situation, we are satisfied in recognizing *A. bifolia* var. *arizonica*, with its corky white bark, at the varietal level, at least until the genetic basis for this character complex is better known.

1. Bark of young and mature trees light gray to creamy white; variably smooth, soft and springy .... **var. *arizonica***
1′ Bark of mature trees dark gray; rough, hard, not springy (does not compress under moderate pressure) .... **var. *bifolia***

var. *arizonica* (Merriam) O'Kane & K. D. Heil (of Arizona) Corkbark fir. Bark of young and mature trees whitish; springy to the touch. Subalpine forests. ARIZ: Apa (Chuska Mountains); COLO: Arc, Hin, LPl, Min, RGr, SJn; NMEX: SJn; UTAH. 2400–3600 m (8000–11800′).

var. *bifolia* Bark not whitish or springy. Subalpine forests. ARIZ: Apa (Chuska Mountains); COLO: Arc, LPl, Min, Mon, SJn; UTAH. 2400–3600 m (8000–11800′).

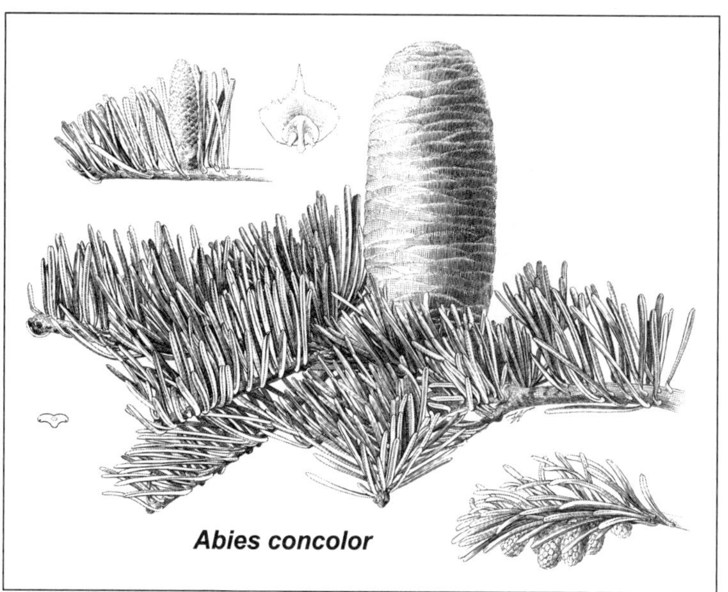
Abies concolor

Abies concolor (Gordon & Glend.) Hildebr. (same color) White fir, Rocky Mountain white fir, pino real blanco, He'kwpa (Hopi). Tree to about 40 m; trunk to 1 m. **BARK** gray and smooth on young trees, becoming thick with age, breaking into deep, longitudinal fissures. **LEAVES** 1.5–6 cm long, pale grayish green, adaxial surface conspicuously glaucous. **SEED CONES** cylindrical, olive-green, 7–12 cm long, 3–5 cm wide. **SEEDS** winged. [*Picea concolor* Gordon & Glend.]. Dry, sunny slopes. ARIZ: Apa; COLO: Arc, Hin, LPl, Min, Mon, SJn; NMEX: RAr, SJn; UTAH. 1900–2600 m (6200–8500′). Colorado, New Mexico, Arizona, Utah, Nevada, California, northern Baja California (Mexico), and southern Oregon and Idaho. *Abies concolor* is probably the most drought-tolerant North American *Abies*, as shown by its scattered but common occurrence throughout the southwestern United States. This fir is also cold-tolerant; it grows (albeit sparingly) on the summit of Navajo Mountain, Utah, at almost 3200 m (10400′). The distributions of *A. concolor* and *Pinus ponderosa* show a strong degree of overlap in our area, with the fir occupying sites that average a little higher, cooler, and moister. The fir is also more shade-tolerant, thrives on north-facing slopes, and maintains itself in the absence of fire. As with *A. bifolia* and *A. lasiocarpa*, the fragmentary distribution of *A. concolor* and other *Abies* taxa in the Southwest has produced a high level of genetic isolation, which in turn has favored genetic differentiation, some of which may rise to the level of taxonomic significance and cause problems of interpretation and classification. Huge quantities of white fir were used by the ancient (Chacoan) people to provide roofing for their ceremonial houses.

Picea A. Dietr. Spruce, C'o (Navajo)

(Latin *pici*, pitch and name of a pitchy pine) Evergreen tree, crown broadly conical to spirelike. **BARK** gray to reddish brown, thin and scaly when young, thick and furrowed with age. **LEAVES** (needles) relatively rigid, borne sessile on a persistent peglike base, apex usually sharp-pointed. **POLLEN CONES** grouped. **SEED CONES** borne mostly on upper branches, maturing in one growing season, pendent, ovoid to cylindric; scales thin, fan-shaped, lacking apophysis and umbo. **SEEDS** winged; cotyledons 5–12. About 35 species, 2 in our flora. Cool-temperate and subarctic portions of Eurasia and North America, with a southern extension into Mexico. Used for timber and pulp, telephone poles, wooden-bodied stringed instruments (violins, cellos, guitars).

1. Branches mostly slender and drooping; young twigs finely pubescent; leaf tips usually acute, but not sharp; seed cone mostly 3–7 cm long, cone scales extending 3–8 mm beyond the seed wing impressions ......***P. engelmannii***
1′ Branches mostly stout and horizontal, young twigs essentially glabrous; leaf tips usually attenuate into rigid, sharp spines; seed cone mostly 6–11 cm, cone scales extending 8–10 mm beyond the seed wing impressions ....***P. pungens***

Picea engelmannii Parry ex Engelm. (named for George Engelmann) Engelmann spruce, white spruce, mountain spruce, silver spruce, epinette, C'o = spruce (Navajo). Tree with stems to 30 m tall, 1 m wide; crown narrowly conical. Young twigs finely pubescent, peglike leaf bases angled distally away from twig at about 15°–45°. **LEAVES** 1.6–3 cm, blue-green, 4-angled in cross section, flexible; apex acute, but not usually very sharp. **SEED CONES** 4–7 (8) cm, widest above middle; scale margin irregularly toothed or erose, tip of scale extending 3–8 mm beyond seed

wing. $2n = 24$. [*P. glauca* (Moench) Voss subsp. *engelmannii* (Parry ex Engelm.) T. M. C. Taylor; *P. columbiana* Lemmon; *P. engelmannii* var. *glabra* Goodman]. Moist upper montane and subalpine forests, primarily on north-facing slopes, often with *Abies bifolia* in dense stands. ARIZ: Apa; COLO: Arc, Con, Hin, LPl, Min, Mon, RGr, SJn; UTAH. 2400–3800 m (8000–12000'). Alberta and British Columbia; California, Colorado, Idaho, Montana, Nevada, New Mexico, Oregon, Utah, Washington, and Wyoming. Our material belongs to **var. *engelmannii***. Some material collected at very high elevations or growing in very wet soil has unusually short needles (± 1 cm) suggestive of the short needles of *P. glauca*. This material probably represents a direct response of *P. engelmannii* to a series of unusual sites, rather than the presence of *P. glauca*. However, since *P. glauca* now occurs naturally in central Wyoming, it quite possibly did occur in this part of Colorado during the glacial stages of the Pleistocene and it would not be surprising to find evidence of past hybridization in modern *P. engelmannii* from this region.

Picea pungens Engelm. (sharp) Colorado blue spruce, blue spruce, Colorado spruce, silver spruce, silvertip fir, pino real, c'o denini = spruce sharp (Navajo). Tree with stems to 25 m tall, 0.6 m wide; crown broadly conical. Young twigs glabrous, peglike leaf bases diverging distally from twigs at about 40°, occasionally to 80°. **LEAVES** 1.6–3 cm, rigid, blue-green to whitish, the latter giving the tree a silvery appearance, 4-angled in cross section, apex spine-tipped. **SEED CONES** (5) 6–11 (12) cm, widest below middle, scale margin erose, tip of scale extending 8–10 mm beyond seed wing impression. $2n = 24$. [*P. parryana* Sarg.]. Usually upper montane to lower subalpine, mostly in our area, often along streams, especially in the lower portion of its altitudinal range. ARIZ: Apa; COLO: Arc, Hin, LPl, Min, Mon, RGr, SJn; NMEX: SJn; UTAH. 2100–3000 (–3400) m (7000–11000'). Arizona, Colorado, Idaho, New Mexico, Utah, Wyoming. Occasional hybridization occurs between *P. pungens* and *P. engelmannii. Picea pungens* is the state tree of both Colorado and Utah.

Pinus L. Pine

(Latin *Pinus*, name for pine) Trees in our flora, some shrubby at timberline; crown often open on mature trees. **BARK** thin, scaly, and continuously shed, or thick, woody, deeply fissured, and persistent. **LEAVES** of 2 kinds: needlelike, borne on deciduous spurs, bound at the base into fascicles (bundles) of (1–) 2–5, reaching 2–18 cm in length, persisting for 2–12 (–16) years; or scalelike leaves subtending needle spurs, lateral buds, and cones. **POLLEN CONES** numerous, elongate, borne mostly on the lower branches, maturing and releasing pollen in one season (spring, adjusted for elevation), then soon shed. **SEED CONES** larger and less numerous than pollen cones, borne mostly on upper branches, ovoid to cylindrical, becoming woody and brown when mature at the end of the second growing season, then usually releasing seeds and falling from the tree intact or with some basal scales remaining attached to branch, or entire cone persisting longer and retaining some or all of the seeds; scales of seed cones each bearing 2 ovules on the upper (adaxial) surface, the end of each scale variously broadened and thickened into a more or less rhombic apophysis, capped by a terminal swelling (umbo), from which a sharp spine may project. **SEEDS** usually winged; cotyledons (3–) 6–10 (–18). (Critchfield and Little 1966; Farjon and Styles 1997; Peattie 1991; Shaw 1914). Approximately 100 species of the Northern Hemisphere (one minor southward extension in Sumatra); tropical highlands to the subarctic; most abundant and diverse in mountainous and semiarid habitats of warm-temperate latitudes. Over half the species are native to southwestern North America. Six species in our flora.

1. Leaves (needles) on mature trees mostly in bundles of 2; seed cones less than 7 cm long ....(2)
1' Leaves on mature trees mostly in bundles of 3 or more; seed cones more than 7 cm long ....(3)

2. Mature trees broad-crowned, usually with major branches near base; leaves mostly upcurved, grayish; scales of seed cone at least 12 mm broad at widest point; seed (or seed impressions in cone scales) more than 7 mm in length .... ***P. edulis***

2′ Mature trees narrow-crowned, usually with a single dominant trunk; leaves mostly straight, dark green; scales of seed cone less than 12 mm broad at widest point; seed (or seed impressions in cone scales) less than 7 mm long .... ***P. contorta***

3. Leaves mostly in bundles of 3 (bundles of 2 locally common on saplings); seed cone leaving a ring of basal scales on branch when falling from tree (basal scales missing from cones on ground) .... ***P. ponderosa***

3′ Leaves mostly in bundles of 5; seed cone falling from tree intact .... (4)

4. Largest scales on seed cones less than 15 mm broad at widest point, each scale terminating in a slender prickle; leaves conspicuously resinous .... ***P. aristata***

4′ Largest scales on seed cones more than 15 mm broad at widest point, lacking terminal prickles; leaves not conspicuously resinous .... (5)

5. Vegetative buds not resinous; seed cones 12–20 (–25) cm long; mature, dry cones quite open, with scales well separated near distal ends, scales attenuate and usually recurved or reflexed at ends; mature trees typically symmetrical, with lower portion clear of branches; in montane forest, usually below 2700 m (8860′) ... ***P. strobiformis***

5′ Vegetative buds resinous; seed cones 6–15 cm long; mature, dry cones compact, with scales overlapping and often touching near distal ends; scales blunt to moderately attenuate at ends, not reflexed; trees typically asymmetrical, often dividing near base; in open, rocky subalpine woodlands, usually above 2700 m (8860′), occasionally as low as 2350 m (7700′) .... ***P. flexilis***

Pinus aristata Engelm. (awned, referring to prickle at end of seed cone scale) Bristlecone pine, Rocky Mountain bristlecone pine, Colorado bristlecone pine, foxtail pine, hickory pine. Tree to 15 (–20) m tall; erect, compact, becoming shorter, shrublike, and twisted toward timberline; trunk diameter to 0.75 (–1) m. **BARK** whitish and smooth on young trees, dark and fissured on old. **LEAVES** 2–4 cm in length; mostly 5 per bundle, curved, dark green, outer surfaces whitened with resin splotches and stomata, persisting 10–16 years. **SEED CONES** dark purplish when young, becoming brown when mature, 6–11 cm long, each scale tipped with a slender incurved prickle to 7 mm in length. [*P. balfouriana* (Balf.) Engelm. var. *aristata* (Engelm.) Engelm.]. Mostly upper subalpine to lower fringes of alpine zone, typically on relatively dry, south-facing slopes. Limited occurrence in the San Juan River drainage; more common east of the Continental Divide. COLO: Min, RGr. 2900–3700 m (9500–12000′). Arizona, Colorado, and New Mexico. *Pinus longaeva* D. K. Bailey (Great Basin bristlecone pine, western bristlecone pine) is closely related and can live to an age of 5000 years. It occurs in the basin ranges of eastern California, Nevada, and Utah.

Pinus contorta Douglas ex Loudon (twisted together, referring to the needles) Lodgepole pine, tamarack pine, black pine. Tree 30–35 m tall, usually narrow-crowned and growing in dense, often pure stands; trunk to 0.8 m in diameter; branches usually shed from lower portion in crowded stands; bark thin, scaly, reddish brown. **LEAVES** 2 per fascicle, (4) 5–8 cm long, stout, often twisted, deep green to yellow-green, persisting 5–8 years. **SEED CONES** maturing, opening and shedding seeds at end of second growing season, or more commonly remaining on tree at least partially closed and shedding seeds over several years (serotinous), 3–7 (8) cm long, more or less globose but somewhat asymmetrical when open; scales narrow (<12 mm broad), thickened at ends, usually tipped by a sharp prickle. **SEEDS** stored in the serotinous cones establish dense, pure stands following fires. [*P. contorta* subsp. *latifolia* (Engelm.) Critchf.]. Subalpine, occasionally approaching timberline; usually on relatively dry slopes. COLO: Hin, LPl, SJn. 2900–3400 m (9500–11200′). Our material belongs to **var. *latifolia*** Engelm. (broad, referring to the needles). Rocky Mountain region to Yukon, Canada. Forest managers have used seeds of this pine for post-fire reforestation beyond its natural range. One example is in San Juan County, Colorado, where seeding was done in the 1800s on the Lime Creek Burn. These stands also represent the southeastern limit of var. *latifolia*. Variety *contorta* extends from the coast of northern California through the Alaskan panhandle, and var. *murrayana* (Balf.) Engelm. ranges from northern Baja California northward along the Sierra Nevada-Cascade axis to Washington State. Used by humans over several millennia. Native people built houses and travois from the wood and also made liquor from the sap. European settlers constructed log cabins and fence posts from the timber, continuing to the present.

Pinus edulis Engelm. (to eat, edible) Pinyon, piñon, two-leaf pinyon, Colorado pinyon pine, nut pine, chá'ol (Navajo), tuvé (Hopi). Tree to 15 m tall, usually shorter, with dense conic crown when young, becoming open and rounded or irregular with age; trunk to 0.8 m wide, often dividing into 2 or more major branches near base and becoming twisted with age. **BARK** grayish to reddish brown, irregularly and shallowly furrowed. **LEAVES** (1) 2 (3) per fascicle, usually upcurved, stout, 2–4 (5) cm long, apex usually sharp to the touch, grayish blue-green, persisting 4–6 years. **SEED CONES** maturing, shedding seeds and usually falling promptly at end of second growing season; resinous, ovoid before opening, (2) 3–5 cm long when open, often as broad or slightly broader, larger scales at least 12 mm wide, thick, blunt. **SEEDS** wingless, large (10–15 mm long), leaving conspicuous depressions in cone scales after being shed. [*Caryopitys edulis* (Engelm.) Small; *P. monophylla* Torr. & Frém. var. *edulis* M. E. Jones; *P. cembroides* Zucc. var. *edulis* (Engelm.) Voss]. Mesas, canyons, mountain slopes, at elevations usually below those of continuous forests, typically forming open woodlands with *Juniperus* spp. and smaller shrubs. ARIZ: Apa, Coc, Nav; COLO: Arc, Dol, LPl, Mon, SMg; NMEX: McK, RAr, San, SJn; UTAH. (1200–) 1500–2100 (–2700) m (4000–9000′). Piñon pine often form xeric "islands" on south-facing slopes within montane forests. This piñon is the most drought-tolerant and widespread species of Pinaceae in the Four Corners region; however, the 2002–2003 drought across southwestern North America caused a massive die off in response to depleted soil water and associated bark beetle infestations (Breshears et al. 2005). *Pinus edulis* is probably the best-known species of the "nut pines," which collectively are distributed from the southwestern United States through central Mexico. Its range overlaps that of *P. cembroides* (three needles per fascicle) in southeastern Arizona and southwestern New Mexico and that of *P. monophylla* (one needle per fascicle) from southeastern California to southwestern Utah, where hybridization apparently is occurring. The nuts (seeds) produced by *P. edulis* are flavorful and highly nutritious. They were (and are) a major item of trade among the native peoples of the region. Nutshells are found at most ancestral pueblo sites, and now the nuts are marketed commercially far beyond their natural range. The wood was used for major construction in the "pit houses" built by ancient Americans circa 400–900 A.D. and was also a major source of fuel, especially for firing ceramic containers. The resin was used for waterproofing baskets, for chewing gum, to cement turquoise in silver, and for making a black dye for wool. *Pinus edulis* is the state tree of New Mexico.

Pinus flexilis E. James (flexible, pliable) Limber pine, Rocky Mountain white pine, gad nezi-tall (Navajo). Tree in our area, narrowly pyramidal when young, crown broadening and opening with age, upper branches often dividing to form several parallel, more or less vertically oriented branches; height 10–15 (–20) m, often shorter; trunk 0.5–1.5 m, often dividing near the base. **BARK** smooth and whitish on younger branches, becoming dark gray and divided into scaly rectangular plates on older trunk(s). **LEAVES** 5 per fascicle, crowded toward the end of the twig, often upcurved, persisting (4) 5–6 years, (3–) 3.5–6 (–9) cm long, 1–1.5 mm wide, apex region smooth, without teeth, slightly yellowish to dark green. **SEED CONES** maturing, releasing seeds and falling at end of second growing season, lance-ovoid to cylindrical before opening, mature cones 6–15 cm, mostly yellowish brown, distal end of scales broadly truncate to moderately attenuate and/or recurved, apophysis thick; stalks 0–12 mm long. **SEEDS** obovoid, body 10–15 mm long, with vestigial wing adhering to scale. [*Apinus flexilis* (E. James) Rydb.]. Mostly subalpine; usually in exposed, windy, rocky, thinly vegetated sites, especially near the upper limits of forest growth, but occasionally much lower. COLO: Arc, Hin, LPl, Min, SJn; UTAH. (2300–) 2450–3200 m (7500–10500′). Where growing conditions allow continuous forest to develop (usually at lower elevations), *P. flexilis* is often replaced by its close relative, *P. strobiformis*, a large, straight-trunked forest tree. The seeds of *P. flexilis* were occasionally eaten by native peoples, but this pine probably never provided more than a meager supply.

1. Seed cone 7–12 cm, very compact, apophysis ± blunt at distal end, not attenuate or recurved .... **var. *flexilis***
1′ Seed cone 10–15 cm, moderately compact to more open, distal portion of apophysis slightly to moderately attenuate and/or recurved .... **var. *reflexa***

var. *flexilis* **LEAVES** 1–1.5 mm broad; white stomatal lines evident on all 3 surfaces, apical region smooth-margined. **SEED CONES** spreading on branch (not pendulous), 7–12 cm, very compact when fully opened, with little space between scales (which may touch adjacent scales at several points); apophysis much-thickened, broad and blunt at distal end, the adaxial (upper) portion marked by a narrow (< 5 mm), light-colored band. Upper elevations on Navajo Mountain, and abundant on the broad summit area. UTAH. 2900–3200 m (9500–10500′). North through the Rocky Mountains to southwestern Alberta, west in the Basin and Range Province across Utah and Nevada to eastern and southern California. Navajo Mountain, Utah, is the only site in the San Juan River drainage where the authors could locate var. *flexilis*, while var. *reflexa* was observed in several localities (including Navajo Mountain) but was usually identified as *P. strobiformis* in herbarium collections.

var. *reflexa* Engelm. (reflexed or turned back) **LEAVES** 0.8–1.2 mm broad, white stomatal lines present on adaxial (inner) surfaces but lacking on abaxial (outer); apical region with scattered minute teeth. **SEED CONES** subpendulous,

10–15 cm, moderately compact with some space between scales, which spread from axis at 50°–70°; apophysis scarcely to moderately attenuate (5–8 mm) and often recurved, but less frequently reflexed. [*P. reflexa* (Engelm.) Engelm.; *P. ayacahuite* C. Ehrenb. ex Schltdl. var. *strobiformis* Lemmon]. COLO: Arc, Hin, LPl, Min, Mon; UTAH (Navajo Mountain). 2135–3350 m (7000–11000′). Arizona, New Mexico, southwest Texas (probably included in *P. strobiformis*). Known in northern Mexico from high elevations in Chihuahua, Coahuila, Nuevo León, and on Cerro Potosí. The name *P. flexilis* var. *reflexa* has not been applied to material north of the United States-Mexico border by most taxonomists working in the United States. They have chosen to include similar United States material in *P. strobiformis*. The Rocky Mountain taxa of subgenus *Strobus* appear to form a "messy" cline from southern Canada well into Mexico. *Pinus flexilis* var. *flexilis* occupies at least the northern half but extends south disjunctly to the San Francisco Peaks in northern Arizona and to the Utah-Arizona border region (Navajo Mountain, Utah). *Pinus strobiformis* and what appears to be *P. flexilis* var. *reflexa* occupy the middle portion (southwestern Colorado to southern Arizona and New Mexico), with the latter occupying mostly high-elevation exposed sites and the former at lower elevations in well-developed mixed conifer forests. South of the Four Corners region *P. strobiformis* becomes more widespread and distinct in southern Arizona and New Mexico, with larger cones than in the populations to the north, and *P. flexilis* var. *reflexa* apparently is even more restricted to high elevations than in southwestern Colorado. This cline continues into southern Mexico, where additional taxa are involved. In our area, the middle portion of the cline described above, typical *P. strobiformis* and *P. flexilis* var. *flexilis* are relatively scarce (the latter observed only at the summit of Navajo Mountain), but material resembling *P. flexilis* var. *reflexa* or suggestive of hybrid combinations of this taxon with *P. flexilis* var. *flexilis* or with *P. strobiformis* is common. We also cannot rule out the possibility that *P. flexilis* var. *reflexa* may even be derived from earlier hybridization between entities resembling *P. flexilis* var. *flexilis* and *P. strobiformis*, since var. *reflexa* is intermediate between them in many of its characters. We have recognized *P. flexilis* var. *reflexa* in our area, even though it can be difficult to decide which individuals to include, because a high proportion of our material makes a better fit with this entity than with *P. strobiformis* or *P. flexilis* var. *flexilis*. For more detailed descriptions of these taxa, see Farjon and Styles (1997).

Pinus ponderosa P. Lawson & C. Lawson (heavy) Ponderosa pine, Rocky Mountain ponderosa pine, western yellow pine, blackjack pine, piño real, piñabete, nídíshchii (Navajo). Tree to about 40 m tall; trunk to 1.5 m in diameter, branches descending to spreading-ascending. **BARK** on mature trees orangish to deep reddish brown, with irregular fissures. **LEAVES** 3 per fascicle (in ours), but young saplings may produce some fascicles with 2 needles, as do older trees to the north and east of our area, dark green, (7–) 10–16 (–18) cm in length. **SEED CONES** broadly conical to egg-shaped, (5–) 7–10 (–12) cm long; apophyses at midcone moderately raised; umbo terminating in a stout, sharp prickle, curving outward or pointed toward the base of the cone. **SEED** body 3–5 mm, portion of wing extending beyond the seed about 3× the length of the seed body. [*Pinus brachyptera* Engelm.; *P. scopulorum* (Engelm.) Lemmon]. Sheltered canyons at its lowest limits to exposed ridges at its highest localities; usually the dominant tree at the lowest forested elevations, forming open, nearly pure stands on warm, south-facing slopes, but extending to much higher elevations as well. ARIZ: Apa, Nav; COLO: Arc, Dol, Hin, LPl, Min, Mon, SJn, SMg; NMEX: McK, RAr, San, SJn; UTAH. (1700–) 1900–2800 (–3000) m (5600–10000′). Our material belongs to **var. *scopulorum*** Engelm. (of the cliffs). British Columbia to northern Mexico, in all the western states and east to Nebraska and South Dakota. Ponderosa pine is adapted to relatively frequent low-intensity fires caused by lightning, which kill most of the understory vegetation but not the mature trees. Remaining saplings of ponderosa pine, which have a high light requirement, are thus able to thrive. The young trees grow rapidly under favorable conditions. Ponderosa pine is second only to Douglas-fir as a lumber-producing tree in North America. Native people used the timber for roof beams, sweathouses, corrals, and fences, and cradleboards were made from the wood. The fragrant branches were used to line floors and as mattresses. The seeds and inner bark were eaten; the resin was chewed as a gum. Fresh sap, rubbed on the skin, was a treatment for

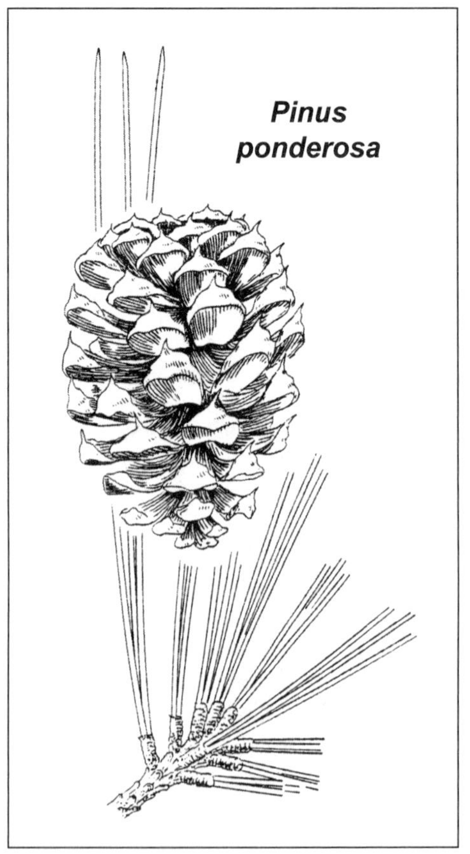

Pinus strobiformis

various dermatological ailments. Pine burls were made into ladles and spoons; red dye can be made from the outer bark, and yellow dye from the pollen. Ponderosa pine is also a tree of important spiritual significance and is used in at least six Navajo ceremonies. Hopi people use the pine to make a pano or prayer stick. It is said that the first Hopi climbed out of the underworld on a pine ladder, so the ladders down to their subterranean ceremonial kivas must be made from pine branches.

Pinus strobiformis Engelm. (plug of lint, shaped like a fir cone) Southwestern white pine, Mexican white pine, border white pine, Arizona white pine, pino emeno. Tree to 30 m tall, conic when young, becoming more rounded and irregular with age; trunk to 1 m in diameter. **BARK** gray and shallowly fissured when young, becoming more deeply furrowed with age, but with reddish brown scaly areas developing between the fissures; resembling somewhat the bark of mature *P. ponderosa*; branches ascending in upper portion of crown, becoming more horizontal in lower portion. **LEAVES** 5 per fascicle, spreading in lower, shaded portion of crown, upcurved in higher, more exposed portion, persisting 3–5 years, 5–9 cm long, 0.6–1 mm wide, apical area usually with a few very fine teeth, green to slightly bluish green; whitish stomatal lines evident on inner (adaxial) needle surface, rarely on outer (abaxial) surface. **SEED CONES** maturing, shedding seeds and falling at end of second growing season; lance-cylindric before opening, pendulous, 12–25 cm long (in ours); apophysis thickened at base, becoming thinner distally, usually conspicuously attenuated, often recurved and/or reflexed as well. **SEEDS** ovoid, body 10–12 mm, wingless or nearly so. Upper montane, but usually not subalpine; in valley bottoms or relatively moist slopes with some soil development and from lower to higher elevations; associates may include *P. ponderosa* (ponderosa pine), *Pseudotsuga menziesii* (Douglas-fir), *Abies concolor* (white fir), *Abies bifolia* (subalpine fir), *Picea pungens* (Colorado blue spruce), and *Picea engelmannii* (Engelmann spruce). COLO: Arc, Hin, Min, Mon, SJn; UTAH. (2200–) 2350–2750 (–3000) m ([7200–] 7800–9000 [–9800′]). Like its eastern [*Pinus strobus* (white pine)] and western relatives [*Pinus monticola* (western white pine) and *Pinus lambertiana* (sugar pine)], *Pinus strobiformis* produces very desirable, high quality timber, although its relative inaccessibility and low frequency have reduced the level of exploitation.

Pseudotsuga Carrière Douglas-fir

(Latin *pseudo*, false, + Japanese *tsuga*, a hemlock) Tree irregularly conic, crown broadening and flattening with age; branches pendulous or not, irregularly whorled. **BARK** gray, smooth with resin blisters when young, becoming dark, very thick, somewhat corky, and deeply fissured in age; buds elongate, not resinous. **LEAVES** borne singly, alternate, short-stalked; persisting about 6–8 years; leaf scars slightly raised proximally, but almost flush with twig distally. **POLLEN CONES** axillary. **SEED CONES** mostly pendent, maturing in one growing season, more or less cylindric, scales persistent, lacking apophysis and umbo, apices rounded; bracts usually exserted and clearly visible between the scales, 3-pronged, with central lobe longer and narrower than the lateral pair. **SEEDS** winged, cotyledons 6–12. $x = 12$ or 13. Five species (three in East Asia, two in western North America, one of these in our flora).

Pseudotsuga menziesii (Mirb.) Franco (for Archibald Menzies, 1754–1842, a Scottish botanist who collected needles and cones on Vancouver Island) Douglas-fir, Rocky Mountain Douglas-fir, piño real Colorado, Douglas-spruce, red fir, yellow fir, false fir, Douglas-tree, ch'o (Navajo), sala'vi (Hopi). Tree to 40 m tall in the study area; trunk to 1.2 m diameter. **LEAVES** 15–30 mm, flattened, radiating from stem in ± brushlike fashion (not 2-ranked), usually bluish green. **SEED CONES** 4–7 cm, bracts spreading, often reflexed. $2n = 26$. [*P. douglasii* (Sabine ex D. Don) Carrière var. *glauca* Mayr]. Mostly in the mountains and occasionally in protected areas at lower elevation. ARIZ: Apa, Nav; COLO: Arc, Dol, Hin, LPl, Min, Mon, RGr, SJn; NMEX: McK, RAr, SJn; UTAH. 1500–3000 m (5000–10000′). A widespread western conifer, native from central British Columbia through all of the western United States and into Mexico. The distribution of Douglas-fir in the San Juan Basin tracks that of ponderosa pine rather closely, and both often grow on the same sites. However, the pine appears to be more tolerant of cold and drought and can grow on dry south-facing slopes and in cold air pockets where Douglas-fir usually does not survive, while the fir may outcompete the pine on moist, shaded north-facing slopes. Our material belongs to **var. *glauca*** (Mayr) Franco (gray-green). Douglas-fir is the most important timber species in North America (ponderosa pine is second). The ancient peoples at Chaco Canyon used the logs as beams in their large structures and traveled great distances to obtain them. *P. menziesii* also is said to have many other medicinal and analgesic properties which the native people put to good use.

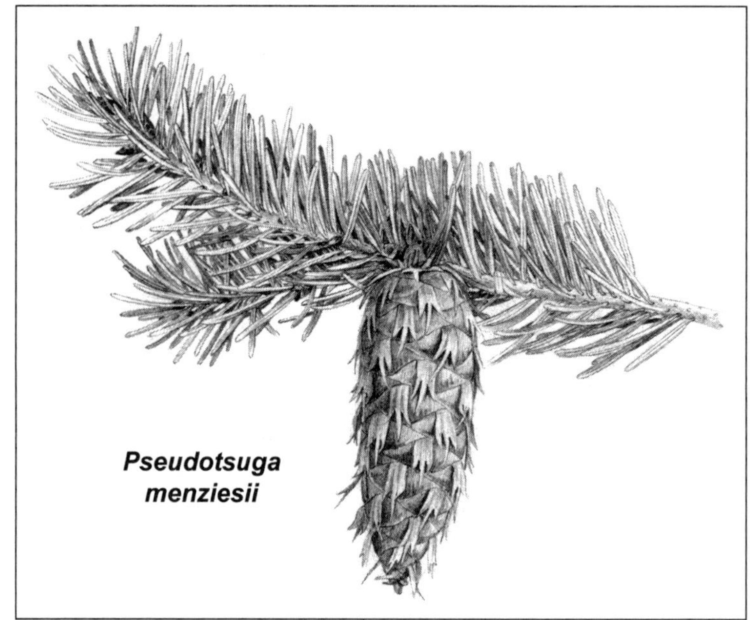
Pseudotsuga menziesii

ANGIOSPERMS

ACERACEAE Juss. MAPLE FAMILY

Linda Mary Reeves

Trees and shrubs with copious sugary sap, monoecious, dioecious, or polygamous. **LEAVES** opposite, simple, unlobed or palmately lobed; palmately or pinnately compound. **INFL** a panicle, raceme, corymb, or umbel. **FLOWERS** perfect or imperfect, small, actinomorphic; sepals generally 4–5, often very reduced or connate; apetalous or with petals, generally 4–5, sometimes absent; stamens 4–10; pistils 1; styles 2, or 1 and deeply divided; ovary superior, 2-celled, 2-lobed, nectary disc often present; ovules 2 per cell. **FRUIT** a samara-like schizocarp, each winged mericarp with 1 seed. Two genera with about 180 species in the north-temperate zone, including the genus *Dipteronia* in China and montane southeast Asian tropical regions (Landrum 1995).

Acer L. Maple

(old name for maple tree; sharp) Deciduous trees or shrubs, monoecious or dioecious. **INFL** a raceme, panicle, umbel, or corymb. **FLOWERS** mostly 5-merous (rarely 4–12); calyx colored, 5-lobed (rarely 4–12); corolla either not present or 5-merous, usually with short claws, inserted on the margin of a perigynous or hypogynous disc; stamens 3–12; styles 2, long, slender, stigmatic surface on the inner side; each carpel bearing a wing, ovary compressed.

1. Leaves on one plant mostly compound, with pinnately veined leaflets .... ***A. negundo***
1' Leaves mostly simple, rarely compound; palmately veined and palmately lobed ....(2)

2. Leaf blades finely pubescent beneath; margin with blunt teeth; outer bud scales brownish ... ***A. grandidentatum***
2' Leaf blades glabrous beneath; margin with sharply pointed teeth; outer bud scales reddish .... ***A. glabrum***

Acer glabrum Torr. (without hairs) Rocky Mountain maple. A small tree (in our area) to 10 m high, essentially glabrous, except for the tomentose inner surface of the bud scales; young twigs dark purplish red to reddish brown, the epidermis of older twigs flaking off in thin whitish sheets; young buds covered with 2 pink to red valvate scales. **LEAVES** petiolate, 1.5–9 cm long; blade suborbicular in outline, 3–7 cm long, 3.5–9 cm wide, palmately 3-lobed, occasionally 5-lobed, rarely compound with 3 leaflets, or a mixture of forms; apex acute or obtuse; base cordate to obtusely cuneate, margin coarsely serrate, the teeth acute to acuminate. **INFL** corymbose. **FLOWERS** 5 mm long, 1.5 mm wide; perianth greenish yellow. **FRUIT** mericarp 2–2.3 cm long, 1 cm wide. Conifer forests, often along streams. [*A. diffusum* Greene; *A. neomexicanum* Greene]. ARIZ: Apa, Nav; COLO: Arc, Hin, LPl, Min, Mon, RGr, SJn; NMEX: McK, RAr, SJn; UTAH. 2000–3900 m (6600–12700′). Flowering: May. Fruit: Jun–Sep. Southeast Alaska, British Columbia to montane elevations of New Mexico, Arizona, and California. Navajos used an infusion of the branches as a treatment for swelling.

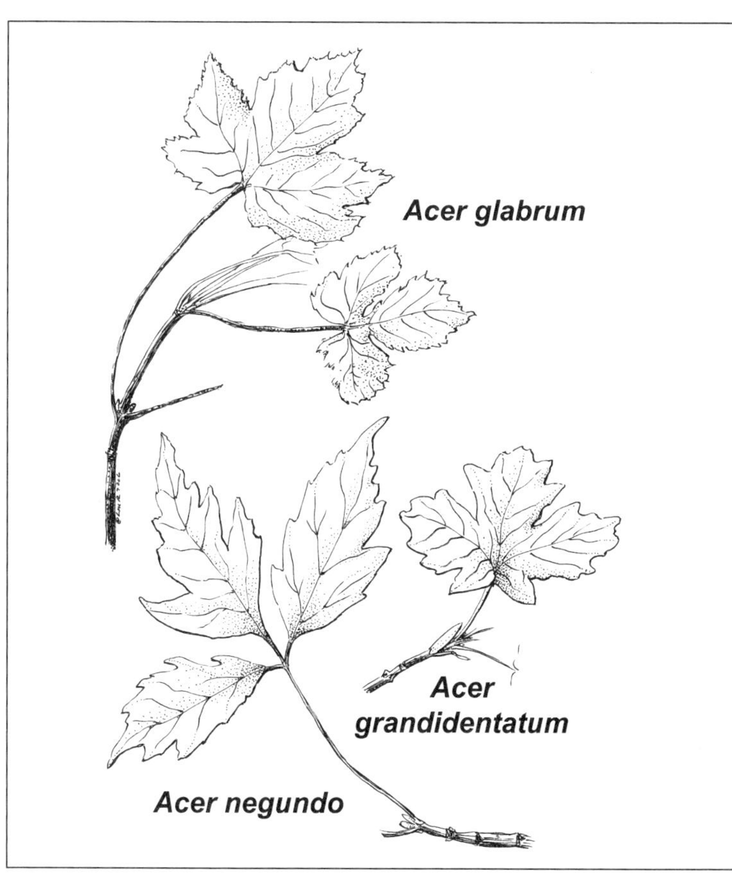

Acer grandidentatum Nutt. (large-toothed) Big-tooth maple, sugar maple, palo de azúcar. Tree to 15 m high, mainly glabrous except for finely pubescent lower leaf surfaces and young growth; young twigs yellowish brown, becoming chestnut-brown and glabrous; buds with 3–4 series of dark brown imbricate scales. **LEAVES** suborbicular in outline, palmately 3–5-lobed, 3–8.5 cm long, 4–12 cm wide; margin sinuate to irregularly dentate; apices bluntly acute; base cordate to subcordate; petiole glabrous or distally pubescent. **INFL** corymbose. **FLOWERS** about 10 mm long, 1.5–3 mm wide at base; pedicel 1–5 cm long; receptacle truncate to obconic; perianth cuplike, greenish yellow, margin sinuate to shallowly lobed. **FRUIT** with samara-like mericarp 2.3–3.6 cm long, wing 0.7–1.5 cm wide; often rose-colored. Conifer forests or riparian areas. COLO: Mon (Mesa Verde National Park); UTAH (Abajo Mountains and Navajo Mountain). 1200–2600 m (4000–8500′). Flowering: Apr–May. Fruit: Jun–Sep. Idaho and Wyoming to the montane Trans-Pecos region and Edwards Plateau of Texas; Sonora, Chihuahua, and Coahuila. Used by some tribes and settlers as a source of sweetener.

Acer negundo L. (name of Native American origin) Boxelder, ash-leafed maple, arce, fresno de guajuco. Generally a small to medium tree, often sparsely to moderately pubescent on young growth and lower leaf surfaces, mostly dioecious. **STEMS** glabrous to densely pubescent when young, often more or less glaucous, greenish or reddish; the older twigs rougher, gray; buds with 2 reddish, tan, or yellowish valvate scales with sparse to dense hairs. **LEAVES** petioled, 2–7 cm long, green or reddish; mainly trifoliolate, occasionally 3-lobed; 3.5–13.5 cm long, 3.5–18 cm wide, the terminal leaflet up to 11 × 8 cm, the lateral leaflets up to 9 × 6 cm; leaflet margin coarsely toothed or lobed; apex of leaflets acute to acuminate; base of leaflets rounded to cuneate, sometimes oblique, sometimes acuminate in terminal leaflets. **INFL** many-flowered; staminate an umbel; pistillate racemose. **FLOWERS** about 5 cm long, less than 1 mm wide; with filiform pedicel blending with receptacle, 1–4 cm long; perianth greenish yellow, with 4 subelliptic segments. **FRUIT** often gray-green, infused with pink or red; mericarp 2.3–3.6 cm long, wing to 1.5 cm wide. [*A. negundo* var. *interius* (Britton) Sarg.; *A. negundo* var. *arizonicum* Sarg.; *A. saccharum* Marshall var. *grandidentatum* (Nutt.) Sudw.; *Negundo aceroides* Moench]. Primarily riparian habitats, especially along the Animas River and other waterways; also cultivated and escaped from cultivation. ARIZ: Apa, Nav; COLO: Arc, Dol, Hin, LPl, Min, Mon; NMEX: McK, RAr, SJn; UTAH. 1500–2400 m (5000–8000′). Flowering: Mar–Jun. Fruit: May–Oct. Widespread in North America from southern Canada to Guatemala, often cultivated and escaped. Apaches used dried inner bark scrapings as winter food. The inner bark was also boiled and the juice crystallized as a sweetener. Navajos used the wood to make tubes for bellows.

ADOXACEAE E. Mey. MOSCHATEL FAMILY

Fred R. Barrie

Perennial herbs, shrubs, or small trees, glabrous or with simple hairs. **LEAVES** basal and alternate or cauline and opposite, simple or compound. **INFL** cymose or paniculate. **FLOWERS** actinomorphic, perfect, sepals 2–5; petals 5, united at the base; stamens 4, 5, or 10, inserted on the corolla; ovary partially or wholly inferior, 3–5-locular with 1 fertile ovule per locule; styles 1 or equal to the number of locules. **FRUIT** a fleshy or dry drupe. **SEEDS** 3–5. Five or six genera with about 175 species, distributed throughout the temperate and montane regions of Europe, Asia, and North America and in the montane regions in tropical South America. Certain species of *Viburnum* are widely planted as ornamentals and the fruits of some species of *Sambucus* are used for preserves and for making wine. Traditionally, this family has included only the five species in *Adoxa*, *Sinadoxa*, and *Tetradoxa* (the latter two endemic to China), but recent morphological and molecular research has demonstrated that *Sambucus* and *Viburnum*, two genera formerly included in Caprifoliaceae, are properly placed here (Donoghue 2001).

1. Rhizomatous herbs, the leaves alternate; stamens 10 .... ***Adoxa***
1′ Shrubs or small trees, the leaves opposite; stamens 5 .... ***Sambucus***

Adoxa L. Moschatel, Muskroot

(Greek *adoxos*, obscure or insignificant) Perennial, rhizomatous herbs, glabrous, somewhat musky-scented. **LEAVES** basal and cauline, the basal leaves alternate, ternate or simple and trifid, cauline leaves opposite, reduced, usually simple, trifid. **INFL** a raceme of 4 to 6 flowers disposed along the rachis or in a terminal cluster. **FLOWERS** with 2–4 sepals, corolla 4- or 5-lobed, rotate, green or yellow; stamens 5, the filaments divided for 1/2 their length, or 10 and paired in the sinuses of the corolla lobes; ovaries partially inferior, locules 3–5, styles 3–5. **FRUIT** a dry drupe. **SEEDS** 3–5. Three species, two endemic to China, the third occurring in temperate and montane zones in Europe, Asia, and North America.

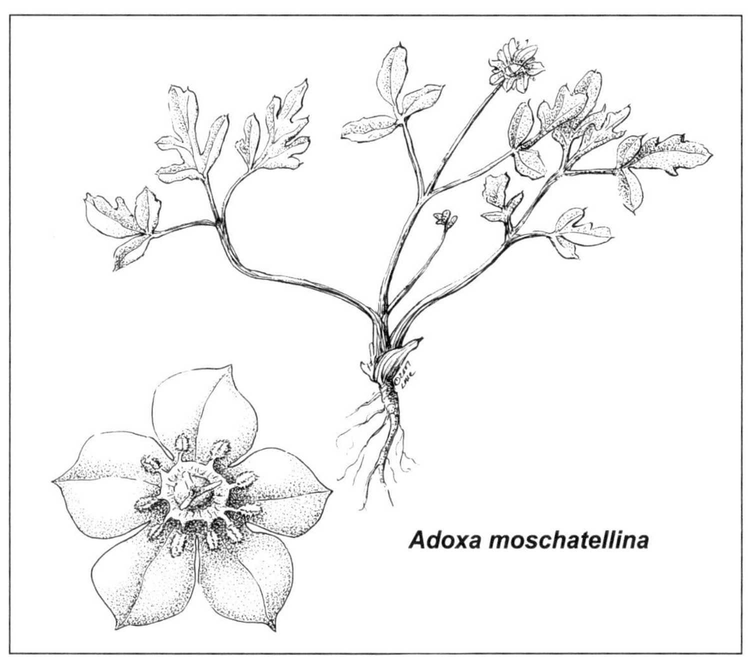

Adoxa moschatellina L. (perfumed with musk) Moschatel, muskroot. **LEAVES** basal and cauline, the basal leaves ternate, 5–15 cm long, the blades ovate, 2–8 cm long, 2–7 cm wide, the leaflets often 3-lobed, the margins crenate; cauline leaves opposite, reduced, usually simple, trifid. **INFL** a terminal cluster of 5 flowers. **FLOWERS** subsessile, the central (terminal) flower of the cluster usually with 2 lanceolate sepals, 1–2 mm long, the 4 surrounding flowers with 3 sepals; corollas green, rotate, 4–8 mm in diameter, the lobes 4 or 5, ellipsoid; stamens 10, paired in the sinuses of the corolla lobes, styles 1 mm long or less. **FRUIT** green, globose, 5 mm in diameter, thickened equatorially, the styles persistent. $2n = 36$. Moist, shady areas in the upper montane, subalpine, and alpine zones, commonly in mossy banks and in the shade of boulders on talus slopes. COLO: Hin, LPl, Min, Mon, SJn. 3000–3800 m (10200–12700′). Flowering: Jun–Aug. Fruit: Jul–Aug. Northern Europe, Asia, Alaska, and Canada east to the Great Lakes region and New England and south along the Rocky Mountains to New Mexico.

Sambucus L. Elder, Elderberry

(classical Latin name for *Sambucus nigra*) Perennial herbs (elsewhere), shrubs, or small trees, glabrous or pubescent. **STEMS** of herbs and young branches with soft, spongy pith. **LEAVES** opposite, compound, pinnate (ours) or bipinnate, a terminal leaflet present. **INFL** a terminal, compound cyme. **FLOWER** sepals 5, corolla of 5 petals united near the base, forming a short tube, white; stamens 5; ovary inferior; style 1, short and thickened, stigma lobes 3–5. **FRUIT** a globose, fleshy drupe, ours red, blue, or purple, elsewhere also white, black, orange, yellow, or green. A genus of 10 to 20 species distributed throughout the temperate regions of Europe, Asia, and North America and in the montane regions of the New World south to Peru.

1. Inflorescence corymbose, up to 20 cm in diameter; fruits blue, waxy .... ***S. cerulea***
1′ Inflorescence domed or conical, 4–8 cm in diameter; fruits red, rarely black, not waxy .... ***S. racemosa***

Sambucus cerulea Raf. (sky-blue) Blue elderberry. Shrubs with 1–several stems, glabrous or puberulent along the leaf midrib and veins; stipules small, bractlike, soon deciduous. **LEAVES** 10–25 cm long, leaflets 7–9 (11), all of similar size, 4–10 cm long, 2–4 cm wide, elliptic, the base often oblique, the apex acuminate, the margins serrate. **INFL** a corymbose, compound cyme, 10–20 cm across in flower. **FLOWERS** rotate, 4–5 mm in diameter. **FRUIT** 4–6 mm in diameter, blue, usually with a waxy bloom. $2n = 38$. [*S. nigra* L. var. *cerulea* (Raf.) Bolli; *S. neomexicana* Wooton; *S. glauca* Nutt.]. Wet areas, stream banks, and valley bottoms in pine and spruce-fir forest zones. ARIZ: Apa; COLO: LPl, Mon; NMEX: SJn; UTAH. 1500–2300 m (5300–7600′). Flowering: May–Aug. Fruit: Jul–Aug. Alberta to British Columbia, south to Arizona, Colorado, New Mexico, Utah, and California.

Sambucus racemosa L. (racemose) Red elderberry. Shrubs or small trees, glabrous or rarely sparsely puberulent; stipules fleshy, glandular, 1–2 mm long, soon deciduous or occasionally persistent through flowering. **LEAVES** 10–25 (35) cm long, leaflets 5–7 (9), all of similar size, 4–10 (18) cm long, 2–4 (8) cm wide, ovate to elliptic, the base oblique, the apex acuminate, the margins serrate. **INFL** convex, domed or conical, 4–8 cm in diameter. **FLOWERS** rotate, 4–5 mm in diameter. **FRUIT** 4–6 mm in diameter, red or rarely black, not waxy. $2n = 36, 38$. [*S. microbotrys* Rydb.; *S. melanocarpa* A. Gray; *S. pubens* Michx.]. Mesic and wet areas in pine and spruce-fir forest zones.

1. Fruit black .... **var. *melanocarpa***
1′ Fruit red .... **var. *microbotrys***

var. *melanocarpa* (A. Gray) McMinn (black-fruited) Black elderberry. ARIZ: Apa; NMEX: SJn; UTAH. 2135–2440 m (7000–8000′). Alberta to British Columbia, south to New Mexico and California.

var. *microbotrys* (Rydb.) Kearney & Peebles (small grape) Red elderberry. ARIZ: Apa; COLO: Arc, Hin, LPl, Min, Mon, RGr, SJn; NMEX: RAr, SJn; UTAH. 2300–3100 m (7700–10500′). Newfoundland to British Columbia, south to Pennsylvania, New Mexico, and Arizona.

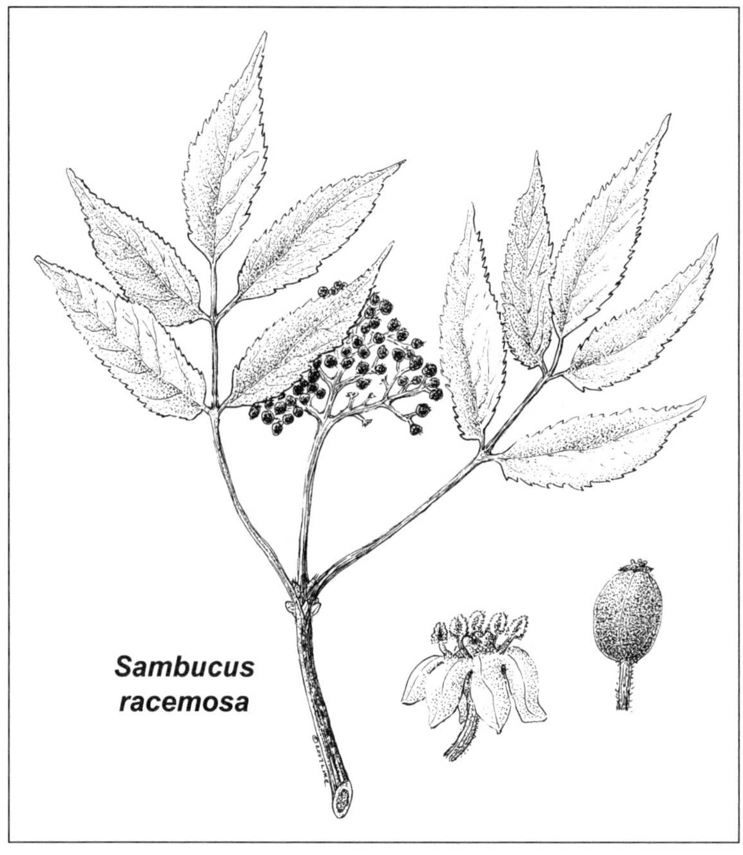
Sambucus racemosa

AGAVACEAE Dumort.
AGAVE FAMILY

William Jennings

Perennials, polycarpic (ours) or monocarpic, succulent to subsucculent, more or less xerophytic, with long-lived leaves, acaulescent to caulescent to arborescent, sometimes suckering at the base to form clumps; or small, slender, erect perennial herbs from perennial rosettes. **STEMS** erect to decumbent or procumbent, thick, simple or branched. **LEAVES** basal or cauline or both, succulent to dry, often large and spine-tipped, often with marginal teeth or with curly fibers, frequently glaucous. **INFL** an elongated raceme or panicle, from small with a few flowers to very large with hundreds of flowers. **FLOWERS** perfect (ours) to polygamous to imperfect; large (ours) to small and frequently showy, mostly white or cream-white (ours) to yellow, orange, or green; tepals of 6 distinct segments or united at the base into a tube, similar or subequal, not differentiated into a distinct calyx and corolla, often fleshy; stamens 6, hypogynous or adnate to the perianth tube; ovary superior or inferior, 3-celled, but often appearing 6-celled, with numerous to few ovules per locule; the slender and often very short style terminated by a capitate or short, 3-lobed stigma. **FRUIT** a loculicidal capsule or berry; membranous to leathery. **SEEDS** 1 to many, with endosperm. About 17 or 18 genera and about 550 species, primarily from arid regions. The circumscription of the Agavaceae is controversial. If the family is not recognized at all, the genera are placed in broadly defined Liliaceae or Amaryllidaceae, depending upon whether the ovary is superior or inferior. If the family is recognized, then the genera *Agave*, *Yucca*, *Furcraea*, *Hesperaloe*, and *Manfreda* are considered to be the core of the family. The treatment of the genera *Sansevieria*, *Cordyline*, *Dracaena*, *Nolina*, and *Dasylirion* has varied, but recent research using DNA sequencing supports the separation of *Dracaena*, *Nolina*, and *Dasylirion* from Agavaceae. Fortunately, in the region of the Four Corners, there is only one agavaceous genus (*Yucca*), which is one of the core genera.

Yucca L. Yucca, Shin Dagger, Soapweed, Soaptree

(Carib Indian name for manihot, erroneously applied) Perennials, shrubby to arborescent, small to large, simple or branched, caulescent or acaulescent, with semiwoody caudices and stems. **LEAVES** stiff, linear to lanceolate, thin to thick and fleshy, with filiferous margins, expanded bases, and spinose tips. **INFL** terminal, annual, many-bracted, few- to many-flowered raceme or panicle. **FLOWERS** large and showy, campanulate to globose, pendulous, cream to white, but also with reddish or purplish highlights; perianth parts 6 (tepals), lanceolate to ovate; stamens 6, hypogynous, filaments broad and flat; ovary superior, 3-locular or falsely 6-locular; style short and thick, with a subglobose or 3-lobed stigma. **FRUIT** a dry-dehiscent erect capsule or a soft, fleshy, spongy, pendulous berry. **SEEDS** obovoid to compressed, black. (McKelvey 1938, 1947; Webber 1953.) Mostly from the dry lands of North America. About 30 species. The taxonomy of *Yucca* has been in chaos since the wildly differing views of McKelvey (1938, 1947) and Webber (1953) were published. Our treatment generally follows that of Reveal in Cronquist et al. (1977). Most of the species treated by Reveal are in the Four Corners region. *Flora of North America* also largely follows the treatment of Reveal for the species in the Colorado Plateau area.

The following key is intended for use as a field key and will be inadequate for most herbarium specimens. When attempting to identify yuccas, it is good to remember that no single character alone will conclusively identify a yucca; the presence of a combination of several features is necessary to make a firm identification. Typical herbarium specimens will consist of a few flowers and/or a few leaves, rather than a whole plant. All too frequently, the herbarium label does not indicate the general aspects of the plant, i.e., general size of the rosette, whether blooming begins above or below the top of the rosette, length of the scape, length of inflorescence, etc. The size and color of the pistil, style, and flowers frequently change on pressing. Collecting yuccas that are not in flower or fruit is a waste of time and resources. Juvenile rosettes of most species of yucca encountered in the Four Corners flora will key immediately as *Yucca harrimaniae*, because of their small size and short leaves. *Intermountain Flora* (v. 6, pp. 527–528) provides very good advice on how to collect and curate a yucca herbarium specimen. *Flora of North America* (v. 26, p. 413) gives essentially the same advice. A series of collections of yuccas, following the procedures in IMF and FNA, would be a worthwhile project.

1. Flower long-campanulate; fruit fleshy, pendulous (shaped like a short, fat banana); leaves over 3 cm wide, very stiff and rigid, very coarsely filiferous on the margins, with a very stiff, sharp terminal spine .... ***Y. baccata***
1' Flower spherically campanulate or globose; fruit a dry capsule, erect; leaves usually less than 2 cm wide, somewhat flexible, finely filiferous on the margins .... (2)

2. Inflorescence starts much above the top of the leaf rosette (long scape), with a long, narrow inflorescence, up to 1 m long; capsules usually constricted (rather dumbbell-shaped) .... ***Y. angustissima***
2' Inflorescence starts below the top of the leaf rosette (short scape), with a rather short inflorescence, typically about 60 cm long; capsules not usually constricted .... ***Y. baileyi***

Yucca angustissima Engelm. ex Trel. (very narrow) Acaulescent or caulescent. **LEAVES** in a rosette near the ground, symmetrical, hemispherical; individual leaves linear, long and very narrow, 25–45 cm long and 0.8 cm wide in typical plants, but up to 60, 75, or even 150 cm and 0.4–1.5 cm wide in some varieties, tapering and long-acuminate, spine-tipped, flexible, with a few fine curly fibers on the margins. **INFL** simple, racemose, occasionally with a few short branches up to 10 cm long, flowering stem including inflorescence 0.5–1.5 m in typical material, but 2–4.5 m in some varieties. **FLOWERS** pendent, campanulate to globose, 4.5–5.5 cm long in typical material, but as small as 3.5 cm in one variety and as large as 6.5 cm in another variety; white to cream or greenish white; style white to pale greenish white. **FRUIT** an erect, dry capsule, commonly with a deep central constriction (dumbbell-shaped), 3.5–5.5 cm long in typical material, but up to 7.5 cm in some varieties. Not recognized by Weber and Wittmann (2001, *Colorado Flora: Western Slope* and 1992, *Catalog of the Colorado Flora*) as occurring in Colorado, but some of the specimens at COLO from Mesa, Montrose, Montezuma, and La Plata counties called *Y. harrimaniae* appear to be this taxon instead. There are several specimens at CS collected by H. D. Harrington and George Kelly from McElmo Canyon west of Cortez, Montezuma County, which are properly identified as *Y. angustissima*. Plants with long, narrow leaves, not heavily filiferous on the margins; a long scape greatly exceeding the leaves; a long, narrow inflorescence; and constricted capsules mark this species. Four varieties are recognized, of which two are in our area. A major food, fiber, and soap plant for most tribes of the region. Young fruits and flowers are roasted or boiled. Leaves are pounded to make cordage, needle and thread, and mats. Roots are source of soap due to saponins. An important ceremonial plant for Navajos.

1. Capsule small, 3.5–5.5 cm, deeply constricted; flowering stem including inflorescence 0.5–1.5 (up to 1.8) m tall; style 10–13 mm .... **var. *angustissima***
1' Capsule larger, 4.5–7.5 cm, moderately constricted, flowering stem including inflorescence 1–2.5 (up to 4.5) m tall; style 3–10 mm .... **var. *toftiae***

var. *angustissima* Mesas and slopes, often in sandy areas or on sandstone outcrops; fairly common in the Four Corners region. ARIZ: Apa, Coc, Nav; COLO: LPl, Mon; NMEX: McK, SJn; UTAH. 1495–2560 m (4900–8400′). Flowering: May–Jul. Southern Utah to central Arizona, eastward to extreme western and southwestern Colorado and northwestern New Mexico.

var. *toftiae* (S. L. Welsh) Reveal (for Catherine Ann Toft, professor of evolution and ecology at University of California, Davis) [*Yucca toftiae* S. L. Welsh]. Sandstone outcrops and hanging gardens along the lower San Juan River and its tributaries. ARIZ: Coc; UTAH. 1065–1220 m (3500–4000′). Just enters the San Juan drainage in the vicinity of the confluence of the Colorado and San Juan rivers, and as far east as the lower slopes of the Abajo Mountains (canyons in the Natural Bridges National Monument vicinity) along the northwest boundary of the study area. The type specimen was taken just outside the San Juan drainage at Three Gardens, one mile north of the confluence of the San Juan and Colorado rivers, San Juan County, Utah (*Welsh, 11935a*, BRY!).

Yucca baccata Torr. (with berries) Shin dagger, Spanish bayonet, datil, banana yucca. **STEMS** acaulescent or nearly so, with a rosette of leaves at or near the ground. **LEAVES** 40–75 cm long, deeply concavo-convex, rigid, dark green to blue-green, glabrous, often glaucous, with a few very coarse fibers on the margins, a very stout, sharp spine on the tip. **INFL** a panicle, 50–90 cm long, mostly hidden by the leaves. **FLOWERS** pendent, oblong-campanulate, 5–15 cm long, but usually about 7–8 cm long, creamy white tinged with reddish brown outside and sometimes inside as well, pendent. **FRUIT** large, fleshy or spongy, 15–20 cm long, tapered when fresh, pendent. Fairly common throughout the Four Corners region on dry, rocky slopes. Mostly in piñon-juniper habitats. ARIZ: Apa, Coc, Nav; COLO: Arc, Dol, LPl, Mon, SMg; NMEX: McK, RAr, San, SJn; UTAH. 1675–2400 m (5500–7875′). Flowering: mid-May–mid-Jun. The very wide, rigid, very painfully pungent leaves, long-campanulate flowers, and the fleshy fruit mark this species well. Anyone who has ever rammed the leaf spine into an ankle or shin will never forget (or misidentify) this species. Important food, fiber, and soap plant for most area peoples.

Yucca baileyi Wooton & Standl. **CP** (for Vernon Bailey, plant collector in the Southwest) Bailey's bouncing yucca. Acaulescent or very short-caulescent. **LEAVES** in a rosette near the ground, symmetrical, hemispherical; individual leaves linear, long, and very narrow, 25–50 cm long and 0.3–0.8 cm wide, tapering and long-acuminate, spine-tipped, flexible, with numerous conspicuous fine curly fibers on the margins. **INFL** racemose, typically 40–60 cm long, but sometimes up to 100 cm, scape short, about 10–15 cm long, the inflorescence thus starting below (usually) to just above the top of the leaf rosette. **FLOWERS** pendent, campanulate to globose, 5–6.5 cm long; white to cream or greenish white, pendent; style 7–9 mm long, white or pale greenish white. **FRUIT** an erect, dry capsule, 4–7 cm long, not constricted or only very slightly so. [*Yucca standleyi* McKelvey; *Yucca navajoa* J. M. Webber]. Dry forest floors, grasslands, sandstone ledges. ARIZ: Apa, Coc, Nav; COLO: Arc, Dol, LPl, Mon; NMEX: McK, RAr, San, SJn; UTAH. 1395–2550 m (4580–8365′). Extreme southern Utah, northern/northeastern Arizona, northwestern New Mexico, southwestern Colorado. The type of *Y. baileyi* was taken in the Tunitcha Mountains, San Juan County, New Mexico (*Standley 7638*). The type of *Y. standleyi* (*McKelvey 4609*) was taken at essentially the same place as the type of *Y. baileyi*. McKelvey discusses at length why she thought the type of *Y. baileyi* did not represent the material that had been called *Y. baileyi* and why that material needed a new name. All subsequent authors have rejected her argument. The type of *Y. navajoa* was taken five miles north of Tohatchi, McKinley County, New Mexico (J. M. Webber in 1944). Plants answering to *Y. navajoa* are subcaulescent, in a dense clump due to branching, and with rather smaller rosettes than typical. These occur, according to J. M. Webber, from three miles north of Tohatchi, extending north for about six miles and west into the Chuska Mountains. Webber later reduced his species to a variety of *Y. baileyi*. Subsequent authors have maintained it as a variety or not recognized it at all. Several specimens of *Y. baileyi* (as *Y. standleyi*) are listed for McKinley County by McKelvey (1947). Neither *Y. baileyi* nor *Y. standleyi* is recognized by Weber and Wittmann as occurring in Colorado. Specimens from the West Slope at COLO are all identified as *Y. harrimaniae*, but *Y. harrimaniae* does not occur in extreme southern Colorado, despite reports to the contrary. Specimens from Montezuma, La Plata, and Archuleta counties identified as *Y. harrimaniae* at COLO are mostly *Y. baileyi* (some are *Y. angustissima*). Older specimens at COLO from the Colorado counties within the range of the *Flora of the Four Corners Region* were annotated by S. D. McKelvey as *Y. baileyi* or *Y. standleyi* (a synonym) and seem to be this taxon, even though they are currently identified as *Y. harrimaniae*. The range of *Y. baileyi* centers on the Four Corners area and, in my opinion, is much more restricted than has been reported in the literature. It does not extend into the San Luis Valley or along the Front Range in Colorado.

ALISMATACEAE Vent. WATER-PLANTAIN, ARROWHEAD FAMILY

C. Barre Hellquist

(ancient Greek name, adopted by Linnaeus from Dioscorides) Annual or perennial herbs. Rhizomatous, stoloniferous, or cormose. Roots septate or nonseptate. **LEAVES** basal, submersed, floating; sessile or petiolate, basal lobes present or absent, veins parallel from base of blade to apex. **INFL** a scapose raceme, panicle, occasionally umbel, erect, floating, or decumbent. **FLOWERS** bisexual or unisexual, if unisexual, plants dioecious; hypogynous; sepals 3, petals

3; stamens (0) 6–9 (–30); pistils 0–6 to 1500; ovules 1–2. **FRUIT** an achene or follicle. **SEEDS** with endosperm absent in mature seed (Hendricks 1957; Bogin 1955; Rataj 1972). Twelve genera, about 80 species worldwide.

1. Pistils and achenes in a single, flat-topped ring, flowers all bisexual, stamens 6 .... ***Alisma***
1' Pistils and achenes in a dense, globose head; flowers bisexual or unisexual, stamens more than 6........ ***Sagittaria***

Alisma L. Water-plantain

(ancient Greek name adopted by Linnaeus from Dioscorides) Perennial herbs, submersed, floating-leaved, emergent; often rhizomatous; stolons, corms, and tubers absent. Roots nonseptate. **LEAVES** sessile or petiolate, petiole triangular, blade linear to ovate, bases attenuate to rounded, apex obtuse to acute. **INFL** a panicle of 2–10 whorls, erect, emergent. **FLOWERS** bisexual, pedicel ascending, a bract subtending the pedicel; receptacle flattened; sepals erect; petals pink or white; stamens 6–9; pistils 15–20 in a single flat-topped, ringed receptacle; ovules 1. **FRUIT** a laterally compressed achene. $x = 7$. Nine species worldwide.

1. Leaves submersed and ribbonlike, or, if emergent, with lanceolate to narrowly elliptic blades; achene with 2 dorsal grooves and a central ridge .... ***A. gramineum***
1' Leaves emergent, with ovate to elliptic blades, occasionally submersed to floating; achenes with a single dorsal groove, no central ridge .... (2)

2. Flower 3–3.5 mm wide, sepals at anthesis 1.5–2.5 mm long; petals 1–3 mm long; fruiting head 2–4 mm in diameter; achene 1.5–2 mm long .... ***A. subcordatum***
2' Flower 7–13 mm wide, sepals at anthesis 3–4 (–6) mm long; petals 3.5–6 mm long; fruiting head 4–7 mm in diameter; achene 2.2–3 mm long .... ***A. triviale***

Alisma gramineum Lej. (grasslike) Grass-leaved water-plantain. Perennial to 50 cm. **LEAVES** submersed, floating, or emergent; if submersed, sessile and linear, 0.2–2 (–3) mm wide, blade present or absent; if emergent, petiolate, rarely sessile, blade linear-lanceolate, lanceolate, or narrowly elliptic, 0.4–1.5 cm wide. **FLOWERS** purplish white, 2–4 mm wide. **FRUIT** in a head. **ACHENES** 3–6 mm diameter, each with 2 dorsal grooves and a central ridge, beak erect. $2n = 14$. [*Alisma geyeri* Torr.; *A. gramineum* var. *angustissimum* (DC.) A. J. Hendricks; *A. gramineum* var. *wahlenbergii* (Holmb.) Raymond & Kucyn.]. Calcareous and brackish waters of shallow and muddy shores. ARIZ: Apa; COLO: Arc, Mon, LPl; NMEX: McK, RAr; UTAH. 2470–3575 m (5100–9450'). Flowering: Jul. Fruit: Jul–Aug. British Columbia and Alberta east to Ontario, Quebec, and Vermont, south to California, New Mexico, and South Dakota; Eurasia; Africa.

Alisma subcordatum Raf. (almost heart-shaped) Southern water-plantain. Perennial herb to 60 cm tall. **LEAVES** emergent, petiolate; blade ovate to elliptic. **INFL** to 1 m. **FLOWERS** with sepals 1.5–2.5 mm long; petals white, 1–3 mm long. **FRUIT** in a head. **ACHENES** 2–4 mm diameter, each obliquely ovoid, with 1 dorsal groove, 1.5–2.2 mm; beak erect. $2n = 14$. [*Alisma parviflorum* Pursh; *A. plantago-aquatica* L. var. *parviflorum* (Pursh) Torrey]. Shallow ponds, stream margins, marshes, and ditches. COLO: Arc, LPl. 1550–2750 m (6050–8000'). Flowering: Jun–Jul. Fruit: Jul–Aug. North Dakota east to Ontario, Quebec, and New Brunswick, south to Arizona, New Mexico, Mississippi, and Georgia; Mexico.

Alisma triviale Pursh (ordinary) Northern water-plantain. Perennial herbs to 1 m. **LEAVES** emergent, petiolate; blade linear-lanceolate to broadly elliptic or oval. **INFL** to 1 m. **FLOWERS** with 3 sepals, 3–4 (–6) mm long; 3 petals, white, 3.5–6 mm. **FRUIT** in a head. **ACHENES** 4–7 mm diameter, ovoid, with 1 dorsal groove, 2.1–3 mm, beak erect or nearly erect. $2n = 28$. [*Alisma plantago-aquatica* L. var. *americanum* Schult. & Schult. f.]. Shallow ponds, stream margins, marshes, and ditches. ARIZ: Apa; COLO: Arc, LPl; NMEX: RAr, SJn. 2350–2700 m (7800–8950'). Flowering: Jun–Jul. Fruit: Jul–Aug. Alaska and British Columbia east across Canada to Newfoundland, south to California, New Mexico, Oklahoma, Illinois, Ohio, and Connecticut. The bulbous base has a strong taste when fresh but becomes a palatable starchy vegetable when dried.

Sagittaria L. Arrowhead

(Latin *sagitta*, arrow) Perennial herbs, rarely annual, submersed, floating-leaved, or emersed; rhizomes usually present, often with tubers or stolons. Roots septate. **LEAVES** sessile or petiolate; petiole triangular to terete; blades sagittate, hastate, cuneate, lanceolate, elliptic, or obovate, apex acute to obtuse. **INFL** a raceme or panicle, rarely an umbel, of 1–17 whorls, erect, emergent or floating; bract apex obtuse to acute. **FLOWERS** unisexual (rarely bisexual); staminate flowers pedicellate, distal to pistillate flowers; pistillate flowers mostly pedicellate, occasionally sessile; a bract subtending the pedicel, pedicel ascending to recurved; receptacle convex; sepals recurved in staminate flowers, recurved to erect in pistillate flowers; petals white, rarely with pink spot or tinge, stamens 7–30; filament linear to dilated, glabrous or pubescent; pistils 1500. **FRUIT** compressed, ribbed or winged with a short, erect or divergent persistent beak. x = 11. About 30 species. North America, South America, Eurasia.

1. Achene with beak ascending or erect; bracts 7–40 mm long; staminate plant with 10–24 stamens; plants submersed rosettes with floating leaves, with flaccid linear leaves, or emergent with recurved petioles...... ***S. cuneata***
1' Achene with beak horizontal; bracts 3–8 mm or less long; staminate plant with 21–40 stamens; plants emergent, usually erect .... ***S. latifolia***

Sagittaria cuneata Sheldon (wedge-shaped) Northern arrowhead. Perennial herbs to 40 cm tall, 80 cm submersed; stolons present, corms present. **LEAVES** emergent, submersed rosettes often with floating leaves, or submersed flaccid, flattened, linear, often with floating leaves; petiole triangular, blade cordate, sagittate, or linear. **INFL** a raceme, rarely a panicle of 2–10 whorls, emergent; peduncles 10–50 cm; bracts connate 1/4 of the total length, 7–40 mm; fruiting pedicels ascending. **FLOWERS** to 2.5 cm diameter; sepals recurved, 10–24 stamens. **FRUIT** in a head. **ACHENES** 0.8–1.5 cm diameter, obovoid, 1.8–2.6 × 1.3–2.5 mm, beak apical, erect, 0.1–0.4 mm. $2n$ = 22. [*Sagittaria arifolia* Nutt. ex J. G. Sm.]. Calcareous waters of rivers, shores of lakes, ponds, wet pastures, and ditches. COLO: Dol, LPl, Mon; NMEX: RAr, SJn; UTAH. 1630–2150 m (5350–7050′). Flowering: Jun–Jul. Fruit: Jul–Aug. Alaska east to Quebec and Newfoundland, south to California, Texas, Illinois, Pennsylvania, and Connecticut. This is a highly variable species that may form submersed, sterile rosettes with or without floating leaves. Plants in deep, flowing rivers may produce broad, flattened, linear leaves and may also produce floating leaves. Emergent plants often form leaf petioles that bend over toward the ground. The corms of this species were eaten by Native Americans.

Sagittaria latifolia Willd. (broad-leaved) Common arrowhead, wapato, duck potato. Perennial herbs to 150 cm tall; rhizomes absent, stolons present bearing corms. **LEAVES** emergent; petiole triangular, erect; blade sagittate, hastate, or lanceolate. **INFL** a raceme, rarely a panicle, of 3–90 whorls, emergent; peduncles 10–59 cm; bracts connate for at least 1/4 of the total length, 3–8 mm. **FLOWERS** to 4 cm diameter; sepals recurved to spreading; stamens 21–40. **FRUIT** in a head 1–1.7 cm diameter; fruiting pedicels spreading. **ACHENES** oblanceolate, 2.5–3.5 × to 2 mm, beak lateral, horizontal, 1–2 mm long. $2n$ = 22. [*Sagittaria latifolia* var. *obtusa* (Muhl. ex Willd.) B. L. Rob.; *S. latifolia* var. *pubescens* (Muhl.) J. G. Sm; *S. ornithorhyncha* Small; *S. planipes* Fernald; *S. pubescens* Muhl; *S. viscosa* C. Mohr]. Pond and lake shores, streams, sloughs, and marshes. COLO: LPl, Mon; NMEX: RAr, SJn. 1525–2285 m (5000–7500′). Flowering: Jul–Aug. Fruit: Jul–Sep. British Columbia east across Canada to Nova Scotia, south to California; New Mexico, Louisiana, and Florida; Mexico; Central America; South America. The leaves are highly variable and different forms have been observed on the same plant. Native Americans ate the tubers roasted or broiled.

AMARANTHACEAE Juss. AMARANTH OR PIGWEED FAMILY

Kenneth R. Robertson

(Greek *amarantos*, unfading, referring to the unwithering sepals) Annual or perennial herbs. **LEAVES** petioled, exstipulate, alternate or opposite, simple, usually entire. **INFL** bracteate cymules arranged in terminal and/or axillary spikes, panicles, or glomerules. **FLOWERS** small or minute, actinomorphic, perfect or imperfect and the plants then monoecious or dioecious; sepals mostly 4 or 5, sometimes fewer, usually dry and scarious or chartaceous; petals absent; stamens 2–5, basally connate or rarely free, the filaments alternating with pseudostaminodia or not, the anthers with 2 or 4 locules; pistil 2- or 3-carpellate; ovary superior, 1-locular, placentation basal, ovules 1 or several, stigmas 2 or 3. **FRUIT** a utricle or pyxis, circumscissile or indehiscent. **SEEDS** small, lenticular, subglobose, or reniform, embryo curved, peripheral, surrounding the mealy perisperm. x = 8, 9, 16, 17 (Robertson and Clemants 2003). Sixty-five genera, 900 species, nearly worldwide, especially in tropical and warm-temperate regions, absent from Arctic and alpine regions. *Amaranthus* cultivated as ornamentals and pseudocereals.

1. Leaves alternate, flowers unisexual, the plants monoecious or dioecious .... ***Amaranthus***
1' Leaves opposite, flowers bisexual .... (2)

2. Indumentum of simple trichomes .... ***Guilleminea***
2' Indumentum of stellate trichomes .... ***Tidestromia***

Amaranthus L. Amaranth, Pigweed, Quelite, Bledo

(Greek *amarantos*, unfading) Monoecious or dioecious annual herbs. **STEMS** simple to much-branched, erect or rarely prostrate, glabrous or variously pubescent with simple hairs. **LEAVES** alternate. **INFL** dense terminal and/or axillary cymules in spikes, panicles, or glomerules, the infl units often subtended by reduced leaves, each cymule subtended by a persistent bract. **FLOWERS** imperfect, greenish or reddish; sepals distinct, persistent, 3–5 or rudimentary to absent in pistillate flowers, glabrous; stamens 3–5, absent in pistillate flowers, filaments free to base, pseudostaminodia absent, anthers 4-locular; pistils with 2–3 stigmas, absent in staminate flowers, ovule 1. **FRUIT** a utricle, 2–3-beaked, membranaceous, usually circumscissile, sometimes irregularly dehiscent or indehiscent. **SEEDS** lenticular to subglobose, smooth, shining or dull, usually black or dark reddish brown. [*Acnida* L. and *Acanthochiton* Torr.]. Specimens of this genus are often difficult to identify by someone not familiar with the group. When using the key, look closely at the tips of pistillate inflorescence branches for staminate flowers to determine whether the plant is monoecious or dioecious. Also, pistillate plants of dioecious species are required for positive identification (Robertson and Clemants 2003; Sauer 1955, 1967, 1972). About 70 species, mostly in tropical, subtropical, and warm-temperate regions, most species originating in the New World, several species weedy and found in agricultural areas and other disturbed habitats nearly worldwide, a few species cultivated as ornamentals or pseudocereal crops. The species below that are not in bold are to be expected in the Four Corners region.

1. Plants dioecious ..... (2)
1′ Plants monoecious ..... (3)

2. Leaves narrowly lanceolate or linear, 2–8 cm long, 0.2–1.2 cm wide; bracts of pistillate flowers deltate, wider than long, accrescent and indurate ..... ***A. acanthochiton***
2′ Leaves ovate, rhomboid, or lanceolate, 4–8 cm long, 2–3 cm wide; bracts of pistillate flowers scarious, much narrower than long ..... ***A. palmeri***

3. Inflorescences terminal panicles or spikes and sometimes also axillary spikes or clusters ..... (4)
3′ Inflorescences strictly axillary clusters, stems then appearing leafy to tips ..... (7)

4. Sepals of pistillate flowers acute or acuminate ..... (5)
4′ Sepals of pistillate flowers obtuse or rounded ..... (6)

5. Inflorescence bracts 3–4 mm long ..... *A. hybridus*
5′ Inflorescence bracts 4–7 mm long ..... ***A. powellii***

6. Plants conspicuously pubescent; inflorescences much-branched, leafless at tip, thick ..... ***A. retroflexus***
6′ Plants glabrous; inflorescences slender ..... ***A. wrightii***

7. Plants erect; sepals of pistillate flowers 3, bracts twice as long as sepals ..... ***A. albus***
7′ Plants prostrate; sepals of pistillate flowers either 1 or 5, bracts equaling sepals ..... (8)

8. Pistillate flowers with only 1 well-developed sepal; seeds 0.7–1 mm in diameter, dull ..... *A. californicus*
8′ Pistillate flowers with 4–5 sepals; seeds 1.3–1.6 mm in diameter, shiny ..... ***A. blitoides***

Amaranthus acanthochiton J. D. Sauer (spiny covering) Greenstripe amaranth. Erect, dioecious herbs, mainly 0.2–0.8 m tall, glabrous or glabrescent. **STEMS** much-branched, with ascending branches, pale white-green to reddish with vertical green stripes. **LEAVES** with petioles shorter than blades; blades narrowly lanceolate to linear, 2–8 cm long, 0.2–1.2 cm wide, apices acute to subobtuse with long mucro, bases narrowly cuneate to narrowly decurrent, margins narrowly hyaline or pale, erose, crispate, or irregularly undulate, prominently white-veined below. **INFL** terminal spikes, erect, stiff; bracts of staminate flowers thin, lanceolate, 2–3 mm long, with moderately heavy, excurrent midribs; bracts of pistillate flowers deltoid, nearly twice as wide as tall, accrescent, recurved at first, then completely enclosing flowers, indurate, conspicuously reticulate, 5 mm long, longer than sepals, up to 10 mm wide, midrib extremely heavy, extending to spinose apex, margins crenate or erose, hyaline or pale. **FLOWERS** (staminate) with 5 sepals, slightly unequal, 2.5–4 mm long; pistillate with 5 sepals, innermost rudimentary, others 4–5 mm long, broadly spatulate, margins crenate, conspicuously veined, styles 3, spreading. **UTRICLES** 2 mm long, elliptic, slightly rugose to smooth. **SEEDS** dark reddish brown, 1–1.3 mm in diameter, lenticular, shiny. [*Acanthochiton wrightii* Torr. not *Amaranthus wrightii* S. Watson]. Sand areas, such as dunes and riverbanks. ARIZ: Nav; NMEX: McK, SJn. 1200–2000 m (3900–6500′). Southern Utah to New Mexico, Arizona, and Trans-Pecos Texas and adjacent Chihuahua, Mexico. Cooked as greens by

Hopis. In herbaria, many specimens of *A. acanthochiton* are misfiled under *A. wrightii*, although the two species are quite distinct.

Amaranthus albus L. (white) Tumbleweed amaranth, quelite. Erect, monoecious, annual herbs, to 1 m tall; large plants forming tumbleweeds. **STEMS** erect, somewhat bushy, many-branched, glabrous, glabrescent, or viscid-pubescent, whitish. **LEAVES** with petioles about 1/2 length of blades, the blades narrowly spatulate or ovate, small, mostly 0.5–3 cm long, 0.5–1.5 cm wide but early leaves to 8 cm long, apices obtuse with a whitish, subspinescent mucro, bases tapering. **INFL** in axillary clusters, staminate and pistillate flowers mixed; bracts of pistillate flowers 2–3 mm, twice as long as sepals, narrow, with stout, long-acuminate, spinescent apices. **FLOWER** sepals 3, staminate with 3 stamens, pistillate with sepals slightly unequal, 1–1.5 mm long, thin, narrowly ovate to linear, apices acute; stigmas 3, erect. **UTRICLES** ellipsoid-ovoid, 1.5 mm long, exceeding sepals, smooth below, coarsely rugose above, circumscissile. **SEEDS** dark reddish brown to black, shiny, 0.6–1 mm in diameter. $2n = 32$. [*Amaranthus graecizans* L., name misapplied to both *A. albus* and *A. blitoides* in older floras.] Weedy plant of cultivated lands, sandy and saline areas, roadsides. ARIZ: Apa, Nav; COLO: Arc, LPl, Mon; NMEX: McK, SJn; UTAH. 1100–2290 m (3600–7500′). Flowering and fruiting: Jun–Sep. Probably originally native to central and western North America, but now widespread throughout the continent and also most other continents. Chiricahua and Mescalero Apache, Navajo, and Hopi used seed ground into flour; young plants also used as greens. Navajo used it with many other plants for ceremonial smoke. Plants from southwestern North America often have viscid pubescence and distinctly crisped leaf margins; these have been separated as **var.** ***pubescens*** (Uline & W. L. Bray) Fernald [*A. pubescens* (Uline & W. L. Bray) Rydb.], but this taxon is not widely recognized today.

Amaranthus blitoides S. Watson (resembling the Old World *A. blitum*) Prostrate or mat amaranth. Prostrate, monoecious, annual herbs. **STEMS** much-branched, 0.2–0.6 m long, spreading from a taproot, glabrous. **LEAVES** with petioles 1/2 as long as blades, the blades ovate, elliptic, obovate, or spatulate, 0.5–2 (4) cm long, 0.5–1 cm wide, apices obtuse, mucronate, bases tapering. **INFL** axillary clusters; bracts of pistillate flowers more or less equaling or slightly exceeding sepals, thin, narrow, apices long-tapering. **FLOWERS** with staminate flowers mixed with pistillate ones, with 3 sepals and 3 anthers; sepals of pistillate flowers mostly 4–5, unequal, 1.5–3 mm long, more or less equaling utricles, thin, narrowly ovate to linear, apices acute or acuminate; stigmas 3, recurved. **UTRICLES** broadly ovoid, 1.7–2.5 mm long, equaling sepals, mostly smooth, circumscissile. **SEEDS** black, rather dull, 1.3–1.6 mm in diameter. $2n = 32$. Disturbed habitats, such as roadsides, riverbanks, railroads, fields. [*Amaranthus graecizans* L, name misapplied to both *A. albus* and *A. blitoides* in older floras.] Weedy species of cultivated areas, roadsides, sandy or hard clay soils, dry lakes, dunes, arroyos. ARIZ: Apa, Coc, Nav; COLO: Dol, LPl, Mon, SMg; NMEX: McK, RAr, San, SJn; UTAH. 1370–2500 m (4500–8200′). Flowering and fruiting: Jul–Oct. Probably native to central North America and now widely distributed throughout temperate regions of the continent and many warm-temperate and subtropical regions around the world. Uses similar to those of *A. albus.*

Amaranthus palmeri S. Watson (for Edward Palmer) Palmer's amaranth, quelite. Coarse, dioecious, annual herb, 1–2 (3) m tall. **STEMS** erect, unbranched or with numerous ascending branches, glabrous. **LEAVES** with petioles 3/4–1 1/2 the length of blades, the blades ovate or rhomboid, 2–6 cm long, 1–3.5 cm wide, apices acute or acuminate with rounded, mucronate tip, bases cuneate to tapering. **INFL** terminal spikes or thyrses, erect or drooping, the terminal unit very elongate; bracts of pistillate flowers 4–6 mm long. **FLOWERS** (staminate) with 5 sepals and stamens, sepals unequal, the outer 3.5–4 mm long, acuminate with long-excurrent midveins; sepals of pistillate flowers 5, recurved, spatulate, unequal, the outer 3–4 mm long, acute with midvein excurrent as a rigid point, stigmas 2 or rarely 3, divergent. **UTRICLES** subglobose, 1.5–2 mm long, shorter than sepals, thin, slightly rugose, circumscissile. **SEEDS** dark reddish brown, shiny, 1–1.2 mm in diameter. $2n = 34$. Waste ground, roadsides, fields, arroyos, river bottoms. ARIZ: Coc; COLO: Mon; NMEX: SJn; UTAH. 1370–1700 m (4500–5500′). Flowering and fruiting: Jul–Oct. Originally native from southern California to Texas and northern Mexico, now a recent introduction throughout much of the eastern United States.

Amaranthus powellii S. Watson (named for John Wesley Powell) Powell's amaranth. Coarse, monoecious, annual herb, 0.3–2 m tall. **STEMS** erect, mostly simple or sometimes many-branched, villous toward apices. **LEAVES** with petioles equaling or exceeding blades, the blades ovate, rhomboid, or lanceolate, 4–8 cm long, 2–3 cm wide, apices tapering to an obtuse, mucronate tip, bases tapering. **INFL** few, long, stiff spikes aggregated into terminal panicles and also from the upper axils; bracts of pistillate flowers lanceolate, rigid, 4–7 mm long, 2–3 times longer than sepals, apices spinose. **FLOWERS** (staminate) clustered at tips of inflorescence branches, sepals 3–5, stamens 3–5; sepals of pistillate flowers 5, unequal, 1.5–3.5 mm long, the outer narrowly elliptic or ovate with aristate apices, stigmas 3, recurved. **UTRICLES** broadly ovoid, compressed, 2–3 mm long, equaling or shorter than sepals, smooth below, the lid smooth or slightly verrucose, circumscissile. **SEEDS** black, shiny, 1–1.4 mm in diameter. $2n = 32, 34$. [*A. bracteosus* Uline & W. L. Bray]. Waste ground, roadsides, sandy and rocky soil, mine spoils, flats, washes, open dry slopes. ARIZ: Apa, Nav; COLO: Arc, Dol, LPl, Min, Mon, SJn, SMg; NMEX: McK, RAr, San, SJn; UTAH. 1580–2700 m (5200–8900′). Flowering: Jun–Oct. Originally native to the American Southwest, but now widespread throughout North America.

Amaranthus retroflexus L. (bent backward) Redroot amaranth. Coarse, monoecious, annual herb, 0.3–1.5 (2+) m tall. **STEMS** erect, unbranched to branched, often reddish toward base, densely short-villous above. **LEAVES** with petioles 1/2 as long to as long as blades, the blades ovate or rhomboid, 4–11 cm long, 1.5–7 cm wide, apices tapering to an obtuse or retuse, mucronate tip, bases tapering. **INFL** numerous, short, thick spikes aggregated into terminal panicles and also from the upper axils; bracts of pistillate flowers 4–6 mm long, twice as long as sepals, apices acuminate, with short-excurrent midrib. **FLOWERS** (staminate) few, at tips of inflorescences, sepals and stamens 5; sepals of pistillate flowers 5, unequal, 2.5–4 mm long, the outer narrowly spatulate, apices mucronate and rounded, truncate, or retuse, stigmas 3, erect or slightly recurved. **UTRICLES** broadly ovoid, 2–2.5 mm long, shorter than sepals, smooth below, slightly rugose above, circumscissile. **SEEDS** black or dark reddish brown, shiny, 1–1.3 mm in diameter. $2n = 32, 34$. Disturbed areas, sandy soil, gravel, alluvial areas, roadsides. ARIZ: Nav; NMEX: McK, San, SJn. 1600–2300 m (5300–7500′). Flowering and fruiting: Jul–Sep. Many specimens identified as this species are actually *A. powellii.* Originally a riverbank pioneer of central and eastern North America, now a common agricultural weed in disturbed habitats.

Amaranthus wrightii S. Watson (for Charles Wright) Wright's amaranth. Plants slender, monoecious annuals, 0.2–1 m tall. **STEMS** erect, simple or with a few ascending branches, glabrous, often whitish or tinged with red. **LEAVES** with petioles shorter than to ± equaling blades, the blades rhomboid-ovate to elliptic-lanceolate, 1.5–6 cm long and 0.5–3 cm wide, apices obtuse to subacute and shortly mucronate, bases broadly to narrowly cuneate, margins entire, mostly plane or slightly undulate. **INFL** short spikes terminating main stem and shorter lateral branches, also axillary; bracts up to twice as long as sepals, apices spinescent. **FLOWERS** (staminate) few, at tips of inflorescences, sepals 5, stamens 4–5; sepals of pistillate flowers spatulate-linear, not clawed, subequal, 1.5–2 mm long, membranaceous, apices emarginate or retuse to obtuse, outer rarely subacute; stigmas 3. **UTRICLES** subglobose to broadly obovoid, 1.3–2 mm in diameter, subequal with sepals, smooth or slightly rugose, circumscissile. **SEEDS** dark reddish brown, 1 mm in diameter, smooth, shiny. Waste ground weed, stream banks, canyons, semideserts. ARIZ: Apa; NMEX: SJn. 1220–2000 m (4000–6560′). Flowering and fruiting: Jul–Sep. Texas to Utah and Colorado. In herbaria, many specimens of *A. acanthochiton* are misfiled under *A. wrightii*, although the two species are quite distinct.

Guilleminea Kunth Matweed

(for French botanist Antoine Guillemin) Perennial herbs. **STEMS** much-branched, radiating from a stout taproot, hairs simple. **LEAVES** opposite, petiolate, large in a basal rosette, moderate-sized and cauline, and small, subtending each inflorescence. **INFL** condensed spikes in axillary glomerules, bracts membranaceous. **FLOWERS** perfect, sepals 5, lower 1/2 connate, free parts 1-veined, stamens 5, filaments basally connate and adnate to calyx, pseudostaminodia absent, anthers 2-locular. **FRUIT** indehiscent, membranous. Five species, south-central and southwestern North America, Mexico, West Indies, and South America (Clemants 2003b; Henrickson 1987; Mears 1967).

Guilleminea densa (Humb. & Bonpl. ex Schult.) Moq. (dense) Small matweed. Plants mostly 20–40 cm in diameter. **STEMS** prostrate, villous, forming mats. **LEAVES** of rosettes early-deciduous, blades 10–40 mm long, leaves of stems very unequal in size, blades 0.4–1.5 cm long and 0.1–0.8 cm wide, becoming smaller below the inflorescences, ovate to oblong or rhomboid, apices acute, bases cuneate to the petiole, glabrous above, villous or lanate below. **FLOWERS** with sepals 1.7–2.8 mm long, ovary ovoid, laterally compressed, ovule 1. **SEEDS** reddish brown, shiny, 0.8–0.9 mm in diameter. [*Brayulinea densa* (Humb. & Bonpl. ex Schult.) Small]. Dry, open, gravelly, rocky flats, ridges, and waste areas. ARIZ: Nav. 1795 m (5600′). Flowering and fruiting: Jun–Oct. Eastern Arizona east to New Mexico and central Texas, south to Oaxaca, Mexico. Henrickson (1987) cites a specimen from the *Flora* area (five miles north of El Capitan [Agathla Peak], 13 May 1935. *Maquire 11332* [NY]). Our material belongs to **var.** ***aggregata*** Uline & W. L. Bray (dense).

Tidestromia Standl. Tidestromia, Honeysweet

(for Ivar T. Tidestrom, Swedish-born American botanist) Annual or perennial herbs, pubescence stellate. **STEMS** ascending, procumbent, or prostrate, much-branched from a caudex. **LEAVES** opposite, petiolate. **INFL** axillary glomerules. **FLOWERS** perfect, sepals 5, distinct; stamens 5, anthers 2-loculed, filaments connate below into cup, pseudostaminodia present or absent; ovary ovoid, stigma 1, capitate, ovule 1. **UTRICLES** indehiscent. About six species of southwestern North America, often gypsophiles (Sánchez del Pino and Clemants 2003).

Tidestromia lanuginosa (Nutt.) Standl. (woolly) Woolly tidestromia. Plants annual low herbs, densely dendroid-stellate pubescent, silvery and whitish to gray-green, somewhat glabrate. **STEMS** prostrate to decumbent or ascending, 10–15 cm tall and up to 1 m in diameter. **LEAVES** with petioles about equaling blades, the blades orbicular to rhomboid, 1–3 cm long, decreasing in size toward the tips of the branches, apices rounded or obtuse, bases cuneate to rounded, bases of uppermost leaves short-petiolate to connate, indurate, forming an involucre. **INFL** few-flowered axillary glomerules surrounded by involucre; bracts broadly ovate, 1–1.5 mm long, shorter than the tepals, stellate-lanuginose. **FLOWERS** yellowish green, sepals 1–2 mm long, mostly glabrous, membranaceous; staminal cup 0.5–1 mm long; pseudostaminodia absent or to 0.2 mm long. **UTRICLES** 1.3–1.6 mm long. **SEEDS** subglobose, reddish brown, 1–1.4 mm in diameter. Roadsides, sandy soil, arroyos, salt desert shrub community. ARIZ: Apa; COLO: Mon; NMEX: SJn; UTAH. 1340–1920 m (4400–6300′). Flowering and fruiting: Jun–Oct. Kansas and Texas west to Utah and New Mexico.

ANACARDIACEAE R. Br. SUMAC FAMILY

John R. Spence

Small to large shrubs, small trees, or woody vines. **STEMS** weakly to strongly branched, sometimes producing aerial roots, bark with resin ducts. **LEAVES** alternate, deciduous or sometimes evergreen, pinnately compound or sometimes simple, stipules usually absent. **INFL** terminal to axillary, racemose to cymose. **FLOWERS** small, perfect or unisexual, hypogynous, actinomorphic, 5-merous; sepals separate or connate below, sometimes persistent in fruit; petals separate, sometimes absent; stamens as many as petals, alternate, sometimes twice as many, attached to outside of or on a 5-lobed nectary disc; ovary of 3 united carpels, 1 to many locules per carpel, styles distinct, ovules 1 per locule. **FRUIT** a smooth, often waxy or resinous drupe. A family of about 650 species and 80 genera in primarily tropical to subtropical regions, with a few north-temperate genera. The sumac family is of considerable economic importance and includes among many tropical fruits such familiar ones as pistachio nuts, mangoes, and cashews. Many members of the family produce allergic reactions in people, from exudates of the resin glands. *Pistacia atlantica* Desf., pistachio nut tree, is commonly cultivated in warmer areas of the southwestern United States and sometimes escapes.

1\. Leaflets more than 3, or, if fewer, then either simple and lobed, or terminal leaflet sessile; fruit red, pubescent .... ***Rhus***

1′ Leaflets 3, shiny, terminal leaflet petiolate; fruit white, smooth .... ***Toxicodendron***

Rhus L. Sumac

(classical Latin and Greek name for the European *Rhus coriaria* L.) Often malodorous, strongly to weakly branched shrubs or small trees, sometimes rhizomatous, aerial roots lacking. **STEMS** smooth, bark not flaky, red-brown to gray, pubescent or glabrous, resinous, not producing aerial roots. **LEAVES** deciduous or evergreen, pinnately compound or simple, petiolate. **INFL** a dense, crowded raceme or umbel, or flowers rarely solitary, terminal or lateral on previous years' twigs. **FLOWERS** polygamous or dioecious; 5-merous, petals pubescent, style short, 3-lobed, ovary with 3 carpels, each with 1 locule. **FRUIT** a small, dry or somewhat fleshy red drupe, glandular-pubescent. A genus of about 150 species widespread in warmer regions of the world. Although many species of *Rhus* are malodorous, few produce the allergic reactions that *Toxicodendron* does.

1\. Leaves 3-lobed to 3-foliate, foliage aromatic .... ***R. aromatica***

1′ Leaves pinnately compound with 7 or more leaflets, foliage not aromatic .... ***R. glabra***

Rhus aromatica Aiton (for the aromatic foliage) Skunkbush, lemonberry. Freely branched, somewhat sprawling shrubs to 3 m. **STEMS** densely pubescent, becoming gray with age, young branches brown. **LEAVES** appearing after the flowers, simple to palmately 3-lobed to 3-foliate, thin to somewhat coriaceous, petiole pubescent, blades glabrous to finely pubescent or ciliate, central leaflet from 1–10 cm long, typically less than 6 cm, lateral lobes or leaflets typically smaller, leaflet margins serrate to nearly entire, leaves turning red in fall. **INFL** appearing before leaves, on short lateral branches, of dense spikes or compact spicate panicles, elliptic to cylindrical, sessile. **FLOWERS** small, subtended by a single bract, with 2 bractlets below, petals 2–3 mm, yellowish, sepals 1–2 mm, not exceeding the petals. **FRUIT**

subglobose, 5–9 mm, dull to bright red or orange-red, pubescent with short, reddish, glandular hairs. A morphologically variable species consisting of numerous mostly ill-defined varieties, with two somewhat distinct varieties occurring in the Four Corners region. These varieties intergrade extensively in leaf and fruit characters, although they are distinct at their extremes. More study of the western taxa is needed. *Rhus aromatica* is widely used by southwestern tribes for food, medicine, dyes, ceremonies, and basketry. Portions of the plants are used in ceremonies such as the Navajo Evil Way and Night Way. The Hopi use it to make ceremonial twigs. The fruit is widely used to make a lemonade-like drink as well as a dye mordant and an orange-red dye. The Navajo use the leaves as a contraceptive as well as an analgesic to soothe upset stomachs.

1. Leaves mostly simple and 3-lobed, lateral lobes much smaller than terminal, terminal lobe wider than long .... **var. *simplicifolia***
1' Leaves mostly compound and 3-pinnate, lateral leaflets not much smaller than terminal, terminal leaflet longer than wide .... **var. *trilobata***

var. *simplicifolia* (Greene) Cronquist (simple-leaved) Dry, rocky slopes and mesas, washes, and riparian zones. ARIZ: Apa, Nav; COLO: Dol, LPl, Mon; NMEX: SJn; UTAH. 1130–2200 m (3700–6900′). Flowering: Mar–May. Fruit: Jun–Jul. Widespread and common in the southwestern United States and northern Mexico, from California to New Mexico, north to Utah and Colorado.

var. *trilobata* (Nutt.) A. Gray ex S. Watson (three-lobed) Stream banks, shaded riparian zones, moist rocky slopes, and around springs. ARIZ: Apa, Nav; COLO: Arc, Dol, Hin, LPl, Mon, SMg; NMEX: McK, RAr, SJn; UTAH. 1200–2200 m (4000–7200′). Flowering: Apr–Jun. Fruit: Jun–Sep. Widespread on the Great Plains and the Rocky Mountains, south to Texas and west to California and Nevada.

Rhus glabra L. (for the smooth branches and twigs) Smooth sumac. Unbranched to weakly branched rhizomatous shrub, 1–2 m. **STEMS** red-brown or gray with age, young twigs glabrous throughout. **LEAVES** appearing with or before the flowers, pinnately compound, petiole glabrous, with (7) 10–20 leaflets, leaflets lanceolate to elliptic, 3–8 cm long, 1–3 cm wide, margins serrate, dark green above, paler below, red in fall. **INFL** terminal, dense and pyramidal, paniculate, 5–16 cm long. **FLOWERS** small, subtended by a small caducous bract, petals yellowish green, 2.5–4 mm, sepals somewhat shorter than petals, 1.5–3 mm. **FRUIT** elliptic to ovate, slightly compressed, 4–6 mm, glandular-pubescent with short reddish hairs. Rare in sunny, open, moist sites with rich soils along stream banks and in rocky canyons. COLO: Arc, LPl, Mon. 1500–1800 m (5000–6000′). Flowering: May–Jun. Fruit: July–Aug. Widespread in temperate North America, west to eastern Washington and Oregon, south to northern Mexico, rare in the Intermountain region. At least on the Colorado Plateau, smooth sumac appears to be relictual in widely scattered locations in moist, rich soil. Because of its regional rarity, little is known about indigenous uses of this species. Elsewhere in North America it was widely used for food and medicine. Unlike *R. trilobata*, the stems are brittle and are not used in making basketry. The staghorn sumac, *R. typhina* L., is often used in landscaping and can escape. This species differs from smooth sumac by its much larger size (to 5 m or more) and its densely hairy twigs.

Toxicodendron Mill. Poison Ivy, Poison Oak

(Greek *toxikos*, poisonous, and *dendron*, tree) Allergenic, strongly to weakly branched shrubs or sometimes weak clambering vines, strongly rhizomatous, aerial roots often present (lacking in ours). **STEMS** smooth, bark not flaky, red-brown, pubescent, resinous, often producing aerial roots. **LEAVES** deciduous, 3-foliate or pinnately compound, terminal leaflet petiolate, mostly near top of stems, shiny bright green, turning red in fall. **INFL** axillary, an open raceme or panicle. **FLOWERS** dioecious; 5-merous, petals glabrous, style short, 3-lobed, ovary with 3 carpels, each with 1 locule. **FRUIT** a fleshy, white to greenish white or yellow drupe, glabrous or sometimes pubescent but not glandular-pubescent. A genus of about 15 species, primarily Neotropical and Southeast Asian. Species of *Toxicodendron* are well named as they can cause severe dermatitis in susceptible people.

Toxicodendron rydbergii (Small ex Rydb.) Greene (for P. A. Rydberg) Rhizomatous shrubs to rarely weakly climbing vines. **STEMS** pubescent, weakly branched, lacking aerial roots. **LEAVES** 3-foliate (rarely 4- or 5-), leaflets shiny green, glabrous or sometimes minutely strigose below along midvein, ovate, 2–12 cm long, acute to acuminate, lateral leaflets smaller than terminal leaflet, margins mostly entire or with a few teeth. **INFL** unbranched or weakly branched raceme, mostly less than 25 flowers. **FLOWERS** small, petals 2–3 mm, yellowish green with dark veins, sepals 1–1.5 mm. **FRUIT** globose, shiny white with yellow or green tints, 4–6 mm. Moist, shaded canyons, stream banks, and around springs. ARIZ: Apa, Nav; COLO: Arc, Dol, Hin, LPl, Mon; NMEX: McK, RAr, SJn; UTAH. 1100–2600 m (3500–8500′). Flowering:

Apr–Jun. Fruit: Jul–Aug. Widespread in the northern and western United States and southern Canada, from Washington and Oregon east to New England, south in the Rocky Mountains to Arizona, New Mexico, and west Texas. Western poison ivy has at times been considered part of the eastern species *T. radicans* (L.) Kuntze, but that species produces abundant aerial roots and is much more prone to climbing and becoming vinelike than *T. rydbergii*. All *Toxicodendron* species produce severe allergic reactions in susceptible people.

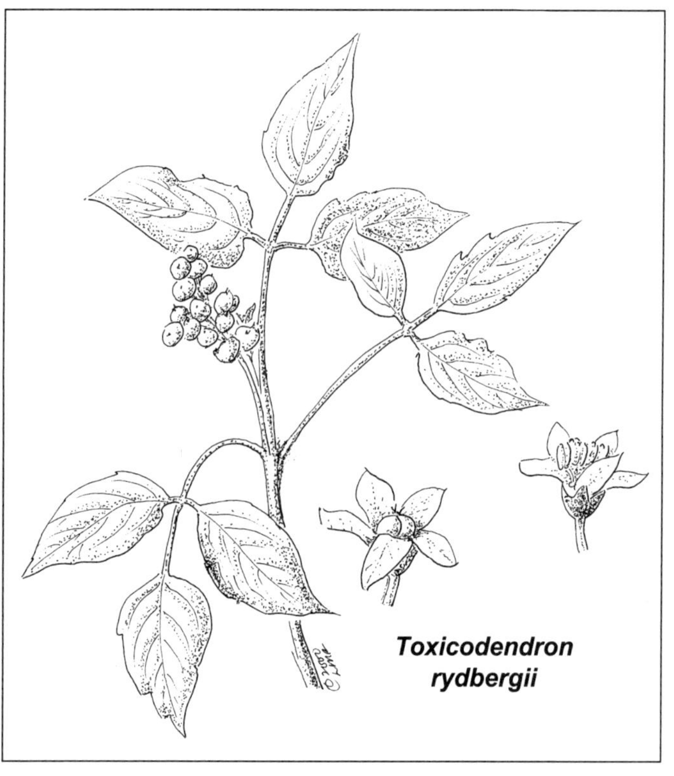
Toxicodendron rydbergii

APIACEAE Juss. (UMBELLIFERAE) PARSLEY FAMILY

Ronald L. Hartman, Sherel Goodrich, and Kenneth D. Heil

Annual, biennial, or perennial acaulescent or caulescent herbs from taproots, fibrous or tuberous roots, or rhizomes; caudex simple to much-branched. **LEAVES** simple to decompound, petioles typically sheathing at base or along petiole or the upper leaves reduced to dilated sheaths. **INFL** of compound umbels, the primary umbels with or without a subtending involucre of bracts, the secondary umbels (umbellets) with or without a subtending involucel of bractlets. **FLOWERS** mostly regular, perfect or often the outer flowers of the umbellet staminate (andromonoecious) or sterile; sepals 5 or absent; petals 5, small, usually inflexed at the tip, white, yellow, or purple; stamens 5, small, alternate with the petals; pistil 1, the ovary inferior, bicarpellate, 2-loculed, with 1 ovule per locule, the 2 styles with or without a conical base (stylopodium). **FRUIT** a schizocarp of 2 mericarps united by their faces (the commissure), terete to dorsally or laterally compressed; mericarps separating at maturity, apically attached to and pendulous from a wirelike, entire or bifid at least in part, carpophore (vascular tissue) or remaining adherent to the commissural face and not apparent, each mericarp usually 5-nerved, 3 of the nerves dorsal and 2 on the lateral margins, the nerves filiform to winged, or obscure or absent, the internerve areas commonly with 1 or more oil tubes, the commissural faces often with 2 or more oil tubes. $x = 4$–12. Nearly 300 genera, about 3000 species, widely distributed. A few genera consist of species that are very poisonous: *Cicuta* and *Conium*. The tuberous roots of *Cymopterus*, *Lomatium*, and *Orogenia* were gathered and eaten by American Indians. Other species were cultivated for spice and food, *Anethum graveolens* L., dill; *Apium graveolens* L., celery; *Carum carvi* L., caraway; *Coriandrum sativum* L., coriander; *Daucus carota* L., carrot; *Pastinaca sativa* L., parsnip; and *Petroselinum crispum* (Mill.) Fuss, parsley (Mathias and Constance 1944; Cronquist, Holmgren, and Holmgren 1997a). The western North American group of umbels, here including *Cymopterus* and *Lomatium*, appear to be extremely polyphyletic based on nuclear ribosomal and chloroplast DNA; thus considerable further work will be necessary to resolve the relationships and develop a new generic classification (Sun and Downie 2004; Sun, Downie, and Hartman 2004). Thus the traditional circumscriptions of genera are here followed, with the exception of *Aletes*.

1. Leaves simple, entire, linear to narrowly oblong, veins 3–5, parallel, prominent .... ***Bupleurum***
1' Leaves deeply cleft to compound, leaflets toothed or lobed, broadly oblong to ovate in outline, veins pinnate or palmate .... (2)

2. Axils of upper leaves bearing bulblets; umbels rarely bearing fruit; plants of wet places .... ***Cicuta***
2' Axils of leaves without bulblets; umbels bearing fruit; plants not restricted to wet places .... (3)

3. Leaves palmately cleft into 5–9 toothed segments; flowers yellow; fruits covered with stout, hooked prickles .... ***Sanicula***
3' Leaves various; fruits not covered by hooked prickles, if so, then flowers white (*Daucus*) .... (4)

4. Plants caulescent; pseudoscape absent; peduncles few to several, mostly shorter than the leafy stem; styles rarely over 1 mm long; stylopodium present; petals white ....(5)
4′ Plants acaulescent, the leaves sometimes whorled atop a pseudoscape or, if subcaulescent, the usually solitary peduncle longer than the short leafy stem, and lateral umbels (if any) typically borne on the lower 1/3 of the plant; styles often over 1 mm long; stylopodium absent or present; petals yellow, white, or purple .... (6)

5. Leaves pinnate or ternate; leaflets mostly sessile....Key 1
5′ Leaves various; leaflets usually petiolulate, at least the primary one ....Key 2

6. Leaves ternate or biternate with 3–9 leaflets or rarely a few simple, usually only 2–3 per plant; petals white ....***Orogenia***
6′ Leaves and leaflets not as above or, if so, the plants mostly taller and/or petals yellow....Key 3

Key 1

Plants caulescent; peduncles and umbels mostly shorter than the stem; stylopodium present (except *Cymopterus*); leaves pinnate or ternate; leaflets sessile

1. Leaflets entire, linear or linear-elliptic, mostly 2–5 cm long; petals yellow when fresh; fruit 6–8 mm long ....***Cymopterus***
1′ Leaflets toothed and/or lobed, not linear; petals yellow or white; fruit various ....(2)

2. Leaves ternate, the upper ones sometimes simple, the leaflets 8–36 cm long, about as wide; plants 1–2 m tall or taller, villous-woolly at least on some of the nodes; petals 4–8.5 mm long (at least some) ....***Heracleum***
2′ Leaves pinnate, the leaflets less than 8 cm long and much narrower; plants shorter or not villous-woolly; petals smaller....(3)

3. Involucre and involucels well developed, sometimes spreading or deflexed, the bractlets (2) 4–12; fruit 1.5–3 mm long, the ribs not winged; plants of very wet places; often growing in water, from fibrous roots....(4)
3′ Involucre lacking or infrequently of 1 or 2 bracts; involucels often lacking; fruit often 3 mm long or else the ribs winged; plants of various habitats, from taproot or tuberous root ....(5)

4. Stems often sprawling, sometimes stoloniferous; leaves with (3) 5–15 opposite pairs of leaflets, these 0.3–4 (6.5) cm long, margins, especially of the upper ones, usually irregularly toothed or cleft; rays 4–16; ribs of the fruit obscure ....***Berula***
4′ Stems erect, not stoloniferous; leaves with 4–6 opposite pairs of leaflets, these 2–8 (15) cm long, margins mostly evenly serrate; rays 11–24; ribs of the fruit prominently corky....***Sium***

5. Umbels often more than 7 per stem; fruit strongly flattened dorsally, 5–8 mm long, 3–6 mm wide, the lateral ribs slightly winged, the dorsal ones filiform; petals greenish yellow or reddish; plants adventive or cultivated ....***Pastinaca***
5′ Umbels fewer than 7 per stem; fruit not strongly flattened dorsally or, if so, 3–5 mm long; petals white or greenish; plants native ....(6)

6. Fruit linear to narrowly clavate, over 10 mm long; leaves 1- to 3-ternate or ternate-pinnate; peduncles mostly not subtended by dilated, bladeless sheaths, or these greatly reduced....***Osmorhiza***
6′ Fruit elliptic to broadly oblong, 3–6 mm long; leaves mostly once-pinnate; peduncles often with subtending dilated sheaths....(7)

7. Fruit strongly flattened, the dorsal ribs filiform, the lateral ribs conspicuously winged; plants with tuberous roots ....***Oxypolis***
7′ Fruit cross section rounded in outline, the dorsal and lateral ribs prominently thickened wings; plants from taproots ....***Angelica***

Key 2

Plants caulescent; peduncles and umbels mostly shorter than the stems; stylopodium present (except *Lomatium*); leaves more than once-compound; primary leaflets not sessile

1. Ultimate leaf segments, in part, over 2 cm long, toothed or lobed, but not entire or pinnatifid ....(2)
1′ Ultimate leaf segments less than 2 cm long or, if longer, entire or pinnatifid....(4)

2. Involucels of mostly 6 bractlets, 1–4 mm long; umbels 6–20 or more per stem, the rays 15–26, 1.5–4 cm long; fruit 1.5–4 mm long, the ribs corky .... ***Cicuta***
2′ Involucels mostly absent, umbels often fewer than 6 per stem and/or the rays either fewer or longer than above or both; fruit 4–25 mm long, the ribs various .... (3)

3. Fruit linear to clavate, (10) 12–25 mm long, bristly hispid (except in *Osmorhiza occidentalis*), the dorsal ribs not prominent; leaflets often hirtellous; dilated sheaths seldom subtending the peduncles .... ***Osmorhiza***
3′ Fruit oblong to elliptic, 4–5 mm long, not bristly hispid, the dorsal ribs with small wings; leaflets glabrous; peduncles often subtended by dilated bladeless or nearly bladeless sheaths .... ***Angelica***

4. Fruits and ovaries with bristly hairs; involucre often of pinnatifid or compound bracts; plants annual .... ***Daucus***
4′ Fruits and ovaries without bristly hairs; involucre mostly of entire bracts; biennial or perennial .... (5)

5. Involucel and involucre absent .... (6)
5′ Involucel and involucre present .... (8)

6. Petals yellow; plants introduced; ultimate segments of leaves filiform, 1–40 mm long, about 0.5 mm wide .... ***Foeniculum***
6′ Petals white or yellow (in *Lomatium*) and the plants native; ultimate segments various, often over 0.5 mm wide .... (7)

7. Petals white; fruit 3–8 mm long, rounded, the dorsal and lateral ribs narrowly winged; stylopodium low-conic .... ***Ligusticum***
7′ Petals yellow or white (*Lomatium nevadense*); fruit 6–20 mm long, dorsally flattened, the dorsal ribs filiform, the lateral ribs winged; stylopodium absent .... ***Lomatium***

8. Stems usually purple-spotted, usually much-branched, mostly with 10–30 or more umbels; plants 5–30 dm tall; involucre of 2–6 bracts, 2–6 (15) mm long; naturalized, weedy in disturbed mesic sites .... ***Conium***
8′ Stems not purple-spotted, few-branched, with (1) 3–7 (12) umbels; plants to 10 dm tall, involucre absent or seldom as above; native, often montane .... (9)

9. Involucels usually with more than 3 bractlets; fruit compressed dorsally; root crown mostly simple without marcescent petiole bases .... ***Conioselinum***
9′ Involucels with 2, rarely 3 bractlets; fruit terete or slightly compressed laterally; root crown simple or branched, usually with fibrous marcescent petiole bases .... ***Ligusticum***

Key 3

Plants typically acaulescent; styles often over 1 mm long; stylopodium absent (except *Podistera*)

1. Fruit strongly flattened dorsally, dorsal ribs filiform, not winged, the lateral ribs winged, body 8–18 (20) mm long or, if shorter, usually pubescent; involucre absent .... ***Lomatium***
1′ Fruit not strongly flattened or, if so, the dorsal ribs winged, at least in part, body usually less than 8 mm long, the wings sometimes to 12 (15) mm long and 2–2.5 mm tall, especially in plants with conspicuous involucres; plants glabrous to hirtellous .... (2)

2. Stylopodium conic; leaves once pinnately compound with palmatifid leaflets; bractlets of involucels, in part, with 2 or 3 or more apical teeth .... ***Podistera***
2′ Stylopodium absent; leaves either more than once compound or leaflets not palmatifid; bractlets entire .... ***Cymopterus***

Angelica L. Angelica

(*angelus*, referring to the medicinal properties of some species) Perennial, caulescent, single-stemmed herbs from a stout taproot. **LEAVES** pinnately to ternately 1–3 times compound, with broad leaflets; lower blades on elongate petioles, the middle ones often arising directly from a dilated sheath, the upper ones often much reduced or absent and the leaves reduced to a dilated sheath. **INFL** an open umbel; involucre and involucel absent or of narrow, scarious or foliaceous bracts or bractlets. **FLOWERS** with calyx teeth minute or obsolete; petals white, seldom pink or yellow; stylopodium broadly conic; carpophore divided to the base. **FRUIT** elliptic-oblong to orbicular, strongly compressed dorsally, the lateral and dorsal ribs with small but obvious wings, or the ribs all corky-thickened and scarcely winged. Fifty species in the Northern Hemisphere. The Paiute boiled the roots and applied a poultice to sores and swellings, especially venereal.

1. Ovaries and immature fruit glabrous; bractlets of involucels linear-lanceolate to lanceolate, usually over 1 mm wide; flowers usually purplish brown .... ***A. grayi***
1' Ovaries and mature fruit scabrous to hispidulous; bractlets of involucels absent, filiform, or narrowly linear, not over 1 mm wide; flowers white to pink .... ***A. pinnata***

Angelica grayi J. M. Coult. & Rose (for Asa Gray, distinguished professor of botany at Harvard) Gray's angelica. **STEMS** 20–60 cm tall, stout, mostly over 1 cm thick at the base. **LEAVES** pinnate to bipinnate or ternate-pinnate, the middle division larger; leaflets 1–5 cm long, ovate to lanceolate, sessile or nearly so, serrate to sometimes lobed; cauline leaves with conspicuously dilated sheaths. **INFL** an open umbel, involucre wanting or of foliaceous bracts; involucel of bractlets 5–18 mm long, usually over 1 mm wide, linear-lanceolate to lanceolate; rays many, spreading-ascending, the whole umbel rather flat on top; pedicels 2–6 mm long, spreading-ascending. **FLOWERS** purplish brown. **FRUIT** 4–5 mm long, glabrous even when young, oval, dorsal ribs narrowly winged, the laterals broader winged; oil tubes solitary in the intervals. Alpine scree slopes. COLO: Arc, Hin, LPl, Min, Mon, RGr, SJn. 3485–4090 m (11500–13500′). Flowering: Jul–Sep. Fruit: Aug–Sep. Wyoming south to Colorado and New Mexico.

Angelica grayi

Angelica pinnata S. Watson (the featherlike arrangement of the leaf) Small-leaved angelica. **STEMS** 4.5–10 (15) dm tall, glabrous or nearly so, except scabrous to hirtellous in the inflorescence, without persistent leaf bases, from a taproot and sometimes branched crown. **LEAVES** pinnate or partly bipinnate with 3 (4) opposite pairs of leaflets, the lowest pair sometimes bipinnate or partly bipinnate, the upper ones pinnate, lower petioles 5–26 cm long, gradually expanded into a dilated, partly sheathing base, reduced and the blades sometimes sessile, lanceolate, elliptic, or ovate, serrate, the margins with 3–7 teeth per cm. **INFL** with peduncles 3.5–14 cm long; umbels (1) 2–5; involucre absent; rays 7–14, 2–8.5 cm long, scabrous to hirtellous; involucels absent or very rarely of 1 or more green to scarious, linear or nearly linear bractlets 3–13 mm long; pedicels 3–7 mm long, glabrous or scabrous. **FLOWERS** with petals white; styles to 1 mm long; ovary glabrous to hirtellous. **FRUIT** 4–5 mm long, glabrous or sparsely hirtellous, the lateral wings about 1 mm wide, the dorsal wings about 0.5 mm wide. [*A. leporina* S. Watson]. Tall forb, oak, maple, aspen, Douglas-fir, spruce-fir, willow, and wet meadow communities, very often along streams or around seeps. ARIZ: Apa; COLO: Arc, Hin, LPl, Min, Mon; NMEX: SJn; UTAH. 2425–3635 m (8000–12000′). Flowering: Jun–Aug. Fruit: Jul–Sep. Wyoming, southwestern Montana, eastern Idaho, south to Colorado, Utah, and New Mexico.

Berula W. D. J. Koch Water Parsnip

(Latin name for an umbellifer) Perennial, acaulescent, glabrous herbs from fibrous roots, often stoloniferous. **LEAVES** pinnately compound or the submerged ones sometimes with filiform-dissected blades. **INFL** an open umbel; involucre and involucel usually well developed. **FLOWERS** with calyx teeth minute or obsolete; stylopodium conic; carpophore divided to the base, inconspicuous, adnate to the mericarps. **FRUIT** elliptic to orbicular, somewhat compressed laterally, glabrous, the ribs inconspicuous, oil tubes numerous. One species of northern temperate regions. The leaves and blossoms were used for food and medicinal purposes by the White Mountain Apache. The Zuni made an ingredient called "schumaakwe cakes," used externally for rheumatism, swelling, rashes, and athlete's foot. The roots are poisonous.

Berula erecta (Huds.) Coville (upright) Cutleaf water parsnip. **STEMS** 5–10 dm long or longer, from numerous fibrous roots. **LEAVES** pinnate with (3) 5–15 opposite pairs of lateral leaflets, or the submerged leaves (if present) often with filiform-dissected blades; petioles to 32 cm long or upper blades sessile on a dilated sheath; blades 2–31 cm long; leaflets 0.3–4 (6.5) cm long, sessile, nearly linear to lanceolate or ovate in outline, toothed to incised or occasionally a few entire. **INFL** with peduncles 1.5–8 cm long; umbels 3–20 or more; bracts of the involucre 1–6, 2–15 (25) mm long, linear or elliptic, entire, toothed, or rarely pinnatifid; rays 4–16, 0.5–2.5 (4) cm long; bractlets of the involucels 4–7, 1–7 mm long, linear or elliptic, entire; pedicels 2–7 mm long. **FLOWERS** with petals white; stamens white; styles less than 1 mm long. **FRUIT** mostly 2 mm long, the ribs obscure. $2n = 18$. In mud and water of streams, seeps, springs, marshes, swamps, margins of ponds and lakes, and in wet hanging gardens. ARIZ: Apa; NMEX: SJn. 1515–1820 m (5000–6000′). Flowering: Jun–Sep. Fruit: Jul–Oct. Widespread in Europe, Mediterranean regions, and North America.

Bupleurum L. Bupleurum, Thorow Wax, Hare's Ear

(Greek, from ox and side) Caulescent perennials from a branching caudex and taproot, the caudex sheathed with dark brown leaf bases, the dead leaves often coiled upon drying. **LEAVES** simple, the basal ones petiolate, the blades entire, parallel-veined, the cauline ones sessile and clasping. **INFL** an open umbel; involucre of conspicuous foliose bracts; involucel of foliose, often connate bractlets; rays few to several; pedicels short. **FLOWERS** yellow, greenish, or purplish; calyx teeth absent; styles short, the stylopodium low-conic. **FRUIT** oblong to orbicular, somewhat laterally flattened, glabrous or roughened; ribs filiform. Seventy species from Eurasia, North Africa, Canary Islands, and Arctic North America.

Bupleurum triradiatum Adams ex Hoffm. (bearing three rays) Thorough-wort, American thorow-wax. **STEMS** few to several, mostly simple, 1.5–5 (7) dm tall. **LEAVES** simple, basal and cauline, 2–25 cm long, narrowly oblong to linear, with 3–5 prominent, parallel veins. **INFL** with rays 1–8 (14), 0.5–5 cm long; involucre of 1–several unequal, lanceolate to ovate bracts; involucel of 5–8 ovate to lanceolate bractlets. **FLOWERS** with petals yellow, greenish, or purple; ovaries glabrous. **FRUIT** glabrous, 3–4 mm long, the ribs raised but wingless. [*Bupleurum americanum* J. M. Coult. & Rose]. Arctic willow community on rocky granite slopes. COLO: LPl. 3875 m (12795′). Flowering: Jul. Fruit: Aug. Alaska and western Yukon; south to Montana, Idaho, Wyoming, and Colorado. Known from a single collection made in 1978 near Trimble Pass in the upper Vallecito Basin. Our material belongs to **subsp. *arcticum*** (Regel) Hultén (the name of a plant taken from *arctos*, "bear").

Cicuta L. Water Hemlock

(hemlock) Perennial, caulescent, glabrous, violently poisonous herbs, from clusters of fibrous roots, some of these commonly tuberous-thickened. **STEMS** thickened at the base with hollow chambers separated by transverse septa; internodes of stems hollow. **LEAVES** 1–3 times pinnate or ternate-pinnate, with well-developed leaflets. **INFL** of several open umbels; involucre wanting or of a few inconspicuous narrow bracts; involucel of several narrow bractlets or rarely absent. **FLOWERS** with petals white or greenish; calyx teeth evident; stylopodium depressed or low-conic; carpophore divided to the base, deciduous. **FRUIT** ovate or orbicular, compressed laterally, the ribs usually prominent and corky. Eight species of northern temperate regions. Considered to be one of the most toxic plants in North America to people, horses, cattle, sheep, and other animals (Mulligan 1980).

1. Axils of upper leaves bearing bulblets; leaflets with narrowly linear segments, most less than 5 mm wide; umbels frequently absent, if present, then rarely bearing fruit .... ***C. bulbifera***

1′ Axils of leaves without bulblets; leaflets with segments usually wider than 5 mm; umbels bearing fruit .... ***C. maculata***

Cicuta bulbifera L. (bearing bulbs) Bulbous water hemlock. **STEMS** 3–10 dm tall, usually with slender, fibrous roots, sometimes becoming thickened; stems 2–8 mm in diameter. **LEAVES** with lower and middle ones more or less dissected into few to many narrow, linear segments, entire to obscurely to prominently few-toothed, segments mostly 0.5–1.5 mm wide and 0.5–4 mm long, the upper reduced, often simple and filiform with dilated bases with fewer segments, or undivided, many of them bearing 1 to several axillary bulblets, these vegetative propagules 2–4.5 mm long, ovoid. **INFL** with peduncles 1–5 cm long; umbel frequently absent, if present, frequently 1, often not maturing fruit, involucre of 0–5 lanceolate ovate bractlets, 1.5–2 mm long, the rays mostly 1–2 cm long, involucel with 0–5 bractlets, ovate to lanceolate; pedicels 2–4 mm long. **FLOWERS** with calyx teeth 0.1–0.4 mm long, pale green to somewhat scarious; petals white; stamens white; styles 0.5–0.6 mm long. **FRUIT** 1.5–2 mm long, orbicular, the ribs low, more or less corky, tan, wider than the green interval. $2n = 11$. Wet shores, marshes, streamsides, and wet meadows. Only known

Cicuta bulbifera

from Scout and Bryce lakes. COLO: SJn. 2660–2860 m (8730–9380′). Flowering: Jul–Aug. Fruit: Aug–Sep. Canada south to Oregon, Montana, Wyoming, and Nebraska.

Cicuta maculata L. (spotted or blotched) Spotted water hemlock, spotted cowbane, beaver-poison. **STEMS** 6–21 dm tall or taller, with clusters of fibrous roots surmounted by a thickened crown; stems 5–15 mm or more in diameter. **LEAVES** pinnate or ternate-pinnate with 4–7 opposite pairs of lateral primary leaflets, the lower ones again pinnate, the upper once pinnate and sessile, the lower petioles 5–40 cm long, the upper ones reduced and the blades often sessile on dilated sheaths, the lowest pair of petiolules 1–3 cm long, leaflets 2–11 cm long, 3–25 mm wide, lanceolate to narrowly so, or linear, finely to coarsely serrate. **INFL** with peduncles (2) 4–15 cm long; umbels 6–30 or more; involucre absent or of 1 or few linear bracts to 1 cm long; rays 15–26, 1.5–4 cm long; bractlets of the involucels about 6, 1–4 mm long, linear or narrowly deltoid, pale yellow-green or purplish, scarious-margined; pedicels 3–10 mm long. **FLOWERS** with calyx teeth about 0.5 mm long, often pale green with whitish margins; petals white; stamens white; styles 0.5–1 mm long. **FRUIT** 2–4 mm long, oval to globose, the ribs prominent, more or less corky, green, often wider than the darker (often purple) intervals. [*C. douglasii* (DC.) J. M. Coult. & Rose, misapplied]. $2n = 11, 22$. Streamsides, rivers, ditches, canals, margins of ponds and lakes, in wet meadows and marshes. ARIZ: Apa; COLO: Arc, LPl, Mon; NMEX: McK, RAr, SJn. 1515–2270 m (5000–7500′). Flowering: Jun–Sep. Fruit: Jul–Oct. Widespread in North America. Our specimens belong to **var.** ***angustifolia*** Hook. (having narrow foliage). This plant is extremely poisonous to both humans and to livestock, due to a yellow-orange resinol, cicutoxin, concentrated in the chambered root crown and less concentrated elsewhere in the plant. Deaths of humans from eating even small portions of the plant have been reported in the western United States in recent years. Losses of cattle due to consumption of this plant are relatively common.

Conioselinum Hoffm. Hemlockparsley

(resemblance to *Conium* and *Selinum*) Perennial, more or less caulescent herbs from a taproot or cluster of fleshy-fibrous roots, sometimes with a caudex. **LEAVES** pinnately or ternate-pinnately decompound. **INFL** open umbel; involucre absent or of a few narrow or leafy bracts; involucels of well-developed, narrow, often scarious bractlets. **FLOWERS** with calyx teeth obsolete; petals white; stylopodium conic; carpophore divided to the base or nearly so. **FRUIT** elliptic or elliptic-oblong, slightly dorsally compressed, glabrous, the lateral ribs evidently thin-winged, the dorsal ribs less so and corky. About 12 species of North America and Eurasia. Navajo uses include blood purifier, smoked for catarrh, snakebite remedy, and snake repellent. The leaves are cooked with meat and used for food.

Conioselinum scopulorum (A. Gray) J. M. Coult. & Rose (rock-loving) Rock lovage, Rocky Mountain hemlockparsley. Plants perennial, 3–10 dm tall, glabrous except in the inflorescence, from a fusiform taproot with simple or very sparingly branched crown, without persistent leaf bases or these few and weakly persisting. **LEAVES** pinnate or ternate-pinnate with (3) 4–5 opposite pairs of lateral primary leaflets, the lower ones 2–3 times pinnate and petiolulate, the upper pinnate, pinnatifid, and sessile or nearly so; petioles 3–23 cm long; blades 3.5–19 cm long, ovate in outline, the lowest pair of primary leaflets 1/2–2/3 as long as the leaf blade, on petiolules (0.5) 1–3.5 cm long, the ultimate segments 2–15 mm long, 1–5 mm wide. **INFL** with peduncles 3–21 cm long, often subtended by a dilated sheath, this usually with a reduced sessile blade; umbels 1–3; involucre absent or of 1 or few linear bracts to 1 cm long; rays 9–15, 1.5–5 cm long; involucels of 3–6 linear or linear-filiform bractlets 2–8 mm long; pedicels 4–12 mm long. **FLOWERS** with petals white; stamens white; styles to 1.3 mm long. **FRUIT** 4–6 mm long, with lateral ribs narrowly corky-winged, the dorsal ones not winged. [*Ligusticum scopulorum* A. Gray]. $2n = 22$. Streamsides and meadows in the mountains. ARIZ: Apa; COLO: Arc, Hin, Min, Mon, SJn; NMEX: RAr, SJn; UTAH. 2250–3200 m (7500–10500′). Flowering: Jul–Aug. Fruit: Aug–Sep. Oregon and Wyoming south to Utah, Arizona, and New Mexico. Plants of *C. scopulorum* are often confused with *Ligusticum porteri*. The two taxa differ in the following, often subtle ways, with features of *L. porteri* in parentheses;

fruit dorsally flattened (nearly terete); bractlets often 3 or more (0–2, rarely more); terminal umbel solitary or subtended by alternate lateral umbels (often subtended by opposite or whorled umbels); and plants from a taproot, with a mostly simple crown and with few if any persisting fibrous leaf bases (the crown simple or branched and usually with numerous persistent fibrous leaf bases). In addition, the rays average shorter and the ultimate segments of the leaves are less conspicuously veined than those of *L. porteri*.

Conium L. Poison Hemlock

(*koneion*, the ancient Greek name of *Conium maculatum*) Biennial, caulescent, glabrous herbs from stout taproots with purple-spotted, freely branching hollow stems. **LEAVES** pinnately or ternate-pinnately dissected. **INFL** an open umbel, several to numerous; involucre and involucels of small, lanceolate to ovate bracts or bractlets. **FLOWER** calyx teeth obsolete; petals white; stylopodium depressed-conic; carpophore entire. **FRUIT** broadly ovoid, somewhat laterally compressed, with prominent, raised, often wavy, slightly winged ribs. Two species, one Eurasian, the other South African. All plant parts are very poisonous.

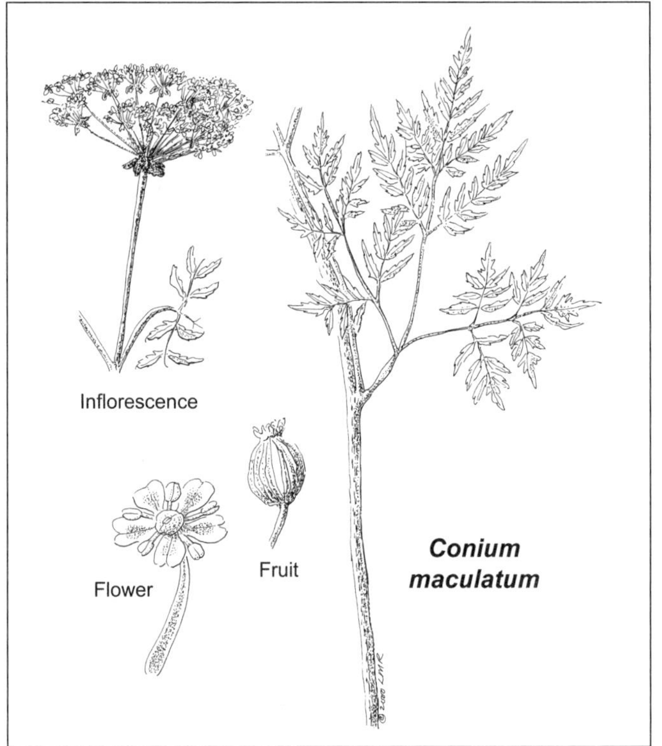

Conium maculatum L. (spotted or speckled) Poison hemlock. **STEMS** 5–30 dm tall, glabrous. **LEAVES** pinnate or ternate-pinnately decompound with 6–9 opposite pairs of lateral primary leaflets, the lower ones usually twice or more pinnate and then pinnatifid, petiolulate, the upper once pinnate, pinnatifid, and sessile; petioles of larger leaves 4–18 cm long; larger leaf blades to 30 cm long, reduced upward and sessile on dilated sheaths, ovate in outline; lowest pair of primary leaflets less than 1/2–2/3 as long as the leaf blade, on petiolules 1–5.5 mm long or shorter upward; ultimate leaflets pinnatifid, the lobes entire or toothed, the widest confluent portions 2–5 (10) mm wide. **INFL** with peduncles 2–7.5 cm long; umbels many; involucral bracts 2–6, 2–6 (15) mm long, entire and ovate or deltoid, caudate to cuspidate, green with scarious margins, or rarely pinnatifid; rays 9–16, 1–4 cm long; bractlets of the involucels 4–6, 1–3 mm long, shaped like the involucral bracts; pedicels 2–6 mm long. **FLOWERS** with petals white; stamens white; styles about 0.5 mm long. **FRUIT** 2–2.5 mm long, the ribs prominently ridged, narrower than the intervals. $2n = 22$. Along ditches, streams, rivers, roadsides, and fence lines, in wet and boggy meadows and moist waste places. COLO: Arc, LPl. 2090 m (6900′). Flowering: Jun–Sep. Fruit: Jul–Oct. Introduced from Eurasia, now widespread in North America. This plant is deadly poisonous due to alkaloids that can cause paralysis of respiratory muscles. It is the classical "poison hemlock" of antiquity.

Cymopterus Raf. Spring Parsley, Indian Turnip

(referring to the winged fruits) Perennial, acaulescent or subcaulescent, glabrous or scabrous herbs from slender to greatly enlarged and tuberlike taproots to branching woody caudices. **LEAVES** all basal (these sometimes elevated on an aerial pseudoscape) or basal and 1 to few cauline mostly on the lower 1/2 of the stems, ternate to pinnate or ternate-pinnately compound, rarely simple and ternately cleft. **INFL** an open to congested umbel, solitary to several, open or reduced to globose heads; involucel of separate or united bractlets; rays few to several, spreading to reflexed; involucral bracts spreading to reflexed or absent; pedicel obsolete or developed. **FLOWERS** with calyx teeth obsolete to conspicuous; petals white, yellow, or purple; stylopodium absent; carpophore absent, inconspicuous and adhering to the inner faces (commissure) of the mericarps, or present, persistent on the pedicel, and divided to the base. **FRUIT** ovoid to oblong, terete to somewhat flattened dorsally, the lateral and usually 1 or more of the dorsal ribs with corky-thickened to papery wings, usually prominent. About 45 species from western and central North America. [*Aletes* J. M. Coult. & Rose in part; *Oreoxis* Raf.; *Phellopterus* (Nutt. ex Torr. & A. Gray) J. M. Coult. & Rose; *Pseudocymopterus* J. M. Coult. & Rose; *Pteryxia* (Nutt. ex Torr. & A. Gray) J. M. Coult. & Rose].

1. Bractlets of the involucel large, showy, scarious, white to purple, and more or less connate below, tending to form a cup around the umbellet; bracts of the involucre similar to the bractlets of the involucel, or smaller and less conspicuous (*Phellopterus*) ....(2)
1′ Bractlets of the involucel (and bracts of the involucre, if present) relatively small and inconspicuous, or larger but herbaceous or coriaceous, the involucel generally asymmetrical ....(4)

2. Umbels in fruit tightly globose, the rays 1–4 (8) mm long, the pedicels 1–4 mm long; carpophores absent ....***C. purpurascens***
2′ Umbels in fruit relatively open, more or less flat-topped, the rays 10–50 mm long, the pedicels 5–12 mm long; carpophores well developed, bipartite, persistent ....(3)

3. Bractlets connate to 1/3 of length, the free portion gradually expanding distally, obovate to spatulate, with mostly 3 veins arising from base, ± parallel below, gradually flaring distally, equal or nearly so ....***C. constancei***
3′ Bractlets connate for 1/3 to 2/3 or more of length, the free portion usually abruptly enlarged distally, broadly ovate to orbicular, with mostly 1 vein, occasionally with 1 or 2 pairs of shorter lateral veins, parallel to divergent or branched ....***C. bulbosus***

4. Plants low, mat-forming perennial herbs with a taproot and often rhizomatous, caudex branching and basal leaves only; herbage glabrous to hirtellous-scabrous, not glandular; alpine and subalpine (*Oreoxis*) ....(5)
4′ Plants of various habit, sometimes densely tufted but not mat-forming ....(6)

5. Bractlets obovate, toothed at the apex, usually purplish ....***C. bakeri***
5′ Bractlets linear or narrowly elliptic, entire, acute to acuminate ....***C. alpinus***

6. Stems evidently scabrous-hirtellous just below the umbel, sometimes also just below the nodes, otherwise glabrous or nearly so; plants more or less leafy-stemmed when well developed, but acaulescent when young (*Pseudocymopterus*) ....***C. lemmonii***
6′ Stems or scapes glabrous or sometimes short-hairy or scabrous, but not with the pattern of pubescence described above; leaves chiefly or all basal or low-cauline, or on pseudoscapes ....(7)

7. Leaves merely pinnatifid to subbipinnatifid (*Aletes*) ....(8)
7′ Leaves mostly 2–3+ times pinnately or ternately dissected ....(9)

8. Fruit rather narrowly winged, the dorsal wings narrower than the lateral ones and papery; flowering stem usually branched ....***C. beckii***
8′ Fruit with thick, corky wings, mostly uniformly developed; flowering stem not branched ....***C. sessiliflorus***

9. Plants from a branched, more or less woody caudex, mostly clothed at the base with marcescent leaf bases, often of rocky places; sepals 0.5–2 mm long (*Pteryxia*) ....(10)
9′ Plants from fibrous taproots with simple or sparingly branched crowns, without or with few marcescent leaf bases, not specifically of rocky places; sepals 0.1–0.4 mm long ....(11)

10. Lowest pair of primary leaflets (1/4) 1/2–3/4 or more the length of the leaf blade, mostly 3–9 cm long, several times longer than the upper pairs, on petiolules 2–4 cm long; plants mostly of lower elevations ....***C. terebinthinus***
10′ Lowest pair of primary leaflets 1/4 or less the length of the leaf blade, to 2.7 cm long, often not more than twice as long as some of the upper pairs, sessile or on petiolules to 1 cm long; plants of mostly 2500 m (9500′) or more ....***C. longilobus***

11. Involucels rarely wholly green, not foliose, often scarious-margined and/or the bractlets linear or narrowly elliptic and not over 1.5 mm wide; plants not viscid and not with adhering sand grains ....***C. purpureus***
11′ Involucels green and foliose, seldom scarious-margined, the bractlets 1.5–4 mm wide; plants obscurely viscid and with adhering grains of sand (*Cymopterus*) ....(11)

12. Leaves once ternate, the 3 leaflets ternately lobed or cleft, the blades with confluent portions 5–35 mm wide; outer rays 1–3.3 cm long; bractlets of involucel entire or rarely tridentate; pseudoscape absent ....***C. newberryi***
12′ Leaves 2–3 times pinnate with 2 (3) opposite pairs of lateral primary leaflets, some rarely ternate, the blades with confluent portions 1–7 (12) mm wide, rays to 1.3 cm long; bractlets of the involucel often with 2–3 teeth; pseudoscape often present; plants widespread ....***C. glomeratus***

Cymopterus alpinus A. Gray (growing in the alpine) Alpine oreoxis. Plants acaulescent, mat-forming to tufted, weakly or not aromatic; taproot 0.2–1.5 cm in diameter near summit (deep-seated, lower portion absent from specimens). **STEM** caudex branched, clothed with persistent leaf bases, pseudoscape absent, rhizomes usually present, often elongate. **LEAVES** herbaceous, sparsely to densely puberulent with horizontal peglike hairs or margins scabrous, not viscid, green; petioles 0.3–5 cm long; blades lanceolate to narrowly ovate in outline, 1–4 cm long, 0.3–2 cm wide, mostly pinnate-pinnatifid to bipinnate-pinnatifid below, with 2–6 opposite pairs of lateral leaflets, leaflets sessile with distinct midribs or not, ultimate leaf segments 0.3–9 mm long, mostly 0.5–1.2 mm wide, elliptic and entire, often overlapping, terminal leaflet variously pinnatifid into oblong to elliptic segments, apices apiculate. **INFL** of 1–15+ open to congested umbels, in fruit subglobose, 0.8–1.5 cm wide; peduncle 1–15 cm tall, at least moderately puberulent like the petioles; involucre absent; rays 4–11, 0.5–4 mm long, lengthening little in fruit; involucel of 5–9 bractlets, linear to ovate, entire, 0.5–6 mm long, free to fused in lower 10%–20%, green to purple, margin narrowly scarious; pedicels 0.3–1 mm long, lengthening little in fruit. **FLOWERS** with calyx teeth 0.2–1 mm long, lanceolate to ovate; petals yellow; styles 1.7–3 mm long. **FRUIT** 4–6 mm long, broadly elliptic, tan to purple, wings 3–5, 0.5–1 mm tall, straight, smooth, corky, oil tubes 1 per interval, 2–4 on commissure; carpophore occasionally present. [*Oreoxis alpina* (A. Gray) J. M. Coult. & Rose]. $2n$ = 60. Forb-grass, spruce, and alpine communities, and raw escarpments and barren ridge communities. COLO: Arc, Con, Hin, LPl, Min, Mon, RGr, SJn; UTAH. 3030–4090 m (10000–13500′). Flowering: Jun–Jul. Fruit: Jul–Aug. Wyoming to New Mexico. Erroneously reported from Arizona (based on *C. breviradiatus*).

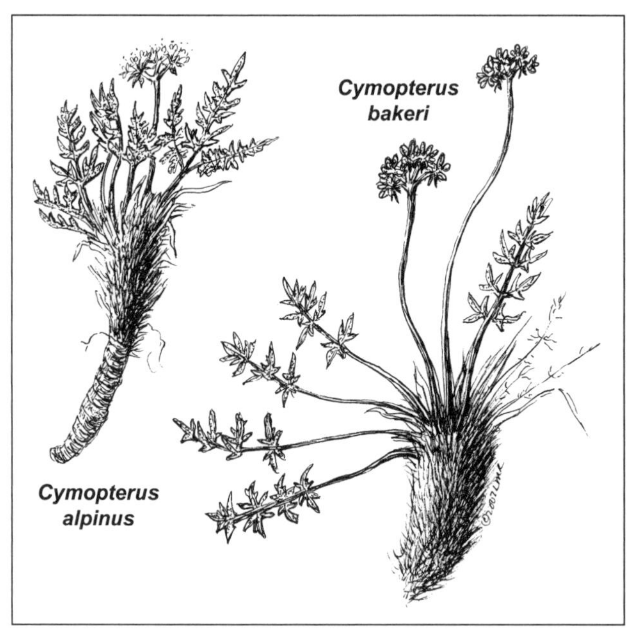

Cymopterus bakeri (J. M. Coult. & Rose) M. E. Jones (for Charles F. Baker) Baker's oreoxis. Plants acaulescent, caespitose to tufted, weakly or not aromatic; taproot 0.2–1 cm in diameter near summit (deep-seated, lower portion absent from specimens). **STEM** caudex branched, clothed with persistent leaf bases, pseudoscape absent, rhizomes absent. **LEAVES** herbaceous; glabrous, not viscid, green to purplish; petioles 1–6 cm long; blades lanceolate, occasionally narrowly ovate in outline, 1–5 cm long, 0.5–3 cm wide, mostly pinnate-pinnatifid to bipinnate below, with 3–7 opposite pairs of lateral leaflets or some alternate, leaflets sessile, usually with distinct midribs, ultimate leaf segments 2–8 mm long, mostly 0.5–1.2 mm wide, linear to lanceolate or oblanceolate and entire, often overlapping, terminal leaflet variously pinnatifid into lanceolate to linear segments, apices usually apiculate. **INFL** of 1–4 umbels, in fruit dense, subglobose, 1–2 cm wide; peduncle 2–14 cm tall, glabrous except slightly to densely puberulent near apex with horizontal peglike hairs; involucre absent; rays 5–14, 2–8 mm long, lengthening little in fruit; involucel of 4–7 bractlets, lanceolate and entire to at least in part obovate and pinnately 3–5 (6)-lobed, 3.5–5 mm long, essentially free to fused in lower 20%, green to purple, margin thinly scarious; pedicels 0.3–1 mm long, to 3 mm in fruit. **FLOWERS** with calyx teeth 0.2–0.4 mm long, narrowly triangular to lanceolate; petals yellow; styles 1–1.3 mm long. **FRUIT** 3.5–5 mm long, broadly elliptic, tan to purple, wings usually 5, 0.2–1 mm tall, straight, smooth, corky, oil tubes 2–3 per interval, 4–6 on commissure; carpophore often present. [*Oreoxis bakeri* J. M. Coult. & Rose]. Alpine communities, and raw escarpments and barren ridge communities. COLO: Arc, Con, Hin, LPl, Min, Mon, RGr, SJn. 3030–4090 m (10000–13500′). Flowering: Jun–Jul. Fruit: Jul–Aug. La Sal Mountains, Utah; Colorado, and New Mexico.

Cymopterus beckii S. L. Welsh & Goodrich (for D. E. Beck, Dept. of Zoology, Brigham Young University) Beck's aletes. Plants caulescent, tufted, weakly or not aromatic; taproot 0.2–0.5 cm in diameter near summit (deep-seated, at least lower portion absent from specimens). **STEM** caudex simple or branched, clothed with persistent leaf bases, pseudoscape absent, rhizomes absent, aerial stems mostly 10–40 cm tall, glabrous. **LEAVES** herbaceous, glabrous, not viscid, green; petioles 3–15+ cm long; blades lanceolate to narrowly ovate in outline, basal and lower pair of cauline leaves 5–10 cm long, 1.5–7 cm wide, pinnate to bipinnate, with 2–3 opposite pairs of lateral leaflets sessile with distinct midribs, ultimate leaf segments 0.5–4 cm long, 0.5–10 mm wide, linear, linear-elliptic to ovate or oblong, seldom overlapping, terminal leaflets usually tripartite into linear segments, apices apiculate; upper 1 (2) pair of cauline leaves similar but usually greatly reduced. **INFL** of 1–10+ flowering stems, commonly each with a terminal and 1 (2) axillary umbel, in fruit open, convex to globose, 1.5–6 cm wide; peduncles 6–18 cm long, glabrous; involucre absent; rays 5–10, 5–10 mm long, to 25 mm long in fruit; involucel of 4–6 bractlets, mostly linear, entire, (1.8) 4–6 mm long, distinct, mostly green, margin thinly scarious; pedicels 1–3 mm long, to 5 mm long in fruit. **FLOWERS** with calyx teeth 0.2–0.7 mm long, triangular to lanceolate; petals yellow; styles 1–1.5 mm long. **FRUIT** 5–8 mm long, narrowly oblong, tan, wings usually 5, 0.2–0.7 mm tall, straight, smooth, membranous, dorsal ones less well developed, oil tubes 1 per interval, 2 on commissure; carpophore present. [*Aletes macdougalii* J. M. Coult. & Rose; *Cymopterus macdougalii* (J. M. Coult. & Rose) Tidestr.; *Oreoxis macdougalii* J. M. Coult. & Rose]. Ponderosa pine and piñon-juniper slopes, sandy stream banks, and sandstone crevices and cliffs. ARIZ: Nav; UTAH. 1750–2600 m (5700–8650′). Flowering: May–Jun. Fruit: Jun–Jul. *Cymopterus beckii* shares many morphological characters with *C. macdougalii* and the two may prove conspecific. This similarity is reinforced thanks to the number of collections of each that have been amassed in recent years. DNA sequence data (Sun and Downie 2004; Sun et al. 2004) indicate that *C. beckii*, *C. macdougalii*, *C.* (*Pteryxia*) *davidsonii* (J. M. Coult. & Rose) R. L. Hartm., and *C. lemmonii* (J. M. Coult. & Rose) Dorn (*Pseudocymopterus montanus* J. M. Coult. & Rose) form a monophyletic group. The rarely collected *C. davidsonii* (southern Arizona and adjacent New Mexico) strongly resembles *C. macdougalii* morphologically.

Cymopterus bulbosus A. Nelson (referring to the bulblike root) Onion spring parsley, Indian turnip, bulbous spring parsley. Plants acaulescent, tufted, weakly or not aromatic; taproot 8–20+ cm long, 0.8–4 cm in diameter, enlarged variously, especially toward base. **STEM** pseudoscapes usually 1 or 2, often conspicuous, each arising 1–7 cm belowground (1–10 cm long) among remnants of old leaf sheaths, 1–3 bladeless scarious sheaths, and often 1 or 2 leaves. **LEAVES** somewhat fleshy, thus often minutely wrinkled on drying, glabrous or margins rarely scaberulous, not viscid, glaucous; petioles 1–8 cm long; blades lanceolate to broadly ovate in outline, 2–8 cm long, 1.5–5 cm wide, pinnate-pinnatifid to bipinnate-pinnatifid below, with 3–6 usually opposite pairs of lateral leaflets, leaflets sessile to petiolulate with distinct midribs, ultimate leaf segments 0.3–5 mm long, mostly 0.1–2.5 mm wide, oblong to elliptic, often overlapping, terminal leaflet variously pinnatifid to pinnate-pinnatifid into oblong to elliptic segments, apices rounded to apiculate. **INFL** of 1–8+ umbels, in fruit loose to somewhat dense, rounded, 1–5 cm wide; peduncles 3–15 cm tall, glabrous; involucre of 6–8 bracts 3–10 mm long, bracts ovate to broadly so, often fused into a cup, white, scarious with 1 green to brown vein; rays 5–9, 2–10 mm long, to 35 mm long in fruit; involucel of 6–8 bractlets, ovate to orbicular, usually rounded and notched, 4–6 mm long, fused in lower 30%–70%, white, scarious with usually 1 green vein arising from base, or with 1 or 2 pair of shorter lateral veins, parallel to divergent or branched; pedicels 1–3 mm long, to 10 mm long in fruit. **FLOWERS** with calyx teeth 0.2–0.6 mm long or obsolete, lanceolate to ovate; petals white or cream to purple; styles 1.5–2 mm long. **FRUIT** 6–11 mm long, broadly elliptic to oblong, tan to purplish, wings usually 5, 2–4 mm tall, usually straight, smooth, membranous, oil tubes 3–4 per interval, 4–7 on commissure; carpophore present. [*C. utahensis* var. *eastwoodiae* M. E. Jones]. Desert shrub and juniper communities. ARIZ: Apa, Nav; COLO: Arc, LPl, Mon; NMEX: McK, RAr, San, SJn; UTAH. 1220–2005 m (4025–6615′). Flowering: Apr–May. Fruit: May–Jun. Utah, Colorado, Arizona, and New Mexico south to northern Mexico. The Acoma, Cochiti, and Laguna eat spring parsley like celery; the Navajo would eat the root raw or roast it in ashes.

Cymopterus constancei R. L. Hartm. **CP** (for Lincoln Constance, University of California, Berkeley) Constance's spring parsley. Plants acaulescent, tufted, weakly or not aromatic; taproot 3–11+ cm long, 0.4–2 cm in diameter, enlarged toward base. **STEM** pseudoscapes usually 1–3, conspicuous, each arising 2–10 cm belowground (3–18 cm long) among remnants of old leaf sheaths, 1–3 bladeless scarious sheaths, and 1 or 2 leaves; rhizomes absent. **LEAVES** somewhat fleshy, thus often minutely wrinkled on drying, usually glabrous or margins sometimes scaberulous, not viscid, often glaucous; petioles 5–10 cm long; blades lanceolate to broadly ovate in outline, 2.5–8 cm long, 1.5–3 cm wide, bipinnate-pinnatifid to tripinnate below, with 3–5 usually opposite pairs of lateral leaflets, leaflets sessile to petiolulate with distinct midribs, ultimate leaf segments 0.2–2.5 mm long, mostly 0.5–1 mm wide, oblong to elliptic, often overlapping, terminal leaflet variously pinnatifid to bipinnatifid into oblong to elliptic segments, apices round to obtuse or apiculate. **INFL** of 1–8+ umbels, in fruit loose, convex to rounded, 2–5 cm wide; peduncle 1–12 cm tall, glabrous; involucre of 1–8 bracts 4–10 mm long, bracts oblong to obovate, often variously lobed, white, scarious with 1–4 purple veins; rays 3–6, 3–5 mm long, to 30 mm long in fruit; involucel of 4–6 bractlets, obovate to spatulate, broadly rounded to truncate, entire or irregularly toothed or lobed, sometimes cleft, 4.5–7 mm long, fused in lower 20%–30%, white, scarious with usually 3 dark green to purple veins arising from base, parallel below, gradually flaring distally, equal or nearly so or lateral pair somewhat shorter; pedicels 1–3 mm long, to 12 mm long in fruit. **FLOWERS** with calyx teeth 0.2–0.7 mm long or obsolete, triangular to ovate; petals white or cream to purple; styles 2–3 mm long.

FRUIT 7–14 mm long, broadly elliptic to suborbicular, tan to purplish, wings 4–5, mostly 3–4 mm tall, usually straight, smooth, membranous, oil tubes 2–3 per interval, 4–7 on commissure; carpophore present. Grasslands, sagebrush, piñon-juniper, and ponderosa pine communities. ARIZ: Apa, Nav; COLO: Arc, LPl, Mon; NMEX: McK, RAr, SJn. 1515–2180 m (5000–7200′). Flowering: Apr–May. Fruit: May–Jun. Southwestern Wyoming south to eastern Utah, western Colorado, and central and northwestern New Mexico.

Cymopterus glomeratus (Nutt.) DC. (stemless or apparently so) Plains spring parsley. **STEMS** 7–18 (27) cm tall from a simple or rarely branched, deep-seated, nearly linear or slightly to much enlarged fibrous taproot; herbage often more or less viscid and dotted with sand grains; pseudoscapes 1–2 (3) per plant, 0.5–5.5 cm long, often partly or wholly subterranean. **LEAVES** basal, or more often whorled and subtending the peduncles atop the pseudoscape, occasionally 1 or 2 on a stem, 2–3 times pinnate, with 1–3 opposite pairs of lateral primary leaflets; petioles 2–8 (11) cm long, blades up to 7 cm long, the confluent portions 1–7 (12) mm wide, oblong, ovate, to nearly linear in outline; primary leaflets 5–35 mm long, gradually reduced upward, pinnate to bipinnatifid with few to several rounded to narrow lobes, the ultimate teeth or lobes to 10 (16) mm long and 2 mm wide. **INFL** peduncles 1–14, equaling or exceeding the leaves, to 14 (19) cm long; involucres absent; rays 6–9, 1–13 mm long; bractlets of the involucel 3–8 (11) mm long, about 1.5–4 mm wide, more or less united at the base, entire or with 2–3 teeth or lobes, green or purple in age, of the texture of the leaves; pedicels to 2 mm long. **FLOWERS** with calyx teeth about 0.2 mm long, greenish; petals yellow when fresh, sooner or later fading to white or cream when dried; styles 2.5 mm long; carpophore absent. **FRUIT** straight or slightly wavy, mostly entire or obscurely erose, to 10 mm long, slightly corky. [*Selinum acaule* Pursh; *Cymopterus acaulis* (Pursh) Raf.; *Cymopterus fendleri* A. Gray; *C. decipiens* M. E. Jones]. $2n = 22$. Desert shrub, blackbrush, sagebrush, and piñon-juniper communities, often on sandy soil. ARIZ: Apa, Coc, Nav; COLO: Arc, Dol, LPl, Mon; NMEX: McK, RAr, San, SJn; UTAH. 1360–2270 m (4500–7500′). Flowering: Apr–May. Fruit: May–Jun. Utah, Colorado, Arizona, and New Mexico south to northern Mexico. Our material belongs to **var. *fendleri*** (A. Gray) R. L. Hartm. (for Augustus Fendler, 1813–1883, German-American botanical collector). *Cymopterus fendleri* has long been treated as distinct from the related *C. glomeratus* (*C. acaulis*) or as a variety of it (Cronquist in Cronquist, Holmgren, and Holmgren 1997a; Goodrich in Welsh et al. 2003). A recent study (Sun, Levin, and Downie 2005) using principal component analyses failed to discriminate among the five varieties recognized by Goodrich in Welsh et al. (2003). Despite this lack of resolution, var. *fendleri* is here recognized. Navajo uses include: dried leaves for food and as a seasoning for cornmeal mush and boiled meat.

Cymopterus lemmonii (J. M. Coult. & Rose) Dorn (for John Gill Lemmon, California botanist) Lemmon's spring parsley. Plants acaulescent to caulescent, tufted, weakly or not aromatic; taproot 7–12+ cm long, 0.4–1.7 cm in diameter, (lower portion absent from specimens), cylindrical to variously enlarged. **STEM** caudex simple or branched, clothed with persistent leaf bases, pseudoscape absent, rhizomes absent, aerial stems 2–50 cm tall, glabrous. **LEAVES** basal and sometimes 1 or 2 cauline ones on the lower 1/3 of the stem or 1 above the middle, herbaceous, glabrous, or margins scaberulous, not viscid, green; petioles 1–13 cm long; blades narrowly lanceolate to broadly ovate or triangular in outline (extremely polymorphic vegetatively and otherwise), 2–15 cm long, 1.5–12 cm wide, pinnate-pinnatifid to tripinnate, in part, with 2–6 mostly opposite pairs of lateral leaflets, leaflets sessile to petiolulate with distinct midribs, ultimate leaf segments mostly 2–40 mm long, 0.5–4 mm wide, linear to lanceolate, seldom overlapping, terminal leaflets variously pinnatifid or bipinnatifid into linear to lanceolate segments, apices rounded to acuminate or apiculate. **INFL** of 1–5+ umbels or sometimes with an axillary umbel from axil of stem leaf, in fruit loose to somewhat dense, rounded, 2–5 cm wide; peduncle 2–40 cm tall, glabrous but scabrous to hirtellous at base of umbel; involucre absent or bracts 1 or 2, 1–3 mm long; rays 8–18, 5–12 mm long, to 35 mm long in fruit; involucel of 5–10 bractlets, linear to lanceolate, entire, acute to acuminate, 3–12 mm long, free or fused in lower 10%, green to purplish, margin often thinly scarious; pedicels 1–6 mm long, lengthening little in fruit. **FLOWERS** with calyx teeth 0.2–0.5 mm long, broadly lanceolate to deltoid; petals bright yellow when fresh, fading or becoming purplish; styles 1.5–2 mm long. **FRUIT** 3–6 mm long, elliptic to broadly so, tan to purplish, wings 2–5, lateral wings 0.8–1.5 mm tall, dorsal wings equally developed to riblike or virtually absent, wings where present straight, smooth, oil tubes 1–5 per interval, 2–6 on commissure; carpophore usually present. [*Thaspium montanum* A. Gray, not *C. montanus*; *Pseudocymopterus montanus* var. *purpureus* J. M. Coult. & Rose; *P. tidestromii* J. M. Coult. & Rose; *P. versicolor* Rydb.; *Peucedanum lemmonii* J. M. Coult. & Rose]. Grass-forb, aspen, Douglas-fir, and spruce-fir communities, and windswept ridges and raw escarpments. ARIZ: Apa; COLO: Arc, Con, Dol, Hin, LPl, Min, Mon, RGr, SJn, SMg; NMEX: McK, RAr, SJn; UTAH. 2270–3485 m (7500–11500′). Flowering: Jul–Aug. Fruit: Aug–Sep. Wyoming south to Utah, Colorado, Arizona, New Mexico, west Texas; northern Mexico. Morphologically, this taxon is highly variable. The Navajo used this species as a ceremonial emetic. The ground root is cooked with meat and the leaves boiled with cornmeal. The Isleta boil the leaves and stems to make a beverage.

Cymopterus longilobus (Rydb.) W. A. Weber **CP** (referring to long leaf segments) Wavewing. Plants acaulescent, tufted, strongly aromatic; taproot 0.4–3.5+ cm in diameter near summit (deep-seated, lower portion absent from specimens), cylindrical. **STEMS** with a branched caudex, clothed with (often long-) persistent leaf and peduncle bases. **LEAVES** leathery, glabrous, not viscid, green to pale green; petioles (1) 2–14 cm long; blades oblong to lanceolate or ovate in outline, (1) 1–10 cm long, 0.8–2.5 cm wide, mostly bipinnate or tripinnate in part, with 5–10 opposite or offset pairs of lateral primary leaflets sessile with distinct midribs (and lateral ribs: fine, reddish to brown oil tubes), ultimate segments 1–12 mm long, 0.3–1.4 mm wide, linear to lanceolate, the widest undivided parts of the blade seldom overlapping, terminal leaflet usually pinnatifid into linear or lanceolate segments, apices acute with usually a whitish tiny mucro. **INFL** of 1–13+ umbels, in fruit open to dense, flat to rounded, mostly 1–5 (6) cm wide; peduncles 7–30 cm tall, glabrous; involucre absent or rarely represented by 1–3 linear to filiform bracts; rays 6–16, 0.5–2.4 cm long (inner ones the shortest, often abortive), lengthening somewhat in fruit; involucel of 2–6 bractlets, linear, entire, 2–10 mm long, distinct, green to brownish (narrow to broad oil tube), margin green; pedicels 1–3 mm long, to 6 mm long in fruit. **FLOWERS** with calyx teeth 0.5–2 mm long, deltoid to lanceolate; petals yellow, fading often to whitish; style 1.5–2 (2.5) mm long. **FRUIT** 4–8 mm long, broadly elliptic, tan, wings 3–5, lateral wings 0.3–1.3 mm tall, dorsal equally to much less well developed, straight, smooth, leathery, oil tubes 1–5 per interval, 3–8 on commissure; carpophore present. [*C. hendersonii* (J. M. Coult. & Rose) Cronquist, name misapplied, restricted to Idaho]. Gravelly slopes, rock outcrops, and alpine meadows. UTAH. 2900–3300 m (9500–10900′). Flowering: May–Jun. Fruit: Jun–Jul. Extreme southeastern Idaho south to Utah and New Mexico.

Cymopterus newberryi (S. Watson) M. E. Jones (for John Strong Newberry, noted geologist) Sweetroot spring parsley. Plants acaulescent, tufted, weakly or not aromatic; taproot 3–6+ cm long, 0.3–1.8 cm in diameter near summit (lower portion usually absent from specimens), variously cylindrical to tapering. **STEM** caudex mostly simple, clothed with persistent leaf bases and 1 to several leaves, pseudoscape absent, rhizomes absent. **LEAVES** usually leathery, glabrous or viscid (sand, debris adherent), green to purplish; petioles 2–12 cm long; blades broadly ovate to triangular in outline, 1–4 (5.5) cm long, (1) 1.5–3.5 (6) cm wide, pinnate-pinnatifid to bipinnatifid with mostly 1 opposite pair of lateral leaflets, leaflets sessile with distinct midribs, ultimate leaf segments 1–5 long, 0.5–5 mm wide, elliptic to ovate or orbicular (in outline), not overlapping, terminal leaflet variously pinnatifid to bipinnatifid into oblong to elliptic segments, apices mostly rounded, often abruptly apiculate. **INFL** of 1–10 umbels, in fruit open, convex, 2–5 cm wide; peduncles 3–17 cm tall, glabrous to viscid; involucre absent or sometimes a poorly developed, usually entire to lobed cup 0.5–1 mm long; rays 5–16 (central ones often greatly reduced or obsolete), lateral rays 2–15 mm long, to 35 mm long in fruit; involucel of 5–9 bractlets, narrowly lanceolate to triangular or elliptic, entire, rounded to apiculate, 3–12 mm long, fused in lower 10%–60%, green to purplish, 1–3 often parallel veins present, margin firm, herbaceous; pedicels 0–3 mm long, lengthening little in fruit. **FLOWERS** with calyx teeth 0.2–0.5 mm long or obsolete, lanceolate to ovate; petals yellow, fading to cream or greenish; styles 2–3 mm long. **FRUIT** 5–8 mm long, broadly elliptic to orbicular, greenish to tan or purplish, wings 2, lateral, sometimes 1 or more dorsal ones maturing at least in part, 1–2.5 mm tall, straight, smooth to irregular, corky, oil tubes 4–6 per interval, 10–12 on commissure; carpophore absent. [*Peucedanum newberryi* S. Watson; *Coloptera jonesii* J. M. Coult. & Rose]. Desert shrub, blackbrush, sand sagebrush, desert grassland, and juniper communities. UTAH. 1180–2180 m (3900–6615′). Flowering: Apr–May. Fruit: May–Jun. Southern Utah and northeastern Arizona. The Navajo use an infusion of the plant as a lotion for wounds and eat the greens with meat. Hopi children would peel the roots and eat them.

Cymopterus purpurascens (A. Gray) M. E. Jones (purple flowers) Wide-wing spring parsley. Plants acaulescent, tufted, weakly or not aromatic; taproot 5–18+ cm long, 0.3–3+ cm in diameter, enlarging variously, especially toward base. **STEM** pseudoscapes 1 or 2, sometimes conspicuous, each arising 1–5 cm belowground (1–7 cm long) among remnants of old leaf sheaths, 1–3 bladeless scarious sheaths, and often 1 to 5 leaves; rhizomes absent. **LEAVES** somewhat fleshy, thus often minutely wrinkled on drying, glabrous or margins rarely scaberulous, not viscid, glaucous; petioles 1–7 cm long; blades lanceolate to broadly ovate in outline, 1.2–7 cm long, 1.5–5 cm wide, pinnate-pinnatifid to bipinnate-pinnatifid below, with 3–6 opposite pairs of lateral leaflets, leaflets sessile to petiolulate with distinct midribs, ultimate leaf segments 0.1–5 mm long, mostly 0.1–2.8 mm wide, oblong to elliptic, often overlapping, terminal leaflet variously pinnatifid to bipinnatifid into lanceolate to ovate segments, apices mostly rounded. **INFL** of 1–8 obscurely distinct umbels, in fruit dense, usually globose, mostly 3–6 cm wide; peduncles 2–14 cm tall, glabrous; involucre usually of 8–10 bracts 8–15 mm long, bracts fused into a lobed to variously parted cup, white, scarious with

1–4 purple veins; rays 0–8, 1–8 mm long, lengthing little in fruit; involucels of 4–6 bractlets, often obscured by the involucre or fruit, oblong to elliptic, usually rounded, 4–7 mm long, fused in lower 30%–60%, white, scarious with 1–4 dark green to purple veins arising from base, equal or lateral veins to 1/2 as long; pedicels 0–5 mm long, lengthening little in fruit. **FLOWERS** with calyx teeth 0.2–0.5 mm long, lanceolate to rounded; petals white or purplish; styles 1.5–2 mm long. **FRUIT** 7–15 mm long, broadly elliptic to suborbicular, tan to purplish, wings 5, 2–4 mm tall, straight to wavy, smooth, membranous, oil tubes 3–4 per interval, 4–7 on commissure; carpophore absent. [*C. montanus* Torr. & A. Gray var. *purpurascens* A. Gray; *C. utahensis* M. E. Jones var. *monocephalus* M. E. Jones]. Desert shrub, sagebrush, piñon-juniper, ponderosa pine communities, on eolian sand to heavy clay. ARIZ: Coc, Nav; UTAH. 1180–2425 m (3900–8000′). Flowering: Mar–May. Fruit: Apr–Jun. Southeastern Idaho to southeastern California and northern Arizona. No authentic material of *C. purpurascens* has been seen by Hartman from Colorado or New Mexico. Sheets so named have thus far been *C. bulbosus* or more commonly *C. constancei.* Navajo uses include an analgesic for backache, a gastrointestinal aid to settle stomachs, and paint for prayer sticks.

Cymopterus purpureus S. Watson (purple flowers) Variable spring parsley, Indian turnip, purple spring parsley. Plants acaulescent, tufted, weakly or not aromatic; taproot 0.4–1.5+ cm in diameter near summit (deep-seated, lower portion absent from specimens), cylindrical to variously enlarged. **STEM** caudex simple or branched, clothed with persistent leaf bases and sheaths of current year's leaves (some often attached above ground level), pseudoscape absent; rhizomes absent. **LEAVES** leathery to somewhat fleshy, thus minutely wrinkled on drying, glabrous, not viscid, green; petioles 2–9 cm long; blades broadly ovate to triangular in outline, 1.5–13 cm long, 1.5–10 cm wide, pinnate-pinnatifid to bipinnate-pinnatifid or ternate, with 2–4 opposite pairs of lateral leaflets or branches, leaflets sessile or petiolulate (petiolules to 32 mm long) with distinct midribs, ultimate segments 2–25 mm long, mostly 1–15 mm wide, linear to ovate, not overlapping, terminal leaflet variously pinnatifid to bipinnatifid into oblong to ovate or elliptic segments, apices rounded to acute or apiculate. **INFL** of 1–6 umbels, in fruit open, rounded, 4–12 cm wide; peduncles 5–25+ cm tall, glabrous; involucre absent; rays 5–22, 5–15 mm long, to 95 mm long in fruit, involucel of (1) 4–8 bractlets, linear to lanceolate or triangular, entire, acute to acuminate, 2–4 mm long, free or fused in lower 20%, green to purplish, margin thinly scarious; pedicels 1–10 mm long, lengthening little in fruit. **FLOWERS** with calyx teeth 0.1–0.5 mm long or obsolete, lanceolate to ovate; petals yellow, sometimes becoming purplish; styles 2–3 mm long. **FRUIT** 4–8 mm long, broadly oblong or elliptic to orbicular, tan to purple, wings usually 5, 1.5–4 mm tall, usually wavy, smooth, membranous, oil tubes 2–7 per interval, 3–10 on commissure; carpophore present. Desert shrub, sagebrush, piñon-juniper, mountain brush, ponderosa pine, and rarely aspen-fir communities in sandy to heavy clay soils. ARIZ: Apa, Coc, Nav; COLO: Arc, Dol, LPl, Mon, SMg; NMEX: RAr, San, SJn; UTAH. 1180–2575 m (3900–8500′). Flowering: Mar–May. Fruit: Apr–Jun. Eastern and southern Utah, western Colorado, northern Arizona, and northwestern New Mexico. Navajo used the plant as a potherb in seasoning mush and soup.

Cymopterus sessiliflorus (W. L. Theob. & C. C. Tseng) R. L. Hartm. (stalkless flower) Sessile-flower Indian parsley. Plants acaulescent, tufted to caespitose, weakly or not aromatic; taproot 0.2–1.4 cm in diameter near summit (deep-seated, lower portion absent from specimens). **STEM** caudex branched, clothed with persistent leaf bases, pseudoscape absent, rhizomes absent. **LEAVES** herbaceous to leathery, glabrous or margins scaberulous to scabrous, not viscid, green; petioles 0.5–11 cm long; blades lanceolate to broadly ovate in outline, 1.5–8 cm long, 8–3.5 cm wide, pinnate to pinnate-pinnatifid or nearly bipinnatifid with 2–6 opposite pairs of lateral leaflets, leaflets sessile, 3–35 mm long, often with distinct midribs, mostly 1–3 mm wide, linear to narrowly elliptic and entire or with 1–3 (5) teeth or lateral lobes, seldom overlapping, terminal leaflet simple to variously pinnatifid into linear to broadly elliptic segments, apices apiculate. **INFL** of 1–5+ umbels, in fruit open to dense, rounded to subglobose or convex, 1.5–3 cm wide; peduncle 3–25 cm tall, glabrous throughout or slightly scabrous at summit; involucre absent; rays 4–9, 2–15 mm long, lengthening little in fruit; involucel of 3–8 bractlets, linear to elliptic, entire to rarely 3-lobed, 2–5 mm long, fused at base or in lower 1/3, green, margin often narrowly scarious; pedicels 0–2 mm long, to 2.5 mm long in fruit. **FLOWERS** with calyx teeth 0.5–1.2 mm long, linear to deltoid; petals yellow when fresh; styles 2–2.5 mm long. **FRUIT** 5–6 mm long, broadly elliptic, tan, wings 3–5, 0.3–1.5 mm tall, straight, smooth, membranous to somewhat corky, oil tubes 1 per interval, 2 on commissure; carpophores present. [*Aletes macdougalii* J. M. Coult. & Rose subsp. *breviradiatus* W. L. Theob. & C. C. Tseng; *Cymopterus breviradiatus* (W. L. Theob. & C. C. Tseng) R. L. Hartm.]. Rocky slopes, ledges, and crevices of badlands and canyonlands, and sandy ground in piñon-juniper communities. ARIZ: Apa, Nav; COLO: Arc, LPl, Mon; NMEX: McK, RAr, San, SJn; UTAH. 1515–2180 m (5000–7200′). Flowering: Apr–May. Fruit: May–Jun. Endemic to northwestern New Mexico, southwestern Colorado, northeastern Arizona, southeastern Utah. The genus *Aletes* sensu stricto is defined, in part, by somewhat laterally flattened fruit (versus terete to dorsally flattened in *Cymopterus*). This lateral compression is prominent in the type species, *A. acaulis* (Torr.) J. M. Coult. & Rose, and its presumed closest relative, *A. humilis* J. M. Coult. & Rose. Neither of the taxa of *Aletes* for which nomenclatural innovations recently have been made (*A. macdougalii* subsp. *breviradiatus*, *A. sessiliflorus*) are flattened laterally (Hartman 2006). The former has fruit with corky-thickened wings similar to those of *C. sessiliflorus* whereas *C. macdougalii* has weakly developed, thin papery wings. *Cymopterus sessiliflorus* is distinguished from *C. breviradiatus* in that the former is more robust and is said to have an additional vitta in each wing (sporadically seen in transactions [Theobald, Tseng, and Mathias 1964]). *Cymopterus sessiliflorus* was based on a single collection near Cuba, New Mexico, whereas the type locality for *C. breviradiatus* is in Mesa Verde National Park (holotype and three paratypes), Colorado. In addition, two paratypes were cited from adjacent San Juan County, Utah. The number of collections of this species pair has increased manyfold since the late 1970s. It has been concluded that *C. breviradiatus* forms a continuum with *C. sessiliflorus* and thus should be treated as a synonym of it. The material representing *C. sessiliflorus* used in DNA studies (as *Aletes s.*; Sun and Downie 2004; Sun, Downie, and Hartman 2004) is from Black Mesa, New Mexico, some 70 miles east of the type locality of *C. sessiliflorus*. When this population was first brought to the attention of Professor Richard Spellenberg (New Mexico State University), it was suggested that it might represent a new taxon. During the recent floristic inventory of Carson National Forest and adjacent Bureau of Land Management lands (Jill Larsen, graduate student, B. E. Nelson, and R. L. Hartman in 2005 and 2006), it has become apparent that Spellenberg may be correct, and we in fact did not have *C. sessiliflorus* represented in the above cited molecular studies.

Cymopterus terebinthinus (Hook.) Torr. & A. Gray (referring to turpentine, probably the aromatic nature of the plant) Rock-parsley, rock-loving parsley, wavewing. Plants acaulescent, tufted, strongly aromatic; taproot 0.3–1+ cm in diameter near summit (deep-seated, lower portion absent from specimens), cylindrical. **STEM** caudex simple or branched, clothed with persistent leaf bases and sheaths of current and previous year's leaves (some often attached above ground level), pseudoscape absent; rhizomes absent. **LEAVES** leathery, glabrous, not viscid, green; petioles 2–13 cm long; blades lanceolate to ovate in outline, 1.5–14 cm long, 2–10 cm wide, mostly pinnate to pinnate-pinnatifid above, bipinnate-pinnatifid or nearly ternate in lower portion, with 4–10+ pairs of opposite or offset lateral leaflets (elongate terminal segment and lateral branches very narrow, most less than 8 mm wide), leaflets sessile to petiolulate (petiolules to 40 mm long) with distinct midribs, ultimate segments finely and completely divided, 0.2–7 mm long, mostly 0.2–1.5 mm wide, mostly linear to oblong, seldom overlapping, terminal leaflet usually pinnatifid into linear segments, apices obtuse to usually apiculate. **INFL** of 1–7+ umbels, in fruit sparse, flat to rounded, mostly 5–10 cm wide; peduncles 10–35 cm tall (among remnants of previous year's peduncles), glabrous throughout; involucre absent; rays 7–13, 0.7–2.5 cm long (central ones often elongating little), lateral rays to 8 cm long in fruit; involucel of (0) 1–5 bractlets, linear to linear-subulate, entire, 2–5 mm long, distinct, green, margin thinly scarious; pedicels 0.5–2 mm long, to 8 (10) mm long in fruit. **FLOWERS** with calyx teeth 0.5–0.9 mm long, linear to lanceolate; petals bright yellow, fading often to whitish; styles 1.5–4 mm long. **FRUIT** 5–8 mm long, ovoid-oblong, tan, wings 2–5, some of the dorsal ones often reduced or obsolete, 0.5–1.5 (2.5) mm tall, usually wavy, smooth to irregular, membranous, oil tubes 4–6 per interval, 8–12 on commissure; carpophore present. Desert shrub, blackbrush, and piñon-juniper communities, often in talus, colluvium, crevices of rock outcrops, and in sandy to clayey soil. ARIZ: Apa; COLO: LPl; NMEX: McK, SJn; UTAH. 1400–2075 m (4600–6850′). Flowering: Apr–May. Fruit: May–Jun. Eastern Utah, Nevada, Idaho, and Arizona. Our material belongs to **var.** ***petraeus*** (M. E. Jones) Goodrich (rock lover), which up until recently was recognized at the species level.

Daucus L. Carrot

(to burn) Biennial, caulescent herbs from taproots. **LEAVES** pinnately dissected. **INFL** open umbels; involucre of pinnatifid bracts; involucel of toothed or entire bracts or absent. **FLOWERS** with calyx teeth evident to obsolete; petals white or those of the central flower of the umbel or umbellet often purple or rarely all the flowers pink or yellow; stylopodium conic; carpophore entire or bifid at the apex. **FRUIT** oblong to ovoid, slightly compressed and evidently ribbed dorsally, with 2 ribs on the commissure, beset with stout, spreading bristles, in part somewhat uncinate on ribs. About 22 species widely distributed, one native to the United States.

Daucus carota L. (referring to the red color of the root) Wild carrot, Queen Anne's lace. **STEMS** 6–10 dm tall, from a taproot; herbarge glabrous or hirsute. **LEAVES** in rosettes and cauline, mostly 1–2 times pinnate and then pinnatifid, with 4–9 opposite or offset pairs of lateral primary leaflets, basal and lower cauline petioles to 15 cm long, basal and lower blades 5–15 cm long or more, the upper ones reduced and sessile on dilated sheaths, the lowest pair of primary leaflets 1/3–1/2 as long as the leaf blade, on petiolules 4–15 mm long, ultimate segments 1–10 mm long, 0.5–2 mm wide, elliptic, narrowly deltoid, or linear, often acute. **INFL** peduncles mostly 8–30 cm long; umbels 4–10 or more; involucre of pinnatifid bracts, 1–5 cm long, the segments linear and narrow; rays 15–60 or more, (0.5) 1–6 cm long; involucels similar to the involucre but smaller, or the bractlets entire, 2–16 mm long. **FLOWERS** white to yellowish, the central flower of each umbellet normally pinkish or purple. **FRUIT** 3–4 mm long, bristly hirsute in rows, the hairs or bristles 2 mm long, minutely glochidiate apically, the intervals often with shorter simple hairs. $2n = 18$. Along roadsides, irrigation ditches, and other moist places. COLO: LPl, Mon; NMEX: SJn. 1575–1970 m (5200–6500′). Flowering: Jul–Aug. Fruit: Aug–Sep. Widespread; introduced from Eurasia. The wild plants [**subsp.** ***carota***] differ from the cultivated ones [**subsp.** ***sativus*** (Hoffm.) Arcang.] primarily in the size and flavor of the root.

Foeniculum Mill. Fennel

(*foenum*, hay, referring to the odor) Biennial or perennial, caulescent herbs with strong odor of anise; glabrous, glaucous, from a taproot. **LEAVES** pinnately dissected with filiform ultimate segments. **INFL** of open umbels; involucre and involucel absent. **FLOWERS** with calyx teeth obsolete; petals yellow; stylopodium conic; carpophore divided to the base. **FRUIT** oblong, subterete, or slightly compressed laterally, with prominent ribs. One species found in Europe and the Mediterranean.

Foeniculum vulgare Mill. (common) Sweet fennel. A short-lived perennial herb. **STEMS** 0.5–2 m tall, from a taproot, solitary, branched above. **LEAVES** to 3 times ternate-pinnately compound with 6–9 opposite pairs of lateral primary leaflets; petioles to 15 cm long, rather abruptly expanded into a dilated sheathing base or absent and blades arising directly from the sheath; larger blades to 30 or 40 cm long, ovate in outline, finely and completely dissected, the elongated filiform ultimate segments 4–40 mm long and less than 1 mm wide, the lowest pair of primary leaflets on petiolules often over 2 cm long. **INFL** peduncles 1.5–6.5 cm long; umbels several; rays 10–40, 2–8 cm long. **FLOWERS** with sepals obsolete, petals yellow; styles 0.3–0.4 mm long. **FRUIT** 3.5–4 mm long. $n = 8, 11, 13, 15, 22$. Roadsides and waste places. Known from one collection in the study area near Kayenta. ARIZ: Apa. 1820 m (6000′). Flowering: Jul–Sep. Fruit: Aug–Sep. Widespread in much of the United States. The plant has been used as a substitute for tobacco by the Hopi.

Heracleum L. Cowparsnip, Hogweed

(for Hercules, whose healing powers were attributed to the herb) Biennial or perennial herbs from a taproot or fascicled fibrous roots. **LEAVES** ternately or pinnately compound, with broad, toothed or cleft leaflets. **INFL** of open umbels; involucel absent or of slender bractlets; flowers of the marginal umbellets generally irregular, the outer petals enlarged and often deeply bilobed. **FLOWERS** with calyx teeth obsolete or minute; stylopodium conic; carpophore divided to the base. **FRUIT** orbicular to obovate or elliptic, strongly flattened dorsally, usually pubescent, the dorsal ribs narrow, the lateral ribs broadly winged. Sixty species, circumboreal, one native to North America.

Heracleum lanatum Michx. (woolly) Common cowparsnip. Stout, single-stemmed perennial herbs. **STEMS** 8–25 dm tall, from a taproot or cluster of fibrous roots, glabrate or thinly to densely villous or villous-hirsute below to villous-woolly above, especially on the nodes. **LEAVES** ternate or the upper ones simple, petioles to 25 cm long or longer, or absent on upper leaves with the petiolules and rachis arising directly from a dilated sheath, blades to 40 cm long or longer, ovate to orbicular; leaflets 8–36 cm long or longer, ovate to orbicular, usually with 3 major lobes that are again lobed and coarsely toothed. **INFL** with peduncles 5–24 cm long; involucre absent or of few, mostly linear, entire bracts to 2 cm long; rays 12–25, 3.5–12 cm long; involucels of 3–5 linear, subulate, or caudate bractlets to 15 mm long; pedicels 6–26 mm long. **FLOWER** petals white, (2) 4–8.5 mm long, at least some deeply bilobed; filaments white, the anthers whitish to dark green or yellow with pollen; styles about 1 mm long, the stigmas incurved. **FRUIT** 8–12 mm long, obovate to obcordate, strongly flattened, the lateral ribs with wings 1–1.5 mm wide, the dorsal ribs filiform. $2n = 22$. [*H. sphondylium* L. subsp. *lanatum* (Michaux) Á. Löve & D. Löve]. Aspen, tall forb, fir, oak-maple, willow, streamside, and wet meadow communities. ARIZ: Apa; COLO: Arc, Hin, LPl, Min, Mon, RGr, SJn; NMEX: RAr; UTAH. 2270–2940 m (7500–9700′). Flowering: Jun–Aug. Fruit: Jul–Sep. Eurasia and across much of North America.

Ligusticum L. Licorice-root, Oshá

(refers to the abundance of the plant in the region of Liguria, Italy) Perennial, caulescent or acaulescent herbs from taproots. **LEAVES** ternately or ternate-pinnately compound or dissected, the lower ones with well-developed petioles, the upper ones with blades arising directly from dilated sheaths. **INFL** of open umbels; involucre and involucel absent or of a few narrow bracts or bractlets. **FLOWERS** with calyx teeth evident or obscure; petals white; stamens white; stylopodium low-conic; carpophore divided to the base. **FRUIT** oblong to ovate or suborbicular, subterete or slightly compressed laterally, the ribs evident, often winged. About 20 species in both the Western and Eastern Hemispheres, mostly in north-temperate regions.

Ligusticum porteri J. M. Coult. & Rose (for Thomas C. Porter) Southern ligusticum, oshá, Porter's licorice-root, Porter's lovage. Robust plants with 1 to several stems from base, mostly 5–12 dm tall. **STEMS** usually branched above. **LEAVES** basal, long-petiolate, and cauline petiolate to sessile, similar to the basal ones; blades orbicular in outline, ternate-pinnately dissected with broader ultimate segments, these (1.5) 3–8 mm wide. **INFL** a terminal umbel often subtended by a whorl of 3–8 lateral umbels, and occasionally with up to 12 or more umbels, but sometimes with the lateral umbels only 2 and opposite but not alternate. **FLOWERS** with calyx teeth evident to obsolete; petals white; stamens white; stylopodium low-conic. **FRUIT** oblong, 5–8 mm long. $n = 11$. [*L. brevilobum* Rydb.]. Sagebrush, oak, aspen, Douglas-fir, spruce-fir, and occasionally in open forb-grass, and hanging garden communities. ARIZ: Apa; COLO: Arc, Con, Hin, LPl, Min, Mon, RGr, SJn; NMEX: McK, RAr, SJn; UTAH. 2120–2940 m (7000–9700'). Flowering: Jun–Aug. Fruit: Jul–Sep. Southern Wyoming to northern Mexico, west to Idaho and Arizona. Plants of this taxon are often mistaken for *Conioselinum scopulorum.* The reader is referred to *C. scopulorum* for discussion of several of the distinguishing features. The Zuni uses include: infusion of root for body aches, root chewed by medicine man and patient during curing ceremonies for various illnesses, and the crushed root and water used as wash and taken for sore throats. Used extensively by northern New Mexico Hispanics for many conditions including colds, flu, and other respiratory ailments.

Lomatium Raf. Desert Parsley, Biscuit-root

(*loma*, border, referring to the winged fruit) Perennial, acaulescent or caulescent, occasionally with a short pseudoscape, glabrous or pubescent, from a slender taproot with sometimes 1 or more tuberlike segments, or from a thickened, woody, branching caudex, sometimes clothed at the base with marcescent leaf bases. **STEMS** simple or rarely branched. **LEAVES** pinnate or pinnately to ternate-pinnately compound, the sheaths often dilated; petioles developed, and distinct or confluent with and poorly differentiated from the sheath, or absent and the petiolules arising directly from the sheath, the ultimate segments extremely variable. **INFL** of open umbels; involucre absent or inconspicuous; rays few to many, spreading to ascending, the central ones often shorter and sterile; involucel mostly separate or partly united bractlets, rarely absent; pedicels slender or stout, the central ones often shorter and sterile; petals small, yellow, white, greenish yellow, or purplish. **FLOWERS** with calyx teeth obsolete or small, or conspicuous in some species; styles slender, often curved or coiled; stylopodium absent; carpophore divided to the base. **FRUIT** oblong to orbicular or obovate, flattened dorsally, glabrous or pubescent, dorsal ribs filiform or obsolete or occasionally with rudimentary wings at the base. About 70 species of western and central North America. The genus is closely related to *Cymopterus*, and the filiform wingless dorsal ribs of the fruit seem to be the only consistent difference from *Cymopterus*. The dependability of this separation is somewhat weakened by the tendency for lack of dorsal wings in some species of *Cymopterus*.

1. Leaf blades ternate- to biternate-pinnately compound, with mostly 9–21 pairs of linear to lanceolate leaflets often 3–13 cm long, 1–6 mm wide............ ***L. triternatum***

1′ Leaves more than once compound, the ultimate segments not over 15 mm long and mostly more than 50 per leaf or, if a few ultimate segments over 15 mm long or fewer than 50 per leaf, the plants pubescent and petals white ....(2)

2. Larger mature leaves with blades (10) 15–30 cm long, ternate-pinnately compound, the larger ultimate segments 2–3 mm wide; plants 3–13 dm tall; peduncles fistulose, (3) 4–6 (10) mm thick at the base .... ***L. dissectum***

2′ Larger mature leaves with blades 2–11 cm long or, if longer, either not at all ternate or with ultimate segments not over 1 mm wide; plants rarely over 50 cm tall; peduncles fistulose or not, often less than 4 mm thick..........(3)

3. Plants pubescent; petals white.......... ***L. nevadense***

3′ Plants glabrous or at most scabrous; petals yellow when fresh, fading whitish when dried..........(4)

4. Lowest pair of primary leaflets less than 1/3 as long as the leaf blade; Navajo Mountain region .... ***L. parryi***

4′ Lowest pair of primary leaflets 1/3–3/4 as long as the leaf blade; widespread.......... ***L. grayi***

Lomatium dissectum (Nutt.) Mathias & Constance (deeply divided, referring to the leaves) Giant lomatium, fern-leaf biscuit-root. Plants mostly short-caulescent, bushy (30–130 cm tall), somewhat aromatic, puberulent or rarely glabrous; taproot 25+ cm long, 1–3 cm in diameter (woody), variously thickened (lower portion absent from specimens). **STEM** caudex with few or no marcescent leaf bases, these soon shredding, deciduous. **LEAVES** herbaceous, green; petioles 3–20 cm long, often absent on cauline leaves (blades sessile on a dilated sheath); blades ovate to orbicular in outline, 10–30 cm long, 6–30 cm wide, or cauline leaves reduced, ternate- to biternate- or pinnate-pinnatifid or decompound, with 5–9 opposite or subopposite pairs of primary leaflets; petiolules 2.5–12 cm long, ultimate segments numerous, 1–12 mm long, 0.5–3 mm wide, rounded to acute. **INFL** of 1–4 umbels, in fruit open, flat to rounded, 10–20 cm wide; peduncles glabrous (often glaucous), 15–50 (90) cm tall; involucre absent or occasionally of 1–3 deciduous, linear to foliaceous bracts; rays 9–27, 2–5 cm long, to 12 cm long in fruit; involucel of 3–7 bractlets, 3–6 mm long or occasionally much longer and foliaceous, linear to lanceolate, entire, green with thinly scarious margin, reflexed; pedicels 3–4 mm long, to 10 mm long in fruit. **FLOWERS** with calyx teeth 0.1–0.3 mm long or obsolete, rounded; petals yellow, yellow-green, or purplish; anthers yellow or yellow-green; styles 1.5–1.8 mm long. **FRUIT** broadly elliptic, 9–15 (20) mm long, 6–10 mm wide, glabrous, lateral wings 1–2 mm wide, corky, the dorsal ribs filiform; oil tubes obscure; carpophores present. [*Leptotaenia dissecta* Nutt.]. Sagebrush, piñon-juniper, oak-maple, aspen-fir, riparian, and rarely greasewood–desert shrub communities. COLO: Arc, Dol, LPl, Mon, SMg; NMEX: RAr; UTAH. 1820–2635 m (6000–8700′). Flowering: Apr–Jun. Fruit: May–Jul. British Columbia and Alberta south to California, Arizona, Colorado, and New Mexico. Our material belongs to **var. *eatonii*** (J. M. Coult. & Rose) Cronquist [*Leptotaenia eatonii* J. M. Coult. & Rose]. Sometimes the leaves are mistaken for those of *Ligusticum porteri*, but the mostly solitary umbel is strikingly different from the usually opposite or whorled lateral umbels in addition to the terminal one in the *Ligusticum*.

Lomatium grayi (J. M. Coult. & Rose) J. M. Coult. & Rose (for Asa Gray, distinguished professor of botany) Milfoil lomatium, Gray's biscuit-root. Plants acaulescent or subcaulescent, tufted, strongly aromatic, glabrous; taproot 20+ cm long, 0.3–3+ cm in diameter (woody), variously thickened (lower portion absent from specimens). **STEM** caudex simple or much-branched, often clothed at the base with thick layer of marcescent, mostly shredded, fibrous leaf bases. **LEAVES** herbaceous, green; petioles 3–14 cm long including dilated sheath or blades arising directly from dilated sheath; blades ovate to lanceolate in outline, 7–16 cm long, 3–5 cm wide, ternate-bipinnately to tripinnately dissected, with 7–10 opposite pairs of lateral primary leaflets; petiolules 1–7.5 cm long, ultimate segments very numerous, extremely fine, 1–3 (6) mm long, 0.2–0.3 mm wide, acute to blunt. **INFL** of 1–8+ umbels, in fruit open to congested, rounded, 2–15 cm wide; peduncles glabrous, 10–45 (70) cm tall; involucre absent; rays 10–26, 0.5–1.5 cm long, to 8 cm long in fruit; involucels of 3–6 bractlets, 3–5 mm long, linear to lanceolate, entire, green, erect to spreading; pedicels 2–5 mm long, to 18 mm long in fruit. **FLOWERS** with calyx teeth 0.2–0.3 mm long or obsolete, ovate; petals yellow fading to whitish with age; anthers yellow; styles 1.5–2.5 mm long. **FRUIT** broadly elliptic to obovate-oblong, 8–12 mm long, 5–8 mm wide, glabrous, lateral wings 1–2 mm wide, membranous, dorsal ribs filiform; oil tubes 1 per interval, 2–4 (6) on commissure; carpophore present. [*Peucedanum millefolium* S. Watson; *L. millefolium* (S. Watson) Macbride; *Peucedanum grayi* J. M. Coult. & Rose; *Cogswellia grayi* (J. M. Coult. & Rose) J. M. Coult. & Rose]. $n = 11$. Sagebrush, piñon-juniper, mountain brush, ponderosa pine, and Douglas-fir communities. COLO: Arc, Dol, LPl, Mon, SMg; NMEX: RAr, SJn; UTAH. 1820–2425 m (6000–8000′). Flowering: Apr–May. Fruit: May–Jun. Washington to Nevada and east to Idaho, Colorado, and New Mexico. Our material belongs to **var. *grayi***.

Lomatium nevadense (S. Watson) J. M. Coult. & Rose (of Nevada) Nevada lomatium, Nevada biscuit-root. Plants acaulescent or subcaulescent, dense bushy clumps, somewhat aromatic, glabrous; taproot 15+ cm long, 1–3+ cm in diameter (lower portion absent from specimens). **STEM** caudex with many branches, branches clothed at base with thick layer of conspicuous, persistent, terete leaf bases. **LEAVES** thickened, herbaceous or coriaceous, green (glaucous); petioles 2–7.5 cm long; blades lanceolate to oblong, often narrowly so, in outline, 7–24 cm long, 1.5–5 cm wide, pinnate-bipinnatifid or partly tripinnatifid, with 7–9 opposite pairs of primary leaflets or the upper leaflets simple; sessile or with petiolules to 1.2 cm long, ultimate segments mostly 50–150, 1–15 mm long, 1–2 mm wide, acute. **INFL** of 10–20+ umbels, in fruit open to congested, flat to rounded, 5–13 cm wide; peduncle glabrous, 5–32 cm long, involucre absent; rays 8–13, 1–5 cm long, to 6 cm long in fruit; involucel of 3–8 bractlets, 3–10 mm long, linear to lanceolate, entire, tridentate, or pinnatifid, green to purplish, spreading to reflexed; pedicels 0.5–1 cm long, to 2 cm long in fruit. **FLOWERS** with calyx teeth 0.3–0.8 mm long, lanceolate; petals yellow, fading to white with age, anthers yellow-green; styles 2–4 mm long. **FRUIT** broadly elliptic to elliptic-oblong, 6–20 mm long, 5–10 mm wide, glabrous, lateral wings 1–3 mm wide, thickened, the dorsal ribs filiform; oil tubes 2–3 per interval, 4 on commissure; carpophore present. [*Peucedanum nevadense* S. Watson]. $2n = 22$. Sagebrush, piñon-juniper, mountain brush, ponderosa pine communities. ARIZ: Apa; NMEX: SJn; UTAH. 2120–2425 m (7000–8000′). Flowering: Mar–May. Fruit: Apr–Jun. Oregon and California east to Utah, Arizona, and Colorado. Material from the San Juan drainage has glabrous fruit and bulbose roots as found in **var. *parishii*** (J. M. Coult. & Rose) Jeps., but the leaf morphology resembles that of **var. *nevadense***. The Paiute would peel the roots and eat them like radishes.

Lomatium parryi (S. Watson) J. F. Macbr. (for Charles Christopher Parry, noted botanist and explorer) Parry's lomatium, Utah desert parsley. **STEMS** 8–40 cm tall, acaulescent, glabrous, from a branched caudex clothed at the base with persistent leaf bases. **LEAVES** bi- or partly tripinnatifid, with mostly 7–9 opposite pairs of primary leaflets or the upper leaflets simple; petioles 3–16 cm long, terete, often persisting for some years without shredding; blades 7–24 cm long, the lowest pair of lateral primary leaflets less than 1/4 as long as the leaf blade, sessile or with petiolules to 1.2 cm long, the ultimate segments mostly 50–150, 1–15 mm long, 1–2 mm wide, acute. **INFL** with peduncles 5–32 cm long; rays of the umbel 8–13, 1–5 cm long; bractlets of the involucels 3–10 mm long, entire, tridentate or rarely pinnatifid, spreading to reflexed in age; pedicels 1–2 cm long. **FLOWERS** with petals yellow, turning white in age; styles 2–4 mm long. **FRUIT** 6–20 mm long, 5–10 mm wide, the lateral wings 1–3 mm wide, the dorsal ribs filiform. [*Peucedanum parryi* S. Watson; *Cogswellia cottamii* M. E. Jones]. Desert shrub, blackbrush, piñon-juniper, mountain brush, ponderosa pine–manzanita communities. ARIZ: Coc, Nav; UTAH. 1820–2120 m (6000–7000′). Flowering: Apr–May. Fruit: May–Jun. Colorado and Utah to eastern California.

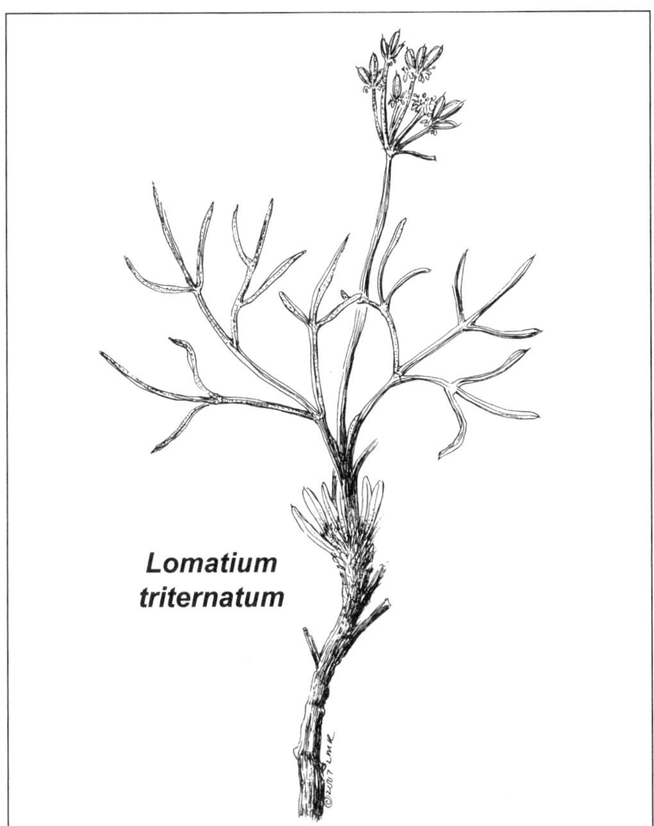
Lomatium triternatum

Lomatium triternatum (Pursh) J. M. Coult. & Rose (divided into three) Nineleaf biscuit-root. Plants acaulescent or subcaulescent (rarely with a cauline leaf), tufted, somewhat aromatic, glabrous (esp. leaves) to hirtellous; taproot 30+ cm long, 0.5–2 cm in diameter, tapering to base (lower portion absent from specimens). **STEM** caudex simple to sparingly branched, usually clothed with persistent leaf bases. **LEAVES** herbaceous, green; petioles 1–25 cm long including the dilated sheathing base, or the sheath only; blades ovate to broadly obovate in outline, 4–20 cm long, 2–10 cm wide, ternate- to biternate-pinnately compound, with (3) 9–21 opposite pairs of primary leaflets or segments, ultimate leaflets or segments linear to lanceolate, mostly 2–13 cm long, 1–6 (10) mm wide, acute to apiculate. **INFL** of 1–6 umbels, in fruit open to congested, flat to rounded, 5–13 cm wide; peduncles hirtellous, 15–55 cm tall; involucre absent; rays 4–20, 0.5–2 cm long, to 10 cm long in fruit;

involucel of 6–10 bractlets, 1.5–8 mm long, linear to lanceolate, entire, green to purplish, erect to spreading; pedicels 0.5–1 mm long, to 10 mm long in fruit. **FLOWERS** with calyx teeth obsolete; petals bright yellow, fading to white with age; anthers yellow; styles 1–1.5 mm long. **FRUIT** broadly elliptic, 8–15 mm long, 4–11 mm wide, glabrous, lateral wings 2–2.5 (4) mm wide, membranous, dorsal ribs filiform; oil tubes 1 per interval, 2 on commissure; carpophore present. Sagebrush-grass, piñon-juniper woodland, mountain brush, ponderosa pine, and dry meadow communities. COLO: Arc, Dol, LPl, Mon, SMg; NMEX: McK, RAr, SJn; UTAH. 1675–2440 m (5500–8000′). Our material is ***subsp. platycarpum*** (Torr.) Cronquist (wide seed).

Orogenia S. Watson Indian Potato

(mountain race, referring to the habitat) Perennial plants from tuber, low, glabrous, acaulescent or very short-caulescent. **LEAVES** once or twice ternate or rarely simple, the leaflets narrow, elongate, mostly entire. **INFL** a loose umbel; involucre absent; involucel of a few minute, narrow bractlets or absent; rays few, unequal, spreading; pedicels short or obsolete. **FLOWERS** with calyx teeth obsolete; petals white; stylopodium absent, styles short and spreading. **FRUIT** oblong to oval, nearly round in cross section, glabrous, dorsal ribs filiform and rather prominent, a corky rib running down middle of commissure; carpophore absent. Two species in the western United States.

Orogenia linearifolia S. Watson (linear-leaved) Turkey pea, Great Basin Indian potato. Root up to about 1.5 cm thick, extending 2–7 cm belowground. **STEMS** ascending a few cm above the ground. **LEAVES** mostly 2 or 3, the blade only slightly elevated above the ground level, the ultimate segments mostly 1–4.5 cm long, 0.5–4 mm wide. **INFL** compact at anthesis, 1–2 cm long, somewhat less wide, becoming more open in fruit, pedicels less than 2 mm long. **FLOWERS** with calyx teeth obsolete, petals white; anthers dark purplish. **FRUIT** 3–4 mm long, oblong-elliptic, with evident dorsal ribs. Blooms as soon as the snow recedes in mountain brush and ponderosa pine communities. COLO: Arc, Dol, LPl, Mon, SMg; NMEX: RAr. 2120–2425 m (6500–8500′). Flowering: Mar–May. Fruit: Apr–Jun. Washington to Montana, south to Colorado and Utah. The root was eaten by many tribes and was available at most times of the year.

Osmorhiza Raf. Sweetroot, Sweet Cicely

(from *osme*, odor, + *rhiza*, root) Perennial, caulescent, usually pubescent herbs from taproots with simple or branched crowns. **LEAVES** ternately or pinnately 1–3 times compound with well marked leaflets. **INFL** open or lose umbel; involucre absent or of 1 to few narrow foliaceous bracts; involucel absent or of several foliaceous reflexed bractlets. **FLOWERS** with calyx teeth obsolete; petals and stamens white, greenish white, yellow, pink, or purple; stylopodium conic to depressed; carpophore bifid less than 1/2 its length. **FRUIT** linear or clavate, somewhat compressed laterally, bristly hispid to glabrous, the ribs narrow. About 10 species of North America, South America, and eastern Asia.

1. Ovaries and fruit glabrous, generally obtuse at both ends; petals and stamens yellow or greenish yellow; leaves (1) 2 times pinnately or ternate-pinnately compound; plants strongly aromatic, usually with more than 2 stems .... ***O. occidentalis***

1′ Ovaries and fruit bristly hispid, with long, pointed, bristly hispid tails; petals white or greenish white; leaves biternate; plants not strongly aromatic, often with solitary stems .... (2)

2. Mature fruit including tails mostly 16–25 mm long, the apex concavely pointed into a beak 1–2 mm long; fruiting pedicels mostly ascending-spreading .... ***O. chilensis***

2′ Mature fruit including tails mostly 13–18 mm long, the apex convex and obtuse; fruiting pedicels horizontally divergently spreading to sometimes ascending........ ***O. depauperata***

Osmorhiza chilensis Hook. & Arn. (of Chile) Chile oreoxis, sweet cicely. **STEMS** often solitary, 18–75 cm tall, from a taproot, without marcescent leaf bases; herbage not strongly aromatic. **LEAVES** basal and 2–3 cauline, biternate, usually with 9 distinct leaflets; petioles 3–16 cm long or cauline leaves sessile; blades 5–15 cm long, the lateral primary leaflets nearly as long as the central one, with petiolules (1) 2–5.5 cm long; blades of leaflets 1–4 (5.5) cm long, elliptic to ovate, lobed to cleft, and toothed, ciliate, and often pubescent on nerves below and sometimes scattered pubescent between the nerves. **INFL** with peduncles 5–34 cm long; umbels 1–5; involucre absent; rays 3–7, 2.5–9 (13) cm long, ascending or spreading-ascending, glabrous to hirtellous; involucels absent; pedicels 5–22 (30) mm long, ascending. **FLOWERS** with petals and stamens greenish white; styles less than 0.5 mm long. **FRUIT** including the tails 16–25 mm long, linear-clavate, bristly hispid, the beak concavely pointed, 1–2 mm long, the concave beak usually evident in young fruits. [*O. nuda* Torr.]. $n = 11$. Mountain brush, aspen, Douglas-fir, white fir, narrowleaf cottonwood, and riparian communities. COLO: Arc, Hin, LPl; UTAH. 1970–2695 m (6500–8900′). Flowering: May–Jul. Fruit: Jun–Aug. Alaska to California, east to Alberta and Arizona; also in Great Lakes region; Argentina and Chile.

Osmorhiza depauperata Phil. (of the poor) Blunt-fruit sweet cicely, blunt-seed sweetroot. **STEMS** mostly solitary, 14–63 (77) cm tall, often with a slight ring of hairs at the nodes, from a taproot, without persisting leaf bases; herbage not strongly aromatic. **LEAVES** basal and 1–3 cauline, biternate, usually with 9 distinct leaflets, or the upper cauline ones once ternate; petioles (1) 3–17 cm long, often with dilated, ciliate bases; blades (2) 4–11 cm long, the lateral primary leaflets almost equal to the central one or a little shorter, with petiolules (0.5) 1–4 cm long, blades of leaflets 1–4 (5.5) cm long, elliptic to ovate, lobed to cleft and toothed, ciliate and often pubescent on nerves below and sometimes scattered pubescent between them. **INFL** with peduncles 3.5–15 (22.5) cm long; umbels 3–6; involucre absent, or rarely of a solitary bract to 12 mm long; rays 3–5, 1.5–8.5 cm long, spreading to divaricate; involucels absent or infrequently of 1 or 2 separate ciliolate bractlets to 3 mm long; pedicels 5–20 mm long, spreading to divaricate. **FLOWERS** with petals greenish white; styles about 0.2 mm long. **FRUIT** including thc tails (11) 13–18 mm long, linear-clavate, bristly hispid, the beak convex-obtuse. [*O. berterii* DC.; *Washingtonia obtusa* J. M. Coult. & Rose]. $n = 11$. Mountain brush, aspen, ponderosa pine, Douglas-fir, spruce-fir, and riparian communities. ARIZ: Apa; COLO: Arc, Hin, LPl, Min, Mon, RGr, SJn; NMEX: McK, RAr, SJn; UTAH. 1980–2695 m (6500–8900'). Flowering: Jun–Jul. Fruit: Jul–Aug. Alaska to California, east to South Dakota and New Mexico; also in Great Lakes region; Argentina and Chile.

Osmorhiza occidentalis (Nutt.) Torr. (western) Western sweet cicely, western sweetroot. **STEMS** mostly 6–13 dm tall, from a taproot, with few or no persistent leaf bases, strongly aromatic. **LEAVES** (1) 2 times pinnate or the upper cauline ones ternate-pinnately compound, with 3–4 pairs of opposite lateral primary leaflets; petioles of lower leaves 4–30 cm long or longer, the upper ones much reduced, the lowest pair of primary leaflets usually again pinnate, usually over 1/2 as long as the leaf blade, with petiolules obovate to lance-elliptic or ovate, coarsely toothed and some often lobed. **INFL** peduncles 6–20 cm long; umbels 3–5; involucre absent or occasionally of 1–2 linear or filiform bracts to 16 mm long; rays 7–13, 2–6.5 cm long; involucels absent; pedicels 2–7 mm long; calyx obsolete; **FLOWERS** with petals greenish white or greenish yellow, 1–2 mm long; stylopodium low; styles 0.7–1 mm long; carpophore divided to the base. **FRUIT** 16–20 mm long, 2–3 mm wide, linear, glabrous. [*Glycosma occidentale* Nutt. ex Torr. & A. Gray; *G. maxima* Rydb.]. $2n = 22$. Tall forb, aspen, oak-maple, spruce-fir, and riparian communities. COLO: Arc, Min, Mon, RGr; UTAH. 1970–2695 m (6500–8900'). Flowering: May–Jul. Fruit: Jun–Aug. British Columbia and Alberta south to California and Colorado. The Paiute used the root for stomachaches, gas pains, and as a dip to kill chicken lice.

Oxypolis Raf. Cowbane, Hog-fennel

(*oxy*, sharp, + *polios*, white, referring to the subulate involucel and white petals) Perennial, caulescent herbs from fascicled tuberous roots. **LEAVES** pinnate. **INFL** an open umbel; involucre absent; rays ascending. **FLOWERS** with calyx teeth conspicuous; petals white to purple; stylopodium conic; carpophore divided to the base. **FRUIT** oblong to oval, strongly flattened dorsally, with dorsal ribs filiform and lateral ribs broadly winged. About seven species of western North America.

Oxypolis fendleri (A. Gray) A. Heller (for Augustus Fendler, German-American botanist) Fendler's cowbane. **STEMS** 6–8 dm tall, without persistent leaf bases. **LEAVES** pinnate with 2–5 pairs of opposite lateral leaflets, the upper ones sometimes reduced to bladeless or nearly bladeless sheaths, the petioles (3) 5–15 cm long or the upper blades sessile on a dilated sheath; blades 7–17 cm long, oblong in outline, the leaflets sessile, 2–5 cm long, ovate to orbicular, shallowly to deeply crenate-dentate or serrate or rarely incised, or those of the upper leaves lanceolate to linear and sometimes entire. **INFL** with peduncles (1) 4–20 cm long; umbels usually 4 or more per stem; involucre absent; rays 5–14, 1–5 (7) cm long, ascending; involucels absent; pedicels 3–10 mm long. **FLOWERS** with petals and stamens white; styles mostly less than 1 mm long. **FRUIT** 3–5 mm long. [*Archemora fendleri* A. Gray]. Stream banks. ARIZ: Apa; COLO: Arc, Con, Hin, LPl, Min, Mon, RGr, SJn; NMEX: SJn; UTAH. 2455–2760 m (8050–9060'). Flowering: Jul–Aug. Fruit: Aug–Sep. Wyoming south to New Mexico.

Pastinaca L. Parsnip

(ancient Latin name for parsnip) Biennial or perennial caulescent herbs from large taproots. **LEAVES** pinnately compound, with broad toothed to pinnatifid leaflets. **INFL** open umbel. **FLOWER** involucre and involucel usually absent; calyx teeth obsolete; petals yellow or red; stylopodium depressed-conic; carpophore divided to the base. **FRUIT** elliptic to obovate, strongly flattened dorsally, the dorsal ribs filiform, the lateral ones narrowly winged. About 14 species native to Eurasia.

Pastinaca sativa L. (cultivated) Garden parsnip, wild parsnip. **STEMS** 8–15 dm tall, caulescent, aromatic, from a taproot. **LEAVES** pinnate or partly bipinnate in some of the lower leaflets, with 3–6 opposite or offset pairs of lateral leaflets; petioles 3–15 mm long or absent and the blade sessile on a dilated sheath; blades 12–35 cm long or longer, oblong in outline; leaflets sessile and sometimes confluent or the lower ones sometimes on petiolules to 1.7 cm long, the blades 2.5–12 cm long, lanceolate to ovate, coarsely serrate, and often lobed. **INFL** a compound umbel, umbels 6–15 or more, the terminal one sessile or pedunculate but shorter than the 2 immediately lateral ones, the lateral umbels alternate or opposite or on opposite branches supporting 2 or more umbels; involucre absent or of 1 to few linear, entire or occasionally toothed or lobed bracts to 2 (4) cm long; rays 9–25, 0.8–8.5 cm long; involucels absent or infrequently of 1–2 linear bractlets to 2 mm long; pedicels 4–20 mm long. **FLOWERS** with petals greenish yellow or reddish; styles 0.4–1 mm long. **FRUIT** 5–8 mm long, 3–6 mm wide, broadly elliptic to orbicular or obovate, strongly flattened dorsally, the dorsal ribs filiform and lateral ones slightly winged. $2n = 22$. Ditch banks, roadsides, fence lines, gardens, fields, margins of ponds and lakes, and moist floodplains. COLO: Arc. 2135 m (7000′). Flowering: May–Jul. Fruit: Jun–Sep. Introduced from Europe, mostly widely established in North America, however, apparently rare in the Four Corners. The cultivated plants, subsp. *sativa*, differ from the wild plants, **subsp. *sylvestris*** Roug & Camus [*P. sylvestris* Mill.], in having larger roots. Some of the apparently wild plants might be recent escapes from cultivation. The Paiute used the root as a medicine for tuberculosis.

Podistera S. Watson Podistera, Woodroot

(*podos*, foot, + *stereos*, solid; probably in regard to the compact habit) Perennial, acaulescent, glabrous plants from taproots or branched caudices. **LEAVES** pinnate with deeply lobed leaflets. **INFL** a solitary, open to congested umbel; involucre wanting; involucel of toothed bractlets. **FLOWER** calyx teeth conspicuous, ovate; petals greenish yellow; stylopodium conic; carpophore stout, undivided. **FRUIT** oval, slightly flattened laterally, the ribs filiform to prominent. Three species native to western North America.

Podistera eastwoodiae (Rose ex Eastw.) Mathias & Constance (for Alice Eastwood, botanist and curator of the herbarium at the California Academy of Sciences) Eastwood's podistera. Plants acaulescent, tufted, somewhat aromatic; taproot 5–15+ cm long, 0.2–1 cm in diameter, enlarging variously especially near base. **STEM** caudex simple or branched, often clothed with persistent leaf bases, pseudoscape absent, rhizomes absent. **LEAVES** herbaceous, often delicately so, glabrous, green, petioles 1–12 cm long; blades narrowly to broadly oblong in outline, 2–9 cm long, 0.8–3 cm wide, pinnate with 4–6 opposite (sometimes appearing whorled) pairs of lateral leaflets, leaflets sessile, distinctly veined, 5–20 mm long, mostly 5–25 mm wide, ovate to obovate in outline, ternately or palmately lobed or cleft, ultimate segments linear to narrowly lanceolate; terminal leaflet deeply pinnatifid to ternately divided similar to lateral leaflets, apices acute to apiculate. **INFL** of 1–10+ umbels, in fruit moderately dense, rounded, 1–2.5 cm wide; peduncles 1.5–35 cm tall, glabrous; involucre absent; rays 5–8, 2–8 mm long, lengthening little in fruit; involucel of 3–6 bractlets, linear to broadly spatulate (outline), entire but usually 3 (5–7)-lobed apically, 4–6 mm long, usually surpassing the flowers, free or fused at very base, green, veins obvious; pedicels 1–2 mm long, lengthening little in fruit. **FLOWERS** with calyx teeth 0.3–0.5 mm long, lanceolate to ovate; petals greenish yellow, often purplish with age; styles 0.8–1.2 mm long. **FRUIT** 3–4 mm long, broadly ovoid, ribs prominent, not winged, stylopodium subconical, prominent, oil tubes 3–4 per interval, 4–6 on commissure; carpophore present. Alpine meadows and spruce-fir forest communities. COLO: Arc, Con, Hin, LPl, Min, Mon, SJn. 3200–3810 m (10500–12500′). Flowering: Jun–Jul. Fruit: Jul–Sep. La Sal Mountains, Utah; Colorado south to northern New Mexico.

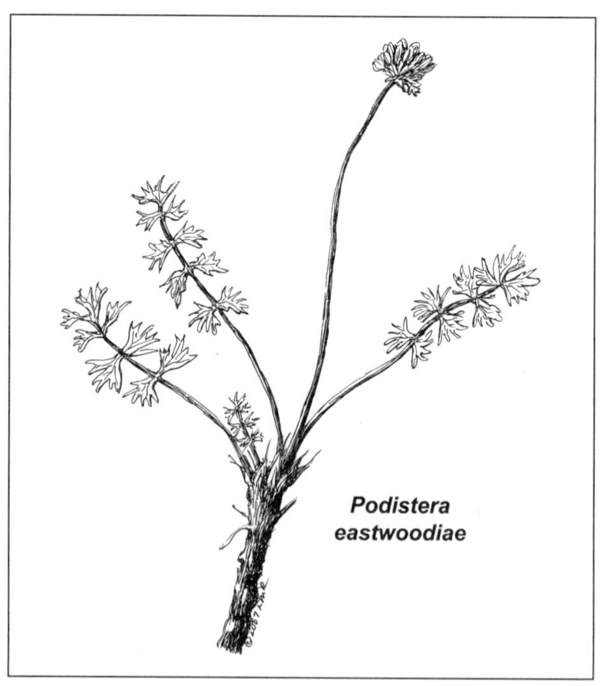

Podistera eastwoodiae

Sanicula L. Black Snakeroot, Sanicle

(to heal) Biennial or perennial herbs. **LEAVES** cleft or dissected. **INFL** of several compact, headlike umbels; involucre more or less foliaceous, often appearing as opposite sessile leaves; involucel of several prominent or inconspicuous bractlets. **FLOWERS** with calyx teeth well developed, mostly connate; petals white, greenish white, yellow, purple, or blue; stylopodium flattened and disclike. **FRUIT** oblong-ovoid to globose, beset with prickles or tubercles, mericarps varying from terete to laterally or dorsally somewhat compressed, the ribs usually obsolete. Nearly 40 species, mostly cosmopolitan in distribution and usually found in temperate regions and mountainous parts of the tropics.

Sanicula marilandica L. (of Maryland) Maryland sanicle. **STEMS** 4–12 dm tall, single, often branching above. **LEAVES** alternate, the lowermost cauline leaves well developed, basal and cauline ones long-petiolate, the blade 6–16 cm wide, palmately 5–7-parted or palmately compound, the segments or leaflets sharply toothed and sometimes shallowly lobed, or some of them often more or less deeply bifid; cauline leaves usually several, gradually reduced upward and becoming sessile. **INFL** ultimate umbels about 1 cm wide or less at anthesis, greenish white, subtended by a few minute narrow bractlets, mostly 15–25-flowered, the staminate flowers more numerous than the erect ones, or some umbellets all staminate. **FLOWERS** with calyx lobes firm, lance-triangular, attenuate, slightly connate at the base; styles elongate, often persistent and longer than the prickles in fruit. **FRUIT** ovoid, 4–6 mm long, 3–5 mm wide, covered with numerous uncinate prickles, the lower ones rudimentary; mericarps subterete in cross section. $2n = 16$. [*S. nevadensis* S. Watson; *S. septentrionalis* Greene]. Lightly wooded slopes and flats from foothills to moderate elevations in the mountains. COLO: Arc; NMEX: RAr. 2135–2440 m (7000–8000′). Flowering: May–Jul. Fruit: Jun–Aug. Washington east to Maine, and south to Florida; most of the United States except much of the Southwest.

Sium L. Water Parsnip, Skirret

(water plant) Perennial, caulescent herbs from fascicles of fibrous roots. **LEAVES** mostly pinnately compound or decompound, with well-marked, toothed to pinnatifid leaflets. **INFL** an open umbel; involucre of entire or incised, often reflexed bracts; involucel of narrow bractlets. **FLOWERS** with calyx teeth minute or obsolete; petals white; stylopodium depressed or rarely conic; carpophore divided to the base (but threadlike and adnate to the faces of the mericarps). **FRUIT** elliptic to orbicular, slightly compressed laterally and somewhat constricted at the commissure, the subequal ribs prominent and corky but hardly winged. Eight species native to northern temperate regions and to South Africa.

Sium suave Walter (sweet-smelling) Hemlock water parsnip. **STEMS** 5–10 dm tall. **LEAVES** pinnate or occasionally partly bipinnate, with 4–6 opposite pairs of sessile lateral leaflets, the lower petioles to 25 cm long, often septate, the upper ones smaller and sometimes reduced to a dilated sheath; lower blades 14–32 cm long, the upper ones reduced; leaflets 2–8 (15) cm long, (1) 3–8 (20) mm wide, linear to lanceolate, sharply and uniformly serrate to pinnatifid with linear segments. **INFL** with peduncles 4–10 cm long; umbels 3–11 or more per stem; involucre of 1–6 separate, often reflexed bracts 2–9 mm long; rays 11–24, 1.5–3 cm long; involucels of (2) 5–12 separate bractlets 2–5 mm long; pedicels 2–8 mm long. **FLOWER** petals and stamens white; styles about 1 mm long. **FRUIT** 2–3 mm long, the ribs prominent. $2n = 6, 12, 22$. Mudflats, marshlands, wet meadows, along streams and shorelines, and in ponds and lakes. COLO: Arc, LPl, SJn. 2285–3200 m (7500–10500′). Flowering: Jul–Aug. Fruit: Aug–Sep. British Columbia to Newfoundland, south to California and Virginia.

APOCYNACEAE Juss. DOGBANE FAMILY

Carolyn Crawford

Perennial herbs, vines, or trees in some tropical genera, most with milky latex. **LEAVES** mostly opposite, some whorled or alternate, without stipules, simple, entire. **INFL** a corymbose cyme, panicle, or sometimes a raceme or solitary flowers in the leaf axils. **FLOWERS** perfect, actinomorphic or sometimes weakly zygomorphic, 5-merous, sometimes 4-merous, except for the pistils which are mostly 2, hypogynous; calyx 5-lobed, usually deeply parted; corolla sympetalous, tubular, campanulate, funnelform, salverform, or urn-shaped, often pubescent within; nectaries 5, usually surrounding the ovaries, sometimes fewer or absent; stamens 5, epipetalous, alternate with the corolla lobes, the anthers often produced at the base into a sterile appendage, often connivent around and sometimes adherent to the stigmatic cap, 2-locular, introrse; pistils of 2 distinct carpels united by a common style or of 2 united carpels forming a single-celled ovary, each ovary superior to subinferior, the style simple, sometimes divided, the stigma simple or enlarged into a stigmatic cap. **FRUIT** a pair of follicles, or sometimes a capsule, berry, or drupe. **SEEDS** few to many, usually flat, with or without a coma. A nearly cosmopolitan family, but mostly tropical or subtropical, about 150 genera and 1000 species. Many species are cultivated as ornamentals: *Mandevilla*, *Nerium* (oleander), *Vinca* (periwinkle), *Plumeria* (frangipani). A number yield useful drugs and alkaloids. Most species are poisonous if ingested. Modern molecular studies strongly support merging Apocynaceae and Asclepiadaceae into a monophyletic family.

1. Leaves alternate; stamens borne near the summit of the corolla tube, the anthers free from the stigma and each other; flowers large; seeds lacking a coma .... ***Amsonia***
1' Leaves opposite; stamens borne near the base of the corolla tube, anthers connivent around the anther cap and adnate to it; flowers small; seeds with a coma .... ***Apocynum***

Amsonia Walter Blue-star

(for Dr. Charles Amson, American physician of Virginia) Perennial herbs, erect, with milky latex. **LEAVES** alternate, irregularly scattered to subverticillate, entire, sessile to petiolate. **INFL** a thyrsoid or corymbose cyme, terminal or lateral in some species, few-flowered. **FLOWER** calyx deeply 5-parted, slightly imbricate; corolla salverform, the tube often narrowly funnelform, somewhat inflated at the insertion of the stamens, constricted at the orifice, pubescent within with reflexed hairs, lobes lanceolate, from 1/5 to as long as the tube, nectaries lacking; stamens included, epipetalous, borne near the summit of the corolla tube, anthers separate; pistils 2, but united by a common filiform style, the stigmatic cap cylindric, subtended by a membranous collar and bilobed above. **FRUIT** 2 erect, slender follicles, cylindric or torulose. **SEEDS** numerous, corky, lacking a coma (Woodson 1928). North America and Japan, about 15 species.

1. Calyx 4–7 mm long; corolla lobes about 1/2 as long as tube; follicles constricted between the seeds (torulose); leaves narrowly lanceolate to linear .... ***A. eastwoodiana***
1' Calyx 1–3 mm long; corolla lobes nearly as long as the tube; follicles cylindrical, not constricted; leaves ovate-lanceolate .... ***A. jonesii***

Amsonia eastwoodiana Rydb. (for Alice Eastwood, longtime herbarium curator at the California Academy of Sciences, who retired in 1949 at the age of 90) Perennial herb. **STEMS** 15–45 cm tall, erect, simple or branched, with several stems clustered on a much-branched caudex at the summit of a taproot; gray-tomentose or glabrous. **LEAVES** alternate, 2–7 cm × 3–8 mm, narrowly lanceolate to linear, more or less sessile. **INFL** a loose thyrse usually surpassed by the foliage; bracts much reduced. **FLOWER** pedicel 1.5–3 cm long; calyx 4–7 mm long, the lobes narrowly lanceolate, attenuate, tomentose or glabrous; corolla salverform, tube 9–12 mm long, constricted at the orifice and inflated at the insertion of the stamens, lobes 4–8 mm long, about 1/2 as long as the tube, whitish or pale bluish, slightly greenish inside the tube, glabrous externally, pubescent internally with stiff, retrorse hairs; anthers included; stigma included and positioned below the anthers, cylindric, bilobed at the apex. **FRUIT** a follicle, erect, 3–7 cm long, constricted between the seeds, with a long beaklike tip. **SEEDS** 1.3 mm long, ellipsoid but obliquely truncate at one or both ends. [*A. tomentosa* Torr. & Frém. var. *stenophylla* Kearney & Peebles]. Sand dunes, deep sand, or gravelly soil, usually in *Coleogyne* communities. ARIZ: Nav; UTAH. 1220–1675 m (4000–5500′). Flowering: late Apr–mid-May. Southeastern Utah and northern Arizona.

Amsonia jonesii Woodson (for Marcus E. Jones, plant collector and taxonomist who worked mostly in Utah) Perennial herb. **STEMS** 20–50 cm tall, more or less erect, several stems clustered on a much-branched caudex at the summit of a taproot, sparingly puberulent to glabrous. **LEAVES** alternate, crowded, overlapping, 3–6.5 cm × 12–27 mm, ovate-lanceolate, acuminate, more or less sessile, glabrous, rather thick and leathery. **INFL** a condensed corymb; bracts much reduced. **FLOWER** pedicel 1.5–4 mm; calyx 1–3 mm long, the lobes lanceolate to narrowly lanceolate, glabrous; the corolla salverform, the tube 6–9.5 mm long, slightly constricted at the orifice, the lobes 4.5–8 mm long, usually just slightly shorter than the tube, whitish or powder blue or darker bluish, often with a patch of hairs at the base of the outside of the lobe, densely pubescent inside, with stiff, retrorse hairs; anthers included; stigma included, positioned below the anthers, cylindric, bilobed at the apex. **FRUIT** a follicle, erect, 6–9 cm long, narrowly cylindric, not constricted between the seeds. **SEEDS** 7–8 mm long, narrowly cylindric but obliquely truncate at both ends. [*A. latifolia* M. E. Jones, not Michx.]. Sandy, gravelly, loamy, or clay soils in sagebrush and piñon-juniper communities. COLO: Mon; NMEX: SJn; UTAH. 1220–1830 m (4000–6000′). Flowering: late Apr–May. Western Colorado, northwestern New Mexico, eastern Utah, and northern Arizona.

Apocynum L. Dogbane, Indian Hemp

(Greek *apo*, away, + *kyon*, dog, the classical name of dogbane) Perennial herbs, erect, with milky latex, rhizomatous. **LEAVES** opposite, petiolate, entire, glabrous to pubescent. **INFL** cymose, often compound, terminal or occasionally lateral, calyx deeply 5-parted. **FLOWERS** small, corolla tubular, campanulate or urceolate, the lobes oblong-lanceolate to ovate, from about 1/2 to about as long as the tube, erect to spreading or reflexed, the tube with 5 small appendages alternating with the stamens, the nectaries of the 5 lobes alternating with the stamens; stamens included, epipetalous, borne near the base of the corolla tube, the filaments short and broad, incurved, usually sparsely puberulent, anthers sagittate, adnate to the stigmatic cap, basal 1/2 sterile; pistils 2, united by a common short style, the stigmatic cap ovoid, obscurely bilobed above, and stigmatic below. **FRUIT** follicles in pairs, usually pendulous, sometimes erect, slender, round, elongate. **SEEDS** numerous, with a long coma. About 70 species in temperate and tropical regions worldwide; in North America, 3–4 species with a conservative interpretation of the genus. The fiber and bark were used by native peoples for cordage, resulting in the common name of Indian hemp. The plants are toxic and the toxin remains in dried material. There is considerable variability in *Apocynum* and at one time 23 species with numerous varieties were recognized in North America. This was reduced to seven, then to two–four in more modern treatments. Hybridization is rampant, accounting for many of the forms recognized at one time. In the Four Corners region, the two species are usually easy to distinguish, but hybrids are known and not all plants will key well. The key is constructed to describe the two species definitely present. Hybrid plants (*A. ×floribundum*; *A. ×medium*) will have intermediate features and will key directly to neither species. When the two species are in proximity, expect hybrids. Both of our species are strongly rhizomatous, and clumps or patches of plants reproducing clonally are common. These could be either of the two species or the hybrid. Such populations will preserve features that may not be typical, leading one to the false conclusion that another species is present. In such situations, survey the whole area inside and outside the clone, make multiple specimens, and take copious notes.

1. Corolla usually 6–10 mm (including the spreading and reflexed lobes), about 3 times as long as the calyx; leaves usually drooping (may be merely spreading); inflorescence congested, the terminal cyme well developed and not surpassed by the leafy lateral cymes; flowers white with pink veins; follicles 5–11 cm long .... ***A. androsaemifolium***
1′ Corolla usually 2.5–5.5 mm long (including the spreading and reflexed lobes), less than 2 times as long as the calyx; leaves usually ascending (may be merely spreading); inflorescence diffuse, the terminal cyme surpassed by the leafy lateral cymes; flowers greenish white to cream; follicles 4–20 cm long .... ***A. cannabinum***

Apocynum androsaemifolium L. (with leaves like *Androsaemum*, probably a reference to *Hypericum androsaemum* L., a shrub native to the Old World, used as a medicinal since antiquity) Spreading dogbane. Perennial herbs, 20–100 cm tall. **STEMS** ascending, mostly alternately branched, herbage glabrous to variously pubescent. **LEAVES** drooping, 3–6 cm long, 2–3.5 cm wide, ovate to broadly lanceolate, obtuse and mucronate at the tip, rounded to subcordate at the base, petiole 2–5 mm long. **INFL** terminal, branching trichotomously, sometimes with a few reduced cymes in the upper leaf axils. **FLOWER** calyx lobes 1.6–3.2 mm long, broadly lanceolate, about 1/3 as long as the corolla, corolla 5–10 mm long, including the lobes, campanulate to broadly urn-shaped, the throat 2.7–6.5 mm broad, lobes reflexed, 1.8–3.5 mm long, white with pink veins or pink with darker red veins; anthers about 2.5 mm long. **FRUIT** follicle usually pendulous (ascending in var. *pumilum*), 6–15 cm long. **SEEDS** about 2 mm long with a 1.5 mm long coma. [*A. ambigens* Greene; *A. occidentale* Rydb. ex Bég. & Belosersky; *A. macranthum* Rydb.]. Moist areas such as meadows, openings in woods, pastures, roadsides, and ditch banks. ARIZ: Nav; COLO: Arc, Dol, Hin, LPl, Min, SJn; NMEX: RAr, SJn, UTAH. 1525–2745 m (5000–9000′). Flowering: mid-Jun–early Aug. Most of North America, widespread and common, usually at higher elevations than *A. cannabinum*. If recognized, var. *pumilum* A. Gray with smaller flowers and smaller, ascending follicles, is to the west of our area.

Apocynum cannabinum L. **CP** (like hemp [*Cannabis*], alluding to the usage as a source of fiber, not to the drug usage) Common dogbane, Indian hemp. Perennial herb, 30–120 cm tall. **STEMS** erect, mostly oppositely branched, herbage glabrous or occasionally pubescent. **LEAVES** ascending or slightly spreading, the petiole 2–6 mm long, 5.5–10 cm long, 2–4 cm wide, blade ovate to lanceolate (0.5–2 cm wide and narrowly lanceolate in var *angustifolium*), acute and apiculate at the apex, cuneate to narrowly rounded basally. **INFL** terminal, branching trichotomously and surpassed by 2 lateral leafy cymes. **FLOWER** calyx lobes 1.4–2.5 mm long, corolla 2.5–4 mm long, cylindric or urn-shaped, the throat 1–2.5 mm broad, the lobes 0.8–1.8 mm long, erect to slightly spreading, greenish white to cream; anthers 1.2–2.4 mm long. **FRUIT** a follicle, 12–16 cm long, pendulous. **SEEDS** about 4 mm long, with a coma 2–2.5 cm long. Moist disturbed areas, such as roadsides or ditch banks. ARIZ: Apa, Nav; COLO: Arc, LPl, Mon, SMg; NMEX: McK, RAr, SJn; UTAH. Below 2285 m (7500′). Flowering: mid-Jun–early Aug. Most of North America, widespread and common. If recognized, the narrow-leaved var. *angustifolium* (Wooton) N. H. Holmgren may be in the southern portions of our area.

Apocynum ×medium Greene, as the species (*Apocynum ×floribundum* Greene, as the species). *Apocynum ×medium* is the name most commonly given to the hybrids, but *Intermountain Flora* has used *A. ×floribundum*. Characteristics are intermediate between the two species. ARIZ: Nav; COLO: Arc.

ARACEAE Juss. ARUM FAMILY

including **LEMNACEAE** DUCKWEED FAMILY

C. Barre Hellquist

Perennial, few annual; terrestrial, wetland, emergent, or floating, sometimes reduced to small floating green leaves, usually with milky or watery latex in terrestrial species. Rhizomes, corms, and stolons present in nonfloating species, or with short roots from floating fronds. **STEMS** absent. **LEAVES** or fronds distinct, mostly alternate or clustered, some not differentiated into petiole. **INFL** a spadix subtended by a spathe, or solitary in *Lemna*. **FLOWERS** monoecious or unisexual, perianth absent, stamens 1–12. Fruit a berry or follicle. **SEEDS** 1–40. One hundred nine genera, about 3,337 species worldwide (Daubs 1965; Jacobs 1947; Landoldt 2000). *Lemna* and *Spirodela* commonly reproduce vegetatively. Flowers are rarely seen.

1. Plants lacking roots; fronds (leaves) boat-shaped, lacking veins .... ***Wolffia***
1′ Plants with roots; fronds (leaves) ovate or obovate, flat or gibbous, 1–21-veined .... (2)

2. Roots 1 per leaf; fronds (leaves) with 1–5 (–7) veins .... ***Lemna***
2′ Roots 2–22 per leaf, leaves with (3–) 5–21 veins .... (3)

3. Fronds (leaves) 7–16 (–21)-veined, 1–1.5 times as long as wide; roots (5–) 7–21 .... ***Spirodela***
3′ Fronds (leaves) (3–) 5–7-veined, 1.5–3 times as long as wide; roots (1–) 2–7 (–12) .... ***Landoltia***

Landoltia Les & D. J. Crawford Giant Duckweed

(for Elias Landolt) Perennial floating aquatic; roots (1–) 2–7 (–12), surrounded by a tubular sheath at the base. **FRONDS** (leaves) floating, with 1–10 leaves cohering together; (3–) 5–7-veined; 1.5 times as long as wide. **FLOWERS** surrounded by small utricular, membraneous scale with a slit on one side; stamens 2. **FRUIT** 4-locular. **SEEDS** 1 (–2), longitudinally ribbed (Les and Crawford 1999). This genus contains only one species, *Landoltia punctata.* Worldwide.

Landoltia punctata (G. Mey.) Les & D. J. Crawford (dotted) Giant duckweed. Roots 0.5–3 cm long. **FRONDS** obovate to elliptic, 1.5–8 mm long, flat or gibbous, often with papillae on upper surface along midvein. **FLOWERS** with 1 or 2 ovules. **FRUIT** laterally winged toward top. **SEEDS** with 10–15 distinct ribs. $2n$ = 40, 46, 50. [*Lemna punctata* G. Mey., *Spirodela oligorrhiza* (Kurz) Hegelm., *Spirodela punctata* (G. Mey.) C. H. Thomps]. Quiet waters. COLO: Arc. 2067 m (6820′). Flowering and fruiting unobserved. Washington east to Pennsylvania, south to California, Arizona, Colorado, Texas, and Florida; worldwide.

Lemna L. Duckweed

(Greek for name of a water plant) Perennial, aquatic. Roots 1 per leaf. **FRONDS** (leaves) floating or submersed, 1–20, often in clusters, lanceolate-ovate, flat or gibbous, 1–15 mm long, upper surfaces often with papillae along the veins, veins 1–5 (–7). **FLOWERS** 1 (–2) per frond, surrounded by a small inflated scale, stamens 2. **SEEDS** 1–4, longitudinally ribbed. x = 10, 21, 22 (Hartog and van der Plas 1970; Landolt 1986; Landolt and Kandeler 1987). Twelve species, worldwide.

1. Plants submersed, fronds long-petiolate, 6–15 mm long, with lateral fronds usually remaining attached to the parent plant .... ***L. trisulca***
1′ Plants floating, fronds elliptic to linear-oblong, 1–6 mm long, forming single plants or in clusters .... (2)

2. Leaves 1 (–3)-veined .... ***L. minuta***
2′ Leaves 3–5 (–7)-veined .... (3)

3. Root sheath winged at base; root usually with acute tip; fronds with 1 or more distinct papillae near the apex .... (4)
3′ Root sheath not winged at base; root usually rounded at tip; fronds with or without distinct papillae near the apex .... (5)

4. Fronds 1–1.7 times as long as broad; root sheath wing 2–3 times as long as wide .... ***L. perpusilla***
4′ Fronds 1–3 times as long as broad; root sheath wing 1–2.5 times as long as wide .... ***L. aequinoctialis***

5. Fronds not reddish on the lower surface .... ***L. minor***
5′ Fronds often reddish on lower surface .... (6)

6. Fronds flat, with papillae on midline of upper surface .... ***L. turionifera***
6′ Fronds usually gibbous, with papillae above node and near apex on upper surface, not between node and apex .... ***L. obscura***

Lemna aequinoctialis Welw. (from regions near the equator) Roots to 3 mm, tip acute; sheath winged at base, wing 1–2.5 times as long as broad. **FRONDS** floating, 1–2 to few in clusters, ovate-lanceolate, flat, 1–6 mm, 1–3 times as long as wide, veins 3; 1 distinct papilla near apex on upper surface and 1 above node. **FLOWERS** with 1-ovulate ovary. **FRUIT** 0.5–0.8 mm, not winged. **SEEDS** 8–26-ribbed. $2n$ = 40, 42, 50, 60, 80, 84. ARIZ: Apa. 2630 m (8680′). Flowering and fruiting unobserved. Nebraska east to Illinois, Kentucky, and North Carolina, south to California, Arizona, New Mexico, Texas, and Florida; worldwide.

Lemna minor L. (smaller) Roots to 15 cm, tip mostly rounded, wingless. **FRONDS** floating, 1 or 2–5 or more in clusters, ovate, slightly gibbous, flat, 1–8 mm, veins 3 (–5); greatest distance between lateral veins near or proximal to the middle; papillae not always distinct. **FLOWERS** imperfect, ovary 1-ovulate. **FRUIT** 0.8–1 mm, laterally winged toward apex. **SEEDS** 8–15-ribbed. $2n$ = 40, 42, 50, 63, 126. Quiet waters. ARIZ: Apa; COLO: Arc; NMEX: SJn. 5450–7050m (1650–2150′). Flowering and fruiting unobserved. British Columbia east to Saskatchewan, Ontario, and Quebec, south to California, Arizona, New Mexico, Oklahoma, Alabama, and Florida; Eurasia; Africa; introduced to Australia and New Zealand.

Lemna minuta Kunth (diminutive) Roots to 15 cm, tip rounded or pointed; sheaths wingless. **FRONDS** floating, 1–2 often clustered, obovate, flat, 0.8–4 mm, longer than broad; veins 1. **FLOWERS** imperfect, ovary 1-ovulate. **FRUIT** 0.6–1 mm, wingless. **SEEDS** 12–15-ribbed. $2n$ = 36, 40, 42. [*Lemna minima* Phil.; *Lemna minuscula* Herter]. Quiet waters. ARIZ: Apa; COLO: Arc; NMEX: RAr, SJn; UTAH. Elevation unknown. Flowering and fruiting unobserved. Washington east to Wyoming, Nebraska, Indiana, Ohio, and West Virginia, south to California, Arizona, New Mexico, Louisiana, and Florida; West Indies; Central America; South America; introduced to Eurasia.

Lemna obscura (Austin) Daubs (hidden) Roots to 15 cm, tip usually rounded; sheath wingless. **FRONDS** floating, 1 or 2–5 or more in clusters, obovate, flat or gibbous, with papillae above node or near apex on upper surface, not between node and apex, 1–3.5 mm, veins 3, greatest distance between lateral veins near middle; papillae near apex, lower surface often slightly red-colored, upper surface often red-spotted, veins 3. **FLOWERS** imperfect, ovary 1-ovulate. **FRUIT** 0.5–0.7 mm, wingless. **SEEDS** 10–16-ribbed. $2n$ = 40, 42, 50. [*Lemna minor* L. var. *obscura* Austin]. Quiet waters. ARIZ: Apa. Elevation unknown. Flowering and fruiting unobserved. Arizona east to South Dakota, Wisconsin, Illinois, and Ohio, south to Texas, Mississippi, and Florida, north along the Atlantic coast to New York; Mexico; South America.

Lemna perpusilla Torr. (very tiny) Roots to 3.5 cm, tip usually sharp-pointed; sheath winged at base, 2–3 times as long as wide. **FRONDS** floating, 1 or 2–few in clusters, ovate-obovate, flat, 1–4 mm, 1–1.7 times as long as wide, 1 distinct papilla near apex of upper surface, veins 3. **FLOWERS** with 1-ovulate ovaries, utricular scale on 1 side. **FRUIT** 0.7–1 mm, wingless. **SEEDS** 35–70-ribbed. $2n$ = 40, 42. Quiet waters. NMEX: SJn. Elevation unknown. Flowering and fruiting unobserved. Nebraska east to Massachusetts, south to New Mexico, Tennessee, and North Carolina.

Lemna trisulca L. (three-furrowed) Star duckweed. Roots lacking or to 2.5 cm, tips pointed; sheaths wingless. **FRONDS** with vegetative submersed, fruiting or flowering floating, usually attached, 3–25 mm long, often connected by 2–20 mm stalk; veins 1 (–3), papillae absent. **FLOWERS** imperfect, ovary 1-ovulate. **FRUIT** 0.6–0.9 mm, laterally winged near tip. **SEEDS** 12–18-ribbed. $2n$ = 40, 42, 44, 60, 63, 80. Still, often alkaline waters. ARIZ: Apa; COLO: Arc. 2590 m (8545′). Flowering and fruiting unobserved. Alaska east to Northwest Territories, Manitoba, Ontario, Quebec, and Nova Scotia, south to California, Arizona, New Mexico, Kansas, Arkansas, Indiana, West Virginia, and Virginia; nearly worldwide but not in South America.

Lemna turionifera Landolt (with turions) Roots shorter than 15 cm, tip usually rounded, sheath wingless. **FRONDS** floating, 1–2 often clustered, obovate, slightly gibbous to flat, with papillae on midline of upper surface, 1–4 mm; veins 3, turions sometimes formed. **FLOWERS** imperfect, ovary 1-ovulate. **FRUIT** 0.5–0.6 mm, wingless. **SEEDS** 30–60-ribbed. $2n$ = 40, 42, 50, 80. Quiet waters. ARIZ: Apa; COLO: Arc, LPl, Mon. 1955 m (5865′). Flowering and fruiting unobserved. Alaska, Yukon, and Northwest Territories east to Manitoba, Ontario, and Nova Scotia, south to California, Arizona, New Mexico, Oklahoma, Alabama, and Georgia.

Spirodela Schleiden Great Duckweed

(Greek *speira*, winding, and *delos*, distinct) Perennial, floating aquatic; roots from lower surface of frond, surrounded at base by tubular sheath. **FRONDS** (leaves) floating, with 1–20 leaves cohering together, often with red spot in center of upper surface, often red beneath; 7–16 (–21)-veined; 1–1.5 times as long as wide; turions present or absent, smaller than the growing frond. **FLOWER** 1 per frond, surrounded by a small membranous scale; stamens 2. **SEEDS** 1–3, longitudinally ribbed. n = 10, 18, 23. Two species.

Spirodela polyrhiza (L.) Schleid. (many-rooted) Giant duckweed, duck-meat, water-flaxseed. Roots 7–21, to 3 cm long. **FRONDS** round-obovate, 3–10 mm in diameter, 1–1.5 times as long as wide, upper surface with red spot in center, lower surface reddish purple; (5–) 7–21 veins, flat, rarely gibbous. **FLOWERS** with 1 or 2 ovules. **FRUIT** 1–1.5 mm long, laterally winged. **SEEDS** with 12–20 ribs. $2n$ = 30, 32, 38, 40, 50, 80. Quiet waters. COLO: Arc. Elevation unknown. Flowering and fruiting unobserved. British Columbia east to Manitoba, Ontario, Quebec, and Prince Edward Island, south throughout the United States; worldwide.

Wolffia Horkel ex Schleid. Water-meal

(for Johann Friedrich Wolff, 1778–1806, German physician) Perennial, floating and submersed; roots absent. **FRONDS** (leaves) 1 or 2, coherent, globular, ovoid, or boat-shaped, smaller than 1.6 mm; veins 0; turions light green, globular, smaller than fronds. **FLOWER** 1 per frond; stamen 1. **FRUIT** 2-locular. **SEED** 1, nearly smooth. x = 10, 20, 21, 22, 23. Eleven species, worldwide.

Wolffia brasiliensis Wedd. (Brazilian) **FRONDS** boat-shaped, 0.5–2.6 mm, rounded at apex, papilla prominent in center of upper surface; upper surface bright green, with 50–100 stomates. $2n$ = 20, 40, 42, 50, 60, 80. [*Wolffia papulifera* C. H. Thomps., *Wolffia punctata* Griesb.]. Quiet waters. COLO: SJn. 2675 m (8825′). Flowering and fruiting unknown. Washington east to Wisconsin and Massachusetts, south to California, Colorado, Texas, and Florida; Mexico; West Indies; Central America; South America.

ARALIACEAE Juss. GINSENG FAMILY

John R. Spence

Woody or herbaceous perennials, with or without prickles. **LEAVES** stipitate, alternate, often very large, simple to compound, often palmately lobed; petiole base broad and sheathing. **INFL** flowers solitary or in a paniculate umbel. **FLOWERS** actinomorphic or rarely zygomorphic, epigynous, perfect or unisexual, usually 5-merous; sepals 5, free, reduced or nearly absent; petals free or united at base; stamens alternate; pistil with 2–5 carpels; ovary inferior, plurilocular; styles distinct or united, often with swollen base; stylopodium attached to nectary disc. **FRUIT** a drupe, often multiseeded, or a berry. x =11, 12 or higher. A family of 75 genera, 700–800 species, worldwide in distribution, but concentrated in the tropics and subtropics. The common English ivy, *Hedera helix*, is a member of this family, as is ginseng.

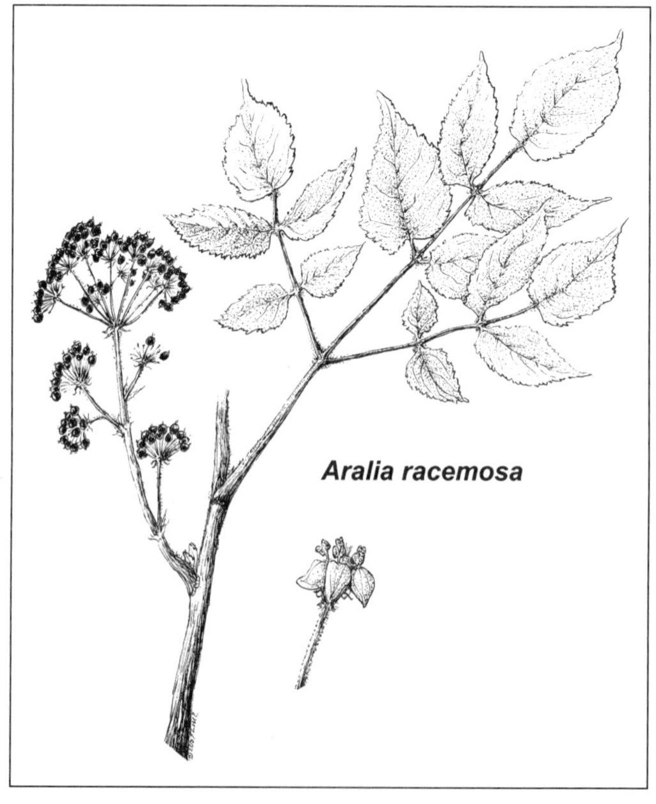

Aralia racemosa

Aralia L. Spikenard

(Latin, of a French-Canadian or American Indian name "aralie") Perennial herbs, shrubs, or small trees, sometimes with prickles. **LEAVES** pinnately or ternately compound, often large, glabrous or puberulent. **INFL** a single or compound umbel; sometimes polygamomonoecious. **FLOWERS** 5-merous; petals white to green, generally imbricate; styles distinct or connate below, persistent in fruit. **FRUIT** a berrylike drupe. A genus of 30–40 species, native to temperate east Asia and North America, Indo-Malaysia (Cronquist, Holmgren, and Holmgren 1997a).

Aralia racemosa L. (for the racemose inflorescence, misapplied) American spikenard. Large unarmed perennial herb from a woody rootstock. **STEMS** 1–2 m tall, somewhat sprawling or clambering. **LEAVES** few, large, 10–100 cm long and wide; primary leaflets pinnately compound; secondary leaflets ovate to orbicular, highly variable on same and different leaves, up to 15–20 cm long; margins serrate to doubly serrate; acuminate; base somewhat obliquely cordate. **INFL** a panicle of several to many umbels, each with 5–20 or more flowers. **FLOWERS** pedicellate; with a persistent cuplike calyx below the ovary, calyx lobes small; petals small, whitish, about 1 mm long, quickly deciduous; styles connate in lower 1/2. **FRUIT** a small, lobed, purple, berrylike drupe, 3–5 mm long. **SEEDS** typically 5 per fruit, large, elliptic, and flattened. $2n = 24$. [*A. bicrenata* Wooton & Standl.; *A. nudicaulis* L. in part]. Moist woodlands, springs, seeps, or riparian zones, in rich soil. ARIZ: Apa, Nav; COLO: LPl, Min, SJn; NMEX: SJn. 1980–2400 m (6500–8000'). Flowering: Jun–Aug. Fruit: Jul–Sep. Eastern North America west across the southern United States to Arizona and Utah. American spikenard is a rare species in the west and may be relictual in our area. A tea made from the roots of *A. racemosa* has been long used as a remedy for colds with coughs, bronchitis, and other chest ailments. Our plants are **subsp. *bicrenata*** (Wooton & Standl.) S. L. Welsh & N. D. Atwood (two notches) [*A. bicrenata* Wooton & Standl.].

ASCLEPIADACEAE Borkh. MILKWEED FAMILY

Carolyn Crawford

Perennial herbs, vines, or shrubs, often rhizomatous, usually with milky latex. **LEAVES** opposite, approximate, whorled, or rarely alternate, simple, usually entire; stipules lacking or small and early-deciduous. **INFL** cymose, usually umbellate. **FLOWERS** perfect, actinomorphic, 5-merous except for the gynoecium, hypogynous; calyx 5-parted or 5-lobed; corolla sympetalous, tubular, campanulate, funnelform or rotate, the lobes often reflexed, 5-lobed to 5-cleft; gynostegium forming a composite structure of filaments, anthers, and style; corona 5-parted, the segments free or coherent and diversely developed as hoods or vesicles, absent in some genera, doubled in others, but usually in one cycle, and each segment with a ligule, inwardly curved appendage, or horn; stamens 5, attached to the base of the corolla tube, the filaments united to form a staminal column around the ovaries, rarely distinct; the anthers basifixed, connivent around the stigma forming an anther head, each introrsely 2-celled, or 4-celled in some, the sacs appendaged or tipped with a scarious, erect, or inflexed appendage, sometimes also appendaged at the base; pollen of each anther sac united into a waxy mass (pollinium), the right-hand pollinium of one anther connected with the left-hand pollinium of the adjacent anther by a translator arm and attached by a gland (corpusculum), so that the 5 pollinia pairs are discharged intact at pollination; gynoecium of 2 distinct carpels, connected at the apex by the stigma head; ovules numerous. **FRUIT** a pair of follicles, usually 1 aborting. **SEEDS** many, compressed, often winged, usually with a coma. Widespread in tropical and subtropical regions, especially in Africa; about 250 genera and 2000 species. Some plants are cultivated: *Asclepias tuberosa* (butterfly-plant), *Hoya*, *Huernia*, *Stapelia*.

1. Plants erect, decumbent, or prostrate herbs or subshrubs .... ***Asclepias***
1' Plants twining vines .... ***Sarcostemma***

Asclepias L. Milkweed, Silkweed

(Latin Aesculapius, from the Greek god of medicine) Perennial herbs (ours), sometimes annual, sometimes shrubby or subshrubs, most with milky latex. **STEMS** erect or decumbent, never twining. **LEAVES** usually decussate, sometimes whorled or irregularly approximate, entire. **INFL** either terminal or lateral or interpetiolar umbellate cymes, occasionally reduced to a single flower. **FLOWERS** of very complex form; calyx deeply 5-parted, the lobes equal, usually bearing few to many small, glandular, chaffy or scalelike appendages at the base; corolla rotate, the lobes either sharply reflexed and concealing the calyx, spreading, or erect; corona of 5 hoods attached to the staminal column, each surrounding and opposite the anthers that are in close contact, but not united by tissue, the hoods of the corona (cowl) shoe- or club-shaped, more or less stipitate to sessile, and deeply saccate at the basal attachment to the column, usually bearing a slender or awl-shaped horn or crest, which often curves upward and over the gynostegium; stamens coherent, the anthers 2-celled, tipped with a triangular to oval hyaline appendage inflexed over the stigma head, with more or less prominent, frequently corneous marginal wings which are nearly face-to-face with those of neighboring anthers, the wings framing the entrances to the 5 stigmatic chambers; pollinia more or less asymmetrically pear-shaped, paired and pendulous from the translator arms, flat, and uniformly fertile; pistils 2, united by a common shield-shaped stigma head, the head adnate to the anthers. **FRUIT** a linear, spindle-shaped, or broadly ovoid follicle, usually erect, but occasionally pendulous. **SEEDS** many, compressed, usually tipped with a coma, rarely naked.

(Woodson 1954). Native to the Americas and Africa, about 120 species. Milkweed flowers are generally considered actinomorphic, except for the two carpels, which can't be seen except by dissection of the flower. A milkweed flower consists of five sepals, usually entire to the base, five corolla lobes, normally reflexed but sometimes ascending, and five hoods, each usually but not always with a protruding horn. However, what makes a milkweed flower unusual is that the staminate and pistillate parts of the flower are "fused" together into one central column, known as the gynostegium, rather than being composed of separate stamens and pistils, as in a conventional flower. The staminate portion of the gynostegium consists of five anther wings, with a pocket at each side enclosing half of a detachable pollination unit called a pollinarium, the whole pollinarium resembling a small wishbone. The pollinarium consists of a single exposed, usually shiny, brownish black "clip" or corpusculum at the center, with a vertical groove that is narrower at the top than at the bottom, and two translator arms, each one extending from either side of the corpusculum. Each arm terminates in a pollen sac, consisting of a large aggregation of pollen grains. The five pollinarium units are "interlinked" with each other by the five anther wings. One pollen sac of each pollinarium is held in the chamber of one anther wing and the other pollen sac of the same unit is held in the chamber of the adjacent anther wing. When the pollinarium is placed in the gynostegium, only the unit's corpusculum is visible, positioned directly above one of the five stigmatic slits. When an insect carries away the pollinarium unit, the anther wings remain attached to the gynostegium. The stigmatic tissue of the gynostegium is mostly located within its central portion, except where it is exposed in the five stigmatic slits, which protrude from between the anther wings. Located at the base of the stigmatic tissue are the two ovaries, or carpels, each capable of producing a follicle with many seeds, each seed usually somewhat flattened and usually with a coma for wind dispersal. The mechanism by which milkweed pollination occurs involves a strong insect, usually a bee (Hymenoptera) or butterfly (Lepidoptera) visiting the flower. While feeding on the copious nectar in the flower's cuplike hoods, one of the hairs on the insect's hind legs often becomes caught in the slit of the corpusculum of one of the pollinaria. Unable to free itself from this entanglement, it struggles until it pulls the whole pollinarium unit from the gynostegium. The insect will then fly around for a brief period, after which the translator arms of the pollinarium unit, which is still caught on its leg hairs, will dry and subsequently rotate ninety degrees, thus positioning the pollen sacs so they can be more easily inserted into the stigmatic slit of another flower. Considering that most asclepiads are obligate outcrossers but are also rhizomatous and thus clonal, the time required to elapse before pollen sac rotation is effected makes it more probable that the insect will visit a flower of a different clone. When the insect then visits another flower, presumably in another population, one of the pollen sacs of the unit will then become inserted into one of the stigmatic slits of the flower being visited. If there is a resident pollinarium unit still in this part of the visited flower's gynostegium, it will impede the removal of the unit being carried by the insect. One of the translator arms of the unit being carried will then get caught in the corpusculum groove of the resident pollinarium. Upon tugging as it tries to free itself, the insect usually manages to break the translator arm of the unit it is carrying, leaving half of the unit (one pollen sac) in the flower's stigmatic slit, thus effecting pollination. Now the insect has a "chain" of half the original unit and a new unit from the pollinated flower hanging from its leg hair. Sometimes chains can consist of three to four partial or whole pollinarium units as the insect travels from plant to plant. The seemingly low probability of this whole procedure occurring is compensated for by the prolific number of wind-borne seeds resulting from a single pollination. The author has observed that a strong insect is necessary to remove the pollinaria from the flowers of larger milkweed species. Many times, weaker honeybees find their legs caught and, unable to pull free, will die in the flowers of showy milkweed (*Asclepias speciosa*).

1. Horns on flowers absent (may require dissection of hood) ....(2)
1' Horns on flowers present (may require dissection of hood) ....(5)

2. Hoods purple, burgundy, or maroon ....(3)
2' Hoods greenish or whitish, clasping the gynostegium ....(4)

3. Hoods curving outward and upward, away from gynostegium, clavate-falciform in shape; corolla lobes green, erect or spreading, not reflexed; follicles erect on deflexed pedicels .... ***A. asperula***
3' Hoods clasping the gynostegium; corolla lobes creamy white; leaves orbicular, glaucous, somewhat bluish; follicles erect on deflexed or spreading pedicels .... ***A. cryptoceras***

4. Hoods scallop-shaped; plants sprawling; leaves usually lanceolate, but can be a variety of shapes and sizes, often crisped on the margins; flowers green; corolla lobes strongly reflexed, giving flower an almost cylindrical look; follicles erect on deflexed pedicels; Rio Arriba County, New Mexico .... ***A. viridiflora***

4′ Hoods truncate or notched at the middle of the apex; plants erect; leaves narrowly linear; corolla white or yellowish green; gynostegium conical, tapering noticeably and extending above the apex of the hood; hoods sometimes with a small "nubbin" inside near the base; follicles erect on deflexed pedicels; plants more widespread .... ***A. rusbyi***

5. Horns present and visible, but only at the tip (may require dissection of hood) ....(6)

5′ Horns clearly visible through most of length (will not require dissection of hood to see all or nearly all of horn) ....(10)

6. Hoods purple, burgundy, or maroon ....(7)

6′ Hoods white, often with a reddish blush ....(8)

7. Hoods curving outward and upward, away from gynostegium, clavate-falciform in shape; corolla lobes green; follicles erect on deflexed pedicels .... ***A. asperula***

7′ Hoods clasping the gynostegium; corolla lobes creamy white; leaves orbicular, glaucous, somewhat bluish; follicles erect on deflexed or spreading pedicels .... ***A. cryptoceras***

8. Plants erect, but less than 20 cm tall, very slender and weak; leaves narrowly linear; follicle pendulous to spreading on a decurved or laterally spreading pedicel .... ***A. cutleri***

8′ Plants sprawling; leaves narrowly lanceolate to broadly ovate; follicle erect on a deflexed pedicel ....(9)

9. Leaves broadly lanceolate to broadly ovate; auricles on hoods slightly below summit of anther head; plants of Utah and Arizona .... ***A. ruthiae***

9′ Leaves lanceolate and narrowly acute at apex (longer and narrower than the preceding); auricles on hoods slightly above the summit of the anther head; plants of New Mexico .... ***A. sanjuanensis***

10. Flowers bright orange or orange-yellow; plants lacking milky latex .... ***A. tuberosa***

10′ Flowers red, pink, creamy, or white, but not orange or orange-yellow; plants with milky latex ....(11)

11. Hoods widely divergent, flaring widely outward and away from gynostegium ....(12)

11′ Hoods more cuplike or extending straight up, but not flaring outward, or only slightly flaring..........(13)

12. Plants erect; flowers large; corolla pink or rose-purple; leaves oblong to lance-ovate, almost twice as long as broad; fruits warty or knobby, not smooth; erect on deflexed pedicel .... ***A. speciosa***

12′ Plants branched from the base in a spherical, basketball-sized shrubby mass, reminiscent of *Ephedra*; flowers small, greenish to pale purple; leaves filiform and revolute; fruits smooth; only one collection in the study area (Pueblo Bonito, San Juan County, New Mexico). A species to the south and southeast of the Four Corners region .... ***A. macrotis***

13. Leaves in whorls, narrow, filiform, less than 4 mm wide, often revolute; follicles erect on an erect pedicel; widespread in the region, probably our commonest milkweed .... ***A. subverticillata***

13′ Leaves wider; linear, linear-lanceolate, lanceolate, or even orbicular, but not in whorls and not filiform and revolute ....(14)

14. Plants small, stems less than 25 cm long; decumbent or prostrate; leaves dimorphic, small ....(15)

14′ Plants large, stems at least 30 cm long; erect; leaves of only one shape, large ....(16)

15. Leaves densely tomentose, dimorphic, ovate below but ovate-lanceolate above, margins somewhat crisped (like cooked bacon), with conspicuous white margins; hoods bright to dull yellow; seeds very large, 10–14 mm .... ***A. macrosperma***

15′ Leaves inconspicuously pilosulous, particularly beneath and on leaf margins, dimorphic, wider below, but narrowly lanceolate near inflorescence, margins not crisped; inflorescence subtended closely by several long, narrow leaves; hoods white to greenish white, sometimes yellowish; seeds smaller than the previous, less than 8 mm long.......... ***A. involucrata***

16. Leaves lanceolate to narrowly lanceolate; follicle pendulous on a spreading pedicel; in the Four Corners region only near the confluence of the San Juan and Colorado rivers .... ***A. labriformis***

16′ Leaves orbicular or elliptic ....(17)

17. Leaves orbicular, often with a notch at the apex (retuse); inflorescence white-tomentulose; follicle erect on a deflexed pedicel, smooth; in a variety of habitats, widespread .... ***A. latifolia***
17' Leaves elliptic; inflorescence white-lanate; follicle spreading to pendulous on a spreading pedicel, with warty processes, not smooth; endemic to sand dunes in Kane County, Utah, and Coconino, Apache, and Navajo counties, Arizona .... ***A. welshii***

Asclepias asperula (Decne.) Woodson (rough, an allusion to the texture of the stem and leaves) Antelope horns. Low herbaceous perennial from a very stout rootstock. **STEMS** several from the rootstock, ascending or decumbent, simple, 20–40 cm tall, minutely and rather roughly pilosulous. **LEAVES** irregularly approximate, with short petioles, lanceolate to linear-lanceolate, with an acuminate tip, base acute to obtuse, 10–20 cm long, 1–3 cm broad, usually folded (conduplicate), with the same pubescence as the stems. **INFL** terminal, umbelliform, many-flowered, crowded, sessile to long-pedunculate, up to 10 cm, the pedicels 1.5–2.5 cm long. **FLOWERS** fragrant; calyx lobes ovate to broadly lanceolate, 4–5 mm; corolla rotate, 9–12 mm, pale yellowish green within and flushed with purple without; gynostegium sessile, the hoods broadly clavate-falciform, spreading with upcurved tips, 8–10 mm, reddish violet to purplish on the back, white within, horn absent; anther head 2 mm high and 3–4.5 mm in diameter, dark purple-brown with a green midsection. **FOLLICLES** erect on deflexed pedicels, rather narrowly spindle-shaped, gradually attenuate to the apex, 6–10 cm long. **SEEDS** oval, 7–8 mm, coma about 3 cm long. [*Acerates asperula* Decne.; *Asclepias decumbens* L.; *Asclepias capricornu* Woodson]. Flats and swales, sandy and rocky hillsides in piñon-juniper or oak woodlands, occasionally in open ponderosa pine woods. ARIZ: Apa, Coc, Nav; COLO: Arc, Dol, LPl, Mon; NMEX: McK, RAr, SJn; UTAH. 1525–2440 m (5000–8000'). Flowering: mid-May–Jul. Southeastern Idaho and eastern Nevada, across Utah and Colorado to western Kansas, Oklahoma, Texas, New Mexico, Arizona, southeastern California; Mexico. Four Corners material belongs to **var. *asperula***. On the southern Great Plains, (Kansas, Oklahoma, and Texas), var. *decumbens* (Nutt.) Shinners replaces the typical form of the species. Used as a snuff ceremonial emetic and as a lotion for mad dog or coyote bite.

Asclepias cryptoceras S. Watson (hidden horn) Pallid milkweed. Perennial herbs. **STEMS** decumbent, 8–35 cm long, 1 or more stems from a thick, woody subtuberous taproot, usually simple or sometimes branched at the base, glabrous, glaucous. **LEAVES** opposite, shortly petiolate, broadly oval or orbicular, 4–9 cm long, glabrous, glaucous. **INFL** terminal in an umbelliform cyme, also in the uppermost leaf axils, few- to several-flowered. **FLOWER** calyx lobes narrowly lanceolate, 6–7 mm long, reflexed; corolla reflexed-rotate, pale yellow or greenish yellow, 10–15 mm long; gynostegium sessile, pale rose; the hoods deeply saccate and decurrent on the column, conspicuously 2-apiculate, 6–9 mm long, reddish violet, horn very inconspicuous or absent; anther head 2.5–3.5 mm high and 4–5 mm broad. **FOLLICLES** erect on erect pedicels, shortly apiculate, 5–7 cm long, smooth, glabrous. **SEEDS** broadly oval, 6–8 mm long, with a coma 1.5–2.5 cm long. Dry, sandy, shaly, or clay hillsides and wash bottoms and arid plains mostly in piñon-juniper habitats but also in sagebrush and shadscale communities. ARIZ: Nav; COLO: Mon; UTAH. 1370–2135 m (4500–7000'). Flowering: late Apr–early Jul. Southwestern Wyoming, western Colorado, Utah, northern Arizona, Nevada; Mono County, California. Our plant is the typical form of the species, **var. *cryptoceras***. Variety *davisii* (Woodson) W. H. Baker is to the north of our area, mostly in eastern Oregon, southern Idaho, and northern Nevada, at the northern edge of the range of the species. Paiutes used a decoction of the root as a wash for headaches and used the latex for ringworm.

Asclepias cutleri Woodson (for Hugh C. Cutler, who collected the type specimen) Cutler's milkweed. Perennial herb, appearing annual, small, with a slender taproot. **STEMS** very slender and weak, simple, branching, if any, from the rootstock only, 5–17 cm tall. **LEAVES** irregularly approximate, sessile, filiform, 3–8 cm long, spreading. **INFL** an umbelliform cyme, terminal, 3–7-flowered. **FLOWER** calyx lobes lanceolate, 2–3 mm long; corolla reflexed-rotate, lobes about 5 mm long, pale greenish rose to pinkish cream, puberulent beneath, column short and thick, hoods 1.5–2 mm long, shortly saccate, with prominent marginal auricles, horn slightly longer than the hood, anther head about 1 mm long, 1.5 mm broad. **FOLLICLES** on pendulous peduncles, narrowly spindle-shaped, gradually attenuate, 3–5 cm long, glabrous to finely puberulent. **SEEDS** broadly oval, 10 mm long, with a white coma 1.5 cm long. Sand dunes or deep sand, sometimes in gravelly places. ARIZ: Apa, Nav; NMEX: SJn; UTAH. 1220–1675 m (4000–5500'). Flowering: late Apr–May. Southeastern Utah and northeastern Arizona; a near endemic to the Four Corners region.

Asclepias involucrata Engelm. ex Torr. (with an involucre, alluding to the whorl of leaves subtending the inflorescence) Perennial herbs, from woody rootstocks. **STEMS** clustered from the rootstock, decumbent to sprawling, branching, up to 25 cm long, puberulent. **LEAVES** irregulary approximate, sessile to subsessile, narrowly lanceolate, apex narrowly acuminate, base acute to obtuse, 1–12 cm long, 3–10 mm wide, usually folded to some degree, pilosulous beneath. **INFL** terminal, and usually also from a few of the uppermost nodes, sessile, rather few-flowered. **FLOWER** calyx lobes ovate-lanceolate, 3–4 mm long; corolla reflexed-rotate, pale green to pale yellowish, tinged with purple without, lobes 3–7 mm long; gynostegium shortly stipitate, the column obconic, 1–1.5 mm long, hoods ovate, acute, 3–4 mm long, horns incurved or ascending, about as long as the hood, the anther head truncately conic, about 2 mm long and 3 mm wide. **FOLLICLES** erect on deflexed pedicels, stoutly spindle-shaped, shortly apiculate, 4–7 cm long, pilosulous to glabrate. **SEEDS** oval, 6–8 mm long, coma 2–3 cm long. Dry, gravelly hills, arroyo bottoms, and flats. ARIZ: Apa, Nav; NMEX: McK, RAr, San, SJn; UTAH. 1220–2285 m (4000–7500′). Flowering: early May–early Jun. Southwestern Kansas, southeastern Colorado, Oklahoma panhandle, Texas panhandle, northeastern and western New Mexico, much of Arizona; Mexico. Does not intergrade with *A. macrosperma*, despite reports to the contrary. Navajos used a poultice of heated roots for toothaches. A favorite jackrabbit food.

Asclepias labriformis M. E. Jones (liplike) Dwarf milkweed. Perennial herbs from a woody taproot. **STEMS** slender, usually simple, 20–50 cm tall, erect, puberulent when young, becoming glabrate. **LEAVES** irregularly approximate, lanceolate to narrowly lanceolate, apex acute to acuminate, base acutely cuneate, 5–15 cm long, 0.6–2.5 cm wide, with a patch of coarse hairs near the base of the midrib. **INFL** umbelliform cymes, lateral in the upper 1/2 of the stem, not terminal, rather few-flowered. **FLOWER** calyx lobes lanceolate, 3–5 mm long, creamy yellow; corolla pale yellowish green, lobes 6–8 mm long, gynostegium shortly stipitate, the column narrowly cylindrical, 1–1.5 mm high, the hoods subquadrate, truncate, 3–4 mm long, horns incurved, somewhat longer than the hood, the anther head truncately conic, 1.5–2 mm. **FOLLICLES** pendulous, ovoid to broadly spindle-shaped, shortly apiculate, 5–7 cm long, glabrous, smooth. **SEEDS** very broadly oval, 1–2 cm, coma 1.5–1.7 cm. Dry, sandy soil with moisture beneath, in washes, sandstone canyons, and other sandy areas or roadsides in blackbrush, rabbitbrush, sagebrush, and piñon-juniper communities. UTAH. 1065–1980 m (3500–6500′). Flowering: late May–early Jul. Endemic to eastern Utah.

Asclepias latifolia (Torr.) Raf. (broad leaves) Broadleaf milkweed. Perennial herb, from a thick woody rootstock. **STEMS** erect, stout, usually simple, 20–60 cm tall, tomentulose when young, becoming glabrate. **LEAVES** opposite, shortly petiolate, broadly ovate, obovate, or elliptic, often retuse with a mucronate tip apically, 4–16 cm long, 4–13 cm wide, somewhat glaucous, with a patch of coarse hairs on the midrib near the base. **INFL** umbelliform cymes, lateral at several of the upper nodes, many-flowered. **FLOWER** calyx lobes ovate-lanceolate, 3–5 mm long, corolla reflexed-rotate, pale green to greenish white, the lobes 8–12 mm long; gynostegium shortly stipitate, the column broadly obconic, 1–2 mm high; hoods subquadrate, 2.5–4 mm long, greenish yellow and yellowing in age, horn sharply incurved, somewhat longer than the hood; anther head 3–4 mm high, 3.4–4 mm wide, truncately conic. **FOLLICLES** erect on deflexed pedicels, broadly spindle-shaped, puberulent to glabrous, smooth. **SEEDS** about 7 mm long, with a coma about 2 cm long. High plains, prairies, dry washes, canyon bottoms, open desert, and roadsides. ARIZ: Apa, Coc, Nav; NMEX: SJn; UTAH. 1390–2070 m (4000–6800′). Flowering: May–Jul. Common in the western half of the southern Great Plains from southern Nebraska to Texas, scattered and uncommon westward in New Mexico, Arizona, and Utah. Becomes weedy in overgrazed rangeland.

Asclepias macrosperma Eastw. (big-seeded, a very appropriate name) Eastwood's milkweed. Perennial herbs, from a woody rootstalk. **STEMS** clustered from the rootstalk, more or less prostrate, simple or occasionally branched, 4–20 cm long, densely white-tomentulose. **LEAVES** irregularly approximate, shortly petiolate, lanceolate to broadly lanceolate, 3–6 cm long, 8–22 mm broad, the margins crenate, crisped (like fried bacon). **INFL** umbelliform cyme, terminal, densely tomentulose, several- to many-flowered. **FLOWER** calyx lobes 2–3 mm long, reflexed, narrowly oblong to lanceolate, tomentulose, pale green; corolla reflexed-rotate, lobes 4–6 mm long, pale yellow or yellowish green, gynostegium very shortly stipitate, less than 1 mm high; hoods broadly ovate, 2.5–3.5 mm long, yellow, the horn shorter than the hood; anther head 1.2–1.5 mm high and 2–2.6 mm wide. **FOLLICLES** erect on deflexed pedicels, 4–7 cm long, broadly spindle-shaped, shortly apiculate, puberulent, smooth. **SEEDS** broadly oval, 10–14 mm long, coma 13–20 mm long. [*Asclepias involucrata* Engelm. ex Torr. var. *tomentosa* Eastw.]. In deep sand or sandy places and along washes or arroyos, usually in blackbrush communities. ARIZ: Apa, Nav; COLO: Mon; NMEX: SJn; UTAH. 1065–1830 m (3500–6000′). Flowering: late Apr–early Jun. Southeastern Utah, extreme southwestern Colorado, northeastern Arizona, and extreme northwestern New Mexico. Woodson (1954) considered the ranges of *A. macrosperma* and *A. involucrata* to overlap in southeastern Utah and extreme northern Arizona. Neither we nor the authors of *Intermountain Flora* have found this to be the case. We have examined about 50 specimens from San Juan County, Utah, alone (28 at BRY), and they were all *A. macrosperma*. Reduction of *A. macrosperma* to synonymy under *A. involucrata* is unjustified (Sundell 1990). The species do not intergrade. At least a dozen specimens at ARIZ identified as *A. involucrata* are really *A. macrosperma*.

Asclepias macrotis Torr. (big ears, alluding to the spreading hoods, large for the flower) Long hood milkweed. Perennial subshrub. **STEMS** densely clustered from the rootstalk, repeatedly branching, very slender and twiggy (resembling a small *Ephedra*), 10–30 cm tall, more or less hemispherical. **LEAVES** opposite, sessile, filiform to acicular, 3.5–8 cm long, 1–4 mm wide, revolute, glabrous. **INFL** solitary and lateral from the upper nodes, few- to several-flowered, surpassed by the foliage (can be hard to see). **FLOWER** calyx lobes ovate, 2–3 mm long; corolla pale greenish yellow, tinged with purple without, the lobes about 5 mm long; gynostegium subsessile, cream or yellowish, the column about 0.5 mm long; hoods very narrowly acuminate, spreading, sometimes with ciliate margins or a couple of marginal teeth, horn much shorter than the hood; the anther head truncately conic, about 1 mm long and 2 mm wide. **FOLLICLES** erect on deflexed pedicels, very narrowly spindle-shaped, long-apiculate, 4–7 cm long, smooth, glabrate. **SEEDS** oval, about 6 mm long, coma 2.5–3.5 cm long. Dry hills and mesas. NMEX: SJn. One collection from the study area (Pueblo Bonito, July 19, 1931, *Castetter 826* UNM, RM). 1340–2010 m (4400–6600′). Flowering: Jun–early Jul. Southeastern Colorado, Oklahoma panhandle, northwestern Texas panhandle and Trans-Pecos Texas, eastern and southwestern New Mexico, southern Arizona; Mexico.

Asclepias rusbyi (Vail) Woodson (for H. H. Rusby, dean of the New York College of Pharmacy, who collected the type specimen) Rusby's milkweed. Perennial herbs. **STEMS** simple, 30–100 cm tall, slender, glabrous, glaucous. **LEAVES** irregularly approximate, sessile, linear, 7–15 cm long, 2–5 mm wide, laxly spreading to reflexed, glabrous, glaucous. **INFL** umbelliform cymes, lateral from a few to several of the upper nodes, pilose. **FLOWER** calyx lobes ovate-lanceolate, 3–4.5 cm long; corolla rotate, pale green or yellowish, with a purplish tinge beneath, lobes 4.5–6.5 mm long; gynostegium sessile to subsessile, hoods 2–2.6 mm long, deeply saccate, truncate, pinkish white or yellowish, auriculate at the base, with well developed winglike auricles, horn, if present, a small crest within the hood; anther head truncately conic, 2–2.5 mm high, 2.8–3.3 mm in diameter, with green wings bent into a small spur at the base. **FOLLICLES** erect on deflexed pedicels, 7–10 cm long, spindle-shaped. **SEEDS** 6–9 mm long with a coma. [*Acerates rusbyi* Vail; *Asclepias engelmanniana* Woodson var. *rusbyi* (Vail) Kearney]. Rocky soil in sagebrush-oak, piñon-juniper, and ponderosa pine–Gambel's oak communities. ARIZ: Apa, Nav; COLO: Arc, LPl, Mon; NMEX: RAr, SJn; UTAH. 1220–2285 m (4000–7500′). Flowering: late May–mid Jul. Western Colorado, southeastern and southwestern Utah, southeastern Nevada, Arizona, western New Mexico. Sundell (1990) focused on the horn morphology and recognized the Arizona plants as *A. engelmanniana* var. *rusbyi*. We find the morphology of the gynostegium to be more significant and the differences consistent between the two species. *Asclepias rusbyi* has a conic anther head; *A. engelmanniana* has a depressed, spheroidal anther head. *Asclepias engelmanniana* is a species of the southern Great Plains, eastern New Mexico, western Texas, and northern Mexico. The two species come in contact only in the Silver City, New Mexico, area. All specimens taken in Arizona are *A. rusbyi*, not *A. engelmanniana*.

Asclepias ruthiae Maguire (for Ruth R. Maguire, Bassett Maguire's wife, to whom he attributed the discovery of the plant) Ruth's milkweed. Perennial herbs, small, low. **STEMS** prostrate or decumbent, 5–15 cm long, simple, or branching belowground from a woody taproot, densely white-tomentulose. **LEAVES** approximate to opposite, petiolate, broadly lanceolate to broadly ovate, acute to acuminate apically, broadly obtuse to rounded basally, 2–4 cm long, 1–2 cm wide, generally white-tomentulose, especially on the leaf margins. **INFL** umbelliform cymes, terminal or in the axils of the uppermost leaves and mostly surpassed by the leaves, few-flowered, densely tomentulose. **FLOWER** calyx lobes ovate-lanceolate, reflexed, 2–3.5 mm long, purple; corolla reflexed-rotate, the lobes 5–6 mm long, pale violet with lighter margins; gynostegium subsessile, greenish; hoods saccate, truncate, 1.5–2 mm long, reddish violet with cream margins and marginal auricles; horn included to barely exserted; anther head 1–1.5 mm high, 1.8–2.2 mm diameter. **FOLLICLES** erect on deflexed pedicels, broadly spindle-shaped, abruptly apiculate, 3–5 cm long, puberulent to nearly glabrous, smooth. **SEEDS** about 10 mm long, coma about 2 cm long. Sandy and hard-packed loamy soils, clay hills, usually in minor drainage channels, desert scrub to piñon-juniper communities. ARIZ: Apa, Nav; NMEX: SJn; UTAH. 1065–1920 m (3500–6300′). Flowering: late Apr–early Jun. Southeast quarter of Utah and adjacent northern edge of Arizona. The small milkweeds (*Asclepiodella* group) are difficult to evaluate, but the group seems to consist of a number of reasonably well-defined taxa with separate ranges. Reduction to a variety of *Asclepias uncialis* is unjustified (Sundell 1990).

Asclepias sanjuanensis K. D. Heil, J. M. Porter & S. L. Welsh (from the region of San Juan, alluding to the river, the county, and the college; the type specimen was taken in a natural area on the San Juan College grounds) San Juan milkweed. Perennial herb, small, low. **STEMS** prostrate or decumbent, 4–8.5 cm long, simple or branching belowground from a woody taproot, minutely tomentulose. **LEAVES** approximate to opposite, petiolate, 2–4 cm long, 4–25 mm wide, oblong-lanceolate, narrowly acute, white-tomentulose on leaf margins and midrib of abaxial leaf surface. **INFL**

umbelliform cymes, usually terminal, sometimes in the axils of the uppermost leaves, few-flowered, sparsely pilosulous. **FLOWER** calyx lobes lanceolate, reflexed, 1.8–3.2 mm long; corolla reflexed-rotate, the lobes 3.5–6 mm long, pale violet; gynostegium subsessile, reddish green; hoods saccate, truncate, 1.5–2.5 mm long, reddish violet with cream margins, with erect marginal auricles, horn included to barely exserted; anther head 1.9–3 mm high, 1.3–2 mm in diameter. **FOLLICLES** erect on deflexed pedicels, broadly spindle-shaped, apiculate, 3.5–6.5 mm long, puberulent, smooth. **SEEDS** about 10 mm long, with a coma. Sandy or gravelly soils, often in small runoff channels, in piñon-juniper habitats. NMEX: McK, SJn. 1525–1950 m (5000–6400′). Flowering: late Apr–late May. A Four Corners regional endemic. Reduction to synonymy under *A. uncialis* is unjustified (Sundell 1990). *Asclepias sanjuanensis* is most similar to *A. ruthiae*.

Asclepias speciosa Torr. (showy, good-looking; the English word "specious" comes from the same root but has a negative tone of falsity in modern usage. Did Torrey imply that the flowers were falsely showy?) Showy milkweed. Perennial herbs, from spreading rootstocks. **STEMS** simple, stout, 60–120 cm tall, generally densely white-tomentose. **LEAVES** opposite, petiolate, broadly ovate to narrowly oblong, apex broadly obtuse or rounded, sometimes acute, base obtuse to rounded, 6–20 cm long, 3–14 cm wide, densely white-tomentose beneath and glabrate above. **INFL** umbelliform cymes in the upper nodes, many-flowered, densely white-tomentose. **FLOWERS** large for the genus, showy; calyx lobes lanceolate, 5–6 mm long, reflexed; corolla purplish, rose, or pink, the lobes 10–15 mm long, broadly lanceolate, gynostegium subsessile, pale rose or pinkish; hoods narrowly ovate-lanceolate, widely spreading, acuminate, pinkish; horn much shorter than the hood; anther head broadly truncate-conic, 2.5–3 mm high, 3–4.5 mm diameter, the wings broad at the base. **FOLLICLES** erect on deflexed pedicels, broadly spindle-shaped, gradually attenuate, with soft thornlike processes, 9–12 cm long, densely white-tomentose. **SEEDS** oval, 6–9 mm long, with a coma 3–4 cm long. Moist or moderately moist places such as meadows, pastures, ditch banks, or along streams and roadsides. ARIZ: Apa, Nav; COLO: Arc, Hin, LPl, Min, Mon; NMEX: McK, RAr, SJn; UTAH. 1370–2500 m (4500–8200′). Flowering: late Jun–Jul. Seemingly ubiquitous from the Missouri River to the Pacific Coast and from southern Canada to central California, northern Arizona, northern New Mexico, and the Texas panhandle, but due to weedy and clonal tendencies, to be expected beyond its natural range wherever there is suitable habitat. Plant used in Eagleway, Female Shootingway, Beautyway, and Beadway as a ceremonial emetic by Navajos. Paiutes and Acoma used a root decoction as a wash for rheumatism. Apaches used the latex as chewing gum.

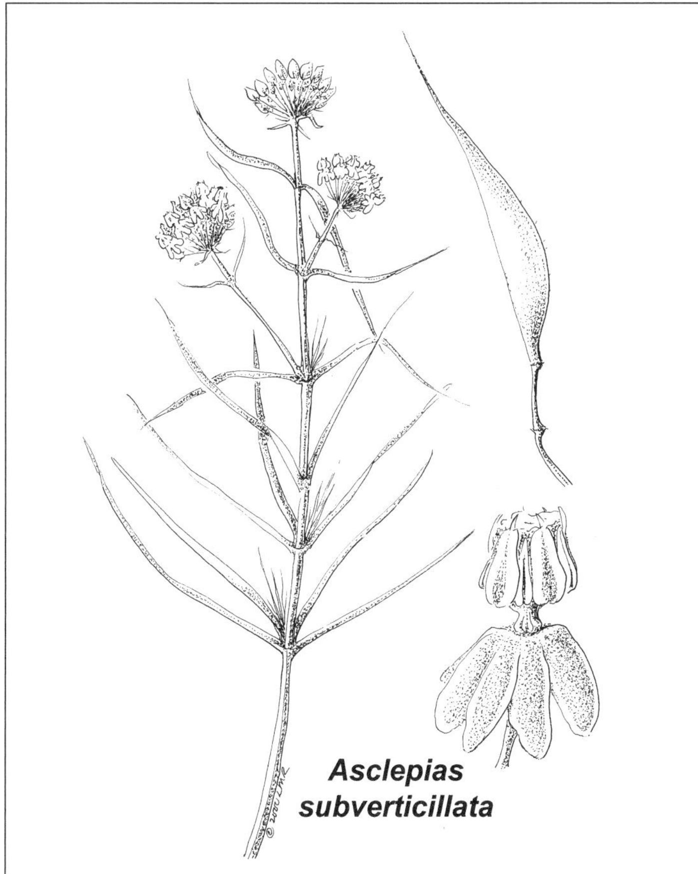
Asclepias subverticillata

Asclepias subverticillata (A. Gray) Vail (not quite verticillate; the leaves on the stems are not all whorled, to distinguish this species from *A. verticillata*) Whorled milkweed. Perennial herbs, from a stout woody rootstock. **STEMS** 30–60 cm tall, occasionally taller, erect, usually several arising from the rootstock, and with short, small-leaved, lateral nonflowering branches, glabrous or the nodes puberulent or puberulent in lines from the nodes. **LEAVES** mostly whorled, usually 3 or 4, sometimes 5 or 6 at a node, the uppermost leaves usually opposite, short-petiolate, linear, 4–10 cm long, 1–4 mm wide, glabrous, or sometimes pilosulous. **INFL** umbelliform cymes in the upper nodes, several- to many-flowered, the flowers fragrant. **FLOWER** calyx lobes lanceolate, reflexed, 1.5–2.5 mm long, pilosulous to glabrate; corolla reflexed-rotate, 3–5 mm long, white, sometimes yellowish or creamy or with a purple tinge; gynostegium narrowly stipitate, 0.6–1.1 mm high, white; hoods cucullate (hood- or cowl-shaped), ovate, 1.2–1.8 mm long, horns narrowly acicular, 2–2.7 mm long, longer than the hoods, incurved, and arching over the anther head so the tips of the horns nearly touch one another; anther head cylindric, 1.3–1.6 mm high and as wide. **FOLLICLES** erect on erect pedicels, narrowly spindle-shaped, 5–9 cm long, 6–8 mm thick, smooth, glabrous or pilosulous. **SEEDS** oval, 6–8 mm long, with a white coma 2 cm long. [*Asclepias verticillata* L. var. *subverticillata* A. Gray]. Moist, sandy places such as roadside ditches and along streams, in sagebrush, piñon-juniper, and ponderosa pine habitats; weedy. ARIZ: Apa, Coc, Nav; COLO: Arc, Dol, Hin, LPl, Mon, SMg; NMEX: McK, RAr, San, SJn; UTAH. 1220–2285 m (4000–7500′). Flowering: mid-Jun–mid-Sep. Southern half of Colorado, southern half of Utah, Arizona, New Mexico, western Texas; Mexico. Weedy and reproducing clonally from the rootstock, it is to be expected occasionally beyond its natural range. *Intermountain Flora* reports that the range is spreading northward in Utah; we have made the same observation in Colorado. The similar *A. verticillata* occurs from the 100th meridian eastward to the Atlantic and Gulf of Mexico. It, too, seems to be spreading as a weed, particularly in western Nebraska. In Colorado, numerous *Pepsis* wasps (tarantula hawks) have been observed on *A. subverticillata*. Used by Hopi mothers to produce a flow of milk. Zunis and Navajos spun seed hair used on prayer sticks. Zunis spun fiber for white dance kilts and women's belts.

Asclepias tuberosa L. (with a tuberous root) Orange milkweed. Perennial herbs; without milky latex. **STEMS** erect or ascending, 20–50 cm tall, occasionally to 90 cm, usually several stems clustered from a woody rootstock, usually branching only at the inflorescence, conspicuously spreading-hispid. **LEAVES** irregularly approximate, 4–8.5 cm long, 4–14 mm wide, shortly petiolate, variable in shape from narrowly lanceolate to broadly oblanceolate, acuminate or rounded at the apex, cuneate to cordate at the base, the margins often revolute. **INFL** umbelliform cymes, in terminal and subterminal helicoid branches, several- to many-flowered. **FLOWER** calyx lobes lanceolate, reflexed, 2–3 mm long; corolla reflexed-rotate, lobes lanceolate, 6–8 mm long, bright orange, yellow-orange, yellow, or reddish; gynostegium 0.6–2 mm high, orange; hoods lanceolate, 4–6 mm long, erect, yellow to orange; horn as long as or slightly longer than the hood, incurved and arching over the anther head; anther heads cylindrical, 1.6–2.2 mm high and as wide. **FOLLICLES** erect on deflexed pedicels, narrowly spindle-shaped, 7–10 cm long, hirsute, smooth. **SEEDS** oval, 6–9 mm, the white coma 3–4 cm long. Moist to moderately moist sandy or gravelly soils in ponderosa pine, oak, and piñon-juniper communities. ARIZ: Apa, Nav; COLO: Arc, LPl; UTAH. 1830–2285 m (6000–7500′). Flowering: late May–Aug. Throughout the eastern United States, sporadic west of the Great Plains in Colorado, Utah, Arizona, New Mexico, and Mexico. *Asclepias tuberosa* is our only milkweed lacking milky latex. Decoction used by Navajos ceremonially and for dog or coyote bites.

Asclepias viridiflora Raf. (green- flowered) Green comet milkweed. Perennial herbs. **STEMS** usually simple, sometimes branching at the base, erect to decumbent, 15–40 cm, inconspicuously puberulent to glabrate. **LEAVES** opposite to irregularly approximate, short-petiolate to subsessile, extremely variable in shape, from suborbicular to linear, 4–13 cm long, 1–6 cm wide, inconspicuously puberulent to glabrate. **INFL** subterminal and lateral from some of the upper nodes, many-flowered, crowded and hemispherical. **FLOWER** calyx lobes ovate-lanceolate, 3–4 mm long, minutely puberulent; corolla reflexed-rotate, pale green, the lobes 6–7 mm long and reflexed so sharply they hug the pedicel, giving an individual flower a narrowly cylindrical appearance; gynostegium sessile, pale green; the hoods deeply saccate, oblongoid, 4–5 mm long, nearly equaling the anther head, lacking horns; anther head 3–4 mm long. **FOLLICLES** erect on deflexed pedicels, narrowly spindle-shaped, long-attenuate, 7–15 cm long, finely puberulent to glabrate. **SEEDS** oval, 6–7 mm long, the coma 3–5 cm long. Prairies, plains, and rocky hillsides. COLO: LPl; NMEX: RAr. 1830–2135 m (6000–7000′). Flowering: mid-Jun–Jul. Great Plains and Midwest, with scattered populations to the Atlantic and Gulf of Mexico, west to Wyoming, eastern Colorado, eastern New Mexico, western Texas, and Mexico, with almost all known occurrences east of the Continental Divide. The populations in Rio Arriba and La Plata counties are the westernmost known.

Asclepias welshii N. H. Holmgren and P. K. Holmgren (for S. L. Welsh, longtime botanist, now emeritus, at Brigham Young University) Welsh's milkweed. Perennial herbs, from a system of vertical taproots and horizontal runners connecting clusters of stems. **STEMS** simple, usually erect, 25–100 cm tall. **LEAVES** opposite, leathery; the upper ones broadly elliptic, ovate, or obovate, 7–9 cm long, 3.5–6 cm wide, rounded to truncate and mucronate apically, rounded to subcordate basally, short-petiolate; the lower leaves smaller, ovate, with a sharply acuminate apex; all leaves densely white-woolly initially, but later becoming glabrous, especially beneath. **INFL** umbelliform cymes lateral at the upper nodes, compactly many-flowered and spherical, white-lanate with cottony pubescent pedicels. **FLOWER** calyx 6–7 mm long, reflexed, linear, lanate beneath; corolla reflexed, but often forced upward due to the compactness of the inflorescence, lobes 5–8 mm long, ovate, cream with a rose-tinged middle, tomentulose beneath; gynostegium subsessile, pale green; hoods broadly truncate and open above, 2.5–3.3 mm long, cream, horns longer than the hoods and curved over the anther head; anther head 1.2–1.8 mm high, 1.5–3.4 mm diameter, cream, with the anther sacs violet-tinged, wings broadest near the middle. **FOLLICLES** pendulous, 4–7 cm long, bearing soft, subulate processes. **SEEDS** 1.2 cm long, coma not specified. Pink to reddish sand dunes with few other plants. ARIZ: Apa, Nav. 1615–1920 m (5300–6300′). Flowering: Jun–Aug. Kane County, Utah (Coral Pink Sand Dunes); Coconino, Apache, and Navajo counties, Arizona.

Sarcostemma R. Br. Climbing Milkweed

(Greek *sarx*, flesh, + *stemma*, crown, alluding to the fleshy inner corona) Perennial, somewhat shrubby vines from a slightly thickened slender taproot. **STEMS** twining, simple to much-branched. **LEAVES** opposite, usually entire, sometimes much reduced and scalelike in some species, cordate, usually with a patch of glands on the midrib near the base, on the top surface of the leaf. **INFL** cymose, racemiform, or umbelliform (ours), in the axils of the leaves. **FLOWER** calyx 5-parted to near the base, where it is adnate to the corolla; corolla rotate to campanulate or salverform; corona complex, consisting of a double ring: an entire ring at the base of the corolla and 5 distinct, swollen, bladderlike structures (corona vescicles) adnate to the corolla or the staminal column directly below the anthers; the anthers 2-celled, the membranous dorsal appendage ovate to deltoid; pollinia pendulous from the translator arms; pistils 2, united by a common stigma head, the head adnate to the anthers. **FRUIT** a follicle, spindle-shaped or obclavate. **SEEDS** unequally biconvex or flattened, tipped with a long coma (Holm 1950; Liede 1996). A genus of wide distribution (mostly the Americas and Africa, but also Asia and Australia), about 60–65 species. The circumscription of *Sarcostemma* is controversial. Holm advocated a broadly drawn circumscription, including in it species that had been assigned at one time or another to several genera (*Funastrum*, *Philibertia*, *Pentacyphus*, *Oxystelma*). Holm's key makes it clear that the differences are not great between the sections of *Sarcostemma* he recognizes. Liede, based on a cladistic analysis of morphological features, recognized all these as separate genera, with our species becoming *Funastrum cynanchoides* (Decne.) Schltr. She provides neither a dichotomous key nor text descriptions of the genera. The floral similarity of the New World species assigned to *Funastrum* and the Old World species remaining in *Sarcostemma* she interprets as parallel evolution, but without specific evidence. Her conclusions as to which species are primitive and which are highly derived are the opposite of Holm's conclusions. More work is necessary to resolve these issues, and the author believes it is premature to unquestioningly accept Liede's interpretation as superior to that of Holm.

Sarcostemma cynanchoides Decne. (like a *Cynanchum*, a similar genus of vining milkweeds) Climbing milkweed. Perennial vine. **STEMS** trailing on the ground or climbing on shrubs, twining, about 2–4 m long, but can be much more in well-developed plants, much-branched, glabrous to sparsely puberulent. **LEAVES** opposite, petiolate, 3–6 cm long, 7–11 mm wide at the broadest basal portion, narrowly lanceolate with an attenuate apex and hastate to cordate at the base, or ovate with an obtuse apex and truncate basally; a patch of coarse hairs near the base of the midrib, the petiole 7–15 mm long. **INFL** an axillary umbel, 8–20-flowered, the peduncle 1–5 cm long. **FLOWER** calyx 2–3 mm long, the segments broadly lanceolate, pilosulous; corolla rotate, the lobes ovate, 5–7 mm long, puberulent to glabrous on the surface, densely ciliate on the margins, white with a median purple streak or pinkish or purplish throughout, the corona ring at the base of the corolla and not adnate to the base of the corona vescicles; the corona vescicles appendaged to the gynostegium, vesicles 1.2–1.8 mm long, suboblongoid, gynostegium subsessile; anther head 1.1–1.4 mm high, 1.5–2.5 mm diameter, with an apical appendage 0.3 mm long, suborbicular. **FOLLICLES** slender, spindle-shaped, pendulous, 7–8 cm long. **SEEDS** about 5 mm long, oblanceolate, coma about 3.5 cm long. Along desert washes, sandy or rocky soil, climbing over whatever shrubby vegetation is present. ARIZ: Nav; UTAH. Below about 1220 m (4000′) in our area, at higher elevations well to the south of us. Flowering: Apr–Sep. Southern Utah, southern Nevada, southeastern California, Arizona below the Mogollon Rim, southern New Mexico, western Texas, southwestern Oklahoma, and southward well into Mexico. Four Corners material belongs to **var.** ***hartwegii*** (Vail) Shinners (for K. T. Hartweg, German botanist who collected in Mexico). It is to be sought along the San Juan River just about anywhere downstream from the Farmington area, perhaps introduced to the river bottomland or along irrigation ditches as a weed. Seek the vines during the winter, when the trees and shrubs are leafless, mark the location, and return during the summer to determine if the vine is *Sarcostemma* or something else. The typical form of the species, var. *cynanchoides*, occurs in southern New Mexico, southeastern Arizona, Oklahoma, and Texas.

ASPARAGACEAE Juss. ASPARAGUS FAMILY

William Jennings

Perennial, tall, fernlike, arising from matted rhizomes or tuberous roots. **STEMS** erect or climbing, freely branched, with filiform to flattened branchlets (which effectively function as leaves), borne in the axils of the true leaves. **LEAVES** alternate, scalelike, dry, thin. **FLOWERS** perfect (ours) or functionally dioecious, borne singly or in pairs along the stem; small, greenish white or greenish yellow, tepals 6, all alike, divided nearly to the base or mostly connate; stamens 6; style with 3 stigmas; ovary superior, 3-celled. **FRUIT** a globose berry. The position of *Asparagus* is controversial. Many taxonomists agree that it should not be placed in the Liliaceae, its traditional location due to the superior ovary. However, recent floras continue to place it in the Liliaceae pending a more general consensus on the dismemberment of the Liliaceae. As here defined it consists solely of the genus *Asparagus*, a large and complex genus of 100 to 300 species, ranging from Siberia to South America, with many species in the Mediterranean region. There is one introduced species in the Four Corners region.

Asparagus L.

(the classical Greek name for the plant) Description of the genus is the same as for the family.

Asparagus officinalis L. (as used in medicine, meaning found in an herb shop) Asparagus. Tall, much-branched, finely dissected perennial herbs, arising from a dense mat of rhizomes. **STEMS** (new) stout and simple with scale leaves up to 5 mm long (the succulent shoots of commercial asparagus); the mature stems freely branching and up to 2 m tall, rather like an overgrown fern. **LEAVES** dry, thin, scalelike, not green, 1–2 mm long on mature plants, with clusters of axillary, filiform to needlelike branchlets 1–2 cm long. **FLOWERS** small, nonshowy, campanulate, 3–7 mm long, greenish white to greenish yellow, cernuous, on filiform jointed pedicels up to 2 cm long. **FRUIT** a berry, 6–8 mm in diameter, red. In cultivation or escaped as a weed along ditches and fencerows, where it is spread (apparently) by droppings from birds that have fed on the fruits. COLO: Arc, Dol, LPl, Mon; NMEX: SJn; UTAH. 1370–1980 m (4500–6500′). Edible portions appear in May and June. Flowering: Jul–Aug. Native to the Old World.

ASTERACEAE Bercht. & J. Presl (COMPOSITAE) SUNFLOWER FAMILY

Stanley L. Welsh, Guy Nesom, and Kenneth D. Heil

Annual, biennial, or perennial herbs or shrubs. **LEAVES** alternate, opposite, or whorled, simple, pinnatifid, or compound. **INFL** of involucrate heads, these solitary or several in corymbose, racemose, paniculate, or cymose clusters. **HEADS** few to numerous on a common receptacle, surrounded by green bracts forming a cup-shaped, cylindrical, or urn-shaped involucre enclosing the flowers in bud; heads entirely of tubular (disc) flowers, of ligulate (ray) flowers, or with tubular corollas forming a central disc and an outer radiating series of ligulate corollas. **RECEPTACLE** flat, convex, conic, or cylindric, naked or bearing chaffy bracts, scales, or hairs. **FLORETS** with calyx lacking or crowning the summit of the ovary and modified as a pappus of capillary bristles, scales, or awns; stamens alternate with corolla lobes; filaments free (rarely connate); anthers united and forming a tube (rarely separate); ovary inferior, of 2 carpels, 1-loculed and with a single ovule; styles 1, 2-cleft, exserted through the anther tube. **FRUIT** a cypsela, an "achene" produced by an inferior ovary. x = 2–19+. A huge cosmopolitan family with about 1550 genera (Welsh 1983).

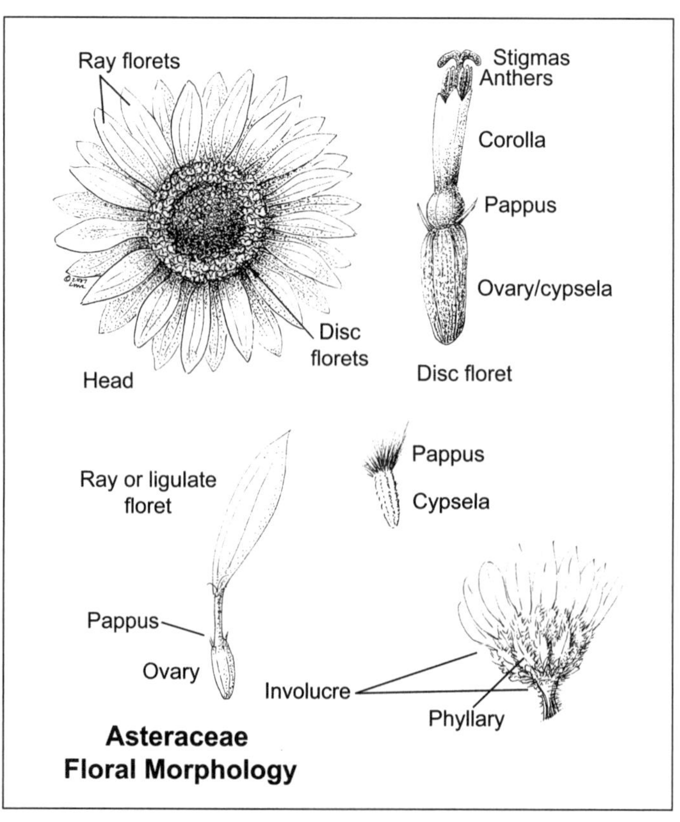

Asteraceae Floral Morphology

1. Corollas all raylike; plants usually with milky juice .... Key 1
1′ Corollas not all raylike, some or all of them tubular; juice seldom if ever milky .... (2)

2. Heads discoid; corollas all tubular; no ray flowers present, or the rays minute and inconspicuous .... Key 2
2′ Heads radiate; corollas not all tubular; ray flowers present and conspicuous .... (3)

3. Pappus of capillary bristles, at least in part .... Key 3
3′ Pappus of awns or scales, or lacking .... (4)

4. Pappus lacking .... Key 4
4′ Pappus present, of awns or scales .... Key 5

Key 1

Corollas all raylike; plants usually with milky juice

Gary Baird

1. Pappus, at least in part, of plumose bristles .... (2)
1′ Pappus of simple bristles, awns, or scales .... (4)

2. Cypselae beaked; leaves long, grasslike; peduncles swollen .... ***Tragopogon***
2′ Cypselae beakless; short, not grasslike; peduncles not swollen .... (3)

3. Involucres mostly 6–12 mm diameter; stems mostly 1; florets 30–100 .... ***Scorzonera***
3′ Involucres mostly 2–5 mm diameter; stems 1–8; florets 5–16 .... ***Stephanomeria***

4. Pappus of 1–3 series of unawned or awned scales .... (5)
4′ Pappus of capillary bristles .... (6)

5. Pappus of 2 or 3 series of unawned scales; corollas blue, closing by midmorning .... ***Cichorium***
5′ Pappus scales in a single series, awned; corollas yellow, not closing by midmorning .... ***Microseris***

6. Achenes more or less flattened; stems leafy; heads in panicles or in corymbose or umbellate clusters .... (7)
6′ Achenes not flattened; stems leafy or scapose; heads solitary or variously disposed .... (9)

7. Involucres cylindric or ovoid-cylindric; achenes beaked .... ***Lactuca***
7′ Involucres broadly campanulate to hemispheric; achenes not beaked .... (8)

8. Corollas yellow .... ***Sonchus***
8′ Corollas blue .... ***Mulgedium***

9. Corollas pink or purplish .... (10)
9′ Corollas yellow or yellowish, or white or cream-colored .... (11)

10. Plants annual; heads mainly 5–7 mm long (from base of involucre to tip of pappus) .... ***Prenanthella***
10′ Plants perennial; heads mainly 8–20 mm long or more .... ***Lygodesmia***

11. Leaves all basal; heads solitary on scapose peduncles .... (12)
11′ Leaves not all basal, the stems leafy; heads not on scapose peduncles .... (15)

12. Cypselae not beaked, truncate; pappus bristles barbellate .... ***Microseris***
12′ Cypselae beaked or not beaked and tapering to apex; pappus not of barbellate bristles .... (13)

13. Cypselae beakless .... ***Uropappus***
13′ Cypselae beaked, beaks 1–10 mm long .... (14)

14. Achenes 10-ribbed or 10-nerved, not spinulose; involucral bracts usually imbricate in several series .... ***Agoseris***
14′ Achenes 4- to 5-ribbed, spinulose, especially near the apex; principal bracts in a single series, the outer much shorter .... ***Taraxacum***

15. Achenes ridged or tuberculate between the angles; leaves crustaceous-margined; plants of southeastern Utah, rare in our study area .... ***Glyptopleura***
15′ Achenes striate between the angles; leaves otherwise; plants widely distributed .... (16)

16. Pappus bristles early-deciduous, more or less united below and falling together, only a few of the stout outer ones may be persistent .... ***Malacothrix***
16′ Pappus bristles persistent or tardily deciduous, and then falling separately .... (17)

17. Pappus tan to brown; involucral bracts not thickened .... ***Hieracium***
17′ Pappus white or whitish; involucral bracts somewhat thickened at base or on midrib .... ***Crepis***

Key 2

Corollas all tubular; no ray florets present, or the ray florets present but with laminae minute and inconspicuous

1. Heads unisexual, the pistillate heads with 1–4 flowers enclosed in involucre; involucre burlike or nutlike; only style tips exserted .... (2)
1′ Heads perfect or unisexual; involucre not burlike or nutlike .... (3)

2. Involucral bracts of the staminate heads separate; fruiting involucres burlike, bearing hooked appendages .... ***Xanthium***
2′ Involucral bracts of the staminate heads united; fruiting involucres various, if burlike, lacking hooked appendages .... ***Ambrosia***

3. Stamens not united by their anthers; florets always unisexual, the pistillate corollas none or much reduced .... (4)
3′ Stamens with united anthers or rarely not united in some species with perfect flowers, at least some flowers usually perfect .... (7)

4. Achenes long-villous; leaves or their lobes linear-filiform .... ***Oxytenia***
4′ Achenes not long-villous; leaves or their lobes not linear-filiform .... (5)

5. Pistillate florets subtended by large, chaffy scales simulating inner involucral bracts; achenes with pectinate or winged margins .... ***Dicoria***
5′ Pistillate florets subtended by chaffy scales or these lacking; achenes without pectinate or toothed wings .... (6)

6. Heads in bracteate racemiform or spiciform arrays .... ***Iva***
6′ Heads in ± ebracteate paniculiform arrays .... ***Cyclachaena***

7. Pappus of capillary bristles, at least in part, these smooth, scabrous, barbellate, or plumose .... (8)
7′ Pappus lacking, or, if present, not of capillary bristles .... (38)

8. Leaves (some or all cauline) opposite or whorled .... (9)
8′ Leaves (at least basally) alternate .... (11)

9. Corollas yellow; involucral bracts in 1 or 2 series, but all equal in length .... ***Arnica***
9′ Corollas white to cream; involucral bracts in 2 to several series .... (10)

10. Achenes 5-angled or 5-ribbed; involucral bracts subequal or in 2 series .... ***Ageratina***
10′ Achenes 10-angled or 10-ribbed; involucral bracts imbricate in several series of different lengths .... ***Brickellia***

11. Leaves spinescent, usually with spiny teeth or lobes, rarely entire but then with spine-tipped apex, mostly thistlelike .... (12)
11′ Leaves entire, denticulate or variously lobed, lacking spines, not thistlelike .... (14)

12. Pappus bristles plumose (rarely some otherwise); receptacle densely bristly .... ***Cirsium***
12′ Pappus bristles merely barbellate .... (13)

13. Receptacle densely bristly, not fleshy or honeycombed; heads nodding .... ***Carduus***
13′ Receptacle with dense bristles or narrow, chaffy scales between disc florets, or the achenes minutely glandular .... ***Onopordum***

14. Plants tomentose to woolly-tomentose .... (15)
14′ Plants glabrous to variously hairy but not tomentose or woolly-tomentose .... (19)

15. Receptacles paleate, outer (pistillate) florets each subtended by a saccate palea, more or less enclosing the achene; outer (fertile) cypselae epappose .... ***Stylocline***

15′ Receptacles epaleate; all florets pappose ..... (16)

16. Plants unisexual or nearly so; heads with either staminate or pistillate florets, or (in *Anaphalis*) predominantly pistillate heads rarely with a few central, functionally staminate florets, and/or predominantly staminate heads rarely with a few peripheral, pistillate florets. ..... (17)
16′ Plants bisexual; each head with pistillate and bisexual or functionally staminate florets. ..... (18)

17. Plants mostly 20–80 (–120+) cm tall; cauline leaves clasping-decurrent, basal usually withering before flowering; phyllaries in 8–12 series ..... ***Anaphalis***
17′ Plants (0.2–) 4–25 (–70) cm tall; cauline leaves not clasping or decurrent, basal usually present at flowering; phyllaries in 3–6 series ..... ***Antennaria***

18. Annuals, (1–) 3–30 cm; heads usually in ± capitate axillary clusters, sometimes in spiciform glomerules; cypselae glabrous or minutely papillate ..... ***Gnaphalium***
18′ Annuals, biennials, or perennials, 15–90 (–100) cm; heads usually in glomerules in corymbiform or paniculiform arrays, sometimes in terminal clusters; cypselae usually smooth or papillate-roughened (papillate in *P. luteoalbum*) ..... ***Pseudognaphalium***

19. Plants shrubs or small trees ..... (20)
19′ Plants herbs or subshrubs, sometimes woody at the base ..... (25)

20. Heads unisexual (staminate flowers may have styles but ovary does not develop); plants dioecious ..... ***Baccharis***
20′ Heads bisexual; plants monoclinous ..... (21)

21. Stems and leaves sparsely to densely silvery sericeous, eglandular; central (staminate) flowers functionally staminate ..... ***Pluchea***
21′ Stems and leaves variously pubescent, not silvery sericeous; central (staminate) flowers fully fertile ..... (22)

22. Phyllaries 4–6 in 2 series, ± equal ..... ***Tetradymia***
22′ Phyllaries 7–60+ in 3–5 series, mostly unequal ..... (23)

23. Phyllaries disposed in spirals and equal or unequal ..... ***Ericameria***
23′ Phyllaries disposed in vertical ranks and unequal ..... (24)

24. Heads in congested, cymiform to corymbiform arrays; disc florets 4–15; cypselae oblong to obconic ..... ***Lorandersonia***
24′ Heads borne singly or in condensed, cymiform clusters grouped in paniculiform or corymbiform arrays; disc florets (2–) 5–6 (–40); cypselae subcylindric ..... ***Chrysothamnus***

25. Plants annual ..... (26)
25′ Plants perennial ..... (30)

26. Plants low, depressed, often moundlike scurfy pubescent herbs; leaves broadly ovate or roundish, entire or toothed ..... ***Psathyrotes***
26′ Plants not as above ..... (27)

27. Involucral bracts in 1 series only (a few short bracts may be present); ray florets absent ..... ***Senecio***
27′ Involucral bracts in 2–5 equal to subequal series; ray florets present but ligules absent or inconspicuous ..... (28)

28. Stems and leaves glabrous; outer phyllaries with foliaceous green zones ..... ***Symphyotrichum***
28′ Stems and leaves variously pubescent; phyllaries without foliaceous green zones ..... (29)

29. Leaf faces often stipitate-glandular or gland-dotted; phyllaries lacking orange to brown midnerves; cypselae densely sericeous, ± strigillose, or glabrous, often stipitate-glandular and/or gland-dotted ..... ***Laennecia***
29′ Leaf faces eglandular; phyllaries with orange to brownish midnerves; cypselae glabrous or sparsely strigillose, eglandular ..... ***Conyza***

30. Involucral bracts in more or less distinct vertical rows ..... (31)
30′ Involucral bracts not in vertical rows ..... (33)

31. Phyllaries disposed in spirals and equal or unequal ..... ***Ericameria***
31′ Phyllaries disposed in vertical ranks and unequal ..... (32)

32. Heads in congested, cymiform to corymbiform arrays; disc florets 4–15; cypselae oblong to obconic .... ***Lorandersonia***
32' Heads borne singly or in condensed, cymiform clusters grouped in paniculiform or corymbiform arrays; disc florets (2–) 5–6 (–40); cypselae subcylindric .... ***Chrysothamnus***

33. Heads in dense spikes; florets rose-purple; Archuleta County, Colorado .... ***Liatris***
33' Heads variously arranged, but if in spikes, the florets not purple; widespread .... (34)

34. Leaves with prominent, whitish, spinescent teeth .... ***Xanthisma***
34' Leaves without spinescent teeth .... (35)

35. Plants herbs; ray florets present but ligules absent or inconspicuous .... ***Erigeron***
35' Plants subshrubs; ray florets absent .... (36)

36. Involucral bracts not longitudinally striate; flowers yellow with dark orange, resinous veins .... ***Isocoma***
36' Involucral bracts longitudinally striate; flowers cream to off-white .... (37)

37. Leaves opposite; achenes 5-angled or 5-ribbed; involucral bracts subequal or in 2 series .... ***Ageratina***
37' Leaves mostly alternate; achenes 10-angled or 10-ribbed; involucral bracts imbricate in several series of different lengths .... ***Brickellia***

38. Receptacle with dense bristles or chaffy with thickened scales between the disc florets .... (39)
38' Receptacle naked or at most short-hairy, never with dense bristles or scales .... (47)

39. Involucral bracts with hooked spines; lower leaves large (resembling rhubarb), cordate at base .... ***Arctium***
39' Involucral bracts without hooked spines; lower leaves not large or cordate at base .... (40)

40. Receptacles bristly .... (41)
40' Receptacles chaffy .... (42)

41. Phyllary appendages dentate or fringed; pappus of persistent, nonplumose bristles .... ***Centaurea***
41' Phyllary appendages entire or lacerate, not fringed; pappus of ± caducous, distally plumose bristles .... ***Acroptilon***

42. Involucral bracts (at least the outer) with foliose, green tips and mainly spinescent teeth; florets yellow to yellow-orange; plants cultivated and escaping .... ***Carthamnus***
42' Involucral bracts seldom, if ever, with tips both foliose and spinescent; flowers variously colored but typically not yellow to yellow-orange .... (43)

43. Involucral bracts in 2 distinct sets: the outer herbaceous, the inner differing in shape and texture; leaves opposite, at least below, or alternate .... (44)
43' Involucral bracts not in 2 unlike sets; leaves alternate or basal .... (45)

44. Leaves alternate throughout; outer involucral bracts about 5, spreading, herbaceous, the inner (1–3 subtending pistillate flowers) larger and broader, becoming strongly accrescent and hooded in fruit .... ***Dicoria***
44' Leaves opposite, at least below; outer involucral bracts various, but not as above, not accrescent and hooded in fruit .... ***Thelesperma***

45. Involucral bracts in 1 series, boat-shaped, each bract enclosing a marginal floret .... ***Dicoria***
45' Involucral bracts in 1 or more series, not boat-shaped and enclosing marginal florets .... (46)

46. Receptacles high-conical; plants herbaceous .... ***Rudbeckia***
46' Receptacles not high-conical; plants woody shrubs .... ***Encelia***

47. Pappus none .... (48)
47' Pappus present .... (52)

48. Leaves opposite, some cauline, somewhat connate at base (to be expected) .... *Flaveria*
48' Leaves alternate or basal .... (49)

49. Heads numerous, in spikes, racemes, or panicles; anthers with acute tips; receptacles flat; plants woody or herbaceous .... (50)
49' Heads solitary on ends of stems, or sometimes corymbose or capitate; anthers with rounded tips; receptacles convex or conic; plants herbaceous, or woody only at base .... (51)

50. Thorny subshrubs or shrubs; disc florets functionally staminate; corollas ± villous .... ***Picrothamnus***
50′ Nonthorny annuals, perennials, subshrubs, or shrubs; disc florets bisexual and fertile, or functionally staminate; corollas usually glabrous, sometimes hairy, not villous .... ***Artemisia***

51. Plants annual; heads solitary or paniculately arranged; leaves green and glabrous .... ***Matricaria***
51′ Plants perennial; heads corymbose or capitate; leaves usually silvery canescent .... ***Chrysanthemum***

52. Pappus of 2–8 caducous awns; plants usually strongly glutinous .... ***Grindelia***
52′ Pappus various, but not of 2–8 caducous awns .... (53)

53. Leaves and involucre conspicuously punctate with translucent oil glands .... ***Dyssodia***
53′ Leaves and involucre sometimes impressed-punctate, but without translucent oil glands .... (54)

54. Pappus of 12 or more scale- or bristle-like segments, these nearly or as long as the achene .... (55)
54′ Pappus of fewer than 12 scalelike segments or else much shorter than the achene .... (56)

55. Pappus of 12–16 linear, acuminate awns; involucres glutinous; leaves 3- to 5-nerved; plants of dunes and other sandy sites in southeastern Utah and northeastern Arizona .... ***Chrysothamnus*** (*Vanclevea*)
55′ Pappus with about 35 flattened, silvery scales and bristles of different widths; involucres not glutinous; leaves 1-nerved; plants not sand-loving .... ***Acamptopappus***

56. Achenes strongly compressed; pappus of 0 or 1 or 2 slender awns .... (57)
56′ Achenes not compressed or, if so, the pappus not of 1 or 2 slender awns .... (58)

57. Involucral bracts in 1 series; leaves long-acuminate or caudate .... ***Pericome***
57′ Involucral bracts in 2 or more series; leaves not long-acuminate or caudate .... ***Perityle***

58. Leaves and/or phyllaries dotted or streaked with pellucid glands containing strong-scented oils .... ***Tagetes***
58′ Leaves and/or phyllaries rarely dotted or streaked (never with pellucid glands containing strong-scented oils); plants sometimes with sessile or stipitate, surface glands and sometimes otherwise strong-scented .... (59)

59. Pappus a crown with margins entire or of short scales united into a crown .... ***Tanacetum***
59′ Pappus not as above .... (60)

60. Involucral bracts with a thin, scarious, white, yellow, or purplish margin and tip .... ***Hymenopappus***
60′ Involucral bracts without a scarious, colored margin and tip .... (61)

61. Leaves all or mostly opposite; annuals; ray florets absent or 1–2, yellow to white .... ***Schkuhria***
61′ Leaves alternate; annuals or perennials; ray florets absent .... (62)

62. Plants scapose; leaves roundish, entire or crenate .... ***Chamaechaenactis***
62′ Plants leafy-stemmed; leaves not roundish and entire or subentire .... ***Chaenactis***

Key 3

Corollas not all tubular; ray florets present; pappus of capillary bristles

1. Rays white, pink, rose, violet, or purple, not yellow .... (2)
1′ Rays yellow or orange-yellow .... (16)

2. Pappus, at least of disc florets, of several to many rigid bristles; achenes pubescent with 2-forked hairs or the hairs barbed at apex .... ***Townsendia***
2′ Pappus, at least of disc florets, of many capillary bristles, at least in part; achenes glabrous or pubescent with simple hairs .... (3)

3. Phyllaries subequal, rarely somewhat graduated; rays usually narrow; style tips short, triangular, rounded, or obtuse; achenes flattened, 2 (–4)-nerved (4–8 or 10–14 in a few species) .... (4)
3′ Phyllaries usually strongly graduated; rays comparatively broad; style tips ovate and acute to subulate, usually lanceolate; achenes fusiform to terete, 1–5-nerved .... (5)

4. Involucres (5–) 7–22 (–25) mm wide; disc florets 25–150; rays usually conspicuous; annual, biennial, or perennial .... ***Erigeron***
4′ Involucres 1–1.5 (–3) mm wide; disc florets 8–30; rays short and inconspicuous; annual .... ***Conyza***

5. Plants perennial and rhizomatous, or annual, or, if from a caudex, then ordinarily less than 10 cm tall ....(6)
5′ Plants perennial from a caudex or taproot ....(10)

6. Low, white-rayed perennial herbs from creeping underground rhizomes and caudexlike stems with scale leaves; cauline leaves coriaceous, densely overlapping; flowering in spring; arid sites .... ***Chaetopappa***
6′ Low to tall, white- to pink- or purple-rayed annual or perennial herbs or shrubs from rhizomes or fibrous roots; cauline leaves herbaceous, not overlapping; mainly flowering in summer and autumn; sites various ....(7)

7. Plants suffrutescent, rushlike, commonly with axillary or subaxillary thorns, from a deep-seated rhizome; southeastern Utah .... ***Chloracantha***
7′ Plants herbaceous, annual or perennial, unarmed, from a taproot or rhizome; distribution various ....(8)

8. Heads in a distinctly corymbiform arrangement; phyllaries rounded, ciliate-fringed, with a green, often basally truncate apical patch; cypselae cylindric, 3.8–4.8 mm, 7–10-nerved; pappus bristles in 2 series of equal length, apically thickened .... ***Eurybia***
8′ Heads in a paniculiform or loosely corymbiform arrangement; phyllaries usually not rounded or ciliate-fringed, green apical patch basally acute; cypselae fusiform, 7–10-nerved; pappus bristles in a single series, subequal in length ....(9)

9. Phyllaries usually unequal, sometimes subequal, proximally indurate, distally with defined green zones, sometimes distally herbaceous or outer wholly foliaceous, sometimes stipitate-glandular; cypselae usually obovoid or obconic, sometimes fusiform, ± compressed, (2–) 3–5 (–10)-nerved .... ***Symphyotrichum***
9′ Phyllaries subequal, herbaceous, without definite distal green zones, not foliaceous, stipitate-glandular; cypselae fusiform, terete, 7–10-nerved .... ***Almutaster***

10. Plants more or less woody, from a ligneous caudex, perennial; heads usually solitary and large; mostly selenophytes .... ***Xylorhiza***
10′ Plants herbaceous, from a taproot, winter-annual, biennial, or perennial; heads usually several to numerous; soil types various ....(11)

11. Annuals or biennials ....(12)
11′ Perennials ....(14)

12. Leaves deeply 1–2-pinnatifid (some or all lobes or teeth acute and bristle-tipped) .... ***Machaeranthera***
12′ Leaves usually entire, toothed, or lobed (if 1–2-pinnatifid, teeth or lobes often rounded, sometimes apiculate, mostly not bristle-tipped) ....(13)

13. Plants hairy, sometimes stipitate-glandular; leaves entire or toothed; ray pappus of 40–50 bristles .... ***Dieteria***
13′ Plants glabrous and leaves entire or toothed (ciliate or teeth bristle-tipped or apiculate), or plants hairy, sometimes stipitate-glandular, and leaves 1–2-pinnatifid; ray pappus usually 0 (if 20–30 bristles, leaves 1–2-pinnatifid) .... ***Arida***

14. Cypselae monomorphic; pappus of relatively fine bristles .... ***Dieteria***
14′ Cypselae ± dimorphic (ray 3-sided and rounded abaxially, disc ± compressed); pappus of relatively coarse (± flattened) bristles ....(15)

15. Stems simple; leaves mostly basal, margins serrate or serrulate; involucres depressed-hemispheric; cypselae 3–9-ribbed per face; pappus bristles coarsely barbellate .... ***Xanthisma***
15′ Stems usually branched; leaves basal (persistent or withering by flowering) and cauline (distally ± reduced) or mostly cauline, margins pinnately lobed or pinnatifid, toothed, or entire; involucres turbinate-campanulate, or hemispheric; cypselae 8–13-nerved per face; pappus bristles barbellulate .... ***Arida***

16. Leaves opposite, at least below ....(17)
16′ Leaves alternate throughout ....(18)

17. Leaves with stiff marginal bristles; involucre and leaves with conspicuous oil glands; plants annual .... ***Pectis***
17′ Leaves without stiff marginal bristles; involucre and leaves without oil glands; plants perennial .... ***Arnica***

18. Pappus of 2–8 stiff, caducous bristles; plants usually glutinous .... ***Grindelia***
18′ Pappus of numerous, usually soft, persistent bristles ....(19)

19. Pappus elements of 2 distinct lengths, the inner of capillary bristles, the outer much shorter, of bristles, setae, and/or scales .... ***Heterotheca***
19′ Pappus elements all relatively long, subequal capillary bristles .... (20)

20. Involucral bracts in distinct vertical ranks .... (21)
20′ Involucral bracts not in distinct vertical ranks .... (23)

21. Outer involucral bracts without loose herbaceous tips; erect stems annual; plants herbaceous; leaves persistent .... ***Petradoria***
21′ Outer involucral bracts with loose herbaceous tips; erect stems perennial; plants shrubs; leaves deciduous ... (22)

22. Phyllaries disposed in vertical ranks and unequal .... ***Chrysothamnus***
22′ Phyllaries disposed in spirals and equal or unequal .... ***Ericameria***

23. Involucral bracts in 1 series, frequently with some smaller bracts at base; style branches truncate apically ..... (24)
23′ Involucral bracts in (2–) 3–6 series, without smaller bracts at base; style branches with distinctly elongate tips .... (25)

24. Roots often fleshy and seldom branched; leaf margins with relatively many callous denticles .... ***Senecio***
24′ Roots seldom fleshy, often branched; leaf margins with relatively few or no callous denticles .... ***Packera***

25. Heads usually numerous and in densely paniculiform arrays, rarely racemiform or corymbiform; involucres usually less than 6 mm long, plants rhizomatous .... (26)
25′ Heads 1 or 2–4, not densely arranged; involucres usually more than 6 mm long; plants taprooted .... (28)

26. Cauline leaves clasping; involucres stipitate-glandular; heads in distinctly flat-topped, tightly corymbiform arrays; involucres 10–11 mm high .... ***Oreochrysum***
26′ Cauline leaves not clasping; involucres sometimes slightly resinous but not distinctly stipitate-glandular; heads in corymbiform to paniculiform or racemiform arrays; involucres 3–6 mm high .... (27)

27. Leaves not punctate .... ***Solidago***
27′ Leaves resinous-punctate .... ***Euthamia***

28. Leaves mostly in a basal rosette or restricted to proximal 1/2 of stem, with 3 parallel, raised nerves; phyllaries not foliaceous, outer similar to inner .... ***Stenotus***
28′ Leaves basal and cauline, nerves not parallel or raised; at least outer phyllaries foliaceous, sometimes grading into immediately subtending foliar bracts .... (29)

29. Cauline leaves clasping or subclasping, upper not closely subtending heads and not grading into phyllaries .... ***Pyrrocoma***
29′ Cauline leaves not clasping or subclasping, upper bracteate, closely subtending heads, and grading into phyllaries .... ***Tonestus***

Key 4

Corollas not all tubular; ray florets present; pappus lacking

1. Rays white, pink, or pink-purple, sometimes yellow .... (2)
1′ Rays yellow, sometimes partly purplish or maroon .... (6)

2. Receptacle naked .... (3)
2′ Receptacle with chaffy scales .... (4)

3. Receptacle broad and flattish; involucral bracts with a dark brown submarginal line .... ***Chrysanthemum***
3′ Receptacle convex, conic, or hemispheric; involucral bracts without a dark brown submarginal line ..... ***Matricaria***

4. Heads small, numerous, in dense flattish or rounded cymose panicles; plants perennial .... ***Achillea***
4′ Heads comparatively large to very large, solitary or few; plants annual or perennial .... (5)

5. Native; perennial with prominent rosettes; widespread .... ***Hymenopappus***
5′ Introduced or native; annual or perennial, without rosettes; uncommon .... (6)

6. Leaves once- or twice-pinnatifid; plants weedy; ray flowers white; San Juan County, New Mexico .... ***Anthemis***
6′ Leaves simple, ovate to lanceolate; plants native; ray flowers yellow; McKinley County, New Mexico .... ***Zinnia***

7. Receptacles naked .... (8)
7′ Receptacles chaffy, at least toward the margin .... (14)

8. Heads 1- or 2-flowered, in dense glomerate clusters, sessile in the forks of the stem, or terminal and leafy-involucrate, (to be expected) .... *Flaveria*
8′ Heads several- to many-flowered, solitary or on terminal peduncles .... (9)

9. Plants woolly .... (10)
9′ Plants not woolly .... (11)

10. Rays persistent, becoming papery .... ***Baileya***
10′ Rays not persistent .... ***Eriophyllum***

11. Involucre and leaves with translucent oil glands .... ***Pectis***
11′ Involucre without translucent oil glands (*Tanacetum* has gland-dotted or glandular-punctate leaves) .... (12)

12. Rays minute; involucral bracts obtuse, with scarious margins .... ***Tanacetum***
12′ Rays conspicuous; involucral bracts acuminate, without scarious margins .... (13)

13. Biennial; stems 10–80 cm tall; leaves mostly alternate .... ***Bahia***
13′ Perennial; stems 3–20 cm tall; leaves mostly opposite .... ***Picradeniopsis***

14. Ray achenes partly or wholly enfolded by their involucral bracts; plants annual, glandular-viscid above; strongly aromatic .... ***Madia***
14′ Ray achenes not conspicuously enfolded by their involucral bracts, or if so, the plants perennial; plants perennial, or if annual, not glandular above; not strongly aromatic .... (15)

15. Involucre distinctly double, the outer bracts herbaceous, the inner ones broader and united to about the middle .... ***Thelesperma***
15′ Involucre not double, the bracts distinct to the base .... (16)

16. Plants subscapose; leaves variously dissected or sagittate; heads broad .... ***Balsamorhiza***
16′ Plants with stems definitely leafy; leaves usually not dissected or sagittate .... (17)

17. Plants shrubby; achenes conspicuously ciliate on the margins, notched at the apex, very flat; plants of the lower San Juan River drainage .... ***Encelia***
17′ Plants herbaceous; achenes not conspicuously ciliate on the margins; plants widespread .... (17)

18. Leaves doubly pinnately dissected; heads numerous in corymbose cymes .... ***Achillea***
18′ Leaves simple, entire or toothed to lobed; heads few to several .... (19)

19. Achenes 2-winged; discs 15–25 mm wide; leaves white-strigose beneath, green above; plants frequently of sandy, disturbed sites; annual .... ***Verbesina***
19′ Achenes not 2-winged; discs 6–15 mm wide; leaves green on both sides; plants commonly of saline clays and commonly in sagebrush and aspen uplands; perennial .... ***Heliomeris***

Key 5

Corollas not all tubular; ray florets present; pappus of awns or scales

1. Receptacle chaffy .... (2)
1′ Receptacle not chaffy, either naked or bristly .... (19)

2. Receptacle bearing a row of chaffy scales between the ray florets and the outer disc florets, otherwise naked; pappus of 10–20 slender setiform scales (ray florets lack a pappus) .... ***Layia***
2′ Receptacle chaffy throughout; pappus not of 10–20 slender setiform scales .... (3)

3. Pappus of awns only, without scales .... (4)
3′ Pappus of scales (at least in part) .... (11)

4. Ray and disc achenes of different compression and texture and often differing in pappus composition and number; McKinley County, New Mexico .... ***Zinnia***
4′ Ray and disc achenes not especially different in compression and texture (or if so, not differing in pappus composition and number) .... (5)

5. Achenes flat and obcompressed .... (6)
5′ Achenes not obcompressed .... (10)

6. Calyculi of (3–) 5–12 bractlets .... (7)
6′ Calyculi none .... (8)

7. Perennial; rays yellow or lacking; achenes not beaked .... ***Bidens***
7′ Annual; rays pink; achenes beaked .... ***Cosmos***

8. Ray florets usually 5–21; annuals .... ***Sanvitalia***
8′ Ray florets usually (2–) 5–35; perennials .... (9)

9. Disc florets functionally staminate (only ray florets produce cypselae) .... (10)
9′ Disc florets bisexual, fertile .... (11)

10. Ray florets 8–35 (in 1–3 series); cypselae shed without accessory structures; achenes winged .... ***Silphium***
10′ Ray florets (2–) 8 (–13); cypselae shed together with subtending phyllary and 2–4 adjacent paleae and disc florets; achenes not winged .... ***Engelmannia***

11. Achenes plump; pappus of 2 to several caducous awns .... ***Helianthus***
11′ Achenes flat, very strongly compressed; pappus various .... ***Encelia***

12. Achenes circular or 4-angled in cross section, with gland-dotted furrows; leaves finely divided; annuals .... ***Anthemis***
12′ Achenes very flat, strongly compressed or thickened; glabrous to hairy, not gland-dotted; leaves entire or coarsely toothed; annuals, biennials, or perennials .... (13)

13. Achenes very flat, strongly compressed .... (14)
13′ Achenes not very flat, usually much thickened .... (16)

14. Receptacle flat or only slightly convex .... ***Helianthella***
14′ Receptacle elongate-cylindric .... (15)

15. Receptacle elongate-cylindric; rays often partly maroon .... ***Ratibida***
15′ Receptacle merely conic; rays mostly pure yellow .... ***Rudbeckia***

16. Pappus caducous (of 2 awns and rarely some scales) .... ***Helianthus***
16′ Pappus persistent .... (17)

17. Inner involucral bracts united to middle into a cup .... ***Thelesperma***
17′ Inner involucral bracts not united into a cup .... (18)

18. Leaves mostly cauline, blades narrowly oblong to linear, 5–20 mm wide .... ***Scabrethia***
18′ Leaves basal and cauline, blades mostly 30–100 mm wide .... ***Wyethia***

19. Rays white or purple .... (20)
19′ Rays yellow, sometimes marked with purple or orange .... (23)

20. Pappus a short crown .... (21)
20′ Pappus of awns or scales .... (22)

21. Leaves entire or pinnately divided .... ***Chrysanthemum***
21′ Leaves irregularly 2–3 times pinnately dissected .... ***Matricaria***

22. Plants dwarf woolly annuals .... ***Eriophyllum***
22′ Plants annual or perennial, not woolly .... ***Townsendia***

23. Receptacle densely bristly or hairy .... (24)
23′ Receptacle naked .... (25)

24. Heads very small; involucres less than 10 mm wide .... *Gutierrezia*
24' Heads medium-sized; involucres more than 10 mm wide .... *Gaillardia*

25. Involucral bracts in 1 series, united almost throughout into a tube or cup .... *Tagetes*
25' Involucral bracts not as above .... (26)

26. Pappus a mere crown or of caducous awns .... (27)
26' Pappus persistent, of awns or scales .... (29)

27. Pappus of 2–8 caducous awns; plants glutinous .... *Grindelia*
27' Pappus a short crown; plants seldom if ever glutinous .... (28)

28. Leaves entire, bristly-margined basally .... *Pectis*
28' Leaves 2- or 3-pinnate .... *Tanacetum*

29. Pappus of several more or less united scales; rays broad, papery, and long-persistent following anthesis .... *Psilostrophe*
29' Pappus not of united scales; rays not papery and persistent .... (30)

30. Leaves and involucre with conspicuous oil glands .... (31)
30' Leaves and involucre without conspicuous oil glands .... (32)

31. Plants annual; leaves bipinnatisect; stems villous .... *Dyssodia*
31' Plants perennial, low shrubs; leaves simple or merely pinnatisect; stems hispidulous .... *Thymophylla*

32. Achenes stout, oblong or obovoid .... *Platyschkuhria*
32' Achenes slender, elongate-clavate .... (33)

33. Involucral bracts spreading or reflexed; receptacle convex to subglobose; cauline leaves decurrent or strongly clasping .... *Helenium*
33' Involucral bracts appressed; receptacle almost flat; leaves not decurrent .... (34)

34. Pappus of numerous scales; stems leafy; leaves linear or linear-spatulate, entire, 2.5 mm wide or less .... *Gutierrezia*
34' Pappus of 5–8 scales; leaves lobed, or if entire, broader and mostly or entirely basal .... (35)

35. Leaf blades simple or 1–2-pinnately lobed, ultimate margins entire or toothed, usually ± gland-dotted (often in pits) .... *Hymenoxys*
35' Leaf blades mostly oblanceolate to linear or filiform, sometimes lobed; eglandular or ± gland-dotted ... *Tetraneuris*

Acamptopappus (A. Gray) A. Gray Goldenhead
G. L. Nesom

(Greek *akamptos*, stiff or unbending, + *pappus*, alluding to thick pappus elements) Deciduous shrubs; taproots woody. **STEMS** erect, branched, spinescent with age (older portions gray, usually with shredding bark), glabrous or hirtellous. **LEAVES** alternate, sometimes in axillary fascicles, sessile to petiolate; blades narrowly lanceolate or narrowly obovate or spatulate, margins entire, 1-nerved. **INFL** a radiate or discoid head, pedunculate, borne singly or in loose corymbiform arrays. **INVOLUCRES** campanulate to hemispheric or globose. **PHYLLARIES** in 3 series, broadly ovate to ovate-elliptic, margins broad, lacerate-hyaline. **RECEPTACLES** flat to slightly convex, deeply alveolate, alveola margins divided into narrowly lanceolate, sharply acute extensions, epaleate. **RAY FLORETS** 5–14 or 0; corollas yellow. **DISC FLORETS** bisexual, fertile. **CYPSELAE** obconic, compressed, nervation not evident, densely sericeous to villous, outer trichomes ascending-spreading, inner trichomes closest to cypsela wall appressed and tortuous. **PAPPUS** of 18–22 persistent, variably thick, broadly flattened, subequal, marginally barbellate to smooth or barbellulate, apically attenuate (ray) to spatulate (disk), bristlelike scales in 1–2 series, plus a few much shorter bristles or setae. $x = 9$. Species two, southwestern United States. The genus is recognized as low, white-barked desert shrubs with pedunculate heads solitary or in loose, corymbiform arrays, few yellow rays, ovate phyllaries with broad hyaline margins, sericeous to villous-cypselate, and thick pappus members.

Acamptopappus sphaerocephalus (Harv. & A. Gray) A. Gray (spherical head) Rayless goldenhead. Plants mostly 20–40 cm. **LEAVES** with glabrous faces, margins sometimes minutely hirtellous. **HEADS** discoid (ray florets 0), in loose, corymbiform arrays. **INVOLUCRES** hemispheric to globose, 7–10 mm wide. $2n$ = 18. [*Haplopappus sphaerocephalus* Harv. & A. Gray]. Rocky hillsides, sandy ridges. UTAH. 1125–1400 m (3700–4595′). Flowering: Mar–Jun. California to Arizona and Utah. Four Corners plants are **var. *sphaerocephalus***.

Achillea L. Yarrow, Milfoil
Brian A. Elliott

(for Achilles, ancient Greek warrior) Perennial, rhizomatous, aromatic herbs. **STEMS** ascending to erect. **LEAVES** alternate, commonly 1–3-pinnately dissected, rarely entire. **INFL** a radiate head or rarely discoid, borne in dense terminal corymbs. **INVOLUCRES** campanulate to narrowly hemispheric. **PHYLLARIES** imbricate in 3–4 series, midrib green, margins scarious or hyaline. **RECEPTACLES** flat to convex or conical, chaffy throughout. **RAY FLORETS** pistillate and fertile; ligules 3–12, white, pink, or yellow. **DISC FLORETS** perfect and fertile, more than 10 per head; corolla white or rarely pink. **CYPSELAE** compressed, glabrous with a callous margin. **PAPPUS** lacking. About 80 species, native to North America, Eurasia, and North Africa.

1. Ray flowers yellow .... ***A. filipendulina***
1′ Ray flowers white, pink, red, or purplish .... ***A. millefolium***

Achillea filipendulina Lam. (resembling the genus *Filipendula*, meadow sweet) Fernleaf yarrow. **STEMS** stiffly erect, to 15 dm high or more. **LEAVES** 4–35 cm long, ultimate segments linear-lanceolate, toothed, often with 1 larger lobe on the upper side. **INFL** head radiate, numerous. **RAY FLORETS** with yellow ligules, to 1 mm long. $2n$ = 18, 36. A cultivated ornamental introduced from Asia, rarely escaping on disturbed ground or riparian areas. Known from one collection. COLO: LPl. 1990 m (6540′). Reportedly escaped in Utah, Michigan, New York, and Vermont. The collection *Heil 22192* (SJNM) is the first documented case of the species escaping in Colorado.

Achillea millefolium L. (referring to the finely dissected leaves) Yarrow, common yarrow, milfoil yarrow, Hazéíyiltsee'í. **STEMS** to 6 (20) dm high. **LEAVES** petiolate (lower) or sessile (upper); ultimate segments narrow and spinulose-tipped. **INFL** head radiate, numerous. **RAY FLORETS** with white to pink, red, or purple ligules, ovate to round, 2.5–4 mm long. A polyploid complex based on x = 9. [*Achillea gracilis* Raf.; *A. lanulosa* Nutt.; *A. subalpina* Greene]. A species of wide ecological amplitude, growing in many habitats. ARIZ: Apa; COLO: Arc, Con, Dol, Hin, LPl, Min, Mon, RGr, SJn, SMg; NMEX: McK, RAr, SJn; UTAH. 1800–3580 m (5950–11750′). Flowering: May–Sep. Throughout North America. *Achillea millefolium* is a complicated polyploid complex in North America, representing several ploidy levels. The foliage has had extensive medicinal use for a variety of ailments, including use as a cold remedy, poultice, analgesic, skin remedy, for headaches, toothaches, and numerous other maladies. Dried leaves can be steeped for tea, and fresh leaves have been used raw as a condiment or sparingly in salad. The flowering tops have also been used sparingly as a condiment. Yarrow should be used in moderation, since it contains many secondary plant products and can cause photosensitization of the skin if used in excess.

Acourtia D. Don Desert-peony
J. Mark Porter

(for Mrs. A'Court, a British amateur botanist) Perennials, (2.5–) 5–50 (–150) cm tall, caudices brown-woolly. **STEMS** glabrate or resinous-punctate. **LEAVES** basal and cauline; shortly petiolate or sessile; blades elliptic-oblong, lanceolate, oblong, oblong-lanceolate, oblong-oblanceolate, orbiculate, ovate, ovate-elliptic, or rhombic-orbiculate, thin and chartaceous to thick and coriaceous, bases cuneate to cordate or clasping, margins entire or lobed or pinnately divided, dentate, or serrate, surfaces usually minutely stipitate-glandular or hirtellous. **INFL** borne singly or in a panicle or corymb. **HEADS** semiradiate or discoid (of 1 floral type). **INVOLUCRES** turbinate or obconic to campanulate, 6–18 mm long. **PHYLLARIES** in 1–7 series, lanceolate to oblanceolate or linear, unequal, rigid, margins scarious, apices obtuse to acute, acuminate, or mucronate. **RECEPTACLES** concave, flat, or convex, usually foveolate, alveolate, or reticulate, pubescent, sometimes paleate, the paleae apically pubescent. **FLORETS** 3–25 (–80), bisexual, fertile; corollas pink to lavender or white to yellowish, zygomorphic, 2-lipped; outer lip liguliform, 3-toothed, inner usually smaller, 2-lobed, lobes often curled; anther basal appendages entire, elongate, rounded, apical appendages lanceolate; style branches relatively short, apices blunt. **CYPSELAE** fusiform or terete to cylindric, 4–10 mm, not beaked, usually ribbed, faces glabrous or stipitate-glandular. **PAPPUS** of 40–60 (–80) tan or white, barbellate to nearly smooth bristles in 1–3 (–9) series. x = 27. About 41 species of the warm regions of North America, Mexico, and Central America. Many previous accounts include *Acourtia* species as members of *Perezia*; however, a recircumscription of *Acourtia* (Reveal and

King 1973; Turner 1978) transfers several former species of *Perezia* to *Acourtia*, a change supported by chloroplast *ndhF* DNA sequence data (Kim et al. 2002). These data indicate *Acourtia* is most closely related to *Proustia* and *Trixis* rather than to *Perezia*. (Bacigalupi 1931; Cabrera 1992, 2001; Kim, Loockerman, and Jansen 2002; Reveal and King 1973; Turner 1978, 1993.)

Acourtia wrightii (A. Gray) Reveal & R. M. King (honoring Charles Wright, 1811–1885, botanist and explorer) Brownfoot, Wright's perezia. Plants herbaceous perennials. **STEMS** 30–120 cm; often purplish with dense red-brown hairs at base. **LEAVES** cauline; petiolate below, sessile and clasping above; blades lance-oblong to elliptic-oblong, 2.5–13 cm long, bases sagittate or clasping, margins spinulose-dentate to denticulate, surfaces minutely stipitate-glandular and hirtellous; vasculature reticulate. **HEADS** in corymbiform arrangement. **INVOLUCRES** turbinate, 5–8 mm. **PHYLLARIES** in 2–3 series, linear to lanceolate, margins fimbrillate-glandular, apices obtuse to acute, abaxial faces glabrous or glandular-hairy. **RECEPTACLES** reticulate, glandular. **FLORETS** 8–12; corollas pink or purple, 9–20 mm. **CYPSELAE** linear-fusiform, 2.8–6 mm long, glandular-puberulent. **PAPPUS** bright white, 9–12 mm. $2n = 54$. Gravel, caliche, or sandy loamy soils in open arid regions of desert shrub and juniper woodlands. ARIZ: Nav; UTAH. 1220–1400 m (4000–4700′). Flowering: Jun–Nov. Fruit: Jul–Nov. Widespread in the deserts of the United States including Arizona, New Mexico, Texas, Utah; Mexico. The Kayenta Navajo have used *A. wrightii* as an aid for difficult labor, and as a postpartum medicine.

Acroptilon Cass.

Patricia Barlow-Irick

(Greek *akron*, tip, and *ptilon*, feather, describing the pappus bristles) Perennials, 30–100 cm, not spiny. **STEMS** erect, branched throughout, branches ascending. **LEAVES** sessile or petiolate; basal and lower cauline blade margins coarsely dentate or 1–2-pinnately lobed, margins of middle and distal cauline dentate or entire, faces glabrate to tomentose and resin gland–dotted. **HEADS** discoid, in leafy-bracted paniculate arrays. **INVOLUCRES** ovoid to urn-shaped, constricted near tip. **PHYLLARIES** many, in 6–8 series, outermost round to ovate, bases tightly appressed, margins entire, apices widely scarious, obtuse or acute, inner lanceolate, margins entire, apices acute to acuminate, innermost bristly-ciliate or -plumose. **RECEPTACLES** flat, bearing flattened bristles. **FLORETS** all fertile; corollas blue, pink, or white. **CYPSELAE** obovoid, slightly compressed, smooth or with indistinct ribs, glabrous, attachment scars subbasal. **PAPPUS** ± falling, of many unequal flattened bristles, proximally barbed, longer ones distally plumose. One species introduced from Eurasia.

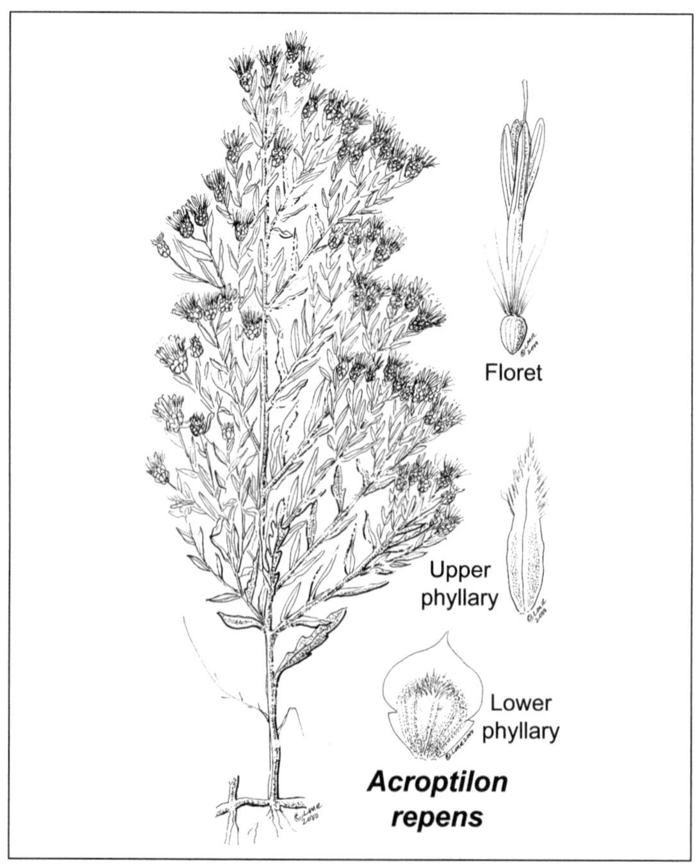

Acroptilon repens (L.) DC. (creeping) Russian knapweed, creeping knapweed, Turkestan thistle, hardheads, mountain bluet. Long-lived perennial with mostly dark brown or black roots; with scaly adventitious buds. **STEMS** ± cobwebby-tomentose, erect, 20–100 cm tall. **LEAVES** (basal) deciduous, blades spatulate, entire to toothed, 4–15 cm; upper leaves linear to linear-lanceolate or oblong, 1–7 cm. **INVOLUCRES** 1–2 cm long, loosely cobwebby. **PHYLLARY** apices of inner acute or acuminate, not spiny, with broad translucent tip, densely short-pilose. **FLORETS** with corollas pink or purplish, 11–14 mm long, tubes 6.5–7.5 mm long, throats 2–3.5 mm long, lobes 3–3.5 mm long. **CYPSELAE** ivory to brown, 2–4 mm long, pappus bristles white, 6–11 mm long. [*Centaurea repens* L.; *Acroptilon picris* (Pall. ex Willd.) C. A. Mey.; *C. picris* Pall. ex Willd.]. Weedy; fields, roadsides, riverbanks, ditch banks, clearcuts, and cultivated ground. ARIZ: Apa, Nav; COLO: Arc, Dol,

LPl, Mon, SMg; NMEX: McK, RAr, San, SJn; UTAH. 1404–2165 m (4605–7100′). Flowering: May–Sep. Introduced from Eurasia; adventive in all of the western and midwestern United States. Russian knapweed is a noxious weed, spreading from seeds and runner roots. It is very poisonous to horses, causing neurological symptoms called "chewing disease." This bitter herb is usually avoided by grazing animals, and consequently it tends to spread under intensive grazing. Stands have been known to persist for 100 years.

Ageratina Spach Snakeroot
G. L. Nesom

(diminutive of genus *Ageratum*) Perennial herbs, subshrubs, or shrubs. **STEMS** usually erect, rarely scandent, sparsely to densely branched. **LEAVES** cauline, mostly opposite or distal, sometimes alternate, petiolate, blades 3 (–5)-nerved from bases, usually deltate, lanceolate, ovate, rhombic, or triangular, sometimes orbiculate, margins entire or toothed, faces glabrous or hispidulous, pilose, or puberulent, sometimes gland-dotted. **HEADS** discoid, usually in compact, terminal and axillary corymbiform arrays, sometimes borne singly. **INVOLUCRES** campanulate. **PHYLLARIES** persistent, in 2 (–3) series, 0- or 2-nerved, lanceolate to linear, ± equal. **RECEPTACLES** convex, epaleate. **FLORETS** 10–60; corollas white or lavender, throats obconic to campanulate; styles with bases sometimes enlarged, glabrous, branches linear, seldom distally dilated. **CYPSELAE** prismatic or ± fusiform, usually 5-ribbed, scabrellous and/or gland-dotted. **PAPPUS** usually persistent, of 5–40 barbellulate bristles in 1 series. $x = 17$. Genus of about 250 species, from the southern United States to Mexico, Central America, and Andean South America.

Ageratina herbacea (A. Gray) R. King & H. Rob. (herbaceous) Fragrant snakeroot. Perennial herbs or subshrubs with woody crowns and woody rhizomes, (20–) 30–60 (–80) cm. **STEMS** minutely puberulous. **LEAVES** opposite; petioles 10–25 mm; blades triangular to lanceolate-ovate or ovate, 2–5 (–7) × 1.5–3.5 (–4.5) cm, bases truncate to shallowly cordate, margins dentate to serrate-dentate, abaxial faces minutely puberulous or sparsely hispidulous to glabrate, eglandular. **HEADS** clustered on peduncles 4–15 mm. **INVOLUCRES** 4–5 mm high; corollas white, glabrous. $2n = 34$. [*Eupatorium herbaceum* (A. Gray) Greene]. Rocks along streams, cliffs and ledges, open woods, burns, sandy washes, juniper, piñon-juniper, ponderosa pine, pine–Gambel's oak, pine–Douglas-fir, sagebrush, lodgepole pine, white pine, aspen. ARIZ: Apa, Nav; COLO: Arc, Hin, LPl, Min, Mon; NMEX: McK, RAr, SJn; UTAH. (1600–) 1850–2700 m [(5250–) (6075–8860′)]. Flowering: Aug–Oct. California to Colorado, south to Texas, Arizona, New Mexico, and Mexico (Baja California, Chihuahua, Coahuila, Sonora). *Ageratina herbacea* is recognized by its yellow-green to grayish yellow-green leaves with triangular to ovate blades, white flowers, relatively small, granular-puberulent involucres, and woody rhizomes. Used by the Ramah Navajo as a cold infusion and as a lotion for headache or fever.

Agoseris Raf. Mountain-dandelion, False-dandelion, Agoseris
Gary I. Baird

(Greek *agos*, chief, + *seris*, chicory, the allusion obscure) Perennials (ours), taprooted, herbage glabrous and glaucous to hairy, sap milky. **STEMS** lacking. **LEAVES** all basal, sessile or obscurely petiolate, erect to decumbent; blades oblanceolate to linear-lanceolate, simple, entire to pinnately, laciniately, or runcinately lobed; lobes linear-lanceolate to deltate, often lobulate; lobules 1, basal, acroscopic, or lacking. **HEADS** liguliflorous, borne singly, erect; peduncles scapose, ebracteate. **INVOLUCRES** cylindric to conic. **PHYLLARIES** mostly herbaceous, subequal to imbricate, spreading (outer) to erect (inner). **RECEPTACLES** flat, epaleate (mostly) or paleate. **RAY FLORETS** matutinal; corolla yellow (outer often abaxially rosy-purple, this still evident when dried) or orange (often drying purplish), rarely pinkish or reddish, ligule often pilose proximally. **DISC FLORETS** lacking. **CYPSELAE** monomorphic or dimorphic (outer differing from inner), light to dark brown or purplish, usually fusiform or narrowly conical, glabrous or scaberulous to hirsutulous, usually 10-ribbed, short- to long-beaked (beak up to 2× length of achene body in ours). **PAPPUS** of bristles (sometimes flattened), persistent, white, ± barbellulate. $x = 9, 18$ (Baird 1996). Ten species of (mostly) western North America and one species of temperate, southern South America.

1. Corollas orange, pink, red, or purple (all of these usually drying purple) .... ***A. aurantiaca***
1′ Corollas yellow (outer usually with a broad abaxial purple stripe that remains evident on drying) .... (2)

2. Mature achene beak 1–4 mm long (usually less than 0.5× length of achene body) .... ***A. glauca***
2′ Mature achene beak over 5 mm long (usually 1–2× length of achene body), rarely less .... (3)

3. Ligules 4–12 mm; leaves mostly entire or with 2–4 pairs of spreading to antrorse lobes; usually of mesic habitats (montane forests and meadows to alpine tundra) .... ***A. aurantiaca***

3′ Ligules 10–20 mm; leaves mostly with (3–) 5–8 pairs of spreading to retrorse lobes, uncommonly entire; usually of xeric habitats (sagebrush, grass, piñon-juniper, or ponderosa pine communities) .... ***A. parviflora***

Agoseris aurantiaca (Hook.) Greene (orange) Orange agoseris. Perennials 10–40 cm, base purplish. **LEAVES** with blades linear-lanceolate to oblanceolate, 10–30 cm × 5–15 mm (excluding lobes), bases ± clasping, ciliate, and purple; margins entire to laciniately pinnatifid, apex acute to obtuse, faces glabrous to sparsely villous, sometimes glaucous; lobes (when present) usually 2–4 pairs, linear to lanceolate, spreading to antrorse; lobules inconspicuous to subequaling lobe, rarely lacking. **HEADS** solitary; peduncles villous to woolly (rarely glabrous), especially distally (at base of head), nonglandular, often much elongating between flowering and fruiting. **INVOLUCRES** 15–30 mm. **PHYLLARIES** subequal to imbricate (rarely outer much longer and foliaceous), green to rosy-purple (except for margin), sometimes with a darker midstripe, often black-purple spotted or blotched; outer linear-lanceolate to ovate or obovate, not elongating with fruit, margins ciliate to pubescent, apex obtuse to acuminate, erect to spreading (adaxial surface often pubescent), faces glabrous to villous; inner phyllaries linear to lanceolate, erect, elongating with fruit, glabrous to (distally) pubescent. **FLORETS** 15–50, outer ± equal to involucre; corolla orange (mostly) or yellow (occasional), rarely pinkish or purplish, tube 5–10 mm, ligule 5–10 × 1–3 mm; anthers 2–4 mm. **CYPSELAE** body fusiform to obconic, 6–9 mm, ribs often thickened distally, apex abruptly to gradually tapered; beak ± slender, smooth, 5–8 mm long (generally equal in length to achene body). **PAPPUS** 9–15 mm. $2n$ = 18, 36. Alaska, Yukon, and Northwest Territories, south to California, Arizona, and New Mexico, disjunct in Quebec. Ramah Navajo uses include: a ceremonial emetic; cold infusion taken and used as a lotion for arrow or bullet wounds; wet leaves rubbed on swollen arms; cold infusion taken and used as lotion for protection from witches.

1. Phyllaries mostly lanceolate, margins ciliate, especially proximally, faces mostly villous; corollas mostly orange; achene apex often abruptly tapered, ribs often thickened distally .... **var. *aurantiaca***

1′ Phyllaries mostly ovate or obovate, margins ciliate, especially distally, faces usually glabrous; corollas orange or yellow (sometimes pink); achene apex usually gradually tapered, ribs mostly uniform .... **var. *purpurea***

var. *aurantiaca* **PEDUNCLES** usually shorter than leaves at flowering, villous to tomentose when young, ± glabrate with age but remaining sparsely to densely pubescent (rarely glabrous) beneath head. **PHYLLARIES** (outer) mostly lanceolate, margins ciliate (especially proximally), apices acute to attenuate, faces sparsely to densely pubescent (rarely glabrous). **FLORETS** mostly orange (rarely yellow or pinkish); ligules mostly less than 9 mm long; anthers 2–4 mm. **CYPSELAE** ribs often thickened distally, glabrous to (mostly) minutely pubescent, apex often abruptly tapered to beak. $2n$ = 18. [*A. gracilens* (A. Gray) Greene; *A. greenei* (A. Gray) Rydb.; *A. lackschewitzii* Douglass M. Hend. & R. K. Moseley; *A. nana* Rydb.]. Alpine and subalpine communities. COLO: Con, Hin, Min, Mon, SJn; UTAH. 3200–3900 m (10500–12800′). Flowering: Jul–Aug. Alaska, Yukon, and Northwest Territories, south to California, Arizona, and New Mexico, disjunct in Quebec. Known in the study area only from the highest elevations in the Abajo and San Juan mountains, which represent part of the southern limits of this variety (only collections from the Sangre de Cristo Mountains in New Mexico are further south). All specimens seen from the study area appear to be (more or less) intermediate with var. *purpurea* and are not as distinct as more typical (and northern) var. *aurantiaca*.

var. *purpurea* (A. Gray) Cronquist (purple, perhaps alluding to the color of the dried florets) Colorado Plateau agoseris. **PEDUNCLES** usually equal to or longer than leaves at flowering, tomentose to woolly when young, glabrate with age but remaining densely (usually) pubescent to woolly beneath head. **PHYLLARIES** (outer) mostly ovate to obovate, margins ciliate to pubescent (especially distally), apices obtuse to acute, faces glabrous to sparsely pubescent. **FLORETS** orange or yellow (rarely pinkish); ligules often over 9 mm long (especially yellow corollas); anthers 3–5 mm. **CYPSELAE** ribs usually uniform, glabrous (mostly) to minutely scabrous, apex mostly gradually tapered to beak. $2n$ = 18, 34, 36. [*A. arachnoidea* Rydb.; *A. arizonica* (Greene) Greene; *A. attenuata* Rydb.; *A. glauca* var. *cronquistii* S. L. Welsh; *A. graminifolia* Greene; *A. frondifera* Osterh.; *A. roseata* Rydb.; *A. rostrata* Rydb.]. Montane and alpine meadows, and forests. ARIZ: Apa; COLO: Arc, LPl, Min, Mon, SJn; NMEX: McK, RAr, SJn; UTAH. 1800–3600 m (6000–11800′). Flowering: Jun–Sep. Arizona, Utah, Colorado, and New Mexico, just reaching eastern Nevada and southern Wyoming. This is the common variety found midmontane and higher on the Colorado Plateau and southern Rocky Mountains. It exhibits a higher frequency of yellow-flowered populations than does var. *aurantiaca*, especially westward in its range (i.e., the Wasatch and Mogollon plateaus). These yellow-flowered populations are often misidentified as *A. glauca* or *A. parviflora* and historically constituted part of the nebulous *A. arizonica*. Variety *purpurea* forms hybrids with both *A. glauca* and *A. parviflora*. *Agoseris* ×*montana* Osterh. (*A. aurantiaca* var. *purpurea* × *A. glauca* var. *dasycephala*) is a mostly yellow-flowered hybrid (polyploid?) that occurs generally above 3000 m (10000′) in the mountains of Colorado and southern Wyoming. Specimens of *A.* ×*montana* show varying degrees of intermediacy between the parental taxa and may be found where the two occur together. It occurs uncommonly in the study area in the San Juan Mountains (COLO: SJn).

Agoseris glauca (Pursh) Raf. (bluish or grayish, due to a waxy coating) Prairie agoseris. Perennials 10–50 cm tall, base not (or rarely) purplish. **LEAVES** with blades lanceolate to oblanceolate, 5–25 cm × 2–50 mm wide (excluding lobes), base ± clasping, not usually ciliate or purple; margins entire (mostly), toothed, or pinnately lobed, apex bluntly obtuse to acuminate, faces glabrous and glaucous to densely villous, sometimes glandular; lobes (when present) mostly 2–3 pairs, lanceolate, spreading; lobules mostly lacking. **HEADS** solitary; peduncles glabrous to densely pubescent, especially distally (at base of head), glandular or not, usually little or not elongating between flowering and fruiting. **INVOLUCRES** 1–3 cm tall. **PHYLLARIES** subequal to imbricate, green or rosy-purple (except for margin), sometimes with a darker midstripe, often black-purple spotted, speckled, and/or tipped; outer lanceolate to ovate or obovate, not elongating with fruit, margins glabrous or ciliate, apex obtuse to acuminate, erect to spreading (adaxial surface glabrous to densely pubescent), faces glabrous to villous, glandular or not; inner phyllaries lanceolate, erect, not (or little) elongating with fruit, glabrous to (distally) pubescent. **FLORETS** 15–150, outer much exceeding involucre; corolla yellow (outer usually with adaxial reddish purple stripe), tube 4–18 mm long, ligule 6–24 × 2–5 mm; anthers 3–6 mm. **CYPSELA** body fusiform to narrowly conic, 5–9 mm, ribs uniform, apex gradually tapered; beak stout, usually ribbed, mostly 1–4 mm long (generally less than 1/2 length of achene body), or lacking. **PAPPUS** 8–18 mm. $2n$ = 18, 36. Mostly in wet to mesic meadows, montane forests, and alpine tundra. British Columbia and southern Northwest Territories, east to Manitoba and western Minnesota, south to California, Arizona, and New Mexico, disjunct in Ontario, Michigan, and the Canadian Arctic.

1. Plants mostly pubescent, especially so at base of head; leaves entire to (commonly) toothed or lobed; involucres ± pubescent, sometimes glandular-pubescent; phyllaries mostly ovate to obovate or oblong, the margins ± wavy or undulate; mostly of alpine or subalpine habitats .... **var. *dasycephala***

1′ Plants mostly glabrous and glaucous, or slightly pubescent; leaves mostly entire, rarely toothed or lobed; involucres glabrous (rarely slightly pubescent proximally), nonglandular; phyllaries mostly lanceolate, the margins ± straight; mostly of wet meadows, prairies, and fields in valleys and foothills .... **var. *glauca***

var. *dasycephala* (Torr. & A. Gray) Jeps. (shaggy head, referring to the involucres) Arctic agoseris. **LEAVES** with blades narrowly to broadly oblanceolate, mostly 5–20 cm tall, margins entire to toothed or lobed, apices bluntly obtuse to acute, rarely acuminate, faces sparsely to densely pubescent, uncommonly glabrous or glaucous. **HEADS** with peduncles generally 5–40 cm tall, glabrate but remaining (usually densely) pubescent at base of head, pubescence sometimes glandular. **PHYLLARIES** with outer mostly ovate to obovate or oblong, usually purplish (to blackish), especially proximally, or densely speckled, sparsely to densely villous, pubescence often yellowish translucent, sometimes glandular, margins wavy, undulate, or recurved. **CYPSELAE** usually glabrous, sometimes scabrous distally. $2n$ = 36. [*A. aspera* (Rydb.) Rydb.; *A. villosa* Rydb.; *A. pubescens* Rydb.; *A. maculata* Rydb.]. Montane meadows and forests to alpine tundra. COLO: Hin, Mon, SJn. 2900–3600 m (9600–12000′). Flowering: Jul–Aug. British Columbia and Alberta, south to Washington, Utah, and New Mexico, with disjunct populations in the Northwest Territories (bordering the Arctic Ocean). This variety typically is found at much higher elevations than var. *glauca*. Its occurrence in the San Juan Mountains represents the southernmost extent of its range. This variety consists of a series of quasi-distinct entities, each with a particular geographic range, that grade into each other along their boundaries, forming a step-clinal distribution for the variety as a whole. The material from Colorado and south-central Wyoming constitutes the *maculata* phase (based on *A. maculata*). This is one of the more morphologically distinct phases and appears to be geographically isolated from its neighboring phases (i.e., the *isomeris* phase of the Uinta Mountains in Utah and the *aspera* phase of the Rocky Mountains in Idaho, Montana, and northwestern Wyoming). Variety *dasycephala* hybridizes with *A. aurantiaca* var. *purpurea* (see discussion of *A.* ×*montana* under *A. aurantiaca* var. *purpurea*).

var. *glauca* **LEAVES** with blades lanceolate to oblanceolate, mostly 10–30 cm long, margins entire (rarely toothed or lobed), apices acuminate to attenuate, rarely acute or obtuse, faces mostly glabrous and glaucous, sometimes slightly pubescent. **HEADS** with peduncles generally 20–60 cm tall, glabrous (rarely slightly pubescent at base of head). **PHYLLARIES** with outer mostly lanceolate, green (mostly) to rosy-purple (except for margin), often black-purple spotted, speckled, and/or tipped, glabrous (rarely slightly pubescent proximally), pubescence (if any) usually white-opaque and nonglandular, margins straight. **CYPSELAE** usually scabrous distally, sometimes glabrous. $2n$ = 18. [*A. longissima* Greene]. Low-elevation (often wet or alkaline) meadows. COLO: Arc. 2200 m (7300′). Flowering: Aug. British Columbia to Manitoba, south to California, Utah, and Colorado, with disjunct populations in Arizona, Michigan, and Ontario. The study area represents the southern limits of var. *glauca*, where it is apparently uncommon (only a disjunct population in eastern Arizona is farther south). This variety typically occurs at much lower elevations than the next. Where the two occur together, they intergrade. *Agoseris* ×*agrestis* Osterh. (= *A. glauca* var. *glauca* × *A. parviflora*) occurs in the foothills of the Rocky Mountains in Colorado and adjacent New Mexico and Wyoming. It is the basis for *A. glauca* var. *agrestis* (Osterh.) Q. Jones ex Cronquist of recent authors but most specimens so identified are more accurately assigned to *A. glauca* var. *dasycephala* (q.v.). In general appearance, *A.* ×*agrestis* approaches *A. glauca* var. *glauca*, except that its leaves are often toothed or laciniate, the peduncle is usually villous, especially beneath the head (in this feature it approaches *A. glauca* var. *dasycephala*, but it is not glandular-hairy, as is usually the case in var. *dasycephala*), and the phyllaries are usually ciliate (as in *A. parviflora*). It also grows at lower elevations, more typical of the parents, than observed for *A. glauca* var. *dasycephala*. It has been found just outside the study area, in the San Juan Mountains in Conejos County.

Agoseris parviflora (Nutt.) D. Dietr. (small flower) Steppe agoseris. Plants mostly 10–25 cm tall, base rarely purplish, herbage glabrous, sometimes glaucous, to (uncommonly) densely villous, nonglandular. **LEAVES** linear-lanceolate to oblanceolate, 4–15 cm × 1–10 mm (excluding lobes), base ± clasping, rarely purple, blade pinnately lobed or toothed, rarely entire (sometimes variable on same plant), apex acuminate to attenuate (rarely obtuse); lobes mostly (3–) 5–8 pairs, linear to lanceolate, retrose to spreading; lobules often present. **HEADS** solitary; peduncles glabrate to villous, base of head mostly remaining villous, not significantly elongating after flowering. **INVOLUCRES** 2–3 cm long. **PHYLLARIES** usually imbricate, rosy-purple (except for margin), sometimes with a darker midstripe, rarely spotted or all green; outer lanceolate to ovate, not elongating with fruit, margins ciliate to lanate, apex bluntly to attenuately acuminate, usually spreading (adaxial surface glabrous to pubescent), faces glabrous to loosely pubescent, not glandular; inner lanceolate, glabrous to pubescent (distally), erect, not or little elongating with fruit. **RECEPTACLES** epaleate. **FLORETS** 30–50, outer exceeding involucre; corolla yellow (outer usually with adaxial reddish purple stripe), tube 6–15 mm, ligule 10–20 mm × 2–4 mm; anthers mostly 3–5 mm. **CYPSELA** body fusiform to narrowly obconic, 5–9 mm, ribs mostly uniform, apex often abruptly tapered; beak slender, mostly 5–10 mm long (generally 1–2× length of achene body), rarely shorter. **PAPPUS** 10–20 mm. $2n = 18$. [*A. glauca* var. *laciniata* (D. C. Eaton) Kuntze]. Mostly in sagebrush, piñon-juniper, and ponderosa pine communities. ARIZ: Apa, Coc; COLO: Arc, Dol, Hin, LPl, Mon; NMEX: McK, RAr, SJn; UTAH. 2100–2700 m (7000–8800′). Flowering: Apr–Aug. Oregon to Montana and South Dakota, south to California, Arizona, and New Mexico. This species has been treated as a variety of *A. glauca* by all recent authors but, in reality, the two are rarely confused and only occasionally form hybrids, the most notable being *A.* ×*agrestis* (see discussion above under *A. glauca* var. *glauca*).

Almutaster Á. Löve & D. Löve Marsh Alkali Aster

G. L. Nesom

(for contemporary American *Aster* sensu lato expert Almut G. Jones) Perennial herbs 1.5–11 dm; fibrous roots fleshy, creeping rhizomes slender, uppermost stems, leaves, and phyllaries stipitate-glandular, lower parts glabrous or glabrate. **LEAVES** basal and cauline, thick, linear-lanceolate to nearly filiform, with 3–5 parallel nerves, sessile and subclasping, 2–12 (–23) cm long, margins entire, often slightly revolute. **INFL** a radiate head, on bracteate peduncles; solitary or 2–10 (–30) in a loosely corymboid or paniculate inflorescence. **INVOLUCRES** turbinate to hemispheric, 4.5–8 mm high. **PHYLLARIES** in 3–4 (–5) subequal series, herbaceous, lacking a white, indurate basal zone. **RECEPTACLES** flat, epaleate. **RAY FLORETS** 15–30 (–45) in 1 series, pistillate, fertile; corollas white or sometimes lilac-tinged. **DISC FLORETS** bisexual, fertile. **CYPSELAE** fusiform, terete, with 7–10 thin nerves, sparsely strigose. **PAPPUS** a single series of subequal barbellate bristles. $x = 9$. A monotypic genus, distributed from south-central Canada (Manitoba) through the western United States into south-central Mexico.

Almutaster pauciflorus (Nutt.) Á. Löve & D. Löve (few-flowered) Marsh alkali aster. Plants laxly colonial. **STEMS** sometimes reddish at base. **LEAVES** 2–12 (–23) cm long, cauline gradually reduced upward, bracteate in the inflorescence, margins sometimes ciliate or (distal) stipitate-glandular; basal sometimes marcescent, petiole bases widened, clasping. **RAY FLORET** laminae 5–8 mm. $2n = 18$. [*Aster pauciflorus* Nutt.]. Seeps, saline or alkaline marshes, wetlands, river bottoms and floodplains, riparian (cottonwood, saltcedar, Russian olive, willow, saltgrass), piñon-juniper, ponderosa pine–juniper. ARIZ: Apa, Nav; COLO: LPl; NMEX: RAr, SJn. 1500–2100 m (4920–6990′). Flowering: (May–) Jul–Sep (–Oct). Canada (Northwest Territories and Alberta to Manitoba); California, Nevada, and North Dakota, south to Arizona, New Mexico, and Texas; Mexico.

Ambrosia L. Ragweed

Gary I. Baird

(in Greek mythology, the name for the food of the gods; an ancient plant name used by both Dioscorides and Pliny) Annuals, perennials, or shrubs, taprooted, sometimes rhizomatous, herbage pubescent or glabrate, usually gland-dotted or stipitate-glandular, sap clear. **STEMS** 1 or several, erect, decumbent, or prostrate, branched (proximally, distally, or both). **LEAVES** cauline, simple, opposite or alternate (often proximally opposite, distally alternate), sessile or petiolate; blades deltate, ovate, rhombic, obovate, elliptic, lanceolate, linear, or filiform, usually lobed to pinnatifid, ultimate margins entire or dentate. **INFL** in terminal or axillary, ebracteate, racemiform to spiciform clusters (pistillate proximal, staminate distal, or intermixed, or pistillate solitary to clustered in axils), erect (mostly pistillate) or nodding (mostly staminate); sessile or pedunculate. **HEADS** discoid, unisexual, calyculi lacking. **PISTILLATE HEADS** with involucres globose, obovoid, conic, obconic, or fusiform; outer phyllaries 5–8 in approximately 1 series, herbaceous, distinct or connate (at least partially); inner phyllaries (actually paleae?) about 5–80 in 1–many series, ± indurate, connate proximally, distal ends forming spines, tubercles, or wings, or none, the whole forming a bur enclosing the cypselae (sometimes smooth and nutlike); florets usually 1 (–5), corolla lacking. **INVOLUCRES** campanulate to rotate. **PHYLLARIES** about 5–15 in 1 series, connate. **RECEPTACLES** flat to convex, epaleate or paleate; paleae (if present) spatulate to linear, membranous, glabrous or pubescent, occasionally gland-dotted or stipitate-glandular. **FLORETS** usually 5–60+, corolla whitish or purplish, funnelform, often gland-dotted; filaments ± connate, anthers distinct or proximally coherent. **CYPSELAE** monomorphic, blackish, ± ovoid, glabrous. **PAPPUS** lacking. $x = 18$. About 40 species of North and South America, some introduced in various regions around the globe. All species are wind-pollinated and some contribute significantly to seasonal allergies.

1. Plants annual from a taproot ....(2)
1′ Plants perennial from a rhizome ....(4)

2. Leaves palmately 3–5-lobed (the most distal sometimes unlobed), all opposite (the most distal sometimes alternate); burs with about 5 (or fewer) distal, short spines (less than 1 mm) plus the apical beak .... ***A. trifida***
2′ Leaves 1–2-pinnatifid (the most distal pinnatifid or sometimes entire), opposite proximally, alternate distally; burs distinctly spiny (spines greater than 2 mm, flattened) or tuberculate, or these lacking (except for apical beak) ......(3)

3. Mature burs mostly 3–5 mm, with about 8–18 flattened spines in 2+ series (usually with 1 proximal series and 1 distal series; spines sometimes reduced or absent); staminate involucres about 3–5 mm and with usually 3 (1–5) thick, black nerves; plants common and widespread in the study area .... ***A. acanthicarpa***
3′ Mature burs mostly 2–3 mm, with about 5 (or fewer) apical teeth (in addition to beak) in 1 series (either distal or medial); staminate involucres mostly 2–3 mm long (sometimes larger), lacking (or rarely) black nerves; plants apparently uncommon in the study area .... ***A. artemisiifolia***

4. Pistillate involucres and burs with a single floret and 1 apical beak; leaves ± uniformly pubescent and gland-dotted .... ***A. psilostachya***
4′ Pistillate involucres and burs with 2 florets and 2 apical beaks; leaves ± greenish adaxially, white abaxially (densely pubescent) .... ***A. tomentosa***

Ambrosia acanthicarpa Hook. (spiny-seeded, referring to the burs) Bur ragweed. Annuals 10–80 cm, herbage coarsely strigose to hispid or sericeous-strigillose, also gland-dotted. **STEMS** 1, erect, or basal branches often divergently ascending, striate, ± strigose (proximally glabrate). **LEAVES** opposite (proximal) to alternate (distal); petioles 5–20 (–35) mm; blades ovate to deltate, 10–70 × 10–60 mm, 1–2-pinnatifid to pinnatisect, base cuneate, ultimate margins entire or sparsely denticulate, adaxial faces strigillose to subglabrous, gland-dotted (or not), abaxial faces strigose to scabrous, usually also scattered hispid, gland-dotted. **PISTILLATE HEADS** clustered (sometimes forming glomerules); florets 1. **STAMINATE HEADS** with peduncles about 1–2 mm. **INVOLUCRES** openly funnelform to nearly flat, ± oblique, 2–3 mm wide, lobed (1–3+ of the outward-facing lobes each with a thickened, black nerve), ± strigose and gland-dotted. **FLORETS** 4–20, corollas sometimes smudged dark or black (especially on the lobes). **BURS** with bodies fusiform to obpyramidal, 3–5 mm, ± smooth to rugose, ± prominently ridged (horizontally and vertically) from base of spines, glabrous to sparsely puberulent, sometimes gland-dotted; spines 8–16+, in ± 2 series (1 proximal, 1 distal), about 2–4 mm, stramineous to purplish, divergent, frequently flattened, sometimes reduced or lacking (often mixed within the same cluster), apices occasionally hooked; beak 1, slender, spinelike. [*Franseria acanthicarpa* (Hook.) Coville]. $2n = 36$. Usually in sandy soils, often in disturbed habitats; desert shrub, sagebrush, piñon-juniper, and ponderosa pine–oak communities. ARIZ: Apa, Coc, Nav; COLO: Arc, LPl, Mon, SMg; NMEX: McK, RAr, San, SJn; UTAH. 1400–2300 m

(4500–7600′). Flowering: Aug–Sep. California to Colorado and New Mexico, south to northern Mexico. Leaf ash used as Evilway blackening by the Ramah Navajo.

Ambrosia artemisiifolia L. (*Artemisia*-like leaves) Annual ragweed. Annuals 10–60 cm, herbage pilose to strigillose, also gland-dotted. **STEMS** 1, erect, branched distally (sometimes throughout), branches ascending, ± pilose (to subglabrous). **LEAVES** opposite (proximal) to alternate (distal); petioles 25–35+ mm; blades ovate to elliptic or lanceolate, 30–50+ × 20–30+ mm, 1–2-pinnatifid, base cuneate, ultimate margins entire or denticulate, adaxial faces strigillose, gland-dotted, abaxial faces pilosulous to strigillose, gland-dotted. **PISTILLATE HEADS** clustered; florets 1. **STAMINATE HEADS** with peduncles about 1 mm long. **INVOLUCRES** cupulate, ± oblique, 3–5 mm wide, shallowly to obscurely lobed (rarely with thickened, black nerves), hispidulous to pilosulous (or glabrous), ± gland-dotted. **FLORETS** 12–20, corollas usually darkened on the lobe margins. **BURS** with bodies ± obovoid or pyriform, about 2–4 mm, ± smooth to reticulate, glabrous or sparsely puberulent; spines 3–5 in 1 distal (or medial) series, less than 1 mm, erect, ± conical; beak narrowly conical. $2n = 34, 36$. [*A. artemisiifolia* var. *elatior* (L.) Descourt.; *A. elatior* L.; *A. glandulosa* Scheele; *A. monophylla* (Walter) Rydb.]. A ruderal weed usually in disturbed habitats. ARIZ: Apa. 1375–1725 m (4510–5660′). Flowering: mostly Aug–Sep. Probably native to the eastern and central regions of North America, introduced westward and now established throughout North America; also introduced in Europe.

Ambrosia psilostachya DC. (a smooth spike, alluding to the inflorescences) Western ragweed. Perennials 10–40+ cm, rhizomatous, herbage strigose to hispid and gland-dotted throughout. **STEMS** 1 (or more), erect, simple or branched (proximally or throughout). **LEAVES** opposite (proximal) to alternate (distal), sessile or petiolate (petioles to 20 mm); blades ovate to lanceolate, about 15–50+ × 8–35+ mm, pinnately toothed or lobed to 1–2-pinnatifid, bases cuneate to truncate, ultimate margins entire or few dentate, faces strigose to hispid (abaxially), gland-dotted. **PISTILLATE HEADS** clustered; florets 1. **STAMINATE HEADS** with peduncles about 1–2 mm. **INVOLUCRES** cupulate, ± oblique, 2–3 mm wide, shallowly to obscurely lobed, papillose-strigillose to hispidulous (-ciliate), gland-dotted. **FLORETS** 8–24, corollas not darkened (rarely slightly purplish) or with black nerves. **BURS** with bodies obovoid to subglobose, 2–4 mm, stramineous to brownish, ± reticulate, ± puberulent and gland-dotted; spines lacking or 1–6 in ± 1 distal series, to 1 mm long, erect; beak 1, conical. $2n = 18, 27, 36, 45, 54, 63, 72, 100–104, 108, 144$. Most commonly in moist riparian or meadow communities, occasional in grasslands or open woodlands, often in disturbed or weedy areas. COLO: Arc, Min; NMEX: RAr, San, SJn; UTAH. 1100–2200 m (3700–7300′). Flowering: Jun–Sep. Possibly native to central prairies but now established throughout North America from southern Canada to northern Mexico.

Ambrosia tomentosa Nutt. (tomentose) Skeleton-leaf ragweed. Perennials 10–30 cm, rhizomatous, herbage sparsely to densely strigillose, obscurely gland-dotted. **STEMS** 1–several, erect, mostly simple (or branched). **LEAVES** alternate; petioles about 10–25 mm; blades obovate to oblanceolate, about 45–60+ × 15–40+ mm, 2–3-pinnately lobed; lobes alternate to opposite; bases obscurely cuneate, ultimate margins deltate-dentate (often also between lobes or downward on petiole), adaxial faces ± strigillose (greenish), abaxial faces densely strigillose (whitish). **PISTILLATE HEADS** clustered; florets 2. **STAMINATE HEADS** with peduncles about 1–2+ mm. **INVOLUCRES** openly funnelform, ± oblique, about 3 mm wide, lobed (several black-nerved), strigillose (sometimes densely). **FLORETS** 25–40, corollas with a black nerve between each lobe (sometimes also purplish and/or lobes darkened or blackened). **BURS** with bodies obovoid, 2–4 mm, ± brownish, ribbed, puberulent and gland-dotted; spines about 10, ± irregularly arranged, about 1 mm long, narrowly conical, spreading; beaks 2, stoutly conical and ± divergent. $2n = 24$. [*Franseria discolor* Nutt.; *F. tomentosa* A. Gray]. Sagebrush to ponderosa pine–Gambel's oak communities. ARIZ: Apa; COLO: LPl, Mon; NMEX: SJn. 2200–2300 m (7200–7400′). Flowering: Jun–Aug. Northern Great Plains, south to Arizona, New Mexico, and Texas.

Ambrosia trifida L. (three-cleft) Giant ragweed. Annuals 30–150+ cm, herbage papillose-strigose to hispid, often glabrate proximally, gland-dotted (or not). **STEMS** 1, erect, branched distally (mostly). **LEAVES** opposite (rarely the most distal alternate); petioles about 10–30 mm, hispid-ciliate; blades ovate to elliptic, 40–100+ × 30–70+ mm, 3 (-5)-palmately lobed (unlobed in depauperate forms), base cuneate to truncate, margins serrate (rarely entire), faces ± papillose-strigillose, gland-dotted (sometimes obscure or lacking). **PISTILLATE HEADS** clustered; florets 1. **STAMINATE HEADS** with peduncles about 1–3 mm. **INVOLUCRES** openly funnelform to nearly flat, ± oblique, 2–4 mm wide, shallowly (to obscurely) lobed (1–3 outward-facing lobes each with a black nerve), ± strigillose to hispid, sometimes gland-dotted; florets 3–25, corollas with a black nerve between each lobe. **BURS** with bodies ± obovoid or obpyramidal, 3–5 mm, ± ribbed from base of spines, glabrous (or sparsely puberulent); spines 4–5 in 1 distal series, less than 1 mm long, stramineous, erect, ± conical; beak 1, conical. $2n = 24, 48$. [*Ambrosia aptera* DC.]. Roadside habitat. COLO: Arc. 1900 m (6200′). Flowering: Jul. Native to eastern and central North America, occasionally introduced westward, now occurring across North America.

Anaphalis DC. Pearly Everlasting

G. L. Nesom

(an ancient name, or perhaps derived from generic name *Gnaphalium*) Perennial herbs (–subshrubs), dioecious or subdioecious, 20–80 (–120+) cm; fibrous-rooted, rhizomatous, not stoloniferous. **STEMS** usually 1, usually erect. **LEAVES** basal and cauline; alternate; petiolate or sessile; blades oblanceolate or lanceolate to linear, bases ± cuneate, margins entire, faces usually bicolored (concolored), abaxial usually white to gray and tomentose, adaxial usually greenish and glabrate or glabrous, sometimes grayish and sparsely arachnose. **INFL** a head, usually discoid (unisexual or nearly so) or disciform, in glomerules in corymbiform or paniculiform arrays. **INVOLUCRES** subglobose. **PHYLLARIES** in 8–12 series, bright white, opaque, at least toward tips, often proximally woolly; stereomes not glandular, unequal, more or less papery at least toward tips. **PERIPHERAL FLORETS** pistillate, 50–150, more numerous than staminate; sometimes a few pistillate florets peripheral in predominantly staminate heads or 1–9 staminate florets central in predominantly pistillate heads; corollas yellowish. **INNER FLORETS** functionally staminate, 30–55; corollas yellowish. **CYPSELAE** oblong (–obclavate, ovoid, or cylindric), 2-nerved, faces ± scabrous, hairs clavate, not myxogenic. **PAPPUS** usually of 10–20, distinct or basally connate, readily falling barbellate bristles, tips more or less clavate in bisexual or functionally staminate florets. $x = 14$. Species about 100, mostly central Asia and India, one in North America.

Anaphalis margaritacea (L.) Benth. & Hook. f. **CP** (resembling a pearl, alluding to the involucre) Pearly everlasting. Perennials; rhizomes relatively slender. **STEMS** white, densely and closely tomentose, not glandular. **LEAVES** with blades 1–3-nerved, 3–10 (–15) cm, bases subclasping, decurrent, margins revolute, abaxial faces tomentose or glabrescent (proximal leaves), not glandular or very sparsely and inconspicuously glandular, adaxial faces green, glabrate. **INVOLUCRES** 5–7 × 6–8 (–10) mm long. **PHYLLARIES** ovate to nearly linear (innermost), subequal to unequal, apices white, opaque. **CYPSELAE** 0.5–1 mm, bases constricted into stipiform carpopodia. $2n = 28$. [*Gnaphalium margaritaceum* L.; *Anaphalis margaritacea* var. *occidentalis* Greene; *A. margaritacea* var. *subalpina* (A. Gray) A. Gray]. Along trails, roadsides, lake margins, ridges; ponderosa pine, spruce-fir, and spruce communities. COLO: Arc, Con, Hin, LPl, Min, RGr, SJn. 2300–3450 m (7545–11320′). Flowering: Jul–Oct. Alaska and Yukon, across Canada, northern and eastern United States, Washington to Dakotas, south through western states (including Arizona, Colorado, New Mexico, and Utah) to Mexico (Baja California); naturalized in Europe and Asia. *Anaphalis margaritacea* has the aspect of *Pseudognaphalium* but is subdioecious and has a distinctive cypselar vestiture. It is further recognized by its combination of rhizomatous habit; subclasping-decurrent, bicolored, revolute leaves; large heads; and distally white phyllaries.

Antennaria Gaertn. Pussytoes, Everlasting

R. J. Bayer and G. L. Nesom

(Latin *antenna* and *-aria*, possession of, alluding to similarity of clavate pappus bristles in staminate florets to insect antennae) Perennials or subshrubs; mostly gynoecious (all plants of a population pistillate) or dioecious (staminate plants in equal frequency as pistillates); sometimes caespitose, sometimes stoloniferous, sometimes rhizomatous. **LEAVES** basal and cauline; alternate; petiolate or sessile; blades 1–7-nerved, mostly cuneate, elliptic, lanceolate, linear, oblanceolate, or spatulate, apices sometimes "flagged" (apices with flat, linear, scarious appendages similar to phyllary tips), margins entire, abaxial faces usually tomentose, adaxial glabrous or more or less tomentose to sericeous or glabrescent. **INFL** a discoid head, unisexual, borne singly or in corymbiform, paniculiform, racemiform, or subcapitate arrays. **INVOLUCRES** (staminate) campanulate to hemispheric, 2–6+ mm diameter; pistillate turbinate or campanulate to cylindric, 3–7 (–9+) mm diameter. **PHYLLARIES** in 3–6+ series, unequal, proximally papery or membranous; distally more or less scarious, often black, brown, castaneous, cream, gray, green, olivaceous, pink, red, white, or yellow. **RECEPTACLES** epaleate. **RAY FLORETS** absent. **DISC FLORETS** mostly 20–100+, functionally staminate or pistillate; staminate corollas white, yellow, or red, narrowly funnelform or tubular; pistillate corollas white, yellow, or red, narrowly tubular to filiform. **CYPSELAE** mostly ellipsoid to ovoid, faces usually glabrous, often papillate (stout, myxogenic twin-hairs). **PAPPUS** of basally connate or coherent caducous bristles, shed together in rings or in groups; staminate usually more or less clavate. $x = 14$ (Bayer 1990; Bayer, Soltis, and Soltis 1996; Bayer and Stebbins 1987, 1993). Species 45; temperate and Arctic/alpine regions, North America, Mexico, South America, Eurasia. Morphologically discrete sexual diploids are recognized at specific rank. Polyploids (nonhybrid- or auto-polyploid) that are morphologically identical with sexual diploid taxa, whether they are agamospermous or amphimictic, are treated as conspecific with their sexual diploids. Sexual and asexual polyploids of hybrid origin (segmental and genomic allopolyploids) are recognized as species because their genetic composition is not attributable to any single diploid origin. Polyploid complexes have been defined primarily by assessing their genomic composition through the use of genetic markers and morphological studies.

1. Heads usually 1, rarely in clusters of 2 or 3, subsessile or pedunculate .... (2)
1' Heads usually in clusters of (2–) 3–15 (–110+), rarely borne singly, pedunculate .... (3)

2. Heads subsessile among basal leaves; plants 0.2–1.5 (–2) cm high; basal leaves oblanceolate to spatulate, 6.5–13 × 2–5 mm, densely and closely silvery gray-sericeous; stolons present, plants forming mats .... ***A. rosulata***
2' Heads short-pedunculate; plants 0.5–4 cm high; basal leaves linear to narrowly spatulate, 8–11 × 1–1.2 mm, gray-tomentose; stolons absent, plants caespitose, not forming mats .... ***A. dimorpha***

3. Stolons none; plants not forming mats; lower cauline leaves 25–150 (–200) mm, 3–5-nerved; outer phyllaries usually each with a dark spot at base .... ***A. anaphaloides***
3' Stolons mostly 1–5 (–18) cm; plants forming mats; basal and lower cauline leaves 6–35 (–45) mm, mostly 1-nerved; outer phyllaries with or without a dark spot at base. .... (4)

4. Basal leaves strongly bicolored, the adaxial surface green, usually stipitate-glandular but otherwise glabrous, margins white-woolly and contrasting with adaxial face. .... ***A. marginata***
4' Basal leaves concolored (or in *A. parvifolia* strongly to weakly bicolored, but the adaxial surface pubescent, sometimes glabrate with age, eglandular), margins not distinctly contrasting with adaxial face .... (5)

5. Pistillate involucres 8–10 (–15) mm; leaves spatulate, commonly weakly to strongly bicolored .... ***A. parvifolia***
5' Pistillate involucres 4–8 mm; leaves narrowly oblanceolate to spatulate, concolored. .... (6)

6. Phyllaries distally mostly dark to light brown, black, or olivaceous (inner sometimes whitish) .... ***A. media***
6' Phyllaries distally mostly whitish to pink or rosy .... (7)

7. Phyllaries with a dark spot at the base of each; dioecious .... ***A. corymbosa***
7' Phyllaries without a basal dark spot; dioecious or gynoecious .... (8)

8. Phyllaries distally white; dioecious .... ***A. microphylla***
8' Phyllaries distally usually pink to red; gynoecious (staminate plants uncommon) .... ***A. rosea***

Antennaria anaphaloides Rydb. (for a resemblance to *Anaphalis margaritacea*) Pearly pussytoes. Dioecious. Plants 15–35 (–50) cm. **STOLONS** none. **LEAVES** (basal) ephemeral, 3–5-nerved, oblanceolate to linear-oblanceolate, 25–150 (–200) × 4–20 (–25) mm, tips mucronate, faces gray-pubescent; cauline oblanceolate or linear, 10–80 mm, usually flagged. **HEADS** 8–30 (–50+) in corymbiform arrays. **INVOLUCRES:** staminate (4–) 5–6.5 mm; pistillate 4.5–7 mm. **PHYLLARIES** distally white or cream, sometimes suffused pink to rose, with a distinct dark brown or blackish spot at the base or middle of each. **COROLLAS:** staminate 2.5–4 mm; pistillate 3–4.5 mm. **PAPPUS** bristles: staminate 3–4.5 mm; pistillate 3.5–4.5 (–5.5) mm. $2n = 28$. [*Antennaria pulcherrima* Greene subsp. *anaphaloides* (Rydb.) W. A. Weber; *A. pulcherrima* var. *anaphaloides* (Rydb.) G. W. Douglas]. Alpine and subalpine meadows, peaty wetlands of alpine lakes. COLO: SJn. 3000–3400 m (9840–11155'). Flowering: Jul–Aug. Canada (Alberta to Saskatchewan), Washington to Montana, south to Nevada, Utah, and Colorado.

Antennaria corymbosa E. E. Nelson (in a flat-topped cluster, in reference to the inflorescence) Flat-top pussytoes, meadow pussytoes. Dioecious. Plants 6–15 cm. **STOLONS** 1–10 cm. **LEAVES** (basal) 1-nerved, spatulate, 18–45 × 2–4 mm, tips mucronate, faces ± gray-tomentose; cauline linear, 8–13 mm, apices acuminate, not flagged. **HEADS** 3–7 in corymbiform arrays. **INVOLUCRES:** staminate 4–5.3 mm; pistillate 4–5 mm. **PHYLLARIES** distally white or light brown, with a distinct dark brown or blackish spot at the base of each. **COROLLAS:** staminate 2–3.2 mm; pistillate 2.5–3.5 mm. **PAPPUS** bristles: staminate 2.5–3.5 mm; pistillate 3.5–4.5 mm. $2n = 28$. [*Antennaria acuta* Rydb.; *A. dioica* Gaertn. var. *corymbosa* (E. E. Nelson) Jeps.; *A. hygrophila* Greene; *A. nardina* Greene]. Alpine tundra and alluvium, fell-fields with Engelmann spruce, moist willow thickets often with *Potentilla fruticosa*, openings in spruce-fir, roadsides. COLO: Con, Hin, LPl, Min, Mon, RGr, SJn. (2950–) 3240–3680 m (9680–12075'). Flowering: Jun–Aug. Washington to Montana, south to California, Utah, and New Mexico.

Antennaria dimorpha (Nutt.) Torr. & A. Gray. (two-shaped) Cushion pussytoes. Dioecious. Plants 0.5–4 cm. **STOLONS** none. **LEAVES** (basal) 1-nerved, linear to narrowly oblanceolate or narrowly spatulate, 8–11 × 1–1.2 mm, tips acute, faces ± gray-tomentose; cauline linear or oblanceolate, 7–12 mm, apices acute, not flagged. **HEADS** borne singly. **INVOLUCRES:** staminate 6–8 mm; pistillate 10–11 mm. **PHYLLARIES** distally brown, apices acute-acuminate.

COROLLAS: staminate 3–5 mm; pistillate 8–10 mm. **PAPPUS** bristles: staminate 4.5–6 mm; pistillate 10–12 mm. 2*n* = 28, 56. [*Gnaphalium dimorphum* Nutt.; *A. macrocephala* (D.C. Eaton) Rydb.]. Sandy or gravelly slopes, roadsides; piñon-juniper-sagebrush, Gambel's oak, sagebrush–Gambel's oak. COLO: Arc, Dol, LPl, Mon, SMg; NMEX: SJn; UTAH. 1800–2400 m (5905–7875′). Flowering: Apr–May (–Jun). Canada (Alberta to Saskatchewan), Washington to Montana and Nebraska, south to California, Nevada, and New Mexico.

Antennaria marginata Greene (margined, referring to the leaves) Whitemargin pussytoes. Dioecious or gynoecious. Plants 5–20 cm. **STOLONS** 2–7 cm. **LEAVES** (basal) 1–3-nerved, spatulate, 15–20 × 4–6 mm, tips mucronate, abaxial faces gray-tomentose, adaxial green, stipitate-glandular at least on the petiolar region, otherwise glabrous, margins white-woolly; cauline linear, 7–16 mm, apices acute, not flagged. **HEADS** 5–8 in corymbiform arrays. **INVOLUCRES:** staminate 4.5–7 mm; pistillate 5–7 (–9) mm. **PHYLLARIES** distally white, apices acuminate. **COROLLAS:** staminate 3–5 mm; pistillate 4.5–6.5 mm. **PAPPUS** bristles: staminate 3.5–5.5 mm; pistillate 5.5–8.5 mm. 2*n* = 28, 56, 84, 112, 140. [*Antennaria dioica* Gaertn. var. *marginata* (Greene) Jeps.; *A. fendleri* Greene]. Slopes, ridges, benches, roadsides; ponderosa pine, ponderosa pine–Douglas-fir–Gambel's oak, Douglas-fir–white fir, spruce-fir, and aspen communities. ARIZ: Apa; COLO: Arc, Hin, LPl, Mon, SJn; NMEX: RAr, SJn. (1750–) 2200–2750 m (5740–9020′). Flowering: May–Jun (–Jul). California to Colorado, New Mexico, and Texas; Mexico (Chihuahua, Coahuila).

Antennaria media Greene (middle, perhaps for its "intermediate" morphology) Rocky Mountain pussytoes. Gynoecious (in our area). Plants 5–13 cm. **STOLONS** 1–4 cm. **LEAVES** (basal) 1-nerved, spatulate to oblanceolate, 6–19 × 2.5–6 mm, tips mucronate, faces gray-pubescent; cauline linear, 5–20 mm, apices acute, not flagged. **HEADS** 2–5 (–9) in corymbiform arrays. **INVOLUCRES:** staminate (3.5–) 4.5–6.5 mm; pistillate 4–8 mm. **PHYLLARIES** distally dark brown, black, or olivaceous. **COROLLAS:** staminate 2.5–4.5 mm; pistillate 3–4.5 mm. **PAPPUS** bristles: staminate 2.5–4.5 mm; pistillate 4–5.5 mm. 2*n* = 56, 98, 112. [*Antennaria alpina* Gaertn. var. *media* (Greene) Jeps.]. Alpine tundra and talus, subalpine meadows, and spruce-fir communities. COLO: Arc, Con, Hin, LPl, Min, Mon, RGr, SJn. 3450–3950 m (11320–12960′). Flowering: (Jun–) Jul–Sep. Alaska, Yukon, and Northwest Territories, south to California, Nevada, Arizona, Utah, and Colorado. The distinction between *A. media* and *A. umbrinella* Rydb. in the *Flora* region needs to be examined.

Antennaria microphylla Rydb. (small, relating to leaves) Littleleaf pussytoes. Dioecious. Plants 9–30 cm, stems sometimes stipitate-glandular distally. **STOLONS** 1–5 cm, often at least slightly woody. **LEAVES** (basal) 1-nerved, spatulate, 6–16 × 2–6 mm, tips mucronate, faces silvery gray-pubescent; cauline linear, 5–25 mm, apices acute, not flagged. **HEADS** 6–13 in corymbiform arrays. **INVOLUCRES:** staminate 5–6.5 mm; pistillate 5.5–7 mm. **PHYLLARIES** distally bright white to tawny. **COROLLAS:** staminate 2.5–3 mm; pistillate 3–4.3 mm. **PAPPUS** bristles: staminate 3–4 mm; pistillate 3–5 mm. 2*n* = 28. [*Antennaria nitida* Greene; *A. rosea* Greene var. *nitida* (Greene) Breitung]. Rocky washes and slopes, ridges, meadows, roadsides; juniper-shrub, ponderosa pine–piñon pine, and aspen-pine-spruce communities. ARIZ: Apa; COLO: Con, Hin, LPl, Min, SJn; NMEX: SJn; UTAH. 2250–2950 (–3500) m (7380–11480′). Flowering: May–Aug (–Sep). Alaska, Yukon, Northwest Territories, and Nunavut, south to British Columbia and Ontario, further to Nebraska, California, Arizona, and New Mexico. *Antennaria microphylla* is a primary sexual progenitor of the *A. rosea* polyploid agamic complex. The two species can be identified by the dioecy vs. gynoecy character (presence of staminate plants almost always indicates *A. microphylla*; presence of only pistillate plants is not indicative). *Antennaria microphylla* also tends to produce more cauline leaves than *A. rosea*, leaves more silvery-white with a greater tendency to have a distinct petiole, and it tends to occur on floodplains. Reliance on phyllary coloration provides a useful sort but may give artificial identifications, especially for *A. rosea*. Carlquist's circumscription of *A. microphylla* (1955) included *A. rosea*, in recognition of their close and intergrading morphologies, and the apparent reticulate hybridity of these taxa may make it too difficult to separate them in the herbarium.

Antennaria parvifolia Nutt. (small leaves) Nuttall's pussytoes. Dioecious or gynoecious. Plants 2–8 (–15) cm. **STOLONS** 1–6 cm. **LEAVES** (basal) 1-nerved, narrowly spatulate to spatulate or oblanceolate, 8–35 × 2–15 mm, tips mucronate, faces gray-tomentose; cauline linear to narrowly oblanceolate, 8–20 mm, apices acute, not flagged. **HEADS** 2–7 in corymbiform arrays. **INVOLUCRES:** staminate 5.5–7.5 mm; pistillate 8–10 (–15) mm (gynoecious), 7–7.2 mm (dioecious). **PHYLLARIES** distally white to pink. **COROLLAS:** staminate 3.5–4.5 mm; pistillate 5–8 mm. **PAPPUS** bristles: staminate 4–5.5 mm; pistillate 6.5–9 mm. 2*n* = 56, 84, 112, 140. [*Antennaria aprica* Greene; *A. dioica* Gaertn. var. *parvifolia* (Nutt.) Torr. & A. Gray]. Meadows, openings, roadsides; piñon-juniper, Gambel's oak, ponderosa pine, pine-fir-aspen, aspen and spruce communities. ARIZ: Apa; COLO: Arc, Dol, Hin, LPl, Min, Mon, SJn; NMEX: McK, RAr, SJn; UTAH. (1850–) 2150–2850 (–3250) m (6070–10660′). Flowering: (Apr–) May–Jul. Canada (British Columbia to Ontario), Washington to Michigan, south to Iowa, Nevada, Arizona, New Mexico, and Texas; Mexico (Chihuahua, Nuevo

León). Recognized by its relatively woody rhizomes, spatulate and bicolored leaves, and large heads with white to pinkish phyllaries.

Antennaria rosea Greene (referring to the rose-colored phyllaries of some individuals) Rosy pussytoes. Gynoecious (staminate plants uncommon). Plants 19–30 cm, stems sometimes stipitate-glandular distally. **STOLONS** 1.5–4.5 cm. **LEAVES** (basal) 1-nerved, 10–20 × 2–10 mm, spatulate to narrowly cuneate, tips mucronate, faces usually gray-pubescent; cauline linear, 9–26 mm, apices subulate, sometimes with flat, lanceolate scarious appendages (flags). **HEADS** usually 6–12 in corymbiform arrays. **INVOLUCRES:** staminate unknown; pistillate 6.5–8 mm. **PHYLLARIES** distally pink to red, apices acute or erose-obtuse. **COROLLAS:** staminate unknown; pistillate 3.5–6 mm. **PAPPUS** bristles: staminate unknown; pistillate 5–6 mm. 2*n* = 42, 56, (70). [*Antennaria arida* E. E. Nelson; *A. scariosa* E. E. Nelson; *A. viscidula* (E. E. Nelson) E. E. Nelson ex Rydb.]. Openings and meadows, roadsides, ponderosa pine, often with oak, Douglas-fir–Gambel's oak, ponderosa pine–blue spruce, spruce-fir, alpine talus. ARIZ: Apa; COLO: Con, Hin, LPl, Min, Mon, SJn; NMEX: SJn; UTAH. 1800–3800 m (5905–12465′). Flowering: May–Aug. Alaska, Yukon, British Columbia to Newfoundland and Maine, south to Nevada, Arizona, and New Mexico. Four Corners plants are **subsp. *arida*** (E. E. Nelson) R. J. Bayer (dry).

Antennaria rosulata Rydb. (in rosettes, in reference to the plant aspect) Kaibab pussytoes. Dioecious. Plants 0.2–1.5 (–2) cm. **STOLONS** 1–2 (–3.5) cm. **LEAVES** (basal) 1-nerved, spatulate, spatulate-obovate, or oblanceolate, 6.5–13 × 2–5 mm, tips mucronate, faces densely and closely silvery gray-pubescent (often obscurely stipitate-glandular); cauline linear, 2–9 mm, apices acute, not flagged. **HEADS** 1 (rarely –3), subsessile among basal leaves. **INVOLUCRES:** staminate 5–7.5 mm; pistillate 6–10 mm. **PHYLLARIES** distally white. **COROLLAS:** staminate 2.5–4.5 mm; pistillate 3.5–5.5 mm. **PAPPUS** bristles: staminate 3.5–5 mm; pistillate 5.5–6.5 mm. 2*n* = 28. [*Antennaria sierrae-blancae* Rydb.]. Meadows, open slopes, sagebrush, ponderosa pine (often with Gambel's oak, sagebrush, or other shrubs) and ponderosa pine–spruce communities. ARIZ: Apa; COLO: Arc, Hin, LPl, Min, Mon; NMEX: McK, RAr; UTAH. 2200–2800 (–3000) m (7215–9840′). Flowering: May–Jun (–Jul). Arizona, Utah, Colorado, and New Mexico.

Anthemis L. Chamomile, Dog-fennel
Brian Elliott

(ancient Greek name for chamomile) Annual, biennial, or perennial taprooted herbs, usually aromatic. **STEMS** erect, typically branched. **LEAVES** alternate, rarely incised-dentate to more commonly 1–3-pinnately dissected with ultimate segments narrowly linear or filiform and cuspidate. **INFL** heads radiate or rarely discoid, pedunculate and solitary. **INVOLUCRES** hemispheric. **PHYLLARIES** imbricate in several series, margins scarious or hyaline. **RECEPTACLES** convex to conical, often naked at base but chaffy toward middle and apex; chaff scales linear to narrow. **RAY FLORETS** pistillate and fertile or neutral; ligules white or yellow. **DISC FLORETS** perfect and fertile, numerous; corolla yellow. **CYPSELAE** terete to 4–5-angled or slightly compressed, glabrous. **PAPPUS** lacking or rarely a short crown. Over 100 species native to Europe, east Asia, and north Africa, now naturalized and often weedy in the United States.

Anthemis cotula L. (a small cup, referring to a hollow at the base of the leaf) Mayweed, dog-daisy, dog-fennel, stinkweed, stinking chamomile. Annual, unpleasantly aromatic herbs. **STEMS** usually 1–6 (9) dm tall, subglabrate. **LEAVES** 2–6 cm long, 2–3-pinnatifid, ultimate segments linear, sparsely villous and gland-dotted. **INFL** heads radiate. **PHYLLARIES** oblong to oblong-oblanceolate, midrib green or yellow, margins white-scarious or hyaline. **RECEPTACLES** chaffy; chaff scales subulate. **RAY FLORETS** sterile; ligules 10–16 (20), white, 5–11 mm long, reflexed in age. **CYPSELAE** subterete, approximately 10-ribbed, ribs glandular-tuberculate. **PAPPUS** lacking. 2*n* = 18. [*Maruta cotula* (L.) DC.]. Inhabiting roadsides, disturbed ground, streamsides, and ditches. NMEX: SJn. 1520–2750 m (5000–9000′). May–Aug. Found throughout the United States including Alaska and Hawaii. *Anthemis cotula* is easily confused with *Matricaria* but can be distinguished by the receptacle, which is chaffy at the middle and apex in *Anthemis* and naked in *Matricaria.* The species has had extensive medicinal use as a cold remedy, for fevers, diarrhea, as an ear wash, and for rheumatism. The flowers have been used to prevent fleas and bedbugs, and the leaves have been used as a general insecticide. It should not be used for food or tea as it is known to irritate the mucous membranes.

Arctium L. Burdock
Patricia Irick-Barlow

(Greek, from the name "arktion" from *arktos* or bearclotbur, burr dock) Biennials, 50–300 cm. **STEMS** coarsely branched, branches ascending. **LEAVES** simple, entire to remotely denticulate, the lower ovate-cordate and petiolate, not spiny, underside resin gland–dotted, upper side often tomentose. **HEADS** subglobose, in leafy-bracted, loosely branching arrays, caducous in anthesis. **INVOLUCRES** spheric to ovoid. **PHYLLARIES** very numerous, stiff, subulate, strongly hooked at apex, appressed at the base, with entire margins. **RECEPTACLES** ± flat, densely scaly. **FLORET** corollas purple, pink, rarely white. **CYPSELAE** obovoid to narrowly ellipsoid, compressed, striate or ribbed, glabrous, attachment scars basal. **PAPPUS** scabrid, yellowish hairs, free from base. Introduced from Eurasia and North Africa, now worldwide. The dry heads disperse by clinging to passing animals. The roots and leaves of several species are used for medicinal herbs, mostly as a blood purifying agent.

1. Heads commonly over 2.5 mm thick, arranged in corymbose clusters, especially the terminal.................***A. lappa***
1' Heads mostly 1.5–2.5 cm thick, arranged in racemelike axillary clusters; the terminal also racemelike ....***A. minus***

Arctium lappa L. (burlike) Great burdock, grande bardane. Plants erect, well-branched, to 2 m tall. **LEAVES:** basal glabrous, short-hairy or thinly cobwebby; blades to 80 cm long on solid petioles almost 1/2 as long, coarsely dentate to subentire. **INFL** usually in corymbose clusters, with long peduncles 2.5–5 cm. **INVOLUCRES** 20–30 × 24–47 mm. **PHYLLARIES** linear, glabrous to sparsely pubescent, exceeding the corolla by 1–5 mm. **FLORETS** purple (rarely white) corollas 9–14 mm, glabrous or glandular-pubescent distally. **CYPSELAE** light brown, 6–7.5 mm. **PAPPUS** bristles 2–5 mm long. Waysides, field borders, wood clearings, and waste places. COLO: SMg. 1980–2135 m (6500–7000′). Flowering: Jul–Oct. Native to Eurasia. Cultivated for its edible roots, and persisting.

Arctium minus (Hill) Bernh. (minor) Common or lesser burdock, petite bardane, cibourroche, chou bourache. Biennial, producing a large-leaved rosette in the first year and a tall (about 1.5 m), erect, much-branched stem in the second year, greenish to reddish purple, sparsely arachnoid-pubescent to glabrous, from stout taproot. **LEAVES** alternate, petiole sparsely arachnoid-pubescent, with adaxial groove, to 40+ cm long, typically hollow but sometimes only at the base; basal very large, to 60+ cm long, cordate; cauline reduced upward, truncate or cordate at base, ovate, with arachnoid pubescence below, puberulent or glabrous above; margins sinuous to undulate or commonly crisped. **INFL** in a loose pedunculate cyme of 1–4 flowers, sessile in the leaf axils or pedunculate at the ends of branches, falling in anthesis as a spiny bur; peduncles 0–9.5 cm, glabrous to densely pubescent. **INVOLUCRES** urceolate,15–40 mm diameter, becoming globose in fruit. **PHYLLARIES** linear-attenuate, uncinate and reddish at apex, to 1.5 cm long, mostly glabrous but strigillose on margins near base; outer phyllaries strongly hooked and inner without hooks. **FLORETS** all discoid, glabrous, usually exceeding the phyllaries; corolla tube to 1 cm long, constricted and white below, lavender to pink above; lobes acute, 1.3–1.9 mm long; inflated throat 2–4 mm long; style tips bifurcate; anthers lavender to white, with tiny mucronate tips. **CYPSELAE** angled, dark brown or with darker spots, 4–8 mm long. **PAPPUS** barbellate bristles 1–3.5 mm long. Waste places, roadsides, fields, forest clearings. ARIZ: Apa; COLO: Arc, LPl, Mon; NMEX: RAr, San, SJn; UTAH. 1525–2285 m (5000–7500′). Flowering: Jul–Oct. Introduced from the Old World. Considered a undesirable or noxious weed because the hooked spines on the bur can cause mechanical injury to animals or tangle animal fibers. Hybrids between *A. lappa* and *A. minus* are known from Europe, showing intermediate characters and reduced achene fertility.

Arida (R. L. Hartm.) D. R. Morgan & R. L. Hartm. Desert Tansy-aster

G. L. Nesom

(Latin *aridus*, alluding to the xeric habitat typical of members of the genus) Annuals, perennials, or subshrubs, usually taprooted, sometimes rhizomatous. **STEMS** glabrous to hairy, sometimes stipitate-glandular. **LEAVES** basal and cauline, alternate, simple, petiolate (basal) or sessile (cauline), entire to dentate, lacerate or deeply 1–2-pinnatifid, with apiculate to spinulose teeth or lobes, basal often petiolate, distal commonly clasping. **INFL** a radiate or discoid head, borne singly or in cymiform or corymbiform arrays. **INVOLUCRES** turbinate to depressed-hemispheric. **PHYLLARIES** in 3–8 graduated series. **RECEPTACLES** convex, shallowly alveolate. **RAY FLORETS** pistillate, fertile, in 1 series, lamina light to dark blue, absent in one species. **DISC FLORETS** bisexual, fertile. **CYPSELAE** narrowly oblong, 8–13 nerves per face, sparsely to densely sericeous. **PAPPUS** of barbellate bristles in 2–3 series, sometimes absent on ray. $x = 5$. [*Machaeranthera* sect. *Arida* R. L. Hartm.]. Species nine, southwestern United States and Mexico.

Arida parviflora (A. Gray) D. R. Morgan & R. L. Hartm. (small-flowered) Annuals or short-lived perennials; involucres, peduncles, and sometimes leaves and midstems stipitate-glandular, without nonglandular hairs. **STEMS** mostly 5–15 cm. **LEAVES** mostly cauline, 1–2-pinnatifid. **INVOLUCRES** hemispheric, 3–5 × 4–6 mm. **PHYLLARIES** in 3–4 series. **RAY FLORETS** epappose. $2n = 10$. [*Machaeranthera parviflora* A. Gray]. Roadsides, wetlands, seeps, along washes and canals, sand to alkaline silt and silty clay, riparian, desert shrub. ARIZ: Apa; NMEX: SJn; UTAH. 1400–2200 m (4595–7215′). Flowering: Jul–Oct. Utah, Colorado, Arizona, New Mexico, Texas; Mexico.

Arnica L. Arnica

Brian Elliott

(ancient name for plants of the genus, derivation obscure) Perennial herbs from rhizomes or caudices with fibrous roots; densely pubescent to glabrous, often glandular. **STEMS** simple or branched, herbaceous, 1–8.5 dm tall. **LEAVES** opposite or basal, rarely alternate above, simple, 1–12 pairs, entire to toothed, sessile to broadly petiolate, uppermost sessile and reduced. **INFL** heads radiate or discoid, solitary to many, cymose or corymbose. **INVOLUCRES** broadly hemispheric to turbinate. **PHYLLARIES** biseriate in 2 equal series, but sometimes appearing uniseriate, connivent, herbaceous, green. **RECEPTACLES** flat or convex, naked, fimbrillate, or hirsute. **RAY FLORETS** pistillate and fertile, rarely with 5 staminodes; ligules (0) 6–21, yellow to orange, 1–4 cm long, 0.4–1.2 cm wide. **DISC FLORETS** perfect and fertile, many; corolla yellow to orange, tubular to goblet-shaped, 5-lobed, lobes triangular-lanceolate. **CYPSELAE** cylindrical to fusiform, 3–10 mm long, 5–10-nerved, glabrous, hirsute, or glandular, with a conspicuous white annulus at the base. **PAPPUS** of numerous barbellate to plumose capillary bristles about equaling the disc corolla, white to tawny or brown. About 30 species, circumboreal, with most diversity found in the western North American mountains. The taxonomy of the genus is complicated by apomixis and polyploidy (Downie and Denford 1988; Maguire 1943; Wolf and Denford 1984).

1. Heads discoid, nodding in bud ..... ***A. parryi***
1' Heads radiate, erect or nearly so in bud ..... (2)

2. Stem leaves (2) 4–10 pairs ..... ***A. chamissonis***
2' Stem leaves 1–4 (5) pairs ..... (3)

3. Pappus subplumose, tawny or brown; apex of phyllaries conspicuously acuminate and reddish-tinged.. ***A. mollis***
3' Pappus barbellate, white; apex of phyllaries obtuse to acuminate, green ..... (4)

4. Leaf blades narrow, the largest 3–10 times as long as wide; basal leaves often densely tufted; rhizomes clothed with old leaf bases ..... ***A. rydbergii***
4' Leaf blades wide, the largest 1–3 times as long as wide; basal leaves not densely tufted; rhizomes naked or nearly so ..... (5)

5. Achenes uniformly (sometimes sparsely so) hirsute; cauline leaves usually petiolate; cauline and basal leaves usually all cordate or truncate; plants usually growing in woods or thickets ..... ***A. cordifolia***
5' Achenes glabrate below, sometimes sparsely short-hirsute above; cauline leaves usually sessile; cauline leaves and basal leaves of short offshoots from the main stem seldom cordate; plants usually growing in meadows or open places ..... ***A. latifolia***

Arnica chamissonis Less. (for Adelbert von Chamisso, 1781–1838, botanist who visited California in 1816) Chamisso arnica, meadow arnica, leafy arnica. Rhizomes long, naked except for reduced scales at the nodes. **STEMS** arising singly, branched above, 1–9 dm tall. **LEAVES**: cauline leaf pairs 4–10; basal leaves smaller than cauline and often withering by anthesis, middle leaves largest, upper somewhat reduced; lower leaves narrowed into petioles connate below into an ocrea, sessile above. **INFL** heads radiate, (1) 3–15; peduncle moderately to densely villous at apex. **PHYLLARIES** villous with septate attenuate trichomes; apex obtuse to acute and pilose. **RAY FLORET** ligules 12–18, yellow, 1–2 cm long, apex shallowly toothed. **DISC FLORETS** 7.5–9 mm long, short-villous and sparsely short-stipitate glandular. **CYPSELAE** 4–6 mm long, short-hirsute to glandular or glabrous. **PAPPUS** barbellate, rarely subplumose, dirty white to tawny. $2n$ = 38, 57, 60, 106–108. [*A. celsa* A. Nelson; *A. denudata* Greene; *A. foliosa* Nutt.; *A. lanulosa* Greene; *A. macilenta* Greene; *A. rhizomata* A. Nelson; *A. stricta* A. Nelson; *A. tomentulosa* Rydb.; many other synonyms]. Montane forests, meadows, riparian areas, and lakeshores, often in moist ground. ARIZ: Apa; COLO: SJn; NMEX: McK, SJn. 2300–2750 m (7600–9000'). Flowering: Jun–Aug. Widespread in western North America from Alaska south to Arizona and New Mexico. A highly polymorphic group with many local variants. Ours is **subsp.** ***foliosa*** (Nutt.) Maguire. The species may be toxic.

Arnica cordifolia Hook. (heart leaf, referring to the heart-shaped leaves) Heartleaf arnica. Rhizomes long, nearly naked, often apically branched and giving rise to several stems in a loose caudex. **STEMS** usually unbranched, sometimes branched, 1–7 dm tall. **LEAVES**: cauline leaf pairs 2 (–6), largest below the middle; lower leaves petiolate, the petiole equaling or exceeding the blade; upper 1–2 leaf pairs reduced, sessile. **INFL** heads radiate, 1–5 (10); peduncle villous and densely stipitate-glandular at apex. **PHYLLARIES** usually glandular, pilose at the base, puberulent above;

apex acuminate or acute and moderately pubescent. **RAY FLORET** ligules rarely absent, usually 9–16, yellow, 1.5–3 cm long, apex subentire to 3-dentate, teeth to 3.5 mm long. **DISC FLORETS** 8–12 mm long, moderately pilosulous and sometimes sparsely glandular. **CYPSELAE** 6–11 mm, uniformly but thinly hirsute, occasionally stipitate-glandular. **PAPPUS** barbellate, white to stramineous. $2n$ = 38, 57, 76, 95, 114. [*A. chionophila* Greene; *A. hardinae* St. John; *A. humilis* Rydb.; *A. paniculata* A. Nelson; *A. parvifolia* Greene; *A. pumila* Rydb.; *A. whitneyi* Fernald; many other synonyms]. Montane forests and meadows to subalpine. ARIZ: Apa; COLO: Arc, Con, Dol, Hin, LPl, Min, Mon, RGr, SJn; NMEX: RAr; UTAH. 2485–4050 m (8150–13280′). Flowering: May–Aug. Widespread in western North America from the Yukon south to Arizona and New Mexico. Apomixis and polyploidy are common in the species, and numerous subspecies and varieties have been named. A poultice or tincture of the plant has been used for inflammation, bruises, cuts, and tuberculosis. The species should only be used externally, as it is not safe for tea.

Arnica latifolia Bong. (broad or wide leaf, referring to the broad leaves) Broadleaf arnica. Rhizomes long or short, giving rise to several basal rosettes, naked except for thin brown scales at the nodes and persistent leaf bases of the previous year at the summit. **STEMS** solitary or clustered, simple or branched above, 8–60 cm tall. **LEAVES**: cauline leaf pairs 2–6, largest below or at the middle (lower leaves sometimes reduced), reduced above; lower leaves abruptly narrowed to a winged petiole, upper leaves sessile or subsessile. **INFL** heads radiate, 1–5 (9); peduncle sparsely to moderately villous at apex. **PHYLLARIES** glandular, sparsely villous to nearly glabrous, often ciliolate; apex short-acute or acuminate to caudate-acute, lacking an apical tuft of hair. **RAY FLORET** ligules 8–15, yellow, 10–25 mm long, apex shallowly 3-toothed, teeth 0.5–1 (1.5) mm long. **DISC FLORETS** 6–10 mm long, sparsely villous or short-pilose. **CYPSELAE** 5–9 mm long, glandular and/or sparsely pubescent above, glabrate at base. **PAPPUS** barbellate, white. $2n$ = 38, 57, 76, 112. [*A. jonesii* Rydb.; *A. latifolia* Bong. var. *angustifolia* Herder; *A. menziesii* Hook.; *A. ovalifolia* Greene; *A. platyphylla* A. Nelson; many others]. Moist montane forests and meadows to the alpine zone. COLO: Con, Hin, Min, RGr, SJn. 2770–3350 m (9100–11000′). Flowering: Jul–Aug. Alaska to northern California, Utah, and Colorado. The species is a mix of sexual diploids and apomictic polyploids.

Arnica mollis Hook. (soft, referring to the pubescence) Rhizomes short and branched, often forming a loose caudex, conspicuously clothed with short roots. **STEMS** 1–several, little-branched. **LEAVES**: cauline leaf pairs 2–5 exclusive of the basal cluster, largest below the middle; petiolate below with a connate-sheathing base, sessile above; basal leaves usually withering by anthesis. **INFL** heads radiate, 1–7; peduncle stipitate-glandular and pubescent with septate hairs at apex. **PHYLLARIES** stipitate-glandular, sparsely to moderately moniliform-pilose at base; apex usually conspicuously acuminate and reddish-tinged. **RAY FLORET** ligules 9–18, yellow, 14–24 mm long, apex 3-toothed, teeth 0.5–1.5 mm long. **DISC FLORETS** 8–10 mm long, sometimes glandular, moderately to sparingly short-hispid pilose with septate hairs. **CYPSELAE** 5–8 mm long, hirsute, usually stipitate-glandular. **PAPPUS** subplumose to plumose, tawny to brownish. $2n$ = 38, 57, 74, 76, 152. [*A. coloradensis* Rydb.; *A. mollis* Hook. var. *scaberrima* Smiley; *A. ovata* Greene; *A. subplumosa* Greene; *A. sylvatica* Greene; many others]. Meadows, stream banks, and forests to the alpine zone. COLO: Con, Dol, Hin, Min, Mon, SJn. 2990–3840 m (9800–12600′). Flowering: Jul–Aug. Alberta and British Columbia south to California, Nevada, Utah, and Colorado.

Arnica parryi A. Gray (for Dr. Charles Christopher Parry, 1823–1890, English-born American botanist and explorer) Rhizomes short, conspicuously clothed with roots or nearly naked. **STEMS** usually solitary, branched above, 2–6 dm tall. **LEAVES**: cauline leaf pairs 2–4, largest below, reduced above; basal leaves petiolate below with narrow-winged and basally connate-sheathing petioles, sessile above. **INFL** heads discoid (in ours), rarely radiate, (1) 3–9 (16), nodding in bud, peduncle pilose and glandular-hairy at apex. **PHYLLARIES** glandular, puberulent; apex ciliolate but lacking a terminal tuft of hair. **RAY FLORETS** usually lacking. **DISC FLORETS** 7.5–9 mm long, pilose. **CYPSELAE** 4–6 mm long, glabrous to hirsute, often stipitate-glandular. **PAPPUS** barbellate to subplumose, stramineous to tawny. $2n$ = 38, 57, 76 and higher. [*A. angustifolia* Vahl var. *eradiata* A. Gray; *A. foliosa* Nutt. var. *sonnei* (Greene) Jeps.; *A. parryi* subsp. *sonnei* (Greene) Maguire; *A. parryi* var. *sonnei* (Greene) Cronquist; *A. sonnei* Greene]. Parks, meadows, moist slopes, and open woods in montane forests. COLO: Dol, Hin, LPl, Min, Mon, SJn, SMg. 3050–3600 m (10000–11800′). Flowering: Jun–Sep. Alberta and British Columbia to California, Nevada, Utah, and Colorado. Ours is **var.** ***parryi***, with discoid heads. A variety with radiate heads, var. *sonnei* (Greene) Cronquist, is known from Nevada, California, and Oregon.

Arnica rydbergii Greene (for Dr. Per Axel Rydberg, 1860–1931, prolific Rocky Mountain botanist) Rydberg's arnica. Rhizomes moderately long, clothed with scales of old leaf bases, branched. **STEMS** clustered on short-branched rhizome tips, forming an approximate crown with caespitose clumps, unbranched, 1–3 dm tall. **LEAVES**: cauline leaf pairs 3–5, reduced above, largest at the middle; basal leaves narrowed to a short, broadly winged petiole, sessile above. **INFL** heads radiate, 1–5; peduncle pilose and glandular-hairy at apex. **PHYLLARIES** glabrate to thinly subpilose and glandular; apex obtuse to moderately acute and pubescent within. **RAY FLORET** ligules 6–10, yellow to yellow-orange, 1–2 (3) cm long, apex entire to minutely subcallously 3-toothed, teeth 0.1–0.5 mm long. **DISC FLORETS** 6–9 mm long, densely pilose, stipitate-glandular. **CYPSELAE** 4–7 mm long, densely short-villous with apical hairs longest. **PAPPUS** barbellate, white. $2n = 38, 76$. [*A. caespitosa* A. Nelson; *A. cascadensis* St. John; *A. lasiosperma* Greene; *A. sulcata* Rydb.; *A. ovalis* Rydb.; *A. tenuis* Rydb.; and others]. Dry meadows and slopes in the subalpine and alpine zones. COLO: Con, Hin. 3050–3765 m (10000–12350′). Flowering: Jul–Aug. Alberta and British Columbia south to northern California, Utah, and Colorado.

Artemisia L. Sagebrush, Wormwood, Felon-herb, Mugwort, Sailor's-tobacco, Chamiso, Hediondo

J. Mark Porter and Arnold Clifford

(Greek *Artemis*, goddess of the hunt and namesake of Artemisia, queen of Anatolia) Annuals, biennials, perennials, subshrubs, or shrubs, 3–350 cm tall, generally aromatic, rarely not. **STEMS** 1–10, generally erect and branched, glabrous or hairy, trichomes basifixed or medifixed. **LEAVES** basal or basal and cauline; alternate; petiolate or sessile; blades filiform, linear, lanceolate, ovate, elliptic, oblong, oblanceolate, obovate, cuneate, flabellate, or spatulate, usually pinnately or palmately lobed, sometimes apically 3-lobed, -toothed, or entire, surfaces glabrous or hairy. **INFL** in relatively broad paniculate, or more narrow racemose or spicate capitulescenses. **HEADS** usually discoid, rarely disciform (e.g., subradiate in *A. bigelovii*). **INVOLUCRES** campanulate, globose, ovoid, or turbinate, 1.5–8 mm diameter. **PHYLLARIES** persistent, 2–20, in 4–7 series, distinct, usually green to whitish green, rarely stramineous, ovate to lanceolate, unequal, margins and apices usually green or white, rarely dark brown or black, scarious, abaxial surfaces glabrous or hairy. **RECEPTACLES** flat, convex, or conic, glabrous or hairy, lacking paleae in ours. **RAY FLORETS** absent. **PERIPHERAL PISTILLATE FLORETS** in disciform heads usually 1–20, and corollas filiform; corollas of 1–3 pistillate florets in capitula of *A. bigelovii* sometimes 2-lobed and weakly raylike. **DISC FLORETS** 2–20 (–32) per head, bisexual and fertile, or functionally staminate; corollas usually pale yellow, rarely red, glabrous or hirtellous, tubes cylindric, throats subglobose or funnelform, lobes 5, deltate. **CYPSELAE** fusiform, brown, ribs absent or 2–5, finely striate, surfaces glabrous or hairy, but not villous, often gland-dotted. **PAPPUS** absent in ours. $x = 9$. About 350–500 species, mostly Northern Hemisphere, some in South America and Africa. *Artemisia* is taxonomically difficult and nomenclaturally controversial. In recent years several segregate genera, e.g., *Seriphidium* and *Picrothamnus*, are commonly recognized in floras. Although these groups can be segregated on the basis of morphology, they are neither phylogenetically distinct nor monophyletic (Vallès et al. 2003; Watson et al. 2002; Vallès and McArthur 2001). Indeed, removal of either of these groups renders *Artemisia* paraphyletic (Hall and Clements 1923; Ling 1982, 1995; McArthur, Welch, and Sanderson 1988; Vallès et al. 2003; Torrell et al. 1999; Vallès and McArthur 2001; Watson et al. 2002).

1. Plants shrubs or subshrubs, woody well above the base ....(2)
1′ Plants herbaceous ....(10)

2. Heads with both ray and disc flowers, the ray flowers 2-lipped; branchlets of inflorescence spreading to reflexed; plants mainly of rimrock areas .... ***A. bigelovii***
2′ Heads with disc flowers only; branchlets of inflorescence variously disposed; plants seldom of rimrock, the distribution various ....(3)

3. Leaves 1- to 3-pinnately or ternately dissected, the segments linear ....(4)
3′ Leaves entire or toothed or, if lobed, the lobes oblong or broader or, if linear, tall shrubs of sandy areas at low elevations ....(7)

4. Plants silvery canescent; receptacle hairy; commonly growing on windswept ridges, but not always so restricted .... ***A. frigida***
4′ Plants green to gray-green; receptacle glabrous or, if hairy, plants of low elevations ....(5)

5. Branches spreading, spinescent; flowering in spring .... ***A. spinescens***
5′ Branches erect or ascending, not spinescent; flowering in late summer and autumn ....(6)

6. Plants commonly 0.5–2 dm tall; leaves 0.3–1 cm long; on clay soils with igneous, calcareous, or dolomitic gravels .... ***A. pygmaea***
6′ Plants usually 5–15 dm tall; leaves mainly 0.6–8 cm long .... ***A. filifolia***

7. Leaves linear-filiform, less than 1 mm wide, entire, or 3-pinnately divided; tall plants of sandy low-elevation sites .... ***A. filifolia***

7′ Leaves broader, entire, or segments broader than 1 mm wide; habitat and elevation various ....(8)

8. Persistent leaves entire or with 1 or 2 teeth (the ephemeral ones often tridentate); heads borne in slender panicles; plants of high elevations .... ***A. cana***
8′ Persistent and ephemeral leaves toothed or lobed at the apex; heads borne in slender-spicate to broad panicles; plants of low to moderate elevations .... (9)

9. Plants usually less than 3 dm tall; leaves usually less than 1 cm long; foliage dull yellowish to lead gray or rarely silvery .... ***A. nova***
9′ Plants mainly more than 3 dm tall; leaves usually more than 1 cm long (at least some); foliage silvery canescent .... ***A. tridentata***

10. Leaves all entire, or the lower ones toothed or lobed, glabrous and green above and beneath, or white-hairy on both surfaces (see also *A. carruthii* and *A. michauxiana*) .... (11)
10′ Leaves deeply incised, pinnatifid, or ternately divided, variously hairy .... (12)

11. Leaves green above and beneath; central flowers of heads with normal ovaries .... ***A. dracunculus***
11′ Leaves white-hairy above and beneath or green above; central flowers of head with abortive ovaries .... ***A. ludoviciana***

12. Plants biennial from a taproot; leaves green, essentially glabrous; naturalized .... ***A. biennis***
12′ Plants perennial from a caudex or rhizomatous; leaves tomentose, strigose, pilose, or bicolored (hairy abaxially, glabrous adaxially) .... (13)

13. Cauline leaves reduced upward, the largest leaves in a basal rosette, silvery villous to strigulose, scarcely tomentose and uniformly colored above and beneath (except *A. franserioides*); plants from caudices, only occasionally rhizomatous .... (14)
13′ Cauline leaves not especially reduced upward, seldom with a basal rosette, variously tomentose and often bicolored; plants often rhizomatous .... (19)

14. Abaxial pubescence of leaves loosely villous or sericeous to glabrous; corollas hairy, the receptacle glabrous; plants of high elevations .... (15)
14′ Abaxial pubescence of leaves tomentose, appressed-strigose or villosulose; corollas glabrous, glandular or hairy but, if hairy, then receptacle long-villous; plants variously distributed .... (17)

15. Involucres 4–5.3 mm high; peduncles to 50 mm .... ***A. norvegica***
15′ Involucres 3–4 mm high; peduncles 10 mm or less, sometimes sessile .... (16)

16. Florets yellow-orange, rarely red-tinged, 2.2–3 mm long, glabrous; disc florets functionally staminate, not setting fruit; plants from a branched caudex .... ***A. borealis***
16′ Florets yellowish or yellow- to red-tinged, 1.5–2 mm long, hairy; disc florets bisexual, setting fruit; plants rhizomatous (to be expected) .... *A. laciniata*

17. Capitulescence a spicate raceme; receptacle and corolla long-villous .... ***A. scopulorum***
17′ Capitulescence a slender panicle; receptacle and corolla glabrous .... (18)

18. Disc florets functionally staminate, not setting fruit; plants from a branched caudex; heads erect; involucres 2–3 mm high .... ***A. campestris***
18′ Disc florets bisexual, setting fruit; plants from a slender rhizome; heads nodding; involucres 3–5 mm high .... ***A. franserioides***

19. Leaves entire or with entire lobes; plants of moderate elevations, mostly below 1525 m (7500′) .... ***A. carruthii***
19′ Leaves bipinnatifid, the lobes again toothed; plants of high elevations, mostly greater than 1525 m (7500′) .... ***A. michauxiana***

Artemisia biennis Willd. (biennial) Biennial wormwood. Annuals or biennials, (10–) 30–80 (–150) cm tall, not aromatic. **STEMS** 1, erect, often reddish, finely striate, glabrous. **LEAVES** cauline, sessile, green or yellow-green; blades broadly lanceolate to ovate, 4–10 (–13) cm long, 1.5–4 cm wide, 1–2-pinnately lobed, terminal lobes coarsely toothed, glabrous. **HEADS** erect, subsessile, in leafy, paniculate to spicate capitulescences 12–35 (–40) cm long, 2–4 cm wide. **INVOLUCRES** globose, 2–4 mm high, (1.5–) 2–4 mm wide. **PHYLLARIES** broadly elliptic to obovate, glabrous, green. **FLORETS**: pistillate 6–25; bisexual 15–40; corollas pale yellow, 1.8–2.2 mm long, glabrous. **CYPSELAE** ellipsoid,

0.2–0.9 mm long, 4–5-nerved, glabrous. $2n$ = 18. Disturbed habitats, often clay soils, logging roads, lake beds, pond margins, mudflats. ARIZ: Apa; COLO: Arc, Dol, Hin, LPl, Min, Mon, RGr, SMg; NMEX: RAr, SJn; UTAH. 1525–2710 m (5000–8900′). Flowering: late Jul–Sep. Fruit: Sep–Oct. Widespread throughout North America; introduced in Europe and New Zealand. *Artemisia biennis* is probably native to the northwest United States and may be introduced in other parts of its range. Used in the blackening ceremony by the Ramah Navajo.

Artemisia bigelovii A. Gray (for John M. Bigelow, 1804–1878, botanist and U.S. Army surgeon) Bigelow sagebrush. Shrubs, 20–40 (–60) cm tall, branched from bases, rounded, mildly aromatic. **STEMS** silvery, canescent, bark gray-brown. **LEAVES** persistent, light gray-green; blades narrowly cuneate, 0.5–3 cm long, 0.2–0.5 cm wide, entire or 3 (–5)-lobed, the lobes acute, 1.5–2 mm long, less than 1/3 blade length, silvery canescent. **HEADS** nodding, capitulescences 6–25 cm long, 1–4 cm wide, branches erect, slightly curved. **INVOLUCRES** globose, 2–3 mm high, 1.5–2.5 mm wide. **PHYLLARIES** 8–15, ovate, canescent or tomentose. **FLORETS**: pistillate 0–2, raylike, laminae to 1 mm long; bisexual 1–3; corollas 1–1.5 mm long, style branches of ray florets elongate, exserted, tips acute, lacking papillae; style branches of disc florets short, truncate, papillate. **CYPSELAE** ellipsoid, 0.8–1 mm long, 5-ribbed, glabrous. $2n$ = 18, 36, 72. [*A. petrophila* Wooton & Standl.; *Seriphidium bigelovii* (A. Gray) K. Bremer & Humphries]. Deserts, sandy or alkaline soils, rimrock, desert shrub, and piñon-juniper woodlands. ARIZ: Apa, Coc, Nav.; COLO: LPl, Mon; NMEX: McK, RAr, San, SJn; UTAH. 1340–2300 m (4400–7550′). Flowering: Jun–Oct. Fruit: Jul–Nov. California and Nevada east to Colorado, New Mexico, and Texas.

Artemisia borealis Pall. (northern) Boreal sage. Perennials, 10–20 (–40) cm tall (caespitose), mildly aromatic; taprooted, caudices branched. **STEMS** (1–) 2–5, gray-green, tomentose. **LEAVES** persistent, basal rosettes persistent, gray-green to white; blades ovate, 2–4 cm tall, 0.5–1 cm wide, 2–3-pinnately or -ternately lobed, lobes linear to narrowly oblong, with acute apices, villous-tomentose, glabrate, or glabrous. **HEADS** sessile below, pedunculate above, in leafy spiciform capitulescences, 4–9 (–12) cm long, (0.5–) 1–5 cm wide. **INVOLUCRES** hemispheric, 3–4 mm high, 3.5–4 mm wide. **PHYLLARIES** obscurely scarious, densely tomentose-villous. **FLORETS**: pistillate 8–10; functionally staminate 15–30; corollas with lobes yellow-orange, rarely red-tinged, 2.2–3 mm long. **CYPSELAE** oblong-lanceoloid, somewhat compressed, 0.4–1 mm long, faintly nerved, glabrous. $2n$ = 18, 36. [*A. campestris* L. subsp. *borealis* (Pall.) H. M. Hall & Clem.]. Open meadows, usually on well-drained soils. COLO: Hin, SJn; NMEX: McK, SJn. 3765–3810 m (12350–12500′). Flowering: Jul–late Aug. Fruit: Jul–Sep. Widely scattered in the mountains of western North America: Alberta and Yukon south to Oregon, east to Maine and south to Colorado; Eurasia. Our plants are **subsp. *borealis***.

Artemisia campestris L. (of the plains) Western sagewort, field sagewort, sand wormwood. Biennials or perennials, 30–100 cm tall, faintly aromatic; taprooted, caudices branched. **STEMS** 2–5, turning reddish brown, tomentose or glabrous. **LEAVES** of ours with persistent basal rosette; basal blades 4–12 cm, cauline gradually reduced, 2–4 cm long, 0.5–1.5 cm wide, 2–3-pinnately lobed, lobes linear to narrowly oblong, apices acute, surfaces green and glabrous or gray-green and sparsely hairy. **HEADS** pedunculate, in mostly leafless panicles, 10–22 cm tall, 1–3 (–7) cm wide. **INVOLUCRES** turbinate, 2–3 mm long, 2 (–3) mm wide. **PHYLLARIES** scarious-margined, glabrous or villous-tomentose. **FLORETS**: pistillate 5–20; functionally staminate 12–30; corollas pale yellow, sparsely hairy or glabrous. **CYPSELAE** oblong-lanceoloid, somewhat compressed, 0.8–1 mm long, faintly nerved, glabrous. [*A. campestris* var. *scouleriana* (Besser) Cronquist]. Sandy soils, arid regions; dunes, desert scrub, piñon-juniper, open meadows, montane. ARIZ: Apa, Coc, Nav; COLO: Arc, LPl, Mon; NMEX: McK, RAr, San, SJn; UTAH. 1220–2710 m (4000–8900′). Flowering: Jul–Oct. Our material belongs to **subsp. *pacifica*** (Nutt.) H. M. Hall & Clem. (of the Pacific Ocean or peace-making). Canada; Alaska south to California and New Mexico, east to Montana. Used by the Kayenta Navajo as a ceremonial fumigant ingredient and the seeds made into mush and used for food.

Artemisia cana Pursh (grayish white) Silver sagebrush, silver wormwood. Shrubs, 50–90 (–150) cm tall, with well-developed trunks and erect branches from base, aromatic. **STEMS** light brown to gray-green, leafy, persistently canescent to glabrescent, ours white and sparsely tomentose or brown and glabrous. **LEAVES** deciduous, whitish gray, ours bright green or dark gray-green; blades narrowly elliptic to lanceolate, ours with blades linear to narrowly lanceolate, (1.5–) 2–3 cm long, 0.2–0.4 cm wide, usually entire, ours irregularly lobed, sparsely hairy or glabrescent, viscid. **HEADS** congested, 2–3 per branch, erect, sessile, in sparsely leafy paniculate capitulescences, 12–20 cm long, 1–2 cm wide. **INVOLUCRES** narrowly to broadly campanulate, 3–4 mm long, 2–3 (–4) mm wide, subtended by green, leaflike bracts. **PHYLLARIES** ovate or lanceolate, densely canescent. **FLORETS** 4–8 (–20); corollas 2–3 mm long, resinous; style branches ellipsoid, to 2.3 mm long, exserted, gland-dotted. **CYPSELAE** 1–2.3 mm long, resinous, light brown. $2n$ = 18, 36, 72. Wet mountain meadows, stream banks, rocky areas with late-lying snows. COLO: Arc, Hin, LPl; NMEX: McK,

RAr, SJn. 2000–3300 m (6600–10800′). Flowering: Jul–Sep. Fruit: Aug–Oct. Four Corners material belongs to **subsp. *viscidula*** (Osterh.) Beetle (a little sticky). British Columbia to Saskatchewan, south to California, Nevada, and northern Mexico.

Artemisia carruthii Alph. Wood ex Carruth (for James H. Carruth) Carruth wormwood. Perennials, 15–40 (–70) cm, faintly aromatic, rhizomatous. **STEMS** mostly 3–8, ascending, brown to gray-green, bases somewhat woody, sparsely to densely tomentose. **LEAVES** cauline, bicolor; blades narrowly elliptic, 0.1–2.5 (–3) cm long, 0.5–1 cm wide, gradually reduced above, deeply pinnatifid, with 3–5 lobes, densely tomentose abaxially, sparsely hairy adaxially. **HEADS** usually nodding, in leafy, paniculate capitulescences, 10–30 cm long, 3–9 cm wide, lateral branches erect. **INVOLUCRES** campanulate, 2–2.5 (3) mm high, 1.5–3 mm wide. **PHYLLARIES** lanceolate, gray-tomentose. **FLORETS**: pistillate 1–5; bisexual 7–25; corollas pale yellow, 1–2 mm long, glandular-pubescent. **CYPSELAE** light brown, cylindro-elliptic, about 0.5 mm long, glabrous. [*A. bakeri* Greene; *A. coloradensis* Osterh.; *A. kansana* Britton]. Open sandy soils, in sagebrush, aspen forests, and spruce-fir forests. ARIZ: Apa, Nav; COLO: Arc, LPl, Mon, SJn; NMEX: McK, RAr, SJn; UTAH. 2155–2750 m (7080–9030′). Flowering: Jul–Sep. Fruit: Aug–Oct. Utah east to Kansas, south to New Mexico and Texas; Mexico. *Artemisia carruthii* is closely related to *A. ludoviciana*, with which it may be confused and occasionally may hybridize. Ramah Navajo uses include: plant ash used as Evilway and Hand Trembling blackenings, branches used in Beautyway garment ceremony; leaves taken for cough; cold infusion used as lotion for sores; used in sweat bath medicines; and root used as a "life medicine."

Artemisia dracunculus L. (little dragon) Wild tarragon. Perennials, often subshrubs, 50–120 (–150) cm tall, strongly tarragon-scented or not aromatic; rhizomatous, caudices coarse. **STEMS** relatively numerous, erect, green to brown or reddish brown, somewhat woody, glabrous. **LEAVES**: proximal blades bright green and glabrous or gray-green and sparsely hairy, 5–8 cm; cauline blades bright green, linear, lanceolate, or oblong, 1–7 cm long, 0.1–0.5 (–0.9) cm wide, mostly entire, sometimes irregularly lobed, acute, usually glabrous, rarely glabrescent. **HEADS** in terminal or lateral, leafy, paniculate capitulescences, 15–45 cm tall, 6–30 cm wide; spheric on slender, often nodding peduncles. **INVOLUCRES** globose, 2–3 mm high, 2–3.5 (–6) mm wide. **PHYLLARIES** broadly lanceolate, light brown, margins broadly hyaline, glabrous, membranous. **FLORETS**: pistillate 6–25; functionally staminate 8–20; corollas pale yellow, 1.8–2 mm long, eglandular or sparsely glandular. **CYPSELAE** oblong, 0.5–0.8 mm long, faintly nerved, glabrous. $2n$ = 18. [*A. aromatica* A. Nelson; *A dracunculina* S. Watson; *A. dracunculoides* Pursh; *A. glauca* Pall. ex Willd.]. Open meadows and fields, desert scrub, moist drainages, roadsides. ARIZ: Apa, Coc, Nav; COLO: Arc, Dol, Hin, LPl, Min, Mon, RGr, SJn, SMg; NMEX: McK, RAr, SJn; UTAH. 1200–3000 m (4000–11000′). Flowering: late Jun–Nov. Fruit: Jul–Nov. Widespread in western North America. Yukon southeast to Illinois and south to Mexico.

Artemisia filifolia Torr. (thread leaf) Sand sage. Shrubs, rounded, 60–180 cm tall, faintly aromatic. **STEMS** green or gray-green, wandlike, slender, curved, rarely stout and stunted, glabrous or sparsely hairy. **LEAVES** gray-green; blades linear if entire, obovate in outline if lobed, 1.5–5 (–6) cm long, 0.1–2.5 cm wide, entire to 3-lobed, lobes filiform, less than 1 mm wide, with acute apices, glabrous or sparsely hairy. **HEADS** mostly sessile, in paniculate capitulescences 8–15 (–17) cm tall, 2–4 (–5) cm wide, branches erect or recurved. **INVOLUCRES** globose, 1.5–2 mm high, 1.5–2 mm wide. **PHYLLARIES** ovate, inconspicuous, margins scarious, densely hairy. **FLORETS**: pistillate 1–4; functionally staminate 3–6; corollas pale yellow, 1–1.5 mm long, glabrous. **CYPSELAE** oblong, distally incurved-falcate and oblique, 0.2–0.5 mm long, obscurely nerved, glabrous. $2n$ = 18. [*A. plattensis* Nutt.; *Oligosporus filifolius* (Torr.) Poljakov]. Open prairies, dunes, sandy soils. ARIZ: Apa, Nav; COLO: Mon; NMEX: McK, San, SJn; UTAH. 1400–2050 m (4600–6725′). Flowering: late Jul–Nov. Fruit: Aug–Nov. Widespread in the western United States. South Dakota south to Colorado, Arizona, New Mexico, and Texas; Mexico.

Artemisia franserioides Greene (like *Franseria*) Bursage mugwort. Biennials or perennials, 30–100 cm tall, faintly aromatic, rhizomatous. **STEMS** 1–3, erect, reddish brown, leafy, glabrous or glabrate. **LEAVES** in basal rosettes, and cauline, petiolate, bicolored, abaxially white and adaxially green; blades ovate, 3–7 (–20) cm long, 2–4 (–6) cm wide, 2–3-pinnately lobed, lobes elliptic, 2–6 mm wide; cauline leaves sessile, reduced, tomentose abaxially or glabrous or glabrescent adaxially, glandular. **HEADS** nodding, peduncles 0–2 mm, in paniculate to racemose capitulescences, 10–35 cm tall, 2–4 cm wide, often secund. **INVOLUCRES** broadly ovate, 3–5 mm high, 4–5 (–6) mm wide. **PHYLLARIES** broadly ovate, sparsely hairy. **FLORETS**: pistillate 4–5 (–13), 1–1.5 mm long; bisexual 25–35, corollas yellow, 1.5–2 mm long, glabrous. **CYPSELAE** elliptic, 0.5–0.8 mm long, glabrous. Open coniferous forests, middle to upper montane, talus slopes and cliffs; openings in spruce-fir forests. COLO: Arc, Hin, LPl, Min, SJn. 2530–3350 m (8300–11000′). Flowering: Aug–Sep. Fruit: late Sep–Nov. Arizona, Colorado, and New Mexico; Mexico.

Artemisia frigida Willd. (cold) Fringed sage, prairie sagewort, Armoise douce. Perennials, 10–40 cm tall, forming silvery mats or mounds, strongly aromatic. **STEMS** gray-green or brown, glabrescent. **LEAVES** persistent, silver-gray; blades ovate, 0.5–1.5 (–2.5) cm long, 1–2-ternately lobed, the lobes 0.2–0.5 mm wide, surfaces densely whitish pubescent. **HEADS** in leafy, paniculate capitulescences, 0.5–2 (–4) cm long, 4–15 (–20) cm wide. **INVOLUCRES** globose, (3–) 5 mm long, (2–) 5–6 mm wide. **PHYLLARIES** gray-green, margins ± brownish, densely tomentose. **FLORETS**: pistillate 10–17; bisexual 20–50; corollas 1.5–2 mm long, glabrous. **CYPSELAE** 1–1.5 mm, glabrous. $2n = 18$. [*A. frigida* var. *gmeliniana* (Besser) Besser; *A. frigida* var. *williamsiae* S. L. Welsh]. Flowering: summer–fall. Rocky, well-drained soils, in desert scrub, piñon-juniper woodlands, and ponderosa pine forests. ARIZ: Apa, Coc, Nav; COLO: Arc, Dol, Hin, LPl, Mon, SJn, SMg; NMEX: McK, RAr, San, SJn; UTAH. 1645–3165 m (5400–10390′). Flowering: Jul–Oct. Fruit: Aug–Nov. Alaska to Quebec, south to Arizona and Kansas; Eurasia. Plant ash applied before painting Witcheryway prayer sticks by the Ramah Navajo.

Artemisia ludoviciana Nutt. (from Louisiana, referring to the Louisiana Purchase) Silver wormwood, white sage, silver sage. Perennials, 20–80 cm tall, rhizomatous, aromatic. **STEMS** few to numerous, erect, gray-green, simple or branched, hairy. **LEAVES** cauline, uniformly gray-green, green, white, or bicolor (white abaxially, green adaxially); blades linear to broadly elliptic, 1.5–11 cm long, 0.5–4 cm wide, entire, lobed or deeply pinnatifid, hairy. **HEADS** erect or nodding, sessile or peduncles 2–5 mm long, in congested to open paniculate capitulescences. **INVOLUCRES** campanulate or turbinate, (1–) 2–5 mm high, 2–5 (–8) mm wide. **PHYLLARIES** gray-green, lanceolate to ovate or obovate, margins narrowly hyaline, densely tomentose. **FLORETS**: pistillate 5–12; bisexual 6–45; corollas yellow, sometimes red-tinged, 1.5–2.8 mm long, glabrous. **CYPSELAE** ellipsoid, 0.4–0.6 mm long, obscurely nerved, glabrous. $2n = 18, 36, 54$. Seven subspecies ranging from Canada and the United States to Mexico.

1. Leaves usually deeply lobed, nearly to midrib, lower leaves sometimes entire; involucres 3–5 mm wide; mountain meadows and slopes .... (2)
1′ Leaves entire or shallowly lobed (lobes to 1/3 width); involucres 2–3 (–4) mm wide; desert valleys and mountains .... (3)

2. Involucres 4–5 × 4–8 mm .... **subsp. *candicans***
2′ Involucres 3–4 × 3–5 mm .... **subsp. *incompta***

3. Capitulescense paniculate, (4–) 8–30 cm wide; leaves mostly 1.5–2 cm .... **subsp. *albula***
3′ Capitulescense narrow-paniculate or racemose, 1–6 cm wide; leaves 1.5–11 cm .... (4)

4. Leaf margins plane, leaves gray, tomentose on both surfaces .... **subsp. *ludoviciana***
4′ Leaf margins revolute, leaves gray-green, strongly bicolored, tomentose abaxially, glabrescent adaxially .... **subsp. *mexicana***

subsp. *albula* (Wooton) D. D. Keck (white) White wormwood. **STEMS** 30–80 cm tall, open-branched, tomentose or glabrous. **LEAVES** uniformly whitish green; blades linear-lanceolate, entire, or obovate-elliptic, with antrorse lobes, to 1/3 blade lengths, usually 1–2 cm, margins revolute, tomentose. **HEADS** in open, paniculate capitulescences, (9–) 15–40 cm long, (4–) 8–30 cm wide. **INVOLUCRES** 1–2 mm wide, 2–3 mm wide. **FLORETS**: pistillate 8–11; bisexual 8–13; corollas 1–1.5 mm long. [*Artemisia albula* Wooton; *A. microcephala* Wooton]. Sandy soils of riparian areas and drainages. ARIZ: Apa, Coc, Nav; COLO: Mon; NMEX: McK, RAr, SJn; UTAH. 1460–2560 m (4785–8400′). Flowering: Jun–Sep. Fruit: Jul–Oct. California and Nevada east to New Mexico and south to Texas; Mexico.

subsp. *candicans* (Rydb.) D. D. Keck (glossy white) **STEMS** 30–50 (–80) cm tall, mostly simple, sparsely tomentose. **LEAVES** white; blades broadly obovate to oblong, 4–10 cm long, 1.5–4 cm wide, apex deeply lobed, lobes lanceolate, 1/3 or more blade lengths, acute, tomentose. **HEADS** in racemose capitulescences, 4–7 cm long, 2–4 cm wide. **INVOLUCRES** broadly campanulate, 4–5 mm high, 4–8 mm wide. **FLORETS**: pistillate 8–10; bisexual 20–30; corollas 1.5–2 mm long. [*A. candicans* Rydb.; *A. ludoviciana* var. *latiloba* Nutt.]. Alcoves, piñon-juniper woodlands, ponderosa pine, Douglas-fir, and aspen communities. ARIZ: Apa, Coc, Nav; COLO: LPl, Min; NMEX: SJn; UTAH. 1570–2590 m (5160–8500′). Flowering: Jul–early Sep. Fruit: Aug–Oct. Alberta, British Columbia, California, Idaho, Montana, Nevada, Oregon, Utah, Washington, and Wyoming.

subsp. *incompta* (Nutt.) D. D. Keck (unadorned) Mountain wormwood. **STEMS** 20–50 (–80) cm tall, mostly simple, hairy. **LEAVES** bicolor, white or gray-green abaxially and green adaxially; blades narrowly to broadly lanceolate, 1.5–11 cm long, 1–1.5 cm wide, irregularly lobed, the lobes usually 1/3 or more the blade width, abaxially hairy, glabrous or glabrescent. **HEADS** in paniculate capitulescences, (10–) 15–25 (–35) cm long, (2–) 3–10 cm wide. **INVOLUCRES** broadly campanulate, 3–4 mm high, 3–5 mm wide. **FLORETS**: pistillate 5–12; bisexual 6–45; corollas yellow, sometimes red-tinged, 1.5–2 mm long. $2n$ = 36, 54. [*A. incompta* Nutt.; *A. ludoviciana* var. *incompta* (Nutt.) Cronquist]. Open meadows, mountain slopes, riparian areas, aspen and spruce-fir woodlands. COLO: Arc, Hin, LPl, Mon; UTAH. 2105–3050 m (6900–10000′). Flowering: Jul–Oct. Fruit: Aug–Nov. Oregon and Montana, south to California, Nevada, and Colorado.

subsp. *ludoviciana* (from Louisiana) Armoise de l'ouest. **STEMS** gray to white, mostly simple, 30–80 cm tall, tomentose. **LEAVES** gray; blades linear to narrowly elliptic, 3–11 cm long, margins plane, lower leaves entire or lobed at apex, lobes to 1/3 blade length; cauline 1.5–11 cm long, 1–1.5 cm wide, entire, lobed or pinnatifid, densely tomentose. **HEADS** in congested, paniculate or racemose capitulescences, 5–30 cm long, 1–4 cm wide. **INVOLUCRES** 3–4 mm high, 2–4 mm wide. **FLORETS**: pistillate 5–12; bisexual 6–30; corollas 1.9–2.8 mm long. $2n$ = 18, 36. Disturbed sites, meadows, and rocky slopes, in sagebrush, piñon-juniper woodlands, and ponderosa pine forests. ARIZ: Apa; COLO: Arc, Dol, Hin, LPl, Min, Mon, SJn, SMg; NMEX: McK, RAr, San, SJn; UTAH. 1530–3050 m (5025–10000′). Flowering: Jul–Oct. Fruit: Aug–Nov. British Columbia to Ontario, south to California, New Mexico, Texas, and east to Indiana. Subspecies *ludoviciana* is widespread, common, and the most morphologically variable in North America, occurring in many habitats.

subsp. *mexicana* (Willd. ex Spreng.) D. D. Keck (from Mexico) Mexican wormwood. **STEMS** 30–80 (–100) cm tall, openly branched, sparsely hairy or glabrous. **LEAVES** strongly bicolor, white or gray-green abaxially and green adaxially, narrowly lanceolate, 4–10 cm long, 0.5–1 cm wide, lobed; abaxially tomentose or glabrescent. **HEADS** in paniculate or racemose capitulescences 1–8 cm long, 2–5 cm wide. **INVOLUCRES** campanulate, 3–4 mm high, 2–3 mm wide. **FLORETS**: pistillate (5–) 8–10; bisexual (6–) 10–20; corollas 1.3–1.7 mm long. $2n$ = 18, 36. [*A. mexicana* Willd. ex Spreng.; *A. neomexicana* Greene ex Rydb.; *A. revoluta* Rydb.]. Mountain slopes, rocky soils, and riparian habitat, in piñon-juniper woodlands, ponderosa pine forests, and hanging gardens. ARIZ: Apa, Coc, Nav; COLO: LPl, Min, Mon, SMg; NMEX: McK, RAr, SJn; UTAH. 1430–2770 m (4700–9090′). Flowering: late Jul–Oct. Fruit: Aug–Nov. Colorado east to Missouri, and south to Mexico. Plant ash used in blackening ceremonies by the Ramah Navajo.

Artemisia michauxiana Besser (for André Michaux) Lemon sagewort. Perennials, 30–100 cm tall, rhizomatous; strongly aromatic, lemon-scented. **STEMS** relatively many, erect, green, simple, glabrous. **LEAVES** cauline, green; blades broadly lanceolate to narrowly elliptic, 1.5–11 cm long, 1–1.5 cm wide, 2-pinnately lobed, ultimate lobes toothed, abaxial surfaces white-tomentose, adaxial surfaces glabrous, yellow gland-dotted. **HEADS** sessile or pedicelled, erect to nodding, peduncles 0–10 mm, in paniculate or spicate capitulescences, 8–15 cm tall, 1–1.5 cm wide. **INVOLUCRES** campanulate, 3–4 mm long, 2–5.5 mm wide. **PHYLLARIES** yellow-green, rarely purplish, broadly ovate, glabrous or sparsely hairy, yellow gland-dotted. **FLORETS**: pistillate 9–12; bisexual 15–35; corollas yellow, 1–1.5 mm long, glandular. **CYPSELAE** yellow to light brown, ellipsoid, 0.4–0.6 mm long, glabrous or glandular. $2n$ = 18, 36. [*A. discolor* Douglas ex DC.; *A. vulgaris* L. subsp. *michauxiana* (Besser) H. St. John]. Talus or scree slopes, alpine and subalpine drainages, spruce-fir to tundra. COLO: RGr. 3810 m (12500′). Flowering: Jul–Sep. Fruit: Aug–Oct. Alberta, British Columbia, Yukon; California and Oregon east to Montana, south to Utah and Colorado. *Artemisia michauxiana* is very similar to some races of *A. ludoviciana* but the former can be distinguished by its glabrous, bright green to yellow-green leaves and lemon fragrance.

Artemisia norvegica Fr. (of Norway) Arctic sagebrush. Perennials, 25–40 (–60) cm tall, mildly aromatic, roots often horizontal, woody. **STEMS** 1–3, erect to ascending, green or reddish, glabrous or sparsely tomentose. **LEAVES** petiolate, mostly in basal rosettes, bright green; blades of basal leaves broadly lanceolate, 5–8 (–10) cm long, 2–3 (–4) cm wide, 1–3-pinnately lobed, terminal lobes 1–7 mm long, 1.5–3 mm wide; midcauline leaves sessile, pinnately lobed; on flowering stems, sessile, linear, entire, glabrous or hairy. **HEADS** nodding, proximal ones with peduncles to 50 mm long, in racemose capitulescences, 10–17 cm long, 1–2 cm wide. **INVOLUCRES** globose, (4–) 5–8 mm high, 4–10 mm wide. **PHYLLARIES** ovate-lanceolate to elliptic, margins dark brown to black, sparsely hairy to villous. **FLORETS**: pistillate 6–20; bisexual (30–) 50–70; corollas yellow or red-tinged, 1.5–2.5 (–3.5) mm long, long-hairy. **CYPSELAE** ovoid-oblong, angular, 2.2–2.7 mm long, glabrous or villous. $2n$ = 18, 36. [*A. arctica* Less. var. *saxatilis* (Besser) Y. R. Ling]. Subalpine to alpine habitats, moist soils, talus slopes and rock outcrops. COLO: RGr, SJn. 2900–3800 m (9500–12500′). Flowering: Jul–Aug. Fruit: Aug–Oct. Alaska and Yukon south to California, east to Utah and Colorado; eastern Asia. Our material is **subsp. *saxatilis*** (Besser) H. M. Hall & Clem. (rock-dwelling).

Artemisia nova A. Nelson (new) Black sagebrush, black sage. Shrubs, 10–30 (–50) cm tall, trunks relatively short, widely branched, aromatic. **STEMS** brown, glabrescent; bark dark gray, exfoliating with age. **LEAVES** persistent, usually bright green to dark green, sometimes gray-green; blades cuneate, 0.5–2 cm long, 0.2–1 cm wide, apex rounded, 3-lobed, the lobes to 1/3 blade lengths, surfaces sparsely hairy, gland-dotted. **HEADS** arranged in paniculate capitulescences, 4–10 cm long, 0.5–3 cm wide. **INVOLUCRES** narrowly turbinate, 2–3 mm long, 2 mm wide. **PHYLLARIES** straw-colored or light green, ovate to elliptic, margins hyaline, resinous, sparsely hairy or glabrous. **FLORETS** 2–6; corollas 2–3 mm long, glabrous, style branches slightly exserted. **CYPSELAE** 0.8–1.5 mm long, ribbed, glabrous or resinous. $2n$ = 18, 36. [*A. arbuscula* Nutt. subsp. *nova* (A. Nelson) G. H. Ward; *A. tridentata* Nutt. subsp. *nova* (A. Nelson) H. M. Hall & Clem.; *Seriphidium novum* (A. Nelson) W. A. Weber]. Shallow soils, desert valleys, exposed mountain slopes, shadscale, juniper, rabbitbrush, piñon-juniper, and mountain brush communities. ARIZ: Apa, Nav; COLO: Arc, Dol, LPl, Mon, SMg; NMEX: McK, RAr, San, SJn; UTAH. 1735–2480 m (5700–8145′). Flowering: Jul–Oct. Oregon to Montana, south to California, Arizona, and New Mexico.

Artemisia pygmaea A. Gray (dwarf) Pygmy sage. Shrubs, 5–10 cm tall, slightly aromatic, caudices branched, woody. **STEMS** stiffly erect, pale to light brown, sparsely tomentose, leaves erect-appressed. **LEAVES** sessile, rigid, bright green, persistent; blades oblong to ovate, 0.3–0.5 cm long, 0.2–0.3 cm wide, pinnately lobed nearly to midribs, lobes 3–7, divergent, glabrous or sparsely tomentose, resinous. **HEADS** sessile, erect, in paniculate or racemose capitulescences, (1–) 2–3 cm long, 0.5–1 cm wide. **INVOLUCRES** narrowly turbinate, 2–3 mm long, 3–4 mm wide. **PHYLLARIES** green, narrowly lanceolate, midribs prominent, glabrous or sparsely tomentose. **FLORETS** 2–6; corollas 2.5–3 mm long, glandular; style branches flat, erose, exserted. **CYPSELAE** prismatic, 0.4–0.5 mm long, glabrous, resinous. $2n$ = 18. [*Seriphidium pygmaeum* (A. Gray) W. A. Weber]. Fine-textured soils of gypsum or shale, sagebrush and piñon-juniper. ARIZ: Apa; COLO: Mon; NMEX: SJn. 1500–2210 m (4920–7255′). Flowering: Jul–Sep. Fruit: Aug–Oct. Arizona, Colorado, Nevada, New Mexico, and Utah.

Artemisia scopulorum A. Gray (of rocks or crags) Alpine sagebrush. Perennials, 10–25 cm tall, caespitose, mildly aromatic, caudices slender. **STEMS** gray-green, glabrate. **LEAVES** persistent, gray-green; blades of basal leaves oblanceolate, 2–7 cm long, 0.1 cm wide, 2-pinnately lobed, the lobes linear or oblanceolate; cauline leaf blades smaller, 1–2-pinnate or entire, silky canescent. **HEADS** 5–22 in spicate capitulescences, 5–9 cm long, 1–1.5 cm wide. **INVOLUCRES** globose or subglobose, 3.7–4.4 mm high, 4–7 mm wide. **PHYLLARIES** green, the margins black or dark brown, densely villous. **FLORETS**: pistillate 6–13; bisexual 15–30; corollas 1.5–2.5 mm long, lobes hairy. **CYPSELAE** 0.8–1 mm long, glabrous. $2n$ = 18. Meadows, bases of rocks, subalpine and alpine communities. COLO: Arc, Con, Hin, LPl, Min, Mon, RGr, SJn. 3445–4030 m (11300–13220′). Flowering: Jul–early Sep. Fruit: Aug–Oct. Higher mountains of Montana and Wyoming, south to New Mexico.

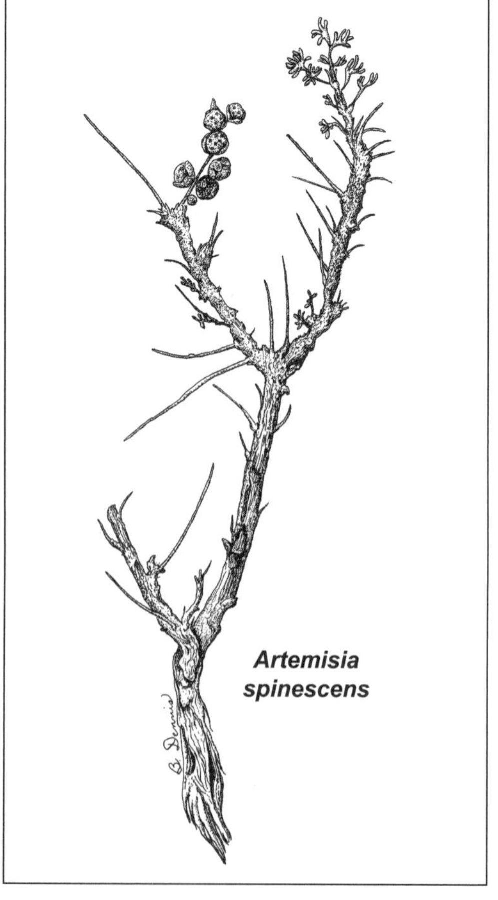
Artemisia spinescens

Artemisia spinescens D. C. Eaton (bearing spines) Budsage, bud sagebrush. Aromatic subshrubs or shrubs, 5–30 (–50) cm tall. **STEMS** 1–10, usually erect, diffusely branched from bases and throughout, some lateral branches persistent, forming thorns, villous to cobwebby with medifixed hairs. **LEAVES** mostly cauline, alternate, petiolate or sessile; 15–20 mm long, 1–5 (–20) wide; blades orbiculate, simple or 1–2-pedately lobed, the lobes orbiculate, spatulate, or linear, ultimate margins entire, surfaces villous and gland-dotted. **HEADS** 2–12, discoid, usually in leafy, racemose or spicate capitulescences, rarely borne singly. **INVOLUCRES** obconic, 2–3 (–5) mm diameter. **PHYLLARIES** persistent and distinct, 5–8 in 2 series, obovate, subequal, margins and apices narrowly scarious, hyaline. **RECEPTACLES** convex, glabrous, lacking paleae. **RAY FLORETS** absent; peripheral pistillate florets 2–8; corollas pale yellow, filiform, villous. **DISC FLORETS** 5–15, functionally staminate; corollas pale yellow, villous, tubes cylindric, throats campanulate, lobes 5, deltate. **CYPSELAE** brown, 1–1.5 mm long, obovoid to ellipsoid, surfaces villous and obscurely nerved. $2n$ = 18, 36. [*Picrothamnus* *desertorum* Nutt.]. Arid slopes and valleys, sand, clay, occasionally saline soils, associated with greasewood, shadscale, black sagebrush, and other desert scrub communities. ARIZ: Apa; COLO: LPl, Mon; NMEX: McK, SJn; UTAH. 1160–2070 m (3800–6795'). Flowering: Apr–Jun. Fruit: May–Jul. Oregon to Montana, south to California and New Mexico. *Artemisia spinescens* is distinguished from the remainder of *Artemisia* by its spinescent branches and large heads.

Artemisia tridentata Nutt. (three teeth) Big sagebrush, big sage, chamiso, hediondo. Shrubs, 40–200 (–300) cm tall, herbage canescent, aromatic; trunks thick. **STEMS** glabrate, bark gray, exfoliating into strips. **LEAVES** persistent, gray-green; blades cuneate, (0.4–) 0.5–3.5 cm long, 0.1–0.7 cm wide, 3-lobed, surfaces densely hairy. **HEADS** usually erect, on slender peduncles, arranged in paniculate capitulescences, 5–30 cm long, 1–6 cm wide. **INVOLUCRES** lanceolate, (1–) 1.5–4 mm long, 1–3 mm wide. **PHYLLARIES** oblanceolate to widely obovate, densely tomentose. **FLORETS** 3–8; corollas 1.5–2.5 mm long, glabrous. **CYPSELAE** 1–2 mm long, hairy or glabrous, glandular. [*Seriphidium tridentatum* (Nutt.) W. A. Weber]. Western North America and northwest Mexico. Navajo uses include: compound of plants used for headaches; plant used for religious and medicinal ceremonies; used for colds and stomachaches; infusion of plants taken by women as an aid for delivery and lotion for snakebites; wood used between the poles of the sweathouse to prevent the sand from sifting through, and for brooms. A major fiber plant used for sandals, mats, etc. by Great Basin tribes. Circumscription of subspecies of *Artemisia tridentata* has been the subject of considerable taxonomic revision and controversy.

1. Shrubs, 100–200 (–300) cm; heads in relatively broad, paniculate capitulescences; involucres 1.5–2.5 mm long, 1–2 mm wide; deep, well-drained (usually sandy) soils in valley bottoms, lower montane slopes along drainages .... **subsp. *tridentata***

1' Shrubs, 30–150 cm; heads in relatively narrow, paniculate capitulescences .... (2)

2. Shrubs, 60–80 (–150) cm; crowns flat-topped; capitulescences 10–15 cm; involucres 2–3 mm long, 1.5–3 mm wide; mountains .... **subsp. *vaseyana***

2' Shrubs, 30–50 (–150) cm; crowns rounded; capitulescences 2–6 (–8) cm; involucres (1–) 1.5–2 mm long, 1.5–2 mm wide; usually cold-desert basins and high plateaus, sometimes foothills .... **subsp. *wyomingensis***

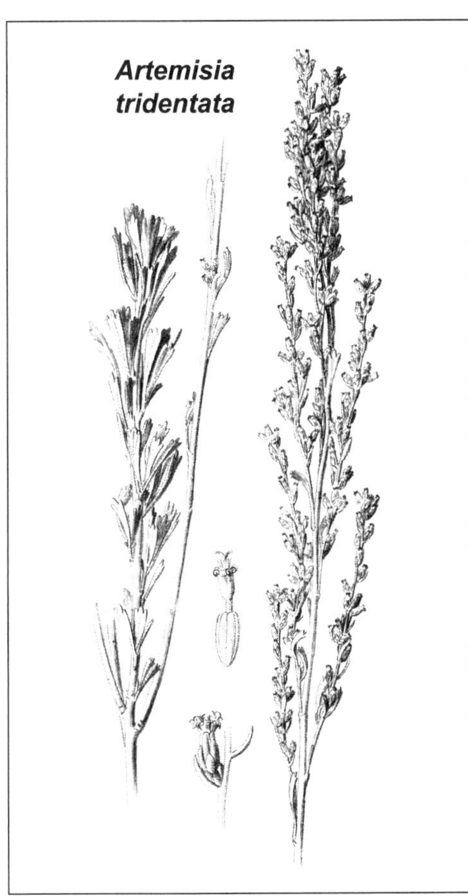

subsp. *tridentata* (three teeth) Great Basin sagebrush, big sage. Shrubs, 10–20 (–30) dm tall. **STEMS** with vegetative branches subequal to flowering branches. **LEAVES** cuneate or lanceolate, 0.5–1.2 (–2.5) cm long, 0.2–0.3 (–0.6) cm wide, 3-lobed. **HEADS** in paniculate capitulescences, 5–15 (–20) cm long, (1.5–) 5–6 cm wide. **INVOLUCRES** 1.5–2.5 mm long, 1–2 mm wide. **FLORETS** 4–6. **CYPSELAE** glabrous. $2n = 18, 36$. Deep, well-drained (usually sandy) soils in valley bottoms, lower montane slopes, along drainages. ARIZ: Apa, Nav; COLO: Arc, Dol, LPl, Mon, SMg; NMEX: McK, RAr, San, SJn; UTAH. 1665–2030 m (5460–6655'). Flowering: Jul–Oct. Fruit: Aug–Nov. Washington and Montana south to California, Arizona, and New Mexico.

subsp. *vaseyana* (Rydb.) Beetle (for George Vasey) Mountain sagebrush. Shrubs, 60–80 (–150) cm tall, plants strongly aromatic, crowns flat-topped. **STEMS** with vegetative branches subequal. **LEAVES** broadly cuneate, 1.2–3.5 cm long, 0.3–0.7 cm wide, regularly 3-lobed to irregularly toothed. **HEADS** in paniculate capitulescences, 10–15 cm long, 2–4 cm wide. **INVOLUCRES** 2–3 mm long, 1.5–3 mm wide. **FLORETS** 3–9. **CYPSELAE** glabrous. $2n = 18, 36$. [*A. tridentata* var. *pauciflora* Winward & Goodrich]. Montane meadows, usually in rocky soils, sometimes in forested areas. COLO: Arc, LPl, Mon; UTAH. 2000–2800 m (6560–9185'). Flowering: Jul–Oct. Fruit: Aug–Nov. Washington to the Dakotas, south to California and Colorado. The most common of the *A. tridentata* subspecies on mountain slopes.

subsp. *wyomingensis* Beetle & A. M. Young (of Wyoming) Wyoming sagebrush. Shrubs, 3–5 (–150) dm tall, rounded in outline. **STEMS** with vegetative branches stiffly spreading, often persisting, scattered among flowering stems. **LEAVES** narrowly to broadly cuneate, (0.4–) 0.7–1.1 (–2) cm long, (0.1–) 0.2–0.3 cm wide, 3-lobed. **HEADS** in paniculate capitulescences, 2–6 (–8) cm long, 1–3 cm wide. **INVOLUCRES** (1–) 1.5–2 mm long, 1.5–2 mm wide. **FLORETS** 4–8. **CYPSELAE** glabrous. $2n = 36, 54$. Rocky or fine-grained soils, cold-desert basins to high plateaus, foothills. ARIZ: Apa, Coc, Nav; COLO: Arc, Dol, LPl, Mon, SMg; NMEX: McK, RAr, San, SJn; UTAH. 1870–2285 m (6135–7500′). Flowering: Jul–Oct. Fruit: Aug–Nov. Wyoming and Idaho south to New Mexico. Subspecies *wyomingensis* is characterized by its shorter leaves, usually shorter stature, and its shorter flowering branches that are retained from year to year.

Baccharis L. Seepwillow
G. L. Nesom

(for Bacchus, the Roman god of wine, the allusion to the genus unclear) Small trees or shrubs, subshrubs, or perennial herbs, dioecious (rarely monoecious), commonly glabrous. **LEAVES** alternate or rarely opposite, simple, linear to ovate or obovate or spatulate, 1-nerved to 3-nerved from the base, sessile or petiolate, entire to toothed, usually punctate-glandular and resinous, commonly coriaceous or subcoriaceous, sometimes reduced to thorns, scales, or wings, or absent. **INFL** a discoid head, unisexual, sessile or short-pedunculate, in thyrsoid panicles or corymboid to paniculate-corymboid aggregations, less often racemoid, spicate, or solitary. **INVOLUCRES** cylindric, campanulate, to hemispheric. **PHYLLARIES** in 3–8 graduated series. **RECEPTACLES** nearly flat to convex, epaleate or paleate in one group. **STAMINATE FLORETS** morphologically bisexual but with sterile ovaries, corollas funnelform with narrow, spreading-reflexing lobes cut nearly to the base of the limb, mostly whitish to purplish. **PISTILLATE FLORETS** with filiform-tubular corollas with short teeth or lobes. **CYPSELAE** oblong to narrowly fusiform, 5–10 (–12)-nerved. **PAPPUS** of 1–3 series of barbellate bristles, staminate often apically dilated, pistillate usually elongating at maturity well past the involucral bracts, basally caducous in sect. *Baccharis*. $x = 9$ (Mahler and Waterfall 1964). A New World genus of 350–450 species, about 90% of which are centered in South America.

1. Perennial herbs or subshrubs, much-branched from bases. .... ***B. wrightii***
1′ Shrubs, branched above the base .... (2)

2. Leaves lanceolate to elliptic or narrowly oblanceolate, margins evenly serrate (early-season forms with elliptic-oblong, entire leaves); cypsela 5-nerved, pappus bristles 4–6 mm; heads in terminal, corymbiform arrays (early-season forms with heads in clusters on short, lateral branches) .... ***B. salicifolia***
2′ Leaves oblong to oblanceolate, margins entire or with 1–4 pairs of teeth on the distal 1/2 or 2/3; cypsela 8–10-nerved, pappus bristles 8–12 mm; heads in a relatively open panicle .... ***B. salicina***

Baccharis salicifolia (Ruiz & Pav.) Pers. (willow leaf) Mule fat, seepwillow, water wally. Shrubs 1.5–3.5 m tall, glabrous, punctate-glutinous or sometimes densely and minutely papillate but not glutinous. **LEAVES** lanceolate to elliptic or narrowly oblanceolate, 5–11 (–14) cm × 5–18 mm, 3-nerved, margins evenly serrate to denticulate from near base (early-season plants sometimes with leaves elliptic, entire, 10–15 mm). **HEADS** short-pedicellate, aggregated in dense, terminal, corymbiform clusters (early-season forms with heads in small clusters on short, lateral branches); pistillate involucre 3–6 mm long. **CYPSELAE** 0.8–1.5 mm, 5 (–7)-nerved; mature pistillate pappus 4–6 mm, silvery white. $2n = 18, 36$. [*Baccharis glutinosa* Pers.; *Baccharis viminea* DC.]. River terraces, washes, alkali seeps, disturbed sites, riparian (saltcedar, salsola). ARIZ: Coc; NMEX: SJn; UTAH. 1125–2400 m (3700–7875′). Flowering: Mar–Aug. Fruit: through Sep. California and Nevada east to southwestern Colorado, New Mexico, and Texas; widespread in Mexico; disjunct to South America. Plants often identified as *B. viminea* DC., with strongly woody stems, relatively small, entire leaves, and small clusters of heads at the ends of numerous, short, lateral branches are early-season forms of *B. salicifolia* (Wilken 1971).

Baccharis salicina Torr. & A. Gray (willow) Willow-baccharis, Great Plains false willow. Shrubs 1–3 m tall, glabrous, glutinous. **LEAVES** oblong to narrowly elliptic or oblanceolate, 2–8 cm × 3–14 (–17) mm, 3-nerved, margins entire or with 1–4 pairs of teeth on the distal 1/2 or 2/3. **HEADS** short-pedicellate or sessile in few-headed clusters, aggregated in a relatively open panicle; pistillate involucre 5–6 (–7) mm. **CYPSELAE** 1.2–2 mm, (5–) 8–12-nerved; mature pistillate pappus 8–12 mm, silvery white. Hanging gardens and seeps, river terraces and floodplains, roadsides. COLO: Mon; NMEX: RAr, SJn; UTAH. 1125–1615 m (3700–5300′). Flowering: May–Sep. Fruit: through Oct. Utah and Arizona east to Kansas, Oklahoma, and Texas; northwestern Mexico (Chihuahua, Durango, Coahuila). *Baccharis salicina* is intermediate between *B. emoryi* A. Gray and *B. neglecta* Britton and intergrades with both, though the region of intergradation with *B. emoryi* is broader. The complex is an east-west transition (*neglecta-salicina-emoryi*) from narrow leaves and small heads to broader leaves and larger heads, but the three entities apparently are distinct. Brief characterizations provide guides to their identification. *B. emoryi*: pistillate capitulescence conspicuously open, with few heads in distinctly terminal glomerules; leaves few, spreading and obovate to narrowly elliptic (–2 mm wide), if narrow in the capitulescence then abruptly wider immediately downstem; pistillate involucre (5–) 6–7 (–8) mm. *B. salicina*: pistillate capitulescence elongate and more dense, with more, smaller heads; leaves more, erect-ascending and mostly narrowly elliptic, continuing narrow downstem below the heads; pistillate involucre 5–6 (–7) mm. *B. neglecta*: pistillate capitulescence elongate and dense with numerous small heads; leaves numerous, linear, and erect; pistillate involucres (3–) 4–5 mm.

Baccharis wrightii A. Gray (honoring Charles Wright, 1811–1885, botanical explorer of the southwestern United States, Cuba, and China) Wright's false willow. Perennial herbs or subshrubs 0.2–0.5 m tall, stems with numerous stiff branches, glabrous, nonglutinous. **LEAVES** linear-oblong, 4–12 mm, usually bractlike and widely spaced, 1-nerved, or rarely obovate, 3-nerved, and up to 10 mm wide, minutely papillate along the midvein, margins entire. **HEADS** solitary; pistillate involucre 8–12 mm long. **CYPSELAE** 4.5–5 mm, 5–9-ribbed, moderately to densely papillate; mature pistillate pappus 15–20 mm, brownish. $2n = 18$. Talus slopes, sandy flats, alluvium, riparian (cottonwood, saltcedar), desert scrub (*Isocoma*, *Ephedra*, *Chrysothamnus*, resinbush), juniper-rabbitbrush-snakeweed communities. ARIZ: Apa, Nav; COLO: Dol; NMEX: McK, SJn; UTAH. 1550–2050 m (5085–6725′). Flowering: Apr–Jun (–Jul). Utah and Arizona east to Kansas, Oklahoma, and Texas; northwestern Mexico (Sonora, Chihuahua, and Durango).

Bahia Lag. Bahia

Brian A. Elliott

(for Juan Francisco Bahi, professor of botany at Barcelona, Spain) Annual, biennial, or perennial taprooted herbs, sometimes suffrutescent. **STEMS** erect to ascending. **LEAVES** alternate or opposite, entire to bipinnatifid, sessile or petiolate. **INFL** heads radiate, rarely discoid, few to many and corymbose; involucre campanulate to hemispheric. **PHYLLARIES** free in 1–3 series, equal or subequal, often reflexed in age, broadest above the middle. **RECEPTACLES** flat to slightly convex, naked or alveolate. **RAY FLORETS** pistillate and fertile; ligules 5–20, yellow or white, entire to 3-lobed. **DISC FLORETS** perfect and fertile, numerous; corolla yellow. **CYPSELAE** narrowly obpyramidal, quadrangular, pubescent or glabrous. **PAPPUS** of scales or none (in ours). Twelve species of the western United States, Mexico, and Chile. The genus has traditionally been treated with three subgenera that are now accepted as separate genera: *Bahia*, *Picradeniopsis*, and *Platyschkuhria* (Ellison 1964).

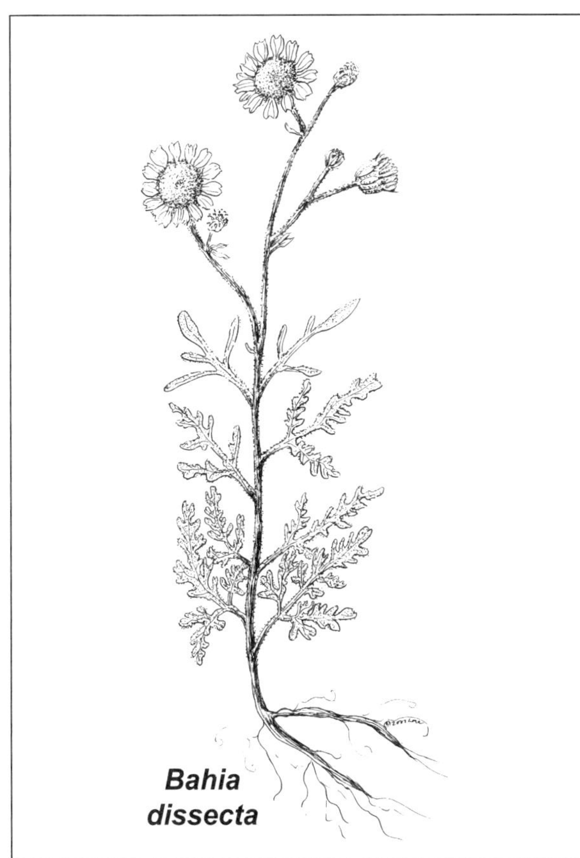

Bahia dissecta (A. Gray) Britton **CP** (apart + cut, referring to the numerous leaf segments) Ragleaf bahia, cutleaf bahia, yellow ragweed. **STEMS** openly branched above, to 10 dm tall, sometimes reddish, hirsutulous to puberulent, also stipitate-glandular above. **LEAVES** alternate, 1–several ternately divided, ultimate segments linear or oblong and 1–5 mm wide, petiolate below, sessile and reduced above, strigillose or puberulent. **INFL** heads radiate, peduncles 1–7 cm long and densely glandular-pubescent. **INVOLUCRES** 4–10 mm high. **PHYLLARIES** 12–24 in 2–3 series, oblong-oblanceolate to oblong-obovate, herbaceous, glandular-pubescent, apex abruptly caudate; outer bracts 3-nerved. **DISC FLORETS** 2.5–4.5 mm long, stipitate-glandular on the basal tube. **CYPSELAE** 3–4.5 mm long, black, striate, glabrous, short-hairy, or glandular. **PAPPUS** none. $2n = 36$. [*Amauriopsis dissecta* Rydb.; *Chrysanthemoides* A. Gray; *Eriophyllum chrysanthemoides* Kuntze; *Villanova chrysanthemoides* A. Gray; *Villanova dissecta* Rydb.]. Sagebrush, montane shrubland, and meadows and open sites from piñon-juniper woodlands to spruce-fir communities. ARIZ: Apa, Nav; COLO: Arc, Hin, LPl, Min, Mon, SJn; NMEX: McK, RAr, SJn; UTAH. 1980–3050 m (6500–10000′). Flowering: Jul–Oct. California to southern Wyoming and south to Arizona, New

Mexico, Texas, and adjacent Mexico. The species has been used medicinally as a cathartic, emetic, for pain, and as a contraceptive.

Baileya Harv. & A. Gray ex Torr. Desert Marigold
Brian A. Elliott

(for Jacob Whitman Bailey, early American microscopist and researcher of algae and diatoms) Annual or biennial to perennial taprooted herbs; white-woolly tomentose to floccose. **STEMS** erect, 1–several from the base. **LEAVES** basal or cauline, often reduced above, alternate, usually petiolate below, sessile above; simple, entire to deeply pinnately lobed or pinnatifid, linear-lanceolate to broadly ovate. **INFL** a radiate head, solitary or cymose, short- to long-pedunculate. **INVOLUCRES** campanulate to hemispheric. **PHYLLARIES** in 1–2 subequal or equal series, linear-lanceolate, woolly-pubescent. **RECEPTACLES** flat to convex, naked, alveolate. **RAY FLORETS** pistillate and fertile; ligules 4–numerous, yellow, apex 3-toothed to 3-lobed or entire, reflexed in age, persistent on achene. **DISC FLORETS** perfect and fertile, many; corolla yellow, gland-dotted, lobes triangular, pubescent; corolla tube shorter than the throat. **CYPSELAE** linear to oblong or clavate, striate to weakly angled and ribbed, glandular-pubescent. **PAPPUS** lacking. Three or four species of the southwestern United States and Mexico. Members of the genus are toxic to stock, particularly on overgrazed rangeland (Turner 1993).

Baileya pleniradiata Harv. & A. Gray (full + bearing ray flowers, referring to the numerous ray flowers) Annual baileya. Annual, winter-annual, or short-lived perennial herbs. **STEMS** freely branching to the middle or above, 0.5–5 dm tall. **LEAVES** basal and cauline, the basal often withering by anthesis, spatulate or oblanceolate to oblong or linear, 1–12 cm long, entire to 1–2-pinnately lobed, petiolate below, sessile and reduced above. **INFL** heads solitary, peduncle 3–12 cm long. **INVOLUCRES** hemispheric, 4–10 mm high. **PHYLLARIES** 20–34, 3–6 mm long. **RAY FLORETS** 20–60; ligule 7–11 mm long, elliptic to obovate, apex shallowly 3-lobed or -toothed. **DISC FLORETS** 3 mm long, tube much shorter than the expanded throat. **CYPSELAE** cylindric, 3–4 mm long; pale, 3–5-angled, ribs most prominent on the angles. $2n = 32$. [*Baileya nervosa* M. E. Jones; *Baileya perennis* Rydb.; *B. thurberi* Rydb.]. Dry, open, sandy and gravelly deserts, plains, mesas, and roadsides. ARIZ: Apa, Nav. Up to 2000 m (6500′). Flowering: Mar–Jun. Sometimes with a second bloom in Oct–Nov. California to west Texas and northern Mexico. Poisonous to livestock.

Balsamorhiza Hook. Balsamroot
Gary I. Baird

(Greek *balsamon*, an aromatic tree resin, + *rhiza*, root) Perennials, taproots slender to massive (some thick-barked), rhizomes lacking or sometimes present (deep), caudices simple or multibranched, herbage variously pubescent and/or glandular, rarely glabrous, sap clear. **STEMS** 1 to many, erect to lax, simple or branched near base. **LEAVES** mostly basal, simple (some deeply pinnatifid), petiolate, cauline reduced, alternate (mostly) or opposite; petioles short to longer than blades, bases persisting as fibers; basal blades deltate-triangular (ours) to elliptic-ovate or lanceolate, bases cordate to sagittate or truncate, margins entire or crenate to deeply pinnatifid or lobed, apices ± acute. **HEADS** radiate, usually borne singly (rarely 2 or 3), erect; peduncles scapiform, often with 2–3 bractlike leaves. **INVOLUCRES** campanulate to hemispheric. **PHYLLARIES** about 10–20 in 2–3 series, herbaceous, outer subequaling to surpassing inner. **RECEPTACLES** flat to convex, paleate; paleae persistent, chartaceous; calyculi lacking. **RAY FLORETS** 5–20, pistillate, fertile; corollas yellow, fugacious. **DISC FLORETS** about 50–150+, bisexual, fertile; corollas yellow. **CYPSELAE** monomorphic, 3–4-angled, glabrous. **PAPPUS** lacking. x = 19. Twelve species of western North America.

Balsamorhiza sagittata (Pursh) Nutt. (shaped like an arrowhead, alluding to the leaves) Arrowleaf balsamroot. Perennials, about 20–40 cm tall, taproots massive, rhizomes lacking, caudices multibranched. **LEAVES**: petioles longer than blades; basal blades lance-deltate, 5–25 × 3–15 cm, bases cordate-sagittate, margins entire, faces ± silvery-tomentose (especially abaxially), often glabrate (especially adaxially); cauline usually at midstem. **HEADS** ± lanate proximally. **INVOLUCRES** about 15–25 mm diameter. **PHYLLARIES** tomentose; outer lanceolate to oblanceolate or linear, apices acute to acuminate. **RAY FLORETS** mostly 8–15; laminae about 20–40 mm. $2n$ = 38. Sagebrush, piñon-juniper, mountain brush, and ponderosa pine–Gambel's oak forest communities. ARIZ: Apa; COLO: Dol, LPl, Mon, SMg; UTAH. 2000–2700 m (6500–8800′). Flowering: Apr–May. British Columbia and Alberta south to California, Arizona, and Colorado.

Bidens L. Beggar-ticks
G. L. Nesom

[Latin *bis*, two, + *dens*, tooth, alluding to the commonly two-awned pappus] Annual to perennial herbs, glabrous to sparsely hairy. **LEAVES** opposite, simple to ternately or pinnately compound or dissected, petiolate to sessile. **HEADS** few to numerous, terminal, sometimes in a subcorymboid arrangement, radiate or discoid, immediately subtended by a series of foliaceous calyculus bracts. **INVOLUCRES** cupulate-campanulate. **PHYLLARIES** linear-lanceolate to ovate, striate with oil ducts. **RECEPTACLES** flat to convex, receptacular bracts flat, striate. **RAY FLORETS** usually about 8 or sometimes absent, sterile, laminae yellow or white. **DISC FLORETS** fertile, yellow to orange, goblet-shaped to tubular. **CYPSELAE** usually strongly flattened, sometimes narrow and 4-angled to subterete. **PAPPUS** of (1–) 2–4 antrorsely or retrorsely barbed awns, sometimes nearly absent. x = 12. About 200 species; widespread, especially in subtropical, tropical, and warm-temperate North America and South America (Sherff 1937).

1. Leaves simple, sessile and usually clasping to subclasping. .... ***B. cernua***
1′ Leaves pinnately compound, with 3 or 5 distinct leaflets or pinnately dissected, usually petiolate and not at all clasping .... (2)

2. Leaves pinnately compound, with broad, serrate leaflets; calyculus bracts distinctly longer than the phyllaries; achenes strongly flattened to weakly 4-angled. .... ***B. frondosa***
2′ Leaves pinnately dissected, with narrow, entire to coarsely toothed segments; calyculus bracts shorter than phyllaries to about the same length; achenes slender and subterete or obscurely 4-angled .... ***B. tenuisecta***

Bidens cernua L. (drooping) Nodding bur-marigold. Annuals, (2–) 20–100 cm. **LEAVES** simple, mostly lanceolate to oblanceolate, 4–10 cm, sessile and usually clasping to subclasping, margins coarsely serrate to dentate, occasionally entire or nearly so. **HEADS** mostly nodding in age. **CALYCULUS** bracts (3–) 5–8 (–10), mostly 8–12 mm, loosely spreading or reflexed, longer than the phyllaries. **RAY FLORETS** 6–8 or sometimes 0, laminae 2–15 mm. **CYPSELAE** strongly flattened to weakly 4-angled, outer (3–) 5–6 mm. **PAPPUS** of (2–) 4 erect, retrorsely barbed awns. $2n$ = 24, 48. Lake and stream shores, floodplains. ARIZ: Apa; COLO: Arc, LPl, Min; NMEX: RAr, SJn. 1615–2640 m (5300–8665′). Flowering: Jul–Sep. Widespread in North America, especially in northern two-thirds of the United States. *Bidens tripartita* L. (including *B. comosa* (A. Gray) Wiegand) is very similar to *B. cernua* but usually has slightly smaller heads with greater tendency to remain erect; the disc corollas are more tubular and 4-lobed (vs. goblet-shaped and 5-lobed in *B. cernua*), the anthers do not protrude from the corolla (vs. short-exserted in *B. cernua*), and the receptacular bracts are narrower and lanceolate (vs. broad and elliptic-ovate in *B. cernua*). *Heil et al. 19854* (SJNM) from Archuleta County, Colorado, has technical characters of disc corollas and receptacular bracts similar to *B. tripartita* but the overall appearance is *B. cernua*. *Bidens tripartita* is common in eastern North America but very scattered in the West. The two species occur in similar habitats.

Bidens frondosa L. (leaf-bearing, alluding to the foliar bracts subtending the heads) Devil's-pitchfork. Annuals, (10–) 20–60 cm. **LEAVES** pinnately compound, leaflets 3 or 5, mostly 15–25 mm wide, serrate, usually petiolate and not at all clasping. **HEADS** erect. **CALYCULUS** bracts (5–) 8 (–10), 6–15 mm, spreading to ascending, usually distinctly longer than the phyllaries. **RAY FLORETS** 0 or 1–3, laminae 2–3.5 mm. **CYPSELAE** strongly flattened to weakly 4-angled, 6–8 mm. **PAPPUS** of 2 erect to spreading, antrorsely or retrorsely barbed awns. $2n$ = 24, 48, 72. Lake and stream sides, sandbars, wetlands. COLO: Arc, LPl, Mon; NMEX: SJn; UTAH. 1340–1920 m (4400–6300′). Flowering: Aug–Sep. Widespread in North America, in every U.S. state.

Bidens tenuisecta A. Gray (slender-cut, alluding to the deeply cut leaves) Slim-lobe beggarticks. Annuals, (10–) 20–40 cm. **LEAVES** pinnately compound, pinnately dissected, segments mostly 2–2.5 wide, entire to coarsely toothed, usually petiolate and not at all clasping. **HEADS** erect. **CALYCULUS** bracts 6–12, 5–7 mm, erect-appressed, shorter than phyllaries or about the same length. **RAY FLORETS** 0 or 3–6, laminae 4–6 mm. **CYPSELAE** slender and subterete to obscurely 4-angled, 6–8 mm. **PAPPUS** of 2 (–3) spreading to divergent, retrorsely barbed awns. Roadsides, roadcuts, ditches, disturbed sites. COLO: Arc, LPl; NMEX: SJn. 1965–2285 m (6450–7500′). Flowering: Jul–Sep. Utah, Colorado, New Mexico, and Arizona; Mexico.

Brickellia Elliott Brickellbush

Randall W. Scott

(for John Brickell, 1749–1809, botanist and physician of Georgia) Ours perennials, subshrubs or shrubs, (12–) 30–120 (–200) cm tall. **STEMS** mostly erect, often much-branched (sometimes virgate, often striate). **LEAVES** cauline; opposite or alternate; petiolate or sessile; blades usually 3-nerved from bases, deltate, lance-elliptic, lance-linear, lanceolate, suborbiculate, margins mostly crenate, dentate, entire, serrate, or toothed, faces glabrous (sometimes shiny) or glandular-puberulent, strigose, or tomentose, gland-dotted. **INFL** of heads usually in corymbiform, sometimes cymiform, paniculiform or racemiform arrays, rarely borne singly. **INVOLUCRES** cylindric to obconic or campanulate, 5.5–12 mm diameter. **PHYLLARIES** persistent, (10–) 21–35 (–48) in (3–) 4–8 series, usually 3–5 (–16)-striate or -nerved, linear or lanceolate to oblanceolate or oblong, usually unequal (usually chartaceous, sometimes herbaceous). **RECEPTACLES** flat to convex, epaleate. **DISC FLORETS** (3–) 8–45 (–90); corollas usually white or whitish to cream, sometimes greenish, purplish, or yellowish, throats mostly cylindric to narrowly funnelform (lengths 3–5 times diameter); style bases enlarged, hairy, branches narrowly clavate (± dilated distally). **CYPSELAE** narrowly prismatic, 10-ribbed, glabrous or hairy to glabrate, often gland-dotted. **PAPPUS** persistent, of 10–80 usually smooth or barbellulate to barbellate, sometimes plumose or subplumose, bristles in 1 series.

1. Pappus usually of plumose or subplumose, sometimes barbellate, bristles ....(2)
1′ Pappus usually of smooth or barbellulate to barbellate, sometimes subplumose, bristles ....(3)

2. Phyllaries puberulent, often densely gland-dotted as well .... ***B. eupatorioides***
2′ Phyllaries glabrous or puberulent (not gland-dotted) .... ***B. brachyphylla***

3. Florets (40–) 45–90 ....(4)
3′ Florets 3–30 (–50) ....(5)

4. Shrubs; leaf margins usually sharply dentate or dentate-serrate, rarely entire; outer phyllaries ovate, lance-ovate, or lance-linear .... ***B. atractyloides***
4′ Perennials (caudices woody); leaf margins crenate, crenate-dentate, dentate, or serrate; outer phyllaries lance-ovate to narrowly lanceolate .... ***B. grandiflora***

5. Florets 3–7 .... ***B. longifolia***
5′ Florets 8–30 (–50) ....(6)

6. Petioles 4–70 mm ....(7)
6′ Petioles 0–5 mm ....(8)

7. Leaf apices acuminate or long-acuminate to attenuate .... ***B. grandiflora***
7′ Leaf apices acute to obtuse or rounded .... ***B. californica***

8. Florets 25–50; leaf blades elliptic, oblong, or lance-linear, margins entire .... ***B. oblongifolia***
8′ Florets 8–28 (–34); leaf blades ovate to suborbiculate, margins entire or dentate .... ***B. microphylla***

Brickellia atractyloides A. Gray (like the genus *Atractylis*) Spearleaf brickellbush. Shrubs, 20–50 cm tall. **STEMS** densely branched, glandular-puberulent. **LEAVES** opposite or alternate; petioles 0–3 mm long; blades deltate, lanceolate, or ovate, 3- or 4-nerved from bases; 10–50 × 5–25 mm, bases acute to truncate or cordate; margins sharply dentate or entire; apices acuminate; faces glabrous or minutely glandular-puberulent. **HEADS** in open, paniculiform arrays; peduncles 10–70 mm long, hispid to hispidulous and stipitate-glandular. **INVOLUCRES** cylindric to broadly campanulate, 10–15 mm long. **PHYLLARIES** 24–33 in 3–4 series, green, 3–16-striate, subequal or unequal, margins narrowly scarious (apices acute to acuminate); outer (often bright green, 7–16-striate) broadly ovate, inner (pale green, 3–4-striate) linear or narrowly lanceolate (often chartaceous, scabrellous, often glandular). **FLORETS** 40–90; corollas pale yellow-green, 6–8 mm. **CYPSELAE** 3–5.5 mm long, scabrellous. **PAPPUS** of 18–25 smooth or barbellulate bristles. $2n = 18$. Rock crevices, cliff faces, talus slopes, and outwash fans; desert scrub and piñon-juniper communities. UTAH. 1110–1525 m (3640–5000′). Flowering: Mar–Oct. Our material belongs to **var. *atractyloides***. Arizona, California, Colorado, Nevada, and Utah.

Brickellia brachyphylla (A. Gray) A. Gray (short-leaved) Plumed brickellbush. Perennials, 30–100 cm tall from a woody caudex. **STEMS** branched, pubescent. **LEAVES** mostly alternate (sometimes subopposite); petioles 0–3 mm long; blades 3-nerved from bases, lanceolate or lance-ovate, 10–50 × 4–20 mm, bases acute to obtuse, margins serrate or entire, apices acute, faces sparsely to densely pubescent, often gland-dotted or stipitate-glandular. **HEADS** usually in open, racemiform or paniculiform arrays, rarely borne singly; peduncles 4–20 mm long, pubescent. **INVOLUCRES** cylindric to campanulate, 8–11 mm long. **PHYLLARIES** 15–20 in 4–5 series, greenish, often purple-tinged, 5–9-striate, unequal, margins narrowly scarious (often ciliate, apices acute to acuminate or subaristate); outer lance-ovate (often puberulent), inner narrowly lanceolate (glabrous). **FLORETS** 9–12; corollas pale yellow-green, often purple-tinged, 4.5–6 mm. **CYPSELAE** 2.5–5.3 mm long, mostly velutinous, sometimes pubescent. **PAPPUS** of 27–32 white, usually plumose, rarely barbellate, bristles. $2n = 18$. [*Clavigera brachyphylla* A. Gray]. Rocky ridges, canyon walls, and hillsides; piñon-juniper woodland, Gambel's oak, and ponderosa pine communities. ARIZ: Apa; COLO: Arc, LPl, Mon; NMEX: McK, RAr, SJn. 1730–2440 m (5675–8000′). Flowering: Jul–Oct. Arizona, Colorado, Kansas, New Mexico, and Texas.

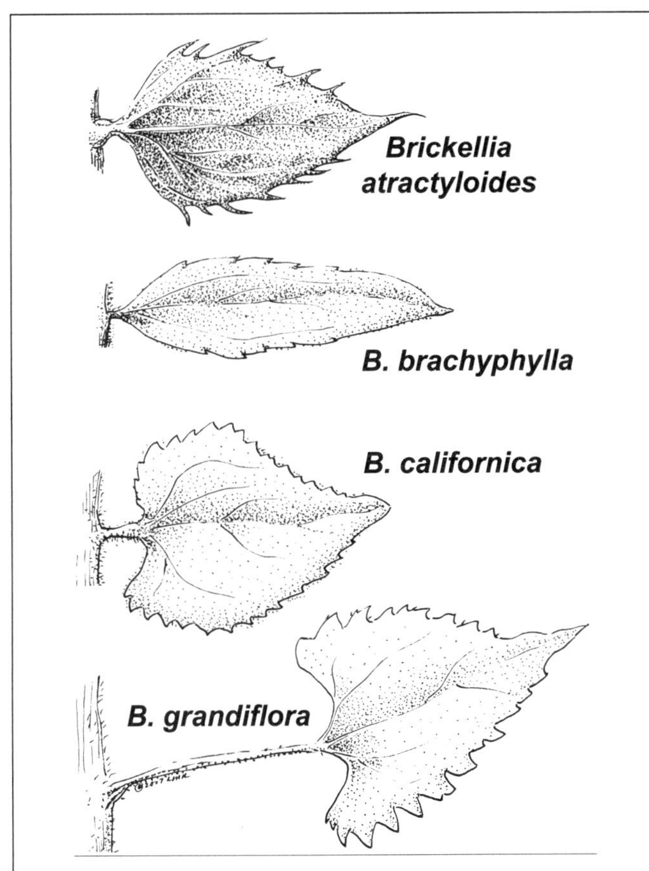

Brickellia californica (Torr. & A. Gray) A. Gray (of California) California brickellbush. Shrubs, 50–200 cm tall. **STEMS** branched from near the base, glandular-pubescent. **LEAVES** alternate; petioles 4–25 mm long; blades 3-nerved from bases, ovate to deltate, 1.7–5.2 × 1.3–4.5 cm, crenate to serrate, apices acute to rounded, faces puberulent to glabrate, often gland-dotted. **HEADS** borne in leafy paniculiform arrays; peduncles 1–5 mm, glandular-pubescent. **INVOLUCRES** cylindric to obconic, 5.5–8 mm long. **PHYLLARIES** 21–35 in 5–6 series, greenish, often purple-tinged, 3–4-striate, unequal, margins scarious; outer ovate to lance-ovate (glabrous or sparsely glandular-pubescent, apices acute to acuminate). **FLORETS** 8–12; corollas pale yellow-green, 5.5–8 mm. **CYPSELAE** 2.5–3.5 mm long, puberulent. **PAPPUS** of 24–30 white, barbellate bristles. $2n = 18$. [*Bulbostylis californica* Torr. & A. Gray]. Dry, rocky outcrops, hillsides, arroyos, and canyons. ARIZ: Apa, Coc; COLO: Arc, LPl, Mon; NMEX: McK, RAr, SJn; UTAH. 1855–2655 m (6090–8720′). Flowering: Jul–Oct. Colorado and New Mexico west to California. Used by the Kayenta Navajo as a ceremonial emetic following clan incest and as a lotion on infant sores caused by prenatal infection. The Ramah Navajo used an infusion of leaves as a lotion for cough or fever.

Brickellia eupatorioides (L.) Shinners (like *Eupatoria*; from the Greek name Mithridates Eupator, king of Pontus) False boneset. Perennials, 30–150 cm tall. **STEMS** branched, pubescent. **LEAVES** opposite; petioles 0–1 mm long; blades 1-nerved from bases, lance-linear to linear, 5–90 × 1.5–9 mm, bases acute, margins entire, apices obtuse to acuminate, faces glandular-pubescent. **HEADS** in paniculiform arrays; peduncles 10–100 mm long. **INVOLUCRES** cylindric to narrowly campanulate, 8–13 mm long. **PHYLLARIES** 22–26 in 4–6 series, greenish, sometimes purple-tinged, 3–7-striate, unequal, margins scarious (often ciliate); outer ovate to lance-ovate (puberulent, often densely gland-dotted, apices obtuse to acuminate), inner lanceolate (± gland-dotted, apices obtuse to aristate). **FLORETS** 16–30; corollas pale yellow-green to pinkish lavender, 4.5–6 mm long. **CYPSELAE** 4–5 mm long, glabrous or strigose, sometimes hispidulous or velutinous and/or gland-dotted. **PAPPUS** of 20–28 white or tawny, usually plumose or subplumose, sometimes barbellate bristles. $2n = 18$. [*Kuhnia eupatorioides* L.]. Hillsides, cliff faces, and rocky ledges; piñon-juniper woodlands, Gambel's oak, montane meadows, and ponderosa pine communities. ARIZ: Apa; COLO: Arc, Hin, LPl, Mon; NMEX: McK, RAr; UTAH. 1730–2695 m (5675–8840′). Flowering: May–Oct. Our material belongs to **var. *chlorolepis*** (Wooton & Standl.) B. L. Turner (green-scaled). [*Kuhnia chlorolepis* Wooton & Standl.; *Brickellia rosmarinifolia* (Vent.) W. A. Weber subsp. *chlorolepis* (Wooton & Standl.) W. A. Weber; *K. eupatorioides* L. var. *chlorolepis* (Wooton & Standl.) Cronquist]. Arizona, Colorado, New Mexico, Texas, Utah; Mexico. Decoction of root taken for an old injury or cough by the Ramah Navajo.

Brickellia grandiflora (Hook.) Nutt. (large-flowered) Tasselflower brickellbush. Perennial, 30–95 cm tall (taproots thickened). **STEMS** branched, puberulent. **LEAVES** opposite or alternate; petioles 10–70 mm long; blades 3-nerved from bases, deltate-ovate, lance-ovate, or subcordate, 1.5–11 cm × 0.6–6.5 cm, bases acute, truncate, or subcordate, margins crenate, dentate, or serrate, apices attenuate, faces puberulent and gland-dotted. **HEADS** nodding in flower and fruit, in loose, corymbiform or paniculiform arrays; peduncles 4–30 mm long, pubescent. **INVOLUCRES** cylindric or obconic, 5.5–8 mm long. **PHYLLARIES** 30–40 in 5–7 series, greenish, 4–5-striate, unequal, margins scarious; outer lance-ovate to lanceolate (pubescent, margins ciliate, apices long-acuminate), inner lanceolate to lance-linear (glabrous, apices acute to acuminate). **FLORETS** mostly 20–40 (–70); corollas pale yellow-green, 6.5–7.5 mm. **CYPSELAE** 4–5 mm long, hispidulous to hirtellous. **PAPPUS** of 20–30 white, barbellate bristles. $2n = 18$. [*Eupatorium grandiflorum* Hook.]. Rocky hillsides, shaded forests, dry slopes, and canyons; piñon-juniper, mountain brush, ponderosa pine, aspen, Douglas-fir, and white fir communities. ARIZ: Apa; COLO: Arc, Hin, LPl, Min, Mon; NMEX: McK, SJn; UTAH. 2315–3130 m (7600–10265′). Flowering: Jul–Oct. Washington south to Baja California, east to Montana, Nebraska, Missouri, and Arkansas. Used by the Navajo as a ceremonial liniment for the "Female Shooting Life Chant." Used by the Ramah Navajo as a cold infusion of dried leaves for headache, as a ceremonial emetic, and for influenza.

Brickellia longifolia S. Watson (long-leaved) Longleaf brickellbush. Shrubs, 20–150 cm tall. **STEMS** branched from above the base, ± glabrous. **LEAVES** alternate; petioles 0–5 mm long; blades 3-nerved from bases, lance-elliptic to lance-linear or linear, 1.2–13.5 cm × 3–8 mm, lengths 8–25 times widths, flat (not folded or falcate), bases tapering, margins entire or nearly so, apices acuminate, faces gland-dotted, often puberulent (shiny). **HEADS** in leafy paniculiform arrays; peduncles 0–3 mm, glabrous or glutinous. **INVOLUCRES** cylindric, 5–7 mm long. **PHYLLARIES** 10–24 in 6–8 series, pale green to stramineous, ovate (apices acute), inner lanceolate (apices obtuse). **FLORETS** 3–7; corollas cream, 3.5–4.5 mm long. **CYPSELAE** 1.8–2.5 mm long, scabrous. **PAPPUS** of 30–40 barbellulate bristles. Washes, stream margins, seeps, and hanging gardens. ARIZ: Apa, Nav; UTAH. 1120–1710 m (3675–5620′). Flowering: Sep–Nov. Four Corners material belongs to **var. *longifolia***. Arizona and Colorado west to Nevada and Utah.

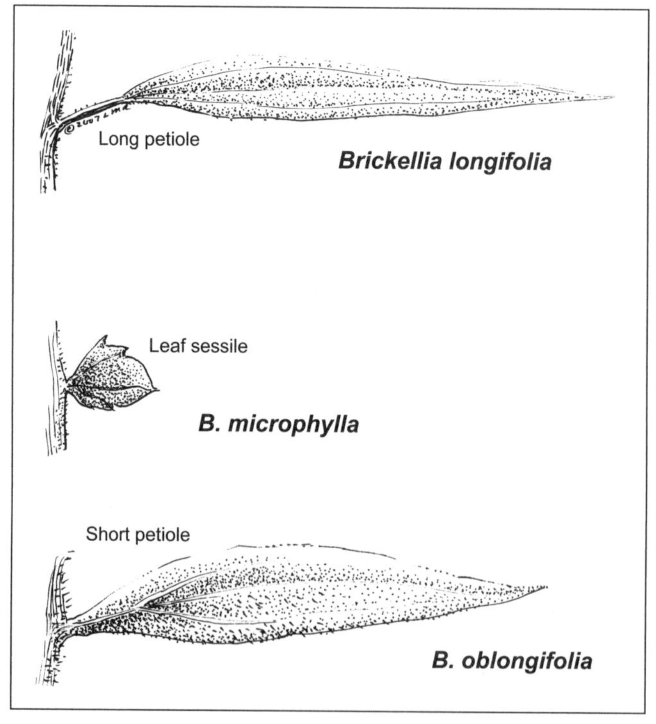

Brickellia microphylla (Nutt.) A. Gray (small-leaved) Littleleaf brickellbush. Shrubs, 30–70 cm tall. **STEMS** much-branched, pubescent, glandular-viscid. **LEAVES** alternate; petioles 0–3 mm long; blades 3-nerved from bases, ovate to suborbiculate, 3–14 (20) × 1–9 (12), bases acute to obtuse or rounded, margins entire, coarsely dentate or serrate, apices rounded to acute, faces glandular-villous or hispidulous. **HEADS** in loose, paniculiform arrays (often clustered at ends of branches); peduncles (bracteate) 2–10 mm long, glandular-viscid. **INVOLUCRES** cylindric to narrowly campanulate, 7–12 mm long. **PHYLLARIES** 30–48 in 5–8 series, greenish, often purple-tinged, 3–5-striate, (recurved or spreading) unequal, margins scarious (apices acute to acuminate); outer obovate to suborbicular (glandular-pubescent; middle sometimes 3-toothed with middle tooth elongated), inner linear-oblong, (glabrous or sparsely gland-dotted). **FLORETS** 8–13; corollas pale yellow, often purple-tinged, 5.5–7 mm long. **CYPSELAE** 3.5–4.7 mm long, glabrous or hirtellous. **PAPPUS** of 18–24 white, barbellulate bristles. $2n = 18$. Dry, rocky places, canyon walls, washes, and hanging gardens, mostly on sandstone outcrops; blackbrush, rabbitbrush, sagebrush, shadscale, greasewood, and piñon-juniper woodland communities. ARIZ: Apa, Coc, Nav; COLO: Dol, LPl, Mon; NMEX: McK, RAr, SJn; UTAH. 1485–2180 m (4870–7160′). Flowering: Jul–Oct. Our material belongs to **var.** ***scabra*** A. Gray (rough). [*B. microphylla* subsp. *scabra* (A. Gray) W. A. Weber; *B. scabra* (A. Gray) A. Nelson ex B. L. Rob.]. Nevada east to Wyoming, Utah, Colorado, Arizona, and New Mexico.

Brickellia oblongifolia Nutt. (oblong-leaved) Narrowleaf brickellbush. Perennials or subshrubs, 10–55 cm tall (caudices woody). **STEMS** branched, stipitate-glandular to glandular-pubescent. **LEAVES** mostly alternate (sometimes subopposite); petioles 0; blades obscurely 3-nerved from bases, elliptic, lance-linear or oblong, 9–40 × 1–15 mm, bases acute to attenuate, margins entire, apices acute or obtuse, faces pubescent to villous, often stipitate-glandular. **HEADS** borne singly or in corymbiform arrays; peduncles 2–50 mm long, glandular-pubescent. **INVOLUCRES** cylindric to campanulate, 10–20 mm long. **PHYLLARIES** 25–35 in 4–6 series, greenish, 4–5-striate, unequal, margins scarious (sometimes ciliate, apices acute to acuminate); outer ovate to lanceolate (gland-dotted to glandular-puberulent), inner linear-lanceolate to linear (glabrous). **FLORETS** 25–50; corollas pale yellow-green or cream, often purple-tinged, 5–10 mm long. **CYPSELAE** 3–7 mm long, setose-hispidulous, mostly lacking glands. **PAPPUS** of 18–25 white, barbellate to subplumose bristles. $2n = 18$. Dry, rocky hillsides; desert grasslands, saltbush, rabbitbrush, blackbrush, piñon-juniper woodland, and ponderosa pine communities. ARIZ: Apa, Coc, Nav; COLO: LPl, Mon; NMEX: RAr, SJn; UTAH. 1345–2085 m (4420–6850′). Flowering: May–Jul. Four Corners material belongs to **var.** ***linifolia*** (D. C. Eaton) B. L. Rob. (narrow-leaved). [*B. linifolia* D.C. Eaton; *B. oblongifolia* subsp. *linifolia* (D. C. Eaton) Cronquist]. British Columbia south to Montana, California, Arizona, and New Mexico. Used by the Kayenta Navajo as a lotion on infant ear and finger sores caused by prenatal infection.

Carduus L. Plumeless Thistle, Chardon
Patricia Barlow-Irick

(Latin *carduus*, ancient classical name for thistle) Annual or biennial, 3–40 dm tall, prickly, ± tomentose, sometimes glabrate. **STEMS** 1–several from the base, spiny-winged from decurrent leaf bases. **LEAVES** alternate, decurrent, pinnatifid with spiny serrate margins. **HEADS** solitary, on long peduncles or clustered at the ends of branches. **INVOLUCRES** cylindric to spheric. **PHYLLARIES** in 5–10+ series, lanceolate, entire, acute, spine-tipped. **RECEPTACLES** flat, epaleate, bearing soft hairlike bristles. **FLORETS** numerous, all tubular; corollas white to pink or purple, subequally 5-lobed, tubes long, slender, throats short, campanulate, abruptly expanded from tubes, lobes linear; anther bases sharply short-tailed, apical appendages oblong; fused portions of style branches with minutely puberulent, slightly swollen basal nodes, distally papillate or glabrous, distinct portions very short. **CYPSELAE** ovoid, slightly compressed, faces smooth, glabrous, attachment scars slightly lateral. **PAPPUS** deciduous as a basally joined ring of many minutely barbed bristles. $x = 8, 9, 10, 11, 13$. About 100 species, Eurasia; naturalized species one.

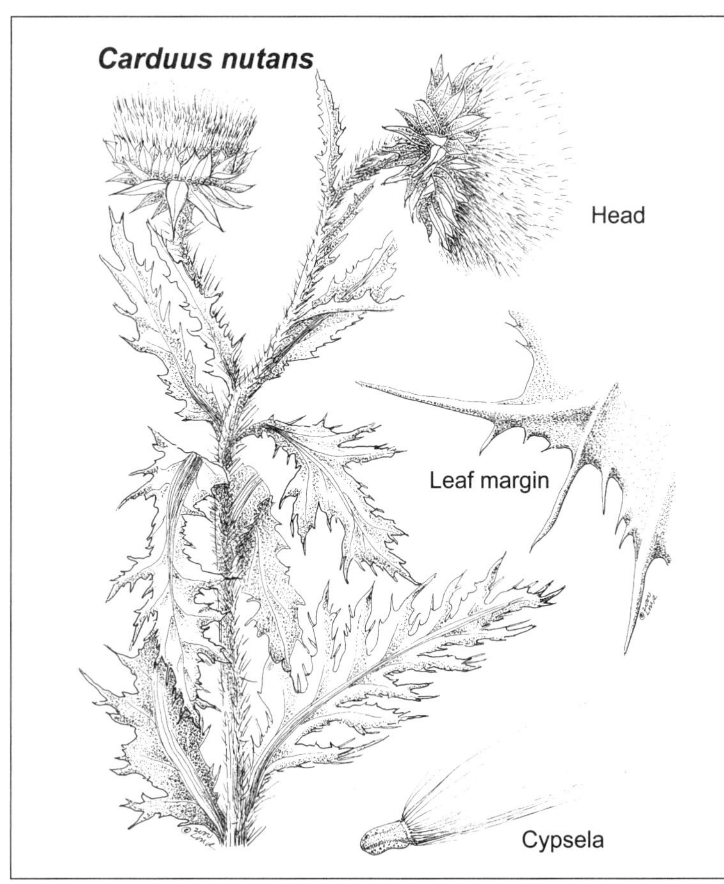

Carduus nutans L. (nodding) Musk thistle, bristle thistle, nodding thistle. Annual or biennial, to 2 m tall. **LEAVES** pinnatifid, lobes coarsely spiny toothed; mostly decurrent; upper leaves much reduced. **HEADS** solitary on naked peduncles or leafy branches, 3–5 cm wide. **PHYLLARIES** green to deep purple, or purple-tipped, outer phyllaries with strong central whitish vein. **COROLLAS**

white, lavender, or purple, 18–25 (30) mm, divided into 8–14 (16) mm tube, 3–6 mm throat, and 6–9 mm lobes; style tip above the node 2–3.5 mm long. **PAPPUS** bristles 14–20 mm long. Disturbed sites along roadsides and in fields; spreading to sagebrush, piñon-juniper woodlands, mountain brush, and aspen communities. ARIZ: Apa, Nav; COLO: Arc, Dol, Hin, LPl, Min, Mon, SJn, SMg; NMEX: McK, RAr, San, SJn; UTAH. 1525–2590 m (5000–8500′). Flowering: Jul–Aug. Origin Eurasian and North African and now widely distributed throughout the United States. An aggressive weed, often brought in on truck tires.

Carthamnus L.

Kenneth D. Heil

Annual herbs with taproots, the juice watery. **STEMS** erect, often branched. **LEAVES** alternate, spinose-denticulate. **HEADS** few, terminating the branches, discoid. **PHYLLARIES** imbricate in several series, the outer at least with prominent foliose, herbaceous tips spinescent along the margin, the body and the inner arachnoid-pubescent and glandular. **RECEPTACLES** densely scaly with slender, flat scales. **FLORETS** all tubular, perfect, corolla yellow to yellow-orange. **PAPPUS** lacking on outer achenes, the inner achenes with persistent slender scales. **CYPSELAE** obpyramidal, 4-angled, glabrous. Fourteen species, Mediterranean, Europe, and Far East (Welsh et al. 2003).

Carthamnus tinctorius L. (used in dyeing) Safflower. Subglabrous annual, mainly 2–6 dm tall. **LEAVES** cauline, sessile, more or less clasping basally, 2–10 cm long, 0.5–2.5 (4) cm wide, oblong-lanceolate to ovate, sinuately dentate, the teeth armed with pale, sharp spines. **HEADS** 1.5–3.5 (4.5) cm wide, campanulate to ovoid. **PHYLLARIES** imbricate, the outer with prominent foliose tips, the inner merely acute to acuminate, broad, flat, tomentose, connivent in bud. **FLORET** corollas yellow to orange. **CYPSELAE** about 4–5 mm long, pale. Cultivated oil plant, escaping and persisting. COLO: Dol, LPl; UTAH. 1830–2285 (6000–7500′). Flowering: May–Aug. Apparently native to the eastern Mediterranean. Safflower is cultivated as an oil seed and is used as a source of vegetable dye. It is also used in birdseed and as an ornamental.

Centaurea L. Knapweed, Star Thistle, Cornflower

Patricia Barlow-Irick

(Greek *kentaurieon*, ancient plant name associated with Chiron, the wise centaur, famous for knowledge of medicinal plants) Annuals, biennials, or perennials, 20–300 cm tall, glabrous or tomentose. **STEMS** erect, ascending, or spreading, simple or branched. **LEAVES** basal and cauline; petiolate or sessile; proximal blade margins often ± deeply lobed, (spiny in *C. benedicta*), distal ± smaller, often entire, faces glabrous or ± tomentose, sometimes also villous, strigose, or puberulent, often glandular-punctate. **HEADS** discoid, disciform, or radiate, borne singly or in corymbiform arrays. **INVOLUCRES** cylindric or ovoid to hemispheric, waisted. **PHYLLARIES** many, in 6–many series, unequal, proximal parts appressed, body margins entire, distal parts expanded into erect to spreading, usually ± dentate or fringed, linear to ovate appendages, spine-tipped or spineless. **RECEPTACLES** flat, epaleate, bristly. **FLORETS** 10–many per head, usually sterile, corollas slender and inconspicuous to much expanded, ± bilateral; inner fertile, corollas white to blue, pink, purple, or yellow, bilateral or radial, often bent at junction of tubes and throats, lobes linear-oblong, acute; anther bases tailed, apical appendages oblong; fused portions of style branches with minutely hairy nodes, distinct portions minute. **CYPSELAE** ± barrel-shaped, ± compressed, smooth or ribbed, apices entire (denticulate in *C. benedicta*), glabrous or with fine, 1-celled hairs, attachment scar lateral (with or without elaiosomes). **PAPPUS** 0 or ± persistent, of 1–3 series of smooth or minutely barbed, stiff bristles or narrow scales. x = 8, 9, 10, 11, 12, 13, 15. About 500 species; Eurasia, northern Africa; widely introduced worldwide.

1. Corollas yellow ..... ***C. solstitialis***
1′ Corollas white to pink, blue, or purple ..... (2)

2. Principal phyllaries spine-tipped; corollas white, rarely pink or purple ..... ***C. diffusa***
2′ Principal phyllaries not spine-tipped; corollas blue, pink, or purple (rarely white) ..... (3)

3. Annuals; corollas mostly blue; a commonly cultivated garden ornamental ..... ***C. cyanus***
3′ Perennials; corollas pink to purple, rarely white; a common weed of roadsides ..... ***C. stoebe***

Centaurea cyanus L. (blue) Bachelor's-button, garden cornflower, cornflower, bluebottle, bluebonnets, corn pinks. Annual from taptroots, 2–10 dm tall. **STEMS** usually 1, erect, ± openly branched distally, loosely tomentose especially when young. **LEAVES** ± loosely gray-tomentose; apices not spiny; basal leaf blades linear-lanceolate, 3–10 cm, entire or few-lobed, cauline linear, slightly smaller, usually entire. **HEADS** in open arrays, pedunculate. **INVOLUCRES**

campanulate, 1–2 cm long. **PHYLLARIES** green, ovate (outer) to oblong (inner), tomentose or becoming glabrous, margins scarious, colored or white, and fringed. **FLORETS** blue (white to purple), those on the periphery of the receptacle sterile and raylike, enlarged, with lobes 7–8 mm long; central florets fertile, discoid, with lobes 4–5 mm long. **CYPSELAE** stramineous, brown to black or bluish, 4–5 mm, finely hairy. **PAPPUS** of many unequal stiff bristles, 3–4 mm long. $2n = 24$. [*Leucacantha cyanus* (L.) Nieuwl. & Lunell]. A cultivated ornamental and now established in disturbed areas from naturalized plants or broadcast with wildflower seed mixes. COLO: LPl, SJn. 2325–2925 m (7630–9600′). Flowering: May–Sep. Native to Europe.

Centaurea diffusa Lam. (spreading) Diffuse knapweed, tumble knapweed, spreading knapweed, white knapweed. Annuals or perennials, 10–60 cm from a deep taproot; may remain a low rosette for first years. **STEMS** 1–several, much-branched throughout, puberulent along the ridges, eventually breaking off and rolling in the wind. **LEAVES** hispidulous and ± short-tomentose; basal and lower cauline petiolate, deciduous, blades 3–20 cm, margins bipinnately dissected into narrow lobes; midcauline sessile, bipinnately dissected; becoming bractlike near the flower heads. **HEADS** disciform, in open panicles. **INVOLUCRES** narrowly urn-shaped, tightly constricted below tips, 8–13 mm long. **PHYLLARIES** with bodies pale green, ovate to lanceolate, glabrous or finely tomentose, with a few prominent parallel veins, margins and erect appendages fringed, each phyllary tipped by spine 1–3 mm, stramineous on margins, sometimes purplish at apex. **FLORETS** cream-white (rarely pink or pale purple), 10–16 mm long, discoid, peripheral florets sterile with longer lobes; free tips of anthers white, 0.3–0.5 mm long; corolla lobes 3–4 mm in central florets, to 5.5 in peripheral florets; style tips above node 1–1.5 mm long. **CYPSELAE** tan to black, about 2.5 mm long. **PAPPUS** lacking or rudimentary, to 1 mm long. [*Acosta diffusa* (Lam.) Soják]. Roadsides and other disturbed sites. ARIZ: Apa; COLO: Arc, Dol, LPl, Min, Mon; NMEX: SJn; UTAH. 1600–2350 m (5250–7710′). Flowering: Jun–Aug. Native to southeastern Europe; widespread throughout the United States.

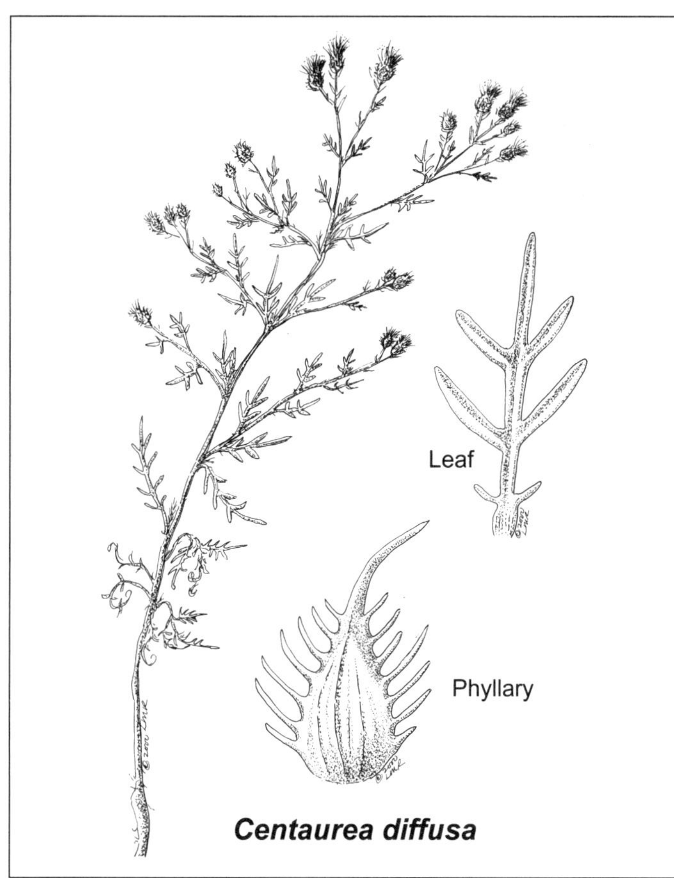

Centaurea diffusa

Centaurea solstitialis L. (relating to midsummer) Yellow star thistle. Annuals. **STEMS** simple or branched from the base, 10–100 cm tall, gray-tomentose. **LEAVES** gray-tomentose, short-bristly; cauline petiolate or tapered to the base; blades 5–15 cm long, margins pinnately lobed or dissected. **HEADS** disciform, borne singly or in open leafy arrays. **INVOLUCRES** ovoid, 13–17 mm long, loosely cobwebby-tomentose or becoming glabrous. **PHYLLARIES** with bodies pale green, ovate, appendages stramineous to brown, each with palmately radiating cluster of spines, stout central spine 10–25 mm long; inner with scarious appendages, obtuse or abruptly spine-tipped. **FLORETS** many; corollas yellow, all ±equal, 13–20 mm long. **CYPSELAE** dimorphic, 2–3 mm long, glabrous, outer dark brown, without pappi, inner white or light brown, mottled. **PAPPUS** of many white unequal bristles 2–4 mm long, fine. $2n = 16$. Roadsides and abandoned fields. NMEX: SJn. 1980 m (6500′). Flowering: Jun–Oct. Widespread in the western United States; native to Europe. Known from one collection within the study area. It is easily identified by its long yellowish spines and has caused poisoning in horses.

Centaurea stoebe L. (stuffing, apparently used for packing wine bottles, making brooms, and bedding) Spotted knapweed, spotted star thistle. Perennial, sometimes short-lived, from a stout taproot. **STEMS** erect, 1–several from the base, branched throughout, ridged, loosely cobwebby, 30–100 cm tall. **LEAVES** glabrous or tomentose, basal spatula-shaped in outline, usually deeply divided, 5–15 cm long; cauline 10–20 cm long, bipinnately dissected into narrow lobes 1–3 mm wide, becoming bractlike near the flower heads. **HEADS** solitary, at the ends of the branches.

INVOLUCRES urn-shaped and somewhat constricted at tips, 11–15 mm long. **PHYLLARIES** with prominent veins below the dark-spotted tip (often colorless in white-flowered forms), the tips fringed, not spiny. **FLORETS** pink to purple, sometimes white, all discoid, the peripheral ones enlarged and sterile; anther tips lavender to purple, 0.3–0.8 mm long; corolla lobes 6–9 mm long in peripheral florets, 3–5 mm long in central florets; style tips 1–1.7 mm long. **CYPSELAE** dark, with lines, 2.5–3.5 mm long. **PAPPUS** of stiff bristles 2–3 mm long or absent. Disturbed areas, often along roadsides, waste places, fields. [*Centaurea maculosa* Lam. subsp. *micranthos* S. G. Gmel. ex Gugler; *C. biebersteinii* DC.]. COLO: Arc, Dol, LPl, Mon, SJn; NMEX: McK, SJn; UTAH. 1645–2590 m (5400–8500′). Flowering: Jun–Oct. Native to Europe, widespread throughout the United States. Spotted knapweed is a serious pest of rangelands, pastures, and open fields in many areas in the northern United States. It may cause dermatitis. Our material belongs to **subsp.** ***micranthos*** (S. G. Gmel. ex Gugler) Hayek (small flower).

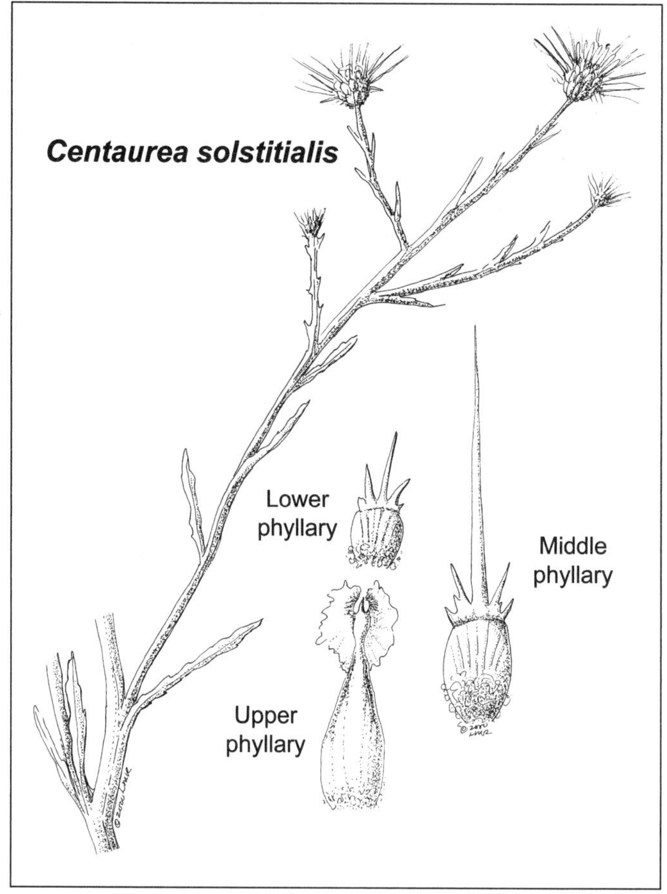

Chaenactis DC. Pincushion, Dusty-maidens
J. Mark Porter

(Greek *chaino*, to gape, + *aktis*, ray) Annuals, biennials, perennials, or subshrubs, (2–) 5–70 (–200) cm, from a taproot. **STEMS** erect or prostrate, generally branched. **LEAVES** basal and cauline; alternate; reduced in size distally; petiolate; blades deltate, elliptic, linear, oblanceolate, or ovate, 1–4-pinnately lobed, glabrous or hairy, often stipitate-glandular or gland-dotted. **INFL** borne singly or cymosely. **HEADS** discoid (sometimes appearing radiate). **INVOLUCRES** hemispheric to obconic or broadly cylindric, (3–) 5–15 (–25) mm diameter. **PHYLLARIES** 5–25 in 1–2 (–3) subequal to unequal series. **RECEPTACLES** convex to flat, pitted or knobby, usually lacking paleae (paleae 3–12 in *C. carphoclinia*). **RAY FLORETS** absent (sometimes simulated by enlarged peripheral disc corollas). **DISC FLORETS** 8–75, bisexual, fertile, diurnal with anthers exserted (nocturnal in *C. macrantha*); corollas white, pinkish, cream, or yellow, tubes shorter than cylindric or funnelform throats, lobes 5, deltate to lanceolate, sometimes enlarged, unequal; style branch appendages blunt, obscure. **CYPSELAE** clavate to cylindric or compressed, obscurely 8–20-angled, faces scabrous and strigose to densely sericeous (usually eglandular). **PAPPUS** usually persistent, of (1–) 4–20 distinct, erose scales in 1–4 equal or unequal series, outer series shorter, scales usually fewer and shorter on peripheral cypselae, sometimes absent or coroniform. n = (5), 6, (7), 8, $2n$= 13, 14, 15, 16, 17, 18, 25, 26, 27, 28, 30, 36, 38. A genus of 18 species (3 in the *Flora*) of western North America and northwest Mexico. *Chaenactis* appears to be most closely related to *Dimeresia* and *Orochaenactis*, which Baldwin et al. (2002) treated together as a narrowly circumscribed tribe, Chaenactideae (Baldwin, Wessa, and Panero 2002; Mooring 1965, 1980; Stockwell 1940).

1. Annuals; hairs of the lower portions mostly farinose, not arachnoid; largest leaf blades (2–) 3–4-pinnately lobed; receptacles bearing paleae .... ***C. carphoclinia***

1′ Annuals, biennials, or perennials; hairs of the lower portions mostly arachnoid, lanuginose, pannose, stipitate-glandular, or glabrescent, but not farinose; largest leaf blades 1–2-pinnately or subpalmately lobed; receptacles lacking paleae .... (2)

2. Biennials or perennials (rarely flowering first year); pappus of (8–) 10–20 scales in 2–4 equal or gradually unequal series; leaf blades gland-dotted beneath hairs .... ***C. douglasii***

2′ Annuals; pappus usually of (1–) 4–8 (–14) scales in 1, 2 abruptly unequal, or 2–3 gradually unequal series; leaf blades not gland-dotted (except *C. macrantha*) .... (3)

3. Peripheral corollas zygomorphic, enlarged; pappus of (1–) 4 (–5) scales, usually in 1 series or with partial outer series; peduncles usually stipitate-glandular distally; corollas 4.5–6.5 mm long .... ***C. stevioides***
3′ Peripheral corollas actinomorphic, not or scarcely enlarged; pappus of 8 scales in 2 abruptly unequal series; peduncles arachnoid to lanuginose distally, not stipitate-glandular; corollas 9–12 (–15) mm long ...***C. macrantha***

Chaenactis carphoclinia A. Gray. (small, dry bed, referring to the drying basal leaves) Pebble pincushion, straw-bed pincushion. Plants (5–) 10–30 (–60) cm tall. **LEAVES** basal (withering) and cauline, 1–6 (–8) cm; sometimes somewhat succulent; primary lobes mostly 2–7 (–10) pairs, linear, terete; peduncles 2–6 cm. **INOVLUCRES** obconic to cylindric or hemispheric, mostly 5–10 mm diameter; longest **phyllaries** 7–10 mm long, granular-glandular and villous; apices acuminate, aristate, terete, erect to incurved. **RECEPTACLES** paleate, (0–) 3–10 paleae, persistent, phyllary-like, apices visible among mature floret buds. **DISC FLORETS** with corollas 4–6 mm. **CYPSELAE** terete, 3–4.5 mm. **PAPPUS** usually of 4 (–5) scales, longest 3–5 mm. $2n$ = 16. Open, rocky, gravelly, or sandy arid slopes and flats, in blackbrush and saltbush communities. UTAH. 1200–1800 m (4000–6000′). Flowering: May–Jun. Fruit: Jun–Jul. California, Nevada, and Arizona; Mexico (Baja California, Sonora). Our material belongs to **var.** ***carphoclinia***.

Chaenactis douglasii (Hook.) Hook. & Arn. (honoring David Douglas, 1798–1834, Scottish botanical explorer) Hoary pincushion, Douglas' dusty-maidens. Biennials or perennials, (2–) 5–50 (–60) cm tall, taprooted; sometimes caespitose or matted; hairs of the lower portions thinning with age, grayish, mostly arachnoid-sericeous to thinly lanuginose. **STEMS** 1–25, erect to spreading. **LEAVES** basal, or basal and cauline, (1–) 2–12 (–15) cm long; largest blades elliptic or slightly lanceolate to ovate, usually 2-pinnately lobed; primary lobes (4–) 5–9 (–12) pairs. **HEADS** 1–27 per stem, opening diurnally; peduncles mostly ascending to erect, 1–10 cm. **INVOLUCRES** obconic to hemispheric; longest **phyllaries** 9–15 (–17) mm, outer usually stipitate-glandular (rarely eglandular), often arachnoid, lanuginose, or sparsely villous; apices usually squarrose, pliant. **DISC FLORETS** with corollas 5–8 mm. **CYPSELAE** 5–8 mm, usually sparsely glandular among other hairs. **PAPPUS** longest scales 3–6 mm.

1. Leaves strictly basal; plants caespitose or matted; stems (1–) 10–25; heads 1 (–2) per stem ....**var. *alpina***
1′ Leaves basal (sometimes withering) and cauline; plants not or scarcely caespitose, not matted; stems usually 1–5 (–12); heads (1–) 2–25 per stem ....**var. *douglasii***

var. ***alpina*** A. Gray (of the alpine) Alpine dusty-maidens, alpine pincushion. Perennials, mostly (2–) 5–10 (–20) cm tall, caespitose to matted. **STEMS** (1–) 10–25. **LEAVES** strictly basal, to (1–) 2–6 cm, hairs persistent or glabrate. **HEADS** 1 (–2) per stem; longest **phyllaries** 9–12 mm; outer stipitate-glandular, rarely lacking glands, and often arachnoid to lanuginose. $2n$ = 12. [*Chaenactis alpina* (A. Gray) M. E. Jones]. Rocky or gravelly alpine ridges, talus, scree, and fell-fields. COLO: Arc, Con, Hin, LPl, RGr, SJn. 2700–4000 m (9000–13500′). Flowering: Jul–Sep. Fruit: Aug–Oct. Oregon to Montana, California, and Colorado.

var. ***douglasii*** False yarrow, hoary pincushion, Douglas' dusty-maidens. Biennials or perennials, (3–) 8–50 (–60) cm tall, rarely slightly woody or flowering first year, not caespitose or matted. **STEMS** usually 1–5 (–12). **LEAVES** basal and cauline, 1.5–12 (–15) cm, hairs persistent. **HEADS** (1–) 2–25 per stem; longest **phyllaries** 9–15 (–17) mm; outer stipitate-glandular and arachnoid, lanuginose, or sparsely villous. $2n$ = 12, 14, 15, 24, 26, 30, 36. Open sandy, gravelly, or rocky soils in sagebrush, piñon-juniper, and ponderosa pine communities. ARIZ: Apa; COLO: Arc, LPl, Mon; NMEX: RAr, SJn; UTAH. 1200–3500 m (4000–11500′). Flowering: May–Sep. Fruit: Jun–Sep. British Columbia to Montana, south to California, Arizona, and Colorado.

Chaenactis macrantha D. C. Eaton (large flower) Showy dusty-maidens, bighead dusty-maidens, Mojave pincushion. Plants 5–25 (–35) cm tall; hairs of the lower portions grayish, arachnoid-sericeous or lanuginose, sometimes ± glabrescent. **STEMS** mostly 1–5; branching below. **LEAVES** basal and cauline, basal often withering, 1.5–7 cm; largest blades elliptic to ovate, not succulent, 1 (–2)-pinnately lobed, gland-dotted beneath hairs; primary lobes mostly 2–5 pairs. **HEADS** opening nocturnally, mostly 1–5 (–7) per stem, nodding in bud; peduncles 1.5–8 cm, arachnoid-sericeous to thinly lanuginose distally, not stipitate-glandular. **INVOLUCRES** obconic to broadly cylindric; longest **phyllaries** 12–18 mm; outer arachnoid-sericeous to thinly lanuginose in fruit, not stipitate-glandular. **DISC FLORETS** with corollas white to pinkish or cream, 9–12 (–15) mm; peripheral corollas nocturnally spreading, actinomorphic, scarcely enlarged. **CYPSELAE** 5–6 (–7) mm. **PAPPUS** of 8 scales in 2 abruptly unequal series, longest 5–7 mm. $2n$ = 12. Open, loose, light-colored, silty, usually calcareous or alkaline, desert soils, associated with shadscale, blackbrush, and piñon-juniper woodlands. ARIZ: Nav; UTAH. 1200–2140 m (4000–7000′). Flowering: Mar–early Jul. Fruit: May–Jul. California, Nevada, Utah, and Arizona.

Chaenactis stevioides Hook. & Arn. (origin unclear, perhaps named for Pedro J. Esteve, or meaning "similar to *Stevia*," a genus of Asteraceae) Desert pincushion, Esteve pincushion, broad-flower pincushion, Steve's pincushion. Plants 5–30 (–45) cm tall; hairs of the lower portions grayish, arachnoid-sericeous, glabrescent at nodes. **STEMS** 1–12, erect or occasionally decumbent; branching from the base or above. **LEAVES** basal and cauline, the basal usually withering, 1–8 (–10) cm; largest blades elliptic, usually not succulent, mostly 1–2-pinnately lobed; primary lobes 4–8 pairs. **HEADS** mostly 3–23 per stem, opening diurnally; peduncles 1–5 (–10) cm, usually stipitate-glandular distally, often arachnoid. **INVOLUCRES** hemispheric to obconic, bases green, rounded in fruit; longest **phyllaries** 5.5–8 (–10) mm, outer stipitate-glandular and/or arachnoid in fruit. **DISC FLORETS** with corollas white to pinkish, cream, or pale yellow, 4.5–6.5 mm; peripheral corollas spreading, zygomorphic, enlarged. **CYPSELAE** (3–) 4–6.5 mm. **PAPPUS** of (1–) 4 (–5) scales, usually in 1 series, rarely with partial outer, abruptly unequal series, longest scales 1.5–6 mm. $2n = 10$. Open, arid or semiarid, sandy or gravelly slopes and flats of desert scrub and juniper woodlands. ARIZ: Apa, Coc, Nav; COLO: LPl, Mon; NMEX: SJn; UTAH. 1220–1900 m (4000–6200′). Flowering: Feb–Jun. Fruit: Apr–Jul. Wyoming to Nevada, west to California; Mexico (Baja California, Sonora). *Chaenactis stevioides* has been employed by the Kayenta Navajo to mend broken ceremonial items, the juice being used as glue.

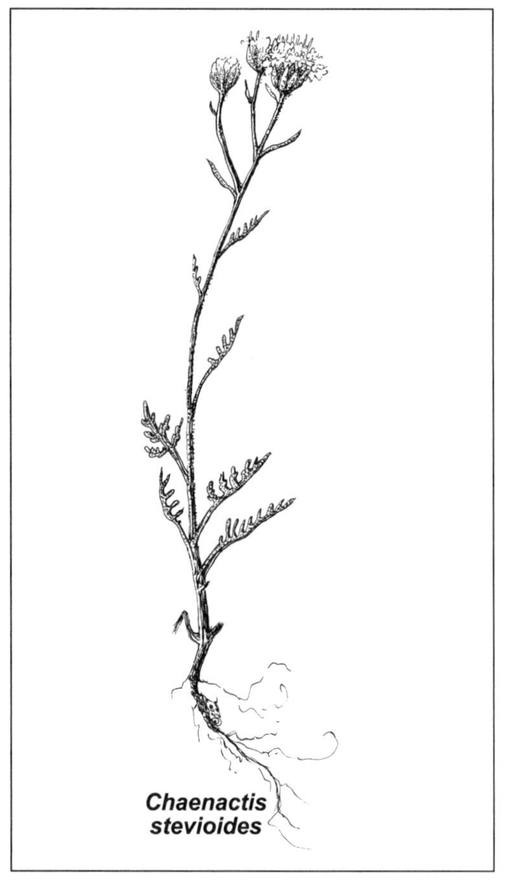

Chaenactis stevioides

Chaetopappa DC. Rose-heath

G. L. Nesom

(Greek *chaite*, long hair, + pappus) Annuals or perennials, 5–30 cm; taprooted or rhizomatous, glandular or eglandular. Stems erect or decumbent. **LEAVES** basal (sometimes not persistent to flowering) and cauline, alternate, blades linear to oblanceolate-obovate or spatulate, margins entire, glabrous to strigose, hispidulous, or hispido-pilose, sometimes stipitate-glandular. **HEADS** radiate, borne singly. **INVOLUCRES** cylindro-turbinate to hemispheric, 3.8–6.5 × 2–10 mm. **PHYLLARIES** in 2–6 series, unequal, elliptic to linear-lanceolate, margins prominently whitish scarious, midveins not evident or indurate, strigose to glabrate. **RECEPTACLES** flat or slightly convex, epaleate. **RAY FLORETS** 6–24 (–30) in 1 series, pistillate, fertile; corollas usually white to bluish, coiling at maturity. **DISC FLORETS** bisexual and fertile or functionally staminate. **CYPSELAE** cylindric to narrowly obovoid, terete to compressed, 2–5-, 8-, or 10-nerved, strigose, sometimes gland-dotted. **PAPPUS** persistent, of hyaline crowns, hyaline scales, awnlike bristles, or combination of alternating scales and bristles, members usually 5 (–30, close to a multiple of 5), or reduced to minute crowns and essentially absent. $x = 8$. [*Leucelene* Greene] (Nesom 1988; Shinners 1946). Species 11, south-central and southwestern United States, adjacent Mexico.

Chaetopappa ericoides (Torr.) G. L. Nesom (similar to *Erica*, alluding to the leaves) Rose-heath. Perennial herbs 6–12 cm, densely stipitate-glandular to eglandular; taprooted (but usually not evident in collections), forming beds or clumps of large, separate tufts connected by creeping underground rhizomes and caudexlike stems with scale leaves. **LEAVES** even-sized, densely overlapping; blades linear-oblanceolate to lanceolate, 5–11 (–20) × 0.5–2 mm, coriaceous, not clasping, midnerves strongly raised, longitudinally 2-grooved adaxially, spinulose-tipped, often densely, minutely, orangish stipitate-glandular. **HEADS** with involucre turbinate to hemispheric, (3.5–) 4.5–6 (–7) × 4–8 mm. **RAY COROLLAS** white. **DISC FLORETS** 12–24, bisexual. **CYPSELAE** 5-nerved, strigose. **PAPPUS** of (20–) 24–26 (–30) barbellate bristles in 1 series. $2n = 16, 32$. [*Aster arenosus* (A. Heller) S. F. Blake; *Leucelene ericoides* (Torr.) Greene]. Open sites, commonly sandy, roadsides, grassland, talus, dunes, washes, desert shrub, sagebrush, juniper to piñon-juniper woodland, ponderosa, and pine-spruce communities. ARIZ: Apa, Coc, Nav; COLO: Arc, Dol, LPl, Mon, SMg; NMEX: McK, RAr, San, SJn; UTAH. 1550–2250 (–2400) m (5085–7380′). Flowering: Apr–Jul (–Oct). Southwestern United States, California and Arizona to Kansas, Nebraska, Oklahoma, and Texas; widely distributed in northern Mexico. *Chaetopappa ericoides* is variable in the type and amount of pubescence as well as other features.

Chamaechaenactis Rydb. Fullstem

J. Mark Porter

(Greek *chamae-*, creeping, low, on the ground, + generic name *Chaenactis*) Perennials, 2–7 (–9) cm, to 10–24 cm across. **STEMS** mostly subterranean, caudices relatively thick, branched, pilose; aerial stems essentially peduncles. **LEAVES** mostly basal; alternate; petiolate; blades cordate, elliptic, ovate, or rounded, margins entire or distally crenate, often revolute; strigose and gland-dotted abaxially, sometimes glabrescent adaxially; 1- or 3-nerved. **INFL** solitary. **HEADS** discoid. **INVOLUCRES** obconic, 6–15 mm diameter. **PHYLLARIES** 11–15 in 2 series, erect, sometimes spreading in senescence, oblong to oblanceolate, unequal, herbaceous, abaxially densely villous, obscurely glandular. **RECEPTACLES** convex, knobby, lacking paleae. **RAY FLORETS** absent. **DISC FLORETS** 10–30, bisexual, fertile; corollas white to pinkish, tubes shorter than cylindric throats, lobes 5, deltate; style branches stigmatic in 2 lines, appendages linear-oblong, blunt. **CYPSELAE** clavate, quadrangular, with 8–12 obscure nerves, densely pilose-strigose, eglandular. **PAPPUS** persistent, of 7–11 distinct, oblanceolate to narrowly spatulate, erose scales in 2 series, midnerves prominent. $x = 16$. One species of western North America. *Chamaechaenactis* has often been considered a close relative of *Chaenactis*; however, nuclear rDNA evidence from B. G. Baldwin et al. suggests closer relationship to *Bartlettia* and *Hymenopappus*, members of Bahiinae (Baldwin, Wessa, and Panero 2002; Preece and Turner 1953).

Chamaechaenactis scaposa

Chamaechaenactis scaposa (Eastw.) Rydb. (bearing a scape) Eastwood-plant, fullstem. Plants densely caespitose, pulvinate; taproots deep; caudices thickly branched. **LEAVES** basal, old leaf bases marcescent; longest petioles 3–40 mm, proximally dilated and chartaceous; larger blades 4–18 mm long, 3–13 (–15) mm wide, coriaceous. **HEADS** (1–) 5–30 or more, ascending to erect; peduncles (0–) 1–5 (–7) cm long, strigose, usually villous distally; longest **phyllaries** 9–17 mm; outer shorter than inner. **DISC FLORETS** with corollas 5–9 mm. **CYPSELAE** 5–8 mm, black. **PAPPUS** with longest scales 4–7 mm. $2n = 32$. Dry, open, relatively barren silty to clay soils from shale, sandstone, marl, or limestone, often rocky or gravelly, in desert scrub, mountain brush, piñon-juniper woodlands, and ponderosa pine forest openings. ARIZ: Apa; COLO: Mon; NMEX: SJn; UTAH. 1500–2600 m (4800–8500′). Flowering: late Apr–early Jul. Fruit: May–Jul. Restricted to the Green, San Juan, and upper Colorado River drainges. Arizona, New Mexico, Utah, Wyoming, and Colorado.

Chloracantha G. L. Nesom, Y. B. Suh, D. R. Morgan, S. D. Sundb. & B. B. Simpson Spiny Aster

G. L. Nesom

(Greek *chloros*, green, + *akantha*, thorn) Perennials or subshrubs (commonly appearing herbaceous), forming large clones, sometimes glaucous, glabrous or glabrate; stoutly rhizomatous. **STEMS** strictly erect, lateral branches commonly modified to thorns. **LEAVES** cauline, early-deciduous, alternate, sessile, blades oblanceolate, 1-nerved, margins entire or rarely with 1–2 pairs of small teeth. **HEADS** radiate, borne singly in loosely corymbo-paniculiform arrays. **INVOLUCRES** broadly turbinate to hemispheric. **PHYLLARIES** 4–5 series, unequal, oblong-elliptic to lanceolate, margins hyaline, veins (1–) 3 (–5), parallel, orange-resinous, apices rounded to lanceolate, glabrous. **RECEPTACLES** shallowly convex, smooth, epaleate. **RAY FLORETS** 1 (–2) series, pistillate, fertile; corolla white, coiling at maturity. **DISC FLORETS** bisexual, fertile; corolla yellow. **CYPSELAE** fusiform-cylindric, 5 (–6)-nerved, glabrous. **PAPPUS** of persistent barbellate bristles in 1–2 series, usually with a few much shorter outer setae. $x = 9$. [*Aster* L. sect. *Spinosi* (Alexander) A. G. Jones] (Sundberg 1991; Nesom et al. 1991). One species. Southwestern United States, western Mexico, Central America. *Chloracantha spinosa* produces no terminal resting buds and its permanently green stems without periderm are herbaceous in aspect, but it behaves more like a subshrub in its perennial stems (alive for up to about four growing seasons) with a quickly developed vascular cambium and its production of axillary buds with bud scales. All or almost all of the leaves are usually shed by flowering and the colonies become masses of erect, green stems with loosely paniculiform arrays of small, terminal, white-rayed heads.

Chloracantha spinosa (Benth.) G. L. Nesom (spiny) Spiny aster. **STEMS** 50–120 (–150) cm, usually thorny proximally but armament of distal portions variable among populations, with some producing only thornless, wandlike branches. **LEAVES** mostly 10–40 mm. **HEADS** with involucre 4–5.5 (–6) mm. **RAY FLORET** laminae 3.5–5 (–7) mm. $2n = 18$. [*Aster spinosus* Benth.; *Erigeron ortegae* S. F. Blake; *Leucosyris spinosa* (Benth.) Greene]. Sandy alluvium, riparian (saltcedar, goldenweed), desert scrub. UTAH. 1125–1220 m (3700–4000′). Flowering: Jun–Oct. California, Nevada, Utah, and Arizona, eastward to Oklahoma and Louisiana; northern Mexico. Four Corners plants are **var. *spinosa.*** Three other varieties occur in western Mexico and Central America.

Chrysanthemum L. Shasta Daisy, Oxeye-daisy

Kenneth D. Heil

(Greek *chrysanthemon*, golden flower) Perennial herbaceous plants. **STEMS** leafy. **LEAVES** alternate, entire or toothed to coarsely incised or pinnatifid. **HEADS** long-pedunculate, few or solitary and terminal on stems or long branches or corymbose. **INVOLUCRES** mostly saucer-shaped; bracts imbricate in several series, appressed, scarious-margined. **RECEPTACLES** flat or convex, naked. **RAY FLORETS** pistillate and fertile, rays conspicuous, white. **DISC FLORETS** many, yellow; anthers united, not caudate at the base. **CYPSELAE** 5–10-ribbed or angled. **PAPPUS** none or rarely of a short crown. About 75 species, worldwide.

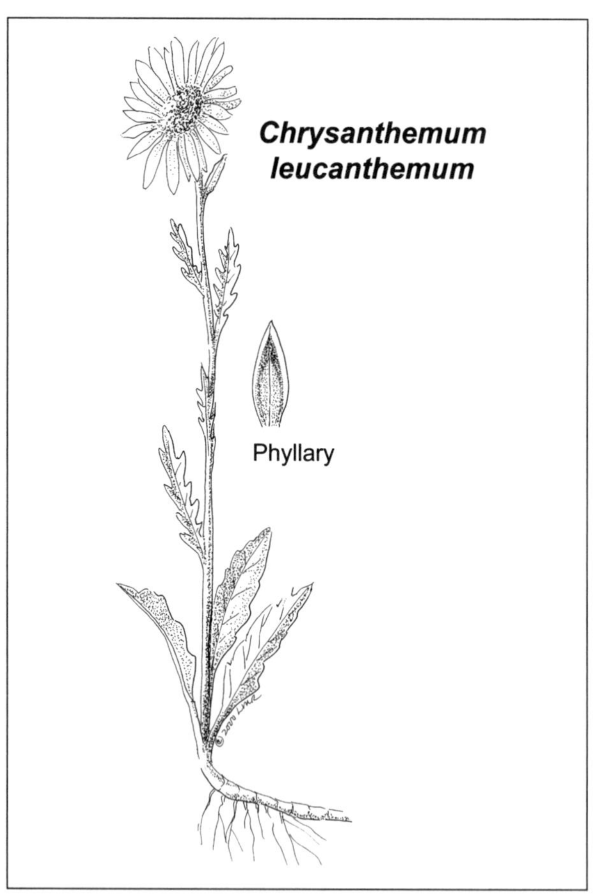

Chrysanthemum leucanthemum L. (white-flowered) Shasta daisy, oxeye-daisy. Perennial. **STEMS** 25–80 cm tall, glabrous or nearly so. **LEAVES** (basal) obovate to spatulate, coarsely and irregularly toothed or incised; stem leaves oblanceolate to linear, toothed to subpinnatifid, base lacerate, sessile at least above. **HEADS** on long peduncles. **INVOLUCRES** 12–35 mm diameter, bracts lanceolate or oblong-lanceolate, light green with a darker green midrib, narrowly scarious on margins. **RAYS** mostly 15–35, white, 1–2 cm long. **CYPSELAE** 10-ribbed. **PAPPUS** none. $2n = 18, 36, 54$. [*Leucanthemum vulgare* Lam.; *L. leucanthemum* Rydb.; *Chrysanthemum leucanthemum* var. *pinnatifidum* Lecoq & Lamotte]. Mountain meadows, roadsides, and disturbed sites. COLO: Arc, Hin, LPl, Min, Mon, SJn; NMEX: RAr, SJn. 2100–3000 m (6890–9840′). Flowering: Jun–Sep. Native of Europe and naturalized throughout North America. Oxeye-daisy has infested many meadows throughout southwestern Colorado and northwestern New Mexico. Infestations are usually not grazed due to a disagreeable taste and can produce unwanted flavor in milk.

Chrysothamnus Nutt. Rabbitbrush

R. P. Roberts

(Latin *chryseus*, golden, + *thamnus*, a bush, hence goldenbush) Shrubs or subshrubs, often rounded or compact. **STEMS** to 1.5 m tall, woody and highly branched, ascending to erect or spreading, bark whitish tan, becoming gray with maturity, sometimes fibrous or flaky, young twigs mostly green, glabrous to pubescent or glandular, often resinous, usually ridged from leaf bases. **LEAVES** alternate, mostly ascending or spreading, rarely deflexed or recurved, blades filiform, lanceolate, elliptic, to obovate, 7–80 mm long, 0.5–25 mm wide, planar to sulcate, twisted in most, sessile or short-petiolate, glabrous to puberulent, sometimes punctate, often resinous, margins entire, ciliate in some, apices usually acute, may be apiculate, midvein usually evident, 0–2 pairs of fainter collateral veins present. **INFL** solitary or in condensed cymiform clusters, these forming paniculiform, corymbiform, or rarely racemiform arrays. **HEADS** discoid, short-pedunculate. **INVOLUCRES** turbinate, obconic, cylindric, sometimes hemispheric or campanulate, 5–15 mm long, 1.5–15 mm wide. **PHYLLARIES** 12–60+, usually 3–7-seriate but sometimes 1-seriate, vertically or spirally ranked, often keeled, tan, linear to elliptic, lanceolate to ovate or obovate to spatulate, apices acute to acuminate or rounded, glabrous or pubescent, often resinous. **DISC FLORETS** usually 5–6, sometimes to 40+, corollas yellow; 3.5–11 mm long; style branches 2–3.8 mm long. **CYPSELAE** tan to reddish brown, turbinate to elliptic or cylindric, often ribbed and angled, 2.5–6.5 mm long, glabrous to densely pubescent, sometimes glandular. **PAPPUS** 12–50+, tan, stramineous, or white barbellate bristles, or scales. $x = 9$. Nine species in North America.

1. Leaves planar or falcate; florets 10–40 per head ....(2)
1′ Leaves often twisted; florets 3–9 (14) ....(3)

2. Leaves ascending to spreading, flat; phyllary apices acute to rounded, erect; faces not resinous; florets 10–20 ....***C. scopulorum***
2′ Leaves spreading to deflexed, usually falcate; phyllary apices acute to acuminate, often recurved, faces often glutinous; florets 30–40 ....***C. stylosus***

3. Cypselae glabrous or with a few glandular hairs ....(4)
3′ Cypselae densely pubescent ....(5)

4. Involucres 5–7 mm long; phyllaries weakly aligned vertically and weakly keeled ....***C. vaseyi***
4′ Involucres 9 mm or more long; phyllaries strongly aligned vertically and keeled ....***C. depressus***

5. Leaves 1–2 mm wide; phyllary apices acuminate to cuspidate; florets 4–5 per head ....***C. greenei***
5′ Leaves 1–10 mm wide; phyllary apices acute to obtuse; florets 3–14 (most often 5) per head..........***C. viscidiflorus***

Chrysothamnus depressus Nutt. (low) Long-flowered rabbitbrush. **STEMS** to 0.5 m, irregularly branched, lower branches decumbent, densely puberulent, tan to gray, becoming flaky with maturity. **LEAVES** erect or closely ascending, sessile, blades linear to oblanceolate or narrowly oblong, 7–30 mm long, 1.5–7 mm wide, planar to keeled, glabrous, puberulent or sparsely stipitate-glandular, midvein evident. **HEADS** densely cymiform, not overtopped by distal leaves. **INVOLUCRES** obconic, 9–15 mm long, 3–5 mm wide. **PHYLLARIES** 20–25, 4–6-seriate, in 5 vertical ranks, tan, often with green and/or purplish markings, keeled, lanceolate to elliptic, 3–8 mm long, 0.5–1.5 mm wide, apices acute to acuminate, midvein evident distally, outer herbaceous, inner scarious. **DISC FLORETS** 5–6, corollas 7–11 mm long; style branches 2.4–3.3 mm long. **CYPSELAE** tan, subcylindric, 5–6.5 mm long, mostly glabrous, sparsely glandular distally. **PAPPUS** whitish tan, 5.5–7.5 mm long. $2n = 18$. [*Ericameria depressa* (Nutt.) L. C. Anderson]. Dry canyons and rocky crevices; sagebrush, salt desert scrub, piñon-juniper woodland, mountain brush, and ponderosa pine communities. ARIZ: Apa, Nav; COLO: Arc, Dol, LPl, Mon, SMg; NMEX: McK, RAr, San, SJn; UTAH. 1125–2700 m (3700–8860′). Flowering: Sep–Oct. Arizona, California, Colorado, Nevada, New Mexico, and Utah. A decoction of the plant was used by the Ramah Navajo to facilitate labor and delivery of the placenta. Used as prayer stick decorations by the Hopi.

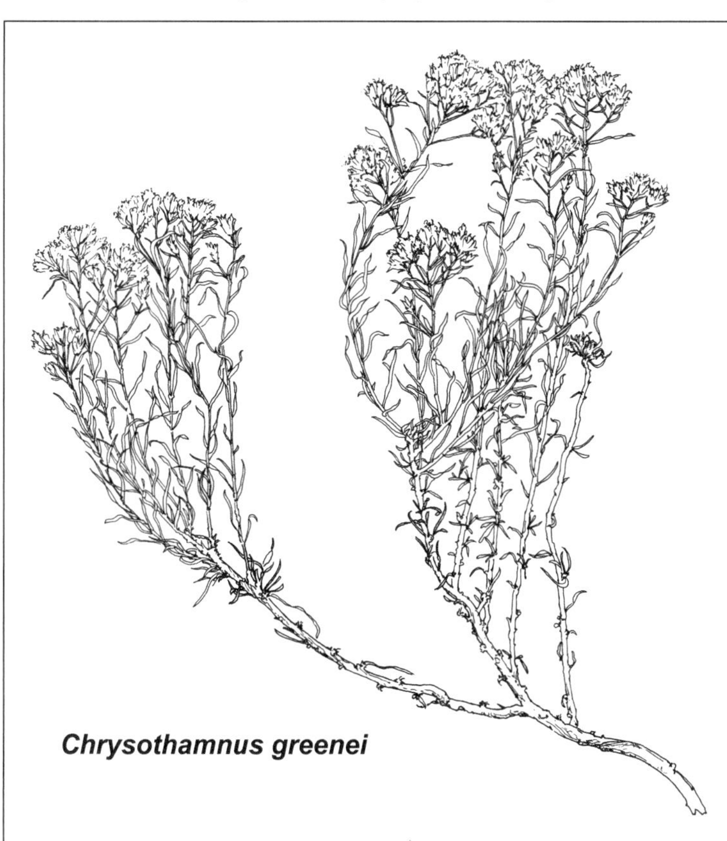
Chrysothamnus greenei

Chrysothamnus greenei (A. Gray) Greene (for Edward L. Greene 1843–1915, botanist, editor, and critic) Greene's rabbitbrush. **STEMS** to 0.5 m tall, moderately branched, bark becoming flaky to fibrous with maturity, twigs ascending, green, becoming whitish tan, glabrous, resinous. **LEAVES** ascending to spreading, linear-filiform, 10–40 mm long, less than 2.5 mm wide, planar to sulcate, sometimes twisted, blades glabrous, midvein faintly evident, margins glabrous or edged with short conical hairs, apices acute to apiculate. **HEADS** in dense cymiform clusters, distal leaves reaching but not exceeding clusters. **INVOLUCRES** cylindric to turbinate, 5–8 mm long, 1.5–2.5 mm wide. **PHYLLARIES** 15–20, in 3–4 vertically aligned or spiral series, mostly tan, greenish apically, ovate or oblong to elliptic, 1.5–5 mm long, apices acuminate to cuspidate, cusp often recurved or falcate, mostly chartaceous, midvein rarely visible. **DISC FLORETS** 4–5, corollas 3.7–5.5 mm long; style branches 2–2.4 mm, exserted beyond corolla lobes. **CYPSELAE** turbinate, reddish brown, 3–4 mm, densely pubescent. **PAPPUS** tan, 3.7–5 mm long. $2n = 18$. [*Bigelowia greenei* A. Gray; *Chrysothamnus filifolius* Rydb.; *Ericameria filifolia* (Rydb.) L. C. Anderson]. Sandy washes, dry open places; black sagebrush, shadscale, sagebrush, and piñon-juniper woodland communities. ARIZ: Apa, Coc, Nav; COLO: LPl, Mon; NMEX: McK, San, SJn. Flowering: Aug–Oct. 1300–2000 m (4265–6560′). Arizona, California, Colorado, Nevada, New Mexico, Utah, and Wyoming. An infusion of plant tops used as a wash for chicken pox and measles eruptions by the Navajo. Used as prayer stick decorations by the Hopi.

Chrysothamnus scopulorum (M. E. Jones) Urbatsch, R. P. Roberts & Neubig (of the cliffs) Grand Canyon glowweed. Shrubs up to 1 m tall. **STEMS** with lower portions woody, branched, glabrous, bark becoming tan to gray and flaky with maturity. **LEAVES** ascending to spreading, sessile; blades usually 5-nerved, linear to narrowly elliptic or lanceolate, 7–80 mm long, 1–12 mm wide, planar, glabrous to minutely puberulent, margins often ciliolate, apices attenuate to spinulose. **HEADS** in cymiform to corymbiform, rarely racemiform clusters. **INVOLUCRES** obconic to subcylindric, 6.5–10 mm long, 3–5 mm wide. **PHYLLARIES** 50–60+, more or less spirally arranged, usually in 5–6 but rarely 7 series, tan, unequal, oblong to elliptic, apices acute to rounded, mostly chartaceous, midvein raised. **DISC FLORETS** usually 10–16, sometimes to 20, corollas 5.5–8 mm long; style branches 2.8–3.7 mm long. **CYPSELAE** reddish brown, cylindric, 4–6 mm, pubescent. **PAPPUS** tan, 6–7.5 mm long. $2n = 18$ [*Bigelowia menziesii* (Hook. & Arn.) A. Gray var. *scopulorum* M. E. Jones; *Isocoma scopulorum* (M. E. Jones) Rydb.; *Haplopappus scopulorum* (M. E. Jones) S. F. Blake]. Mountain slopes; piñon-juniper, mountain brush, rabbitbrush, and ponderosa pine communities. ARIZ: Apa, Nav; UTAH. 1200–2200 m (3935–7215′). Flowering: Aug–Oct. Arizona and Utah.

Chrysothamnus stylosus (Eastw.) Urbatsch, R. P. Roberts & Neubig (small, pillarlike, referent unclear) Resinbush. **STEMS** mostly 40–120 cm tall, woody, green when young, aging to tan or white to grayish with age, branched, resinous. **LEAVES** ascending to spreading, becoming deflexed, sessile, lanceolate to linear-elliptic, 6–60 mm long, to 9 mm wide, often folded, apices acute, faces glabrous, gland-dotted, resinous. **HEADS** dense cymiform arrays, not overtopped by distal leaves. **INVOLUCRES** hemispheric, 8–12 mm long, up to 15 mm wide. **PHYLLARIES** 40–60, in 3–5 series, in spirals, mostly tan, ovate to lanceolate, 3–10 mm long, up to 1.8 mm wide, greenish distally, apices acute or acuminate. **DISC FLORETS** mostly 30–40, corollas 6–8.5 mm long, style branches 2.8–3.5 mm. **CYPSELAE** tan, 4–5 mm long, narrowly cylindric, glabrous or sparsely hairy, resinous. **PAPPUS** 12–15 white or stramineous, lanceolate to lance-linear scales 2–4 mm long. $2n = 18$. [*Grindelia stylosa* Eastw.; *Vanclevea stylosa* (Eastw.) Greene]. Sand dune communities; often with ephedra, four-winged saltbush, sand dropseed, and blackbrush. ARIZ: Apa, Coc, Nav; UTAH. 1190–1700 m (3900–5575′). Arizona and Utah.

Chrysothamnus stylosus

Chrysothamnus vaseyi (A. Gray) Greene (for George Vasey, 1822–1893, botanist) Vasey's rabbitbrush. **STEMS** to 0.3 m tall, branched, bark tan or dark gray, becoming fibrous with maturity, young twigs ascending, green, becoming tan with age, glabrous to puberulent, resin-dotted to resinous. **LEAVES** ascending to spreading, sessile, blades linear to oblanceolate, 10–40 mm long, to 2.5 mm wide, planar or sulcate, rarely twisted, glabrous to puberulent, sometimes gland-dotted, often resinous, apices acute to apiculate, midvein faintly evident. **HEADS** in dense cymiform clusters, sometimes overtopped by distal leaves. **INVOLUCRES** cylindric to obconic, sometimes 5 mm but usually 6–8 mm long, 2–4 mm wide. **PHYLLARIES** 12–18, weakly aligned vertically in 3–4 series, graduated, mostly tan, greenish apically, ovate to elliptic, apices acute to obtuse-rounded, chartaceous, midvein slightly raised. **DISC FLORETS** 5–7, corollas 4.5–6.5 mm long; style branches 3–3.8 mm, appendages 0.8–1.2 mm. **CYPSELAE** 4–5 mm long, cylindric to turbinate, reddish brown, glabrous, distinctly nerved. **PAPPUS** tan, 3.5–5 mm long. $2n = 18$. [*Bigelowia vaseyi* A. Gray; *Ericameria vaseyi* (A. Gray) L. C. Anderson]. Open woods and dry meadows; sagebrush, mountain brush, and ponderosa pine communities. ARIZ: Nav; COLO: Arc, Min, Mon, SMg; NMEX: RAr. 1700–2900 m (5575–9515′). Flowering: Aug–Oct. Colorado, Nevada, New Mexico, Utah, and Wyoming.

Chrysothamnus viscidiflorus (Hook.) Nutt. (sticky-flowered) Yellow rabbitbrush. **STEMS** to 1.5 m tall, branched, bark whitish tan, becoming gray, flaky, and fibrous with maturity, twigs ascending, green, becoming tan at maturity, glabrous to puberulent, often resinous. **LEAVES** ascending, spreading, or deflexed, sessile, blades linear to lanceolate, to 10 mm wide, planar to sulcate, usually twisted, adaxially glabrous to puberulent, margins often undulate and ciliate, apices acute to apiculate, midvein visible, usually flanked by smaller collateral veins. **HEADS** in dense, rounded cymiform clusters. **INVOLUCRES** cylindric to obconic or campanulate, 4–7.5 mm long, 1.5–2.5 mm wide. **PHYLLARIES** 12–24, in 3–5 spirals or weak vertical arrays, graduated, mostly tan, linear-oblong, lanceolate to elliptic, or obovate to spatulate, adaxially glabrous or puberulent, apices acute to obtuse or rounded, sometimes apiculate, chartaceous, costa generally evident at least distally, margins scarious, eciliate or ciliolate to erose-ciliolate. **DISC FLORETS** 3–14, most frequently 4–5, corollas 3.5–6.5 mm long; style branches 2.2–3.2 mm long, extending beyond corolla at maturity. **CYPSELAE** tan to reddish brown, turbinate, 2.5–4.2 mm long, pubescent. **PAPPUS** tan, 3.5–6 mm. $2n = 18, 36, 54$. Arizona, California, Colorado, Idaho, Montana, Nebraska, Nevada, New Mexico, Oregon, South Dakota, Utah, Washington, and Wyoming. Hopi uses include: poultice of chewed plant tips applied to boils, used as a spice, used as a sand break to protect young corn and melons, yellow blossoms used as a yellow dye for wools and cotton yarn, and plant used for roasting corn. Navajo uses include: thatch to prevent the sand on top of the sweathouse from sifting through; orange flowers boiled with roasted alum and used as a light orange dye for leather, wool, and basketry; yellow flowers boiled with roasted alum and used as a yellow dye for leather, wool, and basketry; plant used as an emetic to make a sick person vomit.

1. Upper stems, and frequently the leaves, puberulent ..... (2)
1′ Upper stems glabrous and leaves glabrous ..... (3)

2. Stems and leaves densely puberulent near heads; leaves 1–2 (–4) mm wide ..... **subsp. *puberulus***
2′ Stems and leaves sparsely puberulent near heads; leaves over 2 mm wide ..... **subsp. *lanceolatus***

3. Leaves about 1 mm wide; florets 3–4 (–5) per head; involucres somewhat turbinate ..... **subsp. *axillaris***
3′ Leaves 1–10 mm wide; if 1 mm wide, then florets 4 or more and involucres narrowly cylindric.. **subsp. *viscidiflorus***

subsp. *axillaris* (D. D. Keck) L. C. Anderson (in leaf axils) Inyo rabbitbrush. **STEMS** to 0.5 m tall, glabrous. **LEAVES** green, filiform, less than 2 mm wide, 1-nerved, involute or terete, apices acicular, adaxially pubescent. **HEADS** in rounded cymiform clusters. **INVOLUCRES** turbinate to obconic, 5–6 mm long. **PHYLLARIES** 14–24, 4–5-seriate, forming 4–5 weak vertical ranks, broadly linear to linear-oblong, apices sharply acute or apiculate, spreading at maturity, margins scarious, eciliate, or distally erose-ciliolate, with a green to brown subapical patch, adaxially glabrous. **DISC FLORETS** 3–5; corollas 3.5–5 mm long. $2n = 18$. [*Bigelowia douglasii* A. Gray var. *stenophylla* A. Gray; C. *viscidiflorus* subsp. *stenophyllus* (A. Gray) H. M. Hall & Clem.; *Ericameria viscidiflora* (Hook.) L. C. Anderson var. *stenophylla* (A. Gray) L. C. Anderson; *C. axillaris* D. D. Keck; *E. viscidiflora* subsp. *axillaris* (D. D. Keck) L. C. Anderson]. Desert slopes in granitic sand; blackbrush, sagebrush, galleta grass, shadscale, and piñon-juniper woodland communities. ARIZ: Apa; COLO: Mon; NMEX: SJn. 1300–2000 m (4265–6560′). Flowering: Aug–Oct. Arizona, California, Colorado, Nevada, and Utah.

subsp. *lanceolatus* (Nutt.) H. M. Hall & Clem. (lance-shaped) Yellow rabbitbrush. **STEMS** to 0.5 m tall, puberulent on young twigs near heads. **LEAVES** bright green, linear to lanceolate, 2–6 mm wide, usually not twisted, puberulent abaxially, apices abruptly acute, 3- or 5-nerved. **HEADS** in small, compact, cymiform clusters. **INVOLUCRES** turbinate, 5–6.5 mm long. **PHYLLARIES** 14–18, in 3–4 series, spirally or vertically arranged, lacking subapical green patch, costa evident distally, oblong, unequal, apices often rounded, apiculate, adaxially puberulent. **DISC FLORETS** 5; corollas 5.5–6 mm. $2n = 18, 36$. [*C. lanceolatus* Nutt.; *Ericameria viscidiflora* (Hook.) L. C. Anderson subsp. *lanceolata* (Nutt.) L. C. Anderson]. Juniper/sagebrush savannas. ARIZ: Apa, Coc; COLO: Mon, SMg; NMEX: SJn. 1200–2500 m (3940–8200′). Flowering: Aug–Oct. Arizona, California, Colorado, Idaho, Montana, Nevada, New Mexico, Oregon, South Dakota, Utah, Washington, and Wyoming.

subsp. *puberulus* (D. C. Eaton) H. M. Hall & Clem. (with tiny hairs) Yellow rabbitbrush. **STEMS** to 0.5 m tall, gray-green, densely puberulent. **LEAVES** pale grayish green, narrowly linear, 1–2 mm long, blades often twisted, margins flat or involute, puberulent, apices acute, usually 1-nerved, sometimes proximally 3-nerved, densely puberulent adaxially.

HEADS in small, compact, cymiform clusters. **INVOLUCRES** turbinate, 5–7 mm long. **PHYLLARIES** 14–16, 3–4-seriate, arranged in weakly aligned vertical ranks, costa obscure or distally evident, oblong, unequal, margins scarious, entire or ciliolate, sometimes with subapical green patch, apices acute to mostly obtuse, adaxially puberulent. **DISC FLORETS** mostly 5, corollas 5–6 mm long. [*Linosyris viscidiflora* (Hook.) Torr. & A. Gray var. *puberula* D.C. Eaton; *C. viscidiflorus* subsp. *puberulus* (D. C. Eaton) H. M. Hall & Clem.; *C. marianus* Rydb.]. Rabbitbrush, black sagebrush, shadscale, piñon-juniper woodland, and ponderosa pine communities. ARIZ: Coc, Nav; COLO: Min; NMEX: SJn. 1500–3000 m (4920–9840′). Flowering: Aug–Sep. Arizona, California, Idaho, Nevada, Oregon, and Utah.

subsp. *viscidiflorus* Yellow rabbitbrush. **STEMS** to 1.2 m tall, glabrous. **LEAVES** spreading to deflexed, green, yellowish green or bluish green, linear to lanceolate, 1–10 mm wide, usually twisted, planar to sulcate, glabrous, margins often ciliate, apices acute, often apiculate. **HEADS** in congested cymiform clusters. **INVOLUCRES** obconic to narrowly cylindric, 5.5–7 mm long. **PHYLLARIES** 12–18, in 3 or less often 2–4 series, arranged in weak vertical ranks, often with subapical green patches, lanceolate to elliptic, costa evident or obscured, margins scarious, eciliate or ciliolate, adaxially glabrous. **DISC FLORETS** 4–14, corollas 5–6.5 mm long. $2n = 18, 27, 36, 45, 54$. Rabbitbrush, shadscale, sagebrush, piñon-juniper woodland, mountain brush, ponderosa pine, white fir, and aspen communities. ARIZ: Apa, Coc, Nav; COLO: Dol, LPl, Mon, SMg; NMEX: SJn; UTAH. 1450–3000 m (4755–9840′). Flowering: Aug–Oct. Arizona, California, Colorado, Idaho, Montana, Nebraska, Nevada, New Mexico, Oregon, South Dakota, Utah, Washington, and Wyoming.

Cichorium L. Chicory

Gary I. Baird

(Greek *kichoreia*, an ancient plant name of Arabic origin) Perennials (ours), taprooted, herbage glabrous to sparsely hispidulous, sap milky. **STEMS** usually 1, erect. **LEAVES** basal and cauline, alternate, sessile or petiolate, simple. **INFL** branched, spicate or simple, made up of numerous heads. **HEADS** liguliflorous. **INVOLUCRES** cylindric. **PHYLLARIES** mostly herbaceous, imbricate in 2 series (outer shorter), erect. **RECEPTACLES** flat, epaleate. **RAY FLORETS** matutinal; corolla blue or purple (rarely white). **DISC FLORETS** lacking. **PAPPUS** of truncate scales in 1 (2) series. **CYPSELAE** brownish, obconic, 3–5-angled, glabrous, striate-nerved, beakless. $x = 9$. Six species native to Eurasia and northern Africa, mostly in the Mediterranean region.

Cichorium intybus L. (purportedly from Coptic *Tobi*, the first month of the growing season in ancient Egypt, when this crop was traditionally eaten) **STEMS** to 100 cm tall, usually branched distally. **LEAVES** with blades oblanceolate to linear-lanceolate, basal 10–25 cm × 1–7 cm, margins runcinate to dentate or entire, cauline much reduced upward; lobes retrorsely deltate, denticulate on acroscopic margin (mostly); denticles few to many. **HEADS** (sub)sessile, borne 1–3 in axillary glomerules. **INVOLUCRES** about 10 mm long. **PHYLLARIES** green to purplish, margins usually scabrous-hispidulous, often with a few abaxial glandular setae (usually on distal midrib); outer narrowly ovate to lanceolate, basally cartilaginous; inner linear-lanceolate. **RAY FLORETS** about 10–20, outer exceeding involucre; corolla blue, tube 2–3 mm, ligule 10–15 × 4–5 mm; anthers 4–5 mm. **CYPSELAE** 2–3 mm. **PAPPUS** minute, less than 0.5 mm in length. $2n = 18$. Roadsides, ditch banks, disturbed places, and other ruderal habitats, often in moist, sandy soils. ARIZ: Apa, Nav; COLO: Arc, Dol, LPl, Min, Mon, SMg; NMEX: McK, RAr, SJn; UTAH. 1600–2370 m (5200–7770′). Flowering: May–Oct. Native to Europe and Asia, widely introduced in North America, South America, and Africa. Chicory has been grown since antiquity for its root, which is used as a coffee additive, and its leaves, which are used as salad greens and commonly called radicchio or, sometimes, endive. True endive, *C. endivia* L., is widely grown in North America but rarely escapes cultivation; it differs from *C. intybus* in being an annual-biennial plant with glomerules often 5-flowered, corollas purplish, achenes slightly smaller, and pappus less than 0.5 mm. There is currently significant interest in chicory root as a commercial source of inulin, which is used in the food industry as a source of soluble fiber and artificial sweeteners.

Cirsium Mill. Thistle

Patricia Barlow-Irick

(Greek *kirsion*, thistle) Biennial or perennial spiny herbs. **LEAVES** alternate, dentate or 1–3 times pinnatifid, lobes and leaf margins spine-tipped. **HEADS** discoid, solitary, racemose, or in small terminal clusters; involucral bracts in several series, some or all of them spine-tipped, and in many species with a glutinous dorsal ridge when the capitula are young. **RECEPTACLES** flat to subconic, densely soft-bristly. **FLORETS** tubular and perfect (rarely dioecious), corollas with a slender basal tube that expands with floret maturity, a more expanded throat, and elongate lobes divided by unequal sinuses; filaments usually papillose, anthers and filaments distinct, shaggy hyaline bases subtending the anther and opaque, free tips distal to the pollen-forming section of tube; style branches mostly adnate to each other and marked at their base by a thickened and papillose ring. **CYPSELAE** basifixed, glabrous, with an apical crown on the distal end and an angled attachment scar on the basal end. **PAPPUS** of numerous plumose bristles, deciduous in a ring. About 200 species in North America, Eurasia, and North Africa. The taxonomy of *Cirsium* is complicated by hybridization, infraspecific variation, and morphological changes through the growing season. Vegetative features like leaf lobing, spine size, and vestiture are generally too variable at the infraspecific level to be taxonomically useful. Overall branching patterns change over the season. Ants eat the glutinous dorsal ridges on phyllaries and apical rings on seeds. Therefore, only the floral characters have any phenotypic reliability. Two characters, style tip length and anther tip length, are important. *Cirsium* descriptors: style tip length is measured from the distal tip of the style to the thickened hairy node at the base of the branches; the anther tip length is from the distal free tips of the anther tube to the adnate sections of the opaque anther tip.

1. Plants either dioecious or leaf surfaces covered with tiny stiff prickles; invasive weeds ....(2)
1′ Native plants; not as above ....(3)

2. Dioecious; capitula less than 2 cm diameter; upper leaf surface glabrous to tomentose .... ***C. arvense***
2′ Heads perfect; capitula greater than 2 cm diameter; upper leaf surface with tiny prickles .... ***C. vulgare***

3. Middle phyllary margins fimbriate, erose, dilated with scarious edges, or obscured in dense trichomes ....(4)
3′ Outer and middle phyllaries entire (innermost may be scarious-erose) ....(8)

4. Corolla throats longer than 6 mm ....(5)
4′ Corolla throats shorter than 6 mm ....(6)

5. Pappus longer than 16 mm; corollas mostly longer than 22 mm .... ***C. scariosum***
5′ Pappus shorter than 16 mm; corollas mostly shorter than 22 mm .... ***C. centaureae***

6. Capitula in a thick and spikelike "inflorescence"; stem strongly ribbed .... ***C. scopulorum***
6′ Capitula separated from each other by stems and peduncles; stem without strong ribbing ....(7)

7. Corollas less than 20 mm long, white to yellow; lobes less than 5 mm long .... ***C. parryi***
7′ Corollas greater than 20 mm long, white to purple; lobes greater than 5 mm long .... ***C. wheeleri***

8. Corolla lobes less than 6 mm in length; plants of hanging gardens and other moist sites .... ***C. rydbergii***
8′ Corolla lobes more than 6 mm in length; plants of dry places ....(9)

9. Style tips longer than 3 mm and corolla lobes less than 13 mm ....(10)
9′ Style tips shorter than 3 mm and corolla lobes greater than 13 mm ....(14)

10. Phyllaries long-attenuate, linear, or narrowly lanceolate .... ***C. neomexicanum***
10′ Phyllary bases ovate to lanceolate ....(11)

11. Free anther tips (above the tube) less than 1 mm; corolla lobes shorter than 8 mm ....(12)
11′ Free anther tips longer than 1 mm; corolla lobes longer than 8 mm ....(13)

12. Corolla throat longer than 5 mm .... ***C. tracyi***
12′ Corolla throat shorter than 5 mm .... ***C. wheeleri***

13. Lower leaves decurrent, at least 5 mm; phyllary spines longer than 5 mm .... ***C. ochrocentrum***
13′ Lower leaves not decurrent; phyllary spines shorter than 5 mm .... ***C. undulatum***

14. Corollas bright red or carmine .... ***C. arizonicum***
14′ Corollas pink to purple ....(15)

15. Leaves with 15–25 sinuses between major lobes; stem and leaf midvein of underside with multicellular hairs .... ***C. chellyense***
15′ Leaves with less than 15 sinuses between major lobes; multicellular hairs absent ....(16)

16. Plants from below 1300 m (4265′) elevation; corolla lobes mostly longer than 13 mm .... ***C. calcareum***
16′ Plants from above 1300 m (4265′) elevation; corolla lobes mostly shorter than 13 mm .... ***C. bipinnatum***

Cirsium arizonicum (A. Gray) Petr. (of Arizona) Arizona thistle. Perennial. **STEMS** 1 to several, 5–15 dm tall from a caudex or short taproot; glabrous to tomentose, usually without septate trichomes. **LEAVES** to 45 cm long, 4–8 cm wide, unlobed to deeply divided, usually tomentose, at least abaxially, usually without septate trichomes; principal marginal spines 3–15 mm long, often stout; cauline leaves narrowed at base to truncate or clasping, but not or only slightly decurrent. **INFL** with several heads on short, often leafy-bracteate peduncular branches near the summit, sometimes crowded. **INVOLUCRES** cylindric maturing to campanulate, mostly 1.5–3 cm wide, 2–3.2 cm high. **PHYLLARIES** well imbricate in several series, firm, stramineous or partly greenish, or anthocyanic, glabrous or often inconspicuously tomentulose along the margins, spines of middle phyllaries 2–11 (15) mm, slender to very stout. **FLORETS** with corolla usually red or crimson, 24–34 mm; throat 2–5 mm; lobes 13–18 mm; style tips 1–2.5 mm.; anther tips 0.6–0.9 mm. **CYPSELAE** 3.5–6 mm long, brown. **PAPPUS** 17–25 (30) mm. $2n = 30$. [*C. nidulum* (M. E. Jones) Petr.; *C. arizonicum* (A. Gray) Petr. var. *nidulum* (M. E. Jones) S. L. Welsh]. Hanging garden, salt desert scrub, and pine-oak-juniper woodland communities. UTAH. 1125–1370 m (3700–4500′). Flowering: May–Oct. California and Nevada east to Utah and New Mexico.

Cirsium arvense (L.) Scop. (of cultivated fields) Canada thistle, creeping thistle. Perennial, dioecious, 5–12 dm, from creeping rootstocks. **STEMS** 1 to many, erect, glabrous to appressed gray-tomentose; leaves generally oblong in outline, 3–30 cm long, 1–6 cm wide, the principal spines 1–7 mm long, the abaxial surface glabrous to densely gray-tomentose, the adaxial surface green, glabrous to thinly tomentose; cauline leaves gradually reduced; upper cauline leaves becoming bractlike. **INFL** with heads in corymbiform or paniculiform arrangements. **INVOLUCRES** ovoid in flower, campanulate in fruit, 1–2 cm long, 1–2 cm diameter. **PHYLLARIES** strongly graduated in 6–8 series, outermost ovate, with a narrow glutinous ridge, innermost chartaceous and usually purple-tinged, the spine tips fine, 0–1 mm. **FLORETS** with corolla purple (white or pink); staminate florets 12–18 mm, surpassing the pappus; pistillate 14–20 mm, surpassed by pappus in fruit; staminate throats 1–1.8 mm; pistillate 0.5–1 mm; staminate lobes 3–5 mm; pistillate 2–3 mm; style tips 1–2 mm (pistillate plants may show gynoecium but have a vestigial ovary). **CYPSELAE** 2–5 mm long, brown. **PAPPUS** 13–32 mm long. $2n = 34$. [*Serratula arvensis* L.; *Breea arvensis* (L.) Less.; *Carduus arvensis* (L.) Robson; *Cirsium incanum* (S. G. Gmel.) Fisch. ex M. Bieb.; *C. setosum* (Willd.) Besser ex M. Bieb.; *C. arvense* (L.) Scop. var. *argenteum* (Peyer ex Vest) Fiori; *C. arvense* (L.) Scop. var. *horridum* Wimm. & Grab.; *C. arvense* (L.) Scop. var. *integrifolium* Wimm. & Grab.; *C. arvense* (L.) Scop. var. *mite* Wimm. & Grab.; *C. arvense* (L.) Scop. var. *vestitum* Wimm. & Grab.] Disturbed sites, fields, pastures, roadsides. ARIZ: Apa; COLO: Arc, Dol, Hin, LPl, Min, Mon, RGr, SJn, SMg; NMEX: RAr, SJn; UTAH. 1280–2600 m (4200–8530′). Flowering: Jun–Oct. Adventive from Eurasia.

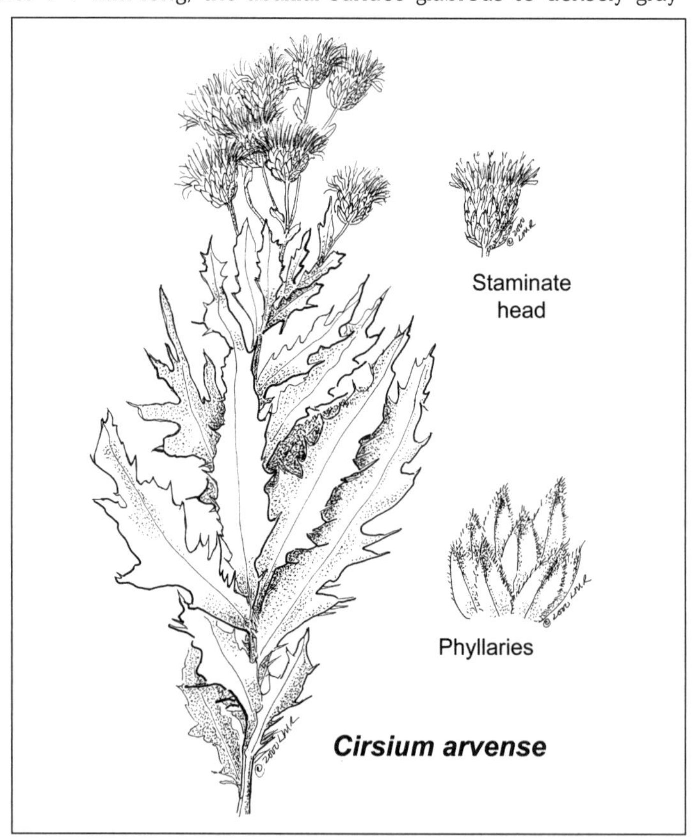

Cirsium arvense

Cirsium bipinnatum (Eastw.) Rydb. (doubly pinnate or feathered) Canyon thistle. Perennials, 5–15 dm tall, from taprooted caudex or runner roots. **STEMS** 1 to several, erect or ascending, glabrous to thinly arachnoid-tomentose with fine trichomes. **LEAVES** oblong-elliptic in outline, 3–40 cm long, 1–13 cm wide, the principal spines 3–7 mm long; the bases sessile, sometimes decurrent as spiny wings to 1.5 cm long or clasping; uppermost cauline leaves reduced or bractlike. **HEADS** 1–20+; peduncles 0–15 cm long. **INVOLUCRES** cylindric maturing to campanulate, 2.3–3 cm long, 1.5–3 cm wide. **PHYLLARIES** in 7–9 series, green, stramineous, or the inner reddish to rich reddish purple, sometimes inconspicuously tomentulose along the margin, spine tips 1–6 mm long, innermost tapering to a slender tip. **FLORETS** with corolla lavender or purple (white), 22–34 mm long; throat 5–9 mm long, not sharply differentiated from the tube; lobes 8–15 mm long; style tips 2–4.5 mm long; anther tips 0.6–0.9 mm long. **CYPSELAE** 3.5–6 mm long, brown. **PAPPUS** 18–25 mm long. [*Cnicus drummondii* (Torr. & A. Gray) A. Gray var. *bipinnatus* Eastw.; *Cirsium pulchellum* (Greene ex Rydb.) Wooton & Standl. subsp. *bipinnatum* (Eastw.) Petr.; *C. calcareum* (M. E. Jones) Wooton & Standl. var. *bipinnatum* (Eastw.) S. L. Welsh; *C. pulchellum* (Greene ex Rydb.) Wooton & Standl. subsp. *diffusum* (Eastw.) Petr.; *C. diffusum* (Eastw.) Rydb.; *C. pulchellum* (Greene ex Rydb.) Wooton & Standl.; *C. calcareum* (M. E. Jones) Wooton and Standl. var. *pulchellum* (Greene ex Rydb.) S. L. Welsh]. Canyons and rocky slopes in desert scrub, piñon pine–Gambel's oak–Utah juniper woodlands, and openings in ponderosa pine communities. ARIZ: Apa, Coc; COLO: Arc, Dol, LPl, Mon; NMEX: McK, RAr, SJn; UTAH. 1460–2620 m (4790–8595'). Flowering: Jul–Sep. Utah, Colorado, Arizona, and New Mexico. This taxon has been treated as *C. pulchellum* (Greene ex Rydb.) Wooton & Standl., but this epithet has priority. There is a population near Navajo Moutain, Utah, that is somewhat similar to *C. bipinnatum* but needs to be studied.

Cirsium calcareum (M. E. Jones) Wooton & Standl. (chalky) Calcareous thistle. Perennial with 1 to several stems, 5–15 dm tall from caudex or short taproot. **STEMS** glabrous to tomentose, usually without septate trichomes. **LEAVES** unlobed to deeply divided, glabrous to tomentose on 1 or both surfaces; principal marginal spines 3–10 mm long, slender to stout; cauline leaves often shortly decurrent as spiny wings to 10 mm long. **HEADS** several on short, often leafy-bracteate, peduncular branches. **INVOLUCRES** cylindric or ovoid maturing to campanulate; mostly 1.5–3 cm wide. 2.2–3.3 cm tall. **PHYLLARIES** with spines in middle 1–6 (9) mm, slender to very stout. **FLORETS** with corolla lavender to purple, 24–32 mm long; throat 4–7 mm long, not sharply differentiated from the tube; lobes 13–18 mm long; style tips 1.2–3 mm long; anther tips 0.7–0.9 mm long. **CYPSELAE** 3.5–6 mm long, brown. **PAPPUS** 19–28 mm long. [*Carduus calcareus* (M. E. Jones) A. Heller; *Cnicus calcareus* M. E. Jones]. Canyons and rocky slopes in desert scrub, piñon pine–Gambel's oak–Utah juniper woodlands, and openings in ponderosa pine forests. ARIZ: Apa, Nav; COLO: Arc, Mon; NMEX: RAr, SJn; UTAH. 1220–2440 m (4000–8000'). Flowering: Jun–Oct. Utah, Colorado, Arizona, and New Mexico. Used by the Ramah Navajo as a wash for sore eyes.

Cirsium centaureae (Rydb.) K. Schum. (a reference to the centaur Chiron, who was supposed to have discovered the medicinal uses of a plant in Greece called *Centaurea* and *Centaurium*) Rocky Mountain fringed thistle. Perennial herb from a simple caudex and taproot. **STEMS** 3–12 dm tall, arachnoid or glabrous. **LEAVES** pinnately or bipinnately divided, tomentose below; cauline leaves short-decurrent (less than 15 mm). **INFL** in a cymose arrangement, each head with more than 60 florets. **INVOLUCRES** ovoid maturing to campanulate, 1.5–3 cm, glabrous or thinly arachnoid-tomentose. **PHYLLARIES** strongly graduated to subequal in 5–6 series, linear to ovate, green or with a maroon to dark brown subapical patch or appendage, outer and middle phyllaries commonly dilated, scarious, and erose-toothed to fringed; spine tips 1–6 mm; apices of inner phyllaries narrow, dilated, scarious, and fimbriate, sometimes flexuous or reflexed. **FLORETS** with corolla white, purple, or lavender, 18–22 mm long; throat 6.5–7.5 mm long; lobes 4–5 mm long; style tips 4–5 mm long. **CYPSELAE** unknown. **PAPPUS** 14.5–15.5 mm long. $2n = 34, 36$. [*Cnicus carlinoides* Schrank var. *americanus* A. Gray; *Cirsium griseum* (Rydb.) K. Schum.; *C. laterifolium* (Osterh.) Petr.; *C. modestum* (Osterh.) Cockerell; *C. scapanolepis* Petr.; *C. spathulifolium* Rydb.]. Gambel's oak, sagebrush, grasslands, piñon-juniper woodlands, aspen, and openings in montane coniferous forests. ARIZ: Apa; COLO: Con, Hin, LPl, Min, RGr, SMg, SJn; NMEX: RAr; UTAH. 1900–2900 m (6235–9515'). Flowering: Jun–Sep. Arizona and New Mexico north to Utah and Wyoming. The correct epithet is in question; this is the most common and least ambiguous. The epithet "*americanum*" was used for varieties of *Cnicus* and *Carduus*, all of which were transferred to *Cirsium*.

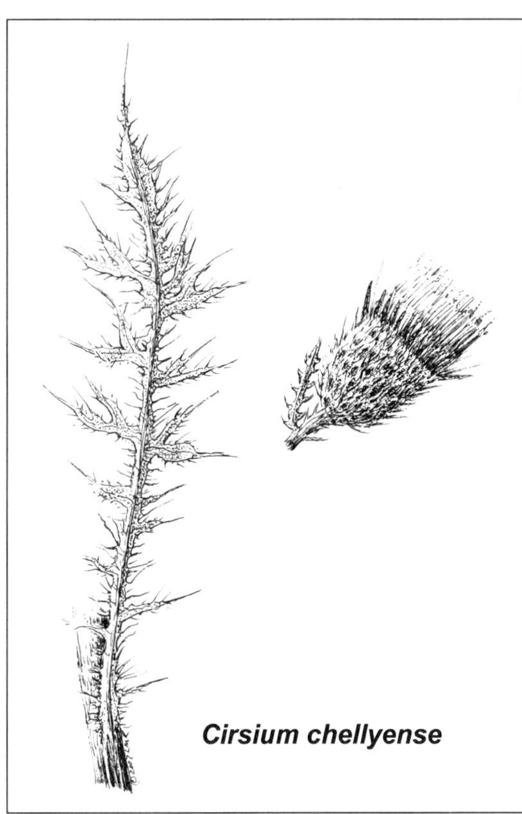
Cirsium chellyense

Cirsium chellyense R. J. Moore & Frankton (of Canyon de Chelly) Navajo thistle. Perennial with 1–several stems 5–15 dm tall from caudex or short taproot, may spread by creeping roots. **STEMS** glabrous or villous to tomentose with multicellular trichomes. **LEAVES** deeply divided into (10) 15–25 narrow lobes with narrow webbing between lobes, glabrous or villous below with multicellular trichomes; principal marginal spines 5–12 mm long, usually very slender; cauline leaves usually decurrent as spiny wings 12–30 mm.

HEADS 1–20+, pedunculate. **INVOLUCRES** cylindric to narrowly ovoid, 2.5–2.7 cm tall. **PHYLLARIES** with spines 6–12 mm long. **FLORETS** with corolla usually reddish purple, 24–32 mm long; throat 4–7.5 mm long; lobes 12–16 mm long; style tips 2.5–3.5 mm long; anther tips about 0.7 mm long. **CYPSELAE** 3.5 mm long, brown. **PAPPUS** 20–27 mm long. $2n = 34$. [*Cirsium chuskaense* R. J. Moore & Frankton; *C. navajoense* R. J. Moore & Frankton; *C. arizonicum* (A. Gray) Petr. var. *chellyense* (R. J. Moore & Frankton) D. J. Keil]. Canyon bottoms and sandstone slopes in desert scrub, desert grassland, piñon-juniper woodland, mountain brush–Gambel's oak, and ponderosa pine communities. ARIZ: Apa, Nav; NMEX: McK, SJn. 1650–2750 m (5415–9020′). Flowering: Jun–Sep. Arizona and New Mexico.

Cirsium neomexicanum A. Gray (of New Mexico) New Mexican thistle. Biennial, 4–29 dm, from taproot. **STEMS** usually 1, erect, thinly gray-tomentose; branches ascending. **LEAVES** oblong-elliptic to oblanceolate, 6–35 cm long, 1.5–7 cm wide, shallowly to deeply pinnatifid, the lobes generally rigidly spreading, undivided or with 1–2 pairs of coarse teeth or lobes, the principal spines 5–15 mm long, both surfaces gray-tomentose, sometimes glabrate; cauline leaves much reduced upward, the bases decurrent as spiny wings; uppermost scarcely more than a bract with long spines. **HEADS** 1 to many on large individuals; peduncles 5–30 cm, with bracts. **INVOLUCRES** shallowly hemispheric maturing to campanulate, 2–3 cm, 2.5–5 cm diameter, arachnoid to tomentose. **PHYLLARIES** in 7–10 series, linear to narrowly lanceolate, spine tips 4–15 mm, spreading to reflexed, without glutinous ridge. **FLORETS** with corolla white to pale lavender or pink, 18–27 mm long; throat 4–7 mm long; lobes 5–9 mm long; style tips 4–6 mm long; anther tips 0.6–1 mm long. **CYPSELAE** 5–6 mm long, dark brown. **PAPPUS** 15–20 mm long. $2n = 30, 32, 30+1$ [*C. utahense* Petr.; *C. neomexicanum* A. Gray var. *utahense* (Petr.) S. L. Welsh; *C. humboldtense* Rydb.; *C. arcuum* A. Nelson; *C. undulatum* (Nutt.) Spreng. var. *albescens* D. C. Eaton]. Blackbrush, sagebrush, and piñon-juniper woodland communities. ARIZ: Apa, Coc, Nav; COLO: Arc, Dol, LPl, Mon, SMg; NMEX: McK, RAr, SJn; UTAH. 1125–2100 m (3700–6890′). Flowering: Apr–Jul. California and Nevada east to Utah, Colorado, New Mexico, Arizona, and Texas; Mexico (Sonora). Plant used for chills and fever by the Navajo. Ramah Navajo uses include infusion of root used as a wash for eye diseases and for livestock with eye diseases; root used as a "life medicine."

Cirsium ochrocentrum A. Gray (with an ocher-colored center) Yellowspine thistle. Perennials, 3–9 dm tall. **STEMS** 1–5+ from base, erect or ascending, densely gray-tomentose. **LEAVES** oblong to narrowly elliptic in outline, 10–30 cm long, 2–8 cm wide, adaxially thinly gray-tomentose, abaxially densely white-tomentose, strongly undulate, the principal spines 5–20 mm, straw-colored; basal leaves usually present at anthesis, wing-petioled; cauline leaves progressively reduced up the stem, sessile, auriculate, or long-decurrent as spiny wings. **HEADS** 1 to few, terminal on branches or in leaf axils later in the season, peduncles 0–4 cm. **INVOLUCRES** ovoid maturing to broadly campanulate, 2.5–4.5 cm, often smaller in later capitula, 2.5–4.5 cm diameter, loosely arachnoid on phyllary margins or glabrate. **PHYLLARIES** graduated in 5–10 series, spines yellowish, spine tips 4–12 mm, spreading; apices of inner phyllaries often flexuous, scabrid-margined, or sometimes erose. **FLORETS** with corolla white or pale lavender to purple, pink, or red, 25–45 mm long; throat 8–11 mm long; lobes 7.5–13 mm long; style tips 4–6.5 mm long; anther tips 1–2.2 mm long. **CYPSELAE** 6–9 mm long, light brown. **PAPPUS** 20–40 mm long, white or tawny. $2n = 15, 16, 17$. [*Carduus undulatus* Nutt. var. *ochrocentrus* (A. Gray) Rydb.; *Carduus ochrocentrus* (A. Gray) Greene; *Cnicus ochrocentrus* A. Gray; *Cnicus undulatus* A. Gray var. *ochrocentrus* A. Gray]. Often along roadsides in desert grasslands, sagebrush, and piñon-juniper woodland communities. NMEX: McK, RAr, San. 1830–2200 m (6000–7220′). Flowering: Jun–Sep. Arizona, New Mexico, and Texas, northeast to Nebraska.

Cirsium parryi (A. Gray) Petr. (for Charles C. Parry, 1823–1890, botanist and explorer) Parry thistle. Biennial, 5–20+ dm tall, from taproot. **STEMS** 1, erect, puberulent to pilose with jointed trichomes, sometimes also thinly arachnoid; branches 0–many, ascending, often nodding at tips. **LEAVES** oblong to lanceolate or oblanceolate, 10–30 cm long, 2–5 cm wide, often only coarsely dentate or else shallowly to deeply pinnatifid, the lobes well separated, the principal spines slender to stout, 1–15 mm long, one or both surfaces thinly pilose, thinly arachnoid, or glabrate at maturity; cauline leaves only moderately reduced upward, the base wing-petioled or merely sessile in the basal leaves, sessile and auriculate-clasping to slightly decurrent in the upper leaves. **HEADS** 1 to many at tip of main stem and branches, often also in upper leaf axils, peduncles 0–4 cm, less than 60 florets per head. **INVOLUCRES** hemispheric to subspheric, 1.5–2.5 cm long, 1.5–3 cm diameter, glabrous to finely arachnoid and/or pilose, often long pilose-ciliate with arachnoid trichomes connecting adjacent phyllaries. **PHYLLARIES** in 5–8 series, linear to narrowly lanceolate, the outer often nearly as long as the inner, proximally greenish, distally darker, becoming brownish, the bodies entire to spiny-ciliate or with expanded, scarious, pectinately fringed terminal appendages, the spine tips 2–6 mm long, straight. **FLORETS** with corolla ochroleucous to yellow, 11–17 mm long; throat 2–4 mm long; lobes 3–5 mm long; style tips

2–4 mm long; anther tips 0.3–0.5 mm long. **CYPSELAE** 4–6 mm long, tan to dark brown, the apical collar narrow, not differently colored. **PAPPUS** 9–15 mm long. 2*n* = 34. [*Cnicus parryi* A. Gray; *Cirsium gilense* (Wooton & Standl.) Wooton & Standl.; *C. inornatum* (Wooton & Standl.) Wooton & Standl.; *C. pallidum* (Wooton & Standl.) Wooton & Standl.; *C. parryi* (A. Gray) Petr. subsp. *mogollonicum* Schaack and G. A. Goodwin]. Stream banks and other sites with damp soil; meadows in ponderosa pine, Douglas-fir, and spruce-fir communities. ARIZ: Apa; COLO: Arc, Hin, LPl, Min, Mon, RGr, SJn; NMEX: RAr, SJn. 2100–3700 m (6890–12140′). Flowering: Jul–Oct. Colorado south to Arizona and New Mexico.

Cirsium rydbergii Petr. (for Per Axel Rydberg, 1860–1931, botanist of the western United States) Rydberg's thistle, alcove thistle. Perennial, 10–30 dm tall, from caudex and taproot, spreading by creeping roots. **STEMS** 1 to several, erect or ascending to lax and hanging, glabrous or thinly tomentose; branches 0 or few, ascending. **LEAVES** elliptic, 30–90+ cm long, 10–40 cm wide, 1 to 2 times pinnately lobed, the lobes linear to ovate, strongly undulate, the principal spines slender, 5–15 mm long, both surfaces glabrous or thinly tomentose and soon glabrate, often glaucous; basal leaves present at anthesis, petioled or wing-petioled; lower cauline leaves wing-petioled, the middle much reduced, less deeply lobed, the base sessile, clasping, short-decurrent, the upper linear or lanceolate, bractlike, very spiny. **HEADS** few to many, erect or nodding in clusters at tips of upper branches; peduncles 0.5–6 cm. **INVOLUCRES** hemispheric, 1.4–2 cm, 1–2 cm diameter, glabrous or margins thinly tomentose. **PHYLLARIES** strongly graduated in 5–8 series, ovate to lance-oblong, the outer and middle with appressed bases and elongated, flattened, spreading or reflexed tips, green to brownish, the abaxial surface with or without a poorly developed glutinous ridge, the spine tips 3–25 mm long, slender; apices of inner phyllaries entire, straight. **FLORETS** with corolla dull white to pink or purple, 16–20 mm long; throat 4–6.5 mm long; lobes 4.5–6 mm long; style tips 2.5 mm long. **CYPSELAE** 3.7–4.5 mm long, gray or brown, the apical collar not differentiated. **PAPPUS** 10–15 mm long. 2*n* = 34. [*C. lactucinum* Rydb.]. Hanging gardens and seeps; rarely in canyons below them. ARIZ: Apa, Nav; UTAH. 1125–1525 m (3700–5005′). Flowering: May–Sep. Arizona and Utah.

Cirsium scariosum Nutt. (scarious) Elk thistle. Plants acaulescent to erect, 0–20 dm tall. **STEMS** 1 to few, usually very stout, leafy, glabrous or villous to tomentose with septate trichomes. **LEAVES** linear to oblong, oblanceolate, or narrowly elliptic, pinnately lobed or often unlobed, the upper narrow, often unpigmented toward base or tinged pink or purplish; adaxial surface glabrous or villous with septate trichomes, abaxial surface glabrous to gray-tomentose. **HEADS** 1–10+, sessile or short-peduncled, crowded at stem tips, usually subtended or overtopped by crowded upper leaves. **INVOLUCRES** 2–3.5 cm. **PHYLLARIES** with outer and middle lanceolate to ovate, the spine tips 1–8 mm long, slender to stout; tips of inner phyllaries acuminate and entire or abruptly expanded into a scarious, erose-toothed appendage. **FLORETS** with corolla white to purple, 20–36 mm long; throat 6–10 mm long; lobes 3.5–9 mm long; style tips 3–6.5 mm long. **CYPSELAE** 4.5–6.5 mm. **PAPPUS** 17–30 mm long. 2*n* = 34, 36. [*Carduus scariosus* (Nutt.) A. Heller; *Ca. butleri* Rydb.; *Ca. magnificus* A. Nelson; *Ca. drummondii* (Torr. & A. Gray) Coville var. *acaulescens* (A. Gray) Coville; *Ca. acaulescens* (A. Gray) Rydb., not Greene, 1893; *Ca. longissimus* A. Heller; *Ca. americanus* (A. Gray) Greene; *Ca. kelseyi* Rydb.; *Ca. erosus* Rydb.; *Ca. lacerus* Rydb.; *Cirsium hookerianum* Nutt. var. *scariosum* (Nutt.) B. Boivin; *Ci. butleri* (Rydb.) Petr.; *Ci. magnificum* (A. Nelson) Petr.; *Ci. minganense* Vict.; *Ci. acaulescens* (A. Gray) K. Schum.; *Ci. drummondii* Torr. & A. Gray var. *acaulescens* (A. Gray) J. F. Macbr.; *Ci. coloradense* (Rydb.) Cockerell ex Daniels subsp. *longissimum* (A. Heller) Petr.; *Ci. americanum* (A. Gray) K. Schum.; *Ci. acaule* Scop. var. *americanum* A. Gray; *Ci. kelseyi* Petr.; *Ci. erosum* (Rydb.) K. Schum.; *Ci. lacerum* Petr.; *Cnicus scariosus* A. Gray; *Cn. drummondi* (Torr. & A. Gray) A. Gray var. *acaulescens* A. Gray]. Moist, sometimes saline soils, meadows, ditches, stream banks, forest openings, sagebrush to spruce-fir communities. ARIZ: Apa; COLO: Arc, Con, Hin, LPl, Min, RGr, SJn; NMEX: RAr. 1310–2800 m (4300–9185′). Flowering: Jun–Sep. British Columbia to Montana and Idaho, south to California, Arizona, and Colorado. This taxon can occur in many forms and hybridizes readily with other taxa. Sometimes its upper leaves will be quite curly and feathery. Many authors recognize the acaulescent plants as a separate taxon, but both forms appear in single populations. The stems are edible.

Cirsium scopulorum (Greene) Cockerell **CP** (of the cliffs) Alpine thistle, horse's ice cream cone. Plants erect or nodding, 5–15 dm tall. **STEMS** strongly ribbed and mostly hollow; leaves oblong-lanceolate, to 15 cm long, glabrous or nearly so or abaxially finely arachnoid-tomentose and/or villous to tomentose with septate trichomes on one or both surfaces. **INFL** in a tight spikelike capitulescence, heads usually sessile, sometimes also sessile in upper leaf axils. **INVOLUCRES** 2–2.5 cm long, densely tomentose with septate trichomes on margins. **PHYLLARIES** 3–4 mm broad at the base, 25–35 mm long, tapering to a strong yellow spine, spines equaling or surpassing the corollas. **FLORETS** with corolla white, yellow, or pink to purple, 14–21 mm long; throat 3.5–5 mm long; lobes 3.5–7 mm long; style tips 4–5 mm long. **CYPSELAE** 5.5 mm long, dark brown. **PAPPUS** 8–17 mm long. [*Cnicus eriocephalus* A. Gray; *Carduus* *hookerianus* (Nutt.) A. Heller var. *eriocephalus* (A. Gray) A. Nelson; *Cirsium eriocephalum* A. Gray; *Cn. hesperius* Eastw.; *Ci. hesperium* (Eastw.) Petr.]. Rocky slopes; subalpine and alpine meadows. COLO: Arc, Con, Hin, LPl, Min, RGr, SJn. 2700–3810 m (8860–12500′). Utah, Colorado, and New Mexico.

Cirsium tracyi (Rydb.) Petr. (for Samuel Mills Tracy, 1847–1920, botanist and agronomist with the U.S. Dept. of Agriculture) Tracy's thistle. Perennial, 5–20+ dm tall, from taproot. **STEMS** 1 to several from base, erect or ascending, thinly gray-tomentose. **LEAVES** elliptic to oblong in outline, 8–40 cm, 1–12 cm wide, thinly tomentose above, densely gray-tomentose below, weakly to strongly undulate, the principal spines 2.5–7+ mm; cauline leaves progressively reduced up the stem, becoming sessile, auriculate-clasping to short-decurrent. **HEADS** 1 to many, terminal on branches and often in leaf axils, the capitulescence leafy; peduncles 0–10+ cm. **INVOLUCRES** ovoid maturing to broadly campanulate, 2–3 cm, 1.7–3.5 cm diameter, loosely arachnoid on phyllary margins or glabrate. **PHYLLARIES** graduated in 6–10 series, spine tips 2–6 mm; apices usually entire. **FLORETS** with corolla white to lavender or pink-purple, 23–30 mm long; throat 5.5–10.7 mm long; lobes 5.5–9.7 mm long; style tips 4–7 mm long; anther tips 0.5–1 mm long. **CYPSELAE** 6–7 mm long, light to dark brown. **PAPPUS** 20–23 mm long, white or tan. $2n = 24$. [*Carduus tracyi* Rydb.; *Cirsium acuatum* (Osterh.) Petr.; *Ci. floccosum* (Rydb.) Petr.; *Ci. undulatum* (Nutt.) Spreng. var. *tracyi* (Rydb.) S. L. Welsh]. Often in disturbed sites; sagebrush, mountain brush, piñon-juniper woodland, aspen, and Douglas-fir communities. COLO: Arc, Dol, LPl, Mon, SMg; NMEX: RAr, SJn; UTAH. 1400–2900 m (4595–9515′). Utah, Colorado, and New Mexico.

Cirsium undulatum (Nutt.) Spreng. (wavy-edged) Wavy-leaf thistle, gray thistle, prairie thistle. Short-lived perennials, 2–10 dm tall. **STEMS** 1 to several from base, erect or ascending, densely gray-tomentose. **LEAVES** elliptic to oblong or ovate in outline, 10–40 cm, 1–10 cm wide, adaxially thinly tomentose, abaxially densely gray-tomentose, strongly undulate, the principal spines 2–12+ mm long, yellowish; basal leaves sometimes present at anthesis, wing-petioled; cauline leaves progressively reduced up the stem. **HEADS** 1–10+, terminal on branches and in leaf axils; peduncles 0–25+ cm. **INVOLUCRES** ovoid maturing to broadly campanulate, 2.5–4.5 cm long, 1.5–4.5 cm diameter, loosely arachnoid on phyllary margins or glabrate. **PHYLLARIES** graduated in 8–12 series, ovate to lanceolate (outer) to linear-lanceolate (inner), spine tips 27 mm long, spreading; apices of inner phyllaries narrow, flat; entire. **FLORETS** with corolla lavender to pink, purple, or white, 24–40 mm long; throat 3.5–11 mm long; lobes 6.5–13 mm long; style tips 4–8.5 mm long; anther tips 0.75–1.75 mm long. **CYPSELAE** 6–7 mm long, light to dark brown. **PAPPUS** 25–38 mm, white or tawny. $2n = 26$. [*Carduus undulatus* Nutt.; *Cirsium megacephalum* (A. Gray) Cockerell; *Ci. undulatum* var. *megacephalum* (A. Gray) Fernald]. Desert scrub, sagebrush, piñon-juniper woodland, mountain brush, ponderosa pine, and aspen communities. COLO: Arc, LPl, Mon, SMg; NMEX: RAr, SJn; UTAH. 1400–2800 m (4595–9185′). Flowering: Jun–Oct. British Columbia to Minnesota, south to Missouri, New Mexico, and Arizona. Ramah Navajo uses include: decoction of root used for gonorrhea, cold infusion of root used as a wash for eye diseases and for livestock with eye diseases; root used as a "life medicine."

Cirsium vulgare (Savi) Ten. (common) Bull thistle, spear thistle. Biennial, 6–15 dm tall, from taproot. **STEMS** 1 to many, erect or ascending, villous with septate trichomes. **LEAVES** oblong-lanceolate to obovate, 15–40 cm, 6–15 cm wide, the principal spines 2–10 mm, abaxial surface villous with septate trichomes along the veins, adaxial surface green, covered with short appressed prickles; leaf bases decurrent, wings 20–35 mm long. **INFL** with few to many heads; peduncles 1–6 cm. **INVOLUCRES** hemispheric to campanulate, 3–4 cm long, 2–4 cm diameter, loosely arachnoid-tomentose, nonglandular. **PHYLLARIES** in 10–12 series, linear-lanceolate (outer) to linear (inner), with spines 2–5 mm long. **FLORETS** with corolla purple (rarely white), 25–35 mm long; throat 5–6 mm long; lobes 5–7 mm long; style tips 3.5–6 mm long. **CYPSELAE** 3–4.5 mm long, light brown with darker streaks. **PAPPUS** 25–28 mm long. $2n = 68$. [*Carduus vulgaris* Savi; *Ca. lanceolatus* L.; *Cirsium lanceolatum* (L.) Scop., non Hill]. Moderately invasive weed of disturbed sites, pastures, meadows, forest openings, roadsides. ARIZ: Apa, Nav; COLO: Arc, Dol, Hin, LPl, Min, Mon, SJn, SMg; NMEX: McK, RAr, San, SJn; UTAH. 1340–2200 m (4395–7220′). Flowering: Jun–Sep. A European introduction that is now widespread.

Cirsium wheeleri (A. Gray) Petr. (for George M. Wheeler, 1842–1905, soldier and topographical engineer) Wheeler's thistle. Perennial, slender, 1.5–6 dm. **STEMS** 1 to few, erect, closely gray-tomentose. **LEAVES** lanceolate to narrowly elliptic, 10–25 cm, 1–4 cm wide, variously lobed, the principal spines slender, 2–5 mm, abaxial surface gray-tomentose, adaxial surface green, glabrous to thinly tomentose; cauline leaves progressively reduced upward, not or scarcely decurrent, sometimes the upper weakly clasping; uppermost leaves often reduced to bracts. **HEADS** 1–6; peduncles 0–10 cm. **INVOLUCRES** hemispheric to subcylindric, 1.5–2.2 cm long, 1.5–2.5 cm diameter, thinly floccose-tomentose or glabrate. **PHYLLARIES** in 6–9 series, lanceolate (outer) to linear-lanceolate (inner), pale green with darker tips, brownish when dry, the abaxial surface with a narrow glutinous ridge, spines 3–7 mm, slender; apices of inner phyllaries scarious, often purplish, flexuous, tapered or expanded, erose to pectinate-fringed. **FLORETS** with corolla white or pink to pale purple, 20–28 mm long; throat 3–5 mm long; lobes 5–10 mm long; style tips 2.5–6 mm long; anther tips 0.5 mm long. **CYPSELAE** 6.5–7 mm long, straw-colored with brownish streaks, the apical collar colored like the body. **PAPPUS** 15–20 mm long. $2n = 28$. [*Cnicus wheeleri* A. Gray; *Cirsium blumeri* Petr.; *Ci. olivescens* (Rydb.) Petr.; *Ci. perennans* (Greene) Wooton & Standl.]. Mountain brush, piñon-juniper woodland, white fir, aspen, and spruce-fir communities. ARIZ: Apa; COLO: Arc, LPl, Mon; NMEX: McK, RAr, SJn; UTAH. 2100–2900 m (6890–9515′). Flowering: Jul–Oct. Utah and Colorado south to Arizona, New Mexico, and Texas; Chihuahua, Mexico.

Conyza Less. Horseweed

G. L. Nesom

(Greek *konis*, dust, from the use of powdered plants to repel insects) Annual or perennial herbs, nearly glabrous to coarsely hispid-pilose. **LEAVES** linear to oblanceolate, entire to toothed or pinnatifid. **INFL** a broadly ellipsoid to columnar head, sometimes somewhat flat-topped. **PHYLLARIES** in 2–4 subequal to slightly graduate series, forming a hypanthium-like cup and appearing inserted on it, the outer usually with 3 orange-resinous nerves. **RAY FLORETS** pistillate, fertile, in numerous series, the corollas eligulate or with a lamina that barely exceeds phyllary length. **DISC FLORETS** perfect, fertile. **CYPSELAE** oblong, flattened, with 2 thin, marginal nerves, glabrous to sparsely strigose, eglandular. **PAPPUS** a single series of barbellate bristles, uncommonly with a short outer series of setae. $x = 9$. Sixty to 100 species, primarily from tropical and subtropical America, including a number of species that are cosmopolitan colonizers (Noyes 2000).

Conyza canadensis (L.) Cronquist (from Canada) Horseweed. Plants erect, (3–) 50–200 (–350) cm. **LEAVES** usually evenly and densely arranged on larger plants; margins, faces, and veins sparsely hispid to strigose-hispid; proximal blades oblanceolate to linear, 20–50 (–100+) × 4–10 (–15) mm, toothed to entire, distal similar, smaller, entire. **HEADS** usually in paniculiform, densely cylindric to pyramidal arrays, corymbiform and more open when smaller. **PHYLLARIES** usually glabrous, sometimes sparsely strigose, with a broad, median, orange resin duct. **RECEPTACLES** 1–1.5 (–3) mm diameter in fruit. **PISTILLATE FLORETS** 20–30 (–45+); corollas ± equaling or surpassing styles, laminae 0.3–1 mm. **PAPPUS** of whitish bristles 2–3 mm. $2n = 18$. [*Erigeron canadensis* L.]. Stream and lake shores, wetlands, seeps, washes, dunes, alluvium in canyon bottoms, roadsides and other disturbed sites, riparian (cottonwood, saltcedar, cattail), mountain brush, piñon-juniper, ponderosa pine–Douglas-fir, and aspen–white fir communities. ARIZ: Apa, Nav; COLO: Arc, Dol, Hin, LPl, Min, Mon; NMEX: McK, RAr, San, SJn; UTAH. 1500–2600 m (4920–8530′). Flowering: Jun–Oct (–Nov). Canada (British Columbia to Quebec and Prince Edward Island); widespread throughout the United States, Mexico, and Central America; native to North America, introduced and often extremely abundant in South America, Europe, Asia, and Africa. Plants with glabrous stems and red-tipped phyllaries are sometimes treated as **var. *pusilla*** (Nutt.) Cronquist; similar plants with glabrous stems and stramineous (not red-tipped) phyllaries are sometimes treated as **var. *glabrata*** (A. Gray) Cronquist. Very young and small plants (5–10 cm) may flower anytime in the season, bearing only 1–5 heads.

Cosmos Cav. Cosmos

Kenneth D. Heil

(Greek *kosmos*, harmoniously ordered universe, or *kosmo*, ornament) Annual (ours), up to 25 dm tall. **STEMS** mostly 1, leafy, erect or ascending, branched distally or ± throughout. **LEAVES** opposite, mostly cauline, petiolate or sessile; blade mostly 1–3-pinnately lobed, margins usually entire, glabrous, hispid or puberulent. **HEADS** radiate, borne singly or in corymbiform arrays. **CALYCULI** of 5–8 basally connate, ± linear to subulate bractlets. **INVOLUCRES** hemispheric, 3–15 mm diameter. **PHYLLARIES** persistent, 5–8 in ± 2 series, membranous or herbaceous margins ± scarious. **RECEPTACLES** flat, paleate. **RAY FLORETS** 0–8, corollas white to pink, purple, or yellow to red-orange. **DISC FLORETS** 10–20 (–80+), bixexual, fertile; corollas yellow or orange. **CYPSELAE** dark brown to black, slender, quadrangular-cylindric or fusiform. **PAPPUS** persistent, of mostly 2–4 retrorsely barbed awns. $x = 12$ (Flora of North America Editorial Committee 2006). About 26 species widely introduced throughout the United States; tropical and subtropical America, especially Mexico.

Cosmos parviflorus (Jacq.) Pers. (small-flowered) Southwestern cosmos. **STEMS** mostly 30–80 cm, glabrous to sparsely pubescent. **LEAVES** with petioles about 0.5 cm long; blades 2.5–6.5 cm long, lobes to 1 mm wide, margins usually spinulose-ciliate, apices acute to obtuse. **PEDUNCLES** 10–30 cm long. **CALYCULI** with spreading to reflexed, linear-oblong to narrowly lanceolate bractlets, 6–9 mm long, apices acute. **INVOLUCRES** mostly 6–7 mm diameter. **PHYLLARIES** erect, oblong, 5–8 mm long, apices acute. **RAY FLORETS** with white to rose-pink or violet corollas, 5–9 mm long. **DISC FLORETS** with corollas 4–5 mm long. **CYPSELAE** 9–16 mm long. **PAPPUS** of 2–4 erect awns, 2–3 mm long. $2n = 24$. Disturbed and cultivated areas; piñon pine and ponderosa pine communities. One collection along highway between Bayfield and Ignacio, Colorado. COLO: LPl. 2070 m (6800′). Flowering: Jul–Sep. Colorado south to New Mexico, Arizona, and Texas; Mexico. A cold infusion of dried leaves used as a ceremonial chant lotion by the Ramah Navajo.

Crepis L. Hawksbeard

Gary I. Baird

(Greek *krepis*, base or foundation, a plant name used by Theophrastus, the application unclear) Perennials (ours), taprooted, often with a caudex, herbage more or less glabrous to pubescent, stipitate-glandular or not, sap milky. **STEMS** generally 1–3, branching distally, erect, usually striate. **LEAVES** basal (rosulate) and cauline (reduced), simple; basal petiolate, clasping, blade oval-elliptic to oblanceolate, margins entire (rare) or dentate to pinnately or runcinately lobed, apices acute to acuminate; cauline alternate, (sub)sessile, nonclasping, blade narrowly elliptic to linear-lanceolate. **INFL** 2–60 heads, borne in corymbiform to paniculiform clusters, erect. **HEADS** liguliflorous, peduncles naked. **INVOLUCRES** calyculate. **PHYLLARIES** herbaceous, greenish to blackish, erect, outer much reduced, inner subequal in 1–2 series, midrib often keeled in fruit. **RECEPTACLES** flat, epaleate. **RAY FLORETS** matutinal to diurnal; corolla ligulate, yellow, not (or rarely) reddish abaxially. **DISC FLORETS** lacking. **CYPSELAE** monomorphic (ours), yellow, light to dark brown, or black, fusiform to columnar, usually curved, glabrous, often roughened, strongly 10–20-ribbed, apex tapered, beakless (ours). **PAPPUS** of numerous bristles, persistent, white, ± barbellulate. $x = 11$. About 200 species in Eurasia, North America, and Africa (Babcock 1947a and 1947b). As a whole, *Crepis* is considered a taxonomically challenging genus. Our native species generally consist of sexually reproducing diploid populations whose boundaries are compromised by asexually reproducing (apomictic) polyploid hybrids. The line separating some of the apomicts in these species complexes sometimes appears both artificial and arbitrary, as they can combine various characteristics of the different parental taxa, making the identification of individual specimens difficult.

1. Basal leaves glabrous, often glaucous, entire to dentate or shallowly lobed; inner phyllaries mostly 12–16; florets 20–40 .... ***C. runcinata***

1′ Basal leaves pubescent, not glaucous, pinnately to runcinately lobed; inner phyllaries mostly 6–12; florets 8–20 .... (2)

2. Basal leaves runcinately lobed, lobes dentate or lobulate; inner phyllaries villous-tomentose, stipitate-glandular; outer phyllaries to 1/2 length of inner .... ***C. occidentalis***

2′ Basal leaves pinnately lobed, lobes entire or weakly dentate; inner phyllaries mostly glabrous to tomentose, setae lacking or, if present, nonglandular; outer phyllaries mostly less than 1/4 length of inner .... (3)

3. Involucres usually villous-tomentulose and with at least some blackish, nonglandular setae; lobes of basal leaves often denticulate or lobate; cauline leaves often pinnately lobed .... ***C. ×intermedia*** (see under ***C. acuminata***)

3′ Involucres villous-tomentulose only on margins or tips of phyllaries, and setae lacking; lobes of basal leaves mostly entire; cauline leaves mostly entire .... ***C. acuminata***

Crepis acuminata Nutt. (tapered or pointed, referring to the leaf tip) Longleaf hawksbeard. Plant bases usually pale, not yellow; caudex more or less shallow, not clothed with long-persistent leaf bases. **STEMS** 30–65 cm, more or less villous-tomentulose proximally, usually glabrous or glabrate distally, branched distally (first branch usually midstem or above). **LEAVES**: basal blades elliptic, 10–30 cm × 10–20 mm (excluding lobes), margins pinnately lobed, apices acuminate, faces villous-tomentulose; lobes 5–8 pairs (ours), linear-lanceolate, to 35 mm long (the middle pairs the longest, proximal and distal pairs progressively shorter), spreading, margins entire (rarely weakly dentate); cauline leaves progressively reduced upward, elliptic-lanceloate, margins often entire. **HEADS** about 10–30 (or more) per stem, lower peduncles and branches much exceeding terminal head. **INVOLUCRES** cylindric-campanulate, 10–12 mm tall. **PHYLLARIES**: outer less than 1/4 length of inner (rarely to 1/2 length of inner), deltate, margins villous-tomentulose; inner about 6–8, lanceolate, green or slightly blackish distally, mostly glabrous except villous-tomentulose apex, margins scarious, midrib keeled in fruit. **RAY FLORETS** about 10 per head (ours), outer exceeding involucre; corolla tube 5–6 mm, scaberulous, ligule about 10 mm × 2–3 mm, glabrous; anthers about 3 mm. **CYPSELAE** 6–9 mm long, pale yellowish brown to tan, 10–12-ribbed.

PAPPUS bristles numerous, 6–9 mm, white. $2n$ = 22, 33, 44, 55, 88. [*Psilochenia acuminata* (Nutt.) W. A. Weber]. Piñon-juniper, mountain brush, oak, ponderosa pine, and aspen communities. COLO: Mon, SMg; UTAH. 2400–2600 m (7800–8500′). Flowering: Jun–Aug. Washington to Iowa, south to California, Arizona, and New Mexico. The polyploid, apomictic hybrid ***C. ×intermedia*** A. Gray exhibits a mix of features from both *C. acuminata* and *C. occidentalis*. In general, it will key here to *C. acuminata* but can be (often arbitrarily) separated from that species in the following ways: leaf lobes usually denticulate or lobate (vs. entire); cauline leaves usually pinnately lobed (vs. usually entire); inner phyllaries more or less villous-tomentulose (vs. only at tip) and with (at least a few) blackish, non-glandular setae distally on midrib (vs. lacking); and fruits yellow to golden (vs. pale yellowish brown to tan).

Crepis occidentalis Nutt. (western) Western hawksbeard. Plant bases often yellow; caudex often deep-seated, clothed with long-persistent leaf bases. **STEMS** 10–30 cm tall, villous-tomentose and usually stiptate-glandular (sometimes glands restricted to heads only), branched throughout (first branch often well below midstem). **LEAVES**: basal blades elliptic, 9–18 cm × 3–22 mm (excluding lobes), margins runcinately lobed to pinnatifid, apices acuminate, faces villous-tomentose and stipitate-glandular (rarely nonglandular); lobes about 8–10 (or more) pairs, lance-deltate, to 17 mm long (middle pairs the longest, proximal and distal pairs progressively shorter), mostly retrorse, strongly dentate to lobate on acroscopic margin; cauline leaves progressively reduced upward, elliptic, pinnately lobed or dentate (uppermost may be entire). **HEADS** 3–15 per stem, lower peduncles and branches less than to just exceeding terminal head. **INVOLUCRES** cylindric, 11–16 mm tall. **PHYLLARIES**: outer to 1/2 length of inner, deltate, tomentose; inner about 7–9, lanceolate, blackish green, villous-tomentose and long stipitate-glandular (stipes blackish and glands yellowish, especially on midrib), margins scarious, midrib keeled (especially proximally). **RAY FLORETS** about 15–20 per head (ours), outer exceeding involucre; corolla tube 5.5–7.5 mm long, scaberulous, ligules 12–14 mm × 2–3 mm, glabrous; anthers about 6 mm long. **CYPSELAE** about 7 mm long, gold-yellow to black-brown, 10–14-ribbed. **PAPPUS** bristles numerous, 8–11 mm, white. $2n$ = 22, 33, 44, 55, 66, 77, 88. [*Psilochenia occidentalis* (Nutt.) Nutt.]. Sagebrush, saltbush-greasewood, grassland, piñon-juniper, and ponderosa pine–oak woodland communities. ARIZ: Apa, Coc; COLO: Arc, LPl, Mon; NMEX: RAr, SJn; UTAH. 1900–2400 m (6100–8000′). Flowering: May–Jun. British Columbia to Saskatchewan, south to California, Arizona, and New Mexico. This is the most common species in our range, where it apparently consists only of polyploid, apomictic populations. Four subspecies (or varieties) are traditionally recognized but their separation is difficult and often arbitrary. All of our material appears assignable (more or less) to **subsp. *costata*** (A. Gray) Babc. & Stebbins (ribbed). Uncommon, eglandular (but with a few blackish setae on the phyllaries) specimens have been found within our area (COLO: Dol, SMg); these may represent transitional forms with *C. ×intermedia*.

Crepis runcinata (E. James) Torr. & A. Gray (sawlike, alluding to the toothed leaf margins) Naked-stem hawksbeard. Plant bases pale or green, not yellow; caudex shallow, not clothed with long-persistent leaf bases. **STEMS** 15–50 cm tall, glabrous or glabrate (remaining sparsely puberulent at nodes) or rarely villous-pubescent distally, scapiform, branched distally (first branch usually midstem or above). **LEAVES**: basal blades oblanceolate, obovate, or spatulate, 2–18 cm × 3–30 mm (including lobes), margins entire to weakly dentate or lobate, occasionally runcinate, apices acute to round-obtuse; lobes (or teeth) mostly 3–5 pairs, often remote, deltate to linear-lanceolate, to 7 mm long (the middle or upper pairs the longest, or all subequal), mostly spreading, entire; cauline leaves much reduced (usually bract-like and inconspicuous), linear, entire. **HEADS** 2–14 per stem, lower peduncles and branches less than to just exceeding terminal head. **INVOLUCRES** turbinate-campanulate, 8–10 mm tall. **PHYLLARIES**: outer to 3/4 length of inner (usually much shorter), lance-deltate, glabrous or villous-ciliate; inner about 12–16, lanceolate, green or blackish green, glabrous or villous-ciliate distally, eglandular or stipitate-glandular (stipes and glands pale), margins scarious, midrib keeled proximally. **RAY FLORETS** about 20–50 per head, outer exceeding involucre; corolla tube 3–6 mm, glabrous, ligules 7–8 mm × 1–2 mm, glabrous; anthers about 4 mm long. **CYPSELAE** 4–8 mm, mostly yellow-brown, 10–12-ribbed. **PAPPUS** bristles numerous, 4–8 mm, white. $2n$ = 22. [*Hieracium runcinatum* E. James]. Seeps, springs, river bottoms, and other wet (or drying) meadow and riparian habitats, commonly in saline or alkaline soils. Flowering: May–Jun. British Columbia to Manitoba, south to California, Texas, and Chihuahua. Apparently not common in our area. A number of poorly defined varieties (sometimes treated as subspecies) have been proposed. The following key can be used to separate the three variants occurring in our area.

1. Phyllaries stipitate-glandular .... **var. *runcinata***
1′ Phyllaries eglandular .... (2)

2. Leaves obovate to spatulate, 5–10 cm long (or longer), 1.5–3 cm wide (or wider) .... **var. *glauca***
2′ Leaves narrowly oblanceolate, less than 5 cm long, less than 1 cm wide .... **var. *barberi***

var. *barberi* (Greenm.) B. L. Turner (for Charles M. Barber, an American botanist, 1876–?) Barber's hawksbeard. **LEAVES**: blades narrowly oblanceolate, 2–4 cm × 3–10 mm, margins entire to runcinately lobed, faces glaucous. **HEADS** 2–4 per stem. **PHYLLARIES**: outer much shorter than inner; inner glabrous (except tip), eglandular. **RAY FLORETS** about 20 per head. **CYPSELAE** 5–7 mm long. **PAPPUS** 7–8 mm long. Saline seeps near the San Juan River. NMEX: SJn. 1500 m (5000′). Southern Nevada to New Mexico, south to Chihuahua. This variant appears most similar to var. *glauca* but is usually distinguished by its narrower leaves, fewer heads per stem, and the features of its fruit.

var. *glauca* (Nutt.) B. Boivin (bluish or grayish, alluding to the foliage) Smooth hawksbeard. **LEAVES**: blades oblanceolate to spatulate, 6–14 cm × 15–30 mm, margins entire to runcinately lobed, faces glaucous. **HEADS** 4–7 per stem. **PHYLLARIES**: outer less than 1/2 length of inner; inner glabrous (except tip), eglandular. **RAY FLORETS** 20–30 (or more) per head. **CYPSELAE** 4–6 mm long. **PAPPUS** 5–6 mm long. [*Crepidium glaucum* Nutt.; *Psilochenia runcinata* (E. James) Á. Löve & D. Löve subsp. *glauca* (Nutt.) Á. Löve & D. Löve]. Saline seeps and springs. COLO: Arc; NMEX: SJn. 1500–1700 m (5100–5500′). Alberta to Manitoba, south to Arizona, New Mexico, and Texas.

var. *runcinata* **LEAVES**: blades oblanceolate, 5–8 cm × 7–15 mm, margins entire to dentate, faces mostly glabrous, not glaucous. **HEADS** 3–4 per stem. **PHYLLARIES**: outer less than 1/2 length of inner; inner stipitate-glandular. **RAY FLORETS** 20–30 (or more) per head. **CYPSELAE** 4–8 mm long. **PAPPUS** 4–6 mm long. [*Crepis aculeolata* Greene]. Moist to dry (seasonal?) meadows. COLO: Arc, LPl; NMEX: RAr. 2500 m (7600′). British Columbia to Manitoba, south to Nevada, New Mexico, and Nebraska.

Cyclachaena Fresen. Sumpweed, False Ragweed

Gary I. Baird

(Greek *cyclo*, circular, + *achenium*, achene, perhaps referring to the position of the cypselae in the head) Annuals, taprooted, rhizomes lacking, herbage scabrellous-strigillose to sericeous and gland-dotted throughout (or proximally glabrate), sap clear. **STEMS** 1, erect, branched distally. **LEAVES** cauline, simple, proximally opposite (sometimes nearly all), distally alternate, long-petiolate; blades ± ovate to deltate, rhombic, or suborbicular, 3- or 5-nerved (rarely larger leaves 3–5-lobed), bases cuneate to shallowly angular-cordate (larger leaves), margins coarsely dentate-serrate (often double), apices acute to acuminate. **INFL** in ebracteate paniculate clusters (with 1–3 heads per node), nodding; sessile or obscurely pedunculate. **HEADS** disciform. **INVOLUCRES** broadly campanulate to subrotate; receptacle ± conic, paleate; paleae spatulate to linear, sometimes lacking, membranous; calyculi lacking. **PHYLLARIES** 10–12 in ± 2 aligned series, distinct, ± subequal, strigillose (especially outer) and gland-dotted (especially inner), outer herbaceous, inner ± membranous, clasping the adjacent cypselae. **PISTILLATE FLORETS** 5; corollas whitish, tubular, inconspicuous, or lacking. **STAMINATE FLORETS** about 5–20; corollas whitish (lobes often blackish), funnelform, gland-dotted; filaments connate, anthers distinct. **CYPSELAE** monomorphic, blackish or brownish, obovoid, ± obcompressed, glabrous (rarely sparsely pubescent and/or gland-dotted distally), minutely striate. **PAPPUS** lacking. $x = 18$. One species of North America, introduced in Europe.

Cyclachaena xanthiifolia (Nutt.) Fresen. (*Xanthium*-like leaves) Giant sumpweed. Perennials 30–200 cm. **LEAVES** with petioles 1–10 cm long; blades about 5–15 × 2–15 mm, reduced upward. **HEADS** with peduncles rarely to 5+ mm; paleae about 2 mm. **INVOLUCRES** 2–3 mm long. **PHYLLARIES** obovate to suborbicular, apices acuminate to cuspidate. **CYPSELAE** 2–3 mm. $2n = 36$. [*Iva xanthiifolia* Nutt.]. Commonly of ruderal sites, roadsides, cultivated or abandoned fields, or other disturbed habitats. COLO: Arc, Dol, LPl, Mon, SJn; NMEX: RAr, San, SJn; UTAH. 1600–2700 m (5200–9000′). Flowering: Aug–Oct. Possibly native to the central prairies, now established over much of North America.

Dicoria Torr. & A. Gray Twinbugs

Gary I. Baird

(Greek *di*, two, + *koris*, bug, alluding to the cypselae) Annuals or perennials (or subshrubs), taprooted, rhizomes lacking, herbage sparsely to densely strigillose (and hispidulous), usually also gland-dotted or stipitate-glandular throughout, sap clear. **STEMS** 1, erect, ± intricately branched, sometimes becoming ± woody proximally. **LEAVES** cauline, simple, alternate (proximal opposite), petiolate; blades lanceolate to lance-linear, lance-ovate, or ± elliptic, usually 3-nerved, bases cuneate, margins entire or dentate, apices acute to acuminate. **INFL** in ± racemiform to paniculiform clusters (sometimes 2–3 heads occur at the same node), nodding; sessile or peduncles obscure. **HEADS** disciform or discoid. **INVOLUCRES** campanulate to rotate. **PHYLLARIES** about 5–10 in 2 series, distinct, outer herbaceous, persistent, inner membranous, each subtending a pistillate floret, usually accrescent and tardily deciduous. **RECEPTACLES** convex, paleate or epaleate; paleae linear, membranous, usually pubescent distally, or lacking; calyculi lacking. **PISTILLATE**

FLORETS 1–4 (or lacking), marginal in head, corollas lacking. **STAMINATE FLORETS** 5–15, central in head, corollas whitish. **CYPSELAE** monomorphic, blackish purple (or brownish), ± lustrous, obcompressed and often slightly involute (or revolute), obovate to elliptic, margins corky-winged and/or toothed (wing ± lacerated or divided), faces smooth or warty, ± sparsely strigillose, gland-dotted (especially distally), apices often with a white tuft of pubescence. **PAPPUS** lacking. $x = 18$. Two species of western North America and northwestern Mexico.

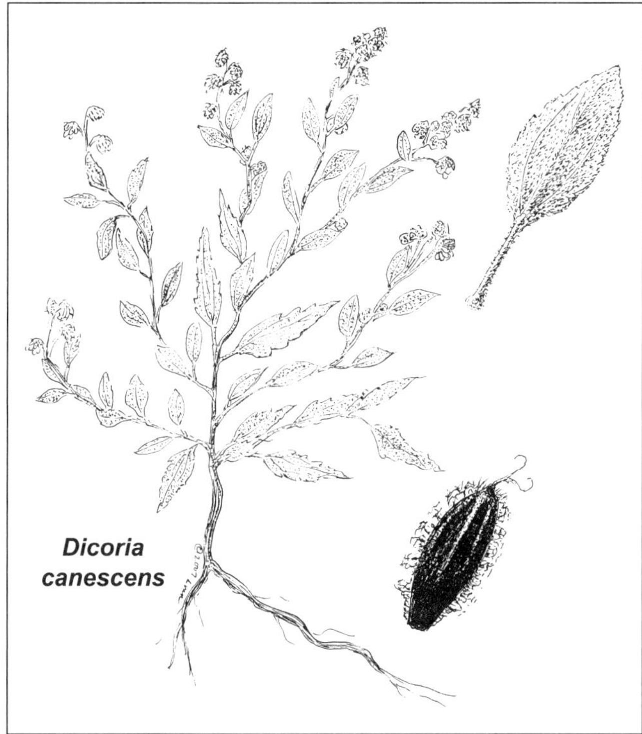

Dicoria canescens A. Gray (hoary or grayish white) Desert twinbugs. Plants 10–90 cm tall (sometimes wider than tall). **STEMS** striate. **LEAVES** with petioles 5–20 mm long; blades about 10–30 × 3–20 mm. **INVOLUCRES** 3–5 mm diameter. **PHYLLARIES** with outer ovate-oblong to lanceolate, faces strigillose and gland-dotted; inner (at maturity) ± obovate to suborbicular, usually saccate or cucullate and ± surrounding adjacent cypsela, gland-dotted or stipitate-glandular, sometimes sparsely strigillose or distally ciliate. **CYPSELAE** 3–8 mm. $2n = 36$. Sand dune communities or in (sometimes alkaline) sandy soils of washes, canyons, or flats. ARIZ: Apa, Nav; COLO: Mon; NMEX: McK, San, SJn; UTAH. 1400–1900 m (4500–6200′). Flowering: Aug–Oct. California to Colorado and New Mexico, south to northern Mexico. This species is generally segregated into two geographically and morphologically overlapping subspecies (each containing a number of varieties), the separation of which seems largely arbitrary. All of the material from the study area is ± assignable to **subsp. *brandegeei*** (A. Gray) Cronquist (for Townshend S. Brandegee, 1843–1925, botanist).

Dieteria Nutt.

G. L. Nesom

(Greek, *di-*, two, and *etos*, year, alluding to the biennial duration of the plants first named by Nuttall) Annuals, biennials, perennials, or subshrubs, taprooted. **STEMS** glabrous to puberulent, stipitate-glandular or eglandular. **LEAVES** basal and cauline, alternate, simple, sessile or petiolate, entire to irregularly serrate or dentate, with spinulose teeth, clasping to subclasping or not at all. **INFL** a radiate head, discoid in 1 variety, borne in loosely corymbiform to paniculate arrays. **INVOLUCRES** turbinate, campanulate, or hemispheric. **PHYLLARIES** in 3–12 graduate series. **RECEPTACLES** convex, shallowly alveolate. **RAY FLORETS** pistillate, fertile (sterile or 0 in 1 variety), in 1 series, lamina white to blue, lavender, or purple. **DISC FLORETS** bisexual, fertile. **CYPSELAE** linear to obovoid, flattened, smooth or 8–12-nerved, glabrous to strigose. **PAPPUS** of barbellate bristles in 1–3 series. $x = 4$. [*Machaeranthera* subg. *Dieteria* (Nutt.) Greene]. Species three, western United States and Mexico.

1. Middle and distal cauline leaves conspicuously auriculate-clasping or subclasping, involucres 10–15 mm .... ***D. bigelovii***

1′ Distal cauline leaves not conspicuously clasping or subclasping; involucres 5–10 mm .... (2)

2. Stems and peduncles conspicuously stipitate-glandular, essentially without nonglandular vestiture; proximal and middle cauline leaves usually serrrate to spinulose-dentate; ray corollas 10–20 mm .... ***D. asteroides***

2′ Stems and peduncles puberulent and eglandular or inconspicuously glandular, glabrous and eglandular in var. *aristata*; proximal and middle cauline leaves usually entire; ray corollas 6–12 mm .... ***D. canescens***

Dieteria asteroides Torr. (resembling *Aster*) **STEMS** and peduncles conspicuously stipitate-glandular, essentially without nonglandular vestiture. **LEAVES**: proximal and middle cauline linear to linear-oblanceolate, usually spreading and at least slightly falcate, usually serrate to spinulose-dentate, not clasping. **INVOLUCRES** 7–10 mm. **PHYLLARIES** triangular-subulate to linear-lanceolate, tips erect to slightly spreading or reflexed, glandular, usually without nonglandular vestiture. **RAY COROLLAS** 10–20 mm, lavender. **CYPSELAE** strigose. [*Machaeranthera asteroides* (Torr.) Greene].

1. Middle and upper cauline leaves mostly 2.5–4.5 cm, faces puberulent to sessile-glandular; phyllary apices linear-lanceolate, reflexed .... **var. *asteroides***
1' Middle and upper cauline leaves mostly 0.5–2 mm, faces sessile- to stipitate-glandular; phyllary apices triangular-subulate, mostly erect .... **var. *glandulosa***

var. *asteroides* Middle and upper cauline leaves mostly 2.5–4.5 cm, faces puberulent to sessile-glandular; phyllary apices linear-lanceolate, reflexed. $2n = 8$. Piñon pine, Douglas-fir, ponderosa pine–Gambel's oak communities. ARIZ: Apa; COLO: Arc; NMEX: RAr. 2050–2550 m (6725–8365'). Flowering: Aug–Oct. California, Nevada, Arizona, New Mexico; Mexico.

var. *glandulosa* (B. L. Turner) D. R. Morgan & R. L. Hartm. (glandular) Middle and upper cauline leaves mostly 0.5–2 mm, faces sessile- to stipitate-glandular; phyllary apices triangular-subulate, mostly erect. $2n = 8$. [*Machaeranthera asteroides* (Torr.) Greene var. *glandulosa* B. L. Turner]. Dunes and sandy soil, alluvium, roadsides and other disturbed sites, less commonly in gypseous soil, canyon bottoms, washes, benches; cottonwood-willow-saltcedar, desert shrub, rabbitbrush, sagebrush, piñon-juniper, and piñon–Gambel's oak communities. ARIZ: Apa, Coc, Nav; COLO: Dol, LPl, Mon, SMg; NMEX: McK, San, SJn; UTAH. 1125–2150 (–2450) m (3700–8040'). Flowering: Jun–Sep (–Oct). Arizona, Utah, New Mexico. Variety *glandulosa* is the more common expression of the species in the Four Corners area, but plants examined of var. *asteroides* appear to be distinct, even though apparently sympatric with var. *glandulosa*.

Dieteria bigelovii (A. Gray) D. R. Morgan & R. L. Hartm. (for John M. Bigelow, 1804–1878, botanist for several surveys and expeditions in the western U.S.) **STEMS** and peduncles densely stipitate-glandular, loosely strigose to stiffly puberulent or without nonglandular vestiture. **LEAVES** cauline, ovate to narrowly ovate, lanceolate, or linear-oblong, usually spreading-ascending, usually widely serrate, auriculate-clasping or subclasping. **INVOLUCRES** 10–15 mm. **PHYLLARIES** linear-lanceolate, tips spreading to reflexed, glandular, usually without nonglandular vestiture. **RAY COROLLAS** 10–25 mm, blue to purple. **CYPSELAE** sparsely strigose. $2n = 8$. [*Machaeranthera bigelovii* (A. Gray) Greene]. Roadsides and other disturbed sites, alluvium, rocky slopes, cliffs, wet grasslands; Gambel's oak, piñon pine–Gambel's oak, ponderosa pine, aspen, fir-aspen, and spruce-fir communities. COLO: Arc, Dol, Hin, LPl, Min, Mon, RGr, SJn; NMEX: McK, RAr. 1950–3000 m (6400–9840'). Flowering: Jul–Sep (–Oct). New Mexico and Colorado, rare in Arizona and Wyoming. Four Corners plants are **var. *bigelovii***.

Dieteria canescens (Pursh) Nutt. (becoming gray) **STEMS** and peduncles puberulent and eglandular or inconspicuously glandular, glabrous and eglandular in var. *aristata*. **LEAVES**: proximal and middle cauline linear to linear-oblanceolate, usually spreading and at least slightly falcate, usually entire, not clasping. **INVOLUCRES** 5–9 mm. **PHYLLARIES** glandular, usually without nonglandular vestiture, tips erect to slightly spreading or reflexed. **RAY COROLLAS** 6–12 mm (absent), blue to purple. **CYPSELAE** strigose or glabrous. [*Machaeranthera canescens* (Pursh) A. Gray].

1. Involucres 7–9 mm; phyllaries eglandular, puberulent, tips appressed; cypselae glabrous .... **var. *ambigua***
1' Involucres 5–7 mm; phyllaries glandular, usually without nonglandular vestiture, tips erect to slightly spreading or reflexed; cypselae strigose .... **var. *aristata***

var. *ambigua* (B. L. Turner) D. R. Morgan & R. L. Hartm. (doubtful or uncertain) **INVOLUCRES** 7–9 mm. **PHYLLARIES** eglandular, puberulent, tips appressed. **CYPSELAE** glabrous. [*Machaeranthera canescens* (Pursh) A. Gray var. *ambigua* B. L. Turner]. $2n = 8$. Roadsides, open slopes and flats, washes, sand; Gambel's oak–sage, piñon-juniper–Gambel's oak, and ponderosa pine communities. ARIZ: Apa, Nav; NMEX: SJn. 2050–2450 m (6725–8040'). Flowering: Aug–Oct. Utah, Colorado, Arizona, New Mexico.

var. *aristata* (Eastw.) D. R. Morgan & R. L. Hartm. (awned) **INVOLUCRES** 5–7 mm. **PHYLLARIES** glandular, usually without nonglandular vestiture, tips erect to slightly spreading or reflexed. **CYPSELAE** strigose. [*Machaeranthera canescens* (Pursh) A. Gray var. *aristata* (Eastw.) B. L. Turner]. $2n = 8$. Roadsides, open areas, sand, riparian; desert grassland, sagebrush, greasewood-sagebrush, cottonwood-fir, and piñon-juniper woodland communities. ARIZ: Apa; COLO: Arc, LPl; NMEX: McK, RAr, San, SJn. 1500–2200 m (4920–7215'). Flowering: (May–) Jun–Sep. Colorado, Arizona, New Mexico. For the most part, var. *aristata* and var. *ambigua* seem to be geographically distinct. As cited and mapped by Turner (1987), var. *aristata* occurs sympatrically with *D. canescens* var. *glabra* A. Gray in the Four Corners area, but it does not seem possible to make an unarbitrary distinction among collections from our area. The cauline glandularity is variable in these plants, but all are identified here as var. *aristata* (midstems stipitate-glandular, puberulent; vs. var. *glabra*, midstems essentially eglandular, otherwise glabrous to puberulent). Intermediates between *D. canescens* var. *aristata* and *D. asteroides* apparently are relatively common.

Dyssodia Cav. Dogweed
Brian A. Elliott

(probably derived from the Greek *dys*, bad, + *osma*, odor, referring to the rank aroma) Annual to perennial, taprooted, ill-scented herbs. **STEMS** spreading to erect, branched. **LEAVES** opposite, sometimes alternate or alternate above, pinnately parted with narrow lobes, sessile or petiolate, glandular with translucent, embedded oil glands. **INFL** heads radiate, rarely discoid, terminating the stems and branches, solitary or in paniculate to cymose clusters. **INVOLUCRES** turbinate to hemispheric, 5–10 mm high. **PHYLLARIES** 4–8 (16) in 2–3 series, free to the base or nearly so, chartaceous to coriaceous; outer bracts linear, not overlapping, rarely lacking; inner bracts wider, overlapping; a calyculus of 1–9 deltate to linear bracts sometimes present, 1/2 to equal the length of the phyllaries. **RECEPTACLES** flat to convex, naked to minutely fimbrillate or finely setose. **RAY FLORETS** pistillate and fertile, 5–8 (12), rarely lacking; ligule yellow to orange. **DISC FLORETS** perfect and fertile, 20–100; corolla yellow-orange, 2–5 mm long, cylindric or nearly so, pubescent on tube and throat. **CYPSELAE** obconic to obpyramidal, 2–4 mm long, glabrous to sericeous. **PAPPUS** of 12–20 scales, each cleft into 5–10 bristles; equaling the disc corolla. Seven species of North and South America. *Dyssodia* has been more broadly circumscribed in the past but was split into seven segregate genera (including *Thymophylla* and *Adenophyllum*) by J. L. Strother in 1986 (Strother 1969, 1986).

Dyssodia papposa (Vent.) Hitch. (with a well-developed pappus) Alkali yellowtops, fetid marigold, pappose glandweed. Annual taprooted herbs. **STEMS** much-branched at maturity, 1–4 (7) dm tall. **LEAVES** pinnatisect into 11–15 lobes 5–20 mm long and 1–3 mm wide, ultimate segments spinulose-tipped, glabrous to sparsely pubescent, irregularly gland-dotted. **INFL** heads radiate, 1–few in numerous dense cymes, subsessile or short-pedunculate. **PHYLLARIES**: principal bracts (6) 8 (12), scarious at the margins and apex, fimbrillate, gland-dotted, distally anthocyanic; the inner wider, overlapping, chartaceous; calyculate bracts 4–9, linear, green. **RECEPTACLES** slightly convex, fimbriate. **RAY FLORETS** with inconspicuous ligules, 1.5–2 mm long, scarcely exceeding the disc flowers, sparsely puberulent, apex 2–3-denticulate. **DISC FLORETS** 12–50; corolla pale yellow, 3 mm long; tube narrowly cylindric and puberulent. **CYPSELAE** obpyramidal, black, 3–5-angled, sericeous. **PAPPUS** about 3 mm long. $2n = 26$. [*Boebera chrysanthemoides* Willd.; *Boebera papposa* Rydb.; *Dyssodia chrysanthemoides* Lag.; *Tagetes papposa* Vent.]. Disturbed sites, roadsides, desert shrubland, rolling hills, and piñon-juniper woodland communities. ARIZ: Apa, Nav; COLO: Arc, LPl, Mon; NMEX: McK, RAr, SJn. 1600–2250 m (5300–7300′). Flowering: Aug–Oct. Native to Mexico and the Great Plains, now weedy in much of the United States. Used medicinally as a poultice for ant bites; smoked for epileptic fits; breathed in for headache; and an infusion of the plant has been used as a rub for fever. It has also been used to treat coughing in horses. Seeds have been dried, ground, and used for flour, and the tops used for greens or seasoning.

Encelia Adans. Brittlebush
Gary I. Baird

(for Christoph Entzelt, 1517–1583, German naturalist) Shrubs (ours), subshrubs, or perennial herbs, taprooted, rhizomes lacking, herbage glabrous or variously pubescent, often gland-dotted, sap clear. **STEMS** many, erect, usually proximally branched. **LEAVES** mostly cauline (basal in some), simple, petiolate, alternate; petiolate (obscure in some); blades deltate, ovate, or rhombic to lanceolate or linear, bases cuneate, margins entire (rarely toothed), apices obtuse to acute. **INFL** borne singly or in open paniculiform clusters, erect or nodding; peduncles long, ± scapiform. **HEADS** radiate or discoid. **INVOLUCRES** hemispheric. **PHYLLARIES** about 20–30+ in 2–3 series, herbaceous, subequal or unequal (outer shorter). **RECEPTACLES** flat to convex, paleate; paleae conduplicate and enfolding achenes, deciduous, ± membranous; calyculi lacking. **RAY FLORETS** 8–25+, or lacking, neuter; corollas yellow; laminae fugacious. **DISC FLORETS** mostly 80–100+, bisexual, fertile; corollas yellow or brownish purple. **CYPSELAE** monomorphic, strongly compressed, obovate to cuneate, margins narrow, white, ciliate, apices usually notched, faces glabrous or pubescent. **PAPPUS** lacking, or of 2 bristlelike awns (persistent or readily deciduous). $x = 18$. Perhaps 13 species of southwestern North America and western South America.

Encelia frutescens (A. Gray) A. Gray (becoming shrubby) Button brittlebush. Shrubs 40–150 cm tall (often wider than high). **STEMS** with slender, intricate branches, glabrous or glabrate. **LEAVES** cauline (mostly proximal on current season's branches); petioles 2–7 mm long; blades ovate or elliptic to lanceolate, 10–25 cm long, apices acute (ours) to obtuse, faces strigose, gland-dotted (ours) or not. **HEADS** borne singly; peduncles long, leafless, strigose to glabrate.

INVOLUCRES 5–12 mm diameter. **PHYLLARIES** lanceolate. **RAY FLORETS** lacking or (ours) 8–13; laminae 8–12 mm. **DISC FLORETS** with corollas 5–6 mm, yellow. **CYPSELAE** 5–9 mm. **PAPPUS** lacking (ours) or of 2 bristlelike scales (often obscure). $2n = 36$. [*E. resinifera* C. Clark]. Blackbrush, sagebrush, piñon-juniper, mountain brush, ponderosa pine–oak forest communities. ARIZ: Apa, Nav; UTAH. 1400–1700 m (4500–5500′). Flowering: May–Jul. California, Nevada, Utah, and Arizona. All material from the study area belongs to **var.** ***resinosa*** M. E. Jones ex S. F. Blake (resin-bearing), which is sometimes recognized as a separate species.

Engelmannia A. Gray ex Nutt. Engelmann's Daisy, Cutleaf Daisy

Gary I. Baird

(for George Engelmann, German-American botanist, 1809–1884) Perennials, taprooted, rhizomes lacking, caudices woody, herbage strigose or pustulose-strigose, often pilose-hispid also, sap clear. **STEMS** usually 1, erect, branched (proximally, distally, or both). **LEAVES** basal and cauline, simple, alternate; petiolate (basal and proximal cauline) to sessile (distal cauline); blades oblong to lanceolate, 1–2-pinnately lobed, bases cuneate, ultimate margins entire, mid-veins often whitish, apices bluntly acute. **INFL** ± corymbiform, erect; peduncles ± ebracteate. **HEADS** radiate, 20–100+ (proliferating with time). **INVOLUCRES** hemispheric. **PHYLLARIES** about 18–24 in about 3 series, imbricate, deciduous, innermost like palea (clasping adjacent floret). **RECEPTACLES** flat, paleate; paleae enfolding adjacent floret, deciduous, chartaceous, the outer becoming thickened; calyculi lacking; **RAY FLORETS** 8–10, pistillate, fertile; corollas yellow, fugacious. **DISC FLORETS** about 30–60, abortive (functionally staminate); corollas yellow. **CYPSELAE** monomorphic, grayish black, obovate, ± compressed, faces keeled, strigose, enclosed by subtending phyllary, falling together with adjacent disc florets and palea. **PAPPUS** of 2–4 bractlike scales, strigulose, ciliate, persistent or tardily deciduous (ray florets), or much reduced and obsolete (disc florets). $x = 9$. One species of south-central to southwestern North America.

Engelmannia peristenia (Raf.) Goodman & C. A. Lawson (very narrow, the allusion unclear) Perennials mostly 20–50 cm tall. **STEMS** densely strigose and often pilose-hispid also (especially when young). **LEAVES**: petioles shorter than blades; basal blades 10–30 × 8–9 cm, faces strigose to pustulose-strigose; cauline reduced distally. **INVOLUCRES** about 6–10 mm diameter. **PHYLLARIES** with bases ovate to orbicular, chartaceous, margins membranous, ciliate, faces subglabrous to strigose, apices abruptly linear-oblong, herbaceous, hispid-strigose, shorter than to surpassing bases. **RAY FLORET** laminae about 10–15 mm long, abaxially sparsely strigose. $2n = 18$. [*E. pinnatifida* Nutt.]. Roadsides in desert grassland communities. NMEX: SJn. 1800 m (6000′). Flowering: Jun–Jul (Oct). South Dakota south to Arizona, Texas, Arkansas, and Mexico. Primarily a plains species, it occurs sporadically westward. Currently known from one locality within the study area.

Ericameria Nutt. Rabbitbrush, Goldenbush

Roland P. Roberts

(Greek *ereike*, heath, + *meris*, part; alluding to the heathlike leaves of the genus) Shrubs to 5 m tall. **STEMS** mostly erect to ascending, rarely prostrate, fastigiately branched; bark typically tan to reddish brown, becoming gray with maturity; twigs usually green to gray or yellowish, glabrous to densely pubescent to tomentose, often glandular, resin-coated in some. **LEAVES** alternate, simple, filiform to laminar, linear or lanceolate to spatulate, entire, sessile or short-petiolate; mostly evergreen, cauline often crowded, mostly ascending to spreading, sometimes recurved, green to grayish, blades adaxially sulcate, concave, or planar, sometimes undulate or crisped, apices acute to rounded or retuse, glabrous to densely tomentose, often glandular, sometimes punctate, often resinous; midvein obscure to prominent, usually bounded by a pair of collateral veins; axillary leaf fascicles sometimes present. **INFL** in solitary, cymiform, or racemiform, sometimes in highly branched and paniculiform or thyrsiform clusters. **HEADS** discoid or radiate. **INVOLUCRES** cylindric, obconic, campanulate, or hemispheric, usually 4–19+ mm long, 2–18 mm wide. **PHYLLARIES** 8–60, in 2–7 strongly graduated or subequal series, usually in vertical ranks, ovate, lanceolate, or elliptic, sometimes glandular and often resinous, outermost often herbaceous or herbaceous-tipped, otherwise mostly chartaceous, 1-nerved. **RAY FLORETS** 0–18, carpellate, fertile, lamina various shades of yellow or white in some, elliptic to oblong, apices shallowly notched or toothed. **DISC FLORETS** 4–70, perfect, fertile, corollas yellow or white, tubes shorter than the narrowly funnelform or campanulate throats, lobes 5, erect to spreading or reflexed, deltate to triangular; style appendages lanceolate to subulate. **CYPSELAE** tan to reddish brown, usually prismatic, sometimes cylindric, ellipsoid, turbinate, or obconic, 5–12-nerved, glabrous to densely pubescent, sometimes gland-dotted. **PAPPUS** of 20–60 barbellate bristles in 1 series, persistent or late-deciduous, whitish or tan to reddish, subequal, apically attenuate. $x = 9$. Thirty-four species in North America.

1. Phyllaries unequal, obtuse to acute, if attenuate then strongly keeled, chartaceous .... ***E. nauseosa***
1′ Phyllaries subequal, attenuate, weakly keeled, usually membranous .... ***E. parryi***

Ericameria nauseosa (Pall. ex Pursh) G. L. Nesom & G. I. Baird (very sickening) Rubber rabbitbrush. **STEMS** to 2.5 m tall, fastigiately branched, usually erect or ascending to spreading, white to green, tomentose. **LEAVES** usually crowded, ascending to spreading, blades filiform to narrowly oblanceolate, 10–70 mm long, 0.5–10 mm wide, mostly adaxially sulcate to concave, adaxially glabrous to tomentose, often gland-dotted, apices acute; midvein mostly evident, axillary leaf fascicles absent. **INFL** in rounded to flat-topped cymiform clusters. **HEADS** discoid. **INVOLUCRES** obconic to subcylindric, 6–16 mm long, 2–4 mm wide. **PHYLLARIES** 10–30, in 3–5 series, tan, strongly graduated and usually in vertical ranks, mostly keeled, ovate to lanceolate, mostly chartaceous, apices acute to obtuse, abaxially resinous; adaxially glabrous or pubescent, costa raised and expanded apically. **DISC FLORETS** usually 5, sometimes 4–6, corollas 6–12 mm long, lobes glabrous to sparsely pubescent; style appendages 1.5–4 mm long, appendages subulate, 1–2.5 mm long. **CYPSELAE** tan, turbinate to cylindric or oblanceoloid, 3–8 mm long, glabrous or pubescent. **PAPPUS** whitish, 3–13 mm long. $2n$ = 18. Arizona, California, Colorado, Idaho, Montana, Nevada, New Mexico, Oregon, Utah, Wyoming; Mexico (Baja California). This species is widespread and represented in North America by 21 varieties.

1. Cypselae glabrous .... (2)
1′ Cypselae pubescent .... (6)

2. Corolla lobes villous .... (3)
2′ Corolla lobes glabrous .... (4)

3. Plants 60 cm or more tall; involucres 10–12.5 mm long; phyllary apices acute to obtuse .... **var. *nitida*** (in part)
3′ Plants less than 60 cm tall; involucres 11.2–16 mm long; phyllary apices acute to short-acuminate .... **var. *turbinata*** (in part)

4. Phyllaries abaxially glabrous .... **var. *leiosperma***
4′ Phyllaries abaxially tomentulose to scurfy-tomentulose .... (5)

5. Stems nearly leafless at flowering; leaves 15–30 mm long; phyllary apices acute to acuminate .... **var. *bigelovii***
5′ Stems leafy at flowering; leaves 30–50 mm long; phyllary apices acute to obtuse .... **var. *nitida*** (in part)

6. Style appendages shorter than stigmatic portions .... (7)
6′ Style appendages longer than stigmatic portions .... (9)

7. Corolla lobes 1.5–2.5 mm long .... **var. *oreophila*** (in part)
7′ Corolla lobes 0.5–1 mm long .... (8)

8. Phyllaries tomentose .... **var. *hololeuca***
8′ Outer phyllaries tomentulose, inner glabrous .... **var. *latisquamea*** (in part)

9. Corolla lobes villous .... (10)
9′ Corolla lobes glabrous .... (12)

10. Stems usually leafless at flowering; corolla tubes 7–8.5 mm long .... **var. *juncea***
10′ Stems usually leafy; corolla tubes 9.5–11.8 mm long .... (11)

11. Leaf blades 30–50 mm long; surfaces glabrate; phyllary apices acute to obtuse; corolla lobes glabrous or villous .... **var. *nitida*** (in part)
11′ Leaf blades 10–20 mm long; surfaces tomentulose to densely tomentose; phyllary apices acute to acuminate; corolla lobes villous .... **var. *turbinata*** (in part)

12. Involucres 16–19 mm; phyllary apices more or less recurved .... **var. *arenaria***
12′ Involucres 6–14.5 mm long; phyllary apices erect .... (13)

13. Phyllaries usually glabrous .... (14)
13′ Phyllaries usually pubescent .... (15)

14. Corolla lobes 0.6 to 1.5 mm long; corolla tubes puberulent or glabrous .... **var. *graveolens***
14′ Corolla lobes 1.5 to 2.5 mm long; corolla tubes glabrous .... **var. *oreophila*** (in part)

15. Stems grayish white to white .... (16)
15' Stems yellowish green to green .... (18)

16. Outer phyllaries densely tomentose, inner glabrous .... **var. *latisquamea*** (in part)
16' All phyllaries tomentose, sometimes sparsely .... (17)

17. Corolla tubes puberulent, usually 6.5–9 mm long .... **var. *nauseosa***
17' Corolla tubes tomentose or glabrous, usually 8.7–13 mm long .... **var. *speciosa***

18. Involucres 10–12.5 mm long; corollas 9.5–11 mm long; pappus 9–10 mm long .... **var. *nitida*** (in part)
18' Involucres 6.5–10 mm long; corollas 6–9 mm long; pappus 4.2–7.6 mm long .... **var. *oreophila*** (in part)

var. *arenaria* (L. C. Anderson) G. L. Nesom & G. I. Baird (sand) Sand rabbitbrush. **STEMS** to 1.2 m tall, grayish green, moderately leafy and tomentose. **LEAVES** grayish green, tomentose, blades linear, 40–75 mm long, 0.8–1.1 mm wide, 1-nerved. **INVOLUCRES** 16–19 mm long. **PHYLLARIES** 22–27, sometimes 19, in strong vertical ranks, apices cuspidate-aristate, somewhat recurved, abaxially glabrous to glabrescent. **COROLLAS** 12–14.2 mm long, tubes glabrous, lobes 1.2–2 mm long, glabrous; style appendages longer than stigmatic portion. **CYPSELAE** pubescent. **PAPPUS** 9.6–12.7 mm long. $2n = 18$. [*Chrysothamnus nauseosus* (Pall. ex Pursh) Britton subsp. *arenarius* L. C. Anderson; *C. nauseosus* var. *arenarius* (L. C. Anderson) S. L. Welsh]. Sandhills and sandstone cliffs. ARIZ: Apa, Nav; NMEX: RAr, San, SJn; UTAH. 1500–2000 (4920–6560′). Flowering: Aug–Oct. Arizona, New Mexico, Utah.

var. *bigelovii* (A. Gray) G. L. Nesom & G. I. Baird (for John M. Bigelow, 1804–1878, botanist with the Mexican Boundary Survey) Bigelow's rabbitbrush. **STEMS** to 1.2 m tall, whitish, tomentose, may be nearly leafless at flowering. **LEAVES** mostly grayish white, tomentulose, blades filiform, 15–30 mm long, 0.5–1 mm wide, 1-nerved. **INVOLUCRES** 10.5–12.5 mm long. **PHYLLARIES** 14–18, sometimes to 25, apices acute to acuminate, erect, abaxially tomentulose. **COROLLAS** 9–11 mm long, tubes glabrous or puberulent, lobes 0.8–1.5 mm long, glabrous; style appendages longer than stigmatic portion. **CYPSELAE** glabrous. **PAPPUS** 7.2–11 mm long. $2n = 18$. [*Chrysothamnus nauseosus* (Pall. ex Pursh) Britton var. *bigelowii* (A. Gray) H. M. Hall; *C. nauseosus* subsp. *bigelowii* (A. Gray) H. M. Hall & Clem.]. ARIZ: Apa, Nav; COLO: Mon; NMEX: McK, RAr, SJn; UTAH. Rimrock and crevices; *Grayia* and piñon-juniper woodland communities. 1300–2200 m (4265–7215′). Flowering: Aug–Oct. Arizona, Colorado, New Mexico, Utah. Plant used as a ceremonial emetic by the Kayenta Navajo. The Ramah Navajo uses include: leaves made into a lotion and used for headache; decoction of root used for menstrual pain; strong decoction of root taken for colds; white galls from plants hung around babies' necks to stop dribbling; branches used to carpet the sweathouse floor; yellow flowers used as a yellow dye for wool; branches used to make Enemyway prayer stick.

var. *graveolens* (Nutt.) Reveal & Schuyler (strong-smelling) Pungent rabbitbrush. **STEMS** to 1.6 m tall, yellowish green to nearly white, leafy, tomentum compact and smooth. **LEAVES** greenish, often tomentulose, blades broadly linear, 30–90 mm long, 1–3 mm wide, some 3–5-nerved. **INVOLUCRES** 6–8 mm long. **PHYLLARIES** 14–18, apices acute, erect, usually glabrous abaxially, outer sometimes sparsely pubescent. **COROLLAS** 7–9 mm or sometimes longer, tubes puberulent or glabrous, lobes 0.6–1.5 mm long, glabrous; style appendages longer than stigmatic portion. **CYPSELAE** densely pubescent. **PAPPUS** 4.9–6.5 mm long. $2n = 18$. [*Chrysocoma graveolens* Nutt.; *Bigelowia graveolens* (Nutt.) A. Gray var. *glabrata* A. Gray; *Chrysothamnus nauseosus* (Pall. ex Pursh) Britton var. *graveolens* (Nutt.) H. M. Hall; *C. nauseosus* var. *glabratus* (A. Gray) Cronquist; *C. nauseosus* subsp. *graveolens* (Nutt.) Piper]. Slightly alkaline soils; willow-cottonwood, saltcedar, sagebrush, shadscale, mountain brush, and ponderosa pine communities. ARIZ: Apa, Nav; COLO: Arc, LPl, Mon; NMEX: McK, RAr, SJn; UTAH. 1125–2450 m (3700–8040′). Flowering: Aug–Oct. Arizona, Colorado, Idaho, Kansas, Montana, Nebraska, Nevada, New Mexico, North Dakota, Oklahoma, South Dakota, Texas, Utah, Wyoming.

var. *hololeuca* (A. Gray) G. L. Nesom & G. I. Baird (wholly white) White rabbitbrush. **STEMS** to 1.2 m tall, whitish, densely tomentose and leafy. **LEAVES** whitish, tomentose, blades filiform to linear, 10–35 mm long, 1–2 mm wide, 1-nerved. **INVOLUCRES** 7–9 mm long. **PHYLLARIES** 14–16, apices obtuse, erect, abaxially tomentose. **COROLLAS** 7–9.5 mm long, tubes puberulent to cobwebby, lobes 0.5–1 mm long, glabrous; style appendages shorter than stigmatic portion. **CYPSELAE** densely pubescent. **PAPPUS** 5.2–7.2 mm long. $2n = 18$. [*Bigelowia graveolens* (Nutt.) A. Gray var. *hololeuca* A. Gray; *Chrysothamnus nauseosus* (Pall. ex Pursh) Britton var. *hololeucus* (A. Gray) H. M. Hall]. Gravelly or sandy slopes, well-drained; shadscale, rabbitbrush, sagebrush, and piñon-juniper woodland communities. ARIZ: Apa, Nav; NMEX: RAr, SJn; UTAH. 1125–2500 m (3700–8200′). Flowering: Aug–Oct. Arizona, California, Idaho, Nevada, New Mexico, Oregon, Utah.

var. ***juncea*** (Greene) G. L. Nesom & G. I. Baird (rushlike) Rush rabbitbrush. **STEMS** to 0.7 m tall, yellowish green, tomentum compact, often leafless at flowering. **LEAVES** yellowish green, compactly tomentose, blades filiform, 10–30 mm long, 0.7–1 mm wide, 1-nerved. **INVOLUCRES** 9.5–12 mm long. **PHYLLARIES** 22–25, apices acute, erect, abaxially tomentose or glabrate. **COROLLAS** 7–8.5 mm long, tubes pubescent, lobes 0.7–0.9 mm long, villous; style appendages longer than stigmatic portion. **CYPSELAE** pubescent. **PAPPUS** 5.6–7.6 mm long. $2n = 18$. [*Bigelowia juncea* Greene; *Chrysothamnus nauseosus* (Pall. ex Pursh) Britton var. *junceus* (Greene) H. M. Hall]. Blackbrush, matchweed, shadscale, and piñon-juniper communities. ARIZ: Apa, Nav; UTAH. 1200–1800 m (3935–5905′). Flowering: Aug–Oct. Arizona, Nevada, Utah.

var. ***latisquamea*** (A. Gray) G. L. Nesom & G. I. Baird (broad-scaled) Broadscale rabbitbrush. **STEMS** to 2 m tall, white, tomentose, leafy. **LEAVES** whitish, loosely tomentose, blades filiform, 15–45 mm long, 1–2 mm wide, 1-nerved. **INVOLUCRES** 7.5–10.5 mm long. **PHYLLARIES** 15–17, apices obtuse, erect, outer abaxially densely tomentulose, inner glabrous. **COROLLAS** 7–9.5 mm long, tubes glabrous or puberulent, lobes 0.6–1 mm long, glabrous; style appendages shorter than or equaling stigmatic portion. **CYPSELAE** densely pubescent. **PAPPUS** 4.8–7.6 mm long. $2n = 18$. [*Bigelowia graveolens* (Nutt.) A. Gray var. *latisquamea* A. Gray; *Chrysothamnus nauseosus* (Pall. ex Pursh) Britton subsp. *latisquameus* (A. Gray) H. M. Hall & Clem.]. Dry streambeds and arroyos; big sagebrush, piñon-juniper woodland, and ponderosa pine communities. ARIZ: Nav; NMEX: McK, San. 1675–2200 m (5500–7215′). Flowering: Sep–Oct. Arizona, New Mexico. The yellow twigs and flowers used as a yellow dye for wool by the Navajo.

var. ***leiosperma*** (A. Gray) G. L. Nesom & G. I. Baird (smooth-seeded) Smooth-fruit rabbitbrush. **STEMS** to 0.6 m tall, greenish yellow, tomentum compact; often leafless at flowering. **LEAVES** yellowish green, surfaces glabrate, blades filiform, 10–30 mm long, 0.8–1 mm wide, 1-nerved. **INVOLUCRES** 8–11.5 mm long. **PHYLLARIES** 17–31, apices acute to obtuse, erect, abaxially glabrous. **COROLLAS** 5–8.5 mm long, tubes mostly glabrous, lobes 0.5–1.1 mm, sometimes incurved, glabrous; style appendages longer than stigmatic portions. **CYPSELAE** glabrous. **PAPPUS** 3–6.6 mm long. $2n = 18$. [*Bigelowia leiosperma* A. Gray; *Chrysothamnus nauseosus* (Pall. ex Pursh) Britton subsp. *leiospermus* (A. Gray) H. M. Hall & Clem.]. Gravelly and sandy soils, rocky crevices; resinbush, blackbrush, shadscale, black sagebrush, piñon-juniper woodland, and ponderosa pine communities. ARIZ: Apa, Nav; COLO: Mon; UTAH. 1370–2440 m (4500–8000′). Flowering: Aug–Oct. Arizona, California, Colorado, Nevada, Utah.

var. ***nauseosa*** Rubber rabbitbrush. **STEMS** to 0.6 mm tall, grayish to white, tomentose, leafy. **LEAVES** grayish green to white, surfaces tomentose, blades narrowly linear, 10–50 mm long, 1–1.5 mm wide, 1-nerved. **INVOLUCRES** 7–10.5 mm long. **PHYLLARIES** 10–18, apices usually obtuse, erect, abaxially densely to sparingly tomentose. **COROLLAS** 6.5–9 mm long, tubes usually puberulent, lobes 1–2 mm long, glabrous; style appendages longer than stigmatic portions. **CYPSELAE** densely pubescent. **PAPPUS** 5.2–7.2 mm long. $2n = 18$. Plains and hillsides, often in barren alkaline soils; desert grassland, juniper savanna, and piñon-juniper woodland communities. ARIZ: Nav; NMEX: McK, RAr, San. 1525–2000 m (5000–6560′). Flowering: Aug–Oct. Arizona, Colorado, Idaho, Montana, Nebraska, New Mexico, North Dakota, South Dakota, Wyoming.

var. ***nitida*** (L. C. Anderson) G. L. Nesom & G. I. Baird (shining or lustrous, referring to the bracts) Shiny-bract rabbitbrush. **STEMS** to 1.5 m tall, yellowish green, densely tomentose, leafy at flowering. **LEAVES** yellowish green, blades glabrate, linear, 30–50 mm long, 1–1.5 mm wide, 1-nerved. **INVOLUCRES** 10–12.5 mm long. **PHYLLARIES** 13–19, outer weakly keeled, apices acute to obtuse, erect, lowermost abaxially scurfy-tomentulose. **COROLLAS** 9.5–11 mm long, tubes mostly glabrous, lobes 0.7–1 mm long, glabrous or villous; style appendages longer than stigmatic portion. **CYPSELAE** usually glabrous, sometimes pubescent. **PAPPUS** 9–10 mm long. $2n = 18$. [*Chrysothamnus nauseosus* (Pall.ex Pursh) Britton subsp. *nitidus* L. C. Anderson; *C. nauseosus* var. *nitidus* (L. C. Anderson) S. L. Welsh]. Sandy gravels of dry creek beds; resinbush-*Ephedra* communities. ARIZ: Apa, Nav; NMEX: RAr, SJn. 1200–1800 m (3935–5905′). Flowering: Aug–Oct. Arizona, New Mexico, Utah.

var. ***oreophila*** (A. Nelson) G. L. Nesom & G. I. Baird (mountain-loving) Great Basin rabbitbrush. **STEMS** to 2.5 m tall, greenish, tomentum compact, usually leafy. **LEAVES** yellowish green, surfaces glabrate, blades filiform to linear, 20–50 mm long, 0.8–1 mm wide, 1-nerved. **INVOLUCRES** 6.5–10 mm long. **PHYLLARIES** 10–20, apices acute, erect, glabrous. **COROLLAS** 6–9 mm, tubes glabrous, lobes 1.5–2.5 mm long, glabrous; style appendages shorter than or equaling stigmatic portions. **CYPSELAE** pubescent. **PAPPUS** 4.2–7.6 mm long. $2n = 18$. [*Chrysothamnus oreophilus* A. Nelson; *C. nauseosus* (Pall. ex Pursh) Britton var. *artus* (A. Nelson) Cronquist; *C. nauseosus* subsp. *consimilis* (Greene) H. M. Hall & Clem.; *C. nauseosus* var. *consimilis* (Greene) H. M. Hall; *C. oreophilus* var. *artus* A. Nelson; *Ericameria nauseosa* var. *arta* (A. Nelson) G. L. Nesom & G. I. Baird; *E. nauseosa* subsp. *consimilis* (Greene) G. L. Nesom & G. I. Baird].

Alkaline valleys and plains; saline meadows, riparian zones in saltgrass–alkali sacaton, shadscale, sagebrush, mountain brush, piñon-juniper, and ponderosa pine communities. ARIZ: Apa, Coc, Nav; COLO: SMg; NMEX: McK, RAr, SJn. 1200–3000 m (3945–9840′). Flowering: Aug–Oct. Arizona, California, Colorado, Idaho, Montana, Nevada, New Mexico, Oregon, Utah, Wyoming.

var. *speciosa* (Nutt.) G. L. Nesom & G. I. Baird (showy) Whitestem rabbitbrush. **STEMS** to 2 m tall, whitish, loosely tomentose, leafy. **LEAVES** dark green to grayish white, surfaces tomentose, blades linear to linear-oblanceolate, 30–70 mm long, usually 0.3–1.5 or sometimes to 2.5 mm wide, 1-nerved. **INVOLUCRES** 7.5–13.5 mm long. **PHYLLARIES** 12–28, apices acute, erect, all but especially the lower abaxially sparsely to densely tomentose. **COROLLAS** 8.7–13 mm long, tubes tomentose or glabrous, lobes 1.1–2.1 mm long, glabrous; style appendages longer than stigmatic portion. **CYPSELAE** densely pubescent. **PAPPUS** 6–11.3 mm long. [*Chrysothamnus speciosus* Nutt.; *C. nauseosus* (Pall. ex Pursh) Britton subsp. *albicaulis* (Nutt.) H. M. Hall & Clem.; *C. nauseosus* subsp. *speciosus* (Nutt.) H. M. Hall & Clem.]. Various dry habitats including juniper-sage, ponderosa pine, and piñon communities. ARIZ: Apa; COLO: Arc; NMEX: RAr, San, SJn; UTAH. 1370–2285 m (4500–7500′). Flowering: Aug–Oct. California, Colorado, Idaho, Montana, Nevada, Oregon, Utah, Washington, Wyoming.

var. *turbinata* (M. E. Jones) G. L. Nesom & G. I. Baird (shaped like a spinning top, reference unknown) Dune rabbitbrush. **STEMS** to 0.6 m tall, yellowish green, compactly tomentose, moderately leafy at flowering. **LEAVES** grayish green, surfaces tomentulose to densely tomentose, blades linear, 10–20 mm long, 0.5–1.5 mm wide, 1-nerved. **INVOLUCRES** 11.2–16 mm long. **PHYLLARIES** 16–25, apices acute to short-acuminate, erect, lower abaxially tomentose or glabrous, inner usually glabrous. **COROLLAS** 9.8–11.8 mm long, tubes puberulent or glabrous, lobes 0.7–0.9 mm long, villous; style appendages longer than stigmatic portion. **CYPSELAE** glabrous to densely pubescent. **PAPPUS** 9.5–12 mm long. $2n = 18$. [*Bigelowia turbinata* M. E. Jones; *Chrysothamnus nauseosus* (Pall. ex Pursh) Britton var. *turbinatus* (M. E. Jones) S. F. Blake; *Ericameria nauseosa* var. *turbinata* (M. E. Jones) G. L. Nesom & G. I. Baird]. Dunes and deep sands, saltbush, ephedra, juniper, and greasewood communities. ARIZ: Apa, Coc, Nav; NMEX: SJn; UTAH. 1300–1700 m (4265–5575′). Flowering: Aug–Oct. Arizona, New Mexico, Nevada, Utah.

Ericameria parryi (A. Gray) G. L. Nesom & G. I. Baird (for Charles C. Parry, 1823–1890, botanist and explorer) Parry's rabbitbrush. Shrubs to 1 m tall. **STEMS** ascending to erect, frequently fastigiately branched; densely tomentose, eglandular, greenish when young, becoming tan or gray with maturity. **LEAVES** sparse to crowded, usually ascending to spreading, planar to concave, blades linear to spatulate, 10–80 mm long, 0.5–8 mm or sometimes to 14 mm wide, glabrous to tomentose or viscid, sometimes glandular-dotted or with stipitate glands, apices acute, 1–3-nerved; axillary leaf fascicles absent. **INFL** solitary, or in racemiform or cymiform clusters, sometimes in paniculiform or thyrsiform arrays. **HEADS** discoid. **INVOLUCRES** subcylindric, 9–18 mm long, 4–8 mm wide, peduncles to 10+ mm long. **PHYLLARIES** 10–20, in 3–6 series, subequal, tan, ovate to lanceolate or elliptic, 5–11+ mm long, 0.7–2 mm wide, chartaceous or herbaceous, generally tomentose, apices acuminate to attenuate, sometimes herbaceous-tipped, may be resinous, midvein mostly evident. **DISC FLORETS** 5–20, corollas 8–12.5 mm long, tubes glabrous to sparsely pubescent, lobes 0.4–2.5 mm; style appendages subulate, longer than stigmatic portion. **CYPSELAE** ellipsoid to subturbinate, tan, 3–8 mm long, sericeous. **PAPPUS** whitish to brown, 3.3–7.5 mm. $2n = 18$. Widespread, often abundant, and variable. Arizona, California, Colorado, Idaho, Nebraska, Nevada, New Mexico, Oregon, Utah, Wyoming.

1. Uppermost leaves not exceeding heads ..... (2)
1′ Uppermost leaves exceeding heads ..... (3)

2. Phyllaries erect; corolla lobes 1.5–2 mm long, tubes glabrous; involucres 11–12.5 mm long ..... **var. *attenuata***
2′ Phyllaries recurved; corolla lobes 1–1.5 mm long, tubes glabrous or sparsely pubescent; involucres 11–15 mm long ..... **var. *nevadensis*** (in part)

3. Phyllary apices acuminate; leaf surfaces eglandular ..... **var. *howardii***
3′ Phyllary apices attenuate; leaf surfaces glandular ..... (4)

4. Phyllaries 24–28; florets 4–6 per head; leaf surfaces densely tomentose ..... **var. *nevadensis*** (in part)
4′ Phyllaries 10–15; florets 8–20 per head; leaf surfaces glabrous or puberulent ..... **var. *parryi***

var. *attenuata* (M. E. Jones) G. L. Nesom & G. I. Baird (narrowed to a point, referring to the bracts) Narrow-bract rabbitbrush. **STEMS** to 0.6 m tall. **LEAVES** green, surfaces glabrous or sparsely pubescent, eglandular, blades linear, 20–40 mm long, mostly 1 mm wide, 1-nerved, somewhat viscid, uppermost not exceeding the heads. **HEADS** 5–10, in compact, racemiform clusters. **INVOLUCRES** 11–12.5 mm long. **PHYLLARIES** 13–22, apices attenuate, erect. **FLORETS** 5–7. **COROLLAS** clear yellow, 10–11 mm long, tubes glabrous, throats gradually flaring, lobes 1.5–2 mm long. $2n = 18$. [*Bigelowia howardii* (Parry ex A. Gray) A. Gray var. *attenuata* M. E. Jones; *Chrysothamnus parryi* (A. Gray) Greene subsp. *attenuatus* (M. E. Jones) H. M. Hall & Clem.]. Mountain slopes, dry, stony soils; sagebrush, piñon-juniper, ponderosa pine, and aspen communities. ARIZ: Apa, Nav; COLO: Min, Mon; NMEX: McK, RAr, SJn; UTAH. 1700–3000 m (5575–9840′). Flowering: Aug–Oct. Arizona, Colorado, Idaho, New Mexico, Utah.

var. *howardii* (Parry ex A. Gray) G. L. Nesom & G. I. Baird (probably for Winslow J. Howard, New Mexico botanist and collector in the Rocky Mountains) Howard's rabbitbrush. **STEMS** to 0.6 m tall. **LEAVES** gray, surfaces tomentose, eglandular, blades linear, 20–40 mm long, about 1 mm wide, 1-nerved, uppermost exceeding heads. **HEADS** usually 5–10, sometimes more, in compact racemiform clusters or terminal glomerules. **INVOLUCRES** 9.5–12 mm long. **PHYLLARIES** 12–20, apices acuminate, spreading, chartaceous. **FLORETS** 5–7. **COROLLAS** pale yellow, 8–10.7 mm long, tubes pubescent, throats gradually flaring, lobes 1.2–1.7 mm. $2n = 18$. [*Linosyris howardii* Parry ex A. Gray; *Chrysothamnus parryi* (A. Gray) Greene subsp. *howardii* (Parry ex A. Gray) H. M. Hall & Clem.]. Upland slopes and tablelands; sagebrush, rabbitbrush, juniper, piñon-juniper, ponderosa pine, and Douglas-fir communities. ARIZ: Apa; NMEX: McK, RAr, San, SJn. 1900–2900 m (6235–9515′). Flowering: Aug–Oct. Arizona, Colorado, Idaho, Nebraska, New Mexico, South Dakota, Utah, Wyoming.

var. ***nevadensis*** (A. Gray) G. L. Nesom & G. I. Baird (of Nevada) Nevada rabbitbrush. **STEMS** to 0.6 m tall. **LEAVES** green to gray, surfaces densely tomentose, glandular, resinous, blades linear, 15–40 mm long, 0.5–3 mm wide, 1-nerved, uppermost seldom exceeding heads. **HEADS** usually 5–10, sometimes more, in racemiform to narrowly paniculiform clusters. **INVOLUCRES** 11–15 mm long. **PHYLLARIES** 24–28, apices attenuate, tips tan or greenish and slightly recurved. **FLORETS** 4–6. **COROLLAS** clear yellow, 9–10.5 mm long, tubes glabrous to sparsely pubescent, throats gradually flaring, lobes 1–1.5 mm long. $2n = 18$. [*Linosyris howardii* Parry ex A. Gray var. *nevadensis* A. Gray; *Chrysothamnus parryi* (A. Gray) Greene subsp. *nevadensis* (A. Gray) H. M. Hall & Clem.]. Big sagebrush, juniper, piñon-juniper, mountain brush, and ponderosa pine communities. ARIZ: Apa, Coc, Nav; UTAH. 1850–2550 m (6070–8365′). Flowering: Aug–Oct. Arizona, California, Nevada, Oregon, Utah.

var. *parryi* Parry's rabbitbrush. **STEMS** to 1 m tall. **LEAVES** green, surfaces glabrous or puberulent, glandular, somewhat resinous, blades linear, 30–80 mm long, 2–3 mm wide, 3-nerved, midvein conspicuous, uppermost exceeding heads. **HEADS** usually 5–12, sometimes more, in racemiform clusters. **INVOLUCRES** 9–12 mm long. **PHYLLARIES** 10–15, apices attenuate, erect, mostly chartaceous, outer sometimes herbaceous-tipped. **FLORETS** 8–20. **COROLLAS** yellow, 7.9–10 mm long, tubes sparsely pubescent, throats gradually flaring, lobes 1.4–2.1 mm long. $2n = 18$. Open dry hillsides and plains; ponderosa pine, Douglas-fir, and spruce-fir communities. COLO: Dol, LPl, Min, Mon; NMEX: RAr, SJn. 2000–2900 m (6560–9515′). Flowering: Aug–Oct. Colorado, Nevada, New Mexico, Utah, Wyoming.

Erigeron L. Fleabane

Guy Nesom

(Greek *eri*, early, or *erio*, woolly, + *geron*, old man, perhaps alluding to the pappus, which becomes gray and accrescent in some species, or to the solitary woolly heads of some of the alpine species) Herbs (less frequently subshrubs, shrubs, and trees), annual, biennial, or perennial, glabrous or hairy, commonly glandular; taprooted, fibrous-rooted, or rhizomatous and fibrous-rooted, sometimes stoloniferous. **LEAVES** basal and/or cauline, alternate, sessile to petiolate. **HEADS** usually radiate, sometimes discoid or disciform, borne singly or in a loose, corymbiform or paniculiform array. **INVOLUCRES** turbinate to hemispheric. **PHYLLARIES** in 2–5 subequal series. **RECEPTACLES** flat to conic, shallowly alveolate, epaleate. **RAY FLORETS** pistillate, fertile, in 1 (–2+) series, sometimes 0, corollas usually white to bluish or purplish to pink, less commonly yellow or the laminae absent. **DISC FLORETS** bisexual, fertile. **CYPSELAE** oblong to oblong-obovoid, compressed to flattened, ribs 2 (–4), or subterete, ribs 5–14, glabrous or strigose or sericeous, eglandular. **PAPPUS** of persistent or deciduous barbellate bristles, usually with outer setae or scales, sometimes only on ray or only on disc cypselae, sometimes absent. $x = 9$. [*Achaetogeron* A. Gray; *Trimorpha* Cass.]. About 390 species, widespread, mostly in temperate regions.

1. Heads disciform (ray/pistillate florets present, but corollas tubular or laminae absent or shorter than the involucres) ....(2)

1′ Heads radiate (corollas of pistillate florets bearing laminae, laminae sometimes filiform and hardly longer than the involucres) ....(3)

2. Leaves ternately dissected or lobed .... ***E. compositus***
2′ Leaves linear-oblanceolate to spatulate, entire .... ***E. aphanactis***

3. Plants with leafy runners or slender scale-leaved rhizomes .... (4)
3′ Plants without leafy runners or scale-leaved rhizomes .... (6)

4. Cauline leaf bases clasping to subclasping; stems minutely glandular to eglandular, sometimes sparsely hirsute-pilose proximal to heads, otherwise glabrous; phyllaries minutely glandular, otherwise glabrous or sometimes sparsely villous at bases .... ***E. eximius***
4′ Cauline leaf bases not clasping .... (5)

5. Stems evenly hirsutulous with spreading-deflexed hairs; stoloniform branches usually without terminal plantlets .... ***E. tracyi***
5′ Stems strigose, often sparsely so, with antrorsely appressed to closely ascending hairs; stoloniform branches usually with terminal plantlets .... ***E. flagellaris***

6. Leaves ternately or pinnately lobed or dissected .... (7)
6′ Leaves mostly entire or dentate, rarely with 1–2 pairs of coarse lobes .... (10)

7. Annuals .... (8)
7′ Perennials .... (9)

8. Stems densely and evenly puberulous-hirsutulous, hairs spreading to spreading-descending or -ascending, often crinkly, minutely glandular at least near heads, usually over whole stem; involucres 3–4 × (5–) 7–11 mm; ray florets 75–150 .... ***E. divergens***
8′ Stems usually sparsely strigose, sometimes sparsely hirsutulous (then only distally), sometimes minutely glandular near heads; involucres 2–3.5 × 5.5–7.5 mm; ray florets 37–85 .... ***E. religiosus***

9. Leaves ternately lobed or dissected .... ***E. compositus***
9′ Leaves pinnately lobed or dissected .... ***E. pinnatisectus***

10. Annuals, biennials, or short-lived perennials .... (11)
10′ Perennials .... (20)

11. Ray corollas with laminae filiform, erect .... (12)
11′ Ray corollas strap-shaped, usually spreading .... (13)

12. Heads usually in loosely racemiform arrays; pistillate florets in 1 series, all with laminae filiform .... ***E. lonchophyllus***
12′ Heads usually in corymbiform arrays; pistillate florets in 2 zones, outer with laminae nearly filiform, inner tubular and essentially elaminate .... ***E. nivalis***

13. Plants fibrous-rooted .... (14)
13′ Plants taprooted .... (16)

14. Cauline leaves clasping to auriculate-clasping, basal (15–) 30–110 (–150) × 10–25 (–40) mm, margins shallowly crenate to coarsely serrate or pinnately lobed; heads 1 (earliest season) or usually 3–50 (–100+); ray florets 125–250 .... ***E. philadelphicus***
14′ Cauline leaves not clasping, basal 10–55 × 3–9 (–12) mm, margins mostly entire or dentate; heads mostly 1, sometimes to 3; ray florets 40–130 .... (15)

15. Stems hirsutulous, hairs evenly spreading-deflexed .... ***E. tracyi***
15′ Stems strigose, hairs antrorsely appressed .... ***E. flagellaris***

16. Stems strigose from bases to tips .... (17)
16′ Stems hispid-pilose, hirsute, hirsutulous (hairs upcurved), or puberulous-hirsutulous .... (18)

17. Plants 3–15 cm; stems erect, usually 1 and simple; heads usually 1 .... ***E. flagellaris***
17′ Plants 6–30 (–40) cm; stems decumbent-ascending, often multiple from bases, commonly branching from midstem; heads initially 1, sometimes more from axillary branches .... ***E. religiosus***

18. Stems densely hirsutulous with evenly spreading-deflexed hairs; heads 1 (–3) .... ***E. tracyi***
18′ Stems hirsutulous, hirsute, puberulous-hirsutulous, or spreading-hairy, hairs not all spreading-deflexed and of relatively even lengths; heads usually (1–) 5–100 .... (19)

19. Stems hirsutulous with upcurved hairs, eglandular; outer pappus a cartilaginous crown, inner of bristles; some ray florets positioned among inner phyllaries .... ***E. bellidiastrum***
19′ Stems puberulous-hirsutulous with spreading, spreading-descending or -ascending, sometimes crinkly hairs, usually minutely glandular; outer pappus of setae or scales, inner of bristles; all ray florets completely interior to inner phyllaries .... ***E. divergens***

20. Plants fibrous-rooted, usually rhizomatous, or without an evident taproot (taproot weakly developed or not collected because of extensive rhizome or caudex system) .... (21)
20′ Plants usually with an evident taproot, sometimes also with branched caudices .... (36)

21. Rhizomes or caudex branches slender, without apparent well-defined central axes .... (22)
21′ Rhizomes thickened, usually evident as single central axes .... (25)

22. Leaves all or mostly cauline (basal relatively small or withering before flowering); cypselae 5–6-nerved, glabrous .... ***E. rhizomatus***
22′ Leaves basal or basal and cauline; cypselae 2-nerved, sparsely strigose .... (23)

23. Leaves all basal, basal leaves 5–12 mm long; ray florets 10–20 .... ***E. scopulinus***
23′ Leaves basal and cauline, basal leaves 15–120 mm long; ray florets 15–100 .... (24)

24. Taprooted (often not evident or not collected); basal leaves oblanceolate to obovate or spatulate; ray laminae reflexing .... ***E. leiomerus***
24′ Rhizomatous; basal leaves narrowly oblanceolate to oblong; ray laminae not reflexing or coiling or sometimes tardily coiling .... ***E. ursinus***

25. Plants 2–25 (–35) cm; leaves mostly basal, bases of cauline not clasping or subclasping; heads 1–3 .... (26)
25′ Plants (5–) 15–90 cm; leaves basal and cauline or mostly cauline, bases of cauline usually clasping to subclasping; heads 1–21 .... (28)

26. Hairs on phyllaries with clear to reddish or reddish purple cross walls; stems often stipitate-glandular over all or part .... ***E. grandiflorus***
26′ Hairs of phyllaries with blackish purple cross walls; stems minutely glandular near heads .... (27)

27. Basal leaves oblanceolate; heads barely protruding past basal rosette; involucres densely villous; phyllaries purple; ray corollas 4–6 mm, laminae erect, not coiling or reflexing; habitats at 3800–4150 m (12465–13615′) .... ***E. humilis***
27′ Basal leaves spatulate; heads held well above leaves; involucres sparsely villous-sericeous; phyllaries green; ray corollas 7–11 mm, laminae spreading, tardily coiling; habitats at (2700–) 3400–3800 m (8860–12465′) .... ***E. melanocephalus***

28. Phyllaries villous-lanate to hirsute-villous, at least basal part of hairs with black or reddish to purple cross walls. .... (29)
28′ Phyllaries glandular or variously hairy, hairs without distinctly colored cross walls .... (30)

29. Phyllaries hirsute-villous, hairs with black cross walls .... ***E. coulteri***
29′ Phyllaries densely villous-lanate, hairs with reddish to purple cross walls .... ***E. elatior***

30. Phyllaries eglandular or glands essentially obscured by relatively dense nonglandular hairs .... ***E. formosissimus***
30′ Phyllaries glandular, if also sparsely to moderately hirsute to villous, then nonglandular hairs not obscuring glandularity .... (31)

31. Primary rhizomes slender, producing scale-leaved runners or stoloniform rhizomes. .... ***E. eximius***
31′ Primary rhizomes relatively thick, sometimes ligneous, fibrous-rooted, without scale-leaved runners or stoloniform rhizomes .... (32)

32. Distal portion of stems glabrous or strigillose with crinkly hairs, eglandular or nearly so; phyllaries densely glandular, otherwise glabrous or nonglandular hairs few .... (33)
32′ Stems hirsute distally or distally and proximally, glandular or eglandular, phyllaries glandular, sometimes also hirsute to villous .... (34)

33. Stems glabrous at least on distal portions; ray corollas about 1 mm wide; plants 30–80 (–100) cm; ray florets 75–150 .... ***E. speciosus***

33′ Stems strigillose from base to apex, more densely so distally; ray corollas 1.5–3 mm wide; plants 5–55 (–70) cm; ray florets 30–80 .... ***E. glacialis***

34. Cauline leaves gradually becoming smaller distally; stems usually ascending at base .... ***E. formosissimus***
34′ Cauline leaves nearly even-sized or sometimes largest at midstem; stems usually erect from the base .... (35)

35. Stems moderately to densely hirsute, eglandular; leaves evenly hirsute to strigose-hirsute on laminae and veins, usually eglandular .... ***E. subtrinervis***
35′ Stems glabrous to sparsely hirsute-pilose, often minutely glandular distally; leaves glabrous or the distal sparsely minutely glandular .... ***E. speciosus***

36. Short-lived perennials; caudices usually simple; stems and leaves arising from near roots .... (37)
36′ Perennials; caudices usually branched .... (38)

37. Stems densely and evenly puberulous-hirsutulous, hairs spreading to spreading-descending or -ascending, often crinkly, minutely glandular at least near heads, usually over whole stem; involucres 3–4 × (5–) 7–11 mm; ray florets 75–150 .... ***E. divergens***
37′ Stems usually sparsely strigose, sometimes sparsely hirsutulous (then only distally), sometimes minutely glandular near heads; involucres 2–3.5 × 5.5–7.5 mm; ray florets 37–85 .... ***E. religiosus***

38. Petioles prominently ciliate with thick-based, spreading hairs; leaves basal and cauline or sometimes mostly basal, mostly linear to narrowly oblanceolate .... (39)
38′ Petioles not prominently spreading-ciliate (or if so, hairs thin-based or ascending to loosely appressed); leaves linear to obovate or spatulate .... (41)

39. Disc corolla throats not indurate or inflated; midvein region of phyllaries greenish; inner pappus of 18–25 bristles; heads 1; stems and leaves usually sparsely hirsute or almost without nonglandular hairs, usually minutely glandular .... ***E. vetensis***
39′ Disc corolla throats indurate and inflated; midvein region of phyllaries orange to yellowish; inner pappus of (7–) 10–14 (–15) bristles; heads 1–5, rarely more; stems and leaves moderately to densely hispid to hirsute, minutely glandular .... (40)

40. Disc corollas slightly puberulent with glandular-viscid, blunt-tipped hairs; outer pappus of setae 0.1–0.3 mm, inner of 15–27 bristles .... ***E. pumilus***
40′ Disc corollas hirsute-strigose with sharp-pointed hairs; outer pappus of scales 0.2–0.5 mm, inner of (7–) 10–14 (–15) bristles .... ***E. concinnus***

41. Stems and leaves glabrous or glabrate; phyllaries minutely glandular, rarely also sparsely strigose or hirsute-villous (sometimes glabrous), commonly purplish; ray corolla laminae reflexing .... (42)
41′ Stems and/or leaves strigose or sericeous to hirsute or villous; phyllaries strigose to hirsute, sometimes glandular, rarely purplish; ray corolla laminae coiling or straight and spreading (reflexing in *E. canus*) .... (44)

42. Leaves all basal, 0.5–1.2 cm × 1–3.5 mm; stems essentially scapose; phyllaries in 3–4 (–5) series .... ***E. scopulinus***
42′ Leaves basal and reduced cauline, 1.3–7 cm × 2–13 (–15) mm; phyllaries in 2–3 series .... (43)

43. Caudex branches variable, short to relatively long and slender; taproots often not evident; heads 1–4; involucres 3.2–4 × 5–6 mm; ray florets 10–15 .... ***E. kachinensis***
43′ Caudex branches relatively long and thick to slender and lignescent; taproots usually evident; heads 1; involucres 4–6 × 7–13 mm; ray florets 15–60 .... ***E. leiomerus***

44. Caudices diffuse with extensive system of relatively long, slender, rhizomelike branches .... ***E. vagus***
44′ Caudices compact and simple or short-branched .... (45)

45. Stems basally ascending to decumbent-ascending, sometimes purplish at bases; stems and leaves strigose to hirtellous .... (46)
45′ Stems erect from the base, usually not purplish at bases; stems and leaves strigose to sericeous .... (48)

46. Stems greenish proximally; basal leaves narrowly obovate to oblanceolate, usually apically rounded to obtuse, hirtellous to loosely strigose; cauline leaves densely arranged; ray corollas reflexing at maturity .... ***E. abajoensis***

46′ Stems often purplish proximally; basal leaves linear to narrowly lanceolate or oblanceolate, apically sharply acute, sparsely, closely strigose to nearly glabrous; cauline leaves relatively few; ray corollas relatively straight at maturity, not reflexing ....(47)

47. Stems 5–11 (–14) cm; basal leaves linear to narrowly linear-oblanceolate, 3–6 cm × 1–2 (–3) mm, 1-nerved. ***E. canaani***
47′ Stems (10–) 12–32 cm; basal leaves narrowly lanceolate to oblanceolate, 7–12 cm × 2–8 mm, 3-nerved ... ***E. eatonii***

48. Leaves mostly spatulate (to be expected) ....*E. acomanus* Spellenb. & P. J. Knight
48′ Leaves mostly linear to oblanceolate........(49)

49. Leaves basal and cauline, the basal mostly withered or deciduous before flowering, not forming a conspicuous tuft........(50)
49′ Leaves mostly basal or basal and cauline, the basal usually persistent in a dense tuft .... (51)

50. Cauline leaves reduced to linear bracts relatively even-sized beyond midstem and continuing to immediately proximal to heads; heads (1–) 3–10; involucres 3–5 × 5–8 mm; ray florets 10–14 (–20), corollas 4–8 mm; disc corollas viscid-puberulent (hairs multicellular, blunt); flowering Jun–Sep........***E. sparsifolius***
50′ Cauline leaves gradually reduced distally, ending well proximal to heads; heads 1–3 (–5); involucres 5–7 × 7–12 (–15 mm); ray florets 28–40, corollas 10–18 (–20) mm; disc corollas sparsely strigose-villous (hairs needlelike); flowering mid-Apr–Jun (–Jul)........***E. utahensis***

51. Basal leaves narrowly oblanceolate........(52)
51′ Basal leaves linear ....(55)

52. Basal leaves linear-oblanceolate to oblanceolate, 2–5 (–7) mm wide, ray corolla laminae reflexing at maturity, sometimes tardily, not coiling; cypselae (8–) 10–14-nerved, glabrous........***E. canus***
52′ Basal leaves spatulate to oblanceolate or narrowly oblanceolate, 1–6 mm wide; ray corolla laminae coiling at maturity; cypselae (2–) 4–8-nerved, strigose to strigose-pilose ....(53)

53. Heads 1–3 (–5), stems branched at middle or above; basal leaves mostly absent by flowering, sometimes persistent but not forming a prominent tuft; involucres 8–11 mm wide; achenes mostly 6–8-nerved ....***E. utahensis***
53′ Heads 1, stems unbranched; basal leaves persistent in a tuft; involucres 10–16 (–20) mm wide; achenes (2–) 4 (–6)-nerved ....(54)

54. Basal leaves mostly narrowly oblanceolate to oblanceolate; stems and leaves silvery ....***E. argentatus***
54′ Basal leaves linear to narrowly oblanceolate; stems and leaves gray-green........***E. pulcherrimus***

55. Cypselae (2–) 4 (–5)-nerved, faces and margins densely strigose-sericeous; pappus bristles 32–50. ***E. pulcherrimus***
55′ Cypselae 2 (–3)-nerved, faces glabrous, margins ciliate to nearly glabrous; pappus bristles 15–27 (–35)........(56)

56. Basal leaves 4–20 mm long; cauline leaves 0 or restricted to proximal 1/4 (–1/2) of stems; stems, leaves, and phyllaries gray-green, densely strigose ....***E. consimilis***
56′ Basal leaves 12–30 mm long; cauline leaves often little reduced on at least proximal 1/2 of stem; stems, leaves, and phyllaries greenish, sparsely strigose........***E. sivinskii***

Erigeron abajoensis Cronquist **CP** (from the Abajo Mountains) Abajo fleabane. Perennial 5–15 (–24) cm tall; taprooted, caudices with relatively thick and short branches, retaining old leaf bases. **STEMS** ascending to decumbent, usually loosely strigose with ascending hairs, uncommonly hirsute or hirtellous with spreading to deflexed hairs, eglandular. **LEAVES** basal and cauline; basal persistent, narrowly obovate to oblanceolate, (15–) 30–70 × 2–6 (–8) mm, mostly 1-nerved, entire; cauline linear to linear-oblong or narrowly lanceolate-oblong, gradually reduced distally; faces strigose, eglandular. **HEADS** 1–4. **INVOLUCRES** 3.7–5.2 × (5–) 10–11 mm. **PHYLLARIES** in 2–3 (–4) series, strigose to strigulose or finely hirsute-villous, minutely glandular. **RAY FLORETS** 35–60; corollas white to light purple, 5–6 mm, laminae reflexing. **CYPSELAE** 2-nerved, strigose. **PAPPUS**: outer of setae, inner of 12–20 bristles. [*E. awapensis* S. L. Welsh]. Rocky or gravelly slopes, talus, cliffs, crevices, sandstone outcrops, Gambel's oak, sagebrush, piñon-juniper, ponderosa pine. ARIZ: Apa; COLO: Mon; NMEX: SJn; UTAH. 2000–2950 (–3250) m (6560–10660′). Flowering: (Jun–) Jul–Aug. Arizona, New Mexico, Colorado, and Utah. *Erigeron abajoensis* is distinguished from *E. caespitosus* Nutt. by its tendency for strigose (versus hirsute) stems and leaves and for 1-nerved leaves. Orientation of cauline vestiture, however, varies in both taxa. Other mostly consistent differences from *E. caespitosus* are the reflexing rays of

E. abajoensis and the densely and minutely glandular (without other hairs) phyllary apices. The Four Corners population system, including a few southward extensions, is identified here as *E. abajoensis*, with the caveat that the zone in central Utah where the ranges of *E. caespitosus* and *E. abajoensis* apparently meet, and from where *E. awapensis* has been described, needs to be studied.

Erigeron aphanactis (A. Gray) Greene (obscure or unseen rays) Perennial, 2–20 (–30) cm tall; taprooted, caudices branched. **STEMS** erect to slightly decumbent-ascending, canescent-hirsute, densely stipitate-glandular, leafy. **LEAVES** basal (persistent) and cauline; basal linear-oblanceolate to spatulate, 20–80 × 2–7 mm, petioles prominently ciliate, at least on proximal portions, hairs thick-based, spreading; cauline gradually or quickly reduced distally, sometimes bractlike (stems scapiform), margins entire, 1-nerved, canescent-hirsute, densely stipitate-glandular. **HEADS** 2–3, disciform. **INVOLUCRES** 4–6 × 7–15 mm. **PHYLLARIES** in 2–3 series, coarsely hirsute, densely minutely glandular. **RAY (PISTILLATE) FLORETS** 30–45; corollas usually tubular, lacking laminae, or laminae shorter than involucres. **DISC COROLLAS** white-indurate and inflated, conspicuously puberulent. **CYPSELAE** 2-nerved, sparsely strigose (carpopodia whitish). **PAPPUS**: outer of scales or setae, inner of 7–20 bristles. $2n = 18$. [*E. concinnus* (Hook. & Arn.) Torr. & A. Gray var. *aphanactis* A. Gray]. Clay, sandy, or rocky soil, talus, washes, riparian, desert shrub, sagebrush, juniper, piñon-juniper, ponderosa pine. ARIZ: Apa; COLO: Dol, Mon, SMg; NMEX: SJn; UTAH. 1600–2350 m (5250–7710′). Flowering: (Apr–) May–Sep. Oregon, California, and Arizona, east to Colorado and New Mexico. Four Corners plants are **var. *aphanactis***. *Erigeron aphanactis* is similiar in habit and vestiture to *E. concinnus*, which has conspicuous rays, usually white to pinkish. The two occur sympatrically without hybrids over a significant part of their ranges.

Erigeron argentatus A. Gray (silvery) Silver fleabane. Perennials, (8–) 15–30 (–40) cm tall; taprooted, caudices branched, retaining old leaf bases. **STEMS** erect, densely gray-green to silvery strigose, hairs white, closely appressed, eglandular. **LEAVES** basal and cauline; basal persistent, spatulate to oblanceolate or narrowly oblanceolate, (15–) 20–50 (–70) × 1–4 (–6) mm, margins entire, 1-nerved, faces silvery strigose, eglandular; cauline linear to linear-oblanceolate, much reduced distally. **HEADS** 1. **INVOLUCRES** 5.5–9 × (10–) 12–22 mm. **PHYLLARIES** in 3–4 series, silvery strigose with closely appressed hairs, minutely glandular. **RAY FLORETS** 20–50 (–75); corollas blue to lavender, less commonly pink or white, 9–15 mm, laminae coiling. **CYPSELAE** 6–8-nerved, strigose-pilose. **PAPPUS**: outer of setae or lanceolate scales, inner of 25–40 bristles. $2n = 18$. Ridges and slopes in dry, sandy or gravelly soil, desert shrub, sagebrush, juniper, and piñon-juniper. ARIZ: Apa; COLO: Mon; NMEX: SJn; UTAH. 1750–1900 m (5740–6235′). Flowering: May–Jul. California to Nevada, Arizona, Utah, and Colorado.

Erigeron bellidiastrum Nutt. (beautiful daisy, *aster*, star or aster) Sand fleabane. Annual (or biennial?), 3.5–30 (–50) cm tall; taprooted, caudex simple. **STEMS** erect to ascending, hirsutulous with upcurved hairs, usually eglandular. **LEAVES** basal (sometimes persistent) and cauline or mostly cauline; linear to linear-oblanceolate, 10–15 (–30) mm, margins entire or rarely with 1 pair of shallow teeth, 1-nerved, faces sparsely strigose, eglandular. **HEADS** 1–12, from branches beyond midstem or sometimes clustered toward stem tips, usually in a diffuse array. **INVOLUCRES** 3–5 × 5–7 (–11) mm. **PHYLLARIES** in 2–3 (–4) series, hispidulous, minutely glandular. **RAY FLORETS** 22–70, some positioned among the inner phyllaries; corollas white, often with abaxial lilac midstripe, drying white to bluish, 4–7.5 mm, laminae not coiling or reflexing. **DISC COROLLA** throats indurate and inflated. **CYPSELAE** 2-nerved, sparsely strigose. **PAPPUS**: outer evidently reduced to a cartilaginous crown, inner of 15–18 bristles. $2n = 18, 36$. Dunes, loose sand, creek sides, washes, floodplains; desert shrub, sagebrush, juniper, piñon-juniper. ARIZ: Apa; COLO: Mon; NMEX: San, SJn; UTAH. 1350–1750 m. Flowering: May–Sep. Montana to South Dakota, south to Arizona and Texas; Mexico. Our material belongs to **var. *bellidiastrum***. *Erigeron bellidiastrum* is recognized by its annual duration, upcurved hairs of the stem, relatively few rays, 1-seriate pappus, and by some ray florets consistently produced between the phyllaries, the mature cypselae of these held in place as the phyllaries reflex at maturity. A cold infusion of dried leaves was used as a ceremonial chant lotion by the Ramah Navajo.

Erigeron canaani S. L. Welsh **CP** (of Canaan Mountain, Utah, the type locality) Canaan fleabane. Perennial (5–) 10–25 (–30) cm tall; taprooted, caudices simple or branched. **STEMS** ascending to decumbent or nearly prostrate, sparsely strigose to glabrate, eglandular. **LEAVES** basal (persistent, old leaf bases persistent, chaffy, fibrous) and cauline; basal linear to narrowly linear-oblanceolate, 50–150 × 1–2 (–3) mm, cauline gradually or abruptly reduced distally, margins entire, 1-nerved, faces glabrate to sparsely strigose, eglandular. **HEADS** 1 (–3). **INVOLUCRES** 5–7 × 10–13 mm. **PHYLLARIES** in 2–3 (–4) series, hirsute, minutely glandular. **RAY FLORETS** 15–22; corollas white to purplish, 3.5–6 mm, laminae not coiling or reflexing. **CYPSELAE** 2-nerved, sparsely strigose. **PAPPUS**: outer of setae, inner of about 20 bristles. Ponderosa pine, aspen, Douglas-fir, limber pine, spruce-fir-aspen, rocky soil, pine duff. UTAH. 2550–3100 m (8365–10170′). Flowering: May–Aug. Utah and Arizona.

Erigeron canus A. Gray (gray) Gray fleabane. Perennial, 5–35 cm; taprooted, caudices with relatively thick branches, usually retaining old leaf bases. **STEMS** erect, densely white strigose-canescent, eglandular. **LEAVES** mostly basal, persistent, ± erect, linear-oblanceolate to oblanceolate, 20–100 × 2–5 (–7) mm, sharply reduced in size on stems, margins entire, 1-nerved, faces densely white strigose-canescent, eglandular. **HEADS** 1 (–4). **INVOLUCRES** 5–7 × 9–16 mm. **PHYLLARIES** in 3–4 series, densely hirsute to strigose-hirsute, minutely glandular. **RAY FLORETS** 20–50 (–70); corollas white to light lavender, 7–12 mm, laminae reflexing, sometimes tardily. **CYPSELAE** (8–) 10–14-nerved, glabrous. **PAPPUS**: outer of setae or bristles, inner of 24–36 bristles. $2n = 18$. Dry hills and grasslands, washes, gravelly, silty, or shaley soil, sagebrush, piñon-juniper, Gambel's oak, ponderosa pine–oak, Douglas-fir. ARIZ: Apa; NMEX: McK, SJn. 1900–2800 m (6235–9185′). Flowering: May–Jul. Utah, Wyoming, South Dakota, south to Arizona, New Mexico, and Oklahoma; Mexico (Chihuahua). The Ramah Navajo used the plant in ceremonial chant lotion and for "deer infection."

Erigeron compositus Pursh **CP** (made up of parts, alluding to the leaves) Dwarf mountain fleabane. Perennial, 5–15 (–25 cm); taprooted, caudex branches usually relatively thick and short, rarely slender and rhizomelike. **STEMS** erect, sparsely hispid-pilose, minutely glandular. **LEAVES** mostly basal, persistent, spatulate to obovate-spatulate, 5–50 (–70) × (2–) 4–12 mm, cauline reduced, bractlike, margins (1–) 2–3 (–4)-ternately lobed or dissected, cauline mostly entire, 1-nerved, faces minutely glandular, densely hispidulous-puberulent to glabrate. **HEADS** 1. **INVOLUCRES** 5–10 × 8–20 mm. **PHYLLARIES** in 2–3 series, purple-tipped, hirsute with spreading hairs, minutely glandular. **RAY FLORETS** 20–60; corollas white to pink or blue, usually 6–12 mm, often reduced to tubes (heads then disciform), laminae not coiling or reflexing. **CYPSELAE** 2-nerved, sparsely strigose-hirsute. **PAPPUS**: outer usually of setae, sometimes 0, inner of 12–20 bristles. $2n = 18, 36, 45, 54$. [*E. compositus* var. *discoideus* A. Gray; *E. compositus* var. *glabratus* Macoun]. Talus, among boulders, riparian, aspen to aspen-fir, spruce, and fir, subalpine meadows, alpine tundra. COLO: Arc, Con, Hin, LPl, Min, Mon, SJn; UTAH. 2900–3850 m (9515–12630′). Flowering: May–Aug. Alaska to Quebec and Greenland, Washington to North Dakota, south to California, Arizona, and Colorado; eastern Asia (Russian Far East).

Erigeron concinnus (Hook. & Arn.) Torr. & A. Gray (well-arranged, striking) Navajo fleabane. Perennial, mostly 10–30 cm; taprooted, caudices simple or branched, branches sometimes rhizomelike. **STEMS** ascending to erect, sparsely to densely hispid-pilose to glabrate, minutely glandular. **LEAVES** usually basal and cauline (petioles prominently ciliate, hairs spreading, thick-based); blades narrowly oblanceolate to linear-oblong, 10–50 (–80) × 1–4 mm, margins entire, usually ciliate, 1-nerved, faces usually hirsute to hirsute-villous, sometimes substrigose to glabrate, eglandular; cauline unreduced or gradually reduced distally. **HEADS** (1–) 2–3 (–5). **INVOLUCRES** 4–7 × 7–12 (–15) mm. **PHYLLARIES** in 2–4 series, hirsute to hirsute-villous, ± minutely glandular. **RAY FLORETS** 50–100 (–125); corollas white to lavender, 6–15 mm, laminae reflexing. **DISC COROLLAS** with throats indurate and inflated, hirsute-strigose with biseriate, sharply pointed hairs. **CYPSELAE** 2-nerved, sparsely strigose-hirsute. **PAPPUS**: outer of scales (0.2–0.5 mm), inner of (7–) 10–14 (–15) bristles. $2n = 18$. [*E. pumilus* Nutt. var. *concinnoides* Cronquist; *E. pumilus* var. *concinnus* (Hook. & Arn.) Dorn]. *Erigeron concinnus* has been treated within *E. pumilus*; Nesom (1983) found that these taxa approach each other closely in geographic range without intergradation. Gravelly or rocky sites, sand, sandy clay, washes, flats, desert scrub (including blackbrush, *Atriplex*), sagebrush, piñon-juniper, Gambel's oak, piñon–mountain shrub. ARIZ: Apa, Coc, Nav; COLO: Dol, LPl, Mon, SMg; NMEX: RAr, SJn; UTAH. 1450–2450 m (4755–8040′). Flowering: Apr–Jun (–Jul). California to Idaho, Wyoming, Colorado, and New Mexico. Four Corners plants belong to **var. *concinnus***. Used by the Ramah Navajo as a cold infusion for general body pain and for "antelope infection."

Erigeron consimilis Cronquist (similar in all respects) San Rafael fleabane. Perennial, 3–10 cm; taprooted, caudices branched. **STEMS** erect, scapiform, finely strigose, eglandular. **LEAVES** mostly basal, persistent, linear, 4–20 × 0.7–1.5 mm, cauline bractlike or absent, margins entire, 1-nerved, faces densely and finely strigose. **HEADS** 1. **INVOLUCRES** 6–8.5 × 11–22 mm. **PHYLLARIES** in 2–3 (–4) series, hispid-hirsutulous with relatively thick-based hairs, inconspicuously glandular. **RAY FLORETS** 30–55; corollas white or pinkish, sometimes with an abaxial lilac midstripe or drying purplish, 7–11 mm, tubes and bases of laminae densely strigose-hirsute, laminae coiling, sometimes tardily. **CYPSELAE** 2-nerved, margins velutinous-ciliate, faces glabrous. **PAPPUS**: outer of setae, inner of 15–25 (–35) bristles. [*E. compactus* S. F. Blake var. *consimilis* (Cronquist) S. F. Blake]. Sandy and silty sites, shale outcrops, desert shrub, piñon-juniper. ARIZ: Apa, Nav; NMEX: SJn. 1700–2100 m (5575–6890′). Flowering: May–Jun. Utah, Arizona, and Colorado.

Erigeron coulteri Porter (for John M. Coulter, 1851–1928, author of *Manual of Rocky Mountain Botany*, first editor of *Botanical Gazette*) Coulter's fleabane. Perennial, 10–70 cm; rhizomatous, fibrous-rooted, with branched caudices and scale-leaved stolons. **STEMS** erect, sparsely hispid-villous, hairs usually with black cross walls, often glabrate, eglandular. **LEAVES** basal (persistent or not) and cauline; proximal broadly oblanceolate to elliptic or oblong-lanceolate, 40–120 (–150) × 7–25 mm, margins entire or with 1–5 pairs of shallow teeth, 1-nerved, faces sparsely strigose to strigose-villous; cauline gradually reduced distally and becoming elliptic-ovate to lanceolate, usually clasping. **HEADS** 1 (–4). **INVOLUCRES** 7–10 × 10–16 mm. **PHYLLARIES** in 2 (–3) series, hirsute-villous, hairs with black cross walls, minutely glandular. **RAY FLORETS** 45–140; corollas 9–25 mm, laminae coiling, white. **CYPSELAE** 2-nerved, sparsely strigose. **PAPPUS**: outer of setae, inner of 20–25 bristles. $2n = 18$. Moist coniferous woods, moist to wet meadows, open areas along creeks and lakes, rocky slopes, talus, poplar-willow, ponderosa pine–Douglas-fir, aspen, spruce-fir, spruce, alpine tundra. COLO: Arc, Con, Hin, LPl, Min, Mon, RGr, SJn; NMEX: SJn. 2400–3700 m (7875–12140′). Flowering: Jun–Sep. Oregon, Idaho, and Montana, south to California and New Mexico.

Erigeron divergens Torr. & A. Gray. (diverging, probably alluding to the branching pattern) Spreading fleabane. Annual or short-lived perennial, (7–) 12–40 (–70) cm; taprooted, caudex simple or less commonly branched. **STEMS** erect to ascending, single or multiple from base, densely and evenly puberulous-hirsutulous, hairs spreading to spreading-descending or -ascending, often crinkly, bases not thickened, minutely glandular at least near heads, usually over whole stem. **LEAVES** basal (usually deciduous) and cauline; basal obovate-spatulate, 10–70 × 4–14 mm, cauline gradually reduced distally, margins entire or with 2–3 pairs of teeth or lobes, 1-nerved, faces hirsute to loosely strigose-hirsute, sometimes sparsely glandular. **HEADS** 1+ in early season, usually 5–100+. **INVOLUCRES** 3–4 × (5–) 7–11 mm. **PHYLLARIES** in 3–4 series, hirsute, minutely glandular. **RAY FLORETS** 75–150; corollas white to light lavender, drying lilac, without abaxial midstripe, (2–) 4–9.5 mm, laminae not coiling or reflexing. **DISC COROLLAS** with throats indurate and slightly inflated. **CYPSELAE** 2-nerved, sparsely strigose. **PAPPUS**: outer of setae or scales, inner of 6–9 (–12) bristles. $2n = 18, 27, 36$. [*Erigeron accedens* Greene]. Gravelly or sandy flats, sand, roadsides and other disturbed sites, desert grasslands, riparian, desert shrub (including saltbush, blackbrush), sagebrush, piñon-junper, juniper-oak, Gambel's oak, ponderosa pine, and fir-aspen. ARIZ: Apa, Coc; COLO: Arc, Dol, Hin, LPl, Min, Mon, SJn, SMg; NMEX: McK, RAr, San, SJn; UTAH. (1150–) 1550–2600 m (3770–8530′). Flowering: Apr–Sep (–Oct). Canada (southern British Columbia and Alberta), Washington to Montana and North Dakota, south to California and Texas; Mexico. Polyploidy and agamospermy apparently are common in *E. divergens* and contribute to the variability and, probably to some extent, the polymorphism characteristic of this species. An infusion of the plant was taken by Navajo women as an aid for delivery. The Kayenta Navajo used the plant as a snuff for headaches. The Ramah Navajo used the plant ceremonially in several ways; as a lotion for "lightning infection"; as an eyewash; and the root was used as a "life medicine."

Erigeron eatonii A. Gray (for Daniel C. Eaton, 1834–1895, Yale botanist and collector with the King Expedition in Utah) Eaton's fleabane. Perennial, (10–) 12–32 cm; taprooted, caudices simple or branched. **STEMS** erect to ascending or decumbent, strigose, rarely hirtellous, sometimes minutely glandular. **LEAVES** basal (persistent) and cauline; basal linear to oblanceolate, 7–12 cm × 2–8 mm, cauline gradually reduced distally, margins entire, 3-nerved, faces loosely strigose to sparsely hirsute-villous, eglandular. **HEADS** 1–2 (–7), held well beyond basal leaves. **INVOLUCRES** 4.5–6 × 10–13 (–16) mm. **PHYLLARIES** in 2–3 (–4) series, sparsely villous, minutely glandular. **RAY FLORETS** 16–42; corollas white or pink to bluish or purple, 5–8 (–9) mm, laminae not coiling or reflexing. **CYPSELAE** 2-nerved, sparsely strigose. **PAPPUS**: outer of setae, inner of 16–20 bristles. $2n = 18$. Piñon-juniper, Gambel's oak, ponderosa pine, Douglas-fir, aspen, riparian. ARIZ: Apa, Nav; COLO: LPl, Mon; NMEX: SJn; UTAH. 1800–2750 m (5905–9020′). Flowering: May–Jul (–Aug). Idaho and Montana south to Nevada, Arizona, Colorado, and New Mexico. Four Corners plants belong to **var. *eatonii***.

Erigeron elatior (A. Gray) Greene (tall or high, alluding to the habit) Tall fleabane. Perennial, 20–60 cm; fibrous-rooted, from ligneous, relatively thick rhizomes. **STEMS** erect, sparsely to moderately villous, minutely glandular. **LEAVES** mostly cauline, basal not persistent; proximal obovate, becoming ovate to ovate-lanceolate distally, (20–) 30–90 × 5–25 mm, nearly equal-sized distally and evenly distributed along stem, clasping, margins entire, 1-nerved, faces sparsely to moderately hirsute-villous. **HEADS** 1 (–6). **INVOLUCRES** 7–12 × 13–20 mm. **PHYLLARIES** in 2–4 series, apices linear, commonly spreading-reflexing, densely villous-lanate, hairs with reddish to purple cross walls, minutely glandular. **RAY FLORETS** 75–150; corollas pink to rose-purple, 10–20 mm, laminae coiling. **CYPSELAE** 2-nerved, sparsely strigose. **PAPPUS**: outer of setae or setiform scales, inner of 15–20 bristles. $2n = 18$. Mountain brush, alpine and subalpine meadows, openings in spruce-fir and aspen, talus. COLO: Arc, Dol, Hin, LPl, Min, Mon, SJn. 2600–3450 m (8530–11320′). Flowering: Jul–Sep. Utah, Wyoming, Colorado, and New Mexico.

Erigeron eximius Greene (exceptional, uncommon, allusion unknown) Spruce-fir fleabane. Perennial, 15–60 cm; fibrous-rooted, caudices usually simple, primary rhizomes slender, simple or branched, usually producing slender, scale-leaved stolons bearing terminal tufts of leaves. **STEMS** erect to ascending, usually glabrous, usually sparsely hirsute-villous immediately beneath heads, densely minutely glandular to nearly eglandular proximally. **LEAVES** basal and cauline; basal persistent, spatulate to elliptic-spatulate to oblanceolate-obovate, 30–150 × 9–35 mm, cauline gradually reduced distally, usually clasping to subclasping, margins entire or serrulate to mucronulate with 3–5 pairs of teeth, 1-nerved, faces glabrous, distal often glandular. **HEADS** 1–5 (–15). **INVOLUCRES** 7–9 × 11–19 mm. **PHYLLARIES** in 3–4 series, minutely glandular, otherwise glabrous except sparsely hirsute-pilose at very base. **RAY FLORETS** 40–80; corollas white to bluish or lavender, 12–20 mm, laminae coiling. **CYPSELAE** 2 (–4)-nerved, sparsely strigose. **PAPPUS**: outer of setae, inner of 20–30 bristles. $2n$ = 18, 36. [*E. superbus* Greene ex Rydb.]. Margins and openings, ponderosa pine, Douglas-fir, aspen, maple-fir, spruce-fir, subalpine meadows. COLO: Arc, Hin, LPl, Min, Mon; NMEX: RAr, SJn. 2500–3350 m (8200–10990′). Flowering: Jul–Oct. Utah and Wyoming, south to Arizona and Texas. *Erigeron eximius* is characterized by a simple caudex producing scale-leaved stolons, clasping, widely spaced cauline leaves tending toward a spatulate shape, and lack of nonglandular hairs. The Ramah Navajo used the plant ceremonially as a lotion for various ills such as a coughing, fever, protection in warfare or hunting, influenza, and for protection from witches.

Erigeron flagellaris A. Gray (alluding to the long, slender, flagella-like leafy runners) Trailing fleabane. Biennial or short-lived perennial, 3–15 cm tall; usually fibrous-rooted, sometimes taprooted, caudices lignescent, rarely branched. **STEMS** first erect, then producing herbaceous, leafy, prostrate runners, usually with rooting plantlets at tips, sparsely strigose with antrorsely appressed hairs consistent in orientation, sometimes slightly glandular near heads. **LEAVES** basal (often persistent) and cauline; basal broadly oblanceolate to elliptic, 20–55 × 3–9 mm, cauline quickly reduced distally, margins entire or dentate, 1-nerved, faces strigose, eglandular. **HEADS** 1 (–3, on proximal branches). **INVOLUCRES** 3–5 × 6–13 mm. **PHYLLARIES** in 2–3 series, strigose to loosely hirsute, minutely glandular. **RAY FLORETS** 40–125; corollas white, often with an abaxial midstripe, often drying lilac, 4–10 mm, laminae not coiling or reflexing. **CYPSELAE** 2-nerved, sparsely strigose. **PAPPUS**: outer of setae, inner of 10–17 bristles. $2n$ = 18, 27, 36, 45, 54. [*E. nudiflorus* Buckley]. Streamsides, meadows, and grassy slopes, often moist, open areas in grasslands, hanging gardens, outcrops, riparian, sagebrush, piñon pine, ponderosa pine–Gambel's oak, aspen, spruce-fir, spruce, bristlecone pine, subalpine meadows. ARIZ: Apa, Nav; COLO: Arc, Dol, Hin, LPl, Min, Mon, SJn, SMg; NMEX: McK, RAr, SJn; UTAH. 1750–3200 m (5740–10500′). Flowering: May–Aug (–Sep). Southern British Columbia and Alberta south to Arizona, New Mexico, and Texas; Mexico. Early-season forms of *E. flagellaris* may consist of a basal rosette and a single, erect, scapiform, monocephalous stem; leafy runners usually develop quickly. The Ramah Navajo used the leaves ceremonially as a medicine and as a fumigant. A poultice of chewed leaves was applied to spider bites, used for "lightning infection"; poultice of plant was applied to snakebite, and infusion of leaves was used as eyewash for livestock.

Erigeron formosissimus Greene (exceptionally beautiful) Beautiful fleabane. Perennial, 10–40 (–55) cm tall; fibrous-rooted, from variably thickened rhizomes. **STEMS** ascending, densely hirsute to hirsutulous or glabrous, minutely glandular to stipitate-glandular. **LEAVES** basal (persistent) and cauline; basal oblanceolate to oblanceolate-spatulate, 20–100 (–150) × 4–10 (–15) mm, entire, closely ciliate, 1-nerved, faces glabrous or sparsely hirsute, sometimes sparsely glandular; cauline gradually reduced distally, becoming ovate to lanceolate, clasping. **HEADS** 1–6. **INVOLUCRES** 5–8 × 10–20 mm. **PHYLLARIES** in 2–3 series, glabrous or hirsute-villous, densely minutely glandular to stipitate-glandular (in var. *formosissimus*, glandularity sometimes obscured by nonglandular hairs). **RAY FLORETS** 75–150; corollas blue to purple, rarely pink to white, 8–15 mm, laminae coiling at tips or not at all (mostly 1 mm wide). **CYPSELAE** 2-nerved, sparsely strigose. **PAPPUS**: outer of setae, inner of 15–25 bristles.

1. Involucre moderately to densely hirsute, also minutely glandular; distal leaves hirsute-villous, eglandular or sometimes also sparsely glandular .... **var. *formosissimus***
1′ Involucre densely minutely glandular, otherwise glabrous to sparsely hirsute-villous; distal leaves minutely glandular, sometimes also sparsely villous-hirsute .... **var. *viscidus***

var. *formosissimus* **STEMS** densely hirsute to hirsutulous with deflexed hairs. **DISTAL LEAVES** hirsute-villous, eglandular or sometimes also sparsely glandular. **INVOLUCRES** minutely glandular (glands usually inconspicuous under nonglandular vestiture), otherwise moderately to densely hirsute. Meadows, margins, ponderosa pine, aspen-spruce-fir. ARIZ: Apa; COLO: Hin. 2250–2500 m (7380–8200′). Flowering: Jul–Sep. Utah and Arizona to Colorado and New Mexico. Used by the Ramah Navajo as a lotion for good luck in hunting.

var. *viscidus* (Rydb.) Cronquist (viscid, sticky) **STEMS** glandular, sometimes sparsely so, less commonly sparsely hirsute. **DISTAL LEAVES** usually glabrous or sometimes sparsely hirsute-villous, minutely glandular. **INVOLUCRES** densely minutely glandular, otherwise glabrous to sparsely hirsute-villous. [*Erigeron viscidus* Rydb.]. Roadsides, moist meadows, streamsides, crevices, aspen, ponderosa pine, ponderosa pine–fir, bristlecone pine, limber pine, spruce. COLO: Arc, Hin, LPl, Min, SJn; NMEX: RAr, SJn. 2300–3200 m (7545–10500′). Flowering: Jun–Sep. Arizona to Colorado,

New Mexico, Wyoming, and South Dakota. Intermediates between the varietal expressions (e.g., *Heil et al. 19781* [SJNM], San Juan County, Colorado, with primarily glandular phyllaries and densely deflexed-hirsute stems) are less common than the extremes.

Erigeron glacialis (Nutt.) A. Nelson (of the glaciers, alluding to the snowy habitat) Subalpine fleabane. Perennial, 5–55 (–70) cm tall; fibrous-rooted, rhizomatous, caudices usually simple, thick. **STEMS** erect to ascending, hirsute or hirsute-villous to densely strigillose with loosely appressed, slightly crinkled hairs, eglandular. **LEAVES** basal (usually persistent) and cauline; basal linear-oblanceolate to broadly lanceolate or spatulate, (20–) 30–160 (–200) × 7–45 mm, cauline gradually reduced distally, usually distinctly subclasping (except when greatly reduced), margins entire, 1-nerved, faces glabrous or villous, eglandular. **HEADS** 1 (–8) on densely strigillose peduncles. **INVOLUCRES** 6–9 (–12) × 10–22 (–25) mm. **PHYLLARIES** in 2–3 (–4) series, faces and margins usually glabrous, rarely sparsely villous, densely and evenly stipitate-glandular. **RAY FLORETS** 30–80; corollas mostly blue to purple or pink, sometimes white to pale blue, 8–16 (–25) mm, laminae mostly 1.5–3 mm wide, coiling. **CYPSELAE** (4–) 5 (–7)-nerved, sparsely strigose. **PAPPUS**: outer of setae, inner of 20–30 bristles. [*Aster glacialis* Nutt.; *Erigeron salsuginosus* (Richardson ex R. Br.) A. Gray var. *glacialis* (Nutt.) A. Gray; *E. callianthemus* Greene; *E. peregrinus* (Banks ex Pursh) Greene var. *callianthemus* (Greene) Cronquist; *E. peregrinus* var. *scaposus* (Torr. & A. Gray) Cronquist]. Roadsides, talus, glacial moraines, sometimes dry slopes, openings in aspen, fir, spruce-fir, moist or wet subalpine meadows, alpine tundra. COLO: Arc, Con, Hin, LPl, Min, Mon, RGr, SJn. 3300–3800 m (10825–12465′). Flowering: Jul–Sep. Alaska, Yukon to British Columbia and Alberta, Washington to Montana, south to California, Nevada, Utah, and New Mexico. Four Corners plants belong to **var. *glacialis***.

Erigeron grandiflorus Hook. (large-flowered) Rocky Mountain alpine fleabane. Perennial, 2–25 cm tall; fibrous-rooted, rhizomatous, caudices or rhizomes crownlike or with relatively short and thick branches. **STEMS** erect to decumbent-ascending, sparsely to moderately pilose to villous-hirsute, often stipitate-glandular over all or part. **LEAVES** basal (persistent) and cauline; oblanceolate to obovate or spatulate, 10–60 (–90) × 3–8 (–14) mm, cauline quickly or gradually reduced distally, margins entire, 1-nerved, faces sparsely hirsutulous or villous to sparsely strigose or glabrate, sometimes sparsely glandular. **HEADS** 1. **INVOLUCRES** 5–8 (–10) × 8–20 mm. **PHYLLARIES** in 2–3 series, moderately to densely woolly-villous, hairs sometimes with reddish cross walls, minutely glandular at least near tips. **RAY FLORETS** 50–130; corollas blue to pink or purplish, rarely white, 7–11 (–15) mm, laminae coiling. **CYPSELAE** 2-nerved, strigose. **PAPPUS**: outer of setae, inner of (7–) 10–18 (–22) bristles. $2n = 18, 27$. [*Erigeron simplex* Greene]. Rocky slopes, talus, meadows, near timberline, alpine, tundra. COLO: Arc, Con, Hin, LPl, Min, Mon, RGr, SJn. 3450–3900 m (11320–12795′). Flowering: Jul–Sep. Southern British Columbia and Alberta, Oregon, Idaho, and Montana, south to Arizona and New Mexico. Triploid populations identified as *E. grandiflorus* (sensu stricto) differ from diploid populations (*E. simplex* sensu stricto) only in slight and overlapping quantitative features.

Erigeron humilis Graham (low, on the ground) Arctic alpine fleabane. Perennial 1–3 cm tall; apparently taprooted, caudices simple or branched. **STEMS** erect, villous-hirsute, minutely glandular (conspicuously so proximal to heads). **LEAVES** basal (persistent) and cauline; blades oblanceolate to spatulate-oblanceolate, 10–25 × 2–5 mm, cauline becoming linear, reduced or not distally, margins entire, 1-nerved, faces sparsely villous, sometimes sparsely minutely glandular. **HEADS** 1. **INVOLUCRES** 5–8 × 8–12 mm. **PHYLLARIES** in (1–) 2 (–3) series, dark purple, villous with hairs with dark reddish to blackish purple cross walls, minutely glandular. **RAY FLORETS** 50–80; corollas white to whitish pink, 4–6 mm, laminae erect, not coiling or reflexing. **CYPSELAE** 2-nerved, finely strigose-hirsute. **PAPPUS**: outer of setae (inconspicuous), inner of 20–30 bristles. $2n = 36$. Alpine tundra, ridges. COLO: Hin, LPl, SJn. 3800–4150 m (12465–13615′). Flowering: Jun–Aug. Alaska and Yukon through Canada, Idaho, and Montana south to Utah and Colorado; Greenland and northern Eurasia. Similar in aspect to *E. melanocephalus* but the heads barely protruding past basal rosettes; basal leaves more oblanceolate than spatulate; phyllaries purple (vs. green), involucres more densely villous; and individual cells of involucral villous hairs more elongate.

Erigeron kachinensis S. L. Welsh & Glen Moore (from Kachina Natural Bridge in Natural Bridges National Monument, Utah) Kachina fleabane. Perennial, 6–18 cm tall; taprooted, caudex branches relatively short or long, sometimes relatively thick. **STEMS** usually decumbent to ascending, glabrous, eglandular. **LEAVES** basal (persistent) and cauline; basal blades oblanceolate to obovate or spatulate, often folding, 13–50 × 2–13 mm, cauline reduced distally, margins entire, 1-nerved, faces glabrous, eglandular. **HEADS** 1–4. **INVOLUCRES** 3.2–4 × 5–6 mm. **PHYLLARIES** in 2–3 series, often purplish (at least at tips), minutely glandular. **RAY FLORETS** 10–15; corollas white or pinkish, 3.5–5.5 mm, laminae reflexing. **CYPSELAE** 2-nerved, sparsely strigose. **PAPPUS**: outer of setae, inner of 12–14 bristles. Wet sandy soil and sandstone crevices, alluvial benches, hanging gardens, piñon-juniper, ponderosa pine, aspen. UTAH. (1200–) 1450–2400 m (3935–7875′). Flowering: May–Aug. Colorado and Utah.

Erigeron leiomerus A. Gray (smooth parts, alluding to the smooth, glabrous caudex branches) Rockslide fleabane. Perennial, 4–12 (–15) cm tall; taprooted (taproot often not evident or not collected), caudices with diffuse system of relatively long and slender, rhizomelike branches. **STEMS** usually purplish near bases, sometimes greenish, decumbent to ascending or erect, glabrous or sparsely strigillose (hairs closely to loosely appressed), eglandular. **LEAVES** basal and cauline; basal oblanceolate to obovate or spatulate, 15–70 × 2–11 (–15) mm, cauline reduced distally, bases sometimes purplish and enlarged, margins entire, 1-nerved, faces glabrous or sparsely strigose, eglandular. **HEADS** 1. **INVOLUCRES** 4–6 × 7–13 mm. **PHYLLARIES** in 2–3 series, often purplish, glabrous or sparsely strigose, minutely glandular. **RAY FLORETS** 15–60; corollas white to blue or purple, 6–11 mm, laminae reflexing. **CYPSELAE** 2-nerved, sparsely strigose. **PAPPUS**: outer of setae, inner of 15–25 bristles. $2n$ = 18. Talus, boulder fields, ridges, cliffs, meadows, lodgepole pine, spruce, spruce-fir, alpine tundra. COLO: Arc, Hin, LPl, Min, Mon, SJn. 2550–3650 m (8365–11975′). Flowering: Jun–Aug. Colorado, Idaho, and Montana, south to Nevada, Utah, and New Mexico.

Erigeron lonchophyllus Hook. (lancelike or spearlike leaves) Short-ray fleabane. Biennial or short-lived perennial, 2–45 (–60) cm tall; fibrous-rooted, sometimes appearing annual, caudices not branched. **STEMS** erect or basally ascending, sparsely to densely hirsute, eglandular. **LEAVES** basal and cauline; basal persistent, oblanceolate to spatulate, 13–80 (–150) × 1.5–5 (–12) mm, margins entire, usually spreading-ciliate, 1-nerved; cauline mostly linear, sometimes longer than basal, usually erect or nearly so; faces sparsely to moderately hirsute to glabrate, eglandular. **HEADS** 1 or 3–12, usually in loosely racemiform arrays, from erect peduncles distal to midstem, sometimes on proximal 1/3. **INVOLUCRES** 4–9 × 7–17 mm. **PHYLLARIES** in 2–3 series, hirsute to strigose-hirsute, eglandular. **RAY (PISTILLATE) FLORETS** 70–130 in 1 zone; corollas white to light pink, 2–3 mm, laminae filiform, erect, not coiling or reflexing, not surpassing involucres. **CYPSELAE** 2-nerved, sparsely strigose. **PAPPUS**: outer of setae, inner of 20–30 nonaccrescent bristles. $2n$ = 18. [*E. racemosus* Nutt.; *Trimorpha lonchophylla* (Hook.) G. L. Nesom]. Ditch banks, roadsides, moist edges of streams and lakes, ponderosa pine–Gambel's oak, Douglas-fir–aspen, spruce-fir. ARIZ: Apa; COLO: Arc, LPl, SJn; NMEX: SJn. 1950–2900 m (6400–9515′). Flowering: Jul–Sep. Alaska and Northwest Territories to Quebec, south to Nebraska, California, Arizona, and New Mexico.

Erigeron melanocephalus (A. Nelson) A. Nelson (black-headed) Black-head fleabane. Perennial, 3–12 (–21 at lower elevations) cm tall; rhizomatous, fibrous-rooted, caudices decumbent, often branched, rhizomelike, sometimes appearing taprootlike. **STEMS** erect, distally villous, hairs with black cross walls, minutely glandular near heads. **LEAVES** mostly basal (persistent), some cauline; spatulate to oblanceolate, (10–) 20–50 (–150) × 4–6 (–15) mm, cauline linear and bractlike, margins entire, 1-nerved, sparsely hirsute or glabrous, eglandular. **HEADS** 1. **INVOLUCRES** 6–9 × 10–14 mm. **PHYLLARIES** in 2 (–3) series, sparsely villous-sericeous, hairs flattened with black cross walls, imparting black color to involucre, glandular. **RAY FLORETS** 45–74; corollas white to purple, 7–11 mm, laminae tardily coiling. **CYPSELAE** 2-nerved, strigose-hirsute. **PAPPUS**: outer of setae, inner of 15–30 bristles. $2n$ = 18. Rocky slopes, talus, outcrops, drainage channels, subalpine and alpine meadows, spruce-fir, tundra. COLO: Arc, Con, Hin, LPl, Min, Mon, RGr, SJn; UTAH. (2700–) 3400–3800 m (8860–12465′). Flowering: Jul–Sep. Wyoming and Utah to Colorado and New Mexico.

Erigeron nivalis Nutt. (snowy, alluding to the habitat of the type collection) Snowy fleabane. Biennial or short-lived perennial, 5–25 (–35) cm tall; fibrous-rooted, caudices simple or branched, sometimes apparently weakly short-rhizomatous, less commonly taprooted. **STEMS** erect to basally ascending, sometimes sparsely hirsute-villous, minutely glandular. **LEAVES** basal (persistent) and cauline; basal oblanceolate to spatulate, 20–60 × 2–6 (–10) mm, cauline gradually reduced distally, margins entire or rarely with 1–2 pairs of shallow teeth, 1-nerved, faces sparsely hirsute-strigose, eglandular. **HEADS** 1–6 (–8), in corymbiform arrays, on curved-ascending peduncles. **INVOLUCRES** 5–6 × 8–11 mm. **PHYLLARIES** in 2–3 (–4) series, minutely glandular, sparsely hirsute-villous or without nonglandular hairs. **RAY (PISTILLATE) FLORETS** in 2 zones; outer 40–70, corollas white to pinkish, erect, filiform, 5.5–7 mm, laminae not coiling or reflexing; inner many fewer than those of outer zone, elaminate. **DISC FLORETS** 4.4–5.5 mm. **CYPSELAE** 2–2.3 mm, 2-nerved, sparsely strigose. **PAPPUS**: outer of setae, inner of (12–) 14–21 accrescent bristles. $2n$ = 18. [*E. acris* L. var. *debilis* A. Gray; *E. debilis* (A. Gray) Rydb.; *Trimorpha acris* (L.) Gray var. *debilis* (A. Gray) G. L. Nesom]. Meadows, open woods, spruce. COLO: LPl, Min, SJn. 2900–3200 m (9515–10500′). Flowering: Jun–Aug. Alaska to British Columbia and Alberta, Washington to Montana, south to California and Colorado, rare in north-central New Mexico.

Erigeron philadelphicus L. (related to Philadelphia, from where it was originally described) Philadelphia fleabane. Annual or biennial, 4–70 cm tall; fibrous-rooted, caudices simple. **STEMS** erect, leafy to heads, loosely strigose to sparsely hirsute, hirsute-villous, or villous, minutely glandular. **LEAVES** basal (commonly withering before flowering) and cauline; basal oblanceolate to obovate, (15–) 30–110 (–150) × 10–25 (–40) mm, margins shallowly crenate to coarsely serrate or pinnately lobed, 1-nerved; cauline oblong-oblanceolate to lanceolate, gradually reduced distally, clasping to auriculate-clasping; faces sparsely hirsute to villous, eglandular. **HEADS** (1–) 3–35, usually in corymbiform arrays, ultimate branches arising near stem tips. **INVOLUCRES** 4–6 × 6–15 mm. **PHYLLARIES** in 2–3 series, sometimes basally connate, hirsute-villous to sparsely hirsute or glabrous, sometimes minutely glandular. **RAY FLORETS** 150–250 (–400); corollas usually white, sometimes pinkish, 5–10 mm, laminae not coiling or tardily coiling. **CYPSELAE** 2-nerved, sparsely strigose. **PAPPUS**: outer of setae, inner of 15–20 (–30) bristles. $2n$ = 18. Canal and river margins, ditch banks, willow, birch, cottonwood. COLO: Arc; NMEX: SJn. 1600–2900 m (5250–9515′). Flowering: May–Jul (–Aug). Alaska, Canada, apparently native to eastern North America, widely naturalized in the West; introduced in Europe, Asia. Four Corners plants belong to **var. *philadelphicus***.

Erigeron pinnatisectus (A. Gray) A. Nelson (divided in a featherlike fashion) Feather-leaf fleabane. Perennial, 4–11 cm; taprooted, caudices usually with relatively thick, woody branches. **STEMS** erect, sparsely to moderately hispidulous and minutely glandular, sometimes merely glandular. **LEAVES** mostly basal and proximal, persistent; basal pinnatifid with linear lobes, often folding, 20–40 (–50) × 5–10 mm, 1-nerved; cauline mostly linear, abruptly smaller distally or nearly absent, margins entire, 1-nerved, usually coarsely ciliate; faces usually minutely glandular, sometimes sparsely hirsute or glabrate. **HEADS** 1. **INVOLUCRES** 5.5–8 × 9–13 mm. **PHYLLARIES** in 3–5 series, sparsely hispidulous to hirsute-villous, minutely glandular. **RAY FLORETS** 40–70; corollas usually light blue to lavender, drying bluish or fading lighter, 7–12 mm, laminae reflexing. **CYPSELAE** 2-nerved, sparsely strigose. **PAPPUS**: outer of setae, inner of 25–30 bristles. [*E. compositus* Pursh var. *pinnatisectus* A. Gray]. Talus, ridges, subalpine and alpine meadows, tundra. COLO: Arc, Con, Hin, LPl, Min, RGr, SJn. 2700–3900 m (8860–12795′). Flowering: late Jun–Aug. Wyoming, Colorado, and New Mexico.

Erigeron pulcherrimus A. Heller (extremely beautiful) Basin fleabane. Perennial, (5–) 7–30 (–35) cm tall; taprooted, caudices with relatively short and thick branches. **STEMS** erect, moderately to densely gray-green–strigose, eglandular. **LEAVES** basal (persistent) and cauline; basal linear to narrowly oblanceolate, 10–70 × 1–3 (–5) mm, cauline little reduced for about 1/2–3/4 of stem, margins entire, 1-nerved, faces moderately to densely strigose with white, stiff hairs, eglandular. **HEADS** 1. **INVOLUCRES** 5–7 (–9) × 10–16 (–20) mm. **PHYLLARIES** in 2–3 (–4) series, hirsute to villous-hirsute, minutely glandular. **RAY FLORETS** 25–60; corollas white to pink or bluish, 8–15 mm, laminae coiling. **CYPSELAE** (2–) 4 (–5)-nerved, densely strigose-sericeous. **PAPPUS**: outer sometimes of setae or bristles (to 1/2 lengths of inner), inner of 32–50 bristles. [*E. bistiensis* G. L. Nesom & Hevron]. Clay, clay-silt, sandy, and gravelly soil, saline, gypseous, and seleniferous, desert shrub, juniper-rabbitbrush, piñon-juniper. ARIZ: Apa; NMEX: McK, RAr, SJn. 1700–2100 m (5575–6890′). Flowering: (Apr–) May–Jun. Wyoming south to Arizona and New Mexico.

Erigeron pumilus Nutt. (diminutive) Shaggy fleabane. Perennial, 5–30 (–50) cm tall; taprooted, caudices with relatively short and thick branches. **STEMS** erect, hirsute to hispid-hirsute, often with slightly deflexed hairs, minutely to stipitate-glandular. **LEAVES** basal (persistent) and cauline; petioles prominently ciliate, hairs thick-based, spreading; basal oblanceolate to narrowly oblanceolate, 20–80 × 1–4 (–5) mm, margins entire, 1-nerved; cauline on distal 1/2–3/4 of stems, little reduced distally, becoming linear-lanceolate; faces hispid to hispid-hirsute, little, if at all, glandular. **HEADS** 1–5 (–50). **INVOLUCRES** 4–7 × 7–15 mm. **PHYLLARIES** in 2–4 series, hirsute to hispid-hirsute, minutely glandular. **RAY FLORETS** 50–100; corollas white, 6–15 mm, laminae reflexing. **DISC FLORETS** with throats distinctly indurate and inflated, glabrous or sparsely puberulent with glandular-viscid, blunt-tipped hairs. **CYPSELAE** 2-nerved, sparsely strigose. **PAPPUS**: outer of setae, inner of 15–27 bristles. $2n$ = 18, 36. Sand, silty clay, silt; desert shrub, sagebrush, piñon-juniper, ponderosa pine–Gambel's oak communities. ARIZ: Apa; NMEX: McK, RAr, San, SJn. 1550–2350 m (5085–7710′). Flowering: May–Jul. Canada (southern Alberta and Saskatchewan), Washington, Montana, and North Dakota, south to Kansas, Arizona, and New Mexico. Four Corners plants belong to **var. *pumilus***.

Erigeron religiosus Cronquist (alluding to its locality centered around Zion National Park) Clear Creek fleabane. Annual or short-lived perennial, 6–30 (–40) cm tall; taprooted, caudices simple or branched. **STEMS** decumbent-ascending to erect, usually sparsely strigose, hairs rarely ascending-spreading, sometimes sparsely hirsutulous distally, sometimes minutely glandular near heads, otherwise sometimes sparsely glandular. **LEAVES** basal (persistent) and cauline; basal oblanceolate-spatulate to spatulate, 10–70 × 2–8 (–13) mm, margins entire, or uncommonly dentate or pinnately divided, 1-nerved, faces sparsely strigose, eglandular; cauline linear to oblanceolate, reduced. **HEADS** 1 to about 50, first 1 per branch, later more from axillary branches, forming diffuse arrays. **INVOLUCRES** 2–3.5 × 5.5–7.5 mm. **PHYLLARIES** in 3–5 series, sparsely to moderately hirtellous, minutely glandular. **RAY FLORETS** 37–85; corollas

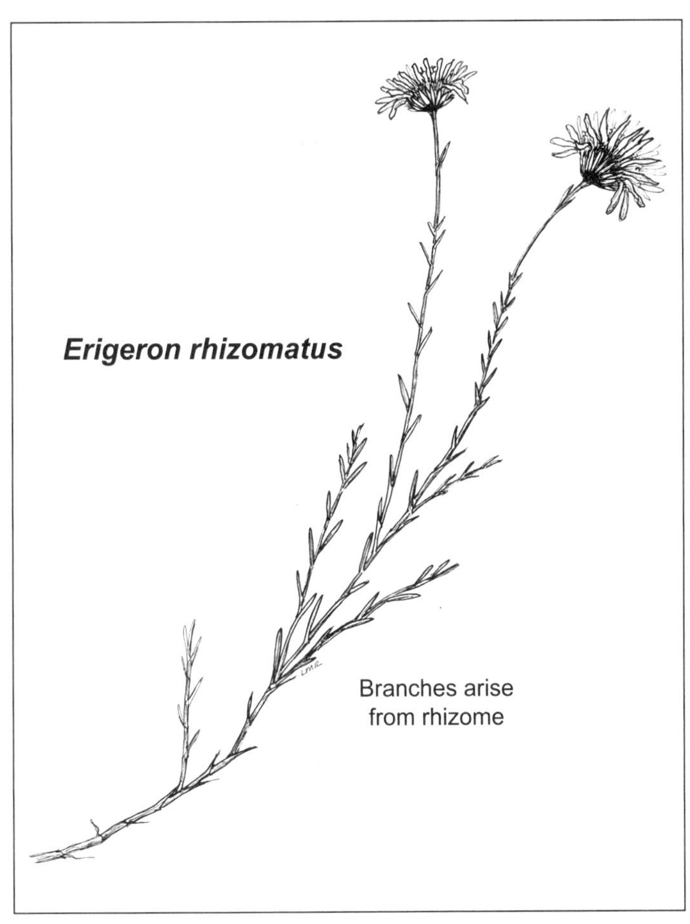

white, drying lilac, with abaxial lilac midstripe, 3.5–7 mm; laminae not coiling or reflexing. **DISC FLORETS** with throats slightly indurate and inflated. **CYPSELAE** 2-nerved, sparsely strigose. **PAPPUS**: outer of setae, inner of 6–12 bristles. $2n$ = 18, 27. Deep sand, desert scrub; riparian. UTAH. 1135–1200 m (3700–3935′). Flowering: May–Sep. Arizona, Utah.

Erigeron rhizomatus Cronquist (characterized by rhizomes) Zuni fleabane. Perennial, 25–45 cm tall; rhizomatous, fibrous-rooted, roots clustered, relatively thick, sometimes forming clumps to about 30 cm diameter; caudex branches rhizomelike, creeping-ascending, slender, scale-leaved, without well-defined central axes. **STEMS** erect, simple or 1–3-branched from near bases, sparsely strigose to strigose-hirsutulous, sometimes sparsely minutely glandular. **LEAVES** cauline, more densely leafy on secondary stems; proximal narrowly oblong to oblong-oblanceolate, quickly linear, 50–100 × 1–3 mm, relatively even-sized distally, margins entire, ciliate, 1-nerved, faces glabrous, eglandular. **HEADS** 1 (rarely 2–3 from proximal branches). **INVOLUCRES** 6–7 × 13–16 mm. **PHYLLARIES** in 4–5 series, sparsely strigose, sometimes sparsely minutely glandular. **RAY FLORETS** 25–45; corollas white with abaxial lilac midstripe, 6–7 mm, laminae not coiling or reflexing. **CYPSELAE** 5–6-nerved, glabrous. **PAPPUS**: outer of setae, inner of 25–35 bristles. $2n$ = 18. Nearly barren detrital clay hillsides or benches, roadcuts, soils derived from shales (often seleniferous) of the Chinle Formation, usually north- or east-facing slopes, piñon-juniper, ponderosa pine, and Douglas-fir communities. ARIZ: Apa; NMEX: SJn. 2000–2500 m (6560–8200′). Flowering: May–Jul. Arizona and New Mexico.

Erigeron scopulinus G. L. Nesom & V. D. Roth (belonging to cliffs or ledges) Rock fleabane. Perennial, 0.5–3.5 cm tall; taprooted (taproots weakly developed, often not evident), forming systems of relatively slender, basal offsets and slender, ligneous rhizomes 1–15 cm. **STEMS**: vegetative decumbent to ascending; flowering erect, scapiform, sparsely strigose, eglandular. **LEAVES** basal, persistent, spatulate, often folding, 5–12 × 1–3.5 mm, margins entire, 1-nerved, faces glabrous or sparsely strigose, eglandular. **HEADS** 1. **INVOLUCRES** 4–4.5 × 4–7 mm. **PHYLLARIES** in 3–4 (–5) series, glabrous or sparsely strigose, sometimes minutely glandular. **RAY FLORETS** 10–20; corollas white, 5.5–9 mm, laminae reflexing. **DISC FLORETS** with throats slightly indurate and inflated. **CYPSELAE** 2 (–3)-nerved, sparsely strigose. **PAPPUS**: outer of setae, inner of 13–18 bristles. $2n$ = 18. Rocky pavement, talus and grass areas, alpine community with *Geum*, *Packera*, *Poa*. COLO: SJn. 3840 m (12600′). Flowering: Jul–Aug. Arizona, New Mexico, Colorado.

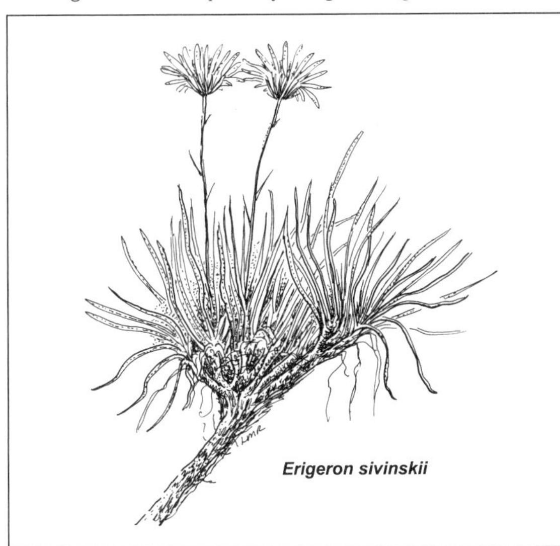

Erigeron sivinskii G. L. Nesom (for its initial collector, Robert Sivinski, New Mexico botanist) Sivinski's fleabane. Perennial, 5–8 cm tall; taprooted, caudices with relatively short, thick branches, retaining old leaf bases. **STEMS** erect, closely strigose with white hairs of even lengths, eglandular. **LEAVES** basal (persistent, persistent portions relatively short and broad) and cauline; basal erect, linear, 12–30 × 0.5–0.8 mm, cauline unreduced for ± 1/2 stem length, margins entire, eciliate, 1-nerved, faces sparsely strigose, eglandular. **HEADS** 1. **INVOLUCRES** 5–6 × (8–) 10–14 mm. **PHYLLARIES** in 2–3 series, (relatively thin-herbaceous), sparsely hirsute-pilose along midribs, minutely glandular. **RAY FLORETS** 21–33; corollas white, with abaxial lilac midstripe, 7–9 mm, laminae coiling. **CYPSELAE** 2 (–3)-nerved, margins sparsely ciliate, faces glabrous. **PAPPUS**: outer of setae, inner of 21–27 bristles. Red clay slopes of the Summerville Formation and eroded, shale slopes of the Chinle Formation in desert shrub and piñon-juniper communities. ARIZ: Apa, Nav. 1700–2350 m (5575–7710′). Flowering: May–Jun. Arizona and New Mexico.

Erigeron sparsifolius Eastw. (few leaves) Bracted Utah fleabane. Perennial, 10–55 cm tall; taprooted, caudices branching. **STEMS** erect, sparsely strigose, eglandular. **LEAVES** basal (usually withering before flowering, not forming conspicuous tufts) and cauline; basal and the most proximal cauline oblanceolate-spatulate, 20–50 × 2–4 mm, margins entire, 1-nerved, faces densely and closely strigose, eglandular; cauline quickly reduced to erect linear bracts 3–20 × 0.5 mm and relatively even-sized, at least beyond midstem to immediately proximal to heads. **HEADS** (1–) 3–10, from branches well beyond midstem. **INVOLUCRES** 3–5 × 5–8 mm. **PHYLLARIES** in 2–3 series, sparsely strigose or without nonglandular hairs, densely minutely glandular. **RAY FLORETS** 10–14 (–20); corollas white to blue, 4–8 mm, laminae coiling. **DISC FLORETS** viscid-puberulent with multicellular, blunt hairs. **CYPSELAE** 3–4-orange-nerved, sparsely strigose. **PAPPUS**: outer of inconspicuous setae, inner of 20–25 bristles. [*E. utahensis* A. Gray var. *sparsifolius* (Eastw.) Cronquist]. Rocky or sandy soil, soil pockets and crevices in sandstone, hanging gardens, canyon bottoms, river terraces. ARIZ: Apa, Nav; COLO: Mon; NMEX: RAr, SJn; UTAH. 1250–1700 m (4100–5575′). Flowering: Jun–Sep. Arizona, Colorado, Utah, and New Mexico. *Erigeron sparsifolius* is partially sympatric with *E. utahensis*. They are distinct in morphology and phenology and intermediates are not common.

Erigeron speciosus (Lindl.) DC. (showy) Showy fleabane. Perennial, 30–80 (–100) cm tall; rhizomatous, fibrous-rooted, caudices relatively thick. **STEMS** erect, glabrous or sparsely hirsute-pilose, hairs 0.5–1 mm, often minutely glandular distally. **LEAVES** basal (usually withering before flowering) and cauline; basal oblanceolate-spatulate, 30–80 (–150) × 4–18 (–28) mm, margins entire, often ciliate, 1-nerved, faces glabrous or distal sparsely minutely glandular; cauline ovate to ovate-lanceolate, oblong-lanceolate, or lanceolate, nearly even-sized distally or sometimes middle the largest, usually clasping to subclasping. **HEADS** (2–) 4–20, in corymbiform arrays. **INVOLUCRES** 6–9 × 11–22 mm. **PHYLLARIES** in 2–3 (–4) series, usually glabrous, sometimes sparsely hirsute-pilose, minutely glandular. **RAY FLORETS** 75–150; corollas blue to lavender, rarely whitish, 8–16 mm, laminae (mostly 1 mm wide) slightly coiling at least at tips. **CYPSELAE** 2 (–4)-nerved, sparsely strigose. **PAPPUS**: outer of setae, inner of 20–30 bristles. [*E. speciosus* var. *conspicuus* (Rydb.) Breitung; *E. speciosus* var. *macranthus* (Nutt.) Cronquist; *E. subtrinervis* Rydb. ex Porter & Britton subsp. *conspicuus* (Rydb.) Cronquist; *E. subtrinervis* var. *conspicuus* (Rydb.) Cronquist]. Rocky slopes, washes, streamsides, roadsides, meadows, piñon-juniper, cottonwood, Gambel's oak, ponderosa pine, pine-fir, spruce-fir, aspen, and aspen-spruce communities. ARIZ: Apa; COLO: Arc, Dol, Hin, LPl, Min, Mon, SJn, SMg; NMEX: McK, RAr, SJn; UTAH. 2100–2950 (–3200) m (6890–10500′). Flowering: Jul–Oct. Canada (British Columbia, Alberta), Washington to Montana and South Dakota, south to Oregon, Arizona, and New Mexico; Mexico (Baja California). Plants completely lacking nonglandular hairs on the phyllaries, stems, and leaves have been recognized as *E. speciosus* var. *macranthus*. Intermediates are encountered between *E. speciosus* and *E. subtrinervis*; *E. subtrinervis* var. *mollis* (A. Gray) S. L. Welsh may be a recurrent hybrid; it is identified here within *E. subtrinervis*. Used by the Ramah Navajo for menstrual pain and as a contraceptive.

Erigeron subtrinervis Rydb. ex Porter & Britton (weakly 3-nerved) Perennial, 15–90 cm tall; rhizomatous to subrhizomatous, fibrous-rooted, caudices usually branched, woody, thick. **STEMS** erect, moderately to densely hirsute, hairs 0.5–0.8 mm, eglandular. **LEAVES** basal (usually withering before flowering) and cauline; basal oblanceolate-spatulate, 30–80 × 6–20 (–27) mm, margins entire, 1-nerved, faces evenly hirsute to strigose-hirsute on laminae and veins, usually eglandular; cauline lanceolate to oblong, oblong-ovate, or broadly ovate, nearly even-sized distally or sometimes middle the largest, clasping to subclasping. **HEADS** 1–6 (–21), in corymbiform arrays. **INVOLUCRES** 6–9 × 13–20 mm. **PHYLLARIES** in 2–3 (–4) series, moderately to densely hirsute, minutely glandular. **RAY FLORETS** 100–150; corollas blue to lavender, 7–18 mm, laminae coiling at tips. **CYPSELAE** 2 (–4)-nerved, sparsely strigose. **PAPPUS**: outer of setae, inner of 20–30 bristles. [*E. speciosus* (Lindl.) DC. var. *mollis* (A. Gray) S. L. Welsh]. Openings and margins, brushy slopes, roadsides, Gambel's oak, ponderosa pine, pine-fir, aspen, mixed conifer, and bristlecone pine communities. COLO: Arc, Hin, Min; NMEX: RAr; UTAH. 1950–3150 m (6395–10335′). Flowering: (Jun–) Jul–Sep. Idaho to North Dakota, south to Utah, New Mexico, and Nebraska. *Erigeron subtrinervis* is variable in vestiture, perhaps reflecting gene exchange with *E. speciosus*.

Erigeron tracyi Greene (for Samuel M. Tracy, 1847–1920, botanist primarily of Missouri and the Gulf states) Running fleabane. Annual, biennial, or short-lived perennial, 2.5–8 (–12, 18) cm tall; taprooted, less commonly fibrous-rooted, caudices simple or branched. **STEMS** first erect, then producing herbaceous, leafy, prostrate runners, sometimes with rooting plantlets at tips, densely hirsutulous with spreading-deflexed hairs of relatively even length and orientation, sparsely minutely glandular. **LEAVES** mostly basal (persistent in early season); oblanceolate to spatulate with obovate-elliptic blades, 10–30 (–60) × 3–6 (–12) mm, cauline quickly reduced distally, margins entire, dentate, or lobed, 1-nerved, faces densely hirsute, eglandular. **HEADS** 1 (–3 rarely, from midstem or proximal branches). **INVOLUCRES** 3.5–4.5 (–6) × 6–9 (–12) mm. **PHYLLARIES** in 3–4 series, sparsely to moderately hirsute, minutely glandular. **RAY FLORETS** 60–130; corollas white, often purplish abaxially, sometimes with an abaxial midstripe, 5–9 mm, laminae not coiling or reflexing. **DISC FLORETS** with throats indurate and slightly inflated. **CYPSELAE** 2-nerved, sparsely strigose. **PAPPUS**: outer of setae, inner of 12–16 bristles. $2n$ = 27. [*E. cinereus* A. Gray; *E. colomexicanus* A. Nelson; *E. commixtus* Greene; *E. divergens* Torr. & A. Gray var. *cinereus* A. Gray]. Roadsides, lake shores, washes, openings, desert shrub, grassy slopes, oak chaparral, piñon-juniper, and Douglas-fir–ponderosa pine communities. ARIZ: Apa, Nav; COLO: Arc, Dol, LPl, Mon, SMg; NMEX: McK, RAr, SJn. 1650–2350 m (5415–7710′). Flowering: Apr–Jun (–Jul). Nevada, Utah, and Arizona east to Kansas, Oklahoma, and Texas; Mexico. In March through June plants produce leaves in a basal rosette usually with a single, erect, monocephalous stem. Stoloniform branches are soon formed, and prostrate runners are usually evident by the end of the season. *Erigeron tracyi* is similar in habit to *E. flagellaris*, but the stem pubescence is different, the stolons much less commonly produce rooting plantlets at the tips, and the plants tend to be perennial with woody or lignescent caudices.

Erigeron ursinus D. C. Eaton (bearlike, alluding to the type locality) Bear River fleabane. Perennial, 3–20 cm tall; fibrous-rooted, forming diffuse systems of slender, rhizomelike caudex branches. **STEMS** ascending, bases usually purplish, glabrous or sparsely strigose, less commonly subvillous, sometimes glandular near heads. **LEAVES** basal (persistent) and cauline; basal and proximal narrowly oblanceolate to oblong, 20–120 × 2–11 mm, cauline reduced distally, margins entire, 1-nerved, faces glabrous or loosely strigose, eglandular. **HEADS** 1 (–3). **INVOLUCRES** 5–7 × 9–19 mm. **PHYLLARIES** in 2–3 (–4) series, often purplish at margins and tips, sparsely to moderately villous to hirsute-villous, densely minutely glandular. **RAY FLORETS** 30–100; corollas 6–15 mm, laminae not reflexing or coiling or sometimes tardily coiling, pink to bluish purple. **DISC FLORETS** 3.2–4.7 mm. **CYPSELAE** 2-nerved, sparsely strigose. **PAPPUS**: outer of setae, inner of 10–20 bristles. $2n$ = 18. [*Erigeron ursinus* var. *meyerae* S. L. Welsh]. Meadows and grassy openings. COLO: Hin, LPl. 3000–3600 m (9840–11810′). Flowering: Jul–Sep. Idaho and Montana south to Nevada, Arizona, and New Mexico.

Erigeron utahensis A. Gray **CP** (from Utah) Utah fleabane. Perennial, 10–60 cm tall; taprooted, caudices branching. **STEMS** erect, densely and closely strigose, eglandular. **LEAVES** basal (mostly withered before flowering, not forming conspicuous tufts) and cauline; basal and proximal cauline linear-oblanceolate to narrowly spatulate, 15–100 × 1–6 mm, cauline gradually reduced distally, ending well proximal to heads, margins entire, 1-nerved, faces densely and closely strigose, eglandular. **HEADS** 1–3 (–5), from branches from midstem or beyond. **INVOLUCRES** 5–7 × 7–12 (–15) mm. **PHYLLARIES** in 3–4 series, loosely hirsute-strigose, often minutely glandular. **RAY FLORETS** 28–40; corollas white, pink, or blue, 10–18 (–20) mm, laminae coiling. **DISC FLORETS** with sparsely strigose-villous, needlelike hairs. **CYPSELAE** 4 (–6)-nerved, sparsely strigose. **PAPPUS**: outer of inconspicuous setae, inner of 20–35 bristles. $2n$ = 18, 36. [*E. utahensis* var. *tetrapleuris* (A. Gray) Cronquist]. Rocky slopes, cliff bases, ledges, and crevices, sandstone outcrops and terraces, sandy soil, gravelly limestone, shale, mountain brush, piñon-juniper. ARIZ: Nav; COLO: Mon; NMEX: RAr, SJn; UTAH. 1650–2100 m (5415–6890′). Flowering: mid-Apr–Jun (–Jul). California and Arizona to Colorado and New Mexico.

Erigeron vagus Payson (wandering, perhaps alluding to the system of long, rhizomelike caudex branches) Rambling fleabane. Perennial, 2–4 (–5.5) cm tall; taprooted, caudices diffuse, divided into a system of relatively long and slender, rhizomelike branches. **STEMS** erect, sparsely hirsute, sparsely to densely minutely glandular. **LEAVES** basal and cauline; basal persistent, spatulate, (5–) 10–25 (–30) × 4–10 mm, distal cauline bractlike, margins 3–5-lobed (lobes rounded, ultimately entire), 1-nerved, faces finely puberulent, minutely glandular. **HEADS** 1. **INVOLUCRES** 5–7 × 8–16 mm. **PHYLLARIES** in 2–3 series, purplish, hirsute to villous-hirsute, minutely glandular. **RAY FLORETS** 25–40; corollas white to pink, 4–7 mm, laminae coiling. **CYPSELAE** 2-nerved, finely strigose-hirsute. **PAPPUS**: outer 0 (or inconspicuous), inner of about 20 bristles. $2n$ = 18. Ridges, talus, alpine tundra, ponderosa pine, and bristlecone pine communities. COLO: Arc, Hin, LPl, SJn. 3500–4150 m (11480–13615′). Flowering: late Jun–Aug. Oregon and California east through Nevada and Utah to Colorado.

Erigeron vetensis Rydb. (from La Veta Pass, Colorado, the type locality) Early blue-top fleabane. Perennial, 5–25 cm tall; taprooted, caudices with relatively short, thick branches, often retaining old leaf bases. **STEMS** erect, densely glandular, otherwise glabrous or sparsely hirsute to villous, very rarely eglandular or nearly so. **LEAVES** basal (persistent) and cauline; linear to narrowly oblanceolate, 20–90 (–150) × 2–4 (–7) mm, cauline on proximal 1/3–2/3 of stems, gradually reduced distally, margins entire, 1-nerved, faces minutely glandular, otherwise glabrous to sparsely hispidulous, petioles prominently ciliate, hairs thick-based, spreading. **HEADS** 1. **INVOLUCRES** 4–8 × 7–15 mm. **PHYLLARIES** in 3–4 series, minutely and densely glandular, otherwise sparsely to moderately hispid or hispid-villous. **RAY FLORETS** 30–90; corollas usually bluish to purplish, sometimes white, drying pinkish, 6–16 mm, laminae reflexing. **DISC FLORETS** with throats not indurate or inflated. **CYPSELAE** 2-nerved, sparsely strigose. **PAPPUS**: outer of inconspicuous, fine setae, inner of 18–25 bristles. [*E. glandulosus* Porter; *E. porteri* S. F. Blake]. Rocky slopes, outcrops, cliffs, ponderosa pine, and spruce-fir communities. COLO: Hin, Min. 2600–2950 m (8530–9680′). Flowering: May–Jul. Wyoming, Nebraska, Colorado, and New Mexico.

Eriophyllum Lag. Woollyleaf

S. L. Welsh

(*erio*, woolly, + *phyllum*, leaf) Annual or perennial woolly herbs. **LEAVES** alternate, entire or toothed to lobed. **INFL** solitary or corymbosely clustered. **HEADS** radiate. **INVOLUCRES** campanulate or hemispheric; bracts (apparently 2)-seriate, firm, erect. **RECEPTACLES** flat to low-conic, naked. **RAY FLORETS** few, pistillate and fertile, yellow or white. **DISC FLORETS** perfect, fertile, the tube glandular or hairy; pappus of firm nerveless chaffy scales; style branches flattened. **CYPSELAE** 4-angled (Constance 1937).

Eriophyllum lanosum (A. Gray) A. Gray (woolly) Gray's woollyleaf. Annual, floccose-tomentose, diminutive herbs. **STEMS** mainly 2–13 cm tall or long, simple and erect or divaricately branching from the base. **LEAVES** 0.5–1.8 cm long, 1–2 mm wide, linear to linear-oblanceolate, entire or essentially so. **HEADS** turbinate, solitary on naked peduncles, 0.5–5 cm long. **INVOLUCRES** 5–6.5 mm wide, 5–7 mm high; bracts 8–10, oblong, acute, distinct or nearly so. **RAYS** 5–10, white, 3–5 mm long, 2.5–3.5 mm wide. **CYPSELAE** 2.5–4.5 mm long, slender, sparsely strigulose. **PAPPUS** of about 5 slender, hyaline, awn-tipped scales. $n = 4$. Blackbrush communities. Only known from one collection in Betatakin Canyon. ARIZ: Nav. 2070 m (6800′). Flowering: May. Washington County, Utah; California, Nevada, and Arizona.

Eurybia (Cass.) Cass. Aster

G. L. Nesom

(Greek *eurys*, wide or widespread, + *baios*, few, perhaps referring to the few wide-spreading ray florets; Cassini did not explain the derivation) Perennial herbs, producing short, thick rhizomes and/or long, slender, scale-leaved rhizomes, stems and leaves mostly glabrate, stipitate-glandular in a few. **LEAVES** linear to obovate or cordate, usually with 3–5 veins entering in parallel from the base, sessile, entire to serrate or spinulose. **INFL** a radiate head, in a compact to loose, corymboid inflorescence, rarely reduced to 1–several heads. **INVOLUCRES** turbinate to campanulate, 4–12 mm. **PHYLLARIES** in 5–7 graduate series, basally indurate, usually with a distinctly demarcated green apical patch, usually distinctly low-keeled, the margins usually minutely ciliate-fringed. **RECEPTACLES** foveolate, epaleate. **RAY FLORETS** pistillate, fertile; corolla blue and coiling at maturity or white and little coiling. **DISC FLORETS** perfect, fertile. **CYPSELAE** cylindric or subcylindric, 8–12 (–18)-nerved, eglandular. **PAPPUS** (1–) 2-seriate, of barbellate bristles, these often stiff and flattened, often apically dilated.

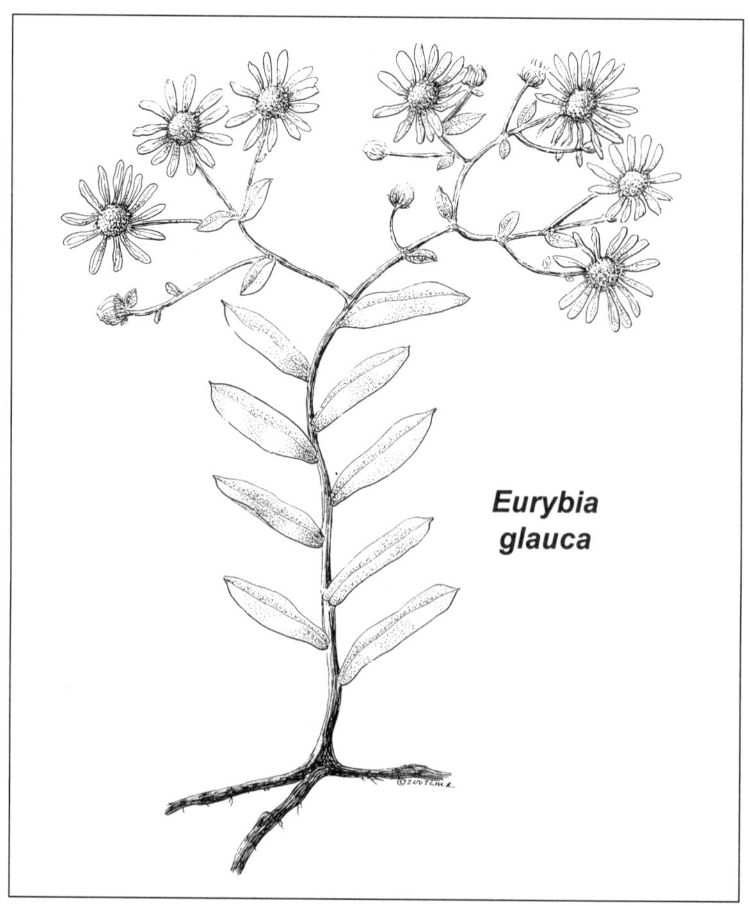

Eurybia glauca

$x = 9$. [*Aster* subg. *Eurybia* Cass.; *Herrickia* Wooton & Standl.; *Weberaster* Á. Löve & D. Löve]. Twenty-eight species consolidated from a number of species groups within North American *Aster* sensu lato (Nesom 1994); eastern and western North America, with one species reaching northward into Alaska, across the Aleutians, and into northeastern Asia. *Eurybia* is distinctive in its combination of corymboid inflorescence; ciliate-fringed, rounded phyllaries with a green, often basally truncate apical patch; linear-lanceolate disc style appendages; cylindric and multinerved cypselae; 2-seriate pappus of equal length; apically thickened bristles; and base chromosome number of $x = 9$. Brouillet et al. (2005) treated *Eurybia* sect. *Herrickia* (Wooton & Standl.) G. L. Nesom as the separate genus *Herrickia*, including *E. glauca* and three other related species.

Eurybia glauca (Nutt.) G. L. Nesom (waxy) Gray's aster. Perennial herbs, 20–70 cm, glaucous, mostly glabrous, minutely stipitate-glandular distally or eglandular; rhizomes woody, elongate and creeping, or short, erect, branched. **STEMS** glabrous to sometimes thinly scabridulous distally. **LEAVES** cauline, proximal reduced and deciduous by flowering, sessile, blades oblong or elliptic-oblong to lanceolate, mostly 40–120 × 5–25 mm, bases subclasping, sometimes cuneate and not clasping, margins entire, faces glabrous. **INVOLUCRES** campanulate, 6–9 mm. **PHYLLARIES** ovate or oblong (outer) to lanceolate (inner), unequal, green zones in distal 1/3–1/5 or less (outer), reduced to midnerve, or none (inner). **RAY COROLLAS** lavender. **CYPSELAE** fusiform, ± compressed, 3.8–4.8 mm, ribs 7–10, faces glabrous or sparsely strigillose. $2n = 18$. [*Aster glaucus* (Nutt.) Torr. & A. Gray; *Aster glaucodes* S. F. Blake; *Herrickia glauca* (Nutt.) Brouillet].

1. Distal stems, peduncles, and phyllaries eglandular .... **var. *glauca***
1′ Distal stems, peduncles, and phyllaries minutely stipitate-glandular .... **var. *pulchra***

var. *glauca* Gray's aster. Distal stems, peduncles, and phyllaries eglandular. $2n = 18$. Rocky slopes, roadsides; sagebrush, mountain shrub, piñon-juniper, ponderosa pine–Douglas-fir, and white fir–Gambel's oak communities. ARIZ: Apa; COLO: Arc, Dol, Hin, LPl, Min, Mon; NMEX: McK, RAr, SJn; UTAH. 1600–2750 m (5250–9020′). Flowering: (Jun–) Jul–Sep. Idaho and Montana, south to Arizona and New Mexico.

var. *pulchra* (S. F. Blake) Brouillet (beautiful) Beautiful aster. Distal stems, peduncles, and phyllaries minutely stipitate-glandular. [*Aster glaucodes* S. F. Blake subsp. *pulcher* S. F. Blake; *Eurybia pulchra* (S. F. Blake) G. L. Nesom; *A. wasatchensis* (M. E. Jones) S. F. Blake var. *pulcher* (S. F. Blake) S. L. Welsh; *Herrickia glauca* (Nutt.) Brouillet var. *pulchra* (S. F. Blake) Brouillet]. Seeps in wash, riparian with cottonwood. UTAH. 1150–1700 m (3770–5575′). Flowering: Aug–Oct. Utah and Arizona.

Euthamia (Nutt.) Cass. Goldentop
G. L. Nesom

(Greek *eu*, well, + *thamees*, crowded, "in allusion to the crowding of the flower heads." – Nuttall protolog, 1818) Perennial herbs, rhizomatous, glabrous or sometimes scabridulous in the inflorescence and leaf margins and veins of lower surfaces, stems branched distally. **LEAVES** alternate, linear to linear-lanceolate, 1–3 (–5)-veined, entire, all cauline, mostly even-sized, punctate and usually at least slightly resinous. **INFL** a short-pedunculate to sessile head, often in glomerate clusters, the whole inflorescence a rounded to distinctly flat, corymboid arrangement. **INVOLUCRES** turbinate or campanulate. **PHYLLARIES** in 3–5 graduate series, with a restricted, greenish, thickened apical region. **RECEPTACLES** deeply foveolate, at least somewhat fimbrillate. **RAY FLORETS** pistillate, fertile, corolla yellow. **DISC FLORETS** perfect, fertile. **CYPSELAE** oblong to narrowly turbinate, subterete, with 2–4 very faint nerves, strigose. **PAPPUS** a single series of barbellate bristles. $x = 9$. [*Solidago* subg. *Euthamia* Nutt.] (Sieren 1981). Species about eight; mostly central and eastern United States, one in western North America (USA; southwestern Canada; Baja California, Mexico). *Euthamia* is recognized by its rhizomatous habit; narrow, punctate, cauline leaves; glomerate or short-pedicellate clusters of small, yellow-rayed heads in distinct corymbs; and small, terete cypselae with a pappus of barbellate bristles.

Euthamia occidentalis Nutt. (western) Western goldentop. **STEMS** 40–150 cm. **LEAVES** 5–10 (–12) cm, mostly 3-veined, margins scabrous. **HEADS** in an elongate to broadly rounded inflorescence. **INVOLUCRES** turbinate, 3.4–5 mm high. **RAY FLORETS** 15–25. **DISC FLORETS** 7–15. Creek and river terraces, sandbars, roadside swales, riparian vegetation. ARIZ: Apa; COLO: LPl, Mon; NMEX: SJn; UTAH. 1200–1800 m (3935–5905′). Flowering: Jun–Sep. Western North America; British Columbia and Alberta south to California, Arizona, and New Mexico; Mexico.

Gaillardia Foug. Indian Blanket, Blanketflower

Brian Elliott

(for Gaillard de Charentonneau, French patron of botany) Annual, biennial, or perennial taprooted herbs. **STEMS** herbaceous, rarely suffrutescent in a few species, scapose or caulescent, mostly 1–8 dm tall, often pubescent with short moniliform hairs and striate. **LEAVES** basal or cauline and alternate, entire to toothed or pinnatifid, petiolate or sessile. **INFL** heads radiate (in ours) or rarely discoid, solitary, usually long-pedunculate. **PHYLLARIES** in 2–3 series, herbaceous, reflexed in age. **RECEPTACLES** convex to subglobose, bristly with long setae, alveolate or merely rugose and fimbrillate. **RAY FLORETS** neutral (in ours), occasionally pistillate and sterile; ligules red to purple or yellow, showy and conspicuous, apex 3-toothed or 3-lobed. **DISC FLORETS** perfect and fertile; corolla yellow or purple, lobes usually with moniliform hairs. **CYPSELAE** obpyramidal, usually villous. **PAPPUS** of 5–10 scarious scales, usually with the midrib exserted as a prominent apical awn as long as or longer than the body (lacking in *G. arizonica*). Fifteen species, fourteen from the western United States and one disjunct in South America (Biddulph 1944).

1. Annual or short-lived perennial; ligules usually red or purple to the middle and yellow at the tips; phyllaries chartaceous at the base, herbaceous above; cypselae pubescent at the base, glabrous above; pappus scales with an awn equal to the scale .... ***G. pulchella***
1′ Perennial; ligules yellow, red or purple only at the base; phyllaries herbaceous throughout; cypselae pubescent throughout; pappus scales with an awn either 1/3 to 1/2 or twice the length of the scale .... (2)

2. Leaves mostly or all deeply and narrowly pinnately cleft; lobes of the disc corolla about 1 mm long, broadly triangular and acute; pappus scales with an awn 1/3 to 1/2 the length of the scale .... ***G. pinnatifida***
2′ Leaves entire, toothed, or broadly pinnately cleft; lobes of the disc corolla about 2 mm long, attenuate or acuminate; pappus scales with an awn twice the length of the scale .... ***G. aristata***

Gaillardia aristata Pursh (Latin, awn or bristle, referring to the pappus) Blanketflower, common blanketflower. Perennial herbs; hirsute with moniliform hairs. **STEMS** 2–8 dm tall. **LEAVES** basal and cauline, 5–15 (20) cm long, linear-lanceolate to oblanceolate or oblong, entire to coarsely toothed, rarely subpinnatifid, often glandular or glandular-punctate. **INFL** phyllaries ovate to lanceolate, villous with moniliform hairs; apex acuminate to long-acuminate or attenuate. **RECEPTACLES** bristly; setae subulate, equaling or exceeding the fruit. **RAY FLORET** ligules yellow, often purplish or reddish at the base, 10–35 cm long, apex 3-lobed. **DISC FLORET** corollas yellow below and purple above (rarely all yellow), lobes triangular-acuminate. **CYPSELAE** 3–4 mm long; densely silky-pubescent throughout. **PAPPUS** of scarious scales, acuminate, midrib exserted as an awn twice the length of the scale. $2n = 34$, 36, 68, 72. [*Gaillardia bicolor* Lam. var. *aristata* (Pursh) Nutt.; *Virgilia grandiflora* Nutt.]. Open plains and prairies, roadsides and disturbed ground, piñon-juniper woodland, ponderosa pine, and spruce-fir communities. COLO: LPl, Min; NMEX: RAr, SJn. 2100–2450 m (7000–8100′). Flowering: Jun–Sep. Western Canada and the northern Great Plains south to New Mexico and California. The species is widely cultivated and has seen extensive medicinal use including: as a poultice of roots for skin disorders and backache, as an infusion for eyewash, gastroenteritis, and cancer, as a decoction for kidney troubles, headache, and tuberculosis, and in baths for venereal disease.

Gaillardia pinnatifida Torr. (pinnately cleft, referring to the leaves) Hopi blanketflower, cut-leaved blanketflower, red-dome blanketflower, pinnate-leaved gaillardia. Perennial herbs, but often appearing biennial or annual; pubescent with moniliform hairs. **STEMS** to 6 dm. **LEAVES** confined to the basal 1/2 of the stem, to 10 cm long, oblanceolate to spatulate in outline, entire to toothed or more commonly pinnatifid, glandular-punctate. **INFL** phyllaries lanceolate, villous with moniliform hairs; apex acute to acuminate. **RECEPTACLES** bristly; setae subulate, shorter than or about equaling the fruit. **RAY FLORET** ligules yellow with purple veins, sometimes reddish or pinkish at the base, 10–20 mm long, apex 3-lobed. **DISC FLORET** corollas yellow below and purple above, lobes triangular. **CYPSELAE** 2–3 mm long, silky-pubescent throughout. **PAPPUS** of lanceolate to oblanceolate scarious scales, acuminate, midrib exserted as an awn 1/3 to 1/2 the length of the scale. $2n = 34$, 36. [*G. crassa* Rydb.; *G. crassifolia* A. Nelson & J. F. Macbr.; *G. globosa* A. Nelson; *G. gracilis* A. Nelson; *G. linearis* Rydb.; *G. mearnsii* Rydb.; *G. straminea* A. Nelson]. Dry, open plains and hills, desert shrub, piñon-juniper woodland, disturbed ground, and roadsides. ARIZ: Apa, Nav; COLO: LPl, Mon; NMEX: McK, SJn; UTAH. 1200–1950 m (4000–6400′). Flowering: Apr–Aug. Known from Utah, Arizona, Colorado, New Mexico, Texas, and northern Mexico. The most common species in our range. Used medicinally as a diuretic, a poultice for gout, a snuff to relieve congestion, and as a decoction for heartburn and nausea.

Gaillardia pulchella Foug. (somewhat beautiful) Indian blanketflower, firewheel, rose-ring gaillardia. Annual or short-lived perennial herbs; villous with short, often appressed moniliform hairs, resin-dotted. **STEMS** 1–6 dm tall. **LEAVES**

lanceolate to oblanceolate or oblong, 2–8 cm long, entire to coarsely toothed, lobed, or subpinnatifid, mostly sessile or the lower occasionally petiolate, often clasping stem. **INFL** phyllaries lanceolate, chartaceous below and green above, reflexed in flower and fruit, ciliate with moniliform hairs; apex acute to acuminate. **RECEPTACLES** bristly; setae subulate, equaling or reaching twice the length of the fruit. **RAY FLORET** ligules red or purple, yellow-tipped, or rarely all red or all yellow, 8–20 mm long; apex 3-toothed. **DISC FLORET** corollas yellow below, purple above, lobes triangular-acuminate. **CYPSELAE** 2–2.5 mm long, hirsute at base, trichomes equaling the achene, glabrous above but often obscured by long basal hairs. **PAPPUS** of lanceolate scales gradually tapering to an awn equaling the basal scale. 2*n* = 34, 36. [*G. drummondii* DC.; *G. neomexicana* A. Nelson]. Open plains, hills, and disturbed roadsides. ARIZ: Apa, Nav; NMEX: McK, RAr, SJn. 1875–2450 m (6150–8000′). Flowering: May–Sep. Ranging from the southeastern United States to the southern Great Plains, Colorado, New Mexico, and Arizona.

Glyptopleura D. C. Eaton Holly Dandelion
Gary I. Baird

(Greek *glyptos*, sculpture, + *pleura*, rib, in reference to the fruit) Taprooted annuals, herbage glabrous, sap milky. **STEMS** obscure, simple to much-branched, prostrate. **LEAVES**: basal rosulate, cauline alternate, progressively much-reduced upward, petiolate; blades simple, margins mostly pinnately lobed; lobes conspicuously white-denticulate and sometimes pinnately lobulate. **HEADS** liguliflorous, mostly terminal, erect; peduncles bracteate. **RECEPTACLES** flat, epaleate. **PHYLLARIES** herbaceous, imbricate to subequal in about 2 (+) series, outer appearing as much-reduced leaves. **RAY FLORETS** matutinal; corollas ligulate, white, often purplish abaxially, fading pink. **DISC FLORETS** lacking. **CYPSELAE** light to dark brown, narrowly obconic, curved, obtusely 5-angled, transcorrugated with about 12 rounded ribs, the most distal forming an apical cupule, glabrous, short-beaked; beak stout, obtusely 5-ribbed, arising from inside the apical cupule. **PAPPUS** of numerous fine bristles, easily deciduous in a fragile ring, white, ± smooth. *x* = 9. Two species of the Great Basin and Mojave deserts.

Glyptopleura marginata D. C. Eaton (margined, referring to the white-denticulate margins of the leaves) Plants 2–3 cm tall (ours). **LEAVES** long-petiolate; petioles broadly flattened, grading into blade, basal 12–24 mm × about 2 mm (up to 3× blade length); blades orbicular to oval, basal 8–18 mm × 7–9 mm; lobes 3–5 pairs, subopposite, oval-obovate, rounded, lobulate; lobules 0–2 pairs, denticulate; denticles white, sometimes basally purplish, rarely branched; cauline proportionally reduced and grading into outer phyllaries. **HEADS** solitary at ends of stems. **INVOLUCRES** 10–15 mm tall. **PHYLLARIES** in about 2 series; outer spatulate, margins white-denticulate, erect or spreading; inner linear-lanceolate, margins entire and scarious, erect, proximally green, distally purple. **RAY FLORETS** subequal or just exceeding involucres; corolla tube 6–7 mm, ligule 7–10 mm × about 1.5 mm; anthers 2–3 mm. **CYPSELAE** about 5 mm long (including beak); beak mostly 1 mm long. **PAPPUS** about 8 mm long. 2*n* = 18. Shadscale, blackbrush, and other desert shrub communities, in mostly sandy soils. UTAH. 1200 m (3900′). Flowering: Apr–May. California, Nevada, Utah, and southern Idaho. This species is found primarily in the Great Basin, but a localized disjunct population occurs in our area along the lower San Juan River.

Gnaphalium L. Cudweed
G. L. Nesom

(Greek *gnaphalion*, a downy plant, the name anciently applied to these or similar plants) Annuals, (1–) 3–30 cm; usually taprooted, sometimes fibrous-rooted. **STEMS** usually 1, erect (often with decumbent-ascending branches from bases; more or less woolly-tomentose, not glandular). **LEAVES** mostly cauline; alternate; more or less sessile; blades oblanceolate to spatulate or linear, bases ± cuneate, margins entire, faces concolor, gray and tomentose. **INFL** a disciform head, usually in more or less capitate clusters (in axils of leaves or bracts), sometimes in spiciform glomerules. **INVOLUCRES** narrowly to broadly campanulate, 2.5–4 mm. **PHYLLARIES** in 3–5 series, usually white or tawny to brown (opaque or hyaline, often shiny; stereomes usually glandular distally), ± equal to unequal, chartaceous toward tips (inner phyllaries narrowly oblong, usually white-tipped and protruding distal to outer). **RECEPTACLES** flat, smooth, epaleate. **PERIPHERAL FLORETS** pistillate, 40–80 (more numerous than bisexual); corollas purplish or whitish. **INNER FLORETS** bisexual, 4–7; corollas purplish or whitish. **CYPSELAE** oblong, faces usually glabrous, sometimes minutely papillate (hairs more or less papilliform, not myxogenic). **PAPPUS** of 8–12 readily falling barbellate bristles in 1 series. *x* = 7. About 38 species, mostly in the Old World, our 2 native.

1. Leaf blades linear, 0.5–3 mm wide; heads in spiciform arrays of spikelike, axillary glomerules; bracts subtending heads linear, 10–25 × 0.5–1 mm, surpassing glomerules; inner phyllaries narrowly triangular, apices acute .... ***G. exilifolium***

1′ Leaf blades spatulate to oblanceolate-oblong, 3–8 (–10) mm wide; heads in capitate clusters; bracts subtending heads oblanceolate to obovate, longest 4–12 × 1.5–4 mm, shorter than or equaling to slightly surpassing glomerules; inner phyllaries narrowly oblong, apices blunt .... ***G. palustre***

Gnaphalium exilifolium A. Nelson (slender leaf, presumably alluding to the linear leaves) Slender cudweed. Annuals, 3–15 (–25) cm; taprooted or fibrous-rooted. **STEMS** commonly branched from bases, erect to ascending, tomentose. **LEAVES** with linear blades, 0.4–5 cm × 0.5–3 mm; linear bracts subtending heads, 10–25 × 0.5–1 mm, surpassing glomerules. **HEADS** in spiciform glomerules (along distal 1/3–2/3 of main stems, sometimes appearing loosely spiciform); involucres 2.5–3.5 mm. **PHYLLARIES** brownish, bases woolly, inner narrowly triangular with whitish, acute apices. $2n = 14$. [*Gnaphalium angustifolium* A. Nelson; *G. grayi* A. Nelson & J. F. Macbr.; *G. strictum* A. Gray]. Lake and pond margins, streamsides, seeps, intermittent ponds, moist meadows, in areas of sagebrush-grassland, piñon-juniper, ponderosa pine–aspen, Douglas-fir, cottonwood-riparian. ARIZ: Apa; COLO: Arc, Dol, Hin, LPl, Min, Mon, SJn; NMEX: McK, SJn; UTAH. 1850–2700 m (6070–8860′). Flowering: (Jul–) Aug–Oct. South Dakota and Wyoming to Utah, Arizona, Colorado, and New Mexico; Mexico (Chihuahua).

Gnaphalium palustre Nutt. (marshy) Western marsh cudweed. Annuals, (1–) 3–15 (–30) cm; taprooted or fibrous-rooted. **STEMS** commonly with decumbent branches produced from bases, densely or loosely and persistently woolly-tomentose. **LEAVES** with spatulate to oblanceolate-oblong blades, 1–3.5 cm × 3–8 (–10) mm; oblanceolate to obovate bracts subtending heads, 4–12 × 1.5–4 mm, shorter than or surpassing glomerules. **HEADS** in capitate glomerules (at stem tips and in the most distal axils). **INVOLUCRES** 2.5–4 mm. **PHYLLARIES** brownish, bases woolly, the inner narrowly oblong with white (opaque), blunt apices. $2n = 14$. Arroyos, rocky and sandy streambeds, pond edges, potholes, other moist, open sites, often heavily grazed. ARIZ: Apa, Nav; COLO: LPl, Mon; NMEX: McK; UTAH. 1700–1800 m (5575–5905′). Flowering: Jun–Sep. Western Canada south through the Dakotas, Montana, and Washington, further through Arizona, Colorado, New Mexico, and Utah to northwestern Mexico.

Grindelia Willd. Gumweed, Resinweed
Guy Nesom

(for David Hieronymus Grindel, 1776–1836, Latvian botanist) Annual, biennial, or perennial herbs, or subshrubs, 15–250+ cm; taprooted (rhizomatous in *G. oölepis*). **STEMS** usually erect, sometimes decumbent to prostrate, glabrous or hairy. **LEAVES** basal and cauline or mostly cauline, alternate, petiolate (proximal) or sessile (distal), cauline blades mostly oblong, obovate, oblanceolate, or spatulate, bases commonly clasping to subclasping, margins usually serrate to dentate, sometimes entire, crenate, or pinnatifid (especially proximal), faces usually glabrous and gland-dotted, sometimes pubescent and/or stipitate-glandular. **INFL** a radiate or discoid head, in corymbiform to paniculiform arrays or borne singly. **INVOLUCRES** mostly globose to hemispheric or cupulate. **PHYLLARIES** mostly linear to lanceolate, unequal to subequal, bases usually ± chartaceous, apices ± herbaceous, looped, hooked, recurved, or straight, usually glabrous and ± resinous. **RECEPTACLES** flat or convex, ± alveolate, epaleate. **RAY FLORETS** absent or 5–60+, pistillate, fertile; corollas yellow to orange. **DISC FLORETS** bisexual and fertile (all or outer) or functionally staminate; corollas yellow. **CYPSELAE** ellipsoid to obovoid, ± compressed, sometimes ± 3–4-angled, faces smooth, striate, ribbed, furrowed, or rugose, glabrous. **PAPPUS** usually of (1–) 2–8 (–15) easily deciduous (persistent or tardily falling in *G. ciliata*), straight or contorted to curled, smooth or barbellulate to barbellate, subulate scales, setiform awns, or bristles. $x = 6$ (Nesom 1990a; Steyermark 1934, 1937). Species about 30, mostly in central and western North America, Mexico, and South America; introduced in the Old World.

1. Ray florets absent .... ***G. nuda***
1′ Ray florets present .... (2)

2. Phyllaries narrowly lanceolate, apices straight to slightly recurved; leaf margins usually entire, leaf bases not at all clasping; tips of foliar teeth sharp and indurate, eglandular .... ***G. arizonica***
2′ Phyllaries linear to lance-linear or lance-subulate, apices usually looped to hooked; leaf margins usually crenate to serrate, leaf bases clasping to subclasping; tips of foliar teeth blunt and gland-tipped .... ***G. squarrosa***

Grindelia arizonica A. Gray (alluding to Arizona) Arizona gumweed. Perennial herbs, (10–) 25–70 cm. **STEMS** stramineous to reddish, glabrous. **LEAVES** with cauline blades oblong to oblanceolate, 15–45 (–85) mm, lengths mostly 3–8 times widths, bases ± cuneate, not clasping, margins mostly entire, less commonly serrate, or (proximal leaves) pinnately lobed to coarsely toothed, teeth sharp and indurate-apiculate (eglandular), faces glabrous and sparsely or not at all gland-dotted. **PHYLLARIES** linear to lanceolate, apices subulate to deltate, spreading to slightly reflexed, slightly to strongly resinous. **CYPSELAE** monomorphic, faces striate to furrowed. **PAPPUS** of 2–4 straight or weakly contorted awns shorter than disc corollas. $2n = 12$. [*G. laciniata* Rydb.]. Clay hills, rocky and shaley slopes, pond edges, roadsides and other disturbed sites, sometimes in alkaline soil, cottonwood, riparian, mountain shrub, rabbitbrush, juniper, piñon-juniper woodland, ponderosa pine, and white fir communities; commonly with Gambel's oak. ARIZ: Apa, Nav; COLO: Arc, LPl, Mon, SMg; NMEX: RAr, SJn. 1700–2750 m (5575–9020′). Flowering: (Jun–) Jul–Sep (–Oct). Utah, Arizona, Colorado, New Mexico, southwestern Texas, and Mexico (Chihuahua, Coahuila). Four Corners plants are **var. *arizonica*.** *Grindelia arizonica* is distinguished by its glabrous, often reddish stems, glabrous and mostly eglandular leaves, spinulose-indurate foliar teeth, and monomorphic cypselae with longitudinal furrows at maturity. *Grindelia arizonica* var. *neomexicana* (Wooton & Standl.) G. L. Nesom, with apices of outer phyllaries linear to lanceolate-acuminate and erect to slightly spreading, occurs mostly in Catron and Grant counties, New Mexico. Plants of *G. arizonica* with leaf margins pinnately lobed to laciniate or coarsely dentate have been called *G. laciniata*—these occur mainly in the Four Corners region; the inclusion here of *G. laciniata* within *G. arizonica* (var. *arizonica*) follows the treatment of Strother and Wetter (2006). *Grindelia decumbens* Greene (type from Cimarron, Montrose County, Colorado) almost certainly occurs in the Four Corners region (Colorado and New Mexico), as inferred from Strother and Wetter (2006), but no specimens of it have been seen in this study. It is similar to *G. arizonica* in perennial duration, radiate heads, glabrous stems and leaves, and leaves with apiculate, non–glandular-resinous foliar teeth; the two species would be distinguished by the following contrasts.

1. Tips of phyllaries mostly slightly incurved, straight, or slightly recurved; leaf bases not at all clasping; leaf faces sparsely, or not at all, resin gland–dotted .... ***G. arizonica***

1′ Tips of phyllaries (most or at least the outer) looped to hooked or spreading (inner may be recurved, straight, or incurved); leaf bases ± cuneate or clasping; leaf faces moderately resin gland–dotted .... ***G. decumbens***

Grindelia nuda Alph. Wood (bare or naked, alluding to the lack of ray florets) Rayless gumweed. Perennial herbs, 15–90 cm. **STEMS** green or commonly reddish, glabrous. **LEAVES** with cauline blades oblanceolate to obovate, subclasping, largest midcauline (10–) 15–65 mm × (2–) 4–13 mm, 4–8 (–10) times longer than wide, margins serrate (rarely nearly entire or with narrow, shallow lobes), teeth blunt and subglandular (not resinous-glandular like *G. squarrosa*), faces glabrous, dark-dotted (but not resinous-punctate like *G. squarrosa*). **PHYLLARIES** lanceolate-linear or lanceolate-subulate, apices subterete to subulate, recurved to barely looped-resinous. **CYPSELAE** monomorphic, often deeply furrowed at maturity. **PAPPUS** of 2–3 (–8) straight awns shorter than disc corollas. $2n = 24$. [*Grindelia aphanactis* Rydb.]. Roadsides and other disturbed sites, washes, meadows; saltcedar, cottonwood-riparian, juniper-shrub, piñon-juniper, Gambel's oak–sage, ponderosa pine–Gambel's oak, and ponderosa pine–Douglas-fir communities. ARIZ: Apa, Nav; COLO: Arc, LPl, Min, Mon; NMEX: McK, RAr, San, SJn; UTAH. (1200–) 1500–2500 m (3935–8200′). Flowering: Jul–Oct. Southeastern Utah to southwestern Kansas, south to Arizona and central Texas; Mexico (Chihuahua). Four Corners plants are **var. *aphanactis*** (Rydb.) G. L. Nesom (Greek *aphanes*, invisible or obscure, + *aktis*, ray). Strother and Wetter (2006) treated *G. nuda* (including var. *aphanactis* and var. *nuda*, all with discoid heads) within the otherwise radiate *G. squarrosa*. The sympatry and abundant occurrence of both species within the Four Corners area, however, provide evidence that they are distinct. Cypselae of *G. squarrosa* are dimorphic (ray and outer disc smooth and compressed but slightly 3–4-angled, inner disc longer, strongly compressed, and 2-angled), while cypselae of *G. nuda* are monomorphic (all smooth or developing shallow furrows late in maturation).

Grindelia squarrosa (Pursh) Dunal (rough with stiff scales, alluding to the heads with imbricate and recurving phyllary tips) Curly gumweed. Biennials or perennials, perhaps flowering first year and appearing annual, (10–) 14–100 cm. **STEMS** usually whitish, sometimes reddish, glabrous. **LEAVES** with cauline blades ovate or obovate to oblong, oblanceolate, or lanceolate, (10–) 15–70 mm, lengths 2–5 (–10) times widths, bases clasping, margins usually crenate to serrate (rarely entire), teeth rounded to obtuse and resinous-glandular, faces glabrous, strongly resinous-punctate. **PHYLLARIES** reflexed to spreading or appressed, filiform or linear to lance-linear or lance-subulate, apices subterete to subulate, usually looped to hooked, sometimes recurved to nearly straight, moderately to strongly resinous. **CYPSELAE** dimorphic, faces smooth, striate, or ± furrowed. **PAPPUS** of 2–3 (–8), straight awns shorter than disc corolla. $2n = 12$. [*G. serrulata* Rydb.; *G. squarrosa* var. *serrulata* (Rydb.) Steyerm.]. Roadsides and other disturbed sites, sand and alluvium, hills, clay flats, along streams, subalkaline soils; saltcedar–Russian olive, juniper–Gambel's oak–sage–rabbitbrush, ponderosa pine–piñon–juniper, and ponderosa pine–aspen communities. ARIZ: Apa, Nav; COLO: Arc, Dol, LPl, Mon, SMg; NMEX: RAr, SJn; UTAH. 1300–2300 (–2550) m (4265–8365′). Flowering: (Jun–) Jul–Sep

(–Oct). Canada (British Columbia to Quebec), south to Virginia, Maryland, Illinois, Indiana, and Kentucky, to California, Arizona, New Mexico, Texas, and Arkansas; Mexico (Chihuahua); introduced in Ukraine.

Gutierrezia Lag. Snakeweed
G. L. Nesom

(possibly for Pedro Gutierrez, Spanish nobleman, not specified by Lagasca) Annuals, perennials, or subshrubs. Stems glabrous or papillate-scabrous to minutely hispidulous or scabrate-hirtellous. **LEAVES** basal and cauline, alternate, sessile to petiolate, blades linear to lanceolate or spatulate, margins entire, sometimes scabroso-ciliate, faces gland-dotted (sometimes obscurely), glutinous. **INFL** a radiate (discoid in 1 species) head, borne singly or in clusters of 3–6. **INVOLUCRES** cylindric to campanulate. **PHYLLARIES** in 2–4 series, imbricate, ovate to lanceolate, bases white-indurate, margins narrowly scarious, flat or less commonly strongly convex or keeled. **RECEPTACLES** flat to conic, foveolate, with numerous 1-seriate, swollen, apically hooked trichomes, epaleate. **RAY FLORETS** pistillate, fertile; corollas yellow or white, coiling at maturity. **DISC FLORETS** bisexual, fertile; corollas concolorous with rays. **CYPSELAE** clavate or cylindric, 5–8-nerved, hairy (essentially glabrous in 1 species), hairs white, usually arising primarily from between ribs, appearing to occur in longitudinal lines, usually obscuring surface. **PAPPUS** of 5–10 whitish, irregular, often erose-margined, persistent scales in 1–2 series, sometimes connate or reduced to coroniform in some species, usually longer in disc than ray. x = 4 (Lane 1985). Species 28, western North America and western South America.

1. Involucre cylindric, 1–1.5 mm wide; ray florets 1 (–2); disc florets 1 (–2), sterile (ovaries vestigial), corollas obdeltate-funnelform (throats widely flaring), lobes long (1/3 the corolla length), recurved-coiling..***G. microcephala***

1′ Involucre cuneate-campanulate to cylindric, 1.5–2 (–3) mm wide; ray florets (2–) 3–8; disc florets (1–) 3–9, fertile, corollas tubular-funnelform, lobes deltate and erect to spreading or recurved ....***G. sarothrae***

Gutierrezia microcephala (DC.) A. Gray (Greek *micro*, small, + *cephal*, head) Small-head snakeweed. Subshrubs 20–140 cm. **STEMS** glabrous to minutely hispidulous. **LEAVES**: basal and proximal absent at flowering; cauline blades linear or filiform to narowly oblanceolate or lanceolate, 0.5–2.2 (–4) mm wide, little reduced distally, 1-nerved. **HEADS** sessile to subsessile, in compact glomerules forming distinctly flat-topped arrays. **INVOLUCRES** cylindric, 1–1.5 mm wide. **RAY FLORETS** 1 (–2), each enclosed by a conduplicate inner phyllary. **DISC FLORETS** 1 (–2), sterile (ovaries vestigial), corollas broadly obdeltate-funnelform, throats widely flaring, lobes recurved-coiling. **CYPSELAE** densely strigose-sericeous. **PAPPUS** (ray) of basally caducous, narrowly lanceolate-oblong scales. $2n$ = 8, 16, 24, 32. Along streams and benches, alluvium, dunes, hanging gardens, disturbed sites; desert shrub, riparian, juniper–mixed shrub, and piñon-juniper communites. ARIZ: Apa, Coc; Nav; COLO: LPl, Mon; NMEX: SJn; UTAH. 1125–1950 m (3700–6400′). Flowering: (Jul–) Aug–Oct. California, Nevada, and Arizona, east to Colorado and Texas; northern Mexico.

Gutierrezia sarothrae (Pursh) Britton & Rusby (Greek *saron*, a broom) Broom snakeweed. Subshrubs 10–60 (–100) cm. **STEMS** minutely hispidulous. **LEAVES**: basal and proximal absent at flowering; cauline blades linear to lanceolate, sometimes filiform and fascicled, 1.5–2 (–3) mm wide, little reduced distally, 1 (–3)-nerved. **HEADS** sessile to subsessile, in compact glomerules forming distinctly flat-topped, dense corymbiform arrays. **INVOLUCRES** cylindric to cuneate-campanulate, 1.5–2 (–3) mm wide. **RAY FLORETS** (1–) 3–8. **DISC FLORETS** (2–) 3–9, fertile, rarely sterile, corolla tubular-funnelform, lobes erect to spreading or recurved. **CYPSELAE** densely strigose-sericeous. **PAPPUS** of basally caducous scales. $2n$ = 8, 16, 32. Sandstone outcrops, dunes, clay hills, rocky slopes, gypseous soil, alluvium, road-sides and other disturbed sites; desert grasslands (often abundant in overgrazed pastures), desert shrub, sagebrush, juniper-sage, piñon-juniper, and ponderosa pine–piñon–juniper–Gambel's oak communities. ARIZ: Apa, Coc, Nav; COLO: Arc, Dol, LPl, Min, Mon, SMg; NMEX: McK, RAr, San, SJn; UTAH. 1450–2400 (–2700) m (4755–8860′). Flowering: Jul–Sep (–Oct). Canada (Alberta to Manitoba), through Washington, Montana, North Dakota, and Minnesota, south to California, Arizona, New Mexico, and Texas; widespread in northern Mexico. Few-flowered forms of *G. sarothrae* can be distinguished from *G. microcephala* by their fertile disc ovaries and tubular-funnelform disc corollas.

Helenium L. Sneezeweed
Brian Elliott

(ancient Greek name, reportedly for Helen of Troy) Annual, biennial, or perennial herbs. **STEMS** 1–several, erect, simple or branched, leafy and often winged. **LEAVES** alternate, simple or pinnatisect, linear to lanceolate, elliptic, oblong, or spatulate, entire or toothed, glandular-punctate, often decurrent as wings on the stem. **INFL** heads radiate or discoid, solitary to numerous and corymbose, pedunculate. **PHYLLARIES** in 2–3 equal or subequal series, linear to lanceolate or subulate, herbaceous, spreading or becoming reflexed, free or fused. **RECEPTACLES** convex to conic or globose, alveolate. **RAY FLORETS** pistillate or neuter, fertile or sterile; ligules conspicuous, yellow, golden yellow, or orange, fan-shaped; apex 3-lobed. **DISC FLORETS** perfect and fertile, numerous; corolla yellow to red or purplish brown; lobes 4–5, glandular-hairy. **CYPSELAE** turbinate to obpyramidal, 4–5-angled with as many intermediate ribs, pubescent on the angles and ribs or glabrous. **PAPPUS** absent or of 5–10 scarious or hyaline scales, often awn-tipped or fimbriate. Approximately 40 species of North and South America. All species are toxic to livestock, and some are cultivated as ornamentals (Bierner 1972).

Helenium autumnale L. (pertaining to autumn, presumably referring to the late summer flowers) Common sneezeweed. Perennial, fibrous-rooted herbs. **STEMS** usually solitary, branched above, 1.5–12 dm high, puberulent and glandular, narrowly angled or winged from decurrent leaf bases. **LEAVES** mostly cauline, 1.5–15 cm long, 0.5–4 cm wide, narrowly elliptic to lanceolate, oblanceolate, or ovate-lanceolate, shallowly toothed to denticulate or subentire, glabrous to sparsely hairy or glandular-punctate, lower leaves soon deciduous; leaf base decurrent on stem, lower leaves with short, winged petioles, upper leaves sessile. **INFL** heads 3–many in a terminal corymb; peduncle 3–10 cm, short-hairy. **INVOLUCRES** 7–10 mm high. **PHYLLARIES** narrow, soon reflexed. **RAY FLORETS** 10–20 per head; ligules golden yellow to orange, 5–25 mm long, often over 5 mm wide. **DISC FLORETS** yellow, 3–3.5 mm long, 5-lobed, the lobes pubescent. **CYPSELAE** 1.5–2 mm long, hirsute and glandular on the angles. **PAPPUS** scales lance-ovate, usually lacerate on the margins and awn-tipped, awn equaling the body; scales 1–3 mm long, much shorter than the disc corolla. $2n = 32, 34, 36$. [*H. montanum* Nutt.]. Generally growing in damp meadows and bottomlands, marshes, stream banks, and ditches. ARIZ: Apa; COLO: Hin; SJn. 1400–2590 m (4600–8500′). Flowering: Jun–Oct. Widely distributed in North America and expected throughout our area. The species is split into three varieties, ours being **var. *montanum*** (Nutt.) Fernald (of the mountains). Although the species is toxic to livestock, it has been used medicinally as a gynecological aid, for treating fever, as an analgesic, and for colds.

Helianthella Torr. & A. Gray Little Sunflower, Helianthella

Gary I. Baird

(diminutive form of *Helianthus*) Perennials, taprooted, rhizomes lacking, caudices branched, woody or not, herbage glabrous to pubescent, obscurely to conspicuously gland-dotted or eglandular, sap clear. **STEMS** usually many, erect, simple or branched distally. **LEAVES** basal and cauline, opposite (at least proximally) or alternate, simple, petiolate or distally sessile, reduced upward; petioles shorter than blades; blades elliptic, ovate, lanceolate, linear, to oblanceolate or spatulate, bases cuneate, margins entire, apices acute to rounded. **HEADS** radiate, often borne singly or 2–15 in corymbiform or racemiform clusters, erect or nodding; peduncles bracteate to ebracteate. **INVOLUCRES** campanulate to hemispheric. **PHYLLARIES** about 20–30+ in mostly 3 series, herbaceous, subequal or unequal, outer sometimes foliaceous and surpassing inner. **RECEPTACLES** convex, paleate; paleae scarious to chartaceous, ± conduplicate and enfolding cypselae, deciduous, ± striate-nerved, apices truncate (to obtuse), erose-ciliate, subcucullate; calyculi lacking. **RAY FLORETS** 8–20, neuter; corollas yellow; laminae fugacious. **DISC FLORETS** 30–200+, bisexual, fertile; corollas yellow or purplish to brownish, lobes pubescent abaxially. **CYPSELAE** monomorphic, brownish, ± compressed, obovate to obcordate, margins sometimes narrowly winged, usually ciliate, faces glabrous or pubescent. **PAPPUS** lacking (rare), or of 2 persistent, subulate scales, plus a few shorter, lacerate scales (all these sometimes connate). $x = 15$. Nine species of western North America.

1. Heads usually several (5–20) per stem; ray laminae less than 15 mm (about equal to width of disc); disc corollas dark purplish, rarely yellow ..... ***H. microcephala***

1′ Heads usually few (1–3) per stem; ray laminae mostly 15–45 mm (usually surpassing width of disc); disc corollas yellow ..... (2)

2. Plants less than 50 cm tall; disc less than 15 mm wide; phyllaries mostly lanceolate; rays pale yellow ..... ***H. parryi***

2′ Plants 50–150 cm tall; disc 25–50 mm wide; phyllaries often ovate; rays ± bright yellow ..... ***H. quinquenervis***

Helianthella microcephala (A. Gray) A. Gray (small-headed) Purpledisc little sunflower. Perennials about 20–80 cm tall, taproot and caudex woody, herbage scabrous to hispidulous or coarsely strigose, often subglabrate, ± gland-dotted (sometimes obscurely so). **LEAVES** mostly basal; blades oblanceolate to linear-oblanceolate or elliptic, 6–25 × 1–3 cm, usually evidently 3-nerved, margins strigose, apices acute to obtuse or rounded; cauline usually several, proximally opposite and distally alternate, or sometimes nearly alternate throughout, progressively reduced upward. **HEADS** about (3–) 5–20 per stem in loose or compact corymbiform or racemiform clusters, erect; paleae subequal to florets, chartaceous, often purplish or blackish (especially distally) and usually gland-dotted, sometimes vernicose, margins membranous to hyaline, apices ± truncate, ciliate (sometimes densely) to erose. **INVOLUCRES** turbinate to campanulate or hemispheric, mostly 10–15 mm diameter. **PHYLLARIES** 20–30+ in 3–4 series, linear-oblong to linear-deltate, sometimes blackish, margins densely strigose-ciliate, faces glabrous or sparsely strigose, gland-dotted (sometimes obscurely so), apices acute to obtuse, usually blackish purple (sometimes margins also). **RAY FLORETS** 5–13; corollas yellow, laminae 6–14 × 2–3 mm, glabrous or ± strigose abaxially, eglandular or obscurely gland-dotted (usually abaxially), uncommonly vernicose abaxially. **DISC FLORETS** about 30–50+; corollas blackish purple distally, uncommonly yellow; anthers blackish purple, apical appendages yellow, eglandular. **CYPSELAE** ± compressed, narrowly obovoid, 7–8 mm, dark brown, densely sericeous. **PAPPUS** of 2 lacerate awnlike scales, 2–3 mm, usually with 1–4 smaller scales (less than 1 mm), these sometimes lacking. $2n = 30$. Mostly in piñon-juniper and ponderosa pine–oak forest communities. ARIZ: Apa; COLO: LPl, Mon, SMg; NMEX: McK, SJn; UTAH. 1500–2600 m (5000–8600′). Flowering: Jun–Sep. Primarily of the Colorado Plateau region of Arizona, Colorado, New Mexico, and Utah.

Helianthella parryi A. Gray (for Charles C. Parry, 1823–1890, American botanist and explorer of the Rocky Mountains) Parry's little sunflower. Perennials, mostly 20–50 cm tall, herbage strigose to strigillose (densely so beneath heads). **LEAVES** mostly basal; blades oblanceolate to spatulate, 6–10 cm × 4–8 mm, evidently 3–5-nerved, margins ± strigose, apices acute; cauline few (about 2–3 pairs), proximally opposite and distally alternate, progressively reduced upward. **HEADS** borne singly or perhaps 2–3 per stem, ± nodding; paleae scarious, stramineous, eglandular, margins membranous, apices ± truncate, densely pubescent-ciliate. **INVOLUCRES** ± hemispheric, 15–20 mm diameter. **PHYLLARIES** 20–30+ in 3 series, lanceolate to lance-linear, herbaceous, base somewhat thickened and chartaceous, margins densely pilose-ciliate, faces glabrous, eglandular, apices acuminate. **RAY FLORETS** 8–14; corollas pale yellow, laminae 25–30 × 5–7 mm, strigillose on abaxial veins, eglandular. **DISC FLORETS** 30–60; corollas pale yellow; anthers blackish, apical appendages yellow, eglandular. **CYPSELAE** compressed, narrowly obovate, about 7–8 mm, black, densely sericeous. **PAPPUS** of 2 lacerate awnlike scales, about 2–3 mm, usually with 1–4 smaller scales (less than 1 mm), these sometimes lacking. Bristlecone pine and spruce-fir communities. COLO: Arc, Min, SJn. 2590–3050 m (8500–10000′). Flowering: Jul. Primarily of the southern Rocky Mountains in Arizona, Colorado, and New Mexico.

Helianthella quinquenervis (Hook.) A. Gray (five-nerved or -veined) Five-nerve little sunflower. Perennials, mostly 50–150 cm tall, herbage glabrous to pilose-strigose (densely so beneath heads). **LEAVES** mostly cauline; proximal blades largest, elliptic-ovate to lance-ovate or lanceolate, 10–30 × 1–4 cm, evidently 3–5-nerved, margins ± strigose, apices acute to acuminate; cauline about 3–5 pairs, mostly opposite (rarely 3-whorled), the most distal sometimes alternate, progressively reduced upward. **HEADS** borne singly, or occasionally 2–3 per stem, ± nodding; paleae scarious, stramineous, often grayish black distally, eglandular, margins scarious, apices ± truncate, densely pubescent to merely ciliate. **INVOLUCRES** ± hemispheric, 25–50 mm diameter. **PHYLLARIES** about 20–30 in about 3 series, ovate to lanceolate, herbaceous, base somewhat thickened, margins densely pilose-ciliate, faces ± glabrous, eglandular, apices acuminate. **RAY FLORETS** 13–21 (sometimes fewer); corollas yellow, laminae 25–40 × 5–12 mm, obscurely strigillose abaxially, eglandular. **DISC FLORETS** numerous (200+); corollas yellow; anthers blackish, apical appendages yellow, eglandular. **CYPSELAE** strongly compressed, obovate, mostly 8–10 mm, blackish brown, margins pilose-ciliate, faces glabrous or sparsely strigillose. **PAPPUS** of 2 bristlelike scales, about 3–4 mm, usually with 2–4 smaller scales (less than 2 mm). Ponderosa pine, spruce-fir, and alpine communities. COLO: Arc, Hin, LPl, Min, Mon, SJn; UTAH. 2400–3500 m (7800–11400′). Flowering: Jun–Sep. Oregon, Idaho, and Montana to Arizona, New Mexico, Colorado, and South Dakota, south into northern Mexico.

Helianthus L. Sunflower

Gary I. Baird

(Greek *helios*, sun, + *anthos*, flower) Annuals or perennials, taprooted (slender to thickened) or rhizomatous, some tuberous, caudices simple or branched, herbage glabrous or pubescent, often gland-dotted, sap clear. **STEMS** mostly 1, erect to procumbent, usually branched distally, or simple. **LEAVES** mostly cauline, simple, petiolate or sessile, opposite or alternate (often proximal opposite, distal alternate); blades deltate to ± lanceolate or linear, usually evidently 3-nerved, bases cordate to narrowly cuneate, margins entire or coarsely serrate or crenate, rarely lobed, apices acute to acuminate. **HEADS** radiate (rarely discoid), borne singly or few to many in various arrays, erect or nodding; peduncles usually ebracteate. **INVOLUCRES** campanulate to hemispheric. **PHYLLARIES** mostly 10–40+ in 2–4 series, herbaceous, subequal to unequal. **RECEPTACLES** flat to ± convex, paleate; paleae persistent, conduplicate and enfolding cypselae, apices ± 3-toothed or (sometimes) entire, often reddish purple (distally) or sometimes blackish;

calyculi lacking. **RAY FLORETS** 5–30+, neuter; corollas yellow; laminae fugacious. **DISC FLORETS** mostly 30–150+, bisexual, fertile; corollas yellow or reddish purple (at least distally), throat abruptly expanded from tube and usually proximally bulbous and pubescent. **PAPPUS** mostly of 2 primary ± subulate scales (to 5 mm), often with some shorter secondary scales (to 2 mm), readily deciduous. **CYPSELAE** monomorphic, mostly blackish or brownish, often mottled or striped, ± compressed to subterete, ± 4-angled, obpyramidal to turbinate, glabrous, glabrate, or pubescent. x = 17. About 50 species of North America, introduced and established in many regions of the world. Species identification is notoriously problematic, especially if the specimen is fragmentary or was growing in unusual circumstances. It is also complicated by polyploidy, hybridization, and phenotypic plasticity due to environmental conditions. The result is that not all specimens are readily assignable.

1. Plants annual from a taproot; leaves mostly alternate (or proximal opposite); paleae prominently 3-toothed (middle tooth much longer than wide); disc corollas reddish purple (at least distally), rarely yellowish ....(2)
1′ Plants perennial from rhizomes or widely creeping roots; leaves opposite or alternate; paleae weakly 3-toothed (middle about as long as wide); disc corollas yellow or reddish purple ....(4)

2. Phyllaries linear, much surpassing disc; pappus of 2 linear scales plus some ± bristlelike scales.......***H. anomalus***
2′ Phyllaries ovate or lanceolate, usually equaling or less than disc; pappus of 2 lanceolate or awnlike scales, occasionally with a few shorter scales also ....(3)

3. Phyllaries ± lanceolate, tapering gradually to apex; paleae strigillose to glabrate, eglandular, the central ones conspicuously white hispid-ciliate (becoming less evident with age)............................***H. petiolaris***
3′ Phyllaries ovate to oblong-ovate, abruptly narrowed to acuminate apex; paleae ± scabrous-hispid and often gland-dotted, not especially ciliate, the central ones ± the same as the marginal ones............................***H. annuus***

4. Plants often over 1 m tall (to 3+ m); leaves not strongly reduced upward; disc corollas yellowish; anther appendages yellow ....(5)
4′ Plants mostly less than 1 m tall; leaves usually strongly reduced upward; disc corollas reddish purplish or yellow; anther appendages blackish or grayish............................(7)

5. Stems densely scabrous-hispid; roots not tuberous; leaves 1-nerved and ± folded; flowers in racemiform or spiciform clusters, usually blooming in autumn (Sep–Oct)............................***H. maximiliani***
5′ Stems glabrous and glaucous, sometimes scattered hispid, or scabrous-hispid; roots or rhizomes ± tuberous; leaves 3-nerved and flat; flowers mostly in corymbiform or paniculiform clusters (sometimes racemiform), usually blooming in late summer (Aug–Sep) ....(6)

6. Stems mostly glabrous and glaucous, occasionally also sparsely hispid; plants native and usually found in wet meadows, streamsides, or other moist habitats ....***H. nuttallii***
6′ Stems mostly scabrous-hispid; plants widely cultivated and sometimes escaping, but not usually found in wet habitats; Jerusalem artichoke (not known from the study area but to be expected) ....*H. tuberosus* L.

7. Plants mostly glabrous and glaucous (occasionally sparsely hispid); leaf margins often undulate; disc corollas usually reddish, sometimes yellow; primary pappus scales 1–2 mm, secondary scales lacking; cypselae 3–4 mm, glabrous (or glabrate) ....***H. ciliaris***
7′ Plants mostly scabrous-hispid (stems sparsely so); leaf margins mostly flat or revolute; disc corollas yellow; primary pappus scales 4–5 mm, secondary scales sometimes present; cypselae 5–6 mm, puberulent (especially distally) ....***H. pauciflorus***

Helianthus annuus L. (annual) Common sunflower. Annuals mostly 1–3 m tall (less in depauperate specimens), taprooted, herbage often coarsely scabrous-hispid throughout. **STEMS** 1, erect, usually branched (simple in depauperate individuals or some cultivated forms). **LEAVES** mostly alternate (proximal opposite), petioles 2–20 cm long; blades broadly ovate to lanceolate, 5–40 × 1–40 cm, bases cordate to cuneate, margins ± serrate-crenate (at least on larger leaves), faces ± scabrous-hispid, abaxially gland-dotted. **HEADS** 1–10+ (proliferating with time) in loose, ± paniculiform clusters; peduncles 2–20 cm, hispid; paleae about 10 mm, apices 3-toothed (middle tooth long-acuminate), ± uniformly hispid to hispidulous distally, sometimes gland-dotted. **INVOLUCRES** ± hemispheric (or broadly so), disc about 20–40 mm diameter (more in cultivated forms, less in depauperate individuals). **PHYLLARIES** about 20–30 (more in cultivated forms), ovate to oblong-ovate or lance-ovate, mostly 15–25 × 5–8 mm, margins hispid-ciliate, abaxial faces ± hispid (rarely glabrous or glabrate), ± gland-dotted, apices abruptly contracted, long-acuminate. **RAY FLORETS** 10–30 (more in cultivated forms, less in depauperate individuals), laminae (15–) 25–50 mm, sparsely pubescent abaxially, or glabrous. **DISC FLORETS** 100+ (much more in some cultivated forms, less in depauperate individuals); corolla throats proximally densely pubescent and distinctly bulbous, lobes mostly reddish purple (uncommonly yellow) and abaxially pubescent to strigillose; anther appendages purplish or yellow, usually evidently gland-dotted. **CYPSELAE** about 4–5 mm long (more in cultivated forms), sericeous-hispidulous (especially distally), sometimes glabrate. **PAPPUS** of 2 lanceolate scales (about 2–4 mm), smaller scales usually lacking, or 1–4 (less than 1 mm). $2n = 34$. Floodplains, wetlands, roadsides, ditch banks, open fields, and other disturbed sites in many plant communities. ARIZ: Apa, Nav; COLO: Arc, Dol, Hin, LPl, Min, Mon, SJn, SMg; NMEX: McK, RAr, San, SJn; UTAH. 1500–2700 m (4900–8800′). Flowering: May–Oct. Throughout almost all of North America (northern Canada to Mexico); introduced almost worldwide. This plant was originally cultivated by native Americans for its edible seed and is today widely grown for the same purpose.

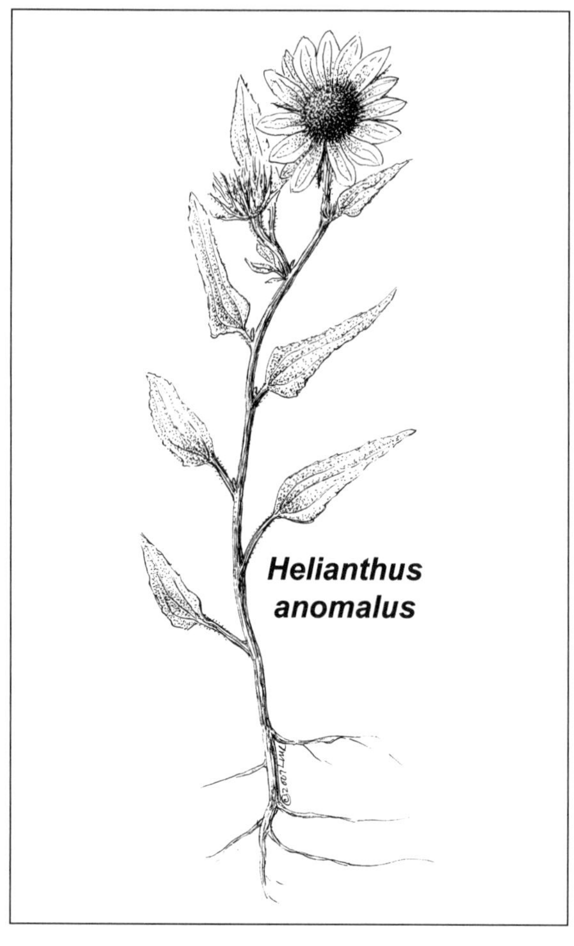

Helianthus anomalus S. F. Blake (anomalous) Western sunflower. Annuals, mostly 10–60 cm tall, taprooted, herbage pubescent; indumentum a mix of very coarse, white, multicellular, ± pustular hairs, often drying blue-green proximally ("white-hispid"), minute ± crinkly yellow hairs ("yellow-pubescent"), and minute glandular hairs ("gland-dotted"). **STEMS** 1, erect, often branched, ± glabrous or sparsely white-hispid. **LEAVES** mostly alternate (proximal opposite), petioles 1–5 cm, ± white-hispid (sometimes very sparse); blades linear-lanceolate to broadly lance-ovate, 20–80 × 7–65 mm, bases cuneate to truncate (occasionally weakly cordate), margins entire, faces white-hispid, sometimes gland-dotted (usually more so adaxially than abaxially) or sparsely yellow-pubescent. **HEADS** 1–10+ (proliferating with time) in loose, ± paniculiform clusters; peduncles 1–9 cm, ± white-hispid and usually yellow-pubescent (at least beneath head); paleae 7–9 mm, apices 3-toothed (middle tooth largest, long-acuminate, lateral teeth sometimes obscure), ± hispidulous distally (rarely central ones white hispid-ciliate). **INVOLUCRES** ± hemispheric, disc about 15–30 mm diameter. **PHYLLARIES** mostly 10–20 in about 2 series, linear, about 10–20 × 2–3 mm, usually much surpassing disc, often with 1 larger (30–35 mm) foliaceous bract subtending head, margins ± white-hispid, abaxial faces ± white-hispid, usually also yellow-pubescent, rarely gland-dotted, midvein sometimes glandular-thickened (sometimes lateral veins also), apices acuminate. **RAY FLORETS** 7–12, laminae 15–30 mm, glabrous or (occasionally) obscurely gland-dotted abaxially. **DISC FLORETS** 25–100+; corolla throats proximally pubescent, ± bulbous, lobes reddish purple and abaxially pubescent; anther appendages purplish, eglandular to evidently gland-dotted. **CYPSELAE** narrowly turbinate to ± compressed, 4–9 mm, sericeous-pilose, ± mottled, often with whitish stripes (especially on margins). **PAPPUS** of 2 main linear scales (about 2–4 mm), and some (up to 10) smaller ± bristlelike scales (about 1–3 mm), all readily deciduous. $2n = 34$. [*H. deserticola* Heiser]. Sand dune communities, or in sandy soils. ARIZ: Apa, Coc, Nav; UTAH. 1400–1600 m (4500–5400′). Flowering: Jun (mostly) –Oct. Utah, Arizona, and Nevada. The segregation of *H. deserticola* from *H. anomalus* is problematic and seems unrealistic. Both were derived through hybridization between *H. annuus* and *H. petiolaris* and in general are sympatric on the western Colorado Plateau. The features used to separate these hybrid taxa overlap to such a degree that they ultimately fail and separation becomes arbitrary.

Helianthus ciliaris DC. (ciliate, alluding to the leaf and phyllary margins) Blueweed, blueweed sunflower. Perennials, 15–70 cm tall, rhizomatous (or from rhizomelike roots), herbage mostly glabrous and glaucous, occasionally scattered hispid. **STEMS** 1 (often clustered), erect to decumbent, simple or few-branched. **LEAVES** opposite (proximal) to alternate (distal), strongly reduced upward, ± sessile; proximal blades oblanceolate to lanceolate or linear, proximal about 40–75 × 5–20 mm, bases cuneate, margins entire, often undulate, faces glabrous and glaucous (sometimes sparsely hispid). **HEADS** borne singly or 2–5; peduncles 2–15 cm; paleae about 7 mm, apices entire to obscurely 3-toothed (middle tooth largest), acute, puberulent and ± gland-dotted distally. **INVOLUCRES** hemispheric, disc 12–25 mm diameter. **PHYLLARIES** mostly 15–20 in about 3 series, ovate to lance-ovate, 3–8 × 2–3 mm, margins ciliate, abaxial faces glabrous or ± pubescent, eglandular, apices acute. **RAY FLORETS** 10–18, laminae 8–10 mm (or less), glabrous. **DISC FLORETS** about 30; corolla throats proximally ± pubescent, not or obscurely bulbous, lobes reddish or yellow, glabrous; anther appendages ± blackish or grayish, gland-dotted. **CYPSELAE** 3–4 mm long, glabrous. **PAPPUS** of 2 scales (about 1–2 mm), additional scales lacking. $2n$ = 68, 102. Roadsides, ditch banks, cultivated fields, or other disturbed habitats. ARIZ: Apa; COLO: Arc, Hin, Min, Mon; NMEX: RAr. 1400–2590 m (4500–8500′). Flowering: Jun–Aug. California and Nevada, eastward to the central plains, southward into Mexico, adventive in Idaho and Washington (and probably elsewhere). This species usually forms large colonies and is easily propagated from pieces of the rhizome. It can be an aggressive weed in disturbed habitats and is considered a noxious weed by some states.

Helianthus maximiliani Schrad. (for Alexander Philipp Maximilian, Prince of Wied-Neuwied, 1782–1867, German naturalist and explorer) Maximilian sunflower. Perennials, 50–300 cm tall, rhizomatous, herbage scabrous to scabrous-hispid and gland-dotted. **STEMS** 1 (usually in groups), erect, branched (branches relatively short), usually densely scabrous-hispid. **LEAVES** mostly alternate (proximal opposite), sessile or (proximal) short-petiolate, petiole to 2 cm long; blades linear-lanceolate, about 10–30 × 2–5 cm, ± conduplicate, 1-nerved bases cuneate, margins entire to serrulate, faces scabrous, gland-dotted. **HEADS** 1–25+, in racemiform or spiciform clusters (proliferating with age); peduncles 1–10 cm, often bracteate (few); paleae 7–11 mm, apices entire and acute, to obscurely 3-toothed (middle tooth largest), sometimes purplish, abaxially pubescent to canescent, gland-dotted. **INVOLUCRES** hemispheric, disc about 15–30 mm diameter. **PHYLLARIES** mostly 30–40 in 3–4 series, linear-lanceolate, 15–20 × 2–3 mm, margins ciliate, abaxial faces scabrous, gland-dotted, apices long-acuminate to attenuate, spreading-ascending. **RAY FLORETS** 10–25, laminae 25–40 mm long, sparsely puberulent and gland-dotted abaxially. **DISC FLORETS** 75–150+; corolla throats proximally pubescent, not or obscurely bulbous, lobes yellow, glabrous or sparsely puberulent; anther appendages yellow, usually gland-dotted. **CYPSELAE** 3–4 mm long, glabrous (or glabrate), usually mottled. **PAPPUS** of 2 scales, about 3–4 mm long, additional scales lacking. $2n$ = 34. Mostly of prairie habitats, commonly found in disturbed areas (roadsides, railroads, etc.). NMEX: McK, SJn. 1800–2000 m (5900–6800′). Flowering: late Sep–Oct. Probably native to the prairie regions of central North America, it is widely cultivated for its horticultural value and is now scattered throughout much of the continent from southern Canada to Mexico.

Helianthus nuttallii Torr. & A. Gray (for Thomas Nuttall, 1786–1859, British botanist and early explorer of western North America) Nuttall's sunflower, New Mexican sunflower. Perennial, 60–300 cm tall, rhizomatous (and with tuberous roots), herbage mostly scabrous or hispid and gland-dotted. **STEMS** 1, erect, branched (distally), glabrous and glaucous, sometimes scattered hispid also. **LEAVES** opposite (proximal or all) to alternate (distal or most), petiole about 5–15 mm, hispid-ciliate; blades lanceolate (to lance-ovate), mostly 5–15 × 1–3 cm, bases cuneate, margins entire or serrulate, ± revolute (or flat), faces adaxially scabrous, abaxially scabrous-hispid and gland-dotted. **HEADS** 1–7+ (proliferating with age) in loose, ± paniculiform clusters; peduncles 2–8 cm, often with a foliaceous bract closely subtending head; paleae 8–12 mm long, apices obscurely 3-toothed (middle tooth acute), abaxially pubescent. **INVOLUCRES** hemispheric, disc about 10–20 mm diameter. **PHYLLARIES** mostly 30–40 in 3–4 series, linear-lanceolate, 8–16 × 1–3 mm, margins ciliate proximally, outer ± revolute, abaxial faces strigose, ± gland-dotted distally, apices long-acuminate to attenuate, ± spreading. **RAY FLORETS** 10–20, laminae 15–30 mm long, sparsely puberulent abaxially, not gland-dotted. **DISC FLORETS** 60–70+; corolla throats proximally glabrous, not or obscurely bulbous, lobes yellow, glabrous; anther appendages yellow, usually gland-dotted. **CYPSELAE** 3–5 mm long, glabrous. **PAPPUS** of 2 scales (about 2–4 mm), smaller scales lacking or 1–2 (less than 1 mm). $2n$ = 34. Riparian communities, wet meadows, streamsides, ditch banks, and other moist sites. COLO: Arc, LPl, Mon; NMEX: RAr. 2000–2100 m (6600–6900′). Flowering: Aug–Sep. British Columbia to Manitoba, south to California, Arizona, New Mexico, Oklahoma, and Arkansas. All of the material from the study area belongs to **subsp. *nuttallii***.

Helianthus pauciflorus Nutt. (few-flowered) Stiff sunflower. Perennial, 30–75+ cm tall, rhizomatous, herbage mostly scabrous-hispid and gland-dotted. **STEMS** 1, erect, mostly simple (or few-branched distally), ± hispid (sometimes sparsely), often reddish. **LEAVES** all opposite (rarely distal alternate), strongly reduced upward; short-petiolate (proximal) or sessile (distal), petiole to 1 cm, ± hispid-ciliate; proximal blades lanceolate to lance-ovate or subrhombic, 50–100 × 15–30 mm, bases cuneate, margins serrulate, faces scabrous to sparsely hispid, abaxially sparsely gland-dotted (or lacking). **HEADS** mostly borne singly, or 1–3+ in loose, cymiform clusters; peduncles 2–23 cm long; paleae 9–10 mm long, often blackish, margins (at least distal) ciliate, apices obscurely 3-toothed (middle tooth obtuse to acute), faces scaberulous. **INVOLUCRES** hemispheric, disc about 15–20 mm diameter. **PHYLLARIES** mostly 25–35 in

3–4 series, green or black, ovate to obovate, 8–16 × 1–3 mm, margins ciliate, abaxial faces glabrous, apices acute to obtuse. **RAY FLORETS** 10–20, laminae 20–35 mm long, glabrous (or obscurely puberulent and/or gland-dotted). **DISC FLORETS** 50–75+; corolla throats proximally puberulent, not or obscurely bulbous, lobes reddish purple (rarely yellow), glabrous; anther appendages blackish, gland-dotted. **CYPSELAE** 5–6 mm long, ± puberulent (especially distally). **PAPPUS** of 2 lanceolate, erose scales (about 4–5 mm), sometimes with 2–4 smaller scales (less than 2 mm). $2n = 102$. [*H. rigidus* (Cass.) Desf.; *H. subrhomboideus* Rydb.]. Ponderosa pine–oak forest, usually dry soils, sometimes in disturbed habitats. COLO: Arc, Hin, LPl, Mon; NMEX: RAr. 2000–2500 m (6600–8200′). Flowering: Jul–Sep. Southern Canada, south through the central and eastern United States to Texas and Georgia. All of the material from the study area belongs to **subsp. *subrhomboideus*** (Rydb.) O. Spring & E. E. Schill. (almost rhombus-shaped).

Helianthus petiolaris Nutt. (with a petiole) Prairie sunflower. Annuals, mostly 30–100 cm tall (or less in depauperate specimens), taprooted, herbage strigose to hispid, often canescent distally (new growth and beneath heads). **STEMS** 1, erect, usually branched distally (sometimes throughout), ± strigose, often hispid also (or strictly hispid). **LEAVES** mostly alternate (proximal opposite), petioles 1–4 cm long, usually hispid; blades lanceolate to lance-ovate, 30–80 × 5–30 mm, bases mostly cuneate, occasionally ± truncate (rarely cordate), margins entire or ± serrulate (rarely undulate), faces strigose, ± glandular-pustulose, not (or obscurely) gland-dotted. **HEADS** mostly ±1–5 (proliferating with age) in loose, ± paniculiform clusters; peduncles 1–15 cm long, strigose to (often) canescent, often with 1 foliaceous bract just beneath head, this much surpassing involucre; paleae about 5–8 mm, apices 3-toothed (middle tooth acuminate), strigillose and conspicuously whitish hispid-ciliate (at least those toward center of head), ± glabrate (when head is fully mature). **INVOLUCRES** ± hemispheric, disc about 10–25 mm diameter. **PHYLLARIES** mostly 15–25, lanceolate to lance-linear or lance-ovate, about 10–15 × 1–5 mm, margins ± strigillose (rarely ciliate), abaxial faces ± strigillose to hispidulous (sometimes glabrate), eglandular, apices gradually tapered, acute-acuminate. **RAY FLORETS** about 7–15, laminae 15–25 mm, sparsely pubescent abaxially (sometimes minutely gland-dotted), or glabrous. **DISC FLORETS** mostly 50–100+; corolla throats proximally densely pubescent, ± bulbous, lobes mostly reddish purple (uncommonly yellowish or greenish) and abaxially pubescent to strigillose; anther appendages purplish, usually inconspicuously gland-dotted or glabrous. **CYPSELAE** about 3–5 mm long, sericeous-villous. **PAPPUS** of 2 lanceolate scales (about 2–3 mm), smaller scales usually lacking, or 1–2 (less than 0.5 mm). $2n = 34$. Riparian, grassland, and desert shrub communities, often along roadsides and other disturbed sites. ARIZ: Apa, Coc, Nav; COLO: Dol, LPl, Mon; NMEX: McK, RAr, San, SJn; UTAH. 1800–2000 m (5800–6800′). Flowering: Jun–Aug. Native to central and western North America, adventive eastward and now found throughout much of the United States and southern Canada. Two weakly defined subspecies are often recognized. All material from the study area ± fits within **subsp. *fallax*** Heiser, but subsp. *petiolaris* is to be expected. The following key may help to separate them.

1. Peduncles usually with a foliaceous bract subtending head; phyllaries mostly 2–3.5 mm wide; stems ± hispid; throat of disc corollas gradually narrowing above ± pubescent basal bulge....................................**subsp. *fallax***

1′ Peduncles usually bractless; phyllaries mostly 3–5 mm wide; stems strigose to hispidulous; throat of disc corollas abruptly constricted above densely pubescent basal bulge..**subsp. *petiolaris***

Heliomeris Nutt. Goldeneye

Gary I. Baird

(Greek *helios*, sun, + *meros*, division or part, possibly alluding to the yellow rays) Annuals or perennials, taprooted or ± fibrous-rooted, rhizomes lacking, caudices branched, herbage strigose to strigillose or hispid, often gland-dotted, sap clear. **STEMS** 1 to many, slender, erect, branched distally or throughout. **LEAVES** cauline, simple, ± sessile, opposite or alternate (distally), reduced distally; blades lanceolate to ovate, rhombic-elliptic, lance-linear, or linear, usually 3-nerved (sometimes obscurely so), bases ± cuneate, margins entire (most) to serrate, flat to revolute, apices ± acute. **HEADS** radiate, usually borne singly or several in cymiform to paniculiform clusters, erect; peduncles obscure to long, ebracteate. **INVOLUCRES** hemispheric to rotate. **PHYLLARIES** mostly 15–25 in 2–3 series, herbaceous, mostly lance-linear, outer less than to surpassing inner. **RECEPTACLES** convex to conic, paleate; paleae chartaceous, conduplicate and enfolding cypselae, deciduous, striate-nerved, apices cuspidate to acuminate, often gland-dotted; calyculi lacking. **RAY FLORETS** 5–15, neuter; corollas yellow (usually distal 1/2 pale, distinctly demarcated, at least when dry); laminae fugacious. **DISC FLORETS** about 25–50+, bisexual, fertile; corollas yellow, tube much shorter than throat and abruptly expanded to base of throat, tube and base of throat ± chartaceous, stipitate-glandular. **CYPSELAE** monomorphic, black or brownish, usually mottled or striate, slightly 4-angled, narrowly turbinate, glabrous. **PAPPUS** lacking. $x = 8$. Five species of southwestern North America.

1. Plants annual from a taproot .... *H. longifolia*
1′ Plants perennial from a woody caudex .... *H. multiflora*

Heliomeris longifolia (B. L. Rob. & Greenm.) Cockerell (long leaf) Longleaf goldeneye. Annuals 7–15 cm tall, taproots usually slender, herbage ± strigose. **STEMS** 1, simple to freely branched. **LEAVES** with blades lance-linear to linear, 10–40+ × 1–5+ mm, margins entire, often revolute, obscurely ciliate proximally, faces gland-dotted abaxially. **HEADS** few to many, proliferating with age; peduncles often obscure (usually less than 5 mm), occasionally to 6 cm, densely strigose distally (beneath head); paleae faces glabrous (or very sparsely strigillose), distally ± gland-dotted, margins scarious, ciliate-fringed distally, apices acute to cuspidate, not especially darkened. **INVOLUCRES** about 5–14 mm diameter. **PHYLLARIES** mostly 20–30 in 2–3 subequal series, strigose, ± gland-dotted. **RAY FLORETS** 12–15; laminae elliptic to elliptic-oblong, 5–17 mm, sparsely puberulent and gland-dotted abaxially, sometimes glabrous. **DISC FLORET** anther appendages eglandular. **CYPSELAE** about 2 mm long. $2n = 16$. [*Viguiera annua* (M. E. Jones) S. F. Blake; *V. longifolia* (B. L. Rob. & Greenm.) S. F. Blake]. Sagebrush, piñon-juniper, and desert shrub communities. ARIZ: Apa, Nav; NMEX: SJn. 1800–2200 m (6000–7200′). Flowering: Aug–Sep. Extreme southern Utah and Nevada to Arizona, New Mexico, Texas, and Mexico. Our material all belongs to the more northern **var. *annua*** (M. E. Jones) W. F. Yates (annual).

Heliomeris multiflora Nutt. (many-flowered) Showy goldeneye. Perennials 20–100+ cm tall, taproots surmounted by a fibrous-rooted, woody caudex, herbage mostly strigose to puberulent, or stems glabrate. **STEMS** 1 to several, usually branched distally. **LEAVES** with blades elliptic, ovate, lanceolate, lance-linear, to linear, 10–90 × 2–20 mm, margins entire to denticulate or (rarely) serrate, flat to revolute, obscurely ciliate proximally, faces gland-dotted abaxially. **HEADS** few to many, proliferating with age; peduncles 1–15 cm, densely strigose to hispid distally (beneath head); paleae faces puberulent, gland-dotted distally, margins membranous, ciliate-fringed distally, apices cuspidate to acuminate, often blackened or black-speckled. **INVOLUCRES** mostly 6–16 mm diameter. **PHYLLARIES** 15–20 in 2 subequal series, strigose, ± gland-dotted. **RAY FLORETS** 5–15; laminae oval to oblong, 7–20 mm, puberulent and gland-dotted abaxially. **DISC FLORET** anther appendages gland-dotted abaxially. **CYPSELAE** 1–3 mm long. $2n = 16, 32$. [*Viguiera multiflora* (Nutt.) S. F. Blake]. Riparian, meadow, or other moist sites in sagebrush, piñon-juniper, ponderosa pine–oak, Douglas-fir, aspen, and spruce-fir communities. ARIZ: Apa, Nav; COLO: Arc, Dol, Hin, LPl, Min, Mon, RGr, SJn, SMg; NMEX: McK, RAr, SJn; UTAH. 1800–3000 m (6000–10000′). Flowering: Jun–Sep. Idaho, Montana, and Wyoming, south to California, Arizona, New Mexico, Texas, and Mexico. Two traditionally recognized but confluent varieties occur within the study area. Their separation seems incomplete and ± arbitrary (var. *multiflora* mostly occurs in moist sites at higher elevations, while var. *nevadensis* mostly occurs in drier sites at lower elevations).

1. Main leaves lance-linear to lance-ovate, mostly 8–25 mm wide, margins flat .... **var. *multiflora***
1′ Main leaves lance-linear to linear, mostly 2–8 mm wide, margins revolute .... **var. *nevadensis***

var. *multiflora* **LEAVES** with blades lance-ovate to lanceolate or lance-linear, 8–20+ mm wide, margins flat and often denticulate (rarely serrate). $2n = 16$. Ponderosa pine–oak, aspen, spruce-fir forests, and montane meadows. COLO: Arc, Hin, LPl, Mon, Min, SJn; UTAH. Mostly over 2500 m (ca 8500′). Montana south to Arizona, Texas, and Mexico.

var. *nevadensis* (A. Nelson) W. F. Yates (for the state of Nevada) Nevada goldeneye. **LEAVES** with blades lance-linear to linear, 2–8 mm wide, margins revolute, entire or obscurely denticulate. $2n = 16, 32$. Sagebrush, piñon-juniper, ponderosa pine–oak, and Douglas-fir forests. ARIZ: Apa, Nav; COLO: Arc, Dol, LPl, Min, Mon, SMg, SJn; NMEX: McK, RAr, SJn; UTAH. Mostly under 2500 m (about 8500′) but as high as 2800 m (about 9000′).

Heterotheca Cass. Goldenaster
G. L. Nesom

(Greek *heteros*, different, + *theke*, case or container, in reference to the dissimilar cypselae of ray and disc) Perennial or annual herbs, taprooted or rhizomatous, often aromatic, stems, leaves, and phyllaries often viscid with sessile or short-stipitate glands, also usually with stiff, nonglandular hairs. **LEAVES** alternate, simple, ovate to lanceolate, entire or with a few shallow teeth, basal often petiolate, upper commonly clasping. **INFL** a solitary head or few or in open, corymboid clusters. **PHYLLARIES** in 4–8 graduate series, lanceolate, rigid. **RECEPTACLES** flat to convex, shallowly alveolate. **RAY FLORETS** pistillate, fertile, lamina yellow to orange-yellow, coiling with maturity, absent in 2 species. **DISC FLORETS** perfect, fertile. **CYPSELAE** cylindrical to slightly compressed, 8–14-nerved or sometimes with a few thicker ribs, sparsely short-strigose to densely sericeous; in sect. *Heterotheca*, disc and ray dimorphic. **PAPPUS** of barbellate bristles somewhat uneven in length, with a shorter, outer series of lanceolate scales or bristlelike squamellae, pappus lacking or merely a short crown or a few much shorter bristles in ray cypselae of species of sect. *Heterotheca*. $x = 9$ (Semple 1996; Semple, Blok, and Heiman 1980; Nesom 1997, 2006). Apparent intermediates are often encountered (whether through phenotypic plasticity, hybridization, or incomplete differentiation is not clear) and significant variability may make unequivocal identifications difficult; the common occurrence of tetraploids also underlies the taxonomic difficulties.

1. Cauline leaves usually held at nearly right angles to stem, often contorted (margins undulate) upon drying; leaf surfaces densely glandular, the glands conspicuous through the sparsely hispid to hispidulous (nonglandular) vestiture .... ***H. horrida***
1′ Cauline leaves mostly ascending, usually not contorted upon drying; leaf surfaces eglandular to glandular, little to partially or mostly obscured by the sparsely to loosely strigose or villous-strigose (nonglandular) vestiture.......(2)

2. Leaves moderately to densely glandular, sparsely strigose, with nonglandular hairs usually restricted to midvein area (middle 1/3) of adaxial surfaces, nonglandular hairs sometimes nearly absent on both surfaces; phyllaries glabrous to very sparsely strigose, sparsely to densely glandular at least on distal 1/2 .... ***H. polothrix***
2′ Leaves glandular to eglandular, sparsely to densely strigose, with hairs relatively evenly distributed on adaxial surfaces; phyllaries strigose to strigose-villous, eglandular to sparsely glandular.......... (3)

3. Cauline leaves oblanceolate to broadly oblanceolate or obovate, 3–10 mm wide, largest at midstem or above, upper usually closely or immediately subtending heads and commonly longer than the involucres; mostly subalpine in habitat .... ***H. pumila***
3′ Cauline leaves oblanceolate to narrowly oblanceolate, oblong-lanceolate, or narrowly obovate, 2–5 mm wide, relatively even-sized, shorter than lower leaves, upper not closely subtending and longer than involucres; mostly lower elevations .... (4)

4. Stems 40–100 cm, strictly erect from the base; stems and leaves silvery to silver-gray sericeous to densely strigose; cauline leaves relatively even-sized upward, sessile.......... ***H. zionensis***
4′ Stems 15–40 cm, ascending from the base; stems and leaves sparsely to moderately gray to gray-green strigose or strigose-villous; cauline leaves diminishing in size upward, usually narrowed to a petiolar base.......... ***H. villosa***

Heterotheca horrida (Rydb.) V. L. Harms (bristly or rough) Spreadleaf goldenaster. Perennial herbs, from a strongly developed taproot. **STEMS** 15–50 cm, hispid-strigose to hispidulous, hairs ascending to spreading or deflexed, minutely puberulent at the surface, glandular to eglandular. **LEAVES** cauline, commonly oriented at nearly right angles to stem, often drying contorted (margins undulate), midcauline mostly lanceolate to lanceolate-oblong, (5–) 10–20 (–30) mm, hispid-strigose to hispid, sometimes nearly without nonglandular hairs, moderately to densely glandular. **HEADS** 1–4, commonly loosely aggregated in an open corymbiform array. **PHYLLARIES** sparsely strigose to nearly glabrous, usually glandular at least distally. $2n = 18, 36$. [*H. villosa* (Pursh) Shinners var. *horrida* (Rydb.) Semple; *H. villosa* var. *nana* (A. Gray) Semple]. Rocky slopes, washes, sandy pockets, roadsides, ledges, meadows; desert grassland, piñon pine, piñon-juniper communities. ARIZ: Apa; COLO: Arc, LPl, Min; NMEX: McK, RAr, SJn. 1600–2350 m (5250–7710′). Flowering: May–Sep (–Oct). Nevada, Arizona, and southern Idaho east to New Mexico, Nebraska, and Oklahoma. Semple (1996) treated *H. villosa* in the Four Corners region to include var. *nana* and var. *scabra*, treated here as *H. horrida* and *H. polothrix*, respectively (Nesom 2006), both of which are morphologically distinct from *H. villosa* and sympatric with it in the Four Corners region.

Heterotheca polothrix G. L. Nesom (pony hairs, alluding to a resemblance of the adaxial foliar vestiture to a pony's mane) Great Basin goldenaster. Perennial herbs, from a strongly developed taproot. **STEMS** 15–40 cm, very sparsely and loosely strigose to sparsely villous (hairs usually ascending, sometimes spreading), minutely puberulent at the surface, densely (less commonly sparsely) glandular. **LEAVES** oblanceolate to oblong-lanceolate, 15–30 × 3–6 mm, ascending, not drying contorted, nearly glabrous to sparsely strigose, hairs of adaxial surfaces usually restricted to midvein area (middle 1/3 of blade), more evenly distributed abaxially, moderately to densely glandular. **HEADS** 1 or loosely 2–4 on leafy peduncles 3–6 cm. **PHYLLARIES**: longest 6–8 mm, glabrous to very sparsely strigose (coarse hairs along midrib, finer hairs laterally, if present), sparsely to densely glandular at least on distal 1/2. $2n = 18, 36$. [*Heterotheca horrida* (Rydb.) V. L. Harms subsp. *cinerascens* (S. F. Blake) Semple; *H. villosa* (Pursh) Shinners var. *scabra* (Eastw.) Semple]. Canyons, washes, roadsides, rocky slopes; rabbitbrush, sagebrush, juniper-sage, piñon-juniper, ponderosa pine–Gambel's oak communities. ARIZ: Apa, Coc, Nav; NMEX: SJn; UTAH. 1550–2500 m (5085–8200′). Flowering: May–Aug. Nevada and south-central Idaho to Utah and northern Arizona, barely into western Colorado and New Mexico. Recognized especially by the distinctive vestiture of leaves and phyllaries.

Heterotheca pumila (Greene) Semple (diminutive) Subalpine goldenaster. Perennial herbs, rhizomelike caudex branches slender, apparently without a strongly developed taproot; stems and leaves eglandular, strigose. **STEMS** 10–25 cm, loosely strigose to strigose-villous, eglandular. **LEAVES** oblanceolate to broadly oblanceolate or obovate, 15–35 × 3–10 mm, largest at midstem or above, uppermost cauline usually closely or immediately subtending heads and commonly longer than the involucres, moderately strigose to villous-strigose, eglandular or minutely and inconspicuosuly glandular. **HEADS** 1 (–3), hardly aggregated. **PHYLLARIES**: longest 8–10 mm, loosely strigose to sparsely strigose-villous, eglandular to sparsely glandular. $2n$ = 18, 36. Woods edges and clearings, meadows, talus slopes, roadsides, subalpine to alpine vegetation, lodgepole pine–fir, pine-spruce, aspen–spruce–white fir, and ponderosa pine–Gambel's oak communities. COLO: Arc, Dol, Hin, LPl, Mon, SJn, SMg; UTAH. (2400–) 2650–3300 m (7875–10825′). Flowering: Jul–Sep. South-central Wyoming through central Colorado and southeastern Utah. The Utah plants are from the Abajo Mountains and grow in ponderosa pine–Gambel's oak communities at slightly lower elevations, 2400–2650 m (7875–8695′), than those from southwestern Colorado, which characteristically occur in subalpine communities. Similar plants from the Abajo and La Sal Mountains of Utah, with slightly smaller upper cauline leaves than typical for *H. pumila*, were cited and mapped by Semple (1996) as *Heterotheca* "aff. *pumila*." Recognized by its low habit, slender stems and caudex branches, distally increasing size of cauline leaves, solitary and relatively large heads closely subtended by upper cauline leaves, subvillous phyllaries, and habitats at relatively high elevation.

Heterotheca villosa (Pursh) Shinners (hairy) Hairy goldenaster. Perennial herbs, from a strongly developed taproot. **STEMS** ascending from the base, 15–40 cm, sparsely to moderately strigose or strigose-villous, eglandular to sparsely glandular. **LEAVES** oblanceolate to narrowly oblanceolate, oblong-lanceolate, or narrowly obovate, (10–) 15–40 × 2–5 mm, ascending, moderately to densely strigose, the vestiture commonly imparting a gray-green to silvery gray color and often (when dense) obscuring the surface, eglandular or sparsely to densely glandular; basal often persistent at flowering. **HEADS** (1–) 3–7 (–12), ebracteate, in distinctly clustered corymbiform arrays, peduncles 1–3 (–4) cm. **PHYLLARIES**: longest (5–) 6–7 (–8) mm, sparsely strigose, eglandular to moderately glandular. $2n$ = 18, 36. [*Heterotheca villosa* var. *pedunculata* (Greene) V. L. Harms ex Semple; *H. villosa* var. *minor* (Hook.) Semple; *H. villosa* var. *hispida* (Hook.) Harms]. Rocky slopes, talus, alluvium, sandbars, meadows, roadsides, disturbed sites, riparian (cottonwood, willow, saltcedar, Russian olive), desert shrub, sagebrush-rabbitbrush, piñon-juniper, ponderosa pine with Gambel's oak, white pine, and/or aspen, spruce-fir communities. ARIZ: Apa, Nav; COLO: Arc, Dol, Hin, LPl, Min, Mon, SJn, SMg; NMEX: McK, RAr, San, SJn; UTAH. 1125–2800 m (3700–9185′). Flowering: late May–Oct. Canada (British Columbia to Manitoba), Washington to Minnesota and Wisconsin, south to California, Arizona, New Mexico, Texas, and Oklahoma. According to Semple (1996), *H. villosa* var. *minor* (upper cauline leaves sparsely to densely glandular) and *H. villosa* var. *pedunculata* (upper cauline leaves eglandular to very sparsely glandular) occur in complete sympatry in the Four Corners area. Only a single entity is recognized here, as unarbitrary morphological distinctions between these two taxa in the Four Corners area are not apparent (nor is the distinction of var. *minor* from typical *H. villosa* (var. *villosa* sensu Semple).

Heterotheca zionensis Semple (from the Zion National Park area) Silvery goldenaster. Perennial herbs, from a strongly developed taproot. **STEMS** strictly erect from the base, 40–100 cm (grazed or damaged plants may be smaller), often 10–20 stems per clump, silvery to silver-gray sericeous to densely strigose with thin-based, closely antrorsely appressed to ascending nonglandular hairs, eglandular. **LEAVES:** cauline mostly oblanceolate-obovate, 15–35 mm × 4–7 mm, relatively even-sized up the stem, spreading to ascending, silvery to silver-gray sericeous to densely strigose, eglandular; basal absent by flowering. **HEADS** 4–22, ebracteate, in a loosely subcorymboid arrangement, peduncles 1–3 cm. **PHYLLARIES**: longest 6–9 mm, sparsely to moderately strigose, eglandular to moderately glandular. $2n$ = 18. Rocky and sandy slopes and flats, roadsides, piñon-juniper. ARIZ: Apa, Nav; UTAH. 1650–2100 m (5500–7000′). Flowering: late June–Sep. Idaho, Utah, Arizona, New Mexico, and Texas.

Hieracium L. Hawkweed

Gary I. Baird

(Greek *hierakion*, an ancient plant name, purportedly derived from *hierax*, hawk) Perennials, taprooted (usually surmounted by a shortened, fibrous-rooted, branched or unbranched root crown), herbage mostly pubescent; indumentum a complex mix of long, setaceous hairs ("pilose-hispid"), short, branched hairs ("dendritic-stellate"), and/or short, glandular hairs ("stipitate-glandular"); sap milky. **STEMS** mostly 1, erect, branched distally. **LEAVES** basal and/or cauline, simple, alternate; petiolate or sessile, not auriculate; blades mostly elliptic to oblanceolate or lanceolate, margins entire or denticulate, apices acute to obtuse, faces glabrous and glaucous to pubescent. **INFL** borne in corymbiform to racemiform clusters, erect; peduncles ebracteate. **HEADS** liguliflorous, glabrous to pubescent. **INVOLUCRES** cylindric to campanulate. **PHYLLARIES** mostly imbricate in 3–4 series, herbaceous, erect, margins not or narrowly membranous, outer often reduced and calyculate. **RECEPTACLES** flat, epaleate. **RAY FLORETS** matutinal or diurnal; corollas ligulate, mostly yellow, sometimes whitish or orangish. **DISC FLORETS** lacking. **CYPSELAE** monomorphic, reddish brown to black, mostly columnar, sometimes tapering distally (ours), glabrous, about 10-ribbed, beakless (or obscurely beaked). **PAPPUS** of bristles, persistent, white to ochroleucous, barbellulate. $x = 9$. Perhaps 1000 species (or more) in Eurasia, Africa, North America, and South America. The Eurasian members form polyploid-apomictic (asexually reproducing) complexes consisting of poorly defined "species," which account for the bulk of the named taxa in the genus. Our native members all appear to be diploid-gametic (sexually reproducing) species that are more readily delineated.

1. Achenes mostly 5 mm or more long, ± tapered apically; involucres mostly more than 10 mm tall, usually bearing a few long, yellowish (or blackish at base) setaceous hairs .... ***H. fendleri***

1' Achenes mostly less than 5 mm long, not tapered apically; involucres 10 mm or less tall, usually bearing many long, black setaceous hairs .... ***H. triste***

Hieracium fendleri Sch. Bip. (for Augustus Fendler, German botanist, 1813–1883) Yellow hawkweed. **STEMS** mostly 10–30 cm, more or less glaucous, proximally yellowish pilose-hispid (hairs to 5 mm long), distally pilose-hispid and stipitate-glandular, occasionally also (sparsely) dendritic-stellate. **LEAVES** all basal or with 1–2 proximal-cauline; blades elliptic-obovate, 4–7 cm × 8–12 mm, margins entire, faces pilose-hispid, ± glaucous. **HEADS** 3–5; peduncles sparsely yellowish pilose-hispid and stipitate-glandular. **INVOLUCRES** 12–15 mm tall. **PHYLLARIES** about 15–25, lanceolate, erect, sparsely yellowish pilose-hispid, sparsely dendritic-stellate (or lacking), and sparsely stipitate-glandular. **RAY FLORETS** mostly 15–30, equaling or just surpassing involucre; corolla tube mostly 5–8 mm, ligule mostly 3–4 mm, yellow. **CYPSELAE** 5–7 mm, apices ± tapered. **PAPPUS** bristles 6–9 mm, ochroleucous. $2n = 18$. Ponderosa pine, aspen, and Douglas-fir communities. ARIZ: Apa; COLO: Arc, Hin, LPl, Min, Mon, SJn; NMEX: McK, SJn; UTAH. 2135–2745 m (7000–9000′). Flowering: Jun–Jul. Nevada to Wyoming, south to Texas, Mexico, and Central America.

Hieracium triste Willd. ex Spreng. (confident or trusty, the allusion unclear) Woolly hawkweed. **STEMS** mostly 10–35 cm, ± dendritic-stellate, usually also black pilose-hispid and/or stipitate-glandular (pubescence increasing distally). **LEAVES** all basal or with about 1–3 (or more) reduced cauline; blades oblanceolate to spatulate, about 2–4+ cm × about 5–15+ mm, margins entire or denticulate, faces glabrous or sometimes stipitate-glandular and/or scabrous. **HEADS** (1) 2–9; peduncles densely dendritic-stellate and long black pilose-hispid, often also stipitate-glandular. **INVOLUCRES** 7–12 mm tall. **PHYLLARIES** mostly 20–40, lanceolate, erect, densely long black pilose-hispid, densely (or sparsely) dendritic-stellate, and ± stipitate-glandular. **RAY FLORETS** 20–60, equaling or just surpassing involucre; corolla tube 4–6 mm, ligule 1–3 mm, yellow; anthers about 1 mm. **CYPSELAE** 2–4 mm, columnar, apices not tapered. **PAPPUS** bristles 5–7 mm, sordid to stramineous (especially distally). $2n = 18$. [*H. gracile* Hook.; *H. triste* var. *gracile* (Hook.) A. Gray]. Spruce-fir and alpine communities. COLO: Arc, Con, Hin, LPl, Min, Mon, RGr, SJn. 3050–3810 m (10000–12500′). Flowering: Jul–Aug. Alaska, Yukon, and Northwest Territories, south (in mountains) to California, Utah, Colorado, and New Mexico.

Hymenopappus L'Hér. Hymenopappus, White Ragweed

Brian Elliott

(Greek *hymen*, membrane, + *pappos*, seed down, referring to the membranous pappus) Aromatic biennial or perennial herbs; taprooted, often with a woody, branched crown. **STEMS** erect, solitary to diffusely branched, subscapose to leafy-stemmed, angled and sulcate, often pithy. **LEAVES** mostly basal, reduced above, alternate, simple to 2-pinnately dissected, often with an inrolled margin, punctate-glandular. **INFL** heads discoid or occasionally radiate, few to many, cymose or corymbose; peduncles slender, conspicuous. **INVOLUCRE** hemispheric to campanulate. **PHYLLARIES** 6–14, subequal in 2–3 series, membranous at the apex or throughout, margins scarious. **RECEPTACLES** flat to convex, usually naked, rarely chaffy. **RAY FLORETS** pistillate and fertile; ligules white, yellow, or reddish; apex 2–3-lobed. **DISC FLORETS** perfect, fertile, 10–many per head; corolla yellow to white or rarely reddish; tube narrow, slender, glandular; throat abruptly flared, campanulate to funnelform; lobes triangular to ovate, reflexed after anthesis. **CYPSELAE** narrowly obpyramidal, 3–4-angled, laterally compressed; faces 0–3-nerved, glabrous to densely pubescent. **PAPPUS** absent or of 12–22 linear-oblong or broadly ovate, obtuse, hyaline scales, usually with a midrib, often obscured by pubescence of the achene. Ten species of the western United States and adjacent Mexico. A taxonomically complex group with intergradation between the species and varieties. *Hymenopappus tenuifolius* Pursh is expected from our area and has been collected near Gallup, New Mexico (Turner 1956).

1. Heads radiate with conspicuous white rays .... ***H. newberryi***
1′ Heads discoid .... (2)

2. Plants biennial with a single, unbranched crown; stem leaves many (10–50) .... ***H. flavescens***
2′ Plants perennial with a branched caudex; stem leaves few (0–12) .... (3)

3. Leaves bipinnately dissected with narrow, linear-filiform ultimate segments; achenes pubescent .... ***H. filifolius***
3′ Leaves simple or once-pinnate with ovate to lanceolate ultimate segments; achenes glabrate .... ***H. mexicanus***

Hymenopappus filifolius Hook. (thread leaf, referring to the narrow leaf segments) Fineleaf hymenopappus. Perennial herbs from a branching caudex; densely tomentose to nearly glabrate. **STEMS** 5–100 cm high. **LEAVES** mostly basal; cauline leaves 0–12, usually much reduced above; basal leaves 3–20 cm long, 1–2-pinnatifid, ultimate segments linear to filiform, 2–50 mm long; leaves minutely impressed-punctate. **INFL** heads discoid, 1–60 per stem, corymbiform; ultimate peduncle 0.5–16 cm long, slender. **INVOLUCRES** subturbinate to broadly campanulate, 4–12 mm high. **PHYLLARIES** 3–14 mm long, 2–5 mm wide; apex acute to obtuse, membranous, white to yellowish or rarely pink-tinged. **RECEPTACLES** naked or rarely with 6–10 chaffy scales. **RAY FLORETS** absent. **DISC FLORETS** 10–80 per head; corolla yellow or white, 2–7 mm long; tube 1–2.5 mm long, densely glandular to nearly glabrous; throat campanulate. **CYPSELAE** 3–7 mm long, densely pubescent with trichomes 0.2–3 mm long. **PAPPUS** of 12–22 linear-oblong scales, 0.1–3 mm long. $2n = 34$ ($2n = 68$ in var. *lugens*). [*Rothia filifolia* (Hook.) Kuntze]. Washington to North Dakota and south to California and Texas, barely entering Canada and Mexico. A highly variable species composed of 13 confluent varieties, 5 found within our range. The plant was reportedly used as an emetic, and roots were mixed with lard and applied to swellings.

1. Basal leaf axils glabrous, subglabrous, or sparsely tomentose, lacking a dense tomentum; cauline leaves 0–2; plants to 45 cm high .... **var. *parvulus***
1′ Basal leaf axils conspicuously woolly-tomentose; cauline leaves 0–12; plants to 100 cm high .... (2)

2. Disc corolla 2–3 mm long; florets 10–30 per head; phyllaries 3–6 mm long; leaves of the basal rosette with a tip 1–6 mm long .... **var. *pauciflorus***
2′ Disc corolla 3–7 mm long; florets 25–60 per head; phyllaries 6–14 mm long; leaves of the basal rosette with a tip 3–30 mm long .... (3)

3. Disc corolla 3–4.5 mm long; pappus 0.2–3 mm long; involucre 6–9 mm high; throat 1.5–2 mm long; anthers 2–3 mm long .... **var. *cinereus***
3′ Disc corolla 4–7 mm long; pappus 1.2–3 mm long; involucre 9–12 mm high; throat 2–5 mm long; anthers 3–4 mm long .... (4)

4. Cauline leaves 1–7, often well developed; leaves of the basal rosette with a tip 5–30 mm long; phyllaries 8–14 mm long; plants of desert areas, usually below 1525 m (5000′) .... **var. *megacephalus***
4′ Cauline leaves 0–3, usually reduced and inconspicuous; leaves of the basal rosette with a tip 3–15 mm long; phyllaries 6–10 mm long; plants of juniper, oak, or pine woodlands, usually at 1525–2750 m (5000–9000′) ... **var. *lugens***

var. *cinereus* (Rydb.) I. M. Johnst. (ash-gray, referring to the grayish tomentose leaves) Plants sparsely grayish green tomentose. **STEMS** to 40 cm high. **LEAVES** mostly basal; basal leaves 5–14 cm long, bipinnately dissected with linear-filiform ultimate divisions 0.5–1 mm wide, conspicuously impressed-punctate; cauline leaves (0) 2–4, much reduced above. **INFL** heads 1–6 per stem; ultimate peduncles 1–6 cm long. **PHYLLARIES** 6–9 mm long, 2–4 mm wide; apex yellow- or rarely white-membranous. **DISC FLORETS** 25–40 per head; corolla yellow, rarely white, 3–4.5 mm long; tube 1.5–2.5 mm long, moderately glandular; throat 1.5–2.5 mm long, campanulate. **CYPSELAE** 4–6 mm long, evenly pubescent, trichomes conspicuous and 1–3 mm long. **PAPPUS** of 14–18 scales, 1–2 mm long. $2n = 34$. [*H. arenosus* A. Heller; *H. cinereus* Rydb.; *H. ochroleucus* Greene]. Dry hills and plains in open exposed areas. ARIZ: Apa, Nav; COLO: Arc, Dol, LPl, Mon, SMg; NMEX: McK, RAr, San, SJn; UTAH. 1690–2820 m (5550–8600′). Flowering: May–Sep. Colorado and eastern Utah to New Mexico and Arizona, with scattered populations in western Texas and Oklahoma. The subspecies has been used to treat coughs.

var. *lugens* (Greene) Jeps. (possibly from the Latin *lugere*, to mourn, the application here uncertain) Plants densely grayish tomentose to greenish glabrate. **STEMS** (20) 30–60 cm high. **LEAVES** mostly basal; basal leaves 5–14 cm long, bipinnately dissected with linear ultimate segments 3–10 mm long, 1–3 mm wide; cauline leaves 0–2 (3), much reduced. **INFL** heads 3–8 per stem; ultimate peduncles 2–12 cm long. **PHYLLARIES** 6–10 mm long, 3–4 mm wide; apex conspicuously reddish- to yellow-membranous. **DISC FLORETS** 20–72 per head; corolla usually bright yellow, sometimes white, 4–6 mm long; tube 1.5–2 mm long, sparsely glandular; throat campanulate-tubular, 2.2–3 mm long. **CYPSELAE** 5–6 mm long, pubescent, trichomes 1–2 mm long. **PAPPUS** of 16–18 scales, 1.2–2.5 mm long. $2n = 34$, 68. [*H. gloriosus* A. Heller; *H. lugens* Greene; *H. macroglottis* Rydb.; *H. nudatus* Wooton & Standl.; *H. scaposus* Rydb.]. Mountains and dry, rocky slopes, often associated with juniper, pine, or oak. ARIZ: Apa; NMEX: McK, San, SJn. 1220–2440 m (4000–8000′). Flowering: May–Sep. Southern California and Nevada to New Mexico and adjacent Mexico. According to Turner, "this is the most variable and perplexing variety in the *H. filifolius* complex." The subspecies has been used medicinally as an emetic, a poultice, for toothache, and to treat blood poisoning.

var. *megacephalus* B. L. Turner (referring to the large heads) Plants uniformly tomentose to nearly glabrate. **STEMS** 30–70 cm tall. **LEAVES** mostly basal; basal leaves 8–20 (30) cm long, bipinnately dissected with coarse, flattened linear divisions 8–30 mm long, 1–2 mm wide; cauline leaves 2–6, reduced above. **INFL** heads 3–14 per stem; ultimate peduncle 2–10 cm long. **PHYLLARIES** 8–14 mm long, 2–5 mm wide; apex yellow- to reddish-membranous. **DISC FLORETS** (20) 30–60 per head; corolla yellow to pale yellow, 4–7 mm long; tube 2–3 mm long, glandular; throat campanulate-tubular, 2–5 mm long. **CYPSELAE** 5–7 mm long, pubescent, trichomes 1–3 mm long. **PAPPUS** of 14–18 scales, 1–3 mm long. On sandy or gravelly desert soils. ARIZ: Apa, Coc, Nav; COLO: Arc, Dol, LPl, Mon; NMEX: McK, SJn; UTAH. 1690–2560 m (5550–8400′). Flowering: May–Nov. California to Colorado, and south to Arizona and New Mexico.

var. *parvulus* (Greene) B. L. Turner (very small, referring to the heads) Plants sparsely grayish green tomentose to reddish glabrate. **STEMS** 10–25 cm high. **LEAVES** mostly basal; basal leaves 3–10 cm long, usually 1-pinnate with a few scattered secondary linear divisions 5–20 mm long and 0.5–1 mm wide. **INFL** heads 3–10 (15) per stem; ultimate peduncle very short, 5–15 mm long. **PHYLLARIES** 4–8 mm long, 2–3 mm wide; apex yellow-membranous, rarely reddish tinged. **DISC FLORETS** 10–18 per head; corolla yellow, 2–3 mm long; tube 1–1.5 mm long, densely glandular; throat campanulate-tubular, 1.5–1.8 mm long. **CYPSELAE** 3.5–4 mm long, pubescent, trichomes approximately 1 mm long. **PAPPUS** of 14–16 scales, 1–1.8 mm long. $2n = 34$. [*H. parvulus* Greene]. Open, rocky or sandy foothills and slopes, often in barren sites. COLO: Dol, LPl, Mon. 1675–2620 m (5500–8600′). Flowering: Jun–Aug. Endemic to Colorado. According to Turner, this is one of the most distinct varieties of the group.

var. *pauciflorus* (I. M. Johnst.) B. L. Turner (few-flowered, probably referring to the disc florets, but application here uncertain) Plants tomentose to nearly glabrate. **STEMS** 20–35 cm high. **LEAVES**: basal 5–7 cm long, bipinnately dissected, ultimate segments 1–5 mm long, 0.5–1 mm wide; cauline leaves 3–9, reduced above. **INFL** heads 2–15 (30) per stem; ultimate peduncles 2–6 cm long, slender. **PHYLLARIES** 3–6 mm long, 2–3 mm wide; apex yellow-membranous. **DISC FLORETS** 10–30 per head; corolla yellow, 2–3 mm long; tube 1–2 mm long, densely glandular; throat 1–1.6 mm long. **CYPSELAE** 3–4.5 mm long, evenly pubescent, trichomes 1–2 mm long. **PAPPUS** of 14–18 scales, 0.5–1.5 mm long. $2n = 34$. [*H. pauciflorus* I. M. Johnst.]. Dry mesas and slopes, often on sandy soils. ARIZ: Apa, Coc, Nav; COLO: LPl, Mon, SJn, SMg; UTAH. 1370–2040 m (4500–6700′). Flowering: May–Oct. Known from the Four Corners area of Utah, Arizona, and Colorado. Although not yet documented, the variety can be expected in northwestern New Mexico.

Hymenopappus flavescens A. Gray (becoming yellow, referring to the florets) College-flower. Taprooted biennials, glabrate to tomentose. **STEMS** to 90 cm high. **LEAVES** basal and cauline; basal leaves to 15 cm long, 4 cm wide, evenly tomentose or canescent on both sides, bipinnately dissected, ultimate segments linear, 1–2 mm wide; cauline leaves 10–50, reduced above. **INFL** heads discoid, 15–100 per stem; ultimate peduncles 1–6 cm long. **PHYLLARIES** 4–8 mm long, 2–4 mm wide; apex yellow-membranous. **RECEPTACLES** naked. **DISC FLORETS** 30–90 per head; corolla yellow, 2.5–3.5 mm long; tube 1.5–2 mm long, glandular; throat 1–1.5 mm long, abruptly campanulate. **CYPSELAE** 3–4.5 mm long, evenly pubescent, trichomes 0.5–1.5 mm long. **PAPPUS** of 16–22 linear-oblong scales, 1–1.5 mm long. $2n = 34$. [*H. fisheri* Wooton & Standl.; *H. flavescens* A. Gray; *H. robustus* Greene; *Rothia flavescens* (A. Gray) Kuntze]. Gravelly, rocky, open, or disturbed sites, often on limestone or sandy soils. ARIZ: Apa; NMEX: McK, SJn. 1500–2285 m (4900–7500′). Flowering: May–Aug. The species is known from Arizona, New Mexico, Colorado, Texas, Oklahoma, and Kansas. The two varieties (var. *flavescens* and var. *canotomentosus* A. Gray) meet in eastern New Mexico, where they intergrade. Ours is the western **var. *canotomentosus*** A. Gray.

Hymenopappus mexicanus A. Gray (of or from Mexico) Mexican woollywhite. Perennial herbs, greenish glabrate to densely white-tomentose. **STEMS** 20–90 cm high. **LEAVES** primarily in a basal rosette, cauline leaves reduced or absent; basal leaves to 20 cm long and 2.5 cm wide, simple to 1-pinnate, lanceolate to spatulate, nearly glabrous to densely tomentose, obscurely impressed-punctate; lobes (if present) broad, ovate to lance-linear, 1–7 mm wide. **INFL** heads discoid, several to numerous in somewhat flat, cymose panicles; peduncle 0.5–10 cm long, slender or short-thickened. **PHYLLARIES** 7–9 mm long; apex membranous, yellow to white or rarely reddish-tinged. **RECEPTACLES** naked. **DISC FLORETS** 20–40 per head; corolla yellow, 3–4.5 mm long; tube 2–2.5 mm long, densely glandular; throat campanulate, 1–2.5 mm long. **CYPSELAE** 4–6 mm long, faces 2–3-nerved, rarely rugose, glabrous to sparsely puberulent. **PAPPUS** obsolete, minute, or of 12–20 short, obtuse to spatulate, laciniate scales, up to 0.4 mm long. $2n = 34$. [*H. integer* Greene; *H. obtusifolius* A. Heller; *H. petaloideus* Rydb.; *Rothia mexicana* Kuntze]. Open woods amongst conifers and aspen. ARIZ: Apa; NMEX: McK. 2160–2620 m (7100–8600′). Flowering: Jun–Oct. Known from New Mexico, Arizona, and adjacent Mexico. A distinct species, but quite variable in leaf shape.

Hymenopappus newberryi

Hymenopappus newberryi (A. Gray) I. M. Johnst. (in honor of John Strong Newberry, 1822–1892, American physician, geologist, paleontologist, botanist, and collector of the type specimen). Wild cosmos, Newberry's hymenopappus. Perennial herbs, sparsely tomentose to nearly glabrate. **STEMS** 20–60 cm tall. **LEAVES** mostly basal; basal leaves 12–25 cm long, 3–5 cm wide, bipinnately dissected, evenly and sparingly canescent on both surfaces to nearly glabrate, inconspicuously impressed-punctate, ultimate segments linear, flattened; cauline leaves 1–3 (5), much reduced. **INFL** heads radiate, 3–8 per stem, corymbose; peduncle 6–15 cm long, stout. **PHYLLARIES** subequal in 2–3 series, 8–10 mm long, 4–7 mm wide; apex obtuse, white or yellow-membranous. **RECEPTACLES** chaffy; chaff scales deciduous, 5–9 mm long, 2–5 mm wide, partly to completely enclosing marginal achenes, minutely glandular to glabrous, yellow-membranous above. **RAY FLORET** ligules 8, corollas white, 10–20 mm long, 8–15 mm wide, conspicuous; apex 3-cleft, sinuses 2–3 mm deep. **DISC FLORETS** with yellow corollas, 3.5–4 mm long; tube 1.5–2 mm long, sparsely glandular to glabrate; throat campanulate, 1.5–2 mm long. **CYPSELAE** strongly incurved, 3.5–4 mm long, glabrous. **PAPPUS** minute (less than 0.1 mm) or absent. $2n = 34$. [*Leucampyx newberryi* A. Gray ex Porter & J. M. Coult.]. Along roadsides or in open areas of the mountains amongst aspen, pine, or spruce woods. COLO: Arc, Hin, Mon. 1980–3050 m (6500–10000′). Flowering: Jun–Sep. Known only from southern Colorado and northern New Mexico.

Hymenoxys Cass. Rubberweed

Brian Elliott

(Greek *hymen*, membrane, + *oxys*, sharp, referring to the aristate or awned pappus scales) Annual, biennial, or perennial taprooted herbs; caulescent or scapose. **STEMS** simple to branched. **LEAVES** basal or cauline, alternate, simple and entire or pinnately to ternately divided, sessile or petiolate, often impressed glandular-punctate. **INFL** a radiate head (in ours) or rarely discoid, solitary to corymbose or cymose, pedunculate. **INVOLUCRES** campanulate to hemispheric. **PHYLLARIES** in 2–3 similar or dissimilar series; outer series free or basally connate 1/4–1/2 of length; inner series similar or dissimilar to outer, free to base. **RECEPTACLES** flat to convex, conic, or hemispheric, naked. **RAY FLORETS** pistillate, fertile; ligules yellow to yellow-orange, showy, often reflexed in age. **DISC FLORETS** perfect, fertile, numerous; corolla yellow. **CYPSELAE** turbinate to obpyramidal, 5-angled, pubescent. **PAPPUS** of hyaline scales, ovate to narrowly lanceolate, apex obtuse to acuminate, often aristate or awned. Recently revised to include 28 species in

8 subgenera, mostly North American but a few species in South America. Members of the genus are considered toxic to livestock (Bierner 1994, 2001, 2004, 2005; Bierner and Janson 1998).

1. Plants annual .... ***H. odorata***
1′ Plants biennial or perennial .... (2)

2. Phyllaries in 2–4 similar and subequal series .... (3)
2′ Phyllaries in 2 dissimilar and unequal series .... (4)

3. Leaves entire or toothed; outer phyllaries spreading to reflexed in fruit .... ***H. hoopesii***
3′ Leaves usually dissected; outer phyllaries not spreading to reflexed in fruit .... ***H. grandiflora***

4. Inner phyllaries lanceolate, apices awn-tipped; pappus scales 10 .... ***H. bigelovii***
4′ Inner phyllaries narrowly obovate to obovate, apices acuminate to mucronate; pappus scales 8 or fewer .... (5)

5. Plants biennial from a basal rosette and taproot; plants monocarpic; stems 1–3 .... ***H. cooperi***
5′ Plants perennial from a simple or branched caudex surmounting a taproot; plants polycarpic; stems 3–10 or more .... (6)

6. Basal leaves densely woolly in the axils; ultimate leaf segments 0.5–2.5 mm wide .... ***H. richardsonii***
6′ Basal leaves glabrous or pubescent in the axils but not woolly; ultimate leaf segments 2–11 mm wide .... ***H. helenioides***

Hymenoxys bigelovii (A. Gray) K. F. Parker (in honor of Dr. John Milton Bigelow, professor of botany at Detroit Medical College, botanist on the Pacific Railroad survey of 1853–1854, and prolific plant collector) Bigelow's rubberweed. Perennial herbs from a sparingly branched caudex; caudex clothed by midveins of decayed leaves. **STEMS** 1–5, usually unbranched, 15–70 cm high, sparsely to densely pubescent, often tomentose at the base. **LEAVES** basal and cauline, simple and entire (blades rarely divided into 3 segments), narrowly linear, glabrous to densely pubescent, eglandular or sparsely gland-dotted; leaf base expanded, clasping, persistent, sparsely to densely long-woolly villous. **INFL** a head, 1–5 per plant, solitary or in paniculate arrays; peduncle (1.5) 6–20 (29) cm long, expanded and densely tomentose below the involucre. **INVOLUCRES** 13–20 mm high. **PHYLLARIES** in 2 series; outer phyllaries basally connate about 1/5 of length, apex acute to acuminate; inner phyllaries scalelike, exceeding the outer series, apex aristate. **RAY FLORETS** 13–15; ligules 13–26 mm long, apex 3-lobed. **DISC FLORET** corollas 5.7–7.4 mm long. **CYPSELAE** 4.2–4.7 mm long. **PAPPUS** of 10 aristate scales, 4.7–7.3 mm long, nearly equaling the disc corolla. $2n = 30$. [*Actinea bigelovii* (A. Gray) A. Nelson; *Actinea gaillardia* A. Nelson; *Actinella bigelovii* A. Gray; *Macdougalia bigelovii* (A. Gray) A. Heller]. Roadsides, dry slopes, and pine forests. One collection in our area. ARIZ: Apa. 2350 m (7700′). Flowering: Apr–Jul. Known only from eastern Arizona and western New Mexico. Used medicinally as a stimulant, a cathartic, and for rheumatism.

Hymenoxys cooperi (A. Gray) Cockerell (in honor of Dr. James Graham Cooper, 1830–1902, California physician, naturalist, and plant collector) Cooper's rubberweed, Cooper's hymenoxys, ragged rustlers. Biennial or perennial taprooted herbs, caudex few-branched or a simple crown. **STEMS** 1–few, branched above, to 10 dm high, often reddish, glabrate to villous or canescent. **LEAVES** basal and cauline, reduced above, simple or pinnately (rarely bipinnately) 3–9-cleft, ultimate segments linear, 0.5–2 mm wide; puberulent glandular-punctate. **INFL** a head, 1 to many, in open panicles or corymbs; peduncle to 13 cm long, sparsely to densely pubescent. **INVOLUCRE** 5–8 mm high. **PHYLLARIES** in 2 series; outer phyllaries basally connate for 1/3–1/2 their length, greener and more keeled than inner series, apex acute to acuminate; inner phyllaries usually exceeding the outer series, apex acute to mucronate. **RAY FLORETS** 7–14; ligules 8–17 mm long, apex 3 (4)-lobed. **DISC FLORET** corollas 2.7–4.8 mm long. **CYPSELAE** 1.7–3.7 mm long. **PAPPUS** of 5–6 (8) obovate, aristate scales, 1.3–3.3 mm long; scales conspicuously shorter than disc corolla. $2n = 30$. [Numerous synonyms exist in *Actinella*, *Actinea*, *Picradenia*, and *Hymenoxys*]. Dry slopes, valleys, and foothills, often associated with piñon-juniper or sagebrush communities. ARIZ: Apa. 2135 m (7000′). Flowering: May–Sep. Ranging from southern Oregon and Idaho south to Arizona. Although previously split into several varieties, the most recent treatment (Bierner 2001) accepts no varieties under *H. cooperi.* The species has reportedly been used as a dye.

Hymenoxys grandiflora (Torr. & A. Gray) K. F. Parker **CP** (large or great flower, referring to the spectacular flowers) Old-man-of-the-mountain, graylocks, tundra hymenoxys. Low-growing, perennial, taprooted herbs, caudex simple or few-branched. **STEMS** 1–10 per caudex, simple or branched below, 8–30 cm high, densely floccose-woolly or villous. **LEAVES** cauline and basal, 2–12 cm long, simple or 1–2-pinnatifid or ternate, ultimate segments 3–15, linear, villous to glabrate. **INFL** with 1–10 heads per plant, usually few, borne singly, showy; peduncle 0.5–10 cm long, densely tomentose below. **INVOLUCRES** 15–25 mm high. **PHYLLARIES** subequal in 2–3 series; outer phyllaries free or basally connate 1/5–1/4 their length, apex acute; inner phyllaries free, shorter than or equaling the outer series, apex acute to acuminate. **RAY FLORETS** 15–50; ligules 15–35 mm long. **DISC FLORET** corollas 5–6 mm long. **CYPSELAE** 3–5 mm long. **PAPPUS** of 5–8 lanceolate-aristate scales, 4.5–5.3 mm long, nearly equaling or equaling the disc corolla. $2n = 30$. [*Actinea grandiflora* (Torr. & A. Gray) Kuntze; *Actinella grandiflora* Torr. & A. Gray; *Ptilepida grandiflora* (Torr. & A. Gray) Rose; *Rydbergia grandiflora* (Torr. & A. Gray) Greene; *Tetraneuris grandiflora* (Torr. & A. Gray) K. F. Parker]. Subalpine meadows, talus slopes, and rocky places, high peaks and alpine communities. COLO: Arc, Con, Hin, LPl, Min, Mon, RGr, SJn. 3100–3840 m (10200–12600′). Flowering: Jun–Sep. Rocky Mountains from southern Colorado north to Idaho and Montana.

Hymenoxys helenioides (Rydb.) Cockerell (resembling *Helenium*) Intermountain rubberweed, sneezeweed hymenoxys. Perennial herbs, caudex simple or branched, woody. **STEMS** 1–5 (to 10 or more), simple below, branched above, 15–60 cm high, subglabrate to puberulent or villosulous. **LEAVES** basal and cauline, simple or pinnately divided into 3–5 segments, linear-oblanceolate; lower leaves largest, mostly 5–20 cm long and 2–10 mm wide. **INFL** with 2–many heads, paniculate or corymbiform; peduncle to 8 (12) cm long, moderately to densely pubescent. **INVOLUCRES** 5–8 mm high. **PHYLLARIES** in 2 subequal but dissimilar series; outer phyllaries basally connate for 1/4–1/3 their length, apex acuminate; inner phyllaries about equaling the outer, apex acuminate to mucronate. **RAY FLORETS** 10–16; ligules 15–31 mm long, apex 3-lobed. **DISC FLORET** corollas 3.5–5.5 mm. **CYPSELAE** 2.5–3.5 mm long. **PAPPUS** of 5–7 obovate to lanceolate-aristate scales, 2.5–4 mm long; 1/2 as long as to nearly equaling the disc corolla. $2n = 30$. [*Actinea helenioides* (Rydb.) S. F. Blake; *Dugaldia helenioides* (Rydb.) A. Nelson; *Hymenoxys* × *helenioides* (Rydb.) Cockerell; *Picradenia helenioides* Rydb.]. Open meadows, rocky ridges, edges of forests, and along creeks in the mountains. ARIZ: Apa; NMEX: SJn. 2290–2590 m (7500–8500′). Flowering: Jun–Aug. Known from eastern Utah, southwestern Colorado, and northeastern Arizona. This taxon appears to be a hybrid of *H. hoopesii* and *H. richardsonii* var. *floribunda.* Bierner (2001) recognizes it as a species because it is not known whether F1 hybrids have given rise to breeding populations.

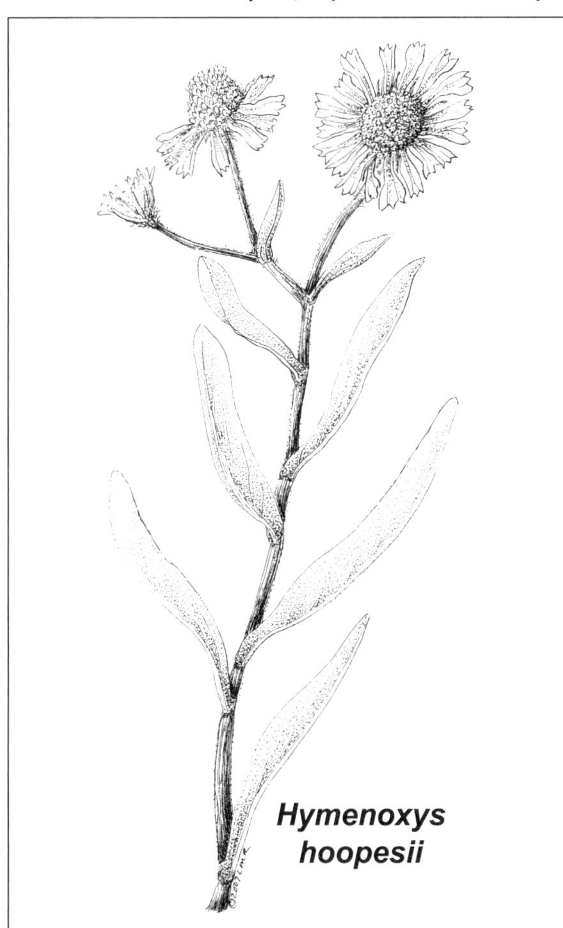
Hymenoxys hoopesii

Hymenoxys hoopesii (A. Gray) Bierner (for Thomas Hoopes, businessman and collector of the plant in the 1850s) Orange sneezeweed, owl's claws. Stout perennial herbs from a short rhizome or compactly branched caudex. **STEMS** 1–several, 2–10 dm high, branched above, glabrous to densely pubescent, striate but lacking wings. **LEAVES** basal and cauline, simple and entire; basal leaves 2–30 cm long, oblong to oblanceolate, usually tapering to a winged petiole or clasping base; upper leaves reduced, linear to oblanceloate or lanceolate, sessile. **INFL** with 2–15 heads, corymbiform or rarely solitary; peduncle 3–20 cm long, lanate below the involucre. **INVOLUCRE** 5–10 mm high. **PHYLLARIES** in 2 series; outer phyllaries basally connate, initially appressed, often reflexed with age, apex acute; inner bracts free, apex acuminate. **RAY FLORETS** 13–26; ligules 15–35 mm long, apex shallowly 3–4-lobed, reflexed in age. **DISC FLORET** corollas 4.5–5.4 mm long. **CYPSELAE** 3.5–4.5 mm long. **PAPPUS** of 5–7 hyaline scales, lanceolate, apex attenuate-acuminate to awned, 2.5–4.1 mm long, 1/2 as long as to nearly equaling the disc corolla. $2n = 30, 34$. [*Dugaldia hoopesii* (A. Gray) Rydb.; *Heleniastrum hoopesii* (A. Gray) Kuntze; *Helenium hoopesii* A. Gray]. Meadows, streamsides, open woods, and moist sites. ARIZ: Apa; COLO: Arc, Dol, Hin, LPl, Min, Mon, RGr, SJn; NMEX: McK, RAr, SJn. 2440–3630 m (8000–11900′). Flowering: May–Oct. Rocky Mountain states from Montana to New Mexico and west to California and Oregon. The species contains glycosides toxic to sheep, causing a disease called "spewing sickness." The plant has been used to inhibit vomiting, and the flowers have been used as a yellow dye.

Hymenoxys odorata DC. (fragrant, referring to the aromatic foliage) Bitterweed, bitter rubberweed. Annual taprooted herbs; bushy and aromatic. **STEMS** 1–several, diffusely branched, 15–60 cm high, slender. **LEAVES** basal and cauline; basal leaves withering by anthesis; cauline leaves 1–10 cm long, pinnatifid into 3–15 narrow filiform segments. **INFL** of many small heads, solitary on axillary or terminal peduncles forming flat-topped corymbs; peduncle 2–15 cm long, expanded apically, sparsely to densely pubescent. **INVOLUCRES** 3–6 mm high. **PHYLLARIES** in 2 dissimilar series; outer bracts basally connate 1/4–1/3 their length, thickened basally, apex acute; inner bracts exceeding the outer, pressed inward in fruit, convergent at maturity, apex acute. **RAY FLORETS** 6–13; ligules 5–11 mm long, persistent on achenes, apex 3-lobed. **DISC FLORET** corollas 3–4 mm long. **CYPSELAE** 1.5–2 mm long. **PAPPUS** of 5 (rarely 6) lanceolate to ovate scales, attenuate or aristate, 1.5–2.3 mm long, 1/2 as long as to nearly equaling the disc corolla. $2n$ = 22, 24, 28, 30. [*H. chrysanthemoides* DC. subsp. *osterhoutii* (Cockerell) Cockerell; *H. multiflora* Rydb.; *Phileozera multiflora* Buckley; *Picradenia multiflora* (Buckley) Greene; *Picradenia odorata* (DC.) Britton]. Plains and valleys, common on overgrazed or disturbed rangeland. NMEX: McK, San, SJn. 1520–1770 m (5000–5800′). Flowering: Apr–Aug. Known from California, Arizona, New Mexico, Colorado, Kansas, Oklahoma, and Texas. Toxic to livestock, particularly sheep.

Hymenoxys richardsonii (Hook.) Cockerell (in honor of Sir John Richardson, 1787–1865, English naturalist and Arctic explorer) Perennial caespitose herbs with a many-branched woody caudex surmounting a taproot. **STEMS** 1–many, simple below, sparingly to much-branched above, 6–50 cm high, sparsely hairy. **LEAVES** basal and cauline, 2–15 cm long, simple or pinnately divided into 3–7 linear-filiform segments up to 2 mm wide. **INFL** with 1–many heads, corymbiform; peduncle up to 17 cm long, nearly glabrous to densely short-pubescent. **INVOLUCRES** 5–11 mm high. **PHYLLARIES** in 2 dissimilar series; outer bracts connate basally 1/2 their length, apex acute to acuminate; inner bracts free, equaling or slightly exceeding the outer bracts, apex mucronate. **RAY FLORETS** 5–14; ligules 7–23 mm long, apex 3 (4)-lobed. **DISC FLORET** corollas 3–5 mm long. **CYPSELAE** 2–4 mm long. **PAPPUS** of 4–8 obovate to lanceolate-aristate scales, 1.5–4.5 mm long, shorter than or nearly equaling disc corolla. $2n$ = 30, 56. [Numerous synonyms exist in *Actinea*, *Actinella*, *Hymenopappus*, *Hymenoxys*, *Picradenia*, and *Ptilepida*]. Open areas and at the edges of forests. ARIZ: Apa, Nav; COLO: Arc, LPl, Mon; NMEX: McK, RAr, San, SJn. 1760–2750 m (5780–9000′). Flowering: May–Oct. The species ranges from New Mexico and Arizona north to Montana and adjacent Canada. Ours is the southern, more montane **var.** ***floribunda*** (A. Gray) K. F. Parker (profusely flowered) of Arizona, New Mexico, Colorado, and Utah. Variety *floribunda* intergrades with var. *richardsonii* where their ranges overlap in Colorado and Utah. The species is toxic to livestock, and the root latex has been investigated as a commercial source of rubber. The yellow flowers have been used as a dye.

Isocoma Nutt. Jimmyweed, Goldenweed

G. L. Nesom

(Greek *isos*, equal, + *kome*, hair of the head; "so called from its equal flowers" – Nuttall, protolog) Subshrubs or sometimes perennial herbs, (4–) 20–120 (–150) cm, bases often woody. **LEAVES** mostly cauline, alternate, sessile, blades linear to oblanceolate or obovate, margins entire to toothed or pinnatifid, teeth or lobes often spinulose-tipped, 1-nerved, faces glabrous or hispidulous, villous, or tomentose, usually gland-dotted (in pits). **INFL** a discoid, sessile to subsessile head in compact clusters, borne in terminal corymbiform arrays. **INVOLUCRES** obconic to turbinate or campanulate. **PHYLLARIES** in (3–) 4–6 series, oblong- to elliptic-lanceolate, unequal, margins narrowly scarious, flat to convex, apices sometimes green, midveins usually barely evident, glabrous or tomentose, sometimes gland-dotted, sometimes with resin pockets. **RECEPTACLES** flat, foveolate, epaleate. **DISC FLORETS** bisexual, fertile; corollas yellow with dark orange-resinous veins, tubes elevating corolla at flowering and outer corollas prominently leaning outward. **CYPSELAE** terete or subterete, 5–11-ribbed (sometimes thick and resinous), sericeous. **PAPPUS** of 40–50 persistent, unevenly thick, barbellate bristles in 2 (–3) series. x = 6. [*Haplopappus* sect. *Isocoma* (Nutt.) H. M. Hall]. Species 16. Southwestern United States and Mexico (Nesom 1991). *Isocoma pluriflora* (Torr. & A. Gray) Greene occurs just to the south of our area.

1. Leaves pinnatifid; cypselar ribs not forming apical horns .... ***I. azteca***
1′ Leaves entire to shallowly toothed or lobed, not pinnatifid; cypselar ribs forming apical horns .... ***I. rusbyi***

Isocoma azteca G. L. Nesom (alluding to the ancient Indian tribe of central Mexico) Apache jimmyweed. Herbage usually glabrous, sometimes sparsely stipitate-glandular, not resinous. **LEAVES** with narrowly oblong to narrowly oblanceolate blades, 20–50 mm long, margins shallowly to deeply pinnatifid, lobes in 3–8 evenly arranged pairs, apically aristate. **INVOLUCRES** 7–8 × 5–7.5 mm. **PHYLLARY** apex green to greenish yellow, not aristate, barely to prominently gland-dotted, without resin pockets. **FLORETS** 18–25; corolla 5–6 mm. **CYPSELAE** ribs not forming apical horns.

Slopes, river and pond edges, washes, roadsides, sandy to clay soil, gypseous or saline, desert shrub, badlands; piñon-juniper woodlands and ponderosa pine communities; commonly with *Atriplex*. ARIZ: Apa; NMEX: McK, RAr, San, SJn. 1500–1950 (–2200) m (4920–7215′). Flowering: Jun–Sep. Arizona and New Mexico. *Isocoma azteca* and *I. rusbyi* are essentially contiguous in range along the New Mexico-Arizona border.

Isocoma rusbyi Greene (honoring botanist H. H. Rusby, 1855–1940) Rusby's jimmyweed. Herbage glabrous, not resinous. **LEAVES** with narrowly elliptic-oblong to elliptic-obovate blades, 20–50 mm, margins entire. **INVOLUCRES** (5.5–) 6–9.5 × 5–7.5 mm. **PHYLLARY** apex yellowish to greenish yellow, not aristate, weakly or not gland-dotted, without resin pockets. **FLORETS** 19–25; corolla 5–6.5 mm. **CYPSELAE** ribs extending slightly past apices, forming hornlike extensions. $2n = 12$. [*Haplopappus rusbyi* (Greene) Cronquist]. Rocky or sandy soil, less commonly in clay, usually saline, dunes, sandy benches, washes, often disturbed sites; desert shrub communities, commonly with *Atriplex* and *Salsola*, riparian with cottonwood, Russian olive, and saltcedar, sometimes with scattered junipers. ARIZ: Apa, Coc, Nav; COLO: Mon; NMEX: SJn; UTAH. (1200–) 1400–1650 (–1900) m (3935–6235′). Flowering: (Jun–) Jul–Sep (–Nov). Arizona, Colorado, New Mexico, and Utah.

Iva L. Marsh-elder

Gary I. Baird

(derivation uncertain) Annuals, perennials, or shrubs, taprooted or rhizomatous, herbage glabrous or scabrous, usually gland-dotted, sap clear. **STEMS** 1 (or more), erect (or decumbent to spreading), usually branched. **LEAVES** cauline, simple, mostly opposite (distal sometimes alternate), sessile to petiolate; blades spatulate to linear or filiform, 1- or 3-nerved, bases cuneate to attenuate, margins entire or dentate, sometimes undulate, apices obtuse to acuminate. **INFL** in bracteate racemiform to spiciform clusters (with 1–2 heads per node), nodding, bracts sometimes foliaceous; sessile or pedunculate. **HEADS** discoid or disciform. **INVOLUCRES** hemispheric, campanulate, or turbinate. **PHYLLARIES** 3–15 in 1–3 series, outer distinct or connate, herbaceous or inner membranous and palealike. **RECEPTACLES** ± flat to hemispheric, paleate; paleae spatulate to linear or setiform, sometimes lacking (in whole or in part), membranous; calyculi lacking. **PISTILLATE FLORETS** 1–8 (or lacking), corollas whitish, tubular, inconspicuous. **STAMINATE FLORETS** 3–20, corollas whitish to pinkish, funnelform; filaments ± connate, anthers connate or distinct. **CYPSELAE** monomorphic, blackish or brownish, ± obcompressed, obovoid to pyriform, sometimes pubescent and/or gland-dotted. **PAPPUS** lacking. $x = 18$. About nine species of North America (mostly eastern).

Iva axillaris Pursh (growing from the axil, referring to the position of the heads) Povertyweed. Perennials 10–60 cm tall, rhizomatous, herbage ± scabrous-strigose and gland-dotted throughout. **STEMS** erect. **LEAVES** opposite (proximal) to alternate (distal), sessile or obscurely petiolate (to 3 mm); blades oblong to obovate or spatulate, mostly 10–30 × 3–10 mm, margins entire. **HEADS** mostly 1 per node in racemiform clusters; peduncles 1–2 mm; bracts foliaceous and only slightly reduced upward; paleae linear, about 2 mm. **INVOLUCRES** hemispheric, 2–3 mm. **PHYLLARIES**: outer about 3–5 in 1 series, connate (at least proximally), or with 1 separate and proximal to the others. **PISTILLATE FLORETS** 3–8. **STAMINATE FLORETS** 4–20; anthers distinct. **CYPSELAE** about 3 mm long, greenish (proximally) to blackish purple (distally), conspicuously gland-dotted. $2n = 36, 54$. Riparian, lacustrine, or palustrine communities, usually where seasonally moist, often in disturbed areas, sometimes in saline soils. ARIZ: Apa; COLO: Arc, Dol, LPl, Mon, SJn; NMEX: SJn; UTAH. 1700–2100 m (5600–6900′). Flowering: Jun–Sep. British Columbia to Manitoba, south to California and Texas. An invasive weed of disturbed sites, it is considered noxious by some.

Lactuca L. Lettuce

Gary I. Baird

(Latin *lactugo*, milky, in reference to the sap) Annuals (ours), taprooted, herbage glabrous or pubescent, usually prickly-setose, sap milky. **STEMS** usually 1, erect, branched distally. **LEAVES** basal and cauline or mostly cauline (at flowering), simple, alternate, mostly sessile, more or less auriculate; blades obovate to lanceolate, margins entire to pinnately or runcinately lobed, ultimate margins prickly-denticulate, apices acute; lobes opposite to subopposite. **INFL** of heads borne in paniculiform clusters. **HEADS** liguliflorous, usually numerous, borne in paniculiform clusters, erect; peduncles bracteate, glabrous. **INVOLUCRES** lacking calyculi, small, mostly cylindrical to narrowly obconic (in fruit). **PHYLLARIES** imbricate, herbaceous, green, erect. **RECEPTACLES** flat, epaleate. **RAY FLORETS** matutinal; corolla ligulate, blue or yellow (ours), outermost rosy-purple abaxially, often drying bluish. **DISC FLORETS** lacking. **CYPSELAE** monomorphic, brownish or grayish to black, often mottled, more or less flattened, ellipsoid to oblanceoloid, glabrous or scaberulous to hispidulous distally, ribbed, marginal ribs somewhat winged; beak lacking, stout (not much different from body), or slender (much different from body), shorter or longer than body. **PAPPUS** of bristles, persistent, white to brownish, ± barbellulate. $x = 9$. About 75 species in Africa, Eurasia, and North America. Both *Lactuca canadensis* L. and *L. sativa* L. have been found near the study area and are to be expected.

1. Corolla blue or white, rarely yellowish; cypselae beakless or gradually tapered to a stout beak less than 1 mm long; pappi sordid or brownish .... ***L. biennis***

1′ Corolla yellow, rarely bluish or white (sometimes drying bluish); cypselae abruptly tapered at apex into a slender beak 1–4 mm long; pappi white .... (2)

2. Plants biennial; achene body about 5–6 mm, faces 1 (3)-nerved; herbage not prickly (reported from the study area and to be expected, but no verified species seen) .... *L. canadensis* L.

2′ Plants annual or biennial; achene body about 3–4 mm, faces (3) 5–9-nerved; herbage prickly or not .... (3)

3. Plants ± annual; leaf blades oblong to oblanceolate, pinnately lobed or (sometimes) entire, twisted at base so blade is held in a vertical plane, margins and lower midvein prickly, rarely otherwise (common weed throughout the study area) .... ***L. serriola***

3′ Plants annual or biennial; leaf blades ovate to orbicular, entire to dentate, not twisted at base or held in a vertical plane, margins and lower midvein not (or rarely) prickly (common garden lettuce but rarely, if ever, escaping in the study area; purportedly it readily hybridizes with the previous species) .... *L. sativa* L.

Lactuca biennis (Moench) Fernald (lasting for two years) Tall blue lettuce. Annuals or biennials 75–200 cm (or more). **STEMS** glabrous or sparsely pilose-setose proximally, often glaucous. **LEAVES** with blades oblanceolate to lanceolate, lower and middle cauline about 10–40 cm × 30–200 mm (including lobes), not twisted at base or held in a vertical plane, margins pinnately lobed, ultimate margins dentate, faces glabrous or sparsely pilose-setose abaxially on veins; auricles obscure, 1–2 mm, straight, bluntly acute; lobes 3–4 pairs, ± deltate, about 25–100 mm long, spreading or slightly recurved; upper cauline progressively reduced and becoming bractlike, ± lanceolate, pinnately lobed to entire. **HEADS** numerous (to 100+, proliferating with age). **INVOLUCRES** 8–14 mm tall. **PHYLLARIES** 15–25 in about 4 series, lanceolate, green to blackish purple, glabrous. **RAY FLORETS** 20–30, outer exceeding involucre; corolla bluish or whitish, sometimes yellowish, tube 5–6 mm, ligule 4–5 mm × about 1 mm, glabrous; anthers about 2 mm. **CYPSELAE** body ± ellipsoid, flattened, about 4–5 mm, mottled brown, tapered distally; ribs 5–6, rugulose; beak lacking or less than 1 mm, stout, pale. **PAPPUS** 4–6 mm, brownish. $2n = 34$. Riverbanks and other moist areas. COLO: LPl, SJn. 2500 m (8300′). Flowering: Aug. Yukon and British Columbia, eastward to Newfoundland, and southward to California, New Mexico, Iowa, Tennessee, and South Carolina.

Lactuca serriola L. (sawlike, alluding to the leaf margins) Prickly lettuce. Annuals 15–90 cm tall. **STEMS** glabrous or (densely) prickly-hirsute proximally, stramineous. **LEAVES** with blades mostly oblanceolate, lower and middle cauline 1–12 cm × 6–60 mm (including lobes), usually twisted at base and held in a vertical plane, margins entire to pinnately lobed, ultimate margins prickly-hirsute, faces prickly-hirsute along midrib (at least abaxially); auricles straight to recurved (less than 90°), acute; lobes 1–3 pairs, lanceolate to deltate, 5–20 mm long, mostly retrorsely curved; upper cauline (distal 1/2 to 1/4 of stem) much reduced, bractlike, ± lanceolate, mostly entire. **HEADS** numerous (to 500+, proliferating with age). **INVOLUCRES** about 10 mm tall. **PHYLLARIES** 12–16 in mostly 4 series, lanceolate, mostly green, sometimes rosy-purple or spotted, glabrous. **RAY FLORETS** 15–20, outer (just) exceeding involucre; corolla yellow, tube 4–7 mm, ligule 5–6 mm × about 1 mm, pilose proximally; anthers 2–3 mm. **CYPSELAE** about 7 mm long (including beak); body oblanceoloid, about 3 mm, grayish brown, mottled darker, abruptly tapered to beak; ribs 5–7, muricate-rugose, scaberulous distally; beak 3–5 mm, flexuous, whitish. **PAPPUS** 4–5 mm, white. $2n = 18$. Moist to dry habitats, often disturbed sites and roadsides, in riparian, sagebrush, piñon-juniper, ponderosa pine, oak, Douglas-fir, aspen, and white fir communities. ARIZ: Apa, Nav; COLO: Arc, Dol, Hin, LPl, Min, Mon; NMEX: McK, RAr, San, SJn; UTAH. 1600–2700 m (5300–8700′). Flowering: Jul–Sep. Europe; widely established in North America from southern Canada to Mexico and introduced nearly worldwide. Cultivated lettuce, *L. sativa* L., is occasionally found as a garden escapee, although it has not been found in the study area. It purportedly hybridizes readily with *L. serriola*, where the two come in close contact.

Laennecia Cass. Woolwort
G. L. Nesom

(Greek *lenos*, woolly hair, in reference to the characteristic vestiture) Annuals, biennials, or short-lived perennials, mostly taprooted, leaves, stems, and phyllaries white-tomentose or cottony, coarsely hairy in 2 species, glandular with sessile or stipitate resin glands, eglandular in some species. **LEAVES** alternate, lanceolate or oblanceolate to oblong or elliptic, toothed to pinnately lobed, rarely entire, epetiolate, clasping or nonclasping. **INFL** a spicate or racemose to loosely paniculate or corymboid inflorescence of heads. **PHYLLARIES** in 2–5 subequal to graduate series. **RECEPTACLES** flat to shallowly convex, epaleate. **PISTILLATE FLORETS** numerous in several series, fertile, white, filiform-tubular, eligulate, or some species with a lamina 0.2–2.5 mm long. **DISC FLORETS** perfect, fertile. **CYPSELAE** narrowly oblanceolate-elliptic to obovate in outline, with sessile or short-stipitate resin glands on the faces, rarely eglandular. **PAPPUS** of 1–2 series of slender, barbellate, often easily caducous bristles, sometimes elongating at maturity past the disc corollas and involucres, with or without an outer series of much shorter setae, bristles, or scales. $x = 9$. [*Conyza* sect. *Laennecia* (Cass.) Cuatrec.] (Nesom 1990b). Seventeen species; South America to Mexico and the southwestern United States. Similar to *Conyza* in production of eligulate or short-ligulate pistillate florets in numerous series but distinguished by phyllaries with a single, greenish (vs. orange-resinous) midvein, light-veined (vs. orange-veined) disc corollas with lanceolate (vs. deltate) lobes, presence of achenial glands, and accrescent pappus bristles. Most *Laennecia* species have herbage with a resinous-glandular, woolly-tomentose vestiture.

1. Ray corollas eligulate; pubescence coarsely villous to villous-pilose, leaves 5–15 mm wide; phyllary margins narrow, barely hyaline.. ***L. coulteri***
1' Ray corollas ligulate; pubescence arachnoid to floccose; leaves 2–8 mm wide; phyllary margins broad and distinctly hyaline.. ***L. schiedeana***

Laennecia coulteri (A. Gray) G. L. Nesom (honoring North American botanist John M. Coulter, 1859–1928) Coulter's woolwort. Taprooted annuals, 3–15 cm, moderately to densely coarsely villous to villous-pilose with jointed, viscid hairs of varying lengths, sessile- and stipitate-glandular. **LEAVES** mostly cauline, oblong to spatulate-oblong, 2–5 cm long, 5–15 mm wide, regularly toothed to shallowly lobed, clasping to subclasping. **HEADS** in a panicle, cylindric to broader, more elliptic or nearly corymboid in shape. **PHYLLARY** margin narrow, barely hyaline. **PISTILLATE** corollas eligulate. $n = 18$. [*Conyza coulteri* A. Gray]. Roadsides, ditch banks, washes, moist canyons, disturbed sites. COLO: LPl; UTAH. 1200–2400 m (3935–7875'). Flowering: Jul–Oct. California and Nevada east to Colorado and Texas; Mexico.

Laennecia schiedeana (Less.) G. L. Nesom (honoring German botanist Christian J. W. Schiede, 1798–1836) Pineland woolwort. Taprooted annuals, 2–5 (–10) cm, glandular, moderately to densely woolly-arachnoid or floccose with jointed, nonviscid hairs protracted into long, minutely filiform, crisped apices, sessile-glandular. **LEAVES** mostly cauline, lanceolate-oblong to oblong or oblong-spatulate, 1.5–5 cm long, 2–8 mm wide, entire to shallowly toothed, rarely shallowly lobed, clasping. **HEADS** in a narrowly cylindric panicle. **PHYLLARY** margin broad and distinctly hyaline. **PISTILLATE** corollas with ligules 0.3–1.4 mm. $n = 18$. [*Conyza schiedeana* (Less.) Cronquist]. Talus and cliff bottoms, trails, roadsides; ponderosa pine–white pine, and Douglas-fir communities. COLO: Hin, LPl. 2500–3000 m (8200–9840'). Flowering: Aug–Sep. Arizona, Colorado, and New Mexico; Mexico; Central America.

Layia Hook. & Arn. ex DC. Tidytips
Brian A. Elliott

(for George Tradescant Lay, 19th-century plant collector, botanist on Beechey's 1825–1828 voyage) Annual taprooted herbs. **STEMS** ascending to erect, simple or branched. **LEAVES** basal and cauline, alternate above but sometimes opposite below, linear-lanceolate to oblanceolate or oblong, entire to toothed or pinnatifid, sessile. **INFL** a radiate head (discoid in 1 species), solitary at branch tips or in an open corymb. **INVOLUCRES** campanulate to hemispheric. **PHYLLARIES** uniseriate but with chaff bracts sometimes exposed and mimicking a second row, herbaceous, flattened on the back, folding about the ray achene and falling with the fruit. **RECEPTACLES** flat to slightly convex, marginally chaffy to chaffy throughout; chaff scales free. **RAY FLORETS** pistillate and fertile; ligules white, yellow, or yellow with a white tip, broad, tube pubescent, apex 3-lobed or 3-toothed. **DISC FLORETS** perfect and fertile, numerous; corolla yellow, style branches bristly at the tip. **CYPSELAE** with ray achenes usually glabrous; disc achenes pubescent. **PAPPUS** lacking on ray achenes; of bristles, awns, or scales on disc achenes. Fourteen species of western North America. A report of *Layia platyglossa* (Fisch. & C. A. Mey.) A. Gray from San Juan County, Utah, is due to confusion in labeling.

Layia glandulosa (Hook.) Hook. & Arn. (an abundance of glands) Whitedaisy tidytips, white tidytips, white layia. **STEMS** to 3(6) dm high, often reddish, hispid with multicellular hairs and stipitate-glandular hairs. **LEAVES** 1–8 cm long, 1–10 (15) mm wide, often reduced above, lower leaves irregularly toothed to pinnately lobed, upper leaves often entire, hispid with multicellular hairs, often strigose on the surfaces. **INVOLUCRES** 6–9 mm. **PHYLLARIES** 4–10 mm long, short-hispid and usually with stalked, purple- or black-tipped glandular hairs. **RECEPTACLES** chaffy; chaff scales in a single row between ray and disc flowers. **RAY FLORETS** with 1–8 (14) ligules, white, generally 4–18 mm long. **CYPSELAE** with ray achenes glabrous; disc achenes appressed-hairy. **PAPPUS** on disc achenes of 10–15 slightly flattened bristles, scabrous above, plumose with capillary hairs at the base to middle and white-woolly within at the base. $2n = 16$. [*Layia glandulosa* subsp. *glandulosa*; *L. glandulosa* subsp. *lutea* D. D. Keck]. Dry slopes, sagebrush communities, piñon-juniper, sandy canyon bottoms, and roadsides. ARIZ: Coc, Nav; UTAH. 1550–1920 m (5100–6300′). Flowering: Apr–Jun. British Columbia, the western United States, and northwest Mexico. The parched seeds have been ground for flour or meal, particularly in California.

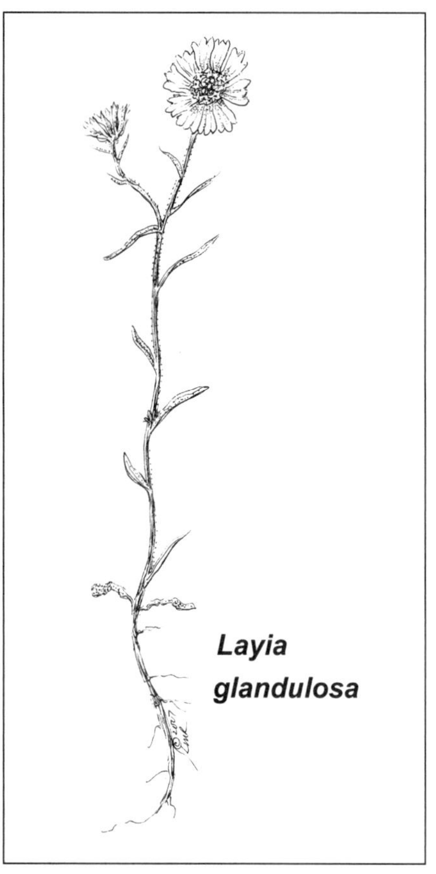
Layia glandulosa

Liatris Gaertn. ex Schreb. Gayfeather, Blazing Star
G. L. Nesom

(derivation unknown) Perennial herbs; corms globose to depressed-ovoid or napiform, sometimes elongated, becoming rhizomes, roots all or mostly adventitious. **STEMS** simple or basally branched. **LEAVES** basal and cauline, alternate, petiolate (basal) or sessile, 1- or 3–5-nerved, mostly linear to ovate-lanceolate, margins entire, faces often gland-dotted. **HEADS** discoid, in corymbiform, cymiform, racemiform, or spiciform arrays. **INVOLUCRES** mostly campanulate to hemispheric or turbinate-cylindric. **PHYLLARIES** in (2–) 3–7 series, herbaceous to petaloid, ovate to elliptic or lanceolate, usually unequal. **RECEPTACLES** flat, epaleate. **RAY FLORETS** absent. **DISC FLORETS** 3–85; corollas usually lavender to pinkish purple. **CYPSELAE** prismatic, 8–11-ribbed, usually hirsutulous to hirtellous-pilose, usually gland-dotted. **PAPPUS** of coarsely barbellate to plumose bristles. $x = 10$. Species 37; North America including Mexico and the Bahamas (Gaiser 1946; Nesom 2005).

Liatris ligulistylis (A. Nelson) K. Schum. (referring to the flattened, exserted style branches, superficially suggesting the slender rays of some *Erigeron*) Northern Plains gayfeather. Plants 20–100 cm tall; corms subglobose, often knotty, densely fibrous. **STEMS** sparsely to densely puberulent, puberulent-villous, or strigose-puberulent. **LEAVES**: basal and lower cauline narrowly oblanceolate to spatulate-lanceolate, 90–150 (–220) × 4–17 (–24) mm, gradually reduced distally to near midstem, then abruptly reduced, sparsely to densely puberulent. **HEADS** 4–21, pedunculate in open, racemiform arrays. **INVOLUCRES** campanulate to turbinate-campanulate, 10–15 mm wide. **PHYLLARIES** oblong-obovate to oblong-spatulate, strongly unequal, apices broadly rounded to truncate. $2n = 20$. Lakeshores, low roadsides, ditches. COLO: Arc. 2000–2450 m (6560–8040′). Flowering: Jul–Oct. Canada (southern Alberta to Manitoba) through Montana, North Dakota, Minnesota, and Wisconsin, south to South Dakota, Iowa, Wyoming, Colorado, and New Mexico.

Lorandersonia Urbatsch, R. P. Roberts & Neubig Rabbitbrush
R. P. Roberts

(for Loran C. Anderson, b. 1936, enthusiast of *Chrysothamnus* and related groups) Shrubs or subshrubs to 3.5 m tall. **STEMS** erect to ascending, often intricately branched; bark typically tan, becoming white or gray with maturity; young twigs usually green, glabrous to scabrous, often resinous or punctate. **LEAVES** alternate, mostly evergreen, cauline, often crowded, ascending to spreading or rarely deflexed, sessile or short-petiolate, glabrous to scabrous, blades linear to oblong or lanceolate to narrowly oblanceolate, planar to concave, sometimes punctate, often resin-coated or resin-dotted, midvein prominent, sometimes bounded by 1–2 pairs of collateral veins, margins entire or edged with trichomes, apices acute. **INFL** with 4–22 florets, usually in congested, rounded cymiform to corymbiform clusters, occasionally racemose. **HEADS** discoid or radiate. **INVOLUCRES** cylindric to obconic or hemispheric, 4–15 mm long. **PHYLLARIES** 13–30, in 3–6 series, imbricate to strongly aligned vertically, green to tan, ovate to oblong or lanceolate to oblanceolate, apices acute, acuminate, cuspidate, obtuse, erect or slightly spreading, often resinous; midvein obscure to conspicuous, sometimes enlarged subapically and glandular; lowermost sometimes herbaceous or herbaceous-tipped, otherwise mostly chartaceous. **RAY FLORETS** 6–8, rarely 1, carpellate, fertile, ranging from pale to dark yellow; laminae elliptic to obovate, 3.5–5 mm long. **DISC FLORETS** 4–15, bisexual, corollas same color as ray corollas, 3.5–14 mm long, lobes erect to spreading or reflexed; style branches 1.7–4.6 mm long. **CYPSELAE** mostly tan to brownish, usually prismatic, oblong to obconic, 1.5–7 mm long, glabrous to densely pubescent. **PAPPUS** similar in ray and disc flowers, whitish tan, 20–80+ subequal, barbellate bristles in 1 series. $x = 9$. Seven species distributed throughout the southern and central Rocky Mountains and southern plains of North America.

1. Cauline leaves mostly less than 3 mm wide; capitula with vertically aligned phyllaries; involucres 10–15 mm long .... ***L. baileyi***
1' Cauline leaves mostly more than 3 mm wide; capitula with phyllaries ± imbricate, not strongly vertically aligned; involucres less than 7 mm long .... ***L. linifolia***

Lorandersonia baileyi (Wooton & Standl.) Urbatsch, R. P. Roberts & Neubig (for William W. Bailey, 1843–1914, botanist) Bailey's rabbitbrush. **STEMS** to 1.4 m tall, greenish when young, becoming whitish to gray with maturity, glabrous to puberulent. **LEAVES** ascending to appressed, linear or linear-oblong, 3 mm or less wide, planar or adaxially concave, margins ciliolate, midvein conspicuous. **HEADS** cymose to corymbose, 0.5–7 cm wide. **INVOLUCRES** obconic to cylindric, 10–15 mm long. **PHYLLARIES** 20–25, in 4–5 series, keeled, ovate to oblong or lanceolate, glabrous, apices acute to attenuate or acuminate, mostly chartaceous, shiny. **DISC FLORETS** 5, corollas 9–14 mm; style branches long-exserted, 4.1–4.6 mm long. **CYPSELAE** turbinate, 5–7 mm long, usually sparsely pubescent or rarely densely pubescent. **PAPPUS** yellowish tan, 9–12 mm. $2n = 18$. [*Bigelowia pulchella* A. Gray; *Chrysothamnus pulchellus* (A. Gray) Green; *Ericameria pulchella* (A. Gray) L. C. Anderson]. Especially on sandy sites; shadscale, blackbrush, piñon-juniper woodland, and ponderosa pine communities. ARIZ: Apa, Nav; NMEX: McK, SJn. 1300–2400 m (4265–7875′). Flowering: Aug–Oct. Arizona, Colorado, Kansas, New Mexico, Oklahoma, Texas, Utah.

Lorandersonia linifolia (Greene) Urbatsch, R. P. Roberts & Neubig (linear-leaved) Spearleaf rabbitbrush. Shrubs spreading by creeping rhizomes. **STEMS** to 3.5 m tall, greenish when young, becoming tan to gray with maturity, glabrous. **LEAVES** ascending, broadly linear to lanceolate or oblong, 3 mm or more wide, planar, somewhat punctate, margins minutely scabrous, midvein and 2 fainter collateral veins evident. **HEADS** corymbiform, to 3–12 cm wide. **INVOLUCRES** obconic, 4.5–7 mm long. **PHYLLARIES** 15–18, in 3–4 series, somewhat keeled, elliptic to lanceolate, glabrous or sparsely scabrous, mostly tan, apices acute, short-acuminate, or rounded, outer often thickened and herbaceous subapically. **DISC FLORETS** 4–6, corollas 4–5.5 mm; style branches 1.9–2.3 mm. **CYPSELAE** subcylindric, 2.5–3.5 mm, densely pubescent. **PAPPUS** whitish tan, 4.5–7 mm. $2n = 18$. [*Chrysothamnus linifolius* Greene; *C. viscidiflorus* (Hook.) Nutt. subsp. *linifolius* (Greene) H. M. Hall & Clem.]. Riparian communities; irrigation canals, stream banks, seeps, and springs. ARIZ: Apa; COLO: Arc, Dol, Mon; NMEX: McK, RAr, SJn; UTAH. 1200–2400 m (3935–7875′). Flowering: Aug–Oct. Arizona, Colorado, Montana, New Mexico, Utah, Wyoming.

Lygodesmia D. Don Skeletonplant, Rushpink

Gary I. Baird

(Greek *lygos*, twig, + *desme*, bundle, alluding to the clustered, almost leafless stems) Perennials, deeply rhizomatous, herbage mostly glabrous, sap milky. **STEMS** 1, erect to decumbent, few- to many-branched, often stiff (rarely subwoody proximally), more or less striate. **LEAVES** basal and cauline, simple, alternate, sessile, not auriculate; blades linear-subulate to linear-lanceolate, or reduced and scalelike, margins entire (ours). **HEADS** liguliflorous, borne singly at ends of branches (obscurely corymbiform), erect; peduncle more or less bracteate, glabrous. **INVOLUCRES** cylindric, calyculate; calyculi about 8, much reduced, lanceolate to deltate, unequal, erect. **PHYLLARIES** about 5–12 in mostly 1 series, subequal, linear-lanceolate, herbaceous, erect. **RECEPTACLES** flat, epaleate. **RAY FLORETS** matutinal; corolla ligulate, pink, lavender, or white. **DISC FLORETS** lacking. **CYPSELAE** monomorphic, pale green to tan, more or less columnar, sometimes curved, sometimes tapered distally, glabrous, sometimes muricate-rugose, obscurely to evidently ribbed, beakless. **PAPPUS** of bristles, persistent, white to tawny, smooth to barbellulate. $x = 9$. About five species in North America (Tomb 1980).

1. Stems not very ramose (simply branched); leaves mostly over 2 cm long (at least lower), not usually bractlike or withered at flowering; phyllaries mostly with apical appendages; ligules 15–25 mm × 5–10 mm; cypselae over 10 mm long; pappus over 10 mm long .... ***L. grandiflora***
1' Stems very ramose (highly branched); leaves mostly reduced and bractlike, usually much less than 2 cm long, often withered at flowering; phyllaries lacking apical appendages; ligules 7–9 mm × 3–5 mm; cypselae less than 10 mm long; pappus less than 10 mm long .... ***L. juncea***

Lygodesmia grandiflora (Nutt.) Torr. & A. Gray (large-flowered) Large flower skeletonplant. **STEMS** 2–45 cm tall, erect, not highly ramose (usually with primary branches only), obscurely striate, mostly glabrous. **LEAVES** basal (not in rosettes) and cauline, present at flowering; blades linear to subulate or linear-lanceolate, to 15 cm long × 1–4 mm wide, glabrous, progressively reduced upward (uppermost sometimes reduced and bractlike), margins entire, whitish, smooth to muriculate. **HEADS** borne singly (at the apex of each branch). **INVOLUCRES** 15–25 mm tall. **CALYCULI** of about 8 reduced bractlets (all much less than phyllaries proper or graduating into phyllaries proper), deltate to lanceolate, erect. **PHYLLARIES** mostly 5–12 in about 1 series, linear-lanceolate, glabrous to pubescent, sometimes lanate-tomentose distally, margins scarious, entire to erosulate, apices mostly appendaged, midrib keeled distally. **RAY FLORETS** 5–12, exceeding involucre; corolla glabrous, tube 10–15 mm long, ligule 15–25 mm long × 5–10 mm wide (often deeply lobed); anthers 5–7 mm long. **CYPSELAE** 10–15 mm long, obscurely prismatic, smooth to rugose, stramineous; ribs about 5–6; apices with a transverse sulcus (directly below pappus). **PAPPUS** 10–13 mm long, tawny. $2n$ = 18, 27. Idaho and Wyoming, south to Arizona and New Mexico. Five variants are recognized (some have been treated as species); three of these occur within the study area, and one occurs just to the north of the study area along the Dolores River in Colorado, var. *doloresensis* (Tomb) S. L. Welsh.

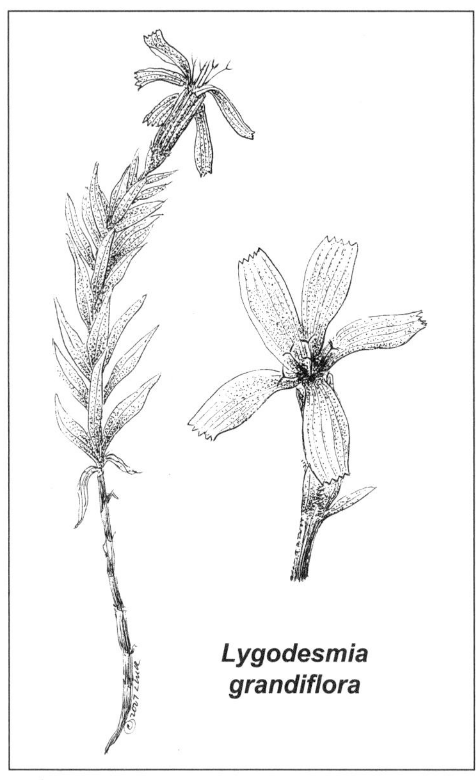
Lygodesmia grandiflora

1. Phyllaries and florets commonly 8–12, rarely fewer; plants often found on alluvial or clay soils rather than sand .... **var. *grandiflora***
1′ Phyllaries and florets mostly 5 or 6; plants usually found on loose sand or sandy soils .... (2)

2. Plants mostly less than 10 cm tall (occasionally up to 15 cm); stems not woody near base, branching throughout, branches short and compact; leaves not stiff and spreading; corollas pink .... **var. *arizonica***
2′ Plants mostly over 20 cm tall; stems more or less woody near base, mostly branching proximally, branches long and loose; leaves stiff and spreading; corollas white .... **var. *entrada***

var. *arizonica* (Tomb) S. L. Welsh (of Arizona) Arizona skeletonplant. Plants mostly low and compact, 2–15 cm tall. **STEMS** mostly herbaceous proximally, internodes mostly less than 10 mm (rarely to 20 mm), branched at base or throughout. **LEAVES** 2–6 cm long, lax (mostly curved and spreading), uppermost often exceeding involucres (rarely shorter and bractlike). **CALYCULI** much shorter than phyllaries (rarely up to 1/4 length of phyllaries). **PHYLLARIES** mostly 5, glabrous to puberulent (especially distally), margins entire, apices more or less keeled (but not usually appendaged). **RAY FLORETS** mostly 5; corollas mostly pink (sometimes pale) or white. $2n$ = 18. Mostly in sand or sandy soils. ARIZ: Apa, Nav; COLO: Mon; NMEX: McK, RAr, SJn; UTAH. 1500–1700 m (4800–6000′). Flowering: May–Jun. Arizona, Colorado, New Mexico, and Utah.

var. *entrada* (S. L. Welsh & Goodrich) S. L. Welsh (for the Entrada Formation) Arches skeletonplant. Plants mostly tall and loose, 20–45 cm tall. **STEMS** more or less woody proximally, internodes mostly 20–70 mm (rarely less), branched at base. **LEAVES** mostly 1–8 cm long, stiff (mostly straight and spreading), uppermost often reduced and bractlike, not exceeding involucres. **CALYCULI** graduating into the phyllaries (the two not significantly distinct). **PHYLLARIES** mostly 6, mostly lanate-tomentose (especially distally), margins entire to (often) erosulate, apices mostly appendaged. **RAY FLORETS** mostly 6; corollas white (drying pinkish). $2n$ = 27. Mostly in sandy soils of the Entrada Sandstone Formation. UTAH. 1400–1500 m (4600–4800′). Flowering: Jun. Restricted to eastern Utah, where it is of conservation concern. This variety reportedly rarely (if ever) sets seeds; it appears to be an asexually reproducing triploid.

var. *grandiflora* Plants low to tall, compact to loose, 10–40 cm tall. **STEMS** mostly herbaceous proximally, internodes 10 (or less) –80 mm, mostly few-branched distally. **LEAVES** 1–10 cm long, somewhat lax (mostly straight and erect), uppermost often reduced and bractlike, not exceeding involucres. **CALYCULI** much shorter than phyllaries (rarely up to 1/4 length of phyllaries). **PHYLLARIES** mostly 8–12 (rarely less), glabrous to puberulent (especially distally), margins entire, apices keeled to appendaged. **RAY FLORETS** mostly 8–12 (rarely less); corollas pink to lavender. $2n$ = 18.

Alluvium, clay, or sandy soil. ARIZ: Apa; COLO: LPl, Mon; NMEX: McK, RAr, SJn. 1600–2300 m (5300–7700′). Flowering: May–Jun. Wyoming, south to Utah, Colorado, and New Mexico.

Lygodesmia juncea (Pursh) D. Don ex Hook. (rushlike, alluding to the clustered, often leafless stems) Rush skeleton-plant. **STEMS** 20–35 cm tall, highly ramose with densely clustered branches, strongly sulcate-striate, glabrous. **LEAVES** cauline, mostly withered or absent at flowering; blades linear-subulate, to 2 cm long, mostly less than 1 mm wide, margins entire, obscurely whitish and muriculate, faces glabrous, progressively reduced upward or all reduced and scalelike. **HEADS** borne singly (at the apex of each branch); peduncle densely bracteate below heads. **INVOLUCRES** 7–14 mm tall. **CALYCULI** of reduced bractlets (all much shorter than phyllaries proper and grading into the scalelike bracts of peduncles), linear-subulate, keeled. **PHYLLARIES** mostly 5 in 1 series, lance-linear, glabrous, margins scarious, entire, apices not appendaged, midrib keeled. **RAY FLORETS** mostly 5, exceeding involucre; corolla glabrous, tube 6–10 mm, ligule 7–9 × 3–5 mm (often deeply lobed); anthers 4–5 mm. **CYPSELAE** 6–10 mm long, weakly ribbed (often aborted); apices transversely sulcate (directly below pappus). **PAPPUS** 6–10 mm, tawny. $2n = 18$. Ponderosa pine–oak–juniper and cottonwood communities, often disturbed sites. ARIZ: Apa; COLO: Arc, Mon; NMEX: McK, RAr, SJn. 1900–2900 m (6300–9500′). Flowering: Jul–Sep. British Columbia to Manitoba and Indiana, south to Nevada, Arizona, New Mexico, and Texas. This is primarily a prairie species that extends sporadically westward into the Rocky Mountain region.

Machaeranthera Nees Tansy-aster
G. L. Nesom

(Latin *machaera*, sword, + *anthera*, anther, alluding to the curved, sword-shaped anther appendages) Annuals or biennials, taprooted. **STEMS** usually stipitate-glandular, otherwise glabrous to puberulent or villous. **LEAVES** basal (withering by flowering) and cauline, alternate, simple, sessile, deeply 2-pinnatifid, with apiculate to spinulose teeth or lobes, not clasping. **INFL** a radiate head, borne singly on leafy peduncles. **INVOLUCRES** broadly turbinate to hemispheric. **PHYLLARIES** in 3–6 graduate series. **RECEPTACLES** flat to convex, shallowly alveolate. **RAY FLORETS** pistillate, fertile, in 1 series, lamina blue, violet, or purple. **DISC FLORETS** bisexual, fertile. **CYPSELAE** narrowly to broadly obovoid, 8–18-nerved, sparsely to densely strigose-sericeous. **PAPPUS** of persistent barbellate bristles in 1–3 series. $x = 4$. Two species, southwestern United States and Mexico. Most species generally identified as *Machaeranthera* over the last 50 years are mostly treated here within the segregate genera *Arida*, *Dieteria*, and *Xanthisma*, following Morgan and Hartman (2003) and related treatments in the *Flora of North America* volumes. Also see *Psilactis* and *Xylorhiza*.

Machaeranthera tanacetifolia (Kunth) Nees (*Tanacetum*-leaved) **STEMS** 5–100 cm tall, minutely glandular, usually also sparsely puberulent to villous-puberulent. **LEAVES** pinnatifid, sparsely stipitate-glandular, basal usually withered by flowering. **INVOLUCRES** 6–11 mm. **PHYLLARIES** linear-lanceolate to linear, apices long-acuminate, spreading to reflexed. $2n = 8$. Roadsides, grass and sagebrush flats, desert scrub, riparian, and ponderosa pine–cliffrose communities. COLO: Arc; NMEX: McK, RAr, SJn; UTAH. 1300–2100 m (4265–6890′). Flowering: (Jun–) Jul–Sep (–Oct). Montana south to California, Arizona, New Mexico, and Texas.

Madia Molina Tarweed
Brian A. Elliott

(Chilean name for members of the genus) Usually annual, occasionally biennial or perennial taprooted herbs; usually glandular and tar-scented. **STEMS** 1–several, often simple below, branched above. **LEAVES** cauline, opposite below, alternate above, simple, narrowly linear to lanceolate or oblong, entire to denticulate. **INFL** a radiate head or rarely discoid; solitary or racemose to cymose or corymbose, pedunculate. **INVOLUCRES** campanulate to ovoid, appearing angled due to keeled phyllaries. **PHYLLARIES** 1–20, free in 1 equal series, strongly carinate, completely enfolding ray achenes and falling with them. **RECEPTACLES** flat to convex, naked, hairy, or chaffy with a single series of bracts between ray and disc flowers. **RAY FLORETS** pistillate and fertile, rarely lacking; ligules yellow or purplish, sometimes minute and inconspicuous, cuneate, 2–3-lobed. **DISC FLORETS** perfect and fertile or staminate; corolla yellow. **CYPSELAE** with ray achene obovoid or clavate, laterally compressed with a sharp ventral angle, rounded dorsally; disc achene similar or abortive. **PAPPUS** none, a short crown, or a few scales. Eighteen species native to western North America, Chile, and Argentina.

Madia glomerata Hook. (a ball, referring to the shape of the heads) Tarweed. Annual taprooted herbs; tar-scented; densely glandular and hispid or villous, usually denser above. **STEMS** simple or ascending-branched above, 1.5–8

(12) dm tall. **LEAVES** linear to lance-linear, 2–9 cm long, 1.5–7 mm wide, loosely strigose and often bristly-ciliate with setose multicellular hairs, often with fascicles of leaves in the axils. **INFL** heads discoid or inconspicuously radiate, in dense to open paniculate clusters or cymes. **INVOLUCRES** fusiform to ovoid, 5.5–9 mm high, 1–5 mm wide. **RECEPTACLES** chaffy; chaff scales resembling phyllaries, borne in a single row between ray and disc flowers. **RAY FLORETS** with ligules 1–3.5 mm long, inconspicuous. **DISC FLORETS** fertile, rarely sterile, 1–12; corolla 3–4.5 mm long. **CYPSELAE** oblanceolate and curved, laterally compressed, 4–6 mm long, black, 5-nerved, glabrous. **PAPPUS** none. $2n = 28$. Roadsides, disturbed sites, and openings in sagebrush, montane shrubland, montane forests, and alpine meadows. ARIZ: Apa; COLO: Arc, Dol, Hin, LPl, Min, Mon, SJn. NMEX: McK, RAr. 2250–3000 m (7500–9600′). Flowering: Jul–Sep. Ranging from Alaska south to California, northern Arizona, and northern New Mexico. The seeds have been eaten raw, roasted, dried, or ground to meal. Edible oil has reportedly been made from the pressed seed. The stems and leaves have been used in steam baths as a treatment for venereal disease.

Malacothrix DC. Desert Dandelion

Gary I. Baird

(Greek *malakos*, soft, + *thrix*, hair, possibly alluding to the pappus) Annuals (ours) or perennials, taprooted, herbage glabrous, sometimes glaucous or pubescent, sap milky. **STEMS** 1–15, erect to reclining, branched. **LEAVES** basal (rosulate) and cauline, simple, alternate, sessile; blades obovate to spatulate or lanceolate, usually pinnately lobed, ultimate margins entire or dentate; lobes opposite, deltate to filiform. **INFL** borne singly (uncommon) or in corymbiform to paniculiform clusters, erect or nodding; peduncles bracteate, glabrous. **HEADS** liguliflorous. **INVOLUCRES** campanulate to hemispheric, calyculate or calyculi lacking. **PHYLLARIES** imbricate or subequal in about 2–6 series, herbaceous, erect. **RECEPTACLES** flat to convex, epaleate but sometimes bristly. **RAY FLORETS** matutinal or diurnal; corollas ligulate, yellow or whitish, sometimes outer reddish abaxially. **DISC FLORETS** lacking. **CYPSELAE** monomorphic, stramineous to purplish brown, ± cylindrical to fusiform (sometimes angled), glabrous, about 15-ribbed (sometimes 5 more pronounced and apically projected into a small crown around the pappus); beak lacking. **PAPPUS** lacking or of 1 series of basally connate and readily deciduous bristles, sometimes also an outer series of 0–6 persistent bristlelike scales and/or crown of teeth, white, barbellulate (ours) to plumose (proximally). $x = 7, 9$. About 20 species in western and southwestern North America, introduced into South America.

1. Basal leaves ± erect, pinnately lobed, lobes filiform, ultimate margins entire; persistent outer pappi of minute teeth plus 1–5 bristlelike scales ..... ***M. glabrata***

1′ Basal leaves ± prostrate, pinnately lobed, lobes deltate, ultimate margins denticulate; persistent outer pappi of minute teeth but lacking any longer, bristlelike scales ..... ***M. sonchoides***

Malacothrix glabrata (A. Gray ex D. C. Eaton) A. Gray (smooth or hairless) Smooth desert dandelion. Annuals, mostly 10–40 cm tall. **STEMS** glabrous or sparsely arachnose-pubescent proximally. **LEAVES**: basal and lower cauline blades pinnatifid, 5–10 × 2–8 cm, mostly erect, ultimate margins entire, faces glabrous or glabrate; lobes 2–6 pairs, filiform, 10–30 mm long, spreading; upper cauline reduced, pinnatifid to entire. **CALYCULI** lacking. **INVOLUCRES** mostly 10–15 mm tall. **PHYLLARIES** 20–25 in 2–3 series, lance-linear to linear, margins membranous, usually glabrous (sometimes pubescent). **RAY FLORETS** 30–100+, outer exceeding involucre; corolla yellow, tube about 8 mm, ± pubescent distally, ligule 8–12 mm × 2–3 mm, glabrous; anthers about 2 mm long. **CYPSELAE** body ± cylindrical (or somewhat 5-angled), 2–3 mm long; ribs uniform or 5 major plus about 10 minor, ± apically forming a low crown. **PAPPUS** about 8 mm long, persistent outer scales usually 2–4 (rarely lacking) plus several smaller teeth (or not). $2n = 14$. Blackbrush and piñon-juniper communities, often in sandy soils. Occurring on the border of the study area in ARIZ: Coc, Nav; UTAH. 1800 m (6000′). Flowering: Mar–Jul. Oregon and Idaho, south to California, Arizona, New Mexico, and northern Mexico.

Malacothrix sonchoides (Nutt.) Torr. & A. Gray (*Sonchus*-like) Sowthistle desert dandelion. Annuals, mostly 10–25 cm tall. **STEMS** glabrous or sparsely stipitate-glandular. **LEAVES**: basal and lower cauline blades oblong to elliptic, pinnately lobed, about 3–10 × 1–3 cm, mostly prostrate, ultimate margins denticulate, faces glabrous; lobes 3–9 pairs, deltate or rounded, about 2–3 mm long, spreading; upper cauline reduced, dentate to subentire. **CALYCULI** of 8–12 bractlets in about 2 series, lance-ovate. **INVOLUCRES** 7–10 mm tall. **PHYLLARIES** 12–14 in 1–2 series, lanceolate, margins membranous, often denticulate and/or sparsely stipitate-glandular, glabrous. **RAY FLORETS** 75–100+, outer exceeding involucre; corolla yellow, tube about 4–5 mm, glabrous, ligule 5–10 mm × about 1–2 mm, glabrous; anthers about 2 mm long. **CYPSELAE** body ± cylindrical (or somewhat 5-angled), 2–3 mm; ribs uniform or 5 major plus about

10 minor, apically forming a low crown. **PAPPUS** about 5 mm, persistent outer series an obscure, low crown (longer, persistent scales lacking). $2n = 14$. Sand dunes or in deep sandy soils. ARIZ: Apa, Nav; COLO: Mon; NMEX: McK, San, SJn; UTAH. 1300–1500 m (4300–4800′). Flowering: May–Jun. California eastward to Wyoming, Colorado, and New Mexico.

Matricaria L. False Chamomile, Mayweed
Brian Elliott

(Latin *matrix*, womb, for medicinal use of the plant in medieval times to treat afflictions of the uterus) Annual, biennial, or perennial taprooted and often aromatic herbs. **STEMS** erect to decumbent, branched and leafy. **LEAVES** alternate, 2–3-finely pinnately dissected, ultimate segments linear to linear-filiform, petiole short or none. **INFL** heads radiate or discoid, solitary at branch tips or corymbose. **PHYLLARIES** imbricate in 2–3 series, oblong to oval, margins broadly hyaline to scarious, apex obtuse. **RECEPTACLES** naked, hemispheric to conical, hollow. **RAY FLORETS** pistillate and fertile; ligules white or lacking. **DISC FLORETS** perfect and fertile, numerous; corolla green or yellow; 4–6-toothed, tube swollen in fruit. **CYPSELAE** cylindrical, 3–5-nerved on the margins and ventrally, nerveless dorsally, glabrous or rugose. **PAPPUS** a narrow crown or none. Approximately 35 species from North America, Europe, and South Africa. Several species have been introduced and are often weedy in North America. Members of the genus are sometimes placed in *Tripleurospermum* based on the number of seed ribs.

1. Head discoid; disc corolla 4-toothed; receptacle conic; achenes rugose, ribs neither thickened nor winglike .... ***M. discoidea***
1′ Head radiate with white rays; disc corolla 5-toothed; receptacle hemispheric; achenes smooth, ribs callous-thickened and winglike .... ***M. maritima***

Matricaria discoidea DC. (discoid, referring to the head) Rayless chamomile, disc mayweed. Annual herbs; pineapple-scented. **STEMS** erect, branched below, 0.5–5 dm tall. **LEAVES** 1–3-pinnatifid, 1–5 cm long. **INFL** heads discoid. **RECEPTACLES** strongly conic, pointed at apex, 4–10 mm wide, hollow. **RAY FLORETS** lacking. **DISC FLORET** corollas 4-toothed. **CYPSELAE** with truncate tip, 3–5-veined with 2 well developed marginal ribs and 1–3 sometimes weak ventral ribs with linear brown glands extending nearly the full length; surface smooth. **PAPPUS** a short crown, minute, or none. $2n = 18$. [*Artemisia matricarioides* Less.; *Chamomilla discoidea* J. Gay ex A. Braun; *Chamomilla suaveolens* Rydb.; *Chrysanthemum suaveolens* Asch.; *Cotula matricarioides* Bong.; *Lepidanthus suaveolens* Nutt.; *M. matricarioides* (Less.) Porter; *M. suaveolens* (Pursh) Buchenau non L.; *Santolina suaveolens* Pursh; *Tanacetum suaveolens* Hook.]. Disturbed ground and roadsides, often on moist or sandy ground. COLO: LPl, SJn. 2100–2750 m (7000–9000′). Flowering: May–Sep. Native in western North America, and introduced and naturalized in eastern North America. The species has seen extensive medicinal use as a laxative, cold remedy, cure-all tonic, and for stomach pain, fevers, and diarrhea. Pulverized, it has been used as a preservative in meat and berries. The flower heads have been eaten raw, dried for tea, or used as seasoning. Its pleasant aroma has led to use as a perfume, lining of baby cradles, and insect repellent.

Matricaria maritima L. (of the sea, referring to the typical European habitat of the species) Wild chamomile, scentless chamomile. Annual, biennial, or short-lived perennial herbs; scentless. **STEMS** 1–8 dm tall, glabrous or sparsely pubescent. **LEAVES** bipinnatifid, 1–8 cm long. **INFL** heads radiate. **RECEPTACLES** rounded to hemispheric, 8–15 mm wide. **RAY FLORETS** 10–25 per head; ligules 6–13 mm long. **DISC FLORETS** 5-toothed. **CYPSELAE** with 2 marginal and 1 ventral rib, callous-thickened and winglike; surface rugose on the back and between ribs. **PAPPUS** a low crown. $2n = 18, 36$. [*Chamomilla maritima* Rydb.; *Chrysanthemum inodorum* L.; *M. chamomilla* L.; *M. inodora* L.; *M. perforata* Mérat; *Tripleurospermum inodorum* (L.) Sch. Bip.; *Tripleurospermum maritimum* (L.) W. D. J. Koch subsp. *phaeocephalum* (Rupr.) Hämet-Ahti]. Roadsides, weedy sites, ditch banks, riparian areas, and floodplains. COLO: Arc, LPl. 1830–2350 m (6000–7700′). Flowering: Jun–Sep. Native to Eurasia, introduced and now widespread in North America. The flower heads have been eaten raw, dried for tea, or used as seasoning.

Microseris D. Don Silverpuffs
Gary I. Baird

(Greek *micros*, small, + *seris*, chicory) Annuals or perennials, taprooted, herbage glabrous or scurfy-pubescent, sap milky. **STEMS** 1 to many, simple or branched, erect. **LEAVES** mostly basal, cauline reduced or lacking, alternate, simple, petiolate; blades linear to lanceolate or oblanceolate, margins entire to dentate or pinnately lobed, apices acuminate to acute or obtuse. **INFL** borne singly, nodding or inclined (in bud) to erect (in flower and fruit); peduncles ebracteate or leafy. **HEADS** liguliflorous. **PHYLLARIES** deltate to lanceolate, imbricate in 3–4 series, herbaceous. **RECEPTACLES** flat, epaleate; calyculi lacking. **RAY FLORETS** matutinal; corolla ligulate, yellow (outer often abaxially rosy-purple). **DISC FLORETS** lacking. **CYPSELAE** ± monomorphic, gray to brown or purplish, columnar to fusiform or obconic, ribs 10–15, glabrous or scabrous, beak lacking. **PAPPUS** of aristate scales, persistent, silvery white to ochroleucous or brownish; aristae barbellulate to plumose. $x = 9$. Fourteen species in western North America, South America, Australia, and New Zealand (Chambers 1955).

Microseris nutans (Hook.) Sch. Bip. (nodding) Nodding silverpuffs. Perennials 10–60 cm tall. **STEMS** simple or branched, erect to decumbent. **LEAVES** mostly basal or lower cauline, blades linear to oblanceolate, 5–30 cm, margins entire, remotely dentate, or pinnately lobed, apices acuminate, faces glabrous or sparsely scurfy-pubescent; cauline reduced, often sessile and clasping; lobes linear. **PEDUNCLES** ebracteate or leafy, glabrous or scurfy-pubescent distally (beneath head). **INVOLUCRES** mostly 10–20 mm tall, nodding in bud, erect in flower and fruit. **PHYLLARIES** imbricate, glabrous or scurfy-puberulent and usually also black-villous; outer lanceolate to deltate or linear, apices acute to acuminate; inner lanceolate, apices acuminate. **RAY FLORETS** 10–75, outer surpassing involucre. **CYPSELAE** ± columnar, 4–8 mm. **PAPPUS** of 15–30 aristate scales, silvery white; scale bodies linear to lanceolate, mostly 1–3 mm; aristae plumose. $2n = 18$. Disturbed area in sagebrush communites. COLO: Mon. 1800 m (5900′). Flowering: May. British Columbia and Alberta south to California, Nevada, Utah, and Colorado. Known in the study area by a single collection taken near Yucca House National Monument.

Mulgedium Cass. Blue Lettuce

Gary I. Baird

(Latin *mulgere*, to milk, referring to the milky sap) Perennials, rhizomatous, herbage glabrous, sap milky. **STEMS** usually 1, erect, branched distally. **LEAVES** basal and (mostly) cauline, simple, alternate; basal petiolate, cauline sessile, not auriculate; blades linear to lanceolate, or linear-oblanceolate, margins entire to dentate or pinnately lobed, apices acute to acuminate; lobes opposite. **INFL** borne in paniculiform clusters, erect. **HEADS** liguliflorous, borne in paniculiform clusters, erect; peduncles bracteate, glabrous. **INVOLUCRES** cylindric. **PHYLLARIES** strongly graduated (outermost more or less calyculate), herbaceous, erect. **RECEPTACLES** flat, epaleate. **RAY FLORETS** matutinal; corolla ligulate, blue (ours). **DISC FLORETS** lacking. **CYPSELAE** monophorphic, light brown to red-brown, more or less flattened, ellipsoid to lanceoloid, glabrous, ribbed, beakless or shortly beaked. **PAPPUS** of bristles, persistent, white, more or less barbellulate. $x = 9$. About 15 species in Asia, Europe, and North America.

Mulgedium pulchellum (Pursh) G. Don (beautiful) **STEMS** 20–100 cm tall, glabrous. Blue lettuce. **LEAVES** with blades mostly linear to lanceolate, lower and middle cauline 5–15 cm × 4–12 mm (excluding lobes), not much reduced upward until inflorescence, margins of basal and lower cauline mostly dentate to pinnately lobed, upper cauline mostly entire, faces glabrous and (usually) glaucous; lobes 3–4 pairs, linear-lanceolate, 1–14 mm long, retrorse. **HEADS** 2–30 (or more, proliferating with age). **INVOLUCRES** 10–15 mm tall. **PHYLLARIES**: outer 8–12, lance-deltate, erect, less than 1/2 length of inner and ± calyculate or completely grading into inner; inner 9–12 in 2 series, lanceolate, green proximally, purplish distally, glabrous. **RAY FLORETS** about 17–50, outer exceeding involucre; corolla tube 5–8 mm, pilose distally with often purplish hairs, ligule 9–12 mm × 2–3 mm, glabrous; anthers 4–5 mm. **CYPSELAE** 4–7 mm long (including beak), somewhat flattened, more or less ellipsoid, tapered to stout beak; body 3–5 mm long, tan to reddish brown; ribs 4–6 per face, muricate-rugose; beak 1–2 mm (less than 1/2 length of body), stout, whitish. **PAPPUS** 7–9 mm. $2n = 18$. [*Lactuca pulchella* (Pursh) DC.; *L. tatarica* (L.) C. A. Mey. subsp. *pulchella* (Pursh) Stebbins]. Moist to dry habitats, often disturbed sites and roadsides, in riparian, piñon-juniper, to spruce-fir communities. ARIZ: Nav; COLO: Arc, Hin, LPl, Min, Mon, SJn; NMEX: RAr, SJn; UTAH. 1600–2500 m (5200–8200′). Flowering: Jun–Sep. Alaska to Maine, south to California, Texas, Louisiana, Alabama, and Pennsylvania (absent from southeastern United States). Traditionally included in *Lactuca*.

Onopordum L. Scotch Thistle

Patricia Barlow-Irick

(Greek *onopordon*, name for cotton thistle) Biennial, 5–40+ dm, coarse, prickly, usually cottony, woolly, or felted, sometimes short and glandular. **STEMS** 1–several from the base, spiny-winged from decurrent leaf bases. **LEAVES** alternate; winged-petiolate (basal) or sessile (cauline); subentire to pinnately lobed; lobes and teeth spine-tipped. **HEADS** solitary or clustered at the ends of branches. **INVOLUCRES** hemispheric to ovoid or spheric. **PHYLLARIES** in 8–10+

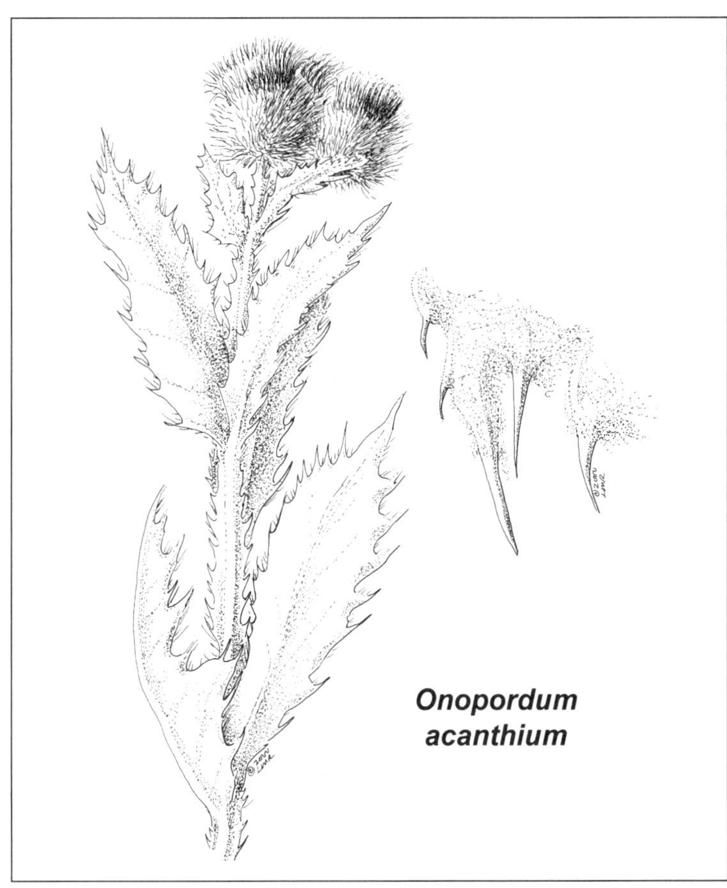

series, linear to ovate, entire, tapered to stiff spines, middle and outer often spreading or reflexed. **RECEPTACLES** epaleate, not bristly, enclosing bases of achenes in pits with dentate margins. **DISC FLORETS** many, all tubular; corollas white or purple, style tips with minutely hairy nodes scarcely divided. **CYPSELAE** ± cylindric, 4–5-angled, usually ± transversely roughened, glabrous, attachment scars basal. **PAPPUS** many barbed or plumose bristles united at base into a deciduous ring. x = 17. About 40 species, Europe, North Africa, West Asia.

Onopordum acanthium L. (spiny) Scotch or heraldic thistle. Taprooted biennial. **STEMS** branched above, covered with dense, white felted tomentum, 0.5–3 m tall, with broad, dentate, spiny wings 2–3 (5) cm wide. **LEAVES** deltoid to broadly elliptic, pinnatifid with triangular lobes, coarsely dentate, gray-green, basal leaves to 50 cm long, cauline leaves much smaller, with dense, fine felted tomentum above and beneath; prickles marginal, yellowish, spreading, 5–10 mm long; capitula broadly ovoid, erect, 2–4 (–5) cm wide, solitary; peduncles elongating just before anthesis. **PHYLLARIES** linear-subulate, 2–3 mm wide at base, the outer and middle with dense cobwebby tomentum; apex spinose, spreading to suberect. **FLORETS** with corolla purple or white, 20–24 mm long; divided into an 8–12 mm long tube, 3.5–4.5 mm long throat, and 5–7 mm long distinctly unequal lobes; style tip 3.5–5 mm beyond node. **CYPSELAE** dark, narrowly obovoid to clavate, weakly transversely flattened and 4-angled, rugose. **PAPPUS** with scabrid bristles, 5–10 mm long. Disturbed sites, roadsides, fields, on fertile soils, and natural areas. ARIZ: Apa, Nav; COLO: LPl, Mon; NMEX: RAr, SJn. 1645–2135 m (5400–7000′). Flowering: Jun–Jul. Native to Europe and now worldwide. Survives in dry conditions as a small herb.

Oreochrysum Rydb. Parry's Goldenweed
G. L. Nesom

(Greek *oreios*, of mountains, + *chrysos*, gold) Perennial herbs 15–60 (–100) cm; rhizomes long, slender, scale-leaved, thickening, lignescent. Stems minutely puberulous or hirtellous, with mixture of stipitate-glandular and nonglandular hairs. **LEAVES** basal and cauline (basal and proximal cauline persistent), alternate, petiolate to subpetiolate, basal and proximal cauline blades spatulate-oblanceolate, middle and distal elliptic to broadly ovate-lanceolate or oblanceolate, margins entire, minutely short-stipitate-glandular or gland-dotted, viscid. **INFL** a radiate head, in distinctly flat-topped, tightly corymbiform arrays. **INVOLUCRES** campanulate to hemispheric, 10–11 × 6–8 mm. **PHYLLARIES** in 3–4 series, strongly unequal to subequal, mostly appressed, outer lanceolate to ovate, inner broadly lanceolate-oblong, herbaceous, apices green-tipped and erect to reflexing, minutely stipitate-glandular, glabrous or hirtellous. **RECEPTACLES** flat, alveolate, epaleate. **RAY FLORETS** 12–20, pistillate, fertile; corolla yellow. **DISC FLORETS** bisexual, fertile. **CYPSELAE** fusiform, plump but distinctly compressed, nerves 12–16, whitish, raised, glabrous. **PAPPUS** of 40–60 equal, persistent, barbellate bristles in 2 (–3) series. x = 9. [*Haplopappus* sect. *Oreochrysum* (Rydb.) H. M. Hall; *Solidago* subg. *Oreochrysum* (Rydb.) Semple]. Species one; western North America.

Oreochrysum parryi (A. Gray) Rydb. (for Charles C. Parry, 1823–1890, surgeon-naturalist and prolific botanical collector with the Mexican Boundary Survey during the years 1849–1852) Parry's goldenweed. **STEMS** often purple proximally. **LEAVES** (3–) 6–15 cm long, sometimes slightly succulent, clasping or subclasping, less commonly nonclasping, relatively unreduced up to capitulescence, distal sometimes grading into phyllaries. **RAY FLORET** corollas 6–10 mm. **DISC**

FLORET corollas 7–9 mm. $2n = 18$. [*Aplopappus parryi* A. Gray; *Solidago parryi* (A. Gray) Greene]. Moist to dry meadows, tundra, clearings, roadsides, wooded slopes, open woods, often in partially shaded understory; ponderosa pine–Gambel's oak, spruce-fir often with aspen, white fir–Douglas-fir, and bristlecone pine communities. ARIZ: Apa; COLO: Arc, Hin, LPl, Min, Mon, RGr, SJn; NMEX: RAr, SJn. (2250–) 2500–3450 (–3900) m (7380–12795′). Flowering: Jul–Sep (–Oct). Arizona, Colorado, Utah, and Nevada to New Mexico and Wyoming; Mexico (Chihuahua).

Oxytenia Nutt. Copperweed
Gary I. Baird

(Greek *oxy*, pointed or sharp, + *tenia*, band or ribbon, referring to the leaves) Perennials (becoming woody proximally), taprooted, rhizomes lacking, herbage white-strigillose (or glabrate), usually gland-dotted, sap clear. **STEMS** usually several, erect, simple (only branched distally in the inflorescence), striate. **LEAVES** cauline, simple, alternate, petiolate or sessile; blades pinnatifid (proximal) to unlobed (distal); lobes (or unlobed leaves) linear to filiform, ultimate margins entire. **INFL** in paniculiform clusters (with 1–5 heads per node), nodding; sessile or obscurely pedunculate. **HEADS** disciform, ± ebracteate. **INVOLUCRES** hemispheric. **PHYLLARIES** about 10–15 in ± 3 series, distinct, outer herbaceous, ± gland-dotted, inner membranous and palealike. **RECEPTACLES** convex, paleate; paleae spatulate to cuneiform, membranous, ± pubescent (distally), sometimes reduced and obscure; calyculi lacking. **PISTILLATE FLORETS** 5, corollas lacking, ovaries densely white-villous (hairs longer than corollas and conspicuous at flowering). **STAMINATE FLORETS** 10–25, corollas whitish or pale yellowish, gland-dotted; filaments connate, anthers distinct (or weakly coherent). **CYPSELAE** monomorphic, blackish, obovoid, ± obcompressed and angled (angles ribbed), smooth, villous, ± eglandular. **PAPPUS** lacking. $x = 18$. One species of western North America.

Oxytenia acerosa Nutt. (rigid or needlelike) Plants 50–200 cm tall. **LEAVES** with petioles to 20+ mm; blades or lobes 20–150 × 1–2 mm. **HEADS** with paleae about 2 mm long. **INVOLUCRES** 2–3 mm. **PHYLLARIES**: outer ovate, apices acute to short-acuminate, sometimes 1–2 outermost reduced and proximal to the others. **CYPSELAE** about 2 mm long. $2n = 36$. [*Iva acerosa* (Nutt.) R. C. Jacks.]. Riparian communities, often in alkaline soils. ARIZ: Apa, Nav; COLO: Mon; NMEX: SJn; UTAH. 1300–1500 m (4400–4900′). Flowering: Sep. California eastward to Colorado and New Mexico.

Packera Á. Löve & D. Löve Ragwort, Groundsel
Debra Trock

(for John G. Packer, Canadian botanist) Herbs, annual, biennial, or perennial from rhizomatous or taprooted caudices. **STEMS** erect, single or several loosely clustered from caudices. **LEAVES** mostly from basal rosettes, petiolate to rarely sessile, blades simple, margins entire to pinnatifid or lobed; cauline leaves few, alternate, gradually to abruptly reduced distally, short-petiolate becoming sessile, blades variously dissected to lobed. **INFL** compact to open compound corymbs or cymes of heads, occasionally a single head; heads radiate or eradiate. **INVOLUCRES** campanulate to cylindrical. **PHYLLARIES** 8, 13, or 21 in a single series, green or reddish, free to the base, with scarious margins. **RAY FLORETS** 5–13, pistillate and fertile; corollas pale yellow to deep orange-red, tubes 1–4 mm, laminae 5–15 mm. **DISC FLORETS** 25–80+, perfect, corollas pale yellow to deep orange-red, tubes 1.5–5 mm gradually expanded to campanulate limbs 1.5–6 mm long, with 5 shallow lobes. **CYPSELAE** cylindrical, 0.75–3.5 mm long, glabrous or short-hirsute along the angles. **PAPPUS** of fine white bristles. Sixty-four species; North America from the Arctic to the Gulf of Tehuantepec in southern Mexico, one or two species extending into eastern Siberia.

1. Heads with ray florets absent or with laminae less than 4 mm long .....(2)
1′ Heads with ray florets 5–13, laminae 5–10+ mm long .....(4)

2. Plants 2–5 dm tall; inflorescence of 6–20+ heads ..... ***P. debilis***
2′ Plants 0.5–1.5 dm tall; inflorescence a single large head or 2–6 much smaller ones .....(3)

3. Stems 3–6+ cm tall; heads eradiate, phyllaries purple to deep reddish; cypselae 3–3.5 mm long, hirtellous; plants of middle elevations ..... ***P. spellenbergii***
3′ Stems 7–15+ cm; heads normally radiate, phyllaries green or with reddish tips; cypselae 1.5–2 mm long, glabrous; plants near or above timberline ..... ***P. werneriifolia***

4. Stem and leaves glabrous or at most with tufts of pubescence at the base of the stem or in the axils of the leaves .....(5)
4′ Stem and leaves distinctly pubescent or becoming glabrous on upper surfaces with age .....(14)

5. Basal leaves with margins lobed or pinnatifid .... (6)
5′ Basal leaves with margins entire to dentate, crenate, or serrate .... (8)

6. Leaf margins weakly lobed .... ***P. streptanthifolia***
6′ Leaf margins distinctly pinnatifid .... (7)

7. Plants with fibrous-rooted rhizomes; leaf blades with margins pinnate, 2–4 pairs of lateral lobes; terminal lobe large, ovate to reniform, base abruptly contracted; damp meadows in spruce-aspen forests ... ***P. sanguisorboides***
7′ Plants taprooted; leaf blades with margins pinnatifid or lyrate, 5–10 pairs of lateral lobes; terminal lobe normally smaller than lateral lobes, or if larger, base gradually tapering; dry, rocky or sandy soils in sagebrush or ponderosa pine forests .... ***P. multilobata***

8. Plants less than 15 cm tall; cauline leaves absent, plants scapose, or cauline leaves reduced to bracts; inflorescence a single large terminal head or at most 2–6 heads .... (9)
8′ Plants usually more than 15 cm tall; cauline leaves gradually reduced upward; inflorescence 5–20+ heads ..... (10)

9. Basal leaves petiolate, with blades distinctly reniform, margins crenate, bases abruptly truncate to cordate, lower surface reddish; caudices with long, slender, branching rhizomes .... ***P. porteri***
9′ Basal leaves sessile or short-petiolate, with blades elliptic or ovate, margins entire, wavy or dentate toward the apex, bases attenuate, lower surfaces not reddish; caudices branching and subrhizomatous or woody .... ***P. werneriifolia***

10. Upper cauline leaves sessile and weakly clasping to distinctly clasping and auriculate; ray florets deep yellow to orange or brick-red .... (11)
10′ Cauline leaves petiolate or sessile, but never clasping or auriculate; ray florets pale to bright yellow .... (12)

11. Plants normally <3 dm tall; leaf blades thick and turgid; lower, middle, and upper cauline leaves sessile, conspicuously clasping and auriculate; upper leaves as large or larger than basal leaves; inflorescence congested, usually fewer than 8 heads; plants mostly collected at over 2740 m (9500′) .... ***P. dimorphophylla***
11′ Plants normally >3 dm tall; leaf blades thin, not turgid; lower and middle leaves sessile or petiolate, may be weakly clasping; upper leaves smaller than basal leaves; inflorescence an open corymb of more than 8 heads; plants mostly collected below 2740 m (9500′) .... ***P. crocata***

12. Basal leaves thick and turgid, blades ovate, obovate, or orbicular; phyllaries dark green, often with reddish tips; conspicuous bracts present at the base of phyllaries .... ***P. streptanthifolia***
12′ Basal leaves not noticeably thickened, blades elliptic to broadly lanceolate; phyllaries light green; bracts on phyllaries absent or small and inconspicuous .... (13)

13. Leaf blades broadly lanceolate to ovate, base of leaf blades truncate to subcordate or rarely obtuse; lower cauline leaves sublyrate to subentire; middle cauline leaves conspicuous, with margins deeply incised .... ***P. pseudaurea***
13′ Leaf blades lanceolate to narrowly elliptic, base of leaf blades tapering to cuneate or obtuse, but never subcordate; lower cauline leaves often sublyrate or with margins variously dissected, but rarely subentire; middle and upper cauline leaves mere sessile bracts .... ***P. paupercula***

14. Stems mostly shorter than 15 cm; inflorescence of 1–6 heads .... (15)
14′ Stems mostly taller than 15 cm; inflorescence of 6–25+ heads .... (16)

15. Cauline leaves absent, plants distinctly scapose; basal leaves lightly to heavily pubescent; inflorescence of 1–6 heads .... ***P. werneriifolia***
15′ Cauline leaves much reduced but present; basal leaves sparsely to densely woolly at least on the lower surface; inflorescence of 6–25+ heads .... ***P. cana***

16. Leaf margins distinctly pinnatifid or lobed .... (17)
16′ Leaf margins entire to serrate or dentate .... (18)

17. Caudices taprooted; stems often clustered into groups of 2–5, base of stems frequently reddish; leaf blades with margins dinstinctly lobed or lyrate, petioles and lower surface of leaf blades frequently reddish, terminal lobe usually larger than lateral lobes .... ***P. multilobata***

17′ Caudices rhizomatous with lateral branches; stems normally single, reddish color not present on stems or leaves; leaf blades with margins regularly and evenly pinnatifid or merely wavy, terminal lobe not distinctly larger .... ***P. fendleri***

18. Caudices subligneous; stems with arachnoid hairs in leaf axils; leaves thick and leathery, narrowly lanceolate or oblanceolate; cauline leaves large and numerous .... ***P. cynthioides***

18′ Caudices taprooted or fibrous-rooted, rarely woody; stems with simple hairs in leaf axils; leaves not at all leathery, blades lanceolate to elliptic or ovate; cauline leaves progressively reduced up the stem .... (19)

19. Plants densely canescent (grayish in color from a covering of hairs); margins of basal leaves entire to wavy; achenes 2.5–3.5 mm long and glabrous .... ***P. cana***

19′ Plants lightly to densely woolly, upper leaf surface becoming glabrous with age; margins of basal leaves serrate to dentate; achenes less than 2.5 cm long, glabrous or hirtellous .... (20)

20. Caudices taprooted; stems and leaf blades densely woolly or glabrous with age, lower surface of leaves frequently reddish; small cymes often arising from upper leaf axils; achenes hirtellous, pappus >5 mm long... ***P. neomexicana***

20′ Caudices fibrous-rooted or stoloniferous; stems and leaf blades lightly woolly, lower surface never reddish; inflorescences without cymes from leaf axils; achenes glabrous, pappus 3.5–4.5 mm long .... ***P. hartiana***

Packera cana (Hook.) W. A. Weber & Á. Löve (grayish white) Woolly groundsel. Perennial, caudices stout, branching-rhizomatous, suberect. **STEMS** 8–40 cm tall; single from basal rosette, or clustered rosettes; woolly. **LEAVES** basal, petiolate, blades ovate to elliptic or lanceolate, 2.5–5 cm long, 1–3 cm wide, bases attenuate, margins entire or irregularly undulate to weakly dentate, persistently woolly on lower surface, woolly to subglabrous on upper surface; cauline leaves petiolate to sessile and weakly clasping, progressively reduced distally, elliptic to lanceolate, margins entire to weakly dentate. **INFL** a corymbiform cyme of 4–14 (20+) heads; heads radiate, 8–12 (14) mm long and 6–11 mm wide, pedicels densely woolly. **PHYLLARIES** 13 or 21, green, 5–8+ mm long, woolly. **RAY FLORETS** 8–10 or occasionally 13, laminae 8–10 mm long. **DISC FLORETS** 35–50+; corolla tube 2.5–3.5 mm long, limb 3.5–4.5 mm long. **CYPSELAE** 2.5–3.5 mm long, glabrous. **PAPPUS** 4.5–7 mm long. $2n = 46$. [*Senecio canus* Hook.; *S. purshianus* Nutt.; *S. howellii* Greene]. Open high plains and in sagebrush associations, dry, rocky slopes and crevices on limestone- and granitic-derived soils. COLO: Arc, Con, Hin, LPl, Min, Mon, SJn; UTAH. 1200–3700 m (3900–12000′). Flowering: Jun–Jul. British Columbia east to Manitoba, south to Nebraska and west to California. Found in a wide variety of habitats throughout its range. Alpine specimens are noticeably dwarfed and often mistaken for *P. werneriifolia*, which is distinguished by its scapose aspect.

Packera crocata (Rydb.) W. A. Weber & Á. Löve (saffron yellow) Saffron ragwort. Perennial, caudices long, stout, horizontal to ascending. **STEMS** 2–6 dm tall; single or rarely 2–3 clustered, glabrous. **LEAVES** basal, thin-petiolate, blades narrowly lanceolate or ovate to obovate, 2–6 cm long, 1–4 cm wide, bases attenuate to occasionally rounded, margins entire to crenate-dentate, glabrous; cauline leaves petiolate to sessile and weakly clasping, progressively reduced distally, lanceolate to oblong, margins sublyrate or lobed. **INFL** an open corymb of 7–15+ heads; heads radiate, 7–9 mm long and 7–10 mm wide, pedicels glabrous. **PHYLLARIES** 13 or 21, light green or with reddish tips, 4–8 mm long, glabrous. **RAY FLORETS** 8 or 13, laminae 6–8 mm long, deep yellow to orange-red. **DISC FLORETS** 60–80+; corolla tube 4.5–5.5 mm long, limbs 2.5–3.5 mm long. **CYPSELAE** 1–1.5 mm long, glabrous. **PAPPUS** 3–5 mm long. $2n = 46$. [*Senecio crocatus* Rydb.]. Wet meadows in grass/sedge/willow associations, along trails and in rocky outcrops. COLO: Arc, Con, Dol, Hin, LPl, Min, RGr, SJn, SMg; NMEX: RAr, San; UTAH. 2100–3000 m (6900–9850′). Flowering: Jul–Aug. Idaho east to Wyoming, south to northern New Mexico and west through Utah and northeastern Nevada.

Packera cynthioides (Greene) W. A. Weber & Á. Löve (doglike) White Mountain ragwort. Perennial, caudices subligneous, rhizomatous, horizontal to erect. **STEMS** 2–5 dm tall, single or 2–3 clustered, densely woolly or canescent with tufts of arachnoid hairs in leaf axils, occasionally becoming glabrous with age. **LEAVES** basal, petiolate, blades thick and leathery, narrowly lanceolate or oblanceolate, 2.5–10 cm long, 0.5–2 cm wide, bases attenuate, margins entire, subentire, dentate, or wavy, upper surface early-glabrate; cauline leaves abundant and only gradually reduced proximally, sessile and numerous, lanceolate to oblanceolate, margins entire to wavy. **INFL** a congested or rarely open cyme of 10–30+ heads; heads radiate, 8–10 mm long and 8–10 mm wide, pedicels woolly. **PHYLLARIES** 13 or occasionally 8, green with red tips, 3–6 mm long, glabrous distally. **RAY FLORETS** 8 or rarely 5, laminae 8–10 mm long. **DISC FLORETS** 35–45+; corolla tube 2.5–3.5 mm long, limb 3.5–4.5 mm long. **CYPSELAE** 1–1.5 mm long, glabrous. **PAPPUS** 5–6 mm long. $2n = 46$. [*Senecio cynthioides* Greene; *S. wrightii* Greenm.]. Subalpine pine and juniper forests in rocky soils on steep slopes. NMEX: SJn. 2200–2900 m (7200–9500′). Flowering: Jul–Sep. Narrowly endemic to higher elevations in west-central New Mexico. Blooming later in the season than other *Packera* species at this latitude. Noted by collectors as usually growing on north- or west-facing slopes in limestone-derived soils.

Packera debilis (Nutt.) W. A. Weber & Á. Löve (weak, feeble) Weak groundsel. Perennial, caudices short, weakly branching. **STEMS** 2–5 dm tall, single or 2–4 clustered, base of stem and axils of leaves woolly. **LEAVES** basal, petiolate and turgid, blades elliptic to ovate or subreniform, 2–4 cm long and 1.5–3 cm wide, bases attenuate to rounded, margins entire to crenate-dentate; cauline leaves progressively reduced distally, sessile, blades pinnate-lobed, sinuses deep and rounded, lobes with entire to subentire margins. **INFL** open to slightly congested corymbiform cymes of 6–20 heads; heads eradiate, 7–9 mm long and 7–9 mm wide, pedicels glabrous or lightly pubescent. **PHYLLARIES** 13 or 21, green with reddish tips, 6–8 mm long, glabrous. **RAY FLORETS** absent. **DISC FLORETS** 45–65+; corolla tubes 3–4 mm long, limb 2–3 mm long. **CYPSELAE** 1–2 mm long, glabrous. **PAPPUS** 4–5 (–5.5) mm long. $2n = 46$. [*Senecio debilis* Nutt.; *S. fedifolius* Rydb.]. Open meadows, alkaline soils at middle to subalpine elevations. COLO: Con. 1700–3000 m (5600–9850′). Flowering: Jun–Aug. Southern Rocky Mountains from Montana and Idaho through Wyoming and Colorado. Usually collected in wet alkaline meadows.

Packera dimorphophylla (Greene) W. A. Weber & Á. Löve (two leaf shapes) Splitleaf groundsel. Perennial, caudices short, stout, fibrous-rooted, horizontal to erect. **STEMS** 1–3+ dm tall, single or rarely 2–3 clustered, glabrous. **LEAVES** basal, petiolate, thick and turgid, blades varying from ovate to subreniform or occasionally oblanceolate, bases tapering or subcordate to abruptly contracted, petioles distinctly winged or thin and delicate, margins entire to crenate; lower and middle cauline leaves sessile, conspicuously clasping and auriculate, blades as large or larger than basal leaves, oblanceolate, obovate, or lyrate, margins subentire to bluntly and irregularly dissected; upper cauline leaves reduced to sessile bracts, or rarely large and conspicuous. **INFL** a congested corymbiform cyme of 1–8+ heads; heads radiate, 1–8 mm long and 8–10 mm wide, pedicels glabrous. **PHYLLARIES** 13 or 21, rarely fewer, green with reddish tips, 5–7 mm long, glabrous. **RAY FLORETS** 8 or 13, laminae 5–8 mm long, deep yellow to orange. **DISC FLORETS** 45–60+, corolla tube 2–3 mm long, limb 2.5–3.5 mm long. **CYPSELAE** 0.75–1.5 mm long, glabrous. **PAPPUS** 3–4 mm long. $2n = 46$.

1. Plants normally less than 3 dm tall; basal leaves ovate to oblong-lanceolate, tapering to a broad-winged petiole .... **var. *dimorphophylla***

1′ Plants more than 3 dm tall; basal leaves suborbicular, ovate-rotund or reniform, subcordate to abruptly contracted at the base .... **var. *intermedia***

var. *dimorphophylla* **STEMS** 1–3 dm tall, single or 2–3 clustered, simple or branched. **LEAVES** basal, ovate to subreniform or occasionally oblanceolate, 1–4 cm long and 1–4 cm wide, bases tapering to a broad-winged petiole; cauline leaves clasping and conspicuously auriculate. [*Senecio dimorphophyllus* Greene var. *dimorphophyllus* W. A. Weber & A. Löve]. Wet or drying open alpine and subalpine meadows and steep, wet slopes. COLO: Arc, Con, Hin, LPl, Min, Mon, RGr, SJn; UTAH. 2800–3800 m (9200–12450′). Flowering: Jul–Aug. From Montana and Idaho, south to north-central New Mexico and west into Utah.

var. *intermedia* (T. M. Barkley) Trock & T. M. Barkley (intermediate) **STEMS** 3–5 dm tall, single and unbranched. **LEAVES** basal, suborbicular to ovate-rounded, 1–5 cm long and 1–4 cm wide, bases rounded to truncate, petiole slender; cauline leaves weakly clasping. [*Senecio dimorphophyllus* Greene var. *intermedius* T. M. Barkley]. Rocky slopes and ridges in the mountains. COLO: LPl, Mon; UTAH. 2600–3200 m (8530–10500′). Flowering: Jul–Aug. This variety has been described as "intermediate" between *P. dimorphophylla* and *P. crocata*. Fieldwork is needed to resolve this relationship.

Packera fendleri (A. Gray) W. A. Weber & Á. Löve (for Augustus Fendler, German botanist) Fendler's ragwort. Perennial, caudices rhizomatous with prominent lateral branches, horizontal to suberect. **STEMS** 1–4 dm tall, single or caespitose in some populations, woolly or becoming glabrous with age. **LEAVES** basal, petiolate, blades lanceolate to oblanceolate, 3–6 cm long and 1–3 cm wide, bases tapering, margins shallowly but evenly pinnatifid to pinnatisect or merely wavy; cauline leaves sessile, gradually reduced distally, margins pinnatisect or wavy. **INFL** open or compact corymbiform cymes of 6–25+ heads; heads radiate, 6–9 mm long and 5–8 mm wide, pedicels heavily to irregularly woolly. **PHYLLARIES** 13, green, 5–7 mm long, woolly proximally to glabrous distally. **RAY FLORETS** 6–8+, laminae 5–7 mm long. **DISC FLORETS** 30–40+, corolla tube 2.5–3 mm long, limb 2.5–3.5 mm long. **CYPSELAE** 2.5–3 mm long, glabrous. **PAPPUS** 4–5 mm long. $2n = 46$. [*Senecio fendleri* A. Gray; *S. nelsonii* Rydb.; *S. canovirens* Rydb.]. Steep slopes in loose, dry, rocky or gravelly soil, along streams and in open forested areas. Also abundant in disturbed sites such as roadsides, picnic areas, and campsites. COLO: Arc, SJn; NMEX: RAr, San. 1600–3200 m (5250–10500′). Flowering: May–Oct. Central Wyoming south through New Mexico. Abundant and almost weedy in the southern Rocky Mountains.

Packera hartiana (A. Heller) W. A. Weber & Á. Löve. (for J. H. Hart) Hart's ragwort. Perennial, caudices fibrous-rooted or stoloniferous, ascending. **STEMS** 4–7 dm tall, single or 2–3 clustered, base of stem and leaf axils woolly, or occasionally glabrate. **LEAVES** basal, petiolate, blades ovate to obovate or rounded, 3–8 cm long and 2–4 cm wide, bases obtuse or tapering, margins serrate-dentate to weakly crenate, abaxial surface sometimes sparsely woolly; cauline leaves: lower leaves petiolate, sublyrate to weakly pinnatisect; middle and upper leaves sessile, pinnatisect to nearly entire. **INFL** an open corymb of 3–10 heads; heads radiate, 7–9 mm long and 7–9 mm wide, pedicels loosely woolly or rarely glabrous. **PHYLLARIES** 21 or rarely 13, green, 5–7 mm long, glabrous or sparsely woolly. **RAY FLORETS** 8 or 13, laminae 5–8 mm long. **DISC FLORETS** 50–65+, corolla tube 2–4 mm long, limb 2.5–3.5 mm long. **CYPSELAE** 1–2 mm long, glabrous. **PAPPUS** 3.5–4.5 mm long. $2n = 44$. [*Senecio hartianus* A. Heller; *S. quaerens* Greene; *Packera quaerens* (Greene) W. A. Weber & Á. Löve]. Meadows or open areas in woodlands and along streams. ARIZ: Apa. 1600–2600 m (5250–8550′). Flowering: May–Jul. A few specimens of this species have also been collected in the Trans-Pecos area of Texas.

Packera multilobata (Torr. & A. Gray ex A. Gray) W. A. Weber & Á. Löve (many-lobed) Lobeleaf groundsel. Perennial or biennial, caudices taprooted, weakly branching, ascending to erect. **STEMS** 2–4+ dm tall, single or 2–5 loosely clustered, glabrous or occasionally lightly woolly throughout, with axils of leaves densely woolly; base of stems frequently reddish. **LEAVES** basal, petiolate, blades obovate, oblanceolate, or spatulate in outline, 4–8 cm long and 1–3 cm wide, bases attenuate, margins deeply pinnatifid or lyrate, leaf segments variously toothed and rounded; cauline leaves progressively reduced distally, sessile. **INFL** an open corymbiform cyme of 10–30+ heads; heads radiate, 8–10 mm long and 7–10 mm wide, pedicels glabrous or with tufts of woolly hairs at the base. **PHYLLARIES** 13–21, green, often with yellow tips, 4–9+ mm long, glabrous to sparsely woolly at the base. **RAY FLORETS** 8–13, corolla laminae 7–10 mm. **DISC FLORETS** 40–50+, corolla tube 4–5 mm long, limb 3–4 mm long. **CYPSELAE** 2–3 mm long, glabrous or hirtellous on the angles. **PAPPUS** 5–6 mm long. $2n = 46, 92$. [*Senecio multilobatus* Torr. & A. Gray ex A. Gray; *S. lynceus* Greene; *S. uintahensis* Greenm.; *S. stygius* Greene; *S. thornberi* Greenm.]. Dry, rocky or sandy substrates in sagebrush, woodland, and subalpine areas. ARIZ: Apa, Coc, Nav; COLO: Dol, LPl, Mon, SMg; NMEX: McK, RAr, San, SJn; UTAH. 1200–2900 m (3950–9500′). Flowering: May–Jul. From Wyoming, southward on the west side of the Continental Divide, westward into California. In colder parts of its range, plants are shorter, the caudex is more developed with several clustered stems, and leaf lobes are large. In desert habitats, basal leaves are narrower and more finely lobed, the tomentum is often persistent, and the stem and leaves are often reddish. The Ramah Navajo used a decoction of the plant for menstrual pain, and as a ceremonial medicine.

Packera neomexicana (A. Gray) W. A. Weber & Á. Löve (referring to New Mexico) New Mexico groundsel. Perennial, biennial, or winter-annual, caudex a taprooted rhizhome, branched, horizontal or ascending to erect. **STEMS** 2–5+ dm tall, single or 2–5 clustered, densely woolly or becoming glabrous with age. **LEAVES** basal, petiolate, blades ovate to lanceolate or narrowly lanceolate, 2–6 cm long and 1–3 cm wide, bases abruptly contracted to attenuate, margins subentire to denticulate or serrate, lower surface usually reddish and woolly, upper surface often becoming glabrous; cauline leaves progressively reduced distally; lower ones petiolate and similar to basal leaves; middle and upper leaves becoming sessile, blades lanceolate, margins entire. **INFL** an open or compact corymbiform cyme of 3–20+ heads, often subtended by smaller cymes arising from leaf axils; heads radiate, 10–14 mm long and 6–10 mm wide, pedicels normally pubescent. **PHYLLARIES** 13 or 21, green or yellowish, 4–7+ mm, woolly to nearly glabrous. **RAY FLORETS** 8 or 13, rarely 5, corolla laminae 4–10 mm. **DISC FLORETS** 40–60+, corolla tubes 1.5–2.5 mm, limb 3.5–4.5 mm. **CYPSELAE** 1.5–2.5 mm long, hirtellous on the angles or infrequently glabrous. **PAPPUS** 5–6+ mm. $2n = 44, 46, 92$. Plant used as an antidote for narcotics, and a poultice of the plant was applied and used as a lotion for burns by the Kayenta Navajo. The Ramah Navajo used the plant as a hunting medicine and a cold infusion for good luck.

1. Perennial from a branching caudex; plants permanently but loosely woolly or occasionally glabrate; basal leaves narrowly lanceolate, margins subentire or irregularly dentate .... **var. *mutabilis***

1′ Perennial, biennial, or winter-annual from a taprooted caudex; plants permanently and densely woolly; basal leaves ovate or broadly lanceolate, margins dentate to deeply dentate or dissected .... **var. *neomexicana***

var. *mutabilis* (Greene) W. A. Weber & Á. Löve (changeable) Perennial from a branching or weakly spreading caudex. Permanently but loosely woolly throughout. **STEMS** single or 2–3. **LEAVES** basal, narrow, bases tapering, margins subentire to dentate. [*Senecio neomexicanus* A. Gray var. *mutabilis* (Greene) T. M. Barkley; *S. mutabilis* Greene; *S. neomexicanus* var. *metcalfei* (Greene ex Wooton & Standl.) T. M. Barkley; *S. metcalfei* Greene ex Wooton & Standl.]. Rocky, well-drained soils along roadsides, in open meadows and coniferous woodlands at middle to high altitudes. ARIZ: Apa, Nav; COLO: Arc, Con, Dol, Hin, LPl, Min, Mon, RGr, SJn, SMg; NMEX: McK, RAr, SJn; UTAH. 1800–3000 m (5900–9850′). Flowering: Apr–Jul. Abundant throughout the Colorado Plateau. May form hybrids with *P. tridenticulata* (Rydb.) W. A. Weber & Á. Löve, *P. streptanthifolia*, and *P. fendleri*.

var. *neomexicana* Winter-annual, biennial, or perennial from taprooted caudex; plant densely woolly throughout, rarely becoming glabrate. **STEMS** normally single. **LEAVES** basal, with blades ovate to obovate, margins deeply dentate or dissected; cauline leaves conspicuous and well developed. Rocky soils in oak-conifer or chaparral associations. ARIZ: Apa, Coc, Nav; NMEX: SJn. 1200–2400 m (3950–7850′). Flowering: Apr–Jun. Most abundant through southern Arizona and southwestern New Mexico.

Packera paupercula (Michx.) Á. Löve & D. Löve (poor little one) Balsam groundsel. Perennial, caudices subrhizomatous, weakly branching, ascending to erect. **STEMS** 2–4 dm tall, single or 2–4 loosely clustered, glabrous or sparsely woolly near the base. **LEAVES** basal, petiolate, blades lanceolate to narrowly elliptic or oblanceolate, 3–6 cm long and 1–2 cm wide, bases attenuate to obtuse, margins subentire to variously dentate or serrate; cauline leaves progressively reduced distally; lower ones petiolate, blades sublyrate; middle ones sessile, blades lanceolate, margins variously dissected; upper ones sessile bracts. **INFL** a loose or compact corymbiform cyme of 2–12 heads; heads radiate, 6–8 mm long and 7–9 mm wide, pedicels glabrous. **PHYLLARIES** 13 or 21, green, 5–8 mm long, glabrous. **RAY FLORETS** 8 or 13, or occasionally absent, corolla laminae 5–10 mm long. **DISC FLORETS** 50–65+, corolla tube 2–3 mm, limb 2–3 mm. **CYPSELAE** 1–2 mm long, glabrous or rarely hispid on the angles. **PAPPUS** 3.5–4.5 mm long. $2n = 44, 46, 92$. [*Senecio pauperculus* Michx.; *S. balsamitae* Muhl. ex Willd.; *S. flavovirens* Rydb.; *S. crawfordii* Britton]. Wet meadows, open woodlands, along stream banks and rocky outcrops. COLO: Hin, LPl, Min. 2500–3000 m (8200–9850′). Flowering: May–Jun. Alaska east to New Brunswick, south to Georgia and west to northern California. Not present in the central Great Plains.

Packera porteri (Greene) C. Jeffrey (for T. C. Porter) Porter's groundsel. Perennial, caudices with long, slender, branching rhizomes. **STEMS** 3–10+ cm tall, single, often reddish, glabrous. **LEAVES** basal, petiolate and turgid, blades reniform, 0.5–1.5 cm long and 0.5–2.5 cm wide, bases abruptly truncate to cordate, margins distinctly crenate or occasionally wavy, lower surface distinctly reddish; cauline leaves abruptly reduced distally to 1–4 sessile bracts with entire margins. **INFL** a single large terminal head; heads radiate, 10–15 mm long and 10–13 mm wide. **PHYLLARIES** 13 or 21, deep red, 8–10+ mm long, glabrous. **RAY FLORETS** 8 or 13, corolla laminae 8–12 mm long. **DISC FLORETS** 40–50+, corolla tube 2.5–3.5 mm long, limb 2.5–3.5 mm long. **CYPSELAE** 1.5–2.5 mm long, glabrous. **PAPPUS** 4–5.5 mm long. [*Senecio porteri* Greene]. Steep talus slopes in alpine habitats. COLO: Hin, SJn. 2800–3900 m (9200–12800′). Flowering: Jul–Aug. This species has a very restricted range in Colorado and is known only from the Sawatch Range westward into Gunnison County and southward into the highest peaks of San Juan and Hinsdale counties. There are two disjunct populations of this species in Washington and Oregon. The Oregon collection was made nearly 100 years ago, and the species has not been collected since then.

Packera pseudaurea (Rydb.) W. A. Weber & Á. Löve (false golden yellow) Falsegold groundsel. Perennial, caudices simple or branched, fibrous-rooted, horizontal to erect. **STEMS** 2–7 dm tall, single or occasionally 2–4 clustered, glabrous or sparsely pubescent at the base. **LEAVES** basal, petiolate, blades broadly lanceolate to ovate, 2–5 cm long and 2–4 cm wide, bases truncate to rounded or obtuse, margins serrate, dentate, or crenate; cauline leaves petiolate, becoming sessile distally, blades sublyrate, laciniate, or rarely subentire. **INFL** an open or congested cyme of 5–20 heads; heads radiate, 6–9 mm long and 7–10 mm wide; pedicels glabrous. **PHYLLARIES** 21, occasionally 13, rarely 30 or more, light green, 5–8 mm long, glabrous. **RAY FLORETS** 8 or 13, corolla laminae 6–10 mm long, or occasionally absent. **DISC FLORETS** 70–80+, corolla tube 2.5–3.5 mm long, limb 2–3 mm long. **CYPSELAE** 1–1.5 mm long, glabrous. **PAPPUS** 4.5–5.5 mm long. $2n = 40, 44, 46, 80$. [*Senecio pseudaureus* Rydb. var. *flavulus* (Greene) Greenm.; *S. flavulus* Greene]. Damp soils along stream banks in woodlands or open meadows. COLO: Arc, Con, Hin, LPl, Min, Mon, RGr; NMEX: McK, RAr. 1800–2700 m (5900–8850′). Flowering: May–Jul. Southern Wyoming south into northern New Mexico. Our plants belong to **var. *flavula*** (Greene) W. A. Weber & Á. Löve (a little yellow).

Packera sanguisorboides (Rydb.) W. A. Weber & Á. Löve (refers to the superficial resemblance to the genus *Sanguisorba*) Burnet ragwort. Perennial or biennial, caudices creeping, fibrous-rooted, ascending to erect. **STEMS** 3–5+ dm tall, single or 2–3 clustered, glabrous or with dense tufts of pubescence in leaf axils. **LEAVES** basal, petiolate, blades broadly oblanceolate, 6–12 cm long and 2–6 cm wide, margins pinnate, 2–3+ pairs of lateral lobes, bases of lobes contracted, lobe margins irregularly dentate or crenate; terminal lobe large, ovate to reniform, base abruptly contracted to unwinged midrib; cauline leaves gradually reduced distally, short-petiolate to sessile, blades lyrate, terminal lobes less distinct, lobe margins shallowly dentate and midrib winged. **INFL** a subumbelliform or compound cyme of 2–6 cymules with 2–5+ heads each; heads radiate, 3–12, 6–9 mm long and 5–8 mm wide; pedicels glabrous or with tufts of pubescence at the base. **PHYLLARIES** 13, bright green with light green to yellow tips, 4–7 mm long, glabrous. **RAY FLORETS** 8, corolla laminae 6–12 mm long. **DISC FLORETS** 35–50+, corolla tube 2.5–3.5 mm long, limb 2–3 mm long. **CYPSELAE** 1.5–2 mm long, glabrous. **PAPPUS** 4.5–5.5 mm long. $2n = 46$. [*Senecio sanguisorboides* Rydb.]. Damp, open meadows in spruce-aspen forest associations. NMEX: RAr. 2700–3700 m (8850–12150′). Flowering: Jul–Sep. Restricted to the San Juan, Sangre de Cristo, Magdalena, and Sacramento mountains of New Mexico.

Packera spellenbergii (T. M. Barkley) C. Jeffrey (for Richard Spellenberg, New Mexico botanist) Carrizo Creek ragwort. Perennial, caudices coarse, weakly creeping or suberect, fibrous-rooted, rhizomatous. **STEMS** 3–6+ cm, single or occasionally 2, dense white tomentum exfoliating with age, upper stem becoming nearly glabrous. **LEAVES** basal, sessile, coriaceous, blades linear and strongly revolute, 10–15+ mm long, 1–2 mm wide, margins entire; cauline leaves abruptly reduced to linear bracts. **INFL** 1 or rarely 2 large heads; heads eradiate, 10–12 mm long, 10–15+ mm wide, pedicels glabrous; phyllaries 13, purple to deep reddish purple, 6–9+ mm, pubescent to glabrescent. **RAY FLORETS** normally absent, if present, 5 or 8, corolla laminae 4–7 mm long. **DISC FLORETS** 30–40+, corolla tubes 2.5–3.5 mm long, limbs 3.5–4 mm long. **CYPSELAE** 3–3.5 mm long, hirtellous. **PAPPUS** 5–6 mm long. [*Senecio spellenbergii* T. M. Barkley]. Calcareous soil in sparsely vegetated areas of grasslands or piñon-juniper woodland communities. ARIZ: Apa; NMEX: McK. 1600–2300 m (5250–7545′). Flowering: mid-Apr–May. New Mexico, south-central Utah, and northwestern Arizona. This plant is semisucculent and has deeply anthocyanic herbage. More fieldwork needs to be done to determine if these populations are indeed disjunct, or if the plant is so small and rare as to be overlooked in other places where it may occur.

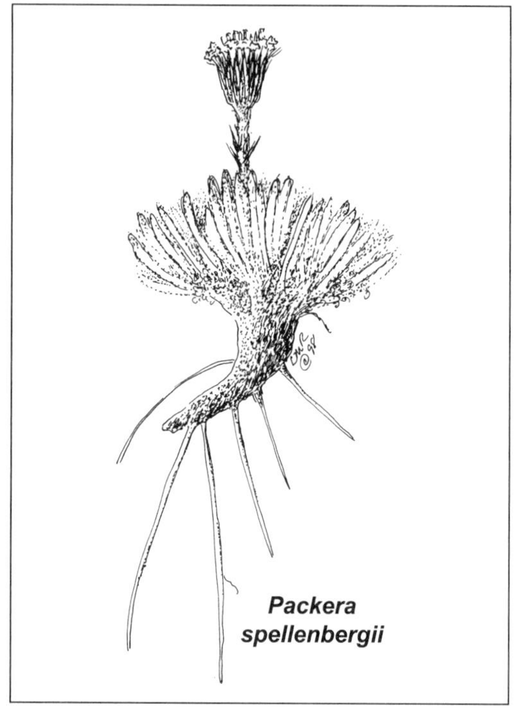
Packera spellenbergii

Packera streptanthifolia (Greene) W. A. Weber & Á. Löve (twisted leaf, refers to the extreme morphological variation in the leaves of this species) Rocky Mountain groundsel. Perennial, caudices weak to stout, fibrous-rooted, horizontal to suberect. **STEMS** 2–4+ dm tall, single or 2–5 clustered, glabrous or occasionally lightly pubescent at the base of the stem and in the axils of the leaves. **LEAVES** basal, petiolate, thick and turgid, blades varying from spatulate to oblanceolate, or ovate to orbicular, 24+ cm long and 1–3 cm wide, bases attenuate to abruptly contracted or rarely subcordate, margins entire, crenate, dentate, or weakly lobed, normally glabrous, rarely persistently pubescent; cauline leaves gradually to abruptly reduced distally, becoming sessile, margins entire to subentire. **INFL** a loose corymbiform cyme of 4–20 heads; heads radiate, 5–10 mm long and 8–11 mm wide; pedicels glabrous or rarely lightly pubescent. **PHYLLARIES** 13 or 21, rarely 8, green or with reddish tips, 4–7+ mm long, glabrous. **RAY FLORETS** 8 or 13, corolla laminae 5–10 mm long, ranging from light yellow to orange-yellow. **DISC FLORETS** 35–60+, corolla tube 2–4 mm long, limb 2.5–4 mm long. **CYPSELAE** 1–2.5 mm long, glabrous. **PAPPUS** 3–6 mm long. $2n = 46, 92$. [*Senecio streptanthifolius* Greene; *S. cymbalarioides* Nutt.; *S. laetiflorus* Greene; *S. platylobus* Rydb.; *S. rubricaulis* Greene]. Forested areas and open meadows and valleys in dry to damp and loamy soils. COLO: Arc, Con, Hin, LPl, Min, Mon, SJn, SMg; NMEX: RAr. 1800–3400 m (5900–11150′). Flowering: May–Jul. From British Columbia east to Saskatchewan, south through the Rocky Mountains, west to California and north to Washington.

Packera werneriifolia (A. Gray) W. A. Weber & Á. Löve (for W. C. Werner) Hoary groundsel. Perennial; caudices branching, subrhizomatous, sometimes several densely crowded and caespitose. **STEMS** 7–15+ cm, single or 3–5 clustered, densely pubescent, becoming glabrous with age, or in some populations, essentially glabrous throughout. **LEAVES** basal, sessile or with very short petioles, blades narrowly lanceolate to elliptic, 1.5–4 cm long and 0.5–2.5 cm wide, bases attenuate, margins entire or dentate toward the apex, frequently revolute; or in some populations leaves petiolate with blades ovate to orbicular, 1–2 cm long and 0.5–1.5 cm wide, bases attenuate to abruptly contracted, margins entire or wavy, occasionally dentate toward the apex; cauline leaves reduced to mere bracts, plant with a scapose aspect. **INFL** a single large head or 2–6 much smaller heads; heads radiate, rarely eradiate, 10–12 mm long and 10–14 mm wide, or if smaller, more numerous heads 8–10 mm long and 7–9 mm wide; pedicels glabrous or densely pubescent if stem is. **PHYLLARIES** 13 or 21, green or with reddish tips, 4–10 mm long, glabrous to pubescent. **RAY FLORETS** 8 or 13, corolla laminae 5–10 mm long. **DISC FLORETS** 30–50+, corolla tube 2.5–3.5 mm long, limb 3–4 mm long. **CYPSELAE** 1.5–2 mm long, glabrous. **PAPPUS** 5–6 mm long. $2n = 44, 46$. [*Senecio werneriifolius* A. Gray; *S. saxosus* Klatt; *S. petrocallis* Greene; *S. muirii* Greenm.; *S. molinarius* Greenm.]. Rocky talus slopes or in sandy soil in forest openings near or above timberline. COLO: Arc, Con, Hin, LPl, Min, Mon, RGr, SJn. 2400–3700 m (7850–12150′). Flowering: Jun–Aug. Idaho south through the southern Rockies, west to California.

Pectis L. Chinchweed, Fetid Marigold
Brian Elliott

(Latin *pectis*, comb, referring to the marginal cilia of the petiole) Annual or perennial, taprooted herbs; aromatic. **STEMS** low, prostrate to erect, slender. **LEAVES** opposite, simple, narrow, entire, sessile, gland-dotted, ciliate with a few stiff bristles at the base. **INFL** heads radiate; solitary or in leafy cymes, pedunculate. **INVOLUCRES** turbinate to cylindric or campanulate. **PHYLLARIES** 3–12 in 1 equal series, free, enclosing ray achenes. **RECEPTACLES** naked. **RAY FLORETS** perfect and fertile, equaling the number of phyllaries; ligules yellow, occasionally purple-tinged; apex entire to 2–3-toothed or -lobed. **DISC FLORETS** perfect and fertile, few–many per head; corolla yellow, 4–5-lobed, often bilabiate with 1–2 deep slits. **CYPSELAE** linear to linear-clavate, terete to mildly angled, 2–4.5 mm long, black. **PAPPUS** of bristles, scales, awns, or a low crown. Approximately 100 species native to North and South America (Keil 1977).

Pectis angustifolia Torr. (referring to the narrow leaves) Narrow-leaf pectis, lemonscent. Annual herbs; bushy and lemon-scented. **STEMS** erect, dichotomously branched, 3–20 cm tall, densely leafy at branch tips. **LEAVES** linear, 1–5 dm long, 1–3 mm wide, with 1–5 pairs of subulate setae at the conspicuously flared and scarious-margined base, submarginal glands conspicuous. **INFL** heads clustered at branch tips in leafy cymes; peduncles 1–10 mm long, 1–4-bracteate, often concealed by subtending leaves. **INVOLUCRES** turbinate. **PHYLLARIES** cartilaginous, keeled, gibbous at base; apex abruptly truncate. **CYPSELAE** sparsely silky-villous. **PAPPUS** coroniform (in ours), 0.1–0.3 mm high, rarely with 1–3 awns on the scales. $2n = 24$. [*Pectidopsis angustifolia* DC.]. Dry, sandy or gravelly plains and hills, piñon-juniper, and desert shrublands. ARIZ: Coc, Nav; COLO: LPl; NMEX: RAr, San, SJn; UTAH. 1600–2000 m (5300–6500′). Flowering: Jun–Oct. Ranging from Nebraska, Kansas, and west Texas to eastern Colorado, New Mexico, Arizona, and northern Mexico. Three varieties are currently recognized. Ours, with flared leaf bases and a coroniform pappus, is **var. *angustifolia***. The plants have been used medicinally for stomach ailments and as an emetic. The foliage has been eaten raw or cooked, used for seasoning, and also makes a palatable tea.

Pericome A. Gray Pericome
Brian A. Elliott

(Greek *peri*, around, + *coma*, hair, referring to ciliate margin of the achene) Perennial, suffrutescent herbs; aromatic. **STEMS** many-branched from the base, striate and puberulent. **LEAVES** opposite, simple, reduced above, triangular-hastate to triangular-cordate or cordate-ovate, puberulent and gland-dotted. **INFL** a discoid head, numerous in terminal cymose or corymbose clusters. **INVOLUCRES** turbinate or cylindric to campanulate. **PHYLLARIES** in 1 equal series, linear, connate by hyaline margins but separating in age, midrib thickened. **RECEPTACLES** convex to conical, naked. **RAY FLORETS** lacking. **DISC FLORETS** perfect and fertile, numerous; corolla yellow, 4-toothed, filiform below and flaring above, glandular throughout but denser at the base. **CYPSELAE** linear-elliptic or oblong to oblanceolate, strongly compressed laterally, margins callous-thickened and ciliate, surfaces dark brown or black and nerveless, puberulent. **PAPPUS** a crown of low laciniate scales, connate below, sometimes with 1–2 awns. Two species of the southwestern United States and Mexico.

Pericome caudata A. Gray (tail, referring to the long-attenuate leaf apex) Pericome. **STEMS** widely spreading, plants to 1.5 m tall and nearly as wide. **LEAVES** opposite but upper occasionally alternate, 3–15 cm long, petiole 1–5 cm long; truncate at base, margins entire, sinuate, or irregularly toothed. **INVOLUCRES** 5–8 mm high. **PHYLLARIES** 16–25, 4.5–7 mm long, woolly-puberulent, apices free. **DISC FLORETS** with corollas 3–5 mm long, exceeding the involucre. **CYPSELAE** 3.5–5 mm long. **PAPPUS** a low crown 1 mm high, awns 1–4.5 mm long. $2n = 36$. [*P. caudata*

A. Gray var. *glandulosa* (Goodman) H. D. Harr.]. Dry canyons, rocky slopes, cliffs and talus amongst piñon-juniper, ponderosa pine, and Douglas-fir. ARIZ: Apa; COLO: Arc, Min; NMEX: McK, SJn. 2450–2950 m (8000–9700′). Flowering: Jul–Oct. California to Colorado and south to Mexico. Medicinal uses of the root include poultices for toothache, as an analgesic, and for coughs. Leaves have been used to soothe fevers and influenza.

Perityle Benth. Rock Daisy
J. Mark Porter

(Greek *peri*, around, + *tyle*, a callus, referring to callous cypsela margin) Annual or perennial herbs, subshrubs, or shrubs, 2–45 (–75) cm, glabrous or hairy, sometimes glandular. **STEMS** erect, spreading, or pendent (ours). **LEAVES** mostly cauline; often proximally opposite, distally alternate, petiolate or sessile, blades usually 3-lobed, ultimate margins entire, toothed, or lobed. **INFL** borne singly or in corymbs. **HEADS** radiate or discoid. **INVOLUCRES** campanulate, cylindrical, funnelform, or hemispheric, 3–15 mm diameter. **PHYLLARIES** persistent, 5–28 in (1–) 2 (–3) series, distinct, linear to ovate, equal or subequal, flat or keeled, glabrous or hairy, apices obtuse, acute, or attenuate. **RECEPTACLES** flat or convex, pitted, naked. **RAY FLORETS** absent (in ours), or (1–) 3–18, pistillate, fertile; corollas cream, yellow, or white, showy or rudimentary. **DISC FLORETS** 5–200, bisexual, fertile; corollas cream, yellow, or white, tubes shorter than or nearly equal to cylindric, funnelform, or campanulate throats, lobes 4, deltate, acute; stamens 4; style branches flattened, linear, usually tapering to fine, minutely hairy tips. **CYPSELAE** black, flattened to subcylindric, linear to oblanceolate or obovate, margins callous and glabrous, hairy, or ciliate, faces usually hairy, sometimes glabrous. **PAPPUS** absent, or persistent or falling, of 1–35 bristles plus callous crowns or hyaline scales. x = 17, 19; $2n$ = 32, 34, 38, 64, 68, 72. With 66 species (1 in our flora), *Perityle* occurs in the United States, Mexico, and South America. Most species are rock-dwelling, found in the eroded mountain and basin region of the southwestern United States (Niles 1970; Powell 1973).

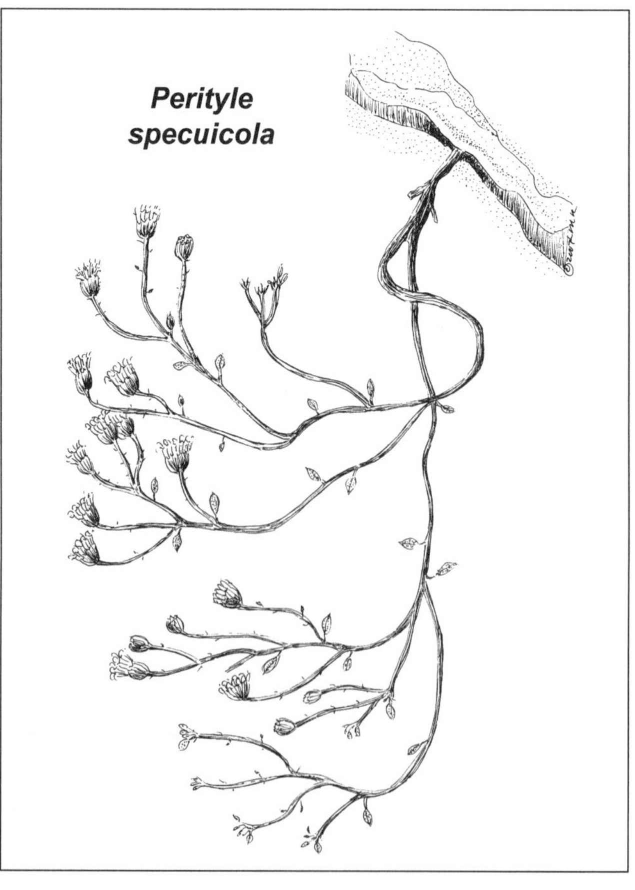

Perityle specuicola S. L. Welsh & Neese (cave-inhabiting) Alcove rock daisy, hanging-garden rock daisy. Perennial herbs or subshrubs with a woody base. **STEMS** 30–75 cm, sprawling or pendulous, much-branched, sparsely leafy; hispidulous or scabous. **LEAVES** mostly alternate, small and inconspicuous, petioles 1–3 (–8) mm long, blades lance-ovate to elliptic, (2–) 4–8 mm long, 1–5 mm wide, margins entire. **HEADS** borne singly or in corymbs, 3.5–5 mm high, 5–6 mm wide; peduncles 4–70 mm long. **INVOLUCRES** campanulate, puberulent. **PHYLLARIES** 11–16, lanceolate to oblanceolate, 3.5–5 mm long, 0.5–1 mm wide. **RAY FLORETS** absent. **DISC FLORETS** 30–60, corollas yellow, tubes 0.5–0.8 mm, throats broadly tubular, 1.3–1.8 mm, lobes 0.4–0.6 mm, triangular-acute. **CYPSELAE** narrowly oblanceolate,

3–3.8 mm, faces glabrous, margins thin-callous, relatively long-ciliate. **PAPPUS** usually of 1–3 (–4) unequal, barbellate bristles 1–2.5 mm, rarely absent. Rock crevices and faces along the Colorado and San Juan rivers; associated with seeps and hanging gardens. UTAH. 1100–1300 m (3700–4200′). Flowering: May–Oct. Fruit: Jul–Oct. Endemic to Grand and San Juan counties, Utah.

Petradoria Greene Rock Goldenrod
G. L. Nesom

(Greek *petros*, rock, + *doria*, an early name for goldenrods) Perennial, suffrutescent herbs; taproots stout, caudices woody. **STEMS** becoming whitish with age. **LEAVES** basal and cauline; basal persistent, linear to lanceolate or oblanceolate, coriaceous, cauline similar, alternate, bases acute to attenuate, expanded laterally and abaxially, clasping, especially proximally, margins entire, veins (1–) 3–5, parallel, conspicuously raised, both faces punctate, resinous, otherwise glabrous to scabrate. **INFL** a radiate head, borne in densely corymbiform arrays. **INVOLUCRES** cylindric to turbinate. **PHYLLARIES** in 3–6 strongly unequal series, imbricate to nearly vertically aligned, ovate to oblong, usually weakly keeled, glabrous, resinous. **RECEPTACLES** foveolate, paleae sporadically present, short, bristlelike. **RAY FLORETS** (0–) 1–2 (–3), pistillate, fertile; corollas yellow. **DISC FLORETS** (1–) 2–4 (–9), functionally staminate, ovary abortive; corollas yellow. **CYPSELAE** (ray) cylindric to slightly compressed, glabrous, 6–9-nerved. **PAPPUS** of somewhat unequal, barbellate bristles. $x = 9$. One species; western United States (Anderson 1963).

Petradoria pumila (Nutt.) Greene (dwarf) Rock goldenrod. **STEMS** 8–30 cm. **LEAVES** 20–120 × 1–12 mm, cauline well developed or reduced distally from basal. **HEADS** with (2–) 4–5 (–8) florets. **INVOLUCRES** 5–9.5 × 1.3–3 mm. **PHYLLARIES** 1–7 × 0.5–1 mm. **RAY FLORETS** (0–) 1–2. **DISC FLORETS** (1–) 2–4 (–5). **CYPSELAE** (ray) 4–5 mm. $2n = 18, 36$. [*P. graminea* Wooton & Standl.; *P. pumila* subsp. *graminea* (Wooton & Standl.) L. C. Anderson; *P. pumila* var. *graminea* (Wooton & Standl.) S. L. Welsh]. Sand, clay, cobble, ridges, talus, benches, crevices, alluvium; desert shrub, sagebrush, juniper-sage, mountain shrub–Gambel's oak, piñon-juniper, and ponderosa pine–Rocky Mountain juniper communities. ARIZ: Apa, Nav; COLO: Dol, LPl, Mon, SMg; NMEX: McK, RAr, San, SJn; UTAH. 1125–2550 m (3700–8365′). Flowering: Jun–Sep (–Oct). California, Nevada, Utah, Idaho, and Wyoming south to Arizona and New Mexico. Variety *graminea* usually has been recognized (e.g., Urbatsch, Roberts, and Neubig [2005], who essentially adopted the concept and measurements of Anderson [1963]), but this grass-leaved form (blades about 1 mm wide, often 1 (–3)-veined), is here interpreted to be a populational variant within the species. Of a total of 51 plants counted from the Four Corners region, disc florets averaged 3.7 in number; ray florets averaged 1.2. Rayless heads occur on plants with most heads radiate. Only plants from Elko, Eureka, and White Pine counties, Nevada, seem distinctively different: they have consistently broad leaves and disc florets range 5–9 in number.

Picradeniopsis Rydb. Bahia
Brian Elliott

(referring to the similarity to *Picradenia*) Perennial herbs with creeping rootstocks, sometimes suffrutescent; strigose and punctate-glandular. **STEMS** arising singly or in clusters, branched below, 8–25 cm tall, strigose. **LEAVES** opposite, palmately 1–3-parted, the uppermost entire and reduced, ultimate segments linear, 1–3 mm wide, sometimes toothed. **INFL** heads radiate, in a terminal corymb. **INVOLUCRES** campanulate, 4–8 mm high. **PHYLLARIES** subequal in 2 series; outer series keeled and canescent. **RECEPTACLES** flat, pitted, naked. **RAY FLORETS** pistillate and fertile; ligules few, yellow, 2–4 mm long, oval to oblong. **DISC FLORETS** perfect and fertile; corolla yellow; tube glandular. **CYPSELAE** narrowly obpyramidal, angled, hirsutulous or glandular. **PAPPUS** a crown of 8–9 lanceolate to ovate or obovate scales. A genus of two species ranging from the northern Great Plains to the southwestern United States and northern Mexico. The genus has previously been included in *Bahia* as a subgenus.

Picradeniopsis woodhousei (A. Gray) Rydb. (for Samuel Washington Woodhouse, 1821–1904, physician, naturalist, and explorer) Woodhouse's bahia. Perennial herbs, sometimes weakly suffrutescent. **STEMS** 5–20 cm tall. **PHYLLARIES** in 1–2 series, puberulent, keeled with a prominent midrib, apex obtuse. **RAY FLORET** ligules oblong. **DISC FLORETS** 25–40. **CYPSELAE** 4-angled, hispid to hispidulous. **PAPPUS** of narrowly lanceolate scales, apex acute to acuminate, midrib reaching apex and often excurrent as an awn. $2n = 24$. [*Bahia woodhousii* A. Gray]. Plains, hills, piñon-juniper woodland, and sagebrush. ARIZ: Apa. NMEX: SJn. 1800–1980 m (5900–6500′). Flowering: Jun–Sep. Ranging from the southern Great Plains to Texas, New Mexico, Arizona, and northern Mexico. The species has been used medicinally as an emetic and for stomachaches. A poultice of the roots has been used to soothe skin rashes and sores.

Platyschkuhria (A. Gray) Rydb. Basin Daisy

J. Mark Porter

(Greek *platys*, broad, + genus *Schkuhria*) Perennials, 10–55 cm tall. **STEMS** erect, from woody-based, branched caudices. **LEAVES** basal and cauline; mostly alternate; petiolate; blades mostly lanceolate, sometimes ovate, margins entire, surface sparsely to densely scabrellous, and gland-dotted. **HEADS** radiate, borne singly or 3–11 in loose, corymbosely or paniculately arranged inflorescences. **INVOLUCRES** campanulate to hemispheric, 12–25 mm diameter. **PHYLLARIES** 9–21 in 2 series, distinct, subequal, oblong or elliptic to lanceolate, herbaceous, membranous-margined, persistent, reflexed in fruit. **RECEPTACLES** convex, pitted, lacking paleae. **RAY FLORETS** 6–12, pistillate, fertile; corollas yellow. **DISC FLORETS** 25–85, bisexual, fertile; corollas yellow to orange (gland-dotted), tubes about equaling funnelform to campanulate throats, lobes 5, deltate to lance-deltate. **CYPSELAE** narrowly obpyramidal, 4-angled, finely nerved, hirsutulous, at least on angles, not gland-dotted. **PAPPUS** of 8–16 scales in 1 series, distinct, lance-elliptic to lance-subulate, basally and medially thickened, distally and laterally scarious, weakly, if at all, aristate. $x = 12$; $2n = 24$, 25, 28, 48, 50, 60, 72. One species of the western United States (Brown 1983; Ellison 1971).

Platyschkuhria integrifolia (A. Gray) Rydb. (leaves entire) Oblongleaf basin daisy, naked-stem bahia. **STEMS** solitary to several, white-strigulose, slightly or not at all stipitate-glandular; leafy nearly to the apex. **LEAVES** basal and cauline; petiolate, blades 2–10 cm long, 5–40 mm wide; cauline leaves reduced in size above, ultimately bractlike. **HEADS** (1–) 2–10. **INVOLUCRES** 9–13 mm long, 12–27 mm wide. **RAY FLORETS** 7–11, corolla 6–16 mm long. **DISC FLORET** corollas 3–6 (–7) mm. **CYPSELAE** 3–8 mm long. **PAPPUS** 0.6–3 mm long. $2n = 24$. Seleniferous clays, shaley slopes, in desert shrub communities. ARIZ: Apa, Nav; COLO: Mon; NMEX: McK, San, SJn; UTAH. 1220–2200 m (4000–7000′). Flowering: May–Jul. Fruit: Jun–Aug. Our material belongs to **var. *oblongifolia*** (A. Gray) W. L. Ellison (oblong leaves).

Pluchea Cass. Arrowweed

G. L. Nesom

(for Abbé N. A. Pluche, 1688–1761, French naturalist) Annuals, perennials, subshrubs, shrubs, or trees (usually fetid-aromatic), (20–) 50–200 (–500) cm; taprooted or fibrous-rooted. **STEMS** erect, simple or branched (seldom winged); usually puberulent to tomentose and stipitate- or sessile-glandular, sometimes glabrous. **LEAVES** cauline; alternate; petiolate or sessile; blades mostly elliptic, lanceolate, oblanceolate, oblong, obovate, or ovate, bases clasping or not, margins entire or dentate, abaxial faces mostly arachnose, puberulent, sericeous, strigose, or villous and/or stipitate- or sessile-glandular, adaxial similar or glabrate or glabrous. **INFL** a disciform head in corymbiform or paniculiform arrays. **INVOLUCRES** mostly campanulate, cupulate, cylindric, hemispheric, or turbinate. **PHYLLARIES** persistent or falling, in 3–6+ series, mostly ovate to lanceolate or linear, unequal. **RECEPTACLES** flat, epaleate. **PERIPHERAL FLORETS** pistillate, in 3–10+ series, fertile; corolla creamy white, whitish, yellowish, pinkish, lavender, purplish, or rosy. **INNER FLORETS** functionally staminate, 2–40+; corollas creamy white, whitish, yellowish, pinkish, lavender, purplish, or rosy, lobes (4–) 5. **CYPSELAE** oblong-cylindric, ribs 4–8, faces strigillose and/or minutely sessile-glandular or glabrous. **PAPPUS** 1 series of barbellate bristles, persistent or tardily falling, distinct or basally connate. $x = 10$ (Robinson and Cuatrecasas 1973; Nesom 1989). Species 40–60, tropical and warm-temperate regions of North America, West Indies, South America, southeast Asia, Africa, Australia, and Pacific Islands.

Pluchea sericea (Nutt.) Coville (silky) Arrowweed. Shrubs or trees, 150–300 (–500) cm. **STEMS** densely leafy, sericeous, not glandular. **LEAVES** sessile; blades lanceolate to narrowly lanceolate or narrowly oblanceolate, 1–5 × 0.2–1 cm, margins entire, faces sparsely to densely silvery sericeous, not glandular, minutely punctate. **HEADS** in cymiform clusters. **INVOLUCRES** ± campanulate, 4–6 × 3–5 mm. **PHYLLARIES** pink to purplish, tomentose to villosulous (outer) to arachnose-ciliate or glabrate (inner); corolla pink to purplish. **CYPSELAE** glabrous. **PAPPUS** persistent, bristles distinct, distally dilated in functionally staminate florets. $2n = 20$. [*Tessaria sericea* (Nutt.) Shinners]. Creek and lake shores, dunes, hanging gardens. UTAH. 1000–1200 m (3280–3935′). Flowering: mostly May–Jul. California and Nevada to Utah, Arizona, New Mexico, and Texas; Mexico (Baja California, Chihuahua, Sonora).

Prenanthella Rydb. Brightwhite

Gary I. Baird

(diminutive form of the genus name *Prenanthes*) Annuals, taprooted, herbage glandular-puberulent, sap milky. **STEMS** mostly 1, much-branched distally (sometimes throughout). **LEAVES** basal (mostly) and cauline, simple, alternate; sessile to petiolate, not auriculate; blades more or less oblanceolate, margins coarsely dentate (sometimes irregularly so), apices obtuse to acute. **INFL** of numerous heads in dense paniculiform clusters. **HEADS** liguliflorous, terminal,

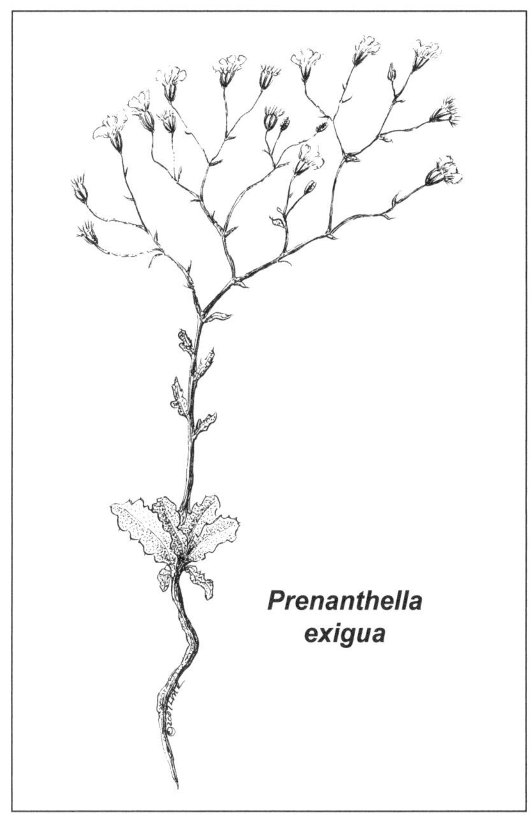
Prenanthella exigua

numerous in dense paniculiform clusters, erect; peduncles bracteate. **INVOLUCRES** cylindric to narrowly campanulate; calyculi of 2–3 bractlets, deltate, unequal. **PHYLLARIES** 4–5 in 1 series, herbaceous, margins scarious. **RECEPTACLES** flat, epaleate. **RAY FLORETS** matutinal; corolla ligulate, white or pinkish. **DISC FLORETS** lacking. **CYPSELAE** monomorphic or dimorphic (occasionally 1 darker, longer, more rugose, more persistent, and pappus early-deciduous), white (mostly) to brown, columnar (slightly thickened at base and apex), glabrous, minutely rugulose, 5-ribbed and -angled, each face longitudinally grooved, beakless. **PAPPUS** of bristles, persistent to deciduous, white, smooth. $x = 7$. One species in western North America.

Prenanthella exigua (A. Gray) Rydb. (very small or minute, alluding to the heads) Brightwhite. Plants 7–20 cm, often about as wide as tall. **STEMS** stramineous and stiff proximally (or with age). **LEAVES** 9–32 mm × 6–12 mm, reduced distally to minute bracts. **INVOLUCRES** 4–5 mm tall. **PHYLLARIES** linear-lanceolate, equal, erect, glabrous to glandular-puberulent, green to purple (at least medially). **RAY FLORETS** mostly 4, just exceeding involucre; corollas glabrous, tube 1–2 mm, ligule 2–3 mm × about 1 mm; anthers 1–1.5 mm. **CYPSELAE** 3–5 mm. **PAPPUS** 2–3 mm. $2n = 14$. Sandy, rocky, or clay soils; sagebrush, blackbrush, shadscale, and piñon-juniper communities, ARIZ: Apa, Nav; COLO: Mon; NMEX: SJn; UTAH. 1500–1700 m (4800–5700′). Flowering: May–Jun. Eastern California to western Colorado, south to Arizona and northern Mexico. This species is found primarily in the Great Basin, Colorado Basin, and Colorado Plateau regions.

Psathyrotes A. Gray Turtleback

Brian A. Elliott

(Greek *psathy*, crumbling, referring to the brittle stems) Annual or short-lived perennial taprooted herbs; aromatic and divaricately branching. **STEMS** leafy, brittle. **LEAVES** cauline, alternate, simple, petiolate, entire to toothed, scurfy-puberulent to floccose-tomentose. **INFL** a discoid head, solitary or in subumbellate clusters. **INVOLUCRES** hemispheric to narrowly cylindric. **PHYLLARIES** 8–24 in 2–3 series, lanceolate to oblong-lanceolate; outer bracts herbaceous, inner bracts often scarious on the margin. **RECEPTACLES** flat, naked, alveolate. **RAY FLORETS** none. **DISC FLORETS** perfect and fertile, 8–32; corolla yellow, reddish purple at apex in age, tube much shorter than the cylindrical throat, lobes 5, distally stipitate-glandular and villous. **CYPSELAE** cylindrical to oblong or obpyramidal, weakly 10-ribbed, subglabrous to villous or stipitate-glandular. **PAPPUS** of numerous smooth or barbellate capillary bristles in 1–4 series. Three species of the western United States and Mexico (Strother and Pilz 1975).

Psathyrotes pilifera A. Gray (bearing hairs, referring to the soft pubescence) Piliferous turtleback, hairybeast turtleback. Annual herbs. **STEMS** freely branched, forming compact mounds to 1.5 dm high and 3 dm wide. **LEAVES** broadly ovate to oval-elliptic or subrotund, 5–15 mm long, 4–16 mm wide, entire, petiole equaling or exceeding the blade, conspicuously ciliate with straight multicellular hairs 5–6 mm long on the margins and veins below, scurfy. **INFL** peduncle 4–20 mm long. **INVOLUCRES** narrowly cylindric, 7–10 mm long. **PHYLLARIES** in 2 series; outer series of 5 bracts, 6–11 mm long, sometimes constricted at the middle and dilated at the tip, glandular-puberulent, ciliate with long multicellular hairs about 5 mm long; inner bracts 8, chartaceous except for the herbaceous tip, deciduous. **DISC FLORETS** 8–14; corolla 6–6.5 mm long, throat conspicuously longer than tube. **CYPSELAE** narrowly cylindric, 4–5 mm long, short stipitate-glandular and subscabrid. **PAPPUS** of numerous barbellate capillary bristles in 2–4 series, basally connate, about 5 mm long, white. $2n = 34$. Piñon-juniper and desert shrub communities, often on alkaline or gypsum soils. UTAH. 1500 m (5000′). Flowering: Jul–Oct. Known only from Utah, Nevada, and Arizona. The plant has been used medicinally as a lotion for chilblain (swellings or sores resulting from exposure to cold), and as a ceremonial emetic.

Pseudognaphalium Kirp. Rabbit Tobacco, Cudweed

G. L. Nesom

(Greek *pseudo*, deceptively similar, + genus name *Gnaphalium*, alluding to resemblance) Annuals, biennials, or perennials, (4–) 15–150 (–200) cm (usually taprooted, sometimes fibrous-rooted). **STEMS** 1+, usually erect, sometimes decumbent to procumbent, more or less woolly-tomentose, sometimes stipitate- or sessile-glandular. **LEAVES** basal and cauline or mostly cauline; alternate; usually sessile; blades mostly narrowly lanceolate to oblanceolate, bases often clasping and/or decurrent, margins entire, faces bicolor or concolor, abaxial white to gray and tomentose to velutinous, adaxial usually greenish and glabrous or glabrescent, sometimes grayish and loosely arachnose (sometimes stipitate- or sessile-glandular). **INFL** a disciform head, usually in glomerules in corymbiform or paniculiform arrays, sometimes in terminal clusters. **INVOLUCRES** mostly campanulate to cylindric, (3–) 4–7 mm. **PHYLLARIES** in (2–) 3–7 (–10) series, whitish, rosy, tawny, or brownish (opaque or hyaline, dull or shiny; stereomes usually green, usually sessile-glandular distally), unequal, usually chartaceous toward tips. **RECEPTACLES** epaleate. **PERIPHERAL FLORETS** pistillate, (15–) 25–250+ (more numerous than bisexual). **INNER FLORETS** bisexual, (1–) 5–20 (–40+). **CYPSELAE** oblong-compressed or cylindric, faces usually smooth, sometimes papillate-roughened and/or with 4–6 longitudinal ridges, usually glabrous (papilliform hairs in *P. luteoalbum*). **PAPPUS** of 10–12 distinct (basally coherent in *P. luteoalbum* and *P. stramineum*), caducous, barbellate bristles in 1 series. $x = 7$ (Nesom 2004). About 100 species, mostly in temperate New World regions, some Old World.

1. Heads in terminal glomerules; pappus bristles loosely coherent basally, released in clusters or easily fragmented rings .... (2)
1′ Heads in corymbiform arrays; pappus bristles basally distinct, released individually .... (3)

2. Involucres 3–4 mm; bisexual florets 5–10, corollas red-tipped; cypselae papillate .... ***P. luteoalbum***
2′ Involucres 4–6 mm; bisexual florets mostly 18–28, corollas evenly yellowish, not red-tipped; cypselae glabrous .... ***P. stramineum***

3. Stems eglandular; leaves concolored .... ***P. jaliscense***
3′ Stems stipitate-glandular; leaves strongly to weakly bicolored .... ***P. macounii***

Pseudognaphalium jaliscense (Greenm.) Anderb. (from Jalisco, Mexico, the type locality) Jalisco rabbit tobacco. Annuals or biennials, 30–70 cm; taprooted. **STEMS** (branched among heads) densely and persistently loosely woolly-tomentose-sericeous, not glandular. **LEAVES** with narrowly lanceolate to nearly linear blades, 3–10 cm × 3–6 mm, bases not clasping, decurrent 4–8 mm, margins flat or slightly revolute, faces concolored, tomentose-sericeous (bases of hairs enlarged), sessile-glandular beneath tomentum. **HEADS** in corymbiform arrays. **INVOLUCRES** campanulate, 5–6 mm. **PHYLLARIES** in 5–6 (–7) series, white (opaque, dull), ovate or elliptic, keeled, apiculate, glabrous. **PISTILLATE FLORETS** (80–) 90–115. **BISEXUAL FLORETS** (6–) 8–12. **CYPSELAE** weakly ridged, papillate-roughened or smooth. [*Gnaphalium jaliscense* Greenm.]. Grassy meadow with scattered ponderosa pine and aspen. COLO: Hin. Aproximately 2300 m (7545′). Flowering: Jul–Sep. Arizona, Colorado, Nebraska, New Mexico, and Texas; the major part of the range in western Mexico. *Pseudognaphalium jaliscense* is recognized by its relatively long, narrow, concolored to weakly bicolored leaves with nonclasping, short-decurrent bases; relatively large heads with white, opaque, dull phyllaries; and relatively large numbers of pistillate and bisexual florets.

Pseudognaphalium luteoalbum (L.) Hilliard & B. L. Burtt (yellow-white) Red-tip rabbit tobacco. Annuals. **STEMS** 15–40 cm; loosely white-tomentose, not glandular; taprooted or fibrous-rooted. **LEAVES** crowded, internodes 1–5, with narrowly obovate to subspatulate blades, 1–3 (–6) cm × 2–8 mm (distal smaller, oblanceolate to narrowly oblong or linear), bases subclasping, usually decurrent 1–2 mm, margins weakly revolute, faces mostly concolor to weakly bicolor, abaxial gray-tomentose, adaxial usually gray-tomentose, sometimes glabrescent, neither glandular. **HEADS** in terminal glomerules (1–2 cm diameter). **INVOLUCRES** broadly campanulate, 3–4 mm. **PHYLLARIES** in 3–4 series, silvery gray to yellowish (hyaline), ovate to ovate-oblong, glabrous. **PISTILLATE FLORETS** 135–160. **BISEXUAL FLORETS** 5–10; corolla red-tipped. **CYPSELAE** not evidently ridged, conspicuously dotted with whitish, papilliform hairs. **PAPPUS** bristles loosely coherent basally, released in clusters or easily fragmented rings. $2n = 14, 16, 28$. [*Gnaphalium luteoalbum* L.]. Ditches, stream banks, washes, seasonal ponds, commonly heavily grazed or disturbed. ARIZ: Nav; UTAH. 1100–1800 m (3610–5905′). Flowering: Apr–Oct. Native to Eurasia and adventive on most continents, scattered in the United States in western and southeastern states, including New Mexico, Arizona, and Utah; Mexico.

Pseudognaphalium macounii (Greene) Kartesz (for John Macoun, 1831–1920, Canadian botanist) Macoun's rabbit tobacco. Annuals or biennials (often sweetly fragrant), 40–90 cm; taprooted. **STEMS** stipitate-glandular throughout (usually persistently lightly white-tomentose distally). **LEAVES** not crowded, internodes mostly 5+ mm, with lanceolate to oblanceolate blades, 3–10 cm × 3–13 mm (distal linear), bases not clasping, decurrent 5–10 mm, margins flat to slightly revolute, faces weakly bicolor, abaxial tomentose, adaxial stipitate-glandular, otherwise glabrescent or glabrous. **HEADS** in corymbiform arrays. **INVOLUCRES** campanulo-subglobose, 4.5–5.5 mm. **PHYLLARIES** in 4–5 series, stramineous to creamy (hyaline, shiny), ovate to ovate-oblong, glabrous. **PISTILLATE FLORETS** 47–101 (–156). **BISEXUAL FLORETS** 5–12 (–21). **CYPSELAE** not ridged, ± papillate-roughened. [*Gnaphalium macounii* Greene, sometimes misidentified as *Gnaphalium viscosum* Kunth]. Roadsides, depressions in rocks; mountain brush, ponderosa pine, piñon pine, and juniper communities. ARIZ: Apa; COLO: LPl; NMEX: McK, SJn. 1800–2600 m (5905–8530′). Flowering: Jul–Oct. Across Canada, south into the northeastern United States, Washington and Montana south to Mexico.

Pseudognaphalium stramineum (Kunth) Anderb. (of straw, probably alluding to the color of the phyllaries) Cotton-batting-plant. Annuals or biennials, 30–60 (–80) cm; taprooted. **STEMS** 1+ from base, erect to ascending, loosely tomentose, not glandular. **LEAVES** crowded, internodes usually 1–5, sometimes to 10 mm, with oblong to narrowly oblanceolate or subspatulate blades, 2–8 (–9.5) cm × 2–5 (–10) mm (smaller distally, narrowly lanceolate to linear), bases subclasping, usually not decurrent, sometimes decurrent 1–2 mm, margins flat or slightly revolute, faces concolor, loosely and persistently gray-tomentose, not glandular. **HEADS** in terminal glomerules (1–2 cm diameter). **INVOLUCRES** subglobose, 4–6 mm. **PHYLLARIES** in 4–5 series, whitish (often yellowish with age, hyaline, shiny), ovate to oblong-obovate, glabrous. **PISTILLATE FLORETS** 160–200. **BISEXUAL FLORETS** (8–)18–28; **CYPSELAE** weakly, if at all, ridged (otherwise smooth or papillate-roughened, glabrous, without papilliform hairs). **PAPPUS** bristles loosely coherent basally, released in clusters or easily fragmented rings. $2n = 28$. [*Gnaphalium stramineum* Kunth; *Gnaphalium chilense* Spreng.]. Streamsides, washes, chaparral slopes, cliff walls. ARIZ: Apa; NMEX: SJn. 1950–2200 m (6400–7215′). Flowering: Jun–Sep. Canada (British Columbia) south to California, Arizona, New Mexico, and west Texas, disjunct adventives along the Atlantic coast; Mexico.

Psilostrophe DC. Paperflower, Paper-daisy

Brian Elliott

(Greek *psilos*, naked, + *strophe*, to turn or twist, derivation obscure) Perennial or biennial, woody-taprooted herbs or low suffrutescent shrubs, often pilose to woolly or floccose. **STEMS** erect, simple below and branched above, 5–60 cm high, glabrous to villous, tomentose or tomentulose. **LEAVES** rosulate and cauline or all cauline, villous to nearly glabrous; lower leaves linear to ovate or spatulate, entire to pinnately lobed or pinnatifid; upper leaves reduced, entire. **INFL** heads radiate, solitary or few to many in corymbose clusters, subsessile to long-pedunculate. **INVOLUCRES** cylindric to campanulate. **PHYLLARIES** in 1–2 series; outer series 4–13, linear-oblong to lanceolate, villous, connivent; inner series (if present) 1–7, smaller than the outer bracts, scarious or with membranous margins; rarely with a calyculus subtending the outer bracts. **RECEPTACLES** flat, naked. **RAY FLORETS** pistillate, fertile; ligules 1–8, yellow to orange, broad and ovate, becoming papery and persistent on achene; apex 3-lobed. **DISC FLORETS** perfect, fertile, 5–25 per head; corolla yellow, 5-lobed, the lobes glandular-pubescent. **CYPSELAE** sublinear or linear, terete to slightly angled, conspicuously striate, glabrate to villous or sparsely glandular. **PAPPUS** of 4–6 hyaline, laciniate or entire, subequal scales. Seven species of the western United States and adjacent Mexico, previously included in *Riddellia.* Toxic to livestock and often weedy (Brown 1978; Heiser 1944).

1. Stems thinly and inconspicuously hairy above; ligules tightly reflexed against involucre at maturity ..... ***P. sparsiflora***
1′ Stems loosely villous to white-woolly; ligules horizontal at maturity ..... (2)

2. Pappus scales ovate-oblong and obtuse, less than 1/2 the length of disc corolla; ray flowers 5–8; disc flowers 10–20; involucre 7–10 mm ..... ***P. bakeri***
2′ Pappus scales lanceolate to subulate and acute, 1/2 to equaling the disc corolla; ray flowers 2–6; disc flowers 5–12; involucre 4.5–7 mm ..... ***P. tagetina***

Psilostrophe bakeri Greene (in honor of the collector of the type specimen) Baker's paperflower, Colorado paperflower. Perennial, tomentose to villous or floccose herbs; caudex branched, woody. **STEMS** 1–several, branched, 0.5–3 dm high. **LEAVES** cauline and basal, simple, loosely to densely villous or floccose; lower leaves to 10 cm long, oblanceolate to spatulate, entire or rarely 3–5-pinnately lobed, petiolate or sessile; upper leaves sessile. **INFL** heads loosely corymbose at branch tips; peduncle to 5 cm. **INVOLUCRES** cylindric, 7–10 mm high, lightly villous. **RAY**

FLORETS 4–8 per head; ligules yellow-orange, 8–17 mm long, to 10 mm broad, not reflexed at maturity. **DISC FLORETS** 10–20 per head; corolla 4–6 mm long. **CYPSELAE** striate, glabrous. **PAPPUS** scales ovate to broadly oblong, rounded or obtuse, short-erose or denticulate, 1–2 mm long, nearly transparent, 1/3 to nearly 1/2 the length of disc corolla. $2n = 32$. [*Riddellia tagetina* Nutt. var. *pumila* M. E. Jones]. Dry plains and hills, often in alkaline soils. COLO: LPl; NMEX: San, RAr. 1980–2165 m (6500–7100′). Flowering: May–Sep. Southern Idaho to Utah and Colorado. The type specimen is from Montrose County, Colorado, but outside the boundary of this flora.

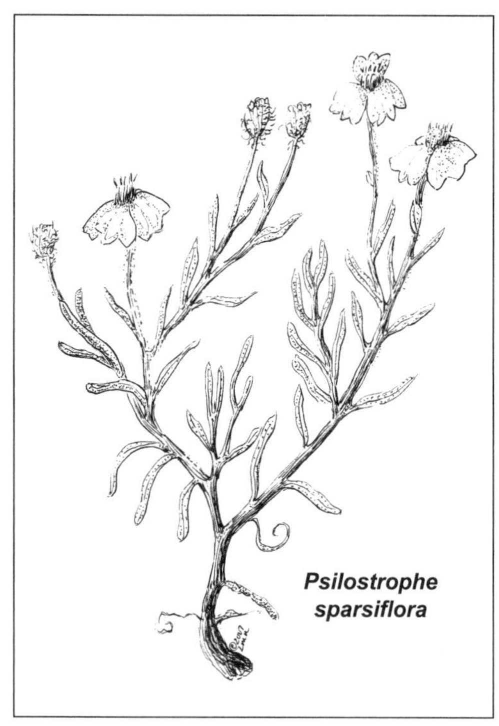
Psilostrophe sparsiflora

Psilostrophe sparsiflora (A. Gray) A. Nelson (probably referring to the relatively few ray flowers) Greenstem paperflower. Perennial herbs from a usually branched caudex; often woolly at the crown, otherwise inconspicuously hairy and often atomiferous-glandular above. **STEMS** single or clustered, freely and divergently branched above, 1–6 dm high, sparsely pilose to glabrate, often appearing twisted or zigzag at the nodes. **LEAVES** rosulate, reduced above; basal leaves to 14 cm long, linear to spatulate, entire or rarely pinnately lobed, petiolate, villous or glabrate with age, often withering with age; cauline leaves smaller than basal, linear to narrowly oblanceolate, sessile, sparsely villous to glabrate, dotted with atomiferous glands above. **INFL** heads in loose corymbs of 3–6; peduncle 0.5–3.5 cm long, slender. **INVOLUCRES** cylindric, 4–6 mm high, loosely woolly. **RAY FLORETS** 1–4 per head; ligules yellow, 5–11 mm long, broader than long, sharply reflexed at maturity. **DISC FLORETS** 5–12 per head; corolla 3–5 mm long. **CYPSELAE** 2.5–3 mm long, glabrous to sparsely sessile-glandular. **PAPPUS** scales lanceolate to linear, often erose, mostly acute, 1.5–3 mm long, 1/2 to 2/3 the length of disc corolla. $2n = 32$. [*P. divaricata* Rydb.; *Riddellia tagetina* Nutt. var. *sparsiflora* A. Gray]. Dry slopes, open woods, sagebrush and desert shrublands. ARIZ: Apa, Coc, Nav; NMEX: McK, SJn; UTAH. 1460–2300 m (4800–7600′). Flowering: Apr–Oct. Known from Utah, Arizona, and New Mexico. Although toxic, the species has been used medicinally as a blood purifier, as a poultice for wounds, and to treat diarrhea.

Psilostrophe tagetina (Nutt.) Greene (resembling *Tagetes*, the marigold) Woolly paperflower. Perennial herbs from a caudex; white-woolly below, loosely tomentose above. **STEMS** clustered at the crown, usually branched above, 1–5 dm high, loosely to densely villous, becoming less so or glabrate with age. **LEAVES** rosulate, reduced above; basal leaves to 15 cm long, ovate to oblanceolate or spatulate, entire to lobed or pinnatisect, loosely to densely white-villous; upper leaves reduced, 1–7 cm long, linear to oblanceolate or spatulate, lightly villous. **INFL** heads numerous in dense to open corymbs terminating the branches; peduncle 0.5–4 cm long. **INVOLUCRES** 4.5–7 mm high, densely white-villous or tomentose. **RAY FLORETS** 2–6 per head; ligules yellow, 5–14 mm long, usually broader than long. **DISC FLORETS** 5–12 per head; corolla 3–5 mm long. **CYPSELAE** 2–2.5 mm long, striate, glabrous or sparsely hairy. **PAPPUS** scales narrowly to broadly lanceolate or oblong, entire to erose, rounded or obtuse to acute, 2 mm long, 1/2 as long as to equaling the disc corolla. $2n = 32$. [*P. grandiflora* Rydb.; *P. hartmanii* Rydb.; *P. tagetina* var. *lanata* A. Nelson; *Riddellia tagetina* Nutt.]. Desert shrubland, dry, open plains and mesas, and pine forests. ARIZ: Coc; NMEX: McK, RAr, San, SJn. 1280–2250m (6900–7425′). Flowering: May–Oct. Southwestern United States from Utah and Arizona eastward to Texas and Kansas and adjacent Mexico. A morphologically diverse species with subspecies historically named on the basis of highly variable characters such as pubescence or size of ligules and peduncles. It is often mistaken for *P. bakeri.* The species is reportedly toxic to sheep but has been used medicinally as an analgesic, a cathartic, a snakebite remedy, and to treat stomachaches and sore throats. The bright yellow flowers have been used for a dye or paint.

Pyrrocoma Hook. Goldenweed
G. L. Nesom

(Greek *pyrrhos*, reddish or tawny, + *kome*, hair of the head, in reference to the reddish pappus in some species) Perennial herbs 5–20 cm high, from a woody taproot and often a short, branched caudex, with decumbent to ascending or erect, often red-tinged stems, leaves, and phyllaries, sericeous, tomentose, or glabrous, sometimes sessile- or stipitate-glandular. **LEAVES** cauline or basal and cauline, basal oblanceolate to elliptic or nearly linear, 3–25 (–45) cm, uncommonly with a short petiole, entire to spinulose-dentate or -serrate or shallowly laciniate, cauline clasping to subclasping, reduced upward. **INFL** a radiate, solitary head, 2–5 (–15) in a racemoid to spiciform or loosely corymboid inflorescence. **INVOLUCRES** hemispheric to narrowly campanulate, 5–15 (–20) mm. **PHYLLARIES** in 2–6 equal to graduate series, herbaceous and yellow-green throughout or with a distinct, green, apical patch, or less commonly the lower 2/3 white-indurate. **RAY FLORETS** 10–80 in 1 series, pistillate, fertile, corollas yellow to orange. **DISC FLORETS** bisexual, fertile. **CYPSELAE** subcylindric-fusiform, terete to somewhat flattened, 3–4-angled, with several nerves between the angles, sericeous to strigose or glabrous. **PAPPUS** several series of brownish, rigid, barbellate bristles of unequal length. $x = 6$. [*Haplopappus* sect. *Pyrrocoma* (Hook.) H. M. Hall] (Mayes 1976). A genus of about 14 species (some with infraspecific variants) in the western United States and Canada.

1. Stems mostly 6–18 cm, sparsely villous to tomentose; basal leaves narrowly lanceolate, faces glabrous, margins ciliate; phyllaries obovate, in 3–4 graduate series, villous, ciliate .... ***P. clementis***
1′ Stems mostly 30–60 cm, glabrous, sometimes tomentulose near the inflorescence; basal leaves oblanceolate to spatulate or narrowly elliptic, faces and margins glabrous; phyllaries oblong to spatulate, in 2–3 equal to subequal series, glabrous .... ***P. crocea***

Pyrrocoma clementis Rydb. (honoring Frederick E. Clements, 1874–1945, ecologist and botanist, collector of the type) Clements' goldenweed. **STEMS** 6–18 (–40) cm, sparsely villous to tomentose. **LEAVES**: basal narrowly lanceolate, 5–14 × 1–1.6 cm, faces glabrous, margins ciliate, cauline smaller, clasping, relatively even-sized or gradually reduced distally. **INVOLUCRES** 20–34 mm wide. **PHYLLARIES** narrowly obovate, acute, in 3–4 graduate series, villous, ciliate. $2n = 12$. [*Haplopappus clementis* (Rydb.) S. F. Blake]. Subalpine and alpine meadows, rocky slopes, openings in aspen and spruce-fir communities. COLO: Hin, LPl, Min. 3100–3750 m (10170–12305′). Flowering: Jun–Sep. Wyoming, Colorado, Utah. Four Corners plants are **var. *clementis***.

Pyrrocoma crocea (A. Gray) Greene **CP** (of saffron, the yellow color) Curly-head goldenweed. **STEMS** 30–60 (–80) cm, glabrous, often finely tomentose near the inflorescence. **LEAVES**: basal spatulate with oblanceolate to elliptic blades, 10–45 × 2–6 cm, faces and margins glabrous, cauline smaller, clasping, relatively even-sized or gradually reduced distally. **INVOLUCRES** 15–60 mm wide. **PHYLLARIES** oblong to spatulate, in 2–3 equal to subequal series, glabrous. $2n = 12$. [*Haplopappus croceus* A. Gray]. Clay hills and flats, roadsides and cuts, disturbed sites; sagebrush and grassland communities. COLO: Arc, Hin, LPl, Min, Mon, SJn; NMEX: RAr. 2150–2350 m (7055–7710′). Flowering: Jul–Sep. Wyoming, Utah, Colorado, New Mexico. Four Corners plants are **var. *crocea***.

Ratibida Raf. Prairie Coneflower, Mexican Hat

Gary I. Baird

(derivation unknown) Perennials, taprooted or fibrous-rooted, or rhizomatous, caudex semiwoody, branched, herbage strigose-hirsute, often gland-dotted, sap clear. **STEMS** 1 to many, erect, branched (at least distally). **LEAVES** basal and cauline, simple, petiolate to (distally) subsessile, cauline ± reduced, alternate; blades ovate to lanceolate, obovate, oblong, or oblanceolate, pinnatifid to bipinnatifid, bases cuneate to attenuate, margins entire to serrate, apices bluntly acute. **HEADS** radiate, borne singly or in corymbiform clusters, erect; peduncles ebracteate. **INVOLUCRES** spreading-campanulate to rotate, sometimes obscure. **PHYLLARIES** 5–15 in 2 series, herbaceous, ± unequal (outer sometimes surpassing inner), spreading to reflexed. **RECEPTACLES** ovoid to columnar, paleate; paleae ± cucullate-conduplicate, clasping adjacent floret, chartaceous, lateral faces each with a purple-black gland, margins membranous, apices herbaceous, densely strigose-velutinous and gland-dotted, deciduous; calyculi lacking. **RAY FLORETS** 3–15, neuter; corollas yellow, reddish brown, purplish brown, or bicolored; laminae reflexed, persistent. **DISC FLORETS** 50–400+, bisexual, fertile; corollas green to yellowish green, lobes spreading to strongly reflexed, sometimes purplish adaxially and/or glandular abaxially; anthers black-purple, glandular. **CYPSELAE** monomorphic, blackish, strongly compressed, linear-oblanceolate to oblong, margins ciliate-erose, faces glabrous or sparsely hairy, apices truncate-oblique. **PAPPUS** lacking, or of 1–2 toothlike projections, or coroniform. $x = 16$. Seven species mostly of central and southern North America.

1. Leaves pinnatifid; peduncles long (to 40+ cm), heads much surpassing the leaves; discs columnar, to 50 mm tall, mostly greenish gray (before flowers open); ray corollas mostly 10–35 mm long .... ***R. columnifera***
1′ Leaves pinnatifid to bipinnatifid; peduncles short (to 7 cm), heads not much surpassing the leaves; discs ± ovoid, to 15 mm tall, purplish gray (before flowers open); ray corollas mostly less than 10 mm long .... ***R. tagetes***

Ratibida columnifera (Nutt.) Wooton & Standl. (column-bearing, alluding to the elongated receptacle) Upright prairie coneflower, Mexican hat. Perennials 20–100+ cm tall. **LEAVES** with blades pinnatifid (rarely some bipinnatifid); lobes 5–13, oblanceolate to linear, margins entire to lobulate, apices acute; cauline ± reduced upward, often sessile. **HEADS** cylindrical, 10–50 × 7–10 mm; peduncles mostly 10–40 cm tall, prominently striate-nerved; paleae white-pubescent apically (uncommonly purple-pubescent), the discs mostly appearing greenish gray before flowers open, lateral glands oval to oblanceolate. **INVOLUCRES** 8–12 mm diameter. **PHYLLARIES** 5–14, linear to linear-lanceolate, about 5–10 mm long, spreading, outer much surpassing inner. **RAY FLORETS** 4–7; corollas yellow to reddish or purplish brown, often bicolored, laminae narrowly ovate to oblong- or elliptic-obovate, 10–35 × 5–15 mm, adaxially strigose and gland-dotted. **DISC FLORETS** 200–400+; corollas greenish, sometimes lobes purplish. **CYPSELAE** ± oblong, mostly 1–3 mm, often as wide as tall, abaxial margin glabrous, adaxial and apical margins ciliate-erose. **PAPPUS** of 1–2 toothlike projections (often unequal), sometimes lacking. $2n = 28$. Sagebrush, piñon-juniper, ponderosa pine–oak communities, often in disturbed sites. ARIZ: Apa, Nav; COLO: Arc, LPl, Mon; NMEX: McK, RAr, SJn; UTAH. 1800–2400 m (6000–7800′). Flowering: Jun–Sep. Southern Canada, south primarily through the Great Plains to northern Mexico. Often cultivated or included in wildflower seed mixes, and now widely introduced outside its native range. Plants with predominantly purplish or reddish brown rays have been treated as f. ***pulcherrima*** (DC.) Fernald (sometimes treated as a variety). Ramah Navajo uses include a cold infusion given to sheep that were "out of their minds."

Ratibida tagetes (E. James) Barnhart (for *Tages*, an Etruscan city) Green prairie coneflower. Perennials 15–40 cm tall. **LEAVES** with blades pinnatifid to often bipinnatifid (basal and proximal cauline sometimes merely lobed, rarely entire); lobes 2–7, linear to linear-lanceolate or lance-ovate, margins entire to lobulate, apices acute; cauline ± reduced upward, sessile. **HEADS** ovoid to ellipsoid, 5–15 × 5–14 mm; peduncles mostly 2–7 cm tall, obscurely nerved; paleae purple-pubescent apically, the discs appearing purplish gray before flowers open, lateral glands linear to oblanceolate. **INVOLUCRES** 4–10 mm diameter. **PHYLLARIES** 10–12, linear-lanceolate, about 2–5 mm long, spreading-reflexed, outer surpassing inner. **RAY FLORETS** 5–10; corollas yellow or reddish or purplish brown, often bicolored, laminae elliptic-oblanceolate, 3–10 × 2–6 mm, adaxially strigose and gland-dotted. **DISC FLORETS** ± 50–150; corollas greenish, lobes often purple. **CYPSELAE** ± oblong, 2–3 mm, sometimes as wide as long, abaxial margin glabrous, adaxial margin narrowly winged and distally ciliate-erose, apical margin obscurely ciliate or glabrous. **PAPPUS** coroniform. $2n = 32$. Disturbed areas in piñon-juniper communities. COLO: Arc, Mon; NMEX: SJn. 1800–2000 m (5600–6600′). Flowering: Aug. Colorado and Kansas, south to Arizona, New Mexico, Texas, and northern Mexico. Ramah Navajo uses include infusion of leaves used for stomachache or as a cathartic; plant used in ceremonial chant lotion; cold infusion of leaves taken for coughs and fever; and plant used as a fumigant for sexual infection.

Rudbeckia L. Coneflower, Black-eyed Susan

Gary I. Baird

(for Olaus J. Rudbeck, 1630–1702, and Olaus O. Rudbeck, 1660–1740, father and son Swedish professors at Uppsala University) Annuals, biennials, or perennials, fibrous-rooted, rhizomatous, or (rarely) taprooted, herbage glabrous or pubescent, sometimes glaucous, sometimes gland-dotted, sap clear. **STEMS** 1 to many, erect, branched distally. **LEAVES** basal and cauline, simple or compound, petiolate or sessile, alternate (occasionally the most proximal opposite); blades lanceolate to ovate, elliptic, oblanceolate, spatulate, or linear, bases generally cuneate, margins entire, dentate, serrate, pinnately lobed, pinnatifid, or bipinnatifid, apices acute to acuminate. **HEADS** radiate or discoid, borne singly or many in corymbiform or paniculiform clusters, erect; peduncles ebracteate. **INVOLUCRES** hemispheric to rotate. **PHYLLARIES** 5–20 in mostly 2–3 series, herbaceous (at least distally), ± subequal, spreading to reflexed. **RECEPTACLES** subspheric to ovoid, conic, or columnar, paleate; paleae ± herbaceous (at least distally), sometimes purplish distally, often clasping adjacent cypsela, often surpassing florets, deciduous; calyculi lacking. **RAY FLORETS** 5–25, or lacking, neuter; corollas yellow to yellow-orange, sometimes brownish purple proximally; laminae spreading to reflexed, fugacious to persistent. **DISC FLORETS** 50–800+, bisexual, fertile; corolla yellow to yellowish green or brownish purple (often bicolored). **CYPSELAE** monomorphic, obpyramidal, 4-angled, glabrous (or pubescent on angles). **PAPPUS** lacking, or of 2–8 scales or coroniform. $x = 16, 18, 19$. Approximately 20–25 species of North America, many species widely cultivated and sometimes adventive (as in Europe). Only one species is native to the study area, but a number of other species are to be expected adventively due to their widespread use in revegetation mixes, wildflower mixes, and as garden ornamentals.

Rudbeckia laciniata L. (cut into strips, alluding to the divided leaves) Cutleaf coneflower, green-eyed susan. Perennials 50–200 cm tall, rhizomatous, herbage mostly glabrous and ± glaucous. **STEMS** 1, usually branched distally. **LEAVES**:

basal long-petiolate, often withered at flowering; cauline petiolate to sessile, reduced upward; blades ovate to lanceolate, (proximal) 10–30 × 5–20 cm (sometimes wider than long), mostly 3–5-pinnately lobed or pinnately compound (the most distal sometimes unlobed), margins entire to serrate, often sparsely strigillose-ciliate, faces glabrous (adaxially) or sparsely strigillose (abaxially); lobes (or leaflets) crowded, often dentate or 1–5-lobulate, sometimes entire. **HEADS** radiate, usually several in a corymbiform cluster; peduncles often long, glabrous or puberulent at base of head. **INVOLUCRES** 15–35 mm diameter, usually rotate. **PHYLLARIES** 10–15, ovate to lanceolate, to 20 mm long, sparsely to densely strigillose-puberulent, or glabrous. **RECEPTACLES** hemispheric to conical, 1–3 cm tall; paleae linear, apices densely puberulent abaxially. **RAY FLORETS** mostly 8–12; laminae elliptic to oblanceolate, 15–50 mm, abaxially strigillose. **DISC FLORETS** 150–300+; corolla yellow to yellowish green. **CYPSELAE** 3–4 mm long. **PAPPUS** of 4 reduced scales, to 1 mm, or coroniform. $2n = 38, 54, 72$. Riparian, meadow, and other moist habitats in ponderosa pine–oak, aspen, Douglas-fir, and spruce-fir communities. ARIZ: Apa; COLO: Arc, Dol, Hin, LPl, Min, Mon, SJn; NMEX: McK, RAr, SJn. 2000–3000 m (6500–10000′). Flowering: Jun–Sep. Eastern North America, westward to the Rocky Mountains. The species contains perhaps four or five varieties. All of the material from the study area is assignable to **var. *ampla*** (A. Nelson) Cronquist (large). Cultivated forms of this species are primarily derived from var. *laciniata*.

Sanvitalia Lam. Creeping Zinnia
Gary I. Baird

(possibly for Federico Sanvitali, an Italian professor, 1704–1761) Annuals (perennials), taprooted, rhizomes lacking, herbage pubescent, sap clear. **STEMS** 1, erect to prostrate, branched from base or throughout. **LEAVES** cauline, simple, petiolate or sessile, opposite; petioles short; blades lanceolate to obovate, spatulate, or linear, usually 3-nerved (at least proximally), bases cuneate to rounded, margins entire (rarely dentate or lobate), apices acute to acuminate. **HEADS** radiate, borne singly, erect; peduncles short to obscure. **INVOLUCRES** hemispheric to rotate. **PHYLLARIES** about 10–20 in 2–3 series, subequal or imbricate (outer shorter). **RECEPTACLES** convex to conical, paleate; paleae persistent, scarious; calyculi lacking. **RAY FLORETS** mostly 5–20, pistillate, fertile; corollas whitish or yellowish; laminae sessile, marcescent, becoming papery. **DISC FLORETS** 15–60, bisexual, fertile; corollas yellow to orange (distally). **CYPSELAE** dimorphic, outer usually angled, inner ± flattened to terete, ribbed, angles thickened or corky or winged, smooth to tuberculate, glabrous or pubescent. **PAPPUS** of 2–4 hornlike awns, or lacking. $x = 8, 11$. Five species of the southwestern United States, south to Central and South America.

Sanvitalia abertii A. Gray (for Lieutenant James W. Abert, a member of John C. Frémont's third expedition, in 1845) Abert's creeping zinnia. Annuals 3–30 cm tall. **STEMS** erect. **LEAVES**: blades lanceolate to linear, 20–40 × 2–5 mm, bases cuneate, margins strigose, otherwise entire, faces sparsely strigose (mostly along veins). **HEADS** broadly campanulate. **INVOLUCRES** mostly 5–10 mm diameter. **PHYLLARIES** 8–10 in about 2 series, subequal, ± herbaceous (at least distally), margins membranous, faces prominently white striate-nerved. **RECEPTACLES** ± conical; paleae linear, hyaline-scarious, prominently white striate-nerved, sometimes purplish distally, surpassing florets, grading into phyllaries, deciduous. **RAY FLORETS** (3) 6–10, each subtended by a phyllary; corollas yellow or whitish, prominently green striate-nerved abaxially, laminae ovate, 2–4 mm long, apices bifid (occasionally trifid). **DISC FLORETS** about 15–25; corollas about 2 mm long, green proximally, whitish distally. **CYPSELAE** about 4 mm; ray cypselae mostly 4-angled, whitish-cartilaginous, sulcate below lateral pappus horns; disc cypselae (outer) ± terete-angled to (inner) somewhat compressed and narrowly winged, tuberculate, blackish green or brownish. **PAPPUS** of (ray florets) 2–4 hornlike projections or knobs, to 1 mm, confluent with cypsela body, or (disc florets) lacking. $2n = 22$. Piñon-juniper and desert grassland communities. ARIZ: Apa; NMEX: McK, San. 1900–2100 m (6200–6800′). Flowering: Aug–Sep. Southern California eastward to trans-Pecos Texas, and south into Mexico.

Scabrethia W. A Weber Whitestem Sunflower
Gary I. Baird

(Latin *scabra*, rough or scratchy, + alteration of *Wyethia*) Perennials, taproots stout, rhizomes lacking, caudices simple or multibranched, somewhat woody, herbage scabrous to hispid, sap clear. **STEMS** often many, erect, simple (rarely branched distally), white. **LEAVES** all cauline, simple, ± sessile, alternate, proximal reduced and scalelike; blades linear to linear-oblong or linear-lanceolate, bases cuneate, margins entire, apices bluntly to acuminately acute. **HEADS** radiate, borne singly, erect; peduncles mostly ebracteate. **INVOLUCRES** ± hemispheric. **PHYLLARIES** mostly 30–50 in 4–6 series, herbaceous, outer often surpassing inner. **RECEPTACLES** flat to convex, paleate; paleae persistent, chartaceous; calyculi lacking. **RAY FLORETS** 10–20, pistillate, fertile; corollas yellow, fugacious. **DISC FLORETS** mostly 40–120+, bisexual, fertile; corollas yellow. **CYPSELAE** monomorphic, 4-angled, glabrous. **PAPPUS** coroniform, persistent. $x = 19$. One species of western North America.

Scabrethia scabra (Hook.) W. A. Weber **CP** (rough or scratchy, alluding to the leaf and stem surfaces) Perennials 20–60 cm tall. **LEAVES**: blades 8–20 cm × 5–20 mm, faces coarsely scabrous or hispid but often glabrate and becoming merely scabrous, midvein prominent, white. **INVOLUCRES** mostly 20–30 mm diameter (usually of equal height). **PHYLLARIES** linear-lanceolate to narrowly deltate or ovate, subequal or unequal (outer surpassing inner), margins ± ciliate, scabrous to densely pilose-hispid, faces scabrous to hispid or subglabrous, sometimes minutely gland-dotted, apices acuminate to attenuately filiform, spreading to reflexed, or erect (inner). **RAY FLORET** laminae 15–50 mm long. $2n = 38$. [*Wyethia scabra* Hook.]. Sandsage, blackbrush, saltbush, and piñon-juniper communities, mostly in sandy soils or dunes, often along the bottom of washes, canyons, or ravines. ARIZ: Apa, Nav; NMEX: McK, SJn; UTAH. 1300–2300 m (4200–7700′). Flowering: May–Sep. Arizona, Colorado, New Mexico, Utah, and Wyoming, generally resticted to the Colorado Plateau. Recognition of intraspecific taxa seems unmerited. Our material belongs to **subsp. *canescens*** (W. A. Weber) W. A. Weber.

Schkuhria Roth False Threadleaf

Kenneth D. Heil

(for Christian Schkuhr, 1741–1811, German botanist) Annual, up to 50+ cm tall. **STEMS** erect, simple or branched. **LEAVES** mostly opposite, mostly cauline; petiolate or sessile; blades linear to filiform or 1–2-pinnately lobed, ultimate margins entire, faces sparsely hairy, glabrescent, usually gland-dotted. **HEADS** radiate or discoid, mostly in corymbiform to paniculiform arrays. **INVOLUCRES** obconic-obpyramidal to ± turbinate or hemispheric, 3–6 mm diameter. **PHYLLARIES** persistent, 4–6+ in 1–2 series. **RAY FLORETS** mostly 1–2 (ours), pistillate, corollas yellow or white. **DISC FLORETS** 2–8 (–30+), bisexual, corollas yellow to yellowish, sometimes red-tipped. **CYPSELAE** narrowly obpyramidal, 4-angled. **PAPPUS** persistent, of 8+ spatulate to lanceolate scales. $x = 11$ (Flora of North America Editorial Committee 2006). Two species in the southwestern United States, Mexico, Central America, and South America.

Schkuhria multiflora Hook. & Arn. (many-flowered) Many-flower false threadleaf. Mostly 3–12 (–25) cm tall. **STEMS** decumbent-ascending to erect. **LEAVES** 1–3 cm long; blades linear or lobed, faces puberulent and gland-dotted. **PEDUNCLES** 5–25 mm long. **PHYLLARIES** 7–10, green to purple, weakly carinate, oblanceolate to obovate, gland-dotted, usually hirsutulous. **RAY FLORETS** 0. **DISC FLORETS** 15–30+; corollas yellowish, 1–2 mm long. **CYPSELAE** blackish to buff, 3 mm long. **PAPPUS** of 8 white to tawny or purplish, obovate-rounded or oblanceolate to quadrate scales, 1–2 mm long. $2n = 22$. [*Bahia neomexicana* (A. Gray) A. Gray; *Schkuhria neomexicana* A. Gray]. Mudflats, cracks of sandstone, and roadsides; ponderosa pine communities. ARIZ: Apa; NMEX: McK, San, SJn. 2105–2280 m (6900–7480′). Flowering: Sep–Oct. California, Arizona, Colorado, New Mexico, and Texas; Mexico. The plant is chewed for 10 minutes for mouth sores by the Ramah Navajo.

Scorzonera L. Vipergrass

Gary I. Baird

(Italian *scorzone* or Spanish *escorzon*, a venomous reptile, alluding to its purported use in treating snakebites) Biennials (mostly), taprooted, herbage sparsely pubescent or glabrate, sap milky. **STEMS** 1, branched at base and/or distally, erect to (frequently) decumbent. **LEAVES** basal (most) and cauline, alternate, simple, sessile or obscurely petiolate, basal clasping, progressively much reduced upward; blade margins entire to pinnatifid. **INFL** terminal, erect; peduncles sparsely bracteate, not inflated. **HEADS** liguliflorous. **PHYLLARIES** herbaceous, imbricate. **RECEPTACLES** flat, epaleate; calyculi lacking. **RAY FLORETS** matutinal; corolla ligulate, yellow, abaxially and distally purplish. **DISC FLORETS** lacking. **CYPSELAE** monomorphic, whitish to light brown, columnar to obclavate (ours), glabrous, not beaked. **PAPPUS** of long, bristlelike, subulate scales in 2–3 series, persistent, ochroleucous to tawny, plumose. $x = 7$. About 30 species in Eurasia.

Scorzonera laciniata L. (cut into narrow strips, alluding to the lobed leaves) Biennials (perennials) 15–45 cm tall. **STEMS** conspicuously white-nerved especially proximally, base often purplish. **LEAVES**: basal and lower cauline blades mostly linear, 6–26 cm × 1–12 mm (excluding lobes), bases prominently 5-nerved, margins entire to pinnately lobed, midrib prominent, white; cauline mostly entire and reduced; lobes 1–3 pairs, subopposite, mostly linear, 2–13 mm long, entire. **INFL** with peduncles pubescent to glabrate, base of heads more or less remaining pubescent. **INVOLUCRES** ± conical, 12–17 mm tall (in flower) to 30 mm tall (in fruit). **PHYLLARIES** imbricate in about 4 series, ovate to lanceolate, erect, margins entire, scarious, mostly purplish, faces glabrous to pubescent, mostly green; outermost often much reduced. **RAY FLORETS** 25–65, subequal or just exceeding involucre; corolla sparsely pubescent at tube-ligule juncture, otherwise glabrous, tube 7–8 mm, ligule 4–7 mm × 1–1.5 mm; anthers 2–3 mm long. **CYPSELAE** 9–12 mm long, 5-angled (especially distally), 10-ribbed, proximal 1/3 hollow and enlarged; ribs white (especially proximally). **PAPPUS** 9–17 mm long. $2n = 14$. [*Podospermum laciniatum* (L.) DC.]. Sagebrush, shadscale, and rabbitbrush communities; on alluvial flats and wash bottoms, mostly in sandy-silty soils, commonly in disturbed areas. COLO: Arc, Mon; NMEX: McK, RAr, SJn. 1500–2200 m (4900–7200′). Flowering: Apr–Jun. Europe, widely introduced into the western United States.

Senecio L. Groundsel, Ragwort, Butterweed

G. L. Nesom and T. M. Barkley

(reputedly from Latin *senex*, old man or woman, alluding to the white pappus bristles resembling the white hair of an elderly person) Annuals, biennials, perennials, subshrubs, or shrubs; herbage glabrous or hairy, often glabrescent at flowering. **LEAVES** basal and/or cauline; alternate; petiolate or sessile, bases sometimes clasping; blades ovate or deltate to oblanceolate, lanceolate, linear, or filiform, rarely suborbiculate, sometimes palmately or pinnately lobed to 2–3-pinnatifid, margins entire or denticulate to serrate or toothed. **INFL** a head, radiate or discoid, sometimes nodding, 1–numerous; calyculi usually of 1–8 bractlets, sometimes absent. **INVOLUCRES** cylindric to turbinate or campanulate. **PHYLLARIES** usually ±5, 8, 13, or 21 in (1–) 2 series, distinct (margins interlocking), mostly oblong to lanceolate or linear, subequal or equal. **RECEPTACLES** epaleate. **RAY FLORETS** usually ±5, 8, 13, or 21, pistillate, fertile, sometimes absent; corollas usually yellow. **DISC FLORETS** bisexual, fertile; corollas usually yellow, tubes shorter than to equaling campanulate throats; style branches stigmatic in 2 lines, apices usually truncate-penicillate. **CYPSELAE** cylindric or prismatic, usually 5-ribbed or 5-angled, glabrous or hairy. **PAPPUS** of 30–80 fragile bristles, usually persistent, sometimes caducous, white, barbellulate or smooth. $x = 10$. Approximately 1000+ species, nearly worldwide, mostly in warm-temperate, subtropical, and tropical regions at middle and upper elevations.

1. Annuals, without prominent caudices or rhizomes; ray florets absent ..... ***S. vulgaris***
1′ Perennials, usually with well-developed caudices or rhizomes; ray florets present and conspicuous (or absent in *S. bigelovii* and *S. pudicus*) ..... (2)

2. Subshrubs or shrubs (if weakly woody, then with upward-branching aspect of shrubs); leaves linear-filiform or pinnately parted into linear-filiform lobes ..... (3)
2′ Perennial herbs, caudices sometimes ± woody; leaves lanceolate to ovate or ovate-orbiculate (rarely sublinear in *S. serra*) ..... (4)

3. Stems and leaves woolly-tomentose to floccose, sometimes unevenly so; calyculus with prominent bractlets, lengths of at least some usually to 1/2 phyllaries; phyllaries ±13 or ±21 ..... ***S. flaccidus***
3′ Stems and leaves glabrous or sparsely tomentose (especially adaxial leaf faces, leaf axils, and among heads), never woolly-tomentose; calyculus with bractlets absent or lengths less than 1/4 phyllaries; phyllaries ±8 (±13) ..... ***S. spartioides***

4. Basal leaves usually withering before flowering, or at least not clustered and hardly distinct from the cauline, cauline ± equal in size, usually evenly distributed or concentrated distally ..... (5)
4′ Basal leaves usually clustered and present at flowering, middle and distal usually progressively smaller, midstem leaves sometimes equaling basal but then distal sharply reduced ..... (8)

5. Stems mostly 25–40 cm, ascending; heads 1–2 (–4); cauline leaves clasping or subclasping, unreduced upward or gradually smaller ..... ***S. fremontii***
5′ Stems mostly 40–120 cm, strictly erect; heads 10–60; cauline leaves not clasping, usually petiolate or subpetiolate, unreduced upward ..... (6)

6. Leaves (at least midcauline) pinnately lobed to irregularly incised or laciniate ..... ***S. eremophilus***
6′ Leaves subentire to dentate or serrate ..... (7)

7. Leaves narrowly lanceolate to sublinear, bases long-attenuate ..... ***S. serra***
7′ Leaves ± triangular, bases cuneate to ± truncate or subhastate ..... ***S. triangularis***

8. Heads usually strongly nodding, at least in bud ..... (9)
8′ Heads usually erect, weakly nodding if at all ..... (14)

9. Heads (3–) 8–16 (–40), distally clustered; basal and lower cauline leaves 8–20 (–28) cm, leaf margins mostly entire; ray florets absent .... ***S. pudicus***
9′ Heads 1–12 (–20), commonly solitary, otherwise loosely clustered; basal leaves (1–) 2–15 (–20) cm, leaf margins subentire to denticulate, dentate, or serrate, or incised to subpinnatifid; ray florets absent or present .... (10)

10. Ray florets absent .... ***S. bigelovii***
10′ Ray florets present .... (11)

11. Principal leaves mostly lower and middle cauline, (5–) 10–20 cm; stems mostly 40–60 cm .... ***S. amplectens***
11′ Principal leaves mostly basal, blades (1–) 2–10 cm, cauline sharply reduced; stems mostly 5–20 cm .... (12)

12. Leaf blades (1–) 2–5 cm, margins dentate to incised or subpinnatifid, abaxial faces lightly but persistently floccose-tomentose. .... ***S. taraxacoides***
12′ Leaf blades 2–10 cm, margins entire to dentate or denticulate, both faces glabrous or nearly so .... (13)

13. Leaf blades elliptic-orbiculate to ovate-orbiculate or broadly ovate, essentially palmately nerved with 1–2 pairs of secondary veins arising from the near the base, margins entire or subentire to weakly dentate or denticulate; stems trailing; ray florets 13–16 mm, slightly longer than involucres .... ***S. soldanella***
13′ Leaf blades ovate to narrowly ovate, ovate-elliptic, broadly elliptic, or elliptic-oblanceolate, pinnately nerved with 3–5 pairs of secondary veins arising regularly from the midvein, margins coarsely to finely dentate; stems erect to ascending; ray florets 17–22 mm, about 2 times longer than involucres .... ***S. holmii***

14. Cauline leaves reduced and bractlike immediately above the basal, not clasping, glaucous; phyllary tips usually greenish .... ***S. wootonii***
14′ Cauline leaves reduced but continuing at least to about midstem, not clasping (*S. atratus*) or clasping, not glaucous; phyllary tips usually black .... (15)

15. Herbage lightly but persistently floccose-tomentose to canescent, sometimes unevenly glabrescent; heads 20–60; phyllaries (±5) ±8; involucres 5–6 mm diameter .... ***S. atratus***
15′ Herbage glabrous or glabrescent but persistently tomentose or villous at least when young; heads 4–15 (–30+); phyllaries (±8) ±13 or ±21; involucres 7–10 (–12) mm diameter .... (16)

16. Cauline leaves reduced but often continuing above midstem, margins sharply dentate or serrate to subentire; herbage glabrous; high elevations .... ***S. crassulus***
16′ Cauline leaves reduced to linear bracts above midstem, margins mostly entire to subentire, less commonly denticulate; herbage persistently tomentose or villous at least when young, at length glabrescent; middle elevations .... ***S. integerrimus***

Senecio amplectens A. Gray (clasping) Showy alpine ragwort. Perennials. **STEMS** (30–) 40–60 cm tall; rhizomatous or with branched caudices; herbage often purplish-tinged, glabrous or sparsely and unevenly hairy, especially near leaf axils and among heads, usually glabrescent. **LEAVES** mostly proximal and midcauline, basal smaller, usually withering before flowering; petiolate, petioles shorter than to equaling blades, often clasping; blades broadly lanceolate or lanceolate to oblanceolate, (5–) 10–20 cm long, 2–4 (–5) cm wide, bases tapered, margins dentate to denticulate, distal leaves smaller, bractlike. **HEADS** corymbose, nodding, 1–5 (–10); calyculi of 2–5 linear to filiform bractlets, lengths to 1/2 phyllaries. **PHYLLARIES** ±13 or ±21, tips often black or purplish, sometimes with scattered black hairs on abaxial faces. **RAY FLORETS** usually ±13, sometimes fewer. $2n = 40$, about 175–180. [*Ligularia amplectens* (A. Gray) W. A. Weber; *Senecio pagosanus* A. Heller]. Meadows, rocky slopes, talus; spruce-willow, spruce-fir, krummholz, and alpine tundra communities. COLO: Arc, Con, Hin, LPl, Min, RGr, SJn. 3100–3750 m (10170–12305′). Flowering: Jul–Aug (–Sep). Colorado and New Mexico. Compared to close relatives *S. holmii* (= *S. amplectens* var. *holmii*), *S. taraxacoides*, and *S. soldanella*, *S. amplectens* sensu stricto has taller stems with prominent cauline leaves (basal leaves usually withered by flowering). The phyllaries of *S. amplectens* are consistently dark purple.

Senecio atratus Greene (black or clothed in black) Black-tip ragwort. Perennials. **STEMS** (20–) 35–70 (–80+) cm tall; rhizomes or caudices branched, erect to weakly creeping; herbage floccose-tomentose to canescent, sometimes unevenly glabrescent. **LEAVES** progressively reduced distally; petiolate; blades oblong-ovate to oblanceolate, (5–) 10–30 cm long, 1.5–4 (–6) cm wide, bases tapered, margins dentate, denticles dark, callous; middle leaves similar, sessile, smaller; distal leaves bractlike. **HEADS** in compact corymbose clusters, 20–60; calyculi of 2–5 linear bractlets, lengths to 1/3 phyllaries. **PHYLLARIES** (±5) ±8, 6–8 mm, tips black. **RAY FLORETS** (±3) ±5. $2n = 40$. [*Senecio* *atratus* var. *milleflorus* (Greene) Greenm.; *S. milleflorus* Greene]. Roadsides, trails, open slopes, creek sides; spruce with fir and aspen, timberline vegetation, and riparian with cottonwood-willow communities. COLO: Arc, Con, Hin, LPl, Min, Mon, RGr, SJn. (2400–) 2750–3600 m ([7875–] 9020–11810′). Flowering: Jul–Sep. Utah and Wyoming to Colorado and New Mexico.

Senecio bigelovii A. Gray (honoring John M. Bigelow, 1804–1878, botanist for several surveys and expeditions in the western U.S.) Bigelow's ragwort. Perennials. **STEMS** (20–) 40–80 (–120) cm tall; caudices fibrous-rooted; herbage glabrous or nearly so, sometimes floccose-tomentose. **LEAVES** progressively reduced distally, blades of proximal leaves with bases attentuate (leaves oblanceolate) to abruptly contracted or at least obtuse (leaves spatulate-obovate), 7–15 cm long, (1–) 2–5 cm wide, margins subentire or serrate to dentate, middle and distal leaves similar, sessile, reduced, often clasping. **HEADS** racemose, nodding, (1–) 3–12 (–20); calyculi of 4–10 linear bractlets, lengths mostly 1/3–1/2 phyllaries, sometimes 1 or 2 equaling phyllaries. **PHYLLARIES** mostly 13, 6–8+ mm, tips green. **RAY FLORETS** absent. $2n = 40$. [*Ligularia bigelovii* (A. Gray) W. A. Weber; *Senecio rusbyi* Greene]. Wet meadows, pond and stream margins, along trails; spruce-fir and aspen communities. COLO: Arc, Hin, LPl, Min, Mon, RGr, SJn; UTAH. 2700–3500 m. Flowering: Jul–Aug. Arizona, Utah, New Mexico, Colorado, and Wyoming. Two varieties in *S. bigelovii* have been recognized. Variety *bigelovii* has leaves contracted to a petiolate base, herbage more consistently floccose-tomentose, especially among heads and on abaxial faces of leaves, and more numerous and longer phyllaries. Variety *bigelovii* is prevalent in most of New Mexico and southeastern Arizona, var. *halli* A. Gray further north in Colorado. Most of ours are more like var. *bigelovii*, but intergrades are common (Hin, Min, RGr) and blur the distinction.

Senecio crassulus A. Gray (thickened) Thick-leaf ragwort. Perennials. **STEMS** (15–) 20–50 (–70) cm tall; rhizomes branched, ± woody; herbage glabrous. **LEAVES** thickish-turgid, progressively reduced distally, petiolate, blades broadly lanceolate to subelliptic, 2.5–15 cm long, 1–5 cm wide, bases tapered, margins sharply dentate to subentire, some teeth callous; middle leaves sometimes larger than proximal; distal leaves sessile, smaller, often clasping. **HEADS** corymbose, (1–) 4–12; calyculi of (1–) 3–6 linear to filiform bractlets, lengths to 1/3 phyllaries. **PHYLLARIES** (±8) ±13 or ±21, 5–9 mm, tips black, villous. **RAY FLORETS** ±8 or ±13. $2n = 40$. [*Senecio lapathifolius* Greene; *S. semiamplexicaulis* Rydb.]. Talus, ridges, open slopes; spruce-fir, near-timberline vegetation, subalpine and alpine meadows, tundra. COLO: Arc, LPl, Min, Mon, SJn. 2800–3750 m (9185–12300′). Flowering: Jul–Sep. Oregon to South Dakota, south to Nevada and New Mexico.

Senecio eremophilus Richardson (solitude-loving) Desert ragwort. Perennials. **STEMS** (20–) 40–80 (–140) cm tall, caudices branched, fibrous-rooted; herbage glabrous or glabrate. **LEAVES** ± evenly distributed, proximal often withering before flowering; petiolate or sessile; blades ovate or lanceolate to narrowly lanceolate, (3–) 6–12 (–20) cm long, (1–) 1.5–5 (–7) cm wide, bases tapered, margins usually pinnate to lacerate, sometimes dentate. **HEADS** corymbose, 20–60; calyculi of 3–5 prominent or inconspicuous bractlets, lengths to 1/2 phyllaries. **PHYLLARIES** (±8) ±13, 3–6 mm, tips green or black. **RAY FLORETS** ±8. $2n = 38, 40, 44$. [*Senecio kingii* Rydb.; *S. ambrosioides* Rydb.; *S. macdougalii* A. Heller; *S. eremophilus* Richardson var. *macdougalii* (A. Heller) Cronquist]. Lake margins, streambeds, springs, disturbed riparian, roadsides, talus, rocky slopes, Gambel's oak, lodgepole pine, ponderosa pine, ponderosa pine–Gambel's oak–Douglas-fir, Douglas-fir–white fir (often with maple), aspen, spruce-fir, and bristlecone pine communities. ARIZ: Apa; COLO: Arc, Dol, Hin, LPl, Min, Mon, RGr, SMg; NMEX: McK, RAr, SJn; UTAH. 2000–3000 (–3300) m (6560–9840′ [–10825′]). Flowering: (Jun–) Jul–Sep. Utah to Wyoming, south to Arizona, Colorado, and New Mexico. Our material belongs to **var.** ***kingii*** Greenm. (honoring Clarence King, 1842–1901, geologist and leader of the geological exploration of the fortieth parallel). Variety *macdougalii*, with less conspicuous calyculus bractlets, slightly smaller heads, and fewer phyllaries, has often been recognized as distinct and is perhaps the most common expression in our area, but intergradation toward var. *kingii* (the earlier name) is so prevalent that many identifications would have to be arbitrary. Variety *eremophilus* occurs north of the southern Rocky Mountains.

Senecio flaccidus Less. (weak or drooping) Thread-leaf ragwort. Subshrubs or shrubs. **STEMS** (30–) 40–120 cm tall; taproots forming woody crowns; herbage ± lanate-tomentose, sometimes unevenly glabrescent. **LEAVES** ± evenly distributed; sessile or obscurely petiolate; blades narrowly linear to filiform, often pinnatifid with linear to filiform segments, 3–10 (–12) cm long, bases ± linear, ultimate margins entire or remotely toothed, fascicles of smaller leaves sometimes borne in axils of larger leaves. **HEADS** in subcorymbose cymes, (1–) 3–10 (–20); involucres weakly campanulate to cylindric; calyculi bractlets minute or absent. **PHYLLARIES** ±13 (±21), 6–10 mm, tips green or minutely black. **RAY FLORETS** usually ±8 or ±13, sometimes ±21. [*Senecio douglasii* DC. var. *jamesii* (Torr. & A. Gray) Ediger ex Correll & M. C. Johnst.; *S. douglasii* var. *longilobus* (Benth.) L. D. Benson; *S. longilobus* Benth.]. Usually in sandy soil, less commonly clay, sandstone outcrops, roadsides, washes, disturbed sites; desert scrub, saltbush, desert grassland, juniper-shrub, piñon-juniper woodland, and ponderosa pine–piñon-juniper woodland communities. ARIZ: Apa, Coc, Nav; COLO: LPl, Mon; NMEX: McK, SJn; UTAH. 1400–2000 m (4595–6560′). Flowering: May–Oct. Utah to Kansas, south to Arizona and Texas; Mexico. Four Corners material belongs to **var.** ***flaccidus***. Used by the Navajo as a ceremonial item. The plant was boiled and taken before the individual entered the sweathouse to get a good voice for the "Night Chant." Also used as a broom to brush spines from cactus fruit by the Hopi and Navajo.

Senecio fremontii Torr. & A. Gray (honoring John C. Frémont, 1813–1890, botanical explorer and surveyor of western North America) Dwarf mountain ragwort. Perennials. **STEMS** (10–) 25–40 cm tall; perennating bases subrhizomatous, spreading, sometimes knotty-woody; herbage often purple-tinged, glabrous. **LEAVES** evenly distributed or smaller and fewer distally, somewhat stiffish-succulent when fresh, the most proximal and the most distal often smaller, bractlike; petiolate; blades ovate or obovate to oblanceolate, 2–7 cm long, 1–3 (–4) cm wide, bases ± truncate to tapered, bases of larger weakly clasping, margins toothed or subentire. **HEADS** 1–2 (–4); calyculi absent or of 1–5 usually lance-deltate to linear, sometimes foliaceous, bractlets, lengths mostly 1/5–1/2 phyllaries. **PHYLLARIES** ±13 or ±21, 10–12 mm, tips green or brownish. **RAY FLORETS** ±8. $2n = 40$, 40+, 80. [*Senecio blitoides* Greene; *S. invenustus* Greene]. Talus, ridges, gravelly slopes; subalpine and alpine meadows, and timberline vegetation. COLO: Arc, Con, Hin, LPl, Min, Mon, RGr, SJn. 3450–3900 m (11320–12795′). Flowering: Jul–Sep. Colorado, Utah, and Wyoming. Our material belongs to **var.** ***blitoides*** (Greene) Cronquist (resembling *Blitum* = *Chenopodium*).

Senecio holmii Greene (honoring Herman T. Holm, 1854–1932, Danish-born botanist, author of *The Vegetation of the Alpine Region of the Rocky Mountains in Colorado*) Holm's ragwort. Perennials. **STEMS** 5–20 cm tall, erect to ascending; caudices branched, fibrous-rooted; herbage glabrous to lightly and inconspicuously strigose-villous, petioles and bases of stems usually purplish. **LEAVES** mostly basal or subbasal; petiolate; blades ovate to narrowly ovate, ovate-elliptic, broadly elliptic, or elliptic-oblanceolate, 2–10 cm long, 2–4.5 cm wide, bases usually gradually tapered, less commonly abruptly tapered to nearly truncate, margins coarsely to finely dentate, cauline leaves abruptly or gradually reduced, bractlike above midstem. **HEADS** nodding, 1–3, usually from branches on proximal 1/2 of primary stem; calyculi of 2–5 linear to filiform bractlets, lengths to 1/2 phyllaries. **PHYLLARIES** 7–12 mm, tips green. **RAY FLORETS** usually ±13, 17–22 mm. [*Senecio amplectens* A. Gray var. *holmii* (Greene) Harr.; *Ligularia holmii* (Greene) W. A. Weber]. Talus, loose shale, rocky meadows; subalpine meadows, spruce-fir, spruce krummholz, and alpine tundra communities. COLO: Arc, Con, Hin, LPl, Min, Mon, RGr, SJn. 3400–4050 m (11155–13285′). Flowering: Jul–Aug (–Sep). Nevada, Utah, Colorado, New Mexico, Wyoming, Montana. Commonly treated as a variety of *S. amplectens*, but the two taxa are sympatric in our area, grow in the same habitats, and have been collected in close proximity. *Senecio holmii* differs from *S. amplectens* sensu stricto in shorter stature, leaves mostly basal and proximal with middle and distal sharply reduced, and phyllaries green or light purplish and glabrous. Weber (1987) and Weber and Wittmann (1990) also have treated them as two distinct species. Plants of *S. holmii* with leaves tending toward orbiculate are sometimes misidentified as *S. soldanella*.

Senecio integerrimus Nutt. (absolutely entire, alluding to the leaf margins) Lamb-tongue ragwort. Perennials (or biennials?). **STEMS** (10–) 20–70 cm tall; caudices buttonlike, roots fleshy-fibrous; herbage copiously to sparsely arachnose, tomentose, or villous at flowering, glabrescent. **LEAVES** progressively reduced distally, basal and proximal cauline usually indistinctly petiolate; blades of cauline mostly elliptic to lanceolate or oblanceolate, sometimes rounded-deltate or suborbiculate, 6–25 cm long, 1–6 cm wide, bases ± tapered or truncate to cordate, margins entire or subentire, less commonly denticulate, distal leaves sessile, bractlike. **HEADS** in a corymbose to subumbellate cyme, 6–15 (–30); calyculi of 1–5 linear to filiform bractlets, seldom more than 2 mm. **PHYLLARIES** usually ±13 or ±21, rarely ±8, ± lanceolate, (4–) 5–10 mm, tips black. **RAY FLORETS** usually ±5, sometimes absent. $2n = 40$, 80. [*Senecio exaltatus* Nutt.; *S. hookeri* Torr. & A. Gray; *S. integerrimus* Nutt. var. *vaseyi* (Greenm.) Cronquist; *S. lugens* Richardson var. *exaltatus* (Nutt.) D. C. Eaton]. Roadsides, outcrops, open slopes, sagebrush, oak-sagebrush, ponderosa pine with aspen, Gambel's oak, and manzanita communities. COLO: Hin, LPl, Min, Mon, SMg; UTAH. 2100–2700 m (6890–8860′). Flowering: May–Jun (–Jul). Southwestern Canada, Washington to Montana, south to California, Nevada, and Colorado. Four Corners material belongs to **var.** ***exaltatus*** (Nutt.) Cronquist (raised high).

Senecio pudicus Greene (modest or bashful) Bashful ragwort. Perennials or biennials (possibly winter-annuals). **STEMS** 50–80 cm tall; caudices fibrous-rooted; herbage glabrous. **LEAVES** progressively reduced distally; petiolate; blades narrowly lanceolate or oblanceolate, becoming linear distally, 8–20 (–28) cm long, (0.5–) 1–3 (–4) cm wide, bases attenuate, margins usually entire, sometimes weakly dentate, middle and distal leaves sessile. **HEADS** nodding, (3–)

8–16 (–40) in racemiform or paniculiform distal clusters; calyculi of 2–6 minute, lance-linear bractlets. **PHYLLARIES** usually ±13, rarely ±8, 5–9 mm, tips green. **RAY FLORETS** absent. [*Senecio cernuus* A. Gray; *Ligularia pudica* (Greene) W. A. Weber]. Roadsides, commonly along streams or in mossy banks in spruce-fir communities. COLO: Hin, Min. 2850–3000 m (9350–9840′). Flowering: Jul–Sep. Colorado, New Mexico, and Utah.

Senecio serra Hook. (saw, alluding to the leaf margins) Tall ragwort. Perennials. **STEMS** 40–100 (–250) cm tall; caudices ligneous, branched; herbage glabrous or lightly floccose-tomentose proximally when young. **LEAVES** evenly distributed, proximal often withering before flowering; petiolate or subsessile; blades lanceolate or narrowly lanceolate to sublinear, 5–15 (–20+) cm long, (1–) 1.5–4 cm wide, bases tapered, margins dentate to subentire, distal leaves smaller, bractlike. **HEADS** corymbose, (12–) 30–50; calyculi of 2–6 linear to filiform bractlets, 0.5–5 mm. **PHYLLARIES** ±13, 7–9 mm, tips usually green, sometimes black. **RAY FLORETS** ±8. 2*n* = 40. [*Senecio admirabilis* Greene]. Roadsides, clearings, along trails, open slopes; aspen, aspen-fir, and spruce-fir communities. COLO: Hin, LPl, Min, Mon, SJn; NMEX: RAr. 2650–3300 m (8695–10825′). Flowering: Jul–Sep. Wyoming, Utah, Colorado, and New Mexico. Four Corners material belongs to **var. *admirabilis*** (Greene) A. Nelson (admirable).

Senecio soldanella A. Gray (a European genus of Primulaceae) Colorado ragwort. Perennials. **STEMS** 5–15 (–20) cm tall, trailing (sharply bent about midstem or below); caudices branched, fibrous-rooted, roots slightly fleshy; herbage slightly succulent, glabrous, petioles and bases of stems purplish. **LEAVES** mostly basal or subbasal; petiolate; blades elliptic-orbiculate or depressed elliptic-orbiculate to ovate-orbiculate or broadly ovate, 2–4.5 (–6) cm long, 1.5–4 (–5) cm wide, bases subcordate to truncate or abruptly tapered, margins subentire to weakly dentate or denticulate, cauline leaves (above primary leaves) much smaller, bractlike. **HEADS** nodding, 1 (–2); calyculi of 3–5 lanceolate to lance-linear bractlets, lengths 1/2–2/3 phyllaries. **PHYLLARIES** ±21, (8–) 10–13 mm, tips green. **RAY FLORETS** ±13, 13–16 mm. [*Ligularia soldanella* (A. Gray) W. A. Weber]. Talus; rocky alpine tundra communities. COLO: LPl, Mon, SJn. 3700–4150 m (12140–13615′). Flowering: Jul–Aug. Colorado and New Mexico. *Senecio soldanella* is distinguished by its slightly succulent, subentire, orbiculate, palmately veined leaves and relatively short ray florets. Plants of *S. holmii* with leaves tending toward orbiculate are sometimes misidentified as *S. soldanella*.

Senecio spartioides Torr. & A. Gray. (resembling a broom) Broom ragwort. Subshrubs. **STEMS** 20–120+ cm tall; taproots forming woody crowns; herbage usually glabrous, sometimes sparsely, unevenly hairy. **LEAVES** evenly distributed, proximal often smaller; sessile or obscurely petiolate; blades narrowly linear to filiform or parted into linear-filiform lobes, 5–10 cm long, 1–6 mm wide, bases ± linear, ultimate margins entire. **HEADS** in branching corymbose cymes, 10–20 (–60); involucres cylindric to narrowly campanulate; calyculi absent or of 1–3 minute, inconspicuous bractlets. **PHYLLARIES** usually ±8, sometimes ±13, (5–) 6–9 (–10) mm, tips green or minutely black. **RAY FLORETS** ±5 (±13). 2*n* = 40. [*Senecio andersonii* Clokey; *S. multicapitatus* Greenm. ex Rydb.; *S. spartioides* var. *multicapitatus* (Greenm. ex Rydb.) S. L. Welsh; *S. spartioides* var. *granularis* Maguire & A. H. Holmgren ex Cronquist; *S. toiyabensis* S. L. Welsh & Goodrich]. Sandy or silty soil, less commonly on shale, dunes, sandstone outcrops, washes, roadsides; desert grassland, desert shrub, cottonwood-willow, big sage, juniper-sage-snakeweed, juniper-rabbitbrush, piñon-juniper woodlands, ponderosa pine–Gambel's oak communities. ARIZ: Apa, Coc, Nav; COLO: Arc, Dol, Mon; NMEX: McK, RAr, San, SJn; UTAH. (1350–) 1600–2300 (–2700) m (4430–8860′). Flowering: (May–) Jul–Sep (–Oct). California to South Dakota, south to Arizona, New Mexico, and Texas. Plants with leaves parted into filiform lobes (seldom more than 1 mm wide) have been recognized as *S. spartioides* var. *multicapitatus* (or as *S. multicapitatus*); widespread occurrence of intermediates makes the distinction arbitrary.

Senecio taraxacoides (A. Gray) Greene (resembling *Taraxacum*) Dandelion ragwort. Perennials. **STEMS** 5–10 (–14) cm tall; caudices branched, fibrous-rooted; herbage sometimes purple-tinged, floccose-tomentose, persistently on abaxial faces of leaves, glabrescent elsewhere. **LEAVES** ± evenly distributed; petiolate; blades oblanceolate to narrowly ovate, (1–) 2–6 cm long, 1–2.5 cm wide, bases tapered, margins dentate or incised to subpinnatifid, distal leaves bractlike. **HEADS** 1 (–5); calyculi of 2–5 lance-linear bractlets, lengths 1/3–2/3 phyllaries. **PHYLLARIES** ±13 or ±21, 7–10 mm, tips green. **RAY FLORETS** usually ±13, sometimes fewer or absent, 8–12 mm. 2*n* = 40. [*Senecio amplectens* A. Gray var. *taraxacoides* A. Gray; *Ligularia taraxacoides* (A. Gray) W. A. Weber]. Alpine tundra communities. 3800–4000 m (12465–13125′). Flowering: Jul–Aug. COLO: Arc, Hin. Colorado and New Mexico.

Senecio triangularis Hook. (triangular) Arrow-leaf ragwort. Perennials. **STEMS** (20–) 50–120 (–200) cm tall; caudices branched, ± woody; herbage glabrous or sparsely floccose-tomentose when young. **LEAVES** evenly distributed; petiolate; blades narrowly triangular, (3–) 4–10 cm long, 2–6 cm wide, bases usually ± truncate, sometimes tapered, margins usually dentate, rarely subentire, distal leaves subsessile, smaller. **HEADS** subcorymbose, 10–30 (–60); calyculi of 2–6 bractlets, rarely more than 2 mm. **PHYLLARIES** (±8) ±13 (±21), 6–10 mm, tips usually green, rarely black. **RAY FLORETS** ±8. $2n$ = 40, 80. [*S. triangularis* var. *angustifolius* G. N. Jones]. Pond and lake margins, seepages, along trails, openings, talus; ponderosa pine, pine-fir, spruce, spruce-fir-aspen, and alpine tundra communities. COLO: Arc, Hin, LPl, Min, Mon, RGr, SJn. 2700–3500 m (8860–11480′). Flowering: Jul–Sep. Alaska, Yukon, and Northwest Territories, south to Washington and Montana, continuing to California, Arizona, and New Mexico.

Senecio vulgaris L. (common) Common groundsel. Annuals. **STEMS** (10–) 20–50 (–60+) cm tall; taprooted or shallowly fibrous-rooted; herbage glabrous or sparsely and unevenly tomentose when young. **LEAVES** evenly distributed; petiolate or distal sessile; blades ovate to oblanceolate, 2–10 cm long, 0.5–2 (–4) cm wide, bases tapered, margins lobulate to dentate, ultimate margins often secondarily dentate to denticulate. **HEADS** 8–20; calyculi of 2–4 (–6) bractlets, black-tipped, lengths about 1/4 phyllaries. **PHYLLARIES** ±21, 4–6 mm, tips usually green, sometimes black. **RAY FLORETS** absent. $2n$ = 40. Along rivers and irrigation canals; riparian vegetation. NMEX: RAr, SJn. 1615–1645 m (5300–5400′). Flowering: Apr–May. Native to Eurasia; widely naturalized in North America.

Senecio wootonii Greene (honoring New Mexico botanist Elmer O. Wooton, 1865–1945) Wooton's ragwort. Perennials. **STEMS** (15–) 20–45 (–60) cm tall; caudices erect or weakly spreading; herbage nearly always glaucous, glabrous. **LEAVES** thickish, progressively reduced distally; petiolate; blades ovate or obovate to lanceolate, 4–9 (–15+) cm long, 1.5–3 (–4+) cm wide, bases tapered, margins wavy or subentire, often with callous denticles; middle and distal leaves sessile, bractlike. **HEADS** (3–) 8–24; calyculi of 1–3 oblong to lance-linear bractlets, less than 3 mm. **PHYLLARIES** ±13 (±21), 6–9 mm, tips green to brownish, not blackened. **RAY FLORETS** 8–10. $2n$ = 40. Creek banks, along trails, meadows, open slopes; spruce, and aspen communities. COLO: Arc, Hin, LPl, SJn, SMg. 2400–3250 m (7875–10660′). Flowering: (May–) Jun–Jul. Arizona and Colorado to New Mexico and Texas; Mexico.

Silphium L. Rosinweed
Kenneth D. Heil

(Greek *silphi*, a plant with medicinal properties) Perennials up to 25 dm tall. **STEMS** usually erect, mostly branched. **LEAVES** basal and cauline; whorled, opposite, or alternate (sometimes all conditions on one plant); petiolate or sessile; blades 1- or 3-nerved, deltate, elliptic, linear, ovate, or rhombic, sometimes 1–2-pinnately lobed or pinnatifid, bases mostly cordate or truncate, margins entire or toothed, faces glabrous or hairy. **INFL** in paniculiform or racemiform arrays. **INVOLUCRES** campanulate to hemispheric, 10–30 mm diameter. **PHYLLARIES** persistent, 11–45 in 2–4 series, outer broader, foliaceous, inner smaller, thinner, each subtending a ray floret. **RECEPTACLES** flat to slightly convex, paleate. **RAY FLORETS** 8–35+ in 1–4 series, pistillate, fertile; corollas yellow or white. **DISC FLORETS** 20–200+, functionally staminate; corollas yellow or white. **CYPSELAE** black to brown, obflattened. **PAPPUS** 0, or persistent, of 2 awns. x = 7. Twelve species in North America (Flora of North America Editorial Committee 2006).

Silphium laciniatum L. (deeply cut) Compassplant. **STEMS** up to 25 dm tall, terete, hirsute, hispid, or scabrous. **LEAVES**: basal persistent, petiolate or sessile; cauline petiolate or sessile; blades lanceolate, linear, or rhombic, 4–60 × 1–30 cm, mostly 1–2-pinnately lobed. **PHYLLARIES** 25–45, in 2 or 3 series, outer reflexed or appressed, apices acuminate to caudate. **RAY FLORETS** 27–38; corolla yellow. **DISC FLORETS** 100–275; corolla yellow. **CYPSELAE** 10–18 × 6–12 mm. **PAPPUS** 1–3 mm. $2n$ = 14. Roadside in a desert grassland community. Known from one collection (*Heil & Clifford 22827*). NMEX: McK. 2090 m (6865′). Flowering: Jul–Aug. New York to South Dakota, south to Alabama and Texas.

Solidago L. Goldenrod
J. C. Semple and G. L. Nesom

(Latin *solidus*, whole, + *ago*, resembling or becoming, probably alluding to healing properties) Plant a perennial, from woody caudices or rhizomes. **STEMS** glabrous to strigose, strigillose, hispid, or short-villous. **LEAVES** basal (persistent or not by flowering) and cauline; petiolate (proximal) or sessile (proximal and distal, latter sometimes subpetiolate); margins often serrate, faces glabrous or densely hairy. **INFL** usually a radiate head, in a corymbiform (flat to round-topped), racemiform (club-shaped or pyramidal), or paniculiform, sometimes secund array, often secund on branches. **RAY FLORETS** (0–) 2–15 (–24), pistillate, fertile; corollas yellow. **DISC FLORETS** 2–35 (–60), bisexual, fertile. **CYPSELAE** narrowly obconic to cylindric, glabrous or moderately strigillose, ribs 8–10. **PAPPUS** of persistent barbellate bristles. x = 9. Species about 100, primarily in North America, some in South America and Eurasia.

1. Heads in corymbiform clusters or in short axillary and terminal clusters and appearing columnar, not secund on branches; rhizomes not creeping. ....(2)
1′ Heads in cone-shaped to pyramidal arrays, usually at least slightly secund on branches; rhizomes creeping.....(6)

2. Heads in short axillary and terminal clusters, appearing columnar overall .... ***S. simplex***
2′ Heads in corymbiform clusters....(3)

3. Cauline leaf blades ovate to rhombiform, 24–50 mm wide, densely hispid to hispid-strigose; plants 30–70 (–90) cm; phyllaries usually striate, outer oblong .... ***S. rigida***
3′ Cauline leaf blades linear-oblanceolate to spatulate, 2–25 mm wide, glabrous or puberulent; plants mostly (5–) 10–50 (–80) cm; phyllaries not striate, outer linear-lanceolate to lanceolate....(4)

4. Stems and leaf faces finely, densely, and closely puberulent.... ***S. nana***
4′ Stems and leaf faces essentially glabrous....(5)

5. Proximal margins of leaves conspicuously ciliate with crinkly hairs.... ***S. multiradiata***
5′ Leaf margins completely glabrous to minutely scabrous .... ***S. simplex***

6. Stems and leaf faces glabrous; stems usually purplish; axillary fascicles of small leaves commonly present and conspicuous.... ***S. missouriensis***
6′ Stems and leaf faces pubescent; stems usually greenish, sometimes purplish near the base; axillary fascicles absent or inconspicuous ....(7)

7. Plant 50–200 cm tall, stems erect from the base; lower cauline leaves usually withered or deciduous by flowering, cauline only slightly or gradually reduced upward, all similar in shape; the most distal leaves, bracts, and/or phyllaries often sparsely to moderately minutely stipitate-glandular (use lens) .... ***S. lepida***
7′ Plant 15–80 (–150) cm tall, stems ascending from the base; lower cauline leaves usually present at flowering, cauline conspicuously reduced upward, the largest borne somewhat above the naked base but below midstem, differently shaped from those more distal; leaves, bracts, and phyllaries eglandular .... ***S. velutina***

Solidago lepida DC. (pleasant, neat) Rocky Mountain goldenrod, Canada goldenrod. Plants 25–150 cm; rhizomes short- to long-creeping. **STEMS** 1–25 (–50+), erect, proximally glabrous or sparsely to moderately villous, distally densely so. **LEAVES**: basal 0; proximal cauline usually deciduous by flowering, cauline sessile, blades narrowly to broadly oblanceolate or elliptic, 100–150 × 15–23 mm, largest near midstem, tapering to bases, margins sharply serrate to subentire, 3-nerved, sometimes obscurely so, apices acute to acuminate, faces sparsely short strigose-villous (more so along veins) to glabrate, the most distal commonly sparsely to moderately minutely stipitate-glandular. **HEADS** often secund on branches, in a pyramidal paniculiform array, usually leafy proximally, proximal branches arching to recurved. **PHYLLARIES** unequal, deltate-lanceolate. **RAY FLORETS** (7–) 10–16 (–22); lamina (0.5–) 0.9–1.6 (–2.2) mm. $2n$ = 18, 36, 54. [*S. serotina* Aiton var. *salebrosa* Piper; *S. canadensis* L. var. *salebrosa* (Piper) M. E. Jones; *S. salebrosa* (Piper) Rydb.]. Meadows, roadsides, along streams and rivers; cottonwood, ponderosa pine, and spruce-fir communities. ARIZ: Apa; COLO: Arc, Dol, Hin, LPl, Min, Mon, SJn; NMEX: RAr, San; UTAH. 1600–2750 m (5250–9020′). Flowering: Jul–Sep (–Oct). Canada (southern Northwest Territories to British Columbia and east to Manitoba), Washington to Montana, south to New Mexico and Arizona. Our material belongs to **var. *salebrosa*** (Piper) Semple (rough), with heads in a pyramidal paniculiform array and at least slightly secund on arching to recurved proximal branches. It is the common expression in the Four Corners region; var. *lepida*, with heads in thyrsiform arrays and proximal branches ascending) is sometimes identified in the Four Corners region, but this likely is a reflection of plasticity within var. *salebrosa*.

Solidago missouriensis Nutt. (from Missouri) Missouri goldenrod. Plants (10–) 30–80 cm; rhizomes short- to long-creeping. **STEMS** erect, glabrous or sometimes sparsely strigose in lines; fascicles of small leaves often present in axils. **LEAVES**: proximal cauline tapering to long, winged petioles, oblanceolate to linear-oblanceolate, 50–100 (–200) × (5–) 10–20 (–30) mm, margins entire to serrulate, usually 3-nerved, apices acute, glabrous; middle to distal cauline sessile, blades lanceolate to linear, 40–60 mm, rapidly reduced distally, margins entire. **HEADS** usually secund on branches, in a paniculiform array, broadly secund-pyramidal, or more rhombic to transversely rhombic, less commonly with nonsecund heads. **PHYLLARIES** strongly unequal, outer ovate to lanceolate, inner linear-ovate to oblong or linear-lanceolate. **RAY FLORETS** 5–14, lamina 1.5–2 (–4) mm. $2n$ = 18, 36. [*S. missouriensis* var. *fasciculata* Holz.; *S. missouriensis* var. *glaberrima* (M. Martens) Rosend. & Cronquist; *S. missouriensis* var. *montana* A. Gray; *S. missouriensis* var. *tenuissima* (Wooton & Standl.) C. E. S. Taylor & R. J. Taylor; *S. missouriensis* var. *tolmieana* (A. Gray) Cronquist]. Open conifer forests, disturbed soils, roadsides. COLO: LPl. 1800–2000 m (5905–6560′). Flowering: Aug–Sep. Canada (British Columbia to Ontario), south to Washington and Oregon, Nevada and New Mexico, Tennessee, Arkansas, and Texas, apparently scattered and uncommon in our area; northern Mexico. Varieties that have been recognized appear to be strongly intergrading and without geographic integrity.

Solidago multiradiata Aiton (many rays, alluding to the ray florets) Rocky Mountain goldenrod. Plants (3–) 10–30 (–80) cm; caudices branched. **STEMS** decumbent to erect, proximally glabrous or sparsely hairy, becoming densely short-hispid-strigose in arrays. **LEAVES**: basal and proximal cauline winged-petiolate, margins ciliate, blades linear-oblanceolate to spatulate, 65–170 × 8–25 mm, serrate to crenate near apices, distal sessile and blades sometimes subclasping, ovate to linear-lanceolate. **HEADS** not secund on branches, in a dense, round-topped, corymbiform array. **PHYLLARIES** unequal to subequal, outer linear-lanceolate to lanceolate. **RAY FLORETS** 12–18, lamina 3–4 mm. $2n = 18$. [*S. multiradiata* var. *scopulorum* A. Gray]. Alpine tundra, slopes, and meadows, talus, peaty wetland of alpine lakes, and subalpine fir–Douglas-fir communities. COLO: Arc, Mon, RGr, SJn; UTAH. 3250–3750 m (10660–12305′). Flowering: Jul–Oct. Alaska, Yukon, and Canada (British Columbia to Newfoundland), south in the western United States through Washington and Montana to California, Arizona, and New Mexico.

Solidago nana Nutt. (dwarf) Dwarf goldenrod. Plants 10–50 cm tall; rhizomes strongly thickened, caudices branched. **STEMS** decumbent to ascending, finely and densely puberulent. **LEAVES**: basal, present at flowering, basal and proximal petiolate, blades mostly spatulate, varying toward oblanceolate or narrowly obovate, 20–100 × 5–20 mm, basal smaller than proximal, margins serrate or entire, somewhat 3-nerved, faces moderately to densely finely and closely puberulent, middle and distal cauline sessile, blades 10–30 × 4–12 mm, greatly reduced distally, 1 prominent vein, margins entire or distally serrate. **HEADS** not secund or sometimes slightly secund, in a rounded-corymbiform array. **PHYLLARIES** strongly unequal, oblong. **RAY FLORETS** (5–) 6–10, lamina about 3 mm. $2n = 18$. Roadsides, meadows, open slopes, woodland openings; ponderosa pine, ponderosa pine–piñon pine–Gambel's oak, and mountain shrub communities. COLO: Arc, Dol, Hin, SMg; NMEX: SJn; UTAH. 2150–2450 m (7055–8040′). Flowering: Jul–Sep. Montana and Idaho south to Nevada, Arizona, and New Mexico.

Solidago rigida L. (stiff) Stiff-leaved goldenrod. Plants 30–70 (–90) cm; caudices branching, woody. **STEMS** erect, stout, densely hispid to hispid-strigose. **LEAVES**: basal and proximal cauline usually present at flowering, abruptly narrowed to long petioles, blades ovate to rhombiform, 80–120 × 24–50 mm, margins entire to crenate, sometimes undulate, faces densely hispid to hispid-strigose; middle to distal cauline sessile, blades lanceolate to ovate, 30–50 × 15–17 mm, greatly reduced distally, margins entire or finely serrate, sometimes undulate. **HEADS** not secund on branches, in a tightly clustered, corymbiform array. **PHYLLARIES** unequal, oblong, conspicuously striate (3–5 pronounced veins), inner often more nearly linear. **RAY FLORETS** 6–13, lamina 1.4–5.4 mm. $2n = 18$. Prairies, grassy clearings. COLO: Arc. About 2100 m (6890′). Flowering: Aug–Sep. Canada (Alberta to Ontario), Montana to Michigan south to New Mexico, Texas, Missouri, and Indiana, apparently adventive and scattered in western states. Four Corners plants are **var. *humilis*** Porter (low). [*Oligoneuron rigidum* (L.) Small var. *humilis* (Porter) G. L. Nesom; *Solidago rigida* subsp. *humilis* (Porter) S. B. Heard & Semple]. We have not seen the single collection of *S. rigida* from within the *Flora* area (COLO: Arc, Pagosa Springs, 30 August 1899, *Baker 722* [COLO]). It is slightly out of range for the species at that locality and likely was an introduction.

Solidago simplex Kunth (single, probably alluding to the number of stems) Mountain goldenrod. Plants 3–40 cm; caudices branching. **STEMS** ascending to erect, proximally glabrous, strigose in arrays. **LEAVES**: basal, sometimes present in rosettes at flowering, basal and proximal cauline spatulate, rarely broadly elliptic, 1-nerved, gradually attenuate to winged petiole, margins weakly crenate, glabrous, middle and distal sessile, similar, blades lanceolate to linear, 5–16, (5–) 12.5–47 (–90) × (0.7–) 2–7 (–13) mm, reduced distally, margins entire to sparsely serrate, sometimes glutinous. **HEADS** not secund on branches, in a terminal corymbiform cluster or in short axillary and terminal clusters and appearing columnar overall. **PHYLLARIES** strongly unequal, outer ovate, inner linear-oblong, viscid-resinous. **RAY FLORETS** 7–16, lamina 2–5 × 0.7–0.9 mm. $2n = 18$. [*S. decumbens* Greene; *S. decumbens* var. *oreophila* (Rydb.) Fernald; *S. multiradiata* Aiton var. *neomexicana* A. Gray; *S. neomexicana* (A. Gray) Wooton & Standl.; *S. oreophila* Rydb.; *S. spathulata* DC. var. *neomexicana* (A. Gray) Cronquist]. Alpine tundra, talus, rocky slopes and ridges, cliffs, subalpine meadows; ponderosa pine–Douglas-fir–aspen, and spruce-fir-aspen communities. ARIZ: Apa, Nav; COLO: Arc, Con, Hin, LPl, Min, Mon, RGr, SJn; NMEX: SJn; UTAH. (2300–) 2650–3750 m (7545–12305′). Flowering: Jul–Sep. Alaska, Canada (Yukon and Northwest Territories south to British Columbia to Quebec), south to Washington and Oregon, Montana, Minnesota, and Michigan, continuing to Arizona, New Mexico; Sierra Madre Oriental in Mexico. Four Corners plants are **var. *simplex***.

Taller plants with more elongate capitulescences occur mostly below 2805 m (9200′). Those from higher elevations characteristically are shorter and bear heads in a single cluster.

Solidago velutina DC. (velvety) Velvety goldenrod. Plants 15–80 (–150) cm; rhizomes creeping, slender. **STEMS** usually ascending at base, less commonly erect, glabrate proximally to sparsely strigose-puberulent distally. **LEAVES**: basal and proximal cauline often persisting to flowering, gradually tapering to winged petioles, blades linear-oblanceolate to oblanceolate, rarely spatulate, 50–120 × 8–30 mm, the most proximal much smaller, margins entire or less commonly shallowly serrate, faces glabrate to moderately scabrose-strigose, middle and distal cauline sessile or subsessile, blades elliptic to oblanceolate or obovate, 10–50 × 3–12 mm, middle tapering to bases, somewhat to strongly 3-nerved, largest, usually much reduced distally, margins entire or sometimes distally serrate, faces sparsely to densely strigose-puberulent. **HEADS** usually slightly to strongly secund on branches, in a narrow to broad, thyrsiform to secund-pyramidal paniculiform array. **PHYLLARIES** strongly unequal, lanceolate to oblong. **RAY FLORETS** 6–12, lamina 2.9–6.3 mm. $2n$ = 18, 36, 54. [*Solidago sparsiflora* A. Gray; *S. arizonica* (A. Gray) Wooton & Standl.; *S. californica* Nutt. var. *nevadensis* A. Gray; *S. canadensis* L. var. *arizonica* A. Gray; *S. velutina* var. *nevadensis* (A. Gray) C. E. S. Taylor & R. J. Taylor]. Riparian alluvium, talus, roadsides, open slopes, meadows, hanging gardens, rocky washes; cottonwood, piñon-juniper, Gambel's oak, ponderosa pine, and fir-aspen communities. ARIZ: Apa, Nav; COLO: Arc, Dol, LPl, Mon, SJn, SMg; NMEX: McK, RAr, SJn; UTAH. (1125–) 1680–2650 (–2950) m (3700–9680′). Flowering: (Jun–) Jul–Sep. Montana and South Dakota, south to California, Arizona, New Mexico, and Texas; widespread in the northern half of Mexico. Our plants belong to **subsp**. ***sparsiflora*** (A. Gray) Semple (few-flowered).

Sonchus L. Sowthistle

Gary I. Baird

(Greek *sonchos*, ancient name of a thistle) Annuals, biennials, or perennials, taprooted or rhizomatous, herbage glabrous and often glaucous, sap milky. **STEMS** usually 1, erect, branched distally, stipitate-glandular distally or eglandular, mostly glabrous (sometimes tomentose-lanate below heads). **LEAVES** basal and cauline, or mostly cauline, simple, alternate; petiolate (basal) to sessile (cauline), more or less winged, base auriculate, margins echinate-dentate; blades lanceolate to oblanceolate or obovate, margins entire to pinnately or runcinately lobed or pinnatifid, ultimate margins echinate-dentate, apices acute to obtuse; lobes lanceolate to deltate, spreading to retrorse. **INFL** of 15–85+ heads, borne in corymbiform, umbelliform, or paniculiform clusters. **HEADS** liguliflorous, erect; peduncles mostly ebracteate, glabrous or pubescent, stipitate-glandular or eglandular. **INVOLUCRES** lacking calyculi. **PHYLLARIES** 30–50, in 3–4 series, unequal (outermost about 1/2 length of inner), herbaceous, green, erect, margins scarious, midribs more or less keeled proximally. **RECEPTACLES** flat to convex, epaleate. **RAY FLORETS** ± matutinal; corolla ligulate, yellow, outermost rosy-purple abaxially. **DISC FLORETS** lacking. **CYPSELAE** monomorphic, stramineous to reddish brown, weakly to strongly compressed, ellipsoid to oblanceoloid, glabrous, muricate (sometimes restricted to margins or ribs), 8–15-ribbed (usually 3–5 on each face), lateral ribs sometimes winged, beak lacking. **PAPPUS** of bristles, more or less persistent, white, smooth to barbellulate. $x = 9$. About 50 species in Eurasia, Africa, and Macaronesia, some species now cosmopolitan (Boulos 1973).

1. Auricles (middle to upper) strongly recurved (usually 1800) and appearing curled or coiled; cypselae strongly compressed, faces smooth, with 3 ribs each, margins winged; ligules shorter than corolla tubes .... ***S. asper***
1′ Auricles (middle to upper) straight or recurved (rarely more than 900); cypselae weakly compressed, faces muricate-rugulose, with usually 5 ribs each, margins not winged; ligules equaling corolla tubes ....(2)

2. Plants perennial from rhizomes or stolons; stem bases usually solid; peduncles strongly (stipitate) glandular, or eglandular .... ***S. arvensis***
2′ Plants annual (or occasionally biennial) from taproots; stem bases usually hollow; peduncles usually eglandular, rarely stipitate-glandular .... ***S. oleraceus***

Sonchus arvensis L. (pertaining to cultivated fields) Field sowthistle. Perennials, deeply rhizomatous or stoloniferous, herbage glabrous, often sparsely to densely tomentose-lanate below heads, and glandular or eglandular; glands yellow, sessile or stipitate, stipes mostly 1–2 mm (rarely less). **STEMS** 12–125 cm tall, basally more or less hardened and solid. **LEAVES**: lower and middle cauline blades oblanceolate to linear-lanceolate, 8–28 cm × 20–80 mm (including lobes), base usually narrowly winged, auriculate, margins mostly pinnately to runcinately lobed, ultimate margins weakly to moderately spinose; auricles straight to recurved (mostly less than 90°), usually rounded; lobes 3–7 pairs (or more), lanceolate to deltate, to 4 cm long, spreading to retrorse; cauline leaves progressively and strongly reduced upward, uppermost linear to linear-lanceolate, entire to lobate. **HEADS** 15–75 (or more, proliferating with age); peduncles naked or 1–2-bracteate, glabrous or sparsely to densely tomentose-lanate, and stipitate- or sessile-glandular, or eglandular. **INVOLUCRES** 10–15 mm tall. **PHYLLARIES** lanceolate, glabrous to (proximally) lanate, eglandular or glandular, glands sessile or stipitate. **RAY FLORETS** numerous (100–200 or more), outer exceeding involucre; corolla tube 7–16 mm, pilose, ligule 7–13 × 1–2 mm, glabrous; anthers 3–4 mm. **CYPSELAE** 2.5–3.5 mm, dark red-brown, ellipsoid to oblanceoloid, slightly compressed, about 5 ribs per face, muricate-rugose. **PAPPUS** bristles numerous, 8–10 mm, white. $2n$ = 36, 54. Riparian communities, moist habitats, often disturbed sites. COLO: LPl, Mon; NMEX: SJn. 1800–2100 m (6000–7000′). Flowering: Jul–Oct. Europe, widely introduced and established throughout North America and elsewhere. This is an aggressive weed that has been listed as noxious by many states and provinces. Two varieties are traditionally recognized (sometimes treated as subspecies).

1. Involucres and peduncles conspicuously stipitate-glandular .... **var. *arvensis***
1′ Involucres and peduncles eglandular or sessile-glandular .... **var. *uliginosus***

var. *arvensis* **HEADS** (and peduncles) long stipitate-glandular. $2n$ = 54. COLO: LPl, SJn; NMEX: SJn. Europe, widespread in North America and elsewhere.

var. *uliginosus* (M. Bieb.) Trautv. (pertaining to marshes) **HEADS** (and peduncles) eglandular or sessile-glandular. $2n$ = 36. COLO: LPL, Mon. Europe, widespread in North America and elsewhere.

Sonchus asper (L.) Hill (rough or harsh, alluding to the often coarsely spiny leaves) Spiny sowthistle. Annuals or biennials, taprooted, herbage glabrous, often sparsely villous to tomentose-lanate distally (especially beneath heads), and glandular or eglandular; glands dark, stipitate, stipes about 1 mm or less. **STEMS** 20–110 cm tall, basally soft and hollow, usually densely stipitate-glandular distally (beneath inflorescence). **LEAVES**: lower and middle cauline blades oblanceolate to obovate, or lanceolate, 6–22 cm × 20–60 mm (including lobes), base narrowly to broadly winged (sometimes broader than blade), auriculate, margins entire to (usually) runcinately lobed or pinnatifid, ultimate margins coarsely and strongly spinose; auricles strongly recurved (usually 180°–270°) and appearing coiled, rounded; lobes 3–7 pairs, lanceolate to deltate or rounded, to 3 cm long, mostly retrorse; cauline leaves ± progressively reduced upward, uppermost lanceolate, mostly unlobed. **HEADS** 3–30 (or more, proliferating with age); peduncles mostly naked (or 1-bracteate), glabrous or sparsely to densely tomentose-lanate, eglandular to densely stipitate-glandular. **INVOLUCRES** 8–15 mm tall. **PHYLLARIES** lanceolate, glabrous to (proximally) tomentose-lanate, eglandular to stipitate-glandular. **RAY FLORETS** numerous (100–150 or more), outer ± equaling involucre; corolla tube 7–10 mm, glabrous to sparsely pilose, ligule 4–6 × 0.5–1 mm, glabrous; anthers 1–1.5 mm. **CYPSELAE** 2–3 mm, stramineous to red-brown, ellipsoid, strongly compressed, mostly 3 (4 or 5) ribs per face, smooth, marginal ribs more or less winged, wings often muricate. **PAPPUS** bristles numerous, 5–9 mm, white. $2n$ = 18. Riparian habitats, seeps, springs, and hanging gardens, often in saline or alkaline soils. ARIZ: Apa, Nav; COLO: Arc, Dol, Mon, SMg; NMEX: SJn; UTAH. 1100–2100 m (3700–6900′). Flowering: Apr–Oct. Europe, widely introduced and established throughout North America and elsewhere (now cosmopolitan).

Sonchus oleraceus L. (a potherb) Common sowthistle. Annuals or biennials, taprooted, herbage glabrous, often villous to tomentose-lanate distally (especially beneath heads), mostly eglandular. **STEMS** 20–60 cm (or more) tall, basally soft and hollow, glabrous (ours) or occasionally stipitate-glandular distally. **LEAVES**: lower and middle cauline blades oblanceolate to obovate, spatulate, or lanceolate, 7–20 cm × 2–7 cm (including lobes), base narrowly to broadly winged (sometimes broader than blade), auriculate, margins entire to runcinately lobed or pinnatifid, ultimate margins weakly to moderately spinose; auricles straight to recurved (less than 90°), acute; lobes 1–4 pairs, lanceolate to deltate, to 3 cm long, mostly retrorse; cauline leaves ± progressively reduced upward, uppermost ± lanceolate, mostly unlobed. **HEADS** 10–15 (or more, proliferating with age); peduncles mostly naked (or 1-bracteate), glabrous or sparsely tomentose-lanate, eglandular (ours). **INVOLUCRES** 10–15 mm tall. **PHYLLARIES** lanceolate, glabrous to (proximally) tomentose-lanate, eglandular (ours). **RAY FLORETS** numerous (100–150 or more), outer ± equalling involucre; corolla tube 5–7 mm, sparsely pilose, ligule 3–6 × 0.5–1 mm wide, glabrous; anthers 1–1.5 mm. **CYPSELAE** 2.5–3.5 mm, stramineous to red-brown, oblanceoloid, somewhat compressed, mostly 5 (often unevenly spaced) ribs per face, muricate-rugulose, marginal ribs not winged. **PAPPUS** bristles numerous, 5–9 mm, white. $2n$ = 32, 36. Ditch banks, cultivated fields, and other mesic habitats, often in disturbed sites. ARIZ: Nav; COLO: Mon; NMEX: SJn. 1700–1800 m (5500–5800′). Flowering: Aug–Sep. Europe, widely introduced and established throughout North America and elsewhere (now cosmopolitan). This and the preceding species are very similar and often confused. Features of the auricles and fruits are most useful in separating them.

Stenotus Nutt. Goldenweed or Mock Goldenweed
G. L. Nesom

(Greek *stenos*, narrow, or *stenotes*, narrowness, alluding to the characteristically narrow leaves) Perennial herbs from woody, multicipital caudices, with few–numerous short, crowded stems, glabrous to sparsely or densely tomentose, not resinous or punctate. **LEAVES** mostly in a basal rosette, entire, linear to linear-oblanceolate or nearly spatulate, with 3 parallel, raised nerves, apparently 1-nerved in narrow leaves. **INFL** a mostly solitary head, the peduncles leafless or less commonly with scattered leaves. **INVOLUCRES** campanulate to hemispheric. **PHYLLARIES** herbaceous with hyaline margins, 3 (–5)-nerved, strongly to weakly graduate in 2–3 (–4) series. **RECEPTACLES** foveolate to nearly smooth. **RAY FLORETS** pistillate, fertile, (absent), corollas yellow. **DISC COROLLAS** bisexual, fertile. **CYPSELAE** subterete to compressed, 2–6-nerved, moderate to densely strigose-sericeous. **PAPPUS** of 1–2 series of basally caducous, subequal, barbellate bristles. $x = 9$. Four species; western North America in the United States and northwestern Mexico.

Stenotus armerioides Nutt. (similar to *Armeria*, a genus of Plumbaginaceae) Thrifty goldenweed. **STEMS** (peduncles) 4–26 cm tall. **LEAVES**: cauline absent or reduced and restricted to proximal 1/2 of stem. **HEADS** 1 (–2); involucres 6–14 mm wide. **RAY FLORETS** 5–15. $2n = 18$. [*Haplopappus armerioides* (Nutt.) A. Gray]. Sand, sandy clay, hanging gardens, sandy benches, rocky ledges; desert scrub, cottonwood-rabbitbrush, piñon-juniper-shrub, and ponderosa pine–Gambel's oak communities. ARIZ: Apa, Coc, Nav; COLO: Dol, LPl, Mon; NMEX: McK, RAr, San, SJn; UTAH. 1550–2250 m (5085–7380′). Flowering: Apr–Jun. Montana and North Dakota south to Arizona and New Mexico. Four Corners plants are **var. *armerioides***.

Stephanomeria Nutt. Wirelettuce, Skeletonweed
Gary I. Baird

(Greek *stephanos*, garland, wreath, or crown, + *meros*, division or part, alluding to the pappus) Annuals, taprooted, or perennials, rhizomatous or sometimes with a woody caudex, herbage (mostly) glabrous or pubescent, sometimes glandular, sap milky. **STEMS** 1 to many, erect, simple to highly branched. **LEAVES** basal and (mostly) cauline, simple, alternate, sessile, not auriculate; basal and lower cauline usually withered by flowering, blades oblanceolate to linear, entire to dentate or pinnatifid; cauline often much reduced and bractlike. **HEADS** liguliflorous, borne singly or clustered, terminal or axillary, few to numerous, in paniculiform or racemiform clusters, erect; peduncles bracteate, glabrous or glabrate. **INVOLUCRES** cylindric-conical; calyculi of 3–7 bractlets, unequal. **PHYLLARIES** 5–8 in 1–2 series, equal, erect, herbaceous, margins scarious to hyaline. **RECEPTACLES** flat, epaleate. **RAY FLORETS** matutinal; corolla ligulate, pink, lavender, or white. **DISC FLORETS** lacking. **CYPSELAE** ± monomorphic, light to dark brown or stramineous, cylindric-prismatic, glabrous, ribs 5, broad and forming corners, separated by a narrow groove on each face (ours), smooth to tuberculate, beakless. **PAPPUS** of bristles, persistent, white to tawny, plumose (at least partly so). $x = 8$. Perhaps 16 species in western North America.

1. Plants annual, taprooted .... ***S. exigua***
1′ Plants perennial, rhizomatous (deeply) and/or with a woody caudex .... (2)

2. Pappus bristles tan or tawny (white only at base), plumose on distal 1/2–3/4; peduncles glabrous .... ***S. pauciflora***
2′ Pappus bristles white, plumose throughout; peduncles glandular-pubescent .... ***S. tenuifolia***

Stephanomeria exigua Nutt. (very small or minute, alluding to the heads) Small wirelettuce. Annuals from a slender to stout taproot, herbage glabrous, often glaucous, to sparsely pubescent or glandular-pubescent distally. **STEMS** mostly 1, 8–64 cm tall, erect, green or gray-white (or purplish) with age (especially proximally), usually diffusely branched distally (occasionally throughout), branches divaricate and intricate. **LEAVES** withered and absent at flowering, basal and lower cauline blades linear to narrowly oblanceolate, 2–9 cm × 1–14 mm, margins dentate to laciniately pinnatifid or subentire; cauline strongly reduced and bractlike, often entire. **HEADS** terminal or axillary, pedunculate, borne in more or less paniculiform clusters. **INVOLUCRES** 7–9 mm tall. **CALYCULI** 3–4, deltate-ovate, 1–2.5 mm, glabrous to sparsely (glandular) pubescent, margins scarious, white, entire to minutely ciliate. **PHYLLARIES** 5–7 in 2 series, linear-lanceolate, glabrous to sparsely puberulent (or glandular), green or purple (especially medially and distally), midrib keeled, thickened proximally, margins scarious and white (outer) or broadly hyaline (inner). **RAY FLORETS** 5–6, exceeding involucre; corollas glabrous, tube 4–5 mm, ligule 8–11 × 3–5 mm wide; anthers 3–4 mm. **CYPSELAE** 3–4 mm, light to dark tan or stramineous; ribs broad, keeled (forming angles of achene), tuberculate, separated by a narrow groove (on each face). **PAPPUS** 5–6 mm, white to stramineous (in part or wholly), sometimes abruptly widened proximally and scalelike, plumose on distal 1/2 (sometimes more) but not throughout. $2n = 16$. Sagebrush, desert shrub, and piñon-juniper communities, often in sandy soils. ARIZ: Apa, Coc, Nav; COLO: LPl, Mon; NMEX: McK, SJn; UTAH. 1100–1900 m (3600–6300′). Flowering: May–Oct. Oregon and Idaho, south to Baja California, Arizona, and Texas. Our material is all assignable to **var.** ***exigua***.

Stephanomeria pauciflora (Torr.) A. Nelson (few-flowered) Prairie wirelettuce. Perennial from a rhizome, root crown usually deeply branched and slender, herbage glabrous, often glaucous, or (rarely) puberulent (not glandular) distally. **STEMS** 1–5 (or more), 10–50 cm tall, erect, green or becoming grayish with age (especially proximally), diffusely branched distally (or throughout), branches mostly divaricate and intricate. **LEAVES** mostly withered and absent at flowering, basal and lower cauline blades linear, 2–6 cm × 2–8 mm (including lobes), margins entire to dentate or irregular-pinnately lobed; cauline reduced distally; teeth or lobes to 3 mm, deltate to linear. **HEADS** terminal and axillary, pedunculate to subsessile, borne in more or less paniculiform clusters. **INVOLUCRES** 6–10 mm tall. **CALYCULI** 3–7 (grading into peduncular bracts), lanceolate, 1–4 mm, glabrous to pubescent, margins narrowly scarious, white, entire. **PHYLLARIES** 5–8 in 2 series, linear-lanceolate, glabrous, green or purple (especially medially and distally), midrib keeled, thickened proximally, margins narrowly scarious and white (outer) or narrowly hyaline (inner). **RAY FLORETS** 5–9, exceeding involucre; corollas glabrous, tube 4–5 mm, ligule 7–10 × 2–5 mm wide; anthers 3–5 mm. **CYPSELAE** 3–5 mm, stramineous; ribs broad, keeled (forming angles of achene), rugulose, separated by a very narrow to obscure groove (on each face). **PAPPUS** 5–7 mm, not abruptly widened proximally, (mostly) proximal 1 mm white and not plumose, or stramineous and plumose throughout. $2n = 16$. [*Prenanthes pauciflora* Torr.; *Lygodesmia pauciflora* (Torr.) Shinners]. Blackbrush, juniper, and desert shrub communities, in sandy to clay soils, often in wash bottoms and alkaline habitats. ARIZ: Apa, Nav; COLO: Arc, Dol, LPl, Mon; NMEX: McK, SJn; UTAH. 1100–2000 m (3700–6700′). Flowering: May–Oct. California, Wyoming, and Kansas, south to Texas and Mexico.

Stephanomeria tenuifolia (Raf.) H. M. Hall (slender-leaved) Slender wirelettuce. Perennials from a rhizome, root crown usually thick and woody, herbage mostly glabrous, glandular-puberulent distally. **STEMS** numerous, slender, persistent (at least the bases), 30–75 cm tall, erect, green (sometimes purplish proximally), becoming stramineous with age (especially proximally), usually striate (especially proximally), branched proximally or throughout, branches ascending to erect. **LEAVES** withered and absent at flowering (at least lower), basal and lower cauline blades linear to filiform, 2–10 cm × 1 mm or less (excluding lobes), margins entire or dentate to lobed; cauline much reduced and scalelike; teeth or lobes less than 0.5 mm, linear. **HEADS** terminal and axillary, pedunculate, borne in racemiform or paniculiform clusters; peduncles glandular-puberulent. **INVOLUCRES** 7–11 mm tall. **CALYCULI** 4–7 (grading into peduncular bracts), linear-lanceolate, 1–4 mm, glandular-puberulent, margins narrowly scarious, white, entire to minutely ciliate. **PHYLLARIES** 4–5 in 1–2 series, linear-lanceolate, (glandular-) puberulent medially and/or distally, rarely glabrous, green and usually purple medially and/or distally, midrib keeled (not especially thickened proximally), margins scarious and white (inner sometimes broadly so; not hyaline). **RAY FLORETS** 4–5, exceeding involucre; corollas glabrous, tube 4–5 mm, ligule 7–9 × 2–3 mm, sometimes laciniately lobed; anthers 4–6 mm. **CYPSELAE** 3–4 mm, stramineous, minutely to obscurely mottled; ribs keeled to rounded, minutely muricate-rugulose to nearly smooth, separated by a very narrow to obscure groove (on each face). **PAPPUS** 5–6 mm, white, not abruptly widened proximally, plumose throughout or proximal 1 mm (or less) with reduced barbs. $2n = 16$. [*Prenanthes tenuifolia* Torr.]. Riparian, hanging garden, and cliff habitats in piñon-juniper, ponderosa pine, and Douglas-fir communities, often in sandy soils, rock outcrops, or canyon bottoms. ARIZ: Apa, Coc, Nav; COLO: Arc, Mon; NMEX: SJn; UTAH. 1700–2700 m (5600–8700′). Flowering: Jul–Sep. British Columbia and Saskatchewan, south to Texas and Mexico. Although widespread and morphologically variable, no taxonomic recognition of intraspecific variation appears merited.

Stylocline Nutt. Nest-straw

G. L. Nesom

(Greek *stylos*, column, + *cline*, bed, alluding to the narrowly cylindric receptacles of the type species) Low, diffuse-branched spring annuals, densely grayish white with a woolly vestiture, stems usually basally ascending. **LEAVES** alternate, entire, narrowly lanceolate to oblanceolate or oblong, woolly on both surfaces. **HEADS** ovoid to subglobose, disciform, in compact clusters immediately subtended by several leaves; true involucral bracts none, vestigial, or 1–3, irregular, grading into receptacular bracts. **RECEPTACLES** columnar to clavate. **PISTILLATE FLORETS** numerous, peripheral, fertile; corollas filiform-tubular, each subtended (at least the inner enclosed) by a dorsally woolly bract with a terminal/marginal hyaline appendage. **FUNCTIONALLY STAMINATE FLORETS** mostly 3–5, central, bracteate; ovaries vestigial or abortive. **CYPSELAE** (fertile) glabrous, terete, nerveless, brown, epappose. **PAPPUS** 0–few thin, caducous bristles. $x = 14$.

Stylocline micropoides A. Gray (resembling the genus *Micropus*) Woollyhead. Annual herbs, taprooted, 2–10 cm. **STEMS** erect to ascending, simple or branched from the base. **LEAVES**: lower lanceolate, longest 8–20 mm. **HEADS** clustered at branch tips, subglobose, 5–8 mm wide. **RECEPTACLES** narrowly cylindric; longest bracts of pistillate flowers 3.3–4.5 mm. **FUNCTIONALLY STAMINATE FLORETS** 3–6, corollas 1.2–1.9 mm long. **CYPSELAE** (fertile) 1–1.4 mm long. **PAPPUS** of 2–5 (–10) fragile bristles 1–2 mm long. $2n = 28$. Steep talus slopes with *Ephedra*, *Yucca*, *Atriplex*. UTAH. 1400–1500 m (4595–4920′). Flowering: Apr–May. Southern California and Nevada to Utah and southwestern New Mexico; Mexico (Baja California, Sonora, Chihuahua).

Symphyotrichum Nees Aster

G. L. Nesom and L. Brouillet

(Greek *symphysis*, junction, + *trichos*, hair, perhaps alluding to a perceived basal connation of bristles in the European cultivar used by Nees as the type) Perennials or annuals, from short to elongate rhizomes, woody caudices, or taproots (annuals). **STEMS** often glabrous, sometimes hairy, usually hairy in decurrent lines distally. **LEAVES** basal and cauline, petiolate, subpetiolate, or sessile; blades (basal) cordiform to elliptic, oblanceolate, or spatulate, (cauline) ovate to lanceolate, oblanceolate, or linear, usually reduced distally, margins serrate or entire, faces glabrous or hairy. **HEADS** radiate or disciform, borne usually in paniculiform, sometimes in racemiform or subcorymbiform arrays. **INVOLUCRES** cylindric or campanulate to hemispheric. **PHYLLARIES** in (3–) 4–6 (–7) series, unequal to subequal, 1-nerved, apices usually with a well-defined green zone, sometimes ± leafy. **RECEPTACLES** flat to slightly convex, epaleate. **RAY FLORETS** in 1 (rarely 3–5) series, pistillate, fertile; corollas white, pink, blue, or purple, sometimes reduced to tubes. **DISC FLORETS** bisexual, fertile; corolla yellow to white, becoming purplish to reddish or pinkish at maturity. **CYPSELAE** usually obovoid or obconic, sometimes fusiform, ± compressed, (2–) 3–5 (–10)-nerved. **PAPPUS** of slender, barbellate bristles. $x = 4–8, 13, 18\ (21)$. [*Aster* subg. *Symphyotrichum* (Nees) A. G. Jones]. Species about 90, mostly in North America, Central America, West Indies, and South America, one species circumboreal; introduced in Europe and elsewhere.

1. Annual (occasionally weakly perennial in *S. frondosum*), taprooted; heads inconspicuously radiate (ray corollas with minute ligules) or disciform (ray corollas reduced to tubes), ray florets 75–100+ in 3–5 series; pappus bristles longer at maturity than disc corollas ..... (2)
1′ Perennial, rhizomatous; heads conspicuously radiate, ray florets (10–) 15–60 in 1 series; pappus bristles at maturity about equal disc corolla length ..... (3)

2. Heads disciform; ray corollas reduced to tubes, ligules absent ..... ***S. ciliatum***
2′ Heads short-radiate; ray corolla ligules 4.5–5 × 0.1–0.2 mm, pink to white ..... ***S. frondosum***

3. Distal leaves, peduncles, and phyllaries stipitate-glandular ..... ***S. campestre***
3′ Leaves, peduncles, and phyllaries eglandular ..... (4)

4. Midstems densely and uniformly hairy ..... (5)
4′ Midstems glabrous or hairy only in narrow vertical lines ..... (7)

5. Middle and upper cauline leaves clasping to auriculate-clasping ..... ***S. foliaceum***
5′ Cauline leaves not at all clasping ..... (6)

6. Stems strigose; phyllary apices erect, obtuse to rounded and not spine-tipped, faces densely strigose; ray florets violet ..... ***S. ascendens***
6′ Stems hispidulous to loosely strigose; phyllary apices spreading to reflexed, spine-tipped, faces sparsely to moderately hispid-strigose; ray florets white or rarely light blue ..... ***S. falcatum***

7. Middle and upper cauline leaves auriculate-clasping or subclasping ..... (8)
7′ Middle and upper cauline leaves not clasping or subclasping, not auriculate ..... (12)

8. Middle and upper cauline leaves subclasping, not strongly auriculate ..... (9)
8′ Middle and upper cauline leaves distinctly auriculate-clasping ..... (11)

9. Plants 20–50 (–60) cm; basal leaves persistent to flowering; heads few, 1–5 (–10), usually on bracteate peduncles, peduncles evenly puberulent ..... ***S. spathulatum***
9′ Plants 30–150 (–200) cm; basal leaves withering by flowering; heads numerous, usually 20–100+, on leafy-bracted peduncles, peduncles puberulent evenly or in lines ..... (10)

10. Phyllaries subequal in length, outermost foliaceous and sometimes slightly longer than inner; peduncles and distal stems sometimes evenly puberulent; heads in racemiform to narrow, paniculiform arrays .... ***S. eatonii***
10' Phyllaries graduate in length, outermost foliaceous or with a green apical patch, shorter than inner; peduncles and distal stems puberulent in lines; heads usually in open, paniculiform arrays.................................... ***S. lanceolatum***

11. Heads relatively few (1–3 [–6]), usually closely subtended by well-developed cauline leaves; outer phyllaries foliaceous; stems pubescent at least near heads... ***S. foliaceum***
11' Heads relatively numerous (mostly 15–50+) in a bracteate array above the leaves; outer phyllaries with an oblanceolate-triangular, green apical zone; stems and leaves glabrous, glaucous.. ***S. laeve***

12. Phyllaries subequal in length, outermost foliaceous and sometimes longer than inner; peduncles and distal stems minutely and evenly puberulent; heads in racemiform to narrow, paniculiform arrays .... ***S. eatonii***
12' Phyllaries graduate in length, outermost with a green apical patch or foliaceous, shorter than inner; peduncles and distal stems puberulent in lines; heads usually in open, paniculiform arrays.................................... ***S. lanceolatum***

Symphyotrichum ascendens (Lindl.) G. L. Nesom (rising, perhaps alluding to the ascending flowering branches) Long-leaved aster, Intermountain aster. Perennials 20–60 cm, long-rhizomatous. **STEMS** densely and uniformly strigose, especially distally. **LEAVES**: basal usually persistent, blades oblanceolate, 50–150 × 5–12 (–15) mm, petiolate; cauline sessile or subpetiolate, not clasping, oblong to narrowly lanceolate or narrowly obovate, 50–120 mm, margins entire, apices acute, faces glabrous or strigose. **HEADS** in paniculiform arrays, on ascending branches; peduncles uniformly strigose. **INVOLUCRES** 4–7 mm long. **PHYLLARIES** narrowly oblanceolate or linear (outer) to linear (inner), unequal, bases indurate, green zones obovate to elliptic, faces densely strigose. **RAY FLORETS** (10–) 15–40; corollas violet. $2n = 26, 36, 52$. [*Aster ascendens* Lindl.; *A. chilensis* Nees subsp. *ascendens* (Lindl.) Cronquist]. Riparian, often disturbed sites, springs, along drainages, hanging gardens, dry lake beds, meadows, sagebrush flats, ponderosa pine, Gambel's oak, piñon pine, piñon-juniper woodland, and cottonwood communities. ARIZ: Apa; COLO: Arc, Dol, Hin, LPl, Min, Mon, SJn, SMg; NMEX: RAr; UTAH. (1125–) 1800–2700 (–2850) m (3700–9350'). Flowering: (Jun–) Aug–Oct. Southwest Canada (British Columbia to Saskatchewan), Washington and California east to Nevada and Arizona, North Dakota, Colorado, and New Mexico. *Symphyotrichum ascendens* is an allopolyploid derived from the hybrid between *S. spathulatum* ($x = 8$) and *S. falcatum* ($x = 5$). Backcrosses to both parental species or hybrids with related taxa sometimes occur where ranges overlap.

Symphyotrichum campestre (Nutt.) G. L. Nesom (pertaining to plains or flat areas) Western meadow aster. Perennials 10–40 cm, long-rhizomatous. **STEMS** proximally glabrous, distally stipitate-glandular and sparsely strigose. **LEAVES**: basal and proximal cauline often withering by flowering, blades (1–3-nerved) narrowly oblanceolate to lanceolate with entire margins, mucronulate or white-spinulose, bases sessile to subclasping, cauline mostly 20–80, 2–8 mm long, faces glabrate to sparsely scabrous or short-strigose, distal cauline stipitate-glandular. **HEADS** borne singly or in paniculiform arrays, branches ascending; peduncles stipitate-glandular. **INVOLUCRES** 5.5–8 mm. **PHYLLARIES** linear to lanceolate with acute to acuminate apices, subequal to unequal, spreading to reflexed, bases ± indurate, green zones covering distal portion, outer ± foliaceous, faces glabrate, sparsely to densely stipitate-glandular. **RAY FLORETS** 15–31; corollas violet. $2n = 10$. [*S. campestre* var. *bloomeri* (A. Gray) G. L. Nesom]. Alkaline shoreline, riparian vegetation. NMEX: McK, SJn. 2200 m (7215'). Flowering: Aug–Oct. Southern British Columbia and Alberta, south to California, Nevada, Colorado, and New Mexico. Known in the study area only from the shoreline of Black Lake, five miles northwest of Crystal, New Mexico.

Symphyotrichum ciliatum (Ledeb.) G. L. Nesom (ciliate) Rayless annual aster, rayless alkali aster. Annuals 7–70+ cm, taprooted. **STEMS** often red-tinged, ± succulent, glabrous. **LEAVES**: basal withering by flowering, blades spatulate, 15–205 × 1.5–9 mm; cauline linear-oblanceolate, (10–) 30–80 (–150), reduced distally, sessile, bases subclasping to rounded; margins usually entire, sometimes serrulate, apices acute to short-acuminate, faces glabrous. **HEADS** disciform, in ± dense, narrow to pyramidal, paniculiform to racemiform arrays; peduncles glabrous, bracts crowding heads. **INVOLUCRES** 5–7 (–11) mm. **PHYLLARIES** linear to narrowly oblanceolate with acute apices, subequal or outer sometimes longer, bases scarious, green zones foliaceous (outer and middle) to lanceolate (inner), faces glabrous. **RAY FLORETS** 75–95+ in 4–5+ series; corolla laminae 0, tubes shorter than style branches. $2n = 14$. [*Aster angustus* (Lindl.) Torr. & A. Gray; *A. brachyactis* S. F. Blake; *Brachyactis angusta* (Lindl.) Britton; *B. ciliata* Ledeb. subsp. *angusta* (Lindl.) A. G. Jones]. Along creeks and washes, alkaline seeps, dry lake beds, riparian. ARIZ: Apa; Nav; COLO: LPl, Mon; NMEX: RAr, SJn; UTAH. 1300–2200 m (4265–7215'). Flowering: Aug–Oct. Canada, Alaska, Washington to New York and Pennsylvania, south to Arizona, New Mexico, Oklahoma, Illinois, Indiana; northern Eurasia west to Romania.

Symphyotrichum eatonii (A. Gray) G. L. Nesom (for Daniel C. Eaton, 1834–1895, Yale botanist and collector with the King Expedition in Utah) Eaton's aster. Perennials 40–100 cm, short-rhizomatous. **STEMS** glabrous or sparsely and evenly puberulent. **LEAVES**: basal withering by flowering, blades narrowly lanceolate, 60–200 × 10–20 mm, petiolate, cauline sessile, narrowly lanceolate to linear, 20–150 mm, bases cuneate, upper sometimes subclasping; margins entire or sometimes serrate, apices acute; faces glabrous or sparsely puberulent. **HEADS** usually 20–100+ in racemiform to narrow, paniculiform arrays, branches usually to 10 cm; peduncles leafy, sparsely, minutely, and evenly puberulent. **INVOLUCRES** 5–8 mm. **PHYLLARIES** often spreading, oblanceolate (outer) to linear (inner), subequal, outer indurate less than 1/2 their length, bases variable, outermost foliaceous and sometimes longer than inner, inner scarious, green zones oblanceolate to elliptic, faces glabrous or sparsely puberulent. **RAY FLORETS** 20–40; corolla laminae white to pink. $2n = 16, 32, 48, 64$. [*Aster bracteolatus* Nutt.; *A. eatonii* (A. Gray) Howell; *Symphyotrichum bracteolatum* (Nutt.) G. L. Nesom]. Moist meadows, creek borders, lakeshores, washes, cliff faces; ponderosa pine, spruce-fir, and cottonwood communities. ARIZ: Nav; COLO: Arc, LPl, Min, Mon, RGr, SJn. 1850–2950 m (6070–9680′). Flowering: Aug–Sep (–Oct). Canada (British Columbia to Saskatchewan), Washington to California and Arizona, east to Montana, Wyoming, Colorado, and New Mexico.

Symphyotrichum falcatum (Lindl.) G. L. Nesom (sickle-shaped or hooked) White prairie aster. Perennials 10–80 cm, long- or short-rhizomatous. **STEMS** moderately to densely hispidulous to loosely strigose, eglandular. **LEAVES**: basal withering by flowering, blades oblanceolate, 10–40 mm; cauline linear-oblanceolate to oblong, 10–40 (–60) mm, reduced distally, sessile, not clasping; margins entire, often coarsely ciliate proximally, apices acute to obtuse, rounded to mucronulate-spinulose, faces glabrate or moderately to densely strigose to hispid-strigose. **HEADS** (1–) 10–200+ in racemiform to diffuse-paniculiform arrays; peduncles densely strigose. **INVOLUCRES** (4.5–) 5–8 mm. **PHYLLARIES** strongly unequal, outer oblanceolate to spatulate, inner linear-lanceolate, basal 1/2–3/4 indurate, green zones (in distal 1/2–1/4) diamond-shaped, apices spine-tipped, spreading to reflexed, faces sparsely to moderately hispid-strigose. **RAY FLORETS** (15–) 20–35; corollas usually white. $2n = 30$. [*Aster commutatus* (Torr. & A. Gray) A. Gray; *A. crassulus* Rydb.; *A. ericoides* L. var. *commutatus* (Torr. & A. Gray) B. Boivin; *A. falcatus* Lindl. subsp. *commutatus* (Torr. & A. Gray) A. G. Jones; *A. falcatus* var. *commutatus* (Torr. & A. Gray) A. G. Jones; *A. falcatus* var. *crassulus* (Rydb.) Cronquist; *Symphyotrichum falcatum* subsp. *commutatum* (Torr. & A. Gray) Semple; *S. falcatum* var. *crassulum* (Rydb.) G. L. Nesom]. Roadsides and other disturbed sites, openings, canyon bottoms, floodplains, riparian, ponderosa pine, Gambel's oak, and piñon-juniper woodland communities. ARIZ: Apa, Coc; COLO: Arc, Dol, LPl, Mon; NMEX: McK, RAr, SJn; UTAH. (1125–) 1600–2400 (–2500, very rarely 2900) m (3700–9515′). Flowering: (Jul–) Aug–Oct. Southern Canada (Alberta to Manitoba), Arizona and Utah, east to Texas, Kansas, North Dakota, Minnesota; Mexico (Chihuahua, Coahuila, Durango, Jalisco, Nuevo León, Sonora). Four Corners plants are **var. *commutatum*** (Torr. & A. Gray) G. L. Nesom (changed or exchanged). Variety *commutatum* is introduced in Ontario, Illinois, Missouri, and Wisconsin, and possibly other eastern states.

Symphyotrichum foliaceum (Lindl. ex DC.) G. L. Nesom (conspicuously leafy) Leafy or leafy-bracted aster. Perennials 5–60 cm; from long, slender rhizomes. **STEMS** glabrous to strigose in lines or evenly puberulent. **LEAVES**: basal usually persistent, blades broadly elliptic to obovate, 30–200 × 8–25 (–30) mm, petiolate to subpetiolate; cauline petiolate to subpetiolate or sessile, oblanceolate to obovate-oblanceolate to obovate, 35–120 mm, commonly little reduced upward, clasping to auriculate-clasping; margins entire or sometimes serrate, faces usually glabrous, sometimes sparsely hairy. **HEADS** 1–3 (–6), borne singly or in paniculiform arrays, usually closely subtended by well-developed cauline leaves; peduncles evenly puberulent or strigose in lines. **INVOLUCRES** 6–16 (–20) mm. **PHYLLARIES** oblanceolate or oblong (outer) to lanceolate or linear (inner), subequal or unequal (outer exceeding inner), outer foliaceous, inner with elongate, elliptic to lanceolate green zones, faces glabrous or puberulent. **RAY FLORETS** 15–60; corollas violet to purple. [*Aster foliaceus* Lindl. ex DC.].

1. Plants 5–20 cm; stems finely and evenly puberulent from base to apex; heads 1 (–3); at least inner phyllaries purple or purple-tinged; alpine or subalpine habitats .... **var. *apricum***
1′ Plants 10–60 cm; stems strigose in lines, sometimes evenly puberulent immediately below heads; heads 1–12; phyllaries green; subalpine or lower habitats .... (2)

2. Phyllaries unequal, outer exceeding inner, oblanceolate and foliaceous .... **var. *canbyi***
2′ Phyllaries subequal, narrowly lanceolate, outer scarcely foliaceous .... **var. *parryi***

var. *apricum* (A. Gray) G. L. Nesom (lying open, exposed to the sun) Plants 5–20 cm. **STEMS** evenly and finely puberulent from base to apex. **CAULINE LEAF BASES** cuneate. **HEADS** 1 (–3). **PHYLLARIES** (at least inner) purple or purple-tinged, oblanceolate or narrowly oblong (outer) to linear (inner), subequal, outer scarcely foliaceous, apices acute. $2n$ = 16, 32. [*Aster foliaceus* Lindl. ex DC. var. *apricus* A. Gray]. Alpine and subalpine meadows, spruce-fir, timberline. COLO: Arc, LPl, Min, RGr, SJn, SMg; NMEX: SJn. 2800–3150 m (9185–10335′). Flowering: Jul–Oct. Canada (British Columbia), Washington, California, and Arizona, east to Montana, Wyoming, Colorado, and New Mexico. Variety *apricum* apparently is more strongly differentiated than var. *canbyi* and var. *parryi* appear to be between themselves. All three varieties are broadly sympatric in the western U.S.

var. ***canbyi*** (A. Gray) G. L. Nesom (for William M. Canby, 1831–1904, Delaware businessman, philanthropist, and avid botanist) Plants 10–60 cm. **STEMS** strigose in lines, sometimes evenly puberulent immediately below heads. **CAULINE LEAF BASES** truncate, auriculate. **HEADS** (1–) 5–15. **PHYLLARIES** green, broadly oblanceolate (outer) to lanceolate or linear (inner), unequal, outer exceeding inner and wider, often foliaceous, apices obtuse to rounded, inner narrowly ovate, apices acute to obtuse. $2n$ = 16. [*Aster foliaceus* Lindl. ex DC. var. *canbyi* A. Gray]. Lakesides, subalpine meadows, steep slopes, riparian, roadsides, oak-juniper, ponderosa pine, ponderosa pine–Gambel's oak, pine-fir, fir-aspen, and spruce-fir communities. ARIZ: Apa, Nav; COLO: Arc, Dol, Hin, LPl, Min, RGr; NMEX: RAr. (1850–) 1900–3000 (–3550) m (6070–11645′). Flowering: Aug–Oct. Washington to Montana and Wyoming, Nevada, Arizona, Colorado, and New Mexico.

var. ***parryi*** (D. C. Eaton) G. L. Nesom (for Charles C. Parry, 1823–1890, surgeon-naturalist and prolific botanical collector with the Mexican Boundary Survey during the years 1849–1852) Plants 10–50 cm. **STEMS** strigose in lines, sometimes evenly puberulent immediately below heads. **CAULINE LEAF BASES** cuneate to truncate. **HEADS** 1–12. **PHYLLARIES** green, oblanceolate or oblong (outer) to linear (inner), subequal, outer scarcely foliaceous, apices acute. $2n$ = 16, 32, 64. [*Aster ascendens* Lindl. var. *parryi* D. C. Eaton; *Aster foliaceus* Lindl. ex DC. var. *parryi* (D. C. Eaton) A. Gray]. Riparian, spruce-cottonwood, and ponderosa pine–Douglas-fir communities. ARIZ: Apa; COLO: Mon; NMEX: McK, SJn. 2200–2650 m (7215–8695′). Flowering: Aug–Oct. Canada (British Columbia, Alberta), Washington, California, east to Montana, Colorado, and New Mexico.

Symphyotrichum frondosum (Nutt.) G. L. Nesom (leafy-branched) Short-rayed alkali aster. Annuals or sometimes perennials 5–140 cm, taprooted. **STEMS** glabrous. **LEAVES**: basal withering by flowering, blades oblanceolate to spatulate, 20–115 × 2–15 mm, petiolate, cauline oblanceolate to linear, subpetiolate or sessile, reduced distally; margins entire (basal sometimes serrulate), sometimes ciliate or remotely scabrous, acute to ± obtuse, faces glabrous. **HEADS** in narrow, paniculiform to spiciform arrays, branches often in axils of nearly every leaf, ascending; peduncles glabrous. **INVOLUCRES** 5–9 mm. **PHYLLARIES** subequal to ± unequal, oblong-oblanceolate or -lanceolate to obovate, linear-lanceolate or linear (innermost), bases scarious, green zones (outer) foliaceous, (inner) lanceolate, faces glabrous. **RAY FLORETS** 90–110 in 4–5+ series; corolla laminae 1.5–2 × 0.1–0.2 mm, pink to pinkish white. $2n$ = 14. [*Aster frondosus* (Nutt.) Torr. & A. Gray; *Brachyactis frondosa* (Nutt.) A. Gray]. River, stream, lake, and pond margins, wet areas, mudflats, riparian, often disturbed sites. ARIZ: Apa, Nav; COLO: LPl, Min; NMEX: McK, RAr, SJn; UTAH. 1750–2700 m (5740–8860′). Flowering: Aug–Oct. Southwest Canada (British Columbia), Washington, California, and Arizona, east to Wyoming, Colorado, and New Mexico; Mexico (Baja California).

Symphyotrichum laeve (L.) Á. Löve & D. Löve (smooth) Perennials (15–) 20–70 (–120) cm, with thick, woody caudices or a few long rhizomes. **STEMS** glabrous, glaucous. **LEAVES**: basal usually withering by flowering, petiolate, blades spatulate or oblong to ovate or lanceolate-ovate, 30–200 × 10–25 (–30) mm, bases attenuate or cuneate to rounded; cauline petiolate or subsessile or sessile, ovate to lanceolate to linear, reduced upward (abruptly so in arrays), bases auriculate-clasping; margins entire; glaucous, thick, margins crenate-serrate or -serrulate to entire, faces glabrous. **HEADS** mostly 15–50+ in broad, sometimes ± flat-topped, paniculiform arrays, branches stiffly ascending (rarely arching); peduncles glaucous, glabrous or very sparsely puberulent in lines, bracts densely spaced, grading into phyllaries. **INVOLUCRES** (4.2–) 5–7 (–8) mm. **PHYLLARIES** unequal to subequal, subulate or lanceolate (outer) to oblong-lanceolate or linear-lancolate or -oblanceolate, basal 1/2–3/4 indurate, green apical zones oblanceolate-triangular, faces glabrous. **RAY FLORETS** (11–) 13–23 (–34); corollas usually pale to dark blue or purple, seldom white. $2n$ = 48. [*Aster laevis* L. var. *geyeri* A. Gray; *A. laevis* subsp. *geyeri* (A. Gray) Piper; *A. laevis* var. *guadalupensis* A. G. Jones]. Pond edges, wet areas, riparian, meadows, talus, clearings and edges, roadsides, oak-juniper, ponderosa pine–Gambel's oak, spruce-fir-aspen, and cottonwood communities. COLO: Arc, LPl, Min, Mon; NMEX: McK, RAr, SJn.1900–2750 m (6235–9020′). Flowering: mid-Jul–Sep. Canada (Yukon and British Columbia to Ontario), Washington to Nevada, east to North Dakota, South Dakota, Colorado, New Mexico, and west Texas; Mexico (Coahuila). Our plants are **var.** ***geyeri*** (A. Gray) G. L. Nesom (for Carl A. Geyer, 1809–1853, German botanist who collected in the western U.S.), Geyer's smooth aster.

Symphyotrichum lanceolatum (Willd.) G. L. Nesom (lance-shaped) Lance-leaved aster, white-panicled aster. Perennials 30–150 (–200) cm, long-rhizomatous. **STEMS** usually puberulent in lines. **LEAVES**: basal withering by flowering, petiolate to subpetiolate, blades elliptic-oblanceolate or obovate to suborbiculate, 10–80 × 5–20 mm, margins crenate-serrate; cauline sessile or subsessile, lanceolate to narrowly lanceolate or oblanceolate, (40–) 50–150 mm, reduced distally, bases subclasping or clasping, margins serrate or middle to distal entire; faces glabrous or adaxial sparsely scabrous. **HEADS** borne singly or numerous, usually 20–100+, congested at ends of lateral branches, usually subtended by foliaceous bracts, on ascending branches, branch leaves often longer than pedicels; peduncles puberulent in lines, bracts foliaceous. **INVOLUCRES** 4–7.2 mm. **PHYLLARIES** graduate in length, linear-lanceolate, basal 1/4–1/2 indurate, green zones lanceolate to linear-lanceolate, outer sometimes foliaceous, abaxial faces glabrous. **RAY FLORETS** 18–45; corollas usually pale to dark purplish blue, sometimes white. $2n = 64$. [*Aster hesperius* A. Gray; *A. hesperius* var. *wootonii* Greene; *A. lanceolatus* Willd. subsp. *hesperius* (A. Gray) Semple & Chmiel.; *Symphyotrichum hesperium* (A. Gray) Á. Löve and D. Löve]. (Semple and Chmielewski 1987.) Pond and creek margins, floodplains, wet areas, riparian, meadows, roadsides, uncommonly in piñon-juniper. ARIZ: Apa, Nav; COLO: Arc, LPl, SJn; NMEX: SJn; UTAH. 1450–2450 m (4755–8040′). Flowering: mid-Aug–Sep (–Oct). Canada (British Columbia and Northwest Territories to Quebec), Washington, Oregon, California, and Arizona, east to Texas, Oklahoma, Nebraska, Iowa, North Dakota, Wisconsin, and Minnesota; Mexico (Baja California, Chihuahua, Sonora). Our plants are **var. *hesperium*** (A. Gray) G. L. Nesom (western).

Symphyotrichum spathulatum (Lindl.) G. L. Nesom (spatula-shaped) Western mountain aster. Perennials 20–50 (–60) cm, from long, slender rhizomes. **STEMS** glabrous or sparsely puberulent. **LEAVES**: basal persistent, blades narrowly elliptic to oblanceolate, 50–150 mm, petiolate; cauline linear to narrowly oblong-lanceolate or oblanceolate, 30–150 mm, reduced distally, sessile, subclasping to clasping but not auriculate; margins entire, apices acute, faces glabrous or sparsely puberulent. **HEADS** few, 1–5 (–10), in corymbiform to paniculiform arrays; peduncles bracteate, evenly puberulent. **INVOLUCRES** 5–10 mm. **PHYLLARIES** unequal, narrowly oblong or linear (outer) to linear (inner), bases indurate, green zones elliptic to lanceolate, apices usually acute, sometimes obtuse, faces glabrous or ± puberulent. **RAY FLORETS** 15–40; corollas violet. $2n = 16, 32, 48, 64$. [*Aster spathulatus* Lindl.; *Aster occidentalis* (Nutt.) Torr. & A. Gray]. Meadows with ponderosa pine, wet areas. COLO: Hin; NMEX: SJn. 2350–3000 m (7710–9840′). Flowering: Jul–Aug. Canada (Northwest Territories, British Columbia, Alberta), Washington to California, east to Montana, Wyoming, Colorado, New Mexico; Mexico (Baja California). Four Corners area plants are **var. *spathulatum***.

Tagetes L. Marigold

Brian Elliott

(for Tages, an Etruscan god said to have sprung from the earth as it was being plowed) Annual or perennial herbs, sometimes suffrutescent or shrubby; conspicuously gland-dotted and strong-scented. **STEMS** erect, simple or diffusely branched, to 2 m tall, glabrous. **LEAVES** opposite, rarely alternate above, simple to pinnately divided or compound, margins of blades or lobes entire to serrate, sessile or petiolate, glabrous to pilosulous. **INFL** heads radiate or discoid, solitary at branch tips or in leafy terminal cymes, peduncles often clavate and fistulose. **INVOLUCRES** fusiform, cylindric, or campanulate. **PHYLLARIES** 3–21 or more in 1 equal or subequal series, connate to near apex, gland-dotted or streaked. **RECEPTACLES** flat to convex or conic, naked, alveolate. **RAY FLORETS** pistillate and fertile, rarely neuter; ligules (0) 1–many; white, yellow, orange, or brown, persistent. **DISC FLORETS** perfect and fertile, (0) 3–many; corolla yellow-orange, tube narrowly cylindric, short. **CYPSELAE** elongate, clavate to cylindric, black, finely striate, 4–5-angled, glabrous or pubescent. **PAPPUS** of 2–6 dissimilar scales, free or connate, sometimes awned; rarely absent. Approximately 50 species of the southwestern United States, Mexico, Central and South America. Many members of the genus are cultivated, becoming naturalized and weedy outside their native range.

Tagetes micrantha Cav. (referring to the small flowers) Licorice marigold. Dwarf, annual, taprooted herbs; anise-scented. **STEMS** erect, usually branched above, 1–3 dm tall, glabrous or sparingly pilose in the axils. **LEAVES** with uppermost bracts alternate; cauline leaves opposite, shortly connate at base, linear, 1–3 cm long, entire or parted into linear divisions, ultimate segments about 1 mm wide, cuspidate, minutely glandular. **INFL** heads terminal on leafy branches, peduncles 0.5–2.5 cm long. **INVOLUCRES** fusiform to subcylindrical, 6–12 mm long, about 1.5 mm wide. **PHYLLARIES** 4–5, linear, connate full length, minutely glandular, margins between phyllaries hyaline; apex with a subterminal dorsal cusp 0.5 mm or less long. **RECEPTACLES** flat to convex, less than 1 mm wide. **RAY FLORETS** pistillate and fertile; ligules 0–3; white or pale yellow, 1–2 mm long and barely exceeding phyllaries, oblong, apex 2–3-toothed. **DISC FLORETS** (4) 5–7 (12); corolla yellow, 3–3.5 mm long, 5-toothed. **CYPSELAE** elongate-linear, slightly flattened, 5–7 mm long, black, finely striate or costellate, strigose on base and margin, becoming glabrate. **PAPPUS** of 2 awns and 2 scales, awns exceeding the scales. $2n = 24$ [*Tagetes fragrantissima* Sessé & Moc.]. Dry, open ground. ARIZ: Apa; NMEX: McK, SJn. 1800–2600 m (6000–8500′). Flowering: Aug–Sep. Ranging from west Texas to Arizona, and south to central Mexico. The species has been used medicinally as a cold remedy, for fevers, stomach complaints, and diarrhea.

Tanacetum L. Tansy
Brian Elliott

(derivation of name uncertain, possibly from the Greek *athanasia* or Latin *tanazita*, meaning immortality, referring to the long-lived flowers) Annual or perennial aromatic herbs; often rhizomatous and suffrutescent. **STEMS** erect or decumbent; branched above or below. **LEAVES** basal and/or cauline, alternate, rarely entire and toothed to more commonly 1–3-pinnately dissected, often resin-dotted or glandular-punctate. **INFL** corymbose, heads rarely solitary. **INVOLUCRES** hemispheric. **PHYLLARIES** imbricate in several series, margins often brown to black; margins and tips of inner bracts usually scarious. **RECEPTACLES** flat to low-convex, naked. **RAY FLORETS** pistillate and fertile or neuter; ligules white or yellow, many, reduced and inconspicuous or expanded and less than 1 cm long, or absent. **DISC FLORETS** perfect and fertile, numerous; corolla yellow, tubular, 5-lobed at tip. **CYPSELAE** subcylindric, truncate, irregularly 3–12-angled or -ribbed; faces usually gland-dotted. **PAPPUS** absent or a short crown. Approximately 160 species, mostly Eurasian, with several native to North America. Closely related to *Chrysanthemum* and difficult to separate taxonomically. Several members are widely cultivated.

Tanacetum vulgare L. (common) Common tansy, tansy. Aromatic perennial herbs from stout, creeping rhizomes; glabrous or nearly so throughout. **STEMS** 3–15 dm tall. **LEAVES** 6–20 cm long, 1–3-pinnatifid, sessile or petiolate, ultimate margins dentate; blade glandular-punctate; rachis winged. **INFL** heads discoid, numerous. **INVOLUCRES** 4–6 mm high. **MARGINAL FLORETS** pistillate; ligule lacking, corolla 3–4-lobed. **DISC FLORETS** 5-lobed. **CYPSELAE** unequally 3–5-angled or -ribbed, 1–2 mm long. **PAPPUS** coroniform, minute. $2n = 18$. [*Chrysanthemum vulgare* (L.) Bernh.; *Tanacetum boreale* Fisch. ex DC.; *T. officinarum* Crantz; *T. vulgare* L. var. *crispum* L.]. A cultivated species escaping to ditches, roadsides, disturbed and waste ground. COLO: LPl. 1980 m (6500′). Flowering: Jul–Sep. Found throughout the United States and adjacent Canada. Known from a single collection in our area, but likely more common. Tansy is quite toxic and has caused human fatalities, mostly from medicinal use or using foliage for tea. It has been used medicinally to kill intestinal worms and for abortion. Dried and crushed leaves have also been used in minute quantities as a spice but are fatal if used in greater amounts.

Taraxacum F. H. Wigg. Dandelion
Gary I. Baird

(Arabic *tarakh shaqn*, ancient name for this or some other bitter herb) Perennials, taprooted, sometimes with a branched, nonwoody caudex, acaulescent, herbage glabrous, sometimes glaucous, to pubescent, sap milky. **STEMS** lacking. **LEAVES** all basal, rosulate, erect to prostrate, simple, petiolate or sessile; petioles usually narrowly winged; blades oblanceolate to obovate, entire to dentate or runcinately pinnatifid; lobes lanceolate to dentate, mostly retrorse, sometimes spreading, often denticulate or lobate; denticles or lobules 0–7 (or more), mostly on acroscopic margin, sometimes irregular. **HEADS** liguliflorous, borne singly, erect; peduncle scapose, ebracteate, glabrous to villous (mostly distally). **INVOLUCRES** more or less campanulate; calyculi of 6–18 (or more) bractlets in 2–3 series, subequal to unequal, erect to squarrose, margins entire, often scarious, apices acute to acuminate. **PHYLLARIES** 8–24 (or more) in 1–2 series, equal, erect (reflexed when fruit mature), herbaceous, glabrous, sometimes glaucous, margins more or less scarious to hyaline (usually folded, interlocked, and weakly connate in bud), apices acuminate (calyculi and phyllaries sometimes corniculate with subterminal protuberances). **RECEPTACLES** flat, epaleate. **RAY FLORETS** ± matutinal; corolla ligulate, yellow (outer usually with purplish or grayish stripe abaxially, this still evident when dried), rarely pinkish or whitish. **DISC FLORETS** lacking. **CYPSELAE** monomorphic, stramineous, olivaceous, brownish, reddish, or grayish, mostly obovoid, coarsely muricate (especially distally), mostly 5–15-ribbed, apically narrowed into a cone that tapers into a slender beak. **PAPPUS** of numerous bristles in 1 series, persistent, white to sordid, barbellulate. $x = 8$. An undetermined number of species of Eurasia, North America, and South America; some species widely introduced and now cosmopolitan (most abundantly in temperate, alpine, and polar climates). The taxonomy and nomenclature within the genus are chaotic and without consensus. Delimitation of species is complicated by hybridization, polyploidy, and apomixis. As many as 2000 asexually reproducing, apomictic microspecies have been recognized. In reality, the number of actual species is much lower, but most characters used to separate the species ultimately fail. A generally accepted approach to separating our North American material is employed here.

1. Phyllaries and calyculi bractlets more or less corniculate (protuberances subterminal) ....(2)
1′ Phyllaries and calyculi bractlets not corniculate ....(3)

2. Leaves usually less than 10, mostly less than 5 cm long, subentire to dentate; bractlets of calyculi erect to spreading; fruit stramineous to brownish or olivaceous; plants of high elevations, over 3000 m (9850′).......***T. ceratophorum***
2′ Leaves usually more than 10, mostly more than 5 cm long, usually lacerately lobed or pinnatifid; bractlets of calyculi soon recurved or reflexed; fruit red to brownish red; plants of lower elevations, mostly below 2500 m (8200′), rarely higher..***T. erythrospermum***

3. Plants less than 5 cm tall; leaves mostly 10 or fewer, dentate, the teeth spreading and with entire margins; calyculus bractlets 10 or fewer, ovate to deltate, erect to spreading (ultimately reflexed) ....***T. scopulorum***
3′ Plants more than 5 cm tall; leaves mostly more than 10, subentire to pinnatifid, the teeth or lobes retrorse and often denticulate or lobulate; calyculus bractlets more than 10, lanceolate, soon recurved or reflexed ....(4)

4. Leaves generally lacerately pinnatifid with often lanceolate or linear lobes; body of fruit red to brownish red ....***T. erythrospermum***
4′ Leaves generally dentate or lobed to pinnatifid, with mostly deltate lobes; body of fruit stramineous, olivaceous, tan, or grayish ....***T. officinale***

Taraxacum ceratophorum (Ledeb.) DC. (horn-bearing, alluding to the projections on the phyllaries and calyculi) Horned dandelion. Perennials 5–6 cm (or more). **LEAVES**: blades 3–5 cm × 10–12 mm (including lobes), margins subentire to dentate (or more or less runcinately lobed); teeth (or lobes) deltate to lanceolate, mostly retrorse, often denticulate; denticles 0–2 per tooth, lanceolate to linear, acroscopic. **INVOLUCRES** 8–12 mm tall. **CALYCULI** of 10–12 bractlets in about 2 series, 6–8 mm, ovate to lanceolate, erect to spreading, pale, margins often purplish, apices acute, mostly corniculate (protuberances subterminal). **PHYLLARIES** 12–14 in 1–2 series, 8–12 mm, lanceolate, blackish green (especially distally), margins scarious, pale, apices acuminate, mostly corniculate (protuberances subterminal). **RAY FLORETS** about 50, outer exceeding involucre; corolla tube about 5 mm, ligule 8–11 × 1–2 mm; anthers about 4 mm. **CYPSELAE** 10–11 mm, stramineous to brownish or olivaceous; body 4–5 mm; ribs 5 (prominent) and often with 10–15 smaller ribs; beak 5–14 mm. **PAPPUS** 5–7 mm. $2n$ = 16, 32, 40, 48. [*Taraxacum eriophorum* Rydb.; *T. ovinum* Greene]. Subalpine to alpine meadows and slopes. COLO: Hin, LPl, Min, Mon, SJn; UTAH. 3500–3870 m (11500–12700′). Flowering: Jul. Alaska to Greenland, south to California, Utah, and New Mexico; Eurasia. Just entering the study area at the highest elevations of the San Juan Mountains.

Taraxacum erythrospermum Andrz. (red-seeded) Red-seeded dandelion. Perennials 3–24 cm tall. **LEAVES**: blades oblanceolate, 5–11 cm × 5–35 mm (including lobes), margins runcinately lobed or pinnatifid; lobes 3–6, deltate to linear-lanceolate, mostly retrorse, lobulate to denticulate; lobules (0) 3–7 (or more) per lobe (at least on middle lobes), lanceolate to linear, acroscopic (rarely basiscopic also), occasionally yet again denticulate with 1–2 denticles. **INVOLUCRES** 10–15 mm tall. **CALYCULI** of 10–16 bractlets in about 2 series, ovate to lanceolate, recurved to reflexed, pale, often glaucous and/or purplish, margins narrowly scarious, apices acute to acuminate, with or without corniculi (protuberances, when present, subterminal). **PHYLLARIES** 12–18 in 1–2 series, 10–15 mm, lanceolate, green to blackish green, margins scarious, pale, apices acuminate, with or without corniculi (protuberances, when present, subterminal). **RAY FLORETS** about 75–150, outer ± exceeding involucre; corolla sparsely pilose, tube 2–4 mm, ligule 5–9 × 1–2 mm; anthers 2–3 mm. **CYPSELAE** 9–13 mm, reddish or reddish brown; body about 3 mm, ribs about 15; beak 5–7 mm. **PAPPUS** 5–7 mm. $2n$ = 16, 24, 32. [*Taraxacum laevigatum* (Willd.) DC. var. *erythrospermum* (Andrz.) J. Weiss; *T. officinale* F. H. Wigg. var. *erythrospermum* (Andrz.) Bab.]. Meadows, hanging gardens, sagebrush, piñon-juniper, ponderosa pine, and aspen communities, often in disturbed sites. ARIZ: Apa, Nav; NMEX: McK, San, SJn. 1800–2700 m (5800–9000′). Flowering: May–Sep. Eurasia, introduced and now widely established in North America and elsewhere.

Taraxacum officinale F. H. Wigg. (storeroom or workplace, alluding to its remedial or medicinal uses) Common dandelion. Perennials 6–30 cm tall. **LEAVES**: blades oblanceolate, 5–30 cm × 10–60 mm (including lobes), margins runcinately dentate, lobed, or pinnatifid; lobes 3–5, deltate, retrorse, entire to lobulate; lobules 0–4 (or more), lanceolate to linear, acroscopic (rarely basiscopic also), occasionally yet again denticulate with mostly 1 denticle. **INVOLUCRES** 10–20 mm tall. **CALYCULI** of 12–24 bractlets in about 2–3 series, ovate to lanceolate, margins sometimes denticulate, recurved to reflexed, pale, often glaucous, sometimes purplish, margins narrowly scarious, apices acute to acuminate, lacking protuberances. **PHYLLARIES** 12–22 in 1–2 series, 10–20 mm, lanceolate, green to blackish green, often glaucous, margins scarious, pale, apices acuminate, lacking protuberances. **RAY FLORETS** about 50–150, outer exceeding involucre; corolla sparsely to densely pilose, tube 3–4 mm, ligule 7–12 × 1–2 mm; anthers 3–4 mm. **CYPSELAE** 7–13 mm, olivaceous, stramineous, or brownish; body about 3 mm, ribs about 5 (prominent) with about 10 smaller ribs; beak 5–8 mm. **PAPPUS** 5–7 mm. $2n$ = 24, 40. [*Leontodon taraxacum* L.; *Taraxacum officinale* var. *palustre* Benth.; *T. sylvanicum* R. Doll]. Riparian, hanging garden, grassland, sagebrush, piñon-juniper, ponderosa pine, aspen, spruce, and alpine communities, often in disturbed or cultivated areas. ARIZ: Apa, Coc, Nav; COLO: Arc, Con, Dol, Hin, LPl, Min, Mon, RGr, SJn, SMg; NMEX: McK, RAr, San, SJn; UTAH. 1800–3800 m (5800–12500′). Flowering: Apr–Oct. Eurasia, introduced and now widely established in North America and elsewhere.

Taraxacum scopulorum (A. Gray) Rydb. (of cliffs or crags) Alpine dandelion. Perennials 1–4 cm tall. **LEAVES**: blades oblanceolate, 2–5 cm × 5–6 mm (including lobes), margins entire to dentate-lobed; lobes 0–5, deltate, spreading (or slightly retrorse), entire. **INVOLUCRES** 6–8 mm tall. **CALYCULI** of about 8 bractlets in 2–3 series, ovate, erect, ± glaucous, often purplish (especially distally), margins green or (rarely) scarious, apices acute to acuminate, lacking protuberances. **PHYLLARIES** 8–12 in 1–2 series, 6–8 mm, lanceolate, green to blackish green, ± glaucous, margins scarious, pale, apices acuminate, lacking protuberances. **RAY FLORETS** 25–50, outer just exceeding involucre; corollas glabrous, tube 2–4 mm, ligule 5–6 × about 1 mm; anthers 1–3 mm. **CYPSELAE** 7–8 mm, reddish brown to stramineous; body about 3 mm, ribs about 5 (prominent) with about 10 smaller ribs; beak 3–4 mm. **PAPPUS** 4–5 mm. $2n$ = 16? [*T. lyratum* (Ledeb.) DC. misapplied]. Alpine tundra. COLO: Con, LPl, Mon, SJn. 3600–3900 m (12000–13000′). Flowering: Jul. High elevations in the Rocky Mountains of British Columbia, Alberta, Idaho, Wyoming, Utah, and Colorado, also at low elevations in the Canadian Arctic and Greenland.

Tetradymia DC. Horsebrush, Cottonthorn

Brian Elliott

(Greek *tetra*, four, + *dymo*, together, referring to the four-flowered head of the type species) Perennial woody shrubs, spinescent or unarmed, floccose-woolly especially when young. **STEMS** erect to spreading, freely and stiffly branched, to 2 dm high, pannose-tomentose. **LEAVES**: primary leaves borne on elongated shoots, alternate, linear to subulate, lanceolate or spatulate, sometimes modified to straight or recurved spines, entire, sessile, tomentose to sericeous or glabrescent; secondary leaves (when present) short-lived, soon withering, fasciculate in the axils of primary leaves, entire, sessile, glabrous to sericeous. **INFL** heads discoid, axillary, racemose or corymbose; peduncles ebracteate or bearing 1–5 diminutive, foliagelike bracts. **INVOLUCRES** cylindric to turbinate or hemispheric, 4–12 mm high. **PHYLLARIES** 4–6, in 1–2 equal or subequal series, oval-elliptic to lanceolate, canescent-tomentose to glabrate, firm; margins imbricate, faces keeled, base often enlarged and thickened. **RECEPTACLES** flat, naked. **RAY FLORETS** absent. **DISC FLORETS** perfect and fertile, 4–8 per head; corolla bright yellow to cream-colored; tube long, slender; throat short; lobes 5, linear-lanceolate, relatively long and spreading. **CYPSELAE** obconic to terete, striate, often angled, glabrous to densely long-hairy. **PAPPUS** none or of many barbellate or penicillate capillary bristles or of barbellate scales, whitish to tawny. Ten species of the western United States and adjacent Canada and Mexico. Toxic to sheep, causing big-head disease due to swelling of the head (Strother 1974).

1. Young stems unevenly white-pannose with longitudinal glabrate streaks below the nodes; primary leaves not spiny; peduncles ebracteate; disc flowers 4 (5); pappus of 100–150 capillary bristles, the bristles not obscured by the pubescence of the cypsela ..... ***T. canescens***

1′ Young stems evenly white-pannose; primary leaves maturing into spines; peduncles bracteate; disc flowers 5–9; pappus of about 25 slender bristles, often somewhat obscured by the pubescence of the cypsela ..... ***T. spinosa***

Tetradymia canescens DC. (grayish white, usually caused by pubescence overlying a green surface, referring to the stems and leaves) Common horsebrush, spineless horsebrush. Unarmed shrubs. **STEMS** much-branched, the branches spreading, 1–8 dm tall and nearly as wide; young stems unevenly white-pannose with green glabrescent streaks below nodes of primary leaves; older stems glabrescent. **LEAVES**: primary leaves 5–40 mm long, 1–6 mm wide, linear to lanceolate, oblanceolate or spatulate, sometimes spinulose-tipped, sparsely tomentose to sericeous; secondary leaves (when present) fascicled, axillary. **INFL** heads 3–6 (8), in small cymose clusters terminating the branches; peduncles 5–15 (25) mm long, ebracteate, tomentose. **INVOLUCRES** 6–15 mm high. **PHYLLARIES** 4 (rarely 5), subequal, oblong to ovate or lanceolate, tomentose, often carinate. **DISC FLORETS** 4 (rarely 5) per head; corolla bright yellow to cream,

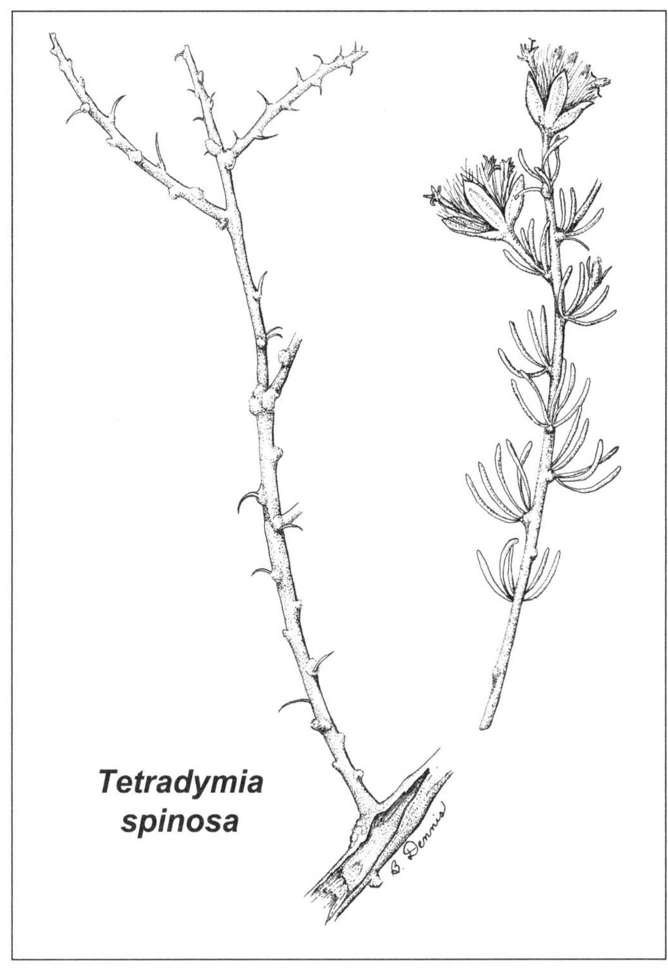

Tetradymia spinosa

7–15 mm long, lobes 2.5–5 mm long. **CYPSELAE** 2.5–5 mm long, glabrous to villous. **PAPPUS** of 100–150 white to tawny capillary bristles, 6–11 mm long, free to the base or united below, some bristles penicillate. 2*n* = 60, 62, 90, or 120. [*T. inermis* Nutt.; *T. linearis* Rydb.; *T. multicaulis* A. Nelson]. Dry, open places in the plains, foothills, and amongst piñon-juniper, ponderosa pine, mixed conifer, and aspen. ARIZ: Apa, Nav; COLO: Arc, Dol, LPl, Mon, SMg; NMEX: McK, RAr, San, SJn; UTAH. 1700–2380 m (5600–7800′). Flowering: Jun–Sep. Southern British Columbia to Montana and south to New Mexico, Arizona, and California. Although toxic, the plant has been used medicinally as a cold remedy, cough medicine, emetic, disinfectant, for fever and stomach-ache, and as a treatment for venereal disease. The yellow flowers, when mixed with other species, have been used as a yellow dye.

Tetradymia spinosa Hook. & Arn. (bearing well-developed spines) Cottonthorn, spiny horsebrush, shortspine horsebrush, thorny horsebrush, catclaw horsebrush. Compact or spreading, spinescent shrubs. **STEMS** intricately branched, the branches straight or arching, to 1.2 m high; young stems densely and evenly white-pannose. **LEAVES**: primary leaves 5–20 (25) mm long, linear, tomentose but becoming glabrate, maturing into harsh recurved spines; secondary leaves (when present) 3–15 (25) mm long, linear-filiform to clavate, fleshy, glabrous or glabrescent. **INFL** heads solitary or in pairs on axillary peduncles from nodes of the previous season; peduncles 5–30 mm long, bracteate, pannose to tomentose. **INVOLUCRES** 8–12 mm high. **PHYLLARIES** 5 (4–6), subequal, oblong to ovate, densely tomentose to glabrous, weakly carinate. **DISC FLORETS** 5–9 per head; corolla pale to bright yellow, 6–10 mm long; throat conspicuously dilated, lobes lance-linear, about 3 mm long. **CYPSELAE** 6–8 mm long, densely pubescent with erect white trichomes 9–12 mm long. **PAPPUS** of 25 slender bristles, 6–9 mm long, often obscured by the pubescence of the cypsela. 2*n* = 60. [*Lagothamnus ambiguus* Nutt.; *Lagothamnus microphyllus* Nutt.]. Dry, open places in the foothills and plains, often on sandy soils of desert scrub or sagebrush. COLO: LPl, Mon; NMEX: RAr, San, SJn; UTAH. 1520–2130 m (4980–7000′). Flowering: May–Jun. A plant of the Great Basin, ranging from Oregon to Montana and south to Colorado and New Mexico, but apparently not known from Arizona.

Tetraneuris Greene Bitterweed

Brian Elliott

(Greek *tetra*, four, + *neuris*, nerved, referring to the venation of ray corolla) Annual or perennial herbs with a branched caudex. **STEMS** 1–many, erect or decumbent, unbranched or branched above. **LEAVES** all basal or basal and cauline, alternate, entire to toothed or lobed, glabrous or sparsely to densely pubescent, eglandular or sparsely to densely impressed gland-dotted, midrib distinct, base of basal leaves often expanded and clasping. **INFL** heads radiate or discoid, solitary, paniculate, or corymbose; peduncle expanded apically, sparsely to densely pubescent. **INVOLUCRES** hemispheric to campanulate. **PHYLLARIES** in 3 series, herbaceous, green or tinged with red or purple, sparsely to densely pubescent; outer series with or without scarious margins; middle series alternate with the outer series, usually with scarious margins; inner series narrower, margins scarious. **RECEPTACLES** hemispheric to conic, naked. **RAY FLORETS** pistillate, fertile, absent or 7–27 per head; ligules yellow, fan-shaped to oblanceolate, glabrous to pubescent, eglandular or sparsely to densely dotted with sessile glands; apex usually 3-lobed. **DISC FLORETS** perfect, fertile, 20–250 or more per head; corolla yellow basally, yellow to purple or red apically; throat cylindric to campanulate, lobes 5. **CYPSELAE** obpyramidal to narrowly obpyramidal, sparsely to densely pubescent. **PAPPUS** of scales, usually aristate. Nine species of the western United States and adjacent Canada and Mexico, with one species disjunct (and possibly extinct) in the upper Midwest. The genus has mostly been treated as a subgenus of *Hymenoxys* but is now considered distinct based on DNA restriction site data. *Tetraneuris scaposa* (DC.) Greene var. *scaposa* may be expected in northwestern New Mexico or southwestern Colorado. (Bierner and Turner 2003.)

1. Leaves all basal; heads 1 or rarely 2 per stem ..... ***T. acaulis***
1' Leaves basal and cauline; heads 1 to 5 (rarely 7) per stem ..... (2)

2. Leaf blades sparsely villous to glabrate; plants bright gray-green ..... ***T. ivesiana***
2' Leaf blades moderately to usually densely strigose-canescent (tightly appressed, often silvery hairs relatively short, about 1–1.5 mm long); plants often silvery from dense pubescence ..... ***T. argentea***

Tetraneuris acaulis (Pursh) Greene (without a stalk or stem, referring to the scapose habit) Stemless woollybase, stemless four-nerve daisy, stemless hymenoxys. Perennial herbs from a short, thick, compactly branched caudex; long-woolly pubescent at the base, sericeous to villous or glabrate above. **STEMS** lacking. **LEAVES** all basal (in ours), tightly clustered, linear to oblanceolate or spatulate, entire, leaf base usually densely woolly or long-pubescent, blade densely villous to glabrous. **INFL** heads mostly radiate, rarely discoid, solitary atop a naked scape, rarely with 2 heads per scape; scape erect, unbranched; 0.5–50 cm, densely villous to glabrous. **PHYLLARIES**: outer and middle often purple-tinged on apex and margin or sometimes throughout, margins sometimes scarious. **RAY FLORETS** 8–15 (21) per head; ligules 5–20 mm long, 3–8 mm wide, broad and showy. **DISC FLORETS** 25–200 or more per head; corolla 2.7–4.3 mm long. **CYPSELAE** 2–4.5 mm long. **PAPPUS** of 5–9 scales, ovate to obovate or lanceolate, apex acute or short-awned, 2–4 mm long including the awn. A complex group composed of four recognized varieties. The varieties are apparently ecologically and geographically significant, but morphologically confluent, and the synonomy is extensive. Note that var. *nana* of southeastern Utah has been reduced to synonymy under var. *arizonica.* A collection of var. *epunctata* at the southern edge of its range from just outside our area in Dolores County is noted in Bierner and Turner (2003). Although not described below, it is included in the key.

1. Leaf blade epunctate or nearly so; in our area from southern Utah to western Colorado ..... var. *epunctata*
1' Leaf blade impressed-glandular punctate; widespread ..... (2)

2. Leaf blade impressed-glandular punctate, but not densely so, glabrous to sparsely or densely pubescent, usually densely strigose-canescent (tightly appressed, often silvery pubescence, hairs relatively short, about 1–1.5 mm long); mostly east of the Continental Divide ..... **var. *acaulis***
2' Leaf blade densely and conspicuously impressed-glandular punctate, glabrous or sparsely to densely pubescent but not strigose-canescent; mostly west of the Continental Divide ..... **var. *arizonica***

var. *acaulis* Stemless four-nerve daisy. **LEAVES** sparsely to moderately pubescent, usually densely strigose-canescent (tightly appressed, often silvery pubescence, hairs relatively short, about 1–1.5 mm long). $2n = 28, 30, 56$, or 60. [Numerous synonyms have been named in several genera, including *Actinella*, *Actinea*, *Cephalophora*, *Gaillardia*, *Ptilepida*, *Tetraneuris*, and *Hymenoxys*]. Plains, grasslands, slopes, and at the edge of woods. COLO: Arc; NMEX: RAr. 1370–2930 m (4500–9600′). Flowering: Apr–Oct. Great Plains and east slope of the Rocky Mountains from Montana to New Mexico. Fully intergradient with var. *arizonica* and closely related to *T. argentea.*

var. *arizonica* (Greene) K. F. Parker (of Arizona) Arizona woollybase, Arizona four-nerve daisy. **LEAVES** glabrate to villous but not strigose-canescent, densely and conspicuously impressed-glandular punctate. $2n = 28, 30, 56, 60$. [Several synonyms have been named in *Actinea*, *Tetraneuris*, and *Hymenoxys*. Also, *T. acaulis* (Pursh) Greene var. *nana* (S. L. Welsh) Kartesz & Gandhi is included here]. Mountains and plains, often in grasslands, sagebrush, desert shrublands, and pine, fir, or aspen forests. ARIZ: Apa; COLO: Arc, LPl, Mon, SMg,; NMEX: McK, RAr, San, SJn; UTAH. 1700–2740 m (5600–9000′). Flowering: Apr–Sep. West of the Continental Divide from New Mexico to Arizona and California, and north to Nevada, Utah, Wyoming, and Idaho. Fully intergradient with var. *acaulis.* The subspecies has been used medicinally as a stimulant and in a poultice for pain.

Tetraneuris argentea (A. Gray) Greene (silvered or silvery, referring to the pubescence) Perennial herbs with a stout, compactly branched caudex surmounting a taproot; woolly at base, silvery silky-villous above. **STEMS** erect, simple or branched, 10–40 cm high. **LEAVES** basal and cauline, progressively reduced above, linear above to lanceolate, oblanceolate, or spatulate below, entire, sericeous and often silvery to glabrous, impressed-punctate, this often hidden by pubescence; basal leaves densely crowded, base woolly or long-pubescent. **INFL** heads radiate, 1–3 (7) per stem, solitary, paniculate or corymbiform; peduncle 2–14 cm long, moderately to densely pubescent and gland-dotted. **PHYLLARIES**: outer series obovate to ovate or lanceolate, margin usually scarious, apex often tinged with red or purple; middle series with margin usually scarious, apex often tinged with red or purple. **RAY FLORETS** 8–21 per head; ligules 5–20 mm long, 4–7 mm wide. **DISC FLORETS** 25–75 or to 100 or more per head; corolla 2.8–3.8 mm long. **CYPSELAE** 2.4–3.1 mm long. **PAPPUS** of 5–6 scales, obovate, aristate. $2n$ = 30, 60. [*Actinea argentea* (A. Gray) Kuntze; *A. formosa* (Greene ex Wooton & Standl.) A. Nelson; *A. leptoclada* (A. Gray) Kuntze; *A. scaposa* (DC.) Kuntze var. *trinervata* (Greene) Kittell; *Actinella argentea* A. Gray; *A. leptoclada* A. Gray; *Hymenoxys argentea* (A. Gray) K. F. Parker; *Ptilepida argentea* (A. Gray) A. Heller; *T. leptoclada* (A. Gray) Greene; *T. formosa* Greene ex Wooton & Standl.]. Dry plains and hills, often amongst piñon-juniper and in sandy soils. Known from one collection in our area. NMEX: RAr. 1800 m (5900′). Flowering: Apr–Oct. Restricted to eastern Arizona and northern New Mexico. Uncertainty regarding the proper generic placement of *Tetraneuris* has led to numerous synonyms. The plant has been used medicinally for heartburn and in lotion for eczema.

Tetraneuris ivesiana Greene (for J. C. Ives, 1829–1868, assistant surveyor with the U.S. Corps of Topographical Engineers on two trips to Colorado searching for a southern railroad route) Ives' woollybase. Perennial herbs from a branched caudex; sparsely to moderately pubescent and gland-dotted. **STEMS** erect, simple or few-branched above, 9–26 cm high. **LEAVES** basal and cauline, the cauline sometimes reduced to a single small leaf; linear-oblanceolate, entire, villous to glabrate and conspicuously impressed-punctate; basal leaves tightly clustered, base woolly or villous. **INFL** heads radiate, 1–2 (5) per stem, solitary, paniculate, or corymbose; peduncle 5–19 cm long, apex often densely pubescent. **PHYLLARIES**: outer series obovate to ovate or lanceolate, margin usually scarious, apex often tinged with red or purple; middle series with margin usually scarious, apex and margins often tinged with red or purple. **RAY FLORETS** 7–10 per head; ligule 6–20 mm long, 4–9 mm wide. **DISC FLORETS** (40) 80–150 or more per head; corolla 3.3–4.5 mm long, glabrous. **CYPSELAE** 3–4.1 mm long. **PAPPUS** of 5–7 scales, obovate to oblanceolate or sometimes lanceolate-aristate. $2n$ = 28, 30, 56, 60. [*Actinea leptoclada* (A. Gray) Kuntze var. *ivesiana* (Greene) J. F. Macbr.; *Hymenoxys acaulis* (Pursh) K. F. Parker var. *ivesiana* (Greene) K. F. Parker; *Hymenoxys argentea* (A. Gray) K. F. Parker var. *ivesiana* (Greene) Cronquist; *Hymenoxys ivesiana* (Greene) K. F. Parker; *T. intermedia* Greene; *T. mancosensis* A. Nelson; *T. pilosa* Greene]. Open areas and roadsides, often in open piñon-juniper woodland or pine forests. ARIZ: Apa, Coc, Nav; COLO: Arc, Dol, LPl, Mon, SMg; NMEX: McK, RAr, San, SJn; UTAH. 1200–2680 m (3950–8880′). Flowering: Apr–Oct. Known only from the Four Corners region of Utah, Colorado, Arizona, and New Mexico. The species is very similar to *T. acaulis* var. *arizonica*.

Thelesperma Less. Greenthread, Navajo Tea

J. Mark Porter

(Greek *thele*, nipple, + *sperma*, seed, alluding to papillate cypselae of some species) Annuals, perennials, or herbaceous subshrubs, 10–75 cm. **STEMS** usually 1, erect, branched distally or throughout. **LEAVES** mostly basal, basal and cauline, or mostly cauline; mostly opposite; blades 1–2 (–3)-pinnately lobed (ultimate lobes oblanceolate to filiform), usually glabrous. **INFL** borne singly or in loose corymbs. **HEADS** radiate or discoid. **CALYCULI** (outer, lower involucral bracts) of 3–8 distinct, usually spreading or reflexed, linear to subulate, herbaceous bractlets. **INVOLUCRES** hemispheric to urceolate, 4–15 mm diameter. **PHYLLARIES** 5–8 in 2 series, persistent, connate 1/5–7/8 their lengths, lanceolate to ovate, subequal, leathery to membranous, margins of free portion of phyllaries scarious. **RECEPTACLES** flat to convex; paleae whitish, red-brown striate, obovate to oblong, scarious, appressed to cypsela, falling. **RAY FLORETS** absent or approximately 8, sterile; corollas yellow or red-brown, or bicolored (yellow and red-brown). **DISC FLORETS** 20–100, bisexual, fertile; corollas yellow (with red-brown nerves) or red-brown, either with throats equal to or longer than 5 deltate, equal lobes, or with throats shorter than 5 lance-linear, unequal lobes (in either form, the abaxial sinus usually deeper than others). **CYPSELAE** dark red-brown or yellow-brown, outer often arcuate, shorter, inner more columnar, usually some or all obcompressed (each usually shed together with its subtending palea), faces smooth or papillate to tuberculate or verrucate, margins sometimes winged. **PAPPUS** absent, or persistent, of 2 retrorsely ciliate, subulate scales or awns. x = 12. Species 10+ (3 in our flora) in western North America, Mexico, and South America (Hansen, Allphin, and Windham 2002; Greer 1997; Strother 2006).

1. Throats of disc corollas equal to or longer than lobes; pappus usually absent, or of 2 awns 0.1–0.3 (–0.5) mm long ..... ***T. subnudum***

1′ Throats of disc corollas shorter than lobes; pappus usually of 2 awns or scales (0.5–) 1–3 mm, rarely absent ...(2)

2. Perennials (sometimes flowering first year); calyculi (outer, lower involucral bracts) ovate to oblong, 1–3 mm; ray florets 0; cypselae 5–8 mm .... ***T. megapotamicum***
2' Annuals (sometimes persisting); calyculi (outer, lower involucral bracts) linear to narrowly triangular, (2–) 4–8 mm; ray florets 8; cypselae 3.5–4.5 mm .... ***T. filifolium***

Thelesperma filifolium (Hook.) A. Gray (thread-leaf) Plains Navajo tea, greenthread, stiff greenthread. Annuals, rarely persisting a second year, 10–40 (–65) cm. **STEMS** 1 to several, loosely branching, internodes mostly 10–35 mm. **LEAVES** mostly cauline, scattered over lower 1/2–3/4 of stems; bipinnately divided, lobes mostly linear to filiform, sometimes oblanceolate, 5–30 (–57) mm long, 0.5–1 (–3.3) mm wide. **HEADS** 1–few, long-pedunculate, terminating branches. **CALYCULI** of 7–8 (–9) linear to narrowly triangular bractlets (2–) 4–8.5 mm. **INVOLUCRES** 6–9 mm long, connate 1/4–1/2 their length, the free portion white scarious-margined. **RAY FLORETS** 8; laminae yellow to golden yellow, sometimes suffused with red-brown, 12–22 mm long. **DISC FLORETS** yellow with red-brown nerves, throats shorter than lobes. **CYPSELAE** 3.5–4.5 mm. **PAPPUS** of 2 subulate awns, 0.5–1 (–2.3) mm long. $2n = 16, 18$. Disturbed locations on dry clays or sandy soils, rocky slopes, plains, and hills. NMEX: SJn. 1525–1980 m (5000–6500′). Flowering: Mar–Aug (–Oct). Fruit: May–Oct. A widespread species, *T. filifolium* occurs in Arkansas, Colorado, Kansas, Louisiana, Mississippi, Missouri, Nebraska, New Mexico, Oklahoma, South Dakota, Texas, and Wyoming; Mexico (Nuevo León). In our area, this species is represented by **var. *intermedium*** (Rydb.) Shinners (intermediate). Used as a tea by many tribes.

Thelesperma megapotamicum (Spreng.) Kuntze (big river) cota (Spanish), ch'il ahwéhé (Diné), molanawe (Zuni), ho hoísi (Hopi), Hopi tea, greenthread, Navajo tea, Indian tea. Perennials or subshrubs, (20–) 30–85 cm tall, with a caudex or creeping rhizomes. **STEMS** several, glabrous and glaucous, internodes mostly 40–100 mm. **LEAVES** mostly cauline, scattered over lower 1/2–3/4 of stems; once- to twice-pinnately divided, lobes mostly linear to filiform, sometimes oblanceolate, 20–40 (–54) mm long, 0.5–1 (–2.5) mm wide, margins of petiole often scabrous-ciliolate. **HEADS** 1–few, long-pedunculate, terminating branches. **CALYCULI** of 3–5 (–6) ovate to oblong bractlets 1–2 (–3) mm. **INVOLUCRES** 6–9 mm long, connate 1/3–2/3 their length, the free portion white scarious-margined. **RAY FLORETS** absent. **DISC FLORETS** yellow, often with red-brown nerves, throats shorter than lobes. **CYPSELAE** 5–8 mm long, thickened, narrowly oblong, warty roughened. **PAPPUS** 2 stout, barbed awn-scales, 1–2 (–3) mm, sometimes with a pair of additional, shorter awns. $2n = 22, 44$. Disturbed places on sands or clays in desert scrub, openings in piñon-juniper, Gambel's oak, and ponderosa pine communities. ARIZ: Apa, Nav; COLO: LPl, Mon; NMEX: McK, RAr, San, SJn; UTAH. 1220–2500 m (4000–8000′). Flowering: (Apr–) May–Oct. Fruit: late May–Nov. Wyoming to Nebraska, south to Arizona and Texas; Mexico (Chihuahua, Coahuila), and South America. *Thelesperma megapotamicum* has been used by the Navajo as a stimulant (tea from leaves and stems taken as a "nervous stimulant"), toothache remedy (tea from leaves and stems), and beverage (tea from leaves and flowers, with or without sugar). Ramah Navajo have used boiled roots as the source of an orange-yellow dye for wool, and the Hopi have used flowers as the source for a reddish brown dye for basket-making yucca fibers.

Thelesperma subnudum A. Gray (nearly naked) Navajo tea, scapose greenthread, sand fringedpod. Perennials, 10–30 (–40) cm, caudices present and often woody. **STEMS** solitary or few, glabrous and ± glaucous, internodes mostly 1–5, 5–35 mm. **LEAVES** basal or restricted to the lower 1/2 of stem; cauline leaves 3–8 cm long, irregularly once- to twice-pinnately divided, lobes mostly oblanceolate to linear, 5–45 mm long, 1–3 (–5) mm wide. **HEADS** 1–few, long-pedunculate, terminating branches. **CALYCULI** of 7–9 deltate to linear-lanceolate bractlets 2–4 mm. **INVOLUCRES** 8–12 mm long, connate for 1/3–2/3 their length, the free portion white scarious-margined. **RAY FLORETS** 8, rarely absent, yellow, 12–25 mm long, 9–15 mm wide. **DISC FLORETS** yellow, sometimes with red-brown nerves, throats equal to or longer than lobes. **CYPSELAE** 4.5–7 mm long, thickened, incurved. **PAPPUS** absent, or a pair of short scales, 0.1–0.5 mm long. $2n = 22, 24$. Desert scrub or openings in piñon-juniper and ponderosa pine forests. ARIZ: Apa, Nav; COLO: Mon; NMEX: McK, San, SJn; UTAH. 1370–2000 m (4500–6500′). Flowering: May–Sep. Fruit: Jul–Sep. Colorado, Utah, Arizona, and New Mexico. In our area, material is assignable to **var. *subnudum***. Dried flowers and tips of young leaves are boiled and used to make a tea beverage by the Hopi and Navajo. Navajo have used the leaves, stems, and flowers as the source of an orange dye for wool; the Hopi have used the flowers to make a reddish brown dye for basketry and textiles.

Thymophylla Lag. Pricklyleaf
Brian A. Elliott

(*thymo*, thyme, + *phyllon*, leaf, referring to the similarity of the narrow leaves to thyme) Annual or perennial herbs, occasionally suffrutescent subshrubs; aromatic and conspicuously gland-dotted. **STEMS** under 3 dm high. **LEAVES** opposite or alternate, simple or pinnately divided. **INFL** a radiate head, rarely discoid, solitary or in few-headed cymes, pedunculate or sessile. **INVOLUCRES** hemispheric to campanulate, 3–7 mm high. **PHYLLARIES** 8–13 (22) in 2–3 series; inner bracts adnate 2/3 of length or more to inner surface of outer bracts; calyculus of 0–8 free, deltate to subulate bracts sometimes present, usually much shorter than phyllaries. **RECEPTACLES** flat, convex, or conic, naked to puberulent or sparsely and minutely fimbrillate. **RAY FLORETS** pistillate and fertile; ligules 0–few, yellow or yellow-orange, rarely white. **DISC FLORETS** perfect and fertile, many; corolla yellow or yellow-orange. **CYPSELAE** cylindric to obpyramidal. **PAPPUS** of 10 (–20) scales, each awn-tipped or cleft into bristles. Approximately 10 species, native to the southwestern United States and Mexico and disjunct in Argentina (Strother 1969, 1986).

1. Leaves 1–2 cm long, simple, entire or toothed but not pinnatisect; heads sessile or subsessile, not exceeding the foliage; plants glabrous, ciliate, or minutely pubescent .... ***T. acerosa***
1′ Leaves 4–9 cm long, pinnatisect into 3–15 linear lobes; heads pedunculate, exceeding the foliage; plants short-hispid .... ***T. pentachaeta***

Thymophylla acerosa (DC.) Strother (referring to the sharp, prickly leaves) Dogweed, pricklyleaf dogweed. Suffruticose perennial with a strong, woody taproot, forming compact clumps. **STEMS** erect to spreading, densely branched, glabrous or villosulous, often with fascicles of leaves in the axils. **LEAVES** opposite, sometimes alternate above, simple, linear to filiform and acerose, 1–2 cm long, 0.5–2 mm wide, sparsely scaberulous and occasionally short-ciliate at the base. **INFL** heads radiate, sessile or subsessile. **PHYLLARIES** 8–13, connate nearly to the scarious apex, conspicuously gland-dotted in the upper 1/2; calyculate bracts about 5. **RAY FLORETS** with 7–8 ligules, yellow. **DISC FLORETS** 18–25; corolla yellow, 3–4 mm long. **CYPSELAE** cylindric, 3–3.5 mm long, strigillose. **PAPPUS** of about 20 scales, each cleft into 3–5 bristles. $2n = 16, 24, 26$. [*Dyssodia acerosa* DC.; *D. fusca* A. Nelson]. Rocky slopes and mesas, low rolling hills, desert shrub communities, and piñon-juniper, often on sandstone or limestone soils. NMEX: San; UTAH. 1600–2950 m (5200–9700′). Flowering: May. Utah, Arizona, New Mexico, and west Texas to Mexico. Leaves have been used as a tobacco substitute and in baths for fever.

Thymophylla pentachaeta (DC.) Small (five + bristle, referring to the pappus scales) Scale glandweed, five-needle pricklyleaf, five-needle thymophylla. Perennial, woody-taprooted subshrubs; mildly aromatic. **STEMS** erect, densely branched, 8–28 cm tall and nearly as wide, densely puberulent. **LEAVES** opposite, often crowded, pinnately divided into 3–5 (7) stiff, linear-filiform, spinulose segments, lobes less than 1 mm wide; sparsely hirtellous and conspicuously gland-dotted. **INFL** a radiate head, solitary on slender peduncles 2–5 cm long, peduncles 2–4-bracteate. **PHYLLARIES** 13–15 in 2 subequal series; outer bracts free to base, margins ciliolate; inner bracts adnate to inner surface of outer bracts; calyculate bracts 3–5, short-triangular, margins ciliolate. **RAY FLORETS** with (10–) 13 (–21) ligules, yellow. **DISC FLORETS** 50–70; corolla yellow, 2.5–3 mm long. **CYPSELAE** obpyramidal, 2.2–3 mm long, hispidulous or glabrous. **PAPPUS** usually of 10 scales, each cleft into 3 awns; sometimes of 10 scales in 2 series, the inner 1–3 awn-tipped, the outer short and awnless; never with both forms on the same plant. $2n = 24, 26, 32$. [*Dyssodia belenidium* Macloskie; *D. cupulata* A. Nelson; *D. pentachaeta* B. L. Rob.; *D. thurberi* (A. Gray) A. Nelson; *Hymenatherum belenidium* DC.; *H. thurberi* A. Gray; *Tagetes belenidium* Kuntze; *Thymophylla belenidium* (DC.) Cabrera; *T. thurberi* (A. Gray) Wooton & Standl.]. Dry hills and mesas, desert shrub communities, and rocky places. ARIZ: Apa; UTAH. 1200–1600 m (3900–5300′). Flowering: Apr–Oct. Ranging from California to Texas, south to Mexico, and disjunct in Argentina. Ours is **var. *belenidium*** (DC.) Strother, one of four varieties. A decoction of the root has been used as a purgative.

Tonestus A. Nelson Serpentweed
G. L. Nesom

(an anagram of *Stenotus*, alluding to a resemblance) Perennial herbs, from thick, woody caudex branches or less commonly with a distinct taproot (slender rhizomes in *T. lyallii*). **LEAVES** 1–5-veined from the base, mostly obovate to oblanceolate, margins toothed-spinulose or entire, subclasping but not auriculate, often persistent in a rosette but the stems leafy, upper cauline grading into phyllaries. **INFL** a discoid or radiate head, sometimes few and loosely corymboid but most commonly solitary. **INVOLUCRES** turbinate to hemispheric or campanulate. **PHYLLARIES** in 3–4 subequal series, at least the outer foliaceous. **RECEPTACLES** convex, foveolate, epaleate. **RAY FLORETS** 0 or present and pistillate, fertile, corollas yellow. **DISC FLORETS** bisexual, fertile. **CYPSELAE** narrowly oblong to slightly fusiform, terete to somewhat compressed, 4–12-nerved, glabrous to strigose or villous. **PAPPUS** of barbellate, apically attenuate bristles in (1–) 2 series. $x = 9$. [*Haplopappus* sect. *Tonestus* (A. Nelson) H. M. Hall].

Tonestus pygmaeus (Torr. & A. Gray) A. Nelson (dwarf) Pygmy serpentweed. **STEMS** (1–) 2–8 cm, minutely puberulent, minutely stipitate-glandular. **LEAVES** basal and cauline, oblanceolate to narrowly oblanceolate or lanceolate. **INVOLUCRES** 7–12 mm wide. **PHYLLARIES**: outer foliaceous, sometimes grading into immediately subtending foliar bracts. **RAY FLORETS** 10–20. **CYPSELAE** 2–5 mm, 8–9-nerved. $2n = 18$. [*Haplopappus pygmaeus* (Torr. & A. Gray) A. Gray]. Rocky flats, slopes, and ridges, talus; alpine tundra communities. COLO: Arc, Con, Hin, LPl, Min, Mon, RGr, SJn. 3450–3750 m (11320–12305′). Flowering: Jul–Aug. Montana (extinct), Wyoming, Colorado, New Mexico.

Tonestus pygmaeus

Townsendia Hook. Townsend Aster or Townsend Daisy
Timothy K. Lowrey

(for David Townsend, amateur botanist of West Chester, Pennsylvania) Plants perennial or biennial, occasionally annual, caulescent or acaulescent; erect, suberect, decumbent, or rosulate. **LEAVES** alternate, spatulate to linear, glabrate to densely pubescent. **INFL** a head, radiate, pedunculate or sessile, sunken in leaf tufts. **PHYLLARIES** in 2–7 series. **RAY FLORETS** female, ligules blue or whitish to pink. **DISC FLORETS** bisexual, corollas yellow. **CYPSELAE** glabrate or frequently pubescent with duplex hairs. **PAPPUS** of barbellate bristles or squamellae, ray and disc pappus similar or ray shorter than disc. Mostly western North America but *T. exscapa* extending to Nebraska and Kansas, and one species in central Mexico.

1. Plants acaulescent, rosulate or caespitose, decumbent ....(2)
1′ Plants caulescent ....(4)

2. Phyllaries obovate, ovate, or lanceolate .... ***T. rothrockii***
2′ Phyllaries linear to narrowly lanceolate ....(3)

3. Leaves involute, midveins usually conspicuous; disc pappus more than 6.5 mm long .... ***T. leptotes***
3′ Leaves not involute, midveins inconspicuous; disc pappus less than 6.5 mm long .... ***T. exscapa***

4. Leaves glabrous, shiny; heads long-pedunculate from axils of upper leaves, peduncles up to 7 cm long..... ***T. glabella***
4′ Leaves pubescent, strigose to canescent, never shiny; heads terminal or short-pedunculate, peduncles less than 2 cm long ....(5)

5. Disc pappus longer than disc corollas; stem densely canescent .... ***T. incana***
5′ Disc pappus shorter than disc corollas; stem strigose to sparsely canescent ....(6)

6. Plants annual; phyllary apices obtuse .... ***T. annua***
6′ Plants perennial; phyllary apices acute .... ***T. fendleri***

Townsendia annua Beaman (annual, referring to annual habit) Annual Townsend daisy. Plants caulescent, prostrate to decumbent annuals, branching at the base. **STEMS** 0.2–2.5 dm long, strigose. **LEAVES** oblanceolate or spatulate, entire, up to 3 cm long, sparsely strigose, basal leaves soon deciduous. **INFL** a terminal head, on the stems or on very short peduncles. **INVOLUCRES** 8–16 mm wide, 4.5–7.5 mm high. **RAY FLORETS** 12–30 per head, corollas white to pink or light lavender, 5–9 mm long. **DISC FLORETS** yellow, sometimes pink- or purple-tipped, 2.3–3.7 mm long. **CYPSELAE** pubescent, 2–2.8 mm long. **RAY PAPPUS** 0.6–1 mm long. **DISC PAPPUS** 1.8–3 mm long. Sandy flats, arroyo bottoms, disturbed ground. ARIZ: Apa, Coc, Nav; COLO: LPl, Mon; NMEX: McK, RAr, San, SJn; UTAH. 1219–2100 m (4000–7000′). Flowering: Apr–Sep. Colorado, Utah, Arizona, New Mexico, and Texas. Quite variable in plant size, which is related to precipitation. Closely related to *T. strigosa* Nutt. and often confused with it.

Townsendia exscapa (Richardson) Porter (without a stem or scape) Easter daisy. Plants acaulescent, rosulate, perennial; caudex well developed, becoming woody. **STEMS** absent. **LEAVES** oblanceolate or narrowly spatulate, entire, strigose or strigose-sericeous, 3–8 cm long, 2–6 mm wide. **INFL** a head, sessile or short-pedunculate, peduncles up to 3 cm long. **INVOLUCRES** 1.3–3.7 cm wide, 1–2.2 cm high. **PHYLLARIES** in 4–7 series, lanceolate or linear, scarious-ciliate margined. **RAY FLORETS** 20–40 per head, corollas white or pinkish, often with darker pink stripe on the back, 12–22 mm long. **DISC FLORETS** yellow, often pink- or purple-tipped, 6–11 mm long. **CYPSELAE** moderately to heavily pubescent, fertile achenes orange at maturity, 3.7–6.5 mm long. **RAY AND DISC PAPPUS** similar, 6–13 mm long. [*Aster exscapus* Richardson; *T. sericea* Hook.; *T. wilcoxiana* Wood; *T. intermedia* Rydb. ex Britton]. Ponderosa pine, piñon-juniper, and sagebrush communities. ARIZ: Apa, Nav; COLO: Arc, Hin, LPl, Min, Mon; NMEX: McK, RAr, San, SJn. 1525–2500 m (5000–8200′). Flowering: Mar–Aug. Canada south to Nebraska, Kansas, Nevada, Arizona, New Mexico, Texas, and Mexico. This is the most variable and widespread species of the genus. It has both sexual and asexual (apomictic) populations and hybridizes with *T. leptotes* in Colorado and New Mexico.

Townsendia fendleri A. Gray (for Augustus Fendler, botanical collector who botanized in Santa Fe in 1846) Fendler's Townsend daisy. Plants caulescent perennials or rarely biennial, decumbent or erect. **STEMS** branched, strigose to canescent. **LEAVES** narrowly oblanceolate to linear, entire, involute, strigose, up to 3.5 cm long and 0.35 cm wide. **INFL** a terminal head. **INVOLUCRES** campanulate, 6.4–13 mm wide, 5–8.5 mm high. **PHYLLARIES** in 4–5 series, lanceolate or ovate-lanceolate, strigose, scarious-margined. **RAY FLORETS** 10–25 per head, corollas white above, often with pinkish or bluish stripe on back. **DISC FLORETS** yellow, 2–3.5 mm long. **CYPSELAE** pubescent, 2.1–3.2 mm long. **RAY PAPPUS** of connate-coroniform squamellae or short bristles up to 1.5 mm long. **DISC PAPPUS** of bristles, 1.9–3.4 mm long, shorter than disc corollas. Piñon-juniper woodland communities, often in gypsiferous soils. ARIZ: Apa; COLO: Dol, LPl, SMg; NMEX: SJn. 1165–1840 m (5500–6000′). Flowering: May–Sep. North-central New Mexico and south-central Colorado. Closely related to *T. gypsophila* Lowrey & P. J. Knight, which occurs only on gypsum outcrops in Sandoval County, New Mexico.

Townsendia glabella A. Gray **CP** (glabrous, smooth) Gray's Townsend daisy or Easter daisy. Plants caulescent perennials, decumbent to caespitose, caudex well developed and woody. **STEMS** unbranched, glabrous, up to 5.5 cm long. **LEAVES** spatulate or oblanceolate, entire, glabrous, dark green and shiny, 4–6.5 cm long, 4–11 mm wide. **INFL** a head, on peduncles arising from axils of upper stem leaves, peduncles up to 7 cm long, naked or with 1 small bract. **INVOLUCRES** 8–24 mm wide and 7.8–12.3 mm high. **PHYLLARIES** in 3–6 but mostly 4 series, lanceolate or oblanceolate, glabrous, scarious-margined. **RAY FLORETS** 12–35, ray corollas white to light blue-pink, 8–14 mm long. **DISC FLORETS** yellow, 3.6–5.3 mm long. **CYPSELAE** oblanceolate, glabrous or minutely papillose. **RAY PAPPUS** of bristles, up to 1.8 mm long. **DISC PAPPUS** bristles 4–7 mm long. [*Townsendia bakeri* Greene]. Piñon-juniper woodland and ponderosa pine communities; foothills and mesa slopes in soils derived from the Mancos Shale Formation. COLO: Arc, LPl, Mon. 1950–2600 m (6400–8500′). Flowering: May–Aug. Known only from southwestern Colorado; a Four Corners regional endemic.

Townsendia incana Nutt. **CP** (silvery white pubescence, referring to hairs on the stems) Silvery Townsend daisy. Plants caulescent, pulvinate or suberect perennials; caudex developed and much-branched. **STEMS** canescent. **LEAVES** spatulate or oblanceolate, entire, strigose, up to 4.5 cm long and 0.5 cm wide. **INFL** a head, terminal, or occasionally pedunculate. **INVOLUCRES** campanulate, 8–19 mm wide, 7–14 mm high. **PHYLLARIES** in 3–4 series, lanceolate, scarious-margined. **RAY FLORETS** 10–30 per head, corollas white on upper surface, often with pink stripe on back of ligule, 7–13 mm long. **DISC FLORETS** yellow, and often pink-tinged, 3.7–6.5 mm long. **CYPSELAE** pubescent, 3–4.7 mm long. **RAY PAPPUS** 0.3–6.3 mm long. **DISC PAPPUS** 3.8–6.6 mm long, longer than disc corollas. [*Townsendia fremontii* Torr. & A. Gray; *T. arizonica* A. Gray; *T. incana* Nutt. var. *ambigua* M. E. Jones; *T. diversa* Osterh.]. Piñon-juniper woodland, mixed desert shrubland, and sagebrush communities. ARIZ: Apa, Coc, Nav; COLO: Arc, Dol, LPl, Mon; NMEX: McK, RAr, San, SJn; UTAH. 1320–2200 m (4325–7215′). Flowering: Apr–Sep. Wyoming to Nevada, Arizona, and New Mexico. A widespread and variable species due to apomixis and occasional hybridization with related species.

Townsendia leptotes (A. Gray) Osterh. (Greek *lepto*, thin or narrow, referring to leaves) Slender Townsend daisy. Plants acaulescent perennials, rosulate-pulvinate. **LEAVES** linear, oblanceolate, or narrowly spatulate, entire, thickish, involute, glabrous to strigose-sericeous, up to 60 mm long, 3.5 mm long. **INFL** a head, sessile or pedunculate, peduncles up to 2.8 cm. **INVOLUCRES** campanulate, 0.8–2.3 cm wide, 0.5–1.5 mm high. **PHYLLARIES** in (3) 4–7 series, lanceolate to linear, glabrous or slightly strigose, scarious-margined. **RAY FLORETS** 15–40 per head, corollas white, cream, pink, or blue, 8–14 mm long. **DISC FLORETS** yellow, 3.2–7 mm long. **CYPSELAE** oblanceolate, glabrous to moderately pubescent. **DISC AND RAY PAPPUS** similar, of bristles 3–8 mm long, or ray pappus bristles absent. [*Towsendia sericea* Hook. var. *leptotes* A. Gray]. Montane sagebrush and open grassland communities; slopes, ridges, and plateau margins. COLO: Arc, LPl, Mon; NMEX: RAr; SJn. 2135–3050 m (7000–10000′). Flowering: May–Aug. Western Montana, Idaho, Wyoming, Utah, Nevada, New Mexico, Colorado, California. Extremely variable, with sexual and asexual propulations. Hybridizes with *T. exscapa.*

Townsendia rothrockii A. Gray ex Rothr. (after Joseph Trimble Rothrock, professor of botany, Pennsylvania) Rothrock's Townsend daisy. Plants acaulescent perennials, rosulate, caudex well developed and woody. **LEAVES** spatulate-oblanceolate, entire, conspicuously thickened, glabrous, 10–35 mm long and 2–7 mm wide. **INFL** a head, nearly sessile or pedunculate, peduncles naked or with 1 bract near the head, up to 2.7 cm long. **INVOLUCRES** 12–28 mm wide, 8–12 mm high. **PHYLLARIES** in 4–6 series, elliptic, ovate, obovate, or lanceolate, glabrous, scarious-margined. **RAY FLORETS** 18–40 per head, blue, 8–16 mm long. **DISC FLORETS** yellow, 3.3–4.8 mm long. **CYPSELAE** oblanceolate; ray achenes papillose and moderately pubescent; disc achenes lightly pubescent to glabrous, not papilose. **RAY PAPPUS** of coroniform squamellae or bristles up to 1.5 mm. **DISC PAPPUS** of bristles 3.2–6 mm long. Alpine meadows, fell-fields, open spruce forests, and limestone outcrops. COLO: Arc, Hin, LPl, SJn. 2600–3800 m (8500–12400′). Flowering: May–August. A rare species endemic to montane habitats in southwestern Colorado.

Tragopogon L. Goatsbeard, Salsify

Gary I. Baird

(Greek *tragos*, goat, + *pogon*, beard, probably alluding to the pappus) Biennials (mostly), stoutly taprooted, herbage glabrous or tomentose to loosely lanate, sap milky. **STEMS** 1, solitary or branched near base and/or distally, erect, ± striate. **LEAVES** basal (1st year mostly) and cauline (2nd year), alternate, simple, progressively reduced upward, sessile; blades linear to ovate-lanceolate, base often dilated, clasping, prominently 5–13-nerved, margins entire. **HEADS** liguliflorous, terminal, erect; peduncles mostly ebracteate, glabrous or tomentose to lanate, especially distally, sometimes becoming inflated and fistulose. **PHYLLARIES** 6–14, ± lanceolate, herbaceous, equal in about 2 series, margins herbaceous to hyaline, apices acuminate, often loosely tomentulose, faces often keeled, sparsely to densely black-puberulent (sometimes with a few white hairs also), or glabrous. **RECEPTACLES** flat, epaleate; calyculi lacking. **RAY FLORETS** matutinal; corolla ligulate, yellow and/or purple. **DISC FLORETS** lacking. **CYPSELAE** ± monomorphic or somewhat dimorphic, dark brown to pale tan, stramineous, or whitish, ± fusiform to cylindrical, ± 5-angled, 10-ribbed, muricate, tomentulose immediately below pappus, otherwise glabrous, gradually or abruptly tapered to a long, often pale beak. **PAPPUS** of long, bristlelike, subulate scales in 1 series, persistent, whitish to brownish, plumose. $x = 6$. About 150 species of Eurasia and Africa, some species introduced to North America, Australia, and the Pacific.

1. Corollas purple ..... ***T. porrifolius***
1′ Corollas yellow ..... (2)

2. Peduncles inflated and fistulose distally; corollas usually pale yellow, the outer ones mostly much shorter than the involucre; leaf apices erect ..... ***T. dubius***
2′ Peduncles not inflated or fistulose; corollas usually bright yellow, the outer ones mostly surpassing the involucre; (at least upper) leaf apices ± recurved or coiled ..... ***T. pratensis***

Tragopogon dubius Scop. (doubtful, perhaps alluding to its placement) Yellow salsify, yellow goatsbeard. Biennials 15–80 cm tall. **STEMS** glabrous or loosely tomentose-lanate when young, quickly becoming glabrous, usually with some residual tomentum in leaf axils. **LEAVES**: basal and lower cauline blades mostly linear, 10–40 cm × 2–4 mm, erect to ascending, ± glabrous, apices erect or straight; cauline blades usually lanceolate, becoming progressively shorter but with wider (to 10 mm) bases; peduncles glabrous or pubescent distally, inflated and fistulose. **INVOLUCRES** cylindrical to narrowly conical, 20–40 mm tall (in flower) to 30–60 mm tall (in fruit). **PHYLLARIES** linear-lanceolate, margins green (outer) or membranous (inner), apices often tomentulose, faces glabrous or occasionally sparsely black-puberulent. **RAY FLORETS** 10–100+, outer much shorter to just equaling the involucre; corolla pale yellow, sparsely pilose at tube-ligule juncture, tube 7–9 mm, ligule 5–20 × 1–3 mm; anthers 3–4 mm, black. **CYPSELAE** 20–30 mm long; beak about equal to body in length. **PAPPUS** scales 20–30 mm, sordid. $2n = 12$. Most commonly found along roadways, ditch banks, or other disturbed sites; riparian, marsh, meadow, saltbush, sagebrush, piñon-juniper, ponderosa pine, Douglas-fir, aspen, and spruce-fir communities. ARIZ: Apa, Nav; COLO: Arc, Dol, Hin, LPl, Min, Mon, SJn, SMg; NMEX: McK, RAr, San, SJn; UTAH. 1400–2900 m (4700–9600′). Flowering: Apr–Sep. Europe, introduced; widely established in North America and elsewhere. Occasionally forms sterile diploid hybrids with the other species. *T.* ×*miscellus* G. B. Ownbey (*T. dubius* × *T. pratensis*) and *T.* ×*mirus* G. B. Ownbey (*T. dubius* × *T. porrifolius*) are allotetraploid hybrids that have arisen in North America. Neither of these hybrid species are known from the study area.

Tragopogon porrifolius L. (leek-leaved) Garden salsify, oyster plant. Biennials 30–60+ cm tall. **STEMS** glabrous or loosely tomentose-lanate and ± glabrate in age, usually with some residual tomentum below nodes and heads. **LEAVES**: basal and lower cauline lanceolate to linear-lanceolate, 15–30 cm × 5–15 mm, erect to ascending, ± glabrous, apices erect or straight; cauline blades broadly lanceolate to ovate-lanceolate, becoming progressively shorter but not always wider (about 10–15 mm) at base; peduncles ± pubescent to tomentose, especially distally, inflated and fistulose. **INVOLUCRES** ± conical, 35–45 mm tall (in flower) to 40–70 mm tall (in fruit). **PHYLLARIES** lanceolate, margins green (outer) or narrowly membranous (inner), apices often tomentulose, faces sparsely black-puberulent and sometimes sparsely and loosely tomentose or lanate proximally. **RAY FLORETS** to 100+, outer shorter than to equaling the involucre; corolla purple, setose-pilose at tube-ligule juncture, tube about 10 mm, ligule 10–25 mm × 1–4 mm; anthers 4–6 mm, black. **CYPSELAE** 25–40 mm long; body abruptly tapered to beak; beak ± longer than body. **PAPPUS** scales about 30 mm, brownish. $2n = 12$. Disturbed sites, usually in more mesic habitats. COLO: Arc, LPl; NMEX: RAr, SJn. 2000–2200 m (6800–7200′). Flowering: May–Jun. Europe, introduced and widely established in North America and elsewhere. Widely grown as a garden vegetable for its edible root.

Tragopogon pratensis L. (pertaining to meadows or prairies) Meadow salsify. Biennials 30–80+ cm tall. **STEMS** glabrous. **LEAVES**: basal and lower cauline lanceolate to ovate-lanceolate, 10–30 cm × 5–20 mm, erect to ascending, glabrous, apices arched to recurved or coiled (at least the upper leaves); cauline blades broadly lanceolate to ovate-lanceolate, becoming progressively shorter but not always wider (about 5–15 mm) at base; peduncles glabrous or sparsely black-puberulent at base of head, uniform throughout, not inflated or fistulose. **INVOLUCRES** ± conical, 10–25 mm tall (in flower) to 20–35 mm tall (in fruit). **PHYLLARIES** lanceolate to narrowly ovate-lanceolate, margins narrowly purple (outer) to hyaline-membranous (inner), apices ± glabrous, faces glabrous or sparsely black-puberulent. **RAY FLORETS** 20–100, outer equaling or surpassing the involucre; corolla bright yellow, pilose at tube-ligule juncture, tube about 5 mm, ligule 5–20 mm × 1–2 mm; anthers about 3 mm, blackish. **CYPSELAE** 15–25 mm long; body tapered to beak; beak shorter than body. **PAPPUS** scales about 15 mm, whitish. $2n = 12$. Roadside in a ponderosa pine–Douglas-fir community. ARIZ: Apa, Nav; COLO: Arc, Dol, Hin, LPl, Min, Mon; NMEX: McK, RAr, SJn. 1585–2590 m (5200–8500′). Flowering: Aug. Europe, introduced and widely established in North America and elsewhere.

Uropappus Nutt. Silverpuffs

Gary I. Baird

(Greek *ura*, tail, and *pappos*, tuft of white hair, alluding to the bristle terminating the scales of the pappus) Annuals, taprooted, usually shortly caulescent, herbage glabrous or villous-puberulent to sparsely scurfy, sap milky. **STEMS** obscure, usually branched near base, erect. **LEAVES** all basal (or appearing so), simple; petioles obscure, clasping; blades linear to linear-lanceolate, often villous-ciliate proximally, margins entire to pinnately lobed, apices acuminate. **HEADS** liguliflorous, borne singly, erect; peduncles ebracteate, scapose, distally inflated. **PHYLLARIES** herbaceous, imbricate in about 4 series. **RECEPTACLES** more or less flat, epaleate. **RAY FLORETS** matutinal; corollas ligulate, pale yellow, abaxially rosy-purple. **DISK FLORETS** lacking. **CYPSELAE** black-purple, columnar-fusiform, approximately 10-ribbed, scaberulous (especially distally), gradually tapered apically to form an indistinct beak. **PAPPUS** of 5 linear-lanceolate scales, apically notched and with a bristle arising from the notch, white or ochroleucous, lustrous, deciduous. $x = 9$. One species of western North America.

Uropappus lindleyi (DC.) Nutt. (for John Lindley, British botanist, 1799–1865) **STEMS** to 20 cm tall, purplish. **LEAVES** with blades less than 10 cm long, puberulent, villous-ciliate proximally, margins entire to pinnately lobed; lobes 1–5 pairs, subopposite, linear-lanceolate, 3–14 mm long, mostly spreading, margins entire. **HEADS** solitary; peduncles much exceeding leaves. **INVOLUCRES** 15–25 mm tall. **PHYLLARIES** lanceolate, erect, glabrous, green or rosy-purple, margins entire, scarious; inner elongating with fruit. **RAY FLORETS** subequal to involucre; corolla sparsely pilose, tube 3–4 mm, ligule 4–6 × approximately 1 mm; anthers 1–2 mm. **CYPSELAE** 10 mm. **PAPPUS** scales approximately 10 mm, notch about 2 mm, bristles about 5 mm. $2n = 18$. [*Microseris lindleyi* (DC.) A. Gray]. Piñon-juniper community in moist, sandy soil. UTAH. 1200 m (4000′). Flowering May–Jun. British Columbia south to Baja California, Sonora, and Texas (mostly west of the Rocky Mountain axis). In the study area known from a single collection taken on north side of Navajo Mountain near Lake Powell.

Verbesina L. Crownbeard

Gary I. Baird

(derivation unknown) Annuals or perennials (some woody), taprooted or rhizomatous, herbage glabrous to densely pubescent, sap clear. **STEMS** 1 to many, usually erect, often branched, internodes sometimes winged. **LEAVES** basal and/or cauline, simple, petiolate or sessile, opposite or alternate (rarely whorled); blades deltate, ovate, elliptic, rhombic, lanceolate, or lance-linear, bases cuneate to cordate, margins entire to pinnately or palmately lobed, apices acute to acuminate. **HEADS** radiate or discoid, borne singly or in various arrays, erect. **INVOLUCRES** campanulate to hemispheric or rotate. **PHYLLARIES** about 10–30 in about 1–4 series, herbaceous to chartaceous, persistent, subequal to unequal (outer shorter than or surpassing inner). **RECEPTACLES** flat to convex or somewhat conic, paleate; paleae herbaceous to scarious, often enfolding achenes, sometimes linear; calyculi lacking. **RAY FLORETS** 0–30, pistillate, fertile or sterile, or neuter; corollas yellow to orange or ochroleucous, ± fugacious. **DISC FLORETS** 10–150+, bisexual, fertile; corollas same color as rays. **CYPSELAE** monomorphic, flattened, orbiculate to oblanceolate or elliptic, usually winged, glabrous to pubescent. **PAPPUS** of usually 2 scales or awns, usually persistent, or lacking. $x = 17$ (18?). Perhaps 200 species mostly of the Neotropics.

Verbesina encelioides (Cav.) Benth. & Hook. f. ex A. Gray (resembling *Encelia*) Golden crownbeard. Annuals 20–100 cm tall, internodes not winged. **STEMS** usually 1, often branched throughout, sometimes striate or nerved (especially proximally). **LEAVES** cauline, usually alternate, sometimes opposite (proximally, or all in young or depauperate plants), petiolate (proximal) to sessile (distal); petioles shorter than blades; blades ovate-deltate to lanceolate or rhombic, 15–70 × 5–40 mm, bases ± cuneate, margins coarsely dentate to lobate, or subentire, faces strigose to sericeous (adaxial often sparsely pubescent and green, with abaxial often densely pubescent and silvery; cauline reduced distally and becoming bractlike. **HEADS** solitary (mostly in young or depauperate plants) to many (proliferating with age), cymiform to corymbiform. **INVOLUCRES** hemispheric to nearly flat, 10–20 mm diameter; palea tardily deciduous. **PHYLLARIES** 12–18 in 2–3 series, subequal or outer just surpassing inner, erect to spreading, strigose; outer lance-ovate to lance-linear or linear, green or blackish. **RAY FLORETS** 10–15, pistillate, sterile (mostly); laminae about 10–20 mm, often broad and deeply 3-lobed. **DISC FLORETS** 40–150+, corollas yellow. **CYPSELA** bodies brownish to blackish, obovate, compressed, margins broadly winged, faces strigillose (at least distally), midrib ± keeled; wings corky, whitish, strigillose (at least distally), apically auriculate. **PAPPUS** of 2 bristlelike awns, 1–2 mm. $2n = 34$. Greasewood, saltcedar, willow, sagebrush, grassland, and piñon-juniper communities, sandy to silty or rocky soils, often in disturbed or weedy habitats. ARIZ: Apa, Coc, Nav; COLO: Arc, Dol, LPl, Mon, SMg; NMEX: McK, RAr, San, SJn; UTAH. 1500–2700 m (4900–8700′). Flowering: May–Oct. Widespread in the central, western, and southeastern United States, to Mexico, the West Indies, and South America; introduced into Asia, Australia, and the Pacific, where it has become an aggressive weed. All of our material is assignable to **var. *exauriculata*** B. L. Rob. & Greenm. (without auricles).

Wyethia Nutt. Mules-ears

Gary I. Baird

(for Nathaniel J. Wyeth, early American explorer, 1802–1856) Perennials, taproots stout to massive, rhizomes lacking, caudices simple or few-branched, sometimes woody, herbage glabrous or pubescent, nonglandular or gland-dotted or finely stipitate-glandular, sometimes vernicose, sap clear. **STEMS** 1 to many, erect, simple or branched near base. **LEAVES** basal and cauline, simple, petiolate (basal) to subsessile (cauline); bases of proximal petioles sometimes expanded; basal blades elliptic-deltate to linear-elliptic or lanceolate, bases mostly cuneate, margins entire, apices acute to acuminate; cauline ± reduced distally, subsessile, alternate. **INFL** borne singly, or 2–5, corymbiform to racemiform, erect; peduncles ebracteate to leafy. **HEADS** radiate. **INVOLUCRES** campanulate to hemispheric. **PHYLLARIES** about 10–30 in 2–3 series, herbaceous, outer subequaling to surpassing inner. **RECEPTACLES** flat to convex, paleate; paleae persistent, chartaceous; calyculi lacking. **RAY FLORETS** 5–25, pistillate, fertile; corollas yellow to white, fugacious. **DISC FLORETS** 35–150+, bisexual, fertile; corollas yellow. **CYPSELAE** monomorphic, 3–4-angled, glabrous or pubescent. **PAPPUS** lacking, or coroniform, persistent. $x = 19$. Eight species of western North America.

Wyethia arizonica A. Gray (of Arizona) Arizona mules-ears. Perennials 10–65 cm tall. **LEAVES**: petioles shorter than to equaling blades; basal blades elliptic to lance-elliptic or lanceolate, 10–30 × 3–10 cm, faces subglabrous to densely hirsutulous, gland-dotted and sometimes vernicose; cauline ± reduced proximally (lowermost often bractlike with an expanded petiole but no blade) and distally (sometimes only slightly). **HEADS** mostly borne singly, sometimes 2–3; peduncles short to long, mostly ebracteate, sparsely to (usually) densely hirsutulous, and finely but densely stipitate-glandular (glands orangish). **INVOLUCRES** about 20–40 mm diameter. **PHYLLARIES**: outer subequal to or surpassing inner, oblong to obovate or ovate, rarely lanceolate, hirsutulous to subglabrous, gland-dotted to finely stipitate-glandular, margins sparsely to densely ciliate, apices acute to obtuse. **RAY FLORETS** mostly 10–15; laminae about 35–50 mm long, usually strigillose on abaxial veins and finely stipitate-glandular abaxially. $2n = 38$. Common in ponderosa pine–Gambel's oak forests, occasional in piñon-juniper or spruce forest communities. ARIZ: Apa; COLO: Arc, Dol, Hin, LPl, Min, Mon, SMg; NMEX: McK, RAr, SJn; UTAH. 2100–2500 m (7000–8200′). Flowering: May–Jul. Arizona, Colorado, New Mexico, and Utah, generally restricted to the southern Colorado Plateau. Intergrades with *W. amplexicaulis* (Nutt.) Nutt., a species that occurs far to the north of the study area. These hybrids are represented by *W.* ×*magna* A. Nelson ex W. A. Weber, which does not appear to occur within the study area.

Xanthisma DC. Sleepy Daisy, Goldenweed

G. L. Nesom

(Greek *xanthos*, yellow, + *-ismos*, condition or quality) Annuals, biennials, perennials, or subshrubs, usually taprooted, sometimes rhizomatous. **STEMS** glabrous to hispid or hispidulous, villous, or stipitate-glandular. **LEAVES** basal and cauline, alternate, simple, petiolate (basal) or sessile (cauline), entire to dentate, or 1–2-pinnatifid, with spinulose teeth or lobes, bases sessile to clasping. **INFL** a radiate or discoid head, borne singly or in corymbiform arrays. **INVOLUCRES** turbinate, campanulate, or hemispheric. **PHYLLARIES** in 2–8 graduate series. **RECEPTACLES** flat to convex, shallowly alveolate. **RAY FLORETS** pistillate, fertile, in 1 series, lamina white, pink, purple, or yellow, absent in 3 species. **DISC FLORETS** bisexual, fertile. **CYPSELAE** ellipsoid to obovoid, oblong, 6–18-nerved, sparsely to densely sericeous. **PAPPUS** of persistent, basally flattened, barbellate bristles in 2–4 series. $x = (2, 3, 4)\ 5$. [*Machaeranthera* sect. *Blepharodon* (DC.) R. L. Hartm.; *Machaeranthera* sect. *Havardii* (R. C. Jacks.) R. L. Hartm.; *Machaeranthera* sect. *Sideranthus* (Nutt. ex Nees) R. L. Hartm.]. Species 17, western United States and Mexico.

1. Ray florets 0; leaves coarsely serrate to serrulate ..... ***X. grindelioides***
1′ Ray florets present and conspicuous ..... (2)

2. Ray florets pink to purple; leaves serrate to crenate-serrate ..... ***X. coloradoense***
2′ Ray florets yellow; leaves pinnatifid to 2-pinnatifid ..... (3)

3. Annuals, caudex simple, not woody ..... ***X. gracile***
3′ Perennials, caudex usually branched, woody ..... ***X. spinulosum***

Xanthisma coloradoense (A. Gray) D. R. Morgan & R. L. Hartm. (from Colorado) Colorado tansy-aster. Perennial herbs, 2.5–14 cm tall, caudices long-branched (often collected without the taproot). **LEAVES**: basal persistent, oblanceolate to spatulate, serrate to crenate-serrate. **HEADS** radiate, solitary, involucres 10–25 mm wide. **RAY FLORETS** 20–35, laminae pink to purple or deep reddish lavender. $2n = 8, 16$. [*Machaeranthera coloradoensis* (A. Gray) Osterh.; *M. coloradoensis* var. *brandegeei* (Rydb.) T. J. Watson ex R. L. Hartm.]. Subalpine outcrops with spruce, ridges, alpine tundra. COLO: Arc, Hin, LPl, Min, RGr, SJn. 3150–3700 m (10335–12140′). Flowering: Jul–Aug (–Sep). South-central Wyoming to southwestern Colorado.

Xanthisma gracile (Nutt.) D. R. Morgan & R. L. Hartm. (slender) Slender goldenweed. Annual herbs, caudices simple. **LEAVES**: basal often persistent, basal and cauline coarsely serrate to serrulate. **HEADS** radiate. **RAY FLORETS** 12–26. $2n = 4, 6, 8$. [*Machaeranthera gracilis* (Nutt.) Shinners; *Haplopappus ravenii* R. C. Jacks.]. Roadsides and other disturbed sites, dunes, sandy soil; sagebrush, piñon-juniper woodlands, and ponderosa pine–Rocky Mountain juniper communities. ARIZ: Nav; COLO: Arc, LPl, Mon; NMEX: McK, RAr, SJn. 1600–2450 m (5250–8040′). Flowering: Aug–Sep (–Oct). California and Nevada to Colorado, New Mexico, and Texas.

Xanthisma grindelioides (Nutt.) D. R. Morgan & R. L. Hartm. (resembling *Grindelia*) Gumweed aster. Subshrubs, caudices much-branched. **LEAVES**: basal often persistent, basal and cauline coarsely serrate to serrulate. **HEADS** discoid. **RAY FLORETS** 0. $2n = 8$. [*Machaeranthera grindelioides* (Nutt.) Shinners]. Sand, sandy clay, rocky and gravelly sites, benches, steep slopes, badlands; desert shrub, sagebrush, juniper, Gambel's oak, and piñon-juniper woodland communities. Sometimes with scattered ponderosa pine, and riparian sites with cottonwood and saltcedar. ARIZ: Apa, Coc, Nav; COLO: Dol, Mon; NMEX: RAr, SJn; UTAH. 1650–2300 m (5415–7545′). Flowering: May–Jun (–Sep). Montana and North Dakota south to Nevada, Arizona, and New Mexico. Four Corners plants are **var. *grindelioides***.

Xanthisma spinulosum (Pursh) D. R. Morgan & R. L. Hartm. (spinulose) Spiny goldenweed. Subshrubs, caudices much-branched. **LEAVES**: basal persistent or deciduous, basal and cauline pinnatifid. **HEADS** radiate. **RAY FLORETS** 14–60. [*Haplopappus spinulosus* (Pursh) DC.].

1. Stems 6–15 cm; basal leaves persistent and dense, cauline mostly on proximal 1/2 of stems; heads on naked or bracteate peduncles 1–4 cm; involucres 15–25 mm .... **var. *paradoxum***
1' Stems (10–) 15–30 cm; basal leaves mostly deciduous by flowering, cauline relatively even-sized upward to near heads; heads on bracteate peduncles 0.5–2 (–3) cm; involucres 8–12 mm wide .... **var. *spinulosum***

var. *paradoxum* (B. L. Turner & R. L. Hartm.) D. R. Morgan & R. L. Hartm. (strange) **STEMS** 6–15 cm. **LEAVES**: basal persistent and dense, cauline mostly on proximal 1/2 of stems. **HEADS** on naked or bracteate peduncles 1–4 cm. **INVOLUCRES** 15–25 mm wide. [*Machaeranthera pinnatifida* (Hook.) Shinners var. *paradoxa* B. L. Turner & R. L. Hartm.]. Washes, clay hills, sandstone, disturbed sites; desert scrub, and piñon-juniper communities. COLO: Mon; UTAH. 1400–1800 m (4595–5905'). Flowering: (Mar–) Apr–Oct. Colorado, New Mexico, Utah.

var. *spinulosum* **STEMS** (10–) 15–30 cm. **LEAVES**: basal mostly deciduous by flowering, cauline relatively even-sized upward to near heads. **HEADS** on bracteate peduncles 0.5–2 (–3) cm. **INVOLUCRES** 8–12 mm wide. $2n = 8$, 16. [*Machaeranthera pinnatifida* (Hook.) Shinners]. Roadsides, bare hills, riparian, sand, clay; desert shrub, shadscale-snakeweed, sagebrush, piñon-juniper woodlands, and ponderosa pine communities. ARIZ: Apa; COLO: Arc, Mon; NMEX: McK, RAr, San, SJn; UTAH. 1650–2350 m (5415–7710'). Flowering: (May–) Jun–Oct. South-central Canada (Alberta to Manitoba), Montana to Minnesota, south to New Mexico, Oklahoma, and Texas; Mexico.

Xanthium L. Cocklebur

Gary I. Baird

(Greek *xanthos*, yellow, the allusion unclear) Annuals, taprooted, rhizomes lacking, herbage strigose to hirtellous, and often gland-dotted, sap clear. **STEMS** 1, erect, branched, sometimes spiny (at nodes). **LEAVES** cauline, simple, alternate (proximal ± opposite), petiolate; blades pentagonal-suborbicular to ± deltate, lanceolate, or linear-lanceolate, usually palmately 3-lobed (or more) to pinnately lobed or pinnatifid, bases cordate (most) to cuneate, ultimate margins coarsely dentate, apices ± acute. **INFL** usually in terminal or axillary, racemiform to spiciform clusters (pistillate heads proximal, staminate heads distal), occasionally borne singly in axils, ± erect; sessile or peduncles obscure. **HEADS** discoid, unisexual. **RECEPTACLES** ± conic, paleate (at least staminate heads); calyculi lacking. **PISTILLATE HEADS** with involucres ± ellipsoid, forming a bur; outer phyllaries mostly 5–15, herbaceous, distinct (outermost sometimes foliaceous); inner phyllaries (actually paleae?) 20–400 in few to many series, indurate, proximally connate, distally distinct and hooked (the most distal 1–3 conspicuously larger, not hooked), forming a hardened bur enclosing the cypselae; florets 2, corollas lacking. **STAMINATE HEADS** with involucres ± rotate (or lacking); phyllaries about 5–15 (or lacking) in 1–2 series, herbaceous, distinct; paleae spatulate to linear, membranous, pubescent, gland-dotted; florets 20–50+, corollas whitish, anthers distinct. **CYPSELAE** monomorphic, black, ± fusiform, glabrous. **PAPPUS** lacking. $x = 18$. About two to three species of North and South America, introduced worldwide.

1. Plants lacking nodal spines; leaves mostly angular-suborbicular to deltate .... ***X. strumarium***
1' Plants with usually 3 nodal spines; leaves lanceolate to lance-linear (not known from the study area but to be expected) .... *X. spinosum*

Xanthium strumarium L. (a place of swelling or enlarged tumor, the allusion unclear) Common cocklebur. Plants mostly 10–80 (up to about 200) cm tall. **STEMS** lacking spines. **LEAVES** long-petiolate; petioles 2–10+ cm; blades suborbicular (often pentagonal) to ± deltate, or ovate-lanceolate, shallowly 3–5-lobed, base angular-cordate to ± truncate, or cuneate (usually in depauperate plants), ultimate margins coarsely dentate, faces ± uniformly hirtellous-strigose and gland-dotted, apices ± acute. **BURS** ellipsoid to cylindric-ellipsoid or subglobose, about 10–35 mm long, brownish or yellowish; prickles hispid-pilose and gland-dotted (or stipitate-glandular) proximally, glabrous and eglandular distally, apices hooked (the 1–3 most distal prickles much stouter, ± conic, incurved or straight). $2n = 36$. [*Xanthium americanum* Walter; *X. strumarium* var. *canadense* (Mill.) Torr. & A. Gray; *X. strumarium* var. *glabratum* (DC.) Cronquist; *X. wootonii* Cockerell]. Riparian, hanging garden, sagebrush, and piñon-juniper communities, usually in disturbed sites. ARIZ: Apa, Nav; COLO: Arc, Dol, LPl, Mon; NMEX: McK, RAr, San, SJn; UTAH. 1100–2200 m (3700–7100'). Flowering: Aug–Oct. Throughout subarctic North America to South America, widely introduced worldwide. Various sub-specific (and specific) taxa have been proposed (mostly based on subtle differences in the burs) but their segregation seems arbitrary and untenable. Cocklebur is listed as a noxious weed and the young seedlings are toxic.

Xylorhiza Nutt. Woody-aster
G. L. Nesom

(Greek *xylon*, wood, + *rhiza*, root) Shrubs, subshrubs, or perennials, from woody taproots and branching, persistent caudices. **STEMS** often white-barked, glabrous to villous or tomentose, sometimes stipitate-glandular. **LEAVES** cauline, alternate, sessile to subpetiolate, blades simple, linear to lanceolate or oblanceolate, oblong, or narrowly elliptic, margins entire to dentate-spinulose. **INFL** a radiate head, borne singly, terminal on naked peduncles. **INVOLUCRES** campanulate to hemispheric, 7–20 mm. **PHYLLARIES** in 3–6 series, graduated, white-indurate toward the base with a green tip, narrowly lanceolate, keeled. **RECEPTACLES** flat, foveolate to alveolate, epaleate. **RAY FLORETS** pistillate, fertile; corolla white to light blue or purple, coiling at maturity. **DISC FLORETS** bisexual, fertile. **CYPSELAE** fusiform to linear or ovate, ribs 4, sericeous with long, silky, subappressed hairs. **PAPPUS** of stout, often flattened, unequal, barbellate bristles in 2–3 series. $x = 6$. [*Machaeranthera* sect. *Xylorhiza* (Nutt.) Cronquist & D. D. Keck]. Species 10; southwestern United States and Mexico (Watson 1977; Nesom 2002). Contemporary botanists have recognized *Xylorhiza* as a genus distinct from *Machaeranthera*, characterized by the following: perennial herbs, subshrubs, and shrubs from large, woody taproots with woody, branching, persistent caudices, large, solitary heads, long phyllaries with erect or spreading apices, white or bluish rays, large cypselae, vernal flowering, and base chromosome number of $x = 6$ (vs. $x = 4$ or 5 in *Machaeranthera*).

1. Leaves 3–20 (–25) mm wide, margins (or many of them), sharply spinulose-toothed .... ***X. tortifolia***
1′ Leaves mostly 2–15 mm wide, margins all or nearly all entire ....(2)

2. Leaves linear-oblong to linear-lanceolate, 2–5 mm wide, truncate or rounded-auriculate at the base .... ***X. linearifolia***
2′ Leaves narrowly oblanceolate to oblong-spatulate, 2–15 mm wide, tapering at the base .... ***X. venusta***

Xylorhiza linearifolia (T. J. Watson) G. L. Nesom (linear-leaved) Moab woody-aster. Subshrubs. **STEMS** 15–35 cm, puberulent to coarsely stipitate-glandular. **LEAVES** linear-oblong to linear-lanceolate, 2–5 mm wide, bases truncate or rounded-auriculate, margins entire (rarely few-toothed), faces puberulent to coarsely stipitate-glandular. **RAY FLORETS** 13–21; corollas light blue to white. $2n = 12, 24$. [*Xylorhiza glabriuscula* Nutt. var. *linearifolia* T. J. Watson; *Machaeranthera linearifolia* (T. J. Watson) Cronquist]. Canyons, sands and clays; desert scrub communities. UTAH. 1200–1400 m (3935–4595′). Flowering: Apr–May (–Jun). Utah endemic.

Xylorhiza tortifolia (Torr. & A. Gray) Greene (twisted leaf) Smooth Mojave woody-aster. Subshrubs. **STEMS** 20–60 (–80) cm, stipitate-glandular, otherwise glabrous except for the sparsely pilose-puberulent leaf margins. **LEAVES** lanceolate to elliptic-oblong or oblanceolate, 3–20 (–25) mm wide, bases often subclasping or sometimes attenuate and not clasping, margins sharply spinulose-toothed, faces stipitate-glandular, otherwise glabrous. **RAY FLORETS** (15–) 18–60 (–85); corollas lavender or seldom white. $2n = 12$. [*Machaeranthera tortifolia* (Torr. & A. Gray) Cronquist & D. D. Keck var. *imberbis* Cronquist]. Sandy or gravelly slopes and flats, washes; juniper, piñon-juniper, sagebrush, and blackbrush communities. ARIZ: Nav; UTAH. 1300–1600 m (4265–5250′). Flowering: Apr–May (–Jun). Nevada, Arizona, and Utah. Four Corners plants are **var. *imberbis*** (Cronquist) T. J. Watson (beardless, alluding to the lack of nonglandular hairs).

Xylorhiza venusta (M. E. Jones) A. Heller (Latin, charming, lovely) Charming woody-aster. Perennials to subshrubs. **STEMS** 10–40 cm, densely villous-puberulent to glabrate, eglandular. **LEAVES** narrowly oblanceolate to oblong-spatulate, 2–15 mm wide, bases attenuate, not clasping, margins entire, faces densely villous-puberulent to glabrate. **RAY FLORETS** 14–34; corollas white to pale lavender. $2n = 12, 24$. [*Machaeranthera venusta* (M. E. Jones) Cronquist & D. D. Keck]. Barren, open sites, often selenious or gypseous, alkaline, dominated by saltbush and shadscale. COLO: Mon; UTAH. 1300–1700 m (4265–5575′). Flowering: Apr–Jun. Colorado, Utah, and north-central Arizona.

Zinnia L. Zinnia
Gary I. Baird

(for Johann G. Zinn, German botanist in Mexico, 1727–1759) Annuals or suffruticose perennials, taprooted and sometimes rhizomatous, caudices woody, herbage pubescent and often gland-dotted, sap clear. **STEMS** few to many, erect to prostrate, often much-branched. **LEAVES** cauline, simple, ± sessile, opposite; blades ovate, oblong, elliptic, lanceolate, linear, or acerose, prominently 1–3-veined, bases cuneate to rounded, margins entire, apices acute to acuminate. **HEADS** radiate (rarely discoid), borne singly, erect; peduncles short to obscure. **INVOLUCRES** campanulate to cylindric or hemispheric. **PHYLLARIES** 12–30 in 3–4 series, imbricate (outer shorter). **RECEPTACLES** conic, paleate; paleae persistent, chartaceous to scarious; calyculi lacking. **RAY FLORETS** 5–20, pistillate, fertile; corollas yellow, orange, red, purple, or white; laminae marcescent, becoming papery. **DISC FLORETS** about 20–150+, bisexual, fertile; corolla yellow, orange, or reddish, lobes usually villous or velutinous adaxially. **CYPSELAE** monomorphic, 3-angled (ray florets) or compressed (disc florets), not winged. **PAPPUS** lacking, or of 1–3 short awns or scales. x = (10–) 12 (–11). Seventeen species of southwestern North America, Central America, and South America.

Zinnia grandiflora Nutt. (large-flowered) Plains zinnia. Perennials 8–22 cm tall (often about as broad), ± suffruticose from a branched and woody caudex, herbage strigose. **STEMS** branched, erect or decumbent. **LEAVES**: blades linear or linear-lanceolate, 10–30 × 2–3 mm, faces coarsely strigose and gland-dotted, often 3-nerved proximally, not much reduced distally. **INVOLUCRES** about 5–8 mm tall, cylindric to narrowly campanulate (about as wide as tall). **PHYLLARIES** oblong, scarious, glabrous or strigillose distally, apices erose-ciliate and often tinted red. **RAY FLORETS** 3–6; corollas bright yellow, laminae 10–18 mm long, ovate to orbiculate, obscurely lobed. **DISC FLORETS** 18–24; corollas red (or green). **CYPSELAE** 4–5 mm, scabrellous. **PAPPUS** lacking, or of usually 2 unequal awns. $2n$ = 42. Sagebrush, juniper, desert shrub, and grassland communities. NMEX: McK. 1800–2000 m (6000–6500′). Flowering: Jun–Sep. Colorado and Kansas, south to Arizona, Texas, and Mexico.

BERBERIDACEAE Juss. BARBERRY FAMILY

John R. Spence

Shrubs or subshrubs, sprawling to erect. **STEMS** weakly to strongly branched, sometimes rhizomatous or stoloniferous, with or without spines. **LEAVES** alternate, petiolate, evergreen to deciduous, simple to trifoliate or odd-pinnate, often spiny-margined, stipules lacking. **INFL** terminal to axillary, either racemes, panicles, umbels, or spikes, or flowers sometimes solitary, ultimate branch of inflorescence arising from the axis of a bract, bractlets sometimes present subtending or well below flower. **FLOWERS** actinomorphic, hypogynous, perfect, often showy; perianth 3–5-merous, sometimes absent; bracts present, sepaloid, grading into sepals; sepals petaloid, free, (4) 6 or sometimes many, in 1–4 series; petals free, (4) 6–9 in 2–3 series, often nectariferous; stamens as many as the petals, typically 6, opposite, in 1–2 series, anthers longitudinally dehiscent or with laterally hinged valves; pistil of 2–3 carpels, united, with 1 locule. **FRUIT** a follicle, berry, or utricle; seeds 1–50, arillate or not. A family of 15 genera and about 650 species, widespread especially in the Northern Hemisphere. The barberry family is of some horticultural importance and is of considerable economic significance as it includes taxa that act as alternate hosts to grass rusts. Some North American genera have a long fossil record, with many apparently relictual distributions with temperate East Asian taxa. Although *Berberis* and *Mahonia* are often combined, they are quite distinct in western North America and are kept separate for the *Flora* following traditional practice. More studies are needed to resolve the relationships between the two genera. Both our genera serve as alternate hosts to *Puccinia graminis*, a rust that infests grains, including wheat.

1. Leaves simple, deciduous; stems spiny .... ***Berberis***
1′ Leaves compound, evergreen; stems lacking spines .... ***Mahonia***

Berberis L. Barberry

(classical Arabic for the fruit, *berberys*) Erect to sprawling shrubs, 1–3 m. **STEMS** dimorphic, composed of elongate primary shoots and short secondary shoots, bark glabrous, gray to red or brown, spines simple or 1–5-branched. **LEAVES** deciduous, simple, clustered on secondary shoots, thin, distinctly 1-veined, margins smooth to serrate and bristle-tipped. **INFL** a lax, open raceme or umbel. **FLOWERS** 3-merous, subtended by bracts; sepals 3 or 6, yellowish; petals 6, yellow, with 2 nectar glands near base; stamens 6, filaments with or without 2 teeth near apex. **FRUIT** a juicy red berry. A large genus of 450 species widespread in Eurasia and extending south through Africa, with only 2 native species in North America. *Berberis* is cultivated for edible fruit, and the wood contains berberidine, a yellow alkaloid, which is widely used to make a dye. Some species are cultivated as ornamentals, as they form dense, strong hedges.

1. Racemes 3–15-flowered; leaf margins smooth or with 3–12 teeth; older bark red-brown to purple .... ***B. fendleri***
1′ Racemes 10–20-flowered; leaf margins typically with 15–30 spinose teeth; older (on second-year stems) bark gray .... ***B. vulgaris***

Berberis fendleri A. Gray (for botanist A. Fendler) Fendler barberry. Erect to sprawling shrubs, 1–2 m. **STEMS** strongly dimorphic, bark glabrous, red-brown to purple on older stems, spines simple or 1–2-branched. **LEAVES** deciduous, simple, thin, lanceolate, oblanceolate to narrowly elliptic from a long-attenuate base, 1–6 cm long, 0.5–1.7 cm wide, margins smooth to rarely somewhat bristle-tipped, bristles to 1.5 mm, apex acute to obtuse, blades dull or glossy but not glaucous. **INFL** a lax raceme of 5–12 flowers. **FLOWER** filaments lacking, 2 teeth near apex. **FRUIT** red, ovoid, 5–8 mm. Rare in mesic to riparian woodlands and around springs, in high-elevation piñon-juniper woodlands to coniferous forest communities. COLO: Arc, Hin, LPl, Min, Mon; NMEX: McK, RAr, SJn; UTAH. 1800–2600 m (6400–8600′). Flowering: Apr–Jun. Fruit: Jun–Jul. Fendler barberry is a relatively uncommon species that appears to be relictual in the Four Corners and was apparently much more common and widespread during the Pleistocene glaciations.

Berberis vulgaris L. (common) Common barberry. Erect shrubs, 1–3 m. **STEMS** strongly dimorphic, bark glabrous, gray, spines simple or 3-branched. **LEAVES** deciduous, simple, thin, obovate to oblanceolate or sometimes elliptic, from a short to long-attenuate base, 2–7 cm long, 0.6–3 cm wide, margins strongly bristle-tipped, bristles to 1.2 mm, apex rounded-obtuse, blades dull or glossy, glaucous below. **INFL** a somewhat open raceme of 10–20 flowers. **FLOWER** filaments lacking 2 teeth near apex. **FRUIT** purple to red, ellipsoid, 9–11 mm. Often cultivated as an ornamental, and sometimes escapes locally but never becomes common or invasive in the Southwest. NMEX: SJn. 1585 m (5200′). Flowering: May–June. Fruit: July. Native to Eurasia. Common barberry is one of the most important alternate hosts for *Puccinia graminis*. The species has ethnobotanical properties similar to the native *Mahonia* species.

Mahonia Nutt. Oregon Grape, Grape-holly, Barberry

(for B. MacMahon, friend and colleague of Nuttall) Erect to sprawling or creeping shrubs, sometimes stoloniferous, 0.1–3 m. **STEMS** monomorphic or weakly dimorphic, composed of elongate primary shoots and short secondary shoots, bark glabrous, gray, brown to purple-brown, lacking spines. **LEAVES** evergreen, pinnately compound, not clustered on secondary shoots, thick, rigid, leaflets 3–9, distinctly or obscurely 1–3-veined, margins with numerous spinose teeth. **INFL** a lax to dense raceme or subumbellate corymb, of 3–many flowers. **FLOWERS** 3-merous, subtended by sepalose bracts; sepals 3–6, yellowish; petals 6, yellow, with 2 nectar glands near base; stamens 6, filaments with 2 recurved teeth near apex. **FRUIT** a juicy to somewhat dry purple, red-brown, blue to blue-black berry. A genus of about 70 species, widely distributed in North and Central America and southeast Asia to Indonesia. *Mahonia* is often utilized or cultivated for ornamentals, for the yellow dye in its roots and stems, and for its edible fruit.

1. Plants tall, erect shrubs; inflorescence open, lax, of 10 or fewer flowers .... ***M. fremontii***
1′ Plants low, creeping, stoloniferous shrubs; inflorescence dense, with more than 10 flowers .... ***M. repens***

Mahonia fremontii (Torr.) Fedde (for John C. Frémont) Frémont barberry. Erect to sprawling shrubs to 3 m, not stoloniferous. **STEMS** weakly dimorphic, with short to elongate secondary shoots, bark on second-year stems gray-brown to brown-purple, glabrous. **LEAVES** odd-pinnate, 2–10 cm long, leaflets (3) 5–9 (11), rigid, ovate to lanceolate, glaucous, dull gray-green, 10–30 mm long, 5–15 mm wide, obscurely 1- to 3-veined, margins serrate with 3–6 pairs of sharp teeth, 2–6 mm long. **INFL** racemose or umbellate, open, 3–10-flowered, up to 8 cm long, 2 smaller red bractlets present below flower, 1 subtending the flower. **FLOWER** petals 3–4 mm long, apices entire, glands marginal at base. **FRUIT** a yellowish, brown-red to purple ellipsoid berry, 2–4 mm, not glaucous. Widespread on rocky slopes and washes in desert scrub, riparian zones, and piñon-juniper woodlands. UTAH. 1135–2200 m (3700–7000′). Flowering: Mar–May. Fruit: May–Jul. A widespread species on the Colorado Plateau. This species is often used in Navajo ceremonies, such as the Evil Way, as well as for various ailments and as a yellow dye. Both the Hopi and Zuni used the plant in various ceremonies and also used the berries for purple dyes and food. Frémont barberry is an alternate host for *Puccinia graminis*. A related species, *M. aquifolium* (Pursh) Nutt., native to the Pacific Northwest, is commonly cultivated as an ornamental. It can be distinguished from *M. fremontii* by its very glossy, dark green leaves with large leaflets that are 3–9 cm long.

Mahonia fremontii

Mahonia repens (Lindl.) G. Don (creeping) Oregon grape, tsinyaachéch'il. Low, sprawling or creeping shrubs to 0.3 m, stoloniferous. **STEMS** monomorphic, mostly lacking short secondary shoots, bark of second-year stems gray-purple, glabrous. **LEAVES** odd-pinnate, 10–30 cm long, dull dark green above, paler green below, leaflets (3) 5–7, ovate, 18–70 mm long, prominently 1- to sometimes 3-veined, margin serrate with 6–10 pairs of sharp teeth 0.5–3 mm long. **INFL** racemose, dense, 4–8 cm long, 10–30-flowered, bractlets absent. **FLOWER** petals 5–7 mm long, cleft at apex and bilobed. **FRUIT** a dark blue to blue or purple-black ovoid berry, 6–10 mm, glaucous. In shaded sites under trees on mountain slopes, in ponderosa pine, mixed conifer, and subalpine forest communities. 1900–3800 m (6500–11000′). ARIZ: Apa, Nav; COLO: Arc, Dol, Hin, LPl, Min, Mon, SJn, SMg; NMEX: McK, RAr, SJn; UTAH. Flowering: Apr–Jun. Fruit: Jun–Aug. Oregon grape is known from a few low-elevation sites around springs in the Great Bend area of Lake Powell. It occurs under dense Gambel's oak in shaded sites and is probably relictual. Used as both a ceremonial and medicinal plant by many tribes. The Navajo use it to treat insect bites, and internally as a cure-all. The Hopi use the yellow roots as a dye in the Home Way. This species is resistant to *Puccinia*.

BETULACEAE Gray BIRCH/ALDER FAMILY

Kenneth D. Heil

Monoecious, deciduous trees or shrubs. **LEAVES** alternate, simple, stipulate, serrate to doubly serrate. **INFL** a catkin; staminate catkin elongated and pendulous; pistillate catkin conelike. **FLOWERS**: staminate with bracts subtending 2 or 3 flowers, calyx 4-parted or perianth lacking; pistillate without or with a minute perianth; ovary 2-loculed; styles 2. **FRUIT** a 1-loculed, 1-seeded nutlet or nut. $x = 8, 14$ (Welsh et al. 2003; Brasher 2001).

1. Nutlet wingless, enveloped by enlarged inflated involucre; plants of alcoves or hanging gardens .... ***Ostrya***
1′ Nutlet mostly winged, lacking an involucre; plants of riparian areas and moist areas beneath alcoves and hanging gardens; widespread .... (2)

2. Pistillate aments 1–several, racemose, bracts persistent, conelike; pith of twigs triangular in cross section. ***Alnus***
2′ Pistillate aments solitary, bracts deciduous with the nutlets; pith of twigs round in cross section .... ***Betula***

Alnus Mill. Alder

(Latin *alnus*, alder) Trees or shrubs. **LEAVES** alternate, simple, serrate to lobed. **STAMINATE CATKINS** 1–3, bracts subtending 3–6 flowers. **FLOWERS** with 2–4 stamens. **PISTILLATE CATKINS** conelike, woody bracts subtending 2 flowers. **FRUIT** a nutlet. About 35 species widely distributed. Useful for timber; roots with nitrogen-fixing bacteria; and often used as an ornamental.

1. Leaves ovate to oblong-ovate, rounded to cordate or subcordate at the base; stamens 4; widespread ..... ***A. incana***
1′ Leaves elliptic or ovate-oblong, acute or short-cuneate at the base; stamens mostly (1–) 2 (–3); Chuska Mountains in New Mexico and Arizona .... ***A. oblongifolia***

Alnus incana (L.) Moench (gray) Thinleaf alder. Large shrubs or small trees up to 10 m tall; bark thin, red; twigs pubescent and glandular. **LEAVES** with petioles 5–30 mm long; blades 2–11 cm long, 1–10 cm wide, ovate or oblong-ovate, obtuse, acute or short-acuminate at apex, rounded or broadly cuneate to subcordate at base, doubly serrate, usually hairy along the veins beneath. $n = 14$. [*Betula alnus* L. var. *incana* L.; *A. tenuifolia* Nutt.]. Riparian areas; stream banks, seeps, and springs. ARIZ: Apa; COLO: Arc, Hin, LPl, Min, Mon, RGr, SJn; NMEX: McK, RAr, SJn; UTAH. 1980–3280 m (6500–10770′). Flowering: Apr–May. Our material belongs to **subsp. *tenuifolia*** (Nutt.) Breitung (slender-leaved). Alaska and Yukon south to New Mexico and California. The brown or reddish powdered bark used as a tan or reddish dye by the Navajo.

Alnus oblongifolia Torr. (oblong leaves) New Mexico alder, Arizona alder. Large shrubs or trees up to 30 m tall; bark of older trees grayish brown. **LEAVES** with blades ovate or oblong-ovate to lanceolate, mostly 3–9 cm long, 1.5–7 cm wide, the bases acutish to short-cuneate; margins doubly serrate. **STAMINATE FLOWERS** with (1–) 2 (–3) stamens. In mountain canyons along streams. ARIZ: Apa; NMEX: McK. 1830–2340 m (6000–7680′). Flowering: May. New Mexico and Arizona.

Betula L. Birch

(Latin *betula*, birch) Trees and shrubs, outer bark smooth, often separable in sheets. **LEAVES** alternate, simple, serrate to crenate or doubly serrate. **STAMINATE CATKINS** 1–4 per bud, pendulous or spreading in flower; staminate flowers in clusters of 3; stamens 2. **PISTILLATE CATKINS** usually solitary, erect in flower; pistillate flowers 2–3 per bract. **FRUIT** a samara. Approximately 60 species in the Northern Hemisphere, 4 in Europe. Timber for plywood, and sap for sweetening, fermenting, and shampoo.

1. Low to moderately sized shrub, 0.5–1.2 m tall; leaf blades 0.5–2.5 cm long, oval to orbicular, usually 10 or fewer teeth per side .... ***B. glandulosa***
1′ Tall shrubs or trees, mostly 4–8 m tall; leaf blades 1.5–8 cm long, ovate to deltoid; usually 10–40 teeth per side .... ***B. occidentalis***

Betula glandulosa Michx. (glandular) Glandular birch, swamp birch, bog birch. Low shrubs, 0.5–1.2 m tall; twigs and branches glabrous, densely resinous-glandular. **LEAVES** with petioles 2–10 mm long; blades 1–2.5 cm long, suborbicular to obovate, apex rounded, base rounded or broadly cuneate, margin crenate-serrate, 10 or fewer teeth per side, not hairy in lower vein axils. **PISTILLATE CATKINS** 7–20 mm long, 3–8 mm thick, bracts commonly glabrous dorsally, ciliate. **FRUIT** a samara with wings narrower than 1/2 the body width. $2n = 28$. Beaver ponds, fens, and iron bogs in subalpine communities. COLO: SJn. 2710–3130 m (8900–10275′). Flowering: May–Jun. Greenland to Alaska, south to Maine, Oregon, and New Mexico.

Betula occidentalis Hook. (of the west) Water birch. Shrubs or small trees, mostly 3–6 m tall, with several trunks, to 2.5 dm thick or more. Bark not exfoliating, reddish or yellowish brown, shining, with pale horizontal lenticels; twigs pubescent to glabrous with reddish crystalline resin glands. **LEAVES** with petioles 5–15 mm long; blades 1–5 cm long, 0.7–4 cm broad, ovate, acute or abruptly acuminate apically, obtuse to rounded, sharply and often doubly serrate, mostly 15–25 teeth per side, not hairy in lower vein axils, minutely hairy to glabrous on margins near the base. **PISTILLATE CATKINS** 15–40 mm thick, the bracts puberulent and ciliate. **FRUIT** a samara with wings subequal to width of the nutlet. Riparian communities; along streams, seeps, and springs. ARIZ: Apa, Nav; COLO: LPl, Mon; NMEX: SJn. Alaska and Yukon east to Northwest Territories, south to South Dakota, California, Colorado, Arizona, and New Mexico.

Ostyra Scop. Hop-hornbeam

(Greek *ostryos*, scale, referring to the scaly catkins) Small trees to 18 m tall, trunks 1 to several, bark brownish gray to light brown; lenticels inconspicuous. **LEAVES** alternate, simple, doubly serrate. **STAMINATE CATKINS** pendulous, the scales abruptly acuminate. **FLOWERS** consisting of 3–6 stamens, each in a bract axil; filaments short, each forked apically and the branches with pilose half-anthers. **PISTILLATE CATKINS** slender, loosely flowered, the ovate hairy bracts caducous, subtending 2 flowers, each enclosed within an ovoid pouch composed of united bracts and bractlets; calyx minute; bracts accrescent and inflated in fruit, hoplike. **FRUIT** a compressed ovoid nutlet. Five species, mostly in northern temperate zones.

Ostrya knowltonii Coville (for F. H. Knowlton, 1860–1926, botanist and paleontologist) Western hop-hornbeam. Small trees, mostly 2–6 m tall, trunks 3–18 cm thick; branchlets spreading-hairy, more or less stipitate-glandular, becoming glabrous. **LEAVES** with blades 0.8–8 cm long, 0.8–5 cm wide, ovate to lance-ovate or elliptic, doubly serrate, acute apically, rounded to obtuse basally. **STAMINATE CATKINS** 1.5–3 cm long. **PISTILLATE CATKINS** 0.7–1 cm long, in fruit to 4.5 cm long, about as wide. **FRUIT** with aments 2–5 cm long, the individual sacs 10–25 mm long, greenish white to brownish. Sandstone areas; shaded defiles and hanging garden communities. UTAH. 1125–1370 m (3700–4500′). Flowering: May–Jun. Utah, Arizona, New Mexico, and Texas. Along with the western redbud, this is one of the principal small trees at Lake Powell. Potential as an ornamental.

BIGNONIACEAE Juss. CATALPA FAMILY

Linda Mary Reeves

Primarily lianas or vines, but many trees and shrubs, a few herbs. **LEAVES** usually opposite, without stipules, usually compound; glands often present at petiole base. **INFL** generally cymose or paniculate, with inconcpicuous bracts. **FLOWERS** showy, zygomorphic or tubular; calyx tubular with 5 lobes; corolla campanulate, funnelform, often irregular with 5 lobes, often frilled and strongly marked; stamens 5, often epipetalous or reduced to staminodes; ovary 1, superior, of 2 carpels, usually bilocular, occasionally unilocular, usually with axile placentation, stigma bifid, nectar disc present at base of ovary. **FRUIT** a loculicidal capsule, occasionally a fleshy fruit. **SEEDS** generally winged, samara-like. $2n = 14$. About 120 genera with about 800 species worldwide, primarily tropical and subtropical regions, especially northern South America. Many species (*Catalpa*, *Tabebuia*) used for timber; many ornamentals, particularly in the tropics, such as *Chilopsis*, *Jacaranda*, *Campsis*, *Doxantha*, and *Tecomaria*. In our area, *Chilopsis linearis* is grown successfully as an ornamental. Future surveys may find this species as an escape.

1. Plants vining, leaves pinnately compound, perianth orange or red-orange .... ***Campsis***
1′ Tree, leaves simple, corollas primarily pale pink or white .... ***Catalpa***

Campsis Lour. Trumpet Creeper, Hummingbird Trumpet

(Greek for bending) Vining shrub or vine with aerial rootlets. **LEAVES** opposite, odd-pinnate; leaflets 9–11, with tufts of sericeous hairs on leaf rachis between petiolules. **INFL** a terminal panicle. **FLOWERS** weakly zygomorphic; calyx campanulate, yellow to reddish, brownish, or greenish; corolla fleshy to papery, tubular-funnelform, yellow to orange to red, about 3–4 times as long as the calyx, the 5 lobes often triangular; anthers not exserted, filaments flaring from top of corolla tube. **FRUIT** a cylindric-oblong, beaked capsule, with keeled sutures, loculicidally dehiscent. **SEEDS** numerous, compressed, with 2 large, papery wings. Two species, one widespread in North America, the other in China and eastern Asia (*Campsis chinensis* (Lam.) Pichon). Some of our escaped specimens may represent the natural and cultivated hybrid between these two species, *Campsis* ×*tagliabuana* (Vis.) Rehder, usually the cultivar 'Madame Galen' or other combinations with various expressions of genetic introgression. The hybrids generally have a brighter red, papery corolla, are more shrubby, very vigorous, and thus likely to escape.

Campsis radicans (L.) Seem. (root) Trumpet honeysuckle, trumpet creeper, hummingbird trumpet, cow-itch vine. Deciduous vine climbing to 10 m or more, often developing multiple strong woody stems. **LEAVES** with leaflets broadly ovate to ovate-oblong to elliptic-lanceolate; apex acuminate; base broadly rounded to cuneate; to 9 cm long and 4 cm wide; margin serrate to dentate. **FLOWERS** with calyx yellowish brown to orange, occasionally red-orange, lobes 5, triangular; corolla tubular, dull reddish to bright orange-red, the interior floral tube often brighter and a more saturated color than exterior or corolla lobes, 4–9 cm long; anthers 4 mm long. **FRUIT** 8–12 cm long. [*Tecoma radicans* (L.) Juss.; *Bignonia radicans* L.] Escaped cultivated ornamentals around abandoned dwellings and town lots or agricultural homesteads. Becoming an invasive weedy species in the eastern U.S. Flowering: May–Oct, but more typically in our area: Jul–Sep. Fruit: Jul–Aug. ARIZ: Apa; COLO: Mon; NMEX: SJn; UTAH. Native from New York to Ohio and Iowa, south to Florida and Texas, but occurring as a weed in New England, the upper Midwest, Colorado, and Utah. Very attractive to hummingbirds. Crows, ravens, and jays often eat the young flower buds. Woody stems sometimes damage fences and buildings. Contact with the plant can produce a mild urticating rash in livestock and some people.

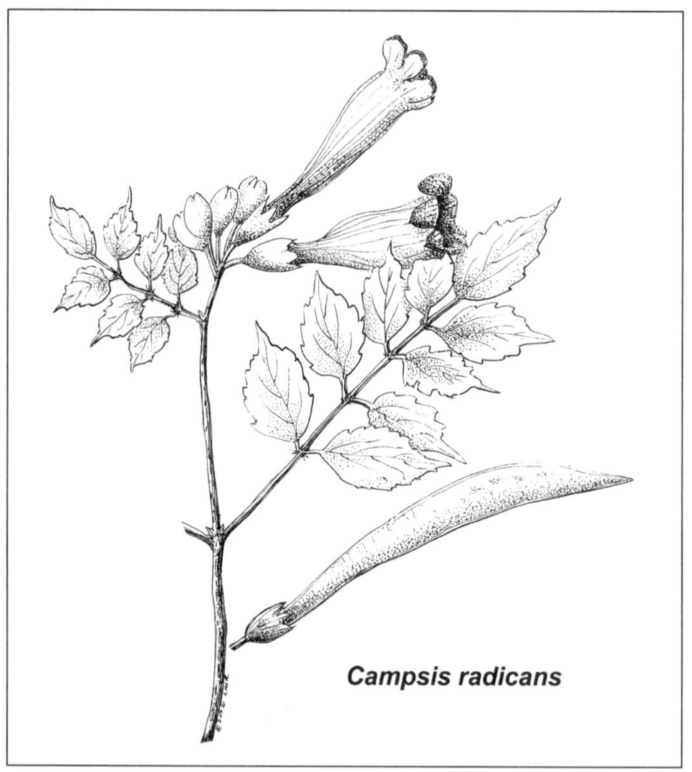
Campsis radicans

Catalpa Scop. Catalpa, Catawba

(from the Indian name for an Asian species) Trees. **LEAVES** opposite, occasionally whorled, long-petioled, large, simple, entire or lobed, often pungent when crushed. **INFL** a large panicle. **FLOWER** calyx often 2-lipped; corolla very showy, white, pink, yellowish or spotted and/or lined, zygomorphic, campanulate, 2-lipped, each spreading lip split into 2 upper lobes and 3 lower lobes; fertile stamens 2; style bifid. **FRUIT** a cylindrical capsule, dehiscent in 2 parts at maturity. **SEEDS** small, compressed, winged, with white pubescent tuft at each end. About 10 species in North America and Asia.

Catalpa speciosa Warder ex Engelm. (beautiful) Northern catalpa, western catalpa, showy catalpa, Indian bean, catawba, cigartree. Trees to 15 m high, bark red-brown, texture scaly. **LEAVES** long-petiolate to 15 cm; blade 15–30 cm

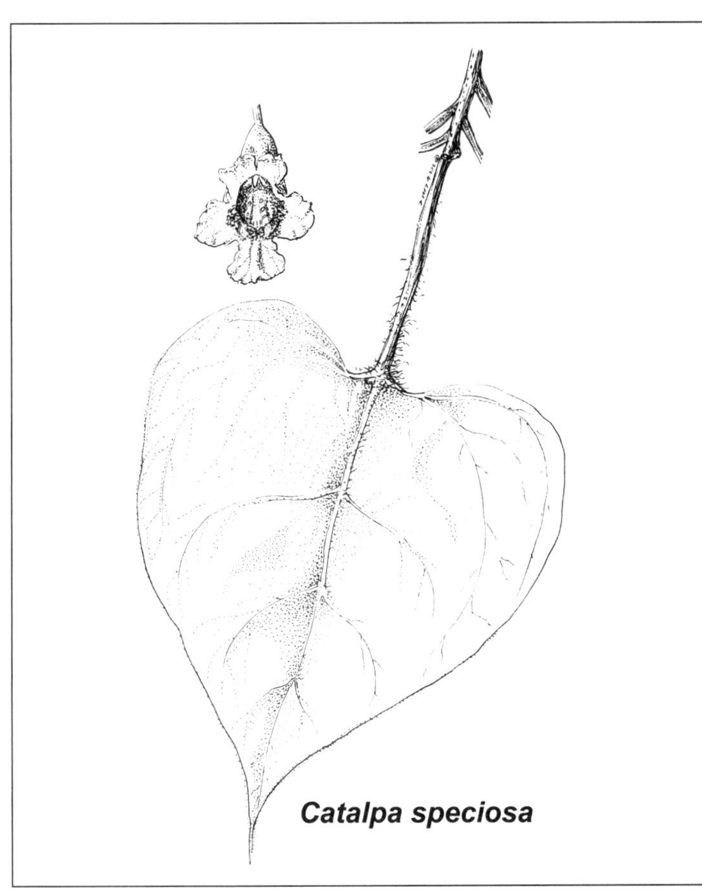
Catalpa speciosa

long, cordate to ovate-oblong; apex long-acuminate, base truncate to cordate, bright green, somewhat glabrous to scabrous above, pubescent beneath. **INFL** few-flowered, about 15 cm long, not as broad as long. **FLOWERS** 3–5 cm long, 3–4 cm wide; lower corolla lobe emarginate, white, with yellow stripes and purple-brown stripes on inner surface. **FRUIT** a capsule, 20–50 cm long, 1–1.5 cm wide. **SEEDS** 4–5 mm wide. Escaped cultivated ornamentals, primarily in towns or abandoned homesteads, trading posts, and former townsites, or along watercourses, springs, and acequias. ARIZ: Apa; COLO: Mon; NMEX: SJn. Below 1950 m (below 6400′). Flowering: May–Jul. Fruit: Jun–Aug. Native to the midwestern United States from southern Illinois and southern Indiana south to east Texas; escaped in many western states.

BORAGINACEAE Juss. BORAGE FAMILY

Robert C. Sivinski

Annual and perennial herbs in our flora. **LEAVES** alternate or sometimes opposite at the stem base, simple, entire, without stipules and usually rough-hairy. **INFL** usually of cymosely disposed helicoid racemes or spikes, short cymules, or rarely solitary, axillary flowers. **FLOWERS** sympetalous, usually actinomorphic, rarely zygomorphic, 5-merous as to the calyx, corolla, and stamens, sometimes cleistogamous or heterostylous; sepals distinct or connate at the base below the middle; corolla salverform or funnelform, usually with 5 protruding fornices at the juncture of the tube and limb; stamens inserted on the corolla tube and alternate with the lobes; ovary superior, 2-carpellate, each carpel usually 2-ovulate with a secondary partition; style solitary, simple or bifid; stigma capitate or geminate, sometimes disc-shaped. **FRUIT** a schizocarp (in our species), separating into 4 hard mericarps (nutlets), often fewer by abortion. **NUTLETS** smooth or variously ornamented, rarely pubescent, each one basally or ventrally attached to a nearly flat receptacle or a central conical gynobase. $x = 6, 8$ (Higgins 1979). Worldwide about 120 genera and 2000 species in Arctic, temperate, and tropical regions, but especially well developed in western North America, the Mediterranean, and other arid regions. Several European and Asian genera (*Echium*, *Myosotis*, *Symphytum*, etc.) are occasionally planted in domestic gardens and may escape into our flora.

1. Nutlets armed on the surfaces or margins with stout, hooked or barbed prickles or bristles ....(2)
1′ Nutlets unarmed, sometimes ornamented with minute setae ....(5)

2. Nutlets radially spreading at maturity; bristles of the nutlets merely hooked at the tips, not glochidiate with several barbs .... ***Pectocarya***
2′ Nutlets erect or spreading; bristles of the nutlets glochidiate with 2 or more barbs ....(3)

3. Nutlets covered with numerous short, glochidiate prickles on both dorsal and ventral surfaces .... ***Cynoglossum***
3′ Nutlet prickles only on the dorsal surface or margins, not on ventral surface ....(4)

4. Annual; pedicels erect in fruit .... ***Lappula***
4′ Biennial or perennial; pedicels deflexed in fruit .... ***Hackelia***

5. Style attached to a receptacle or central gynobase, arising between the lobes of the ovary; ovary deeply 4-lobed ....(6)
5′ Style, or a sessile stigma, attached to the apex of the ovary; ovary entire or shallowly 2- or 4-grooved ....(11)

6. Nutlets basally and broadly attached to a broad receptacle or low gynobase; nutlet attachment scar basal, large, greater than 1/2 the nutlet width, sometimes surrounded by an evident rim; stigmatic surface geminate, usually slightly or conspicuously bilobed .... (7)
6′ Nutlets variously attached to the gynobase, if basal, the attachment scar always small, less than 1/2 the nutlet width; stigma entire .... (8)

7. Calyx cleft to near the base; corolla yellow or yellow-green in our species; base of nutlet with a sharp rim around the attachment scar, or without an evident rim .... ***Lithospermum***
7′ Calyx cleft to near, or slightly below, the middle; corolla blue; base of nutlet a broad, thickened rim .... ***Anchusa***

8. Corolla limb at anthesis campanulate, or subcylindric-erect and appearing to be a slightly wider continuation of the tube; corolla tube greater than 3 mm long .... ***Mertensia***
8′ Corolla limb at anthesis abruptly flared or rotate, or if erect, then the corolla tube less than 3 mm long; corolla tube 1–15 mm long .... (9)

9. Corolla blue, rarely white in occasional individuals; plants of alpine habitats .... ***Eritrichium***
9′ Corolla white or yellow; plants of lower elevations .... (10)

10. Nutlets with a closed or open groove scar (formed by the nonfusion of the pericarp) running most of the length of the ventral surface, this often expanded or forked at the base .... ***Cryptantha***
10′ Nutlets with a ventral keel of fused pericarp terminating at a small, nearly basal attachment scar, or the scar near the middle of the nutlet and raised on the ventral keel .... ***Plagiobothrys***

11. Style evidently bifid, branched well below the capitate stigmas .... ***Tiquilia***
11′ Style undivided; stigma a tuft of coarse hairs subtended by a swollen ring, or conic on a peltate disc ***Heliotropium***

Anchusa L. Alkanet

(from the Greek *anchousa*, ancient name of a related plant, *Alkanna tinctoria*) Annual or perennial herbs. **LEAVES** alternate, often pungently hirsute. **INFL** a panicle of helicoid racemes. **FLOWERS** actinomorphic; calyx lobes narrow, corolla funnelform or salverform, throat obscured by fornices, limb widely spreading; stamens included; style slender; stigma capitate; ovules 4. **NUTLETS** with a stipelike basal attachment that surrounds the attachment scar with a prominent thickened rim that detaches from a pit on the low gynobase. An Old World genus of about 40 species, centering in the Mediterranean region.

Anchusa officinalis L. (on official medicinal list) Bugloss. Perennial herb. **STEMS** 1–several from a taproot, erect, branching from near the base when well developed, 3–10 dm tall, coarsely hirsute. **LEAVES** strigose-hirsute; basal leaves petiolate, oblanceolate, 6–20 cm long; cauline leaves lanceolate, gradually reduced and becoming sessile above. **INFL** a panicle of racemes on the upper stem. **FLOWERS** nearly sessile; calyx 5–8 mm long at anthesis, scarcely longer in fruit, lobes narrowly triangular; corolla blue, tube nearly equal the length of the calyx, limb 6–9 mm wide; fornices prominent, papillate. **NUTLETS** oblique, rugose or granulate, 3–4 mm long, the tip directed inward. $2n = 16$, 32. Occasional weed on roadsides and other disturbed habitats at 1970 m (6500′). COLO: LPl. Flowering: May–Jul. Fruit: Jun–Aug. Introduced from Eurasia.

Cryptantha Lehm. ex G. Don Cryptantha, Cat's-eye, Hidden-flower

(Greek *kryptos*, hidden, + *anthos*, flower, for the cleistogamous flowers of the first-described South American species) Annual, biennial, or perennial herbs, strigose, rarely glabrate, usually hirsute or hispid-setose. **LEAVES** narrowly oblanceolate, spatulate, or occasionally linear. **INFL**, in most species, a series of helicoid, naked or bracteate modified cymes appearing as racemes or spikes, usually elongating with maturity or often short and aggregated into a terminal thyrse or capitate cluster. **FLOWERS** actinomorphic, calyx cleft to the base or nearly so; corolla white, ochroleucous, or yellow, salverform with a spreading or rotate limb and fornices at the throat; anthers included; usually homostylous, but heterostylous in some perennial species; stigma capitate; ovules usually 4, rarely 2. **NUTLETS** 1–4, triangular, ovate to lanceolate, smooth to variously roughened, heteromorphic within a fruit or all similar, affixed to an elongate gynobase, the ventral scar either closed, narrowly open, or forming a triangular areola (Higgins 1971; Johnston 1925). A diverse genus of about 150 species in the western half of North America and arid regions of South America. *Cryptantha* is especially interesting for its complex reproductive patterns ranging from autogamy to outcrossing reinforced by heterostyly. Various forms of nutlet abortion are evident to some degree in almost all our species. Abortion rates can be relatively mild and random, or taken to an extreme, where only one nutlet at a genetically fixed location is matured in each flower. This has led to nutlet heteromorphy in some annual species where a single, always matured nutlet is larger and/or more firmly attached to the gynobase, while the three consimilar nutlets are deciduous and randomly matured at various rates, or always aborted. The persistent odd nutlet is dispersed as a unit with the calyx.

1. Plants biennial or perennial; corolla conspicuous, limb 4 mm or greater in diameter .... (2)
1′ Plants annual, rarely woody at the stem bases; corolla inconspicuous, limb less than 3 mm wide .... (13)

2. Dorsal (facing away from the center of the flower) surfaces of mature nutlets smooth and shiny .... (3)
2′ Dorsal nutlet surfaces variously roughened with tubercles, murications, or wrinkles .... (5)

3. Corolla limb white; fornices and corolla tube often yellowish .... ***C. cinerea***
3′ Corolla entirely yellow .... (4)

4. Mature nutlets usually 1 per calyx (older flower with shed corolla), sometimes 2; inflorescence an elongate, somewhat cylindrical arrangement of alternating cymules of nearly equal size .... ***C. flava***
4′ Mature nutlets usually 4 per calyx, seldom fewer; inflorescence a capitate arrangement of a few terminal cymules, sometimes subtended by 1 or 2 smaller cymules .... ***C. confertiflora***

5. Corolla tube elongate, distinctly surpassing the calyx lobes (usually by 2 mm or more); flowers heterostylic .... (6)
5′ Corolla tube short, about equal to the tips of the calyx lobes; flowers not heterostylic .... (10)

6. Nutlets finely muricate-papillate (the murications sometimes setulose-tipped), usually maturing 1, sometime 2, per calyx .... ***C. fulvocanescens***
6′ Nutlets rugose or tuberculate, usually maturing 4 per calyx, rarely fewer by abortion .... (7)

7. Leaves conspicuously pustulate on both surfaces; corolla tube 12–17 mm long; calyx segments 7–13 mm long at anthesis .... ***C. longiflora***
7′ Leaves only conspicuously pustulate on the lower surface; corolla tube 5.5–12 mm long; calyx segments 3.5–8 mm at anthesis .... (8)

8. Plants low, less than 1.5 dm tall; inflorescence subcapitate, less than 5 cm long .... ***C. paradoxa***
8′ Plants usually taller than 1.5 dm; inflorescence more elongate, 5–25 cm long .... (9)

9. Corolla limb widely spreading, rotate; fornices prominent, erect, 1–2 mm long .... ***C. flavoculata***
9′ Corolla limb funnelform, never opening out flat; fornices low, broad, less than 1 mm long .... ***C. tenuis***

10. Nutlet margins conspicuously papery-winged; plants coarse biennials; 4–10 dm tall .... ***C. setosissima***
10′ Nutlet margins rounded or sharply angled, never winged; plants biennial or perennial; less than 4 dm tall .... (11)

11. Nutlets finely papillate with some of the papillae coalescing into low ridges; nutlet scar without raised margins .... ***C. humilis***
11′ Nutlets coarsely rugose-tuberculate; nutlet scar with elevated margins .... (12)

12. Biennial or short-lived perennial; nutlet scar closed or nearly so .... ***C. bakeri***
12′ Perennial; nutlet scar narrowly open, constricted below the middle .... ***C. osterhoutii***

13. Nutlet margins conspicuously winged .... ***C. pterocarya***
13′ Nutlet margins rounded or sharply angled to a knifelike edge .... (14)

14. Fruiting calyx circumscissile a little below the middle .... ***C. circumscissa***
14′ Fruiting calyx not circumscissile .... (15)

15. Taproot strongly charged with red-purple dye; gynobase elongate, surpassing the nutlets and terminated by a sessile stigma, without a differentiated style .... ***C. micrantha***
15′ Taproot usually without red dye (sometimes weakly red-dyed in *C. recurvata*); gynobase inconspicuous or shorter than the nutlets and topped by a definite style that may or may not surpass the nutlets .... (16)

16. Nutlet usually solitary per mature calyx .... (17)
16′ Nutlets normally 4 per calyx, sometimes fewer by abortion .... (18)

17. Calyx lobes and nutlet conspicuously recurved or deflexed .... ***C. recurvata***
17′ Calyx lobes and nutlet not curved or bent .... ***C. gracilis***

18. Nutlets in each calyx all smooth-surfaced .... ***C. fendleri***
18′ Nutlets rough-surfaced, or at least some so .... (19)

19. Nutlets in each calyx all alike in size and surface ornamentation .... ***C. nevadensis***
19′ Nutlets in each calyx conspicuously heteromorphic, with 1 long nutlet and 3 shorter, ± similar nutlets .... (20)

20. Flowers pedicellate; pedicels 1–4 mm long .... ***C. racemosa***
20′ Flowers sessile or pedicels less than 1 mm long .... (21)

21. Nutlet margins sharply angled to a knifelike edge .... ***C. inaequata***
21′ Nutlet margins rounded .... (22)

22. Odd nutlet surface usually with minute murications; attachment scars of consimilar nutlets deeply impressed; midrib of fruiting calyx lobes conspicuously thickened and bony .... ***C. crassisepala***
22′ Odd nutlet finely granulate or nearly smooth; attachment scars of consimilar nutlets shallowly impressed; midrib of fruiting calyx lobes moderately thickened .... ***C. kelseyana***

Cryptantha bakeri (Greene) Payson (for Charles F. Baker, Colorado botanist) Baker's cryptantha. Perennial. **STEMS** simple, 1–several from a branching caudex or simple tuft of basal leaves, 1–4 dm tall, spreading-hispid and softer, finer under-pubescence. **LEAVES** oblanceolate or spatulate, spreading-hirsute and sericeous-strigose or long-spreading villous, pustulate; cauline leaves ascendingly reduced. **INFL** elongate-cylindric in flower, lower part interrupted, 0.6–2.6 dm long, cymules somewhat spreading in fruit. **FLOWERS** homostylic; fruiting calyx strigose-hirsute, 6–9 mm long; corolla tube white or pale yellow, equal or subequal to the apex of the calyx, fornices yellow or ochroleucous; corolla limb 7–10 mm wide. **NUTLETS** ovate-lanceolate, usually all 4 maturing, rugose-tuberculate on both surfaces; scar closed with conspicuously raised margins. $n = 12$. [*Oreocarya bakeri* Greene]. Sandstones or sandy clay soils in sagebrush, piñon-juniper, oak brush, and ponderosa pine communities. ARIZ: Apa, Nav; COLO: Arc, Dol, LPl, Mon, SMg; NMEX: McK, RAr, SJn; UTAH. 1350–2650 m (4500–8700′). Flowering: May–Jun. Fruit: Jun–Jul. A Colorado Plateau endemic in Colorado, Utah, and the Four Corners region of Arizona and New Mexico.

Cryptantha cinerea (Greene) Cronquist (ash-gray, for the appearance of the surface pubescence) Bow-nut cryptantha. Perennial (rarely appearing biennial). **STEMS** 1–several from a branching, sometimes woody caudex, simple or branched, strigose, sometimes loosely villous, with or without varying amounts of spreading-bristly hairs, or glabrate. **LEAVES** oblanceolate to lance-linear, obtuse to acute, strigose to villous-puberulent or glabrate, usually pustulate on lower surface. **INFL** narrow or somewhat broad, cymules often elongating at maturity. **FLOWERS** homostylic; pedicels 1–3 mm long; calyx segments ovate-lanceolate, tomentose to strigose-hirsute, 5–7 mm long in fruit; corolla tube white or pale yellow, equal or subequal to the apex of the calyx; fornices yellow or sometimes ochroleucous, corolla limb 4–8 mm wide. **NUTLETS** 1.8–2.5 mm long, smooth and glossy, sometimes minutely papillate-velvety on the ventral surfaces, deeper than wide, dorsal surface narrowly ovate, bowed outward from the base and inward to the tip; scar closed and discontinuous with a small basal pocket, 4 or fewer maturing. An extremely polymorphic species with many named and unnamed variations. In our flora, var. *cinerea* and var. *jamesii* are neither geographically nor ecologically distinct. Intermediate forms are as prevalent as those with the characteristics of these poorly delimited infraspecific taxa. I can recognize only two varieties in our flora.

1. Upper leaf surface pubescent, at least sparsely so .... **var. *cinerea***
1′ Upper leaf surface glabrous .... **var. *pustulosa***

var. *cinerea* A distinctly heterogeneous taxon as to pubescence, branching of the stems, leaf arrangement, and disposition or length of the cymes in the inflorescence. $n=12$. [*Oreocarya cinerea* Greene; *C. jamesii* (Torr.) Payson var. *cinerea* (Greene) Payson; *C. jamesii* var. *setosa* (M. E. Jones) I. M. Johnst. ex Tidestr.; *O. multicaulis* (Torr.) Greene; *C. jamesii* var. *multicaulis* (Torr.) Payson; *C. cinerea* var. *jamesii* Cronquist]. Desert scrub and sagebrush, piñon-juniper, and ponderosa pine communities. ARIZ: Apa, Coc, Nav; COLO: Arc, Dol, LPl, Mon; NMEX: McK, RAr, San, SJn; UTAH. 1500–2600 m (4900–8600′). Flowering: Apr–Sep. Fruit: Jun–Oct. South Dakota, Oklahoma to Utah, south to Arizona, Texas, and Chihuahua. Used as a lotion for snakebites by the Kayenta Navajo. Used by the Ramah Navajo as a ceremonial medicine and a poultice for sores.

var. *pustulosa* (Rydb.) L. C. Higgins (full of blisters, for the siliceous discs on the herbage) A unique variety with dark green herbage that is nearly glabrous, except in the inflorescence. Often lacks a basal leaf tuft. $n = 12$. [*C. pustulosa*

(Rydb.) Payson; *C. jamesii* (Torr.) Payson var. *pustulosa* (Rydb.) H. D. Harr.; *Oreocarya pustulosa* Rydb.]. Sandy soils in desert scrub, sagebrush, and piñon-juniper communities. ARIZ: Apa; NMEX: SJn; UTAH. 1400–2000 m (4600–6600′). Flowering: Jun–Sep. Fruit: Jul–Oct. Arizona, Colorado, New Mexico, and Utah. The sandy habitats of the Colorado Plateau populations differ from those in central New Mexico, which are restricted to gypsum substrates.

Cryptantha circumscissa (Hook. & Arn.) I. M. Johnst. (split, referring to the circumscissile calyx) Cushion cryptantha. Annual. **STEMS** branched from the base and above, forming a hemispherical mass 2–10 cm high. **LEAVES** linear, 5–13 mm long, strigose or short-hispid, densely crowded toward the branch ends, lower leaves more remotely arranged. **INFL** of indefinite racemelike clusters with flowers in the axils of leafy bracts and forks of branches. **FLOWERS** sessile; fruiting calyx 2–4 mm long, circumscissile a little below the middle, basal part persistent, cupulate and white-scarious, upper lobes herbaceous and deciduous; corolla inconspicuous, 1–2 mm broad; style equaling or barely exceeding the nutlets. **NUTLETS** 4, homomorphous or nearly so, smooth or minutely muricate, narrowly triangular-ovate, 1.2–1.7 mm long, back flattened, especially near the apex, margins angled; scar closed and forked at the base. $2n = 24, 36$. [*Lithospermum circumscissum* Hook. & Arn.; *C. depressa* A. Nelson]. Sandy soils in desert scrub. ARIZ: Coc; UTAH. Up to 1680 m (5500′). Flowering: Apr–Jun. Fruit: May–Jul. Wyoming to Washington, south to Arizona and Baja California, also amphitropically to Chile and Argentina. Used as a dermatological lotion for itching by the Kayenta Navajo.

Cryptantha confertiflora (Greene) Payson (crowded flowers, referring to the nearly capitate inflorescence) Basin yellow cryptantha. Caespitose perennial. **STEMS** few to several from the branches of a woody caudex, 1.5–3 dm tall, tomentose at the base, setose and strigose above. **LEAVES** in basal clusters and cauline, narrowly oblanceolate, acute, strigose, pustulate only on the lower surface. **INFL** subcapitate, consisting of a relatively large, terminal cluster of cymules often subtended by 1 or more remote, smaller cymules. **FLOWERS** heterostylic; calyx segments linear-lanceolate, 6–8 mm long at anthesis, 10–14 mm long in fruit; corolla yellow throughout, tube 9–13 mm long, surpassing the calyx at anthesis, limb 8–12 mm wide. **NUTLETS** triangular to ovate, margins sharply angled to nearly winged, surfaces smooth and glossy, 3.5–4 mm long, 2.5–3 mm wide; scar closed and lacking elevated margins, usually 4 maturing. [*Oreocarya confertiflora* Greene; *O. alata* A. Nelson; *O. lutea* Greene ex Brand]. Desert scrub, sagebrush, and piñon-juniper communities. ARIZ: Apa, Coc, Nav; UTAH. 1200–2000 m (3900–6600′). Flowering: Apr–Jun. Fruit: May–Aug. Arizona and Utah, west to California.

Cryptantha crassisepala (Torr. & A. Gray) Greene (thick sepals, referring to the thickened midribs of the fruiting calyx) Thicksepal cryptantha. Annual. **STEMS** branched from the base and above, 0.5–1.5 dm long, spreading or erect without a strong central axis, spreading-hirsute. **LEAVES** linear to narrowly oblanceolate, finely strigose and stiffly hispid, pustulate. **INFL** terminating the branches, cymes naked or rarely 1–2 leafy bracts at the base, elongating at maturity. **FLOWERS** nearly sessile; fruiting calyx 4–6.5 mm long, segments narrowly lanceolate, hirsute and spreading pungent-hispid, midribs becoming thick and bony at maturity; corolla white, limb 1–2.5 mm wide. **NUTLETS** lance-ovate, usually 4 maturing, heteromorphic, 1 evidently larger, 2–2.5 mm long and more firmly attached to the gynobase, minutely granulate or muricate; the 1–3 consimilar nutlets readily deciduous, 1.2–1.8 mm long and granulate-tuberculate; scar open and usually excavated. [*Eritrichium crassisepalum* Torr. & A. Gray; *Krynitzkia crassisepala* (Torr. & A. Gray) A. Gray]. Sandy soil in desert scrub, sagebrush, and piñon-juniper communities. ARIZ: Apa, Coc, Nav; COLO: Arc, Dol, LPl, Mon; NMEX: McK, RAr, San, SJn; UTAH. 1200–2000 m (3900–6600′). Flowering: Apr–Jun. Fruit: May–Jul. Utah and Colorado, south to Texas and Chihuahua. Our plants belong to the small-flowered **var. *elachantha*** I. M. Johnst. (disgrace flower, probably in reference to the small flower).

Cryptantha fendleri (A. Gray) Greene (for Augustus Fendler, first resident botanist in New Mexico) Fendler's cryptantha. Annual. **STEMS** solitary or occasionally branching from the base, 1–4 dm tall, erect and forming a central axis that is paniculately branched above with rigid ascending branches, strigose and spreading-hirsute. **LEAVES** mostly cauline, linear-acute, hirsute and spreading-hispid, pustulate. **INFL** broad, cymes naked, terminating the stem and branches. **FLOWERS** sessile or nearly so; fruiting calyx 4–6 mm long, strigose and pustulate-hispid; corolla inconspicuous, 1 mm wide, white. **NUTLETS** all alike, usually 4 maturing, narrowly lanceolate, smooth and glossy; scar closed except at the small basal areola. [*Krynitzkia fendleri* A. Gray; *C. pattersonii* (A. Gray) Greene]. Sandy soils in sagebrush, piñon-juniper, and ponderosa pine communities. ARIZ: Apa, Nav; COLO: Mon; NMEX: McK, RAr, SJn; UTAH. 1520–2380 m (5000–7800′). Flowering: May–Jul. Fruit: Jun–Aug. Alberta to Washington, south to Arizona and New Mexico. Decoction of plant taken for coughs, and plant used for sheep feed by the Ramah Navajo.

Cryptantha flava (A. Nelson) Payson (yellow, referring to the yellow corolla) Yellow cryptantha. Caespitose perennial. **STEMS** few to several from the branches of a woody caudex, 1–4 dm tall, spreading-hirsute. **LEAVES** linear-oblanceolate to spatulate, mostly basal and silvery strigose, but cauline leaves usually well developed and spreading-hispid. **INFL** typically elongate and thyrsoid-cylindric, conspicuously yellow-setose, foliar bracts inconspicuous above. **FLOWERS** heterostylic; calyx 8–10 mm long at anthesis, 9–12 mm long in fruit, lobes linear; corolla yellow throughout, tube 9–12 mm long and surpassing the calyx at anthesis, limb 7–11 mm wide. **NUTLETS** lance-ovate, usually only 1, sometimes 2, maturing, 3.4–4.2 mm long, smooth and glossy; scar closed and lacking an elevated margin. $n = 12$. [*Oreocarya flava* A. Nelson; *O. lutescens* Greene]. Sandy soils in desert scrub, sagebrush, and piñon-juniper communities. ARIZ: Apa, Coc, Nav; COLO: Arc, Dol, LPl, Mon; NMEX: McK, RAr, San, SJn; UTAH. 1280–2300 m (4200–7500′). Flowering: Apr–Jun. Fruit: May–Jul. Wyoming and Utah, south to Arizona and New Mexico. Similar in appearance to *C. confertiflora*, but *C. flava* is distinguished by its fuller inflorescence and solitary nutlets. Used for cancer treatment by the Hopi. The Kayenta Navajo used the plant as an eye medicine, gastrointestinal aid for inflammation, and gynecological aid for postpartum purification.

Cryptantha flavoculata (A. Nelson) Payson (yellow eye, referring to the yellow fornices centered in the white corolla limb) Plateau cryptantha. Caespitose perennial. **STEMS** few to several from a branching caudex, 0.5–2.5 dm tall, strigose and spreading-setose. **LEAVES** oblanceolate to spatulate, obtuse, rarely acute, coarsely strigose to sericeous-puberulent, lower surface conspicuously pustulate, pustules fewer or lacking on upper surface. **INFL** narrow-cylindric, rarely somewhat open and spreading, 0.5–3 dm long, cymules not much elongating in fruit. **FLOWERS** heterostylic; calyx segments lanceolate to ovate, 5–6 mm long at anthesis, 8–10 mm long in fruit; corolla tube white or pale yellow, 7–10 (12) mm long, surpassing the calyx at anthesis; fornices yellow, conspicuous, 1–2 mm long; corolla limb white, rotate, 8–12 mm wide. **NUTLETS** lanceolate to lance-ovate, usually all 4 maturing, 2.5–3.8 mm long, tuberculate-muricate with conspicuous ridges formed by coalescence of tuberculations; scar open, constricted near the middle and surrounded by highly elevated margins. [*Oreocarya flavoculata* A. Nelson]. Sagebrush and piñon-juniper communities. COLO: Mon, SMg; NMEX: SJn; UTAH. 1350–2100 m (4500–7000′). Flowering: Apr–Jun. Fruit: May–Jul. Wyoming and Colorado west to California and south to Arizona, barely reaching northwestern New Mexico. Many of the Colorado plants in our flora belong to an unnamed variation with atypical 10–12 mm long corolla tubes. Their corolla tubes resemble those of *C. longiflora*, but they are typical of *C. flavoculata* in all other respects.

Cryptantha fulvocanescens (S. Watson) Payson (tawny grayish) Plateau cryptantha. Caespitose perennial. **STEMS** few–several from the branches of a woody caudex, 0.5–3 dm tall, spreading-hispid above. **LEAVES** mostly basal, narrowly oblanceolate, acute or obtuse, uniformly silvery strigose, lower surfaces pustulate, cauline leaves usually narrower and spreading-hispid. **INFL** narrow-cylindric to somewhat open at maturity, rarely subcapitate, conspicuously tawny-setose or silvery strigose. **FLOWERS** heterostylic; pedicels 1–10 mm long at maturity; fruiting calyx 6–13 mm long, lobes linear, hispid to strigose; corolla tube white or pale yellow, 6–12 mm long and surpassing the calyx at anthesis; fornices usually yellow, conspicuous, 1–1.5 mm long; limb white, 6–9 mm wide, reflexed after anthesis. **NUTLETS** lance-ovate, 3.1–4.4 mm long, usually only 1 or sometimes 2 maturing per calyx, both surfaces muricate, often with sharp, setose tips terminating some or all of the murications; scar closed or only slightly open, lacking elevated margins. $n = 12$. Decoction of plants taken at childbirth by the Navajo. Used as a snakebite and toothache remedy by the Ramah Navajo.

1. Calyx densely hispid-strigose; interior calyx lobe faces strigulose, the green surface partly visible .... **var. *fulvocanescens***
1′ Calyx densely strigose and sparsely hispid, interior calyx lobe faces obscured by dense, silvery pubescence .... **var. *nitida***

var. *fulvocanescens* CP [*Oreocarya fulvocanescens* (A. Gray) Greene; *O. echinoides* (M. E. Jones) J. F. Macbr.; *C. echinoides* (M. E. Jones) Payson; *C. fulvocanescens* var. *echinoides* (M. E. Jones) L. C. Higgins]. Shale, clayey sand, or gypseous soils in saltbush, sagebrush, and piñon-juniper communities. ARIZ: Apa, Nav; COLO: LPl; NMEX: McK, RAr, San, SJn; UTAH. 1500–2300 m (4900–7500′). Flowering: Apr–Jun. Fruit: May–Jul. Utah south to Arizona and New Mexico, barely reaching southwestern Colorado.

var. *nitida* (Greene) Sivinski (shining, for its silvery pubescence) The inflorescence of this variety is more silvery strigose and less hispid than var. *fulvocanescens*. On average, it also has shorter pedicels, longer calyx, more flowers per cymule, and is less likely to have setose tips on the nutlet murications. [*Oreocarya nitida* Greene]. Alkaline sandstone or sandy gypsum in sagebrush, mixed desert scrub, and piñon-juniper communities. ARIZ: Apa, Nav: COLO: Dol,

Mon; NMEX: SJn; UTAH. 1600–2140 m (5200–7000′). Flowering: Apr–Jun. Fruit: May–Jul. A Colorado Plateau endemic in Colorado, Utah, and the Four Corners region of Arizona and New Mexico.

Cryptantha gracilis Osterh. (slender, presumably for the erect, slender growth form) Slender cryptantha. Annual. **STEM** usually solitary, but branching from the base and above when well developed, 1–2.5 dm tall, strigose and spreading-hirsute. **LEAVES** basal and scattered along the stem, linear or narrowly spatulate, hispid-hirsute, pustulate on lower surface. **INFL** of few to several naked, compact cymes terminating the branches, not much elongating at maturity. **FLOWERS** nearly sessile; fruiting calyx 2–3 mm long, densely white or tawny appressed hispid-villous, segments lanceolate; corolla white, inconspicuous, 1 mm wide; mature nutlet surpassing the style. **NUTLETS** lanceolate, 1.4–2 mm long, smooth and glossy, usually 1, rarely 2–3, maturing per calyx; scar closed except for small basal areola. [*C. hillmanii* A. Nelson & Kenn.; *C. gracilis* var. *hillmanii* (A. Nelson & Kenn.) Munz & I. M. Johnst.]. Sandstone ledges or sandy soils in piñon-juniper and sagebrush communities. ARIZ: Apa, Nav; COLO: Arc, Dol, Mon, LPl; NMEX: RAr, SJn; UTAH. 1100–2300 m (3600–7550′). Flowering: Apr–Jun. Fruit: May–Jul. Colorado and New Mexico, west to California and Oregon.

Cryptantha humilis (Greene) Payson (low-growing, referring to the caespitose habit) Round-spike cryptantha. Caespitose perennial. **STEMS** several, from a branching caudex, 0.5–2 dm tall, finely strigose and spreading-setose. **LEAVES** mostly basal, oblanceolate to spatulate, strigose-puberulent and appressed-setose; cauline leaves similar, but fewer and smaller, usually setose-ciliate. **INFL** narrowly cylindrical, opening out somewhat at maturity. **FLOWERS** homostylic; calyx 3–5 mm in anthesis, 5–9 mm long in fruit; corolla tube white or pale yellow, equal or subequal with the apex of the calyx; fornices yellow, low and rounded, 0.5 mm long; corolla limb white, 6–10 mm wide. **NUTLETS** lance-ovate, 1–4 maturing, both surfaces muricate-rugulose; scar open and lacking elevated margins. [*Oreocarya humilis* Greene; *O. nana* Eastw.; *C. nana* (Eastw). Payson; *C. humilis* var. *nana* (Eastw.) L. C. Higgins]. Shaley soils in desert scrub. COLO: Mon. Up to about 1800 m (6000′). Flowering: Apr–Jun. Fruit: May–July. Idaho and Oregon, south to California, Utah, and Colorado.

Cryptantha inaequata I. M. Johnst. (uneven, referring to the larger odd nutlet in each fruit) Unequal cryptantha. Annual. **STEMS** several from the base, also branching above when well developed, strigose and hispid, 2–4 dm tall. **LEAVES** mostly cauline, linear or narrowly lanceolate, hispid-pustulate, 1–3 cm long. **INFL** terminating the branches, cymes usually naked or 1–few leafy bracts scattered throughout, especially near the base, elongating at maturity. **FLOWERS** sessile or pedicel less than 1 mm; calyx 2.5–3 mm long in fruit, lobes lanceolate, villous on the margins, midribs thickened and hispid; corolla inconspicuous, 1–2.5 mm wide. **NUTLETS** heteromorphic, usually 4 maturing, triangular-ovate with small, pale tuberculations, margins sharply angled to a knifelike edge, scar closed above and enlarging to a small triangular areola near the base; odd nutlet larger and more firmly attached to the gynobase, about 1.7 mm long; consimilar nutlets deciduous, about 1.3 mm long. Desert scrub and blackbrush communities. UTAH. Up to 1200 m (4000′). Flowering: Apr–May. Fruit: May–Jun. Arizona, Utah, and Nevada.

Cryptantha kelseyana Greene (for Francis D. Kelsey, Montana botanist) Kelsey's cryptantha. Annual. **STEMS** branched from the base and above, 0.5–2.5 dm long, spreading or erect without a strong central axis, spreading-hirsute. **LEAVES** linear to narrowly oblanceolate, finely strigose and stiffly hispid, pustulate. **INFL** terminating the branches, cymes naked or rarely 1–2 leafy bracts at the base, elongating at maturity. **FLOWERS** nearly sessile; fruiting calyx 5–7 mm long, segments narrowly lanceolate, hirsute and spreading pungent-hispid, midribs somewhat thickened at maturity; corolla white, limb 1–2 mm wide. **NUTLETS** lance-ovate, usually 4 maturing, heteromorphic, 1 evidently larger, 2–2.6 mm long and more firmly attached to the gynobase, minutely granulate and appearing dull and nearly smooth; the 1–3 consimilar nutlets readily deciduous, 1.2–1.8 mm long and granulate-tuberculate; scar narrow and shallowly impressed, not excavated. Sandy or gravelly soils in sagebrush and piñon-juniper communities. UTAH. Up to about 2300 m (7500′). Flowering: May–Jul. Fruit: Jun–Aug. Saskatchewan to Idaho, south to Colorado and Utah. Barely entering our flora, where often confused with *C. crassisepala*, which is distinguished by thicker fruiting sepals, excavated nutlet scars, and, usually, minutely muricate odd nutlets.

Cryptantha longiflora (A. Nelson) Payson (long flowers, referring to the long corolla tubes) Long-flower cryptantha. Short-lived perennial or biennial. **STEMS** 1 to few from a simple or branching caudex, 0.5–2.5 dm tall, strigose and spreading-setose. **LEAVES** oblanceolate to spatulate, obtuse, rarely acute, both surfaces strigose and pustulate-hirsute. **INFL** broad and open, 0.7–3 dm long, cymules often elongating in fruit. **FLOWERS** heterostylic; calyx segments linear-lanceolate, 7–10 mm long at anthesis, 10–16 mm long in fruit; corolla tube white or pale yellow, 12–17 mm long, surpassing the calyx at anthesis; fornices yellow, conspicuous, 0.5–1 mm long; corolla limb white, rotate, 9–11 mm wide. **NUTLETS** lanceolate to lance-ovate, 2–4 maturing, 3–4 mm long, tuberculate with low ridges formed by coalescence of tuberculations; scar closed to narrowly open, lacking conspicuously elevated margins. [*Oreocarya longiflora* A. Nelson]. Desert scrub, sagebrush, and piñon-juniper communities. UTAH. 1450–1800 m (4500–5900′). Flowering: Apr–Jun. Fruit: May–Jul. Endemic in Colorado and Utah.

Cryptantha micrantha (Torr.) I. M. Johnst. (small flowers, referring to the inconspicuous corollas) Redroot cryptantha. Annual. **STEMS** slender and wiry from a red taproot, branching at base and diffusely above, 3–15 cm tall, strigose. **LEAVES** mostly cauline, small, linear, strigose to ciliate-hirsute near the base. **INFL** of many small, biseriate, bracteate cymes terminating the branches. **FLOWERS** nearly sessile; fruiting calyx 1.5–2.5 mm long, and tardily deciduous, segments lanceolate, hirsute; corolla white, inconspicuous, less than 1 mm wide; gynobase capped by sessile stigma and evidently surpassing the nutlets. **NUTLETS** narrowly lanceolate, smooth or finely tuberculate, usually 4 maturing, all alike or somewhat heteromorphic with an odd nutlet that is slightly larger and more firmly attached to the gynobase; scar narrowly open or closed and slightly dilated at the base. $2n = 24$. [*Eritrichium micranthum* Torr.; *Krynitzkia micrantha* (Torr.) A. Gray; *Eremocarya micrantha* (Torr.) Greene]. Desert scrub and blackbrush communities. ARIZ: Apa, Nav; NMEX: SJn; UTAH. Up to about 1550 m (5100′). Flowering: Mar–May. Fruit: Apr–Jun. Oregon, California, and Baja California, east to New Mexico and Texas.

Cryptantha nevadensis A. Nelson & P. B. Kenn. (of Nevada) Nevada cryptantha. Annual. **STEMS** branching from base and above, 1–4 dm tall, erect or lax, branches often curved or flexuous and spreading, strigose, spreading hairs few or lacking. **LEAVES** linear or nearly so, strigose to sparsely spreading-hispid, pustulate. **INFL** terminating the branches with naked cymes, usually paired, sometimes ternate or single. **FLOWERS** nearly sessile; fruiting calyx 6–10 mm long, segments lance-linear, subequal with recurved tips, white ciliate-margined, midrib pungent-hispid; corolla white, inconspicuous, less than 2 mm wide. **NUTLETS** homomorphic, 1–4 matured per calyx, lanceolate, 2–2.4 mm long, densely verrucose-muricate; scar closed above, gradually dilated toward the basal triangular areola. [*C. leptophylla* Rydb.]. Sandy or gravelly soils in desert scrub or blackbrush communities. UTAH. Up to 1550 m (5100′). Flowering: Mar–May. Fruit: Apr–Jun. California, Nevada, and Utah, south to New Mexico, Arizona, and Baja California.

Cryptantha osterhoutii (Payson) Payson (for George Osterhout, Colorado and Utah botanist) Osterhout's cryptantha. Caespitose perennial. **STEMS** several from a branching caudex, slender or lax, 0.5–1.5 dm tall, spreading-setose and finely strigose. **LEAVES** mostly basal, oblanceolate or spatulate, obtuse, strigose and appressed-setose, pustulate on the lower surface, cauline leaves similar, but fewer and less crowded. **INFL** capitate or shortly interrupted-cylindric, 3–8 cm long, mature cymules not much elongating. **FLOWERS** homostylic; calyx 2.5–4 mm long at anthesis, 6–9 mm long in fruit; corolla tube white or pale yellow, equal or subequal to the apex of the calyx, fornices yellow, about 0.5 mm long; corolla limb rotate, 5–8 mm wide. **NUTLETS** lance-ovate, 1–4 maturing per calyx, 2.7–3.2 mm long, rugose and tuberculate on both surfaces, carinate on dorsal surface; scar open, constricted above the base, margins slightly elevated. [*Oreocarya osterhoutii* Payson]. Sandstone and sandy shale in blackbrush and mixed desert scrub communities from 1520–2000 m (5000–6600′). ARIZ: Apa, Nav; UTAH. Flowering: May–Jun. Fruit: Jun–Jul. Endemic in Arizona, Colorado, and Utah.

Cryptantha paradoxa (A. Nelson) Payson **CP** (for the Paradox Valley in western Colorado) Paradox cryptantha. Caespitose perennial. **STEMS** several from a branching caudex, 0.5–1.5 dm tall, spreading-hirsute. **LEAVES** mostly basal with fine strigose pubescence and appressed pustulate bristles on the lower surface, cauline leaves reduced, more loosely villous. **INFL** subcapitate, 1–4 cm long. **FLOWERS** heterostylic; calyx 5–6 mm long in anthesis, 6–8 mm long in fruit; corolla tube yellow, 9–12 mm long, surpassing the apex of the calyx; fornices yellow, about 0.5 mm long; corolla limb white, rotate, 9–12 mm wide. **NUTLETS** lance-ovate, usually 4 maturing, 2–2.8 mm long, rugose-tuberculate; scar narrowly open, constricted below the middle, margins elevated. [*Oreocarya paradoxa* A. Nelson]. Gypseous or alkaline soils in saltbush, sagebrush, or piñon-juniper communities. COLO: Mon; NMEX: SJn; UTAH. 1400–2000 m (4600–6600′). Flowering: May–Jun. Fruit: Jun–Jul. Colorado Plateau endemic in Colorado and Utah, barely reaching New Mexico.

Cryptantha pterocarya (Torr.) Greene (wing nut, referring to the winged nutlets) Wing-nut cryptantha. Annual. **STEMS** freely branching from the base and above when well developed, erect to spreading, 1–4 dm tall, strigose. **LEAVES** scattered, linear or nearly so, hirsute, pustulate. **INFL** terminating branches, cymes naked, paired or rarely solitary or ternate, elongating in fruit. **FLOWERS** nearly sessile; fruiting calyx 4–5 mm long, segments lanceolate-ovate, strigose and often sparsely hispid; corolla white, inconspicuous, 0.5–2 mm wide. **NUTLETS** all alike or heteromorphic, usually

4 maturing, 2.2–3.2 mm long, all, or only 3, conspicuously wing-margined with thin, entire to erose wings, body surface muricate-tuberculate, scar narrowly open above and dilated below to an excavated areola.

1. Nutlets alike, all 4 winged .... **var. *cycloptera***
1′ Nutlets heteromorphic, consisting of 3 winged consimilar nutlets and 1 wingless, lanceolate nutlet... **var. *pterocarya***

var. *cycloptera* (Greene) J. F. Macbr. (circular wing, for the fact that all 4 nutlets in each fruit are winged) [*C. cycloptera* Greene; *Krynitzkia cycloptera* Greene]. Sandy soils in desert scrub, sagebrush, and piñon-juniper communities. ARIZ: Apa; NMEX: SJn; UTAH. Up to 1650 m (5400′). Flowering: Apr–Jun. Fruit: May–Jul. California and Nevada, east to Texas and south to Mexico.

var. *pterocarya* The winged nutlets are deciduous, while the wingless nutlet is more firmly attached to the gynobase and dispersed with the calyx. [*Eritrichium pterocaryum* Torr.; *Krynitzkia pterocarya* (Torr.) A. Gray]. Sandy soils in desert scrub, sagebrush, and piñon-juniper communities. ARIZ: Apa, Coc, Nav; COLO: Mon, LPl, SMg; NMEX: SJn; UTAH. Up to 2040 m (6700′). Flowering: Apr–Jun. Fruit: May–Jul. Idaho and Washington, south to California, Arizona, and New Mexico.

Cryptantha racemosa Greene (racemose, referring to pedicellate flowers on racemelike branchlets) Bushy cryptantha. Long-lived annual. **STEMS** diffusely slender-branched, becoming woody near the base where the epidermis exfoliates, 1–10 dm tall, strigose and sometimes spreading-hirsute. **LEAVES** linear or narrowly oblanceolate, hirsute, pustulate, early-deciduous ones 3–6 cm long, later more numerous ones 1–4 cm long. **INFL** of short, paniculately or cymosely disposed racemes, irregularly bracteate, not much elongating in fruit. **FLOWER** pedicels 1–4 mm long; fruiting calyx 2–4 mm long, corolla white, inconspicuous, about 1 mm wide. **NUTLETS** heteromorphic, usually 4 maturing, triangular-ovate, margins sharply angled to a knifelike edge, muriculate or pale-tuberculate; attachment scar gradually broadening into a basal, triangular areola; odd nutlet 1–2 mm long and only slightly longer than the 3 consimilar nutlets. [*Eritrichium racemosum* S. Watson ex A. Gray]. Rocky or gravelly slopes or canyons in desert scrub and blackbrush communities. UTAH. Up to 1100 m (3700′). Flowering: Apr–Oct. Fruit: May–Nov. California and Baja California, east to Arizona, Nevada, and Utah.

Cryptantha recurvata Coville (bent over and downward, referring to the recurved nutlets) Bent-nut cryptantha. Annual. **STEMS** freely branching from the base and above when well developed, lax and spreading or ascending, 1–4 dm long, strigose. **LEAVES** scattered, linear to lance-oblong, appressed-hispid and minutely pustulate. **INFL** terminating branches, cymes naked, usually paired, elongating at maturity. **FLOWERS** sessile; fruiting calyx 2.5–3.5 mm long, asymmetrical, bent and recurved, strigose and usually sparsely spreading-hispid; corolla white, inconspicuous, about 1 mm wide. **NUTLET** lanceolate, only 1 matured per calyx, somewhat recurved-bent in alignment with the recurved calyx, finely granulate-muricate, scar closed or narrowly open. Sandy or shaley soils in desert scrub, sagebrush, and piñon-juniper communities. ARIZ: Apa; COLO: Mon; NMEX: SJn; UTAH. Up to 1920 m (6300′). Flowering: Apr–Jun. Fruit: May–Jul. Oregon and California, east to Colorado and New Mexico.

Cryptantha setosissima (A. Gray) Payson (bristly-hairy, referring to the very setose indument) Bristly cryptantha. Biennial. **STEM** simple and solitary from a stout taproot, 4–10 dm tall, finely puberulent and coarsely spreading-setose with 2–4 mm bristles. **LEAVES** basal and cauline, oblanceolate, villous-tomentose with pustulate subappressed setae, gradually reduced above. **INFL** densely setose, narrow-cylindric and interrupted below when in flower, open in fruit with cymules elongating at maturity. **FLOWERS** homostylic; calyx 4–6 mm at anthesis, 6–13 mm long in fruit, segments broadly lanceolate, hispid; corolla tube equal with calyx apex; corolla limb rotate, white, 7–10 mm wide; fornices yellow. **NUTLETS** usually 4 maturing, 4.5–6 mm long, margins decidedly papery-winged, dorsal body surface finely muricate; scar narrowly open. [*Eritrichium setosissimum* A. Gray; *Oreocarya setosissima* (A. Gray) Greene]. Sedimentary and igneous soils in oak brush, ponderosa pine, and mixed conifer communities. ARIZ: Apa, Nav; NMEX: McK, SJn; UTAH. 2000–2700 m (6600–8850′). Flowering: Jul–Aug. Fruit: Aug–Sep. Arizona, New Mexico, Utah, and Nevada.

Cryptantha tenuis (Eastw.) Payson (thin, referring to the relatively slender stems) Canyon cryptantha. Caespitose perennial. **STEMS** slender, 1 to several, 1–3 dm tall from a branching, somewhat woody caudex, strigose. **LEAVES** mostly basal, fewer cauline, narrowly oblanceolate to linear-spatulate, strigose and weakly spreading-setose, lower surface pustulate. **INFL** narrow, interrupted, 6–14 cm long; cymule not much elongated at maturity. **FLOWERS** heterostylic; calyx 5–6 mm long in anthesis, 7–9 mm long in fruit, segments linear-lanceolate; corolla white, funnelform, tube barely surpassing the calyx; corolla limb campanulate, never rotate, 5–8 mm wide; fornices pale yellow, broad,

0.5 mm long. **NUTLETS** lanceolate, usually all 4 maturing, 3–4 mm long, dorsal surface carinate, both surfaces deeply rugose-tuberculate; scar open, constricted above the base, and with elevated margins. [*Oreocarya tenuis* Eastw.]. Sandstone and sandy clay soils in desert scrub, blackbrush, sagebrush, and piñon-juniper communities. UTAH. 1300–2000 m (4300–6600′). Flowering: May–Jun. Fruit: Jun–Jul. A Colorado Plateau endemic of Utah.

Cynoglossum L. Hound's Tongue

(Greek *kynos*, dog, + *glossa*, tongue, possibly for the shape of the leaves) Taprooted, robust biennial or perennial herbs. **LEAVES** simple, alternate, basal ones long-petioled. **INFL** of several false racemes that elongate at maturity; racemes naked or with 1 or more leafy bracts at the base. **FLOWERS** actinomorphic, calyx cleft to below the middle; corolla funnelform or salverform, tube short, limb broad and spreading, fornices well developed and often exserted; stamens 5, filaments inserted near the corolla throat; stigma capitate. **NUTLETS** widely spreading at maturity and attached to the gynobase by the apical end, all surfaces covered with short, stout, glochidiate prickles; scar broad, not extending below the middle. A cosmopolitan genus of about 80 species.

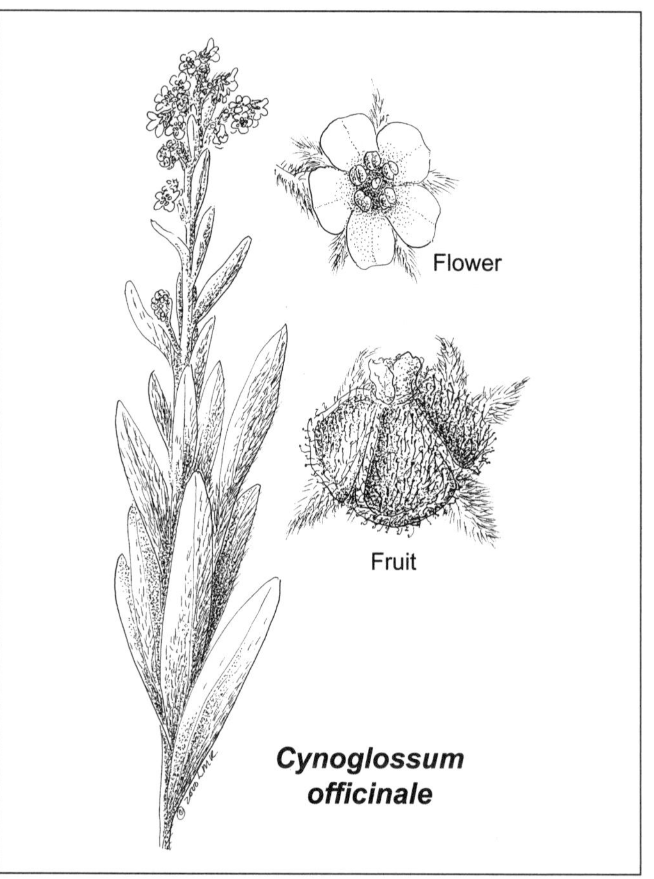

Cynoglossum officinale

Cynoglossum officinale L. (used in medicine) Hound's tongue. Coarse biennial. **STEM** solitary, stout, leafy to the top, 4–10 dm tall. **LEAVES** villous-tomentose; basal leaves long-petioled, oblong to oblanceolate, 1–3 dm long; cauline leaves gradually reduced above, lanceolate, sessile to somewhat clasping. **INFL** of several false racemes in upper axils or terminating short branches. **FLOWERS** with mature pedicels curved-spreading; sepals broad, obtuse, 5–8 mm in fruit; corolla reddish purple, tube 3–5 mm long, limb 6–8 mm wide. **NUTLETS** ovate, 5–7 mm long, descending-spreading from an apical attachment to the gynobase, upper surface flattened and margined. $2n$ = 24, 48. Valley bottoms and disturbed areas in piñon-juniper, oak brush, mountain riparian, ponderosa pine, and spruce-fir communities. COLO: Arc, Dol, LPl, Mon, SJn; NMEX: RAr, SJn. 1700–3000 m (5600–9800′). Flowering: May–Jul. Fruit: Jul–Aug. Introduced from Europe.

Eritrichium Schrad. ex Gaudin Alpine Forget-me-not

(Greek *erion*, wool, + *trichos*, hair, for the villous indument of the herbage) Dwarf, pulvinate-caespitose perennials. **LEAVES** small, densely crowded on the branch ends of a branching caudex and imbricate on short stems. **INFL** a racemelike cluster terminating the stems. **FLOWERS** actinomorphic, calyx cleft to near the base; corolla tube short; fornices conspicuous; anthers included; stigma entire. **NUTLETS** smooth, pubescent or glabrous, apex obliquely truncate with entire or toothed margins, ventral surface carinate; basilaterally attached to a broad, low gynobase; attachment scar small and less than 1/2 the width of the nutlet. A circumboreal genus of about four species in Eurasia and North America.

Eritrichium nanum (Vill.) Schrad. ex Gaudin **CP** (dwarf, referring to the low, pulvinate habit) Alpine forget-me-not. Cushionlike, long-lived perennial. **STEMS** short and slender, up to 1 dm tall, or acaulescent. **LEAVES** silvery villous, narrowly ovate to oblong, 5–10 mm long. **FLOWERS** with blue, rarely white, corollas; corolla tube equal to the calyx; corolla limb rotate, 4–8 mm wide. **NUTLETS** 1–4 maturing, margins entire or with a few short teeth, glabrous. $2n$ = 24. [*Myosotis nana* Vill.]. Open, rocky places in spruce-fir forest and alpine tundra. COLO: Con, Min. 3350–4270 m (11000–14000′). Flowering: Jul–Aug. Fruit: Aug–Sep. Europe and Asia to Alaska, south to New Mexico. The plants in our flora belong to the more densely hairy Rocky Mountain **var. *elongatum*** (Rydb.) Cronquist (elongated).

Hackelia Opiz Stickseed

(for Joseph Hackel, Czech botanist) Coarse biennial or perennial herbs. **LEAVES** alternate, basal and cauline. **INFL** a panicle of racemelike sympodial cymes terminating the stem and branch ends. **FLOWERS** actinomorphic; pedicel deflexed or spreading in fruit; calyx cleft to the base; corolla blue, pink, or white, tube short; fornices conspicuous; anthers included; stigma capitate. **NUTLETS** erect on a pyramidal gynobase, margins armed with glochidiate prickles, prickles sometimes extending into the intramarginal area of the dorsal surface (Gentry and Carr 1976). A genus of about 45 species in North America, South America, and Eurasia.

1. Cymes naked; leaves not pustulate .... ***H. floribunda***
1′ Cymes bracteate with small, 1–6 mm leaflike bracts subtending most flowers; leaves pustulate .... (2)

2 Corolla conspicuous, 5–8 mm wide; dorsal nutlet surface with 1 or more short intramarginal prickles ... ***H. gracilenta***
2′ Corolla inconspicuous, 1.5–3 mm wide; dorsal nutlet surface lacking intramarginal prickles .... ***H. besseyi***

Hackelia besseyi (Rydb.) J. L. Gentry (for Charles E. Bessey, midwestern U.S. botanist) Bessey's stickseed. Coarse biennial herb. **STEMS** 1–few, usually solitary, erect, strigose or villous-hirsute, 3–10 dm tall. **LEAVES** spreading-hirsute, pustulate; basal leaves oblanceolate, 2–10 cm long; cauline leaves gradually reduced above, lanceolate to ovate-acute. **INFL** of paniculately disposed cymes; cymes bracteate, at least in the lower 1/2. **FLOWERS** small; corolla blue, limb 1.5–3 mm wide, tube not exceeding the calyx. **NUTLETS** 2.5–3.5 mm long, marginal prickles distinct or somewhat confluent near the base, dorsal surface muriculate-hispidulous, intramarginal prickles absent; usually 4 maturing. [*Lappula besseyi* Rydb.; *H. grisea* (Wooton & Standl.) I. M. Johnst.]. Piñon-juniper, oak brush, ponderosa pine, and pine-fir communities. Presently known in our flora from two collections, COLO: SJn; NMEX: McK. 2100–2750 m (7000–9000′). Flowering: Jul–Sep. Fruit: Aug–Oct. Colorado, New Mexico, and Texas.

Hackelia floribunda (Lehm.) I. M. Johnst. (many flowers) Many-flowered stickseed. Robust biennial or short-lived perennial. **STEMS** 1–few, often solitary, erect, strigose, 4–12 dm tall. **LEAVES** strigose or appressed-hirsute; basal leaves petiolate, up to 15 cm long, oblanceolate with some spreading hairs; upper cauline leaves strigose, sessile, lanceolate or linear-oblong. **INFL** branches short and strict; cymes naked, or with 1 small bract at the base, densely flowered. **FLOWERS** with blue or white corollas, tube not exceeding the calyx, limb 3–7 mm wide. **NUTLETS** 3–4 mm long, marginal prickles distinct or somewhat confluent near the base, dorsal surface muriculate-hispidulous, intramarginal prickles absent; usually 4 maturing. [*Echinospermum floribundum* Lehm.; *Lappula floribunda* (Lehm.) Greene]. Valley bottoms and forest openings in sagebrush, ponderosa pine, aspen, and spruce-fir communities. ARIZ: Apa; COLO: Arc, Hin, LPl, Min, RGr, SMg; NMEX: RAr, SJn; UTAH. 1800–3300 m (5900–10800′). Flowering: Jun–Aug. Fruit: Jul–Sep. British Columbia and Saskatchewan, south to California, Arizona, and New Mexico. The Ramah Navajo used this plant for serious injury such as fracture.

Hackelia gracilenta

Hackelia gracilenta (Eastw.) I. M. Johnst. (fully slender, referring to the relatively slender stems and elongate inflorescence branches when in fruit) Mesa Verde stickseed. Coarse biennial or short-lived perennial. **STEMS** few–several, erect or spreading, branching from the base and above when well developed, spreading-hirsute, 2.5–7.5 dm tall. **LEAVES** spreading-hirsute, pustulate; basal leaves petiolate, up to 7 cm long, ovate to oblong-spatulate; cauline leaves sessile, lanceolate or linear-oblong. **INFL** of paniculately disposed cymes on the stems, cymes bracteate. **FLOWERS** with blue or very pale blue corollas, tube not exceeding the calyx, limb 5–8 mm wide. **NUTLETS** 4–5 mm long, marginal prickles distinct or somewhat confluent near the base, dorsal surface muriculate-hispidulous, intramarginal prickles few and short; usually 4 maturing. [*Lappula gracilenta* Eastw.]. Sandy soils in oak brush and piñon-juniper communities. COLO: Mon. 2100–2400 m (7000–8000′). Flowering: May–Jun. Fruit: Jun–Jul. Endemic to Mesa Verde National Park and immediate vicinity, in southwestern Colorado.

Heliotropium L. Heliotrope

(Greek name for some of the species, from *helios*, sun, + *tropos*, turn) Annual or perennial herbs in our flora. **LEAVES** alternate, mostly cauline. **INFL** cymose to scorpioid, terminating the branches or from the axils. **FLOWERS** with white corolla limbs in our species; anthers included; ovary entire or shallowly 2- or 4-grooved; style present or absent, attached to the ovary apex; stigma conic or tufted on a broad, swollen or discoid base. **NUTLETS** usually 4, but sometimes failing to break apart and producing 2 coherent, 2-seeded nutlets, glabrous or pubescent. A large and widespread genus of more than 200 species, mostly in tropical and warm-arid regions.

1. Plants hairy, not succulent; stigma a swollen ring with a tuft of hairs .... ***H. convolvulaceum***
1′ Plants glabrous, succulent; stigma conic on a peltate disc .... ***H. curassavicum***

Heliotropium convolvulaceum (Nutt.) A. Gray **CP** (resembling *Convolvulus*, a genus of morning glory) Showy heliotrope. Annual. **STEMS** freely branching from near the base and above, well developed, 1–3 dm tall, strigose. **LEAVES** mostly cauline, petiolate, lanceolate to ovate, 10–25 mm long, coarsely strigose, pustulate. **INFL** extra-axillary, appearing at the leaf nodes of the elongating branches. **FLOWERS** short-pedicellate; calyx 4–6 mm long at anthesis, 6–9 mm in fruit, lobes linear and unequal; corolla white, yellow at the throat, fragrant, tube strigose and well surpassing the calyx, swollen at the summit where the anthers are inserted; corolla limb widely funnelform with 5 points marking the petal tips, 12–22 mm wide; style slender, 3–5 mm long; stigma an expanded ring of receptive area topped by a short, hispid cone. **NUTLETS** hairy, 2–2.5 mm long. $2n = 42$. [*Euploca convolvulacea* Nutt.]. Sand dunes and sandy arroyos in desert scrub. ARIZ: Apa, Coc, Nav; UTAH. Up to 1950 m (6400′). Flowering: May–Sep. Fruit: Jun–Sep. California and Nevada, east to Kansas and Texas, south to Chihuahua. The seeds were made into a mush and used for food by the Kayenta Navajo.

Heliotropium curassavicum L. (for the Caribbean island of Curaçao) Seaside heliotrope. Fleshy-succulent annual or short-lived perennial. **STEMS** branched from the base and above, prostrate to ascending, 1–6 dm long, glabrous. **LEAVES** all cauline, glabrous, glaucous, oblanceolate or spatulate-obovate, up to 8 cm long and 18 mm wide. **INFL** cymes naked, scorpioid, usually 2–5 at the branch ends. **FLOWERS** nearly sessile; calyx segments ovate-lanceolate, 2–3 mm long; corolla white with a purplish or yellowish throat; style lacking; stigma sessile on the ovary apex, peltate-discoid with a short, conic top. **NUTLETS** 2–3 mm long, glabrous. $2n = 26$.

1. Corolla limb 5–10 mm wide, white or at most purplish-tinged or yellow at the throat; leaves spatulate or obovate, 10–18 mm wide .... **var. *obovatum***
1′ Corolla limb 3–5 mm wide, white with an initially yellow throat becoming purple with age; leaves linear-oblanceolate or spatulate, usually less than 10 mm wide .... **var. *oculatum***

var. *obovatum* DC. (obovate, for the leaf shape) Saline soils in low-lying places that are wet or seasonally wet in saltgrass, rush, and greasewood communities. Presently known in our flora from a single collection at Black Lake in the Chuska Mountains. NMEX: SJn. Up to 2250 m (7500′). Flowering: May–Oct. Fruit: Jun–Oct. Saskatchewan to Washington, south to Arizona and Chihuahua.

var. *oculatum* (A. Heller) I. M. Johnst. ex Tidestr. (eye, referring to the purple throat in a white corolla limb) [*H. oculatum* A. Heller]. Saline soils in low-lying places that are wet or seasonally wet in saltgrass, saltbush, and greasewood communities. ARIZ: Apa; NMEX: McK, SJn; UTAH. Up to 1800 m (6000′). Flowering: May–Oct. Fruit: Jun–Oct. Nevada, California, and Baja California, east to Utah, New Mexico, and Texas.

Lappula Moench Stickseed, Sheepbur

(Latin diminutive of Greek *lappa*, a bur) Taprooted annuals. **LEAVES** alternate, basal and cauline. **INFL** of bracteate cymes terminating the branches. **FLOWERS** on ascending-erect pedicels; calyx cleft to near the base; corolla actinomorphic, small and inconspicuous, funnelform with definite fornices at the throat; stamens included; stigma capitate. **NUTLETS** armed with glochidiate prickles on the margins, attached to the gynobase along the length of the ventral keel, usually 4 maturing. A genus of about a dozen species, mostly in the Northern Hemisphere.

1. Marginal prickles of the nutlet in a single row around the nutlet, sometimes confluent and swollen at the base .... ***L. occidentalis***
1′ Marginal prickles of the nutlet in 2 or 3 rows, not confluent at the base .... ***L. squarrosa***

Lappula occidentalis (S. Watson) Greene (western) Spiny sheepbur. Annual. **STEMS** 1–several, branching from the base and above, 1.5–4 dm long, erect or spreading, hirsute to strigose. **LEAVES** pustulate-hirsute; basal leaves oblanceolate to spatulate, 1–3 cm long; cauline leaves linear-oblong or narrowly oblanceolate, 1–4 cm long, gradually reduced above. **INFL** of solitary or geminate, bracteate cymes terminating the stems and branches, greatly elongating in fruit. **FLOWER** pedicels 1–3 mm long, ascending to erect in fruit; corolla blue or white, limb 1–2.5 mm wide. **NUTLETS** lanceolate with a single row of glochidiate prickles on the margins, 2–2.5 mm long, dorsal surface tuberculate. The very similar *L. redowskii* (Hornem.) Greene is treated here as an entirely Old World species that does not occur in North America. Often considered a nuisance because the prickly nutlets tenaciously adhere to clothing and sheep's wool.

1. Marginal prickles of most nutlets distinctly confluent into a swollen base to form a cupulate margin; stems usually several, ascending from the base .... **var. *cupulata***
1′ Marginal prickles of the nutlets distinct to the base or slightly confluent, but not swollen; stems solitary or few, usually erect .... **var. *occidentalis***

var. *cupulata* (Gray) Higgins (like a cup, referring to the connate-swollen marginal prickles on the nutlets) [*L. redowskii* (Hornem.) Greene var. *cupulata* (A. Gray) M. E. Jones; *L. redowskii* var. *texana* (Scheele) Brand]. Saltbush, sagebrush, and piñon-juniper communities. ARIZ: Apa; COLO: LPl, Mon; NMEX: McK, RAr, San, SJn; UTAH. 1400–1700 m (4600–6400′). Flowering: May–Jun. Fruit: May–Jul. Western North America to Mexico. The Navajo used parts of the plant for nosebleeds. The Kayenta Navajo used it as a lotion for itching.

var. *occidentalis* (of the west) [*L. redowskii* (Hornem.) Greene var. *occidentalis* Rydb.]. Desert scrub, sagebrush, piñon-juniper, and ponderosa pine communities. ARIZ: Apa, Coc, Nav; COLO: Arc, Dol, LPl, Mon, SMg; NMEX: McK, RAr, San, SJn; UTAH. Up to 2450 m (8000′). Flowering: Apr–Jun. Fruit: May–Jul. Western North America to Mexico. The Kayenta Navajo used the plant as a poultice on sores caused by insects. The Ramah Navajo used it as a lotion for sores or swellings and as fodder for sheep feed.

Lappula squarrosa (Retz.) Dumort. (rough with scales, referring to the multiple rows of marginal prickles on the nutlets) Bristly sheepbur. Annual. **STEMS** 1–few, branching from the base and above, erect, 2–8 dm tall, villous-hirsute and strigose. **LEAVES** villous-hirsute; basal leaves linear-lanceolate or oblong, 2–5 cm long; cauline leaves narrowly oblong, sessile, gradually reduced above. **INFL** of geminate, bracteate cymes terminating the stems and branches, elongating in fruit. **FLOWER** pedicels 1–4 mm long, ascending-erect in fruit; corolla blue, limb 2–4 mm wide. **NUTLETS** lanceolate with at least 2 rows of marginal prickles; prickles distinct to the base and sometimes scattered across the ventral surface of the nutlets; dorsal surface muricate or verrucose. $2n = 48$. [*L. echinata* Gilib.]. Piñon-juniper, oak brush, ponderosa pine, aspen, and mixed conifer communities. COLO: Arc, SMg. 2100–2600 m (6800–8500′). Flowering: Jun–Aug. Fruit: Jun–Sep. Introduced from Eurasia.

Lithospermum L. Stoneseed, Gromwell

(Greek *lithos*, stone, + *sperma*, seed, referring to the usually gray, bony nutlets) Annual or perennial herbs. **LEAVES** alternate, usually cauline, sometimes also a basal tuft. **INFL** of cymes, or solitary flowers in the leaf axils. **FLOWERS** pedicellate, calyx cleft to near the base, corolla actinomorphic, funnelform or salverform, fornices evident or obscure; anthers included; style filiform, homostylic or heterostylic; stigma geminate. **NUTLETS** 4 or fewer by random abortion, erect, ovoid or angular, smooth, pitted or verrucose, attachment scar basal and broad. (Johnston 1952.) A genus of about 60 species in North America, Eurasia, and Africa.

1. Corolla tubes 15–35 mm long on chasmogamous spring flowers; corolla limb margins erose; summer flowers much smaller, usually cleistogamous .... ***L. incisum***
1′ Corolla tube 4–13 mm long; corolla limb margins entire; flowers all chasmogamous .... (2)

2. Corolla bright yellow or yellow-orange; corolla tube strigose-pubescent on exterior surface; flowers heterostylic .... ***L. multiflorum***
2′ Corolla pale yellow, often tinged with green; corolla tube glabrous or sparsely pubescent; flowers homostylic .... ***L. ruderale***

Lithospermum incisum Lehm. (deeply cut, referring to the erose margins of the petals) Plains stoneseed. Perennial from a woody taproot. **STEMS** 1–several, erect or ascending, branching from the base and above, 0.5–3 dm long. **LEAVES** strigose, linear-lanceolate to linear, 1–5 cm long, crowded in the inflorescence, basal leaves deciduous before flowering. **INFL** of leafy racemes with the flowers in the leaf axils. **FLOWERS** pedicellate; calyx 6–10 mm long; corolla salverform, yellow; tube 15–35 mm long, sparsely strigose or glabrous outside; limb 12–25 mm wide, petal margins erose to nearly entire; fornices small but evident; long-styled, chasmogamous spring flowers often infertile; smaller, cleistogamous, short-styled summer flowers are highly fertile and occur further down the stem on recurved pedicels. **NUTLETS** 3–3.5 mm long, white or gray, glossy, smooth or slightly pitted, ventral keel evident, attachment scar basal and surrounded by a prominent collar. $2n$ = 24. Desert scrub, sagebrush, piñon-juniper, mountain brush, and ponderosa pine communities. ARIZ: Apa, Nav; COLO: Arc, Hin, LPl, Mon; NMEX: McK, RAr, San, SJn; UTAH. 1370–2400 m (4500–8000′). Flowering: Apr–Jul. Fruit: Jun–Sep. Central United States, Canada, and Mexico, west to Nevada.

Lithospermum multiflorum Torr. ex A. Gray **CP** (many flowers) Purple gromwell. Perennial from a woody, purple dye-charged taproot. **STEMS** usually several, erect, simple or branched above, 1.5–6 dm tall, strigose-hispid. **LEAVES** cauline, appressed-strigose above, hirsute below, linear-lanceolate to narrowly oblong, 2–6 cm long. **INFL** of bracteate helicoid cymes, elongating in fruit. **FLOWERS** heterostylic, short-pedicellate, corolla bright yellow or yellow-orange, tubular-funnelform; tube 8–13 mm long, strigose on the outside; limb ill-defined, 5–11 mm wide, petal margins entire or slightly erose; fornices reduced to a small papillate patch at the base of each petal. **NUTLETS** about 3 mm long, white, glossy, smooth or slightly pitted, ventral keel evident, attachment scar not much depressed and without a raised collar, often only 1 maturing per calyx. Piñon-juniper, oak brush, ponderosa pine, aspen, and mixed conifer communities. ARIZ: Apa, Nav; COLO: Arc, Dol, Hin, LPl, Min, Mon, SJn, SMg; NMEX: McK, RAr, SJn; UTAH. 1850–3500 m (6000–11500′). Flowering: Jun–Aug. Fruit: Jul–Sep. Colorado and Utah south to Texas and Mexico. Indigenous people prepare a purple dye from the roots. Used as a "life medicine" and "big medicine" by the Ramah Navajo.

Lithospermum ruderale Douglas ex Lehm. (growing among rubbish) Western stoneseed. Perennial. **STEMS** usually several, erect or decumbent, strigose, villous near the base, simple or branched above, 2–5 dm tall. **LEAVES** cauline, crowded above, strigose, scabrous on the margins, lanceolate to linear, 3–10 cm long. **INFL** of short leafy racemes with flowers in the leaf axils. **FLOWERS** homostylic, pedicel short and stout; calyx lobe subulate in fruit; corolla funnelform or wide salverform, pale greenish yellow; tube glabrous, 4–7 mm long; limb 8–13 mm wide; fornices lacking. **NUTLETS** 4–6 mm long, gray, smooth, glossy, ventral keel usually evident, often only 1 or 2 maturing per calyx. $2n$ = 34. Piñon-juniper, sagebrush, oak brush, ponderosa pine, aspen, and mixed conifer communities. COLO: Dol, Mon, SMg; UTAH. 1720–2800 m (5650–9200′). Flowering: Apr–Jun. Fruit: May–Jul. Southwestern Canada, south to California, Utah, and Colorado. Used as an oral birth control by the Navajo.

Mertensia Roth Bluebells

Larry C. Higgins and Kenneth D. Heil

(for F. C. Mertens, a German botanist) Glabrous or pubescent, caulescent perennial herbs with fleshy, fusiform, rhizomelike roots. **STEMS** 1 to many from each root, decumbent to erect, usually branched below the inflorescence. **LEAVES** sessile or petiolate, entire, linear to cordate, alternate. **INFL** a lax or congested, ebracteate, unilateral, modified scorpioid cyme, or with the lowest flowers often single and subtended by leaves, often becoming panicled in age. **FLOWER** calyx 5-parted, occasionally campanulate, the expanded limb shorter or longer than the tube, with or without fornices in the throat; corolla blue, occasionally white or pink; filaments attached below the throat; anthers exserted or included; style shorter or longer than the corolla, in some di- or trimorphic; stigma entire or slightly lobed; ovary 2-loculed, each locule 2-lobed. **NUTLETS** 4, attached laterally to the gynobase, usually rugose or pectinately rugose, coriaceous or smooth and shining.

1. Plants relatively tall and robust, 4–15 dm tall when fully developed; with evident lateral veins in the cauline leaves; flowering in late spring and summer ....(2)

1′ Plants smaller, seldom as much as 4 dm tall; usually without evident lateral veins in the cauline leaves; flowering as soon as snow and temperature permit ....(3)

2. Leaves strigose at least on the upper surface; calyx lobes acute, often hairy on the back and ciliate-margined .... ***M. franciscana***

2′ Leaves glabrous on both surfaces, or pustulate on the upper surface, but lacking hairs; calyx lobes ciliate but usually not hairy on the back .... ***M. ciliata***

3. Upper surface of the leaves with loosely appressed hairs directed toward the margins of the leaves; stems 1–3 (5); plants of middle to lower elevations .... ***M. fusiformis***

3′ Upper surface of the leaves glabrous or with strigose, forward-pointing hairs; stems few to many; plants of middle and upper elevations .... ***M. lanceolata***

Mertensia ciliata (James ex Torr.) G. Don (fringed with hairs) Mountain bluebell. Plants erect or ascending, 1–12 dm tall, usually with many stems from each rootstock. **LEAVES** basal, petioles longer or shorter than the blades; variable, oblong to ovate, or lanceolate, subcordate, 4–15 cm long, 3–10 cm broad, ciliate on the margins, often papillate on the upper surface; cauline leaves lanceolate to ovate, acute, acuminate or obtuse at the apex, attenuate to subcordate at the base, the lowermost papillate on the upper surface, often quite glaucous, thin in texture. **INFL** from the axils of leaves, the peduncles elongated in mature or well-developed plants, in young plants the flowers aggregated at the top of the plant, each peduncle terminated in a modified, ebracteate, scorpioid cyme, or occasionally subumbellate. **FLOWER** calyx lobes 1.5–3 mm long, glabrous on the back, ciliate to papillate on the margins, more or less strigose within, obtuse or rarely enlarged in fruit; corolla tube 6–8 mm long, glabrous or with crisped hairs within; corolla limb 4–10 mm long, sometimes longer than the tube, moderately expanded; anthers 1–2.5 mm long, as long as or shorter and narrower than the expanded part of the filament; fornices prominent, glabrous, papillate, or pubescent; style about as long as the corolla or exceeding it. **NUTLETS** rugose or mammillate. $2n = 24, 48$. [*Pulmonaria ciliata* James ex Torr.; *M. stomatechoides* Kellogg]. Aspen, spruce-fir, and alpine communities. COLO: Arc, Con, Hin, LPl, Min, Mon, RGr, SJn. 2745–3750 m (9000–12300′). Flowering: late May–Sep. Montana to Colorado, New Mexico, and west to Oregon and California.

Mertensia franciscana A. Heller (for the San Francisco Peaks) Flagstaff bluebell. Plants with erect or ascending stems, 1–10 (17) dm tall, usually several from each rootstock. **LEAVES** basal, petioles longer or shorter than the blade; oblong-elliptic to elliptic, 6–20 cm long, 5–9 cm broad, the base subcordate to obtuse, the apex acuminate, acute or obtuse, short-strigillose above, glabrous or with spreading pubescence beneath; cauline leaves elliptic to narrowly ovate, 4–14 cm long, 1–5 cm broad, obtuse to acuminate, the lowermost petiolate, becoming glabrous to densely pubescent with spreading hairs below. **INFL** paniculately disposed in a bracteate, modified, scorpioid cyme, the branches elongating in age; pedicels strigose, 1–20 mm long. **FLOWER** calyx 2.5–5 mm long, divided, almost acute, rarely obtuse, glabrous or pubescent within; corolla limb 4–6 (9) mm long, subequal to or slightly shorter or longer than the corolla tube, moderately expanded; anthers 2.5–3 mm long, longer than the filaments; filaments 2–2.5 mm long, glabrous or with spreading hairs; fornices prominent, usually pubescent; style 9–20 mm long, usually shorter than the corolla, sometimes exceeding it. **NUTLETS** rugose and papilliferous. [*M. toyabensis* J. F. Macbr. var. *subnuda* J. F. Macbr.; *M. arizonica* Greene var. *subnuda* (J. F. Macbr.) L. O. Williams]. Moist canyons and streamsides or wet meadows in mountain brush, ponderosa pine, aspen, Douglas-fir, willow, and spruce-fir communities. ARIZ: Apa; COLO: Arc, Hin, LPl, Min, Mon, SJn; NMEX: RAr, SJn; UTAH. 2260–3750 m (6500–12300′). Flowering: late May–Sep. Mountains of Colorado, Arizona, New Mexico, and southeast Utah; Snake Range in Nevada.

Mertensia fusiformis Greene (swollen at the middle and tapering to each end like a spindle; referring to the root) Spindle bluebell. Plants with erect stems or nearly so, 1–3 dm tall, glabrous or sparingly pubescent, 1–few from each rootstock, this typically rather large and fusiform. **LEAVES** basal, petiole 7–12 cm long; elliptic to oblong-ovate, 4–12 cm long, 1.5–3 cm broad, usually densely strigose above, glabrous below; cauline leaves short-petiolate, linear-oblong to ovate-oblong, 1.5–10 cm long, 0.4–3 cm broad, sessile or the lowermost more or less densely strigose above, glabrous below, usually quite obtuse, rarely somewhat acute. **INFL** usually congested, sometimes paniculate; pedicels 1–15 mm long, densely strigose. **FLOWER** calyx 3–6 mm long, slightly accrescent, the lobes lanceolate to lance-ovate, 2–5 mm long, acute, ciliate, usually pubescent on the backs, occasionally nearly glabrous, not divided to the base; corolla 4–7 mm long, with a ring of crisp hairs within at the base; corolla limb 5–7 mm long, moderately expanded, usually subequal to or shorter than the tube, but sometimes longer; anthers 1.5–2.5 mm long; filaments 1–3 mm long; fornices present but usually not conspicuous, glabrous or nearly so; style usually surpassing the anthers, sometimes shorter. **NUTLETS** rugose, about 3 mm long. [*M. papillosa* Greene var. *fusiformis* A. Nelson]. Sagebrush, piñon-juniper woodland, mixed mountain brush, aspen, ponderosa pine, Douglas-fir, and spruce-fir communities. COLO: Arc, Dol, Hin, LPl, Min, Mon, SJn; NMEX: RAr, SJn; UTAH. 2135–2995 m (7000–9830′). Flowering: late Apr–Jun. Wyoming, Utah, Colorado, and New Mexico.

Mertensia lanceolata (Pursh) DC. ex A. DC. (lance-shaped, referring to the leaves) Lanceleaf bluebell. **STEMS** 1 to many, 1–4.5 dm tall, erect or ascending, canescent to glabrous. **LEAVES** basal, sessile or with the petioles longer than the blade; ovate-lanceolate, glabrous to densely canescent on both surfaces; cauline leaves lanceolate to oblong-lanceolate, moderately reduced toward the inflorescence, mostly sessile. **INFL** congested to loosely paniculate, especially in age; bracts only near the base. **FLOWER** calyx 2–5 (9) mm long in fruit, divided to below the middle and to the base, or near, the segments lanceolate to ovate-trianglular, glabrous to strigose; corolla tube 3–7 mm long, with a ring of dense hairs near the base, the limb 3–9 mm broad, moderately expanded; fornices conspicuous, glabrous to pubescent; anthers 1–2 mm long, well exserted from the tube, straight or curved; filaments longer or shorter than their anthers; style longer or shorter than the corolla tube. **NUTLETS** 2–3 mm long, rugose.

1. Calyx not divided quite to the base; filaments longer than the anthers, anthers usually curved; styles reaching or surpassing the anthers .... ***var. lanceolata***
1′ Calyx divided to or near the base; filaments shorter than the anthers, anthers straight; styles usually not reaching the anthers .... (2)

2. Leaves pubescent at least on one surface; plants usually occurring above 3000 m (9840′) .... ***var. nivalis***
2′ Leaves glabrous on both surfaces; plants mostly below 3000 m (9840′) .... ***var. coriacea***

var. *coriacea* (A. Nelson) L. C. Higgins & S. L. Welsh (leathery) Prairie bluebell. [*M. coriacea* A. Nelson var. *dilata* A. Nelson]. Open sites in sagebrush, piñon-juniper woodland, mountain mahogany, mountain shrub, and ponderosa pine communities. COLO: SMg; NMEX: SJn. 2490–2725 m (8175–8950′). Flowering: May–Jul. Wyoming, Utah, Colorado, New Mexico.

var. *lanceolata* Lance-leaf bluebell. Upper piñon-juniper woodland, Gambel's oak, ponderosa pine, and Douglas-fir communities. COLO: Hin, LPl, Min, SJn; NMEX: McK, RAr, SJn. 1885–2770 m (5750–9085′). Flowering: May–Jul. Saskatchewan south to Montana, the Dakotas, Nebraska, Wyoming, Colorado, and New Mexico.

var. *nivalis* (S. Watson) L. C. Higgins **CP** (of the snow) Snow bluebell. [*M. paniculata* (Aiton) G. Don var. *nivalis* S. Watson; *M. lanceolata* var. *viridis* A. Nelson; *M. viridis* (A. Nelson) A. Nelson; *M. canescens* Rydb.]. Fell-fields, talus and open rocky slopes in mountain meadows, spruce-fir, krummholz, and alpine tundra communities. COLO: Arc, Con, Dol, Hin, LPl, Min, Mon, RGr, SJn; NMEX: RAr; UTAH. 2285–3960 m (7500–13000′). Flowering: May–Sep. Saskatchewan, Montana, and North Dakota, south through Idaho, Wyoming, Colorado, and northern New Mexico.

Pectocarya DC. ex Meisn. Combseed

(Greek *pektos*, combed, + *karyon*, nut, referring to a row of prickles on the nutlet margins) Annual taprooted herbs. **LEAVES** alternate or sometimes the lower cauline ones opposite, small, narrow. **INFL** in terminal, naked or irregularly bracteate racemelike cymes. **FLOWERS** pedicellate; calyx cleft to near the base, sepals all alike or more or less dissimilar; corolla actinomorphic, small and inconspicuous, white; fornices nearly closing the throat; stamens included; stigma capitate. **NUTLETS** flattened, radially spreading at maturity, uncinate-bristly, at least toward the apex, margins often raised or winged, usually 4 maturing, sometimes in dissimilar pairs. A genus of about 15 species in western North America and temperate South America.

Pectocarya heterocarpa I. M. Johnst. (different fruit, referring to the different nutlet morphologies within each fruit) Chuckwalla combseed. **STEMS** several from the base, prostrate to ascending, strigose, 3–15 cm long. **LEAVES** linear, hispid-strigose, 1–3 cm long. **INFL** of indeterminate, cymosely branching racemes that make up the entire length of the stems, irregularly bracteate with leafy bracts. **FLOWERS** on strongly recurving pedicels in fruit; calyx lobes asymmetrical and unequal in fruit; corolla inconspicuous, white, limb 1–1.5 mm wide. **NUTLETS** heteromorphic, 1.8–3 mm long, uncinate-bristly on the margins, at least toward the apex, sometimes in dissimilar pairs, 2 or 3 of them strongly wing-margined and 1 or 2 with a narrow or no winged margin. $2n = 24$. Desert scrub community. UTAH. Up to 1200 m (3900′). Flowering: Apr–May. Fruit: May. California, Nevada, and Utah, east to New Mexico, south to Baja California and Sonora.

Plagiobothrys Fisch. & C. A. Mey. Popcorn-flower

(Greek *plagios*, oblique, + *bothros*, pit, referring to basilateral attachment scar on the nutlets of some species) Annual or perennial herbs. **LEAVES** alternate above, often opposite on the lower stem. **INFL** a series of helicoid, naked or bracteate modified cymes appearing as racemes or spikes, usually elongating at maturity. **FLOWERS** actinomorphic; calyx cleft to below the middle; corolla white, limb usually rotate; fornices evident at the throat; stamens included, stigma capitate. **NUTLETS** homomorphic, 4 or fewer by abortion; ventral keel usually evident; attachment scar elevated, caruncle-like, basilateral or near the middle of the ventral keel (Johnston 1923). A genus of about 50 species in western North America and South America.

1. Plants hispid, densely spreading hirsute; nutlets broadly lanceolate, attachment scar lateral and near the middle of the ventral surface .... ***P. jonesii***
1′ Plants sparsely strigose-hirsute; nutlets narrowly lanceolate, attachment scar basilateral .... ***P. scouleri***

Plagiobothrys jonesii A. Gray (for Marcus E. Jones, western botanist, especially in Utah) Jones' popcorn-flower. **STEMS** branching from base and divergent above, erect to ascending, without a central axis, spreading-hispid throughout and also retrorsely strigose, 1–3 dm tall. **LEAVES** basal and cauline, spreading-hispid; basal leaves narrowly oblanceolate, up to 10 cm long; cauline leaves shorter and sessile, lanceolate to lance-ovate. **INFL** of helicoid racemes terminating the branches, leafy-bracteate at the base, naked above, somewhat elongating in fruit. **FLOWERS** short-pedicellate; calyx 6–8 mm long in fruit, lobes subulate-linear; corolla inconspicuous, limb 1–2 mm wide. **NUTLETS** ovate, 2.5–3 mm long, abruptly pointed at the apex, incurved, ventral and dorsal keels evident, surface tessellate; attachment scar narrow and elongate at the base of the ventral keel and near the middle of the nutlet. Desert scrub community. UTAH. Up to 1330 m (4350′). Flowering: Apr–May. Fruit: May–Jun. Arizona, California, Nevada, and Utah.

Plagiobothrys scouleri (Hook. & Arn.) I. M. Johnst. (for John Scouler, Scottish botanist) Scouler's popcorn-flower. **STEMS** several, branching from the base, prostrate to ascending, appressed-setose, 0.5–2 dm long. **LEAVES** essentially all cauline, lower 1–4 pairs opposite, alternate above, linear to narrowly oblanceolate, strigose, 1–7 cm long. **INFL** of helicoid racemes terminating the branches, irregularly bracteate with leaflike bracts, elongating in fruit. **FLOWERS** short-pedicellate, calyx 2–4 mm long in fruit; corolla inconspicuous, limb 1–2 mm wide. **NUTLETS** lance-ovate, 1.5–2 mm long, variously roughened with rugae or tuberculations, with or without short setose projections; attachment scar small, basilateral. $2n = 54$. Our plants belong to the smaller flowered, more prostrate **var. *hispidulus*** (Greene) Dorn (with stiff hairs). [*P. scouleri* var. *penicillatus* (Greene) Cronquist]. Moist soils or drying mud in low areas in piñon-juniper, sagebrush, ponderosa pine, aspen, and mixed conifer communities. COLO: Arc, Dol, Hin, LPl, Mon, SJn, SMg. 1850–3500 m (6100–11500′). Flowering: May–Aug. Fruit: Jun–Sep. Manitoba to British Columbia, south to California and New Mexico.

Tiquilia Pers. Tiquilia, Crinklemat

(derived from the Quechua vernacular name *tiquil-tiquil* for the Peruvian species *T. dichotoma*) Annual or perennial herbs or suffruticose shrubs. **LEAVES** alternate, entire, all cauline. **INFL** in axillary clusters or solitary and axillary. **FLOWERS** essentially sessile; corolla broadly tubular-funnelform, limb usually spreading, fornices lacking or inconspicuous; stamens included; style terminal at the apex of a 4-grooved ovary, bifid to well below the 2 capitate stigmas. **NUTLETS** 4 or fewer by abortion, each with a long ventral scar indicating attachment to adjacent nutlets (Richardson 1977). A genus of about 25 species in arid habitats of western North America, Mexico, and western South America.

Tiquilia latior (I. M. Johnst.) A. T. Richardson (the wider, in comparison to *T. hispidissima*, a similar species with narrower leaves) Matted tiquilia. Perennial from a woody taproot. **STEMS** several to numerous, dichotomously branched, spreading along the ground forming mats 2–6 dm in diameter. **LEAVES** firm, linear, revolute, narrowed to a somewhat coriaceous base, margins hispid-ciliate, 5–15 mm long, crowded on the short, terminal branchlets. **INFL** of solitary flowers in the leaf axils. **FLOWERS** sessile; calyx 3–3.5 mm long; corolla pink, tube 4–7 mm long; style cleft for only about 0.5 mm below the stigmas. **NUTLETS** lance-ovate, ventrally convex, 1.1–1.7 mm long, minutely papillate-granular, usually only 1 or 2 maturing per calyx. [*Coldenia hispidissima* (Torr. & A. Gray) A. Gray var. *latior* I. M. Johnst.]. Saltbush, desert scrub, and piñon-juniper communities. ARIZ: Apa, Nav; UTAH. 1000–1800 m (3300–5900′). Flowering: May–Aug. Fruit: Jun–Sep. Arizona, Nevada, and Utah. Used for gastrointestinal disease by the Kayenta Navajo.

BRASSICACEAE Burnett (CRUCIFERAE) MUSTARD or CRUCIFER FAMILY

Steve L. O'Kane, Jr.

(named for the genus *Brassica*, Latin for cabbage; the classical alternative name, Cruciferae, refers to the cross-shaped flowers of four petals) Annual, biennial, or perennial, herbs to subshrubs or small shrubs (in ours), containing pungent-smelling and -tasting oils (glucosinolates); glabrous, glaucous, or variously pubescent with simple, 2–many-branched, glandular, stellate, or peltate trichomes. **LEAVES** alternate (rarely opposite), exstipulate, cauline, basal (often forming a rosette), or both. **INFL** indeterminate racemes or rarely spikes, sometimes these arranged into panicles, loosely to densely flowered and often greatly elongating in fruit, usually ebracteate. **FLOWERS** hypogynous, actinomorphic to somewhat bilateral, perfect or rarely imperfect; sepals 4, distinct, usually caducous, lateral (inner) pair often saccate or gibbous at the base; petals 4 or rarely absent, clawed (usually) or not, distinct, usually entire; stamens 6 (rarely 2 or 4), the outer 2 usually shorter than the inner 4 (tetradynamous), rarely of equal length or in 3 pairs of different length; nectar glands various and arranged around the filament bases; gynoecium 2-carpellate, syncarpous, sessile, substipitate, or stipitate, 2-locular with a false septum spanning a thickened rim (replum). **FRUIT** usually a bivalved capsule (referred to as a silique when 3 or more times longer than broad or a silicle when less than 3 times longer than broad), usually dehiscent, variously terete and uncompressed, quadrangular in cross section, or compressed parallel (latiseptate)

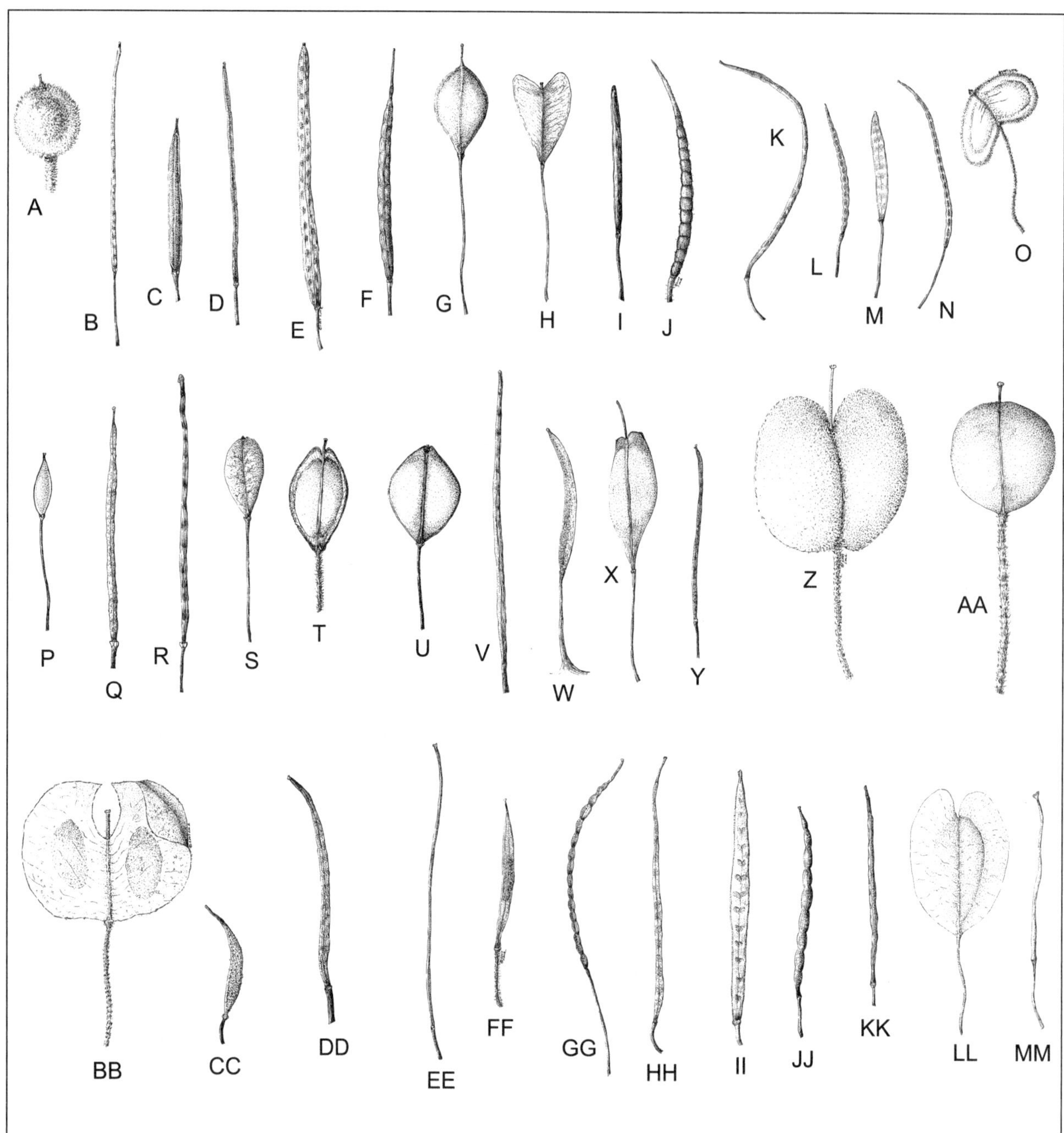

Brassicaceae fruits: Bars are 1 cm unless otherwise noted. **A.** *Alyssum simplex.* **B.** *Arabis pycnocarpa.* **C.** *Barbarea orthoceras.* **D.** *Boechera fendleri.* **E.** *Boechera stricta.* **F.** *Brassica tournefortii.* **G.** *Camelina microcarpa.* **H.** *Capsella bursa-pastoris.* **I.** *Cardamine cordifolia.* **J.** *Chorispora tenella.* **K.** *Conringia orientalis.* **L.** *Descurainia incisa.* **M.** *Descurainia pinnata.* **N.** *Descurainia sophia.* **O.** *Dimorphocarpa wislizenii.* **P.** *Draba fladnizensis.* **Q.** *Erysimum capitatum.* **R.** *Hesperidanthus linearifolius.* **S.** *Hornungia procumbens.* **T.** *Lepidium campestre.* **U.** *Lepidium perfoliatum.* **V.** *Malcomia africana.* **W.** *Nasturtium officinale.* **X.** *Noccaea fendleri.* **Y.** *Pennellia micrantha.* **Z.** *Physaria acutifolia.* **AA.** *Physaria fendleri.* **BB.** *Physaria newberryi.* **CC.** *Rorippa sinuata.* **DD.** *Sinapis arvensis.* **EE.** *Sisymbrium altissimum.* **FF.** *Smelowskia americana.* **GG.** *Stanleya pinnata.* **HH.** *Streptanthella longirostris.* **II.** *Streptanthus cordatus.* **JJ.** *Thelypodium integrifolium.* **KK.** *Thelypodiopsis purpusii.* **LL.** *Thlaspi arvense.* **MM.** *Turritis glabra.*

or perpendicular (angustiseptate) to the false septum, inflated or not. **SEEDS** winged or wingless, often exuding mucilage when wetted. $x = 5–12+$. Three hundred thirty-eight genera and 3712 species. A worldwide family centered in temperate regions of the Northern Hemisphere (Appel and Al-Shehbaz 2003; Holmgren, Holmgren, and Cronquist 2005; Rollins 1993; Tai-yien et al. 2001; Warwick and Al-Shehbaz 2006; Welsh et al 2003). Warm, arid areas, alpine habitats, and regions with Mediterranean climates are especially rich in members of this family, which also contains important crops like *Brassica oleracea* (cabbage, broccoli, cauliflower, Brussels sprouts, kale, kohlrabi, savoy), *B. nigra* (black mustard), *B. napus* (rutabaga), *B. juncea* (Indian mustard), *B. rapa* (rapeseed or canola oil, turnip, bok choy, Chinese mustard, Chinese cabbage), *Lepidium sativum* (cress), *Armoracia rusticana* (horseradish), *Raphanus sativus* (radish), *Sinapis alba* (table mustard), *Nasturtium officinale* (watercress), and *Eutrema wasabi* (wasabi). Ornamental species include *Erysimum cheiri* (wallflower), *Hesperis matronalis* (dame's rocket), and *Lunaria annua* (money plant, honesty). In addition to these many useful species, the family is also known for a large number of weedy members, especially in arid and semiarid regions.

The generic descriptions below are based on the family-wide treatment of Appel and Al-Shehbaz (2003). The family as a whole has rather stereotyped flowers and, therefore, the following will be assumed in descriptions: 4 sepals, 4 petals, 6 tetradynamous stamens, and fruit dehiscent. The presence of mucilage can be detected by soaking mature seeds in water; if present, the mucilage will form a clear, sticky halo around the seed. This halo can help cement seeds in place when on precarious slopes and can aid in dispersal by sticking to the feet, hair, or plumage of animals. Mature fruits are nearly always needed to positively identify members of the family because of the homogeneity of the flowers.

1. Fruit less than 3× longer than broad (a silicle) .... (2)
1′ Fruit more than 3× longer than broad (a silique) .... (13)

2. Petals deeply bilobed, white; stellate-pubescent; annual .... ***Berteroa***
2′ Petals entire to emarginate, not deeply bilobed, color various; plants glabrous or pubescent (stellate-pubescent or not); annual, biennial, or perennial .... (3)

3. Silicle strongly flattened parallel to the septum (latiseptate) .... (4)
3′ Silicle terete or only very slightly flattened, flattened perpendicular to the septum (angustiseptate), or the replum perpendicularly compressed and the fruit valves papery-inflated .... (5)

4. Silicle nearly round in face view; seeds 2–4 per fruit .... ***Alyssum***
4′ Silicle distinctly longer than broad; seeds more than 4 per fruit .... ***Draba***

5. Silicle didymous and strongly flattened, resembling a pair of spectacles .... ***Dimorphocarpa***
5′ Silicle not didymous, or if didymous then papery and strongly inflated .... (6)

6. Trichomes (look carefully at 10–20×) absent or, if present, all of them simple .... (7)
6′ Trichomes present, at least some 2-forked, dendritic, or stellate .... (10)

7. Delicate, slender-stemmed annuals with small, erect to ascending silicles on ascending pedicels .... ***Hornungia***
7′ Plants not especially delicate; annual, biennial, or perennial; silicles various .... (8)

8. Silicle elliptic to oval or suborbicular in outline; 1 seed per locule .... ***Lepidium***
8′ Silicle triangular-obovate or obcordate in outline or, if orbicular, with 2 or more seeds per locule and over 1 cm in length .... (9)

9. Plants annual; silicle orbicular in outline, over 1 cm long .... ***Thlaspi***
9′ Plants perennial; silicle triangular-obovate or obcordate in outline, less than 1 cm long .... ***Noccaea***

10. All cauline leaves petiolate or sessile, never auriculate-clasping .... (11)
10′ At least some cauline leaves auriculate-clasping .... (12)

11. Plants very delicate; slender-stemmed annuals with small, erect to ascending silicles on ascending pedicels, nearly glabrous or with scattered irregularly branched trichomes .... ***Hornungia***
11′ Plants not at all delicate; short-lived to long-lived perennials with various kinds of silicles; plants densely covered with stellate to stellate-peltate trichomes, usually silvery .... ***Physaria***

12. Silicle obovoid, ± acute at apex; petals pale yellow to almost white .... ***Camelina***
12′ Silicle cuneate-obcordate, truncate to truncate-emarginate at apex; petals white .... ***Capsella***

13. Silique on a long stalk (gynophore) 10–25 mm long .... ***Stanleya***
13′ Silique sessile or on a short gynophore less than 6 mm long .... (14)

14. Silique flattened parallel to the septum (latiseptate) (because the flattening may not be pronounced, roll the fruit between your thumb and forefinger to be sure) .... (15)
14′ Silique terete to 4-angled or 4-ribbed .... (21)

15. Petals narrow, ± involute with unexpanded blades, greenish yellow, brownish purple, or reddish purple .... (16)
15′ Petals flat, with a broadened blade, white, pink, or lavender .... (17)

16. Petals greenish yellow, often with purple veins, rarely almost white .... ***Streptanthella***
16′ Petals brownish purple or reddish purple .... ***Streptanthus***

17. Valves (silique walls) veinless (not counting the replum rim) .... (18)
17′ Valves veined .... (19)

18. Leaves pinnately compound or simple and cordate; silique elastically dehiscent .... ***Cardamine***
18′ Leaves simple and usually lanceolate, not cordate; silique not elastically dehiscent .... ***Draba***

19. Basal leaves green at flowering time; plants perennial; silique strongly flattened, or if long and slender (and flattening hard to discern), not stiffly erect .... ***Boechera***
19′ Basal leaves withering by flowering time; plants monocarpic (dying following first flowering); silique only slightly flattened, long and slender, stiffly erect .... (20)

20. Cauline leaves glabrous, glaucous, and auriculate; petals pale yellow .... ***Turritis***
20′ Cauline leaves at least somewhat hairy, neither glaucous nor auriculate; petals white .... ***Arabis***

21. Plants with at least some malpighian, branched, 3-forked, or stellate trichomes .... (22)
21′ Plants either glabrous or with simple trichomes only (these may be stipitate-glandular) .... (27)

22. Leaves pinnately compound or deeply divided .... (23)
22′ Leaves simple and entire, toothed, or sinuate-dentate .... (24)

23. Petals yellow; leaves pinnately compound; of various habitats .... ***Descurainia***
23′ Petals white, cream, or pale lavender; leaves deeply divided; alpine habitats .... ***Smelowskia***

24. Trichomes mostly malpighian (sometimes 3- or 4-branched), appressed to leaf surface and with principal branches parallel to the midvein; petals yellow to burnt-orange, sometimes lavender or red-purple at high elevations .... ***Erysimum***
24′ Trichomes variously branched, but not malpighian, appressed, or with branches parallel to the midvein; petals white, pink, lavender, or purple .... (25)

25. Petals barely exceeding the sepals .... ***Pennellia***
25′ Petals well exceeding the sepals .... (26)

26. Siliques sharply pointed, very stiff; pedicels nearly as thick as the silique .... ***Malcolmia***
26′ Siliques not especially pointed, not especially stiff; pedicels distinctly narrower than the silique .... ***Hesperis***

27. Plants with gland-tipped trichomes; petals purplish blue; when fully mature, the silique transversely jointed between the seeds, essentially indehiscent but breaking into seed-bearing segments .... ***Chorispora***
27′ Plants entirely lacking stipitate-glandular trichomes; petals variously colored; silique not transversely jointed, dehiscent .... (28)

28. Silique strongly beaked with a prominent sterile portion .... (29)
28′ Silique not or only slightly beaked .... (30)

29. Valves of the silique prominently or obscurely 1-nerved; sepals erect or ascending .... ***Brassica***
29′ Valves of the silique 3–7-veined; sepals spreading .... ***Sinapis***

30. Mature fruit less than 18 mm long; plants of wet and moist places, streamsides, permanent and temporary pond margins, drying mud, or marshes; at least some cauline leaves with auriculate blades or petioles (except *Rorippa alpina*, *R. tenerrima*, and some *R. sphaerocarpa*) .... (31)
30′ Mature fruit greater than 18 mm long; plants of drier habitats; leaves auriculate or not .... (33)

31. Petals white; leaves succulent .... ***Nasturtium***
31' Petals yellow; leaves not especially succulent .... (32)

32. Silique somewhat 4-angled at least at the base, elongate, with a short beak, not torulose .... ***Barbarea***
32' Silique terete, often club-shaped but sometimes elongate or nearly spherical, torulose or not .... ***Rorippa***

33. At least some cauline leaves auriculate .... (34)
33' All leaves sessile or petiolate, never auriculate .... (35)

34. Leaves ovate to broadly elliptic, glaucous, entire, rounded at apex and clasping the stem .... ***Conringia***
34' Leaves not as above .... ***Thelypodiopsis***

35. Cauline leaves deeply and regularly pinnatifid; petals yellow .... ***Sisymbrium***
35' Cauline leaves not deeply and regularly pinnatifid; petals yellow, white, lavender, or purple .... (36)

36. Silique torulose; plants biennial; petals white to purple .... ***Thelypodium***
36' Silique not torulose; plants perennial; petals purple .... ***Hesperidanthus***

Alyssum L. Alyssum, Madwort

(Greek for without madness, apparently referring to medicinal uses, especially to treat the symptoms of rabies) Annual, biennial, and perennial herbs, pubescent with stellate or dendritic hairs, these sometimes mixed with simple ones. **STEMS** simple or much-branched. **LEAVES** mostly cauline, petiolate to subsessile, entire, oblanceolate or narrower. **FLOWER** sepals ascending, connivent or distinct, deciduous or more rarely persisting into fruit (e.g., *A. alyssoides*); petals yellow to cream. **FRUIT** a silicle, elliptic to orbicular, latiseptate, ovules usually 2–4. **SEEDS** 2 rows per locule, mucilaginous or not. $x = 7, 8$. One hundred ninety-five species mostly of Europe and adjacent southwestern Asia; several species are widespread weeds.

1. Silicles glabrous .... ***A. desertorum***
1' Silicles pubescent with stellate hairs .... (2)

2. Sepals persistent in fruit; silicles 3–4 mm wide .... ***A. alyssoides***
2' Sepals deciduous after anthesis; silicles 4–5.2 mm wide .... ***A. simplex***

Alyssum alyssoides (L.) L. (resembling *Alyssum*) Pale madwort, pale alyssum. Annual herb, gray-green and covered with appressed-stellate trichomes. **STEMS** simple or much-branched from near or at the base. **LEAVES** entire, spatulate to narrowly oblanceolate, narrowing to a sessile base, up to 3 (4) cm long. **INFL** elongated in age with divergent fruiting pedicels 2–3 mm long. **FLOWER** sepals persistent; petals pale yellow and fading to cream or white, 2.5–4 mm; filaments slender, neither winged nor appendaged. **FRUIT**: silicle ovate to orbicular, inflated in the center with a flattened margin, 3–4 mm wide, covered with minute stellate trichomes, emarginate at the apex. **SEEDS** 2 per locule, ovoid-obovoid, plump, 1.2–2 mm long, narrowly winged. $n = 16$, $2n = 32$. [*Clypeola alyssoides* L.; *Alyssum calycinum* L.]. Disturbed sites, roadsides, wash bottoms, Gambel's oak, and piñon-juniper communities. ARIZ: Apa; COLO: SJn; NMEX: SJn. 1900–2750 m (6250–9000'). Flowering: late April–early June. Native to Europe and Asia, now widely distributed and weedy in North America.

Alyssum desertorum Stapf (of the desert) Desert madwort, desert alyssum. Annual herb, gray-green and covered with appressed-stellate trichomes. **STEMS** simple or much-branched from near or at the base. **LEAVES** entire, spatulate to narrowly oblanceolate, narrowing to a sessile base, up to 2.5 (3.5) cm long. **INFL** elongated in age with divergent fruiting pedicels 1.5–4 mm long. **FLOWER** sepals deciduous; petals pale yellow and fading to cream or white, 1.8–3 mm; median pairs of filaments narrowly winged at base, lateral filaments with a broadly winged appendage apically notched into 2 teeth. **FRUIT**: silicle orbicular, inflated in the center with a flattened margin, 3–4 mm wide, glabrous (rarely with a few scattered stellate trichomes), emarginate at the apex. **SEEDS** 2 per locule, broadly ovate, 1.2–1.5 mm long, narrowly winged. $n = 16$, $2n = 32$. Disturbed sites, ponderosa pine, and shrub communities. ARIZ: Apa; COLO: Mon; NMEX: SJn. 1900–2250 m (6200–7350'). Flowering: Apr–May. Native of Europe, now weedy and widely distributed in the Rocky Mountains and the Great Basin.

Alyssum simplex Rudolphi (simple) Wild alyssum, small alyssum, field alyssum. Annual herb, gray-green and covered with appressed-stellate trichomes. **STEMS** simple or much-branched from near or at the base. **LEAVES** entire, spatulate to narrowly oblanceolate, narrowing to a sessile base, 4–25 mm long, the basal and lower cauline leaves early-deciduous. **INFL** elongated in age, with divergent fruiting pedicels 2.5–5 mm long. **FLOWER** sepals deciduous; petals pale yellow, aging cream or white, 2.7–3.6 mm long; median pairs of filaments broadly winged and apically 1- or 2-

toothed, lateral filaments with a broadly winged appendage apically. **FRUIT**: silicle broadly obovate to suborbicular, inflated in the center with a flattened margin, 4–5.2 mm wide, covered with coarse, stellate trichomes, emarginate at the apex. **SEEDS** 2 per locule, broadly ovate, 1.8–2.1 mm wide, narrowly winged. n = 8, $2n$ = 16. [*A. micranthum* C. A. Mey.; *A. minus* (L.) Rothm.; *A. minus* var. *micranthum* (C. A. Mey.) T. R. Dudley; *A. parviflorum* Fisch. ex M. Bieb.]. Disturbed sites, roadsides, wash bottoms, sagebrush, rabbitbrush, and piñon-juniper communities. ARIZ: Apa, Nav; COLO: Arc, Dol, LPl, Min, Mon, SMg; NMEX: McK, RAr, SJn; UTAH. 1525–2350 m (5000–7750′). Flowering: mid-Apr–early Jun. Montana to Colorado and west to California.

Arabis L. Rockcress
Michael D. Windham

(for the Arabian region) Biennials or short-lived perennials, rarely annuals. **HAIRS** mostly simple or forked (in ours), occasionally dendritic. **STEMS** simple or branched distally, lower portion hirsute with simple and 2-rayed hairs. **LEAVES** basal and cauline, the former often deciduous in late anthesis; basal leaves petiolate, rosulate, broadly oblanceolate to obovate, (5–) 10–25 mm wide, thin, simple, entire to dentate or rarely lyrate-pinnatifid, hirsute with simple or forked hairs, 3-rayed hairs rarely present; upper cauline leaves sessile, with or without auricles, entire or dentate, sparsely to copiously ciliate, often hirsute. **INFL** ebracteate or with bracts at lower nodes, racemose to narrowly paniculate, often elongate in fruit; fruiting pedicels erect (in ours), ascending, or divaricate. **FLOWERS** ascending; sepals ovate or oblong, erect or ascending, caducous, bases of lateral pair saccate or not, margins membranous; petals white (in ours), lavender, or purple, 3–5.5 mm long (in ours); blade spatulate, oblong, or oblanceolate, apex obtuse or emarginate; claw shorter than sepals; stamens 6, tetradynamous; filaments usually not dilated at base; ovules 52–70 per ovary (in ours). **FRUIT** linear, distinctly flattened parallel to the septum, straight, sessile or short-stipitate, unsegmented, 0.8–1.2 mm wide (in ours) at dehiscence, erect (ours) to ascending or divaricate, not secund, glabrous; valves papery, with an obscure or prominent midvein; septa complete, membranous, veinless; styles obsolete or cylindric; stigmas capitate, entire or slightly 2-lobed. **SEEDS** in 1 (ours) or rarely 2 rows per locule, flattened, oblong or orbicular, winged or margined; seed coats smooth or minutely reticulate. x = 8. About 70 species: Asia; Europe; Africa; North America. As delimited by Rollins (1993) and most local floras, *Arabis* included a heterogeneous assemblage of species here distributed among *Arabis*, *Boechera*, and *Turritis*. Recent molecular studies (e.g., Koch, Haubold, and Mitchell-Olds 2001) demonstrate that these three genera are not closely related.

Arabis pycnocarpa M. Hopkins (dense fruit) Hopkins' rockcress. Plants biennials or short-lived perennials. **STEMS** 1 (rarely more), arising from center of basal rosette, 2–5.5 dm tall, often branched distally, lower portions hirsute with simple and forked hairs 0.2–1 mm long. **LEAVES** rosulate at base of stem, oblanceolate or obovate, the largest 11–17 mm wide, with obtuse apex and entire to dentate margins, sparsely hirsute with simple and stalked, forked hairs; cauline leaves 5–23, often overlapping proximally but rarely concealing stem, entire or dentate; upper cauline leaves lanceolate or nearly linear, without auricles or with small auricles to 0.5 mm long, sparsely ciliate, sporadically hirsute to nearly glabrous. **INFL** usually branched; fruiting pedicels erect to erect-ascending, 4–13 mm long, glabrous. **FLOWERS** inconspicuous; sepals greenish to beige, 2.5–4 mm long; petals white, 3–5.5 mm long. **FRUIT** erect to erect-ascending, often appressed to rachis, glabrous; style 0.2–0.5 mm long, cylindric. **SEEDS** in 1 row in each locule. $2n$ = 32. [*A. hirsuta* (L.) Scop. var. *glabrata* Torr. & A. Gray misappl.; *A. hirsuta* var. *pycnocarpa* (M. Hopkins) Rollins]. Riparian habitats and open forest floor in mountainous areas. ARIZ: Apa; NMEX: SJn. 2300–2650 m (7500–8700′). Flowering: Jun–Sep. Throughout most of the United States except for the Gulf Coast region. Our plants belong to **subsp**. ***pycnocarpa***.

Barbarea W. T. Aiton Yellowrocket

(for St. Barbara, Catholic martyr, 3rd or 4th century, protector from thunderstorms) Biennial or perennial herbs, glabrous or sparsely pubescent with simple trichomes. **STEMS** erect, branched, usually stiff, angled. **LEAVES**: basal pinnatifid or pinnately compound; cauline auriculate at base, entire to pinnatifid. **FLOWERS** rather showy; sepals ascending, lateral pair often gibbous-based; petals yellow or yellowish, obovate to oblong. **FRUIT** a silique, linear, terete, quadrangular or latiseptate, often torulose, erect to divaricate. **SEEDS** 1 row per locule, nonmucilaginous. x = 8, 9. Twenty-nine species, nearly worldwide.

1. Venation of upper leaves pinnate; upper leaves pinnatifid at least to some degree; siliques relatively thick, 1.1–1.5 mm thick ..... ***B. orthoceras***

1′ Venation of upper leaves palmate-pinnate; upper leaves coarsely dentate; siliques narrow, 0.5–1.1 mm thick ..... ***B. vulgaris***

Barbarea orthoceras Ledeb. (straight + horn, referring to the straight beak of the fruit) American yellowrocket. Biennial or perennial from a taproot, caudex usually simple, herbage glabrous or sparsely hirsute. **STEMS** stiff, erect, angled, single but freely branched, 2–6 (7.5) dm tall. **LEAVES** basal and cauline, the basal ones often withering, long-petioled, 3–10 (15) cm long, reduced upward, auriculate-clasping, blades (oblong-elliptic to) lyrate-pinnatifid to pinnate, terminal lobe ovate, margins nearly entire to crenate or dentate, lower petiole bases sometimes ciliate with simple hairs. **INFL** an ebracteate raceme. **FLOWER** sepals broadly lanceolate, 2.2–3.5 mm long; petals bright to pale yellow, 3–6 mm long, oblanceolate or spatulate. **FRUIT**: silique 1.5–4 cm long, 1.1–1.5 mm thick, slightly flattened parallel to the septum but 4-angled, ascending to divaricately ascending, ± linear, acute at apex, somewhat torulose. **SEEDS** 1 row per locule, 1.3–1.6 mm long. $n = 8$, $2n = 16$. [*B. americana* Rydb.; *Campe americana* Cockerell]. Damp or moist places, especially when disturbed, of riverbanks, aspen groves, lake margins, cottonwood groves, valley bottoms, vacant lots, and meadows. COLO: Arc, Hin, LPl, Min, Mon, SJn; NMEX: SJn; UTAH. 1875–2750 m (6150–9050′). Flowering: May–early Oct. Circumboreal and south to a line from New Hampshire to California.

Barbarea vulgaris W. T. Aiton (common) Garden yellowrocket. Biennial (or perennial) from a taproot, caudex usually simple, herbage glabrous or petiole bases sometimes ciliate with simple hairs. **STEMS** erect, usually branched above, 1.5–10 dm tall. **LEAVES** basal and cauline, the basal ones often withering, petiolate, 5–13 cm long, blade oblanceolate in outline, lyrate-pinnatifid, (entire or) crenate-dentate, the terminal lobe larger and broadly ovate to suborbicular, leaves reduced upward, auriculate-clasping. **INFL** ebracteate raceme, dense in flower but elongating in fruit. **FLOWER** sepals broadly lanceolate, 2.7–4.5 mm long; petals bright yellow, 4.5–6.5 (8) mm long, oblanceolate or spatulate. **FRUIT**: silique 1.5–3 cm long and 0.5–1.1 mm thick, ± terete, weakly torulose, divaricately ascending, prominently nerved. **SEEDS** 1 row per locule, 1–1.5 mm long. $n = 8$, $2n = 16$. [*Erysimum barbarea* L.; *E. arcuatum* Opiz ex J. Presl & C. Presl]. Damp or most places, especially when disturbed. COLO: LPl, Mon; UTAH. 2125–2300 m (7000–7600′). Flowering: mid-May–Jul. Native of Europe and Asia, now weedy throughout North America.

Berteroa DC. Berteroa, False Madwort

(for Carlo Giuseppe Bertero, 1789–1831, Italian botanist and physician) Annual or perennial herbs, densely pubescent with mostly stellate trichomes. **STEMS** erect, branched. **LEAVES**: basal leaves forming a rosette, entire to pinnatifid; cauline leaves usually entire and sessile. **FLOWER** sepals ascending to spreading; petals bilobed at apex, white. **FRUIT**: silicles elliptic to ovate or orbicular, latiseptate, ± inflated, erect to divaricate. **SEEDS** 2 rows per locule, nonmucilaginous, winged or margined. $x = 8$. Five species of Europe and the Middle East.

Berteroa incana (L.) DC. (hoary) Hoary false madwort, hoary alyssum. Annual or perennial, grayish green with fine, dense, appressed-stellate trichomes. **STEMS** erect, little branched at the base, more freely branched above, up to 10 dm tall. **LEAVES**: basal ones usually forming a rosette, entire to pinnatifid, slenderly petiolate, 3–5 cm long; the cauline ones few to numerous, appressed-ascending, entire (dentate), obtuse apically, 2–5.5 cm long, 3–7.5 cm wide, somewhat petiolate below and becoming sessile and reduced above, basal and lower cauline leaves typically withering by anthesis. **INFL** dense, narrow, ebracteate, elongating in fruit. **FLOWER** sepals early-deciduous, 2–3 mm long; petals bilobed, yellow (more rarely white, cream, purplish, or white with purple veins), clawed or not, entire to emarginate, 4–6 mm long. **FRUIT**: silicle 4–8 mm long, pubescent, somewhat inflated, valves with 1 faint vein near the base, erect. **SEEDS** 3–7 per locule, 2 rows per locule, orbicular, wing-margined, the wing thick and 0.1–0.2 mm wide. $n = 8$, $2n = 16$. [*Alyssum incanum* L.]. Roadsides and disturbed areas (one record not seen). UTAH. Native of Europe and now established and weedy mainly in the northern and central states and in southern Canada.

Boechera Á. Löve & D. Löve Rockcress

Michael D. Windham

(for Tyge Böcher, 1909–1983) Perennial, rarely biennial, with simple, malpighiaceous, forked, dendritic, or rarely absent, never cruciform or stellate, trichomes. **STEMS** simple or branched distally, lower portion dendritic-pubescent to hirsute or glabrous. **LEAVES** basal and cauline, basal leaves persistent or deciduous in late anthesis, petiolate, rosulate or not, oblanceolate (rarely obovate) to linear, 2–10 (–14) mm wide, somewhat leathery, simple, entire or dentate, dendritic-pubescent to hirsute or nearly glabrous, 3-rayed hairs present in some species; upper cauline leaves sessile, with or without auricles, entire or dentate, rarely ciliate, sparsely pubescent to glabrous. **INFL** ebracteate, usually racemose, often elongate in fruit; fruiting pedicels reflexed to divaricate, ascending or occasionally erect. **FLOWERS** ascending or downcurved in late anthesis; sepals ovate or oblong, erect or ascending, caducous, bases of lateral pair slightly saccate or not, margins membranous; petals white, pale lavender, or purple, (4–) 6–18 mm long; blade spatulate or oblanceolate, apex obtuse; claw shorter than sepals or undifferentiated from blade; stamens 6, tetradynamous;

filaments not dilated at base; ovules 26–216 per ovary. **FRUIT** linear, distinctly flattened parallel to the septum, straight or curved, sessile, unsegmented, (1–) 1.5–3.5 mm wide at dehiscence, pendent to loosely ascending or erect, secund or not, glabrous or pubescent; valves papery, with an obscure or prominent midvein; septa complete, membranous, veinless; styles obsolete or cylindric; stigmas entire. **SEEDS** in 1 or 2 rows per locule, flattened, oblong or orbicular, winged to nearly wingless; seed coats smooth or minutely reticulate. $x = 7$. At least 70 diploid species (8 in the flora) and numerous allopolyploid, apomictic hybrids: United States, Canada, and Mexico; 1 in Russian Far East (Windham and Al-Shehbaz 2006). Traditionally, this species complex has been included in the genus *Arabis.* Recent molecular studies (e.g., Koch, Haubold, and Mitchell-Olds 2001) demonstrate that these genera are not closely related. Two species of *Arabis* (now *Boechera*) were used by the Navajo. Plants identified as *A. perennans* were reported to have anticonvulsive properties, and specimens called *A. holboellii* were used as ceremonial items. Because neither of these taxa actually occurs in the Four Corners region, the true identity of the *Boechera* species used by the Navajo is uncertain. The taxonomic complexity of *Arabis* sensu lato is legendary and, when the genus is split as outlined above, all of the most problematic groups come to reside in *Boechera*. A rare confluence of hybridization, polyploidy, and apomixis make this one of the most difficult genera in the North American flora. The sexual diploid species are relatively distinct from one another morphologically, but they hybridize whenever they come into contact. Through polyploidy and apomixis, these hybrids become stable, self-propagating lineages that have the potential to outcompete the sexual diploids that gave rise to them. Most of the hybrid derivatives in *Boechera* are triploids, but apomictic diploids are known as well. Thus, for any pair of sexual diploid species (e.g., AA and BB), this process can yield several different intermediates, including AB apomicts and two possible apomictic triploids (AAB and ABB). Under these circumstances, even the most distinctive sexual diploid progenitors can become lost in a seemingly continuous cline of morphological variability. In the following treatment, only the sexual diploid species of the Four Corners region are described and included in the key. These represent the morphological extremes of the complex, and all have validly published binomials. Although commonly encountered in nature and the herbarium, apomictic hybrids are not formally treated here for two reasons. First, the hybrids are difficult to circumscribe, representing almost every possible combination of diploid genomes. Second, very few of these hybrids have been named, and none of these have yet been transferred to *Boechera*. Given our current state of knowledge, the best way to identify a possible hybrid is to provide a formula name based on the hypothesized parentage (e.g., *Boechera fendleri* × *B. stricta* or *B. fendleri* hybrid). This requires an accurate understanding of the sexual diploids, which we hope the following keys and descriptions will provide. In all cases, measurements of stem length are taken from fruiting plants, those of basal leaves from the largest in the basal rosette, for the fruiting pedicels from the longest in the infructescence, and for the stem hairs from the largest near the base. Descriptions of the pedicels, flowers, and fruits, as well as the number of flowers, are taken from the main inflorescence rather than its lateral branches; number of seed rows per locule is determined near the middle of the fruit.

1. Fruits pubescent throughout; upper cauline leaves densely pubescent and lacking auricles; petals usually more than 10 mm long .... ***B. formosa***

1′ Fruits glabrous; upper cauline leaves glabrous to sparsely pubescent or, if densely pubescent, with well-developed auricles; petals less than 10 mm long (except in some *B. stricta*) .... (2)

2. Plants nearly glabrous except for scattered, sessile, 2-rayed hairs on margins (and often surfaces) of basal leaves; fruits erect, with 2 rows of seeds in each locule .... ***B. stricta***

2′ Plants pubescent (at least proximally) with simple or branched hairs, the latter raised above the leaf surface on a short but definite stalk; fruits pendent to loosely ascending, with 1 or 2 rows of seeds in each locule .... (3)

3. Basal leaves coarsely ciliate and hirsute with simple or 2- to 3-rayed hairs, occasionally almost glabrous .... (4)

3′ Basal leaves minutely pubescent with dendritic hairs, rarely ciliate (except on petioles) and never glabrous .... (6)

4. Upper cauline leaves with prominent auricles, the largest more than 0.8 mm long; stems of mature plants usually more than 2.6 dm tall, arising from center of basal rosette or laterally below sterile caudex shoot; longest fruiting pedicels more than 9 mm long .... ***B. fendleri***

4′ Upper cauline leaves without auricles or the auricles less than 0.8 mm long; stems of mature plants usually less than 2.6 dm tall, arising laterally below sterile caudex shoot; longest fruiting pedicels less than 9 mm long .... (5)

5. Basal leaves oblanceolate to obovate, the largest more than 2.5 mm wide; lower portion of stems hirsute with mostly simple hairs, the largest more than 0.4 mm long; upper cauline leaves often with small auricles; seeds in 2 rows in each locule; species barely entering our region from the northwest .... ***B. pendulina***

5′ Basal leaves narrowly oblanceolate to linear, the largest less than 2.5 mm wide; lower portion of stems glabrous or pubescent with mostly 2-rayed hairs less than 0.4 mm long; upper cauline leaves without auricles; seeds in 1 row in each locule; species barely entering our region from the northeast .... ***B. oxylobula***

6. Lower portion of stems glabrous or very sparsely pubescent; fruits ascending or rarely almost perpendicular to rachises, secund; fruiting pedicels usually less than 6 mm long; sepals dark purple, at least in part; species confined to high montane and alpine habitats .... ***B. lemmonii***
6′ Lower portion of stems usually densely pubescent; fruits pendent to loosely ascending, not obviously secund (except in some *B. retrofracta*); fruiting pedicels usually more than 6 mm long; sepals greenish to dull lavender; species found in desert to midmontane habitats .... (7)

7. Plants with a single dominant flowering stem arising from center of basal rosette; inflorescences usually with more than 15 flowers or fruits; pedicels turning sharply downward in late anthesis, the young flowers ascending and the older flowers reflexed; fruits usually appressed-pendent .... ***B. retrofracta***
7′ Plants usually with several flowering stems arising laterally below sterile caudex shoot; inflorescences with 5–15 (rarely more) flowers or fruits; pedicels remaining upright or curving gently downward in fruit, the orientation of young and old flowers generally similar; fruits ascending to widely pendent .... (8)

8. Mature fruits loosely ascending or occasionally perpendicular to rachises; basal leaves oblanceolate to obovate with the largest blades usually more than 6 mm wide, often shallowly dentate, pubescent with slightly overlapping hairs .... ***B. pallidifolia***
8′ Mature fruits widely pendent; basal leaves narrowly oblanceolate to linear with blades usually less than 6 mm wide, entire, pubescent (often canescent) with densely overlapping hairs .... ***B. lignifera***

Boechera fendleri (S. Watson) W. A. Weber (for Augustus Fendler) Fendler's rockcress. Perennial (sometimes biennial), hirsute proximally, nearly glabrous distally. **STEMS** 1 to several (up to 7) per caudex branch, arising from center of basal rosette or laterally below sterile caudex shoot, 2.6–8 dm tall, lower portions hirsute with simple or spurred hairs 0.5–0.9 mm long. **LEAVES** at base of stem oblanceolate, 4–10 (–14) mm wide, often dentate, strongly ciliate (at least on the petioles), blade surfaces nearly glabrous or hirsute with coarse, simple and stalked, 2-rayed hairs mostly 0.4–1 mm long; upper cauline leaves glabrous, with auricles 0.8–3 mm long. **INFL** 6–40 (–74)-flowered; fruiting pedicels 9–23 mm long, perpendicular to rachises or slightly ascending, curved or angled downward, glabrous. **FLOWERS** ascending prior to fruit development; sepals greenish to dull lavender (rarely purple), sparsely hirsute; petals whitish to lavender, 5–9 mm long. **FRUIT** 1.3–2.2 mm wide at maturity, widely pendent, not secund, glabrous. **SEEDS** in 2 rows in each locule. $2n = 14$. [*Arabis fendleri* (S. Watson) Greene; *A. holboellii* Hornem. var. *fendleri* S. Watson]. Meadows, open forest floor, and roadsides in mountainous areas. ARIZ: Apa; COLO: LPl, Mon; NMEX: McK, SJn; UTAH. 2000–2800 m (6600–9100′). Flowering: May–Jul (Sep). Although reported from surrounding areas, typical diploid *B. fendleri* apparently is confined to the Four Corners states. Local plants attributed (incorrectly) to var. *spatifolia* (Rydb.) Dorn are apomictic triploid hybrids with *B. oxylobula*.

Boechera formosa (Greene) Windham & Al-Shehbaz (beautiful) Pretty or pallid rockcress. Long-lived perennials (often with an elevated, woody base), pubescent throughout. **STEMS** usually 1 (–3) per caudex branch, arising terminally, 2–5 dm tall, pubescent with stalked, dendritic hairs 0.1–0.3 mm long. **LEAVES** at base of stem linear or rarely linear-oblanceolate, 2–3 (–4) mm wide, entire, not ciliate, blade surfaces pubescent like the stems; upper cauline leaves densely pubescent, without auricles. **INFL** (6–) 9–22-flowered; fruiting pedicels 4–10 mm long, slightly to strongly downcurved, uniformly pubescent. **FLOWERS** ascending early in anthesis but divaricate or even downcurved later; sepals dull lavender to greenish, pubescent; petals white to pale lavender, (8–) 10–18 mm long. **FRUIT** 1.6–2.7 mm wide at maturity, pendent and occasionally appressed, often somewhat secund, pubescent throughout. **SEEDS** in 2 rows in each locule. $2n = 14$. [*Arabis formosa* Greene; *A. pulchra* M. E. Jones ex S. Watson var. *pallens* M. E. Jones; *Boechera pulchra* (M. E. Jones ex S. Watson) W. A. Weber subsp. *pallens* (M. E. Jones) W. A. Weber]. In sandy soil, usually among blackbrush, sagebrush, or piñon-juniper. Confined to sandy or rocky habitats in the Four Corners states. ARIZ: Apa, Nav; COLO: Mon; NMEX: SJn; UTAH. 1300–1900 m (4300–6200′). Flowering: Apr–Jun. Although rarely confused with other species, *B. formosa* has hybridized with several other taxa. The resulting apomictic diploids and triploids are immediately recognizable by the sporadic and incomplete pubescence of ovaries and fruits.

Boechera lemmonii (S. Watson) W. A. Weber (for John Lemmon, 1832–1908) Lemmon's rockcress. Plants usually long-lived, somewhat matted perennials, pubescent proximally, nearly glabrous distally. **STEMS** usually 1 (–3) per caudex branch, arising from center of basal rosette or laterally below sterile caudex shoot, 0.5–2 dm tall, lower portions glabrous or very sparsely pubescent with stalked, dendritic hairs 0.1–0.2 mm long. **LEAVES** at base of stem oblanceolate to obovate, 2.5–5 mm wide, entire (rarely dentate), often ciliate on petioles with branched hairs, blade surfaces sparsely to densely pubescent with dendritic hairs 0.1–0.2 mm long; upper cauline leaves glabrous to sparsely pubescent, without auricles (rarely) or the auricles 0.1–0.5 (–1) mm long. **INFL** 5–14-flowered; fruiting pedicels 2.5–6 mm long, ascending or rarely perpendicular to rachises, usually curved, glabrous. **FLOWERS** ascending prior to fruit development; sepals purple or purplish green, glabrous or sparsely pubescent; petals lavender to purple, 4–6 mm long. **FRUIT** 1.5–2 mm wide at maturity, ascending or rarely almost perpendicular to rachises, secund, glabrous. **SEEDS** in 1 row in each locule. 2*n* = 14. [*Arabis lemmonii* S. Watson]. Rocky habitats above tree line. COLO: Hin, SJn. 3800–4050 m (12500–13300′). Flowering: Jun–Aug. Alaska and Alberta south to Colorado and California. Local plants have more ascending fruits and slightly larger hairs than typical *B. lemmonii*; they closely match the type of *Arabis egglestonii* Rydb. from Gunnison County, Colorado.

Boechera lignifera (A. Nelson.) W. A. Weber (wood-bearing) Nelson's rockcress. Short-lived perennials, pubescent proximally, nearly glabrous distally. **STEMS** usually 2–5 (–9) per caudex branch, arising laterally below sterile caudex shoot, 1.2–3 (–4) dm tall, the lower portions finely pubescent with dendritic hairs 0.1–0.2 mm long. **LEAVES** at base of stem narrowly oblanceolate, 2–5 (–7) mm wide, entire, rarely ciliate at base of petiole, blade surfaces pubescent with stalked, densely overlapping, dendritic hairs to 0.1–0.2 mm long; upper cauline leaves sparsely pubescent, with auricles (0.5–) 1–2 mm long. **INFL** 6–15-flowered; fruiting pedicels (5–) 7–16 mm long, divaricate, gently downcurved, sparsely pubescent to glabrous. **FLOWERS** all ascending prior to fruit development; sepals greenish to dull lavender, sparsely pubescent; petals whitish but often aging pale lavender, 5–7 mm long. **FRUIT** 1.2–2 mm wide at maturity, widely pendent, not secund, glabrous. **SEEDS** in 1 row in each locule. 2*n* = 14. Sandy soil, usually under piñon or juniper trees. ARIZ: Apa, Nav; COLO: Mon; NMEX: McK, RAr, San, SJn. 1900–2300 m (6200–7500′). Flowering: Apr–early Jun. Idaho and Wyoming south to Arizona and New Mexico. Occasionally misidentified as *B. perennans* (S. Watson) W. A. Weber, a species with predominantly 2-rayed stem hairs, that does not occur in the Four Corners region.

Boechera oxylobula (Greene) W. A. Weber (sharp pod or capsule) Low rockcress. Perennial, usually sparsely pubescent proximally, often glabrous distally. **STEMS** (1–) 3–7 per caudex branch, arising laterally below sterile caudex shoot, 0.4–1.7 dm long, lower portion glabrous or pubescent with mostly 2-rayed hairs 0.1–0.4 mm long. **LEAVES** at base of stem narrowly oblanceolate to linear, 1–2.5 mm wide, entire or rarely dentate, ciliate with simple, subsetose hairs 0.3–0.7 mm long, blade surfaces glabrous or sparsely pubescent with stalked, 2- and 3-rayed hairs 0.1–0.4 mm long; upper cauline leaves sparsely pubescent or glabrous, without auricles. **INFL** (2–) 4–12-flowered; fruiting pedicels 3–6 mm long, divaricate, slightly to strongly downcurved, glabrous or with occasional isolated hairs. **FLOWERS** ascending prior to fruit development; sepals greenish to dull lavender, glabrous or sparsely pubescent; petals white to pale lavender, 4–5 mm long. **FRUIT** 1.2–2 mm wide at maturity, pendent, rarely somewhat secund, glabrous. **SEEDS** in 1 row in each locule. 2*n* = 14. [*Arabis demissa* Greene; *A. oxylobula* Greene; *A. rugocarpa* Osterh.; *Boechera demissa* (Greene) W. A. Weber]. Rocky habitats in the mountains. COLO: Min. 3600 m (11800′). Flowering: May–Jul. Endemic to the mountains of western Colorado. The name *B. demissa* (Greene) W. A. Weber has been widely misapplied (Windham and Al-Shehbaz 2006); the type specimen closely resembles *B. oxylobula* and is here placed in synonymy under the older name. Our plants represent the more pubescent extreme, typified by *Arabis rugocarpa* Osterh. from Lake County, Colorado.

Boechera pallidifolia (Rollins) W. A. Weber (pale-leaved) Four Corners rockcress. Perennial, pubescent proximally, nearly glabrous distally. **STEMS** usually 2–5 (–9) per caudex branch, arising laterally below sterile caudex shoots, (0.5–) 1.5–3.7 dm tall, lower portion finely pubescent with stalked, dendritic hairs 0.1–0.3 mm long. **LEAVES** at base of stem oblanceolate to obovate, 5–11 (–13) mm wide, shallowly dentate or occasionally entire, often somewhat ciliate on petioles, blade surfaces pubescent with stalked, slightly overlapping, dendritic hairs 0.1–0.3 mm long; upper cauline leaves sparsely pubescent, with auricles (0.5–) 1–2 mm long. **INFL** 4–15 (–20)-flowered; fruiting pedicels (5–) 7–15 mm long, ascending, straight or slightly recurved, glabrous or sparsely pubescent. **FLOWERS** ascending prior to fruit development; sepals greenish to dull lavender (rarely purplish), sparsely pubescent; petals pale lavender to lavender or (rarely) whitish, 5–9 mm long. **FRUIT** 1–2 mm wide at maturity, ascending or rarely perpendicular to rachises, not secund, glabrous. **SEEDS** in 1 row in each locule. 2*n* = 14. [*Arabis pallidifolia* Rollins; *A. thompsonii* S. L. Welsh]. Sandy soil, usually under piñon or juniper trees. ARIZ: Apa, Coc, Nav; COLO: LPl, Mon; NMEX: McK, RAr, San, SJn; UTAH. 1750–2300 m (5700–7500′). Flowering: Apr–Jun. Wyoming and Utah south to Arizona and New Mexico. This species has hybridized extensively with *B. fendleri*, producing a confusing array of apomictic triploids. Such plants often are misidentified as *B. perennans* (S. Watson) W. A. Weber, a low desert species that does not enter our region. The type specimen of *Arabis selbyi* Rydb. is one such triploid hybrid.

Boechera pendulina (Greene) W. A. Weber (hanging down) Pendulous rockcress. Perennial, hirsute proximally, nearly glabrous distally. **STEMS** usually 2–6 per caudex branch, arising laterally below sterile caudex shoots, 0.6–2.6 (–3) dm tall, lower portion sparsely to moderately hirsute with mostly simple hairs 0.3–0.8 mm long. **LEAVES** at base of stem oblanceolate to obovate, 2.5–6 mm wide, entire (rarely dentate), ciliate with setose, mostly simple hairs 0.4–1 mm long, blade surfaces glabrous or sparsely hirsute with simple and stalked, 2-rayed hairs 0.3–0.8 mm long; upper cauline leaves sparsely ciliate or hirsute to glabrous, without auricles or with small auricles to 0.7 mm long. **INFL** 6–14-flowered; fruiting pedicels 4–7 (–9) mm, perpendicular to rachises or slightly ascending, usually curved or angled downward, glabrous (very rarely sparsely pubescent). **FLOWERS** ascending prior to fruit development; sepals green to dull lavender, sparsely pubescent (rarely glabrous); petals whitish to pale lavender, 4–6 mm long. **FRUIT** 1.2–2.1 mm wide at maturity, widely pendent, not secund, glabrous. **SEEDS** in 2 rows in each locule. $2n$ = 14. [*Arabis setulosa* Greene; *Boechera demissa* (Greene) W. A. Weber var. *pendulina* (Greene) N. H. Holmgren]. Rocky and/or sandy soil in piñon-juniper woodland or sagebrush areas. UTAH. 2050–2700 m (6700–8800′). Flowering: Apr–Jun. Wyoming to Nevada south to Arizona and Colorado. Specimens of *B. pendulina* usually are misidentified as *Arabis demissa* Greene. As indicated by Windham and Al-Shehbaz (2006), the latter name is a synonym of *B. oxylobula*, a taxon that is quite distinct from *B. pendulina*. Although rare in the flora area, *B. pendulina* has hybridized with *B. fendleri* to form apomictic triploids known from both Arizona and Utah.

Boechera retrofracta (Graham) Á. Löve & D. Löve (backwards and broken) Reflexed rockcress. Biennial or short-lived perennial, densely pubescent proximally, sparsely pubescent to nearly glabrous distally. **STEMS** usually 1 per caudex branch, arising from center of basal rosette, 3–7 dm tall, lower portion finely pubescent with stalked, dendritic hairs 0.1–0.3 mm long. **LEAVES** at base of stem oblanceolate, 2.5–6 mm wide, entire or shallowly dentate, not cilate, blade surfaces pubescent with stalked, densely overlapping, dendritic hairs 0.05–0.2 mm long; upper cauline leaves pubescent (especially on abaxial surfaces), with auricles 0.5–2.5 mm long. **INFL** 15–60-flowered; fruiting pedicels 7–12 mm long, descending, abruptly angled downward near base, sparsely (rarely densely) pubescent. **FLOWERS** ascending in early anthesis but strongly downcurved later; sepals dull lavender to greenish, pubescent; petals whitish to lavender, 4–7 mm long. **FRUIT** 0.9–1.5 mm wide at maturity, pendent and usually appressed to rachises, often somewhat secund, glabrous to sparsely pubescent with dendritic hairs. **SEEDS** usually in 2 irregular rows in each locule. $2n$ = 14. [*Arabis holboellii* Hornem.; *A. holboellii* var. *retrofracta* (Graham) Rydb.; *A. holboellii* var. *secunda* (Howell) Jeps.; *A. retrofracta* Graham; *Boechera holboellii* (Hornem) Á. Löve & D. Löve var. *secunda* (Howell) Dorn]. Typical diploid *B. retrofracta* has a more northerly distribution and is not yet known from the *Flora* area. However, this species has contributed the dominant genome to several apomictic triploid hybrids found in the Four Corners region. The name *Arabis consanguinea* Greene [*A. holboellii* var. *consanguinea* (Greene) G. A. Mulligan] is based on one of these hybrids. Most common is *B. retrofracta* × *B. fendleri*. Piñon-juniper woodland and ponderosa pine communities. COLO: Arc, Dol, LPl, Mon, SMg; NMEX: RAr, SJn; UTAH. 1640–2460 m (5000–7500′). Flowering: May–Jun. Hybrids with a single genome of *B. retrofracta* are commonly misidentified as *B. holboellii* (Hornem.) Á. Löve & D. Löve var. *pinetorum* (Tidestr.) Dorn, a distantly related taxon confined to the central Sierra Nevada.

Boechera stricta (Graham) Al-Shehbaz (very upright) Drummond's rockcress. Perennial (sometimes biennial), glabrous except for sporadic hairs near base. **STEMS** usually 1 (up to 7) per caudex branch, arising from center of basal rosette or (rarely) below sterile caudex shoot, (1.5–) 2–8 dm tall, lower portion glabrous or with a few malpighiaceous (rarely simple) hairs 0.3–0.7 mm long. **LEAVES** at base of stem oblanceolate, 2–8 (–14) mm wide, entire or rarely dentate, often ciliate with short simple hairs on petiole, blade surfaces with sessile, 2-rayed hairs 0.3–0.7 mm long usually concentrated along margins and on abaxial surfaces, rarely almost glabrous; upper cauline leaves glabrous, with auricles (0.5–) 1–3 mm long. **INFL** 8–35-flowered; fruiting pedicels (5–) 8–18 mm long, erect and appressed to rachis, glabrous. **FLOWERS** erect-ascending; sepals green to dull lavender, usually glabrous; petals white (but often aging pale lavender), 6–11 mm long. **FRUIT** 1.5–3.1 mm at maturity, erect and often appressed to rachis, not secund, glabrous. **SEEDS** in 2 rows in each locule. $2n$ = 14. [*Arabis drummondii* A. Gray; *A. connexa* Greene; *Boechera drummondii* (A. Gray) Á. Löve & D. Löve; *Turritis stricta* Graham]. Meadows, open forest floor, and roadsides in mountainous areas. ARIZ: Apa; COLO: Arc, Hin, LPl, Min, Mon, RGr, SJn; NMEX: McK, RAr, SJn; UTAH. 2350–3900 m (7700–12800′). Flowering: May–Aug. Alaska and Northwest Territories south to New Mexico and California. Apomictic triploid hybrids between *B. stricta* and *B. fendleri*, differing from the former in their more divergent fruits and minutely stalked, 2-rayed hairs, are common at middle elevations in the *Flora* area. Such specimens often are misidentified as *B.* ×*divaricarpa*, a more northerly taxon with a different hybrid parentage.

Brassica L. Mustard, Turnip, Cabbage, Broccoli

(the Latin name for cabbage) Annual (ours), biennial or perennial, often glaucous, glabrous or pubescent with simple hairs, herbs (ours) to shrubs. **STEM** usually single and freely branched above. **LEAVES**: basal usually rosulate, petiolate, pinnate or lyrate-pinnatifid; cauline short-petiolate or sessile, sometimes auriculate to amplexicaul. **FLOWER** sepals erect or ascending, lateral pair usually gibbous-based; petals yellow, broadly to narrowly obovate. **FRUIT**: silique oblong to linear, terete to latiseptate or rarely quadrangular, erect to divaricate, often divided into 2 dissimilar portions, a usually seedless beak and a basal seed-bearing segment. **SEEDS** 1 row per locule, mucilaginous or not. x = 7–11. Thirty-nine species centered in Macaronesia and the Mediterranean. Several species are important crops (see discussion in family description) and weeds. The genus as a whole is polyphyletic and will likely be split in the future into segregate genera.

1. Upper cauline leaves auriculate-clasping, with a wide blade .... ***B. rapa***
1′ Upper cauline leaves cuneate at base and scalelike .... ***B. tournefortii***

Brassica rapa L. (of fields) Field mustard, rape mustard, turnip. Annual, taprooted, glabrous to sparsely pubescent below with simple hairs. **STEM** usually single but freely branched above, up to 10 dm tall. **LEAVES**: basal and lower cauline ones petiolate, ± lyrate-pinnatifid with 2–4 lateral lobes, the terminal lobe ovate; rosulate leaves withering by flowering time; cauline leaves becoming progressively reduced in size and less lobed, the petioles also becoming shorter until from middle stem to the apex the leaves are auriculate-clasping. **INFL** corymbose raceme in flower, elongating in fruit. **FLOWER** sepals lanceolate to narrowly oblong, 4–6 mm long; petals yellow, 6–10 mm long with narrow claws and oblanceolate blades possessing darker veins. **FRUIT**: silique 3–7 cm long, flattened, weakly torulose at most, divaricately ascending, clearly differentiated into a body and a beak 8–15 mm long, the beak tapering into a slender style. **SEEDS** 1 row per locule, spherical, about 1.5 mm long, reddish brown. n = 10, 2n = 20. [*Brassica campestris* L.]. One record from a parking area. ARIZ: Nav. 1760 m (5775′). Flowering: late Apr–early May. Native to Europe but now established in much of North and Central America. The source of canola (rapeseed) oil, bok choy, and turnip.

Brassica tournefortii Gouan (named for French botanist Joseph Pitton de Tournefort, 1656–1708) Asian mustard, African mustard, Sahara mustard. Annual, covered with stiff white trichomes, sometimes glabrescent above. **STEM** single but freely branched from the base to the apex, up to 6 (9) dm tall. **LEAVES**: basal in a dense rosette, persistent, on short, broad petioles, pinnately lobed to lyrate with 8–14 pairs of lobes, dentate, terminal lobe large; cauline leaves rapidly reduced upward and mostly bractlike above. **INFL** paniculately branched and greatly elongating in fruit. **FLOWER** sepals with hyaline margins, lanceolate to narrowly oblong, 2.6–4 mm long; petals 4–6.5 mm long, pale yellow, often with some violet in the throat, rarely white. **FRUIT**: silique divaricately spreading, torulose, differentiated into a body 2.7–4.5 cm long and a tapering terete beak 12–19 mm long. **SEEDS** 1 row per locule, spherical, mucilaginous when wet, dark brown or orange-brown. n = 10, 2n = 20. [*Coincya tournefortii* (Gouan) Alcaraz, T. E. Díaz, Rivas Mart. & Sánchez-Gómez]. Only recently discovered in our area; spreading into weedy areas near the Colorado and San Juan rivers around and near Lake Powell on sandy shores and on sand dunes; its spread is facilitated by oscillating lake levels. UTAH. 1100–1150 m (3600–3800′). Flowering: Apr–early May. Native to the Mediterranean area and now found in California, Arizona, southern Utah, Texas, and Mexico.

Camelina Crantz False Flax

(Greek *chamai*, on the ground, + *linon*, flax, apparently referring to the plant as a weed in flax fields) Annual or biennial, usually pubescent with dendritic to stellate (rarely simple) hairs. **STEM** single and simple or branched above. **LEAVES** both basal and cauline, simple, entire to dentate, cauline leaves sessile, sagittate or amplexicaul, lanceolate to linear. **FLOWER** sepals erect to ascending, the inner slightly gibbous-based; petals yellow to white, obovate or spatulate. **FRUIT** a silique or silicle, pyriform to clavate or oblong, generally a little inflated but ± latiseptate and reticulate, ascending to divaricate or erect and appressed to the rachis, valves firm, 1-nerved. **SEEDS** copiously mucilaginous. x = 6, 8, 10, 13. Eleven species native to southeast Europe and the Middle East; four species are widespread weeds.

1. Silicle 7–10+ mm long; plants ± glabrous to sparsely pubescent .... ***C. sativa***
1′ Silicle less than 7 mm long; plants densely pubescent, especially below .... (2)

2. Petals less than 5 mm long, yellow; basal leaves withering by flowering time .... ***C. microcarpa***
2′ Petals 6–9 mm long, white outside, pale yellow inside; basal leaves present at flowering time .... ***C. rumelica***

Camelina microcarpa Andrz. ex DC (small fruit) Littlepod false flax. Annual, hirsute with simple (rarely forked) trichomes with a lower level of smaller stellate trichomes below, sometimes glabrous above. **STEM** single and occasionally branched, 3–10 dm tall. **LEAVES** primarily cauline, the basal ones withering by anthesis, short-petiolate to sessile, becoming sagittate-auriculate above, 2–8 cm long, lanceolate to oblanceolate, entire or nearly so. **INFL** becoming elongated and often compound, with fruiting pedicels spreading-ascending. **FLOWER** sepals oblong, 2–3 mm long with rounded apices; petals pale yellow, fading white, 3–6 mm long. **FRUIT** a silicle, obovoid, obtuse at apex, 5–7 mm long, inflated but firm, glabrous. **SEEDS** plump, 0.9–1.4 mm long, narrowly ovoid. $n = 20$, $2n = 40$. [*C. sativa* (L.) Crantz subsp. *microcarpa* (Andrz. ex DC.) Em. Schmid]. Weedy areas of desert shrub, piñon-juniper, slickrock, roadsides, irrigation ditches, mountain shrub, sagebrush, and pond margins. ARIZ: Apa, Nav; COLO: Arc, Dol, LPl, Min, Mon, SMg; NMEX: RAr, SJn; UTAH. 1525–2425 m (5000–7950′). Flowering: mid-May–early Jul. Eurasian, widely introduced and weedy in North America. Seeds used by the Chiricahua and Mescalero Apache people to make bread.

Camelina rumelica Velen. (for Rumelia, Bulgaria) Graceful false flax. Annual, hirsute with ± dense, simple, long white trichomes, nearly glabrous above, especially in the inflorescence. **STEM** single, simple or branched from the base, 2–6 dm tall. **LEAVES**: basal leaves forming a loose rosette, lanceolate to oblong, entire to irregularly dentate; cauline leaves oblong, irregularly toothed, with acute auricles. **INFL** with fruiting pedicels horizontally spreading. **FLOWER** sepals with long simple trichomes; petals pale yellow-white to white, oblong-spatulate with a very short claw, 6–9 mm long. **FRUIT** a silicle, obovate, 5–7 (–8) mm long. **SEEDS** 1–1.2 mm long, plump, reddish brown or brown. $n = 6, 12$, $2n = 12, 26$. [*C. sylvestris* Wallr. var. *albiflora* Boiss.]. Disturbed areas of piñon-juniper communities. COLO: Arc, Mon; NMEX: SJn. 1850–2225 m (6100–7300′). Flowering: May–Jul. Eurasian, a weedy species expanding its range in the American West.

Camelina sativa (L.) Crantz (sown, cultivated) Gold-of-pleasure. Annual (biennial), ± glabrous, if trichomes present, these simple and branched. **STEM** single, usually branched above. **LEAVES**: basal leaves oblong-lanceolate, sinuately and irregularly toothed; cauline leaves linear-oblong, base sagittate or strongly auriculate, margins entire or remotely denticulate. **INFL** with fruiting pedicels ascending. **FLOWER** sepals 2–3 mm long; petals yellow, spatulate with a narrow claw. **FRUIT** a silicle, obovate, 7–10 mm long, typically truncate apically. **SEEDS** oblong, 1.5–3 mm long, plump, dark brown. $n = 20$, $2n = 26, 28, 40$. Disturbed areas, especially roadsides. NMEX: McK, SJn. 1850–1925 m (6100–6300′). Flowering: May. Eurasian, now widely introduced in temperate North America.

Capsella Medik., nom. conserv. Shepherd's Purse

(Latin *capsella*, a little box) Annual or biennial herbs with dendritic or stellate trichomes mixed with simple trichomes. **STEMS** erect or ascending, 1 to several from base. **LEAVES**: basal rosulate, petiolate, entire to pinnatifid; cauline sessile, auriculate or sagittate at base. **FLOWER** sepals ascending; petals white, obovate to spatulate, sometimes very short or lacking. **FRUIT** a silicle, obdeltate to obcordate, strongly angustiseptate, ascending or divaricate, valves keeled. **SEEDS** mucilaginous. $x = 8$. A genus of five species. Some authors have recognized a single, very variable species (*C. bursa-pastoris*), while others have recognized as many as 200 species! Native to Eurasia.

Capsella bursa-pastoris (L.) Medik. (purse + shepherd, referring to the shape of the fruit) Shepherd's purse. Annual or winter-annual, pubescent with a mix of simple and stellate trichomes but ± glabrous above. **STEMS** simple or branched, 1–5 dm tall. **LEAVES**: basal leaves strongly rosette-forming, oblanceolate, subentire to toothed to lyrate-pinnatifid, 3–6 cm long on a winged petiole; lowest cauline leaves ± like the rosette leaves but becoming lanceolate-oblanceolate, remotely serrate-dentate, sessile and auriculate-clasping above. **INFL** a many-flowered raceme or panicle. **FLOWER** sepals ovate and obtuse, 1–2 mm long, scarious-margined and sometimes white at the tip; petals oblanceolate to spatulate, clawed, 2–3 mm long, white. **FRUIT** a silicle, markedly triangular-obcordate with a cuneate base, strongly flattened perpendicular to the septum, midnerve forming a keel, 4.5–7 mm long. **SEEDS** oblong, brown, 0.8–1 mm long, mucilaginous when wetted. $n = 8, 16$, $2n = 16, 32$. [*Thlaspi bursa-pastoris* L.; *Bursa bursa-pastoris* (L.) Britton]. Riparian areas, meadows, disturbed weedy areas, wash bottoms, roadsides, grassy ponderosa pine forests, piñon-juniper, and oak brush communities. ARIZ: Apa, Nav; COLO: Arc, Dol, Hin, LPl, Min, Mon, SJn, SMg; NMEX: McK, RAr, SJn; UTAH. 1850–3050 m (6100–10000′). Flowering: late Apr–early Sep. Native to Eurasia but now the second most common weed on earth. Seeds used by the Chiricahua and Mescalero Apache people to make bread. Used in the Far East as a vegetable and as a treatment for eye disease and dysentery.

Cardamine L. Bittercress

(*kardamon*, a Greek name used by Dioscorides for a cresslike plant) Annual, biennial, or perennial herbs, often rhizomatous and these sometime tuberous, glabrous or pubescent with simple hairs. **STEMS** erect to ascending, simple or branched. **LEAVES**: basal pinnately compound, trifoliate, pinnatifid or entire; cauline petiolate or sessile, sometimes auriculate, simple or compound. **FLOWER** sepals erect or ascending; petals white to purplish, obovate to spatulate. **FRUIT** a silique, linear to subulate, somewhat latiseptate to subterete, with coiling, elastically dehiscent valves. **SEEDS** 1 row per locule, mucilaginous or not. $x = 6, 7, 8, 10$. [*Dentaria* L.]. One hundred ninety-seven species worldwide, especially in temperate zones.

1. Leaves all simple (a few may be slightly pinnatifid); petals showy, 7–12 mm long .... ***C. cordifolia***
1' Leaves all pinnate (some may be pinnatifid); petals small, 1.2–2.1 mm long .... ***C. pensylvanica***

Cardamine cordifolia A. Gray (heart-shaped leaf) Heartleaf bittercress, large-mountain bittercress. Perennial with extensively spreading slender rhizomes, mostly glabrous but the stems sometimes densely pubescent near the base with simple, short, spreading trichomes. **STEMS** mostly simple, erect, to 6 (–8) dm tall. **LEAVES** all cauline, those near the base with petioles up to 8 cm long, the petioles decreasing in size upward, blades reniform to ovate-cordate to deltate-cordate, simple, somewhat fleshy, the lowermost sometimes with a pair of small ovoid leaflets below the larger terminal leaflet. **INFL** a simple raceme. **FLOWER** sepals elliptic to oblong, with white tips, the inner pair gibbous at the base; petals relatively showy, white, with a broad obovate blade emarginate or truncate-rounded at the tip, 7–12 mm long. **FRUIT** a silique, linear, straight, very slightly flattened parallel to the septum, 2–3.7 mm long. **SEEDS** in 1 row per locule, oblong, dark brown, 1.4–1.6 mm long. $2n = 24$. [*C. cardiophylla* Rydb.; *C. uintahensis* F. J. Herm.]. Wet meadows and riparian areas in Douglas-fir, spruce, ponderosa pine, cottonwood, and willow communities. ARIZ: Apa; COLO: Arc, Con, Hin, LPl, Min, Mon, SJn; UTAH. 2100–3725 m (6900–12200′). Flowering: mid-June–mid-Aug. Western North America. The species can only with difficulty be separated into three very weakly differentiated varieties that are not here described. Densely pubescent plants have been referred to var. *incana* A. Gray; plants with leaf blades longer than wide are var. *cordifolia*; and plants with leaf blades wider than long are var. *lyallii* (S. Watson) A. Nelson & J. F. Macbr.

Cardamine pensylvanica Muhl. ex Willd. (of Pennsylvania) Pennsylvania bittercress. Biennial or short-lived perennial from a fibrous root system, glabrous but sometimes hirsute near the base. **STEMS** one or few, erect to decumbent, branched from the base or throughout. **LEAVES** pinnate (–pinnatifid) with 5–11 (–19) leaflets, lateral leaflets broadly elliptic to obovate, sessile, cuneate at base; terminal leaflets more broadly obovate and slightly larger; leaflets (and hence leaves) becoming smaller distally; margins entire to dentate. **INFL** a raceme elongating in fruit. **FLOWER** sepals lanceolate, 1.2–2.1 mm long; petals white, oblanceolate to narrowly spatulate, not clawed, 2–4 mm long. **FRUIT** a silique, slender, 5–10 mm long, about 1 mm wide, terete. **SEEDS** in 1 row per locule, oblong, brown, 0.7–1 mm long. $n = 16$, $2n = 32$. [*C. flexuosa* With. subsp. *pensylvanica* (Muhl. ex Willd.) O. E. Schulz; *C. multifolia* Rydb.; *C. oregana* Piper]. Along stream margins and irrigation canals. NMEX: SJn. 1625–1650 m (5300–5400′). Flowering: May. Northern Alaska and Canada south to the northern and eastern portions of the contiguous United States.

Chorispora R. Br. ex DC., nom. conserv. Blue Mustard

(Greek *choris*, separate, + *spora*, seed, referring to the one-seeded disarticulating segments of the fruit) Annual (rarely perennial) herbs with simple trichomes and often also with multicellular glandular hairs. **STEMS** rather stout, usually single but branched throughout. **LEAVES**: basal rosulate, pinnately lobed or dentate; cauline petiolate (absent in perennial species), entire to pinnatifid. **FLOWER** sepals erect, lateral pair strongly gibbous-based; petals lavender to purple, yellow, or white, spatulate to oblong. **FRUIT** a silique, linear-terete, torulose to moniliform, indehiscent and breaking into 1-seeded segments, divaricate, beaked. **SEEDS** 1 row per locule, nonmucilaginous. $x = 7, 9$. Eleven species native to central Asia and the Middle East.

Chorispora tenella (Pall.) DC. (tender, soft) Blue mustard, purple mustard, musk mustard, crossflower, chorispora. Annual, from a taproot, with stipitate-glandular trichomes and often a few simple nonglandular trichomes. **STEMS** branched throughout, up to 5 dm tall, lower branches often decumbent. **LEAVES** elliptic-oblong to lanceolate (to oblanceolate), deeply sinuate-dentate, lower leaves with relatively long petioles, these reduced apically and leaves becoming sessile, main leaves with blades 3–8 cm long. **INFL** an elongated raceme, the lower flowers subtended by leaflike bracts. **FLOWER** sepals linear, obtuse to rounded, 4–7 mm long; petals blue-purple or magenta with darker veins (rarely white), often with a yellowish throat, long-clawed, oblanceolate, 9–12 mm long. **FRUIT** a silique, woody, with a stout, tapering beak, torulose, terete, 3–4.5 mm long including the beak, curved upward to less commonly nearly straight, indehiscent and breaking into 1-seeded segments. **SEEDS** greenish yellow, 1.2–1.5 mm long, narrowly ovoid. $n = 7$, $2n = 14, 28$. Disturbed sites of canyon bottoms, roadsides, and sagebrush, piñon-juniper, cottonwood, and ponderosa pine communities. ARIZ: Apa, Nav; COLO: Arc, LPl, Mon, SMg; NMEX: McK, RAr, San, SJn; UTAH. 1525–2125 m (5000–7000′). Flowering: late Mar–May. Native of south-central Eurasia, weedy and now abundant in the Great Plains and Intermountain West of the United States and southern Canada.

Conringia Heist. ex Fabr. Hare's Ear

(for Hermann Conring, 1606–1681, professor of medicine, University of Helmstedt, Germany) Annual or rarely perennial herbs, glabrous and usually glaucous. **STEMS** erect, simple. **LEAVES**: basal petiolate with somewhat sheathing petioles, entire; cauline sessile, amplexicaul, elliptic to oblong or ovate, entire to slightly wavy. **FLOWER** sepals erect or ascending, lateral pair subgibbous-based; petals white or yellow, oblanceolate to suborbicular, long-clawed. **FRUIT** a silique, linear, quadrangular, terete or latiseptate, often torulose. **SEEDS** 1 row per locule, strongly mucilaginous. $x =$ 7, 9. Six species of Europe and the Middle East.

Conringia orientalis (L.) Dumort (of the east) Hare's-ear mustard. Annual or winter annual, glabrous and glaucous. **STEMS** simple or a little branched above, 3–6 (9) dm tall including the overtopping fruits. **LEAVES**: basal obovate to oblanceolate and tapering to a winged petiole, (sub)entire, 5–9 cm long, cauline leaves oblong-lanceolate, sessile, cordate-clasping, entire. **INFL** a corymbiform raceme, few-flowered. **FLOWER** sepals linear to lanceolate, acute apically, 5–8 mm long; petals cream-white to very pale yellow, long-clawed, narrowly oblanceolate, 9–12 mm long. **FRUIT** a silique, ascending to nearly erect, linear-quadrangular but sometimes nearly terete, somewhat torulose, tapering to a slender tip with a thick style, 8–13 mm long. **SEEDS** brown, slightly winged, 1 row per locule, 2–2.5 mm long. $n = 7$, $2n = 14, 28$. [*Conringia perfoliata* (Crantz) Link]. Usually disturbed sites of roadsides, canyon bottoms, and in piñon-juniper and grass-dominated communities. ARIZ: Apa; COLO: Arc, Mon; NMEX: RAr, SJn; UTAH. 1850–2200 m (6100–7150′). Flowering: Jun. Native of Eurasia now abundant in the plains, but found throughout North America and almost worldwide.

Descurainia Webb & Berthel. Tansymustard

(for François Descurain, 1658–1740, French pharmacist and botanist) Annual, biennial, or less frequently perennial herbs (rarely shrubs) with dendritic trichomes, these sometimes mixed with unicellular glandular or simple trichomes. **STEMS** usually single, erect, usually branched. **LEAVES** petiolate, basal leaves rosulate, withering early and often not seen; cauline 2- or 3-pinnatisect to pinnate, bipinnate, or tripinnate, often very finely divided. **FLOWER** sepals erect, ascending, or rarely spreading or reflexed, not saccate; petals yellow to more rarely cream-white, spatulate or obovate to oblong, weakly clawed. **FRUIT** a silique or silicle, linear, oblong, clavate, elliptic, or fusiform, terete or quadrangular, the false septum sometimes perforated or lacking. **SEEDS** 1 or 2 rows per locule, mucilaginous (in ours). $x = 7$. Forty to 45 species, principally of North and South America and Macaronesia (Detling 1939). *Descurainia* is a difficult genus that is, apparently, rife with hybridization between both species and subspecies. Difficulties are exacerbated by a great deal of intrataxon morphological variation and by the weedy nature of the species, which occupy disturbed sites where two or more taxa may be colonists. A definitive monograph of the genus is needed. Fully mature fruits must be available for accurate identification. The keys below will work for most specimens, but intermediate morphologies are to be expected, especially between *D. incisa* and *D. pinnata* and among their subspecies.

1. Plants diminutive, < 1.5 cm tall; short-lived perennial; known only from Stony Pass (San Juan County, Colorado) above 3760 m (12300′) .... ***D. kenheilii***

1′ Plants larger; annual or biennial (short-lived perennial in the rare *D. torulosa*); distribution various .... (2)

2. Siliques fusiform, short and plump, obviously tapered at both ends, 2.5–4.5 mm long; styles relatively prominent, > (0.3) 0.4 mm long .... ***D. californica***

2′ Siliques linear, or clavate to obovate, sometimes tapered at both ends, (4) 5–28 mm long; styles obsolete or < 0.3 mm long .... (3)

3. Mature pedicels short, 1–2.5 (3) mm long; fruits torulose; growing at high elevations near 3658 m (12000′); stems and roots fine; short-lived perennial; very rare, Archuleta County, Colorado .... ***D. torulosa***

3′ Mature pedicels > 3 mm long, (some *D. incana* may be shorter); fruits not or only somewhat torulose; generally growing at lower elevations, (but up to 3590 m [11775′] in a few *D. incana*); stems and roots not especially fine; annual or biennial; common and widespread .... (4)

4. Siliques (13) 15–28 mm long; lower leaves bi- or tripinnate .... ***D. sophia***

4′ Siliques 4–15 (20) mm long; lower leaves all or mostly pinnate, sometimes a few bipinnate ones present (be sure to differentiate between pinnate-bipinnatifid and bipinnate) .... (5)

5. Siliques rounded to obtuse at apex, clavate or subclavate to narrowly obovate; seeds mostly in 2 rows per locule (note that all species have funicles on both sides of the replum, but in this species the seeds appear to be in 2 rows rather than crowded into 1 row); style obsolete or nearly so .... ***D. pinnata***

5′ Siliques acute or acuminate at apex, linear to narrowly oblong; seeds in 1 row per locule; style short but evident .... (6)

6. Siliques crowded and ± appressed to the rachis of the raceme, 1-nerved; pedicels erect .... ***D. incana***
6' Siliques not appressed to the rachis of the raceme, weakly 1-nerved; pedicels divaricately to widely spreading .... ***D. incisa***

Descurainia californica (A. Gray) O. E. Schulz (of California) Sierra tansymustard. Annual, winter-annual, or biennial, pubescent with small dendritic trichomes. **STEMS** usually single, erect, branched above, (3) 4–8 (10) dm tall. **LEAVES** obovate to oblanceolate, pinnate, the leaflets pinnatifid, crenate, or ± entire, upper leaves often simply pinnatifid. **INFL** an elongated raceme, with ascending to divaricately ascending pedicels. **FLOWER** sepals oblong, spreading, 0.9–1.4 mm long; petals bright yellow, oblanceolate, 1.1–1.7 mm long. **FRUIT**: silicles fusiform, relatively thick and plump, acutely tapered at both ends, 2.5–4.5 mm long. **SEEDS** light brown, few, 1–3 per locule, in 1 row. $n = 7$, $2n = 14$. [*Sisymbrium californicum* (A. Gray) S. Watson; *Sophia leptostylis* Rydb.]. Piñon-juniper, mountain shrub, Gambel's oak, spruce-fir, and white fir forests, especially where disturbed. COLO: LPl, Mon, SJn; NMEX: McK; UTAH. (1400) 1950–3400 m [(4600) 6400–11125′]. Flowering: late Jun–Aug. Wyoming, Colorado, and New Mexico west to California and Oregon. The short, plump, few-seeded, and fusiform fruits are distinctive in the genus.

Descurainia kenheilii Al-Shehbaz (for Ken Heil) Heil's tansymustard. Short-lived perennial, sparsely pubescent throughout with dendritic trichomes. **STEMS** 1–1.5 cm tall, erect, single from base, simple or rarely few-branched above. **LEAVES**: basal and lowermost cauline pinnate, oblanceolate in outline, petiolate; upper cauline leaves subsessile, smaller, sparsely pubescent. **INFL** a raceme, not elongated in fruit, fruiting pedicels erect to ascending, not appressed to rachis. **FLOWER** sepals yellowish, ovate, 1–1.4 mm long, ascending; petals yellowish, narrowly oblanceolate. **FRUIT**: silique linear, 6–10 × 1–1.3 mm, erect, not appressed to rachis, straight, terete, glabrous, torulose, with a distinct midvein; septum with a prominent midvein. **SEEDS** reddish brown, uniseriate, oblong, 4–8 per fruit. Thin soil and talus slopes in alpine tundra. COLO: SJn. 3764 m (12350′). Flowering: late Aug.–mid-Sep. Endemic to the San Juan Mountains, but only recently discovered and known only from the type location at the top of Stony Pass.

Descurainia incana (Bernh. ex Fisch. & C. A. Mey.) Dorn (hoary) Mountain tansymustard. Biennial, pubescent with fine dendritic trichomes, canescent to greenish, sometimes with some glandular trichomes. **STEMS** usually single, erect, branched above or throughout, (1.3) 3–12 dm tall. **LEAVES** ovate to broadly lanceolate or oblanceolate, pinnate, the leaflets of lower leaves pinnatifid, those of the upper entire. **INFL** an elongated raceme, with strongly ascending to erect pedicels, rather crowded. **FLOWER** sepals oblong, erect, 1–1.5 mm long; petals yellow, oblanceolate, 1.2–1.8 mm long. **FRUIT**: siliques ascending to erect and appressed, crowded, linear, acutely tapering at both ends or with the apex obtuse, 4–10 (12) mm long. **SEEDS** reddish brown, 4–8 per locule, in 1 row. $n = 7$, $2n = 14$. [*Sysymbrium incanum* Bernh. ex Fisch. & C. A. Mey.; *S. richardsonii* Sweet; *Descurainia richardsonii* O. E. Schulz nom. illeg.; *Sophia richardsonii* Rydb.]. Openings in spruce-fir forests, wetlands, meadows, riparian areas, often in disturbed sites. COLO: Arc, Min, RGr, SJn. 2925–3590 m (9600–11775′). Flowering: Jul–mid-Aug. Across Canada and Alaska, south to Maine in the east and Texas and southern California in the west, but apparently not present in Arizona, Oregon, and Washington.

Descurainia incisa (Engelm. ex A. Gray) Britton (cut, referring to the incised leaves) Mountain tansymustard. Annual, densely to sparsely pubescent (rarely subglabrous) with small dendritic trichomes, glandular hairs present in subsp. *viscosa*. **STEMS** single, erect, branched above or throughout, 2–7.5 dm tall. **LEAVES**: lower leaves obovate, pinnate with pinnatifid leaflets, the upper leaves reduced and less divided. **INFL** elongated racemes, with ascending to widely spreading pedicels. **FLOWER** sepals oblong, ovate, or lanceolate, erect, 1–2 mm long; petals yellow, spatulate to oblanceolate, 1.5–3 mm long. **FRUIT**: siliques linear to very narrowly elliptic, acute to acuminate above and below, 7–18 mm long. **SEEDS** light brown, few to several per locule, in 1 row. A difficult and variable species that cannot always be reliably distinguished from *D. pinnata* and its subspecies. Four subspecies in our flora.

1. Pedicels equaling or shorter than the mature siliques, ± the same length throughout the inflorescence .... (2)
1' Pedicels longer than the mature siliques, lower pedicels longer than the upper .... (3)

2. Rachis and branches of the raceme with stipitate-glandular hairs; petals bright yellow .... **subsp. *viscosa***
2' Rachis and branches of the raceme lacking glands; petals pale yellow .... **subsp. *incisa***

3. Herbage canescent; pedicels long, often > 15 mm .... **subsp. *paysonii***
3' Herbage sparsely to densely pubescent, but not canescent; pedicels < 15 mm .... **subsp. *filipes***

subsp. *filipes* (A. Gray) Rollins (thread foot) Mountain tansymustard. Herbage sparsely to moderately pubescent. **INFL** rachis and branches lacking stipitate-glandular hairs; pedicels longer than the mature siliques, lower ones longer than the upper. **FRUIT** on pedicels (10) 12–20 mm long. $2n = 14, 28, 42$. [*Sisymbrium incisum* Engelm. ex A. Gray var. *filipes* A. Gray; *Descurainia pinnata* (Walter) Britton var. *filipes* (A. Gray) M. Peck; *D. incisa* var. *filipes* (A. Gray) N. H. Holmgren]. Piñon-junper, ponderosa pine, Gambel's oak, spruce-fir, Douglas-fir, riparian shrub, and meadow communities. ARIZ: Apa, Coc; COLO: Arc, Dol, LPl, Mon; NMEX: SJn; UTAH. 1800–2850 (3535) m [5900–9350 (11600′)]. Flowering: May–mid-Aug. Idaho and western Wyoming south to New Mexico, west to California, and north to Washington.

subsp. *incisa* Mountain tansymustard. Herbage subglabrous to moderately pubescent. **INFL** rachis and branches lacking stipitate-glandular hairs; pedicels equaling or shorter than the mature siliques. **FRUIT** on pedicels 6–20 mm long. $2n = 42$. [*Sisymbrium incisum* Engelm. ex A. Gray]. Waste areas, alluvium, ponderosa pine, riparian areas, spruce-fir, dunes, juniper, and mountain shrub. ARIZ: Apa, Nav; COLO: Hin, Mon, SJn; NMEX: San, SJn. (1250) 1675–3450 m [(4100) 5500–11300′]. Flowering: May–Aug. Alberta to New Mexico, west to California, north to British Columbia.

subsp. *paysonii* (Detling) Rollins (named for Edwin Blake Payson, 1893–1927, botanist at the Missouri Botanical Garden and first monographer of the genus *Lesquerella*) Payson's tansymustard. Herbage canescent. **INFL** rachis and branches lacking stipitate-glandular hairs; pedicels longer than the mature siliques, lower ones longer than the upper. **FRUIT** on pedicels (15) 18–23 mm long. [*D. pinnata* (Walter) Britton subsp. *paysonii* Detling; *D. pinnata* var. *paysonii* (Detling) S. L. Welsh & Reveal; *D. incisa* var. *paysonii* (Detling) N. H. Holmgren]. Piñon-juniper, sagebrush, desert shrub, and bitterbrush communities. COLO: Mon; NMEX: RAr, SJn; UTAH. 1750–1890 m (5740–6200′). Flowering: late Mar–May. Southern Wyoming to southeastern Idaho, southeast to Colorado and New Mexico.

subsp. *viscosa* (Rydb.) Rollins (sticky) Herbage typically sparsely pubescent. **INFL** rachis and branches with stipitate-glandular hairs; pedicels equaling or shorter than the mature siliques. **FRUIT** on pedicels 5–10 mm long. $n = 7$, $2n = 14$. [*Sisymbrium viscosum* (Rydb.) Blank.; *Descurainia richardsonii* O. E. Schulz nom. illeg. subsp. *viscosa* (Rydb.) Detling; *D. incana* (Bernh. ex Fisch. & C. A. Mey.) Dorn var. *viscosa* (Rydb.) Dorn]. Utah juniper, Gambel's oak, ponderosa pine, and piñon-juniper communities. ARIZ: Apa; NMEX: SJn. 1580–2350 m (5200–7700′). Flowering: May–Aug. Texas to Nevada, Oregon, and Montana.

Descurainia pinnata (Walter) Britton (feathered, referring to the pinnate leaves) Western tansymustard. Annual, sparsely to densely pubescent with small, dendritic trichomes, eglandular or glandular above (subsp. *halictorum*). **STEMS** single, erect, branched starting at the base, sometimes only above, rarely unbranched, 0.8–5 (7) dm tall. **LEAVES** ovate to lanceolate or oblanceolate, pinnate (bipinnate sometimes below), the leaflets pinnatifid to bipinnatifid, progressively reduced upward. **INFL** elongated raceme with ascending, divaricately ascending to widely spreading pedicels. **FLOWER** sepals oblong, ovate, or lanceolate, erect, often scarious, 1–2 mm long; petals bright yellow to almost white; oblanceolate, 1.3–2.5 mm long. **FRUIT**: siliques clavate to broadly linear, acute to acuminate at the base, obtuse or rounded at the apex, 4–15 (20) mm long. **SEEDS** light brown, 5–20 per locule, usually in 2 evident rows but sometimes crowded and imperfectly so. Hopis used this species both as a spice in cooking and cooked alone as a green. Navajos ground the seeds to make bread. A difficult and variable species that cannot always be reliably distinguished from *D. incana* and its subspecies. Three subspecies in our flora.

1. Fruiting pedicels spreading about 75°, sometimes wider; leaves ± canescent; inflorescence branches often with at least some glandular hairs .... **subsp. *halictorum***
1′ Fruiting pedicels spreading about 45°; leaves usually pubescent, but not canescent .... (2)

2. Siliques < 8 mm long; flowers small; petals < 1.5 mm; fruiting pedicels < 6 mm .... **subsp. *nelsonii***
2′ Siliques > 8 mm long; flowers conspicuous; petals ≥ 2 mm; fruiting pedicels > 6 mm .... **subsp. *intermedia***

subsp. *halictorum* (Cockerell) Detling (pertaining to the sea, referring to saline habitats) Western tansymustard. Herbage usually canescent. **INFL** fruiting pedicels spreading about 75° or more. **FRUIT**: siliques (5) 6–15 (20) mm long. [*Sophia halictorum* Cockerell; *Sisymbrium halictorum* (Cockerell) K. Schum.; *Descurainia pinnata* var. *halictorum* (Cockerell) M. Peck; *D. pinnata* var. *osmiarum* (Cockerell) Shinners]. Sandy areas of dunes, piñon-juniper, washes, mountain shrub, desert shrub, grass communities, and ledges and slickrock. ARIZ: Apa, Coc, Nav; COLO: Dol, LPl, Mon; NMEX: McK, RAr, SJn; UTAH. (1125) 1550–2130 (2310) m [(3700) 5100–7000 (7575′)]. Flowering: late Apr–early Jun. Arkansas west through the southern Rocky Mountains to Oregon and California, south to northern Mexico and Baja California. Plants sometimes become greatly enlarged and multibranched when growing on sand dunes. This subspecies is especially variable in morphology.

subsp. *intermedia* (Rydb.) Detling (intermediate, reference uncertain) Western tansymustard. Herbage pubescent, but not canescent. **INFL** fruiting pedicels spreading about 45°. **FRUIT**: siliques 8–12 mm long. $n = 7$, $2n = 14, 28$,

42. [*Sophia intermedia* Rydb.; *Descurainia pinnata* var. *intermedia* (Rydb.) C. L. Hitchc.]. Piñon-juniper, mountain shrub, Gambel's oak, Douglas-fir, desert shrub, alcoves and hanging gardens. ARIZ: Coc, Nav; COLO: Arc, Dol, LPl, Mon; NMEX: McK, RAr, SJn; UTAH. (1120) 1250–2375 m [(3665) 4100–7760′)]. Flowering: Apr–early Jun. British Columbia and Alberta south to southern Colorado and northern Arizona.

subsp. *nelsonii* (Rydb.) Detling (for Aven Nelson, 1859–1952, prolific botanist of the University of Wyoming) Nelson's tansymustard. Herbage pubescent, but not canescent. **INFL** fruiting pedicels spreading about 45°. **FRUIT**: siliques (4) 5–8 (10) mm long. $2n = 28$. [*Sophia nelsonii* Rydb.; *Descurainia pinnata* var. *nelsonii* (Rydb.) M. Peck]. Piñon-juniper, desert shrub, blackbrush, washes. COLO: Mon; NMEX: McK, SJn; UTAH. 1310–1875 m (4300–6130′). Flowering: late Apr–early Jun. Montana to New Mexico, west to Nevada, Oregon, and Washington.

Descurainia sophia (L.) Webb ex Prantl (wise or wisdom) Northern tansymustard. Annual or biennial, sparsely to densely pubescent with small dendritic trichomes, sometimes mixed with simple trichomes. **STEMS** usually single, erect, short-branched above, 2.5–8 dm tall. **LEAVES** broadly ovate or obovate to oblanceolate, bi- or tripinnate. **INFL** elongated racemes with divaricately ascending pedicels. **FLOWER** sepals narrowly oblong, spreading, 1.6–2.8 mm long; petals pale yellow, narrowly oblong, 2–3 mm long. **FRUIT**: siliques linear, straight to incurved, 14–28 (30) mm long. **SEEDS** light brown, 10–20 per locule, in 1 row. $n = 10, 14$, $2n = 20, 28, 38$. [*Sisymbrium sophia* L.]. Typically in disturbed areas of sagebrush, saltbush, mountain shrub, ponderosa pine, spruce-fir, and Douglas-fir, riparian areas, wash bottoms. ARIZ: Apa, Nav; COLO: Arc, Dol, LPl, Min, Mon, SMg; NMEX: McK, RAr, San, SJn; UTAH. (1120) 1600–2370 (2830) m [(3670) 5300–7770 (9300′)]. Flowering: late Mar–mid-Sep. Native of Eurasia, weedy and now throughout temperate North America. Navajos used a poultice of plant parts to treat toothache and used the seeds to make cakes.

Descurainia torulosa Rollins (referring to the alternating swellings and contractions along the fruit) Wind River tansymustard. Short-lived perennial from fine, elongating roots and/or rhizomes, sparsely pubescent with minute dendritic trichomes. **STEMS** single, branching throughout, weak and typically drooping, up to 1.5 dm long. **LEAVES** pinnate to deeply pinnatifid, the segments broad and rounded, not or few-lobed. **INFL** elongating slightly in fruit, with erect to closely ascending pedicels. **FLOWER** sepals elliptic, 1–1.8 mm long; petals pale yellow, narrowly oblanceolate, slightly longer than the sepals. **FRUIT**: siliques linear, straight to more usually outcurved, 6–8 mm long, torulose, erect to ascending, on pedicels 1–1.5 (2.5) mm long. **SEEDS** light brown, 2–3 per locule, in 1 row. Growing in moist, shaded cracks of lower cliff faces in the alpine. COLO: Arc. 3658 m (12000′). Flowering: Late Jul–early Aug. Northwestern and east-southwestern Wyoming, where it is rare and occupies similar habitats. This is the only known report of the species outside of Wyoming. There is debate as to whether this taxon should best be considered a subspecies of *D. incana*, or if it should be recognized at all (Bricker, Brown, and Patts Lewis 2000). The molecular data of Bricker et al. (2000), however, clearly indicate that the Wyoming material is monophyletic. It is unclear if recognizing the species will cause other *Descurainia* taxa to be paraphyletic. Torulose fruits, very short pedicels, and high-elevation habitat seems diagnostic, at least in our material. It is possible that our material represents an undescribed taxon, as it differs somewhat from the original description of the species and from typical material from Wyoming.

Dimorphocarpa Rollins Spectacle-pod

(two kinds of fruit, referring to the double spectacle-like silicles) Annual, biennial, or perennial herbs, pubescent with dendritic trichomes. **STEMS** erect or decumbent, generally single, branched or not. **LEAVES**: basal and cauline petiolate (or sessile above), entire, dentate, or lobed. **FLOWER** sepals spreading to reflexed; petals white to lavender, obovate. **FRUIT** a silicle, strongly didymous and angustiseptate, strongly flattened, spectacle-shaped, divaricate, valves winged and dehiscing with the enclosed seed. **SEEDS** 1 per locule, nonmucilaginous. $x = 9$. Four species of arid, usually sandy regions of the southwestern United States and adjacent Mexico.

Dimorphocarpa wislizenii (Engelm.) Rollins (for Friedrich Adolph Wislizenus, 1810–1889, a St. Louis physician who collected plants in the West) Spectacle-pod, touristplant. Annual (sometimes short-lived perennial), gray-white with dense, branched trichomes. **STEMS** erect, branched from the base upward (rarely simple), 2–6 dm tall. **LEAVES** basal and cauline, the basal and lower cauline ones soon withering; lower leaves petiolate, oblanceolate, 2.5–7 (10) cm long, 3–12 mm wide, entire to dentate to somewhat lobed or pinnatifid; upper leaves sessile, lanceolate, entire to somewhat dentate. **INFL** flowers crowded but elongating in fruit. **FLOWER** sepals narrowly oblong, densely pubescent, spreading and soon deciduous, 2.5–4 mm long; petals white or tinged purple, spreading, broadly ovate with an abruptly narrowed claw, 4–8 mm long, the claw expanded at the base and sometimes toothed; stigmas triangular and capitate.

FRUIT a silique, strongly didymous, flattened parallel to the replum and appearing spectacle-like, ± reniform in outline, cordate basally, 11–18 mm wide, densely pubescent to glabrous, sometimes trichomes present on the interior. **SEEDS** 1 per locule, 2.5–3 mm long, broadly oblong to suborbicular, brown. n = 9, $2n$ = 18. [*Dithyrea wislizenii* Engelm.; *Dithyrea griffithsii* Wooton & Standl.; *Dimorphocarpa wislizenii* var. *griffithsii* (Wooton & Standl.) Payson]. Sandy areas of hanging gardens, sand dunes, sandy pockets in slickrock, piñon-juniper communities, and disturbed areas. ARIZ: Apa, Nav; COLO: Dol, Mon; NMEX: McK, RAr, San, SJn; UTAH. 1275–2150 m (4200–7000′). Flowering: Apr–Jun, occasionally with a second flowering period in late Aug–Sep. Western Texas, Arizona, New Mexico, southern Utah, southeastern Nevada, and southwestern Colorado. Navajo uses include internal and external treatment of venereal disease; treatment of itchy or irritated skin; an infusion of the plant as a lotion for insect bites; and ceremonially in the Beauty Way, Water Way, Hail Way, and Mountaintop Way. Hopi uses include powdered leaves to treat abrasions.

Draba L. Draba, Whitlow-grass

Ihsan A. Al-Shehbaz and Michael D. Windham

(Greek *drabe*, acrid) Plants annual, biennial, or perennial, with or without caudices; trichomes simple, forked, stellate, malpighiaceous, or dendritic, stalked or sessile, often more than 1 kind present. **STEMS** erect to ascending, sometimes prostrate, simple or branched basally and/or apically, leafy or leafless. **LEAVES** (basal) petiolate or rarely sessile, most commonly rosulate, simple, entire or toothed, rarely pinnately lobed; cauline leaves sessile (in ours) or petiolate, cuneate (in ours) or auriculate at base, sometimes absent, entire or dentate. **INFL** a raceme, few- to many-flowered, bracteate or ebracteate, corymbose, elongated or not in fruit; fruiting pedicels slender, erect, ascending, or divaricate. **FLOWER** sepals ovate, oblong, or elliptic, free, deciduous or rarely persistent, erect, ascending, or spreading, base of inner pair not saccate or subsaccate; petals yellow, white, pink, purple, orange, rarely longer or shorter than sepals; blade obovate, spatulate, oblong, oblanceolate, orbicular, or linear, apex obtuse, rounded, or rarely deeply 2-lobed; claw obscurely to well differentiated from blade; stamens 6, slightly to strongly tetradynamous; filaments wingless, unappendaged, glabrous, free, dilated or not at base; anthers ovate or oblong, not apiculate at apex; nectar glands 1, 2, or 4, distinct or confluent and subtending bases of all stamens; median nectaries present or absent; ovules 4 to numerous per ovary. **FRUIT** a silicle or silique (ours), dehiscent, ovate, lanceolate, elliptic, oblong, linear, orbicular, ovoid, or globose, flattened parallel to the septum (in ours) or terete, plane or sometimes spirally twisted, sessile, unsegmented; valves papery, usually with a distinct midvein and obscurely to prominently anastomosing lateral veins, glabrous or variously pubescent, not keeled, smooth, wingless, unappendaged; replum rounded; septum complete; style distinct or obsolete; stigma capitate, entire or slightly 2-lobed. **SEEDS** biseriate, wingless (in ours) or rarely winged, oblong, ovate, or orbicular, flattened; seed coat minutely reticulate, not mucilaginous when wetted; cotyledons accumbent. Three hundred fifty to 400 species; native or introduced to every continent except Antarctica. *Draba* species often hybridize where their ranges overlap, forming both sterile and fertile (polyploid) intermediates. Such hybrids are common in the flora, making current species circumscriptions somewhat arbitrary.

1. Plants annuals, biennials, or short-lived perennials, taprooted and lacking caudices ....(2)
1′ Plants definite perennials, usually with well-developed, branched caudices ....(8)

2. Petals 5–7 mm long; styles 1.2–3.5 mm long; fruits often twisted .... ***D. helleriana***
2′ Petals 1.2–5 mm long; styles 0.01–0.12 mm long; fruits not twisted ....(3)

3. Rachises and fruiting pedicels pubescent ....(4)
3′ Rachises and fruiting pedicels glabrous (rarely with a few scattered trichomes) ....(5)

4. Petals 1–2 mm wide, white (lacking in late-season flowers); cauline leaves 0–3 (–6); below 2500 m (8200′) elevation .... ***D. cuneifolia***
4′ Petals 0.5–1 mm wide, yellow at anthesis (and present in all flowers); cauline leaves 3–10 (–17); above 2500 m (8200′) elevation .... ***D. rectifructa***

5. Fruiting pedicels (1.5–) 2–7× longer than fruits; cauline leaves 4–12 .... ***D. nemorosa***
5′ Fruiting pedicels usually shorter (rarely up to 1.5× longer) than fruits; cauline leaves 0–3 (–5) ....(6)

6. Petals white (lacking in late-season flowers); inflorescences usually subumbellate; seeds 0.5–0.8 mm long; below 2500 m (8200′) elevation .... ***D. reptans***
6′ Petals yellow to cream (but often fading white); inflorescences usually elongate in fruit; seeds 0.7–1.1 mm long; above 2500 m (8200′) elevation ....(7)

7. Lower leaf surfaces with 2–4-rayed trichomes .... ***D. albertina***
7′ Lower surfaces of basal leaves glabrous or rarely with a few simple and 2-rayed trichomes .... ***D. crassifolia***

8. Surfaces of basal leaves glabrous .... (9)
8′ Surfaces of basal leaves (at least the lower surfaces) pubescent .... (11)

9. Plants not caespitose; cauline leaves absent or rarely 1; petals 1.5–2.5 (–3) × 0.5–1 mm .... ***D. crassifolia***
9′ Plants caespitose; cauline leaves 2–12; petals 3–6 × 1.5–4 mm .... (10)

10. Stems glabrous near base; inflorescences ebracteate; basal leaves oblanceolate, 2–10 mm wide .... ***D. crassa***
10′ Stems pubescent throughout; inflorescences bracteate basally; basal leaves linear to linear-oblanceolate, 0.3–2 (–3) mm wide .... ***D. graminea***

11. Rachises and fruiting pedicels glabrous .... (12)
11′ Rachises and fruiting pedicels pubescent .... (16)

12. Styles (0.5–) 1–2.7 mm long; petals 4–6.5 mm long; leaf trichomes sessile or subsessile, cruciform or malpighiaceous .... ***D. spectabilis***
12′ Styles 0.01–0.35 mm long; petals 1.5–3.4 mm long; leaf trichomes stalked, 2–8-rayed (often simple in *D. crassifolia* and *D. fladnizensis*) .... (13)

13. Lower surfaces of basal leaves pubescent with (2–) 3–8-rayed trichomes .... (14)
13′ Lower surfaces of basal leaves glabrous or rarely with a few simple and 2-rayed trichomes .... (15)

14. Lower leaf surfaces with 2–4-rayed trichomes; petals yellow; ovules and seeds (20–) 24–44 per fruit .... ***D. albertina***
14′ Lower leaf surfaces with 4–8-rayed trichomes; petals white; ovules and seeds 16–24 (–28) per fruit .... ***D. lonchocarpa***

15. Petals yellow or cream (but often fading white); styles 0.01–0.1 mm long; plants rarely caespitose .... ***D. crassifolia***
15′ Petals white; styles (0.05–) 0.1–0.3 mm long; plants often caespitose .... ***D. fladnizensis***

16. Basal leaves pubescent with sessile to subsessile cruciform or malpighiaceous trichomes .... ***D. spectabilis***
16′ Basal leaves pubescent with simple or stalked, 2–12-rayed trichomes .... (17)

17. Trichomes of basal leaves usually concentrated near leaf margins, the rays twisted and contorted .... ***D. streptobrachia***
17′ Trichomes of basal leaves uniformly distributed, the rays straight .... (18)

18. Styles 0.05–0.6 mm long; petals white .... (19)
18′ Styles (0.5) 0.6–3.5 mm long; petals yellow or cream (white in *D. smithii* and occasional faded specimens) .... (20)

19. Fruits pubescent with 3–7-rayed trichomes; inflorescences bracteate below; fruiting pedicels often appressed to rachises; ovules and seeds 28–48 per fruit .... ***D. cana***
19′ Fruits glabrous or pubescent with simple and 2-rayed trichomes; inflorescences ebracteate; fruiting pedicels not appressed to rachises; ovules and seeds 16–24 (–28) per fruit .... ***D. lonchocarpa***

20. Stem trichomes minute (0.05–0.2 mm long) and highly branched (4–12-rayed), simple trichomes lacking; caudex branches many, slender, forming extensive, prostrate mats .... ***D. smithii***
20′ Stem trichomes generally larger (0.1–1.8 mm long) and less branched (up to 6-rayed), at least some simple trichomes present; caudex branches few, thick, not forming prostrate mats .... (21)

21. Petals 3.5–5 mm long; styles 0.5–1.2 (–1.5) mm long; inflorescences bracteate below; seeds and ovules 28–38 (–44) per fruit .... ***D. aurea***
21′ Petals 5–7 mm long; styles 1.2–3.5 mm long; inflorescences ebracteate, very rarely the lowermost flowers bracteate; ovules and seeds 14–28 per fruit .... ***D. helleriana***

Draba albertina Greene (from Alberta, Canada) Slender draba or whitlow-grass, Alberta draba. Short-lived perennials or biennials, not caespitose. **STEMS** 3–30 cm tall, simple or branched above, erect, pubescent proximally with simple and occasionally 2-rayed trichomes 0.1–0.9 mm long, glabrous distally. **LEAVES** at stem bases obovate to linear-lanceolate, 7–18 × 2–6 (–9) mm, ciliate with simple trichomes 0.4–1 mm long; lower surfaces uniformly pubescent with stalked, 2–4-rayed trichomes 0.05–0.4 (–0.5) mm long, rarely with simple trichomes along midvein; upper surfaces with simple and 2-rayed trichomes 0.07–0.4 mm long, rarely glabrous; cauline leaves 1–3 (–5) or rarely absent, elliptic to ovate-lanceolate, entire to denticulate, pubescent like basal leaves. **INFL** ebracteate, (2–) 6–30-flowered, elongated in fruit; rachises glabrous; lowermost fruiting pedicels 3–14 mm, divaricate-ascending or horizontal, glabrous. **FLOWER** sepals 1.4–2.1 mm long, glabrous or with few simple trichomes; petals yellow, 2–3.2 × 0.7–1.2 mm; anthers 0.15–0.25 mm long. **FRUIT** a narrowly elliptic to linear silique, 4–15 × 1–2.1 mm, not twisted, glabrous or rarely puberulent with simple trichomes 0.05–0.1 mm long; style 0.01–0.12 mm long; ovules and seeds (20–) 24–44 per fruit. **SEEDS** 0.7–1 × 0.4–0.5 mm. $2n = 24$. [*D. stenoloba* Ledeb. var. *nana* (O. E. Schulz) C. L. Hitchc.]. Meadows, open forest floor, and roadsides in mountainous areas. COLO: Hin, LPl, Min, SJn; NMEX: SJn. 2740–3660 m (9000–12000′). Flowering: Jun–Aug. From Alaska and Nunavut south to New Mexico and California. The single collection seen from New Mexico is rather typical of the species, but plants from southwestern Colorado appear to intergrade with *D. crassifolia*.

Draba aurea Vahl ex Hornem. (gold) Golden draba or whitlow-grass. Perennials, with simple or branched caudices. **STEMS** 5–35 cm tall, simple or branched above, erect, pubescent throughout with a mixture of simple (0.4–1.3 mm long) and stalked, 3–6-rayed (0.1–0.5 mm long) trichomes. **LEAVES** at stem bases oblanceolate to obovate, (4–) 7–22 × 2–7 (–10) mm, ciliate with simple trichomes to 0.8 mm; both surfaces uniformly pubescent with stalked, (2–) 4–8-rayed trichomes 0.2–0.6 mm long; cauline leaves 5–26, oblong to lanceolate or ovate, dentate to entire, pubescent like basal leaves or sometimes with simple trichomes on upper surfaces. **INFL** bracteate proximally (very rarely ebracteate), 10–52-flowered, elongated in fruit; rachises pubescent like stem; lowermost fruiting pedicels 3–10 (–14) mm long, ascending to suberect, pubescent with simple and branched trichomes. **FLOWER** sepals 2.2–3 mm long, with simple and branched trichomes; petals yellow, 3.5–5 × 1.5–2.5 mm; anthers 0.4–0.5 mm long. **FRUIT** a lanceolate to linear-lanceolate silique, 6–17 × 2–3.5 mm, slightly twisted or plane, pubescent with a mixture of simple and short-stalked, 2–4-rayed trichomes 0.05–0.3 mm long; style 0.5–1.2 (–1.5) mm long; ovules and seeds 28–38 (–44) per fruit. **SEEDS** 0.9–1.3 × 0.5–0.7 mm. $2n = 64$, about 72, 74, 82. [*D. aureiformis* Rydb.; *D. bakeri* Greene; *D. decumbens* Rydb.; *D. luteola* Greene; *D. mccallae* Rydb.]. Mountain meadows, subalpine conifer woodlands, rocky ledges and crevices, and alpine tundra. COLO: Arc, Hin, LPl, Min, Mon, RGr, SJn; UTAH. 2500–3840m (8200–12600′). Flowering: Jun–Aug. Alaska to Greenland, south to Ontario, New Mexico, and Arizona. Highly variable, as indicated by the number of species-level synonyms and the reported variation in chromosome number. Preliminary evidence suggests that populations assigned to *D. aurea* represent a polyploid species complex formed by hybridization among several local species. *Draba streptocarpa* A. Gray is one of the parents, and specimens assigned to that species in the Four Corners catalog (Heil and O'Kane 2005) are here included within *D. aurea*. Both *D. cana* and *D. helleriana* have contributed genes to the complex as well, and *D. aurea* is sometimes difficult to distinguish from both.

Draba cana Rydb. (grayish) Hoary draba, lanceleaf draba. Perennials, with simple or few-branched caudices. **STEMS** (4–) 6–20 cm tall, simple or branched above, erect, hirsute proximally with simple trichomes 0.5–1 mm long, pubescent throughout with stalked, 4–10-rayed trichomes 0.05–0.2 mm long. **LEAVES** at stem bases linear-lanceolate to oblong, (5–) 8–20 × 3–8 (–11) mm, ciliate with simple trichomes 0.3–0.8 mm long; both surfaces uniformly pubescent with short-stalked, 4–12-rayed trichomes 0.1–0.3 mm long; cauline leaves 3–12 (–17), ovate-lanceolate to oblong, dentate (rarely entire), pubescent like basal leaves but with simple and forked trichomes near leaf base on upper surfaces. **INFL** basally bracteate, 10–47-flowered, elongated in fruit; rachises pubescent like stems; lowermost fruiting pedicels 2–5 (–10) mm long, usually erect and subappressed, pubescent like rachises. **FLOWER** sepals 1.5–2 mm long, with simple and few-rayed trichomes; petals white, 2.3–4.5 × 0.7–1.7 mm; anthers 0.1–0.2 mm long. **FRUIT** a narrowly lanceolate to linear silique, 5–10 × 1.5–2.5 mm, slightly twisted or plane, pubescent with short-stalked, 3–7-rayed trichomes 0.05–0.3 mm; style 0.1–0.6 mm; ovules and seeds 28–48 per fruit. **SEEDS** 0.5–0.9 × 0.3–0.5 mm. $2n = 32$. [*D. lanceolata* Royle misapplied]. Rock crevices and talus mostly in alpine habitats. COLO: LPl, SJn. 3050–4115 m (10000–13500′). Flowering: Jun–Aug. Alaska to Greenland, south to New Hampshire, Wisconsin, New Mexico, and California. Listed in the Four Corners catalog (Heil and O'Kane 2005) as *Draba breweri* S. Watson var. *cana* (Rydb.) Rollins, this taxon is distinguished from typical *D. breweri* (endemic to the Sierra Nevada and vicinity) by its noncaespitose habit, taller, more commonly branched stems, and mostly dentate cauline leaves. Specimens from the Four Corners region are rare, and there is evidence of intergradation with *D. aurea* (a polyploid species of hybrid origin that has *D. cana* as one of its parents).

Draba crassa Rydb. (thick) Thickleaf draba or whitlow-grass. Perennials, caespitose, with thick, usually branched caudices. **STEMS** (4–) 6–15 cm tall, simple, decumbent to ascending, glabrous proximally, pubescent distally with twisted, simple and stalked, 2-rayed trichomes 0.1–0.8 mm long. **LEAVES** at stem bases oblanceolate, 20–60 (–70) × 2–8 (–10) mm, ciliate with primarily simple trichomes 0.3–0.8 mm long; both surfaces glabrous; cauline leaves 2–4 (–6), ovate to oblong, entire, ciliate with simple and stalked, 2-rayed trichomes. **INFL** ebracteate, 4–25-flowered, somewhat elongated in fruit; rachises pubescent with twisted, simple and short-stalked, 2-rayed trichomes 0.1–0.8 mm long; lowermost fruiting pedicels 5–10 (–15) mm long, divaricate-ascending, pubescent like rachises. **FLOWER** sepals 2–3.3

mm long, pubescent with simple and 2-rayed trichomes; petals yellow, 3.5–6 × 2–4 mm; anthers 0.5–0.6 mm long. **FRUIT** a lanceolate to ovate-lanceolate silique, 7–14 × 3–5 mm, slightly twisted, glabrous; style 0.4–1.5 mm long; ovules and seeds 16–20 per fruit. **SEEDS** 1.2–1.7 × 0.8–1.1 mm. $2n = 24$. Cliffs and talus in alpine and subalpine habitats. COLO: Con, Min, LPl, SJn. 3780–4270 m (12400–14000′). Flowering: Jun–Aug. Montana to Wyoming, south to New Mexico and Utah. The specific epithet could apply equally well to the caudex or the leaves, both of which are thick and fleshy. These features, combined with the decumbent stems and glabrous leaf surfaces, make this a rather distinctive species.

Draba crassifolia Graham (thick leaves) Snowbed draba or whitlow-grass, hairy draba. Short-lived perennials, rarely with poorly developed caudices. **STEMS** 2–15 cm tall, usually simple, erect, glabrous or rarely pubescent proximally with simple trichomes 0.3–0.7 mm long, glabrous distally. **LEAVES** at stem bases oblanceolate to obovate, 5–17 × 2–6 mm, with a few simple cilia or occasionally eciliate; both surfaces glabrous or with a few simple and 2-rayed trichomes 0.3–0.9 mm long; cauline leaves absent or rarely 1, oblong to ovate, entire, glabrous. **INFL** ebracteate, 2–25-flowered, elongated in fruit; rachises glabrous; lowermost fruiting pedicels 1–8 mm long, horizontal to divaricate-ascending, glabrous. **FLOWER** sepals 1–2 mm long, glabrous; petals yellow or cream, often fading white, 1.5–3 × 0.5–1 mm; anthers 0.15–0.25 mm long. **FRUIT** a narrowly elliptic to linear-lanceolate silique, 3–10 × 1.5–2.5 mm, not twisted, glabrous; style 0.02–0.1 mm long; ovules and seeds 8–30 per fruit. **SEEDS** 0.8–1.1 × 0.4–0.6 mm. $2n = 40$. [*D. parryi* Rydb.]. Alpine tundra and talus slopes. COLO: Con, Hin, LPl, Min, Mon, RGr, SJn. 3260–4270 m (10700–14000′). Flowering: Jun–Sep. Alaska to Greenland, south to Quebec, New Mexico, and Arizona. Apparently formed by ancient hybridization between *D. albertina* and *D. fladnizensis*, this allopolyploid species can be confused with either. The yellow to cream-colored petals of *D. crassifolia* become whitish with age, and herbarium specimens that don't provide information on petal color before pressing can be nearly impossible to distinguish from *D. fladnizensis*.

Draba cuneifolia Nutt. ex Torr. & A. Gray (cuneate or wedge-shaped leaves) Wedgeleaf draba or whitlow-grass. Annuals, taprooted and lacking caudices. **STEMS** 2–27 cm tall, simple or branched at base, erect to ascending, hirsute proximally with simple or spurred trichomes 0.5–1.2 mm long and pubescent throughout with stalked, 2–4 (–5)-rayed trichomes 0.05–0.4 mm long. **LEAVES** at stem bases spatulate or broadly obovate, 8–35 × 5–18 mm, often ciliate proximally with simple trichomes 0.4–0.7 mm long; both surfaces uniformly pubescent with stalked, 2–4 (–5)-rayed trichomes 0.1–0.7 mm long; upper surfaces usually with a few simple trichomes; cauline leaves absent or 1–6 along lower 1/3 of stem, often dentate on distal 1/2, pubescent like basal leaves. **INFL** ebracteate, 10–50-flowered, elongated in fruit; rachises densely pubescent with 2–4-rayed trichomes; lowermost fruiting pedicels 1–10 mm long, horizontal to divaricate-ascending, pubescent like rachises. **FLOWER** sepals 1.5–2.5 mm long, glabrous or pubescent with simple trichomes; petals white (lacking in late-season flowers), 2–5 × 1–2 mm; anthers 0.1–0.4 mm long. **FRUIT** an oblong to linear silique, 6–16 × 1.7–3 mm, not twisted, antrorsely puberulent with simple (rarely some 2-rayed) trichomes 0.1–0.3 mm long or glabrous; style 0.01–0.1 mm long; ovules and seeds 24–66 (–72) per fruit. **SEEDS** 0.5–0.7 × 0.4–0.5 mm. $2n = 30, 32$. In soil (often disturbed) or on rocky slopes in desert scrub and piñon-juniper woodlands. ARIZ: Apa, Nav; COLO: Arc, Dol, LPl, Mon; NMEX: McK, RAr, SJn; UTAH. 1100–2350 m (3600–7700′). Flowering: Mar–May. South Dakota and Pennsylvania south to Florida and California; Mexico. Collections from the Four Corners region represent **var. *cuneifolia***, distinguished from other varieties by having simple (rarely 2-rayed) vs. branched trichomes on the fruits. *Draba cuneifolia* is occasionally confused with *D. reptans* (see comments under that species).

Draba fladnizensis Wulfen (from Fladnitz, Austria) Austrian draba or whitlow-grass, Patterson's draba, Arctic draba, Fladnitz draba. Perennials, often somewhat caespitose from branched caudices. **STEMS** 1.5–13 cm tall, simple, erect, glabrous or very rarely sparsely pubescent proximally with simple trichomes 0.1–0.4 mm long. **LEAVES** at stem bases linear to narrowly obovate, 3–12 (–16) × 1–4 mm, ciliate with primarily simple trichomes 0.25–0.6 mm long; lower surfaces with simple trichomes and (rarely) a few short-stalked, 2-rayed trichomes; upper surfaces usually glabrous; cauline leaves absent or 1–2, oblong to ovate, entire, ciliate. **INFL** ebracteate (rarely lowermost flower bracteate), 2–14-flowered, elongated in fruit; rachises glabrous; lowermost fruiting pedicels 1–6 mm long, divaricate-ascending, glabrous. **FLOWER** sepals 1.2–2.2 mm long, glabrous or with few simple trichomes; petals white, 2–3.5 × 0.8–1.5 mm; anthers 0.2–0.25 mm long. **FRUIT** an elliptic-lanceolate to oblong silique, 3–9 × 1.5–2 mm, not twisted, glabrous; style (0.05) 0.1–0.3 mm long; ovules and seeds 12–24 per fruit. **SEEDS** 0.8–1 × 0.5–0.6 mm. $2n = 16$. [*D. pattersonii* O. E. Schulz]. Alpine cliffs, talus, and meadows. COLO: LPl, Min, SJn. 3500–3840 m (11500–12600′). Flowering: Jun–Aug. Alaska to Greenland, south to Quebec, Colorado, and Utah; Eurasia. Local collections have been identified as **var. *pattersonii*** (O. E. Schulz) Rollins, distinguished from the primarily Arctic var. *fladnizensis* by shorter stems, larger basal leaves, and broader fruits. Separation from *D. crassifolia* can be problematic (see comments under that species). Plants identified as *D. porsildii* G. A. Mulligan in the Four Corners catalog (Heil and O'Kane 2005) do not match that species morphologically and are tentatively included here in *D. fladnizensis*.

Draba graminea Greene **CP** (grasslike) Rocky Mountain draba. Perennial; caespitose, with well-developed, branched caudices. **STEMS** 1–8 cm tall, simple, ascending to decumbent, pubescent throughout with twisted, simple, spurred, and very few, subsessile, 2-rayed trichomes 0.1–0.5 mm long, very rarely glabrous. **LEAVES** at stem bases linear to linear-oblanceolate, (5–) 10–40 × 0.3–2 (–3) mm, ciliate with simple (rarely 2-rayed) trichomes 0.2–0.8 mm long; both surfaces glabrous; cauline leaves 2–12, all as floral bracts, linear to oblanceolate, entire, ciliate proximally like basal leaves. **INFL** bracteate, 3–15-flowered, somewhat elongated in fruit; rachises pubescent like stems (rarely glabrous); lowermost fruiting pedicels 3–10 (–15) mm long, divaricate-ascending, pubescent like rachises. **FLOWER** sepals 1.5–2.5 mm long, glabrous or with simple and 2-rayed trichomes; petals yellow, 3–5 × 1.5–3 mm; anthers 0.3–0.5 mm long. **FRUIT** an ovate-elliptic to lanceolate silique, 5–11 × 2.5–5 mm, often slightly twisted, glabrous; style 0.2–0.7 mm long; ovules and seeds 8–16 per fruit. **SEEDS** 1.2–1.5 × 0.7–1 mm. $2n = 18$. Cliff ledges, talus, and gravel bars in alpine habitats. COLO: Hin, LPl, Min, SJn. 3660–4120 m (12000–13500′). Flowering: Jul–Sep. Endemic to southwest Colorado. A very distinctive species, marked by the combination of long, linear leaves forming grasslike tufts, pale yellowish green foliage, and strongly bracteate inflorescences.

Draba helleriana Greene (for Amos Arthur Heller) Heller's draba. Perennials or biennials, with branched or simple caudices. **STEMS** 10–34 cm tall, usually branched, erect, moderately to densely hirsute proximally with simple and 2-rayed trichomes 0.4–1.8 mm long, these mixed with short-stalked to subsessile, 3–5-rayed trichomes 0.2–0.4 mm long and somewhat coarser 2-rayed ones, glabrous or pubescent distally. **LEAVES** at stem bases oblanceolate to obovate, 9–41 (–52) × 2–7 (–10) mm, usually ciliate proximally with simple or 2-rayed trichomes; both surfaces uniformly pubescent with stalked, 3–5-rayed trichomes 0.1–0.6 mm long; upper surfaces sometimes with a few (rarely exclusively) simple or 2-rayed trichomes 0.4–1 mm long; cauline leaves (8–) 12–30, ovate-lanceolate to oblong, entire to dentate, pubescent with stalked, mostly 4-rayed trichomes mixed with a smaller number of simple or 2-rayed trichomes to 1.3 mm long. **INFL** ebracteate (rarely lowermost 1–3 flowers bracteate), 10–52-flowered, elongated in fruit; rachises pubescent with short-stalked to subsessile, 3–5-rayed trichomes with or without simple trichomes; lowermost fruiting pedicels 4–13 mm long, horizontal to divaricate-ascending, pubescent abaxially. **FLOWER** sepals 2.5–4 mm long, with simple and short-stalked, 2-rayed trichomes to 0.9 mm long; petals yellow, 5–7 × 1.5–2.2 mm; anthers 0.5–0.8 mm long. **FRUIT** an oblong-lanceolate to ovate silique, 5–15 × 2–3.5 mm, slightly to strongly twisted or plane, puberulent with simple and subsessile, 2 (–4)-rayed trichomes 0.03–0.25 (–0.8) mm long; style 1.2–3.5 mm long; ovules and seeds 14–28 per fruit. **SEEDS** 1–1.3 × 0.6–0.8 mm. $2n = 18$. [*D. aurea* Vahl ex Hornem. var. *stylosa* A. Gray; *D. neomexicana* Greene]. Rocky areas in montane forests and meadows. COLO: Arc, Con, RGr; NMEX: McK, RAr, SJn. 2100–3500 m (6900–11500′). Flowering: May–Sep. Colorado south to New Mexico and Arizona; northern Mexico. Our plants are **var. *helleriana***, distinguished by having upper leaf surfaces covered with stalked cruciform trichomes (mostly simple or forked in other varieties). As a genome donor to several polyploid hybrids, this species can be difficult to distinguish from both *D. aurea* and *D. spectabilis*. It had a wide variety of uses among the Ramah Navajo. A decoction of the leaves was used to treat bad coughs, kidney problems, and gonorrhea; the whole plant was used as a ceremonial emetic, and a cold infusion of the leaves provided a ceremonial eyewash and a lotion to protect against witches.

Draba lonchocarpa Rydb. (spear fruit) Lancepod draba or whitlow-grass. Perennials, caespitose (rarely mat-forming) with branched caudices. **STEMS** 1–11 cm tall, simple, erect, pubescent proximally or throughout (rarely glabrous) with minutely stalked, 4–8-rayed trichomes 0.05–0.2 mm long. **LEAVES** at stem bases oblanceolate to obovate, 3–10 × 1–3 mm, ciliate proximally with simple and 2-rayed trichomes 0.1–0.8 mm long; both surfaces uniformly pubescent with short-stalked, 4–8-rayed trichomes 0.2–0.5 mm long; cauline leaves absent or 1 (–4), ovate or oblong, entire, pubescent like basal leaves. **INFL** ebracteate (rarely lowermost flowers bracteate), 3–9-flowered, elongated in fruit; rachises glabrous or pubescent like stem; lowermost fruiting pedicels 2–9 mm long, ascending, glabrous or pubescent like stem. **FLOWER** sepals 1.5–2 mm long, glabrous or pubescent with simple and short-stalked, 2–5-rayed trichomes; petals white, 2–3.4 × 1–1.5 mm; anthers 0.2–0.3 mm long. **FRUIT** a linear to narrowly oblong-lanceolate silique, 6–15 × 1–2.2 mm, slightly twisted or plane, glabrous or sparsely puberulent with simple and minutely stalked, 2-rayed trichomes 0.1–0.2 mm long; style 0.05–0.35 mm long; ovules and seeds 16–24 (–28) per fruit. **SEEDS** 0.7–1 × 0.5–0.6 mm. $2n = 16$. [*D. nivalis* Lilj. var. *exigua* (O. E. Schulz) C. L. Hitchc.]. Rock outcrops and talus slopes in alpine habitats. COLO: SJn. 3750–4025 m (12200–13200′). Flowering: Jun–Jul. Alaska to Northwest Territories, south to Colorado and California. Rarely encountered in the Four Corners region. Our plants are **var. *lonchocarpa***.

Draba nemorosa L. (of woods) Woodland draba or whitlow-grass. Annuals, taprooted and lacking caudices. **STEMS** 3–45 cm tall, simple or branched, erect to ascending, densely pubescent proximally with a mixture of simple (to 1.3 mm) and stalked, 2–4-rayed trichomes (to 0.5 mm long), glabrous distally. **LEAVES** at stem bases oblong-obovate or oblanceolate, 8–25 × 5–12 mm, obscurely ciliate proximally or eciliate; both surfaces uniformly pubescent with stalked, 2–4-rayed trichomes 0.1–0.5 mm long; cauline leaves 4–12, broadly ovate to oblong, dentate or denticulate, lower surfaces pubescent like basal leaves, upper surfaces with primarily simple (some 2–3-rayed) trichomes. **INFL** ebracteate, 10–60-flowered, elongated in fruit; rachises glabrous; lowermost fruiting pedicels 0.7–2.8 cm long, divaricate, glabrous. **FLOWER** sepals 0.7–1.6 mm long, with few simple trichomes; petals yellow, 1.2–2.5 × 0.4–1 mm long, deeply notched; anthers 0.1–0.2 mm long. **FRUIT** an elliptic to oblong-oblanceolate silique, 3–10 × 1.5–2.5 mm, not twisted, glabrous or puberulent with simple, antrorse trichomes 0.05–0.2 mm long; style 0.01–0.1 mm long; seeds and ovules 30–72 per fruit. **SEEDS** 0.5–0.8 × 0.3–0.5 mm. $2n$ = 16. In disturbed soil, mostly sagebrush communities. ARIZ: Apa; COLO: Hin. 1830–2440 m (6000–8000′). Flowering: Apr–Jul. Alaska to Nunavut, south to Ontario, Colorado, and Arizona; native of Asia and Europe, naturalized in North America and Australia. Not included in the Four Corners catalog (Heil and O'Kane 2005), this species is a relatively recent introduction to the region.

Draba rectifructa C. L. Hitchc. (straight fruit) Mountain draba or whitlow-grass. Annuals, taprooted and lacking caudices. **STEMS** 5–30 cm tall, simple or few-branched above, erect to ascending, hirsute proximally with simple and stalked, 2-rayed trichomes 0.4–1.3 mm long and pubescent throughout with stalked, 2–4-rayed trichomes 0.05–0.2 mm long. **LEAVES** at stem bases oblanceolate to obovate, 10–20 (–30) × 2–7 (–10) mm, ciliate with simple trichomes to 1.5 mm long; lower surface uniformly pubescent with a mixture of simple trichomes 0.6–1.3 mm long and stalked, 2–4-rayed trichomes 0.1–0.8 mm long; upper surface with simple and stalked, 2-rayed trichomes 0.4–1.2 mm long; cauline leaves 3–10 (–17), oblong to ovate-lanceolate, entire, pubescent like basal leaves. **INFL** ebracteate, 15–50-flowered, elongated in fruit; rachises pubescent like stem; lowermost fruiting pedicels 2–10 mm long, horizontal to divaricate-ascending, pubescent with simple and stalked, 2–4-rayed trichomes. **FLOWER** sepals 1.3–2 mm long, with simple trichomes; petals yellow, sometimes fading to white, 1.5–3 × 0.5–1 mm; anthers 0.2–0.3 mm long. **FRUIT** a lanceolate silique, 6–11 × 1.3–2.3 mm, not twisted, puberulent with antrorse simple trichomes 0.1–0.2 mm long; style 0.01–0.1 mm long; ovules and seeds 32–60 per fruit. **SEEDS** 0.5–0.7 × 0.4–0.5 mm. In soil or on rocky slopes in meadows in Douglas-fir and spruce-fir communities. COLO: Hin, LPl, Min, SJn. 2680–3230 m (8800–10600′). Flowering: Apr–Aug. Endemic to the Four Corners states. Local collections are rather uniform and similar to the type collection. A chromosome count of $2n$ = 24 has been attributed to *D. rectifructa*; it was derived from a morphologically distinctive metapopulation in north-central Utah that almost certainly represents a different species. An infusion of this plant was reportedly used by the Navajo as a diuretic.

Draba reptans (Lam.) Fernald (creeping) Carolina draba or whitlow-grass. Annuals, taprooted and lacking caudices. **STEMS** 1–12 cm tall, often branched at base, erect to ascending, sparsely to densely pubescent proximally with 2–3-rayed trichomes 0.1–0.6 mm long, these sometimes mixed with simple or spurred trichomes to 0.9 mm long, usually glabrous distally. **LEAVES** at stem bases elliptic to suborbicular, 5–23 × 4–8 (–13) mm, often ciliate proximally with simple trichomes 0.4–1 mm long; lower surfaces uniformly pubescent with stalked, 2–4-rayed trichomes 0.1–0.5 mm long; upper surfaces with larger (0.6–1 mm long) simple and smaller (to 0.7 mm long) 2-rayed trichomes; cauline leaves absent or rarely 1–3 just above the base, entire, pubescent like basal leaves. **INFL** ebracteate, 3–16-flowered, usually subumbellate; rachises glabrous or rarely with a few trichomes; lowermost fruiting pedicels 1–9 mm long, horizontal to divaricate-ascending, glabrous or rarely with a few trichomes. **FLOWER** sepals 1.5–2.3 mm long, with simple trichomes; petals white (or lacking in late-season flowers), 2–4.5 × 1–1.5 mm; anthers 0.4–0.5 mm long. **FRUIT** a linear to linear-oblong silique, 5–20 × 1.2–2.3 mm, not twisted, glabrous or with antrorse, simple (rarely spurred or 2-rayed) trichomes 0.1–0.3 mm long; style 0.01–0.1 mm long; ovules and seeds 32–88 per fruit. **SEEDS** 0.5–0.8 × 0.3–0.5 mm. $2n$ = 16, 30, 32. [*Arabis reptans* Lam.; *Draba coloradensis* Rydb.; *D. reptans* var. *stellifera* (O. E. Schulz) C. L. Hitchc.]. In sandy soil (often disturbed) or rocky slopes in desert scrub and piñon-juniper woodland habitats. ARIZ: Apa, Coc, Nav; COLO: Arc, Dol, LPl, Mon; NMEX: McK, SJn; UTAH. 1650–2250 m (5400–7400′). Flowering: Feb–Jun (–Aug). British Columbia to Ontario and Massachusetts, south to Georgia, Texas, and California. Occasionally confused with *D. cuneifolia*, with which it often grows. There is no evidence of hybridization, and *D. reptans* is easily distinguished from *D. cuneifolia* by the glabrous rachises and fruiting pedicels and usually subumbellate inflorescences. Reportedly used by the Ramah Navajo as a dermatological aid; a poultice of crushed leaves was applied to sores.

Draba smithii Gilg & O. E. Schulz (for E. C. Smith) Smith's draba. Perennials, the caudex branches many, slender, prostrate, and forming extensive mats. **STEMS** 5–30 cm tall, usually simple, ascending to erect, pubescent throughout with stalked, dendritic, 4–12-rayed trichomes 0.05–0.2 mm long. **LEAVES** at stem bases obovate to narrowly oblanceolate,

5–13 × 2–4 (–7) mm, not ciliate or base very rarely with simple or spurred trichomes to 0.25 mm long; both surfaces uniformly pubescent with short-stalked, dendritic, 5–12-rayed trichomes 0.05–0.3 mm long; cauline leaves (2–) 3–8, oblong to lanceolate, entire or rarely denticulate, pubescent like basal leaves or rarely with long-stalked trichomes with fewer rays on upper surfaces. **INFL** ebracteate, 12–28-flowered, elongated in fruit; rachises pubescent like stems; lowermost fruiting pedicels 3–13 mm long, divaricate, pubescent like stems. **FLOWER** sepals 2–2.5 mm long, pubescent with trichomes similar to leaves; petals white, 4–6 × 2–3 mm; anthers 0.5–0.6 mm long. **FRUIT** an ovate-lanceolate silique, 5–9 × 2–3 mm, twisted, pubescent with simple and short-stalked, 2–4-rayed trichomes 0.05–0.2 mm long; style 0.7–2.3 mm long; ovules and seeds 16–20 per fruit. **SEEDS** 0.8–1.2 × 0.5–0.7 mm. $2n = 32$. Cliff ledges and talus slopes. COLO: Arc, Hin. 2350–2470 m (7700–8100′). Flowering: May–Jul. Endemic to the mountains of south-central Colorado. A very distinctive species marked by its intricately branched, mat-forming caudices, flowers with white petals and long styles (an unusual combination), and highly branched (5–12-rayed) leaf trichomes.

Draba spectabilis Greene (showy) Showy draba or whitlow-grass, splendid draba or whitlow-grass. Perennials, with simple or few-branched caudices. **STEMS** 7–53 cm tall, simple, erect to ascending, pubescent proximally with sessile, malpighiaceous cruciform (simple in var. *spectabilis*) trichomes 0.15–0.5 mm long, usually glabrous or sparsely pubescent distally. **LEAVES** at stem bases oblanceolate to obovate, 10–60 × 5–12 (–15) mm, sometimes ciliate with simple trichomes; both surfaces uniformly pubescent with sessile or subsessile, malpighiaceous or crucifrom trichomes 0.1–0.6 mm long; cauline leaves 3–12 (–16), broadly ovate to lanceolate or oblong, entire or dentate, ciliate or not, pubescent like basal leaves or upper surfaces also with simple and stalked, 2-rayed trichomes. **INFL** ebracteate, 10–61-flowered, elongated in fruit; rachises glabrous or pubescent with malpighiaceous, cruciform and/or simple trichomes; lowermost fruiting pedicels 5–26 mm long, horizontal to divaricate-ascending, glabrous or pubescent abaxially like rachises. **FLOWER** sepals 2.2–4 mm long, glabrous or with simple and 2-rayed trichomes; petals yellow, 4–6.5 × 1–2.5 mm; anthers 0.5–0.7 mm long. **FRUIT** a lanceolate to elliptic silique, 6–13 × 2–3.5 mm, not twisted, glabrous or puberulent with simple and sessile, 2-rayed trichomes 0.03–0.3 mm long; style (0.5–) 1–2.7 mm long; ovules and seeds 12–24 per fruit. **SEEDS** 1–1.4 × 0.7–0.9 mm.

1. Lower stems pubescent with sessile to subsessile, malpighiaceous or cruciform trichomes; cauline leaves often not ciliate; fruits glabrous or puberulent with simple and 2-rayed trichomes .... **var. *oxyloba***
1′ Lower stems hirsute with simple trichomes; cauline leaves ciliate; fruits glabrous .... **var. *spectabilis***

var. *oxyloba* (Greene). Gilg & O. E. Schulz (sharp lobe) **STEMS** pubescent proximally with sessile, malpighiaceous or cruciform trichomes, usually the 2 rays parallel to stem axis longer. **CAULINE LEAVES** often not ciliate, with sessile, malpighiaceous or cruciform trichomes only. **FRUIT** glabrous or puberulent with simple and 2-rayed trichomes. $2n = 20, \pm 32, 40$. Meadows, talus, and margins of spruce-fir forests. COLO: Arc, LPl, SJn. 2960–3750 m (9700–12300′); Flowering: Jun–Aug. Wyoming south to Colorado, Utah, and New Mexico. The two varieties of *D. spectabilis* form discrete populations in some regions but intergrade extensively in southwestern Colorado.

var. *spectabilis* **STEMS** hirsute proximally with simple trichomes. **CAULINE LEAVES** ciliate. **FRUIT** glabrous. $2n = 20$. Aspen, Douglas-fir, spruce-fir, and alpine communities. ARIZ: Apa; COLO: Arc, Hin, LPl, Mon, SJn; NMEX: McK, SJn; UTAH. 2840–3960 m (9300–13000′). Flowering: Jun–Aug. Wyoming south to Colorado, Utah, and Arizona. Superficially similar to the Arctic/boreal species *D. borealis* DC. There is no evidence that true *D. borealis* occurs in the region; the report in the Four Corners catalog (Heil and O'Kane 2005) is based on misidentification of local hybrids involving *D. spectabilis*.

Draba streptobrachia R. A. Price (twisted arms) Alpine tundra draba. Perennials from well-developed, few- to many-branched caudices. **STEMS** 1–13 cm tall, simple, decumbent to ascending, pubescent throughout with subsessile, 3–5-rayed trichomes 0.03–0.25 (–0.4) mm long, the rays twisted and contorted. **LEAVES** at stem bases oblanceolate to linear-oblanceolate, 4–30 (–40) × 1–5 mm, rarely sparsely ciliate with simple trichomes to 0.6 mm long; both surfaces unevenly pubescent with trichomes concentrated near leaf margins, the trichomes short-stalked, 3–8-rayed and 0.05–0.4 mm long, with the rays twisted and contorted; cauline leaves 1–5, oblong to linear, entire, pubescent like basal leaves. **INFL** ebracteate, 4–10 (–18)-flowered, elongated in fruit; rachises pubescent like stems; lowermost fruiting pedicels 2–12 mm long, ascending, pubescent like stems. **FLOWER** sepals 2–3 mm long, pubescent with simple and short-stalked, 2–4-rayed trichomes; petals yellow, 3–5 × 1.5–3 mm; anthers 0.25–0.4 mm long. **FRUIT** elliptic to ovate-lanceolate, 3–10 × 2–4 mm, slightly twisted or plane, pubescent with simple and minutely stalked, 2–4-rayed trichomes 0.03–0.25 mm long, or occasionally glabrous; style 0.3–0.8 (–1.2) mm long; ovules and seeds 10–16 (–18) per fruit. **SEEDS** 1–1.6 × 0.6–1 mm. $2n =$ about 64. [*Draba spectabilis* Greene var. *dasycarpa* (O. E. Schulz) C. L. Hitchc.].

Cliffs and talus slopes in alpine habitats. COLO: Arc, Con, Hin, LPl, Min, SJn. 3660–3960 m (12000–13000′). Flowering: Jun–Aug. Endemic to western Colorado. An apomictic species of hybrid origin; shared morphological features suggest *D. spectabilis* and *D. crassa* as possible parents. The report of *D. oligosperma* Hook. in the Four Corners catalog (Heil and O'Kane 2005) was based on misidentified specimens of *D. streptobrachia*.

Erysimum L. Wallflower

(Greek *eryomai*, to help or save, referring to purported medicinal properties) Annual, biennial, or perennial, herbs or rarely subshrubs or shrubs, pubescent with 2–few-branched malpighian or stellate trichomes. **STEMS** ribbed, single to multiple from base, branched or not. **LEAVES**: lower petiolate, entire to dentate, rarely pinnatifid; upper short-petiolate or sessile. **FLOWER** sepals erect, the lateral pair gibbous-based or not; petals yellow to orange, less frequently white, magenta, or purple, orbicular or obovate to oblong, long-clawed. **FRUIT** a silique, linear (rarely oblong), plain or torulose, terete or 4-angled (rarely a little compressed), valves sometimes ± keeled. **SEEDS** 1 row per locule, nonmucilaginous. $x = 6–17$. Two hundred twenty-three species of the Northern Hemisphere and especially speciose in eastern Europe and the Middle East.

1. Petals greater than 12.5 mm long, showy; style mostly greater than 1.5 mm long; plants biennial or perennial .... ***E. capitatum***
1′ Petals less than 12 mm long, not especially showy; style mostly less than 1.5 mm long; plants annual, biennial, or perennial .... (2)

2. Siliques widely spreading and strongly torulose at maturity; pedicels thick, ± the same diameter as the siliques .... ***E. repandum***
2′ Siliques ascending to erect, not noticeably torulose at maturity; pedicels slender, narrower than the siliques ...(3)

3. Plants annual; petals less than 5 (6) mm long; siliques ± quadrangular at maturity, the "facets" flat; flower buds not especially notched at apex; seeds twisted .... ***E. cheiranthoides***
3′ Plants biennial or short-lived perennial; petals (4–) 5–7 mm long; siliques subterete (but may be 4-ribbed at maturity, and then the "facets" rounded); flower buds noticeably notched at apex; seeds not twisted ...... ***E. inconspicuum***

Erysimum capitatum (Douglas ex Hook.) Greene (a head, referring to the relatively dense inflorescence) Western wallflower, sand dune wallflower. Biennial or perennial from a thick, leaf-covered caudex, variously pubescent with 2-branched trichomes with short stalks, upper leaf surfaces sometimes with 3 (4–7)-branched trichomes mixed in. **STEMS** 1 to several (more generally at higher elevations), up to 8 (10) dm tall. **LEAVES**: basal leaves linear-lanceolate to spatulate, entire to dentate, acute (rarely obtuse) at apex, gradually narrowing to the base. **INFL** raceme or panicle, congested in flower but elongating in fruit. **FLOWER** sepals 7–14 mm long, elliptic to oblong, gibbous at base; petals 13–25 (–32) mm long, the claw exceeding the blade, showy, yellow, orange-yellow, cream-yellow, or lavender (high elevations). **FRUIT** a silique, divaricately ascending to ascending, quadrangular to slightly flattened in cross section, 3–15 mm long, 1–3 mm wide, tapering to the style. **SEEDS** 1 row per locule, 1.7–2 mm long, narrowly oblong or narrowly ovoid. $2n = 36$. [*E. arkansanum* Nutt.; *E. elatum* Nutt.; *E. pumilum* Nutt.; *E. wheeleri* Rothr.; several varieties of *E. asperum* (Nutt.) DC.; *E. californicum* Greene; *E. argillosum* (Greene) Rydb.; *E. aridum* (A. Nelson) A. Nelson; *E. oblanceolatum* Rydb.; *E. radicatum* Rydb.; *E. moniliforme* Eastw.] Hopis used the plant for advanced cases of tuberculosis; Navajos smelled crushed leaves for headache and used the plant as an emetic in ceremonials. The long list of synonyms illustrates the confusion that is present in this taxon. Various ways have been attempted to parse out the infraspecific variation. As delineated here, var. *purshii* is the common and widespread entity at generally lower elevations. Variety *nivale* is removed from var. *purshii* (sensu Rollins 1993) and is found at higher elevations. Variety *capitatum* is apparently rare in our area. Each of the varieties has an extensive synonymy of its own which is not presented here. The Great Plains species *E. asperum* (Nutt.) DC. differs from *E. capitatum* in that its siliques are horizontally spreading and are stiff with a striped appearance; *E. asperum* is possibly to be expected in grassy areas of the easternmost portions of the *Flora* area.

1. Subalpine to alpine areas, generally above 2800 m (9200′); stems several to many from the base, caespitose, up to 20 (25) cm tall; petals yellow-orange to lavender or even red-violet; siliques somewhat torulose; long-lived perennial .... **var. *nivale***
1′ Low elevations up to lower subalpine areas, generally below 2450 m (8000′); stems 1 to several, not appearing especially caespitose, up to 50 cm or more tall; petals yellow, yellow-orange, burnt-orange, or copper; siliques not noticeably torulose; biennial to short-lived perennial .... (2)

2. Upper leaf surfaces with mostly 3-rayed trichomes; some plants up to 7 dm or more tall; seeds greater than 1.2 mm wide, often distally winged .... **var. *capitatum***
2′ Upper leaf surfaces with mostly 2-rayed trichomes; some plants up to 5 (6) dm tall; seeds less than 1.2 mm wide, essentially wingless .... **var. *purshii***

var. *capitatum* Wallflower. Biennial to short-lived perennial from a typically unbranched caudex. **STEMS** 1 to several, branched or unbranched, up to 7 or even 10 cm tall. **FLOWER** sepals green; petals yellow, yellow-orange, sometimes fading to orange. **FRUIT** not especially torulose. **SEEDS** typically greater than 1.2 mm wide. One record in our area from a canyon bottom with Douglas-fir and ponderosa pine. ARIZ: Apa. 2042 m (6700′). Flowering: late May. Texas to Saskatchewan to the Pacific states and central Mexico, disjunct in the East to Ohio, Tennessee, and Arkansas.

var. *nivale* (Greene) N. H. Holmgren (of the snow) Wallflower. Long-lived perennial, caespitose from a usually branched and at least partly underground caudex which is ± densely clothed in old leaf bases. **STEMS** several to many, typically up to 20 (25) cm tall. **FLOWER** sepals purple to green with purple tips; petals deep yellow-orange or lavender. **FRUIT** somewhat torulose. **SEEDS** typically less than 1.2 mm wide. Alpine and subalpine areas in meadows, fell-fields, talus slopes, upper spruce forests, krummholz, tundra, and drainages with willows. COLO: Hin, LPl, Min, Mon, RGr, SJn; UTAH. (2750–) 3200–3975 m [(9000–) 10500–13000′]. Flowering: late May–late Jul, occasionally through late Aug. Utah and Colorado.

var. *purshii* (Durand) Rollins (for Frederick Pursh, botanist who worked with the Lewis and Clark collections) Wallflower. Biennial to short-lived perennial from a typically unbranched caudex. **STEMS** 1 to several, branched or unbranched, up to 50 (60) cm tall. **FLOWER** sepals green but sometimes fading purple at upper elevations; petals typically yellow or yellow-orange but occasionally copper or burnt-orange especially when fading. **FRUIT** not especially torulose. **SEEDS** typically less than 1.2 mm wide. Sagebrush flats, sandy terraces, piñon-juniper, ponderosa pine with Gambel's oak, mixed mountain shrubs, and roadsides. ARIZ: Apa, Nav; COLO: Arc, Dol, LPl, Mon, SMg; NMEX: McK, RAr, San, SJn; UTAH. 1275–2450 m (4200–8000′). Flowering: late Apr–early Aug, occasionally flowering again in Sep with adequate moisture. Rocky Mountain states to the Sierras and to central Mexico. Based on the taxonomy adopted here, our most common variety.

Erysimum cheiranthoides L. (resembling *Cheiranthus*) Wormseed wallflower. Annual (or biennial), pubescent with mostly 2-branched trichomes. **STEMS** usually single, often branched from the base or more commonly above, up to 10 dm tall. **LEAVES** linear to narrowly lanceolate, lanceolate or oblanceolate, 3–11 cm long and up to 2.2 cm wide, entire to denticulate. **INFL** loose and greatly elongated. **FLOWER** sepals narrowly elliptic to narrowly lanceolate, 1.8–3.3 mm long; petals pale yellow, 3.5–5 mm long. **FRUIT**: siliques divaricately ascending to erect, 1.5–3 cm long and about 1 mm wide, terete-quadrangular, pubescent with 3- or 4-branched trichomes. **SEEDS** 1 row per locule, 1–1.4 mm long, twisted, orangish. $n = 8$, $2n = 16$. Piñon-juniper communities, clay hills, wash bottoms with tamarisk. ARIZ: Apa; COLO: Arc, LPl, Mon; NMEX: McK, RAr, SJn; UTAH. 1850–2175 m (6100–7150′). Flowering: late Mar–May. Introduced from Eurasia but some populations in the north of its North American range and in the Rocky Mountains possibly indigenous; now growing from Alaska to northern California, east across the Rocky Mountains and plains to Newfoundland and south to Florida.

Erysimum inconspicuum (S. Watson) MacMill. (not or less + visible, referring to the small flowers) Shy wallflower, lesser wallflower. Perennial (or possibly biennial), pubescent with mostly 2-branched trichomes except leaf surfaces sometimes have 3-branched trichomes. **STEMS** usually single from an often swollen caudex, rarely branched, up to 6 dm tall. **LEAVES**: basal linear to linear-oblanceolate, entire to sparsely dentate, petiolate, 3–8 cm long, 2–6 mm wide; leaves becoming reduced and sessile upward. **INFL** dense and elongating. **FLOWERS** notched at apex in bud; sepals narrowly lanceolate, 4.5–8 mm long, the outer pair gibbous-based and hooded at the apex, the inner pair with a scarious apex; petals 5–12 mm long, pale yellow. **FRUIT**: siliques divaricately ascending to ascending, 3–6 dm long, 1–1.7 mm wide, subterete, pubescent. **SEEDS** 1 row per locule, 1.3–2 mm long, orangish. $n = 27$, $2n = 54$. [*E. parviflorum* Nutt.; *E. asperum* (Nutt.) DC. var. *inconspicuum* S. Watson; *E. syrticolum* E. Sheld.]. Roadsides and ponderosa pine, piñon-juniper, Gambel's oak, and mountain shrub communities. ARIZ: Apa, Nav; COLO: Arc, LPl, Mon, SMg; NMEX: McK, RAr, San, SJn; UTAH. 1750–2450 m (5750–8000′). Flowering: late Apr–early Jun. Canada, Pacific Northwest east of the Cascades, Rocky Mountain states, north-central United States. Hopis used the plant as a treatment of tuberculosis. Our plants are **var. *inconspicuum***.

Erysimum repandum L. (turned up or bent back, meaning sinuate and referring to the leaf margins) Spreading wallflower, treacle wallflower. Annual, pubescent with 2- or 3-branched trichomes. **STEMS** usually single and branched throughout, up to 3.5 (5) dm tall. **LEAVES** narrowly oblanceolate to linear, sinuate-dentate to almost entire, 1–6 (–10) cm long, petioled below, sessile above. **INFL** relatively few-flowered, elongating. **FLOWER** sepals 4–6 mm long, narrowly lanceolate, the outer pair gibbous-based and hooded at the apex, petals 6–8 mm long, pale yellow. **FRUIT**: siliques horizontally spreading to slightly spreading, 4–8 cm long, 1.1–1.5 mm wide, on pedicels essentially the same diameter as the fruit, torulose. **SEEDS** 1 row per locule, 1–1.4 mm long, yellow-orange. n = 7, 8, $2n$ = 14, 16. [*E. pygmaeum* (Adams) J. Gay]. Disturbed areas, tamarisk, saltbush, desert shrub, roadsides, saline areas. ARIZ: Apa; COLO: LPl, Mon; NMEX: McK, San, SJn. 1650–2250 m (5400–7400′). Flowering: late Apr–Jun. Native of Eurasia, now distributed from central Canada to North Carolina and west to California.

Hesperidanthus (B. L. Rob.) Rydb. Plains-mustard

(the genus *Hesperis* + Greek *anthos*, flower, referring to the showy purple flower of the genus) Perennial herbs from a woody crown, glabrous and glaucous. **STEMS** single to few, erect to somewhat decumbent, branched above. **LEAVES** all cauline, lowermost sometimes obovate and petiolate, but leaves generally sessile, linear, lanceolate, or oblanceolate, entire to ± pinnatifid. **FLOWER** sepals with inner pair gibbous-based, outer pair apically cucullate; petals showy, clawed, blade obovate to suborbicular, the apex sometimes ± erose or crenulate, lavender to purple. **FRUIT** a silique, linear, terete. **SEEDS** 1 row per locule. x = 11. Five species of western North America and Mexico. Members of the genus have variously been assigned to *Schoenocrambe*, *Streptanthus*, *Thelypodium*, and *Sisymbrium*.

Hesperidanthus linearifolius (A. Gray) Rydb. (linear-leaved) Slimleaf plains-mustard. Perennial, glabrous and glaucous. **STEMS** 1 or a few, erect, branched upward, up to 15 dm tall. **LEAVES** a rosette on first-year plants, petiolate, obovate, dentate to entire, early-deciduous; cauline leaves oblanceolate and becoming linear or linear-lanceolate upward, thick, short-petiolate to cuneate, entire to weakly dentate. **INFL** few-flowered, elongating some in fruit. **FLOWER** sepals purplish, inner pair gibbous-based, outer pair with cucullate apex, 4.5–7 mm long; petals showy, lavender to purplish with darker veins, 9–16 (22) mm long, claw linear, blade obovate to suborbicular and spreading at right angle to the claw, apically usually finely crenulate. **FRUIT**: siliques on pedicels divaricately ascending to spreading, straight, 3.5–9 mm long, terete. **SEEDS** 1 row per locule, oblong, 1.5 mm long. n = 11, ±20, $2n$ = 22, 40, 44. [*Schoenocrambe linearifolia* (A. Gray) Rollins; *Streptanthus linearifolius* A. Gray; *Thelypodium linearifolium* (A. Gray) S. Watson; *Sisymbrium linearifolium* (A. Gray) Payson; *Thelypodiopsis linearifolia* (A. Gray) Al-Shehbaz]. Ponderosa pine communities variously containing juniper, piñon pine, Gambel's oak, and cliffrose. NMEX: McK, SJn. 1975–2400 m (6500–7900′). Flowering: Jun–early Sep. Colorado, Texas, New Mexico, Arizona south to mid-Mexico. This species, as the synonymy shows, has moved among several genera.

Hesperis L. Rocket

(from Greek *hesperos*, evening, referring to the time of maximum floral fragrance) Biennial or perennial herbs, pubescent with forked or simple mixed with forked trichomes, uniseriate multicellular glands sometimes present. **STEMS** 1–few, simple or branched. **LEAVES**: basal usually rosulate, entire to pinnately lobed; cauline petiolate or sessile, sometimes auriculate or clasping. **INFL** racemose or paniculate. **FLOWER** pedicels with 2 minute glands; sepals erect, the lateral pair saccate; petals white to purple or yellow to brownish or green, obovate to oblong, clawed. **FRUIT**: silique linear, terete, quadrangular or latiseptate, often torulose; valves with a prominent midvein; tardily dehiscent. **SEEDS** 1 row per locule, nonmucilaginous. x = 6, 7, 8, 10. Forty-six species mostly of the Middle East and Europe, a few in northern Africa and east and central Asia.

Hesperis matronalis L. (of a matron or dame) Dame's rocket, sweet rocket, damask violet, mother-of-the-evening. Biennial or perennial, hirsute with simple and forked trichomes, sometimes with shorter forked ones mixed in. **STEMS** 1 (–several), erect, simple or sparingly branched. **LEAVES** essentially all cauline (if near-basal ones present, these withering early), petiolate, the lower on long petioles, lanceolate to ovate-lanceolate, ± serrate to dentate, cuneate basally, generally 7–12 cm long, larger below, smaller above. **INFL** terminal and axillary racemes, not greatly elongating in fruit. **FLOWER** sepals oblong, obtuse to rounded, sometimes long-pilose near the apex, 5.5–9 mm long; petals lilac to purple (rarely white), blade broadly ovate, 17–23 mm long including the long claw. **FRUIT** a silique, terete, torulose, glabrous, longitudinally ribbed, 4–10 (14) cm long, narrow. **SEEDS** 1 row per locule, reddish brown, narrowly oblong or ellipsoid, 3–4 mm long. n = 8, 12, 14, 16, 24, $2n$ = 12, 14, 24, 26, 28. Riparian area near canal. COLO: LPl. 2100 m (6890′). Flowering: May–Jul. Native of Europe and escaping from gardens where it is grown as an ornamental. So far, known from one self–propagating population in our study area and escaped from cultivation. To be expected elsewhere, especially near abandoned towns and mining camps. The flowers are unusual for the family in that they are quite fragrant, especially at night, hence the common name mother-of-the-evening.

Hornungia Rchb. Hutchinsia, Ovalpurse

(for Ernst G. Hornung, 1795–1862, German botanist) Annual or perennial herbs, pubescent (some individuals nearly glabrous in ours) with 2-forked, dendritic or stellate trichomes, sometimes mixed with simple ones. **STEMS** few to several from base, usually branching. **LEAVES** basal or all cauline, petiolate, entire to pinnatifid. **FLOWER** sepals spreading; petals exceeding or subequaling the sepals, white, obovate or oblong to cuneiform. **FRUIT**: silicle elliptic to ovate or lanceolate, angustiseptate, ascending to divaricate, valves at least somewhat reticulate. **SEEDS** 2 rows per locule (aseriate when few), mucilaginous or not. $x = 6$. [*Hutchinsia* R. Br., nom. illeg.; *Pritzelago* Kuntze; *Hymenolobus* Nutt. ex Torr. & A. Gray]. Three species of Europe, Southeast Asia, and possibly North America.

Hornungia procumbens (L.) Hayek (to become prostrate) Ovalpurse. Delicate annual, glabrous or sparsely pubescent with simple or branched trichomes. **STEMS** freely branched, generally diminutive and up to 15 cm tall. **LEAVES** 1–3 cm long, obovate to lyrate-pinnatifid below, reduced in size upward and becoming obovate to narrowly lanceolate, cuneate at base. **INFL** loose and elongate. **FLOWER** sepals less than 1.1 mm long; petals white, spatulate, about equaling the sepals. **FRUIT** a silique, elliptic to slightly obovate, 2.3–4.3 mm long, 1.4–2 (2.4) mm wide, valves reticulate-veined, standing erect on divaricate pedicels. **SEEDS** broadly oblong or broadly oval, yellowish brown, mucilaginous when wetted. $n = 6$, $2n = 12, 24$. [*Hutchinsia procumbens* (L.) Desv.; this species has historically been placed in *Lepidium*, *Thlaspi*, *Capsella*, *Noccaea*, *Hymenolobus*, and *Bursa*.]. Seeps, moist areas in and below hanging gardens, shaded areas on benches above wash bottoms. ARIZ: Apa, Nav; NMEX: SJn; UTAH. 1350–1700 m (4400–5600′). Flowering: May–Apr. Native of Europe (and North America), widespread but infrequent in Canada and the western United States.

Lepidium L. Peppergrass, Pepperwort, Peppercress

Steve L. O'Kane, Jr. and Ihsan A. Al-Shehbaz

(Greek *lepidion* or *lepidos*, scale, referring to the fruit appearance) Herbs, annual, biennial, or perennial, sometimes subshrubs, rarely shrubs or lianas; trichomes absent or simple. **STEMS** erect or ascending, sometimes creeping, simple or branched basally and/or apically. **LEAVES**: basal leaves rosulate or not, simple, entire or variously toothed or divided and 1–3-pinnatisect; cauline leaves petiolate or sessile, base cuneate, attenuate, auriculate, sagittate, or amplexicaul, margin entire, dentate, or dissected. **FLOWER** sepals ovate or oblong, rarely orbicular, base of lateral pair not saccate; petals white, yellow, pink, or purple, erect or spreading, sometimes rudimentary or absent; blade obovate, spatulate, oblong, oblanceolate, orbicular, linear, or filiform, apex entire or emarginate, claw absent or distinct; stamens 2 or 4 and equal in length, lateral or median, or 6 and usually tetradynamous. **FRUIT** a silicle, dehiscent, schizocarpic, or indehiscent, strongly flattened parallel to the septum or inflated and terete, sessile, unsegmented; valves veinless or prominently veined, keeled or rounded, apically winged or wingless, glabrous or pubescent, thin or strongly thickened and ornamented, readily releasing or enclosing seed; septum complete or perforated; style absent, obsolete, or distinct, included or exserted from apical notch of fruit. **SEEDS** 1 per locule, oblong or obovate; usually copiously mucilaginous when wetted. $x = 4, 8$. Two hundred thirty-one species present on all continents except Antarctica (Al-Shehbaz, Mummenhoff, and Appel 2002; Bowman et al. 1999; Hitchcock 1936, 1945, 1950; Mulligan and Frankton 1962; Mummenhoff, Brüggemann, and Bowman 2001). Young plants of several species have been boiled, fried, and eaten by many tribes. While the species limits of *Lepidium* are reasonably well defined and marked, infraspecific taxonomy is currently in a state of disarray (e.g., see *L. montanum* below). The seeds of most species of *Lepidium* can easily be transported by animals, especially migratory birds, for long distances due to copious amounts of mucilage released upon wetting. The genus here includes species formerly contained in *Cardaria* (Al-Shehbaz et al. 2002; Bowman et al. 1999; Mummenhoff, Brüggemann, and Bowman 2001). When collecting specimens, collect basal material, including leaves, as well as fruits and flowers. If the caudex cannot be collected, be sure to note whether the base of the plant is woody or not and if the caudex is elevated above the ground or not. In the descriptions below, **INFL** refers to the infructescence.

1. Silicles indehiscent, entire at apex, inflated, somewhat flattened or not, with rounded or obtuse apex; plants rhizomatous, perennial (genus *Cardaria* of floras) ....(2)
1′ Silicles dehiscent, usually notched at the apex (sometimes barely so), distinctly flattened with a carinate or winged apex; plants from a taproot or fibrous-rooted (but rhizomatous in *L. latifolium*), annual, biennial, or perennial and then sometimes woody at the base (genus *Lepidium* sensu stricto of floras) ....(4)

2. Silicles and sepals pubescent .... ***L. appelianum***
2′ Silicles and sepals glabrous (sepals sometimes sparsely hairy in *L. chalepense*) ....(3)

3. Silicles cordate at base, the apex obtuse .... *L. draba*
3′ Silicles rounded, truncate, or cuneate at base, the apex rounded .... *L. chalepense*

4. Leaves (at least the upper ones) with clasping or amplexicaul, auriculate bases .... (5)
4′ Leaves cuneate or petiolate .... (6)

5. Upper and lower cauline leaves distinctly different, the lower bi- or tripinnatifid, the upper entire ..... *L. perfoliatum*
5′ Upper and lower cauline leaves essentially the same, entire to serrate .... *L. campestre*

6. Styles either lacking or included within the apical notch of the silicle; stamens 2 .... (7)
6′ Styles exceeding the apical notch of the silicle; stamens 4 or 6 (rarely 2 in *L. integrifolium*) .... (10)

7. Silicles pubescent (at least slightly so and noticeable), usually slightly reticulate-veiny; stems several from the base .... *L. lasiocarpum*
7′ Silicles essentially glabrous, not reticulate-veiny; stem single from the base (but may branch above) .... (8)

8. Petals present, up to 1.5 mm long; pedicels longer than the silicle .... *L. virginicum*
8′ Petals either absent or very short and rudimentary, less than 1 mm long; pedicels usually as long as or shorter than the silicle .... (9)

9. Stem branched throughout, the branches shorter than the main stem; seeds lacking a narrow margin; silicles sparsely pubescent or glabrous .... *L. ramosissimum*
9′ Stem unbranched or branched above and the branches almost the same length as the main stem; seeds with a narrow margin, at least apically on one side; silicles glabrous .... *L. densiflorum*

10. Plants rhizomatous; leaf blades leathery .... *L. latifolium*
10′ Plants with a taproot or fibrous-rooted; leaf blades not especially leathery .... (11)

11. Plants annual, biennial, or perennial, herbaceous throughout .... (12)
11′ Plants perennial, typically subshrubs, but at least woody at the base, which is often elevated above the ground ..(14)

12. Basal and lowermost leaves entire or nearly so .... *L. integrifolium*
12′ Basal and lowermost leaves all or mostly pinnatifid to pinnatisect to (bi-)pinnate .... (13)

13. Plants 4.5–16 (or more) dm tall; upper cauline leaves mostly greater than 4 mm wide ...... *L. eastwoodiae* (in part)
13′ Plants less than 4.4 dm tall; upper cauline leaves mostly less than 4 mm wide .... *L. montanum* (in part)

14. Basal leaves from short sterile shoots, crenate or serrate-crenate .... *L. crenatum*
14′ Basal leaves not from short sterile shoots, all or mostly pinnatifid to pinnatisect to (bi-)pinnate .... (15)

15. Plants 4.5–16 (or more) dm tall; upper cauline leaves mostly greater than 4 mm wide ...... *L. eastwoodiae* (in part)
15′ Plants less than 4.4 dm tall; upper cauline leaves mostly less than 4 mm wide .... (16)

16. Leaf lobes (especially the terminal one) narrowly oblanceolate to oblong, often with a rounded apex, mostly more than 2 mm wide .... *L. montanum* (in part)
16′ Leaf lobes tapering to a point from the base to the apex, mostly less than 2 mm wide .... *L. alyssoides*

Lepidium alyssoides A. Gray (resembling *Alyssum*) Mesa pepperwort. Subshrubs or herbs, perennial with woody base often elevated aboveground, glabrous or minutely puberulent. **STEMS** 1–4.8 (–6.1) dm, erect to ascending, few to many from base, branched throughout. **LEAVES**: basal often not rosulate; petiole 1–6 cm, blade 1.5–8 (–11) × (0.5–) 1–3.5 cm, pinnately lobed, lobes entire or denticulate; middle cauline leaves sessile, blade linear, (0.8–) 1.3–7 (–9.5) cm × 1–2 (–3) mm, base attenuate, not auriculate, margin entire. **INFL** raceme, elongated; rachis puberulent or glabrous; fruiting pedicel 3.5–8 (–11) mm, terete, straight or recurved to somewhat sigmoid, divaricate to horizontal, glabrous or puberulent adaxially. **FLOWER** sepals ovate to oblong, 1–2 mm long, deciduous; petals white, suborbicular, 2–3 mm long, claw 0.5–1.5 mm; stamens 6. **FRUIT** dehiscent, broadly ovate, 2–3.7 (–4.3) × 1.8–2.9 (–3.4) mm, apically winged, apical notch 0.1–0.3 mm deep; valves thin, smooth, glabrous, not veined; style 0.2–0.6 mm, exserted beyond apical notch. **SEEDS** brown, ovate, 1.5–1.8 mm long. $2n = 32$. [*L. montanum Nutt.* subsp. *angustifolium* (C. L. Hitchc.) C. L. Hitchc.; *L. montanum* var. *angustifolium* C. L. Hitchc.; *L. tortum* L. O. Williams]. Piñon-juniper and sagebrush communities, often in sandy or sandy and wet areas. ARIZ: Apa, Coc; NMEX: RAr, SJn. 2100–2550 m (6900–8350′). Flowering: late May–Jul. Arizona, Colorado, Nevada, New Mexico, southwestern Texas, Utah; Mexico.

Lepidium appelianum Al-Shehbaz (for O. Appel, contemporary student of the Brassicaceae) Hairy whitetop, globe-podded hoarycress. Herbs, perennial, rhizomatous, often densely hirsute. **STEMS** 1.5–3.5 (–5) dm, erect or ascending, 1 to several from base, branched above. **LEAVES**: basal not rosulate, often withered by anthesis; petiole 0.5–1.5 cm; blade obovate to oblanceolate, 2–6 (–7) × 0.3–2 cm, margin dentate to sinuate; middle cauline leaves sessile, blade oblong or lanceolate, 1–5 (–8) × (0.3–) 0.5–1.5 (–3) cm, pubescent, base sagittate, margin dentate or subentire. **INFL** corymbose or rarely racemose panicle, not or rarely elongated; rachis pubescent with often curved trichomes; fruiting pedicel 3–9 (–12) mm, terete, straight or slightly curved, divaricate to ascending, pubescent all around. **FLOWER** sepals oblong, 1.4–2 mm long, deciduous; petals white, broadly obovate, (2.2–) 2.8–4 mm long, claw 1–1.4 mm long; stamens 6. **FRUIT** indehiscent, globose to subglobose, (2–) 3–4.4 (–5) mm in diameter, inflated, apically wingless, apical notch absent; valves thin, smooth, densely puberulent, not veined; style 0.5–1.5 mm. **SEEDS** brown or dark brown, ovoid, 1.5–2 × 1–1.5 mm. 2*n* = 16. [*Cardaria pubescens* (C. A. Mey.) Jarm.]. Roadsides, sagebrush communities, alkaline meadows, waste ground, ditch and stream sides, fields, pastures. COLO: Arc, LPl, Mon; NMEX: SJn; UTAH. 1750–2075 m (5750–6800′). Flowering: May–Jul (–Sep). Native of central Asia, naturalized elsewhere in Asia and North and South America. The species has become a noxious weed in most of its range in North America.

Lepidium campestre (L.) W. T. Aiton (of fields) Fieldcress, field pepperwort. Herbs, annual, densely hirsute. **STEMS** 1.2–5 (–6.3) dm, erect, simple or branched above. **LEAVES**: basal rosulate; petiole (0.5–) 1.5–6 cm; blade oblanceolate or oblong, (1–) 2–6 (–8) × 0.5–1.5 cm, margin entire, lyrate, or pinnatifid; middle cauline leaves oblong, lanceolate, or narrowly deltoid-lanceolate, (0.7–) 1–4 (–6.5) cm × (2–) 5–10 (–15) mm, base sagittate or auriculate, margin dentate or subentire. **INFL** raceme, much elongated; rachis hirsute with straight, spreading trichomes; fruiting pedicel (3–) 4–8 (–10) mm, terete, straight or slightly recurved, horizontal, hirsute all around. **FLOWER** sepals oblong, (1–) 1.3–1.8 mm long, deciduous; petals white, spatulate, (1.5–) 1.8–2.5 mm long, claw 0.6–1 mm; stamens 6. **FRUIT** dehiscent, broadly oblong to ovate, (4–) 5–6 (–6.5) × (3–) 4–5 mm, curved adaxially, apically broadly winged, apical notch (0.2–) 0.4–0.6 mm deep; valves thin, papillate except for wing, not veined; style 0.2–0.5 (–0.7) mm, slightly exserted beyond to included in apical notch. **SEEDS** dark brown, ovoid, 2–2.3 (–2.8) mm. 2*n* = 16. [*Thlaspi campestre* L.; *Neolepia campestris* (L.) W. A. Weber]. Roadsides, pastures, gardens, open flats, piñon-juniper and Gambel's oak communities, waste ground, disturbed areas, meadows, fields, washes, barren shale slopes. COLO: Arc, Dol, LPl, Min, Mon; NMEX: RAr, SJn; UTAH. 1625–2850 m (5350–9350′). Flowering: Jun–Aug. Native of Europe and Asia, naturalized in South Africa and South America and throughout North America.

Lepidium chalepense L. (after Aleppo, a city in Syria) Whitetop. Herbs, perennial, rhizomatous, densely hirsute to subglabrous. **STEMS** (0.8–) 2.1–6.6 (–9.2) dm, erect or decumbent basally, many-branched above. **LEAVES**: basal not rosulate, withered early; petiole 0.9–4.4 cm; blade obovate, spatulate, or ovate, (1.8–) 2.5–8.6 (–14) × 1–3.7 cm, margin subentire or dentate; middle cauline leaves sessile; blade obovate to oblong or lanceolate to oblanceolate, (1.5–) 2.6–9.3 (–13.2) × 1.2–3.1 (–4.5) cm, pubescent or glabrous, base sagittate-amplexicaul or auriculate, margin dentate or entire. **INFL** a corymbose panicle, elongated; rachis glabrous or puberulent with straight or curved cylindrical trichomes; fruiting pedicel 5–16 (–19) mm, terete, straight, ascending to horizontal, glabrous or sparsely puberulent adaxially. **FLOWER** sepals oblong to ovate, 1.7–3 mm long, deciduous; petals white, obovate, 3–5 mm long, claw 1.2–2 mm; stamens 6. **FRUIT** indehiscent, obovoid to subglobose or obcompressed globose, 3.5–5.8 (–7) × (3.5–) 4–6.2 (–7) mm, apex wingless, apical notch absent; valves thin, smooth, glabrous, often not veined; style (0.8–) 1.2–2 (–2.3) mm. **SEEDS** dark reddish brown, ovate, 1.5–2.3 mm. 2*n* = 48, 80, 128. [*Cardaria boissieri* (N. Busch) Soó; *Cardaria chalepensis* (L.) Hand.-Mazz.; *C. draba* (L.) Desv. subsp. *chalepensis* (L.) O. E. Schulz; *L. draba* L. (several subsp. including var. *repens*. (Schrenk) Thell.); *Thlaspi chalepense* (L.) Poir.]. Disturbed and waste areas. COLO: LPl; NMEX: SJn, Mon. About 1830 m (6000′). Flowering: May–Jun. Native of Asia, naturalized in Europe, South America, and the western United States and Canada.

Lepidium crenatum (Greene) Rydb. (a notch, referring to a leaf margin) Alkaline pepperwort. Subshrubs, perennial with woody base elevated aboveground, puberulent. **STEMS** (2–) 3–8 (–11) dm, erect, single from base, branched above. **LEAVES**: basal rosulate on sterile shoots; petiole (1.5–) 2.5–8 (–10) cm; blade oblanceolate to spatulate, (2–) 3–7 (–9) × 0.5–2.3 (–3.2) cm, margin crenate to serrate-crenate; middle cauline leaves shortly petiolate to sessile, blade oblong to oblanceolate, 1–3.5 × 0.4–1.5 cm, base cuneate, not auriculate, margin entire. **INFL** subcorymbose panicle, slightly elongated; rachis puberulent with straight trichomes; fruiting pedicel 3–6 (–8) mm, terete, straight, divaricate-ascending to horizontal, puberulent adaxially. **FLOWER** sepals oblong to ovate, 1.3–1.8 mm long, deciduous; petals white, suborbicular to broadly obovate, 2–3 mm long, claw 0.5–1 mm; stamens 6. **FRUIT** dehiscent, broadly ovate, (2.5–) 3–4 × (2–) 2.3–2.8 mm, apically winged, apical notch 0.1–0.2 mm deep; valves thin, smooth, glabrous, not veined; style

0.2–0.6 mm, exserted beyond apical notch. **SEEDS** brown, ovate, 1.5–2 × 1–1.3 mm. [*L. montanum* Nutt. subsp. *spatulatum* (B. L. Rob.) C. L. Hitchc.; *L. montanum* Nutt. var. *spatulatum* (B. L. Rob.) C. L. Hitchc.; *L. scopulorum* M. E. Jones var. *spatulatum* B. L. Rob.]. Piñon-juniper and sagebrush communities; clay bluffs of sandstone mesas, arroyo banks. COLO: LPl, Mon; NMEX: SJn. 1830–2285 m (6000–7500'). Flowering: Jun–Aug. Western Colorado and northern New Mexico. This species was reported from Utah by Hitchcock (1936) and Rollins (1993), but material has not been seen that unequivocally proves that.

Lepidium densiflorum Schrad. (dense flowers) Prairie pepperwort, common peppergrass. Herbs, annual or biennial, puberulent or glabrous. **STEMS** (1–) 2.5–5 (–6.5) cm, erect, often single, branched above. **LEAVES**: basal rosulate, withered early; petiole 0.5–1.5 (–2) cm; blade oblanceolate, spatulate, or oblong, (1.5–) 2.5–8 (–11) cm × 5–10 (–20) mm, margin coarsely serrate or pinnatifid; middle cauline leaves shortly petiolate; blade narrowly oblanceolate or linear, (0.7–) 1.3–6.2 (–8) cm × (0.5–) 1.5–10 (–18) mm, base attenuate to cuneate, not auriculate, margin entire or irregularly serrate to dentate or rarely pinnatifid. **INFL** raceme, much elongated; rachis puberulent with straight, slender to subclavate trichomes; fruiting pedicel (1.5–) 2–3.5 (–4) mm, terete, straight or slightly recurved, divaricate-ascending to horizontal, puberulent adaxially. **FLOWER** sepals oblong, 0.5–1 mm wide, deciduous; petals absent or rudimentary and white, filiform, 0.3–0.9 mm, claw absent; stamens 2. **FRUIT** dehiscent, obovate to obovate-suborbicular, (2–) 2.5–3 (–3.5) × 1.5–2.5 (–3) mm, widest above middle, apically winged, apical notch 0.2–0.4 mm deep; valves thin, smooth, glabrous or sometimes sparsely puberulent at least on margin, not veined; style 0.1–0.2 mm, included in apical notch. **SEEDS** brown, ovate, 1.1–1.3 mm. $2n = 32$. [*L. densiflorum* var. *elongatum* (Rydb.) Thell., var. *macrocarpum* G. A. Mulligan, var. *pubicarpum* (A. Nelson) Thell., and var. *ramosum* (A. Nelson) Thell.; *L. elongatum* Rydb.]. Waste places, disturbed sites, fields, pastures, grasslands, sagebrush flats, floodplains, gravelly hillsides, rock crevices, and roadsides from desert scrub up to aspen and spruce-fir communities. ARIZ: Apa, Nav; COLO: Arc, Hin, LPl, Mon, SJn; NMEX: McK, RAr, SJn; UTAH. 1530–2750 m (5020–9031'). Flowering: May–Jul (–Sep). Native of and present throughout North America, naturalized in Asia and Europe. Navajos used the plant to treat the effects of swallowing an ant and as an infant sedative by rubbing the plant on an infant's face. The number and limits of recognized varieties of this species have varied wildly among authors, and Rollins (1993: 554) admits that infraspecific taxa are "very weak at best." Variation very rarely correlates with geography, and, therefore, the recognition of varieties in this species is neither practical nor very useful. Of the recognized varieties, perhaps var. *pubicarpum* (with puberulent silicles and including var. *elongatum*) might merit recognition (records from NMEX: SJn and UTAH). It is distributed in almost all of the Rocky Mountain and Pacific states and is distinguished from the other varieties solely by the presence of trichomes or minute papillae on the fruit valves.

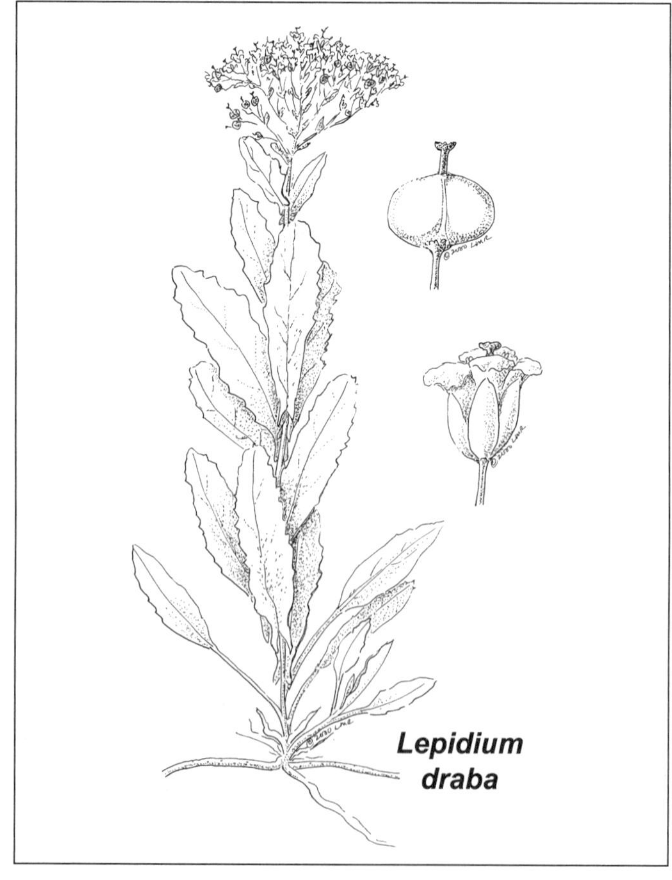
Lepidium draba

Lepidium draba L. (apparently named for the genus *Draba*) Hoarycress, whitetop. Herbs, perennial, rhizomatous, hirsute or subglabrous. **STEMS** (0.8–) 2–6.5 (–9) dm, erect or decumbent basally, often single, many-branched above. **LEAVES**: basal not rosulate, withered early; petiole 1–4 cm; blade obovate, spatulate, or ovate, (1.5–) 3–10 (–15) × 1–4 cm, margin sinuate to dentate or entire; middle cauline leaves sessile; blade ovate, elliptic, oblong, or lanceolate, oblanceolate, or obovate, (1–) 3–9 (–15) × 1–2 (–5) cm, pubescent or glabrous, base sagittate-amplexicaul or auriculate, margin dentate or entire. **INFL** corymbose panicle, slightly or much elongated; rachis glabrous or puberulent with straight or curved cylindrical trichomes; fruiting pedicel 5–10 (–15) mm, terete, straight, ascending to horizontal, glabrous or sparsely puberulent adaxially. **FLOWER** sepals oblong to ovate, 1.5–2 mm long, deciduous; petals white, obovate, 3–4 (–4.5) mm long, claw 1–1.7 mm; stamens 6. **FRUIT** indehiscent, cordate to subreniform, (2–) 2.5–3.7 (–4.3) × 3.7–5 (–5.6) mm, apex obtuse to subacute, wingless, apical notch absent; valves thin, smooth, glabrous, reticulate-veined; style 1–1.8 (–2) mm. **SEEDS** brown, ovate, 1.5–2.3 mm long. 2*n* = 32, 64. [*Cardaria draba* (L.) Desv.]. Roadsides, fields, agricultural lands, riversides, disturbed ground, pastures, waste areas. COLO: Arc, Dol, LPl, Mon; NMEX: RAr. 1825–2150 m (6000–7000′). Flowering: late Apr–Jul. Native of southwest Asia and perhaps southern Europe, naturalized in South Africa, Australia, and North and South America. The species has become a noxious weed in several western states.

Lepidium eastwoodiae Wooton (for Alice Eastwood, premier western botanist and curator at California Academy of Sciences herbarium) Eastwood's tall pepperwort. Herbs, annual, biennial, or perennial with woody base, glabrous or pubescent. **STEMS** (3.5–) 4.5–18 dm, erect, simple, branched above. **LEAVES**: basal not rosulate, soon deciduous; petiole (1–) 2–5.5 (–7.5) cm; blade (2–) 3–6.8 (–9) × 1–3 cm, pinnatifid, lobes dentate to serrate; middle cauline leaves short-petiolate or sessile, blade narrowly lanceolate or oblanceolate to linear, 3–7 cm × (2.5–) 4–10 mm, gradually reduced in size upward, base attenuate to cuneate, not auriculate, margin entire or rarely dentate. **INFL** raceme, elongated; rachis puberulent with straight or curved trichomes; fruiting pedicel (3–) 3.5–7.5 (–8) mm, wingless, slightly recurved or somewhat sigmoid, rarely straight, divaricate-ascending to horizontal, puberulent adaxially. **FLOWER** sepals suborbicular to oblong, 0.8–1.5 mm long, deciduous; petals white, suborbicular, 2.2–3.5 mm long, claw 0.7–1.5 mm; stamens 6. **FRUIT** dehiscent, broadly ovate, 2–3.5 × 1.8–2.6 mm, apically winged, apical notch 0.1–0.2 mm deep; valves thin, smooth, glabrous, not veined; style 0.3–0.6 mm, exserted beyond apical notch. **SEEDS** dark brown, ovate, 1.4–1.8 mm. [*L. alyssoides* A. Gray var. *eastwoodiae* (Wooton) Rollins; *L. moabense* S. L. Welsh; *L. montanum* Nutt. var. *eastwoodiae* (Wooton) C. L. Hitchc.]. Piñon-juniper, sagebrush, or mixed desert shrub communities. COLO: Arc, Dol, Mon. 1675–2650 m (5550–8725′). Flowering: Jun–early Sep. Arizona, southern and western Colorado, New Mexico, and Utah. Although it was previously considered a variety of either *L. montanum* or *L. alyssoides*, the differences in morphology and flowering periods support its recognition as an independent species. Further, Holmgren et al. (2005) followed in reducing *L. moabense* to synonymy under *L. eastwoodiae*.

Lepidium integrifolium Nutt. (with entire leaves) Entire-leaf pepperwort, thick-leaf pepperwort. Herbs, perennial, often with a thick, unbranched caudex covered with petiolar remains, puberulent. **STEMS** (1–) 1.5–3.5 (–4) dm, ascending, several from caudex, branched above. **LEAVES**: basal rosulate; petiole (0.5–) 1.5–6 (–7.5) cm; blade oblanceolate to obovate, (1.5–) 2.5–7 (–9) × 1.5–2.5 cm, margin entire or rarely subapically denticulate; middle cauline leaves shortly petiolate to sessile, blade narrowly lanceolate to broadly oblanceolate, 1–5 cm × 2–9 (–12) mm, base cuneate, not auriculate, margin entire or rarely subapically denticulate. **INFL** raceme, elongated; rachis puberulent with straight, sometimes clavate trichomes; fruiting pedicel (4–) 5–10 mm, wingless, straight, divaricate-ascending to horizontal, puberulent adaxially. **FLOWER** sepals oblong-obovate, 1.8–2.5 mm long, deciduous; petals white, obovate, 2.5–3.6 mm long, claw 0.5–1 mm; stamens (2 or) 4 (or 6), median and lateral when 4. **FRUIT** dehiscent, ovate, 3.2–4 × 2–3.5 mm, apically winged, apical notch 0.1–0.3 mm deep; valves thin, smooth, glabrous, not veined; style 0.5–0.8 (–1) mm, exserted beyond apical notch. **SEEDS** brown, ovate, 1.8–2 mm. [*L. heterophyllum* (S. Watson) M. E. Jones; *L. montanum* Nutt. subsp. *heterophyllum* (S. Watson) C. L. Hitchc.; *L. montanum* subsp. *integrifolium* (Nutt.) C. L. Hitchc.; *L. montanum* var. *integrifolium* (Nutt.) C. L. Hitchc.; *L. utahense* M. E. Jones; *L. zionis* A. Nelson]. Piñon and sagebrush communities and margins of washes. NMEX: SJn; UTAH. Approximately 2075 m (6800′). Flowering: Jun–Jul. Central Arizona, east-central Nevada, Utah, southwest Wyoming.

Lepidium lasiocarpum Nutt. (hairy fruit) Hairy-pod pepperwort. Herbs, annual, hirsute or hispid. **STEMS** (0.15–) 0.6–3 (–3.8) dm, erect to ascending or outer ones decumbent, few to several or rarely 1 from base, branched above. **LEAVES**: basal not rosulate, later withered; petiole (0.4–) 1–3.5 (–5); blade spatulate to oblanceolate, 1.5–4.5 (–7.5) × 1.2–2 (–3) cm, lyrate-pinnatifid, pinnatisect, or 2-pinnatifid, very rarely dentate, lobes entire or dentate; middle cauline leaves subsessile or with petioles 0.8–2.2 cm, blade lanceolate to oblanceolate, (0.7–) 1.2–3.3 (–5) cm × (2–) 4–12 mm, base cuneate, not auriculate, margin subentire to dentate. **INFL** a raceme, often much elongated; rachis hirsute or hispid with straight cylindrical trichomes; fruiting pedicel 2–4 (–4.6) mm, often strongly flattened, straight or slightly curved, divaricate-ascending to horizontal, hirsute to hispid all around or adaxially. **FLOWER** sepals oblong, 1–1.3 mm long, deciduous; petals absent or present and white, oblanceolate to linear, (0.3–) 0.6–1.5 mm, claw absent; stamens 2, median. **FRUIT** dehiscent, ovate to ovate-orbicular, 2.8–4 (–4.6) × 2.4–3.6 mm, base broadly cuneate to rounded, apically winged, apical notch 0.3–0.6 mm deep; valves thin, smooth, hirsute to hispid on surface or along margin, not veined; style obsolete or to 0.1 mm, included in apical notch. **SEEDS** brown, ovate, 1.4–2.2 mm. 2*n* = 32. [*L. georginum* Rydb.; *L. lasiocarpum* subsp. *georginum* (Rydb.) Thell.; *L. lasiocarpum* var. *georginum* (Rydb.) C. L. Hitchc.; *L. lasiocarpum* var. *pubescens* Thell.; *L. lasiocarpum* var. *rosulatum* C. L. Hitchc.; *L. ruderale* L. var. *lasiocarpum* (Nutt.) Engelm.

ex A. Gray]. Desert scrub, blackbrush, sagebrush, and piñon-juniper woodlands; dry washes and flats, waste places, streambeds, roadsides, sandy areas, rock slides, stony slopes. ARIZ: Coc, Nav; COLO: Mon; NMEX: SJn; UTAH. 1100–1700 m (3600–5600′). Flowering: Mar–early Jun. Our material is **subsp. *lasiocarpum*.** Arizona, southern and east-central California, southwest Colorado, southern Nevada, New Mexico, western Texas, southern Utah, and Mexico. There is considerable variation in the density and location of trichomes on the fruit valve.

Lepidium latifolium L. (broad leaf) Perennial pepperweed, perennial pepperwort, broadleaf pepperwort, tall whitetop. Herbs, perennial, with thick rhizomes, glabrous or pubescent. **STEMS** (2–) 3.5–12 (–15) dm, erect, single, branched above. **LEAVES**: basal not rosulate; petiole 1–9 (–14) cm; blade leathery, elliptic-ovate to oblong, (2–) 3.5–15 (–25) × (0.5–) 1.5–5 (–8) cm, margin entire or serrate; middle cauline leaves sessile or short-petiolate, blade oblong to elliptic-ovate or lanceolate, 2–9 (–12) × 0.3–4.5 cm, base cuneate, not auriculate, margin serrate or entire. **INFL** subcorymbose panicles, slightly elongated or not; rachis glabrous or sparsely puberulent with straight cylindrical trichomes; fruiting pedicel 2–5 (–6) mm, terete, straight or slightly ascending to divaricate, glabrous or puberulent adaxially. **FLOWER** sepals suborbicular to ovate, 1–1.4 mm long, deciduous; petals white, obovate, 1.8–2.5 mm long, claw 0.7–1 mm; stamens 6. **FRUIT** dehiscent, oblong-elliptic to broadly ovate or suborbicular, 1.8–2.4 (–2.7) × 1.3–1.8 mm, apically wingless, apical notch absent or rarely to 0.1 mm; valves thin, smooth, glabrous or sparsely pilose, not veined; style 0.05–0.15 mm, exserted beyond (when present) apical notch. **SEEDS** brown, oblong, (0.8–) 1–1.2 mm. $2n = 24$. [*Cardaria latifolia* (L.) Spach]. Piñon-juniper woodlands, sagebrush, mountain shrub, and ponderosa pine communities. ARIZ: Apa; COLO: Arc, Hin, Mon; NMEX: McK, RAr, SJn; UTAH. 1525–2350 m (5000–7725′). Flowering: Jun–Sep. Native to northern Africa, Asia, southern Europe, and Australia, now naturalized in southern Canada and much of the continental United States.

Lepidium latifolium

Lepidium montanum Nutt. (of the mountains) Mountain pepperwort. Herbs, annual, biennial, or perennial, caespitose or not, woody or not basally, glabrous or pubescent. **STEMS** 0.4–5 (–7) dm, erect to ascending, 1 to many from base, often much-branched above. **LEAVES**: basal rosulate or not; petiole (0.5–) 1.2–5.3 (–7.6) cm; blade 1- or 2-pinnatifid to pinnatisect, rarely undivided, 1.5–4 (–6) cm, lobes entire or dentate; middle cauline leaves shortly petiolate, blade similar to basal leaves or undivided and linear, gradually reduced in size upward, base cuneate to attenuate, not auriculate. **INFL** raceme, often much elongated; rachis puberulent with straight or curved, cylindrical trichomes, rarely glabrous; fruiting pedicel (2.7–) 3.3–8.5 (–10) mm, terete, slightly recurved or somewhat sigmoid, divaricate to horizontal, sparsely to densely puberulent adaxially. **FLOWER** sepals oblong to broadly ovate, 1.2–1.8 (–2.1) mm long, deciduous; petals white, spatulate to oblanceolate, 2.2–3.7 (–4.3) mm long, claw 1–1.4 mm; stamens 6. **FRUIT** dehiscent, ovate to suborbicular or rarely oblong, 2–4.3 (–5) × 1.8–3.6 (–4) mm, apically winged, apical notch 0.1–0.3 mm deep; valves thin, smooth, glabrous or rarely puberulent, not veined; style 0.2–0.7 (–0.9) mm, exserted beyond or rarely subequaling apical notch. **SEEDS** brown, oblong, 1.2–1.8 mm. $2n = 32$. [*L. albiflorum* A. Nelson & P. B. Kennedy; *L. alyssoides* A. Gray var. *jonesii* (Rydb.) Thell.; *L. alyssoides* var. *stenocarpum* Thell.; *L. corymbosum* Hook. & Arn.; *L. crandallii* Rydb.; *L. heterophyllum* (S. Watson) M. E. Jones; *L. integrifolium* Nutt. var. *heterophyllum* S. Watson; *L. jonesii* Rydb.; *L. montanum* subsp. *alpinum* (S. Watson) C. L. Hitchc.; *L. montanum* var. *alpinum* S. Watson; *L. montanum* subsp. *canescens* (Thell.) C. L. Hitchc.; *L. montanum* var. *canescens* (Thell.) C. L. Hitchc.; *L. montanum* subsp. *cinereum* (C. L. Hitchc.) C. L. Hitchc.; *L. montanum* var. *cinereum* (C. L. Hitchc.) Rollins; *L. montanum* var. *claronense* S. L. Welsh; *L. montanum* subsp. *glabrum*

(C. L. Hitchc.) C. L. Hitchc.; *L. montanum* var. *glabrum* C. L. Hitchc.; *L. montanum* subsp. *heterophyllum* (S. Watson) C. L. Hitchc.; *L. montanum* var. *heterophyllum* (S. Watson) C. L. Hitchc.; *L. montanum* subsp. *jonesii* (Rydb.) C. L. Hitchc.; *L. montanum* var. *jonesii* (Rydb.) C. L. Hitchc.; *L. montanum* var. *neeseae* S. L. Welsh & Reveal; *L. montanum* var. *nevadense* Rollins; *L. montanum* var. *soliarborense* S. L. Welsh; *L. montanum* var. *stellae* S. L. Welsh & Reveal; *L. montanum* var. *stenocarpum* Thell.; *L. montanum* var. *wyomingense* (C. L. Hitchc.) C. L. Hitchc.; *L. philonitrum* A. Nelson & J. F. Macbr.; *L. scopulorum* M. E. Jones; *L. utahense* M. E. Jones]. Blackbrush, shadscale, sagebrush, ponderosa pine, and purple sage communities, often in silty or clay soils. ARIZ: Apa, Coc, Nav; COLO: Arc, LPl, Mon; NMEX: RAr, SJn; UTAH. 1250–2300 m (4100–7500′). Flowering: Apr–Aug (–Sep). Native to the western United States. Navajos used the species as a gastrointestinal aid to treat biliousness and other disorders; also used for palpitations and dizziness. Although *L. alyssoides*, *L. barnebyanum*, *L. crenatum*, *L. davisii*, *L. eastwoodiae*, *L. integrifolium*, and *L. papilliferum* are not included in the species, which has reduced some of the heterogeneity in *L. montanum*, a thorough study is needed to determine the number and range of infraspecific taxa, to understand the nature of variation, and to determine how distinct the species is from the seven species just mentioned. Some of the varieties (e.g., *alpinum*, *coloradense*, *montanum*, *neeseae*, and *stellae*) recognized by both Rollins (1993) and Holmgren et al. (2005), are distinct enough to merit recognition at some rank. If these taxa, however, were to be recognized here, it is unclear where the vast amount of remaining variation should be placed. Until a thorough study of the species and its relatives is available, we choose not to formally recognize infraspecific taxa. Furthermore, construction of a workable key given our current understanding of the species is impossible. This is far from an ideal situation. A glance at the partial synonymy for *L. montanum* given above reveals something of the complexity of this issue.

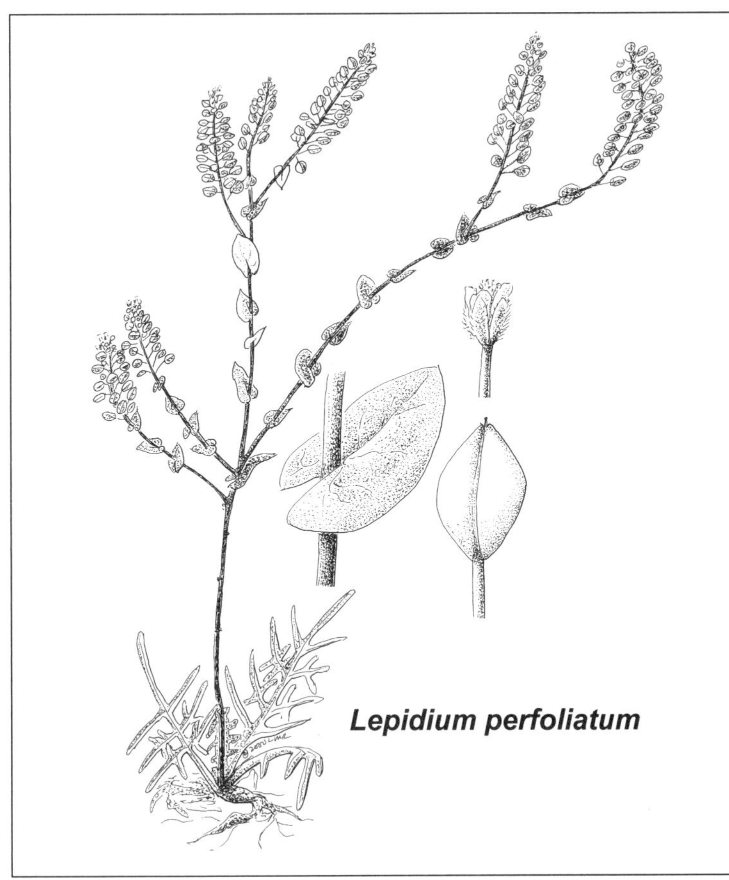
Lepidium perfoliatum

Lepidium perfoliatum L. (through + leaf, referring to the stem going through the leaf) Clasping pepperwort. Herbs, annual or occasionally biennial, glaucous, glabrous or sparsely pubescent below. **STEMS** (0.7–) 1.5–4.3 (–5.6) dm, erect, single at base, branched above. **LEAVES**: basal rosulate; petiole (0.5–) 1–2 (–4) cm, 2- or 3-pinnatifid or pinnatisect, 3–8 (–15) cm, lobes linear to oblong, entire; middle and upper cauline leaves sessile, ovate to cordate or suborbicular, 1–3 (–4) × 1–2.5 (–3.5) cm, base deeply cordate-amplexicaul, margin entire. **INFL** a raceme, much elongated; rachis glabrous; fruiting pedicel 3–6 (–7) mm, terete, straight, divaricate-ascending to horizontal, glabrous. **FLOWER** sepals oblong, 0.8–1 (–1.3) mm long, deciduous; petals pale yellow, narrowly spatulate, 1–1.5 (–1.9) mm long, claw 0.5–1 mm; stamens 6. **FRUIT** dehiscent, orbicular to rhombic or broadly obovate, 3–4.5 (–5) × 3–4.1 mm, apically winged, apical notch 0.1–0.3 mm deep; valves thin, smooth, glabrous, not veined; style 0.1–0.4 mm, subequaling or slightly exserted beyond apical notch. **SEEDS** dark brown, ovate, 1.6–2 (–2.3) mm. $2n = 16$. Waste places, dry sandy slopes, piñon-juniper woodlands, sagebrush flats, open deserts, roadsides, pastures, meadows, open grasslands, alkaline flats and sinks, fields, disturbed sites. ARIZ: Apa, Nav; COLO: Arc, LPl, Mon; NMEX: RAr, SJn; UTAH. 1460–2050 m (4800–6700′). Flowering: Apr–Jun (–Jul). Native of northern Africa, Asia, and Europe; naturalized in Australia, South America, and much of the United States and southern Canada.

Lepidium ramosissimum A. Nelson (many-branched) Branched pepperwort. Herbs, biennial, puberulent. **STEMS** 1–5.3 (–7.7) dm, erect, single from base, much-branched above. **LEAVES**: basal not rosulate, soon withering; petiole 1–4 cm; blade oblanceolate, pinnatifid, 2–5 × 0.8–1.5 cm, lobes entire, serrate, or dentate; middle cauline leaves shortly petiolate to sessile, blade oblanceolate, 1.2–4.8 (–6) cm × 1–8 (–10) mm, base attenuate to cuneate, not auriculate, margin dentate or rarely lobed, uppermost leaves linear, entire. **INFL** a raceme, slightly elongated; rachis puberulent with curved cylindrical to subclavate trichomes; fruiting pedicel (1.6–) 2–3.8 (–5) mm, terete, straight or often recurved, divaricate-ascending to horizontal, puberulent adaxially or rarely all around. **FLOWER** sepals oblong, 0.6–1 mm long, deciduous; petals absent or rudimentary and white, linear, 0.2–0.8 (–1) mm long, claw absent; stamens 2. **FRUIT** dehiscent, elliptic, 2.2–3.2 × 1.7–2.1 mm, apically winged, apical notch 0.1–0.3 mm deep; valves thin, smooth, glabrous or puberulent at least along margin, not veined; style obsolete or rarely to 0.1 mm, included in apical notch. **SEEDS** brown, oblong, 1.2–1.6 mm. 2*n* = 32, 64. [*L. divergens* Osterh.; *L. fletcheri* Rydb.; *L. ramosissimum* var. *bourgeauanum* (Thell.) Rollins; *L. ramosissimum* var. *divergens* (Osterh.) Rollins; *L. ramosissimum* var. *robustum* Thell.]. Sagebrush, ponderosa pine, and Gambel's oak communities; waste ground, roadsides, railroad embankments, alkaline flats, abandoned fields. ARIZ: Apa; COLO: Hin, Min, Mon, RGr, SJn; NMEX: McK, RAr, SJn. 2250–3075 m (7400–10100′). Flowering: Jun–Aug (–Sep). Canada, northern Mexico, and western United States. As noted by Rollins (1993:581), the varieties of *L. ramosissimum* are "weak at best" and are not here recognized. *Lepidium divergens* is included in the synonymy of *L. ramosissimum* pending further study.

Lepidium virginicum L. (of Virginia) Virginia cress, poor-man's pepper, wild cress. Herbs, annual, puberulent. **STEMS** (0.6–) 1.5–5.5 (–7) dm, erect, single from base, branched above. **LEAVES**: basal not rosulate, withered by anthesis; petiole 0.5–3.5 cm; blade obovate, spatulate, to oblanceolate, (1–) 2.5–10 (–15) × 0.5–3 (–5) cm, pinnatifid to lyrate or dentate; middle cauline leaves shortly petiolate, oblanceolate or linear, 1–6 cm × 3–10 mm, base attenuate to subcuneate, not auriculate, margin serrate or entire. **INFL** a raceme, much elongated; rachis puberulent with curved cylindrical trichomes, rarely glabrous; fruiting pedicel 2.5–4 (–6) mm, slender, terete or flattened, straight or slightly recurved, divaricate-ascending to nearly horizontal, puberulent adaxially or rarely all around or glabrous. **FLOWER** sepals oblong to ovate, 0.7–1 mm long, deciduous; petals white, spatulate to oblanceolate, 1–2 (–2.5) mm long, rarely rudimentary, claw undifferentiated or to 0.8 mm; stamens 2, median. **FRUIT** dehiscent, orbicular or nearly so, 2.5–3.5 mm in diameter, apically winged, apical notch 0.2–0.5 mm; valves thin, smooth, glabrous, not veined; style 0.1–0.2 mm, included in apical notch. **SEEDS** brown, ovate, 1.3–1.9 (–2.1) mm. 2*n* = 32. [*Lepidium menziesii* DC.; *L. bernardinum* Abrams; *L. californicum* Nutt.; *L. hirsutum* Rydb.; *L. glaucum* Greene; *L. idahoense* A. Heller; *L. intermedium* A. Gray; *L. intermedium* var. *pubescens* Greene; *L. medium* Greene; *L. occidentale* Howell; *L. robinsonii* Thell.; *L. simile* A. Heller]. Roadsides, bottomlands, gravelly and sandy shores, waste ground, riverbanks, grassy meadows, dry flats and creek beds, abandoned fields; piñon-juniper up to Douglas-fir communities. COLO: LPl, Mon; NMEX: SJn. 2350–2775 m (7750–9125′). Flowering: May–Jul (–Sep). Southwest Canada, western United States and Mexico. Our material is **subsp. *menziesii*** (DC.) Thell. (for Archibald Menzies, 1754–1842, British surgeon-naturalist), which has flattened pedicels (rather than terete in subsp. *virginicum*). *Lepidium virginicum* is easily distinguished from the similar *L. densiflorum* by having well-developed or rarely rudimentary petals, accumbent cotyledons, orbicular fruits, and raceme rachises with curved cylindrical trichomes. By contrast, *L. densiflorum* has rudimentary or often no petals, incumbent cotyledons, obovate fruits, and raceme rachises with straight, often subclavate trichomes.

Malcolmia W. T. Aiton African Mustard

(for William Malcolm, 1769–1820, British horticulturist) Annual herbs with simple, forked, or malpighian trichomes. **STEMS** several from base, at least the central one branched. **LEAVES** simple to bipinnatifid, sessile or petiolate and occasionally auriculate. **FLOWER** sepals erect, lateral pair mostly gibbous-based; petals lavender, purple, violet, or rarely white, obcordate or obovate to oblong. **FRUIT** a silique, linear, terete or quadrangular, often torulose, erect to recurved. **SEEDS** 1 row per locule, nonmucilaginous. x = 7, 8, 10, 12. Thirty-two species of Eurasia and northern Africa; some species naturalized and weedy in North America and Australia.

Malcolmia africana (L.) Botsch. (of Africa) African mustard, African adder's-mouth. Annual, densely pubescent with small, coarse, freely branched trichomes. **STEMS** simple to branched, but the central branch erect, the outer prostrate or decumbent, 1.5–4 dm tall. **LEAVES** oblanceolate, petiolate, remotely and coarsely dentate, 3–7 cm long, becoming somewhat smaller and sessile. **INFL** few-flowered, elongate, lower flowers in leaf axils. **FLOWER** sepals linear to narrowly lanceolate, 3.5–6 mm long; petals pink to rose-violet, clawed with an obovate blade, 6–10 mm long. **FRUIT** a silique, terete to subquadrangular, straight, slightly torulose, divaricately ascending to divaricate, 4–6.5 cm long on stout pedicels about the same thickness as the silique. **SEEDS** 1 row per locule, oblong, 1–1.3 mm long. *n* = 7, 2*n* = 14, 28. [*Hesperis africana* L.]. Desert scrub, saltbush, piñon-juniper communities; roadsides, disturbed areas. ARIZ: Apa, Nav; COLO: LPl, Mon; NMEX: SJn; UTAH. 1250–2050 m (4150–6700′). Flowering: Apr–May. Native to Mediterranean North Africa, now in southern Canada, the Intermountain region, New Mexico, Colorado, and Arizona.

Nasturtium Mill. Watercress

(Pliny's name, meaning twisted nose, and referring to the pungent odor) Perennial, rhizomatous, aquatic or semiaquatic herbs, glabrous or pubescent with simple trichomes. **STEMS** rooting at lower nodes, 1–many, branched or unbranched, typically prostrate or decumbent, less frequently erect. **LEAVES** all cauline, petiolate (–sessile), usually pinnate with 1–6 (12) pairs of lateral leaflets. **FLOWER** sepals ascending to erect, the lateral pair subgibbous-based or not; petals white, obovate or narrowly spatulate, not clawed. **FRUIT** a silique, linear, terete, smooth or somewhat torulose, obscurely veined. **SEEDS** 1 or 2 rows per locule, nonmucilaginous. $x = 7, 8$. Five species, Eurasia, North Africa, and North America (two species native).

Nasturtium officinale R. Br. (Latin *officin*, herb pharmacy; plants so named were once on an official herbal medicine list) Watercress. Perennial, aquatic or semiaquatic, from creeping stolons (or rhizomes), prostrate or prostrate-decumbent, glabrous. **STEMS** leafy throughout, creeping and often decumbent. **LEAVES** petiolate and the base of the petiole often with small auricles, pinnate (occasionally simple ones present) with 3–7 ovate to broadly oblong, sessile, ± entire leaflets, terminal leaflet larger and often broader. **INFL** axillary and terminal, loose, ebracteate above but often subtended by small leafy bracts below. **FLOWER** sepals oblong, inner pair gibbous-based, 2–3 mm long; petals spatulate, white and sometimes tinged with purple, 3–5.5 mm long. **FRUIT** a silique, on widely spreading pedicels, very narrowly oblong, straight or somewhat upcurved, slightly flattened parallel to the septum. **SEEDS** 2 rows per locule, 0.8–1.1 mm long, suborbicular. $n = 16$, $2n = 32$. [*Sisymbrium nasturtium-aquaticum* L.; *Rorippa nasturtium-aquaticum* (L.) Hayek]. Riparian areas, seeps, willow thickets, irrigation ditches, hanging gardens. ARIZ: Apa; COLO: LPl, Mon, SJn; NMEX: SJn; UTAH. 1700–2525 m (5600–8300′). Flowering: May–Jul. From southern Canada to Panama, native (probably) to Europe and western Asia but now naturalized throughout the world. Watercress is a popular, nutritious, and widely planted peppery salad green. Caution should be used when eating wild-grown plants as they can harbor parasites and diseases, especially when grown in polluted water or water found near livestock.

Noccaea Moench Pennycress

(for Domenico Nocca, 1758–1841, Italian professor of botany, clergyman, and botanical garden director) Biennial or perennial (annual) herbs, glabrous or rarely pubescent with simple hairs. **STEMS** few to several from the base, these not or little-branched. **LEAVES** simple, entire to dentate, basal often forming a rosette and petiolate; cauline (at least above the very base) sessile and auriculate-clasping. **FLOWER** sepals erect to spreading, not gibbous-based; petals white to violet, narrowly oblanceolate, clawed. **FRUIT**: silicle angustiseptate, obovoid to orbicular with a rounded or notched apex. **SEEDS** 1 row per locule, mucilaginous or not. $x = 7$. Seventy-seven species of the Northern Hemisphere but only four to six in North America. Traditionally this genus was included in a much larger *Thlaspi*.

Noccaea fendleri (A. Gray) Holub **CP** (for Augustus Fendler, 1813–1883, botanical collector in Texas and Mexico) Wild candytuft, mountain cress, alpine pennycress. Perennial, glabrous, often glaucous, herbage green to purple-tinged. **STEMS** single to few (to many) from the base, these simple or branched above, 0.1–2.5 (4) dm tall. **LEAVES**: basal rosulate, often numerous, petiolate, linear, oblong, oblanceolate, ovate, to spatulate, 0.4–3 cm long, cuneate to attenuate at base, margin entire, denticulate, or dentate, apex obtuse to acute; cauline leaves 2–21 per stem, sessile, ovate or suboblong, 0.4–2.8 cm long, auriculate to subamplexicaul at base, entire or dentate, apex obtuse to subacute. **INFL** congested or elongated. **FLOWER** sepals ovate or oblong, 1.6–5.3 mm long, margin membranous; petals white to pinkish purple, spatulate, (3.4) 4.2–13 mm long, apex obtuse. **FRUIT** a silicle, divaricately ascending to horizontal-spreading, obovate to obcordate, cuneate at base, winged (rarely wingless), 5–10 mm long, 2.5–4.5 (6) mm wide. **SEEDS** 2–4 per locule, 1.1–2 mm long, ovoid, brown. $2n = 14, 28$. [*Thlaspi fendleri* A. Gray]. Mountain brush, Gambel's oak, ponderosa pine, spruce and fir, aspen, subalpine meadows, alpine tundra. ARIZ: Apa, Nav; COLO: Arc, Con, Dol, Hin, LPl, Min, Mon, RGr, SJn, SMg; NMEX: McK, RAr, SJn; UTAH. Flowering: May–Aug. 1900–4000 m (6250–13100′). Western United States and northern Mexico. Navajo people used the plant as a ceremonial chant lotion; cold infusions for internal and external itching and as a preventive from deer injury; a cold infusion or lotion as protection from witches. All or nearly all of our material belongs to **subsp. *glauca*** (A. Nelson) Al-Shehbaz & M. Koch (not having hairs). Possibly subsp. *fendleri* makes it north as far as our area. Subspecies *fendleri* has petals that are usually pinkish purple, styles 2.5–3.5 mm in fruit, silicles 7–12 mm long, and compact infructescences. Subspecies *glauca* has petals usually white, styles 1–2.2 mm in fruit, silicles 5–8 mm long, and lax infructescences. North American members of the species have traditionally been placed in *Noccaea montana* (L.) F. K. Mey. (*Thlaspi montanum* L.), which is here interpreted to be European.

Pennellia Nieuwl. Mock Thelypody

(for Francis W. Pennell, 1886–1952, Scrophulariaceae expert and curator of the herbarium at the Academy of Natural Sciences, Philadelphia) Biennial or perennial herbs, pubescent with 2-forked or dendritic hairs or rarely glabrous. **STEMS** single, erect, simple or branched. **LEAVES**: basal rosulate, entire or sinuate-dentate to shallowly lobed; cauline petiolate to sessile, entire or dentate to shallowly lobed. **INFL** a lax raceme. **FLOWER** sepals erect to ascending, the lateral pair subgibbous-based or not; petals purple to white, subequaling or rarely exceeding the sepals. **FRUIT**: silique linear, terete or latiseptate, erect to pendent. **SEEDS** 1 (± 2) rows per locule, mucilaginous or not. $x = 8$. Nine species centered in Mexico, two of which make it into the United States.

1. Siliques pendent on downward arching pedicels .... ***P. longifolia***
1′ Siliques erect or ascending on erect or divaricately ascending pedicels .... ***P. macrantha***

Pennellia longifolia (Benth.) Rollins (long leaf) Long-leaf mock thelypody. Biennial or short-lived perennial, pubescent with coarse, simple or forked spreading trichomes below. **STEMS** mostly single, branched above. **LEAVES**: basal rosulate, usually withering before flowering, oblanceolate, entire to sinuate-dentate, apically obtuse; cauline leaves linear, few, entire or lower shallowly dentate, densely pubescent. **INFL** strongly elongated in fruit, secund. **FLOWER** sepals oblong, reddish purple; petals purplish, often white at the base, veins darker, narrowly oblanceolate, 4.5–6 mm long. **FRUIT** a silique, pendent, straight, linear, slightly latiseptate, 5–9 cm long. **SEEDS** crowded, plump, angled. $n = 8$, $2n = 16$. [*Streptanthus longifolius* Benth.; *Thelypodium longifolium* (Benth.) S. Watson]. Ponderosa pine–Gambel's oak community. ARIZ: Apa. 2450 m (8050′). Flowering: Aug. Western Texas to Arizona, New Mexico, and Colorado south to central Mexico.

Pennellia micrantha (A. Gray) Nieuwl. (small flower) Mountain mock thelypody. Perennial or biennial, glabrous above and pubescent below with forked or dendritic trichomes. **STEMS** single or more rarely, several, each usually simple but branched in the inflorescence, 5–10 dm tall. **LEAVES**: basal rosulate, oblanceolate, sinuate-dentate to shallowly lobed (rarely entire), obtuse at apex, densely pubescent with dendritic trichomes; cauline leaves ± like the basal ones except fewer, becoming much reduced and glabrous above. **INFL** narrow and elongated. **FLOWER** sepals oblong, often purplish; petals white or more rarely purplish, spatulate to narrowly ligulate, gradually narrowing to base, 3–8 mm long. **FRUIT** a silique, erect or nearly so, slender, terete, 2–4 cm long. **SEEDS** crowded, plump, angled. [*Streptanthus micranthus* A. Gray; *Thelypodium micranthum* (A. Gray) S. Watson]. Slopes and talus in spruce-fir-aspen zone. COLO: LPl, Mon. 2700–2925 m (8850–9600′). Flowering: Jul–early Aug. West Texas to southern Arizona to Colorado; mountains of northern Mexico. In our area the species is limited to the La Plata Mountains. Navajos used a decoction of the root to hasten birth; a poultice of the crushed, heated roots was used for toothache.

Physaria (Nutt. ex Torr. & A. Gray) A. Gray Bladderpod

(Greek *physa*, a bladder, referring to the inflated fruits of some species) Annual, biennial, or perennial herbs, pubescent with stellate to webbed-stellate to lepidote trichomes (rarely a few simple ones present), which often give the plants a gray or silvery appearance. **STEMS** 1 to several, erect to prostrate, often from an enlarged or branching caudex. **LEAVES**: basal leaves usually rosulate, entire to repand to pinnatifid, petiolate or blade tapering into the petiole; cauline leaves entire to repand to dentate, petiolate, often with the blade tapering into the petiole or ± sessile. **INFL** racemose. **FLOWER** sepals erect or spreading, pubescent, subgibbous or gibbous at the base, often cucullate at the apex; petals typically yellow but sometimes cream-yellow, white, white with purple veins, or purplish; in some species the petals turn lavender or maroon upon drying. **FRUIT** a silicle, sometimes short-stipitate, basically of two forms: either didymous, latiseptate, and with inflated, usually papery or leathery valves (*Physaria* sensu stricto of floras), or not didymous, terete to latiseptate or angustiseptate, with valves not independently inflated, ± chartaceous or firm (*Lesquerella* of floras), the false septum occasionally perforated. **SEEDS** 4–28 (40) per locule, mucilaginous or not. $x = 4, 5, 7–10, 12, 15$. [*Lesquerella* S. Watson except species of the southeastern United States now in *Paysonia* O'Kane & Al-Shehbaz]. One hundred seven species, mostly of western North America and a few in Argentina and immediately adjacent Bolivia (Al-Shehbaz and O'Kane 2002; Payson 1921; Rollins 1939; Rollins and Shaw 1973). Seeds contain hydroxy fatty acids, and some species, notably *P. fendleri*, are being intensively studied as a source of specialized lubricants. Most of the genus *Lesquerella* was recently united with *Physaria*, making *Physaria* much larger, but monophyletic and morphologically coherent (Al-Shehbaz and O'Kane 2002).

1. Silicles not or only slightly inflated; rounded, acute, or obtuse apically (rarely slightly notched), not or only slightly (in *P. scrotiformis*) didymous (formerly *Lesquerella*, the bladderpods) .... (2)

1′ Silicles highly inflated and notched apically (sometimes basally as well), strongly didymous (formerly *Physaria* sensu stricto, the double bladderpods) ....(11)

2. Silicles glabrous, usually maturing, at least in part, to reddish or coppery brown ....(3)
2′ Silicles pubescent, typically not becoming reddish or coppery brown ....(5)

3. Trichomes with rays fused (webbed) to about the middle or a little more (view at 10× or more) ....***P. fendleri***
3′ Trichomes with rays distinct (or only little fused at the base and visible with high magnification) ....(4)

4. Plants forming dense, low cushions from an intricately branched underground caudex; leaves linear-oblanceolate, usually less than 8 mm long; inflorescence and infructescence few-flowered and barely exceeding the leaves; limited to windswept exposures of the nearly white Todilto Limestone ....***P. navajoensis***
4′ Plants upright to decumbent from a ± simple caudex; leaves with blades suborbicular or obovate to rhombic and up to 8 cm long; inflorescence dense and elongating in fruit; limited to gray clay soils derived from the Mancos Shale ....***P. pruinosa***

5. Leaves all linear to narrowly oblanceolate, with no distinction between blade and petiole, the blades generally less than 3 (4) mm wide ....(6)
5′ Leaves wider, the blade ± distinct from the petiole, at least some (usually all or most) of the blades greater than 3 mm wide ....(8)

6. Plants arising from a branched, often underground caudex, the caudex branches clothed with old leaf bases; infructescence not elongating [*Physaria pulvinata* O'Kane & Reveal would key here; although not yet found in the study area, it is to be expected, especially in the southwest corner of Colorado or the southeast corner of Utah. The species is currently known from argillaceous shale outcrops just outside the study area in Dolores and San Miguel Counties, Colorado. *P. pulvinata* differs from *P. intermedia* in its shorter leaves (up to only 1.5 cm) and dense, mound-forming habit.] ....***P. intermedia***
6′ Plants arising from a unbranched, essentially ground-level caudex, not especially clothed with old leaf bases; infructescence elongated ....(7)

7. Leaves very narrow, linear, mostly involute; silicles pendulous on recurved pedicels ....***P. ludoviciana***
7′ Leaves broader, with flat blades; silicles on spreading or sigmoid pedicels, not (except for rarely a lower few) pendulous ....(10)

8. Silicles ellipsoid, ± densely pubescent; leaves with blade at least somewhat distinct from the petiole, basal leaves not forming a distinctive erect tuft; common in our area ....***P. rectipes*** (in part)
8′ Silicles ovoid, obovoid, or obpyriform, sparsely to moderately pubescent; leaves with blade tapering into the petiole, the basal leaves typically forming a central erect tuft; rare in our area ....(9)

9. Plants above 3500 m (11500′) elevation; silicles slightly didymous, obovoid or obpyriform; fruiting stems prostrate; stem leaves all less than 5 mm long ....***P. scrotiformis***
9′ Plants below 2440 m (8000′) elevation; silicles not at all didymous, ovoid; fruiting stems prostrate to erect; stem leaves greater than 9 mm long ....***P. montana***

10. Silicle bodies ± terete, not at all flattened (the apex may be slightly flattened); leaves tapering into the petiole, the basal leaves typically forming a central erect tuft; common ....***P. rectipes*** (in part)
10′ Silicles at least somewhat angustiseptate; leaf blades ± distinct from the petiole, basal leaves not forming a central erect tuft; infrequent in Apache County, Arizona, and adjacent Utah ....***P. kingii***

11. Silicle valves angled and the sides concave, the style shorter than the valves; leaf blades ± tapering into the petiole ....***P. newberryi***
11′ Silicle valves rounded, the style longer than the valves; leaf blades distinct from the petiole ....***P. acutifolia***

Physaria acutifolia Rydb. **CP** (sharp-pointed leaf, a bit of a misnomer as most plants have rounded leaves) Rydberg's twinpod, sharpleaf twinpod (a misnomer). Perennial, caespitose, silvery from the dense, stellate, many-rayed trichomes. **STEMS** numerous, somewhat decumbent, arising laterally, 5–15 cm long. **LEAVES**: basal numerous in a rosette, entire or with a few scattered teeth, blade obovate to orbicular, usually obtuse, on a slender narrowly winged petiole, to 8 cm long; cauline entire, spatulate to oblanceolate, obtuse at apex, 1–3 cm long. **INFL** racemose and elongating in fruit. **FLOWER** sepals linear-oblong; petals erect to ascending, yellow, narrowly oblong to oblanceolate, 6–11 mm long.

FRUIT a silicle, didymous, inflated, apical and basal sinuses deep, the basal sometimes a little less so, valves suborbicular, pubescent. **SEEDS** (1) 2 per locule, 2–3 mm long, brown. *n* = 5, 8, 12, 2*n* = 10, 16, 24. [*P. stylosa* Rollins; *P. didymocarpa* (Hook.) A. Gray var. *australis* Payson]. Piñon-juniper, ponderosa pine, sagebrush, rabbitbrush, blackbrush, and saltbush communities, especially in clayey substrates. ARIZ: Apa; COLO: Dol, LPl, Mon; NMEX: RAr, SJn; UTAH. 1475–2300 m (4850–7500′). Flowering: Apr–May. Southeast Montana south to New Mexico and west to Arizona, Utah, and Idaho. Our plants are **var. *acutifolia***.

Physaria fendleri (A. Gray) O'Kane & Al-Shehbaz (for Augustus Fendler, 1813–1883, botanical collector in Texas and Mexico) Fendler's bladderpod. Perennial from a usually branched caudex, densely pubescent with stellate, many-rayed trichomes webbed to about 1/2 their length. **STEMS** several, each usually unbranched, erect or the lateral ones decumbent, 0.5–2.5 (4) dm long. **LEAVES** mostly cauline, basal rosette of ± similar leaves rarely present, linear to more commonly narrowly oblanceolate, entire, rarely somewhat involute, main ones 1.5–3 (4.5) cm long. **INFL** racemose, elongating slightly in fruit. **FLOWER** sepals narrowly elliptic or oblong; petals yellow, often with darker, orangish veins especially at the junction of blade and claw, 5–10 mm long, blade oblanceolate. **FRUIT** a silicle, globose or more frequently broadly ellipsoid or ovoid and acute, glabrous, 4–8 mm long, short-stipitate. **SEEDS** 2–4 per locule, 2 mm long. *n* = 6, 7, 12, 2*n* = 12. [*Lesquerella fendleri* (A. Gray) S. Watson]. Piñon-juniper, piñon-juniper-shrub, and desert shrub communities; slickrock. ARIZ: Apa; COLO: Mon; NMEX: McK, SJn. 1450–2200 m (4750–7250′). Flowering: Apr–early Jun. Texas to southern Utah and southern Colorado and south to northern Mexico. This species is under active investigation for use as an agronomic crop, because the seeds are a good source of a hydroxy fatty acid that yields a high quality industrial lubricant.

Physaria intermedia (S. Watson) O'Kane & Al-Shehbaz (intermediate) Mid bladderpod, Watson's bladderpod. Perennial, caespitose from a usually branched underground caudex clothed in old leaf bases, silvery or gray from dense crust of stellate trichomes with forked or bifurcate rays. **STEMS** several, usually 1 per caudex branch. **LEAVES**: basal clustered on stem base, entire, linear to linear-oblanceolate, usually involute, sometimes somewhat flattened, 2–5 cm long; cauline (upper) remote, linear-oblanceolate to linear, usually involute, 1–4 cm long. **INFL** dense raceme, corymbiform, not or barely elongating in fruit. **FLOWER** sepals erect, yellowish or greenish yellow; petals yellow, 6.5–10.5 mm long, blade spreading, oblanceolate. **FRUIT** a silicle, subglobose to slightly ovoid, rarely a little compressed, acute at apex, sparsely pubescent, 3.5–5.5 mm long. **SEEDS** 2–4 per locule, plump, about 1.5 mm long. *n* = 10, 18, 2*n* = 18. [*Lesquerella intermedia* (S. Watson) A. Heller; *L. alpina* (Nutt.) S. Watson var. *intermedia* S. Watson]. Piñon-juniper, sometimes with scattered Douglas-fir. ARIZ: Nav. 1750–2450 m (5700–8000′). Flowering: Jun. Southern Utah, northern Arizona, and reported from New Mexico. Hopis used an infusion of the roots as a ceremonial emetic; rubbed the plant on the abdomen when the uterus failed to contract after childbirth; and ate the roots and made a poultice of chewed roots to treat snakebite. Navajos used a root poultice to treat sore eyes, and the plant is used in the Night Way.

Physaria kingii (S. Watson) O'Kane & Al-Shehbaz (for Clarence King, 1842–1901, geologist and mining engineer) King bladderpod. Perennial from a simple caudex, densely pubescent with stellate trichomes with branched rays. **STEMS** few to several, prostrate to decumbent, each usually simple, arising from beneath the basal tuft of leaves. **LEAVES**: basal petiolate, outer ones spreading to ascending, oblanceolate, elliptic, obovate, rhomboidal (or suborbicular), 5–16 mm wide, abruptly narrowed to petiole, the inner ones erect, with narrower blades; cauline oblanceolate to spatulate, becoming subsessile above, entire. **INFL** congested in flower, elongating in fruit. **FLOWER** sepals erect to spreading, lanceolate; petals yellow (rarely cream or cream-yellow and then to the west of our range on the Kaibab Plateau), obovate to oblanceolate, 6–10 mm long. **FRUIT** a silicle, obovoid, ellipsoid (ours), or subglobose, slightly inflated, slightly angustiseptate, pubescent, 3.2–5 (7.5) mm long. **SEEDS** 4–8 per locule, slightly flattened, 1.8–2.5 mm long. 2*n* = 10, 12. [*P. wardii* (S. Watson) O'Kane & Al-Shehbaz; *Lesquerella wardii* S. Watson; *L. latifolia* A. Nelson; *L. barnebyi* Maguire; *L. occidentalis* (S. Watson) S. Watson var. *parvifolia* Maguire & A. H. Holmgren]. Piñon-juniper communities, often with big sagebrush or manzanita. ARIZ: Apa; UTAH. 2125–2225 m (6950–7300′). Flowering: late Apr–Jun. Utah, northern Arizona, southern Nevada, and southern California. Our plants are **subsp. *latifolia*** (A. Nelson) O'Kane & Al-Shehbaz.

Physaria ludoviciana (Nutt.) O'Kane & Al-Shehbaz (of Louisiana, referring to the Louisiana Purchase) Silver bladderpod, foothill bladderpod. Perennial herb from a simple or short-branched caudex, densely pubescent with 4- to 7-rayed stellate trichomes, the rays usually bifurcate. **STEMS** few to several, each unbranched, outer ones decumbent, inner ones erect. **LEAVES**: basal forming a tuft, basal and cauline linear, entire (rarely shallowly dentate), some simple hairs present on leaf bases, no differentiation between blade and petiole, flat or often involute, up to 8 (12) cm long. **INFL** dense in flower and greatly elongating. **FLOWER** sepals narrowly oblong, erect; petals yellow, erect, linear or narrowly oblong to lanceolate, 6.5–8.2 mm long. **FRUIT** a silicle, subglobose or shortly obovoid, terete to slightly latiseptate, densely pubescent, pendulous on recurved pedicels. **SEEDS** 2–6 per locule, ovoid, orangish brown, 1.7–2.5 mm long. n = 5, 10, 15, $2n$ = 10, 30. [*Lesquerella ludoviciana* (Nutt.) S. Watson; *L. argentea* (Pursh) MacMill.]. Piñon-juniper communities in deep sand and wash bottoms. ARIZ: Coc, Nav; UTAH. 1525–2325 m (5000–7600′). Flowering: May–Jun. Wisconsin and Illinois, west to Wyoming, southern Nevada, and northern Arizona.

Physaria montana (A. Gray) Greene (mountains) Mountain bladderpod. Perennial herb, from a simple, branched, sometimes enlarged and somewhat woody caudex, pubescent with stellate trichomes with 4–7 forked or bifurcate rays. **STEMS** several, prostrate to decumbent to erect. **LEAVES**: basal tufted or rosette-forming, petiolate, blades suborbicular or obovate to elliptic, entire to sinuate or shallowly dentate; cauline often secund, linear to obovate or rhombic, entire or shallowly dentate, short-petiolate below, becoming sessile above. **INFL** dense and compact, elongating a little in fruit. **FLOWER** sepals elliptic, the median ones with thickened, cucullate tips; petals yellow to orange-yellow and sometimes fading purple, narrowly spatulate, 7.5–12 mm long. **FRUIT** a silicle, ellipsoid or ovoid, slightly angustiseptate, densely pubescent, 6–12 mm long. **SEEDS** generally 6–10 per locule. n = 5, $2n$ = 10. [*Lesquerella montana* (A. Gray) S. Watson; *L. shearis* Rydb.]. Piñon-juniper communities, sometimes with Gambel's oak. COLO: Arc, LPl, Mon. 1950–2300 m (6400–7550′). Flowering: May–Jun. Southwestern South Dakota to southeast Wyoming to northern New Mexico, mostly on the east side of the Rocky Mountains, and sporadically west of them to about the Utah border. In our area these plants seem to be somewhat intermediate with *P. rectipes*. The *rectipes/montana* "complex" is in need of further study.

Physaria navajoensis (O'Kane) O'Kane & Al-Shehbaz (of the Navajo) Navajo bladderpod. Perennial herb, pulvinate-caespitose and soboliferous from a much-branched caudex, forming hemispherical cushions, silvery gray from a dense crust of stellate trichomes, each with 5 rays each twice bifurcate. **STEMS** numerous, crowded, not exceeding the leaves, basal portions covered with old leaf bases. **LEAVES** essentially all cauline, linear-oblanceolate, tapering to an indistinct petiole, entire, 3–8 (13) mm long. **INFL** dense, few-flowered, corymbose, not or barely exceeding the leaves. **FLOWER** sepals yellow-green, elliptic; petals spatulate, deep yellow, faintly orange at junction of blade and claw, 5.2–6.5 mm long. **FRUIT** a silicle, becoming reddish or copper-colored at maturity, glabrous, ovate, often a little compressed at margins apically, acute, 3–4.9 mm long. **SEEDS** usually 1 per locule, plump and a little flattened, strongly mucilaginous. [*Lesquerella navajoensis* O'Kane]. Piñon-juniper communities on limestone. NMEX: McK, SJn. 2225–2300 m (7300–7500′). Flowering: May–early Jun. Limited to the vicinity of Thoreau, New Mexico, and two small populations near the southern end of the Chuska Mountains in New Mexico near the Arizona state line. A very rare, mat- and mound-forming species apparently limited to windswept outcrops of Todilto Limestone.

Physaria newberryi A. Gray (for John Strong Newberry, 1822–1892, American botanist, physician, geologist, paleontologist, and professor at Columbia University School of Mines) Newberry's twinpod. Perennial, caespitose herb from a simple or branched caudex, densely pubescent with silvery stellate trichomes. **STEMS** few to many, usually unbranched, arising laterally from the basal leaves. **LEAVES**: basal usually a dense tuft or rosette, obovate to oblanceolate, petiolate, tapering into the petiole, the apex obtuse or rounded, veins prominent on underside, entire or few-toothed, up to 8 (12) cm long; cauline entire, oblanceolate, few, smaller than the basal, entire, lanceolate to narrowly lanceolate. **INFL** not elongating to elongating in fruit (a variable character). **FLOWER** sepals narrowly lanceolate, gibbous-based, 7–8.5 mm long; petals yellow, erect to ascending, narrowly oblanceolate, tapering to base, 9.5–12 mm long. **FRUIT**: silicle strongly didymous, inflated but with angular valves with concave sides, the apical sinus deep and V-shaped, the base nor or barely notched, style included in the apical notch. **SEEDS** 2–4 per locule. n = 8, $2n$ = 16. [*P. didymocarpa* (Hook.) A. Gray var. *newberryi* (A. Gray) M. E. Jones]. Piñon-juniper, saltbush, desert shrub, and wash bottom communities. ARIZ: Apa; NMEX: McK, SJn. 1725–2325 m (5650–7600′). Flowering: Apr–May. Southern Utah, Arizona, and New Mexico. Rollins (1993) recognized var. *racemosa* as well as var. *newberryi* from in or near our area, but a complete continuum of intermediates is present. Therefore, only **var. *newberryi*** is recognized and var. *racemosa* is relegated to synonymy. Variety *yesicola* Sivinski was recently discovered and named from central New Mexico; it has a long, filiform style that exceeds the apical notch of the fruit.

Physaria pruinosa (Greene) O'Kane & Al-Shehbaz (frosty, referring to the frosted appearance of the leaves) Frosty bladderpod, Pagosa Springs bladderpod. Perennial herbs from a simple or branched caudex basally covered with old leaf bases, grayish green, densely pubescent with 4- to 7-rayed trichomes with forked or bifurcate rays a little fused (webbed) toward their bases. **STEMS** few to several, these usually unbranched, decumbent to erect. **LEAVES**: basal petiolate with suborbicular or obovate to rhombic blades, entire to sinuate or shallowly dentate, abruptly narrowed to the petiole, 4–8 cm long; cauline obovate to rhombic, entire to shallowly toothed, petiolate below, becoming sessile above. **INFL** dense and slightly elongating in fruit. **FLOWER** sepals elliptic to oblong, cucullate at apex; petals yellow, spatulate, about 9 mm long. **FRUIT** a silicle, subglobose or ellipsoid, inflated and thin-walled, glabrous, often becoming coppery at maturity, terete, 6–9 mm long. **SEEDS** usually 2–4 per locule, somewhat flattened. [*Lesquerella pruinosa* Greene]. Shale hills with ponderosa pine and Gambel's oak or sagebrush and grasses surrounded by ponderosa pine. COLO: Arc, Hin; NMEX: RAr. 2300–2400 m (7550–7875'). Flowering: May–early Jun. Limited to a small area around Pagosa Springs, Colorado, and barely extending into New Mexico. Restricted to gray clay soils derived from Mancos Shale.

Physaria rectipes (Wooton & Standl.) O'Kane & Al-Shehbaz (straight foot, possibly referring to the straight stems) Straight bladderpod. Perennial herbs from a simple or branched caudex, herbage gray-green and often scabrous, pubescent with 4- to 6-rayed trichomes, the rays forked or bifurcate. **STEMS** few to several (to many in some extreme forms), prostrate, decumbent or erect, simple or branched. **LEAVES**: basal ± rosulate, especially the outer ones, these withering early, tapering to the petiole, entire (to weakly toothed), oblanceolate on the outside and often becoming narrower toward the middle and often involute when young; cauline oblanceolate, obtuse, ± entire. **INFL** ± dense in flower and somewhat elongating in fruit. **FLOWER** sepals lance-elliptic, 4.3–6.5 mm long; petals yellow, erect to spreading, oblanceolate, 6.5–10 mm long. **FRUIT** a silicle, ovoid to ellipsoid to subglobose, pubescent, terete or a little latiseptate at the apex, 4–7 mm long. **SEEDS** usually (4) 6–8 per locule, somewhat flattened. $n = 5+1, 9, 10, \pm 20$, $2n = 18$. [*Lesquerella rectipes* Wooton & Standl.]. Wash bottoms and piñon-juniper, piñon-juniper–Douglas-fir, ponderosa pine, and Gambel's oak communities. ARIZ: Apa, Nav; COLO: Arc, Dol, LPl, Mon, SMg; NMEX: McK, RAr, San, SJn; UTAH. 1600–2650 m (5200–8700'). Flowering: May–early Jul. Northern New Mexico, adjacent Colorado to south-central Utah and central Arizona. Large sprawling forms found in Arizona (Apa), Colorado (Mon), and Utah, which flower in early July at elevations ranging from 1615–1920 m (5300–6300'), may represent an undescribed species. *Physaria pulvinata* O'Kane & Reveal, a mound-forming plant from just north of our area in Colorado, has recently been segregated from *P. rectipes*. *Physaria rectipes* is extremely variable and is in need of detailed study.

Physaria scrotiformis O'Kane **CP** (saclike, referring to the scrotum-shaped fruit) West Silver bladderpod. Long-lived perennials from a simple or sparsely branched underground caudex with a distal short thatch of old leaf bases, silvery gray-green to silvery purplish, the silver color due to the dense covering of stellate trichomes. **STEMS** (0) 1–5 per plant, prostrate, arising laterally, up to 3 cm long. **LEAVES**: basal entire, in an erect or ascending tuft, oblanceolate, elliptic or rhombic, mostly flat, sometimes somewhat folded, attenuate at base and tapering to a slightly winged petiole, apex rounded to rounded-acute, 0.6–2.7 cm long and 2–5 mm wide; cauline entire, elliptic to oblanceolate, short-petiolate to essentially entire, 3–7 per stem, 3–5 mm long and 1.2–2 mm wide. **INFL** racemose and few-flowered. **FLOWERS** not yet observed. **FRUIT** a silicle, ovoid to obpyriform, apex rounded, flattened, or even slightly emarginate; base rounded-obtuse; wider than long; 3–4.5 mm long and 3.7–5 mm wide, the valves inflated and a little wider than the replum, making the fruits slightly didymous; pubescent with scattered trichomes; becoming purplish or greenish purple at maturity. **SEEDS** (1) 2–3 per locule, 1.9–2.2 mm long, apparently not mucilaginous. Subalpine-alpine boundary in a matrix of Engelmann spruce islands and tundra on cobbly limestone. COLO: LPl. 3535–3625 m (11600–11850'). Flowering: probably late May–early Jul, but not yet observed in flower. Limited to a single large population on West Silver Mesa, San Juan Mountains, Colorado, growing on nearly barren exposures of Leadville Limestone. A very rare species discovered while conducting fieldwork for this publication.

Rorippa Scop. Yellowcress

(from *rorippen*, a Saxon common name) Annual, biennial, or perennial herbs, rarely rhizomatous, glabrous or pubescent with simple trichomes. **STEMS** 1 to many, erect to prostrate, branched to much-branched. **LEAVES**: basal, if present, usually rosulate, short-petiolate or sessile, simple or 1–3-pinnatisect; cauline petiolate or auriculate to amplexicaul, progressively reduced in size upward, entire to pinnatisect. **FLOWERS** on pedicels that often have 2 minute basal glands; sepals erect to spreading, the lateral pair rarely subgibbous-based; petals subequaling or longer that the sepals, rarely absent, yellow, white, or purplish, ovate to oblanceolate, rarely clawed. **FRUIT** a silique or silicle, 2 (rarely 3–6)-loculed and -valved (very unusual in the family), globose, obovoid, oblong to linear, terete or somewhat latiseptate, plain or torulose, the false septum occasionally perforated. **SEEDS** (1) 2 rows per locule, mucilaginous or not. $x = 5, 6, 7, 8$ (Stuckey 1972). Eighty-six species worldwide, generally of wet to moderately moist locations, but a few weeds in drier situations.

1. Fruit a silicle, globose to broadly ovoid, about as wide as long .... ***R. sphaerocarpa***
1' Fruit an elongate silicle or silique, definitely longer than wide .... (2)

2. Plants pubescent with short, vesicular hairs ..... ***R. sinuata***
2′ Plants glabrous or pubescent, but lacking vesicular hairs ..... (3)

3. Fruit with papillate hairs on the valves; stems tending to be light-colored ..... ***R. tenerrima***
3′ Fruit glabrous (or very nearly so) and lacking papillate hairs on the valves; stems of typical color ..... (4)

4. Plants perennial from a simple or branched caudex; petioles of cauline leaves not at all auriculate; high elevations ..... ***R. alpina***
4′ Plants annual, biennial, or short-lived perennial from a taproot; petioles of cauline leaves auriculate or auriculate-clasping; low to high elevations ..... (5)

5. Petioles of cauline leaves short-auriculate; petals shorter than the sepals, 0.7–1.3 mm long; stems few (rarely single), prostrate, decumbent, or ascending ..... ***R. curvipes***
5′ Petioles of cauline leaves auriculate-clasping; petals shorter or longer than the sepals, 0.8–2.7 mm long; stems usually single, stout, erect or sometimes decumbent at the base ..... ***R. palustris***

Rorippa alpina (S. Watson) Rydb. (of the alpine) Alpine yellowcress. Perennial herbs from a simple or usually branched caudex, herbage glabrous. **STEMS** slender, weak, several to many, prostrate to decumbent, freely branching. **LEAVES** basal and cauline, somewhat reduced in size upward, basal ones withering early, lanceolate to obovate, crenate to pinnatifid to some nearly entire, blade tapering to the petiole. **INFL** narrow, elongating in fruit, terminal. **FLOWER** sepals oblong to ovate, 1–2 mm long; petals yellow, fading to pale yellow, 2–2.5 mm long, spatulate to ovate. **FRUIT** a silique (or silicle), oblong, tapering toward apex, 3–5 (–8) mm long, straight or slightly curved upward, terete, usually slightly torulose, glabrous, styles 0.5–1 mm long. **SEEDS** orbicular, brown. $2n = 16$. [*Nasturtium obtusum* Nutt. var. *alpinum* S. Watson; *Rorippa obtusa* (Nutt.) Britton var. *alpina* (S. Watson) Britton; *Radicula alpina* (S. Watson) Greene; *Rorippa curvipes* Greene var. *alpina* (S. Watson) Stuckey]. Lake, pond, and stream margins, alpine meadows, talus slopes, moist open forests. ARIZ: Apa; COLO: Arc, Con, Hin, LPl, Min, RGr, SJn; NMEX: McK, RAr, SJn. 2375–3850 m (7800–12600′). Flowering: Jul–Sep. Central and southern Rocky Mountain states. Navajos made a tonic to be taken after childbirth.

Rorippa curvipes Greene (bent foot, probably referring the habit of the plant) Bluntleaf yellowcress. Annual or short-lived perennial from a taproot, glabrous or sparsely pubescent with simple, short, retrorse trichomes. **STEMS** single to several, prostrate to erect, branched. **LEAVES** basal and cauline, basal ones withering early, entire to pinnately divided, margins entire or slightly and irregularly toothed, cauline auriculate or not, oblong to oblanceolate, pinnatifid to near the midrib, the terminal lobe similar to the others, margins entire to crenate. **INFL** axillary and terminal, elongating in fruit. **FLOWER** sepals oblong, 1–1.5 mm long; petals spatulate to narrowly oblanceolate, yellow, 0.5–1.5 (–2.8) mm long, shorter than the sepals. **FRUIT** a silique (or silicle), pyriform to short-cylindrical, usually slightly curved upward, (1.4–) 2.5–8 mm long, glabrous. **SEEDS** orbicular, brownish. [*R. underwoodii* Rydb.; *R. integra* Rydb.; *R. sinuata* (Nutt.) Hitchc.]. Disturbed sites of lakeshores and floodplains. ARIZ: Apa; COLO: Arc, LPl, Mon; NMEX: SJn. 1900–2350 m (6200–7750′). Flowering: Aug–Sep. Western United States and Mexico. Our plants seem to be all **var. *curvipes***. Variety *truncata* (Jeps.) Rollins, which is to be expected in our area, has longer and narrower fruits (4.5–8 mm long vs. 2.5–4.7 mm long).

Rorippa palustris (L.) Besser (swamp, marsh, referring to moist habitat) Bog yellowcress. Annual, biennial, or short-lived perennial, glabrous or sparingly to densely hirsute below and becoming sparingly hirsute to glabrous above, trichomes simple. **STEMS** usually single, stout to more rarely decumbent at the base or even prostrate, branched. **LEAVES** basal and cauline, short-petiolate (the uppermost sometimes sessile and clasping), petiole auriculate (at least slightly so), oblong to oblanceolate, deeply pinnatifid, the lobes irregularly dentate to crenate-sinuate, less frequently entire, the terminal lobe larger, ovate, rounded to obtuse apically. **INFL** axillary and terminal, elongating in fruit. **FLOWER** sepals 1.2–2.5 mm long, ovate or oblong, slightly gibbous-based or not; petals oblanceolate or spatulate, yellow, 0.8–2.7 mm long. **FRUIT** a silicle or silique, (2.2–) 3–11 (–14) mm long, obtuse or rounded on each end, the shortest fruits sometimes 3- or 4-valved (very unusual for members of the family), globose to elongate-oblong, glabrous. **SEEDS** suborbicular, light brown. $2n = 32$. [*R. islandica* (Oeder ex Murray) Borbás misapplied].

1. Plants mostly greater than 4 dm tall, often reddish; leaves thick-textured, on winged petioles ..... **var. *fernaldiana***
1′ Plants mostly 1–4 dm tall, often purplish; leaves thin-textured, nearly sessile ..... **var. *palustris***

var. *fernaldiana* (Butters & Abbe) Stuckey (for Merritt Lyndon Fernald, 1873–1950, American botanist and editor of the 8th and last edition of *Gray's Manual of Botany*) Fernald's yellowcress. Shallow water, pond margins, springs, and wetlands. COLO: Arc, LPl; NMEX: RAr, SJn. 1600–2850 m (5300–9400′). Flowering: May–Jul. North America, Mexico, and Central America. Navajos used the plant as a ceremonial eye wash.

var. *palustris* Bog yellowcress. Lake, pond, and steam margins and wet meadows. ARIZ: Apa; COLO: Arc, Con, Hin, LPl, Min, SJn; NMEX: McK, RAr, SJn. (1850) 2750–3625 m [(6100) 9000–11900′]. Flowering: Jun–Jul. Northern North America and possibly introduced from Europe in some areas.

Rorippa sinuata (Nutt.) Hitchc. (bend or curve, probably referring to the leaf margin) Spreading yellowcress. Perennial from a taproot, often with creeping underground roots and adventitious shoots, sparsely to densely pubescent with simple vesicular trichomes. **STEMS** few to many from the base, prostrate to decumbent, rarely erect and then rhizomatous. **LEAVES** all cauline, somewhat fleshy, short-petiolate below and becoming sessile and slightly auriculate above, deeply pinnatifid, the lobes rounded, entire to coarsely sinuate or crenate to toothed, upper surfaces glabrous. **INFL** axillary and terminal. **FLOWER** sepals ovate, the outer 2 gibbous-based, green with a pale hyaline margin, (2–) 2.5–3.2 mm long; petals oblanceolate, yellow, fading to pale yellow, (2.5–) 3–4.5 (–6) mm long. **FRUIT** a silique or silicle, short- to long-cylindrical or oblong to lanceolate, often curved upward, terete, glabrous to densely covered with vesicular trichomes. **SEEDS** angular, light brown. [*Nasturtium sinuatum* Nutt.; *Rorippa trachycarpa* (A. Gray) Greene]. Lake, pond, and stream margins, irrigation ditches, wash bottoms, and wet meadows. ARIZ: Apa; COLO: Hin, LPl, Mon, SJn; NMEX: McK, RAr, SJn; UTAH. 1500–3450 m (4900–11300′). Flowering: Jun–Aug. Western two-thirds of North America, with a few outliers on the East Coast. Navajos soaked watermelon seeds in a cold infusion of this plant to increase plant productivity.

Rorippa sphaerocarpa (A. Gray) Britton (spherical fruit) Roundfruit yellowcress. Biennial from a taproot, sometimes rooting at the nodes in wet ground, glabrous or hirsute below (rarely hirsute throughout). **STEMS** single to many from the base, decumbent to erect, usually branching throughout. **LEAVES** basal and cauline, short-petiolate, basal ones withering by flowering time, very slightly auriculate to nonauriculate, oblong to oblanceolate, pinnately divided to the midrib, the lobes irregularly serrate, terminal lobe larger than the others. **INFL** axillary and terminal, elongating in fruit. **FLOWER** sepals ovate or oblong, sometimes slightly gibbous at base, 0.8–1.2 mm long; petals obovate or spatulate, often shorter than the sepals, 0.6–1.2 mm long, yellow. **FRUIT** a silicle, globose to ovoid, less than 1.5 times as long as wide, 1–2.5 (–3) mm long, terete. **SEEDS** suborbicular, brown. [*Nasturtium sphaerocarpum* A. Gray; *Rorippa obtusa* (Nutt.) Britton var. *sphaerocarpa* (A. Gray) Cory]. Ephemeral creek banks, lake margins, marshy areas, hanging gardens, ephemeral ponds. ARIZ: Apa; COLO: LPl; NMEX: McK, SJn; UTAH. 1125–2725 m (3675–8950′). Flowering: May–Jun. Texas, Colorado, and Wyoming west to California, south to northern Mexico.

Rorippa tenerrima Greene (smooth, round) Southern marsh yellowcress. Annual from a taproot, glabrous. **STEMS** few to many from the root crown, freely branched, prostrate to decumbent, noticeably light-colored. **LEAVES** basal and cauline, basal ones withering early, short-petiolate, deeply and very regularly pinnatifid, narrowly oblong to narrowly elliptic, lobes entire or sinuate-margined, the terminal lobe larger. **INFL** axillary and terminal, elongating in fruit. **FLOWER** sepals lanceolate or oblong, purplish, 0.8–1.2 mm long; petals oblanceolate or spatulate, 0.5–0.8 mm long, yellow, shorter than the sepals. **FRUIT** a silique (or silicle), sparsely to densely papillate-pubescent, cylindrical to narrowly lanceolate, (2–) 3–7 (–9) mm long. **SEEDS** orbicular, brown. Wash bottoms, moist disturbed areas, sagebrush, ephemeral ponds, floodplains, hanging gardens. NMEX: McK, RAr, SJn; UTAH. 1125–2075 m (3700–6800′). Flowering: Jun–Aug. Western North America south to Baja California.

Sinapis L. Charlock

(Latin for mustard) Annual (rarely perennial) herbs or subshrubs, glabrous or pubescent with simple trichomes. **STEMS** erect, usually branched. **LEAVES** usually all cauline, petiolate, lyrate-pinnatifid or pinnatisect, sometimes dentate, progressively reduced in size upward, ± sessile and entire (except lobing). **INFL** racemose. **FLOWER** sepals widely spreading or reflexed; petals yellow (rarely white), obovate. **FRUIT** a silique, oblong to linear, often divided into a body and a beak, terete or strongly latiseptate, sometimes ± corky, often torulose. **SEEDS** 1 row per locule, mucilaginous or not. $x = 7–9, 12$. Seven species of southern Europe, northern Africa, and the Middle East; two species weedy and naturalized worldwide.

Sinapis arvensis L. (of the field) Charlock, wild mustard, corn-mustard. Annual from a taproot, herbage hispid, at least below, with simple hairs. **STEMS** simple, single, erect, branched above. **LEAVES** all cauline, petiolate below, sessile above, lyrate-pinnatifid with coarsely toothed lobes, the terminal lobe broadly ovate. **INFL** dense in flower and greatly elongating in fruit. **FLOWER** sepals narrowly oblong, 4–6 mm long; petals with a slender claw and a broadly obovate to suborbicular blade, 7–11 mm long, yellow. **FRUIT** a silique, 2–3.5 cm long, differentiated into a body and a beak, the body 1.5–2.5 cm long, linear, subterete, torulose, the valves distinctly 5–7-nerved, the beak 0.7–1.6 cm long, narrowly conical, slightly flattened distally. **SEEDS** spherical, finely reticulate, brown. $n = 9$, $2n = 18, 24$. [*Brassica arvensis* L.; *B. kaber* (DC.) L. C. Wheeler]. Moist disturbed areas of alkali seeps, pond margins, ponderosa pine, and meadows. COLO: Arc, Hin, LPl, Mon, SJn; UTAH. 1100–2650 m (3650–8700′). Flowering: late Apr–early Oct. Eurasia; highlands of Mexico and Central America; nearly cosmopolitan in temperate North America. Used by the Navajo as a ceremonial emetic to treat "deer infection."

Sisymbrium L. Tumblemustard, Hedgemustard

(a name for a mustard mentioned by both Dioscorides and Pliny) Annual or perennial herbs, glabrous or pubescent with simple, rarely branched, trichomes. **STEMS** usually erect and branched. **LEAVES** cauline, petiolate, runcinate, lyrate-pinnatisect, or pinnatifid, the upper progressively reduced upward and becoming short-petiolate to sessile. **INFL** racemose. **FLOWER** sepals erect (rarely ascending), rarely somewhat connivent at base, the lateral pair sometimes subgibbous-based; petals yellow, rarely white or lavender, clawed. **FRUIT** a silique, linear or lanceolate, terete or rarely ± latiseptate, plain or torulose, straight to slightly arched. **SEEDS** 1 row per locule, nonmucilaginous. $x = 7, 9, 10, 13$ (Warwick et al. 2002). Ninety-five species nearly worldwide, but the genus likely to be divided.

1. Upper leaves entire; perennials; glabrous and usually glaucous .... ***S. linifolium***
1′ Upper leaves all or mostly entire; annuals; glabrous or pubescent .... (2)

2. Upper and lower leaves distinctly different, the lower with oblong and dentate lobes, the upper with narrow linear or filiform lobes; fruiting pedicels thick and stout, almost as thick as the siliques .... ***S. altissimum***
2′ Upper and lower leaves essentially the same, differing only in size; fruiting pedicels slender, narrower than the siliques .... (3)

3. Flowers showy, sepals 3–4 mm long, petals 5.2–6.5 mm long, bright yellow .... ***S. loeselii***
3′ Flowers not especially showy, sepals 1.7–2.5 mm long, petals 2–4 mm long, pale yellow .... ***S. irio***

Sisymbrium altissimum L. (very high, referring to the tallness of the plant compared to most other members of the genus) Tall tumblemustard. Annual from a taproot, herbage hispid below, with simple hairs, usually glabrous above. **STEMS** usually single, erect, stout, branched. **LEAVES** petiolate, broadly lanceolate to oblong or oblanceolate, up to 20 cm long, the lower ones runcinate to pinnatifid, the upper pinnately divided into narrowly linear segments. **INFL** dense in flower and elongating in fruit. **FLOWER** sepals lanceolate or narrowly oblong, the outer 2 with a short horn at the apex, 3–5.5 mm long; petals oblanceolate, 5.5–8.5 mm long, pale yellow, fading to white. **FRUIT** a silique, linear, about the same diameter as the pedicel, terete, 5.5–8.5 cm long, valves 3-nerved. **SEEDS** oblong, flattened, cream to yellow, slightly mucilaginous when wetted. $n = 7$, $2n = 14$. [*Hesperis altissima* (L.) Kuntze]. Disturbed areas of desert grassland, washes, roadsides, townsites, piñon-juniper, desert shrub, and sagebrush communities. ARIZ: Apa, Nav; COLO: Arc, Dol, LPl, Min, Mon, SMg; NMEX: McK, RAr, San, SJn; UTAH. 1575–2240 m (5150–7350′). Flowering: late May–early Sep. Native of Eurasia, weedy in arid areas from Central America to northern North America. Navajos made a porridge from the seeds cooked in goat's milk.

Sisymbrium irio L. (possibly from Latin *iri*, the iris of the eye) London rocket. Annual from a taproot, herbage glabrous or sparsely hispid above with simple trichomes. **STEMS** usually single, erect, stout, often much-branched throughout. **LEAVES** petiolate, pinnately lobed with a larger hastate terminal lobe, becoming reduced and sometimes simple above. **INFL** dense in flower and not greatly elongating. **FLOWER** sepals narrowly lanceolate, 1.7–2.5 mm long; petals oblanceolate, 2–4 mm long, pale yellow. **FRUIT** a silique, straight or slightly incurved, terete, glabrous to sparsely pubescent, slender, 3–5 cm long, valves 3-nerved. **SEEDS** oblong, yellow. $n = 7, 14$, $2n = 14, 28$. [*Descurainia irio* (L.) Webb & Berthel.; *Erysimum irio* (L.) Farw.]. Disturbed areas of piñon-juniper communities, wash bottoms, roadsides, and campgrounds. COLO: Mon; NMEX: RAr, SJn; UTAH. 1100–1850 m (3600–6100′). Flowering: late Mar–early Jun. Native of Europe, now in southwestern United States and adjacent Mexico and eastern United States.

Sisymbrium linifolium (Nutt.) Nutt. ex Torr. & A. Gray (linear-leaved) Flaxleaf plainsmustard. Perennial, rhizomatous, glabrous and glaucous (rarely sparsely pilose with simple trichomes below). **STEMS** few to several from a short, branched caudex, erect and often decumbent at the base. **LEAVES** all cauline, the lower ones up to 7 cm long, somewhat fleshy, linear to linear-oblanceolate, entire to pinnatifid, becoming entire, narrower, and linear above as well as nearly terete. **INFL** elongating in fruit. **FLOWER** sepals narrowly elliptic, yellow, 4–6.5 mm long; petals narrowly oblanceolate or spatulate, claw lacking, 7–12 mm long, yellow. **FRUIT** a silique, linear, terete, 4–7 cm long, wider than the pedicel. **SEEDS** oblong, plump. $n = 7$, $2n = 14$. [*Schoenocrambe linifolia* (Nutt.) Greene; *Sisymbrium pygmaeum* Nutt.]. Wash bottoms and hanging gardens. ARIZ: Apa, Coc, Nav; COLO: Arc, Dol, LPl, Mon; NMEX: RAr; UTAH. 1750–1850 m (5800–6100′). Flowering: May–Jun. Western North America to British Columbia.

Sisymbrium loeselii L. (for J. A. Loiseleur des Longchamps) Small tumbleweed mustard. Annual from a taproot, herbage sparsely to densely hispid below with simple spreading to retrorse trichomes, usually glabrous above. **STEMS** usually single, erect, stout, usually much-branched, especially above. **LEAVES** short-petiolate, up to 15 cm long below, smaller above, broadly deltoid-lanceolate to lanceolate, runcinate-pinnatifid, typically with a large, acuminate, irregularly serrate to dentate terminal lobe. **INFL** dense at first, then rapidly elongating. **FLOWER** sepals narrowly lanceolate, 3–4 mm long; petals obovate, 5.2–6.5 mm long, long-clawed, bright yellow. **FRUIT** a silique, straight or slightly incurved, terete, valves prominently 3-nerved, 2–3.4 (–4) cm long, evidently wider than the pedicel. **SEEDS** oblong, orange, mucilaginous when wetted. $n = 7$, $2n = 14$. [*Turritis loeselii* (L.) W. T. Aiton; *Erysimum loeselii* (L.) Farw.]. Piñon-juniper communities; disturbed areas. COLO: Mon; NMEX: SJn; UTAH. 1950–2100 m (6400–6950′). Flowering: Jul–early Nov. Native of Eurasia, now scattered throughout North America.

Smelowskia C. A. Mey. Smelowskia, False Candytuft

(for Timotheus Smelowsky, 1769–1815, Russian pharmacist and botanist) Perennial, often pulvinate, pubescent with simple and 2-forked or dendritic trichomes or subglabrous. **STEMS** numerous, branched. **LEAVES**: basal rosulate, petiolate, pinnate or pinnatifid; cauline similar but reduced, becoming sessile upward. **INFL** racemose. **FLOWER** sepals ascending to spreading, often persistent; petals white to cream or pink to purple, spatulate to orbicular. **FRUIT** a silique or silicle, ovate, pyriform, or ellipsoid to linear, terete, angustiseptate or slightly 4-angled, ascending to erect. **SEEDS** 1 row per locule, nonmucilaginous. $x = 6, 10, 11, 12$ (Warwick et al. 2004). Twenty-five species of mountainous or high-latitude regions of central Asia, eastern Siberia, and western and northwestern North America.

Smelowskia americana Rydb. (of America) Alpine smelowskia, American false candytuft. Perennial, caespitose, from a branched caudex clothed with persistent leaf bases, herbage grayish green due to dense covering of fine, soft, freely branching trichomes. **STEMS** several to many from the caudex, unbranched or sometimes branched above, erect or ascending. **LEAVES** basal and cauline, (ob)ovate, reduced in size upward, the basal petiolate and ± rosulate, cauline sessile, all pinnately (to bipinnate in robust specimens) divided (rarely a few nearly entire), lobes oblong to cuneate, more nearly linear above. **INFL** dense in flower, elongating in fruit, the lower pedicels sometimes bracteate. **FLOWER** sepals (ob)ovate, obtuse with scarious margins, cream to violet, 1.8–3.4 mm long; petals obovate to orbicular, 3.5–5.5 mm long, cream to pinkish and occasionally lavender. **FRUIT** a silique, fusiform to linear, tapered at both ends, slightly latiseptate, usually glabrous, 4–8 (–12) mm long. **SEEDS** ellipsoidal. $2n = 44$. [*S. calycina* (Stephan) C. A. Mey. var. *americana* (Regel & Herder) W. H. Drury & Rollins]. Alpine tundra and krummholz, often among rocks, and high-elevation talus slopes. COLO: Arc, Con, Hin, LPl, Min, Mon, RGr, SJn. 3200–4125 m (10500–13500′). Flowering: Jul–mid-Aug. British Columbia and Alberta south to Nevada, Utah, and Colorado. Western United States material has traditionally been recognized as *S. calycina* var. *americana*. Recent molecular studies (Warwick et al. 2004), however, show that the varieties of *S. calycina* should be recognized at the species level.

Stanleya Nutt. Prince's Plume

(for Lord Edward Smith Stanley, Earl of Derby, 1775–1851, British statesman and ornithologist) Annual, biennial, or perennial herbs, often woody at the base, glabrous or pubescent with simple (rarely some branched in upper portions of the plant) trichomes, often glaucous. **STEMS** erect or ascending, branched or not. **LEAVES** often with a basal rosette of petiolate, entire, toothed, or pinnatifid leaves; cauline leaves entire, toothed, or pinnatifid, petiolate or occasionally auriculate or amplexicaul. **INFL** racemose. **FLOWER** sepals spreading to reflexed; petals yellow (less frequently cream or white), ovate, elliptic, or rarely oblong to linear, papillose on the claw (in ours); stamens not tetradynamous, all ± the same length and coiling as they dehisce from the top downward. **FRUIT** a silique on a long stipe of 1–3 cm at maturity, linear, subterete or latiseptate, ± torulose. **SEEDS** 1 row per locule, mucilaginous. $x = 6, 7$. Six species of the western United States. The plants are an indicator of selenium in the soil and have a distinctive strong odor like the selenium-accumulating species of *Astragalus.* The accumulation of selenium can lead to livestock poisoning in areas where animals are forced to eat the unpalatable plants for lack of better forage. Populations of *Stanleya pinnata* were sought during the uranium mining boom of the 1950s and 1960s as a possible indicator of uranium ore.

1. Cauline leaves, at least above, sessile and auriculate at the base .... ***S. viridiflora***
1′ Cauline leaves petiolate, the blades cuneate and not auriculate at the base .... (2)

2. Petal blade 2.5–5 mm broad, white to pale yellow, obovate to orbicular, hairy only at the inner summit of the claw .... ***S. albescens***
2′ Petal blade 1–2.5 mm broad, bright yellow, oblanceolate to obovate, the claw hairy nearly its full length on the inner side .... ***S. pinnata***

Stanleya albescens M. E. Jones (becoming white, referring to the white petals) White prince's plume. Biennial from a taproot, glabrous and glaucous. **STEMS** single to few, branching from the base and above. **LEAVES** ± fleshy, petiolate, essentially all cauline, lyrate-pinnatifid, runcinate or the uppermost rarely entire, to 20 cm long. **INFL** relatively dense, but less so than in other species. **FLOWER** sepals linear, reflexed, 9–13 mm long; petals hairy on inner side near the junction of blade and claw, 2.5–5 mm long, blade obovate to orbicular, white to cream-white (rarely very pale yellow). **FRUIT** a silique, on usually widely spreading pedicels, glabrous, erect to spreading, (sub)terete, gently curved upward, 3–6 cm long, on a long stipe 15–26 mm long. **SEEDS** oblong to ovoid, brown. Desert scrub of saline soils. ARIZ: Apa, Nav. 1850–1975 m (6100–6500′). Flowering: late May–August. Southwest Colorado, northwest New Mexico, and northeast Arizona. Hopis ate the boiled leaves.

Stanleya pinnata (Pursh) Britton (feathered or pinnate, referring to the leaves) Prince's plume. Perennial subshrub, glabrous or sparsely pubescent with simple hairs, ± glaucous. **STEMS** few to many, erect or ascending. **LEAVES** all cauline, petiolate, lower leaves entire to deeply pinnately lobed, broadly lanceolate, up to 15 cm long and 5 cm wide, upper leaves linear-lanceolate to ovate, entire or pinnatifid, smaller than lower leaves. **INFL** densely flowered, elongating somewhat in fruit. **FLOWER** sepals linear, with a somewhat thickened tip, yellowish, (8–) 10–16 mm long; petals hairy nearly the entire length of the inner side, 1–2 cm long, blade oblanceolate to obovate, white to yellow. **FRUIT** a silique, on stout, widely spreading pedicels, glabrous, (sub)terete, arcuate to nearly straight, 4–8 (–12) cm long, on a long stipe 12–21 mm long. **SEEDS** oblong, brown, 0.8–1 mm long, mucilaginous when wetted.

1. Leaves all entire (sometimes some lower ones dentate or shallowly pinnatifid); plants infrequent .... **var. *integrifolia***
1′ Leaves (at least the lower ones) pinnately divided; plants common .... **var. *pinnata***

var. *integrifolia* (E. James) Rollins (entire leaf) Golden prince's plume. $n = 12, 14$, about 24. [*S. integrifolia* E. James; *S. pinnatifida* Nutt. var. *integrifolia* (E. James) B. L. Rob.]. Piñon-juniper, piñon-shrub, desert scrub, clay hills, and shaley slopes. COLO: LPl, Mon; NMEX: SJn. 1500–2075 m (4950–6800′). Flowering: Jul–Sep. Nevada, Utah, Colorado, Arizona, New Mexico, Wyoming, and Kansas.

var. *pinnata* (Pursh) Britton **CP** Desert prince's plume. $n = 12, 14, 2n = 24$. [*Cleome pinnata* Pursh; *Stanleya pinnatifida* Nutt.; *S. arcuata* Rydb.; *S. canescens* Rydb.]. Selenium, and often uranium-bearing soils of desert scrub, piñon-juniper, sagebrush, badlands, and (rarely) hanging gardens. ARIZ: Apa, Coc, Nav; COLO: Dol, LPl, Mon; NMEX: McK, RAr, SJn; UTAH. 1125–2150 m (3700–7050′). Flowering: May–Sep. Western continental United States except, apparently, Washington. Navajos used a poultice of this plant to treat glandular swelling.

Stanleya viridiflora Nutt. (green flower) Green prince's plume. Perennial from a short-branched caudex covered with old leaf bases surmounting a thick, woody taproot, glabrous, glaucous, the leaves ± fleshy. **STEMS** single to few, erect, unbranched or branched above. **LEAVES** ± fleshy, petiolate, basal and cauline, the basal rosette well developed, lanceolate to ovate, narrowly cuneate at the base, entire or occasionally coarsely toothed or lyrate-pinnatifid, cauline leaves crowded. **INFL** loosely flowered and little-elongating in fruit. **FLOWER** sepals narrowly elliptic, spreading to reflexed, pale greenish yellow to yellow; petals linear-oblong, erose at apex, blade tapering into the claw, 13–21 mm long, lemon yellow to pale yellow to nearly white. **FRUIT** a silique, on stout, widely spreading pedicels, glabrous, arcuate, (sub)terete, 3–7 cm long, on a long stipe 10–20 (–25) mm long. **SEEDS** oblong to ovoid, dark brown. $n = 12$, $2n = 24, 28$. [*S. collina* M. E. Jones]. Clayey soils of mat saltbush and desert scrub communities. COLO: Mon; NMEX: SJn; UTAH. 1225–1700 m (4000–5600′). Flowering: Apr–Jun. Southern Montana south to northern New Mexico and west to California except (apparently) Arizona and Washington.

Streptanthella Rydb. Fiddle-mustard, Streptanthella

(diminutive of *Streptanthus*, another genus in the family) A monotypic genus.

Streptanthella longirostris (S. Watson) Rydb. (long bird's beak) Longbeak fiddle-mustard, longbeak twistflower, longbeak streptanthella. Annual herbs, glabrous and usually glaucous. **STEMS** usually single, erect or ascending, usually branched, 1–4 (5) dm tall. **LEAVES** all cauline (basal ones, if present, very early-deciduous), shortly petiolate, narrowly (ob)lanceolate, obtuse apically, entire, dentate, or rarely pinnatifid, becoming linear and entire above. **INFL** elongated at tips of branches. **FLOWER** sepals ascending, the lateral pair subgibbous-based, 2.5–5.5 mm long; petals white or yellowish, sometimes purple-tinged, narrowly spatulate to narrowly oblanceolate, crispate, 3.5–6.5 mm long. **FRUIT** a silique, linear, latiseptate, narrow, 3–5 cm long, pendent or pendent-appressed, on strongly reflexed pedicels. **SEEDS** 1 row per locule, nonmucilaginous, brownish, membranous-winged, 2–2.5 mm long. $x = 7$. Piñon-juniper, juniper-shrub, slickrock, desert scrub, alcoves, and hanging gardens. ARIZ: Apa, Coc, Nav; COLO: LPl, Mon; NMEX: McK, RAr, San, SJn; UTAH. 1275–2125 m (4200–7000′). Flowering: late Mar–early May. Montana to Idaho and westward in dry areas to Washington and California; Sonora, Mexico.

Streptanthus Nutt. Jewelflower, Twistflower

(Greek *streptos*, twisted, + *anthos*, flower, referring to the often wavy petal margins) Annual or rarely perennial herbs, often glaucous, glabrous to pubescent with simple trichomes. **STEMS** single to several from base, unbranched or branched above. **LEAVES**: basal petiolate; cauline sessile and auriculate or amplexicaul, entire or dentate (rarely pinnatifid). **INFL** racemose. **FLOWER** sepals erect or ascending, distinct or connivent, lateral pair gibbous-based or not; petals lavender to purple or magenta, obovate, elliptic to linear or lanceolate, crispate or involute-rolled blades. **FRUIT**: silique sessile or stipitate, linear, latiseptate or subterete, plain or torulose. **SEEDS** 1 row per locule, nonmucilaginous, winged. x = 12, 14. Thirty-four species (in a broadly defined genus) limited to the western and southwestern United States and northern Mexico.

Streptanthus cordatus Nutt. (heart) Heartleaf twistflower. Perennial, from a simple or closely branched caudex, sometimes rhizomatous, glaucous, glabrous except the ciliate margins of some lower leaves. **STEMS** single or few, mostly simple. **LEAVES**: basal and lower cauline leaves on winged petioles, spatulate-obovate, dentate, up to 10 cm long; cauline leaves broadly oblong to nearly orbicular, auriculate and sagittate at the base, entire to dentate. **INFL** lax and elongate in fruit. **FLOWER** calyx flask-shaped, the sepals greenish brown to purple, the outer pair strongly gibbous-based, scarious-margined, 7.5–12 mm long; petals narrowly oblong, involute along margins, dark purple to brownish, often with white margins, 9.5–15 (–16) mm long. **FRUIT** a silique, linear to curved, strongly latiseptate, 5–10 cm long and up to 7 mm wide. **SEEDS** orbicular, with a broad membranous wing, brown. n = 12, 14. [*S. crassifolius* Greene; *S. coloradensis* A. Nelson]. Piñon-juniper, piñon-juniper-shrub, piñon–Douglas-fir, and Douglas-fir–ponderosa pine communities, often in wash bottoms. ARIZ: Apa, Coc, Nav; COLO: Arc, Dol, LPl, Mon; NMEX: RAr, San, SJn; UTAH. 1575–2600 m (5150–8500′). Flowering: late Apr–May. Wyoming, Colorado, and New Mexico west to California and Oregon. Used as vegetable greens by the Navajo. Our plants are **var. *cordatus***.

Thelypodiopsis Rydb. Tumblemustard

(diminutive of *Thelypodium*, another genus in the family) Annual, biennial, or rarely perennial herbs, often glaucous, glabrous or pubescent with simple trichomes. **STEMS** erect, single to several from the base, usually branched above. **LEAVES** all cauline or both basal and cauline (the basal rosette often soon deciduous), petiolate to sessile, the lower ones sometimes petiolate, but all leaves usually auriculate to amplexicaul, entire, toothed to pinnatifid. **INFL** racemose. **FLOWER** sepals erect or rarely spreading, the lateral pair subgibbous-based or not; petals white, lavender, purple, or yellow, differentiated into a blade and claw, the blade oblong to obovate. **FRUIT** a silique, sessile or stipitate, linear, ± terete, occasionally torulose. **SEEDS** 1 row per locule, nonmucilaginous, not winged in ours. x = 10, 11. Seventeen species of western North America, south to Guatemala.

1. Petals yellow; siliques on stipes 2–8 mm long ..... ***T. aurea***
1′ Petals white to pale lavender; siliques on stipes less than 0.3 mm long ..... ***T. purpusii***

Thelypodiopsis aurea (Eastw.) Rydb. (golden, referring to the flower color) Durango tumblemustard. Biennial or short-lived perennial from a taproot, glabrous or rarely sparsely pubescent with simple hairs at the stem base. **STEMS** single or few, erect. **LEAVES** somewhat fleshy, basal leaves oblanceolate, tapering to a winged petiole, irregularly dentate; cauline leaves sessile, auriculate, oblong to lanceolate, acute-tipped. **INFL** elongating in fruit. **FLOWER** sepals oblong, scarious-margined, 5–7 (–9) mm long; petals narrowly oblong with a slight constriction at the junction of the blade and claw, 7–13 mm long, yellow. **FRUIT** a silique, erect to slightly divaricate, slender, nearly straight to more commonly incurved, somewhat torulose, stipe 2–8 mm long, body 4.5–9 mm long. **SEEDS** oblong, plump. n = 11. [*Thelypodium aureum* Eastw.; *Sisymbrium aureum* (Eastw.) Payson]. Ledges, alkali flats, clay flats and hills, desert shrub, piñon-juniper, badlands, saltbush communities. ARIZ: Apa; COLO: Dol, Mon; NMEX: San, SJn; UTAH. 1475–2100 m (4800–6900′). Flowering: Apr–May. Endemic to the general Four Corners region.

Thelypodiopsis purpusii (Brandegee) Rollins (named for Carl Albert Purpus, western plant collector) Purpus' tumblemustard. Annual from a taproot, glabrous and leaves glaucous. **STEMS** single, often branched above, erect. **LEAVES** somewhat fleshy, basal leaves few and early-withering, these tapering to a winged petiole, pinnatifid to shallowly sinuate-dentate; cauline leaves sessile, auriculate, ovate to elliptic-lanceolate, entire. **INFL** loose, few-flowered, elongating in fruit. **FLOWER** sepals oblong, scarious-margined, 3.5–4.5 mm long, white to lavender or purplish. **FRUIT** a silique, ascending, terete, narrowly linear, sessile to subsessile on a short stipe less than 0.3 mm long, body 4–5.5 cm long. **SEEDS** oblong, brown. [*Thelypodium purpusii* Brandegee; *Sisymbrium purpusii* (Brandegee) O. E. Schulz]. Growing in duff beneath trees and shrubs in piñon-juniper and mountain shrub communities. ARIZ: Apa; COLO: LPl; NMEX: McK, SJn. 1900–2200 m (6250–7220′). Flowering: May–early Jun. New Mexico, southwest Colorado, west Texas, and northeast Arizona.

Thelypodium Endl. Thelypody

(Greek *thelys*, female, + *podion*, foot, referring to the "little foot," or gynophore, of the ovary) Annual, biennial, or rarely short-lived perennial herbs, glaucous and glabrous or more rarely pubescent with simple trichomes. **STEMS** erect to few from the base, usually branched above. **LEAVES**: basal ± rosulate, usually withering early, petiolate, relatively broad, entire, toothed or pinnately lobed; cauline petiolate to sessile, often auriculate to amplexicaul, entire, toothed, or pinnately lobed. **INFL** racemose. **FLOWER** sepals erect to reflexed, lateral pair not gibbous-based; petals white to purple, spatulate or linear to obovate. **FRUIT** a silique, on an indistinct to long stipe, linear, terete or ± latiseptate, usually torulose. **SEEDS** 1 row per locule, nonmucilaginous. $x = 13$. Nineteen species of central and western North America.

1. Cauline leaves sessile ..... ***T. integrifolium***
1′ Cauline leaves petiolate ..... (2)

2. Sepals ascending; siliques 2.5–6 cm long, strongly torulose; plants pubescent or not below ..... ***T. laxiflorum***
2′ Sepals widely spreading to reflexed; siliques 5–9 cm long, moderately torulose; plants glabrous ..... ***T. wrightii***

Thelypodium integrifolium (Nutt.) Endl. (entire leaf) Entire-leaved thelypody, tall thelypody. Biennial from a taproot, glabrous, glaucous and somewhat fleshy. **STEMS** usually single, unbranched or branched near the top. **LEAVES** somewhat thick and fleshy, basal rosulate, petiolate, blade oblong, oblanceolate, or spatulate, entire, less frequently denticulate or repand, up to 7 cm long; cauline leaves oblanceolate to lanceolate below and becoming linear-lanceolate to linear above, sessile, entire. **INFL** dense and little-elongating. **FLOWER** sepals lanceolate to linear, pale green, white, or lavender to purplish, 2.3–4.8 mm long; petals with a long, slender claw, blades oblong to obovate, white to lavender to purplish, 4.5–10 mm long. **FRUIT** a silique straight to strongly incurved, terete to a little flattened, torulose to submoniliform, divaricately ascending to horizontal (rarely reflexed), 1–3.5 (–6) cm long on a stipe 0.3–2.2 (5.5) mm long. **SEEDS** ovoid or oblong.

1. Fruiting pedicel stout, strongly flattened at base; infructescence moderately congested, the raceme of the main stem 5.5–30 cm long ..... **subsp. *gracilipes***
1′ Fruiting pedicel slender and somewhat flattened at base; infructescence densely congested, the raceme of the main stem 1–5 cm long ..... **subsp. *integrifolium***

subsp. *gracilipes* (B. L. Rob.) Al-Shehbaz (thin foot) $n = 13$. Moist areas of canyon bottoms, floodplains, cottonwood groves, alcoves, hanging gardens, piñon-juniper communities, and willow and shrub communities. ARIZ: Nav; COLO: LPl, Mon; NMEX: McK, RAr, SJn; UTAH. 1125–2225 m (3700–7300′). Flowering: Jun–Oct. Limited to the Four Corners states.

subsp. *integrifolium* $n = 13$. [*T. lilacinum* Greene]. Ponderosa pine–Gambel's oak community. UTAH. 2400 m (7845′). Flowering: early Jul. One record on a roadcut.

Thelypodium laxiflorum Al-Shehbaz (relaxed, loose flower) Droopflower thelypody. Biennial from a taproot, herbage glabrous or sparsely to densely hirsute below. **STEMS** usually single, erect, unbranched or branched above. **LEAVES**: basal petiolate, oblanceolate, often pinnately to lyrately lobed (rarely laciniate or sinuate, somewhat glaucous (rarely sparsely pubescent); cauline leaves short-petioled, linear to linear-lanceolate or lanceolate, sometime oblanceolate or oblong especially below, sinuate to dentate. **INFL** elongating and lax in fruit. **FLOWER** sepals oblong or oblanceolate, 3.2–6 mm long; petals clawed, the blade spatulate or (ob)lanceolate to narrowly oblong, white to rarely lavender. **FRUIT** a silique, slender, straight or somewhat variously curved, terete, divaricate to somewhat reflexed, submoniliform to strongly torulose, (2.5) 3–6 cm long on a stipe 0.3–3 mm long. **SEEDS** oblong. Wet areas in canyon bottoms and Gambel's oak communities. COLO: Arc, LPl, Mon; NMEX: McK, SJn. 1800–2500 m (5900–8200′). Flowering: Jul–Aug. Western Colorado, Utah, and Nevada.

Thelypodium wrightii A. Gray (for Charles Wright, American botanical explorer, 1811–1885) Wright's thelypody. Biennial from a taproot, glabrous and glaucous. **STEMS** usually single, erect, unbranched or branched above. **LEAVES**:

basal rosulate, these and lowermost cauline leaves early-withering, oblanceolate, pinnatifid to coarsely toothed; cauline leaves reduced in size upward, lanceolate to narrow-lanceolate, tapering to the petiole, entire (usually) to dentate to sinuate. **INFL** dense and becoming strongly elongated. **FLOWER** sepals narrowly oblong, 3.5–5.5 mm long; petals long, slender-clawed, the blade (ob)lanceolate, 4.5–6 (–7.5) mm long, white to rarely lavender. **FRUIT**: silique latiseptate to subterete, straight to somewhat curved, torulose, horizontal to reflexed, 5–7 (–9) cm long. **SEEDS** ovoid or oblong. [*Stanleya wrightii* (A. Gray) Rydb.]. Talus and rock outcrops. NMEX: McK, RAr. 1825–2125 m (6000–7000′). Flowering: Jul–Aug. New Mexico, Arizona, Utah, Oklahoma, and Texas. Navajos rubbed the ashes of this plant on eyelids for eye disease; plants were tied to the cradle bow to help babies sleep. Various Pueblo tribes used the plant as greens. Our plants are **subsp. *wrightii***.

Thlaspi L. Pennycress

(Greek name for some members of the family) Annual, biennial, or perennial herbs, often glaucous, glabrous or pubescent with simple trichomes. **STEMS** usually single, branched or unbranched. **LEAVES** all cauline, sessile (lowermost sometimes petiolate), auriculate, entire or rarely dentate. **INFL** racemose. **FLOWER** sepals erect to ascending, not gibbous-based; petals white, narrow-clawed with an expanded blade, subequaling or exceeding the sepals. **FRUIT** a silicle (rarely a silique), obcordate to suborbicular, rarely elliptic or linear, angustiseptate, apex rounded or deeply emarginate, valves keeled and usually broadly winged. **SEEDS** few per locule, mucilaginous or not. $x = 7$. Six species (after segregate genera such as *Noccaea* are removed) of the Old World.

Thlaspi arvense L. (field) Field pennycress. Annual or winter-annual from a taproot, glabrous. **STEMS** single, unbranched to branched throughout, erect. **LEAVES** basal and cauline, the basal ones early-deciduous and not usually seen, decreasing in size upward, oblanceolate, irregularly dentate to sinuate, obtuse to rounded apically, petiolate below, becoming sessile and sagittate-clasping above, up to 4.5 (–6) cm long. **INFL** elongating in fruit. **FLOWER** sepals oblong, broadly rounded, 1.4–2.5 mm long; petals spatulate, 2.5–3.7 mm long, white. **FRUIT** a silicle, strongly angustiseptate, oval to oblong-obcordate, wing-margined, 7–10 (–17) mm long. **SEEDS** ovoid, brown. $n = 7$, $2n = 14$. Disturbed areas of roadsides, canal banks, stock ponds, sagebrush, piñon-juniper communities, and spruce. COLO: Arc, Dol, Hin, LPl, Mon, SJn; NMEX: SJn. 1650–9200 m (5380–9200′). Flowering: May–Jul. Native of Europe, now weedy throughout temperate areas in Central America, Mexico, and North America.

Turritis L. Tower Mustard, Tower Cress

Michael D. Windham

(Latin *turris*, tower) Plants biennials, rarely short-lived perennials. **HAIRS** simple and forked to subcruciform. **STEMS** simple or branched distally, lower portions hirsute with simple and 2–3-rayed hairs. **LEAVES** basal and cauline, the former often deciduous in late anthesis; basal leaves petiolate, rosulate, oblanceolate to oblong, (7–) 10–30 mm wide, thin, simple, crenate to repand (rarely entire), hirsute with 2–4-rayed (and a few simple) hairs, 3-rayed hairs abundant; upper cauline leaves sessile, with prominent auricles, entire, without cilia, glabrous and glaucous. **INFL** ebracteate, usually racemose, elongate in fruit; fruiting pedicels erect (in ours) or divaricate. **FLOWERS** ascending; sepals oblong, erect, caducous, bases of lateral pair not saccate, margins membranous; petals pale yellow to cream-colored (rarely aging lilac); blade spatulate to narrowly oblanceolate, apex obtuse; claw undifferentiated from blade; stamens 6, tetradynamous; filaments not dilated at base; ovules 130–200 per ovary. **FRUIT** linear, subterete-quadrangular, sessile, unsegmented, 0.7–1.5 mm wide at dehiscence, glabrous; valves leathery, with a prominent midvein; septa complete, membranous, veinless; styles obconic, the upper portion gradually expanded and distinctly wider than the lower; stigmas often slightly 2-lobed. **SEEDS** in 2 rows in each locule, flattened, elliptic or orbicular, wingless or rarely narrowly winged. $x = 6, 8$. A genus of two species: North Africa, Eurasia, North America, naturalized elsewhere. Traditionally, this taxon has been included in the genus *Arabis*. Recent molecular studies (e.g., Koch, Haubold, and Mitchell-Olds 2001) demonstrate that the two genera are not closely related.

Turritis glabra L. (hairless) Tower mustard, tower cress. Biennial, rarely short-lived perennial. **STEMS** 1 (rarely more) arising from center of basal rosette, 4–6 (–10) dm tall, often branched distally, lower portions hirsute with simple and stalked, 2–3-rayed hairs 0.2–1.2 mm long, upper portions glabrous and glaucous. **LEAVES** rosulate at base of stem, oblanceolate or oblong, the largest 10–30 mm wide, with obtuse to acute apex and crenate to repand (rarely entire) margins, hirsute with 2–4-rayed (and a few simple) hairs; cauline leaves 15–27, overlapping and often concealing stem proximally, entire or dentate; upper cauline leaves lanceolate, with prominent auricles to 4 mm long, glabrous. **INFL** simple or branched; fruiting pedicels erect, 5–12 mm long, appressed to rachis, glabrous. **FLOWERS** inconspicuous;

sepals greenish to beige, 2.5–5 mm long, glabrous; petals pale yellow to cream-colored (sometimes aging lilac), 3–5.5 mm long. **FRUIT** erect, appressed to rachis, glabrous; styles obconic, 0.5–1.1 mm long. $2n$ = 12, 16, 32. [*Arabis glabra* (L.) Bernh.]. Meadows, forest margins, and riverbanks, often in disturbed habitats. COLO: Hin, Min. 2400 m (7800′). Flowering: Jun–Aug. Throughout most of the United States except for the Gulf Coast region; North Africa, southwest Asia, Europe, naturalized in Australia.

CACTACEAE Juss. CACTUS FAMILY

Kenneth D. Heil and J. M. Porter

Perennial stem succulents, some becoming woody. **STEMS** globose, cylindroid, columnar, or flattened. **AREOLES** (nodes) producing spines and sometimes glochids. **SPINES** variable, may be alike or diverse, long or short, straight or curved. **INFL** produced on either new or old growth, located near or below apex of stem. **FLOWERS** epigynous, with hypanthium forming a floral tube, perianth of numerous segments (tepals) grading from sepaloids to petaloids; stamens numerous; carpels 3–24, fused; style 1, stigma lobes free; ovary inferior. **FRUIT** fleshy or dry at maturity, a many-seeded berry. x = 11 (Benson 1982; Anderson 2001). Eighty-six to 104 genera, 800–2000 species. New World natives except for *Rhipsalis baccifera* (tropical Africa). British Columbia and Alberta south to the Argentine portion of Patagonia, and introduced to Australia, Asia, Africa, and the Mediterranean. Sea level to 4500 m (15000′). Most cacti are found in arid and semiarid areas; however, they are also found in a diversity of habitats, including rainforests. Uses include: horticulture, food for humans and livestock, emergency water source, dye, and ceremonial purposes, including hallucinogens.

1. Areoles bearing minute, sharp-pointed, barbed bristles (glochids) as well as other spines; seeds bony; stems made up of a series of cylindroid, club-shaped, or flattened joints ....(2)
1′ Areoles not bearing glochids; seeds not bony; stems generally without a series of joints ....(4)

2. Joints of the stem flattened, not circular in cross section; spines with no sheath (pricklypears) .... ***Opuntia***
2′ Joints of the stem club-shaped or cylindroid or circular in cross section; spines with or without a thin paperlike sheath ....(3)

3. Joints of the stem cylindroid or circular in cross section; spine forming a thin paperlike sheath (cholla) .... ***Cylindropuntia***
3′ Joints of the stem clavate or club-shaped; spine with no sheath .... ***Corynopuntia***

4. Flowers and fruits always appearing below the current stem apex on stems that are 1 or more years old (hedgehog) .... ***Echinocereus***
4′ Flowers and fruits produced on new growth of the current season ....(5)

5. Stems without ribs; tubercles grooved when mature; fruit fleshy, green or red, naked .... ***Escobaria***
5′ Stems with either ribs or separate tubercles, if stems with tubercles, the tubercles not grooved; fruit dry, green and turning pink, reddish, or brown, often with a few scales ....(6)

6. Stems not ribbed, bearing separate tubercles; fruit dehiscent by a vertical slit .... ***Pediocactus***
6′ Stems strongly ribbed, the ribs coalescent with the bases of the tubercles; fruit indehiscent .... ***Sclerocactus***

Corynopuntia F. M. Knuth Club-cholla

Donald J. Pinkava

(*coryne* for club + *Opuntia*) Shrubs trailing, forming mats or clumps. **STEM** segments firmly attached (in ours) to easily dislodged, subequal in length, cylindric and usually clavate, curving upward from near bases, to subspheric, glabrous. **SPINES** with epidermis sheath deciduous only at apices, exposing yellow spine tips; at least 1 of the major spines angular-flattened to ribbonlike. **GLOCHIDS** commonly increasing greatly in number as areoles age. **FLOWERS** with ovary and floral tube bearing large tufts of white to tan wool. **FRUIT** narrowly obconic to ellipsoid, smooth, fleshy at first but soon drying, sometimes spiny. **SEEDS** pale yellow to brownish; discoid, the girdle smooth. Twelve species. [Genera *Marenopuntia* Backeb.; section *Clavatae* Engelm.; series *Clavatae* (Engelm.) K. Schum.]. North American deserts.

Corynopuntia clavata (Engelm.) F. M. Knuth (club, referring to shape of stem segments) Club-cholla. Shrubs mat-forming, wide-spreading, 5–15 cm tall. **STEM** segments short-clavate, strongly narrowed at base, 2.5–5 (7.5) cm long, 1.5–3 cm wide. **AREOLES** circular, 3 mm diameter; wool white to gray. **SPINES** 7–15, mostly in distal areoles; major 1–3 adaxial spines ascending, white to tan, angularly flattened to subterete; major 3–5 spines deflexed, white, flattened, the longest daggerlike, 12–35 mm long, at least 1.5 mm wide basally. **GLOCHIDS** yellow, in adaxial 1/4 of areole, ± 4 mm long. **FLOWERS** with inner tepals bright yellow, to 2.5 cm long; filaments pale yellow-green to white; anthers yellow; style and stigma lobes white. **FRUIT** yellow, barrel-shaped to ellipsoid, fleshy, 3–4.5 cm long, 1.2–2.5 cm diameter; spineless but densely yellow-glochidiate; areoles 35–55. **SEEDS** yellow-white, smooth, 4.5 mm long, 4 mm wide. $2n = 22$. [*Opuntia clavata* Engelm.]. Desert grassland communities. NMEX: McK, San, SJn. 1700–1750 m (5600–5700′). Flowering: Jun–Jul. Fruit: Jul–Aug. A northwestern New Mexico endemic. If following a morphologic, geographic, and DNA study by Griffith (2002), which relegates the name *Grusonia* to one species (in Mexico), this taxon becomes *Corynopuntia clavata* (Engelm.) F. M. Knuth.

Cylindropuntia (Engelm.) F. M. Knuth Cholla

Donald J. Pinkava

(*cylindrus* for cylinder + *Opuntia*; possibly a Greek town, Opus, where a cactuslike plant perhaps grew) Trees or shrubs, erect, rarely creeping, much-branched. **STEM** segments firmly attached to easily dislodged, of varied lengths, cylindric to slightly clavate, straight, glabrous. **SPINES** with whole epidermis sheath deciduous; major spines not or only basally angularly flattened. **FLOWERS** with inner tepals yellow-green, yellow to bronze, red to magenta, spatulate, emarginate-apiculate; outer tepals green with margins tinged color of inner tepals. **FRUIT** fleshy and green to yellow to scarlet, sometimes tinged red to purple, or dry and tan to brown, cylindric to subspherical, sometimes clavate, spineless or spiny. **SEEDS** pale yellow to tan to gray, flattened, angular to squarish or circular, often warped, each commonly bearing 1–4 large depressions due to pressures from adjacent developing seeds, the girdle smooth or as a low marginal ridge. [*Opuntia* subgenus *Cylindropuntia* Engelm.]. Thirty-two species: Great Basin, Chihuahuan, Mojave, and Sonoran deserts; United States, Mexico, and Caribbean.

1. Trees medium-sized to large, to 3 m tall, widely branching; inner tepals rose to magenta; spines of 1 kind, subequal or, if some longer, not arranged in a cross .... ***C. imbricata***

1′ Trees small, much-branched, grading into low, compact shrubs, to 1.3 m tall; inner tepals pale yellow to yellow; spines dimorphic, the smaller radial spines surrounding (1–) 4 larger central spines, these spreading into a cross .... ***C. whipplei***

Cylindropuntia imbricata (Haw.) F. M. Knuth (gutter-tiled) Tree cholla, coyonostyle. Trees, shrubs with short trunks, or large shrubs, (1) 1.5–2.5 (3) m tall. **STEM** segments gray-green, cylindric to weakly clavate, 8–25 cm long, 1.5–4 cm in diameter; tubercles very prominent, widely spaced, (1.5) 2–5 cm long. **AREOLES** elliptic, 5–8 mm long, 3–4 mm wide; wool yellow-tan, aging gray to black. **SPINES** usually at most areoles, silver to yellow to usually reddish to tan to brown, not obscuring the stem, stout, terete or sometimes flattened basally, spreading, straight or slightly curved, (0) 8–15 (30) per areole, 8–30 (40) mm long; sheaths silver to yellow to usually tan to dirty-white, yellow-tipped. **GLOCHIDS** pale yellow, in dense apical tuft, 0.5–3 mm long. **FLOWERS** with inner tepals dark pink to magenta to red-magenta, obovate, apiculate, 1.5–3.5 cm long; filaments green basally to pink to magenta apically; anthers yellow; style light green basally to pink to red-magenta apically; stigma lobes green to cream. **FRUIT** yellow, fleshy, obovoid, spineless, 2.4–4.5 cm long, 2–4 cm in diameter; tubercles subequal, prominent but occasionally smooth at maturity; umbilicus 7–14 mm deep; areoles 18–30. **SEEDS** 2.5–4 mm diameter, yellow-tan, subcircular to angled, warped, sides smooth to slightly lumpy, the girdle usually narrow, not protruding, 2.5–4 mm diameter. $2n = 22$. [*Cactus imbricatus* Lem.; *Opuntia imbricata* (Haw.) DC.]. Sandy to gravelly soils, grasslands, piñon-juniper communities. COLO: Arc; NMEX: McK, SJn. 1900–2200 m (6300–7300′). Flowering: Jun–Jul. Fruit: Jul–Aug. Arizona, Colorado, Kansas, New Mexico, Oklahoma, Texas; Chihuahua, Mexico. Our material belongs to **var. *imbricata*** [*Opuntia arborescens* Engelm.]. *Cylindropuntia imbricata* var. *imbricata* is the wide-ranging, aspect-dominant cholla of the Chihuahuan Desert and barely enters into the Four Corners region. It hybridizes with *C. whipplei* and occurs in scattered localities (perhaps it or its parents transported by Native Americans) in Arizona (Nav), New Mexico (SJn), and Colorado (Arc). This hybrid, ***C. ×viridiflora*** (Britton & Rose) F. M. Knuth, has a low, bushy habit, ripe cantaloupe–colored inner tepals, and an irregular nondimorphic spine pattern, $2n = 22$. Often used as an ornamental.

Cylindropuntia whipplei (Engelm. & J. M. Bigelow) F. M. Knuth (for A. W. Whipple, leader of railroad expedition, Mississippi River to Pacific Ocean) Whipple cholla. Trees or shrubs, low to upright, sparingly to very densely branched with whorled to subwhorled branchlets, 0.5–1.3 m tall. **STEM** segments green, 3–9 (15) cm long, 0.5–1.5 (2.2) cm in diameter; tubercles very prominent and short, 5–10 mm long. **AREOLES** 2–6 mm long, 1.5–4 mm wide; wool pale yellow to white, aging gray, oval to obdeltoid. **SPINES** in all but basalmost areoles, best developed toward apex, whitish or pale yellow, pink, pale red-brown, sometimes yellow-tipped, (1) 3–8 (10) per areole, interlacing, dimorphic, the sheaths whitish to pale yellow (rarely golden) throughout or yellow- to gold-tipped; central spines stout, subterete, usually 4 (–6) spreading into a cross, 2–3.4 (4.5) cm long; radial spines slender, flattened basally, deflexed, usually 4, 5–8 mm long; also 0–2 bristly spines. **GLOCHIDS** yellow, in apical tuft, 1–3 mm long. **FLOWERS** with inner tepals pale yellow, yellow to green-yellow, spatulate, apiculate, 1.5–2.5 (3) cm long; filaments yellowish to yellow-green; anthers yellow; styles white to yellowish; stigma lobes whitish, yellowish to pale green, rarely pink-tinged. **FRUIT** yellow to greenish yellow, pulpy-fleshy, broadly cylindric to subspheric, spineless, 18–30 (35) mm long, 15–22 (32) mm in diameter; tubercles subequal or uppermost longest, usually prominent; umbilicus 7–8 mm deep; areoles 36–62. **SEEDS** 3–3.5 mm long, 2.5–3.5 mm wide, pale yellow, subcircular to slightly angular in outline, nearly flat to warped, the sides smooth or with 1–3 depressions, the girdle smooth. $2n = 22, 44$. [*Opuntia whipplei* Engelm. & J. M. Bigelow]. Sandy, clay, gravelly soils to rocky slopes, sage, grasslands, piñon-juniper communities. ARIZ: Apa, Coc, Nav; COLO: Arc, Dol, LPl, Mon; NMEX: McK, RAr, San, SJn; UTAH. 1700–2300 m (5600–7500′). Flowering: May–Jul. Fruit: Jul–Aug. Arizona, Colorado, New Mexico, Utah. Hybridizes with *C. imbricata* as *C.* ×*viridiflora* (Britton & Rose) F. M. Knuth. See *C. imbricata*. This is the common cholla found in the Four Corners region. Used as an antidiarrheal by the Hopi. The Ramah Navajo used it to make cactus prayer sticks. The branches were made into a wand and used in the Red Antway.

Echinocereus Engelm. Hedgehog Cactus, Pitayitá

David J. Ferguson

(*echinos* for spiny + *cereus* for candle) Perennial stem succulents. **STEMS** single or caespitose, cylindric, ribbed, mostly under 10 cm in diameter and 40 cm tall; ribs raised, longitudinal. **AREOLES** round to elliptic, woolly when young. **SPINES** borne from areoles, never hooked. **FLOWERS** funnelform, from areoles over 1 year old, buds often rupturing stem below areole upon emergence; ovary with lanceolate foliaceous scales, bearing wool, spines, and hairlike wool from axils; perianth segments numerous, linear-lanceolate to cuneate, entire; outer segments typically less brightly pigmented and shorter than inner segments. **FRUIT** an ovoid to spherical berry, at least somewhat juicy, bearing slender spines that are deciduous at maturity; indehiscent or dehiscent by vertical slit(s). **SEEDS** many, black, about 1 mm long; testa cells convex, rugose, with pits at margins; micropyle included within hilum. Western United States and northern Mexico. Mostly in gravelly to rocky habitats in temperate to subtropical semiarid to arid regions. About 44 species.

1. Flowers small, usually less than 2.5 cm diameter, green, yellow, or brownish, highly fragrant; fruit usually under 1.5 cm; stems usually single and small, exceptionally to 7.5 × 15 cm; ribs usually 10 or more; radial spines usually 12 or more, slender, pectinate, usually not reaching adjacent ribs .... ***E. viridiflorus***
1′ Flowers large, usually greater than 2.5 cm diameter, orange, red, pink, or purple, not highly fragrant; fruit usually over 1.5 cm; stems mostly larger; ribs usually 13 or fewer; radial spines usually 12 or fewer, mostly stout, not pectinate, usually reaching adjacent ribs .... (2)

2. Flowers usually over 6 cm wide, pink to purple, closing at night; tepals pointed; stems mostly not branching, or with only a few branches; spines usually pale, with a dark stripe down length of outward-facing side .... (3)
2′ Flowers usually under 5 cm, orange to red, open day and night; tepals rounded apically; stems mostly freely branching, forming dense mounds; spines not colored as above .... (4)

3. Stems mostly single, usually conical apically; ribs 7–10; radial spines usually fewer than 10; central spine 0–1, porrect, terete, sometimes a second (smaller, below main porrect central) .... ***E. fendleri***
3′ Stems often clustering, usually rounded apically; ribs 9–13; radial spines usually more than 9; central spines mostly 3 or more, main one spreading downward and subulate to flattened .... ***E. engelmannii***

4. Stems mostly with 10 or fewer ribs; spines usually angular in cross section; central spines usually 0–2 .... ***E. triglochidiatus***
4′ Stems mostly with 9 or more ribs; spines not angular in cross section, terete or lower centrals subulate; central spines usually more than 2 (rarely 1–2) .... ***E. coccineus***

Echinocereus coccineus Engelm. (deep red) Beehive claretcup cactus. **STEMS** cylindrical, rounded apically, 5–10 cm in diameter; ribs (8) 9–12 (14); branching from base, caespitose, forming large, mostly hemispheric clusters of stems to 1 m across and 45 cm tall. **AREOLES** mostly 6–12 mm apart on ribs. **CENTRAL SPINES** (0) 3– 4 (6) per areole, 1–5.5 cm long, to 0.8 mm basal diameter (above swelling), terete or longest flattened-subulate, bulbous basally; of similar colors as the radial spines, but often darker than radials. **RADIAL SPINES** 8–12 per areole, 5–30 mm long, to about 0.5 mm diameter (above swelling), upper shortest, lower longest; terete, bulbous basally; white to yellow, reddish, brown, or nearly black, often varicolored and/or different colors in same areole. **FLOWERS** salverform, 2.5–4 cm diameter; functionally dioecious or rarely perfect; flower size often reduced, with anthers reduced and nonfunctional in pistillate plants; stigmas somewhat reduced and ovules abortive in staminate plants; tepals entire, mostly cuneate, obtuse to rounded apically, of firm texture; red, sometimes orange, rarely pink or white, never purplish, often paler basally; stigma lobes 6–12, green, thick, linear, erect to slightly spreading. **FRUIT** subglobose, juicy, 2–4 cm, usually brownish or reddish, sometimes green; bearing clusters of slender spines deciduous at maturity; mostly indehiscent but sometimes dehiscent by vertical slit(s). $2n = 44$. [*Cereus coccineus* Salm-Dyck ex DC. var. *cylindricus* Engelm.; *Cereus coccineus* var. *melanacanthus* Engelm.; *Cereus conoideus* Engelm. & J. M. Bigelow; *Echinocereus phoeniceus* (Engelm.) Rümpler; *Mammillaria aggregata* Engelm. ex S. Watson; *Echinocereus triglochidiatus* Engelm. var. *melanacanthus* (Engelm.) L. D. Benson]. Exposed, sunny, rocky slopes and ledges in blackbrush, sagebrush, salt desert scrub, and piñon-juniper communities. ARIZ: Apa; COLO: Mon; NMEX: McK, RAr, San. 1830–2135 m (6000–7000′). Flowering: late Apr–Jun. Fruit: Jun–Jul. Four Corners material is **var. *coccineus***. New Mexico to Colorado and Arizona. Distribution peripheral in our area. This taxon is confused in the literature with *E. triglochidiatus* var. *mojavensis*, but distribution within the study area is allopatric; variety *coccineus* has the stem apex usually more broadly rounded; ribs average more numerous; spines average more numerous, terete to subulate and rarely angular in cross section; dioecious; tetraploid. This species occurs sympatrically with *E. triglochidiatus* var. *triglochidiatus*, which is easily distinguished by fewer ribs (5–7) and fewer, much stouter angular spines on more widely separated areoles. Flowers are not quantifiably distinguishable from those of *E. triglochidiatus*, but tend to open less widely, be darker in color, and if formed normally should always have both good pollen and ovules upon dissection. *E. coccineus* usually flowers slightly earlier than *E. triglochidiatus* in any given location.

Echinocereus engelmannii (Parry ex Engelm.) Lem. **CP** (for George Engelmann, 19th-century physician and botanist) Engelmann hedgehog cactus, variable-spine hedgehog cactus, strawberry cactus. **STEMS** cylindrical, mostly rounded apically, 4–7.5 cm in diameter, to 40 cm tall; ribs 9–13; mostly single, sometimes branching from base to form clusters of up to 6 or more unequal stems. **AREOLES** about 1 cm apart on ribs. **SPINES** often 2-toned and then darker apically and/or with darker color as stripe along length of spine on side facing away from plant body; light-colored ones usually white, yellow, pale brown, or gray; dark-colored ones highly variable, yellowish, reddish, purplish, brown, gray, or black; upper centrals usually darkest. **CENTRAL SPINES** (2) 4 (6) per areole, 1–7 cm long, to 1.5 mm basal diameter (above swelling); upper centrals subulate to nearly terete, usually shorter, porrect or spreading; lower main central(s) flattened and deflexed, longer and wider than uppers, often curved. **RADIAL SPINES** 8–13 per areole, up to 0.5 mm diameter, 1–2 cm long, upper shortest, lower longest; nearly terete, sometimes slightly angular in cross section or lowest somewhat flattened, often curved, bulbous basally. **FLOWERS** perfect, large, 6–10 cm diameter; tepals with gaps between bases, entire, mostly narrowly cuneate, acute apically, of delicate texture; pink to magenta, always purplish, usually darker toward base and often green basally; stigma lobes about 10, green, thick, linear, ascending to spreading. **FRUIT** broadly elliptic to subglobose, white-pulpy, becoming clear-juicy in age, 2.5–4 cm long, usually red, sometimes brownish; bearing clusters of slender spines deciduous at maturity; dehiscent by vertical slit(s). $2n = 44$. Mostly rocky slopes and ledges in salt desert scrub, blackbrush, or piñon-juniper communities. ARIZ: Apa, Nav; UTAH. 1125–1980 m (3700–6500′). Flowering: late Apr–early Jun. Fruit: May–Jul. Utah and Arizona. Four Corners material belongs to **var. *variegatus*** (Engelm. & J. M. Bigelow) Rümpler (varied spine coloration). Closely related to *E. fendleri*, but separable by spination, less tapered stem, often more ribs, larger size, and polyploidy. The two are mostly allopatric, with *E. fendleri* occurring in cooler environments, but occasionally they are sympatric. Occasional confusing but apparently sterile hybrids have been found further south in Arizona, New Mexico, and Sonora. Such hybrids might be expected in our area but have not been seen. Distinction from *E. engelmannii* var. *chrysocentrus* (Engelm. & J. M. Bigelow) Rümpler, which occurs in the Mojave Desert and Great Basin regions, is dubious. The name *E. engelmannii* var. *variegatus* has been maintained until the range of variability seen throughout the Colorado Plateau, Great Basin, and Mojave deserts can be studied more thoroughly. Plants in our area with fewer than four central spines resemble *E. engelmannii* var. *rectispinus* (Engelm. ex S. Watson) Blum, Lange & Rutow; however, in our area such plants occur as individuals among normal populations, and that variety is not recognized as occurring in our area. Used in xeric landscaping.

Echinocereus fendleri (Engelm.) Sencke ex J. N. Haage (for Augustus Fendler, 19th-century plant collector) Fendler hedgehog cactus, strawberry cactus. **STEMS** cylindrical, mostly narrowing apically, 4–7.5 cm in diameter, to 25 cm tall; ribs (5) 7–10 (12); mostly single, sometimes branching from base to form clusters of up to 6 or more unequal stems. **AREOLES** mostly approximately 1 cm apart on ribs. **SPINES** often 2-toned and then darker apically and/or with darker color as stripe along length of spine on side facing away from plant body; lighter colored usually white, yellowish, pale brown, or gray; darker colored usually brown to near black, sometimes yellowish or purplish; main centrals usually darkest. **CENTRAL SPINES** 0–1 (2) per areole, 1–60 mm long, to 1 mm basal diameter (above swelling); main one nearly terete, long and porrect, usually curving upward, usually dark; second one, if present, usually much smaller, pale in color, often flattened. **RADIAL SPINES** 5–10 per areole, up to 0.5 mm diameter, 5–20 mm long, upper shortest, lower longest; nearly terete, sometimes slightly angular in cross section or lowest somewhat flattened, often curved, bulbous basally. **FLOWERS** perfect, large, 5–10 cm diameter; tepals with gaps between bases, entire, mostly narrowly cuneate, acute apically, of delicate texture; pink to magenta, always purplish, usually darker toward base and often green basally; stigma lobes mostly 10, green, thick, linear, ascending to spreading. **FRUIT** broadly elliptic to subglobose, white-pulpy, becoming clear-juicy in age, 2.5–4 cm long, usually red, sometimes greenish or brownish; bearing clusters of slender spines that are deciduous at maturity; dehiscent by vertical slit(s). $2n = 22$. [*Cereus fendleri* Engelm. var. *pauperculus* Engelm.]. Mostly gravelly slopes and hilltops, desert in grassland or open piñon-juniper communities. ARIZ: Apa; COLO: Arc, LPl, Mon; NMEX: McK, RAr, SJn. 1645–2285 m (5400–7500′). Flowering: late Apr–early Jun. Fruit: Jun–Jul. Colorado and Utah to Texas and northern Mexico. Four Corners material belongs to **var. *fendleri***. Used in xeric landscaping.

Echinocereus triglochidiatus Engelm. (three-glochidiate, bearing barbs) Claretcup cactus. **STEMS** cylindrical, mostly narrowing apically; branching from base, caespitose, forming irregular to hemispheric clusters of stems. **SPINES** mostly angular in cross section, grayish or brownish, sometimes near white, yellowish, reddish, or near black, especially when young. **FLOWERS** perfect, 2.5–5 cm diameter; tepals entire, mostly cuneate, obtuse to rounded apically, of firm texture; red, never purplish, sometimes paler basally; stigma lobes 6–12, green, thick, linear, erect to slightly spreading. **FRUIT** subglobose, juicy, 2–5 cm, usually brownish or reddish, sometimes green; bearing clusters of slender spines deciduous at maturity; mostly indehiscent but sometimes dehiscent by vertical slit(s). $2n = 22$.

1. Stems with 7 or more (most often 9) ribs; areoles more closely spaced and spines often obscuring green of stem; spines more slender, usually under 1 mm thick, less angular in cross section, and usually nearly terete; radials usually more than 8 .... **var. *mojavensis***

1′ Stems with 5 to 7 ribs; areoles widely spaced so that green of stem is not obscured; spines stout and thick, often over 1.5 mm thick, prominently angular in cross section; radial spines usually 8 or fewer ....**var. *triglochidiatus***

var. *mojavensis* (Engelm. & J. M. Bigelow) L. D. Benson (of the Mojave Desert or River) Mojave claretcup cactus. **STEMS** 2.5–7.5 cm in diameter; ribs 7–12 (usually 9); clusters to 60 cm and 30 cm tall. **AREOLES** mostly 6–12 mm apart on ribs. **CENTRAL SPINES** 0–2 (4) per areole, 1–5 cm long, to 1 mm basal diameter (above swelling), nearly terete but usually angular in cross section, often curved and twisted, bulbous basally; of similar colors as the radial spines. **RADIAL SPINES** 5–12 per areole, 5–30 mm long, to about 0.5 mm diameter (above swelling), lowermost one often shortest, and uppermost one often longest; nearly terete but usually angular in cross section, often some curved, bulbous basally. [*Cereus mojavensis* Engelm. & J. M. Bigelow var. *zuniensis* Engelm.; *Cereus bigelovii* Engelm.; *Echinocereus sandersii* Orcutt; *Echinocereus krausei* de Smet]. Blackbrush, ephedra, sagebrush, and piñon-juniper communities. ARIZ: Apa, Coc, Nav; COLO: Mon, LPl; NMEX: San, SJn; UTAH. 1125–2440 m (3700–8000′). Flowering: Apr–May. Fruit: Jun–Jul. New Mexico, Wyoming, California, and Baja California.

var. *triglochidiatus* CP Thick-spine claretcup cactus. **STEMS** 5–12 cm in diameter; ribs 5–7 (8); clusters of stems to 90 cm across and 60 cm tall. **AREOLES** mostly 1.5 cm or more apart on ribs. **CENTRAL SPINES** 0–1 (2) per areole, 1–5 cm long, to 1.8 mm basal diameter (above swelling), angular in cross section, often curved and twisted, bulbous basally; of similar colors as the radial spines. **RADIAL SPINES** (0) 5–12 per areole, 5–30 mm long, to 1.5 mm diameter (above swelling), lowermost one often shortest, and uppermost one often longest; angular in cross section, often curved and twisted, bulbous basally; mostly gray to light brownish, but variable and sometimes white to yellow, reddish, brown, or nearly black. [*Echinocereus triglochidiatus* var. *gonacanthus* (Engelm. & J. M. Bigelow) Boissev.]. Sagebrush, piñon-juniper, and mountain brush communities. ARIZ: Apa; COLO: Arc, LPl, Mon, SMg; NMEX: McK, RAr, San, SJn. 1525–2285 m (5000–7500′). Flowering: May. Fruit: Jun–Jul. New Mexico to Colorado and Arizona. This variety blends with var. *mojavensis* to the northwest, and plants of the Four Corners region may be intermediate in character. Used in xeric landscaping.

Echinocereus viridiflorus Engelm. **CP** (green flower) Green-flowered hedgehog cactus, hen-and-chicks cactus. **STEMS** cylindrical, mostly narrowing apically, 1.5–7.5 cm in diameter, to 15 cm tall; ribs 8–12 (20); mostly single, sometimes branching from base and forming small clusters of stems. **AREOLES** small, elliptic, mostly 5 mm apart on ribs. **CENTRAL SPINES** 0–1 (4) per areole, arranged vertically, 1–15 mm long, mostly 0.6 mm basal diameter (above swelling), terete, bulbous basally, at first porrect and often curved upward slightly, becoming more deflexed in age; of similar colors as the radial spines. **RADIAL SPINES** 8–12 (20) per areole, about 4 mm basal diameter, (above swelling), 3–10 mm long, pectinate, tending to obscure plant body from view; lowermost and uppermost often shortest; terete, bulbous basally; mostly white to purplish or reddish, but sometimes brown or yellow, often tipped dark, plant typically bearing spines in alternating bands of two different colors. **FLOWERS** strongly sweetly fragrant; perfect, rotate, 2–2.5 cm diameter; tepals entire, mostly linear-lanceolate, rounded to more often acute, of delicate texture; pale yellow to yellow-green or sometimes light brown; 5–10 green to brownish, capitate to slightly spreading stigma lobes. **FRUIT** ovoid, sticky and a little juicy, 1–1.5 cm, green to brownish; bearing clusters of slender white spines deciduous at maturity; dehiscent by vertical slits; often drying on plant. $2n = 22$. [*Echinocereus viridiflorus* var. *minor* Engelm.]. Sunny, gravelly to rocky slopes, hilltops, and ledges, often in grassland, shrubland, or open woodland. NMEX: RAr, San. 1830–2440 m (6000–8000′). Flowering: late March–early June. Fruit: May–July. South Dakota to Texas and New Mexico. Four Corners material is **var. *viridiflorus***.

Escobaria Britton & Rose Pincushion Cactus, Nipple Cactus

David J. Ferguson

(for Rómulo and Numa Escobar, Mexico) Perennial stem succulent. **STEMS** single or caespitose, globose, conic, or cylindric, tuberculate, mostly under 10 cm in diameter; tubercles conic to cylindric. **AREOLES** elongate, forming groove on upper side of tubercles on mature plants, on juveniles may be restricted to apex of tubercles at top of stem. **SPINES** borne at apex of tubercles, never hooked. **FLOWERS** from base of groove between tubercles; ovary naked or bearing a few minute membranous scales; perianth segments numerous, linear-lanceolate, outer perianth segments fimbriate, typically less brightly pigmented and shorter than inner segments. **FRUIT** an indehiscent juicy berry; clavate to fusiform, naked or with few minute scales. **SEEDS** many, brown to black, approximately 1–2 mm long; testa cells bordered by fine raised line and centrally pitted; hilum excluded from micropyle. Western United States, northern Mexico, and Cuba. Mostly in gravelly to rocky habitats in temperate to subtropical semiarid to arid regions. About 22 species.

1\. Flowers usually under 2.5 cm diameter, translucent pale green, yellow, brownish, or rarely pinkish; stigma lobes green. Fruit usually not exceeding spines when mature, bright red (occasionally orange); seeds globose, black .... ***E. missouriensis***

1′ Flowers usually over 2.5 cm diameter, pink to magenta, rarely white; stigma lobes white to pink. Fruit usually protruding beyond spines when mature, greenish to brownish; seeds reniform, brown .... ***E. vivipara***

Escobaria missouriensis (Sweet) D. R. Hunt **CP** (of the Missouri region) Missouri pincushion cactus. **STEMS** hemispheric to flattened above, up to 10 cm in diameter, usually less; mostly solitary, sometimes caespitose, forming flat to hemispheric clusters of stems to 15 cm. **AREOLES** densely white-woolly. **SPINES** translucent, usually some pubescent, white to pale yellowish, becoming gray in age. **CENTRAL SPINES** 0–4 per areole, 1–2 cm long. **RADIAL SPINES** 10–20 (25) per areole, 1–1.5 cm long, about 0.2 mm diameter. **FLOWERS** 1.5–2.5 cm diameter, usually pale translucent yellowish, sometimes greenish, brownish, or pinkish, with 3–6 spreading, linear, green stigma lobes. **FRUIT** bright red (occasionally orange), mostly under 1.5 cm long. $2n = 22$. [*Mammillaria nuttallii* Engelm.; *M. notesteinii* Britton; *Coryphantha marstonii* Clover; *Escobaria missouriensis* (Sweet) D. R. Hunt var. *marstonii* (Clover) D. R. Hunt]. Mostly exposed gravelly to rocky slopes, ledges, and hilltops, most often in open piñon-juniper woodland communities. COLO: Mon. 1830–2285 m (6000–7500′). Flowering: Apr–Jun. Fruit: May–Sep. Minnesota to Idaho, Arizona, New Mexico, and Nebraska. Other varieties from southern Nebraska and Missouri to Nuevo León, Mexico.

Escobaria vivipara (Nutt.) Buxb. **CP** (germinating or sprouting while still attached to parent) Plains pincushion cactus, nipple cactus, spiny stars, ball cactus, cushion cactus. **STEMS** hemispheric to conic, to 10 cm in diameter, usually less; usually solitary, sometimes caespitose, forming large clusters of stems to 30 cm, usually less. **CENTRAL SPINES** (3) 5–7 (10) per areole on mature plants, 1–2 cm long; 0–3 and much shorter on immature plants; mostly darker than radials. **RADIAL SPINES** (10) 15–20 (25) per areole, 1–1.5 cm long, about 0.2 mm diameter; white or pale pinkish to pale brown (rarely pale yellow), often darker brown apically. **FLOWERS** 2.5–5 cm diameter, magenta, sometimes partly or entirely pink or rarely white; with 5–10 spreading linear stigma lobes white to occasionally pale pink, rarely darker. **FRUIT** 1.5–3 cm, usually green, varying to brown. $2n = 22$. [*Coryphantha vivipara* (Nutt.) Britton & Rose var. *kaibabensis* P. C. Fisch.]. Mostly exposed gravelly to rocky slopes and hilltops, but widespread in other well-drained sunny habitats. ARIZ: Apa, Nav; COLO: Arc, Dol, LPl, Mon, SMg; NMEX: McK, RAr, San, SJn; UTAH. 1125–2440 m (3700–8000′). Flowering: Apr–Jun. Fruit: Jul–Oct. Colorado to New Mexico, Arizona, and Utah. Other varieties from Manitoba to Alberta, Texas, northern Mexico, and California. All material examined from the Four Corners region is referable to **var. *arizonica*** (Engelm.) D. R. Hunt (of Arizona), Arizona pincushion cactus, Navajo pincushion cactus. The name *E. vivipara* var. *borealis* Engelm., often misapplied to San Juan drainage material, was described from North Dakota and northeast New Mexico and is synonymous with *E. vivipara* var. *vivipara*. This latter name is also often misapplied to our plants; it differs in central spines not reduced in juveniles, and stigmas pink to magenta. This taxon is from east of the Rockies and is not found in the San Juan River drainage.

Opuntia Mill. Pricklypear
Donald J. Pinkava

(derivation uncertain, presumably based on Opus, a Greek town where a cactuslike plant perhaps grew) Trees or shrubs, erect, decumbent to trailing, or forming clumps much-branched, these segmented. **STEM** segments strongly flattened or, in a few species, subcylindric to subspheric, of varied lengths and widths. **AREOLES** bearing glochids and a dense wool. **SPINES** with epidermis intact, not sheathing; major spines flattened to cylindric. **FLOWERS** with inner tepals pale yellow to orange, pink to red or magenta, rarely white, or with bases of a different color, oblong to spatulate, emarginate-apiculate; outer tepals green to yellow with margins tinged color of inner tepals. **FRUIT** fleshy to juicy (bleeding), green, yellow, or red to purple, usually spineless; or dry, tan to gray, clavate to cylindric to subspheric, and usually spiny. **SEEDS** glabrous, pale yellow to tan to gray; 3–10 mm long, generally circular to reniform, flattened (discoid) to subspheric, angular to squarish, sometimes warped, commonly bearing 1–4 large, shallow depressions due to pressures from adjacent developing seeds; the girdle smooth to protruding, ridged to strongly winged. About 180 species. Canada to southern Argentina, West Indies, Galápagos Islands. Many introduced to the Old World; many cultivated (Parfitt 1980, 1998). The Navajo used the plant as a dermatological aid for boils, and *Opuntia* is also used in the design of sand paintings of the Cactus People for the Wind Chant. The Hopi remove the spines, boil the pads, and eat them. The Isleta and Jemez eat the fruit.

When collecting herbarium specimens, be sure to obtain at least two or three consecutive stem segments plus fresh flowers and/or mature fruits, with detailed descriptions as to color, shape, and size. Characters of spines, unless otherwise stated, are based on those in well-developed areoles, usually in the distal portions of stem segments. Opuntioideae (also including *Cylindropuntia* and *Corynopuntia*) is taxonomically difficult at the genus and species levels. Evolution, in ours, has been complicated by interspecific hybridization, facultative apomixis, vegetative propagation, and polyploidy, all promoted by the perennial habit. Often hybrids set seeds and have a rather high percentage of pollen stainability (estimated viability). Many hybrids have been named as species and varieties.

1. Fruits fleshy or juicy ("bleeding") when ripe, dull red or purple, spineless; inner tepals yellow with red bases ...(2)
1′ Fruits dry at maturity, sometimes tardily so, tan to gray, usually bearing spines and an apical flange; inner tepals yellow to magenta throughout, rarely with reddish bases ....(3)

2. Major spines 1–4 in upper (particularly marginal) areoles, terete or 1 flattened, erect and/or reflexed; stem segments 5–11 cm long, 3.5–7.5 cm broad, flabby, cross-wrinkled; stigma lobes white to yellow-cream (rarely tinted pale green) .... ***O. macrorhiza***
2′ Major spines (0) 2–8 in areoles of upper 3/4 of stem segment, most flattened, divergent or reflexed, the lower spiny areoles usually with 1 (–3) deflexed gray to tan spines; stem segments 10–25 cm long, 7–20 cm broad, firm; stigma lobes green .... ***O. phaeacantha***

3. Depressed spines 0–3 at lower edge of areole, 1–3 mm long; longest spine per areole 0.8–3.5 cm; stem segments 1.5–5.5 cm long, 1.5–3 cm broad, subspheric, cylindric to flattened, elliptic-obovate; areoles 3–5 in a diagonal row across midstem segment .... ***O. fragilis***
3′ Depressed spines 5–11 at lower edge of areole, 4–20 mm long; longest spine per areole 2–15 cm long; stem segments firmly attached, 6–27 cm long, 5–18 cm broad, flattened, obovate to circular (or, if elliptic, then stem segments longer than 9 cm); areoles 4–14 in a diagonal row across midstem segment .... ***O. polyacantha***

Opuntia fragilis (Nutt.) Haw. (easily detached) Brittle pricklypear, little pricklypear. Shrubs low, mat-forming, 2–10 cm tall. **STEM** segments subspheric to cylindric to flattened, dark green, glabrous, the terminal ones easily detached, elliptic-obovate, (1.5) 2–5.5 cm long, (1) 1.5–3 cm broad; tubercles low (but becoming pronounced when dried). **AREOLES** 3–5 in a diagonal row across midstem segments, oval, 3 mm long, 2.5 mm wide; wool white. **SPINES** in most areoles, gray with brown tips, terete, straight, spreading, 3–8 per areole, the longest to 3.5 cm long. **GLOCHIDS** inconspicuous, tan to brown, in apical (adaxial) crescent in areole. **FLOWERS** with inner tepals yellow, sometimes basally red, 2–2.6 cm long; filaments white or red; style white; stigma lobes green. **FRUIT** tan, dry, 1–3 cm long, 0.8–1.5 cm in diameter; areoles 19–21, the uppermost ones bearing 1–6 short spines. **SEEDS** 5–6 mm in diameter; girdle protruding 1–1.5 mm. 2*n* = 66. [*Cactus fragilis* Nutt.; *Opuntia brachyarthra* Engelm. & J. M. Bigelow]. Sandy or gravelly soils or on outcrops, barren areas in grasslands, piñon-juniper, and ponderosa pine communities. ARIZ: Apa, Nav; COLO: Arc, LPl, Dol, Mon, SMg; NMEX: McK, RAr, SJn; UTAH. 1850–2400 m (6000–7800′). Flowering: May–Jun. Fruit: Jun–Jul. British Columbia to Ontario, Canada; south to Arizona and New Mexico.

Opuntia macrorhiza Engelm. **CP** (large root) Western pricklypear, plains pricklypear. Plants forming clumps, to 35 cm tall, sometimes from tuberlike rootstocks. **STEM** segments flattened, dark green, glabrous, usually glaucous, fleshy to flabby, often cross-wrinkled with stress, obovate to subcircular, 5–11 cm long, 3.5–7.5 cm broad. **AREOLES** 5–7 (8) in diagonal row across midstem segment, oval to subcircular, 2–4 mm in diameter; wool tan. **SPINES** mostly in pads' upper areoles, white to red-brown, the major ones straight, rather stout (more or less 0.5 mm in diameter near base), erect and/or deflexed, (0) 1–4 per areole, the longest 2–6 cm long. **GLOCHIDS** pale yellow, tan to red-brown, aging brown, forming a well-developed, dense apical tuft, to 6 mm long. **FLOWERS** with inner tepals yellow with reddish bases, 2.5–4 cm long; filaments pale yellow; style white; stigma lobes yellow. **FRUIT** green, yellowish, to dull red, elongate-obovoid, long-stipitate, spineless, fleshy, 2.5–4 long, 1.5–2.8 cm diameter; areoles 12–26. **SEEDS** tan, subcircular, 4–5 mm in diameter, thickish, warped; girdle broad, protruding to 0.5 mm. 2*n* = 44. [*Opuntia mesacantha* Raf. var. *macrorhiza* (Engelm.) J. M. Coult.; *O. compressa* J. F. Macbr. var. *macrorhiza* (Engelm.) L. D. Benson]. Clay, sandy to rocky soils, slopes and mesas, chaparral, grassy woodlands, and coniferous forests. ARIZ: Apa, Coc, Nav; COLO: Arc, Dol, LPl, Mon; NMEX: McK, RAr, San, SJn; UTAH. 1970–2800 m (6000–8500′). Flowering: May–Jun. Fruit: Jun–Aug. Utah to Oklahoma, Missouri, south to Arizona, Texas; Chihuahua and Sonora, Mexico.

Opuntia phaeacantha Engelm. **CP** (dark spine) Brown-spined pricklypear. Shrubs decumbent to commonly trailing, 0.3–1 m tall. **STEM** segments flattened, green to dark green, glabrous, obovate to circular, 10–25 cm long, 7–20 cm broad. **AREOLES** 5–7 in a diagonal row across midstem segments, obovate to elliptic to circular, 10–25 mm long, 7–20 mm broad; wool tan to brown, aging grayish. **SPINES** at most areoles to only distal 1/4 of pad surface, to essentially absent, brown to white, (0–) 2–8 per areole; major central spines straight or curved, brown to red-brown (to blackish), partly to wholly chalky white, subulate, usually flattened at base, 3.5–7 cm long; basal spines usually 1 (–3), to 2 cm long, shorter ± white, deflexed, flattened. **GLOCHIDS** tan to red-brown, dense in an apical crescent and a subapical tuft, to 8 mm long. **FLOWERS** with inner tepals yellow with basal portions red, rarely entirely pink to red, 3–4 cm long; filaments greenish below, pale yellow to white above; style white; stigma lobes green to yellow-green. **FRUIT** wine-red to purple, with ± greenish flesh (± juicy in age), obovate to barrel-shaped, not long-stipitate, spineless, 3–5 cm long, 2–3 cm in diameter; areoles 15–40. **SEEDS** tan, subcircular, 4–5 mm in diameter, notched, warped; girdle protruding about 1 mm. 2*n* = 66. [*Opuntia angustata* Engelm. & J. M. Bigelow; *O. phaeacantha* var. *brunnea* Engelm. and var. *camanchica* (Engelm. & J. M. Bigelow) L. D. Benson]. Sandy to rocky soils, sagebrush and piñon-juniper communities. ARIZ: Apa, Coc, Nav; COLO: Arc, Dol, LPl, Mon; NMEX: McK, RAr, San, SJn; UTAH. 1215–2295 m (3700–7000′). Flowering: May–Jul. Fruit: Jul–Aug. Utah to Colorado, south to California, Texas; Chihuahua, Mexico. One population (ARIZ: Nav., e.g., *Heil & Clifford s.n.* [ASU, SJNM]) approaches *O.* ×*curvispina* Griffiths (*O. chloorotica* × *O. phaeacantha*), with narrow, large-seeded fruit, but more study is needed. *Opuntia phaeacantha* may represent more than one species.

Opuntia polyacantha Haw. (many spines) Shrubs low, with more or less prostrate branches, 5–50 cm tall. **STEM** segments flattened, green, glabrous, the terminal ones not easily detached, elliptic to narrowly to broadly obovate to circular, 6–27 cm long, 5–18 cm broad. **AREOLES** 4–14 in a diagonal row across midstem segment, subcircular, 3–6 mm in diameter; wool tan to brown. **SPINES** at all to only distal areoles of stem segment, either all as 1 kind of basic spine: yellow to dark-brown to black turning gray, pink-gray to gray-brown, (0) 1–18 per areole, the largest (3.5) 4–9 (18.5) cm long and spreading and curling in all directions to more or less straight, erect, ascending to deflexed; or of ± 2 kinds: major spines yellow-brown to brown to gray, reflexed to porrect, 1–5, the longest 2–15 cm long; minor spines white to white-gray, deflexed, 5–11, the largest 4–16 mm long. **GLOCHIDS** yellow, aging brown, in crescent at areole apex,

1–5 mm long. **FLOWERS** with inner tepals yellow to magenta throughout, 2.5–4 cm long; filaments white, yellow to magenta; style white to pale pink; stigma lobes green. **FRUIT** tan to brown, dry at maturity, 1.5–4.5 cm long, 1.2–2.5 cm in diameter; areoles 10–33, each or only the uppermost bearing 3–16 spines, 4–20 mm long, sometimes burlike. **SEEDS** tan to gray, oblong to subcircular, 3–7 mm long, 2–4 mm wide; girdle protruding 1–2 mm.

1. Spines of 1 kind, grading in length, from ascending to deflexed, commonly dark brown, 7–18 per areole, the longest spines porrect to ascending, shorter ones similar, 5–8 cm long, deflexed to porrect ....**var. *hystricina***
1' Spines of 2 kinds: major spines 1–3 per areole, 2–4 cm long, mostly deflexed, yellow-brown to gray, sometimes ascending at stem segment apex; minor spines 1–5 per areole, deflexed, white, subtending major spines ....**var. *polyacantha***

var. *hystricina* (Engelm. & J. M. Bigelow) B. D. Parfitt **CP** (porcupine) Porcupine pricklypear. **STEM** segments obovate, 8–10 cm long, 5–8 cm broad. **AREOLES** about 10–17 mm apart; (6) 8–10 in a diagonal row across midstem segments. **SPINES** in most areoles, grading in size and orientation, the longest (yellow-gray to) brown to black, slightly descending, porrect, ascending near apex, (4) 5–8 cm long, (1) 2–6 per areole, the smaller spines gray-white, reflexed, 4–6 per areole. **FRUIT** stout; areoles 11–21, most bearing 4–8 spines, 4–18 mm long. 2*n* = 44, 66. [*Opuntia hystricina* Engelm. & J. M. Bigelow; *O. rhodantha* K. Schum.; *O. xanthostemma* K. Schum.]. Clay, sandy or gravelly soils, desert scrub, grasslands to piñon-juniper communities. ARIZ: Apa, Nav: COLO: Mon, LPl; NMEX: RAr, San, SJn; UTAH. 1215–2560 m (3700–7800′). Flowering: May–Jun. Fruit: Jun–Jul. Nevada, eastern California to Colorado, south to Arizona, New Mexico.

var. *polyacantha* Starvation pricklypear. **STEM** segments broadly obovate to circular, (4–) 8.5–12 cm long, (4) 5.5–11 cm broad. **AREOLES** 6–13 mm apart; 6–11 in a diagonal row across midstem segments. **SPINES** in most areoles, of 2 kinds: major spines yellow-brown to gray, deflexed to reflexed or porrect at pad apex, 1–3 (5) per areole, 2–3.5 (4) cm long; minor spines white, deflexed, 0–5 per areole, 5–10 (16) mm long. **FRUIT** stout; areoles 12–28, most bearing (4) 6–15 spines, 5–10 mm long. 2*n* = 22, 44. [*Opuntia heacockiae* Arp; *O. juniperina* Britton & Rose; *O. missouriensis* DC. var. *microsperma* Engelm.; *O. polyacantha* var. *rufispina* (Engelm. & J. M. Bigelow) L. D. Benson; *O. polyacantha* var. *trichophora* (Engelm. & J. M. Bigelow) J. M. Coult.]. Clay, sandy or gravelly soils, grasslands, sage, piñon-juniper communities. ARIZ: Apa, Coc, Nav; COLO: Arc, Dol, LPl, Mon, SMg; NMEX: McK, RAr, San, SJn; UTAH. 1215–2625 (2985) m [3700–8000′ (9100′)]. Flowering: May–Jul. Fruit: Jul–Aug. Western Canada; Idaho south to Arizona and Texas; Chihuahua, Mexico.

Pediocactus Britton & Rose Pincushion Cactus

Kenneth D. Heil and J. M. Porter

(*pedio* for plains, referring to the Great Plains, habitat of *P. simpsonii*) Plants mostly single or occasionally clumped. **STEMS** globular to broadly obovoid to globose, 1–7.5 cm long, 1–5.5 cm diameter. **SPINES** differentiated into 1 or 2 types. **CENTRAL SPINES** 0–8, 7.5–13.5 cm long, 0.3–0.7 mm wide, red to reddish brown. **RADIAL SPINES** 20–28, 1–7 mm long. **FLOWER** petaloids pink to magenta, white, or yellow, 1–4 cm long, 2–5 cm wide. **FRUIT** green to tan, 4–9 mm long, 3–7 mm broad. **SEEDS** gray to black, 1.5 mm long, 1–2 mm wide (Heil, Armstrong, and Schleser 1981). Eight species limited to North America.

1. Central spines lacking; stems up to 3.8 cm high, mostly less than 2.5 cm wide; rare, San Juan County, New Mexico ....***P. knowltonii***
1' Central spines of flowering individuals 5–8; stems 2.5–7.5 cm long, 2.5–5.5 cm wide; widespread in the mountain foothills ....***P. simpsonii***

Pediocactus knowltonii L. D. Benson **CP** (honoring Fred Knowlton, discoverer) Knowlton cactus. **STEMS** solitary or rarely clustered, barely protruding above ground level, up to 3.8 cm long, 1–2.5 cm wide. **CENTRAL SPINES** 0. **RADIAL SPINES** 18–23, reddish tan, 1–1.4 mm long,. **FLOWER** petaloids pink, mostly 10 mm long, 1.5–2.5 mm diameter. **FRUIT** green to tan, mostly 4 mm long, 3 mm wide. **SEEDS** dark gray to black, 1.5 mm long, 1–1.2 mm wide. Gravelly hills in piñon-juniper-sagebrush communities at 1900 m (6200–6300′). NMEX: SJn (near the Los Piños River and to be sought in Colorado). Flowering: May. Fruit: Jun. Endemic to the San Juan River drainage; our rarest and most endangered plant. Heavily collected after its initial discovery and still a favorite in the cactus trade.

Pediocactus simpsonii (Engelm.) Britton & Rose **CP** (honoring James H. Simpson, explorer and topographic engineer) Simpson's mountain cactus. Stems mostly solitary, 2.5–7.5 cm long, 2.5–5.5 cm wide. **SPINES** differentiated into 2 types. **CENTRAL SPINES** 5–8, reddish brown, 7.5–13.5 mm long, 0.3 mm wide. **RADIAL SPINES** 20–28, whitish, up to 6 mm long. **FLOWER** petaloids pink to magenta, white, or yellow, 1.2–4 cm long, 1.2–5 cm wide. **FRUIT** green, 6–9 mm long, 4.5–7.5 mm wide. **SEEDS** mostly 1.5 mm long, up to 2 mm wide. Rocky soils in Rocky Mountain montane forests, parklands, and piñon-juniper-sagebrush communities at 1970–2760 m (6500–9100′). ARIZ: Apa; COLO: Dol, Mon, SMg; NMEX: SJn; UTAH. Flowering: May. Fruit: Jun. Eastern Oregon, southern Idaho, Montana, to southern Wyoming, west-central Nevada, northeastern Arizona, northern New Mexico, and westernmost South Dakota. This hardy cactus is usually situated on warm, sunny mountain slopes up to an elevation of 3350 m (11000′).

Sclerocactus **Britton & Rose Devil's-claw, Kah-bes-zi**
Kenneth D. Heil and J. M. Porter

(*sclerao* for cruel or obstinate, referring to the hooked spines) Plants mostly single or occasionally clumped. **STEMS** depressed-globose, globose, obvoid, cylindroid to elongate-cylindroid, mostly ribbed, 3–30 cm long, 3.8–15 cm wide. **SPINES** differentiated into 2 or more types. **CENTRAL SPINES** 0–9, the lower usually hooked, 0.7–7.2 cm long, 0.5–2 mm wide. **LATERAL SPINES** 0–8, usually not hooked, 1.4–6.2 cm long. **RADIAL SPINES** 2–17, 0.5–3.6 cm long. **FLOWERS** cylindric to funnelform, petaloids rose to pink, purple, yellow, cream, or white, 1–5.7 cm long, 1–5.5 cm wide. **FRUIT** green or tan, occasionally turning pink, 0.8–2.5 cm long, 1–1.5 mm wide. **SEEDS** brown to black, 1.5–3 mm long, 2–4 mm wide (Heil and Porter 1994). Thirteen species limited to North America.

1. Central spine 0–1; perianth cream or pink; clay soils; in southwestern Colorado and northwestern New Mexico .... ***S. mesae-verdae***
1′ Central spines 4–9; perianth color various; on various soil types; more widely distributed .... (2)

2. Flower 3–5.7 cm long, 2.5–5.5 cm wide, funnelform, pericarpel with large papillae, surface granular; upper central spine flat or angled .... ***S. parviflorus***
2′ Flower 2.2–3.2 cm long, 1.5–2 cm wide, turbinate, pericarpel with small papillae, smooth; upper central spine flat and ribbonlike .... (3)

3. Petaloids (inner tepals) yellow; central spines 4; northeastern Arizona and San Juan County, Utah ...... ***S. whipplei***
3′ Petaloids purple; central spines 6–9; southwestern Colorado and northwestern New Mexico .... ***S. cloveriae***

Sclerocactus cloveriae K. D. Heil & J. M. Porter (for Elzada Clover, field botanist) Clover's fishhook cactus, kah-bes-zi (Navajo). **STEMS** ovoid to elongate-cylindrical, 3–30 cm long, 1.8–15 cm wide. **CENTRAL SPINES** 4–9, the lower one hooked or absent, 1.5–4.6 cm long, 1.5 mm wide. **LATERAL SPINES** 3–8, mostly a bit shorter than the central spines, usually not hooked. **RADIAL SPINES** 2–8, mostly 0.5–2.5 cm long. **FLOWERS** purple, mostly 2.3–3 mm long, mostly 1.5–2.5 mm wide. **FRUIT** green to tan, 7–15 mm long, 5–12 mm wide. **SEEDS** brown or black, 1.2–2.5 mm long, 1.9–3.5 mm wide. [*S. whipplei* (Engelm. & J. M. Bigelow) Britton & Rose var. *heilii* Castetter, P. Pierce & K. H. Schwer.; *S. whipplei* var. *reevesii* Castetter, P. Pierce & K. H. Schwer.].

1. Stems mostly 3–7.5 cm long; central spines mostly 4 (5), the lower often absent; San Juan County, New Mexico .... **subsp. *brackii***
1′ Stems mostly 5–25 cm long; central spines mostly 8 (6–9), the lower hooked; widespread .... **subsp. *cloveriae***

subsp. *brackii* K. D. Heil & J. M. Porter (for Steven Brack, student of *Sclerocactus*) **STEMS** 2.9–7.5 cm long, 1.8–7 cm wide. **CENTRAL SPINES** mostly 4, the lower one often absent. **LATERAL SPINES** 3. **RADIAL SPINES** mostly 6–7. **FLOWERS** 2.3–3.5 mm long. Sandy clay soils of the Nacimiento Formation. Desert scrub with scattered Utah juniper. NMEX: SJn. 1500–1800 m (5000–6000′). Flowering: late Apr–May. Fruit: late May–Jun. Endemic to the San Juan River drainage.

subsp. *cloveriae* CP STEMS 3.9–30 cm long, 4.8–15 cm wide. **CENTRAL SPINES** 6–9, mostly 8, the lower one hooked. **LATERAL SPINES** 5–8. **RADIAL SPINES** 2–6, mostly 4. **FLOWERS** 2.5–4 cm long. Sandy or clay soils, desert grasslands, desert scrub and piñon-juniper to ponderosa pine communities. COLO: LPl; NMEX: RAr, San, SJn. 1515–2180 m (5000–7200′). Flowering: May–Jun. Fruit: Jun. Scattered from south of Albuquerque, New Mexico, northward along the San Pedro, Rio Puerco, and San Juan river valleys to near Waterflow and northward into southern Colorado along the Los Piños, La Plata, and Animas rivers.

Sclerocactus mesae-verdae (Boissev. & C. Davidson) L. D. Benson **CP** (for the Mesa Verde Plateau) Mesa Verde cactus. **STEMS** mostly pale green, depressed-globose to ovoid, 3.2–11 cm long, 3.8–8 cm wide. **CENTRAL SPINES** 0–1, rarely hooked, 7–15 mm long. **RADIAL SPINES** 7–13, 6–13 mm long. **FLOWERS** 1–3.5 cm long, 1–3 cm wide, yellow to cream, rarely pink. **FRUIT** green, becoming tan at maturity, 8–10 mm long. **SEEDS** black, 2.5–3 mm long, 3–4 mm wide. [*Coloradoa mesae-verdae* Boissev. & C. Davidson; *Echinocactus mesae-verdae* (Boissev. & C. Davidson) L. D. Benson; *Pediocactus mesae-verdae* (Boissev. & C. Davidson) Arp]. Clay hills, desert scrub communities. COLO: Mon; NMEX: SJn. 1480–1660 m (4900–5500′). Flowering: late Apr–early May. Fruit: Jun. Endemic to the San Juan River drainage. During summer and winter months the stems lose moisture and shrivel down into the soil. This cactus grows in some of the most arid and alkaline conditions imaginable.

Sclerocactus parviflorus Clover & Jotter (small flower) Fishhook cactus. **STEMS** solitary or clustered, depressed-globose, globose, cylindroid to elongate-cylindroid, 5–27 mm long, 4.5–13 cm wide. **CENTRAL SPINES** 4–6, the lower central spine hooked, 1.5–7.2 cm long, 0.6–1 mm wide. **LATERAL SPINES** similar to the lower, usually shorter, rarely hooked. **RADIAL SPINES** 3–17, 0.6–3.6 cm long. **FLOWER** petaloids rose to purple, pink, yellow, rarely white, 3–5.7 cm long, 2.5–5.5 cm in wide. **FRUIT** green, turning reddish pink, 1–2.5 cm long, 1–1.5 cm wide. **SEEDS** dark brown to black, 1.5–3 mm long, 2.5–3.5 mm wide.

1. Flower petaloids (inner tepals) yellow; upper central spines 1.5 mm wide; upper piñon-juniper woodland communities and in sagebrush communities .... **subsp. *terrae-canyonae***
1′ Flower petaloids purple to pink-purple, rarely white; central spines 1–2 mm wide; Navajoan Desert to lower piñon-juniper woodland communities .... (2)

2. Upper central spine mostly 1 mm wide; lower Navajoan Desert .... **subsp. *parviflorus***
2′ Upper central spine mostly 1.5–2 mm wide; upper Navajoan Desert and lower piñon-juniper woodland communities .... **subsp. *intermedius***

subsp. *intermedius* (Peebles) K. D. Heil & J. M. Porter (intermediate spines) Intermediate fishhook cactus. **LOWER CENTRAL SPINE** 1.7–5.9 cm long, 0.5–1 mm wide. **UPPER CENTRAL SPINE** mostly angled, rarely flat or rhombic, 1.9–5.7 cm long, mostly 1.5–2 mm wide. **FLOWER** petaloids purple, rose, pink, or rarely white. [*S. intermedius* Peebles; *S. whipplei* (Englem. & J. M. Bigelow) Britton & Rose var. *intermedius* (Peebles) L. D. Benson; *S. parviflorus* var. *blessingae* W. Earle]. Gravelly and sandy hills, mesas; upper Navajoan Desert and piñon-juniper woodland communities. ARIZ: Apa, Coc, Nav; COLO: Dol, Mon; NMEX: McK, SJn; UTAH. 1350–1800 m (4500–6500′). Flowering: Apr–May. Fruit: May–Jun. Eastern Utah and adjacent Colorado, south to northern Arizona and northwestern New Mexico.

subsp. *parviflorus* CP LOWER CENTRAL SPINE 2–6.3 cm long, 0.6–1 mm wide. **UPPER CENTRAL SPINE** mostly angled, rarely flat or rhombic, 1.7–6.2 cm long, mostly 1 mm wide. **FLOWER** petaloids rose or purple. [*Echinocactus parviflorus* (Clover & Jotter) L. D. Benson; *S. havasupaiensis* Clover; *S. whipplei* (Engelm. & J. M. Bigelow) Britton & Rose var. *roseus* (Clover) L. D. Benson; *S. contortus* K. D. Heil]. Along major river drainages in mostly sandy soils of the lower Navajoan Desert. ARIZ: Nav; UTAH. 1050–1500 m (3500–5000′). Flowering: Apr–May. Fruit: May–Jun. Southeastern Utah, adjacent western Colorado, and northern Arizona.

subsp. *terrae-canyonae* (K. D. Heil) K. D. Heil & J. M. Porter (canyonlands) Canyonlands fishhook cactus. **LOWER CENTRAL SPINE** 2.9–8.2 cm long, 0.75–2 mm wide. **UPPER CENTRAL SPINE** flat, rhombic or rounded, 2.6–6.2 cm long, mostly 1.5 mm wide. **FLOWER** petaloids yellow. [*S. terrae-canyonae* K. D. Heil]. Sandy soils in upper piñon-juniper woodlands and sagebrush communities. ARIZ: Nav, Coc; UTAH. 1800–2300 m (6500–7500′). Flowering: May. Fruit: Jun. Endemic to the San Juan River drainage.

Sclerocactus whipplei (Engelm. & J. M. Bigelow) Britton & Rose (for Lt. Amiel W. Whipple, explorer and astronomer) Whipple's fishhook cactus, kah-bes-zi (Navajo). **STEMS** depressed-globose, or globose to elongate-cylindroid, 3–14 cm long, 4–11 cm wide. **CENTRAL SPINES** 4, the lower one hooked, 1.6–4.5 cm long, 0.5–1 mm in wide. **LATERAL SPINES** 2, 1.4–4.5 cm long, mostly 1 mm wide. **RADIAL SPINES** 5–12, 0.6–2.4 cm long. **FLOWER** petaloids yellow, 2.2–3.2 cm long, 1.5–2 cm wide. **FRUIT** green or tan, usually reddish at maturity, 0.8–2.5 cm long, 0.6–1.5 mm wide. **SEEDS** black, 2 mm long, 1.5 mm wide. [*Echinocactus whipplei* Engelm. & J. M. Bigelow; *Pediocactus whipplei* (Engelm. & J. M. Bigelow) Arp]. Gravelly and sandy hills, canyon rims, and mesas; desert grasslands and piñon-juniper woodlands. ARIZ: Apa, Nav; UTAH. 1500–1970 m (5000–6500′). Flowering: late Apr–May. Fruit: May–Jun. Northwestern Arizona and southeastern Utah (Bluff). Widespread throughout much of the Navajo and Hopi Indian Reservations. From a distance these cacti often look like mounds of dried grass and are very difficult to spot. Occasionally used as an ornamental.

CAMPANULACEAE Juss. BELLFLOWER FAMILY

Nancy R. Morin and Tina J. Ayers

Perennial herbs in our flora. **LEAVES** estipulate, alternate, simple. **INFL** a panicle or raceme. **FLOWERS** actinomorphic or zygomorphic, sometimes inverted (resupinate), perfect; sepals 5; petals 5, united proximally; stamens typically 5, free or filaments and anthers united; pistil 1-, 2-, or 3-carpellate; ovary inferior, style 1, stigma 2- or 3-lobed. **FRUIT** a capsule, 2- or 3-chambered, dehiscing by lateral pores or apical valves. $x = 6–17$. Approximately 90 genera, 2000+ species. Worldwide. Most members of the family exude a milky juice.

1. Flowers regular .... ***Campanula***
1′ Flowers irregular .... (2)

2. Flowers minute, 5 mm long or less; anthers all alike .... ***Nemacladus***
2′ Flowers 10 mm long or more; anthers united, 3 larger than the other .... ***Lobelia***

Campanula L. Bellflower

(Latin, little bell, from corolla shape) **LEAVES** mostly basal; cauline leaves bractlike. **FLOWERS** not resupinate, actinomorphic, petals blue, fused 2/3–3/4; stamens 5, free; ovary hemispheric or obconate, 3-chambered. **FRUIT** a capsule, hemispheric or obconate, spreading or pendulous, opening by lateral pores. **SEEDS** many. Mostly Old World. Three hundred to 400 species primarily in temperate areas.

1. Basal leaves nearly orbicular to oblanceolate, base obtuse, conspicuously petiolate; flowers and capsules generally nodding, corolla campanulate; capsules nodding, pores basal .... ***C. rotundifolia***
1′ Basal leaves oblanceolate to linear, narrowed to a petiole; flowers erect or horizontal; capsules mostly erect, pores subapical .... (2)

2. Flowers mostly erect (sometimes slightly nodding); corolla usually broadly campanulate, 8–20 mm long; stigma lobes 1/3 length of style .... ***C. parryi***
2′ Flowers erect or held horizontally; corolla usually narrowly funnelform, 5–8 mm long; stigma lobes 1/4 length of style .... ***C. uniflora***

Campanula parryi A. Gray **CP** (after North American plant collector C. C. Parry) Rocky Mountain bellflower. Plants erect, herbage glabrous but with marginal cilia. **STEMS** 0.5–1.5 dm tall. **LEAVES**: basal 10–35 mm × 3–10 mm, narrowed to a winged petiole, blade oblanceolate to linear; cauline leaves sessile, increasingly narrowed toward apex of stem. **INFL** usually solitary or 2–3 terminating axillary shoots. **FLOWER** calyx tube obconic, lobes long-triangular, 5–15 mm long, margins toothed; corolla broadly campanulate, 8–20 mm long, 10–25 mm wide, lobes erect, 1/3 length of corolla; stigma lobes about 1/3 length of style. **CAPSULES** 5–15 mm long, obconic, erect, pores near apex. $2n = 34$. Alpine meadows, riparian areas, local moist spots. ARIZ: Apa; COLO: Arc, Hin, LPl, Min, RGr, SJn; NMEX: McK, RAr, SJn. 2285–3460 m (7500–11360′). Flowering: Jun–Sep. Washington to Montana, south to Arizona and New Mexico. Used by Kayenta Navajo as a gynecological aid.

Campanula rotundifolia L. (round leaf) Harebell. Plants erect or ascending, herbage glabrous or slightly hairy on lower stems and leaves. **STEMS** 10–60 cm tall. **LEAVES**: basal 5–40 mm long, conspicuously petiolate, blades orbicular to oblanceolate, base obtuse; cauline leaves lanceolate to linear. **INFL** several in a raceme. **FLOWER** calyx tube hemispheric, lobes narrowly triangular to nearly linear, 3–9 mm long, margins entire; corolla narrowly campanulate, 10–20 mm long, 8–15 mm wide, lobes reflexed, 1/4 length of corolla; stigma lobes about 1/8–1/4 length of style. **CAPSULES** 3–8 mm long, hemispheric, nodding, pores near base. $2n = 34$. Drying meadows, roadsides, rocky areas; ponderosa pine, aspen, spruce-fir, and alpine communities. ARIZ: Apa; COLO: Arc, Con, Hin, LPl, Min, Mon, RGr, SJn; NMEX: McK, RAr, SJn. 2285–3810 m (7500–12500′). Flowering: Jun–Sep. Panboreal; Alaska, Canada, south to northern California, east to New York. Used ceremonially by Ramah Navajo, also as an analgesic, disinfectant, eye medicine, and hunting medicine.

Campanula uniflora L. (single flower) Arctic harebell. Plants erect, glabrous. **STEMS** 4–6 cm tall. **LEAVES**: basal 15–20 mm long, narrowed to a petiole, blades narrowly ovate to oblanceolate, cauline leaves sessile, linear-lanceolate. **INFL** solitary. **FLOWER** calyx tube narrowly obconic, lobes linear, 2.5–5 mm long, margins entire; corolla narrowly funnelform, 5–8 mm long, 4–6 mm wide, lobes erect, 1/3 length of corolla; stigma lobes about 1/4 length of style.

CAPSULES 0.8–1.5 cm long, obconic, erect, pores near apex. Alpine meadows, often in talus. COLO: SJn. 3655–3960 m (12000–13000′). Flowering: Jul–Aug. Panboreal; Rocky Mountains; Alaska and Yukon south to Colorado; Eurasia.

Lobelia L. Lobelia

(for Matthias de l'Obel, Flemish botanist, 1538–1616) **LEAVES** mostly basal or all cauline. **FLOWERS** resupinate, zygomorphic, petals red, fused about halfway, limb strongly 2-lipped, 2 lobes of upper lip shorter than 3 of lower; stamens 5, fused, 2 smaller anthers each with terminal tuft of bristles; stigma 2-lobed; ovary ± spheric, 2-chambered. **FRUIT** a capsule, spheric, opening by apical valves. **SEEDS** many. Worldwide. Three hundred fifty species.

Lobelia cardinalis L. (red) Cardinal flower. **STEMS** 4–20 dm tall, purple-red. **LEAVES** sessile, linear-lanceolate to elliptic, 5–15 mm wide, margins with small, gland-tipped teeth. **INFL** a raceme. **FLOWER** tube 15–20 mm, anther tube 3.5–4.5 mm, triangular bristles at tips of 2 shorter anthers 0. Stream bottoms, seeps, and hanging gardens. UTAH. 1100–1160 m (3600–3800′). Flowering: Aug–Oct. California east to Quebec and Florida and south to Panama.

Nemacladus Nutt.

(branched thread) Annuals. **STEMS** much-branched, diffuse. **LEAVES** basal in a compact rosette, cauline largely reduced to bracts. **INFL** a panicle. **FLOWERS** minute; sepals 5, triangular; corolla irregular, more or less bilabiate, the lower lip 3-lobed, the upper 2-lobed; ovary with 3 flattened, rounded glands near the base; stamens monadelphous above, the 2 filaments that arise between the glands with small, stipelike appendages and with 1 or more transparent, terminal, rodlike cells. **FRUIT** a bilocular capsule, dehiscing by valves.

Nemacladus glanduliferous Jeps. (bearing small glands) Glandular threadplant. Diminutive, puberulent or glabrous annuals. **STEMS** diffusely branched, 5–15 cm tall; pedicels spreading-ascending, 3–10 mm long, capillary, bracts 2–7 mm long. **FLOWERS** 1.5–2.5 mm long, white, lobes with purple tips; sepals 0.8–1.2 mm long; anthers 0.2–0.3 mm long. **FRUIT** a capsule 1.5–3 mm long, 1/2 inferior. $n = 9$. Mixed desert shrub communities. UTAH. 1220–1310 m (4000–4300′). Flowering: Mar–May. California, Nevada, Utah, Arizona, New Mexico, and Mexico. Our material belongs to **var**. ***orientalis*** McVaugh.

CANNABACEAE Martinov HEMP FAMILY

Linda Mary Reeves

Erect, pubescent or glandular herbs, occasionally weakly vining, annual or perennial, dioecious or monoecious. **LEAVES** opposite, occasionally alternate, palmately compound or lobed, with margins serrate or serrulate, occasionally simple, with margins entire. **INFL** a cyme, panicle, or spike; axillary or terminal; staminate inflorescences not congested, many-flowered; pistillate inflorescences few-flowered, in compact spikelike clusters. **FLOWERS** small, bracteate; staminate flowers pedicellate, sepals 5, petals absent, stamens 5; pistillate flowers sessile or subsessile, often paired; sepals 5, tubular; petals absent; ovary superior, 2-carpelled; style short; stigmas 2, filiform. **FRUIT** an achene, often with papery bracts. $x = 8, 10$. Two genera, about four species, primarily in the north-temperate zone (Mason 1999).

1. Herbs erect; leaves palmately compound .... ***Cannabis***
1′ Herbs weakly vining; leaves simple, palmately lobed or unlobed .... ***Humulus***

Cannabis L. Hemp, Marijuana

(Greek for hemp) Annual erect herbs, dioecious, rarely monoecious, to 6 m, ours considerably shorter. **LEAVES** alternate to opposite, often congested on a stem; palmately compound with 3–12 linear to linear-lanceolate leaflets, 3–15 cm long, 0.3–1.5 cm wide; scabrous, abaxial side more densely puberulent, margins serrate, teeth often with hooked apices. **INFL** numerous; staminate inflorescence a leafy axillary cyme or panicle; pistillate inflorescence an erect or spreading congested spike on leafy branches. **FLOWERS** small, inconspicuous; staminate flowers with pedicel 0.5–3 mm long, sepals ovate to lanceolate, 2.5–4 mm long; pistillate flowers subsessile, enclosed in a bracteole and subtended by a bract. **FRUIT** white, mottled with purple, lenticulate, enclosed by persistent calyx. $x = 10$. One species, widely distributed in North America, native originally to Eurasia.

Cannabis sativa L. (cultivated) Hemp, marijuana, hashish, pot, grass, maryjane, weed. Characters of the genus. $2n = 20$. Mostly a weedy species of roadsides and fields. No doubt cultivated widely. Not well established in our area, but occasionally seen in wet years. COLO: Mon; NMEX: RAr, SJn. 1450–1980 m (4800–6500′). Flowering: Jun–Sep. Fruit: Jul–Oct. *Cannabis sativa* is cultivated as a fiber (hemp) and is grown as a medicinal and recreational drug.

Humulus L. Hops

(Latin for hop plant) Perennial, dioecious, herbaceous vines with forked hairs. **LEAVES** opposite, simple, 3–5- palmately lobed or unlobed; abaxial surface glandular. **INFL**: staminate an axillary panicle; pistillate a short spike, with flowers in pairs. **FLOWERS** with pistillate pairs subtended by a foliaceous bract; calyx membranous, unlobed, closely covering the ovary; stigmas 2, elongate. **FRUIT** lenticular, enclosed by the persistent calyx and bracts. $x = 8, 10$. Two species in Europe, Asia, and North America.

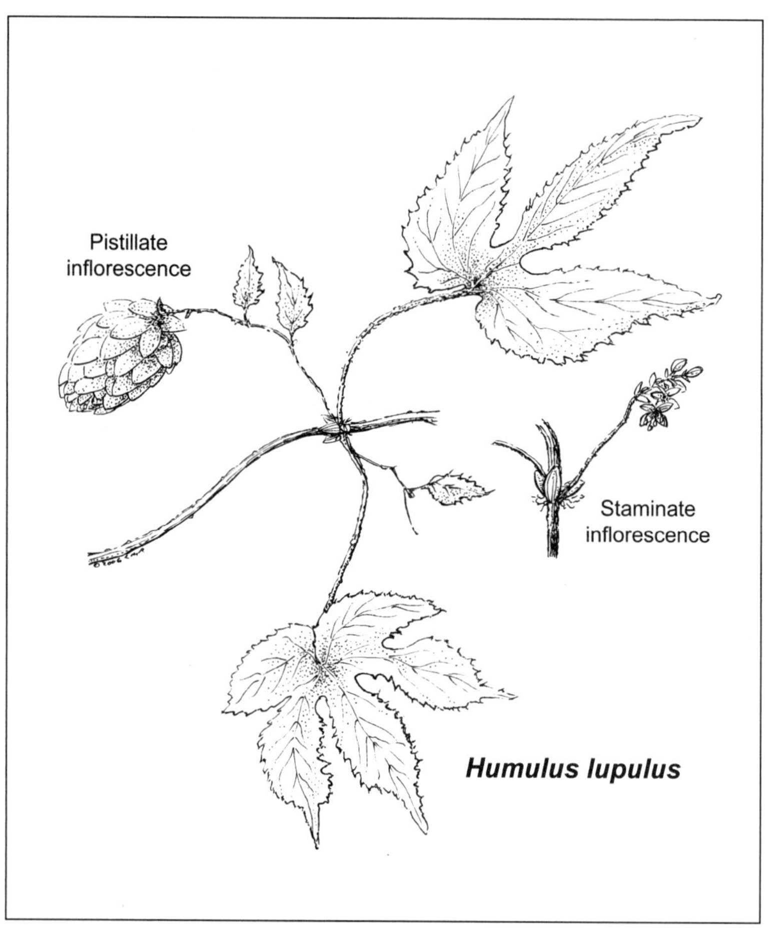

Humulus lupulus L. (hop plant) Viny herbs to 10 m or more long. **STEMS** pubescent at nodes. **LEAVES** long-petiolate, but usually shorter than the blade; blade ovate, unlobed or 3–7-lobed, 10 cm or more long. **INFL**: staminate panicle 7–15 cm long; pistillate inflorescence 2–5 cm long. $2n = 20, 40$. Four varieties. Ours is **var. *neomexicanus*** A. Nelson and Cockerell. Wooded areas and stream banks. ARIZ: Apa, Nav; COLO: Arc, Hin, LPl, Min, Mon; NMEX: McK, SJn. UTAH. 1450–3000 m (4700–9900′). Flowering: Jun–Oct. Fruit: Jul–Oct. Southern British Columbia to Manitoba, south to California, east to Texas and northern Mexico. *Humulus lupulus* is cultivated for use in brewing beer and gives the product a bitter taste. Plant used by Navajos for coughs and protection against witches. Flowers and fruit also used by Apaches and Navajos as a flavoring agent.

CAPRIFOLIACEAE Juss. HONEYSUCKLE FAMILY

Fred R. Barrie

(Latin, goat leaf) Shrubs, woody vines, or prostrate, woody perennials, glabrous or variously pubescent with simple or glandular hairs. **LEAVES** opposite, simple, estipulate. **INFL** a terminal or axillary cyme or the flowers solitary or paired in the leaf axils. **FLOWERS** actinomorphic or zygomorphic; perfect; 4- or 5-merous; the sepals free or connate at the base, or reduced; corollas sympetalous, occasionally bilabiate, the tube often lobed or spurred at the base; stamens attached within the tube, alternate with the corolla lobes, included to weakly or strongly exserted; ovary inferior, 2–5-locular, ovules 1–several per locule or some of the locules sterile; style elongate, stigma capitate or lobed. **FRUIT** a fleshy drupe with 1–several seeds. Ten to 12 genera, 250–300 species. Distributed throughout the north-temperate and boreal regions, and in montane areas in the tropical zones. Several of the genera, e.g., *Abelia* and *Lonicera*, include popular ornamental species that are commonly planted. Some of these taxa, notably in *Lonicera*, readily naturalize and may be found far outside their natural range. Recent morphological and molecular research has demonstrated that *Sambucus* and *Viburnum*, two genera formerly included in Caprifoliaceae, are properly placed in Adoxaceae (Donoghue et al. 2001).

1. Prostrate, evergreen perennials with creeping stems ..... ***Linnaea***
1′ Shrubs or vines ..... (2)

2. Corollas actinomorphic or nearly so; fruits white ..... ***Symphoricarpos***
2′ Corollas zygomorphic; fruits red, yellow, or black ..... ***Lonicera***

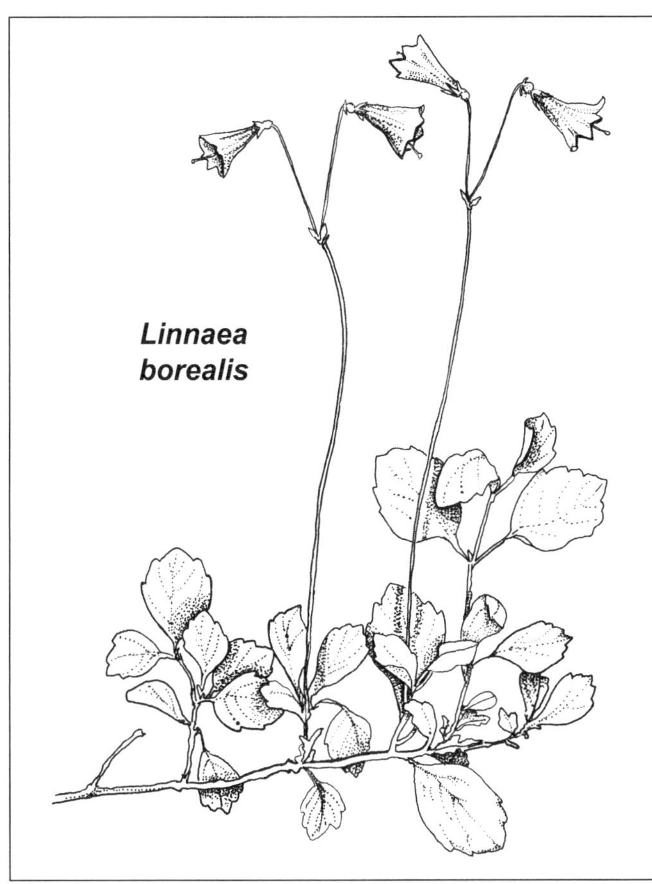

Linnaea L. Twinflower
(named for Carolus Linnaeus, 1707–1778, Swedish botanist) A genus with one recognized species.

Linnaea borealis L. (northern) Evergreen perennials. **STEMS** creeping, with numerous lateral branches 2–10 cm long, usually sparsely sericeous. **LEAVES** simple, 8–20 mm long, the blades elliptic to widely elliptic or orbicular, 6–15 mm long, 5–12 mm wide, the margins entire or with 2–6 broad shallow teeth, the apex widely acute or rounded, glabrous or with scattered hairs along the petiole. **INFL** erect, terminal on the lateral branches, 3–10 cm long, bearing 2 (rarely 4 or 6) paired, nodding, terminal flowers. **FLOWERS** actinomorphic, perfect, the pedicels 1–2 cm long; the ovary pubescent with a mix of simple and glandular hairs; sepals 5, ovate, 1–2 mm long; corolla 7–10 mm long, campanulate, pink, pubescent within, lobes 5, rounded, 1/4–1/3 the length of the tube; stamens 4, 3–4 mm long, inserted in the corolla tube, style 5–12 mm long. **FRUIT** dry and indehiscent, ellipsoid, 2–3 mm long, 3-locular, 2 sterile, the other with a single seed; pubescent with a mixture of simple and glandular hairs. $2n = 32$. [*L. americana* J. Forbes; *L. borealis* var. *americana* (J. Forbes) Rehder]. Moist sites; Douglas-fir, aspen, and spruce-fir communities. COLO: Arc, Hin, LPl, Min, SJn; NMEX: SJn. 2300–3300 m (8500–11000′). Flowering: Jun–Aug. Fruit: Jul–Sep. Northern Europe and Asia; Alaska and Canada south to Washington and New England, and along the Rocky Mountains to New Mexico and Arizona. *Linnaea borealis* is often divided into two varieties, var. *longiflora* Torr., with corollas 10–17 mm long and a well-defined tube 1–3 mm long at the base of the corolla, and **var. *borealis*** (ours), with corollas 7–10 mm long and no basal tube.

Lonicera L. Honeysuckle
(named for Adam Lonitzer, 1528–1586, German botanist) Shrubs or woody vines, the stems, leaves, and inflorescences variously pubescent, the hairs simple or glandular. **LEAVES** opposite, simple, all distinct (ours) or the distal 1–2 pairs connate-perfoliate, the margins entire. **INFL** axillary, pedunculate pairs of sessile flowers (ours) or terminal cymes. **FLOWERS** actinomorphic or zygomorphic, perfect; calyx 5-lobed or reduced and ring-shaped, the lobes obsolete; corolla tubular or funnel-shaped, typically with a glandular nectary near the base, with 5 regular lobes or bilabiate, the upper lip with 4 lobes, the lower with 1; stamens 5, inserted on the corolla, presented at the mouth of the corolla or strongly exserted; ovary 2- or 3-locular, ovules 2 per locule; style 1, exserted, glabrous or pubescent, the stigma capitate. **FRUIT** a red, pink, yellow, or black globose berry, the seeds few. One hundred eighty to 200 species, distributed throughout Europe, Asia, and North America.

1. Trailing or climbing vines, the flowers strongly bilabiate .... ***L. japonica***
1′ Shrubs, the flowers zygomorphic, but not strongly bilabiate .... (2)

2. Inflorescence bracts prominent, involucrate, red or purple .... ***L. involucrata***
2′ Inflorescence bracts linear, less than 15 mm long, green .... (3)

3. Plants glabrous, the inflorescence bracts 2–3 mm long, the corolla lobes 1/4 to 1/3 the length of the tube .... ***L. utahensis***
3′ Plants sparsely pubescent, the inflorescence bracts 5–15 mm long, the corolla lobes equal to or longer than the tube .... ***L. ×bella***

Lonicera ×bella Zabel (lovely) Showy fly honeysuckle. Shrubs to about 3 m tall. **STEMS** soft, hollow, the new growth usually sparsely pubescent, the hairs simple. **LEAVES** ovate-oblong, the blades 3–6 cm long, 1.5–3.5 cm wide, the base rounded to subcordate, the apex acute to obtuse, sparsely pubescent on the underside and along the petiole. **INFL** with the peduncle 15–20 mm long, glabrous or sparsely pubescent, the bracts linear, 5–15 mm long. **FLOWERS** zygomorphic, the bractlets ovate, about 1 mm long, glabrous, sepals ovate, about 0.5 mm long, corollas pink or white, becoming yellow, 15–20 mm long, the tube pubescent on the inner surface, the lobes equal to or longer than the tube, lobed at the base; stamens exserted; ovary 2–3 mm long, glabrous; style pubescent. **FRUIT** red or yellow, about 1 cm in diameter. [*L. morrowii* sensu Weber and Wittmann 1996]. COLO: LPl; NMEX: SJn. 1800–2000 m (5400–6500′). Flowering: May–June. Fruit: Jul. Introduced; widely planted as an ornamental and naturalized locally. A hybrid between two Asian species, *L. tatarica* L., which is glabrous, and *L. morrowii* A. Gray, which is pubescent on the inflorescences and the undersides of the leaves. This plant is known from two collections, one from a floodplain along the Animas River in Durango, the other from along an irrigation canal in Farmington. Listed by the USDA as an invasive plant.

Lonicera involucrata (Richardson) Banks ex Spreng. (involucrate, referring to the inflorescence bracts) Bearberry honeysuckle; twinberry. Shrubs 0.5–3 m tall. **LEAVES** with petioles 5–10 mm long, blades ovate to elliptic or obovate, 5–12 cm long, 2–6 cm wide, the midvein pubescent on the upper surface, the lower surface uniformly pubescent, the margins ciliate; base cuneate to truncate. **INFL** with the peduncle 2–5 cm long; bracts leafy, widely ovate, 1–25 mm long, forming a showy involucre surrounding the flowers and fruits, red or purple in flower, becoming dark purple in fruit, glandular-pubescent. **FLOWERS** with bractlets similar to the bracts and forming part of the involucre, calyx reduced, corolla yellow or with a touch of red, 1–1.5 cm long, the outer surface glandular-pubescent, the lobes shorter than the tube, spurred at the base; stamens presented at the mouth of the corolla or weakly exserted; ovary 4–5 mm long; style glabrous. **FRUIT** glossy black, about 1 cm in diameter. $2n$ = 18. Streamsides and other moist sites; riparian, aspen, Douglas-fir, and spruce-fir communities. ARIZ: Apa; COLO: Arc, Hin, LPl, Min, Mon, RGr, SJn; NMEX: SJn; UTAH. 2100–2900 m (8000–11000′). Flowering: May–Jul. Fruit: Jul–Aug. Alaska and Alberta south to California, Arizona, and Chihuahua, Mexico; east to Northern Michigan and Quebec.

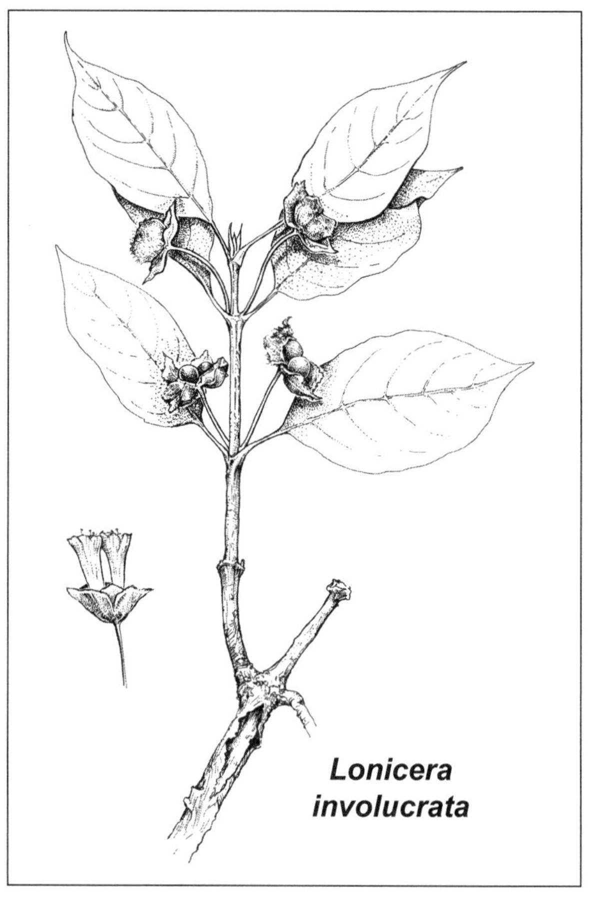
Lonicera involucrata

Lonicera japonica Thunb. ex Murray (of Japan) Japanese honeysuckle. Trailing or climbing vines. Stems, petioles, and inflorescences pubescent, the hairs simple or glandular, up to 1 mm long. **LEAVES** with petioles 5–10 mm long, blades ovate to elliptic, 3–8 cm long, 1.5–4 cm wide, pubescent along the midvein, the margins ciliate, the base truncate or rounded, the apex acute. **INFL** with the peduncle 3–5 mm long, the bracts similar in form to the leaves, up to 1 cm long. **FLOWERS** zygomorphic, sepals lanceolate, about 1 mm long, corolla white or yellow, bilabiate, 2–5 cm long, the lobes equal to the length of the tube; stamens strongly exserted, ovary 2–3 mm long, style glabrous. **FRUIT** black, 1–1.5 cm in diameter. $2n$ = 18. Introduced; known from a single collection made along a forest trail at 2500 m (8500′). COLO: Arc. Flowering: Jun. East Asia; introduced, widely planted as an ornamental, and locally naturalized. Once established, *L. japonica*, perhaps even more so than *L.* ×*bella*, can be an aggressive colonizer. Listed by the USDA as an invasive plant.

Lonicera utahensis S. Watson (of Utah) Utah honeysuckle. Shrubs 0.5–2 m tall, all parts glabrous or with a few hairs on the petioles and leaf margins. **LEAVES** oblong or ovate, the blades 1.5–5 cm long, 1–3 cm wide, the base truncate to rounded or subcordate, the apex obtuse or rounded. **INFL** peduncle 8–15 mm long, the bracts linear, 2–3 mm long. **FLOWERS** with bractlets that are obsolete or minute, less than 1 mm long; sepals reduced and nearly obsolete; corolla white or yellow, 1–2 cm long, the lobes 1/4–1/3 the length of the tube, spurred at the base; stamens weakly exserted; ovary 2–3 mm long. **FRUIT** red or salmon-pink, about 1 cm in diameter. [*L. ebractulata* Rydb.]. Spruce-fir forests. UTAH. 2800 m (9500′). Flowering: Jun. Fruit: July. Southeast Oregon east to Wyoming, south to Baja California, Mexico, in the west and Utah in the east. Known from a single collection at the northwestern margin of our range, west of Monticello.

Symphoricarpos Duhamel Snowberry

(Greek *symphorein*, to bear together, + *karpos*, fruits, a reference to the clustered fruits) **LEAVES** simple, the margins entire, sinuate, or dentate. **INFL** terminal or axillary, racemose, with tightly clustered flowers or the flowers solitary or paired in the leaf axils. **FLOWERS** actinomorphic, perfect, 4- or 5-merous, corollas pink or white, tubular, bell-shaped or funnelform; stamens with short filaments attached near the mouth of the tube, included to weakly exserted; ovary 4-locular with 2 sterile locules, 2 fertile, each bearing a single ovule; style 1, the stigma capitate or lobed. **FRUIT** a drupe, typically 2-seeded, white (ours), coral, red, or black. A genus of 9–16 species, one in China (*S. sinensis* Rehder), the remaining endemic to North America (Jones 1940).

1. Leaves elliptic or ovate, the margins entire; corollas glabrous, the style pubescent .... ***S. longiflorus***

1′ Leaf shape variable, the margins entire to sinuate or lobed; corollas pubescent on the inner surface, the style glabrous .... ***S. oreophilus***

Symphoricarpos longiflorus A. Gray (long-flowered) Desert snowberry. Shrubs 0.5–2 m tall. **STEMS** often intricately branched, the smaller twigs persistent, giving the plant a thorny appearance; the bark on older branches shredding in strips. **LEAVES** with petioles about 1 mm long, blade elliptic or ovate, 5–20 cm long, 2–8 cm wide, uniformly pubescent, the margins entire, the base and apex cuneate. **INFL** terminal, with 5–6 flowers clustered at the end of the branches, or more commonly 1–2 flowers in the leaf axils; bracts ovate, 2 mm long, the apex acute, glabrous or pubescent. **FLOWERS** with the calyx lobes triangular, connate, 1–2 mm long, the apex acute; corollas tubular, light pink, 10–15 mm long, the lobes 1/4–1/3 the length of the tube, glabrous; stamens with sessile anthers attached at or just below the mouth of the corolla; style 4–7 mm long, pubescent. **FRUIT** white, ellipsoid, 6–10 mm long, nutlets usually 2, about 5 mm long. Piñon-juniper woodlands and ponderosa pine communities. ARIZ: Apa, Nav; NMEX: SJn; UTAH. 1525–2285 m (5000–7500′). Flowering: May–Jun. Fruit: Jun–Jul. Oregon south to California, Arizona, New Mexico, and Texas; Mexico.

Symphoricarpos oreophilus A. Gray **CP** (round-leaved) Mountain snowberry, roundleaf snowberry. Shrubs 0.5–2 m tall, the vestiture variable on all parts, from glabrous or nearly so to puberulent with straight or recurved hairs or short-pilose. **STEMS** red or tan, the bark shredding on older branches. **LEAVES** with petioles 2–4 mm long, blade elliptic to ovate or broadly ovate, the margins entire to sinuate or irregularly dentate, the base acute to truncate, the apex acute. **INFL** terminal, few-flowered clusters or the flowers in drooping, axillary pairs, the peduncle 1–3 mm long; bracts ovate, connate at the base, the apices acute to acuminate. **FLOWERS** with the calyx lobes triangular, connate, up to 1 mm long, the apex acute; corollas funnel-shaped, pink or white, 7–14 mm long, the lobes about 1.3× the length of the tube, pubescent on the inner surface; stamens with filaments 1–3 mm long; style glabrous. **FRUIT** white, ellipsoid, 6–15 mm long. [*S. rotundifolius* A. Gray; *S. utahensis* Rydb.; *S. vaccinioides* Rydb.]. Gambel's oak woodlands, ponderosa pine, Douglas-fir, and spruce-fir communities. ARIZ: Apa, Coc, Nav; COLO: Arc, Dol, Hin, LPl, Min, Mon, RGr, SJn, SMg; NMEX: McK, RAr, SJn; UTAH. 1900–3400 m (6500–11500′). Flowering: Jun–Jul. Fruit: Jul–Sep. An extremely variable species with respect to vestiture, leaf shape, and corolla length. The leaf form may vary on the same plant, with the leaves on the flowering branches differing from the leaves on the sterile branches. Jones (1940) recognized four species in what is here considered one. His concepts, however, fail to hold up, as his defining characters combine independently across population boundaries and no well-demarcated groups can be seen.

CARYOPHYLLACEAE Juss. PINK FAMILY

Ronald L. Hartman and Richard K. Rabeler

Annual to perennial herbs. **LEAVES** simple, stipulate or not, opposite, whorled, or appearing so, nodes often swollen. **INFL** of cymes or flowers solitary; bracts foliaceous, scarious, or absent. **FLOWERS** actinomorphic, usually perfect; hypogynous or perigynous; sepals 4–5, distinct or connate below, apex sometimes hooded or with apical or subapical spine; petals 4–5 or absent, distinct, often clawed, sometimes with auricles and coronal appendages, blade apex entire, notched, or 2 (–4)-fid, sometimes dentate or laciniate; stamens (0) 1–10, arising from base of ovary, nectariferous disc, or hypanthium rim, ovary 1, superior, 1- or 3–5-locular, placentation free-central, basal, or axile in proximal 1/2; styles (0) 1–5, distinct or connate proximally; stigmas linear along adaxial surface of styles (or style branches), subcapitate, or terminal. **FRUIT** a capsule with carpel valves splitting apically to throughout or, more rarely, an indehiscent utricle. **SEEDS** reniform to triangular, often laterally compressed, horizontal wing sometimes present, spongy appendage (strophiole) rarely present. Eighty-three or 89 genera. About 3000 species. Worldwide, especially in north-temperate, montane and alpine, and Mediterranean areas (Rabeler and Hartman 2005).

1. Leaves with stipules ovate to triangular, mostly scarious, connate .... (2)
1′ Leaves with stipules absent .... (4)

2. Fruit a utricle; seed 1; petals absent; sepals with subapical awns; staminodes filiform .... ***Paronychia***
2′ Fruit a capsule; seeds 3–150+; petals rarely absent; sepals without awns; staminodes absent .... (3)

3. Petal blade apex entire; styles 3, distinct .... ***Spergularia***
3′ Petal blade apex divided into 2 or 4 lobes; styles 3, connate proximally .... ***Drymaria***

4. Sepals connate 1/2 + of length, usually forming a tube; petals white to pink, red, or purple; perianth hypogynous .... (5)
4′ Sepals distinct or essentially so; petals usually white; perianth hypogynous or sometimes perigynous .... (7)

5. Styles 3 (4) or 5 (absent in staminate flowers); fruit valves 3–5 or splitting into 6–10 teeth; flowers perfect (except *S. latifolia*) .... ***Silene***
5′ Styles 2 (3); fruit valves usually 4; plants with perfect flowers .... (6)

6. Plant perennial; calyx tube terete; coronal appendages 2 .... ***Saponaria***
6′ Plant annual; calyx tube 5-angled or -keeled; coronal appendages absent .... ***Vaccaria***

7. Petals absent .... (8)
7′ Petals present .... (9)

8. Capsule cylindric, often curved, opening by 8 or 10 teeth .... ***Cerastium nutans***
8′ Capsule ovoid to globose, symmetrical, opening by 6 (8 or 10) valves or teeth (*S. calycantha*, *S. media*, *S. obtusa*, and *S. umbellata*) .... ***Stellaria***

9. Petal apices 2-lobed, often divided nearly to base .... (10)
9′ Petal apices entire, emarginate, or notched .... (12)

10. Petals 2-lobed for 1/10–1/5 their length; fruit spheric, opening by 6 2–3× recoiled valves; seeds 1–3; rhizomes usually with tuberous thickenings .... ***Pseudostellaria***
10′ Petals 2-lobed or 2-fid nearly to base; fruit ovoid or globose to cylindrical, opening by teeth or valves, valves not recoiled; seeds 5–30+; rhizomes, when present, without tuberous thickenings .... (11)

11. Capsule cylindric, often curved, opening by 10 teeth .... ***Cerastium***
11′ Capsule ovoid to globose, opening by 6 (8 or 10) valves or teeth .... ***Stellaria***

12. Capsule valves and styles equal in number .... (13)
12′ Capsule valves or teeth 2× number of styles .... (14)

13. Sepals (4) 5; styles (4) 5; capsule valves 4 or 5 .... ***Sagina***
13′ Sepals 5; styles 3; capsule valves or teeth 3 .... ***Minuartia***

14. Styles 5; capsule cylindric, opening by 10 teeth .... ***Cerastium***
14′ Styles 3; capsule ovoid to urceolate or globose, opening by 6 valves or teeth .... (15)

15. Leaf blade filiform to subulate, congested at or near base of flowering stem, apex narrowly blunt to usually apiculate or spinose .... ***Eremogone***
15′ Leaf blade ovate to lanceolate (sometimes narrowly so), not congested at or near base of flowering stem, apex acute to obtuse .... (16)

16. Seed 1–2.2 mm with appendage present, elliptic, white, spongy; plant perennial .... ***Moehringia***
16′ Seed 0.4–0.8 mm with appendage absent; plant annual or perennial .... ***Arenaria***

Arenaria L. Sandwort

(sand, a common habitat) Annual or perennial herbs with taproot or slender rhizomes. **STEMS** prostrate to ascending or erect, sometimes densely matted, simple or branched, terete to ellipsoid. **LEAVES** not congested at or near base of flowering stem, without stipules, blades linear-lanceolate to broadly ovate or rarely orbiculate. **INFL** an open cyme or flowers solitary. **FLOWERS** hypogynous; sepals 5, distinct or nearly so, apex acute to acuminate, not hooded, awn absent; petals 5 or sometimes absent, white, not clawed, auricles absent, coronal appendages absent, blade apex entire; stamens 10, arising from ovary base, staminodes absent; ovary 1-locular; styles 3. **FRUIT** a broadly ellipsoid or ovoid capsule, opening by 6 ascending to recurved teeth. **SEEDS** 5–35, brown to black, shiny, smooth, marginal wing absent, appendage absent. About 210 species, temperate Americas, Eurasia, and Asia Minor.

1. Leaf blades 1-veined, linear-lanceolate to oblanceolate; sepals often prominently keeled proximally, glabrous; seeds black, suborbiculate, slightly compressed, 0.7–0.8 mm, shiny, smooth .... ***A. lanuginosa***
1' Leaf blades 3–5-veined, elliptic to broadly ovate or rarely orbiculate; sepals not keeled, stipitate-glandular; seeds ashy black, reniform, plump, 0.5–0.6 mm, not shiny, with prominent tubercles .... ***A. serpyllifolia***

Arenaria lanuginosa (Michx.) Rohrb. (wool-like surface) Perennial herbs, sometimes blooming first year. **STEM** internodes retrorsely puberulent throughout or in lines. **LEAVES** 1-veined, linear-lanceolate to narrowly elliptic or oblanceolate. **INFL** pedicels retrorsely pubescent, proliferating, 1–80+-flowered cymes. **FLOWER** sepals often appearing prominently keeled proximally, to 5.5 mm long in fruit, not pustulate, glabrous; petals narrowly spatulate to obovate, sometimes absent. **FRUIT** ± loosely to tightly enclosed by calyx, ovoid, 3–6 mm, 4/5–1 1/2× sepal length. **SEEDS** black, suborbiculate, slightly compressed, 0.7–0.8 mm long, shiny, smooth. $2n = 40, 44$. [*A. saxosa* A. Gray; *A. confusa* Rydb.; *A. lanuginosa* subsp. *saxosa* (A. Gray) Maguire; *A. saxosa* A. Gray var. *cinerascens* B. L. Rob.; *A. saxosa* A. Gray var. *mearnsii* (Wooton & Standl.) Kearney & Peebles; *Spergulastrum lanuginosum* Michx. subsp. *saxosum* (A. Gray) W. A. Weber]. Understory of diverse shrub and conifer communities, stream banks, alpine meadows and talus slopes; disturbed habitats. ARIZ: Apa; COLO: Arc, Hin, LPl, Min, Mon, RGr, SJn; NMEX: McK, RAr, SJn; UTAH. 2100–4000 m (7000–13000′). Flowering: Jun–Sep. Colorado and Utah south to western Texas, northern Mexico, and southern California. Our material belongs to **var. *saxosa*** (A. Gray) Zarucchi, R. L. Hartm. & Rabeler (rocky places). The varietal combination *A. lanuginosa* var. *cinerascens* (B. L. Rob.) Shinners, often applied to this taxon, cannot be used because the earlier autonym *A. saxosa* var. *saxosa* has priority. Used as a lotion for headaches and for pimples by the Ramah Navajo.

Arenaria serpyllifolia L. (thymelike leaves) Thymeleaf sandwort. Annual herbs. **STEMS** uniformly puberulent. **LEAVES** 3–5-veined, elliptic to broadly ovate or rarely orbiculate. **INFL** a terminal, 3–50+-flowered cyme. **FLOWER** sepals not keeled, to 4 mm in fruit, ± minutely pustulate, stipitate-glandular; petals present, oblong. **FRUIT** loosely enclosed by the calyx, ovoid to cylindric-ovoid, 3–3.5 mm, 4/5–1 1/5× sepal length. **SEEDS** ashy black, reniform, plump, 0.5–0.6 mm long, not shiny, with low-elongate, prominent tubercles. $2n = 40, 44$. Roadsides, open, sandy, rocky, or disturbed sites. ARIZ: Apa. 1700 m (5500′). Flowering: May–Jul. Introduced throughout North America. Our material belongs to **var. *serpyllifolia***.

Cerastium L. Mouse-ear Chickweed

(Greek *ceras*, horn, alluding to the shape of the capsule) Annual or perennial herb with taproot or slender rhizomes. **STEMS** ascending to erect or decumbent, simple or branched, terete. **LEAVES** not congested at or near base of flowering stem, without stipules, blade linear to oblanceolate or elliptic. **INFL** an open cyme. **FLOWERS** hypogynous or weakly perigynous; sepals 5, distinct or nearly so, apex acute, acuminate, to obtuse, not hooded, awn absent; petals 5 or rarely absent, white, clawed, auricles absent, coronal appendages absent, blade apex bilobed 1/5–1/2 of length; stamens usually 10, arising from ovary base, staminodes absent; ovary 1-locular; styles 5. **FRUIT** a cylindric, usually curved capsule, opening by 10 erect or spreading, convolute or revolute teeth. **SEEDS** 25–50, orange to brown, papillate-tuberculate, marginal wing absent, appendage absent. About 100 species worldwide, but mainly north-temperate regions (Morton 2005a).

1. Plant annual, without matted basal branches; mature capsule 2–3× calyx length .... ***C. nutans***
1' Plant biennial or perennial, with matted basal branches; mature capsule 1–2× calyx length .... (2)

2. Petals about equaling the sepals .... ***C. fontanum***
2' Petals usually at least 1.5× sepal length .... (3)

3. Bracts herbaceous or only the most distal slightly scarious-margined; leaves of flowering stems narrowly oblong to oblong-lanceolate or oblanceolate, apex usually obtuse, sterile axillary clusters of leaves absent; plants strictly alpine .... ***C. beeringianum***
3' Bracts scarious-margined, or only the most proximal completely herbaceous; leaves of flowering stems linear to narrowly lanceolate, apex usually acute, sterile axillary clusters of leaves present; plants general in distribution .... ***C. arvense***

Cerastium arvense L. (of cultivated fields) Field or prairie mouse-ear. Perennial herb, clumped and taprooted, or mat-forming and long-creeping rhizomatous. **STEMS** flowering, ascending to erect, 5–30 cm long, pilose-subglabrous proximally, glandular-pubescent distally; nonflowering shoots present. **LEAVES** of flowering stem lanceolate or oblanceolate to linear, apex acute, rarely obtuse, sterile axillary clusters of leaves present proximally. **INFL** bract margins narrowly scarious, glandular-pubescent. **FLOWER** sepal margins narrowly scarious, softly pubescent; petals obovate, 7.5–12.5 mm long, about 2× calyx length, apex bifid; stamens 10. **FRUIT** less than 1.5× calyx length. **SEEDS** brown, 0.6–1.1 mm long. $2n = 36$. Roadsides and other disturbed areas, meadows and riparian areas, coniferous and deciduous forests and woodlands, alpine meadows and talus slopes. ARIZ: Apa; COLO: Arc, Hin, LPl, Min, Mon, RGr; NMEX: RAr, SJn; UTAH. 2000–3800 m (6500–12500′). Flowering: May–Sep. Throughout much of North America except the Southeast. Our material belongs to **subsp. *strictum*** (L.) Gaudin (upright).

Cerastium beeringianum Cham. & Schltdl. (for Graf von Beering) Beering mouse-ear chickweed. Perennial herbs, taprooted, with short, prostrate sterile shoots, rarely rhizomatous. **STEMS**: flowering usually erect, 10–25 cm long, glabrous to sparsely pubescent proximally but more so distally, hairs widely spreading to slightly deflexed, often glandular; nonflowering shoots present. **LEAVES** of flowering stems narrowly oblong to oblong-lanceolate or oblanceolate, apex obtuse, sterile axillary clusters of leaves absent. **INFL** bract margins herbaceous, only the most distal sometimes slightly scarious, pubescence glandular and eglandular. **FLOWER** sepal margins broad, densely glandular-pubescent; petals broadly oblanceolate, 6–12 mm long, usually equaling calyx length (rarely 2×), apex deeply bifid; stamens 10. **FRUIT** about 2× calyx length. **SEEDS** pale to dark brown, 0.7–1.1 mm long. $2n = 72$. Subalpine riparian areas and meadows, but normally alpine meadows and talus slopes. COLO: Arc, Con, Hin, LPl, Min, Mon, RGr, SJn. UTAH. 3500–4250 m (11400–14000′). Flowering: May–Sep. Canada; Oregon to Montana, south to California and New Mexico. A cold infusion of the plant used for sheep or horses with eye troubles by the Ramah Navajo.

Cerastium fontanum Baumg. (of a spring) Common mouse-ear chickweed. Perennial herbs, usually tufted to mat-forming, often rhizomatous. **STEMS**: flowering erect, 10–45 cm long, softly pubescent with eglandular, straight hairs; nonflowering shoots sometimes present, decumbent. **LEAVES** of flowering stems elliptic to ovate-oblong, apex subacute, sterile axillary clusters of leaves absent. **INFL** bract margins herbaceous, often narrowly scarious distally, eglandular-pubescent. **FLOWER** sepal margins narrow, pubescent with eglandular, rarely glandular hairs; petals oblanceolate, 5–7 mm long, 1–1.5× calyx length, apex deeply bifid; stamens 10, occasionally 5. **FRUIT** about 2× calyx length. **SEEDS** reddish brown, 0.4–1.2 mm long. $2n =$ usually 144. [*C. vulgare* Hartm.; *C. fontanum* subsp. *triviale* (Link) Jalas]. Riparian and other wet areas in meadows or in understory of forests and woodlands. ARIZ: Apa; COLO: Arc, Hin, LPl, Min, Mon, SJn. 2200–3200 m (7200–10500′). Flowering: May–Sep. Introduced throughout North America. Our material belongs to **var. *vulgare*** (Hartm.) M. B. Wyse Jacks. (common).

Cerastium nutans Raf. (nodding) Nodding mouse-ear chickweed. Annual herbs, slender, with slender taproot. **STEMS**: flowering erect, 10–50 cm long, softly pubescent, often with a few long, woolly hairs at proximal nodes, glandular distally; nonflowering shoots absent. **LEAVES** of flowering stems lanceolate to spatulate proximally, lanceolate to narrowly elliptic distally, apex mostly acute, sterile axillary clusters of leaves absent. **INFL** bract margins herbaceous, glandular-pubescent. **FLOWER** sepal margins herbaceous or narrow (outer) or as wide as herbaceous center (inner), glandular-puberulent; petals oblanceolate, sometimes absent, mostly 3–6 mm long, less than 1.5× calyx length, apex bifid, stamens 10. **FRUIT** 2–3× calyx length. **SEEDS** golden brown, 0.5–0.8 mm long. $2n = 36$. Meadows and disturbed areas. NMEX: SJn; UTAH. 2100–2500 (6800–8100′). Flowering: May–Aug. Throughout North America except California, Nevada, and Utah. Our material belongs to **var. *nutans*.**

Drymaria Willd. ex Schult. Drymary

(Greek *drymos*, forest, alluding to the habitat of at least one species) Annual herbs with taproot. **STEMS** ascending to erect, simple to often branched distally, terete. **LEAVES** not congested at or near base of flowering stem, with stipules subulate, herbaceous to indurate, blades orbiculate to spatulate or linear to oblong. **INFL** terminal cymes. **FLOWERS** hypogynous; sepals 5, distinct, apex acute to acuminate or blunt to rounded, hooded, awn absent; petals 5, white, clawed, auricles absent, coronal appendages absent, blade apex divided into 2 or 4 lobes; stamens 5, arising from base of ovary, staminodes absent; ovary 1-locular; styles 3. **FRUIT** an ellipsoid to globose capsule, opening by 3 spreading to recurved valves. **SEEDS** 3–25, tan, reddish brown, or dark brown to purplish, tuberculate, marginal wing absent, appendage absent. About 48 species, mostly tropical and arid America and Africa, Australia, and Indonesia (Hartman 2005).

1. Cauline leaves appearing whorled, at least in part; inflorescences initially a terminal cyme but becoming elongate and racemose; petals 4-lobed; seeds 0.8–0.9 mm long .... ***D. molluginea***

1′ Cauline leaves opposite; inflorescences a congested to open terminal cyme; petals 2-lobed; seeds 0.5–0.7 mm long .... (2)

2. Herbaceous portion of sepals ± oblong, apex blunt or rounded, veins ± parallel, apically confluent; stipules entire .... ***D. depressa***
2' Herbaceous portion of sepals lanceolate, apex acute to acuminate, veins with lateral pair distinctly arcing outward; stipules divided into 2 segments .... ***D. leptophylla***

Drymaria depressa Greene (low as to stature) Annual herbs, glabrous to minutely puberulent, 0.5–5 cm tall. **STEMS** ascending to erect, generally branching at base. **LEAVES** opposite, stipules entire, subulate, 0.5–1.2 mm long, ± decidious, blades (basal ones orbicular to spatulate) oblong, 0.3–1 cm long, 0.02–0.3 cm wide, base obtuse, apex rounded to acute. **INFL** 3–25+ flowers in a congested to open terminal cyme. **FLOWER** sepals ± equal, lanceolate, oblong, or ovate (herbaceous portion ± oblong), 1.8–2.3 mm long, nerves (ribs) 3, prominent, ± parallel, apically confluent, apex blunt to rounded, glabrous; petals 1.5–2.8 mm long, 2-lobed. **SEEDS** light reddish brown to tan, snail shell- to teardrop-shaped, 0.5–0.6 mm long. Rocky depressions of Twilight Gneiss in montane woodland communities. COLO: LPl, SJn. 2520–2790 m (8270–9160′). Flowering: Aug–Sep. Arizona, Colorado, New Mexico; Mexico and Central America. This species has been treated as a variety of *Drymaria effusa* A. Gray, from which it differs in a number of respects as detailed in the key in Hartman (2005).

Drymaria leptophylla (Cham. & Schltdl.) Fenzl ex Rohrb. (slender-leaved) Annual herbs, glabrous, 8–25 cm tall. **STEMS** erect, sparingly branched. **LEAVES** mostly opposite, stipules divided into 2 segments, segments filiform, 0.3–1 mm long, ± deciduous, blade linear to narrowly oblong, 0.5–2.5 cm long, 0.02–0.12 cm wide, base gradually tapered, apex rounded to apiculate. **INFL** of (5–) 15–75+ flowers in open terminal cymes. **FLOWER** sepals: outer 2 often shorter (± 0.5 mm) than inner, lanceolate to ovate (herbaceous portion lanceolate), 1.8–3.5 mm long, nerves (ribs) 3, often prominent, apically confluent, apex acute to acuminate (herbaceous portion similar), glabrous or with a few sessile glands; petals 1.3–2.6 mm long, 2-lobed. **SEEDS** tan, snail shell-shaped, 0.6–0.7 mm long. $2n = 36$. Rocky or gravelly flats or slopes or disturbed areas in woodlands. NMEX: McK. 1700–2400 m (5575–7875′). Flowering: Aug–Oct. Colorado, Arizona to Texas, and Mexico. Our material belongs to **var.** ***leptophylla***; the other two varieties are restricted to Mexico.

Drymaria molluginea (Ser.) Didr. (resemblance to the genus called carpet-weed) Annual herbs, glabrous or glandular, 3–25 (–30) cm tall. **STEMS** erect, unbranched or dichotomously branched. **LEAVES** mostly appearing whorled, stipules entire, filiform to subulate, 0.5–2 mm long, ± persistent, blade linear, 1–3 (–3.5) cm long, 0.05–0.15 (–0.2) cm wide, base gradually tapered, apex rounded, sometimes apiculate. **INFL** of 3–30+ flowers in open terminal cymes or racemose cymes. **FLOWER** sepals subequal, broadly ovate to ± orbicular (herbaceous portion elliptic to oblong or lanceolate), 2–3.5 mm long, midnerve often prominent, the lateral pair less so, distinct, apex obtuse (herbaceous portion similar), hood formed in part by the scarious margin, glabrous; petals 1.7–2.5 mm long, 4-lobed. **SEEDS** dark brown to purplish, horseshoe-shaped, 0.8–0.9 mm long. Rocky to sandy soil from 2100–2400 m (6890–7875′). NMEX: McK. Flowering: Aug–Oct. Arizona to Texas, and northern Mexico.

Eremogone Fenzl Sandwort

(Greek *eremo*, solitary or deserted, + *gone*, seed or offspring, allusion uncertain) Perennials usually with branched, woody base. **STEMS** ascending to erect, simple or branched, terete. **LEAVES** congested at or near base of flowering stems, without stipules; blades needlelike or filiform to subulate. **INFL** an open to congested cyme. **FLOWERS** weakly perigynous, hypanthium shallowly cup-shaped; sepals 5, distinct or nearly so, apex acute or acuminate, not hooded, awn absent; petals 5, white, clawed or not, auricles absent, coronal appendages absent, blade apex entire or emarginate; stamens 10, arising from hypanthium, staminodes absent; ovary 1-locular; styles 3. **FRUIT** a capsule, ovoid, opening by 6 ascending to recurved teeth. **SEEDS** 2–5, dark reddish to black, smooth to tuberculate, marginal wing absent, appendage absent. About 90 species, temperate western North America and Eurasia.

1. Flowers in 1 or more dense capitate umbellate cymes .... ***E. congesta***
1' Flowers in open cymes .... (2)

2. Sepals glabrous throughout or essentially so .... ***E. eastwoodiae***
2' Sepals moderately to densely stipitate-glandular .... ***E. fendleri***

Eremogone congesta (Nutt.) Ikonn. (congested) Ballhead sandwort. Perennial herbs, tufted, green. **STEMS** 3–40 cm long, mostly glabrous. **LEAVES** with basal blades erect-ascending to arcuate-spreading, subulate or needlelike to filiform, 2–11 cm long, flexuous or rigid, apex obtuse to sharply acute or spinose, glabrous, sometimes glaucous. **INFL**

6–35+-flowered, of 1 or more dense capitate umbellate cymes; pedicels 0.1–3 mm long, usually glabrous. **FLOWER** sepals ovate to lanceolate, margins narrowly scarious, apex obtuse or acute, glabrous; petals white, oblong, 1.5–2× sepal length; nectaries as lateral and abaxial mound with crescent-shaped groove at base of filaments, 0.3 mm long. **SEEDS** reddish brown to black, 1.4–3 mm long, tuberculate; tubercles low, rounded, often elongate. $2n = 22$. [*Arenaria congesta* Nutt.]. Meadows, sagebrush slopes, oak thickets, and ponderosa pine woodlands. ARIZ: Apa; COLO: Arc, Dol, LPl, Mon, SMg; UTAH. 2250–3000 (7400–9800′). Flowering: May–Aug. Washington to Montana, south to California and Colorado. Four Corners material belongs to **var. *congesta***.

Eremogone eastwoodiae (Rydb.) Ikonn. (for Alice Eastwood, 1859–1953, botanical collector in the Four Corners region, and curator of botany, California Academy of Sciences) Eastwood's sandwort. Perennial herbs, densely tufted, green. **STEMS** 10–25 cm long, glabrous or stipitate-glandular. **LEAVES** with basal blades spreading to recurved, needlelike, 1–3 cm long, flexuous to rigid, apex spinose, glabrous to puberulent, not glaucous. **INFL** 3–17-flowered, mostly an open cyme; pedicels 3–30 mm long, glabrous or stipitate-glandular. **FLOWER** sepals lanceolate to ovate-lanceolate, margins broadly scarious, apex narrowly acute to acuminate, glabrous or stipitate-glandular; petals yellowish white or sometimes brownish to reddish pink, broadly oblong-elliptic to oblanceolate, 0.9–1.1× sepal length, nectaries narrowly longitudinally rectangular, apically cleft or emarginate, adjacent to filaments, 1–2 mm long. **SEEDS** brown, 1.2–1.7 mm long, papillate, subechinate; tubercles conical. Plant used as an emetic for the stomach by the Hopi.

1. Stem and pedicel stipitate-glandular .... **var. *adenophora***

1′ Stem and pedicel glabrous .... **var. *eastwoodiae***

var. *adenophora* (Kearney & Peebles) R. L. Hartm. & Rabeler (gland-bearing) Stem and pedicel stipitate-glandular. [*Arenaria eastwoodiae* Rydb. var. *adenophora* Kearney & Peebles]. Sandy washes and badlands, desert shrub and sagebrush communities, and ponderosa pine savannas. ARIZ: Apa, Coc, Nav; COLO: Arc, LPl; NMEX: RAr, San, SJn; UTAH. 1200–2560 m (4000–7600′). Flowering: May–Jul. Utah, southern Colorado, northern Arizona, and New Mexico.

var. *eastwoodiae* Stem and pedicel glabrous. [*Arenaria eastwoodiae* Rydb.; *A. fendleri* A. Gray var. *eastwoodiae* (Rydb.) H. D. Harrington]. $2n = 22$. Sandy washes, piñon-juniper and oak woodlands, and conifer savannas. ARIZ: Apa, Coc, Nav; COLO: Mon, SMg; NMEX: RAr, San, SJn; UTAH. 2150–2500 m (7100–8200′). Flowering: May–Aug. Southern Wyoming, Utah, western Colorado, northern Arizona, and New Mexico.

Eremogone fendleri (A. Gray) Ikonn. (for Augustus Fendler, 1813–1883, botanical collector in Venezuela, Panama, Trinidad, and New Mexico) Fendler's sandwort. Perennial herbs, mostly tufted, bluish green. **STEMS** 10–30 (–40) cm long, densely stipitate-glandular. **LEAVES** with basal blades ascending or recurved, filiform, 1–10 cm long, flexuous, apex apiculate to spinose, glabrous to puberulent, not glaucous. **INFL** 3–35-flowered, ± an open cyme; pedicel 3–25 mm long, stipitate-glandular. **FLOWER** sepals linear-lanceolate, margins broadly scarious, apex acuminate, moderately to densely stipitate-glandular on herbaceous portion; petals white, oblong-elliptic to spatulate, 0.9–1.3× sepal length, nectaries as lateral and abaxial rounding of base of filaments, 0.2 mm long. **SEEDS** black, 1.5–1.9 mm long, tuberculate; tubercles rounded, elongate to rounded-conical. $2n = 44$. [*Arenaria fendleri* A. Gray; *A. fendleri* var. *brevifolia* (Maguire) Maguire; *A. fendleri* var. *porteri* Rydb.; *A. fendleri* var. *tweedyi* (Rydb.) Maguire]. Grasslands and sagebrush slopes, ponderosa and oak woodlands, rocky slopes, and alpine meadows. ARIZ: Apa, Nav; COLO: Arc, Con, Dol, Hin, LPl, Min, Mon, RGr, SJn; NMEX: McK, RAr, San, SJn; UTAH. 2200–4000 m (7200–13000′). Flowering: May–Sep. Utah, Colorado, and southern Wyoming south to Arizona and Texas. The root is used only in the summer as a "life medicine," and the powdered root used as a snuff to cause sneezing for a "congested nose" by the Ramah Navajo.

Minuartia L. Starwort

(for J. Minuart, 18th-century Spanish botanist) Annual or perennial herbs with taproot, branched caudex, or trailing rhizomes. **STEMS** ascending to erect or prostrate, sometimes mat-forming, simple or branched, terete or nearly so. **LEAVES** not congested at or near base of flowering stem, without stipules, blade filiform to subulate, lanceolate or oblanceolate. **INFL** an open cyme or flowers solitary. **FLOWERS** perigynous, hypanthium usually disc- or cup-shaped; sepals 5, distinct, apex rounded or obtuse to acute or acuminate, sometimes hooded, awn absent; petals 5, white, not clawed, auricles absent, coronal appendages absent, blade apex entire to notched; stamens 10, arising from hypanthium, staminodes absent; ovary 1-locular; styles 3. **FRUIT** an ovoid capsule, opening by 3 erect to recurved valves. **SEEDS** 5–20, reddish brown to brown or black, smooth, tuberculate, or with marginal papillae, marginal wing absent, appendage absent. About 175 species, temperate and Arctic Northern Hemisphere, North Africa, Asia Minor.

1. Sepal apices obtuse or rounded, hooded .... ***M. obtusiloba***
1′ Sepal apices acute to acuminate or spinescent, not hooded .... (2)

2. Pedicels stipitate-glandular; seeds reddish brown, 0.4–0.5 mm long; capsules 4.5–5 mm long, longer than sepals; sepals 2.5–3.2 mm long, not enlarging in fruit .... ***M. rubella***
2′ Pedicels glabrous; seeds black, 0.7–1 mm long; capsules 3–3.8 mm long, shorter than sepals; sepals ovate to lanceolate, 3.5–5 mm long, to 5.5 mm in fruit .... ***M. macrantha***

Minuartia macrantha (Rydb.) House (large-flowered) House's stitchwort, large-flower sandwort. Perennial herbs, glabrous. **LEAVES** flat to 3-angled distally, subulate to linear, apex rounded, glabrous. **INFL** 1- or 2–5-flowered; bracts broadly subulate, herbaceous or scarious-margined proximally. **FLOWER** sepals ovate to lanceolate, 3.5–5 mm long, to 5.5 mm in fruit, apex sharply acute to acuminate, not hooded, glabrous; petals oblong to obovate. **FRUIT** 3–3.8 mm long, shorter than sepals. **SEEDS** black, 0.7–1 mm long, tuberculate; tubercles low, rounded (50×). [*Alsinanthe macrantha* (Rydb.) W. A. Weber; *Arenaria filiorum* Maguire; *A. rubella* (Wahlenb.) Sm. var. *filiorum* (Maguire) S. L. Welsh; *Minuartia filiorum* (Maguire) McNeill]. Spruce-fir forests, talus slopes, and alpine meadows. COLO: Arc, Con, Hin, LPl, Mon, RGr, SJn. 3200–4000 m (10600–13000′). Flowering: Jun–Sep. Nevada to northwest Wyoming, south to Arizona and New Mexico.

Minuartia obtusiloba (Rydb.) House **CP** (blunt-lobed) Twin-flower sandwort, alpine stitchwort. Perennial herbs, stipitate-glandular. **LEAVES** 3-angled, needlelike to subulate, apex rounded to acute, glabrous. **INFL** 1- or occasionally 2–3-flowered; bracts subulate, herbaceous. **FLOWER** sepals narrowly ovate to oblong, 2.9–6.5 mm long, not enlarging in fruit, apex narrowly rounded, hooded, stipitate-glandular; petals ovate to spatulate. **FRUIT** 3.5–6 mm long, equaling sepals. **SEEDS** reddish tan, 0.6–0.7 mm long, obscurely sculptured (50×). $2n = 26$, approximately 52, 78. [*Arenaria obtusiloba* (Rydb.) Fernald; *Lidia obtusiloba* (Rydb.) Á. Löve & D. Löve]. Alpine meadows and rocky areas including talus slopes. COLO: Arc, Con, Hin, LPl, Min, Mon, RGr, SJn. 3350–4200 m (11000–13800′). Flowering: May–Sep. Northwestern North America; Washington to Montana, south to California and New Mexico.

Minuartia rubella (Wahlenb.) Hiern (slightly red) Beautiful or reddish sandwort, boreal stitchwort. Perennial herbs, usually moderately to densely stipitate-glandular. **LEAVES** flat to 3-angled, subulate, apex acute to apiculate, often stipitate-glandular. **INFL** 3–7- or rarely 1-flowered; bracts herbaceous, scarious-margined. **FLOWER** sepals ovate to lanceolate, 2.5–3.2 mm long, not enlarging in fruit, apex acute to acuminate, not hooded, stipitate-glandular; petals elliptic. **FRUIT** 4.5–5 mm long, longer than sepals. **SEEDS** reddish brown, 0.4–0.5 mm long, tuberculate; tubercles low, elongate, usually rounded (50×). $2n = 24$. [*Arenaria rubella* (Wahlenb.) Sm.; *Tryphane rubella* (Wahlenb.) Rchb.]. Alpine meadows and rock fields. COLO: Arc, Con, Hin, LPl, Min, Mon, SJn. 3450–3750 m (11400–13400′). Flowering: May–Sep. Canada; Washington to Montana and South Dakota, south to Arizona and New Mexico.

Moehringia L. Sandwort

(for P. H. G. Moehring, 18th-century Danzig naturalist) Perennial herbs with slender rhizomes. **STEMS** prostrate or ascending to erect, simple or branched, terete or angled. **LEAVES** not congested at or near base of flowering stem, without stipules, blades oblanceolate to elliptic. **INFL** an open cyme or flowers solitary. **FLOWERS** weakly perigynous, hypanthium minute, disc-shaped; sepals 5, distinct, apex obtuse to acute or acuminate, not hooded, awn absent; petals 5, white, not clawed, auricles absent, coronal appendages absent, blade apex entire; stamens 10, arising from hypanthium, staminodes absent; ovary 1-locular; styles 3. **FRUIT** an ovoid to subglobose capsule, opening by 6 revolute teeth. **SEEDS** 3–6, reddish brown to blackish, shiny, smooth to minutely tuberculate, marginal wing absent, appendage white, elliptic, spongy. About 25 species, temperate North America and Eurasia.

1. Stem pubsecence retrorse; sepals (herbaceous portion) oblong to elliptic, apex mostly rounded or obtuse, 1.7–2.8 (–3) mm; petals about 2× sepal length .... ***M. lateriflora***
1′ Stem pubescence peglike, spreading; sepals (herbaceous portion) lanceolate, apex acute to acuminate, (2.8) 3–6 mm; petals 3/4–1 1/2× sepal length .... ***M. macrophylla***

Moehringia lateriflora (L.) Fenzl (lateral flowers) Grove or blunt-leaf sandwort. Perennial herbs. **STEMS** ascending or decumbent, terete, 5–30 cm long, uniformly retrorsely pubescent. **LEAVES** broadly elliptic to oblong-elliptic or oblanceolate, blades 6–35 mm long, margins granular to minutely serrulate-ciliate, apex obtuse or rounded. **FLOWER** sepals not keeled, ovate or obovate, herbaceous portion oblong to elliptic, 1.7–3 mm long, apex mostly obtuse or rounded; petals 1.7–2.8 (–3) mm long, about 2× sepal length. **FRUIT** subglobose, 3–5 mm long, 1 1/2–2× sepal length. **SEEDS** reniform, 1 mm long, smooth. $2n = 48$. [*Arenaria lateriflora* L.]. Wooded areas; ponderosa pine and Douglas-fir communities. ARIZ: Apa; COLO: Arc, LPl, Mon. 2450–2650 m (8040–8695′). Flowering: May–Aug. Throughout much of North America excluding Texas, Kansas, Oklahoma, and the Southeast.

Moehringia macrophylla (Hook.) Fenzl (large-leaved) Bigleaf sandwort. Perennial herbs. **STEMS** ascending to erect, angled or grooved, 2–18 cm long, hairs minute, spreading, peglike. **LEAVES** lanceolate to elliptic, blades 15–70 mm long, margins smooth to minutely granular, often ciliate in proximal 1/2, apex acute. **FLOWER** sepals at least somewhat keeled, ovate, herbaceous portion lanceolate, 2.8–6 mm, apex acute to acuminate; petals (2.8–) 3–6 mm long, 3/4–1 1/2× sepal length. **FRUIT** ovoid, 5 mm long, about equaling sepals. **SEEDS** oval, 1.5–2.2 mm long, tuberculate. $2n = 48$. [*Arenaria macrophylla* Hook.]. Mostly coniferous woodland communities. COLO: Arc, Hin, LPl, Mon, SJn. 2600–3350 m (8500–10000′). Flowering: Jun–Aug. Canada and adjacent northeastern United States; Washington to Montana, south to California and New Mexico (excluding Nevada).

Paronychia Mill. Nailwort, Whitlow-wort

(Greek *para*, beside, + *onychos*, fingernail, alluding to use for treating whitlow, a disease of fingernails) Perennial herbs with branched, woody base. **STEMS** prostrate, ascending, or erect, simple or branched, terete or angled. **LEAVES** not congested at or near base of flowering stem, with stipules ovate or lanceolate to subulate, scarious, blades linear-subulate or narrowly elliptic or oblong. **INFL** an open cyme or flowers solitary. **FLOWERS** perigynous, hypanthium cup-shaped, expanded distally; sepals 5, connate basally, apex obtuse to rounded, hooded, awned subapically on abaxial surface, awn filiform to stout (often thickened-conic proximally, spinose distally); corolla 0, stamens 5, arising from hypanthium rim opposite sepals, staminodes 5; ovary 1-locular; style 1. **FRUIT** an ovoid utricle. **SEEDS** 1, brown, smooth, marginal wing absent, appendage absent. About 110 species, mostly warm-temperate Americas, Eurasia, and Africa.

1\. Leaf blades narrowly elliptic or oblong, indistinctly nerved, fleshy, not spinulose-tipped, the stipules broadly ovate, entire; plants growing above 3050 m (10000′) .... ***P. pulvinata***
1′ Leaf blades linear-subulate, distinctly nerved, rigid, spinulose-tipped, the stipules lanceolate to subulate, cleft; plants usually growing below 2950 m (9000′) .... ***P. sessiliflora***

Paronychia pulvinata A. Gray (cushionlike) Rocky Mountain nailwort. Perennial herbs, cushion-forming. **STEMS** prostrate, 5–10 cm long. **LEAVES** with ovate stipules, 3–6 mm long, apex subobtuse, entire; blade narrowly elliptic or oblong, 2–5 mm long, fleshy, indistinctly nerved, apex obtuse to subacute, not spinulose-tipped. **INFL** with mostly solitary flowers among the leaves and terminating shoots. **FLOWERS** elliptic-oblong, calyx straight to tapering distally, 2.5–2.8 mm; sepals whitish to green, veins indistinct, narrowly oblong to ovate-oblong, 1.5–1.7 mm long, papery to herbaceous, margins white, 0.2–0.3 mm wide, papery, awn subterminal, hood ascending as continuation of sepal, broadly rounded to notched, awn erect, 0.3–1 mm long, white, ± glabrous spine; style 0.8–1 mm long. **FRUIT** glabrous. $2n = 32$. [*P. pulvinata* var. *longiaristata* Chaudhri; *P. sessiliflora* Nutt. subsp. *pulvinata* (A. Gray) W. A. Weber]. Alpine meadows and rocky areas. COLO: Con, Hin, Min. 3650–4250 m (12000–14000′). Flowering: Jun–Aug. Southern Wyoming, eastern Utah, Colorado, and northern New Mexico.

Paronychia sessiliflora Nutt. (stalkless flower) Creeping nailwort. Perennial herbs, mat-forming. **STEMS** erect to ascending, 5–25 cm long. **LEAVES** with lanceolate to subulate stipules, 2–3 mm long, apex long-attenuate, often deeply cleft; blade linear-subulate, 4–7.5 mm long, leathery, distinctly nerved, apex acute or shortly cuspidate-mucronate. **INFL** mostly in a congested 3–6-flowered terminal cyme. **FLOWER** calyx ovoid, narrowing distally, 3.6–5 mm long; sepals green to red-brown, midrib and lateral pair of veins prominent, lanceolate-oblong, 1.5–2 mm long, leathery, margins whitish to translucent, 0.1–0.2 mm wide, scarious, awn terminal, hood ± obscure, narrowly rounded, awn erect to somewhat spreading, 1–2 mm long, narrowly conic in proximal 1/2 with white, scabrous spine; style 1.4–1.5 mm long. **FRUIT** densely pubescent in distal 1/2. $2n = 64$. Sagebrush slopes, piñon-juniper woodland, and clay badlands communities. NMEX: SJn; UTAH. 1800–2000 m (6000–6500′). Flowering: May–Jun. Southwest Canada south to Nevada, Utah, New Mexico, and Texas.

Pseudostellaria Pax Sticky Starwort

(Greek *pseudo*, false, referring to a resemblance to *Stellaria*) Perennial herb, rhizomes with spherical or elongate, tuberous thickenings. **STEMS** ascending to erect, simple or branched, 4-angled. **LEAVES** not congested at or near base of flowering stem, without stipules, blades linear to broadly lanceolate. **INFL** an open cyme or flowers solitary. **FLOWERS** hypogynous; sepals 5, distinct, apex obtuse to acute, not hooded, awn absent; petals 5, white, not clawed, auricles absent, coronal appendages absent, blade apex bilobed 1/8–1/5; stamens 5, arising from ovary base, staminodes absent; ovary 1-locular; styles 3. **FRUIT** a capsule, opening by 6 2–3× recoiled valves. **SEEDS** 1–3, brown, tuberculate, marginal wing absent, appendage absent. About 21 species, western United States, southeast Europe, and eastern Asia.

Pseudostellaria jamesiana (Torr.) W. A. Weber & R. L. Hartm. (for Edwin James, 1797–1861, the first botanical collector of the central Rockies) Perennial herbs from rhizomes with spherical or elongate tuberous thickenings 0.5–2.5 cm broad. **STEMS** 4-angled, 12–60 cm long, glabrous proximally or stipitate-glandular throughout or at least in inflorescence, often densely so. **LEAVES** with blades mostly 2–10 cm long, margins flat to briefly revolute, smooth or granular to serrulate, glabrous or stipitate-glandular. **INFL** an open cyme, flowers often proliferating with age; pedicels recurved to reflexed in fruit, uniformly stipitate-glandular. **FLOWER** sepals lanceolate to narrowly ovate, 3–6 mm long, stipitate-glandular, often densely so; petals 7–9.5 mm long, apex with V-shaped notch 1–2 mm deep, lobes broadly rounded; anthers 10, purple; styles 3.5–4.5 mm long; stigmas terminal. **FRUIT** 4.5–5 mm long. **SEEDS** 1–3, reddish brown, broadly elliptic, ± plump, 2–3.4 mm long; tubercles conical to elongate, rounded. $2n = 96$. [*Stellaria jamesiana* Torr.]. Meadows, shrublands, and coniferous forests. ARIZ: Apa, Nav; COLO: Arc, LPl, Mon, SJn; NMEX: McK, RAr, SJn; UTAH. 2100–3200 m (7000–10500′). Flowering: May–Aug. Washington to Montana, south to California, Arizona, New Mexico, and Texas. A poultice of the plant is applied to hailstone injuries and the plant is chewed for corral dance by the Kayenta Navajo.

Sagina L. Pearlwort

(Latin *sagina*, ancient name for *Spergula*, which was once considered congeneric) Annual or perennial herbs with taproot. **STEMS** ascending to decumbent, simple or branched, terete to somewhat angled. **LEAVES** not congested at or near base of flowering stem, without stipules, blades linear. **INFL** a cyme or flowers solitary. **FLOWERS** hypogynous; sepals usually 5, distinct, apex obtuse or rounded to somewhat acute, not hooded, awn absent; petals usually 5, white, not clawed, auricles absent, coronal appendages absent, blade apex entire; stamens usually 5, arising from ovary base, staminodes absent; ovary 1-locular; styles 5, rarely 4. **FRUIT** an ovoid capsule, opening by 4 or 5 recurved valves. **SEEDS** 5–24, tan to brown, smooth to tuberculate, marginal wing absent, appendage absent. Fifteen to 20 species, chiefly cold-temperate Northern Hemisphere.

Sagina saginoides (L.) H. Karst. (fattening) Perennial herbs, tufted to caespitose, glabrous. **STEMS** ascending or sometimes procumbent, branched. **LEAVES** with blades 10–20 mm long, not succulent, apex usually apiculate, glabrous. **INFL** with frequently recurved pedicels after flowering, erect in fruit, filiform, glabrous. **FLOWERS** mostly 5-merous, sepals elliptic, 2–2.5 mm long, glabrous, margins hyaline, white to purple, apex obtuse to rounded, remaining appressed to capsule; petals elliptic, 1.5–2 mm long; stamens mostly 10. **FRUIT** 2.5–3.5 mm long, 1.5–2× sepal length, dehiscing to base. **SEEDS** brown, obliquely triangular with distinct adaxial groove, 0.3–0.4 mm long, smooth to slightly roughened. $2n = 22$. Meadows, stream margins, disturbed areas, coniferous forests, alpine meadows and talus slopes. COLO: Arc, Con, Hin, LPl, Min, Mon, SJn; NMEX: SJn; UTAH. 2150–3700 m (7100–12400′). Flowering: May–Sep. Northern North America; Washington to Montana, south to California, Arizona, and New Mexico.

Saponaria L. Soapwort, Bouncing Bet

(Latin *saponis*, soap, + *aria*, pertaining to, referring to the sap) Perennial herbs with slender to stout rhizomes. **STEMS** erect to spreading, simple or branched, terete. **LEAVES** not congested at or near base of flowering stem, without stipules, blades ovate to elliptic. **INFL** a dense to open cyme. **FLOWERS** hypogynous; sepals 5, connate proximally into tube, terete, apex acute to acuminate, not hooded, awn absent; petals 5, pink to white, clawed, auricles absent, coronal appendages 2, blade apex entire or emarginate; stamens 10, adnate with petals to carpophore, staminodes absent; ovary 1-locular; styles 2. **FRUIT** a cylindrical to ovoid capsule, opening by 4 ascending or recurved teeth. **SEEDS** 20–75, dark brown, papillose, marginal wing absent, appendage absent. About 40 species, Eurasia and Africa.

Saponaria officinalis L. (medicinal) Bouncing bet. Plants perennial, colonial. **STEMS** erect, simple or branched distally, 30–90 cm long. **LEAVES** with an often absent or winged petiole 0.1–1.5 cm long; blade strongly 3 (–5)-veined, 3–15 cm long. **INFL** a dense to open cyme. **FLOWERS** sometimes double; calyx green or reddish, often cleft, 15–25 mm long; petals pink to white, often drying to dull purple, blade 8–15 mm long. **FRUIT** 15–20 mm long. **SEEDS** 1.6–2 mm wide. $2n = 28$. Roadsides, floodplains, and weedy areas. COLO: Arc, LPl, Mon; NMEX: McK, RAr, SJn. 1800–2100 m (6000–6900′). Flowering: Jun–Sep. Introduced throughout most of North America.

Silene L. Campion, Catchfly

(Greek *seilenos*, Silenus, intoxicated foster father of Greek god Bacchus, who was described as covered with foam, perhaps alluding to the viscid nature of many species) Annual or perennial herbs with taproot or branched, woody caudex. **STEMS** ascending to erect, simple or branched, terete to somewhat angled. **LEAVES** not congested at or near base of flowering stem, without stipules, blades linear to lanceolate or broadly elliptic. **INFL** with congested cymes, or frequently flowers few or solitary. **FLOWERS** hypogynous; sepals 5, connate proximally into tube, terete, apex triangular to linear-acuminate, not hooded, awn absent; petals 5, white, pink, or scarlet, clawed, auricles 2, coronal appendages 2, blade apex 2–4-lobed, stamens 10, adnate with petals to carpophore, staminodes absent; ovary 1- or 3–5-locular; styles usually 3 or 5. **FRUIT** an ovoid to cylindric capsule, opening by 3–5 valves, frequently splitting into 6–10 teeth. **SEEDS** 20–60, reddish to black, tuberculate or papillate, marginal wing present (2 species), appendage absent. About 700 species, mainly Northern Hemisphere (Morton 2005b; Hitchcock and Maguire 1947).

1. Calyx glabrous .... (2)
1′ Calyx variously pubescent .... (4)

2. Plants densely matted and forming a cushion; stems 4–10 cm tall; leaves linear, less than 1.5 mm wide; flowers solitary, bright pink or rarely white in some individuals; plants of alpine areas .... ***S. acaulis***
2′ Plants not densely matted; stems 15–80+ cm tall; leaves lanceolate to oblanceolate, most 3–35 mm wide; flowers mostly 2–many in open cymes, variously colored; plants mostly of lower elevations .... (3)

3. Calyx 10-ribbed (in fruit), with obvious reticulations of secondary veins absent, ovoid to ellipsoid, 4–6 mm wide, base tapering into pedicel, lobes (teeth) narrowly triangular, 1–2 mm long, carpophore less than 1 mm long; petals in flower 1.1–1.2× calyx length; stamens included; styles included; upper internodes often with a glutinous band .... ***S. antirrhina***
3′ Calyx 20-nerved (in fruit), with distinctive reticulations of secondary veins, campanulate, 10–15 mm wide, base depressed at attachment to pedicel, lobes (teeth) broadly triangular, 3–4 mm long, carpophore 2–3 mm long; petals in flower 2× calyx length; stamens exserted 2–4 mm; styles 2× calyx length; upper internodes glabrous .... ***S. vulgaris***

4. Petals scarlet, lobes 4–6, linear to oblong, central pair of lobes 4–10+ mm long; calyx (in flower) 1.9–2.3 cm long, cylindrical .... ***S. laciniata***
4′ Petals white to greenish white or pink, if red or purple then calyx (in flower) 0.9–1.4 cm long and ovoid to ellipsoid, lobes 2, rounded or blade emarginate .... (5)

5. Plants 2–20 cm tall; cauline leaves in 1–3 pairs, most shorter than the internodes; seed winged, wing 0.25–1× width of body of seed; plants of alpine areas .... (6)
5′ Plants mostly 20–80 cm tall, if less than 10 cm (*S. menziesii*), then cauline leaves in 5–10+ pairs and closely overlapping; seed not winged; plants mostly of lower elevations .... (7)

6. Flower nodding, calyx broadly elliptic to campanulate (pressed), 12–14 mm long, 6–10 mm wide (to 16 mm, 15 mm, respectively, in fruit), thin and membranous between nerves; petals purple-red; seed 1.5–2 mm long, marginal wing nearly the width of body of seed .... ***S. uralensis***
6′ Flower erect, calyx elliptic (pressed), 6–12 mm long, 5–8 mm wide, enlarging little in fruit, thicker in texture; petals white to pink; seed 0.7–1.2 mm long, marginal wing about 1/4 the width of body of seed .... ***S. hitchguirei***

7. Plants annual or perennial; calyx usually 15–25 cm long; corolla white or pinkish, petal blades 7–15 mm long; introduced, of disturbed places .... (8)
7′ Plants perennial; calyx usually 5–16 mm long; corolla white to pink or purple, petal blades 1–6 mm long; native, of undisturbed areas .... (9)

8. Flowers perfect; calyx teeth 6–13 mm long, linear-lanceolate to subulate, nerves 10, prominently anastomosing in mature flowers; styles 3; capsule teeth 6 .... ***S. noctiflora***
8′ Flowers unisexual, plants usually dioecious; calyx teeth 3–5 (6) mm long, narrowly to broadly triangular, nerves 10 in staminate flowers, 20 in pistillate flowers, without prominent anastomoses; styles 5; capsule teeth mostly 5 or 10 .... ***S. latifolia***

9. Calyx in flower 5–7 mm long, nerves obscure; cauline leaves 5–20, of relatively uniform size and distribution .... ***S. menziesii***

9′ Calyx in flower usually 8–16 mm long, nerves prominent; cauline leaves mostly 3–5, usually progressively and markedly reduced above ....(10)

10. Inflorescence with 2–8+ flowers per node, flowering nodes 4–6+, flowering progressing up the inflorescence; styles mostly 3; capsule teeth 3 or 6 .... ***S. scouleri***
10′ Inflorescence with 1 or 2 flowers (occasionally flowering branches) per node, flowering nodes mostly 2 or 3, flowering beginning from apex as determinate cymes or only central flower developing; styles mostly 5; capsule teeth mostly 5 .... ***S. drummondii***

Silene acaulis L. (stalkless) Moss campion. Perennial herbs, densely matted and forming cushions. **STEMS** 1–3 cm tall, internodes glabrous. **LEAVES** 3–6 pairs per stem, 0.4–1.5 cm long, 0.5–1.2 mm wide, linear, bases closely enveloping internodes. **INFL** a solitary flower, pedicel erect or ascending. **FLOWERS** perfect and unisexual; calyx tubular to campanulate (pressed), 6–9 mm long, 2–3 mm wide (in fruit to 13 mm and 5 mm, respectively), 10-nerved, not anastomosing, glabrous, lobes lanceolate to ovate, 1–1.5 mm long, base becoming depressed at pedicel; petals bright pink, rarely white in some individuals, 1.2–1.5× calyx length, rounded to bilobed, lobes rounded, appendages poorly developed; stamens exserted 0–1 mm in staminate flowers; styles 3, exserted 2–3 mm. **FRUIT** opening by 6 teeth; carpophore 1 mm long. **SEEDS** light brown, 0.8–1 mm long, margins neither prominently papillate nor winged. $2n = 24$. [*S. acaulis* subsp. *subacaulescens* (F. N. Williams) Hultén]. Subalpine and alpine meadows and talus slopes. COLO: Arc, Con, Hin, LPl, Min, Mon, RGr, SJn. 3650–4250 m (12000–14000′). Flowering: Jun–Sep. Canada; Washington to Montana, south to Arizona and New Mexico.

Silene antirrhina L. (with leaves like *Antirrhinum*) Sleepy catchfly. Annual herbs (ours), caudex 1–few-branched. **STEMS** 15–70 cm tall, upper internode(s) with a glutinous band accumulating debris. **LEAVES** 7 (11) pairs per stem, 1–7 cm long, 2–12 mm wide, narrowly oblanceolate to linear, much shorter to occasionally 1.3× internode length. **INFL** an open, 1–25+-flowered cyme, pedicel erect to ascending. **FLOWERS** perfect; calyx ovoid to ellipsoid (pressed), 5–6 mm long, 2–3 mm wide (in fruit to 8 mm and 5 mm, respectively), 10-nerved, not anastomosing, glabrous, lobes (teeth) narrowly triangular, 1–2 mm long, base tapered at pedicel; petals mostly dark red throughout or in part, 1.1–1.2× calyx length, bilobed, lobes rounded, appendages 2, 0.1–0.4 mm long; stamens included; styles 3, included. **FRUIT** opening by 6 teeth; carpophore 0.8–1 mm long. **SEEDS** gray-brown, 0.5–0.8 mm long, margins neither prominently papillate nor winged. $2n = 24$. Desert shrub and piñon-juniper woodland communities; grasslands, riparian areas, and disturbed sites. ARIZ: Apa; COLO: Dol, LPl, Mon; NMEX: McK, SJn; UTAH. 1100–2400 m (3600–6800′). Flowering: Mar–Jun. Introduced throughout most of North America.

Silene drummondii Hook. (for Thomas Drummond, 1780–1835, Scottish botanical explorer and naturalist) Drummond's catchfly, forked catchfly. Perennial herbs, caudex 1–few-branched. **STEMS** 20–70 cm tall, internodes uniformly stipitate-glandular, without glutinous bands. **LEAVES** 2–5 pairs per stem, 3–10 cm long, 4–12 mm wide, narrowly lanceolate to oblanceolate or linear, most shorter than internodes. **INFL** with 1 or 2 flowers (occasionally flowering branches) per node, flowering nodes mostly 2 or 3, flowering beginning from apex as determinate cymes or only central flower developing. **FLOWERS** perfect; calyx broadly tubular to ellipsoid (pressed), 12–18 mm long, 4–8 mm wide (in fruit to 14–18 mm and 5–8 mm, respectively), 10-nerved, not anastomosing, glandular-pubescent, lobes (teeth) triangular, 1–1.5 mm long, base tapered at pedicel, petals white, often dark red at least in part, slightly exceeding calyx, bilobed, lobes rounded, appendages 2, 0.1–0.4 mm; stamens included; styles mostly 5, included. **FRUIT** opening by mostly 5 teeth. **SEEDS** dark brown, 0.7–1 mm long, margins minutely papillate, not winged. Root used as a "life medicine" by the Ramah Navajo.

1. Petals not exceeding calyx, white; fruiting calyces mostly broadly ovoid (pressed), 2.25–3× as long as broad; seeds 0.7–0.8 mm long; inflorescence mostly 3–10-flowered .... **var. *drummondii***
1′ Petals exserted 4–6 mm, dark red to purple apically; fruiting calyces narrowly ovoid to ellipsoid (pressed), 2× as long as broad; seeds 1–1.2 mm long; inflorescence mostly 1–4-flowered .... **var. *striata***

var. *drummondii* **INFL** mostly 3–10-flowered. **CALYX** in fruit mostly broadly ovoid (pressed), 2.25–3× as long as broad. **PETALS** not exceeding calyx, white. Oak thickets, ponderosa pine, aspen forests, riparian areas, and subalpine and alpine meadows and talus slopes. ARIZ: Apa; COLO: Arc, LPl, Min, Mon, SJn; NMEX: McK, SJn; UTAH. 2135–3810 m (7000–12500′). Flowering: May–Aug. British Columbia to Manitoba, south to Arizona, New Mexico, and Nebraska.

var. *striata* (Rydb.) Bocquet (striped) **INFL** typically 1–4 (–8)-flowered. **CALYX** in fruit narrowly ovoid to ellipsoid (pressed), 2× as long as broad. **PETALS** exserted 4–6 mm, dark red to purple apically. [*S. drummondii* subsp. *striata*

(Rydb.) J. K. Morton]. Ponderosa pine and spruce-fir forests, subalpine and alpine meadows and rocky areas. ARIZ: Apa; COLO: Hin, LPl, Min, RGr, SJn; NMEX: McK, SJn; UTAH. 2500–4000 m (8200–13000′). Flowering: Jun–Aug. Saskatchewan south to Nevada, Arizona, and New Mexico.

Silene hitchguirei Bocquet (contraction of C. L. Hitchcock and B. Maguire) Mountain campion. Perennial herbs, caudex 1–few-branched, with tuft of basal leaves. **STEMS** 2–10 cm tall, internodes uniformly stipitate-glandular, septa purple. **LEAVES** 1–3 pairs per stem, 5–8 cm long, 5–10 mm wide, narrowly oblanceolate and long-spatulate, most shorter than internodes. **INFL** mostly 1–4-flowered. **FLOWERS** perfect; calyx elliptic (pressed), 6–12 mm long, 5–8 mm wide, enlarged little in fruit, 10-nerved, anastomosing apically, densely pubescent, hairs purple-septate, lobes triangular, 1.5–4 mm long, base tapered at pedicel, petals white or pink, 1–1.5× calyx, emarginate, appendages 2, 0.2–0.3 mm long; stamens equaling calyx; styles 5, equaling calyx. **FRUIT** opening by 5 or 10 teeth; carpophore 0.5–1 mm long. **SEEDS** brown, 0.7–1.2 mm long, margin not papillate, wing less than 1/4 width of body of seed. $2n = 24$. [*Lychnis apetala* L. var. *montana* (S. Watson) C. L. Hitchc.; *Silene uralensis* (Rupr.) Bocquet subsp. *montana* (S. Watson) McNeill]. Subalpine and alpine meadows; rocky areas. COLO: Hin, Min, SJn. 3400–4200 m (11200–13850′). Flowering: Jun–Aug. Alberta south to Utah and Colorado.

Silene laciniata Cav. (deeply cut) Gregg's Mexican pink or campion. Perennial herbs, caudex 1–few-branched. **STEMS** 20–60 cm tall, internodes uniformly stipitate-glandular, without glutinous bands. **LEAVES** 4–7 pairs per stem, 3–6 cm long, 8–30 mm wide, lanceolate or elliptic to oblanceolate, shorter than to occasionally 1.5× internode length. **INFL** a (1) 3–9-flowered cyme, pedicel mostly ascending to erect. **FLOWERS** perfect; calyx cylindrical (pressed), 19–23 mm long, 5–6 mm wide, (in fruit to 25 mm and 8 mm, respectively), 10-nerved, anastomosing near lobes, stipitate-glandular with clear cross walls, lobes 3–3.5 mm long, triangular, bases becoming depressed at pedicel; petals scarlet, 1.4–1.6× calyx length, 4–6-lobed, lobes linear to oblong, appendages 2, 1–2 mm long; stamens exserted 8–10 mm; styles 3, exserted 9–12 mm. **FRUIT** opening by 6 teeth; carpophore 2–4 mm long. **SEEDS** reddish brown, 1.2–2.5 mm long, margin long-papillate, not winged. $2n = 48$. Oak woodlands, ponderosa pine and spruce forests. ARIZ: Apa; NMEX: McK, SJn. 2100–2700 m (6900–8900′). Flowering: Jul–Sep. Arizona, New Mexico, and Texas; Mexico. Our material belongs to **var. *greggii*** (A. Gray) S. Watson (for Josiah Gregg, 1806–1850, merchant and explorer who collected plants in the region around Santa Fe, New Mexico).

Silene latifolia Poir. (broad-leaved) White campion or cockle. Annual or short-lived perennial herb, caudex 1–few-branched, plants of disturbed areas. **STEMS** 30–80+ cm tall, internodes uniformly pubescent, often stipitate-glandular. **LEAVES** 3–8 pairs per stem, 2–12 cm long, 6–25 mm wide, lanceolate to elliptic, much shorter than to 2× internode length. **INFL** open, 3–30-flowered cymes, pedicels erect to widely spreading. **FLOWERS** unisexual; calyx in staminate flowers tubular (pressed), 15–20 mm long, 5–8 mm wide (in fruit to 22 mm and 9 mm, respectively), 10-nerved, not anastomosing, in pistillate flowers elliptic, 15–25 mm long, 5–9 mm wide (in fruit to 23 mm and 15 mm, respectively), 20-nerved, not anastomosing, lobes narrowly to broadly triangular, 3–6 mm long, tapering at base, often abruptly; petals white, 0.4–0.6× calyx length, mostly deeply bilobed, lobes rounded, appendages 2, 1–1.5 mm long; stamens exserted 5–7 mm; styles mostly 5, exserted 7–10 mm. **FRUIT** opening by mostly 5 or 10 teeth; carpophore 1–2 mm long. **SEEDS** dark gray-brown, 1.2–1.5 mm long, margins neither prominently papillate nor winged. [*Lychnis alba* Mill.; *Silene pratensis* (Raf.) Gren. & Godr.]. Roadsides, alluvial flats, other disturbed sites. COLO: Arc, Hin, Min. 1800–2900 m (6000–9500′). Flowering: May–Aug. Introduced throughout North America.

Silene menziesii Hook. (for Archibald Menzies, 1754–1842, British physician, gardener, botanist, and plant collector in northwest North America) Menzies' catchfly. Perennial herbs, shoots decumbent to matted at base. **STEMS** mostly 5–30 cm tall, internodes uniformly stipitate-glandular, without glutinous bands. **LEAVES** 5–20 pairs per stem, 2–4 (6) cm long, 3–20 mm wide, broadly elliptic to oblanceolate, mostly 1–3× internode length. **INFL** of solitary, terminal flowers or 3–7+ in a leafy cyme. **FLOWERS** functionally perfect or unisexual; calyx campanulate (pressed), 5–7 mm long, 3–6 mm wide, enlarged little in fruit, obscurely 10-nerved, not anastomosing, puberulent to stipitate-glandular, lobes 1.5–3 mm long, bases tapered at pedicel; petals white, 1–1.7× calyx length, 2-lobed, lobes oblong, appendages 2, 0.1–0.3 mm long; stamens equaling calyx in staminate flowers; styles mostly 3, exserted 1–3 mm. **FRUIT** opening by 3 or 6 teeth; carpophore 0.5–1 mm long. **SEEDS** brown, 0.6–1 mm long, margins neither prominently papillate nor winged. $2n = 24, 48$. [*S. menziesii* var. *viscosa* (Greene) C. L. Hitchc. & Maguire]. Oak and aspen woodlands, ponderosa, Douglas-fir, and spruce forests. ARIZ: Apa; COLO: Arc, Hin, LPl, Min, Mon, SJn; NMEX: McK, RAr, SJn. 2225–3350 m (7350–11000′). Flowering: May–Aug. Northwestern North America; Washington to Montana, Nebraska, south to Arizona and New Mexico.

Silene noctiflora L. (night-flowering) Night-flowering catchfly. Annual herbs, caudex 1–few-branched, plants of disturbed areas. **STEMS** 30–70+ cm tall, internodes uniformly stipitate-glandular, without glutinous bands. **LEAVES** 5–9 pairs per stem, 1–12 cm long, 3–35 mm wide, broadly elliptic to lanceolate, much shorter to 2× internode length. **INFL** a 3–9-flowered cyme, pedicels erect to widely spreading. **FLOWERS** perfect; calyx ovate-elliptic (pressed), 17–20 mm long, 5–8 mm wide (in fruit to 22 cm and 12 mm, respectively), 10-nerved with prominent anastomoses, densely pilose, often stipitate-glandular, lobes 6–13 mm long, linear-lanceolate to subulate, base tapered, often abruptly, at pedicel; petals white or pinkish, 1.2–1.3× calyx length, deeply bilobed, lobes usually oblong, appendages 2, 0.5–1.5 mm long; stamens exserted 1–3 mm; styles 3, exserted 3–4 mm. **FRUIT** opening by 6 teeth; carpophore 1–3 mm long. **SEEDS** grayish brown, 1–1.6 mm long, neither winged nor papillate. $2n = 24$. Disturbed areas. COLO: Arc. 2300 m (7500′). Flowering: Jun–Sep. Introduced throughout North America. A poultice of leaves applied to a prairie dog bite by the Ramah Navajo.

Silene scouleri Hook. (for John Scouler, a Scottish geologist and botanist, naturalist on the voyage to the Columbia River with David Douglas in 1824–1825) Hall's catchfly. Perennial herbs, caudex 1–several-branched. **STEMS** mostly 25–70 cm tall, internodes uniformly stipitate-glandular, without glutinous bands. **LEAVES** 1–12 pairs per stem, 6–25 cm long, 4–30 mm wide, ovate-lanceolate to lanceolate, oblanceolate, or rarely linear, much shorter than to 2× internodes. **INFL** 2–8+ flowers per node, flowering nodes 4–6+, flowering progressing up the inflorescence. **FLOWERS** perfect; calyx campanulate (pressed), 8–13 mm long, 3–7 mm wide (in fruit to 18 mm and 8 mm, respectively), 10-nerved, anastomosing apically, densely stipitate-glandular, lobes lanceolate, 2–5 mm long, base becoming depressed at pedicel; petals white, greenish white, or pink, 1.2–1.3× calyx length, deeply 2-lobed, lobes broadly oblong, appendages 2, 1–3 mm long; stamens equaling calyx; styles mostly 3, exserted 4–6 mm. **FRUIT** opening by mostly 6 teeth; carpophore 1.5–4 mm long. **SEEDS** grayish brown, 0.9–1.5 mm long, margins neither prominently papillate nor winged. $2n = 48$, 96. Oak and juniper thickets, meadows, subalpine forests, and alpine meadows and rocky areas. ARIZ: Apa; COLO: Arc, Hin, LPl, Min, Mon, SJn; NMEX: SJn; UTAH. 1800–3600 m (6000–11800′). Flowering: May–Sep. British Columbia and Alberta south to Oregon, Idaho, and New Mexico. Our material belongs to subsp. ***hallii*** (S. Watson) C. L. Hitchc. & Maguire (for Elihu Hall, 1820–1882, botanist, farmer; with J. P. Harbour he collected plants in the mountains of central Colorado in 1862).

Silene uralensis (Rupr.) Bocquet (of the Ural Mountains) Nodding campion. Perennial herbs, caudex 1–few-branched, with tuft of basal leaves. **STEMS** 8–20 cm tall, internodes uniformly stipitate- and nonstipitate-glandular, septa purple. **LEAVES** 1–3 pairs per stem, 0.5–4 cm long, 2–4 mm wide, linear to linear-lanceolate, most shorter than internodes. **INFL** mostly 1 (3)-flowered per stem, pedicels recurved, becoming erect in fruit. **FLOWERS** perfect; calyx broadly elliptic to campanulate (pressed), 12–14 mm long, 6–10 mm wide (in fruit to 16 mm and 15 mm, respectively), 10-nerved, anastomosing apically, densely pubescent, hairs purple-septate, lobes triangular to oblong, 2.5–4 mm long, base tapered or abruptly narrowed at pedicel; petals dingy pink to purple, 1–1.3× calyx, emarginate to 2-lobed, lobes rounded, appendages 2, 0.2–0.3 mm long; stamens equaling or slightly exceeding calyx; styles 5, equaling or slightly exceeding calyx. **FRUIT** opening by 5 or 10 teeth; carpophore 1–2 mm long. **SEEDS** brown, 1.5–2 mm long, margin not prominently papillate, wing about equaling width of body of seed. $2n = 24$. [*Lychnis apetala* L. var. *attenuata* (Farr) C. L. Hitchc.]. Alpine meadows and rocky areas. COLO: Hin, SJn. 3800–3900 m (12500–12800′). Flowering: Jun–Aug. Northern North America; Montana to Utah and Colorado. Our material belongs to var. ***uralensis***.

Silene vulgaris (Moench) Garcke (common) Bladder campion. Perennial short-lived herbs, caudex 1–few-branched. **STEMS** 20–45 cm tall, upper internodes glabrous. **LEAVES** 7–12 pairs per stem, 2–8 cm long, 5–30 mm wide, broadly oblong to oblanceolate or lanceolate, much shorter to 1.5× internodes. **INFL** an open 5–50-flowered cyme, pedicel erect to spreading. **FLOWERS** perfect; calyx campanulate (pressed), 10–16 mm long, 7–11 mm wide (in fruit to 20 mm and 15 mm, respectively), in fruit 20-nerved with distinctive reticulations of secondary veins, glabrous, base depressed at pedicel, lobes (teeth) broadly triangular, 2–3 mm long; petals white, 1.3–1.5× calyx length, mostly bilobed, lobes broadly rounded, appendages absent or essentially so; stamens exserted 1–3 mm; styles 3, exserted 5–12 mm. **FRUIT** opening by 6 teeth; carpophore 2–3 mm long. **SEEDS** brown, 1–1.3 mm long, margins neither prominently papillate nor winged. [*Silene cucubalus* Wibel]. Disturbed areas. COLO: Arc. 1800–2400 m (6000–8000′). Flowering: Jun–Aug. Introduced throughout much of North America.

Spergularia (Pers.) J. Presl & C. Presl Sand-spurry

(derivative of the genus *Spergula*) Annual or perennial herbs with taproot. **STEMS** erect to sprawling, simple to freely branched distally or throughout, terete. **LEAVES** not congested at or near base of flowering stem, with stipules lanceolate to triangular, scarious, blades threadlike to linear. **INFL** symmetrical or 1-sided cymes or flowers solitary. **FLOWERS** briefly perigynous; sepals 5, briefly connate, apex acute to obtuse, not hooded, awn absent; petals 5, white to pink, not clawed, auricles absent, coronal appendages absent, blade apex entire; stamens 1–10, arising from rim of hypanthial disc, staminodes absent; ovary 1-locular; styles 3. **FRUIT** an ovoid capsule, opening by 3 spreading valves with recurved tips. **SEEDS** 30–90, dark or reddish brown, smooth to papillate, submarginally grooved, marginal wing present or absent, appendage absent. About 60 species, mostly temperate and arid America and Mediterranean region of Europe and Africa (Rossbach 1940).

1. Stipules conspicuous, shiny white, lanceolate; axillary leaves 2–4+ per cluster; seed sculpturing of parallel, wavy lines; stamens 6–10 .... ***S. rubra***
1′ Stipules inconspicuous, dull white, broadly triangular; axillary leaf clusters mostly absent; seeds ± smooth; stamens 1–3 (5) .... ***S. salina***

Spergularia rubra (L.) J. Presl & C. Presl (red) Red sand-spurry. Annual or short-lived perennial herbs, delicate, 4–25 cm tall, with stalked glands distally. **STEMS** with primary ones 0.3–0.5 mm diameter proximally. **LEAVES** with conspicuous stipules, shiny white, lanceolate, 3.5–5 mm long, apex long-acuminate; blade filiform to linear, 0.4–1.5 cm long, scarcely fleshy, apex apiculate to spine-tipped; axillary leaves 2–4+ per cluster. **INFL** a cyme, simple or 3 or more times compound or flowers solitary and axillary. **FLOWER** sepal lobes lanceolate, 2–3.2 mm long, to 4 mm in fruit, apex obtuse to acute; petals pink, obovate to ovate; stamens 6–10. **FRUIT** 3.5–5 mm long, equaling to 1.2× sepal length. **SEEDS** red-brown to dark brown, broadly ovate or ± truncate, angular at broad end, plump, 0.4–0.6 mm long, sculpturing of parallel, wavy lines, margins with peglike papillae (30×); wing absent. $2n = 18, 27, 36, 54$. Along trails and other disturbed sites. COLO: LPl, Min, Mon, SJn. 2700–3300 m (8900–10800′). Flowering: May–Sep. Alaska south through Washington and Montana to California and New Mexico (excluding Arizona); eastern North America.

Spergularia salina J. Presl & C. Presl (salt) Salt-marsh sand-spurry, lesser sea-spurry. Annual herbs, delicate, 8–30 cm tall, with stalked glands at least distally. **STEMS** with primary ones 0.6–2 (–3) mm diameter proximally. **LEAVES** with inconspicuous stipules, dull white, broadly triangular, 1.2–3.5 mm long, apex acute to short-acuminate; blade linear, (0.8–) 1.5–4 cm long, fleshy, apex blunt to apiculate; axillary leaf clusters usually absent. **INFL** a cyme, simple or 3 or more times compound or flowers solitary and axillary. **FLOWER** sepal lobes ovate to elliptic, 2.5–4.5 mm long, to 4.8 mm in fruit, apex rounded to acute; petals white or pink to rosy, ovate to elliptic-oblong; stamens 2–3, rarely 5. **FRUIT** 2.8–6.4 mm long, equaling to 1.5× sepal length. **SEEDS** light brown to reddish brown, broadly ovate, mostly plump, 0.5–0.7 (–0.8) mm long, dull, ± smooth, often with gland-tipped papillae (30×); wing usually absent or incomplete. $2n = 18, 36$. [*S. marina* (L.) Griseb.]. Alkali seeps. UTAH. 1100–1380 m (3700–4525′). Flowering: Apr–Jun. Throughout much of North America except the highest latitudes.

Stellaria L. Chickweed, Starwort

(Latin *stella*, star, + *aria*, pertaining to, referring to the shape of the flower) Annual or perennial herbs with taproot or slender rhizomes. **STEMS** prostrate to ascending or erect, simple or branched, terete or 4-angled. **LEAVES** not congested at or near base of flowering stem, without stipules, blades linear or lanceolate to broadly ovate or elliptic. **INFL** an open cyme or umbellate or flower solitary. **FLOWERS** hypogynous or weakly perigynous; sepals 4 or 5, distinct, apex acute, acuminate, or obtuse, not hooded, awn absent; petals 4 or 5, white (sometimes translucent in *S. borealis*), not clawed, auricles absent, coronal appendages absent, blade apex bifid for 2/3–4/5 its length; stamens 4, 5, or 10, arising from nectariferous disc (prominent in *S irrigua*), staminodes absent; ovary 1-locular; styles usually 3. **FRUIT** an ovoid to globose capsule, opening by 6 (8 or 10) ascending to recurved valves. **SEEDS** 5–35, yellow-brown to dark brown, papillate or rugose, marginal wing absent, appendage absent. About 120 species worldwide, mainly north-temperate regions (Morton 2005c).

1. Plant annual, taprooted; stem terete, villous in a line between nodes; at least basal leaves with petiole 5–20+ mm long .... ***S. media***
1′ Plant perennial, rhizomatous; stem 4-angled (except *S. obtusa*), glabrous or uniformly pubescent; leaves sessile or petiole less than 2 mm long .... (2)

2. Bracts scarious or largely so, 0.1–0.4 cm long; flowers in a terminal cyme .... (3)
2′ Bracts herbaceous or foliaceous, mostly 0.5–3 cm long; flower in a leafy terminal cyme or solitary in upper leaf axil .... (5)

3. Cyme umbelloid with 3–5+ pedicels appearing to radiate from one point; pedicels deflexed below, often curved apically in fruit; petals absent .... ***S. umbellata***
3′ Cyme not umbelloid; pedicels straight to arcuate; petals present .... (4)

4. Leaf margins minutely tuberculate-scaberulous as viewed under a strong (20×) lens, dull; cymes axillary or terminating axillary shoots with divaricately branched pedicels; petals bifid to near base .... ***S. longifolia***
4′ Leaf margins smooth, lustrous, often coriaceous; cymes terminal with erect or ascending pedicels; petals deeply notched .... ***S. longipes***

5. Sepals 3.5–5 mm long in flower; leaf margins smooth, lustrous, often coriaceous; petals deeply notched, 1.1–1.6× sepal length .... ***S. longipes***
5′ Sepals 1.3–3 mm long in flower; leaf margins herbaceous or scarious and transparent to white or purple; petals bifid to near base or absent, less than 1× sepal length (to 1.5× longer in *S. crassifolia*) .... (6)

6. Mature plant purplish throughout; leaves fleshy, midrib prominent abaxially, margin transparent to white or purple; flower solitary, axillary, often concealed by upper leaves .... ***S. irrigua***
6′ Mature plant green; leaves herbaceous (somewhat fleshy in *S. crassifolia*), midrib not prominent, margins herbaceous; flower solitary, axillary, but not concealed by leaves, or 3–many in a leafy cyme .... (7)

7. Flower solitary, axillary; bracts absent; plant prostrate or creeping .... (8)
7′ Flowers 3–many in leafy cymes; bracts leaflike; plant mostly erect, sometimes prostrate .... (9)

8. Petals 5; sepals 5, 2–3 mm long in flower, narrowly triangular-lanceolate, acute, veins 3; stems 4-angled (most obvious in proximal 1/2) .... ***S. crassifolia***
8′ Petals 0; sepals 4 (5), 1.3–1.7 mm long in flower, ± ovate, ± obtuse, veins obscure; stems ± terete .... ***S. obtusa***

9. Mature capsule dark purple, ± globose; styles usually 0.5–0.9 mm long, thick, curved outward; leaves (primary ones) ovate to broadly elliptic, 0.5–2.5 cm long; sepals obscurely veined, 1.2–1.5 mm long in flower, to 2.5 mm long in fruit .... ***S. calycantha***
9′ Mature capsule light green to tan, ovate-elongate; styles usually 0.9–1.6 mm long, slender, straight (often contorted when pressed); leaves (primary ones) elliptic-lanceolate to narrowly elliptic or linear-lanceolate, 2–4 cm long; sepals 1–3-veined, 2–2.5 mm long in flower, to 3 mm long in fruit; peripheral to the Four Corners region .... *S. borealis*

Stellaria calycantha (Ledeb.) Bong. (cupped flower) Northern starwort. Perennial herbs, forming clumps, green, rhizomes slender. **STEMS** mostly erect, sometimes trailing, branched, 4-angled, 15–35 cm tall, glabrous or pilose. **LEAVES** with petiole absent or essentially so, blades ovate to elliptic, 0.5–2.5 cm long (widest over lower 1/2), thin, margins glabrous or rarely ciliate, herbaceous to thinly scarious, midrib moderately prominent on lower surface, apex acute. **INFL** mostly a 3–many-flowered terminal cyme; bracts foliaceous, 0.5–3 cm long; pedicels ascending in fruit, 5–35 mm long. **FLOWER** sepals 5, veins obscure, ovate, 1.2–1.5 mm long in flower, to 2.5 mm long in fruit, margins broadly scarious, apex broadly acute, glabrous; petals 5 or 0, bilobed to near base, 0.4–0.6× sepal length; stamens 5; styles 3–5, usually 0.5–0.9 mm long, thick, curved outward. **FRUIT** dark purple, ± globose, 3–5 mm long, valves 6, 8, or 10. **SEEDS** brown, ovate, 0.5–0.9 mm long, smooth or shallowly tuberculate. $2n = 26$. [*S. simcoei* (Howell) C. L. Hitchc.]. Meadows, willow bogs, and spruce-fir communities; scree slopes in alpine communities. COLO: LPl, SJn. 3290–3900 m (10800–12800′). Flowering: Jul–Aug. Alaska to the Atlantic, south to California, Wyoming, Michigan, and New York; Eurasia.

Stellaria crassifolia Ehrh. (thick-leaved) Thick-leaved starwort. Perennial herbs, delicate, mat-forming, green, rhizomes slender. **STEMS** prostrate or creeping, diffusely branched, 4-angled (most obvious in lower 1/2), 3–30 cm long, glabrous. **LEAVES** with petiole absent, blades broadly elliptic-lanceolate to linear-lanceolate, 0.2–1.5 cm long, somewhat fleshy, margins glabrous, transparent to thinly scarious, midrib thinly evident on lower surface, apex acute to acuminate. **INFL** with a solitary flower, axillary but not concealed; bracts foliaceous, 0.5–1 cm long; pedicels sharply curved at apex in fruit, 3–40 mm long. **FLOWER** sepals 5, veins 1–3, narrowly triangular-lanceolate, 2–2.5 mm long in flower, to 3 mm long in fruit, margins narrowly scarious, apex acute, mostly glabrous; petals 5, bilobed to near base, 0.9–1.3× sepal length; stamens 5–10; styles 3. **FRUIT** tan, conic to ellipsoid, 4–5 mm long, valves 6. **SEEDS** tan to reddish brown, reniform to round, 0.7–1 mm long, rugose. $2n = 26$. Riparian areas. COLO: Mon. 3500 m (11700′). Flowering: Jun–Aug. Canada and adjacent areas; North Dakota, South Dakota, Wyoming, Colorado, and Utah. This species is documented in the flora based solely on one collection; it may prove to be far more common.

Stellaria irrigua Bunge (well-watered or wet) Altai chickweed, Altai starwort. Perennial herbs, forming mats or low cushions, mature plant purplish throughout, rhizomes elongate. **STEMS** ascending to spreading, somewhat branched, 4-angled, 2–10 cm tall, glabrous. **LEAVES** with petiole absent, blades elliptic or lanceolate to oblanceolate, 0.1–1 cm long, fleshy, margins glabrous, transparent, white to purple, midrib prominent on lower surface, apex acute. **INFL** a solitary flower, often concealed by upper leaves; pedicels curved to deflexed in fruit, 3–15 mm long. **FLOWER** sepals 5, veins 3, lanceolate, 2.5–3 mm long in flower, to 4 mm long in fruit, margins membranous, apex acute, glabrous; petals 5, bilobed to near base, 0.4–0.5× sepal length; stamens 5; styles 3. **FRUIT** green to tan, ovoid-obtuse, 2.5–3 mm long, teeth 6. **SEEDS** pale brown, reniform, 1–1.2 mm long, not glossy, sides smooth to shallowly rugose, margins thickened with shallow longitudinal ridges. Rocky areas in alpine communities. COLO: Arc, Hin, LPl, Min, Mon, RGr, SJn. 3600–4250 m (11700–14000′). Flowering: Jun–Aug. Siberia; Colorado and New Mexico. The type specimen for this species is from the Altai, Siberia. If *S. irrigua* is conspecific with material from southern Colorado and adjacent New Mexico, it represents a geographical disjunction of more than 12000 km (7500 miles).

Stellaria longifolia Muhl. ex Willd. (long-leaved) Long-leaved starwort. Perennial herbs, forming loose clumps, green, rhizomes elongate. **STEMS** straggling to erect, diffusely branched, 4-angled, 10–35 cm tall, glabrous except angles minutely roughened. **LEAVES** with petiole absent, blades linear to very narrowly elliptic, 0.8–4 cm long, thickish, margins ciliate near base, minutely tuberculate-scaberulous (20×), dull, midrib prominent on lower surface, apex acuminate to acute. **INFL** a 2–many-flowered, divaricately branched cyme, axillary or terminating axillary shoots, visible; bracts mostly scarious, 1–4 mm long; pedicels straight to arcuate in fruit, 3–30 mm long. **FLOWER** sepals 5, obscure veins 3, ovate-elliptic, 2.5–3 mm long in flower, to 4 mm long in fruit, margins broadly scarious, apex acute, glabrous; petals 5, bilobed to near base, 0.9–1.3× sepal length; stamens 5–10; styles 3. **FRUIT** blackish purple or tan, ovoid-conic, 3–6 mm long, valves 6. **SEEDS** brown, broadly reniform, 0.7–0.8 mm long, slightly rugose. $2n = 26$. Most common in moist meadows. COLO: Arc, Con, Hin, LPl, Min, Mon, RGr, SJn; NMEX: SJn. 1950–3050 m (6400–10000′). Flowering: Jun–Aug. Alaska to Newfoundland, south to California and east to South Carolina; northern Mexico.

Stellaria longipes Goldie **CP** (long-stalked) Goldie's starwort. Perennial herbs, forming small to large clumps or mats, or diffuse, green, rhizomes slender. **STEMS** ascending to erect, often branched, 4-angled, 3–32 cm tall, glabrous or softly pubescent. **LEAVES** with petioles absent, blades linear-lanceolate to ovate-triangular, 0.8–4 cm long, margins glabrous or ciliate, smooth, lustrous, often coriaceous, midrib prominent on lower surface, apex acute to acuminate. **INFL** a 1–20-flowered terminal cyme, visible; bracts foliaceous or scarious and 1–3 mm long; pedicels ascending to erect in fruit, 5–30 mm long. **FLOWER** sepals 5, veins 3, midrib prominent, lanceolate to ovate-lanceolate, 3.5–4 mm long in flower, to 5 mm long in fruit, margins narrowly scarious, apex acute, glabrous or pubescent; petals 5, deeply notched, 1.1–1.6× sepal length; stamens 5–10; styles 3. **FRUIT** purplish black, ovoid to lanceoloid, 4–6 mm long, valves 6. **SEEDS** brown, reniform to globose, 0.6–0.9 mm long., shallowly tuberculate to smooth. $2n =$ mostly 52, 78, or 104. [*S. longipes* var. *altocaulis* (Hultén) C. L. Hitchc.; *S. longipes* var. *monantha* (Hultén) S. L. Welsh; *S. longipes* var. *laeta* (Richardson) S. Watson]. Riparian areas, coniferous forests, alpine and subalpine meadows and talus slopes. ARIZ: Apa; COLO: Arc, Con, Hin, LPl, Min, Mon, SJn; NMEX: McK, RAr, SJn; UTAH. 2400–4000 m (7900–13200′). Flowering: May–Sep. Canada and northern United States south to California, Arizona, and New Mexico. Our material belongs to **var**. ***longipes***.

Stellaria media (L.) Vill. (intermediate) Common chickweed. Annual herb, green, taprooted. **STEMS** creeping to ascending, diffusely branched, terete, 5–40 cm long, villous in a line between nodes. **LEAVES** with petiole 5–20+ mm long at least below, blades ovate to broadly elliptic, 0.5–4 cm long, thickish, margins often ciliate, herbaceous, midrib often prominent on lower surface, apex acute or shortly acuminate. **INFL** a 5–many-flowered terminal cyme, visible; bracts foliaceous, 0.5–4 cm long; pedicels ascending, deflexed at base in fruit, 3–40 mm long. **FLOWER** sepals 5, veins obscure, ovate-lanceolate, 4–5 mm long in flower, to 6 mm long in fruit, margins narrowly scarious, apex obtuse, usually glandular; petals 5 or 0, 0.5–0.9× sepal length, bilobed to near base, stamens usually 3–5; styles 3. **FRUIT** green to tan, ovoid-oblong, 3–5 mm long, valves 6. **SEEDS** reddish brown, broadly reniform to round, 0.9–1.3 mm long, with rounded or flat-topped tubercles. $2n = 40, 42, 44$. Disturbed areas. COLO: LPl; NMEX: SJn. 1800–2400 m (6000–8000′). Flowering: Apr–Aug. Introduced throughout North America.

Stellaria obtusa Engelm. (blunt) Blunt-sepaled starwort, Rocky Mountain starwort. Perennial herbs, creeping, mat-forming, green, rhizomatous. **STEMS** prostrate, branched, ± terete, 3–23 cm long, mostly glabrous. **LEAVES** with petioles to 2 mm long at least below, blades broadly ovate to elliptic, 0.2–1.2 cm long, thin, margins often ciliate near base, herbaceous, midrib faint on lower surface, apex acute. **INFL** a solitary flower, axillary but not concealed; bracts absent; pedicels spreading, 3–12 mm long. **FLOWER** sepals 4–5, veins obscure, ± ovate, 1.3–1.7 mm long in flower, to 3 mm long in fruit, margins narrowly scarious, apex ± obtuse, glabrous; petals 0; stamens 10 or fewer; styles 3 (4). **FRUIT** green to tan, globose to broadly ovoid, 2.3–3.5 mm long, valves 6. **SEEDS** reddish to grayish black, broadly elliptic, 0.5–0.7 mm long, finely reticulate. 2n = 26, 52, about 65, about 78. Riparian areas. COLO: Hin, LPl, Min. 2400–2500 m (8000–8200′). Flowering: May–Aug. Southwestern Canada; Washington to Montana, south to California, Arizona, and New Mexico. This species is likely more common than represented by specimens.

Stellaria umbellata Turcz. (in umbels) Umbellate starwort. Perennial herbs, clumped or matted, green, rhizomes slender. **STEMS** erect, branched at base, 4-angled, 5–30 cm tall, glabrous. **LEAVES** with petiole ± absent, blades elliptic to lanceolate, 3–9 cm long, somewhat fleshy, margins glabrous, herbaceous, midrib prominent on lower surface, apex acute. **INFL** a (2)- 5–many-flowered terminal or subterminal umbelloid (3–5+ pedicels appearing to radiate from one point) cyme; bracts mostly scarious, 1–4 mm long; pedicels deflexed below, often curved apically in fruit, 7–20 mm long. **FLOWER** sepals 5, veins 3, lanceolate, 1.5–2 mm long in flower, to 3 mm long in fruit, margins narrowly scarious, apex obtuse, glabrous; petals 0; stamens 5; styles 3. **FRUIT** tan, conic, 3–4.5 mm long, valves 6. **SEEDS** brownish, round, 0.5–0.7 mm long, shallowly rugose. $2n = 26$. [*S. weberi* B. Boivin]. Spruce-fir forests, stream banks, and alpine meadows and talus slopes. COLO: Arc, Con, Hin, LPl, Min, Mon, SJn; UTAH. 2400–4000 m (7875–13125′). Flowering: Jun–Sep. Canada and Alaska; Washington to Montana, south to California, Arizona, and New Mexico.

Vaccaria Wolf Cow Herb, Cow Soupwort

(Latin *vacca*, cow, + *aria*, pertaining to, referring to its value as fodder) Annual herbs with taproots. **STEMS** ascending to erect, simple proximally, branched distally, terete. **LEAVES** not congested at or near base of flowering stem, without stipules, blades lanceolate to broadly so. **INFL** a dense to open cyme. **FLOWERS** hypogynous; sepals 5, connate proximally into 5-angled, 5-winged tube, apex acute to acuminate, not hooded, awn absent; petals 5, pink to purplish, clawed, auricles absent, coronal appendages absent, blade apex entire or briefly bilobed; stamens 10, adnate with petals, staminodes absent; ovary 1-locular or 2-locular proximally; styles usually 2. **FRUIT** an oblong capsule, opening by usually 4 slightly spreading teeth. **SEEDS** 5–10, black, papillose, marginal wing absent, appendage absent. One or four species; Eurasia, introduced into South America, Africa, Australia.

Vaccaria hispanica (Mill.) Rauschert (Spanish) Cow herb, cowcockle. Plants glabrous, glaucous. **STEMS** 20–100 cm tall. **LEAVES** with blades 2–10 cm long, base cuneate to cordate. **INFL** an open cyme, 16–50 (–100)-flowered. **FLOWER** calyx 9–17 mm long; petals with claw 8–14 mm long, blade 3–8 mm long. **FRUIT** included in calyx tube. **SEEDS** 2–2.5 mm wide. 2n = 30. [*V. pyramidata* Medik.]. Roadsides, floodplains, and weedy areas. COLO: Arc. 1800–2100 m (6000–6900′). Flowering: Jun–Sep. Introduced throughout much of North America.

CELASTRACEAE R. Br. STAFF-TREE or BITTERSWEET FAMILY

John R. Spence

Shrubs or trees, sometimes climbing or twining. **LEAVES** persistent, stipitate or stipules sometimes lacking, alternate to opposite, sessile or pedicellate. **INFL** axillary to terminal, cymose, racemose, or flowers solitary. **FLOWERS** perfect, actinomorphic, hypogynous or somewhat perigynous, 4–5-merous; sepals small, connate at base or distinct; petals small, distinct, often spreading, sometimes lacking; stamens alternate, on or just oustide nectary disc; ovary of 2–5 united carpels, superior; style single, terminal, with a variously lobed or capitate stigma. **FRUIT** a capsule, samara, berry, achene, or drupe. **SEEDS** typically encased in an aril. $x = 8, 12, 14$. A family of 95 genera and 1300 species. Worldwide, but concentrated in the tropics and Southern Hemisphere, rare in the Northern Hemisphere. The genus *Euonymus* (burning bush) has many cultivated species. The genus *Catha* is the source of the stimulant drink khat.

Paxistima Raf. Mountain-lover, Box-leaf Myrtle, Dinasts' óóz

(unexplained, perhaps Greek *pachy*, thick, + *stima*, stigma) Low-growing to decumbent evergreen shrub from a creeping fibrous root base, stems often arranged in flat sprays, twigs glabrous when young, roughened-scurfy when older. **LEAVES** small, opposite, persistent, rather coriaceous, glossy, margins smooth to serrate or crenate; stipules small, caducous. **INFL** of short axillary cymes or flowers solitary. **FLOWERS** 4-merous; petals green or maroon, free, spreading; stamens alternate, attached to base of nectary disc; style single, stigma capitate; carpels 2, united into compound 2-locular ovary. **FRUIT** a 2-valved capsule. **SEEDS** partly enclosed in a white aril. A genus of two species, one in western North America, and the second in the southern Appalachian Mountains. This genus is considered relictual from the Tertiary (Navaro and Blackwell 1990).

Paxistima myrsinites (Pursh) Raf. (for superficial similarity to *Myrsine*) Low-growing shrub from creeping fibrous root base, branches splayed to erect. **LEAVES** closely crowded, glossy, elliptic to oblong or elliptic-oblanceolate, 1–4.5 cm long, to 1.5 cm wide, serrate above middle or sometimes lower, petiole very short, 1–2 mm long. **INFL** of solitary axillary flowers, or short axillary cymes of few flowers. **FLOWER** sepals very small, greenish, petals maroon to brown-red, 1–2 mm long, spreading, stamen filaments short, usually less than 1 mm. **FRUIT** capsule valves 4–5 mm long. **SEEDS** 2/3 covered by the white aril. [*Pachystima m.*, incorrect spelling; *Paxistima myrtifolia* L. C. Wheeler]. Moist forests and slopes, talus, rocky outcrops, from the lower montane zone to the lower subalpine zone. ARIZ: Apa; COLO: Arc, Dol, Hin, LPl, Min, Mon, RGr, SJn; NMEX: McK, RAr, SJn. UTAH. 2400–3400 m (6500–8000′). Flowering: May–Jul. Fruit: Jul–Sep. Widespread in the mountains of western North America from Alberta and British Columbia south to California, Arizona, and New Mexico. This plant is important in Navajo ceremonies as an emetic in the Holy Way, Hand-trembling Way, and others. Elsewhere a tea has been used for colds, tuberculosis, and as a pain killer. The fruits (seeds) have sometimes been eaten in times of scarcity. It is an attractive, low-growing ornamental shrub.

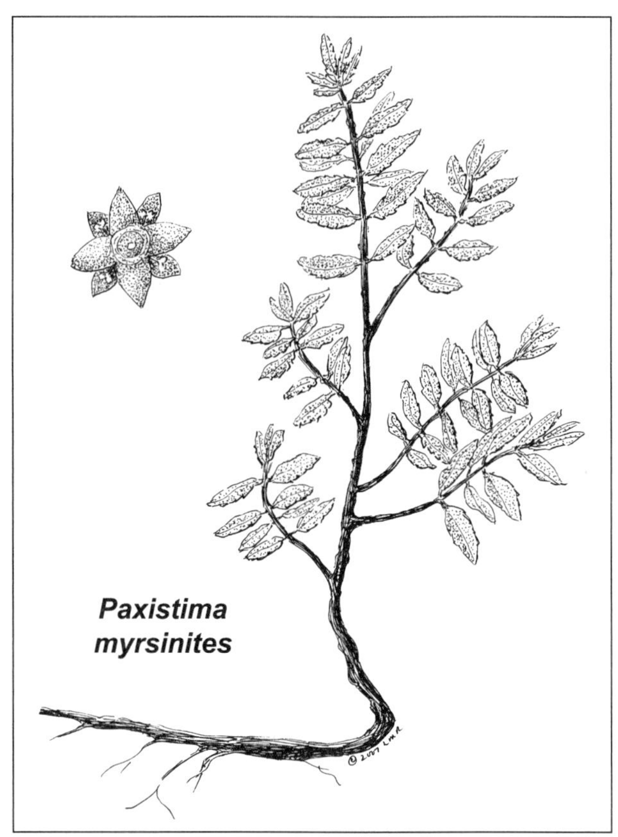

CERATOPHYLLACEAE Gray HORNWORT FAMILY

C. Barre Hellquist

Perennial, rootless, aquatic. **STEMS** slender, with lateral branches. **LEAVES** whorled, sessile, finely dissected. **FLOWERS** small, unisexual, lacking a perianth, solitary, sessile in leaf axil, subtended by an 8–14-cleft involucre replacing the calyx; stamens 12–20; filaments short, the large anthers terminating in 2–3 sharp points; pistillate flowers consisting of a simple 1-celled ovary. **FRUIT** a 1-seeded, ovoid-oblong achene (Les 1986, 1988; Lowden 1978). One genus, six species, worldwide. Seeds and foliage may provide food for ducks.

Ceratophyllum L. Hornwort

(Greek *ceras*, a horn, and *phyllum*, leaf) **STEMS** submersed, floating, olive-green, lacking rhizomes; fragments generate into new plants. **LEAVES** 3–11 per whorl, usually 1–4-dichotomously dissected into filiform to narrowly linear divisions, completely submersed. **INFL** extra-axillary, alternating with leaves. **FLOWERS** with basal connate bracts; pedicel less than 1 mm, almost absent. **FRUIT** an ellipsoidal, moderately compressed achene with marginal or basal spines. x = 12, 19, 20.

Ceratophyllum demersum L. (submerged) Hornwort, coontail. **STEMS** elongate, branching, brittle, cordlike. **LEAVES** finely dissected into capillary to linear and flattened serrate divisions, as many as 12 in a verticil; leaf divisions toothed, about 15 mm long. **FLOWERS** monoecious, occasionally dioecious; staminate in pairs on opposite sides of the stem, short-stalked, with a calyxlike involucre of 8–12 short, stout, oblong segments, each 2–3-toothed at apex; pistillate usually solitary, with a unilocular ovary and long style. **FRUIT** ellipsoid, flattened, 4–6 mm long, with a terminal persistent style 4–6 mm long, with 2 recurved basal, lateral spines. Still water of lakes, ponds, and streams. ARIZ: Apa; COLO: SJn; NMEX: SJn. 1850–2750 m (6100–9400′). Flowering and fruiting: Jun–Sep. Alaska east to Prince Edward Island and Nova Scotia, south to California, Arizona, Texas, and Florida; Mexico; West Indies; Central America. $2n$ = 24, 38, 40, 48. The main means of reproduction is asexual due to fragmentation.

CHENOPODIACEAE Vent. GOOSEFOOT FAMILY

S. L. Welsh & N. D. Atwood

(*chen*, goose, + *pous*, foot, referring to the leaf shape) Herbs, subshrubs, or shrubs, often succulent or scurfy. **LEAVES** simple, alternate or opposite, estipulate. **FLOWERS** inconspicuous, monoecious, dioecious, polygamous, or perfect; calyx persistent, 1- to 5-lobed, enclosing the fruit, or lacking in some pistillate flowers; corolla none; stamens opposite the calyx lobes and as many or fewer; pistil 1, the ovary superior, with 1–3 stigmas, 1-loculed and 1-ovuled. **FRUIT** a utricle. $x = 6–9$ (Flores and Davis 2001; Jacobs 2001; Welsh 1984). One hundred genera, 1300 species worldwide, especially in deserts and saline or alkaline soils. *Beta* (beet, sugar beet, chard) and *Chenopodium* (quinoa) cultivated for food.

1. Leaves scalelike; stems fleshy; plants of saline pans and other salty sites .... ***Salicornia***
1′ Leaves well developed, not scalelike; stems not fleshy; plants of various habitats .... (2)

2. Leaves or bracts of inflorescence tipped with a spine or a spinelike bristle .... (3)
2′ Leaves and floral bracts not bristle- or spine-tipped .... (4)

3. Leaves linear to subulate; bracts of inflorescence ovate-lanceolate, spine-tipped; fruiting sepals winged on the back; flowers not embedded in hairs .... ***Salsola***
3′ Leaves sausagelike, abruptly spine-tipped; bracts of inflorescence not different from the leaves; sepals ending in wings; flowers embedded in hairs .... ***Halogeton***

4. Leaves sub- or semicylindric to linear, usually fleshy .... (5)
4′ Leaves with flattened blades, not especially fleshy .... (7)

5. Shrubs, armed with thorny branchlets; staminate flowers in spikes, pistillate flowers solitary and axillary .... ***Sarcobatus***
5′ Shrubs or herbs, not armed; flowers perfect or both perfect and pistillate .... (6)

6. Herbage villous-tomentose; plants low subshrubs .... ***Bassia***
6′ Herbage glabrous and glaucous, or puberulent; plants annual or perennial, or if subshrubs, then tall .... ***Suaeda***

7. Plants densely white-hairy with at least some dendritic hairs, these becoming golden brown in age; shrubs of broad distribution .... ***Krascheninnikovia***
7′ Plants variously hairy or glabrous, but not as above; shrubs or herbs of various distribution .... (8)

8. Flowers imperfect, the pistillate enclosed in 2 accrescent or connate bracteoles .... (9)
8′ Flowers perfect or some also pistillate, all with sepals and not enclosed by paired bracteoles .... (12)

9. Leaf blades orbicular or suborbicular, flabellate to broadly obtuse at the base; fruit flask-shaped in outline .... ***Suckleya***
9′ Leaves mainly ovate to lanceolate, acute to obtuse basally .... (10)

10. Bracteoles dorsally compressed, variously tuberculate, smooth, or winged; pubescence of inflated hairs or none; plant a shrub or perennial or annual herb; axillary rounded buds lacking .... ***Atriplex***
10′ Bracteoles laterally compressed or 6- to 8-ribbed, lacking appendages; pubescence of simple or branched hairs; plants shrubby; axillary rounded buds present .... (11)

11. Shrubs with divaricate, often thorny branches; bracteoles with margins thickened, spongy within; pubescence of branched hairs .... ***Grayia***
11′ Shrubs with erect, nonthorny branches; bracteoles with margins not spongy-thickened, either obcompressed or dorsiventrally compressed and 6-ribbed; pubescence of scurfy or moniliform hairs .... ***Zuckia***

12. Plants more or less tomentulose; calyx transversely winged in fruit .... ***Cycloloma***
12′ Plants glabrous, scurfy, pilose, or otherwise pubescent, but not or seldom as above; calyx not transversely winged in fruit, except in *Bassia*, and then not tomentulose .... (13)

13. Perianth developing conspicuous horizontal, scarious wings or armed with curved or uncinate spines .... ***Bassia***
13′ Perianth lobes rounded or keeled on the back, lacking wings or spines .... (14)

14. Calyx lobes 5, largely concealing to exposing the fruit; stamens usually 5; herbage glabrous or scurfy .***Chenopodium***
14′ Calyx lobes 1–3, the fruit largely exposed; stamens 1–3; plants not scurfy ....(15)

15. Leaves hastately lobed, or if entire, 4–12 mm long, the blades mainly 2–8 mm broad or more; calyx 1-lobed; stamens 1; plants widespread, of many habitats .... ***Monolepis***
15′ Leaves linear, 0.8–6 cm long, entire, 1–2 mm wide; calyx 1- to 3-lobed; stamens 1–3; plants of sandy, low-elevation sites..***Corispermum***

Atriplex L. Saltbush

(ancient Latin name) Monoecious or dioecious herbs or shrubs, often with scurfy (mealy) collapsed hairs. **LEAVES** alternate or opposite. **INFL** an axillary cluster, glomerule, or spicate panicle. **FLOWERS** small, inconspicuous; staminate flowers with calyx 3- to 5-parted, bractless; stamens 3–5; pistillate flowers without a perianth, pistil naked or rarely with a perianth, commonly enclosed within a pair of foliaceous bracteoles, enlarged in fruit, variously thickened and appendaged; styles 2. **FRUIT** with the pericarp free. **SEEDS** flattened, mainly erect. The genus is complex both taxonomically and nomenclaturally. It consists of both native and introduced herbs and shrubs. Indigenous shrubby species form hybrids with all or most of their constituent taxa, wherever they come in contact. The resulting plasticity allows these remarkable plants to occupy numerous habitats but poses problems that preclude a "neat" taxonomic treatment. The following keys are tentative at best. Several species of *Atriplex* have been demonstrated to be poisonous to livestock due to secondary or facultative selenium accumulation. Kranz anatomy, related to the C_4 photosynthetic pathway, has become a useful tool in recognition of species and species groups not only within *Atriplex*, but within the family as a whole (see Jacobs 2001). Chloroplasts are aligned along veinlets, a feature easily seen by scraping off the epidermis and extraneous coverings (Hanson 1962; Stutz and Sanderson 1983; Thorne 1977). About 250 species in temperate to subtropical saline or alkaline soils, sometimes accummulating selenium. Many weedy.

1. Plants herbaceous annuals..KEY 1
1′ Plants woody, at least below; perennial ....KEY 2

KEY 1

PLANTS HERBACEOUS ANNUALS

1. Seeds of 2 types, black and brown; plants mainly introduced .... (2)
1′ Seeds all alike, either black or brown; plants indigenous...(7)

2. Fruiting bracteoles orbicular or nearly so, the dorsal surfaces smooth, entire ....(3)
2′ Fruiting bracteoles triangular to ovate or rhombic, the dorsal surfaces usually tubercled, denticulate to entire .(4)

3. Black seeds horizontal, enclosed in a membranous calyx; brown seeds vertical between very large bracteoles ....***A. hortensis***
3′ Black seeds vertical between small bracteoles; brown seeds vertical between large bracteoles ......***A. heterosperma***

4. Lowermost leaves ovate, dentate, sessile or short-petiolate; leaves with Kranz anatomy.......................... ***A. rosea***
4′ Lowermost leaves lanceolate, rhombic, triangular, or hastate, dentate or entire, petiolate; leaves lacking Kranz anatomy ....(5)

5. Lowermost leaves lanceolate to linear-lanceolate; plants ruderal, not typically of saline sites ....***A. patula***
5′ Lowermost leaves (sometimes all) triangular or hastate-triangular in outline, or if lanceolate to linear-lanceolate, the plants of moist saline sites ....(6)

6. Leaves thick-textured, more or less scurfy, mostly longer than broad; fruiting bracteoles often spongy-thickened at the base; plants indigenous in moist saline sites...***A. dioica***
6′ Leaves thin-textured, typically glabrous, at least some often as broad as long or broader; fruiting bracteoles thin-textured throughout; plants introduced in moist saline sites ....***A. prostrata***

7. Plants mainly dioecious; widely distributed ....***A. powellii***
7′ Plants mainly monoecious, of various distribution ....(8)

8. Fruiting bracteoles all stipitate, or some of them sessile to subsessile, cuneate, and entire on the sides, when stipitate usually prominently tubercled; leaves mainly oval-ovate; plants forming low rounded clumps on saline substrates .... ***A. saccaria***

8′ Fruiting bracteoles all sessile, subsessile, or short- to long-stipitate, variously tubercled, or if definitely stipitate, as in *A. graciliflora*, *A. pleiantha*, and some *A. argentea*, the surfaces smooth or the leaves definitely triangular-ovate; plants slender or clump-forming, variously distributed ....(9)

9. Fruiting bracteoles 1.5–2.5 mm long, 1–2.5 mm wide, ovate-oblong or oval to obovate, sometimes constricted below, giving an overall violin shape .... ***A. powellii***
9′ Fruiting bracteoles of various length and width, but if as above then differing otherwise ....(10)

10. Fruiting bracteoles with margins dentate, foliaceous well below the apex, the surfaces sometimes with appendages .... ***A. argentea***
10′ Fruiting bracteoles samara-like and orbicular or triangular-ovate, the surfaces smooth ....(11)

11. Fruiting bracteoles samara-like and orbicular, enclosing a single flower .... ***A. graciliflora***
11′ Fruiting bracteoles triangular-ovate, enclosing 2 to 6 flowers .... ***A. pleiantha***

KEY 2

PLANTS WOODY, AT LEAST BELOW; PERENNIAL

1. Bracteoles with 4 lateral wings or 4 rows of teeth; plants unarmed ....(2)
1′ Bracteoles without lateral wings, merely tuberculate or smooth dorsally; plants sometimes thorny ....(4)

2. Leaves more than 8 mm wide; bract tip with or without lateral teeth; lower canyons of the San Juan River .... ***A. garrettii***
2′ Leaves less than 8 mm wide; bract tip without lateral teeth; distribution various .... (3)

3. Bracteoles more than 9 mm wide, with tips not exceeding the wings; staminate flowers yellow; shrubs to 2 m tall; widely distributed .... ***A. canescens***
3′ Bracteoles less than 9 mm wide, with tips exceeding the wings; staminate flowers mostly brown; plants mainly 5–10 dm tall; clay soils .... ***A. gardneri***

4. Plants with thorny branches; bracteoles foliose, united only basally, the surfaces smooth; staminate flowers yellow; widely distributed .... ***A. confertifolia***
4′ Plants lacking thorny branches; bracteoles not foliose, at least 1/3 united and the surfaces typically with appendages; staminate flowers yellow or brown ....(5)

5. Leaves 2–4 mm wide; bracteoles with appendages on lower 1/3; staminate flowers in spikes; plants prostrate .... ***A. corrugata***
5′ Leaves often more than 4 mm wide; bracteoles with appendages various; staminate flowers mainly in panicles; plants not or seldom prostrate ....(6)

6. Leaves oblong-ovate to orbicular, more than 10 mm wide, the lowermost alternate; stems stiffly erect; staminate glomerules very numerous .... ***A. obovata***
6′ Leaves linear to oblong, mainly less than 10 mm wide, or if wider, the lowermost opposite; stems mainly prostrate to ascending from the root-crown, less commonly erect; staminate glomerules numerous .... ***A. gardneri***

Atriplex argentea Nutt. (silvery) Silver orach. Plants annual, monoecious. **STEMS** simple or freely branched. **LEAVES** petiolate or the upper ones subsessile, with Kranz anatomy, the blades 0.5–6 cm long, 0.4–5 cm wide, elliptic, lance-ovate, lanceolate, deltoid, or cordate, runcinate to subhastate or acute basally, obtuse to acute apically, entire or repand-denticulate, scurfy (glabrous). **INFL** with staminate glomerules borne in terminal spikes or panicles, or terminal on lateral branches. **FLOWERS**: staminate with 5-parted calyx. **FRUIT** bracteoles sessile, subsessile, or short- (rarely long-) stipitate, 4–11.2 mm long, 4–10 (–14) mm wide, the margin foliaceous below the apex, dentate to laciniate, the face smooth, tubercled, or crested. **SEEDS** about 2 mm wide, brown. $2n = 18$. Valuable for salt concentration, used as a forage and fodder plant by Navajos. Ramah Navajos used the leaves as a fumigant for pain, as an infusion for colds, and as a poultice for spider bites. Acomas used fruits for food. Leaves were also boiled with fat or meat and eaten as greens by Hopis. Seeds were parched, ground into a flour, and made into a mush by Paiutes.

1. Leaf blades triangular-ovate to oval, the base broadly obtuse to acute or less commonly acute ....**var. *argentea***
1′ Leaf blades elliptic to oval, attenuate to a cuneate base....**var. *rydbergii***

var. *argentea* **STEMS** 3–8 dm tall, erect, the branches ascending. **LEAVES** with blades 0.6–5 cm long, 0.4–4 cm wide, ovate or ovate-elliptic to deltoid, the upper short-petioled, entire or repand-dentate, grayish scurfy when young; fruiting bracteoles sessile or some stipitate (the stipe 0.5–15 mm long), cuneate to obovate to suborbicular in profile, 3.8–11.2 mm long, 4–8.5 (–14) mm wide, usually compressed, united nearly to the apex, the green free margins extending nearly to the base, dentate, the sides smooth or sparsely to densely tuberculate or cristate. $2n$ = 18. Mat-atriplex, shadscale, and greasewood communities. ARIZ: Apa, Nav; COLO: LPl, Mon; NMEX: RAr, SJn; UTAH. 1515–1820 m (5000–6000′) Flowering/Fruit: Jun–Sep. Widespread in western United States and Mexico.

var. *rydbergii* (Standl.) S. L. Welsh (for P. A. Rydberg, Colorado botanist) Rydberg's orach. **STEMS** (0.2) 1–4 dm tall, erect, the branches ascending. **LEAVES** with blades (0.4) 0.6–3.5 cm long, (2) 4–25 mm wide, elliptic to ovate or rhombic, acute to cuneate at the base, rounded to obtuse or acute at the apex, entire or rarely toothed, grayish scurfy. **INFL** of staminate flowers in axillary glomerules or terminal spikes. **FRUIT** bracteoles short-stipitate (the stipe to 4 mm long) or subsessile, compressed, flabelliform to obovate or suborbicular, 2.5–7 (8.4) mm long (including the stipe length), (2) 2.7–6 (7.5) mm wide, widest near or above the middle, united nearly to the middle, the free portion scurfy, shallowly or deeply and coarsely dentate, the sides smooth or with 1 to few thickened processes, and these sometimes again appendaged. [*Atriplex rydbergii* Standl.; *A. pachypoda* Stutz & G. L. Chu]. Saline, fine-textured substrates derived mainly from Cretaceous Mancos Shale and Morrison formations and clay or silty alluvium, growing in salt desert shrub and floodplain communities. ARIZ: Apa; COLO: LPl, Mon; NMEX: SJn; UTAH. 1100–2070 m (3600–6830′). Flowering/Fruit: May–Sep. A Four Corners region endemic.

Atriplex canescens (Pursh) Nutt. (grayish) Four-wing saltbush. Dioecious or rarely monoecious shrubs, mainly 8–20 dm tall, not especially armed. **LEAVES** persistent, alternate, sessile or nearly so, 10–40 mm long, 2–8 mm wide, linear to oblanceolate, oblong, or obovate, entire, retuse to obtuse apically. **INFL** of staminate flowers in clusters 2–3 mm wide, borne in panicles or axillary spikes; pistillate flowers in terminal panicles, 5–40 cm long. **FLOWERS**: staminate yellow (rarely brown). **FRUIT** bracteoles 9–25 mm long and as wide, on a pedicel 1–8 mm long, with 4 prominent wings extending the bract length, united throughout, the surface of wings and body smooth or reticulate; wings dentate to entire, the apex toothed. **SEEDS** 1.5–2.5 mm wide. $2n$ = 18, 36, or higher. [*Calligonum canescens* Pursh; *Atriplex nuttallii* S. Watson; *A. tetraptera* (Benth.) Rydb.; *Pterochiton occidentale* Torr. & Frém; *A. occidentalis* (Torr. & Frém.) D. Dietr.; *A. canescens* var. *occidentalis* (Torr. & Frém.) S. L. Welsh & Stutz]. Sandy, commonly nonsaline, sites in blackbrush, greasewood, salt desert shrub, sagebrush, mountain brush, and piñon-juniper communities. ARIZ: Apa, Coc, Nav; COLO: Arc, Dol, LPl, Mon; NMEX: McK, RAr, San, SJn; UTAH. 1060–2170 m (3500–7160′). Flowering: Apr–Jul. Fruit: Jun–Oct. Washington to Alberta and South Dakota, south to Mexico and Texas. This species forms hybrids with *A. confertifolia* and *A. gardneri* varieties. It is an important browse plant for both wildlife and domestic livestock. It is used in reclamation projects and might be found established at sites beyond its usual range and habitat latitude. Navajos used a decoction of tops or roots as an emetic, used the leaves for cough, and a poultice of leaves for ant bites. Leaves and twigs were used to make a yellow dye and to intensify red dyes. Hopis used the plant for kiva fires, twigs as prayer sticks, and ashes as a baking soda and to maintain the blue in blue cornmeal. Zunis used the twigs in prayer plumes.

Atriplex confertifolia (Torr. & Frém.) S. Watson (crowded leaves) Shadscale. Dioecious spinescent shrubs, 3–8 dm tall. **LEAVES** persistent, alternate, with petioles 1–4 mm long; blades 9–25 mm long, 4–20 mm wide, orbicular to ovate, elliptic, or oval, entire, obtuse apically. **INFL** of staminate flowers in axillary clusters 2–4 mm wide or in spikes to 1 cm long; pistillate flowers paniculate, 3–15 cm long. **FLOWERS**: staminate yellow. **FRUIT** bracteoles sessile or subsessile, suborbicular to rhombic or elliptic, 4–12 mm long and wide, the surface smooth, lacking appendages; terminal teeth distinct, foliaceous, shorter than the bracteoles, entire or toothed below, spreading at maturity. **SEEDS** 1.5–2 mm broad. $2n$ = 18, 36, 54, or higher. [*Obione confertifolia* Torr. & Frém.; *A. collina* Wooton & Standl.]. Gravelly to fine-textured soils in greasewood, mat-atriplex, shadscale, sagebrush, and piñon-juniper communities. ARIZ: Apa, Nav; COLO: LPl, Mon; NMEX: McK, RAr, San, SJn; UTAH. 1160–2150 m (3830–7100′). Flowering: Apr–Jul. Fruit: Aug–Oct. Oregon east to North Dakota, south to California, Arizona, New Mexico, and Texas. Shadscale forms hybrids with *A. canescens*, *A. garrettii*, *A. corrugata*, and *A. gardneri* varieties. It is a valuable browse plant for wildlife and livestock, especially sheep. Hopis used the burned plant as an anticonvulsive; the leaves were also used in puddings and boiled as a green. Navajos rubbed the plant on horses to repel gnats and Paiutes used the plant as a cold and sore muscle remedy.

Atriplex corrugata S. Watson (wrinkled leaves) Mat-saltbush, mat-atriplex. Dioecious, low, spreading shrubs (often appearing as if prostrate), mainly 3–15 cm tall and 3–15 dm broad. **LEAVES** persistent, sessile, opposite below, alter-

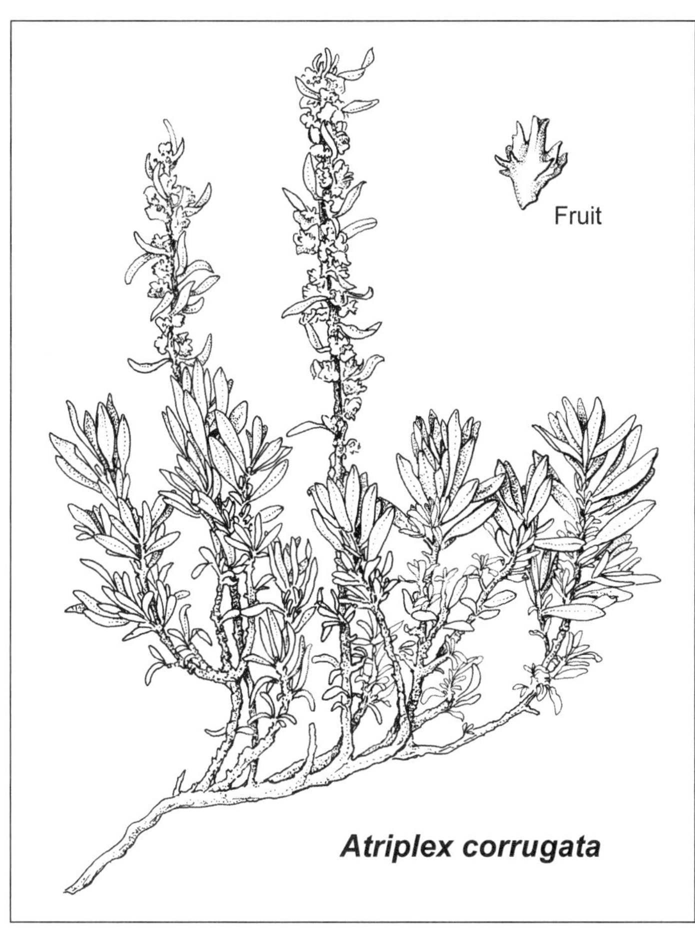

Atriplex corrugata

nate above, 3–18 mm long, 1–6 mm wide, linear to linear-oblanceolate or oblong, entire, obtuse apically. **INFL**: staminate in clusters 3–6 mm wide, borne in spikes 1–8 cm long; pistillate in leafy bracteate spikes 5–15 cm long. **FLOWERS**: staminate yellow to brownish. **FRUIT** bracteoles sessile or subsessile, 3–5 mm long, 4–6 mm wide, densely tuberculate (or smooth), entire or undulate, rounded to acute apically. **SEEDS** about 1.5 mm wide. $2n = 36$. Saline, usually fine-textured substrates derived from Mancos Shale, Morrison, and other similar formations in mat-atriplex and Castle Valley saltbush communities. COLO: Mon; NMEX: SJn; UTAH. 1455–2150 m (4800–7100′). Flowering: Apr–Jul. Fruit: May–Oct. Mat-saltbush is known to form intermediates with both *A. confertifolia* and *A. gardneri* var. *cuneata*. This saltbush is a valuable browse plant on the sparsely vegetated clays and silts of the Four Corners region, especially on Mancos Shale exposures, where it is often the only woody vegetation present and the only thing available for grazing animals to eat.

Atriplex dioica Raf. (dioecious) Thickleaf orach. Monoecious annual herb, 3–15 dm tall, typically erect and often branching, the stems angular, green or striped. **LEAVES** without Kranz anatomy, opposite or subopposite at least below, the petiole 1–3 cm long, the blades 3–12.5 cm long, 2.5–6 (8) cm wide, thickened, lanceolate to linear-lanceolate or often triangular-ovate or lance-ovate, the basal or subbasal lobes spreading to mainly antrorse, obtuse, the blade otherwise entire to sparingly dentate, typically scurfy and strongly 3-veined from near the base. **INFL** in terminal clusters or terminal on lateral branches, in spiciform spikes 2–9 cm long, naked except at the base. **FRUIT** bracteoles sessile, thick, green, blackening in age, 3–10 mm long, broadly triangular to triangular-ovate, usually longer than wide, the apex acute, the base truncate to obtuse, the margins united at the base, the lateral angles entire or with short, sharp teeth, the faces smooth or with 1 or more tubercles, with an inner spongy layer, rarely smooth, veined or the veins obscure. **SEEDS** dimorphic, brown ones 1.5–3 mm wide, wider than long, and flattened at base, and black ones shiny, 1–2 mm wide, wider than long, the radicle inferior; $2n = 36, 54$. [*Chenopodium subspicatum* Nutt.; *Atriplex subspicata* (Nutt.) Rydb.; *A. lapathifolia* Rydb.; *A. carnosa* A. Nelson]. Roadsides, irrigation canals, stream banks, salt flats, and other typically moist saline sites, with saltgrass, pickleweed, and rush. ARIZ: Apa; COLO: Arc, Mon; NMEX: SJn; UTAH. 1525–2120 m (4620–7000′). Flowering: Jul–Sep. Fruit: Aug–Oct. Coastal eastern North America, and west of the 95th meridian from Yukon and the Northwest Territories, south to southern California, northern Arizona, northern New Mexico, and Oklahoma. The habitats are mainly saline, and the species is apparently indigenous. It has long been mistaken for the closely similar *A. prostrata*.

Atriplex gardneri (Moq.) D. Dietr. (for Alexander Gordon, botanical collector; his name misspelled by Moquin) Gardner's saltbush. Dioecious or sparingly monoecious subshrubs, mostly 1–4.5 dm tall and 4–20 dm wide, with decumbent branches. **LEAVES** opposite to subalternate below, alternate above, with petioles 2–12 mm long, the blades 10–50 mm long, 5–25 mm wide, about 1.5 to 4.5× longer than wide, obovate to spatulate, orbicular or oblong, entire or rarely remotely dentate. **INFL**: staminate in glomerules 3–4 mm thick, borne in terminal panicles 2–11 cm long; pistillate in rather leafy, unbranched spikes or spicate panicles 5–23 cm long. **FLOWERS**: staminate dark brown or rarely yellow. **FRUIT** bracteoles sessile or on stipes to 9 mm long, (4) 6–9 mm long and wide, globose to flattened, ovate, obovate, or elliptic to cuneate in profile, the surface densely tuberculate to smooth and free of tubercles, the terminal tooth to 2 mm long, subtended by 2–4 lateral teeth. **SEEDS** 2–2.5 mm broad, brown, the radicle superior. $2n = 18, 27, 36$, and higher. [*Obione gardneri* Moq.; *Atriplex nuttallii* S. Watson, misapplied]. Plants of this complex occur commonly in fine-textured saline substrates in much of the western Great Plains and in the Intermountain region.

Diploids, triploids, tetraploids, and hexaploids (and higher polyploids, all multiples of the base number 9) are known within the complex, and hybrids are known between the constituents and the other woody species which they contact, i.e., *A. canescens*, *A. confertifolia*, and *A. corrugata*. The treatment essentially follows the alignment of taxa suggested by Hanson (1962), with the exception that they are reduced to varietal status and var. *bonnevillensis* (C. A. Hanson) S. L. Welsh is placed within the *gardneri* phase and not with *A. canescens*. The name *A. nuttallii* S. Watson has been used for this complex, but it is based squarely on *A. canescens* (Pursh) Nutt. and is illegitimate. Within Utah there are six morphologically intergrading entities that seem worthy of taxonomic recognition. Our material belongs to **var. *cuneata*** (A. Nelson) S. L. Welsh (referring to the wedge-shaped bracteoles), Castle Valley saltbush. [*Atriplex cuneata* A. Nelson]. Saline fine-textured substrates on Mancos Shale, Morrison, and other formations of similar texture and salinity, but also in alluvium, in greasewood, shadscale, matchweed, horsebrush, juniper, and mat-atriplex communities. ARIZ: Apa, Nav; COLO: Mon; NMEX: McK, RAr, SJn; UTAH. 1120–1740 m (3700–5750′). Flowering: Mar–Aug. Fruit: May–Oct. The hybrids between var. *cuneata* and *A. confertifolia* have been described as *A. neomexicana* Standl. They occur where the ranges of the species overlap. Possibly both of the described intermediates warrant taxonomic recognition, but no nomenclatural combination is intended or implied herein.

Atriplex garrettii Rydb. (for Albert O. Garrett, teacher and botanist in Salt Lake City, Utah) Garrett's saltbush. Dioecious (rarely monoecious) shrubs or subshrubs, mainly 2–6 dm tall, unarmed. **LEAVES** opposite or subopposite below, petiolate, the blades 8–55 mm long, 6–32 mm wide, ovate to obovate, lanceolate, elliptic, or orbicular, yellow-green, sparingly scurfy, entire or repand-dentate, obtuse to cuneate basally, rounded or emarginate to acute apically. **INFL**: staminate brown to tan (rarely yellow), in clusters 2–4 mm wide on panicles 2–8 cm long; pistillate in spikes or spicate panicles 4–30 cm long. **FLOWERS**: staminate brown to tan (rarely yellow). **FRUIT** bracteoles 6–10 mm long and wide, 4-winged, the surface smooth, reticulate, or with flattened processes, toothed apically. **SEEDS** about 2 mm wide, brown. $2n = 18$. [*Atriplex canescens* (Pursh) Nutt. subsp. *garrettii* (Rydb.) H. M. Hall & Clem.; *A. canescens* var. *garrettii* (Rydb.) L. D. Benson]. Shadscale, ephedra, eriogonum, blackbrush, and mixed shrub-grass communities on talus slopes of canyons of the lower San Juan River. UTAH. 1125–1210 m (3410–4000′). Flowering: Mar–Oct. Fruit: May–Oct. Endemic to southeast Utah. This distinctive plant has been regarded as a portion of the variation within an expanded *A. canescens*, with which it shares the feature of four-winged pistillate fruiting bracteoles, but it is possibly more closely allied with *A. confertifolia*, with which it hybridizes. The hybrid intermediates are distinguished by their armed branches and four-winged fruiting bracteoles. Utah plants belong to **var. *garrettii***; those from the vicinity of Navajo Bridge in Coconino County, Arizona, belong to var. *navajoensis* (C. A. Hanson) S. L. Welsh & Crompton. The whole plant has been used with alum to produce a yellow dye by Great Basin tribes.

Atriplex graciliflora M. E. Jones (slender flower) Blue Valley orach. Monoecious annual herbs, mainly 1–4 (5) dm tall, branching from the base, forming clumps 0.5–6 dm wide. **STEMS** ascending to erect, sometimes suffused with red. **LEAVES** alternate, petiolate, the blades 8–15 (23) mm long and 5–19 mm wide, ovate to orbicular, reniform, cordate, or deltoid, truncate to cordate basally, rounded to obtuse or acute apically, entire. **INFL**: staminate in panicles overtopping the foliage; pistillate axillary. **FLOWERS**: staminate with the perianth 5-lobed. **FRUIT** bracteoles samara-like, 6–16 mm wide, stipitate, compressed, orbicular, oblong or cordate in outline, winged, the wings undulate or entire, the surfaces smooth. **SEEDS** about 3 mm wide, dull white. Saltbush, seepweed, greasewood, rabbitbrush, and tamarix communities on saline, often salt-encrusted, soils. COLO: LPl; UTAH. 1210–1820 m (4000–6000′). Flowering: May–Jul. The plants grow on Mancos Shale at the type locality, on Tropic Shale in eastern Kane County, Utah, and on the same or other similar formations (e.g., Entrada) throughout its range. Less commonly it occurs on saline alluvium. Blue Valley orach is easily distinguished by its flattened, samara-like fruiting bracteoles.

Atriplex heterosperma Bunge (different seed) Russian atriplex; two-seed orach. Monoecious annual herbs, mainly 1.5–14 dm tall, erect, branched from below the middle or above. **LEAVES** opposite or subopposite below and sometimes almost throughout, commonly alternate above and petiolate, the blades mainly 15–80 mm long and as wide or wider, hastately lobed, triangular, the main lobe entire or sharply dentate, palmately veined from near the base, the base truncate to cordate or obtuse, acute apically. **FLOWERS**: staminate with 5 sepals. **FRUIT** bracteoles 2–7 mm long, orbicular to suborbicular or ovate, entire, the surfaces smooth, dimorphic, the larger with a pale brown vertical seed 2–3 mm wide, the smaller with a shiny black vertical seed about 1 mm wide. $2n = 36$. Riparian and palustrine (less commonly ruderal) habitats in greasewood, saltgrass, cocklebur, tamarix, cottonwood, and rush-cattail communities. COLO: Arc, Dol, LPl, Mon; NMEX: RAr, SJn; UTAH. 1615–2260 m (5330–7460′). Flowering: Jul–Sep. Fruit: Aug–Oct. Widespread in North America south of the Arctic; adventive from Eurasia. This is a handsome vigorous annual that appears to be invading saline lowland but also occurs in other disturbed, nonsaline areas throughout the regions occupied.

Atriplex hortensis L. (refers to garden) Garden orach. Monoecious annual herbs, mainly 5–15 (25) dm tall, erect, branching from the middle or above. **LEAVES** without Kranz anatomy, opposite or subopposite below, alternate above, petiolate, the blades commonly 1.5–13.5 cm long and 1–13 cm wide, ovate to lanceolate, not especially hastate, the base acute to cordate, acute to rounded apically. **FLOWERS** shortly pedicellate; staminate with 3–5 sepals; pistillate dimorphic, the pistil vertical and enclosed in bracteoles or the pistil horizontal and enclosed in a 4- or 5-lobed calyx. **FRUIT** bracteoles 8–18 mm long, orbicular to ovate, entire, the surfaces smooth, greenish or reddish. **SEEDS** dimorphic, either 2–4 mm wide and brown or about 1 mm wide and black. Disturbed sites in riparian and ruderal habitats. COLO: LPl, Mon; NMEX: SJn. 1425–1820 m (4700–6000′). Flowering: Jul–Sep. Practically cosmopolitan, introduced from Eurasia. This plant is grown as a potherb and is to be expected practically anywhere. It persists and escapes following cultivation.

Atriplex obovata Moq. (inversely ovate-shaped leaves) New Mexico saltbush. Dioecious shrubs, mainly 2–8 dm tall. **LEAVES** tardily deciduous, alternate, shortly petiolate, the blades 8–30 mm long, 6–20 mm wide, obovate to elliptic or orbicular, gray-green, entire or rarely crenate-dentate, rounded to retuse or obtuse apically. **INFL**: staminate in clusters 2–3 mm wide, borne in panicles 6–30 cm long, to 1/2 or more of plant height. **FLOWERS**: staminate yellow. **FRUIT** bracteoles 4–5 mm long, 5–9 mm wide, sessile, broadly cuneate, the surfaces smooth or rarely tubercled, the margins entire, the apical tooth subtended by 2–6 equal or smaller teeth. **SEEDS** 1–1.5 mm wide, brownish. Salt desert shrub and lower piñon-juniper communities, mainly on Mancos Shale. ARIZ: Apa, Nav; COLO: LPl, Mon; NMEX: McK, RAr, San, SJn; UTAH. 1435–2105 m (4735–6945′). Flowering: May–Aug. Fruit: Aug–Oct. Arizona, New Mexico, and Mexico. Plant burned by Hopis as an anticonvulsive and also used as greens or a spice.

Atriplex patula L. (open or spreading) Fat-hen saltplant. Monoecious annual herbs, mainly 1.5–10 dm tall, prostrate-ascending or erect, simple or branched. **LEAVES** lacking Kranz anatomy, alternate or some or all opposite, petiolate, the blades mainly 1–12 cm long, 1–5 cm wide or more, ovate, deltoid, lance-ovate, or lance-linear, cordate to hastate, truncate, or acute to cuneate basally, rounded to obtuse or acute apically, entire to dentate or hastate, thin or thick, green, glabrous or scurfy. **INFL** in paniculate clusters. **FLOWERS**: staminate with 4 or 5 sepals. **FRUIT** bracteoles sessile, subsessile, or rarely stipitate, 2–12 mm long, 3–9 mm wide, deltoid to ovate or rhombic, sometimes spongy-thickened, the margin entire or denticulate, the face smooth, roughened, or tuberculate. **SEEDS** vertical, dimorphic, either black and 1–2 mm wide or brown and 1–3 mm wide. $n = 9, 18, 36$. Roadsides, ditch banks, gardens, and sedge-reed, willow, and cottonwood communities, and other nonsaline ruderal habitats. COLO: Arc, Dol, Hin, LPl, Min, Mon, SMg; UTAH. 1940 m (6400′). Flowering: Jul–Aug. Fruit: Sep. Widespread in North America; native to Eurasia. Materials tentatively assigned here are thin-leaved, but bracteoles vary in outline from rhombic to ovate or narrowly oblong and the surfaces from smooth to tubercled.

Atriplex pleiantha

Atriplex pleiantha W. A. Weber (more flower) Four Corners orach. Monoecious, glabrous or sparingly scurfy annual herbs, mainly 0.5–1.5 dm tall, branching from the base. **LEAVES** alternate to subopposite, petiolate, the blades 5–18 (20) mm long and about as wide, ovate to suborbicular, obtuse to acute apically, entire. **INFL**: staminate in short terminal spikes. **FRUIT** bracteoles 3–7 mm wide and about as long, triangular-ovate, short-stipitate, compressed, entire, enclosing 2–5 flowers, these with perianth well-developed, consisting of 5 hyaline, sparsely ciliate scales 1–1.2 mm long. **SEEDS** black, smooth and shining, about 1.5 mm long, falling at maturity. [*Proatriplex pleiantha* (W. A. Weber) Stutz & G. L. Chu]. Salt desert shrub community, Morrison and Mancos Shale formations. COLO: Mon; NMEX: SJn; UTAH. 1495–1665 m (4530–5500′). Flowering: May–Jun. Fruit: Jul. A Four Corners endemic. Those who wish to recognize this entity at the generic level overlook the similarity of this species with other species of *Atriplex* in the Colorado Plateau, especially *A. saccaria*. The presence of apparently primitive inflorescences within the flowering bracteoles is probably a derived, not primitive, feature.

Atriplex powellii S. Watson (for John Wesley Powell, geologist and naturalist) Powell's orach. Dioecious (or sparingly to often monoecious) annual herbs. **STEMS** slender to stout, mainly 1–5 (7) dm tall, branching almost throughout;

herbage pubescent with scurfy and arachnoid hairs. **LEAVES** with Kranz anatomy, alternate, the lowermost with petioles (0.3) 0.5–3 cm long, becoming short-petiolate or sessile upward, the blades 0.4–5 cm long, 0.2–3 cm wide, ovate to rhombic, orbicular, or elliptic, entire, rounded to cuneate basally, acute to obtuse apically, prominently 3-veined. **FLOWERS**: staminate with calyx 4- or 5-lobed. **FRUIT** bracteoles sessile, 1.5–5.5 mm long, 1.5–5 mm wide, thick, united to the apex, ovate to oblong or broadly cuneate, truncate to cuspidate apically, the surfaces with thickened processes or rarely smooth. **SEEDS** about 2 mm long, greenish or yellowish brown, the radicle superior. $2n = 18$. Saline, usually fine-textured substrates, in greasewood, rabbitbrush, shadscale, seepweed, mat-atriplex, piñon-juniper, and riparian communities. ARIZ: Apa, Nav; COLO: Arc, LPl, Mon; NMEX: McK, RAr, SJn; UTAH. 1495–1980 m (4530–6535′). Flowering: May–Sep. Fruit: Aug–Oct. Montana and South Dakota to Arizona and New Mexico. This is the only annual atriplex in the study area that approaches being truly dioecious, but flowers of the opposite gender occur also, resulting in monoecious individuals. Hopis and Navajos boiled the leaves and ate them with fat and as a salt substitute. Zunis mixed the seeds with ground corn to make a porridge.

Atriplex prostrata Boucher ex DC. (procumbent branching) Thinleaf orach; hastate saltplant. Monoecious annual herb, 1–10 dm tall, erect, decumbent or procumbent, branching. **STEMS** subangular to angular, green or striped. **LEAVES** without Kranz anatomy, opposite or subopposite at least below and often far upward, the petiole (0) 1–3 (4) cm long, the blades 2–10 cm long and almost as wide or wider, triangular-hastate, the lobes spreading, entire, serrate, dentate, or merely irregularly toothed, thin and green (turning reddish or purplish in autumn), subpalmately veined from near the base, the base truncate or cordate, the apex acute to obtuse. **INFL** spiciform, the naked spikes 2–9 cm long, sometimes forming terminal panicles. **FRUIT** bracteoles 3–5 mm long, tringular-hastate to tringular-ovate, green, becoming brown to black at maturity, the apex acute, the base truncate to obtuse, the margins united at the base, the lateral angles mostly entire, the faces smooth or with 2 tubercles, veined or the veins obscure. **SEEDS** dimorphic, brown ones 1–2.5 mm wide, and black ones 1–1.5 mm wide, the radicle subbasal, obliquely antrorse to spreading. $2n = 18$. [*Atriplex triangularis* Willd.; *A. patula* L. var. *triangularis* (Willd.) K. H. Thorne & S. L. Welsh; *A. hastata* L.]. Saltgrass, sedge-rush, rush-cattail, and other palustrine and riparian habitats, usually in saline mucky soils. ARIZ: Apa; NMEX: SJn. 1535–1800 m (5065–5940′). Flowering: Jun–Aug. Fruit: Aug–Sep. Widely distributed in North America; Europe. Though mainly thin-leaved, there are specimens with at least somewhat thickened leaves included here also. Thus, not all specimens will fit neatly here or with the closely similar *A. dioica*, for which the thinleaf orach has long been mistaken. The problem of typification of *A. hastata* L. was reviewed by Taschereau (1972). That name evidently replaces the long-established European *A. calotheca* (Rafn) Fr., a plant not known from New Mexico.

Atriplex rosea L. (for the rosy-colored stems and leaves) Tumbling orach. Monoecious, coarse, annual herbs. **STEMS** simple or more commonly branching throughout, (1) 2–10 (20) dm tall; herbage scurfy to glabrate. **LEAVES** alternate, petiolate, with Kranz anatomy, the blades mainly 1.2–8 cm long, 0.6–5 cm wide, ovate to lanceolate, acute to obtuse apically, irregularly dentate and often subhastately lobed or rarely entire, prominently 3-veined. **INFL** pistillate in axillary glomerules of 5–10. **FLOWERS**: staminate with 4 or 5 sepals. **FRUIT** bracteoles sessile or short-stipitate, (3) 4–6 (10) mm long and as wide, ovate to rhombic, united to the middle, conspicuously dentate, sharply tuberculate to almost smooth on the surfaces, prominently 3–5-nerved. **SEEDS** dimorphic, brown and 2–2.5 mm wide; or black and 1–2 mm wide, the radicle inferior. $n = 9$. [*Atriplex spatiosa* A. Nelson; *A. virgata* Osterh.]. Widely established weedy species of disturbed sites, often in riparian habitats, pond margins, or in barnyards or on animal bedgrounds, in numerous plant community types. ARIZ: Apa, Nav; COLO: Arc, Mon, SMg; NMEX: RAr, SJn; UTAH. 1515–2140 m (4970–7020′). Flowering: Jun–Aug. Fruit: Aug–Oct. Widespread in North America; native to Eurasia. Used for sheep and horse feed, as porridge, and as a black dye by Ramah Navajos.

Atriplex saccaria S. Watson (bag-shaped or pouched, referring to the bracteoles) Stalked orach, medusa-head orach. Low monoecious herbs, forming rounded clumps, mainly 0.5–4 (5) dm tall. **STEMS** usually branched from the base, often suffused red-purple when young; herbage scurfy. **LEAVES** alternate or the lowermost subopposite, with petioles 2–5 mm long or the upper ones sessile, with Kranz anatomy, the blades mainly 0.6–4 cm long, 0.4–3 cm wide, cordate-ovate or subreniform to ovate or deltoid-ovate to oval, entire or in some the base subhastately lobed, truncate to subcordate or broadly cuneate basally, acute to rounded apically. **INFL**: staminate borne in the upper axils or in short, naked, terminal (early-deciduous) spikes or panicles; pistillate usually in fascicles of 1–3 in the lower axils. **FLOWERS**: staminate with 5-parted perianth. **FRUIT** bracteoles monomorphic or typically dimorphic, the larger ones on stipes (2) 4–10 (15) mm long, the others sessile, united to the base, round-triangular or suborbicular, 4–6 mm long, irregularly and coarsely dentate and more or less densely beset with flat, cristate, or hornlike appendages, the smaller bracteoles in the same axils, oblong to cuneate, 3–4 mm long, truncate at the apex, dentate only at the summit and the faces smooth. **SEEDS** 1.5–2.3 mm wide, brownish to whitish, the radicle superior. $n = 9$. Young tender leaves cooked and eaten as greens by Hopis.

1. Leaves oval to rhombic, the bases truncate to cuneate-attenuate; fruiting bracteoles with stipes of various length .... **var. *cornuta***
1' Leaves mainly ovate to cordate in profile, the uppermost bracteate ones (and sometimes all) often broadly so in age; fruiting bracteoles with stipes seldom exceeding 10 mm .... **var. *saccaria***

var. *cornuta* (M. E. Jones) S. L. Welsh (horned) Blade orach. **LEAVES** with petioles mainly 1–3 cm long, gradually shortened upward, the blades 1–2.5 cm long and as broad or broader (to 3.7 cm), rhombic to deltoid or ovate, acute to rounded apically, broadly to narrowly cuneate basally. **INFL**: staminate in small sessile glomerules axillary in the uppermost bracteate leaves, less commonly terminal and spicate or paniculate. **FRUIT** bracteoles dimorphic, those on stipes (2) 4–15 mm long or longer, with surfaces covered with flattened processes, the sessile ones cuneate as in the type variety. [*Atriplex cornuta* M. E. Jones; *A. argentea* Nutt. var. *cornuta* (M. E. Jones) M. E. Jones]. Salt desert shrub communities, on fine-textured, saline substrates associated with the Mancos Shale and other fine-textured formations. ARIZ: Apa; NMEX: RAr, SJn; UTAH. 1425–2045 m (4700–6750'). Flowering: May–Aug. Fruit: Aug–Oct. Perhaps this phase represents nothing more than the extreme ends of variation of the next variety and possibly does not deserve taxonomic status. Certainly the tendency to flattened, sometimes acute and sometimes flabellate to rounded processes is compelling.

var. *saccaria* **LEAVES** all petiolate or the upper ones subsessile, the blades 1–3 cm long and 1–2.5 cm wide, cordate-ovate or subreniform to oval or ovate, the base cordate or some only broadly truncate or less commonly rather abruptly cuneate, entire or slightly hastate. **INFL**: staminate in glomerules in upper axils and in terminal deciduous panicles. **FRUIT** bracteoles dimorphic, those with subglobose bodies bearing flattened or cristate processes and borne on stipes to 10 mm long or more, and those mainly lacking processes, cuneate in outline, and sessile. [*Atriplex caput-medusae* Eastw.; *A. argentea* Nutt. var. *caput-medusae* (Eastw.) Fosberg; *A. saccaria* S. Watson var. *caput-medusae* (Eastw.) S. L. Welsh]. Mat-atriplex, shadscale, greasewood, and piñon-juniper communities, on Mancos Shale, Morrison, and other fine-textured, saline substrates. ARIZ: Apa, Nav; COLO: LPl, Mon; NMEX: McK, San, SJn; UTAH. 1450–2215 m (4800–7310'). Flowering: Mar–Jul. Fruit: Jun–Oct. Wyoming, Utah, Colorado south to northern Arizona and New Mexico.

Bassia All. Summer Cypress

(for Ferdinando Bassi, 18th-century Italian botanist) Annual herbs or subshrubs. **LEAVES** alternate or some opposite, linear to narrowly lanceolate, in some fleshy and terete, entire, sessile; herbage pilose or tomentose, at least in inflorescence. **INFL** glomerate or solitary in leaf axils and in short axillary spikes. **FLOWERS** perfect and pistillate, bracteate; calyx 5-lobed, depressed-globose, enclosing the fruit and usually prominently armed with a curved or hooked spine on the dorsal surface of each lobe, or the lobes keeled and horizontally winged; stamens mostly 5, hypogynous; styles 1, with 2 (3) stigmas. **FRUIT** compressed. **SEEDS** horizontal [*Kochia* Roth] (Scott 1978). About 20 species in western North America and Eurasia. Used as a drug for venereal disease by Navajos.

1. Calyx in fruit armed with hooked spines .... ***B. hyssopifolia***
1' Calyx in fruit with horizontal wings or wartlike tubercles .... (2)

2. Plants annual, introduced weeds of disturbed habitats .... ***B. scoparia***
2' Plants perennial, woody at the base, indigenous in saline habitats .... ***B.americana***

Bassia americana (S. Watson) A. J. Scott (America) Gray Molly. Plants mainly 5–30 (45) cm tall, with erect branches from a woody base; herbage villous-pilose to glabrous. **LEAVES** 5–25 mm long, 1–2 mm wide, linear, semiterete and fleshy. **INFL** often more than 1/2 the branch length; solitary flowers, or 2–5, sessile in axils of scarcely reduced leaves. **FLOWER** perianth segments pubescent, at least apically, 1–1.5 mm long, hooded above, somewhat enlarged in fruit, ultimately keeled and with a membranous, striate wing to 2 mm long and 3 mm wide. [*Kochia americana* S. Watson; *K. americana* var. *vestita* S. Watson; *K. vestita* (S. Watson) Rydb.]. Greasewood, seepweed, saltbush, saltgrass, matchweed, horsebrush, and piñon-juniper communities. ARIZ: Apa; COLO: Mon; NMEX: SJn; UTAH. 1450–1985 m (4790–6550'). Flowering/Fruit: May–Sep. Oregon to Montana, south to California, Arizona, and New Mexico.

Bassia hyssopifolia (Pall.) Kuntze (aromatic herb) Bassia. Annual herbs, the main stem erect, the lower lateral ones often decumbent, 2–10 dm tall or more; herbage more or less lanate, especially in the inflorescence. **LEAVES** 4–40 mm long,

1–5 mm wide, linear to oblong or narrowly oblanceolate. **INFL** clustered in terminal or lateral spikes, or solitary in leaf axils. **FLOWER** bracteoles reduced; pistillate and sterile flowers mixed with perfect ones. **FRUIT** calyx about 2 mm wide, each lobe with a stout, curved to uncinate spine; pericarp membranous, planoconvex. $2n = 18$. [*Salsola hyssopifolia* Pall.; *Echinopsilon hyssopifolium* (Pall.) Moq.; *Kochia hyssopifolia* (Pall.) Schrad.]. Commonly on saline substrates, often in riparian, lacustrine, or palustrine habitats, in saltgrass, greasewood, horsebrush, shadscale, and cottonwood-tamarix communities. ARIZ: Apa, Nav; COLO: Arc, LPl, Mon; NMEX: RAr, SJn; UTAH. 1585–2380 m (5230–7855′). Flowering/Fruit: Jul–Oct. Adventive from Eurasia. This species forms apparent hybrids with *B. scoparia*, from which it differs in having spines on the sepals instead of horizontal flattened processes.

Bassia scoparia (L.) A. J. Scott (form of a broom) Summer cypress. Annual herbs, mainly 3–12 (15) dm tall, green, or suffused with red in autumn; simple or branched from the base, villous and often finely lanate to glabrous. **LEAVES** 0.8–4.5 (6) cm long, 1–4 mm wide, lanceolate to oblanceolate, elliptic or linear, usually 3- to 5-veined, glabrous or softly pilose below (and above) or glabrous above, generally ciliate, acute. **INFL** spicate, interrupted. **FLOWERS**: perfect ones with fruiting perianth glabrous dorsally, ciliate, mostly transversely keeled, tubercled or sometimes horizontally winged from middle of the keel, often lacking a keel. **SEEDS** ovate in outline, 1.5–2 mm long. $2n = 18$. [*Chenopodium scoparium* L.; *Kochia scoparia* (L.) Schrad.]. Disturbed roadsides, canal banks, field margins, and other waste places in salt marsh, sedge-rush, sagebrush, mountain brush, and piñon-juniper communities. ARIZ: Apa, Nav; COLO: Arc, Dol, Hin, LPl, Min, Mon, SJn, SMg; NMEX: McK, RAr, San, SJn; UTAH. 1425–2695 m (4700–8890′). Flowering/Fruit: Jul–Oct. Widespread in North America; adventive from Eurasia. This species is known to form apparent hybrids with *B. hyssopifolia*. The plants are intermediate in every regard, and vegetative specimens are difficult if not impossible to distinguish to species. Because of this, and other considerations, the writer somewhat hesitantly agrees with the A. J. Scott treatment wherein the two genera were combined.

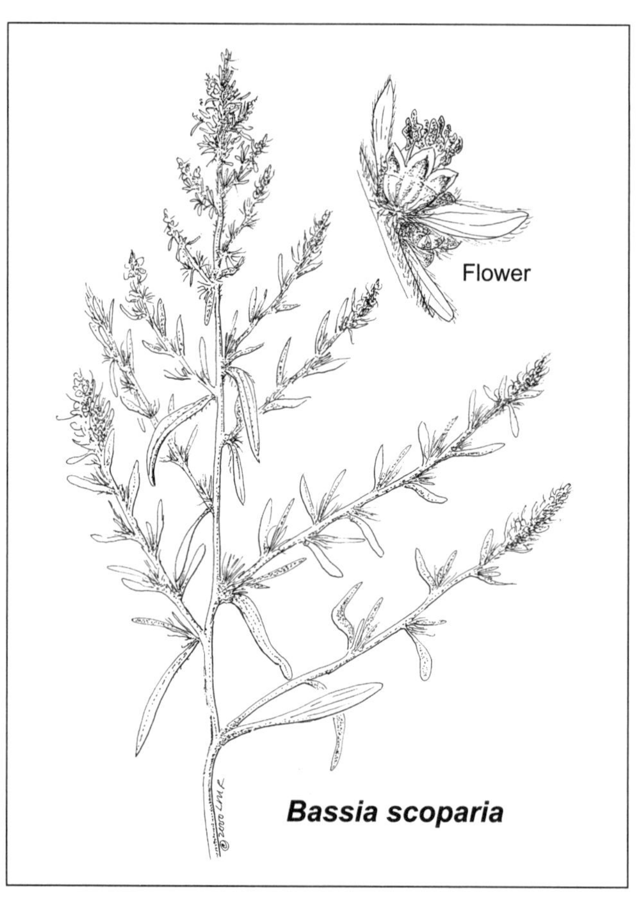

Bassia scoparia

Chenopodium L. Goosefoot, Pigweed, Quelitos

(*chen*, goose, and *pous*, foot, referring to the leaf shape) Annual herbs, glabrous, pubescent, glandular, or farinose (mealy). **LEAVES** alternate, flat, entire, toothed, or lobed. **INFL** usually in cymes, variously arranged in spicate or paniculate inflorescences. **FLOWERS** perfect or some pistillate only, ebracteate; calyx segments usually 4 or 5, persistent, flat or keeled, more or less covering the fruit; stamens commonly 5; styles 2 (3). **FRUIT** rarely becoming fleshy. **SEEDS** lenticular, horizontal or vertical. The genus is notoriously complex for several reasons. The floral features are greatly reduced and diagnostic characteristics are often based on either vegetative structures or on minutiae of the calyx, pericarp, and seed coat, which are often subject to interpretation and might be demonstrated ultimately as trivial. Nomenclature is tangled within both the native and introduced entities, leading to taxonomic treatments that do not satisfactorily circumscribe the taxa as represented by actual specimens. Further, there is variability within the diagnostic features, leading to contradictory statements in taxonomic treatments; e.g., with regard to such characters as adherent versus nonadherent pericarps. Thus, the treatment presented below attempts to provide names for the taxa recognizable in the Four Corners region based on examination of actual specimens. The entities seem to be real, but the names might be misapplied in some cases (Bassett and Crompton 1982; Wahl 1954). About 150 temperate species; some cultivated for greens (*Beta*) or grain (*Chenopodium*).

1. Herbage with yellow resinous glands or glandular hairs, not farinose, aromatic .... (2)
1′ Herbage glabrous or farinose, eglandular, not aromatic .... (4)

2. Flowers in small sessile clusters, borne in bracteate panicles; plants rare .... ***C. ambrosioides***
2' Flowers solitary or loosely clustered, borne in small dichotomous cymes, these spreading-recurved along the axis of an elongate panicle; plants common or rare .... (3)

3. Plants densely glandular-pubescent; keel calyx lobes smooth, not tuberculate; plants common and widespread .... ***C. botrys***
3' Plants sparsely puberulent, the lower leaves and calyx with yellow resinous dots; keel of calyx lobes tuberculate .... ***C. graveolens***

4. Seeds all, or at least some, vertical in the flowers (except sometimes in C. *glaucum*, keyed both ways) .... (5)
4' Seeds usually all horizontal .... (9)

5. Leaves mainly 0.5–2 cm long, 2–7 mm wide, irregularly dentate, glaucous-farinose beneath .... ***C. glaucum***
5' Leaves mainly larger and often hastately lobed, or, if as small as above, green or reddish beneath .... (6)

6. Plants with prostrate to spreading-ascending stems, growing on beaches and in draw-down areas of reservoirs, ponds, and lakes .... ***C. humile***
6' Plants with erect to ascending stems, or if these prostrate (as in phases of *C. capitatum*), the plants not or seldom of palustrine sites .... (7)

7. Flowers in elongate axillary clusters, these forming erect or steeply ascending compact panicles, or in compact axillary glomerules mainly less than 4 (5) mm wide; plants commonly palustrine .... ***C. rubrum***
7' Flowers in subglobose axillary clusters, these forming bracteate spikes; plants usually montane .... (8)

8. Glomerules subtended by foliose bracts throughout the inflorescence; flowers maturing from base to plant apex .... ***C. foliosum***
8' Glomerules not subtended by leafllike bracts in distal 1/2 of spike; flowers maturing uniformly from apex to plant base .... ***C. capitatum***

9. Leaf blades 0.5–2 cm long, 2–7 mm wide, sinuate-dentate, glaucous-farinose beneath .... ***C. glaucum***
9' Leaf blades various, but not simultaneously as above .... (10)

10. Larger cauline leaves with blades more or less cordate to truncate basally, often over 4 cm wide, thin and glabrous; sepals not keeled dorsally; panicles large and open .... ***C. simplex***
10' Larger cauline leaves with bases various, but seldom as above, and less than 4 cm wide, often farinose beneath; sepals usually keeled .... (11)

11. Larger cauline leaf blades mainly 3–5 or more times longer than wide, entire or with a pair of basal lobes ...... (12)
11' Larger cauline leaf blades mainly 1–3 times longer than broad, hastately lobed, toothed, or entire .... (13)

12. Leaves 1-veined, linear, entire; pericarp adherent to the seed, this typically less than 1 mm wide ... ***C. leptophyllum***
12' Leaves 3-veined (at least near the base in larger ones), distinctly broader than linear, often lanceolate to oblong or elliptic, entire or with 2 basal lobes; pericarp not adherent to the seed .... ***C. desiccatum***

13. Leaf blades hastately lobed (the lobe sometimes again lobed or toothed), or oval to elliptic and entire, or rarely with 1 or more teeth on the apical larger lobe; calyx lobes obscurely or narrowly membranous-margined; plants indigenous .... (14)
13' Leaf blades sinuate-dentate, ovate to lanceolate, or entire; calyx lobes with broad scarious margins; plants adventive .... (15)

14. Leaf blades not hastately lobed; pericarp usually not adherent; seeds black, shining, 0.8–1.3 mm wide; plants usually of upper middle to higher elevations .... ***C. atrovirens***
14' Leaf blades often hastately lobed; pericarp not adherent; seeds dark reddish brown to black, shining, 1–1.3 mm wide; plants of wide altitudinal range .... ***C. fremontii***

15. Fruits sharply angled on the margin; seed coat with minute rounded pits .... ***C. murale***
15' Fruits rounded to obtuse on the margin; seed coat smooth to sculptured; plants common or uncommon, of various distribution .... (16)

16. Seeds sculptured, alveolate-reticulate, or reticulate .... ***C. berlandieri***
16' Seeds smooth or faintly striate .... ***C. album***

Chenopodium album L. (white) Lamb's quarters, pigweed, quelitos. Erect annual herbs. **STEMS** red-striate, 1–10 dm tall or more, simple or more commonly branched; herbage more or less farinose, at least when young. **LEAVES** petiolate, the blades 1–6.5 cm long, 0.5–5.6 cm wide, ovate to rhombic-ovate or lanceolate, sinuate-dentate and often subhastately lobed or the upper (rarely all) entire. **INFL** in dense glomerules, these spicate in upper axils. **FLOWER** calyx with keeled lobes, enclosing the fruit. **FRUIT** pericarp adherent. **SEEDS** horizontal, rounded marginally, smooth to sculptured, black, 1–1.5 mm wide. $2n$ = 18, 36, 54, 108. Weedy species of disturbed habitats. ARIZ: Apa; COLO: Arc, Hin, LPl, Min, Mon; NMEX: McK, RAr, San. 1515–2265 m (5000–7475′). Flowering/Fruit: Jul–Sep. Widespread in North America; adventive from Eurasia. This taxon has been confused with *C. fremontii*, q.v., but in those plants having mature fruits the adherent pericarps are diagnostic. Our material belongs to **var. *album***. This was and is a very useful plant to many Native Americans and Hispanics. Navajos used the plant as a burn dressing and for deflecting snakes and snakebite. Seeds were dried and used like corn by Navajos, especially in the Nightway chant, and also by Hopis. The young greens were used, both raw and cooked, by almost every tribe as well as Hispanics.

Chenopodium ambrosioides L. (food of the gods) Mexican-tea. Aromatic annual herbs. **STEMS** erect or ascending, mainly 4–10 dm tall; herbage pubescent and with sessile glands. **LEAVES** short-petioled, the blades 2–10 cm long, 3–30 mm wide (or more), usually lanceolate, dentate to laciniate. **INFL** paniculate, the cymes sessile on ultimate branches, usually bracteate. **FLOWER** calyx united to the middle or above. **SEEDS** 0.5–0.8 mm wide. $2n$ = 32, 36, 48–64. [*Teloxys ambrosioides* (L.) W. A. Weber; *Dysphania ambrosioides* (L.) Mosyakin & Clemants]. Ruderal and garden weeds. ARIZ: Apa. Flowering/Fruit: Jul–Sep. Widespread in tropical and temperate New World; adventive from Mexico.

Chenopodium atrovirens Rydb. (terrible, green) Mountain goosefoot. Plants mainly 2–75 cm tall. **STEMS** erect or steeply ascending, usually branched; herbage sparingly scurfy to glabrous. **LEAVES** petiolate, the blades 0.6–4 cm long, 2–23 mm wide, lanceolate to ovate, entire or obscurely hastately lobed, otherwise entire, obtuse basally. **INFL** clustered in leaf axils or in interrupted terminal spikes, the lower ones subtended by foliose bracts, becoming ebracteate upward. **FLOWER** perianth lobes free to below the middle, keeled dorsally. **FRUIT** pericarp adherent or not adherent to the horizontal, rugulose to smooth, obtusely margined seed. **SEEDS** about 0.8–1.3 (1.5) mm wide. $2n$ = 18. [*Chenopodium fremontii* S. Watson var. *atrovirens* (Rydb.) Fosberg; *C. incognitum* Wahl]. Sagebrush, piñon-juniper, mountain brush, ponderosa pine, Douglas-fir, aspen–tall forb, and spruce-fir communities. ARIZ: Apa; COLO: Arc, Hin, LPl, Min, Mon, SJn; NMEX: McK, RAr, SJn; UTAH. 1890–3050 m (6240–10065′). May–Oct. British Columbia to Saskatchewan south to California, Nevada, Colorado, and Iowa. Relationship of this species probably lies closer to the *leptophyllum* end of the spectrum. The occasional specimens with free pericarps might indicate intermediacy with *C. fremontii*, however. There has been little agreement between previous authors as to whether the pericarp was adherent or not, but in the specimens examined from Utah, the pericarps are usually adherent.

Chenopodium berlandieri Moq. (for Jean L. Berlandier, Swiss botanist who collected in Texas) Berlandier's pigweed. Ill-scented, mostly glabrous annual. **STEMS** up to 1 m or more tall with slender, ascending-arched branches. **LEAVES** thin, oblong, ovate, or rhombic, 2–4 cm long, sinuate to dentate, pale green and glabrate on the upper surface, glabrous to farinose beneath. **INFL** in glomerules borne on slender-branched panicles. **FLOWER** perianth segments sharply carinate, slightly farinose. **FRUIT** pericarp adherent to the seed, minutely pitted. **SEEDS** horizontal, minutely pitted. [*Chenopodium album* L. var. *berlandieri* (Moq.) Mack. & Bush]. Weedy or pioneer plants of disturbed substrates in various plant communities in saline to nonsaline substrates. ARIZ: Apa; COLO: Arc, Dol, LPl, Min, Mon; NMEX: McK, RAr, San, SJn. 1640–2410 m (5410–7955′) Flowering/Fruit: Jun–Sep. Widespread in North America. Plants from New Mexico belong to the widespread and common **var. *zschackei*** (Murr) Murr ex Graebn. n = 18. [*C. zschackei* Murr; C. *berlandieri* subsp. *zschackei* (Murr.) A. Zobel]. I have been unable to distinguish from among the relatively small sample of specimens available to me the var. *sinuatum* (Murr) Wahl, which is mapped in *Flora of North America* (Clemants 2003a) to include most of New Mexico. It differs in an intangible feature of the style base possessing a yellow area.

Chenopodium botrys L. (like a bunch of grapes) Jerusalem-oak. Aromatic annual herbs. **STEMS** commonly 1–5 dm tall, erect or ascending, usually branched; herbage glandular-villous. **LEAVES** petiolate, the blades sinuate-pinnatifid, the lobes again toothed or lobed, oblong to oval in outline. **INFL** an erect panicle of loosely spreading-recurved cymes, mainly shortly bracteate. **FLOWER** sepals about 1 mm long. **SEEDS** horizontal or vertical, 0.5–0.8 mm wide, dull, dark. $2n$ = 18, 36. [*Teloxys botrys* (L.) W. A. Weber; *Dysphania botrys* (L.) Mosyakin & Clemants]. Widespread ruderal weedy species, established locally in indigenous communities, especially in gravelly washes. ARIZ: Apa; COLO: LPl; NMEX: McK, RAr, San, SJn. 1665–2105 m (5500–6945′). Flowering/Fruit: Aug–Sep. Widely distributed in the United States; adventive from Eurasia.

Chenopodium capitatum (L.) Ambrosi (knoblike head) Strawberry-spinach. Plants mainly 1–4 dm tall. **STEMS** erect or decumbent-ascending, simple or more commonly branched from the base; herbage glabrous. **LEAVES** petiolate, the blades (1) 1.5–10 cm long, 1–10 cm wide, triangular-hastate to lanceolate, shallowly to deeply toothed or subentire, hastately lobed or the upper entire, acute to obtuse apically, often turning reddish. **INFL** axillary capitate clusters, these borne in spikes, the lower clusters subtended by foliose bracts, the upper ones ebracteate or with reduced bracts. **FLOWER** perianth lobes free to below the middle, not mealy, dry and greenish or becoming fleshy and reddish, shorter than the fruit. **FRUIT** pericarp adherent to the erect (or less commonly horizontal) seed. **SEEDS** about 1 mm long. $2n = 18$. [*Blitum capitatum* L.; *C. overi* Aellen; *C. chenopodioides* (L.) Aellen]. Plant used as a lotion for head bruises and for black eyes by Kayenta Navajo.

1. Flower clusters often over 6 (to 12) mm wide, the calyx becoming red and fleshy at maturity; plants uncommon .... **var. *capitatum***
1' Flower clusters commonly less than 6 (to 8) mm wide, the calyx not fleshy, though sometimes reddish at maturity; plants common .... **var. *parvicapitatum***

var. *capitatum* $2n = 18$. Gravelly soil or sandy soils in meadows and clearings in ponderosa pine, Douglas-fir–aspen, and spruce-fir forests. COLO: Arc, Min; NMEX: McK, RAr. 2440–2685 m (8050–8860′). Flowering/Fruit: Aug–Oct. Alaska to Quebec, south to California and New England. The entire plant becomes anthocyanic late in its life cycle. Northward the species is abundant in Alaska and Yukon.

var. *parvicapitatum* S. L. Welsh (smallhead) Smallhead chenopod [*Blitum hastatum* Rydb.; *Chenopodium overi* Aellen]. Mountain brush, ponderosa pine, aspen, and spruce-fir communities. ARIZ: Apa; COLO: Arc, LPl, Min, Mon, SJn; NMEX: McK, RAr, SJn. 1860–3050 m (6140–10065′). Flowering/Fruit: Jun–Sep. British Columbia to Saskatchewan, south to California, Nevada, Wyoming, and Colorado. At specific level the correct name would be *C. overi* Aellen. This common plant of montane habitats has been identified variously as *C. capitatum* or *C. rubrum*, and more recently, as *C. chenopodioides* (L.) Aellen. The latter plant is a portion of the flora of Russia and might represent nothing more than phases of *C. rubrum* sensu lato. Certainly the description provided for that entity (Shishkin 1970) is not of our specimens, and neither is the habitat cited (i.e., "wet solonchaks"). The name given here is for the purpose of providing an unequivocal epithet for this montane western American phase of *C. capitatum*. According to *Flora of North America* (2003: 276), *C. chenopodioides* is allied to *C. rubrum*, differing in the perianth segments being connate almost to the tip in the former, and "mostly" near the very base in the latter, and both of them segregated on elusive and overlapping characteristics from *C. capitatum* (and *C. foliosum*), i.e., glomerules of *chenopodioides–rubrum* being 2–5 mm (not 3–10), and presence of both horizontal and vertical seeds (not all vertical). The other characteristics noted are inconclusive and subject to interpretation. *Chenopodium foliosum* (Moench) Asch. [*Morocarpus foliosus* Moench, 1794 Methodus 342] is closely allied morphologically to *C. capitatum* var. *parvicapitatum*, differing in the foliose bracts throughout the inflorescence and in the sepals that turn fleshy-red at maturity as in var. *capitatium* (see *Flora of North America* 2003: 276, wherein it is mapped for San Juan County, Utah). None of the specimens seen bearing the name *C. foliosum* from the study area are indeed that species; they are all var. *parvicapitatum*.

Chenopodium desiccatum A. Nelson (dried up, for its desert habitat) Desert goosefoot. Plants mainly 3–8 dm tall. **STEMS** erect, simple or branched; herbage commonly more or less scurfy. **LEAVES** petiolate, the blades mostly 1.3–6 cm long and 2–10 (15) mm wide, linear to narrowly lanceolate or elliptic, entire or less commonly hastately lobed, cuneate basally. **INFL** clustered in terminal or axillary spicate panicles. **FLOWER** perianth lobes free to below the middle, keeled dorsally. **FRUIT** pericarp not adherent to the horizontal, smooth to rugulose, obtusely margined seed. **SEEDS** 0.9–1.2 mm wide. $2n = 18$. [*Chenopodium pratericola* Rydb.; *C. petiolare* Kunth var. *leptophylloides* Murr.; *C. leptophyllum* (Moq.) Nutt. ex S. Watson var. *desiccatum* (A. Nelson) Aellen; *C. leptophyllum* var. *oblongifolium* S. Watson; *C. pratericola* var. *oblongifolium* (S. Watson) Wahl]. Shadscale, hopsage, rabbitbrush, tamarix-poplar, sagebrush, and piñon-juniper communities. ARIZ: Apa, Nav; COLO: Arc, Dol, Hin, Mon, Smg; NMEX: RAr, San. 1635–2565 m (5400–8420′). Flowering/Fruit: Aug–Sep. Yukon to Manitoba, south to California, New Mexico, and Nebraska. Material assigned to *C. desiccatum* in a strict sense, as distinct from *C. pratericola* in a strict sense, has perianth lobes covering the mature fruit, as opposed to having the fruit not covered. The distinction is not great. Reduction of this taxon to *C. leptophyllum* (as var. *oblongifolium*) begs the question of a probably nearer relationship to *C. fremontii*, with which it shares nonadherent pericarp and broader, more veined leaves. Furthermore, apparent intermediates between *C. fremontii* and *C. desiccatum* exist. The entire complex is in need of monographic study.

Chenopodium foliosum (Moench) Asch. (leafy) Leafy goosefoot. Annual, mostly 0.5–6 dm tall. **STEMS** mostly branching, glabrous to minutely scaly. **LEAVES** with blades 7–40 mm long, entire to irregularly toothed, glabrous to minutely scaly; lower lanceolate to deltate, base 2-lobed to hastate; upper oblong to lanceolate, base tapered. **INFL** 3–5 mm diameter, axillary, mostly spheric, leafy-bracted. **FLOWER** sepals mostly 3, smooth, mostly glabrous, often becoming reddish; stamens 3–4. **FRUIT** 1 mm diameter; wall mostly adherent to seed. **SEEDS** usually vertical. $2n = 16, 18$. Rocky soils; riparian areas in ponderosa pine and subalpine communities. COLO: Arc, LPl, Mon, SJn; NMEX: SJn. 2440–3050 m (8000–10000′). Flowering: (May) Jun–Sep. Western Canada south to New Mexico. Native to Europe.

Chenopodium fremontii S. Watson (for John C. Frémont, explorer and naturalist) Frémont's goosefoot. Plants mainly 1–8 (12) dm tall. **STEMS** erect or ascending, usually branched; herbage more or less scurfy to glabrous. **LEAVES** petiolate, the blades 0.6–5 (6) cm long, and about as broad, less commonly 2–3 times longer than broad, triangular-ovate to ovate or lanceolate, commonly hastately lobed, the lobes often again lobed or toothed, otherwise entire or rarely with 1 or few teeth on the main apical lobe, broadly cuneate to subcordate basally. **INFL** clustered in large, interrupted terminal and smaller lateral spikes, often in naked or near naked panicles. **FLOWERS** scurfy; perianth lobes free to below the middle, keeled dorsally. **FRUIT** pericarp not adherent to the horizontal, smooth to rugulose, obtusely margined seed. **SEEDS** 0.9–1.2 mm wide. $n = 9$. This closely interrelated complex of forms involves the linear-leaved *C. leptophyllum* (q.v.) at one end of the spectrum and the broad-leaved phases of *C. fremontii* at the other end (and with both *C. atrovirens* and *C. desiccatum*, inter alia, between the extremes). The intervening plants have been regarded as species or some of them have been placed within expanded species concepts at both ends of the series. The course followed herein is a compromise between having one all-inclusive species, with numerous varieties, and recognizing all of the named entities at specific level. The proposed treatment attempts to represent the major taxa as they occur in the Four Corners region; the synonymy might not be properly applied in all cases. Hopis used the seeds and leaves as food, flavoring, and packing around yucca fruits when baked. Navajos used the seeds to make bread.

1. Plants 0.5–8 dm tall, variously branched but, if as below, the lateral branches much shorter than the main stem; leaves white-farinose to glabrous and green (at least above), becoming anthocyanic in age, at least some typically over 15 mm long .... **var. *fremontii***

1′ Plants mainly less than 2.5 dm tall, branching from the base, the curved ascending branches subequal to the main stem; leaves more or less white-farinose, at least beneath and sometimes on both sides, mainly 4–15 (20) mm long and typically as broad .... **var. *incanum***

var. *fremontii* $2n = 18$. Warm desert shrub, salt desert shrub, sagebrush, mountain brush, piñon-juniper, aspen, Douglas-fir, ponderosa pine, and spruce-fir communities. ARIZ: Apa, Nav; COLO: Arc, LPl, Min, Mon, SJn, SMg; NMEX: McK, RAr, San, SJn; UTAH. 1650–2770 m (5445–9140′). Flowering/Fruit: Jul–Oct. British Columbia to Manitoba, south to California and Mexico.

var. *incanum* S. Watson (hoary, for the white hairs) Silvery goosefoot. $2n = 18$. [*Chenopodium incanum* (S. Watson) A. Heller; *C. incanum* var. *occidentale* D. J. Crawford; *C. watsonii* A. Nelson]. Blackbrush, salt desert shrub, sagebrush, piñon-juniper, mountain brush, and ponderosa pine communities. ARIZ: Apa; COLO: Arc, Hin, LPl, Mon, SMg; NMEX: RAr, San, SJn; UTAH. 1850–2780 m (6105–9175′). Flowering/Fruit: May–Sep. Nevada to Nebraska, south to Texas and Mexico. There are numerous intermediates that form a transition to var. *fremontii*. Specimens assigned here are not uniform with regard to leaf shape, plant height, and openness of the inflorescence. The plant is treated at species level in *Flora of North America* (2003: 291).

Chenopodium glaucum L. (whitish covering of the leaves) Oakleaf goosefoot. Plants mainly 3–30 (exceptionally to 70) cm long. **STEMS** prostrate to ascending or erect, usually branched; herbage farinose, especially on lower leaf surfaces. **LEAVES** with short to rather elongate petioles, the blades 4–25 (30) mm long, 2–10 (18) mm wide, lanceolate to oblong or ovate, coarsely sinuate-dentate. **INFL** in clusters in numerous short, bracteate or ebracteate, axillary spikes and a terminal spicate panicle. **FLOWER** perianth cleft almost to base, not enclosing the fruit. **FRUIT** pericarp not adherent to the seed. **SEEDS** horizontal or vertical, 0.8–1.3 mm wide, smooth. $2n = 18, 36$. Often in saline substrates on pond and lake shores and stream banks, in sedge-rush, tamarix-sedge, rabbitbrush, piñon-juniper, and aspen to spruce-fir communities. ARIZ: Apa; COLO: Arc, Dol, Hin, LPl, Min, Mon, SMg; NMEX: McK, RAr, SJn; UTAH. 1090–2470 m (3600–8150′). Flowering/Fruit: May–Oct. Widespread in the United States and Canada; Eurasia. Our material is assignable to **var. *salinum*** (Standl.) B. Boivin (salty) on the basis of its larger fruits (0.8–1.3 mm, not 0.6–0.9 mm). [*Chenopodium salinum* Standl.]

Chenopodium graveolens Willd. (strong-smelling) New Mexico goosefoot, foetid goosefoot. Plants annual, strongly scented. **STEMS** erect, 1.5–4 (6) dm tall, simple or paniculately branched from the base, sparsely puberulent to glabrate, often suffused red. **LEAVES** 2–6.5 cm long, 1.5–3 cm wide, lanceolate to narrowly oblong or deltoid-ovate, sinuate- or laciniate-pinnatifid, the lobes triangular, bright green, glabrous or viscid-villous above, covered with yellow glands beneath, obtuse to acuminate apically, truncate to cuneate at the base. **INFL** of numerous loosely few-flowered axillary cymes, these aggregated into narrow, elongate naked panicles. **FLOWERS** sessile in branch axils and at ends of slender, lateral branches, pedicellate flowers usually abortive; calyx deeply cleft, the lobes cornately appendaged, bearing yellow or orange-yellow sessile glands, incompletely enclosing the fruit. **SEEDS** horizontal, 0.5–0.8 mm wide, dark brown, the pericarp adherent. [*Teloxys graveolens* (Willd.) W. A. Weber; *Dysphania graveolens* (Willd.) Mosyakin & Clemants]. Disturbed sites, pond margins, and roadsides in various indigenous plant communities. ARIZ: Apa, Nav; COLO: LPl, Mon; NMEX: McK, SJn. 2200–2745 m (7260–9060′) Flowering/Fruit: Aug–Sep. Arizona and New Mexico to Texas (and elsewhere in the U.S.); Central and South America; Africa. Our material has been assigned to **var**. ***neomexicanum*** (Aellen) Aellen. [*Chenopodium incisum* Poir. var. *neomexicanum* Aellen]. Zunis steeped the plant in water and inhaled the vapor for headaches. Hopis ground the seeds, mixed them with cornmeal, and cooked the dumplings in corn husks. Ramah Navajos made a cold infusion to be taken as protection in warfare.

Chenopodium humile Hook. (low) Low goosefoot. **STEMS** mainly 5–15 cm long, at least the lowermost branches prostrate-ascending; herbage glabrous or nearly so. **LEAVES** petiolate, the blades 0.5–3 cm long, 0.2–2.5 cm wide, triangular to ovate, lanceolate, or elliptic, entire or somewhat lobed, some often subhastate, fleshy and often suffused with red. **INFL** sessile, the clusters borne in simple or branched axillary glomerules and terminal spicate panicles. **FLOWER** perianth lobes cleft nearly to the base. **FRUIT** pericarp not adherent. **SEEDS** nearly always vertical, oval in outline, 0.8–1 (1.2) mm broad. Often, but not exclusively, on moist, saline substrates in palustrine and riparian habitats and in pond or reservoir draw-down areas. COLO: LPl, Min, Mon, SMg; NMEX: RAr, SJn; UTAH. 2380–2525 m (7855–8330′). Flowering/Fruit: Jun–Sep. Widespread in the United States and Canada; Eurasia. [*Chenopodium rubrum* L. var. *humile* (Hook.) S. Watson]. This plant has been mistaken previously for *C. capitatum* var. *parvicapitatum* and for *C. rubrum*, with which it is sometimes placed at varietal level. In some floras similar plants are keyed to *C. chenopodioides* (L.) Aellen (Correll and Johnston 1979a), which is distinguished by the "perianth segments connate to near the tip." Careful examination is necessary to distinguish the proper course through the key presented above. The apparent restriction to lacustrine and/or palustrine habitats, the tendency to entire or merely hastate leaves, and the prostrate to prostrate-ascending habit of these dwarf plants in combination are definitive for this entity. In both habit and habitat the low goosefoot approaches *C. glaucum* but differs in mainly (but not always) lacking scurf on the undersurface of the leaves and the leaves are not as distinctly dentate. If draw-down areas of reservoirs, ponds, and other similar habitats were searched in late summer, the species would probably be found to be much more common than indicated by the few collections examined.

Chenopodium leptophyllum (Moq.) Nutt. ex S. Watson (narrow leaf) Narrowleaf goosefoot. Plants mainly 12–70 cm tall, erect or the branches ascending, simple or branched. **LEAVES** short-petiolate, the blades mainly 0.7–4 cm long, 1–5 (7) mm wide, linear to narrowly oblong or narrowly lanceolate, 1-veined, cuneate to acute basally, entire. **INFL** in loose to compact cymes aggregated into terminal or axillary spicate panicles. **FLOWER** perianth lobes cleft to well below the middle, keeled dorsally. **FRUIT** pericarp adherent. **SEEDS** horizontal, 0.9–1.1 mm wide, black, finely rugulose to smooth. $2n = 18$. [*Chenopodium album* L. var. *leptophyllum* Moq.]. Shadscale, greasewood, rabbitbrush, tamarix, sagebrush, fringed sagebrush, mountain brush, and aspen communities. ARIZ: Apa, Nav; COLO: LPl, Mon; NMEX: McK, RAr, San, SJn; UTAH. 1525–2260 m (4620–7460′). Flowering/Fruit: May–Oct. British Columbia to Saskatchewan, south to California and Mexico. The one-veined, narrow leaf blades and adherent pericarps are apparently definitive for this plant. The young leaves were cooked by Apaches and Hispanics as greens. The sprouts were also boiled with meat. Seeds were considered important food by Zunis, Apaches, Navajos, and Hopis, often ground and put into cakes or steamed balls. Indeed, Zunis considered this one of the most important food plants.

Chenopodium murale L. (growing on walls) Nettleleaf goosefoot. Plants mainly 2–8 dm tall. **STEMS** erect or with branches ascending; herbage glabrous or sparingly farinose, especially in inflorescences. **LEAVES** petiolate, the blades 1–5 (7) cm long and as broad or nearly so, ovate to oval or lanceolate, irregularly sinuate-dentate and some often subhastate, cuneate to subcordate basally. **INFL** sessile and solitary to clustered in axillary or terminal panicles not much, if at all, surpassing the leaves. **FLOWER** perianth lobes free to below the middle, keeled dorsally. **FRUIT** pericarp adherent to the horizontal, rugulose to smooth, sharply margined seed. **SEEDS** horizontal, rugulose to smooth, sharply margined; 1–1.5 mm long. $2n = 18$. Ruderal weeds. ARIZ: Apa. 2170 m (7160′). Flowering/Fruit: Aug–Sep.

Widespread in the United States and Canada; adventive from Eurasia. This is a robust, weedy species with large, irregularly sinuate-dentate leaves, and both lateral and terminal paniculate inflorescences.

Chenopodium rubrum L. (red) Red goosefoot. Plants mainly 0.5–10 dm tall, erect, simple or with steeply ascending branches; herbage glabrous or somewhat villous in inflorescence. **LEAVES** petiolate, the blades 0.7–9 (13) cm long, 0.4–7 (10) cm wide, triangular to ovate, lanceolate, or elliptic, sinuate-dentate or -lobed, some often subhastate, fleshy and often suffused with red. **INFL** sessile, the clusters borne in simple or branched axillary and terminal spicate panicles. **FLOWER** perianth lobes cleft to the middle or below, rounded or sometimes keeled dorsally. **FRUIT** pericarp not adherent. **SEEDS** nearly always vertical, oval in outline, 0.6–1 mm long. $2n = 18, 36$. Saline, moist substrates in lacustrine, palustrine, and riparian habitats. ARIZ: Apa, Nav; COLO: Arc, Hin, LPl, Mon; NMEX: RAr, SJn; UTAH. 1520–2680 m (5025–8850′). Flowering/Fruit: Aug–Sep. Widespread in the U.S. and Canada; Eurasia. The plants turn anthocyanic in autumn.

Chenopodium simplex (Torr.) Raf. (simple) Mapleleaf goosefoot. Plants mainly 2–10 dm tall. **STEMS** erect, simple or branched; herbage glabrous, except in inflorescence. **LEAVES** alternate or the lower often opposite, long-petiolate, the blades commonly 1.7–10 cm long, 1.2–10 cm wide, ovate to deltoid-ovate, sinuate-dentate to lobate, with 2–4 teeth or lobes, cordate to truncate or obtuse basally. **INFL** in small cymes, these arranged in large terminal (to 30 cm long or more) and smaller axillary panicles, more or less farinose and often sparingly glandular. **FLOWER** perianth cleft nearly to the base, not strongly keeled dorsally. **FRUIT** pericarp not or moderately adherent. **SEEDS** 1.2–1.9 mm wide, with obtuse margin, smooth or somewhat sculptured. $2n = 18, 36$. [*Chenopodium hybridum* L. var. *simplex* Torr.; *C. gigantospermum* Aellen; *C. hybridum* var. *gigantospermum* (Aellen) Rouleau]. Sagebrush, piñon-juniper, mountain brush, ponderosa pine, and aspen communities, less commonly in riparian or palustrine habitats. COLO: LPl, Mon; NMEX: RAr. 2680 m (8850′). Flowering/Fruit: Jul–Sep. Widely distributed in the United States and Canada; Europe. A case can be made for inclusion of our material as a variety of *C. hybridum* L. Certainly they are more than mere look-alikes. However, on the basis of the seed being less sculptured than in typical European material, it is herein regarded as specifically distinct. The name taken up here is one long overlooked by students of the genus, but noted by Robert Dorn in his treatment of the vascular plants of Wyoming.

Corispermum L. Bugseed

(*coris*, bedbug, + *sperma*, seed) Annual herbs, often pubescent with stellate hairs. **LEAVES** alternate, sessile, entire, 1-veined. **INFL** solitary or clustered in bract axils, arranged in dense or lax spikes. **FLOWERS** perfect; perianth segments 1–3, minute, unequal, the posterior 1 largest, erect, 1-nerved, scarious; stamens 1–3 (5); stigmas 2, connate basally. **FRUIT** an achene, strongly flattened, plano-convex, indurate, the margin winged or acute (Mosyakin 1995). Sixty north-temperate species.

Corispermum americanum (Nutt.) Nutt. (America) American bugseed, tickseed. **STEMS** mainly 8–50 cm tall, commonly branched throughout, the lower branches often curved-ascending to ascending, usually reddish or purplish; herbage glabrous or sparsely to densely pubescent with soft, branched hairs. **LEAVES** alternate, sessile, semicylindric and more or less involute or subulate, mostly 0.8–6 cm long, 0.5–3 mm wide, apiculate. **INFL** typically slender and elongate, the bracts broadly scarious-margined, with the lower ones narrower or broader than the fruit, the upper ones usually broader than the fruit. **FLOWER** perianth mostly consisting of a single posterior erose segment (rarely with 2 additional small anterior-lateral segments); stamens 3–5. **SEEDS** 2.9–3.5 mm long, 1.6–2.6 mm wide, oval to suborbicular, smooth, glabrous, brownish; wing 0.2–0.3 mm wide, opaque, stramineous, 1/8–1/5 as wide as the body. $2n = 18$. [*Coriospermum hyssopifolium* L. *var. americanum* Nutt.; C. *hyssopifolium* L.; *C. nitidum* Kit.]. Usually growing on sandy substrates, in ephedra, four-wing saltbush, rabbitbrush, scurfpea, and piñon-juniper communities. ARIZ: Apa, Nav; NMEX: McK, RAr, San, SJn; UTAH. 1605–2395 m (5290–7900′). Flower/Fruit: Jul–Sep. Montana and North Dakota south to Arizona, New Mexico, and Texas. The American bugseed species have long been mistaken as introductions from the Old World and have been regarded mainly as either *C. hyssopifolium* and/or *C. nitidum* Kit. ex Schult. The problem was discussed in *A Utah Flora* (Welsh et al. 1993). Mosyakin recognizes a rather large set of intergrading native American taxa, which require careful examination prior to their acceptance. The plants become anthocyanic in autumn and at least in some years form beautiful assemblages of pink, orange, and purple clumps on sands, especially on bars in the various river drainages.

Cycloloma Moq. Winged Pigweed

(*cyclos*, a circle, + *loma*, a border, from the encircling calyx wing) Annual herbs. **STEMS** commonly branched and forming rounded clumps. **LEAVES** alternate, sinuate-dentate; herbage more or less pubescent, eglandular. **INFL** a panicle of interrupted spikes. **FLOWERS** sessile; perianth 5-lobed, stamens 5. **FRUIT** perianth segments with a transverse wing on the back, the wings connate and completely encircling the fruit. **SEEDS** depressed; horizontal. One species native to central North America.

Cycloloma atriplicifolium (Spreng.) J. M. Coult. (leaves resemble *Atriplex*) Winged pigweed. Plants mainly 0.8–5 dm tall (rarely more), divaricately branched, and forming clumps about as broad as tall or broader. **STEMS** striate, loosely and sparingly tomentulose, becoming glabrate. **LEAVES** short-petioled to sessile, the blades 1–8 cm long, 2–15 mm wide, coarsely serrate-dentate, acute apically, cuneate basally. **FLOWERS** perfect and pistillate; sepals 5, keeled, the perianth developing into a horizontal wing; wings white-hyaline, lobed or toothed, 4–5 mm in diameter, often red or purple at maturity; ovary tomentulose; styles 2 or 3. **FRUIT** enclosed in calyx. **SEEDS** about 1.5 mm wide, black, smooth; pericarp not adherent. $2n = 36$. [*Salsola atriplicifolia* Spreng.]. Sandy habitats in blackbrush, mixed desert shrub, and juniper communities. ARIZ: Apa, Nav; NMEX: RAr, SJn; UTAH. 1370–1890 m (4500–6220′). Flowering/Fruit: May–Oct. Manitoba and Indiana, south to Arizona and Texas; adventive in Europe. Hopis used the plant to treat headaches, for fever and rheumatism; the red seeds were used to produce a pink dye. Apaches, Hopis, and Zunis used the seeds in porridge or steamed cakes. The blossoms were chewed and rubbed on the hands for protection against arrows.

Grayia Hook. & Arn. Hopsage

(for Asa Gray) Dioecious or less commonly monoecious shrubs or subshrubs; branches more or less thorny; axillary buds subglobose, prominent. **LEAVES** alternate, entire; herbage pubescent with simple and stellate hairs. **INFL** in terminal and axillary spicate panicles; staminate flowers 2–5 in clusters in bract axils, not separately bracteolate; pistillate flowers 1–several per bract. **FLOWERS** imperfect; staminate flowers with the perianth 4- or 5-lobed, subequal to the 4 or 5 stamens; pistillate flowers often vestigial, each enclosed by 2 connate bracteoles, the more or less accrescent bracteoles obcompressed, the margins thickened and spongy within; stigmas 2. **FRUIT** vertical. One species.

Grayia spinosa (Hook.) Moq. (spiny) Hopsage. Shrubs, mainly 5–12 (15) dm tall; branches gray-brown; branchlets often spinose-persistent, the bark exfoliating in long strips, pubescent with scurfy and stellate hairs when young. **LEAVES** 5–30 mm long (or more), mainly 2–12 mm wide, spatulate to oblanceolate, entire, tapering to a short petiole. **INFL** with pistillate flowers in short spicate inflorescences, the subtending bracts reduced, enclosed by paired, accrescent, obcompressed bracts, orbicular or cordate, the wings thickened and spongy within, 6–15 mm wide, greenish, straw-colored, or suffused with red. **FLOWERS**: staminate with usually 4-lobed perianth, enclosing the 4 stamens, 1.5–2 mm long. $2n = 36$. [*Chenopodium spinosum* Hook.; *Grayia polygaloides* Hook. & Arn.]. Blackbrush, other warm desert shrub, shadscale, horsebrush, rabbitbrush, sagebrush, and piñon-juniper communities. ARIZ: Apa, Coc; COLO: Mon; NMEX: SJn; UTAH. 1130–1790 m (3425–5905′). Flowering/Fruit: Apr–May. Washington to Montana, south to California, Arizona, and New Mexico. This is a valuable browse plant for livestock, especially for sheep. Locally it is called "applebush" because of its palatability. For consideration of relationship to *G. brandegeei* A. Gray, see discussion under *Zuckia*.

Halogeton C. A. Mey. Halogeton

(salt neighbor) Annual herbs. **LEAVES** alternate, fleshy and sausagelike, bearing an apical slender spine. **FLOWERS** perfect or partially pistillate, usually bracteolate; perianth of 5 segments, free nearly to the base, embedded in white hair, the segments gibbous, winged in fruit; stamens 2–5, connate basally into a glandular, hypogynous disc; stigmas 2. **SEEDS** vertical, laterally flattened, adherent to the pericarp. A mostly Mediterranean genus of nine species.

Halogeton glomeratus (M. Bieb.) C. A. Mey. (close together into a head) Halogeton. Plants mainly 3–45 cm tall and as broad, glaucous, usually branched from the base, with curved-ascending branches. **LEAVES** mainly 3–15 mm long and 1–2 mm thick, terete, dilated and semiamplexicaul basally, obtuse and terminating in a deciduous slender spine about 1–1.5 mm long, bearing a tuft of hairs and fascicled leaves in the axils; bracteoles ovate. **FLOWER** perianth segments membranous, ovate to oblong, 1-veined, with lustrous, membranous, fanlike, veiny wings 2–3 mm long and 3–4 mm wide; stamens united into 2 clusters of 2 or 3, with 1 anther per cluster. **FRUIT** oval to obovate, 1.2–1.8 mm long, with an erect cusp on 1 or both sides. $2n = 18$. [*Anabasis glomerata* M. Bieb.]. Mainly in disturbed sites in cheatgrass, Russian thistle, saltgrass, mixed desert shrub, salt desert shrub, and piñon-juniper communities. ARIZ: Apa, Nav; COLO: LPl, Mon; NMEX: McK, RAr, San, SJn; UTAH. 1220–1985 m (4025–6550′). Flowering/Fruit: Jul–Sep. Widely distributed in the western United States; adventive from Eurasia. This plant was introduced into northern Nevada in the early 1930s (first collected in 1934), possibly for use in grazing experiments. It spread quickly into the lower-elevation desert grazing lands of Nevada and western Utah, and subsequently into eastern Utah and other states. The plant is rich in oxalates and poses a serious threat to grazing animals, especially to sheep, which have suffered thousands of death losses for several decades. As many as 1250 sheep of a herd of 1500 succumbed following ingestion of this plant in Antelope Valley, Millard County, Utah, in the 1970s.

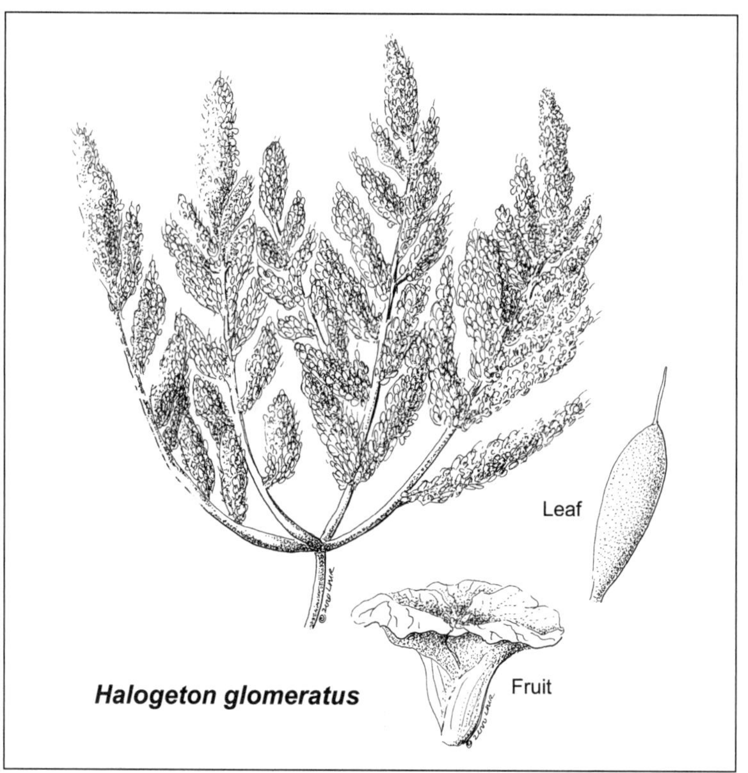

Halogeton glomeratus

Krascheninnikovia Gueldenst.
Winterfat

(for S. P. Krascheninnikov, Russian botanist of the 18th century) Monoecious tomentose shrubs. **LEAVES** alternate, entire. **FLOWERS**: staminate ebracteate, with calyx 4-lobed; stamens 4; pistillate lacking a perianth, enclosed in 2 villous-pilose, partially connate bracteoles, the tips divergent and hornlike; styles 2, slender. **FRUIT** with pericarp thin, free from the seed. [*Eurotia* Adans. in 1768 and *Ceratoides* Gagnebin in 1772, both misapplied] (Howell 1971). About eight species of the northern Mediterranean, temperate Asia, and western North America.

Krascheninnikovia lanata (Pursh) A. Meeuse & A. Smit (with woolly hairs) Winterfat, white-sage, eurotia. Shrubs, woody for 0.2–8 dm aboveground (or more), and with numerous annual branchlets mainly 0.5–3 (5) dm long; herbage stellate-hairy, commonly with longer straight hairs intermixed, the hairs white or becoming yellowish in age. **LEAVES** 1–4.5 cm long, 1–6.5 mm wide, linear to narrowly lanceolate, entire, revolute to almost flat, sessile above, short-petiolate below. **INFL** borne in dense axillary clusters or more or less spicate along branch tips; pistillate flowers 2–4 per axil; staminate flowers in spicate axillary clusters. **FLOWERS**: staminate with perianth segments 1.5–2 mm long; fruiting bracts 3–6 mm long, obscured by the long covering hair. $2n = 18$. [*Diotis lanata* Pursh; *Eurotia lanata* (Pursh) Moq.; *Ceratoides lanata* (Pursh) J. T. Howell]. Shadscale, black sagebrush, sagebrush, and piñon-juniper communities. ARIZ: Apa, Coc, Nav; COLO: LPl, Mon; NMEX: McK, RAr, San, SJn; UTAH. 1550–2840 m (5115–9370′). Flowering/Fruit: Jun–Oct. Yukon to Saskatchewan, south to California, New Mexico, and Texas. Name changes for this common and widespread American species can be traced to interpretations of the International Code of Botanical Nomenclature, whose rules are finally, and reluctantly, bowed to by this botanist. Apologies are owed to all those who have to learn the impossibly long and difficult name dictated by stipulations of the Code. This is an instance where common names have greater staying power than the scientific ones, but the search for stability in nomenclature continues. Cold infusion of plant taken by Ramah Navajos as an antidote for *Datura* poisoning. Chewed leaves used for poison ivy rash. Plant is a "Life Medicine" and a major forage plant for sheep. Also used in the sweathouse for Mountain Chant. Used by Hopis in ceremonials to produce steam. Zunis made a poultice of ground roots for burns.

Monolepis Schrad. Poverty-weed

(single scale) Annual, polygamo-monoecious herbs. **LEAVES** simple, hastately lobed or entire, alternate, mealy to subglabrous, fleshy. **INFL** borne in axillary clusters. **FLOWERS** unisexual, inconspicuous; perianth consisting of 1 bract-like scale (rarely 2 or 3, or lacking), not enclosing the fruit; stigmas 2. **FRUIT** with pericarp reticulately patterned or warty, adherent to the erect seed. Three species of western North America.

Monolepis nuttalliana (Schult.) Greene (for Thomas Nuttall, Harvard botanist) Poverty-weed. Plants mainly 4–30 cm tall. **STEMS** prostrate or ascending to erect, simple or much-branched from the base, mealy to subglabrous; leaves 5–50 mm long, the blades 1–15 mm wide, lanceolate to elliptic or oblong, with 1 pair of lateral lobes near the middle, reduced upward and sometimes entire, the petiole 1–20 mm long. **INFL** borne in dense, sessile, axillary clusters. **FLOWER** perianth segments 1–2 mm long, more or less acute apically; pericarp pitted, usually pale. **FRUIT** 0.9–1.5 mm broad. $2n = 18$. [*Blitum nuttallianum* Schult., based on *B. chenopodioides* L.; *Monolepis chenopodioides* Moq.]. Pioneer plant of open sites in blackbrush, shadscale, mat saltbush, sagebrush, piñon-juniper, mountain brush, ponderosa pine,

Douglas-fir, and aspen communities. ARIZ: Apa, Nav; COLO: Arc, Hin, LPl, Mon, SMg; NMEX: McK, RAr, San, SJn; UTAH. 1515–2550 m (5000–8415′). Flowering/Fruit: Apr–Jul. Alaska and Yukon to California and New Mexico, east to Manitoba and Missouri. The plants are often anthocyanic in whole or in part, even when very young. Plant used by Ramah Navajos as a ceremonial emetic, poultice for skin abrasions, and as a hunting medicine. Hopis used the seeds to make mush. Navajos used the plants as sheep fodder.

Salicornia L. Pickleweed

(*sal*, salt, + *cornu*, horn) Annual herbs from taproots. **LEAVES** simple, scalelike, opposite, connate, glabrous. **INFL** with flowers in opposite groups of 3, sunken in depressions of thickened, terminal spikes. **FLOWERS** perfect, borne sessile; subtended by scalelike bracts; perianth consisting of 4 connate segments free at the tip around a slitlike opening; stigmas commonly 2. **FRUIT** with perianth enclosing; pericarp thin, free from the erect, retrorsely pubescent seed. About 13 species worldwide.

Salicornia europaea L. (Europe) Annual samphire. Plants annual, mainly 9–30 cm tall, from slender taproots. **STEMS** fleshy, erect or ascending, commonly branched, often reddish at maturity. **LEAVES** scalelike, often with a scarious margin; spikes 0.5–5 cm long, 2–3 mm thick, the joints 2–4 mm long. **INFL** with central flowers much above the lateral ones. **FRUIT** dehiscent, the seeds falling separately. $2n = 18, 36$. [*Salicornia rubra* A. Nelson; *S. europaea* L. subsp. *rubra* (A. Nelson) Breitung; *S. rubra* var. *prona* Lunell; *S. europaea* var. *prona* (Lunell) B. Boivin]. Saline, typically fine-textured substrates in salt marsh, seepweed, poverty weed, alkali sacaton, and saltgrass communities. COLO: Mon. 2000 m (6600′). Flowering/Fruit: Jul–Sep. Widespread in North America and Eurasia. Our specimens are supposed to differ from the more coastal ones in having slender spikes with joints about as thick as long, but no such correlation is apparent. Recognition of our material at infraspecific level seems moot.

Salsola L. Tumbleweed

(an Arabic name) Annual herbs. **LEAVES** alternate, entire, commonly spinulose. **INFL** spicate; solitary or clustered in axils of spiny bracts, each with 2 smaller bracteoles. **FLOWERS** perfect, 5-merous; stamens 5, usually inserted at the margin of a lobed disc; styles 2 or 3. **FRUIT** closely enveloped in the persistent calyx; fruiting perianth with winglike, mostly horizontally spreading ridges. **SEEDS** horizontal to oblique (Beatley 1973). About 100 species worldwide of maritime and other saline habitats.

1. Bracts appressed and imbricate along slender inflorescence branches, gradually attenuate to the subulate, spinulose apex; spikes rather dense, continuous; perianth segments wingless or the wing narrow and erose; stems usually erect ..... ***S. collina***
1′ Bracts reflexed, not imbricate when mature, usually rather abruptly narrowed to the spinose or mucronulate apex; spikes interrupted at maturity, at least in the lower portion; perianth segments winged; stems erect or ascending ..... ***S. tragus***

Salsola collina Pall. (found on a hill) Pallas' tumbleweed. **STEMS** simple or much-branched, forming rounded clumps that tumble freely when mature; green or streaked with red, glabrous or somewhat hirsute. **LEAVES** mainly 2–6 cm long and about 1 mm wide, narrowly linear to filiform, modified upward as spinescent bracts with expanded bases and scarious margins. **INFL** elongate-spicate, mainly 15–40 cm long, the tip frequently nodding. **FLOWER** bracts ovate, 4–8 mm long, appressed, the apex straight or slightly recurved, weakly spinose, the margins entire to crenulate, the bracteoles somewhat shorter and narrower; perianth segments 2.5–3.5 mm long, ovate to oval below the somewhat inconspicuous protuberance, above the protuberance ovate to narrowly deltoid, the apex acute, the margins entire to crenulate; stamens included to exserted, the anthers about 0.6 mm long. **FRUIT** ovoid and horizontally ridged at the apex, 1.5–2.5 mm in diameter. **SEEDS** blackish, shiny, cochleate, about 1.5 mm wide, smooth. $n = 9$. Weedy areas with cockleburs, chenopods, and sunflowers. COLO: Arc, LPl; NMEX: SJn. 1875–1985 m (6190–6550′). Flowering/Fruit: Jun–Sep. Rather widespread in North America; adventive from Asia.

Salsola tragus L. (goat) Russian-thistle, tumbleweed. **STEMS** simple or freely branched, clump-forming, taller than wide to wider than tall, 1–15 dm tall and wide or even larger; with red-purple, longitudinal striations, glabrous or pubescent. **LEAVES** mainly 1.5–4 (6) cm long, 0.3–0.8 mm wide, narrowly linear or filiform, spinulose apically, modified upward as spinescent bracts with expanded bases and scarious margins. **INFL** short-spicate, mainly 1–10 cm long. **FLOWER** bracts spreading and often recurved, ovate to narrowly deltoid, 3–8 mm long, glabrous or pubescent basally, strongly spinose apically; perianth segments distinct, 2.5–3.5 mm long, ovate to oval or oblong below the median protuberance, above the protuberance narrowly deltoid, acute apically, entire to crenulate, membranous in anthesis, in fruit becoming transversely winged, the wings mainly 1–2 mm long. **SEEDS** horizontal, blackish, shiny, about 1.5 mm wide, smooth. *n* = 18, 36. [*Salsola pestifer* A. Nelson; *S. iberica* Sennen & Pau; *S. kali* L.]. Weedy species of disturbed habitats. ARIZ: Apa, Coc, Nav; COLO: Arc, Dol, Hin, LPl, Mon, SJn, SMg; NMEX: McK, RAr, San, SJn; UTAH. 1060–2380 m (3500–7855′). Flowering/Fruit: Jun–Oct. Widespread in North America; adventive from Asia. Russian thistle was first introduced with flaxseed from Russia into South Dakota circa 1873. In the following few decades it spread over the American West. This Russian thistle species forms intermediates with *S. paulsenii* Litv., which differs in largely intangible ways, i.e., in three of the perianth segments being attenuate into spinose tips and forming a column at maturity. In specimens examined the spinose development is one of degree and is not absolute. Nevertheless, I have been unable to discern specimens of *S. paulsenii* from the Four Corners region. Russian thistle is indeed a pestiferous weed that grows vigorously in dryland grain fields and elsewhere. The rounded dead plants with their interlaced branches blow in the wind and stack against fences, fill ravines, clog irrigation ditches, and are generally a nuisance. However, it produces considerable biomass on arid rangelands and is eaten by livestock in winter, especially when wet following a storm. Some livestock operators consider it as security for their livestock when other more palatable plants are not available. Chewed plants used as a poultice for bee and ant stings by Navajos; ashes also used internally and externally for smallpox. Roasted seeds and young sprouts used as food or for horse feed.

Sarcobatus Nees Greasewood

(fleshy bush) Thorny shrubs. **LEAVES** mostly alternate, linear, fleshy, sessile. **INFL** borne in axillary spikes. **FLOWERS** imperfect, the staminate ones spirally arranged, ebracteate, and lacking a perianth; stamens 2 or 3, borne beneath stalked peltate scales; pistillate flowers sessile, 1 or 2, in the axils of scarcely reduced leaflike bracts, the pistil surrounded by a cuplike, shallowly lobed to subentire perianth. **FRUIT** with perianth accrescent and adherent to the fruit base, its upper portion flaring to form a broad, winglike border. **SEEDS** erect, flattened, orbicular. One species in western North America.

Sarcobatus vermiculatus (Hook.) Torr. (worm-eaten) Greasewood. Shrubs, mainly 10–20 dm tall or more; branches rigid, spreading, often modified as thorns. **LEAVES** 0.3–4.5 cm long, 1–3 mm wide, semicylindric, linear. **INFL** with staminate spikes catkinlike, 1–4 cm long; pistillate flowers fewer than the staminate ones. **FLOWERS**: pistillate with the perianth about 1 mm long; calyx wing 2–6 mm long. **FRUIT** 4–5 mm long, cup-shaped below the wing. **SEEDS** brown, about 2 mm long. *n* = 18. [*Batis vermiculata* Hook.; *Fremontia vermiculata* (Hook.) Torr.]. Greasewood, seepweed, saltbush, and other plant communities of saline substrates. ARIZ: Apa, Nav; COLO: Arc, Dol, LPl, Mon; NMEX: McK, RAr, San, SJn; UTAH. 1325–1960 m (4350–6470′). Flowering/Fruit: May–Sep. Washington and Alberta to North Dakota, south to California, Arizona, New Mexico, and Texas. Four Corners plants are referable to **var. *vermiculatus***. The plant is characteristic of fine-textured saline valley bottoms and slopes through much of the American West. This is an important browse species for cattle and sheep, even though potentially poisonous due to oxalate salts of sodium and potassium, and oxalic acid. The wood is tough and durable and was used both as tools and as firewood by prehistoric Indians. Hopis use the plant as kiva fuel because it gives a sparkling fire; also used to make furniture and clothes hooks. Used by Navajos for insect bites and by Paiutes as an antidiarrheal. Wood used for cooking tools by many tribes.

Suaeda Forssk. ex Scop Sea Blite, Seepweed

(an Arabic name) Annual or perennial herbs or shrubs. **LEAVES** alternate, entire, terete or flattened, often succulent. **INFL** solitary or clustered in leaf axils. **FLOWERS** inconspicuous, mostly perfect, bracteate; calyx 5-lobed, fleshy, the lobes equal and unappendaged or unequal and some more or less corniculately appendaged; stamens 5, the filaments short; ovary subglobose or depressed. **SEEDS** horizontal or vertical. One hundred fifteen species worldwide in saline and alkaline soils.

1. Plants suffrutescent or definitely shrubby; leaves abruptly short-petiolate; calyx lobes equal, not appendaged, smooth dorsally; herbage glabrous or puberulent; seeds vertical or horizontal ....(2)
1′ Plants annual; leaves sessile; calyx lobes unequal in fruit, horned; herbage glabrous; seeds horizontal ....(3)

2. Leaves 10–30 mm long; glaucous .... ***S. moquinii***
2′ Leaves 3–5 mm long; rarely glaucous .... ***S. nigra***

3. Plants often over 3 dm tall, erect, not clump-forming, the branches stiffly erect-ascending; flowers mostly 3–7 per axil .... ***S. calceoliformis***
3′ Plants mainly 0.5–3 dm tall, forming depressed rounded clumps, the branches spreading, more or less flexuous; flowers 1–3 per axil .... ***S. occidentalis***

Suaeda calceoliformis (Hook.) Moq. (like a little shoe) Broom seepweed. Plants glabrous, often glaucous, erect, simple or with erect-ascending branches and broomlike, 1–5 (8) dm tall. **LEAVES** mainly 1–4 cm long, 1–2 mm wide, linear or tapering from base to apex, semiterete, intergrading with floral bracts upward. **INFL** spikes slender; flowers sessile, mostly in clusters of 3–7. **FLOWERS** with calyx lobes unequal, about 1.5 mm long, at least some conspicuously horned. **FRUIT** horizontal. **SEEDS** smooth, 1–1.5 mm wide, dark brown. $2n = 54$. [*Chenopodium calceoliforme* Hook.; *Suaeda depressa* (Pursh) S. Watson; *S. depressa* (Pursh) S. Watson var. *erecta* S. Watson]. Saline palustrine or riparian sites in saltgrass, greasewood, seepweed, alkali sacaton, and cattail-sedge communities. ARIZ: Apa, Nav; COLO: LPl; NMEX: McK, RAr, SJn. 1420–2260 m (4680–7460′). Flowering/Fruit: Aug–Oct. British Columbia to Saskatchewan, south to California, Arizona, New Mexico, and Texas. Seeds were eaten as food by Paiutes.

Suaeda moquinii (Torr.) Greene (for C. Moquin-Tandon, a French chenopod expert) Bush seepweed. Subshrub or shrub up to 15 dm tall, glabrous or hairy, glaucous. **STEMS** spreading or erect, the base usually woody, yellow-brown; branches spreading. **LEAVES** ascending or widely spreading, petiole 1 mm; blade 10–30 mm, subcylindric to flat, linear to narrowly lanceolate. **INFL** open; clusters confined to the upper stems, 1–12 per cluster, bracts usually shorter than the leaves. **FLOWERS** bisexual or lateral pistillate; calyx lobes rounded; stigmas 3. **SEEDS** horizontal or vertical, 0.5–1 mm, shiny, black. [*Suaeda torreyana* S. Watson incl. var. *ramosissima* (Standl.) Munz; *S. fruticosa* Forssk. ex J. F. Gmel.]. Alkaline and saline places. COLO: LPl, Mon; NMEX: RAr. 1780–1980 m (5800–6535′). Flowering/Fruit: Aug–Oct. Western Canada south to Texas and Mexico. Hopis used a poultice of dry leaves on sores and as a medicinal bath. Kayenta Navajos used the plant for bleeding bowels and Paiutes used it for chicken pox. Navajos boiled the seeds into a gruel.

Suaeda nigra (Raf.) J. F. Macbr. (black, dark) Torrey's seepweed. Plants glabrous or pubescent, sometimes glaucous, suffrutescent or definitely shrubby, 1–12 (15) dm tall or more and as broad or broader, with slender ascending to spreading branches. **LEAVES** 0.5–3.5 cm long, 1–3 mm thick, subterete to flattened, abruptly short-petiolate, intergrading with floral bracts upward. **INFL** with flowers 1–8 or more per axil. **FLOWER** calyx lobes equal, about 1.5–2 mm long, the lobes merely rounded dorsally, not horned or tuberculate. **FRUIT** horizontal or vertical. **SEEDS** 0.8–1.2 mm wide, black, shiny. $n = 9, 18$. [*Chenopodium nigrum* Raf.; *Suaeda intermedia* S. Watson; S. *nigrescens* I. M. Johnst.; *S. suffrutescens* S. Watson; *S. torreyana* S. Watson]. Greasewood, seepweed, saltgrass, and other salt desert shrub communities, often in riparian or palustrine habitats. ARIZ: Apa, Nav; COLO: Mon; NMEX: RAr, SJn; UTAH. 1320–2075 m (4350–6850′). Flowering/Fruit: Jun–Oct. California, Nevada, Wyoming, Arizona, and Mexico.

Suaeda occidentalis (S. Watson) S. Watson (western) Western seepweed. Plants glabrous, often glaucous, forming depressed rounded clumps, not broomlike, but sometimes slender and more or less unbranched, mainly 0.5–3 dm tall. **LEAVES** mostly 0.5–2 (3) cm long, linear-oblong, semicylindric, intergrading with floral bracts upward. **INFL** a slender spike, mostly 1–3 flowers per cluster. **FLOWER** calyx lobes unequal, about 1.5 mm long, at least some conspicuously horned. **FRUIT** horizontal. **SEEDS** smooth, 1–1.5 mm wide, dark brown. [*Schoberia occidentalis* S. Watson]. Saline palustrine or riparian habitats in greasewood, saltgrass, seepweed, and other such communities. ARIZ: Apa, Nav; NMEX: McK. 1985–2220 m (6550–7320′). Flowering/Fruit: Jun–Sep. Washington to Wyoming, south to Nevada, Utah, and Colorado.

Suckleya A. Gray Suckleya

(for Dr. George Suckley, surgeon and naturalist) Monoecious annual. **STEMS** prostrate or ascending, branched, terete, not jointed, unarmed, not fleshy. **LEAVES** alternate, long-petiolate, the blade rhombic-obovate to suborbicular or flabellate, abruptly short-cuneate, repand-dentate, the apex rounded or acute, sparingly scurfy. **INFL** borne in axillary clusters of nearly all leaves. **FLOWERS** small, inconspicuous; staminate flowers with 4-parted calyx, stamens usually 4; pistillate flowers with perianth segments becoming connate and 4-lobed; pistil enclosed by enlarged, obcompressed perianth; stigmas 2. **FRUIT** with pericarp free. **SEEDS** vertical, brown at maturity. A genus of a single species in the Rocky Mountain reigon.

Suckleya suckleyana (Torr.) Rydb. Poison suckleya. **STEMS** stout, terete, much-branched, 2–5 dm long, prostrate or ascending, sparsely scurfy or glabrate. **LEAVES** numerous, the petioles equaling or exceeding the blades, the blades suborbicular to rhombic or rhombic-ovate, 1–2.5 cm long, rounded at the apex, abruptly short-cuneate at the base, repand-dentate with short, triangular, acute or obtuse teeth, green, sparsely scurfy when young, soon glabrate. **INFL** with flowers in dense clusters in the axils of nearly all the leaves, the staminate in the upper axils. **FRUIT** bracteoles ovate-rhombic, often subhastate, 5–6 mm long, dorsiventrally 4-ribbed, laterally 2-winged, bidentate at the apex, glabrous or nearly so, green. **SEEDS** filling the cavity, ovate, about 3–3.3 mm long, reddish brown, the radicle superior.

[*Atriplex suckleyana* (Torr.) S. Watson; *Obione suckleyana* Torr.; *Suckleya petiolaris* A. Gray]. Around drying pond margins and on dry lake beds with ruderal plants. NMEX: McK, RAr, San, SJn. 1650–2230 m (5445–7360′). Flowering/Fruit: Jun–Sep. Montana south to Colorado and Texas. The genus *Suckleya* was early treated within *Atriplex*, but the obcompressed completely connate fruiting bracteoles (or connate subligneous perianth parts) are quite unlike anything in *Atriplex*, and the plants have long been recognized within a stand-alone genus.

Zuckia Standl. Siltbush

(for Ethyl Zuck, a collector of *Zuckia*) Dioecious or less commonly monoecious shrubs or subshrubs; branches not thorny; axillary buds subglobose, prominent. **LEAVES** alternate, entire or more or less lobed; herbage more or less scurfy. **INFL** with staminate flowers 2–5 in clusters in bract axils, not separately bracteolate; pistillate flowers 1 to several per bract, often some vestigial. **FLOWERS**: staminate with perianth 4- or 5-lobed, subequal to the 4 or 5 stamens; pistillate each enclosed by 2 bracteoles, these dorsiventrally flattened and unequally 6-keeled or obcompressed and thin-margined, often subtended by a single filiform bractlet; stigmas 2. **FRUIT** vertical or horizontal.

Zuckia brandegeei (A. Gray) S. L. Welsh & Stutz (for Townshend S. Brandegee, botanist and civil engineer) Siltbush. Plants mainly 1–5 dm tall; branching from a persistent woody base about 0.5–2 dm tall. **STEMS** erect or ascending; herbage more or less scurfy and less commonly with some moniliform hairs in the inflorescence. **LEAVES** subsessile or tapering to a short petiole, 13–80 mm long, 15–42 mm wide, linear or narrowly oblanceolate-spatulate to elliptic, ovate, obovate, or orbicular, entire or rarely hastately lobed. **FLOWERS**: staminate with a 4- or 5-lobed stramineous perianth, cleft to the middle or below, about 1.5–1.8 mm long; pistillate flower bracts obcompressed or dorsiventrally compressed. **FRUIT** bracts vertical or horizontal, respectively, but not exclusively, when mature either flattened and 4–8 mm wide or 6-keeled (with 4 small and 2 large keels) and 2–4 (5) mm wide; fruit included within the bracts. $2n = 36$. [*Grayia brandegeei* A. Gray]. Bract and bud differences in *Zuckia* are correlated, apparently, with the C_3 type of photosynthesis and its attendant foliar morphology, while the shrubby atriplexes have the C_4 type of photosynthesis with its Kranz-type morphology.

1. Leaves mainly less than 6 mm broad .... **var. *brandegeei***
1′ Leaves mainly over 6 mm broad (to 25 mm or more) .... **var. *plummeri***

var. *brandegeei* $2n = 18$. Fine-textured, often saline and seleniferous substrates on the Jurassic Morrison, Triassic Chinle, and Permian Cutler formations (and probably others). ARIZ: Apa, Nav; COLO: Mon; UTAH. 1250–2210 m (4125–7295′). Flowering/Fruit: May–Oct. Utah, Colorado, and Arizona.

var. *plummeri* (Stutz & S. C. Sand.) Dorn (for Perry Plummer, longtime Utah botanist with the intermountain Forest Service) Plummer's siltbush. $2n = 36$. [*Grayia brandegeei* var. *plummeri* Stutz & S. C. Sand. Fine-textured, often saline and seleniferous substrates on Mancos Shale Formation. ARIZ: Coc; NMEX: SJn. 1635–1705 m (5400–5630′). Flowering/Fruit: May–Sep. New Mexico, Wyoming, and Colorado.

CLEOMACEAE Bercht. & J. Presl CLEOME FAMILY

Steve L. O'Kane, Jr.

Annual to rarely perennial in our flora, ill-smelling herbs. **LEAVES** stipulate, alternate, palmately compound; leaflets 3 to 7, rarely simple above. **INFL** a bracteate raceme. **FLOWERS** actinomorphic or zygomorphic by displacement of petals to one side, perfect; sepals 4, free; petals 4, free, ± clawed; stamens typically 6 but up to 20 in *Polanisia*; pistil 1, 2-carpellate; ovary superior, with a replum (but no false septum as in Brassicaceae), stipitate on an elongated gynophore or subsessile, locule 1, placentae 2, parietal, style 1, stigma 2-lobed. **FRUIT** a 2-valved septicidal capsule or didymous and forming a pair of 1–3-seeded nutlets. $x = 8–17$ (Ernst 1963; Iltis 1957, 1958). Thirty-seven to 50 genera, 800–900 species. Worldwide in tropical and subtropical regions, with few species in warm-temperate, primarily arid regions. Many members of the family have an unpleasant odor, hence common names like stinkflower, stinkweed, and skunkweed. Clammyweed is another common name, referring to the sticky stems and foliage of some species. Cleomaceae is here recognized as distinct from Capparaceae, which lacks a replum and has a woody habit (Hall, Sytsma, and Iltis 2002).

1. Stamens 8–20; plants sticky, glandular-puberulent; petals white to cream; ovary subsessile .... ***Polanisia***
1′ Stamens 6; plants glabrous or at least not glandular; petals colored, not or very rarely white or cream; ovary stipitate .... (2)

2. Fruit didymous, indehiscent, forming 2 nutlets .... ***Wislizenia***
2′ Fruit a dehiscent capsule .... (3)

3. Fruit short and broad, as wide as long or wider, 2–12-seeded, not constricted between the seeds .... ***Cleomella***
3′ Fruit clearly longer than broad, many-seeded, somewhat constricted between the seeds .... ***Cleome***

Cleome L. Beeplant, Spiderplant, Waa' (Navajo), Tu'mi (Hopi)

(derivation obscure, an early European name for a mustardlike plant) **LEAVES** palmately compound, leaflets 3–7. **FLOWERS** zygomorphic but occasionally appearing actinomorphic, petals yellow or pink to pinkish purple (rarely white in a few); stamens 6; ovary stipitate with a basal gland. **FRUIT** a capsule, longer than wide, linear to oblong, spreading or pendulous, somewhat constricted between the seeds. **SEEDS** many. Mostly tropical and subtropical America and Africa. One hundred fifty to 200 species primarily in warm tropical areas.

1. Petals yellow; leaflets 3–7 .... ***C. lutea***
1′ Petals pink to purplish (rarely white); leaflets 3 .... ***C. serrulata***

Cleome lutea Hook. (yellow) Yellow beeplant, waa' (Navajo). Plants erect, herbage glabrous or slightly hairy. **STEMS** mainly 3–15 dm tall, branched or simple. **LEAFLETS** 3–5 (7), oblong to lanceolate or elliptic, 0.8–5 cm long, 2–10 mm wide. **INFL** 2–10 cm long. **FLOWERS** with pedicels 3–10 mm long; sepals and petals yellow; stamens yellow. **FRUIT** 1–3 (4) cm long, surface striate, spreading, on stipes 6–12 mm long. $2n = 32, 34$. [*Peritoma lutea* (Hook.) Raf.] Clay flats, desert scrub, piñon-juniper to ponderosa pine communities. ARIZ: Apa, Coc, Nav; COLO: Mon; NMEX: SJn; UTAH. 900–2400 m (2950–7900′). Flowering: May–Jul. Washington to Nebraska, south to California and New Mexico. The seeds were used as a food source prior to the introduction of the Spanish horse (Ute Mountain Ute).

Cleome serrulata Pursh (saw, referring to the minutely serrulate sepals) Rocky Mountain beeplant, Rocky Mountain beeweed, blue Colorado beeplant, spiderplant, clammyweed, stinkflower, stinkweed, guaco, waa' (Navajo). Plants erect, herbage glabrous or slightly hairy. **STEMS** mainly 3–20 dm tall or more, simple or branched above. **LEAFLETS** 3, lanceolate, elliptic, or oblanceolate, 1.5–7 cm long, 4–14 mm wide, entire. **INFL** 5–25 cm long or more. **FLOWERS** with pedicels 12–23 mm long; petals lavender; sepals purple to green; stamens with purple filaments. **FRUIT** 2.5–7 cm long, surface smooth, descending to pendulous, on stipes 1–25 mm long. $2n = 32, 34, 60$. [*C. inornata* Greene; *C. integrifolia* Torr. & A. Gray; *C. serrulata* var. *angusta* Tidestr.; *C. serrulata* forma *inornata* (Greene) W. A. Weber; *Peritoma inornata* Greene; *P. serrulata* (Pursh) DC.]. Disturbed sites and often along roads and stream banks. ARIZ: Apa, Nav; COLO: Arc, LPl, Mon; NMEX: McK, RAr, San, SJn; UTAH. 900–2800 m (2950–9200′). Flowering: May–Aug. Washington to Saskatchewan, south to California and Arizona. Young plants used among the Navajo as a potherb (its foul odor dissipates upon cooking); seeds ground for meal when mixed with corn; plants used in making prayer-sticks (paho) for the Powamu ceremony (Hopi). Boiled stems produce an important yellow dye used in Anasazi pottery. This dye, when mixed with other ingredients, produces a permanent black color when the pots are fired. The plant is also reported as a treatment for stomach disorders, fever, and worms, and as a general tonic.

Cleomella DC. Stinkweed, Cleomella

(diminutive of *Cleome*) Plants generally ascending to erect, usually branched from the base, glabrous, often red-tinged. **STEMS** 6–30 cm tall. **LEAVES** palmately compound, leaflets 3. **FLOWERS** actinomorphic, petals yellow to yellow-orange, upper 2 often recurved, stamens 6, anthers coiled when dry, ovary stipitate. **FRUIT** obdeltoid, rhomboidal, deltoid, or ovoid, usually wider than long, seeds 2–12 (Payson 1922). Ten species limited to arid Mexico and the southwestern United States.

1. Leaflets about 3.5 times longer than wide (2–7 mm wide); foliage bracts reduced to setae .... ***C. palmeriana***
1′ Leaflets about 5.5 times longer than wide (1–3 mm wide); foliage bracts linear to 3-foliate .... ***C. plocasperma***

Cleomella palmeriana M. E. Jones (for Civil War general William J. Palmer) Palmer cleomella. Plants erect-ascending. **STEMS** branching from the base, 6–30 cm tall. **LEAFLETS** 9–20 mm long, 2–9 mm wide, elliptic to oblong or lance-oblong; upper bracts reduced to setae. **FLOWERS** on pedicels 5–6 mm long; sepals acute; petals 3–4 mm long and tipped with red in bud; filaments to about twice as long as petals. **FRUIT** obtuse apically, base triangular-acute, 2–5 mm long and 5–9 mm wide on an often recurved stipe 3–7 mm long. **SEEDS** few to several. [*C. nana* Eastw.; *C. cornuta* Rydb.; *C. montrosae* Payson; *C. oöcarpa* A. Gray]. Mancos Shale in desert shrub communities. ARIZ: Apa; COLO: Mon; NMEX: SJn. 1200–1850 m (3900–6100′). Flowering: May–Jul. A Colorado Plateau endemic limited to the Four Corners region except for var. *goodrichii* Welsh with conspicuously horned fruits which is endemic to the Uinta Basin. Our plants are **var. *palmeriana***.

Cleomella plocasperma S. Watson (twisted seed, referring to the odd-shaped fruits) Watson's cleomella. Plants erect-ascending. **STEMS** several, diffuse, generally branching from upper nodes, 15–40 cm tall. **LEAFLETS** 7–18 mm long, 1– 3 mm wide, linear to narrowly oblong; bracts linear or lowermost 3-foliate. **FLOWERS** on pedicels 9–15 mm long; sepals lanceolate, 0.9–2.2 mm long; petals 3–5 mm long. **FRUIT** ovoid, rhomboidal, or obovoid, 1.5–3 mm long and 3–6 mm wide on a stipe 2–7 mm long. **SEEDS** several. Moist or seasonally moist saline-alkaline areas with saltgrass and other salt-tolerant species. ARIZ: Apa; COLO: Mon; UTAH. 1400–1525 m (4670–5000′). Flowering: Jun–Aug. Mostly limited to the Great Basin in Utah, Oregon, Idaho, and California; apparently disjunct in our area.

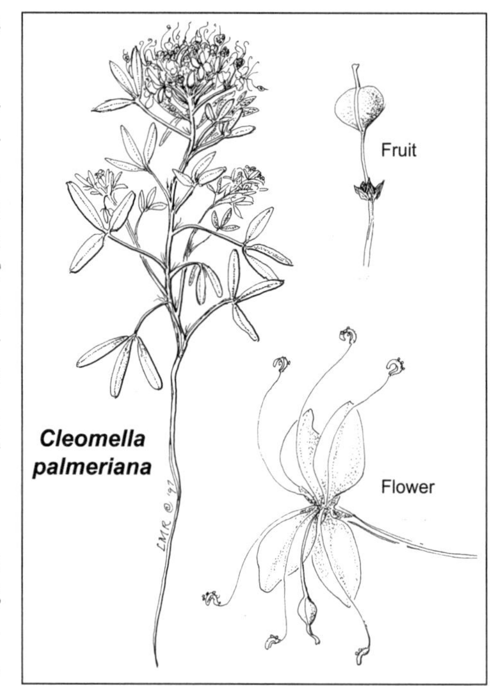

Cleomella palmeriana

Polanisia Raf. Clammyweed

(Greek *polys*, many, + *anisos*, unequal, referring to the many unequal stamens) Plants viscid-puberulent and strongly rank-smelling. **LEAVES** palmately compound, leaflets 3. **INFL** a dense terminal raceme. **FLOWERS** zygomorphic, petals white, stamens 8–20, long-exserted, purple, and unequal. **FRUIT** capsular, elongate, erect, subsessile, glandular, somewhat flattened, valves persistent and dehiscing apically, seeds many (Iltis 1954, 1958, 1966). Four to six species limited to North America.

Polanisia dodecandra (L.) DC. (12 men, referring to 12 stamens) Clammyweed. Plants glabrous. **STEMS** 1.5–8 dm tall. **LEAFLETS** 0.8–4.5 cm long, 3–18 mm wide, obovate to oblanceolate or elliptic. **FLOWERS** on pedicels 10–21 mm long; sepals purplish; petals white to cream, emarginate; stamens with filaments purple, long-exserted. [*P. graveolens* Raf.; *P. trachysperma* Torr. & A. Gray; *P. uniglandulosa* DC.]. Desert shrub and juniper communities, often in wash bottoms and often growing on soil derived from Mancos Shale. COLO: Arc; NMEX: SJn. 1060–2000 m (3480–6560′). Flowering: May–Oct. British Columbia east to South Dakota, south to Texas and Mexico and west to California. Our plants belong to **var. *trachysperma*** (Torr. & A. Gray) H. H. Iltis.

Wislizenia Engelm. Jackass-clover

(named for Friedrich Adolph Wislizenus, 19th-century naturalist-physician and plant collector) (Keller 1979). A genus of one species.

Wislizenia refracta Engelm. (bent abruptly backward, referring to the reflexed stalk of the fruits) Jackass-clover. Plants annual or perennial, glabrous to puberulent. **STEMS** branched from the base, 0.5–24 dm tall. **LEAVES** palmately compound with usually 3 leaflets, petiole 3–25 mm long. **INFL** a dense, terminal raceme up to 20 cm in fruit. **FLOWERS** actinomorphic; pedicels 5–10 mm long; sepals about 2 mm long, green; petals elliptic, yellow, tapered to the base; stamens 8–14 mm long, yellow; ovary of 2 nearly separate lobes, each usually with 1 ovule. **FRUIT** of 2 nutlets with deciduous valves on a reflexed, stalklike receptacle. **SEEDS** usually 1 per nutlet. $2n = 40$. Alkaline soils in valleys and washes. ARIZ: Nav; UTAH. Up to 2000 m (6560′). Flowering: May–Jun. California to Texas and northwest Mexico. Plants are reported to be toxic to humans but in California are considered an important honey plant. Our plants belong to **var. *refracta***.

CLUSIACEAE Lindl. ROSE APPLE, ST. JOHNSWORT FAMILY

(GUTTIFERAE, HYPERICACEAE)

Linda Mary Reeves

Primarily trees and shrubs, some epiphytic; most green plant parts produce glandular secretions in canals. **LEAVES** usually opposite, without stipules, entire, sometimes fleshy, as in *Clusia*. **INFL** a cyme or thyrse. **FLOWERS** perfect or imperfect and dioecious; perianth in 4–5 parts, sepals and petals usually distinct, sometimes fleshy; petals often contorted when in bud; androecium, when present, of 2 whorls, each with 5 stamen bundles, filaments free nearly to base, the outer whorl often sterile or absent, inner whorl always fertile, except in imperfect female flowers; stamens may be variously modified by fusion or reduction, such as reduction of a stamen bundle to a single stamen; ovary superior and of 2–13, usually 3–5, united carpels and many locules; ovules 1 to many per locule; placentation axile or parietal; styles free, united, or lacking. **FRUIT** usually a capsule, but sometimes a fleshy berry or drupe. **SEEDS** often with an aril or wing. x = 7, 8, 9, 10. About 50 genera with about 1000 species worldwide, primarily tropical and subtropical. The Clusiaceae contains many economically useful plants, such as timber sources (*Cratoxylum*, *Mesua*, etc.), gums and resins (*Clusia*, *Garcinia*), drugs (*Hypericum*), and fruits such as mangosteen (*Garcinia mangostana*) and mammey apple (*Mammea americana*). *Hypericum anagalloides*, a small, peach-flowered herb, is known from Arizona and should be searched for in our area.

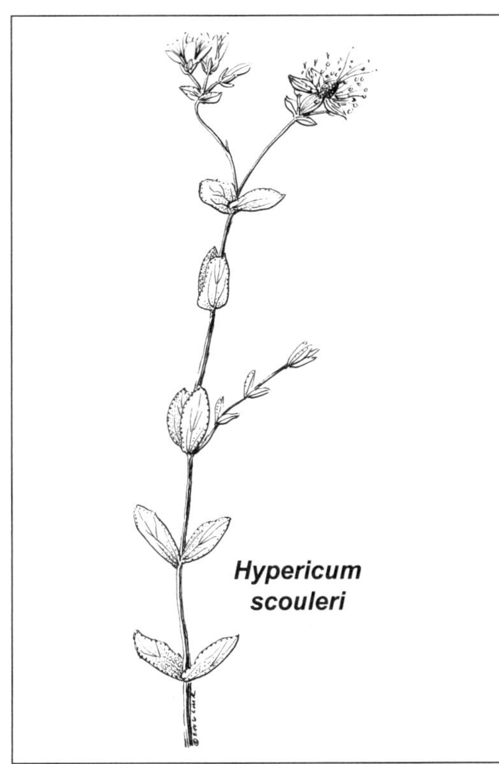

Hypericum L. St. Johnswort

(Greek name for St. Johnswort) Perennial glabrous herbs or shrubs. **LEAVES** opposite; simple; sessile; entire; glandular-punctate. **INFL** a cyme or solitary flower. **FLOWERS** perfect; actinomorphic; perianth 5-merous; corolla yellow, pinkish, reddish, or purplish; stamens numerous, clustered in 3–5 fascicles, filaments elongate; ovary 3-loculed, stigmas minute. **FRUIT** a capsule or berry, many-seeded. **SEEDS** short-cylindric, reticulate. About 300 species in cosmopolitan distribution.

Hypericum scouleri Hook. (after John Scouler) Scouler's St. Johnswort. Herbaceous perennial. **STEMS** erect from creeping rootstocks, 20–70 cm tall, occasionally branched. **LEAVES** 1–3.5 cm long, ovate to elliptic or oblong; apex obtuse or acuminate, black punctate-glandular. **INFL** a branching cyme with few to many flowers. **FLOWERS** actinomorphic; sepals ovate, 2–5 mm long, punctate-glandular; petals 6–15 mm long, yellow; stamens numerous. **FRUIT** a capsule, 8 mm long. $2n$ = 16. [*Hypericum formosum* Kunth var. *scouleri* (Hook.) J. M. Coult.]. Banks of montane watercourses and meadows in ponderosa pine and aspen communities. ARIZ: Apa; COLO: Arc, Hin, LPl, Min, Mon, SJn; NMEX: McK, SJn, RAr; UTAH. 1830–2760 m (6000–9000′). Flowering: late Jun–Aug. Fruit: Jul–Sep. Washington, Oregon, and California to Montana, Wyoming, Colorado, and New Mexico. Our material belongs to **subsp. *scouleri*.**

COMMELINACEAE Mirb. SPIDERWORT FAMILY

William Jennings

Perennial (ours) or annual herbs, usually somewhat succulent. **LEAVES** basal or cauline, with sheathing bases, alternate, simple, margins entire, venation parallel. **INFL** terminal or terminal and axillary, cymose, thyrsiform, or umbel-like. **FLOWERS** actinomorphic or zygomorphic, perfect or both perfect and staminate on the same plant; sepals 3, petals 3, stamens 6, all fertile or some staminoidal or some absent; ovary superior, 2–3-locular, stigma 1, simple. **FRUIT** a loculicidal capsule, seeds 1 to several per locule. Forty genera, about 630 species. Pantropical and nearly pantemperate. The petals are ephemeral and do not press well.

1. Flowers zygomorphic; stamens of 2 types, the filaments naked; inflorescence subtended by 1 spathelike bract .... ***Commelina***

1′ Flowers actinomorphic; stamens all alike, filaments bearded below; inflorescence subtended by several leaflike bracts ..... ***Tradescantia***

Commelina L. Dayflower, Widow's-tears

(for the brothers J. and K. Commelijn, Dutch botanists) **LEAVES** 2-ranked or spirally arranged, not glaucous; blade sessile or petiolate. **INFL** terminal, leaf-opposed; cymes 1 or 2, enclosed in spathes. **FLOWERS** both perfect and staminate, zygomorphic; sepals distinct or proximal 2 connate; petals distinct, petals all the same color or the proximal petal a different color and smaller than the distal 2; stamens 5 or 6, 3 fertile, 2 or 3 sterile (staminoidal), filaments glabrous. **FRUIT** 2- or 3-valved, 2- or 3-locular. **SEEDS** 1 or 2 per locule. Mainly tropical, but almost worldwide, about 170 species.

Commelina dianthifolia Delile **CP** (leaves like a *Dianthus*) Perennial plants with tuberous-thickened roots. **STEMS** simple or branched, erect or decumbent. **LEAVES** linear-lanceolate, acuminate, 4–15 × 0.4–1 cm. **INFL** a distal cyme, usually 1-flowered, exserted, spathe 3 to 6 cm, not connate at base, very long-acuminate or caudate, with the tip as long as or longer than the body, glabrous or short-pubescent. **FLOWER** petals blue, 10–12 mm long, proximal petal smaller. **FRUIT** capsule 3-locular, 2 valves, 5 seeds, brown, 2 to 3 mm, rugose, pitted. Rocky soils in shaded or open ground, ponderosa and oak habitat; uncommon in our area at the northern limit of its range. ARIZ: Apa; COLO: LPl, SJn; NMEX: McK, SJn. 1675–2670 m (5500–8765′). Flowering: Aug. Southern Colorado, Arizona, and New Mexico to Texas; Mexico. Infusion given to livestock as an aphrodisiac by Ramah Navajos.

Tradescantia L. Spiderwort

(for John Tradescant, gardener to King Charles I of England) **LEAVES** arranged spirally or 2-ranked, sessile, linear to oblong, produced in a perfoliated sheath. **INFL** terminal or terminal and axillary pairs of cymes, cymes sessile, umbel-like, subtended by a spathelike bract, either similar to leaves or different. **FLOWERS** perfect, actinomorphic, sepals distinct, petals distinct, white to pink, blue, or violet, stamens 6, all fertile, filaments bearded (ours) or glabrous; ovary with 3 locules. **FRUIT** a dry loculicidal capsule, 3-valved, 3-locular. **SEEDS** usually 2 per locule. About 70 species of the temperate and tropical regions of the New World (including *Tradescantia*, *Rhoeo*, *Setcreasea*, and *Zebrina*).

1. Roots thick, but not tuberlike, rootstock none; stems stout, often branched; sheaths glabrous; corolla usually more than 2 cm in diameter .... ***T. occidentalis***

1′ Roots partly tuberous-thickened, fascicled at base of stem or borne on a creeping rootstock; stems slender; usually unbranched; sheaths pubescent or puberulent, especially on the margins; corolla not more than 2 cm in diameter (all flower parts smaller than above) .... ***T. pinetorum***

Tradescantia occidentalis (Britton) Smyth (of the West) Western spiderwort. **STEMS** typically 30–60 cm tall, erect, branching, glabrous. **LEAVES** linear-lanceolate, long-acuminate, glabrous, 0.2–2 cm wide × 5–30 cm long. **INFL** an umbel-like cyme, terminal or axillary; pedicels erect to spreading or reflexed, glandular-puberulent. **FLOWERS** broadly campanulate, sepals elliptic, acute to acuminate, 4–10 mm long, greenish, or suffused with rose or purple, glandular-puberulent; petals broadly ovate, 7–16 mm long, magenta, purple, rose-purple, or bluish; stamens erect, filaments densely pilose. Rather dry, often sandy or rocky habitats, frequently found at roadsides. ARIZ: Apa, Nav; NMEX: McK, San, SJn; UTAH. 1735–2060 m (5700–6755′). Flowering: late May–early Jul. North Dakota to Texas, west to eastern Montana, central Wyoming, eastern Colorado, across New Mexico and Arizona into southern Utah. Our material belongs to **var. *occidentalis***. *Tradescantia occidentalis* var. *scopulorum* (Rose) E. S. Anderson & Woodson, with glabrous pedicels and sepals, occurs to the south of the San Juan drainage. Tender shoots eaten by Puebloans. Decoction of root taken for internal injuries by Navajos.

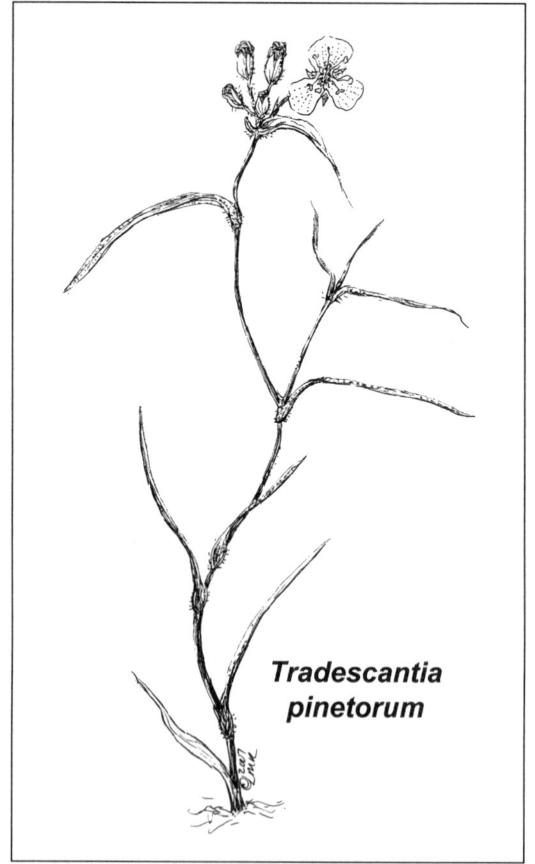

Tradescantia pinetorum

Tradescantia pinetorum Greene (of the pine woods) **STEMS** slender, sparsely branched, 8–39 cm, usually scabridulous. **LEAVES** linear-lanceolate, glabrous, glaucous, 0.15–0.8 cm wide × 1–10 cm long. **INFL** terminal, solitary or with axillary inflorescences from distal nodes, pedicels glandular-puberulent. **FLOWER** sepals 4–6 mm, frequently suffused with red or purple, glandular-puberulent; petals 9–12 mm long, magenta, purple, rose-purple, or bluish; filaments pilose. Moist canyons and stream banks, usually in pine woods. ARIZ: Apa; NMEX: McK, SJn. 1735–2490 m (5700–8170′). The specimen from San Juan County (*Heil & Hyder 7082*, SJNM!) is a significant range extension northeastward from the next-nearest populations in the Chuska Mountains, which are on the southern edge of the San Juan drainage. Flowering: Jul–Sep. Arizona, New Mexico; Mexico. Used by Ramah Navajos as a livestock aphrodisiac.

CONVOLVULACEAE Juss. MORNING GLORY FAMILY

Daniel F. Austin

Herbs, shrubs, or vines, the sap milky in some species; rootstocks sometimes large, otherwise fibrous. **LEAVES** simple, pinnately lobed to pedate or pectinate, or palmately compound; exstipulate. **INFL** axillary, dichasial, solitary, racemose, or paniculate. **FLOWERS** perfect in ours, actinomorphic or slightly zygomorphic, hypogynous, small and inconspicuous to large and showy, but mostly wilting quickly; sepals 5, free, imbricate, equal or unequal, persistent, occasionally accrescent in fruit; petals 5, united; stamens 5, distinct, epipetalous; ovary of 2–3 (4) carpels, usually 1–3-locular; style 1–2; stigmas variable. **FRUIT** 1–4-locular, dehiscent capsule or utricle, or indehiscent and baccate or nutlike. **SEEDS** 1–4, glabrous or hairy. $x = 6–15$ (Austin 1990, 1991, 1998). Fifty-five genera and 1200 species. These plants are most frequent in the tropics and subtropics. Includes the sweet potato (*Ipomoea batatas*) and many ornamentals.

1. Yellow-orange parasitic plants; leaves reduced, minute, scalelike .... ***Cuscuta***
1′ Green plants with normally developed leaves .... (2)

2. Leaf bases obtuse to acute .... (3)
2′ Leaf bases truncate, cordate to hastate .... (4)

3. Flowers white and salverform, or white to blue and rotate to broadly campanulate; styles 2 .... ***Evolvulus***
3′ Flowers purple and funnelform, or white and salverform; styles 1 .... ***Ipomoea***

4. Flower limb lavender, blue, red, or white, throat purple to purple-red .... ***Ipomoea***
4′ Flower limb white, with or without tinges of lavender to pink, throat white .... (5)

5. Calyx enclosed by 2 foliaceous bracts; stigmas oblong, flattened when fresh .... ***Calystegia***
5′ Calyx not enclosed by bracts .... ***Convolvulus***

Calystegia R. Br. Hedge Bindweed

(Greek *caly*, beautiful, + *stegia*, roof, in reference to the bracts) Herbs. **STEMS** twining to erect, glabrous or pubescent. **LEAVES** glabrous or pubescent, ovate, the base sagittate or hastate. **INFL** axillary, often solitary, or in few-flowered cyme. **FLOWER** sepals subequal, ovate, ovate-lanceolate to oblong, usually glabrous; corolla funnelform, white or pink, 3–6 cm long, glabrous or puberulent on the margin of the limb, especially the apex of the petal lobes, stamens included, subequal, the filaments sometimes with glandular trichomes basally; ovary ovoid, glabrous, 2-locular basally and 1-locular apically, 4-ovulate, the style 1, unlobed, glabrous, the 2 stigma lobes oblong, flattened. **FRUIT** capsular, 4-valved, surrounded by enlarged sepals and (usually) bracts. **SEEDS** 1–4, glabrous, black to dark brown, smooth to verrucose (Brummitt 1965; Austin, Diggs, and Lipscomb 1997). About 12 species in the United States and Eurasia.

1. Blade of leaf basally rounded; plants normally pubescent on all vegetative parts .... ***C. macounii***
1′ Blade of leaf basally 2-angled; plants normally glabrous or with a few trichomes on petioles .... ***C. sepium***

Calystegia macounii (Greene) Brummitt (named for John Macoun) Herbs. **STEMS** erect or sparsely twining, these simple or branched, usually from near base, finely pubescent. **LEAVES** with petioles 0.5–40 mm long, blade ovate to ovate-lanceolate, finely pubescent, 2–6 cm long, 1–5 cm wide, basally cordate to subsagittate, the lobes rounded, less often angled, apically obtuse to rarely acute, the margins entire. **INFL** often fewer than 4 per plant, frequently arising from lower few axils. **FLOWERS** on a peduncle 3–5 cm long, rarely longer; pedicel absent; bracts surrounding the calyx, ovate to ovate-mucronate, mostly obtuse, sometimes acute; the sepals elliptic to ovate, subequal, 10–12 mm long, 5–7 mm wide, thin, transparent at least on margins, acute to acuminate, mucronate, glabrous or ciliate; corolla funnelform, white, 4–5 cm long; stamens subequal, 25–28 mm long, filaments basally glandular-pubescent, the anthers 4–5 mm long, basally sagittate; the style 20–23 mm long. **FRUIT** globose to ovoid, brown. **SEEDS** not seen. [*Calystegia interior* House; *C. sepium* (L.) R. Br.]. Moist habitats, near lakes and streams. ARIZ: Apa, Nav. 1950–2050 m (6400–6600′). Flowering: July. Western Minnesota and Iowa to northwestern Missouri and Oklahoma, west to Montana, Wyoming, and Colorado. This is a Great Plains species, ranging from Canada south to Texas. The Arizona plants were introduced to the region during an early logging period; the currently known plants, and former locations, are mostly along the old logging railway and roads.

Calystegia sepium (L.) R. Br. (of hedges) Hedge bindweed. Herbs with rhizomatous, twining stems, these cylindrical or angulate, glabrous. **LEAVES** with petioles 2–7 cm long, blade ovate to ovate-lanceolate, glabrous, 2–15 cm long, 1–9 cm wide, basally cordate-sagittate to hastate, 5-nerved, the auricles obtuse to acute or 2- or 3-dentate, rarely 2-lobed, apically acute to acuminate, the margins entire or undulate. **INFL** of solitary flowers. **FLOWERS** on a peduncle 3–13 cm long; pedicels absent; bracts surrounding calyx angular, ovate, convex, glabrous or ciliate, foliaceous, borders sometimes pinkish, 14–26 mm long, 10–18 mm wide, mucronate, mostly acute; sepals elliptic to ovate-lanceolate, more or less equal, 11–15 mm long, 4–6 mm wide, thin, transparent, acute to more or less obtuse, mucronate, apically ciliate; corollas funnelform, white or tinged at least on limb with rose or pink, 4.5–5.8 cm long; stamens subequal, 23–29 mm long, filaments basally glandular-pubescent, anthers 4.5–6 mm long, basally sagittate; style 20–23 mm long. **FRUIT** ovoid to globose, brown. **SEEDS** 4.5–5 mm long, black, glabrous, smooth or granulose. $2n = 22$. [*Convolvulus sepium* L.; *C. americanus* (Sims) Greene]. Thickets and fence rows. COLO: Mon; NMEX: RAr, SJn. 1525–2195 m (5000–7200′). Flowering: Jun–Aug. Canada from British Columbia to New Brunswick; Washington south to Utah and New Mexico, and from Vermont and New Hampshire south to North Carolina. This is the most common western variant of a highly polymorphic species that has been variously interpreted in the past. Four Corners material is **subsp.** ***angulata*** Brummitt (angles, referring to the teeth or shallow lobes near the base of the leaf blades).

Convolvulus L. Bindweed

(to entwine) Woody or herbaceous vines or shrubs. **LEAVES** petiolate, rarely sessile, blade herbaceous to coriaceous, linear to ovate or elliptic with subtruncate, cordate, sagittate, or hastate bases, glabrous or hairy, the margins usually undulate to crenate or irregularly lobed or laciniate. **INFL** of solitary flowers or in cymose groups. **FLOWER** pedicels mostly 1–3 cm long, bracts and bracteoles linear, elliptic or ovate; sepals subequal, the inner 3 often somewhat longer, suborbicular, elliptic to ovate, hairy or glabrous, obtuse to acute, usually mucronate; corollas white to rose or purple or blue on the limb and white or purplish within the tube, funnelform, the limb 5-angulate to 5-lobed, midpetaline bands glabrous or hairy; stamens included, unequal, with glandular trichomes on the filament base, anthers oblong, basally auriculate, introrse; disc usually lobed; ovary 2-locular, 4-ovulate, ovoid to subglobose, glabrous or hairy; style 1; with 2 filiform, papillose stigmas. **FRUIT** a capsule, 4-valved, mostly brown, chartaceous, ovoid to conical-ovoid, glabrous or hairy. **SEEDS** 1–4, trigonous or rounded, smooth or verrucose, black to dark brown, glabrous (Sa'ad 1967; Austin 2000). This basically Mediterranean genus probably has about 100 species, many of which are found only in Europe and Asia. A few species are native in the Americas, particularly in temperate latitudes or elevations.

1. Leaf blades almost as broad as long; calyx 3–5 mm long; perennials from deep creeping root, forming large patches .... ***C. arvensis***

1′ Leaf blades usually much longer than broad; calyx 6–12 mm long; perennials from taproot, sometimes divided at apex but not forming large, creeping patches .... ***C. equitans***

Convolvulus arvensis L. (of the field) Field bindweed, ch'il na'atlo'ii, naa-xóoyaaih (Navajo). Widely spreading rhizomatous herbs with branched, decumbent or twining stems. **LEAVES** with petioles 3–40 mm long, blade variable, often ovate, ovate-lanceolate to elliptic, 1–10 cm long, 0.3–6 cm wide, entire or with the margin somewhat undulate, basally cordate to subtruncate, hastate, or sagittate, the lobes obtuse or acute, entire or with 2–3 teeth, glabrous or inconspicuously puberulent. **INFL** cymose or 2 or 3 flowers solitary on a peduncle 3–3.5 mm long; bracts elliptic, linear or obovate, 2–3 (9) mm long, the bracteoles linear, 2–4 mm long. **FLOWER** pedicels 5–18 (35) mm long, reflexed in fruit, usually glabrous; sepals obtuse, or less often truncate or emarginate, mucronate, ciliate, the outer elliptic, 3–4.5 mm long, 2–3 mm wide, glabrous or tomentose, the inner suborbicular to obovate, 3.5–5 mm long, 3–5 mm wide; corollas campanulate, white or tinged with pink, 1.2–2.5 cm long; stamens 8–13 mm long, anthers 2–3.5 mm long; ovary ovoid, glabrous, the style 7–10 mm long. **FRUIT** subglobose to ovoid, 5–7 mm wide. **SEEDS** 1–4, 3–4 mm long, black to dark brown, glabrous, tuberculate. $2n = 48, 50$. Cultivated fields, gardens, roadsides. ARIZ: Apa, Nav; COLO: Arc, Dol, Hin, LPl, Mon, SJn, SMg; NMEX: McK, RAr, San, SJn; UTAH. 1125–2350 m (3700–7700′). Flowering: Apr–Oct. Throughout the temperate United States and nearby Canada; naturalized from Europe. Used as a medicine against spider bites (Ramah Navajo).

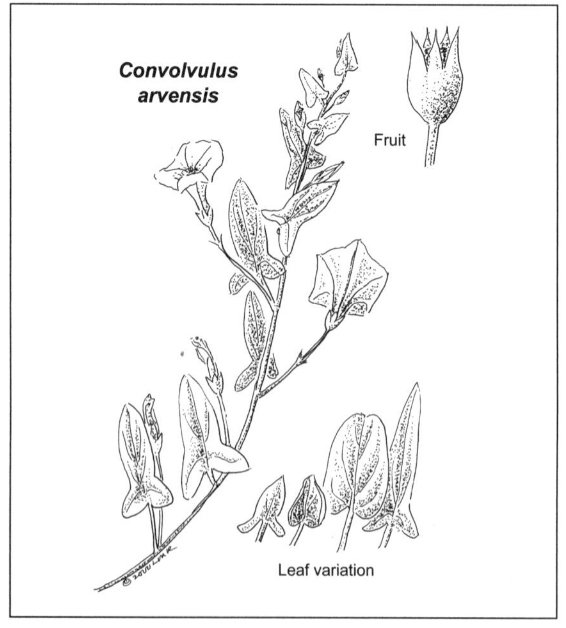

Convolvulus equitans Benth. (overlapping in two ranks) Silver bindweed, desert bindweed. Herbs with branched, prostrate or decumbent stems arising from a taproot, densely hairy. **LEAVES** with petioles 0.25–5 cm long, blade variable, ovate-elliptic to triangular-lanceolate or narrowly oblong with projecting basal lobes; blades most often deeply indented basally, 1–7 cm long, 0.2–4 cm wide, densely hairy on both surfaces with loosely appressed indumentum, margins toothed or lobed or both, rarely entire. **INFL** usually 1-flowered, less often 2–3 and cymose, peduncle 0.5–10.5 cm long; bracts and bracteoles scalelike, hairy like leaves. **FLOWER** pedicels 5–24 mm long; sepals oblong to ovate, obtuse to weakly retuse apically, 6–12 mm long, 3–6 mm wide, appressed-sericeous, the margins membranaceous, subcordate with age; corolla white to pink, at times with a reddish center, (1.5) 2.5–3 cm long, sericeous on the petal lobes; ovary ovoid, glabrous. **FRUIT** subglobose, 7–8 mm wide, glabrous. **SEEDS** 1–4, 4–4.5 mm long, black, granulate, glabrous. [*Convolvulus hermannioides* A. Gray; *C. incanus* Vahl]. Dry plains and hills. ARIZ: Apa, Nav; NMEX: McK, San, SJn. 1525–2000 m (5000–6500′). Flowering: Mar–Nov. Arizona, New Mexico, Colorado, Texas, Oklahoma, Kansas, Nebraska, disjunct into Arkansas and Alabama; Chihuahua, Coahuila to Nuevo León, Mexico, and South America.

Cuscuta L. Dodder

Mihai Costea

(from the Aramaic and Hebrew triradical root of the verb K-S-Y [Kaph, Shin, Yodh], which means "to cover") Holoparasitic, rootless herbs. **STEMS** filiform, yellow or orange, glabrous, trailing or twining and attached to the host by numerous small haustoria. **LEAVES** reduced to alternate, minute scales. **INFL** loose to dense, monochasial clusters grouped in glomerulate, paniculiform, or umbelliform cymes, often confluent in larger inflorescences. **FLOWERS** commonly 5-merous, but often some 4-merous on the same plant; fleshy or thin-membranous, white or white-creamy; laticifers often visible in the calyx, corolla, ovary/capsules, pellucid, yellow or orange, isolated or arranged in rows especially in the midveins of calyx and corolla lobes, ovoid to elongate; calyx fused at least at the base, but distally lobes distinct for various lengths, calyx tube cylindric, cupulate, or campanulate; corolla tube cupulate to cylindric with lobes erect, spreading, or reflexed; stamens inserted near the base of the sinuses; infrastaminal scales present (in the flora), fringed or fimbriate scalelike appendages fused with the corolla tube base and forming a corona alternating with the corolla lobes; ovary superior, 2-locular, each locule with 2 anatropous ovules; styles 2, terminal, stigma spherical (ours). **FRUIT** a capsule, indehiscent to irregularly dehiscent or circumscissile along a transverse line near the base. **SEEDS** 1–4 per capsule; endosperm nuclear; embryo uniformly slender, 1–3-coiled (except globose-enlarged at the base in *C. denticulata*), without a radicle and cotyledons. About 180 species, nearly cosmopolitan in distribution. Molecular studies have shown that *Cuscuta* is nested within Convolvulaceae. All the species in the flora belong to the subgenus *Grammica*. Identification of most *Cuscuta* species is a lengthy process because rehydration of flowers, dissection, and examination under a microscope are necessary. Color and texture of calyx were noted only on dried material. Observation of papillae and laticifers requires magnifications of at least 100× (Costea, Nesom, and Tardif 2005; Costea, Nesom, and Stefanović 2006). Common hosts include *Ambrosia*, *Artemisia*, *Chrysothamnus*, *Grindelia*, *Gutierrezia*, and *Polygonum*.

1. Capsules circumscissile; inflorescence umbelliform ..... ***C. umbellata***
1′ Capsules indehiscent or irregularly dehiscent; inflorescences glomerulate, paniculiform, or corymbiform ..... (2)

2. Calyx lobes obovate-orbicular, margins denticulate; corolla lobes with straight apices and denticulate margins; seeds 1 per capsule ..... ***C. denticulata***
2′ Calyx lobes triangular-ovate to lanceolate, margins entire; corolla lobes with inflexed apices and entire margins; seeds usually 2–4 per capsule ..... (3)

3. Flowers fleshy, papillate; calyx 1/2–3/4 of the corolla tube, calyx lobes acute ..... ***C. indecora***
3′ Flowers membranous, not papillate; calyx equaling corolla tube, calyx lobes obtuse to rounded ..... ***C. campestris***

Cuscuta campestris Yunck. (of the field) Field dodder. **INFL** dense, corymbiform or glomerulate; pedicels 0.3–2.5 (–3.5) mm. **FLOWERS** (1.9–) 2.1–3.6 mm long, membranous, white-creamy when fresh, creamy or golden yellow when dried; papillae absent; laticifers evident in the calyx and less obvious in the corolla and ovary/capsule, isolated or arranged in rows, rotund, ovoid, or elongated; calyx yellow, reticulate, shiny, cupulate, about as long as corolla tube, divided 2/5–3/5 the length; lobes overlapping, ovate-triangular, margins entire, apex obtuse to rounded; corolla tube campanulate, (1.1–) 1.5–1.9 mm long; lobes spreading, triangular-lanceolate, about as long as the tube, margins entire, apex acute to acuminate, inflexed; stamens exserted, anthers broadly elliptic, (0.3–) 0.4–0.5 mm; infrastaminal scales equaling or slightly exceeding corolla tube, oblong-ovate to spatulate, rounded, uniformly dense-fimbriate. **CAPSULES**

indehiscent to irregularly dehiscent, globose-depressed to depressed; not thickened or risen around the large interstylar aperture, sometimes translucent, persistent corolla enveloping 1/3 or less of the capsule base. **SEEDS** 4 per capsule, angled, subrotund to broadly elliptic, 1.12–1.54 × 0.9–1.1 mm. 2*n* = 56. [*C. arvensis* Beyr. ex Engelm. var. *calycina* Engelm.; *C. pentagona* Engelm. var. *calycina* Engelm.]. On numerous hosts growing in waste and disturbed places, margins of roads, and agricultural fields. The most common species in the Four Corners region. Worldwide dispersal has been through contaminated seeds of previously infested forage legumes. ARIZ: Nav; COLO: Arc, LPl, Mon; NMEX: McK, RAr, SJn; UTAH. 1125–1900 m (3700–6235′) Flowering/Fruit: Jul–Sep. Native to North America, very common, and perhaps the most successful and widespread *Cuscuta* weed species worldwide.

Cuscuta denticulata Engelm. (from the small-toothed calyx and corolla lobes) Small-tooth dodder. **INFL** dense, glomerulate; pedicels (0–) 0.5–2.2 mm. **FLOWERS** 1.8–3.1 mm long, membranous, white when fresh, straw-yellow when dried; papillae absent; laticifers visible in the midveins of calyx lobes, arranged in rows, elongated; calyx straw-yellow, reticulate and shiny, campanulate to urceolate, as long as the corolla tube or nearly so, lobes basally overlapping, obovate-orbicular, margins denticulate, apex rounded; corolla tube campanulate, 0.6–1.5 mm; lobes reflexed, about equaling the tube, ovate to broadly elliptic, margins irregularly denticulate, apex rounded, straight; stamens included to slightly exserted, anthers 0.25–0.4 mm, subrotund to elliptic; infrastaminal scales reaching the filament bases, oblong-ovate, rounded, uniformly denticulate or fringed. **CAPSULES** indehiscent to irregularly dehiscent, globose-ovoid, 1.3–2.1 × 1–2 mm, not thickened or risen around the inconspicuous interstylar aperture, translucent, capped by the withered corolla. **SEEDS** 1 per capsule, globose to globose-ovoid, 0.85–1.1 × 0.82–1.1 mm; embryo globose-enlarged at the base. 2*n* = 30. Hosts from desert scrubs or dry, sandy places. COLO: Mon. 1370–2000 m (4500–6560′). Flowering/Fruit: Jul–Sep. Native to western North America, from Washington to Mexico (Baja California).

Cuscuta indecora Choisy (unfit or improper) Large-seed dodder. **INFL** loose to dense, paniculiform or corymbiform; pedicels 0.5–6 mm long. **FLOWERS** 2–5.3 mm long, fleshy, translucent white when fresh, creamy yellow to dark brown when dried; perianth cells convex, domelike, papillae usually present on the pedicels, perianth, and ovary; laticifers visible in the perianth along the midveins and in ovary/capsules, isolated or in longitudinal rows, ovoid to elongated; calyx creamy yellow to brownish, not reticulate or shiny, cupulate, 1/2–3/4 of the corolla tube, divided 1/3–2/3 of the length, lobes overlapping or not at the base, triangular-ovate to lanceolate, margins entire, apex acute to attenuate; corolla tube campanulate, campanulate-cylindric, subglobose, or suburceolate, 1–3.2 mm long, lobes suberect to erect, 1/3 to equaling the tube, triangular-ovate, margins entire, apex acute, inflexed; stamens barely exserted or enclosed, anthers ovate-elliptic to oblong, 0.3–0.5 mm long; infrastaminal scales reaching the filament bases, more or less spatulate, rounded, rarely truncate, uniformly dense-fimbriate. **CAPSULES** indehiscent to irregularly dehiscent, globose to subglobose, 2–3.5 × 1.9–4 (–5) mm, thickened and risen around the interstylar aperture, translucent, surrounded or capped by the withered corolla. **SEEDS** 2–4 per capsule, broadly elliptic to transversely oblique, 1.4–1.85 × 1.25–1.6 mm. 2*n* = 30. Wide range of herbaceous and woody host species. COLO: LPl, Mon; NMEX: SJn; UTAH. 1370–1800 m (4500–5905′). Flowering/Fruit: Jul–Aug. Common in North America and South America. Our material belongs to **var.** ***indecora***.

Cuscuta umbellata Kunth (from the umbrella-like appearance of the inflorescence) Flat-globe dodder. **INFL** dense to loose, umbelliform with up to 300 pedicellate flowers; pedicels 2–10 mm long. **FLOWERS** 2–4 mm, membranous, creamy white when fresh, straw-yellow when dried; papillae usually absent, laticifers evident in the calyx and corolla, isolated, ovoid; calyx straw-yellow, not or finely reticulate, slightly shiny, campanulate, equaling or somewhat longer than the corolla tube, divided 1/2–3/5 the length, lobes not basally overlapping, triangular-ovate margins entire, apex acute to acuminate; corolla tube campanulate, 1.5–2.2 mm, lobes erect to reflexed, as long as or somewhat longer than the tube, lanceolate, margins entire, apex acute to acuminate, straight; stamens exserted, anthers oblong, 0.6–0.9 mm; infrastaminal scales reaching the filament bases, oblong-spatulate, rounded, uniformly dense-fringed to fimbriate. **CAPSULES** circumscissile, depressed-globose, 1.6–2.3 × 1.6–3 mm, thickened and slightly risen around the inconspicuous interstylar aperture, translucent, surrounded by the withered corolla. **SEEDS** 4 per capsule, ovate to broadly elliptic or subrotund, 0.92–1.1 × 0.85–1 mm. Numerous herbaceous hosts. NMEX: RAr. 1900 m (6235′). Flowering/Fruit: Jul–Sep. Common from southern United States and Mexico to South America. Our material belongs to **var.** ***umbellata***.

Evolvulus L. Evolvulus, Dwarf Morning Glory

(to unroll) Herbs or small suffrutescent shrubs, annual or perennial, not twining but sometimes creeping. **LEAVES** usually small, ovate to almost linear, entire. **INFL** in axillary, pedunculate, 1- to several-flowered dichasia. **FLOWER** pedicels about as long as calyx or apparently absent; sepals equal or subequal; corolla conspicuous, blue, or inconspicuous, faded pale bluish white, rotate, funnelform or salverform, the limb plicate, mostly subentire, the lobes pilose externally; stamens with filiform filaments, the anthers ovate to oblong or linear; ovary 2-locular, each locule 2-ovulate, sometimes 1-locular and 4-ovulate; styles 2, free or partially united at the base, each style deeply bifid for at least 1/2 its length into long, terete, filiform to subclavate stigmas. **FRUIT** globose to ovoid, 4-valved. **SEEDS** 1–4, small, smooth or minutely verrucose (Ooststroom 1934). About 100 American species; mostly tropical.

Evolvulus nuttallianus Roem. & Schult. **CP** (for Thomas Nuttall, 1786–1859) Evolvulus, dwarf morning glory. Suffrutescent herbs, erect to ascending. **STEMS** 10–15 cm tall, densely spreading-pilose with an indumentum of ferruginous, brown, fulvous, or gray color. **LEAVES** with short or absent petioles, blades linear-oblong, narrow-lanceolate to narrow-oblanceolate or rarely oblong, 8–20 mm long, 1.5–5 mm wide, entire, attenuate basally, acute to obtuse apically, densely pilose on both surfaces. **INFL** solitary, in axils over whole length of stem. **FLOWER** pedicels 3–4 mm long, reflexed in fruit; bracteoles subulate, 1–4 mm long; sepals lanceolate to narrowly lanceolate, long-acuminate, 4–5 mm long, spreading-villous; corollas rotate to broadly campanulate, 8–12 mm wide, subentire, purple or blue; anthers 1–2 mm long, oblong, basally auriculate, filaments twice as long as the oblong anthers; ovary subglobose, glabrous. **FRUIT** ovoid, about as long as sepals, glabrous. **SEEDS** (1) 2, brown, smooth. [*Evolvulus pilosus* Nutt.]. Sandy and rocky prairies and plains and piñon-juniper woodland communities. ARIZ: Apa, Nav; COLO: LPl; NMEX: RAr, San, SJn; UTAH. 1125–2450 m (3700–8000′). Flowering: May–Jul. Montana, North Dakota south to Missouri, Arkansas, Kansas, and Texas, west to Utah and Arizona, disjunct to Illinois and Tennessee; Chihuahua and Coahuila, Mexico. The Kayenta Navajo use the plants as a snuff for itching of the nose and sneezing.

Ipomoea L. Morning Glory

(a reference to the twining habit) Annual or perennial vines, shrubs, or trees. **STEMS** usually twining, sometimes prostrate or floating, glabrous or hairy. **LEAVES** petiolate, blade variable in shape and size, simple, lobed, divided, or less often compound. **INFL** mostly axillary, 1 to many flowers, in cymes, rarely paniculate. **FLOWERS** on long or short pedicels, the bracts scalelike to foliose; sepals herbaceous to subcoriaceous, ovate to oblong or lanceolate, glabrous to hairy, often somewhat enlarged in fruit but not markedly accrescent; corolla purple, red, pink, white, or less often yellow, regular or rarely slightly zygomorphic, mostly funnelform, less often campanulate, tubular, or salverform, the limb shallowly or rarely deeply lobed, the midpetaline bands well defined by 2 distinct nerves; stamens included or less often exserted, the filaments filiform, often triangular-dilated at the base, mostly unequal in length; ovary usually 2–4-locular, 4-ovulate, less often 3-locular, 6-ovulate; style simple, filiform, included or less often exserted; stigmas capitate, entire or 2–3-lobed, globose. **FRUIT** a globose to ovoid capsule, mostly 4–6-valved or splitting irregularly. **SEEDS** 1–4 (6). $x = 15, 30, 45$ (Austin 1986; Austin and Huáman 1996; Austin and Bianchini 1998). Perhaps over 600 species spread through the tropics and subtropics of the world.

1. Leaves simple, linear to lanceolate; stems erect to decumbent, somewhat fleshy .... ***I. leptophylla***
1′ Leaves simple to pedately dissected, cordate to orbicular in outline; stems often twining, less often decumbent, rarely fleshy although herbaceous .... (2)

2. Leaves pedatisect .... ***I. costellata***
2′ Leaves simple or toothed to shallowly or deeply lobed, usually cordate, rarely acute .... (3)

3. Corollas salverform, scarlet, orange, or yellow; bird flowers .... ***I. cristulata***
3′ Corollas funnelform, lavender to white or purple; bee or moth flowers .... ***I. purpurea***

Ipomoea costellata Torr. (having ribs, on the leaves) Crestrib morning glory. Low annuals from a slender taproot. **STEMS** erect at first, in age trailing or twining at the tips. **LEAVES** with blades subsessile or on petioles 1–3 cm long, deeply pedatisect, the segments 5–9, linear or linear-lanceolate, 7–25 mm long. **INFL** mostly solitary; peduncles 1–3 (7) cm long. **FLOWER** pedicels 15–25 mm long, erect in fruit; bracts subulate, to 1 mm long; sepals slightly unequal, the outer 3–5 mm long, 1–2 mm wide, the inner 4–6 mm long, 2–3 mm wide, oblong-lanceolate, acute, mucronulate, scarious-margined, at least the inner slightly rugose along the veins; corolla pale lavender to pink, 1–1.2 cm long; stamens 3–5 mm long, anthers 1 mm long, with white trichomes along filaments from base to anther; ovary ovoid, 1 mm long, 2-locular, cream, glabrous; styles 4 mm long, green. **FRUIT** subglobose to ellipsoid-globose, 4–5 mm wide, with a 1–2 mm apiculum which is caducous, tan. **SEEDS** 3 mm long, ovoid, black, glabrous. [*Ipomoea pusilla* Brandegee; *I. futilis* A. Nelson]. Rocky sites in the ponderosa pine zone. ARIZ: Apa, Nav. 1980–2285 m (6500–7500′). Flowering: Jul–Oct. Arizona, New Mexico, Texas; Baja California, Chihuahua, Coahuila south to Chiapas, Mexico; introduced into South America.

Ipomoea cristulata Hallier f. (having a crest, on the sepals) Star glory, Trans Pecos morning glory. Annual. **STEMS** twining, glabrous or pilose on the nodes. **LEAVES** with petioles 2–9 cm long, blades ovate, 1.5–10 cm long, 1–7 cm wide, typically the lower entire and the upper 3–5-parted or palmate, or all palmately parted or lobed, glabrous or pilose below, the margins irregularly dentate, the base cordate to subtruncate, the lobes rounded to acute, apically acute to acuminate, rarely obtuse, mucronate. **INFL** of 3–7 flowers, cymose or rarely solitary; peduncles 3–6 (25) cm long. **FLOWER** pedicels 5–14 mm long, reflexed in fruit; bracts linear-lanceolate to ovate, aristate, 1–3.5 mm long, bracteoles ovate to lanceolate, 1–2 mm long; sepals unequal, the outer oblong, 3–3.5 mm long, 2–2.5 mm wide, obtuse and rounded to subtruncate apically, muricate or smooth, with a subterminal arista 3–5 mm long, glabrous, the inner oblong, 4–5.5 mm long, 3–3.5 mm wide, apically truncate, with a subterminal arista 2.5–3.5 mm long; corolla salverform, 1.8–2.6 cm long, red or red-orange, glabrous, the limb 1–1.5 cm wide; stamens 21–23 mm long, exserted, anthers 1.5 mm long; ovary ovoid, 1 mm long, 4-locular, glabrous; styles 18–20 mm long. **FRUIT** subglobose, 7–8 mm wide, with an apiculum 2 mm long. **SEEDS** 1–4, 3.5–5 mm long, ovoid, black to dark brown, finely tomentose. [*Quamoclit gracilis* Hallier f.; *Ipomoea coccinea* L. var. *hederifolia* (L.) A. Gray; *I. coccinea* L.]. Ponderosa pine zone. ARIZ: Apa, Nav; NMEX: RAr, San, SJn. 1980–2800 m (6500–9100'). Flowering: May–Oct. Arizona, New Mexico, and Texas, disjunct and possibly introduced.

Ipomoea leptophylla Torr. (fine, slender leaf) Bush morning glory. Herbs, perennial, from an enlarged taproot. **STEMS** erect to decumbent, 0.3–1.2 m tall or sometimes slightly taller, glabrous. **LEAVES** with petioles 1–7 mm long, blade linear-lanceolate to linear, 3–15 cm long, 2–8 mm wide, entire, basally acute to attenuate, acute, glabrous. **INFL** of 1–3 flowers, rarely more in axillary cymes; peduncles 7–10. **FLOWER** pedicels 5–10 mm long; sepals unequal, the inner longer and wider than outer, 5–10 mm long, ovate to elliptic to orbicular-ovate, obtuse; corollas funnelform, 5–9 cm long, purple-red to lavender-pink with a darker throat. **FRUIT** ovoid, 1–1.5 cm long, brown, glabrous. **SEEDS** 1–4, oblong-ellipsoid, covered with dense, short, brown indumentum. $2n = 30$. Plains and prairies. NMEX: RAr, San, SJn. 1675–1980 m (5500–6500'). Flowering: May–Sep. New Mexico to Wyoming and South Dakota, south to Texas and Oklahoma. Endemic to the Great Plains (Keeler 1980). The enlarged roots of these plants were used, at least in time of famine, by several groups of people on the Great Plains in spite of being laxative. The Western Keres were among those using the plant, but it was used as far north as the Lakota and Pawnee. Plants were also used as medicine.

Ipomoea purpurea (L.) Roth (purple, in reference to the flowers) Tall morning glory. Annual, loosely hairy to tomentose with short, appressed trichomes, retrorse and often large trichomes, also with antrorse, oblique to erect trichomes up to 4 mm long. **STEMS** twining, branched to simple. **LEAVES** with petioles 1–14 cm long, blade ovate, subtrilobate, trilobate, or rarely 5-lobed, also unlobed, 1–11 cm long, 1–12 cm wide, basally cordate, apically acute to acuminate, rarely obtuse, mucronate. **INFL** cymose, with (1) 2–5 flowers. **FLOWER** pedicels 5–16 mm, erect in flower, reflexed and enlarged in fruit, reaching 25 mm long; bracts linear to lanceolate, 1.3–9 mm long, bracteoles similar to bracts, 4.5 mm long; sepals subequal, the outer narrowly ovate-lanceolate to elliptic, 8–15 mm long, (1.5) 2.5–4.5 mm wide, acute to abruptly acuminate apically, more hairy near base, the inner ovate-lanceolate, 8–15 mm long, 2.5–3 mm wide, acute to abruptly acuminate; corolla funnelform, 2.5–4.3 (5) cm long, blue (white to purple in cultivated plants), white within tube, glabrous, the limb 2.4–4.8 (7) cm wide; stamens 8–10 (14) mm, the anthers 1.5–2 (3) mm long; ovary ovoid to conic, 1.5–2 mm long, green, 3-locular, 6-ovulate, glabrous; style 14–22 (30) mm long; stigmas 3, globose. **FRUIT** subglobose to ovoid, 7–8 (10) mm wide, with an apiculum 2–4 mm long, 6-valvate. **SEEDS** 3–6, 4–5 mm long, ovoid, black to dark brown, finely tomentose. $2n = 30$. [*Ipomoea hirsutula* J. Jacq.]. Disturbed sites, cultivated fields, and in cultivation. ARIZ: Apa, Nav; NMEX: San, SJn. 1525–2300 m (5000–7500'). Flowering: Jul–Nov. Pantropical, widespread in North America, from Maine to North Dakota, south to Utah, Nevada, and Arizona, California to Washington along the coast; probably naturalized from Mexico. This is an unusually variable species, at least in part due to cultivation. Cultivated forms are always larger than wild forms, but the size of flowers and sepals may vary even in the wild plants. Throughout much of its pantropical range, the seeds are valued as a laxative.

CORNACEAE Bercht. & J. Presl DOGWOOD FAMILY

Kenneth D. Heil and J. M. Porter

Shrubs or herbaceous perennial plants. **LEAVES** opposite or whorled, simple, entire. **INFL** cymose or capitate. **FLOWERS** perfect, regular, small; sepals 4–5; petals 4–5, distinct; stamens the same number as the petals and alternate to them; pistil 1, the ovary inferior, mostly 2-loculed, 1 ovule per locule; style 1. **FRUIT** a 1–2-seeded drupe. x = 8–13, 19. Twelve genera, 90 species of the north-temperate zone, rare in tropics and south-temperate zones.

Cornus L. Dogwood

(*cornus*, horn, carved from the hard wood) **LEAVES** deciduous, 3–9 cm long, lanceolate, elliptic, ovate to obovate, acute to short-acuminate. **INFL** in a terminal cyme with many flowers or capitate; sepals 0.4–0.5 mm long; petals 1.5–3 mm long, yellow to purple or white. **FRUIT** 7–9 mm in diameter. Forty-five species of the north-temperate zone. Many species are cultivated ornamentals with large, showy bracts. Fruit used in jam and syrup.

1. Plants herbaceous, not over 25 cm tall; leaves 4–6 in a single whorl; inflorescence capitate, subtended by 4 petaloid bracts; fruit red .... ***C. canadensis***

1′ Woody shrubs over 25 cm tall; leaves many and opposite; inflorescence a flat-topped cyme, petaloid bracts not present; fruit white .... ***C. sericea***

Cornus canadensis L. (of Canada) Bunchberry. Herbaceous plants from woody rootstocks. **STEMS** 5–20 cm tall. **LEAVES** 4–6, whorled in 1 series near apex, often a pair of smaller or scalelike leaves below; 3–6 cm long, ovate to obovate, acute to short-acuminate, cuneate at base, sparsely strigose above, glabrous and lighter in color below. **INFL** a capitate cyme with many flowers subtended by 4 large white or yellowish bracts, these about 10–15 mm long, nearly as wide. **FRUIT** red. $n = 22$, $2n = 44$. Spruce-fir and aspen communities. COLO: Arc. 3330 m (11000′). Flowering: Aug. Fruit: Sep. Greenland to Alaska, south to New Jersey, Pennsylvania, Minnesota, and West Virginia to California; eastern Asia and India. Uses include drugs, food, and ceremonial.

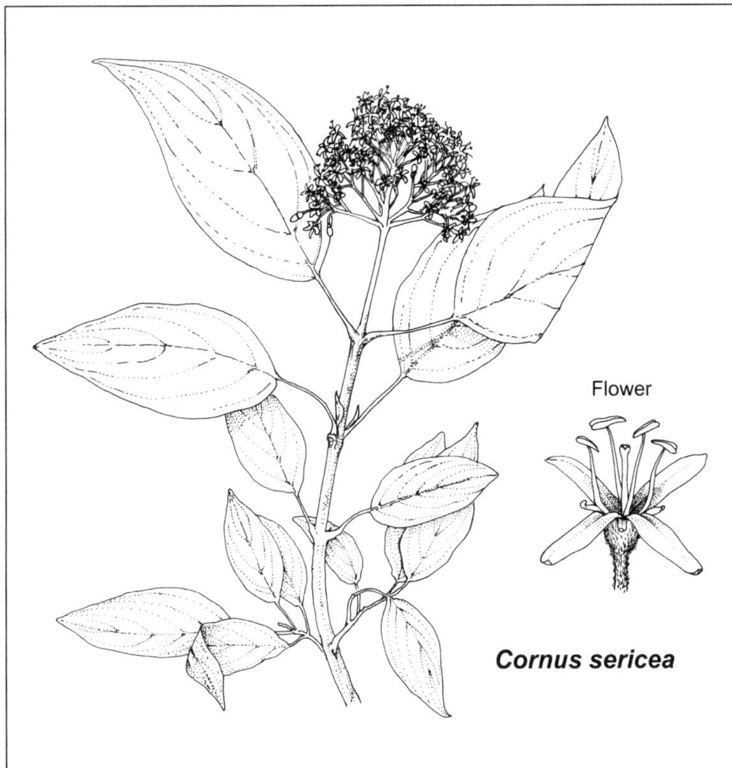

Cornus sericea

Cornus sericea L. (with silky hairs) Red-osier dogwood, kinnikinnik. Woody shrubs up to 4 m tall, and often as broad. Branches red to purplish or yellowish, subglabrous to strigulose, older stems grayish green, mostly glabrous. **LEAVES** mostly 5–9 cm long, lanceolate, elliptic to ovate, acute to acuminate, cuneate, glabrous or nearly so above, strigillose below with spreading hairs along the veins. **INFL** a flat-topped cyme with many flowers. **FLOWERS** with white petals. **FRUIT** white. $2n = 22$. Stream banks and other moist sites. ARIZ: Apa, Nav; COLO: Arc, Hin, LPl, Min, Mon, RGr, SJn; NMEX: McK, RAr, SJn; UTAH. 1515–3030 m (5000–10500′). Flowering: May–Aug. Fruit: Jun–Sep. Newfoundland to Alaska, south to Mexico. Uses include numerous drugs, fiber, dye, and ceremonial items; fruit used as food. Dried parts are a component of a common smoking mixture.

CRASSULACEAE J. St.-Hil. STONECROP or ORPINE FAMILY

Steve L. O'Kane, Jr.

(Latin *crass*, thick, referring to the typically thick, succulent leaves) Annual to perennial, shrubs, a few trees, and herbs (ours), succulent, glabrous and often glaucous. **LEAVES** alternate, opposite, or whorled, entire to less frequently toothed, usually succulent, a basal rosette often present, often with Kranz anatomy. **INFL** a branched cyme, sometimes flowers solitary, terminal or axillary. **FLOWERS** usually bisexual, sometimes unisexual and the plants dioecious or polygamodioecious; (3–) 4–20-merous; calyx and corolla both distinct or somewhat fused at the base; stamens 1–2× the number of sepals or petals, free or sometimes the antisepalous ones epipetalous; carpels distinct, typically the same number as the sepals and petals, superior, placentation marginal; nectaries scalelike and at the base of and opposite the carpels. **FRUIT** a follicle, rarely capsular. **SEEDS** with oily or proteinaceous endosperm, (1–) few or more commonly numerous in each carpel. $x = 4–22+$. The family is nearly cosmopolitan (except in Australia and Polynesia) and is especially speciose in arid- and warm-temperate regions; about 33 genera and 1500 species (Gontcharova, Artyukova, and Gontcharov 2006; Mayuzumi and Ohba 2004). Like the Cactaceae, the family consists almost entirely of succulent plants and the genera are often difficult to distinguish because of a stereotyped morphology. The genus *Sedum* is the largest in the family and has, over the years, been variously divided into segregate genera. Recent molecular work (e.g., refs. above) shows that our material is assignable to *Sedum* and the segregate genus *Rhodiola*. Members of the Crassulaceae are commonly cultivated in rock gardens.

1. Flowers pink (sometimes very pale and nearly white), red, or purple; shoots typically greater than 1.2 dm tall and ± erect; leaves flattened; cauline leaves persistent .... ***Rhodiola***
1′ Flowers white, cream, or yellow; shoots usually less than 1.2 dm tall, sprawling or erect; leaves subterete or flattened; cauline leaves deciduous after flowering .... ***Sedum***

Rhodiola L. Stonecrop

Perennial herbs, glabrous and often glaucous, from scaly, short, thick, branched rootstocks (ours). **STEMS** clustered, erect to ascending. **LEAVES** alternate, succulent, entire to weakly toothed or crenate, flattened. **FLOWERS** 5-merous; bisexual or unisexual and the plants then dioecious or polygamodioecious, pink, purple, or sometimes nearly white; sepals nearly distinct or united at the base; petals distinct; stamens 10; carpels the same number as the petals. **FRUIT** follicles erect, the style slightly divergent. **SEEDS** numerous, minute, pointed at each end, finely longitudinally ribbed. $x = 7, 8$. Thirty-six species in north-temperate areas. Our species are often frequent and beautiful members of moist subalpine and alpine areas

1. Flowers red or purple, unisexual, in a flat-topped inflorescence; leaf midrib inconspicuous .... ***R. integrifolia***
1′ Flowers pink or nearly white, perfect, in capitate clusters; leaf midrib prominent .... ***R. rhodantha***

Rhodiola integrifolia Raf. (entire leaf) King's crown, roseroot, ledge stonecrop. Perennial, succulent, usually glaucous. **STEMS** 3–20 (35) cm tall, stout, clustered, but each stem usually unbranched, erect or ascending. **LEAVES** all cauline, persistent, rhombic, oblanceolate, obovate, sometimes even wider, flattened, succulent, spreading to ascending, irregularly toothed, crenate, or entire. **INFL** a dense, terminal, corymbose cyme. **FLOWERS** often imperfect, the plants dioecious or polygamodioecious; sepals (4–) 5, narrowly lanceolate to lanceolate, 1.5–3 mm long, acute; petals (4–) 5, (ob)lanceolate or a little wider, red to purple; stamens (8–) 10, vestigial or lacking in pistillate flowers. **FRUIT** follicles distinct, erect with a slightly divergent style. **SEEDS** brown. $2n = 22, 33, 36$. [*Sedum integrifolium* (Raf.) A. Nelson; *S. rosea* (L.) Scop.; *S. roseum* (L.) Scop.; *Tolmachevia integrifolia* (Raf.) Á. Löve & D. Löve]. Aspen, spruce-fir, and alpine communities. COLO: Arc, Con, Hin, LPl, Min, Mon, RGr, SJn. 2470–3890 m (8100–12760′). Flowering: Jun–Sep. Our material is **subsp. *integrifolia***. Western half of North America but apparently not yet found in Arizona. Used by native Alaskans as a tea (the plant tops), for sores in the mouth, as an analgesic, and a gastrointestinal aid.

Rhodiola rhodantha (A. Gray) H. Jacobsen (rose flower) Queen's crown, rose crown, redpod stonecrop. Perennial, succulent. **STEMS** 8–30 (–40) cm tall, stout, clustered, but each stem usually unbranched, erect or ascending. **LEAVES** all cauline, persistent, narrowly oblong, narrowly elliptic to (ob)lanceolate, flattened, succulent, spreading, entire or weakly toothed, becoming somewhat involute. **INFL** terminal, racemiform or paniculiform, bracteate, often headlike. **FLOWER** sepals 5, narrowly lanceolate, 4–7 mm long, obtuse, rounded, or acute; petals 5, lanceolate, pink to almost white; stamens 10, filaments pink, anthers dark purple; carpels 5. **FRUIT** follicles distinct, erect except for divergent tip. **SEEDS** pale brown. $2n = 14$. [*Clementsia rhodantha* (A. Gray) Rose; *Sedum rhodanthum* A. Gray]. Aspen, spruce-fir, alpine communities. COLO: Arc, Con, Hin, LPl, Min, Mon, RGr, SJn. 3230–3690 m (10600–12100′). Flowering: Jun–Sep. Arizona, Colorado, New Mexico, Utah, and Montana.

Sedum L. Stonecrop

(Latin name for stonecrop) Perennial (–annual) herbs, sometimes slightly suffrutescent, glabrous and often glaucous, often rhizomatous or stoloniferous. **STEMS** erect to spreading or decumbent. **LEAVES** alternate, rarely opposite, succulent, entire, denticulate, or toothed, sometimes imbricate, often a basal rosette present, terete to flattened. **FLOWERS** 4- or 5- (rarely 3, 6, or 7)-merous; bisexual (ours) or rarely unisexual, yellow or white (ours), pink or purple; sepals distinct or united at the base; petals distinct or united at the base; stamens twice the number of petals, the ones opposite the petals epipetalous at the base of the petal; carpels the same number as the petals. **FRUIT** follicles erect to widely divergent, with a slender or short and tapered style. **SEEDS** numerous, minute, pointed at each end, finely longitudinally ribbed. $x = 7, 8, 9$. More than 600 species mostly in warm or arid north-temperate areas. Many species are common rock garden or house plants.

1. Flowers white or cream; leaves flattened .... ***S. cockerellii***
1' Flowers yellow; leaves subterete or the largest ones slightly flattened .... ***S. lanceolatum***

Sedum cockerellii Britton (for University of Colorado entomologist T. D. A. "Theo" Cockerell, 1866–1948) Cockerell's stonecrop. Perennial, succulent. **STEMS** several, growing in clumps, individually weak, 4–20 cm tall, decumbent, ascending or erect. **LEAVES** sessile, flattened, not or little overlapping, spreading to ascending, at least the basal ones minutely papillose, linear, spatulate, narrowly elliptic to narrowly obovate, deciduous after flowering. **INFL** a cyme, terminal or in the axils of the apical leaves, bracteate. **FLOWER** sepals (4–) 5, obtuse, 4–5 mm long; petals (4–) 5, linear-lanceolate, white or white and pink-tinged; longer than the sepals; stamens (8 or) 10; carpels (4–) 5. **FRUIT** follicles erect and often spreading at the tip. **SEEDS** brownish. [*Cockerellia cockerellii* (Britton) Á. Löve & D. Löve; *Sedum griffithsii* Rose; *S. wootonii* Britton]. Shady, rocky sites; piñon-juniper woodlands, Gambel's oak, ponderosa pine, Douglas-fir, and spruce-fir communities. NMEX: McK, RAr. 2135–3505 m (7000–11500′). Flowering: Jun–Sep. Arizona, New Mexico, Texas; northern Mexico.

Sedum lanceolatum Torr. **CP** (lanceolate) Common stonecrop, spearleaf stonecrop. Perennial, succulent, much-branched below with fragile sterile shoots. **STEMS** 5–15 (–20) cm tall, erect, often with a decumbent base. **LEAVES** sessile, terete or the larger ones somewhat flattened, ascending to ± appressed to the stem, narrow-lanceolate to elliptic-ovate; deciduous after flowering. **INFL** a terminal cyme. **FLOWER** sepals (4–) 5, rounded to obtuse, 2–4 mm long; petals (4–) 5, lanceolate to ovate, yellow, longer than the sepals; stamens (8 or) 10; carpels (4–) 5. **FRUIT** follicles somewhat connate at the base, erect, styles erect or spreading. **SEEDS** brownish. $n = 8$. [*Amerosedum lanceolatum* (Torr.) Á. Löve & D. Löve; *Sedum stenopetalum* Pursh, misapplied]. Rocky outcrops; sagebrush, piñon-juniper woodlands, ponderosa pine, aspen, Douglas-fir, and spruce-fir communities. ARIZ: Apa, Nav; COLO: Arc, Con, Dol, Hin, LPl, Min, Mon, RGr, SJn, SMg; NMEX: McK, SJn; UTAH. 2365–3900 m (7765–12800′). Flowering: Jun–Aug. Western half of North America. Our material is **subsp. *lanceolatum***. Leaves are edible and can be used as an addition to salads.

CROSSOSOMATACEAE Engl. CROSSOSOMA FAMILY

John R. Spence

Small to medium-sized shrubs, often spinescent and strongly branched, branches and stems often ridged or with hyaline to black trichomes. **LEAVES** microphyllous, deciduous or persistent, alternate to opposite, entire or tridentate, single or in fascicles, stipules present and small or absent. **INFL** axillary to terminal, flowers single or in axillary clusters. **FLOWERS** perfect or sometimes unisexual, actinomorphic, perigynous, sometimes with a nectary disc present on the hypanthium or at base; sepals 3–6, equal to unequal, imbricate and persistent; petals 3–6, typically 4–5, imbricate and distinct, longer than sepals; stamens 4–50, of equal or unequal lengths in 1–4 whorls, sometimes attached to nectary disc, alternate; ovary with 2–many ovules per carpel; style short, thick, with capitate stigma. **FRUIT** a coriaceous follicle that dehisces ventrally. **SEEDS** black to brown, with white or yellow entire to fimbriate aril. $x = 12$. A family of three genera and about seven to eight species mostly limited to arid habitats of the southwestern United States and northern Mexico.

Glossopetalon A. Gray Greasebush

(Greek *glosso*, tongue-shaped, + *petalon*, petal) Intricately branched spinescent shrubs, 1–2 m tall, branches ribbed by decurrent lines from the nodes. **STEMS** ribbed, young stems green, bark becoming orange-tan to gray. **LEAVES** small, alternate, entire, deciduous, sometimes absent most of the year; petioles short, decurrent; stipules minute, lanceolate. **INFL** of short axillary cymes or flowers solitary. **FLOWERS** perfect or sometimes unisexual; sepals 4–6, persistent, white; petals 3–5, free, spreading, white, inserted under a fleshy lobed nectary disc; stamens 4–10, attached to base of nectary disc; style lacking, stigma sessile; carpels 1–2, ovary 1-celled, ovules 1–2. **FRUIT** an asymmetrical, coriaceous, striate follicle, dehiscing ventrally. **SEEDS** small, partially enclosed by a white aril. A genus of five species, distributed primarily in arid regions of the southwestern United States and adjacent Mexico, but extending north to southern Washington, Idaho, and Wyoming. *Glossopetalon* (as *Forsellesia*) was placed in the Celastraceae until recently, but recent DNA studies have shown convincingly that it belongs in the Crossosomataceae (Sosa and Chase 2003; Holmgren 1988).

Glossopetalon spinescens A. Gray (for the spinescent branches) Low-growing, intricately branched, spinescent shrub, rarely to 2 m. **STEMS** ribbed, young stems green, bark becoming orange-tan to gray with age. **LEAVES** grayish green, somewhat glaucous, young leaves glabrous to somewhat puberulent, 3–15 mm long, 1–4 mm wide, with petioles 0.5–2.5 mm long, the blade oblanceolate with an apiculate tip, cuneate basally, typically appearing when the plant begins flowering. **INFL** with flower solitary at nodes. **FLOWERS** subtended by a fascicle of bracts; pedicel 1–5 mm long; sepals (3) 5, white, rounded, 1–2 mm long, hyaline-margined; petals (3) 5, white, oblanceolate to linear, 2–7 mm long. **FRUIT** a single follicle, 3–6 mm long, longitudinally ribbed, green to brown, beaked, the stigma discoid, oblique. **SEEDS** 1, more or less globose, 2–3 mm wide, white to brown and shiny-smooth, with a fimbriate aril. [*Forsellesia meionandra* (Koehne) A. Heller]. Rocky outcrops and slopes, sandy washes in desert scrub, and piñon-juniper woodlands. COLO: Mon; UTAH. 1100–2000 m (3500–6800′). Flowering: Mar–May. Fruit: Jun–Aug. Widespread in the Intermountain region from southeastern Washington, central Idaho, and southwestern Wyoming, Nevada, and Arizona south to Mexico, east of the Oregon and California mountains; also on western Great Plains of Oklahoma, New Mexico, and Texas. Locally there are no known uses of this shrub, but the Shoshoni of the Great Basin apparently used a decoction of the plant for tuberculosis. *Glossopetalon spinescens* is a highly variable shrub with several described varieties. Most of our material can be referred to **var. *meionandrum*** (Koehne) Trel. Because of its variability, some plants in the Four Corners region seem to resemble varieties *aridum* M. E. Jones and *microphyllum* N. H. Holmgren from other parts of the species range to the west and north. This species is similar in appearance to other small, highly branched desert shrubs of the Four Corners region, such as *Fendlerella utahensis* (S. Watson) A. Heller, *Coleogyne ramosissima* Torr., *Cercocarpus intricatus* S. Watson, and *Menodora spinescens* A. Gray.

CUCURBITACEAE Juss. GOURD, SQUASH, CUCUMBER FAMILY

Linda Mary Reeves

Typically perennial, monoecious or dioecious, climbing herbaceous vines with some shrubs, succulent lianas, and a few annuals. Most have a swollen tuberous hypocotyl which develops into a large subterranean rootstock. **LEAVES** alternate; simple; palmately lobed and/or compound with 3 or more leaflets; palmately veined; with spiraling solitary or branched tendrils that arise from the base of the petiole or leaf base. The tendril's tip curls around any nearby plant or object and the remainder of the tendril coils up, springlike, which draws the stem close to the support. **INFL** with flower solitary, a small cyme or raceme. **FLOWERS** imperfect; perianth 5-merous, borne at the top of a hypanthium; petals more or less united at the base; stamens 3; filaments and anthers somewhat united; ovary inferior. **FRUIT** a berry, pepo, or fleshy, leathery, or dry indehiscent capsule. **SEEDS** without endosperm, usually large. $x = 7–14$. About 90 genera with about 700 species. Primarily in the tropics with some desert and temperate distributions. A major food family including cucumbers (*Cucumis sativus*), squashes, chayote, pumpkins (*Cucurbita*), melons (*Cucumis*), watermelon (*Citrullus lanatus*), calabash (*Lagenaria siceraria*), as well as succulent ornamentals (*Xerosicyos*, *Cyclantheropsis*).

1. Stems clambering, vining; seeds less than 10 per fruit or, if more than 10 seeds per fruit, fruit with spines or prickles .... ***Echinocystis***

1′ Stems not clambering, trailing; seeds numerous in each fruit or, if not numerous, fruit spineless .... (2)

2. Corolla 5-lobed, united from middle to base .... ***Cucurbita***

2′ Corolla 5-parted to near or at the base .... (3)

3. Tendrils simple, not branched .... ***Cucumis***

3′ Tendrils 2–3-branched .... ***Citrullus***

Citrullus Schrad. Watermelon, Sandia, Citron

(Greek, little lemon) Mostly pubescent, vining annuals. **LEAVES** ovate to palmately lobed. **FLOWERS** small, solitary; corolla rotate; 5-lobed. **FRUIT** a pepo; globular or oblong, fleshy; generally green- or yellow-skinned, glabrous or finely pubescent-skinned with juicy or hard white, red, pink, yellow, or green flesh. **SEEDS** numerous, black, white, or red; flat. Several species from tropical and warm-temperate Africa and Asia.

Citrullus lanatus (Thunb.) Matsum. & Nakai (woolly) Watermelon, sandia. **LEAVES** ovate to ovate-oblong in outline, to 20 cm long, base cordate; pinnatifid lobes divided into 3–4 pairs, which are again lobed and/or toothed, apices broad; tendrils 2–3-forked. **INFL** with axillary solitary flower. **FLOWERS** monoecious; small; corolla 4 cm wide. **FRUIT** usually greenish, variously striped, rind nondurable; flesh generally reddish or yellowish. **SEEDS** generally black or whitish; smooth, to 15 mm long. Native to Africa. Escaped in our area around dwellings and agricultural fields. $2n = 22$. ARIZ: Apa. 1400–2000 m (4500–6500′). Flowering: May–Sep. Fruit: Aug–Oct. Our specimens are **var. *lanatus***. Flesh used as an important staple food by Hopis and Navajos (Diné), both fresh and sun-dried in strips for winter use. Ground seeds used by Hopi to grease piki bread stones and often parched and mixed with corn for piki bread.

Cucumis L. Melon, Cucumber, Gherkin

(Latin, cucumber) Annual or perennial trailing or climbing pubescent vines. **LEAVES** entire or partially dissected; tendrils simple. **INFL** usually with axillary, solitary flower; staminate often with more than 1 per axil. **FLOWERS** usually monoecious, corolla yellow; campanulate to rotate, deeply 5-lobed; anthers free; stigmas 3–5, styles short; ovules many. **FRUIT** fleshy and usually indehiscent, globular to elongated, glabrous, pubescent or echinate. About 25 species in tropical and warm-temperate regions, Africa and Middle East, western Asia.

Cucumis melo L. (fruit) Muskmelon, honeydew, cantaloupe. Trailing, pubescent vine with striate or angled stems. **LEAVES** orbicular-ovate to subreniform, to 13 cm wide, angled but usually not distinctly lobed, pubescent, scabrous; apex rounded; margins sinuate-dentate. **FLOWERS** about 25 mm across; corolla lobes obtuse. **FRUIT** various, mostly globular or ellipsoid; sometimes furrowed, pubescent at first, becoming glabrous or with raised ribs at maturity; fragrance musky, flesh orange, yellow, or green. **SEEDS** numerous, white, slender, about 12 mm long. $2n = 24, 48$. Escaped near agricultural fields and oil/gas rigs. NMEX: SJn. 2000 m (6500′). Flowering: May–Sep. Fruit: Jul–Sep. Native to Africa and western Asia. A food staple cultivated by many tribes.

Cucurbita L. Gourd, Squash, Pumpkin, Calabacita

(Latin, gourd) Annual or perennial with long, prostrate or climbing stems. **LEAVES** entire or lobed; usually scabrous or pubescent; tendrils branched. **FLOWERS** axillary, solitary; large; staminate flowers long-pedicelled; corolla yellow; lobed to about the middle or above; anthers united; stigmas 3–5, bilobed; ovary 1-celled. **FRUIT** mostly glabrous or slightly hairy, indehiscent, mostly fleshy or with a hard rind. **SEEDS** usually numerous, ovate to oblong-ovate, flat, white, tan, or black. About 15 species from the Western Hemisphere, primarily tropical and warm-temperate or desert habitats.

1. Leaves typically strongly triangular, longer than broad .... ***C. foetidissima***
1′ Leaves not as above .... (2)

2. Fruit globose; orange .... ***C. pepo***
2′ Fruit various; yellow, green, or pale orange .... (3)

3. Leaves broadly ovate to weakly triangular .... ***C. moschata***
3′ Leaves circular to reniform .... ***C. maxima***

Cucurbita foetidissima Kunth (evil-smelling) Buffalo gourd, Missouri gourd, calabacilla loca, wild pumpkin, chichicoyota. Rank-smelling, spreading perennial; stems often numerous, to 6 m or more long; root often large and fusiform. **LEAVES** coarse, thick; triangular-ovate; often shallowly angulate-lobed; base broadly rounded to cordate; apex acute to acuminate; to at least 30 cm long; ill-smelling when bruised; grayish green; scabrous. **FLOWER** corollas to 10 cm long, flaring above the middle. **FRUIT** globose; mostly green with lighter stripes; yellowish and dryish when ripe; diameter to 75 mm [*C. perennis* (E. James) A. Gray]. In sandy or gravelly roadsides or disturbed areas. ARIZ: Apa; COLO: Arc, LPl, Mon; NMEX: SJn. 1650–2000 m (5400–6500′). Flowering: Jun–Sep. Fruit: Aug–Oct. From Missouri and Nebraska south and west to Texas, Arizona, California, and Mexico. Mashed stems, leaves, and roots used as a veterinary poultice and as a green paint by Apaches. Used as a skin poultice and for chest pains by Puebloans and Zunis. Used as a treatment to constrict blood vessels and to hasten childbirth by Apaches. Gourds used as dancing rattles and containers by many tribes. Flowers eaten as food by many tribes. Roots and fruits contain saponins and a bleaching agent used for washing clothing. Gourds often used in Mexico to wean babies from the breast because of the distasteful, foul-smelling cucurbitins, thus chichicoyotas, or "trickster breasts."

Cucurbita maxima Duchesne (greatest) Fall squash, winter squash. Stems short or long-trailing, with soft pubescence. **LEAVES** lax; orbicular; not lobed; base cordate; margins dentate. **FLOWERS** solitary; corolla with broad, reflexed lobes; pedicel short, sometimes enlarged at the middle. **FRUIT** glaucous or glabrous; variously colored and shaped. **SEEDS** inflated; margin obtuse. Escaped near fields or from agricultural runoff in canyons. ARIZ: Apa. 2450 m (8000′). Flowering: May–Sep. Fruit: Aug–Oct. Native of Central America. Blossoms baked in cakes by Hopis and Apaches.

Cucurbita moschata Duchesne (musk) Butternut, winter crookneck, cushaw. **STEMS** long-trailing, soft. **LEAVES** velvety, limp, broadly ovate to somewhat triangular-orbicular; not lobed; margin crenate to denticular-dentate. **FLOWERS** with calyx lobes often long; corolla tube broad at base, lobes crinkly, widely spreading. **SEEDS** thin, margin hyaline in fresh material. Escaped from cultivation in Canyon de Chelly. ARIZ: Apa. 1753 m (5750′) Flowering: Jun–Sep. Fruit: Sep–Oct. Native to Mexico and Central America, but commonly escaped near agricultural fields, old homesteads, and roadsides.

Cucurbita pepo L. (ripe) Field pumpkin. Trailing, spreading annual. Stems with stiff, translucent hairs. **LEAVES** large, stiff, rigid, scabrous; triangular or ovate-triangular, 10–35 cm long, lobed; apex acute and apiculate; margins irregularly serrate. **FLOWERS** solitary; corolla with erect or spreading pointed lobes; pedicel angled. **FRUIT** large, highly variable, usually globose or ovoid, orange, furrowed. $n = 10$. Escaped from agricultural fields. ARIZ: Nav; NMEX: SJn. 1600–1800 m (5300–6100′). Flowering: May–Sep. Fruit: Aug–Oct. Native of Central America. A food staple of Apaches, who also baked blossoms in cakes. Ramah Navajos used the blossoms as seasoning for soups. Many tribes used the dried fruit as a winter food.

Echinocystis Torr. & A. Gray Wild Mock-cucumber

(Greek *echinos*, spiny, + *cystis*, bladder, for the prickly fruit) Annual; clambering, vinaceous, usually thin-stemmed plants with branched tendrils. **LEAVES** lobed or angled. **INFL** monoecious; staminate flowers borne on a raceme or panicle; pistillate flowers solitary. **FLOWERS** small; corolla white, whitish, or cream, rotate, 5- or 6-lobed; filaments connate, anthers connate or distinct; ovary 2-loculed; ovules 2 per locule. **FRUIT** fleshy or dry, often echinate. Monotypic genus in North America.

Echinocystis lobata Torr. & A. Gray (lobed) Smooth wild cucumber. High-climbing annual, mostly glabrous. **LEAVES** suborbicular-ovate; to 12 cm long, to 12 cm wide; with 5 triangular serrate lobes; tendrils forked. **INFL** a long raceme or panicle in staminate flowers; pistillate flowers solitary or in small clusters, axillary; short-pedicelled; both staminate and pistillate inflorescences often from the same axil. **FLOWERS** 5–6-merous; greenish to white; corolla rotate, about 1 cm wide, lobes narrowly lanceolate and acuminate; stamens 3, filaments united into a column; ovary 2-celled; stigma broad-lobed. **FRUIT** ovoid; 3–5 cm long, to 25 mm diameter; inflated; beaked and with leathery glabrous prickles to about 6 mm long. **SEEDS** flat, dark. $2n = 32$. [*Sicyos lobatus* Michx.]. In moist soil along acequias in shade. COLO: LPl, Mon; NMEX: SJn. 1000–2100 m (4600–6900′). Flowering: Jul–Sep. Fruit: Sep–Oct. Saskatchewan south to Florida and west to Texas, Colorado, and Arizona. Escaped from cultivation and sporadic in the West.

CYPERACEAE Juss. SEDGE FAMILY

Sherel Goodrich

Perennial or annual, graminoids. **CULMS** (stems) triangular, rounded, or flattened in cross section, not jointed. **LEAVES** simple, linear, entire, 3-ranked, sometimes reduced to bladeless sheaths, these mostly closed. **INFL** spicate, racemose, or umbellate, of 1 to several spikes or spikelets (small spikes). **FLOWERS** much reduced, imperfect or perfect, sessile or nearly so, subtended by a very small bract referred to as a scale; perianth lacking or reduced to bristles; stamens usually 3, rarely 1 or 2, exserted at anthesis, the anther basifixed; ovary superior; stigmas 2 or 3. **FRUIT** a lenticular or trigonous achene.

1. Achene enclosed or folded within a closed or open sac or bract as well as subtended by a scale; perianth bristles lacking; flowers imperfect only ....(2)
1′ Achenes exposed, neither enveloped by nor enfolded in a sac or bract, merely subtended by a scale and sometimes by perianth bristles; flowers mostly perfect (except in *Cladium*) ....(3)

2. Achene completely enclosed in a sac (perigynium), this closed except at the apex, through which the stigmas are exserted, concealing the attachment of the style to the achene .... ***Carex***
2′ Upper part of the achene exposed, the subtending bract with open margins, exposing the attachment of the style to the achene .... ***Kobresia***

3. Perianth bristles numerous, long, cottonlike, concealing other parts of the inflorescence .... ***Eriophorum***
3′ Perianth bristles lacking to several, but not cottonlike, not concealing other parts of the inflorescence..........(4)

4. Achene subtended by perianth bristles ....(5)
4′ Perianth bristles lacking ....(7)

5. Fertile flowers or achenes solitary (rarely 2) in each spikelet, the lower scales of the spikelet empty.......... ***Rhynchospora***
5′ Fertile flowers or achenes few to several in each spikelet (rarely solitary), the lower scales usually fertile (except in *Cladium*, in which the lower scales are empty) ....(6)

6. Spikelets solitary and terminal and without subtending leafy bracts; style thickened at the base, persistent on and forming a point or cap on the achene .... ***Eleocharis***
6′ Spikelets commonly 2 or more, often subtended by 1 or more leafy bracts .... ***Scirpus***

7. Plants 1–2 m tall, with leafy stems and foliose bracts; lower scales empty; flowers in each spike all staminate except the terminal perfect one .... ***Cladium***
7′ Plants various but rarely as large as above; scales typically all subtending flowers, these mostly perfect .... (8)

8. Scales of spikes strongly 2-ranked .... ***Cyperus***
8′ Scales of spikes not 2-ranked .... ***Fimbristylis***

Carex L. Sedge

(Latin name for the genus) Plants perennial, monoecious or occasionally dioecious; stems tufted or arising singly or few together from creeping rhizomes. **INFL** of 1, few, or several spikes; spikes solitary, few, or several, sessile or pedunculate, staminate, pistillate, androgynous. **STAMINATE FLOWERS** borne above the pistillate ones, or gynaecandrous. **PISTILLATE FLOWERS** with few to many imperfect, sessile or short-stipitate flowers; without perianth, the staminate ones reduced to 2 or 3 stamens, the pistillate ones consisting of a pistil, this enveloped in a closed scale or saclike perigynium or perigynia (plural) with the stigmas exserted through the apex (Hermann 1970; Hurd et al. 1998; Johnston 2001). About 1000 species. Cosmopolitan, especially temperate and cold wet places.

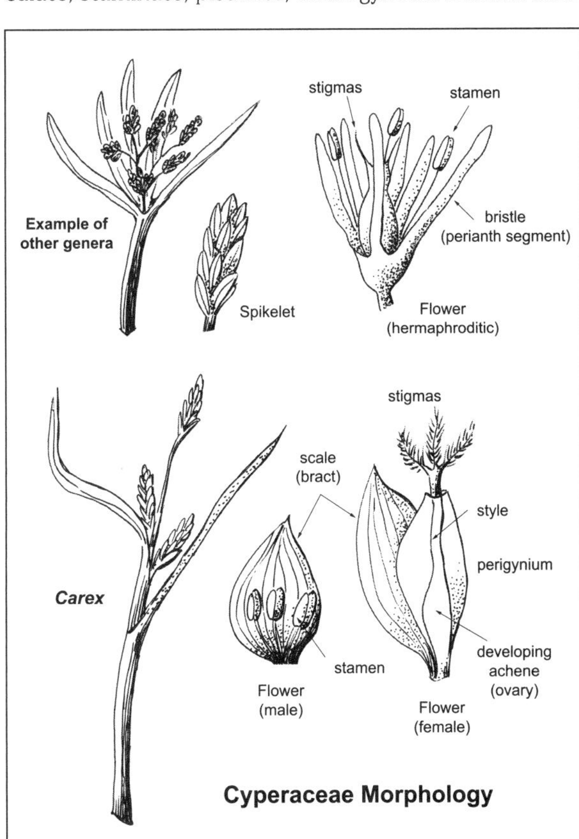

Cyperaceae Morphology

Munz and Keck, in *A California Flora* (1973), noted that hybridization in this genus has been recognized as common in parts of its range, but they also noted that instances of hybridization in California have been rarely observed or reported. Examination of numerous specimens for *A Utah Flora* (Welsh et al. 2003) and for this *Flora of the Four Corners* indicates hybridization is uncommon. The taxa treated are quite distinct. However, due to the large number of species and small differences between some of the species, identification can be difficult and hybridization assumed. Within some parts of the genus, including section *Ovales*, distinction of species can be particularly troublesome. There appears to be no way to construct a definitive key for separation in all cases. The interpretation of relative size of scales and perigynia and the shape of the beak of perigynia commonly used extensively in section *Ovales* appears to be more an art than a science. However, the taxa appear distinct when considering a composite of features that are sometimes difficult to encompass or combine in a key. Throughout the genus, slight morphological differences are often well supported by ecological and geographical differences (Hermann 1970). Distribution maps of *Carex* provided by Johnston (2001) were used extensively in preparation of this treatment. His maps (based on specimens) were considered valid documentation for inclusion in this flora.

1. Spike solitary; perigynia attached directly to the rachis .... **Key 1**
1′ Spikes more than 1; perigynia attached to a rachilla .... (2)

2. Spikes all sessile, aggregated into a headlike or spikelike inflorescence .... (3)
2′ Lowest spike borne on a peduncle, other spikes sometimes also on peduncles .... (6)

3. Stigmas 3; achenes trigonous; inflorescence with 3–5 spikes; terminal spike gynaecandrous, the lateral ones all pistillate .... **Key 5**
3′ Stigmas 2; achenes lenticular; inflorescence commonly with more than 5 spikes; spikes typically all androgynous or all gynaecandrous or all staminate or all pistillate .... (4)

4. Spikes typically androgynous or the plants with well-developed, creeping rhizomes .... **Key 2**
4' Spikes gynaecandrous; plants caespitose; rhizomes lacking or short .... (5)

5. Perigynia round-margined, not winged, not conspicuously flattened, mostly less than 3.5 mm long; scales pale green to brown; inflorescence commonly less than 2 cm long and/or less than 1 cm wide; plants mostly of wet places .... **Key 3**
5' Perigynia wing-margined, often conspicuously flattened, (2.5) 3.5–8 mm long; scales commonly brownish to dark brown, often with a green midrib; inflorescence often longer or wider than above; plants of dry or wet places (section *Ovales*) .... **Key 4**

6. Terminal spike gynaecandrous, the lateral ones all pistillate or gynaecandrous with few staminate flowers; scales generally blackish .... **Key 5**
6' Terminal spike and sometimes 1 or more lateral spikes staminate or androgynous; scales variously colored .... (7)

7. Stigmas 2; achenes lenticular; scales often black or black-purple or with blackish lines flanking a greenish or pale midstripe, often contrasting with greenish or straw-colored perigynia .... **Key 6**
7' Stigmas 3; achenes trigonous; scales greenish or brownish, or if blackish, then usually about the same color as the perigynia .... **Key 7**

KEY 1

1. Spikes all staminate or all pistillate; perigynia pubescent .... (2)
1' Spike androgynous; perigynia glabrous or puberulent (in *C. filifolia*) .... (3)

2. Plants caespitose, 20–45 cm tall, often in hanging gardens in canyons below 1400 m (4600') .... ***C. curatorum***
2' Culms arising singly or few together from rhizomes; plants 5–25 (30) cm tall, alpine above 3000 m (9840') .... ***C. scirpoidea***

3. Stigmas 2; achenes lenticular .... ***C. capitata***
3' Stigmas 3; achenes trigonous .... (4)

4. Plants densely caespitose, often fasciculate; rhizomes lacking or nearly so; leaf blades rarely over 1 mm wide .... (5)
4' Stems arising singly or few together from creeping rhizomes .... (8)

5. Rachilla lacking; perigynia slightly spreading or widely so in age .... ***C. pyrenaica***
5' Rachilla about equaling or longer than the achene; perigynia appressed or nearly so .... (6)

6. Plants known from below 2300 m (7550') elevation; perigynia hirtellous above; spikes 1–4 cm long...... ***C. filifolia***
6' Plants alpine or subalpine, known from above 3600 m (11800'); spikes 0.5–1.8 cm long .... (7)

7. Perigynia with flattened margins; spike equaling or included in the leaves, the staminate part inconspicuous, 2–3 mm long .... ***C. nardina***
7' Perigynia with rounded, somewhat involute margins; spike often exceeding the leaves, the staminate part 3–7 mm long .... ***C. elynoides***

8. Perigynia 1 or 2 per spike, 5–6 mm long; staminate part of spike 1–2 (3) cm long .... ***C. geyeri***
8' Perigynia either more numerous or smaller than above; staminate part of spike usually less than 1 cm long .... (9)

9. Perigynia ascending, 1–6 per spike, these and scales persistent in age .... ***C. obtusata***
9' Perigynia strongly spreading to reflexed at maturity, these and scales deciduous in age .... (10)

10. Rachilla exserted beyond the orifice of the perigynium; scales and perigynia greenish or pale brown ***C. microglochin***
10' Rachilla obsolete; scales and perigynia blackish, at least in part .... ***C. nigricans***

KEY 2

1. Inflorescence globose, with tightly compacted spikelets; plants known from above 3500 m (11500') elevation.... (2)
1' Inflorescence spicate, not globose; plants known mostly from below 3500 m (11500') elevation .... (3)

2. Perigynia ovate-elliptic, inflated; leaf blades to 1.5 mm wide .... ***C. perglobosa***
2' Perigynia lanceolate, not inflated; leaf blades 2–4 mm wide .... ***C. vernacula***

3. Culms arising singly or few together from creeping rhizomes ....(4)
3′ Plants caespitose; rhizomes lacking or short ....(9)

4. Spikes mostly well separated from each other, about 5 mm long, each with 1–3 staminate and pistillate flowers; scales and perigynia green ....***C. disperma***
4′ Spikes congested into a headlike or continuous spikelike inflorescence, mostly with more flowers and longer than above; scales and perigynia brownish to blackish ....(5)

5. Perigynia winged; plants of coniferous forests and openings in forests ....***C. siccata***
5′ Perigynia not winged or but slightly so; plants mostly not of forests ....(6)

6. Perigynia 1.7–2.6 mm long, the beaks 0.2–0.5 mm long; plants usually of boggy meadows ....***C. simulata***
6′ Perigynia or beaks longer than above; plants various ....(7)

7. Rhizomes averaging 2–4 mm wide, blackish or dark brown; lower leaves reduced to bladeless sheaths, these blackish or dark brown; plants 10–70 cm tall ....***C. praegracilis***
7′ Rhizomes averaging less than 2 mm wide, brownish; lower leaves with or without blades, the sheaths light brown or greenish; plants 8–28 cm tall ....(8)

8. Plants unisexual, the spikes all staminate or pistillate; inflorescence 1.5–5 cm long ....***C. douglasii***
8′ Plants bisexual, the spikes androgynous; inflorescence 0.8–2 cm long ....***C. duriuscula***

9. Inflorescence simple, the perigynia attached to a simple rachilla which is attached to the rachis ....(10)
9′ Inflorescence compound ....(11)

10. Perigynia smooth and wingless; spikes with 3–7 perigynia ....***C. vallicola***
10′ Perigynia usually minutely serrulate or slightly wing-margined in the upper 1/2; spikes with 4–12 perigynia ....***C. occidentalis***

11. Scales with awns 1–5 mm long ....***C. vulpinoidea***
11′ Scales without awns ....(12)

12. Perigynia 3.6–5.2 mm long ....***C. stipata***
12′ Perigynia 2–3 mm long ....***C. diandra***

KEY 3

1. Inflorescence an ovoid head with internodes not greater than 1 mm long and concealed in the dense spikes; perigynia and scales dark brown to blackish in age ....***C. illota***
1′ Inflorescence spikelike with at least the lower internodes generally over 1 mm long and conspicuous; perigynia greenish to dark brown ....(2)

2. Perigynia widely spreading at maturity, the lower ones sometimes reflexed ....(3)
2′ Perigynia appressed ....(4)

3. Beak of perigynium 1/4–1/3 (1/2) the length of the body, inconspicuously bidentate, with broad, short teeth; perigynia mostly 2.2–3.2 mm long ....***C. interior***
3′ Beak of perigynium 1/2 or more the length of the body, conspicuously bidentate, the teeth narrow; perigynia mostly 2.8–3.5 (4) mm long ....***C. echinata***

4. Perigynia averaging over 3.5 mm long; inflorescence 4–6.5 cm long ....***C. deweyana***
4′ Perigynia averaging less than 3.5 mm long; inflorescence various, but often shorter than above ....(5)

5. Dorsal suture of perigynium 0.7–1.5 mm long, extending onto the distal part of the body of the perigynium; perigynia 2–3.5 mm long; pistillate scales as large as and covering the perigynia, brownish; spikes 2–4; plants mostly alpine or nearly so ....***C. bipartita***
5′ Dorsal suture of perigynium mostly less than 0.7 mm long, mostly confined to the beak of the perigynium; perigynia 1.7–2.5 mm long; pistillate scales shorter and narrower than the perigynia, variously colored; spikes 4–10; plants mostly subalpine ....(6)

6. Spikes 2–4, brownish or greenish brown; dorsal suture of perigynium extending the full length of the beak ....***C. praeceptorum***

6′ Spikes generally 4–8, greenish, silvery green, pale gray, or sometimes turning brownish in age; dorsal suture of perigynium less than the length of the beak .... ***C. canescens***

KEY 4

1. Inflorescence spicate, with 2–7 spikes; at least the lower spikes evident without teasing the inflorescence apart ....(2)
1′ Inflorescence capitate, with 3–20 spikes, the spikes mostly closely aggregated such that individual spikes at least toward the center of the head are not distinguishable without teasing the head apart ....(5)

2. At least the lower 2–3 spikes only 1–2 times longer than the associated internode; inflorescence rather flexuous, sometimes over 3 cm long; plants 40–80 cm tall .... ***C. praticola***
2′ Spikes mostly at least 2 times as long as the associated internodes; inflorescence mostly strict-erect, mostly less than 2.5 cm long; plants various ....(3)

3. Plants with short, stout rhizomes .... ***C. tahoensis***
3′ Plants densely caespitose, without rhizomes ....(4)

4. Plants montane in big sagebrush and aspen belts .... ***C. petasata***
4′ Plants subalpine and alpine in the San Juan Mountains .... ***C. phaeocephala***

5. Perigynia 5.5–7.5 mm long and 2.5–3.8 mm wide; inflorescence with 3–8 spikes, these ovate or nearly globose and 9–15 mm long .... ***C. egglestonii***
5′ Perigynia shorter or narrower than above or both; inflorescence with about 3–20 spikes, these of various shape and size, but often shorter than above or not ovate or globose ....(6)

6. Lowest bract exceeding the inflorescence .... ***C. athrostachya***
6′ Lowest bract inconspicuous or at least much shorter than the inflorescence ....(7)

7. Beak of perigynia flattened ....(8)
7′ Beak of perigynia terete ....(9)

8. Perigynia 2.3–3.7 mm long .... ***C. bebbii***
8′ Perigynia 4.5–5.5 mm long .... ***C. arapahoensis***

9. Perigynia sharp-edged but not wing-margined, entire, 2.5–3 mm long; inflorescence 0.8–1.4 cm long, dark brown to blackish .... ***C. illota***
9′ Perigynia wing-margined, mostly 3–7.1 mm long; inflorescence sometimes longer than above ....(10)

10. Spikes averaging over 10 mm long, the lower 1 or 2 commonly 11–13 and up to 15 mm long, elliptic or lanceolate, smooth with the perigynia appressed; inflorescence often cuneate at the base; perigynia 5.2–7.1 mm long, about 1–2 mm wide .... ***C. ebenea***
10′ Spikes averaging less than 10 mm long, the larger ones rarely over 11 mm long, ovate to nearly globose; inflorescence various; perigynia often shorter or wider than above or both ....(11)

11. Perigynia usually averaging over 2.2 mm wide, up to 3 mm wide, the body brown to dark brown, the winged margins sometimes greenish; scales about as dark as the perigynia; inflorescence usually not conspicuously bicolored; plants mostly alpine .... ***C. haydeniana***
11′ Perigynia averaging less than 2.2 mm wide, up to 2.4 mm wide, the body often greenish; scales often darker than the perigynia until after maturity; inflorescence commonly strongly bicolored; plants commonly midmontane in the aspen and coniferous forest belts ....(12)

12. Perigynia readily deciduous upon maturity, plano-convex, the perigynial cavity nearly filled by the plump achene .... ***C. pachystachya***
12′ Perigynia persistent well after maturity, strongly flattened except where extended by the relatively small achene .... ***C. microptera***

KEY 5

1. Spikes congested into a dense head, the internodes of the rachis obsolete or short and hidden in the closely aggregated spikes; plants generally high montane to alpine ....(2)
1′ At least the lowest spike separated from the others by a conspicuous rachis internode; plants various ....(3)

2. Perigynia 1.1–1.8 mm wide, often somewhat inflated, often without flattened margins or these inconspicuous, the width nearly filled by the mature achene .... ***C. nelsonii***
2′ Perigynia 2–3 mm wide, strongly flattened, with conspicuous flattened margins, conspicuously wider than the mature achene .... ***C. nova***

3. Pistillate scales aristate, the awn-point 1–2 mm long .... ***C. buxbaumii***
3′ Scales not aristate .... (4)

4. Culms arising singly or few together from long-creeping rhizomes, these covered with a yellowish feltlike tomentum; scales greenish brown; spikes spreading or drooping on slender peduncles .... ***C. magellanica***
4′ Culms tufted; rhizomes lacking or short; scales usually blackish; spikes various .... (5)

5. Spikes not over 1.3 cm long; perigynia 2–2.8 mm long, green at maturity and strongly contrasting with the dark scales .... ***C. norvegica***
5′ At least one spike generally over 1.3 cm long and/or the perigynia over 2.8 mm long, the perigynia variously colored .... (6)

6. Lateral spikes erect or nearly so; plants of hanging gardens of canyons .... (7)
6′ Lateral spikes spreading to pendulous; plants mostly montane .... (8)

7. Achenes mostly lenticular; stigmas mostly 2; lateral spikes more or less of equal length; leaves comparatively lax; culms various but often less than 1 mm thick, also often lax compared to those of the following taxon .... ***C. specuicola***
7′ Achenes mostly trigonous; stigmas mostly 2; lateral spikes commonly unequal; leaves and culms comparatively rigid, the culms commonly over 1 mm thick .... ***C. parryana***

8. Perigynia whitish green, rarely with red markings, strongly contrasting with the shorter, dark scales; lateral spikes ascending, spreading, or drooping at maturity, each usually on a slender peduncle, at least some in each inflorescence commonly gynaecandrous .... ***C. bella***
8′ Perigynia olive-green or commonly marked with dark red, purple, or black, often about as dark as the scales distally; lateral spikes ascending to erect, the upper ones commonly sessile or short-pedunculate, all pistillate or infrequently gynaecandrous in *C. heteroneura* .... (9)

9. Perigynia 2.7–3.5 mm long, 1.6–2.1 mm wide, consistently papillate, sometimes completely black or black-purple at maturity; lowest rachis internode 0.2–1 cm long; lowest spike sessile or on a peduncle to about 0.3 cm long, the upper ones sessile or subsessile; plants 10–28 cm tall .... ***C. albonigra***
9′ Perigynia mostly more than 3.5 mm long or over 2.1 mm wide, rarely papillate, sometimes greenish at maturity; lowest internode of the rachis about 0.3–7 cm long; lowest spike sessile or on a peduncle to 4.5 cm long; plants (2) 2.5–10 dm tall .... (10)

10. Perigynia flattened, broadly elliptic, ovate, or obovate, mostly suffused with dark purple or black, mostly wider than the mature achene; plants widespread .... ***C. heteroneura***
10′ Perigynia not flattened, lanceolate or narrowly elliptic, olive-green, becoming yellow-brown at maturity, seldom suffused with dark purple or black except the beak, not much wider than the mature achene .... ***C. atrosquama***

KEY 6

1. Scales pale green; lowest peduncle enveloped at the base by a closed sheath 2–6 mm long, often originating on the lower 1/2 of the culm .... ***C. aurea***
1′ Scales darker or with at least a dark midrib; lowest peduncle not in a closed sheath, or if so, the sheath less than 2 mm long, mostly originating from the upper 1/2 of the culm .... (2)

2. Pistillate spikes 1 or 2, 6–15 mm long; plants with mostly gynaecandrous terminal spikes, these occasionally all staminate; plants mostly of hanging gardens .... ***C. specuicola***
2′ Pistillate spikes generally more numerous or some usually over 15 mm long; habitat various .... (3)

3. Style strongly bent at maturity and persistent on the achene, mature perigynia about as dark as the scales, slightly inflated; stigmas 2 and 3 .... ***C. saxatilis***
3′ Style not strongly bent, deciduous; perigynia various but often greenish or at least paler than and strongly contrasting with the dark scales (except in *C. scopulorum*); stigmas 2 .... (4)

4. Plants tufted, with numerous fibrous roots, without rhizomes, the larger roots covered with yellowish or yellow-brown feltlike hairs; pistillate spikes 3–4 mm wide; perigynia rather quickly deciduous after maturity .... ***C. lenticularis***
4′ Culms arising singly or few together from long-creeping rhizomes; fibrous roots comparatively few, lacking feltlike hairs or these whitish; pistillate spikes various but often wider than above; perigynia persistent or deciduous ....(5)

5. Midrib of pistillate scales conspicuous throughout the length of the scale, commonly projected beyond the scale as a mucronate tip 0.5–1 mm long; perigynia nerved on both sides as well as on the margins; plants of valleys to midmontane .... ***C. nebrascensis***
5′ Midrib of pistillate scales commonly faint toward the apex of the scale, not or rarely projected into a mucronate tip; perigynia nerveless except on the margins; elevation various .... (6)

6. Lowest bract generally shorter than the inflorescence; mature pistillate scales dark brown or blackish throughout or with a slender pale midvein; mature perigynia usually about as dark as the scales at least where exposed beyond the scales; pistillate spikes often 5 mm wide or wider .... ***C. scopulorum***
6′ Lowest bract usually equaling or exceeding the inflorescence; pistillate scales usually with conspicuous, greenish or lighter midveins; mature perigynia greenish or suffused with red or red-brown but usually not as dark as the outer portions of the scales; pistillate spikes 3–5 mm wide .... (7)

7. Ligule acute, elongate; light central area of pistillate scales usually much narrower than the outer dark portions on either side; perigynia persistent through the summer; beak of perigynium often with a reddish or dark ring at the apex; plants midmontane to high montane .... ***C. aquatilis***
7′ Ligule truncate or broadly rounded; light central area of pistillate scales as wide or wider than the outer dark portions; perigynia often deciduous in early summer; beak of perigynium often white throughout, occasionally darker at the apex; plants of lower elevations, often along ditches .... ***C. emoryi***

KEY 7

1. Perigynia pubescent .... (2)
1′ Perigynia glabrous .... (5)

2. Fertile culms alike, bearing both staminate and pistillate spikes; basal spikes absent; plants with creeping rhizomes .... (3)
2′ Fertile culms of 2 types, some short and hidden in the leaf bases and bearing only pistillate spikes, others elongate and equaling or exceeding the leaves and bearing a terminal staminate spike as well as laterally pistillate spikes; plants densely caespitose .... (4)

3. Plants to 23 cm tall, of well-drained uplands; pistillate spikes 7–10 mm long, with 5–15 perigynia .... ***C. inops***
3′ Plants mostly over 23 cm tall, riparian or near riparian; pistillate spikes over 10 mm long, with more than 15 perigynia .... ***C. pellita***

4. Bract of lowest nonbasal pistillate spike leaflike, commonly exceeding the inflorescence .... ***C. rossii***
4′ Bract of lowest nonbasal pistillate spike commonly not exceeding the inflorescence .... ***C. geophila***

5. Style continuous with, persistent on, and of the same firm texture as the achene, not withering; staminate spikes more than 1 and/or usually 2.1–8.5 cm long .... (6)
5′ Style jointed to, deciduous from, and of softer texture than the achene, withering; staminate spikes mostly 1, 0.3–2.1 cm long .... (11)

6. Leaf sheaths pubescent; style straight .... ***C. atherodes***
6′ Leaf sheaths glabrous; style curved to strongly bent in age .... (7)

7. Pistillate scales narrowed to a serrulate-ciliolate awn .... ***C. hystericina***
7′ Pistillate scales not awned .... (8)

8. Pistillate scales soon purple-black or black except for the hyaline acute tip; styles 2 or 3; achenes lenticular or trigonous .... ***C. saxatilis***
8′ Pistillate scales green, stramineous, reddish brown, or if (rarely) dark brown then usually acuminate or caudate-acuminate; styles 3; achenes trigonous .... (9)

9. Inflorescence 5–13 cm long; spikes crowded, the pistillate ones longer than the rachis internodes; lowest bract 2–3 times longer than the inflorescence; perigynia 7–10 mm long .... ***C. retrorsa***
9′ Inflorescence commonly over 13 cm long; spikes not crowded, pistillate spikes from shorter than to longer than the rachis internodes; lowest bract 1–2 times longer than the inflorescence; perigynia 4–7 mm long ....(10)

10. Perigynia spreading at maturity, the ellipsoid to subglobose body more or less abruptly contracted to a conspicuous beak; culms arising singly or few together from long-creeping rhizomes .... ***C. utriculata***
10′ Perigynia appressed or slightly ascending, the lanceolate to ovate body gradually tapering to the often poorly defined beak; culms tufted, rhizomes lacking or short (north of the Four Corners) .... *C. vesicaria*

11. Peduncle of the lowest and often of the upper pistillate spikes enveloped in a closed sheath 0.4–2 (4) cm long ....(12)
11′ Peduncle of the lowest and other spikes not in a closed sheath or this less than 0.4 cm long ....(13)

12. Pistillate spikes sessile or on erect or ascending peduncles, all crowded or the lowest one sometimes separate; inflorescence 2–5 (10) cm long; staminate spike 0.7–2.1 cm long .... ***C. viridula***
12′ Pistillate spikes borne on capillary, spreading or drooping peduncles, remote; inflorescence 5–30 cm long; staminate spike 0.3–0.9 cm long .... ***C. capillaris***

13. Pistillate spikes 4–6 mm long, sessile; perigynia widely spreading, greenish; scales greenish or pale; plants caespitose .... ***C. interior***
13′ Pistillate spikes longer than 6 mm and/or at least one on a conspicuous peduncle; perigynia or scales or both brownish or blackish ....(14)

14. Pistillate spikes sessile or on erect or strongly ascending peduncles; culms various .... ***C. parryana***
14′ Pistillate spikes borne on slender, spreading to drooping peduncles; culms arising singly or few together from long-creeping, slender rhizomes; roots and rhizomes covered with a feltlike tomentum..(15)

15. Terminal spike 4–12 mm long, occasionally with a few perigynia at the base; pistillate scales narrower than and longer than the perigynia; leaves mostly with well-developed blades .... ***C. magellanica***
15′ Terminal spike (10) 15–21 (27) mm long, staminate only; pistillate scales as wide as the perigynia and barely exceeding them in length; lower leaves commonly bladeless .... ***C. limosa***

Carex albonigra Mack. (white and black) White and black sedge. Plants (10) 15–28 cm tall, the culms tufted; rhizomes lacking or short. **LEAVES** basal and on the lower culms, the blades 1–5 mm wide, bract shorter than the inflorescence. **INFL** (1) 1.5–3 cm long, surpassing the leaves, the lowest internode 2–10 mm long; spikes 3 (2–4), 8–15 mm long, 3–6 mm wide, elliptic to narrowly ovate, the terminal gynaecandrous, often a little larger than the lateral pistillate ones, the lowest one sessile on a peduncle to 3 mm long, the others sessile or subsessile; pistillate scales subequal to or shorter than the perigynia, black or blackish purple, with white hyaline margins, the midrib sometimes greenish or pale. **PERIGYNIA** 2.7–3.5 mm long, 1.6–2.1 mm wide, elliptic to nearly orbicular, papillate (at least apically), black or blackish purple throughout; stigmas 3; achenes trigonous, sessile or nearly so. High mountains in fell-fields, dry and wet meadows, and alpine tundra. COLO: LPl. 3740 m (12265′). Flowering: Jul–Aug. Alaska and Northwest Territories south to California and Arizona. Known from one collection west of Trimble Pass near the head of Missouri Gulch, *O'Kane and Jamieson 7964*. *Carex albonigra* appears to be allied to *C. heteroneura* L., but it is consistently smaller, always of high elevations, and does not show the variation found in *C. heteroneura*. The perigynia are always papillate, but only occasionally and randomly papillate in *C. heteroneura*. Although apparently a distinct taxon, *C. albonigra* seems to pass into both *C. heteroneura* and *C. nova*. *Carex heteroneura* is much more common than *C. albonigra* in the Four Corners region.

Carex aquatilis Wahlenb. (living in water) Water sedge. Plants (0.6) 1.5–9.5 dm tall, from thick, long rhizomes; culms tufted or arising singly or few together. **LEAVES** on the lower 1/2 of the culms, the blades 1.5–5.5 mm wide, shorter than to exceeding the culms; bracts leaflike, the lowest shorter than, subequal to, or surpassing the inflorescence. **INFL** 3–19 cm long with 1–2 (3) staminate spikes above 2–3 (5) pistillate spikes, the transitional ones often androgynous; terminal staminate spike 1–3.5 cm long, the first lateral one 0.8–1.8 cm long, the second lateral one, if present, mostly larger than the first; pistillate spikes (0.7) 1.5–4.5 (7) cm long, 3–5 mm wide; pistillate scales blackish or black-purple, shorter to longer and narrower than the perigynia, lanceolate to ovate, acute to acuminate or rather rounded at the tip, entire, the midrib green or pale or colored like the scale, not raised, not excurrent. **PERIGYNIA** numerous, 2–3.3 mm long including the beak, elliptic to obovate, more or less flattened, marginally nerved only, light green or stramineous in age, often speckled or suffused with reddish brown, minutely granular-muricate, the beak 0.1–0.3 mm, entire or oblique but not bidentate or ciliolate; stigmas 2. **ACHENES** lenticular, shorter and usually narrower than the perigynia. $2n = 76$. Wet and boggy meadows, along streams, and margins of ponds and lakes. ARIZ: Apa; COLO: Arc, Con, Hin, LPl, Min, Mon, RGr, SJn. 2500–3750 m (8200–12300′). Flowering: Jun–Aug. Circumboreal, south in North America to California and New Jersey. Highly selected forage species by livestock, elk, and likely moose.

Carex arapahoensis Clokey (of the Arapaho) Arapaho sedge. Plants 12–30 cm tall, densely caespitose; rhizomes lacking or very short. **LEAVES** basal and on the lower culms, the blades 1.5–4 mm wide, flat or nearly so; lowest bract lacking. **INFL** 1–2 cm long; capitate, equaling or surpassing the leaves, the lowest internode of the rachis about 2–6 mm long; spikes usually 3–6, gynaecandrous, 8–12 mm long, sessile, closely aggregated but more or less distinct; pistillate scales equaling or longer than and as wide or wider than the perigynia, brown, with a pale midrib and hyaline margins. **PERIGYNIA** 4.5–5.5 mm long, 2–2.5 mm wide, more or less elliptic to ovate or elliptic-obovate, flattened, with thin winged margins, the wings extending onto the beak, serrulate at least in the distal 1/4–1/2, finely nerved on both sides, dark reddish brown; stigmas 2. **ACHENES** lenticular. Alpine communities. COLO: LPl, Mon, SJn. 3350–3720 m (11000–12200′). Flowering: Jul–Aug. Wyoming, Colorado, and eastern Utah.

Carex atherodes Spreng. (a beard or awn) Awned sedge, wheat sedge. Plants 4–15 dm tall; culms arising singly or few together from robust, creeping rhizomes. **LEAVES** distributed along the culms, the sheaths villous-hirsute, or sometimes only toward the ventral summit, the blades 4–12 mm wide. **INFL** to 45 cm long with well-spaced spikes mostly shorter than the internodes or the upper ones longer; upper 2–4 spikes staminate, 2.5–5 cm long, sessile or nearly so, except the terminal on a peduncle to 3 cm long; lower spikes pistillate or the intermediate ones sometimes androgynous, cylindrical, 5.5–9 cm long, about 10 mm wide, on peduncles 1–11 cm long, the lower ones enveloped in a closed sheath up to about 9 cm long; pistillate scales aristate-awned, the awn 1–5 mm long and subequal to the perigynia, the body pale, rather scarious, narrower than the perigynia. **PERIGYNIA** 7–10 mm long, lanceolate or lance-ovate, inflated below, conspicuously ribbed, ascending, with a bidentate beak, the teeth 1.5–3 mm long and often more or less divergent; stigmas 3. **ACHENES** trigonous, style continuous with, persistent on, and of the same firm texture as the achene, more or less contorted in age. $2n = 74$. Wet, low ground. The one specimen seen (*Heil 13399B* SJNM) was from along the Piedra River Trail. COLO: Arc, Hin. 2590–3375 m (8500–11080′). Flowering: Jun–Jul. Circumboreal, south in America to New York, Colorado, and Oregon.

Carex athrostachya Olney (of a single spike) Slender-beak sedge. Plants 10–52 cm tall, densely caespitose; rhizomes lacking or very short. **LEAVES** basal and on the lower culms, the blades 1–3 mm wide, flat or channeled; lowest bract (at least) subequal to or surpassing the inflorescence, but not uncommonly shorter, or broken and appearing shorter. **INFL** 1–2.5 cm long, 8–15 mm wide, capitate, usually exserted beyond the leaves, the lowest internode 1–3.5 mm long; spikes about 3–15, gynaecandrous, 5–10 mm long, about 5 mm wide, sessile, tightly aggregated, much longer than and concealing the internodes of the rachis, not, or hardly, evident without dissecting the inflorescence or the lower ones conspicuous; pistillate scales subequal to and almost as wide as the perigynia, brown with a green or pale midstripe and with or without narrow hyaline margins. **PERIGYNIA** 3.5–4.5 mm long, 1–1.5 mm wide, lanceolate or narrowly ovate, flattened, with winged margins, slightly nerved dorsally and sometimes ventrally, greenish, serrulate at least in the distal 1/2, tapered into a more or less terete, often brownish beak; stigmas 2. **ACHENES** lenticular. Drying mud of ephemeral pools, margins of ponds and lakes, and aspen-spruce, forb-grass, and wet meadow communities. ARIZ: Apa; COLO: Arc, Hin, Min, SJn; NMEX: McK, SJn; UTAH. 2470–2805 m (8100–9200′). Flowering: Jun–Aug. Alaska to Saskatchewan, south to California and Colorado.

Carex atrosquama Mack. (dark scale) Dark-scale sedge. Similar to *C. heteroneura*, but with **PERIGYNIA** lanceolate or narrowly elliptic, not flattened, olive-green, becoming yellow-brown at maturity, seldom suffused with dark color except for the beak, not much wider than the mature achene. [*C. atrata* L. subsp. *atrosquama* (Mack.) Hultén]. Mountain meadows and alpine slopes. COLO: LPl, SJn. 2525–3635 m (8000–12000′). Flowering: Jul–Aug. British Columbia to Montana and south to Colorado.

Carex aurea Nutt. (golden) Golden sedge. Plants 4–40 (65) cm tall; culms tufted or arising singly or few together from slender pale rhizomes. **LEAVES** borne on the lower culms, the blades 1–4 mm wide, flat, surpassing or shorter than the inflorescence; bracts leaflike, with sheathing bases, some usually equaling or exceeding the inflorescence, the lowest with a closed sheath 2–6 mm long or longer. **INFL** with terminal spike staminate or often gynaecandrous, rarely androgynous, 0.4–1.5 (2) cm long, 2–3 mm wide, with peduncles 0.4–2 cm long; lateral spikelets 1–3 (4), pistillate, 0.5–2.5 cm long, 3–5 mm wide, usually well spaced and pedunculate; peduncles filiform or capillary, commonly 0.3–7 cm long, or the lowest one arising at or near the culm base and to 11 cm long, this often enclosed in a long sheath; pistillate scales shorter than the perigynia, greenish or brown with a green or pale center. **PERIGYNIA** 4–18, 1.7–3 mm long, ellipsoid to obovoid-globose, pale green and whitish papillate when young, sometimes turning golden and somewhat fleshy in age; stigmas 2, rarely 3. **ACHENES** lenticular. 2*n* = 52. [*C. garberi* Fernald var. *bifaria* Fernald; *C. hassei* L. H. Bailey]. Around seeps and springs, along streams, wet meadows, and hanging gardens. ARIZ: Apa, Nav; COLO: Arc, Hin, LPl, Mon, SJn; NMEX: McK, RAr, SJn; UTAH. 1110–3445 m (3650–11300′). Flowering: Mar–Jul. Alaska to Newfoundland south to New Mexico and Pennsylvania. Tall specimens from the Canyonlands of the Colorado River are somewhat similar in appearance to *C. aquatilis*. Such specimens might be referred to as *C. hassei*. However, the numerous specimens seen show a continuum in size from small to quite large, and the color (golden versus green) of the perigynia seems poorly correlated with any other feature including habitat (Cronquist et al. 1977).

Carex bebbii (L. H. Bailey) Olney ex Fernald (for M. S. Bebb) Bebb sedge. Plants 3–7 dm tall, densely tufted; rhizomes lacking. **LEAVES** borne on the lower culms, the lower reduced to bladeless sheaths or with short blades, the upper ones with blades 2–4 mm wide. **INFL** 1–3 cm long, 5–12 mm wide, subequal to or surpassing the leaves, without a subtending bract, the lowest internode of the rachis 3–4 mm long; spikes usually 4–12, gynaecandrous, 5–9 mm long, sessile, closely aggregated but more or less evident without dissecting the inflorescence, equal to or to 3 times as long as the rachis internodes; pistillate scales shorter and narrower than the perigynia, stramineous to light brown, with a greenish or pale midrib, the margins hyaline or not. **PERIGYNIA** 2.3–3.7 mm long, 1–1.2 mm wide, pale green to stramineous, flattened, with thin winged margins, the wings extending to near the ill-defined, flattened beak apex, serrulate, densely crowded in the spike and spreading-ascending; stigmas 2. **ACHENES** lenticular. 2*n* = 68. Local in moist, swampy meadows. COLO: Hin, LPl, Min, SJn. 2440–3050 m (8000–10000′). Flowering: Jun–Jul. British Columbia to Newfoundland and south to Oregon, Colorado, and New Jersey.

Carex bella L. H. Bailey (beautiful) Beautiful sedge. Plants 20–68 cm tall, tufted, rhizomes lacking or short. **LEAVES** basal and on the lower culms, the blades 1–5 mm wide, flat; lower bract usually leaflike, typically shorter than or subequal to the inflorescence, or infrequently to 2.5 cm longer than the inflorescence, with a sheathing base about 3–4 mm long, the scarious margins of the sheath free or nearly so. **INFL** somewhat flexuous and tending to nod, 4–7 cm long or longer; spikes (2) 3–4, commonly gynaecandrous, pedunculate, more or less cylindrical, 1–3.2 cm long, 3–6 mm wide, the uppermost usually with a conspicuous but short staminate portion, subsessile to peduncled; the lateral ones usually with only a few stamens and infrequently all pistillate; peduncles slender-flexuous, the lower ones shorter to longer than the spikes, the lowest one rarely arising near the base of the culm; pistillate scales shorter than the perigynia, blackish or black-purple, the pale midrib obscure or slightly keeled, especially apically and sometimes excurrent as a mucro. **PERIGYNIA** 2.8–4 mm long, broadly oval to oblong-oval, flat but swollen by the ripening achene, green and strongly contrasting with the dark scales; stigmas 3. **ACHENES** trigonous. Krummholz meadows, subalpine meadows, stream sides in spruce-fir forest, clearcuts and other openings in forests. COLO: Arc, Con, Hin, LPl, Min, Mon, RGr, SJn; UTAH. 2620–3610 m (8600–11850′). Flowering: Jul–Aug. Colorado and South Dakota, south to Arizona, New Mexico, and Mexico.

Carex bipartita All. (two parts) Bipartite sedge. Plants 5–20 cm tall, tufted, rhizomes lacking or short. **LEAVES** basal and on the lower culm, the blades 0.5–2 mm wide; lowest bract lacking or shorter than the first spike. **INFL** 1–1.5 cm long, 0.5–1 cm wide, surpassing the leaves; spikes 2–4, gynaecandrous, 0.5–1 cm long, 3–5 mm wide, sessile, approximate, longer than the internodes of the rachis, but evident in the inflorescence; pistillate scales subequal to and as wide as the perigynia, largely concealing them, except for the beak, brown to blackish purple-brown, with hyaline margins and usually a pale midstripe. **PERIGYNIA** 2.4–3.4 mm long, elliptic to obovate, more or less concavely tapered to a short beak to 0.7 mm long, the dorsal suture extending from the beak to the distal portion of the body, more or less flattened but with rounded margins, finely several-nerved on each side, yellow-green to brown-green with the nerves often brown-red, finely granular-roughened, ascending, about 10–20 per spike; stigmas 2. **ACHENES** lenticular. [*C. lachenalii* Schkuhr]. One specimen seen (*C. Holmes 877a*, SJNM) is from Hermosa Creek, Cascade Divide. Meadows and swamps from spruce-fir to alpine communities. COLO: LPl, SJn. 3595–3810 m (11800–12500′). Flowering: Jul–Aug. Alaska to Greenland south to Utah and Colorado in the Rocky Mountains and to New Brunswick. With brownish scales, reddish-nerved perigynia, approximate spikes, and alpine habitat, plants of *C. bipartita* are more like those of *C. praeceptorum* than they are like those of *C. canescens*.

Carex buxbaumii Wahlenb. (for Johann Christian Buxbaum, 1693–1730, professor of botany) Buxbaum sedge. Plants 3–5.5 dm tall, the culms arising singly or 2–3 together and well spaced on long-creeping rhizomes. **LEAVES** all cauline, borne on the lower culms, the lower ones reduced to bladeless sheaths, the blades 1–3 mm wide; lowest bract leaflike, shorter than to exceeding the inflorescence by 2 cm. **INFL** 3.5–6.5 cm long, mostly surpassing the leaves; spikes 3–4, 0.7–2.2 cm long, 4–8 mm wide, ovoid to nearly cylindrical, erect or strongly ascending, the terminal one gynaecandrous, longer than its peduncle, the lower ones pistillate, sessile or longer than their peduncles, each slightly shorter or exceeding the internode it subtends; pistillate scales conspicuously aristate, the awn-point about 1–2 mm long, the body usually shorter than the perigynia, but with the awn exceeding it, brownish purple or blackish purple, with a greenish midrib. **PERIGYNIA** 2.5–4 mm long, elliptic or elliptic-obovate, faintly nerved; stigmas 3. **ACHENES** trigonous. The two specimens seen (*K. Heil 14073, S. L. O'Kane 4952*, both SJNM) were from peaty wetlands at Andrews Lake. COLO: SJn. 3280 m (10760′). Flowering: Jul–Aug. Circumboreal, south in America to California and North Carolina.

Carex canescens L. (grayed) Silvery sedge. Plants 18–50 cm tall, densely caespitose; rhizomes lacking or very short. **LEAVES** basal and on the lower culms, the blades 1.5–3.1 mm wide; lowest bract lacking or mostly shorter than the first spike and never exceeding the second one. **INFL** 1.4–5.2 cm long, 0.5–1 cm wide, spikelike, equaling or surpassing the leaves; spikes 4–8, androgynous, 0.5–1.4 cm long, 2.5–5 mm wide, sessile, more or less cylindrical, greenish or silver, approximate or the lower ones spaced, equal to or longer than the internodes or the lowermost rarely shorter; pistillate scales usually conspicuously shorter than the perigynia, pale green or silvery, mostly hyaline except for the green midstripe, sometimes pale brown in age. **PERIGYNIA** 1.8–2.5 mm long, 1–1.2 mm wide, elliptic to elliptic-ovate, more or less compressed but with rounded (wingless) margins, greenish, yellow-green, or grayish green, sometimes brownish toward the beak, finely nerved on both sides, the nerves green or rarely pale brown, (10) 15–32 per spike, quite noticeably granular-roughened distally, very short-beaked, the beak granular-roughened and often serrulate (use high magnification), the dorsal suture evident only on the beak, if at all; stigmas 2. **ACHENES** lenticular. $n = 28, 36$. Wet meadows and other riparian or lacustrine communities. COLO: Arc, Con, Hin, LPl, SJn. 2925–3735 m (9600–12250′). Flowering: Jul–Aug. Circumboreal, south in America to California and Virginia.

Carex capillaris L. (hairlike) Hair sedge. Plants 5–70 cm tall, densely caespitose from short rhizomes; culms slender, rather weak. **LEAVES** basal and on the lower culm, or 1 above the middle, the blades 1–3 mm wide; lowest bract foliose, shorter than the inflorescence, strongly sheathing, the sheath closed for 9–20 mm. **INFL** 5–30 cm long, surpassing the leaves; terminal spike staminate, rarely gynaecandrous, 3–9 mm long, borne on a capillary peduncle 1–5 cm long; lateral spikes pistillate, 6–11 mm long, with 5–25 perigynia, borne on capillary spreading to drooping peduncles to 3.5 cm long, each peduncle enveloped basally by the closed sheath of the subtending bract, the lowest spike and peduncle together much shorter than the accompanying internode of the rachis; pistillate scales 1/2–3/4 as long as the perigynia, with broad hyaline margins and greenish centers or greenish midstripe flanked by light reddish brown stripes. **PERIGYNIA** 2.4–3.3 mm long, more or less elliptic or lance-ovate, more or less rounded, the marginal nerves rather prominent, green at first, turning a glossy brown; stigmas 3. **ACHENES** trigonous, not filling the distal part of the perigynia. $2n = 50, 52, 54$. Moist and wet places, often in shade. COLO: LPl, SJn. 2670–3655 m (8760–12000′). Flowering: Jul–Aug. Circumboreal and south in America to New Mexico and New York.

Carex capitata L. (flowers form in a headlike cluster) Capitate sedge. Plants 5–20 cm tall, with culms closely spaced on short rhizomes. **LEAVES** borne on the lower 1/4 of the culms, the lower ones often reduced to bladeless sheaths, the upper ones with blades to about 1 mm wide, involute, filiform; bract lacking. **INFL** a solitary spike, erect, androgynous, 4–10 mm long, 3–7 mm wide, with 6–25 perigynia, the staminate short or elongate; pistillate scales shorter and narrower than the perigynia, dark brown with broad hyaline margins. **PERIGYNIA** 2–3.5 mm long, ovate to suborbicular, conspicuously beaked; rachilla about reaching the beak of the perigynia; stigmas 3. **ACHENES** trigonous, not filling the perigynia. Subalpine and alpine peat fens. COLO: SJn. 3625 m (11900′). Flowering: Jul–Aug. The single specimen seen (*G. Rink 3668*, SJNM) is from the north side of Vestal Peak. British Columbia and Alberta south to Utah and Colorado.

Carex curatorum Stacey (reference uncertain) Canyonlands sedge. Plants caespitose, shortly if at all rhizomatous, dioecious, 2–4.5 dm tall. **LEAVES** with sheaths puberulent, hyaline, purplish black to chestnut at culm base, lower leaves reduced to bladeless sheaths, upper leaves with blades 1–4 mm wide, flat; lowermost bract of inflorescence mostly lacking or rarely to 1.5 cm long, the sheath short, open. **INFL** a solitary spike, erect or nearly so, the staminate 2.5–3 cm long, 3 mm wide, the pistillate 2.5–4.5 cm long, 3–5 mm wide; pistillate scales oblong or lance-ovate, shorter than the perigynia, reddish purple, with lighter center and narrow hyaline margins. **PERIGYNIA** 3.2–4.5 mm long, (1.8) 2–2.3 (2.7) mm wide, broadly elliptic or obovate to nearly orbicular, abruptly contracted into a bidentate hyaline beak, blackish purple distally, greenish below, conspicuously pubescent with translucent hairs; stigmas 3. **ACHENES** trigonous. Hanging gardens and canyons along the Colorado and San Juan rivers. UTAH. 1100–1340 m (3600–4400′). Flowering: Jun–Jul. Southern Utah and northern Arizona.

Carex deweyana Schwein. (for Chester Dewey) Dewey sedge. Culms 40–70 cm tall, tufted; rhizomes lacking or short. **LEAVES** borne on the lower 1/3 of culms, the blades 1–3 mm wide, flat; lowest bract lacking or present and shorter than or subequal to the inflorescence. **INFL** 4–6.5 cm long, about 0.5–1 cm wide, spikelike, equaling the leaves or surpassing them; spikes 4–6 (10), gynaecandrous, 0.7–1.6 cm long, 4–6 mm wide, pale green, sessile, the lowest one usually remote, subequal to or to 2.5 times shorter than the internode, the upper ones remote or approximate, but individually evident without dissecting the inflorescence; staminate flowers few, their scales rather closely appressed to a peduncle-like portion of the rachis; pistillate scales as long as the body of the perigynia, pale green or slightly stramineous, often largely hyaline, with a darker green midrib, this sometimes excurrent. **PERIGYNIA** 3.8–4.6 mm long, 1–1.2 mm wide, lanceolate to lance-elliptic, pale green, the margins not winged or scarcely so distally, the marginal nerves prominent, raised, sometimes slightly extended onto the ventral surface, the dorsal ribs conspicuous, the ventral lacking to obscure, the body tapering to a serrulate bidentate beak with teeth to 0.4 mm long; stigmas 2. **ACHENES** lenticular, 1.5–2 mm long. *Populus-Salix* and other riparian communities. COLO: Arc, Con, Hin, Min. 1830–3385 m (6000–11100′). Flowering: Jun–Aug. Alaska to Newfoundland, south to California, Colorado, and Pennsylvania.

Carex diandra Schrank (with 2 stamens) Lesser panicled sedge. Plants 3–10 cm tall, caespitose, with short rhizomes. **LEAVES** borne on the lower culm, the lowermost reduced to bladeless sheaths, the upper with elongate blades 1–2.5 (3) mm wide, these subequal to or surpassing the inflorescence, flat or nearly so. **INFL** 1–6 cm long, 7–10 mm wide, compound but spikelike; spikes crowded, scarcely evident without dissecting the head, about 3–5 mm long, few-flowered, androgynous, pistillate scales rather hyaline, brownish or stramineous, the midrib firmer and sometimes projected as an aristate awn. **PERIGYNIA** 2.4–3 mm long, lance-ovate or lance-truncate, short-stipitate, gradually to abruptly tapered to a serrulate beak, the dorsal suture sometimes extended into a membranous keel, this usually conspicuous toward the apex, the body thick-walled, dark brown, shiny; stigmas 2. **ACHENES** lenticular. The one specimen seen (*Heil 19507* SJNM) is from a riparian community at Pastorius Reservoir. COLO: LPl. 2105 m (6900′). Flowering: Jun–Aug. Circumboreal, south in America to California and Pennsylvania.

Carex disperma Dewey (two-seeded) Softleaved sedge. Culms 10–45 cm long, densely tufted, from short slender rhizomes, rather flaccid or flexuous. **LEAVES** basal and on the lower culms, the blades 0.5–2 mm wide. **INFL** 0.8–4 cm long, 3–5 mm wide, usually interrupted, spikelike, sometimes subtended by a narrow bract, this shorter than the inflorescence; spikes sessile, 3–5, to about 5 mm long, about as wide as long, androgynous, with 1–2 (3) staminate flowers and 1–3 perigynia, the staminate flowers inconspicuous and the spikes sometimes appearing pistillate; pistillate scales shorter than to equaling the perigynia, pale green, scarious. **PERIGYNIA** 2–3 mm long, elliptic, abruptly contracted to a very short beak; stigmas 2. **ACHENES** lenticular, filling the perigynium; style semipersistent on the achene. $2n = 70$. Seeps and springs and along streams, mostly in shade of woods. COLO: Arc, Hin, LPl, Min, Mon, SJn. 3080–3340 m (10100–10950′). Flowering: Jun–Jul. Circumboreal, south in America to New Mexico and New Jersey.

Carex douglasii Boott (for David Douglas, botanical explorer) Douglas sedge, dioecious sedge. Plants 8–28 cm tall, mostly dioecious; culms arising singly or few together from long, slender, brownish rhizomes less than 2 mm in diameter. **LEAVES** basal and on the lower culms, the blades 0.5–2 mm wide; lowest bract lacking, or sometimes present and foliose. **INFL** 1.5–5.5 cm long, (5) 10–20 mm wide, spicate or capitate, exceeded by the leaves or surpassing them; spikes about 10–25, generally all staminate or all pistillate, rarely androgynous, densely crowded, over twice as long as the internodes of the rachis, more or less evident without dissecting the head or the upper ones sometimes not, the lower 0.8–2 cm long, ovoid or more commonly the larger ones elongate and nearly linear or linear-elliptic; pistillate scales longer than the perigynia and concealing them, these and usually the staminate ones with a sometimes excurrent, light green midstripe, this flanked on both sides by a narrow stramineous or pale brown stripe and then the broad hyaline margins very conspicuous in the spikes. **PERIGYNIA** 2.2–4.7 mm long, 1.1–1.8 mm wide, ovate to elliptic, brownish, the marginal nerves prominent and distended toward the ventral surface, otherwise obscurely nerved; beak of the perigynium 1–1.5 mm long, obliquely cleft, often hyaline, the dorsal suture often conspicuous and with a hyaline flap, this sometimes encroaching onto the body of the perigynium; stigmas 2. **ACHENES** lenticular. $2n = 60$. Alkaline sites, foothills and lower mountains. COLO: Arc, LPl, Min, SJn. 1830–2440 m (6000–8000′). Flowering: May–Jun. British Columbia to Manitoba and south to California and Iowa.

Carex duriuscula C. A. Mey. (somewhat hard) Narrowleaf sedge. Plants 8–28 cm tall; culms arising singly or few together from slender, brownish or pale-colored rhizomes less than 2 mm in diameter, and with numerous fibrous roots. **LEAVES** basal and on the lower culm, the blades 0.5–1.5 mm wide; lowest bract lacking. **INFL** 0.8–2 cm long, 5–10 mm wide, spicate or capitate; spikes about 3–10, sessile, androgynous or rarely all unisexual, 4–9 mm long, longer than the internodes of the rachis and closely aggregated, discernible or not without dissecting the inflorescence;

pistillate scales subequal to or slightly longer than the perigynia, brownish, with a dark green or brownish often keeled midrib, and with narrow or broad scarious margins. **PERIGYNIA** 2.6–3.5 mm long, rather narrowly fusiform-elliptic to broadly elliptic-ovate, brownish, the marginal nerves prominent, otherwise obscurely nerved; beak of the perigynium obliquely cleft, 0.5–1.2 mm long, the dorsal suture inconspicuous; stigmas 2. **ACHENES** lenticular. $2n = 60$. [*C. eleocharis* L. H. Bailey; *C. stenophylla* Wahlenb.]. Piñon-juniper, ponderosa pine, and Gambel's oak communities. ARIZ: Apa; COLO: Arc, SJn; NMEX: McK, SJn; UTAH. 2255–2725 m (7400–8940′). Flowering: May–Jun. Mostly of the Great Plains, but from Iowa to California and south to Arizona.

Carex ebenea Rydb. (of ebony) Ebony sedge. Plants 19–59 cm tall, densely caespitose; rhizomes lacking or short. **LEAVES** basal and on the lower culms, the blades 1–4.5 mm wide, flat or nearly so; lowest bract usually lacking, or if present, mostly much shorter than the inflorescence. **INFL** a head 1.3–2.8 cm long and about as wide, or more commonly slightly narrower (in unpressed specimens), ovate or broadly elliptic, truncate or more commonly cuneate at the base, the internodes of the rachis 0.5–2 mm long; spikes (3) 6–12, gynaecandrous, commonly 9–15 mm long, averaging over 10 mm long, sessile, densely congested, usually slightly more discernible than in *C. microptera*, about 5–10 times longer than the internodes of the rachis; pistillate scales slightly shorter and usually narrower (at least distally) than the perigynia, dark brown to blackish brown, with or without hyaline margins. **PERIGYNIA** (4.8) 5.2–7.1 mm long, 0.9–2 mm wide, the better developed ones commonly 4–6.4 times longer than wide, brown to dark brown in the center and commonly with narrow greenish wings, faintly nerved on both sides, serrulate except near the apex of the prominent slender beak; stigmas 2. **ACHENES** lenticular, 1.5–1.8 mm long. Subalpine and alpine meadows, riparian communities and talus. COLO: Arc, Con, Hin, LPl, Min, Mon, RGr, SJn; UTAH. 2530–3800 m (8300–12450′). Flowering: Jul–Aug. Wyoming to Arizona and New Mexico. Plants of *C. ebenea* are more similar to *C. haydeniana* than to *C. microptera* (q.v.) especially in the darker-colored heads, fewer spikes on the average, and somewhat larger perigynia. They are also more common at higher elevations than is *C. microptera*. However, the perigynia in *C. microptera* are sometimes quite narrow and, except for mostly being shorter, could be mistaken for those of *C. ebenea*. Although intergrading specimens are sometimes encountered, most specimens seem reasonably assigned to one of the taxa.

Carex echinata Murray (spiny) Boreal sedge, star sedge. Plants 1–3 dm tall, densely caespitose; rhizomes lacking or very short. **LEAVES** borne on the lower 1/3 of the culms, the blades 1–2.2 mm wide, flat; lowest bract lacking or usually shorter than the first spike. **INFL** 1–2.3 cm long, about 7 mm wide, exceeded by the leaves or surpassing them; terminal spike gynaecandrous, usually with more staminate flowers than pistillate ones or sometimes staminate, 7–10 mm long, often clavate with a slender staminate portion and wider pistillate portion; lateral spikes (1) 2–3, gynaecandrous, with only a few staminate flowers and 5–10 perigynia, 5–7 mm long and nearly as wide, subequal to or a little longer than the internodes, rather closely spaced but readily evident in the spike; pistillate scales shorter than the perigynia, pale green or pale stramineous with a green midstripe and broad hyaline margins. **PERIGYNIA** widely spreading and conspicuously beaked and giving a bristly appearance to the spikes, 2.8–3.5 (4) mm long, including the prominent serrulate beak 1.1–1.6 mm long, this 1/2 or more as long as the body, neither winged nor thin-margined but the marginal nerves prominent and sometimes ridged, these displaced toward the flattened ventral surface, the dorsal surface convex; stigmas 2. **ACHENES** lenticular. [*Carex muricata* L.; *C. angustior* Mack.]. Three specimens were seen from the edge of the San Juan drainage. One (*Heil 10346*, SJNM) is from 1.25 miles northwest of Elwood Pass. Another (*Heil, Mietty, King 13446*, SJNM) is from Little Sand Creek. The third location is near a small lake on Granite Bench (*Heil, Mietty 22316*, SJNM). Infrequent in swampy meadows. COLO: Arc, Hin, RGr. 2680–3320 m (8800–10900′). Flowering: Jun–Jul. British Columbia to Newfoundland and south to California and North Carolina.

Carex egglestonii Mack. (for Willard W. Eggleston, agriculturalist and economic botanist) Eggleston sedge. Plants (1.5) 3–9 dm tall, caespitose; rhizomes lacking or very short. **LEAVES** borne on the lower culms, the blades 1.5–5 mm wide, flat. **INFL** 1.5–2.7 cm long, 1–3 cm wide, capitate, with the lowest internode to about 5 mm long; surpassing the leaves; without a subtending bract, or this shorter than or rarely equaling the inflorescence; spikes 3–6, gynaecandrous, 10–17 mm long, 7–12 mm wide, ovate to nearly globose, sessile, closely aggregated, but more or less evident without dissecting the head, to about 3 or more times longer than the rachis internodes, with up to 45 or more perigynia; pistillate scales conspicuously shorter and narrower than the perigynia, brownish, with a green or pale brown midstripe and hyaline margins, acute. **PERIGYNIA** (5) 6–8 mm long, 2.5–3.5 mm wide, ovate, elliptic, or lanceolate, greenish or brownish, strongly flattened, with thin, broadly winged margins, finely serrulate distally and frequently below, gradually or rather abruptly tapered to a bidentate, more or less flattened, winged beak; anthers 2.5–3.7 mm long; stigmas 2. **ACHENES** lenticular. Engelmann spruce parklands. COLO: Hin, LPl, Mon, SJn; UTAH. 2880–2895 m (9450–10960′). Flowering: Jun–Aug. Wyoming, Utah, and Colorado. Apparently rarely collected in the Four Corners region.

Carex elynoides Holm (reference uncertain) Kobresia sedge, blackroot sedge. Plants 5–16 cm tall, densely caespitose, lacking rhizomes, the culms and leaves fasciculate. **LEAVES** basal and on the lower 1/4 of the rounded culms, mostly equal to, or surpassed by, the inflorescence, the blades 0.3–0.5 mm wide, strongly folded or involute and nearly terete; bract lacking. **INFL** a solitary spike, erect, androgynous, 1–1.8 cm long, the staminate part 3–7 mm long, shorter to longer than the pistillate part; pistillate scales wider and mostly longer than the perigynia, brown to dark brown, the centers usually not noticeably green or pale, the margins hyaline. **PERIGYNIA** usually 4–12, 2.5–4.5 mm long, oblong or ellipsoid, filled by the mature achene, glabrous or occasionally very sparingly serrulate-ciliolate near the beak; rachilla subequal to the achene; stigmas 3. **ACHENES** trigonous. Alpine tundra and fell-fields. COLO: Arc, Con, Hin, Min, Mon, SJn. 3550–3900 m (11700–12800′). Flowering: Jul–Aug. Montana to Nevada and Colorado. *Carex elynoides* is closely allied to *C. filifolia*, and except for the glabrous or nearly glabrous perigynia and different habitat, it would be almost impossible to distinguish them. Plants of *C. elynoides* are easily mistaken for *Kobresia myosuroides*, which sometimes grows in the same habitat.

Carex emoryi Dewey (for William H. Emory, engineer and major in the U.S. Army) Early sedge, Emory sedge. Plants 30–80 cm tall, with culms arising singly or few together from long-creeping rhizomes. **LEAVES** with reddish brown sheaths, the blades 3–6 mm wide; ligule truncate or broadly rounded; bracts, at least the lower, more or less equal to the inflorescence. **INFL** of 1–7 linear spikes, the terminal spike staminate, about 2–4 cm long, the lateral spikes pistillate or the upper ones androgynous or sometimes staminate, 2–10 cm long; pistillate scales with a light central area as wide or wider than the outer dark portions. **PERIGYNIA** about 2–3 mm long, ellipsoid, papillose, often deciduous in early summer; beak of perigynium often white throughout, occasionally darker at the apex; stigmas 2. **ACHENES** lenticular. Irrigation canal banks, roadsides, margins of ponds and lakes. COLO: Arc; NMEX: RAr, SJn. 1615–2105 m (5300–6900′). Flowering: Apr–Jun. Wyoming to Ontario and south to New Mexico, Texas, and Virginia.

Carex filifolia Nutt. (threadlike leaves) Threadleaf sedge. Plants 8–38 cm tall, densely caespitose, without rhizomes, the culms and leaves fasciculate. **LEAVES** basal or on the lower 1/4 of the rounded culms, equaling or exceeded by the inflorescence, the blades 0.3–0.5 mm wide, strongly folded to involute and nearly terete; bract lacking. **INFL** a solitary spike, erect, androgynous, 1.3–2.5 cm long, the staminate part usually equaling or longer than the pistillate part, commonly 8–13 mm long; scales light brown or brown with broad white hyaline margins, the pistillate ones wider and longer than the perigynia. **PERIGYNIA** usually 5–15, 3–4.5 mm long, obovoid to ellipsoid, filled by the mature achene, minutely hirtellous at least on the upper 1/2; rachilla subequal to the achene; stigmas 3. **ACHENES** trigonous. $2n = 50$. *Cowania*-juniper, piñon-juniper, ponderosa pine–oak communities. NMEX: McK, SJn. 1920–2225 m (6300–7300′). Flowering: Apr–Jul. Yukon to Manitoba south to California and Texas.

Carex geophila Mack. (earth-loving) White Mountain sedge. Similar to *C. rossii*, with the following differences: **LEAVES** appearing more rigid; base of plants brownish, without or with some reddish color; old sheaths perhaps becoming more fibrillose at the base. **INFL** with bract subtending the upper pistillate and staminate spikes shorter than or slightly longer than the spikes (this feature does not apply to the basally borne spikes); pistillate scales commonly with 1–3 perigynia. Ponderosa pine–oak communities. COLO: LPl; NMEX: McK, RAr, SJn; UTAH. 1890–2650 m (6200–8720′). Flowering: May–Jul. Colorado to New Mexico, Arizona, and Mexico.

Carex geyeri Boott (for Charles A. Geyer, Austrian botanical collector) Elk sedge. Plants 15–30 (50) cm tall, more or less caespitose and with short to elongate rhizomes. **LEAVES** basal and cauline, the lower ones often reduced to bladeless sheaths, the blades 1–3 mm wide, flat or nearly so, evergreen, the tips often turning yellow or brown but the lower parts remaining green through at least one winter; bract lacking or in robust specimens to 1.5 cm long. **INFL** a solitary spike, erect, androgynous, the staminate part 1–2 (3) cm long, linear, the pistillate part usually with 1 or 2 plump perigynia, these proximal to or somewhat remote from the staminate part; scales light brown, with broad hyaline margins, the pistillate ones usually equaling or to 4 mm longer than the perigynia. **PERIGYNIA** 5–6 mm long, 2-ribbed, greenish or light brown, ellipsoid or obovoid, solitary or, if 2, rather remote, with the internode subequal to or longer than the perigynia; rachilla obsolete or less than 1/2 as long as the achene; stigmas 3. **ACHENES** trigonous, filling the perigynium. Ponderosa pine, Douglas-fir–oak, grass-forb, and spruce-fir communities. ARIZ: Apa; COLO: Arc, Dol, Hin, LPl, Min, Mon, SJn, SMg; NMEX: RAr, SJn; UTAH. 2250–3400 m (7380–11200′). Flowering: Jun–Jul. Alberta and British Columbia south to California and Colorado. This winter-green sedge is highly selected by elk and perhaps other ungulates in the late fall and winter.

Carex haydeniana Olney (for Ferdinand V. Hayden, geologist and leader of the Hayden Survey) Cloud sedge. Plants 9–33 cm tall, densely caespitose; rhizomes lacking or very short. **LEAVES** basal and on the lower culm, the blades 1–4 mm wide; lowest bract lacking or, if present, not over 1/2 as long as the inflorescence. **INFL** a head 1–1.5 cm long, ovoid to nearly orbicular in outline, with internodes of the rachis about 0.5–1 mm long, usually truncate at the base, surpassing the leaves; spikes 3–8, gynaecandrous, 7–11 mm long, ovate or nearly so, sessile, densely congested, hardly if at all discernible without dissecting the head, 7–10 times longer than the internodes of the rachis; pistillate scales shorter and narrower than the perigynia, dark brown to blackish, with a lighter brown or green midrib, the margins narrowly hyaline or not. **PERIGYNIA** 4–6 (6.3) mm long, (1.7) 2.2–3 mm wide, 1.6–2.9 times longer than wide, ovate or lance-ovate to elliptic, brown, dark brown, or blackish, nearly as dark as the scales or somewhat lighter and sometimes the margins green, faintly nerved dorsally, nerveless or faintly nerved ventrally, strongly flattened except where distended by the achene, the margins thin and winged, serrulate except on the subterete beak tip; stigmas 2. **ACHENES** lenticular, commonly 1.5–2 mm long. The two specimens seen are from alpine communities. COLO: Arc, LPl, SJn. 3595–4265 m (11800–14000′). Flowering: Jul–Aug. British Columbia and Alberta south to California and Colorado. This species is similar to and probably transitional with *C. microptera*.

Carex heteroneura W. Boott (variously nerved) Blackened sedge. Plants 2.5–10 dm tall; culms tufted; rhizomes lacking or short. **LEAVES** basal and on the lower culms, the blades 1.5–10 mm wide, flat; lowest bract mostly subequal to or shorter than the inflorescence, often leaflike. **INFL** 2–10 cm long; first internode of the rachis (0.5) 1–7 cm long or almost obsolete, shorter than to surpassing the spike; second internode about 1–8 mm long; spikes 3–5, 0.8–2.5 cm long, 4–8 mm wide, narrowly oblong to linear, the terminal one gynaecandrous, the lateral ones pistillate or infrequently gynaecandrous, the lowest one typically on a peduncle to 4.5 cm long, the upper pedunculate or sessile; pistillate scales shorter than to much longer than the perigynia, blackish, dark brownish purple, or blackish purple. **PERIGYNIA** 3–5 mm long, 1.7–4 mm wide, oblong, ovate, or nearly orbicular, olive-green, yellow-brown, or greenish marginally but often suffused with dark purple or purple-black; stigmas 3. **ACHENES** trigonous, commonly short-stipitate. [*C. atrata* L.]. Alpine tundra, wet meadows, krummholz, fell-fields, and talus. COLO: Arc, Con, Hin, LPl, Min, Mon, SJn; UTAH. 3445–4240 m (11300–13900′). Flowering: Jul–Aug. Montana to Nevada and New Mexico. Based on the number of specimens seen (37) this is one of the more common sedges of the mountains of the Colorado part of the San Juan River drainage. All mature specimens seen were rather well placed with **var. *chalciolepis*** (Holm) F. J. Herm. [*C. chalciolepis* Holm; *C. atrata L.* var. *chalciolepis* (Holm) Kük.]. Several specimens from the San Juan Mountains had been identified as *C. albonigra*. For the most part these were immature specimens that were sympatric with mature specimens of *C. heteroneura* var. *chalciolepis*.

Carex hystericina Muhl. ex Willd. (from a porcupine) Bottlebrush sedge. Plants 3–6 (10) dm tall; culms clustered on rhizomes. **LEAVES** basal and on the lower culm, the lowermost leaves sometimes reduced to bladeless sheaths, the upper ones with flat blades 2–9 mm wide. **INFL** 6–10 cm long; terminal spike staminate or rarely gynaecandrous, 2–3 cm long; lateral spikes 1–3, pistillate, (1) 1.5–3 cm long, 12–15 mm wide, ascending to strongly spreading, on slender peduncles to 5 cm long or the upper one sessile or subsessile, the lowest subtended by a leafy bract usually exceeding the inflorescence, this usually sheathless or the sheath closed for up to 14 mm; pistillate scales equaling or much shorter than the perigynia not counting the awn, pale green or straw-colored in age, aristate-awned, the awn 2–6 mm long and serrulate-ciliolate. **PERIGYNIA** 5–7 mm long, pale green, prominently nerved, lanceolate or lance-ovate, slightly inflated, with a conspicuous narrow beak about 2 mm long, strongly spreading and giving a bristly appearance to the spikes; stigmas 3. **ACHENES** trigonous, much smaller than the perigynia; style continuous with, persistent on, and of the same firm texture as the achene, straight or bent in age. $2n = 58$. Moist and wet places. COLO: Arc, LPl, Min; UTAH. 1125–2590 m (3700–8500′). Flowering: Jun–Aug. Washington to New Brunswick and south to California and Virginia.

Carex illota L. H. Bailey (dirty or unwashed) Sheep sedge. Plants 10–42 cm tall, densely tufted; rhizomes lacking or short. **LEAVES** basal and on the lower culm, the blades 0.5–2.8 mm wide, flat; bracts not much larger than the scales. **INFL** 0.8–1.4 cm long, nearly or fully as wide as long, capitate, much surpassing the leaves; spikes 3–6, gynaecandrous, about 4–6 mm long, sessile, closely aggregated and scarcely evident from one another in the dense head, with 5–15 ascending or spreading perigynia; pistillate scales about 1/2–3/4 as long as the perigynia, blackish or black-brown, often with a greenish or yellow-brown midstripe and with or nearly without hyaline margins. **PERIGYNIA** 2.5–3 mm long, 1–1.4 mm wide, lanceolate to narrowly ovate, yellowish or greenish brown to blackish, dorsally and sometimes ventrally nerved, the nerves sometimes reddish, the margins entire, rounded or rather sharp, sometimes with very small wings in the distal portion, gradually to rather abruptly tapered to an ill-defined or conspicuous beak, the beak usually blackish, obliquely cleft, the dorsal suture conspicuous through the length of the beak and onto the distal part of the body, usually with an overlapping involute flap; stigmas 2. **ACHENES** lenticular. Wet and boggy meadows, flooded pond and lake margins, evaporated tarns, and along streams. COLO: Arc, Con, Hin, LPl, Min, RGr, SJn. 2925–3730 m (9600–12250′). Flowering: Jul–Aug. British Columbia to Montana, south to California and Colorado.

Carex inops L. H. Bailey (weak) Sun sedge. Plants 8–23 cm tall from thick rhizomes, reddish or brownish at the base, old leaf bases persistent and more or less forming a thatch. **LEAVES** basal or borne on the lower 1/5 of the culm, the blades equaling or exceeding the inflorescence or somewhat shorter, 1–2 mm wide; lowest bract obsolete to about 1/2 as long as the inflorescence. **INFL** terminal, 1.5–3 cm long, with a terminal staminate spike 10–18 mm long with peduncle 1.2–4.5 cm long, and 1 or 2 sessile, lateral, pistillate spikes 7–10 mm long, with 5–15 perigynial scales reddish brown with a pale, raised midrib, the lower pistillate scales acute or long-acuminate to aristate, the upper ones acute. **PERIGYNIA** pubescent, 2.8–4.6 mm long; stigmas 3. **ACHENES** trigonous. Oak, ponderosa pine, and subalpine communities. COLO: Arc, Hin, LPl, Mon; NMEX: McK, RAr, SJn. 2010–2775 m (6600–9100′). Flowering: May–Jul. British Columbia to Ontario south to New Mexico, Missouri, and Indiana. Plants of the Four Corners belong to **subsp. *heliophila*** (Mack.) Crins (sun-lover) [*C. heliophila* Mack.; C. *pensylvanica* Lam. subsp. *heliophila* (Mack.) W. A. Weber]. Used by the Ramah Navajo as a disinfectant for "eagle infections," and as a gastrointestinal aid to relieve discomfort from overeating.

Carex interior L. H. Bailey (inner) Inland sedge. Plants 17–45 cm tall, densely caespitose; rhizomes lacking. **LEAVES** borne on the lower 1/3 of the culms, the blades 1–2 mm wide. **INFL** 0.8–2 cm long, 4–6 mm wide, included in or exserted beyond the leaves, without a subtending bract; spikes (2) 3–4, sessile, gynaecandrous, the terminal one often clavate with a conspicuous slender staminate portion, rarely all staminate, 7–10 mm long, the lateral ones about 4–6 mm long and about as wide, with only a few staminate flowers and (1) 5–10 perigynia, rather closely spaced but readily discerned, subequal to or a little longer than the internodes of the rachis; pistillate scales shorter than the perigynia, pale green, stramineous or marked with brown, with broad hyaline margins. **PERIGYNIA** widely spreading, beaked but not conspicuously so as in *C. echinata* and the spikes not so bristly, 2.2–3.2 mm long including the shallowly bidentate serrulate beak, this 1/4–1/2 as long as the body, not winged or thin-margined, the marginal nerves prominent and slightly ridged, these displaced toward the flattened ventral surface, the dorsal surface convex; stigmas 2. **ACHENES** lenticular. $2n = 54$. Wet and moist sites in mountains. ARIZ: Apa; COLO: Arc, LPl, SJn. 2135–3350 m (7000–11000′). Flowering: Jun–Aug. British Columbia to Labrador and south to Mexico and Pennsylvania.

Carex lenticularis Michx. (lens-shaped) Kellogg sedge. Plants 15–61 cm tall, caespitose from fibrous roots, the larger roots covered with feltlike yellowish or yellow-brown hairs; rhizomes lacking or short. **LEAVES** mostly on the lower 1/4 of the culms, the blades 1–2.6 mm wide; lowest bract shorter than or exceeding the inflorescence, leaflike. **INFL** 4–13 cm long, with a solitary (or 2) staminate spike(s)1–3 cm long above 2–4 cylindrical pistillate ones, these 0.6–4 cm long and 3–4 mm wide, the uppermost pistillate one sometimes reduced; pistillate scales shorter and narrower than the perigynia, blackish or black-purple with a green midrib, this not extending to the tip of the scale and neither excurrent nor ridged. **PERIGYNIA** 2–3 mm long, lanceolate, elliptic to ovate, more or less faintly nerved dorsally and ventrally, deciduous at maturity, usually with a well-defined stipe 0.1–0.4 mm long, green except for the often blackish beak, this 0.1–0.4 mm long, entire; stigmas 2. **ACHENES** lenticular. $n = 34, 44$. [*C. kelloggii* W. Boott]. Riparian communities. COLO; Arc, Mon, SJn. 1830–2440 m (6000–8000′). Flowering: Jun–Aug. Alaska to Labrador, south to California and Michigan. Our plants belong to **var. *lipocarpa*** (Holm) L. A. Standl. (fat-fruited).

Carex limosa L. (of mud) Mud sedge, quaking bog sedge. Plants 13–26 cm tall, often glaucous; culms arising singly or 2–3 together from slender rhizomes; roots covered with a yellowish or yellow-brown feltlike tomentum. **LEAVES** mostly on the lower 1/2 of the culms, the lower ones usually reduced to bladeless sheaths, the upper ones with blades 0.5–1.5 mm wide, more or less flat, the midrib strongly keeled below, often sharply serrulate. **INFL** mostly 3.5–6.5 cm long, shorter or longer than the subtending lowest bract; terminal spike staminate, (1) 1.5–2.1 cm long; lateral spikes 1 or 2, pistillate or occasionally with 1–2 apical staminate flowers, 8–18 mm long, spreading or drooping, on flexuous peduncles 5–22 mm long, or rarely the lowest spike on an elongate peduncle originating near the culm base; pistillate scales shorter than to slightly longer than and nearly as wide as the perigynia, greenish brown to dark reddish brown, with a green center, the margins not or scarcely hyaline, obtuse or acute to cuspidate. **PERIGYNIA** 2.9–3.5 mm long, 1.2–1.9 mm wide, substipitate, with a very short, more or less conspicuous beak, pale green and densely white-granular (the grains visible at high magnification); stigmas 3. **ACHENES** trigonous. $2n = 56, 64$. Sphagnum bogs and wet meadows. COLO: SJn (Johnston 2001). 3415–3535 m (11200–11600′). Flowering: Jul–Aug. Circumboreal, south in America to California and Colorado.

Carex magellanica Lam. (from Magellan) Poor sedge, boreal bog sedge. Plants 15–30 cm tall. Culms arising singly or few together from short to long rhizomes; roots covered with a yellowish or yellow-brown feltlike tomentum. **LEAVES** basal and cauline, mostly bearing blades, the blades usually 1–2 mm wide, flat, sometimes serrulate-ciliolate, especially apically, the midrib more or less keeled below, this sometimes serrulate; lowest bract shorter than to longer than the inflorescence, this 2–5 cm long; terminal spike staminate or sometimes with 1 to several perigynia at or near the apex, 0.9–1.2 cm long; lateral spikes 1–3, pistillate or often with 1–3 staminate flowers at the base, 0.7–1.6 cm long, spreading-ascending to drooping, on slender flexible peduncles 1–2.5 cm long; pistillate scales longer and slightly narrower than the perigynia, acuminate, brown or dark purple-brown. **PERIGYNIA** 2.2–2.9 (3.5) mm long, 1.2–1.7 mm wide, sessile or substipitate, with the beak nearly obsolete, densely white-granular on a pale green background; stigmas 3. **ACHENES** trigonous. n = 29. [*C. paupercula* Michx.]. Infrequent to rare in bogs and wet meadows. COLO: Hin, LPl, SJn (Johnston 2001). 3320–3550 m (10890–11650′). Flowering: Jul–Aug. Circumboreal, south in America in the Rocky Mountains to Utah and Colorado and to the New England states. *Carex magellanica* is similar to *C. limosa*.

Carex microglochin Wahlenb. (small barb) Subulate sedge, few-seeded bog sedge. Plants (5) 15–25 cm tall; culms arising few together from slender rhizomes, the few leaves mostly toward the base. **LEAVES** with blades about 0.5–0.8 mm wide, channeled or nearly terete; bract lacking. **INFL** a solitary spike, androgynous, about 0.5–1 cm long, 3–5 mm wide, erect; scales brownish or tinged with purple, the pistillate ones almost 1/2 as long as the perigynia, quickly deciduous. **PERIGYNIA** awl-shaped or linear-lanceolate, with the rachilla conspicuously exserted from the orifice, 3–5 mm long including the rachilla, about 1 mm wide at the base, strongly spreading, soon deflexed and deciduous; stigmas 3. **ACHENES** elongate, nearly cylindrical, or obtusely trigonous. $2n$ = 58. Calcareous habitats in the alpine. COLO: SJn. 3290 m (10800′). Flowering: Jul–Aug. Greenland to Alaska and south in the Rocky Mountains to Utah and Colorado, and to Quebec.

Carex microptera Mack. (small wing) Small-wing sedge. Plants 2–8 dm tall, densely caespitose; rhizomes lacking or very short. **LEAVES** basal and on the lower culm, the blades 1–5 mm wide, flat or nearly so. **INFL** a head, 1.2–2.5 cm long, and as wide, orbicular or nearly so, usually truncate at the base, the internodes of the rachis about 0.5–2 mm long or the lower one to 3 (4) mm long, without a subtending leafy bract or this rarely to about 1/2 as long as the inflorescence; spikes (2) 5–15 (21), gynaecandrous, commonly 5–11 mm long, sessile, densely congested, hardly if at all discernible without dissecting the head, much longer than the internodes of the rachis; pistillate scales shorter than and narrower than the perigynia, brownish. **PERIGYNIA** (2.5) 3–5 mm long, 1.3–2.2 mm wide, 2–3 (4) times longer than wide, greenish or brownish to dark brown in age, with a green or pale midstripe, lightly or obscurely nerved on both sides, strongly flattened, except where distended by the mature achene, the margins thin, winged, serrulate except on the more or less terete beak-tip, not hyaline or very narrowly so; stigmas 2. **ACHENES** lenticular, commonly 1.1–1.5 mm long. [*C. limnophila* F. J. Herm.; *C. macloviana* d'Urv. var. *microptera* (Mack.) B. Boivin; *C. festivella* Mack.]. Moist and wet places in many montane plant communities. ARIZ: Apa; COLO: Arc, Dol, Hin, LPl, Min, Mon, SJn; NMEX: McK, RAr, SJn; UTAH. 2050–3400 m (6760–11150′). Flowering: Jun–Aug. British Columbia to Saskatchewan, south to California and New Mexico. This is one of the most common and widespread sedges in the area. It is often confused with some other members of section *Ovales*. *Carex microptera* forms a series of intergrades with *C. ebenea*, *C. haydeniana*, and *C. pachystachya*, and all of the above listed species except *C. ebenea* have been reduced previously to varietal status under *C. macloviana* d'Urv. (Cronquist et al.1977). Specimens with small heads and small perigynia with transverse folds ventrally have been recognized as *C. limnophila*. However, such specimens seem to occur at random through much of the range of *C. microptera*. This intergrading phase seems reasonably included in an expanded concept of *C. microptera*. Used as a ceremonial medicine and emetic by the Ramah Navajo.

Carex nardina Fr. (spikenard) Spikenard sedge. Plants 3–14 cm tall, densely caespitose, without rhizomes, the culms and leaves fasciculate. **LEAVES** basal and on the lower rounded culm, mostly surpassing the inflorescence, the blades 0.3–0.7 mm wide, folded to involute and nearly terete; bract lacking. **INFL** a solitary spike, erect, androgynous, 0.5–1 cm long, the staminate part shorter than the pistillate part, exserted beyond the perigynia by 2–3 mm; scales brown with green or pale center and hyaline margins, the pistillate ones as wide or a little wider than the perigynia. **PERIGYNIA** about 5–15, 3.5–4.5 mm long, elliptic, lanceolate or oblong-obovate, with conspicuous, more or less flattened margins, the margins usually serrulate-ciliolate, not wholly filled by the achene; rachilla subequal to the achene; stigmas 2 or 3. **ACHENES** accordingly lenticular or trigonous. $2n$ = 68. [*C. hepburnii* Boott] Dry alpine communities. COLO: Hin, LPl, SJn. 3725 to 4265 m (12220–14000′). Flowering: Jul–Aug. Circumboreal, south to Washington and Colorado.

Carex nebrascensis Dewey (of Nebraska) Nebraska sedge. Plants (1) 2–11.5 dm tall or taller, from robust, long-creeping, scaly rhizomes; culms somewhat tufted or more often arising singly or 2–3 together. **LEAVES** borne on the lower 1/4 of the culms, the blades 3–11 mm wide. **INFL** 4–25 cm long, mostly exceeding the leaves, but often surpassed by some of the foliose bracts, with 1–3 staminate spikes above 2–4 pistillate spikes, the transitional ones occasionally androgynous; terminal staminate spike 1.5–5 cm long, lateral staminate spikes, if present, to 2.5 cm long; pistillate spikes (1.5) 2–6 (10) cm long, 5–8 mm wide, cylindrical, the lower one pedunculate, the upper sessile or short-pedunculate; pistillate scales mostly longer and narrower than the perigynia, lanceolate or narrowly awl-shaped, glabrous or occasionally ciliolate apically, blackish or black-purple, with a narrow green or at least pale line in the center, this extending with the midrib to the apex; midrib prominent, often slightly ridged dorsally and excurrent as a mucro or aristate point, this glabrous or occasionally antrorsely ciliolate. **PERIGYNIA** 2.6–4.1 mm long including the beak, elliptic to obovate, with some conspicuous dorsal and ventral nerves plus the marginal nerves, not as flattened as in *C. aquatilis*, greenish at first, yellowish brown in age, sessile or nearly so, persistent, the beak (0.2) 0.4–0.6 mm long, more or less bidentate and ciliolate; stigmas 2. **ACHENES** lenticular. Two specimens seen (*R. Fleming 1214*, *O'Kane and Heil 5647*, both SJNM) from near Mancos. Swamps, wet meadows, and ditches. ARIZ: Apa; COLO: Mon, SJn; UTAH. 2135–2390 m (7000–7850'). Flowering: Jun–Aug. Washington to South Dakota and south to California and New Mexico. The distribution map provided by Johnston (2001) demonstrates much of the San Juan River drainage as a void in the distribution of this sedge that is widespread and common in much of the western United States.

Carex nelsonii Mack. (for Aven Nelson) Nelson sedge. Plants 10–35 cm tall, caespitose; rhizomes lacking or short. **LEAVES** basal and on the lower culm, the blades not over 12 cm long, 1–2.5 mm wide, flat. **INFL** capitate, 8–15 mm long, 5–12 mm wide, mostly much surpassing the leaves and bract if present; spikes 2 or 3, sessile, congested and sometimes appearing solitary, 5–10 mm long, 4–8 mm wide, ovate to oblong, the terminal one gynaecandrous, usually larger than the lower pistillate ones; pistillate scales shorter than or equal to the perigynia, blackish or black-purple, occasionally with a pale midrib, the margins sometimes narrowly hyaline. **PERIGYNIA** 3–3.8 mm long, 1.1–1.8 mm wide, elliptic, narrowly ovate or obovate, subsessile or with a stipe to 1 mm long, rather gradually tapered to a beak 0.5 mm long, cellular-striate on the faces and minutely papillate, at least apically; stigmas 3. **ACHENES** sharply trigonous, the angles more or less nerved, the nerves slightly ridged. Alpine communities. COLO: Con, Hin, LPl, SJn. 3350–3990 m (11100–13100'). Flowering: Jul–Aug. Montana to Utah and Colorado.

Carex nigricans C. A. Mey. (blackish) Black alpine sedge Plants 7–21 cm tall; culms arising few together from short to elongate rhizomes, with few to several leaves mostly borne toward the base; leaf blades 1–3 mm wide, flat or slightly channeled; bract lacking. **INFL** a solitary spike, erect, 0.8–1.8 cm long, 5–8 mm wide, androgynous, the staminate portion about 1/4–1/2 as long as the pistillate; scales blackish purple with scarious margins, the staminate ones erect, persistent, the pistillate ones spreading with the perigynia and soon deciduous. **PERIGYNIA** 3–4.5 mm long, lanceolate, often greenish except blackish purple on the beak, soon strongly spreading to reflexed, often deciduous after maturity, stipitate, the stipe about 0.5 mm long; rachilla obsolete; stigmas 3. **ACHENES** trigonous. $2n = 72$. Subalpine and alpine communities, often of moist to wet places. COLO: Arc, Con, Hin, LPl, Min, RGr, SJn. 3125–3900 m (10255–12800'). Flowering: Jul–Aug. Alaska and the Siberian coast and south to California and Colorado. *Carex nigricans* is an indicator of snowbeds.

Carex norvegica Retz. (Norwegian) Scandinavian sedge. Plants 12–68 cm tall, densely caespitose, rhizomes lacking or very short. **LEAVES** basal and on the lower culm, the blades 1–4 mm wide, flat. **INFL** (1) 1.5–3.5 (6) cm long, mostly surpassing the leaves and exceeding or exceeded by a leaflike bract; spikes mostly 3 (2–4), 0.5–1.3 cm long, about 4 mm wide, ovoid to nearly cylindrical, erect or strongly ascending, the terminal one gynaecandrous, equal to or longer than its peduncle, the lower spikes pistillate, mostly longer than their slender peduncles, or the lowest 1 rarely on an elongate peduncle arising on the lower 1/3 of the culm, the intermediate 1 or 2 often nearly or quite sessile, crowded toward the terminal one, and longer than their internodes, with the lowest shorter to longer than its internode; pistillate scales shorter than the perigynia, blackish or blackish purple, the midrib inconspicuous and rarely pale, sometimes papillate or ciliate, the margins often white-hyaline. **PERIGYNIA** 2–2.8 mm long, including the short beak, elliptic to narrowly obovate, green, except for the dark beak, strongly contrasting with the dark scales, becoming greenish brown in age, cellular-striate; stigmas 3. **ACHENES** trigonous, the angles slightly rounded, not strongly nerved. $2n =$ 54, 56. [*C. media* R. Br.] Wet areas. COLO: Arc, Hin, LPl, RGr, SJn; NMEX: RAr. 2750–3655 m (9000–12000'). Flowering: Jul–Aug. Circumboreal, south in America to Quebec and New Mexico.

Carex nova L. H. Bailey (new) Blackhead sedge. Plants (6) 16–72 cm tall, densely caespitose; rhizomes lacking or very short; culms erect to somewhat flexuous and nodding. **LEAVES** basal and on the lower culm, the blades 1–5.5 mm wide; bract lacking or more often present and immediately subtending the inflorescence or to 3 cm below it. **INFL** a head, 0.9–2.5 cm long, 9–20 mm wide, with 3 or 4 spikes, usually surpassing the leaves; spikes mostly sessile, or on peduncles to 3 mm long, 5–15 mm long, 3–8 mm wide, ovoid to oblong, the terminal one gynaecandrous, usually slightly larger than the lateral pistillate ones; pistillate scales shorter than or equaling the perigynia, and much narrower, black or purple-black, rather glossy. **PERIGYNIA** 3.5–4.9 mm long, 2–3 mm wide, broadly elliptic, obovate, to ovate-orbicular, flattened, with broad margins, conspicuously wider than the achene, greenish at first, but soon blackish or blackish purple, more or less glossy, glabrous or the upper margins ciliolate; stigmas 3. **ACHENES** trigonous. [*C. pelocarpa* F. J. Herm.] Riparian communities within spruce-fir forests, alpine tundra, willow carrs, ledges, and scree. ARIZ; Apa; COLO: Arc, Con, Dol, Hin, LPl, Min, Mon, RGr, SJn. 2710–4250 m (8900–13900′). Flowering: Jul–Aug. Oregon to Montana and south to Nevada and New Mexico. Plants with flexuous, nodding culms and smooth-margined perigynia have been recognized as *C. nova* var. *pelocarpa* (F. J. Herm.) Dorn.

Carex obtusata Lilj. (blunt) Blunt sedge. Plants 7–20 cm tall; culms usually arising singly from and well spaced (commonly 1–5 cm apart) along very dark brown to purplish black creeping rhizomes. **LEAVES** basal or nearly so, the blades about 1–1.5 mm wide, flat or loosely folded. **INFL** a solitary spike, erect, androgynous, 5–17 mm long, linear to nearly ovoid, the staminate portion longer or shorter than the pistillate portion; scales brownish with hyaline margins, the pistillate ones mostly shorter and narrower than the perigynia. **PERIGYNIA** usually 1–6, 3–4 mm long, narrowly ovoid or ellipsoid, thick-walled, brown or blackish brown, glossy, obscurely to conspicuously ribbed, conspicuously beaked; rachilla subequal to or longer than the achene, often with a flattened, hyaline, apical appendage; stigmas 3. **ACHENES** trigonous, filling the perigynia. $2n = 52$. Dry plains, ridges, and rocky, open slopes. ARIZ: Apa; COLO: Mon. 2135–3350 m (7000–11000′). Flowering: Jun–Jul. Yukon to Manitoba, south to New Mexico and South Dakota; also in Eurasia.

Carex occidentalis L. H. Bailey (of the west) Western sedge. Plants 2–8 dm tall, densely caespitose; rhizomes lacking or very short. **LEAVES** basal and on the lower culm, the blades 1–3 mm wide; lowest bract lacking or less than 1/2 as long as the inflorescence. **INFL** (1) 1.5–4.5 cm long, about 5–10 mm wide, spicate or capitate, oblong to linear-oblong or linear, at least the lower spikes usually separated, the first internode of the rachis about 2–10 mm long; spikes about 4–10, androgynous, not over 5 mm long, sessile, with about 4–12 perigynia. Pistillate scales equal to and nearly concealing the perigynia, brownish with a green midstripe and hyaline margins. **PERIGYNIA** 3–3.8 mm long, 1.6–2 mm wide, elliptic to ovate or ovate-orbicular, the body greenish or brownish with greenish, slightly winged, serrulate (use 20×) margins at least in the upper 1/2 and beak, this bidentate, the teeth about 0.3 mm long, the marginal nerves on the winged edges or scarcely displaced to the ventral surface, the faces faintly nerved or essentially nerveless; stigmas 2. **ACHENES** lenticular. Piñon-juniper, ponderosa pine–oak communities, aspen and spruce-fir parklands. ARIZ: Apa; COLO: Arc, Hin, LPl, Min, Mon, SJn; NMEX: McK, RAr, SJn; UTAH. 2040–2985 m (6700–9800′). Flowering: Jun–Jul. Montana and North Dakota south to California and Texas. This is one of the more frequently collected sedges of the Four Corners. Due to abundance, western sedge is likely an important forage species for ungulates. However, it can be expected to be of low or moderate palatability.

Carex pachystachya Cham. ex Steud. (thick spike) Chamisso sedge. Plants 3–10 dm tall, densely caespitose; rhizomes lacking or very short. **LEAVES** borne on the lower 1/2 of the culms, the lower ones reduced to bladeless sheaths, the upper ones with essentially flat blades 2–6 mm wide; lowest bract lacking. **INFL** 15–23 cm long, 13–16 mm wide, usually oblong, but sometimes ovoid, surpassing the leaves or sometimes subequal to or surpassed by them, the internodes of the rachis concealed by the dense spikes or apparent and 2–4 mm long; spikes 4–12, gynaecandrous, 6–9 mm long, closely aggregated, hardly evident without dissecting the head but usually more conspicuous than in *C. microptera*; pistillate scales shorter and narrower than the perigynia, brown, with or without white hyaline margins to 0.1 mm wide. **PERIGYNIA** 3.5–4.3 (5) mm long, 1.1–2 mm wide, ovate, plano-convex to concavo-convex, lightly nerved dorsally, nerveless or lightly nerved and infrequently with a transverse fold ventrally, greenish and soon stramineous or brownish to copper-colored, usually widely spreading at maturity and often deciduous, the winged margins rather narrow, serrulate, rather abruptly narrowed into a short, darkened, more or less terete, smooth beak; stigmas 2. **ACHENES** lenticular, 1.5–2 mm long, nearly filling the perigynial cavity. [*C. macloviana* d'Urv. var. *pachystachya* (Cham. ex Steud.) Hultén]. Douglas-fir, ponderosa pine–dogwood, and spruce-fir communities, usually in moist or wet soil. COLO: Arc, Hin, LPl, Min, Mon, SJn. 2925–3400 m (9600–11140′). Flowering: Jul–Aug. Alaska, Yukon, Saskatchewan south to California and Colorado. Plants of this taxon are similar to those of *C. microptera*. They differ in general from *C. microptera* in being less common and restricted to wetter places. The perigynia of *C. pachystachya* are plano-convex with the achene more nearly filling the cavity, and they are not so strongly flattened as in *C. microptera*. Also, the perigynia are more widely spreading to deflexed and much sooner deciduous, and the achenes are commonly larger (1.6–2 mm long versus 1.1–1.5 mm).

Carex parryana Dewey (for Charles C. Parry, British-American botanist and explorer) Parry sedge. Plants 18–50 cm tall; culms densely tufted or arising a few together from short to rather elongate rhizomes. **LEAVES** basal and on the lower culm, the blades 1–4 mm wide; lowest bract shorter than or subequal to the inflorescence. **INFL** 1.8–6 cm long, 5–10 mm wide, with (1) 2–4 (6) spikes, the lowest internode of the rachis about 0.5–2 times as long as the lowest spike, the upper ones usually equal to or shorter than the spikes; terminal spike gynaecandrous with a short staminate portion or occasionally nearly or entirely staminate or all pistillate, 1.2–3 cm long, 3–5 mm wide, subsessile or on a peduncle to 1.5 cm long; lateral spikes pistillate, 0.7–2.4 cm long, about 4–5 mm wide, ovoid to cylindrical, sessile or essentially so, or the lowest on a peduncle to 4 cm long; pistillate scales shorter than or subequal to the perigynia and narrower apically, blackish or dark brown-purple, with greenish or pale midrib, sometimes ciliolate apically. **PERIGYNIA** 1.9–3 (3.6) mm long, more or less obovate, glabrous to papillate, or occasionally ciliolate to strigose-hirtellous, the dorsal surface greenish marginally and basally and usually blackish purple in the center near the beak; stigmas mostly 3 but sometimes 2. **ACHENES** mostly trigonous. $2n = 54$. Hanging gardens, and seep and spring communities mostly in canyons of the Colorado River. COLO: Hin; UTAH. 1150–1860 m (3780–6100′). Flowering: May–Jul. Alaska to Ontario and south to Nevada and Colorado. Most specimens are from the canyons of the Colorado River, but *L. Lundquist et al. 269* SJNM is montane. Some plants have at least some flowers with two stigmas and lenticular achenes. These are features of *C. specuicola* J. T. Howell, and some collections from Utah have been identified as *C. specuicola*. However, all mature specimens I have seen from Utah seem to fit better with *C. parryana* than with *C. specuicola*. See also the discussion under *C. specuicola*.

Carex pellita Willd. (covered with skin) Woolly sedge. Plants 24–80 cm tall or taller; culms arising singly or more often few together or in small tufts from thick, creeping rhizomes. **LEAVES** mostly borne on the lower 1/2 of the culms, the lower ones often reduced to bladeless sheaths, the upper ones with flat blades 2–6 mm wide. **INFL** 8–21 cm long, leafy-bracteate, the first and sometimes the second bracts equaling to much surpassing the inflorescence, mostly not sheathing but the lowest one sometimes with a closed sheath to 17 mm long; terminal spike staminate, 1.7–4.5 cm long, cylindrical, sessile or subsessile or on peduncles to 7 cm long, often subtended by 1–2 smaller staminate spikes; lateral pistillate spikes 1–3 (4), 1–4.8 cm long, 5–7 mm wide, the lower ones well spaced and mostly shorter than their internodes, sessile or nearly so, or on peduncles up to 8 cm long, with many perigynia; pistillate scales usually shorter and narrower than the perigynia, acute or awn-tipped, brownish or purplish, typically with a green or pale midstripe. **PERIGYNIA** densely pubescent, 3.3–5 mm long, ovoid to subglobose, with an abrupt, conspicuous, bidentate beak, many-ribbed, thick-walled; stigmas 3. **ACHENES** trigonous, loosely filling the perigynium. $2n = 78$. [*C. lanuginosa* Michx.; *C. lasiocarpa* Ehrh. var. *lanuginosa* (Michx.) Kük.]. Riparian and wetland communities, including hanging gardens, and montane areas. ARIZ: Apa, Nav; COLO: Arc, LPl, Mon; NMEX: RAr, SJn; UTAH. 1125–2590 m (3700–8500′). Flowering: May–Jul. British Columbia to New Brunswick and south to California and Arkansas. Plants of this widespread taxon have passed under the name of *C. lanuginosa* in a number of floras. The type of *C. lanuginosa* is *C. lasiocarpa* Ehrh., leaving *C. pellita* the name of priority.

Carex perglobosa Mack. (spherical) Globe sedge, Mount Baldy sedge. Plants 6–20 cm tall, loosely caespitose from slender, creeping rhizomes. **LEAVES** clustered at the base of the culms, the blades 0.7–1.5 mm wide; bracts lacking. **INFL** a globose head about 1 cm in diameter, usually equaling or surpassing the leaves; spikes 6–15, androgynous, not evident individually in the crowded head, less than 1 cm long, the staminate flowers inconspicuous; pistillate scales subequal to the perigynia, brownish with hyaline margins. **PERIGYNIA** 4–4.8 mm long, 1.7–2.3 mm wide, ovate-elliptic, inflated, gradually tapered to a bidentate beak; stigmas 2. **ACHENES** lenticular. Alpine communities. 3690–4235 m (12100–13900′). Flowering: Jul–Aug. COLO: Con, LPl, Min, SJn. Colorado and the La Sal Mountains of Utah.

Carex petasata Dewey (prepared for a journey or having a cap on) Liddon sedge. Plants 2–7 dm tall, caespitose; rhizomes lacking or very short. **LEAVES** basal and on the lower culm, the blades 1–4 mm wide, flat or channeled; lowest bract lacking. **INFL** 2.2–5 cm long, 8–12 mm wide, spicate, surpassing the leaves, the lowest internode of the rachis 4–7 mm long, the second lowest subequal; spikes 3–6, gynaecandrous, 10–22 mm long, 5–7 mm wide, more or less clavate, sessile, loosely aggregated and evident without dissecting the inflorescence, about 2–3 times longer than the internodes of the rachis; pistillate scales largely concealing the perigynia, stramineous or light reddish brown, with a green or pale midstripe and hyaline margins. **PERIGYNIA** (5.7) 6–7.5 (8) mm long, 1.5–2.8 mm wide, lanceolate, narrowly ovate or nearly elliptic, greenish or light brown, plano-convex, with thin winged margins, serrulate at least in the distal 1/2, gradually tapered to an obliquely cleft beak, conspicuously nerved on both sides; stigmas 2. **ACHENES** lenticular. Montane communities. COLO: Arc, Hin, Mon. The one specimen seen (*Heil et al. 13366*, SJNM) is from Piedra Trail below Ice Cave Ridge. UTAH. 2590–2895 m (8500–9500′). Flowering: Jun–Jul. British Columbia to Saskatchewan and south to California and Arizona. This plant passes into *C. praticola* and perhaps into *C. phaeocephala*. Additional study of this complex is indicated.

Carex phaeocephala Piper (dusky head) Dunhead sedge. Plants (10) 15–30 (55) cm tall, densely caespitose; rhizomes lacking. **LEAVES** basal and on the lower culm, the blades about 1–3 mm wide, flat, channeled, or involute; lowest bract lacking or occasionally present and subequal to the inflorescence or longer. **INFL** 1.4–3 cm long, rarely shorter, 5–15 mm wide, equaling or surpassing the leaves; spikes (1) 2–5 (7), gynaecandrous, 8–15 mm long, 5–7 mm thick, sessile, approximate but mostly evident without dissecting the head, mostly twice or more as long as the internodes; lowest internode of the rachis commonly 4–7 mm long; pistillate scales as long and almost as wide as the perigynia, largely concealing them (sometimes pressed to one side in herbarium specimens), brown to dark brown, with green or pale midstripe and usually conspicuous hyaline margins. **PERIGYNIA** 4–5.5 mm long, 1.5–2.2 mm wide, lanceolate to ovate-elliptic, brownish or stramineous, with greenish stramineous, thin-winged margins, finely but conspicuously nerved dorsally and often ventrally, gradually or rather abruptly tapered to a bidentate beak; stigmas 2. **ACHENES** lenticular. Krummholz spruce, rocky alpine tundra, and talus. COLO: Con, Hin, LPl, Min, Mon, SJn. 2820–3900 m (9260–12800′). Flowering: Jul–Aug. British Columbia to Alberta and south to California and Colorado.

Carex praeceptorum Mack. (teacher) Slope sedge. Plants 10–31 cm tall, densely to loosely tufted, with short or somewhat prolonged rhizomes. **LEAVES** basal and on the lower culm, the blades 1.5–3 mm wide; lowest bract lacking or shorter than the lowest spike. **INFL** 1–2.7 cm long, about 5–8 mm wide, usually equaling or surpassing the leaves; spikes 4–6, gynaecandrous, 5–12 mm long, 3–4 mm wide, sessile, approximate or the lower ones somewhat separate, but usually twice as long as the internodes or occasionally equaling them; pistillate scales shorter than the perigynia, brown to dark brown, with a green midstripe and narrow to rather broad hyaline margins. **PERIGYNIA** 1.7–2.2 (2.5) mm long, 1–1.2 mm wide, elliptic to elliptic-ovate, more or less flattened, but with rounded, wingless margins, yellow-green or brownish green, (10) 15–25 per spike, slightly granular-roughened, infrequently serrulate on the beak, finely nerved on both sides, the nerves frequently reddish or brown-red, sometimes only faintly so, the dorsal suture conspicuous on the beak and extending barely onto the body; stigmas 2. **ACHENES** lenticular. Moist and wet meadows. COLO: Con, Hin, LPl, Min, SJn. 3130–3660 m (10270–12000′). Flowering: Jul–Aug. Washington to Montana and south to Nevada and Colorado. Plants of *C. praeceptorum* are similar to those of *C. canescens*. A combination [*C. canescens* L. var. *praeceptorum* (Mackenzie) Bailey] has been made.

Carex praegracilis W. Boott (very slender) Black creeper sedge. Plants (1) 1.5–7 dm tall, dioecious (and forming clones) or monoecious; culms arising singly or few together or occasionally tightly clustered, from robust blackish or dark brown rhizomes 2–4 mm thick, these covered with scales that often become filamentous in age. **LEAVES** with the lowest reduced to bladeless sheaths, these often dark brown or blackish, the upper leaves borne on the lower 1/4 of the culms, with blades 1–3.5 mm wide. **INFL** 1.5–4.3 cm long, about 5–15 mm wide, simple or compound, capitate or spicate, usually surpassing the leaves, without a subtending leaflike bract or this to about 1/2 as long as the inflorescence; spikes about 6–25, androgynous, staminate or pistillate, to about 1 cm long, closely aggregated, or the lower one somewhat separate, all longer than the rachis internodes, sessile; pistillate scales mostly as long and as wide as the perigynia and largely concealing them, these and the staminate ones brown with a green or paler midrib and hyaline margins. **PERIGYNIA** 2.8–3.5 mm long, 1.1–1.7 mm wide, ovate or lance-ovate to elliptic, usually short-stipitate, the margins rather thin when young, firm in age and slightly turned toward the ventral surface; beak of the perigynium 0.6–1.3 mm long, obliquely cleft, serrulate, the minute teeth often extending onto the body of the perigynium; stigmas 2. **ACHENES** lenticular. $2n = 60$. Riparian and palustrine communities including hanging gardens and irrigated pastures. ARIZ: Apa, Nav; COLO: Arc, Hin, LPl, Mon, SJn; NMEX: RAr, SJn; UTAH. 1400–2665 m (4600–8750′). Flowering: May–Jul. Yukon to Quebec and Maine, south to California and Virginia; Mexico.

Carex praticola Rydb. (meadow dweller) Meadow sedge. Plants 30–80 cm tall, densely caespitose; rhizomes lacking or very short. **LEAVES** basal and on the lower culm, the blades 2–4 mm wide, flat. **INFL** (2.5) 3–4 cm long, 1–1.5 cm wide, spicate, more or less flexuous, sometimes more or less secund; overtopping the leaves and without a leafy bract or this shorter than the inflorescence; the lowest internode about 5–12 mm long, the second one as long or nearly so; spikes (2) 4–7, gynaecandrous, (8) 10–15 mm long, about 5–7 mm wide, usually clavate, sessile, loosely aggregated and evident without dissecting the inflorescence, about 1–2 times as long as the internodes, or the lower one slightly shorter, and the upper ones to about 3 times as long; pistillate scales subequal to and as wide as the perigynia, largely concealing them, dull reddish brown with a green midstripe and broad hyaline margins. **PERIGYNIA** (4) 4.5–5.7 (6) mm long, 1.2–2.1 mm wide, plano-convex, with thin-winged margins, serrulate, gradually tapered to a shallowly bidentate beak, finely nerved dorsally, faintly nerved ventrally, the beak more or less terete and not serrulate at least in the distal portion; stigmas 2. **ACHENES** lenticular. $2n = 64, 70$. Ponderosa pine–Gambel's oak communities. COLO: Arc (*Heil & Mietty 13313* SJNM). 2440 m (8000′). Flowering: Jun–Jul. Alaska to Labrador, south to California and Quebec. Plants of this taxon pass imperceptibly into those of *C. petasata.*

Carex pyrenaica Wahlenb. (of the Pyrenees) Pyrenean sedge. Plants 5–15 cm tall, densely caespitose, without rhizomes, the culms and leaves often fasciculate. **LEAVES** 2–4 per culm, the lower bladeless, the upper with blades 0.5–1.5 mm wide, flat or loosely to strongly channeled; bract lacking. **INFL** a solitary spike, erect, androgynous, 0.7–2 cm long, 4–7 mm thick, the staminate portion about 2–4 mm long; pistillate scales tan to dark brown, the margins hyaline, subequal to the perigynia, deciduous in age. **PERIGYNIA** usually (5) 10 or more, about 2.5–4 mm long, lanceolate, somewhat stipitate and jointed to the rachis, ascending, or sometimes reflexed in age, readily deciduous; rachilla lacking; stigmas 3 (in ours). **ACHENES** trigonous (in ours). Late-melting alpine snowbeds. COLO: Con, Hin, LPl, SJn. 3500–4100 m (11400–13400′). Flowering: Jul–Aug. Alaska to California and Colorado; Eurasia. Our plants are referable to **subsp. *micropoda*** (C. A. Mey.) Hultén (small foot) [*C. crandallii* Gand.]. Most of our specimens appear to have three stigmas and trigonous achenes.

Carex retrorsa Schwein. (twisted or bent) Knotsheath sedge, reflexed sedge. Plants 3–10 dm tall; culms densely clustered on short rhizomes. **LEAVES** mostly borne on the culms, the blades 4–10 mm wide. **INFL** 5–13 cm long, with 1 staminate, gynaecandrous, or androgynous spike above 4–6 pistillate spikes, these 1.5–4.5 cm long, about 1.5–2 cm thick, cylindrical, closely spaced and much longer than the internodes of the rachis or the lowest one widely spaced but still longer than its internode, each subtended by leafy bracts that much surpass the inflorescence; pistillate scales shorter and narrower than the perigynia, greenish or stramineous, narrowly acute, acuminate, or short-awned. **PERIGYNIA** widely spreading, giving a bristly appearance to the spikes, 7–10 mm long, shiny, greenish, conspicuously nerved, the body inflated, ellipsoid to subglobose, narrowed to a conspicuously bidentate beak 2–3 (4) mm long; stigmas 3. **ACHENES** trigonous, much smaller than the perigynia, the style continuous with, persistent on, and of the same firm texture as the achene, contorted in age. $2n = 70$. Infrequent in marshes, wet meadows, or along river bottoms. The two specimens seen are from Opal Lake Trail (*Heil 11453*, SJNM) and upper Price Lakes (*Heil 12668*, SJNM). COLO: Arc, Hin, LPl, Mon. 2650–2745 m (8700–9000′). Flowering: Jun–Aug. British Columbia to Quebec and south to Oregon, Colorado, and New Jersey.

Carex rossii Boott (for Edith A. Ross, botanical collector) Ross sedge. Plants 5–32 cm tall, densely caespitose, with or without short rhizomes, often deep reddish purple basally. **LEAVES** basal and on the lower culm, mostly overtopping the inflorescence, the blades 0.5–3 mm wide. **INFL** essentially the length of the culms, with the lowest spike borne on a slender peduncle arising near the culm base and much removed from the rest of the inflorescence; terminal spike staminate, 5–12 mm long; lateral spikes mostly 2–5, very short, with only 1–6 perigynia, bracteate or leafy-bracteate, the lower bracts (excluding those of the basal spike) usually exceeding the inflorescence; pistillate scales shorter than the perigynia, with a greenish, keeled, often scabrescent-ciliolate midrib, this flanked by stripes of brown or brownish purple and then by broad hyaline margins. **PERIGYNIA** sparsely to densely puberulent, 2.7–4.5 mm long including the oblique beak, this 0.7–1.8 mm long, the body plump, with 2 prominent marginal nerves; stigmas 3. **ACHENES** trigonous, filling the perigynia. $2n = 36$. [*C. brevipes* W. Boott] Mountain brush and piñon-juniper, Douglas-fir, and aspen communities. ARIZ: Nav; COLO: Arc, Dol, Hin, LPl, Mon, SJn, SMg; NMEX: SJn; UTAH. 2040–3540 m (6685–11600′). Flowering: Jun–Jul. Alaska to California and Arizona and east to Michigan. This is a widespread and common to abundant species to the north and west, but it is rarely collected in the Four Corners region.

Carex saxatilis L. (growing among rocks) Russet sedge. Plants 21–40 (56) cm tall; culms arising singly or rather closely spaced on creeping rhizomes. **LEAVES** mostly borne on the lower 1/2 of the culms and sometimes higher, the blades 1.5–3 mm wide, flat. **INFL** 4–9 cm long, with 1–2 (3) staminate spikes 1–3 cm long borne above 1–2 pistillate spikes 1–2.6 cm long and about 5–8 mm wide, the lateral staminate ones well spaced, subequal to or shorter than the rachis internodes, the pistillate ones sessile or on slender, spreading or drooping peduncles to 3 cm long, each subtended by a leafy bract, the lowest bract usually overtopping the inflorescence; pistillate scales slightly to conspicuously shorter than the perigynia, acute, very dark brown, purple-black, or black, sometimes with a pale midstripe, the margins whitish hyaline at least toward the often erose tip. **PERIGYNIA** 3.2–5.2 mm long, 1.5–3.3 mm wide, lanceolate to nearly orbicular, slightly inflated, soon black or blackish purple at least where exposed beyond the scale, with marginal or sometimes dorsal nerves conspicuous, nerveless ventrally, tapering to a short beak, the beak nearly entire or with inconspicuous teeth, these not over 0.5 mm long and easily broken off; stigmas 2 or occasionally 3, sometimes both in the same spike. **ACHENES** accordingly lenticular or trigonous, the style continuous with, persistent on, and of the same firm texture as the achene, often strongly bent or recurved back against the achene in age. 2*n* = 78? 80. [*C. physocarpa* J. Presl & C. Presl] Peaty wetlands of evaporated tarns and other wetlands. COLO: Con, Hin, LPl, SJn. 3290–3720 m (10800–12200′). Flowering: Jul–Aug. Circumboreal; south in America to Utah and Colorado.

Carex scirpoidea Michx. (scirpuslike) Spike-rush sedge. Plants 4–30 cm tall, unisexual; culms arising singly or few together from rather short but conspicuous rhizomes. **LEAVES** basal and on the lower culm, the blades 1.5–3.5 cm long, 3–7 mm wide, linear. **INFL** a solitary spike, wholly staminate or wholly pistillate, about 1.5–3.5 mm long; scales mostly rounded at the apex, blackish purple with hyaline, ciliolate margins, sometimes sparsely pubescent dorsally, the pistillate ones mostly longer and wider than the perigynia; anthers 3–5 mm long. **PERIGYNIA** obovoid, 2–4 mm long including the beak, pubescent; rachilla obsolete.

1. Plants 4–34 cm tall; spike to 4 mm wide when pressed; krummholz and alpine .... **var. *pseudoscirpoidea***
1′ Plants 20–45 cm tall; spike 2.5–3 mm wide when pressed; montane .... **var. *scirpoidea***

var. *pseudoscirpoidea* (Rydb.) Cronquist (false scirpuslike) Krummholz and alpine tundra communites. COLO: Con, Hin, LPl, Mon, SJn. 3500–3900 m (11500–12800′). Flowering: Jul–Aug. Labrador to Alaska, south to New Hampshire and California.

var. *scirpoidea* The one specimen seen is from a calcareous fen north of Haviland Lake (*K. Heil & A. Clifford 26644* SJNM). COLO: LPl. 2620 m (8600′). Flowering: Jul–Aug. Alberta to Manitoba, south to North Dakota, Utah, and Colorado.

Carex scopulorum Holm (of the cliff) Mountain sedge. Plants 22–67 cm long; culms arising singly or few together or somewhat tufted, from robust, scaly rhizomes. **LEAVES** basal and mostly on the lower culm, the blades (2) 3–7 mm wide. **INFL** 2.5–9 cm long, exceeding the lowest foliar bract, with 1 or 2 staminate spikes 1–2.2 cm long above 2–4 cylindrical or ovoid pistillate spikes 1–3 cm long and 5–10 mm wide, the intermediate spikes occasionally androgynous; pistillate scales shorter than or longer than the perigynia and mostly narrower at least toward the apex, black or purple-black. **PERIGYNIA** 1.8–3.3 mm long, elliptic to broadly obovate, nerveless except for the 2 marginal nerves, blackish to dark reddish brown at least distally, sessile or nearly so, the beak nearly obsolete; stigmas 2 or rarely 3. **ACHENES** respectively lenticular or trigonous. Wet meadow, krummholz, and alpine communities. COLO: Arc, Con, Hin, LPl, Min, SJn. 3445–3870 m (11300–12700′). Flowering: Jul–Aug. British Columbia to California and east to Colorado. In addition to the features listed in the key, *C. scopulorum* is distinguished from the similar *C. aquatilis* with comparatively crowded spikes that are shorter on average.

Carex siccata Dewey (dry) Silvertop sedge. Plants 19–36 cm tall; culms arising singly or few together and loosely to rather closely clustered on stout rhizomes. **LEAVES** clustered on the lower 1/4 of the culms, the lower ones often reduced to bladeless sheaths, the upper ones with blades 1–2 mm wide, flat, the longer ones subequal to or shorter than the inflorescence; lowest bract subequal to the lowest spike. **INFL** 1–3 (6) cm long, 5–10 mm wide, linear or clavate or occasionally orbicular; spikes sessile, congested or interrupted, about 5–12 mm long, the lower ones gynaecandrous or pistillate, some (especially the middle ones) staminate, the upper one often pistillate and closely subtended by a staminate spike and thus falsely appearing to be gynaecandrous; scales light brown with broad hyaline margins, the pistillate ones subequal to the body of the perigynia but surpassed by the beak. **PERIGYNIA** 4.5–6.2 mm long including the prominent beak, flattened, serrulate to near the middle and sometimes below, slightly wing-margined below, the body more or less elliptic; stigmas 2. **ACHENES** lenticular, filling the perigynia, sometimes rupturing it at maturity. 2*n* = 70 [*C. foenea* sensu Svenson, not Willd.]. Ponderosa pine–oak, aspen, spruce-fir, and subalpine communities. COLO: Arc, Con, Hin, LPl, Min, RGr, SJn; NMEX: McK, RAr, SJn; UTAH. 1765–3800 m (5800–12400′). Flowering: Jun–Aug. Yukon to Quebec and south to Arizona, New Mexico, and Maine.

Carex simulata Mack. (likeness) Lookalike sedge. Plants 22–45 cm tall, dioecious (and commonly forming unisexual clones) or monoecious, the culms mostly arising singly from, and closely to remotely spaced on long, whitish or light brown rhizomes. **LEAVES** cauline, mostly borne on the lower 1/4 of the culms, the lower ones reduced to bladeless sheaths, the upper ones with blades 1.5–4 (6) mm wide; bracts subequal to the spikes or the lower 1–3 from 1/2 as long to equaling the inflorescence, long acuminate-caudate. **INFL** 1.8–3.5 cm long, 7–25 mm wide, simple or occasionally compound, capitate or spicate, linear to oblong or occasionally ovate, subequal to or surpassing the leaves; spikes commonly 10–20, unisexual or androgynous, crowded and usually concealing the rachis, more or less evident without dissecting the inflorescence; pistillate scales longer than the perigynia, brownish, with narrow to rather broad pale hyaline margins, the midrib sometimes green. **PERIGYNIA** 1.7–2.6 mm long, elliptic-ovate, or nearly orbicular, shining brown, sessile or occasionally very short-stipitate, the margins rounded, not thin, the dorsal surface with a few raised nerves extending to about 1/2 or more to apex or occasionally nerveless, the ventral nerves like the dorsal ones but mostly shorter and more often lacking, the wall thick and coriaceous, the beak 0.2–0.6 mm long, inconspicuously winged and serrulate near the confluence with the body; stigmas 2. **ACHENES** lenticular. Wet meadows and swamps. The one specimen seen (*Rich Fleming 1210*, SJNM) is from west of Puett Reservoir. The other specimen (*Peggy Lyon 114* SJNM), was collected from an iron bog at Chattanooga, Colorado. COLO: LPl, Mon, SJn. 2165–3145 m (7100–10320′). Flowering: Jun–Jul. Washington to Saskatchewan and south to California and New Mexico.

Carex specuicola J. T. Howell (cave-dweller) Navajo sedge, hanging garden sedge. Plants (15) 20–40 (66) cm tall, densely to loosely caespitose with short rhizomes. **LEAVES** basal and borne on the lower 1/5–1/3 of the culm, the blades 0.5–2 (3) mm wide, lax. **INFL** 2–4.2 (10) cm long with 2–3 spikelets, the terminal 6–15 (23) mm long, gynaecandrous or sometimes staminate or with a pistillate apex and base and staminate middle, the lateral spikes 5–15 mm long, often of uniform length, pistillate. **PERIGYNIA** 2.5–3 (3.5) mm long, greenish or straw-colored after maturity; stigmas mostly 2, sometimes 3. **ACHENES** mostly lenticular, sometimes trigonous. Seeps and springs of hanging gardens in sandstone alcoves. ARIZ: Apa, Nav. 1495–1980 m (4900–6500′). Flowering: May–Jul. Arizona. Specimens from hanging gardens of San Juan County, Utah, have also been identified as *C. specuicola*. However, mature specimens examined from San Juan County, Utah, have mostly trigonous achenes. They also tend to have lateral spikes with considerable variability in size. These are features of *C. parryana*. Some immature specimens from Utah might have features of *C. specuicola*. However, these specimens are more or less sympatric with mature specimens that are dominated by features of *C. parryana*. Perhaps *C. parryana* grades into *C. specuicola* in the Four Corners region near the Arizona-Utah border. The transition from three to two stigmas in *C. parryana* begins as far north as Juab County, Utah, where a specimen from Mt. Nebo (*Collins & Harper 937* BRY) has some flowers with two stigmas and others with three stigmas, and in Desolation Canyon, Uintah County, Utah, where a collection with two stigmas (*Atwood 25765* BRY) was made near a collection (*Atwood 24445* BRY) with three stigmas. Occasional presence of two stigmas continues southerly in Emery County, Utah, up to 3110 m (*M. E. Lewis 5125* BRY) and to San Juan County (*A. Clifford 00–216* SJNM). Although some immature specimens from Utah (*A. Clifford 93–125* SJNM) might have features of *C. specuicola*, I am reluctant to include that taxon for Utah without seeing mature specimens that clearly verify its presence there. *Carex parryana* and *C. specuicola* share a number of features including terminal spikes that are gynaecandrous, sometimes staminate, and sometimes pistillate at the apex and base with a staminate center. There appears to be little more than dominance of stigma number and achene shape, and perhaps uniform versus variable length of spikes to serve as diagnostic features. However, several specimens from Arizona consistently show narrower, flaccid leaves, smaller diameter culms, and smaller spikelets than most specimens from Utah. These seem to be additional features that help separate plants of the two taxa. However, a collection (*M. A. Porter 4199* SJNM, BRY) from the type locality (Inscription House, Coconino County, Arizona) has leaves as wide as those in typical *C. parryana*. Much of the interpretation of the *C. specuicola* side of this complex is based on specimens collected by Arnold Clifford, who, apparently, has made more collections of *C. specuicola* than all other collectors combined.

Carex stipata Muhl. ex Willd. (with a stipe) Prickly sedge, owlfruit sedge. Plants 28–75 cm tall or taller, caespitose; rhizomes lacking or very short. **LEAVES** basal and on the lower 1/2 of the culms, the lower culm leaves reduced to bladeless sheaths, the upper ones with blades 4–8 (11) mm wide, flat, the ventral side of the sheath membranous, usually cross-corrugated; lowest bract not much if any longer than the lowest branch of the inflorescence, more or less awnlike, usually broken off in older specimens. **INFL** (2.5) 3–5.5 cm long, 1–2 cm wide, compound but spikelike, with many short, crowded spikes, appearing bristly from the widely spreading, narrowly beaked perigynia; spikes crowded, androgynous, to about 1 cm long and nearly as wide; pistillate scales mostly shorter than the perigynia, pale or brownish, rather scarious, with the more or less greenish or pale midrib sometimes exserted as an awn-tip. **PERIGYNIA** (3.6) 4–5.2 mm long, widely spreading, lance-triangular or lance-ovate, broadest at the sometimes abruptly truncate, somewhat spongy-thickened base, gradually tapered to a long, minutely serrulate beak, conspicuously nerved at least dorsally, subsessile to conspicuously stipitate; stigmas 2. **ACHENES** lenticular. $2n = 52$. Riparian areas including ditches. COLO: Arc, Hin, LPl, Min, Mon; NMEX: SJn. 1615–3050 m (5300–10000′). Flowering: Jun–Jul. Alaska to Newfoundland and south to California and Texas. Our plants are referable to **var. *stipata***.

Carex tahoensis Smiley (of Tahoe) Tahoe sedge. Plants 20–45 cm tall, caespitose, from short, stout rhizomes. **LEAVES** with blades 2–3 mm wide; lowest bract generally lacking. **INFL** 2–4 cm long, 7–15 mm wide, more or less spikelike, usually much surpassing the leaves; internodes of the rachis obscured by the dense spikes or sometimes discernible, the lower 2 collectively 8–18 mm long; spikes 3–6, gynaecandrous, 8–17 mm long, sessile, approximate but usually readily evident without dissecting the inflorescence, about twice as long as the internodes of the rachis; pistillate scales shorter to longer than the perigynia, reddish brown with a greenish midstripe and hyaline margins. **PERIGYNIA** 4.2–7 (7.4) mm long, 1.9–2.8 mm wide, elliptic or ovate, slightly to strongly plano-convex, lightly to evidently many-nerved on both sides, or the ventral side nerveless or with few nerves, wing-margined to the tip of the ill-defined flattened beak, serrulate, greenish, stramineous, or pale brown; stigmas 2. **ACHENES** lenticular. $2n = 68$. Forb-grass communities and open woods. ARIZ: Apa; UTAH. 2500–2600 m (8300–8500′). Flowering: Jun–Jul. Yukon to Alberta and south to California and Arizona. Plants of this taxon have routinely been included in *C. xerantica* L. H. Bailey (Flora of North America Editorial Committee 2002), which is known to the north of our area.

Carex utriculata Boott (with bladders) Bladder-sac sedge. Plants 6–14 dm tall; culms arising singly or 2–3 together from robust, scaly, long-creeping rhizomes, often spongy-thickened and to 1 cm thick near the base. **LEAVES** mostly borne on the lower 1/2 of the culm, the lower sheaths septate-nodulose, the blades 3–12 mm wide, flat; ligule tapering across the ventral surface of the blade at a very rounded angle, not noticeably projected into a point. **INFL** 15–45 cm long, with 2–4 staminate spikes borne above 2–3 (5) pistillate ones, the intermediate ones sometimes androgynous; terminal staminate spike (2) 3–8.5 cm long, the lateral staminate ones somewhat to much shorter, approximate to widely separated; pistillate spikes 2.5–10.5 cm long, 6–13 mm wide, cylindrical, sessile or on peduncles to 3 (7) cm long, rarely compound at the base in large spikes, well spaced and much shorter to longer than the rachis internodes, each subtended by a foliose bract, the bracts often equaling or exceeding the inflorescence; pistillate scales shorter than or subequal to and narrower than the perigynia, greenish, turning red-brown to purple-brown, with a paler median, narrowly acute to acuminate or short-awned. **PERIGYNIA** strongly spreading, glossy, light brown to very dark brown, 4–7 mm long, the body inflated, broadly ellipsoid to subglobose, rather abruptly contracted to a prominent, conspicuously bidentate beak 1–2 mm long, the teeth 0.2–1 mm long; stigmas 3. **ACHENES** trigonous, much smaller than the perigynia, the style continuous with, persistent on, and of the same firm texture as the achene, twisted or abruptly bent. $2n = 70, 72, 76$. [*C. rostrata* Stokes ex Withering] Ponds, lakes, wet meadows, and along streams in ponderosa pine, willow, and spruce-fir communities. ARIZ: Apa; COLO: Arc, Con, Dol, Hin, LPl, Min, Mon, RGr, SJn; NMEX: RAr, SJn; UTAH. 2105–3110 m (6900–10200′). Flowering: Jul–Sep. Circumboreal, south in America to California and Delaware.

Carex vallicola Dewey (living in a valley) Valley sedge. Plants 15–56 cm tall, densely caespitose; rhizomes lacking or very short. **LEAVES** basal and on the lower culms, the blades 0.5–2.5 mm wide. **INFL** 0.8–3 cm long, 5–9 mm wide, slightly to much overtopping the leaves, without a subtending bract or this shorter than the inflorescence, spicate or capitate, shortly interrupted or more often continuous, with the individual spreading perigynia more evident than the individual spikes; spikes 2–10, to 5 mm long, with about 2–7 perigynia, sessile, androgynous, the staminate portion inconspicuous; pistillate scales shorter than the perigynia, with greenish or brownish centers and hyaline margins. **PERIGYNIA** 3.3–3.75 mm long, 1.7–2.3 mm wide, oblong-elliptic, with more or less rounded, wingless margins, pale green or straw-colored, the marginal nerves somewhat displaced onto the ventral surface, otherwise nerveless or nearly so, the body entire, rather abruptly contracted to a very finely serrulate beak 0.6–1 mm long, this slightly or not at all bidentate; stigmas 2. **ACHENES** lenticular, filling the perigynia and sometimes rupturing it at maturity. Gambel's oak and other communities. COLO: Con, Hin, LPl, Min, Mon; NMEX: RAr. 2225–2530 m (7300–9535′). Flowering: Jun–Jul. California to South Dakota, south to Mexico. This taxon is rather easily confused with *C. occidentalis*.

Carex vernacula L. H. Bailey (of the spring) Alpine blackhead sedge. Plants 5–25 cm tall; culms arising singly or 2 or 3 together from rhizomes. **LEAVES** basal and on the lower culms, the blades 2–4 mm wide, flat or nearly so. **INFL** a globose or ovoid head 8–16 mm long, surpassing the leaves, without a subtending bract or this shorter than the inflorescence; spikes usually 10 or more, androgynous, about 4–7 mm long, sessile, very densely crowded and mostly not evident in the head, concealing the very short internodes of the rachis; pistillate scales subequal to the perigynia, dark brown with paler midrib and hyaline margins. **PERIGYNIA** 3.3–4.7 mm long, 1–1.6 mm wide, lanceolate to narrowly elliptic, usually brownish, often with paler greenish margins, flattened, with the marginal nerves prominent, faintly nerved toward the base on both sides, conspicuously beaked; stigmas 2. **ACHENES** lenticular. Alpine communities. COLO: Arc, Con, Hin, LPl, Min, Mon, SJn. 3200–3900 m (10500–12800′). Flowering: Jul–Aug. Washington to Wyoming and south to California and Colorado. This taxon is similar to the European *C. foetida* All., and it is often combined with it. The number of collections from our area indicate that it is not common in the San Juan Mountains.

Carex viridula Michx. (green) Green sedge. Plants 12–33 cm tall, densely caespitose; rhizomes lacking. **LEAVES** basal and on the lower culm, the blades 1–3 mm wide, flat or nearly so. **INFL** 2–5 (10) cm long, exceeded by the leafy bracts; terminal spike 7–21 mm long, staminate or with a few perigynia at the base or intermixed with the staminate flowers;

lateral spikes 2–4, pistillate, 0.5–1 cm long, about 5 mm wide, the upper ones aggregated, closely subtending the terminal spike, and sessile or nearly so, the lowest one approximate and sessile to widely spaced (the internode 5–12 cm long) and on a peduncle to 2.3 cm long, this enclosed in the sheath (to 2 cm long) of the subtending bract; pistillate scales much shorter than the perigynia, pale green, mostly hyaline. **PERIGYNIA** 2.2–3.5 mm long, including the conspicuous beak, pale green or yellow-green, strongly ribbed, obovoid; stigmas 3. **ACHENES** trigonous, filling the lower part of the perigynia only. $2n = 70, 72$. Wet places from foothills to subalpine. COLO: LPl, SJn (Johnston 2001). 1830–2745 m (6000–9000′). Flowering: Jun–Aug. Alaska to Newfoundland and south to California, Arizona, and the New England states; Eurasia, Japan. Our plants are referable to **var.** ***viridula.***

Carex vulpinoidea Michx. (like a fox) Fox sedge. Plants 20–100 cm tall, caespitose from short, tough rootstocks; culms stiff, sharply triangular but roughened on the angles above, usually shorter than the leaves. **LEAVES** flat, 2–6 mm wide, long-attenuate, the sheaths cross-rugulose ventrally. **INFL** compound, spikelike, interrupted with lower branches more or less separate; spikes numerous, subtended by setaceous-elongate bracts, androgynous but the stamens few and inconspicuous; pistillate scales awned. **PERIGYNIA** 1.7–3 mm long, about 1–2 mm wide, with thickened edges and base, tapering gradually into a serrulate beak 1/2 as long as the body or longer. **ACHENES** lenticular. Infrequent in swampy habitats. Three of the four specimens seen are from a riparian community at Sambrito wetlands near Navajo Lake. COLO: Arc, LPl. 1860–2080 m (6100–6820′). Flowering: Jun–Aug. British Columbia to Newfoundland, and south to Oregon, Arizona, and Florida.

Cladium P. Browne Sawgrass

Plants perennial from robust rhizomes, with hollow, leafy stems. **INFL** a compound umbel with numerous spikelets, these few-flowered, scales spirally arranged, the lower ones empty or staminate. **FLOWERS** perfect, solitary at the apex of the spikelet; perianth bristles lacking; stamens 2; style cleft with 2 or 3 branches. **FRUIT** an achene. Two species. North America and cosmopolitan. Used for thatching.

Cladium californicum (S. Watson) O'Neill (of California) California sawgrass. Culms 1–2 m tall, subterete to subtriangular, about 1–1.5 cm thick. **LEAVES** 0.8–1.5 cm wide; flat, serrulate, the teeth cartilaginous. **INFL** of 3–10 or more umbels, with several rays and numerous spikes 3–4 mm long; scales few, reddish brown, the lower ones empty, the middle ones empty of subtending staminate flowers, the terminal one subtending a perfect flower. **ACHENES** ovoid, about 2 mm long, without a tubercle. Hanging gardens along Lake Powell. UTAH. 1125–1160 m (3700–3800′). Flowering: Jun–Jul. Rising water of Lake Powell has drowned most of the sites discovered for this plant. The few surviving relics are perched just above the high water mark of the lake. California to southern Utah and south to Mexico. Used for basketry.

Cyperus L. Flatsedge, Umbrella-sedge

(from *kypeiros*, the ancient Greek name) Plants annual or perennial. **LEAVES** linear, mostly basal and on the lower 1/2 of the culm. **INFL** of few to several clusters of spikelets, the terminal cluster commonly sessile and exceeded by lateral ones, these borne on short to elongate peduncles originating in the axils of the leafy involucral bracts; involucres commonly of well-developed, sheathless, leafy bracts, these sometimes larger than the leaves and commonly exceeding the inflorescence; spikelets with few to several scales subtending flowers, the axis of the spikelet (rachilla) winged or wingless, persistent or deciduous and breaking at the base as a unit or at the nodes as multiple units. **FLOWERS** lacking; stamens 1–3; stigmas 2 or 3. **ACHENES** lenticular or trigonous with a short stylar point but without a tubercle. About 600 species, tropical and warm regions. *Cyperus papyrus* was used as paper in Egypt.

Other species expected include *Cyperus acuminatus* Torr. & Hook., *C. schweinitzii* Torr., *C. strigosus* L., and *C. odoratus* L.

1. Rachis more or less visible between loosely arranged racemose spikelets; larger bracts commonly over 15 cm long and over 3 mm wide ....(2)
1′ Rachis mostly concealed by a dense, capitate cluster of spikelets; bracts various but often shorter than 15 cm and narrower than 3 mm ....(3)

2. Plants annual; spikelets 3–8 (11) mm long ....***C. erythrorhizos***
2′ Plants perennial from elongate stolons; larger spikelets commonly 10–20 mm long or longer ....***C. esculentus***

3. Plants perennial from rhizomes; culms arising singly or few together, commonly over 20 cm tall ..***C. fendlerianus***
3′ Plants annual, culms often clustered from fibrous roots, commonly less than 20 cm tall (in ours) ....(4)

4. Scale with a squarrose awn-point 0.3–1 mm long, with (5) 7–9 nerves .... ***C. squarrosus***
4' Scale entire or with a minute, blunt point, with 2–3 nerves .... ***C. bipartitus***

Cyperus bipartitus Torr. (2 parts) Shining flatsedge, slender flatsedge. Plants tufted, annual, 5–20 (30) cm tall. **LEAVES** on the lower 1/4 of the culms slender, 0.5–2 mm wide; involucral bracts 1–2 (3), similar to the leaves. **INFL** of spicate or headlike clusters of ascending or spreading spikelets, the clusters sessile or nearly so, or some on rays to 2.5 cm long, or rarely longer; spikelets 1–10 per cluster, (3) 7–12 mm long, linear or ovate, persistent on the rachis; rachilla wingless; scales crowded, about 14–18 per spikelet, mostly 2–2.5 mm long, blunt, with a prominent pale midrib and otherwise anthocyanic at maturity, deciduous from the persistent rachilla; stamens 2, rarely 3; style deeply bifid. **ACHENES** lenticular, 1–1.3 mm long, olive or blackish, minutely roughened or cross-ridged at maturity. [*C. rivularis* Kunth]. Shorelines, ditches, often in disturbed places. The one specimen seen (*Rich Fleming 427* SJNM) is from a riparian area at Aztec. NMEX: SJn. 1705 m (5600′). Flowering: Sep. Widespread in the United States, southeastern Canada, and South America.

Cyperus erythrorhizos Muhl. (red root) Redroot flatsedge. Plants annual, 10–73 cm tall; culms sharply triangular, arising singly or few to several together. **LEAVES** crowded toward the base of the culms, the blades elongate, 2–10 mm wide. **INFL** compact, or open and umbel-like, the clusters of spikelets elongate, sometimes cylindrical, 1–3 cm long, the terminal sessile, the others borne on rays to 9 cm long; involucral bracts commonly 4–10, unequal, the larger ones to 40 cm long, sometimes wider than the leaves; spikelets 3–8 (11) mm long, 1–1.5 mm wide, spirally arranged; rachilla narrowly hyaline-winged, the wings readily deciduous in short segments at maturity; scales about 1–1.5 mm long, broadly rounded or blunt apically, with 3–5 faint lateral nerves, closely overlapping, eventually deciduous, the midrib scarcely excurrent as a blunt mucro; stamens 3; styles 3-branched for about 1/4–1/2 their length. **ACHENES** trigonous, pale, shiny, 0.7–1 mm long. Mudflats of reservoirs and other disturbed riparian sites and sometimes associated with agriculture. COLO: Arc, LPl; NMEX: SJn; UTAH. 1100–1850 m (3600–6100′). Flowering: Sep. United States and southern Canada.

Cyperus esculentus L. (edible) Chufa flatsedge. Plants perennial, 30–70 cm tall; culms sharply triangular, usually arising singly from numerous fibrous roots and slender rhizomes which terminate in small tubers. **LEAVES** few to many, clustered toward the base of the culms, the blades elongate, 3–8 mm wide. **INFL** umbel-like and congested to open and elongate with the terminal spikelet sessile, the others on rays to 10 cm long; involucral bracts about 3–6, unequal, from shorter than to much exceeding the inflorescence; spikelets slender, 5–20 mm long, 1–2 mm wide, with about 12–28 flowers; rachilla narrowly hyaline-winged; scales 2–3 (4) mm long, several-nerved, overlapping about 1/2 their length, deciduous at maturity; stamens 3; styles 3-branched for about 1/2–3/4 their length. **ACHENES** unequally trigonous, 1.3–2 mm long. n = 48, 54, 104. Disturbed soils. Only two specimens seen. COLO: LPl; NMEX: SJn. 1705 m (5600′) Flowering: Aug. Widespread in tropical and warm-temperate regions. Used as a ceremonial emetic and ceremonial medicine by the Ramah Navajo.

Cyperus fendlerianus Boeck. (for Augustus Fendler, German-American botanical collector) Fendler flatsedge. Plants perennial from tuberlike rhizomes mostly not over 1 cm long, mostly 20–70 cm tall (about 40 cm in ours). **LEAVES** 2 or 3, commonly 20–40 cm long, about 2 mm wide (in ours). **INFL** capitate, without rays, mostly 2–3 cm long; involucral bracts 2 or 3, unequal, the longer ones up to about 15 cm long, 1–2 mm wide (in ours); spikelets sessile, 5–8 mm long, about 2–3 mm wide; scales 3–6 mm long, several-nerved, overlapping 1/2 or more their length; stamens 3; styles 3-branched. **ACHENES** trigonous. Riparian communities and ephemeral pools of slickrock communities with stunted ponderosa pine and cliffrose. ARIZ: Apa, Coc; COLO: LPl, Min, SJn; NMEX: McK, SJn. 2150–2660 m (7100–8740′). Flowering: Sep. Wyoming, Colorado, and south to Arizona and Texas. The rayless inflorescence coupled with the perennial, tuberous-rhizomatous habit readily sets this species apart from other flatsedges of the area. Flowers and seeds salted and fed to horses by the Chiricahua and Mescalero Apache.

Cyperus squarrosus L. (spreading horizontally) Bearded flatsedge. Plants annual, 2–15 cm tall, tufted, with fibrous roots. **LEAVES** borne near the base of the culms, mostly 0.5–2.5 mm wide. **INFL** with involucral bracts commonly 2–6, foliose, some or all exceeding the inflorescence; spikelets 4–10 mm long, flattened, in capitate clusters, the terminal cluster sessile, the others, if present, on slender rays to 2 cm long; scales about 1–2 mm long, with (5) 7–9 more or less raised nerves, deciduous with age, the narrowed apex awnlike, 0.3–1 mm long, spreading to somewhat recurved; rachilla wingless, persistent; stamen solitary; style 3-branched. **ACHENES** trigonous, 0.6–1 mm long. [*Cyperus aristatus* Rottb.]. Streams, terraces, and potholes in piñon-juniper and ponderosa pine communities. ARIZ: Apa, Coc; NMEX: McK, SJn; UTAH. 1100–2285 m (3600–7500′). Flowering: Jul–Sep. More or less cosmopolitan except at high latitudes.

Eleocharis R. Br. Spike-rush

(from the Greek *helos*, a marsh, and *charis*, grace) Plants annual or perennial. **CULMS** angular, flattened, or terete. **LEAVES** reduced to bladeless sheaths, or the sheaths bristle-tipped, basal or nearly so. **INFL** a solitary, terminal spikelet; involucral bracts lacking; scales spirally arranged, the lower empty or with flowers. **FLOWERS** perfect; perianth lacking or of up to about 10 bristles; stamens 1–3; styles 2 or 3, thickened toward the base, the thickened part persistent on the achene as a tubercle, this confluent with or sharply differentiated from the achene. **ACHENES** lenticular to more or less trigonous.

1. Stigmas 2 .... ***E. palustris***
1′ Stigmas 3 .... (2)

2. Tubercle constricted at the base, more or less forming a cap on the achene, or at least wider than high .... (3)
2′ Tubercle not constricted at the base, confluent with and not forming a cap on the achene, as high as or higher than wide .... (5)

3. Achenes yellowish, brown, or blackish in age, without cross corrugations; spikelets 5–16 cm long, usually with more than 15 flowers .... ***E. parishii***
3′ Achenes whitish or pale gray, longitudinally many-ribbed with cross corrugations forming ladderlike configurations; spikelets 2.5–9 mm long, 3–15 flowers .... (4)

4. Plants perennial, with slender rhizomes; anthers 0.7–1.3 mm long; spikelet 2.5–9 mm long .... ***E. acicularis***
4′ Plants annual; rhizomes lacking or rarely developed; anthers 0.2–0.4 mm long; spikelet 1.5–3 mm long .... ***E. bella***

5. Achenes 0.9–1.3 mm long; scales 1.5–2.5 mm long; spikelets 2.5–6 mm long .... ***E. parvula***
5′ Achenes 1.9–2.8 mm long; scales 2.5–5.5 mm long; spikelets 4–11 mm long .... (6)

6. Plants mostly 40–80 cm tall; from ascending or vertical stout rhizomes; larger culms 1–2 mm thick; spikelets 5–13 mm long; with mostly 10–20 flowers .... ***E. rostellata***
6′ Plants mostly 7–20 cm tall; from slender rhizomes; spikelets 4–8 mm long, with mostly 3–9 flowers .... ***E. quinqueflora***

Eleocharis acicularis (L.) Roem. & Schult. (needlelike) Slender spike-rush, needle spike-rush. Plants perennial, 2–15 (20) cm tall; culms 0.3–0.5 mm thick, tufted, from very slender rhizomes, usually with a solitary sheath about 3–15 mm long. **INFL** a spikelet, 2–8 mm long, with about 4–25 flowers; scales (1.3) 1.5–2.2 mm long, blackish purple, typically with a green midstripe and hyaline margins. **FLOWER** perianth bristles commonly 3–4, equaling or surpassing the achene or sometimes reduced or lacking; anthers 0.7–1.3 mm long; stigmas 3. **ACHENES** white to pale gray, rounded-trigonous, 0.7–1.1 mm long including the low-conic or more or less triangular-conic, basally constricted tubercle, with 8–18 longitudinal ribs, and with numerous cross corrugations connecting the ribs. $2n = 20$. [*Scirpus acicularis* L.]. Ephemeral pools and ponds, margins of permanent ponds and lakes, and riparian communities, often in drying mud. ARIZ: Apa; COLO: Arc, Dol, Hin, LPl, Mon, SJn; NMEX: McK, RAr, SJn; UTAH. 2315–3050 m (7600–10000′). Flowering: Jun–Aug. Circumboreal, south in America to Florida and Mexico.

Eleocharis bella (Piper) Svenson (beautiful) Pretty spike-rush. Plants annual but forming dense tufts and appearing perennial, rarely with slender rhizomes, 1–8 cm tall; culms about 0.2 mm thick, usually with a solitary short sheath. **INFL** a spikelet, 1.5–3 mm long, with 3–15 flowers; scales 1–1.3 mm long, with a green or pale midstripe and hyaline margins and sometimes with dark purplish stripes flanking the green midstripe. **FLOWER** perianth bristles lacking; anthers 0.2–0.4 mm long; stigmas 3. **ACHENES** white or nearly so, rounded-trigonous, with several longitudinal ribs and numerous cross corrugations connecting the ribs, these forming ladderlike configurations, 0.6–0.8 mm long including the low-conic, basally constricted tubercle. [*E. acicularis* (L.) Roem. & Schult. var. *bella* Piper]. Borders of marshes and lakes. An immature specimen (*Reeves 9773* SJNM) from near Berland Lake appears to belong to this taxon. NMEX: SJn. 2660 m (8730′). Flowering: Sep. Washington to Idaho, south to California and New Mexico.

Eleocharis palustris (L.) Roem. & Schult. (from a swamp or marsh) Common spike-rush. Plants perennial, 1–7 (10) dm tall; culms arising singly or few together or more or less tufted, from slender creeping rhizomes, 0.5–3 (4) mm thick, usually with 2 sheaths, at least the lower one usually reddish or dark purple, the upper one extending up to about 2–15 cm on the culm. **INFL** a spikelet, 5–25 mm long, linear to lance-ovate in outline, with several to many flowers; scales 2–4.5 mm long, brownish purple, with a greenish midstripe and hyaline margins, the lower 1 or 2 empty. **FLOWER**

perianth bristles 4 (5 or 6), retrorsely barbellate, a little longer than the achene or sometimes reduced or lacking; anthers 1.3–2.5 mm long (dry); stigmas 2. **ACHENES** lenticular, green-yellow, yellow, or medium brown, finely cellular-roughened, about 1.5–2.5 mm long including the tubercle; tubercle constricted at the base, rather elongate-conic, 0.4–0.7 mm long. $2n = 14, 16, 18, 38$. [*Scirpus palustris* L.; *E. calva* Torr.; *E. erythropoda* Steud.; *E. macrostachya* Britton; *E. uniglumis* (Link) Schult.]. Ephemeral pools, margins of ponds and lakes, other riparian and palustrine communities, and hanging gardens; tolerant of saline conditions. ARIZ: Apa, Nav; COLO: Arc, Dol, Hin, LPl, Min, Mon, SJn, SMg; NMEX: McK, RAr, SJn; UTAH. 1100–2745 m (3600–9000′). Flowering: May–Jul. Widespread in the Northern Hemisphere. A conservative approach for this complex that has been divided into several taxa is followed here. The various taxa of the complex have been separated on small differences, and in most cases the segregates intergrade morphologically and lack clear ecogeographic correlations on anything beyond a local basis (Cronquist et al. 1977).

Eleocharis parishii Britton (for Samuel Bonsall Parish) Plants perennial, 7–40 cm tall; culms to about 1 mm thick, arising singly or few together from slender rhizomes; sheaths usually 2 per culm, reddish or purplish at least in part, the upper one extending to as much as 4.5 cm up the culm, usually at least a few of them ending in a mucro to 0.8 mm long. **INFL** a spike, 5–16 mm long, about 3–7 times longer than wide, with about 20–40 flowers; scales usually 2.4–3 mm long, purplish or brownish purple, usually with a green or pale midstripe and hyaline margins, the lowest one tending to be shorter and broader than the others and empty. **FLOWER** perianth bristles usually 6–7, white or brown, retrorsely barbellate, longer or shorter than the achene; stigmas 3. **ACHENES** plano-convex or unequally trigonous, greenish yellow when young, yellow or brownish to blackish when mature, 1–1.2 mm long, the tubercle more or less elongate-conic, much like that of *E. palustris* but perhaps not quite so noticeably constricted at the base. Wet places in desert regions and moist meadows in the mountains. ARIZ: Apa; COLO: Hin; NMEX: SJn; UTAH. 1125–2550 m (3700–8530′). Flowering: Jun–Sep. Oregon to Utah and south to California and Texas, also in Nebraska.

Eleocharis parvula (Roem. & Schult.) Link ex Bluff, Nees & Schauer (very small) Small spike-rush. Plants perennial, from very slender, inconspicuous rhizomes, 2–7 (10) cm tall; culms about 0.5–1 mm thick, densely tufted, sometimes forming mats, usually with only 1 sheath, this to about 1 cm long. **INFL** a spikelet, 2.5–5 mm long, with 2–9 (20) flowers; scales 1.5–2 (2.5) mm long, the lowest one empty. **FLOWER** stigmas 3. **ACHENES** 0.9–1.3 mm long, usually unequally trigonous; tubercle confluent with the achene, not constricted, very short. Wet, alkaline sites. The one specimen seen (*S. L. O'Kane et al. 6609* SJNM) is from drying mud at Totten Lake shore. COLO: Mon. 2105 m (6900′). Flowering: Jun–Sep. British Columbia to Newfoundland, and south to South America; Europe.

Eleocharis quinqueflora (Hartmann) O. Schwarz (five-flowered) Fewflower spike-rush. Plants perennial, 7–20 (30) cm tall; culms 0.2–1 mm thick, arising singly or few together or occasionally tufted, from slender rhizomes, usually with 1 or 2 sheaths, these entire, pale or reddish, the upper one extending 1–2 cm up the culm. **INFL** a spikelet 4–8 mm long, with about 3–9 flowers; scales 2.5–5.5 mm long, commonly purplish to purple-black, with hyaline margins, often without a green or pale midstripe, the lowest 2 or 3 from 1/2 to 3/4 as long as the spikelet or longer. **FLOWER** perianth bristles equaling or exceeding the achene or sometimes reduced, finely retrorsely barbellate; stigmas 3. **ACHENE** 1.9–2.6 mm long, equally or unequally trigonous or plano-convex, broadest above the middle; tubercle confluent with the achene, not constricted, higher than wide. $n = 68$. [*E. pauciflora* (Lightf.) Link var. *fernaldii* Svenson]. Fens, wet meadows, and seeps. COLO: Arc, Con, Hin, LPl, SJn. 2680–3700 m (8800–12150′). Flowering: Jul–Aug. Circumboreal, south in America to California and New Jersey.

Eleocharis rostellata (Torr.) Torr. (with a small beak) Beaked spike-rush. Plants perennial, (1.5) 4–8 (10) dm tall; culms usually 1–2 mm thick, few together or densely clustered, from a thick, ascending or nearly vertical rhizome, some of the culms often arching and rooting from an apical bulbil (proliferous). **LEAVES** with sheaths usually 2 per culm, brownish or occasionally reddish or purple, the upper one extending about 2–8 cm up the culm. **INFL** a spikelet (5) 8–11 mm long, with (5) 10–20 (25) flowers; scales purplish to black-purple, with broad pale hyaline margins, with or without a green or pale midstripe, about 3–4 mm long; lowest scale a little larger than the others, rarely over 1/2 as long as the spikelet, empty. **FLOWER** perianth bristles somewhat shorter to a little longer than the achene, retrorsely barbellate; stigmas 3. **ACHENES** 1.9–2.8 mm long, rounded-trigonous to plano-convex, smooth or slightly cellular-roughened, greenish, the tubercle confluent with the body of the achene, not constricted, medium to high-conic. [*Scirpus rostellatus* Torr.]. Baltic rush–arrowgrass communities and other lowland saline communities. COLO: Arc, LPl; NMEX: SJn; UTAH. 1615–2500 m (5300–8200′). Flowering: Jul–Aug. British Columbia to Nova Scotia, south to South America. The arching vegetative stems, rooted at both ends, catching one's feet when walking, are unusual features of this taxon. Ceremonial emetic and ceremonial medicine of the Ramah Navajo.

Eriophorum L. Cotton-grass, Cotton-sedge

(bearing wool or cotton) Perennial, grasslike plants with mostly solid culms. **LEAVES** with leaf sheaths closed, the blades grasslike or the upper ones sometimes lacking; bracts subtending the inflorescence scalelike to leaflike. **INFL** a spikelet, 1 to several in a terminal inflorescence; scales of spikelets spirally arranged, not awned. **FLOWER** perianth of numerous persistent bristles, these elongate at least in fruit and forming a cottonlike tuft; stamens 3; styles 3. **ACHENES** unequally trigonous, with a usually slender apiculate tip. Twenty species. North-temperate.

1. Spikes (2) 3–6 (8); leaves with well-developed blades 2–6 mm wide .... ***E. angustifolium***
1' Spikes solitary; leaves with reduced blades, these mostly not over 1 mm wide .... (2)

2. Anthers not longer than 1.5 mm; perianth bristles bright white; spikelets broadly obovoid to subglobose in fruit .... ***E. scheuchzeri***
2' Anthers 1 mm or longer; perianth bristles red-brown to white; spikelets globose in fruit .... ***E. chamissonis***

Eriophorum angustifolium Honck. (narrow-leaved) Manyhead cotton-grass. Plants perennial, 20–60 cm tall; culms mostly arising singly, from long-creeping rhizomes. **LEAVES** with usually well-developed blades, mainly 2–6 mm wide, flat or narrowed and channeled to triangular distally. **INFL** a spikelet (2) 3–6 (8) cm, mostly on slender peduncles to 2.5 cm long, or rarely sessile or nearly so, in an umbellate inflorescence; lowest involucral bract more or less leaflike, 2.5–5.5 cm long, often dark purple, at least toward the base, the upper bracts much smaller; scales greenish, brownish, or purplish, with hyaline margins. **FLOWER** perianth bristles numerous, white, forming cottonlike tufts 2–3.5 cm long. **ACHENES** blackish, 2–3 mm long, oblanceolate to obovate. $2n = 58, 60$. [*Eriophorum polystachion* L. (a rejected name)]. Wet meadows and peaty bogs. COLO: Arc, Hin, LPl, RGr, SJn. 2680–3595 m (8800–11800′). Flowering: Jul–Aug. Circumboreal, south in America to Oregon, New Mexico, and New York. Our plants belong to **var. *angustifolium***.

Eriophorum chamissonis C. A. Mey. (for Ludolf Karl Adalbert von Chamisso, German literature) Rusty cotton-grass. Plants perennial, 20–80 cm tall; colonial from long-creeping rhizomes. **LEAVES** with filiform blades, trigonous-channeled, 3–10 cm long, 1–2 mm wide. **INFL** a solitary spikelet, erect, globose in fruit, 1.5–4 cm; involucral bracts absent; scales blackish to purplish brown, margins white or paler. **FLOWER** perianth bristles 10 or more, red-brown to white, 20–40 mm long. **ACHENES** oblong-obovoid, 2–2.7 mm, as wide as long, apex apiculate. $2n = 58$. Peat bogs and marshes. Two records. Peat fen on Endlich Mesa and one mile southeast of Grizzly Peak. COLO: LPl, SJn. 3525–3810 m (11560–12500′). Flowering: Jul–Aug. Greenland, Alberta, British Columbia south to Idaho and Colorado.

Eriophorum scheuchzeri Hoppe **CP** (after the naturalist Johann Jakob Scheuchzer, 1672–1733) Singlehead cotton-grass, white cotton-grass. Plants perennial, 13–40 cm tall; culms mostly arising singly from long-creeping rhizomes. **LEAVES** with blades to 8 cm long and 1 mm wide, channeled or triangular, often much reduced or lacking, especially the upper one. **INFL** a solitary spike, ovoid to subglobose; subtending involucral bracts scalelike, 3–8 mm long, blackish or black-purple; fertile scales blackish, with broad white hyaline margins, acute. **FLOWER** perianth bristles numerous, white, forming cottonlike tufts 2–3.5 cm long. **ACHENES** brown or blackish, about 2 mm long. Tundra, wet peat, marshy ground, peaty soils, and pond shores. COLO: Hin, LPl, SJn. 3290–3830 m (10800–12560′). Flowering: Jul–Aug. Circumboreal, south in America to Colorado and Newfoundland. *Eriophorum russeolum* Fr. has been reported for the area. Four subspecific names for this taxon are treated as synomyns of *E. chamissonis*, but *E. russeolum* is not formally recognized for North America by the Flora of North America Editorial Committee (2002).

Fimbristylis Vahl Fimbry

(fringed style) Plants annual or perennial; culms leafy at the base. **INFL** a loose umbel or capitate cluster of spikelets, subtended by 1 to several leafy, sheathless involucral bracts; spikelets with spirally arranged scales. **FLOWERS** all perfect; perianth bristles lacking; stamens 1–3; style with 2–3 branches, enlarged toward the base. **FRUIT** an achene. Two hundred fifty species. Warm regions.

Fimbristylis spadicea (L.) Vahl (brown) Fimbry. Plants perennial, 20–90 cm tall. **LEAVES** basal and on the lower 1/5 of the culms, the sheaths more or less closed, glabrous or pubescent, the ventral side membranous, often with minute brownish or reddish spots, the blades 1–4 mm wide. **INFL** simple or more often compound, with the terminal spikelet sessile and subtended by an involucral bract, the lateral spikelets on rays or peduncles mainly 1–7 cm long; spikelets 8–23 mm long, 3–5 mm thick, elliptic-cylindrical, many-flowered; scales 3–5 mm long, usually ciliate to pubescent over the back, reddish brown or grayish, the midrib lighter and prolonged into a mucro or short awn; stamens 3; style 2-lobed about 1/4 the length. **ACHENES** about 1.5 mm long, obovate, minutely apiculate, finely many-ribbed and cross-rugulose. $n = 10$. [*Scirpus spadiceus* L.; *F. puberula* (Michx.) Vahl; *F. thermalis* S. Watson]. Wetlands in valleys and lowlands. The one specimen seen (*Cottam 9546* BRY) is from Bluff Cliffs. UTAH. 1370 m (4500′). Flowering: May–Jun. Tropical America north to California, Ontario, and New York. The name *F. spadicea* is used here to encompass this complex as a broadly defined species.

Kobresia Willd. Kobresia

Plants perennial, densely caespitose, grasslike; culms obtusely triangular, solid. **LEAVES** with closed sheaths and narrow blades. **INFL** of 1 (in ours) to several spikes with few to several spikelets with spirally arranged scales. **FLOWERS** imperfect, without a perianth, staminate and pistillate borne together throughout the spike (unlike *Carex*) or sometimes separately; stamens 3; stigmas 3. **ACHENES** subtended by 2 small bracts, the lower bract more or less corresponding to the lower bracteate scale in *Carex* and the inner bract corresponding to the perigynium but with unsealed margins and exposing the achene. Thirty-five species. North-temperate.

Kobresia myosuroides (Vill.) Fiori (like a mouse) Bellard kobresia. Plants 5–20 cm tall, caespitose. **LEAVES** basal and on the lower stem, the blades tighly involute or channeled, wiry, about 0.4–0.8 mm wide. **INFL** a solitary spike, 1–2 cm long, 2–3 mm wide, with the terminal spikelet staminate; the lateral spikelets mostly with 2 flowers, mostly androgynous, occasionally pistillate; scales 3–4 mm long, brownish, largely scarious. **ACHENES** about 2.5 mm long, brownish with a darker, persistent, apiculate style base. [*Carex bellardii* All.; *C. myosurioides* Vill.; *K. bellardii* (All.) Degl. ex Loisel.]. Moderately windswept alpine benches and convex slopes where snow accumulation is comparatively shallow over the winter. COLO: Arc, Con, Hin, RGr, SJn. 3535–3900 m (11600–12800′). Flowering: Jul–Aug. Circumboreal and south in western North America to Idaho and Colorado. This plant is sometimes confused with *Carex elynoides*. However, the split perigynium and flower arrangement (pistillate and staminate flowers borne throughout the spike) provide strong (although not readily conspicuous) separation of the two.

Rhynchospora Vahl Beak Rush

(nose-seed) Ours perennial, perfect or monoecious, caespitose to shortly rhizomatous. **LEAVES** 3-ranked, with blades short to elongate, the sheaths closed. **INFL** a spikelet aggregated in a dense, headlike cluster. **FLOWERS** solitary, in axils of scales, few to several per spikelet, the lower perfect, the upper imperfect; ovary lacking an enclosed scale; stigmas 2; stamens mostly 3; perianth of 9–15 bristles. **FRUIT** a lens-shaped achene (Welsh 1974). Over 250 species worldwide, mostly in wet, acidic soils.

Rhynchospora alba (L.) Vahl (white, referring to the white scales) Plants caespitose, with short rhizomes. **STEMS** 1.5–3.5 dm tall. **LEAVES** with straw-colored sheaths, persistent, blades 0.3–1 mm broad, linear; lowermost bract 0.5–6 cm long, sheathless or with a closed cylindrical sheath 1–15 mm long. **INFL** a spikelet, aggregated into 1–4 dense, headlike clusters; scales ovate to lance-ovate, whitish, bristles retrorsely barbed, about as long as the brownish tuberculate achene. **ACHENES** 1.5–2.5 mm long, topped by an acuminate tubercle about 1/3–1/2 as long as the achene. Wet, boggy fens. COLO: LPl. 3050 m (8600′). Flowering/Fruit: Jul–Aug (Sep). Alaska disjunctly to Newfoundland and south to California, Idaho, Illinois, Ohio, and North Carolina.

Scirpus L. Bulrush

Plants annual or perennial, rhizomatous or tufted. **LEAVES** with well-developed linear blades or reduced to bladeless sheaths. **INFL** capitate, spicate, paniculate, or umbellate; involucre of 1–several scalelike or leaflike bracts; spikes or spikelets 1 to many; scales spirally arranged, with or without an excurrent awn. **FLOWERS** perfect; perianth of 1–6 bristles, these sometimes obsolete; stamens 3 (rarely fewer); stigmas 2. **ACHENES** lenticular or 3 and trigonous; with or without a stylar apiculus, but without a tubercle. As represented in the Four Corners, the eight species of *Scirpus* could be split into four genera. However, variation and relationships within *Scirpus* as a diverse genus are adequately treated at the sectional level without requiring an expanded binomial nomenclature. Thus a broad interpretation of *Scirpus* is followed in this treatment. About 200 species. Cosmopolitan.

1. Involucre with 2 or more leaflike, spreading bracts ....(2)
1′ Involucre a single leaflike bract, this often erect or nearly so and more or less appearing as a continuation of the culm ....(4)

2. Spikelets mostly 3–25, 10–35 mm long, sessile in 1–5 headlike clusters .... ***S. maritimus***
2′ Spikelets often over 100, mostly 3–6 mm long, in an open umbellate inflorescence ....(3)

3. Stigmas 2; midrib of scales abruptly contracted into a short mucro .... ***S. microcarpus***
3′ Stigmas 3; midrib of scales tapered into a short awn .... ***S. pallidus***

4. Inflorescence with conspicuous branches with more than 20 spikelets; culms terete, 80–300 cm tall; leaves reduced to bladeless sheaths or the blades short and erect ....(5)
4′ Inflorescence spicate with spikes attached directly to the culm; culms subangular to sharply 3-angled, 10–100 cm tall or rarely taller....(6)

5. Scales dull orange or reddish brown, 2–3 (3.5) mm long and equaling or slightly surpassing the achenes ..***S. validus***
5′ Scales dull gray-brown, 3.5–4 mm long and well exceeding the achenes ....***S. acutus***

6. Achenes entire at the apex, not apiculate; scales mostly entire at the apex, the midrib hardly (if at all) exserted; spikelets 1 (in ours) ....***S. nevadensis***
6′ Achenes apiculate; scales notched at the apex, with the midrib extending beyond the notch as a mucro; spikelets mostly more than 1 ....(7)

7. Upper leaf blade shorter than, equaling, or rarely 1.5 times longer than the sheath; apical notch of spikelet scales narrow and mostly exceeding a straight mucro; culms sharply triangular, ridged to nearly winged on the angles, conspicuously concave on the sides (these culm features work best with unpressed specimens) ..***S. americanus***
7′ Upper leaf blade usually over 1.5 times longer than the sheath; apical notch of spikelet scales broad and mostly shorter than a straight or curved mucro or awn; culms subterete to sharply triangular but not concave on the sides ....***S. pungens***

Scirpus acutus Muhl. ex Bigelow (sharpened to a point) Hardstem bulrush, tule. Plants perennial, commonly 10–30 dm tall, from robust rhizomes; culms terete, about 5–20 (30) mm thick. **LEAVES** borne on the lower 1/4 of the culms, reduced to bladeless sheaths, or the upper one with a blade up to 12 cm long. **INFL** compact, umbellate, subtended by a greenish involucral bract that simulates a continuation of the culm, this commonly shorter than or equaling the inflorescence, rarely longer, also with additional inconspicuous bracts; spikelets usually numerous, more or less grayish or gray-brown, commonly 8–15 mm long; lower scales (3) 3.5–4 mm long, mostly gray-hyaline except for the greenish midrib, the gray-hyaline background sharply contrasting with reddish brown short lines, the margins ciliolate or lacerate. **FLOWER** perianth bristles retrorsely barbellate, subequal to or slightly exceeding the achene; styles 2-branched for about 3/4 their length. **ACHENES** 2.2–2.5 mm long, plano-convex, more or less completely hidden by the scales. $2n = 36$. [*Schoenoplectus acutus* (Muhl. ex Bigelow) Á. Löve & D. Löve]. Margins of ponds and lakes, marshes, washes, and floodplains. ARIZ: Apa; COLO: Arc, LPl, Mon; NMEX: RAr, SJn; UTAH. 1000–2650 m (3300–8700′). Flowering: Jun–Aug. Temperate North America. Used as a ceremonial emetic and ceremonial medicine by Ramah Navajo and for basketry by Northern Paiute.

Scirpus americanus Pers. (of America) American threesquare, chairmakers bulrush. Plants to 1 m tall from robust rhizomes; culms very sharply triangular, ridged or winged on the 3 edges, often conspicuously concave on the sides, easily flattened in pressing, mostly thicker than in *S. pungens*, commonly 5–10 mm thick toward the base. **LEAVES** commonly 1–4, borne on the lower 1/3 of the culm, the lower ones often reduced to bladeless sheaths, the upper ones with blades to 10 cm long, rarely longer, the blades usually strongly folded, to 6 mm wide when pressed, gradually tapered. **INFL** spicate or capitate, appearing lateral to a solitary leaflike involucral bract that appears to be a continuation of the culm, this 1–3.5 (5) cm long and commonly 1–3 times longer than the inflorescence; spikelets 2–15, sessile, usually 6–15 mm long; scales yellowish brown, reddish brown, or purplish brown, hyaline, with a narrow apical notch, the midrib firm and commonly exserted into the notch as a straight mucro that is shorter than the notch. **FLOWER** perianth bristles 4 (6), usually retorsely barbellate; styles 2. **ACHENES** 2–3 mm long including the apiculus, 1.4–1.7 mm wide, plano-convex. $n = 39$. [*Schoenoplectus americanus* (Pers.) Volkart ex Schinz & R. Keller]. Often alkaline sites; marshes and wet meadows. The one specimen seen (*D. Loebig 320* SJNM) is from along a stream in Cross Canyon, Colorado. ARIZ: Nav; COLO: Dol, Mon. 1525–1980 m (5000–6500′). Flowering: Jun–Aug. British Columbia to Maine and south to Mexico.

Scirpus maritimus L. (of the sea coast) Alkali bulrush, bulbous bulrush. Plants perennial, 2–15 dm tall, from robust rhizomes, these commonly bearing tubers; culms sharply triangular, often with concave sides, 3–13 mm thick toward the base. **LEAVES** commonly 4–8, borne on the lower 1/2 (3/4) of the culm, mostly all with well-developed folded or flat blades (5) 10–40 cm long or longer and to 1 cm wide, the lower 1 or 2 sometimes reduced to bladeless sheaths. **INFL** of sessile and pedunculate clusters of spikelets, subtended by (1) 2–4 leaflike, unequal involucral bracts, the longest bract up to 34 cm long; scalelike involucral bracts usually few to several, the larger ones commonly with the green midrib exserted as a caudate awn; spikelets 3–25 or rarely more, 10–35 mm long, 5–10 mm thick, all sessile in a compact cluster or 1–4 additional clusters borne on peduncles to 6 cm long; scales tan or light brown, rarely dark brown, scarious, minutely hirtellous on the back, the firm midrib exserted as a mucronate awn, this commonly 1–3 times longer than the apical notch of the scale. **FLOWER** perianth bristles few, minutely retrorse-barbellate, about 1/4–1/2 (3/4) as long as the achenes; styles 2-branched for about 1/3 the length. **ACHENES** lenticular, 2.5–4 mm long, minutely cellular-reticulate. $n = 55$. $2n = 40, 90$. [*S. paludosus* A. Nelson; *Bolboschoenus maritimus* (L.) Palla]. Tolerant of alkaline and saline conditions, riparian and palustrine communities. COLO: Dol, LPl, Mon; NMEX: RAr, SJn; UTAH. 1220–2225 m (4000–7300′). Flowering: May–Jul. Widespread in the Northern Hemisphere. Seeds were parched and ground into flour and made into mush by Northern Paiute.

Scirpus microcarpus C. Presl (small fruits) Panicled bulrush. Plants perennial, (3) 6–15 dm tall, from robust rhizomes; culms obtusely triangular, 6–15 (20) mm thick toward the base. **LEAVES** well distributed on the stems, the sheaths often reddish purple in part, the blades well developed, flat, 15–60 cm long, 6–20 mm wide; leaflike involucral bracts 3–5, to 27 cm long. **INFL** a compound umbellate, terminal cyme, subtended by 3–7 leaflike involucral bracts, the longer bracts to 27 cm long; spikelets very numerous, 3–6 (8) mm long, borne in small clusters of mostly 5–20 at the ends of rather slender branches or some of the clusters sessile; scales with a green midrib and broad scarious margins, flecked with dark purple and appearing greenish black, the midrib hardly, if at all, exserted as a mucro. **FLOWER** perianth bristles 4–6, rather sparsely retrorsely barbellate, subequal to the achene; styles 2-branched for about 1/2–3/4 their length. **ACHENES** lenticular, about 1 mm long. $n = 33$. $2n = 67$. Riparian communities. ARIZ: Apa; COLO: Arc, Dol, Hin, LPl, Min, Mon; NMEX: RAr, SJn; UTAH. 2000–2880 m (6200–9450′). Flowering: Jun–Aug. Canada south to California and West Virginia.

Scirpus nevadensis S. Watson (of Nevada) Nevada bulrush. Plants perennial, 1–2.5 dm tall (in ours), with creeping rhizomes; culms subterete. **LEAVES** few to several, clustered near the base of the culms, the lower ones sometimes reduced to bladeless sheaths, the upper ones with channeled or flat blades to 12 cm long (in ours) and 1–3 mm wide; leaflike involucral bract solitary, about 1 cm long (in ours); scalelike involucral bracts 1 or more, sometimes with the firm midrib exserted as an awn shorter than the body. **INFL** 1 spikelet (in ours), up to 10 elsewhere, sessile, 1–2.2 cm long; scales shining brown, with white, hyaline, sometimes ciliolate margins, the apex entire, the midrib firmer than the body and greenish, not exserted as a mucro. **FLOWER** perianth bristles 1–3 (4), retrorsely barbellate, mostly less than 1/2 as long as the achene; styles 2-branched for about 1/2 their length or less. **ACHENES** plano-convex, about 2 mm long, cellular-reticulate, not at all apiculate. [*Amphiscirpus nevadensis* (S. Watson) Oteng-Yeb.]. Alkaline/saline wet and cienega meadows, with saltgrass and other halophytes. ARIZ: Apa. 1400–1525 m (4600–5000′). Washington to Saskatchewan, south to California and Arizona; Argentina. Flowering: Jun–Jul. The plants of the two collections seen (*S. L. O'Kane et al. 5591* and *A. Clifford et al. 01–713*, both SJNM) all have a single spike with the bract shorter than the inflorescence. Although unusual for the species, these features are not unique, and the achenes clearly belong to *S. nevadensis*.

Scirpus pallidus (Britton) Fernald (pale) Pale bulrush. Plants perennial, 40–150 cm tall, from robust rhizomes; culms triangular, 6–15 (20) mm thick toward the base. **LEAVES** cauline, the sheaths not marked with reddish purple, the blades well developed, flat, commonly 20–60 cm long, 6–20 mm wide. **INFL** a compound umbellate, terminal cyme; involucral bracts leaflike, commonly 3–5, mostly 3–15 cm long or longer; spikelets very numerous, 3–4 mm long, borne in sessile or pedunculate clusters averaging larger than in *S. microcarpus*, about (20) 40 or more per cluster; scales with a green midrib and scarious margins flecked with dark purple and appearing greenish black, the midrib exserted as a short mucro. **FLOWER** perianth bristles mostly 6, minutely retrorse-barbellate above the middle, subequal to or shorter than the achene. **ACHENES** trigonous, with the ventral side the widest, about 1 mm long. Wet, low ground. The only specimen seen (*Heil 19547*, SJNM) is from a riparian community. COLO: Arc, LPl. 1860 m (6100′). Flowering: Jun–Aug. Washington to Minnesota and south to Texas and Missouri. Used as a ceremonial emetic and ceremonial medicine by Ramah Navajo.

Scirpus pungens Vahl (spiny) Common threesquare. Plants perennial, 1.3–11.6 dm tall, from robust rhizomes; culms subterete to sharply triangular, the sides not concave, 2–5 (7) mm thick. **LEAVES** commonly 2–4, borne on the lower 1/3 of the culms, the lower ones often reduced to bladeless sheaths, the upper ones with blades commonly (5) 8–25 (38) cm long, the blades flat to involute, 0.5–4 mm wide, linear, well developed. **INFL** spicate or capitate, appearing lateral to a solitary leaflike involucral bract (rarely 2) that appears to be a continuation of the culm, this 2–11 cm long and commonly 3–7 times longer than the inflorescence, these also accompanied by 1 or 2 smaller, empty, scalelike bract(s), these often blackish purple and usually with an awn from about 1/3 to as long as the body; spikelets 1–6, sessile or essentially so in a compact cluster, commonly 7–20 mm long; scales yellowish brown to reddish brown, some (especially the lower ones) often blackish purple, rather scarious, the midrib prominent and exserted as a straight or curved mucro, the mucro longer than the broad, apical notch of the scale. **FLOWER** perianth bristles 4–6, retrorsely barbellate; stigmas 2 or 3. **ACHENES** lenticular or trigonous, 2.2–3.3 mm long including the conspicuous apiculus, 1.6–2.3 mm wide. $n = 39$. $2n = 74$. [*Schoenoplectus pungens* (Vahl) Palla]. Tolerant of saline and alkaline conditions, along floodplains, ditches, and in seeps, springs, hanging gardens, and other riparian and palustrine communities with saltgrass and other halophytes. ARIZ: Apa, Nav; COLO: Arc, Dol, Mon; NMEX: RAr, SJn; UTAH. 1370–2010 m (4500–6600′). Flowering: May–Jul. Alaska, southern Canada, throughout the United States and to South America, Europe, Australia, and New Zealand. Unfortunately the name *S. americanus* Pers. has generally been misapplied to this species. While plants of the two taxa are apparently closely related, they are strikingly different in the field. Our plants are referable to ***var. longispicatus*** (Britton) Cronquist [*S. americanus* var. *longispicatus* Britton]. Seeds ground lightly into a flour and boiled into a mush by Northern Paiute.

Scirpus validus Vahl (robust) Softstem bulrush, tule. Plants perennial, commonly 8–12 dm tall, from robust rhizomes; culms terete, about 5–10 (15) mm thick. **LEAVES** borne on the lower 1/4 of the culms, reduced to bladeless sheaths or the upper one with a mostly erect blade to 9 cm long and 4 mm wide. **INFL** commonly umbellate, subtended by a greenish involucral bract that simulates a continuation of the culm, this shorter or longer than the inflorescence, also with additional inconspicuous bracts; spikelets usually numerous, more or less orange-brown or reddish brown, commonly 6–10 (15) mm long, borne singly or 2–3 together; scales (2) 2.5–3 (3.5) mm long, with red-brown short lines that hardly, if at all, contrast with the red-brown background color of the scales except on the sometimes pale hyaline margins, the margins entire or shortly fringed-ciliolate to somewhat lacerate. **FLOWER** perianth bristles retrorsely barbellate, subequal to the achenes; styles 2-branched for about 3/4 their length. **ACHENES** about 1.8–2.3 mm long, plano-convex, the margins usually not covered by the scales. [*Schoenoplectus tabernaemontani* (C. C. Gmel.) Palla]. Riparian and palustrine communities. COLO: Arc, Dol, LPl, Mon, SMg; NMEX: Mon, SJn; UTAH. 1705–2350 m (5600–7700′). Flowering: May–Jul. Temperate North America and into tropical America. American plants might be lumped with *Scirpus tabernaemontani* C. C. Gmel. of Europe, which was published in 1805, as was *S. validus*. If lumped, the name of priority might be difficult to determine. The lower end of the stalk eaten raw by Hopi.

DROSERACEAE Salisb. SUNDEW FAMILY

David W. Jamieson

Annual or perennial, insectivorous herbs. **LEAVES** alternate and basal, petioles longer than the variously shaped blades; blades bearing dark-colored digestive glands at the tips of long, red, multiseriate stalks. **INFL** arranged singly or a secund raceme on a long scape. **FLOWERS** regular and hypogynous, or sometimes perigynous; sepals 5, separate or connate at the base; petals 5, white; stamens 4–20; gynoecium of 3–5 carpels united into a single pistil with parietal placentation or 3–5-loculed with axile placentation; styles 3–5 and deeply bifurcate for much of their length. **FRUIT** a loculicidal capsule. **SEEDS** numerous.

Drosera L. Sundew

(Greek *droseros*, dewy) Annual or mostly perennial, insectivorous herbs of bogs and marshes. **LEAVES** basal and alternate and usually clearly differentiated into petioles and blades; petioles longer and narrower than the blades; blade shape variable, but characteristically covered ad- and abaxially by long, multiseriate, glandular hairs with enlarged, terminal glands which secrete insect-digesting fluids. **FLOWERS** as described for the family. *Drosera rotundifolia* L. is thus far known in Colorado only from iron fens northwest of Crested Butte but should be sought in the San Juan Mountains.

Drosera anglica Huds. (English) English sundew. Perennial herbs. **LEAVES** alternate and basal, 1–6 cm long, typically ascending to erect; blades variously obovate to mostly oblanceolate or spatulate, green, bearing abundant long, red, stalked, multicellular hairs with large, terminal, blackish secretory digestive glands; bases slightly expanded and bearing several to many long, marginal and adaxial, multicellular, hyaline hairs, reminiscent of, but devoid of the red pigmentation and terminal digestive glands of hairs on the leaf blades; petioles long and slender, dark red to reddish black, and bearing widely but ± evenly spaced short, hyaline, uniseriate, multicellular, glandular hairs. **INFL** of a single terminal flower, or 2–3-flowered secund racemes, on a scape 3–16 mm tall. **FLOWERS** hypogynous and 5-merous; sepals 4–6 mm long, connate at the base; corolla apopetalous and white, up to 8 mm long, petals narrowly oblanceolate; stamens 5, shorter than the petals; gynoecium of 3 united carpels with parietal placentation; styles 3 and each bifurcate for much of its length. **CAPSULES** loculicidal. **SEEDS** with a dark flattened testa. Peat bogs surrounded by spruce-fir forests. Known from one location near Durango Mountain Ski Resort. COLO: LPl. 2625 m (8610′). Flowering: Jul–Aug. Cordilleran northwestern North America, across Canada south of Hudson Bay from the Maritime Provinces to

Alberta, Great Lakes states. Our location is new for Colorado and is disjunct from the Yellowstone-Teton areas of Wyoming.

ELAEAGNACEAE Juss. OLEASTER FAMILY

Kenneth D. Heil

Shrubs or trees with lepidote or stellate trichomes. **LEAVES** alternate or opposite, simple, entire. **INFL** in axillary clusters. **FLOWERS** perfect or imperfect, regular, mostly 4-merous; perianth 4-lobed; stamens 4–8; pistil 1; ovary superior, 1-loculed; style 1; stigma 1. **FRUIT** a drupe or berrylike, dry, indehiscent achene. **SEEDS** mostly single, little or no endosperm. x = 6, 10, 11, 13. Three genera, 45 species. Northern Hemisphere to tropical Asia and Australia. Uses include drugs, food, fiber, and jewelry. Often planted as ornamentals.

1. Leaves alternate; stamens 4 .... ***Elaeagnus***
1′ Leaves opposite; stamens 8 .... ***Shepherdia***

Elaeagnus L. Oleaster

(Greek *elais*, olive, + *agnos*, the chaste-tree) Trees or shrubs, young twigs with dense lepidote-stellate trichomes. **LEAVES** alternate. **INFL** small axillary clusters on the twigs of the current year. **FLOWERS** perfect or imperfect, actinomorphic, stamens 4. **FRUIT** drupelike, with persistent hypanthium base, dry and mealy. Forty species in Europe, Asia, and North America.

1. Leaves mostly 3–8 times as long as wide; branchlets and leaves with silvery peltate scales only; cultivated and escaped; widespread; mostly thorny .... ***E. angustifolia***
1′ Leaves mostly 1.5–3 times as long as wide; branchlets with both silvery and brown peltate scales; native; rare; no thorns .... ***E. commutata***

Elaeagnus angustifolia L. (narrow-leaved) Russian olive, oleaster. Rapid-growing, mostly thorny, small to large trees, up to 12 m tall, trunks 1–5 dm thick. **LEAVES** lance-linear or narrowly elliptic, mostly 3–10 cm long, 0.5–1.5 cm wide, silvery and densely stellate or lepidote-stellate beneath, less so above, therefore bicolored. **FLOWERS** sickly fragrant, 8–12 mm long, silvery and with yellow corolla lobes, slightly stellate-hairy; stamens 4. **FRUIT** ellipsoid, mostly 1 cm long, at first covered with white scales; however, at full maturity dull orange-yellow with scattered scales. $2n$ = 12, 28. Introduced from Europe and found along roadsides, streams, and other moist sites. ARIZ: Apa, Nav; COLO: Arc, Dol, LPl, Mon, SMg; NMEX: McK, RAr, San, SJn; UTAH. 1180–2145 m (3900–7080′). Flowering: May–Jun. Fruit: Sep–Oct. Widespread in the United States. Russian olive is an example of what can happen when an invasive plant is not controlled. At lower elevations it has found a natural niche along most, if not all, of the Four Corners waterways.

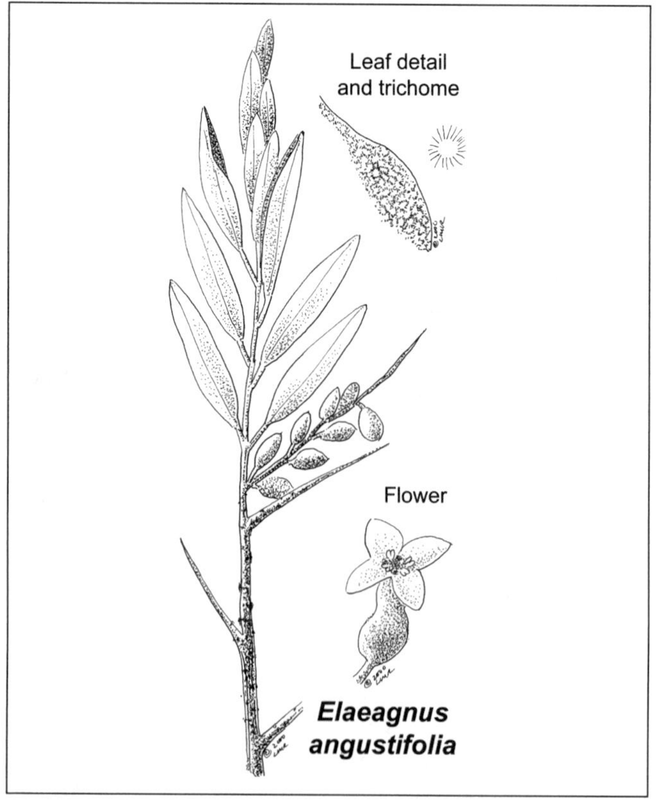

Elaeagnus angustifolia

Elaeagnus commutata Bernh. (commutative, pertaining to *Elaeagnus*) Silverberry. Unarmed shrub, mostly 2–5 m tall; young branches with brownish lepidote scales. **LEAVES** mostly 2–7 cm long, 1–3 cm wide, elliptic to oblanceolate or lanceolate, acute, obtuse, or rounded apically, silvery on both sides but greener above than beneath. **INFL** 1–4 per axil. **FLOWERS** 1–1.5 cm long, corolla lobes yellowish; stamens 4. **FRUIT** ellipsoid, 9–14 mm long, silvery, covered with scales. [*E. argentea* Pursh, not Moench]. $2n$ = 28. Stream banks with willow and Russian olive. COLO: LPl; UTAH. 1950–2230 m (6400–7350′). Flowering: Jun. Fruit: Aug. Quebec west to Alaska, south to Minnesota, North Dakota, Wyoming, and Idaho. A native ornamental.

Shepherdia Nutt. Buffaloberry, Soapberry

(for John Shepherd, 1764–1836, curator of the Liverpool Botanic Gardens and friend of Thomas Nuttall) Shrubs, young twigs densely lepidote-stellate, brownish to silvery. **LEAVES** opposite. **INFL** axillary on twigs of the previous year. **FLOWERS** perfect or imperfect, actinomorphic, sepals persistent in fruit; pistillate flowers with the hypanthium constricted at the summit; staminate flowers with 8 stamens alternating with the nectary glands. **FRUIT** berrylike.

1. Leaves evergreen, blades oval to orbicular, rounded, up to 3.5 cm long, less than twice as long as wide; fruit dry, densely lepidote-stellate .... ***S. rotundifolia***
1′ Leaves deciduous, blades variously shaped, more than twice as long as wide; fruit juicy, red or yellow, with scattered stellate scales .... (2)

2. Leaf blades mostly 3–7 cm long, 1.5–4 cm wide, mostly acute at the base; fruit red, edible; plants often thorny .... ***S. argentea***
2′ Leaf blades mostly 2–5 cm long, 0.5–2.5 cm wide, mostly rounded at the base; fruit yellowish red, unpalatable; plants unarmed .... ***S. canadensis***

Shepherdia argentea (Pursh) Nutt. (silvery) Silver buffaloberry. Often thorny, deciduous shrub or small tree 2–5 m tall; branchlets covered with silvery peltate scales. **LEAVES** short-petiolate, blades 0.5–6 cm long, mostly 3–14 mm wide, oblong or oblong-lanceolate to oblong-elliptic, mostly acute at the base, silvery with lepidote-stellate scales on both sides. **FLOWERS** subsessile, 2.5–4 mm long, corolla lobes yellowish. **FRUIT** 4–7 mm long, red, edible. $2n = 22, 26$. [*Hippophae argentea* Pursh; *Lepargyrea argentea* (Pursh) Greene]. Stream banks and other moist sites. ARIZ: Apa; COLO: Arc, LPl, Mon; NMEX: RAr, SJn; UTAH. 1600–2320 m (5300–7650′). Flowering: May. Fruit: Jun–Jul. British Columbia to Manitoba, Minnesota west to California, Nevada, and North Dakota. The berries are used to make jam and jelly.

Shepherdia canadensis (L.) Nutt. (white) Soapberry. Unarmed, highly branched deciduous shrub, up to 2 m tall; branchlets with brown peltate scales. **LEAVES** short-petiolate, blades 0.5–1 cm long, 1.5–4 cm wide, ovate to lanceolate, rounded apically and basally, green above, brownish lepidote beneath. **INFL** 1 to several per leaf axil. **FLOWERS** 2–3 mm long, the lobes brownish. **FRUIT** ellipsoid, fleshy, bright red or yellow, bitter, 4–7 mm long. $2n = 22$. [*Hippophae canadensis* L.; *Lepargyrea canadensis* (L.) Greene]. Ponderosa pine, Douglas-fir, white fir, aspen, and spruce-fir communities. ARIZ: Apa; COLO: Arc, Hin, LPl, Min, SJn; NMEX: McK, RAr, SJn; UTAH. 2500–3270 m (8250–10800′). Flowering: May–Jul. Fruit: Jul–Sep. Newfoundland to Alaska, south to New York, South Dakota, Oregon, and northern California.

Shepherdia rotundifolia Parry (nearly circular leaves) Buffaloberry. Unarmed, free-branched, evergreen shrub, up to 2 m tall; branchlets stellate-hairy with white or yellow trichomes. **LEAVES** short-petiolate, larger ones 1.5–3.5 cm long, 1–3 cm wide, ovate to orbicular or lanceolate, rounded apically, the base rounded to obtuse, silvery green above, pale beneath. **INFL** 1 to few per axil. **FLOWERS** 3.5–5 mm long, corolla lobes yellowish. **FRUIT** dry, ellipsoid, 5–8 mm long, densely covered with silvery stellate hairs. $n = 15$. [*Lepargyrea rotundifolia* (Parry) Greene; *Elaeagnus rotundifolia* (Parry) A. Nelson]. Blackbrush, sagebrush, and piñon-juniper communities. ARIZ: Apa, Coc, Nav; UTAH. 1180–2450 m (3900–8100′). Flowering Apr–May. Fruit: Jun–Aug. A Colorado Plateau endemic. This beautiful, silvery shrub has excellent potential for xeric landscaping.

ELATINACEAE Dumort. WATERWORT FAMILY

C. Barre Hellquist

Small annual or perennial herbs. **LEAVES** opposite, simple, with membranous stipules. **FLOWERS** small, actinomorphic, axillary; persistent or withering, but persistent sepals and petals imbricated in the bud; stamens the same number as the petals, alternate with them or twice the number; ovary with 2–5 locules, placenta in the axis. **FRUIT** a capsule; valves alternate with ovary partitions. **SEEDS** oblong-cylindric, straight or curved, usually with a reticulate surface pattern, the testa nearly filled by the cylindric embryo (Tucker 1986). Two genera, about 40 species worldwide, especially in temperate regions.

Elatine L. Waterwort

(from a classical name of a low, creeping plant, transferred to this genus) Annual, erect or prostrate, flaccid to succulent plants up to 10 cm long, aquatic, amphibious, or terrestrial dwarf plants. **LEAVES** opposite, sessile or petioled, with hyaline or toothed stipules; glabrous, blades linear-spatulate to oblong or orbicular-obovate, the margin obscurely and remotely crenate. **INFL** nodal, 1 or 2 flowers per node. **FLOWERS** sessile or pedicelled, 2-, 3-, or 4-merous; sepals

2–4, obtuse, or unequal in size, withering, persistent in some species; petals 2–4, hypogynous, usually orbicular in outline; stamens as many as or twice as many as petals or reduced to 1; styles or capitate stigmas 2–4. **FRUIT** a membranous capsule, 2–4-celled, several- to many-seeded, 2–4-valved, partitions left attached to the axis or evanescent. **SEEDS** are needed for a correct identification. The plants often appear quite different when submersed or emergent. About 20 species.

1. Seeds 20 or more per locule; pits 25–35 per row per seed .... ***E. chilensis***
1′ Seeds up to 15 per locule; pits 16–25 per row per seed .... ***E. triandra***

Elatine chilensis Gay (of Chile) Creeping aquatic or terrestrial plants to 10 cm long, rooting at the nodes. **LEAVES** obovate to spatulate, rounded at summit, 3–4 mm long and 1–3 mm wide, narrowed at base to a petiole with hyaline stipules. **FLOWERS** solitary in leaf axils, sessile; sepals 2, sometimes with a third, much reduced, oblong; petals white to pink, orbicular; stamens 3, alternate with the carpels. **SEEDS** 20 or more per cell, borne at the base of the placental axis, slightly curved, with 25–35 pits broader than long. In shallow water or on mud of shores of lakes and ponds. ARIZ: Apa; NMEX: SJn; UTAH. 1685–2760 m (5125–9100′). Flowering: Jul–Sep. Fruit: Jul–Sep. California, Utah, and Arizona; South America.

Elatine triandra Schkuhr (with three stamens) Matted, creeping, aquatic or terrestrial plants. **LEAVES** linear, narrowly lanceolate or narrowly spatulate, mostly truncate or emarginate, to 7 mm long and 3 mm wide. **FLOWERS** sessile, 3-merous. **SEEDS** 15 or fewer per cell, slender-cylindric and curved, with pits broader than long. In shallow water or on mud of ditches, lakes, and ponds. COLO: Hin, LPl; NMEX: McK, SJn. 2000–2735 m (6600–9030′). Flowering: Jul–Aug. Fruit: Jul–Aug. Washington east to Wisconsin, Massachusetts, and Maine, south to California, Colorado, and Texas; Mexico.

ERICACEAE Juss. HEATH FAMILY

John L. Anderson

Shrubs, subshrubs, perennial herbs (in ours), or small trees, some without chlorophyll and mycoheterotrophic. **LEAVES** usually petiolate, lacking stipules, simple, evergreen or deciduous, some scalelike or lacking, alternate or rarely opposite, entire or toothed. **INFL** usually bracteate racemes or corymbs or solitary and terminal or axillary. **FLOWERS** usually perfect, actinomorphic or nearly so and hypogynous, usually pedicellate; sepals 4 or 5, distinct or shortly connate; petals 4 or 5, distinct or connate with short lobes; stamens as many as the corolla lobes and alternate to them or often twice as many; anthers dehiscing by terminal pores or less often by longitudinal slits, often with awnlike appendages called spurs; pollen in monads, dyads, or more often tetrads; ovary superior or sometimes inferior, pistil and style 1, stigma 1 and capitate or lobed, mostly 4- or 5-locular and with axile placentation. **FRUIT** a capsule or sometimes a berry or drupe, seeds small. $x = 6$–23 (Copeland 1947; Haber 1974, 1992; Freudenstein 1999; Kron et al. 2002; Wallace 1975). About 124 genera and 4100 species. Worldwide in cool, temperate, and subtropical regions. The heath family has many important horticultural ornamentals, rhododendrons, and heaths, as well as commercially important edible species (blueberries and cranberries).

1. Plants green, with chlorophyll, autotrophic; with large leaves .... (2)
1′ Plants not green, without chlorophyll, mycoheterotrophic; leaves absent or scalelike .... (3)

2. Petals united; plants woody (suffrutescent in *Gaultheria*) .... (4)
2′ Petals distinct; plants herbaceous (or suffrutescent in *Chimaphila* and *Orthilia*) .... (6)

3. Corolla of united petals; stem over 20 cm tall, red-brown; inflorescence a long raceme .... ***Pterospora***
3′ Corolla of separate petals; stem shorter; plant pale yellow or pinkish; flowers in a nodding terminal cluster .... ***Monotropa***

4. Fruit a mealy drupe with 1–10 seeds .... ***Arctostaphylos***
4′ Fruit a fleshy berry with many seeds .... (5)

5. Plants woody and freely branching .... ***Vaccinium***
5′ Plants herbaceous to slightly woody, creeping and rooting at the nodes .... ***Gaultheria***

6. Flowers solitary .... ***Moneses***
6′ Flowers several, more than one .... (7)

7. Flowers in corymbs; stems leafy; leaves oblanceolate; staminal filaments dilated at base .... ***Chimaphila***
7' Flowers in racemes; stems scapose or nearly so, the leaves mostly basal, ovate to orbicular; staminal filaments slender ....(8)

8. Racemes secund, 1-sided; style straight, 3–5 mm long; 10-lobed nectary disc present at the base of ovary; tubercles present on the adaxial base of petals .... ***Orthilia***
8' Racemes spiral, not 1-sided; style curved, or if straight, 2 mm long; nectary disc absent from the base of ovary; tubercles absent from the adaxial base of petals .... ***Pyrola***

Arctostaphylos Adans. Manzanita, Bearberry

(Greek *arktos*, bear, + *staphyle*, bunch of grapes; fruits eaten by bears) Plants evergreen shrubs. **STEMS** prostrate to erect with bark thin, smooth, exfoliating, and red- to orange-brown. **LEAVES** evergreen, leathery, alternate, margins usually entire, petioles short or the blade subsessile. **INFL** terminal racemes or panicles. **FLOWERS** perfect, actinomorphic; sepals 5, imbricate and distinct; corolla white to pink, urceolate, sympetalous with 5 lobes; stamens 10, included, anthers spurred; filaments dilated, hairy; ovary superior with hypogynous disc at base, 5–8-locular, with 1 ovule. **FRUIT** fleshy, berrylike drupe with 2–10 nutlets. $x = 13$ (Wells 2000). Primarily western North America including Mexico.

1. Shrubs with low, creeping stems, less than 2 dm tall; leaves obovate; berries bright red .... ***A. uva-ursi***
1' Shrubs with tall, erect or spreading stems, more than 1 m tall; leaves orbicular or elliptic; berries dull orange ..(2)

2. Axis of inflorescence white-puberulent; inflorescence a raceme (rarely few-branched) .... ***A. pungens***
2' Axis of inflorescence glandular to glandular-pubescent; inflorescence a dense panicle .... ***A. patula***

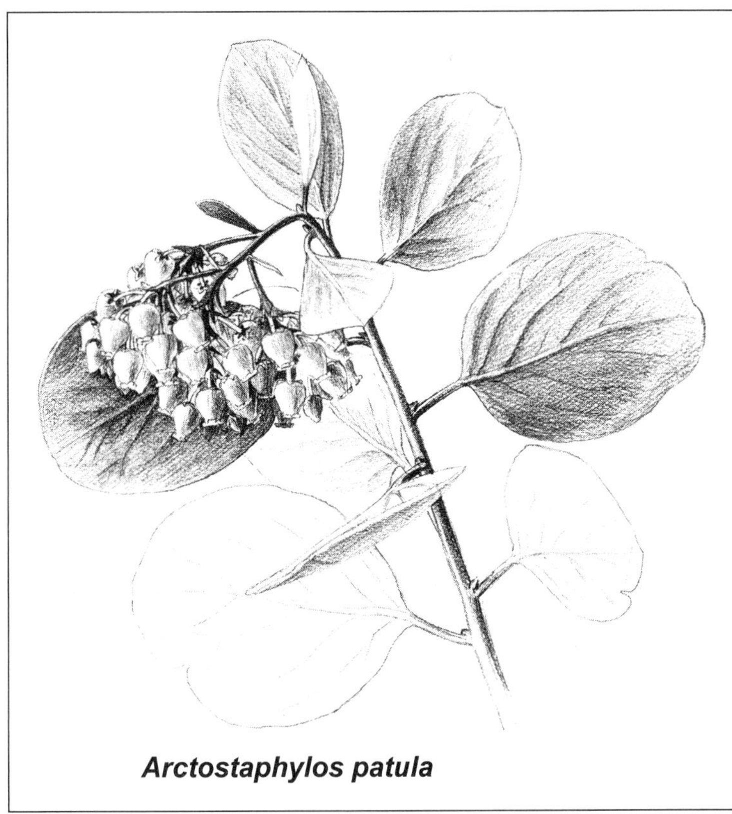

Arctostaphylos patula

Arctostaphylos patula Greene (outspread) Greenleaf manzanita. Broad shrub, 1–2 m tall. **STEMS** with rigid, spreading branches, lower ones at ground level rooting and forming low thickets, bark reddish brown, smooth, branchlets glandular-pubescent. **LEAVES** bright green, shiny, petiole 7–15 mm long, pubescent, blade 1.2–4 cm long, 1.5–4 cm wide, mostly orbicular to elliptic and narrowly elliptic, base rounded to truncate, tip obtuse to acute, margin entire, glabrous. **INFL** a dense panicle, glandular-pubescent, bracts 1–4 mm long, acuminate. **FLOWERS** pedicellate, 2–8 mm long; sepals glabrous, lobes ovate, 2 mm long; corolla white to pink, 5–8 mm long, urceolate; ovary glabrous. **FRUIT** a drupe, 8–12 mm wide, depressed-globose, glabrous, dull orange to brown. $2n = 26$. [*A. platyphylla* (A. Gray) Kuntze; *A. pungens* Kunth var. *platyphylla* A. Gray; *A. parryana* Lemmon var. *pinetorum* (Rollins) Wiesl. & B. Schreib; *A. pinetorum* Rollins]. Open forests, often in ponderosa pine savanna, and also piñon-juniper woodlands. ARIZ: Nav; NMEX: SJn; UTAH. 1890–2710 m (6200–8890'). Flowering: Mar–Jul. Washington and Montana south to Baja California, east to Arizona and Colorado. Greenleaf manzanita berries are edible and can be made into jelly and cider or the seeds ground into flour.

Arctostaphylos pungens Kunth (terminating in a sharp point) Point-leaf manzanita, Mexican manzanita. Erect, broad shrub, 1–2 m tall. **STEMS** with rigid, spreading branches, branchlets densely pubescent, bark reddish brown, smooth. **LEAVES** bright green, shiny, petiole 4–9 mm long, blade 1.5–4 cm long, 0.5–2 cm wide, elliptic to lance-elliptic, base acute to rounded, tip acute and mucronate, margin entire (young leaves may be toothed), glabrous. **INFL** simple or few-branched racemes, densely puberulent, bracts 1.5–4 mm long, acuminate. **FLOWERS** pedicellate, 2.5–6.5 mm long, glabrous; sepals glabrous, lobes 1–2 mm long, reflexed; corolla white to pink, 5–8 mm long, urceolate; ovary glabrous. **FRUIT** a drupe, 5–11 mm wide, depressed-globose, orange to brownish red. $2n = 26$. Rocky slopes in foothills with piñon-juniper woodlands and openings in ponderosa forest. ARIZ: Nav; UTAH. 2210–2805 m (7250–9200′). Flowering: May–June. Mexico north to Texas, New Mexico, Arizona, California, Nevada, and Utah.

Arctostaphylos pungens

Arctostaphylos uva-ursi (L.) Spreng. (bunch of grapes + bear) Bearberry, kinnikinnik. Prostrate shrub 0.1–0.2 m tall. **STEMS** with trailing branches along the ground rooting and forming mats, branchlets glabrous to puberulent. **LEAVES** dark green above, light green below, petiole 2–5 mm long, glandular, blade 1–2.5 cm long, 0.3–1 cm wide, oblanceolate to obovate, base wedge-shaped, tip rounded, not mucronate, margin entire, glabrous. **INFL** simple or few-branched raceme, densely puberulent; bracts acuminate, 1.5–4 mm long. **FLOWERS** pedicellate, 2–4 mm long, glabrous; sepals glabrous, lobes reflexed, 1–2 mm long; corollas white to pink, 4.5–8 mm long, urceolate; ovary glabrous. **FRUIT** a drupe, 6–12 mm wide, depressed-globose, glabrous, bright red. $2n = 26$. [*A. uva-ursi* var. *adenotricha* Fernald & J. F. Macbr.; *A. adenotricha* (Fernald & J. F. Macbr.) Á. Löve, D. Löve & B. M. Kapoor; *Arbutus uva-ursi* L.] Ground cover under coniferous forest in dry and moist sites. ARIZ: Apa; COLO: Arc, Hin, LPl, Min, Mon, SJn; NMEX: McK, RAr, SJn; UTAH. 1900–3200 m (6250–10500′). Flowering: May–Aug. Circumboreal. Bearberry leaves have many reported medicinal uses. A variable species with many taxa named based on pubescence characters that have been shown to have continuous variation (Rosatti 1987). Hybrids with *A. patula* have been reported scattered throughout their common ranges. In our area this hybrid has been collected from Navajo Mountain, San Juan County, Utah (*Holmgren et al. 10,622* UTC).

Chimaphila Pursh Prince's Pine

J. Mark Porter

(Greek *chemim*, winter, + *philos*, loving, referring to the evergreen habit) Perennial, suffruticose, semiherbaceous plants, 10–30 cm tall, from creeping rhizomes. **STEMS** erect to ascending, terete, glabrous. **LEAVES** persistent, evergreen, simple, coriaceous to leathery, alternate to subverticillate or apparently whorled. **INFL** a 1–7-flowered, umbellate corymb. **FLOWERS** actinomorphic, pedunculate; sepals 5, free nearly to the base, persistent; petals 5, distinct, rotate-campanulate, waxy, white to pink; stamens 10, the filaments dilated below, ciliate at the base, anthers subapically attached, 2-celled, opening by short tubelike pores; ovary 5-celled, the styles united, straight or curved, persistent; stigma 5-lobed and peltate. **FRUIT** a depressed-globose capsule, loculicidally dehiscent. **SEEDS** many. $x = 13$. Four to five species, circumboreal, south to Central America (Blake 1917; Freudenstein 1999).

Chimaphila umbellata (L.) W. P. C. Barton (having a simple umbel) Pipsissewa, prince's pine. Perennial plants, evergreen, suffruticose. **STEMS** spreading-ascending, the fertile branches erect, from creeping rhizomes, 10–25 (30) cm tall, glabrous. **LEAVES** bright leathery green, pale green below, simple, subverticillate or whorled, 3–6 per node, 2–6 cm long, 0.5–1.8 cm wide, petioles 3–6 mm long, exstipulate, blade elliptic to oblanceolate, cuneate, the margins sharply serrate and slightly revolute; upper surface glabrous. **INFL** a terminal umbellate corymb, 4–8 flowers, the peduncles erect, 4–7 cm long, glabrous to glandular-puberulent. **FLOWERS** 1–1.6 cm in diameter, perianth 5-merous; sepals 5, pink-purple, basally connate, the lobes 1–2 mm, ovate, erose-ciliate; corolla 4–7 mm long, distinct, white to pink, waxy, concave, broadly ovate with ciliate margin; stamens 10, the filaments dilated below,

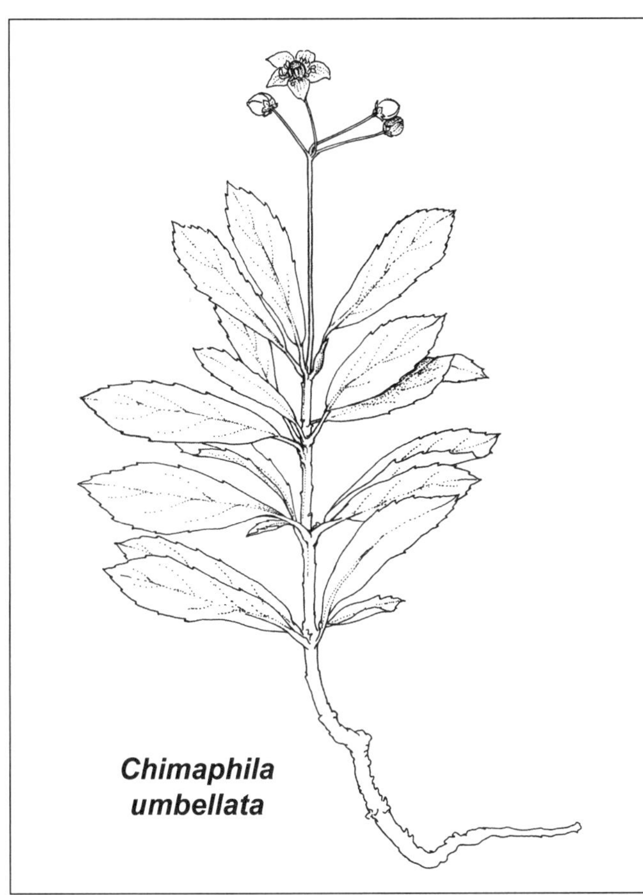

ciliate, abruptly narrowed, the short, glabrous upper portion 2 mm long; anthers plump, 2-celled, opening by 2 wide, basal pores, appearing apical by inversion; ovary 5-celled, the style short; stigma 2 mm wide, peltate, 5-lobed, persistent. **FRUIT** a capsule, 5–6 mm in diameter, loculicidally dehiscent from the apex. $2n = 26$. [*C. occidentalis* Rydb.]. Our material represents **subsp. *occidentalis*** (Rydb.) Hultén. Rich, rocky or sandy substrate in dry, coniferous forests. ARIZ: Apa; COLO: LPl, Min. 2195–3050 m (7200–11650′). Flowering: Jun–Aug. Fruit: Jul–Sep. Circumboreal, in North America south to Central America and the Caribbean.

Gaultheria L. Wintergreen

(named for Jean Francois Gaultier, 18th-century physician and botanist of Quebec) Plants low or creeping, rooting at the nodes and forming mats, slightly woody to herbaceous. **LEAVES** evergreen, alternate, margins serrate or crenate. **INFL** solitary flowers in leaf axils (in ours) or in racemes. **FLOWERS** perfect, actinomorphic, sepals 5, deeply lobed, persistent; petals 5, white, urn-shaped to campanulate; stamens 10, included, filaments dilated, anthers dehiscing by terminal pores, not spurred; ovary superior (ours) or inferior, 4–5-celled. **FRUIT** a many-seeded capsule, loculicidal, enclosed by the expanded fleshy calyx forming a dry berrylike fruit. About 115–200 species in cool and temperate regions worldwide. *Gaultheria* is a natural source of oil of wintergreen used in making tea and flavoring candies, chewing gum, and medicines. Now, commercial oil of wintergreen is synthetically produced or comes from *Betula lenta* L., sweet birch.

Gaultheria humifusa (Graham) Rydb. (spread out over the ground) Alpine wintergreen, creeping wintergreen. Shrublet, suffruticose, mostly herbaceous, less than 2 dm long. **STEMS** creeping and rooting along the ground, forming mats, glabrous or puberulent to sparsely pilose. **LEAVES** leathery, not persistent, with a short petiole, blade 0.8–2 cm long, 5–15 mm wide, oval to elliptic, margin entire at base and setose-serrulate near tip, base and tip rounded. **FLOWER** sepals glabrous; corolla white to pink, 3–4 mm long, campanulate; filament glabrous, anthers awnless (ours); ovary superior (ours). **FRUIT** berrylike, 5–7 mm wide, subglobose, red. [*G. myrsinites* Hook.; *Vaccinium humifusum* Graham]. Acidic, moist, mossy soil, meadows and stream banks, under subalpine forest. COLO: Hin, LPl, Min, Mon, SJn. 2635–3810 m (8640–12500′). Flowering: late Jun–Aug. British Columbia and Alberta south to California and Colorado.

Moneses Salisb. ex Gray Wood Nymph, Wax Flower, Single-delight
J. Mark Porter

(Greek *mono*, single, + *hesis*, delight, referring to the solitary flower) Rhizomatous, herbaceous perennials. **STEMS** short, glabrous. **LEAVES** chlorophyllous, coriaceous, persistent, mainly basal, but sometimes opposite or in whorls. **INFL** composed of a solitary, nodding flower, borne on a long peduncle. **FLOWER** sepals usually 5, persistent; petals usually 5, distinct, spreading; stamens usually 10, the filaments tapering to the apex, the anthers awnless, nodding, opening by means of apparently terminal pores; ovary superior, 5-loculed, the stigma borne on an elongate, glabrous style. **FRUIT** a loculicidal capsule. **SEEDS** many. $n = 13$ (Knudsen and Tollsten 1991; Freudenstein 1999). This monotypic genus has been shown to be closely related (i.e., a sister group) to *Chimaphila*, in spite of the morphological similarity with *Pyrola*. The exclusion of *Moneses* from *Pyrola* is unquestionable. Circumboreal.

Moneses uniflora (L.) A. Gray **CP** (one flower) Wood nymph, wax flower, one-flowered wintergreen, single-delight. Perennial herbs, from slender, creeping rhizomes. **STEMS** solitary, ascending to a slender scape, 3–10 (–12) cm tall, recurved above. **LEAVES** 2–10, opposite or in clusters of 3, crowded, petiole 5–10 mm long, as long as to shorter than the blade; blade 1–2.5 cm long, 6–20 mm wide, ovate-elliptic to obovate, crenate to finely serrate or subentire, retuse or rounded at the apex, tapered decurrently to the petiole. **INFL** composed of a solitary, nodding flower, borne on a long peduncle, 3–15 cm long, 1 (2) ovate scales, 2–3 mm long at or below the recurved portion of peduncle. **FLOWERS** 1.3–2.5 cm in diameter; sepals (4) 5, 15–25 mm long, greenish white to yellowish, the lobes ovate, reflexed, ciliate, the apex rounded, persistent; petals (4) 5, fragrant, waxy-white or tinged with pink, 7–12 mm long, spreading, orbicular or broadly ovate; stamens (8) 10, filaments 5–6 mm long, subulate, dilated below, incurved; anthers 2–3.5 mm long, attached subapically, each theca narrowed above the filament to a tubular pore; style prominent, 2–4 (6) mm long, straight, persistent, thickened distally, stigma peltate, (4) 5-lobed, lobes radiating. **FRUIT** (4) 5-celled, 5–8 mm in diameter, depressed-globose, dehiscing loculicidally from the summit. **SEEDS** numerous. $2n = 22, 24, 26, 32$. [*M. reticulata* Nutt.; *M. uniflora* var. *reticulata* (Nutt.) S. F. Blake; *Pyrola uniflora* L.]. Moist, mossy areas, boggy stream banks under coniferous forests. COLO: Arc, Hin, LPl, Min, Mon, SJn. 2100–3600 m (6900–11900′) Flowering: Jul–Aug. Fruit: late Jul–Sep. Circumboreal.

Monotropa L. Pinesap, Indian-pipe

Gary D. Wallace

(Greek *mono*, one, + *tropos*, direction, referring to the one-sided inflorescence) Plants lacking chlorophyll, white to reddish; arising from a ball-like cluster of matted roots. **STEMS** fleshy, with reduced scalelike leaves. **INFL** a raceme of solitary flowers. **FLOWERS** perfect; sepals 2–5, separate and bractlike; petals 3–6, erect, scalelike, often saccate at base, distinct; stamens 6–12, twice as many as the petals, anthers disclike. **FRUIT** a capsule. **SEEDS** numerous. Two species, circumboreal (Wallace 1975). Recent research indicates that the two species may warrant separation into two distinct genera. Achlorophyllous mycoheterotrophic species that obtain their nutrients through an ectomycorrhizal association with green plants (usually oaks or conifers).

Monotropa hypopitys L. (beneath pines) **INFL** a raceme, rarely scapose, generally under 30 cm tall. **FLOWERS** yellow to orange or reddish, never white; sepals unlike the petals in form; petals yellow to orange or reddish, 4 or 5, 8–17 mm long, broader at apex and narrowly saccate at base, often pubescent within; anthers hippocrepiform, under 1.5 mm long; stigma not funnelform, often subtended by a ring of hairs; style slender, elongate; ovary elongate-ovoid to occasionally ovoid, nectary lobes short and stout. **FRUIT** elongate, 6–10 mm long, 4–8 mm wide, ovoid, walls thin and often deciduous. Coniferous forests. ARIZ: Apa; COLO: Arc, Min; NMEX: McK, SJn. 1800–3350 m (6000–11000′). Flowering: Jul–Sep. Circumboreal with southern extensions in mountains and highlands.

Orthilia Raf. Sidebells

J. Mark Porter

(Greek *orthos*, straight, + *helix*, spiral, referring to the one-sided raceme) Herbaceous perennials or subshrubs with long slender rhizomes. **STEMS** usually solitary, erect, to 5–25 (–30) cm tall. **LEAVES** simple, alternate or subopposite, clustered in (1) 2–4 pseudowhorls at the base of the stem; blades with slightly revolute, crenulate, serrulate, or rarely entire margins, subcoriaceous. **INFL** strongly 1-sided (secund) racemes; bracts (0) 4–7; pedicellate. **FLOWERS** nodding; calyx persistent in fruit; corolla as long as or longer than broad, actinomorphic, petals distinct, broadly ovate, greenish white, yellowish green, or white, with 2 inconspicuous, basal, adaxial tubercles; stamens included or slightly exserted; filaments flattened, tapering smoothly to their bases, glabrous, light brown; anthers oblong, wrinkled and papillate, the pores large and elongate; ovary subtended by a 10-lobed, nectariferous disc; style straight, exserted; stigma pale green, peltate, with 5 prominent, spreading lobes. **FRUIT** a depressed-globose capsule, light greenish brown, pendent, the pedicel and scape eventually becoming erect, the dehiscence incomplete, the margins of the valves cobwebby. **SEEDS** light golden brown, the testa pitted. $x = 19$. Monotypic genus. Circumboreal.

Orthilia secunda (L.) House (having organs, i.e., flowers, toward the same side) One-sided wintergreen. Herbaceous to somewhat woody perennials, rhizomes slender and creeping. **STEMS** solitary, rarely 3, scape 6–19 (21) cm tall, with 1–3 lanceolate bracts; basal bracts below and among the mostly basal leaves. **LEAVES** light green, 2–10, somewhat scattered with conspicuous internodes on the ascending stem, petiole 0.6–2 cm long; blade thin, ovate to elliptic, blades 1.3–4 (–5) cm long, 1–3 cm wide, crenulate-serrulate, rounded to slightly tapering at the base, acute to rounded at the apex, longer than the petiole. **INFL** a secund raceme, 2–15-flowered, pedicels slender, 2–5

(–7) mm long, at first horizontal, then drooping, bractlets ciliate, light green, ascending, lanceolate, 2–4 mm long. **FLOWERS** actinomorphic, 4–6 mm wide; sepals 0.5–1.5 mm long, spreading, persistent, the lobes ciliate-triangular; petals white to greenish yellow, oblong, longer than wide, 3.5–5 (6) mm long, each with 2 small tubercles basally on the inner face, campanulate; stamens included, about as long as the petals, filaments slender, not declined, the anthers 1.5 mm long, not apiculate, the pores sessile; style straight, slender, (2) 3–5 mm long, in fruit to 9 mm long, exserted, lacking a collar but the stigma peltate, 5-lobed; ovary with a 10-lobed, hypogynous disc. **FRUIT** a capsule, 4–5 mm wide, depressed-globose, loculicidally dehiscing from the base. **SEEDS** many. $2n = 38$. [*Pyrola secunda* L.; *Ramischia secunda* (L.) Garcke; *Actinocyclus secundus* (L.) Klotzsch; *P. secunda* var. *obtusata* Turcz.; *Orthilia secunda* var. *obtusata* (Turcz.) House; *P. secunda* var. *pumila* Paine; *P. secunda* forma *eucycla* Fernald]. Moist woods in ponderosa pine, Douglas-fir, aspen, spruce-fir. ARIZ: Apa; COLO: Arc, Hin, LPl, Min, Mon, SJn; UTAH. 2000–3400 m (6600–11300′). Flowering: Jun–Aug. Fruit: Jul–Sep. Circumboreal, south to Mexico.

Pterospora Nutt. Pinedrops
Gary D. Wallace

(Greek *ptero*, wing, + *sporo*, seed, referring to the membranous wing on the seed) Plants without chlorophyll, purplish brown or chestnut-colored. **STEMS** absent. **LEAVES** absent, reduced or scalelike. **INFL** a raceme, 1.5–17 dm long, persistent after seed dispersal. **FLOWERS** perfect, many, pendulous; calyx of 5 slightly united sepals; corolla urceolate, globose, the petals united with short spreading lobes; stamens 10, included, anthers longitudinally dehiscent; style stout. **FRUIT** a 5-lobed capsule. **SEEDS** numerous, broadly winged. $x = 8$ (Wallace 1975). A genus with one species in North America. An achlorophyllous mycoheterotrophic species that obtains its nutrients through an ectomycorrhizal association with green plants (usually oaks or conifers).

Pterospora andromedea Nutt. **CP** (Greek *Andromeda*, daughter of Cepheus and Cassiope) Pinedrops. Achlorophyllous, mycoheterotrophic perennial with inflorescences arising annually from persistent roots. **STEMS** absent. **LEAVES** absent. **INFL** an elongate raceme to 20 dm tall, emerging from soil in an erect position, conspicuously glandular-pubescent, viscid, persistent to next year as a brown stalk. **FLOWERS** actinomorphic, pendulous on recurved pedicel; sepals 5; corolla 6–9 mm long, urceolate, cream to yellowish, sympetalous, 5 lobes reflexed; stamens of alternating lengths, anthers awned on back; stigma 5-lobed; style short, straight; ovary superior, 5-carpellate, nectary inconspicuous at base. **FRUIT** pendulous, capsular, oblate-spheroidal, 7–10 mm wide, persistent, opening from base. **SEEDS** minute, each with a membranous wing several times the size of the seed. $2n = 16, 48$. Various types of coniferous forests, growing in litter. ARIZ: Apa; COLO: Arc, Hin, LPl, Mon, SJn; NMEX: McK, RAr, SJn; UTAH. 2135–2895 m (7000–9500′). Flowering: Jun–Aug. North America especially west of the Rocky Mountains including western Canada, disjunct in the northeast from Michigan to New Hampshire and adjacent eastern Canada, also in scattered areas of Mexico. Used as an emetic or ceremonially smoked in the kiva by Pueblo peoples.

Pyrola L. Wintergreen, Shinleaf
J. Mark Porter

(Latin diminutive of *pyrus*, pear, referring to the leaf shape) Herbaceous, somewhat suffruticose perennials, rhizomatous. **STEMS** scapose, mostly simple, 5–30 cm tall, glabrous. **LEAVES** chlorophyllous, simple, coriaceous, persistent, 6–12, alternate or opposite, basal or apparently basal, rarely absent, petiolate; blades broadly ovate to orbicular, entire to toothed, glabrous. **INFL** scapose, 2–several-flowered bracteate racemes or solitary and terminal, the pedicels 8–12 mm long, spreading or recurved. **FLOWERS** actinomorphic (bell-shaped) or zygomorphic (bowl-shaped); calyx (4) 5-lobed, connate at the base; corolla broadly bowl-shaped, the petals spreading or concave, distinct, white to pink or greenish, deciduous; stamens (8) 10, the filaments narrow or basally dilated, erect or declined; anthers 2-celled, opening by 2 pores or irregular slits; ovary (4) 5-merous, the style straight or declined and curved upward at the end, with 5 spreading stigmatic lobes, persisting in fruit. **FRUIT** a loculicidal capsule. $x = 23$ (Copeland 1947; Haber 1987; Freudenstein 1999). Fifteen species. Circumboreal and extending south to the northern tropics.

1. Styles straight, short, less than 2 mm long, stigmas much wider than the styles; anther pores sessile with no evident subtending tube .... ***P. minor***
1′ Styles declined or bent, longer than 4 mm, stigmas little wider than the styles; anther pores usually borne on short tubes .... (2)

2. Leaves mottled on the upper surface and with pale streaks associated with the main veins; blades broadly ovate, the apex acutely pointed .... ***P. picta***
2′ Leaves uniformly green above, blades elliptic to orbicular .... (3)

3. Petals rose-pink to purple; leaves usually more than 3 cm long or wide .... ***P. asarifolia***
3′ Petals greenish white; leaves usually less than 3 cm long or wide .... ***P. chlorantha***

Pyrola asarifolia Michx. (*Asarum*-like leaves) Round-leaved wintergreen, liverleaf wintergreen. Plants with extensively creeping, scaly rhizomes. **STEM** a scape, 13–30 (40) cm tall, with 2 or 3 widely spaced bracts. **LEAVES** bright green above, lustrous-coriaceous, pale below, petioles mostly longer than the blade, 1–9 cm long, 3–7, blade simple, basal, persistent, 1.3–7.5 cm long, 1.1–7.4 cm wide, oval to reniform-orbicular with a rounded apex, entire to crenulate or serrulate. **INFL** an elongate raceme, (2) 4–12 (20)-flowered, pedicels 3–8 mm long, slightly exceeding the lanceolate bractlets. **FLOWERS** 1–1.5 cm wide; calyx similar in color to the petals, the lobes (1.5) 2–4 mm long, lance-deltoid; corolla campanulate, pink to purplish, the petals broadly ovate, (5) 6–8 mm long, veins prominent; stamens included, the filaments glabrous, declined, anthers 2 mm long, contracted to an apiculate tip at the lower end, dehiscing by tubular pores; style 5–7 mm long, sharply curved downward immediately above the ovary, tapered to a collar, broader and just below the stigma. **FRUIT** a capsule, 4–8 mm in diameter, depressed-globose, loculicidally dehiscing from the base. **SEEDS** many. $2n = 46$. [*P. rotundifolia* L. var. *bracteata* (Hook.) A. Gray; *P. asarifolia* var. *bracteata* (Hook.) Jeps.; *P. rotundifolia* var. *purpurea* Bunge; *P. asarifolia* var. *purpurea* (Bunge) Fernald; *P. incarnata* Fisch.; *P. asarifolia* var. *ovata* Farw.; *P. uliginosa* Torr. & A. Gray ex Torr.; *P. rotundifolia* var. *uliginosa* (Torr. & A. Gray) A. Gray; *P. asarifolia* var. *uliginosa* (Torr. & A. Gray) A. Gray; *P. elata* Nutt.; *P. bracteata* Hook. var. *hillii* J. K. Henry]. Moist, shady woods and boggy stream banks in subalpine forests, mountain brush, aspen, and spruce-fir . COLO: Arc, Hin, LPl, Mon, SJn; NMEX: SJn. 1675–3200 m (5500–10500′). Flowering: Jul–Aug. Fruit: Aug–Sep. Newfoundland and Prince Edward Island, west to Alaska, and south to California, Utah, New Mexico, South Dakota, and New England; also in Asia.

Pyrola chlorantha Sw. (green flower) Green-flowered wintergreen. Herbaceous perennials from widely spreading, slender rhizomes. **STEMS** solitary (occasionally 2), a scape (6) 9–25 cm tall. **LEAVES** thick, coriaceous, pale green, 4–11, in dense basal cluster, petiole 0.8–6 cm long, blade elliptic, reniform to suborbicular, 0.6–3.5 cm long, 0.5–3 cm wide, obscurely crenate, apex rounded, base rounded or broadly cuneate, not decurrent on the petiole. **INFL** a cylindrical raceme, 3–9-flowered, rarely more, with 1 or 2 lance-shaped bracts, pedicels spreading, 3–8 mm long, bractlets almost as long. **FLOWERS** 8–12 mm across; calyx 0.5–1.5 mm long, lobes about 1 mm long, rounded, broader than long, persisting; corolla open-campanulate, greenish but becoming pale yellow-white at anthesis, petals (4) 5–7 mm long, spreading-ascending, oval, obtuse; stamens mostly included, 4 mm long; filaments subulate, dilated below; anthers 2–3 mm long, terminal pores forming elongate tubes; style 5–7 mm long, deflexed and then arched upward toward the apex, a flaring collar below the stigma, persistent. **FRUIT** a capsule, 5–8 mm in diameter, depressed-globose, loculicidally dehiscing from the base. **SEEDS** many. $2n = 46$. [*P. virens* Schreber]. Moist to dry, upland woods; ponderosa pine, aspen, spruce-fir, and subalpine communities. ARIZ: Apa; COLO: Arc, LPl, Min, Mon, RGr, SJn; NMEX: McK; UTAH. 2225–3050 m (7300–10000′). Flowering: Jun–Jul (early Aug). Fruit: Jul–Sep. Ranging from Labrador and Newfoundland west to Alaska, south to Maryland, New Mexico, and Arizona; also in Eurasia. Kayenta Navajo employed *P. chlorantha* to treat bloody diarrhea in infants as well as other blood-related disorders; it is an ingredient for paint used in the Nightway ceremony.

Pyrola minor L. (smaller, lesser) Lesser wintergreen. Herbaceous perennials from slender rhizomes. **STEMS** solitary, rarely 2, scapes 8–25 (30) cm tall, with 1–2 (3) lanceolate, scarious bracts, 6–8 mm long, scattered. **LEAVES** dark green, thin, 3–7, mostly basal in a rosette, petioles shorter than the blade, 0.2–3 cm long, blade (0.4) 1.1–3.3 cm long, (0.6) 0.9–2.5 cm wide, the blades rounded-ovate to suborbicular or elliptic, margins entire to remotely crenate, apex obtuse to rounded, base short-cuneate, tapering decurrently to the petiole. **INFL** a cylindrical raceme, (5) 6–13-flowered, pedicels spreading to arching, 2–3 (5) mm long, the lanceolate bracts scarious, almost equaling the pedicel. **FLOWERS** 1–1.5 cm in diameter, calyx spreading, the lobes oblong-lanceolate, erose to subentire, 1–1.5 mm long, persistent; petals 3.5–4.5 mm long, creamy white to pale pink, fragrant, leathery, rounded-obovate; stamens included, 4–6 mm long, filaments flat, anthers 2–3 mm long, the pores sessile; style straight, short, 1.3–2 mm long, the 5-lobed stigma peltate, persisting. **FRUIT** a capsule, 4.5–6 mm in diameter, depressed-globose, loculicidally dehiscing from the base. **SEEDS** many. $2n = 46$. [*Amelia minor* (L.) Alef.; *Erxlebenia minor* (L.) Rydb.; *P. minor* var. *conferta* Cham. & Schltdl.; *P. conferta* (Cham. & Schltdl.) Fisch. ex Ledeb.]. Moist, mossy sites in montane aspen, spruce, and subalpine forests. COLO: Arc, Con, Hin, LPl, Min, Mon, SJn; NMEX: McK. 2255–3420 m (7400–11200′). Flowering: Jul–Aug. Fruit: Aug–Sep. Circumboreal.

Pyrola picta Sm. **CP** (colored or painted) Picture-leaf wintergreen. Herbaceous perennials from slender, scaly rhizomes. **STEMS** reddish brown, solitary or 2, 10–25 cm tall, with 1–3 lanceolate bracts. **LEAVES** coriaceous, dull green with mottled gray or white areas along the main veins, lavender on the lower surface, 2–several on sterile stems, 0–4 on fertile stems, basal, the petiole shorter than the blade, 0.3–3.5 cm long; blade 2–7 cm long, 1–3.5 cm wide, ovate to elliptic, obtuse to rounded apically, acute at base, the margins thick, entire to denticulate. **INFL** a cylindrical raceme, (2) 5–10 (20)-flowered, the pedicels 3–8 mm long, spreading, with lanceolate bractlets almost as long. **FLOWERS** about 1 cm in diameter; sepals reddish green, the lobes triangular-acute, (1) 1.5–2 mm long; petals 6–8 mm long, ovate, creamy yellow to greenish or purplish; stamens included, curved inward, the anthers contracted below the pores, forming a tubelike pore; style 6–7 mm long, deflexed at the base and arched upward, with a flaring collar below the stigma, stigma small, persistent. **FRUIT** a capsule, 5–7 mm in diameter, depressed-globose, loculicidally dehiscing from the base. **SEEDS** many. $2n = 46$. [*P. pallida* Greene]. Moist woods, in deep shaded ravines of the San Juan Mountains; ponderosa pine, fir, and aspen communities. COLO: Hin, Min; NMEX: RAr. 2530–2930 m (8100–9700′). Flowering: Jul–Aug. Fruit: Aug–Sep. British Columbia to Montana, south to California, Arizona, and Colorado.

Vaccinium L. Blueberry, Cranberry

(Latin *bacca*, berry) Plants shrubs. **STEMS** recumbent to erect. **LEAVES** evergreen or deciduous, alternate, margins entire or serrulate. **FLOWERS** perfect, actinomorphic, solitary or few in axils or terminal racemes; calyx sepals 4–5 with small lobes or fused, persistent; corolla white to pinkish, shallowly 4–5-lobed, urn-shaped to cylindrical or campanulate; stamens 10, twice as many as the corolla lobes, usually included, at the base of the corolla, anthers dehiscing terminally, with a pair of spurs; ovary inferior, 4–5 locules, style slender. **FRUIT** a berry with many seeds. $x = 12$ (ours) (Vander Kloet 1992; Vander Kloet and Dickinson 1999). About 400 species in temperate, subtropical, and tropical montane regions worldwide. *Vaccinium* contains commercially important shrubs with edible fruits: blueberries and cranberries.

1. Plants tall, more than 4 dm; pedicels long (5–16 mm); leaves wide (15–35 mm) .... ***V. membranaceum***
1′ Plants low, less than 4 dm; pedicels short (less than 4 mm); leaves narrow (less than 15 mm) .... (2)

2. Branchlets of current year's growth reddish green or yellowish green, terete or slightly angled .... ***V. caespitosum***
2′ Branchlets of current year's growth bright green, sharply angled .... (3)

3. Berries red; plants with broomlike habit; leaves short (up to 1.5 cm) .... ***V. scoparium***
3′ Berries blue; plants with flexuous branches; leaves long (1–4 cm) .... ***V. myrtillus***

Vaccinium caespitosum Michx. (forming dense patches) Dwarf blueberry, dwarf bilberry, dwarf whortleberry. Low shrubs, 5–30 cm tall. **STEMS** densely branched, forming dense patches from rhizomes, branchlets reddish green or yellowish green, not bright green, puberulent to glabrous, terete or slightly angled. **LEAVES** deciduous, thin, short-petiolate to sessile, usually glabrous or sparsely glandular; blade 15–35 mm long, 4.5–15 mm wide, obovate to elliptic, base tapered to cuneate, tip obtuse to rounded, margins serrulate to near base. **INFL** solitary in leaf axils of current year's growth. **FLOWERS** pedicellate, 1.2–3.5 mm long; sepals glabrous, lobes none to obscure; corollas white to pink, 4–6 mm long, 3–5 mm wide, urn-shaped to globose; filaments glabrous, anthers awned. **FRUIT** a berry, 6–8 mm wide, subglobose, blue-glaucous. $2n = 24$. Mountain slopes and alpine tundra, edges of coniferous woods, meadows, and steamsides. COLO: Hin, LPl, Min, Mon, SJn. 2250–3600 m (7500–12000′). Flowering: Jun–Aug. Across Canada, Alaska south to California and Colorado.

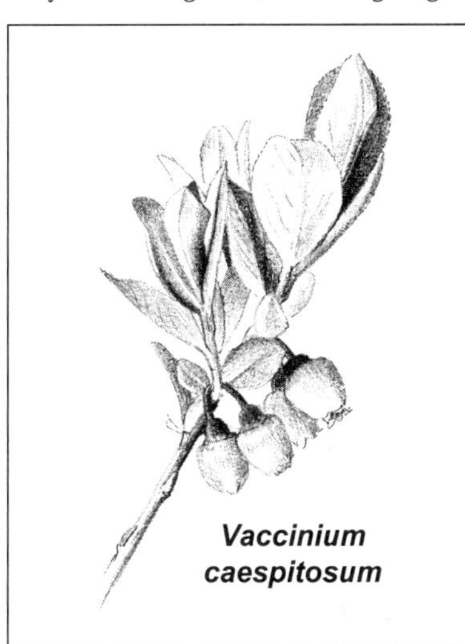
Vaccinium caespitosum

Vaccinium membranaceum Douglas ex Torr. (thin) Mountain bilberry, thinleaf bilberry. Tall shrubs, 4–12 dm. **STEMS** rigid and clump-forming, not rhizomatous, branchlets reddish green, terete to slightly angled, glabrous. **LEAVES** deciduous, very thin, blade 2–5.5 cm long, 1.5–3.5 cm wide, ovate to elliptic, base rounded, tip acute, margins serrate, green, pubescence of scattered hairs.

INFL solitary in leaf axils of current year's growth. **FLOWERS** pedicellate, 5–16 mm long; sepals glabrous, lobes none or obscure; corollas white to pinkish, 3–6 mm long, 4–7 mm wide, urn-shaped to cylindrical; filaments glabrous, anthers awned. **FRUIT** a berry, 6–11 mm wide, purple to black, not glaucous. 2*n* = 48. [*V. globulare* Rydb.]. Mountain slopes, coniferous woods, and wet meadows. COLO: Arc, Hin, Mon. 1600–3100 m (5300–10250′). Flowering: Jun–Jul. Alaska south to California and Colorado, east to Minnesota, Michigan, and Ontario.

Vaccinium myrtillus L. (little myrtle) Blueberry, bilberry, whortleberry. Shrubs 10–40 cm tall. **STEMS** openly branched and forming open colonies from woody rhizomes, branchlets bright green, glabrous, sharply angled, flexuous. **LEAVES** deciduous, thin, short-petiolate; blade 1–4 cm long, 7–16 mm wide, ovate to elliptic, base obtuse, tip acute, margins serrulate to base, green above, glabrous or with scattered glandular hairs. **INFL** solitary in leaf axils of current year's growth. **FLOWERS** pedicellate, 2–4 mm long; sepals glabrous, lobes none to obscure; corolla white or pink, 4–5 mm long, 5–7 mm wide, globose; filaments glabrous, anthers awned. **FRUIT** a berry, 5–9 mm wide, blue or blue-black. 2*n* = 24, 48. [*V. myrtillus* var. *oreophilum* (Rydb.) Dorn; *V. oreophilum* Rydb.]. Mountain slopes, coniferous woods, and openings. ARIZ: Apa; COLO: Arc, Hin, LPl, Min, Mon, RGr, SJn. 2300–3500 m (7600–11500′). Flowering: Jun–Jul. British Columbia and Alberta south to Arizona and New Mexico; Eurasia. Hybrids occur between diploid races in *Vaccinium* section *Myrtillus*. In our area, hybrids between *V. myrtillus* and *V. caespitosum* have been collected in Hinsdale and San Juan counties, Colorado.

Vaccinium scoparium Leiberg ex Coville (in the form of a broom) Grouseberry, dwarf red whortleberry. Erect shrubs, 1–3 dm tall. **STEMS** narrowly branched, with broomlike habit, forming colonies from woody rhizomes, branchlets green, glabrous, sharply angled, rigid. **LEAVES** deciduous, thin, blade 0.5–1.5 cm long, 3–7 mm wide, ovate to elliptic; base obtuse, tip acute, margins serrulate, pale green, glabrous. **INFL** solitary in leaf axils of current year's growth. **FLOWERS** pedicellate, 1.5–3.5 mm long; sepals glabrous, lobes none to obscure; corolla pinkish, 2–4 mm long, 3–4 mm wide, broadly urn-shaped; filaments glabrous, anthers awned. **FRUIT** a berry 3–6 mm wide, bright red. 2*n* = 24. [*V. myrtillus* L. var. *microphyllum* Hook.; *V. microphyllum* (Hook.) Rydb.; *V. erythrococcum* Rydb.]. High mountains, edges of subalpine coniferous forest, and alpine meadows. COLO: Arc, Hin, LPl, Min, Mon, SJn. 2100–3500 m (7000–11500′). Flowering: Jun–Aug. British Columbia and Alberta south to California, Nevada, and Colorado, east to the Black Hills, South Dakota.

EUPHORBIACEAE Juss. SPURGE FAMILY

David L. Bleakly

(*eys*, good, + *phorbe*, feed, pasture, or fodder, from Euphorbus, a 1st-century A.D. physician to Juba, a king of Mauretania, an ancient country in northwestern Africa) Annual or perennial herbs (ours), woody elsewhere, ours tap-rooted or rhizomatous; sap mostly milky; vestiture none, simple, or rarely stellate hairs (*Croton texensis*) or stinging hairs (*Tragia ramosa*). **LEAVES** simple, entire, serrulate, or serrate. **INFL** various, cyathium, raceme, or compact cymules. **FLOWERS** monoecious or rarely dioecious; petals lacking in all of ours; sepals lacking or present; stamens 1 to numerous, distinct (most of ours) or connate; pistils 1, carpels 3, united, styles 3, distinct or united, often bifid or divided. **FRUIT** a capsular schizocarp with 3 locules and 3 lobes. **SEEDS** 1 or rarely 2 per locule, sometimes with a caruncle. About 8100 species, perhaps the sixth largest plant family, essentially cosmopolitan, but most taxa in the tropics and subtropics. Considering the size of the family, relatively few species are of economic value. The most important species are *Manihot esculenta* (manioc, cassava, or yuca; tapioca and Brazilian arrowroot are derivatives), *Hevea brasiliensis* (Pará rubber), *Ricinus communis* (castor bean), *Aleurites* species (tung oil), and *Euphorbia pulcherrima* (Christmas poinsettia). A number of other species are of mostly minor importance as ornamentals and as sources of oils, dyes, fruits, and timber. Many species produce toxic or poisonous proteins and/or alkaloids. Those with milky sap (particularly *Euphorbia*) can cause dermatitis or poisoning in susceptible people (Webster 1994a, 1994b).

1. Herbage with stinging hairs; leaves serrate, 1–4 cm long; plants perennial; ovules 1 per locule; fruit usually with 3 seeds; sap not milky .... ***Tragia***

1′ Herbage glabrous or pubescent, but never with stinging hairs; leaves entire, or if serrate, then usually less than 1 cm long; sap milky or not; plants annual or perennial .... (2)

2. Herbage usually densely (or sparsely) covered with stellate hairs; plants dioecious annuals; leaves entire; ovules 1 per locule; fruit usually with 3 seeds; sap not milky .... ***Croton***

2′ Herbage glabrous or pubescent but never with stellate hairs; plants monoecious; sap milky or not .... (3)

3. Inflorescences axillary to leaves on lateral branches only, composed of cymules, each cymule consisting of 1 pistillate flower and a few staminate flowers, all dark reddish purple; plants glabrous annuals; leaves entire, linear; ovules 2 per locule; fruit usually with 6 seeds; sap not milky .... ***Reverchonia***
3' Inflorescences in stem forks or congested axillary clusters, or terminal, composed of cyathia, cuplike structures containing separate male flowers (which are single stamens) and 1 female flower; plants glabrous or pubescent, annuals or perennials; sap milky .... (4)

4. Leaves all opposite, bases usually distinctly asymmetric; stipules usually present and apparent, sometimes inconspicuous; inflorescences in stem forks or congested axillary clusters .... ***Chamaesyce***
4' Leaves below inflorescence alternate, leaf bases always symmetric; stipules lacking or minute and more or less glandular; inflorescences terminal .... ***Euphorbia***

Chamaesyce Gray Prostrate Spurge, Spurge, Sandmat

(Greek *chamai*, on the ground, + *sykon*, fig, referring to the commonly prostrate habit and the milky sap which also occurs in *Ficus*, and perhaps to the vaguely figlike shape of the fruits of *Chamaesyce*) Annual or perennial taprooted herbs; sap milky. **STEMS** prostrate or erect, rarely winged (*C. serpyllifolia*), annuals usually branching near ground level, sometimes above, perennials often branching below the ground surface. **LEAVES** simple, cauline, short-petiolate, base usually asymmetric, margins entire, serrulate, or serrate, stipules usually apparent. **INFL** composed of cyathia, solitary or congested, in stem forks or leaf axils, often present through most of stem length. **CYATHIUM** solitary, axillary. **GLANDS** 4; appendages usually present. **FLOWERS** pistillate, few to numerous, each consisting of a single stamen. **FRUIT** 3-loculed and 3-angled, glabrous or pubescent, smooth, loculicidal and septicidal, columella persistent. **SEEDS** 1 per locule, smooth, ridged, or with irregular pits or bumps, small, often with white coating, without a caruncle. *Chamaesyce*, which is often considered a subgenus of *Euphorbia*, contains approximately 250 species, mostly in tropical America and Africa, but also more or less cosmopolitan in temperate and subtropical habitats, then often as weeds. Photosynthesis in *Chamaesyce* apparently follows the C_4 photosynthetic pathway, while *Euphorbia* uses the C_3 pathway. In addition to the usual plant parts, a specimen should have seeds for accurate identification (Wheeler 1941).

1. Leaves toothed at least near apex, sometimes very inconspicuously; plants annual .... (2)
1' Leaves entire; plants annual or perennial .... (4)

2. Herbage pubescent .... ***C. serrula***
2' Herbage glabrous .... (3)

3. Seeds usually smooth (occasionally faintly wrinkled or pitted); most internodes (or at least the distal parts of some) almost always narrowly winged and/or appearing flattened and sometimes striate or ridged; leaves variable in size and shape, sometimes with reddish blotch; stamens 5–18; stems prostrate to erect .... ***C. serpyllifolia***
3' Seeds, on all faces, with 4–6 (8) coarse, transverse ridges; internodes round, rarely slightly flattened; leaves never with reddish blotch; stamens usually 4; stems prostrate, rarely ascending .... ***C. glyptosperma***

4. Plants perennial, glabrous; stems prostrate to erect, rather wiry, to 15 cm; leaves orbicular to broadly ovate to lanceolate, <10 mm long .... ***C. fendleri***
4' Plants annual, glabrous; stems erect, ascending, or prostrate, bushy-branching, (5) 10–30+ cm; leaves linear, apex rounded and mucronate, margins sometimes folded or revolute, 1–2.5 (3.5) cm long .... ***C. parryi***

Chamaesyce fendleri (Torr. & A. Gray) Small (for the German Augustus Fendler, 1813–1883, who collected this plant and others in the Santa Fe area in 1847) Fendler's prostrate spurge. Perennial, glabrous herbs from branched, underground woody caudex. **STEMS** several to numerous, sometimes prostrate, usually decumbent or ascending to erect, slender or wiry, often reddish or purplish, to about 15 cm long. **LEAVES** orbicular to broadly ovate to lanceolate, entire, 3–10 mm long; stipules separate, linear. **CYATHIUM** 1–2 mm long; appendages absent, shorter than gland width, or conspicuous, lanceolate or deltoid and 1–3 times as long as gland width; sometimes reddish; margin entire or irregular; stamens (15) 25–35. **FRUIT** glabrous, 2.2–2.5 mm long. **SEEDS** quadrangular, usually smooth, sometimes faintly wrinkled, white, 1.4–2 mm long. $2n = 28$. Sandy or gravelly soil in deserts, shrublands, woodlands, and ponderosa forests. ARIZ: Apa, Nav; COLO: Arc, Dol, LPl, Mon; NMEX: McK, RAr, San, SJn; UTAH. 1525–1860 m (5000–6100'). Flowering: May–Sep. Widespread in the southwestern United States, from the western Great Plains and Texas to southern Nevada, southeastern California, and northern Mexico. In our area, this is the only *Chamaesyce* which is perennial and glabrous. Navajos used an infusion for stomachaches and diarrhea; plant parts were used topically for poison ivy and warts and to stimulate breast milk production; poultices were applied for toothaches and to cuts as a hemostatic. Hopis fed roots to babies whose mother's milk was failing.

Chamaesyce glyptosperma (Engelm.) Small (carved seed, referring to the ridged seeds) Ribseed prostrate spurge. Annual, glabrous, taprooted herbs. **STEMS** prostrate to ascending, 5–40 cm long, much-branched, round and wingless. **LEAVES** mostly at least partly serrate, some may be entire, mostly oblong, 2–12 (15) mm long, often falcate, apex obtuse; stipules subulate, laciniate or dissected, sometimes entire, to about 1 mm long. **CYATHIUM** 0.5–1 mm long; appendages short, to about as long as gland width, entire or erose; stamens usually 4. **FRUIT** glabrous, about 1.5 mm long. **SEEDS** sharply quadrangular, faces and sometimes angles with 4–6 (8) coarse, transverse ridges, 1–1.5 mm long. $2n = 22$. Dry or mesic open places, in deserts, shrublands, woodlands, ponderosa pine forests, sometimes near streams or hanging gardens, sometimes weedy in gardens and roadsides. ARIZ: Apa, Coc, Nav; COLO: Dol, SMg; NMEX: McK, RAr, SJn; UTAH. 1125–1950 m (3700–6400′). Flowering: Jun–Sep. Widespread in North America. Similar to *C. serpyllifolia*, but *C. glyptosperma*'s wingless stems and strongly ridged seeds are distinguishing.

Chamaesyce parryi (Engelm.) Rydb. (for Charles Christopher Parry, 1823–1890) Parry's prostrate spurge. Annual, glabrous, taprooted herbs. **STEMS** erect, ascending, or decumbent, occasionally prostrate, 5–50 cm long, branches on erect plants often bushy. **LEAVES** more or less linear, 1–2.5 (3.5) cm long, 1–3 mm wide, apex rounded and mucronate, margins entire, sometimes folded or revolute; stipules distinct and linear, sometimes entire, to about 1 mm long. **CYATHIUM** 1–2 mm long; appendages typically short and erect (sometimes forming a cup), but sometimes as long as or much longer than the gland is wide and spreading (short and long appendages sometimes on same plant); stamens 35–55. **FRUIT** glabrous, 1.5–2.5 mm long. **SEEDS** plumply ovoid-triangular, smooth or slightly roughened, mottled brown and white, 1.4–1.8 mm long. Sandy places, dunes, along washes, in deserts and shrublands. ARIZ: Apa, Nav; COLO: LPl, Mon; NMEX: RAr, SJn; UTAH. 1490–1890 m (4900–6200′). Flowering: May–Sep. Trans-Pecos Texas, western New Mexico, eastern and northern Arizona, southern Utah, southern Nevada. Very similar to and maybe not distinct from *C. missurica* (Raf.) Shinners, which occurs east of our area. Despite its range of habits, *C. parryi*'s longevity, glabrous foliage, plump, essentially smooth seeds, and linear leaves are unique in our area.

Chamaesyce serpyllifolia (Pers.) Small (from *Thymus serpyllum* + Latin *folium*, leaf; referring to the shape of the leaves) Thymeleaf prostrate spurge. Annual, glabrous, taprooted herbs. **STEMS** prostrate or erect, much-branched, 5–35 cm long; most, some, or at least the distal parts of some or most internodes almost always narrowly winged and/or appearing flattened and sometimes striate or ridged. **LEAVES** variable in size and shape, but oblong to elliptic or elliptic-oblanceolate; usually serrate, at least at rounded or almost truncate apex, 2–12 (15) mm long, often falcate, sometimes with a reddish blotch; stipules distinct, linear, entire or divided, 1–1.5 mm long. **CYATHIUM** 0.5–1.5 mm long; appendages short, white, entire or crenulate; stamens 5–18. **FRUIT** glabrous, 1.5–2 mm long. **SEEDS** quadrangular, usually smooth, sometimes faintly wrinkled or pitted, 1–1.5 mm long. $2n = 22, 28$. Common in dry or moist open places, in many substrates, in shrublands, woodlands, often weedy in gardens and roadsides. ARIZ: Apa, Coc, Nav; COLO: Arc, Hin, LPl, Min, Mon, SJn; NMEX: McK, RAr, San, SJn; UTAH. 1190–2165 m (3900–7100′). Flowering: Jun–Sep. Widespread in the north-central and western United States, from Michigan to Washington, south to California, Texas, and northern Mexico. The combination of winged internodes and more or less smooth seeds, along with toothed leaves and glabrous vestiture, is diagnostic. Navajos used an infusion for stomachaches and diarrhea; plant parts were used topically for poison ivy and warts and to stimulate breast milk production; poultices were applied for toothaches and to cuts to stop bleeding. Hopis fed roots to babies whose mother's milk was failing.

Chamaesyce serrula (Engelm.) Wooton & Standl. (*serra*, saw, + *ula*, little, referring to the serrulate leaves) Sawtooth prostrate spurge. Annual, pilose, taprooted herbs. **STEMS** prostrate to decumbent, much-branched, to 25 cm long. **LEAVES** mostly oblong, but sometimes elliptic or obovate-oblong; serrate, smaller leaves may be entire, often falcate, 5–10 mm long; stipules distinct, deltoid-attenuate, trilobate, 1–2 mm long. **CYATHIUM** about 1 mm long; appendages 1–4 times as long as gland width, entire to crenulate; stamens 7–13. **FRUIT** glabrous, 2–2.5 mm long. **SEEDS** quadrangular, smooth, whitish, 1.5–2 mm long. Grows in gravelly desert soils. NMEX: SJn. 1600 m (5250′). Flowering Aug–Oct. Trans-Pecos Texas, New Mexico, Arizona, south to central Mexico. Apparently rare in our area. This is the only hairy *Chamaesyce* in the study area.

Croton L. Croton

(Greek *kroton*, tick; *Croton* was originally applied to what is now called *Ricinus*, the seeds of which resemble ticks; Linnaeus apparently arbitrarily transferred *Croton* to this group of plants as a replacement for an earlier name, *Ricinoides*) Annual herbs (ours) or elsewhere perennial herbs, shrubs, trees. Hairs usually stellate, but may be lepidote, glandular, and/or tomentose. Plants dioecious or monoecious (ours). Foliage often ill-smelling or sometimes fragrant; sap not milky. **LEAVES** simple, alternate, stipulate or not (ours). **INFL** androgynous (male flowers above a few female flowers) in a terminal spike or raceme in monoecious species; staminate and pistillate flowers on separate plants in dioecious species. **MALE FLOWERS** with sepals only (ours) or sepals and petals, usually 5-lobed; stamens few to numerous. **FEMALE FLOWERS** with sepals but no petals; carpels 1–3, styles usually distinct. **FRUIT** a capsule, usually with persistent columella. **SEEDS** carunculate (ours) or not. About 750 species, most in New World tropics. A few species are used medicinally (especially *C. tiglium* L., croton oil, one of the most purgative substances known), others as scents or flavorings (*C. eluteria* (L.) W. Wright, cascarilla, and *C. niveus* Jacq., copalchi), one as a host plant for lac insects (*C. laccifer* L.), and a few for timber (Ferguson 1901).

Croton texensis (Klotzsch) Müll. Arg. (belonging to Texas) Texas croton, doveweed, skunkweed. Annual, taprooted, malodorous, dioecious herbs with sparse to dense stellate pubescence which may give the plants a grayish or yellowish cast. **STEMS** erect, 2–8 dm tall, often branched above. **LEAVES** linear- to broadly lanceolate, elliptic, or oblong, entire, mostly 1.5–5 cm long and 0.5–1.5 cm wide, petiole length about equal to leaf width. **INFL** a raceme on separate plants. **FLOWER** calyx 2–4 mm wide, sepals 1–2 mm long; male flowers 1–2 cm long at branch ends; stamens 8–12; female flowers 1–4, about 1 cm long; styles 3, about 2 mm long, with many branches. **FRUIT** a capsule 4–6 mm long, stellate-tomentose, usually warty. **SEEDS** 1–3, 3.5–4 mm long, caruncle conspicuous, about 1 mm long. Dry, sandy sites, disturbed areas. ARIZ: Apa, Nav; UTAH. 1220–1830 m (4000–6000′). Flowering: May–Sep. Often common from the central United States to Wyoming, Utah, Arizona, New Mexico, Texas, and into northern Mexico. The habit, leaf shape and size, and degree of hairiness is quite variable. The species is toxic to livestock. The seeds are apparently eaten by doves. The Hopis have used it as an emetic and eyewash. Various western Indian tribes used the plant primarily as a purgative or to relieve gastrointestinal distress.

Euphorbia L. Spurge

(for Euphorbus, a 1st-century A.D. physician to Juba, king of Mauretania, an ancient country in northwestern Africa) Annual or perennial, glabrous or pubescent, taprooted or rarely rhizomatous herbs (ours), woody or succulent shrubs or trees elsewhere; sap milky. **STEMS** erect or ascending (ours), prostrate elsewhere. **LEAVES** cauline, mostly alternate, rarely opposite, entire to dentate; stipules lacking or minute and glandlike. **INFL** composed of cyathia (see glossary for details about this structure); more or less dichotomously divided or umbellate, essentially terminal. **GLANDS** 1 or 4–5; appendages none (or present on *E. marginata*). [*Esula*, *Galarhoeus*, *Poinsettia*, and *Tithymalus*].

1. Appendages white, conspicuous, petaloid, 1.5–2 mm long; glands usually 4; leaves of inflorescence broadly white-margined; plants robust annuals ..... ***E. marginata***
1′ Appendages lacking; glands 1 (rarely 2) or 4–5; annuals or perennials ..... (2)

2. Glands 1 (rarely 2), deeply cupped; inflorescence never branching into symmetrical 3–several-rayed cymes ..... (3)
2′ Glands 4 or 5, entire or crescent-shaped, flat or convex, not cupped; inflorescence branching into 3–several-rayed cymes, which resemble umbels ..... (4)

3. Leaves widest near middle, linear-elliptic to broadly elliptic; petioles often nearly as long as blades at midstem; vestiture stiffly pubescent, hairs ascending, hairs on lower leaf surface stiff, strongly tapered, with a broad basal cell; seeds angular in cross section, unevenly tuberculate, (2) 2.3–2.7 mm long ..... ***E. davidii***
3′ Leaves usually widest below middle, narrowly to broadly lanceolate or ovate to trowel-shaped; petioles usually not more than 1/3 as long as blade at midstem; vestiture mostly short-hairy, more or less retrorse or strigose, some longer and multicellular, hairs on lower leaf surface weak, filiform, lacking broad basal cell; seeds rounded in cross section, evenly tuberculate, 1.7–2 (2.2) mm long ..... ***E. dentata***

4. Plants annual; capsules (and ovaries) verrucose (warty with low bumps), 2–3 mm long ..... ***E. spathulata***
4′ Plants perennial; capsules smooth (never warty), 3–4 mm long ..... (5)

5. Cauline leaves broadly linear or lanceolate-linear to narrowly oblong or narrowly oblanceolate, 3–9 cm long, 2.5–7 (10) mm wide; inflorescence bracts orbicular, starkly contrasting with the narrow cauline leaves, often conspicuously yellowish; capsule finely granular or rugose-tuberculate, glabrous, 3–3.5 mm long; plants strongly rhizomatous, colonial, weedy, invasive ..... ***E. esula***
5′ Cauline leaves variable, oblong-spatulate or elliptic to broadly elliptic or rotund-ovate, not linear, 0.7–2.5 cm long, 3–20 mm wide; inflorescence bracts similar to but slightly larger than cauline leaves, green (not yellowish); capsules glabrous or short-hairy, smooth, about 4 mm long; plants from semiwoody crown, not rhizomatous, widespread native plants ..... ***E. brachycera***

Euphorbia brachycera Engelm. (short-horned, referring to the hornlike extensions of the gland) Horned spurge. Perennial herbs from a robust, semiwoody taproot. **STEMS** usually several, branching below ground level from root crown, 1–4 dm tall, glabrous or velvety-puberulent. **LEAVES** variable, oblong-spatulate or elliptic to broadly elliptic or rotund-ovate, entire, sessile, 0.7–2.5 cm long, 3–20 mm wide; leaves subtending entire inflorescence usually larger and whorled; those subtending distal inflorescence branches paired, green (not yellowish). **INFL** umbellate, the rays 3–5 (8), each ray divided one or more times; 2 foliose bracts subtend each fork. **CYATHIUM** 2–3 mm long; stamens several. **GLANDS** usually 4, yellowish, often with conspicuous horns, but sometimes hornless and the margin crenate, sometimes both types on same plant; appendages none. **FRUIT** glabrous or short-hairy, smooth, about 4 mm long. **SEEDS** grayish, with irregular shallow pitting, 1.8–3 mm long, caruncle conspicuous. $2n = 28$. [*E. lurida* Engelm.; *E. montana* Engelm.; *E. robusta* (Engelm.) Small; *Tithymalus brachycerus* (Engelm.) Small]. Sandy, gravelly, or rocky slopes, shrublands, woodlands, to mixed confer forests. ARIZ: Apa, Nav; COLO: Arc, Hin, LPl, Mon, SJn; NMEX: McK, SJn. 1525–2600 m (5000–8500′). Flowering: May–Aug. The variation in leaf shape throughout its wide range led earlier botanists to recognize several taxa where in fact there is only one. Despite its variation in leaf shape and gland morphology, its perennial, nonrhizomatous growth and smooth capsules make this plant readily identifiable. The Navajo used an infusion of the plant as a purgative and during childbirth; leaves and stems were used as a liniment and topical analgesic; the root sap was rubbed on clothing for good luck.

Euphorbia davidii Subils (for David L. Anderson of Villa Mercedes, Argentina, who collected one of the paratypes) Toothed spurge. Annual herbs from taproots, with stiff, tapered hairs. **STEMS** erect, usually branched, 1–4 (8) dm tall. **LEAVES** mostly opposite, widest near middle, linear-elliptic to broadly elliptic; petioles often nearly as long as blades at midstem, 2–7 cm long, hairs on lower leaf surface stiff, strongly tapered, with a broad basal cell. **INFL** terminal on main stem and branches, congested, bracteal leaves sometimes pale at base. **CYATHIUM** 2–2.5 mm long; stamens many. **GLANDS** 1, cupped, fleshy; appendages none. **FRUIT** smooth, usually glabrous, rarely sparsely hairy, 4–5 mm wide, 2–3 mm long. **SEEDS** angular in cross section, unevenly tuberculate, (2) 2.3–2.7 mm long, caruncle 1–1.4 mm long. $2n = 56$. Dry, open, often disturbed places, in valleys and foothills. COLO: Arc, LPl, Mon; NMEX: SJn. 1525–2150 m (5000–7000′). Flowering: Jul–Sep. Apparently introduced to the Four Corners study area by human activities from the central United States (east of the Rocky Mountains); now fairly widespread in North and South America. Only recently was this species realized to be distinct from what has traditionally been called *E. dentata*. Most plants in our area are *E. davidii*. The general appearance of the two plants is very similar, but they are usually distinguishable if a reasonably complete specimen is available.

Euphorbia dentata Michx. (with toothlike projections, referring to the toothed leaves) Western toothed spurge. Annual herbs from taproots, with retrorse or strigose, weak hairs. **STEMS** erect, 1–4 (6) dm tall, usually branched. **LEAVES** broadest below middle, narrowly to broadly lanceolate or ovate to trowel-shaped, petioles not usually more than 1/3 as long as blade at midstem, 1–9 cm long, hairs on lower leaf surfaces weak, filiform, lacking broad basal cell. **INFL** terminal on main stem and branches, congested, bracteal leaves sometimes pale at base. **CYATHIUM** 2–3 mm long; stamens many. **GLANDS** 1, cupped, fleshy; appendages none. **FRUIT** smooth, glabrous, 4–5 mm wide, 2–3 mm long. **SEEDS** rounded in cross section, evenly tuberculate, 1.7–2.2 mm long, caruncle to about 1 mm long. $2n = 28$. Usually in moister places, often a weed in gardens, etc. NMEX: SJn. 1525–1825 m (5000–6000′). Flowering: Jul–Sep. Apparently introduced to the Four Corners region from the eastern United States by human activities. It is seemingly rare in our area. Until recently, the plants in our area were known by this name, but now it is understood that most are actually *E. davidii*.

Euphorbia esula L. (origin uncertain; two possibilities: Latinized form of Celtic name for sharp, referring to the acrid juice; or from German *esel*, donkey, perhaps referring to the similarity of the long, narrow leaves to donkeys' ears) Leafy spurge, wolf's milk (German: Esels-Wolfsmilch). Perennial, glabrous, rhizomatous herbs, typically colonial and invasive. **STEMS** erect, branching, 3–6 (9) dm tall, arising in clumps from semiwoody crowns connected by often deeply growing rhizomes. **LEAVES** broadly linear or lanceolate-linear to narrowly oblong or narrowly oblanceolate, entire, glabrous, sessile, 3–6 (9) cm long, 2.5–7 mm wide; inflorescence bracts usually broad and short (starkly contrasting with the narrower, longer, cauline leaves) and often conspicuously yellowish; whorled at base of entire inflorescence, paired at base of each dichotomy. **INFL** an umbel with 7–15 rays, each ray repeatedly dichotomous. **CYATHIUM** 2–3 mm long; stamens 12–25. **GLANDS** 4, yellowish green, horned; appendages none. **FRUIT** finely granular to rugose-tuberculate, glabrous, 3–3.5 mm long, sometimes becoming reddish at maturity. **SEEDS** essentially smooth, 2–2.5 mm long. $2n = 16, 20, 48–60, 64$. Weedy and invasive in mesic areas, roadsides, pastures. COLO: Arc, LPl, Mon, SJn; NMEX: SJn. 1525–2285 m (5000–7500′). Flowering: May–Sep. Introduced from Eurasia, initially to New England in the early

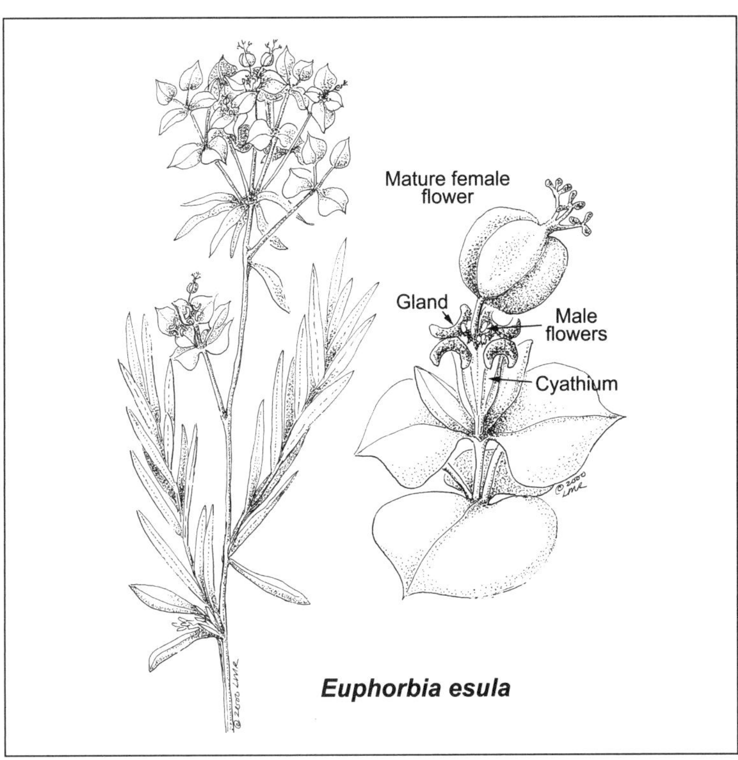

Euphorbia esula

1800s and many times subsequently to other areas in the United States from various Old World populations. Grows from England to Russia, south to Spain and Italy; introduced nearly worldwide, except for Australia; particularly abundant and troublesome in the north-central United States and adjacent Canada. This species is a member of an aggregate of very similar if not conspecific taxa in the Old World (species distinguishable only by leaf shape). Similarly, because of the extensive overlap in features, some recent European authorities do not recognize subspecific designations within *E. esula*. Since it is very difficult to control once it has become established, this plant is considered to be a noxious, invasive weed wherever it is found in the United States and elsewhere outside of its native range. Plants contain the alkaloid euphorbon, which is toxic to cattle and horses; however, goats and sheep are unaffected and are sometimes used to control infestations.

Euphorbia marginata Pursh (having a border, referring to the white leaf margins) Snow on the mountain. Stout, annual, taprooted, typically villous, annual herbs. **STEMS** erect, simple below inflorescence, 3–8 dm tall. **LEAVES** cauline, sessile, alternate, oblong to ovate or elliptic, entire, 3–8 (10) cm long, 1–3 cm wide; inflorescence leaves broadly white-margined. **INFL** umbel-like, usually 3-rayed, each ray branched several times. **CYATHIUM** villous, about 4 mm long; stamens 35–60. **GLANDS** 5, cupped; appendages conspicuously showy, white, petaloid. **FRUIT** hairy, 4–6 mm long. **SEEDS** tuberculate, about 4 mm long. $2n = 56$. Apparently rare in the Four Corners; weedy in disturbed sites, often in calcareous soils. COLO: LPl; NMEX: SJn. 1525–1825 m (5000–6000′). Flowering: Jun–Oct. Introduced from the central United States. Native range from Montana to Minnesota, south to western Texas and eastern New Mexico. Sometimes grown in gardens for showy bracts, occasionally escaping and establishing. This plant's suite of characteristics make it unmistakable.

Euphorbia spathulata Lam. (*spathula* or *spatula*, shoulder blade, referring to the leaf shape) Warty spurge. Glabrous, annual herb. **STEMS** simple or branched, erect, 1–4+ dm tall. **LEAVES** cauline, alternate, obovate to oblong-spatulate, serrulate, 1–4 cm long. **INFL** umbellate, rays usually 3, each repeatedly dichotomous. **CYATHIUM** glabrous, about 1 mm long; stamens 5–10. **GLANDS** 4, minute, appendages absent. **FRUIT** glabrous but warty, 2–3 mm long. **SEEDS** roundish, finely reticulate, 1.5–2 mm long. Apparently uncommon in many habitats, including grasslands, on slopes, stream banks, woodlands, and roadsides. COLO: Arc, Mon; NMEX: McK, SJn; UTAH. 1525–2285 m (5000–7500′). Flowering: May–Aug. Widespread in the central and western United States, Mexico, and South America. The combination of annual growth, serrulate leaves, warty fruit (and ovary), and reticulate seeds is diagnostic.

Reverchonia A. Gray Sand Reverchonia, Sand Spurge

(for the Frenchman Julien Reverchon, 1837–1905, who collected in Texas) A monotypic genus. Glabrous, monoecious, annual herbs from taproots. **LEAVES** alternate, simple, entire, short-petiolate, stipulate. **INFL** in small, dark reddish purple, bracteolate cymules in leaf axils of lateral branches, never on the main stem. **INFL** a cymule with 1 central female flower and 4–6 lateral male flowers. **FLOWERS** pedicellate and having sepals but no petals; staminate calyx 4-lobed, stamens 2; pistillate calyx 6-lobed, pistil with 3 united carpels. **FRUIT** 3-locular, columella usually deciduous. **SEEDS** 2 per locule, grooved along ventral margin (Webster and Miller 1963).

Reverchonia arenaria A. Gray (sand or sandy place; refers to the species' restriction to sandy habitats) Sand spurge, sand reverchonia. **STEMS** erect, 1–3 (5) dm tall, branches numerous and spreading; sap not milky. **LEAVES** narrowly elliptic to linear, 1.5–4 cm long, 2–8 mm wide; stipules lanceolate, acuminate, 1–2 mm long, often dark reddish. **FLOWER** sepals 1.5–2.5 mm long. **FRUIT** subglobose, smooth, 7–10 mm wide, hanging on a curved pedicel 2–5 mm long. **SEEDS** 4–6.5 mm long. $2n = 16$. Restricted to sand dunes and very sandy soils. ARIZ: Apa, Nav; UTAH. 1525–1830 m (5000–6000′). Flowering: Jun–Aug. Western Kansas, Oklahoma, and Texas west to northern Arizona and southern Utah, south to Chihuahua. The color and insertion of the cymules, narrow leaves, and sandy habitat preference make this plant readily recognizable.

Tragia L. Noseburn

[for the German herbalist Hieronymus Bock (German: buck, male deer), 1498–1554, whose name was Latinized after the Greek as Tragus, he-goat] Perennial, monoecious herbs from semiwoody taproots, with stinging and other types of hairs throughout. **LEAVES** alternate, petiolate or sessile, margins entire to serrate (ours) or toothed. **INFL** androgynous (female flowers at lowest 1–2 nodes, male flowers above) racemes, opposite leaves or terminal on lateral branches. **FLOWERS** bracteolate and pedicellate, with sepals, but not petals; male flower stamens 2–4 (10); female flower carpels 3, stigmatic surfaces smooth, roughened, or papillate. **FRUIT** a capsule, explosively dehiscent, columella persistent. **SEEDS** 1 per locule, without a caruncle.

Tragia ramosa Torr. (*ramus*, branch, + *-osa*, full of, referring to the habit) Branched noseburn. **STEMS** usually many, ascending to erect, 1–3 dm long; sap not milky. **LEAVES** narrowly deltoid or linear-lanceolate to ovate, 1–4.5 cm long, 4–20 mm wide, base often truncate or obtuse, margins serrate, petioles 1–5 (10) mm long; stipules lanceolate to ovate, 1–4.5 mm long. **FLOWERS**: staminate calyx lobes 3–4 (6), 1–2 mm long, stamens mostly 3–4; pistillate calyx lobes (5–) 6 (–7), 1.5–2.5 mm long, stigma tips slender, recurved, surfaces not or barely papillate. **FRUIT** 3-lobed, 2–4 mm tall, 6–8 mm wide. **SEEDS** globose, smooth, brownish, 2.5–3 mm long. Rocky, open areas. ARIZ: Apa; COLO: Arc, LPl; NMEX: McK, SJn. 1525–2130 m (5000–7000′). Flowering: May–Aug. Missouri to Texas, west to southeastern California, southern Nevada, and southern Utah; also northern Mexico. Except for *Urtica dioica*, the only plant with stinging hairs in the Four Corners region.

FABACEAE Lindl. LEGUME FAMILY

(LEGUMINOSAE)

S. L. Welsh & N. D. Atwood

Herbs, shrubs, or trees. **LEAVES** alternate, pinnately or palmately compound, or simple, stipulate. **INFL** usually a raceme. **FLOWERS** perfect, irregular or regular, usually borne in racemes; calyx 5-lobed; petals 5 (a banner, 2 wings, and 2 keels) or fewer, less commonly reduced to 1 (banner), or lacking; stamens 10 or 5, or numerous, diadelphous, monadelphous, or distinct; pistil 1, the ovary superior, 1- or 2-loculed, 1-carpelled, the style and stigma 1. **FRUIT** (pod) a legume or loment, sessile, subsessile, stipitate, or with a gynophore, dehiscent or indehiscent. $x = 5–14$. [Leguminosae Juss.]. The family as here constituted consists of some 677 genera worldwide (Isely 1998; Welsh 1978).

1. Flowers regular, in dense heads or compact spicate racemes; stamens 5 or numerous (subfamily Mimosoideae) .... **Key 1**
1′ Flowers irregular (only slightly so in some); stamens 10 or fewer .... (2)

2. Corolla not papilionaceous, sometimes nearly regular, the upper petal enclosed by the others; stamens 10 or fewer, commonly distinct (subfamily Caesalpinoideae) .... **Key 2**
2′ Corolla papilionaceous, the upper petal (banner) enclosing the wing and keel petals in bud, lacking in *Dalea*, *Psorothamnus*, and *Parryella*; stamens 10 or 5 (subfamily Papilionoideae) .... (3)

3. Plants woody; trees, shrubs, or woody vines .... **Key 3**
3′ Plants herbaceous perennials or annuals .... (4)

4. Leaves even-pinnate .... **Key 4**
4′ Leaves odd-pinnate, simple, or palmate .... (5)

5. Leaflets 3 only .... **Key 5**
5′ Leaflets 5 or more, or the leaves simple .... **Key 6**

KEY 1

Flowers regular; stamens 10 or numerous (Mimosoideae)

1. Stamens distinct, not united at the base; plants mainly 3–10 dm tall .... ***Acacia***
1′ Stamens basally connate, the filaments united into a tube; plants mainly 1–2 dm tall .... ***Calliandra***

KEY 2

Corolla not papilionaceous (Caesalpinoideae)

1. Leaves simple, the blades rotund-ovate; flowers pink, appearing before the leaves .... ***Cercis***
1′ Leaves once or twice compound; flowers yellow, white, or greenish, appearing after the leaves ....(2)

2. Herbs; flowers with yellow petals, the stamens exserted or not; plants indigenous ....***Caesalpinia***
2′ Trees; flowers with yellow, white, or greenish yellow petals, the stamens included or not much exserted; distribution broad .... ***Gleditsia***

KEY 3

Trees, shrubs, or woody vines (Papilionoideae)

1. Leaves simple; plants suffruticose, with axillary, spine-tipped, sterile branchlets and raceme axes; flowers red..***Alhagi***
1′ Leaves compound or simple; plants shrubs or trees, but not with axillary spine-tipped sterile branchlets and raceme axes; flowers variously colored but not red ....(2)

2. Plants trees or less commonly low and somewhat shrubby; armed with stipular spines ....***Robinia***
2′ Plants shrubby, unarmed, or armed only with stems modified as thorns ....(3)

3. Petals lacking; leaflets linear .... ***Parryella***
3′ Petals present; leaflets broad .... ***Psorothamnus***

KEY 4

Leaves even-pinnate (Papilionoideae)

1. Style bearded down 1 side; wings of corolla essentially free from the keel .... ***Lathyrus***
1′ Style bearded in a tuft or ring at apex; wings of corolla adherent to the keel .... ***Vicia***

KEY 5

Leaflets 3 (Papilionoideae)

1. Leaves palmate, the terminal leaflet neither stalked nor jointed ....(2)
1′ Leaves pinnate, the terminal leaflet stalked or jointed ....(4)

2. Flowers golden yellow, the banner orbicular, large; legumes narrowly oblong, erect or ascending; staminal filaments distinct ....***Thermopsis***
2′ Flowers ochroleucous to white or pink to pink-purple, the banner not orbicular, moderate to small in size; staminal filaments diadelphous .... (3)

3. Leaflets usually toothed; flowers mostly in heads, commonly pink or white...***Trifolium***
3′ Leaflets entire; flowers not in heads, commonly ochroleucous or pink... ***Astragalus***

4. Herbage glandular-punctate; indigenous plants with usually linear to oblanceolate leaflets .... (5)
4′ Herbage not glandular-punctate; indigenous or cultivated plants with spatulate to obovate or oblanceolate to ovate leaflets...(6)

5. Plants caulescent, with 5 or more developed internodes; pods not included in the calyx at maturity .***Psoralidium***
5′ Plants acaulescent or short-caulescent, usually with fewer than 5 developed internodes; pods included in the calyx at maturity .... ***Pediomelum***

6. Leaflets entire .... (7)
6′ Leaflets toothed (except in some *Trifolium* species)... (8)

7. Flowers in umbels, loosely capitate, or solitary in leaf axils, yellow or suffused with orange......................... ***Lotus***
7′ Flowers in interrupted racemes or panicles, purplish .... ***Phaseolus***

8. Flowers usually in heads; corolla persistent, investing the fruit; fruit straight .... ***Trifolium***
8′ Flowers usually in racemes; corolla not persistent; fruit straight or curved to coiled .... (9)

9. Leaflets toothed along the distal 1/2 or more; racemes elongate, several times longer than broad........ ***Melilotus***
9′ Leaflets toothed along the distal 1/3 only; racemes compact or loose, seldom more than twice longer than broad ....***Medicago***

KEY 6

Leaflets (4) 5 or more, or leaves simple (Papilionoideae)

1. Leaves palmately compound, with usually 5–11 leaflets, long-petiolate .... (2)
1' Leaves pinnately compound or, if rarely palmately compound (as in some *Lotus* species), sessile or with only 4 leaflets or the leaves simple .... (4)

2. Herbage not glandular-dotted; leaflets usually 7–11, variously shaped; stamens monadelphous; pods several-seeded .... ***Lupinus***
2' Herbage glandular-dotted; leaflets usually 5, broadly obovate-spatulate; stamens usually diadelphous; pods 1-seeded .... (3)

3. Plants caulescent, usually with 5 or more developed internodes; pods not included within the calyx at maturity .... ***Psoralidium***
3' Plants subacaulescent to short-caulescent, usually with fewer than 5 developed internodes; pods included within the calyx at maturity .... ***Pediomelum***

4. Herbage glandular-dotted .... (5)
4' Herbage not glandular-dotted .... (6)

5. Racemes spicate; legumes 1-seeded, not bearing appendages; stamens 5; petals (except banner) inserted on staminal tube .... ***Dalea***
5' Racemes not spicate; legumes several-seeded, bearing hooked appendages, or rarely glabrous; stamens 10; petals not inserted on staminal tube .... ***Glycyrrhiza***

6. Leaves simple; plants suffruticose, armed with axillary, thorn-tipped, sterile branchlets and raceme axes; flowers red .... ***Alhagi***
6' Leaves mainly compound or the plants not otherwise as above .... (7)

7. Margin of leaflets toothed; corolla persistent, investing the fruit .... ***Trifolium***
7' Margin of leaflets entire; corolla usually deciduous .... (8)

8. Flowers in umbels, loosely capitate, or solitary in leaf axils; petals yellow, often suffused with orange, or pink .... ***Lotus***
8' Flowers in racemes, loose or rather dense, but not in umbels; petals typically white to pink or various shades of pink-purple .... (9)

9. Keel petals much longer than the wings; fruit a flattened loment .... (10)
9' Keel petals subequal to the wings or shorter; fruit a legume (a terete loment in *Sophora*) .... (11)

10. Fruit 4- to several-seeded, not spiny; plants indigenous .... ***Hedysarum***
10' Fruit 1- to 2-seeded, more or less spiny-toothed; plants adventive, cultivated and escaping .... ***Onobrychis***

11. Staminal filaments distinct; fruit a terete to somewhat flattened loment; plants with blue or white flowers, usually of sandy sites .... ***Sophora***
11' Staminal filaments diadelphous or monadelphous; fruit a legume; plants from a caudex and/or taproot, rarely rhizomatous; habitats various .... (12)

12. Keel with a porrect beak; ventral suture of legume forming a partial or complete partition; plants usually acaulescent .... ***Oxytropis***
12' Keel beakless, or the beak diverging from the floral axis; ventral suture usually not produced internally, the dorsal usually produced in bilocular fruits; plants usually caulescent .... (13)

13. Flowers red-orange when fresh; plants adventive .... ***Sphaerophysa***
13' Flowers pink, pink-purple, lavender, or white to ochroleucous; plants indigenous, or rarely adventive .... ***Astragalus***

Acacia Mill. Acacia

(*akakie*, thorny tree) Armed trees, or in ours suffruticose herbs. **LEAVES** alternate, often clustered on short axillary shoots, bipinnate, petiolate, the pinnae bearing several leaflets; lacking internodal spines; stipules small, slender and soon deciduous. **INFL** borne in numerous flowering capitate spikes or umbels. **FLOWER** calyx 5-lobed; corolla regular, 5-lobed, inconspicuous; stamens numerous, exserted.

Acacia angustissima (Mill.) Kuntze (narrow) Prairie acacia. Sufffrutescent unarmed perennial, arising from a ligneous root crown. **STEMS** sprawling to erect, mainly 3–10 (20) dm tall, hirsute (in ours). **LEAVES** bipinnate, with 4–12 pairs of pinnae; pinnae with 10–24 (30) pairs of leaflets, these oblong and more or less asymmetric, the secondary venation scarcely discernible; stipules inconspicuous, acicular but herbaceous and soon deciduous. **INFL** of pedunculate, capitate umbels mainly 1–1.5 cm thick, arising 1–3 per axil, or many aggregated into a terminal racemose cluster. **FLOWERS** 4–10 (15) per umbel, the sepals whitish, the petals pale greenish or yellowish; stamens numerous, distinct or closely aggregated and appearing connate, whitish (sometimes fading orangish). **PODS** indehiscent, oblong, flat, 3–6 cm long and 6–10 (12) mm wide, straight or sinuate, the valves membranous, glabrous or puberulent. [*Mimosa angustissima* Mill.]. Roadsides in alkaline soils. Evidently established along a field margin east of Shiprock, two miles west of The Hogback (*K. Heil & A. Clifford 19310*, SJNM xerograph BRY). NMEX: SJn. 1525 m (5000′). Flowering: Jun–Jul. Native to the southern United States (Arizona east to Florida); Mexico, Central America. The solitary specimen examined apparently belongs to var. *angustissima*.

Alhagi Gagnebin

(Arabic *al-hajji*, the pilgrim) Much-branched, bushy, suffruticose herbs, armed with sterile axillary thorns and spine-tipped raceme axes. **LEAVES** simple, of 1 petiolulate leaflet, soon deciduous. **INFL** of few-flowered racemes, the flowers minutely bracteate, and with bracteoles. **FLOWER** calyx campanulate, short-toothed; corolla papilionaceous, reddish, the petals all free; stamens diadelphous; style filiform, glabrous, the stigma terminal. **PODS** short-stipitate, linear, incurved, torulose (constricted between the essentially round seeds), indehiscent. $x = 8$.

Alhagi maurorum Medik. (of North Africa) Camel thorn. Suffruticose armed herbs. **STEMS** mainly 3–10 dm tall, from a deep-set rhizome and root system; freely branched, armed at the nodes with stiff, spreading to ascending, leafless thorns, these 1–3.5 cm long; stipules 0.5–3 mm long, subulate to lance-ovate, caducous. **LEAVES** of 1 petiolulate leaflet, oblanceolate to oblong-elliptic, obtuse or apiculate, 3–25 mm long, 1–8 mm wide, pale beneath, minutely red-dotted above. **INFL** a raceme, (1) 2- to 7-flowered, the axis prolonged as a thorn; bracts 0.5–1 mm long, subulate; pedicels 1–2 mm long. **FLOWER** petals reddish purple, 7–9.5 mm long; calyx 2.5–3.2 mm long, glabrous, the margin sometimes puberulent and with low broad teeth 0.2–0.5 mm long; bracteoles 2. **PODS** stipitate, the stipe as long as the persistent calyx, the body with (1 or 2) 3–8 seeds, indehiscent, glabrous. $2n = 16$. [*Hedysarum alhagi* L.; *Alhagi camelorum* Fisch.]. Sandy sites along the San Juan River and its side canyons. ARIZ: Apa; NMEX: SJn; UTAH. 1370 to 1630 m (4495–5350′). Flowering: Jun–Jul. Texas west to Nevada and California. The plant, which is spreading and subdominant along the San Juan River, is armed with nodal thorns, is difficult to eradicate, and is regarded as a noxious weed in Arizona and California. Camel thorn, a provider of valued spring forage in the Near Eastern deserts of the Old World, has become well established in saline meadows, sandbars, and playas in widely scattered portions of the American Southwest. The rhizomes can penetrate several inches of asphalt paving.

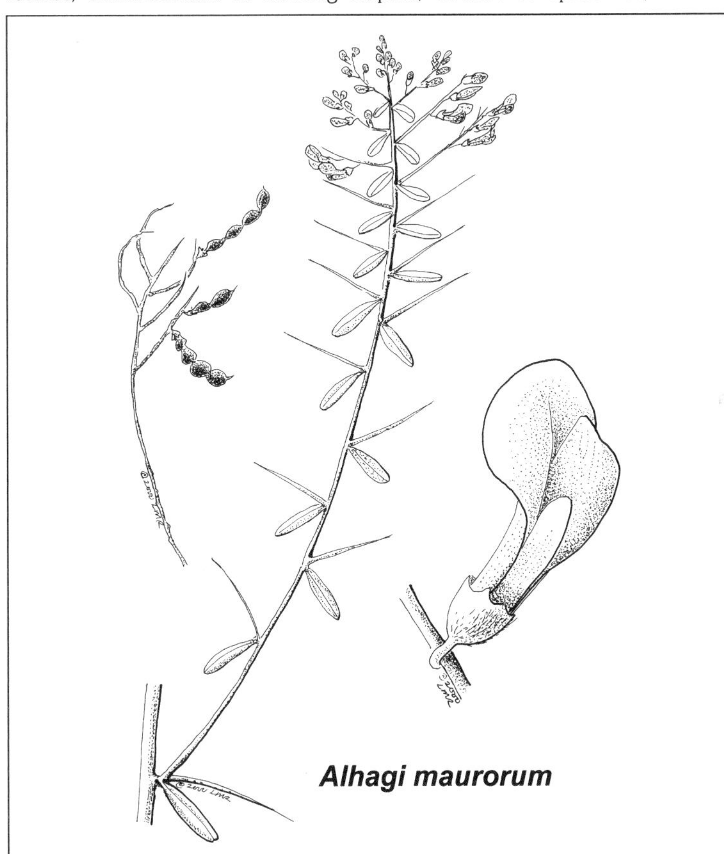
Alhagi maurorum

Astragalus L. Milkvetch

(from the classical Greek word for ankle bone) Milkvetch, locoweed. Acaulescent, subacaulescent, or caulescent annual or perennial herbs, rarely somewhat suffruticose at the base, arising from a taproot and aerial, caespitose, or subterranean caudex. **PUBESCENCE** of basifixed or malpighian (dolabriform) hairs, or glabrous. **LEAVES** alternate, odd-pinnate, digitately 3-foliolate, or unifoliolate, the leaflets entire, petiolulate or sessile, or decurrent; stipules adnate to the petiole or free, or connate opposite the petiole, especially on lowermost reduced leaves; bracts often small, membranaceous, subtending the pedicels; bracteoles 1 or 2, very small or lacking, borne when present on the calyx base. **INFL** borne in racemes or spikes, or rarely subumbellately disposed, rarely solitary; peduncles axillary, or apparently arising from the caudex. **FLOWERS** papilionaceous, violet, purplish, white, pale yellow, or yellow-white to greenish; calyx campanulate to cylindric, with 5 subequal teeth, or the lowermost the longest; petals mostly long-clawed, the banner erect, ovate, oblong, or panduriform, recurved through 20–180°, the wings oblong, usually shorter than the banner and longer than the keel, the keel usually shorter than the wings, suberect, obtuse or acute; stamens diadelphous, 9 connate and 1 (vexillar) free, the alternate anthers of 2 slightly different sizes; ovary sessile or stipitate, or borne on a gynophore, 2 to many-ovulate, the style filiform, straight or incurved, smooth, the small stigma terminal. **PODS** sessile or stipitate, or borne on a gynophore, deciduous or persistent, often deciduous with the calyx and pedicel, the body unilocular or partially or completely bilocular by intrusion of the dorsal inferior suture, varying from linear to globose, or even broader than long, straight or incurved, or more rarely decurved or coiled inward, dehiscent throughout or only at the apex; ovules 2-seriate, 2 to many, borne on slender funicles sometimes united at base by a membrane or funicular flange simulating a septum, but issuing from the seminal suture. **SEEDS** mostly reniform. x = 8, 11, 12. Two thousand species, mostly in western North America.

KEY 1. Key to groups.

1. Leaves with 5–7 spinulose-tipped, decurrent leaflets; plants prostrate to ascending or erect .... ***A. kentrophyta***
1′ Leaves simple, 3-foliolate, or plurifoliolate, but the leaflets not spinulose-tipped; plants of various habit .... (2)

2. Plants with malpighian (dolabriform) pubescence .... **Key 2**
2′ Plants with hairs basifixed .... (3)

3. Plants annual or winter-annual .... **Key 3**
3′ Plants perennial (sometimes short-lived) .... (4)

4. Leaves all, or at least the upper, with terminal leaflet not jointed to the rachis .... **Key 4**
4′ Leaves all with the terminal leaflet jointed to the rachis .... (5)

5. Lowermost stipules and sometimes all connate-sheathing or -clasping .... (6)
5′ Lowermost stipules distinct as are the others .... (7)

6. Pods stipitate or with a gynophore more than 1 mm long .... **Key 5**
6′ Pods sessile or subsessile .... **Key 6**

7. Pods stipitate or with a gynophore 0.5 mm long or longer .... **Key 7**
7′ Pods sessile or subsessile, the stipe if present less than 1 mm long .... (8)

8. Plants acaulescent or subacaulescent, the stem axis aboveground lacking or, when present, shorter than leaves or inflorescence .... **Key 8**
8′ Plants caulescent, the main stem axis longer than the longest leaf or inflorescence .... **Key 9**

KEY 2

Leaves unifoliolate, plurifoliolate, or some 3-foliolate; pubescence malpighian or incipiently so

1. Terminal leaflets of at least some upper leaves continuous with the rachis (the upper leaves often reduced to phyllodia consisting of a subfiliform rachis); plants definitely caulescent .... (2)
1′ Terminal leaflets all jointed, the leaves typically all odd-pinnate .... (3)

2. Raceme and peduncle together surpassing the length of the stem, much surpassing the leaves; plants smelling of selenium, of southeast Utah and northern Arizona .... ***A. moencoppensis***
2′ Raceme and peduncle together not or seldom surpassing the length of the stem, or if so not much surpassing the leaves; plants of various or other distribution .... ***A. ceramicus***

3. Pods fully bilocular, erect or declined; plants distinctly caulescent .... ***A. falcatus***
3′ Pods unilocular, or if bilocular, the plants acaulescent, subacaulescent, or caulescent. .... (4)

4. Wing petals deeply cleft at apex; plants definitely acaulescent, with scapiform peduncles; pods oblong, compressed-triquetrous, fully bilocular .... ***A. calycosus***
4′ Wing petals entire or obscurely emarginate; pods variously unilocular or bilocular .... (5)

5. Stipules, at least at the lowermost 1–3 nodes, connate into a bidentate sheath ....(6)
5′ Stipules all distinct (though fully amplexicaul in some); plants acaulescent or shortly to definitely caulescent ..(9)

6. Plants prostrate-spreading ....(7)
6′ Plants erect or ascending, not prostrate-spreading, when low still more or less tufted and acaulescent or subacaulescent ....(8)

7. Racemes 1- to 3-flowered; flowers pink-purple; leaflets mostly 7–11; plants of sandstone escarpments in southern Utah, southwest Colorado, and northern Arizona .... ***A. sesquiflorus***
7′ Racemes 7- to many-flowered; flowers dull purplish; leaflets mostly 11–17; plants rather widespread in western and southwestern New Mexico, Arizona, southern Nevada, and southern Utah .... ***A. humistratus***

8. Peduncles (3) 5–23 cm long, usually longer than the subtending leaf; plants flowering mainly in springtime (sometimes again in autumn), in southwest Colorado, southeast Utah, northwest New Mexico, and northeast Arizona .... ***A. flavus***
8′ Peduncles 1.5–4 (5.5) cm long, much shorter than the subtending leaf; plants flowering in summer, of northeastern Arizona, northern New Mexico, and La Plata County, Colorado. .... ***A. albulus***

9. Plants mat-forming, pulvinate, or in tiny tufts, of rimrock, northwest New Mexico and southwest Colorado; flowers tiny; leaves tiny; pods ovoid, 3–4.5 mm long; ovules 4–6. .... ***A. humillimus***
9′ Plants definitely caulescent, or subacaulescent to acaulescent and loosely tufted to somewhat mat-forming, but then not with tiny leaves, pods, and flowers; distribution various ....(10)

10. Pods narrowly oblong to oblong-ellipsoid, straight or nearly so, laterally compressed when ripe, bicarinate by the sutures; plants of plains and the eastern foothills of the Rocky Mountains, extending west only in the upper San Juan Basin; southwest Colorado, northwest New Mexico, and adjacent Arizona .... ***A. missouriensis***
10′ Pods plumply ovate to broadly lanceolate in outline, scarcely to moderately incurved, 9–22 mm long, laterally compressed only in the beak; flowers 11–18.5 mm long, whitish or tinged purplish; plants of northern Arizona and southeast Utah ....(11)

11. Pods readily deciduous; plants widespread from Clark and Lincoln counties, Nevada, through southern Utah, and northern Arizona to southwest Colorado and west Texas. .... ***A. amphioxys***
11′ Pods persistent or very tardily deciduous; plants of southwest Colorado and adjacent Utah, northeast Arizona, and northwest New Mexico .... ***A. missouriensis***

KEY 3

Leaves unifoliolate, plurifoliolate, or some 3-foliolate; pubescence basifixed; plants annual (or flowering as annual)

1. Pods stipitate, the body ovoid-ellipsoid, erect-ascending; plants selenophytes, of San Juan County, Utah. ***A. cutleri***
1′ Pods sessile or stipitate, of various shape and habit, but if as above, not of San Juan County, Utah ....(2)

2. Pods obovoid- to oblong-ellipsoid, 10–18 mm long; flowers 4–7 mm long, the flowers recurved through 90°; plants scattered throughout the Four Corners region .... ***A. brandegeei***
2′ Pods either obliquely ovoid to subglobose, half-elliptic, lunately elliptic, or linear-oblong in outline; flowers various, but if of that size, the banner not reflexed so greatly ....(3)

3. Racemes all (or all except some subterminal and obviously depauperate ones) more than 10-flowered ....(4)
3′ Racemes (1) 2- to 10-flowered (seldom some with more flowers) ....(5)

4. Banners reflexed through about 90°; ovules 21–32; plants typical of sandy desert shrublands in southeast Utah, northwest Arizona, and northwest New Mexico .... ***A. fucatus***
4′ Banners recurved through about 50°; ovules 10–21; plants of various or other distribution; pods subsymmetrically ovoid, ovoid-ellipsoid, or subglobose, the beak short and obscure, not well differentiated from the body; plants of southwest Colorado and northwest New Mexico .... ***A. wootonii***

5. Pods villous or hirsute, the hairs 0.7–2.2 mm long; plants of southeast Utah and northwest New Mexico .... ***A. sabulonum***
5′ Pods glabrous or strigulose, if loosely strigulose the hairs not over 0.7 mm long ....(6)

6. Pods loosely strigulose, the hairs spreading, incumbent, or curly .... ***A. sabulonum***
6′ Pods truly strigulose, the hairs straight and appressed .... ***A. nuttallianus***

KEY 4

Plants perennial (sometimes short-lived); pubescence basifixed; leaves all, or at least the upper terminal leaflet, continuous with the rachis

1. Ovaries and pods stipitate, the stipe mostly 2.5–12 mm long .... (2)
1. Ovaries and pods sessile or subsessile .... (5)

2. Flowers lemon-yellow, concolorous, the calyx not strongly contrasting in color with the petals; plants local in Archuleta County, Colorado .... ***A. ripleyi***
2′ Flowers not lemon-yellow, if ochroleucous, as in *A. lonchocarpus*, the calyx contrasting in color with the petals; plants of various distribution .... (3)

3. Pods pendulous, stipitate, the stipe 3.5–6 mm long, the body narrowly oblong, 20–32 mm long, 3.5–4.5 mm thick, dorsoventrally compressed, strigose, unilocular; plants intricately branched, forming birdsnest-like clumps .... ***A. nidularius***
3′ Pods various, but not simultaneously as above; plants not excessively branched, usually with a small number of stems arising from the superficial caudex .... (4)

4. Body of pods laterally compressed, bicarinate, the valves flat or almost so .... ***A. coltonii***
4′ Body of pods dorsiventrally more or less compressed, flattened or openly grooved dorsally, the valves convex .... ***A. lonchocarpus***

5. Flowers 14.5–30 mm long; calyx tubes mostly over 5 mm long, campanulate to cylindric; calyx inflated, densely silky-pilose, enclosing the small, ventrally bisulcate pods; plants of the upper San Juan Valley, New Mexico, and adjacent Colorado. .... ***A. oocalycis***
5′ Flowers mostly 4.3–12.5 mm long; calyx tubes (1.2) 2–5 mm long, campanulate; pods laterally or dorsally compressed, uni- or bilocular .... (6)

6. Flowers 4.3–6.1 mm long; calyx tubes 1.2–1.8 mm long; pods 9–12 mm long, 2.5–3.5 mm wide, strongly laterally compressed; plants of McKinley County, New Mexico .... ***A. cliffordii***
6′ Flowers 5–12.5 mm long; calyx tubes 1.8–5 mm long; pods various, sometimes strongly laterally compressed .. (7)

7. Plants more or less rushlike, the leaflets linear (see also *A. moencoppensis*, next lead), known from southeast Utah, northeast Arizona, and southwest Colorado .... ***A. episcopus***
7′ Plants of various habit, the leaflets often broader than linear, known from various or other distribution .... (8)

8. Stems arising from branched aerial caudex; plants either dwarf and tufted, or moderately tall selenophytes .... ***A. moencoppensis***
8′ Stems arising from subterranean or superficial caudex, often pale and leafless at the base; plants of various habit and distribution .... ***A. wingatanus***

KEY 5

Plants perennial (sometimes short-lived); pubescence basifixed; leaves with terminal leaflet jointed with the rachis; lowermost stipules connate; pods stipitate or on a definite stipelike gynophore

1. Body of pods unilocular and bladdery-inflated, terete or shallowly sulcate along the sutures but not laterally angled .... ***A. hallii***
1′ Body of pods uni- or bilocular, not inflated or if swollen then scarcely bladdery or also angled laterally or bluntly so .... (2)

2. Flowers at once nodding, retrorsely imbricate, and relatively large, (12.5) 14–23.5 mm long; stipe of pods 4–22 mm long .... (3)
2′ Flowers of various orientation, but if nodding then often loosely and openly racemose (or becoming so) and if retrorsely imbricate then smaller; stipe of pods various .... (4)

3. Flowers of moderate size, ochroleucous; body of pods merely triquetrous; plants not with the odor of selenium, montane; widespread throughout the Four Corners .... ***A. scopulorum***
3′ Flowers small, ochroleucous; body of pods depressed ventrally, rounded dorsally, the ventral suture between parallel grooves .... ***A. bisulcatus***

4. Flowers nodding and retrorsely imbricate; body of pods depressed ventrally, the suture between parallel grooves; herbage smelling of selenium. .... ***A. bisulcatus***
4′ Flowers with various orientation, if nodding then openly racemose or becoming so; body of pods not as above; herbage not smelling of selenium; plants variously distributed .... (5)

5. Pods laterally compressed, 2-sided, the faces plane or low-convex, the cavity unilocular .... (6)
5′ Body of pods subterete to obcompressed or trigonous or, if laterally compressed, then swollen and the valves inflexed dorsally as a narrow partial septum .... (7)

6. Stipules and ripe pods commonly black or blackish; racemes often paired in leaf axils; flowers 5.5–10 mm long; body of pods 7–17 mm long, 2.5–4.5 mm wide; plants widespread .... ***A. tenellus***
6′ Stipules and pods not blackening; racemes 1 per leaf axil; plants widespread throughout the Four Corners .... ***A. wingatanus***

7. Flowers 6–7.2 mm long, mostly 12–40 on an axis 4–17 mm long that elongates in fruit; leaflets 7–11; pods 6- to 13-ovulate; plants of northwest New Mexico and southwest Colorado. .... ***A. proximus***
7′ Flowers 7 mm long or more, variously numerous, the axis elongating as above or not; leaflets mostly 11 or more; pods pubescent or glabrous, with 10 or more ovules; plants of various distribution .... (8)

8. Flowers 7–11 mm long, mostly 7–26 on an axis that elongates in fruit; plants of the lower mountains; southwest Colorado, southeast Utah, and northwest New Mexico. .... ***A. flexuosus***
8′ Flowers often more than 11 mm long, various in number, the axis not or much elongating in fruit; plants typically of higher montane sites .... ***A. alpinus***

KEY 6

Plants perennial (sometimes short-lived); pubescence basifixed; leaves with terminal leaflet jointed with the rachis; lowermost stipules connate; pods sessile or the stipe or gynophore less than 1 mm long

1. Flowers at once narrowly ascending or erect and crowded into heads or short, spicate racemes; stems arising from subterranean caudex or rhizomelike caudex branches .... (2)
1′ Flowers variously oriented, but if erect at anthesis then often loosely racemose or becoming so and often spreading or soon declined; stems sometimes as above, but the caudex superficial in many species (see exceptions below) .. (3)

2. Stems mostly less than 2.5 dm tall; flowers purple or lavender; leaflets commonly 13–21; pod ovoid- or oblong-ellipsoid, scarcely swollen, obtusely trigonous, dorsally sulcate; plants of moist sites, often in meadows, widely distributed .... ***A. agrestis***
2′ Stems 3–15 dm tall; petals ochroleucous; leaflets 17–31; pod broadly ovoid or globose, inflated, but firm, sulcate ventrally, black at maturity; plants introduced for reclamation, and to be expected throughout the West .... ***A. cicer***

3. Flowers and pods both erect-ascending and laxly racemose, the axes of the lower peduncle and raceme together longer than the primary stem; raceme axis 3–23 cm long .... ***A. moencoppensis***
3′ Flowers and/or pods finally spreading or declined, the raceme shorter or the axes of the longest peduncle and raceme together shorter than the primary stem; plants of various distribution .... (4)

4. Plants either strongly or definitely caulescent or of different distribution, or both .... (5)
4′ Plants low, tufted, of sandy habitats in southeast Utah, northeast Arizona, and northwest New Mexico .... (6)

5. Calyx 2–3.5 mm long; the tube 1.2–2.5 mm long; flowers 4–6.2 mm long; plants of McKinley County, New Mexico .... ***A. heilii***
5′ Calyx much more than 5 mm long, the tube 3–13.5 mm long; flowers 9–26 mm long; plants of southeast Utah, northeast Arizona, rare in northwest New Mexico .... ***A. zionis***

6. Pods at once laterally flattened and unilocular, bicarinate, narrowly oblong-elliptic or linear-oblanceolate; plants widespread throughout the Four Corners .... ***A. wingatanus***
6′ Pods terete, triquetrous, obcompressed, or bladdery-inflated or bilocular or subbilocular (or both), or if laterally compressed and unilocular, the valves convex or produced into a ridge parallel to the dorsal suture, or not bicarinate .... (7)

7. Plants strongly caulescent, with stems mainly (1.5) 2–9 dm long .... (8)
7′ Plants not strongly caulescent, mainly dwarf clump-formers with stems less than 2 dm long .... (10)

8. Pods (3) 4–12 mm thick, stiffly cartilaginous to coriaceous; plants of northeast Arizona, southwest Colorado, and northwest New Mexico .... ***A. hallii***
8′ Pods linear-oblanceolate or elliptic and 3–5 mm thick, tumid or bladdery and typically more than 5 mm thick and the distribution otherwise .... (9)

9. Pods oblong-oblanceolate or elliptic in outline, 3–5 mm thick; plants from southeast Utah and northeast Arizona eastward and northward .... ***A. flexuosus***
9′ Pods tumid or bladdery, (4) 5–27 (35 when pressed) mm thick; plants typical of sandy desert shrublands; widely scattered throughout the Four Corners. .... ***A. fucatus***

10. Plants rather loosely tufted, or if densely so, not mat-forming; leaves and leaflets various in size and arrangement; plants forming small tufts 5–10 cm tall, known only from Mesa Verde, Colorado. .... ***A. deterior***
10′ Plants forming compact mats, known from slopes of degraded sandstone or bases of sandstone buttes in New Mexico .... (11)

11. Leaves mostly 4–10 (rarely to 20) mm long; leaflets (3) 5–9, subpalmately crowded, mostly 1–4 mm long; known from bases of sandy escarpments in San Juan, Rio Arriba, and McKinley counties, New Mexico .... ***A. micromerius***
11′ Leaves mostly (10) 15–40 mm long; leaflets (7) 9–15, not especially crowded, 0.7–3 (4) mm long; racemes 4- to 10-flowered; plants of the Chuska Mountains; Apache County, Arizona, and McKinley and San Juan counties, New Mexico .... ***A. chuskanus***

KEY 7

Plants perennial (sometimes short-lived); pubescence basifixed; leaves with terminal leaflet jointed with the rachis; lowermost stipules distinct; pods elevated on a stipe or gynophore more than 0.5 mm long

1. Pods elevated on, and jointed to, a stipelike gynophore .... (2)
1′ Pods elevated on, and continuous with, a true stipe or stipelike neck .... (4)

2. Plants typically caulescent, not especially tuft-forming; pods bladdery-inflated .... ***A. oophorus***
2′ Plants subacaulescent, tuft-forming; pods not bladdery-inflated .... (3)

3. Pods sparingly strigulose; plants of southeast Utah, southwest Colorado, and northwest New Mexico .... ***A. naturitensis***
3′ Pods more or less spreading-hirsute; plants of various distribution .... ***A. desperatus***

4. Body of pods laterally compressed, bicarinate by the sutures, with 2 flat or convex faces, unilocular .... (5)
4′ Body of pods either trigonously or dorsiventrally compressed, or subterete (and in some bicarinate), if rarely laterally compressed or appearing so, then fully bilocular or the faces convex or both .... (6)

5. Body of pods erect-ascending, 22–30 mm long, 6–9 mm wide, from an erect-ascending stipe 12–19 mm long, this contorted through almost a half circle, the pods thus inverted; plants from near Towaoc, Montezuma County, Colorado .... ***A. tortipes***
5′ Body of pods descending or pendulous, various in length and width, from a descending or pendulous, noncontorted stipe of various length; plants of other distribution. .... ***A. coltonii***

6. Pods at once strongly inflated or swollen and fully unilocular but not bladdery; plants smelling of selenium .... (7)
6′ Pods either not inflated or, if inflated, then at least incipiently bilocular; plants smelling of selenium or not .... (10)

7. Flowers with both petals and calyx white or the calyx greenish; pods erect and more or less woody at maturity; plants widespread throughout the Four Corners .... ***A. praelongus***
7′ Flowers with petals white to purple, the calyx often purple; pods erect or spreading, merely cartilaginous; plants of various distribution .... (8)

8. Flowers white or pale purple; plants along San Juan River, San Juan County, Utah .... ***A. cutleri***
8′ Flowers purple, often drying dark purple; plants of various distribution, if partially as above, then often flowering as an annual or forming large clumps .... (9)

9. Pods erect or narrowly ascending from the erect peduncle; stems erect and incurved-ascending, 1–4 (4.4) dm tall; plants from northeast Arizona, northwest New Mexico, and southeast Utah .... ***A. preussii***

9′ Pods horizontally spreading, declined, or only weakly ascending, borne on weakly ascending or reclining peduncle; stems various in attitude and height; plants from southeast Utah .... ***A. eastwoodiae***

10. Flowers ochroleucous, the keel tip purple; pod erect, succulent, becoming woody, subbilocular; plants with odor of selenium .... ***A. praelongus***
10′ Flowers variously colored, the keel tip maculate or not; pods otherwise or, if erect, not succulent; plants not smelling of selenium .... (11)

11. Flowers ochroleucous; body of pods (25) 30–40 mm long, triquetrous, the stipe 5–10 (12) mm long; plants local on Mesa Verde, Montezuma County, Colorado. .... ***A. schmolliae***
11′ Flowers ochroleucous, greenish white, lilac, or red or pink-purple; body of pods mostly less than 30 mm long, variously shaped, the stipe length various; plants of other distribution .... ***A. robbinsii***

KEY 8

Plants perennial (sometimes short-lived); pubescence basifixed; leaves with terminal leaflet jointed with the rachis; lowermost stipules distinct; pods sessile or the stipe less than 1 mm long; plants acaulescent or subacaulescent

1. Calyx 2.5–9 mm long; flowers 6.5–16 mm long. .... (2)
1′ Calyx 9.5–16 (19) mm long; flowers 17–25 (26) mm long .... (3)

2. Flowers 11–17 mm long; calyx 6.2–8 mm long; the tube 4.8–6.7 mm long .... ***A. cottamii***
2′ Flowers 8–10 mm long; calyx 3.6–4.5 mm long, the tube 3.1–3.5 mm long .... ***A. monumentalis***

3. Pods bilocular below the beak .... ***A. mollissimus***
3′ Pods unilocular .... (4)

4. Plants strictly acaulescent, the bases thatched with marcescent stipules and leaf bases; pods both hirsute and tomentose; plants widespread in the Four Corners .... ***A. newberryi***
4′ Plants sometimes as above, but mostly with bases not thatched with marcescent stipules and leaf bases, but if so, not strictly acaulescent; pods not at once both hirsute and tomentose .... (5)

5. Pods at once papery-membranous and hirsute with long lustrous hairs; plants of rimrock habitats .... ***A. desperatus***
5′ Pods firm, either leathery or woody, the pubescence various; plants of various substrates and distribution .... (6)

6. Pods strigulose or thinly villosulous, with appressed or incumbent hairs less than 1 mm long; flowers 13–15.5 mm long; pods more or less trigonous, 13–22 mm long .... ***A. naturitensis***
6′ Pod bodies either simply hirsute or at once hirsute and tomentulose, the longer hairs more than 1 mm long, or shaggy-villous, or glabrous, and not otherwise as above. .... (7)

7. Leaflets more than 21 in at least some mature leaves .... ***A. iodopetalus***
7′ Leaflets not more than 21 on any leaves. .... ***A. zionis***

KEY 9

Plants perennial (sometimes short-lived); pubescence basifixed; leaves with terminal leaflet jointed with the rachis; lowermost stipules distinct; pods sessile or the stipe less than 1 mm long; plants definitely caulescent

1. Pods nodding, ellipsoid, 2.5–5 mm thick, slightly decurved, the ventral suture more convex than the dorsal; stems arising from a superficial caudex. .... (2)
1′ Pods of various orientation, either straight or incurved, or more than 5 mm thick, if at once decurved and less than 5 mm thick, the stems from a subterranean caudex .... (3)

2. Pods bilocular or semibilocular, the septum at least 1 mm wide; plants smelling of selenium; plants widespread throughout the Four Corners .... ***A. praelongus***
2′ Pods unilocular or essentially so, the septum, if present, less than 1 mm wide; plants of southwest Colorado and northwest New Mexico .... ***A. wootonii***

3. Pods at once inflated or swollen (ovoid to cylindroid) and bilocular or almost so; plants widespread in the Four Corners .... ***A. lentiginosus***

3′ Pods not or scarcely inflated, unilocular, imperfectly bilocular, or bilocular; compressed-trigonous or subterete and grooved dorsally .... (4)

4. Pods bladdery-inflated, (18) 23–40 mm long, (10) 12–20 mm thick, terete or somewhat dorsiventrally compressed, or openly sulcate ventrally; plants not smelling of selenium .... ***A. wootonii***

4′ Pods not bladdery-inflated, or if so, mainly smaller than above and of other distribution; plants smelling of selenium .... (5)

5. Stems from a subterranean caudex; pods pendulous or spreading-declined; plants sprawling, known from San Juan County, Utah, and Montezuma County, Colorado. .... ***A. cronquistii***

5′ Stems from a superficial caudex; pods humistrate or stiffly erect; plants more widespread .... (6)

6. Petals purple; pods strictly unilocular, the valves stiffly chartaceous; plants uncommon; northeast Arizona, northwest New Mexico .... ***A. preussii***

6′ Petals ochroleucous, the keel tip purple or white; pods initially fleshy, becoming ligneous, the septum narrow or lacking .... (7)

7. Flowers pure white; calyx tube deeply campanulate or short-cylindric, 6–9 mm long, the lobes broadly spreading .... ***A. pattersonii***

7′ Flowers ochroleucous, the keel tip maculate; calyx tube campanulate, mostly 4.5–6.5 mm long, the lobes erect-ascending .... ***A. praelongus***

Astragalus agrestis Douglas ex G. Don (field) Field milkvetch. Perennial, caulescent, 9–43 cm tall, from a subterranean caudex and long, rhizomelike caudex branches. **PUBESCENCE** basifixed, strigulose to villous-pilose. **STEMS** erect to decumbent-clambering; stipules (1) 2–11 mm long, at least the lowermost connate-sheathing. **LEAVES** 2–10 cm long; leaflets 13–23, 4–18 mm long, 2–5 mm wide, narrowly elliptic to lance-oblong, obtuse to retuse or acute, strigulose above and below. **INFL** a raceme, subcapitate, 5- to 15-flowered, the flowers ascending-erect at anthesis, the axis 0.5–2.5 cm long in fruit; bracts 3–7 mm long; pedicels 0.5–1.5 mm long; bracteoles 0; peduncles 1.5–15 cm long. **FLOWERS** 17–24 mm long, pink-purple, ochroleucous, or almost white, the banner recurved through about 25°; calyx 7–12.5 mm long, the tube 5–7.8 mm long, cylindric, villous, the teeth 2.5–5.5 mm long, linear. **PODS** erect, short-stipitate, the stipe 0.3–1 mm long, the body 7–10 mm long, 2.8–4.5 mm thick, oblong-ellipsoid, triquetrous, bilocular, silky-villous; ovules 14–26. $2n = 16$. [*A. hypoglottis* Hook.; *A. dasyglottis* Fisch. ex DC., non *A. dasyglottis* Pallas; *A. hypoglottis* var. *dasyglottis* (DC.) Ledeb.; *A. goniatus* Nutt. ex Torr. & A. Gray; *A. hypoglottis* var. *polyspermus* Torr. & A. Gray; *A. agrestis* G. Don var. *polyspermus* (Torr. & A. Gray) M. E. Jones; *A. virgultulus* E. Sheld.; *A. agrestis* f. *virgultulus* (E. Sheld.) B. Boivin; *A. hypoglottis* var. *bracteatus* Osterh.; *A. agrestis* var. *bracteatus* (Osterh.) M. E. Jones; *A. tarletonis* Rydb.]. Meadows, prairies, stream banks, and openings in sagebrush and aspen. COLO: Arc, LPl; NMEX: RAr. 2285–2535 m (7500–8320′). Flowering: May–early Sep. Southern Yukon east to Manitoba and south to northeast California, northern Nevada, Utah, western Nebraska, northwest Iowa, and central to western Minnesota; widely dispersed in eastern and central Asia.

Astragalus albulus Wooton & Standl. (whitish) Cibola milkvetch. Tall, robust, caulescent perennial, 15–70 cm tall, from a branching caudex. **PUBESCENCE** strigulose to pilose, malpighian. **STEMS** erect or ascending, numerous and forming bushy clumps; stipules 2.5–10 mm long, all connate-sheathing. **LEAVES** 3–10 (14) cm long; leaflets (7) 11–23, (2) 4–27 (45) mm long, linear, linear-oblanceolate, or narrowly elliptic, acute to subobtuse, strigose below, glabrous and green on the upper surface. **INFL** a raceme, 9- to 47-flowered, the flowers ascending at anthesis, the axis (2) 3–15 (20) cm long in fruit; bracts (2) 3–7 mm long; pedicels 0.3–1.2 mm long; bracteoles 0–2; peduncles 1–4 (5.5) cm long. **FLOWERS** 13–17.2 mm long, whitish, the keel tip faintly maculate, the banner recurved through about 45°; calyx (7.2) 7.5–11 mm long, the tube (5.4) 5.7–7.4 mm long, short-cylindric, strigulose to pilose, the teeth (1.5) 1.8–3.6 mm long, lance-subulate. **PODS** declined or pendulous, stipitate, the stipe 0.7–2.5 mm long, the body 9–12 mm long, 3.3–5 mm thick, narrowly ellipsoid or lance-ellipsoid, gently incurved, laterally compressed, keeled ventrally, sulcate dorsally, strigose or glabrous, unilocular. $n = 11$. [*Batidophaca albula* (Wooton & Standl.) Rydb.]. Gullied badlands and sandy clay talus under cliffs, in seleniferous substrates. ARIZ: Apa, Nav; COLO: LPl; NMEX: McK, RAr, San, SJn. 1820–2265 m (5975–7425′). Flowering: Jul–Oct. This mainly New Mexican plant is one of the few species of *Astragalus* that ignore the spring flowering season and flower instead from midsummer onward.

Astragalus alpinus L. (alpine) Alpine milkvetch. Low, caulescent or short-caulescent perennial, 2–30 cm tall, from a subterranean caudex and rhizomatous caudex branches. **PUBESCENCE** strigulose to villosulous, basifixed. **STEMS**

decumbent to ascending, subterranean for a space of 2–15 cm or more; stipules 1.5–8 mm long, at least the lowermost connate-sheathing. **LEAVES** (2) 3–15 cm long; leaflets (11) 15–25 (27), (2) 6–20 (24) mm long, 2–10 mm wide, ovate to elliptic or oblong, retuse to rounded, strigulose above and below. **INFL** a raceme, 5- to 17 (23)-flowered, the flowers erect to declined at anthesis, the axis 0.5–5 (7) cm long in fruit; bracts 1–2.5 mm long; pedicels 0.5–2.3 mm long; bracteoles 0; peduncles 3–15 (17) cm long. **FLOWERS** (6.2) 7–12 (13.1) mm long, pink-purple (rarely white), the banner abruptly incurved through 75°–90°; calyx 3.2–6.5 mm long, the tube 2–4 mm long, campanulate, strigulose, the teeth 1–3.2 mm long, subulate. **PODS** pendulous, stipitate, the stipe 2–5 mm long, the body oblong-lanceolate in outline, 7–17 mm long, 1.5–4 mm thick, strigulose, semibilocular; ovules 5–11. $2n = 16, 32$. [*Phaca astragalina* DC. f. *occidentalis* Gand.; *A. andinus* M. E. Jones; *A. alpinus* var. *parvulus* J. Rousseau; *A. alpinus* subsp. *alaskanus* Hultén; *A. alpinus* f. *lepageanus* J. Rousseau; *A. alpinus* var. *alaskanus* f. *albovestitus* Lepage]. Arctic and alpine tundra, and woods, riverbeds, terraces, or moraines. COLO: Arc, LPl, Min, RGr, SJn; NMEX: SJn. 2530–3810 m (8300–12500′). Flowering: mid-Apr–Sep. Alaska, Yukon, east to Newfoundland, south to Utah, southwest Colorado, South Dakota; Eurasia. San Juan River drainage materials belong to **var. *alpinus***.

Astragalus amphioxys A. Gray (sharp at both ends, referring to the pods) Crescent milkvetch. Perennial (rarely flowering the first year), subacaulescent to shortly caulescent, 2–35 cm tall, from a weak caudex. **PUBESCENCE** malpighian. **STEMS** lacking or up to 20 cm long, the internodes often concealed by stipules; stipules 2–13 mm long, all distinct or the lowermost sometimes connate-sheathing. **LEAVES** 2–13 mm long; leaflets (1) 5–21, 3–20 mm long, 1–9 mm wide, elliptic to obovate or oblanceolate, obtuse to acute, strigose on both sides. **INFL** a raceme, 2–13-flowered, the flowers ascending at anthesis, the axis (0.5) 1–6.5 cm long in fruit; bracts 2.5–8 mm long; pedicels 0.6–2.5 mm long; bracteoles 0–2; peduncles (1) 2–15 (20) cm long. **FLOWERS** 12.8–31 mm long, pink-purple, rarely white, the banner recurved through about 40°; calyx 6.3–14.2 mm long, the tube 5.2–10.5 mm long, cylindric, strigose, usually purplish, the teeth 1.1–4.5 mm long, subulate. **PODS** ascending, sessile, 1.5–5 cm long, 5–12 mm thick, usually curved, mostly dorsiventrally compressed, unilocular, strigose; ovules 42–70. [*A. shortianus* Nutt. var. *minor* A. Gray; *Xylophacos amphioxys* (A. Gray) Rydb.; *A. amphioxys* var. *typicus* Barneby; *A. crescenticarpus* E. Sheld.; *A. selenaeus* Greene; *Xylophacos aragalloides* Rydb.; *A. amphioxys* × *layneae* M. E. Jones; *Xylophacos melanocalyx* Rydb., non *A. melanocalyx* Boiss.]. Used as a snakebite remedy by Zunis.

1. Flower small, the calyx 6.3–9.5 mm, the banner (9.5) 11–16 mm, the keel 8–13 mm long; plants of the Lake Powell region of Utah .... **var. *modestus***
1′ Flower large, the calyx 9.3–14 mm, the banner (16.4) 19–28 mm, the keel (13.2) 14.3–23.7 mm long; range of the species .... (2)

2. Flower variable in size, but the banner less than twice or barely twice the length of the calyx; keel (13.2) 14.3–18.8 mm long; range of the species (if led hither by material from southwest Colorado or southeast Utah, cf. *A. missouriensis* var. *amphibolus*) .... **var. *amphioxys***
2′ Flower always large, the banner 2–2 1/2 times the length of the calyx; keel 19–23.6 mm long;..... **var. *vespertinus***

var. *amphioxys* **FLOWERS** (16.2) 19–24.5 mm long; calyx (5.8) 7–10.5 mm long, 3.2–4.2 (4.7) mm in diameter, the teeth 1.5–3.7 (4.5) mm long. **PODS** as described for the whole species; ovules (42) 44–56, averaging ± 50; $n = 11$. Sandy valleys, plains, gravelly hillsides, sometimes on dunes, rarely on gumbo clay flats, without apparent rock preference (except locally). ARIZ: Apa, Coc, Nav; COLO: Arc, LPl, Mon; NMEX: McK, RAr, SJn; UTAH. 1160–2285 m (3800–7500′). Flowering: late Mar–Jun. North of the Mogollon Escarpment into central New Mexico, south along the Rio Grande into extreme west Texas and adjoining Chihuahua. (If led hither by material from southwest Colorado or southeast Utah, cf. *A. missouriensis* var. *amphibolus*.)

var. *modestus* Barneby (unassuming) **FLOWERS** up to 16 mm long; calyx 6.3–9.5 mm, the teeth 1.1–2.6 mm long. **PODS** 3–4 cm; ovules 44–58. Blackbrush, saltbush, Indian ricegrass, old man sagebrush, and mixed grass communities. UTAH. 1280 m (4200′). Mohave County, Arizona, to southern Utah. Flowering: Mar–May. Known only from valleys affluent to the Colorado River.

var. *vespertinus* (E. Sheld.) M. E. Jones (evening) Evening milkvetch. Habit of var. *amphioxys*. **FLOWERS** a little larger, and the petals less strongly graduated; 23–27 (28) mm long, 7.8–12.7 mm wide; calyx tube 8.8–13.2 mm long, 3.8–5.2 mm in diameter, the teeth 1.3–3 mm long. **PODS** 3–3.5 cm long, gently incurved; ovules 50–70 (averaging ± 60). [*A. vespertinus* E. Sheld.; *Xylophacos vespertinus* (E. Sheld.) Rydb.]. Sandy valley floors, ledges under cliffs, and gravelly hillsides, in sagebrush or juniper forest, mostly on sandstone. COLO: Mon; NMEX: SJn; UTAH. 1150–1820 m (3770–5970′). Flowering: Apr–Jun. Locally plentiful within the Colorado Basin, south of the Tavaputs Escarpment, southwest Colorado and southeast Utah, extending west to the Grand Canyon and Kanab Plateau, northwest Arizona, and northwest New Mexico.

Astragalus bisulcatus (Hook.) A. Gray (two grooved, of the pod) Two-grooved milkvetch, skunkweed. Clump-forming, caulescent perennial, 15–70 cm tall, from a branching caudex. **PUBESCENCE** strigulose, basifixed. **STEMS** erect or ascending, several or numerous; stipules 2.5–10 (12) mm long, at least the lowermost connate-sheathing. **LEAVES** 3–13.5 cm long; leaflets (7) 15–35, 5–33 mm long, 1.5–11 mm wide, lance-oblong to oblong, elliptic, or oblanceolate, glabrous above or nearly so. **INFL** a raceme, 35- to 80-flowered, the axis (4) 5.5–25 cm long in fruit; bracts 2.5–6 mm long; pedicels 1–3.5 mm long; bracteoles 0–2; peduncle 2.5–13 cm long. **FLOWERS** 8–11 mm long, white or whitish, the banner recurved through 45°; calyx typically pallid, the tube 3.1–4 mm long, the teeth 1–2.7 mm long, subulate. **PODS** pendulous, stipitate, stipe 1.4–3 mm long, about equaling the calyx tube, the body (5) 6.5–9.5 mm long, 2–4 mm wide, ellipsoid or oblong-ellipsoid, transversely rugose-reticulate, strigulose; ovules 5–8. n = 12. [*Phaca bisulcata* Hook.]. Fruit used as a ceremonial emetic and as an eye and toothache medicine by Ramah Navajos.

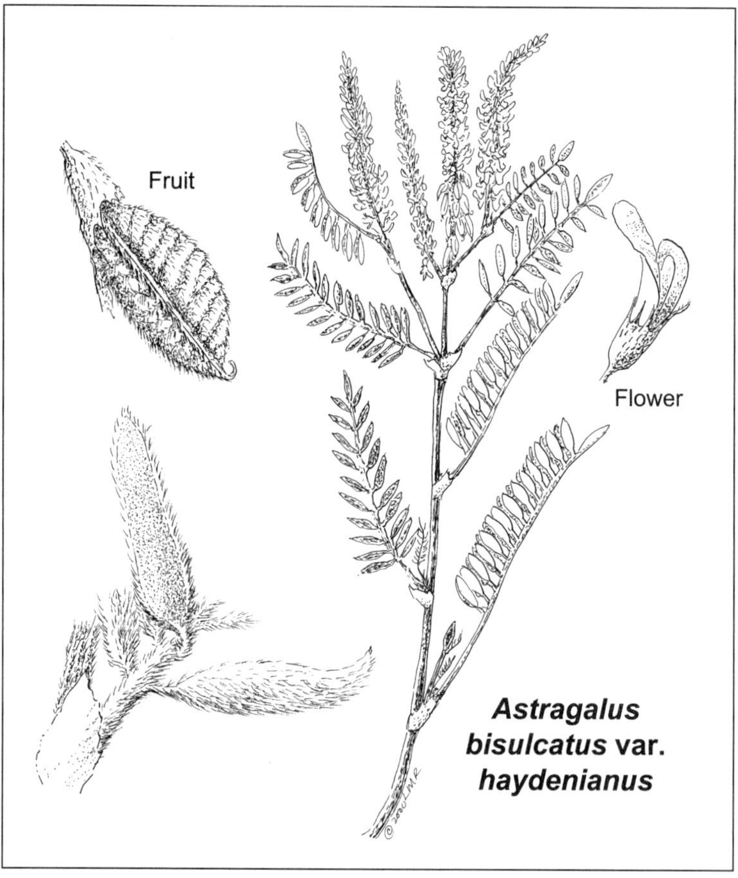

Astragalus bisulcatus var. haydenianus

1. Flowers bright pink-purple; calyx purple; plants of San Juan County, New Mexico; rare ....**var. *bisulcatus***
1′ Flowers ochroleucous or whitish, the keel often purple-tipped; calyx ochroleucous, whitish, or greenish; widespread throughout the Four Corners region ....**var. *haydenianus***

var. *bisulcatus* Plants typically stout, thinly pubescent or with glabrous stems. **LEAVES** with leaflets 17–29, (5) 10–25 (32) mm long. **INFL** a raceme, 25–75-flowered, the axis 5–18 cm long in fruit; bracts (2) 2.5–7 mm long. **FLOWERS** 13–17.5 mm long, pink-purple, pallid and tipped or suffused with purple or lilac, or white or whitish with maculate keel tip; calyx typically red-purple, the tube 3.3–5.7 mm long, the teeth mostly 1.5–4.5 (6) mm long, subulate. **PODS** with stipe 3–5 mm long, the body (8) 10–17 (20) mm long, 2–4.5 mm wide, linear- or narrowly oblong-ellipsoid, smooth or faintly reticulate, glabrous or strigulose; ovules 10–14 (15). n = 11, 12. Fine-textured, saline or seleniferous substrates, in prairies and plains, or in steppes. In our area, along roadsides. NMEX: SJn. 1615 m (5300′). Flowering: May–Aug. Southern Alberta and Saskatchewan to southwest Manitoba, southward through Montana, eastern Idaho, eastern Utah, Wyoming, Colorado, northern New Mexico, North Dakota, western South Dakota, northeast Nebraska, and western Kansas. In addition to containing the element selenium, which is poisonous to livestock, the plants of both vars. *bisulcatus* and *haydenianus* are reported to contain the indolizidine alkaloid swainsonine (Kingsbury 1964).

var. *haydenianus* (A. Gray) Barneby (for F. V. Hayden, leader of the United States Geological and Geographical Survey of the Territories, 1867–1879) Hayden's milkvetch. **STEMS** mostly decumbent and radiating. **LEAVES** with leaflets (13) 21–35, 0.5–2.7 cm long. **INFL** a raceme, 35–80-flowered; bracts (2.5) 3–5 mm long. **FLOWERS** 9–11 mm long, ochroleucous or whitish; calyx ochroleucous, whitish, or greenish, the tube 3.1–4 mm long, the teeth mostly 1–2.7 mm long. **PODS** with stipe 1.4–3 mm long, the body (5) 6.5–9.5 mm long, 2–4 mm wide, ellipsoid or oblong-ellipsoid; ovules 5–8. [*A. haydenianus* A. Gray ex Brandegee; *Tragacantha haydeniana* (A. Gray) Kuntze; *Diholcos haydenianus* (A. Gray) Rydb.; *A. grallator* S. Watson; *Homalobus grallator* (S. Watson) Rydb.]. Sagebrush–mountain brush communities, in fine-textured, often saline, seleniferous substrates. ARIZ: Apa; COLO: Arc, Hin, LPl, Mon; NMEX: McK, RAr, SJn; UTAH. 1935–2500 m (6350–8200′). Flowering: May–Jul. Utah and Wyoming south to northeastern Arizona and northern New Mexico.

Astragalus brandegeei Porter (for Townshend S. Brandegee, botanist) Brandegee's milkvetch. Delicate, low, caulescent perennial, though sometimes flowering as an annual, 5–35 (40) cm tall, from a superficial branching caudex. **PUBESCENCE** strigulose, basifixed. **STEMS** prostrate-spreading, very slender; stipules 1.5–5 mm long, at least the lowermost usually connate-clasping. **LEAVES** (1) 2–11.5 (15) cm long; leaflets 5–15, 5–27 mm long, 0.5–2.6 mm wide, linear-filiform to narrowly oblong, acute to obtuse, strigose beneath, glabrous above. **INFL** a raceme, 1- to 5 (7)-flowered, the flowers ascending at anthesis, the axis 0.5–6 (8) cm long in fruit; bracts 1–2 mm long; pedicels 1.2–4 mm long; bracteoles 2; peduncles 2.5–14 (17) cm long, very slender. **FLOWERS** 4.5–6 (7) mm long, ochroleucous or tinged violet, the banner abruptly curved through 90°; calyx 2.7–4 mm long, the tube 1.8–2.5 mm long, campanulate, black-strigose, the teeth 0.9–2 mm long, subulate. **PODS** pendulous to ascending, sessile or subsessile, the body obovoid to oblong-ellipsoid, 10–18 mm long, 3.5–5 mm thick, slightly dorsoventrally compressed, semibilocular, the septum 1.2–2 mm wide, strigose. $n = 11$. [*Tragacantha brandegeei* (Porter) Kuntze; *Atelophragma brandegeei* (Porter) Rydb.]. Growing on various gravelly substrates in mixed shrublands, piñon-juniper, or oak brush. ARIZ: Apa; NMEX: McK, San, SJn. 1830–2320 m (6000–7610′). Flowering: May–Sep. Utah and in south-central to southwest Colorado, New Mexico, and east-central to central and northern Arizona. This is a cryptic plant that is seldom collected, probably because of its inconspicuous, tiny flowers, slender peduncles, and slender, prostrate stems. When first seen, the small, sausage-like pods seem to float in the air; only careful examination reveals the slender pedicels and peduncles attached to an inconspicuous, small plant.

Astragalus calycosus Torr. ex S. Watson (with a conspicuous calyx) Torrey's milkvetch. Perennial, low, acaulescent herbs, 1–12 cm tall, from a branching caudex. **PUBESCENCE** malpighian. **STEMS** lacking or to 2 cm long, the internodes concealed by stipules; stipules 1.5–6 mm long, all distinct. **LEAVES** 1–8 (12) cm long; leaflets (1) 3–13, 2–19 mm long, 1–7 mm wide, obovate, oblanceolate, or elliptic, obtuse to acute, silvery strigose on both sides. **INFL** a raceme, 1- to 8-flowered, the flowers ascending to spreading at anthesis, the axis 0.2–2.5 cm long in fruit; bracts 0.5–2 mm long; pedicels 0.7–3 mm long; bracteoles lacking or minute; peduncles 0.5–10 cm long, rarely longer. **FLOWERS** 10–16.5 mm long, varicolored, ochroleucous to shades of pink and purple, with white or pale wing tips, the wings bilobed apically, the banner recurved through about 45° or 90°; calyx 5–8.5 mm long, the tube 4–6.7 mm long, campanulate to short-cylindric, strigose, the teeth 1–4.2 mm long, subulate. **PODS** ascending, sessile, narrowly oblong, usually curved, 8–25 mm long, 3–4.5 mm thick, laterally compressed, bilocular, strigose; ovules 13–28. $2n = 22$.

1. Leaves with 3–13 leaflets along a rachis usually less than 1 cm long; scapes ascending or decumbent, 1–7 cm long; raceme axis less than 2 cm long; plants of various distribution .... **var. *calycosus***

1′ Leaves with 5–13 leaflets along a rachis usually more than 1 cm long; scapes erect-ascending, usually over 7 cm long; raceme axis usually over 2 cm long; plants widespread throughout the Four Corners region .... **var. *scaposus***

var. *calycosus* Plants tufted or forming mounds, low. **LEAVES** 1–7 cm long, with 3–7 leaflets; scapes either shorter or longer than the leaves. **INFL** a raceme (1) 2- to 6 (8)-flowered, the axis 2–20 (25) mm long. **FLOWERS** 10–16.5 (20.8) mm long, whitish to pink-purple; calyx 5.2–10.6 mm long, the tube 4–6.4 (6.7) mm long, the teeth (1) 1.5–4.2 mm long. [*Tragacantha calycosa* (Torr.) Kuntze; *Hamosa calycosa* (Torr.) Rydb.; *A. brevicaulis* A. Nelson]. Mixed desert shrublands, sagebrush, piñon-juniper, and ponderosa pine woods. ARIZ: Apa, Coc, Nav; COLO: LPl, Mon; NMEX: McK, SJn; UTAH. 1650–2255 m (5410–7400′). Flowering: May–early Jul. Wyoming, Idaho, and Nevada to California and Arizona.

var. *scaposus* (A. Gray) M. E. Jones (scapose, referring to the peduncles) Scapose milkvetch. Plants tufted. **LEAVES** 1–9 cm long, with (1) 7–13 leaflets; scapes longer than the leaves. **INFL** a raceme, (1) 2- to 6 (8)-flowered, the axis 2–20 (25) mm long. **FLOWERS** 10.5–14 mm long, pink-purple except the whitish wing tips; calyx 5.2–10.6 mm long, the tube 4–6.4 (6.7) mm long, the teeth 1–2 mm long. [*A. scaposus* A. Gray; *Hamosa scaposa* (A. Gray) Rydb.; *A. candicans* Greene]. Juniper, piñon-juniper, and mixed desert shrub communities. ARIZ: Apa; COLO: LPl, Mon; NMEX: McK, SJn; UTAH. 1345–1980 m (4420–6500′). Flowering: Apr–Jun. Nevada to northern Arizona, southwest Colorado, and northwest New Mexico. This is the most showy phase of Torrey's milkvetch. The white wing tips contrast with the rather bright pink-purple of the remainder of the flower. Used as a lotion and poultice applied to injuries from hailstones by the Navajos.

Astragalus ceramicus E. Sheld. (referring to pottery, the painted pod) Painted milkvetch. Slender, caulescent perennial, 3–40 cm tall, from elongate, rhizomelike caudex branches and deeply buried caudex (in sandy sites often extensively branched and pervasive). **PUBESCENCE** malpighian. **STEMS** sprawling to erect, subterranean for a space of 3–40 cm or more. **LEAVES** 2–17 cm long; leaflets 3–13 or only 1, the terminal continuous with the rachis, 3–50 (80) mm long or longer, 0.3–3 mm wide, filiform to narrowly oblong, obtuse to retuse or acute; stipules 1.5–9 mm long, at least some of the lowermost ones connate-sheathing. **INFL** a raceme, 2–15 (25)-flowered (rarely more), the flowers ascending to declined at anthesis, the axis 1–12 cm long in fruit; bracts 1–2.5 mm long; pedicels 0.7–3.1 mm long; bracteoles 0; peduncle 0.7–7.5 cm long. **FLOWERS** 6.3–10.8 mm long, dull purplish to pink or rarely whitish, the banner abruptly recurved through about 85°–90°; calyx 3.1–4.5 (6) mm long, the tube 2.1–3.3 mm long, campanulate, strigose, the teeth 1–2.4 mm long, subulate. **PODS** pendulous, sessile to subsessile or stipitate, the stipe 1–3.3 mm long, the body bladdery-inflated, ellipsoid to glabrous, unilocular; ovules 12–29. $2n = 22$. [*Phaca picta* A. Gray; *A. filifolius* Smyth; *A. pictus* var. *foliolosus* A. Gray; *A. foliolosus* (A. Gray) E. Sheld.; *A. pictus* var. *angustus* M. E. Jones; *A. pictus* var. *magnus* M. E. Jones; *Tragacantha picta* (A. Gray) Kuntze; *A. angustus* var. *pictus* (A. Gray) M. E. Jones; *A. angustus* var. *ceramicus* (E. Sheld.) M. E. Jones; *A. ceramicus* var. *jonesii* E. Sheld.; *A. angustus* (M. E. Jones) M. E. Jones]. Hopi children ate the roots as candy.

1. Vesture of the stems and herbage composed entirely or very largely of malpighian hairs; calyx mostly 3.1–4.2 mm, rarely up to 5.7 mm long; body of the pod (1) 1.5–3 cm long, 5–15 mm in diameter; ovules 12–16, rarely up to 26; plants widespread throughout the Four Corners region .... **var. *ceramicus***

1′ Vesture composed entirely of basifixed hairs (rarely a few minutely spurred at base); calyx mostly 4–6 mm long; body of the pod (2) 3–5 cm long, 1.4–2.6 cm in diameter; ovules (17) 20–29; plants of the Great Plains and east slope of the Rocky Mountains, rare in the Four Corners .... **var. *filifolius***

var. *ceramicus* **LEAVES** 2–12 cm long, at least the lowest and sometimes all bearing 1–6 pairs of lateral leaflets, the uppermost ones quite often reduced to the naked rachis. **INFL** a raceme, 6–15 (25)-flowered, the axis (1) 1.5–8 (15) cm long in fruit. **FLOWERS** 6.3–8.3 (9.5) mm long; calyx 3.1–4.2 mm long, the tube 2.1–2.6 (3.3) mm long, the teeth 1–1.8 mm long. **POD** stipe (1) 1.5–3.3 mm long. Dunes and other sandy sites in piñon-juniper, sagebrush, stream bank, grassland, and mixed desert shrub communities. ARIZ: Apa, Coc, Nav; COLO: Mon; NMEX: McK, San, SJn; UTAH. 1540–2155 m (5050–7065′). Flowering: late Apr–Jul. In some sandy sites along washes, where the plants are easily excavated, the caudex branches have been demonstrated to form extensive subterranean systems that support numerous aboveground stems.

var. *filifolius* (A. Gray) F. J. Herm. (with threadlike leaves) Bradbury's milkvetch. **LEAVES** 2.5–17 cm long, commonly all reduced to the filiform rachis, only some lower ones with 1 or 2 (3) pairs of lateral leaflets. **INFL** a raceme, 2–7-flowered, the axis 1–4.5 (5.5) cm long in fruit. **FLOWERS** 7.4–10.8 mm long; calyx (3.7) 4–6 mm long, the tube (2.3) 2.5–3.5 mm long, the teeth (1.4) 1.6–3 mm long. **POD** stipe 1.5–3 mm long. [based on: *Psoralea longifolia* Pursh; *Orobus longifolius* (Pursh) Nutt.; *Physondra longifolia* (Pursh) Raf.; *Phaca longifolia* (Pursh) Nutt. ex Torr. & A. Gray; *A. ceramicus* var. *imperfectus* E. Sheld., an improper substitute; *A. angustus* (M. E. Jones) M. E. Jones var. *longifolius* (Pursh) M. E. Jones; *A. longifolius* (Pursh) Rydb.; *A. angustus* var. *imperfectus* (E. Sheld.) M. E. Jones; *A. filifolius* Smyth; *A. pictus* var. *filifolius* (A. Gray) A. Gray; *A. mitophyllus* Kearney, a legitimate substitute]. Dunes and sandy hollows in rolling plains, sometimes in sandy fields or on sandbars of intermittent streams. NMEX: SJn. 1820 m (6000′). Flowering: late Apr–Jul. In the higher Great Plains, from western North Dakota and eastern Montana, south through eastern Wyoming and Nebraska to the Arkansas Valley in Colorado and western Kansas, and into the Oklahoma panhandle. A specimen from San Juan County, New Mexico (*A. Clifford 350*, BRY), appears, on the basis of basifixed hairs and large pods, to belong here. However, the calyx is somewhat small for the variety.

Astragalus chuskanus Barneby & Spellenb. **CP** (of the Chuska Mountains) Chuska milkvetch. Prostrate, caulescent, evidently long-lived perennial, forming mats or cushions 8 dm wide, radiating from a branching caudex. **PUBESCENCE** strigulose, basifixed. **STEMS** numerous, 10–35 (40) cm long or more; stipules 2–5.5 mm long, all connate-sheathing. **LEAVES** (1) 1.5–5.5 cm long; leaflets (7) 9–15, 1.5–10 mm long, 0.7–3 (4) mm wide, obovate- or oblong-elliptic, abruptly acuminate, pilosulous on both sides. **INFL** a raceme, compactly 4–10-flowered, the flowers ascending-spreading at anthesis, the axis very short in fruit, 3–15 mm long; bracts 2–3.5 mm long; pedicels 0.9–1.5 mm long; bracteoles 0; peduncle 1.5–3.5 (4) cm long, shorter than the leaf. **FLOWERS** 7.3–8.7 (8–10) mm long, whitish fading ochroleucous, the banner recurved through about 40°; calyx 4.5–5.3 (5.5–6.9) mm long, the tube 2.5–3.2 (3–3.8) mm long, campanulate, pilosulous, the teeth 1.5–2.6 (2–3.1) mm long, subulate. **PODS** humistrately ascending, sessile, obliquely semiovoid, about 6 mm long and 3 mm wide, triquetrously compressed, strigulose; ovules 4–6. $n = 24$. Slopes of degraded Chuska Sandstone, with ponderosa pine, Douglas-fir, and Rocky Mountain juniper; endemic to the Chuska Mountains. ARIZ: Apa; NMEX: McK, SJn. 2135–2865 m (7000–9400′). Flowering: Jun–Jul. There are evidently two rather distinctive morphological phases within the rather numerous specimens now available for examination, which were described so succinctly by Barneby and Spellenberg. One is rather more compact and more obviously gray-hairy, and has the calyx 4.5–5.3 mm long, with tube 2.5–3.2 mm and teeth 1.5–2.6 mm long; the second has the calyx 5.5–6.9

mm long, with tube 3–3.8, and teeth 2–3.1 mm long. Banner length in the first phase is 7.3–8 mm and in the second is 8–10 mm. Perhaps they represent nothing more than responses to differences in ecology, possibly only shade versus open sunlight. Further work is indicated. A Four Corners flora endemic.

Astragalus cicer L. (generic name of the chickpea) Chickpea milkvetch. Perennial, caulescent, 30–90 (110) cm tall, from a subterranean caudex and long, rhizomelike caudex branches. **PUBESCENCE** basifixed, strigose-pilose. **STEMS** at first erect, finally decumbent-clambering. **LEAVES** 2–10 cm long; leaflets 17–29 (31), 5–35 mm long, 2–12 (15) mm wide, lance-elliptic or oblong, obtuse, mucronulate, or acute, strigulose on both sides or glabrescent above; stipules 2–8 mm long, at least the lowermost connate-sheathing. **INFL** a raceme, subcapitate, (6) 10- to 30-flowered, the flowers ascending, the axis (1.5) 2–5 cm long in fruit; bracts 2–6.5 mm long; pedicels 0.3–1.5 mm long; bracteoles 0; peduncle 3.5–11 cm long, shorter than the leaf. **FLOWERS** 12.5–16.5 mm long, ochroleucous, the banner slightly if at all recurved; calyx 6.5–9 mm long, the tube 5–6 mm long, subcylindric, strigulose, the teeth 1.6–3 mm long, triangular-acuminate. **PODS** ascending to spreading, short-stipitate, the stipe to 0.8 mm long, the body 6–14 mm long, 5–10 (12) mm thick, the body obovoid to subglobose, pilose, green becoming black; ovules 14–26. 2*n* = 16, 64. [*Cystium cicer* (L.) Steven]. Mixed desert shrub, sedge-willow, piñon-juniper, sagebrush, aspen communities, and along roadsides. COLO: Mon; NMEX: SJn. 1525–2225 m (5000–7300′). Flowering: Jun–Sep. Southern Manitoba and southern Saskatchewan to western Nevada, Utah, and Colorado. This is a vigorous, introduced European species, now being spread about in reclamation plantings. It is a plant with stems ultimately sprawling, easily distinguished by its ascending ochroleucous flowers and ultimately black, swollen pods.

Astragalus cliffordii S. L. Welsh & N. D. Atwood (for Arnold Clifford, botanist of the Navajo Nation) Clifford's milkvetch. Perennial, caulescent, 35–65 cm tall, from a subterranean caudex. **PUBESCENCE** basifixed. **STEMS** buried for 1–8 cm or more, erect-ascending, forming diffuse clumps. **LEAVES** 3.5–6.5 cm long; leaflets 5–7 (9), 8–28 mm long, 0.4–1 mm wide, linear, acute, strigose; stipules 1.5–4.5 mm long, at least some lower ones connate-sheathing. **INFL** a raceme, very loosely 5–19-flowered, the flowers ascending at anthesis, the axis 3–10.5 (13.5) cm long in fruit; bracts 0.3–0.8 mm long; pedicels 2–2.5 mm long; bracteoles 0; peduncle 1.2–12 cm long. **FLOWERS** 4.3–6.1 mm long, pale, faintly suffused with purple, the keel maculate; calyx 1.8–3 mm long, the tube 1.2–1.8 mm long, campanulate, strigose, the teeth 0.6–1 mm long, triangular. **PODS** declined, subsessile or short-stipitate, the stipe to 0.2 mm long, the body elliptic to oblong in outline, straight or slightly curved, 9–12 mm long, 2.5–3 mm wide, compressed, glabrous, unilocular; ovules 4 or 5. Piñon-juniper and sagebrush communities. NMEX: McK. 1690–2140 m (5545–7015′). Flowering: May–Jun. A Four Corners flora endemic. The alliance of Clifford's milkvetch, as indicated by the smallish, laterally flattened pods, is evidently with *A. wingatanus*, but it differs from that entity in having both filiform peduncles and rachis, a confluent terminal leaflet, and tiny, evidently pale flowers, with lilac pencilled lines on the banner. It is a tall, clump-forming, rushlike plant with numerous stems arising from a subterranean caudex.

Astragalus coltonii M. E. Jones (for W. F. Colton) Colton milkvetch. Perennial, caulescent, 10–75 cm tall, from a branching caudex. **PUBESCENCE** basifixed. **STEMS** arising from a buried caudex, strigulose, cinereous, greenish cinereous, or canescent. **LEAVES** (2) 3–9 cm long; leaflets (5) 9–17 (19), (3) 5–20 mm long, oblong, cuneate-oblong, or ovate, obtuse, truncate, to retuse, brighter green and usually less densely pubescent above than beneath, all jointed or the joint obscure in upper leaves; stipules 1–7 mm long, all distinct. **INFL** a raceme, (6) 10–30-flowered, the flowers spreading-declined at anthesis, the axis 3–20 cm long in fruit; bracts 0.5–3.2 mm long; pedicels 0.8–2.5 mm long; bracteoles 0; peduncle (4) 6.5–21 cm long. **FLOWERS** 12–19 mm long, pink-purple, the banner recurved through about 35°–50°; calyx 4.5–8 mm long, the tube 4–6.7 mm long, cylindric, strigose, purplish, the teeth 0.6–2.3 mm long, broadly subulate. **PODS** pendulous, stipitate, the stipe 5–11 mm long, the body 19–35 mm long, (3) 3.5–6 mm wide, strongly laterally flattened, glabrous, unilocular; ovules 14–20. Our material belongs to ***var. moabensis*** M. E. Jones (for Moab, Utah) Moab milkvetch [*A. coltonii* var. *foliosus* M. E. Jones ex Eastw.; *Homalobus canovirens* Rydb.; *A. canovirens* (Rydb.) Barneby]. Bunchgrass, salt desert shrub, piñon-juniper, and mountain brush communities. ARIZ: Apa, Nav; COLO: Dol, Mon, SMg; NMEX: SJn; UTAH. 1600–2285 m (5250–7500′). Flowering: May–Jun. Disjunct in Uinta County, Wyoming. Despite its obvious relationship to *A. coltonii*, the Moab milkvetch should best be considered at species level. It has a distinctive morphology and geography.

Astragalus cottamii S. L. Welsh (for Walter Pace Cottam) Cottam's milkvetch. Perennial, sometimes flowering the first year, acaulescent or subacaulescent, 1.2–8 cm tall, from a branching caudex. **PUBESCENCE** basifixed. **STEMS** lacking or 0.5–6 cm long. **LEAVES** 1.2–8 cm long; leaflets (5) 9–19 (21), 2–9 mm long, 1–4.2 mm wide, elliptic to oval or oblanceolate, acute to obtuse, strigose on both sides or glabrous above; the internodes mostly obscured by stipules, these 2–6 mm long, all distinct. **INFL** a raceme, 3- to 9-flowered, the flowers ascending at anthesis, the axis 0.5–2 cm long in fruit; bracteoles 0–2; peduncle 0.7–7 cm long. **FLOWERS** 11–17 mm long, pink-purple or bicolored; calyx 6.2–8 mm long, the tube 4.8–6.7 mm long, cylindric, strigulose, purplish, the teeth 1.2–2 mm long, subulate. **PODS** spreading-descending, sessile, curved, oblong to oblong-lanceolate in outline, triquetrous, the dorsal suture sulcate, bilocular, strigose, usually purple-blotched. [*A. monumentalis* Barneby var. *cottamii* (S. L. Welsh) Isely]. Rimrock and ledges of Cedar Mesa, Kayenta, Entrada, and Point Lookout Sandstone formations and in the sandy canyons cut from them. Piñon-juniper and blackbrush communities. ARIZ: Nav; COLO: LPl, Mon; NMEX: SJn; UTAH. 1320–1900 m (4325–6255′). Flowering: Apr–May. A Four Corners endemic. The longer calyx, with cylindric, not campanulate, tube and slightly longer flowers distinguish Cottam's milkvetch from the closely allied rimrock species *A. monumentalis*.

Astragalus cronquistii Barneby (for Arthur John Cronquist, noted botanist) Cronquist's milkvetch. Low, caulescent perennial, 1.5–4 dm tall, from a stout taproot and subterranean caudex. **PUBESCENCE** strigulose, basifixed, with some hairs flattened and scalelike. **STEMS** prostrate to decumbent-ascending, subterranean for a space of 5–14 cm; stipules 2–6 mm long, all distinct. **LEAVES** 1.5–4.5 cm long; leaflets 7–15, 6–23 mm long, 1.5–4 mm wide, oblong to narrowly elliptic, retuse to truncate, strigose beneath, glabrate above. **INFL** a raceme, 6–20-flowered, the flowers declined at anthesis, the axis 1.5–8.5 cm long in fruit; bracts 0.6–1.2 mm long; pedicels 1.5–2.5 mm long; bracteoles 1; peduncle 2–6.5 cm long. **FLOWERS** 8–9 mm long, dull pink-purple, the banner recurved through 90°–100°; calyx 3.8–5.3 mm long, the tube 3.3–4 mm long, campanulate, strigose, the teeth 0.5–1.3 mm long, triangular. **PODS** declined-pendulous, sessile or subsessile, the body narrowly elliptic in outline, 13–30 mm long, 3–4.8 mm wide, trigonous, grooved dorsally, strigose, semibilocular, the septum 0.3–0.6 mm wide. Blackbrush and salt desert shrub on Cutler, Morrison, and Mancos Shale formations. COLO: Mon; UTAH. 1440–1830 m (4720–6000′). Flowering: Apr–Jun. A Four Corners flora endemic. The species occupies two disjunct regions in San Juan County, Utah: Comb Wash (Cutler Formation type vicinity) and the Aneth vicinity (Morrison Formation). The principal distribution of the species, however, is in Montezuma County, Colorado (on Mancos Shale). Its preference for fine-textured, seleniferous substrates is evident.

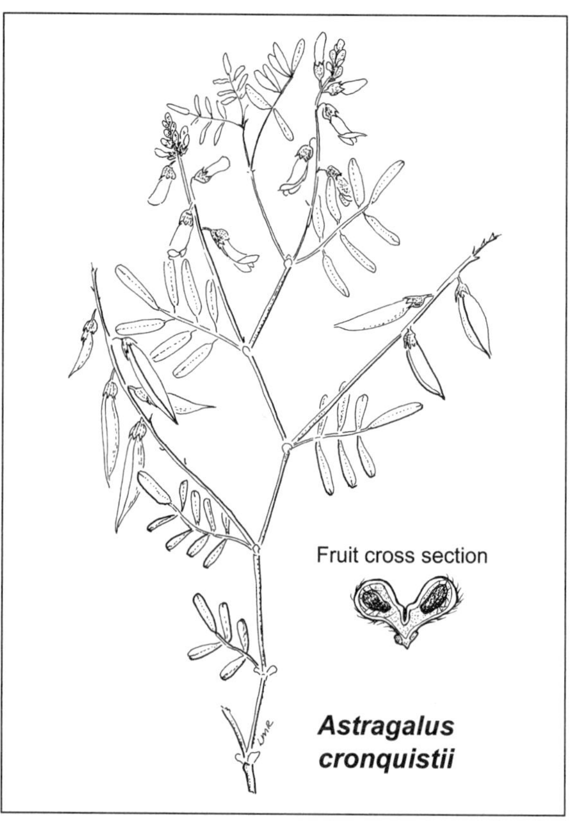

Astragalus cutleri (Barneby) S. L. Welsh (for Hugh Carson Cutler) Cutler's milkvetch. Moderate, caulescent, short-lived perennial, often flowering as an annual, 10–30 (35) cm tall, from a superficial caudex. **PUBESCENCE** sparingly strigulose to subglabrous, basifixed. **STEMS** few to several, ascending to erect, forming bushy clumps. **LEAVES** 3–13 cm long; leaflets 5–17 (19), 3–17 (20) mm long, (3) 5–12 mm broad, elliptic to lanceolate, oblanceolate or obovate, acute to obtuse or mucronulate, strigulose to glabrous below, glabrous above; stipules 2–6.5 mm long, all distinct. **INFL** a raceme, 5- to 9-flowered; pedicels 1.5–2.5 mm long; bracteoles 2; peduncle 2.5–10 cm long; bracts 1.5–2.5 mm long. **FLOWERS** 15–16 mm long, white or tinged (or drying) purplish, the banner recurved through about 40°–45°; calyx (7.3) 7.5–8.5 (9) mm long, the tube 5.9–6.7 mm long, cylindric, pale purple or whitish, drying purplish, sparsely black-strigose, the teeth 1.3–1.7 (2.3) mm long, subulate. **PODS** ascending to erect, stipitate, the stipe 3–3.5 mm long, the inflated body oblong-ellipsoid, 14–18 mm long, 9–11 mm thick, the valves thinly cartilaginous, greenish suffused (sometimes) with purple, often drying stramineous, unilocular, glabrous; ovules 20–38. Saltbush and blackbrush communities on Permian formations. UTAH (Copper Canyon, San Juan arm of Lake Powell). About 1160–1260 m (3810–4125′). Flowering: Apr–May. A Four Corners endemic. This taxon when first characterized was known only from plants flowering as annuals. Later collections demonstrate that the plant is at least a short-lived perennial. The other characteristics hold, however, even if there is more overlap in leaflet number with *A. preussii* var. *latus* (q.v.) than previously known. Also, the pods are of thin texture, approaching *A. eastwoodiae* more so than *A. preussii*, with which it shares features of ascending-erect pods. Cutler's milkvetch differs from *A. preussii* by about the same order of magnitude as does *A. eastwoodiae*.

Astragalus desperatus M. E. Jones (despairing) Rimrock milkvetch. Perennial, acaulescent or subacaulescent, 1–12 cm tall, from a branching caudex. **PUBESCENCE** basifixed. **STEMS** to 8 cm long, the internodes often obscured by stipules. **LEAVES** 1–12 cm long; leaflets (3) 7–17, 2–13 mm long, 1–5 mm wide, elliptic to oblanceolate or obovate, acute to obtuse, strigose on both sides or glabrate above; stipules 1.5–7 mm long, at least the lowermost connate-sheathing. **INFL** a raceme, 3- to 18-flowered, the flowers ascending to declined at anthesis, the axis 0.4–13 cm long in fruit; bracts (1) 1.5–5 mm long; pedicels 0.5–1.4 (2) mm long; bracteoles 0–2; peduncle 0.5–13 cm long. **FLOWERS** 6–9 mm long, pink-purple or bicolored, the banner recurved through about 45°; calyx 3.5–6 mm long, the tube 2.5–4.2 mm long, campanulate, strigose-pilose, the teeth 0.8–2.6 mm long, subulate. **PODS** declined to deflexed, sessile or short-stipitate, the stipe (gynophore) to 1.2 mm long, the body obliquely ovoid to lance-ellipsoid, curved, 6–19 mm long, 3–6 mm thick, hirsute with lustrous hairs, unilocular; ovules 16–28. [*Tium desperatum* (M. E. Jones) Rydb.; *Batidophaca desperata* (M. E. Jones) Rydb.; *A. desperatus* var. *typicus* Barneby]. Mixed desert shrub and piñon-juniper communities, often on rimrock. ARIZ: Apa, Nav; NMEX: SJn; UTAH. 1100–1950 m (3605–6400′). Flowering: Apr–Jun. Utah, Colorado, and Arizona. Our material belongs to **var. *desperatus***.

Astragalus deterior (Barneby) Barneby (meaner, by contrast with the genuine Naturita milkvetch) Cliff Palace milkvetch. Low, slender, subacaulescent or shortly caulescent, taprooted perennial. **PUBESCENCE** strigulose-cinereous, with narrowly ascending straight hairs up to 0.7 mm long. **STEMS** arising together from the root crown or from a shortly forking caudex, usually few, up to 7 cm long. **LEAVES** 2–20 cm long, slender-petioled, with 11–15 (17) linear-elliptic and acute or oblanceolate and obtuse, commonly folded leaflets 3–10 mm long; stipules ovate, papery-scarious, 1–4 mm long. **INFL** a raceme, loosely 2–5-flowered, the flowers ascending; bracts papery-scarious, ovate, 1.5–2.5 mm long; peduncle ascending at anthesis, mostly 2.5–8 cm long. **FLOWERS** ochroleucous, the keel top spotted with dull purple; banner recurved through about 45°; calyx 5–5.5 mm long, strigulose with black and white hairs, the subulate teeth 1.5–2 mm long. **PODS** ascending, gently incurved, 1.2–2 cm long, 3.5–5 mm in diameter, red-mottled; ovules 8–10. [*A. naturitensis* Payson var. *deterior* Barneby]. Sandy soil, bases of cliffs or in sandy pockets along the upper rim. COLO: Mon. Mesa Verde National Park. 1890–1940 m (6210–6365′). Flowering: May–Jun. A Four Corners endemic.

Astragalus eastwoodiae M. E. Jones (for Alice Eastwood, curator of botany at the California Academy of Sciences) Eastwood's milkvetch. Moderate, short-caulescent perennial, 8–20 cm tall, from a superficial caudex. **PUBESCENCE** lacking except on the calyx, strigulose, basifixed. **STEMS** few to several, decumbent to ascending, forming small bushy clumps. **LEAVES** 3–13 cm long; leaflets 13–25, 1–15 mm long, 1–5 mm broad, elliptic to lance-elliptic, oblanceolate or obovate, obtuse to truncate-emarginate, glabrous; stipules 2–6.5 mm long, all distinct. **INFL** a raceme, 3–7-flowered; bracts 1.5–4.5 mm long; pedicels 1.5–3.5 mm long; bracteoles 2; peduncle 2–10.5 cm long. **FLOWERS** 18–22 mm long, pink-purple, the banner recurved through about 45°; calyx 10–12.2 mm long, the tube 8–9.5 mm long, cylindric, purple, sparsely black-strigose, the teeth 1.3–2.7 mm long, subulate. **PODS** spreading to declined, stipitate, the stipe 1.5–4.5 mm long, the inflated body oblong-ellipsoid, 14–26 mm long, 7–14.5 mm thick, the valves papery and straw-colored, unilocular, glabrous; ovules 20–38. $2n = 24, 26$. [*A. preussii* A. Gray var. *sulcatus* M. E. Jones; *Phaca eastwoodiae* (M. E. Jones) Rydb.; *A. preussii* var. *eastwoodiae* (M. E. Jones) M. E. Jones]. Mixed desert shrub and piñon-juniper communities in seleniferous, often fine-textured soils. NMEX: McK; UTAH. 1455 m (4900′). Flowering: late Apr–Jun. Western Colorado, Utah, and New Mexico. The Eastwood milkvetch is closely allied to *A. preussii*; it differs mainly in the shorter stems, smaller leaflets, and spreading-descending, thin-textured pods. A specimen from McKinley County (*S. O'Kane et al. 4846b*, SJNM) looks like this species but is a late specimen with a solitary unattached, damaged pod.

Astragalus episcopus S. Watson (a bishop, a tribute to the collector Capt. F. M. Bishop) Bishop's milkvetch. Perennial, caulescent, rushlike, 20–45 cm tall, arising from a subterranean caudex; pubescence basifixed. **STEMS** erect or ascending. **LEAVES** 2–10 cm long, most of them reduced to the rachis, some with leaflets 3–13 in number, these 1–15 mm long and 0.5–2 mm wide, linear to elliptic or oblong, acute to obtuse or emarginate, strigose on both sides; stipules 2–13 mm long, all distinct. **INFL** a raceme, very loosely 6- to 30-flowered, the flowers ascending at anthesis, the axis 3–30 cm long in fruit; bracts 1.3–3 mm long; pedicels 1.5–3.5 mm long; bracteoles 0–2; peduncle 6–23 cm long. **FLOWERS** 10–15.5 mm long, pale pink or whitish to pink-purple, the banner recurved through about 40°–45°; calyx 4.1–8 (8.5) mm long, the tube 3.4–5.2 (6) mm long, short-cylindric, always much longer than broad, suffused with purple or very pale, white-strigose, the teeth 0.6–2.2 mm long, triangular to subulate. **PODS** pendulous, sessile or subsessile, the body oblong to lance-elliptic in outline, slightly curved to straight, 14–32 mm long, 4–8 mm wide, laterally compressed, glabrous to strigose, straw-colored, or suffused with purple or mottled, unilocular; ovules 16–26. $n = 11$.

[*Tragacantha episcopa* (S. Watson) Kuntze; *Homalobus episcopus* (S. Watson) Rydb.; *A. kaibensis* M. E. Jones; *Lonchophaca kaibensis* (M. E. Jones) Rydb.]. Mixed desert shrub and piñon-juniper communities, often in clay or silty soils derived from Carmel, Entrada, Chinle, Moenkopi, and other exposed, fine-textured strata. ARIZ: Apa, Nav; NMEX: SJn; UTAH. 1370–1555 m (4500–5100′). Flowering: May–Jun. Southeast Utah, Arizona, and New Mexico. Bishop's milkvetch is closely allied to the geographically contiguous *A. lancearius* A. Gray, and a case has been made for its inclusion within this entity at infraspecific level. It differs, however, in a similar manner as other species within sect. *Lonchocarpi*, subsect. *Lancearii*. The laterally flattened, pale pods protruding from short-campanulate calyces are characteristic for the species.

Astragalus falcatus Lam. (curved like a sickle) Russian sickle milkvetch. Perennial, caulescent, 40–90 cm tall from a branching caudex. **PUBESCENCE** malpighian. **STEMS** ascending to erect, forming large clumps. **LEAVES** 5–22 cm long; leaflets 19–37, 6–35 mm long, 1.5–10 mm wide, oblong to elliptic or oblanceolate, acute to apiculate, strigose below, glabrous above, green on both sides; stipules 2–12 mm long, at least some connate-sheathing. **INFL** a raceme, 20- to 70-flowered, the flowers declined at anthesis, the axis 3–20 cm long in fruit; bracts 2–5 mm long; pedicels 32–65 mm long, recurved in fruit; bracteoles 2; peduncle 6–17 cm long. **FLOWERS** 9–11 mm long, greenish white, sometimes suffused with purple; calyx 3.6–4.7 mm long, the tube 3–3.5 mm long, campanulate, strigose, the teeth 0.5–1.2 mm long, triangular. **PODS** decurved, subsessile, curved-oblong, 13–23 mm long, 2.5–4.5 mm wide, triangular, strigose, bilocular; ovules 12–14. $n = 8$. Introduced soil stabilization plant, sparingly naturalized in the western United States. COLO: Mon (Echo Basin Road). 2285 m (7500′). Flowering: Jun–Aug. This is a robust perennial, capable of surviving in harsh, fine-textured soils in mountain brush and piñon-juniper communities. The plant is also suspected of being poisonous to livestock.

Astragalus flavus Nutt. ex Torr. & A. Gray (yellow) Yellow milkvetch. Perennial, caulescent, 5–30 (40) cm tall, from a typically superficial branching caudex. **PUBESCENCE** strigulose, malpighian. **STEMS** decumbent to ascending or erect. **LEAVES** 3–15 (18) cm long; leaflets (5) 9–21, 3–31 mm long, 0.5–6 mm wide, linear, narrowly oblong, or oblanceolate to ovate, obtuse to acute, silvery strigose (greenish) on both sides or glabrate to glabrous above; stipules 2–10 mm long, all connate-sheathing. **INFL** a raceme, 6- to 30-flowered, the flowers ascending at anthesis, the axis 2–12 cm long in fruit; bracts 1.5–5 mm long; pedicels 0.7–1.2 mm long; bracteoles 0; peduncle 3–23 cm long. **FLOWERS** 9–17.8 mm long, yellow to ochroleucous, whitish, lilac, or pink-purple, the banner recurved through 45 (90)°; calyx 5.5–9.5 mm long, the tube 3–5.2 mm long, campanulate, strigose to pilose, the teeth 2–6 mm long. **PODS** erect, sessile, oblong, 7–13 mm long, 3.5–5.5 mm thick, straight, dorsoventrally compressed, strigose, unilocular; ovules 6–17. $2n = 24, 26$. [*Cnemidophacos flavus* (Nutt.) Rydb.; *Tragacantha flaviflora* Kuntze; *A. flaviflorus* (Kuntze) E. Sheld.; *A. confertiflorus* var. *flaviflorus* (Kuntze) M. E. Jones; *A. flavus* var. *candicans* A. Gray; *A. confertiflorus* A. Gray; *Cnemidophacos confertiflorus* (A. Gray) Rydb.]. This species, including its infraspecific taxa, typically occupies fine-textured, seleniferous substrates. It is commonplace on exposures of the Chinle, Morrison, San Jose, and other formations composed of clays, muds, and silts exposed over great expanses of the West.

1. Calyx shaggy long-villous; flowers pink-purple; plants of northeast Arizona and San Juan County, Utah ..... **var. *argillosus***
1′ Calyx strigose to short-villous; flowers yellow to white or tinged purplish, rarely pink-purple ..... (2)

2. Peduncle mainly less than 12 cm long; raceme not or only somewhat elevated much above the leaves; flowers yellow or white; plants widespread ..... **var. *flavus***
2′ Peduncle mainly 12–23 cm long; raceme elevated much above the leaves; flowers white or tinged lilac or bluish; Utah, northern Arizona, and northwest New Mexico ..... **var. *higginsii***

var. *argillosus* (M. E. Jones) Barneby (clay) Clay milkvetch. [*A. argillosus* M. E. Jones]. Mancos Shale and Morrison formations, on saline clays and silts with salt desert shrubs. ARIZ: Apa, Nav; UTAH. 1235–1750 m (4060–5740′). Flowering: Apr–May. Southeastern Utah and adjacent Arizona. The specimens examined from southeastern Utah come from populations where the flowers are pink-purple throughout, not intermixed with the more typical ochroleucous to white-flowered plants of var. *flavus*. However, some specimens from San Juan County, Utah, assigned here have white flowers but bear the long shaggy-villous trichomes on the calyx. There is a collection from Navajo County, Arizona, with evident pink-purple coloring in the petals.

var. *flavus* Habit of the species, ultimately tuft-forming, the caudex branches often impacted by clay. **INFL** a raceme not much elevated above the leaves. **FLOWERS** 11–17.8 mm long, cream-colored to lemon-yellow, rarely suffused with pale pink or purple; calyx 4–7.5 mm long, the tube 3.2–5.2 mm long, campanulate, pilosulous or strigulose, the teeth shorter than the tube; ovules 8–17. Growing on seleniferous substrates composed of saline silts and clays in salt desert

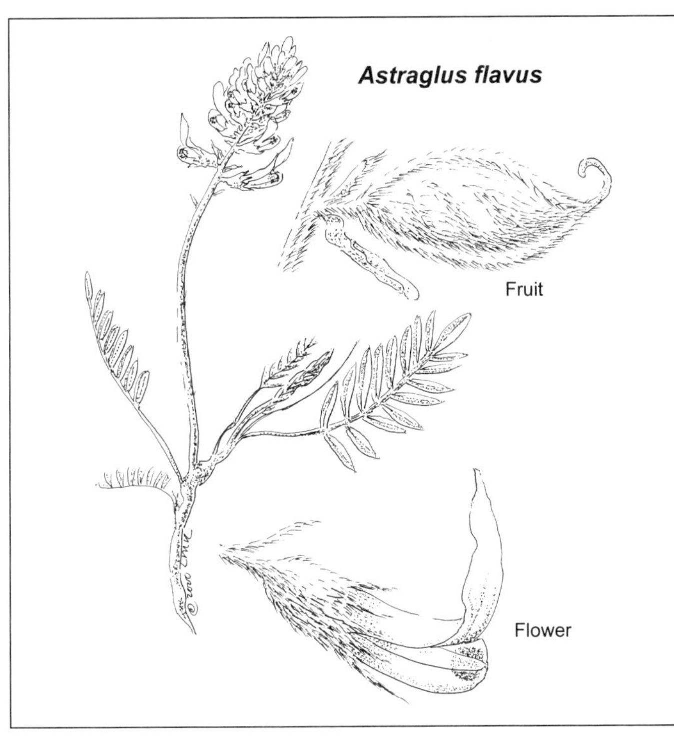

shrub and piñon-juniper communities. ARIZ: Apa; COLO: Arc, Dol, LPl, Mon; NMEX: McK, RAr, San, SJn; UTAH. 1765–2180 m (5800–7160′). Flowering: Apr–Jul (Sep). Southern Wyoming, western Colorado, to San Juan County, Utah; northern Arizona; and northern New Mexico. It was noted by Barneby (1964a: 399) to occur ". . . in places so abundant that it beautifies with it's prolific blossoms extensive tracts of alkaline bottomland and playa otherwise given over to the gray monotony of halophytic shrubs." Most plants of this variety have ochroleucous or indeed yellowish flowers, but plants with pink-purple flowers occur sporadically in its range, never as populations. Sheep poisoning attributable to var. *flavus* is known from the lower-elevation portions of the Uinta Basin in northeastern Utah.

var. *higginsii* S. L. Welsh (for Larry Charles Higgins, botanist) Larry's stinking milkvetch. Habit of the species, but forming smaller clumps when old, sometimes flowering the first season. **INFL** a raceme elevated above the leaves. **FLOWERS** with petals white to cream- or straw-colored; calyx 4–6.7 mm long, the tube 3.7–5.2 mm and 2.5–3.8 mm thick; ovules (6) 8–14. Chinle, Moenkopi, and other similar, fine-textured, seleniferous formations in salt desert scrub and piñon-juniper woodland communities. ARIZ: Apa; COLO: LPl, Mon; NMEX: McK, RAr, SJn; UTAH. 1465–2135 m (4800–7000′). Flowering: Apr–Jun. Southeast Nevada. Relegation of the concept of var. *candicans* to var. *flavus* leaves the Washington County, Utah, specimens, and the pale-flowered materials from adjacent Nevada, southeast Utah, and adjacent Arizona without a varietal designation. Although white or pale, or even lilac-flowered specimens occur elsewhere within the range of var. *flavus*, in a broad sense, they seldom have features of very elongate peduncles and racemes much elevated above the leaves, and they seldom occur as populations. This phase of the stinking milkvetch is commonly taller than elsewhere in the species, frequently more than 30 (to 40) cm tall.

Astragalus flexuosus (Hook.) Douglas ex G. Don (bend) Flexible milkvetch. Perennial, caulescent, 1–6 dm tall, from a branching buried caudex. **PUBESCENCE** strigulose to villosulous, basifixed. **STEMS** decumbent or ascending, buried for a space of 1–20 cm or more. **LEAVES** 1.5–9 cm long; leaflets (5) 9–25, (2) 3–19 mm long, linear or oblong to oblanceolate or obovate, obtuse to truncate or retuse, strigose to glabrate beneath, usually glabrous above; stipules 1–7 mm long, at least the lowermost connate-sheathing. **INFL** a raceme, 7- to 26-flowered, the flowers spreading at anthesis, the axis 0.6–4.5 cm long in fruit; pedicels 0.7–3.5 mm long; bracteoles 0–2; peduncles 1.5–14 cm long. **FLOWERS** 7–11 mm long, pink-purple to dull purplish, the banner recurved through 45°–90°; calyx 3.3–5.8 mm long, the tube 2.4–4.3 mm long, campanulate, strigose, the teeth 0.5–1.7 mm long, subulate. **PODS** descending to spreading, sessile or short-stipitate, the stipe 0.5–1.3 mm long, the body oblong to oblanceolate or elliptic in outline, 8–24 mm long, 2.7–4.8 mm thick, subterete or variously somewhat flattened, strigose to glabrous, unilocular; ovules 12–25. $2n = 22$. [*Phaca flexuosa* Hook.; *Tragacantha flexuosa* (Hook.) Kuntze; *Homalobus flexuosus* (Hook.) Rydb.; *Pisophaca flexuosa* (Hook.) Rydb.; *A. flexuosus* var. *albus* Douglas ex G. Don; *Phaca elongata* Hook.; *Phaca elongata* var. *minor* Hook.; *Phaca fendleri* A. Gray; *Pisophaca sierrae-blancae* Rydb.; *Pisophaca ratonensis* Rydb.; *Pisophaca saundersii* Rydb.].

1. Calyx 3.3–4.1 mm long; banner abruptly recurved through ± 90°; keel 5–5.5 mm long, the half-circular blades very strongly incurved through ±120°; pod 11–15 mm long, 3–4 (4.5) mm in diameter; plants of the Colorado River drainage, eastern Utah, western Colorado, northwest New Mexico, and Arizona; mostly below 1830 m .... **var. *diehlii***

1′ Calyx 3.5–6 mm long; banner at full anthesis recurved through ± 45°–50°, sometimes further in withering; keel 5.3–8.2 mm long, the blades half-obovate, 3.4–5.2 mm long, incurved through 90°–100°; pods various; plants widespread from east-central Arizona and New Mexico but on the west slope in Colorado, mostly above 1830 m elevation .... **var. *flexuosus***

var. *diehlii* (M. E. Jones) Barneby (for I. E. Diehl, a correspondent of M. Jones) Diehl's milkvetch. **STEMS** 1.5–3 dm tall, the herbage greenish cinereous. **LEAFLETS** (7) 11–17, mostly linear or narrowly oblong-oblanceolate. **INFL** a raceme, 12–26-flowered, the axis 2.5–9.5 cm long in fruit; peduncle 2.5–6 cm long. **FLOWERS** 7–9 mm long; calyx 3.3–4.1 mm long, the tube 1.9–2.3 mm long, the teeth 0.7–1.1 (1.4) mm long. **PODS** sessile, the body oblong-ellipsoid, the valves strigulose; ovules 12–18. [*A. diehlii* M. E. Jones; *Pisophaca diehlii* (M. E. Jones) Rydb.]. Salt desert shrub and piñon-juniper communities. NMEX: San, SJn. 1760–1860 m (5775–6105′). Flowering: late Apr–Jun. East-central Utah, western Colorado, northwest New Mexico, and northeast Arizona (Navajo County to Meteor Crater, Coconino County).

var. *flexuosus* **STEMS** 1.5–6 dm tall, the herbage greenish or silky canescent. **LEAFLETS** 11–25 (29), mostly linear or narrowly oblong-oblanceolate. **INFL** a raceme, (7) 12–26 (30)-flowered, the axis 3–13 cm long in fruit; peduncle 4–19 cm long. **FLOWERS** 7.4–11 mm long; calyx 3.5–5.8 mm long, the tube 2.7–4.3 mm long, the teeth 0.5–1.7 mm long. **PODS** stipitate or subsessile, the stipe 0.5–1.3 mm long, concealed by the marcescent calyx, the body linear-oblong, linear-oblanceolate, or narrowly oblong-elliptic in outline, (8) 12–24 mm long, 2.7–4.8 mm thick, the valves finely strigulose, villosulous, or rarely glabrous; ovules 14–20. $2n = 22$. Piñon-juniper, sagebrush-oak, ponderosa pine, and mountain brush communities. ARIZ: Apa; COLO: Arc, Dol, LPl, Mon; NMEX: RAr, SJn; UTAH. 1780–2255 m (5840–7405′). Flowering: May–Aug. British Columbia east to southern Ontario and south to Utah, New Mexico, Nebraska, and Minnesota.

Astragalus fucatus Barneby (painted, referring to the mottled pod) Hopi milkvetch. Moderate, caulescent perennial (sometimes flowering as an annual), 7–45 (70) cm tall, from a subterranean to superficial caudex. **PUBESCENCE** strigulose, basifixed. **STEMS** ascending to erect or sprawling, naked and buried for a space of (0) 1–4 cm. **LEAVES** 2–13 cm long; leaflets 9–17, 3–20 (25) mm long, 0.5–4 mm wide, obtuse to retuse, strigose beneath, glabrous above; stipules 1–5.5 mm long, the lowest connate-sheathing. **INFL** a raceme, 9- to 22-flowered, the flowers ascending to declined at anthesis, the axis 2–11.5 cm long in fruit; bracts 0.8–2 mm long; pedicels 0.7–3.5 mm long; bracteoles 0; peduncle 1–6.5 cm long. **FLOWERS** 6.4–8.7 mm long, pink-purple, the banner recurved through 90°; calyx 3.3–5.4 mm long, the tube 2.3–3.3 mm long, campanulate, strigose, the teeth 0.8–2.2 mm long. **PODS** spreading to declined, sessile, bladdery-inflated, ovoid, ellipsoid, or subglobose, 17–32 mm long, 12–20 mm wide (when pressed), mottled, strigose, unilocular; ovules 21–32. $n = 12$. [*A. subcinereus* A. Gray; *Phaca subcinerea* sensu Rydb.]. Mixed sandy desert shrub (resinbush, blackbrush, juniper) communities. ARIZ: Apa, Nav; COLO: Mon; NMEX: McK, SJn; UTAH. 1315–1830 m (4320–6000′). Flowering: May–Jul (Sep). Widespread throughout the Four Corners region.

Astragalus hallii A. Gray (for Elihu Hall, Colorado botanist) Hall's milkvetch. Perennial, caulescent, 12–50 cm tall, from a subterranean caudex. **PUBESCENCE** strigulose to villosulous, basifixed. **STEMS** decumbent to ascending or erect, subterranean for a space of 1–8 cm. **LEAVES** 2–9 cm long; leaflets 11–27 (31), (1.5) 3–14 (15) mm long, 1–7 mm wide, obovate to oblanceolate or elliptic, retuse to truncate or obtuse, strigulose beneath, sparingly hairy or glabrous above; stipules 1–7 mm long, at least the lowermost connate-sheathing. **INFL** a raceme, (7) 9–28-flowered, the flowers spreading-declined at anthesis, the axis 1–7 cm long in fruit; bracts 1.5–5 mm long; pedicels 1.2–4 mm long; bracteoles 0–2; peduncle 3–9.5 (11) cm long. **FLOWERS** 12.4–15 (18.5) mm long, pink-purple, the banner reflexed through about 45°; calyx (5) 6–7 (9) mm long, the tube 5–6.2 mm long, the teeth 0.5–2.3 mm long, triangular-subulate. **PODS** spreading to declined, sessile or short-stipitate, the stipe 1.5–4 (4.5) mm long, the slightly to greatly inflated body cylindroid to obliquely ovoid-ellipsoid, 17–27 mm long, (3) 4–12 mm thick, strigulose to villosulous, unilocular; ovules 20–34. [*Tragacantha hallii* (A. Gray) Kuntze; *A. gracilentus* (A. Gray) A. Gray var. *hallii* (A. Gray) M. E. Jones; *Homalobus hallii* (A. Gray) Rydb.; *Pisophaca hallii* (A. Gray) Rydb.; *A. shearii* Rydb.; *Atelophragma shearii* (Rydb.) Rydb.]. Hillsides and meadows, nearly always in sagebrush, and in mixed conifer-aspen. ARIZ: Apa; COLO: LPl; NMEX: San, SJn. 2045–2345 m (6705–7700′). Flowering: Jun–Sep. Northeast Arizona, northern New Mexico, and Colorado. Four Corners material belongs to **var. *hallii***.

Astragalus heilii S. L. Welsh & N. D. Atwood **CP** (for Kenneth D. Heil, botanist) Heil's milkvetch. Tufted, low, subacaulescent perennial, about 2–4 cm long, from a superficial branching caudex, this clothed with marcescent leaf bases and peduncles. **PUBESCENCE** basifixed, strigulose. **STEMS** obscured by stipules and leaf bases. **LEAVES** 1–2.5 cm long; leaflets 7–13, 2–3.5 mm long, 1–1.6 mm wide, folded, elliptic, obtuse, strigulose on both sides; stipules 2–3 mm long, merely amplexicaul or the lowermost connate-sheathing. **INFL** a raceme (1) 2- to 4-flowered, the flowers ascending at anthesis, declined in age; peduncle 1–7 cm long, very slender. **FLOWERS** 4–5 mm long, whitish or tinged with violet, the banner recurved through about 45°–60°; calyx 2.3–3 mm long, the tube 1.7–1.9 mm long, campanulate, white- or black-strigose, the teeth 0.7–0.9 mm long, subulate. **PODS** spreading or pendulous, substipitate, subinflated, 9–9.8 mm long, 4.5–4.6 mm thick, slightly dorsiventrally compressed, thin, red-mottled, becoming papery, unilocular; ovules 8–10. Piñon-juniper woodland communities on rocky ledges of the Mesa Verde Formation. NMEX: McK. 1980 m (6500′). A Four Corners endemic.

Astragalus humillimus A. Gray ex Brandegee (smallest of all) Mancos milkvetch. Perennial, tufted or matted, diminutive, forming cushions to 2 dm wide, acaulescent, 0.5–2 cm tall, from a superficial sand-impacted caudex, the branches clothed with a thatch of marcescent stipules and leaf bases (these spinose-persistent). **PUBESCENCE** malpighian, ashy strigulose. **STEMS** obscured by stipules and leaf bases. **LEAVES** 8–28 mm long; leaflets 7–11, 1–5 mm long, 0.3–0.8 mm wide, obovate-cuneate to oblong-elliptic, obtuse or subacute, folded, silvery strigulose on both sides; stipules 1–1.5 mm long, submembranous, all distinct. **INFL** a raceme, 1–3-flowered, the flowers ascending, the axis 1–2 mm long; bracts 0.6–0.8 mm long; pedicels 1–2.5 mm long; bracteoles 0; peduncle 2–4 mm long, shorter than the leaves. **FLOWERS** 9–12 mm long, pink-purple, the keel maculate; calyx 3.3–4.5 mm long, the tube 2.2–3.8 mm long, campanulate to subcylindric, strigulose, the teeth 0.5–1.3 mm long, subulate. **PODS** ascending to spreading, sessile, the body oblong-ellipsoid, 4.5–5.5 mm long, 2–2.5 mm wide, laterally compressed and bicarinate, greenish, densely strigulose, unilocular; ovules about 4. 2*n* = 22. [*Tragacantha humillima* (A. Gray) Kuntze]. Ancient channel sandstone in hogbacks within the Point Lookout Sandstone of the Mesa Verde Group, with Bigelow sagebrush, matchweed, and widely spaced junipers. COLO: Mon; NMEX: SJn. 1545–1835 m (5075–6015′). Flowering: May–Jul. The species has been known since the initial collection by Brandegee in 1875 but was only recently rediscovered after a hiatus of more than a century. A Four Corners endemic.

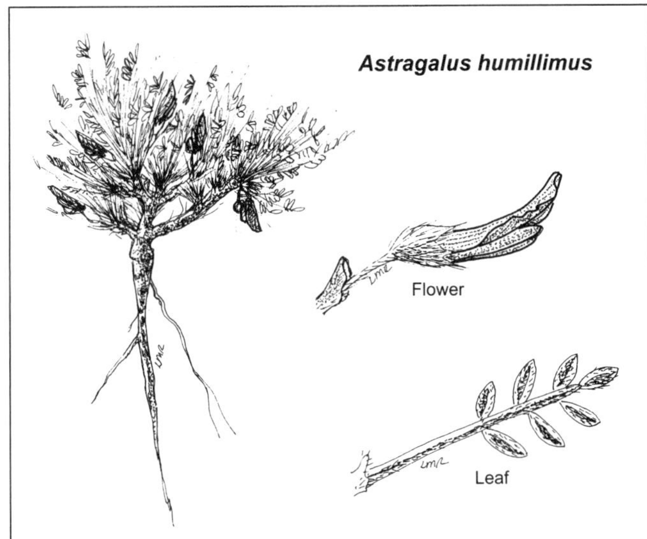

Astragalus humistratus A. Gray (stretched out on the ground) Groundcover milkvetch. Prostrate, spreading, or ascending caulescent perennial, 7–80 cm long, radiating from a superficial caudex. **PUBESCENCE** strigose, malpighian. **STEMS** mostly sprawling. **LEAVES** 1–6 (7.5) cm long; leaflets (5) 9–17 (19), (2) 3–17 (19) mm long, 0.5–6 mm wide, elliptic to oblong or oblanceolate to obovate, acute or obtuse, strigose on both sides or glabrate to glabrous above; stipules 2–10 (12) mm long. **INFL** a raceme, 3- to 30-flowered, the flowers ascending at anthesis, the axis 1–12 (13) cm long in fruit; bracts 1.5–7 mm long; pedicels 0.4–2.2 mm long; bracteoles 0–2; peduncle 1–9 cm long. **FLOWERS** (5.9) 7–11.8 mm long, greenish to ochroleucous, often suffused or veined purplish, or pink-purple, the banner reflexed through 50°–85°; calyx 4.5–8.8 mm long, the tube 2.4–4.1 mm long, campanulate, strigose, the teeth 1.4–5 mm long. **PODS** sessile, ascending to spreading, obliquely ovoid or oblong-ellipsoid, 6–20 mm long, 3.5–5.7 mm wide, variously compressed, strigose, unilocular; ovules 6–26. *n* = 12. [*Tragacantha humistrata* (A. Gray) Kuntze; *Tium huministratum* (A. Gray) Rydb.; *Batidophaca humistrata* (A. Gray) Rydb.; *Pisophaca datilensis* Rydb.; *A. datilensis* (Rydb.) Tidestr.]. Ramah Navajos used the plant as a ceremonial chant lotion, a powder for sores, and as a "life medicine."

1. Pod (10) 13–20 mm long, 3.1–6.5 mm in diameter, 3–4 times longer than wide, dorsally sulcate its whole length up to the base of the beak .... **var. *humistratus***

1′ Pod 6–14 mm long, 2.5–5.7 mm in diameter, 1 1/2–2 1/2 times longer than wide, sulcate dorsally from the middle downward or esulcate .... **var. *humivagans***

var. *humistratus* A. Gray **LEAFLETS** glabrate above. **COROLLA** 9.5–11.5 mm, generally pale, drying purple or pale. **PODS** oblong-lanceoloid, incurved, 15–18 × 4–6 mm, proximally triquetrous, dorsally obcompressed, sulcate entire length. Commonly disturbed sites. Piñon-juniper, ponderosa pine, roadsides. ARIZ: Apa; NMEX: McK, RAr, SJn. 2225–2780 m (7305–9130′). Flowering: May–Sep. To be sought in southwestern Colorado. Disjunct in the Davis Mountains, Texas, and south to Chihuahua.

var. *humivagans* (Rydb.) Barneby (spreading on the ground) **LEAFLETS** pubescent on both sides. **COROLLA** 7–10 mm, generally pale and remaining so, or becoming purple on drying. **PODS** shortly lunate, obcompressed at base;

laterally compressed above, 8–14 mm × 3.5–5 mm, conspicuously beaked; ventral suture conspicuous; dorsal face sulcate only at base. Piñon-juniper to ponderosa pine. ARIZ: Apa; NMEX: SJn; UTAH. 2315–2345 m (7600–7700′). Flowering: May–Sep. Northwest Arizona, Nevada, and Utah.

Astragalus iodopetalus (Rydb.) Barneby. (with violet-colored petals) Violet milkvetch. Caulescent or subacaulescent perennial, from a taproot and superficial root crown. **PUBESCENCE** basifixed, villous-villosulous. **STEMS** 1.5–10 (18) cm long, prostrate and radiating, the internodes either all concealed by stipules, or some of them developed and up to 0.5–2 (2.5) cm long. **LEAVES** (4) 5–15 (20) cm long; leaflets 7–31, 3–17 (20) mm long, oblanceolate, obovate, or elliptic, mostly obtuse or emarginate, flat or loosely folded; stipules submembranous, 2.5–12 cm long, distinct or nearly so. **INFL** a raceme, rather closely (10) 12–20 (25)-flowered, the axis (1.5) 2–8 cm long in fruit; bracts 2.5–8.5 mm long; pedicels 1.3–3.6 mm long; bracteoles 0; peduncle (1.5) 3–10 cm long, shorter than the leaf, prostrate in fruit. **FLOWERS** 17–23.5 mm long, bright reddish violet or merely tinged with violet or almost or quite white, the banner incurved 40°–50°; calyx 10–15 mm long, the tube (6.8) 7.5–10.5 mm long, cylindric to deeply campanulate, thinly villous-villosulous, the teeth 2.5–5.5 mm long, lanceolate to subulate. **PODS** ascending (humistrate), obliquely ovate, lance-elliptic, oblong-elliptic, or lanceolate in outline, (17) 20–30 mm long, 7–10 mm in diameter, straight or nearly so and obcompressed, the green, fleshy, glabrous valves becoming stiffly leathery or subligneous, brownish, or ultimately blackish, subunilocular, the narrow septum to 1.7 mm wide; ovules 30–44. [*Xylophacos iodopetalus* Rydb.; *A. iodopetalus* Greene ex M. E. Jones; *Xylophacos stipularis* sensu Rydb., non *A. arietinus* M. E. Jones var. *stipularis* M. E. Jones]. Dry, stony hillsides and benches, commonly on granite, often about oak thickets, in oak-piñon forest, or among sagebrush. COLO: Arc, Hin, LPl; NMEX: RAr. 2085–2530 m (6850–8300′). Flowering: May–Jul. Locally plentiful and not uncommon around the western and southern slopes of the Rocky Mountains in the valleys of the Dolores, Gunnison, and upper San Juan rivers in southwestern Colorado, extending more rarely southeast to the western tributaries of the Rio Arriba in Taos and Sandoval counties, New Mexico. A specimen, apparently belonging within the concept of *A. iodopetalus*, from Hinsdale County, Colorado (about 1/4 mile north of the Archuleta County line, 8 Jun 1996, *K. Heil 9882*, SJNM!), but with pale flowers and shorter calyx measurements, might represent a phase of the species worthy of taxonomic recognition. The flowers are about 18 mm long, the calyx 6.5–7.2 mm, with the tube 4–4.8 mm, and the teeth 2.2–3.2 mm.

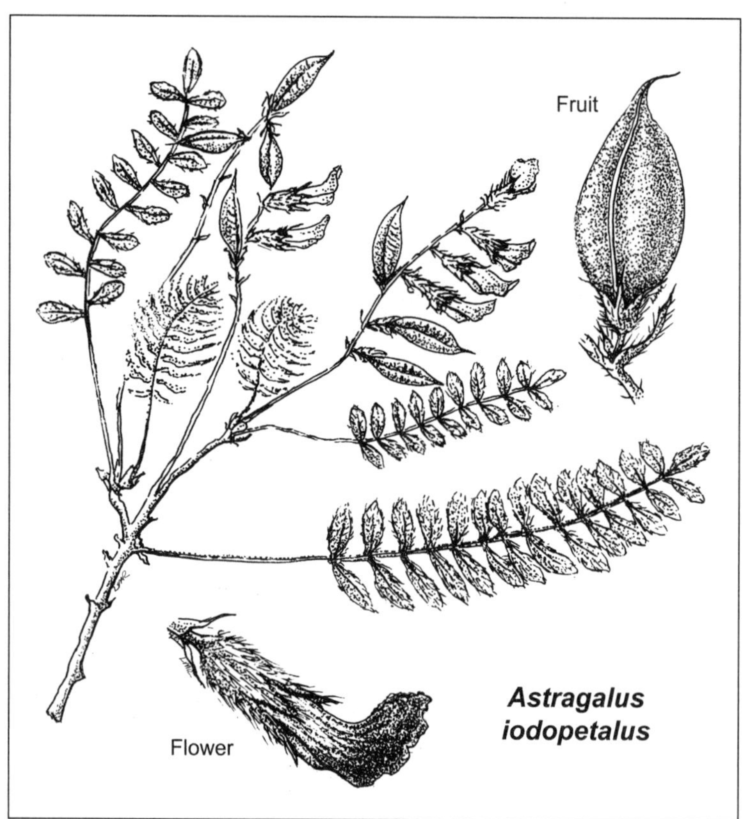

Astragalus iodopetalus

Astragalus kentrophyta A. Gray (spiny growth) Nuttall's kentrophyta. Perennial, caulescent, mat-forming to erect, 15–45 cm long, from a caudex and stolonlike, creeping stems. **PUBESCENCE** basifixed or malpighian. **STEMS** prostrate to erect, compact to elongate. **LEAVES** 0.4–2.6 cm long; leaflets 3–9, (1) 3–13 (17) mm long, 0.5–1.5 mm wide, linear to narrowly elliptic or lanceolate, all continuous with the rachis and spinulose-tipped, strigose on both sides; stipules 1.5–5 mm long, at least some connate-sheathing. **INFL** a raceme, 1- to 3-flowered, the flowers declined at anthesis, the axis to 0.5 cm long in fruit; bracts 0.8–3.5 mm long; pedicels 0.5–2 mm long; bracteoles 0; peduncle to 1.5 (3) cm long. **FLOWERS** 4.5–10 mm long, pink-purple or whitish, ochroleucous or purple-tinged, the banner recurved through about 45°; calyx 2.4–8.3 mm long, campanulate, strigose, the teeth 1.5–5 mm long, subulate. **PODS** declined or spreading, sessile, elliptic to oblong or lance-acuminate in outline, usually curved, 4–10 mm long, 1.3–4 mm wide, strigose, unilocular; ovules 2–8. $2n = 24$. [based on: *Kentrophyta montana* Nutt. 1838, Fl. N. Amer. 1: 353, not *Astragalus montanus* L. 1753; *Tragacantha montana* (Nutt.) Kuntze; *A. montanus* (Nutt.) M. E. Jones; *Kentrophyta viridis*

Nutt. ex Torr. & A. Gray; *A. kentrophyta* var. *viridis* (Nutt. ex Torr. & A. Gray) Hook.; *A. viridis* (Nutt.) E. Sheldon; *Phaca viridis* (Nutt.) Piper; *A. tegetarius* S. Watson var. *viridis* (Nutt. ex Torr. & A. Gray) Barneby]. Used as a ceremonial chant lotion, for rabies, and as a "life medicine" by the Navajos.

1. Pod ovate- or lance-acuminate in profile, gently incurved into a definite beak; calyx 3.4–4.4 mm long, the teeth 1.5–2.4 mm long .... **var. *elatus***
1′ Pod ovoid-lenticular, beakless or nearly so; calyx 2.4–3.3 mm long, the teeth 0.7–1.3 mm long ...**var.*neomexicanus***

var. *elatus* S. Watson (tall) Tall kentrophyta. Suffruticulose and often bushy-branched at base, forming low, prickly bushes 10–45 cm tall. **PUBESCENCE** of stems and herbage strigulose, malpighian, with leaflets pubescent on both sides, or glabrous above. **LEAVES** 10–26 mm long; leaflets (3) 5–7, (2) 5–15 (17) mm long; stipules 1–1.2 mm long. **PEDUNCLES** 1–6 mm long. **FLOWERS** 4.8–6.2 mm long, whitish or faintly purplish-veined or -tinged; calyx 3.4–4.4 mm long, the tube 1.8–2.3 mm long, the teeth 1.5–2.4 mm long, subulate, spinulose. **PODS** (3.5) 4–7 mm long, 1.5–2 mm wide, narrowly ovoid-acuminate; ovules 2–4. [*A. viridis* Bunge var. *impensus* E. Sheld.; *A. viridis* var. *elatus* (S. Watson) Cockerell; *A. kentrophyta* var. *impensus* (E. Sheld.) M. E. Jones; *Kentrophyta impensa* (E. Sheld.) Rydb.; *A. impensus* (E. Sheld.) Wooton & Standl.; *A. montanus* (Nutt.) M. E. Jones var. *impensus* (E. Sheld.) M. E. Jones; *A. tegetarius* S. Watson var. *elatus* (S. Watson) Barneby; *A. kentrophyta* subsp. *elatus* (S. Watson) W. A. Weber]. Mixed desert and salt desert shrub, juniper-piñon, ponderosa pine, and pine-spruce communities, often in floodplains. ARIZ: Apa; NMEX: McK, SJn. 1830–2430 m (6000–8000′). Flowering: Jun–Sep. Inyo County, California; Elko, Nevada; west and south in Nevada; Sweetwater County, Wyoming; and southern and eastern Utah. Both erect and prostrate phases are known, but typically it is a small prickly bush that has borne, with good reason, the name of "barb-wire" kentrophyta. It is common on limestone in the ponderosa pine forests of southern Utah.

var. *neomexicanus* (Barneby) Barneby (of New Mexico) New Mexico spiny milkvetch. Decumbent and forming mats or cushions or erect and bushy-branched at base, and 4–30 (40) cm tall. **PUBESCENCE** of stems and herbage strigulose, malpighian, with leaflets pubescent on both sides. **LEAVES** 8–22 mm long; leaflets mostly 5, 3–13 mm long; stipules 1–1.5 (2.5) mm long. **PEDUNCLES** 1–3 mm long. **FLOWERS** 4.8–5.2 mm long, whitish; calyx 2.4–3.3 mm long, the tube 1.8–2.1 mm long, the teeth 0.7–1.3 mm long, subulate, spinulose. **PODS** 3–4 mm long, 1.8–2.4 mm wide, obliquely ovoid in outline; ovules 2 or 3. [*A. tegetarius* S. Watson var. *neomexicanus* Barneby]. Bluffs, badlands, and dunes, in mixed desert shrub and piñon-juniper communities. ARIZ: Apa; NMEX: McK, San, SJn. 1890–2265 m (6200–7435′). Flowering: Jun–Sep. A northwest New Mexico endemic.

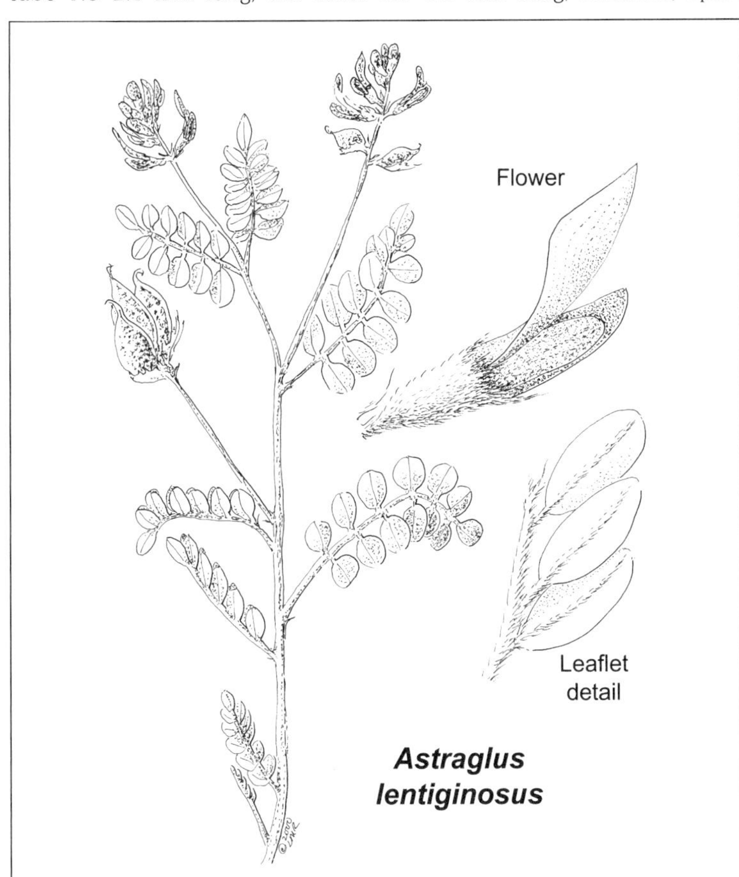

Astraglus lentiginosus

Astragalus lentiginosus Douglas (referring to the freckled or spotted pod) Freckled milkvetch. Perennial, caulescent, mostly 1.5–6 dm tall, from a caudex. **PUBESCENCE** basifixed; stipules 1.5–7 mm long or more, all distinct. **LEAVES** 2.4–15 cm long; leaflets 9–23, 2–23 mm long, 1–13 mm wide, elliptic to ovate or lanceolate, obtuse to rounded, emarginate, or acute, pubescent to glabrous on one or both sides. **INFL** a raceme, (5) 11- to 30-flowered, the flowers ascending to declined at anthesis, the axis 1–18 cm long in fruit; pedicels 1–4 mm long; bracts 1.5–6 mm long; bracteoles 0–2; peduncle 1–14 cm long, sometimes more. **FLOWERS** 8.4–22 mm long, pink-purple, ochroleucous, whitish, or variously suffused with pink or purple; calyx 3.5–11.6 mm long, the tube 3–9 mm long, cylindric to short-cylindric, strigose, the teeth

0.6–2.5 mm long, subulate or triangular. **PODS** ascending to declined, sessile or with a gynophore, variable in outline, either inflated and ovoid (12–26 mm long, 5–20 mm thick) or not inflated and oblong in outline (15–25 mm long, 3–7.5 mm thick), strigose or glabrous, mottled or not, leathery to membranous, bilocular; ovules 16–38. *n* = 11. The freckled milkvetch, considered broadly, has more infraspecific taxa than any others in the genus. Of the great many, perphaps more than 50, varieties known within the species as a whole, 2 are present in the Four Corners region, probably all of which are poisonous to livestock. Indolizidine alkaloids are responsible for the syndrome known as locoism, which is responsible for death and spontaneous abortion in sheep, especially. As in other short-lived perennial species in the genus, populations of freckled milkvetch rise and fall cyclically, dependent on short-term climatic cycles with abundant seasonally critical precipitation followed by other drought periods. Poisoning of livestock occurs when the plants are most abundant. Plant used as a charm by Navajos and pods dried for winter use by Zunis.

1. Pods strongly (-slightly) inflated .... **var. *diphysus***
1' Pods not inflated, nearly straight to hamately incurved .... **var. *palans***

var. *diphysus* (A. Gray) M. E. Jones (double-bladdered, referring to the swollen, bilocular pod) Double bladder freckled milkvetch. Perennial. **STEMS** simple or branched at base, (1) 1.5–3.5 (4) dm long. **LEAVES** 3–10 (14) cm long; leaflets (11) 15–21 (23), 4–20 mm long, oblong-oblanceolate, obovate, rhombic-obovate, elliptic, or ovate-cuneate, truncate-emarginate, flat or loosely folded. **INFL** a raceme, shortly, at first rather densely, (10) 12–24-flowered, the axis not or little elongating, 1–4 (6) cm long in fruit; peduncle erect or incurved, (1.5) 2.5–8.5 cm long, shorter than or equaling the leaf. **FLOWERS** (12.6) 14.5–19 mm long, (5.8) 7–9.8 mm wide, pink-purple; calyx 7–10.4 mm, the cylindric or deeply campanulate tube (5.1) 5.5–8 mm long, (2.5) 2.7–3.5 (4) mm in diameter, the teeth 1.5–3.2 mm long. **PODS** (1) 1.4–2.7 (3) cm long, (6.5) 8–18 mm in diameter, the body mostly plumply ovoid or subglobose, the uni- or bilocular beak (3) 4–10 mm long, green, purple- or red-tinged, or brightly mottled, glabrous, becoming stiffly papery or leathery, semibilocular; ovules 28–35. *n* = 11. [*A. diphysus* A. Gray; *Tragacantha diphysa* (A. Gray) Kuntze; *Cystium diphysum* (A. Gray) Rydb.; *A. diphysus* var. *albiflorus* A. Gray; *A. lentiginosus* var. *albiflorus* (A. Gray) Schoener; *A. macdougalii* E. Sheld.; *A. lentiginosus* var. *macdougalii* (E. Sheld.) M. E. Jones; *Cystium macdougalii* (E. Sheld.) Rydb.]. Yucca-grassland, piñon-juniper forest, and other xeric communities. ARIZ: Apa, Nav; COLO: LPl, Mon; NMEX: McK, RAr, SJn; UTAH. 1645–2285 m (5400–7500'). Flowering: Mar–Jun (Aug). Northern Arizona, south of the Little Colorado River; Henry Mountains, Utah; southwest Colorado; and northwestern and central New Mexico.

var. *palans* (M. E. Jones) M. E. Jones (straggling) Straggling milkvetch. **STEMS** 1.5–5 dm tall, glabrate. **LEAFLETS** broadly obovate to obovate-elliptic, 6–20 mm, sometimes folded. **PODS** turgid or slightly inflated, ovate-acuminate to linear-lanceolate, straight to hamately curved. [*A. palans* M. E. Jones; *Tium palans* (M. E. Jones) Rydb.; *A. bryantii* Barneby]. Salt desert shrub, blackbrush, juniper, piñon-juniper, and mixed desert shrub communities. ARIZ: Coc, Nav; UTAH. 1110–1845 m (3650–6050'). Flowering: Mar–Jun. Southern Utah, northern Arizona, western Colorado, and Nevada. The straggling milkvetch is the common phase of *A. lentiginosus* in the canyons of the lower San Juan River. It is distinctive in having oblong, usually curved pods, seldom more than 7 mm thick, pale pink-purple flowers, and elongate inflorescences. Specimens from along Glen Canyon in San Juan County, Utah, cited previously as *A. bryantii* Barneby, do not differ in any remarkable way from var. *palans*.

Astragalus lonchocarpus Torr. (with long pod) Great rushy milkvetch. Tall, slender, caulescent, clump-forming, evidently long-lived perennial, 30–85 (90) cm tall, from a shallowly subterranean caudex (this seldom collected). **PUBESCENCE** strigulose, basifixed. **STEMS** erect, often in dense clumps, buried for a space of 1–8 cm. **LEAVES** 2–13 cm long, the uppermost and sometimes all simple, the lower with leaflets 3–9 (11), 2–36 mm long, 0.5–4 mm wide, linear to narrowly oblanceolate, obtuse to acute, strigose on both sides or glabrous above; stipules 1–9 (10) mm long, all distinct. **INFL** a raceme, loosely 7- to 40-flowered (or more), the flowers spreading-declined at anthesis, the axis 3.5–45 cm long in fruit; bracts 0.8–2.5 mm long; pedicels 1.3–4.5 mm long; bracteoles 0; peduncle 6–24 cm long. **FLOWERS** 13–20 mm long, ochroleucous to almost white, concolorous, the banner recurved through about 50°; calyx 5.8–10.3 mm long, usually brown (and contrasting with the pale petals), the tube 5–8 mm long, cylindric, gibbous, strigose, the teeth 0.6–2.5 mm long, subulate. **PODS** pendulous, stipitate, the stipe 3–15 mm long, the body elliptic to oblong in outline, 22–50 mm long, 3.3–6.2 (7.5) mm wide, dorsiventrally compressed, the low, concave ventral face, or both faces, carinate by the sutures, the valves often brownish, strigose, unilocular; ovules 12–26. 2*n* = 22. [based on: *Phaca macrocarpa* A. Gray; non *A. macrocarpus* Pall.; *Tragacantha lonchocarpa* (Torr.) Kuntze; *Homalobus macrocarpus* (A. Gray) Rydb.; *Lonchophaca macrocarpa* (A. Gray) Rydb.; *A. macer* A. Nelson; *Lonchophaca macra* (A. Nelson) Rydb.]. Salt desert shrub, blackbrush, and piñon-juniper communities; often on low-quality substrates (i.e., often in saline shales and clays). ARIZ: Apa; COLO: Arc, Dol, LPl, Mon; NMEX: McK, RAr, SJn; UTAH. 1840–2450 m (6040–8045′). Flowering: May–Jul. Eastern Nevada, southern and western Utah. Plant used as an emetic and for goiter by the Navajos.

Astragalus micromerius Barneby (small in all parts) Chaco milkvetch. Prostrate, caulescent, evidently long-lived perennial, forming mats or cushions 1–6 dm wide, radiating from a branching caudex. **PUBESCENCE** villous-hirtellous, basifixed. **STEMS** 5–30 cm long or more, the older, subterranean portions accommodating burial, up to 30 cm long. **LEAVES** 0.4–1 (2) cm long; leaflets (3) 5–9, 1–3.5 (6) mm long, 0.6–2 mm wide, oblong-elliptic to obovate, obtuse, pubescent on both sides; stipules 1–3 mm long, all connate-sheathing. **INFL** a raceme, 1–3 (5)-flowered, the flowers spreading at anthesis, the axis very short in fruit; bracts 1–2 mm long; pedicels less than 1 mm long; bracteoles 0; peduncle 0.3–1 cm long. **FLOWERS** 5.5–6.5 mm long, whitish, the wing tips tinged lilac, the keel purple-tipped, the banner recurved through about 45°; calyx 2.5–3.5 mm long, the tube 1.5–2.5 mm long, obconic-campanulate, the teeth 1–1.3 mm long, subulate. **PODS** spreading, sessile, the body 4–5 mm long, 2.5–3 mm wide, obliquely ovoid in outline, obcompressed and dorsally flattened or shallowly sulcate in part, hirsutulous; ovules 4. Shelving ledges of sandstone cliffs or on steep, boulder-strewn talus, often partly submerged in drifting sand, and on steep slopes of Mancos clay. NMEX: McK, SJn. 2020–2235 m (6635–7340′). Flowering: Jul–Aug. A northwest New Mexico endemic.

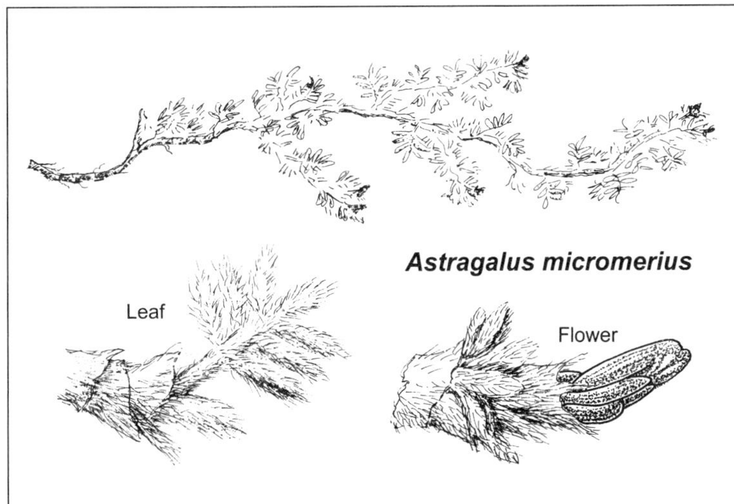

Astragalus missouriensis Nutt. **CP** (of the Missouri River) Missouri milkvetch. Perennial, low, loosely tufted or prostrate, shortly caulescent or subacaulescent, from a taproot and subterranean caudex. **PUBESCENCE** malpighian, strigose to strigulose. **STEMS** up to 1.5 (2) dm long, prostrate and radiating, the internodes often concealed by stipules; stipules (2) 3–9 mm long, distinct. **LEAVES** (2) 4–14 cm long; leaflets (5) 11–17 (21), 3.5–13 (17) mm long, 3–7 (8) mm wide, flat, elliptic to obovate, acute or mucronulate, or obtuse, pubescent on both sides. **INFL** a raceme, (3) 5–15-flowered, the flowers ascending to spreading, the axis (0.5) 1–4 cm long in fruit; bracts 2.5–8 mm long; pedicels 1–3.5 mm long; bracteoles 0–2; peduncles (1.5) 3.5–11 cm long, shorter to longer than the leaf. **FLOWERS** 9.5–22 (24) mm long, pink-purple (or white), the banner recurved through about 45°; calyx 5–12 mm long, the purple-tinged, strigulose tube cylindric or deeply campanulate, (4.1) 6.3–9 (10) mm long, the teeth (0.7) 1.4–3.3 mm long, subulate. **PODS** ascending, sessile and (commonly) persistent on the receptacle, subsymmetrically or obliquely oblong or oblong-elliptic in outline, 1.4–2.7 (3) cm long, (4) 5–9 (10) mm in diameter, variably compressed, straight or slightly incurved, strigulose, brownish and finally black, unilocular or subunilocular; ovules (33) 36–50 (56). [*Tragacantha missouriensis* (Nutt.) Kuntze; *Xylophacos missouriensis* (Nutt.) Rydb.; *A. missouriensis* var. *typicus* Barneby; *A. melanocarpus* Hook.; *A. missouriensis* f. *longipes* Gand.; *A. missouriensis* f. *microphyllus* Gand.; *A. missouriensis* f. *leucophaeus* Gand.]. Only in the Four Corners regions of southwest Colorado and adjacent southeast Utah, west-central to northwest New Mexico, and east-central to northeast Arizona has the species undergone taxonomically significant and geographically correlated morphological differentiation. Otherwise, in its immense range from the Canadian provinces of Alberta, Saskatchewan, and Manitoba southward along the prairies and high plains to Trans-Pecos Texas, it is remarkably uniform.

1. Pod subsymmetrically oblong-elliptic in outline, straight or nearly so, subcylindric when first formed, becoming laterally compressed and bicarinate, the lateral faces transversely dilated and convex; plants widespread east of the Continental Divide from southern Canada to west Texas, extending around the southern foothills of the Rocky Mountains to the upper San Juan River in northwest New Mexico, northeast Arizona, and southwest Colorado .... **var. *missouriensis***

1′ Pod obliquely ellipsoid, lunately incurved, obcompressed below the incurved beak; plants of the western slope of the Rocky Mountains in southwest Colorado, southeast Utah, and northwest New Mexico (Rio Arriba County)...(2)

2. Plants subacaulescent to shortly caulescent, the stems to 10 cm long; racemes with 4–8 flowers; plants from La Plata and Montezuma counties, Colorado; San Juan County, Utah, and San Juan County, New Mexico ... . **var. *amphibolus***
2' Plants rather definitely caulescent, the stems 10–15 (20) cm long, radiating from the root crown; racemes with 9–12 flowers; plants of La Plata and Archuleta counties, Colorado, and adjacent Rio Arriba County, New Mexico .... **var. *humistratus***

var. *amphibolus* Barneby (ambiguous, combining features of *A. missouriensis* and *A. amphioxys*) Missouri milkvetch. **FLOWERS** 15–22 mm long, pink-purple, rarely white; calyx 8.5–13 mm long, the tube 7–10 mm long, cylindric, strigose, the teeth 1.5–3 mm long, subulate. **PODS** ascending to descending, sessile, ellipsoid, (12) 15–25 mm long, 7–9 mm thick, curved, dorsiventrally compressed, strigose, unilocular; ovules 35–55. Piñon-juniper and sagebrush communities, on igneous or sandstone outcrops or substrates derived from them. ARIZ: Apa; COLO: Arc, Dol, LPl, Mon, SMg; NMEX: RAr, SJn; UTAH. 1495–2430 m (4900–7970'). Flowering: May–Jul. On the west slope of the southern Rocky Mountains, Colorado, from Garfield County south through the Gunnison, Mancos, and San Juan River valleys, east to the head of the Rio Grande in Conejos County, west into the foothills of the La Sal and Abajo Mountains, Grand and San Juan counties, Utah.

var. *humistratus* Isely (on the ground) Archuleta milkvetch. **FLOWERS** 19–20.5 mm long, lavender, purple, or almost white, the wing tips often white; calyx 7.8–10 mm long, the tube 7–10 mm long, cylindric, strigose, the teeth 1.5–3 mm long, subulate. **PODS** ascending to descending, sessile, oblong-ellipsoid, 17–20 mm long, 7–9 mm thick, curved, dorsiventrally compressed, glabrous or sparsely pubescent, unilocular; ovules 33–40. Oak brush with scattered ponderosa pine on clay knolls. COLO: Arc, Hin; NMEX: RAr. 2010–2440 m (6600–8000'). Flowering: May–Jul. A Four Corners endemic. The banner much surpasses both the wing and keel petals, with the wings often pale or white or only faintly purplish apically.

var. *missouriensis* **FLOWERS** (14.5) 16–22 (24) mm long, (7.2) 8–12.5 mm wide; calyx 9–12 (14.3) mm long, the tube 6.3–9 (9.3) mm long, 3.3–4.6 (4.9) mm in diameter, the teeth 1.4–3.3 (5.3) mm long. **PODS** 1.5–2.7 (3) cm long, (4) 5–9 (10) mm in diameter, obtuse or sometimes cuneate at base, abruptly contracted distally into a subulate, pungent beak 1.5–4 mm long, at first subterete or a trifle dorsiventrally compressed, when ripe somewhat laterally dilated and obtuse-angled; ovules (33) 40–50 (56). Prairies, in dry, open places, most abundant on limestones, shales, and sandstones, often on gypsum. ARIZ: Apa; COLO: LPl, Mon; NMEX: McK, RAr, SJn; UTAH. 1560–2210 m (5115–7260'). Flowering: late Mar–Jul. Widespread and common over the Great Plains from southern Alberta, Saskatchewan, and southwest Manitoba to northwest and Trans-Pecos Texas, east to the Missouri Valley, western Iowa, central Kansas, and western Oklahoma, west to the upper Missouri, Montana, to Yellowstone Park, and thence south along the Rocky Mountain piedmont to the Rio Grande Valley; west in New Mexico, northeast Arizona (Apache County), and southwest Colorado to the headwaters of the San Juan River. Plants of the Missouri milkvetch are common in the shortgrass prairies of the plains states from Canada south to Texas and New Mexico. Eastern portions of its distribution in western Iowa are along the Missouri River bluffs, which support other plains species in those peculiar habitats. Not uncommonly, white-flowered specimens grow intermixed with the usual bright pink-purple ones.

Astragalus moencoppensis M. E. Jones (Moenkopi Wash) Moenkopi milkvetch. Slender, caulescent perennial, 9–60 cm tall, from a branching caudex. **PUBESCENCE** strigulose, basifixed. **STEMS** erect or ascending, commonly shorter than the longest peduncles, numerous and forming clumps. **LEAVES** 4–16.5 cm long; leaflets 5–15, 2–23 mm long, 0.3–2 mm wide, filiform to linear or narrowly elliptic, acute to obtuse, the terminal often continuous with the rachis, strigose below, glabrous on the involute upper surface; stipules 1.5–7 mm long, at least some connate-sheathing. **INFL** a raceme, 6- to 34-flowered, the flowers ascending at anthesis, the axis 3–25 cm long in fruit; bracts 1.5–3.5 mm long; pedicels 0.3–2 mm long; bracteoles 0–1; peduncle 4–25 cm long. **FLOWERS** 8–11 mm long, pink-purple, the banner recurved through about 45°; calyx 5–7.5 mm long, the tube 3–4 mm long, campanulate, white-pilose, the teeth 1.8–3.5 mm long, lance-subulate. **PODS** ascending-spreading, sessile, ovoid to ellipsoid, 6–7 mm long, 2.3–3.4 mm thick, strigulose, unilocular. [*Cnemidophacos moencoppensis* (M. E. Jones) Rydb.]. Salt desert shrub, mixed desert shrub, and piñon-juniper communities, usually in saline, silty or clay, seleniferous soils. ARIZ: Nav; UTAH. 1340–2165 m (4390–7100'). Flowering: May–Jul. Southern Utah and adjacent Arizona. The Moenkopi milkvetch is a primary selenium indicator of distinctive mien and odor. The very elongate racemes of small pink-purple flowers borne on a sparsely foliose stem is diagnostic for this species. The odor is that of selenium, though it is not as strong as in some other selenophytes.

Astragalus mollissimus Torr. (referring to the vesture being very soft) Woolly locoweed. Perennial, acaulescent to subacaulescent, mainly 10–30 cm tall, from a taproot and superficial root crown. **PUBESCENCE** basifixed, villous-tomentose. **STEMS** when caulescent composed of several short internodes, up to 2 cm long, forming clumps. **LEAVES** (3) 6–26 cm long; leaflets 11–35, 3–45 mm long, suborbicular, ovate, obovate, or rhombic-elliptic, often thick-textured, pubescent on both sides; stipules 3–17 mm long, all distinct. **INFL** a raceme, (5) 7–45-flowered, the axis (0.5) 1–17 cm long in fruit; bracts 2.5–10 (12) mm long; pedicels 0.5–3 mm long; bracteoles commonly 0 (2); peduncle (1.5) 3–25 cm long. **FLOWERS** 12–24.5 mm long, pink-purple, dull pinkish lavender, yellowish suffused with lilac, or creamy white, the banner recurved through about 30°; calyx (6.8) 8–14 (15.3) mm long, the tube (4.5) 5–10 (11.2) mm long, cylindric to deeply campanulate, villous-tomentose, the teeth (1.7) 2–5.5 (6.8) mm long, lanceolate to subulate or subulate-setaceous. **PODS** spreading or ascending, commonly humistrate, the body obliquely ovate, lance-elliptic, lunate, or linear-oblong in outline, 9–25 mm long, 4–13 mm in diameter, sometimes decidedly inflated but never bladdery, terete when narrow, obcompressed and shallowly sulcate along both sutures when broad, nearly straight to incurved through 90° or slightly more, fleshy, stiffly papery, leathery, or subligneous, glabrous, strigose, or villous-tomentose, bilocular; ovules (12) 20–37 (41). The *mollissimus* complex has been variously interpreted as belonging to several species, combined in varietal status under one or more species, or split into species each with segregate varieties, and with peripheral taxa still represented at specific rank (Isely 1998). The present treatment follows Barneby (1964a), wherein the overall similarity of the included taxa seems to be best represented. Leaves used as a ceremonial emetic by Navajos.

1. Raceme (5) 7- to 12-flowered; peduncle mainly recurved in fruit; pod broadly ovoid, 12–18 mm long, 7–13 mm in diameter, shortly conic-beaked; longest hairs of calyx and herbage 1–2 mm long; plants of northwest New Mexico (Santa Fe to McKinley County) .... **var. *matthewsii***

1′ Raceme (10) 12- to many-flowered; peduncle mainly erect or ascending in fruit; if of northwestern New Mexico either the pod less than 7 mm in diameter and the longest hairs of the calyx and herbage up to 2–3.5 mm long, or the pod strongly beaked and the beak unilocular; plants of various or other distribution .... (2)

2. Pod mostly turgid, ovoid, the beak unilocular; plants of the Colorado Basin west through southern Utah to eastern Nevada .... **var. *thompsoniae***

2′ Pod solid or nearly so, narrowly oblong- or lance-ellipsoid, 3–8 mm in diameter, the beak fully bilocular; plants of other distribution .... **var. *bigelovii***

var. *bigelovii* (A. Gray) Barneby ex B. L. Turner (for John Milton Bigelow, botanist with the Whipple Expedition of 1853–1855). Bigelow's woolly locoweed. Shortly caulescent. **LEAVES** 9–26 cm long; leaflets (13) 19–27, 6–25 mm long, ovate, obovate, oval, or broadly elliptic, obtuse or subacute; stipules large and conspicuous, 6–20 mm long. **INFL** an oblong raceme, rather densely (15) 20–45-flowered, the flowers subcontiguous or interrupted proximally, the axis (4) 5–11 cm long in fruit; peduncle (5) 8–22 cm long. **FLOWERS** 17–22.5 mm long, 9–11.5 mm wide, pink-purple; calyx 10.5–13.5 cm long, the tube (8) 8.3–10.3 mm long, (3.2) 4–5.2 mm in diameter, the teeth (1.7) 2.6–4.4 mm long. **PODS** ovoid-acuminate or lance-ellipsoid, gently incurved or nearly straight, 1–1.5 cm long, (4) 4.5–8 mm in diameter, sometimes a little turgid, the stiffly papery or leathery valves densely villous-tomentulose, the longest hairs up to 1–1.6 mm long, the beak bilocular; ovules 20–31. $n = 11$. [*A. bigelovii* A. Gray; *Tragacantha bigelovii* (A. Gray) Kuntze; *A. bigelovii* var. *typicus* Barneby]. Dry plains and foothills, in desert- or mesquite-grassland, sometimes among junipers, apparently most abundant on calcareous soils but also in sandy loams of various origins and occasionally on basalt gravel, especially common on overgrazed and badly eroded cattle ranges. ARIZ: Apa; NMEX: San. 1225–1840 (rarely up to 2305) m (4025–6040′; 7560′). Flowering: Mar–Jun (also as early as Jan). Isely (1998) regards this at species level and includes within it the vars. *marcidus*, *matthewsii*, and *mogollonicus*, primarily on the basis of the completely bilocular pods.

var. *matthewsii* (S. Watson) Barneby (for Washington Matthews, ethnobotanist and archaeologist; U.S. Army surgeon at Fort Wingate) Matthews woolly locoweed. Acaulescent or nearly so. **STEMS** not over 1.5 cm long, concealed by imbricated stipules. **LEAVES** (3) 5–12 cm long; leaflets 11–23, 3–12 mm long, obovate; stipules (3) 4–8 mm long. **INFL** a raceme, (5) 7–12-flowered, the axis (0.5) 1–4.5 cm long in fruit, not or scarcely surpassing the foliage; peduncle scapiform, (1.5) 2.5–8 cm long. **FLOWERS** 18.5–22.5 mm long, pale purple; calyx 10–13 mm long, the tube 7–8.6 mm long, 3.4–4.7 mm in diameter, the teeth 2.4–5.2 mm long. **PODS** broadly and plumply ovoid, widest near the obtuse or truncate base, 12–18 mm long, turgid and 7–13 mm in diameter, gently or rather abruptly incurved distally into the conical, obscurely compressed, fully bilocular beak; ovules 24–31. [*A. matthewsii* S. Watson; *A. bigelovii* A. Gray var. *matthewsii* (S. Watson) M. E. Jones]. Open slopes and hilltops, mostly in ponderosa pine forest, but descending along canyons into the juniper-piñon belt, in light sandy or gravelly, sedimentary, granitic, or volcanic soils. ARIZ: Apa; NMEX: McK, San, SJn. 1735–2285 m (5700–7500′). Flowering: Apr–Jun. Uncommon and scattered in the mountains of the Four Corners region.

var. *thompsoniae* (S. Watson) Barneby (for Ellen P. Thompson, sister of Capt. John W. Powell) Thompson's woolly locoweed. Perennial, acaulescent, 6–45 cm tall, from a caudex. **PUBESCENCE** basifixed. **STEMS** mostly obscured by stipules. **LEAVES** 2–28 cm long; leaflets 15–35, 2–18 mm long, 1–14 mm wide, obovate to suborbicular or elliptic, obtuse to retuse or acute, densely woolly-tomentose on both sides; stipules 4–13 mm long, all distinct. **INFL** a raceme, 7–25-flowered, the flowers ascending at anthesis, the axis 1.5–18 cm long in fruit; bracts 2.5–8 mm long; pedicels 0.5–3 mm long; bracteoles 0–1; peduncles 2.5–24 cm long. **FLOWERS** 18–25 mm long, pink-purple; calyx 11–15.5 mm long, the tube 7.7–13 mm long, cylindric, villous, the teeth 2–4.2 mm long, subulate. **PODS** descending, sessile, ovoid, 11–23 mm long, 6–11 mm thick, curved, densely villous-tomentose, bilocular; ovules 28–38. [*A. thompsoniae* S. Watson; *Tragacantha thompsoniae* (S. Watson) Kuntze; *A. bigelovii* A. Gray var. *thompsoniae* (S. Watson) M. E. Jones; *A. syrticola* E. Sheld.]. Salt desert shrub, mixed desert shrub, grassland, and piñon-juniper communities; typically in sandy substrates. ARIZ: Apa, Coc, Nav; COLO: LPl, Mon; NMEX: McK, RAr, San, SJn; UTAH. (760) 1340–2345 m (4400–7690′). Flowering: Mar–Jun (Dec). The variety apparently is transitional to var. *matthewsii* in northwestern New Mexico. It is regarded at specific rank by Isely (1998).

Astragalus monumentalis Barneby (of Monument Valley) Monument Valley milkvetch. Perennial, acaulescent or subacaulescent, 3–18 cm tall, from a branching caudex. **PUBESCENCE** basifixed or shortly malpighian. **STEMS** 1–6 cm long, ascending, the internodes commonly concealed by stipules. **LEAVES** 1.5–8 (11) cm long; leaflets oblanceolate, strigulose beneath, glabrous or glabrate above; stipules 2–4 mm long, all distinct. **INFL** a raceme, 3- to 9-flowered, the flowers ascending at anthesis, the axis 0.5–7 cm long in fruit; bracts 1.5–5 mm long; pedicels 0.8–2.2 mm long; bracteoles 0; peduncle 1–12 cm long. **FLOWERS** 8–9 mm long, pink-purple; calyx 3.6–4.5 mm long, the tube 3.1–3.5 mm long, campanulate, strigose, purplish, the teeth 0.5–1 mm long. **PODS** ascending, sessile or nearly so, narrowly oblong to lanceolate in outline, 12–21 mm long, 2.3–3 mm wide, straight or curved, triangular in cross section, the dorsal suture sulcate, strigose, bilocular, often mottled. Rimrock and other slickrock sites on Cutler and Cedar Mesa Sandstone formations in mixed desert shrub and piñon-juniper communities. ARIZ: Nav; UTAH. 1215–1900 m (3995–6235′). This is a mirror image congener of *A. cottamii*, differing in its smaller floral parts and overall flower size, and in the incipient malpighian pubescence.

Astragalus naturitensis Payson (of Naturita, Colorado) Naturita milkvetch. Perennial, subacaulescent, 4.5–16 cm tall, from a taproot and superficial caudex, the branches clothed with persistent leaf bases. **PUBESCENCE** basifixed, loosely strigulose. **LEAVES** 1.5–7 cm long; leaflets 9–17, 2–8 mm long, mainly 1–2.5 mm wide, elliptic to obovate or oblanceolate, mostly obtuse, strigulose on both sides or glabrescent above; stipules 2–7 mm long, distinct. **INFL** a raceme, 3–8 (11)-flowered, the axis 0.5–2.5 cm long in fruit; pedicels 0.7–1.8 mm long; bracteoles 0; peduncle 1–6.5 cm long, shorter than the leaves. **FLOWERS** (11.2) 13–15.5 mm long, bicolored, the banner whitish or suffused or lined with lilac, the wing and keel tips purple; banner recurved through about 40°; calyx 5–7.4 mm long, the tube cylindro-campanulate, strigulose, 4–6.2 mm long, the teeth 1–1.5 (2) mm long, triangular-subulate. **PODS** ascending (humistrate), sessile on a gynophore to 0.8 mm long, deciduous from the receptacle, obliquely ellipsoid, 13–22 mm long, 4–6 mm wide, incurved, dorsiventrally compressed to more or less trigonous, the dorsal suture depressed but not inflexed as even a partial partition, usually red-mottled, the valves coriaceous, strigulose, unilocular; ovules 22–31. [*A. arietinus* M. E. Jones var. *stipularis* M. E. Jones; *A. stipularis* M. E. Jones]. Sandstone outcrops in sagebrush and piñon-juniper communities. COLO: Mon; NMEX: McK; UTAH. 1660–2050 m (5445–6730′). Flowering: Apr–Jun. West-central Colorado (valleys of the Colorado and Dolores rivers and McElmo Creek; Mesa, Montrose, and Montezuma counties); San Juan County, Utah, and McKinley County, New Mexico. The flowers of the New Mexico materials average smaller than for the species as a whole but seem not to differ otherwise.

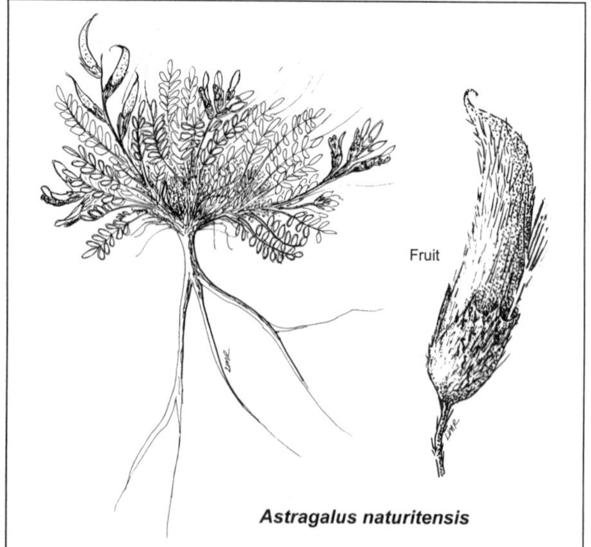

Astragalus naturitensis

Astragalus newberryi A. Gray (for Dr. John Strong Newberry, geologist and naturalist) Newberry's milkvetch. Perennial, acaulescent, 2–12 cm tall, from a taproot and superficial caudex, the branches commonly clothed with a thatch of persistent leaf bases. **PUBESCENCE** basifixed, silky pilose. **LEAVES** 1.5–14 (15) cm long; leaflets (1) 3–15, 3–20 mm long, 2–14 mm wide, obovate to elliptic, oblanceolate, or orbicular, acute to obtuse or retuse, villous-tomentulose on both surfaces; stipules (2.5) 4–11 mm long, all distinct. **INFL** a raceme, 2- to 8-flowered, the flowers ascending, the axis 0.2–2.7 cm long in fruit; bracts 3.5–10 mm long; pedicels 1.4–5 mm long; bracteoles 0–2; peduncle 0.5–11 cm long, shorter than the leaf. **FLOWERS** 14–32 mm long, pink-purple or less commonly whitish with purple keel tip, the banner recurved through about 40°; calyx 11–14.5 mm long, the tube 9–11 mm long, cylindric, villous, the teeth 1.5–4 mm long, subulate to lanceolate. **PODS** ascending (humistrate), obliquely ovoid, curved, (13) 18–28 (36) mm long, 7–13 (17) mm thick, obcompressed, fleshy, thinly hirtellous to densely villous-tomentose, unilocular; ovules 27–40 (46). $n = 11$. [*Xylophacos newberryi* (A. Gray) Rydb.; *A. newberryi* var. *typicus* Barneby; *A. eriocarpus* S. Watson; *Tragacantha watsoniana* Kuntze; *A. newberryi* var. *escalantinus* Barneby]. Sagebrush, rabbitbrush, matchweed, juniper-piñon, shadscale, galleta communities in clays, sands, and gravels. ARIZ: Apa, Nav; COLO: Mon; NMEX: McK, SJn; UTAH. (650) 1525–2350 m (5000–7710′). Flowering: Apr–Jun. Our material belongs to **var. *newberryi***. Utah, Arizona, and New Mexico.

Astragalus nidularius Barneby (like a bird's nest, referring to the many incurved and entangled branches and branchlets) Birds-nest milkvetch. Perennial, caulescent, 15–51 cm tall, from a subterranean caudex; pubescence basifixed. **STEMS** ascending to erect, often branched, subterranean for a space of 3–14 cm or more. **LEAVES** 1.5–7 cm long; leaflets 5–11 (13), 2–20 (25) mm long, 1.3–2 mm wide, linear to oblong, obtuse to emarginate or acute, the terminal leaflet of upper leaves continuous with the rachis, pubescent on both sides or glabrous above stipules 1–6 mm long, at least some connate-sheathing. **INFL** a raceme, (3) 8- to 33-flowered, the flowers rather widely spaced along the rachis, ascending to declined at anthesis, the axis 1.5–28 cm long in fruit; bracts 1.2–2.2 mm long; pedicels 1.2–3 mm long; bracteoles 0–2; peduncle 4–16 (28) cm long. **FLOWERS** 10–15 mm long, pink-purple, the banner reflexed through about 30°–45°; calyx 3.8–7 mm long, the tube 3.3–5.5 mm long, campanulate, strigose, the teeth (0.5) 1–2.2 mm long, subulate. **PODS** pendulous, stipitate, the stipe 3.5–6 mm long, the body narrowly oblong, 20–32 mm long, 3.5–4.5 mm thick, dorsoventrally compressed, strigose, unilocular; ovules (16) 20–26. Piñon-juniper and mixed desert shrub communities. ARIZ: Nav. 1655 m (5425′). Flowering: May–Jun. Utah, in San Juan, Garfield, and Wayne counties. The species is appropriately named, the plants having the appearance of an untidy bird's nest due to the entangled interlocking branches forming rounded clumps overtopped by the flowering racemes.

Astragalus nuttallianus DC. (for Thomas Nuttall, 1786–1859) Small-flowered milkvetch. Annual or winter-annual, caulescent, 3–18 (25) cm long, from a slender taproot and superficial root crown. **PUBESCENCE** basifixed, subglabrous, thinly strigulose, villosulous, or pilose. **STEMS** erect, ascending, prostrate, or decumbent. **LEAVES** 1.5–4.5 (6.5) cm long, dimorphic, those of the lower ones obovate or oblong, retuse or emarginate to obcordate, those of the upper ones mostly longer, oblanceolate, elliptic, or linear-oblong, all or at least the lateral ones obtuse to subacute, the terminal sometimes emarginate; leaflets (5) 7–19 (23), 2–14 (17) mm long, linear-elliptic to obcordate, acute to retuse, flat, pubescent beneath, glabrous or pubescent above; stipules (1) 1.5–6 (9) mm long, mostly herbaceous, all distinct. **INFL** a raceme, (1) 3- to 7-flowered, the axis not over (0) 0.5–2 cm long, the flowers ascending to declined, the axis 0–3 cm long (commonly less than 2 cm); bracts 0.5–2.5 mm long; pedicels 0.4–1.6 mm long; bracteoles 0 or 1; peduncle (1.5) 2.5–5.5 (6.5) cm long, typically surpassing the leaf. **FLOWERS** 3.7–10 (13) mm long, whitish, lilac, or pink-purple, the banner recurved through 40°–45°; calyx 3.4–4.7 mm long, loosely strigulose, the tube 2–2.8 mm long, the teeth (1.2) 1.5–2.2 (2.5) mm long. **PODS** ascending, spreading, or declined, rather hamately curved, the body linear or linear-oblanceolate in profile, (12) 14–20 mm long, 2.1–3.3 mm thick, triquetrous, glabrous, strigulose, or villosulous, bilocular; ovules 12–18. [substitute name for *A. micranthus* Nutt.]. Mixed salt desert and desert shrub, sagebrush, and piñon-juniper communities. ARIZ: Apa, Coc; COLO: Arc, LPl, Mon; NMEX: McK, SJn; UTAH. (1070) 1160–2025 m (3515′; 3800–6650′). Flowering: Apr–Jun. In the south and southeast portions of the Colorado Basin, from Grand Canyon, Arizona (also Apache County), San Juan County, Utah, and southwest Colorado, to the Rio Grande Valley, north-central New Mexico, and to the Arkansas River, southeast Colorado. San Juan River drainage material belongs to **var. *micranthiformis*** Barneby (small-flowered form).

Astragalus oocalycis M. E. Jones (with egg-shaped calyx) Arboles milkvetch. Coarse, caulescent perennial, (15) 25–40 cm tall, from a branching caudex. **PUBESCENCE** strigulose, basifixed. **STEMS** erect or ascending, several to numerous, forming bushy clumps. **LEAVES** 5–15 (17) cm long; leaflets (9) 19–27, 3–35 mm long, linear- or linear-oblong, acute, or obtuse and mucronate, glabrous or glabrate above; stipules 2.5–6 mm long, at least the lowermost connate-sheathing.

INFL a raceme, 35- to 60-flowered, the flowers nodding at anthesis, the axis 4–8 (11) cm long in fruit; bracts 3–8 mm long; pedicels 2–5 mm long; bracteoles 0–2; peduncle 9–17 mm long. **FLOWERS** 14–17 mm long, ochroleucous, the banner recurved about 45°; calyx 10–11 mm long (accrescent to 14 mm long and 11 mm wide in fruit), finally ovoid and contracted apically, hirsute with long straight hairs, the teeth 2–3 mm long, subulate. **PODS** sessile on a stipelike gynophore about 1 mm long, included within the calyx, the body of the pod 6–7 mm long, about 3.5 mm thick, strongly obcompressed, bisulcate, the valves glabrous, unilocular. $2n = 24$. [*Diholcos oocalycis* (M. E. Jones) Rydb.; *Cnemidophacos urceolatus* Rydb.]. Knolls, hillsides, and plains in sagebrush and piñon-juniper communities; usually in seleniferous soils of the San Jose Formation. COLO: Arc, LPl; NMEX: RAr, SJn. 1925–2070 m (6320–6800′). Flowering: May–Jul. A Four Corners endemic. The Arboles milkvetch is one of the most distinctive species of North American *Astragalus*. Its most striking characteristic is its greatly inflated calyx, densely hairy with cream-colored hairs. The numerous flowers are sessile or subsessile, the inflorescence becoming thereby a massive, cylindroid, spiciform head. The greatly inflated, accrescent calyx enclosing the small, few-seeded pod is unmatched in the genus in North America.

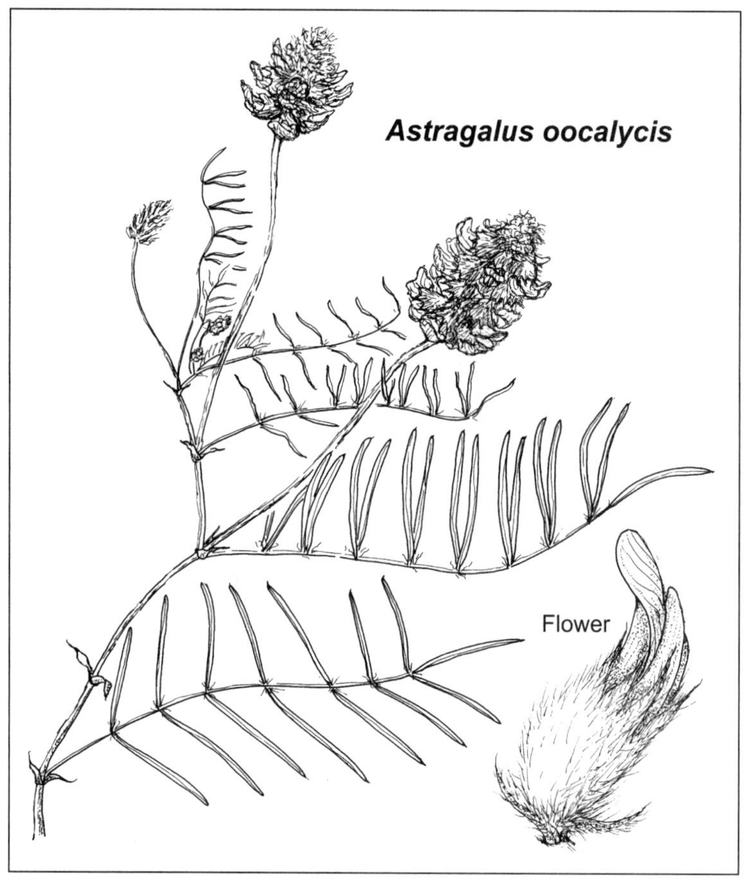

Astraglus oophorus S. Watson (egg-bearing, for the large pods) Egg milkvetch. Perennial, caulescent, 15–30 cm tall, from a caudex; pubescence basifixed. **STEMS** decumbent to ascending, radiating from the caudex; stipules 1.5–7 mm long, all distinct. **LEAVES** 3–21 cm long; leaflets 9–25, 3–20 mm long, 2–11 mm wide, oval to obovate or orbicular, obtuse to retuse or mucronate, glabrous on both surfaces, often ciliate. **INFL** a raceme, 4- to 13-flowered, the flowers spreading at anthesis, the axis 1–8 cm long in fruit; bracts 1.5–5 mm long; pedicels 2–6 mm long; bracteoles 0–1; peduncles 4–13 cm long. **FLOWERS** 17–24 mm long; calyx 6–12 mm long, tube 4–8.5 mm long, cylindric or short-cylindric, glabrous or sparingly strigose, the teeth 2–5 mm long, subulate; petals ochroleucous or concolorous (ours). **PODS** spreading to pendulous, stipitate, the stipe (gynophore) 3.5–12 mm long, the body bladdery-inflated, ellipsoid, 25–55 mm long, 10–30 mm wide (when pressed), glabrous, unilocular, often mottled; ovules 28–54. $2n = 24$. Sagebrush, piñon-juniper, and mountain-brush communities. COLO: Arc, Mon. 1830–2110 m (6000–7000′). Flowering: May–Jun. Fruit: Jun–Jul. Our material belongs to **var. *caulescens*** (M. E. Jones) M. E. Jones (not fully with stem). [*A. megacarpus* (Nutt.) A. Gray var. *caulescens* M. E. Jones; *A. artipes* A. Gray; *Phaca artipes* (A. Gray) Rydb.; *A. oophorus* var. *artipes* (A. Gray) M. E. Jones]. Colorado, Arizona, and Nevada. The bladdery-inflated pods of *A. oophorus* are diagnostic.

Astragalus pattersonii A. Gray (for H. N. Patterson of Oquawka, Illinois, newspaper printer and plant collector) Patterson's milkvetch. Perennial, caulescent, 20–45 (50) cm tall, from a branching caudex. **STEMS** decumbent to ascending or erect. **LEAVES** 5–13 cm long; leaflets 7–15 or more, 6–38 mm long, 3–16 mm wide, elliptic to lanceolate, oblanceolate, or obovate, obtuse to acute, retuse, or mucronate, strigose to glabrous on both sides; stipules 3–8 mm long, all usually distinct. **INFL** a raceme, 6- to 24-flowered, the flowers declined-nodding at anthesis, the axis 2–15 cm long in fruit; bracts 2–8 mm long; pedicels 1–4.5 mm long; bracteoles 2; peduncle 3–18 cm long. **FLOWERS** 14–22 mm long, white, concolorous or the keel tip faintly purplish; calyx 8.8–14.2 mm long, the tube 6–8.8 mm long, cylindric, gibbous, pale tan or whitish, colored like the petals, thinly strigulose, the teeth 2.3–6.5 mm long, subulate. **PODS** erect, sessile, cylindric to ellipsoid or ovoid, 17–35 mm long, 6–10 mm thick, glabrous or puberulent, unilocular; ovules 22–38. $2n = 24$. [*Rydbergiella pattersonii* (A. Gray) Fedde; *Jonesiella pattersonii* (A. Gray) Rydb.]. Piñon-juniper and mixed desert shrub communities, often on fine-textured, seleniferous substrates. COLO: Arc, LPl, Mon; NMEX: SJn; UTAH. 1260–2285 m (4130–7500′). Flowering: May–Jul. Colorado; Uinta Mountains, Utah; and Coconino County, Arizona. Plant used for ear and eye medicine, sore throats, and as an emetic by Navajos.

Astragalus praelongus E. Sheld. (very tall) Stinking milkvetch. Perennial, caulescent, 1–9 dm tall, from a branching superficial caudex. **PUBESCENCE** basifixed. **STEMS** erect or ascending, forming clumps. **LEAVES** 3–22 cm long; leaflets 7–33, 3–50 mm long, 2–24 mm wide, obovate, elliptic, oblong, lanceolate, or oblanceolate, obtuse or retuse to acute, sparingly strigose beneath, glabrous above; stipules 2.5–9 mm long, all distinct. **INFL** a raceme, 10- to 33-flowered, the flowers deflexed at anthesis, the axis 3–16 cm long in fruit; bracts 1–7 mm long; pedicels 1–7 mm long; bracteoles 2; peduncle 4–26 cm long. **FLOWERS** 15–24 mm long, ochroleucous, the keel often faintly purplish-tipped; calyx 5.8–14 mm long, the tube 4.4–7.5 mm long, cylindric, gibbous, glabrous, or thinly strigose, green or yellowish, usually differently colored than the petals, the teeth 0.3–6 mm long, subulate. **PODS** erect to declined, sessile, subsessile, or stipitate, inflated, ellipsoid, ovoid, obovoid, or subglobose, 18–42 mm long, 5–25 mm thick, usually straight, glabrous or puberulent, subunilocular, leathery-woody; ovules 40–75. $n = 12$. [*A. procerus* A. Gray, not Boiss. & Hausskn. 1872; *A. pattersonii* A. Gray var. *praelongus* (E. Sheld.) M. E. Jones; *Phacopsis praelongus* (E. Sheld.) Rydb.; *Rydbergiella praelonga* (E. Sheld.) Fedde & P. Syd.; *Jonesiella praelonga* (E. Sheld.) Rydb.]. Leaves used as a ceremonial emetic by Ramah Navajos.

1. Pods long-stipitate, the stipe 4–8 mm long. .... ***var. lonchopus***
1′ Pods sessile or short-stipitate, the stipe, when present, less than 3 mm long .... (2)

2. Pods narrowly elliptic to oblong, 6–10 mm thick; plants of Rio Arriba and McKinley counties, New Mexico .... ***var. ellisiae***
2′ Pods broadly oblong to elliptic, 10–15 mm thick; plants of Apache County, Arizona, and San Juan County, New Mexico .... ***var. praelongus***

var. *ellisiae* (Rydb.) Barneby (for Charlotte Cortland Ellis, nurse and plant collector) Ellis' milkvetch. [*Jonesiella ellisiae* Rydb.]. Clay soil, commonly on Mancos Shale, Moenkopi, and Chinle formations, but also on alluvial substrates containing selenium, in warm and salt desert shrub and piñon-juniper communities. NMEX: McK, RAr, SJn. 1945–2165 m (6380–7100′). Flowering: Apr–Jul. Relatively widespread from southern Utah, western Colorado, and in New Mexico and adjacent Texas. The Ellis' stinking milkvetch passes by degree to var. *praelongus* especially. Flowering specimens are difficult, if not impossible, to assign to a given variety. The type was collected in the Sandia Mountains, near Tijeras, Bernalillo County, New Mexico, on 30 Apr 1914, by Charlotte Cortland Ellis. Some specimens from adjacent Torrence County (*K. Heil 7585, 7626* SJC!) have very slender pods, tapering to the base (not truly stipitate) and a sulcate dorsal suture. Such specimens were noted briefly by Barneby (1964a: 588), who evidently failed to assign significant taxonomic value.

var. *lonchopus* Barneby (with long stipe) Longstipe milkvetch. Seleniferous fine-textured substrates in blackbrush, mixed desert shrub, and piñon-juniper communities. ARIZ: Apa, Nav; COLO: Mon; NMEX, McK, SJn; UTAH. 1160–2380 m (3800–7805′). Flowering: late Apr–Jul. A Colorado Plateau endemic. Canyons of the Colorado and San Juan rivers.

var. *praelongus* Clay and silt of the Mancos Shale, Moenkopi, and Chinle formations, and other seleniferous soils, in salt desert shrub and piñon-juniper communities. ARIZ: Apa; COLO: Mon; NMEX: McK, RAr, SJn; UTAH. 1630–2455 m (5350–8050′). Flowering: Apr–Jul. New Mexico, Arizona, southwest Colorado, southern Utah, and adjacent Nevada. The type variety of stinking milkvetch demonstrates great variation.

Astragalus preussii A. Gray **CP** (for Charles Preuss, artist with Captain Frémont) Preuss' milkvetch. Perennial or annual, caulescent, mostly 12–45 (50) cm tall, from a woody superficial caudex. **PUBESCENCE** basifixed; stems erect or ascending, forming clumps. **LEAVES** 3.5–13 cm long; leaflets 7–25, 6–28 mm long, 1–6 mm wide, obovate or obcordate to oblong, narrowly elliptic, lanceolate, or linear, emarginate to rounded, obtuse, or acute, glabrous; stipules 2–7 mm long, all distinct. **INFL** a raceme, 3- to 22-flowered, the flowers ascending, the axis 1–20 cm long in fruit; bracts 1.5–4 mm long; pedicels 1–5.5 mm long; bracteoles 2; peduncle 2–15 cm long. **FLOWERS** 14–24 mm long, pink-purple, bicolored, or white; calyx 6.4–12.3 mm long, the tube 5.1–9.7 mm long, cylindric, thinly strigose, purple, the teeth 1.3–2.6 mm long, subulate. **PODS** erect to ascending, stipitate or subsessile, the stipe 2–7 mm long, the body oblong-ellipsoid, inflated, 12–34 mm long, 6–13 mm thick, glabrous or puberulent, stiffly papery to leathery, unilocular; ovules 20–44. $2n = 24$. Clay and silt of the Mancos Shale, Moenkopi, and Chinle formations, and other seleniferous soils, in salt desert shrub and piñon-juniper communities. ARIZ: Apa; NMEX: McK, SJn; UTAH. 1790–2085 m (5840–6850′).

Flowering: Mar–Jun. Northeast Arizona, northwest New Mexico, and southern Utah. Four Corners material belongs to **var. *latus*** M. E. Jones (broad, referring to the pod) Shale milkvetch. [*A. preussii* var. *sulcatus* M. E. Jones].

Astragalus proximus (Rydb.) Wooton & Standl. (referring to the close relationship with *A. flexuosus*) Aztec milkvetch. Slender, caulescent perennial, 1.5–4.5 dm tall, from a branching, buried caudex. **PUBESCENCE** thinly strigulose, basifixed. **STEMS** erect or ascending, subterranean for a space of 1–8 cm, often branched below. **LEAVES** 2–8 cm long, all shortly petioled; leaflets 7–11, 6–22 mm long, linear, linear-oblanceolate, or filiform to narrowly oblong, obtuse, strigulose; stipules 1.5–4.5 mm long, at least the lowermost connate-sheathing. **INFL** a raceme, loosely (7) 12- to 40-flowered, the flowers spreading at anthesis, the axis 4–17 cm long in fruit; pedicels 0.5–1.8 mm long; bracteoles 0; peduncle 3–11 cm long. **FLOWERS** 6.1–7 mm long, whitish, the banner and keel tip lilac-tinged, the banner recurved through 45°–50°; calyx 2.5–3.5 mm long, the tube 1.8–2.5 mm long, campanulate, the teeth 0.6–1.5 mm long, subulate or some deltoid. **PODS** pendulous, stipitate, the stipe (1) 1.2–2 mm long, the body linear-ellipsoid, 10–15 mm long, 2.3–3.2 mm thick, obcompressed, the valves glabrous, unilocular; ovules 6–10. $n = 11$. [*Homalobus proximus* Rydb.; *Pisophaca proxima* (Rydb.) Rydb.]. Sandy, often saline substrates derived from sandstone, among junipers or sagebrush. COLO: Arc, LPl; NMEX: McK, RAr, SJn; UTAH. 1650–2255 m (5410–7400′). Flowering: late Apr–Jul. Northwest New Mexico and adjacent Colorado. The Aztec milkvetch is a close match for the partially sympatric *A. wingatanus* in general aspect and flower and pod size. The pods are, however, glabrous, obcompressed, and definitely stipitate. In *A. wingatanus* the pods are strigulose, laterally compressed, and mainly sessile (or only short-stipitate). They form a juxtaposed species pair.

Astragalus ripleyi Barneby (for Harry Dwight Dillon Ripley) Ripley's milkvetch. Robust, caulescent perennial, 40–70 cm tall, from a branching subterranean caudex. **PUBESCENCE** strigulose, basifixed. **STEMS** erect or ascending, buried for a space of 2–10 cm. **LEAVES** 4–9 (11) cm long; leaflets 13–17 (19), linear or linear-elliptic to subfiliform, obtuse to acutish or truncate, the terminal one nearly always continuous with the rachis; stipules 1–5 mm long, all distinct. **INFL** a raceme, loosely (5) 15–45-flowered, the flowers declined and secund at anthesis, the axis 2–16 cm long in fruit; bracts 1–2 mm long; pedicels 2.3–7 mm long; bracteoles 0; peduncle (3.5) 6–12 (15) cm long. **FLOWERS** 13–17 mm long, pale lemon-yellow, concolorous, the banner recurved through about 35°; calyx 5.5–7 mm long, the tube 5–6.6 mm long, cylindric, strigose, purplish or whitish, the teeth 0.5–1.1 mm long, triangular-subulate. **PODS** pendulous, stipitate, the stipe 8–15 mm long, the body linear-oblong to lanceolate or narrowly elliptic in outline, (14) 20–33 mm long, (3.5) 4–6 mm wide, strongly laterally flattened, straight or only slightly curved, strigulose, subdiaphanous at maturity, unilocular; ovules 11–17. Sagebrush, rabbitbrush, piñon-juniper, ponderosa pine, Douglas-fir, and aspen communities. COLO: Arc. 1980 m (6500′). Flowering: Jun–Jul. Known from one location about 20 miles south of Pagosa Springs. The terminal leaflet in most leaves is continuous with the rachis, and the leaflets are glabrous above. These features, along with the laterally compressed pods and elongate fruiting pedicels, easily distinguish Ripley's milkvetch from the disjunct *A. schmolliae*. The smaller, lemon-yellow flowers and pod differences easily differentiate Ripley's milkvetch from the coarser, larger-flowered *A. lonchocarpus*.

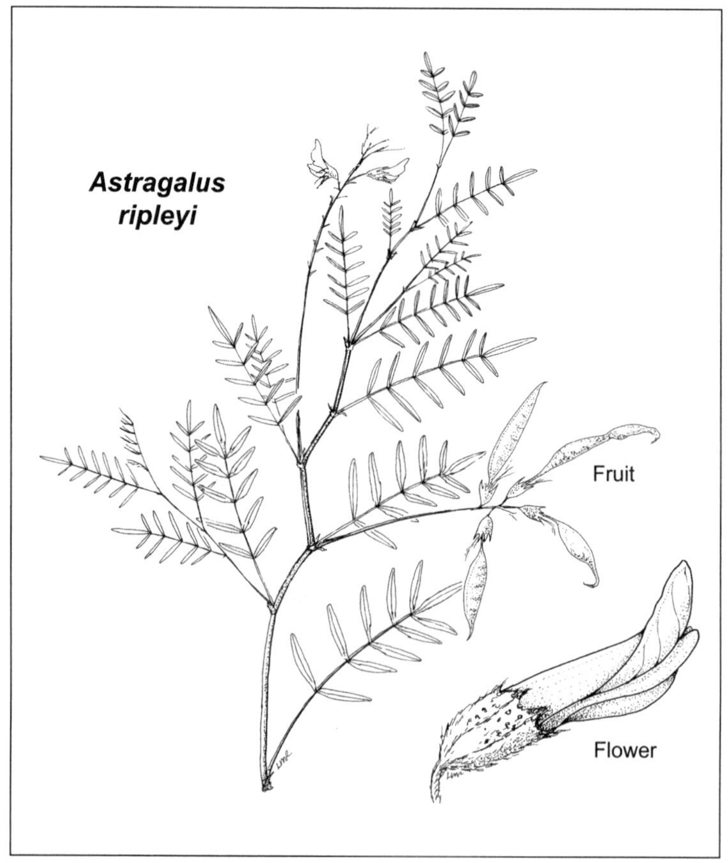

Astragalus robbinsii (Oakes) A. Gray (for James Watson Robbins) Robbins' milkvetch. Slender to robust caulescent perennial, 10–40 (60) cm long, from a superficial or barely subterranean caudex. **PUBESCENCE** basifixed. **STEMS** (1)

1.5–4 (6) dm long, usually ascending, subterranean for 0–2 cm. **LEAVES** 3–12 cm long; leaflets (7) 9–13 (15), (5) 7–25 (32) mm long, strigulose beneath with subappressed or narrowly ascending-incumbent hairs to 0.2–0.6 mm long, or glabrate throughout; stipules 1.5–6 mm long, at least some of the lower ones connate-sheathing. **INFL** a raceme, (5) 7–20 (33)-flowered, the axis (2) 3–18 (20) cm long in fruit; peduncle (3.5) 4.5–16 (23) cm long. **FLOWERS** 5.2–8 mm long, pink-purple or less commonly ochroleucous and tinged with purple, the banner recurved 50°–70°. **POD** stipe (1) 1.5–5 (6.5) mm long, the body ellipsoid, (1) 1.3–2.5 cm long, 3.5–5.5 mm thick, obtusely trigonous, the valves loosely strigulose-pilosulous, the septum 0.2–1 (1.5) mm wide; ovules (6) 7–10. Our material belongs to **var. *minor*** (Hook.) Barneby (small) Small elegant milkvetch. [*Phaca elegans* Hook. var. *minor* Hook.; *A. minor* (Hook.) M. E. Jones; *A. blakei* Eggl.; *A. robbinsii* var. *borealis* Eggl.; *Atelophragma blakei* (Eggl.) Rydb.; *A. robbinsii* var. *blakei* (Eggl.) Barneby ex Gleason; *A. macounii* Rydb.; *Atelophragma macounii* (Rydb.) Rydb.; *Atelophragma collieri* Rydb.; *A. collieri* (Rydb.) A. E. Porsild]. Stream banks, meadows, thickets, and moraines in humus or alluvial soils. COLO: SJn. 2895–3290 m (9500–10800′). Flowering: Jun–Aug. Disjunct from east-central (and disjunctly southeast) Alaska, south-central Yukon, and Northwest Territories, south to northern (Tetsa River Valley) and southern British Columbia, Alberta, eastern Washington, Montana, high mountains of central Idaho, northwest Wyoming, Utah (Summit County), and high mountains of central and southwest Colorado, and east to northern and central Vermont (Mount Mansfield; Willoughby Mountain; Pond Mountain, Wells, Rutland County), and St. John River, northern Maine, and in Cumberland County, Nova Scotia. The materials forming this variety are highly variable not only throughout but on an individual population basis. Variation involves great diversity in stature, in length of peduncle, fruiting raceme, and pods, and in number of flowers. The extremes serve to entice the taxonomist into recognition of subordinate taxa within this variety with its enormous geographical range. Such segregation, even though some of the populations are widely disjunct and more or less distinctive, appears to be unnecessary.

Astragalus sabulonum A. Gray (of coarse sands and gravels) Gravel milkvetch. Annual, winter-annual, or biennial, caulescent, 4–35 (42) cm tall or long, radiating from a root crown. **PUBESCENCE** basifixed, villous-hirsutulous. **STEMS** decumbent to ascending, rarely erect. **LEAVES** 1.5–7 cm long; leaflets (5) 9–15, 2–13 mm long, 1–5 mm wide, oblanceolate to oblong or obovate, retuse to truncate or obtuse, loosely villous on both surfaces or glabrate above; stipules 1–4 mm long, all distinct. **INFL** a raceme, 2–7-flowered, the flowers ascending-spreading at anthesis, the axis 0.3–2.5 cm long in fruit; bracts 1–2.5 mm long; pedicels 0.8–2 mm long; bracteoles 0; peduncle 0.5–4 cm long. **FLOWERS** 6.2–8 mm long, pink-purple or less commonly ochroleucous and tinged with purple; calyx 3.3–6.2 mm long, the tube 1.8–2.5 mm long, campanulate, hirsutulous, the teeth 1.8–3.5 mm long, subulate. **PODS** spreading-declined, sessile, obliquely ovoid, inflated, 9–17 (20) mm long, (4) 5–8 (11) mm thick, curved, thinly fleshy, green or purple-checked or -dotted, white-hirsutulous, unilocular; ovules 10–19. $2n = 24$. [*Phaca sabulonum* (A. Gray) Rydb.; *A. virgineus* E. Sheld.; *Phaca arenicola* Rydb.; *Phaca lerdoensis* Rydb.]. Warm desert shrub, mixed cool desert shrub, salt desert shrub, and lower piñon-juniper communities; often in sandy substrates. ARIZ: Apa, Nav; NMEX: McK, SJn; UTAH. 1095–1870 m (3600–6140′). Flowering: Feb–Jul. In the southeast counties of Utah (Emery County southward); northwest New Mexico; across northern Arizona; Clark, Esmeralda, Lincoln, Mineral, and Nye counties, Nevada; Riverside and Imperial counties, California; and Sonora. Though sometimes mistaken, because of the hirsutulous pod and occasional low growth, for *A. desperatus*, it is easily distinguished by being caulescent.

Astragalus schmolliae Ced. Porter (for Hazel Marguerite Schmoll, who botanized in southwest Colorado) Schmoll's milkvetch. Robust, caulescent perennial, 30–70 cm tall, from a branching, shallowly subterranean caudex. **PUBESCENCE** strigulose, basifixed. **STEMS** erect or ascending, buried for a space of 2–10 cm. **LEAVES** 4–10 (11) cm long; leaflets (7) 11–21, linear or linear-oblong to -elliptic, obtuse to retuse, pubescent on both surfaces, all articulate; stipules 2–7 mm long, all distinct. **INFL** a raceme, loosely (7) 10- to 28-flowered, the flowers declined and secund at anthesis, the axis (2.5) 4.5–20 cm long in fruit; bracts 1.5–3 mm long; pedicels 1–2.5 mm long; bracteoles 0–2; peduncle 9–21 cm long. **FLOWERS** 14.5–18 mm long, ochroleucous, concolorous, the banner recurved through about 45°; calyx 6–7.5 mm long, the tube (5) 5.5–6 mm long, cylindric, strigose, the teeth 1–1.7 mm long, triangular-subulate. **PODS** pendulous, stipitate, the stipe 5–10 (12) mm long, the body linear-oblanceolate in outline, straight or evenly to hamately curved, (25) 30–40 mm long, 3.5–5 mm thick, dorsally compressed, triquetrous, strigulose, unilocular; ovules 18–20. [*A. platycarpus* (Rydb.) Barneby var. *montezumae* Barneby]. Piñon-juniper community, Cliff House Sandstone member of the Mesa Verde Group. COLO: Mon; on Chapin Mesa, Mesa Verde National Park. 1690–2145 m (5545–7045′). Flowering: May–Jun. A Four Corners endemic. Schmoll's milkvetch is locally common on Chapin Mesa in the near vicinity of Mesa Verde National Park headquarters and is easily seen when in full anthesis, growing among the low trees of the piñon-juniper forest.

Astragalus scopulorum Porter (of craggy rocks, symbolizing the Rocky Mountains) Rocky Mountain milkvetch. Moderate, caulescent perennial, 15–48 (65) cm tall, from a subterranean caudex. **PUBESCENCE** strigulose, basifixed. **STEMS** decumbent to ascending, several to many, radiating from the caudex, buried for a space of 3–13 cm. **LEAVES** 1.5–8.5 cm long; leaflets (13) 15–29 (35), 2–19 mm long, 1–8 mm wide, oblong to elliptic or oblanceolate, some narrowly so, acute to obtuse or mucronate, thinly strigose to glabrous beneath, glabrous above, thinly ciliate; stipules 3–9 mm long, at least the lowermost ones connate-sheathing. **INFL** a raceme, 4–22-flowered, the flowers declined to nodding at anthesis, the axis 1–7 cm long in fruit; bracts 1.5–7 mm long; pedicels 1–4 mm long; bracteoles 0–2; peduncle 2–14 cm long. **FLOWERS** 18–24 mm long, ochroleucous, concolorous or the keel faintly purplish, the banner incurved through 80°–95°; calyx 9–11.5 (14) mm long, the tube 6.5–8.5 mm long, subcylindric, strigulose, the teeth 1.5–4 (6) mm long, subulate. **PODS** pendulous, stipitate, the stipe 4–9 mm long, the body oblong, straight or curved, 18–35 mm long, 3–6.5 mm wide, triquetrous, glabrous, bilocular, the septum 2–4 mm wide; ovules 18–25. $2n = 22$. [*Tragacantha scopulorum* (Porter) Kuntze; *Tium scopulorum* (Porter) Rydb.; *A. subcompressus* A. Gray; *A. rasus* E. Sheld.; *Tium stenolobum* Rydb.]. Mountain brush, sagebrush, ponderosa pine, piñon-juniper, and aspen–white fir communities. ARIZ: Apa; COLO: Arc, Dol, LPl, Mon, SMg; NMEX: McK, RAr, San, SJn; UTAH. 1830–2525 m (6010–8290′). Flowering: May–Aug. Utah, Colorado, eastern Arizona, and New Mexico.

Astragalus sesquiflorus S. Watson (1–2-flowered) Sandstone milkvetch. Prostrate, caulescent perennial, forming mats or cushions 1–6 dm wide, radiating from a branching caudex, this more or less clothed with a thatch of old leaf bases. **PUBESCENCE** strigulose, malpighian, or merely subbasally attached. **STEMS** 5–28 (37) cm long or more, arising from a superficial root crown. **LEAVES** 1–4 (6) cm long; leaflets 7–13, 1.5–10 mm long, 0.6–2 mm wide, elliptic to obovate, acute to obtuse, strigose on both sides (commonly involute); stipules 1.5–4.5 mm long, all connate-sheathing. **INFL** a raceme, 1–4-flowered, the flowers ascending at anthesis, the axis very short in fruit; bracts 1.2–3 mm long; pedicels 0.7–4 mm long; bracteoles 0–2; peduncle 0.8–4.5 cm long, filiform. **FLOWERS** 6–8 mm long, pink-purple, the banner reflexed through about 80°; calyx 3.7–5.5 mm long, the tube 1.5–2.8 mm long, campanulate, strigulose, the teeth 1.9–3 mm long, subulate. **PODS** spreading-ascending, sessile or subsessile, the body obliquely oblong in outline, 8–10 mm long, 3–4 mm wide, trigonously compressed, strigulose; ovules 7–10. [*Tragacantha sesquiflora* (S. Watson) Kuntze; *Phaca sesquiflora* (S. Watson) Rydb.; *Batidophaca sesquiflora* (S. Watson) Rydb.; *A. sesquiflorus* var. *brevipes* Barneby]. Mixed desert shrub, piñon-juniper, and ponderosa pine or aspen to spruce-fir communities on sandstone or sandy substrates. ARIZ: Apa, Coc, Nav; UTAH. 1105–3140 m (3635–10310′). Flowering: May–Aug. South-central and southeast Utah, northern Arizona, and Montrose County, Colorado. The plant forms distinctive mats or powder-puff cushions on sand below the sandstone escarpments.

Astragalus tenellus Pursh (rather delicate) Pulse milkvetch. Perennial, caulescent, 10–70 (75) cm tall, from a branching, superficial caudex. **PUBESCENCE** basifixed. **STEMS** erect or ascending, or less commonly decumbent; forming clumps. **LEAVES** 2–9 cm long; leaflets 11–21, 3–24 mm long, 0.4–6 mm wide, narrowly oblong to elliptic, linear, oblanceolate, or obovate, acute to obtuse, mucronate, or emarginate, thinly strigose beneath, glabrous above; stipules 1.5–7 mm long, turning black in drying, at least some connate-sheathing. **INFL** a raceme, (1) 3–23-flowered, the flowers ascending at anthesis, the axis 0.5–11 cm long in fruit; bracts 0.5–2.7 mm long; pedicels 0.7–3.2 mm long; bracteoles 0–2; peduncle 0.2–4 cm long, often paired. **FLOWERS** 6–9 (11) mm long, white to ochroleucous, or less commonly pink-purple, the banner recurved through about 40°; calyx 2.6–5.2 mm long, the tube 2–2.7 mm long, campanulate, strigose, the teeth 0.7–2.5 mm long. **PODS** pendulous, stipitate, the stipe 0.6–5.5 mm long, the body elliptic to oblong in outline, straight or curved, 7–16 mm long, 2.5–4.5 mm wide, laterally flattened, glabrous or less commonly strigose, unilocular; ovules 3–9. $2n = 16, 24$. [*Ervum multiflorus* Pursh, technically superfluous; *Homalobus multiflorus* (Pursh) Nutt.; *A. multiflorus* (Pursh) A. Gray; *Tragacantha multiflora* (Pursh) Kuntze; *Homalobus tenellus* (Pursh) Britton; *Orobus dispar* Nutt.; *Physondra dispar* (Nutt.) Raf.; *Homalobus dispar* (Nutt.) Nutt.; *Phaca nigrescens* Hook.; *Homalobus nigrescens* (Hook.) Nutt. ex Torr. & A. Gray; *A. nigrescens* (Hook.) A. Gray; *Homalobus stipitatus* Rydb.; *Homalobus strigulosus* Rydb.; *A. tenellus* f. *strigulosus* (Rydb.) J. F. Macbr.; *A. tenellus* var. *strigulosus* (Rydb.) F. J. Herm.]. Prairies and plains, mountain brush, sagebrush, piñon-juniper, ponderosa pine, lodgepole pine, aspen-fir, and spruce-fir communities. COLO: Arc, Hin; NMEX: McK, RAr, SJn. 2285–2445 m (7500–8020′). Flowering: May–Aug. Southern Yukon, eastern to northern Manitoba, and south to Nevada, Utah, Nebraska, and Minnesota. Small, laterally compressed pods, peduncles often two per node, and stipules (and often the plants themselves) that blacken on drying are diagnostic for this species. Barneby (1964a) discusses the variation within the species at length, which mainly involves length of pod stipe, pubescence, and flower size and color. He cites five minor variants with at least some geographical and morphological integrity, each of which has received a formal name, mainly by Rydberg, as noted in the synonymy above. Perhaps some, possibly all, should receive formal recognition, but the problems of such recognition are still those outlined by Barneby. Perhaps the most distinctive are those populations with pink-purple flowers that occur in Colorado. These plants were discussed as minor variants by Barneby (1964a), i.e., ". . . , but flowers purplish to bright purple; stipe of pod 1 cm long or less, the body pubescent with black or mixed black and white hairs (*Homalobus clementis*). Local on both sides of the Gunnison-Arkansas Divide, Gunnison and Chaffee Counties, Colorado, south to the Sangre de Cristo Range and the upper Rio Grande in Mineral County." These plants can be known as *A. tenellus* var. *clementis* (Rydb.) J. F. Macbr. These are the plants from Mineral County previously identified as *A. robbinsii* A. Gray var. *minor* (Hook.) Barneby (*K & M Heil 7939* SJNM), which does occur, however, within the San Juan River drainage, in the high mountains of southwestern Colorado.

Astragalus tortipes

Astragalus tortipes J. L. Anderson & J. M. Porter (twisted, referring to the pedicels) Sleeping Ute milkvetch. Tall, robust, caulescent perennial, 30–80 cm tall, arising from a shallowly subterranean caudex. **PUBESCENCE** strigulose, basifixed. **STEMS** few, erect or ascending, buried for a space of 0.5–2.2 cm. **LEAVES** (6) 8–14 (18) cm long; leaflets 7–15, 10–43 mm long, 1–4 mm wide, linear, acute, pubescent on both surfaces; stipules 3–12 mm long, all distinct, the subterranean ones white and chartaceous. **INFL** a raceme, loosely 10- to 25 (30)-flowered, the flowers declined at anthesis, the axis 7–12 cm long in fruit; bracts 1.5–2 mm long; pedicels 1–5 mm long, at first with normal orientation, ultimately twisted through 180°; bracteoles 1–2; peduncle 8–15 cm long. **FLOWERS** (12) 14–18 mm long, lemon-yellow, concolorous, the banner recurved through 80°–110°; calyx (4) 7–9 mm long, the tube 3.5–7 mm long, campanulate, strigose, the teeth (0.5) 1.2–2.5 mm long, lance-subulate. **PODS** inverted due to flexion of the pedicel, ascending to erect, stipitate, the stipe 12–16 (19) mm long, the body oblong-elliptic in outline, 22–30 mm long, 6–9 mm wide, only moderately laterally compressed, bicarinate, glabrous, unilocular; ovules 17–27. Mixed salt desert shrub communities, on Mancos Shale knolls and ridges with pedimental gravels. COLO: Mon. 1710–1750 m (5610–5740′). Flowering: May–Jun. This is a most unusual member of section *Lonchocarpi*; its affinities, name, and description were supplied to its nominal authors by Rupert Barneby. Its erect, stipitate pods are borne on stiff pedicels that are contorted through almost or quite 180°, thus presenting the ventral suture in a dorsal position. It grows intermixed with yet another rarity, *A. cronquistii*, which is abundant at this location, growing as a spreading-prostrate plant on the semi-barrens of the Mancos Shale Formation.

Astragalus wingatanus S. Watson (of Fort Wingate) Fort Wingate milkvetch. Perennial, caulescent, 15–45 (60) cm tall, from a subterranean caudex. **PUBESCENCE** basifixed. **STEMS** buried for 1–8 (16) cm, spreading-ascending, forming diffuse clumps. **LEAVES** 1.5–6.5 cm long; leaflets 7–15 (17), 3–18 mm long, 0.4–3.6 mm wide, linear to narrowly oblong, elliptic, or oblanceolate, acute, obtuse, or retuse, strigose to glabrous beneath, glabrous above, often ciliate; stipules 1.5–5 mm long, at least some connate-sheathing. **INFL** a raceme, very loosely 7- to 35-flowered, the flowers ascending at anthesis, the axis 3–8 cm long in fruit; bracts 0.5–2 mm long; pedicels 0.8–3 mm long; bracteoles 0–2; peduncle 2–14 cm long. **FLOWERS** 5.5–8 mm long, pink-purple, the wing tips white or pale; calyx 2.5–3.7 mm long, the tube 1.5–2.6 mm long, campanulate, strigose, the teeth 0.4–1.4 mm long, triangular-subulate. **PODS** deflexed, subsessile or short-stipitate, the stipe to 1.7 mm long, the body elliptic to oblong in outline, straight or slightly curved, 9–15 mm long, 3–4.5 mm wide, compressed, glabrous, unilocular; ovules 4–8. n = 11. [*Homalobus wingatanus* (S. Watson) Rydb.; *A. dodgeanus* M. E. Jones; *A. wingatanus* var. *dodgeanus* (M. E. Jones) M. E. Jones]. Piñon-juniper, mixed desert shrub, salt desert shrub, and less commonly in mountain brush communities. ARIZ: Apa, Nav; COLO: Arc, Dol, LPl, Mon; NMEX: McK, RAr, SJn; UTAH. 1540–2225 m (5050–7300′). Flowering: May–Jul. Southeastern Utah and adjacent Colorado, southeast Arizona and northwest New Mexico. This species simulates *A. tenellus* (q.v.) in its small flowers and laterally flattened pods. The buried caudex and elongating fruiting racemes are diagnostic for *A. wingatanus* and indicate relationships elsewhere (i.e., with taxa such as *A. flexuosus* var. *diehlii*). Its nearest ally in the near vicinity is *A. cliffordii* (q.v.), which differs in being a tall, clump-forming, rushlike plant with filiform peduncles and leaflets greatly reduced.

Astragalus wootonii E. Sheld. (for Elmer Ottis Wooton, New Mexico botanist) Wooton's milkvetch. Winter-annual or biennial (short-lived perennial), caulescent, (10) 15–50 cm tall, from a taproot and superficial root crown. **PUBESCENCE** basifixed, strigulose. **STEMS** commonly several or numerous, decumbent to incurved-ascending or almost prostrate. **LEAVES** (2) 4–10 (12) cm long; leaflets (7) 11–23, 5–20 mm long, narrowly oblanceolate, linear-oblong, or oblong-obovate, retuse-truncate or obtuse, often callous-mucronulate, folded or rarely flat, pubescent on both sides or glabrous above; stipules submembranous, (1.5) 2.5–7 (10) mm long, all distinct. **INFL** a raceme, 2- to 10 (15)-flowered, the flowers ascending to spreading or declined, the axis (0.5) 1–4 (5) cm long in fruit; bracts 1–3.2 mm long; pedicels 1–3.5 mm long; bracteoles nearly always 2; peduncle (0.5) 1.5–5.5 (7) cm long, shorter than the leaf. **FLOWERS** 4.6–7.5 mm long, whitish, sometimes tinged with pink or lavender, or pale to vivid reddish lilac, the banner recurved through about 50°; calyx 4.3–6.4 mm long, strigulose or villosulous, the tube 2.1–2.9 (3.2) mm long, campanulate or turbinate-campanulate, the teeth 2–3.5 mm long, lance-subulate. **PODS** spreading or declined, commonly humistrate, sessile on the conical receptacle and readily disjointing, broadly and subsymmetrically or somewhat obliquely ovoid, ovoid-ellipsoid, ellipsoid, or subglobose, bladdery-inflated, (10) 15–37 (43) mm long, (8) 12–20 mm in diameter, green or purplish-tinged, rarely lightly mottled, thinly strigulose, subvillosulous, or glabrate, becoming papery, stramineous, lustrous, unilocular; ovules (10) 13–21. $n = 11$. [*Phaca wootonii* (E. Sheld.) Rydb.; *A. wootonii* var. *typicus* Barneby; *A. playanus* M. E. Jones; *A. allochrous* A. Gray var. *playanus* (M. E. Jones) Isely; *Phaca tracyi* Rydb.; *A. tracyi* (Rydb.) Cory; *A. triflorus* sensu A. Gray, non *A. triflorus* (DC.) A. Gray]. Plains, hillsides, and valley floors, or along highways. COLO: Arc, LPl; NMEX: McK, RAr, SJn. 1870–2255 m (6145–7400'). Flowering: Mar–Jul. Sagebrush and piñon-juniper communities. From the upper Pecos across northern Arizona south of the Colorado River to the extreme eastern Mojave Desert in eastern San Bernardino County, California, south and southeast to northeast Sonora, Durango, Chihuahua, and Trans-Pecos Texas.

Astragalus zionis M. E. Jones (of Zion National Park) Zion milkvetch. Perennial, subacaulescent or short-caulescent, 3–23 cm tall, from a branching subterranean caudex, this sometimes clothed with a persistent thatch of leaf bases. **PUBESCENCE** basifixed. **STEMS** buried for 2–15 cm or more, 0–11 cm long, prostrate to ascending, the internodes often concealed by stipules. **LEAVES** mostly 2–15 cm long; leaflets 13–25, 2–16 mm long, 1–6 mm wide, elliptic or ovate, acute or less commonly obtuse, silvery villous on both sides; stipules 1.5–5.5 mm long, all distinct or some shortly connate-sheathing. **INFL** a raceme, 1- to 11-flowered, the flowers ascending at anthesis, the axis 0.3–6 cm long in fruit; bracts 2–5 mm long; pedicels 1–3 mm long; bracteoles 0–2; peduncle 0.5–15 cm long. **FLOWERS** 18–26 mm long, pink-purple or sometimes pale; calyx 8.3–18 mm long, the tube 6.5–12.7 mm long, cylindric, villous, the teeth 1.5–5.7 mm long, subulate. **PODS** ascending, sessile, obliquely ovoid-oblong in outline, 15–28 mm long, 5.5–9 mm wide, usually curved, dorsiventrally compressed, strigose or villosulous, brightly mottled, usually unilocular; ovules 24–30. $2n = 22$. [*Xylophacos zionis* (M. E. Jones) Rydb.]. On sandstone and in sandy and gravelly soils in blackbrush, sagebrush, ephedra, other mixed desert shrub (sometimes salt desert shrub), mountain brush, ponderosa pine, and riparian communities. ARIZ: Apa, Nav; NMEX: SJn; UTAH. 1345–3185 m (4420–10445'). Flowering: Mar–May. Southern Utah and adjacent Arizona. A Mojave-San Juan regional endemic.

Caesalpinia L. Rush-pea

(for Andrea Cesalpino, pioneer botanical systematist) Unarmed shrubs or perennial herbs, glandular-punctate in part. **LEAVES** alternate, odd-bipinnate, petiolate, the pinnae bearing several leaflets; stipules small, persistent. **INFL** a raceme, with several to many flowers; the axis of the raceme densely glandular. **FLOWER** calyx 5-lobed; corolla irregular; petals 5, conspicuous, the uppermost not enclosing the others in bud; stamens 10, long-exserted, brightly colored and showy, distinct; pistils sessile. **PODS** flattened, dehiscent. $x = 12$. One hundred species, mostly tropical.

1. Leaves with 2 or 3 pairs of pinnae; calyx lobes 6–9 mm long; petals 6–9 mm long .... ***C. jamesii***
1' Leaves with 3–7 (9) pairs of pinnae; calyx lobes 8–10.5 mm long; petals 10–12 mm long .... ***C. repens***

Caesalpinia jamesii (Torr. & A. Gray) Fisher (for Edwin James, first botanical collector of the central Rockies) James' rush-pea. Glandular-punctate herbaceous perennial from subterranean caudex and stout taproot. **STEMS** 0.5–3 (5) dm tall, erect or ascending, generally clustered, subterranean for 3–5 cm or more. **LEAVES** about 3–5 cm long, on petioles 1.5–2.5 cm long; stipules subulate, to 3 mm long, soon deciduous; pinnae 2 or 3 pairs, and with a terminal pinna, these with 5–8 (10) pairs of ovate- to elliptic-oblong leaflets 3.5- (7) mm long, obviously black-punctate and puberulent or glabrate. **INFL** a raceme, shorter than or overtopping the leaves, arising opposite the leaves, with 5–15 spreading or declined flowers; pedicels 2–5 mm long. **FLOWER** petals 6–9 mm long, yellow, drying orange to orange-red, punctate, the banner spotted; calyx lobes 6–9 mm long; stamens surpassing the petals. **PODS** horizontal to declined and curved, finally dehiscent, the valves punctate, stellate-pubescent. [*Hoffmannseggia jamesii* Torr. & A. Gray]. Sand dune communities. ARIZ: Apa; NMEX: SJn. 1675–1980 m (5500–6500′). Flowering: Jun. Central and southern Texas to Arizona, north to southwest Nebraska, and eastern Colorado. Zunis gave infusions to sheep to make them prolific.

Caesalpinia repens Eastw. (creeping) Creeping rush-pea. Subacaulescent or shortly caulescent, 5–12.5 cm tall (above-ground), from a deeply subterranean caudex, the branches buried 3–15 cm, pale. **PUBESCENCE** basifixed. **LEAVES** 2.5–9.5 cm long; pinnae 3–7 (9); leaflets 4–14, 3–12 mm long, 1–6 mm wide, asymmetrically obovate-elliptic to oblong, crowded, entire, villosulous. **INFL** a raceme, 7- to 26-flowered, the flowers spreading at anthesis, the axis 3–8 cm long in fruit; bracts 3–7 mm long, caducous; bracteoles 0; pedicels 2–7 mm long; peduncle 1.2–6 cm long. **FLOWERS** opening flat or nearly so; petals yellow, 10–12 mm long, red-spotted near the base, the whole fading pink-orange; calyx 8–10.5 mm long, the tube 2–4.5 mm long, campanulate, retrorsely villosulous, the teeth 6–8.5 mm long, oblong-lanceolate, villosulous. **PODS** pendulous, oblong, 20–50 mm long, 10–20 mm wide, membranous, pilosulous. [*Hoffmannseggia repens* (Eastw.) Cockerell; *Moparia repens* (Eastw.) Britton & Rose]. Sandy deserts with ephedra, Indian ricegrass, and other arenophilous plants. UTAH. 1160–1680 m (3800–5510′). Flowering: early Apr–Jun. A Utah endemic.

Calliandra Benth. Stickpea, Fairy-duster

(beautiful stamens) Herbs (in ours), shrubs, or trees, unarmed. **LEAVES** bipinnate, with numeous leaflets. **INFL** borne in globose heads or capitate clusters. **FLOWER** calyx toothed or deeply cleft, 5-lobed; corolla small, obscured by the numerous, long-exserted, brightly colored staminal filaments. **PODS** straight, the margin thickened, elastically dehiscent. Two hundred species. Tropical America, Madagascar, India.

Calliandra humilis Benth. (low) Dwarf stickpea. Perennial herb, from a subterranean rootstalk and vertical caudex. **STEMS** solitary or clustered, asending, mainly 1–2 dm long. **PUBESCENCE** sparse or abundant, of flexuous hairs about 0.3–1 mm long. **LEAVES** 1.5–7 cm long, on petioles mainly 1–2.5 cm long; stipules 2–3 mm long; pinnae 1–5 (8) pairs; leaflets about 5–14 (20) pairs, imbricate, shortly oblong, oblique, the venation obscure. **INFL** axillary, sessile or in capitate, pedunculate clusters. **FLOWERS** (4) 6–8 (10), 1.8–2.5 cm in diameter, white to pinkish. **PODS** 3–5 cm long, 4–5 mm wide, oblong-oblanceolate, strongly margined, the beak apiculate. Slickrock with scattered, stunted ponderosa pine. NMEX: McK. 2285 m (7500′). Flowering: Jul–Aug. West Texas to Arizona. The lone specimen examined bears only 1–3 pairs of pinnae, each with 5–8 pairs of leaflets, and belongs to **var. *reticulata*** (A. Gray) L. D. Benson (network). [*C. reticulata* A. Gray]. Plant used as a "life medicine" by Navajos and as a powder (roots) for rashes by Zunis.

Cercis L. Redbud

(ancient Greek name for poplar or aspen) Small trees or shrubs. **LEAVES** alternate, simple, palmately veined, cordate-ovate or orbicular; stipules deciduous. **FLOWERS** clustered, appearing before the leaves from spurs on old branches or cauliflorous, pink, showy; calyx turbinate-campanulate, shallowly 5-lobed; corolla irregular, the keel larger than the banner; stamens 10, distinct; ovary subsessile. **PODS** short-stipitate, laterally flattened, the ventral suture somewhat winged, indehiscent or tardily dehiscent. $x = 7$. Six species in North America and one species in Europe.

Cercis occidentalis Torr. ex A. Gray (western) Western redbud. Plant commonly a shrub, mostly 1.5–3.5 m tall, rarely more. **LEAVES** cordate-reniform, cordate basally, rounded to emarginate apically, 2–7 cm long, commonly broader than long, glabrous or puberulent along vein axils beneath. **FLOWERS** appearing before the leaves, cauliflorous; pedicel 8–12 mm long; corolla pink to pink-purple, 12–15 mm long; keel petals 5.5–8 mm wide; calyx asymmetric, 3–4.5 mm long, 5.5–8 mm wide. **PODS** short-stipitate, 4–10 cm long, 13–20 mm wide, winged, the wings 1.5–2.5 mm wide, glabrous. Indigenous or rarely cultivated. In sandstone canyons and alcoves, often in hanging gardens. UTAH. 1125–1220 m (3700–4000′). Flowering: Mar–Apr. California, Nevada, and Arizona. Variation in *Cercis* has been summarized by Isely (1998). Our material was recognized by him as being distinctive as a population when compared to typical California specimens. Plants from Utah, Arizona, and Nevada clearly belong to **var. *orbiculata*** (Greene) Tidestr. (circular). [*C. orbiculata* Greene; *C. canadensis* L. var. *orbiculata* (Greene) Barneby]. The western redbud is placed within an expanded *C. canadensis* by Barneby but is maintained at varietal rank. The problem hinges on the distribution of typically eastern *canadensis* far to the west, in Texas especially. The two main taxonomic units are herein maintained as species rather than varieties. The type of *C. occidentalis* is either from south-central Texas (upper Guadalupe River, by

Lindheimer, s.n.), very close to the distribution of *C. canadensis*, which is regarded as *C. canadensis* var. *texensis* (S. Watson) M. Hopkins, or from California. If the former, it is far removed from the western materials. Thus, the nomenclatural question as to whether var. *orbiculata* should be maintained at specific or infraspecific rank remains unanswered. Barneby (in Cronquist et al. 1989) reports that the strong wood of this species was used by Indians in the Canyonlands region for bows.

Dalea L. Dalea

(for Samuel Dale, English physician and pharmacologist) Perennial herbs or subshrubs, unarmed, trailing or ascending to erect. **LEAVES** alternate; stipules linear to subulate; odd-pinnate with 3 or more leaflets, glandular-dotted. **INFL** in a dense spike; peduncle opposite the leaves. **FLOWER** calyx campanulate, 5- to 10-ribbed and 5-lobed; corolla papilionaceous, the banner petal attached near the rim of the floral cup, the outer 4 variously inserted or all near the rim of the staminal tube; stamens 5, monadelphous; petals white, yellowish, pink, or pink-purple to indigo. **PODS** 1- to 2-seeded, indehiscent, included in the calyx or slightly exceeding it. $x = 7$ or 8. Note: Included herein are the herbaceous species traditionally placed in *Petalostemon*, and excluded are those shrubby species now regarded as *Psorothamnus* (Wemple 1970). One hundred sixty species. Canada to Argentina, especially Mexico and the Andes in desert areas.

1. Plants annual ....(2)
1′ Plants perennial ....(3)

2. Leaves 2–9.5 cm long, short-petiolate or subsessile, the main cauline ones with (7) 21–35 (49) leaflets (those of branchlets mainly only 5–21) .... ***D. leporina***
2′ Leaves 1–2.5 cm long, with petioles 5–15 mm long, often deciduous at anthesis; leaflets 5–9..... ***D. polygonoides***

3. Stems prostrate; spike slender, commonly less than 8 mm thick; leaflets 5–17 ....(4)
3′ Stems decumbent to ascending or erect; spike thicker, commonly 10 mm or more; leaflets 7 (9) or fewer..........(5)

4. Leaflets (5) 7–9; bracts caducous; spike 8–11 mm thick (ignoring petals); calyx definitely glandular..... ***D. scariosa***
4′ Leaflets 5–15; bracts persistent; spike 7–8 mm thick (ignoring petals); calyx inconspicuously glandular...... ***D. lanata***

5. Herbage pilose, pilosulous, or hirsute with lustrous hairs .... ***D. flavescens***
5′ Herbage glabrous..........(6)

6. Petals pink to pink-purple; calyx densely silvery canescent, not long-ciliate.......... ***D. purpurea***
6′ Petals white (fading cream); calyx with teeth ciliate ....(7)

7. Calyx tube glabrous; flowers white, not especially compact in the spike.......... ***D. occidentalis***
7′ Calyx tube definitely pilose, the teeth long plumose-hairy; flowers white to pale pink, definitely compact in the spike.......... ***D. cylindriceps***

Dalea cylindriceps Barneby (inflorescence is cylinder-like) Andean prairie-clover. Perennial, erect or ascending from a root crown and stout taproot, glabrous to the inflorescence. **STEMS** 2–7 dm tall, few to several from the root crown, simple or branched above. **LEAVES** 1–6 cm long, the petiole 3–7 (14) mm long; leaflets 7–9, linear-lanceolate to elliptic-oblong or -lanceolate, 1.2–2.5 cm long, 1–4 mm wide. **INFL** a spike on a peduncle 1–7 cm long, cylindric, 2.5–10 (12) cm long, 9–12 mm thick, very compact, pubescent; bracts caducous, but held by appressed calyces, pilose. **FLOWERS** nonpapilionaceous, the petals 4.7–6.2 mm long, white or less commonly pale pink, the epistemonous petals arising from the apex of the staminal column, soon withering, not concealing the stamens; the calyx 3.4–4.2 mm long, the tube 1.9–2.3 mm long, pilose, the lobes deltoid to lance-caudate; stamens typically 5. [*Petalostemon macrostachyus* Torr.; *D. macrostachya* Moric. 1833]. Dune sands and drainages with sand sagebrush and Indian ricegrass. NMEX: McK, RAr, San, SJn. 1645–2040 m (5400–6700′). Flowering: Jun–Sep. Southern Arizona and east to Texas. The Four Corners specimens represent a significant range extension for this handsome species, with its rather inconspicuous white to pale pink or purplish flowers and definitely plumose-hairy calyx teeth. It is easily distinguished from *D. purpurea* by the typically more numerous leaflets (7–9, not 3–7), and by the characteristics stated above.

Dalea flavescens (S. Watson) S. L. Welsh ex Barneby (yellowish) Kanab prairie-clover. **STEMS** 20–52 cm tall, from a superficial to subterranean caudex, glabrous to strigulose or pilose. **LEAVES** 1.5–4.7 cm long; stipules 1–4 mm long, lance-subulate to linear, persistent; leaflets 3–7, 5–20 mm long, 1–9 mm wide, folded or flat, oblong to oblanceolate or elliptic to linear, lustrous-strigulose or -pilose on both sides, glandular beneath; terminal leaflet petiolulate on a continuation of the rachis or subsessile. **INFL** a spike, 1.5–9 (14) cm long, 10–18 mm wide (when pressed), the rachis spreading-hairy; bracts 4–8 mm long, lance-aristate to lance-subulate or lanceolate, villous to pilose and sometimes glandular; peduncle 1.5–20 cm long, glabrous to sparingly or densely pilose- or villous-hirsute. **FLOWERS** 6.2–11 mm long, the petals white, fading cream; calyx 4–7 mm long, the tube 10-ribbed, not translucent, the teeth 1.5–4 mm long; pistils 7.5–13 mm long, the style 6–9.5 mm long. **PODS** villous. [*Petalostemon flavescens* S. Watson; *Kuhnistera flavescens* (S. Watson) Kuntze]. Infusion of plant taken by Navajos when lightning strikes near a hogan.

1. Spikes rather lax, the longest 5–14 cm long; calyx teeth 2.7–4 mm long; pistils 11.5–13 mm long; styles 8.5–9.5 mm long; plants of central Glen Canyon .... **var. *epica***
1′ Spikes dense, the longest seldom more than 5 cm long; calyx teeth 1.5–2.5 mm long; pistils 7.5–10.3 mm long; styles 6–8 mm long; plants widespread .... **var. *flavescens***

var. *epica* (S. L. Welsh) S. L. Welsh & Chatterley (refers to the epic journey of the Bluff pioneers along the Hole-in-the-Rock trail) [*D. epica* S. L. Welsh]. Hole-in-the-Rock prairie-clover. Sandstone bedrock and sand in blackbrush and mixed desert shrub communities. UTAH. 1320–1410 m (4335–4635′). Flowering: Apr–Jun. Endemic to Utah. This phase of the species intergrades more or less completely with the typical materials and is possibly taxonomically inconsequential.

var. *flavescens* Grasslands, mixed desert shrub, blackbrush, and piñon-juniper communities, commonly in sandy soils. ARIZ: Nav; UTAH. 1220–1640 m (4000–5400′). A Navajo Basin endemic. Flowering: Apr–Jun.

Dalea lanata Spreng. (woolly) Woolly dalea. Perennial herbs from a woody root crown. **STEMS** prostrate, 15–60 cm long or more, from a subterranean to superficial caudex, pilosulous to glabrous. **LEAVES** 0.9–3 cm long; stipules 1–2 mm long, subulate, more or less persistent; leaflets 5–17, 1.5–10 mm long, 1–5.5 mm wide, obovate to cuneate, truncate to emarginate, commonly folded, pilosulous or glabrous. **INFL** a spike, 1.8–7.5 cm long, lax in middle to late anthesis and in fruit; bracts ovate-acuminate, pilosulous and with 1 or more large glands; peduncle 0.5–2.5 cm long. **FLOWERS** 6–7 mm long, the petals indigo to rose-pink; calyx 3.2–4.6 mm long, the tube 2.1–2.7 mm long, glabrous, the teeth 0.9–1.9 mm long, pilosulous dorsally and ciliate; pistils 5–6 mm long, the style 3.5–4.5 mm long. **PODS** villous to glabrous. $2n = 14$. [*Parosela lanata* (Spreng.) Britton]. Stabilized dunes and other sandy sites in salt desert shrub and blackbrush communities. ARIZ: Apa, Nav; NMEX: McK, SJn; UTAH. 1370–1910 m (4495–6260′). Flowering: Jun–Sep. Colorado and Kansas, south to Mexico. Four Corners material belongs to **var. *terminalis*** (M. E. Jones) Barneby (an end) [*D. terminalis* M. E. Jones]. This variety differs from var. *lanata* inter alia in the glabrous, shining, membranous calyx tube. Some plants from San Juan County, Utah (*B. F. Harrison 12194 et al.*, BRY!), are glabrous throughout but seem not to differ otherwise. Scraped roots eaten as candy by Hopis. Poultice used for centipede bites by Navajos.

Dalea leporina (Aiton) Bullock (spotted) Fox-tail prairie-clover. Slender, diminutive to rather robust annual, glabrous to the inflorescence. **STEMS** 1.5–12 (15) dm tall, simple or branched above. **LEAVES** 2–9.5 cm long, short-petiolate or subsessile, the main cauline ones with (8) 10–17 (24) pairs of leaflets (those of branchlets mainly only 4–10 pairs), these oblong-oblanceolate or obovate, emarginate or retuse, or less commonly obtuse, (2) 3–12 mm long, flat and relatively thin or loosely folded and thicker (in xeric habitats). **INFL** a spike, terminal to the main axis and all branches, the latter overtopping the former, the peduncles (1.5) 3–12 (15) cm long, the spike moderately dense, becoming cylindroid, 8–12 (15) mm thick (excluding petals and stamens); bracts caducous. **FLOWER** petals milky white to bluish purple or almost blue, in some races white, eglandular or the banner rarely gland-tipped, the epistemonous ones perched 1.2–2.8 mm below separation of the short filament tassel, readily deciduous; banner (3.4) 4.4–6 mm long; stamens 9 or 10, exserted. **PODS** 2.4–3 mm long, the style base lateral. [*Psoralea leporina* Aiton; *Dalea alopecuroides* Willd.]. Roadsides and other ruderal sites. COLO: Arc; NMEX: SJn. 1495–2135 m (4900–7000′). Flowering: Jul–Sep. Southern Wisconsin to North Dakota, south to Texas and Arizona.

Dalea occidentalis (Rydb.) Riley (western) Western prairie-clover. **STEMS** decumbent to erect, 4–9 dm tall, from a superficial caudex, glabrous. **LEAVES** 1.5–5.2 cm long; stipules 1–4.5 mm long, fragile, often coiled; leaflets 4–9, 5–27 mm long, 1–7 mm wide, oblanceolate to elliptic or oblong, truncate to emarginate, commonly folded, glabrous, glandular beneath. **INFL** a spike, 1–6.8 cm long, 8–12 mm thick, the rachis commonly glabrous; bracts lance-acuminate, caducous; peduncle 1.5–15 cm long, glabrous. **FLOWER** petals loosening with age; petals 6–7 mm, white; calyx strongly 10-ribbed, usually pubescent between the ribs, the tube 2.3–3 mm long, the lobes 1–1.3 mm long; pistils 9–11 mm long, the style 8–10 mm long. **PODS** glabrous or sparingly hairy apically. [*Kuhnistera occidentalis* A. Heller ex Britton & Kearney; *Petalostemon occidentalis* (A. Heller ex Britton & Kearney) Fernald; *Kuhnistera candida* (Willd.) Kuntze var. *occidentalis* Rydb.; *Petalostemon gracilis* Nutt. var. *oligophyllus* Torr.; *Kuhnistera oligophylla* (Torr.) A. Heller; *Petalostemon* *oligophyllum* (Torr.) Rydb.; *Petalostemon candidus* (Willd.) Michx. var. *oligophyllus* (Torr.) F. J. Herm.; *Dalea oligophylla* (Torr.) Shinners; *D. candida* Willd. var. *oligophylla* (Torr.) Shinners]. Sandy drainages and crevices in rimrock in mixed desert shrub, blackbrush, piñon-juniper, and hanging garden communities. ARIZ: Apa, Nav; COLO: Mon; NMEX: McK, RAr, San, SJn; UTAH. 1495–2345 m (4900–7695′). Flowering: Jun–Sep. Alberta to Saskatchewan, south to Arizona, New Mexico, Texas, Mexico, and Iowa. I agree with Isely (1998) that the epithet *occidentalis* has priority at specific rank for this widespread and distinctive species, which, though similar to *D. candida* Willd., is essentially allopatric and differs in several features, "and can be distinguished at a glance" (Isely l.c.).

Dalea polygonoides A. Gray (many angles) Six weeks prairie-clover. Often diminutive, erect or ascending, glabrous or inconspicuously glandular, mainly 0.5–3 dm tall, simple or with spreading to ascending branches. **LEAVES** 1–2.5 cm long, with petioles 5–15 mm long, often deciduous at anthesis; leaflets 5–9, linear to oblong or oblanceolate, 5–15 mm long, 1.5–2.5 mm wide. **INFL** a spike, 1–2.5 cm long, 6–8 mm thick, initially ovoid, finally cylindrical, foliar bracts persistent, pilosulous. **FLOWERS** subpapilionaceous, about 4–6 mm long; corolla circa 4–5 mm, lavender to pale pink-purple, the epistemonous petals arising medially from the staminal column; calyx tube 2–2.5 mm long, slit above, thinly pilose or glabrate, the rib intervals each usually with a single large gland, sometimes 2, the lobes subulate; stamens about 7–9, sometimes some of them nonfunctional. Ground layer in ponderosa pine woodland. NMEX: McK. 2285 m (7500′). Flowering: Aug–Sep. Arizona, New Mexico, west Texas south to Mexico.

Dalea purpurea Vent. (purple) Violet prairie-clover. Perennial, erect or spreading-decumbent from the root crown, villosulous to glabrate, glandular. **STEMS** 2–8 dm tall, clump-forming, very leafy, simple or branched above. **LEAVES** with petioles 1.5–2.5 (3.5) cm long; leaflets typically 5 (3–7), linear-oblanceolate to filiform, 7–24 (28) mm long, folded, tubular or involute. **INFL** a spike, very dense, becoming oblong-cylindroid, mainly 7–12 (13) mm thick (ignoring petals), the axis becoming (1) 1.5–7 cm long; bracts deciduous with the fruiting calyx; peduncle (0) 1–9 (15) cm long, the first spike typically pedunculate, the lateral ones (if any) typically subsessile. **FLOWER** corolla 6–7 mm long, purple to lavender (rarely white); calyx tube 2–3.5 mm long, conspicuously subappressed- or matted-villosulous, with plainly antrorse hairs, eglandular, the lobes deltoid to subulate, subequal, shorter than the tube; epistemonous petals arising from apex of the staminal column; stamens 5. [*Petalostemon purpureus* (Vent.) Rydb.]. With grass. NMEX: McK, San. 2040–2150 m (6700–7050′). Flowering: Jul–Sep. Widespread from Montana east to Minnesota, and south to Texas, Louisiana, Mississippi, and Alabama.

Dalea scariosa S. Watson (thin, dry, membranous texture, not green) La Jolla prairie-clover. Perennial herbs from a woody root crown. **STEMS** prostrate to decumbent-assurgent, 30–60 (80) cm long or more, from a subterranean to superficial caudex, glabrous, glandular. **LEAVES** 1–3 cm long; stipules 1–2 mm long, subulate, more or less persistent; leaflets (5) 7–9, 3–10 mm long, 1–5.5 mm wide, obovate to cuneate, truncate to emarginate, commonly folded, glabrous. **INFL** a spike, 1.5–7.5 (10) cm long, lax in middle to late anthesis and in fruit; bracts caducous; peduncle 0.5–2.5 cm long. **FLOWERS** 9–10 mm long, the petals rose-pink; calyx with tube 3–4 mm long, glabrous, with 2 or more glands between the intervals, the teeth 0.9–1.9 mm long, ciliate; pistils 4–6 mm long, the style less than 3 mm long; stamens 5. **PODS** glabrous. [*Petalostemon scariosus* (S. Watson) Wemple]. Sandy habitats with juniper and among shrubs. NMEX: San. 1635 m (5365′). Flowering: Aug–Sep. (Sandoval Co.; *B. Hutchins 6676*. SJNM); endemic to north-central New Mexico.

Gleditsia L. Locust

Trees, often armed with simple or branched, brown, stout thorns. **LEAVES** alternate, deciduous; stipules minute, caducous; pinnately once to twice compound (with both kinds of leaves often on the same branch, or some intermediate); with 3–5 pinnae, each with (8) 14–36 leaflets, or with 14–36 leaflets on once-pinnate ones. **INFL** spikelike axillary racemes. **FLOWERS** polygamous, almost regular; petals 3–5, very narrow, yellowish, the uppermost internal in bud; inconspicuous, in sepals equal or nearly so; stamens 3–10, distinct, the anthers in pistillate flowers abortive. **PODS** flattened, straplike, indehiscent. Fourteen species. Eastern North America, India, Japan to Philippines.

Gleditsia triacanthos L. (three thorns) Honey locust. Trees to 20 m tall or more; bark smooth. **LEAVES** compound; leaflets 12–35 mm long, the pinnae 3–5 (8) when bipinnate, the leaflets 14–36 on once pinnately compound leaves and on pinnae, 10–42 mm long, 3–14 mm wide, lanceolate to oblong, crenate, obtuse to cuspidate, glabrous above, puberulent along veins beneath. **INFL** a raceme, many-flowered, 3–7 cm long, short-pedunculate or subsessile. **FLOWER** petals 4–5 mm long, greenish; sepals separate. **PODS** sessile, oblong in outline, laterally flattened, 7–35 cm long, 15–30 mm wide, curved, indehiscent. **SEEDS** embedded in tissue. $2n = 28$. Cultivated shade tree, rarely escaping. ARIZ: Apa, Nav; COLO: Arc, LPl, Mon; NMEX: McK, SJn; UTAH. 1220–2135 m (4000–7000′). Flowering: May–Jun. Native to the eastern United States. Honey locust is a popular, very hardy shade tree, represented by several cultivars, including thornless and yellow-leaved varieties. Collections from along the San Juan River, remote from any settlement, appear to be established.

Glycyrrhiza L. Licorice

(sweet root) Perennial, caulescent, from stout, sweet roots. **LEAVES** alternate, odd-pinnate, glandular-punctate; stipules subulate, distinct. **INFL** in axillary racemes, each subtended by a lanceolate, deciduous bract; bracteoles 0. **FLOWERS** papilionaceous; calyx 5-toothed; petals 5, white to cream, the keel shorter than the wings; stamens 10, diadelphous; ovary enclosed in the staminal sheath, the style glabrous. **PODS** sessile, elliptic to oblong in outline, burlike, armed with uncinate appendages, or smooth, indehiscent. **SEEDS** few. $x = 8$. Twenty species. Mostly Eurasia with few in Australia, North America, and South America.

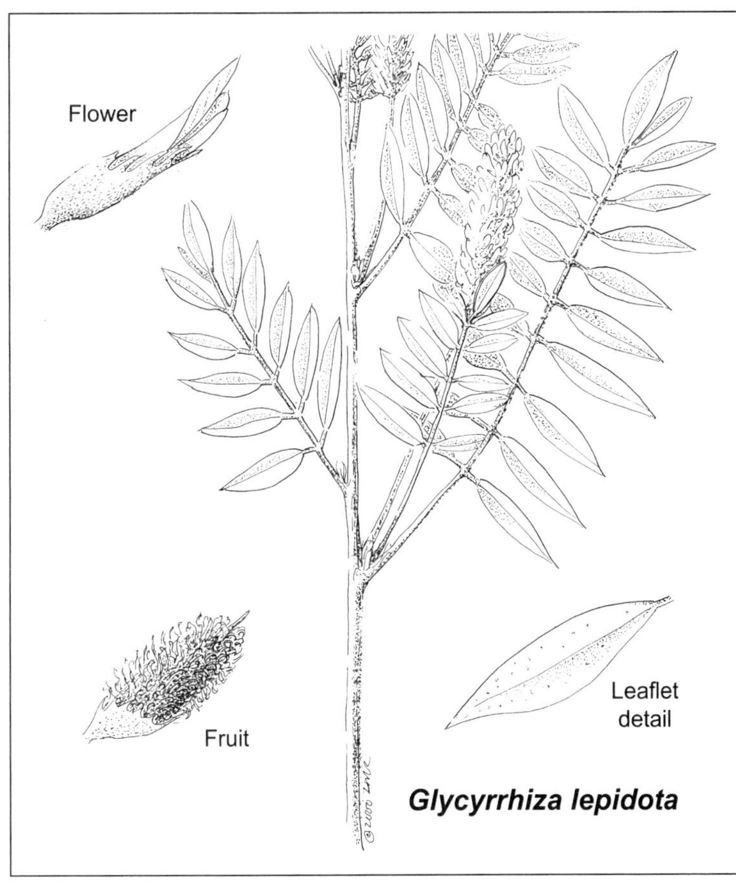

Glycyrrhiza lepidota

Glycyrrhiza lepidota Pursh (covered with small scales) Licorice. Plants 4–12 dm tall, from a deep-seated root; stipules 2–7 mm long, subulate. **LEAVES** 8–19 cm long; leaflets 13–19, 8–53 mm long, 3–15 mm wide, lanceolate to oblong, acute and mucronate, glabrous above, glandular-dotted and puberulent beneath. **INFL** a raceme, 20- to 50-flowered, the flowers ascending at anthesis, the rachis 2.5–9 cm long in fruit; bracts 5–8 mm long, caducous; peduncles often paired, 3–8 cm long. **FLOWERS** 9.1–13 mm long, white to cream; calyx 4.8–6.9 mm long, the tube 2.5–4.9 mm long, campanulate to short-cylindric, stipitate-glandular, the teeth 1.5–3.6 mm long. **PODS** spreading, laterally compressed, oblong, 13–20 mm long, the body 5–7 mm wide, beset with hooked prickles, simulating cockleburs. $2n = 16$. Terraces, streamsides, seeps, and other semimoist sites in greasewood, mixed desert shrub, piñon-juniper, and cottonwood-willow communities. ARIZ: Apa, Nav; COLO: Arc, LPl, Mon; NMEX: McK, RAr, San, SJn; UTAH. 1480–2265 m (4850–7425′). Flowering: May–Aug. Widespread in the United States, except for the southeast. Decoction of root used as an emetic by Navajos, and roots used as a breath freshener by Hopis.

Hedysarum L. Sweetvetch

(*hedys*, sweet, + *saron*, broom) Perennial herbs, caulescent, from a caudex and taproot. **LEAVES** alternate, odd-pinnate; stipules adnate to the petiole base, at least the lowermost connate-sheathing. **INFL** an axillary raceme, each subtended by a bract; bracteoles 2. **FLOWERS** papilionaceous; petals 5, red-purple to pink or pink-purple, the keel much longer than the wings, abruptly bow-shaped; calyx 5-toothed; stamens 10, diadelphous; ovary enclosed in the staminal sheath, the style glabrous. **LOMENTS** with 2–8 segments, prominently reticulate (Northstrom 1974; Northstrom and Welsh 1970; Rollins 1940; Welsh 1995a). One hundred species. North America, Eurasia, and Mediterranean.

1. Leaflets thick, the veins not apparent; fruit segments not winged; calyx lobes subequal, longer than the tube; plants of lower elevation; desert scrub to lower montane .... ***H. boreale***

1′ Leaflets thin, the veins readily apparent; fruit segments winged; calyx lobes unequal, shorter than the tube; plants of higher elevation; upper montane .... ***H. occidentale***

Hedysarum boreale Nutt. (northern) Northern sweetvetch. Perennial, caulescent, 17–70 cm tall, from branching, subterranean to superficial caudex. **PUBESCENCE** basifixed. **STEMS** decumbent to erect. **LEAVES** 3–12 cm long; stipules 2–10 mm long, at least some connate-sheathing; leaflets 5–15, 7–35 mm long, 2–19 mm wide, oblong to elliptic, lance-oblong, or ovate (rarely linear), strigose on both sides or glabrate to glabrous above. **INFL** a raceme, 5- to 45-flowered, the flowers ascending at anthesis, the axis 5–28.5 cm long in fruit; bracts 2–4 mm long; pedicels 0.8–4.5 mm long; bracteoles 2; peduncle 2.8–15 cm long. **FLOWERS** 10–19 mm long, red-purple to pink or pink-purple, less commonly white; calyx 4.5–8 mm long, the tube 2.5–3.5 mm long, campanulate, strigose, the teeth 2–6 mm long, subulate. **LOMENTS** stipitate, pendulous to spreading, with 2–8 segments, not winged, prominently reticulate. $2n = 16$. [*H. canescens* Nutt. ex Torr. & Gray, not L.; *H. cinerascens* Rydb.; *H. boreale* var. *cinerascens* (Rydb.) Rollins; *H. pabulare* A. Nelson; *H. mackenziei* Richardson var. *pabulare* (A. Nelson) Kearney & Peebles; *H. boreale* var. *pabulare* (A. Nelson) Dorn; *H. boreale* var. *obovatum* Rollins]. Mixed desert shrub, piñon-juniper, mountain brush, ponderosa pine, and aspen communities. ARIZ: Apa, Nav; COLO: Dol, LPl, Mon, SMg; NMEX: McK, San, SJn; UTAH. 1180–2845 m (3875–9335′). Flowering: May–Jul. Alberta, and east to Manitoba, and south to Nevada and Texas. Four Corners material belongs to **var. *boreale***. The use of leaf pubescence, or lack thereof, to segregate the var. *boreale* into further taxa is an exercise in frustration, leading to two essentially sympatric phases that might reflect ecology more than genetics. However, there are phases of the species with subtle differences that might be worthy of taxonomic consideration.

Hedysarum occidentale Greene (western) Western sweetvetch. Perennial, caulescent, 30–90 cm tall, from a branching, superficial caudex. **PUBESCENCE** basifixed. **STEMS** ascending to erect. **LEAVES** 8–20 cm long; stipules 10–17 mm long; leaflets 11–19, 9–37 mm long, 4–16 mm wide, ovate to lance-ovate or elliptic, apiculate to emarginate, strigose on both sides or glabrous above. **INFL** a raceme, 10- to 50-flowered, the flowers spreading to declined at anthesis, the axis 6–14 cm long in fruit; bracts 2–8 mm long; bracteoles 2; peduncle 3.7–15 cm long. **FLOWERS** 16–23 mm long, pink to red-purple; calyx 3.5–11 mm long, the tube 2.3–6 mm long, campanulate, glabrous to strigose, the teeth 0.5–2 mm long, triangular. **LOMENTS** stipitate, pendulous, with 1–5 segments, winged. [*H. lancifolium* Rydb.; *H. marginatum* Greene; *H. uintahense* A. Nelson]. Douglas-fir and spruce-fir communities. COLO: Arc, LPl, Min, Mon, SJn. 2955–3505 m (9700–11500′). Flowering: Jul–Aug. British Columbia to Washington, Idaho, Montana, Wyoming, Utah, and Colorado. Four Corners material belongs to **var. *occidentale***.

Lathyrus L. Sweetpea

(from the Greek *lathyros*, a sort of pulse) Annual or perennial herbs, clambering, trailing, or climbing. **LEAVES** alternate, even-pinnately compound, the rachis terminating in a bristle or prehensile tendril; stipules herbaceous, semihastate or semisagittate; leaflets 2–12, very variable. **INFL** an axillary raceme. **FLOWERS** papilionaceous; petals 5, white or cream to pink, purplish, or otherwise (in cultivated types), the wings not adnate to the keel, but fitted together in a groove; calyx 5-toothed, obliquely campanulate; stamens 10, diadelphous; style laterally compressed, bearded along the ventral (upper) edge. **PODS** oblong, several-seeded, the valves coiling upon dehiscence (Hitchcock 1952; Welsh 1965). One hundred fifty species. North-temperate; Europe, East Africa, and temperate South America.

1. Leaflets 2; stems winged; plants introduced .... ***L. latifolius***
1′ Leaflets 4 or more; stems angled but not winged; plants indigenous .... (2)

2. Keel conspicuously shorter than the wings; calyx glabrous or the teeth merely ciliate, the lower tooth usually longer than the tube; stipules large, foliaceous; petals pink-purple, rarely white .... ***L. pauciflorus***
2′ Keel commonly subequal to the wings; calyx often hairy, the lower tooth shorter than the tube; stipules not foliaceous; flowers pink-purple, pale lavender, pinkish violet, cream, or white .... (3)

3. Flowers 8–15 (22) mm long; petals pale lavender tinged to pinkish violet, cream, or white, often polychrome in populations; plants common at middle elevations, especially in aspen, flowering in summer .... ***L. lanszwertii***
3′ Flowers 15–30 mm long; petals bright pink to blue-purple; plants widespread at lower elevations, flowering mainly in springtime .... (4)

4. Plants villous; calyx 5–8 mm long, the tube 3.5–5.5 mm long; pods not stipitate .... ***L. brachycalyx***
4′ Plants glabrous or sparingly villous; calyx 8–12 mm long, the tube 5–7 mm long; pods stipitate, the stipe 4–6 mm long .... ***L. eucosmus***

Lathyrus brachycalyx Rydb. (with a short calyx) Rydberg's sweetpea. Perennial, clambering herbs from a subterranean caudex. **STEMS** buried for a space of 1–15 cm or more, decumbent to erect, 10–50 cm long, the herbage villous or sometimes glabrous above (exceptionally throughout). **LEAVES** 2–9 cm long (excluding tendrils); stipules 6–15 mm long, semisagittate; leaflets 6–12, 5–50 (70) mm long, 2–10 mm wide, linear to elliptic, oblong, lanceolate, or oblanceolate; tendrils simple or branched. **INFL** a raceme, 2- to 5-flowered, the flowers spreading at anthesis; peduncle 4–10 cm long. **FLOWERS** 15–25 mm long, pink to pink-purple (white); calyx tube 3.5–7 mm long, campanulate, the teeth 1.5–4 mm long, triangular to triangular-lanceolate. **PODS** 30–45 mm long, 5–8 mm wide. $2n = 14$. Sandy soils in piñon-juniper, mixed desert shrub, and riparian communities. ARIZ: Apa; COLO: LPl, Mon; NMEX: SJn; UTAH. 1250–2400 m (4100–7870′). Flowering: May–Jul. Southern and southeastern Utah, Colorado, and northwestern New Mexico. Four Corners material belongs to **var. *zionis*** (C. L. Hitchc.) S. L. Welsh (for Zion National Park) Zion sweetpea. [*L. zionis* C. L. Hitchc.]. Zion sweetpea forms extensive stands in sagebrush and piñon-juniper communities. It forms beautiful rounded clumps in full flower, often contrasting with the sandy habitats in which it grows.

Lathyrus eucosmus Butters & H. St. John (well-adorned) Seemly sweetpea. Plants 1.5–4 (6) dm tall. **LEAVES** 3–10 (13) cm long; stipules 8–20 mm long, semisagittate; leaflets (2–4) 6–10, mainly 3–6.5 cm long, 4–12 (16) mm wide, narrowly elliptic to lance-elliptic, coriaceous. **INFL** a raceme, 2- to 5-flowered; peduncle 3–12 cm long. **FLOWERS** 20–30 mm long; calyx 8–12 mm long, the tube 5–7 mm long, prominently veined, the teeth 2–5 mm long. **PODS** stipitate, the stipe 4–6 mm long, the body 3–4.5 cm long, 7–10 mm wide. $2n = 14$. [*L. brachycalyx* subsp. *eucosmus* (Butters & H. St. John) S. L. Welsh; *L. brachycalyx* var. *eucosmus* (Butters & H. St. John) S. L. Welsh]. Clay soils in washes in salt desert shrub communities. ARIZ: Apa; COLO: Arc, Dol, LPl, Mon; NMEX: McK, RAr, SJn; UTAH. 1690–2245 m (5545–7360′). Flowering: Jun–Aug. Colorado, New Mexico, Arizona, and southeast Utah; Mexico. The much larger flowers are diagnostic for distinguishing this species from the previous one. Plant used by Navajos to remove placenta and for "deer infection," or swellings on horses.

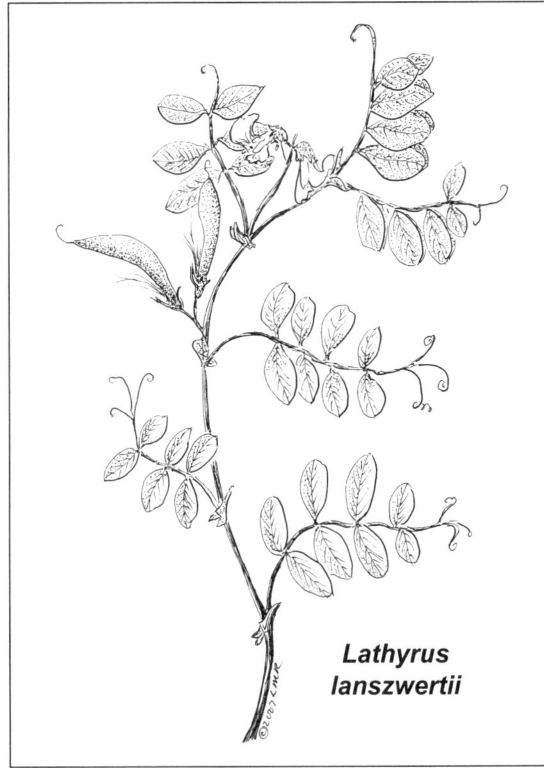
Lathyrus lanszwertii

Lathyrus lanszwertii Kellogg (for Dr. L. Lanszweert, a Belgian physician, early member of the California Academy of Sciences) Lanszwert's sweetpea. Plants clambering, decumbent to erect, 20–60 cm tall, the herbage glabrous to villous. **LEAVES** 2–14 cm long (excluding tendrils); stipules 7–20 mm long, semisagittate; leaflets 4–12, 7–75 mm long, 3–18 (26) mm wide, elliptic to lanceolate, oblanceolate, or oval; tendrils short and simple to more commonly branched and prehensile. **INFL** a raceme, 2- to 5-flowered, the flowers spreading at anthesis; peduncle 2–8.5 cm long. **FLOWERS** 12–22 mm long, pink-purple to white or cream and commonly suffused or veined with pink or purple; calyx tube 3.5–6 mm long, campanulate, the lower lateral teeth 1.8–4.2 mm long, triangular to lanceolate. **PODS** 30–60 mm long, 3–7 mm wide. $2n = 14, 28$. Pods dried and stored in winter by Apaches.

1. Tendrils commonly branched and/or coiled; leaflets often more than 6 .... ***var. laetivirens***
1′ Tendrils reduced to a simple filiform stalk, rarely coiled; leaflets commonly 6 only .... ***var. leucanthus***

var. *laetivirens* (Greene ex Rydb.) S. L. Welsh (richly + greenish white) Largeflower sweetpea. [*L. laetivirens* Greene ex Rydb.; *L. leucanthus* Rydb. var. *laetivirens* (Greene) C. L. Hitchc.]. Riparian, mountain brush, aspen, coniferous forest, and other montane communities. ARIZ: Apa; COLO: Arc, LPl; UTAH. 1840–3150 m (6040–10330′). Flowering: May–Jun. Northern Arizona; Utah; Lincoln County, Nevada; and western Colorado. Mostly the members of this variety are distinctive by their broader leaflets and larger flowers, but they are transitional in every way with var. *leucanthus*.

var. *leucanthus* (Rydb.) Dorn (white-flowered) Whiteflower sweetpea. [based on priorable autonym generated by *L. leucanthus* Rydb. var. *laetivirens* (Greene) C. L. Hitchc.; *L. leucanthus* Rydb.; *L. arizonicus* Britton; *L. lanszwertii* var. *arizonicus* (Britton) S. L. Welsh]. Mountain brush, ponderosa pine, aspen, and spruce-fir communities. ARIZ: Apa, Nav; COLO: Arc, Dol, Hin, LPl, Min, Mon, SJn, SMg; NMEX: McK, RAr, SJn; UTAH. 1830–3200 m (6010–10500′). Flowering: May–Jun. Grand Canyon Plateau of Arizona, Utah, Colorado, and New Mexico. This is the material previously regarded as either *L. arizonicus* or *L. lanszwertii* var. *arizonicus*. Occasional specimens have the tendrils produced into at least three prehensile branches.

Lathyrus latifolius L. (broad-leaved) Perennial sweetpea. Perennial, climbing or clambering vines, 8–20 dm tall, the stems broadly winged, glabrous. **LEAVES** 6–12 cm long (excluding tendrils), the rachis broadly winged; stipules 9–40 mm long, semihastate to semisagittate; leaflets 2, 35–80 (150) mm long, 5–23 (50) mm wide, lance-elliptic to oblong or ovate; tendrils branched, coiled. **INFL** a raceme, 5- to 15-flowered, the flowers spreading at anthesis; peduncle 7–15 cm long. **FLOWERS** 20–25 mm long, pink-purple, pink, or white; calyx tube 5.8–6.2 mm long, campanulate, the lower lateral teeth 3–6 mm long, lanceolate. **PODS** 6–8 (9.5) cm long, 7–10 mm wide, glabrous. $2n = 14$. Cultivated ornamental, persisting, escaping and now established, mainly along canal banks. COLO: Arc, LPl, Mon; NMEX: SJn. 1525–2150 m (5000–7045′). Flowering throughout the summer. Widely grown in the United States; introduced from Europe. Several species of *Lathyrus* have been shown to be toxic in experiments using laboratory animals. Included are *L. latifolius*, *L. odoratus* L., and *L. sylvestris* L., all introduced and grown as ornamentals.

Lathyrus pauciflorus Fernald (few-flowered) Utah sweetpea. Perennial, 2–10 dm tall or more, climbing vines, glabrous. **STEMS** merely angled. **LEAVES** 2–12.5 cm long (excluding tendrils); stipules 8–32 mm long, the larger ones, at least, foliose and toothed; leaflets 8–12, 14–50 mm long, 8–32 mm wide, ovate to elliptic; tendrils well developed, prehensile. **INFL** a raceme, 3- to 10-flowered, the flowers spreading at anthesis; peduncle 3.5–24 cm long. **FLOWERS** 13–27 mm long, pink to pink-purple, with keel usually pale or white; calyx tube 5–7.3 mm long, obliquely campanulate, more or less gibbous, the lower lateral teeth 2.5–7 mm long, often curved and spreading. **PODS** 4–7.5 cm long, 7–11 mm wide. $2n = 14$. Sagebrush, mountain brush, aspen, mixed conifer, and meadow communities. COLO: Arc, LPl, Mon. 2285–2440 m (7500–8000′). Flowering: May–Jul. Oregon, Idaho, Utah, Colorado, and California. Our material belongs to **var. *utahensis*** (M .E. Jones) Peck (of Utah). [*L. utahensis* M. E. Jones]. Apparent hybrids are known between this and varieties *laetivirens* and *lanszwertii* of *L. lanszwertii*.

Lotus L. Trefoil

(Greek *lotos*, a plant name of diverse application in antiquity) Plants annual or perennial herbs or suffrutescent, caulescent, from a taproot and caudex. **LEAVES** alternate, pinnately (or appearing palmately) compound; stipules foliaceous, scarious, or glandlike. **INFL** in axillary pedunculate umbels or solitary; bracts leaflike. **FLOWERS** papilionaceous; calyx 5-toothed; petals 5, yellow or white, sometimes suffused with red, the keel long-attenuate; stamens 10, diadelphous; ovary enclosed in the staminal sheath, the style glabrous. **PODS** flattened or subterete, straight, 1- to several-seeded, dehiscent. $x = 6, 7, 10$ (Ottley 1944). One hundred species. North-temperate.

Lotus wrightii (A. Gray) Greene (for Charles Wright, explorer of the southwestern United States) Wright's trefoil. Perennial, 12–60 cm tall. **STEMS** erect-ascending from a commonly superficial caudex. **LEAVES** petiolate (sometimes shortly so), palmate; stipules reduced to glands; leaflets 3–5, 2–22 mm long, 1–5 mm wide, spatulate to oblanceolate, oblong, linear, obtuse to acute. **INFL** commonly solitary, rarely 2; peduncle 0–2.6 cm long; bracts 1- to 5-foliolate. **FLOWERS** 14–18 mm long, yellow suffused with red; calyx 7.5–9.2 mm long, longer than or subequal to the tube. **PODS** narrowly oblong, 25–34 mm long, 2–2.6 mm wide, strigulose to villosulous, straight. [*Hosackia wrightii* A. Gray; *Anisolotus wrightii* (A. Gray) Rydb.]. Ponderosa pine woods. ARIZ: Apa; COLO: Arc, Dol, LPl, Mon; NMEX: McK, RAr, San, SJn; UTAH. 1925–2725 m (6325–8940′). Flowering: Jun–Aug. Decoction of leaves used as a cathartic for "deer infections" and as a "life medicine" by Navajos. Used as a witchcraft medicine by Zunis.

Lupinus L. Lupine

(wolfish; the plants were once thought to rob the soil of nourishment) Plants annual or perennial herbs. **LEAVES** alternate, palmately compound; stipules slender, persistent. **INFL** a terminal raceme, perfect. **FLOWER** calyx bilabiate, the lips entire or toothed, commonly with bracteoles; petals usually blue or blue-purple, less commonly whitish, yellow, or reddish, the banner variously reflexed, glabrous or variously hairy dorsally, the wings mostly glabrous, the keel glabrous or ciliate on upper (less commonly lower) edges; stamens 10, monadelphous, with 5 long filaments alternating with 5 short ones. **PODS** laterally compressed, 2- to several-seeded. Two hundred species. Andes, Rockies, Mediterranean, Europe, tropical Africa highlands, and South America. Note: The genus is notoriously difficult because of lack of clear diagnostic features. Taxa tend to grade morphologically into each other, probably due to hybridization. The basic chromosome number is $x = 12$, but most of ours are polyploids, and numerous aneuploids are known. Wide-ranging perennial taxa tend to intergrade with all others they contact. Because of these problems, and the likelihood of cleistogamy in some taxa, it is not possible to assign all specimens to described entities. *Lupinus* is the most perplexing genus in the legumes. The following summary treatment is tentative. For a complete list of synonyms involved refer to Barneby (in Cronquist et al. 1989). Several of the species have been implicated in poisoning of livestock. Quinolizidine alkaloids and other kinds of alkaloids have been extracted from the plants. Sheep are the most affected, but horses, cattle, and other animals have also been poisoned (Cox 1970; Dunn 1956, 1957, 1959, 1964; Hess 1969; Fleak and Dunn 1971).

1. Plants annual, the cotyledons commonly persistent .... **Key 1**
1′ Plants perennial, the cotyledons not present at flowering .... **Key 2**

Key 1

Plants annual

1. Axis of the raceme elongate, mainly much over 2.5 cm long (except at early anthesis) .... ***L. pusillus***
1′ Axis of raceme contracted, mainly much less than 2.5 cm long (even in fruit) .... (2)

2. Plants subacaulescent or acaulescent, the internodes seldom to 1 cm long; upper calyx lip 2 mm long or less, entire .... ***L. brevicaulis***
2′ Plants caulescent, with 1 or more elogated internodes, at least some more than 2 cm long; upper calyx lip 3–6 mm long, bilobed .... ***L. kingii***

Key 2

Plants perennial

1. Leaflets glabrous above .... (2)
1′ Leaflets pubescent above .... (3)

2. Leaves mainly basal, the petioles 8–13 cm long or more, coarsely hirsute; stems from a subterranean caudex (or if from a superficial caudex, the plants cultivated); plants indigenous to Grand and San Juan counties, Utah, and cultivated elsewhere .... ***L. polyphyllus***
2′ Leaves mainly well distributed along the stem, the petioles commonly less than 8 cm long, strigose or silvery hairy; stems from a caudex; plants of various distribution .... ***L. argenteus***

3. Banner glabrous dorsally .... (4)
3′ Banner pubescent on the back, at least beneath upper lip of calyx .... (6)

4. Flower 10–13.5 mm long; calyx saccate-gibbous or shortly spurred; leaves mainly basal .... ***L. polyphyllus***
4′ Flower 7–13 mm long; calyx not especially gibbous; leaves distributed along the stem, or all leaves basal .... (5)

5. Plants acaulescent or short-caulescent; leaves essentially all basal .... ***L. lepidus***
5′ Plants caulescent; leaves distributed along the stem .... ***L. argenteus***

6. Calyx with a gibbous-saccate spur at base of upper lip; wings or keel (or both) ciliate below the claws .... ***L. caudatus***
6′ Calyx at most gibbous at base of upper lip; wings and lower edge of keel glabrous .... (7)

7. Leaves mainly basal; plants commonly less than 5 dm tall .... ***L. polyphyllus***
7′ Leaves mainly well distributed along the stems; plants often over 5 dm tall .... (8)

8. Banner reflexed at or below the midpoint, strigose to thinly strigose near the tip or rarely hairy along the crest .... ***L. sericeus***
8′ Banner reflexed beyond the midpoint, strigose on the back beneath the calyx lobe or over much of the back .... ***L. argenteus***

Lupinus argenteus Pursh (silvery) Silvery lupine. Plants perennial, 18–90 cm tall, from a superficial caudex, puberulent to strigose on stems and petioles. **LEAVES** mainly cauline; petioles 1.5–8 cm long; leaflets 6–9, 7–95 mm long, 2–22 mm wide, oblanceolate to spatulate or almost linear, flat or folded, strigulose to strigose on both surfaces or almost or quite glabrous above. **INFL** a raceme, 15- to 92-flowered, 5–24 cm long in anthesis, the axis 6–29 cm long in fruit; pedicels 1.5–6 mm long; peduncle 1.5–14.5 cm long. **FLOWERS** (5–7) 8.5–16 mm long, blue-purple, blue, white, or rarely other hues; calyx gibbous or rounded at base of upper lip; banner with a central yellow or white spot, pubescent or glabrous dorsally, reflexed above the midpoint, the wings and keel glabrous or variously sparingly ciliate; ovules 3–6. The silvery lupine is represented by several more or less distinctive but intergrading varieties. Furthermore, at least some of the phases grade into other taxa, especially into *L. caudatus*, but also into *L. sericeus*. Silvery lupine, along with those species, constitutes the most common and most widespread complex of the perennial lupines. Possibly all of them are poisonous to livestock due to the presence of alkaloids. Ramah Navajos applied a poultice of crushed leaves to poison ivy blisters. The large proportion of the specimens encountered can be segregated by use of the following, admittedly arbitrary key.

1. Leaflets more or less evenly pubescent above, generally folded, the upper surface thus obscured; plants generally of middle and lower elevations .... (2)
1′ Leaflets glabrous above, or with hairs scattered or merely with a few near the margin, commonly flat and green, but sometimes folded; plants generally of middle and upper elevations .... (3)

2. Stems with spreading or retrorse hairs; flower mainly 7–10 mm long; plants of Navajo County, Arizona .... **var. *palmeri***
2′ Stems with appressed or ascending hairs; flower size various; plants of various distribution .... (4)

3. Flower 12–14 mm long; plants of sandy washes in southeastern Utah and northwestern New Mexico .... **var. *moabensis***
3′ Flower 8–12 mm long or more; plants widely distributed in the western United States. .... **var. *argenteus***

4. Flower 5–7 (7.5) mm long; pedicels typically more than 2.5 mm long .... **var. *parviflorus***
4′ Flower typically larger or, if almost as small, the pedicels less than 2.5 mm long .... (5)

5. Flower mainly 6–8 mm long, with pedicels less than 3.5 mm long .... **var. *fulvomaculatus***
5′ Flower mainly 8–11 mm long and with pedicels more than 3.5 mm long .... (6)

6. Stems with the inflorescence solitary and terminal, rarely with 1 or 2 depauperate lateral racemes ..... **var. *rubricaulis***
6′ Stems branched, each with a solitary and terminal raceme, each raceme maturing at about the same time .... **var. *argenteus***

var. *argenteus* [*L. tenellus* Douglas ex G. Don; *L. argenteus* subsp. *argenteus* var. *tenellus* (Douglas) Dunn; *L. alpestris* A. Nelson, as to concept but not as to type; *L. lucidulus* Rydb.; *L. laxus* Rydb.; *L. macounii* Rydb.; *L. stenophyllus* Rydb.; *L. pulcher* Eastw.]. Shadscale, ephedra, grass, sagebrush, piñon-juniper, mountain brush, ponderosa pine, aspen, Douglas-fir, and less commonly spruce-fir communities ARIZ: Apa; COLO: Arc, Dol, LPl, Mon; NMEX: McK, RAr, SJn; UTAH. 1525–2590 m (5000–8495′). Flowering: mid-Apr–May. Alberta and Saskatchewan south to Oregon, California, Nevada, and the Dakotas. Those phases of the species with slightly gibbous calyx base and usually smallish, often compact flowers have been regarded as *L. ×alpestris*. They tend to tie the *argenteus* and *caudatus* complexes together, but as pointed out by Barneby (in Cronquist et al. 1989), they occupy "today a range greater than either of its two parents." Such materials are herein included within var. *argenteus* in a broad sense. However, they might be worthy of some taxonomic recognition. Choice of a name from the more than 80 names involved in the *argenteus-caudatus* complexes that matches these plants, which are not geographically correlated, would require examination of type material of all of those names, a task beyond the scope of this work. The huge number of names is indicative of the complexity of the group, and also of the lack of understanding of the various authors, especially Charles Piper Smith (who contributed about half of the names, mostly if not entirely inconsequential).

var. *fulvomaculatus* (Payson) Barneby (yellow spot) Yellow-spot lupine. [*L. fulvomaculatus* Payson; *L. parviflorus* Nutt. ex Hook. & Arn. var. *fulvomaculatus* (Payson) Harmon]. Mountain brush, aspen, spruce-fir, and subalpine meadow communities. COLO: SJn. About 2745 m (9000′). Flowering: mid-Apr–May. Southwest Montana, southern Idaho, Wyoming, southeast Utah, and Colorado. Flowers are typically lupine blue-purple, but occasional plants have white flowers.

var. *moabensis* S. L. Welsh (of Moab, Utah) Moab lupine. Blackbrush and mixed desert shrub communities. ARIZ: Apa; COLO: LPl; NMEX: SJn; UTAH. 1290–1950 m (4225–6400′). Flowering: mid-Apr–May. In Colorado (Mesa County) and New Mexico; a Navajo Basin endemic. This is the large-flowered, xeromorphic ecotype of lower elevations in the canyon country of the Four Corners. Large-flowered specimens occur elsewhere within var. *argenteus*, but not so consistently as for this variety.

var. *palmeri* (S. Watson) Barneby (for Edward Palmer, naturalist and explorer) Palmer's lupine. [*L. palmeri* S. Watson] Sagebrush and piñon-juniper communities. ARIZ: Nav. About 1980 m (6500′). Flowering: mid-Apr–May. Southwest Utah, Arizona, Nevada, and California.

var. *parviflorus* (Nutt.) C. L. Hitchc. (small-flowered) Small-flower lupine. [*L. parviflorus* Nutt. ex Hook. & Arn.; *L. argenteus* subsp. *parviflorus* (Nutt.) Phillips]. Aspen, lodgepole pine, and spruce-fir communities. NMEX: SJn; UTAH. 2285–3200 m

(7500–10500′). Flowering: mid-Apr–May. Montana and Idaho to Wyoming and South Dakota, and south to New Mexico. Small-flower lupine is matched in its flower size here and there throughout var. *argenteus*.

var. *rubricaulis* (Greene) S. L. Welsh (red stem) Subalpine lupine. [*L. rubricaulis* Greene; *L. caudatus* Kellogg var. *rubricaulis* (Greene) C. P. Sm.; *L. argenteus* subsp. *rubricaulis* (Greene) L. W. Hess & D. B. Dunn, 1970 Rhodora 72:111; *L. argenteus* var. *boreus* (C. P. Sm.) S. L. Welsh, at least as to concept; *L. alpestris* A. Nelson, as to type, but not as to concept; *L. maculatus* Rydb.]. Aspen, meadow, mixed conifer, riparian, and spruce-fir communities. COLO: Arc, Hin; UTAH. 2485–3390 m (8150–11120′). Flowering: mid-Apr–May. Idaho, Montana, and South Dakota, south to Nevada and Colorado (type from Crested Butte). It is with some reluctance that I synonymize *L. maculatus*.

Lupinus brevicaulis S. Watson (short stem) Shortstem lupine. Annual, 4–11 cm tall from a taproot; cotyledons sessile. **STEMS** 0–2 cm long, when developed at all usually obscured by the leaf bases. **LEAVES** in a basal tuft; petioles 0.8–6.5 cm long; leaflets 3–9, 5–18 mm long, 1.5–9 mm wide, oblanceolate, flat or folded, pilose beneath, glabrous above (except marginally in some). **INFL** a raceme, 4- to 12-flowered, 1–2.5 cm long in anthesis, the axis 1.5–3 cm long in fruit; peduncle 0.6–6.5 cm long. **FLOWERS** 5.2–7 mm long, blue-purple or white; pedicel 0.3–0.8 mm long; calyx tapering to the pedicel, the upper lip very short; banner with a central yellow spot, glabrous dorsally, reflexed near the midpoint. **PODS** 7–12 mm long, obliquely ovate in profile; ovules 2 or 3. Blackbrush, salt desert shrub, piñon-juniper, and sagebrush communities. ARIZ: Apa, Nav; COLO: Arc, LPl, Mon; NMEX: McK, RAr, San, SJn; UTAH. 1660–2080 m (5445–6825′). Flowering: early May–Jul. Throughout Utah, and from Oregon to Colorado, south to Arizona and New Mexico. The plant was rubbed on boils as a liniment and used for sterility by Navajos.

Lupinus caudatus Kellogg (ending with a tail-like appendage) Spurred lupine. Perennial, 21–80 cm tall, from a woody caudex. **LEAVES** mainly cauline; petioles 1–12 cm long, commonly 2–8 cm long; leaflets 5–9, 10–45 (60) mm long, 2–14 cm wide, oblanceolate to elliptic or narrowly oblanceolate, pilose on both surfaces or glabrate above. **INFL** a raceme, 10- to 57-flowered, 3–16 cm long in anthesis, the axis 4.5–17 cm long in fruit; peduncle 1–6.5 cm long. **FLOWERS** 8–12.5 (13.5) mm long, blue-purple or less commonly white; pedicel 1–3 (5) mm long; calyx with a gibbous-saccate spur 0.2–1.5 (2) mm long at the base of the upper lip; banner pubescent dorsally, reflexed at or beyond the midpoint; wings commonly ciliate above and near the claws, the keel commonly ciliate above and near the claws; ovules 3–5. Two rather weak varieties are known from the San Juan River drainage. They are separable only arbitrarily but seem to represent at least trends within the variation. Leaves of this plant were used by Ramah Navajos as a ceremonial emetic and a lotion for poison ivy.

1. Leaflets bicolored, green to yellow-green above, dull green to grayish beneath; flowers mainly 10–12 mm long .... **var. *argophyllus***

1′ Leaflets more uniformly colored, either green or gray on both surfaces or only somewhat bicolored..... **var. *utahensis***

var. *argophyllus* (A. Gray) S. L. Welsh (bright or shining leaf) Silverleaf lupine. [*L. decumbens* Torr. var. *argophyllus* A. Gray; *L. argenteus* Pursh var. *argophyllus* (A. Gray) S. Watson; *L. argophyllus* (A. Gray) Cockerell; *L. laxiflorus* Douglas ex Lindl. var. *argophyllus* (A. Gray) M. E. Jones; *L. caudatus* subsp. *argophyllus* (A. Gray) L. Phillips; *L. cutleri* Eastw.; *L. caudatus* var. *cutleri* (Eastw.) S. L. Welsh]. Piñon-juniper, mountain brush, ponderosa pine, and grassland communities. ARIZ: Apa; COLO: Arc, Dol, LPl, Mon; NMEX: McK, RAr, SJn; UTAH. 1580–2445 m (5180–8020′). Flowering: late May–Jul. Montana, Idaho, Wyoming, Utah, and south to Arizona and New Mexico. Flowers of this variety are almost always at least 10–12 mm long, and the calyx definitely though shortly spurred. Specimens previously identified as var. *cutleri* do not differ in any major regard from those of var. *argophyllus*, whose range overlaps the area supposed to bear var. *cutleri*.

var. *utahensis* (S. Watson) S. L. Welsh (of Utah) Utah lupine. [*L. holosericeus* Nutt. var. *utahensis* S. Watson; *L. argenteus* Pursh var. *utahensis* (S. Watson) Barneby; *L. argentinus* Rydb.; *L. leucophyllus* Douglas ex Lindl. var. *lupinus* Rydb.]. Sagebrush, piñon-juniper, mountain brush, ponderosa pine, aspen, mixed conifer, and grassland communities. ARIZ: Apa; COLO: Dol, LPl, Mon, SMg; NMEX: McK, SJn; UTAH. 1480–2680 m (4850–8800′). Flowering: May–Jul. In Oregon, Idaho, and Wyoming, south to California, Nevada, Arizona, New Mexico, and Colorado.

Lupinus kingii S. Watson (for George King) King's lupine. [*L. sileri* S. Watson; *L. capitatus* Greene]. Annual, 6–28 cm tall, from a taproot; cotyledons sessile. **STEMS** with 1 or more apparent internodes more than 2 cm long. **LEAVES** mainly cauline; petioles 0.8–3.5 cm long; leaflets 5–7, 7–23 mm long, 2–6.5 mm wide, oblanceolate, flat or folded, long-pilose on both surfaces or glabrous medially above or overall. **INFL** a raceme, 5- to 12-flowered, 1–2.3 cm long at anthesis, the axis 1.8–3.7 cm long in fruit; peduncle 0–9.5 cm long. **FLOWERS** 7.8–9.2 mm long, blue-purple or less commonly pallid; calyx somewhat gibbous at base of upper lip; banner with a central yellow spot, glabrous dorsally, reflexed below the midpoint; ovules usually 2. Sagebrush, piñon-juniper, mountain brush, ponderosa pine, and grass-forb-sagebrush communities. ARIZ: Apa; COLO: Arc, Dol, LPl, Mon; NMEX: McK, RAr, San, SJn; UTAH. 1840–2375 m (6040–7800′). Flowering: Jun–Aug. Nevada, Utah, Arizona, Colorado, New Mexico. Hopis used the plant as an eye medicine. Navajos made poultices of crushed leaves for skin irritations. Also a Navajo "life medicine."

Lupinus lepidus Douglas ex Lindl. (charming or pleasant) Stemless lupine. Perennial, 2.5–40 cm tall, caespitose from a branching caudex, acaulescent or subacaulescent to short-caulescent. **LEAVES** mainly basal; petioles 1–10 cm long; leaflets 5–9, 3–40 mm long, 1.5–9 mm wide, oblanceolate to elliptic, flat or folded, mucronate, pilose on both surfaces, or less so above. **INFL** a raceme, 12- to 85-flowered, 1–23 cm long at anthesis, the axis 2–30 cm long in fruit; peduncle 0–3 cm long. **FLOWERS** 7–14 mm long, blue-purple or white; pedicel 1–4 mm long; calyx tapering to the petiole or only somewhat gibbous at base of upper lip; banner with a central yellow spot, glabrous dorsally, reflexed at or below the midpoint; ovules 2–4. $2n = 48$. Meadows, open deciduous woodland, mixed conifer, and sagebrush communities. ARIZ: Apa; COLO: Arc, Hin; NMEX: SJn. 2440–2775 m (8000–9100′). Flowering: Jun–Sep. Idaho and Montana, south to California, Nevada, Colorado, and New Mexico. Four Corners material belongs to **var. *utahensis*** (S. Watson) C. L. Hitchc. (of Utah) Watson's lupine. [*L. caespitosus* Nutt.; *L. lepidus* subsp. *caespitosus* (Nutt.) Detling; *L. aridus* Douglas ex Lindl. var. *utahensis* S. Watson; *L. watsonii* A. Heller; *L. caespitosus* var. *utahensis* (S. Watson) B. Cox].

Lupinus polyphyllus Lindl. (many leaves) Showy lupine, meadow lupine, blue pod. Plants perennial, 13–70 cm tall, from a subterranean to superficial caudex; pubescence of stems spreading-hirsute to hirsute-pilose or appressed-ascending. **LEAVES** mainly basal or basal and cauline; petioles of lowermost leaves 3.5–30 cm long; leaflets 8–13, 12–75 mm long, 1.2–7 mm wide, oblanceolate to obovate, plane or folded, pilose beneath, glabrous above or pilose on both sides. **INFL** a raceme, 13- to 68-flowered, 6–25 cm long in anthesis, 8–36 cm long in fruit; peduncle 3.5–13 cm long. **FLOWERS** 10–16 mm long, blue-purple or rarely white; pedicel 3–9 mm long; calyx gibbous or saccate-spurred at base of upper lip; banner with a central yellow or white spot, glabrous dorsally, reflexed near or above the midpoint; wings glabrous; keel sparingly ciliate near the apex; ovules 3–7.

1. Stems from a subterranean caudex; plants widespread throughout the Four Corners region ..... **var. *ammophilus***
1′ Stems from a subterranean caudex; plants of Dolores, La Plata, and San Miguel counties, Colorado .... (2)

2. Leaflets silky strigose on both surfaces; petioles of basal leaves only somewhat (if at all) longer than the cauline ones; plants of La Plata County, Colorado .... **var. *humicola***
2′ Leaflets glabrous above or essentially so; petioles of basal leaves markedly longer than the cauline ones; plants of Dolores, La Plata, and San Miguel counties .... **var. *prunophilus***

var. *ammophilus* (Greene) Barneby (sand-loving) [*L. ammophilus* Greene]. Sand lupine. Sagebrush, piñon-juniper, mountain brush, ponderosa pine, and aspen–Douglas-fir forests, commonly in sandy soils. ARIZ: Apa; COLO: Arc, Dol, LPl, Mon, SMg; NMEX: RAr, San, SJn; UTAH. 1735–2760 m (5700–9060′). Flowering: May–Aug. Southeast Utah; southwest Colorado; Apache County, Arizona; and northwest New Mexico. The long (2 mm plus) spreading hairs and petioles subequal to or longer than the stem proper easily characterize this plant within its area of growth.

var. *humicola* (A. Nelson) Barneby (ground-dweller) Meadow lupine. [*L. humicola* A. Nelson; *L. wyethii* S. Watson]. Wash bottoms and terraces or rimrock in piñon-juniper and riparian communities. COLO: LPl. 2500 m (8195′). Flowering: Jun–Aug. British Columbia and Alberta south to California, Nevada, and Colorado.

var. *prunophilus* (M. E. Jones) L. Phillips (plum-loving) Hairy bigleaf lupine. [*L. prunophilus* M. E. Jones; *L. wyethii* S. Watson var. *prunophilus* (M. E. Jones) C. P. Sm.; *L. arcticus* S. Watson var. *prunophilus* (M. E. Jones) C. P. Sm.; *L. tooelensis* C. P. Sm.]. Sagebrush, piñon-juniper, and mountain brush communities. COLO: Dol, LPl, SMg. 1830–2135 m (6000–7000′). Flowering: Jun–Aug. Washington to Nevada, east to Montana, Wyoming, Utah, and Colorado. Both varieties *ammophilus* and *prunophilus* are sufficiently beautiful as to be of horticultural significance. The flowers are large in both of them, and the basal cluster of long-petiolate leaves is attractive, even when the plants are not in flower.

Lupinus pusillus Pursh (dwarf) Dwarf lupine, rusty lupine. Annual, 3–24 cm tall, from a taproot; cotyledons sessile; pubescence of stems and petioles spreading long-hairy. **LEAVES** mainly cauline; petioles 1–9 cm long; leaflets 3–9 (14), 11–48 mm long, 2–10 mm wide, oblanceolate, flat or folded, long-pilose beneath, glabrous above. **INFL** a raceme, 4- to 38-flowered, 1–17 cm long in anthesis, the axis 4–21 cm long in fruit; peduncle 0.5–3.5 cm long. **FLOWERS** 8.5–12 mm long, blue or variously pink or white; pedicel 1–3.5 mm long; calyx tapering to the pedicel; banner with a central yellow spot, glabrous dorsally, reflexed near the midpoint; ovules 2. **PODS** constricted between the seeds. Blackbrush, mixed desert shrub, piñon-juniper, and mountain brush communities; commonly in sand. ARIZ: Apa, Coc, Nav; COLO: Arc, LPl, Mon; NMEX: McK, RAr, San, SJn; UTAH. 1220–1925 m (4000–6325′). Flowering: May–Jul. Washington to Alberta and Saskatchewan, south to Utah, Arizona, Colorado, New Mexico, and Kansas. Four Corners material belongs to **var. *pusillus***. This variety occurs mainly in the Colorado drainage system. Hopis used the plant as an ear and eye medicine, and the juice as holy water in the Po-wa-mu ceremony.

Lupinus sericeus Pursh (silky) Silky lupine. Perennial, 3–12 dm tall, from a branching caudex; pubescence of stems short-villous to pilose or strigose, sometimes spreading. **LEAVES** with petioles 1.2–9 cm long; leaflets 5–9, 0.7–7.8 cm long, 2–15 mm wide, oblanceolate, commonly flat (at least some), pilose to puberulent on both surfaces or glabrous to glabrate above. **INFL** a raceme, 14–70-flowered, 6–28 cm long in anthesis, the axis 8–37 cm long in fruit; peduncle 1.3–9 (12) cm long. **FLOWERS** 10–16 mm long, blue, blue-purple, pale, or white; pedicels 2–7 mm long; calyx more or less gibbous at the base of the upper lip; banner with yellow or brown eyespot, strigose along the dorsal crest or more widely; ovules 5–7. 2*n* = 48. [*L. flexuosus* Lindley ex J. Agardh; *L. sericeus* var. *flexuosus* (Lindley) C. P. Smith]. Sagebrush, mountain brush, piñon-juniper, ponderosa pine, aspen, spruce-fir, and alpine meadow communities. COLO: Arc, Dol, LPl, Mon, SMg; NMEX: RAr; UTAH. 2135–2895 m (7000–9500′). Flowering: Jun–Aug. British Columbia east to Alberta, south to California, Nevada, Utah, Colorado, Arizona, and New Mexico. Four Corners material belongs to **var. *sericeus***.

Medicago L. Alfalfa, Black Medic

(Medea, source of alfalfa) Annual or perennial herbs, caulescent from a taproot or caudex. **LEAVES** alternate, pinnately trifoliolate, the leaflets serrate in the distal 1/2 or less; stipules herbaceous, often toothed. **INFL** in axillary, pedunculate racemes or heads; bracts subulate. **FLOWERS** papilionaceous, calyx 5-toothed; petals 5, yellow, white, blue, pink, lavender, or purple; stamens 10, diadelphous; ovary enfolded by the staminal sheath, the style subulate. **PODS** curved to spirally coiled, 1- to several-seeded, indehiscent, reticulate or spiny. Fifty-six species. Europe, Mediterranean, Ethiopia, and South Africa.

1. Flowers 2–3 mm long; inflorescence less than 10 mm long in anthesis; pods coiled through a single spiral, 1-seeded, unarmed; plants annual, prostrate to decumbent or rarely erect .... ***M. lupulina***
1′ Flowers 4–10 mm long; inflorescence longer than 10 mm (including flower length), or pods differing from above; plants various .... (2)

2. Flowers yellow (sometimes tinged violet or violet throughout), mainly over 8 mm long; pods merely curved; plants uncommon .... ***M. falcata***
2′ Flowers blue, pink, lavender, purple, or white, mainly 8 mm long or shorter; pods spirally coiled; plants common .... ***M. sativa***

Medicago falcata L. (curved like a sickle) Yellow alfalfa. Perennial (rarely functionally annual), 4–10 dm tall or more, the stems erect or ascending, strigulose; stipules 4–12 mm long, persistent, conspicuously veined. **LEAVES** short-petiolate, the leaflets linear, oblong, oblanceolate, or elliptic, 6–20 mm long, 1–6 (10) mm wide, few-toothed, tridentate, or merely apiculate apically, strigulose beneath. **INFL** a raceme, 6- to 20-flowered, mostly 10–20 mm long; peduncle subequal to the subtending leaves or longer. **FLOWERS** 6–10 mm long, yellow, sometimes suffused with violet; calyx campanulate, the tube 1–2 mm long, the teeth 1.5–3 mm long, lance-subulate. **PODS** 6–10 mm long, merely curved, unarmed, several-seeded. 2*n* = 16, 32. Sparingly cultivated forage plant, escaping and persisting. ARIZ: Apa, Nav; COLO: Arc, Dol, Hin, LPl, Min, Mon, SJn; NMEX: SJn. 1675–2145 m (5500–7045′). Flowering: Jun–Aug. Introduced from the Old World. This plant forms hybrids with *M. sativa* and is sometimes considered a phase of that species. It seldom grows in pure stands; usually it occurs mixed with *M. sativa* and is impressive where grown along interstate highways in the Four Corners region. Mature fruiting pods are diagnostic for the entity.

Medicago lupulina L. (hop plant) Black medic, hop clover. Annual, the stems prostrate to decumbent or sometimes erect, 10–50 cm long or more. **LEAVES** short-petiolate; stipules entire or nearly so, 3–6 mm long, persistent; the leaflets cuneate to obcordate, 4–17 mm long, 2–15 mm wide, toothed in the apical 1/3 (rarely more), pubescent to glabrous. **INFL** a raceme, compactly 6- to 25-flowered, less than 25 mm long in fruit; peduncle mostly equaling or surpassing the subtending leaves. **FLOWERS** 2–3 mm long, yellow; calyx campanulate, about 1 mm long. **PODS** spiral through about 1 coil, unarmed, 1-seeded. 2*n* = 16, 28, 32. Introduced weedy species of lawns, fields, roadsides, other open sites, and in native vegetation. ARIZ: Apa, Nav; COLO: Arc, Dol, Hin, LPl, Min, Mon, SJn, SMg; NMEX: McK, RAr, SJn; UTAH. 1065–3200 m (3500–10500′). Flowering: May–Oct. Introduced from Europe.

Medicago sativa L. (cultivated) Alfalfa, lucerne. Perennial, or functionally annual, the stems 4–12 dm long or more, ascending to erect, finally sprawling, strigulose. **LEAVES** short-petiolate; stipules entire or toothed, 4–12 mm long, persistent; the leaflets elliptic to oblanceolate, 8–40 mm long, 2–15 mm wide, apically few-toothed, pubescent. **INFL** a raceme, 6- to 25-flowered, 10–35 mm long or more; peduncle often surpassing the subtending leaves. **FLOWERS** 6–8 (10) mm long, blue, lavender, pink, purple, or white; calyx campanulate to short-cylindric, the tube 1.5–2.5 mm long, the lance-subulate teeth 2–4 mm long. **PODS** spirally coiled, unarmed, several-seeded. $2n = 16, 32, 64$. Introduced forage plant, escaping and persisting, now almost or quite cosmopolitan. ARIZ: Apa, Nav; COLO: Arc, Dol, Hin, LPl, Min, Mon, SJn; NMEX: McK, RAr, San, SJn; UTAH. 1690–2945 m (5545–9670′). Flowering: May–Oct. Introduced from Europe. This Old World introduction is possibly the most important forage species grown in the Four Corners, as it is in much of the American West.

Melilotus Mill. Sweet Clover

(*mel*, honey, + *lotus*, a plant name of many applications) Plants annual or biennial herbs, caulescent, from a stout taproot. **LEAVES** alternate, pinnately trifoliolate, the leaflets dentate-serrate in the distal 1/2 or more; stipules herbaceous, distinct, subulate, entire or hastately lobed. **INFL** borne in axillary, pedunculate racemes; bracts subulate. **FLOWERS** papilionaceous; petals 5, white or yellow, the keel obtuse; calyx 5-toothed; stamens 10, diadelphous; ovary enfolded by the staminal sheath, the style subulate. **PODS** straight, ovoid, reticulately veined or cross-ribbed, unarmed, glabrous, 1- to 2-seeded, indehiscent. $x = 8$. The generic name *Melilotus* is considered to be feminine, despite its ending; hence, the specific epithets must agree with that gender. Twenty species. Temperate and subtropical; Europe, North Africa, and Ethiopia.

1. Flowers white; pods reticulately veined .... ***M. albus***
1′ Flowers yellow; pods cross-ribbed .... ***M. officinalis***

Melilotus albus Medik. (white) White sweet clover. Annual or biennial, the stems commonly 5–15 dm tall or more, erect, strigulose. **LEAVES** short-petiolate; stipules entire or hastately lobed, mostly 5–10 mm long, persistent; the leaflets obovate to elliptic or oblanceolate, 8–35 mm long, 1–15 mm wide, pubescent or glabrous. **INFL** a raceme, 38- to 115-flowered, 1.8–14.5 cm long or more; peduncle commonly surpassing the subtending leaves. **FLOWERS** 4–5.5 mm long, white; calyx campanulate, the tube 1.2–1.8 mm long, the teeth 1–1.5 mm long, acuminate. **PODS** 2.5–6 mm long, reticulately veined, 1- to 2-seeded. $2n = 16, 24$. Introduced forage plant, now widely established. ARIZ: Apa, Nav; COLO: Arc, Dol, Hin, LPl, Min, Mon, SJn, SMg; NMEX: McK, RAr, San, SJn; UTAH. 1225–2365 m (4025–7755′). Flowering: Jun–Sep. Introduced from Europe. White sweet clover is an excellent source of honey for domestic bees. This species and yellow sweet clover both contain the substance coumarin and are responsible for a hemorrhagic disease and subsequent death in livestock. The substance or one of its isomers, commercially known as coumadin or warfarin, is utilized in medicine as a blood thinner following stroke and other human ailments. It is also the basis for a common rat and mouse poison, inducing fatal hemorrhagic disease in those animals as well.

Melilotus officinalis (L.) Lam. (used in medicine) Yellow sweet clover. Annual, winter-annual, or biennial, the stems 5–15 dm tall or more, erect, strigulose. **LEAVES** shortly petiolate; stipules entire or with 1–3 basal teeth, 3–10 mm long, persistent; the leaflets cuneate to elliptic or oblanceolate, 8–38 mm long, 3–16 mm wide, pubescent or glabrous. **INFL** a raceme, 20- to 65-flowered, 1.8–11 (14) cm long; peduncle shorter to longer than the subtending leaves. **FLOWERS** 4.5–6 (7) mm long, yellow, fading cream; calyx campanulate, the tube 1–1.8 mm long, the teeth 1–1.5 mm long, acuminate. **PODS** 3–5 mm long, cross-ribbed, 1- or 2-seeded. $2n = 16$. [*Trifolium melilotus-officinalis* L.]. Common ruderal weed of almost cosmopolitan distribution in fields, along roadsides, and other open sites, often in native plant communities. ARIZ: Apa, Nav; COLO: Arc, Dol, Hin, LPl, Min, Mon, SJn; NMEX: McK, RAr, San, SJn; UTAH. 1065–2455 m (3500–8050′). Flowering: May–Jul. Introduced from Europe, and widely grown and established in North America. Drought-resistant strains of yellow sweet clover are commonly used in reclamation plantings. It is genetically incompatible with the closely similar *M. alba*. Ramah Navajos used a cold infusion for chills.

Onobrychis Mill. Sainfoin

(*onos*, donkey, + *bruchein*, to bray) Perennial herbs, caulescent, from a caudex and taproot. **LEAVES** alternate, odd-pinnate; stipules adnate to the petiole base, the lowermost amplexicaul but not connate. **INFL** in an axillary raceme, each subtended by a bract; bracteoles 2. **FLOWERS** papilionaceous; petals 5, red-purple to lavender or pink, the keel much longer than the wings, abruptly bow-shaped; calyx 5-toothed; stamens 10, essentially diadelphous; ovary enclosed in the staminal sheath, the style glabrous. **FRUIT** a loment reduced to 1 segment, this armed with prickles. One hundred thirty species. Eurasia and Ethiopia.

Onobrychis viciifolia Scop. (vetch-leaved) Sainfoin, holy clover. Perennial, caulescent, 20–45 cm tall, from a branching, superficial caudex. **STEMS** ascending to erect. **LEAVES** 3–12 mm long; stipules 3–12 mm long, all more or less amplexicaul; leaflets 11–21 (27), 8–25 mm long, 2–7 mm wide, oblong to elliptic or oblanceolate, pilose mainly along veins beneath, glabrous above. **INFL** a raceme, 14- to 39 (50)-flowered, the flowers ascending-spreading at anthesis, the axis 4–14 cm long in fruit; bracts 2.5–4.5 mm long; pedicels 0.2–1.5 mm long; bracteoles 1; peduncle 8–19 cm long. **FLOWERS** 10–13 mm long, red-purple, lavender, or pink; calyx 5.5–6.5 mm long, the tube 2.3–3 mm long, campanulate, the teeth 2.2–4 mm long, subulate. **LOMENTS** sessile, ascending, armed with prickles. $2n = 28$. [*Hedysarum onobrychis* L.; *O. onobrychis* Rydb.; *O. sativa* Lam.]. Introduced forage and reclamation plant, escaping and persisting, roadsides, canal banks, and reclamation areas. COLO: Dol, LPl, Mon, SMg. 1855–2365 m (6095–7760′). Flowering: Jul–Aug. Native to Europe. The genus is closely allied to *Hedysarum* and is equally attractive.

Oxytropis DC. Locoweed

(Greek *oxys*, sharp, + *tropis*, keel) Perennial, caulescent or acaulescent herbs, from a taproot and caudex. **LEAVES** alternate or basal, odd-pinnate; stipules adnate to the petiole base, often connate-sheathing. **INFL** scapose or an axillary raceme, each subtended by a single bract; bracteoles 0 (rarely 2). **FLOWERS** papilionaceous; petals 5, pink, pink-purple, or white, the keel shorter than the wings, the keel tip produced into a porrect beak; calyx 5-toothed; stamens 10, diadelphous; ovary enfolded in the staminal sheath, the style glabrous. **PODS** sessile or stipitate, straight, erect, ascending, or spreading-declined, 1- or 2-loculed, or partially 2-loculed by intrusion of the ventral (upper) suture, dehiscent apically or throughout. $x = 8$ (Barneby 1952; Ralphs, Welsh, and Gardner 2002; Welsh 1991, 1995b, 2001). Three hundred species. Mostly north-temperate.

1. Plants caulescent, with 1 or more internodes apparent, or sometimes acaulescent; pods stipitate, pendulous .... ***O. deflexa***
1′ Plants acaulescent, the internodes not apparent; pods sessile, steeply ascending .... (2)

2. Pubescence of malpighian hairs; corollas typically some shade of pink-purple .... ***O. lambertii***
2′ Pubescence of basifixed hairs; corollas typically white .... ***O. sericea***

Oxytropis deflexa (Pall.) DC. (turned away) Stemmed oxytrope. Perennial, caulescent to subacaulescent, (5) 7–48 cm tall. **PUBESCENCE** basifixed, villous-pilose. **STEMS** few to several from a caudex, with usually 1–7 apparent internodes (sometimes none), flexuous, sometimes acaulescent, more or less villous with spreading or retrorse hairs. **LEAVES** 2–22 cm long; stipules 7–20 mm long, subherbaceous, sometimes connate opposite the petiole, slightly adnate to the petiole, the free blades caudate-acuminate; leaflets (9) 15–41, 3–25 mm long, 1–8 (11) mm wide, not fasciculate, lance-oblong to lanceolate, pilose on both surfaces or glabrous above, quite sessile. **INFL** a raceme, 3- to 30-flowered or more, the flowers ascending to declined at anthesis, the axis 3.5–10 cm long in fruit; bracts pilose; peduncle 3.5–32 (36) cm long, villous-pilose. **FLOWERS** 5–10.5 (12) mm long, whitish, lilac, or blue purple; calyx 3–8 mm long, the tube 2–3.5 (4.5) mm long, campanulate, the teeth 1.5–5 mm long, lance-subulate. **PODS** spreading-declined, subsessile to shortly stipitate, the body oblong to ellipsoid, 8–18 mm long, 3–4.5 mm wide, subunilocular, pilosulous. $2n = 16$. [*Astragalus deflexus* Pall.; *Aragallus deflexus* (Pall.) A. Heller; *Astragalus retroflexus* Pall.] (Boivin 1962).

1. Flowers pink-purple, pink, or less commonly whitish, 2–15 (20) per raceme; racemes hemispheric (or subcapitate) to shortly subcylindrical, 14–20 mm broad in flower (when pressed), not much elongating in fruit (1–10 cm long); herbage variously pubescent; plants often of woods or thickets .... **var. *pulcherrima***
1′ Flowers whitish, pinkish, bluish, or less commonly pink-purple, 10–30 or more per raceme; racemes subcylindrical, 10–13 (14) mm broad in flower (when pressed), much elongating in fruit (to 18 cm long); herbage usually conspicuously long-villous; plants commonly of open sites .... **var. *sericea***

var. *pulcherrima* S. L. Welsh & A. Huber (beautiful or handsome) Beautiful oxytrope. Caulescent or subacaulescent to acaulescent. **INFL** a raceme, rather densely 10- to 20-flowered, 20–27 mm wide when pressed, typically 1.5–6 cm long in fruit. **FLOWERS** bright pink-purple or purple. Alpine tundra, spruce-fir, and aspen communities, on gravels and in meadows COLO: LPl, SJn. 3260–3810 m (10700–12500′). Flowering in summer. Northeast Utah and southwest Colorado. Large-flowered materials from the Uinta Mountains of Utah and from the Rockies of southwestern Colorado have long been identified as *O. deflexa* var. *deflexa*. Examination of critical material, including types from Siberia, has demonstrated that these American materials are distinct from all Siberian phases of the species.

var. *sericea* Torr. & A. Gray **CP** (of silk) Silky oxytrope. [*O. retrorsa* var. *sericea* (Torr. & A. Gray) Fernald; *O. deflexa* subsp. *foliolosa* (Hook.) Cody; *O. retrorsa* Fernald; *O. deflexa* subsp. *retrorsa* (Fernald) Á. Löve & D. Löve; *O. deflexa* var. *culminis*

Jeps.; *O. deflexa* var. *parviflora* B. Boivin]. Typically caulescent, rarely acaulescent. **INFL** a raceme, usually 10–25-flowered, normally much elongated in fruit. **FLOWERS** typically dirty white or variously suffused with purple. $n = 8$ (*Spellenberg 4002*, NMC). Spruce-fir, other conifer, aspen, willow, alder, and meadow communities. COLO: LPl, Min, SJn. 2600–3585 m (8535–11760′). Flowering in summer. Alaska, Yukon, Northwest Territories, British Columbia, Alberta, Saskatchewan, Manitoba; Washington, California, Idaho, Montana, Colorado, New Mexico, Oregon, Utah, and Wyoming.

Oxytropis lambertii

Oxytropis lambertii Pursh (for Aylmer Bourke Lambert) Lambert's locoweed, purple locoweed. Caespitose, acaulescent, (10) 14–50 cm tall. **PUBESCENCE** malpighian, strigose. **LEAVES** 3–24 cm long; stipules pilose; leaflets 7–13, 7–45 mm long, 2–8 mm wide, lanceolate to oblong or linear; scapes 4–28 (45) cm long, strigose. **INFL** a raceme, 8- to 40-flowered, the flowers ascending at anthesis, the axis 3–23 cm long in fruit; bracts strigose. **FLOWERS** 17–25 mm long, pink-purple; calyx 6.5–10 mm long, the cylindric tube 4.5–7.5 mm long, the teeth 1.5–4.5 mm long, subulate. **PODS** sessile or shortly stipitate, erect or ascending at maturity, cylindroid to lance-acuminate in outline, the body 15–27 mm long, 2.5–6 mm thick, bilocular, strigose to strigulose. $n = 16$. Mixed desert shrub, piñon-juniper, sagebrush, and grass communities. ARIZ: Apa, Nav; COLO: SMg; NMEX: McK, RAr, SJn; UTAH. 1675–2285 m (5500–7500′). Flowering: May–Jul. The species ranges from Saskatchewan and Manitoba southward. Our material belongs to the southern montane and steppe **var. *bigelovii*** A. Gray (for John Milton Bigelow, botanist and surgeon). [*Aragallus bigelovii* (A. Gray) Greene; *Astragalus lambertii* Pursh (Spreng.) var. *bigelovii* (A. Gray) Tidestr.; *O. lambertii* subsp. *bigelovii* (A. Gray) W. A. Weber]. This species, like its near congener *O. sericea*, produces locoism in livestock, due to the presence of indolizidine alkaloids in at least some populations.

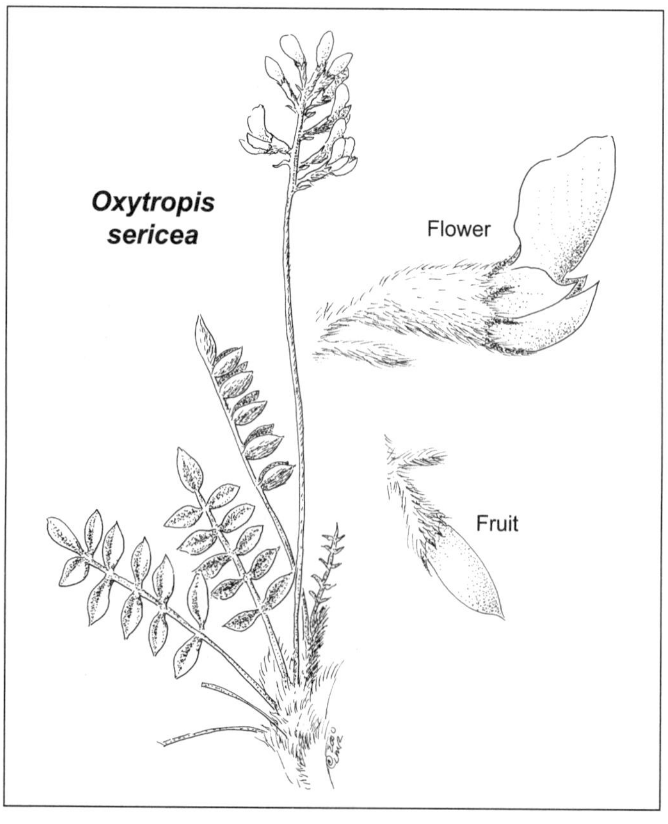

Oxytropis sericea

Oxytropis sericea Nutt. (of silk) Silky or white locoweed. Caespitose, acaulescent, 13–32 cm tall. **PUBESCENCE** basifixed, silky pilose. **LEAVES** 3.5–21 cm long; stipules pilose to subtomentose; leaflets 9–23, 4–32 (40) mm long, 1.5–10 mm wide, lanceolate to oblong, elliptic, or ovate, pilose; scapes 7–26 cm long, pilose. **INFL** a raceme, 6- to 27-flowered, the flowers ascending to spreading, the axis 1.5–12 cm long in fruit; bracts pilose. **FLOWERS** 15–26 mm long, white or tinged with purple; calyx 8–12 mm long, the tube cylindric, the teeth triangular to subulate. **PODS** erect, sessile, the body subcylindric to ovoid-oblong, 10–25 mm long, 4–7.5 mm thick, bilocular or nearly so, strigose or pilosulous. [*O. lambertii* Pursh var. *sericea* (Nutt.) A. Gray; *Spiesia lambertii* (Pursh) Kuntze var. *sericea* (Nutt.) Rydb.;

Aragallus sericeus (Nutt.) Greene; *Aragallus lambertii* (Pursh) Greene var. *sericeus* (Nutt.) A. Nelson; *Aragallus majusculus* Greene]. Sagebrush, piñon-juniper, and grassland (rarely mixed desert shrub) communities. NMEX: McK, RAr. 2135 m (7000′). Flowering: May–Aug. Yukon south to Nevada, New Mexico, and Oklahoma. Our material belongs to **var. *sericea***. The white locoweed contains indolizidine alkaloids associated with so-called locoism in livestock. Another syndrome in cattle, brisket disease, is also symptomatic of poisoning due to this plant. Livestock should not be grazed where the plant is common, or the animals must be closely monitored and removed from the range to other feeding stations free of the plants, should symptoms appear.

Parryella Torr. & A. Gray Dunebroom

(named for Charles Christopher Parry) Unarmed shrubs. **LEAVES** alternate, pinnate, with numerous leaflets, glandular-dotted; stipules subulate, caducous; peduncles opposite the leaves. **INFL** in a loose, spicate raceme or panicle. **FLOWER** calyx turbinate-campanulate, 10-ribbed near the base, 5-lobed; petals lacking; stamens 10, the filaments distinct, inserted on the hypanthium. **PODS** 1-seeded (2-ovuled), indehiscent, obliquely ovoid, glandular-dotted, exserted from the calyx. (Rydberg 1919a.) One species. Southwest United States and Mexico.

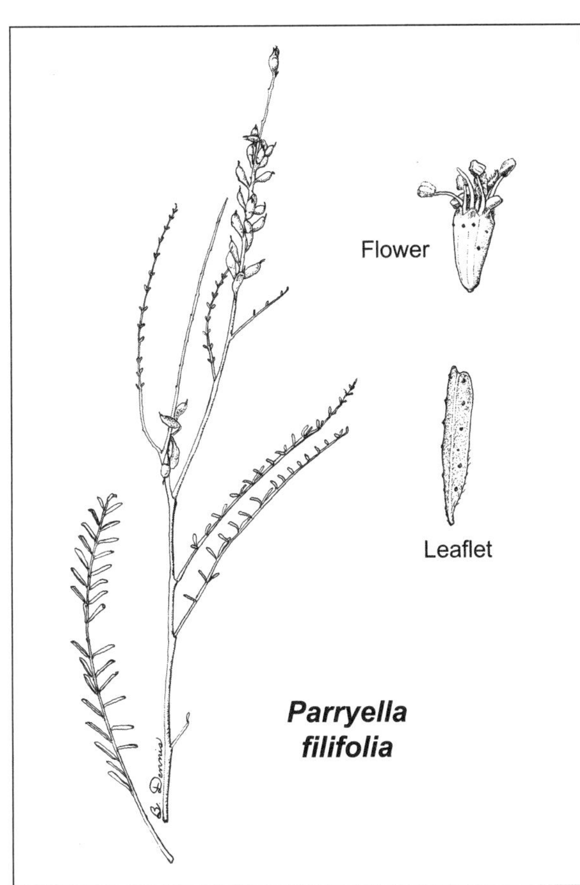

Parryella filifolia Torr. & A. Gray ex A. Gray (referring to the threadlike leaves) Narrow-leaf dunebroom. Shrubs to 15 dm tall or more, often partially buried in sand, the branchlets strigose and with mammiform, glandular protuberances. **LEAVES** 3.5–13 cm long; stipules 1–2 mm long, chestnut-brown, fragile; leaflets 8–40, 1–21 mm long, linear (rarely oblong), all involute, strigose to glabrate and glandular on the visible surface. **INFL** a raceme, 4–10 cm long, the main ones often branched near the base, forming panicles, 35- to 90-flowered, the axis not much elongating in fruit; bracts reduced to gland-tipped vestiges; peduncle 0–1 cm long. **FLOWERS** 5–6.5 mm long in anthesis, calyx 2.5–3.1 mm long, 10-ribbed near the base, the tube opaque, the teeth 0.2–0.5 mm long, ciliate; stamens long-exserted. **PODS** short-stipitate, the body 5–6.5 mm long, 2.5–3 mm wide, the pilose style base persistent, glabrous, glandular-punctate. n = 10. Stabilized dune sands with sand sagebrush, purple sage, blackbrush, Indian ricegrass, and four-wing saltbush, or less commonly in sandstone pockets or talus with skunkbrush, singleleaf ash, cliff rose, and rabbitbrush. ARIZ: Apa; COLO: Mon; NMEX: McK, SJn. 1625–1900 m (5330–6235′). Flowering: Jun–Sep. Grand and San Juan counties, Utah; Montezuma County, Colorado; northeast Arizona; and northwest New Mexico. This plant shows promise for use in reclamation of sandy sites. Hopis used the beans for toothaches, the plant for basketry and kachina masks, and the roots for stable gear.

Pediomelum Rydb. Breadroot

(Greek *pedion*, plain, + *melon*, apple) Perennial herbs, unarmed, from deep-seated, tuberous roots. **LEAVES** alternate; stipules triangular to subulate; blade palmately 3- to 7-foliolate, glandular-dotted. **INFL** borne in an axillary raceme or spikelike raceme. **FLOWER** petals blue to purple, lavender, or white; calyx subcylindric, gibbous at the base on the upper side, 5-lobed, the lowest lobe the longest; stamens 10, diadelphous (rarely all connate). **PODS** 1-seeded, irregularly dehiscent, usually included within the calyx. Note: The name is from the Greek words for prairie and apple, a literal translation of the French "pomme de prairie," a reference to the edible, tuberous roots of most species. The species have been traditionally assigned within an expanded *Psoralea* L., a genus now restricted to South African plants (see also *Psoralidium*) (Rydberg 1919b; Ockendon 1965). Fifty species. Mostly North America.

Pediomelum megalanthum (Wooton & Standl.) Rydb. (referring to the large flower) Large-flowered breadroot. Subacaulescent to caulescent, 4–25 cm tall, from slender, subterranean caudex branches arising from deep-seated tuberous roots. **STEMS** with 0–5 elongated internodes, incurved-strigose or with a few (rarely most) hairs spreading-ascending. **LEAVES** mainly 5 (8)-foliolate; stipules scarious, 2–15 mm long; petioles 1.2–9.5 cm long, pubescent like the stems or more commonly spreading-hairy; leaflets 9–34 mm long, 4–23 mm wide, cuneate-obovate to subrhombic, gray-green, strigulose, and punctate beneath, yellow-green, punctate, and strigose overall above. **INFL** a raceme, mainly 6- to 24-flowered, 2–5 cm long; pedicels 1.5–5 mm long; bracts commonly bidentate, lance-ovate, 3–12 mm long; peduncle 1.4–5 cm long. **FLOWERS** 12.5–21 mm long, the banner, wings, and keel commonly purple or suffused with purple; calyx 12.5–18.5 mm long, the tube 5.5–8 (9) mm long, only somewhat broader than the lateral ones. **PODS** included in the calyx. [*Psoralea megalantha* Wooton & Standl.]. Mixed desert shrub, piñon-juniper woodland, and blackbrush communities. ARIZ: Apa; NMEX: SJn; UTAH. 1750–2160 m (5740–7085′). Flowering: Apr–Jun. Four Corners material belongs to **var. *megalanthum***. Eastern Utah and adjacent Colorado to northwest New Mexico and northeast Arizona.

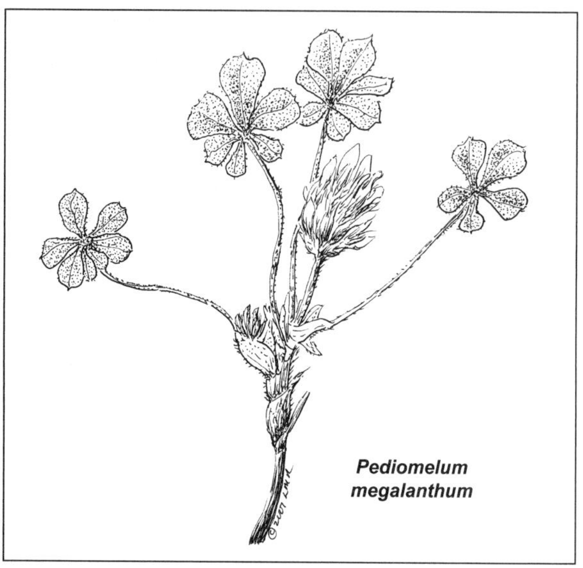
Pediomelum megalanthum

Phaseolus L. Bean

(Greek for an edible bean) Annual or perennial herbs, from a taproot. **LEAVES** alternate; stipules herbaceous, distinct; pinnately trifoliolate. **INFL** axillary or in an axillary raceme, each subtended by a bract; bracteoles 2, attached at base of calyx. **FLOWER** papilionaceous; petals 5, pink to purplish or white; calyx 5-toothed; stamens 10, diadelphous; ovary few- to several-ovuled, the style twisted or coiled in the keel, bearded toward the apex, the stigma oblique. **PODS** linear to oblong, laterally flattened to subterete, the valves coiling upon dehiscence. Fifty species. Tropics and the deserts of America.

Phaseolus angustissimus A. Gray (referring to the narrow leaves) Perennial, decumbent or clambering. **STEMS** 1.5–5 (10) dm long, glabrous, generally clustered and often branched. **LEAVES** mainly 2–6 cm long or more; stipules 2–3 mm long, often reflexed; leaflets 3, deltoid to oblong-lanceolate or linear-lanceolate, often asymmetric and lobed at the base. **INFL** 2–20 cm long, with flower clusters in the upper bracteate axils. **FLOWER** petals 6–10 mm long, lavender to pink; calyx 2–3 mm long. **PODS** oblong-falcate, 1.5–2 cm long, 6–7 mm wide, the beak filiform, about 2 mm long; seeds rugose-warty. [*P. dilatatus* Wooton & Standl.]. Mixed desert shrublands with ephedra, *Gutierrezia*, Bigelow sagebrush, shadscale, and galleta. ARIZ: Apa; NMEX: McK. 1600–1915 m (5245–6285′). Flowering: Jun–Sep. West Texas and west across New Mexico and Arizona.

Psoralidium Rydb. Scurfpea

(diminutive of *Psoralea*) Perennial, unarmed herbs, from rhizomes. **LEAVES** alternate; stipules triangular to subulate; pedately 3-foliolate, glandular-dotted. **INFL** an axillary, pedunculate, and interrupted spikelike raceme. **FLOWER** petals bluish or purplish to lavender, or less commonly white; calyx campanulate, the tube short, 5-lobed, the lowermost lobe typically longer than the others; corolla papilionaceous; stamens 10, diadelphous. **PODS** 1-seeded, indehiscent, usually not included in the calyx (Rydberg 1919c).

1. Peduncle 15–40 cm long or more; leaves few, mostly deciduous by anthesis, the leaflets sharply acuminate; plants of southeast Utah .... ***P. junceum***
1′ Peduncle mainly shorter than 15 cm; leaves numerous, persisting through the season, the leaflets obtuse to rounded or cuspidate; distribution various .... (2)

2. Plants commonly 4–10 dm tall, at least some leaves 4- to 5-foliolate; flowers mainly indigo .... ***P. tenuiflorum***
2′ Plants often less than 4 dm tall, all leaves commonly 3-foliolate; flowers white to purple .... ***P. lanceolatum***

Psoralidium junceum (Eastw.) Rydb. (rushlike) Rush scurfpea. Caulescent, 4.8–9 dm tall, from a rhizome. **STEMS** with 5 or more elongated internodes, strigose. **LEAVES** commonly 3-foliolate; stipules acuminate, strigose; often deciduous by anthesis; petiole 1.4–7 cm long, pubescent like the stems; leaflets 3, 19–44 mm long, 3–7 mm wide, oblanceolate to elliptic, acuminate apically, strigose and glandular on both surfaces, greenish; peduncle (8) 11–48 cm long. **INFL** a raceme, 7- to 20-flowered, 5–11 cm long; bracts lance-acuminate, glabrous dorsally, 1.5–2.5 mm long, deciduous. **FLOWERS** 4.2–5.8 mm long, the petals indigo; calyx 2.7–3.4 mm long, the tube 1.3–2.5 mm long, campanulate, not especially gibbous, the lower tooth 0.7–1.4 mm long, longer than the others. **PODS** 1-seeded, densely silky villous. [*Psoralea juncea* Eastw.]. Stabilized dunes and other sandy sites, often growing with *Amsonia tomentosa*, Indian ricegrass, resinbush, purple sage, sand sagebrush, and blackbrush, and less commonly with juniper. UTAH. 1145–1710 m (3760–5610′). Flowering: Apr–Jun. Southeast Utah; Coconino County, Arizona. A Glen Canyon–San Juan regional endemic.

Psoralidium lanceolatum (Pursh) Rydb. (lance-shaped) Dune scurfpea. Caulescent, 1.5–6.8 dm tall, from a rhizome, clump-forming, the stems with 5 or more elongated internodes, glabrous or strigose. **LEAVES** persistent at flowering; stipules lance-attenuate, 3–16 mm long; petiole 0.8–3 cm long, strigose to glabrate; leaflets 3, 14–50 mm long, 0.5–9 mm wide, oblanceolate below, becoming linear upward, obtuse to acute or cuspidate, sparingly strigose on both sides or glabrous above, yellow-green. **INFL** a raceme, 5- to 41-flowered, 1.2–17 cm long; bracts ovate to elliptic or lanceolate, glabrous dorsally, 1.3–2.8 mm long, persistent; peduncles 3.3–24 cm long. **FLOWERS** 4.8–6.3 mm long, blue, white, or bicolored; calyx 1.6–2.8 mm long, the tube 1.3–2 mm long, campanulate, not especially gibbous, the lower tooth 0.2–0.8 mm long, not much larger than the others. **PODS** 1-seeded, conspicuously glandular. $2n = 22$. [*Psoralea lanceolata* Pursh; *P. elliptica* Pursh; *Lotodes ellipticum* (Pursh) Kuntze; *Psoralea laxiflora* Nutt., not (Pursh) Poir.; *Psoralea micrantha* A. Gray ex Torr.; *Psoralidium micranthum* (A. Gray) Rydb.]. Trends in morphological variation are apparent among Utah specimens; in their extremes they are readily recognized, but whether they represent taxa is open to question. Navajos used the plant as a poultice for itches and sores, for stomachache, venereal disease, menstrual pain, and for protection against witchcraft. Zunis ate fresh flowers for stomachaches.

1. Racemes compact to moderately lax, 1.2–5.5 cm long at anthesis, the flower nodes seldom widely spaced or mainly with more than 3 flowers per node; blue and white flowers variably abundant; plants with a widespread distribution .... **var. *lanceolatum***
1′ Racemes lax, 2.5–17 cm long at anthesis, the flower nodes widely separated, often with 2 or 3 flowers per node; blue flower color predominating; plants of southeastern Utah and Rio Arriba County, New Mexico **var. *stenophyllum***

var. *lanceolatum* Plains scurfpea. Sand dunes and other sandy sites. ARIZ: Apa, Nav; COLO: Arc, LPl, Mon; NMEX: McK, RAr, San, SJn; UTAH. 1100–2130 m (3610–6995′). Flowering: May–Jul. Washington and Saskatchewan, south to California, Arizona, New Mexico, and Oklahoma.

var. *stenophyllum* (Rydb.) S. L. Welsh (narrow-leaved) Slenderleaf scurfpea [*Psoralea stenophylla* Rydb.; *Psoralidium stenophyllum* (Rydb.) Rydb.; *Psoralea lanceolata* Pursh var. *stenophylla* (Rydb.) Toft & S. L. Welsh]. Sandy sites. ARIZ: Apa; NMEX: RAr; UTAH. 1150–2125 m (3775–7000′). Flowering: May–Jul. Southeast Utah, northeast Arizona, and northwest New Mexico.

Psoralidium tenuiflorum (Pursh) Rydb. (slender-flowered) Prairie scurfpea. Caulescent, 4–10 dm tall or more, forming large clumps. **STEMS** with 8 or more elongated internodes, strigose. **LEAVES** with scarious stipules 2–13 mm long, strigose; 3- to 5-foliolate; persistent at flowering; petioles 0.1–2.2 cm long, strigose to strigulose; leaflets 6–40 mm long, 2–16 mm wide, oblanceolate throughout, mainly rounded to a cuspidate apex, gray-green and strigose beneath, yellow-green and glabrous or pubescent only along the veins above. **INFL** a raceme, mainly 7- to 21-flowered, 1–5.9 cm long; bracts ovate-acuminate, glabrous dorsally, 1.5–2.5 mm long, persistent; peduncle 0.5–7 cm long or lacking, sometimes bracteate or some flowers axillary. **FLOWERS** 4.5–6 cm long, the petals indigo; calyx 2.5–3.2 mm long, the tube 1.5–2 mm long, campanulate, not gibbous, the lower tooth 0.8–1.7 mm long, noticeably larger than the others. **PODS** 1-seeded, glabrous, conspicuously glandular. [*Psoralea tenuiflora* Pursh; *Psoralea floribunda* Nutt. ex Torr. & A. Gray; *Psoralidium floribundum* (Nutt. ex Torr. & A. Gray) Rydb.; *Psoralea obtusiloba* Torr. & A. Gray; *Psoralidium bigelovii* Rydb.; *Psoralea bigelovii* (Rydb.) Tidestr.; *Psoralea tenuiflora* var. *bigelovii* (Rydb.) J. F. Macbr.]. Piñon-juniper, sagebrush, mountain brush, and ponderosa pine communities. ARIZ: Apa; COLO: Arc, LPl; NMEX: McK, RAr. 1895–2510 m (6215–8230′). Flowering: Apr–Sep. North Dakota and Montana, south to Arizona, New Mexico, Texas, and Mexico. The segregation of varieties from among our materials seems unwarranted. Our material represents the edge of the range for this species more common on the prairies and plains of the Great Plains. Navajos smoked the leaves for influenza and the Night Chant. Zunis made a poultice for body purification.

Psorothamnus Rydb. Dalea

(scurfy shrub) Shrubs, typically armed. **LEAVES** alternate; stipules subulate or vestigial; odd-pinnate, with 5 or more leaflets, glandular-dotted. **INFL** in a lax raceme; peduncle opposite the leaves. **FLOWERS** papilionaceous; petals mainly indigo; calyx campanulate, 10-ribbed, 5-lobed; the petals all inserted on the hypanthium; stamens 9 or 10, monadelphous. **PODS** 1- to 2-seeded, indehiscent, exserted from the calyx. $x = 10$. Note: Traditional treatments have placed the species of *Psorothamnus* within an expanded *Dalea* (Rydberg 1919d).

1. Leaves unifoliolate (often deciduous by anthesis), oblanceolate to linear .... ***P. scoparius***
1' Leaves plurifoliolate (usually present) ....(2)

2. Branchlets densely reflexed-puberulent, bearing conspicuous yellow to red-orange resinous glands ...***P. thompsoniae***
2' Branchlets merely appressed-strigose, lacking glands or only obscurely glandular .... ***P. fremontii***

Psorothamnus fremonti (Torr.) Barneby (for John Charles Frémont, explorer, general, naturalist) Frémont's dalea. Armed shrubs, 5–15 dm tall or more; branchlets strigose, sparingly if at all glandular. **LEAVES** 1.8–6.5 cm long; stipules 0.3–14 mm long, fragile; leaflets (1) 3–9, 3–14 mm long, 0.8–6 mm wide, glandular beneath, strigose on both sides, linear to oblong or elliptic, obtuse to rounded, the uppermost often confluent with the rachis. **INFL** a raceme, 8- to 41-flowered, 3.5–14.3 cm long, the rachis strigose to strigulose; bracts 0.8–1.8 mm long, lanceolate, glabrous or else foliose and with 1–4 leaflets; peduncle 0–2.2 cm long, or some flowers axillary. **FLOWERS** 8.5–12 mm long, indigo, rarely white; calyx 4.5–6.5 mm long, the tube 2.5–3.3 mm long, obscurely if at all 10-ribbed, appressed-strigose to glabrate, the teeth 1.8–3.2 mm long, strigose, the upper markedly wider than the others. Blackbrush, mixed desert shrub, and (less commonly) piñon-juniper communities. ARIZ: Nav; UTAH. 1110–1490 m (3640–4890′). Flowering: Apr–Jun. Southern Utah, Arizona, Nevada, and California. The plants are strikingly beautiful, contrasting indigo flowers with grayish foliage. Much of the material, if not all, from along Lake Powell shows the influence of *P. arborescens* (Torr. ex A. Gray) Barneby (i.e., very elongate, lance-linear calyx teeth) and might be best assigned to that species. Decoction of root or plant top taken as an antihemorrhagic.

Psorothamnus scoparius (A. Gray) Rydb. (broomlike) Broom dalea. Unarmed shrubs, mainly to about 1 m tall or more; branchlets rather copiously glandular. **LEAVES** 5–20 mm long, soon deciduous; stipules very small or lacking; unifoliolate (rarely some trifoliolate), petiolate to subsessile, the blades linear-oblanceolate to oblong. **INFL** a spike, subcapitate, 6–15-flowered, 8–12 mm wide; pedicels very short; peduncle 0.5–3.5 cm long. **FLOWER** petals 6–9 mm long, blue to violet; calyx 3.5–4.5 mm long, pilosulous, the lobes unequal, shorter than the tube. **PODS** about 4 mm long, somewhat exserted from the calyx. [*Dalea scoparia* A. Gray]. Growing with four-wing saltbush, sand sagebrush, and other psammophytes. ARIZ: Nav; NMEX: San, SJn. 1455 m (4800′). Flowering: Aug–Sep.

Psorothamnus thompsoniae (Vail) S. L. Welsh & N. D. Atwood (for Ellen P. Thompson, sister of Major John Powell) Thompson's indigo-bush or dalea. Armed shrubs, 2.5–8 dm tall or more; branchlets velvety with retrorse, short hairs, conspicuously glandular with yellow to orange-red, mammiform, resinous glands. **LEAVES** 0.7–5 cm long; stipules vestigial or to 0.8 mm long; leaflets 7–17, 1–10 mm long, 0.6–4 mm wide, linear to oblong, oval, or obcordate, glandular and strigose to glabrate beneath, strigose above, the uppermost jointed to the rachis. **INFL** a raceme, mainly 8- to 25-flowered, 2–9 cm long, the rachis retrorsely hairy; bracts 0.5–1.5 mm long, lanceolate, glabrous or hairy, soon deciduous; peduncle 0.4–1.8 cm long, rarely obsolete and some flowers axillary. **FLOWERS** 7.8–10.8 mm long, indigo or purple-pink; calyx 3.7–5 mm long, the tube 2.1–3.2 mm long, conspicuously 10-ribbed, glabrous or villous, ovate to oval, obtuse or only the lowermost acute. **PODS** glandular-dotted. $n = 10$. [*Parosela thompsoniae* Vail]. Two more or less distinctive and easily recognizable varieties are known.

1. Calyx tube glabrous; leaflets oblong to oval or obcordate; plants of San Juan County and elsewhere .... **var. *thompsoniae***
1' Calyx tube villous; leaflets linear to narrowly elliptic or oblong; plants of San Juan County .... **var. *whitingii***

var. *thompsoniae* [*Parosela thompsoniae* Vail; *Dalea thompsoniae* (Vail) L. O. Williams] Blackbrush, shadscale, piñon-juniper, ephedra, galleta, sand sagebrush, resinbush, and *Ephedra* communities, in sand and among boulders. UTAH. 1135–1675 m (3730–5500′). Flowering: May–Aug. Southeastern Utah. The plants grow well on sand, even dunes, but also occur in fine-textured saline soils and talus.

var. *whitingii* (Kearney & Peebles) Barneby (for Alfred F. Whiting, botanist, anthropologist, and ethnobotanist) [*Dalea whitingii* Kearney & Peebles]. Whiting's indigo-bush. Mixed desert shrub community. UTAH. 1410–1520 m (4620–5030′).

Flowering: May–Aug. Monument Valley and west to Navajo Mountain, south of the San Juan River, Utah (San Juan County); Wupatki National Monument, Coconino County, Arizona. Navajo Basin endemic.

Robinia L. Locust

(for Jean and his son Vespasien Robin, 17th-century herbalists and gardeners) Shrubs or trees, often armed. **LEAVES** alternate; stipules setaceous and caducous or persistent as spines; odd-pinnate, the leaflets petiolulate and stipulate. **INFL** an axillary raceme. **FLOWER** corolla white, pinkish, or pink, very showy, often aromatic; papilionaceous; calyx campanulate to turbinate, the 5 teeth triangular to triangular-acuminate; stamens 10, diadelphous; ovary subsessile or sessile. **PODS** oblong, straight, laterally flattened, tardily dehiscent. $x = 10, 11$ (Isely and Peabody 1984). Four to five species found in North America.

1. Uppermost pair of calyx teeth free for about 2/3 of their length; branchlets and often the peduncle hispid .... ***R. neomexicana***
1′ Uppermost pair of calyx teeth connate almost to the apex, forming an emarginate lip; branchlets and peduncle lacking hispid processes .... ***R. pseudoacacia***

Robinia neomexicana A. Gray (of New Mexico) New Mexico locust. Small trees or shrubs, mainly 1–8 m tall; branchlets villosulous, rarely glandular. **LEAVES** 8–20 (28) cm long; stipules 3–10 mm long or more, often spiny; leaflets 9–19, 12–40 mm long, 7–20 mm wide, lance-oblong to oblong, obtuse, cuspidate, sparingly pubescent above and below, finally glabrate on one or both sides; petioles villosulous near the base. **INFL** a raceme, 3- to 14-flowered, 3–8 cm long; peduncle 1–4 cm long, glandular-pubescent to hispid throughout. **FLOWERS** 16–24 mm long, calyx campanulate, the tube 5 mm long, the teeth 3–5 mm long, triangular-acuminate; corolla pale pink or bright pink-purple. **PODS** 40–80 mm long, glandular-pubescent to hispid or glabrous. [*R. neomexicana* var. *luxurians* Dieck; *R. luxurians* (Dieck) Rydb.; *R. breviloba* Rydb.; *R. subvelutina* Rydb.]. Mesic, often shaded talus slopes and river terraces in cottonwood, sagebrush, and piñon-juniper communities. ARIZ: Nav; COLO: Arc, LPl, Mon; NMEX: SJn. 1610–2070 m (5280–6800′). Flowering: Jun. Nevada, Arizona, New Mexico, Texas, and Mexico. Used by Hopis to purify the stomach. Apaches ate the raw pods as food and made high quality bows from the wood.

Robinia pseudoacacia L. (false acacia) Black locust, false acacia. Trees to 25 m tall or more and to 1.5 m in diameter; branchlets puberulent to villosulous, lacking glands. **LEAVES** (6) 8.5–26 cm long; stipules minute or represented by spines; leaflets (5) 11–25, 11–60 mm long, 6–30 (38) mm wide, lance-oblong to oblong, obtuse to retuse or cuspidate, puberulent to glabrate on both sides; petioles villosulous to pilose near the base. **INFL** a raceme, 11- to 27-flowered, 4–13 cm long; peduncle 1.4–4.3 cm long, puberulent to villosulous. **FLOWERS** 12–20 (23) mm long; calyx broadly campanulate to turbinate, the tube 3.5–5.5 mm long, the teeth 1.5–2 mm long, the upper pair connate, the sinus shallow; corolla white (fading cream), or pale pink. **PODS** 4–12 cm long, glabrous. $2n = 22$. Cultivated ornamental, street, and shade trees, long persisting, escaping, and established along fencerows and stream bottoms. ARIZ: Nav; COLO: Dol; NMEX: McK, SJn; UTAH. 1350–2100 m (4425–6895′). Flowering: Jun. Indigenous in the summer deciduous forest of the eastern United States. This is a handsome shade tree with very hard wood, which has been grown in the Four Corners region since earliest pioneer times. The fragrant flowers yield nectar for excellent honey, and the wood has been used for fence posts. The seeds have been implicated as poisonous to livestock. The herbage evidently contains phytotoxins.

Sophora L. Sophora

(Arabic *sofera*, a plant with yellow flowers) Perennial herbs, unarmed. **LEAVES** alternate; stipules obsolete or herbaceous; odd-pinnately compound. **INFL** in a terminal raceme. **FLOWERS** perfect, papilionaceous; calyx 5-toothed; petals all distinct; stamens 10, distinct or essentially so; ovary stipitate. **PODS** spreading to pendulous, subterete, constricted between the seeds, indehiscent or tardily so. $x = 9$. Fifty-two species. Mostly tropical and north-temperate.

1. Flower white to cream; leaflets less than 5 times longer than broad, gray-green .... ***S. nuttalliana***
1′ Flower blue-purple to blue; leaflets more than 5 times longer than broad, silvery hairy .... ***S. stenophylla***

Sophora nuttalliana B. L. Turner (for Thomas Nuttall, English-American naturalist and botanist) Silky sophora, white sophora. Perennial, caulescent, 12–27 (30) cm tall, from a rhizome. **PUBESCENCE** essentially basifixed. **STEMS** ascending to erect. **LEAVES** 2.5–6 (7) cm long; stipules 1.5–4 mm long, distinct, caducous; leaflets 13–23, 2–11 mm long, 0.8–6 mm wide, oblong-obovate to obovate, rounded to retuse, often folded, strigose beneath, glabrous above. **INFL** a raceme, mainly 8- to 52-flowered, 3–9 cm long at anthesis, the axis 3–13 cm long in fruit; bracts 3–7 mm long; pedicels 2–4 mm long; bracteoles 1; peduncle 0.8–3 cm long. **FLOWERS** 14–19 mm long, white to cream (fading cream); calyx 8–10.5 mm long, gibbous, the tube 6.4–8.5 mm long, obliquely short-cylindric, the teeth 2–2.5 mm long, triangular. **PODS** erect-ascending, stipitate, the stipe 6–12 mm long, the body 12–40 mm long, 3–4.5 mm thick, constricted tightly between the usually 1–3 seeds, strigose. [*S. sericea* Nutt., not Duham.]. River bottoms and roadsides in various plant communities. ARIZ: Apa; COLO: Arc, LPl, Mon; NMEX: McK, RAr, SJn. 1760–2310 m (5780–7580′). Flowering: May–Jun. Wyoming and South Dakota, south to Arizona, Colorado, New Mexico, and Texas. The terminal inflorescences immediately distinguish this from an *Astragalus*, with which it is occasionally confused. Used by Navajos as sheep forage.

Sophora stenophylla A. Gray **CP** (narrow-leaved) Silvery sophora. Perennial, caulescent, 13–41 cm tall (aboveground), from deeply seated rhizomes. **PUBESCENCE** basifixed. **STEMS** ascending to erect. **LEAVES** with stipules 3–12 mm long, distinct, caducous, or obsolete; 1.7–5.6 mm wide, linear to narrowly oblong, acute to attenuate, silvery pilosulous, the pubescence fading yellowish in time. **INFL** a raceme, mainly 12- to 39-flowered, 5–17 cm long in anthesis, 6–23.5 cm long in fruit; bracts 3–7 mm long; pedicels 1–8 mm long; bracteoles 0; peduncle 1.7–5 cm long. **FLOWERS** 15–27 mm long, blue-purple to blue; calyx 6.5–10.8 mm long, gibbous, the tube 4.8–7.2 mm long, obliquely short-cylindric, the teeth 1.7–3.6 mm long, ovate-triangular. **PODS** spreading-declined, stipitate, the stipe 8–16 mm long, the body 15–60 mm long, 6–8 mm wide, strongly constricted between the usually 1–5 seeds, strigose. Sand dunes and other sandy sites, in sand sagebrush, eriogonum-ephedra, blackbrush, piñon-juniper, and ponderosa pine (less commonly salt desert shrub) communities. ARIZ: Apa, Nav; NMEX: McK, SJn; UTAH. 1370–1910 m (4500–6260′). Flowering: Apr–May. Southeast Utah, northwest New Mexico, and northeast Arizona. This is a strikingly beautiful plant with silvery foliage and large, blue, scented flowers.

Sphaerophysa DC. Swainsonpea

Perennial, caulescent, from rhizomes. **LEAVES** alternate, odd-pinnate; stipules adnate to the petiole base, all distinct. **INFL** borne in an axillary raceme, each subtended by a single bract; bracteoles 1. **FLOWERS** papilionaceous; petals 5, dull red, drying lavender to brown; calyx 5-toothed; stamens 10, diadelphous; ovary enfolded by the staminal sheath; style glabrous except for a tuft of hair below the stigma. **PODS** stipitate, bladdery-inflated, subunilocular. Two species. Eastern Mediterranean to central Asia.

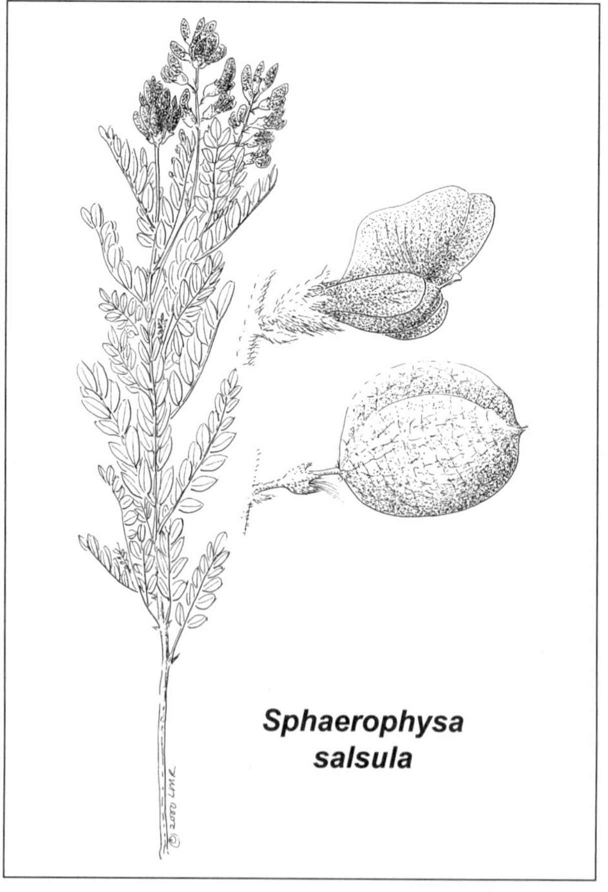

Sphaerophysa salsula (Pall.) DC. (refers to the alkaline habitat in which it grows) Alkali swainsonpea. Perennial, caulescent, 4–7 dm tall, from a deeply placed rhizome. **PUBESCENCE** basifixed. **LEAVES** 3–10 cm long; stipules 1–4 mm long, all distinct; leaflets 15–25, 3–18 mm long, 1–7 mm wide, oblong-obovate to elliptic, retuse to obtuse and apiculate, strigose beneath, glabrous above. **INFL** a raceme, 5- to 17-flowered, the flowers ascending in anthesis, finally nodding, the axis 2.5–9 cm long in fruit; bracts 1–2 mm long; pedicels 2.5–8 mm long; bracteoles 2; peduncle 2.5–9 cm long. **FLOWERS** 12–14 mm long, dull red, fading lavender to brown; calyx 5–6 mm long, the tube 3.8–4.6 mm long, campanulate, the teeth 1.2–2 mm long, triangular. **PODS** ascending to declined, stipitate, the stipe 4–7 mm long, the body bladdery-inflated, ovoid, 13–24 mm long, 9–20 mm wide (when pressed), unilocular, strigulose. [*Phaca salsula* Pall.; *Swainsonia salsula* Taubert]. Introduced weedy species along the San Juan River. NMEX: SJn (B-Square Ranch). 1615 m (5300′). Flowering: May–Jun. Widespread in the western United States; adventive from Asia. The species is locally common in Wyoming, Colorado, New Mexico, and Arizona, and less so, evidently, in Idaho and Nevada. This plant resembles an *Astragalus* species and has been mistaken twice as belonging to previously undescribed and unnamed indigenous species (*A. violaceus* H. St. John; *A. iochrous* Barneby). Generic concepts involving this genus are unresolved, and it seems likely that the plants will ultimately be placed in some earlier-named genus.

Thermopsis R. Br. False Lupine

(Greek *thermos*, lupine, + *opsi*, resemblance) Perennial herbs, caulescent, from rhizomes. **LEAVES** alternate, palmately trifoliolate; stipules foliaceous. **INFL** in a terminal raceme; bracts herbaceous, persistent. **FLOWERS** papilionaceous; calyx 5-toothed; petals 5, yellow or suffused with purple, the keel rounded; stamens 10, distinct; ovary stipitate, the style glabrous. **PODS** narrowly oblong, flattened, many-seeded. $x = 9$.

1. Pod straight, not especially lomentlike, erect or ascending; plants mostly 20–70 cm tall or more ....***T. montana***
1' Pod curved, lomentlike, spreading to recurved; plants mostly 14–40 cm tall ....***T. rhombifolia***

Thermopsis montana

Thermopsis montana Nutt. (of Montana) Golden pea, yellow pea. Caulescent, 2–7.5 (10) dm tall, the stems erect, pilosulous to glabrate. **LEAVES** with stipules foliar, lanceolate to ovate, 13–60 mm long, 3–30 mm wide; petioles 0.8–3.7 cm long; leaflets 3, 21–92 mm long, 5–36 mm wide, elliptic to lanceolate or oblanceolate, acute to rounded, pilosulous beneath, glabrous to glabrate above. **INFL** a raceme, mainly 2- to 23-flowered, 6–25 cm long in anthesis, the axis 9–28 cm long in fruit; peduncle 2.2–13 cm long; bracts 5–11 mm long. **FLOWERS** 20–26 (29) mm long, yellow; calyx 10.2–12.3 mm long, the tube 7–8.3 mm long, obliquely campanulate, the teeth 3.1–4 mm long, ovate-triangular. **PODS** erect or ascending, stipitate, the stipe 2.5–6 mm long, the body 40–54 mm long, 5–7 mm wide, pilose, stramineous or turning black. $2n = 18$. Moist sites along streams, in meadows, around seeps and springs. COLO: Arc, Dol, Hin, LPl, Mon; NMEX: McK, RAr, SJn; UTAH. 2410–2745 m (7910–9000′). Flowering: May–Jun. British Columbia south to Montana, California, Arizona, Colorado, New Mexico, and Utah. When included within an expanded *T. rhombifolia*, our material is perhaps best regarded as *T. rhombifolia* var. *montana* (Nutt.) Isely (Isely 1978). Ramah Navajos used the plant as a fumigant for headaches, sore eyes, and for sickness caused by hunting. The plant was also used as a lotion for protection from witches.

Thermopsis rhombifolia (Nutt. ex Pursh) Richardson (diamond-shaped leaves) Goldenbanner. Caulescent, 15–40 cm tall, the stems erect, glabrate. **LEAVES** with stipules foliose, ovate to lanceolate, 6–30 mm long, 2–22 mm wide; petioles 0.3–2.5 cm long; leaflets 3, 15–47 mm long, 7–25 mm wide, obovate to oblanceolate, obtuse to rounded, glabrous on both sides. **INFL** a raceme, mainly 4- to 30-flowered, 2–10 cm long in anthesis, the axis 2–12 cm long in fruit; peduncle 0.5–5.8 cm long; bracts simple to foliose. **FLOWERS** 18–22 mm long; calyx 7.5–10 mm long, the tube 4.5–6 mm long, the teeth 3–4 mm long, triangular-ovate. **PODS** divaricate, finally recurved, lomentlike, stipitate, the stipe 1.5–4 mm long, the body 25–70 mm long, 5–7 mm wide, pilose to glabrate. $2n = 18$. [*Cytisus rhombifolius* Nutt. ex Pursh]. Sandy and clay soils, mainly where moist, in mixed desert shrub and piñon-juniper communities. NMEX: McK, RAr, SJn. 1510–1810 m (4950–5940′). Flowering: May–Jun. Alberta and Saskatchewan, south to Colorado, New Mexico, and Nebraska.

Trifolium L. Clover

(three-parted leaves) Perennial or short-lived perennial or annual, caulescent or acaulescent, from taproot and caudex, rhizome, or stolon. **LEAVES** alternate; stipules membranous to foliaceous, often connate; palmately to pinnately 3-foliolate, or rarely 4- to 7-foliolate, commonly serrate throughout, rarely entire. **INFL** borne in a terminal or axillary, pedunculate to sessile, subcapitate head or raceme. **FLOWERS** papilionaceous; petals 5, pink, white, or red-purple, withering and persistent, finally investing the pod; calyx 5-toothed; stamens 10, diadelphous. **PODS** usually shorter than the calyx, indehiscent, 1- to several-seeded (Gillett 1965, 1969, 1971, 1972; Hermann 1953; McDermott 1910; Martin 1946).

1. Plants acaulescent or subacaulescent, mainly 1.5–10 cm tall .... (2)
1' Plants caulescent, with 1 or more elongated internodes (see also *T. parryi*), mainly 10–60 cm tall .... (8)

2. Head 1- to 4-flowered, essentially sessile, the flower 15–23 mm long; plants of high elevations .... ***T. nanum***
2' Head several-flowered, sessile or pedunculate, the flower either shorter or the head pedunculate; plants of various distribution .... (3)

3. Plants densely pulvinate-caespitose, matted, known from Navajo Mountain. .... ***T. andinum***
3' Plants typically not densely pulvinate-caespitose, or if so then of other distribution .... (4)

4. Leaflets toothed; calyces villosulous to pilosulous; flower less than 9 mm long; plants of moderate elevation, broadly distributed .... ***T. gymnocarpon***
4' Leaflets entire or, if toothed, the calyces glabrous; flower more than 9 mm long; plants of high elevations .... (5)

5. Leaflets entire .... (6)
5' Leaflets toothed .... (7)

6. Calyces and herbage strigose, less commonly glabrous; banner rounded to acute .... ***T. dasyphyllum***
6' Calyces and herbage villous-pilose; banner attenuate .... ***T. attenuatum***

7. Head subtended by a conspicuous involucre, the bracts often connate at least at the base and over 3 mm long .... ***T. parryi***
7' Head not subtended by an involucre, the bracts if present not at all connate .... ***T. brandegeei***

8. Plants stoloniferous, prostrate and rooting at the nodes; flowers white; calyx not bladdery-inflated; introduced, widespread .... ***T. repens***
8' Plants not stoloniferous (except in some *T. fragiferum*), usually erect or ascending, not or seldom rooting at the nodes; flower mainly pink to red-purple, though sometimes white .... (9)

9. Calyx soon bladdery-inflated and enclosing the corolla; plants introduced .... ***T. fragiferum***
9' Calyx not accrescent or only slightly so, never enclosing the corolla; plants various .... (10)

10. Head sessile or nearly so, commonly immediately subtended by a trifoliolate bract; plants cultivated, escaping, and persisting .... ***T. pratense***
10' Head with well-developed peduncle, not immediately subtended by foliose bracts; plants indigenous or cultivated .... (11)

11. Head subtended by a spinose-toothed involucre .... ***T. wormskjoldii***
11' Head lacking an involucre .... (12)

12. Flower mainly 7–9 mm long; head axillary, from the uppermost nodes; plants cultivated, escaping and persisting .... ***T. hybridum***
12' Flower 10–16 mm long; head terminal, solitary; plants indigenous .... ***T. longipes***

Trifolium andinum Nutt. (of the Andes Mountains) Andean clover. Apparently acaulescent (in reality short-caulescent, the stem with usually 2 few-flowered heads, each subtended by a bracteate leaf), densely pulvinate-caespitose, mat-forming, 0.3–5 cm tall, from woody caudex branches and a thick taproot, the stems obscured by imbricated stipules and persistent leaf bases; petioles 0–2.3 cm long. **LEAVES** with scarious stipules yellowish, glabrous except toward the tip, 3–9 mm long; leaflets 3, 2–12.5 mm long, 1.2–5 mm wide, oblanceolate to obovate, toothed in the apical 1/3, strigose to strigulose on both sides, commonly folded, abruptly acuminate. **INFL** a head, closely subtended by reduced leaflike bracts, these with or without a trifoliolate bract, sessile or appearing pedunculate by elongation of internodes on floriferous branches, the internodes glabrous. **FLOWERS** 7–15, in 2 closely associated heads, the banner violet-purple, the keel and wings ochroleucous, 9–12 mm long, on a pedicel 0.5–1 mm long, these glabrous to strigose; calyx 5–7 (8.5) mm long, the tube 2.2–3.6 (4) mm long, sparingly pilose to glabrous, the teeth 2–3.8 (4.5) mm long, lance-subulate, pilose. **PODS** 4–5 mm long, 2.3–2.7 mm wide. Ponderosa pine, Rocky Mountain juniper, and Douglas-fir–aspen communities. UTAH. 2800–3330 m (9195–10930′) Flowering: May–Jun. Navajo and Abajo mountains. Wyoming, Nevada, and Arizona.

Trifolium attenuatum Greene (weak or meager) Rocky Mountain clover. Acaulescent, densely caespitose, mat-forming or tufted, 8–26 cm tall, from a caudex and thick taproot, the stems obscured by imbricated stipules and marcescent leaf bases; petioles 1.5–10 cm long. **LEAVES** with scarious stipules, glabrous, strongly veined, 5–17 mm long; leaflets 3, 11–45 (60) mm long, 1–7 mm wide, narrowly lanceolate to lance-linear, entire, villous-pilose on both sides (sometimes sparingly so), commonly flat, attenuate or less commonly merely acute, sharply apiculate. **INFL** a head, 20–30 mm wide, terminal, on a peduncle 5–18 (23) cm long, these pilose, the head closely subtended by an inconspicuous involucre of ovate or ovate-attenuate, mainly diminutive scales, distinct almost or quite to the base. **FLOWERS** 12–29, erect, the banner violet or whitish, the keel and wings lavender-purple, 12–15 mm long, the banner attenuate apically; pedicel 0.5–1.5 mm long, finally deflexed; calyx 6.5–7.2 mm long, the tube 2.2–2.5 mm long, villous-pilose, the teeth 3.3–4.7 mm long, linear-subulate, villous-pilose. **PODS** 4–6 mm long. Alpine meadows and rock stripes or spruce-fir communities. COLO: Arc, Con, Hin, LPl, Min, RGr. 3050–3805 m (10000–12485′). Flowering: Jun–Aug. Southern Colorado and northern New Mexico. *Trifolium attenuatum* has been distinguished from *T. dasyphyllum* on the basis of shorter pedicels and flowers that ultimately are reflexed in late anthesis. Neither of these features is diagnostic when applied to most herbarium specimens from the San Juan drainage. However, the pubescence of the herbage, except for the stipules, is villous-pilose, not strigose, and the leaflets are lance-linear and long-attenuate, and evidently the pedicels do reflex at maturity. The type came from "near Pagosa Peak, southern Colorado, 6 Aug. 1899, C. F. Baker, No. 443." Greene (1900) noted that the "silky hairiness," attenuate banner apex, and pedicels that deflex in age are evidently significant and diagnostic, but the latter is difficult to discern except when plants are taken at the close of the flowering season. Other supposed differences noted by Greene, i.e., the more numerous flowers per head in *T. dasyphyllum*, were evidently based only on the solitary plant available to him. Longest leaflets certainly surpass in length those of *A. dasyphyllum*, a feature indicated by Greene.

Trifolium brandegeei S. Watson **CP** (for Townshend Stith Brandegee, botanist) Brandegee's clover. Acaulescent and caespitose, mainly 6–15 cm tall, the leaves and peduncles lax and ascending, the herbage glabrous or sparingly long-hairy. **LEAVES** with petioles elongate, much surpassing the leaflets in length; typically 3 leaflets, mainly 7–30 mm long and 5–15 mm wide, oval, obovate, or elliptic-oblong, entire to sharply denticulate, glabrous or glabrate. **INFL** a head; the involucre very small or lacking; peduncle surpassing the leaves. **FLOWERS** declined, calyx 6–10 mm long, glabrous or with scattered long hairs, green or purple, the teeth lanceolate or ovate-lanceolate, attenuate apically, as long as or longer than the tube; corolla 13–20 mm long, purple. **PODS** 2- to 7-seeded. High mountains in alpine plant communities and with krummholz. COLO: Arc, Con, LPl, Min, Mon, RGr, SJn. 3050–3740 m (10000–12275′). Flowering: Jun–Aug. Southwest Colorado and central New Mexico.

Trifolium dasyphyllum Torr. & A. Gray (hairy leaves) Thickleaf clover. Acaulescent, loosely mat-forming, 2–14 cm tall, from a caudex and thick taproot, the stems obscured by imbricated stipules and leaf bases; petioles 0.3–4 cm long. **LEAVES** with 3 leaflets, 3–28 mm long, 1–5 mm wide, oblanceolate, entire, strigose beneath, strigose to glabrate above, flat or folded, sharply apiculate; stipules scarious, glabrous or strigose, 5–17 mm long. **INFL** a head, closely subtended by a long-lobed involucre (the linear-subulate lobes distinct almost to the base, or rarely all reduced to triangular or ovate-apiculate scales), (3) 11–18 (30) mm wide, terminal, sessile or on a peduncle 0.5–10 (15–23) cm long, these strigose. **FLOWERS** 6–16, erect, the banner violet to ochroleucous, the keel and wings purple, or all purple or rarely white, 11–13 mm long; pedicel 0.5–1.5 mm long; calyx 5.7–9.1 mm long, the tube 2.5–2.9 mm long, strigose to glabrate, the teeth 1.9–6.4 mm long, subulate, strigose to glabrate. **PODS** 4–6 mm long. Alpine meadows and rock stripes. COLO: Arc, Hin, LPl, Min, Mon, SJn. 3350–3900 m (11000–12800′). Flowering: Jun–Sep. Montana, Wyoming, Colorado, and New Mexico. *Trifolium attenuatum* has been distinguished from *T. dasyphyllum* on the basis of shorter pedicels and flowers that ultimately are reflexed in late anthesis.

Trifolium fragiferum L. (strawberry-bearing) Strawberry clover. Caulescent, 5–30 cm long, rhizomatous and sometimes stoloniferous, decumbent to ascending. **LEAVES** with petioles 0.5–13 cm long, obovate, toothed from near the base, truncate to retuse and apiculate, flat, glabrous to glabrate on both sides; stipules scarious, 8–20 mm long. **INFL** a head, involucrate, many-flowered, subglobose, 10–22 mm wide, on a peduncle 2–17 cm long, these glabrous. **FLOWERS** 4–6 mm long, purplish, finally included within the accrescent calyx; calyx finally bladdery-inflated, pilose, reticulately veined around translucent lacunae. $2n = 16$. Meadows, roadsides, and other disturbed sites. NMEX: SJn. 1645 m (5400′). Flowering: Jul–Aug. Utah, Colorado, and New Mexico; native to Europe. Strawberry clover was evidently introduced as a forage and honey plant but is found most commonly as a weedy species in at least mildly saline sites.

Trifolium gymnocarpon Nutt. (naked fruit) Nuttall's clover, dwarf clover. Acaulescent to short-caulescent, 4–16 cm tall, from a caudex and taproot, the stems mainly obscured by imbricated stipules and leaf bases. **LEAVES** with stipules scarious to herbaceous, 6–23 mm long; petioles 1–9 cm long; leaflets 3–5, 6–23 mm long, 2–10 mm wide, elliptic to oblong, ovate, or obovate, toothed from near the base, pilosulous beneath, glabrous to pilosulous above, flat or folded, rounded to acute and apiculate. **INFL** a head, without an involucre, hemispheric, commonly 12–18 mm wide, terminal and axillary, on a peduncle 1–6.5 cm long, these strigulose, erect to bent apically. **FLOWERS** 6–15, the lower ones reflexed in age, the petals pink to lavender or purple, 8.5–11 mm long; pedicel 0.5–1 mm long; calyx 4.4–7 mm long, the tube 2.2–3.3 mm long, strigose, the teeth 1.8–3.7 mm long, subulate, strigose. **PODS** 4–8 mm long, 3–4.5 mm wide. [*T. subcaulescens* A. Gray; *T. gymnocarpon* var. *subcaulescens* (A. Gray) A. Nelson; *T. plummerae* S. Watson; *T. gymnocarpon* f. *plummerae* (S. Watson) McDermott; *T. gymnocarpon* var. *plummerae* (S. Watson) J. S. Martin; *T. gymnocarpon* subsp. *plummerae* (S. Watson) J. M. Gillett; *T. nemorale* Greene]. Mixed desert shrub, sagebrush, piñon-juniper, mountain brush, ponderosa pine, Douglas-fir, and spruce-fir communities. ARIZ: Apa; COLO: Arc, LPl, Mon, SMg; NMEX: McK, RAr, SJn; UTAH. 1865–2440 m (6120–8000'). Flowering: Apr–Jun. Oregon, California, Idaho, Nevada, Wyoming, Colorado, New Mexico, and Arizona. Nuttall's clover is common in the sagebrush communities of the region, where it is often overlooked, due in part to its low stature and to the varicolored or mottled leaves that blend with the background. Attempts at recognition of infraspecific taxa appear to be futile.

Trifolium hybridum L. (hybrid) Alsike clover. Caulescent, 15–70 cm tall, erect or ascending (rarely decumbent), from a caudex and taproot. **LEAVES** with petioles 0–16 cm long; stipules herbaceous, 5–25 mm long; leaflets 3, 5–38 mm long, 3–28 mm wide, oval to lance-elliptic, ovate, or obovate, flat, toothed from near the base, glabrous on both sides, obtuse to retuse and apiculate. **INFL** a head, without an involucre, 12–25 mm wide, terminal and axillary on peduncles 1.5–13 cm long, these glabrous or glabrate, erect. **FLOWERS** many, the lower reflexed in age, the calyx 2.7–4 mm long, the tube 1.2–1.6 mm long, glabrous, scarious, the teeth 1–2.5 mm long, subulate, glabrous; petals white to pink or reddish, fading red-brown, 5–9 mm long. **PODS** 1- to 3-seeded. $2n = 16, 32$. Cultivated, short-lived forage plant, escaping and persisting. COLO: Arc, Dol, Hin, LPl, Min, Mon, RGr, SJn; NMEX: SJn. 1485–3225 m (4870–10590'). Flowering: Jun–Sep. Established throughout much of the temperate world; introduced from Europe. Ingestion of alsike clover has resulted in production of photosensitization in horses, and also in hogs, sheep, and cattle.

Trifolium longipes Nutt. (long-stalked) Rydberg's clover, summer clover. Caulescent (rarely acaulescent), 5–31 (37) cm tall, erect or ascending from a branching caudex and stout to slender taproot, or rhizomatous. **LEAVES** with petioles 1.2–10 cm long; stipules foliaceous, strongly veined, 8–40 mm long; leaflets 3, 5–47 (57) mm long, 3–18 mm wide, narrowly oblong to elliptic, oblanceolate, or obovate, flat, toothed from near the base, pilosulous beneath, glabrous above, acute to obtuse and apiculate. **INFL** an erect head, without an involucre, 17–31 mm long, 15–33 mm wide (when pressed), terminal on peduncles 0.5–17 cm long, these strigulose. **FLOWERS** many, finally reflexed, calyx 4.5–7.8 mm long, the tube 1.6–2.5 mm long, scarious, pilose distally, the teeth 2.9–5.8 mm long, pilose, subulate; the petals whitish to pink or purple; 11–13 mm long. **PODS** 1- to 4-seeded. $2n = 16, 24, 32, 48$.

1. Plants rhizomatous, the roots slender, not much enlarged; leaflets of main leaves commonly more than 5 times longer than broad; plants of Arizona and Colorado. .... **var. *reflexum***

1' Plants from a caudex and tuberous root; leaflets less than 4 times longer than broad; roots tuberous-thickened; plants more widespread.......... **var. *rusbyi***

var. *reflexum* A. Nelson (reflexed) Wind River clover, Rydberg's clover. [*T. longipes* subsp. *reflexum* (A. Nelson) J. M. Gillett; *T. rydbergii* Greene; *T. oreganum* Howell var. *rydbergii* (Greene) McDermott]. Meadows, stream banks, woods, and willow communities. ARIZ: Apa; COLO: Arc, Hin, LPl, Mon; UTAH. 1900–2645 m (6235–8680'). Flowering: Apr–Jun. Wyoming (type from Wind River Mountains, Fremont County), Colorado, Arizona, New Mexico, and Utah.

var. *rusbyi* (Greene) Harr. (for Henry Hurd Rusby, botanist, physician, explorer) Rusby's clover. [*T. rusbyi* Greene; *T. longipes* var. *pygmaeum* A. Gray; *T. longipes* subsp. *pygmaeum* (A. Gray) J. M. Gillett; *T. longipes* var. *brachypus* S. Watson; *T. brachypus* (S. Watson) Blank.; *T. confusum* Rydb.]. Alpine and subalpine meadows, open woods, stream banks, and grasslands with *Geum*, *Pedicularis*, *Smelowskia*, and *Saxifraga*, often in aspen-spruce-fir communities. ARIZ: Apa; COLO: Arc, Hin, LPl, Mon, SJn; NMEX: RAr, SJn; UTAH. 1840–2895 m (6040–9500'). Flowering: Apr–Jun. Southern Utah, southwest Colorado, northwest New Mexico, and northern Arizona.

Trifolium nanum Torr. **CP** (dwarf) Dwarf clover, tundra clover. Acaulescent, pulvinate-caespitose, 2–4 cm tall, from a caudex and taproot, the stems obscured by imbricated stipules and leaf bases. **LEAVES** with petioles 0.3–2 mm long; stipules scarious to herbaceous; leaflets 3, 3–11 mm long, 1–5 mm wide, oblanceolate to obovate, toothed to entire, glabrous or with some hairs on the lower surface, folded or flat, acute to mucronate. **INFL** a head, 1- to 4-flowered, with an involucre of distinct to connate bracts, terminal, sessile or on peduncles 0.3–4 cm long, these glabrate to glabrous. **FLOWERS** 15–23 mm long; pedicel 1–2 mm long; calyx 5–7 mm long, the tube 3.5–4 mm long, scarious, glabrous, the teeth 2.2–2.8 mm long, triangular-subulate, glabrous; pale purplish (fading dark violet), erect. **PODS** 1-

to 4-seeded. 2*n* = 16. Alpine meadows and scree. COLO: Arc, Con, Hin, LPl, Min, Mon, RGr, SJn. 3550–3870 m (11655–12690′). Flowering: Jun–Aug. Montana, Wyoming, Colorado, and New Mexico.

Trifolium parryi A. Gray (for Charles C. Parry, British-American botanist and explorer) Parry's clover. Acaulescent or short-caulescent and with 1 elongate internode, 4–25 cm tall, from a caudex and taproot. **LEAVES** with petioles 0.6–13 cm long; stipules scarious to herbaceous, 6–18 mm long; leaflets 3, 5–43 mm long, 1.5–13 (16) mm wide, oblanceolate or obovate to elliptic or oblong, flat, toothed from near the base, glabrous on both sides, acute to obtuse and apiculate. **INFL** a head, 5- to 20-flowered, subtended by distinct involucral bracts, terminal on peduncles 1.8–22 cm long, these glabrous or sparingly hairy near the apex. **FLOWERS** 12–17 mm long, erect; pedicel 0–1 mm long; calyx 4–7.1 mm long, the tube 2–3.9 mm long, scarious, glabrous, the teeth 2–3.2 mm long, lance-subulate; the petals pale to dark pink-purple (fading dark violet). **PODS** 1- to 4-seeded. 2*n* = 16, 32. Alpine meadows, openings in spruce and other coniferous woods, and on talus slopes. COLO: Arc, Con, Hin, LPl, Min, Mon, SJn. 3350–4025 m (11000–13200′). Flowering: Jul–Aug. Montana, Wyoming, Colorado, and New Mexico. Materials from Mineral County, Colorado, have been assigned to **subsp. *salictorum*** (Greene ex Rydb.) J. M. Gillett (1965 Brittonia 17:132), which is characterized by "thicker, fleshier and more obtuse leaves, more elongate flower heads, thicker peduncles and larger stipules than those of subsp. *parryi*."

Trifolium pratense L. (in meadows) Red clover. Caulescent, short-lived perennial, 18–60 cm tall or more, from a taproot, erect or ascending. **LEAVES** with petioles 0.8–19 cm long; stipules scarious to subherbaceous, 11–24 mm long; leaflets 3, 11–54 mm long, 8–28 mm wide, elliptic to lanceolate, ovate, or obovate, flat, toothed from near the base (the teeth inconspicuous), long-pilose beneath, glabrous above, obtuse to retuse. **INFL** a head, closely subtended by 1 or more foliose bracts, these often 3-foliolate, sessile, or on spreading-hairy peduncles to 3 cm long, many-flowered, 22–36 mm long, 20–34 mm wide, axillary, erect. **FLOWERS** 13–20 mm long, calyx 7.5–9.7 mm long, the tube 3.2–4.1 mm long, strigose, scarious, the teeth 4.3–5.6 mm long, subulate, pilose; corolla deep red. **PODS** 2-seeded. 2*n* = 14, 28, 48. Cultivated forage plant, escaping and at least locally established. COLO: Arc, Dol, Hin, LPl, Min, Mon, SJn; NMEX: RAr, SJn. 1720–2715 m (5640–8910′). Flowering: Jun–Sep. Probably universal; native to Europe. This important agronomic crop plant is a common component of seed mixtures for pasture and hay lands. It is of importance as a forb in meadows cut for wild hay and is commonly found established along roadsides. Bees utilize its nectar to produce fine quality honey. Poisoning of livestock has been reported for this plant, possibly due to the presence of cyanogenetic glycosides. According to Barneby (in Cronquist et al. 1989:214–215), most if not all of our plants have coarse, hollow stems and belong to what has been collectively called **var. *sativum*** Schreb. (von Schreber 1804).

Trifolium repens L. (creeping) White clover. Caulescent, 8–35 cm tall, the stems stoloniferous, creeping and rooting at the nodes, the petioles and peduncles often arising at right angles to the stem axis, radiating from a root crown. **LEAVES** with petioles 1.8–24 cm long; stipules scarious, 3–10 mm long; leaflets 3, 5–22 (38) mm long, 4–18 (30) mm wide, obcordate or obovate to oval or elliptic, flat, toothed from near the base, glabrous on both sides, truncate to emarginate. **INFL** a head, without an involucre, many-flowered, 10–32 mm long, 15–30 mm wide, axillary, on a peduncle 6–33 cm long, these glabrous or sparingly pilose, erect. **FLOWERS** 5–9 (10) mm long, pedicels 1–6.4 mm long; calyx 3.2–5.4 mm long, the tube 2.2–2.7 mm long, scarious, glabrous, the teeth 1–2.7 mm long, subulate, glabrous; white or pinkish, fading brown, the lower reflexed in age. **PODS** 1- to 3-seeded. 2*n* = 16, 22, 30, 32. Commonly grown valuable forage and pasture plant now established both in weedy open sites and in native plant communities. ARIZ: Apa, Nav; COLO: Arc, Dol, Hin, LPl, Min, Mon, SJn; NMEX: McK, RAr, SJn; UTAH. 1645–3445 m (5400–11305′). Flowering: Apr–Oct. Probably universal; native to Europe. White clover has also been implicated in animal poisoning.

Trifolium wormskioldii Lehm. (for Morton Wormskjold, explorer and botanist) Wormskjold's clover, cow clover. Perennial, caulescent, 12–35 cm tall or more, rhizomatous. **STEMS** ascending to erect. **LEAVES** with petioles 1.2–4 cm long; stipules herbaceous, 8–15 mm long, many-toothed; leaflets 3, 6–30 mm long, 3–14 mm wide, oblanceolate to elliptic or obovate, flat, toothed from near the base, glabrous throughout. **INFL** a head, subtended by an involucre, 20–30 mm wide, many-flowered, axillary, erect. **FLOWERS** 10–18 mm long, calyx 7–9 mm long, the tube 2.9–3.7 mm long, glabrous, the teeth 4.1–5.3 mm long, subulate, glabrous; corolla reddish to purple. **PODS** 1- to 4-seeded. 2*n* = 16, 32. [*T. involucratum* Ortega; *T. fimbriatum* Lindl.; *T. involucratum* var. *fimbriatum* (Lindl.) McDermott; *T. willdenovii* Spreng. var. *fimbriatum* (Lindl.) Ewan]. Meadows and streamsides. ARIZ: Apa. 2135–2745 m (7000–9000′). Flowering: May–Sep. British Columbia and Idaho, south to California, Colorado, New Mexico, and Mexico. Specimens from southwestern Colorado have been assigned to both **var. *wormskioldii*** and to **var. *arizonicum*** (Greene) Barneby, 1989

Intermountain Fl. 3B: 228. [*T. arizonicum* Greene, 1895 Erythea 3:18]. The former has flowers mainly 9–13 mm long, the latter 7–9 mm long. The relatively small number of herbarium specimens from the San Juan drainage in herbaria precludes much in the way of analysis of varietal status. Paiutes ate the leaves uncooked as greens.

Vicia L. Vetch

(English vetch) Annual or perennial herbs, clambering, trailing, or climbing. **LEAVES** alternate, even-pinnately compound, the rachis terminating in a usually prehensile tendril; stipules herbaceous, entire to semisagittate; leaflets 4–12 or more, very variable. **INFL** solitary, axillary, or an axillary raceme. **FLOWERS** papilionaceous; calyx 5-toothed, obliquely campanulate to short-cylindric; petals 5, pink to white, the wings adnate to the keel; stamens 10, diadelphous; style filiform, bearded around the circumference below the stigma. **PODS** oblong, 2- to several-seeded, the valves coiling upon dehiscence (Hermann 1960).

1. Flowers 15 or more in dense, secund racemes; introduced plants of cultivated lands and other sites .... ***V. villosa***
1' Flowers 10 or fewer, in secund racemes or otherwise; introduced or indigenous plants .... (2)

2. Flowers 5–8 mm long; plants very slender .... ***V. ludoviciana***
2' Flowers 12–25 mm long or more; plants not very slender, indigenous and widespread .... ***V. americana***

Vicia americana Muhl. ex Willd. (of America) American vetch. Perennial, 1.2–12.7 dm tall, the stems glabrous or pubescent. **LEAVES** (excluding tendrils) 2–3 cm long; stipules 3–10 mm long, semisagittate, deeply toothed in the lower portion; leaflets 8–16, 3–44 mm long, 1–19 mm wide, linear, elliptic, oblong, ovate, lanceolate, oblanceolate, or obovate, glabrous to pubescent, acute to truncate, rounded, or retuse and apiculate, less commonly toothed apically; tendrils branched or simple. **INFL** a raceme, 3- to 7 (10)-flowered, the flowers spreading at anthesis; peduncles 1.8–6.7 cm long. **FLOWERS** 13–22 (25) mm long; calyx 6.2–8.4 mm long, the tube 4.8–6.5 mm long, the lowermost tooth 0.7–1.9 (2.5) mm long, triangular; corolla pink to pink-purple. **PODS** stipitate, the stipe 2.5–4.5 mm long, the body 23–35 mm long, 6–8 mm wide, glabrous. $2n = 14$. [*V. oregana* Nutt.; *V. americana* var. *oregana* (Nutt.) A. Nelson; *V. americana* subsp. *oregana* (Nutt.) Abrams; *V. truncata* Nutt.; *V. americana* var. *truncata* (Nutt.) Brewer]. Sagebrush, piñon-juniper, mountain brush, ponderosa pine, aspen, and spruce-fir communities. ARIZ: Apa; COLO: Arc, Dol, Hin, LPl, Min, Mon, RGr, SJn; NMEX: McK, RAr, SJn; UTAH. 2285–3200 m (7500–10500′). Flowering: May–Aug. British Columbia east to Ontario, Kansas, Missouri, Virginia, and south to Mexico. Infusion of the plant was used by Navajos as an eye wash. The plant is a "life medicine."

Vicia ludoviciana Nutt. ex Torr. & A. Gray (of Louisiana) Louisiana vetch. Annual or winter-annual, 3–8.5 dm tall, the stems glabrous or puberulent. **LEAVES** (excluding tendrils) 1.8–5.5 cm long; stipules semihastate to linear-oblong, 1–4 mm long; leaflets 6–10, 7–28 mm long, 0.6–4 mm wide, linear to oblong or oblanceolate, pilosulous to glabrous, obtuse to acute and mucronate; tendrils branched and prehensile. **INFL** a raceme, 1 (2)-flowered, the flowers ascending to spreading at anthesis; peduncles 0.4–4.8 cm long. **FLOWERS** 6.3–7.4 mm long; calyx 2.2–3.3 mm long, the campanulate tube 1.4–1.9 mm long, the lowermost tooth 0.8–1.4 mm long, lance-subulate; corolla lavender. **PODS** stipitate, the stipe 0.8–1.4 mm long, the body 16–28 mm long, 5–6.2 mm wide, glabrous. $2n = 14$. [*V. exigua* Nutt.; *V. thurberi* S. Watson]. Blackbrush, mixed desert shrub, and piñon-juniper communities. ARIZ: Apa; COLO: LPl, Mon; NMEX: SJn; UTAH. 1665–1910 m (5465–6265′). Flowering: Apr–May. Oregon, California, Nevada, Utah, Arizona, New Mexico, Colorado, and Texas.

Vicia villosa Roth (hairy) Hairy vetch. Annual or biennial, 5–20 dm tall, the stems spreading-hairy. **LEAVES** (excluding tendrils) 2.3–8 cm long; stipules toothed or entire, 5–15 mm long; leaflets 10–18, 8–30 mm long, 1–6 mm wide, linear to oblong or narrowly lanceolate, long-pilose or hirsute on both sides, acute to obtuse and apiculate. **INFL** a raceme, mainly 15- to 25-flowered, the flowers declined at anthesis; peduncles 1.8–7.5 cm long. **FLOWERS** 15–17 mm long; calyx 7–7.8 mm long, the gibbous tube 3.8–4.7 mm long, the teeth 3.1–4.3 mm long, subulate, pilose; corolla pink-purple or reddish violet. **PODS** 20–30 mm long, 7–10 mm wide, glabrous. $2n = 14$. Weedy introduction in cultivated lands and other disturbed sites, often along fencerows. COLO: LPl, Mon. 1975–2110 m (6475–6930′). Flowering: Jun–Jul. Adventive from Europe. Four Corners material belongs to **var. *villosa***. This plant, long grown for forage in agricultural areas, is often conspicuous along fencerows in populated areas.

FAGACEAE Dumort. BEECH, OAK FAMILY

Rodney Myatt

Woody shrubs or trees, evergreen or deciduous, monoecious. **LEAVES** alternate; petiolate; stipules small, deciduous; blade simple, entire, toothed, or lobed. **INFL** with staminate catkins or short, pistillate branches; male catkins pendulous or somewhat erect spikes; female branches with 1–few flowers, involucre cup-shaped or lobed and burlike. **FLOWERS** with staminate calyx lobes 4–6, small; petals 0; stamens 4–12; pistillate calyx 6-lobed; petals 0; style branches 2–3; ovary inferior. **FRUIT** a nut subtended by a scaly involucral cup (acorn) or 1–3 nuts subtended by a spiny, burlike involucral cup. Seven genera and about 900 species worldwide, especially in the Northern Hemisphere. Only *Quercus* occurs naturally in our area. Fruits of most members (*Quercus*, *Fagus*, *Castanea*, *Lithocarpus*, etc.) important as food source for wildlife and indigenous peoples. Trees important as habitat for forest animals. Bark used as cork (*Quercus suber*) or as tannin source (*Lithocarpus* spp.).

Quercus L. Oak

(ancient Latin name for oak) Woody shrubs to trees, evergreen or deciduous, sometimes clonal, with stems rooting at ground; monoecious; often with dense pubescence on young twigs, petioles, leaf blades, and involucral cups, usually more glabrous with age; hairs both single and stellate, with 2–10+ rays; some species with sparse, golden, glandular hairs on leaves and acorn cup. **LEAVES** alternate, deciduous, thin to sclerophyllous, entire to lobed or toothed; lobes rounded, acute, or mucronate; teeth acute, mucronate, or spinose; stipules linear to linear-spatulate. **INFL** catkins; staminate catkins pendulous, with many flowers, occurring early in season, often before leaves are fully developed, and located in axils toward ends of young branches; pistillate catkins short, erect, 2–4 flowers, in axils toward base of young stems. **FLOWERS** with staminate calyx lobes 4–6, ciliate-fringed; pistil with 3 stigmas, short, recurved in flower; ovary subtended by scaly involucral cup, scales with thickened bases. **FRUIT** an acorn with the enlarged involucral cup covering 1/4 to 1/2 of the nut. All of the naturally occurring species in our region are members of the subgenus *Quercus* (*Lepidobalanus*, white oaks), which lack aristate bristles on leaf lobes, have thickened cup scale bases, smooth (nontomentose) inner wall of pericarp, short, thickened styles, and fruits maturing in one year. Species of this subgenus, including ours, hybridize readily and form populations of introgressed individuals and local hybrid swarms, making identification often problematic. Approximately 600 species worldwide, especially in temperate regions of the Northern Hemisphere. An important food source for wildlife and Native Americans. In some regions important as a source of hardwood lumber and firewood.

1. Leaves deciduous, thin, deeply lobed, lobe sinuses generally more than halfway to midrib of leaf; mostly small to medium-size trees or not sclerophyllous, occasionally shrubs .... ***Q. gambelii***

1′ Leaves deciduous or evergreen, but sclerophyllous, entire, toothed, or shallowly lobed; lobe sinuses less than halfway to midrib; shrubs or small trees, may be clonal or not .... (2)

2. Plants semievergreen, mainly 10–30 dm tall or more, or if deciduous, then typically hairy above and densely so beneath; forming clones within and adjacent to stands of *Q. gambelii*, and occurring as scattered individuals where *Q. gambelii*, *Q. turbinella*, or *Q. welshii* coexist; acorns typically less than 5 mm long and less than 10 mm wide, if formed at all .... ***Q. pauciloba*** and ***Q. eastwoodiae*** (hybrids)

2′ Plants not as above .... (3)

3. Leaves evergreen, entire (only occasionally with teeth), grayish, with light gray pubescence on petioles and twigs, mostly small trees .... ***Q. grisea***

3′ Leaves evergreen or deciduous, lobed or toothed, not particularly grayish or with dense gray pubescence .... (4)

4. Leaves deciduous, lobed or shallow, with rounded to acute teeth .... (5)

4′ Leaves evergreen, toothed, with spinose tips on teeth; shrubs, but not clonal; plants usually of more upland, desert chaparral, or woodland habitats .... (6)

5. Leaf lobes mostly acute, mucronate, or apiculate; young leaves densely hairy; plants low shrubs, branch bases at ground usually buried; plants of sandy, low desert habitats in western portion of the Four Corners .... ***Q. welshii***

5′ Leaf lobes mostly rounded to acute; young leaves not so densely hairy; plants shrubs to small trees; some may be semievergreen; plants of more upland, woodland-sagebrush habitats in eastern part of the Four Corners .... ***Q. ×undulata***

6. Leaf spines 1–2.5 mm; young leaves with dense stellate pubescence, rays 8–10+, more or less appressed; golden hairs present on leaves; common .... ***Q. turbinella***
6' Leaf spines 3–5 mm; young leaves sparsely pubescent, stellate hairs with fewer rays; golden hairs lacking; rare, Mesa Verde National Park, Colorado .... ***Q. ajoensis***

Quercus ajoensis C. H. Mull. (from Spanish word for garlic, the origin of name for Ajo Mountains, Arizona) Shrubs to small trees. **LEAVES** evergreen, orbicular to elliptic, toothed, with elongate spines on teeth; slightly pubescent to glabrous, lacking golden hairs. **ACORN** cups very shallow. Growing on shallow, red sand, on gentle, south-facing slope, with piñon-juniper. COLO: Mon. 2135 m (7000′) Flowering: May: There is some question as to whether this population is distinct from *Q. turbinella*. One specimen examined was definitely *Q. ajoensis*; others may be hybrids with *Q. turbinella*. The species occurs in southwest Arizona (type locality is Ajo Mountains, Pima County) and into Mexico, but hybrids with *Q. turbinella* occur in mountains to the north and east in Arizona, and apparently in the Sacramento Mountains, New Mexico. Found in Mesa Verde National Park, Wetherill Mesa (Muller 1954).

Quercus gambelii Nutt. (for William Gambel, naturalist, ornithologist, and botanist) Gambel's oak. Trees to upright shrubs, mostly 2–5 m. **STEMS** single to multiple from ground, sometimes clonal as shrubs. **LEAVES** deciduous, deeply lobed; lobes 3–4 per side, often divided toward tip, rounded to acute at tip, nonaristate; blade 5–12+ cm long, elliptic to obelliptic in outline; base tapered; petiole 0.5–1.5 cm; young leaves with stellate pubescence on both surfaces, especially dense near base and petiole, upper surface glabrous with age; stellate hairs with 2–6 rays, upright; no golden, glandular hairs. **ACORNS** with involucral cup 1–2 cm diameter; covering about 1/3 nut; nut up to 2 cm long. Widespread and variable, occurring in piñon-juniper and upland scrub communities. ARIZ: Apa, Nav; COLO: Arc, Dol, Hin, LPl, Min, Mon, SJn, SMg; NMEX: McK, RAr, San, SJn; UTAH. Flowering: May–Jun. Hybridizes with each of the other species in the region in which it comes in contact. Much of the variation in each species may be the result of past and present introgression of characters from *Q. gambelii.* Some populations or individuals of hybrids are recognized with binomials (cf. *Q.* ×*eastwoodiae* Rydb., *Q.* ×*pauciloba* Rydb., *Q.* ×*undulata* Torr.). Particularly tall, more arborescent trees which occur occasionally in protected, well-watered sites may be recognized as var. *bonina* S. L. Welsh. Based on plants from the type locality (Goodhope Bay, Utah), these plants have larger acorns and grow much faster under similar conditions than var. *gambelii*. Decoction of the root used for postpartum pain, as a cathartic, and as a "life medicine." Acorns were a staple, ground into a meal or eaten whole by Apaches, Acoma, most Puebloans, and Navajos. The red leaf galls were used as dyes and the wood used for baby cradles, tools, ceremonial objects, and shade house construction by Navajos. The bark was also used to tan hides (Tucker 1961).

Quercus grisea Liebm. (gray) Gray oak. Small trees, occasional shrubs; petiole and young twigs with light grayish pubescence. **LEAVES** evergreen, mostly entire, ovate to ovate-elliptic, upper side often grayish green; base rounded to truncate. Hybrids with *Q. gambelii* (*Q.* ×*undulata*) occur in the south and southeastern part of the region. Sagebrush and montane communities. COLO: Arc, Mon. Flowering: May. Occurs mostly in the southern part of Arizona and New Mexico, and into Texas, possibly getting into the montane sagebrush habitats of the southern and eastern parts of the region in Sandoval County, New Mexico. Arizona, New Mexico, Texas, and south to Mexico. The raw fruit was used by Apaches and Navajos as food. Ramah Navajos used the wood to protect new or ceremonial hogans from lightning, ghosts, and witches (Tucker 1961, 1971).

Quercus turbinella Greene (almost top-shaped, referring to the acorns) Shrub live oak. Shrubs, crown spreading, multiple stems, but not clonal. **LEAVES** evergreen, sclerophyllous, elliptic to round-elliptic; petiole 0.5–1 cm; blade 2–4 cm, often slightly grayish green on upper surface, dull green below; base rounded to truncate; blade toothed to shallow, acutely lobed; lobes or teeth spinose-tipped, spines 0.5–1.2 mm; young leaves with dense pubescence, more dense on lower surface, upper usually nearly glabrous with age; stellate hairs with 8–10 rays, more or less appressed to leaf surface; some golden, glandular hairs usually present; veins somewhat raised, lighter in color, evident. Sandy to rocky soils, upland desert scrub, sagebrush, rabbitbrush communities, bordering piñon-juniper. ARIZ: Apa, Nav; COLO: LPl, Mon; NMEX: McK, SJn; UTAH. Flowering: May. Arizona, California, and Nevada, east to Colorado, Utah, New Mexico, and south to Mexico.

Quercus welshii R. A. Denham (for Stanley L. Welsh, Utah botanist) Welsh's oak. Shrubs, with multiple erect stems, usually bases buried by sediment. They appear not to be clonal. **LEAVES** deciduous; blade 4–6 cm, occasionally shorter, elliptic, shallow lobes; upper side slightly darker green than lower; lobes acute, may be mucronate, not spine-tipped; young leaves densely pubescent on both surfaces, petiole, and young stems; stellate hairs variable, 4–10 rays, usually upright; no golden, glandular hairs. Plants of lowland, sandy habitats, mostly in the western part of the region. The multiple-base habit often traps sand into dunelike formations. ARIZ: Apa, Nav; NMEX: SJn; UTAH. 1300–1800 m

(4000–5300′). Flowering: May. Utah, Arizona, New Mexico, Oklahoma, and Texas. There is some question as to whether these plants represent a distinct species or a variety *of Q. havardii* (var. *tuckeri*). Our plants differ from the more eastern *Q. havardii* Rydb. (eastern New Mexico, Texas, Oklahoma) in having somewhat deeper, more acute, often mucronate lobes, characters apparently influenced by introgression with *Q. gambelii* and possibly *Q. turbinella*. [*Quercus havardii* Rydb. var. *tuckeri* S. L. Welsh] (Tucker 1970).

Quercus* ×*eastwoodiae Rydb. (for Alice Eastwood, western botanist) Eastwood's oak. Hybrid (*Quercus gambelii* × *Quercus havardii*). Shrubs, clonal. **LEAVES** deciduous, generally lobed halfway to midrib; lobes acute to mucronate or short-spinose; pubescence dense on leaf surface, petioles, and young twigs; rays upright, variable in number. Sandy to rocky soils in alluvial sites, sandy canyon bottoms, more or less riparian habitats in western part of the region. ARIZ: Nav; UTAH. Arizona and Utah. Flowering: Apr–May (Tucker 1970).

Quercus* ×*pauciloba Rydb. (small-lobed) Hybrid (*Quercus gambelii* × *Quercus turbinella*). Shrubs. **LEAVES** deciduous, shallow to moderately lobed, although lobe sinuses usually 1/2 or less to leaf midrib; lobes acute, mucronate, or short-spinose; pubescence variable, usually not very dense; some stellate hairs with 6–8 rays, upright, some with 8–10 rays, appressed. Most of the specimens examined for the area have characteristics closer to *Q. gambelii* and are closer to *Q. gambelii* populations than *Q. turbinella*. Occurring in upland scrub and piñon-juniper communities, along canyon draws and rocky slopes. ARIZ: Apa, Nav; NMEX: McK, SJn; UTAH. 1200–2000 m (3500–6000′). Flowering: Apr–May. Colorado, Utah, Nevada, Arizona, and New Mexico (Tucker 1961).

Quercus* ×*undulata Torr. (wavy) Wavy-leaved oak. Hybrid (*Quercus gambelii* × *Quercus grisea*). Mostly shrubs to small trees. **LEAVES** shallowly lobed, elliptic to narrowly elliptic, 3–5 cm, a few leaves persistent more than 1 year; leaves often grayish above; lobes acute to rounded, sometimes mucronate; pubescence on leaf base, petiole, and young twigs often grayish silvery. Occurs in upland sagebrush, piñon-juniper communities in eastern part of region. NMEX: McK, RAr, San. Flowering: May. Colorado, Utah, Arizona, and New Mexico (Tucker 1961, 1971).

FRANKENIACEAE Desv. ALKALI HEATH FAMILY

Linda Mary Reeves

Salt-tolerant herbs. **LEAVES** generally opposite, sessile, small, heathlike, often decussate, simple, entire, often with inrolled margins. **INFL** a terminal or axillary cyme, occasionally solitary. **FLOWERS** actinomorphic, perfect, rarely unisexual and monoecious or dioecious; calyx tubular, lobes 4–7; petals free, 4–7, sometimes with a long claw, often with a scale at the base; stamens usually 6, occasionally 24, in 2 whorls, occasionally united at ovary base; ovary superior, with 2–4 united carpels, with a single locule and parietal placentation; style simple; stigma 2- or 3-lobed. **FRUIT** a capsule with persistent calyx. $2n = 20, 30$. About four genera (*Anthobryum*, *Frankenia*, *Hypericopsis*, *Niederleinia*) and 90 species from warm-temperate and subtropical regions, primarily the Mediterranean region, Australia, southern South America, the southwestern U.S., and Mexico. Some ornamental use, especially the St. Helena shrub seaheaths. *Frankenia ericifolia* is used as a fish poison.

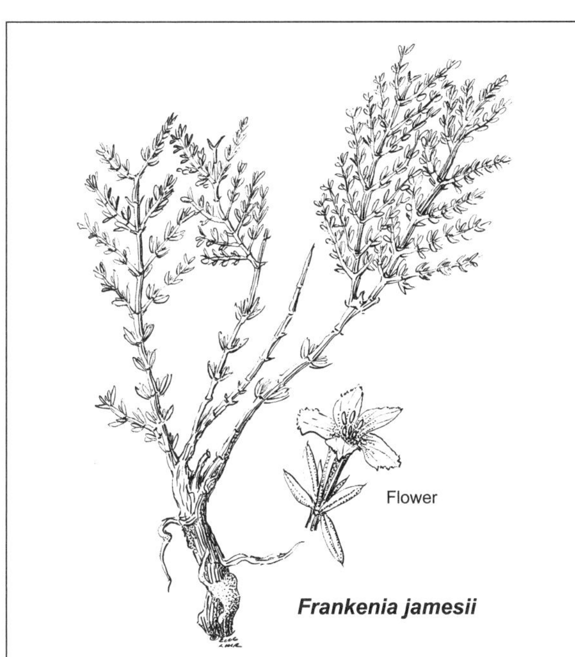

Frankenia jamesii

Frankenia L. Alkali Heath, Seaheath, Wispweed (for Johannes Franckenius, 17th-century Swedish botanist) Subshrubs, usually salt-tolerant, often pubescent. **LEAVES** small, often congested on small lateral branches, pubescent, often salt-encrusted. **FLOWERS** perfect, subtended by a pair of bracts and a pair of bracteoles, often each pair united by a stipular sheath; calyx ribbed, 5-lobed, margin ciliate, persistent; petals 4–6, pink, white, or purplish; stamens 4–6 in 2 whorls, the outer shorter than the inner, filaments often fused at bases; ovary 3-carpellate, placentation basal or parietal. About 80 species of seaside, gypsum, or saline habitats (50 species in Australia). Primarily Mediterranean, Australian, warm-temperate, and arid subtropical distribution.

Frankenia jamesii Torr. ex A. Gray (for Edwin P. James, 1797–1861, an American naturalist and botanical explorer in the Rocky Mountains) James' seaheath. Small shrub with woody base to 30 cm tall. **STEMS** jointed; puberulent. **LEAVES** decussate, appearing whorled; linear to elliptic, somewhat terete, aculeate; somewhat clasping; hispid; margins strongly revolute, often tightly coiled, appearing to be fascicled; base margins ciliate; 3–5 mm long, about 1 mm wide. **INFL** cymose or solitary. **FLOWERS** sessile; calyx tube ribbed, 4–5 mm long, lobes toothlike, 5; petals 5, clawed, about twice as long as calyx tube, white, apex margin somewhat denticulate; stamens 6, 3 long and 3 short; style with 3 branches. **FRUIT** a linear, angled capsule about 5 mm long. Primarily in desert scrub, usually associated with gypsum deposits. Possibly an obligate gypsophile. COLO: Mon; NMEX: SJn, McK. 1500–2000 m (5000–6500′). Flowering: May–Jun. Fruit: Jun–Jul. West Texas and Mexico to New Mexico and Colorado.

FUMARIACEAE Marquis FUMITORY FAMILY

Lynn M. Moore

Winter-annual, biennial, or perennial herbs with watery juice. **LEAVES** alternate or basal, divided or compound; without stipules. **INFL** a raceme or panicle; pedunculate; with bracts. **FLOWERS** irregular; perfect; sepals 2, small and bractlike, peltate, not enclosing the bud, persistent or caducous; petals 4, unequal, outer 2 petals connate, upper petal forming a spurlike saccate base, inner 2 petals coherent, forming a hood over the stigma and anthers; pistil 1; superior; carpels 2; locule 1; style 1; stigma capitate or lobed, wet; stamens 6, diadelphous in 2 bundles of 3, filaments connate at base, the middle anther 2-locular, the lateral anthers 1-locular. **FRUIT** a valvate, dehiscent or indehiscent capsule. **SEEDS** 1 to many. $x = 8$. Four genera and 23 species in North America (19 genera and 450 species worldwide). Temperate, often montane regions of North America, Eurasia, Africa, and Australia. Most botanists, especially those from Europe, treat the Fumariaceae as a subfamily of the Papaveraceae. Intermediate genera do exist, i.e., *Hypecoum* and *Pteridophyllum* (characterized by weakly irregular flowers and four distinct stamens), and cladistic analysis also supports the treatment of the Fumariaceae within the Papaveraceae. Nevertheless, our flora shows distinct morphological differences in the two families and will be treated here as separate to facilitate ease of use in the field (Arber 1931; Judd, Sanders, and Donoghue 1994; Thorne 1974; Zomlefer 1989). One genus in our flora.

Corydalis DC. Corydalis, Fumewort

(Greek *korudallis*, from *korudos*, crested lark, from the shape of the flowers) Plants erect to prostrate herbs. **LEAVES** basal or cauline; compound; 2–4 times pinnate or pinnate-pinnatifid. **INFL** a raceme or panicle. **FLOWER** sepals 2, small and bractlike, peltate, not enclosing the bud, persistent or caducous; petals 4, unequal, outer 2 petals connate, upper petal forming a saccate spur, inner 2 petals coherent, forming a hood over the stigma and anthers. **FRUIT** a dehiscent capsule. **SEEDS** few to many; small; elaisome present (oil-bearing appendage). About 100–300 species in temperate North America, Eurasia, Australia, and Africa.

1. Petals pale to bright yellow; winter-annual or biennial; plants less than 50 cm tall .... ***C. aurea***
1′ Petals white, inner petals red to purple-tipped; perennial; plants tall, up to 2 m .... ***C. caseana***

Corydalis aurea Willd. (with gold) Plants winter-annual or biennial; glaucous; less than 50 cm tall; more or less branched caudex. **STEMS** 10–50; prostrate-ascending; 2–3.5 dm long. **LEAVES** bipinnate; leaflets pinnatifid. **FLOWER** petals pale to bright yellow. **FRUIT** linear capsules, often torulose.

1. Raceme not longer than leaves, slender or weak; spur less than 1/2 the length of the rest of the corolla from the pedicel; capsules pendent or spreading at maturity .... **subsp. *aurea***
1′ Raceme longer than leaves, stout or robust; spur about equal the length of the rest of the corolla from the pedicel; capsules erect at maturity, stout .... **subsp. *occidentalis***

subsp. *aurea* Scrambled eggs, golden corydalis, goldensmoke, mountain corydalis, Tazhii yilchin á ts'íísígíí (little turkeylike odor) **INFL** a raceme, not longer than leaves; slender or weak. **FLOWERS** with upper petal spur less than 1/2 the length of the rest of the corolla from the pedicel. **FRUIT** with capsules pendent or spreading at maturity; slender; 18–24 mm long. **SEEDS** smooth, without marginal ring. $2n = 16$. [*Capnoides aureum* (Willd.) Kuntze; *Corydalis washingtoniana* Fedde]. Hillsides, roadsides, and dry washes in montane, ponderosa pine, sagebrush, and piñon-juniper woodland communities. ARIZ: Apa, Nav; COLO: Arc, Hin, LPl, Min, Mon, SJn; NMEX: McK, RAr, SJn; UTAH. 1980–3275 m (6500–10750′). Flowering: May–Aug. Native across the United States (absent from the Southeast) and Canada. Kayenta and Ramah Navajo drug plant used as an antidiarrheal, dermatological aid, disinfectant, and gynecological aid. Plant sprinkled on livestock for snakebites. Used for sheep feed. Cold infusion used to soak watermelon seeds to increase production.

subsp. *occidentalis* (Engelm. ex A. Gray) G. B. Ownbey (of the west) Curvepod fumewort, curvepod, corydalis. **INFL** a raceme longer than leaves; stout or robust. **FLOWERS** with upper petal spur about equal the length of the rest of the corolla from the pedicel. **FRUIT** capsules erect at maturity; stout; 12–20 mm long. **SEEDS** smooth; with narrow marginal ring. [*Corydalis aurea* Willd. var. *occidentalis* Engelm. ex A. Gray; *C. curvisiliqua* (A. Gray) Engelm. ex A. Gray subsp. *occidentalis* (Engelm. ex A. Gray) W. A. Weber; *C. montana* Engelm. ex A. Gray; *Capnoides montanum* (Engelm. ex A. Gray) Britton]. Hillsides, sandy terraces, and disturbed sites in desert scrub, sagebrush, mountain shrub, piñon-juniper woodland, ponderosa pine, and mixed conifer communities. ARIZ: Apa, Nav; COLO: Arc, Hin, LPl, Min, Mon; NMEX: McK, RAr, SJn; UTAH. 1310–2530 m (4300–8300′). Flowering: Apr–Jul. Native in the United States west of the Mississippi River. Ramah Navajo drug plant used as a rheumatic remedy. Cold infusion taken for stomachache, for sore throat, and as lotion for backache. Compound decoction of plant used for menstrual difficulties, and injuries. Used for sheep feed. Cold infusion used to soak watermelon seeds to increase production.

Corydalis caseana A. Gray **CP** (for Eliphalet Lewis Case, 1843–1925) Brandegee's fumewort, Sierra fumewort, Sierran fumewort, Sierra corydalis, fitweed, cimmarona. Plants perennial; glaucous; up to 2 m tall; large, fleshy roots. **STEMS** 1–several; 10–15 dm long. **LEAVES** 2–4 times pinnate; large, 20–30 cm long. **INFL** a raceme or panicle; 50 or more flowers; bracts linear to linear-elliptic, 10–12 mm long, becoming reduced distally. **FLOWER** petals white, inner petals red to purple-tipped. **FRUIT** capsule ellipsoid; reflexed; 10–15 mm long. **SEEDS** with numerous minute protuberances. [*Corydalis brandegeei* S. Watson]. Wet aspen and spruce-fir forests in the San Juan Mountains. COLO: Arc, Hin, LPl, Min, Mon, RGr, SJn. 2650–3569 m (8700–11800′). Flowering: Jun–Aug. Our material belongs to **subsp. *brandegei*** (S. Watson) G. B. Ownbey (named for Townshend Brandegee, 1843–1925, and Mary Katharine (Layne) Brandegee, pioneer western botanists). Our subspecies is endemic to southern Colorado and adjacent New Mexico. Palatable to both sheep and cattle, although highly toxic to livestock. Significant losses have occurred to both cattle and sheep. Often located along streams of conservation interest.

GENTIANACEAE Juss. GENTIAN FAMILY

John R. Spence

Annual, biennial, or perennial herbs (ours), caespitose to erect, from either a weakly to well-developed taproot, from rhizomes or stolons, or sometimes from fibrous rootlets. **STEMS** branched or not, round to angled, erect, ascending, or decumbent, sometimes winged. **LEAVES** opposite, whorled or sometimes alternate, sessile or rarely petiolate, stipules lacking; blade simple, entire, basal rosettes often present, bases often sheathing-connate, sometimes white-margined and wavy-margined. **INFL** solitary or a simple to complex, thyrsoid cyme. **FLOWERS** actinomorphic, hypogynous, perfect, often showy, 4- to 5-merous; sepals persistent, united, cleft or shallowly lobed, sometimes lobes reduced to absent, inner membranaceous rim often present; petals united, tubular, funnelform, campanulate, rotate, or salverform, plicae sometimes present between lobes, fringed coronae sometimes present on inner surface of lobes below tip, 1 or 2 nectaries present at base of corolla tube or on ovary base; stamens versatile or rarely basifixed, epipetalous, inserted or exserted, alternate with the corolla lobes, sometimes connate, anthers longitudinally dehiscent, sometimes becoming spirally twisted after anthesis, free or rarely connate; ovary of 2 carpels, sessile or stipitate, with 1 or 2 locules each, style obsolete to capitate to filiform, stigmas 2, entire or 2-lobed, often persistent. **FRUIT** a 2-valved septicidal capsule. **SEEDS** typically numerous, small, smooth, reticulate, striate, or pitted, sometimes winged. A family of 85 or more genera and about 1650 species, distributed worldwide, especially common in moist, cool-temperate sites in the Northern Hemisphere, and mountains in the subtropics and tropics. *Lomatogonium* is included in the key and is likely to occur in subalpine-alpine areas of the San Juan Mountains. Many members of the family in the subtropics and tropics are woody shrubs or small trees, lianas, or even root-parasites, but temperate genera are primarily herbaceous. The spectacular Arctic-alpine and Sino-Himalayan perennial gentians are widely cultivated as ornamentals, and many of our local species are among the most attractive wildflowers in the West. Gentians contain unusual bitter substances, such as mangiferin and swertianolin, known as xanthone glycosides, which are reported to have medicinal properties. Many species have been used in traditional herbal remedies. Generic limits, especially in *Gentiana*, *Gentianella*, and *Swertia*, are controversial. While *Gentianella* has been split into several segregate genera, most authors still treat *Gentiana* as one large genus (Ho and Liu 1990; Struwe and Albert 2002). However, including tiny annuals such as *Chondrophylla fremontii* in the same genus as the type, the European *G. lutea*, a meter-tall, coarse, perennial forb with petals that are mostly separate and yellow or red, is inexplicable. The sections of *Gentiana* are strongly marked and morphologically distinctive; thus this treatment follows W. A. Weber and R. C. Wittmann (2001) in taking a relatively narrow view, as they are easily separated in the study area.

1. Corolla rotate, with 1 or 2 fringed glands on the upper surface of each petal lobe; leaves opposite or in whorls, basal rosette typically present ....(2)
1' Corolla tubular, salverform, or campanulate, obvious glands lacking on upper surface of lobes at base although fringed corona sometimes present near base of lobe; leaves opposite, basal rosette present or absent ....(4)

2. Plants slender annuals from a weakly developed taproot; basal leaves few or absent; stigmas decurrent along carpel sutures; flowers white or light blue; subalpine bog plants .... *Lomatogonium*
2' Plants biennials or perennials from a well-developed taproot; basal leaves numerous and well developed; stigmas terminal; flowers white or greenish; habitats and elevations various....................(3)

3. Flowers 5-merous, glands 2 per petal lobe; stem leaves opposite or subopposite; style distinctly short and thickened; high elevations in moist to wet sites.................... ***Swertia***
3' Flowers predominantly 4-merous, glands 1–2 per petal lobe; stem leaves opposite or whorled; style long and slender-filamentous; low to mid-elevations in upland sites.................... ***Frasera***

4. Corolla salverform, pink to rose; anthers becoming spirally twisted after anthesis; in low-elevation wet sites .... ***Zeltnera***
4' Corolla funnelform-tubular to long-campanulate, blue to white; anthers not spirally twisted after anthesis; primarily montane plants in a variety of habitats ....(5)

5. Plants perennial; corolla usually longer than 2.5 cm, corolla lobes with conspicuous plicae between them, lobe margins never fringed ....(6)
5' Plants annual or biennial; lobes with or without plicae, corolla typically shorter than 2.5 cm, if longer or if plants perennial, then corolla lobe margins distinctly fringed ....(7)

6. Flowers white or yellow-white; basal rosette present; low-growing (less than 0.2 m); alpine plants .... ***Gentianodes***
6' Flowers blue to blue-purple; basal rosette lacking; taller (0.3–1 m); montane to subalpine plants........ ***Pneumonanthe***

7. Corolla lobes with plicae, lacking fringed corona below apex; leaves connate into a tubular sheathing base, white-margined; low-growing, often prostrate plants.................... ***Chondrophylla***
7' Corolla lobes lacking plicae, sometimes with 1–2 distinct fringed coronas below tips on inside; leaves not connate or tubular-sheathing at base, not white-margined; plants mostly erect ....(8)

8. Corolla large, often longer than 2.5 cm, dark blue, corolla lobe margins distinctly fringed, fringed corona lacking .... ***Gentianopsis***
8' Corolla shorter than 2.5 cm, pale blue to white, corolla lobes not fringed, fringed corona present just below corolla lobe tips on inside ....(9)

9. Fringed coronae 2 per petal lobe; flowers on long naked peduncles, solitary, longer than subtending internodes; calyx gibbous at base .... ***Comastoma***
9' Fringed corona 1; flowers on short peduncles, shorter than subtending internodes, often clustered; calyx lobes not gibbous at base .... ***Gentianella***

Chondrophylla A. Nelson Moss Gentian

(Latin *chondro*, cartilage, + *phyllum*, leaf, referring to the white-margined leaves) Caespitose, annual or biennial herbs from slender taproot. **STEMS** 1 or several, erect, curved, or prostrate, somewhat angled, simple or branched. **LEAVES** in basal rosettes, stem leaves opposite, sessile, sometimes recurved, connate-sheathing at base, with narrow to wide whitish margins, sometimes obscurely so. **INFL** of terminal solitary flowers (ours). **FLOWERS** predominantly 4-merous, sessile or short-pedicellate; calyx tubular, lobes equal, not gibbous at base, with an intracalycine membrane present, sometimes not continuous; corolla tubular-salverform, blue, purple, green, or white, with symmetrical plicae between lobes, lobes more or less entire, small nectary glands at base on pistil, alternate with filaments, lobes lacking fimbriate scales; stamens free, inserted; pistil on short stipe, elongating in fruit; stigmas 2, sessile, persistent. **FRUIT** an obovate or cylindric capsule, the valves recurved when split open. **SEEDS** small, numerous, ovoid to ellipsoid, finely reticulate or smooth, lacking wings. A large genus of 175 species, distributed almost worldwide in montane habitats, with large concentrations in China and adjacent regions, New Guinea, and reaching Australia and South America. The corollas of some *Chondrophylla* species are temperature- and touch-sensitive, closing when touched or when clouds cover the sun. Species are among the few annuals in Arctic-alpine habitats. The life strategies of our species are not well known but are probably mostly biennial.

1. Capsule obovoid, mostly less than 3 times as long as wide, with the valves deeply split and gaping widely when mature, shortly exserted from corolla on stout stipe; corolla pale blue or white .... ***C. fremontii***
1' Capsule cylindric, more than 4 times longer than wide, with valves only parted halfway down tube or less when mature, exserted or not from corolla on a slender stipe; corolla typically deep blue .... (2)

2. Flowering stems erect; flowers erect; capsule not fully exserted from corolla; plants mostly less than 5 cm tall; corolla distinctly temperature- and touch-sensitive .... ***C. prostrata***
2' Flowering stems arcuate; flowers nodding, not erect; capsule fully exserted from corolla on long stipe; plants typically greater than 5 cm tall; corolla not especially sensitive .... ***C. nutans***

Chondrophylla fremontii (Torr.) A. Nelson (for J. C. Frémont, explorer and naturalist) Caespitose, annual or biennial herbs from slender taproot. **STEMS** 1 or several, ascending-erect to erect, not arcuate, 2–10 cm. **LEAVES**: cauline leaves lanceolate to elliptic, erect, with wide white margins. **FLOWERS** not obviously touch- or temperature-sensitive; calyx 4–8 mm, lobes 2–4 mm, linear to lanceolate, scarious-margined, midrib distinct; corolla white to pale blue, sometimes with greenish or purplish spots, tube 6–10 mm, lobes 2–4 mm, plicae small, 0.1–0.3 mm, mostly entire. **FRUIT** a short, obovate or elliptic capsule, about 2–3 times as long as wide, 2–6 mm, the valves deeply split nearly to fruit base, and strongly recurved on dehiscence, forming an open cup, partially included or barely exserted past corolla on short, stout stipe. [*Gentiana aquatica* L.]. Uncommon in meadows and wetlands in montane and subalpine forests and alpine tundra communities. COLO: LPl, Min, RGr, SJn. 2440–3655 m (8000–12000'). Flowering: Jun–Jul. Fruit: Aug. Alaska and Yukon south through the Rocky Mountains; west to Nevada and California.

Chondrophylla nutans (Bunge) W. A. Weber (nodding, for the nodding flowers) Caespitose, biennial herbs from slender taproot. **STEMS** 1 or several, ascending or curved, arcuate to somewhat nodding, 4–12 cm. **LEAVES** ovate to obovate, somewhat recurved, with green to narrow white margins. **FLOWERS** not obviously touch- or temperature-sensitive; calyx 6–12 mm, lobes 4–6 mm, broadly lanceolate to triangular, not scarious-margined, midrib not distinct; corolla dark blue, tube 8–20 mm, lobes 2–4 mm, plicae large, 0.5–2 mm, split into 2–3 segments. **FRUIT** an elongate, elliptic to fusiform capsule, more than 4 times as long as wide, the valves split about halfway or less to fruit base, somewhat recurved on dehiscence but not forming an obvious cup, on a slender dark brown or blackish stipe, 1–3 cm long, long-exserted well past corolla. Rare in meadows and slopes in alpine tundra. COLO: LPl, Mon. 3700–4150 m (12000–13700'). Flowering: Jun–Aug. Fruit: Aug–Sep. Colorado; Siberia.

Chondrophylla prostrata (Haenke) J. P. Anderson (for the prostrate stems) Caespitose, biennial herbs from slender taproot. **STEMS** 1 or several, ascending, decumbent, or prostrate, somewhat angled, simple or branched, 1–5 cm. **LEAVES** ovate to obovate, recurved, with narrow white to sometimes mostly green margins. **FLOWERS** distinctly touch- or temperature-sensitive; calyx 6–12 mm, lobes 4–6 mm, broadly lanceolate to triangular, not scarious-margined, midrib not distinct; corolla dark blue, tube 8–20 mm, lobes 2–4 mm, plicae large, 0.5–2 mm, split into 2 or 3 segments. **FRUIT** an elongate, elliptic to fusiform capsule, more than 4 times as long as wide, the valves split about halfway or less to fruit base, somewhat recurved on dehiscence but not forming an obvious cup, on a slender, tan to dark brown stipe less than 1 cm long, partially included or barely exserted past corolla. Fairly common in meadows and slopes in subalpine meadows and alpine tundra. COLO: Hin, LPl, RGr, SJn. 3500–4200 m (11500–14000'). Flowering: Jun–Jul. Fruit: Aug–Sep. Alaska south to Alberta, to Colorado and California.

Comastoma Toyok. Lapland Gentian

(Latin *coma*, mane, + *stoma*, opening, for the fringed scales partially occluding the corolla opening) Caespitose to erect annual (ours), biennial, or perennial herbs from fibrous rootlets, taproots, or rhizomes. **STEMS** 1 or several, erect, angled, simple or branched from near the base. **LEAVES** in basal rosettes, stem leaves few, sessile, not connate at base, lacking white margins. **INFL** of terminal solitary flowers (ours). **FLOWERS** predominantly 4-merous, long-pedicellate; sepals united in short tube, unequal, with 2 somewhat larger lobes, lobes gibbous at base, lacking intracalycine membrane; corolla tubular-campanulate, blue or white, lacking plicae between lobes, lobes entire, small nectary glands at base of lobes, alternate with filaments, each lobe with 2 fimbriate scales at base; stamens free, inserted; pistil sessile, stigmas 2, sessile, persistent. **FRUIT** an elliptic, fusiform to ovoid capsule. **SEEDS** small, numerous, smooth, lacking wings. A small genus of 15 species, circumboreal in montane regions, with a concentration in China and adjacent regions. Although included within *Gentianella* by most past authors, recent work has shown that *Comastoma* is more closely related to *Lomatogonium*.

Comastoma tenellum (Rottb.) Toyok. (*tenellus*, delicate, for its small, slender stature) Lapland gentian. Annual herbs from weakly developed, fibrous taproot. **STEMS** 1 to several, simple or branched, curved-ascending, 4–15 cm high. **LEAVES** glabrous, mostly basal, 4–15 mm long, oblanceolate, cauline leaves smaller, 4–10 mm, lanceolate to elliptic, somewhat clasping at base. **INFL** of solitary, terminal or axillary flowers. **FLOWER** pedicel 3–70 mm long; calyx 4–10 mm, the lobes lanceolate to narrowly triangular, parted nearly to base, distinctly gibbous at base; corolla white, sometimes blue-tinged, 5–15 mm long, lobes 2–6 mm long. **FRUIT** a fusiform capsule, 9–16 mm long. [*Gentianella tenella* (Rottb.) Börner]. Rare in damp alpine meadows. COLO: Hin, LPl, Min, RGr, SJn. 3650–4120 m (12000–13500′). Flowering: Jul–Aug. Fruit: Aug–Sep. Alaska to Greenland, south to California, Nevada, Arizona, New Mexico; Eurasia.

Frasera Walter Elk Gentian, Monument Plant, Tc'il peezet–'ool–itsh–íkíih

(for John Fraser, 18th-century surveyor and botanist) Erect biennial or perennial herbs from a well-developed taproot. **STEMS** 1 or several, round, erect, smooth, simple or branched. **LEAVES**: a basal rosette present or absent, cauline leaves opposite or whorled, entire, somewhat thickened, glabrous to puberulent, sometimes white-margined. **INFL** a cyme or loose to dense thyrse or rarely flowers solitary. **FLOWERS** 4- or rarely 5-merous; calyx deeply cleft with linear to lanceolate lobes, lacking intracalycine membrane; corolla rotate or campanulate, white or yellowish white to green, sometimes with blue or purple streaks, lacking plicae between lobes, each lobe with 1 or 2 foveate nectary glands, sometimes with surrounding corona or fringe of hairs or scales, lobes entire; stamens free or connate at base, exserted; pistil sessile, style slender, filiform, with 2-parted stigma. **FRUIT** an ovoid to elliptic capsule. **SEEDS** relatively few, 5–25 per carpel, somewhat flattened, sometimes narrowly winged along margin. A small genus of 15 species, native and endemic to North America, mostly found in the West, in dry to moist upland habitats. The *Four Corners Flora* follows Struwe and Albert (2002) in segregating *Frasera* from *Swertia*, based on molecular and morphological data. All species of *Frasera* studied are biennial (rosettes sometimes live for several years before bolting) and monocarpic, while *Swertia perennis* is a polycarpic, long-lived perennial. Species are widely used as medicinal plants in the Four Corners region.

1. Petal glands 2; leaves whorled, basal ones large, 20–50 cm long, not white-margined, cauline leaves not connate at base .... ***F. speciosa***
1′ Petal glands 1; leaves opposite to whorled, basal leaves less than 20 cm, typically white-margined, cauline leaves connate-sheathing at base .... (2)

2. Glands narrow, 2-forked (lobed) at apex; stem leaves in whorls of 4 .... ***F. albomarginata***
2′ Glands broad, with 2 parallel lobes at base; stem leaves in opposite pairs .... ***F. utahensis***

Frasera albomarginata S. Watson (white-margined, for the leaves) White-margined frasera. Biennial herbs. **STEMS** 1 or several, mostly unbranched, 2–6 dm high, not glaucous. **LEAVES** glabrous to finely puberulent, in whorls of 4, white-margined, plane or undulate, weakly connate-sheathing, basal leaves 3–10 cm, oblanceolate, with a rounded tip, cauline leaves few, abruptly reduced in size. **INFL** a wide panicle of pedunculate cymes, typically 7–25 cm broad. **FLOWER** calyx 2–6 mm, the lobes lanceolate to narrowly ovate; corolla white, sometimes greenish-tinged or with green dots, lobes 6–10 mm long, glands 1 at base of lobe, split above into 2 lobes, fringed with short whitish hairs, crown scales lacking. **FRUIT** an elliptic capsule, 11–17 mm long. [*Swertia albomarginata* Kuntze]. Common in open sandy to clay soils in sagebrush and piñon-juniper woodlands. ARIZ: Apa; COLO: Mon; NMEX: SJn; UTAH. 1670–2200 m (5500–7200′). Flowering: May–Jun. Fruit: Jul–Aug. California and Nevada east to Arizona, New Mexico, and Colorado.

Frasera speciosa Douglas ex Griseb. (showy, in reference to the large size of the plant) Elkweed, p–ih il tjaa'íh. Biennial or short-lived perennial (but monocarpic) herbs. **STEMS** mostly 1, simple, 5–20 dm high, yellowish to somewhat glaucous green. **LEAVES** glabrous or puberulent, in whorls, not white-margined, mostly plane, not or only weakly connate-sheathing, basal leaves 20–50 cm, 4–9 mm wide, spatulate to oblanceolate, with an acute to rounded tip, cauline leaves numerous, 3–6 in whorls, gradually reduced in size. **INFL** a dense, narrow, compact, leafy thyrse of cymes, lower ones pedunculate, upper ones sessile, verticillate. **FLOWER** calyx 9–25 mm, the lobes narrowly lanceolate, not scarious-margined; corolla pale green, with purple blotches, lobes 9–20 mm long, with deeply split corona scales alternating with nectary, nectary glands 2, elliptic, not split, lobes fringed around margins. **FRUIT** an oblong capsule, 18–25 mm long. [*Swertia radiata* (Kellogg) Kuntze]. Common in open to shaded meadows, clearings, and woodlands in piñon-juniper, sagebrush, mountain brush, montane and subalpine conifer forests, aspen, and alpine tundra; rarely in lower-elevation springs and hanging gardens. ARIZ: Apa, Nav; COLO: Arc, Dol, Hin, LPl, Min, Mon, SJn; NMEX: McK, RAr, SJn; UTAH. 1800–3700 m (5900–12000′). Flowering: May–Jul. Fruit: June–Aug. Elkweed, and probably other species of *Frasera*, have a long tradition of use by indigenous tribes. The roots and leaves were used as a general-purpose tonic, while the leaves were mixed with tobacco and smoked to achieve calmness and to clear the mind. Washington to South Dakota, south to California, Nevada, Colorado, Arizona, New Mexico, and Texas; Mexico.

Frasera utahensis M. E. Jones. (of Utah) Utah frasera. Biennial herbs. **STEMS** 1 or several, simple or sometimes branched from base, 5–10 dm high, glaucous. **LEAVES** glabrous, opposite, white-margined, arched, plane or undulate, connate-sheathing, basal leaves 6–12 cm long, lanceolate, with acuminate tips, cauline leaves few, abruptly reduced in size, not always connate. **INFL** a wide, triangular panicle of thyrsoid or racemiform branches of pedunculate cymes, 12–30 cm broad. **FLOWER** calyx 3–6 mm, the lobes lanceolate to narrowly ovate, scarious-margined; corolla white to green, spotted with darker green, lobes 6–10 mm long, glands 1 at base of lobe, broad, rounded, with 2 basal tubes, strongly fimbriate around margins with hairs, crown scales lacking. **FRUIT** an elliptic capsule, 9–16 mm long. [*Swertia utahensis* (M. E. Jones) H. St. John, not *Swertia paniculata* Wall. (*F. paniculata* Torr.)]. Common on sandy to silty or clay soils and slickrock in open shrublands and piñon-juniper woodlands, rarely in lower-elevation springs and hanging gardens. ARIZ: Apa, Nav; NMEX: SJn; UTAH. 1370–2040 m (4500–6700′). Flowering: May–Jul. Fruit: June–Aug. Utah, Arizona, and New Mexico.

Gentianella Moench Little Gentian, Felwort

(diminutive of *Gentiana*) Erect, annual or biennial herbs from a well-developed taproot. **STEMS** 1 or several, erect, angled or sometimes winged, simple or branched. **LEAVES** in basal rosette, stem leaves opposite, sessile, not connate-sheathing, lacking white margins. **INFL** terminal or axillary cymes, or flowers solitary. **FLOWERS** 4 (5)-merous; sepals united in short tube to almost free, lacking intracalycine membrane, lobes equal to often strongly unequal in 2 series; corolla tubular, funnelform-tubular, campanulate, or salverform, blue, pink, white, yellow, or purple, each lobe with 1 (ours) small nectary gland at base alternate with filaments, each lobe with 1 fimbriate scale present on inside, lobes entire; stamens free, inserted; stigmas 2, sessile, persistent. **FRUIT** an ovoid capsule. **SEEDS** small, numerous, smooth, lacking wings. A large genus of about 150–200 species, widespread in the Northern Hemisphere, south in the mountains of Central and South America. The genus is apparently polyphyletic (Struwe and Albert 2002), but our species are related to the type of the genus, characterized by one nectary gland per petal, and thus are likely to remain in *Gentianella* sensu stricto.

1. Calyx lobes subequal, 1–2 slightly longer than others, united at base into a short tube, lobes 4–12 mm long; corolla lobes 3–5 mm long, fimbriae separate to base .... ***G. acuta***
1′ Calyx lobes unequal, 2 outer lobes much larger, foliaceous, 7–20 mm long, covering inner lobes; corolla lobes 4–8 mm long, fimbriae united at base in a scalelike sheet .... ***G. heterosepala***

Gentianella acuta (Michx.) Hultén (acute or narrowed, in reference to the lanceolate, acute calyx lobes) Felwort. Erect, annual or biennial herbs. **STEMS** simple or sometimes branched, 1–3 (4) dm, branches stiffly erect. **LEAVES** 1–6 cm, basal leaves oblanceolate to elliptic, withering early, cauline leaves sessile, lanceolate to ovate. **INFL** several terminal or axillary cymes each of several flowers, on short pedicels 0.5–3 cm. **FLOWERS** 4 (5)-merous; calyx 4–12 mm, the tube short, 2–5 mm, the lobes unequal, 2 slightly longer, 2–8 mm, narrowly lanceolate; corolla tubular-campanulate, 8–17 mm, purplish blue to pale blue, corolla lobes 3–5 mm, fimbriae free, not united in a scale at base, rarely absent in late-flowering plants; anthers 0.5–0.9 mm. **FRUIT** a fusiform capsule, 10–20 mm, slightly exserted, dehiscent at tips. [*G. amarella* (L.) Börner subsp. *acuta* (Michx.) J. M. Gillett]. Fairly common in moist to wet meadows in montane to low alpine areas. ARIZ: Apa; COLO: Arc, Hin, LPl, Min, Mon, RGr, SJn; NMEX: McK, SJn; UTAH. 2500–4000 m (8000–13000′). Flowering: Jul–Aug. Fruit: Aug–Sep. Circumboreal. Alaska to the Atlantic; California, Arizona, New Mexico, and North Dakota.

Gentianella heterosepala (Engelm.) Holub (different sepals, for the distinctly unequal calyx lobes) Felwort. Erect, annual or biennial herbs. **STEMS** usually simple, sometimes branched, 1–5 dm, branching open and somewhat ascending. **LEAVES** 2–6 cm, basal leaves spatulate to obovate, soon withering, cauline leaves sessile from a somewhat cordate base, oblanceolate to elliptic. **INFL** 1 or more few-flowered cymes or flowers sometimes solitary, lower ones at least long-pedicellate, 2–8 cm. **FLOWERS** 4 (5)-merous; calyx 6–20 mm, lobes strongly unequal, with 2 large outer foliaceous ovate lobes, 7–20 mm, enclosing smaller inner lanceolate lobes, 5–10 mm, free almost to base, tube usually absent; corolla tubular-campanulate, 10–25 mm, blue, white, pink, or occasionally yellow, tube 6–16 mm, lobes 4–8 mm, fimbriae bases united into a scale; anthers 0.7–1.6 mm. **FRUIT** a fusiform capsule, 12–18 mm, slightly exserted, dehiscent at tips. [*G. amarella* (L.) Börner subsp. *heterosepala* (Engelm.) J. M. Gillett]. Fairly common in moist to wet meadows in montane to subalpine areas. ARIZ: Apa; COLO: Arc, Con, Hin, LPl, Min, Mon, RGr, SJn; NMEX: RAr, SJn; UTAH. 2500–3500 m (8000–11500'). Flowering: Jun–Jul. Fruit: Jul–Aug. Arizona, Colorado, and New Mexico.

Gentianodes Á. Löve & D. Löve Arctic Gentian

(Latin for Gentius, a king of Illyria, who according to Pliny was the first to discover the medicinal properties of gentians) Caespitose perennial herbs from a well-developed taproot, sometimes with short rhizomes. **STEMS** 1 or several, short, erect or ascending, strongly angled, simple or branched from near the base. **LEAVES** in well-developed basal rosettes, cauline leaves few, sessile, connate-sheathing at base, lacking white margins. **INFL** of terminal flowers, typically 1 but sometimes 2 or 3 in a cluster. **FLOWERS** predominantly 5-merous, sessile, tubular-funnelform; sepals united in long tube, lobes equal, not gibbous at base, with an intracalycine membrane; corolla tubular-campanulate, white or rarely yellow, often with purple-green blotches, symmetrical plicae present between lobes, small nectary glands at base of pistil, alternate with filaments, lobes lacking fimbriate scales, entire; stamens free, inserted; pistil stipitate, style short, stigmas 2, persistent. **FRUIT** an elliptic to oblong capsule. **SEEDS** large, 1–2 mm, with irregular winged margins. A genus of about 100 species, circumpolar in Arctic-alpine regions, with a major concentration in China and the Himalayas, and only one species reaching North America.

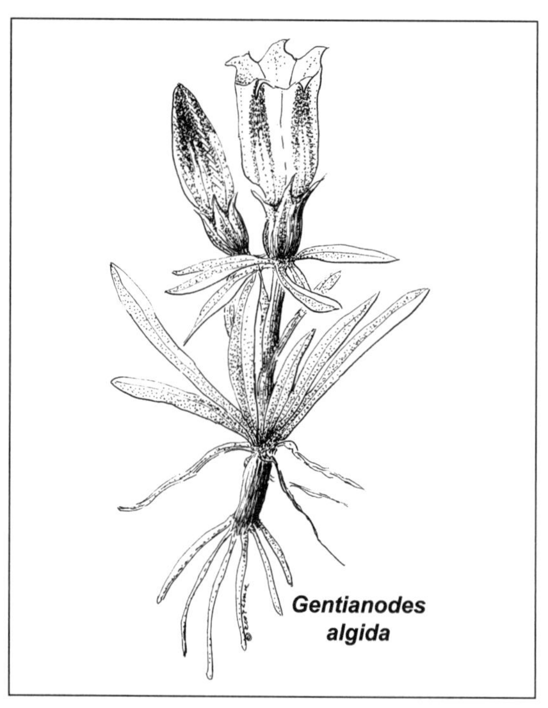
Gentianodes algida

Gentianodes algida (Pall.) Á Löve & D. Löve. (cold, for the tundra habitat) **STEMS** 1 or several, 5–20 cm, ascending to erect, simple. **LEAVES**: basal leaves 4–10 cm, long-lanceolate or linear, cauline leaves 2–6 cm, lanceolate to oblanceolate, in 2–3 pairs below flower bracts. **INFL** typically 1-flowered, sometimes 2–3 present. **FLOWER** calyx tube 9–18 mm, the lobes often of unequal lengths, 6–12 mm; corolla funnelform, tube 30–50 mm, white or pale yellow with green to purple blotches inside, lobes short, 2–6 mm. **FRUIT** a capsule, elliptic-fusiform, short-stipitate, not strongly exserted beyond corolla. [*Gentiana algida* Pall.]. Common in alpine tundra and adjacent subalpine meadows and rocky slopes. COLO: Arc, Con, Hin, LPl, Min, Mon, RGr, SJn. 3300–4200 m (11000–13500'). Flowering: Jul–Aug. Fruit: Aug–Sep. Alaska and Yukon south to New Mexico.

Gentianopsis Ma Fringed Gentian

(Latin for resemblance to *Gentiana*) Erect annual, biennial, or perennial herbs, from a slender taproot or rhizomes. **STEMS** 1 or several, erect, angled, simple or branched from near the base or higher. **LEAVES** in weakly developed basal rosettes, cauline leaves opposite, sessile, somewhat connate-sheathing at base, lacking white margins. **INFL** of terminal solitary flowers, sometimes with axillary flowers below. **FLOWERS** predominantly 4-merous, short- to long-pedicellate; sepals united in long tube, lobes equal or unequal, with 2 larger and 2 smaller lobes, not gibbous at base, with a discontinuous triangular intracalycine membrane often present; corolla tubular-campanulate or funnelform, blue, lacking plicae between lobes, lobes strongly fringed and with tips erose or dentate, small nectary glands at base alternate with filaments, lacking fimbriate scales; stamens free, inserted; pistil stipitate, stipe sometimes elongating in fruit, stigmas 2, sessile, persistent. **FRUIT** a slender, fusiform to elliptic capsule. **SEEDS** small, numerous, angular, papillate, lacking wings. A small, circumboreal genus of 25 species, segregated from *Gentianella*.

1. Perennials; flowers sessile or on short pedicels in axes of 2 leafy bracts; corolla lobes oblong.......... ***G. barbellata***
1' Annuals; flowers on long pedicels, bracts lacking; corolla lobes obovate.......... ***G. thermalis***

Gentianopsis barbellata (Engelm.) H. H. Iltis (for the short, stiff fimbriae of the petals) Erect perennials arising from rhizomes. **STEMS** 5–15 cm, simple or branched above. **LEAVES** in well-developed basal rosettes, 2–8 cm, strongly connate-sheathing at base, stem leaves few, 1–4 cm. **INFL** of terminal solitary flowers. **FLOWERS** sessile or sometimes

Gentianopsis barbellata

short-pedicellate, subtended by 2 bractlike leaves; calyx tube 10–25 mm, lobes 5–12 mm, equal, with an interrupted intracalycine membrane with 1–6 short processes; corolla broadly funnelform, 20–40 mm, lobes 12–24 mm, oblong, fringed in lower 1/2 or to near tip, tips somewhat rounded to acute, erose or dentate; anthers 1.5–2.5 mm. Uncommon in damp mountain and subalpine meadows, rocky slopes, and forest clearings, extending into low alpine tundra. COLO: LPl, RGr, SJn. 3000–3960 m (9800–13000′). Flowering: June–Aug. Fruit: Aug–Sep. Wyoming south to Colorado, Arizona, and New Mexico.

Gentianopsis thermalis (Kuntze) H. H. Iltis (referring to warm springs; the species was first collected around hot springs in Yellowstone National Park) Erect annuals from a slender taproot. **STEMS** 2–6 dm, simple or branched above. **LEAVES** in weakly developed basal rosettes, stem leaves numerous, only weakly or not connate-sheathing at base, 2–7 cm long, oblanceolate, becoming smaller and elliptic above. **INFL** of terminal solitary flowers. **FLOWERS** on a long pedicel, 3–10 cm; calyx tube 12–40 mm, lobes 8–25 mm, unequal, in 2 series, lacking an intracalycine membrane; corolla broadly funnelform, 20–80 mm, lobes 10–20 mm, obovate, fringed in lower 1/2, tips rounded and erose or dentate; anthers 2–4 mm. [*G. detonsa* (Rottb.) Ma]. Common in damp mountain and subalpine meadows, rocky slopes, riparian zones, and forest clearings, extending into low alpine tundra. COLO: Arc, Con, Hin, LPl, Min, Mon, RGr, SJn. 2700–3650 m (8900–12000′). Flowering: June–Aug. Fruit: Aug–Sep. The flowers of *G. thermalis* are among the largest of any native species in the Four Corners region. Some populations have white- or pink-flowered individuals. Circumboreal. Alaska to Newfoundland, south to Mexico.

Pneumonanthe Gled. Bottle Gentian

(Latin *pneumon*, lung, + *anthos*, flower, for the inflated bottlelike flowers) Erect perennial herbs from a well-developed taproot, short, erect, rhizomatous corms sometimes present. **STEMS** 1 or several, erect, angled, simple or branched from near the base. **LEAVES** lacking basal rosettes, cauline leaves numerous, opposite, sessile, connate-sheathing at base, lacking white margins. **INFL** of terminal and lateral bracteate cymes. **FLOWERS** predominantly 5-merous, sessile or pedicellate; calyx tubular, entire or split on 1 side to base, lobes equal or unequal, not gibbous at base, with an intracalycine membrane; corolla funnelform to campanulate, blue or blue-purple, lobes lacking fimbriate scales, entire, with asymmetrical plicae between lobes; small nectary glands at base of pistil, alternate with filaments; stamens free or anthers connate, inserted; pistil sessile or stipitate; style short, slender, stigmas 2, persistent. **FRUIT** an elliptic to ovoid capsule. **SEEDS** small, numerous, reticulate, winged or wingless. A circumtemperate genus of about 40 species, with the majority of the species in North America, extending south into the mountains of Mexico and Guatemala.

1. Flowers arising from below the stem in leaf axils as well as the apex; floral bracts narrowly lanceolate, not scarious or boat-shaped; upper stem leaves lanceolate .... ***P. affinis***
1′ Flowers mostly terminal; floral bracts large, broadly ovate, somewhat scarious, boat-shaped; upper stem leaves below the flowers ovate .... ***P. parryi***

Pneumonanthe affinis (Griseb.) Greene (affinity, in reference to its close relationship to several other species of the genus) Rocky Mountain bottle gentian. Erect to ascending, perennial herbs. **STEMS** 1 or several, 1–7 dm high, puberulent throughout. **LEAVES** few to crowded, elliptic to lanceolate, smaller near base and becoming narrower near tip, 2–4 cm long, 0.7–2 cm wide, margins puberulent, midribs often puberulent. **INFL** of terminal and lateral bracteate cymes, 1–5 per node, lateral flowers usually present; bracts lanceolate, slightly scarious-margined, green, not boat-shaped, shorter than the flowers. **FLOWERS** sessile to short- or long-pedicellate; calyx tube 5–10 mm, often split on 1 side to near base, scarious-margined between lobes, calyx lobes variable, lanceolate to narrowly elliptic, 1–3 mm, equal to strongly subequal; corolla funnelform to campanulate, strongly plicate, often narrower at mouth than in middle, blue or blue-purple, 15–40 mm, with asymmetrical plicae between lobes, tips split 2 (3–5) times, lobes 3–10 mm; anthers free, 1.5–3 mm, inserted; pistil stipitate. **FRUIT** an elliptic to ovoid capsule, 12–30 mm. [*Gentiana affinis* Griseb.]. Uncommon in dry to damp, montane to subalpine meadows and grasslands, forest clearings, and grassy slopes, rarely extending to lower elevations at springs and hanging gardens. ARIZ: Nav; COLO: Arc, LPl, SJn; NMEX: SJn; UTAH. 1800–3200 m (6000–10500′). Flowering: Jun–Jul. Fruit: Jul–Sep. There is some confusion regarding the taxonomy of this and related species. *Pneumonanthe oregana* (Engelm. ex A. Gray) Greene occurs in the Northwest and is replaced by *P. affinis* in the Rocky Mountains and adjacent plains south to Utah and Colorado. *Pneumonanthe bigelovii* (A. Gray) Greene is found in southern Colorado, New Mexico, and Arizona. In Arizona and New Mexico a fourth species is found, *P. rusbyi* (Greene ex Kusn.) Greene. Material in the Four Corners region is interpreted as *P. affinis*, but some collections in the eastern part of the region approach *P. bigelovii*. Plants recently collected in Navajo County, Arizona, on the Navajo Nation approach the southern *P. rusbyi*. More work is needed to sort out this group. British Columbia to Montana, south to California and New Mexico.

1. Corolla broadly funnelform to campanulate, widest at mouth, deep blue-purple; bracts shorter than the corolla; calyx tube typically split down 1 side; Colorado and Utah northward .... ***P. affinis***
1' Corolla funnelform to fusiform, narrowed above, dull to pale blue or sometimes white; bracts variable, exceeding the corolla or shorter; calyx split down 1 side or not split; New Mexico, Arizona, or southeastern Colorado .....(2)

2. Corolla fusiform, closed at maturity; bracts shorter than the corolla; calyx tube often split on 1 side; New Mexico and Arizona .... *P. rusbyi*
2' Corolla funnelform, open at maturity; bracts exceeding the corolla; calyx tube mostly entire; southeastern Colorado, eastern New Mexico, and southern Arizona .... *P. bigelovii*

Pneumonanthe parryi (Engelm.) Greene (for C. C. Parry) Parry's bottle gentian. Erect to ascending, perennial herbs. **STEMS** 1 or several, 1–5 dm, sparsely puberulent. **LEAVES** somewhat crowded, numerous, broadly elliptic to ovate, smaller near base and becoming larger above, numerous, 2–4 cm long, 1–2.5 cm wide, margins somewhat puberulent, midribs glabrous to puberulent. **INFL** of terminal, dense, bracteate cymes, with (1) 2–6 flowers; bracts large, slightly shorter than flowers, boat-shaped, reddish, with scarious margins. **FLOWERS** sessile or short-pedicellate; calyx tube 10–25 mm, usually united but sometimes split on 1 side, scarious-margined between lobes, calyx lobes elliptic, mostly equal to subequal, 3–11 mm; corolla broadly funnelform to campanulate, widest at mouth, blue or blue-purple, 20–50 mm, strongly plicate, with asymmetrical plicae between lobes, tips split 2–5 times, lobes 3–15 mm; anthers free, 3–5 mm, inserted; pistil stipitate. **FRUIT** an elliptic to ovoid capsule, 15–45 mm. [*Gentiana parryi* Engelm.]. Fairly common in damp to wet, montane to subalpine meadows and grasslands, forest clearings, and alpine tundra communities. COLO: Arc, Con, Hin, LPl, Min, Mon, RGr, SJn. 2300–4100 m (7500–13500'). Flowering: Jul–Aug. Fruit: Aug–Sep. Wyoming and Colorado, south to Arizona and New Mexico.

Swertia L. Star Gentian, Felwort

(for Emmanuel Sweerts, Dutch engraver and gardener) Erect to ascending perennial herbs arising from a rhizome. **STEMS** 1 or several, round, smooth, erect to decumbent, simple. **LEAVES** at base alternate, somewhat crowded, cauline leaves subopposite or alternate, entire, somewhat thickened, glabrous, not connate at base, not white-margined. **INFL** a cyme or loose to dense thyrsoid panicle. **FLOWERS** 5-merous; calyx deeply cleft with lanceolate lobes, lacking intracalycine membrane; corolla rotate, deeply cleft with oblong lobes, blue, purple, or rarely white, each lobe with 2 small, foveate, fringed nectary glands at base, lacking plicae between lobes, lobes erose-dentate or entire, but not fringed; stamens free, exserted; pistil sessile, style short or lacking, stigmas 2. **FRUIT** an ovoid capsule. **SEEDS** large, flattened, winged around margin. A large genus of perhaps 140 species, distributed in Eurasia, Africa, and Madagascar, with only one species reaching North America. *Swertia* is polyphyletic as currently circumscribed according to von Hagen and Kadereit (2002). Thus it may be best to restrict the use of the generic name to the type section, of which the type species is *S. perennis*. The elk gentians are not closely related to *Swertia* and are segregated as *Frasera* following Struwe and Albert (2002).

Swertia perennis L. (perennial) Felwort. Erect perennial herbs arising from a short rhizome. **STEMS** typically 1, simple, usually not branched, 2–6 dm. **LEAVES** forming basal rosette, long-petiolate, 4–20 cm long, petiole often as long as the blade, oblanceolate to obovate, cauline leaves few, small, 2–6 cm, lanceolate or narrowly oblanceolate, sessile. **INFL** a few-flowered cyme. **FLOWERS** mostly 5-merous; calyx tube short or nearly absent, lobes lanceolate, 4–8 mm; corolla bluish purple, tube short, lobes 7–15 mm, oblong to elliptic, often erose at tips, each lobe with 2 small, round glands near base, with sparse, long fimbriae. **FRUIT** an ovoid to elliptic capsule, 8–12 mm. Fairly common in wet meadows, marshes, clearings in subalpine forests, stream banks, and bogs. COLO: Arc, Con, Hin, LPl, Min, Mon, RGr, SJn. 2600–4000 m (8600–13000'). Flowering: July. Fruit: Aug. Alaska to California; Eurasia.

Zeltnera G. Mans. American Centaury

(for botanist L. Zeltner, specialist of *Centaurium*) Erect, annual to biennial herbs from a taproot. **STEMS** 1 or several, erect, angled, often winged, simple or branched from near the base. **LEAVES** with or without basal rosettes, stem leaves opposite, somewhat sheathing, not connate at base, lacking white margins. **INFL** a terminal cyme or spike. **FLOWERS** predominantly 5-merous, sessile to pedicellate; calyx deeply cleft into equal linear lobes, sometimes scarious-margined, lobes not gibbous at base, lacking intracalycine membrane; corolla rotate or salverform, pink, rose, or rarely white, corolla tube slender, lacking plicae between lobes, lobes entire, small nectary glands at base alternate with filaments, lacking fimbriate scales; stamens free, exserted; pistil sessile, style simple or 2-lobed, persistent. **FRUIT** a cylindrical-fusiform capsule. **SEEDS** small, numerous, reticulate, lacking wings. A genus of about 25 species, endemic to western and southern North America, Mexico, and Guatemala. The North American species have traditionally been placed in the primarily Eurasian *Centaurium*, but Mansion and Struwe (2004) have shown the genus to be polyphyletic. The North American species are segregated as *Zeltnera* and appear to be most closely related to the Mediterranean genera *Schenkia* and *Exaculum*, and to North American *Sabatia*.

1. Corolla tube 10–20 mm, lobes 7–13 mm, exserted well past the calyx lobes .... ***Z. calycosa***
1′ Corolla tube 6–12 mm, lobes 3–5 mm, barely exserted past the calyx lobes .... ***Z. exaltata***

Zeltnera calycosa (Buckley) G. Mans. **CP** (calyx) Buckley's centaury. Annual to rarely biennial herbs. **STEMS** 5–60 cm. **LEAVES**: basal rosette present, well developed, leaves lanceolate to ovate or often oblanceolate, 2–6 cm long. **FLOWER** calyx 8–14 mm, lobes not scarious-margined; corolla well-exserted past calyx, tube yellowish, 10–20 mm, lobes light pink or white, ovate to obovate, 7–13 mm, anthers 1.5–3.5 mm. **FRUIT** a cylindrical capsule, 8–15 mm. [*Centaurium calycosum* (Buckley) Fernald]. Common in low to mid-elevation wet meadows, riparian zones, springs, and seeps. ARIZ: Apa, Nav; COLO: Mon; NMEX: SJn; UTAH. 1100–1850 m (3600–6000′). Flowering: May–Jul. Fruit: Jul–Sep. California to Texas; Mexico.

Zeltnera exaltata (Griseb.) G. Mans. (very tall) Great Basin centaury. Annual herbs. **STEMS** 2–30 cm, highly variable in height in same population. **LEAVES**: basal rosette usually lacking, occasionally weakly developed in larger plants, leaves linear to lanceolate, 1–3 cm long. **FLOWER** calyx 6–10 mm, lobes often scarious-margined; corolla barely exserted past calyx, tube yellowish, 6–12 mm, lobes pink to rose or rarely white, strap-shaped, 3–5 mm; anthers 0.5–1.5 mm. **FRUIT** a cylindrical capsule, 6–10 mm. [*Centaurium exaltatum* (Griseb.) W. Wight ex Piper]. Uncommon in low-elevation wet meadows, riparian zones, springs, and seeps. COLO: Mon; NMEX: SJn. 1100–1700 m (3600–5600′). Flowering: Jun–Aug. Fruit: Jul–Sep. British Columbia to Idaho and Colorado; Baja California.

GERANIACEAE Juss. GERANIUM FAMILY

Lynn M. Moore

(Greek, a crane, for the long fruit beak) Annual or perennial herbs. **LEAVES** opposite or alternate, stipulate, simple or compound. **INFL** axillary, cymose, sometimes appearing umbellate. **FLOWERS** mostly regular, perfect; sepals 5, distinct; petals 5, distinct; pistil 1, ovary superior, usually 5-loculed, 1–2 ovules per locule, axile placentation, style 1 with 5 stigmatic lobes; stamens 5 or 10, the filaments usually united at the base. **FRUIT** a dry schizocarp, 1 seed per locule, the mericarps separating from the base and coiling at maturity. $x = 7$–14. Eleven to 14 genera, 700–775 species. Widely distributed in temperate to subtropical regions, relatively rare in tropical regions, where it is mostly restricted to higher elevations. *Pelargonium* species are widely cultivated as house plants.

1. Leaves pinnate-pinnatifid or tripartite; stamens 5 .... ***Erodium***
1′ Leaves cordate-palmatifid, 5 (7) segments; stamens usually 10 .... ***Geranium***

Erodium L'Hér. ex Aiton Storksbill, Filaree

(Greek *erodios*, a heron, referring to the long beaks of the fruit) Plants winter-annual or short-lived perennial. **LEAVES** opposite, pinnate-pinnatifid or tripartite, petiolate, stipulate. **INFL** a few-flowered umbel on axillary peduncles. **FLOWERS** with persistent sepals, usually awn-tipped; petals deciduous; fertile stamens 5, alternate with 5 reduced scalelike staminodia, filaments distinct and broadest at the base. **FRUIT** a schizocarp with an elongated style column, styles bearded inside; each mericarp spindle-shaped, becoming free from the tip of the central style column, then spirally coiling, mericarp remaining intact. **SEEDS** smooth. In *Erodium* the beaks of the fruit form a corkscrewlike tail that tightens when dry and loosens when moist. This action drives the sharp, pointed beaks into the soil, thus planting the fruit. This is similar to the seeds of *Stipa* and *Cercocarpus*. Approximately 12 species (20 taxa) in North America. Worldwide about 75 species widespread in temperate and semitropical regions.

1. Leaves pinnate; petals 3–6 (7) mm long .... ***E. cicutarium***
1′ Leaves tripartite; petals 6–12 mm long .... ***E. texanum***

Erodium cicutarium (L.) L'Hér. ex Aiton (for hemlock, which it is said to resemble) Storksbill, filaree, redstem filaree, redstem storksbill, alfilaree, alfilaria, heronsbill, bee food, menstruation medicine, tis'nádáá (Navajo). Plants glandular, villous, strigose. **STEMS** few to several, erect or decumbent at first, becoming prostrate. **LEAVES** basal and cauline, pinnate, 3–14 cm long, 4–8 opposite pairs of pinnatifid lateral leaflets, leaflets 6–13 mm long, stipules lanceolate, acute, fimbriate. **INFL** 2–5 (10)-flowered umbel, glandular-pubescent; pedicels 5–18 (22) mm long, spreading to reflexed at base, recurved at the tip. **FLOWERS** with sepals 2.8–6.4 mm long, lanceolate to ovate, awn mucronate, 0.1–0.4 (1) mm long; petals 5, 3–6 (7) mm long, ciliate at the base, rose-lavender; receptacle and style 2.5–4.5 cm long; style pilose-strigulose; mericarps 4–5.5 mm long, hirsute-ascending; filaments broad at the base, anthers mainly 0.8 mm long. **FRUIT** a schizocarp with a persistent elongated style column; mericarps separating from the tip of the central axis, spirally coiling, bearded, hairs strigose. $2n = 20, 36, 38, 40, 42$. [*Geranium cicutarium* L.]. Piñon-juniper, sagebrush, desert scrub, montane, and spruce-fir communities; frequently disturbed and often dry places, roadsides, farmland, and rangelands. ARIZ: Apa, Coc, Nav; COLO: Arc, Dol, Hin, LPl, Min, Mon, SJn, SMg; NMEX: McK, RAr, San, SJn; UTAH. 1125–2600 m (3700–8500′). Flowering: Apr–Aug. Considered a noxious weed; native of Europe or Asia; introduced by the Spanish in the 16th century into Mexico, then colonized and naturalized north into Arizona, New Mexico, and throughout the U.S. Occasionally grown for forage, it is a serious competitor of early-planted spring crops, facilitated by its ability to emerge and thrive under cool to moderate temperatures. Plants were used medicinally by the Jemez and Zuni. The Navajo used it to treat excessive menstruation and bobcat or mountain lion bites. The Navajo also used it on prayer sticks and as a medicine in many ceremonies.

Erodium texanum A. Gray (of Texas) Texas storksbill, desert heronsbill, desert storksbill, large-flowered storksbill, bull filaree, Texas filaree, Texas fillarie, tufted filaree. Plants eglandular, canescent-strigulose, hairs retrorse on lower stems and petioles, ascending on the leaves, upper stem, and inflorescence. **STEMS** few to several, decumbent to prostrate, reddish. **LEAVES** basal and cauline, 1–3.3 cm long, 0.8–3 cm wide, ovate and tripartite, lobes rounded, crenate, cordate at the base, middle segment sometimes pinnately lobed, veins often reddish, stipules lanceolate to deltate. **INFL** (1) 2–4-flowered umbel, pedicels 6–18 mm long, spreading and abruptly recurved at the tip. **FLOWERS** with sepals 5–11 mm long, lanceolate, broadly oblanceolate to ovate, 5 reddish nerves, margins hyaline and glabrous, awn 0.2–0.8 mm long; petals 5, 6–12 mm long, obovate, broadly rounded, purplish red; receptacle and style 4.5–7 cm long, style pilose-strigulose; mericarps 7–9 mm long, hirsute-ascending from a pustulose base, anthers mainly 1.1–1.2 mm long. **FRUIT** a schizocarp with a persistent elongated style column; mericarps separating from the tip of the central axis, spirally coiling, bearded, hairs pilose-ascending. $2n = 20$. Gravelly or rocky soils in desert scrub communities along the San Juan arm of Lake Powell. UTAH. 1125 m (3700′). Flowering: Feb–Apr. Native to the southwestern United States from California to Texas and Oklahoma, east to Missouri and South Carolina, south to Mexico.

Geranium L. Cranesbill, Geranium

(modern Latin from Greek *geranion*, *geranos*, crane, alluding to the long beak of the fruits) Plants annual or (ours) perennial herbs. **LEAVES** alternate or opposite, cordate-palmatifid, 5 (7) segments, petiolate, stipulate. **INFL** a 2-flowered dichasial cyme. **FLOWERS** with sepals imbricate, persistent, mucronate or awn-tipped; petals deciduous; style column usually beaked; stigmas (5) free; styles nearly glabrous inside; stamens 10, usually all anther-bearing, occasionally 5 fertile and 5 sterile, filaments often unequal, with the inner whorl the longest. **FRUIT** a schizocarp, mericarps separating at maturity below and recurving, remaining attached to the upper column; seeds forcibly expelled. **SEEDS** smooth, reticulate-alveolate, or pitted. Dichotomous keys for this genus are tentative at best. This genus has considerable variability, and diagnostic characters in the past have been based upon pubescence type and its position on the plant (Aedo 2001). A genus of about 300 species of temperate and tropical mountain regions from around the world.

1. Petals white, occasionally light pink to lavender in bud, pilose on bottom adaxial surface for 1/2 or more the length .... (2)
1′ Petals pink to purple, pilose on bottom adaxial surface for 1/4 to 1/2 the length .... (3)

2. Petals 14–20 mm long, with dark purple veins; stems and pedicels hirsute with red- or purple-tipped glandular hairs .... ***G. richardsonii***
2′ Petals 5–10 mm long; stems and pedicels hirsute with yellowish or whitish-tipped glandular hairs .... ***G. lentum***

3. Leaves 6–12 (15) cm wide; petals 14–22 mm long, pilose on bottom adaxial surface for only 1/4 the length .... ***G. viscosissimum***
3′ Leaves 1.5–7 (7.5) cm wide; petals (8) 10–15 (17) mm long, pilose on bottom adaxial surface for 1/4 to 1/2 the length .... ***G. caespitosum***

Geranium caespitosum E. James ex Torr. (caespitose or tufted) Perennial herb from a branched caudex with thick marcescent leaf bases. Plants erect to ascending, sometimes sprawling; usually few stems per caudex base. **LEAVES** palmately lobed, 3–5 segments, each 3–5 (7)-lobed, each ultimate segment obovate or rhombic, 1.5–7 (7.5) cm wide. **FLOWERS** with sepals 8–11 mm long, mucronate tip (0.5) 1–2 mm long; scarious margins; petals pink to purple, (8) 10–15 (17) mm long, pilose at bottom for 1/4 to 1/2 the length; gynoecium 10–13 mm long; mature stylar column 22–31 (40) mm long; beak 2.5–5.5 mm long; stamens 10 (15), the inner whorl sometimes longer. **FRUIT** 22–31 (40) mm long. **SEEDS** reticulate-pitted, dark brown. $n = 26$. Northern Tiwa and western Keres pueblos used this as a treatment for sore throats, rashes, and abrasions along with other plants with high tannin content. The Navajo also used this for overexertion and as a life medicine, and the blossoming plants were dipped in salt water to bring rain. The Jemez used the split fiber to sew moccasins. The *G. caespitosum* complex has been subject to controversial interpretation for over 100 years. Most botanists believe that there is one species. Over the years different varieties have been recognized based upon the glandular versus nonglandular pubescence type. Cronquist, Holmgren, and Holmgren (1997) give a detailed discussion of the taxonomic problems of this species complex; a conservative view is taken here.

1. Stems, petioles, or pedicels nonglandular (sometimes with a few short glandular hairs among the longer nonglandular hairs) .... **subsp. *atropurpureum***
1' Stems, petioles, or pedicels with glandular-tipped hairs .... **var. *caespitosum***

subsp. *atropurpureum* (A. Heller) W. A. Weber (dark purple) Western purple cranesbill, magenta geranium. **STEMS** reddish-tinged, mostly retrorse to spreading nonglandular hairs, occasionally short-hirsute, whitish- to yellowish-tipped glandular hairs present within the longer nonglandular hairs. **INFL** pedicels mostly with retrorse nonglandular hairs, sometimes with a few short yellowish- or whitish-tipped glandular hairs among the longer nonglandular hairs. [*G. atropurpureum* A. Heller; *G. marginale* Rydb. ex Hanks & Small; *G. toquimense* N. H. & A. H. Holmgren]. Coniferous forest, including piñon-juniper, ponderosa pine, spruce-fir, aspen, oak, and montane tall shrub communities. Occasionally along riparian corridors and wetlands. ARIZ: Apa; COLO: Arc, Dol, Hin, LPl, Min, Mon, SJn, SMg; NMEX: McK, RAr, SJn; UTAH. 2100–2900 m (6800–9500'). Flowering: Jun–Aug (Sep). Native to the western United States from Wyoming south to Mexico, west to Nevada, and east to the panhandle of Texas.

var. *caespitosum* Pineywoods geranium, purple cluster geranium, tufted geranium. Perennial herb from a branched caudex with thick marcescent leaf bases. Plants erect, sometimes sprawling, usually few stems per caudex base. **STEMS** with pilose-spreading to retrorse nonglandular hairs and glandular yellowish- or whitish-tipped hairs. **INFL** with pilose pedicels, spreading nonglandular hairs and yellowish- or whitish-tipped glandular hairs. [*G. caespitosum* subsp. *caespitosum*; *G. fremontii* Torr. ex A. Gray; *G. parryi* (Engelm. ex A. Gray) Heller]. Coniferous forest from ponderosa pine to spruce-fir; montane tall shrub and meadow communities. ARIZ: Apa; COLO: Arc, Dol, Hin, LPl, Min, Mon, RGr, SJn; NMEX: McK, RAr, SJn; UTAH. 2000–3000 m (6500–10000'). Flowering: Jun–Aug (Sep). Native to the western United States from Wyoming south to Mexico, west to Nevada, and east to the panhandle of Texas. This variety is less common in the Four Corners region then subsp. *atropurpureum*.

Geranium lentum Wooton & Standl. (pliable) Mogollon geranium, white geranium. Perennial herb from a branched or unbranched woody caudex with wiry marcescent leaf bases. Plants erect, usually few stems per caudex base. **STEMS** with pilose-spreading to retrorse nonglandular and glandular yellowish- or whitish-tipped hairs. **LEAVES** palmately lobed, 5 (7) trifid lobes, ultimate segment obovate or rhombic, sometimes shallowly lobed, 2.5–5 (7) cm wide. **INFL** with densely hirsute pedicels, spreading to retrorse nonglandular hairs and short- to long-hirsute, yellowish- or whitish-tipped glandular hairs. **FLOWER** sepals 6–8 mm long, tip mucronate, 0.5–1 mm long, margins scarious; petals white with cream to light pinkish veins, 5–10 mm long, pilose at bottom for 1/2 to 3/4 the length; gynoecium 7–8 mm long; mature stylar column 22–27 mm long; beak 1–2 (3.5) mm long. **FRUIT** 22–27 mm long. **SEEDS** reticulate. Sandstone canyons; coniferous forests, black sagebrush, aspen, and Gambel's oak communities. ARIZ: Apa. 2225 m (7300'). Flowering: Jun–Sep. Native to the southwestern United States from Arizona to New Mexico and south to Mexico. Navajo applied a poultice of moist leaves and roots to injuries; a decoction of the plant was taken for internal injury and was used as a life medicine.

Geranium richardsonii Fisch. & Trautv. (honors Sir John Richardson, 1787–1865, Scottish naturalist) Richardson's geranium. Perennial herb from a branched or unbranched woody caudex; frequently rhizomatous; covered with marcescent leaf bases. Plants erect, usually solitary. **STEMS** with pilose-spreading to retrorse nonglandular and glandular purple- or red-tipped hairs, lower stems sometimes nearly glabrous. **LEAVES** palmately lobed, 3–5 (7) trifid lobes, these divisions pinnately coarsely toothed, the teeth acute, (4) 6–12 (15) cm wide. **INFL** with dense to sparsely hirsute pedicels, spreading nonglandular hairs and glandular, usually purple- or red-tipped hairs. **FLOWER** sepals 7–8 mm long, tip mucronate, 1–3 mm long, margins scarious; petals usually white, sometimes pink or lavender in bud, with light pink to dark purple veins, 14–20 mm long, pilose at bottom for 1/2 or more the length; gynoecium 8–9 mm long; mature stylar column 19–28 (35) mm long; beak 1–3.5 (4) mm long. **FRUIT** 19–28 (35) mm long. **SEEDS** reticulate-alveolate, dark brown. $2n = 52, 56$. [*G. gracilentum* Greene]. Coniferous forests and mixed coniferous forests from piñon-juniper to subalpine; meadows and tall shrub communities; riparian corridors and alpine. ARIZ: Apa; COLO: Arc, Con, Dol, Hin, LPl, Min, Mon, RGr, SJn, SMg; NMEX: McK, RAr, SJn; UTAH. 2400–4200 m (7800–13800′). Flowering: Jun–Aug. Native to western North America from the Northwest Territories south to California, Arizona, and New Mexico, east throughout the Rocky Mountains to the Black Hills. Occasionally the purple-tipped glandular hairs are undeveloped in small early growth. Northern Tiwa and western Keres pueblos used this plant as a treatment for sore throats, rashes, and abrasions along with other plants with high tannin content. The Navajo used it as a "life medicine," and the Cheyenne used it for nosebleeds.

Geranium richardsonii

Geranium viscosissimum Fisch. & C. A. Mey. (sticky, alluding to the densely glandular pubescence throughout the stems and petioles) Sticky purple geranium, sticky geranium. Perennial herb from an elongate, thick, sometimes branched caudex, often short-rhizomatous; plants erect, usually solitary. **STEMS** with pilose-spreading to retrorse nonglandular and glandular yellowish- or whitish-tipped hairs. **LEAVES** palmately lobed, 5 (7) segments, each 9 (13)-lobed, these divisions pinnately coarsely toothed, the teeth acute, 6–12 (15) cm wide. **INFL** with pubescent pedicels, mostly yellowish- or whitish-tipped glandular hairs and retrorse to spreading nonglandular hairs. **FLOWERS** with sepals 8–11 mm long, tip mucronate, 1–2 mm long, margins scarious; petals usually white, sometimes pink or lavender in bud, with light pink to dark purple veins, 14–20 mm long, pilose at bottom for 1/2 or more the length; gynoecium 11–13 mm long; mature stylar column 28–41 mm long; beak 4–6 (7) mm long. **FRUIT** 19–28 (35) mm long. **SEEDS** reticulate. $2n = 52$. [*G. nervosum* Rydb.; *G. viscosissimum* var. *nervosum* (Rydb.) C. L. Hitchc.; *G. viscosissimum* subsp. *nervosum* (Rydb.) W. A. Weber]. Roadsides, mountain shrub, montane aspen, and spruce-fir communities. COLO: Arc, Dol, Hin, LPl, Min, Mon, RGr, SJn; UTAH. 1600–3200 m (5200–10500′). Flowering: Jun–Aug. Native to northwestern North America from British Columbia south to northern California, east across the mountains to Colorado and the Black Hills. This species is uncommon in the San Juan River drainage. Our material belongs to **var. *incisum*** (Torr. & A. Gray) N. H. Holmgren (deeply cut). It has been reported from New Mexico, but no specimens of *G. viscosissimum* in New Mexico have been verified. It is reported to occur along the Piedra River in Archuleta County, Colorado, and unverifed specimens do exist from other locations in the San Juan Mountains. It is easily confused with *G. caespitosum*. *Geranium viscosissimum* differs in having larger leaves, fruit, and erect, often solitary stems as well as no reddish tinge as in *G. caespitosum* subsp. *atropurpureum*. It was used by the Blackfoot as a cold remedy, gynecological aid, and treatment for sore eyes. It was also thought to work as a love charm or love potion.

GROSSULARIACEAE DC. CURRANT or GOOSEBERRY FAMILY

Roy Taylor

(Latin *grossularia*, gooseberry) Shrubs (ours). **STEMS** spiny. **LEAVES** mostly alternate, rarely opposite, stipules absent or present; simple and sometimes deeply cleft. **INFL** a raceme or panicle, or of solitary flowers in the upper axils. **FLOWERS** perfect, rarely unisexual, regular or nearly so, epigynous with a saucer-shaped to tubular hypanthium; sepals 3–9, sometimes larger than the petals; petals 3–9, alternate with the sepals; stamens as many as and alternate with the petals; nectary disc often present internal to the stamens; gynoecium of 2 or 3 (7) carpels united to form a compound, superior to partly or wholly inferior ovary; ovules numerous. **FRUIT** a berry. **SEEDS** numerous. x = 8, 9, 11, 12, 17, 30. About 340 species of cosmopolitan distribution (Cronquist, Holmgren, and Holmgren 1997a).

Ribes L. Gooseberry or Currant

(Syrian *ribas*, derived from the Persian word for the genus *Ribes*) Shrubs with or without bristles and spines. **LEAVES** alternate; stipules none or adnate to the petiole; palmately lobed, crenate or dentate. **INFL** a raceme or rarely solitary; pedicels subtended by bracts, usually with 2 bractlets about midlength. **FLOWERS** perfect; hypanthium mostly corolla-like; sepals 5, mostly petaloid; petals (4) 5, often smaller than the sepals; stamens (4) 5 (6); ovary inferior, 1-loculed; styles 2, united or distinct. **FRUIT** a berry, crowned by the withered flowers. **SEEDS** several to many. The genus *Ribes* includes both currants and gooseberries. Currants have jointed and disarticulating pedicels, several flowers per raceme, and stipitate-glandular or glabrous berries. The currants in our area are without spines or bristles except for *R. lacustre* and *R. montigenum*. Gooseberries have nonjointed, persistent pedicels, 3–23 flowers per raceme, and no stipitate-glandular hairs on the berries. The gooseberries of the Four Corners region all have nodal spines.

1. Plants armed with nodal spines, often with internodal bristles ....(2)
1′ Plants unarmed ....(5)

2. Hypanthium tubular to narrowly campanulate, 1–6 mm long; racemes 1–5-flowered; pedicel not jointed just below flower ....(3)
2′ Hypanthium shallowly cupped to saucer-shaped, 0.5–1.5 mm long; racemes 3–23-flowered; pedicel jointed just below flower at maturity ....(4)

3. Stamens about equal to petals; berry glabrous, pubescent or glandular-pubescent, not spiny; anthers 0.7–1.6 mm long, not apiculate; style glabrous, connate nearly full length, short-bifid at apex .... ***R.. leptanthum***
3′ Stamens exceeding petals; berry with long, stout spines; anthers longer than 1.6 mm, apiculate.......... ***R. inerme***

4. Leaves densely glandular or glandular-pubescent; racemes 3–11-flowered; pedicels 1–5 mm long; berry bright red .... ***R. montigenum***
4′ Leaves glabrous or sparsely pubescent; racemes 5–23-flowered; pedicels 2.5–7 mm long; berry black or dark purple .... ***R. lacustre***

5. Hypanthium saucer-shaped to shallowly cup-shaped or turbinate, 0.6–3.5 mm long ....(6)
5′ Hypanthium tubular-campanulate to cylindric, 4–13 mm long ....(8)

6. Ovary, berry glabrous .... ***R. inerme***
6′ Ovary, berry with stipitate-glandular hairs..........(7)

7. Racemes loosely flowered, floriferous nearly to base; peduncle absent or to 1.5 mm long; plants with weak decumbent stems; floral bracts 1–3.5 mm long, less than 1/2 as long as pedicels .... ***R. laxiflorum***
7′ Racemes densely flowered, obviously pedunculate, at least 2 cm long; plants erect; floral bracts conspicuous, 2.8–7 mm long, more than 1/2 as long as pedicels.......... ***R. wolfii***

8. Flowers bright yellow, sometimes aging orange or pink to red-purple, nearly always glabrous; berry glabrous; anther without a cuplike gland at tip.......... ***R. aureum***
8′ Flowers white to pinkish white or greenish white, to yellow, pubescent; berry glabrous to glandular-pubescent; anther tipped by small, cup-shaped gland ....(9)

9. Hypanthium narrowly tubular, 1.5–4 mm wide, 6–9.5 mm long, at least twice as long as the sepals; sepals 1.5–3.5 mm long, usually remaining recurved after anthesis; pedicels 0.4–3.4 mm long; leaf blades 0.5–5 cm wide .... ***R. cereum***
9′ Hypanthium broad, usually campanulate, 3.5–7.5 mm wide, 4.5–7 mm long, about the same length as the sepals; sepals 3.5–7 mm long; pedicels 3–17 mm long; leaf blades 2–10.5 cm wide.......... ***R. viscosissimum***

Ribes aureum Pursh (golden) Golden currant, Lewis' currant. Shrubs, 1–3 m tall, unarmed; branchlets glabrous. **LEAVES** with petioles 0.5–2.5 (3) cm long; blades (1) 1.6–4.7 cm long, 1–6.7 cm wide, orbicular, reniform, obovate, cuneate to truncate basally, strongly 3-lobed, the lobes entire or crenate to lobed, glabrous. **INFL** a raceme with (3) 6–9 flowers; bracts 3–12 mm long, entire; pedicels to 3 mm long. **FLOWER** hypanthium cyclindric, yellow, or often reddish in age, corolla-like; sepals mostly 4–6 mm long, yellow, spreading; petals about 2 mm long, yellow, cream, or reddish; stamens subequal to the petals, the anthers longer than the filaments; styles united to near the apex. **FRUIT**

8–12 mm long, black, red, orange, or translucent-golden, glabrous. Riparian, palustrine, and hanging garden habitats; along irrigation canals; greasewood-shadscale, sagebrush, piñon-juniper, serviceberry-hawthorn, ponderosa pine, and Douglas-fir communities. ARIZ: Apa, Nav; COLO: Arc, Dol, LPl, Mon, SMg; NMEX: RAr, SJn; UTAH. 1485–2165 m (4870–7100′). Flowering: Apr–Sep. Washington to Saskatchewan, south to California and New Mexico.

Ribes cereum Douglas (waxen) Wax or squaw currant. Shrubs, (0.2) 0.5–1.5 (2) m tall, unarmed; branchlets pilose-villous and stipitate-glandular. **LEAVES** with petioles 0.4–2.2 (2.9) cm long; blades 0.5–2.5 (3.4) cm long, 0.7–3 (4.4) cm wide, orbicular, reniform, rarely ovate, cordate or truncate basally, with 3–7 shallow lobes, the lobes crenate or dentate, puberulent and stipitate-glandular, or glabrous except on margins and along veins beneath. **INFL** a raceme with 2 or 3 flowers, the axis very short; bracts 2–5 mm long, ciliate, fringed, or lacerate, glandular. **FLOWER** hypanthium 4–11 mm long, pinkish, pilose, sometimes also stipitate-glandular; sepals about 2 mm long, spreading, whitish or pinkish; petals about 1 mm long, whitish; staminal filaments subequal to the anthers; styles united to near the apex; ovaries stipitate-glandular. **FRUIT** 6–8 mm long, reddish, sparingly stipitate-glandular, rarely glabrate. Alcoves and riparian sites; mountain brush, sagebrush, piñon-juniper, ponderosa pine, aspen, spruce-fir communities. ARIZ: Apa, Nav; COLO: Arc, Hin, LPl, Min, Mon, SJn; NMEX: McK, RAr, SJn; UTAH. 1910–3595 m (6260–11800′). Flowering: Apr–Jul. British Columbia to Montana, south to California and New Mexico. The Hopi and Navajo used the berries for food. The Kayenta Navajo used it as an Evilway, Nightway, and Mountaintopway emetic. Also used as a poultice applied to sores and to purify a child who has seen a forbidden sand painting. The Ramah Navajo used the stems to make arrow shafts; the green plant indicated time for plowing and the leafy plant indicated time to plant maize.

Ribes inerme Rydb. (not spiny) Whitestem gooseberry, wine gooseberry. Shrubs 7.5–20 dm tall, branchlets often whitish, glabrous, armed at the nodes with 1 (3) spines, or the spines lacking; internodal bristles mostly lacking or few and sparse. **LEAVES** with petioles (0.3) 0.5–4.5 cm long, sometimes with 1 to few pilose, gland-tipped hairs; blades (0.8) 1.5–9 cm long, orbicular or nearly so, cordate to truncate with 3–5 main lobes, these again lobed and crenate-dentate toothed, the major sinuses cut 1/3–2/3 to the base, glabrous, glabrate with hairs mostly along the veins, or occasionally moderately dense-strigose, not glandular, paler beneath than above. **INFL** a 1–4-flowered raceme, the axis to about 12 cm long; bracts 1–2 mm long, greenish, glabrous or glandular-ciliolate and puberulent; pedicels 2–5 mm long. **FLOWER** hypanthium 2–3.5 mm long, cylindric to narrowly campanulate, greenish or greenish cream, sometimes purplish-tinged, densely pilose to villous-woolly inside; sepals about 3 mm long, colored as the hypanthium; petals about 1–1.5 mm long, obovate to narrowly fan-shaped, white; stamens (1.5) 2–2.5 times longer than the petals, 2.4 (5) mm long; styles subequal to or slightly longer than the stamens, cleft 1/2–2/3 to the base, rather densely pilose on the lower 1/2 or more. **FRUIT** 7–10 mm long, reddish or reddish purple, succulent, more or less edible. [*Grossularia inermis* (Rydb.) Coville & Britton]. Piñon-juniper, mountain brush, aspen, willow, Douglas-fir, spruce-fir communites; often in mountain meadows. ARIZ: Apa, Nav; COLO: Arc, Hin, LPl, Min, Mon, SJn, SMg; NMEX: RAr; UTAH. 1985–3290 m (6520–10800′). Flowering: May–Jun. British Columbia to California, Montana to New Mexico. The berries eaten during the winter by the Navajo.

Ribes inerme

Ribes lacustre (Pers.) Poir. (growing by lakes or ponds) Swamp black gooseberry. Shrubs, 7.5–15 dm tall; branchlets armed with internodal prickles and nodal spines, puberulent, eglandular. **LEAVES** with petioles 0.35 cm long, glabrous or with scattered stipitate-glandular hairs; blades (0.6) 1.5–5.6 cm long, (1) 2–8 cm wide, orbicular in outline, cordate at the base, usually 5-lobed, the lobes again lobed and doubly crenate-dentate, glabrous or sparingly hairy along the veins. **INFL** a raceme, rather loosely 5- to 15-flowered, the axis to 4.5 cm long in fruit, stipitate-glandular with reddish or purplish glands, puberulent; bracts 2–3 mm long, ciliate-glandular; pedicels 3–8 mm long. **FLOWER** hypanthium less than 1 mm long, saucer-shaped, yellow-green, pinkish, or reddish; sepals 2.5–3 mm long, yellow-green, pinkish, or reddish; petals shorter than the sepals, pinkish; stamens subequal to the petals; styles parted to the base. **FRUIT** 6–8 mm long, dark purple, coarsely stipitate-glandular. [*R. oxyacanthoides* L. var. *lacustre* Pers.]. Moist sites, often in conifer and aspen woods. COLO: Arc, Hin, LPl, Min, Mon, SJn; UTAH. 2300–2620 m (7545–8600′). Flowering: May–Jul. Alaska to Newfoundland, south to California, Colorado, South Dakota, and Michigan.

Ribes laxiflorum Pursh (loose flowers) Western or trailing black currant. Shrubs to about 0.7 m tall, the stems sprawling or ascending, unarmed; branchlets and some older branches puberulent. **LEAVES** with petioles (0.5) 2–4.5 cm long, puberulent, sometimes short stipitate-glandular near the blade; blades (1) 2–5 cm long, 1.5–6.5 cm wide, orbicular or nearly so, cordate, with 3–5 primary lobes, these again lobed and crenate-dentate toothed, the major sinuses cut about 1/3–1/2 to the base, glabrate or with some puberulent and stipitate-glandular hairs, especially toward the base on veins beneath, slightly paler beneath with translucent, crystalline, sessile glands. **INFL** a 5–10-flowered raceme, stipitate-glandular and puberulent, the axis 24 cm long; bracts 1–2 mm long, linear or narrowly triangular, greenish; pedicels 4–10 mm long, jointed just below the ovary, some usually persisting at least until fruit is nearly mature. **FLOWER** hypanthium less than 1 mm long; sepals 2–3 mm long, pinkish or purplish; petals about 1 mm long, 1–1.3 mm wide, broadly fan-shaped with concave margins; stamens subequal to the petals; styles cleft 1/3–2/3 their length. **FRUIT** to about 1 cm long, blackish, stipitate-glandular, the glands and stalks mostly purplish. [*R. coloradense* Coville]. Moist sites; aspen and spruce-fir communities. COLO: Arc, Hin, LPl, Min, Mon, SJn. 2530–3290 m (8300–10800′). Flowering: May–Aug. Alaska to Washington and along the coast to southwest California, east to Alberta and northern Idaho.

Ribes leptanthum A. Gray (thin flowers) Trumpet gooseberry. Shrubs, 0.5–2 m tall; branchlets armed at the nodes with 1–3 spines, usually lacking internodal bristles, puberulent. **LEAVES** with petioles 0.2–1.2 cm long; blades 0.5–1.6 cm long, 0.7–2 cm wide, orbicular, cordate basally, mostly 5-lobed, the main lobes again shallowly lobed or toothed, glabrous or less commonly puberulent and rarely glandular. **INFL** a 1–3-flowered raceme, the axis very short; bracts glabrous except glandular-ciliate or -toothed; pedicels about 1 mm long. **FLOWER** hypanthium 4–5.5 mm long, whitish, pilose or short-villous; sepals 4–6 mm long, whitish; petals 2.5–3 mm long, whitish; stamens subequal to the petals; anthers shorter than the filaments; styles glabrous, apically notched. **FRUIT** mostly 6–10 mm long, blackish, glabrous. [*Grossularia leptantha* (A. Gray) Coville & Britton]. Piñon-juniper, mountain brush, ponderosa pine, aspen, spruce-fir, and mountain meadow communities. ARIZ: Apa; Nav; COLO: Mon; NMEX: SJn; UTAH. 2040–2785 m (6700–9135′). Flowering: May–Jun. Colorado, Utah, New Mexico, and Arizona. The Isleta and Jemez Indians eat the fresh berries.

Ribes montigenum McClatchie (mountain-born) Red prickly currant, gooseberry currant. Shrubs, 0.5–2 m tall; branchlets armed at the nodes with 1–3 spines, usually lacking internodal bristles, puberulent. **LEAVES** with petioles 0.2–1.2 cm long; leaf blades 0.5–1.6 cm long, 0.7–2 cm wide, orbicular, cordate basally, mostly 5-lobed, the main lobes again shallowly lobed or toothed, glabrous or less commonly puberulent and rarely glandular. **INFL** a 1–3-flowered raceme, the axis very short; bracts glabrous except glandular- or ciliate-toothed; pedicels about 1 mm long. **FLOWER** hypanthium 4–5.5 mm long, whitish; sepals mostly 2.5–4 mm long, yellowish green, pink, red, or orange; petals 2.5–3 mm long, whitish; stamens subequal to the petals; anthers shorter than the filaments; styles glabrous, apically notched. **FRUIT** about 6–10 mm long, blackish, glabrous. [*R. lacustre* (Pers.) Poir. var. *molle* A. Gray; *R. molle* (A. Gray) Howell; *R. lacustre* var. *lentum* M. E. Jones]. Douglas-fir, spruce-fir, aspen; often in talus and scree slopes. ARIZ: Apa; COLO: Arc, Con, Hin, LPl, Min, Mon, RGr, SJn; UTAH. 2865–3445 m (9400–11305′). Flowering: May–Jul. British Columbia to southern California, east to Montana, and south to New Mexico.

Ribes viscosissimum Pursh (sticky) Sticky currant. Aromatic shrubs, mostly 1–2 m tall, unarmed; branchlets pilose-hirsute and stipitate-glandular. **LEAVES** with petioles 0.6–7 cm long; blades 0.9–6.6 cm long, 1.3–10 cm wide, orbicular, rarely ovate, cordate basally, 3–7-lobed, the main lobes crenate or dentate and sometimes again lobed, glandular-hairy and often pilose or hirsute. **INFL** a 4–12-flowered raceme, the axis 5–30 mm long, glandular; bracts 5–8.5 mm long, entire to toothed, glandular; pedicels 3–17 mm long. **FLOWER** hypanthium 5–9 mm long, whitish or pale green, stipitate-glandular and pilose-hirsute; sepals 3–5.5 mm long, white or yellow-green, or occasionally pinkish; petals 2–3 mm long, whitish; stamens subequal to the petals; filaments longer than the anthers; styles glabrous or nearly so. **FRUIT** 10–13 mm long, black, rather dry, stipitate-glandular. Moist canyon bottoms and slopes in aspen, fir, Douglas-fir, lodgepole pine, and spruce-fir communities. ARIZ: Coc, Nav. 1835–1950 m (6020–6400′). Flowering: May–Jun. British Columbia, south to California, east to Montana and Arizona.

Ribes wolfii Rothr. (for John Wolf, 1820–1897, botanist and naturalist) Rothrock's currant. Shrubs, 0.5–3 m tall, unarmed; branchlets glabrous or puberulent. **LEAVES** with petioles 0.7–4.5 cm long, glabrous or puberulent; blades 1.2–5.7 cm long, 1.2–8 cm wide, orbicular, cordate basally, 3 (5)-lobed, the main lobes again lobed and variously 1–2-crenate or -dentate, glabrous except for sessile, clear, crystalline glands. **INFL** an 8–16-flowered raceme, glandular, the axis about 1–4 cm long; bracts 3–6 mm long, mostly entire; pedicels 1–5 (7) mm long. **FLOWER** hypanthium 0.7–1.5 mm long, green, bowl-shaped, glabrous or puberulent; sepals 2–3 mm long, whitish; petals about 1–1.5 mm long, white; styles free or united below the middle. **FRUIT** 6–10 mm long, blackish, not very fleshy, stipitate-glandular. [*R. mogollonicum* Greene]. Aspen, Douglas-fir, and spruce-fir communities, usually in shade. ARIZ: Apa; COLO: Arc, Hin, LPl, Min, Mon, SJn; NMEX: SJn; UTAH. 1830–3500 m (6005–11480′). Flowering: May–Jul. Colorado, Arizona, New Mexico, north to Washington and Idaho.

HALORAGACEAE R. Br. WATER-MILFOIL FAMILY

C. Barre Hellquist

Rhizomatous, submersed and emergent herbs. **LEAVES** simple, opposite, alternate, or whorled. **INFL** submersed or emergent in leaf or bract axils. **FLOWERS** perfect or imperfect, polypetalous or apetalous; sepals usually developed; petals small or lacking; stamens twice the number of sepals; ovary 1–4-celled; ovules 1 per cell; styles 1–4. **FRUIT** indehiscent, nutlike (Aiken 1981; Crow and Hellquist 1983; Mason 1957; Orchard 1981). Seven genera, about 160 species worldwide.

Myriophyllum L. Water-milfoil

(Greek *myrios*, many, + *phyllon*, leaf) Perennial, aquatic herbs. **LEAVES** whorled or opposite, submersed leaves capillary, emergent; leaves bractlike. **INFL** sessile, in axils of leaves or bracts, the uppermost staminate; the upper flowers often in an emersed terminal spike. **FLOWERS** perfect or imperfect; calyx 4-merous; petals 4 or none; stamens 4–8; ovaries inferior, 4-celled; ovule single in each cell; stigmas 4, recurved. **FRUIT** hard, nutlike, 4-locular, splitting into 4 mericarps. Forty species. *Myriophyllum spicatum* L. is to be looked for in our range.

1. Bracts of emersed leaves entire, shorter than flowers or fruit .... ***M. sibiricum***
1′ Bracts of emersed leaves pectinate, longer than flowers and fruit .... ***M. verticillatum***

Myriophyllum sibiricum Kom. (of Siberia) Shortspike water-milfoil. **STEMS** simple or branching, purple to brownish purple, turning white upon drying. **LEAVES** in whorls of 3 or 4, about 1 cm or more apart, with midstem whorls up to 3 cm apart, simple pinnate with 11 or fewer segments on each side of the rachis, lower leaves often shorter than those toward stem tip. **INFL** a spike, flowers in verticils, lower pistillate, upper staminate; in axils of tiny bracts, shorter than flowers. **FLOWER** petals oblong-ovate, about 2.5 mm long; anthers 8. **FRUIT** subglobose, 2–3 mm long, mericarps rounded on back, smooth or rugulose. [*M. exalbescens* Fernald; *M. magdalenense* Fernald; *M. spicatum* L. var. *exalbescens* (Fernald) Jeps.]. Lakes, ponds, slow-flowing streams, and ditches. ARIZ: Apa; COLO: Arc, Dol, Hin, LPl, Mon, SJn; NMEX: McK, RAr, SJn. 2200–3275 m (7300–10775′). Flowering: Jun–Aug. Fruit: Jul–Aug. Alaska east to Newfoundland, south to California, Arizona, New Mexico, Kansas, Minnesota, Illinois, West Virginia, Maryland, and Connecticut. It has been used as a medicine for poor circulation and as an emetic.

Myriophyllum verticillatum L. (whorled) Northern water-milfoil, whorl-leaf water-milfoil. **STEMS** simple, in the late season producing clavate winter buds. **LEAVES** in whorls of 4 or 5; submersed leaves 0.8–4.5 cm long, with 9–13 opposite or alternate pairs of capillary flaccid divisions; divisions to 28 mm long; emergent leaves and bracts smaller, coarser, or pectinate-pinnate. **INFL** a spike, flowers in whorls or groups of 4–6, lower pistillate, upper staminate; bracteoles palmately 7-lobed, 0.5 mm long. **FLOWER** petals are rudiments in pistillate flowers, spoon-shaped, obtuse, 2.5 mm long; anthers 4 or 8. **FRUIT** subglobose, 2–2.5 mm long, deeply 4-furrowed, mericarps rounded on back, smooth or slightly tuberculate. Lakes and ponds. COLO: Arc, LPl. 2650 m (8700′). Flowering and fruiting unknown.

British Columbia east to Saskatchewan and Newfoundland, south to California, Utah, Colorado, Texas, Michigan, Indiana, Maryland, and Delaware. The plant has been used as a stimulant by eastern Native American tribes.

HYDRANGEACEAE Dumort. HYDRANGEA FAMILY

Roy Taylor

(Greek for water jar, in reference to the cup-shaped fruit) Shrubs (ours). **LEAVES** opposite or alternate, simple, sometimes lobed. **INFL** corymbiform to paniculiform. **FLOWERS** perfect, regular, marginal ones often sterile and irregular; hypogynous or 1/2 to fully epigynous, with a short or prolonged hypanthium; sepals 4 or 5; petals 4 or 5; stamens (1) 2 to several times as many as the petals, sometimes up to 70; nectary disc usually present on top of the ovary; gynoecium of (2) 3–5 (12) carpels united to form a compound, 1/2 to fully inferior, rarely superior ovary; ovules (1) several or numerous. **FRUIT** a capsule (ours). **SEEDS** straight, linear, embryo embedded in the fleshy endosperm. $x = 13–18+$. About 17 genera and 170 species widespread in temperate and subtropical parts of the Northern Hemisphere (Cronquist, Holmgren, and Holmgren 1997a).

1. Stamens 8, the filaments flat, broad; leaves up to 3 cm long, not prominently nerved; flower petals mostly 10–20 mm long, 5.5–12 mm wide .... ***Fendlera***

1′ Stamens more than 20, the filaments terete, subulate; leaves 1–1.5 cm long, 3-nerved beneath; flower petals 5–17 mm long, 3.5–12 mm wide .... ***Philadelphus***

Fendlera Engelm. & A. Gray Fendlerbush

(for Augustus Fendler, 1813–1883, a German-born botanical explorer) Shrubs. **LEAVES** opposite, nearly sessile, deciduous. **FLOWERS** perfect, showy; hypanthium calyxlike; sepals 4; petals 4; stamens 8, filaments flattened, lobed at the apex; ovary inferior at the base, 4-loculed; styles 4. **FRUIT** a capsule, over 1/2 superior. **SEEDS** few in each locule. Two or three species in the southwestern United States (Welsh et al. 2003).

Fendlera rupicola Engelm. & A. Gray **CP** (living near rocks) Cliff fendlerbush. Many-branched shrub to 2 m tall, the bark striate, reddish to straw-colored on the younger twigs, grayish on the older. **LEAVES** opposite, often forming a tight fascicle, blade lanceolate to linear-elliptic and often revolute, 1–4 cm long, 2–7 mm wide, sparingly strigose on both sides, the midrib prominent, grooved above, ridged beneath, nearly sessile leaves without stipules, short-petiolate, deciduous. **FLOWER** hypanthium 2–3 mm long; perfect, showy, solitary or 2–3 together at the ends of short branches or spurs; sepals 3–5 mm long, to 8 mm in fruit, persistent, strigose beneath, tomentose-villous above; petals 13–20 mm long, constricted to a narrow claw, the blade to 11 mm wide, white; staminal filaments about 6–8 mm long, to 2 mm wide at the base, 2-lobed apically, the lobes 2–3 mm long, the anthers shorter or longer than the lobes; styles 4, appearing as 2 at anthesis, glabrous or with multicellular hairs. **FRUIT** a capsule 8–15 mm long. Often on dry, rocky, gravelly slopes; blackbrush, desert scrub, and piñon-juniper woodland communities. ARIZ: Apa, Coc, Nav; COLO: Arc, Dol, LPl, Mon; NMEX: McK, RAr, SJn; UTAH. 1495–2560 m (4900–8400′). Flowering: Apr–Jul. Utah, Colorado, and New Mexico to Arizona and Texas. Navajo uses include: an infusion of inner bark for swallowed ants; the notched stick rubbed with a smooth stick instead of beating a drum in the Mountain Chant Ceremony; used by the Home God in the Mountain Chant Ceremony; wood used to make arrow shafts; used to kill hair lice; used to make weaving forks, planting sticks, and knitting needles. The Kayenta Navajo used the plant as a cathartic and for Plumeway, Nightway, Male Shootingway, and Windway ceremonies; boiled with juniper berries, piñon buds, and cornmeal, and used in mush-eating ceremonies.

Philadelphus L. Mock Orange

(from the Greek *philos*, love, + *adelphos*, brother; said to be named for Ptolemy Philadelphus, king of Egypt, 283–247 B.C.) Shrubs. **LEAVES** opposite, subsessile or on short petioles. **INFL** in few-flowered cymes at the ends of leafy branches. **FLOWER** hypanthium calyxlike; perfect; sepals 4 (5); petals 4 (5), white or nearly white; stamens many, mostly 20–60; ovary at least 2/3 inferior, with 3–5 locules; styles 3–5, distinct or united. **FRUIT** a loculicidal capsule, leathery or woody. **SEEDS** numerous (Welsh et al. 2003).

Philadelphus microphyllus A. Gray (small-leaved) Littleleaf mock orange. Shrubs 8–20 dm tall. **STEM** branchlets appressed, pubescent. **LEAVES** with petioles about 1–3 mm long, blades 10–40 mm long, 4–15 mm wide; elliptic or ovate to lanceolate, base cuneate, tip acute, margins entire, slightly revolute, sparingly to moderately strigose-sericeous. **FLOWER** hypanthium hemispheric, about 2–4 mm long; pubescent; sepals 3–6 mm long, ovate, acuminate, tips often reflexed, densely white-tomentose beneath and on margins, strigose to glabrate on outer surface; petals 5–17 mm long, oblanceolate, entire to erose, white; stamens many, filaments 2–5.8 mm long, somewhat united at base, white with anthers yellowish; ovary complete, inferior at anthesis, styles 4. **FRUIT** a capsule 4–10 mm long, ovoid, woody, 4-valved. **SEEDS** 1–3.5 mm long. Rocky cliffs, sandy soils, talus, and hanging gardens; piñon-juniper woodlands, mountain brush, ponderosa pine, aspen, and spruce-fir communities. [*P. occidentalis* A. Nelson; *P. microphyllus* var. *occidentalis* (A. Nelson) Dorn; *P. microphyllus* subsp. *stramineus* f. *zionensis* C. L. Hitchc.]. ARIZ: Apa, Coc, Nav; COLO: Dol, LPl, Mon; NMEX: McK, SJn; UTAH. 1605–2805 m (5270–9200′). Flowering: May–Jul. Fruit: Sep. Wyoming and Utah to Texas. The fruits were formerly eaten for food by the Isleta.

HYDROCHARITACEAE Juss. TAPEGRASS or FROG-BIT FAMILY

(including **NAJADACEAE** Juss. BUSHY PONDWEED, NAIAD, or WATER-NYMPH FAMILY)

C. Barre Hellquist

Annual or perennial, aquatic, submersed herb; caulescent, with or without evident stem; submersed and floating leaves or submersed leaves, in fresh, brackish, and marine waters. **STEMS** rhizomatous, creeping or rooting at proximal nodes, leafy. **LEAVES** basal, alternate, opposite, or whorled; sessile or petiolate, margin entire, spinulose to serrate. **INFL** axillary, terminal, or scapose, 1-flowered or cymose, subtended by spathe or as involucres. **FLOWERS** unisexual, staminate and pistillate on same or different plants, epigynous, free; mostly 6-parted or with perianth absent; stamens (0–) 2–many. **FRUIT** berrylike or achenelike. **SEEDS** 1–many, fusiform, ellipsoid, ovoid, spheric, or straight (Cook and Urmi-König 1985; St. John 1962). Eighteen genera, about 116 species worldwide.

1. Leaves usually 3, in whorls along stem; blades linear, lacking basal sheaths .... *Elodea*
1′ Leaves subopposite, appearing whorled; blades with distinct basal sheaths .... *Najas*

Elodea Michx. Waterweed

(Greek *helodes*, marshy) Perennial, freshwater herbs. **STEMS** rooted in substrate, branched or unbranched. **LEAVES** submersed and floating, cauline, whorled 3 (–7) at each node, or leaves opposite at base of stems; sessile, blades linear to linear-lanceolate; apex acute, midrib lacking lacunae on either side, margins with fine serrations. **INFL** solitary, sessile. **FLOWERS** unisexual, staminate and pistillate on separate plants, rarely bisexual, usually raised to the surface of the water by elongate floral tube base. **FRUIT** ovoid to lance-ellipsoid, smooth. **SEEDS** cylindric to fusiform. Five species in North America, South America, Europe, introduced into Australia.

1. Staminate pedicel remaining until after anthesis; midstem leaves in whorls of 3 with numerous groups of 2, spreading; pistillate spathe 9–67 mm; seeds 2.8–3 mm .... *E. bifoliata*
1′ Staminate pedicel detached before or during anthesis; midstem leaves in whorls of 3, 2 on the lowest portion of stem, recurved; pistillate spathe 8.3–17.5 mm; seeds 4.5–5.7 mm or longer .... *E. canadensis*

Elodea bifoliata H. St. John **CP** (two-leaved) **LEAVES** with many in 2s, especially on lower stem; linear to narrowly elliptic, 4.7–24.8 × (0.8–) 1.8–4.3 mm. **INFL** with a staminate spathe, 10.2–42 mm; peduncles abscissing after anthesis; pistillate spathe 9–67 mm. **FLOWERS** imperfect; pedicel remaining briefly after anthesis; stamens 7–9, pollen in monads; styles 2.3–3 mm. **SEEDS** 2.8–3 mm. [*E. longivaginata* H. St. John]. Lakes, ponds, reservoirs, and streams. COLO: Hin, LPl; NMEX: SJn. 1700–2750 m (5625–9075′). Flowering and fruiting: unknown. Widely scattered. British Columbia, east to Minnesota, south to California, Utah, Colorado, and Kansas. *Elodea bifoliata* appears to be the more common species throughout the southwestern United States.

Elodea canadensis Michx. (Canadian) **LEAVES** mostly in 3s, spreading or recurved, linear, oblong to ovate, 5–13 × 2–5 mm. **INFL** with staminate spathe 8.2–13.5 mm, peduncle abscissing just before or during anthesis; pistillate spathe 8.3–17.7 mm. **FLOWERS** imperfect; pedicel detaching before anthesis; stamens 7–9 (–18), pollen in tetrads; styles 2.6–4 mm. **SEEDS** 4–5.7 mm or longer. $2n = 24$ (Great Britain). Lakes, ponds, and streams, often alkaline. COLO: Arc. 2050–2750 m (6850–9075′). Flowering and fruiting: unknown. British Columbia and Alberta east across Canada to Quebec, south to California, Utah, Colorado, Arkansas, Alabama, and Virginia; Europe; Asia; Australia. Has been used as a strong emetic by eastern American tribes.

Najas L. Bushy Pondweed, Naiad, Water-nymph

(Greek *naias*, a water nymph) Aquatic submersed annuals. **STEMS** slender, highly branched, rooting at proximal nodes, some with prickles on internodes. **LEAVES** with sheaths variously shaped, margins serrate or dentate; blades linear, 1-veined; margins serrate to minutely serrulate with 50–100 teeth per side; apex acute to acuminate. **INFL** involucres mostly present in staminate flowers, rare in pistillate flowers. **FLOWERS** unisexual; staminate flower subtended by membranous involucre, pistillate flower sessile. **FRUIT** closely enveloping seed. **SEEDS** fusiform to obovoid, areola usually arranged into longitudinal rows. $x = 6$ (Haynes 1979; Lowden 1986). Forty species. *Najas guadalupensis* (Spreng.) Magnus is to be looked for in our study area.

Najas marina L. (of the sea) Spiny naiad. **STEMS** branched, internodes with prickles. **LEAVES** stiff with age; sheaths 2–4.4 mm wide; apex acute; blade 0.4–4.5 mm wide; margins with 8–13 coarse serrations per side. **INFL** with 1 flower per axil; staminate and pistillate on separate plants. **FLOWERS**: staminate 1.7–3 mm wide; pistillate 2.5–5.7 mm wide; styles 1.2–1.7 mm. **SEEDS** 2.2–4.5 mm long, areoles of testa irregularly arranged, not in distinct rows. $2n = 12$ (Europe). [*N. gracilis* Morong]. Lakes and ponds, in brackish water or water of high calcium or sulfur content. NMEX: SJn (Morgan Lake). 1615 m (5450′). Flowering: Jul–Aug. Fruit: Jul–Aug. Widely scattered. California and Nevada east to Minnesota and New York, south to Arizona, Texas, Alabama, and Florida.

HYDROPHYLLACEAE R. Br. WATERLEAF FAMILY

N. Duane Atwood and Stanley L. Welsh

Perennial, biennial, or annual herbs, or shrubs. **LEAVES** simple or pinnatifid. **INFL** commonly in cymes, these mostly scorpioid. **FLOWERS** perfect, regular, 5-merous; calyx lobes 5, similar or dissimilar, sometimes accrescent in fruit; corolla lobes 5; stamens 5, exserted or included; pistil 1–2-carpellate, the ovary superior, 1-loculed or more or less completely 2-loculed; styles 2, or if 1 then 2-cleft. **FRUIT** a longitudinally dehiscent capsule. **SEEDS** 1 to many. $x = 5–13+$.

1. Inflorescence capitate or almost so; stamens long-exserted; plants of mountain brush communities and upward .... ***Hydrophyllum***
1′ Inflorescence definitely scorpioid, thyrsoid, or axillary, or the flowers sometimes solitary; stamens included or exserted .... (2)

2. Ovary unilocular; plants annual with diffusely branched, angled stems. .... ***Ellisia***
2′ Ovary more or less bilocular; plants annual to biennial or perennial, stems not angled .... (3)

3. Stamens unequally inserted on the corolla tube; flowers axillary or solitary in small, dense, leafy clusters; plants low, branching annuals. .... ***Nama***
3′ Stamens equally inserted on the corolla tube; flowers mostly in cymes; plants low to tall, annual, biennial, or perennial .... ***Phacelia***

Ellisia L. Ellisia

(named for John Ellis, English botanist) Annual. **STEMS** simple to diffusely branched and angled. **LEAVES** pinnately divided, the lower opposite, the upper alternate. **FLOWERS** solitary in the upper leaf axils and retrosely hirsute; corolla white to blue, campanulate; calyx divided nearly to the base, enlarging in fruit; stamens included; style divided at the apex less than 1/2 its length. **FRUIT** unilocular, ovules on fleshy placentae attached at the top and bottom of the cell. **SEEDS** usually 4, regularly reticulated.

Ellisia nyctelea L. (night) Aunt Lucy. Annual, 1–4 dm tall. **STEMS** simple or diffusely branched, angled, retrorsely hirsute, 10–40 cm tall. **LEAVES** 2–8 cm long, divisions pinnately divided, 7–13, oblong, acute or obtuse, hispidulous. **FLOWER** calyx lobes 3–5 mm long, lanceolate or ovate-lanceolate; stamens included. **FRUIT** hispid. **SEEDS** 2–3 mm in diameter, globose, dark brown. [*Macrocalyx nyctelea* (L.) Kuntze; *Ellisia nyctelea* var. *coloradensis* Brand]. Valleys and mountains, shaded places. 1220–2745 m (4000–9000′). COLO: Mon. Flowering: May–June. Atlantic states west to Saskatchewan, south to New Mexico.

Hydrophyllum L. Waterleaf

(Greek *hydro*, water, + *phyllon*, leaf) Perennial herbs. **STEMS** erect, succulent, from horizontal rhizomes, these bearing fleshy, fibrous to tuberous roots. **LEAVES** pinnately compound, mostly basal, oblong or oval in outline, petioles slightly dilated and clasping at the base, ciliate. **INFL** composed of 1 to several globose or lax cymes, short-pubescent or strigose and hispid. **FLOWERS** campanulate, purplish to blue, white, or violet; calyx divided nearly to the base, the lobes linear, oblong, or lanceolate; stamens 5, exserted; style 1, exserted 5–10 mm, cleft 1–2 mm; stigma capitate; ovules attached to the front of the 2 large parietal placentae. **SEEDS** 1–3, brown, subglobose, reticulate.

1. Flowers in dense capitate clusters; peduncles shorter than the petioles of the subtending leaves; anthers short-oblong, 0.6–1 mm long .... ***H. capitatum***
1' Flowers in open clusters; peduncles longer than the petioles of the subtending leaves; anthers linear to oblong, 1–2 mm long .... ***H. fendleri***

Hydrophyllum capitatum Douglas ex Benth. (for the headlike flower clusters) Capitate waterleaf. Plants 1–5 dm high, from short rhizomes, these bearing a fascicle of tuberous roots. **STEMS** short. **LEAVES** pinnately compound, ovate to oval in outline, strigose, the blades 2.5–10 cm long, 2–13 cm wide, the primary divisions 5–7, obovate to oblong or lanceolate, the lobes and divisions acute, obtuse, or mucronate. **INFL** of 1 to several globose cymes, the peduncles 1–5 cm long, shorter than the subtending leaves, mostly recurved in fruit. **FLOWER** pedicel 2–5 mm long; sepals obtuse or abruptly acute, 3–4 mm long, 1.5 mm broad or less, ciliate and strigose; petals 5–9 mm long, purplish or white; stamens exserted 5 mm; style exserted 5–10 mm. **SEEDS** normally 2, light brown, 2–3 mm in diameter. $2n = 18$. [*H. capitatum* var. *pumilum* Geyer ex Hook.] Mountain brush, aspen, sagebrush, and mixed conifer communities. COLO: Hin; UTAH. 1340–2745 m (4400–9000′). Flowering: Jun–Jul. Washington to Wyoming, south to Utah and Colorado. Our plants belong to **var.** ***capitatum.***

Hydrophyllum fendleri (A. Gray) A. Heller (for Augustus Fendler, first resident botanist in New Mexico) Fendler's waterleaf. Perennial, 2–9 dm tall from short rhizomes, these bearing fibrous roots. **STEMS** erect, retrorse-hispid. **LEAVES** pinnately compound, ovate or oval in outline, strigose, the blade 6–30 cm long, with 9–13 primary divisions, these ovate to lanceolate, acuminate, coarsely serrate or incised. **INFL** of 1 to several lax cymes, peduncles 3–17 cm long, often branched, mostly longer than the subtending leaves (at least in fruit). **FLOWER** pedicel 2–7 mm long; sepals linear to lanceolate (in fruit), 4–6 mm long , 1–2 mm broad, ciliate and strigose, often hispid dorsally; petals 6–10 mm long, white to violet; stamens exserted 4–6 mm; style exserted 5–7 mm. **SEEDS** 1–3, light brown, 2.5–3 mm in diameter. $2n = 18$. [*H. occidentale* (S. Watson) A. Gray var. *fendleri* A. Gray; *H. albifrons* A. Heller var. *fendleri* (A. Gray) Brand]. Piñon pine, mountain mahogany, oak, ponderosa pine, and spruce communities. COLO: Arc, Hin, LPl, Min, Mon; UTAH. 2500–2970 m (8200–10157′). Flowering: Jun–Jul. British Columbia to Oregon, south to Colorado and southeastern Utah.

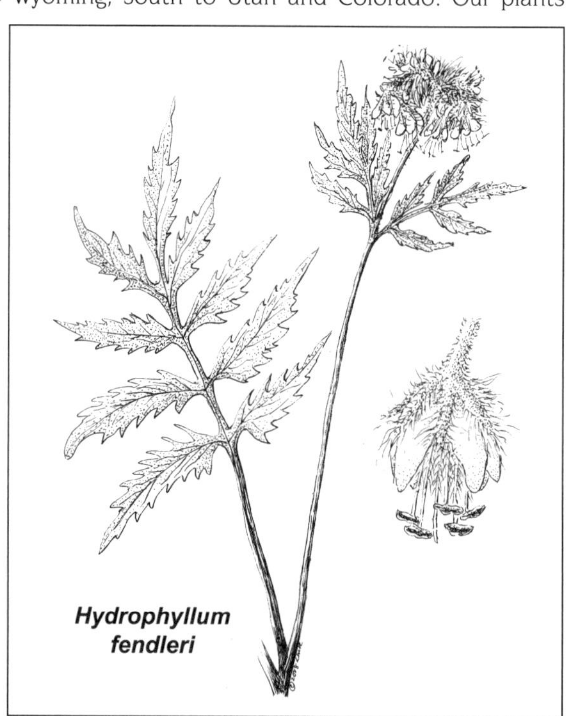
Hydrophyllum fendleri

Nama L. Nama

(Greek, spring) Low, branching annuals. **LEAVES** alternate, entire, hirsutulous to hispid, retrorse to erect. **INFL** of terminal nonscorpioid cymes. **FLOWER** sepals subequal, linear to lanceolate; petals purple or lavender, deciduous, tubular to funnelform; stamens included, borne unequally on the corolla tube; style included, divided to the base or 2-lobed at the apex. **FRUIT** a capsule falsely bilocular by intrusion of the placentae; ovules numerous. **SEEDS** numerous, brown, mostly reticulate.

1. Styles solitary, 2-lobed at the apex; corolla tubular, 3–5 mm long .... ***N. densum***
1' Styles 2, distinct to the base; corolla tubular-campanulate to funnelform-campanulate, 5–16 mm long, or if smaller, stem hairs retrorse (4–7 mm long in *N. retrorsum*) .... (2)

2. Shorter stem pubescence retrorse .... ***N. retrorsum***
2' Shorter stem pubescence spreading to erect .... (3)

3. Corolla relatively showy, purple, 9–15 mm long, funnelform-campanulate, plants relatively coarsely hairy with long and short hairs .... *N. hispidum*
3' Corolla smaller, about 5 mm long, tubular-campanulate, pale lavender, finely glandular and not coarsely hairy .... *N. dichotomum*

Nama densum Lemmon (compact) Compact nama. Annual. **STEMS** dichotomously branched, branches mostly prostrate, spreading-hirsute, 1–10 cm across. **LEAVES** oblanceolate, 0.4–3 cm long, 1–5 mm wide, entire. **FLOWERS** sessile and solitary in the upper leaf axils; calyx lobes linear, 3.8–7 mm long, 1–3 mm wide; corolla funnelform, white to lavender, 3–7.2 mm long, the tube long, the lobes short; style 0.5–2 mm long, 2-lobed at the apex. **FRUIT** 2–3.5 mm long. **SEEDS** 15, 0.5–0.9 mm long, dark brown to blackish, pitted and reticulate. $2n = 14, 28$. [*Conanthus densus* Lemmon ex A. Heller]. Sandy soils or often loose gravel in juniper, sagebrush, and desert shrub communities. ARIZ: Apa; UTAH. 1430–1865 m (4699–6112'). Flowering: May–Aug. Central Washington to eastern California, east through southeastern Oregon and the western part of the Snake River Plains through Nevada to Elko and Nye counties, and disjunct in southwestern Wyoming and eastern Utah.

Nama dichotomum (Ruiz & Pav.) Choisy (having divisions in pairs, referring to the stems) Wishbone fiddleleaf. Annual. **STEMS** erect or ascending, sparsely leafy, often dichotomously branched, glandular and setose, 5–27 cm long. **LEAVES** alternate, linear to narrowly oblanceolate or spatulate, petiolate, 1–3 cm long, 2–5 mm wide. **FLOWERS** pale lavender, with prominent dark guidelines, tubular-campanulate, 3.3–4.5 mm long, borne singly or in pairs in the upper branches, some subtended by a very long leaf or bract, sessile or the lower sometimes with a pedicel to 5 mm long; calyx lobes linear to narrowly spatulate, 5–10 mm long; stamen bases dilated into free-margined scales shorter than the free filament. **FRUIT** oblong, 3.5–4.5 mm long, 1.8–2.3 mm wide, short-pubescent. **SEEDS** 56–62, brown, 0.7 mm long, 0.4 mm wide, oblong, prominently large-pitted. Limestone gravels to sandy soils in piñon-juniper to ponderosa, fir, aspen communities. ARIZ: Apa; COLO: Arc; NMEX: McK, SJn. 1220–2989 m (4000–9800'). Flowering: Aug–Sep. Colorado, New Mexico, Arizona; Mexico and South America.

Nama hispidum A. Gray (referring to the stems' coarse, rigid, erect hairs) Hairy nama. Leafy, branched annuals, 1–3 dm tall. **STEMS** more or less spreading, hispid. **LEAVES** 1–7 cm long, 2–6 mm wide, entire, revolute, finely glandular and with somewhat appressed, nonglandular hairs, oblanceolate to spatulate. **INFL** solitary to several in terminal cymes. **FLOWER** sepals linear to lanceolate, 5–8 mm long; petals purple, broadly funnelform, 8–14 mm long, 7–8 mm wide; style 2–5 mm long, cleft to the base. **SEEDS** numerous, 0.5 mm long, yellowish brown, reticulate. $2n = 14$. Sandy soils. ARIZ: Apa; NMEX: McK, RAr, SJn; UTAH. 1130–1220 m (3700–4000'). Flowering: May–Jun. Western Texas and northern Mexico to southeastern California, north to Kane County, Utah. Our plants belong to what has been referred to as **var.** ***spathulatum*** C. L. Hitchc. Plant used by Navajos as a lotion for spider bites.

Nama retrorsum J. T. Howell (refers to the backward-pointing hairs on the stem) Howell's nama. Leafy, freely branched, erect annuals, pubescent throughout, 1–3 dm tall. **STEMS** erect, fastigiate, hirsute, the shorter stem hairs retrorse. **LEAVES** linear to linear-oblong, sessile, 1–5 cm long, 2–5 mm wide, entire. **FLOWERS** sessile or nearly so and solitary in the upper leaf axils; sepals linear, 5–6 mm long, with narrow, hispid lobes; corolla purple, funnelform, 4–7 mm long, the lobes 1–2 mm; styles 2, distinct to the base. **FRUIT** 3–4 mm long. **SEEDS** numerous, narrowly ellipsoid, shallowly and obscurely pitted, 0.6–0.8 mm long. Sand dunes and other sandy soils in desert shrub–grass communities. ARIZ: Apa, Nav; COLO: Mon; NMEX: RAr, San, SJn; UTAH. 1060–2135 m (3500–7000'). Flowering: Apr–Aug. Grand County, Utah, south to Coconino and Navajo counties, Arizona.

Phacelia Juss. Phacelia

Herbaceous annuals, biennials, or perennials, mostly pubescent and with glandular hairs. **LEAVES** mostly alternate, the lower sometimes opposite, entire to pinnately compound. **INFL** in variously disposed scorpioid cymes, or apparently in lax racemes. **FLOWERS** few to numerous; calyx divided to the base; corolla tubular to broadly campanulate, blue, purplish, violet, or white, mostly deciduous, a few species with a tardily deciduous corolla; stamens included or exserted, equally inserted at the base of the corolla tube, with a pair of scales attached to the base of the corolla and filaments; style included or exserted, bifid, mostly pubescent. **FRUIT** a capsule, unilocular (or nearly bilocular by the intrusion of the placentae). **SEEDS** 1 to numerous, variously roughened, boat-shaped, terete or flattened.

1. Stamens included in the corolla .... (2)
1' Stamens exserted from the corolla more than 2 mm .... (5)

2. Ovules and seeds about 40 .... *P. indecora*
2′ Ovules and seeds 1–20 .... (3)

3. Leaves deeply pinnatifid; flowers white .... *P. ivesiana*
3′ Leaves entire or undulate; flowers lavender to purple .... (4)

4. Leaves oblong to elliptic; style and branches 1.5 mm long; filaments glabrous; flowers in dense sessile clusters; plants of Apache County, Arizona, and McKinley County, New Mexico .... *P. cephalotes*
4′ Leaves broadly ovate to orbicular; style including branches 1.5–4 mm long; filaments sparsely hairy; flowers in racemes, these 1–4 cm long; plants more broadly distributed .... *P. demissa*

5. Seeds excavated on the ventral surface on 1 or both sides of a prominent ridge .... (6)
5′ Seeds not excavated ventrally .... (13)

6. Corolla lobes pubescent and erose-fimbriate or erose-denticulate; leaves strongly dissected .... *P. neomexicana*
6′ Corolla lobes not both pubescent and erose; leaves less dissected, only the lower or none of the sinuses reaching the midrib .... (7)

7. Corolla small, 4 mm long or less, white, blue, or lavender; seeds rather shallowly excavated on both sides of the ventral centrally placed ridge .... *P. alba*
7′ Corolla more than 4 mm long, variously colored or rarely white .... (8)

8. Leaves glabrous or nearly so, except sometimes with hairs on the petiole .... *P. splendens*
8′ Leaves distinctly hairy or glandular .... (9)

9. Plants endemic to the high mountains of Colorado in Archuleta, Mineral, Hinsdale, and Rio Grande counties .... *P. bakeri*
9′ Plants of lower elevations in the San Juan River drainage .... (10)

10. Mature seeds black, 2.5–2.8 mm long, 1–1.2 mm wide; plants restricted to the Moenkopi Formation in San Juan County, Utah .... *P. constancei*
10′ Mature seeds reddish or brown and large in size; plants of sandy or sandy-gravelly sites and more widely distributed .... (11)

11. Corolla rotate to funnelform, the tube white; dorsal surface of mature seeds reddish and smooth; plants restricted to Grand and San Juan counties, Utah .... *P. howelliana*
11′ Corolla tubular-campanulate to campanulate, the tube lavender to bluish; dorsal surface of mature seeds brown or dark brown to black and with transverse ridges or corrugated; plants widespread .... (12)

12. Corolla lavender; seeds lacking ventral corrugations; plants of sand dunes and other sandy sites .... *P. integrifolia*
12′ Corolla blue or purple; seeds corrugated ventrally .... *P. crenulata*

13. Leaves entire or with 1–4 pairs of lobes; stems strigose to hispid, not silvery .... *P. heterophylla*
13′ Leaves pinnatifid, the segments also cleft or entire; stem pubescence silky .... *P. sericea*

Phacelia alba Rydb. (refers to white flowers) White-flowered phacelia. Annual, 0.5–7 dm tall. **STEMS** simple to branched, erect or ascending, leafy, hirsute to setose and stipitate-glandular. **LEAVES** pinnatifid to subpinnatifid, 2–10 cm long, 2–8 cm wide, short-hairy and somewhat glandular, the lowermost long-petiolate, the upper sessile or subsessile. **INFL** of dense, terminal, compound scorpioid cymes, the lower ones often simple, densely glandular and puberulent to hirsute, the cymes 1–2 cm long in flower, to 8 cm long in fruit. **FLOWERS** subsessile, or the pedicel to 1 mm long; sepals linear to oblanceolate, 3.5–4 mm long; petals white to lavender or pale blue, tubular-campanulate; stamens exserted 2–4 mm; style exserted 2–4 mm, divided to below the middle. **FRUIT** ovoid to subglobose, 3–3.3 mm long, glandular-puberulent. **SEEDS** 1–2 (4), light to dark brown when mature, 2.4–3 mm long, uniformly alveolate and cymbiform, the ventral surface shallowly excavated on both sides of the ridge, lacking corrugations, the margins thick and entire. [*P. neomexicana* Thurb. ex Torr. var. *alba* (Rydb.) Brand]. Sagebrush, piñon-juniper, mountain brush, ponderosa pine, aspen, and spruce-fir communities. ARIZ: Apa; NMEX: McK, RAr, San, SJn. 1370–2930 m (4520–9600′). Flowering: late May–Sep. Wyoming, Colorado, Utah, New Mexico, Arizona, and Chihuahua, Mexico.

Phacelia bakeri (Brand) J. F. Macbr. (for C. F. Baker, entomologist and botanist) Baker's phacelia. Annual or biennial, 0.5–4.8 dm tall. **STEMS** with dense, black, multicellular, stipitate glands, pilose-hirsute. **LEAVES** pinnately divided, irregularly crenate-dentate, 2–8 cm long, 0.5–3 cm wide, reduced above. **INFL** of compound terminal and lateral scorpioid cymes. **FLOWER** sepals oblanceolate to narrowly spatulate, pubescent, 4.5–5 mm long; petals campanulate, violet to dark blue, 7–8 mm long, 5–7 mm wide, pubescent on the lobes; stamens exserted 5–9 mm, anthers greenish, filaments bluish; style exserted 5–9 mm, pubescent below. **FRUIT** oblong to oval, 3.5–4 mm long, 3–3.2 mm wide, setose and glandular. **SEEDS** elliptic, brown, 2.7–3 mm long, 1.3–1.6 mm wide, pitted ventrally, the margins thin, the ridge tapering to the margins and lacking excavations, pitted dorsally and flattish, with a shallow central groove. [*P. crenulata* Torr. ex S. Watson var. *bakeri* Brand]. Talus slopes and gravelly soils of open tundra, grassy alpine slopes in sagebrush-grass, spruce, fir, pine, and aspen communities. COLO: Arc, Hin, LPl, Min, RGr. 2150–3722 (7045–12204′). Flowering: Jul–Sep. Colorado endemic.

Phacelia cephalotes A. Gray (refers to the flower head) Chinle phacelia. Annual, 0.5–1.3 dm tall. **STEMS** low, almost prostrate, glandular-villous. **LEAVES** entire, oblong, ovate to elliptic, petiolate, hirsutulous, and glandular, the lower petioles longer than the blade, 0.5–1.8 cm long. **INFL** of compact, sessile or subsessile, scorpioid cymes, these terminal on the branches of the stems. **FLOWER** sepals oblanceolate, 3–4 mm long in flower, to 8 mm long in fruit; petals lavender, inconspicuous, 3–5 mm long, the tube yellowish; stamens and style included. **FRUIT** ovate, 3–4 mm long, shorter than the calyx. **SEEDS** when mature, 8–12, oblong, angular, the angles denticulate, pitted, 1–1.5 mm long. $n = 11$. Salt desert shrub and juniper communities primarily on the Chinle Formation. ARIZ: Apa; NMEX: McK; UTAH. 1060–1525 m (3500–5000′). Flowering: May–Jun. Washington, Kane, and San Juan counties, Utah, east to McKinley County, New Mexico.

Phacelia constancei N. D. Atwood (for Lincoln Constance, a California botanist) Constance's phacelia. Erect biennial herb, 1.5–4.3 dm tall. **STEMS** coarse, 1–several, reddish, hirsutulous to hirsute, and finely glandular. **LEAVES** basal and cauline, linear-lanceolate, revolute, crenate to dentate, the lower short-petiolate or sessile and larger than the upper, the upper sessile. **INFL** of terminal cymes. **FLOWER** sepals elliptic to oblanceolate, 3–4 mm long; petals pale lavender to white, (4) 5–6 mm long, tubular-campanulate; stamens and style exserted 3–4 mm. **SEEDS** 4, black, 2.5–2.8 mm long, 1–1.2 mm wide, elliptic, the margins corrugated, the ventral surface finely pitted, excavated and divided by a prominent ridge, this corrugated along one side, the dorsal surface finely pitted. $n = 11$. Salt desert shrub and juniper communities. COLO: Mon; UTAH. 1250–2013 m (4100–6600′). Flowering: Jun–Sep. Southeastern Utah, northern Arizona, and southwestern Colorado.

Phacelia crenulata Torr. ex S. Watson (having small, rounded teeth) Corrugate phacelia. Annual, 0.3–8.3 dm tall. **STEMS** simple or more commonly branched, stipitate-glandular with short or long trichomes, and usually with some nonglandular hairs intermixed. **LEAVES** oblong-elliptic, subentire to deeply lobed, the lower lobes sometimes almost distinct, the lower leaves the largest, petiolate, the upper mostly reduced and finally sessile. **INFL** scorpioid, terminal on the stems and branches. **FLOWER** sepals elliptic to oblanceolate, 3–10 mm long; petals blue, violet to purple, campanulate, 4–7 mm long; stamens exserted 3–11 mm; style exserted, bifid 1/2 its length, glandular-pubescent below. **FRUIT** globose to subglobose, 2.6–4.1 mm long, 2.8–3.2 mm wide, puberulent to glandular. **SEEDS** 4, elliptic to oblong, 2.8–4 mm long, 1.3–3.2 mm wide, pitted, the ventral surface corrugated on the margins and one side of the ridge. $2n = 22$. Desert shrub of saltbush, blackbrush, budsage, *Ephedra*, *Yucca*, *Hilaria*, *Poliomintha*, and piñon-juniper communities on rocky, sandy, gravelly, silty clay. ARIZ: Apa, Nav; COLO: LPl, Mon; NMEX: McK, SJn; UTAH. 1148–1922 (3742–6292′). Flowering: late Mar–Jun. Eastern Nevada, east through much of Utah to western Colorado, south to northern Arizona and northwestern New Mexico. Our plants belong to **var. *corrugata*** (A. Nelson) Brand. Plant used for horse injuries by Hopis.

Phacelia demissa A. Gray (refers to lowly) Brittle phacelia. Annual, branched from the base, mostly broader than high, 0.3–2 dm tall. **STEMS** brittle and succulent when fresh, densely glandular-villous, usually reddish. **LEAVES** mostly cauline, broadly ovate to orbicular, 1–2.6 cm long, entire to undulate, the lower ones long-petiolate, the upper short-petiolate. **INFL** of terminal or axillary, sessile, scorpioid cymes. **FLOWER** sepals oblong to oblanceolate, 5–7 mm long in fruit; petals lavender to purple with a yellow tube, easily deciduous, 3.5–9 mm long, tubular-campanulate when fully open, sessile to short-pedicellate; stamens included; style included, hairy, shortly bifid. **FRUIT** oblong, 2.5–4 mm long, finely pubescent; ovules 4–20. **SEEDS** ovate to elliptic, 1.1–1.5 mm long, brown to reddish, pitted-reticulate. $2n = 24$. Salt desert shrub, saltcedar, cottonwood, rabbitbrush, and piñon-juniper commuities on Chinle, Dakota, Mancos Shale, Morrison, and Tropic Shale formations. ARIZ: Apa, Nav; COLO: Mon; NMEX: McK, SJn. 1300–2196 m (4266–7200′). Flowering: May to Jun. Eastern, central, and southern Utah, southwest Colorado, northern Arizona, and northwest New Mexico. Our plants belong to **var. *demissa***.

Phacelia heterophylla Pursh (having leaves of more than one form) Varileaf phacelia. Biennial or perennial, 2–6 dm tall. **STEMS** simple, sometimes branched, erect, stout, strigose to hispid, not silvery but sometimes also glandular. **LEAVES** basal and cauline, the lower petiolate, entire or with 1–4 pairs of lobes. **INFL** somewhat virgate, typically branched, densely pilose to spreading-hispid. **FLOWER** sepals lanceolate to oblong, 3–6 mm long, unequal; petals dirty white to purplish, campanulate, 4–7 mm long. **FRUIT** ovoid, 2–3 mm long, pubescent. **SEEDS** 1 or 2, 2–2.5 mm long, brown. $n = 11$. Sagebrush, mountain brush, aspen, and spruce-fir commuities. ARIZ: Apa; COLO: Arc, Hin, LPl, Min, Mon, RGr, SJn, SMg; NMEX: RAr, San, SJn: UTAH. 1370–3340 m (4497–10855′). Flowering: Jun–Sep. Rocky Mountains of Montana, west to Oregon and Washington, east and south to Wyoming, Utah, western Colorado, New Mexico, Arizona to Mexico. Our plants belong to **var. *heterophylla***. Used for greens by Navajos.

Phacelia howelliana N. D. Atwood (for J. T. Howell, botanist) Howell's phacelia. Annual, 0.9–2.3 dm tall. **STEMS** mostly branched and leafy at the base, glandular and hirsute. **LEAVES** oblong to oval, 2–6 cm long, 1–2.5 mm wide, irregularly crenate to lobed, strigose and slightly glandular, the petiole to 5 cm long. **INFL** of branched scorpioid cymes. **FLOWER** pedicels to 2 mm long; sepals linear to narrowly oblanceolate, 3.5–4.5 mm long, 1–1.2 mm wide, glandular and hirsute; petals 5–6 mm long, 6–7 mm wide, rotate to funnelform, the lobes pale violet to blue, the tube white; stamens and style exserted 3–4 mm, the style shorter than the stamens, bifid 3/4 its length, the lower 1/4 pubescent. **FRUIT** oblong to subglobose, glandular and hirsutulous, especially near the apex. **SEEDS** 4, brown, 3.2–4 mm long, 1.4–1.8 mm wide, elliptic, the margins corrugated, involute to flattened, the ventral surface pitted, excavated and divided by a prominent ridge, this sometimes curved to one side and barely corrugated, the dorsal surface reddish brown, smooth and surrounded by a lighter margin. Salt and warm desert shrub and piñon-juniper communities. ARIZ: Apa, Nav; UTAH. 1125–1525 m (3688–5000′). Flowering: late Apr–Jun. Endemic to Utah in Emery, Garfield, Grand, and San Juan counties; adjacent Navajo and Apache counties, Arizona.

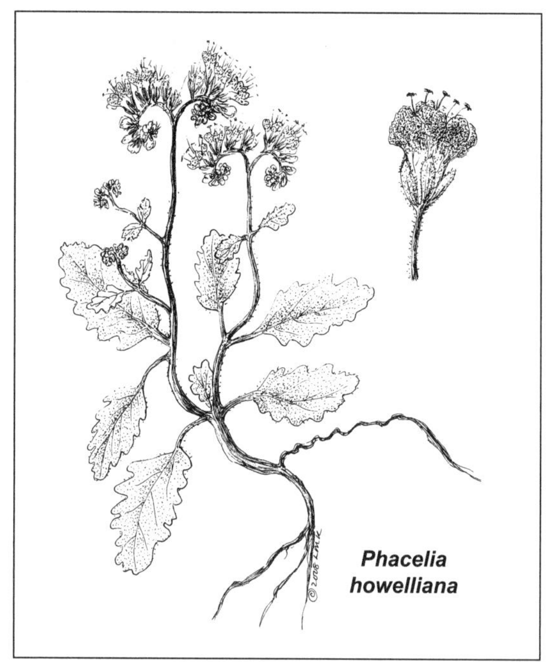
Phacelia howelliana

Phacelia indecora J. T. Howell (without elegance) Bluff phacelia. Annual, 3–14 cm tall. **STEMS** erect to spreading, branched, glandular. **LEAVES** elliptic to oblong, 4–26 mm long, hirsutulous and glandular. **FLOWER** sepals oblanceolate, 3–5 mm long,; petals narrowly campanulate, pale blue, 3–4 mm long, the lobes pubescent, the tube pale yellow and streaked with blue lines. **FRUIT** elliptic, 3–4 mm long. **SEEDS** about 40. Mainly associated with hanging garden communities. UTAH. 1370 m (4488′). Flowering: May–mid-Jun. Endemic to San Juan County, Utah.

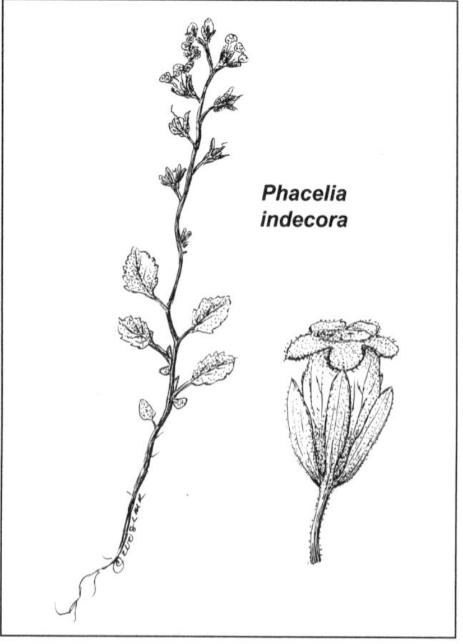
Phacelia indecora

Phacelia integrifolia Torr. (with entire leaves) Torrey's phacelia. Annual or winter-annual, 1–6 dm tall. **STEMS** simple or somewhat branched, densely stipitate-glandular and spreading-hirsute. **LEAVES** simple, crenate to somewhat cleft, oblong to ovate or lanceolate, strigose, finely glandular and with spreading, setose hairs, the lower leaves largest, petiolate, the upper smaller, short-petiolate to sessile. **INFL** terminal, elongate. **FLOWER** sepals oblanceolate to elliptic, 3–4.5 mm long in flower, 4.4–6.5 mm long in fruit; petals lavender to bluish, 5–6.5 mm long and broad, tubular-campanulate; stamens and style exserted 5–6 mm, the style bifid 2/3–3/4 its length, puberulent below. **FRUIT** ovoid to globose, 3.2–5.3 mm long, pubescent. **SEEDS** when mature, 4, oblong to elliptic, dark brown to black, 3.1–4.5 mm long, 1.7–2.2 mm wide, the transverse ridges on the dorsal surface quite distinct, the ventral surface lacking corrugations, the ridge often curved to one side. $2n = 22$. Sandy soil, sometimes on sand dunes in blackbrush and other warm desert shrub, shrub-grass communities. ARIZ: Apa, Nav; COLO: Mon; NMEX: McK, RAr, San, SJn; UTAH. 1190–1710 m (3900–5606′). Flowering: late Mar–Jun. Kane and San Juan counties, Utah, south through central and eastern Arizona, across New Mexico to Trans-Pecos Texas and Mexico, northern Texas, Oklahoma, and southern Kansas.

Phacelia ivesiana Torr. **CP** (for J. C. Ives) Newberry's phacelia. Annual, 0.4–2.7 dm tall, hirsutulous and mostly finely glandular (rarely dark or black). **STEMS** prostrate or ascending, branched. **LEAVES** pinnately divided or lobed, oblong to lanceolate, 1–5 cm long, basal and cauline. **INFL** of laxly flowered, scorpioid cymes. **FLOWER** sepals oblong to oblanceolate, 2–4 mm long in flower, elongating to 7 mm in fruit; petals white, inconspicuous, 2–4 mm long, the tube yellowish; stamens included, the filaments glabrous; style included, divided 1/4 its length, glabrous. **FRUIT** oblong, 3–4.5 mm long, pubescent, especially apically. **SEEDS** 8–15, brown, 1–1.5 mm long, corrugated transversely and alveolate. $n = 22$. Sandy, sandy-clay, silty gravelly soils and occasional on limestone gravels in warm desert shrub, cool desert shrub, sagebrush, rabbitbrush, ephedra, snakeweed, mahogany, and piñon-juniper communities. ARIZ: Apa, Coc, Nav; COLO: Arc, LPl, Mon; NMEX: McK, RAr, San, SJn; UTAH. 1220–2166 m (4000–7100′). Flowering: late Mar–Jul. Sonoran Desert of southeastern California, east through Nevada, Arizona, Utah, Colorado to Wyoming and the Snake River Plains of Idaho.

Phacelia neomexicana Thurb. ex Torr. (for New Mexico) New Mexico phacelia. Annual, 0.8–6.8 dm tall. **STEMS** erect to sparsely branched, setose, hispid and with small stipitate-glandular hairs, often reddish, leafy. **LEAVES** pinnate, the secondary pinnae irregularly incised, 3–8.5 cm long, 1–4.5 cm wide, strigose and stipitate-glandular, petiolate, the petiole 1.5 cm long or less. **INFL** congested, terminal on the main stem and lateral branches (sometimes arising from the axils of the uppermost branches with 1–3 cymes, the cymes up to 1 dm long in fruit. **FLOWERS** short-pedicellate (0.5 mm long); sepals linear to narrowly oblanceolate, 2–3 mm long in flower, to 4 mm in fruit, hispid and densely glandular; petals campanulate, blue, about 4 mm long, 3–3.5 mm wide, the lobes pubescent and erose; stamens and style exserted, the stamens exceeding the corolla 1–2 mm, the filaments bluish, anthers yellow, style exserted 1 mm, bifid 3/4 its length, glandular on the lower 1/4. **FRUIT** oval to elliptic, 4.5–4.7 mm long, 3 mm wide, setose and densely glandular, the raphe oblanceolate. **SEEDS** 4, oblong, brown, 3.2–3.3 mm long, 1.1–1.5 mm wide, alveolate, ventral surface excavated on both sides of the ridge. $n = 11$. Pine and oak woods in canyons and on mountain slopes in rocky to sandy soils. ARIZ: Apa; NMEX: McK, San. 2074–2745 m (6800–9000′). Flowering: late Jul–mid Oct. Eastern Arizona in Apache County, southeastern New Mexico, north through the mountains of central New Mexico, and possibly southern Colorado. Zunis powdered the root and used for rashes.

Phacelia sericea (Graham) A. Gray **CP** (for the silky hairs on the stem) Silky phacelia. Stout perennial, 1–5 dm tall. **STEMS** 1 to several, the pubescence silky, hairs appressed to spreading, or loosely woolly. **LEAVES** pinnatifid, the segments also cleft or entire, the lower well developed, cauline well developed but reduced upward. **INFL** a dense terminal thyrse, usually elongate, the cymes many, compact, short. **FLOWER** sepals linear, pubescent; petals dark purple or blue, persistent, campanulate, 5–7 mm long and as broad, pubescent inside and out; stamens and style long-exserted, the filaments hairy at the base, the style 6–13 mm long, bifid 1/2 its length. **FRUIT** ovoid, 4–6.5 mm long, pubescent. **SEEDS** 8–20, oblong, 1–2 mm long, brown to black, reticulate and grooved longitudinally. $2n = 22$. Open sagebrush slopes, aspen and spruce-fir wooded areas, commonly in rocky soils at middle and upper elevations. COLO: Arc, Hin, LPl, Min, Mon, RGr, SJn; UTAH. 2285–3540 m (7491–11606′). Flowering: Jun–Aug. Southeastern Utah, Colorado, central Nevada, north through the mountains of Wyoming, Montana, central and northeastern Idaho to Washington; British Columbia and Alberta, Canada. Our plants belong to **var.** ***sericea***.

Phacelia splendens Eastw. (brilliant) Eastwood's phacelia. Annual, 0.5–2.7 dm tall. **STEMS** erect, simple or branched, leafy, puberulent and stipitate-glandular. **LEAVES** pinnatifid, 2–7.5 cm long, 0.7–4 cm wide, petiolate, blade essentially glabrous. **INFL** terminal on the main stem and branches, cymes compact. **FLOWER** pedicels short; sepals linear to narrowly oblanceolate, 2.5–3 mm long in flower, 4–4.4 mm long, 0.6–1 mm wide in fruit, hirsute and scattered glandular-hairy; petals campanulate, blue, 4–8 mm long and as broad, glabrous to sparsely pubescent; stamens and style exserted 7–11 mm, filaments blue, anthers yellow, style bifid about 2/3 its length, puberulent and glandular below. **FRUIT** subglobose, 4–4.5 mm long, 3–3.5 mm wide. **SEEDS** 4, finely pitted, ventral surface excavated on both sides of the ridge, ridge corrugated on one side, margins ± revolute. $n = 11$. Mancos Shale Formation. COLO: Mon; NMEX: SJn. 1373–1830 m (4500–6000′). Flowering: late Apr–Jun. Southwestern Colorado, northwestern New Mexico, and southeastern Utah.

IRIDACEAE Juss. IRIS FAMILY

William Jennings

Perennial (ours) or annual herbs herbs from rhizomes, bulbs, or corms. **LEAVES** mostly basal, long and narrow, 2-ranked. **INFL** a raceme or panicle, subtended by 2 spathelike bracts. **FLOWERS** perfect, regular (ours) or irregular, showy, 6 petaloid segments (tepals) in 2 whorls, similar or distinct and dissimilar; stamens 3, usually partly adnate to the outer petaloid segments; style usually 3-branched, branches often flattened, opposite the stamens and outer petaloid segments; ovary inferior with 3 locules. **FRUIT** loculicidally dehiscent capsule. **SEEDS** many, membranous, usually with fleshy endosperm, sometimes arillate. About 60 genera and 1500 species, pantropical, pantemperate. Widely grown as ornamentals (*Iris*, *Crocus*, *Gladiolus*).

1. Style branches large, petal-like; flowers large, over 3 cm wide; sepals and petals dissimilar .... ***Iris***
1' Style branches small, not petaloid; flowers small, less than 3 cm wide; sepals and petals alike .... ***Sisyrinchium***

Iris L. Iris, Flag, Fleur-de-lis, Gladden

(Greek *iris*, rainbow) Perennial, rhizomatous or bulbous. **STEMS** erect, simple or branched. **LEAVES** mostly basal, linear to swordlike; flat (ours), terete, or quadrangular. **INFL** spike, raceme, or panicle bearing 1 to several flowers from paired spathelike bracts. **FLOWERS** showy, brightly colored, petaloid perianth parts differentiated into 2 whorls of 3 each, and united below into a tube; the outer whorl spreading, recurved, or reflexed; the inner whorl erect or arching; stamens 3, alternating with the inner whorl and next to the style branch; style divided into 3 petaloid branches, arched over the outer perianth whorl. **FRUIT** a loculicidal capsule. **SEEDS** numerous, in 2 rows in each locule. A genus of the north-temperate zone; about 150 species. There are numerous cultivated species and hybrids of iris, and sometimes these escape or persist, usually where iris rhizomes have been dumped by gardeners culling their beds. The most commonly seen are *I. germanica* (the common large-flowered and many-colored garden iris) and *I. pseudacorus* (a yellow-flowered species of wet habitats). The iris flower should be familiar to nearly everyone, and our one species of native iris looks like a smaller, blue-flowered version of the large, cultivated garden iris.

Iris missouriensis Nutt. (of Missouri, meaning the river, the type specimen being taken "toward the sources of the Missouri") Western iris, wild iris, blue flag. Rhizomatous perennials. **STEMS** 20–60 cm tall, barely surpassing the leaves. **LEAVES** linear, flat, erect. **FLOWERS** typical of the genus; light blue, occasionally white. **FRUIT** capsule 3–5 cm long, short-cylindrical, 6-ridged. Wetlands such as marshes, moist meadows, stream banks, and pond shores, often in places that dry up as the season progresses. ARIZ: Apa; COLO: Arc, Dol, Hin, LPl, Min, Mon, SJn, SMg; NMEX: McK, RAr, SJn; UTAH. 1525–3050 m (5000–10000′). Flowering: May–Jul. Practically ubiquitous in the United States west of the Great Plains. Used as a ceremonial emetic by Ramah Navajo. Chewed root poultice applied to newborns by Zunis to increase strength.

Sisyrinchium L. Blue-eyed Grass

(Greek *sys*, pig, + *rynchos*, snout, alluding to swine grubbing the roots for food) Perennial (ours) or annual herbs, often caespitose, rhizomatous or not. **STEMS** scapelike or branched, compressed and 2-winged. **LEAVES** basal or basal and cauline, alternate, flat, ensiform, glabrous. **INFL** a terminal cluster from a 2-bracted spathe, bracts approximately equal, or somewhat or greatly unequal. **FLOWERS** actinomorphic, tepals 6, all alike, widely spreading to reflexed, bluish violet to light blue (ours), white, lavender to pink, magenta, purple, or yellow; not fragrant; stamens symmetrically arranged, filaments distinct, connate basally or into tube, tapering to apex; anthers parallel, surrounding but not appressed to style branches; styles 3, erect, connate basally; filiform, extending between stamens and usually beyond anthers. **FRUIT** a capsule, more or less globose. **SEEDS** many, seed coat black, granular to rugulose (Henderson 1976; Cholewa and Henderson 1984). Native to the New World, about 80 species.

Mature flowers and spathes provide the best identification material, but identification is best accomplished on fresh material, as the flowers press poorly. Multiple specimens from the same population are recommended, due to variability of the species. White-flowered plants of a normally blue-flowered species are occasionally found, leading to reports of *S. campestre* E. P. Bicknell far beyond its normal range.

1. Spathe bracts subequal, the inner bract typically 12–20 mm, the outer bract typically 13–24 mm; stems branched, with a leaflike bract subtending 2 or more pedunculate spathes .... ***S. demissum***
1' Spathe bracts unequal, the inner bract typically 13–35 mm, the outer bract typically 20–75 mm, the ratio of the length of the outer bract to the inner bract 1.2 to 2.8; stems simple (unbranched) ....(2)

2. Outer-to-inner bract ratio 1.2 to 1.8; outer bract tapered evenly to an acute apex; outer whorl of perianth parts (sepals) with length-to-width ratio of 2.2 to 3.0 (shorter and broader than the following species), apex usually rounded .... ***S. idahoense***
2' Outer-to-inner bract ratio 1.8 to 2.8; outer bract often widening above the apex of the inner bract before tapering to an acuminate apex; inner bract gibbous-based (bulging out); outer whorl of perianth parts (sepals) with a length-to-width ratio of 2.9 to 4.5 (longer and narrower than the preceding species) .... ***S. montanum***

Sisyrinchium demissum Greene (lowly or humble) Perennial herb, caespitose, to 50 cm tall, often somewhat glaucous. **STEMS** branched, with 1 or 2 nodes. **LEAVES** 1.5–5 mm wide, glabrous, basal, about 1/2 the height of the whole plant. **INFL** borne singly on branched stems; spathes green, wider than supporting branch, glabrous, tapering evenly to apex. **FLOWER** tepals deep bluish violet, bases yellowish, outer tepals 6–15 mm, apex rounded to acute or occasionally emarginate, aristate; filaments connate for most of length. **FRUIT** a capsule, tan when mature, 4–8 mm, pedicel erect to ascending. **SEEDS** 1–2 mm, globose to obconic, granular. Moist areas such as wet meadows, springs, seeps, and stream banks. ARIZ: Apa; COLO: Arc, LPl; NMEX: McK, SJn. 1400–2775 m (4620–9100′). Flowering: Jun–Jul. *Sisyrinchium demissum* is a southwestern species of Arizona, New Mexico, and western Texas, reaching its northern limits in the Four Corners region. It also occurs in Colorado's San Luis Valley.

Sisyrinchium idahoense E. P. Bicknell (of Idaho) Perennial herb, caespitose, to 45 cm tall, not glaucous. **STEMS** simple, winged, glabrous. **LEAVES** glabrous. **INFL** borne singly on an unbranched stem; spathes glabrous, tapering evenly to the apex, outer bract typically 13–16 mm longer than the inner. **FLOWER** tepals light to deep bluish violet, occasionally purple, bases yellowish, outer tepals 8–13 mm, apex rounded, truncate, or emarginate, aristate; filaments connate for most of length, bluish violet. **FRUIT** capsule beige, light or dark brown when mature, 3–6 mm, pedicel erect to ascending. **SEEDS** 1–3 mm, globose to obconic, usually granular. [*S. occidentale* E. P. Bicknell]. In moist montane areas such as wet meadows, springs, seeps, and stream banks. NMEX: RAr. 2250 m (7380′). Flowering: late Jun–Jul. Our material belongs to **var.** ***occidentale*** (E. P. Bicknell) Douglass M. Hend. (of the West) and is found in the Rocky Mountains, reaching its southern limits at the northern edges of the San Juan drainage. Typical var. *idahoense* occurs in the Pacific Northwest, mainly Oregon and Washington

Sisyrinchium montanum Greene **CP** (of the mountains) Perennial herb, caespitose, to 50 cm tall; not glaucous. **STEMS** simple, winged, glabrous. **LEAVES** glabrous. **INFL** borne singly on an unbranched stem; spathes glabrous, outer bract 12–46 mm longer than the inner bract; outer bract often constricted near the apex (bract gradually tapers, then narrows noticeably, then widens again before tapering to apex); inner bract with gibbous base (bulges out noticeably). **FLOWER** tepals dark bluish violet; bases yellowish; outer tepals 9–14.5 mm; apex emarginate to retuse, aristate; filaments connate for nearly entire length. **FRUIT** capsule tan to dark brown, globose, 4–7 mm, pedicel erect to spreading. **SEEDS** globose to obconic, 1–1.5 mm, rugulose. In moist montane areas such as wet meadows, springs, seeps, and stream banks. COLO: Arc, Hin, LPl, San Juan. 2040–2855 m (6700–9365′). Flowering: late May–early Jul. *Sisyrinchium montanum* is transcontinental in distribution, across Canada and the northern tier of states in the United States, extending south in the Rockies to southern Colorado and northern New Mexico; it does not occur in Utah or Arizona. The species reaches its southern limits at the northern edge of the San Juan drainage. Our material is **var.** ***montanum***. Variety *crebrum* Fernald occurs in the northeast United States and Canada.

JUGLANDACEAE DC. ex Perleb WALNUT FAMILY

Linda Mary Reeves

A small family of monoecious, rarely dioecious, aromatic trees and shrubs. **LEAVES** pinnately compound, primarily alternately, rarely oppositely; deciduous or evergreen. **INFL** with staminate pendulous catkins and pistillate flowers clustered or in solitary erect spikes or androgynous panicles. **FLOWERS** imperfect, wind-pollinated; staminate calyx 2–5-lobed; stamens 3–40; pistillate flowers bracteate; sepals 0–4; petals 0, stamens 3–105, sessile; pistil 1; ovary inferior, 2–4-carpelled, generally 2; ovule 1; style short, 1; stigmatic branches 2–4. **FRUIT** a drupe or winged nut with a fibrous husk derived from the involucre and calyx. **SEEDS** solitary, 2-lobed, with large, fleshy, oily cotyledons. x = 16. About 7 genera with about 60 species. North-temperate zone but also subtropical areas of southeast Asia and the Andes. The best known genera include the walnuts and butternuts (*Juglans*), and the hickories and pecans (*Carya*). Both produce fine-grained timber and valuable edible seeds with much ellagic acid (Laferriére 1993; Correll and Johnston 1979b).

1. Staminate and pistillate flowers usually without sepals; husk of fruit valvular; endocarp smooth or with reticulate markings .... ***Carya***
1′ Staminate and pistillate flowers with 4 small sepals; fruit with weakly dehiscent husk and irregularly furrowed endocarp .... ***Juglans***

Carya Nutt. Hickory, Pecan

(Greek, a nut) Trees with durable, hard wood and scaly buds, twigs reddish, with prominent light-colored lenticels. **LEAVES** petiolate; blade odd-pinnately compound; often glandular; leaflets 5–25, margin serrate to entire. **INFL** with staminate catkins usually in fascicles of 3 in the bud scale axils; pistillate flowers 2–10 in a cluster or short spike on a peduncle terminating the seasonal shoot above the leaves. **FLOWERS** (staminate) with 3–8 stamens, adnate to the bract and 2 bracteoles; filaments short or none, free; pistillate with a bract and 3 bracteoles sepal-like in flower, calyx absent; stigmas sessile, 2, often divided, stigmatic disc at base, papillose, persistent. **FRUIT** a drupe with a 4-valved husk, stony endocarp smooth (in ours) or reticulated. About 15 species in eastern North America and eastern Asia. Tasty seeds often eaten as food and cultivated widely in the Southwest, West, and Southeast; wood used for durable purposes and fine furniture.

Carya illinoinensis (Wangenh.) K. Koch (of Illinois) Pecan, nogal morado, nuez encarcelada. Large tree to 50 m high (ours shorter), occasionally developing a massive trunk; thick bark light brown to reddish, often deeply furrowed. Overwintering buds flattened and tomentose. **LEAVES** to 50 cm long (often shorter in ours), petiolate; leaflets 5–15, petiolulate, oblong-lanceolate to lanceolate, laterals falcate, apex acuminate, base often asymmetric, cuneate to rounded, margin serrate to doubly serrate, to 20 cm long and 7.5 cm wide (often smaller in both measurements); yellow-green; glabrous or pilose above; paler and glabrous to pubescent below. **FRUIT** in spikes or clusters of 3–11, cylindric or ellipsoid; apex acute or apiculate; base rounded. **SEEDS** oily, furrowed and edible. [*C. oliviformis* (Michx.) Nutt.; *C. pecan* (Marshall) Engl. & Graebn.; *Hicoria pecan* (Marshall) Britton; *Juglans illinoinensis* Wangenh.]. Scattered, naturalized or escaped specimens from homesteads and cultivation. Known in our area only from Canyon del Muerto, Arizona; escaped from Navajo orchards. ARIZ: Apa. 1900–1950 m (6200–6350′). Flowering: May–Jun. Fruit: Aug. Texas north to Iowa and Illinois, east to Ohio, Maryland, south to Florida. Widely cultivated for food and wood; seeds and wood gathered by many tribes.

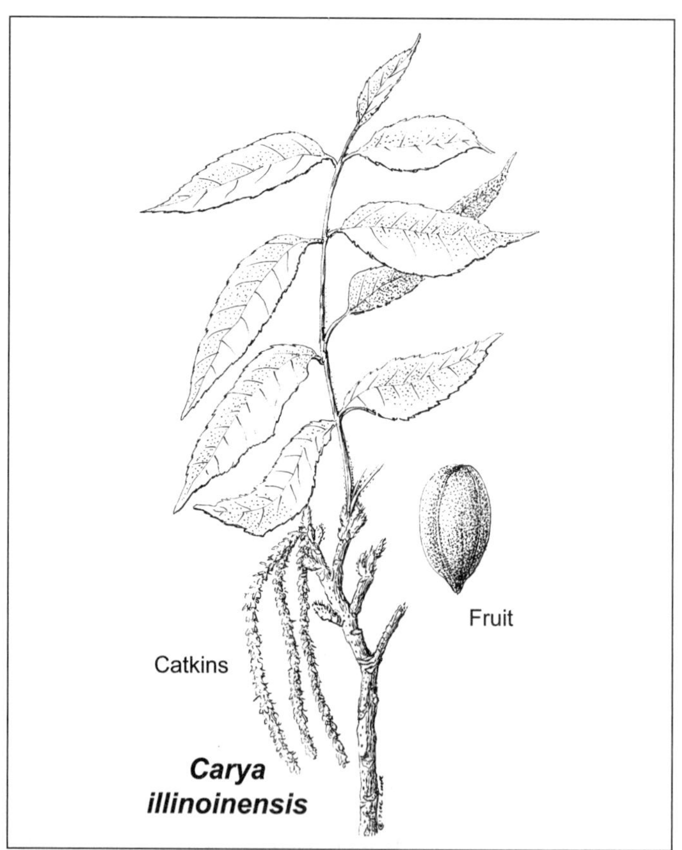

Juglans L. Walnut, Nogal, Butternut

(classical Latin for walnut) Primarily large trees with furrowed, scaly bark and hard, fine-grained wood. Occasionally large shrubs. **LEAVES** oddly pinnately compound, with largest leaflets at midrib; leaflet margins serrate or entire, gland-dotted. **INFL** greenish, with staminate catkins sessile, axillary, pendulous on old wood; pistillate flowers solitary or in clusters, terminal on new leafy growth. **FLOWERS** subtended with 2–6 bracts; sepals 4, occasionally reduced to minute teeth; inner surfaces of stigmatic branches often deeply fringed. **FRUIT** a drupe with fibrous indehiscent exocarp and mesocarp; stony, furrowed endocarp. About 20 species in both the Eastern and Western Hemispheres. Other naturalized *Juglans* species, especially *J. nigra* L., which is probably escaped, should be searched for in riparian areas and around abandoned homesteads.

Juglans major (Torr.) A. Heller (greater) Arizona walnut, nogal, nogal silvestre, ha'atłtse'dii. Tree to about 15 m with trunk to 120 cm diameter; however, often less than 8 m tall in our area and often with several trunks and shrublike.

Bark deeply furrowed and/or ridged; twigs reddish to dark brown, puberulent when young, gray with age. **LEAVES** 15–43 cm long; short-petiolate; leaflets 5–13 (occasionally more or less), lanceolate to falcate-lanceolate, apex acuminate, to 10 cm long, margin dentate-serrate, puberulent when young. **FRUIT** globose to oval, to 30 mm diameter; exocarp and mesocarp light brown, thin, glandular-pubescent; endocarp hard, furrowed. **SEEDS** small, edible. [*J. elaeopyren* Dode; *J. microcarpa* Berland. var. *major* (Torr.) L. D. Benson; *J. rupestris* Engelm. ex Torr. var *major* Torr.] Scattered in streams, washes, and canyons. Known in our area only from Canyon de Chelly and Navajo National Monument, Arizona. ARIZ: Apa, Nav. 1700–2100 m (5700–6800′). Flowering: Apr–May. Fruit: July–Aug. From the Edwards Plateau in Texas, Oklahoma through Utah and southwestern New Mexico, Arizona and south to Guerrero, Mexico. Our specimens probably naturalized from transplants in Navajo orchards. The wood has been used for durable fence posts. The seeds were eaten as winter food by many tribes. A major Navajo dye plant; the fruits are boiled for a golden brown dye, and the young twigs are used for a light brown or reddish dye.

JUNCACEAE Juss. RUSH FAMILY

David W. Jamieson

Annuals or mostly perennials, caespitose or rhizomatous grasslike herbs. **CULMS** terete or flattened, leafless or leafy. **LEAVES** with sheathing bases; sheaths open or closed, often supporting membranous to cartilaginous auricles at the transition to the blade. Blades reduced to bristles, or well developed and terete, or laterally or adaxially-abaxially flattened. **INFL** congested heads or open panicles. **FLOWERS** small, the tepals arranged in 2 series of 3 similar greenish to brown or purplish brown segments; stamens 3 or 6. **FRUIT** a 3-carpellate ovary that matures into a 1- or 3-loculed capsule. **SEEDS** 3 or many (Flora of North America Editorial Committee 2000; Hermann 1975; Cronquist et al. 1977; Hurd, Goodrich, and Shaw 1994; Weber and Wittmann 2001).

1. Leaves glabrous, sheaths open; ovary incompletely to completely 3-loculed; seeds numerous .... ***Juncus***
1′ Leaves and sheaths usually with hairy margins, sheaths closed; ovary 1-loculed; seeds 3 .... ***Luzula***

Juncus L. Rush

(Latin, to join or bind, from use of the stems) Annual or mostly perennial grasslike herbs. **CULMS** caespitose, or arising individually or in clusters from slender to stout rhizomes, terete to sometimes flattened, leafless to leafy. **LEAVES** basal, alternate or 2-ranked; sheaths open basally, the upper margins often scarious and prolonged into membranous or cartilaginous auricles; blades absent, or present as a short bristle or with a well-developed blade, flattened, equitant, inrolled, channeled, or terete, and when so, often hollow or pith-filled and weakly to strongly septate-nodulose. **INFL** terminal or apparently lateral, essentially cymose, flowers arranged rarely singly, or in mostly few- to many-flowered clusters in open, paniculate or corymblike arrangements or in congested, many-flowered heads; subtended by a leaflike bract that is divergent or extends beyond the flowers as a continuation of the culm. **FLOWERS** small, sessile or pedicellate, pedicels and or flowers subtended by single scarious bracts, or the flowers subtended by a pair of variously shaped, scarious bracts called prophylls; perianth regular and hypogynous, in 2 series of 3 small and similar, variously scarious, chartaceous or tawny to green tepals; stamens 3 or 6, anthers shorter or longer than the filaments; ovary 3-carpelled, style single, often persistent in fruit and bearing 3 long stigma lobes. **CAPSULES** loculicidal, often triquetrous, 1, or usually 3-loculed. **SEEDS** tiny and numerous, usually less than 1 mm, variously shaped, apiculate or with one or both ends protruding as parenchymatous tails. About 350 species, Arctic and north-temperate regions.

1. Flowers prophyllate (each subtended by 2 hyaline-scarious bracts)........(2)
1′ Flowers not prophyllate........(9)

2. Plants annual........***J. bufonius***
2′ Plants perennial, densely caespitose or rhizomatous........(3)

3. Plants densely caespitose........(4)
3′ Plants rhizomatous........(8)

4. Leaf sheaths bladeless or bristle-tipped. ....***J. drummondii***
4′ Lower leaf sheaths bladeless or bristle-tipped, but upper leaves clearly bladed........(5)

5. Inflorescence lateral, the inflorescence bract extending beyond the inflorescence like a continuation of the stem (sometimes shorter in *J. hallii*); seeds tailed........(6)
5′ Inflorescence apparently terminal, the inflorescence bract more or less divergent from beneath the inflorescence; seeds not tailed........(7)

6. Capsules acute........***J. parryi***
6′ Capsules retuse........***J. hallii***

7. Capsules obtuse to blunt, incompletely 3-loculed; auricles short, rounded, and cartilaginous........***J. dudleyi***
7′ Capsules clearly retuse, completely 3-loculed; auricles membranous........***J. confusus***

8. Tepals acuminate; leaves reduced to bladeless sheaths, which may be bristle-tipped........***J. arcticus***
8′ Tepals obtuse and short, outer segments cucullate-tipped; leaves clearly bladed........***J. compressus***

9. Plants caespitose........***J. triglumis***
9′ Plants rhizomatous........(10)

10. Leaves equitant, that is, flattened as in *Iris*, or the blade dorsiventrally flattened........(11)
10′ Leaves terete and hollow, with internal septae, blades sometimes channeled toward the blade-sheath transition, or inrolled throughout........(12)

11. Leaf blades dorsiventrally flattened; seeds apiculate........***J. longistylis***
11′ Leaves equitant and ensifoliate, that is, folded and flattened as in *Iris*........***J. ensifolius***

12. Leaf blades largely inrolled, becoming terete and septate only in the tip; auricles absent; seeds long-tailed at both ends........***J. castaneus***
12′ Leaf blades uniformly terete, hollow and septate; auricles present; seeds apiculate........(13)

13. Capsules acuminate........(14)
13′ Capsules rounded-obtuse........(16)

14. Flowers in flowering heads little divergent, mostly erect; inflorescence open, flowering heads on spreading to divaricate branches........***J. articulatus***
14′ Flowers in flowering heads radiating in all directions at maturity........(15)

15. Tepals long-acuminate, subspinulose and slightly reflexed in the upper 1/2; inner series shorter than the outer; inflorescence of mostly 2–10 crowded, dense, many-flowered (25–70+) clusters; leaves diverging from the stems........***J. torreyi***
15′ Tepals merely acuminate, straight; inflorescence open, of many, fewer-flowered (5–10) clusters; leaves erect........***J. nodosus***

16. Flower cluster solitary and terminal; tepals acuminate and exceeding the capsule........***J. mertensianus***
16′ Flower clusters numerous on erect branches in an open inflorescence; tepals acute or obtuse and shorter than the capsule........***J. alpinoarticulatus***

Juncus alpinoarticulatus Chaix (alpine rush) Northern green rush. Tufted perennial from a short rhizome. **CULMS** 5–30 cm tall. **LEAVES** 1–2, mostly toward the base, sheath ending in a short, rather cartilaginous auricle, blade terete, hollow and distantly septate. **INFL** of 5–40 small flower clusters; 2–8 flowers borne on ascending to erect branches in a terminal panicle, 2–8 cm long. **INVOLUCRAL BRACTS** leaflike, shorter than the inflorescence. **FLOWERS**

nonprophyllate; tepals 2–2.5 mm long, outer set obtuse to acute, sometimes apiculate, inner tepals shorter, obtuse; stamens 6. **CAPSULES** slightly longer than the tepals, acute, brown. **SEEDS** elliptic, apiculate, less than 0.5 mm long. Seeps, streams, ponds, and lake margins, often in calcareous sites. COLO: Arc. 2255 m (7400′). Flowering: Jul–Aug. Scattered, boreal North America.

Juncus articulatus L. (divided into joints) Joint-leaf rush. Perennial, arising from a stout rhizome. **CULMS** terete, 15–40 cm tall. **LEAVES** 3–4, cauline. Lower sheaths bristle-tipped, upper sheaths obtuse, membranous-auriculate; upper leaf blades terete, hollow, and conspicuously septate. **INFL** terminal, of 2–25 flower clusters borne on spreading to divaricate branches, clusters 2–10-flowered. **INVOLUCRAL BRACTS** leaflike, septate, shorter than the inflorescence. **FLOWERS** nonprophyllate, subtended by 1 scarious bract; tepals 2 mm long, lanceolate, acute to acuminate, sometimes slightly apiculate; stamens 6. **CAPSULES** ovoid, triquetrous, style persistent as a beak, 3 mm and exceeding the tepals. **SEEDS** less than 0.5 mm long, ovoid-apiculate, striate-reticulate, brown. Roadsides, ditches, stream banks, pond and lake margins, and wet meadows. ARIZ: Apa; COLO: Arc, LPl, Mon, SJn; NMEX: SJn; UTAH. 1525–3170 m (5000–10400′). Flowering: Jul–Sep. Scattered across North America.

Juncus arcticus Willd. (of the Arctic) Arctic rush. Perennial, arising from a moderate to stout rhizome. **CULMS** 20–110 cm tall. **LEAVES** bladeless, sheaths well developed. **INFL** lateral, of 5–50 pedicellate flowers in a congested to a mostly open panicle. **INVOLUCRAL BRACTS** terete, extending well beyond the inflorescence as continuation of the stem. **FLOWERS** prophyllate; tepals light to chestnut brown, 3–6 mm long, lanceolate, acute to acuminate; stamens 6; anthers twice as long as the filaments. **CAPSULES** brown, shorter or longer than the tepals, ovoid to slightly oblong, apex obtuse to mostly acute, style base persistent as a short beak. **SEEDS** less than 0.5 mm, light brown, ovoid to pyriform, obliquely ellipsoid, not tailed. Mostly riparian, but may occur in hanging gardens, grasslands, sagebrush, piñon-juniper to montane forests; often on alkaline or saline soils. ARIZ: Apa; COLO: Arc, Dol, LPl, Mon, SMg; NMEX: McK, RAr, SJn; UTAH. 1525–2590 m (5000–8500′). Flowering: May–Aug. Scattered across boreal North America, northern and western United States. Our material belongs to **var.** ***balticus*** (Willd.) Trautv. (of the Baltic).

Juncus bufonius L. (toad, referring to wet habitats) Toad rush. Plants annual, weakly to strongly caespitose. **CULMS** slender, 2–35 cm tall, few to many and crowded, rarely branched. **LEAVES** few per culm, mostly cauline and on the lower culm, or basal in small plants; sheaths well developed, with membranous margins that disappear in the transition to the flat to inrolled blade, up to 1.5 mm wide. **INFL** of 1–20 flowers, sessile to shortly pedicellate in open panicles. **INVOLUCRAL BRACTS** 1 or 2, lowermost often lanceolate-setaceous, upper mostly ovate to lanceolate, scarious. **FLOWERS** with prophylls ovate and entirely scarious; outer tepals 4–7 mm long, lanceolate-acuminate, midvein green, narrow to broad with narrow to wide, scarious margins, inner tepals similar, but generally shorter, tips variable, obtuse, acute, or acuminate; stamens 6. **CAPSULES** greenish to castaneous, oblong to broadly ovoid, mucronate, incompletely 3-celled. **SEEDS** ovoid, brown, lacking appendages, less that 0.5 mm. Moist to wet meadows, pond margins. ARIZ: Apa, Nav; COLO: Arc, LPl, Mon; NMEX: McK, SJn; UTAH. 1125–2775 m (3700–9100′). Flowering: May–Sep. Cosmopolitan.

Juncus castaneus Smi. (for the chestnut color) Chestnut rush. Rhizomatous perennial. **CULMS** 10–40 cm tall. **LEAVES** mostly toward the culm base; sheath not auriculate, transition to blade gradual, blade flat, but strongly inrolled through most of its length, becoming terete at the tip. **INFL** of 1–4 flower clusters each with 2–7 flowers, flower clusters nearly sessile to pedicellate. **INVOLUCRAL BRACTS** exceeding the inflorescence; sheath somewhat inflated, but tapering to a terete tip. **FLOWERS** nonprophyllate; tepals lanceolate, acute to slightly apiculate, 5–7 mm long, scarious-brown to brown; stamens 6, filaments much longer than the anthers. **CAPSULES** dark brown, well exceeding the tepals, somewhat broadly lanceolate to oblong with a short stylar beak. **SEEDS** long-tailed at both ends, tails longer than the seed, 1.5 mm overall. Wet meadows, ponds, and stream margins in the subalpine and alpine. COLO: Arc, LPl, SJn. 3505–3810 m (11500–12500′). Flowering: Jul–Aug. Circumboreal in the north, cordilleran in the west.

Juncus compressus Jacq. (compressed, but what it refers to is unclear) Compressed rush. Perennial, arising from a slender rhizome. **CULMS** terete, 30 cm tall. **LEAVES** 1–3, cauline, sheath well developed, with opaque auricles, blades flat but broadly channeled; 1 or 2 basal leaf sheaths bristle-tipped. **INFL** open, of 30–40 individually pedicellate flowers. **INVOLUCRAL BRACTS** leaflike and reaching beyond the inflorescence; bract blade channeled. **FLOWERS** prophyllate; tepals 1.5–2.3 mm long; outer tepals longer than the inner and with distinctly cucullate tips; shorter inner tepals with tips rounded-obtuse; stamens 6. **CAPSULES** to 3 mm long, ovoid, dark brown. **SEEDS** ovoid, slightly curved, less than 0.5 mm long, apiculate, golden. Riparian. One collection along the Mancos River. COLO: Mon. 2135 m (7000′). Flowering: Jun. Widely scattered across North America.

Juncus confusus Coville (confusing) Perplexing rush. Caespitose perennial. **CULMS** 10–40 cm tall. **LEAVES** all attached toward the base, uppermost leaves with well-developed, flat to channeled blades, auricles membranous, about 1 mm long. **INFL** of few to 20 congested, short-pedicellate flowers, clusters less than 2 cm in diameter. **INVOLUCRAL BRACTS** slender, well exceeding the inflorescence. **FLOWERS** prophyllate; tepals 3–4.5 mm long, lanceolate to ovate-lanceolate, acute; stamens 6. **CAPSULES** equaling or shorter than the tepals, ovoid-oblong, completely 3-loculate, apex distinctly retuse. **SEEDS** less than 0.5 mm long, golden, striate, apiculate. Roadside ditches, meadows. COLO: Mon, SJn. 2375–2805 m (7800–9200′). Flowering: Jul–Aug. Western North America.

Juncus drummondii E. Mey. (for Thomas Drummond, 1790–1835, Scottish botanical explorer and naturalist) Drummond's rush. Densely caespitose perennials. **CULMS** terete, 15–40 cm tall. **LEAVES** basal, occupying the lower 1/4 of the stem, sheaths 2–6 cm long, blade absent or present as a 1–2 mm long bristle. **INFL** appearing lateral by means of the erect involucral bract surpassing the flowers. **INVOLUCRAL BRACTS** terete beyond scarious auricles, 1–4 cm long. **FLOWERS** prophyllate, in groups of 1–3, all may be sessile or pedicellate, though the lowermost flower tends to be sessile, while upper flowers tend to be short-pedicellate; prophylls scarious, the first prophyll of the lowermost flower tends to be lanceolate-acuminate, but the acumen is often elongated as a 1–2 mm long awn, other prophylls are broader and acute; tepals 5–7 mm long, lanceolate-attenuate, castaneous with scarious margins; stamens 6. **CAPSULES** oblong, truncate and usually retuse, trilocular. **SEEDS** narrow, elliptic-fusiform, both ends long-tailed, reaching 1.5–2 mm. Very common in upper montane meadows, along stream banks, and under forested slopes. COLO: Arc, Con, Hin, LPl, Min, Mon, RGr, SJn; NMEX: RAr; UTAH. 3110–3930 m (10200–12900′). Flowering: July–Aug. Cordilleran.

Juncus dudleyi Wiegand (for William Russel Dudley, 1849–1911, its discoverer) Dudley's rush. Caespitose perennial. **CULMS** 15–50 cm tall. **LEAVES** arising from the lower 1/3 of the culm; sheath margins scarious, ending in short, round, opaque to yelllowish, distinctively cartilaginous auricles; blades narrow, flat, with very slightly ridged margins, shallowly channeled to inrolled. **INFL** 3–60-flowered, flowers pedicellate, congested to lax. **INVOLUCRAL BRACTS** slender, leaflike, 1–12 cm long, exceeding the inflorescence. **FLOWERS** prophyllate; tepals lanceolate-acuminate, 3–5 mm long, scarious-margined, with a broad green midvein; stamens 6. **CAPSULES** ovoid, 1-celled, tip obtuse to slightly retuse with a short stylar beak. **SEEDS** less than 0.5 mm long, apiculate, not tailed, golden brown. In ditches, along streams, around ponds and lakes, wet meadows. ARIZ: Apa; COLO: Arc, LPl, SJn; NMEX: McK, SJn; UTAH. 1125–2620 m (3700–8600′). Flowering: Jun–Sep. Widely scattered throughout North America.

Juncus ensifolius Wickstr. (for the sword-shaped blades) Sword-leaf rush. Rhizomatous perennial. **CULMS** 20–60 cm tall. **LEAVES** 1–3 per culm, upper edges of the sheath scarious-margined, ending in small membranous auricles or transitioning smoothly to the blades, blades equitant and ensifoliate, upper blade irregularly and incompletely septate. **INFL** paniculate, variable, composed in one extreme of 2–6 more or less globose, many-flowered (20+) flower clusters or, in a second extreme, of 10–30 or 40 or more small, few-flowered (10 or fewer) clusters of erect to spreading flowers. **INVOLUCRAL BRACTS** leaflike and ensifoliate, shorter than the inflorescence. **FLOWERS** nonprophyllate; clusters and flowers subtended by single, scarious, acuminate bracts; tepals lanceolate-acuminate, 3–3.5 mm long, greenish brown to dark chestnut; stamens 6, anthers shorter or longer than the filaments. **CAPSULES** shorter than the tepals, ovoid to oblong, rounded toward the apex, style persistent as a short beak, chestnut-brown. **SEEDS** about 0.5 mm long, ellipsoid, apiculate or very short-tailed. Riparian, springs, roadside ditches, seepy cliff crevices. ARIZ: Apa, Nav; COLO: Arc, Dol, LPl, Min, Mon, SJn; NMEX: McK, SJn; UTAH. 1125–2925 m (3700–9600′). Flowering: Apr–Sep. Western North America. Our material belongs to **var. *montanus*** (Engelm.) C. L. Hitchc. (of mountains). Various botanists have recognized two species (*J. saximontanus* and *J. tracyi*) as segregated from *J. ensifolius* and two varieties within *J. ensifolius*, named var. *montanus* and var. *brunnescens*. The segregate species and the two varieties are distinguished in part from *J. ensifolius* by the presence of three stamens in the latter and six in all of the former. Careful dissections reveal that all specimens seen from the study area have six stamens; therefore, only *J. ensifolius* is recognized in the flora. *Juncus saximontanus* and *J. tracyi* are names that elevate to species various expressions of plants also named *J. ensifolius* var. *montanus*. *Juncus ensifolius* var. *montanus* is characterized by Cronquist et al. (1977) as having few (10 or less), many-flowered (10–25) heads and distinguished from var *brunnescens*, which is said to have "heads relatively numerous, mostly 10–50 and few flowered (4–12)." In the study area there are no consistent differences between the two varieties and the two segregate species. Therefore, only *J. ensifolius* var. *montanus* is recognized for the study area (Brooks and Clemants 2000). However, *Juncus* is a sadly undercollected genus in this region, especially the La Plata and San Juan mountains. Additional collections are needed and will shed much needed insight into the difficulties with the *J. ensifolius* complex.

Juncus hallii Engelm. (for Elihu Hall, 1822–1882, botanist) Hall's rush. Caespitose perennials. **CULMS** 10–40 cm tall. **LEAVES** basal or cauline, mostly toward the base, basal leaves bladeless or bristle-tipped, cauline leaves with terete blades. **INFL** appearing terminal, involucral bract said to be deflected to the side, flowers 2–3 (7) per cluster, sessile to short-pedicellate. **INVOLUCRAL BRACTS** glumelike at the base, with scarious margins and tapering into a terete awn that may exceed the inflorescence. **FLOWERS** prophyllate, usually in clusters of 2–3; tepals 4–5 mm long, acute; stamens 6. **CAPSULES** oblong-ovoid, dark brown, 3-celled, tip retuse. **SEEDS** slightly elongate-ovate, striate, short-tailed, about 1 mm with tails. Dry to boggy meadows, around ponds and lakes, and along streams into montane and aspen forests. COLO: Arc, Dol; NMEX: RAr. 2440–3140 m (8000–10300′). Flowering: Jun–Jul. Rocky Mountains, Montana to Colorado. The species is little known in the region and should be sought. The species is often confused with *J. parryi*, which differs in having acuminate tepals and acute capsules.

Juncus longistylis Torr. (long style) Longstyle rush. Perennial from slender, spreading rhizomes. **CULMS** terete (15–45) 60+ cm tall. **LEAVES** mostly basal, 1–3 cauline and remote from each other, sheath margins scarious-hyaline, terminating in rounded, sometimes slightly cartilaginous auricles, blades flat. **INFL** often of 1–3 crowded and terminal heads, occasionally with 2–3 (4) additional heads elevated and slightly separated on peduncles above the others. **INVOLUCRAL BRACTS** with the lowest inflorescence bract scarious, lanceolate to broadly lanceolate, broad midvein often extending as an awn. **FLOWERS** nonprophyllate, arranged in clusters of 2–4, individual flowers subtended individually or in small groups by a scarious, ovate to lanceolate, often acuminate bract; tepals 5–7 mm long, green, brownish to sometimes reddish anthocyanic, lanceolate, acute to slightly acuminate, margins scarious to hyaline; stamens 6, anthers longer than the filaments. **CAPSULES** oblong or ellipsoid, acute to retuse, incompletely trilocular; style 1 mm or longer, stigma long-exserted. **SEEDS** less than 0.5 mm long, apiculate, but not tailed. Wet meadows, canyon-side seeps and springs, streams; occasionally in saline places from desert shrub to spruce forest. ARIZ: Apa, Nav; COLO: Arc, Dol, Hin, LPl, Mon; NMEX: McK, RAr, SJn; UTAH. 1675–3050 m (5500–10000′). Flowering: May–Sep. West across Canada into montane western North America.

Juncus mertensianus Bong. (for C. H. Mertens, 1796–1830, botanist at Sitka, Alaska) Mertens' rush. Perennial, arising from a creeping rhizome. **CULMS** slender, 7–35 cm tall. **LEAVES** 1–3, mostly cauline, sheaths scarious toward the blade, ending in a rounded to acute, membranous, 0.5–1 mm long auricle, blades terete, septate, ending below the inflorescence. **INFL** terminal, of 1 or 2 mostly solitary flower clusters of 10–50+ flowers. **INVOLUCRAL BRACTS** diverging from beneath the inflorescence, leaflike, with well-developed sheath, blade terete, shorter than or shortly exceeding the inflorescence. **FLOWERS** nonprophyllate, each subtended by 1 lanceolate-acuminate bract; tepals lanceolate-acuminate, dark shiny purplish brown to nearly black, 3–5 mm long; stamens 6. **CAPSULES** slightly shorter than or equaling the tepal length, apex oblong, truncate to slightly retuse, color similar to the tepals. **SEEDS** less than 0.5 mm long, brown, elliptic, apiculate, not tailed. Wet meadows, ponds, lakes, and stream banks. COLO: Arc, Con, Hin, LPl, Min, Mon, SJn; NMEX: McK. 2925–3840 m (9600–12600′). Flowering: Jul–Sep. Western North America.

Juncus nodosus L. (jointed, referring to the septate leaves) Nodose rush. Slender perennials arising from slender rhizomes. **CULMS** terete, 20–40 cm tall. **LEAVES**: cauline shorter or longer than the inflorescence; sheath ending in slightly cartilaginous auricles; blades hollow and septate, terete or with a slight adaxial channel. **INFL** of 1–6 open to slightly congested flower clusters, included within or exceeded by the length of the involucral bract; flower clusters 6–11 mm diameter, 5–25-flowered, mostly 15 or fewer and radiating in all directions. **INVOLUCRAL BRACTS** slender, leaflike, short or long, but exceeding the inflorescence. **FLOWERS** nonprophyllate, each subtended by a single scarious, ovate-acuminate bract; tepals 3–4 mm long, more or less equal, narrowly lanceolate, subulate-acuminate, more or less straight; stamens 6. **CAPSULES** exceeding the tepals, trigonous, light to dark brown. **SEEDS** less than 0.5 mm long, ovoid and apiculate, striate, light to dark brown. Riparian; in and along ditches and streams from sagebrush to wet montane meadows. COLO: Arc, LPl, Mon, SJn; NMEX: SJn. 1705–2620 m (5600–8600′). Flowering: Jul–Sep. Scattered across North America.

Juncus parryi Engelm. (for Charles Parry, 1823–1890, botanist and explorer) Densely caespitose perennials. **CULMS** terete, 10–40 cm tall. **LEAVES** essentially basal, lower sheaths bristle-tipped, upper sheaths bearing well-developed terete blades. **INFL** terminal, lowest involucral bract said to reflex away. **INVOLUCRAL BRACTS** with lowest bract 1–3 cm long, glumelike, with scarious basal margins, abruptly narrowed into a terete awn, generally deflected laterally; flowers sessile to short-pedicellate. **FLOWERS** few, 1–3, prophyllate, first prophyll of the only or lowest flower often elongated in a short awn, other prophylls simple hyaline scales; tepals lanceolate-acuminate, 5–7 mm long; stamens 6. **CAPSULES** narrow and acute, brown. **SEEDS** ovoid-elliptic, brown, surface finely striate, to 2 mm long with tails.

Meadows; along streams or on dry, rocky slopes in upper montane forests into the alpine. COLO: Arc, Con, Hin, LPl. 3050–3960 m (10000–13000′). Flowering: Jul–Aug. Cordilleran; British Columbia and Alberta to Colorado. While the capsule is described as having acute tips, occasional specimens are more oblong, with abruptly acute to almost blunt tips; nevertheless, there is no suggestion of their being retuse, and therefore confused with *J. hallii*.

Juncus torreyi Coville (for John Torrey, 1796–1873, botanist) Torrey's rush. Perennials arising from stout, creeping rhizomes. **CULMS** stout, 30–75 cm tall. **LEAVES** 1–4, long cauline-sheathed, with membranous auricles; blades septate thorughout and terete, but the proximal blade is slightly channeled on the adaxial curvature, while completely terete in the extremities. **INFL** terminal, composed of 2–12 (18) generally congested flower clusters, each 10–15 mm in diameter. **INVOLUCRAL BRACTS** equaling or exceeding the inflorescence. **FLOWERS** nonprophyllate, 10–75+ in crowded clusters, radiating in all directions within a cluster; outer tepals dry-grass green to tawny, 4–5 mm long, lanceolate-acuminate, often gently to strongly reflexed in the rigid acumen; inner tepals slightly shorter than the outer; stamens 6, about 1/2 as long as the tepals. **CAPSULES** equal to or slightly exceeding the tepals, narrowly triquetrous and subulate, golden brown. **SEEDS** less than 0.5 mm long, oval, striate, apiculate at either end, golden brown. Seeps and springs, ditches, ponds and lakes, and rivers. COLO: Arc, LPl, Mon; NMEX: RAr, SJn; UTAH. 1125–1980 m (3700–6500′). Flowering: Jun–Sep. Canada to northern Mexico.

Juncus triglumis L. (three flowered) Three-flowered rush. Small, densely caespitose perennials. **CULMS** terete, filiform, 6–12 cm tall. **LEAVES** basal or on the lower culm; sheaths yellow to brownish, well developed and auriculate, blades less than 1/2 the length of the culm, abruptly narrowed, filiform, terete, with callous, septate tips. **INFL** terminal and solitary, 2–3-flowered, 3–5 mm long. **INVOLUCRAL BRACTS** 2, rarely 1, divergent, weakly spathiform, mostly acute, or the lowermost bract rarely tapered into a short (1–2 mm) awn, mostly pale yellow, sometimes pale chestnut at the base. **FLOWERS** nonprophyllate; tepals subequal, 4–5 mm long, oblong-lanceolate, obtuse to acute, mostly pale yellow or colorless, sometimes very pale brown; stamens 6, anthers 1 mm or less long, much shorter that the filaments. **CAPSULES** 3–4 mm long, chestnut, ovoid, mostly acute and apiculate by means of the persistent style, or slightly obtuse. **SEEDS** golden, about 1 mm long, tails white, to about 0.5 mm long. Wet to boggy, peaty meadows and associated lake margins. COLO: LPl, SJn. 3810 m (12500′). Flowering: Jul. North American, Arctic-alpine, Labrador to Alaska and south in the Rocky Mountains to Colorado and New Mexico. Our material belongs to **var. *albescens*** Lange (referring to the white tepals). Inflorescences typically produce two or three flowers that are subtended by two inflorescence bracts that nearly reach the tepal tips. In other individuals one may occasionally find one of the two bracts being awned and reaching beyond the flowers, or just a single awned bract reaching well beyond the inflorescence. In all cases, the tepals are pale.

Luzula DC. Woodrush

(Latin from *gramen luzulae* or *luxae*; *lux*, light; perhaps alluding to the appearance of the heads when covered with dew) Caespitose or rhizomatous, perennial, grasslike herbs. **LEAVES** composed of closed sheaths and flattened blades, sheath collar and lower blade margins bearing few to many, long-pilose hairs, at least in young leaves. **INFL** composed of several to many, few-flowered clusters arranged in open, cymose panicles or in congested, capitate or spikelike heads. **FLOWERS** perfect, subtended by ciliate to ciliate-fimbriate pairs of prophylls; each flower is further subtended by individual secondary bracts with ciliate margins; tepals 6, in 2 similar series of 3, un-petal-like, scarious and chartaceous, hyaline to sometimes greenish, mostly brown, purple-brown, or blackish; stamens 6. **CAPSULES** 3-carpellate, uniloculate, ovules 3 and basal, maturing into 3 seeds. (Cronquist et al. 1977; Hitchcock et al. 1969; Flora of North America Editorial Committee 2000; Weber and Wittmann 2001).

1. Inflorescence an open, regularly drooping, cymoselike panicle, flowers many, arranged in groups of 2–3; leaves mostly glabrous at maturity .... ***L. parviflora***

1′ Inflorescence of several to many clusters of flowers arranged in congested capitate heads or spikes .... (2)

2. Inflorescence elongated, spikelike; primary and secondary involucral bract margins abundantly ciliate, secondary bracts long-acuminate; prophylls ciliate-fimbriate; leaves 1–4 mm wide .... ***L. spicata***

2′ Inflorescence capitate, composed of a few congested flower clusters; primary and secondary involucral bract margins sparsely hairy; leaf blades to 8 mm wide .... ***L. subcapitata***

Luzula parviflora (Ehrh.) Desv. (little flowers) Millet woodrush. Caespitose perennials, rarely with a short rhizome. **CULMS** 20–100 cm tall. **LEAVES** 2–4 per culm, 6–11 mm wide. In young plants upper sheath and blade margins may be sparsely hairy, but glabrous at maturity. **INFL** an open, drooping panicle, flowers sessile to mostly pedicellate; nodes subtended by scarious, hyaline, lanceolate bracts with regularly lacerate margins. **FLOWERS** prophyllate, prophylls scarious-hyaline to brown, ovate with lacerate margins; tepals ovate to broadly lanceolate-acuminate, 2–2.5 mm long, scarious, hyaline to brown; stamens 6, anthers shorter than or equal to the filament length. **CAPSULES** ovoid to ovoid-oblong, slightly shorter or longer than the tepals, green to brown, style remnant persistent as a beak. **SEEDS** ellipsoid, 1–1.4 mm long, brown. Moist meadows, along streams, mixed conifer forest, spruce-fir forest, subalpine to alpine. COLO: Arc, Con, Hin, LPl, Min, Mon, RGr, SJn; NMEX: SJn; UTAH. 2590–3690 m (8500–12100′). Flowering: Jun–Sep. Circumboreal; western North America.

Luzula spicata (L.) DC. (for the spicate inflorescence) Spike woodrush. Caespitose perennial. **CULMS** 3–25 cm tall. **LEAVES** basal and crowded, cauline leaves 1–4 mm wide, leaf margins and especially sheath margins pubescent in age. **INFL** of congested, 1–several-flowered clusters of sessile to very short-pedicellate flowers arranged in an erect or drooping, but mostly nodding spike; secondary bracts strongly ciliate and long-awned. **INVOLUCRAL BRACTS** leaflike, shorter to slightly longer that the inflorescence, generally divergent. Sheath and lower blade margin ciliate. **FLOWERS** subtended by irregular, ovate, hyaline to purplish brown, ciliate-fimbriate prophylls; tepals hyaline to mostly dark, purplish brown, ovate to lanceolate, 2–3.5 mm long, acuminate, acumen to almost 1 mm; stamens 6, anthers slightly shorter than to ± equal to the filaments. **CAPSULES** globose, light to dark brown, slightly shorter than to about equal to the tepals. **SEEDS** brown, ovoid-oblong, 1–1.2 mm long, slight caruncle. Rocky ground; subalpine to mostly alpine meadows. COLO: Arc, Hin, LPl, Mon, SJn. 3350–3960 m (11000–13000′). Flowering: Jun–Aug. Western North America, New England, and eastern Canada.

Luzula subcapitata (Rydb.) H. D. Harr. (for the subcapitate inflorescence) Colorado woodrush. Caespitose perennial. **CULMS** crowded on short rhizomes, 10–35 cm tall. **LEAVES** basal and cauline, crowded at the base, cauline leaves 2–4. Blades up to 8 mm wide and often tinged reddish or reddish brown. **INFL** of 6–10 flower clusters arranged into a congested, irregularly shaped capitate group. **INVOLUCRAL BRACTS** leaflike, but reduced, mostly exceeding the inflorescence. **FLOWERS** prophyllate, prophylls entire to slightly lacerate; flowers essentially sessile; tepals up to 2 mm long, ovate to ovate-lanceolate, sometimes slightly acuminate, scarious, reddish brown, outer series slightly keeled; stamens 6, small; anthers about as long as, or slightly longer than the filaments. **CAPSULES** globose, ± equal to the tepals, dark purplish brown. **SEEDS** ellipsoid, slightly over 1 mm long, brown. Bogs and streamsides in the alpine. COLO: SJn. 3780–3830 m (12400–12560′). Flowering: Aug. A Colorado Rockies endemic.

JUNCAGINACEAE Rich. ARROW-GRASS FAMILY

C. Barre Hellquist

Perennial or annual herbs, rhizomatous. **LEAVES** basal, emergent, sessile, sheath longer than the blade, ligulate, auriculate, blade linear. **INFL** pedunculate, a terminal or axillary scapose spike or spikelike raceme, occasionally with solitary flowers. **FLOWERS** bisexual or unisexual, monoecious, perianth usually present; tepals 1, or 6 in 1–2 series; pistils 1, 3, or 6, if 3 or 6, coherent or weakly connate. **FRUIT** a nutlet or schizocarp. **SEED** 1. Four genera, approximately 15 species worldwide.

Triglochin L. Arrow-grass

(Greek *treis*, three, + *glochis*, a point) Perennial herbs. Roots occasionally with tubers, rhizomes stout. **LEAVES** erect, terete, sheath with ligule. **INFL** a spikelike raceme, scape shorter or longer than leaves. **FLOWERS** bisexual, short-pedicellate; tepals 6, in 2 series; stamens 4 or 6; pistils 6, 3 fertile, 3 sterile, or 6 fertile. **FRUIT** a schizocarp, mericarps 3 or 6. $x = 6$ (Ford and Ball 1988; Löve and Löve 1958). Species around 12 worldwide. The genus is treated as feminine rather than neuter as treated in some floras (Crow and Hellquist 2000).

1. Carpels and stigmas 6; mature fruit 2–3 mm wide, about 2 times as long as wide; no stolons formed .... ***T. maritima***

1′ Carpels and stigmas 3; mature fruit 1 mm wide, about 5–7 times as long as broad; stolons present, bearing small bulbs .... ***T. palustris***

Triglochin maritima L. (of the sea) **LEAVES** erect from stem, 2.2–11.5 cm; sheath 0.7–2.5 cm long; ligules occasionally hoodlike, 2-lobed, apex obtuse to round. **INFL** scape often purple at base, usually exceeding leaves, pedicel 1–4 mm long. **FLOWER** tepals elliptic, apex acute; pistils 6, all fertile. **FRUIT** receptacle lacking wings; schizocarp linear to globose, 2–4.5 mm long; mericarp linear to linear-obovate, 1.5–3.5 mm long, beak 0.2 mm. $2n = 12, 24, 36, 48, 120$. [*T. concinna* Burtt Davy; *T. concinna* var. *debilis* (M. E. Jones) J. T. Howell; *T. debilis* (M. E. Jones) Á. Löve & D. Löve; *T. elata* Nutt.]. Mountain marshes and wet alkaline meadows in our area. ARIZ: Apa; COLO: Arc, LPl, Mon, SJn; NMEX: SJn.

1400–2150 m (4650–8800′). Flowering and fruiting: Jun–Jul. Alaska and Yukon east across Canada to Newfoundland, south to California, New Mexico, Iowa, Indiana, and Rhode Island; Mexico; Central America; Eurasia. *Triglochin concinna* is recognized as synonymous with *T. maritima* following the treatment of Haynes and Hellquist (2000) in the *Flora of North America*, where they also note that *T. maritima* is high in cyanide and is toxic to livestock.

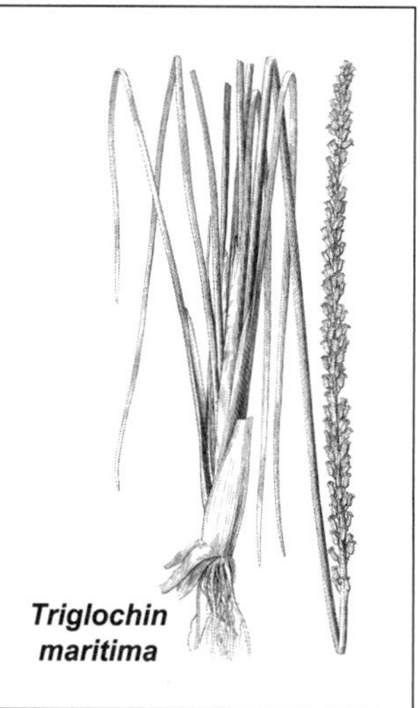
Triglochin maritima

Triglochin palustris L. (of marshes) **LEAVES** erect from sheath, 5–24.5 cm; sheath 3.5–5 cm long; ligule not hooded, unlobed; apex acute. **INFL** scape often purple at base, usually exceeding leaves. **FLOWER** pedicel 0.4–4.5 mm long, tepals elliptic, apex round; pistils 6, 3 fertile, 3 sterile. **FRUIT** a receptacle with wings; schizocarp linear, 7–8.3 mm long; mericarps linear, 6.5–8.5 mm long; beak 0.3 mm. $2n = 24$. Mountain marshes and wet alkaline meadows in our area. ARIZ: Apa; COLO: LPl. 1650–2150 m (5450–7050′). Flowering and fruiting: Jun–Jul. Alaska east across Canada to Newfoundland, south to California, New Mexico, Iowa, Ohio, and Rhode Island; Mexico; South America; Eurasia.

LAMIACEAE Martinov (LABIATAE) MINT FAMILY

Charlotte M. Christy

Plants annual to perennial herbs or shrubs, mostly pubescent and aromatic. **STEMS** typically 4-angled. **LEAVES** opposite (rarely otherwise), simple, rarely entire. **INFL** a cyme, but often congested, forming a dense verticil and these often arranged in a spike, in some forming a head, or flowers 1–few per axil and then forming a raceme or panicle. **FLOWERS** usually perfect; calyx of 5 fused sepals with 2–5 lobes or teeth, actinomorphic to zygomorphic, typically persisting; corolla of 5 fused petals with 4–5 lobes, mostly bilabiate, a few only weakly zygomorphic; stamens 4 or 2 and staminodes 2 or 0; pistil 1; carpels 2; ovary superior, lobes 4, ovules 4; style 1, the stigmas 2, these equal or not, in some appearing only 1. **FRUIT** a schizocarp of 1-seeded nutlets, these spherical, oblong, ovate, or obdeltoid (Cantino 1992). Around 200 genera and 3500–5000 species. Cosmopolitan. The volatile, aromatic oils lead many species to be used as herbal remedies or as culinary herbs; others are cultivated for ornament. Two introduced, naturalized genera (*Lamium* and *Leonurus*) are apparently not known from our area but may perhaps be locally encountered. Both species of *Lamium* (*L. amplexicaule* L. and *L. purpureum* L.) are common weeds of lawns and disturbed areas. *Leonurus cardiaca* L. is most likely to be found in upper-elevation riparian or other moist habitats.

1. Calyx zygomorphic, the teeth or lobes unevenly fused, or markedly different in orientation or size ....(2)
1′ Calyx actinomorphic or nearly so, the teeth or lobes at most slightly differing in fusion, orientation, or size ...(11)

2. Fertile stamens 2, staminodia may be present ....(3)
2′ Fertile stamens 4, staminodia absent ....(4)

3. Upper calyx teeth apparently 0, or 3 and strongly fused; connective markedly elongate, the filament appearing forked or jointed; only 1/2 of each anther functional .... ***Salvia***
3′ Upper calyx teeth 3, not strongly fused; connective not unusually elongate, the filament appearing linear; both anther halves fully developed .... ***Hedeoma***

4. Calyx teeth strongly fused, appearing 2, forming 2 lips, the upper lip crested and falling at maturity; flowers solitary in the axils .... ***Scutellaria***
4′ Calyx teeth evidently more than 2, forming 0–2 lips, the upper lip, if present, lacking a crest and attached at maturity; flowers several to many per axil ....(5)

5. Uppermost calyx lobe markedly larger than the other 4 lobes, forming a lip .... ***Dracocephalum***
5′ Upper calyx sections similar in size, a lip absent or formed by fusion of teeth or lobes ....(6)

6. Upper 3 calyx teeth at least partially fused, forming a lip ....(7)
6′ Upper calyx teeth separate, not forming a lip ....(8)

7. Inflorescence a spike; upper 3 calyx teeth almost completely fused .... ***Prunella***
7′ Inflorescence a single verticil; upper 3 calyx teeth fused approximately halfway .... ***Satureja***

8. Calyx with all teeth equal or the upper teeth longer than the lower ....(9)
8′ Calyx with the lower teeth longer than the upper ....(10)

9. Upper lip of the corolla so deeply lobed as to appear absent; flowers in terminal spikes .... ***Teucrium***
9′ Upper lip of the corolla 2-lobed; flowers in terminal panicles .... ***Trichostema***

10. Plants creeping, sparsely pubescent; axils with 1–few flowers .... ***Glechoma***
10′ Plants erect, tomentose; axils with many-flowered verticils .... ***Nepeta***

11. Stamens 2 ....(12)
11′ Stamens 4 ....(14)

12. Corolla actinomorphic or only slightly zygomorphic, 4-lobed .... ***Lycopus***
12′ Corolla zygomorphic, 5-lobed ....(13)

13. Herbs; flowers in dense verticils at apex or in few axils .... ***Monarda***
13′ Shrubs; flowers 1–3, forming loose clusters in many axils .... ***Poliomintha***

14. Calyx teeth usually 10, their tips hooked; plants tomentose .... ***Marrubium***
14′ Calyx teeth 5, their tips straight; plants glabrous to hairy but not tomentose ....(15)

15. Flowers 1–3 per axil ....(16)
15′ Flowers more than 5 per axil ....(18)

16. Corolla appearing 1-lipped, the upper lip deeply cleft .... ***Teucrium***
16′ Corolla bilabiate, the upper lip entire or 2-lobed ....(17)

17. Stamens included or somewhat exserted; upper corolla lip entire or notched .... ***Stachys***
17′ Stamens long-exserted; upper corolla lip 2-lobed .... ***Trichostema***

18. Corolla weakly zygomorphic, the lobes all similar ....(19)
18′ Corolla zygomorphic, 1 or more lobes markedly different in size ....(20)

19. Verticils many, axillary or a terminal spike .... ***Mentha***
19′ Verticil single, terminal .... ***Monardella***

20. Verticils pedunculate and loose, the flowers obviously pedicellate .... ***Nepeta***
20′ Verticils about sessile, compact, the flowers sessile or subsessile ....(21)

21. Anther sacs parallel .... ***Agastache***
21′ Anther sacs divergent ....(22)

22. Bracts reniform, the edges overlapping and forming a bowl under verticils; corolla red-purple .... ***Lamium***
22′ Bracts elliptic or narrowly triangular, the edges not overlapping; corolla red, orangish, or pink-purple .... ***Stachys***

Agastache Gronov. Giant Hyssop

(*agan*, much, + *stachys*, spikes) Herbs, perennial from rhizomes or a crown, pubescent, aromatic. **LEAVES** petiolate, toothed. **INFL** an axillary verticil forming a dense or interrupted spike; bracts leafy or reduced. **FLOWERS** sessile or short-pedicellate; calyx weakly zygomorphic; teeth 5, their apices not spinose; corolla zygomorphic, 2-lipped, upper lip slightly concave, 2-lobed, lower lip 3-lobed; tube longer than calyx; stamens 4, exserted; anther sacs parallel, none abortive; ovary sessile, lobed to base; stigma evenly 2-lobed. **FRUIT** papillate or hairy. $x = 9$. Twenty-two species in the southwestern United States and northern Mexico.

Agastache pallidiflora (A. Heller) Rydb. (pale-flowered) Bill Williams mountain giant hyssop. Plants rhizomatous. **STEMS** 11–90 cm tall. **LEAVES** with petioles to 20 mm, the blades broadly ovate to triangular or reniform, 1–5 cm long × 1–4 cm wide, the margins coarsely toothed, the base cordate to truncate or slightly attenuate, the apex rounded to acuminate, gland-dotted below. **INFL** 2.5–7 cm long in flower, the lowest verticils ± distinct, the bracts sessile, only those of the lowest verticils leaflike. **FLOWER** calyx 6–9 mm long, base green, distally usually pink to purple, the 2 upper teeth slightly broader than the 3 lower, 1–2 mm; corolla 13–15 mm long, the lobes pink to purple, the tube paler to white; stamens exserted 1–2 mm, the filaments parallel but often curling in age. **FRUIT** 2 mm long, 1.5 mm wide, the apex hairy. $2n = 18$. [*Brittonastrum pallidiflorum* A. Heller]. Riparian or other moist areas in ponderosa pine and

Douglas-fir communities. ARIZ: Apa; COLO: Arc, Hin, LPl, Min, Mon, RGr, SJn; NMEX: McK, RAr, SJn. 2450–3300 m (7700–10200′). Flowering: Jun–Sep. Arizona, Colorado, New Mexico, Texas; Coahuila, Mexico. *Agastache urticifolia* (Benth.) Kuntze has been reported for this area but no specimens have been seen. It differs in being overall a larger plant with the filaments markedly nonparallel: the lower pair of stamens are straight and emerge from under the upper corolla lip, and the upper pair angle strongly downward and emerge alongside the lower corolla lip. *Agastache pallidiflora* was used by the Ramah Navajo as a ceremonial chant lotion and cough medicine; the pulverized root was used as a dusting powder for sores or cankers as a fumigant for "deer infection." The Acoma used the leaves for flavoring.

Dracocephalum L. Dragonhead

(*draco*, dragon, + *cephalus*, head) Annual to perennial taprooted herbs, pubescent. **STEMS** erect, single or branched freely from crown. **LEAVES** petiolate; blades obovate or elliptic to linear-lanceolate; margins entire or toothed. **INFL** an axillary verticil, forming a spike; bracts leafy or reduced. **FLOWERS** sessile; calyx zygomorphic; lobes 5, their apices acute to spinose; corolla zygomorphic, 2-lipped, upper lip slightly concave, notched; lower lip 3-lobed; tube included in calyx; stamens 4, exserted or held under upper lip; anther sacs divaricate, none abortive; ovary sessile, lobed to base; stigma evenly 2-lobed. **FRUIT** glabrous. $x = 9, 10$. Three species in North America and Europe.

Dracocephalum parviflorum Nutt. (small-flowered) American dragonhead. Plants with a taproot. **STEMS** 9–50 cm tall, the more robust frequently much-branched. **LEAVES** petiolate, the blades broadly to narrowly triangular, 1–7 cm long, margins irregularly toothed, the apex long-acuminate, spinose in upper leaves. **INFL** a spike, the lowest verticils often distinct, bracts leafy and with teeth mostly spine-tipped. **FLOWER** calyx 10–13 mm long, the lobes spine-tipped, the upper lobe larger than lower 4; tube at base of sinus with a prominent tuft of long trichomes; corolla about 1 cm long, scarcely exceeding calyx, pale rose-purple, often drying medium blue, the tube paler to white; stamens held under upper corolla lip, the anther sacs placed end-to-end. **FRUIT** 2 mm long, dark brown to black. [*Moldavica parviflora* (Nutt.) Britton]. Sand or clay soils, sage, piñon-juniper woodland, and ponderosa pine communities. ARIZ: Apa, Nav; COLO: Arc, Dol, Hin, LPl, Min, Mon, SJn; NMEX: McK, RAr, SJn; UTAH. 2000–3000 m (6200–9000′). Flowering: Apr–Oct. Mostly in the northern half of the United States; southern Canada. Plant used for infants with diarrhea and as a life medicine by the Kayenta Navajo. Used by the Ramah Navajo for headache; infusion of leaves used as an eyewash and for fever.

Glechoma L. Ground Ivy

Perennial herbs, pubescent. **STEMS** creeping, mat-forming. **LEAVES** long-petiolate; blades round to reniform; margins coarsely crenate. **INFL** axillary, interrupted; bracts leafy. **FLOWERS** sessile; calyx zygomorphic, lobes 5, their apices spinose or aristate; corolla zygomorphic, 2-lipped; upper lip slightly concave, entire; lower lip 3-lobed; tube equaling or shorter than calyx; stamens 4, held alongside upper lip; anther sacs parallel, none abortive, ovary sessile, lobed to base; stigma evenly 2-lobed. **FRUIT** apex densely hairy. $x = 9, 10$. Ten species of temperate Eurasia. The genus is often spelled *Glecoma*.

Glechoma hederacea L. (*Hedera*, the genus of ivy) Creeping Charlie, ground ivy. **STEMS** 5–10 cm tall. **LEAVES** petiolate, reniform, toothed. **INFL** a branching cyme, the flowers 1–few per axil, the pedicels about 3 mm long. **FLOWER** calyx 5–6 mm long, reflexed in fruit; corolla 1.5–2 cm long, long-exserted from calyx, blue with white markings. $2n = 18, 36$. Moist areas in ponderosa pine communities. COLO: Arc, LPl. 2285–2550 m (7500–8000′). Flowering: May–Sep. Native of Europe and now widely naturalized in the United States; however, uncommon in our area. This species is regarded as poisonous to horses.

Hedeoma Pers. Mock-pennyroyal

(*hedys*, sweet, + *osme*, scented) Perennial taprooted herbs, pubescent, aromatic. **STEMS** erect, single or branched freely from crown. **LEAVES** with winged petiole or subsessile; blades obovate or elliptic to linear-lanceolate; margins entire to toothed. **INFL** a few-flowered axillary verticil; bracts reduced or leafy. **FLOWERS** pedicellate; calyx zygomorphic, lobes 5, their apices narrowly acute; corolla zygomorphic, 2-lipped; upper lip slightly concave, entire; lower lip 3-lobed; tube equaling or slightly exceeding calyx; stamens 2, exserted; anther sacs divergent, none abortive; ovary sessile, lobed to base; stigma unevenly 2-lobed. **FRUIT** glabrous. $x = 9, 10$. About 20 species; New World.

1. Calyx closed in fruit, the upper lobes parallel with and appressed to the lower, the lower teeth to 3 mm long .... ***H. drummondii***

1′ Calyx open in fruit, the upper lobes arched upward away from the lower, the lower teeth to 1.5 mm long........ ***H. nana***

Hedeoma drummondii Benth. (for T. Drummond, plant collector) Drummond's false pennyroyal. **STEMS** 6–30 cm tall, often freely branching. **LEAVES** subsessile or petiole to 2 mm, overall 7–13 mm long × 3–8 mm wide, elliptic to oblong (ovate), entire, frequently folded, gland-dotted above and below. **INFL** axillary, 1–5-flowered, the bracts inconspicuous. **FLOWERS** with pedicel 1–3 mm long; calyx 5–7 mm long, sometimes bicolored, the lower lobes to 3 mm long, the upper to 2 mm long, parallel and appressed to the lower in fruit; corolla 6–10 mm long, rose-pink to purple; stamens scarcely exserted, the anther sacs placed end-to-end. **FRUIT** 1–1.5 mm long, brown. $2n = 24, 36$. Sandy or gravelly soils, desert shrub and piñon-juniper woodland communities. ARIZ: Apa, Nav; COLO: Arc, LPl, Mon; NMEX: McK, SJn; UTAH. 1150–2250 m (4600–7000'). Flowering: May–Sep. North Dakota and Montana, south to Arizona, New Mexico to San Luis Potosí, Mexico. Infusion of plant taken in large quantities for influenza by the Ramah Navajo. Plant used for pain by the Navajo.

Hedeoma nana Briq. (dwarf) Dwarf false pennyroyal. **STEMS** 10–23 cm tall, branching at base and below the inflorescence. **LEAVES** with petioles 2–3 mm long; blades 6–15 mm long × 2–5 mm wide, ovate to elliptic or oblong, entire, not folded, gland-dotted. **INFL** axillary, 1–5-flowered, bracts leafy, reduced upward. **FLOWERS** with pedicel 1–4 mm long; calyx 4–6 mm long, often bicolored, the upper side purple, the lower green, the lower teeth to 1.5 mm long, the upper to 1 mm long, angled upward and not appressed to the lower in fruit; corolla 5–10 mm long, rose-purple, the lower lip often white-mottled; stamens scarcely exserted, the anther sacs placed end-to-end. **FRUIT** 1–1.5 mm long, brown. $2n = 36$. [*H. dentata* Torr. var. *nana* Torr.]. Sandy soils, often in crevices of sandstone; piñon-juniper woodlands, and ponderosa pine communities. ARIZ: Apa; NMEX: McK; UTAH. 1150–2400 m (3600–7400'). Flowering: May–Sep. Arizona, California, Colorado, Nevada, New Mexico, Texas; south to central Mexico. The plant used by assistant during the Navajo War Dance. Leaves chewed for the mint flavor by the Isleta.

Lamium L. Dead-nettle

(Greek *lamium*, throat, from the corolla shape) Annual or perennial taprooted herbs, glabrous or somewhat pubescent, aromatic in some. **STEMS** erect or decumbent, usually much branched from base. **LEAVES** petiolate or the upper sessile; blades ovate to orbicular or reniform; margins entire or toothed to incised. **INFL** in axillary verticils; bracts leafy. **FLOWERS** sessile or subsessile; calyx actinomorphic, lobes 5, their apices acuminate or awned; corolla zygomorphic, 2-lipped; upper lip hooded, notched; lower lip 3-lobed; tube exceeding calyx; stamens 4, included in hood; anther sacs divaricate, none abortive; ovary sessile, lobed to base; stigma evenly 2-lobed. **FRUIT** glabrous. $x = 9$. About 40 species in Eurasia and North Africa. Several species are grown for ornament.

Lamium amplexicaule L. (*amplexus*, clasping, + *caulis*, stem) Henbit. Plants annual. **STEMS** 5–40 cm tall. **LEAVES** 15–50 mm long, 5–20 mm wide, triangular-ovate to orbicular, the petiole 10–35 mm long, the base truncate to cordate, the margin irregularly coarsely crenate or incised. **INFL** 10–15 mm broad, the bracts sessile, mostly reniform, horizontal to ascending, much exceeding flowers. **FLOWER** calyx 4–7 mm long; corolla red-purple, with white on lower lip, 10–20 mm long or shorter in cleistogamous flowers. **FRUIT** about 2 mm long, brown mottled with white. $2n = 18, 36$. Lawns and other disturbed areas. COLO: LPl; NMEX: SJn. 1850 m (6000'). Flowering: Mar–Sep. Naturalized nearly throughout the Americas; Europe. Known in our area only from one recent collection in Farmington.

Lycopus Tourn. ex L. Bugleweed, Water Horehound

(*lycos*, wolf, + *apus*, footed) Perennial taprooted herbs, glabrous or somewhat pubescent. **STEMS** erect, simple or much-branched from crown. **LEAVES** sessile or subsessile; blades oblong-lanceolate; margins pinnatifid to serrate, rarely subentire. **INFL** in axillary verticils; bracts leafy. **FLOWERS** sessile; calyx actinomorphic; teeth 4 or 5, their apices aristate; corolla only weakly zygomorphic, 4- or 5-lobed; tube shorter than calyx; stamens 2, exserted or included in corolla tube; anther sacs divaricate, none abortive; ovary sessile, lobed to base; stigma evenly 2-lobed. **FRUIT** glabrous, with corklike margin or crest. $x = 11$. Fourteen species in temperate North America and Eurasia.

1. Leaves pectinate; fruits with corky margin at apex .... ***L. americanus***
1' Leaves serrate; fruits with corky margin extending to base .... ***L. asper***

Lycopus americanus Muhl. ex W. P. C. Barton (of America) American water horehound. **STEMS** 25–50 cm tall, branching below the inflorescence; if robust, hairs conspicuous only at the nodes. **LEAVES** 3–7.5 cm long × 2–3.5 cm wide, broadly to narrowly elliptic, the petiole 4–9 mm long, winged, the base attenuate, the margin pectinately lobed at least near base, lobed or coarsely toothed above, gland-dotted above and below. **INFL** 5–7 mm broad, in axillary clusters, bracts much exceeding the flowers. **FLOWER** calyx 2–3 mm long, the lobes attenuate and somewhat spinose; corolla white, 3–4 mm long; stamens scarcely exserted. **FRUIT** exceeding calyx tube, corky margin forming a crest at the apex.

$2n = 22$. Riparian or other moist to wet areas; piñon-juniper, ponderosa pine, and Douglas-fir communities. ARIZ: Apa; COLO: LPl; NMEX: SJn. 1700–2450 m (5300–7500′). Flowering: Jun–Jul. Throughout the United States; southern Canada, northern Mexico. It should be looked for in other counties in our area.

Lycopus asper Greene (rough) Rough bugleweed. **STEMS** 25–65 cm tall, may branch below the inflorescence, pubescent. **LEAVES** 4.5–7 cm long × 1–2 cm wide, elliptic, sessile below, the upper leaves with a broadly winged petiole to about 2 mm, the base attenuate, the margins serrate, glandular above and below. **INFL** 5–10 mm broad. **FLOWER** calyx about 3 mm long, the lobes attenuate, the tip herbaceous; corolla white, 3–4 mm long; stamens scarcely exserted. **FRUIT** exceeding the calyx tube, corky margin at apex extending along sides to base. $2n = 22$. Riparian or other moist to wet areas; desert shrub and piñon-juniper communities. COLO: LPl; NMEX: SJn. 1550–1800 m (4900–5600′). Flowering: Jun–Sep. Throughout the western United States and east to New York; western Canada.

Marrubium L. Horehound

(Hebrew *marrob*, bitter juice) Biennial or perennial taprooted herbs, often tomentose, aromatic. **STEMS** usually erect and several to many from crown. **LEAVES** mostly petiolate; blades broadly deltoid, ovate, or orbicular; margins toothed. **INFL** in axillary verticils; bracts leafy. **FLOWERS** subsessile; calyx zygomorphic to actinomorphic; teeth 5 or 10, their apices indurate, curled in some; corolla zygomorphic, 2-lipped; upper lip flat, 2-lobed; lower lip 3-lobed; tube shorter than calyx; stamens 4, included; anther sacs divaricate, none abortive; ovary sessile, lobed to base; stigma evenly 2-lobed. **FRUIT** glabrous. Thirty species, Eurasian.

Marrubium vulgare L. (common) Common horehound. Plants with glandular and stellate trichomes, appearing tomentose. **STEMS** 14–100 cm tall. **LEAVES** with petioles to 5 cm long; blades strongly rugose, broadly deltoid to orbicular, to 5 cm long and wide, the upper reduced; margins coarsely crenate to dentate. **INFL** in interrupted spikes of dense verticils, 1–3 cm broad; bracts petiolate, reflexed, much exceeding the flowers. **FLOWER** calyx actinomorphic, 3–6 mm long; teeth usually 10, the apices curled, indurate hooks; corolla white, 5–10 mm long. **FRUIT** about 2 mm long, dark brown. $2n = 34, 36$. Disturbed areas, including roadsides and riparian zones; desert shrub, piñon-juniper woodlands, ponderosa pine, Douglas-fir, and aspen communities. ARIZ: Apa, Coc, Nav; COLO: Arc, Dol, LPl, Mon, SMg; NMEX: McK, RAr, San, SJn; UTAH. 1600–2450 m (5000–7500′). Flowering: May–Oct. Naturalized nearly throughout the United States and much of the New World. The pubescent, nearly orbicular and rugose leaves and curled sepals make this a distinctive plant. Originally introduced for medicinal uses and as flavoring for horehound candy. Infusion of plant taken for sore throats by the Navajo. Ramah Navajo uses include: decoction of plant used for stomachache and influenza, strong infusion used for "lightning infection," root used before and after childbirth, and plant taken for influenza.

Mentha L. Mint

(Greek *Minte*, a nymph's name) Perennial rhizomatous herbs, glabrous or pubescent, aromatic. **STEMS** mostly erect from rhizome. **LEAVES** petiolate to subsessile; blades ovate to oblong; margins toothed. **INFL** verticillate, interrupted; bracts leafy. **FLOWERS** subsessile; calyx actinomorphic; teeth 5, their apices acute; corolla weakly zygomorphic, 2-lipped; upper lip flat, 2-lobed; lower lip 3-lobed; tube shorter than calyx; stamens 4, exserted; anther sacs divaricate, none abortive; ovary sessile, lobed to base; stigma evenly 2-lobed. **FRUIT** glabrous. About 30 species in temperate zones worldwide. Sometimes confused with *Lycopus*, from which it differs by having pedicellate flowers, four stamens, and fruits without an elaborated, corky margin. About 25 species, temperate, one circumboreal, the others in Eurasia and Australia.

1. Inflorescence bracts exceeding flower clusters; verticils along the leafy stem .... ***M. arvensis***
1′ Inflorescence bracts inconspicuous; verticils forming a terminal spike.. ***M. spicata***

Mentha arvensis L. (of fields) Wild mint. Plants with short hairs. **STEMS** 10–100 cm tall. **LEAVES** 2.5–8.5 cm long × 1–3 cm wide, petioles 5–12 mm long, lanceolate to elliptic to somewhat ovate, the margins serrate, base attenuate, glandular above and below. **INFL** 1–2 cm diameter, in the axils of leaves, these reduced upward but always exceeding the verticils. **FLOWERS** short-pedicellate; calyx 2–3.5 mm long, purplish in some; corolla 5–6 mm long, pale purple, some with a red or blue tinge; stamens exserted 1–3 mm beyond corolla. **FRUITS** less than 1 mm long. $2n = 24, 54, 72, 90$. [*M. canadensis* L.; *M. borealis* Michx.]. Riparian or other moist areas; piñon-juniper woodlands, ponderosa pine, Douglas-fir, aspen, and spruce-fir communities. ARIZ: Apa, Nav; COLO: Arc, Dol, Hin, LPl, Mon; NMEX: McK, RAr, SJn; UTAH. 1700–3250 m (5300–10000′). Flowering: Jun–Sep. United States except the southeastern states; Canada and

Europe. Kayenta Navajo uses include: plant used as a lotion for swellings, roots used for prenatal snake infection, and used as flavoring with meats or cornmeal mush. Ramah Navajo uses include: cold infusion taken and used as lotion for fever and influenza, cold infusion given to counteract effects of being struck by a whirlwind.

Mentha spicata L. (bearing a spike) Spearmint. Plants essentially glabrous. **STEMS** to 50 cm tall. **LEAVES** to 4.5 cm long, to 1.5 cm wide, sessile, elliptic, the margins toothed; glandular above and below. **INFL** to 7 mm diameter, forming a terminal spike, the bracts inconspicuous. **FLOWERS** short-pedicellate; calyx 1–2 mm long, green; corolla 2–3 mm long, purple, fading to white in age; stamens included or slightly exserted. **FRUIT** less than 1 mm long. $2n = 36, 48$. Riparian or other moist areas in ponderosa pine communities. ARIZ: Apa; COLO: LPl; NMEX: SJn. 2200–2450 m (6800–7600′). Flowering: Jun–Aug. Naturalized in nearly every state and extensively in the New World; European native.

Monarda L. Bee-balm, Horsemint

(for N. B. Monardes, Spanish physician and botanist) Annual or perennial, rhizomatous or taprooted herbs or shrubs, usually pubescent, aromatic. **STEMS** usually erect, simple or branched. **LEAVES** petiolate to sessile; blades linear, oblong, elliptic, or ovate, margins mostly toothed. **INFL** in dense verticils, these single or forming an interrupted spike, bracts numerous, crowded, large, often colored. **FLOWERS** subsessile; calyx actinomorphic; teeth 5, their apices acute to aristate; corolla zygomorphic, 2-lipped; upper lip hooded, 2-lobed; lower lip 3-lobed; tube longer than calyx; stamens 2, under the hood or exserted; anther sacs divaricate, none abortive; ovary sessile, lobed to base; stigma unevenly 2-lobed. **FRUIT** glabrous. Sixteen species in North America. Several species are cultivated as ornamentals.

1. Verticils of flowers 1; corolla greater than 2 cm long; calyx teeth about 1 mm long .... ***M. fistulosa***
1′ Verticils of flowers 2 to several; corolla less than 2 cm long; calyx teeth about 3 mm long .... ***M. pectinata***

Monarda fistulosa L. **CP** (hollow tube, with the ends closed) Bee-balm. Plants rhizomatous. **STEMS** 30–75 cm tall, sparsely branched, sparsely hairy, especially below. **LEAVES** 3–6.5 cm long × 1–3 cm wide, the petioles to 8 mm long, lanceolate to ovate, the apex attenuate, the base more or less rounded or cordate in broader leaves, the margins shallowly serrate, occasionally approaching entire, glandular above and below. **INFL** a single, terminal verticil, 3–8 cm broad including flowers, the bracts several, green or with purple markings, mostly exceeding sepals. **FLOWER** calyx 8–10 mm long, the teeth deltoid, to 1 mm long; corolla medium red-purple, 2.5–4 cm long, the upper lip straight, the lower curved downward, the tube markedly exceeding calyx; stamens long-exserted. Riparian and other moist areas; ponderosa pine and Douglas-fir communities. ARIZ: Apa; COLO: Arc, Dol, Hin, LPl, Mon; NMEX: RAr, SJn; UTAH. 2200–2800 m (6900–8600′). Flowering: May–Aug. Continental United States except California, Florida. Cold infusion of plant used as a wash for headaches by the Navajo. Cold infusion taken and used as a lotion for gunshot or arrow wounds by the Ramah Navajo.

Monarda pectinata Nutt. (comblike) Plains bee-balm. Plants taprooted. **STEMS** 10–50 cm tall, branched mostly from base. **LEAVES** up to 3.5 cm long, to 1 cm wide, the petioles to 7 mm long, lanceolate to oblong, the base attenuate, the margins entire to shallowly serrate, glandular above and below. **INFL** of 2–4 (5) verticils, 1.5–2.5 cm broad including flowers, the bracts with 1–2 pairs, leaflike, exceeding flowers, often reflexed, plus several pairs of reduced bracts. **FLOWER** calyx 6–8 mm long, the teeth aristate, 2–3 mm long; corolla pale red-purple to white, with or without darker and lighter spots on lower lip, about 1.5 cm long, the upper and lower lips curving, the tube slightly exceeding calyx; stamens under upper lip or slightly exserted. $2n = 18, 36$. Riparian and other moist areas in ponderosa pine and Douglas-fir communities. ARIZ: Apa, Nav; COLO: LPl; NMEX: McK, RAr, SJn; UTAH. 1850–2700 m (5800–8300′). Flowering: May–Aug. Colorado and Nebraska south to Arizona, New Mexico, and Texas. Plant used for stomach disease by the Kayenta Navajo. Ramah Navajo uses include: cold infusion taken and used as poultice for headache, cough, or influenza, used in a ceremonial lotion. Leaves ground for seasoning and mixed with sausage by the Acoma and Laguna.

Monardella Benth. Monardella

(little *Monarda*) Perennial taprooted herbs or subshrubs, pubescent, aromatic. **STEMS** usually erect, often several-branched from crown. **LEAVES** petiolate to subsessile; blades linear to elliptic or oblong, margins finely toothed to entire. **INFL** mostly a terminal verticil subtended by several crowded series of mostly leaflike bracts. **FLOWERS** subsessile; calyx actinomorphic; teeth 5, their apices acute; corolla weakly zygomorphic, weakly 2-lipped; upper lip flat, 2-lobed; lower lip 3-lobed; tube longer than calyx; stamens 4, exserted; anther sacs divaricate, none abortive; ovary sessile, lobed to base; stigma unevenly 2-lobed. **FRUIT** glabrous. $x = 21$. Twenty species in western North America.

Monardella odoratissima Benth. (extremely fragrant) Montana monardella. **STEMS** 8–35 cm tall. **LEAVES** 1–2.5 cm long, 0.3–10 mm wide, lanceolate, the petioles to 4 mm long, the base rounded to attenuate, the apex mostly obtuse, glandular beneath. **INFL** a single verticil, 1.5–2.5 cm broad, the bracts 1 pair green and leaflike and 3+ pairs reduced and purple, closely imbricate, broadly ovate to broadly lanceolate. **FLOWER** calyx 5–7 mm long, teeth less than 1 mm; corolla 10–11 mm long, white to pale pink or occasionally purple, the tube slightly exceeding the calyx, lobes of upper lip slightly wider than lower; stamens slightly to obviously exserted. Sandy to clay soils; Gambel's oak, ponderosa pine, and Douglas-fir communities. NMEX: SJn; UTAH. 2900–3350 m (9100–10300′). Flowering: Jul–Sep. Western United States, mostly in mountainous areas.

Nepeta L. Catnip, Catmint

(ancient Latin name for catnip) Mostly perennial taprooted herbs, mostly pubescent, aromatic. **STEMS** usually erect, mostly much-branched. **LEAVES** petiolate to subsessile; blades ovate to deltoid; margins toothed. **INFL** in an axillary verticil, forming an interrupted spike; bracts leafy or reduced; flowers subsessile or short-petiolate. **FLOWER** calyx zygomorphic to nearly actinomorphic; teeth 5, their apices acute; corolla zygomorphic, 2-lipped; upper lip hooded, 2-lobed; lower lip 3-lobed; tube longer or shorter than calyx; stamens 4, under upper lip or slightly exserted; anther sacs divaricate, none abortive; ovary sessile, lobed to base; stigma evenly 2-lobed. **FRUIT** glabrous or tuberculate. About 250 species native to Eurasia. Many species are attractive to and eaten by cats.

Nepeta cataria L. (of a cat) Catnip. Plants white-tomentose. **STEMS** to 100 cm tall, freely branched in age, mound-forming if robust. **LEAVES** with petiole 4–40 mm long; blades to 70 mm long, to 40 mm wide, cordate, ovate, or deltoid; margin coarsely toothed. **INFL** many-flowered but in somewhat loose verticils, the lowest separate, the upper forming a spike; bracts inconspicuous. **FLOWER** calyx 6–7 mm long, the opening oblique; teeth narrowly triangular, attenuate, the 2 lower smaller than the 3 upper; corolla 8–9 mm long, white with purple spots, the lower lip reflexed and markedly longer than the erect upper lip. **FRUIT** 1.5 mm long, dark brown. Riparian or other moist areas in ponderosa pine communities. ARIZ: Apa; COLO: LPl, Mon; NMEX: SJn; UTAH. 1750–2350 m (5400–7300′). Flowering: Jul–Sep. Naturalized throughout the continental United States except Florida.

Poliomintha A. Gray Rosemary-mint

(Greek for a hoary white mint) Shrubs, densely white-pubescent, aromatic. **STEMS** erect to spreading, often much-branched. **LEAVES** subsessile; blades linear to oblong; margins mostly entire. **INFL** axillary, few-flowered; bracts leafy. **FLOWERS** subsessile to short-pedicellate; calyx nearly actinomorphic; teeth 5, their apices acute; corolla zygomorphic, 2-lipped; upper lip flat, 2-lobed; lower lip 3-lobed; tube exceeding calyx; stamens 2, held under upper lip or slightly exserted, anther sacs divaricate, none abortive; ovary sessile, lobed to base; stigma unevenly 2-lobed. **FRUIT** glabrous. Four species in the southwestern United States; adjacent Mexico. This genus is sometimes lumped with *Hedeoma*.

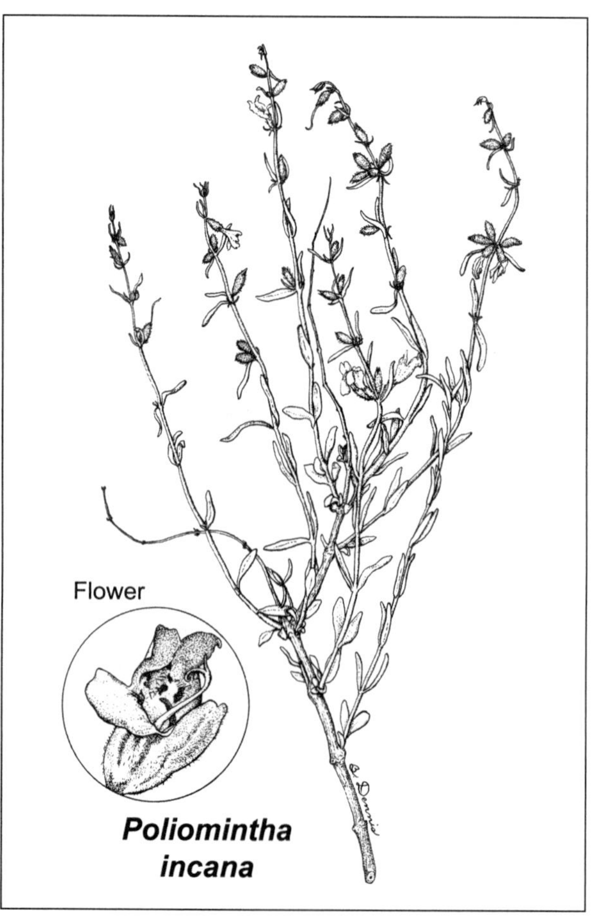

Poliomintha incana

Poliomintha incana A. Gray (grayish white) Purple sage, frosted mint. Plants gray-green. **STEMS** to 40 cm tall. **LEAVES** 5–20 mm long, 1–4 mm wide, ascending, the apex truncate, rounded, or slightly apiculate; glandular above and below. **INFL** of 1–3 flowers. **FLOWER** calyx 4–7 mm long, green or purplish, with dense white hairs about 1 mm long; corolla 8–13 mm long, pale red-purple, the lower lip darker and white-mottled, reflexed and larger than the erect upper lip; stamens with anther sacs slightly separated by enlarged connective, placed end-to-end. [*Hedeoma incana* Torr.]. Sandy soils, often on stabilized dunes; salt desert scrub, blackbrush, rabbitbrush, and piñon-juniper communities. ARIZ: Apa, Nav; COLO: Mon; NMEX: San, SJn; UTAH.

1200–1900 m (3800–5800′). Flowering: Apr–Oct. Arizona, California, Colorado, Texas, Utah. Plant used for sores by the Kayenta Navajo. Plant used for rheumatism and ear trouble by the Hopi.

Prunella L. Self-heal

(early German name of a plant) Perennial herbs, mostly pubescent. **STEMS** erect to prostrate and rooting at nodes. **LEAVES** petiolate; blades oblong; margins entire or weakly toothed. **INFL** verticillate, forming a dense spike; bracts leafy or broad and sheathing. **FLOWERS** short-pedicellate; calyx zygomorphic; teeth 5, their apices cuspidate and acute; corolla zygomorphic, 2-lipped; upper lip hooded, weakly 2-lobed; lower lip 3-lobed; tube exceeding calyx; stamens 4, included in hood; anther sacs strongly divaricate, none abortive; ovary sessile, lobed to base; stigma evenly 2-lobed. **FRUIT** glabrous. Four species, one nearly cosmopolitan, the others native to Europe and Asia.

Prunella vulgaris L. (common) Common self-heal. Plants rhizomatous. **STEMS** 5–40 cm tall, with 0–few branches. **LEAVES**: the lower long-petiolate, the upper subsessile; 2–7 cm long, 1–3 cm wide; blade lanceolate to ovate or oblong; margins entire to irregularly toothed. **INFL** with lowest pair of bracts leafy, the others sessile, reniform, exceeding calyx, often purple-tipped. **FLOWER** calyx 8–9 mm long, upper lip formed from 3 strongly fused lobes, much broader than the partially fused lower 2; corolla 10–15 mm long, purple with white mottling on the lower lip, this reflexed, erose to fringed, the upper lip erect, hooded; stamens with anther sacs placed end-to-end, appearing laterally attached to the branched filament. **FRUIT** 2 mm long, 1 mm wide, dark brown with darker longitudinal lines. $2n = 28, 32$. [*P. parviflora* Gilib.]. Riparian or other moist areas; ponderosa pine, Douglas-fir, aspen, and spruce-fir communities. ARIZ: Apa; COLO: Arc, Hin, LPl, Min, Mon, SJn; NMEX: McK, SJn; UTAH. 2150–3250 m (6700–10100′). Flowering: Jun–Sep. Widespread throughout the United States; circumboreal.

Salvia L. Sage

(to save, as from harm) Annual, biennial, or perennial taprooted herbs or shrubs, pubescent, aromatic. **STEMS** usually erect, often much-branched. **LEAVES** long-petiolate to subsessile; blades ovate to orbicular; margins mostly toothed. **INFL** a few- to many-flowered axillary verticil, in some forming a head or spike; bracts leafy or reduced. **FLOWERS** subsessile to pedicellate; calyx zygomorphic, 2-lipped; teeth 5 or apparently 3, their apices acute; corolla zygomorphic, 2-lipped; upper lip hooded or flat, entire or 2-lobed; lower lip 3-lobed; tube usually longer than calyx; stamens 2, included in hood or exserted; anther sacs separated by strongly developed connective, the filament appearing forked or jointed, 1 sac abortive; ovary sessile, lobed to base; stigma evenly 2-lobed. **FRUIT** glabrous. At least 700 species, worldwide. Seeds are often a part of traditional diets. Many *Salvia* species used as herbs, especially *S. officinalis*, culinary sage, and many also are cultivated.

1. Leaves primarily basal, typically exceeding 10 cm long; plants tomentose .... ***S. aethiopis***
1′ Leaves primarily cauline, rarely exceeding 4 cm long; plants sparsely hairy .... ***S. reflexa***

Salvia aethiopis L. (of Ethiopia) Mediterranean sage. Tomentose biennials. **STEMS** to 60 cm tall, much-branched above, the mature plant often globose. **LEAVES** with petioles to 12 cm long, mostly in a basal rosette, the cauline ones mostly reduced to bracts, blade to 25 cm long, 16 cm wide, the margin coarsely incised or irregularly toothed. **INFL** a panicle with axillary, few-flowered verticils, the bracts broadly ovate to orbicular, spine-tipped. **FLOWER** calyx 10–15 mm long, the lobes spine-tipped, lower 2 longer and broader than upper; corolla 10–20 cm long, yellow, the upper lip strongly arched and mostly including the stamens and stigma. **FRUIT** 2–3 mm long. $2n = 24$. Sandy disturbed soils; piñon-juniper communities. Known from one collection in Mesa Verde National Park. COLO: Mon. 2375 m (7800′). Flowering: Jun–Sep. Arizona, California, Colorado, Idaho, Oregon, Utah, Washington. This is a naturalized invasive, considered a noxious weed in several states, that is sometimes cultivated as an ornamental.

Salvia reflexa Hornem. (bent abruptly backward) Lanceleaf sage. Sparsely hairy perennials. **STEMS** to 50 cm tall, freely branched. **LEAVES** with petioles to 15 mm long, cauline, blade 10–40 mm long, 3–10 mm wide, the margin entire to crenate or serrate. **INFL** a diffuse spike, the flowers solitary in the axils, the bracts strongly reduced, inconspicuous, not spinose. **FLOWER** calyx 4–8 mm long, the teeth not spinose, lower 2 narrower than the apparently single upper one; corolla 6–9 mm long, the lobes pale blue, the tube white, the upper lip hooded and including the stamens, the style slightly exserted. **FRUITS** 2–2.5 mm long, 1.5 mm wide. [*Salvia lanceolata* Brouss.]. Sandy or clay soils; piñon-juniper woodland and ponderosa pine communities. ARIZ: Apa, Nav; NMEX: McK, RAr, SJn. 1650–2300 m (5100–7100′). Flowering: Aug–Sep. Nearly throughout the United States, except the extreme northwest, northeast, and southeast.

Satureja L. Savory

(Latin, of uncertain meaning) Annual or perennial rhizomatous herbs or shrubs, glabrous or pubescent, aromatic. **STEMS** erect, single to several, sparsely or freely branched. **LEAVES** petiolate or sessile; blades linear, lanceolate, ovate, or oblanceolate; margins entire or toothed. **INFL** axillary, few-flowered or in dense verticils; bracts leafy or reduced. **FLOWERS** short-pedicellate; calyx mostly zygomorphic; teeth 5; apices acute to acuminate; corolla zygomorphic, 2-lipped; upper lip flat or cupped; lower lip 3-lobed; tube usually exceeding calyx; stamens 4, held under upper lip; anther sacs divergent, none abortive; ovary sessile, lobed to base; stigma evenly 2-lobed. **FRUIT** glabrous, smooth to reticulate. One hundred fifty species, almost cosmopolitan but mostly Mediterranean. Summer savory (*S. hortensis*) cultivated as an herb.

Satureja vulgaris L. Fritsch (common) Wild basil. Perennial herbs from rhizomes. **STEMS** to 45 cm tall, simple or sparingly branched. **LEAVES** with petioles to 6 mm long; blades ovate to broadly lanceolate, 1–3 cm long, 1–2.5 cm wide; margins entire to toothed; apex obtuse. **INFL** of 1–2 dense verticils with leafy bracts, the floral bracts setaceous. **FLOWER** calyx 5–8 mm long, 2-lipped, the 3 upper teeth shorter and broader than the 2 lower ones; corolla 10–15 mm long, pale red-purple. **FRUIT** 1–1.5 mm long, brown, epidermis smooth. [*Clinopodium vulgare* L.]. Sandy soils; ponderosa pine, Douglas-fir, aspen, and spruce-fir communities. COLO: LPl, Min. 2550–2700 m (8000–8300′). Flowering: Aug–Sep. Northeast to north-central United States and scattered at higher elevations in the arid West. Placed in *Calamintha* by some authors.

Scutellaria Riv. ex L. Skullcap

(like a tray or shield, referring to the calyx shape) Annual or perennial herbs or subshrubs, often rhizomatous or with a woody caudex, pubescent, not aromatic. **STEMS** erect or ascending, simple or branched. **LEAVES** petiolate, the uppermost petiolate or sessile; blade smooth, reniform, ovate, oblong, orbicular, elliptic, or lanceolate; margin serrate, crenate, or entire. **INFL** racemose to paniculate; flowers 1 per axil, on a short branch and pedicellate; bracts leafy or reduced. **FLOWER** calyx strongly zygomorphic, 2-lipped; lips entire; upper with a dome or transverse, arcuate crest or ridge; in fruit markedly enlarging and splitting longitudinally; corolla zygomorphic, 2-lipped; upper lip hooded, entire; lower lip 3-lobed; tube exceeding calyx; stamens 4, all fertile, unequal, included in hood or exserted; anther sacs somewhat divaricate, ciliate, 2 stamens with 1 sac abortive; ovary stalked, deeply lobed; stigma evenly 2-lobed. **FRUIT** variously papillate or tuberculate. Two hundred to 300 species in the Northern Hemisphere, mostly in temperate regions. Some species are cultivated.

Scutellaria galericulata L. (with a helmet) Marsh skullcap. Perennial rhizomatous herbs. **STEMS** 17–50 (–75) cm tall. **LEAVES** 14–55 mm long, 6–20 mm wide; lower leaves with petioles to 5 mm long, the upper subsessile; blade lanceolate to ovate or oblong, the base often slightly cordate; margin shallowly serrate to crenate. **FLOWER** calyx 3–4 mm long, often purple above and green below, crest about 0.8 mm high; in fruit the calyx 5–6 mm long, the crest about 3 mm tall; corolla blue and white, 15–18 mm long, the tube long-exserted, the stamens and style included. **FRUIT** about 1.5 mm long, yellowish brown with an olive cast; epidermis papillate. $2n = 30$. Marshy areas along streams and lakes; ponderosa pine, Douglas-fir, aspen, and spruce-fir communities. ARIZ: Apa; COLO: Arc, LPl, SJn; NMEX: SJn. 1950–3050 m (6100–9500′). Flowering: Jul–Aug. United States except the southeastern states; circumboreal.

Stachys L. Betony, Hedge-nettle

(spike) Annual to perennial herbs, taprooted or rhizomatous, usually pubescent and aromatic. **STEMS** mostly erect, variously branched. **LEAVES** petiolate to sessile; blades lanceolate to ovate or oblong; margins toothed, crenate, or entire. **INFL** in axillary verticils, forming an open or crowded spike; bracts leafy; flowers sessile or subsessile. **FLOWER** calyx zygomorphic or actinomorphic; lobes or teeth 5, their apices acute to spinose; corolla zygomorphic, 2-lipped; upper lip concave or hooded, entire or notched; lower lip usually 3-lobed; tube longer or shorter than calyx; stamens 4, held under upper lip or exserted; anther sacs divaricate, none abortive; ovary sessile, deeply lobed; stigma evenly 2-lobed. **FRUIT** glabrous or roughened. About 300 species, cosmopolitan except Australia, mostly in temperate areas (Epling 1934). Several of the large-flowered species are cultivated ornamentals.

1. Corolla bright red, conspicuously exceeding the calyx; petioles mostly exceeding 10 mm long .... ***S. coccinea***
1′ Corolla pink to purple or brownish orange, only slightly exceeding the calyx; leaves sessile or the petioles less than 7 mm in length .... (2)

2. Corolla pink to purple with darker markings; plants green .... ***S. palustris***
2′ Corolla brownish orange, the markings purplish; plants gray-green .... ***S. rothrockii***

Stachys coccinea Ortega (scarlet) Scarlet hedge-nettle. Perennial, rhizomatous herbs, green; hairs sparse, spreading. **STEMS** 30–80 cm tall. **LEAVES** 2–8 cm long, to 4 cm wide, petioles 1–7 cm long, reduced in upper leaves; blades elliptic to triangular-ovate; margins coarsely serrate or crenate; bases cordate to truncate, attenuate in some. **FLOWER** calyx nearly actinomorphic, 5–10 mm long; corolla scarlet, or in some with an orangish tint, 2–3 cm long, its tube markedly exceeding calyx. **FRUIT** 2 mm long and wide. $2n = 84$. Known from one collection in 1940 by Edward Castetter in Chaco Culture National Historic Park, New Mexico. Canyons and rocky slopes; desert grassland and scattered Utah juniper communities. NMEX: SJn. 1860–1950 m (6100–6400′). Flowering: Jun–Oct. West Texas, New Mexico, and Arizona; Mexico. A very showy wildflower often grown for ornament, with several cultivars available.

Stachys palustris L. (swampy) Marsh hedge-nettle. Perennial, rhizomatous herbs, green; hairs sparse, spreading or appressed. **STEMS** 10–100 cm tall. **LEAVES** to 7 cm long, to 2.5 cm wide, sessile or petioles to 7 mm long; blades broadly triangular-ovate to oblong; margins serrate; bases cordate to truncate. **FLOWER** calyx nearly actinomorphic, 5–10 mm long; corolla pink to red-purple, 1–1.8 cm long, its tube slightly exceeding calyx. **FRUIT** about 2 mm long and wide. $2n = 68, 102$. Riparian or other moist areas, ponderosa pine communities. ARIZ: Apa; COLO: LPl; NMEX: SJn. 1830–2725 m (6000–8945′). Flowering: Jul–Sep. United States except southeastern states; Canada and northern Mexico. This is a widespread and variable circumpolar species. Our material belongs to **subsp**. ***pilosa*** (Nutt.) Epling (hairy), which is considered a full species (*S. pilosa* Nutt.) by some authors. In many specimens the calyx is bicolored, purple above and green below.

Stachys rothrockii A. Gray (for J. Rothrock) Rothrock's hedge-nettle. Perennial, rhizomatous herbs; grayish from appressed pubescence. **STEMS** to 20 cm. **LEAVES** 3–4 cm long, 0.5–1.5 cm wide, subsessile; petioles to 2 mm long; blades oblong to narrowly elliptic; margins finely serrate to entire; bases short-attenuate. **FLOWER** calyx nearly actinomorphic, 5–7 mm long; corolla brownish orange with purple markings, 7–10 mm long, its tube equaling or slightly exceeding calyx. **FRUIT** 1.5–2.5 mm long, 1.5 mm wide. $2n = 34$. Sandy or gravelly soils, grassland, piñon-juniper woodlands, and ponderosa pine communities. ARIZ: Apa, Nav; NMEX: McK, SJn. 2150–2400 m (6700–7400′). Flowering: Jun–Sep. Arizona, New Mexico, and Utah. Ramah Navajo uses include: dried leaves used as a ceremonial medicine and plant used in the chant lotion, used as a foot deodorant, and dried leaves used for "deer infection."

Teucrium L. Germander

(for Teucer, the first king of Troy) Annual or perennial herbs, taprooted or rhizomatous, glabrous or pubescent, aromatic. **STEMS** mostly erect, simple or branched. **LEAVES** petiolate; blades lanceolate, elliptic, or ovate; margins lobed or toothed. **INFL** axillary, few-flowered; bracts leafy or reduced. **FLOWERS** pedicellate; calyx actinomorphic or zygomorphic; teeth or lobes 5, their apices acute to acuminate; corolla zygomorphic, the upper lip deeply lobed and appearing absent, not 2-lipped, 5-lobed; tube mostly shorter than calyx; stamens 4, exserted; anther sacs divaricate, none abortive; ovary sessile, not lobed to base; stigma evenly 2-lobed. **FRUIT** pubescent or glabrous. $x = 5, 8, 13$. About 100 species, cosmopolitan in temperate areas. The herb germander, *T. chamaedrys* L., is often cultivated.

Teucrium canadense L. (of Canada) Canada germander. Perennial from rhizomes, pubescent. **STEMS** 60–150 cm tall, branching above. **LEAVES** elliptic or some lanceolate to narrowly ovate, 4.5–9 cm long, 1–3 cm wide, the margins serrate. **INFL** 1–few flowers per axil, in loose spikes, the lowest with leafy bracts; pedicels 2–4 mm long. **FLOWER** calyx zygomorphic, 5–7 mm long; 2 lower teeth narrower and longer than the 3 upper ones; corolla 8–15 mm long, white with red-purple markings or mostly purplish, the center lobe markedly larger than the 4 lateral ones. **FRUIT** about 2 × 2 mm, brown, glabrous. $2n = 16$. Riparian areas; ponderosa pine, Douglas-fir communities. Known from one collection by Herb Owen in 1964 at Spruce Haven, Colorado. COLO: LPl. 2375 m (7800′). Flowering: Jul–Sep. Continental United States; southern Canada and southern Mexico. This widespread and variable species includes many named variants, none of which are recognized here.

Trichostema L. Blue Curls, Flux Weed, Turpentine Weed

(Greek *thrix*, hairy, + *stema*, stamen) Annuals or perennials, herbs or suffrutescent, taprooted, pubescent, aromatic. **LEAVES** petiolate or subsessile; blades linear, lanceolate, elliptic, ovate, or obovate; margins mostly entire. **INFL** axillary elongate cymes, forming an open panicle; bracts leafy or reduced. **FLOWERS** pedicellate; calyx actinomorphic or somewhat zygomorphic; teeth or lobes 5, their apices acuminate or obtuse; corolla zygomorphic, the upper lip deeply 5-lobed; tube mostly markedly exceeding calyx; stamens 4, often long-exserted and strongly curved; anther sacs slightly to strongly divaricate, none abortive; ovary sessile, lobed in upper 2/3; stigma unevenly 2-lobed. **FRUIT** glabrous or pubescent, often reticulate. $x = 7, 10, 19$. About 16 species, throughout America; southeastern Canada, Mexico.

Trichostema arizonicum A. Gray (of Arizona) Arizona blue curls. Perennial. **STEMS** woody at the base, 20–60 cm tall, bearing short, downward-curling hairs, sometimes glandular. **LEAVES** oblong to ovate, 1–3 cm long, to 12 mm wide, entire or crenate, often glandular, acute or rounded at the base, often subsessile. **FLOWER** calyx actinomorphic, about 2 mm long in flower, to about 6 mm long in fruit, the teeth about as long as the tube; corolla 7–14 mm long, white with blue or violet lower lip, the lower lip 5–10 mm long; stamens strongly arched and longer than corolla, emerging between lobes of upper corolla lip. **FRUIT** about 2.5 mm long, the apex glandular. $2n = 20$. Rocky sites; desert grasslands and scattered Utah juniper communities. Known from one collection in 1940 by Clark in Chaco Culture National Historic Park (Stony Point, New Mexico). NMEX: SJn. 1860–1950 m (6100–6400′). Flowering: Jul–Oct. This plant is much more common in southern New Mexico and Arizona.

LENTIBULARIACEAE Rich. BLADDERWORT FAMILY

C. Barre Hellquist

Perennial or annual herbs; aquatic or terrestrial; many insectivorous. **LEAVES** alternate, cauline or in basal rosettes, simple or dissected. **INFL** with 1–several flowers on an erect scape. **FLOWER** calyx 2-lobed; corolla deeply 2-lipped, the lower 3-lobed, spurred at base; with a conspicuous palate, palate usually branched; stamens 2; ovary free, style short or lacking; stigma 1–2-lipped. **FRUIT** ovoid to globose, 2- or 4-valved capsule. **SEEDS** minute (Crow and Hellquist 1985, 2000). Four genera and about 170 species worldwide.

Utricularia L. Bladderwort

(from *utriculus*, a little bladder) Herbs perennial, aquatic or terrestrial. **STEMS** submersed, creeping, or floating. **LEAVES** dissected, green, usually bearing bladderlike carnivorous traps or these sometimes on separate nonphotosynthetic leaves; bladderlike traps with a valvelike action for trapping microorganisms; producing winter buds in autumn. **INFL** 1–few on erect scapes. **FLOWER** corolla 2-lipped, the lower lip 3-lobed with conspicuous palate, upper lip erect, usually entire; stamens 2; ovary superior. **FRUIT** a 2-valved capsule (Lloyd 1935; Taylor 1989). About 200 species worldwide.

Utricularia macrorhiza J. Le Conte (large-rooted) Common bladderwort. Herbs, perennial, submersed aquatic, floating just below surface. **STEMS** elongate, 1–3 m, 0.5 mm or more in thickness, plumose branches to 12 cm in diameter; the terminal bud becoming a winter bud. **LEAVES** elliptic to ovate, 14–20 mm long, much dissected, with capillary segment bearing bladderlike traps; scape erect, 1–8 dm tall, emergent, bearing 5–20 yellow flowers. **FLOWERS** with corolla 12–18 mm broad, the lips closed, the lower lip slightly longer than the upper one and more or less 3-lobed, palate conspicuous, spur conical. **FRUIT** pedicel reflexed. **SEEDS** brown, shiny, striate-reticulate. [*U. vulgaris* L.; *U. vulgaris* var. *americana* A. Gray; *U. vulgaris* subsp. *macrorhiza* (J. Le Conte) R. T. Clausen]. Ponds, lakes, streams, and ditches. ARIZ: Apa; COLO: Arc, LPl, SJn; NMEX: McK, SJn. 2450–2850 m (8150–9450′). Flowering and fruiting: Jun–Aug. Alaska east to Newfoundland, south to California, New Mexico, Texas, Ohio, and Virginia; Mexico. Taylor (1989) considers the North American *U. macrorhiza* separate from the Eurasian *U. vulgaris* but notes that the differences between the two species appear small, being based mainly on spur shape. The name *macrorhiza* is a misnomer because bladderworts lack roots.

LILIACEAE Juss. LILY FAMILY

William Jennings

(Greek *leirion*, lily) Perennial herbs, sometimes woody. Underground stems scaly or tunicate (fiber- or membranous-coated) bulbs, solid corms, or short or elongate rhizomes. **STEMS** simple or branched, leafy or scapose. **LEAVES** basal or cauline or both, usually alternate, sometimes opposite or whorled, usually narrow, even grasslike, but sometimes broad, usually parallel-veined. **INFL** of various forms (racemose, paniculate, spicate, cymose, umbelliform, or single or paired in leaf axils). **FLOWERS** perfect to polygamous or imperfect, small to large, often showy individually or as an inflorescence; tepals usually 6 (4 or 8 in some genera), distinct or fused into a tube, frequently all 6 alike, but in some genera differentiated into petals and sepals; stamens usually 6, or otherwise in some genera; filaments slender to dilated or united into a crown or an expanded winglike base, anthers of various forms, most often cordate, oblong, linear, or lanceolate and extrose, versatile, or introrse; ovary superior or partly inferior, usually 3-celled with 1 to numerous ovules in each locule; styles 1 to 3, terminated by a capitate or lobed stigma. **FRUIT** a loculicidal or septicidal capsule or a fleshy berry. **SEEDS** 1 to many, with abundant endosperm.

As broadly recognized here, a family of many diverse forms, widely distributed in temperate, subtropical, and tropical regions, up to 280 genera and 4200 species. It seems inevitable that a broadly defined lily family eventually will be broken into several segregate families. Many proposals have been advanced over the years, with more in recent years as molecular and genetic research has progressed. *Flora of North America* indicates that the 70 genera in North America north of Mexico could be divided among as many as 20 families, leaving only *Lilium*, *Tulipa*, *Erythronium*, *Lloydia*, and *Fritillaria* in Liliaceae. The problem is how to do it. As yet, a consensus has not emerged. In the *Flora of the Four Corners*, Agavaceae and Asparagaceae are recognized, with the other traditional liliaceous genera left in the family. Editors' note: If the genera here assigned to Liliaceae were to be assigned to the maximum number of segregate families proposed, their disposition would be as follows (Flora of North America Editorial Committee 2002): Alliaceae (*Allium*); Anthericaceae (*Echeandia*, *Eremocrinum*, *Leucocrinum*); Calochortaceae (*Calochortus*); Convallariaceae (*Maianthemum*); Melanthiaceae (*Veratrum*, *Zigadenus*); Liliaceae s. s. (*Erythronium*, *Fritillaria*, *Lilium*, *Lloydia*); Themidaceae (*Androstephium*); and Tricyrtidaceae (*Prosartes*, *Streptopus*). Most authors do, however, continue to include *Calochortus* in Liliaceae s. s. (e.g., Judd et al. 2008; Simpson 2010). Further, recent research has shown that species here placed in *Zigadenus* are better assigned to *Anticlea* (*A. elegans* and *A. vaginatus*) and *Toxicoscordion* (*T. paniculatum* and *T. venenosum*; Zomlefer and Judd 2002). We have noted in the key below where segregate families appear.

1. Flowers several to many in a terminal, umbel-like inflorescence on a leafless scape, subtended by spathelike bracts ....(2)
1′ Flowers solitary or racemose or paniculate; not in a scapose umbel subtended by spathelike bracts ....(3)

2. Tepals distinct or nearly so; plants with an onion odor and taste [Alliaceae—eds.] ....***Allium***
2′ Tepals united into a basal tube; filaments united into a tubular corona [Themidaceae—eds.] ....***Androstephium***

3. Plants from rhizomes or thickened underground stems; not bulbous ....(4)
3′ Plants from tunicate (coated) or scaly bulbs, or from corms; not from rhizomes ....(10)

4. Leaves all basal or the basal leaves much larger than the reduced cauline leaves [Anthericaceae—eds.] ....(5)
4′ Leaves well distributed on the stem, not all basal ....(7)

5. Plants acaulescent, inflorescence a sessile cluster at ground level atop a tuft of leaves; perianth segments united into a tube; plants of piñon-juniper, sagebrush, and ponderosa pine communities ....***Leucocrinum***
5′ Plants caulescent, inflorescence atop a long, slender scape or stem ....(6)

6. Anthers dorsifixed near the base, tepals reflexed ....***Echeandia***
6′ Anthers basifixed, tepals not reflexed but spreading distally ....***Eremocrinum***

7. Inflorescence a large, densely flowered panicle 20–80 cm long; plants up to 2 m tall; fruit a capsule [Melanthiaceae in part—eds.] ....***Veratrum***
7′ Inflorescence a raceme or panicle less than 20 cm long, of few to many flowers or solitary; fruits ovoid or globose berries ....(8)

8. Flowers in terminal panicles or racemes [Convallariaceae—eds.] ....***Maianthemum***
8′ Flowers solitary or at most 2 or 3, flowers pendulous [Tricyrtidaceae—eds.] ....(9)

9. Flowers terminal, 1 to 3, on stout, pendulous, pubescent pedicels ....***Prosartes***
9′ Flowers on the back side of the leaf, on slender, geniculate (with a joint like a knee), twisted, glabrous pedicels; pendulous ....***Streptopus***

10. Inflorescence a densely flowered panicle or raceme; flowers small, no more than about 10 mm long [Melanthiaceae in part—eds.] ....***Zigadenus***
10′ Flowers solitary or few, if few, in an open raceme; flowers 10 mm long or more [Liliaceae s. s.—eds.] ....(11)

11. Flowers 2–5, nodding, yellow, the tepals spreading and becoming reflexed ....***Erythronium***
11′ Flowers erect or nodding, but if nodding, the flowers brownish or purplish and the perianth parts not reflexed ....(12)

12. Flowers erect, petals and sepals dissimilar, with a large, conspicuous gland (about 5 mm) at the adaxial base of the petal ....***Calochortus***
12′ Flowers erect or nodding, petals and sepals (tepals) in 2 very similar series; a gland, if present, small and indistinct ....(13)

13. Flowers erect, small, white with greenish or purplish veins; leaves basal and cauline; alpine ....***Lloydia***
13′ Flowers erect or nodding, red, orange, purple, or brownish; leaves various; not alpine ....(14)

14. Flowers large, erect, red or orange; uppermost cauline leaves whorled .... ***Lilium***
14′ Flowers small, nodding, brownish or purplish; cauline leaves all alternate .... ***Fritillaria***

Allium L. Wild Onion, Wild Garlic

(the classical name for garlic) Perennial herbs, with an onion or garliclike odor, scapose from a bulb, or terminated on primary or secondary rhizomes. **LEAVES** sheathing, linear, terete to flat, solid or hollow, soon withering and deciduous in some species. **INFL** scapose, terminal, umbelliform; bracts free or partly connate, usually dry and membranous, persistent, 1–several-nerved. **FLOWERS** perfect, perianth parts 6 (tepals), all alike, white, pink, rose, or white with a colored midvein, usually persistent in fruit; or some flowers replaced by bulblets; stamens 6, more or less connate at the base and adnate to the perianth base; ovary 3-celled, 3-lobed, and crested in some species; style 1, capitate to trifid. **FRUIT** a capsule, loculicidal. **SEEDS** usually 6 or fewer, black, alveolate. Widely distributed in the Northern Hemisphere in a variety of habitats, about 500 species (Ownbey 1947; Ownbey and Aase 1955). The important diagnostic features of *Allium* are: bulb coat, bulb shape, or the presence of rootstocks and/or rhizomes; relative length of the scape and leaves; color and shape of the tepals (or if the flowers are replaced by bulblets); and characteristics of the ovary and capsule (presence or absence of crests). When collecting onion specimens, carefully dig a large hole to obtain the bulb, and *gently* break apart the dirt clod to make sure any rhizomes are collected (see *A. bisceptrum* var. *palmeri*). Do not pull the onion from the ground or outer bulb coats and any rootstocks or rhizomes will be left behind. Press the specimen, if possible, between foam sheets; do not mash the bulb flat in a regular press. This makes for a bulky specimen but preserves the diagnostic features.

1. Scape more or less recurved at the apex, umbel at least partially nodding; flowers pale lavender; stamens exserted beyond the perianth .... ***A. cernuum***
1′ Scape straight (more or less erect); umbel erect, not nodding; stamens included .... (2)

2. Bulb elongate and little more than a slightly swelled base of the stem and terminating a short, stout, irislike rhizome; leaves flat and greater than 5 mm wide .... ***A. gooddingii***
2′ Bulb traditional onion shape (more or less spherical or ovoid); leaves narrow, less than 5 mm wide .... (3)

3. Ovary with 6 conspicuous "crests" (like miniature triangular sails; use strong hand lens) .... (4)
3′ Ovary lacking "crests" (top of ovary may have knobs or be rounded or flat, but not with triangular sail-like prominences; *A. textile* has rather prominent knobs on top of the ovary, but in this key it goes with the crestless group) .... (6)

4. Plants with 1 leaf, this much exceeding the short (3–5 cm) scape, often dried up and coiled at the tip ("curlicued" like a pig's tail); Utah .... ***A. nevadense***
4′ Plants with 2 or more leaves, these either shorter or longer than the scape, green at anthesis, not coiled; widespread throughout the Four Corners region .... (5)

5. Crests large, about the same length as the ovary beneath; leaves 2 or more, shorter than the scape; bulb with a reticulate coat; Utah and Arizona .... ***A. bisceptrum***
5′ Crests small, shorter than the ovary beneath; leaves 2, exceeding the scape; bulb with a coarsely fibrous coat; widespread in our area .... ***A. macropetalum***

6. Outer bulb coats thin and membranous to tough and leathery, marked on surface with fine, quadrangular or hexagonal reticulations, but with no coarse fibers present (more or less like commercial onions) .... ***A. acuminatum***
6′ Outer bulb coat traversed by coarse, often interwoven fibers, becoming loose and twinelike or netted (rather like coarse burlap) .... (7)

7. Plants of moist areas in the mountains (primarily montane or subalpine, but sometimes at higher or lower elevations); leaves 3 or more, generally shorter than the scape; flowers pink; tepals with a slightly darker pink midrib .... ***A. geyeri***
7′ Plants of dry areas, at elevations below 2285 m (7500′); leaves 2, rarely 3, about equaling the scape; flowers white; tepals with a red-brown midrib, the midrib color not "bleeding" into the rest of the tepal .... ***A. textile***

Allium acuminatum Hook. (Latin: drawn out at the apex into a tapering point, alluding to the shape of the floral parts) Acuminate onion. **BULBS** ovoid, subglobose, or nearly spherical; outer bulb coat with square or hexagonal reticulations. **LEAVES** 2 to 4, narrow, concavo-convex, shorter than the scape, early-deciduous. **INFL** scape 10–40 cm; bracts 2, lanceolate to ovate, acuminate; umbelliform, flowers erect. **FLOWER** tepals 6, rose or pink, with long, recurved tips, the inner tepals serrate-denticulate on the margins, and shorter than the outer tepals; stamens shorter than the tepals; ovary inconspicuously crested with 3 low, rounded processes; stigma capitate. Dry hillsides and slopes, rocky and bare places, sometimes under ponderosa pine and Gambel's oak. ARIZ: Apa; COLO: Arc, Dol, LPl, Mon; NMEX: RAr, SJn; UTAH. 2065–2655 m (6780–8710′). Flowering: Jun–early Jul. Southern Canada to central California and east to Idaho, Wyoming, western Colorado, northwestern New Mexico, and central Arizona. Bulbs and leaves used by Utes for food.

Allium bisceptrum S. Watson (two scepters, an allusion to the paired crests on the ovary) Palmer's onion. **BULBS** ovoid, commonly producing either a cluster of basal bulblets or rhizomes up to 10 cm long terminated by larger bulblets; some of the outer bulb coats with rectangular reticulations, varying to irregular shapes with sinuous walls. **LEAVES** 2 or more, tapering from a broad base, usually shorter than the scape, green at anthesis. **INFL** scape 10–30 cm; bracts 2, ovate to lanceolate, acuminate, 3–5-nerved; umbelliform, **FLOWER** tepals pale pink to lilac, purplish, or white; stamens shorter than the tepals; ovary conspicuously crested with 6 distinct (3 pairs), erect, papillose-denticulate processes; style included; stigma capitate. **FRUIT** a capsule, the pedicel becoming flexuous or deflexed in fruit. [*A. palmeri* S. Watson]. Meadows, slopes, and aspen groves. ARIZ: Apa, Nav; UTAH. 1220–2950 m (4000–9685′). Flowering: May–Jul. Four Corners material belongs to **var.** ***palmeri*** (S. Watson) Cronquist (for E. J. Palmer, plant collector of the Southwest and Mexico). Central Nevada, central Utah, southward and southeastward to central Arizona and western New Mexico. Paiutes used the roasted bulbs, seed heads, and leaves for food.

Allium cernuum Roth (nodding, alluding to the flowers that face downward) Nodding onion. **BULBS** elongate, often clustered, sometimes short-rhizomatous; outer bulb coats membranous, often tinged reddish or purplish (rather like an elongated miniature red cultivated onion or scallion); the bulb coats minutely striate with elongate cells in regular rows. **LEAVES** several per bulb, concavo-convex to nearly flat, shorter than the scape. **INFL** scape 10–50 cm tall, abruptly recurved near the apex, the inflorescence nodding; bracts acuminate, breaking into 2 separate 3- to 5-nerved, reflexed portions; campanulate. **FLOWER** tepals lavender, elliptic-ovate, obtuse, entire; stamens exserted; style exserted; ovary crested. **FRUIT** a capsule, the pedicels elongating in fruit, bending upward, so that the capsules are held erect. [*Allium cernuum* Roth var. *obtusum* Cockerell ex J. F. Macbr.]. Gambel's oak, sagebrush, ponderosa pine, Douglas-fir, and aspen communities; mostly in cool, usually shaded spots. ARIZ: Apa, Nav; COLO: Arc, Hin, LPl, Min, Mon, SJn; NMEX: McK, RAr, SJn; UTAH. 2015–2720 m (6610–8920′). Flowering: late Jul–Aug. Widespread in northern North America, southward in the mountains to Mexico. Used by Hopis, Navajos, and Apaches to flavor soups and gravies. The leaves used in salads and bulbs stored for winter use.

Allium geyeri S. Watson (for C. A. Geyer, plant collector on the Oregon Trail) **BULBS** ovoid, often clustered; the outer bulb coats persisting as a coarse, open-fibrous, reticulate coat. **LEAVES** usually 3 per bulb, sometimes more; shorter than the scape. **INFL** scape 10–60 cm tall; bract breaking into 2 or 3 acuminate, 1-nerved portions. **FLOWER** tepals pink, occasionally white; or replaced partially or wholly by bulblets; stamens included or about as long as the perianth; style included; ovary with inconspicuous mounds (crests). **FRUIT** a capsule, with the dried tepals investing the capsule. $n = 7$. Apaches and Hopis used the bulbs as flavoring or raw with piki bread.

1. Some or all the flowers replaced by bulblets .... **var. *tenerum***
1′ None of the flowers replaced by bulblets .... (2)

2. Bulb coats many-layered, extending from the base of the bulb nearly to the base of the leaves, loosely wrapping the bulb .... **var. *chatterleyi***
2′ Bulb coats of fewer layers, from the base to the top of the bulb, but not extending far up the stem, more tightly wrapping the bulb .... **var. *geyeri***

var. ***chatterleyi*** S. L. Welsh (for M. Chatterley, associate of S. L. Welsh) Similar to var. *geyeri*, but differing in the many-layered marcescent bulb coats, 4.5–11 cm in length, extending from the base of the bulb to the base of the leaves. UTAH. 2680 m (8800′). The features that purport to separate this from typical *A. geyeri* do not seem to be significant. Specimens seen had large bulbs with thick bulb coats but seemed to be otherwise typical.

var. ***geyeri*** Piñon-juniper, ponderosa pine, aspen, spruce-fir, and alpine communities; mostly in moist sites. ARIZ: Apa; COLO: Arc, Dol, Hin, LPl, Min, Mon, SJn; NMEX: McK, RAr, SJn; UTAH. 2240–3960 m (7355–13000′). Flowering: May–Jul. Mostly in the Rockies from southern Canada to New Mexico and western Texas, and west to portions of Washington, Oregon, Nevada, and Arizona.

var. ***tenerum*** M. E. Jones (tender or delicate) [*Allium rubrum* Osterh.]. Similar to var. *geyeri*, except most of the flowers replaced by bulblets on short pedicels. The flowers that do remain seem to be on pedicels that are overly long, giving the inflorescence an unkempt appearance. Wet meadows, mostly in the subalpine zone. COLO: Arc, Hin, LPl, RGr, SJn. 2775–3650 m (9100–11980′). Flowering: Jun–Jul.

Allium gooddingii Ownbey (for Leslie Goodding, plant collector of southern Arizona, who collected the type specimen) Goodding's onion. **BULBS** elongate, terminating a thick, irislike rhizome; inner bulb coats whitish or pinkish; outer bulb coats brownish, membranous, minutely striate with elongate cells in regular rows; not fibrous-reticulate, but with persistent parallel fibers. **LEAVES** several, plane, obtuse, entire, 4–8 mm broad, shorter than the scape, green at anthesis. **INFL** scape 35–45 cm tall; bracts 2, membranous, withering. **FLOWER** tepals elliptic, obtuse, entire, pink, withering in fruit; stamens about as long as the perianth; style capitate; ovary crestless. **FRUIT** a capsule, broader than long. Aspen and spruce-fir communities; usually moist areas. ARIZ: Apa; NMEX: SJn. 2285–2745 m (7500–9000′). Flowering: mid-Jul–mid-Aug. Arizona and New Mexico. Distribution of this species is scattered. Specimens have been seen from areas to the south of the San Juan drainage area (the high country of the White Mountains area, Arizona; the Mogollon Mountains and the Sierra Blanca areas, New Mexico).

Allium macropetalum Rydb. (big petals, alluding to the size of the flowers, which are a little larger than most other western onions) **BULBS** ovoid, usually clustered, outer coats coarse, fibrous, persistent. **LEAVES** usually 2, surpassing the scape. **INFL** scape 5–20 cm tall; bract membranous, breaking into 2 or 3 separate 3- to 5-nerved portions; perianth campanulate. **FLOWER** tepals usually white or pinkish with a dark red midrib, the color of which "bleeds" away from the midrib, lanceolate, obtuse to acuminate, entire; stamens shorter than the perianth; stigma capitate; ovary crested. **FRUIT** a capsule, the crests on the ovary best developed in fruit; tepals persistent and papery. [*Allium reticulatum* J. Presl & C. Presl var. *deserticola* M. E. Jones; *Allium deserticola* (M. E. Jones) Wooton & Standl.]. Shadscale and other salt desert shrub, and piñon-juniper communities. ARIZ: Apa, Nav; COLO: LPl, Mon; NMEX: McK, RAr, San, SJn; UTAH. 1340–2405 m (4400–7890′). Flowering: mid-Apr–early Jun. Southeastern quarter of Utah, west-central and southwestern Colorado, eastern half of Arizona, western half of New Mexico, western Texas; Mexico. *Allium macropetalum* and *A. textile* are very similar in many ways, and the two species come in contact in a zone along the Colorado-Utah and Colorado-New Mexico state lines. *Allium macropetalum* has leaves surpassing the scape, a 3- to 5-nerved bract, the color of the midrib diffusing into the background color of the tepal, and crests on the ovary. *Allium textile* has a scape surpassing or about equaling the leaves, a 1-nerved bract, a sharp demarcation between the color of the midrib and the rest of the tepal, and only low mounds on the ovary. In close cases, the deciding factor is the number of nerves on the bract.

Allium nevadense S. Watson (of Nevada, where it was first collected) **BULBS** ovoid, sometimes proliferating by stalked basal bulblets; outer bulb coats with contorted cellular reticulations. **LEAVES** 1, at least twice as long as the scape, the proximal portion of the leaf usually green at anthesis, the distal portion coiled or frequently dried up and broken off. **INFL** scape short, 3–5 cm tall; bracts 2 or 3, ovate to lanceolate, acuminate, spreading or reflexed, 3- to 7-nerved. **FLOWER** tepals whitish with a pale green midrib or pinkish with deep pink midrib; stamens shorter than the tepals; stigma almost as long as the stamens, capitate; ovary crested with 6 distinct (3 pairs), thin, entire or toothed processes. **FRUIT** a capsule, with the persistent papery tepals present. Sandy, gravelly, or clayey soils in desert regions. UTAH. 1635 m (5360′). Flowering: Apr–Jun. Southern Idaho and southeastern Oregon, south through extreme western Colorado (Moffat and Mesa counties only), much of Utah, and Nevada, to southeastern California and northwestern Arizona. The very short scape, the single long leaf, typically with the tip coiled, and the very conspicuous crests characterize this species. Four Corners material belongs to **var. *nevadense***.

Allium textile A. Nelson & J. F. Macbr. (woven, alluding to the fibrous bulb coats, which look like burlap) **BULBS** ovoid, usually in a cluster; inner bulb coats whitish, with epidermal cells vertically elongate and regular; outer bulb coats persisting as a fibrous, fine-meshed, open reticulum. **LEAVES** usually 2 per bulb, about equaling the scape, green at anthesis. **INFL** scape 6–30 cm tall; bracts 3, ovate, acuminate, 1-nerved; tepals white, with a reddish brown midrib; stamens shorter than the perianth; stigma capitate; ovary with obscure or obvious mounds, but not the sail-like processes of the conspicuously crested species. **FRUIT** a capsule, invested by the callous-keeled tepals. [*Allium reticulatum* Nutt., not J. Presl & C. Presl; *A. aridum* Rydb.]. Shadscale, mat-atriplex, greasewood, sagebrush, and piñon-juniper communities. COLO: Arc, LPl, Mon; NMEX: RAr, San, SJn. 1865–2040 m (6115–6700′). Flowering: May–Jun. The northern Great Plains, westward into Montana, Idaho, and northern Utah, south to western Nebraska, Colorado, and northern New Mexico. The northern edge of the San Juan River drainage in Colorado and New Mexico is the southern limit of the range of this species.

Androstephium Torr. Funnel-lily

(Greek *andros*, stamens, + *stephanos*, crown, referring to the united filaments) Erect, glabrous, scapose, perennial herbs. **CORMS** ovoid, fibrous-coated. **LEAVES** usually several, linear, channeled. **INFL** scape 10–30 cm, inflorescence umbellate, subtended by 3 bracts, lanceolate to broadly lanceolate. **FLOWER** tepals 6, all alike, joined below to form a funnelform tube, about 1/2 the length of the tepals, narrowly oblong, spreading; stamens 6, the filaments partly united into a tube with erect, bifid lobes between the introrse, basifixed, mostly acute anthers; style slender, persistent, terminated by 3 small stigmas; ovary superior, 3-celled, sessile. **FRUIT** a capsule, subglobose, obtusely 3-angled. **SEEDS** black, flat. Southwestern United States and northern Mexico. Two or three species.

Androstephium breviflorum S. Watson (short flower) Erect perennial herb. **CORMS** ovoid, fibrous-coated. **LEAVES** usually several, but occasionally just 1, equaling or exceeding the scape. **INFL** a scape, 10–30 cm; bracts lanceolate or broadly lanceolate, 8–15 mm long; flowers on ascending pedicels, 1–3 cm long. **FLOWERS** usually light violet-purple to whitish, with a diffuse grayish or brownish stripe down the center of the tepal; tepals united to form a tube; stamens 6, the filaments united and longer than the anthers, the lobes of the crown shorter than the anthers. **FRUIT** a capsule, 3-lobed, 10–15 mm long. **SEEDS** black, 7–9 mm long, flat, 4 to 6 per capsule. [*Brodiaea breviflora* (S. Watson) J. F. Macbr.; *Brodiaea paysonii* A. Nelson]. Blackbrush, shadscale, mat-atriplex, sagebrush, and piñon-juniper communities; usually in dry, sandy places, to rocky soil, often saline and seleniferous. ARIZ: Apa, Nav, Coc; COLO: Arc, LPl, Mon; NMEX: McK, SJn; UTAH. 1220–2135 m (4000–7000′). Flowering: mid-April–May. Southwestern Wyoming, western Colorado, northwestern New Mexico, west across Utah and northern Arizona, and Nevada. Reported for California. Occurrences in the Four Corners region represent the southeastern limit of the species range.

Calochortus Pursh Sego Lily, Mariposa

(beautiful grass) Perennial herbs. **BULBS** deep-seated, tunicate, membranous- or fibrous-coated. **STEMS** simple or branched, erect or somewhat twining, frequently bulbiferous at about ground level or from the lower leaf axils. **LEAVES** few, linear, alternate, the basal one solitary and larger, the cauline ones reduced up the stem. **INFL** solitary or a few, umbelliform, erect or nodding. **FLOWERS** large, showy, white, yellow, red, lavender, or purple and usually tinged with other colors; the sepals and petals distinct and different, the sepals ovate to lanceolate, greenish, brownish, or colored the same as the petals; petals larger and broader, cuneate to clawed, hairy in some species, usually with a depressed gland near the base, which is often spotted, colored, and hairy; stamens 6, in 2 series; ovary superior, 3-celled, contracting or tapering to a trifid, sessile, persistent stigma. **FRUIT** a capsule, linear to orbicular, 3-angled or winged, septicidal, erect or nodding. **SEEDS** numerous, irregular or flattened (Ownbey 1940). Western North America from southern Canada to Central America. About 60 to 70 species, best developed in California, where there are numerous species of different forms (43 species in *The Jepson Manual*). In *Flora of North America*, 56 species are treated; we have 4 in our area. The features of the gland (near the base of the petal) are very important for identification of *Calochortus*. When collecting, be sure that the flower is pressed wide open so the gland can be examined, and attach a note on the label whether the gland is depressed or not (look on the back of the petal for a "bump"). Bulbs tend to be buried deeply and pulling will not yield the bulb; it has to be dug up.

1. Anthers acute to strongly apiculate (a point on the tip); glands on petals oblong, depressed; hairs on faces of petals usually branched, densely hairy on petals above gland; purplish band on petals above gland and purple dot on claw (below gland) .... ***C. gunnisonii***

1′ Anthers obtuse; glands on petals circular to lunate (crescent moon shape), depressed or not; hairs on faces of petals seldom branched, sparsely hairy on petals above gland; marked either above or below gland on petal but not both places .... (2)

2. Stems flexuous (zigzag, not usually fully erect, sometimes twining), not bulbous near the base of the stem; glands on petals not depressed and not surrounded by membranes; flowers a soft lavender above, yellow near the gland, and with purple spot on claw below gland .... ***C. flexuosus***

2′ Stems straight, erect or nearly so, usually bulbous near the base of the stem; glands on petals more or less depressed (look on back of petal for a "bump"), surrounded by membranes; flower color various .... (3)

3. Petals and sepals both a clear, uniform, golden yellow; with a red-brown or maroon band above gland on the petal and a similar mark on the sepal .... ***C. aureus***

3′ Petals variously colored, but most often white, or white suffused with pink, fully pink, or with pale yellowish highlights, and always with a yellow claw and a red-brown or maroon band above the gland; sepals most often whitish with brownish lengthwise striations and red-brown band and yellow base, but may be the same color as the petal .... ***C.nuttallii***

Calochortus aureus S. Watson **CP** (golden yellow) Golden mariposa. Perennial herb, erect. **STEMS** usually simple, 10–30 cm tall. **LEAVES** few, basal leaf largest, up to 15 cm long. **INFL** campanulate. **FLOWER** petals and sepals golden yellow, with a maroon band above the gland on the petal, the sepal similarly marked; sepals broadly lanceolate, acuminate; petals broadly obovate, cuneate, short-acuminate, with a few hairs near the gland; gland circular, depressed, surrounded by a conspicuosly fringed membrane, and covered with short, simple or distally branched prosesses; anthers oblong, obtuse, yellowish to cream, about equaling the length of the filaments; ovary linear, not winged. **FRUIT** a capsule, linear-lanceolate, acuminate, 3-angled, 2.5–5 cm long. [*Calochortus nuttallii* Torr. var. *aureus* (S. Watson) Ownbey]. Blackbrush, shadscale, and piñon-juniper communities; often in dry sandy, clayey, or rocky places. ARIZ: Apa, Coc, Nav; NMEX: McK, SJn; UTAH. 1740–2255 m (5700–7400′). Flowering: mid-May–mid-Jun. Southern Utah, northern Arizona, northwestern New Mexico. This is our only species of sego lily with bright yellow flowers. Plant used in the Hopi flute ceremony. Bulbs, flowers, and root eaten raw by Navajos and Hopis. Bulb is a Navajo life medicine.

Calochortus flexuosus S. Watson (bent alternately in different directions; zigzag; alluding to the twisted nature of the stem) Sinuous mariposa. Perennial, erect or decumbent or twining among other plants or straggling along the ground. **STEMS** usually branched, 20–40 cm long. **LEAVES** linear, sparse. **INFL** campanulate, erect. **FLOWERS** soft lavender or lilac-tinged, each petal with a yellow band at the gland and a purple spot on the claw; sepals lanceolate to obtuse, shorter than the petals, colored similarly to petals; petals obovate, cuneate, rounded to obtuse at the apex, glands transverse-lunate to nearly circular, not depressed and not covered with a membrane, but densely covered by short processes; anthers oblong, about as long as the filaments; ovary linear, not winged. **FRUIT** a capsule, lanceolate, acute, 3-angled, 2.5–3.5 cm long, erect. Blackbrush, shadscale, and piñon-juniper communities; often on dry stony slopes, rocky mesas, and flats. ARIZ: Apa, Nav; COLO: Arc, Mon; NMEX: SJn; UTAH. 1190–1765 m (3900–5800′). Flowering: mid-Apr–mid-May. Southeastern California, southern Utah, southwestern Colorado, northwestern New Mexico, northern and central Arizona.

Calochortus gunnisonii S. Watson (for John W. Gunnison, leader of the Gunnison Expedition) Gunnison's mariposa lily. Perennial, erect. **STEMS** usually simple, 30–60 cm tall. **LEAVES** linear, becoming slightly reduced above, not exceeding the inflorescence. **INFL** campanulate, erect. **FLOWERS** usually white, but plants with purple or yellow flowers are known, usually with a transverse purple band above the gland and a purple spot on the claw; sepals lanceolate to broadly lanceolate, acute, shorter than the petals, colored similarly to petals; petals obovate, cuneate, obtuse to rounded at the apex, with numerous distally branched, gland-tipped hairs on the claw and near the gland; glands transversely oblong, depressed, surrounded by a deeply fringed but indistinct membrane, and densely covered with short, distally branched processes; anthers acute to strongly apiculate, yellowish to cream, longer than the filaments; ovary linear, not winged. **FRUIT** a capsule, linear-oblong, acute, 3-angled, 3–5 cm long, erect. Sagebrush, piñon-juniper, and aspen communities; dry to moist, often grassy, slopes. ARIZ: Apa, Nav; COLO: Arc, Dol, Hin, LPl, Mon, SMg; NMEX: McK, RAr, SJn; UTAH. 2115–2775 m (6940–9110′). Flowering: mid-Jun–mid-July. Montana, Wyoming, South Dakota (Black Hills), Colorado, eastern Utah, northeastern Arizona, northern New Mexico. In northeastern Arizona, *C. gunnisonii* replaces *C. ambiguus* and their ranges do not overlap. Plant used as a ceremonial medicine by Navajos. Also a "life medicine" and used to deliver placentas. Bulbs eaten as food.

Calochortus nuttallii Torr. (for Thomas Nuttall of Harvard University, botanist of the early 19th century) Sego lily. Perennial, erect. **STEMS** usually simple, 20–40 cm tall, often bulbiferous at the base. **LEAVES** few, linear, reduced above, seldom exceeding the inflorescence. **INFL** campanulate, erect. **FLOWERS** most often white, but can be white suffused with pink, fully pink, or magenta, or with pale yellowish highlights, and always with a yellow claw and a red-brown or maroon band above the gland; sepals lanceolate to broadly lanceolate, acuminate, shorter than the petals, often whitish with brownish lengthwise striations and a red-brown band and yellow base, but may be the same color as the petals; petals broadly obovate, cuneate, short-acuminate, 3–4 cm long, with scattered yellowish hairs near the gland; gland circular, depressed, surrounded by a conspicuously fringed membrane and densely covered with short, simple or distally branched processes; anthers oblong, yellowish or pinkish, about equaling the length of the filaments; ovary linear, not winged. **FRUIT** a capsule, linear-lanceolate, acuminate, 3-angled, 3–5 cm long, erect. Dry slopes and flats in shadscale, greasewood, mat-atriplex, sagebrush, piñon-juniper, and ponderosa pine communities. [*Calochortus luteus* Nutt., not Douglas; *C. watsonii* M. E. Jones; *C. rhodothecus* Clokey]. ARIZ: Apa, Nav; COLO: Arc, Dol, LPl, Mon, SMg; NMEX: McK, RAr, San, SJn; UTAH. 1510–2560 m (4960–8400′). Flowering: May–Jun. Eastern Montana, western North Dakota, South Dakota (Black Hills), most of Wyoming, southeastern Idaho, eastern Nevada, much of Utah (except where displaced by *C. aureus* in south-central and southeastern Utah), western Colorado, Kaibab Plateau of

Arizona, and northwestern New Mexico. *Calochortus nuttallii* reaches the southeastern limit of its range in the northeastern quarter of the Four Corners. Bulbs eaten by Navajo children. Paiutes used bulbs as food in spring, eaten roasted or raw. Flower used ceremonially by Hopis.

Echeandia Ortega Amber Lily

(for Pedro Echeandia, Spanish botanist) Perennial herbs, from corms with enlarged storage roots. **LEAVES** both basal and cauline; narrowly linear to narrowly oblong or elliptic, with the base surrounded by marcescent fibrous leaf bases. **INFL** scapose, racemose, or paniculate, 1–4 flowers per node. **FLOWERS** perfect or pseudohermaphroditic, staminate, more or less erect to pendulous, tepals 6, strongly reflexed to spreading, outer whorl distinct from inner whorl, yellow (ours), to orange, cream, or white, 3-veined, more or less elliptic; stamens 6; anthers connate or distinct, yellow; style 1; ovary superior, ellipsoid. **FRUIT** a capsule, broadly to narrowly oblong, dehiscence loculicidal. **SEEDS** black, irregularly compressed and folded. Southwestern United States, south to Argentina, Bolivia, and Peru. About 80 species, with at least 60 of those in Mexico and Central America (Cruden 1981).

Echeandia flavescens (Schult. & Schult. f.) Cruden (turning yellow) Amber lily. Perennial, erect, storage roots enlarged from corms. **LEAVES** basal, 3 to 15, 8–40 cm long, narrowly linear, margins denticulate, ciliate; cauline leaves reduced. **INFL** scape unbranched or branched, very slender, 20–40 cm tall (taller to the south of us), glabrous, paniculate, usually with 3 flowers per node, but occasionally more at the lower nodes and occasionally solitary at the upper nodes; flowers facing away from scape or somewhat upward. **FLOWERS** pale yellow, with 3 to 5 central veins accented in greenish or brownish; tepals elliptic, 9–12 mm long, with those of the outer whorl 2–4 mm wide and those of the inner whorl 4–8.5 mm wide; anthers bright yellow; style about 1/3 longer than the anthers, with a minutely 3-lobed apex; ovary obovate, 2–5 mm. **FRUIT** a capsule, broadly oblong to oblong, 3-lobed, 8–12 mm long. $2n = 16, 32, 48$. [*Anthericum flavescens* Schult. & Schult. f.; *Anthericum torreyi* Baker; *Anthericum stenocarpum* Baker]. Dry grasslands and canyons, piñon-juniper woodlands, and openings in ponderosa pine forests. ARIZ: Apa; NMEX: McK. 2155–2470 m (7065–8100′). Flowering: Jul–Aug. Arizona, New Mexico, western Texas; Mexico. The species reaches the northern limit of its range in the southern San Juan River drainage.

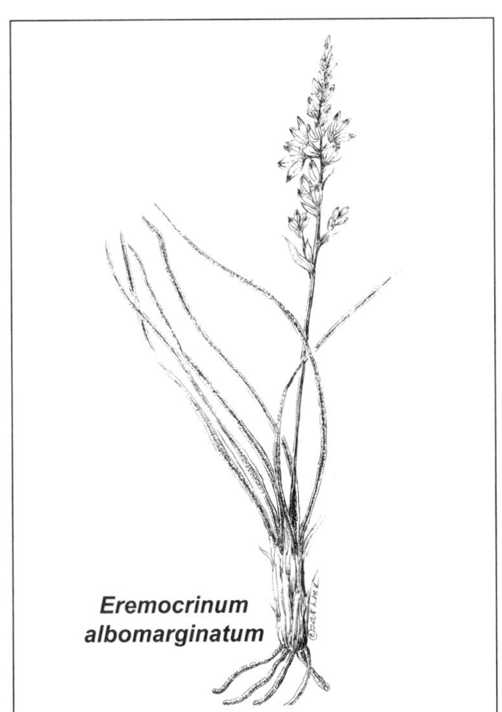
Eremocrinum albomarginatum

Eremocrinum M. E. Jones Sand Lily, Desert Lily

(desert lily) Perennial herbs, from a short rhizome with fleshy roots. **LEAVES** all basal, narrowly linear, 10–40 cm long, 1–3 mm wide, erect or spreading, grasslike, each tuft surrounded by fibrous, marcescent sheaths. **INFL** scape 15–30 cm long, racemose, densely flowered. **FLOWERS** erect or spreading, rather showy but small, on a short pedicel subtended by a scarious bract; tepals 6, 1–2 cm long, oblong, alike, united below into a tubular stipelike base, white to greenish white, the 3 central veins accented with green to brown; stamens 6, shorter than the tepals, anthers lanceolate; style terminated by a discoid stigma; ovary superior, 3-celled. **FRUIT** a loculicidal capsule, 3-lobed, oblong, 4–6 mm long. **SEEDS** black, angled. A monotypic genus of the Southwest. The description for the species is the same as that for the genus.

Eremocrinum albomarginatum (M. E. Jones) M. E. Jones (with white margins) Sand lily. [*Hesperanthes albomarginata* M. E. Jones]. Deep sandy places or sand dunes. ARIZ: Apa, Nav; UTAH. 1220–1660 m (4000–5455′). Flowering: mid-Apr–mid-Jun. Southeastern quarter of Utah and northeastern Arizona. This looks superficially like a *Zigadenus*, but its affinities are with *Echeandia* and *Anthericum*. Plant used for snakebites by Navajos.

Erythronium L. Dogtooth Violet

(red, alluding to the pink or red flowers of some species) Perennial herbs, from deep-seated, elongate corms. **LEAVES** basal, in pairs, opposite, in some species with a reduced leaf at the base of the lowermost flower, petiolate to subsessile, green or mottled. **INFL** solitary or racemose. **FLOWERS** large and showy, pale to deep yellow (ours), or white, deep pink, or violet, nodding; tepals 6, alike or with the inner series broader, lanceolate or acuminate, spreading at anthesis and later becoming reflexed; stamens 6, in 2 slightly unequal sets, filaments slender and erect, tapering slightly from the base or with a broadly dilated base, covering the ovary, and surrounding the style; anthers shrinking by 1/2 after pollen dehiscence; style 1, with a 3-lobed stigma; ovary superior, 3-celled. **FRUIT** a capsule, obovoid to cylindric-clavate, loculicidal. Mostly in North America, about 12 species.

Erythronium grandiflorum Pursh (large-flowered) Dogtooth violet, trout lily, fawn lily, glacier lily, avalanche lily. Perennial herbs, erect, from an elongated corm. **LEAVES** usually 2, narrowly to broadly oblong-elliptic, 10–20 cm long, 1–5 cm wide, acuminate, subsessile, green, not mottled. **INFL** scape 10–30 cm, flowers 2 to 4 on naked peduncles, nodding. **FLOWERS** golden yellow, becoming pale yellow as flowers fade, the base lighter within and streaked with green without; tepals 6, all alike, 2.5–3.5 cm long × 4–7 mm wide, lanceolate, spreading, becoming reflexed; stamens 6, exserted when the tepals reflex, but actually shorter than the tepals, filaments erect, not concealing the ovary or surrounding the style, the anthers yellow, red, somewhat purplish, or white, 10–12 mm long before dehiscence; style projecting the 3 spreading stigmatic lobes beyond the anthers; ovary superior. **FRUIT** a capsule, cylindric-clavate, 3–6 cm long. Moist sites in aspen and spruce-fir communities. COLO: Arc, Hin, LPl, Mon. 2405–3675 m (7900–12060′). Flowering: May–Jul, usually just after snowmelt in areas where snow accumulates. In the Rockies and Cascades from southern Canada to California, Utah, and Colorado. In the region of the Four Corners, known only from the San Juan Mountains. In spite of the common name of "dogtooth violet," this lily does not look anything like a violet (*Viola* spp.). A mass blooming just after snowmelt turns the forest floor yellow, a dramatic sight early in the season, when nothing else is in flower.

Fritillaria L. Fritillary

(dice box, either an allusion to the cubical shape of the capsule or to the checkered markings of some species) Perennial herbs, erect, glabrous, unbranched. **BULBS** small, consisting of a few fleshy scales, and often accompanied by numerous offset bulblets the size of rice grains. **LEAVES** few, alternate to whorled, sessile, linear to lanceolate. **INFL** scapose, solitary or a few flowers in a raceme. **FLOWERS** showy, white, yellow, red, or purplish, and often mottled or striped with white, yellow, red, brown, or purple, usually nodding, campanulate, tepals 6, similar, spreading, each with a gland at the base; stamens 6; style 1, entire or trifid; ovary superior, 3-celled. **FRUIT** a capsule, 6-angled or -winged, loculicidal. **SEEDS** flat, brownish, obovate. Northern Hemisphere, about 50 species.

Fritillaria atropurpurea Nutt. (blackish or very dark purple) Checker lily, chocolate lily. Perennial herb, erect, slender. **BULBS** tan, 1–2 cm in diameter, composed of a few small, thin scales and few, if any, auxiliary bulblets. **LEAVES** several, linear, 6–10 cm long, alternate or whorled, scattered on the upper 1/2 of the stem. **FLOWERS** campanulate, nodding, usually 1 to 4, brown or purple (liver-colored), with yellow or white spots or mottling; tepals 6, alike, oblong or rhombic, tapering abruptly at the base, 10–20 mm long, with the gland as an indistinct brownish yellow spot at the base; stamens with slender filaments, 6–15 mm long; style connate at base only, branches 6–9 mm long. **FRUIT** a capsule, broadly obovoid, acutely angled. Mountain foothills, often among Gambel's oak, sagebrush, or under aspen or conifers. ARIZ: Apa, Nav; COLO: Dol, LPl, Mon; UTAH. 2255–2805 m (7405–9200′). Flowering: May–Jun. North Dakota to Oregon, south to California, Arizona, and New Mexico. Utes regarded a decoction of this plant to be dangerously poisonous, but used as a medicine.

Leucocrinum Nutt. ex A. Gray Star Lily, Sand Lily

(white lily) Perennial herb, acaulescent, tufted, from a short, vertical, deeply buried rootstock. **LEAVES** tufted, narrowly linear, grasslike, spreading, 10–20 cm long × 2–6 mm wide, with each tuft of leaves surrounded at the base by membranous sheathing bracts. **INFL** in a central cluster at ground level, with underground jointed pedicels arising directly from the rootstalk. **FLOWERS** showy; tepals 6, white, all alike, united below the middle into a long, slender tube, separate above, with spreading oblong segments, glandless and clawless; stamens 6, inserted near the apex of the floral tube, exserted, but actually shorter than the spreading tepal segments; style filiform, with 3 short stigmatic lobes; ovary superior. **FRUIT** a capsule, 3-angled, loculicidal, subterranean and thus seldom seen or collected. **SEEDS** few to several, black, angled. A monotypic genus of the western United States. The description of the species is the same as that for the genus.

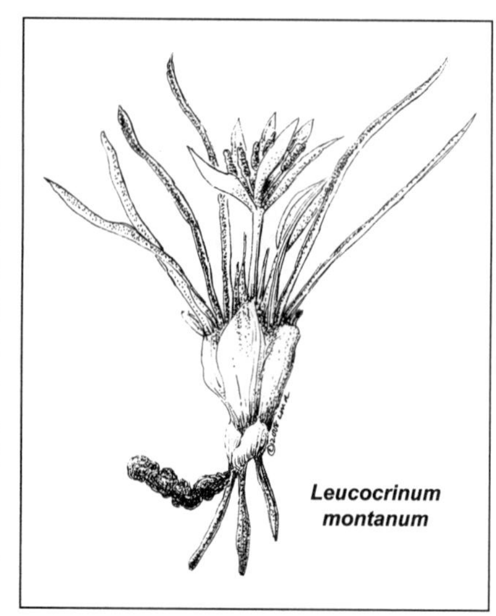
Leucocrinum montanum

Leucocrinum montanum Nutt. ex A. Gray (of the mountains) Common star lily. The description of the species is the same as that for the genus. Sagebrush, piñon-juniper, and ponderosa pine communities. COLO: Arc, LPl; NMEX: RAr. 1875–2440 m (6150–8000′). Flowering: Apr–Jun. Central Oregon, south to central California, Idaho, northern Nevada, Montana, South Dakota (Black Hills), south to southern Utah and northern New Mexico. Paiutes applied a poultice of pulverized roots to sores or swellings.

Lilium L. Lily

(Greek *leirion*, the classical name for lilies) Perennial herbs, tall, erect, glabrous, from fleshy-scaled bulbs or short, scaly rhizomes. **STEMS** unbranched, leafy. **LEAVES** numerous, alternate or whorled, sessile, often lanceolate. **INFL** solitary or 2 or more in a terminal raceme, often leafy-bracted. **FLOWERS** showy, variously colored in the different species (white, pink, orange, or brick-red), frequently spotted with red, brown, or purple; nodding or erect; funnelform to campanulate; tepals 6, all alike, separate, spreading to recurved, often semiclawed and with a gland near the base; stamens 6, borne on the receptacle, filaments slender, anthers usually facing away from the axis; ovary superior, 3-celled; style slender or stout, with 3 short lobes. **FRUIT** a capsule, loculicidal, ovoid to cylindric. **SEEDS** many, flat. North America and Eurasia, about 75 species; well developed in California, where there are 12 species.

Lilium philadelphicum L. (loving brother, possibly an allusion to the proximity of the anthers to one another as the flower is opening) Wood lily. **BULBS** depressed-globose, of thick, fleshy scales. **STEMS** erect, 30–60 cm tall. **LEAVES** alternate, but the uppermost in 1 or 2 whorls, linear to lanceolate, glabrous. **INFL** usually solitary, uncommonly umbelliform with 2 or 3 flowers. **FLOWERS** large, very showy orange-red to red, with purplish red spots at the base; tepals 6, all alike, 5–6 cm long, elliptic-oblong, acute, clawed. **FRUIT** a capsule, cylindric to cylindric-ovoid. [*Lilium umbellatum* Pursh; *Lilium montanum* A. Nelson]. Moist, open woods, often in aspen groves or bordering ponds; rare west of the Continental Divide. COLO: Arc, Hin, LPl. 2315–2515 m (7600–8260′). Flowering: mid-Jun–early Aug. Ohio to Alberta, south to Arkansas and New Mexico, with a population known from the Guadalupe Mountains of west Texas. Wood lily is one of our most beautiful wildflowers, with large, red-orange flowers about the size of the palm of the hand. Even to the novice wildflower lover, the plant is instantly recognizable as a lily, and herbarium specimens are never misidentified. Populations tend to be small and scattered and wood lilies should not be picked for a bouquet.

Lloydia Salisb. ex Rchb. Alp Lily

(for Edward Lloyd, curator of the Oxford Museum) Perennial herbs. Rhizomes oblong, with persistent sheaths, simulating the look of an onion bulb, and terminating a creeping rootstock. **STEMS** slender, erect, less than about 15 cm tall. **LEAVES** 2 to a few, linear, reduced above. **INFL** solitary or a few in a terminal raceme. **FLOWERS** small, white, with colored veins; tepals 6, all alike, separate, spreading, with a small gland near the base; stamens 6; ovary superior, 3-celled; style short, persistent, with 3 short stigmatic lobes. **FRUIT** a capsule, loculicidal, more or less globose. Eurasia and North America, about 12 species.

Lloydia serotina

Lloydia serotina (L.) Salisb. ex Rchb. (late-coming) Alp lily. Perennial herbs, erect, slender, with a short, thick rhizome, the withered stems and leaves of prior years' growth persistent and coating the rhizome, making it resemble a bulb. **LEAVES** both basal and cauline, the basal leaves linear and about as long as the stem, the cauline ones lanceolate and reduced. **FLOWERS** 1 or 2, sometimes as many as 4, erect, broadly turbinate, white to yellowish white, with greenish or brownish coloration on the central veins and usually tinged rose on the back; tepals 6, oblong or broadly oblanceolate, obtuse, 9–12 mm long. **FRUIT** a capsule, obovoid, 6–8 mm long. Occurs mainly at or above timberline. COLO: Hin, LPl, Min, Mon, SJn. 3175–3870 m (10420–12700′). Flowering: Jun–Jul. Eurasia, North America (circumpolar); from Alaska south to Oregon in the Cascades and from Alberta south to northern New Mexico in the Rockies. Utah populations are in the Uinta Mountains.

Maianthemum F. H. Wigg. False Solomon-seal

(Latin *Maius*, month of May, + Greek *anthemon*, flower) Perennial herbs, terrestrial (ours) or aquatic. Rhizomes spreading and filiform or densely clumped, cylindrical, fleshy, zigzag by virtue of one branch of the rhizome growing and the other aborting (sympodial). **STEMS** simple, erect or arching. **LEAVES** cauline, alternate, clasping or short-petiolate, glabrous or puberulent, ovate, with a round or cordiform base, the margins flat or undulate, denticulate or entire, the apex acute or caudate. **INFL** paniculate or racemose. **FLOWERS** with 6 tepals and 6 stamens (ours) or with 4 tepals and 4 stamens, perianth spreading, tepals white, separate, ovate to triangular; stamens inserted at base of tepal; ovary superior, 2- or 3-celled; style short, stigma 2- or 3-lobed. **FRUIT** a berry, variously mottled when immature, bright red at maturity, usually lobed. **SEEDS** 1 to 12, globose (LaFrankie 1986). North America, Central America, northern Europe, eastern Asia, 30 species.

1. Inflorescence paniculate; flowers numerous; tepals 1–2 mm long; stamens exceeding the tepals.... ***M. racemosum***
1′ Inflorescence racemose; flowers few; tepals 4–7 mm long; stamens not exceeding the tepals .... ***M. stellatum***

Maianthemum racemosum (L.) Link **CP** (with the flowers in a raceme; a misnomer, the flowers of this species are in a panicle) Solomon's plume; false Solomon-seal. Perennial herbs, erect to suberect, to arching. Rhizomes rather thick. **STEMS** 30–90 cm, glabrous or finely pubescent. **LEAVES** alternate, sessile or short-petiolate, with a clasping base, flat, broadly ovate to elliptic, 7–20 cm long, the apex acuminate to acute. **INFL** a panicle with dense clusters of flowers, 5–12 cm long. **FLOWER** tepals 6, all alike, white, narrowly oblong, 1.5–2 mm long; stamens 6, the filaments broadly dilated at the base and usually broader than the tepals; style less than 1 mm long. **FRUIT** a berry, red at maturity, speckled with purple, particularly noticeable when immature, globose, 4–6 mm in diameter. [*Smilacina racemosa* (L.) Desf.]. In shaded, cool, moist places, in oak-maple, piñon-juniper, Douglas-fir, aspen, and spruce-fir communities. ARIZ: Apa; COLO: Arc, Dol, Hin, LPl, Min, Mon, SJn; NMEX: McK, RAr, SJn; UTAH. 1985–3230 m (6520–10600′). Flowering: Jun–Jul. Throughout most of North America.

Maianthemum stellatum (L.) Link (with narrow divisions radiating from the center like rays of a star, an allusion to the shape of the flower) Stellate Solomon-seal. Perennial herbs, erect. Rhizome stout. **STEMS** 20–60 cm, glabous or finely pubescent. **LEAVES** alternate or spiraling, sessile, sometimes with a clasping base, flat to folded on the midvein, lanceolate to oblong-lanceolate, 6–15 cm long, the apex acute. **INFL** a raceme, rather few-flowered. **FLOWER** tepals 6, all alike, white, lanceolate to narrowly oblong, 4–7 mm long; stamens 6, erect; style 1–2 mm long. **FRUIT** a berry, greenish yellow when immature, often dark striped, red, or reddish purple at maturity, globose, 5–8 mm in diameter. [*Smilacina stellata* (L.) Desf.]. Moist, shaded or open ground in hanging gardens, piñon-juniper, sagebrush, oak-maple, ponderosa pine, aspen, and spruce-fir communities. ARIZ: Apa, Nav; COLO: Arc, Hin, LPl, Min, Mon, RGr, SJn; NMEX: McK, RAr, SJn; UTAH. 1125–3720 m (3700–12200′). Flowering: Apr–Jun. Alaska to Newfoundland, south to the Ohio Valley, southern Kansas, western Oklahoma, much of New Mexico, eastern and northern Arizona, California; Mexico. Used as a ceremonial emetic by Navajos and in the Fire Dance.

Prosartes D. Don Fairybells

(to append, alluding to the pendulous ovules in the type species) Perennial herbs. Rhizomes slender, knotty, with fibrous roots. **STEMS** branched distally, generally pubescent. **LEAVES** sessile or subsessile, broadly ovate to oblanceolate, with reticulate veining. **INFL** terminal, more or less umbellate, 1 to 4, up to 7 in some species, drooping, pedicellate. **FLOWER** tepals 6, not persistent in fruit, separate, weakly gibbous proximally; stamens hypogynous, basally adnate to tepals, filaments filiform to basally dilated; overy superior, 3-celled; style included or exserted, stigma not lobed or weakly 3-lobed. **FRUIT** a berry, yellowish to red, and more or less fleshy. **SEEDS** light yellow to orange-brown, ellipsoidal to oblong, smooth (Shinwari et al. 1994). North America, five species.

Prosartes trachycarpa S. Watson (rough fruit, alluding to the papillate nature of the surface of the berry) Fairybells. Perennial herbs. Rhizomes slender to thickish. **STEMS** suberect, sparsely branched, crisp-pubescent. **LEAVES** alternate, 4–10 cm long, ovate to ovate-oblong, apex acute, pubescent below, glabrous above, the margins ciliate. **INFL** 1 to 3 flowers on stout, pendulant, pubescent pedicels, flowers facing downward. **FLOWERS** narrowly campanulate, 10–15 mm long, tepals 6, creamy white, narrowly oblanceolate, spreading from near the narrowed base; stamens 6, mostly exserted; ovary broadly ovoid to obovoid, 3-celled; style 9–12 mm long, with short stigmatic lobes up to 1 mm long. **FRUIT** a berry, 7–10 mm long, depressed-globose, strongly papillate, reddish orange when mature. [*Disporum trachycarpum* (S. Watson) Benth. & Hook. f.]. Oak-maple, Douglas-fir, white fir, ponderosa pine, and spruce-fir communities; often in shady, wooded places, often near streams or on north-facing slopes. ARIZ: Apa; COLO: Arc, Hin, LPl, Min, Mon, SJn; NMEX: RAr, SJn; UTAH. 2150–2805 m (7050–9205′). Flowering: Jun–Jul. Ontario to British Columbia, south in the mountains to Utah, Colorado, northern New Mexico, and Arizona. The New World species have traditionally been treated as section *Prosartes* of the genus *Disporum*, which otherwise occurs only in Asia. Recent work suggests that there is sufficient difference to recognize two genera.

Streptopus Michx. Twisted Stalk

(twisted foot, alluding to the bent or twisted peduncles) Perennial herbs. Rhizomes horizontal, extensive. **STEMS** simple to highly branched and leafy. **LEAVES** numerous, alternate, elliptic to ovate, bases sessile to mostly clasping, apex acute to acuminate. **INFL** rotate to campanulate. **FLOWERS** small, white, greenish white, pink, or rose-red, depending upon the species, borne beneath the leaf (extra-axillary), 1 or sometimes 2 together, on a bent or twisted peduncle; tepals 6, alike, mostly oblanceolate to oblong, erect, spreading, or recurved; stamens 6, hypogynous, filaments short and flat; ovary superior, 3-celled; style slender to bulbous, with an entire or short-trifid stigma. **FRUIT** a berry. North America and Eurasia, seven species.

Streptopus amplexifolius (L.) DC. (with leaves clasping the stem) Twisted stalk. Perennial herbs, stout, erect. Rhizomes thick, highly branched. **STEMS** 50–80 cm tall, glabrous to pubescent stems and branches, the thick rigid hairs red at the base. **LEAVES** ovate-oblong to oblong-lanceolate, 5–15 cm long, clasping at the base, apex acuminate. **INFL** campanulate. **FLOWERS** white with a greenish or yellowish tinge, 9–15 mm long, arising extra-axillary (back side of leaf), the arched peduncle 1–3 cm long, bearing a short-stalked gland at the tip, then sharply geniculate (like a knee joint), the twisted pedicel 1–2 cm long, bearing 1 or 2 flowers; tepals narrowly oblong-lanceolate, spreading, recurved at the tip; stamens unequal, outer series 1 mm long, inner series 2–3 mm long, the filaments only slightly less broad than the anthers; ovary elliptic; style 4–5 mm long, with an unlobed or minutely lobed stigma. **FRUIT** a berry, 10–12 mm long, ellipsoid, whitish green when immature, yellowish orange or red when mature. $2n = 16, 32$. [*Uvularia amplexifolia* L.; *Convallaria amplexifolia* (L.) E. H. L. Krause; *Streptopus fassettii* Á. Löve & D. Löve]. Aspen and spruce-fir communities; moist places, especially along streams. COLO: Arc, Hin, LPl, Min, Mon, SJn; UTAH. 2325–3290 m (7630–10790′). Flowering: mid-May–Jul. Eurasia and North America, from Alaska across Canada and the northern United States; in the West, south to Arizona.

Veratrum L. False Hellebore, Cornhusk Lily, Skunk Cabbage

(an ancient name for hellebore, probably from *vere*, true, + *ater*, black, alluding to the black rhizomes of some species) Perennial herbs. Rhizomes short, thick, stout. **STEMS** tall, stout, erect, leafy, unbranched. **LEAVES** numerous, narrow to broad, often with lengthwise plaits, strongly clasping in most species. **INFL** a terminal panicle or compound raceme. **FLOWERS** small, greenish white, yellow, or purple, depending upon the species, broadly spreading, polygamous (upper perfect; lower staminate), on a short pedicel; tepals 6, separate or sometimes with a short hypanthium, generally without a gland, clawed in some, oblong-ovate, obtuse; stamens 6, opposite the tepals, free, almost as long as the tepals, filaments slender, anthers cordate-ovoid; ovary superior, 3-celled, glabrous or woolly; styles 3, terminated by an elongated stigma. **FRUIT** a capsule, 3-lobed, septicidal, membranous, ovoid, and tipped with the persistent styles. **SEEDS** numerous, flat, and surrounded by a broadly winged, loose outer seed coat. Northern Hemisphere, about 30 species.

Veratrum californicum Durand (of California) Skunk cabbage. Perennial herbs, tall, erect, stout. **STEMS** typically 1.5–2 m tall, glabrous below, becoming tomentose in the inflorescence. **LEAVES** numerous, large, oblong-lanceolate to broadly elliptic or ovate, 25–40 cm long, 10–20 cm wide, gradually reduced up the stem, leaves pubescent below and sometimes above on the veins, plicate and strongly nerved, the base sheathing, apex acute. **INFL** a panicle, freely branching, stiffly erect. **FLOWERS** spreading, greenish white to dull white, subsessile or on a short pedicel, gradually reduced upward; tepals 6, lanceolate to oblong-ovate or elliptic, 10–14 mm long, entire, erose, or somewhat denticulate, glabrous or slightly woolly, the margin green, with a gland at the base, V-shaped (2 confluent glands); stamens 6, shorter than the tepals, 6–8 mm long, filaments slender, anthers broader than long; ovary glabrous or with a few hairs; styles 3. **FRUIT** a capsule, narrowly ovoid, 2–3 cm long, glabrous. **SEEDS** flat, winged, 10–12 mm long. [*Veratrum tenuipetalum* A. Heller]. Aspen, mixed conifer, and spruce-fir communities; moist open meadows, along streams, and by lakes. ARIZ: Apa; COLO: Arc, Con, Hin, LPl, Min, Mon, RGr, SJn; NMEX: RAr, SJn. 2345–3650 m (7700–11975′). Flowering: Jun–Aug. Montana to Washington, south to California, Nevada, Utah, Colorado, Arizona, and New Mexico. *Veratrum* has some alkaloids that have been useful for the treatment of hypertension. The plant is also the cause of occasional livestock poisoning. Paiutes applied a poultice of root for rheumatism. Also for burns and bruises. Used as a gynecological aid and gland medicine. Also used to cover berries to keep them fresh.

Zigadenus Michx. Death Camas

(*zygos*, yoke, + *aden*, gland, an allusion to the paired glands of some species) Perennial herbs, erect, simple, leafy, glabrous. **BULBS** deep-seated, coated (ours) or rhizomatous. **LEAVES** linear, grasslike, clasping, mostly basal, cauline leaves reduced. **INFL** terminal raceme or panicle. **FLOWERS** small, greenish to yellowish white or white, rotate to shallowly campanulate, perfect or unisexual, on an erect spreading pedicel, subtended by long-acuminate bracts; tepals 6, all alike, lanceolate to ovate, separate or adnate to the lower part of the ovary, clawed at the base, 1 or 2 yellow to greenish glands near the base; stamens 6, hypogynous or perigynous, filaments basally dilated, anthers basifixed, extrorse, cordate-reniform; ovary superior or partly inferior; 3-celled; styles 3, distinct, stigmatic at tips. **FRUIT** a capsule, 3-lobed, loculicidal. **SEEDS** subrhomboid, irregularly angled. North America and Asia, about 15 species. As recognized here (and following *Flora of North America*), the genus is polyphyletic. Most species of *Zigadenus* are quite poisonous, and the toxicity is said to increase upon drying. *Zigadenus* poisoning can be fatal. Many of the alkaloids found in *Veratrum* are found in *Zigadenus* as well. Persons mistakenly eating *Zigadenus* and then dying, when they thought they had collected the edible bulbs of *Camassia*, has led to the common name of "death camas."

1. Perianth segments no more than 7 mm long; ovary superior; glands of perianth ovate or semicircular, often with an indefinite margin ....(2)
1' Perianth segments at least 6 mm long; ovary partly inferior; glands of perianth obcordate (heart-shaped with the point facing the stem), with a definite margin ....(3)

2. Inflorescence usually a panicle, with at least 2 basal branches, less frequently a raceme; basal leaves 5–15 mm wide; tepals usually clawless or with short claws .... ***Z. paniculatus***
2' Inflorescence usually a raceme, less frequently a near-panicle with 1 basal branch; basal leaves 3–6 mm wide; tepals usually clawed, less frequently clawless .... ***Z. venenosus***

3. Plants 30–60 cm tall, inflorescence a raceme, sometimes a near-panicle with 1 basal branch, usually at elevations above 1980 m (6500′) in the mountains .... ***Z. elegans***
3' Plants 60–100 cm tall or more, inflorescence a large, diffuse panicle, usually at elevations below 1980 m (6500′) in hanging gardens and seeps in canyon country .... ***Z. vaginatus***

Zigadenus elegans Pursh **CP** (elegant, referring to the showy flowers) Mountain death camas. Perennial herbs, 15–70 cm tall. **BULBS** deep-seated, narrowly globose, 1–3 cm long. **LEAVES** basal, linear, 10–25 cm long, 2–15 mm wide, cauline leaves becoming much reduced above. **INFL** a raceme, or sparingly branched below with ascending branches, rather few-flowered. **FLOWERS** rotate, creamy white or greenish white, on a spreading or ascending pedicel, 10–35 mm long; tepals 6, spreading, obovate or oblong, 6–11 mm long, not clawed, with an obcordate gland at the base of each tepal, yellow, yellow-green, or green; stamens 6, perigynous, as long as the tepals, the filament broadly dilated basally, anther reniform; ovary partly inferior, 3 distinct styles, 2.5–3 mm long. **FRUIT** a capsule, 15–20 mm long. [*Anticlea elegans* (Pursh) Rydb.; *Zigadenus chloranthus* Richardson; *Z. longus* Greene; *Z. dilatatus* Greene; *Z. alpinus* Blank.]. Aspen, spruce-fir, and alpine communities; often in moist meadows. ARIZ: Apa; COLO: Arc, Con, Hin, LPl, Min, Mon, RGr, SJn; NMEX: RAr, SJn; UTAH. 2325–3765 m (7630–12350′). Flowering: Jun–Jul. Alaska, south to British Columbia and Alberta, south in the more western mountains to Washington, Oregon, and northern Nevada, and south in the Rockies to Arizona, New Mexico, western Texas, and Mexico. Navajos used infusion of plant for mad coyote bites.

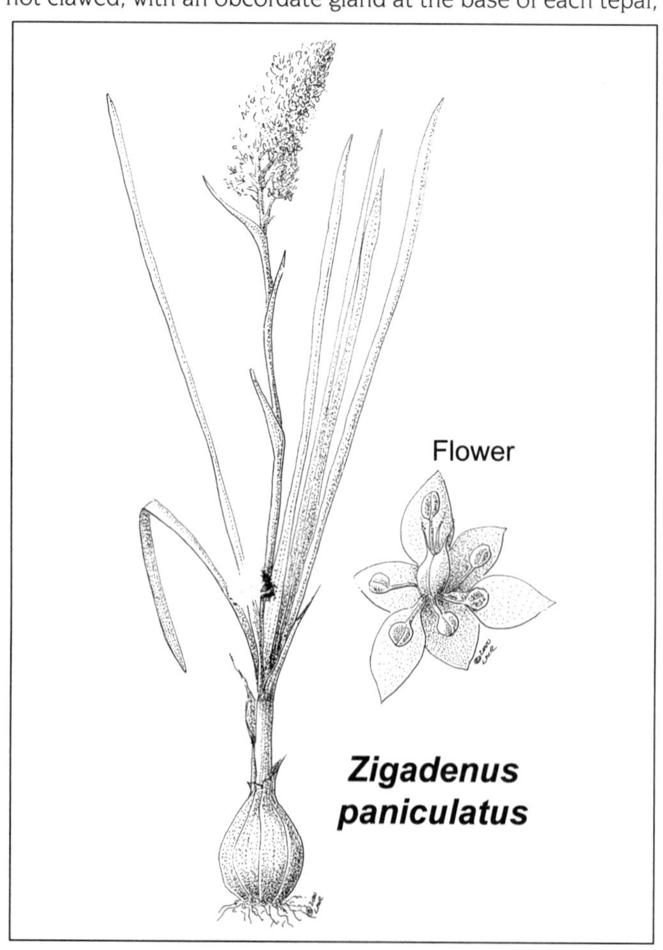

Zigadenus paniculatus

Zigadenus paniculatus (Nutt.) S. Watson (with the flowers in a panicle) Foothills death camas. Perennial herbs, 30–50 cm tall. **BULBS** oblong-ovoid, deep-seated, 3–4 cm long, with persistent bulb coats. **LEAVES** mostly basal, some along the lower stem, linear, 15–40 cm long, 5–15 mm wide, much reduced above. **INFL** a terminal panicle, densely flowered, 10–30 cm long. **FLOWERS** broadly campanulate, perfect or staminate only, white or yellowish white; tepals 6, persistent in fruit, those of the outer series broadly ovate, 3.5–4.5 mm long, scarcely or not clawed, those of the inner series oblong-lanceolate, 4–5 mm long, with a narrow claw less than 1 mm long, gland 1, yellow or yellowish green, broadly ovate or obovate, distal margin evident or indefinite; stamens 6, usually equaling or slightly longer than the tepals, filaments slender, anther reniform or cordate; ovary superior; styles 3, distinct, 2.5–3 mm long. **FRUIT** a capsule, 10–20 mm long. [*Toxicoscordion paniculatum* (Nutt.) Rydb.]. Blackbrush, piñon-juniper, sagebrush, ponderosa pine, and Douglas-fir communities; often in dry, sandy or rocky areas. ARIZ: Apa, Nav; COLO: Arc, Dol, LPl, Mon, SMg; NMEX: McK, RAr, SJn; UTAH. 1820–2500 m (5970–8200′). Flowering: Apr–Jun. Washington south to California, east to Idaho, western Montana, western Wyoming, south through Nevada, Utah, western Colorado to northern Arizona and northwestern New Mexico. Navajos gave the plant to sheep with bloat. Paiutes used the bulb for neuralgia and rheumatism, for swellings, toothache, and as an emetic. Navajos cooked the bulbs with meat and ate the greens.

Zigadenus vaginatus (Rydb.) J. F. Macbr. (sheathed) Alcove death camas. Perennial herbs, 30–100 cm or more, base of stem with marcescent leaf bases. **BULBS** narrowly ovoid, 3–6 cm, with persistent bulb coats. **LEAVES** mostly basal, 20–75 cm long × 6–18 mm wide, sheathing the stem below, reduced up the stem. **INFL** a large, broad, diffuse panicle, up to 45 cm long. **FLOWERS** perfect, rotate, white; tepals 6, obovate to spatulate or elliptic, 6–7 mm long, rounded to obtuse, with a yellow obcordate gland at the base, clawless; stamens 6, shorter than the tepals; ovary partly inferior, styles 3, about 2–3 mm long. **FRUIT** a capsule, 10–15 mm long. [*Anticlea vaginata* Rydb.]. Hanging garden communities in seeps and alcoves. ARIZ: Apa, Nav; UTAH. 1125–1980 m (3700–6500′). Flowering: Aug–Oct. Arizona, Utah, and Colorado.

Zigadenus venenosus S. Watson (very poisonous) Meadow death camas. Perennial herbs, 20–60 cm tall. **BULBS** oblong-ovoid, deep-seated, 1.5–2.5 cm long, with persistent bulb coats. **LEAVES** mostly basal, some along the lower stem, linear, 15–30 cm long, 3–6 mm wide, much reduced above. **INFL** a terminal raceme, sometimes with 1 or 2 basal branches and thus paniculate, densely flowered, up to 20 cm long. **FLOWERS** campanulate, perfect or staminate only on some branches, white or yellowish white; tepals 6, persistent in fruit, those of the outer series ovate, 4.5–5 mm long, clawed (up to 1 mm long) or not clawed, those of the inner series oblong-lanceolate, 5–6 mm long, with a narrow claw 0.7–1.2 mm long, gland 1, yellow or yellowish green, broadly ovate, distal margin evident or indefinite; stamens 6, usually equaling or slightly longer than the tepals, filaments slender, anther reniform or cordate; ovary superior; styles 3, distinct, 2–3 mm long. **FRUIT** a capsule, 10–20 mm long. [*Toxicoscordion venenosum* (S. Watson) Rydb.; *Zigadenus gramineus* Rydb.; *Z. intermedius* Rydb.; *Z. falcatus* Rydb.; *Z. acutus* Rydb.; *T. arenicola* A. Heller; *Z. salinus* A. Nelson]. Sagebrush, piñon-juniper, oak-maple, and white fir communities; often in dry, sandy or rocky areas. ARIZ: Apa, Coc, Nav; COLO: Arc, LPl, Mon, SMg; NMEX: RAr, SJn. 1605–2065 m (5275–6775′). Flowering: Apr–Jun. Southwestern Canada, Pacific Northwest, south to California, east to the Dakotas, south to Nevada, Utah, Colorado. Ours are **var. *gramineus*** (Rydb.) O. S. Walsh ex C. L. Hitchc. Seeds and roots considered deadly poison. Paiutes used the bulb as a poultice for swellings, snakebite, rheumatism, and burns.

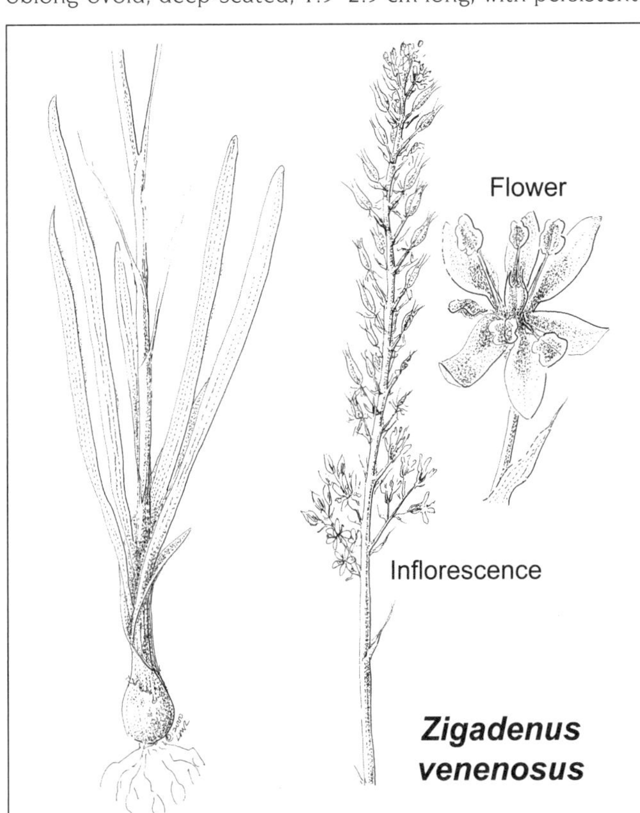

LINACEAE DC. ex Perleb FLAX FAMILY

Kenneth D. Heil and J. Mark Porter

Ours taprooted annual or perennial herbs. **LEAVES** alternate, simple. **INFL** terminal racemes or cymes. **FLOWERS** perfect, actinomorphic, nodding in bud; sepals 5, persistent; petals 5, free, blue, yellow, or orange, soon falling;

stamens 5, alternate the petals, often appendaged, staminodia 0 or 5, alternate the stamens, the filaments united at the base; pistil 1, the ovary superior, of 5 united carpels, locules 2–5, becoming 4–10 by the growth of false septa (secondary partitions); styles 2–5, distinct or united below. **FRUIT** a capsule. **SEEDS** compressed, shining, becoming gelatinous when wet. $x = 6–11+$. Fifteen genera, 300 species, widely distributed.

Linum L. Flax

(flax) Annual or perennial herbs from a taproot; stems 4–80 cm tall, often branched. **LEAVES** 3–30 mm long, 0.5–4 mm wide, appressed-ascending to ascending, stipular glands present or absent. **FLOWERS** with peduncle 0.5–35 mm long; sepals 3.5–9 mm long, lanceolate, acuminate, or elliptic; entire, often with a slender awn tip; petals yellow, orange, blue, or white, 5–23 mm long; style 2.5–10 mm long, often cleft at apex. **FRUIT** ellipsoid, or ovoid to globose, 3–8 mm long. **SEEDS** mucilaginous. More than 90 species of temperate and warm regions. Commercial flax is derived from retting stems of *L. usitatissimum* L. Other uses include linseed oil, stock feed, medical, and ornamental. Flax blossoms open early in the morning and fall by afternoon. Mature petals are seldom preserved on specimens; therefore, care should be taken during collection.

1. Petals blue or rarely white .... ***L. lewisii***
1′ Petals yellow or orange .... (2)

2. Styles distinct to the base; flowers in racemes; pine forests in Apache County, Arizona, and McKinley County, New Mexico .... ***L. neomexicanum***
2′ Styles united nearly to the apex; flowers in terminal cymes; widespread .... (3)

3. Plants short-lived perennials (rarely annual); sepals persistent in fruit; secondary partitions of carpels incomplete, edges hairy; no stipular glands evident; flowers 0.5–6 cm long .... ***L. subteres***
3′ Plants annual; sepals soon deciduous; secondary partitions complete, edges glabrous; stipular glands brown to dark brown; flowers 0.5–2.5 cm long .... (4)

4. Stems densely hirtellous with spreading hairs; petals mostly orange to salmon, the bases reddish .... ***L. puberulum***
4′ Stems glabrous or scaberulous; petals yellow to yellow-orange .... (5)

5. Styles 2–4 mm long; stems scaberulous below, usually simple below the middle; stipular glands brown .... ***L. australe***
5′ Styles 4.5–7 mm long; stems glabrous, usually branched from the base; stipular glands dark brown .... ***L. aristatum***

Linum aristatum Engelm. (an awn or bristle, referring to the awn at the apex of the leaf or sepal) Broom flax. Annual to short-lived perennial, glabrous herbs from taproots. **STEMS** 8–45 cm tall, usually branched from near the base. **LEAVES** 3–15 mm long, 0.5–1 mm wide, appressed-ascending, stipular glands present. **INFL** a cyme, the peduncles 5–35 mm long. **FLOWER** sepals 5–9 mm long; petals yellow to yellow-orange, 7–12 mm long; anthers 0.8–1 mm long; style 4–7 mm long, cleft at apex. **FRUIT** ellipsoid, 3.5–4.5 mm long; secondary partitions hyaline, almost complete. $n = 15$. Sand dunes and other sandy sites in blackbrush, sand sagebrush, resinbush, piñon-juniper, and ponderosa pine communities. ARIZ: Apa, Coc, Nav; COLO: Mon; NMEX: McK, SJn; UTAH. 1180–2000 m (3900–6060′). Flowering: May–Aug. Fruit: Jun–Sep. Texas south to Mexico.

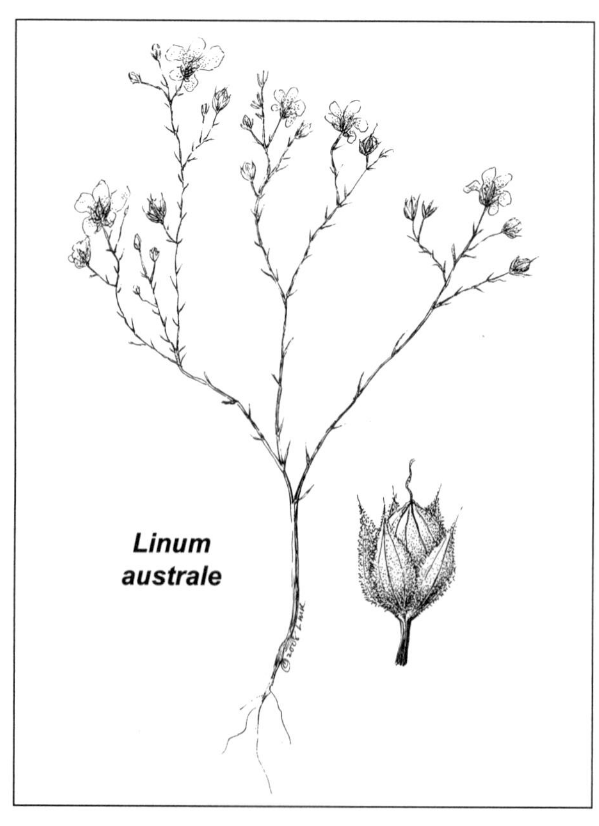
Linum australe

Linum australe A. Heller (southern) Southern flax. Annual, scabrous herbs from taproots. **STEMS** 10–40 cm tall, usually branched above the base. **LEAVES** mostly 6–20 mm long, 0.5–1.5 mm wide, appressed-ascending to ascending, stipular glands present on ours. **INFL** a cyme, the peduncles mostly 3–15 mm long. **FLOWER** sepals 4.5–7 mm long; petals yellow to yellow-orange, 5–9 mm long; anthers 0.4–0.8 mm long; style 2.5–4 mm long, cleft at apex. **FRUIT**

ovoid, 3.5–4.5 mm long; secondary partitions hyaline, almost complete. $n = 15$. Piñon-juniper, sagebrush, and ponderosa pine communities. COLO: Arc, LPl, Mon; NMEX: McK, RAr, SJn. 2000–2750 m (6600–9075′). Flowering: May–Aug. Fruit: Jun–Sep. Montana, Wyoming, Nevada south to Texas and northern Mexico. Used as a drug for stomach disorders by the Hopi, and for kidney disease by the Navajo.

Linum lewisii Pursh (honoring Meriwether Lewis of Lewis and Clark) Blue flax. Perennial, glabrous herbs from a caudex and stout taproot. **STEMS** 15–80 cm tall, usually simple. **LEAVES** 4–30 mm long, 0.5–4 mm wide, ascending to spreading-ascending. **INFL** in cymose clusters, the peduncles 1–3 cm long. **FLOWER** sepals 3.5–6 mm long; petals blue with a whitish or yellowish base or white, 12–23 mm long; staminodia present, about 1 mm long; styles distinct. **FRUIT** ovoid to globose, 6–8 mm long, opening from the top, primary and secondary sutures fringed-ciliate. [*L. perenne* L. subsp. *lewisii* (Pursh) Hultén; *L. perenne* L. var. *lewisii* (Pursh) Eaton & Wright]. $2n = 18$. Frequently found along roadsides; sagebrush, mountain brush, piñon-juniper, and ponderosa pine communities. ARIZ: Apa, Nav; COLO: Arc, Dol, Hin, LPl, Mon, SJn, SMg; NMEX: McK, RAr, SJn; UTAH. 1440–3180 m (4750–10500′). Flowering: May–Sep. Fruit: Jun–Oct. Alaska to Mexico; Asia. Used for fiber, food, and drugs.

Linum neomexicanum Greene (of New Mexico) New Mexico yellow flax. Perennial, glabrous herbs from a caudex. **STEMS** up to 40 cm tall, usually branching above the base. **LEAVES** mostly 5–15 mm long, 1–2 mm wide, appressed-ascending to ascending. **INFL** in racemes, the peduncles 1–3 mm long. **FLOWER** petals yellow, 5–8 mm long; sepals 3–4 mm long; styles distinct to the base. **FRUIT** ovoid, 4–5 mm long, opening from the top, primary partitions ciliate-hairy adaxially, secondary partitions incomplete and sometimes fringed. Rocky hillsides in coniferous forest communities. ARIZ: Apa; NMEX: McK. 2135–2515 m (7050–8300′). Flowering: Jul–Sep. Fruit: Aug–Sep. New Mexico and Arizona south to northern Mexico.

Linum puberulum (Engelm.) A. Heller (minutely downy) Plains flax. Annual, hirtellous herbs from a taproot. **STEMS** 4–20 cm tall, usually branched from the base. **LEAVES** 3–15 mm long, 0.5–1 mm wide, appressed-ascending to ascending, stipular glands present in ours. **INFL** in cymose peduncles mostly 0.5–2.5 cm long. **FLOWER** sepals 4–7 mm long; petals orange to salmon, with a dark base, 8–15 mm long; styles 3–6 mm long, cleft apically. **FRUIT** ovoid, 3–4 mm long, the secondary partitions hyaline, almost complete. $n = 15$. Grassland, salt desert scrub, sagebrush, and piñon-juniper communities. ARIZ: Apa, Nav; COLO: LPl, Mon; NMEX: McK, RAr, SJn; UTAH. 1620–2075 m (5340–6290′). Flowering: May–Aug. Fruit: Jun–Sep. California, Nevada east to Nebraska, south to Texas; Mexico. Used as a drug; eye medicine and heartburn.

Linum subteres (Trel.) J. P. Winkl. (somewhat terete or circular in cross section) Utah yellow flax. Perennial (or annual), glabrous, herbs from taproots. **STEMS** 10–45 cm tall, usually branched above the base. **LEAVES** 4–20 mm long, 0.5–2 mm wide, appressed-ascending to ascending, stipular glands not present in ours. **INFL** cymose on peduncles 0.5–6 cm long. **FLOWER** sepals 4–7 mm long; petals yellow, 7–10 mm long; anthers 1–2 mm long; styles 4.5–9 mm long, cleft apically. **FRUIT** ovoid, 3.5–4.5 mm long, the secondary partitions prominently lacerate-fringed, incomplete. $n = 15$. Sandy soils in salt desert scrub, mountain shrub, rabbitbrush, and piñon-juniper communities. UTAH. 1635–2045 m (5400–6700′). Flowering: Jun–Aug. Fruit: Jul–Sep. Utah to Nevada.

LOASACEAE Juss. STICKLEAF or BLAZING-STAR FAMILY

Charlotte M. Christy

Annuals or perennials, mostly herbs; hairs harsh, often barbed, some stinging. **STEMS** mostly erect, usually brittle. **LEAVES** exstipulate, mostly alternate, simple. **INFL** cymose in ours. **FLOWERS** perfect, actinomorphic, epigynous; sepals 4–5, usually persistent; petals 4–5, ours free; petaloid staminodia present or absent; stamens mostly 10–many, mostly free; pistil 3–5-carpellate, unilocular; placentation usually parietal; style 1, often persistent. **FRUIT** a capsule (ours). **SEEDS** 1 to many (Hufford 1989; Schenk 2009; Schenk and Hufford 2010). Fifteen genera, about 250 species. New World except *Kissenia* (Africa and Arabia) and *Plakothira* (Marquesas Islands). Uses include ornamentals and some herbal remedies.

Mentzelia L. Blazing-star, Stickleaf

(for Christian Mentzel, 17th-century German botanist) Annuals or perennials; ours herbaceous. **STEMS** often with mature epidermis white. **LEAVES** sessile or petioles winged and usually indistinct; blades linear, lanceolate, elliptic, ovate, or oblanceolate; margins sinuate-dentate or pinnately lobed, entire or crenate in some. **FLOWERS** usually bracteate; sepals 5; petals 5, white to yellow or orangish; petaloid staminodia present or absent; outer stamens with filiform to petaloid filaments. **CAPSULES** dehiscent by 3 apical valves, exceptionally more. **SEEDS** mostly 10–many, 1–4 mm long, endosperm present; seed coat sculptured, appearing smooth or reticulate at low magnification, some requiring 40–100× magnification to see structural details. $x = 14$. At least 100 species, New World. Fruit measurements refer to the body (the seed-containing portion). Seeds provide important characters; thus immature specimens may be unidentifiable.

1. Plants biennials or short-lived perennials; seeds 2–4 mm long, the embryo-containing portion ovate, the margin winged; petals exceeding 5 mm long; petaloid stamens or staminodia often present (section *Bartonia*)..............(2)
1′ Plants annual; seeds 1–2 mm long, the embryo-containing portion irregular, often blocky or faceted, wingless; petals rarely exceeding 5 mm long in ours; petaloid stamens or staminodia absent (section *Trachyphytum*).......(9)

2. Outer surface of petals with evident trichomes; seed epidermal cells with wavy adjoining walls......***M. cronquistii***
2′ Outer surface of petals glabrous except for an apical tuft; seed epidermal cells with straight, wavy, or strongly sinuate adjoining walls.......(3)

3. Plants branching only in upper 1/2, the branches strongly ascending; petals nearly erect; adjoining walls of seed epidermal cells strongly sinuate.......***M. rusbyi***
3′ Plants potentially producing branches at most nodes, the branches spreading; petals widely spreading; adjoining walls of seed epidermal cells straight, wavy, or strongly sinuate.......(4)

4. Cauline leaves obovate to oblanceolate, the margins crenate; wing of seed 1 mm broad, the epidermal cells with straight walls.......***M. pterosperma***
4′ Cauline leaves narrowly elliptic to lanceolate or linear, the margins sinuate-dentate to lobed; wing of seed less than 1 mm broad, the epidermal cells with wavy to strongly sinuate walls.......(5)

5. Seed with wing about 0.5 mm broad, the epidermal cells with strongly sinuate walls; cauline leaves with broad to clasping bases, the lobes or teeth usually irregular.......***M. multiflora***
5′ Seed with wing nearly obsolete to 0.3 mm broad, the epidermal cells with wavy to sinuate walls; cauline leaves sessile but not clasping, the lobes or teeth similar in size and spacing, mostly deeply pinnately lobed.............(6)

6. Wing of seed nearly obsolete to 0.25 mm broad, the epidermal cells with wavy to sinuate walls.........***M. laciniata***
6′ Wing of seed about 0.3 mm broad, the epidermal cells with sinuate walls.......(7)

7. Leaves oblanceolate to elliptic, the lobes filiform, 1.4 mm wide or less, up to 17 mm long.......***M. filifolia***
7′ Leaves spatulate, the lobes narrow but not filiform, more than 1 mm wide, less than 13 mm long.......(8)

8. Leaves to 1.2 cm wide; seeds less than 3 mm long.......***M. sivinskii***
8′ Leaves to 3.2 cm wide; seeds more than 3 mm long.......***M. holmgreniorum***

9. Plants rarely exceeding 9 cm tall; leaves entire.......***M. thompsonii***
9′ Plants usually exceeding 20 cm tall unless depauperate; leaves, at least the basal, toothed or lobed...............(10)

10. First-produced fruits bending more than 90°, often at least 180°; seeds rounded, only weakly faceted.......***M. obscura***
10′ First-produced fruits straight or bending no more than 90°; seeds angular, strongly faceted.......(11)

11. Fruits long-tapering to base, the first-produced ones often bending, potentially to 90°; bracts mostly subtending ovary and fruit, rarely white-based.......***M. albicaulis***
11′ Fruits slightly tapering, the first-produced ones straight or rarely with a slight bend; bracts commonly attached to ovary and fruit, mostly white-based.......***M. montana***

Mentzelia albicaulis (Douglas ex Hook.) Douglas ex Torr. & A. Gray (white-stemmed) Western blazing-star. Annual. **STEMS** to 30 (–45) cm tall, branching throughout unless depauperate. **LEAVES** to 12 (–15) cm long, to 1 (2) cm wide, narrowly elliptic to lanceolate; basal leaves petiolate, the margins deeply lobed, rarely entire; cauline leaves sessile, often with fewer lobes or entire, the bases often clasping. **BRACTS** of early flowers leaflike, the others strongly reduced, narrowly lanceolate to ovate, green, sometimes with faintly whitish bases, mostly subtending the ovary; margins entire or few-toothed. **FLOWERS** with calyx lobes 2–3 mm long; petals yellow, spreading, broadly obovate or obcordate, 2–6 mm long, 2–4 mm wide; staminodia and broad-filament stamens absent; ovary 4–10 mm long; style 2–4 mm long. **CAPSULES** sessile, clavate, often long-tapering to base; body 8–28 mm long, 3–4 mm wide at apex, usually straight or the first-formed bent slightly, occasionally as much as 90°. **SEEDS** irregularly blocky, several-faceted, the angles sharp, 1–2 mm long, wingless; epidermal cells with straight adjoining walls, their surface flat to strongly pointed-papillate. $2n$ = 54, 72. [*Bartonia albicaulis* Douglas ex Hook.; *Acrolasia albicaulis* (Douglas ex Hook.) Rydb.; *A. gracilis* Rydb.]. Sand, clay, or gravelly soils, grassland, desert shrub, piñon-juniper woodlands, and ponderosa pine communities. ARIZ: Apa, Coc, Nav; COLO: Arc, Dol, LPl, Mon, SMg; NMEX: McK, RAr, San, SJn; UTAH. 1125–2250 m (3700–7400′). Flowering: Mar–Jun. Fruit: Apr–Jul. Arizona, California, Colorado, Idaho, New Mexico, Nevada, Oregon, Utah, Washington, Wyoming; British Columbia, Canada; Baja California, Sonora, Mexico. A widespread species with many named variations, commonly intergrading with *M. montana* and others of its relatives where they co-occur. The Ramah Navajo formed a poultice of crushed and soaked seeds and applied it for toothaches; a compound of the leaves was used for snakebite. The Hopi used the plant as a substitute for tobacco.

Mentzelia cronquistii H. J. Thomps. & Prigge (for Arthur Cronquist, 20th-century botanist) Cronquist's blazing-star. Perennial. **STEMS** to 60 cm tall, potentially branching throughout. **LEAVES** to 10 cm long, to 2 cm wide, oblanceolate to narrowly elliptic or lanceolate; basal leaves petiolate, sinuate-dentate; cauline leaves sessile, the bases or basal lobes often clasping; margins lobed, toothed, or entire. **BRACTS** of early flowers reduced leaves, the others further reduced, linear-lanceolate, often entire, subtending the ovary. **FLOWERS** with calyx lobes 5–10 mm long; petals yellow, widely spreading at anthesis, 5–10 (–16) mm long, 1.5–5 mm wide, narrowly oblanceolate, the apex acute, with trichomes covering the outer surface; staminodia (0) 3–5, slightly shorter and narrower than petals; outer 1–2 whorls of stamens with broadened filaments; ovary 3–5 mm long; style 4–8 mm long. **CAPSULES** usually pedicellate, cylindrical or subglobose when short, 4–10 (–15) mm long, 3–7 mm wide, straight. **SEEDS** oval, 2–3 mm long, the wing about 0.5 mm wide; epidermal cells with curved to wavy adjoining walls, the surface walls with 15–20 small bumps. $2n$ = 20. [*Nuttallia cronquistii* (H. J. Thomps. & Prigge) W. A. Weber]. Most commonly on sandy soils; desert shrub, sagebrush, and piñon-juniper woodland communities. ARIZ: Apa, Coc; COLO: Dol, Mon; NMEX: SJn; UTAH. 1125–1900 m (3700–6200′). Flowering: Apr–Sept. Fruit: May–Nov. Arizona, Colorado, New Mexico, Utah. This species should be looked for in Navajo County, Arizona. Best recognized by the relatively small flowers and fruits and the trichomes that cover the dorsal surface of its often relatively narrow petals. A somewhat similar species, *M. marginata* (Osterh.) H. J. Thomps. & Prigge, occurs north of our area. It also has trichomes on the dorsal surface of the petals but has straight adjoining walls in the seed coat, somewhat larger flowers, and broader leaves.

Mentzelia filifolia J. J. Schenk & L. Hufford **CP** (narrow leaf) Filiform blazing-star. Biennial. **STEMS** to 75 cm tall, usually branching throughout. **LEAVES** to 11.5 cm long, to 3.5 cm wide; mostly oblanceolate to elliptic, rachis filiform, 8–20 lobes. **FLOWERS** with calyx lobes 6–11 mm long; petals yellow, 14–18.5 mm long, 3.5–6 mm wide, oblanceolate, glabrous on abaxial surface; outer whorl of stamens fertile and staminodial; second whorl of stamens fertile; style 10–14 mm long. **CAPSULES** cylindrical, 11–19 mm long, 5–7.5 mm wide, straight. **SEEDS** gray to light brown, lenticular-ovoid, 2.9–3.2 mm long, winged, seed coat anticlinal cell walls sinuate, with central papillae. n = 10. Loamy and rocky soils. Sagebrush, piñon-juniper, and mixed conifer communities. ARIZ: Apa; NMEX: McK. 2100 m (6960′). Flowering: Jun–Aug. Fruit: Jul–Sep. In the past, this taxon was treated as part of *M. laciniata*.

Mentzelia holmgreniorum J. J. Schenk & L. Hufford (in honor of Noel and Patricia Holmgren) Holmgren blazing-star. Biennial. **STEMS** to 50 cm tall, main stem erect, straight. **LEAVES** to 9 cm long, to 3.2 cm wide; narrowly to broadly spatulate, petiolate, margins pinnatisect with 14–20 lobes. **FLOWERS** with calyx lobes 6.5–9.5 mm long; petals 5, yellow, 13.5–18.5 mm long, mostly 5–6 mm wide, narrowly spatulate, glabrous on abaxial surface; outer whorl of stamens fertile and staminodial; second whorl of stamens fertile; style 8.5–10.5 mm long. **CAPSULES** cylindrical, 13–14.5 mm long, 6–7 mm wide, base tapering. **SEEDS** pale gray with a white wing, lenticular-ovoid, 3.7–3.8 mm long, seed coat anticlinal cell walls sinuate, with central papillae. n = 10. Sandy soils in washes, roadsides, and other disturbed sites. Mixed conifer communities. ARIZ: Apa. 1495–2225 m (6000–7200′). Flowering: Jun–Aug. Fruit: Jul–Sep. This species has broader lamina lobes and rachises than *M. filifolia*. Within our study area, the Holmgren blazing-star is found in the Canyon de Chelly region.

Mentzelia laciniata (Rydb.) J. Darl. (laciniate leaves) Cutleaf blazing-star. Perennial. **STEMS** to 60 cm tall, usually branching throughout. **LEAVES** to 10 cm long, to 3 cm wide; elliptic to lanceolate or linear; basal leaves petiolate, usually deeply and narrowly pinnately lobed, sometimes merely toothed or approaching entire; cauline leaves sessile, usually without clasping bases or basal lobes, narrowly and regularly lobed or toothed, approaching entire only if very narrow. **BRACTS** of early flowers reduced leaves; others linear to lanceolate, entire or toothed, mostly subtending the ovary. **FLOWERS** with calyx lobes 5–6 mm long; petals yellow, widely spreading, 8–15 mm long, 3–5 mm wide, oblanceolate to spatulate, obtuse to acute, with trichomes at apex only; staminodia (0) 3–5, slightly smaller than petals; outer several whorls of stamens with slightly broadened to petaloid filaments; ovary 4–7 mm long; style 7–10 mm long. **CAPSULES** pedicellate, cylindrical, 9–18 mm long, 3–5 mm wide, straight. **SEEDS** oblong to oval, 2–3 mm long, the wing nearly obsolete to about 0.25 mm wide; epidermal cells with wavy to slightly sinuate adjoining walls, the surface walls with 5–10 (–15) small bumps. 2*n* = 20. [*Nuttallia laciniata* (Rydb.) Wooton & Standl.]. Sandy or clay soils, desert shrub, and piñon-juniper communities. ARIZ: Apa; COLO: Arc, LPl, Mon; NMEX: McK, RAr, SJn. 1900–2200 m (6000–7200′). Flowering: Jun–Oct. Fruit: Jul–Oct. Colorado, New Mexico, possibly adjacent Apache County, Arizona. Often misidentified as *M. humilis* (Urb. & Gilg) J. Darl., a species with similar leaves but more broadly winged seeds, that is gypsiferous, and common in southeastern New Mexico and west Texas. An infusion of the flowers used as an eyewash by the Ramah Navajo.

Mentzelia montana (Davidson) Davidson (of mountains) Variegated-bract blazing-star. Annual. **STEMS** to 50 cm tall, potentially branching throughout. **LEAVES** to 10 cm long, to 1.5 cm wide, narrowly elliptic to lanceolate or nearly linear; margins often entire; basal leaves petiolate, toothed to lobed; cauline leaves sessile, mostly entire, occasionally lobed throughout, the bases broad to clasping. **BRACTS** of early flowers leaflike, the others shorter but broader, often ovate or obovate, with whitish bases, commonly on the ovary; margins mostly entire or few-toothed. **FLOWERS** with calyx lobes 1–5 mm long; petals yellow, spreading, broadly obovate or obcordate, 2.5–5 (–10) mm long, 2–5 mm wide; staminodia and broad-filament stamens lacking; ovary 4–10 mm long; style 1.5–4 (–7) mm long. **CAPSULES** sessile, clavate to nearly cylindric, mostly tapering only near base; body 10–16 (–20) mm long, apex 2–3 mm wide, when mature straight, rarely with a slight bend. **SEEDS** irregular, several-faceted, the angles sharp; epidermal cells with adjoining walls straight, the surface walls flat to strongly pointed-papillate. 2*n* = 36. Sand and clay soils, desert shrub, and piñon-juniper communities. ARIZ: Apa, Coc, Nav; COLO: Arc, Dol, LPl, Mon; NMEX: McK, RAr, San, SJn; UTAH. 1350–2350 m (4500–7700′). Flowering: Apr–Aug. Arizona, California, Colorado, New Mexico, Oregon, Texas, Utah; northern Mexico. Immature and depauperate specimens are difficult or impossible to separate from *M. albicaulis*. Best recognized by its shorter, straight fruits that taper only near the base and its white-based bracts.

Mentzelia multiflora (Nutt.) A. Gray (many-flowered) Adonis blazing-star, desert blazing-star. Perennial. **STEMS** to 80 cm tall, potentially producing branches throughout. **LEAVES** to 15 cm long, to 3 cm wide; narrowly elliptic to lanceolate, occasionally oblanceolate; basal leaves petiolate, the margins toothed to lobed; cauline leaves sessile, commonly with clasping bases or basal lobes, the margins deeply and unevenly lobed, toothed or sometimes approaching entire in very narrow leaves. **BRACTS** of early flowers reduced leaves, the others linear-lanceolate and entire, few-toothed or lobed, mostly subtending the ovary. **FLOWERS** with calyx lobes 6–12 mm long; petals yellow, rarely cream or nearly white, widely spreading, 9–23 mm long, 3–10 mm wide, the apex rounded to obtuse, with trichomes at apex only; staminodia (0–) 5, slightly smaller than petals; outer several whorls of stamens with petaloid or broadened filaments; ovary 4–7 mm long; style (7–) 10–14 mm long. **CAPSULES** pedicellate, cylindrical, (6–) 10–20 mm long, 4–6 mm wide, straight. **SEEDS** oval, 2.5–4 mm long, the wing 0.5 mm wide or slightly wider; epidermal cells with sinuate adjoining walls, the surface walls with a finely cobbled dome. 2*n* = 18 (20). [*Bartonia multiflora* Nutt.; *Nuttallia multiflora* (Nutt.) Greene]. Sand, clay, alluvium, shale soils, desert shrub, piñon-juniper woodlands, and ponderosa pine communities. ARIZ: Apa, Coc, Nav; COLO: Arc, Dol, LPl, Mon; NMEX: McK, RAr, San, SJn; UTAH. 1125–2300 m (3700–7500′). Flowering: Apr–Sep. Fruit: May–Oct. Arizona, California, Colorado, Nebraska, New Mexico, Oklahoma, Texas, Utah, Wyoming; northern to central Mexico. A widespread, common species, distinguishable from *M. laciniata* by its wider seed wings and frequently irregular lobes or teeth of its leaves. Often misidentified as *M. pumila* Torr. & A. Gray, due to previous misapplication of that name. This species occurs much further north of the study area and has seeds with straight adjoining walls. The plant is used as an emetic by the Navajo and the seeds are used for food; the leaves are chewed and sprayed with the mouth on offerings before and after making prayer sticks. The plant is used as a fumigant for the collared lizard ceremony by the Kayenta Navajo.

Mentzelia obscura H. J. Thomps. & J. E. Roberts (poorly known) Pacific blazing-star. Annual. **STEMS** to 30 cm tall, potentially branching throughout. **LEAVES** to 8 cm long, to 1 cm wide, elliptic to broadly lanceolate; basal leaves petiolate, the margins usually lobed or toothed; cauline leaves sessile, the bases broad to clasping, the margins entire or toothed. **BRACTS** of early flowers leaflike, usually entire; others smaller and broader, when ovate mostly 2-toothed near base, green, occasionally with whitish bases, mostly subtending the ovary. **FLOWERS** with calyx lobes 1.5–4 mm long; petals yellow, spreading, broadly obovate or obcordate, 2.5–6 mm long, 1.5–3.5 mm wide; staminodia and broad-filament stamens lacking; ovary 3–5 (–9) mm long; style 2–4 mm long. **CAPSULES** sessile, clavate, long-tapering to base; body 11–25 mm long, apex 2–3 mm wide, the earliest usually bent 180°, less commonly S-shaped or bent at least 90°. **SEEDS** irregularly ovate, weakly several-faceted, the angles rounded, 1 mm long, wingless; epidermal cells with straight adjoining walls, the surface walls domed or with a short central papilla. $2n = 36$. Silty to gravelly alluvial soils, riparian and desert shrub communities. UTAH. 1125–1350 m (3700–4500′). Flowering: Mar–May. Fruit: Apr–May. Arizona, California, Nevada, Utah; northern Mexico. Apparently reaching the northeastern limits of its range here, in the low-elevation river channels and canyon bottoms.

Mentzelia pterosperma Eastw. (winged seed) Wingseed blazing-star. Perennial. **STEMS** to 25 cm tall, potentially branching throughout when robust. **LEAVES** to 6 cm long, to 2.5 cm wide; oblanceolate, or obovate to broadly lanceolate; basal leaves long-petiolate, entire, crenate or shallowly toothed, rarely lobed; cauline leaves sessile, often clasping, crenate, toothed or sometimes entire. **BRACTS** of all flowers lanceolate or linear, entire or shallowly toothed, subtending the ovary. **FLOWERS** with calyx lobes 6–10 mm long; petals yellow, widely spreading, 10–15 (–20) mm long, 3–5 mm wide, oblanceolate, the apex acute, with trichomes at apex only; staminodia 0 (–5), smaller than petals; stamens with broad filaments common in outer whorl; ovary 7–10 mm long; style (8–)10–15 mm long. **CAPSULES** pedicellate, cup-shaped to cylindrical, 8–15 mm long, 5–10 mm wide, straight. **SEEDS** nearly orbicular, 3–4 mm long, the wing 1 mm wide; epidermal cells with straight to slightly curved adjoining walls, the surface walls with about 10–15 bumps. $2n = 22$. [*Touterea pterosperma* (Eastw.) Rydb.; *Nuttallia pterosperma* (Eastw.) Greene]. Sand or clay soils, sometimes also on gypsum; desert shrub, sagebrush, and piñon-juniper communities. COLO: Mon; UTAH. 1125–1350 m (3700–4500′). Flowering: May–Sep. Fruit: Jun–Oct. Arizona, California, Colorado, Nevada, Utah. The diminutive stature of many specimens suggests that this species may be facultatively annual.

Mentzelia rusbyi Wooton (for H. H. Rusby, botanist) Rusby's blazing-star. Perennial. **STEMS** to 100 cm tall, branches strongly ascending and restricted to the upper 1/2. **LEAVES** to 15 cm long, to 2 cm wide; narrowly elliptic to lanceolate; basal leaves short-petiolate, shallowly toothed, occasionally subentire; cauline leaves sessile, toothed, the bases clasping. **BRACTS** of early flowers broadly lanceolate, lobed, the others further reduced, often arising from the ovary. **FLOWERS** with calyx lobes 7–10 mm long, petals white or cream but drying pale yellow, erect-spreading, 10–23 mm long, 4–6 mm wide, narrowly oblanceolate or spatulate, the apex acute, with trichomes at apex only; staminodia usually 3, equal to or smaller than petals; stamens with broad filaments few; ovary 4–10 mm long; style 9–12 mm long. **CAPSULES** mostly pedicellate, cylindrical, 14–32 mm long, to 10 mm wide, straight. **SEEDS** oval, 3–4 mm long, the wing about 0.5 mm wide; epidermal cells with strongly sinuate adjoining walls, the surface walls a finely cobbled dome. $2n = 20$. [*M. pumila* Torr. & A. Gray var. *rusbyi* (Wooton) Urb. & Gilg; *M. nuda* (Pursh) Torr. & A. Gray var. *rusbyi* (Wooton) H. D. Harr.; *Nuttallia rusbyi* (Wooton) Rydb.]. On a wide range of substrates; piñon-juniper woodlands, Gambel's oak, and ponderosa pine communities. ARIZ: Apa; COLO: Arc, Dol, LPl, Mon; NMEX: McK, RAr, SJn; UTAH. 2000–2600 m (6500–8500′). Flowering: Jun–Sep. Fruit: Jul–Oct. Arizona, Colorado, New Mexico, Utah. This species is easily recognized by its height, branching pattern, large fruits, and conspicuous bracts.

Mentzelia sivinskii J. J. Schenk & L. Hufford **CP** (in honor of Robert Sivinski, New Mexico state botanist) Sivinski blazing-star. Biennial. **STEMS** to 70 cm tall, main stem erect, straight, epidermis pubescent. **LEAVES** to 11.2 cm long, 3–11.5 mm wide; narrowly to broadly spatulate, petiolate, margins pinnate with 18–24 lobes. **FLOWERS** with calyx lobes 5.5–9.3 mm long; petals 5, light yellow to yellow, 9–14.5 mm long, mostly 3–6.5 mm wide, narrowly spatulate, glabrous on abaxial surface; outer whorl of stamens fertile and staminodial; second whorl of stamens fertile; style 4.5–10 mm long. **CAPSULES** cup-shaped, 8.2–12.5 mm long, 5–7.5 mm wide. **SEEDS** pale gray to light brown with a white wing, lenticular-ovoid, 2.7–2.8 mm long, seed coat anticlinal cell walls sinuate, with central papillae. Sandy, clay soils of the Nacimiento Formation. Desert scrub and Utah juniper communities. COLO: LPl; NMEX: RAr, SJn. 1525–1815 m (5000–5955′). Flowering: Jun–Aug. Fruit: Jul–Sep. Schenk (2009) found *M. sivinskii* to be most closely related to *M. integra* (M. E. Jones) Tidestr. and *M. procera* (Wooton & Standl.) J. J. Schenk & L. Hufford. The Sivinski blazing-star is a narrow endemic in northwest New Mexico.

Mentzelia thompsonii Glad (for H. J. Thompson, botanist) Thompson's mentzelia, Thompson's blazing-star. Annual. **STEMS** to 9 cm tall, branching throughout when robust. **LEAVES** to 5 cm long, 1.5 cm wide, elliptic to lanceolate; basal leaves short-petiolate, entire; cauline leaves sessile to short-petiolate, entire (rarely with 1–few teeth). **BRACTS** of early flowers leaflike, the others reduced, narrowly lanceolate to ovate, green, mostly subtending the ovary; margins entire, rarely few-toothed. **FLOWERS** with calyx lobes 1–2 mm long; petals yellow, broadly obovate or obcordate, 2–3 mm long, about 2 mm wide, with a few trichomes at the apex; petaloid staminodia and broad-filament stamens absent; ovary 4–7 mm long at anthesis; style 1–2 mm long. **CAPSULES** sessile, cylindric but slightly tapering to base, straight (rarely slightly bent); body 6–16 mm long; apex 2–3 mm wide. **SEEDS** rounded to weakly faceted, the angles blunt, 2 mm long, wingless, epidermal cells with straight adjoining walls, the exposed surface flat or with a raised, conic center. Mostly on

Mancos clay soils; salt desert shrub communities. COLO: Mon; NMEX: SJn; UTAH. 1600–1950 m (5100–6400′). Flowering: Apr–May. Fruit: Apr–Jun. Colorado, New Mexico, Utah. Its combination of short stature; relatively long, broad, entire leaves; and small flowers makes this an easy species to recognize.

MALVACEAE Juss. MALLOW FAMILY

S. L. Welsh & N. D. Atwood

Herbs or less commonly shrubs, and some trees, usually pubescent with branched or stellate hairs, annual, biennial, or perennial, with mucilaginous juice. **LEAVES** alternate, stipulate, simple, mostly palmately veined. **INFL** solitary or in thyrsoid cymes or more or less racemose or paniculate, sometimes with an involucel of sepaloid bractlets. **FLOWERS** perfect (or imperfect), regular; sepals 5, more or less persistent; petals 5, separate, adnate to the staminal sheath; stamens numerous, united by the filaments (monadelphous); ovary superior, 3- to many-loculed. **FRUIT** a capsule or schizocarp. $x = 6–17+, 20+$. (Welsh 1980.) Cosmopolitan, especially the tropics; approximately 1550 species.

1. Involucel (epicalyx of sepaloid bractlets) lacking; petals white, pink, or lavender; plants of moist sites, usually at middle and higher elevations .... ***Sidalcea***
1′ Involucel (epicalyx of sepaloid bractlets) present, or if lacking, otherwise different than above; petals variously colored .... (2)

2. Petals orange or rarely purplish pink; plants indigenous perennial herbs mainly of arid habitats at middle and low elevations .... ***Sphaeralcea***
2′ Petals variously colored, but not orange; plants indigenous or introduced perennial, biennial, or annual plants of various distribution .... (3)

3. Flower corolla rose-pink or rarely white; plants indigenous, 7–15 dm tall, perennial, of middle and higher elevations .... ***Iliamna***
3′ Flower corolla white, pink, rose, yellow, or other hues; plants differing in one or more ways from above .... (4)

4. Flower mostly 6–10 cm broad, opening flat; plants tall, adventive or cultivated biennials .... ***Althaea***
4′ Flower less than 6 cm broad or, if broader, the plants shrubby or perennial .... (5)

5. Style branches 5, elongate; fruit a capsule; plants low annual .... ***Hibiscus***
5′ Style branches more than 5, short; fruit a schizocarp; plants annual or biennial .... (6)

6. Style branches filiform, with elongate stigmatic lines; plants common, weedy .... ***Malva***
6′ Style branches with capitate or truncate stigmas; plants rare .... ***Anoda***

Althaea L. Hollyhock

(in Greek mythology, the wife of Oeneus, King of Calydon) Plants herbaceous, biennial, with coarse stellate hairs. **LEAVES** alternate, petiolate, cordate at the base, lobed. **INFL** a solitary flower, (2) or borne in racemiform inflorescence; involucel of 6–9 bractlets, connate at the base. **FLOWER** corolla of various colors; calyx 5-cleft. **FRUIT** flattened, wheel-like, invested by the calyx, the numerous carpels separating at maturity. Europe to Asia, naturalized in the United States; approximately 12 species.

Althaea rosea (L.) Cav. (rose-colored) Hollyhock. Coarse biennials to 20 dm tall or more, the stems erect, stellate-hairy. **LEAVES** (3) 5- to 7-lobed, mostly 3–15 cm long (from sinus to apex) and often much broader. **FLOWERS** shortly pedicellate, 6–12 cm wide or more; calyx lobes triangular, investing the fruit at maturity, the involucel calyxlike; corolla variously colored, often rose to pink or lavender or sometimes white, usually with a dark center; carpels numerous; stellate along the margins, and reticulate on the sides, 5–7 mm long. $2n = 42$. [*Alcea rosea* L.]. Cultivated ornamental, persisting and escaping. ARIZ: Nav; COLO: LPl, Mon; NMEX: SJn. 1615–2160 m (5300–7095′). Flowering: Jul–Aug. Widespread in North America; introduced from China. This is a hardy, robust species with flowers of great charm and beauty. It has been widely grown in the American West since pioneer times. Given any chance at all, it seeds spontaneously and grows in the same place for long periods. It is a not-infrequent roadside waif.

Anoda Cav. Anoda

(Ceylonese name) Herbaceous annuals from taproots, the pubescence sparsely hirsute to densely puberulent or tomentose. **LEAVES** alternate, long-petiolate, usually more or less truncate to subcordate basally, commonly 3–5-

lobed, the midlobe comprising 1/2 or more of the leaf length. **INFL** a solitary flower, axillary; involucel lacking. **FLOWER** calyx 5-cleft. **FRUIT** a schizocarp, the inner layer forming a saclike envelope around the seed or becoming closely adherent to the seed coat, the dorsal wall more persistent. Mostly in tropical America; approximately 10 species.

Anoda cristata (L.) Schltdl. (crested) Crested anoda. Plants to approximately 8 dm tall, branching from near the base, the herbage sparingly hirsute with mostly simple hairs. **LEAVES** long-petiolate, the blades 2–6 cm long, deltoid to triangular-ovate, from longer than broad to about as broad as long, 3–5-lobed, the terminal lobe making up 1/2 or more of the blade length, often marked or margined with purple. **INFL** axillary, a solitary flower, long-pedicellate. **FLOWER** petals purple, 1–2.5 cm long; carpels 15–20, rather conspicuously beaked, hispid, the basal portion entirely thin-scarious and veinless, and with slender midnerve, the partitions wholly obliterated and breaking up in fruit. [*Sida cristata* L.]. The solitary specimen examined from the Four Corners (B-Square Ranch) was collected in a hayfield (*Heil & Bolack 14364* SJNM). NMEX: SJn. 1625 m (5335′). Flowering: Sep–Oct. Texas to Arizona and South America.

Hibiscus L. Rosemallow

(Greek *hibiscos*, mallow, derived from Ibis, an Egyptian deity) Plants herbaceous or woody, annual or perennial, with stellate or simple hairs. **LEAVES** alternate, petiolate, obtuse to truncate or cordate basally, lobed to incised. **INFL** axillary, solitary-flowered; involucel of 5–10 distinct bractlets. **FLOWER** calyx 5-cleft, more or less accrescent in fruit. **FRUIT** a loculicidal capsule, the carpels 5. **SEEDS** several in each locule. Warm-temperate to tropical; approximately 200 species.

Hibiscus trionum L. (three-parted) Flower-of-an-hour. Annual, commonly 1.5–5 dm tall, the lower branches often prostrate, coarsely hispid-stellate to glabrate. **LEAVES** 3-lobed or more commonly 3–5-parted, the main lobes cuneate basally, the middle lobe the largest. **INFL** a solitary flower, axillary, mostly 3–6 cm wide; bractlets usually 10, linear, often coarsely hispid, much shorter than the fruiting calyx. **FLOWER** petals cream-colored to yellowish, with a purple center, closing in shade. $2n = 28, 56$. Weedy species of gardens, open spaces, and cultivated land. ARIZ: Nav; COLO: Mon; NMEX: SJn. 1590–1750 m (5215–5740′). Flowering: Sep–Oct. Widespread in North America; adventive from central Africa.

Iliamna Greene Wild Hollyhock

(possibly from the place name *Iliamna*, for a lake, a volcano, and a town in Alaska) Plants herbaceous, perennial, sparingly and minutely stellate-hairy. **LEAVES** alternate, petiolate, the blade cordate to truncate basally, the margin lobed. **INFL** a thyrsoid panicle; involucel of 3 narrow, persistent bractlets. **FLOWER** calyx 5-cleft. **FRUIT** a loculicidal capsule, the carpels many. **SEEDS** usually 3 in each locule (Wiggins 1936). North American, seven species.

Illiamna grandiflora (Rydb.) Wiggins (large-flowered) Wild hollyhock. Perennial. **STEMS** few to many from a woody caudex, mostly 10–15 dm tall, minutely stellate-puberulent, green. **LEAVES** 5–7-lobed, cordate to truncate basally, 2.5–15 cm long (from petiole apex to tip), 6–10 cm broad, the lobes triangular to lanceolate, crenate-serrate, finely stellate; petioles mostly less than 1 (15) cm long; bracteoles lanceolate, more than 1/2 as long as the calyx. **FLOWER** calyx lobes 5–8 mm long (longer in fruit); petals rose-pink (rarely white), 30–37 mm long; carpels about 10 mm long in fruit, hispid and stellate. [*Sphaeralcea grandiflora* Rydb., not *S. grandiflora* Phil; *S. rydbergii* Tidestr.]. Streamsides and mesic slopes. COLO: LPl, Mon (Mesa Verde National Park); UTAH (Abajo Mountains). 2105–2590 m (6900–8500′). Flowering: Jul–Aug. Colorado, Utah, Arizona, and New Mexico.

Malva L. Mallow

(Greek *malache*, soft) Herbaceous annual, biennial, or perennial, from taproots, the pubescence simple to branched or stellate. **LEAVES** alternate, petiolate, usually more or less cordate basally, commonly lobed. **INFL** in an axillary cluster (sometimes solitary) or in subterminal panicles; involucel of 3 narrow to broad persistent bractlets. **FLOWER** calyx 5-cleft, the petioles to 20 cm long or more; the carpels mostly 10–15. **FRUIT** a schizocarp. European; 30 species.

Malva neglecta Wallr. (overlooked) Cheese mallow. Annual or biennial. **STEMS** prostrate-spreading, commonly 1–6 dm long, stellate-hairy. **LEAF BLADES** reniform-orbicular, 0.6–3 cm long (from sinus to apex) or more, and much broader, crenate and not at all to only shallowly 5- to 7-lobed, the petioles to 20 cm long or more. **INFL** clustered (or solitary) in the axils; bractlets linear. **FLOWER** calyx (3) 4–6 mm long at anthesis, the lobes acuminate; petals white to pink or lilac, about twice as long as the sepals; carpels hairy, rounded on the back. $2n = 42$. Weeds of disturbed sites and cultivated land. ARIZ: Apa, Nav; COLO: Arc, Dol, Hin, LPl, Mon, SJn, SMg; NMEX: McK, RAr, San, SJn; UTAH. 1525–2430 m (5000–8000′). Flowering: May–Oct. Likely universal; widespread in North America; native to Eurasia. This is a persistent, vigorous weed of gardens. The taproot is far-reaching and difficult to pull. Most herbicides do not kill this plant. It seeds heavily, and the seeds persist in the soil for a long time. Immature fruits are edible and have a good flavor; they have been eaten by generations of children.

Sidalcea A. Gray Checker Mallow

Herbaceous perennial from taproots or short rhizomes, usually stellate and somewhat hirsute. **LEAVES** alternate, petiolate, often dimorphic, the lowermost merely palmately lobed, the upper ones commonly cleft and with linear lobes. **INFL** a semispicate raceme, dimorphic, those of plants with perfect flowers the largest; involucel lacking. **FLOWER** calyx 5-cleft; carpels 5–10, 1-seeded, tardily separating (Hitchcock 1957; Roush 1931). Western North America; 20 species.

1. Petals white or merely pinkish-tinged, often drying yellow; anthers bluish pink; plants rhizomatous; stems hirsute below .... ***S. candida***
1′ Petals pink to lavender; anthers usually yellow to white; plants rhizomatous or not; stems hirsute to glabrous or tomentose below........ ***S. neomexicana***

Sidalcea candida A. Gray (very white) White checker. Plants from slender rhizomes. **STEMS** 4–10 dm tall, glabrous to hirsute with simple hairs below, more or less stellate above. **LEAVES** with blades 6–20 cm wide, the basal ones shallowly 5- to 7-lobed and coarsely crenate, the upper ones divided into 3–5 entire segments. **FLOWER** calyx 7–10 mm long, variously stellate-hairy and glandular-puberulent; petals white to pinkish, often drying yellow, 12–20 mm long; carpels about 3 mm long. Stream banks, lake shores, wet meadows, and seeps, mainly in sagebrush, mountain brush, and aspen-spruce-fir communities. COLO: Arc, Hin, LPl, Min, Mon, SJn; NMEX: RAr. 2365–2730 m (7800–9000′). Flowering: June–Sep. Wyoming west to Nevada.

Sidalcea neomexicana A. Gray (of New Mexico) New Mexico checker. Plants from thickened taproot or fascicled roots, and often with a caudex. **STEMS** 2–9 (10) dm tall, hirsute below (or rarely glabrous) with simple or bifurcate hairs. **LEAVES** with blades 1.5–11 cm wide, the basal ones crenate to shallowly 5- to 7-lobed, the cauline ones divided usually into 5 laciniate to entire segments; petioles slender, usually equaling or surpassing the calyx (to 2 cm long or more). **FLOWER** calyx 5–10 mm long, usually with some simple pustulose hairs interspersed with the stellate ones; petals rose-pink (fading blue-purple), 11–19 mm long; carpels 2–3 mm long. $2n = 20$. Wet meadows, stream banks, and seeps. ARIZ: Apa; COLO: Arc, Hin, LPl, Min, Mon; NMEX: McK, RAr, SJn; UTAH. 1980–2680 m (6500–8800′). Flowering: Jun–Sep. Utah, Oregon, Idaho, and Wyoming, south to California, Arizona, and Mexico. The Ramah Navajo used a cold infusion for internal injuries.

Sphaeralcea A. St.-Hil. Globemallow

Herbaceous perennial from taproots or rhizomes, glabrescent to canescent with stellate hairs. **LEAVES** alternate, petiolate, sometimes dimorphic, the lowermost merely toothed or palmately lobed (rarely entire), the upper ones cleft to entire. **INFL** a racemose to thyrsoid cyme; involucel of 3 or fewer filiform bractlets. **FLOWER** calyx 5-cleft; carpels 8–20, the seeds 1 or 2 per carpel. **FRUIT** a schizocarp, the mature fruit segments divided into a basal indehiscent, reticulate portion and an apical dehiscent portion (Atwood and Welsh 2002; Jefferies 1972; Kearney 1935). Flowers of *Sphaeralcea* are characteristically grenadine (a shade of orange), typically the only flowers so colored in vast stretches of the American Southwest; occasional plants and even small populations bear flowers that are pink, which fade mainly a violet color. Deserts of America; approximately 60 species.

1. Inflorescence racemose, rarely with more than 1 flower per node; reticulate part of carpel conspicuously wider than the nonreticulate part, forming 2/3 of the carpel ....(2)
1′ Inflorescence thyrsoid to thyrsoid-glomerate, with usually more than 1 flower per node; reticulate part of carpel not much wider than the dehiscent part, less than 2/3 of the carpel. ....(3)

2. Leaf blades distinctly 2–5-lobed, -parted, or -divided, the uppermost ones entire; hairs of rays radiating in a single plane; carpels 7–9; plants often 1.5 dm tall or more........ ***S. leptophylla***
2′ Leaf blades usually 3–5-lobed, the lobes usually toothed or again lobed, the upper ones not or seldom entire; carpels 10–14; rays of hairs usually radiating in several planes .... ***S. coccinea***

3. Leaf blades 1.5–3.5 times longer than broad, the terminal lobe exaggerated........(4)
3′ Leaf blades as broad as long or almost so ....(5)

4. Calyx lobes much surpassing the capsuloid schizocarp; reticulations of the carpel coarse and prominent .... ***S. subhastata***
4' Calyx lobes shorter than, subequal to, or only slightly surpassing the capsuloid schizocarp; reticulations of the carpel neither especially coarse nor prominent .... ***S. fendleri***

5. Leaves shallowly 3–5-lobed; carpels with well-defined to nearly obscure reticulae on the lower 1/3; plants mainly of western Colorado, east across Utah to eastern California, and south to northwest New Mexico and central Arizona .... ***S. parvifolia***
5' Leaves 3–5-cleft, -parted, or -divided; carpels with well-defined reticulae on less than 1/2; plants of various distribution .... (6)

6. Midlobe of leaf not more than 5 mm wide; inflorescence racemiform or subthyrsoid, few-flowered .... ***S. digitata***
6' Midlobe of leaf often more than 5 mm wide, mainly 10–20 mm; inflorescence interrupted-thyrsoid, many-flowered .. (7)

7. Herbage bright green; leaves thin-textured, often narrowly lobed; plants of the San Juan arm of Lake Powell, San Juan County, Utah .... ***S. moorei***
7' Herbage gray-green to whitish-canescent; leaves often thick-textured, seldom narrowly lobed; plants more widely distributed .... ***S. grossulariifolia***

Sphaeralcea coccinea (Nutt.) Rydb. (scarlet) Common globemallow. **STEMS** solitary or few to many from the apex of a woody caudex and stout taproot, or less commonly from creeping rhizomes, 0.6–4.2 dm tall, white to yellowish-canescent. **LEAVES** with blades 1.1–3.7 cm long, 1.2–5.2 cm wide, usually wider than long, ovate to cordate-ovate in outline, the base often cordate, usually 3–5-lobed, with main divisions cleft almost or quite to the base, the lobes usually again toothed or lobed, varying from linear to broadly spatulate-obovate and more or less confluent. **INFL** racemose, sometimes paniculate, rarely thyrsoid, the flowers mostly solitary at the nodes; pedicels shorter than calyx. **FLOWER** calyx uniformly stellate, the rays or hairs not radiating in a single plane, the lobes lance-acuminate; petals 8–15 mm long, orange; carpels 8–14, 2–3 mm high, the indehiscent part forming 2/3 or more of the carpel, reticulate on the sides and on the back. $n = 5$. [*Malva coccinea* Nutt.; *Cristaria coccinea* (Nutt.) Pursh; *Malva coccinea* Nutt.; *Malvastrum coccineum* (Nutt.) A. Gray; *Sida dissecta* Nutt. ex Torr. & A. Gray; *Malvastrum coccineum* var. *dissectum* (Nutt. ex Torr. & A. Gray) A. Gray; *M. dissectum* Harv.; *Sphaeralcea dissecta* (Nutt. ex Torr. & A. Gray) Rydb.; *S. coccinea* subsp. *dissecta* (Nutt. ex Torr. & A. Gray) Kearney; *S. coccinea* var. *dissecta* (Nutt. ex Torr. & A. Gray) Garrett; *Malvastrum coccineum* var. *elatum* Baker f.; *M. elatum* (Baker f.) A. Nelson; *S. elata* (Baker f.) Rydb.; *S. coccinea* var. *elata* (Baker f.) Kearney; *Malvastrum cockerellii* A. Nelson; *M. micranthum* Wooton & Standl.]. Blackbrush, greasewood, shadscale, mat-atriplex, *Ephedra*, Indian ricegrass, galleta, sagebrush, juniper-piñon, mountain brush, and ponderosa pine communities. ARIZ: Apa, Nav; COLO: Arc, Dol, LPl, Mon, SMg; NMEX: McK, RAr, San, SJn; UTAH. 1555–2390 m (5100–7850′). Flowering: Apr–Sep. Saskatchewan and Alberta south to Arizona, New Mexico, and Texas. Used as a ceremonial medicinal plant, dermatological aid, dietary aid, disinfectant, and beverage by the Kayenta Navajo. The Ramah Navajo used the plant as a "life medicine."

Sphaeralcea digitata (Greene) Rydb. (fingered) Juniper globemallow. Perennial. **STEMS** few to many from a woody caudex, mainly 2–5.5 dm tall, the herbage mainly thinly canescent to rather sparsely stellate-hairy and green. **LEAVES** with blades 1.5–4 cm long, and about as wide, pedately 5-lobed, the lobes distinct to the base, all divisions oblanceolate to narrowly obovate, cuneate at base, obtuse to acute and often mucronulate apically, the midlobe not more than 5 mm wide and not much surpassing the main lateral lobes, the lobes entire or the midlobe irregularly few-toothed or cleft. **INFL** usually 10–20-flowered, narrowly thyrsoid, with often only the lowermost nodes with more than 1 flower per node; pedicels shorter than to much longer than the calyx. **FLOWER** calyx 3.5–7 mm long, uniformly stellate, the rays of hairs not or less commonly radiating in a single plane, the longest trichome rays mainly less than 0.4 mm long except on some calyx bases, the calyx lobes ovate to lance-acuminate; petals 8–14 mm long, orange (grenadine); carpels 9–13, 3–4 mm high, the indehiscent portion forming from 1/3–1/2 of the carpel, not prominently reticulate on the sides, with lacunae opaque or nearly so. [*Malvastrum digitatum* Greene]. Piñon-juniper woodland communities, often along or near drainages. ARIZ: Nav. 1525–2135 m (5000–7000′). (*O'Kane et al. 5277*, 5 Jun 2001, w. branch Piute Creek, SJNM). We have not seen any specimens matching the above description from the Four Corners except that from Navajo County. Flowering: May–Jun. The Ramah Navajo used an infusion of the whole plant for stomachaches and the root as a "life medicine."

Sphaeralcea fendleri A. Gray (for Augustus Fendler, German-American botanical collector) Fendler's globemallow. Perennial. **STEMS** few to several from a woody caudex, 4–14 dm tall or more, the herbage green and sparsely to densely white to gray-canescent. **LEAVES** with blades 2–6 cm long, 1.3–5.2 cm wide, from longer than wide to wider than long, ovate-oblong to very broadly ovate, the base usually strongly cuneate, shallowly to deeply 3-lobed, the lateral lobes about 1/4 as long as the midlobe, the main division sometimes coarsely and irregularly few-toothed or -cleft. **INFL** a very narrow, interrupted, few- to many-flowered thyrse, with lower branches seldom exceeding 3 cm, with usually more than 1 flower per node; pedicels equaling to longer than the calyx. **FLOWER** calyx 4.5–6 mm long, uniformly stellate, the rays of hairs not or less commonly radiating in a single plane, the longest trichome rays mainly less than 0.4 mm long except on some calyx bases, the calyx lobes deltoid to ovate-lanceolate and acuminate; petals 8–13 mm long, orange (grenadine) or less commonly rose-pink and drying violet; carpels 11–15, 4–5 mm high, the indehiscent portion forming from 1/5–1/3 of the carpel, faintly and finely reticulate on the sides, with lacunae opaque or nearly so. [based on *S. miniata* A. Gray, not *S. miniata* (Cav.) Spach; *S. leiocarpa* Wooton & Standl.]. Desert shrublands in sagebrush upward to ponderosa pine and spruce-fir communities. ARIZ: Apa, Nav; COLO: Arc, LPl; NMEX: McK, RAr, San, SJn. 1470–2350 m (4820–7710'). Flowering: Jun–Sep. Texas and south to Mexico. Perhaps the leaf shape of this species most closely simulates *S. subhastata*, which is generally a much smaller plant, mainly much less than 5 dm tall, which is further distinguished by the coarsely and prominently reticulate portion of the fruiting carpel, the reticulate portion forming 1/3–1/2 of the carpel, and by the calyx lobes much surpassing the capsuloid schizocarps. The Kayenta Navajo used the plant for sand cricket bites and for a sore mouth. The Ramah Navajo used the plant for internal injuries, hemorrhage, and as a lotion for external injuries.

Sphaeralcea grossulariifolia (Hook. & Arn.) Rydb. (gooseberry-like leaves) Gooseberry-leaf globemallow. **STEMS** few to many from a woody caudex, 3.5–7.5 dm tall or more, the herbage mainly white to gray- or yellowish-canescent to rather sparsely stellate-hairy and green. **LEAVES** with blades 1.2–5 cm long, 1.3–5.2 cm wide, from longer than wide to wider than long, cuneate-ovate to ovate or cordate-ovate in outline, the base cordate to truncate or obtusely 3–5-lobed, the lobes distinct to the base or confluent to well above the base, the main division usually cleft or parted to irregularly toothed. **INFL** compact-thyrsoid, with usually more than 1 flower per node; pedicels shorter than to much longer than the calyx. **FLOWER** calyx 4.5–9 mm long, uniformly stellate, the rays of hairs not or less commonly radiating in a single plane, the longest trichome rays mainly less than 0.5 mm long except on some calyx bases, the calyx lobes ovate to lance-acuminate; petals 8–18 mm long, orange or rarely rose-pink; carpels 10–14, 2.9–3.7 mm high, the indehiscent portion forming from 2/5–3/5 of the carpel, reticulate on the sides, with lacunae opaque or nearly so. [*Sida grossulariifolia* Hook. & Arn.; *Malvastrum coccineum* (Nutt.) A. Gray var. *grossulariifolium* (Hook. & Arn.) Torr.; *M. grossulariifolium* (Hook. & Arn.) A. Gray; *Sphaeralcea pedata* Torr. ex A. Gray; *S. grossulariifolia* subsp. *pedata* (Torr. ex A. Gray) Kearney]. Salt, warm, and cool desert shrub, piñon-juniper, and ponderosa pine communities. ARIZ: Apa, Coc, Nav; COLO: Mon; NMEX: SJn; UTAH. 1485–1880 m (4870–6160'). Flowering: Apr–Aug. Washington, Oregon, Idaho, California, Nevada, and Arizona. There is much variation within the widely distributed *S. grossulariifolia*, with herbage varying in aspect from gray to gray-green or brownish, to yellow-green, or green. Leaf blades tend not to be cut completely to the base as is often the case in the related *S. moorei*, *S. fumariensis*, *S. rusbyi*, and *S. gierischii*. Most specimens range in size from 3.5–5.5 dm in height, while all of the other closely allied species except *S. fumariensis* commonly tend to exceed that height. The roots have been chewed by the Hopi and boiled with cacti roots for difficult defecation, and boiled and chewed for broken bones.

Sphaeralcea leptophylla (A. Gray) Rydb. **CP** (thin-leaved) Slenderleaf globemallow, rimrock globemallow. **STEMS** few to many from a woody caudex, clump-forming, 2–5.5 dm tall, grayish-canescent throughout, the stems of the season often growing up through those of the previous years. **LEAVES** with blades 10–32 mm long, digitately 3-lobed, the lobes entire, linear to oblanceolate, 1–4 mm wide, or the upper leaves simple and linear. **INFL** racemose, elongate, usually with 1 flower per node; pedicels from much shorter to longer than the calyx. **FLOWER** calyx uniformly stellate, the rays of hairs radiating in a single plane, the lobes lance-attenuate; petals 8–12 mm long, orange; carpels 7–9, 3–3.5 mm high, the indehiscent portion forming 2/3–3/4 of the carpel, coarsely reticulate, ridged or tuberculate on the back. [*Malvastrum leptophyllum* A. Gray]. Shadscale, blackbrush, ephedra, rabbitbrush, piñon-juniper communities; on the Carmel, Morrison, Entrada, Navajo, Kayenta, Chinle, Moenkopi, and Cutler formations. ARIZ: Apa, Nav; COLO: LPl, Mon; NMEX: SJn; UTAH. 1460–1785 m (4790–5855'). Flowering: Apr–Jul. Texas and Mexico.

Sphaeralcea moorei (S. L. Welsh) N. D. Atwood & S. L. Welsh (named for G. Moore, Utah botanist) Moore's globemallow. **STEMS** few to many from a woody caudex, 3–7.5 dm tall or more, the stems typically dark red-purple at least near the base, the herbage green and rather sparsely stellate-pubescent, the trichomes 0.3–0.8 mm wide, the rays radiating in a horizontal plane. **LEAVES** with blades 0.9–3.5 (7.4) cm long, 1–2.6 (15, a rare excption) cm wide, often longer than wide, ovate to cordate-ovate or cordate in outline, the base cordate to truncate or obtuse, digitately 3- to 5-

lobed, the main lobe 0.8–3.5 (7.4) cm long and 0.8–1.5 cm wide, entire or cleft or parted to irregularly toothed, or rarely the blade confluent and merely palmately lobed. **INFL** rather compactly thyrsoid, with usually more than 1 flower per node, or rather compactly glomerate-paniculate with 2–5 flowers on an axillary peduncle; pedicels shorter than to much longer than the calyx; bracteoles linear, sometimes red-purple, but not especially contrasting with the calyx. **FLOWER** calyx 4.5–7.2 (8.5) mm long at anthesis, green, becoming stramineous in fruit, uniformly stellate-hairy or less commonly glabrous externally only near the apex, the rays of hairs mainly radiating in a single plane, the lobes ovate to lance-acuminate; petals 11–15 mm long, orange (grenadine); carpels 10–14, 4.5–5.1 mm high, 2.1–2.3 mm wide, the reticulate portion forming about 1/2 of the carpel, the lacunae opaque. [*S. grossulariifolia* (Hook. & Arn.) Rydb. var. *moorei* S. L. Welsh]. Blackbrush, ephedra, matchweed, resinbush, *Yucca*, piñon-juniper, chaffbush, indigo bush, sagebrush, and hanging garden communities, typically in sandy substrates. UTAH. 1130–1355 m (3710–4440′). Flowering: Apr–Jul. Along the Colorado River in Glen Canyon and its tributaries. This is mainly a low-elevation species of sandy tracts along Glen Canyon and the San Juan River arm of Lake Powell.

Sphaeralcea parvifolia A. Nelson (small-leaved) Small-leaf globemallow. **STEMS** few to many from branching woody caudex. **LEAVES** with blades 1.5–5.5 cm long, 1.2–5.2 cm wide, ovate to orbicular, reniform, or cordate-ovate, the base cordate to truncate or obtuse, usually shallowly 3–5-lobed, the sinuses usually shallow, the lobes crenate-dentate. **INFL** commonly narrowly thyrsoid, typically with more than 1 flower per node; pedicels usually shorter than the calyx. **FLOWER** calyx uniformly stellate, the rays of hairs not in a single plane, the lobes lance-ovate to deltoid; petals 7–15 mm long, orange (grenadine), rarely white; carpels 10–12, 3–4 mm high, the indehiscent part forming from 1/4–1/3 of the carpel, faintly reticulate on the sides. [*S. marginata* York ex Rydb.; *S. arizonica* A. Heller ex Rydb.]. Blackbrush, other warm desert shrub, salt desert shrub, sagebrush, piñon-juniper, and mountain brush communities. ARIZ: Apa, Nav; COLO: Arc, LPl, Mon; NMEX: McK, RAr, San, SJn. 1215–2135 m (3990–7005′). Flowering: May–Sep. Arizona, New Mexico, Colorado, and Utah; west to Nevada and California. The confluent leaf blades are diagnostic for most specimens but should not be taken as indicative in all cases. Color of the herbage varies from distinctly gray-canescent to green or yellow-green. The Hopi used the plant for sores, cuts, and wounds and chewed or boiled the root for broken bones.

Sphaeralcea subhastata J. M. Coult. (somewhat spear-shaped) Spear globemallow. **STEMS** few to many from a rather stout taproot, 1–5 dm tall or long (mainly less than 2.5 dm), the herbage canescent. **LEAVES** with blades 2–5.5 cm long, 1–3 cm wide, longer than wide, oblong-lanceolate to narrowly ovate, acute to short-acuminate at the apex, less densely pubescent above, scarcely lobed but subhastately angled or toothed at base, the margins subentire to irregularly crenate or dentate, the base cuneate, prominently veined beneath. **INFL** racemose, with usually more than 1 flower per node at the lowermost nodes, few-flowered (mostly fewer than 12), usually leafy near the apex; pedicels usually stout and shorter than the calyx, rarely exceeding 1 cm. **FLOWER** calyx 4–11 mm long, uniformly stellate, the calyx lobes ovate-lanceolate to lanceolate; petals 10–18 mm long, orange (grenadine) or rarely rose-pink or drying violet; carpels 10–17, 4–6 mm high, the dehiscent portion usually acute and with cusps to about 2 mm long, the indehiscent portion coarsely and prominently reticulate, often blackish, forming from 1/3–1/2 of the carpel. Greasewood and mixed desert shrubland. ARIZ: Apa, Nav; NMEX: RAr. 1615–2535 m (5300–8320′). Flowering: May–Sep. Arizona and New Mexico south to Texas and Mexico. Subhastate globemallow is evidently uncommon in the Four Corners region and has been confused with *S. fendleri*, which also has elongate leaves. The long-attenuate calyx lobes far surpassing the carpels, the conspicuous reticulum on the carpels, and the small stature of the plant are the features that are diagnostic from that species (Welsh and Erdman 1964).

MENYANTHACEAE Dumort. BOGBEAN FAMILY

C. Barre Hellquist

Perennial aquatic herbs; rhizomatous. **LEAVES** submersed, floating, or erect; basal, alternate, petiolate, simple or trifoliate. **INFL** a raceme or umbel in leaf axils. **FLOWERS** perfect, regular; corolla sympetalous, petals 5, margins valvate, rolled inward in bud, entire or fimbriate; stamens 5, inserted on corolla; styles short or long. **FRUIT** a 2-valved capsule. Five genera and about 70 species, worldwide. Family is sometimes treated in the Gentianaceae.

Menyanthes L. Bogbean, Buckbean

(Greek *mei*, small, + *antheo*, flower). One species, mainly boreal, worldwide.

Menyanthes trifoliata L. **CP** (three-leaved) Bogbean, buckbean. A perennial, glabrous, herbaceous aquatic. Rhizomes thick, creeping, sheathed by membranous bases of long petioles. **LEAVES** with 3 oval or oblong leaflets, narrowed at base and 2–6 cm long, entire; petioles 3–30 cm long. **INFL** 5–20-flowered raceme on a naked peduncle 20–40 cm long, pedicels 3–5 mm long. **FLOWER** calyx 5-lobed, persistent; corolla short-funnelform, the tube as long as the calyx, 1.5–2 cm long, mainly white, slightly reddish or purplish, fimbriate; stamens 5, sympetalous, alternate with corolla lobes; anthers sagittate, longitudinally dehiscent, filaments short; ovary superior. **FRUIT** a 2-valved capsule, irregularly dehiscent, ovoid, 8 mm long. **SEEDS** subglobose. $2n = 54$. [*Menyanthes trifoliata* var. *minor* Fernald]. Ponds, bogs, and marshes. COLO: Hin, LPl, SJn. 2620–3505 m (8600–11500′). Flowering: May–Jul. Alaska east to Labrador and Newfoundland, south to California, Arizona, New Mexico, Colorado, Nebraska, Missouri, Illinois, West Virginia, and North Carolina.

MORACEAE Gaudich. FIG (MULBERRY) FAMILY

Linda Mary Reeves

Primarily trees and shrubs with milky sap, of warm-temperate and tropical regions worldwide. **LEAVES** most often alternate, but also spiral or opposite; stipules present; blade variously shaped, but often elliptic, lanceolate, ovate, or cordate; occasionally fleshy or prominently veined; margins entire, dentate, serrate, or lobed. **INFL** monoecious or dioecious, sometimes cauliflorous on large branches or trunks; rarely solitary, in catkins, aments, or sometimes on flattened, hollowed and/or variously thickened receptacles (syconia). **FLOWERS** imperfect; actinomorphic or tubular; small, often minute; perianth parts 4; corolla generally none; stamens 4, rarely 1 or 2; carpels 2, 1 abortive, with persistent style; ovary superior trending toward inferior in more advanced genera, with intermediate states; ovule single, pendulous. **FRUIT** variable, fleshy, multiple and/or accessory, produced by the receptacle (syconium), occasionally an achene or drupe. About 70 genera with about 3000 species. An important economic and ornamental family, it includes the edible fig *Ficus carica*, *Artocarpus* (breadfruit and jackfruit), and *Morus* (mulberries). Many useful tropical landscape plants (*Ficus lyrata*, *F. elastica*) and timber trees (*Chlorophora*, *Maclura*), as well as ecologically important plants, are represented within this family. Many species of *Ficus* are pollinated by minute female wasps, which use the inflorescence as a brood chamber. Fig species in tropical forests are known to be continuously fruit bearing, providing reliable food between major fruiting seasons.

1. Leaves with entire margins; branches spiny; flowers in heads or clusters; fruit globose, with a crustlike rind .... ***Maclura***
1′ Leaves serrate or dentate, undivided or lobed; branches not spiny; staminate inflorescence an ament; pistillate inflorescence a head or cluster .... ***Morus***

Maclura Nutt. Osage Orange, Bois d'Arc, Bow-wood, Naranjo Chino

(for William Maclure, 1763–1840) Dioecious trees with hard wood. **LEAVES** with stipules small or minute; blade entire. **INFL** staminate, a peduncled head or cyme; pistillate, a dense, globose, axillary head. **FLOWERS** imperfect; perianth 4-merous. **FRUIT** multiple, globose. $x = 14$. About 15 species in temperate and tropical North America, Africa, and Asia.

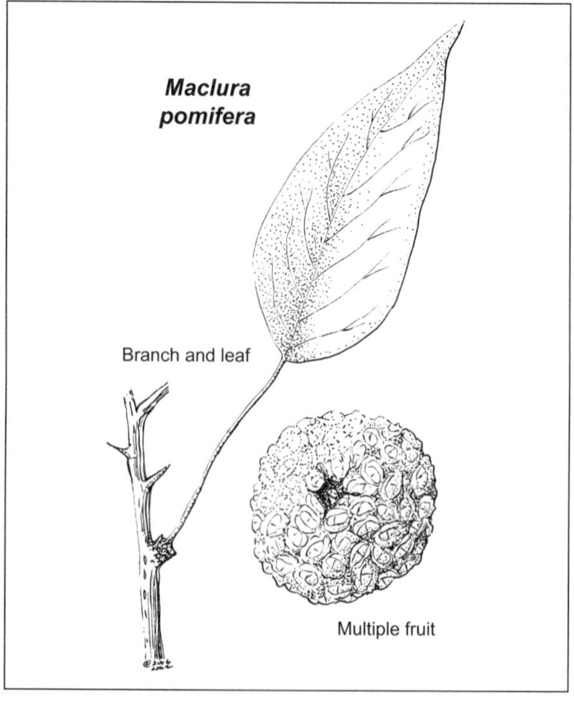

Maclura pomifera (Raf.) C. K. Schneid. (fruit-producing) Tree or shrub to 12 m tall, often thicket-forming; spiny, deciduous, dioecious; open-crowned; wood hard, yellow; sap milky; bark furrowed, yellow-brown; thorns to 25 mm long, stout. **LEAVES** alternate or opposite; stipules minute; petioles to 5 cm long; blade entire, ovate to elliptic-lanceolate, turning bright yellow in fall, to 12 cm long; base broadly cuneate to subcordate; apex acuminate. **INFL**: staminate a loose axillary head or umbel; pistillate a head. **FLOWERS** imperfect; 4-merous; pistillate flowers sessile, 2–2.5 cm across, style long, filiform; staminate flowers pedicelled, 25–35 mm long. **FRUIT** a globose multiple of achenelike fruits to 15 cm diameter; rind wrinkled, hard. **SEEDS** about 1 cm long. [*Toxylon pomiferum* Raf.] Escaped along rivers and at old homesteads. NMEX: SJn. 1525 m (5000′). Flowering: Apr–May. Fruit: Aug–Oct. Known in our area from one riparian location near Shiprock; two large female trees and several shrubby male trees forming a thicket. Eastern Texas, Arkansas, and Oklahoma but widely escaped and cultivated as a thorny hedge throughout the southern United States. The wood was widely used for bows by many tribes, including Apaches. Sometimes the roots were used by several tribes as a yellow dye.

Morus L. Mulberry, Moral

(Latin for mulberry) Winter-deciduous, dioecious or monoecious trees or shrubs; bark scaly and/or thick. **LEAVES** with stipules lanceolate, deciduous; blade undivided or lobed; serrate or dentate; basal nerves 3–5. **INFL** a stalked, axillary, pendent, catkinlike spike or ament, both sexes often present. **FLOWERS** imperfect; calyx 4-parted, sepals involute, partially enclosing filaments; stamens 4; stigmas 2. **FRUIT** an ovoid compressed achene, enclosed by the fleshy white, black, or dark purple calyx, aggregating into a multiple accessory structure, similar in outward appearance to the aggregate fruits of *Rubus*, but with a multiple floral origin. About 12 species in temperate and subtropical Northern Hemisphere regions. Trees often grown for edible fruit, as ornamentals, or as food for silkworm larvae.

Morus alba

Morus alba L. (white) White mulberry, Russian mulberry, silkworm mulberry, moral blanco. Small tree to 15 m high with spreading branches which eventually form a rounded crown; young branches slightly pubescent, becoming glabrous. **LEAVES** with petioles 1–3 cm long; ovate to subcordate; to 20 cm long; apex subobtuse to acuminate; dentate, sometimes lobed; light green and usually glabrous above, hairy in tufts at axils of major veins on underside or glabrous. **INFL** a pendent catkinlike spike; pistillate spikes 5–15 mm long; staminate spikes about twice as long. **FLOWERS** small, greenish or whitish. **FRUIT** a fleshy multiple, ovoid to oblong-elliptic; 1–2.5 cm long; white, pink, violet, or reddish; edible, sweet and/or tart. $2n$ = 28, 30. Along rivers and acequias or escaped from cultivation near habitations. Attracting and spread by migrating fruit-eating birds, such as scrub jays, cedar waxwings, or orioles. ARIZ: Apa; COLO: Mon; NMEX: SJn; UTAH. 1300–2000 m (4300–6500′). Flowering: Mar–May. Fruit: May–Jul. Native of China, but naturalized and invasive in much of Europe and North America.

NYCTAGINACEAE Juss. FOUR-O'CLOCK FAMILY

John R. Spence

Annual or perennial herbs, subshrubs, or shrubs. **STEMS** unarmed (ours), prostrate, procumbent, ascending to erect, often clambering on other plants, sometimes swollen at the nodes, with or without sticky bands in the internodes, variously glabrous or pubescent, sometimes white-pubescent or reddish in color. **LEAVES** deciduous, sessile to petiolate, opposite or rarely alternate, pairs often unequal in size, simple, bases symmetric to strongly unequal, blades variably glabrous or pubescent, often somewhat thick and fleshy, sometimes viscid, stipules lacking. **INFL** terminal or lateral bracteate cymes or racemes, or flowers solitary, bracts sometimes sepaloid, often accrescent in fruit, peduncles sometimes with sticky bands in internodes. **FLOWERS** actinomorphic to sometimes zygomorphic, perigynous, perfect or unisexual, showy to inconspicuous, chasmogamous to cleistogamous; calyx 5-merous, inconspicuous to showy, connate, salverform, funnelform to campanulate, constricted beyond ovary; corolla absent; stamens typically 5, sometimes more, alternate with calyx, filaments free or connate at base, dehiscing longitudinally; nectary disc sometimes present at base of pistil, ovary of 11 carpels, styles fusiform-linear, capitate, or hemispherical, ovules solitary, placentation basal. **FRUIT** an anthocarp consisting of an achene, or utricle enclosed in persistent base of calyx, woody to fleshy, often sticky-glandular, smooth or ribbed to winged. **SEEDS** 1. A family of 35 genera and about 350 species, distributed primarily in the New World, but extending to Africa, Australia, and Eurasia. The four-o'clock family includes the well-known *Bougainvillaea* vine, native to South America, as well as other ornamentals. The family appears to be closest to the Phytolaccaceae. Generic and species limits are controversial, especially in *Mirabilis* (Levin 2000; Flora of North America Editorial Committee 2003).

1. Flowers numerous in capitate umbels, subtended by 5–10 broad bracts; stigmas linear ....(2)
1′ Flowers single or in few-flowered, cymose to umbellate clusters, subtended by 1–5 narrow to broad bracts; stigmas capitate to hemispheric ....(3)

2. Plants annual; fruits large, greater than 1 cm long and wide, with large, conspicuous, papery-translucent wings extending above and below body of fruit .... ***Tripterocalyx***
2′ Plants perennial or rarely annual; fruits small, less than 1 cm, with smaller, less conspicuous, opaque wings that do not extend above or below the fruit .... ***Abronia***

3. Flowers subtended by large, foliaceous, often connate bracts that form an involucre, sometimes enlarging in fruit ....(4)
3′ Flowers ebracteate or subtended by small bracts that are not connate or forming an involucre, not enlarging in fruit ....(5)

4. Prostrate, strongly sticky-glandular herbs; flowers bilaterally symmetrical, in clusters of 3 appearing to be 1 flower, subtended by 3 free papery-translucent bracts .... ***Allionia***
4′ Prostrate to erect, glabrous to glandular herbs; flowers mostly radially symmetric, in clusters of 4 or more or sometimes single, bracts 4–6, free to distinctly connate, green or papery .... ***Mirabilis***

5. Plants perennial; fruits with 3–5 thin, papery-translucent wings; inflorescence stem lacking sticky internodal bands, flowers dull green-white, long-funnelform, 30–50 mm long .... ***Acleisanthes***
5′ Plants annual; fruits lacking distinct wings; inflorescence stems with sticky internodal bands, flowers white to pink, campanulate, small, 1–1.5 mm long .... ***Boerhavia***

Abronia Juss. Sand-verbena, K'íneetlíciit–a'–ipáih

(Greek *abros*, for the delicate and graceful plants) Annual or perennial herbs from taproots or rhizomes. **STEMS** prostrate, ascending to erect, or plants sometimes acaulescent, lacking sticky bands on internodes, glabrous, glandular, or pubescent. **LEAVES** commonly basal with fewer cauline leaves, opposite pairs usually distinctly unequal, petiolate, blades thick and fleshy, bases mostly asymmetric, glabrous or pubescent. **INFL** axillary or scapose, pedunculate, in capitate, cymose to umbellate heads, involucre of 5–10 bracts, persistent, not accrescent, connate, papery or translucent, sometimes green. **FLOWERS** chasmogamous, actinomorphic, perfect, funnelform to salverform, tube often constricted above ovary, abruptly 5-limbed; stamens 5–9, included in tube, styles included, stigmas linear. **FRUIT** a symmetric anthocarp, not distinctly ribbed, turbinate or obovate, usually winged, wings expanded at distal end or not, wall smooth to slightly rugose, not tuberculate, glabrous to pubescent. A North American genus of 20 species, primarily in arid to semiarid habitats in the West and Southwest, and extending into northern Mexico. Many species require mature fruits to identify and are highly variable morphologically, suggesting hybridization. Some species have fragrant flowers. The Navajo name is obscure, usually translated as either stinkbug food or gray.

1. Plants usually acaulescent, most or all leaves basal .... ***A. nana***
1′ Plants usually caulescent, peduncles arising from axils of stem leaves ....(2)

2. Plants with long, slender rhizomes; flowers typically 10–25 per head; fruit wings expanded distally ...... ***A. bolackii***
2′ Plants typically arising from a taproot; flowers typically more than 25 per head; fruit wings expanded distally or not, or fruits wingless ....(3)

3. Fruits wingless, or if winged, then wings not dilated distally .... ***A. fragrans***
3′ Fruit wings conspicuous, dilated distally .... ***A. elliptica***

Abronia bolackii N. D. Atwood, S. L. Welsh & K. D. Heil (named in honor of Tommy Bolack, on whose ranch the type specimen was collected) Bolack's sand-verbena. Perennial herb from rhizomes. **STEMS** short, erect to decumbent, or plants sometimes appearing acaulescent, glandular-pubescent to glabrous. **LEAVES** numerous, mostly basal; petiole 1–4 cm; blade ovate or elliptic, 1–3 cm long, 0.4–1.5 cm wide, margins sometimes weakly sinuate to undulate, glabrous or finely puberulent. **INFL** peduncle 1.6–5.5 cm long, glandular-pubescent, bracts 5, green, becoming paler whitish with age, ovate, 6–10 mm long, 4–10 mm wide, apex broadly acute to rounded. **FLOWERS** 10–25 per head, perianth greenish white, tube 7–11 mm long, limb white, 2–3 mm long. **FRUIT** obovate to obconic, 4.5–7 mm long, puberulent, thin-walled, wings (3) 5, not expanded at distal end or folded together. Rare on gypsiferous clay badlands; saltbush and shadscale shrubland communities. NMEX: SJn. 1660–1730 m (5400–5700′). Flowering: Apr–May. Fruit: May–Jul. *Abronia bolackii* is the only rhizomatous species in the study area and is restricted to the Ojo Alamo Formation, which weathers into a gypsiferous clay soil. It is closest to *A. elliptica*, which is overall a much larger plant that develops from a distinct taproot.

Abronia elliptica A. Nelson (for the elliptical shape of the leaves) Sand-verbena. Perennial herb from stout taproot. **STEMS** erect to decumbent, rarely nearly acaulescent, glandular-pubescent to rarely glabrous. **LEAVES** numerous, basal and cauline; petiole 1–8 cm; blade ovate-elliptic, 1.5–5.5 cm long, 0.6–3 cm wide, glaucous, margins often sinuate to undulate, abaxial surface usually glabrous, adaxial surface glandular-puberulent. **INFL** peduncle 3–8 cm long, glandular-pubescent, bracts 5, scarious, obovate to ovate, 6–20 mm long, 4–8 mm wide, puberulent to villous, apex acute to rounded. **FLOWERS** numerous, (20) 25–70 (75) per head; perianth tube pink or greenish, 12–20 mm long, limb white, 6–8 mm long. **FRUIT** turbinate, 5–12 mm long, glabrous to puberulent, thin-walled, wings (2) 3–5, thin, expanded at distal end, 2 folded together to form a groove. [*A. fragrans* Nutt. ex Hook. var. *elliptica* (A. Nelson) M. E. Jones]. Sandy to rocky soils in open shrublands and barrens, and sand dunes. ARIZ: Apa, Coc, Nav; COLO: LPl, Mon; NMEX: McK, RAr, San, SJn; UTAH. 1125–2285 m (3700–7500′). Flowering: Apr–Sep. Fruit: May–Oct. Although closely related to *A. fragrans*, it is distinct in fruit morphology. The Hopi claim that putting a plant of *A. elliptica* on the head of a child will induce sleepiness.

Abronia fragrans Nutt. ex Hook. (for the fragrant flowers) Fragrant sand-verbena. Perennial herb from stout taproot. **STEMS** erect to decumbent, rarely nearly acaulescent, glandular-pubescent to rarely glabrous. **LEAVES** numerous, basal and cauline; petiole 1–7 cm; blade ovate, triangular, or broadly lanceolate, 3–11 cm long, 1–7 cm wide, green, not glaucous, margins often sinuate to undulate, adaxial and abaxial surfaces glandular-puberulent to densely pubescent, sometimes villous. **INFL** with peduncle 3–8 cm long, glandular-pubescent, bracts 5, scarious, lanceolate to ovate, 8–22 mm long, 3–10 mm wide, puberulent to villous, apex acute to rounded. **FLOWERS** numerous, (25) 30–75 (80) per head; perianth tube green, pink, red, or purple, 12–24 mm long, limb white, 7–10 mm long. **FRUIT** turbinate, 5–12 mm long, glabrous to puberulent, thick-walled, wings absent and fruit grooved, or present and (2) 3–5, thick, stiff, not expanded at distal end, not folded together. Uncommon to locally common on sandy soils in open shrublands and barrens, open piñon-juniper woodlands, and sand dunes, ARIZ: Apa, Coc, Nav; COLO: LPl, Mon; NMEX: McK, RAr, San, SJn; UTAH. 1525–2285 m (5000–7500′). Flowering: Apr–May. Fruit: May–Jul. One of the Navajo names translates as spider medicine. Supposedly the effects of swallowing a spider can be eliminated by taking a medicine prepared from the plant. Many American Indian tribes used the plant as a cathartic and a diaphoretic, as well as for a variety of ailments, including dermatitis, stomach cramps, and cold sores. The roots were sometimes used as food by the Acoma and Laguna Pueblos.

Abronia nana S. Watson (for its low-growing stature) Dwarf sand-verbena. Perennial herb from rhizomes. **STEMS** absent, plants typically acaulescent and caespitose, rarely very short stems present, glandular-pubescent to glabrous, not whitish. **LEAVES** numerous, basal; petiole 1–5 cm; blade lanceolate to elliptic or ovate, 0.5–2.5 cm long, 0.7–1 cm wide, margins entire to undulate, glabrous or finely puberulent. **INFL** with peduncle 3–10 cm long, glandular-puberulent, bracts 5, scarious, lanceolate to narrowly ovate, 5–10 cm long, 3–6 mm wide, apex broadly acute to rounded, glandular-puberulent. **FLOWERS** 12–25 per head; perianth tube pink or rarely white, 10–28 mm long, limb white, 7–10 mm long. **FRUIT** obovate to cordate, 5–10 mm long, wings 5, not expanded distally or folded together. Clayey, gypsiferous soils; desert grassland, rabbitbrush, saltbush, sagebrush, and piñon-juniper woodland communities. ARIZ: Apa; NMEX: SJn. 1615–2135 m (5300–7000′). Flowering: Apr–May. Fruit: May–Jul.

Acleisanthes A. Gray Trumpets, Moonpod

(Greek *acleis*, a lack, + *anthos*, flower, for flowers lacking an involucre) Perennial herbs, sometimes becoming woody, from taproots. **STEMS** procumbent to erect, often clambering on other plants, sometimes swollen at the nodes, lacking sticky bands on internodes, glabrous or pubescent, often whitish. **LEAVES** sessile to petiolate, opposite pairs usually unequal, often strongly so, blades mostly thick and fleshy, ovate, bases symmetric to asymmetric, glabrous or pubescent. **INFL** terminal or axillary, solitary flowers or few-flowered cymes; bracteate or not, if present, involucres subtending 1–25 flowers, bracts 1–3, persistent, connate, not accrescent, remaining green. **FLOWERS** actinomorphic, perfect, sometimes cleistogamous and rounded-closed, chasmogamous flowers showy, funnelform; stamens 2–6, exserted; styles exserted beyond stamens, stigmas peltate. **FRUIT** symmetric, 4–5-ribbed, ellipsoid to oblong, leathery, smooth to finely pubescent, ribbed, wings often present, 3–5, sometimes hyaline, not distinctly tuberculate, glandular at one end. A genus of 17 species distributed primarily in the deserts of the Southwest and northern Mexico, but extending to arid northeast Africa. Our single species was originally placed in *Selinocarpus*, but Levin (2000) showed that it is probably embedded within *Acleisanthes*, which is also supported by morphology.

Acleisanthes diffusa (A. Gray) R. A. Levin (apart, for open branching of plants) Spreading moonpod. Perennial herb to subshrub. **STEMS** prostrate to erect, sometimes clambering on other plants, dichotomously to irregularly branched, densely brown to white-pubescent, sometimes appearing whitish. **LEAVES** numerous, spreading, green, petiolate, opposite pairs usually slightly unequal, blades mostly thick and fleshy, ovate, 12–28 mm long, 5–15 mm wide, with undulate margins, glabrous or pubescent, apex rounded to acute, leaf bases more or less symmetric. **INFL** terminal or axillary, bracteate, involucres subtending 1 (2–3) flowers, bracts 2–3, small, linear, 2–3 mm long. **FLOWERS** actinomorphic, perfect, mostly cleistogamous, small and rounded-closed, occasional chasmogamous flowers present, perianth 30–45 mm long, dull to pale green or yellow-green, sometimes with green stripes, funnelform; stamens 4–5. **FRUIT** symmetric, ellipsoid, pubescent with white hairs, smooth, wings present, 1–2 mm wide, not tuberculate. [*Selinocaprus diffusus* A. Gray]. Rare in arid to semiarid shadscale, saltbush, and blackbrush shrublands; usually on clay soils. UTAH. 1125–1220 m (3700–4000′). Flowering: May–Aug. Fruit: Jun–Oct.

Allionia L. Windmills

(for C. Allioni, Italian botanist) Annual or perennial herbs from slender to stout taproot. **STEMS** prostrate to clambering on other plants, sometimes swollen at the nodes, lacking sticky bands on internodes, glabrous or pubescent, somewhat reddish. **LEAVES** petiolate, opposite pairs unequal, blades thin, broadly lanceolate to ovate or triangular, bases more or less asymmtetric, glabrous or pubescent. **INFL** axillary, pedunculate, a cluster of 3 flowers; involucres distinct, of 3 bracts, persistent, connate, not accrescent, thin and papery or translucent. **FLOWERS** strongly bilaterally symmetrical, chasmogamous, perfect, showy, funnelform, limb short to almost absent; stamens 4–7, exserted; styles exserted beyond stamens, stigmas capitate. **FRUIT** compressed dorsiventrally, with concave and convex faces, obovoid, 5-ribbed, 2 lateral ribs enlarged and curved inward on convex face, ribs on convex face with 1–8 glandular teeth or teeth and glands sometimes lacking, fruit wall smooth or weakly rugose adaxially, not tuberculate, glandular on the concave face. A small genus of two species, widespread from southern North America through South and Central America and the West Indies. The fruit morphology is very complex, and mature fruits are usually necessary for proper identification. The three bilaterally symmetric flowers are closely clustered, giving the appearance of a single radially symmetric flower.

1. Annual; fruit lateral ribs with 4–8 regular teeth; fruit weakly convex adaxially .... ***A. choisyi***
1′ Perennial or rarely annual; fruit lateral ribs with 0–4 short, irregular teeth; fruit strongly convex adaxially .... ***A. incarnata***

Allionia choisyi Standl. (for Jacques Denys Choisy, botanist, clergyman, and philosopher, 1799–1859) Trailing four-o'clock, garapatilla. Annual herbs from slender taproot. **STEMS** sprawling-prostrate, to 2 m, somewhat sticky-pubescent, often with adherent sand grains. **LEAVES** green to glaucous-green, paler below, becoming smaller distally; petiole 4–20 mm, same length or shorter than blade; larger leaf blades 12–26 mm long, 8–20 mm wide, margins undulate or somewhat plane, glabrous or pubescent, sometimes viscid, base asymmetric. **INFL** peduncle 2–10 mm, pubescent, involucres broadly ovate to rounded at maturity, 5–7 mm, margins not ciliate. **FLOWERS** pink, magenta, or rarely white, small, 2–6 mm. **FRUIT** brown at maturity, 3–4 mm long, 2–3.2 mm wide, weakly convex, teeth of lateral wings regularly shaped, 4–8 in number, slender, attenuate, glandular or not, concave side glandular, stalks of glands the same length or longer than head. Disturbed and open sandy or gravelly to rocky sites in desert scrub, black sagebrush, piñon-juniper woodland, and riparian communities. ARIZ: Apa, Nav; NMEX: McK, SJn; UTAH. 1650–1830 m (5420–6000′). Flowering: May–Sep. Fruit: Jun–Oct. The small fruits resemble ticks, thus the Spanish name garapatilla. The fruits of this species are variable and sometimes are more strongly convex with regularly shaped teeth, resembling fruits of *A. incarnata*.

Allionia incarnata L. (flesh-colored, in reference to the flowers) Trailing four-o'clock, allionia, windmills. Annual to more commonly perennial herbs from slender to stout and sometimes woody taproot. **STEMS** sprawling-prostrate, to 1.2 m, somewhat sticky-pubescent to villous, often with adherent sand grains. **LEAVES** green to glaucous-green, paler below, becoming smaller distally, petiole 3–20 mm long, usually shorter than blade, larger leaf blades 22–60 mm long, 12–32 mm wide, margins flat to sometimes undulate, glabrous or pubescent, sometimes viscid, base asymmetric. **INFL** with peduncle 4–26 mm long, pubescent, involucres ovate at maturity, 5–8 mm, margins not ciliate. **FLOWERS** pink to magenta, small, 6–14 mm. **FRUIT** brown at maturity, 3–4.5 mm long, 1.8–2.5 mm wide, strongly and deeply convex, teeth of lateral wings absent, or up to 4 present, broadly triangular to blunt, lacking glands, concave side glandular, stalks of glands the same length or shorter than head. Common in sandy to rocky sites in open desert scrub, blackbrush, shadscale, snakeweed, and piñon-juniper woodland communities. ARIZ: Apa, Nav; COLO: LPl, Mon; NMEX: McK, SJn; UTAH. 1430–1830 m (4700–6000′) Flowering: Apr–Oct. Fruit: May–Nov. The plants in the study area are referred to **var. *incarnata***, while a second variety, var. *villosa* (Standl.) Munz, occurs in the Mojave and Sonoran deserts.

Boerhavia L. Spiderling

(for botanist Herman Boerhaave) Annual or perennial herbs or subshrubs from slender to swollen and tuberous taproots. **STEMS** procumbent, ascending to erect, sometimes swollen at the nodes, with or without sticky bands on internodes, glabrous to glandular-pubescent or villous, not whitish. **LEAVES** petiolate, opposite pairs distinctly unequal, blades thin to sometimes weakly fleshy, margins plane to undulate, narrowly lanceolate to broadly ovate, bases symmetric to asymmetric, glabrous or pubescent. **INFL** terminal or axillary, peduncle short or long, often obscure, highly branched, of a diffuse cyme, raceme, spike, or umbel, rarely borne singly; bracts 1–3, subtending 1–few flowers, persistent or deciduous, weakly connate to distinct, not accrescent in fruit, papery or translucent. **FLOWERS** more or less actinomorphic to sometimes bilaterally symmetric, perfect, chasmogamous flowers funnelform to campanulate, limbs abruptly flared above tube, cleistogamous flowers absent; stamens 2–8, included; styles at same level as stamens or somewhat exserted, stigmas peltate. **FRUIT** symmetric, 4–5-ribbed, fusiform, elliptic, obovoid, or pyramidal, ribs 3–5, angular to smooth or sometimes weakly winged, wall glabrous or puberulent, smooth to rugose, not tuberculate, glabrous to pubescent. A genus of 40 species, distributed in warm regions of the world, including North, Central, and South America, Africa, and Australia. The genus is taxonomically difficult, and mature fruits are usually needed for correct identification.

Boerhavia torreyana (S. Watson) Standl. (for John Torrey, 1796–1873, botanist and colleague of A. Gray) Spiderling. Annual herbs. **STEMS** ascending to erect, to 50 cm, irregularly branched, densely glandular-villous and viscid, at least at base, not whitish. **LEAVES** numerous, mostly basal; petiole 12–25 mm; blades oval to ovate or triangular, 20–40 mm long, 15–25 mm wide, becoming smaller distally, variably glabrous to glandular-pubescent or brown-punctate, paler below, margins smooth to undulate, apex rounded to obtuse. **INFL** terminal and axillary, irregularly branched, sticky internodes present, spicate to racemose; bracts 2, deciduous, lanceolate, 1–2 mm long, apex acute to acuminate. **FLOWERS** actinomorphic, small; perianth 1–1.5 mm, white to pink, campanulate; stamens 2–3. **FRUIT** obovoid, gray to red-brown or tan, 2–3 mm long, ribs 5, rounded, sometimes slightly winged, wall smooth to weakly rugose. Talus slopes of the De Chelly Sandstone. Known from Canyon de Chelly. ARIZ: Apa. 1730 m (5675′). Flowering: Jul–Aug. Fruit: Sep–Oct.

Mirabilis L. Four-o'clock, Umbrellawort, Tshétitéeh

(Latin *mirabilis*, wonderful) Perennial herbs or subshrubs from slender to swollen and tuberous taproots. **STEMS** procumbent, ascending to erect, often clambering on other plants, sometimes swollen at the nodes, lacking sticky bands on internodes, glabrous or pubescent, sometimes whitish. **LEAVES** sessile to petiolate, opposite pairs usually equal, blades thin to distinctly thick and fleshy, narrowly lanceolate to broadly orbicular, glabrous or pubescent, bases more or less equal. **INFL** terminal or axillary, bracteate, involucres subtending 1–15 flowers, persistent, connate, accrescent, remaining green or often becoming papery or translucent. **FLOWERS** more or less actinomorphic, perfect, sometimes cleistogamous and rounded-closed, chasmogamous flowers showy, funnelform to campanulate; stamens 3–6, exserted; styles exserted beyond stamens, stigmas capitate. **FRUIT** symmetric, 4–5-ribbed, obovoid to elliptic or globose, ribs smooth to tuberculate, wall between ribs smooth to often strongly tuberculate, glabrous to pubescent, not distinctly glandular at one end. A genus of about 60 species, primarily New World with a large concentration in the Southwest, Mexico, and into South America. Our species are easily separated into two well-marked groups, section *Quamoclidion* (*M. glandulosa*, *M. multiflora*, *M. oxybaphoides*) and section *Oxybaphus* (*M. albida*, *M. glabra*, *M. linearis*). Although *Oxybaphus* has often been segregated from *Mirabilis* and treated at the generic level, it has been reluctantly decided to not recognize it. In some preliminary phylogenetic DNA work, Levin (2000) showed that *Oxybaphus* may be derived from within section *Quamoclidion*, and both were shown to be distinct from section *Mirabilis*. However, *Quamoclidion* has generally not been recognized as a genus in North America. Clearly more work needs to be done on sectional relationships, but it is likely that the three principal sections may eventually be recognized as distinct genera. The species in *Oxybaphus* are notoriously variable and apparently hybridize where their ranges overlap; thus this treatment is very preliminary, largely following Spellenberg (2003). Fruits are generally needed to distinguish the species. Some species of *Mirabilis* were widely used by American Indian tribes as medicine, usually from concoctions of the root. The generic name used by the Navajo refers to a tendency for plants to force their roots into rocky sites and outcrops. The showy species *M. multiflora* and *M. glandulosa* are easily transplanted from pieces of the root and make outstanding xeric ornamentals.

1. Involucre becoming enlarged, tan and somewhat translucent with conspicuous veins; fruit distinctly 5-ribbed; leaf blades linear to narrowly ovate (*Oxybaphus*) ....(2)
1′ Involucre not enlarging in fruit, remaining green with veins obscure; fruit smooth to weakly 5-ribbed or somewhat 5-angled; leaf blades broadly ovate, deltate, or cordate (*Quamoclidion*) ....(4)

2. Fruit glabrous, grayish brown to green-brown .... *M. glabra*
2′ Fruit pubescent, dark brown to dark olive-brown .... (3)

3. Leaf blades narrowly linear-lanceolate, mostly less than 1 cm wide; fruit ribs with smooth, low tubercles barely raised above fruit surface; stems whitish-pubescent .... *M. linearis*
3′ Leaf blades lanceolate to ovate, at least some greater than 2 cm wide; fruit ribs with distinctly raised tubercles; stems green .... *M. albida*

4. Leaves relatively thin, cordate; flowers 1–3 per involucre; perianth mostly less than 1 cm long .... *M. oxybaphoides*
4′ Leaves thick, fleshy, orbicular to triangular-cordate; flowers typically 6 or more per involucre; perianth large, showy, 2–6 cm long .... (5)

5. Plants generally irregularly sprawling and branched, leaves rounded-cordate to orbicular; bracts with rounded-obtuse tips; inflorescence sticky-glandular; fruit strongly tuberculate, sticky when wet .... *M. glandulosa*
5′ Plants sprawling to erect, branched from base, forming mounds; leaves triangular-cordate; bracts with acute tips; inflorescence weakly glandular; fruit smooth or weakly tuberculate, not or slightly sticky when wet ... *M. multiflora*

Mirabilis albida (Walter) Heimerl (for the white flowers) Umbrellawort. Perennial herbs. **STEMS** erect or procumbent, glabrous or puberulent, sometimes viscid or hirsute and spreading, not whitish. **LEAVES** ascending or spreading, green to glaucous-green or grayish green; petiolate; ovate-lanceolate to mostly ovate or cordate, variable in shape, lanceolate leaves 1–10 cm long, 0.5–2.5 cm wide, ovate leaves to 5 cm wide, blades thin to distinctly thick and fleshy, glabrous or pubescent, sometimes viscid, apex variable, rounded to acute, base symmetric to somewhat asymmetric. **INFL** terminal or axillary, peduncle short, to 2 cm, pubescent, involucres campanulate, 5–7 mm, lobes ovate to triangular, pubescent, margins often ciliate with dark hairs, involucre enlarging and becoming pale tan and translucent in fruit, 6–14 mm long. **FLOWERS** (1) 2–3 per involucre; perianth small, 1–2 cm, white or sometimes pink, short-funnelform, small cleistogamous flowers sometimes present, typically single. **FRUIT** brown, distinctly ribbed, ovoid to obovate, 4–6 mm long, pubescent, sometimes weakly viscid, ribs with tubercles, wall rugose, with low distinct tubercles or rarely almost smooth [*M. comata* (Small) Standl.; *Oxybaphus comatus* (Small) Weath.]. Rocky slopes, fields, and sandy sites. ARIZ: Apa; COLO: Arc, Hin, LPl, Min, SJn; NMEX: McK. 1950–2530 m (6400–8300′). Flowering: Jul–Aug. Fruit: Sep–Oct.

Mirabilis glabra (S. Watson) Standl. (smooth, in reference to fruit) Umbrellawort. Perennial herbs. **STEMS** erect to ascending, glabrous, glandular, puberulent, or pubescent, not whitish. **LEAVES** ascending to erect, glaucous-green or grayish green; sessile or short-petiolate; blade linear to narrowly ovate, 5–9 cm long, 0.5–7 cm wide, somewhat lanceolate to mostly ovate or cordate, variable in shape, lanceolate leaves 1–10 cm long, 0.5–2.5 cm wide, ovate leaves to 5 cm wide, somewhat thick but not fleshy, glabrous or pubescent, apex variable, rounded to acute, base symmetric to somewhat asymmetric. **INFL** terminal or rarely axillary, peduncle short, to 1 cm, pubescent, involucres campanulate, 5–7 mm long, lobes ovate to triangular, glabrous, margins sometimes sparsely ciliate or margins with pale hairs, involucre enlarging and becoming pale tan and translucent in fruit, 8–12 mm long. **FLOWERS** 1–3 per involucre; perianth small, less than 1 cm long, white or sometimes pink, short-funnelform, typically lacking cleistogamous flowers. **FRUIT** pale grayish green to greenish brown, distinctly ribbed, obovate, 4–5 mm long, glabrous or rarely somewhat puberulent, ribs with tubercles, wall not distinctly rugose, with low to often distinct tubercles present [*Oxybaphus glaber* S. Watson]. Rocky slopes, fields, and sandy sites. NMEX: SJn. 1890–2050 m (6200–6725′). Flowering: May–Sep. Fruit: Jun–Oct.

Mirabilis glandulosa (Standl.) W. A. Weber (for the sticky fruit and inflorescence) Showy four-o'clock. Perennial herbs. **STEMS** sprawling to ascending, forming straggling mats, glabrous to pubescent, not whitish. **LEAVES** spreading to erect, green, sometimes glaucous, petiolate at least proximally; blade broadly cordate to nearly orbicular, 4–10 cm wide × 5–8 cm long, thick and somewhat fleshy, glabrous, apex variable, rounded to acute, base asymmetric. **INFL** terminal or axillary, peduncle short or long, 0.5–6.5 cm, strongly glandular-pubescent and viscid, involucres broadly tubular to campanulate, 25–40 mm, lobes ovate, rounded-obtuse, glabrous to pubescent, margins not ciliate, involucre not enlarging in fruit, remaining green. **FLOWERS** 6 per involucre; perianth large, showy, 2–6 cm long, pink to magenta, long-funnelform, typically lacking cleistogamous flowers. **FRUIT** brown to brown-black, ribs present or nearly absent, ovoid, 5–10 mm long, distinctly sticky-mucilaginous when wet, wall distinctly and strongly tuberculate. [*M. multiflora* (Torr.) A. Gray var. *glandulosa* (Standl.) J. F. Macbr.]. (Ours) mostly on clay slopes derived from the Morrison Formation; shadscale, sagebrush, and piñon-juniper woodland communities. ARIZ: Apa, Nav; COLO: Dol, Mon; NMEX: SJn; UTAH. 1525–1980 m (5000–6500′). Flowering: Apr–Jun. Fruit: Jun–Jul. *Mirabilis glandulosa* in its typical phase is quite distinct from *M. multiflora*, although there are intermediates. The very broad, rounded leaves, rounded involucre tips, and sticky inflorescence and fruit generally distinguish it from *M. multiflora*. It also tends to flower earlier and is less likely to flower after the summer monsoons, which *M. multiflora* commonly does.

Mirabilis linearis (Pursh) Heimerl (line, for the slender, linear leaves) Narrowleaf umbrellawort. Perennial herbs. **STEMS** decumbent, ascending to erect, variably glabrous, pubescent, or puberulent, sometimes glandular-pubescent and viscid or densely hirsute, whitish. **LEAVES** ascending to spreading, green to glaucous-green or blue-green or gray; petiolate; blade linear-lanceolate to narrowly lanceolate, variable in shape, 4–12 cm long, 0.1–2 cm wide, thin to somewhat thick and fleshy, glabrous, glandular, or hirsute, apex variable, rounded to acute, base symmetric to somewhat asymmetric. **INFL** terminal and axillary, peduncle short, 4–10 mm, glandular-pubescent to pilose or hirsute, involucres green, campanulate, 4–6 mm, lobes ovate, short, ciliate-pubescent in fruit, involucre enlarging and becoming pale tan and translucent in fruit, 5–12 mm long. **FLOWERS** typically 3 per involucre; perianth small, 0.7–1 cm, white to pink or purple, cleistogamous flowers lacking. **FRUIT** olive to brown, weakly ribbed, obovate, 3–5 mm long, pubescent to hirsute, not viscid, wings not or weakly tuberculate, wall smooth to somewhat rugose, with low or distinct tubercles. [*Oxybaphus linearis* (Pursh) B. L. Rob.]. A variable species with two varieties in the study area. Medicines from the roots of this species were widely used for stomach ailments by the Navajo and Zuni, who also ate the seeds and fruits on occasion. More work is needed to clarify the relationships and distributions of the two varieties, and also to determine whether var. *decipiens* is distinctive enough to be considered a separate species.

1. Leaf blades lanceolate, 0.5–2 cm wide, green; perianth pink to purple .... **var. *decipiens***
1' Leaf blades linear, 0.1–1 cm wide, glaucous-gray or blue-green; perianth white to pink .... **var. *linearis***

var. *decipiens* (Standl.) S. L. Welsh (deceiving) Uncommon in ponderosa pine, Gambel's oak, piñon-juniper woodland, aspen, spruce-fir, and riparian communities. ARIZ: Apa; COLO: Arc, Dol, LPl, Mon; NMEX: McK, RAr, SJn; UTAH. 1900–3050 m (6235–10005'). Flowering: June–Aug. Fruit: Jul–Sep. Generally at higher elevations than var. *linearis*.

var. *linearis* Shadscale, sagebrush, rabbitbrush, piñon-juniper woodland, mountain brush, and ponderosa pine communities. ARIZ: Apa, Coc, Nav; COLO: Arc, Dol, LPl, Mon, SMg; NMEX: McK, RAr, San, SJn; UTAH. 1125–2715 m (3700–8905'). Flowering: May–Sep. Fruit: Jun–Oct.

Mirabilis multiflora (Torr.) A. Gray **CP** (for the numerous flowers) Showy four-o'clock. Perennial herbs. **STEMS** sprawling to ascending, forming hemispherical mounds, glabrous to pubescent, not whitish. **LEAVES** spreading to erect, green, sometimes glaucous; petiolate at least proximally, blade broadly triangular-cordate or ovate, 4–9 cm wide × 4–7 cm long, thick and somewhat fleshy, glabrous, apex variable, rounded to acute, base asymmetric. **INFL** terminal or axillary, peduncle short or long, 0.5–6.5 cm, glabrous or weakly glandular-pubescent, not distinctly viscid, involucres broadly tubular to campanulate, 25–40 mm, lobes triangular-ovate, acute, glabrous to pubescent, margins not ciliate, involucre not enlarging in fruit, remaining green. **FLOWERS** 6 per involucre; perianth large, showy, 2–6 cm long, pink to magenta, long-funnelform, typically lacking cleistogamous flowers. **FRUIT** brown to brown-black, ribs present or nearly absent, ovoid, 5–10 mm, not or only slightly sticky-mucilaginous when wet, wall smooth to weakly tuberculate. Sagebrush, blackbrush, piñon-juniper woodlands, other wooded slopes, canyons, and riparian communities. ARIZ: Apa, Coc, Nav; COLO: Arc, LPl, Mon; NMEX: McK, RAr, San, SJn; UTAH. 1220–2560 m (4000–7800'). Flowering: May–Sep. Fruit: Jun–Oct. This species was widely used in the region as a drug to induce hallucinations (Hopi), to aid digestion (Zuni), for rheumatism (Navajo), as a dermatological aid and cleanser (many tribes), taken internally to decrease appetite (Zuni), and to make a tea (Navajo).

Mirabilis oxybaphoides (A. Gray) A. Gray (for similarity to *Oxybaphus*) Spreading four-o'clock, t–'iicnát'ooh. Perennial herbs. **STEMS** prostrate, ascending, sometimes climbing on and tangled in other plants, glandular-pubsecent to puberulent, not whitish. **LEAVES** sparse, spreading; petiolate; blade broadly ovate to deltate, 2–8 cm long, 1–7 cm wide, thin or sometimes weakly fleshy, glabrous to glandular-pubescent, apex mostly acuminate or rarely rounded. **INFL** a terminal or axillary cyme, involucres clustered at ends of branches, sometimes solitary, peduncle variable in length, pubescent, involucres bell-shaped, 6–8 mm, lobes ovate to triangular, pubescent, margins not distinctly ciliate with dark hairs, not enlarging or becoming pale tan and translucent in fruit. **FLOWERS** 3 per involucre; perianth small, 0.5–1 cm, pale pink to purplish, rarely white, typically lacking cleistogamous flowers. **FRUIT** olive-green or green-brown to dark brown or black, not ribbed although faint grooves sometimes present, obovoid to subspherical, 2–3 mm, not pubescent, wall smooth or only slightly rugose, lacking distinct tubercles. Piñon-juniper and oak woodlands, riparian woodlands, and other shaded sites. ARIZ: Apa, Nav; COLO: Arc, Hin, LPl, Min, Mon; NMEX: McK, RAr, San, SJn. 1645–2565 m (5400–8415'). Flowering: May–Jul. Fruit: Jun–Sep. The Navajo name is usually translated as snake tobacco. They used the powdered leaves and roots to reduce swelling from spider bites, to control dandruff, to aid healing of fractures, and sometimes as food or tea.

Tripterocalyx (Torr.) Hook. Sand-puffs

(Greek *tri*, three, + *pteron*, wing, + *calyx*) Annual herbs from slender, fibrous to somewhat woody taproots. **STEMS** green to often reddish throughout or at nodes, decumbent to erect, viscid-glandular to pubescent, lacking sticky bands on internodes, glabrous or pubescent. **LEAVES** petiolate, opposite pairs unequal; blades thick and fleshy, broadly lanceolate, elliptic, ovate, or obovate, glabrous or pubescent, bases mostly asymmetric. **INFL** axillary, pedunculate, of capitate heads or umbels, bracteate, bracts 5, subtending 5–25 (30) flowers, persistent, weakly connate to nearly distinct, not accrescent, thin, translucent to papery. **FLOWERS** actinomorphic, perfect, chasmogamous, perianth funnelform to salverform, 4–5-lobed, caudocous above the ovary; stamens 3–5, included in tube, styles included, stigmas linear. **FRUIT** symmetric, slender-elliptic to fusiform, sometimes somewhat spongy, ribs extending into 2–4 wings, translucent, large, wall smooth to rugose, not tuberculate, glabrous to pubescent, not distinctly glandular at one end. A small genus of four species endemic to western North America and Mexico.

1. Floral tube relatively short, 6–16 mm, limb 3–5 mm, greenish pink, lobes inconspicuous ....***T. micranthus***
1' Floral tube longer, 13–28 mm, limb 9–12 mm, pink, magenta, or white, lobes distinct...(2)

2. Limb of perianth white, at least adaxially; fruits less than 20 mm long..***T. wootonii***
2' Limb of perianth pink to purple on both sides; fruits mostly greater than 20 mm long...........................***T. carneus***

Tripterocalyx carneus (Greene) L. A. Galloway (flesh-colored, for the showy flowers) Pink sand-puffs. Annual herbs. **STEMS** decumbent, ascending to erect, 10–60 cm, glandular-pubescent to puberulent, often reddish. **LEAVES** spreading, petiole 1.5–7 cm, blades oblong to ovate or broadly lanceolate, 3–9 cm long, 1.2–4 cm wide, somewhat glaucous abaxially, glabrous to glandular-pubescent, margins entire to undulate and ciliate, apex acute to obtuse. **INFL** short- to long-pedunculate, 1.5–10 cm, 10–25-flowered, bracts lanceolate, 6–15 mm long, 3–6 mm wide, margins glandular-ciliate. **FLOWER** perianth tube 13–25 mm, pale pink to magenta, limb 9–12 mm, pale pink, showy. **FRUIT** ovate, 16–30 mm long, lateral ribs weakly developed or lacking; wings 2–4. Blackbrush, *Ephedra*-resinbush, purple sage, and sand sagebrush communities. COLO: Mon; NMEX: McK, SJn. 1615–1890 m (5300–6200′). Flowering: Apr–Sep. Fruit: Jun–Oct. This showy species reaches its northern limits within the study area and is one of the few elements of the Chihuahuan Desert flora in the Four Corners region.

Tripterocalyx micranthus (Torr.) Hook. (for the small flowers) Small-flowered sand-puffs. Annual herbs. **STEMS** decumbent, ascending to erect, 10–50 cm, glandular-pubsecent and viscid, reddish. **LEAVES** spreading, petiole 1.2–3.5 cm, blades broadly lanceolate to elliptic, 1.5–5 cm long, 0.6–2.2 cm wide, green or glaucous abaxially, mostly glandular-pubescent, margins entire to undulate, ciliate, apex acute to obtuse. **INFL** short- to long-pedunculate, 1.5–9 cm, 5–15-flowered, bracts lanceolate to ovate, 3.5–8 mm long, 1–2.5 mm wide, margins glabrous to glandular-pubescent. **FLOWER** perianth tube short, 5–16 mm, greenish pink, limbs short, 3–5 mm, greenish white or pink. **FRUIT** rounded-oval, 11–20 mm long, lateral ribs weakly developed or with poorly developed wings, wings typically 3 (2–4). Sandy sites in desert grasslands and desert shrub communities. ARIZ: Apa, Nav; COLO: LPl; NMEX: McK, RAr, SJn; UTAH. 1125–1860 m (3700–6100′). Flowering: Apr–Sep. Fruit: Jun–Oct. This showy species is most common to the north, reaching its southern limit in the Four Corners region.

Tripterocalyx wootonii Standl. (for Elmer Ottis Wooton, 1865–1945, New Mexico botanist) Wooton's sand-puffs. Annual herbs. **STEMS** decumbent, ascending to erect, 10–70 cm, glandular-pubescent and viscid, green to often reddish at nodes. **LEAVES** spreading; petiole 1–6 cm, blades broadly ovate to oblong, 1.8–7.6 cm long, 0.6–4.5 cm wide, more or less glaucous abaxially, glabrous or glandular-pubescent, margins entire to weakly undulate, scabrid to short-ciliate, apex acute to obtuse. **INFL** short- to long-pedunculate, 1.7–9.5 cm, 10–25-flowered, bracts narrowly lanceolate to ovate, 4.5–13 mm long, 0.5–4.7 mm wide, margins glandular-ciliate or sometimes glabrous. **FLOWER** perianth tube 13–25 mm, pink, limbs distinct, 9–11 mm, white adaxially. **FRUIT** oval, 14–24 mm, lateral ribs present, 1–3 or sometimes absent, if present, lacking wings or wings poorly developed, wings typically (2–) 3 (–4). [*T. carneus* (Greene) L. A. Galloway var. *wootonii* (Standl.) L. A. Galloway]. Sandy soils in desert grasslands, desert scrub, sagebrush, and juniper savanna communities. ARIZ: Apa, Coc, Nav; COLO: Mon; NMEX: McK, RAr, San, SJn; UTAH. 1125–2200 m (3700–7215′). Flowering: Apr–Sep. Fruit: Jun–Oct. *Tripterocalyx wootonii* largely replaces *T. carneus* in the southern portions of the study area and further south.

NYMPHAEACEAE Salisb. WATER-LILY FAMILY

C. Barre Hellquist

Perennial, aquatic, rhizomatous herbs. **STEMS**: rhizomes branched or unbranched, erect or horizontal, some tuberous, some stoloniferous. **LEAVES** arising from rhizomes, alternate, floating, submersed, and emersed; stipules present or absent; petioles long; blades lanceolate to ovate, or orbicular with basal sinus, margins entire, spinose, or dentate. **INFL** axillary or extra-axillary, flowers solitary. **FLOWERS** bisexual, diurnal or nocturnal, floating, emergent, occasionally submersed with long peduncle; involucre absent; perianth often persistent; hypogynous, perigynous, or epigynous; sepals usually 4–9 (–12); petals numerous (rarely absent), often transitional to stamens; stamens numerous; pistil 1, 3–35-carpellate and -locular; ovules numerous; stigma sessile, radiate on stigmatic disc. **FRUIT** berrylike, indehiscent or irregularly dehiscent. **SEEDS** several to numerous; aril present or absent, fleshy (Wood 1959). Six genera, about 50 species worldwide.

Nuphar Sm. Spatterdock, Cow-lily, Yellow Pond-lily

(ancient Arabic or Persian) Rhizomes horizontal, branched. **LEAVES** floating, submersed, and emersed; blades orbicular to linear, basal lobes divergent or overlapping, margins entire; primary venation mostly pinnate, basal section of midrib with parallel veins, light green, some reddish when young. **FLOWERS** floating or emersed, opening diurnally; perianth hypogynous, nearly globose at anthesis; sepals 5–9 (–12), green to yellow on outer surface, yellow, often red-tinged on inner surface, oblong to obovate or orbicular; petals numerous, smaller than sepals, spirally arranged, stamenlike; stamens yellow or red-tinged, inserted below ovary; ovary longer than petals and stamens; stigmatic disc with margin entire, crenate, or dentate. **FRUIT** borne on peduncles. **SEEDS** ovoid, to 6 mm; aril absent. $x = 17$ (Beal 1956; Padgett 1997). Ten to 12 species. North America, Mexico, Europe, Asia.

Nuphar polysepala Engelm. (many-sepaled) Yellow cow-lily, spatterdock. Rhizomes 3–8 cm in diameter or larger. **LEAVES** floating, submersed, and emergent; petioles terete; blades green, widely ovate, 10–40 (–45) × 7–30 cm, sinus 1/3–2/3 length of midrib, lobes divergent to overlapping. **FLOWERS** 5–10 cm in diameter; sepals (6–) 9 (–12), green on undersurface, yellow above, sometimes red-tinged toward base; petals oblong, thick; anthers 3.5–9 mm, shorter than the filaments. **FRUIT** green to yellow, cylindric to ovoid, 4–6 (–9) × 3.5–6 cm, strongly ribbed, slightly constricted below the stigmatic disc; disc green, 20–35 mm diameter. **SEEDS** 3.5–5 mm long. $2n = 34$. [*N. lutea* (L.) Sm. subsp. *polysepala* (Engelm.) E. O. Beal; *Nymphaea polysepala* (Engelm.) Greene]. Ponds, lakes, and streams. COLO: LPl, SJn. 2950–3275 m (9400–10800′). Flowering and fruiting: Jun–Aug. Alaska, Yukon, and Northwest Territories south to California, Arizona, Colorado, and New Mexico. This species has been used by various Native Americans as an antirheumatic medicine, blood medicine, and for various other maladies. The roots may be boiled or roasted and eaten as a vegetable. Ground seeds have been used as flour and for porridge. Dried roasted seeds are similar to popcorn.

OLEACEAE Hoffmanns. & Link OLIVE FAMILY

Kenneth D. Heil

Trees, shrubs, or subshrubs. **LEAVES** opposite (rarely alternate in *Menodora*), simple or pinnately compound, stipulate. **INFL** racemose, paniculate, or thyrsoid. **FLOWERS** perfect or imperfect; calyx mostly 4-lobed or absent; corolla usually united or with distinct petals, or lacking; stamens 2, distinct; pistil 1, the ovary superior, 2-carpelled and 2-loculed; style 1 or lacking, stigmas 1 or 2. **FRUIT** a berry, drupe, loculicidal capsule, circumscissile capsule, or samara. $x = 10$, 11, 13, 14, 23, 24. Approximately 30 genera and about 600 species worldwide, mostly in warmer temperate and subtropical regions. Numerous ornamentals, which include *Chionanthus virginicus* L. (fringe tree), *Forsythia* (golden bells), *Ligustrum* (privet), and *Syringa* (lilac). Several species are used for timber including species in *Olea*, *Fraxinus*, *Nestegis*, and *Notelaea* (Welsh et al. 2003).

1. Fruit a samara; leaves mostly compound (except in *Fraxinus anomala*); large shrubs or trees .... ***Fraxinus***
1′ Fruit a drupe or capsule; leaves simple; spiny shrubs or subshrubs .... (2)

2. Fruit a drupe; corolla lacking; indigenous shrub of river valleys .... ***Forestiera***
2′ Fruit a capsule; corolla well developed; piñon-juniper woodland communities or cultivated shrubs .... (3)

3. Corolla yellow; fruit a membranous circumscissile capsule; indigenous subshrub .... ***Menodora***
3′ Corolla lavender to red, purple, lilac, or white, never yellow; fruit a loculicidal capsule; cultivated shrub ..... ***Syringa***

Forestiera Poir. Desert Olive, New Mexico Privet

(for Charles Le Forestier, French physician and botany teacher of Poiret, the describer) Sprawling shrubs. **LEAVES** opposite, often appearing fascicled at the ends of the branches, serrate or entire. **FLOWERS** very small, polygamo-dioecious, appearing before the leaves; calyx none or minute, unequally 5–6-cleft; corolla none or 1 or 2 small petals; stamens 2 or 4; ovary 2-loculed, with 2 ovules. **FRUIT** a thin-fleshed drupe. **SEEDS** bony, 1. About 15 species, mostly in southwest North America.

Forestiera pubescens Nutt. (becoming hairy) New Mexico privet, New Mexico olive. Shrubs up to 3 m tall, bark gray with spiny branches. **LEAVES** (0.8) 1.5–5.5 cm long, (0.3) 0.5–2 cm wide, oblanceolate to elliptic, entire to serrulate. **INFL** with flowers in fascicles. **FLOWERS**: staminate sessile; pistillate pedicellate. **DRUPES** 5–8 mm long, ellipsoid, blue-black. [*F. neomexicana* A. Gray]. River terraces with Russian olive, saltcedar, skunkbush, and cottonwood. ARIZ: Apa, Nav; COLO: Arc, LPl, Mon; NMEX: McK, RAr, SJn; UTAH. 1360–1905 m (4460–6260'). Flowering: Apr–Jun. Fruit: Jun–Sep. California east to Oklahoma and Texas; Chihuahua, Mexico. Leaves used by the Ramah Navajo as a ceremonial emetic; stem used to make Evilway big hoop. Used to make prayer sticks and as a dye plant for wool by the Navajo. The large shrubs were considered water indicators by the Isleta. The fruit is eaten by numerous birds and mammals, especially foxes and coyotes.

Fraxinus L. Ash

Deciduous trees or shrubs. **LEAVES** opposite, pinnately compound (simple in *F. anomala*). **INFL** a panicle. **FLOWERS** perfect or unisexual, inconspicuous; calyx lacking or 4-cleft; corolla lacking or of 2 or more distinct petals; stamens mostly 2; ovary 2-loculed; style 1; stigmas 1 or 2. **FRUIT** a samara. About 65 species, mostly north-temperate, with a few extending to the tropics. An important timber tree and used for tool handles and sporting goods.

1. Flower with a corolla, in terminal branches on lateral leafy branchlets of the current year; McKinley County, New Mexico .... ***F. cuspidata***

1' Flower without a corolla, in axillary panicles from separate buds in the axils of leaves of the previous year; widespread .... (2)

2. Leaves mostly simple, sometimes 3–7-foliate; twigs quadrangular; wings of the samaras not or seldom continuing to the base of the fruit; blackbrush, shadscale, piñon-juniper, and ponderosa pine communities .... ***F. anomala***

2' Leaflets 3–7; twigs terete or nearly so; wings of the samaras continuing to the base of the fruit; stream courses and flood plains .... ***F. velutina***

Fraxinus anomala Torr. ex S. Watson (abnormal, referring to the single leaf) Singleleaf ash. Shrub or small tree, mostly 2–5 m tall; twigs 4-sided. **LEAVES** glabrous, mostly simple, occasionally 3- to 7-foliolate; the blade ovate to oval, entire or crenulate; 1–5 cm long, 1–7.5 cm wide, acute or subcordate basally, acute to rounded apically. **INFL** a 3–12 cm many-flowered panicle. **FLOWERS** mostly perfect; calyx 1–2 mm long; petals none. **SAMARAS** 12–27 mm long, 5–12 mm wide, obovate-oblanceolate, winged almost to the base. [*Fraxinus anomala* var. *triphylla* M. E. Jones]. Often on slickrock in blackbrush, ephedra, shadscale, Gambel's oak, piñon-juniper, and ponderosa pine communities. ARIZ: Apa, Coc, Nav; COLO: Dol, Mon; NMEX: SJn; UTAH. 1360–2135 m (4460–7000'). Flowering: Apr–May. Fruit: May–Jun. New Mexico, Colorado to California; Mexico. Seeds used in prayer for rain by the Kayenta Navajo. Used for prayer sticks by the Hopi.

Fraxinus cuspidata Torr. (with a stiff point, referring to the leaves that are occasionally cuspidate) Fragrant ash. Shrub or small tree up to 6 m high; trunk up to 2 dm in diameter; bark gray, smooth, becoming fissured into ridges. **LEAVES** petioled with (3) 4–7 (–9) leaflets, leaflets lanceolate to broadly ovate, long-pointed and sometimes cuspidate, 3.5–7 cm long, entire to coarsely toothed, dark green above, paler and pubescent beneath when young. **INFL** a panicle, 7–10 cm long, loose. **FLOWERS** appearing with the leaves; petals 4, about 10 mm long, 1 mm wide, white; stamens shorter than the petals. **SAMARAS** including the wing about 1.2 cm long, oblong-obovate to lanceolate; wing 6 mm wide in upper 1/2 and extending nearly to base of flattened fruit body. Local on rocky slope of canyons. Willow, Gambel's oak, and piñon-juniper communities. NMEX: McK. 2000–2135 m (6560–7000'). Flowering: May–Jun. Fruit: Jun–Jul. Nevada, Arizona, New Mexico, and Texas. The Navajo used the stems to make arrows and bows. The wood is used to make weaving tools.

Fraxinus velutina Torr. (velvety) Velvet ash, Arizona ash. Moderate-size trees; branchlets terete, spreading-hairy to sparingly or nearly glabrous. **LEAVES** with 3–5 leaflets (rarely simple), petiolulate, lanceolate to ovate, elliptic, or orbicular, cuneate to acute basally, acuminate to rounded apically, nearly entire to serrate, hairy or glabrous on the

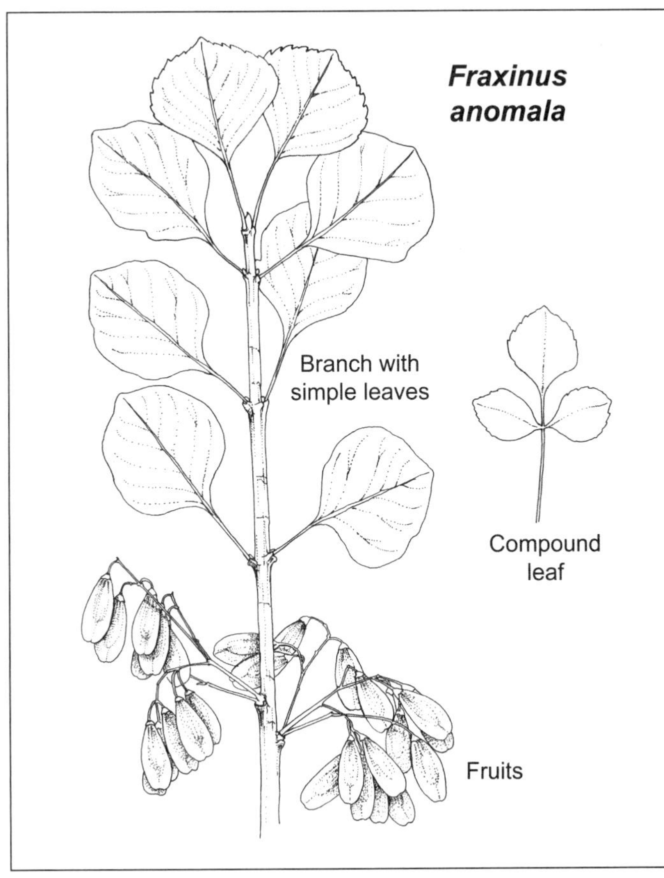

lower surface. **FLOWERS** imperfect; calyx persistent, campanulate; corolla none; anthers apiculate, oblong. **SAMARAS** 16–34 mm long, 4–6 mm wide, the blade decurrent about halfway along the terete body. [*F. pennsylvanica* Marshall subsp. *velutina* (Torr.) G. N. Mill.]. Stream courses and flood plains. A single collection along the Animas River in Farmington. NMEX: SJn. 1645 m (5400′). Flowering: May–Jun. Fruit: Jun–Jul. California, Nevada, Utah, east to Texas.

Menodora Bonpl. Menodora

(Greek *menos*, force, + *doron*, gift) Menodora. Subshrubs. **LEAVES** alternate, lowermost opposite, simple, sessile or subsessile, entire. **INFL** solitary, corymbose, or paniculate. **FLOWERS** showy; calyx 5–15-lobed; corolla yellow, subrotate, 5–6-lobed; stamens 2, inserted on the corolla tube; ovary superior, 2-loculed, 2–4 ovules per locule; style slender, stigma capitate. **FRUIT** a circumscissile capsule.

Menodora scabra A. Gray (rough) Rough menodora. Erect or ascending subshrub, mostly 2–3.5 dm tall, woody at the base only. **LEAVES** 0.5–3 cm long, 2–5 mm wide, narrowly elliptic to oblong or lanceolate, glabrous or scaberulous. **FLOWER** calyx glabrous to scabrous, 7–11 linear lobes, 4–5 mm long; corolla bright yellow, subrotate, lobes 5–9 mm long. **CAPSULE** 5–7 mm long, 8–12 mm thick. **SEEDS** 4–5 mm long, about 3 mm wide. $2n = 22$. Rocky soils. Ponderosa pine, Rocky Mountain juniper, piñon pine, and Gambel's oak communities. NMEX: McK. California, Utah south to Arizona, New Mexico, and Texas; Mexico. Used by the Ramah Navajo in the following ways: decoction of root used for back pain, infusion taken for heartburn, decoction of plant taken to facilitate labor, and plant used as a "life medicine."

Syringa L. Lilac

(Greek *syrinx*, a hollow tube or pipe, referring to the hollow state of the younger shoots in certain species) Shrubs. **LEAVES** petiolate, opposite, simple. **INFL** a terminal or lateral panicle. **FLOWERS** perfect, campanulate, 4-toothed, persistent; corolla tubular, limb 4-lobed and rotate; stamens 2, inserted on the corolla tube; ovary 2-loculed, each locule with usually 2 ovules; style with a 2-lobed stigma. **FRUIT** a loculicidal capsule. About 25 species, mostly Asiatic. Commonly cultivated and with strongly fragrant flowers.

Syringa vulgaris L. (common) Common lilac. Shrubs up to 4 m tall or more, branches mostly erect. **LEAVES** 3–12 cm long, 1.5–8 cm wide, ovate to cordate to rounded, truncate or obtuse basally, acute to acuminate apically, glabrous. **INFL** a panicle, 10–20 cm long. **FLOWERS** purple, lilac, or white; stamens included. $2n = 44, 46$. An ornamental that persists around homesites. Introduced from Europe. COLO: Mon, SMg; NMEX: SJn; UTAH. 1325–2225 m (4350–7300′). Flowering: Apr–Jun. Fruit: Jun–Jul. Widespread throughout the United States.

ONAGRACEAE Juss. EVENING PRIMROSE FAMILY

Warren L. Wagner and Peter C. Hoch

Annual or perennial herbs and shrubs, rarely trees. **LEAVES** usually cauline, often with a basal rosette, rarely basal only, simple, alternate or opposite, occasionally whorled or spirally arranged, entire or toothed to pinnatifid; stipules present, usually small and caducous, or absent in tribes Epilobieae and Onagreae. **FLOWERS** perfect and hermaphroditic or rarely unisexual, actinomorphic or zygomorphic, (2–) 4 (–7)-merous; axillary, in leafy spikes or racemes or solitary, or occasionally in panicles, flower parts distal to the ovary deciduous after anthesis (in ours); floral tube present or occasionally absent, nectariferous within; sepals green or colored; petals as many as sepals or rarely absent, variously colored, occasionally clawed; stamens twice as many as sepals and in 2 series, the antisepalous set usually longer, rarely all equal, or as many as sepals, anthers versatile or sometimes basifixed, dithecal, polysporangiate, opening by longitudinal slits, pollen grains united by viscin threads, (2–) 3 (–5)-aperturate, shed singly or in tetrads or polyads; ovary inferior, consisting of as many carpels and locules as sepals, the septa sometimes thin or absent at maturity, placentation axile or parietal, ovules 1 to numerous per locule, in 1 or several rows or clustered, style 1, stigma with as many lobes as sepals or clavate to globose. **FRUIT** a loculicidal capsule or indehiscent berry or nutlike. **SEEDS** usually small, sometimes with a coma or wings. x = 7, 8, 10, 11, 15, 18. (Wagner, Hoch, and Raven 2007; Munz 1965; Raven 1979; Levin et al. 2004.) Twenty-two genera and 657 species (806 taxa) distributed nearly worldwide, but especially rich in the New World.

1. Stipules present, sometimes deciduous; fruit indehiscent, burlike with hooked hairs; flowers 2-merous [tribe Circaeae] .... ***Circaea***

1′ Stipules absent; fruit a dehiscent capsule or sometimes indehiscent, but not burlike; flowers 4-merous .... (2)

2. Seeds comose (rarely secondarily lost); sepals erect or spreading; stigmas commissural with dry multicellular papillae [tribe Epilobieae] .... (3)

2′ Seeds not comose; sepals reflexed; stigmas not commissural [tribe Onagreae] .... (4)

3. Floral tube absent; petals entire; leaves spirally arranged or rarely subopposite or verticillate; stamens subequal; style and stamens reflexed at maturity .... ***Chamerion***

3′ Floral tube present; petals with apical notch; leaves opposite, at least near base of the stem; stamens in 2 unequal whorls; style and stamens erect .... ***Epilobium***

4. Capsule 2-loculed; stems very slender, hairlike; petals white with 1 or 2 yellow areas at base .... ***Gayophytum***

4′ Capsule 4-loculed; stems usually not very slender; petals yellow or white but not white with yellow areas .... (5)

5. Style with a peltate indusium at base of stigma, at least at younger stages prior to anthesis; stigma 4-lobed, receptive all around (or peltate to discoid or nearly square in sect. *Calylophus*) .... ***Oenothera***

5′ Style without an indusium; stigma hemispherical .... (6)

6. Seeds in 2 rows per locule; capsule pedicellate; leaves ± basal, pinnatifid with a large terminal lobe, in some species either lateral lobes ± reduced or leaf ovate-cordate on a long petiole; lower surface of leaves or leaf margins with conspicuous (brown) oil cells .... ***Chylismia***

6′ Seeds in 1 row per locule; capsule sessile; leaves entire to toothed, not predominantly basal or with oil cells .... ***Camissonia***

Camissonia Link Camissonia, Sun Cup

(named in honor of Adelbert von Chamisso, 1781–1838, German poet, and botanist on the Russian ship *Rurik*, which sailed around the world from 1815 to 1818, including collecting in California in 1816) Annual, usually erect, sometimes prostrate, caulescent herbs, with or usually without a well-defined basal rosette, white or reddish brown epidermis conspicuously exfoliating. **LEAVES** alternate to subfasciculate, subsessile or petiolate, blades linear to narrowly elliptic, margins entire to weakly denticulate; stipules absent. **INFL** a spike, nodding initially and becoming erect in fruit. **FLOWERS** actinomorphic; sepals 4, reflexed singly or in pairs; petals 4, yellow, sometimes with 1 to several red dots near base and fading red; stamens 8 in 2 unequal series, anthers versatile, pollen shed singly; ovary with 4 locules, stigma subentire, subcapitate to subglobose. **FRUIT** a sessile capsule, regularly but sometimes tardily loculicidal, straight to somewhat flexuous, subterete. **SEEDS** many, in 1 row per locule, narrowly obovoid to narrowly oblanceoloid, with the surface smooth, glossy. $2n$ = 14, 28, 42; x = 7. Reproductive features: flowers diurnal; self-compatible or self-incompatible; autogamous (rarely cleistogamous) or outcrossing; pollinated by bees. As now reconfigured, *Camissonia* consists of 12 species (16 taxa). Endemic to western North America, with one species in South America from Peru to Chile and Argentina. This treatment departs significantly from the most recent monograph of *Camissonia* by Raven (1969), in that we have segregated from it the genera *Camissoniopsis*, *Chylismia*, *Chylismiella*, *Eremothera*, *Eulobus*, *Neoholmgrenia*, *Taraxia*, and *Tetrapteron*, based on molecular and morphological analysis (Levin et al. 2004; Wagner, Hoch, and Raven 2007). The more narrowly delimited *Camissonia* is morphologically defined by having capsules subterete; seeds in one row per locule, glossy, triangular in cross section; and plants flowering from the upper nodes only.

Camissonia parvula (Nutt. ex Torr. & A. Gray) P. H. Raven (Latin *parvus*, small or little) Slender annuals, ± strigillose, rarely with scattered spreading hairs, also usually sparsely glandular-puberulent, especially in the inflorescence. **STEMS**

1 or more, wiry, arising from the base, usually (3–) 15–30 cm long, with basal rosette absent. **LEAVES** 10–30 mm, linear, margin subentire. **INFL** nodding. **FLOWERS** with floral tube 1.3–2 mm long; sepals 1.5–2.5 mm long; petals 1.5–3.6 mm long, yellow, fading reddish orange. **FRUIT** 1.5–3 cm long, cylindrical, straight, usually swollen by seeds. **SEEDS** 0.7–0.8 mm, shiny, minutely pitted. $2n = 28$. [*Oenothera parvula* Nutt. ex Torr. & A. Gray]. Sandy soils or dunes, usually in sagebrush shrubland to sagebrush-juniper woodland. ARIZ: Coc, Nav; NMEX: SJn; UTAH. 1130–2100 m (3600–6900′). Flowering: May–Jul. Widespread in the western United States from eastern Washington, Oregon, and California to Montana and Wyoming, western Colorado to northern Arizona, and San Juan County, New Mexico. Self-pollinating.

Chamerion (Raf.) Raf. ex Holub Fireweed, Willow Herb

(probably a contraction of *Chamaenerion*, a pre-Linnaean epithet for *Epilobium angustifolium* meaning dwarf oleander) Erect perennial herbs, pubescent (eglandular) to subglabrous, with shoots from woody caudex or spreading lateral roots. **STEMS** often clumped, simple or rarely branched. **LEAVES** spirally arranged, very rarely subopposite or subverticillate, often subcoriaceous; cauline blades linear to lanceolate, elliptic, or ovate; stipules absent. **INFL** a simple raceme or spike, rarely branched. **FLOWERS** 4-merous, slightly zygomorphic, the lower petals somewhat narrower than upper ones, strongly protandrous, opening on axis nearly perpendicular to stem axis; floral tube absent, producing nectar from raised disc at base of style and stamens; sepals green or reddish green, spreading; petals pink to rose-purple, rarely white, obcordate or obtrullate, entire; stamens 8 in 1 whorl, subequal, erect at onset of anthesis, later deflexed; filament bases slightly bulged to form chamber around nectary disc; anthers versatile; pollen blue or yellow, shed singly; ovary with 4 locules; style deflexed with unopened stigma during anther dehiscence, later straightening and stigma lobes opening, stigma deeply 4-lobed and revolute, receptive only on inner surfaces. **FRUIT** an elongate, slender, loculicidal capsule, terete to quadrangular, splitting to base with intact central column. **SEEDS** numerous, in 1 row per locule, narrowly clavate, with coma of silky hairs at chalazal end. $2n = 36, 72, 108$; $x = 18$. Flowers diurnal, hermaphroditic, protandrous, generally remaining open for 3–5+ days; self-compatible or rarely some populations of *C. angustifolium* self-incompatible; outcrossing and pollinated mainly by bees. This genus of eight species is divided into two sections, section *Rosmarinifolium*, found in Europe and the Caucasus region, and section *Chamerion*, found in eastern Asia, with two species widespread in the Northern Hemisphere. Although sometimes included within *Epilobium*, *Chamerion* forms a well-differentiated sister group to that genus, based on both morphological characters (lack of a floral tube; subequal stamens; zygomorphy involving the petals, stamens, and stigma; and leaves nearly always spirally arranged, rarely subopposite or verticillate) and molecular data. Two species are native to North America, and both are in our flora (Mosquin 1966; Small 1968; Baum, Sytsma, and Hoch 1994).

1. Bracts much smaller than cauline leaves, sublinear; leaves linear to lanceolate, with distinct submarginal vein; petioles 2–7 mm long; sepals 9–19 mm long; petals 14–25 mm long; seeds 1–1.3 mm long....... ***C. angustifolium***

1′ Bracts not much smaller than cauline leaves, lanceolate; leaves narrowly ovate, elliptic to lance-elliptic, lacking submarginal vein; petioles 0–3 mm long; sepals 10–16 mm long; petals 10–32 mm long; seeds 1.2–2.1 mm long .... ***C. latifolium***

Chamerion angustifolium (L.) Holub (Latin *angustus*, narrow, + *folius*, leaf; narrow-leaved) Fireweed. Perennial herbs, forming large clones by vigorous shoots from a woody caudex or from long lateral roots, glabrous to densely strigillose, especially on inflorescence. **STEMS** 30–200 cm tall. **LEAVES** on petioles 2–7 mm long; cauline blades (6–) 9–23 × (0.7–) 1.5–3.4 cm, linear to oblong- or elliptic-lanceolate, green, subglabrous to strigillose, on abaxial side and on abaxial midrib, margins ± denticulate, with prominent submarginal vein. **INFL** bracts much smaller than cauline leaves. **FLOWERS** nodding in bud, suberect at anthesis; sepals 9–19 mm long; petals 14–25 mm long, pale pink to purple or rarely white. **FRUIT** 5–9.5 cm long, densely appressed-canescent, pedicel 1–3 cm long. **SEEDS** 1–1.3 mm long, irregularly reticulate, with indistinct chalazal collar; coma dingy or white, 10–17 mm long, not easily detached. $2n = 72, 108$. Widespread, in open (often disturbed) habitats; common in our area. ARIZ: Apa; COLO: Arc, Con, Hin, LPl, Min, Mon, RGr, SJn; NMEX: RAr, SJn; UTAH. 1130–4200 m (3700–13700′). Flowering: Jul–Sep. Circumboreal, in North America from Greenland, Canada, and Alaska to North Carolina, Ohio, New Mexico, California, and Mexico. Four Corners material belongs to **subsp. *circumvagum*** (Mosquin) Hoch (wandering around). An extremely widespread species in boreal regions worldwide, this species survives and spreads quickly following fires, hence its common name. Used by many cultures as a potherb, and the young leaves are eaten as salads. The flowers are favored by bees, and fireweed honey is highly valued.

Chamerion latifolium (L.) Holub (Latin *latus*, broad, + *folius*, leaf; broad-leaved) Arctic fireweed. Perennial herbs, with thick, woody rhizomes and wiry mass of roots. **STEMS** 12–35 cm tall, glabrous below to sparsely or rarely densely strigillose on upper stem and inflorescence. **LEAVES** on petioles 0–2 mm long; cauline blades 2–5 (–8) × 0.6–1.7 (–2.6) cm, narrowly ovate or elliptic to lance-elliptic, green or pale green, subglabrous or strigillose, especially on veins, margins subentire to remotely punctate-denticulate with 4 to 7 teeth, lacking submarginal vein. **INFL** bracts about 1/2 as long as cauline leaves, foliaceous. **FLOWERS** erect in bud, nodding at early anthesis; sepals 10–16 mm long; petals 10–24 (–32) mm long, rose-purple or pink. **FRUIT** 2.5–8 cm long, strigillose, pedicel 1.2–2.5 cm long. **SEEDS** 1.2–2.1 mm long, surface irregularly low-reticulate, with distinct chalazal collar; coma tawny or dingy, 9–15 mm, not easily detached. $2n = 36, 72$. Widely distributed but uncommon, usually on river gravel and talus slopes. Not common in our area. COLO: Hin, LPl, Min, RGr, SJn. 1130–3700 m (3700–12000′). Flowering: Jun–Aug. Circumpolar, circumboreal, in North America from Greenland and eastern Canada to Alaska and along mountain ranges south to Colorado and California.

Chylismia (Nutt. ex Torr. & A. Gray) Raim. Sun cup

(Greek *chul*, juicy, + *-ism*, suffix to form an adjective, in reference to the succulence of many species in this genus) Robust annual herbs, or sometimes perennial herbs with woody base, stems often slightly succulent, usually branched, usually with well-developed basal rosette. **LEAVES** alternate, long-petiolate, blades usually deeply pinnately divided or rarely simple, veins on lower surface usually with conspicuous brown oil cells, margins sharply dentate to entire; stipules absent. **INFL** an erect or nodding raceme. **FLOWERS** actinomorphic, subsessile to long-pedicellate; floral tube short (in ours) or long, funnelform; sepals 4, reflexed separately; petals 4, yellow or white (often fading orange-red) or lavender, if yellow usually strongly ultraviolet reflective, often with 1 or more red dots near base; stamens 8 in 2 subequal series (in ours) or rarely 4 in 1 series, anthers versatile, long-ciliate or glabrous, pollen shed singly (in ours) or occasionally in tetrads; ovary with 4 locules; stigma entire and capitate or rarely conical-peltate and ± 4-lobed. **FRUIT** a loculicidal capsule, long-pedicellate to subsessile, straight, subterete and clavate or linear, often spreading or deflexed. **SEEDS** numerous, in 2 rows per locule, lenticular to narrowly ovoid, with a more or less pronounced membranous margin when immature, narrowly ovoid when mature, finely lacunose. $2n = 14, 28; x = 7$. Flowers diurnal (in ours) or vespertine; self-incompatible or self-compatible; outcrossing diurnal species pollinated by bees, outcrossing vespertine species pollinated by small moths or hawkmoths, or autogamous. *Chylismia* is a genus of 16 species in two sections (ours all sect. *Chylismia*), distributed in desert regions of western North America (Raven 1969). This genus is clearly distinguished from other species formerly and currently included in *Camissonia* by having straight to arcuate (never twisted or coiled) capsules on distinct pedicels and seeds in 2 rows per locule. Molecular analyses (Levin et al. 2004) showed that *Chylismia* is not in a clade with other *Camissonia* species but rather is a separate clade sister to the strongly supported and realigned genus *Oenothera* (Wagner, Hoch, and Raven 2007).

1. Capsule 0.4–1.2 cm long; inflorescence with intricate, filiform branches, corymbiform .... ***C. parryi***
1′ Capsule 1–5 cm long; inflorescence not with filiform branches, usually unbranched or widely branching, but not forming an intricate corymbiform inflorescence .... (2)

2. Inflorescence erect, elongating in flower and fruit; capsule 1.2–1.8 mm thick .... ***C. walkeri***
2′ Inflorescence nodding, elongating after anthesis; capsule 1.5–2.6 mm thick .... (3)

3. Flowers outcrossing, petals 5.5–9 mm long, stigma exserted beyond anthers at anthesis .... ***C. eastwoodiae***
3′ Flowers self-pollinating, petals 1.5–5.5 mm long, stigma not exserted beyond anthers .... ***C. scapoidea***

Chylismia eastwoodiae (Munz) W. L. Wagner & Hoch (for Alice Eastwood, 1859–1953, who worked on the flora of western North America, especially in California) Succulent annual, glabrous, glandular-puberulent, or villous below. **STEMS** 3–30 cm tall. **LEAVES** in a basal rosette, sometimes cauline, simple, blades oblanceolate to cordate, 0.8–7.5 cm × 0.4–3 cm, margins entire or denticulate. **INFL** nodding, elongating after anthesis. **FLOWERS** with floral tube 2–4.5 mm long, glandular-puberulent or strigose externally, villous near base within; sepals 3–8 mm long, without free tips in bud; petals 5.5–9 mm long, bright yellow with red dots near the base; stigma exserted beyond stamens at anthesis. **FRUIT** 1.8–4 cm long, 1.5–1.9 mm thick, curved, erect, pedicel spreading or slightly deflexed, 0.4–2.8 cm long. **SEEDS** 1.2–1.7 mm long. $2n = 14$. [*Camissonia eastwoodiae* (Munz) P. H. Raven; *Oenothera eastwoodiae* (Munz) P. H. Raven]. Barren clay flats or slopes, especially Mancos clay. ARIZ: Apa; COLO: Mon; UTAH. 1200–1800 m (3900–5900′). Flowering: Apr–Jun. Colorado Plateau endemic from Grand and Emery counties, Utah, to Mesa County, Colorado, south to San Juan County, Utah, east to Montrose County, Colorado. Outcrossing.

Chylismia parryi (S. Watson) Small (for Charles C. Parry, 1823–1890, English-born American botanist and explorer of western North America) Erect and often intricately branched annual herb 5–40 (–80) cm tall, densely villous with white hairs about 2 mm long. **LEAVES** in poorly defined basal rosette, mostly cauline, simple or very rarely with a few small lateral segments, the blade or terminal segment ovate or cordate, 1.4–4 cm × 0.5–2.2 cm, margin weakly dentate, reduced upward. **INFL** nodding, mostly glabrous, with intricate, filiform branches, corymbiform. **FLOWERS** with floral tube 0.5–2 mm long, glabrous or villous externally and within; sepals 1.5–4 mm long, with clusters of light brown oil cells at tip, without free tips in bud; petals 2–7 mm long, bright yellow, often with red dots near base, sometimes fading rose; stigma well exserted beyond anthers at anthesis. **FRUIT** 0.4–1 (–1.2) cm long, 1.2–1.5 mm thick, pedicel 4–20 mm long, filiform, widely spreading or reflexed but capsule held erect. **SEEDS** 0.7–1.2 mm long, lenticular, with narrow cellular rim, few and crowded in 4-septate capsule so as to appear 1-rowed, $2n = 14$. [*Camissonia parryi* (S. Watson) P. H. Raven; *Oenothera parryi* S. Watson] Dry, open sites in clay, gypsiferous, or gravelly soils, slopes or flats, desert shrublands to juniper woodland. UTAH (only Road Canyon). 1800–1900 m (5900–6200′). Flowering: May–Sep. Washington to Beaver counties, Utah; Clark County, Nevada; and Mohave to Coconino counties, Arizona; apparently disjunct to San Juan County, Utah. Outcrossing.

Chylismia scapoidea (Torr. & A. Gray) Small (Latin *scapus*, scape, in reference to the nearly leafless flowers and stems) Annual herb 3–45 cm tall. **STEMS** strigose, villous, or glandular-puberulent below. **LEAVES** mostly in a basal rosette, greatly reduced upward, the blade or terminal segment ovate, narrowly ovate, or elliptic, 1.7–6 × 1–3 cm, ± with several lateral smaller segments. **INFL** nodding, elongating after anthesis, glandular-puberulent, strigose, or subglabrous. **FLOWERS** with floral tube 1–4.5 mm long, glandular-puberulent, strigose, or glabrous externally, sparsely villous to glabrous within; sepals 1.2–5 mm long, ± free tip in bud; petals 1.5–5.5 mm long, bright yellow with fine red dots toward the base; stigma surrounded by 8 dimorphic stamens, the longer usually surpassing the shorter by 1–2 mm. **FRUIT** ascending, weakly clavate, curved to nearly straight, 1–5 cm long, 1.8–2.6 mm thick, the valves sometimes twisted at maturity; pedicel ascending or spreading, 4–20 mm long. **SEEDS** 1–2 mm long. $2n = 14, 28$. [*Camissonia scapoidea* (Torr. & A. Gray) P. H. Raven; *Oenothera scapoidea* Torr. & A. Gray]. Dry, open sites in clay or sandy soils, slopes or flats, desert shrublands to juniper woodland. Flowering: May–Jun. Widespread in the western United States from eastern Oregon to western and central Wyoming and western Montana, south to northern Arizona and northwestern New Mexico. Self-pollinating.

1. Capsule 2.5–5 cm long; petals 1.5–2 mm long .... **subsp. *macrocarpa***
1′ Capsule (1–) 1.5–3 cm long; petals 1.7–5 mm long .... **subsp. *scapoidea***

subsp. *macrocarpa* (P. H. Raven) W. L. Wagner & Hoch (large-fruited) **LEAVES** with ovate blades, often with cordate bases, 1.7–3.5 × 1–1.5 cm. **FLOWERS** with petals 1.5–2 mm long. **FRUIT** 2.5–5 cm long, stout, very large for size of plant, pedicel 2–8 (–14) mm long. [*Camissonia scapoidea* (Torr. & A. Gray) P. H. Raven subsp. *macrocarpa* (P. H. Raven) P. H. Raven] ARIZ: Apa, Nav. 1860–1955 m (6100–6415′). Northeastern Arizona.

subsp. *scapoidea* **LEAVES** with narrowly ovate to ovate blades, often with cordate bases, 2–5.5 × 1–3 cm. **FLOWERS** with petals 1.7–5 mm long. **FRUIT** (1–) 1.5–3 cm long; pedicel 5–20 mm long. $2n = 14, 28$. [*Camissonia scapoidea* (Torr. & A. Gray) P. H. Raven subsp. *scapoidea*]. COLO: Mon; NMEX: SJn; UTAH. 1485–1900 m (4880–6230′). Eastern Oregon to southwestern Montana and western Wyoming, south through Nevada to northern Arizona and northwestern New Mexico.

Chylismia walkeri A. Nelson (after zoologist Ernest Pillsbury Walker, 1891–1969) Slender annual herb 10–70 cm tall. **STEMS** villous below. **LEAVES** in a basal rosette, the cauline blades well developed or absent, the rosette leaves oblong, 3–22 cm long, terminal segment 1–5 × 0.5–3 cm, lateral segments ± developed, villous, often densely so on lower surface, with brown oil cells prominently lining veins below, often purple-dotted, margin doubly serrate. **INFL** erect, branching, elongating in flower and fruit, the buds individually drooping. **FLOWERS** with floral tube 0.5–1.3 mm long, glandular-puberulent or hispid externally, glabrous to sparsely villous within; sepals 1.2–2 mm long, often purple-dotted, with short free tips in bud; petals 1.2–3 mm long, bright yellow; stigma surrounded by anthers at anthesis. **FRUIT** 1.2–4.5 cm long, 1.2–1.8 mm thick, spreading or ascending, the valves often twisted at maturity, straight or slightly curved, pedicel 5–15 mm long. **SEEDS** 0.6–1.2 mm long. $2n = 14, 28$. [*Camissonia walkeri* (A. Nelson) P. H. Raven subsp. *walkeri*; *Oenothera walkeri* (A. Nelson) P. H. Raven subsp. *walkeri*] Dry, open places in sandy, talus, or clay sites, in desert shrublands. ARIZ: Apa, Nav; COLO: Mon; NMEX: SJn; UTAH. 1100–1800 m (3600–5900′). Flowering: Apr–Jun. Nearly confined to the Colorado Plateau, from southeastern Utah, adjacent Colorado, and northern Arizona. Our material belongs to **subsp. *walkeri***. Self-pollinating.

Circaea L. Enchanter's Nightshade

(named for Circe, the enchantress of Greek mythology) Perennial herbs, producing rhizomes, often forming large colonies, sometimes terminated by tubers or with stolons. **LEAVES** opposite and decussate, becoming alternate toward the inflorescence, petiolate; stipules setaceous or glandlike, deciduous or rarely persistent. **INFL** a simple or branched raceme. **FLOWERS** hermaphroditic (protogynous, one anther shedding pollen at anthesis, the other one with delayed dehiscence), bilaterally symmetrical, axillary, pedicellate, erect in bud, spreading to strongly reflexed in fruit; floral tube a mere constriction to very short, subcylindrical to funnelform, with a nectary wholly within and filling lower portion of floral tube or elongated and projecting above the opening of the floral tube as a fleshy, cylindrical or ringlike disc; sepals 2, usually white or pink, spreading or reflexed; petals 2, white or pink, notched at the apex or subentire; stamens 2, anthers versatile, pollen shed in monads; ovary with 2 locules, style filiform, stigma shortly bilobed. **FRUIT** an indehiscent capsule, deciduous with the pedicel at maturity, covered with stiff uncinate hairs. **SEEDS** 1 per locule, smooth, fusiform, or, more commonly, broadly clavoid to slenderly ovoid, adhering ± firmly to the inner ovary wall. $2n = 22$. Flowers diurnal; self-compatible; outcrossing, and pollinated by syrphid flies and small bees, or sometimes autogamous. *Circaea* consists of eight species (14 taxa) and a number of common hybrids, which occur throughout the Northern Hemisphere in moist, temperate, broad-leaved evergreen, deciduous, coniferous, or cool boreal forests at elevations from sea level to 5000 m. Unlike most genera of the Onagraceae, which are restricted to or are most diverse in the New World, *Circaea* is most diverse in eastern Asia, where 12 of the 14 taxa occur. The genus is strikingly distinct within Onagraceae by having 2-merous flowers and indehiscent fruits covered with hooked hairs (Boufford 1982; Boufford et al. 1990).

Circaea pacifica Asch. & Magnus (named for its distribution in western North America) Erect perennial herb, over-wintering by small tubers at the end of slender rhizomes, lower parts retrorsely puberulent, the upper part and inflorescence glandular-puberulent. **STEMS** 1–4 dm tall. **LEAVES** thin, ovate to ovate-cordate, the lowermost smaller, 1.5–7 (–10) × 1–6 (–8) cm, margin subentire to sharply serrate, petioles 0.3–4 cm long. **INFL** a terminal raceme to 17 cm, and also with smaller lateral racemes 0.8–3 cm long, elongating after anthesis. **FLOWERS** with pedicel 0.7–3.5 mm long, elongating to 1.8–5 mm in fruit; floral tube 0.3–0.6 mm long; sepals 1–2.1 mm long; petals white or pink, 0.8–1.6 mm long, deeply notched up to 1/4 to 1/3 their length; stigma capitate to weakly 2-lobed. **FRUIT** 1.5–2.5 mm, clavate, 1-locular by abortion of other carpel. **SEEDS** 1. $2n = 22$. [*C. alpina* L. subsp. *pacifica* (Asch. & Magnus) P. H. Raven]. In forest and shrubland, usually in moist to wet areas along streams and marshy sites, often in partial shade. COLO: LPl, Min. 1500–2900 m (4900–9500′). Flowering: May–Sep. The species occurs in western North America from British Columbia southward and is increasingly disjunct to Arizona and New Mexico. Usually self-pollinating.

Epilobium L. Willow Herb

(Latin *epi*, upon, + *lobium*, pod; probably referring to the inferior ovary of the flower) Perennial herbs with rosettes, fleshy decussate turions, soboles, or stolons, rarely woody at base, or annual herbs with taproot. **STEMS** erect to ascending or decumbent, simple to well branched, strigillose, glandular, villous, or glabrous, often with raised strigillose lines descending from leaf axils. **LEAVES** opposite below inflorescence or only at base, alternate distally, subsessile to petiolate; cauline blades sublinear to lanceolate, oblong, or ovate, lower ones often obovate; stipules absent; bracteoles absent. **INFL** a spike, raceme, or panicle, or flowers solitary in leaf axils. **FLOWERS** actinomorphic or (*E. canum* only) zygomorphic, pedicellate to sessile; floral tube short or elongate, usually with ring of hairs or scales at mouth within, nectary at base of tube, parts distal to ovary deciduous after anthesis; sepals 4, green or rarely colored, erect or rarely spreading; petals 4, rose-purple to white, rarely cream-yellow, or rarely orange-red (then sepals and floral tube also colored), obcordate to obtrullate, apically notched; stamens 8 in 2 unequal series; anthers versatile, rarely basifixed, pollen shed in tetrads or rarely singly; ovary with 4 locules, stigma entire and clavate or capitate to deeply 4-lobed, the lobes commissural, receptive only on inner surfaces. **FRUIT** a loculicidal capsule, usually narrowly cylindrical, terete to sharply quadrangular, splitting to base with intact central column or rarely splitting only on upper 1/3 with central column disintegrating. **SEEDS** numerous or rarely 1–8 per locule, in 1 (2) rows per locule, with ± persistent coma of hairs, or some species (not in our area) lacking coma. $2n = 18, 20, 24, 26, 30, 32, 36, 38, 60$; $x = 18$. Reproductive features: Flowers diurnal, sometimes weakly protandrous, usually self-compatible, and remaining open for more than one day; primarily autogamous, but about 20 species modally outcrossing, pollinated by bees, flower flies, butterflies, or rarely (sect. *Zauschneria*) hummingbirds. *Epilobium* is the largest genus in the Onagraceae, with 165 species distributed on all continents except Antarctica. The genus currently is divided into seven sections, one of which has two subsections. *Epilobium* now includes the former segregate genera *Zauschneria* and *Boisduvalia*. The monophyly of this redefined genus *Epilobium* is very strongly supported by both molecular and morphological characters and forms a sister relationship with *Chamerion*.

1. Floral tube 19–24 mm long; calyx and corolla red-orange, slightly zygomorphic .... ***E. canum***

1′ Floral tube 0.5–12 (–16) mm long; calyx ± green, corolla white to rose-purple, actinomorphic ....(2)

2. Annual with taproot; lower stem peeling; leaves often folded on midrib and clustered at upper nodes .... ***E. brachycarpum***
2′ Perennials from caudex; lower stem not peeling; leaves flat, not clustered at nodes .... (3)

3. Stems erect, with basal rosettes or fleshy turions .... (4)
3′ Stems ascending, clumped, or matted, with soboles or stolons from caudex .... (6)

4. Leaf veins conspicuous; stems (3–) 10–120 (–190) cm tall, with leafy basal rosettes or fleshy turions; seed surface conspicuously ridged .... ***E. ciliatum***
4′ Leaf veins inconspicuous; stems 2–60 cm tall, with basal fleshy turions; seed surface rugose to papillose .... (5)

5. Pedicel in fruit 0.8–3.8 cm long; inflorescence nodding in bud; leaves 0.5–4.7 cm long .... ***E. halleanum***
5′ Pedicel in fruit 0–0.5 cm long; inflorescence suberect; leaves 1–5.5 (–6) cm long .... ***E. saximontanum***

6. Plants caespitose, 3–20 (–25) cm tall; leaves (5.5–) 8–25 mm long; fruit 1.7–4 (–5.6) cm long .... ***E. anagallidifolium***
6′ Plants loosely clumped, 10–50 cm tall; leaves 15–55 mm long; fruit 4–10 cm long .... (7)

7. Pedicel in fruit 0.5–2 cm long; fruit 4–6.5 cm long; petals pink to rose-purple or rarely white; seeds 0.9–1.2 mm long, papillose .... ***E. hornemannii***
7′ Pedicel in fruit 2–4.5 cm long; fruit 5–10 cm long; petals white or rarely flushed pink; seeds 1.1–1.6 mm long, reticulate .... ***E. lactiflorum***

Epilobium anagallidifolium Lam. (with leaves like the genus *Anagallis* [Myrsinaceae]) Alpine willow herb. Low, mat-forming, perennial herb, spreading by thin soboles up to 5 cm long with scattered small leaves. **STEMS** 3–20 (–25) cm tall, ascending basally, often sigmoidally bent or nodding, unbranched, subglabrous or with strigillose lines decurrent from margins of the petioles, rarely strigillose on upper stem with scattered glandular hairs. **LEAVES** often less than length of internodes, attenuate to petioles 1–6 mm long, or subsessile on inflorescence; cauline blades (5.5–) 8–25 × 2.5–10 mm wide, spatulate to oblong basally to elliptic at midstem to lanceolate or sublinear above, subglabrous, margins subentire above to scarcely denticulate below, lateral veins obscure. **INFL** few-flowered, nodding in bud to suberect later. **FLOWERS** erect or nodding; floral tube 0.6–1.2 mm long, glabrous within; sepals 1.5–5 mm long; petals (1.7–) 2.5–6.5 (–8) mm long, pink to rose-purple, rarely white, the notch 0.5–1.2 mm; stigma broadly clavate or subcapitate, at least the longer stamens shedding pollen directly onto stigma at anthesis. **FRUIT** 1.7–4 (–5.6) cm long, slender, often reddish purple, pedicel 0.5–3.5 (–6.8) cm long. **SEEDS** 0.7–1.4 mm long, narrowly obovoid, surface reticulate, sometimes smooth, with inconspicuous chalazal collar; coma dull white, persistent. $2n = 36$. [*E. alpinum* L., nom. rej.] Moist rockslides, talus slopes, and gravelly areas near streams or seeps in alpine and upper montane communities. COLO: Con, Hin, LPl, Min, Mon, RGr, SJn. (1300–) 2100–4100 m ([4300–] 6900–13500′) Flowering: Jun–Sep. Circumboreal. Similar and related to *E. hornemannii* and *E. lactiflorum*, this differs from those species by consistently smaller stature, subentire leaves, and shorter capsules. Also similar to and confused with *E. clavatum* Trel., which occurs in high mountains north of our region and has a more robust habit, more pubescence, wider leaves, and larger seeds.

Epilobium brachycarpum C. Presl (Latin *brachy-*, short, + *-carpus*, fruit; short-fruited) Annual willow herb. Erect annual with taproot. **STEMS** single, 15–200 cm tall, terete, with exfoliating epidermis near base, often paniculate-branched, glabrous below, often strigillose and glandular above. **LEAVES** sparse, often much shorter than internodes, the lower ones often deciduous, opposite below, alternate and gradually reduced in size above, sometimes fascicled, with short axillary shoots of crowded leaves; cauline blades 1–5.5 × 0.1–0.8 cm, linear to linear-lanceolate or narrowly elliptic, often folded along midrib, subglabrous to sparsely strigillose, margins remotely denticulate, lateral veins very obscure, 2–4 on each side of midrib. **INFL** usually an open, indeterminate panicle, subglabrous to strigillose and often glandular. **FLOWERS** erect, the subtending bract sometimes partially fused to the pedicel; floral tube 1–12 (–16) mm long, with ring of hairs near mouth within; sepals 1.2–8 mm long; petals 1.5–15 (–20) mm long, white to deep rose-purple, the notch 0.5–6.5 mm; pollen shed singly or rarely as tetrads; stigma clavate or deeply 4-lobed. **FRUIT** 1.5–3.2 cm long, glabrous to usually slightly glandular-strigillose, pedicels 0.1–1.7 cm long. **SEEDS** 1.5–2.7 mm long, obovoid, surface low-papillose, with a constriction 0.5–0.7 mm from micropylar end, chalazal collar inconspicuous; coma white, easily detached. $2n = 24$. [*E. paniculatum* Nutt. ex Torr. & A. Gray and varieties] Dry woods, grasslands, roadsides. Relatively uncommon in our area. COLO: Arc, Dol, Hin, LPl, Min, Mon, SJn; NMEX: RAr; UTAH. 2165–2745 m (7100–9000′). Flowering: Jun–Sep. Fruit: Jul–Oct. Widespread in western North America from northern Canada to Minnesota, south to California, Arizona, and New Mexico. Naturalized in South America and Europe.

Epilobium canum (Greene) P. H. Raven (Latin *canus*, grayish white, in reference to whitish hairs common on this plant) Perennial herbs or subshrubs, often clumped, with basal shoots from ± woody caudex. **STEMS** 15–40 cm tall, terete, well branched throughout to nearly simple, often scaly basally, grayish to green, generally moderately pubescent, with whitish, spreading-villous hairs and shorter, erect, glandular hairs. **LEAVES** mostly longer than internodes, often with a short axillary shoot and appearing fascicled, subsessile; cauline leaves 20–45 × 5–21 mm wide, ovate to broadly elliptic, moderately pubescent, margins prominently denticulate, veins prominent, 3–7 on each side of the midrib. **INFL** an erect raceme, mainly glandular-pubescent. **FLOWERS** erect to horizontal; zygomorphic, with upper petals slightly flared to right angle with calyx tube, lower ones parallel with it, red-orange or very rarely white; floral tube 19–24 mm long, the base slightly expanded, a ring of 8 irregular scales near base; sepals 4–8 mm long; petals 8–13 mm long, obcordate, the notch 2.2–3.2 mm; stigma 4-lobed, exserted beyond the stamens. **FRUIT** 1.5–2 cm long, glandular-pubescent, sometimes beaked, subsessile or on pedicels to 0.5 cm long. **SEEDS** 1.5–2.6 mm long, narrowly ovoid with constriction 0.6–0.8 mm from micropylar end, surface low-papillose, chalazal collar inconspicuous, coma white, easily detached. $2n = 30$. [*Zauschneria latifolia* (Hook.) Greene var. *garrettii* (A. Nelson) Hilend]. Sandy or rocky soils on steep slopes, rocky hillsides, roadsides, and dry streambeds. Uncommon in our area. UTAH. 1200–2900 (–3400) m (3900–9500′ [11200′]). Flowering: Jun–Oct. Endemic in the western United States, from northwestern Wyoming and southeastern Idaho to northern Nevada and montane areas of north-central to southwestern Utah, disjunct to the Abajo Mountains of San Juan County in southeastern Utah, and to southeastern California. Our material belongs to **subsp.** ***garrettii*** (A. Nelson) P. H. Raven (for A. O. Garrett, 1870–1948, botanist and teacher in Salt Lake City, Utah).

Epilobium ciliatum Raf. (Latin *ciliatus*, ciliate or hairy) Erect perennial herb perennating from leafy basal rosettes or fleshy underground shoots. **STEMS** (3–) 10–120 (–190) cm tall, well branched or simple, terete, lower stem subglabrous with raised lines of strigillose hairs decurrent from margins of the petioles, the upper stem strigillose and usually glandular, or very rarely sericeous. **LEAVES** subglabrous with strigillose margins, or sometimes strigillose throughout, sessile or on petiole 0.1–1 cm long; cauline blades (0.5–) 3–12 × (0.2–) 0.6–5.5 cm, mostly longer than internodes, very narrowly lanceolate to ovate or elliptic, the basal ones often obovate, margins densely serrulate, lateral veins conspicuous, 4–10 on each side of the midrib. **INFL** open or condensed, bracts very reduced or leafy. **FLOWERS** erect; floral tube 0.5–2.6 mm long, with ring of hairs at the mouth within; sepals 2–7.5 mm long, sometimes keeled; petals 2–14 mm long, white to pink or rose-purple, the notch 0.4–2.5 mm; stigma clavate or subcapitate, as long as or rarely exserted beyond anthers. **FRUIT** (1.5–) 3–10 cm long, strigillose and glandular, on pedicel (0–) 0.5–1.5 (–4) cm long. **SEEDS** (0.6–) 0.8–1.6 (–1.9) mm long, narrowly obovoid, surface with conspicuous longitudinal ridges of flattened papillae, rarely reticulate, chalazal collar sometimes conspicuous, coma white or dingy, easily detached, very rarely lacking. $2n = 36$.

1. Petals 2–6 (–9) mm long, white or pink; stem with basal rosettes; leaves generally lanceolate, reduced in open inflorescence .... **subsp. *ciliatum***
1′ Petals 4.5–12 (–14) mm long, deep rose-purple to pink or rarely white; stem with fleshy turions; leaves generally ovate, not much reduced in condensed inflorescence .... **subsp. *glandulosum***

subsp. ***ciliatum*** Variable, weedy, perennial herb, with basal rosettes of subobovate obtuse leaves 10–35 mm long, or rarely with fleshy turions. **STEMS** (3–) 5–120 (–190) cm tall, often well branched. **LEAVES** (0.5–) 3–12 × (0.2–) 0.6–3.7 cm, very narrowly lanceolate to narrowly ovate, or obovate near base, those on the inflorescence narrower. **INFL** open, bracts very reduced. **FLOWERS** with floral tube 0.5–1.8 × 0.9–3 mm long; sepals 2–6 × 0.7–1.6 mm long; petals 2–6 (–9) × 1.3–4 mm, white or sometimes pink; stigma rarely exserted beyond stamens. **FRUIT** (1.5–) 4–10 cm long, pedicel 0.2–1.5 (–4) cm long; frequently 200–500 capsules per plant. **SEEDS** (0.6–) 0.8–1.2 (–1.5) mm long, surface distinctly ridged, coma present or very rarely lacking. $2n = 36$. [*E. adenocaulon* Hausskn. and varieties]. Meadows, stream banks, roadsides, and other moist, often disturbed habitats in full sun to moderate shade. Very common in our area. ARIZ: Apa, Nav; COLO: Arc, Hin, LPl, Min, Mon, SJn, SMg; NMEX: McK, RAr, SJn; UTAH. 1125–4100 m (3700–13500′). Flowering: May–Nov. Very widespread in North America from northern Canada and Alaska to North Carolina, Ohio, Missouri, Texas, New Mexico, California, southern Mexico, and Guatemala; also widespread in South America, Australasia, Europe, and East Asia. This species was used as an analgesic for leg pains by Hopi and as a lotion or poultice of roots for muscular cramps by Navajo. *Epilobium ciliatum* is the most common species in the genus in North America and in our area.

subsp. ***glandulosum*** (Lehm.) Hoch & P. H. Raven (glandular) Perennial herb with basal turion of fleshy scales or leaves just below ground level, or rarely with leafy rosette. **STEMS** 20–110 (–170) cm tall, simple or sparsely branched.

LEAVES 3–10.5 (–15.5) × 1–4.5 (–5.5) cm, typically not much reduced in size onto a crowded inflorescence, narrowly ovate to ovate, rarely lanceolate, often clasping. **INFL** leafy, condensed. **FLOWERS** with floral tube 1–2.6 × 1.4–3.5 mm long; sepals 4.5–7.5 × 1.2–2 mm, sometimes keeled; petals 4.5–12 (–14) × 2.5–6.5 mm, deep rose-purple to pink or rarely white; stigma often exserted beyond stamens. **FRUIT** 4–8.5 cm long, pedicel 0.5–2.5 cm long, or rarely subsessile. **SEEDS** 1.1–1.6 (–1.9) mm long, surface distinctly ridged or rarely reticulate; coma present. $2n = 36$. [*E. brevistylum* Barbey]. Montane, subalpine, and maritime meadows, stream banks, roadsides, and other moist habitats. Uncommon in our area. COLO: Hin; NMEX: SJn. 1500–3400 m (5000–11200'). Flowering: Jun–Sep. Boreal North America from eastern Canada and Alaska to New England, Wisconsin, and south along the mountain ranges to New Mexico, northern Arizona, and California.

Epilobium halleanum Hausskn. (for Elihu Hall, 1820–1882, Colorado collector and farmer) Hall's willow herb. Slender perennial herbs with compact, underground, fleshy turions. **STEMS** 2–60 cm tall, rarely branched, subglabrous below inflorescence with raised lines of strigillose hairs decurrent from margins of the petioles, the upper stem strigillose and glandular, or subglabrous throughout. **LEAVES** subglabrous, with strigillose margins, petioles 0–0.2 cm; cauline blades 0.5–4.7 × 0.25–1.4 cm, mostly subequal to internodes but much reduced in upper pairs, ovate and obtuse basally to lanceolate or narrowly elliptic above, margins subentire below, often denticulate above, lateral veins inconspicuous, 3–6 on each side of midrib. **INFL** nodding in bud, erect later, often bracts very reduced in size. **FLOWERS** erect; floral tube 0.5–1.7 mm long, with a ring of hairs at the mouth within; sepals 1.2–2.8 mm long; petals 1.6–5.5 mm long, white, often fading pink, with apical notch 0.3–1.2 mm; stigma entire, clavate, rarely exserted beyond the anthers. **FRUIT** 2.4–6 cm long, slender, subglabrous to quite pubescent, on pedicel 0.8–3.8 cm long. **SEEDS** 1.1–1.6 mm long, attenuate, narrowly obovoid, surface papillose; coma white, easily detached. $2n = 36$. [*E. pringleanum* Hausskn.]. Shaded or open meadows, stream banks, and seasonally moist roadside areas. Uncommon in our area. COLO: Arc, Hin, LPl, Min, Mon; UTAH. 1200–3950 m (4000–13000'). Flowering: Jun–Sep. Western North America, from British Columbia to Saskatchewan in Canada south through the mountain ranges to California, Arizona, and New Mexico.

Epilobium hornemannii Rchb. (for Jens Hornemann, 1770–1841, Danish botanist) Hornemann's willow herb. Ascending perennial herb, often clumped, with short leafy soboles. **STEMS** 10–45 cm tall, simple or branched from base, subglabrous below inflorescence, except for raised strigillose lines decurrent from the margins of the petioles, inflorescence often ± densely strigillose and glandular. **LEAVES** glabrous except for strigillose hairs on margins, reduced in size on inflorescence, on petioles 0.2–0.8 cm; cauline blades 1.5–5.5 × 0.7–2.9 cm, obovate below to broadly elliptic or narrowly ovate above, often lanceolate on inflorescence, margins denticulate, lateral veins inconspicuous, 3–5 on each side of midrib. **INFL** erect or sometimes nodding. **FLOWERS** erect; floral tube 1–2.2 mm long, glabrous within; sepals 3–6.6 mm long; petals 4–11 mm long, pink to rose-purple or rarely white, the notch 0.7–2.4 mm; stigma clavate to cylindrical, rarely exserted beyond anthers. **FRUIT** 4–6.5 cm long, slender, on pedicel 0.5–2 cm long. **SEEDS** 0.9–1.2 mm long, very narrowly obovoid, surface papillose, coma white to tawny, easily detached. $2n = 36$. Moist alpine and montane meadows, stream banks, and fine scree slopes. Proper habitat restricted in our area, but not uncommon. COLO: Arc, Con, Hin, LPl, Min, Mon, RGr, SJn. 1500–3900 m (5000–12000'). Flowering: Jun–Aug. Nearly circumboreal, in North America from Greenland and northeastern Canada to Alaska, south to New England, Quebec, South Dakota, and along mountain ranges to New Mexico, Arizona, and California. Our material belongs to **subsp**. ***hornemannii***.

Epilobium lactiflorum Hausskn. (white- or milky-flowered) Suberect to ascending perennial herb, often clumped, with leafy soboles. **STEMS** 15–50 cm tall, simple or rarely branched from the base, subglabrous below the inflorescence except for densely strigillose lines decurrent from the margins of the petioles, more generally strigillose and glandular on the inflorescence. **LEAVES** glabrous except for strigillose hairs on margins, on petioles 0.3–1.2 cm; cauline blades 2–5.5 × 0.8–2.4 cm, elliptic or narrowly ovate to narrowly lanceolate, basal ones often spatulate or obovate, denticulate, lateral veins inconspicuous. **INFL** nodding in bud, then erect. **FLOWERS** erect, floral tube 1–2.2 mm long, glabrous within; sepals 3–5.5 mm long, frequently keeled; petals 4–8.8 mm long, white, rarely with red veins or flushed light pink after pollination, the notch 0.7–1.4 mm; stigma clavate, rarely subcapitate and indented at the top, rarely exserted beyond anthers. **FRUIT** 5–10 cm long, slender, pedicel 2–4.5 cm long. **SEEDS** 1.1–1.6 mm long, narrowly obovoid, surface reticulate (smooth) or rarely low-rugose, chalazal neck often conspicuous, coma white, easily detached. $2n = 36$. Montane to subalpine meadows, stream banks, moist ledges, gravel bars, and sometimes roadside ditches. Not uncommon in the proper habitats in our region. COLO: Arc, Hin, LPl, RGr, SJn; NMEX: RAr, SJn. 2590–3595 m (8500–11800'). Flowering: Jun–Sep. Nearly circumboreal; in North America, from eastern Canada to Alaska, south to New England and along the western mountain ranges to New Mexico, Arizona, and California.

Epilobium saximontanum Hausskn. (Latin *saxi-*, rocky, + *montanum*, mountain; Rocky Mountains) Erect perennial herbs, with fleshy underground turions. **STEMS** 4–55 cm tall, simple or well branched, terete, lower stem subglabrous, with raised lines of strigillose hairs decurrent from margins of the petioles, the upper part strigillose and usually glandular. **LEAVES** subglabrous, with strigillose margins, subsessile; cauline blades 10–55 (–60) × 4–20 (–25) mm, lanceolate or narrowly elliptic to ovate, the lower leaves often subobovate, margins faintly denticulate, lateral veins often inconspicuous, 3–6 on each side of the midrib. **INFL** erect. **FLOWERS** erect; floral tube 0.8–1.4 mm, with a ring of hairs at the mouth within; sepals 1.2–3.5 mm long; petals 2.2–5 (–7) mm long, white, infrequently pink to rose-purple, the notch 0.4–1.5 mm; stigma narrowly to broadly clavate or rarely subcapitate, the longer stamens shedding pollen onto stigma. **FRUIT** 2–5.5 (–7) cm long, subsessile or rarely on pedicels 0.1–0.5 cm long, held close to stem. **SEEDS** 1.1–1.6 (–1.8) mm long, very narrowly obovoid, surface rugose to papillose, chalazal collar conspicuous, coma white, easily detached. $2n = 36$. Meadows and stream banks, wet slaty cliffs, often seasonally moist, sometimes disturbed habitats. Relatively uncommon in our area. ARIZ: Apa; COLO: Arc, Con, Hin, LPl, Min, Mon, RGr, SJn. 1400–3500 m (4600–11500′) Flowering: Jul–Sep. Western and northern North America, from eastern Canada (sparse) to western Canada, south mainly in the southern and central Rocky Mountains from Idaho and Montana to Arizona and New Mexico, also barely reaching eastern California (White Mountains and east slope of the Sierra Nevada).

Gayophytum A. Juss. Ground Smoke

(named for Claude Gay, 1800–1873, a French botanist, author of flora of Chile, and the Greek word for plant, *phyton*). Annual herbs, glabrous to strigillose, sometimes villous. **STEMS** very slender, hairlike, usually erect, densely branched or simple, the epidermis usually exfoliating near the base. **LEAVES** alternate or the lowest subopposite, sessile or petiolate, blades narrowly lanceolate, margins entire; stipules absent. **INFL** a panicle or raceme. **FLOWERS** actinomorphic, pedicellate or sessile; floral tube very short, with nectary at base; sepals 4, green, reflexed individually or in pairs; petals 4, oblong, entire, white with 1 or 2 yellow areas at base, fading pink or red; stamens 8, in 2 unequal series, anthers ± basifixed, often adhering to stigma and shedding pollen directly on it; pollen shed singly; ovary 2-locular; stigma entire, globose to hemispheric, the surface wet and nonpapillate. **FRUIT** a loculidical capsule, flattened or subterete and often constricted between seeds, with 4 valves, all becoming free or 2 remaining attached to septum. **SEEDS** many or 2–5 per locule, 1 row per locule, ovoid, glabrous or puberulent, without appendages. $2n = 14, 28$; $x = 7$. Flowers diurnal; self-compatible; one species (*G. heterozygum*) is a permanent translocation heterozygote (PTH); autogamous and sometimes cleistogamous, or some species outcrossing and pollinated by syrphid flies or bees. *Gayophytum* consists of nine species (10 taxa), most in western North America, with one species (*G. micranthum*) endemic to southern South America in Chile and Argentina, and one species (*G. humile*) on both continents (Lewis and Szweykowski 1964).

1. Seeds in each chamber crowded and overlapping ....(2)
1′ Seeds in each chamber not crowded and overlapping ....(3)

2. Petals 0.7–1.5 mm long; pedicel longer than capsule .... ***G. ramosissimum***
2′ Petals 1.5–3 mm long; pedicel shorter than capsule .... ***G. diffusum***

3. Plant branched throughout, usually with 2–8 nodes between branches; capsules usually not much constricted between seeds .... ***G. decipiens***
3′ Plant branched ± at the base, always branched above at every node or with 1–2 nodes between branches; capsules constricted between the seeds .... ***G. diffusum***

Gayophytum decipiens H. Lewis & Szweyk. (Latin for deceiving, used when a species closely resembles another) **STEMS** to 50 cm, glabrous or pubescent; branched throughout, usually with 2–8 nodes between branches; branching usually not dichotomous. **LEAVES** 1–3 cm long, somewhat smaller above; petiole 0–5 mm long. **FLOWERS** appearing 1–5 nodes from base and above; petals 1.1–1.8 mm long; stigma spherical, not exserted beyond stamens. **FRUIT** 6–15 mm long, ascending, longer than pedicel, not conspicuously flattened, somewhat constricted between seeds; pedicels 1–2 (–5) mm long. **SEEDS** all developing, 10–25, subopposite, glabrous or densely puberulent. $2n = 14$. Edges of meadows, moist sites along streams, occasionally on slopes. ARIZ: Apa; COLO: Arc, Hin, Min, SMg; NMEX: SJn; UTAH. 1200–2900 m (3900–9500′). Flowering: Jun–Aug. Widespread in the western United States from Washington to Montana, south to southern California, Arizona, and Colorado. Self-pollinating.

Gayophytum diffusum Torr. & A. Gray (Latin *diffusus*, diffuse, loosely spreading) **STEMS** to 60 cm, glabrous or pubescent; branched or unbranched near base, branched above, usually with 1 or 2 nodes between branches; upper branch-

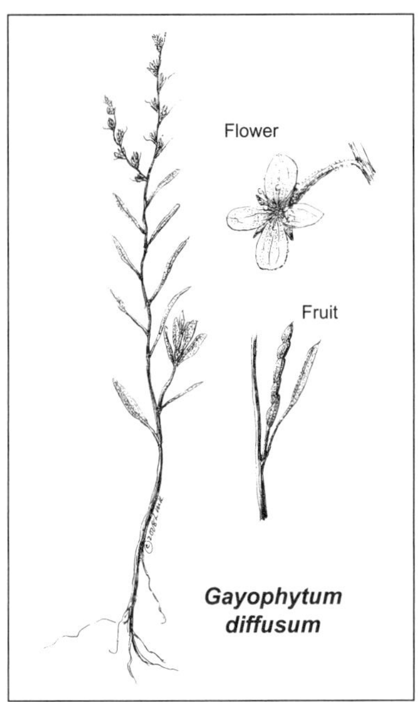

Gayophytum diffusum

ing dichotomous. **LEAVES** 1–6 cm long; petiole 0–10 mm long. **FLOWERS** appearing 1–20 nodes from base and above; petals 1.2–3 mm long; stigma spherical, not exserted beyond stamens. **FRUIT** 3–15 mm long, ascending, spreading, or reflexed, usually longer than pedicel, constricted between seeds; pedicels 2–11 (–15) mm long. **SEEDS** generally all maturing, 3–18, staggered, opposite or crowded and overlapping, glabrous or pubescent. $2n=28$. Occurring on slopes and flats, in dry to moist sites, in open pine forests, meadow margins, or sagebrush shrublands, occasionally in disturbed sites. ARIZ: Apa, Nav; COLO: Arc, Dol, Hin, LPl, Min, Mon, RGr, SJn, SMg; NMEX: McK, RAr, SJn; UTAH. 1130–3000 m (3600–9800′). Flowering: May–August. Extremely variable species widespread in western North America from British Columbia to western South Dakota, south to northern Baja California, Arizona, and New Mexico. Usually self-pollinating. Four Corners material belongs to **subsp.** ***parviflorum*** F. H. Lewis & Szweyk. (small-flowered).

Gayophytum ramosissimum Torr. & A. Gray (Latin for very much branched) **STEMS** to 50 cm, glabrous or nearly so; branched throughout, usually at every other node; branching dichotomous except near base. **LEAVES** 1–4 cm long, much reduced above; petiole 0–10 mm long. **FLOWERS** appearing 5–15 nodes from base and above; petals 0.7–1.5 mm long; stigma spherical, not exserted beyond stamens. **FRUIT** 3–9 mm long, shorter than pedicel, cylindrical; pedicels (3–) 5–12 mm long. **SEEDS** all developing, 10–30, crowded, overlapping, glabrous. $2n=14$. Occurring in open areas, often in sandy soils, usually in piñon-juniper woodland or sagebrush shrubland. ARIZ: Apa; COLO: Arc, LPl, Mon; NMEX: RAr, SJn; UTAH. 1130–2500 m (3600–8200′). Flowering: May–Aug. Widespread in the western United States from Washington to Montana, south to eastern California, Arizona, and New Mexico. Self-pollinating. The Navajo used the plant in a lotion or poultice for cuts and abrasions. The whole plant was used to create the effect of a dream of a spider bite.

Oenothera L. Evening Primrose

Annual, biennial, or perennial herbs, caulescent or acaulescent; stems erect, ascending, or occasionally decumbent, epidermis green or whitish and exfoliating, with a taproot or fibrous roots, occasionally with shoots arising from spreading lateral roots, or with rhizomes. **LEAVES** alternate, usually with a basal rosette present before flowering but often absent later; stipules absent. **INFL** of solitary flowers in leaf axils, or when numerous forming spikes, racemes, or corymbs. **FLOWERS** actinomorphic or (in sect. *Gaura*) zygomorphic and all petals held in the upper 1/2 of the flower, buds erect or recurved; floral tube well developed; sepals (3) 4, reflexed individually, in pairs, or as a unit and reflexed to one side at anthesis, green to tinged or striped red or purple; petals (3) 4, yellow, purple, or white, rarely pink or red, sometimes with a red basal spot, or the base pale green to yellow, usually aging orange, purple, pale yellow, red, or white, usually obcordate or obovate, sometimes (sect. *Gaura*) clawed; stamens (6) 8, subequal or in 2 unequal series, anthers versatile; pollen shed singly; ovary with (3) 4 locules or septa incomplete (sect. *Gaura*) and 1-locular, ovules numerous or (sect. *Gaura*) reduced to 1–8; style glabrous or pubescent, stigma deeply divided into (3) 4 linear lobes or sometimes (sect. *Calylophus*) peltate, discoid to nearly square or obscurely and shallowly 4-lobed, entire surface of lobes receptive, subtended by a more or less conspicuous peltate indusium in early development, persisting to anthesis but often obscured by developing stigma. **FRUIT** a capsule, usually loculicidally dehiscent, sometimes tardily so, rarely an indehiscent, nutlike capsule with hard, woody walls (sects. *Gauropsis*, *Gaura*), straight or curved, terete to angled or winged, ellipsoid to clavate, sometimes tapering to a sterile basal stipe, with (3) 4 locules or sometimes unilocular and the septa incomplete and fragile, not evident at maturity (sect. *Gaura*); sessile or sometimes with sterile basal stipe. **SEEDS** usually numerous, in 1 or 2 (3) rows or in clusters in each locule, or reduced to 1–4 per capsule (sect. *Gaura*). $2n = 14, 28, 42, 56$; $x = 7$. Flowers vespertine or diurnal, usually lasting less than one day; self-incompatible or self-compatible; outcrossing diurnal flowers pollinated by bees, small moths, butterflies, syrphid flies, or hummingbirds (two species), and vespertine flowers pollinated by hawkmoths or sometimes other small moths, rarely wasps or ant lions (sect. *Gaura*), or flowers autogamous and occasionally cleistogamous. As delimited here *Oenothera* is the second largest genus in the Onagraceae, with 145 species divided into 18 sections. The genus is widely distributed in temperate to subtropical areas of North and South America, with a few species in Central America, usually of open, often disturbed habitats, from sea level to nearly 5000 m; several species are widely naturalized. Based on convincing molecular analyses (Levin et al. 2003, 2004; Hoggard et al. 2004; Wagner, Hoch, and Raven 2007) and the consistent synapomorphy of an indusiate style, *Oenothera* has been broadened by including in it *Gaura* (Raven and Gregory 1972) and *Stenosiphon* for the first time, and returning *Calylophus*, which Munz (1965) and others had included in *Oenothera*, but which Raven (1964) and Towner (1977) had excluded. *Gaura*, *Stenosiphon*, and to some degree *Calylophus* each have similar but slightly different variations on the basic lobed stigma. Several of our species have been recently revised (Dietrich and Wagner 1987; Dietrich, Wagner, and Raven 1997; Wagner, Stockhouse, and Klein 1985).

1. Stigma peltate (sect. *Calylophus*) .... (2)
1′ Stigma divided into 4 linear lobes, the lobes very short to long .... (4)

2. Sepals with keeled midrib; stamens biseriate, the antisepalous set conspicuously longer; capsule tardily dehiscent, often slightly recurved (subsect. *Calylophus*) .... ***O. berlandieri***
2′ Sepals flat, without keeled midrib; stamens subequal; capsule promptly dehiscent, usually straight (subsect. *Salpingia*) .... (3)

3. Plants subglabrous; stems sparingly to moderately branched, ascending to suberect .... ***O. hartwegii***
3′ Plants densely strigillose throughout, becoming ± silky-strigose, also sometimes glandular-puberulent on flower parts; stems caespitose, moderately branched, spreading-decumbent to ascending .... ***O. lavandulifolia***

4. Capsule indehiscent; flowers zygomorphic, with all petals held in the upper 1/2 of the flower, or if nearly actinomorphic, then petals very short (0.15–0.3 cm long); floral tube 0.15–1 (–1.3) cm long; seeds 1–4 (sect. *Gaura*) .... (5)
4′ Capsule dehiscent, sometimes tardily so; flowers actinomorphic, petals usually longer than 1 cm [(0.5–)1–6.5 cm long]; floral tube (1–) 1.5–13.5 cm long; seeds numerous .... (7)

5. Flowers nearly actinomorphic; petals 0.15–0.3 cm long; sepals 0.2–0.4 cm long; stigma surrounded by anthers at anthesis (self-pollinating) (subsect. *Schizocarya*) .... ***O. curtiflora***
5′ Flowers zygomorphic; petals 0.3–1.4 cm long; sepals 0.5–1.5 cm long; stigma exserted beyond anthers at anthesis (outcrossing) .... (6)

6. Clumped perennial herbs from a thick taproot, often branching belowground with underground stems, or branching only at surface or not at all; underground stems often becoming horizontal or nearly so and giving rise to new plants; sepals 0.5–1 cm long; capsule pyramidal in upper 1/2, constricted sharply to a terete base (subsect. *Campogaura*) .... ***O. suffrutescens***
6′ Biennial herb from a fleshy taproot, stems several from the base; sepals 1.1–1.5 cm long; capsule ellipsoid, 4-angled (subsect. *Gaura*) .... ***O. coloradensis***

7. Capsule winged; plants subacaulescent; petals yellow (sect. *Lavauxia* subsect. *Lavauxia*) .... ***O. flava***
7′ Capsule not winged, but sometimes valves tuberculate or ridged; plants caulescent, or if acaulescent, then petals white; petals yellow or white .... (8)

8. Capsule valves tuberculate or with a nearly smooth to irregular undulate ridge; petals white (sect. *Pachylophus*) .... ***O. cespitosa***
8′ Capsule valves not tuberculate or ridged; petals yellow or white .... (9)

9. Petals white, fading pink or lavender .... (10)
9′ Petals yellow, fading yellow to orange or red (sect. *Oenothera*) .... (12)

10. Seeds in 1 row per locule; capsule 1.5–2.5 mm in diameter, straight or contorted, tapering to the apex (sect. *Anogra*) .... ***O. pallida***
10′ Seeds in 2 rows per locule; capsule 3–5 mm in diameter, straight or sometimes curved, not tapering throughout to the apex (sect. *Kleinia*) .... (11)

11. Annual herb from a taproot; mouth of floral tube glabrous; leaves subentire or coarsely dentate or pinnatifid .... ***O. albicaulis***
11′ Perennial herb from a taproot and spreading lateral roots; mouth of floral tube densely pubescent with straight white hairs 1–2 mm long; leaves pinnatifid .... ***O. coronopifolia***

12. Flower nodding; capsule cylindrical, 3–4 mm in diameter (subsect. *Nutantigemma*) .... ***O. pubescens***
12′ Flower erect; capsule lanceoloid, 4–9 mm in diameter (subsect. *Oenothera*) .... (13)

13. Petals 0.7–2 cm long; sepals 0.9–1.8 cm long; stigma surrounded by anthers at anthesis; pollen about 50% fertile (2 sizes under 10× magnification, with abortive grains smaller) .... ***O. villosa***
13' Petals 2.5–6.5 cm long; sepals 2.5–5.5 cm long; stigma exserted beyond anthers at anthesis; pollen nearly all fertile (1 size under 10× magnification) .... (14)

14. Floral tube 6–13.5 cm long .... ***O. longissima***
14' Floral tube 2.5–5 (–5.5) cm long .... ***O. elata***

Oenothera albicaulis Pursh (Latin *albus*, white, + *caulis*, stem, in reference to the white stems) Annual herb with taproot and a ± persistent basal rosette, 0.5–3 dm tall, stems usually several from the base, the central one erect, others ascending, strigillose, and also sparsely hirtellous, at least in the upper parts, epidermis white to pink, not exfoliating. **LEAVES** 5–10 × 0.3–2.5 cm, the cauline blades smaller, margin subentire or coarsely dentate or pinnatifid. **INFL** of solitary flowers in leaf axils. **FLOWERS** nodding, actinomorphic; floral tube 1.5–3 cm, cylindrical; sepals 1.5–2.7 cm, without free tips in bud; petals (1.5–) 2–3.5 (–4) cm, white, fading pink; stigma with 4 linear lobes, exserted beyond anthers at anthesis. **FRUIT** 2–4 cm long, 3–4 mm in diameter, straight or sometimes curved, cylindrical, weakly 4-angled, dehiscent nearly to the base. **SEEDS** numerous in 2 rows per locule, about 1 mm long, ellipsoid, pitted-reticulate in rows. $2n = 14$. Dry, open sites, usually sandy flats and slopes. ARIZ: Apa, Coc, Nav; COLO: Arc, Dol, LPl, Mon; NMEX: McK, RAr, San, SJn; UTAH. 1125–2000 m (3700–6600′). Flowering: Mar–Jun. Widespread in western North America from western North Dakota and eastern Montana, southward through the plains and Rocky Mountain region to Arizona, New Mexico, western Texas, and northern Sonora and Chihuahua, Mexico. Probably self-incompatible, outcrossing.

Oenothera berlandieri Walp. (for Jean Louis Berlandier, 1805–1851, Belgian explorer in North America) Sundrops, Texas primrose, Berlandier's sundrops. Perennial from a woody caudex, basal rosette usually not present at anthesis, stems usually several from the base, simple to moderately branched, erect to sprawling, 1–4 dm tall, glabrous to strigillose, especially in upper parts. **LEAVES** 1–4 cm × 0.1–0.6 cm wide, usually not much reduced on upper stem, margin subentire to serrate; fascicles of small leaves often present in nonflowering axils. **INFL** of solitary flowers in axils, but crowded at the stem apex. **FLOWERS** erect, actinomorphic; floral tube 0.5–2 cm long, funnelform; sepals 0.4–1.2, midrib raised or keeled, with free tips 0–2 mm long in bud; petals (0.6–) 1–2.5 cm long, yellow, fading yellow to orange or purple; stigma discoid to nearly square, exserted beyond anthers at anthesis. **FRUIT** 1–3.5 cm long, 1–2 mm in diameter, often slightly recurved, cylindrical, bluntly 4-angled, dehiscent nearly to the base, often tardily so. **SEEDS** numerous in 2 rows per locule, 1–1.8 mm long, sharply angled, truncate at apex. $2n = 14$. [*Calylophus berlandieri* Spach]. Occurring on sandy, gravelly, or limestone soils in dry areas. NMEX: SJn (near Bloomfield). 1650–1700 m (5400–5600′). Flowering: Apr–Jun. Occurs in the high plains from south-central Kansas, Texas panhandle, and scattered localities in eastern to central New Mexico and adjacent southeastern Colorado; with recently discovered disjunct populations near Bloomfield, San Juan County, New Mexico. Four Corners material belongs to **subsp.** ***berlandieri***. Self-incompatible, outcrossing.

Oenothera cespitosa Nutt. **CP** (growing in a tuft) Sand-lily, fragrant, gumbo, white, or tufted evening primrose. Acaulescent or caulescent perennial herbs with definite basal rosette, loosely caespitose to pulvinate from stout taproot, adventitious shoots sometimes produced from slender, spreading, lateral roots. **STEMS** 1 to many, branching from near the base, ascending or rarely nearly erect, 1–4 dm long, epidermis not exfoliating. **LEAVES** in a rosette, 2–25 (–33) × (0.3–) 1–4 (–6.5) cm, margin irregularly sinuate-dentate, pinnatifid, or subentire. **INFL** solitary in leaf axils. **FLOWERS** erect or sometimes nodding, actinomorphic; floral tube (2.8–) 4–14 (–16.5) cm long, cylindrical; sepals (1.5–) 2–4.5 (–5.4) cm long, without free tips in bud; petals (1.6–) 2–4.5 (–5.6) cm long; stigma with 4 linear lobes, exserted beyond stamens. **FRUIT** (1–) 1.5–5 (–6.8) cm long, 5–9 mm in diameter, cylindrical to lanceoloid or ellipsoid, straight to falcate or sigmoid, nearly symmetrical or strongly asymmetrical at base, tapering to a sterile beak 6–8 mm long, the free tips 0.2–4 mm long, erect to recurved, the valves with rows of distinct tubercles to sinuate or nearly smooth ridges along the margin, dehiscing 1/3–7/8 of their length, sessile to long-pedicellate. **SEEDS** numerous, in (1) 2 rows per locule, obovoid, with a seed collar sealed by a thin membrane, this flat or depressed into the raphial cavity, when depressed often splitting and becoming separated from the seed collar. $2n = 14, 28$. [*Oenothera caespitosa* Sims, orth. var.]. Local and colonial in open sites, slopes, and flats, rocky to sandy or clay soils, from grasslands, desert scrub, and piñon-juniper woodland to montane conifer forests. Widespread in western North America, from southeastern Washington and Oregon, to southern Alberta and Saskatchewan, south to southern California and east to North and South Dakota, west of the Missouri River, through much of Colorado, western two-thirds of New Mexico to west Texas and northern Chihuahua and Sonora, Mexico. Self-incompatible, outcrossing. Used by Navajos in various ceremonies, as a dusting powder for chafing, and as a gynecological aid in a poultice of ground-up plants applied for prolapse of the uterus. Used by Hopi for sore eyes, toothache, and in a poultice of the root for swellings. The flowers were used by Hopi ceremonially as "white flower," and the ground plant used as a tobacco substitute. Often cultivated.

1. Capsule oblong-lanceoloid, abruptly contracted to a sterile beak, the dehiscing valves erect or incurved; floral tube 3.5–6 (–7) cm long, often recurved when young; plants shaggy-villous, sometimes densely so.... .... **subsp. *navajoensis***
1′ Capsule cylindrical to lance-cylindrical, tapering gradually to a sterile beak, the dehiscing valves spreading; floral tube (5–) 7–16 cm long, erect; plants hirsute or glabrous .... (2)

2. Capsule somewhat curved, the valves with smooth to irregular ridges; leaves mostly oblanceolate to spatulate, the margins dentate .... **subsp. *macroglottis***
2′ Capsule straight, the valves with tuberculate ridges or rows of nearly distinct tubercles; leaves oblanceolate to lanceolate or elliptic; typically pinnately lobed .... **subsp. *marginata***

subsp. *macroglottis* (Rydb.) W. L. Wagner, Stockh. & W. M. Klein (large glottis, possibly referring to the hypanthium) Acaulescent or rarely with a short congested stem 0.4–0.8 dm tall, hirsute and glandular-puberulent, sometimes glabrous. **LEAVES** (6.8–) 9.5–23 (–32) × (1.3–) 2.4–4.5 (–6.5) cm, oblanceolate to spatulate, margin regularly to irregularly dentate, usually 1–3 small teeth 1–2 mm high between each pair of larger teeth 3–4 mm high, sometimes with a few larger lobes toward the base, rarely the blade coarsely and irregularly pinnately lobed. **FLOWERS** erect; floral tube (4.5–) 7.5–11 (–15.3) cm long; sepals (2.2–) 3–4.5 (–5) cm long; petals (2.1–) 3.5–4.3 (–5) cm long, fading pink to pale rose. **FRUIT** (1.7–) 2.5–4.5 (–5.6) cm long, lance-cylindrical to cylindrical, somewhat curved, terete, symmetrical throughout or occasionally asymmetrical at the base by being slightly flattened on one side, margins of valves with a conspicuous, nearly smooth to irregular undulate ridge about 1.5 mm high, the cross veins visible but not conspicuous, the valves spreading, pedicel 2–7 mm long. **SEEDS** 2.5–3 mm long. $2n$ = 14, 28. [*O. cespitosa* Nutt. var. *macroglottis* (Rydb.) Cronquist]. Igneous soils, piñon-juniper to spruce-fir. COLO: Arc, Dol, Hin, LPl, Min, Mon, SMg; NMEX: RAr, SJn; UTAH. 1850–3000 m (6100–9800′). Flowering: late May–Sep. Endemic to the southern Rocky Mountains, from southern Wyoming through Colorado to southeastern Utah and central New Mexico.

subsp. *marginata* (Nutt. ex Hook. & Arn.) Munz (margined) Acaulescent or with congested stems 1–4 dm long, hirsute and glandular-puberulent. **LEAVES** (2.8–) 10–26 (–36) × (0.6–) 1–3 (–4.5) cm, oblanceolate to narrowly elliptic, rarely lanceolate, coarsely and irregularly pinnately lobed to dentate or rarely serrate, the lobes rarely cut more than halfway to the midrib, sometimes only coarsely erose. **FLOWERS** erect; floral tube (4.1–) 8–14 (–16.5) cm long; sepals (2.2–) 3.4–4.5 (–5.4) cm long; petals (2.4–) 3.5–5 (–6) cm long, fading pink to lavender. **FRUIT** (2.1–) 2.5–5 (–6.8) cm long, cylindrical to sometimes lance-cylindrical, rounded to obtusely quadrangular in cross section, straight and nearly symmetrical and obtuse to cuneate at the base, the margins of the valves with minute to conspicuous tubercles, these sometimes coalesced into a sinuate ridge, the tubercles 0.3–2 mm high, 1–4.5 mm apart, the cross veins prominent to obscure, the valves spreading, pedicel 0–40 (–55) mm long. **SEEDS** 2.2–3.4 mm long. $2n$ = 14. Soils derived from igneous rock, sandstone, or limestone, rarely shale. ARIZ: Apa, Coc, Nav; COLO: Arc, Dol, Mon, SMg; NMEX: McK, RAr, SJn; UTAH. 1200–2300 m (3900–7500′). Flowering: Apr–Jul. Widespread in western North America, from southeastern Washington and Idaho, south to southern California and western Colorado, western two-thirds of New Mexico to west Texas and northern Chihuahua and Sonora, Mexico.

subsp. *navajoensis* W. L. Wagner, Stockh. & W. M. Klein (of the Navajo) Caulescent and loosely caespitose or with stems 1–2.5 dm, crinkly villous, the hairs usually somewhat tangled and glandular-puberulent. **LEAVES** (3.5–) 4–13 (–16) × (0.7–) 1–3.2 cm, oblanceolate to rhombic-obovate, coarsely and irregularly dentate or serrate, sometimes pinnately lobed, often with several larger lobes toward the base of the blade. **FLOWERS** recurved or erect; floral tube (3.5–) 4–7 (–8) cm long; sepals 2.2–2.7 (–3.2) cm long; petals (2.5–) 2.8–3.2 (–3.4) cm long, fading pink to light rose. **FRUIT** 1.3–3.5 (–4) cm long, oblong-lanceoloid, quadrangular or sometimes only obscurely angled in cross section, straight, asymmetrical at the base, usually truncate to rarely obtuse, the margins of the valves with a low, sinuate ridge to 8–15 small, nearly distinct tubercles 0.4–1 mm high, 2–3 mm apart, the cross veins prominent, the valves remaining erect or incurved, pedicel flattened to rounded, 1–3 mm long. **SEEDS** 2.1–2.6 mm long. $2n$ = 14, 28. [*O. cespitosa* Nutt. var. *navajoensis* (W. L. Wagner, Stockh. & W. M. Klein) Cronquist]. Clay and fine sand, desert shrubland to piñon-juniper. ARIZ: Apa, Coc, Nav; COLO: Arc, Dol, LPl, Mon; NMEX: SJn; UTAH. 1130–1800 m (3600–5900′). Flowering: Apr–Jun. Nearly endemic to and widespread in the Colorado Plateau, from eastern Utah, western Colorado, northeastern Arizona, and northwestern New Mexico.

Oenothera coloradensis (Rydb.) W. L. Wagner & Hoch (origin in Colorado) Biennial herb with fleshy taproot and ± persistent basal rosette. **STEMS** several from the base, erect, 5–12.5 dm tall, unbranched to moderately branched, villous and strigillose, epidermis not exfoliating. **LEAVES** 2–13 × 1–4 cm, those in rosette larger, margin subentire to repand-denticulate. **INFL** a spike. **FLOWERS** erect, zygomorphic, with petals in the upper 1/2 and the style and stamens usually in the lower 1/2; floral tube 0.8–1.2 cm long, cylindrical; sepals 1.1–1.5 cm long, without free tips in bud; petals 1.1–1.4 cm long; stigma with 4 short linear lobes, exserted beyond anthers at anthesis. **FRUIT** 0.8–1.1 cm long, 2–5 mm in diameter, narrowly ellipsoid, sharply 4-angled, indehiscent. **SEEDS** 2–4, 2–3 mm long, yellowish to light brown. $2n = 14$. [*Gaura neomexicana* Wooton subsp. *neomexicana*]. Mountain meadows in openings in coniferous forest, growing in heavy soils. COLO: Arc, LPl; NMEX: RAr, SJn. 1800–2800 m (5900–9200′). Flowering: July–Sep. Occurs in two disjunct areas: Sacramento Mountains and Sierra Blanca, Lincoln, and Otero counties (south-central), New Mexico, and San Juan Mountains, near Pagosa Springs, and adjacent areas of New Mexico. Our material belongs to subsp. ***neomexicana*** (Wooton) W. L. Wagner & Hoch (of New Mexico). Self-compatible, outcrossing.

Oenothera coronopifolia Torr. & A. Gray (Latin *corona*, crown, + Greek *opi*, vision or looking like, + Latin *folia*, leaves, in reference to the pinnatifid leaves looking like the top margin of a crown) Perennial herb with a taproot and basal rosette not present by anthesis, forming spreading lateral roots. **STEMS** 1–6 dm tall, usually several, erect, usually branched, strigose and also hirsute, epidermis not exfoliating. **LEAVES** 2–7 × 0.5–1.5 cm, pinnatifid, sometimes the lower ones not as dissected. **INFL** of solitary flowers in leaf axils. **FLOWERS** nodding, actinomorphic; floral tube 1–2.5 cm, cylindrical, densely pubescent at the mouth with straight white hairs 1–2 mm long; sepals 1–2 cm, without free tips in bud; petals 1–1.5 (–2) cm, white, fading pink; stigma with 4 linear lobes, exserted beyond anthers at anthesis. **FRUIT** 1–2 cm long, 3–5 mm in diameter, straight, cylindrical, ± weakly 4-angled, dehiscent nearly to the base. **SEEDS** numerous in 2 rows per locule, about 2 mm long, ellipsoid, minutely tuberculate in rows. $2n = 14, 28$. Dry, open sites, grassy meadows, slopes, along drainages, in foothills and mountains. ARIZ: Apa, Coc, Nav; COLO: Arc, Dol, LPl, Mon, SMg; NMEX: McK, RAr, San, SJn; UTAH. 1500–2800 m (4900–9200′). Flowering: Jun–Sep. Widespread in western North America from western North and South Dakota, eastern Wyoming, southward through the plains and Rocky Mountain region to northern Arizona, New Mexico, and eastern Kansas. Probably self-incompatible, outcrossing. Used by Navajo to improve the taste of tobacco, in a cold infusion of leaves for stomachache, and in a poultice of the plant or root for swellings.

Oenothera curtiflora W. L. Wagner & Hoch (Latin *curtus*, short, in reference to the short flower parts) Rank annual herb with heavy taproot and basal rosette, usually not present at anthesis. **STEMS** 3–20 (–30) dm tall, erect, usually unbranched or branching just below the inflorescence, densely glandular-puberulent and moderately long-villous, sometimes glabrate above. **LEAVES** 2–12.5 × 0.5–4 cm, those in rosette larger, narrowly elliptic to narrowly ovate, margin subentire to sinuate-dentate. **INFL** a dense spike, nodding toward apex. **FLOWERS** erect, nearly actinomorphic; floral tube 0.15–0.5 cm long, cylindrical; sepals 0.2–0.4 cm long, without free tips in bud; petals 0.15–0.3 cm, white, quickly fading pink; stigma with 4 very short, linear lobes, surrounded by anthers at anthesis. **FRUIT** 0.5–1.1 cm long, 1.5–3 mm in diameter, fusiform, reflexed at maturity, tapering ± abruptly toward the base, terete, weakly 4-angled in the upper 1/3, the angles becoming broad and rounded in lower part, indehiscent. **SEEDS** 3 or 4, 2–3 mm long. $2n = 14$. Occurring in disturbed places and along streams. [*G. parviflora* Douglas ex Lehm.; *Gaura mollis* James, nom. rej.]. ARIZ: Apa, Nav; COLO: Arc, LPl, Mon; NMEX: McK, RAr, SJn; UTAH. 1125–2000 m (3700–6600′). Flowering: May–Aug. Widely distributed throughout the western and central United States, from southern Washington and northeastern Oregon south through Idaho, Nevada to Arizona, east to Indiana, and south to Sinaloa and Zacatecas, Mexico; naturalized in Australia, China, Japan, and southern South America. Self-compatible, self-pollinating. The root used by Hopi in a concoction for snakebite. Fresh leaves were used in a headband for cooling in hot weather. The Navajo used the leaves in an infusion for burns or inflammation, as a fumigant, and in a poultice applied postpartum to sore breasts. Sometimes the roots were added to meat stew, and the plant was also used to protect dancers during the fire dance of the Mountain Chant.

Oenothera elata Kunth (Latin *elatus*, tall, in reference to the tall habit) Biennial to short-lived perennial herbs with a heavy taproot and ± persistent basal rosette. **STEMS** 3–25 dm tall, erect, 1 to several, ± branching above, exclusively densely strigillose with some longer appressed hairs or with spreading to erect hairs, some with red-pustulate bases, sometimes also sparsely to densely glandular-puberulent, especially in inflorescence. **LEAVES** 4–25 × 1–2.5 (–4) cm, rosette leaves often much larger, oblanceolate to narrowly lanceolate or very narrowly to narrowly elliptic, margins bluntly dentate or subentire, the teeth sometimes widely spaced, the lower ones sometimes sinuate-dentate toward the base. **INFL** a dense spike, erect. **FLOWERS** erect, actinomorphic; floral tube 2.5–5 (–5.5) cm long, cylindrical; sepals 2.7–5 cm long, with free sepal tips 2–7 mm long; petals 3–5.5 cm long, yellow, fading orange to yellow; stigma with 4 linear lobes, exserted beyond anthers at anthesis. **FRUIT** 2–6.5 cm long, 4–7 mm in diameter, narrowly lanceoloid, bluntly 4-angled, dehiscent nearly to the base. **SEEDS** numerous, in 2 rows per locule, 1–1.9 mm long, angled, reticulate-pitted. 2*n* = 14. [*O. hookeri* Torr. & A. Gray; *O. hookeri* Torr. & A. Gray var. *angustifolia* R. R. Gates; *O. venusta* Bartlett; *O. hewetti* Cockerell]. Occurring along streams, in meadows, or in disturbed sites. ARIZ: Apa, Coc, Nav; COLO: Arc, Dol, Hin, LPl, Min, Mon, SJn; NMEX: McK, RAr, SJn; UTAH. 1130–3000 m (3600–9800′). Flowering: Jun–Sep. Widespread in the western United States, from southern Washington south to California, east and south to Oklahoma, central and western Texas, and Durango, Coahuila, and Baja California in Mexico. Four Corners material belongs to **subsp**. ***hirsutissima*** (A. Gray ex S. Watson) W. Dietr. (shaggy or bristly). Self-compatible, mostly outcrossing. This species has an AA genomic constitution with plastome I (Dietrich, Wagner, and Raven 1997). Used by Navajo as a Plumeway emetic, as a cold remedy, and in a poultice of the root for swellings.

Oenothera flava (A. Nelson) Garrett (Latin *flavus*, yellow, in reference to the yellow petals) Subacaulescent, short-lived perennial or rarely annual herb, with a fleshy taproot and 1 to several densely leafy rosettes. **LEAVES** (3.4–) 6–30 (–36) × (0.5–) 1.5–5 (–7) cm, oblanceolate to linear-oblong in outline, often somewhat fleshy, appearing glabrous but sparsely to moderately strigillose mostly along the margins and sometimes the veins, often mixed with more evenly distributed glandular hairs, margin irregularly and coarsely pinnately lobed, the lobes variable in size and shape from linear to deltoid to coarsely dentate or subentire, sometimes with a large, terminal, lanceolate lobe. **INFL** of solitary flowers from the rosette. **FLOWERS** erect, actinomorphic; floral tube (2.4–) 4–20 (–26.5) cm long, cylindrical; sepals (0.8–) 1.1–4 (–4.2) cm long, with free tips 1–5 (–12) mm long in bud; petals (0.7–) 1–4.5 (–5) cm long, bright yellow, sometimes paler in smaller-flowered plants, fading orange, drying purple; stigma with 4 linear lobes, exserted beyond or surrounded by anthers at anthesis. **FRUIT** (1–) 2–3.5 (–4.3) cm long, 4–8 mm in diameter, narrowly ovoid or ellipsoid, sometimes lanceoloid, becoming hard and leathery in age, the surface usually conspicuously reticulate-veined, gradually constricted to a short beak, the valves with a narrowly oblong wing (2–) 3–5 (–6) mm wide, confined to the upper 2/3 of the capsule, dehiscing 1/4–1/2 the length of the capsule. **SEEDS** numerous in 2 rows per locule, 1.8–2.2 (–2.6) mm long, asymmetrically cuneiform, the surface minutely beaded, narrowly winged at the distal end and along 1 adaxial margin, the raphe raised and conspicuous. 2*n* = 14. [*O. taraxacoides* (Wooton & Standl.) Munz]. Local and colonial, sometimes common in wettish (at least seasonally moist) clay to gravelly sand of swales, desiccating flats and ponds, montane meadows, margins of permanent or seasonal watercourses, and disturbed sites. ARIZ: Apa, Nav; COLO: Arc, Dol, Hin, LPl, Mon, SJn; NMEX: McK, RAr, San, SJn; UTAH. 1130–3000 m (3600–9800′). Flowering: May–Sep. Widespread in western North America from southern Alberta and southwestern Saskatchewan, across the plains of Montana and North Dakota, south to northern Nebraska, throughout the Rocky Mountain region to the Sierra Madre Occidental and adjacent lowlands to Durango, Mexico, and also from disjunct areas: 1) Yakima County, Washington, 2) eastern Oregon and northeastern California, 3) mountains of Nevada, and 4) Jalisco, Guanajuato, and Hidalgo, Mexico. Wagner (1986) recognized two subspecies within the self-compatible *O. flava*. Subsequent study of the complex variation pattern suggests that larger-flowered plants in the southern part of the range evolved on numerous occasions in areas where longer-tongued hawkmoths occur; therefore this species is no longer subdivided. Our plants are the small-flowered (petals 0.7–2.5 cm long), mostly self-pollinating type. The ashes of capsules are used by Navajo for burns, and a poultice of the root for swellings, internal injuries, and throat trouble.

Oenothera hartwegii Benth. (for Karl T. Hartweg, 1812–1871, a German botanist who collected plants in Colombia, Ecuador, Guatemala, Mexico, and California in the United States for the London Horticultural Society) Perennial from a usually stout, woody caudex, basal rosette not present. **STEMS** several to many, sparingly to moderately branched, ascending to suberect, 1.4–4 dm tall, glabrous to sparsely glandular-puberulent. **LEAVES** 1–5 × 0.1–1 cm, linear to oblanceolate to lanceolate, margin subentire, occasionally slightly undulate, rarely with small axillary leaves present. **INFL** of solitary flowers in the upper axils. **FLOWERS** erect, actinomorphic; floral tube 3–5 cm long, narrowly funnelform; sepals 0.9–2.8 cm long, midrib not keeled, with free tips 0.5–3 mm long; petals 1–3 cm long, yellow, usually fading red to purple; stigma peltate, exserted beyond anthers at anthesis. **FRUIT** 1–4 cm long, 2–3 mm in diameter, cylindrical and often narrowed at each end, obtusely 4-angled, dehiscent nearly to the base. **SEEDS** numerous, in 2 rows per locule, 1–1.5 mm long, narrowly obovoid and somewhat angled, surface smooth. 2*n* = 14. [*Calylophus hartwegii* (Benth.) P. H. Raven subsp. *fendleri* (A. Gray) Towner & P. H. Raven; *Oenothera fendleri* A. Gray]. Occasional on clay or gravelly soils, sometimes calcareous soils. NMEX: RAr. 1130–2200 m (3600–7215′). Flowering: May–Aug. Occurring from eastern Arizona, on the Mogollon Rim, through much of New Mexico, to western parts of Oklahoma and Texas, and rare in eastern Chihuahua, Mexico. Our material belongs to **subsp.** ***fendleri*** (A. Gray) W. L. Wagner & Hoch (for Augustus Fendler, 1813–1883, botanical collector in New Mexico, Venezuela, Panama, and Trinidad). Self-incompatible, outcrossing. Used as a drug by the Navajo, especially for internal bleeding. Often cultivated.

Oenothera lavandulifolia Torr. & A. Gray (Latin *lavandulus*, lavender, perhaps in reference to the color of the faded petals) Perennial from a stout, woody caudex, basal rosette not present. **STEMS** caespitose, several to many, moderately branched, spreading-decumbent to ascending, 0.4–2 (–3) dm tall, densely strigillose throughout, becoming ± silky-strigose, also sometimes glandular-puberulent on flower parts. **LEAVES** 0.6–5 × 0.1–0.6 cm, linear to narrowly lanceolate or narrowly oblanceolate, margin subentire, occasionally slightly undulate, infrequently revolute, sometimes with small axillary leaves present. **INFL** of solitary flowers in the upper axils. **FLOWERS** erect, actinomorphic; floral tube 2.5–8 cm long, narrowly funnelform; sepals 0.8–2 cm long, midrib not keeled, with free tips 0.3–3 mm long; petals 1.2–2.8 cm long, yellow, usually fading pink to pale purple; stigma peltate, exserted beyond anthers at anthesis. **FRUIT** 0.6–2.5 cm long, 1–3 mm in diameter, cylindrical and often narrowed at each end, obtusely 4-angled, dehiscent nearly to the base. **SEEDS** numerous, in 2 rows per locule, 1.5–2.5 mm long, narrowly obovoid and somewhat angled, surface smooth. $2n = 14$. [*Calylophus lavandulifolius* (Torr. & A. Gray) P. H. Raven]. Dry, open sites, in sandy to clay soils. ARIZ: Apa, Nav; COLO: Dol, Mon; NMEX: RAr, SJn; UTAH. 1300–2500 m (4300–8200′). Flowering: May–Aug. Occurring from eastern Nevada through Arizona north of the Mogollon Rim, Utah to southeastern Wyoming, adjacent Nebraska, south to western Texas. Self-incompatible, outcrossing. The capsules used by the Apache as a cooked food for children (although identification may refer to *O. hartwegii* as well as or instead of this species).

Oenothera longissima Rydb. (Latin *longissimus*, extremely long, in reference to the much longer floral tubes in this species relative to closely related species) Bridges evening primrose. Biennial to short-lived perennial herbs from a heavy taproot, basal rosette present at anthesis. **STEMS** erect, 1 to several, ± branching above, 6–30 dm tall, densely strigillose or also with spreading to erect hairs, some with red-pustulate bases, and sometimes also sparsely to densely glandular-puberulent, especially in inflorescence. **LEAVES** 5–25 × 0.8–2.5 cm, rosette leaves often much larger, oblanceolate to narrowly lanceolate or very narrowly to narrowly elliptic, margins bluntly dentate or subentire, the teeth sometimes widely spaced, the lower ones sometimes sinuate-dentate toward the base. **INFL** an erect, dense spike. **FLOWERS** erect, actinomorphic; floral tube 6–13.5 cm long, cylindrical; sepals 2.5–5.5 cm long, with free sepal tips 2–6 mm long; petals 2.8–6.5 cm long, yellow, fading orange to yellow; stigma with 4 linear lobes, exserted beyond anthers at anthesis. **FRUIT** 2.5–5.5 cm long, 4–9 mm in diameter, narrowly lanceoloid, bluntly 4-angled, dehiscent nearly to the base. **SEEDS** numerous, in 2 rows per locule, 1–1.9 mm long, angled, reticulate-pitted. $2n = 14$. Occurring in at least seasonally moist sites, usually in sandy or sandy loam soils, sometimes alkaline soils. ARIZ: Apa, Coc, Nav; COLO: Mon; NMEX: SJn; UTAH. 1125–2800 m (3700–9200′). Flowering: Jul–Sep. Ranging from southwestern Colorado, southern Utah, and northwestern New Mexico to eastern Nevada and southern California. Self-compatible, mostly outcrossing.

Oenothera pallida Lindl. (Latin *pallidus*, pale, in reference to often pale leaves) Pale evening primrose. Perennial herbs with taproot and lateral roots, basal rosette ± present at anthesis, or annual herbs from a taproot. **STEMS** 1 to several from the base, 10–50 cm tall, ± branching from near the base, central one erect, subglabrous to sparsely to densely strigillose, also often spreading-villous, at least in the upper parts, epidermis white, exfoliating. **LEAVES** 1–6 × 0.3–1.5 cm, the cauline ones smaller, coarsely dentate or sinuate-dentate. **INFL** of solitary flowers in the upper leaf axils. **FLOWERS** nodding, actinomorphic; floral tube 1.5–3.5 cm long, cylindrical; sepals 1.3–2.5 cm long, with free tips in bud 0.5–2 mm long; petals (1–) 1.5–2.5 (–3) cm long, white, yellow at the base, fading pink or lavender; stigma with 4 linear lobes, exserted beyond anthers at anthesis. **FRUIT** 1.5–6 cm long, 1.5–2.5 mm in diameter, straight or contorted, cylindrical, tapering to the apex, weakly 4-angled, the angles blunt, dehiscent nearly to the base. **SEEDS** in 1 row per locule, 1.5–2.2 mm long, narrowly obovoid, minutely pitted at high magnification. $2n = 14$. Dry, open sites, usually sandy flats and slopes. Widespread in western North America from Washington and northern Oregon to southwestern South Dakota, southward through the plains and Rocky Mountain region to Arizona, New Mexico, western Texas, and northern Chihuahua, Mexico. Self-incompatible, outcrossing. A poorly understood species currently subdivided into four subspecies (Wagner, Hoch, and Raven 2007) that differ largely in aspect, leaf division, capsule configuration, and pubescence. Used by Navajo as a dusting powder for venereal disease sores; a poultice of the plant used for spider bites, and a decoction of roots and leaves for snakebite. It was also used as a Beadway emetic, and as a treatment for livestock with colic. An infusion of the plant was used for kidney disease and for sore throats. Used by the Hopi ceremonially as the white flower associated with the northeast direction.

1. Plants subglabrous, rarely with a few long hairs; leaves often denticulate; capsule usually contorted .... **subsp. *pallida***

1′ Plants sparsely to densely strigillose, also often spreading-villous; leaves usually sinuate-dentate; capsule straight, curved, or contorted .... (2)

2. Upper part of plant conspicuously villous; capsule usually contorted; plant not branching much above base; short-lived perennial to annual herbs ....**subsp. *trichocalyx***
2' Upper plant sparsely to densely strigillose or subglabrous; capsule usually straight or curved; plants usually much-branched above the base; perennial, often becoming a subshrub ....**subsp. *runcinata***

subsp. *pallida* Perennial herbs from taproot and spreading lateral roots, basal rosette not present at anthesis. **STEMS** usually branched throughout, subglabrous, rarely with a few long hairs or sparsely strigillose. **LEAVES** subentire to denticulate, rarely more deeply divided. **FLOWER** sepals 1.2–1.8 cm long, usually red-tinged. **FRUIT** usually contorted. In sandy soils or on dunes, often in disturbed areas, sometimes in alkaline soils. ARIZ: Apa, Coc, Nav; COLO: LPl, Mon; NMEX: McK, RAr, San, SJn; UTAH. 1125–2000 m (3700–6600′). Flowering: May–Sep. Widespread in the Intermountain region of the western United States, from Washington and northern Oregon, through Idaho to central Wyoming, Utah, southern Nevada, northern Arizona, and northwestern New Mexico.

subsp. *runcinata* (Engelm.) Munz & W. M. Klein (pinnatifid or coarsely serrate, in reference to the leaves) Perennial herbs from a taproot and spreading lateral roots, basal rosette not present at anthesis. **STEMS** usually branched throughout, often becoming a subshrub, usually densely strigillose throughout, rarely sparsely so. **LEAVES** sinuate-dentate, rarely some of them merely dentate. **FLOWER** sepals 1–2.5 cm long, usually gray-green. **FRUIT** usually straight or somewhat curved. In sandy soils or on dunes, often in disturbed areas, sometimes in alkaline soils. COLO: Mon; NMEX: McK, SJn. 1130–2300 m (3600–7500′). Flowering: May–Sep.

subsp. *trichocalyx* (Nutt. ex Torr. & A. Gray) Munz & W. M. Klein (hairy calyx) Annual or short-lived perennial from a taproot and sometimes from spreading lateral roots, basal rosette usually present at anthesis. **STEMS** 1 to several from the base, relatively little-branched above, villous, especially in the upper parts, also usually sparsely to moderately strigillose. **LEAVES** sinuate-dentate, rarely some of them merely dentate. **FLOWER** sepals 1–1.8 cm long, often red-tinged. **FRUIT** usually contorted. In sandy, silty, or rocky soils. ARIZ: Apa, Nav; COLO: LPl, Mon; NMEX: SJn; UTAH (Glen Canyon area). 1125–2400 m (3700–7900′). Flowering: May–Jun. Occurring from Wyoming to the eastern half of Utah, western half of Colorado, and San Juan County, New Mexico.

Oenothera pubescens Willd. ex Spreng. (becoming hairy) South American evening primrose. Erect, annual or biennial herbs, basal rosette present at anthesis. **STEMS** erect, simple or with a branched main stem and arcuate to procumbent lateral branches arising from rosette, 0.5–5 (–8) dm tall, densely to sparsely strigillose, sometimes in upper part also villous and glandular-puberulent. **LEAVES** 2–8 × 0.5–2.5 cm, rosette leaves often much larger, narrowly oblanceolate, narrowly oblong to lanceolate or narrowly elliptic, pinnately lobed to subentire. **INFL** of solitary flowers in the upper axils, the apex often curved. **FLOWERS** nodding, actinomorphic; floral tube 1.5–5 cm long, cylindrical; sepals 0.5–2.5 cm long, with free tips 0.1–1 mm long; petals 0.5–2.5 (–3.5) cm long, yellow, fading red; pollen about 50% (–70%) fertile; stigma divided into 4 linear lobes, surrounded by anthers at anthesis. **FRUIT** 2–4.5 cm long, 2–3 mm in diameter, cylindrical, dehiscent nearly to the base. **SEEDS** numerous, in 2 rows per locule, 0.9–1.5 mm long, broadly ellipsoid to subglobose, the surface pitted. $2n = 14$ (ring of 14 in meiosis). Scattered to locally common in open sites in montane communities. NMEX: McK. 2600–2800 m (8500–9200′). Flowering: May–Aug. Widespread from Arizona and New Mexico, south to Guatemala, and Andes of South America in Colombia, Ecuador, and Peru. Self-compatible, self-pollinating, permanent structural heterozygote species.

Oenothera suffrutescens (Ser.) W. L. Wagner & Hoch (Latin, slightly woody, in reference to the woody base) Scarlet gaura, scarlet beeblossom, linda tarde wild honeysuckle, butterfly weed. Clumped perennial herb from a thick taproot, often with branching underground stems, or branching only at the surface or not at all; underground stems often becoming horizontal or nearly so and giving rise to new plants, basal rosette not present at anthesis. **STEMS** 1–12 dm tall, varying from densely branched at the base but little above, to strict and scarcely branched at the base but densely so above, densely strigillose, also often with long spreading hairs near the base, to subglabrous. **LEAVES** 0.7–6.5 × 0.1–1.5 cm, linear to narrowly elliptic, margin entire to remotely and coarsely serrate. **INFL** of erect, dense spikes. **FLOWERS** erect, zygomorphic; floral tube 0.4–1.3 cm long, cylindrical; sepals 0.5–1 cm long, without free tips in bud; petals 0.3–0.8 mm long, white, fading to orange-red to deep maroon; stigma with 4 linear lobes, exserted beyond anthers at anthesis. **FRUIT** 0.4–0.9 cm long, 1–3 mm in diameter, pyramidal in upper 1/2, constricted sharply to a terete base, indehiscent. **SEEDS** (1–) 3–4, 1.5–3 mm long, 1–1.5 mm thick, light to reddish brown. $2n = 14, 28, 42, 56$. [*Gaura coccinea* Pursh]. Occurring in sandy to sometimes clay soils, often calcareous, in desert shrublands to piñon-juniper woodland, disturbed areas, meadows, slopes. ARIZ: Apa; COLO: LPl, Mon; NMEX: McK, RAr, San, SJn; UTAH. 1130–2000 m (3600–6600′). Flowering: May–Aug. Widespread in North America from southern British Columbia to southern Manitoba, south across the plains to western parts of Minnesota, Iowa, and Missouri, and south in the Rocky

Mountain region from Montana, eastern Wyoming, all but northwestern Colorado, northern and eastern Arizona, southern Nevada, southern California, and through much of Mexico in the Sierra Madre Occidental, Sierra Madre Oriental, and the transvolcanic belt. Self-incompatible, outcrossing. The Navajo used a cold infusion of this plant to settle children's upset stomachs after vomiting. Also as a "life medicine," especially for serious internal injuries.

Oenothera villosa Thunb. (Latin *villosus*, villous, in reference to the spreading hairs often present) Biennial evening primrose. Biennial to short-lived perennial herbs from a heavy taproot, basal rosette present at anthesis. **STEMS** erect, 1 to several, ± branching above, 3–20 dm tall, with a mix of 3 hair types, densely strigillose, spreading-villous, some hairs with red-pustulate bases, and sparsely to densely glandular-puberulent, especially in inflorescence. **LEAVES** 5–20 × 1–2.5 (–4) cm, rosette leaves often larger, oblanceolate to narrowly lanceolate or narrowly elliptic, margins bluntly dentate or subentire, the teeth sometimes widely spaced, the lower ones sometimes sinuate-dentate toward the base. **INFL** a dense spike, erect. **FLOWERS** erect, actinomorphic; floral tube 2.3–4.4 cm long; sepals 0.9–1.8 cm long, with free sepal tips 0.5–3 mm long; petals 0.7–2 cm long, yellow, fading orange to yellow; pollen about 50% (–70%) fertile; stigma with 4 linear lobes, surrounded by anthers at anthesis. **FRUIT** 2–4.3 cm long, 4–7 mm in diameter, narrowly lanceoloid, bluntly 4-angled, dehiscent nearly to the base. **SEEDS** numerous, in 2 rows per locule, 1–2 mm long, angled, surface reticulate-pitted. $2n = 14$ (ring of 14 in meiosis). Occurring primarily in open, often wet sites such as streamsides, fields, roadsides. COLO: Arc, Hin, LPl, Min, Mon, SJn; NMEX: McK, RAr, SJn; UTAH. 1130–3150 m (3600–10300′). Flowering: Jun–Aug. Occurring widely in western North America from southern British Columbia eastward to Manitoba, south to northern California, the northern half of Nevada, Utah, Arizona, and New Mexico, and east to Minnesota and Nebraska; possibly naturalized around the Great Lakes. Self-compatible, self-pollinating, permanent structural heterozygote species. Our material is **subsp. *strigosa*** (Rydb.) W. Dietr. & P. H. Raven. Dried leaves were smoked with tobacco for good luck in hunting by the Navajo. Also used in a cold root infusion for "deer infection."

ORCHIDACEAE Juss. ORCHID FAMILY

Linda Mary Reeves

(*orchis*, testis, referring to the underground tubers of some European species) Mostly perennial herbs in ours, vines or shrublike; most epiphytic, but ours terrestrial, sometimes mycotrophic with reduced chlorophyll content and reduced or absent leaves; most completely or partially dependent, at least as seeds and young protocorms, upon mycorrhizal fungi for nutrition. Underground parts rhizomes or tuberous; roots fleshy and few, often with a velamen; sometimes difficult to differentiate histologically from underground stems. **STEMS** commonly erect, sheathed in bracts in most, pendent to prostrate in others, often swollen and succulent, forming a pseudobulb. **LEAVES** if present, often sheathing, cordate, elliptic, ovate, obovate, lanceolate, or linear, occasionally terete, commonly somewhat fleshy and succulent. **INFL** solitary, spicate, corymbose, or racemose. **FLOWERS** complete, mostly perfect, few imperfect (such as *Cycnoches*), minute and inconspicuous to showy, usually zygomorphic and resupinate (twisting 180°), sometimes hypersupinate (twisting more than 180°); perianth of 6 tepals, or more commonly 3 sepals and 3 petals, the lowermost or uppermost often expanded into a lip (labellum) or pouch and often uniquely colored or ornamented with calli, nectar guidelines, or other unusual structures such as dangling wax secretions or pseudostamens, often containing nectaries and a spur or nectar bowl, many without nectaries or food reward for pollinators; stamen(s) completely fused to the style and stigma, or nearly so, to form the column; fertile anthers 3, 2, or 1 (typical in ours); pollen mealy, or more usually agglutinated into 1 or more pollinia, often in derived genera connected to gluelike viscin adherents and/or elastic fibers (the caudicle or stipe) and/or a sticky pad derived from the rostellum (the viscidium), the whole forming the pollinarium or, in some genera, hemipollinaria; pollinia not in pollinaria in some genera, adhering to pollinator by gluelike secretions of the column; stigma forming a sticky concave surface on the lower side of the column tip; ovary inferior, 3-carpellate; ovules numerous, with parietal placentation. **FRUIT** in most a dry, dehiscent capsule, sometimes uncommon. **SEEDS** mostly minute, occasionally substantial as in *Vanilla*, numerous, often as many as a million per fruit, mostly airborne, dry, lacking endosperm or obvious differentiation of the embryo. Seedling initially an undifferentiated callus, later often with poorly differentiated embryonic tissues. Chromosome number various. About 25,000–35,000 species distributed worldwide in almost all habitats with moisture, but most abundant in the tropics. Terrestrial orchids, especially, often inhabiting slightly or infrequently disturbed sites. Pollination systems commonly spectacular and varied; autogamy present in many genera. *Piperia unalascensis* (Spreng.) Rydb. should be searched for in our area as it has been found in Linda Canyon of the Zuni Mountains (McKinley County), New Mexico, as well as in Dolores County, Colorado. Both locations are just outside the study area. *Vanilla planifolia*, source of vanilla "bean" and flavoring; pseudobulbs of some genera (*Ophrys* or *Orchis*) eaten as food in Asia (salepi dondurma, "fox testicle ice cream" in Turkey) or dried and powdered for thickening stews. *Cattleya*, *Laelia*, *Dendrobium*, *Oncidium*, *Cymbidium*, *Phalaenopsis*, *Vanda*, and other genera important in the cut-flower and horticulture trade with many multigeneric hybrids produced (Luer 1975; Dressler 1981, 1993; Coleman 2002).

1. Leaves present at flowering .... (2)
1' Leaves absent at flowering .... (9)

2. Fertile anthers 2; flowers large and showy, yellow and brown or, rarely, yellow and green .... ***Cypripedium***
2' Fertile anther 1; flowers various .... (3)

3. Leaf mainly solitary, basal, slightly to somewhat plicate; flowers showy, pink and white with yellow pseudostamens on saccate lip .... ***Calypso***
3' Leaves various, usually more than 1, flowers various, no pseudostamens, lip not or only slightly saccate .... (4)

4. Leaves primarily basal, not well distributed along the stem; if present, small, few, and bractlike .... (5)
4' Leaves distributed along the stem; if more than 2, longer basally, and smaller, more frequent, and leaflike bracts approaching the inflorescence .... (6)

5. Lip not deeply saccate; inflorescence a congested spiral; leaves entirely green .... ***Spiranthes***
5' Lip deeply saccate; inflorescence not a spiral, but a 1-sided, spikelike raceme; leaves often marked conspicuously with white or silver-white .... ***Goodyera***

6. Leaves 2, opposite, borne more or less midstem, ovate or cordate, margins often undulate; flowers small, green, dull purple, or purplish green .... ***Listera***
6' Leaves more than 2, alternate, often mostly basal, linear, straplike, or tending toward plicate; flowers various .... (7)

7. Leaves lanceolate, tending toward plicate; flowers solitary or in a loose raceme; slightly showy, spreading, flower color variable but often green, pinkish, and yellow .... ***Epipactis***
7' Leaves straplike, elliptic or linear, not plicate; inflorescence generally a congested raceme, occasionally loose; flowers not showy, greenish or rarely white .... (8)

8. Lip apex divided into 2 lobes; flowers green, often with pinkish bases of sepals, petals, lip, and/or column .... ***Coeloglossum***
8' Lip apex not divided obviously into 2 lobes; flowers green, yellowish green, purplish, or white, often with blue-green markings near base of lip .... ***Platanthera***

9. Flower solitary, pink, rarely white; lip saccate, with a tuft of yellow hairs, 2-horned .... ***Calypso***
9' Flower not solitary, brownish purple or dull yellowish, lip not or weakly 3-lobed or auricled toward the base, often many-spotted or -striped .... ***Corallorhiza***

Calypso Salisb. Fairy Slipper

(Greek *Kalypso*, a secluded sea nymph in Homer's *Odyssey*) Small, low herbs, usually with a single plicate leaf and short flowering stalk with 2–4 small sheathing bracts; small globose corm, coralloid rhizome; roots somewhat fleshy, closely intertwined with fungal mycelia. **LEAF** often absent or fading during flowering; glabrous, petiole often long; blade ovate, plicate. **FLOWERS** solitary, showy, deep pink or pinkish lavender, occasionally white or light pink, fragrant, scent reminiscent of lilac (*Syringa*), often appearing while snow is still on the ground; sepals and lateral petals usually deep pink, linear, erect-spreading; lip saccate, 2-horned, pink, white, mottled and/or striped maroon, bladelike, dilated lamina covering the lip apex, often with prominent tufted yellow hairs (pseudostamens); column suborbicular or ellipsoid, broad, deep pink; pollinaria with 2 pairs of pollinia, flat, waxy, with flat, elliptic, detachable viscidium. **FRUIT** erect when mature, ellipsoid, about 2–3 cm long × 1–2 cm wide, often few and uncommon to rare. Prefers sites of occasional disturbance with downed and well-rotted timber, usually fir, on north-facing banks. Colonies eventually die out as dense conifer shade develops. One circumboreal species.

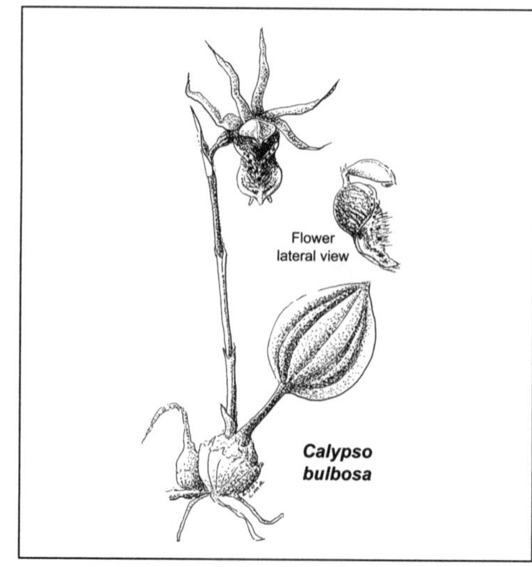

Calypso bulbosa (L.) Oakes (bulblike corm) Fairy slipper, Venus' slipper, calypso, hider-of-the-north, deer head orchid. [*C. americana* R. Br.;

C. borealis (Sw.) Salisb.]. $2n = 28$. Cool, mossy, coniferous forests, usually under fir with considerable wood and bark debris, occasionally under ponderosa pine in logging road ruts or other moist humus soils. Also found at edges of conifer or aspen woods. Populations tend to be large in areas of previous logging, dying out as disturbed forest matures and dense shade develops. Locally common in Chuska and San Juan Mountains. ARIZ: Apa; COLO: Arc, Hin, LPl, Min, RGr, SJn; NMEX: SJn; UTAH. 2405–3200 m (7900–10500′). Flowering: May–Jul. Fruit: Aug–Oct. Northeastern United States; Canada and Alaska south along the higher elevations of the Rockies to Arizona and New Mexico, also northern Europe and Asia. Ours is **var. *americana*** (R. Br.) Luer. The flower attracts pollinators with bright yellow pseudo-stamens on the lip, mimicking a pollen food source. Pollinators are evidently young, naïve queen bees of *Bombus*, *Pyrobombus*, and *Psithyrus*.

Coeloglossum Hartm. Frog Orchid

(Greek *koilos*, hollow, + *glossa*, tongue) Plant an erect herb, leafy, glabrous, 5–30 cm tall, roots annual, forked, fleshy, tuberoid. **LEAVES** 2–5, elliptic, obtuse, elliptic-obovate to oblanceolate, to 18 cm long, 7 cm wide, clasping, medium green. **INFL** a congested, spicate raceme. **FLOWERS** arising from a leaflike floral bract; sepals and lateral petals green, often with pinkish bases, conniving over a broad column with diverging anther cells; the lip white or whitish green, often with pale pink or peach coloring at the base, apex 2-lobed, a small, central, toothlike prominence between the lobes; pollinia 2, granular, each in 1 hemipollinarium, viscidium with a small, thin membrane. **FRUIT** an elongate, semi-erect capsule, 1–1.5 cm long, 0.5 cm wide; few to many. $2n = 40$. One circumboreal, polymorphic species with two varieties. Recent DNA evidence indicates that this clade is embedded in the Asian and European genus *Dactylorhiza*, but its morphological features and lack of genetic divergence are coherent enough to warrant keeping it as a separate genus (Pridgeon et al. 1997). It has been placed previously in the genera *Habenaria* and *Platanthera*, possibly due to convergently evolved features (small, green, translucent flowers) as a response to similar pollinators. Pollinators reported to include two wasp species (*Tenthredopsis* sp., *Cryptus* sp.), a mosquito (*Tipula* sp.), and a beetle (*Cantharis* sp.).

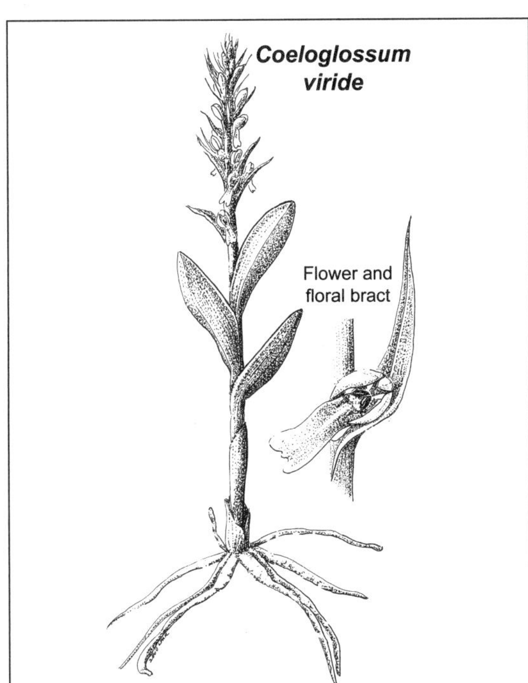

Coeloglossum viride (L.) Hartm. (green) American frog orchid, long-bracted orchid, satyr rein orchid. [*Orchis virescens* Muhl.; *Habenaria virescens* (Muhl.) Spreng.; *Platanthera viridis* (L.) Lindl. var. *bracteata* (Muhl. ex Willd.) Rchb. f.]. Rare, but often misidentified. In our area at high elevations in shady, rich, moist, cool woods under conifers, willow and aspen, or forest edge. COLO: Hin, LPl, Mon, SJn. 2470–3505 m (8100–11500′). Flowering: Jun–Jul. Fruit: Aug–Sep. Alaska through Canada and the northeast coast of Asia to the northeastern United States and Canada, south to the southern Appalachians, west to North Dakota and Washington, south to Arizona (White Mountains) and New Mexico. Ours are assigned to **var. *virescens*** (Muhl. ex Willd.) Luer.

Corallorhiza Gagnebin Coralroot

(Greek *korallion*, coral, + *rhiza*, root) Erect herbs with reduced or absent chlorophyll arising from coralloid rhizomes, mycotrophic, roots absent, usually in association with decaying conifer wood and mycorrhizal fungi. **LEAVES** absent or reduced to small bracts. **INFL** a loose to congested raceme, flowers few to many. **FLOWERS** slightly showy to inconspicuous; perianth parts brownish or yellowish, often cupped, sometimes spreading; lip often, but not always, with brightly colored spots or stripes, margin often undulate, base occasionally auricled; column sometimes with 1 or more spots, compressed; pollinia waxy, 4, connected by a single elastic band to a detachable viscidium in a pollinarium. **FRUIT** ovoid to ellipsoid, pendent, often remaining conspicuously through the winter and following growing season, few to many per flower stalk. About 11 primarily temperate and boreal species of North and Central America, one species circumboreal.

1. Perianth parts with reddish brown, purple, or yellowish brown stripes; lip striped .... ***C. striata***
1′ Perianth parts not prominently striped, lip often spotted..(2)

2. Lip with obvious lateral lobes on each side; often robust stalk; common .... ***C. maculata***
2′ Lip without lateral lobes or lobes small and/or shallow; slender stalk; uncommon or rare .... (3)

3. Lip with shallow lobes, not spotted; flower somewhat spreading; plant pale yellowish green; rare .... ***C. trifida***
3′ Lip without lobes, spotted; flower not spreading, plant bluish to dark reddish brown; uncommon .. ***C. wisteriana***

Corallorhiza maculata (Raf.) Raf. (spotted) Spotted coralroot, large coralroot, many-flowered coralroot. Plant reddish or brownish purple, 10–60 cm tall, without leaves, but with many sheathing bracts. **INFL** a bracteate raceme, 5–35 flowers. **FLOWERS** often spreading at anthesis; sepals and petals reddish to brownish purple, linear to oblanceolate, with lateral petals slightly shorter than sepals; lip white or buff, often with dark red, bright pink, or purple spots, 2 prominent, basal, auriculate, lateral lobes, curved, with apex acute, over 1 mm in length, the margin of the main lip lobe often undulate. **FRUIT** sometimes numerous, 2.5 cm long, 1 cm wide. $2n = 42$. [*C. multiflora* Nutt.; *C. hortensis* Suksd.]. Moist and dry, dappled shady to full sun montane locations under conifers or aspens, often at old logging sites. Locally common to abundant. ARIZ: Apa; COLO: Arc, Hin, LPl, Min, Mon, SJn; NMEX: RAr, SJn; UTAH. 2285–3350 m (7500–11000′). Flowering: late May–Aug. Fruit: Jun–Nov. Alaska to Nova Scotia, south to California and east to Texas, North Carolina; Mexico and Guatemala. The species is reported to be epiparasitic in a relationship with mycorrhizal fungi and adjacent trees. Varied color forms occur in our area, especially forma *flavida*, which lacks the purplish pigments and appears as a striking lemon-yellow, with the lip a pure white. This form is often collected and confused with *C. trifida*, which is much smaller, more delicate, yellowish green, and rare in our area. The species is reported to be pollinated by dance flies (*Empis*) or is self-pollinated.

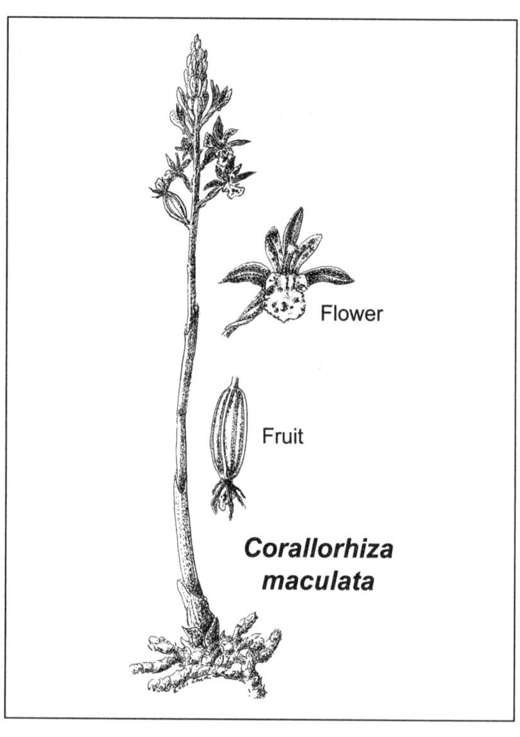

Corallorhiza maculata

Corallorhiza striata Lindl. (striped) Striped coralroot. Plant reddish, brownish, reddish purple, or yellowish brown, 8–40 cm tall, without leaves, but stem often with sheathing bracts. **INFL** a bracteate raceme of usually 2–25 flowers. **FLOWERS** buff to yellowish, often not conspicuous in ours, perianth parts with 3–5 pinkish or purplish to brownish stripes; sepals lanceolate; lateral petals shorter and ovate to obovate; lip elliptic, entire, white to yellowish, strongly to weakly (in ours) striate, stripes merging at the apex, base with 2 median calli. **FRUIT** less frequent per stalk than *C. maculata*, 2 cm long, 1 cm wide, ellipsoidal, pendent. $2n = 42$. [*C. macraei* A. Gray; *Neottia striata* (Lindl.) Kuntze]. Moist to dry, dappled shade to full sun in montane conifer and aspen woods with decayed wood, often at old logging sites. Uncommon to rare, but often overlooked. ARIZ: Apa; COLO: Hin, LPl, Mon; NMEX: RAr. 2405–2590 m (7900–8500′). Flowering: late May–Jul. Fruit: Jun–Sep. British Columbia to Quebec and Michigan, south throughout the western United States to Texas and northern Mexico. Ours, washed out and dull in color, are **var. *vreelandii*** (Rydb.) L. O. Williams, although var. *striata*, which is more colorful and striped, occurs close to the study area in Colorado and central New Mexico. Most likely pollinated by parasitic wasps, *Coccygonemus pedalis*, although ours more likely pollinated by flies. The species may also be self-pollinating.

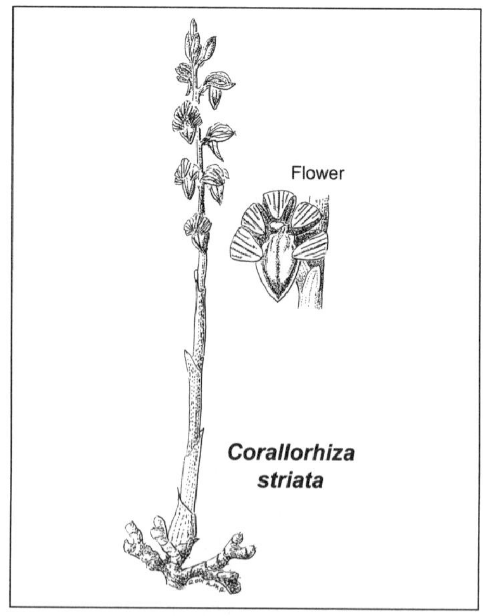

Corallorhiza striata

Corallorhiza trifida Châtel. (triparted, referring to the lip, which is not, in fact, trifid, but instead weakly trilobed) Yellow coralroot, early coralroot, northern coralroot. Plant yellowish, greenish, slender, to 25 cm tall or smaller in our area, without leaves, but often with sheathing bracts. **INFL** a terminal raceme, lax, with up to 20 flowers, usually fewer

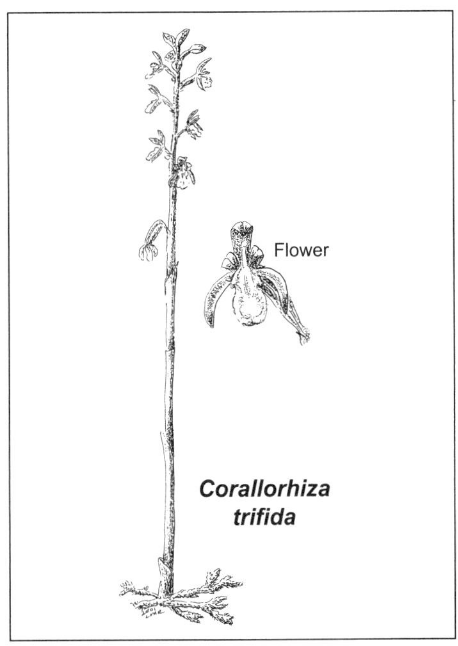

Corallorhiza trifida

in our area. **FLOWER** sepals and lateral petals lanceolate, oblanceolate to falcate, tan to brownish or greenish; lip white, often with a few purplish spots, with 2 small side lobes, margin undulate. **FRUIT** pendent, ellipsoid, about 1 cm long, 0.5 cm wide. Moist, shady locations under conifers. Rare, known only from Little Sand Creek, Colorado, but probably more frequent in high, moist locations. COLO: Hin. 2710 m (8900′). Flowering: Jun. Fruit: Jun–Jul. Circumboreal, south along the Rocky Mountains to New Mexico; Illinois and Missouri. The only *Corallorhiza* species found in Asia and Europe. Although mycotrophic, may not rely as much on fungi for nutrition due to pale green color of entire plant. Reported to be autogamous or pollinated by syrphid flies.

Corallorhiza wisteriana Conrad (named for Charles J. Wister, who first collected the species) Spring coralroot, early coralroot, Wister's coralroot. Plant similar to and easily confused with *C. maculata*, but smaller (less than 25 cm tall), with thinner stems; dark, dull, inky purple rather than reddish purple flowers; earlier blooming. **INFL** a terminal raceme, loosely flowered with up to 25 flowers per scape. **FLOWERS** not spreading; sepals and lateral petals forming a hood over the lip; lip margin undulate to crenulate, without lateral lobes, white, often with 1 to many pale pink to purple spots. **FRUIT** slightly smaller and more globose than those of *C. maculata*, often low frequency per plant, 1 cm long, 0.6 cm wide. $2n = 42$. [*C. ochroleuca* Rydb.]. Moist to dry, dappled shady montane locations in areas with decaying conifer wood, often in association with *C. maculata* and *C. striata*. Uncommon to locally common. ARIZ: Apa; COLO: Arc, Hin, Min, SJn. 2405–2865 m (7900–9400′). Flowering: early May–Jun. Fruit: Jun–Sep. Montana south to Arizona, New Mexico, and western Mexico, South Dakota and southeastern United States. The author has observed flowers visited by small, black bees, although it has been suggested that some populations are self-pollinating.

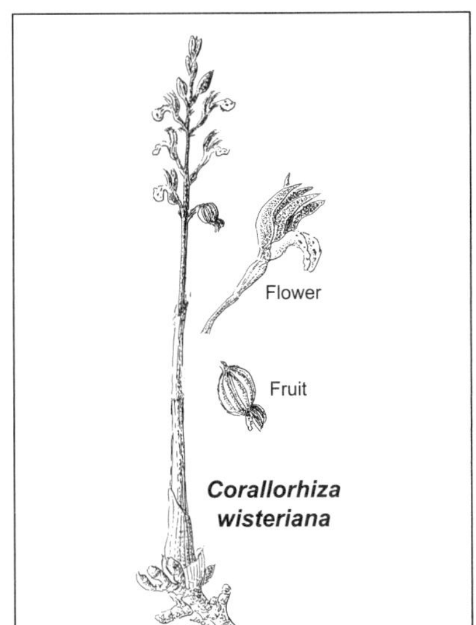

Corallorhiza wisteriana

Cypripedium L. Lady's Slipper, Moccasin Flower

[Greek *Kypris*, Cypriot one (Aphrodite), + *podion* or *pedilon*, little foot or slipper] Leafy erect herbs with rhizomes and many fibrous roots. **LEAVES** sheathing, large, ovate, plicate and/or sessile, often glandular-pubescent; apex acute. **INFL** a raceme, flowers 1 to several, with bracts. **FLOWERS** mostly spreading, often showy and brightly colored; sepals 3, the lower 2 often partially to completely united (synsepalous); lateral petals spreading, occasionally twisted, lip inflated, pouchlike, often puckered with a small dorsal opening; column bearing a large, central, flaplike staminode and 2 fertile anthers on each side; pollen mealy, not in pollinia or pollinaria. **FRUIT** a ribbed, 3-locular, ellipsoid capsule. **SEEDS** numerous, minute, with papery seed coat. About 25 species in North America and Eurasia, with one species in Mexico and Guatemala. Some species are grown or collected for horticultural value. Many possess detaching glandular hairs that may cause a skin rash. Most species are trap-pollinated by bees, as the flowers offer no reward but may serve as sleeping locations for late foraging bumblebees. One species pollinated by fungus gnats (Cribb 1997).

Cypripedium parviflorum Salisb. (small-flowered) Yellow lady's slipper, yellow moccasin flower, whippoorwill shoe. Plant robust, densely pubescent, to 80 cm high. **LEAVES** 3–5, ovate to lanceolate, plicate, bright green. **INFL** with 1 or 2 conspicuous large flowers. **FLOWERS** showy; sepals and lateral petals brown, brownish green, or brownish purple; lip large, inflated, bright yellow. **FRUIT** 2–3 cm long, 1–2 cm wide, often few and rare. $2n = 20$. [*C. calceolus* L. var. *pubescens* (Willd.) Correll; *C. flavescens* DC.; *C. pubescens* Willd.]. Moist, shady locations, often with seeps, sometimes on steep, mossy slopes. Rare and local. COLO: LPl; NMEX: SJn. 2375–2895 m (7800–9500′). Flowering: early Jun–Aug. Fruit: Aug–Oct. Labrador to Alaska, south to all states except Florida, Nevada, and California. Our plants are **var. *pubescens*** (Willd.) O. W. Knight. Originally thought to be a variety of *C. calceolus* L., which is now considered by Cribb (1997) to be a European species that exhibits characters convergent (probably as an adaptation to similar pollinators) with *C. parviflorum.* Pollinator an andrenid bee.

Epipactis Zinn Helleborine

(Greek *epipaktis* or hellebore) Short to tall perennials to 1 m; often forming clumps with several leafy, bracteate, flowering stems from short underground rhizomes with fibrous roots. **LEAVES** alternate, sheathing, often somewhat plicate. **INFL** a loose, bracteate raceme with few flowers. **FLOWERS** somewhat spreading, white, greenish, purplish, or purplish-pink, with open nectaries near the base of the lip; sepals and petals free; lip somewhat saccate near the base, often auricled and somewhat differently colored than the apex of the lip (epichile); column with 1 broad stigma, the central lobe forming a rostellum; stigma lobes often 2 and prominent; pollinia 2 pairs, soft, without caudicles. **FRUIT** a pendent capsule. **SEEDS** numerous, minute. About 25 species in temperate North America, Mexico, North Africa, and Eurasia. One species native to North America, but *E. helleborine* (L.) Crantz has been introduced from Europe and is becoming widely established at moist sites near urban centers, including Albuquerque, New Mexico, where it grows in riparian cottonwood forests. It should be searched for in riparian locations in our area. Most species are pollinated by wasps, bees, or syrphid flies, in some cases female flies attempting to lay eggs on aphidlike warts on the lip.

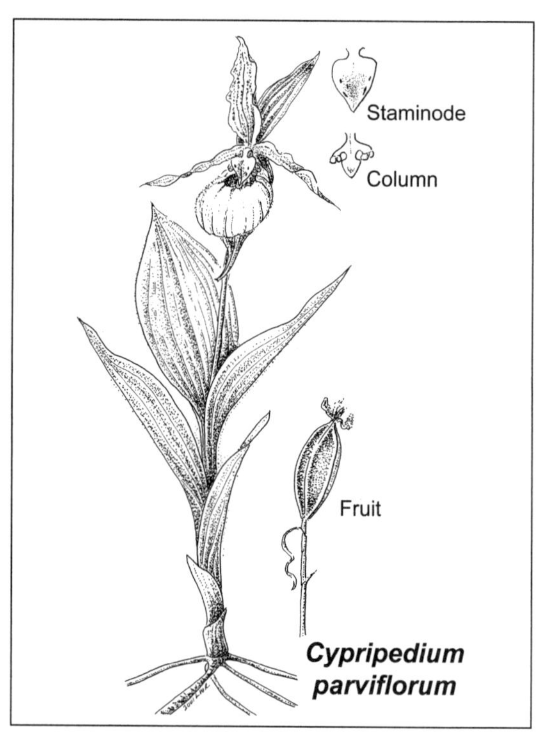

Cypripedium parviflorum

Epipactis gigantea Douglas ex Hook. (gigantic, referring to flowers and possibly, leaves) Chatterbox, giant helleborine, stream orchid. Plant terrestrial, up to 1 m tall, often much shorter in our area, in clumps, especially where mowed or grazed regularly. **STEMS** rhizomatous, with fibrous roots. **LEAVES** alternate, 4–12, ovate-elliptic to lanceolate, plicate, medium green, somewhat clasping, becoming bracteate ascending the flowering stem, 3–16 cm long, 1–7 cm wide. **INFL** a loose raceme with up to 15 flowers with floral bracts. **FLOWERS** 1.5–3 cm wide, often multicolored, pale to intense green, purple, pink, and yellow, spreading; sepals greenish purple, lip and lateral petals often with purple reticulate stripes and shallow, red, warty protuberances and/or orange calli, anthers green; pollinia in 2 pairs, soft. **FRUIT** ellipsoid, 2–3 cm long, 1–2 cm wide, may be sparsely pubescent, often numerous per stem. $2n = 32, 40$. [*Helleborine gigantea* (Douglas) Druce; *Epipactis americana* Lindl.]. Locally abundant near permanent streams, springs, and hanging gardens or at the base of north-facing cliffs in moist humus; our only "desert" orchid. ARIZ: Apa, Nav; COLO: Arc, Mon; NMEX: SJn; UTAH. 1095–2440 m (3600–8000′). Flowering: Apr–Aug. Fruit: May–Sep. British Columbia south to Baja California and central Mexico, east to Texas and South Dakota; India to China and Japan (often identified as *E. royleana* Lindl., a synonym). Pollinator is a syrphid fly.

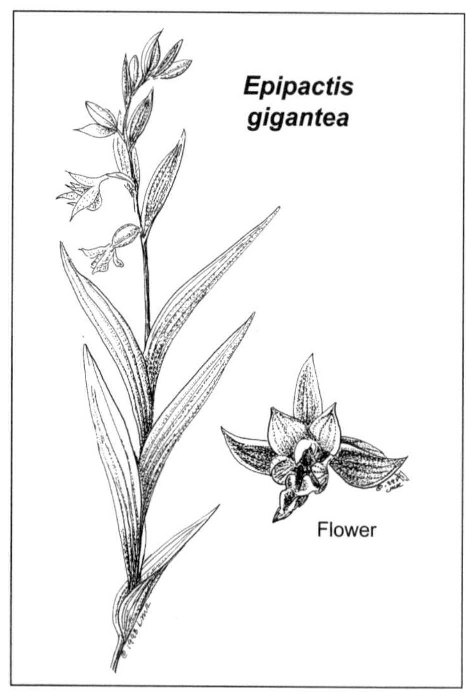

Epipactis gigantea

Goodyera R. Br. Rattlesnake Plantain

(named for Hampshire botanist John Goodyear, 1592–1664) Small to medium herbs, often forming matlike colonies; rhizomes horizontal, creeping, thick; roots slightly succulent. **LEAVES** forming a basal rosette, often with silvery, pale green, or whitish netlike markings. **INFL** a bracteate scapose spike with flowers more common on one side or loosely spiraled. **FLOWERS** small, white, fragrant, often with green veins, glandular-pubescent; upper sepal connivent with lateral petals, forming a tubelike structure over the column; lip deeply saccate; column 2-horned; pollinia 2, granular, in clumps maintained by elastic threads, but not in well-defined pollinaria. **FRUIT** an erect capsule, often numerous per flower stalk. **SEEDS** numerous, minute, papery. Hybrids often form, particularly in the Great Lakes area. A genus of worldwide distribution of about 25 species, with four temperate and boreal North American species. Some species ("jewel orchids") grown for the attractively patterned and sometimes velvety leaves rather than the flowers. The leaves were used in the past for medicinal purposes.

1. Leaves pale to dark green, 3–11 cm long, 2–4 cm wide, often prominently reticulated, with a broad central stripe; flowering plant usually more than 20 cm tall; common .... ***G. oblongifolia***
1' Leaves often pale green, 1–4 cm long, 0.5–3 cm wide, reticulation not present or prominent in ours, no prominent central stripe, flowering plant usually less than 15 cm tall; rare .... ***G. repens***

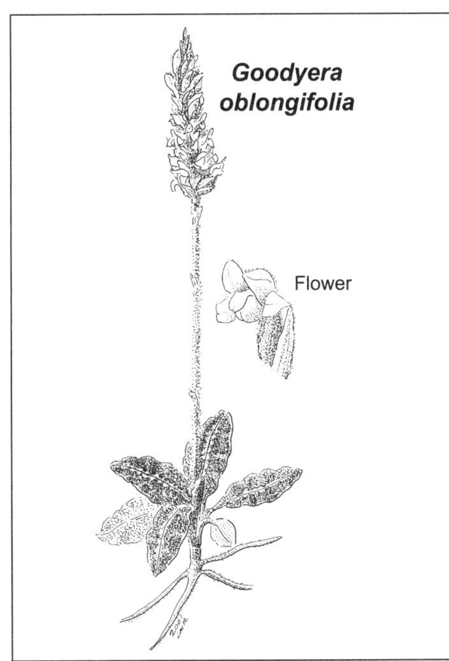

Goodyera oblongifolia Raf. (oblong leaf) Menzies' rattlesnake plantain, rattlesnake plantain. Terrestrial, densely pubescent, mat-forming, sometimes rhizomatous herb, 20–45 cm tall. **LEAVES** 2–many, 2–10 cm long, 1–4 cm wide in a rosette, elliptic, bluish green with prominent broad white or silvery central stripe, often with white, silver, or pale green reticulation. **INFL** a 1-sided, undulate, semisecund, densely flowered spicate raceme with up to 40 flowers. **FLOWERS** white, often with green or pinkish veining, often sweetly fragrant; sepals, ovary, and flowering stalk densely pubescent; lateral petals and dorsal sepal connivent over the column, forming a tube with the lip; lateral sepals spreading; lip bulbous-saccate, white; column short, rostellum pointed; pollinia granular, soft, yellow, viscidium elongate. **FRUIT** ellipsoid, pubescent, about 1 cm × 0.5 cm wide, often many per stem. $2n = 22, 30$. [*G. oblongifolia* var. *reticulata* B. Boivin; *G. decipiens* (Piper) Hubbard]. Common in dry or moist, dark coniferous forests. ARIZ: Apa; COLO: Arc, Hin, LPl, Min, Mon, SJn; NMEX: SJn; UTAH. 2315–3260 m (7600–10700′). Flowering: Jul–Sep. Fruit: Aug–Oct. Southern Canada and Alaska, south to Minnesota, Michigan, and Maine, along the Appalachians to North Carolina and Tennessee; Montana, Idaho, South Dakota, New Mexico, Arizona; Sonora and Tamaulipas, Mexico. Pollinated by bumblebees.

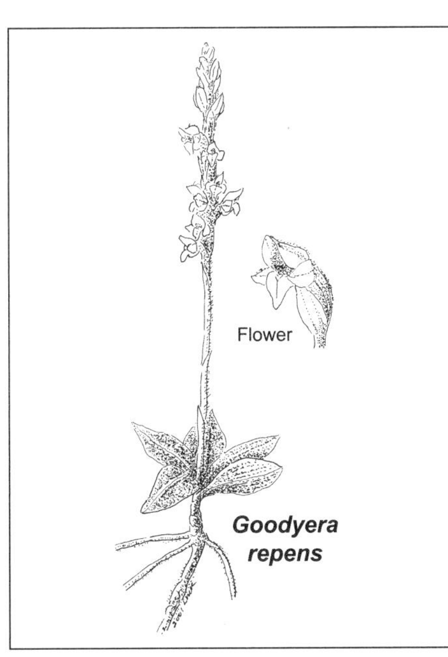

Goodyera repens (L.) R. Br. (creeping) Lesser rattlesnake plantain. Small, pubescent, spicate herb up to 25 cm tall. **LEAVES** 2 to many, 1–4 cm long, 0.5–3 cm wide, ovate, in basal rosette, dark to light green, sometimes marked with light green, white or silvery reticulation without broad, light central stripe. **INFL** a loosely flowered, undulate, semisecund, spicate raceme with up to 20 flowers. **FLOWERS** white, often tinged with green or pink, fragrance "soapy"; sepals pubescent, sepals and lateral petals connivent over the short column, tips of lateral sepals not obviously reflexed; lip globose-saccate, white, often with pink; pollinia blunt, yellow, soft, viscidium orbiculate. **FRUIT** ellipsoid, 0.7 cm long, 0.5 cm wide, few to many. $2n = 22, 30$. [*G. ophiodes* (Fernald) Rydb.]. Uncommon and local in mountain ranges of Colorado in moist, dark, coniferous forests among mosses and elfin flowering plants. COLO: Arc, Hin, LPl, Min. 2405–3080 m (7900–10100′). Flowering: Jul–Aug. Fruit: Jul–Sep. Canada and Alaska, northeastern United States, to western North Carolina and eastern Tennesee, Michigan, Wisconsin, Minnesota, Montana, Idaho, Wyoming, South Dakota, Colorado, New Mexico, and Arizona. Pollinator reported to be *Bombus pratorum* or autopollinated, although the scent and flower size suggest a dipteran.

Listera R. Br. Twayblade

(for Martin Lister, 17th-century English naturalist) Small herbs with slender rhizomes, fibrous roots. The adult plant has the appearance of a dicot seedling. **LEAVES** 2 (rarely 3), broad, opposite, borne midstem. **INFL** a terminal raceme. **FLOWERS** small, inconspicuous, widely spreading, often greenish or purplish and/or translucent; most sepals and lateral petals short, often reflexed; the lip mostly cleft, 1-nerved, at least twice as long as the other petals; column possessing a nectary under the stigma; pollinia in pollinaria 2, soft, tear-shaped, without caudicles, viscidium circular.

FRUIT a small, globose to elliptic capsule. **SEEDS** larger than those of most orchids, but thin and papery. A genus of cool regions in the Northern and Southern Hemispheres with about 25 species. Pollinators generally minute flies, fungus gnats, or wasps.

Listera cordata (L.) R. Br. (heart-shaped) Heart-leaved twayblade. Plant erect, slender, to 15 cm tall (taller in more northerly and wetter areas), usually less in our area. Plant appears insignificant and similar to a just-germinated dicot seedling. Underground base of stem swollen, with several thin, stringy roots. **STEM** base surrounded by sheaths, glabrous below, weakly pubescent above. **LEAVES** 2, opposite, cordate, 2 cm wide × 2 cm long or slightly smaller; light green in ours, reticulate-veined, margins undulate or entire. **INFL** a slender raceme, mostly loosely flowered with up to 25 small, but not minute, flowers. **FLOWER** pedicel 3 mm long; green to light green, often translucent with sepals ovate to elliptic, sepals 2 × 1 mm, lateral sepals oblique; petals similar to sepals; lip somewhat linear, about 5–6 mm long, with 2 slender lobes forking from about midway to the apex, a pair of hornlike protuberances at the base; column short, wide. **FRUIT** somewhat erect, globose to ellipsoid, 0.5 cm long × 0.4 cm wide. $2n = 36, 38$. [*Listera nephrophylla* Rydb.; *Ophrys nephrophylla* Rydb.; *L. cordata* (L.) R. Br. var. *chlorantha* P. Beauv.]. Growing sometimes in great masses in shady moist places among a thick mat of mosses under fir and spruce or aspen, often along the edges of streams or springs. Thought to be relatively uncommon in our area, but may be overlooked and locally abundant. COLO: Arc, Hin, LPl, Min, Mon, SJn. 2710–3290 m (8900–10800′). Flowering: Jun–Jul. Fruit: Jun–Aug. Circumboreal in distribution in the Northern Hemisphere, south to the west coast of North America to northern California, along the Rocky Mountains to New Mexico and south to Japan in the Eastern Hemisphere. Our plants belong to **var. *nephrophylla*** (Rydb.) Hultén, with uniformly light green flowers and the leaves significantly larger compared to var. *cordata*. Pollination studies have shown fungus gnats from two families to be the major pollinators through an elaborate mechanism to ensure cross-pollination.

Platanthera Rich. Fringed Orchid, Bog Orchid

(Greek *platys*, wide, broad, + *anthera*, anther) Plants erect, glabrous; roots stoloniferous tuberoids or fusiform. **LEAVES** bright to dark green, 1 or 2 to many, often basal and cauline, becoming reduced toward the inflorescence. **INFL** a loose to extremely congested spike, bracteate spike, or spicate raceme with 5 or 6 to hundreds of flowers. **FLOWERS** large and showy to small and inconspicuous (generally, in ours), often green or white, but yellow, pink, purple, or gold in many fringed species, often fragrant; sepals and lateral petals spreading, reflexed, or hooded; lip pendent, upcurved, entire, crenulate, lobed, or fringed, spur, if present, often short and saccate to several cm in length; column short and broad, anther sacs widely separated, often opening in front; pollinia consisting of mealy grains to tight masses in 2 viscous hemipollinaria, often with elastic caudicles. **FRUIT** semierect to ascending, ellipsoid. A diverse, polymorphic, temperate, terrestrial genus of about 200 species formerly included in *Habenaria*, a genus which now includes only subtropical and tropical species. Most species, especially the fringed orchids, are probably pollinated by butterflies and moths. However, as our species are mostly green, inconspicuous, and oddly fragrant, pollinators are more likely flies and bees. All of our species are included in the *P. dilatata - P. hyperborea* complex of section *Limnorchis*, which includes many closely related, variable, circumboreal species, subspecies, and varieties, endlessly split, lumped, and subsequently resplit. Each species exhibits a few to many geographical races or named varieties. Most of these taxa are capable of hybridization. *Platanthera stricta* Lindl. does not occur in our area, its range extending farther north and west. *Platanthera saccata* (Greene) Hultén is a synonym for this taxon. *Platanthera hyperborea* (L.) Lindl. is now restricted to Iceland and Greenland and is, therefore, not in our area. No definite voucher specimens of the autopollinating *P. aquilonis* Sheviak have been seen from the study area, the main character being the horizontally positioned pollinia, but this species surely occurs here. It is imperative that collectors of this genus note floral scent and flower color as these are not evident in herbarium specimens and are absolutely essential in identification. See notes under Orchidaceae concerning *Piperia unalascensis* (Spreng.) Rydb.

1. Flowers white; lip obviously dilated at the base .... ***P. dilatata***
1′ Flowers green, yellow-green, whitish green, or purplish, lip not obviously dilated .... (2)

2. Leaves 1–2; scape naked or with 0 or 1 bract (rarely 2) .... ***P. obtusata***
2′ Leaves 1–several; scape with many bracts, gradually reduced distally .... (3)

3. Inflorescence sparsely flowered, the column broad, filling more than 1/2 the length of the dorsal sepal; small to medium-sized plants generally inhabiting streamside or hanging garden locales .... (4)
3′ Inflorescence a loose spike to extremely congested, column varied, but not broad, generally moist seepage or damp locales .... (5)

4. Flowers yellowish green to green, often sparsely flowered, mostly streamside .... *P. sparsiflora*
4′ Flowers dark or medium green, slightly congested, few- to many-flowered, populations restricted primarily to hanging gardens and cool cliff bases .... *P. zothecina*

5. Spur saccate or scrotiform; lip dull yellowish to bluish, sometimes reddish; musty-scented .... *P. purpurascens*
5′ Spur slenderly cylindric; lip whitish or whitish green; sweet or cucumber-scented .... *P. huronensis*

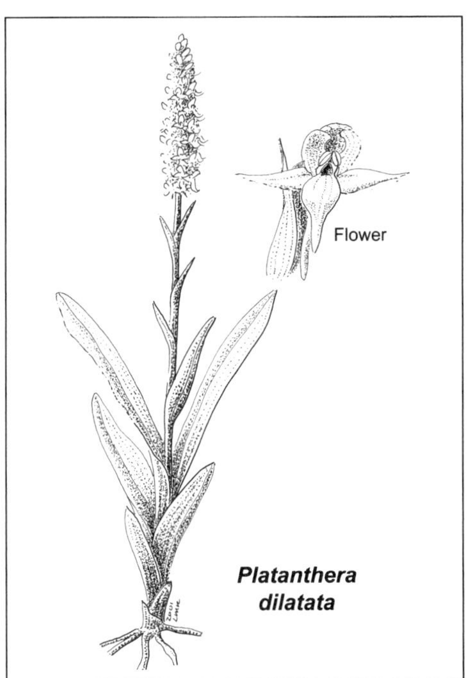

Platanthera dilatata

Platanthera dilatata (Pursh) Lindl. ex L. C. Beck (broadened or expanded, referring to the base of the lip) Tall white bog orchid, tall white northern orchid, fragrant orchid, bog candle. Plants erect, glabrous, slender or stout, to about 1 m tall, roots fleshy, fusiform, elongated. **LEAVES** few–12, linear to lanceolate or oblong, becoming bractlike above, to about 30 cm long × 7 cm wide. **INFL** a many-flowered spike, congested to lax. **FLOWERS** white; dorsal sepal ovate-elliptic, forming a hood with the lateral petals; lateral sepals elliptic-lanceolate, oblique, usually spreading; petals ovate-lanceolate, oblique, falcate; lip linear-lanceolate, dilated at the base; spur cylindric, equal to lip length. **FRUIT** semierect, ellipsoid, about 1–1.5 cm long × 0.6 cm wide. 2*n* = 42. [*Orchis dilatata* Pursh; *Habenaria dilatata* (Pursh) Hook.; *Limnorchis dilatata* (Pursh) Rydb.]. Montane wet meadows, ditches, roadsides, stream banks, damp slopes, and wet roadsides. COLO: LPl; NMEX: RAr. 1830–3230 m (6000–10600′). Flowering: Jun–Aug. Fruit: Jul–Oct. Alaska east to Labrador, south to New York, Illinois, and Colorado. Ours are **var. *dilatata*.** This species has been known to hybridize freely with other green platantheras growing adjacently, often resulting in some plants with greenish flowers. *Platanthera dilatata* often has a strong clove, cinnamon, or cinnamon-vanilla scent, no doubt to attract the many species of small moths and skippers as pollinators.

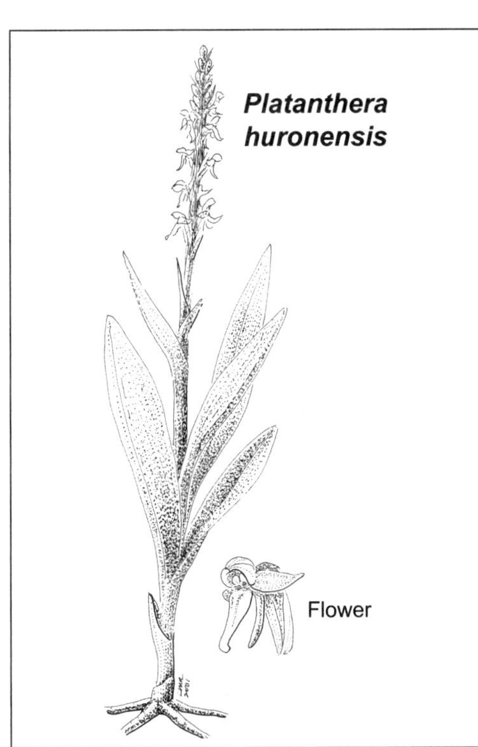

Platanthera huronensis

Platanthera huronensis (Nutt.) Lindl. (of Lake Huron, referring to its northern location) Tall northern green orchid, tall green bog orchid, tall leafy green orchid. Plants erect, to about 40 cm tall. **LEAVES** few to many, gradually reducing to bracts; blade oblong to linear-lanceolate, about 5–25 cm long × 0.8–4 cm wide. **INFL** a spikelike raceme; lax to dense. **FLOWERS** whitish or light green; calyx sometimes darker than corolla; lateral sepals spreading to reflexed, dorsal sepal forming a hood with lateral petals; petals ovate to falcate; lip descending, lanceolate, with no basal thickening; spur slender, cylindric to clavate, often with tapering apex; 5–14 mm; anther cells upright and parallel. **FRUIT** ellipsoidal, erect, 1 cm long × 0.5 cm wide. 2*n* = 84. [*Orchis huronensis* Nutt.; *Habenaria hyperborea* (L.) R. Br. var. *huronensis* (Nutt.) Farw.; *Platanthera hyperborea* (L.) Lindl. var. *huronensis* (Nutt.) Luer]. Montane wet meadows, stream banks, marshes, acequias, seeping slopes, roadsides. COLO: Arc, Hin, LPl, Mon, SJn; NMEX: RAr, SJn. 2285–3445 m (7500–11300′). Flowering: Jun–Sep. Fruit: Jul–Oct. Alaska east to Labrador, south to New Jersey, South Dakota, New Mexico, and Oregon. May hybridize with others in the genus. Intensely fragrant with a sweet, pungent, cucumber-like scent, indicating a different pollinator than that of the sometimes sympatric *P. purpurascens*.

Platanthera obtusata (Banks ex Pursh) Lindl. (blunted, referring to the blunt leaves) Small northern bog orchid, blunt leaf orchid. Plants erect, 5–about 20 cm. **LEAVES** 1 or 2, rarely with 1 small bract; blade obovate to oblanceolate, or elliptic-oblong, about 12 cm long × 4 cm wide, clasping the stem. **INFL** a lax spike with about 15 flowers. **FLOWERS** greenish white to yellowish, corolla often lighter than calyx; lateral sepals reflexed, dorsal sepal forming a hood with lateral petals;

petals falcate; lip descending, linear with median basal thickening; spur slender, conic, about 3–10 mm. **FRUIT** erect, elliptic, to 8.5 cm long × 4 cm wide. $2n = 42, 63$. [*Orchis obtusata* Banks ex Pursh; *Habenaria obtusata* (Banks ex Pursh) Richardson; *Habenaria obtusata* var. *collectanea* Fernald]. Mesic to wet coniferous forest, sphagnum bogs, stream banks. Known in our area only from relatively high, spruce-fir forests. COLO: Hin. 2895–3170 m (9500–10400′). Flowering: Jul–Aug. Fruit: Aug–Sep. Circumpolar in distribution, with an arm traveling down the Rocky Mountains: Alaska east to Labrador, south to Massachusetts, Wisconsin, and Colorado; Eurasia. Mosquitoes, particularly males, are the pollinators of this species and the pollinarium is deposited on one of the large eyes while he feeds from the conical entrance to the nectary.

Platanthera purpurascens (Rydb.) Sheviak & W. F. Jenn. (becoming purple, referring to the occasional purplish coloring of the flowers and vegetative parts) Southwestern green bog orchid. Plants 10–50 cm tall. **LEAVES** few to many, larger basally, scattered and becoming smaller along stem; blade oblong to lanceolate, variable size, about 5–20 cm long × 1–5 cm wide. **INFL** a lax to dense spikelike raceme, flowers sometimes in fascicles. **FLOWERS** green to yellowish green, often with a bluish or reddish tint, especially the buds; lateral sepals spreading, dorsal sepal forming a hood with lateral petals; petals falcate; lip descending to reflexed or projecting, lanceolate to ovate, without basal thickening; spur clavate to scrotiform, 2–3 mm long. **FRUIT** ellipsoidal, about 1 cm long × 0.5 cm wide. $2n = 42$. [*Limnorchis purpurascens* Rydb.; *Platanthera hyperborea* (L.) Lindl. var *purpurascens* (Rydb.) Luer; *Habenaria purpurascens* (Rydb.) Tidestr.]. Common in moist montane habitats, such as seeping, mossy slopes, stream banks, acequias, roadsides, conifer woods, and near old buildings. ARIZ: Apa; COLO: Arc, Hin, LPl, Min, Mon, SJn; NMEX: RAr, SJn. 2345–3655 m (7700–12000′). Flowering: Jun–Sep. Fruit: Jul–Oct. California, southern Wyoming south to New Mexico and eastern Arizona. May grow and hybridize with others of the genus. The flowers are strongly seminiferously scented, sometimes with vanilla overtones, probably indicating a dipteran pollinator, possibly a mosquito.

Platanthera sparsiflora (S. Watson) Schltr. (sparse-flowered) Sparsely flowered bog orchid, canyon habenaria. Plants glabrous, leafy, often slender, sometimes stout and robust, 10–75 cm tall; roots fleshy, elongated, often fusiform. **LEAVES** light to bright green, 10 or fewer, oblong to lanceolate, more frequent basally, clasping the stem and becoming bracteate above, to 30 cm long × 3.5 cm wide, but often much smaller. **INFL** a laxly few- to many-flowered, often elongated spikelike raceme. **FLOWERS** light or yellowish to deep green, often held at a 90° angle to the rachis; dorsal sepal ovate, generally longer and larger than in our other species of *Platanthera*; lateral sepals elliptic-lanceolate, downwardly recurved; lateral petals lanceolate, oblique, forming an acutely pointed hood with the large dorsal sepal; lip linear, the base often with a slightly thickened median ridge, tapering slightly toward the apex, spur cylindrical, as long as or slightly longer than the lip; column large, often filling the entire hood, 2.5–4 mm long × 2–3 mm wide. **FRUIT** semierect, ellipsoid, about 1 cm long × 0.5 cm wide, occasionally heavily produced. [*Habenaria sparsiflora* S. Watson; *Limnorchis ensifolia* Rydb.; *Limnorchis sparsiflora* (S. Watson) Rydb.]. Moist locations, often close to streams and permanent water sources in open or dense woods or enclosed seepages. ARIZ: Apa, Nav; COLO: Arc, Hin, LPl, SJn; NMEX: SJn. 1370–3080 m (4500–10100′). Flowering: Jun–Sep. Fruit: Jul–Oct. Southern Washington south through California, northern Utah south to Arizona, east to Lincoln County, New Mexico, south to Baja California, western and south-central Mexico. May be moth-pollinated. A variable species somewhat resembling other green species of *Platanthera* and occupying a similar range in the West, but easily distinguished from them by the very large column, possibly indicating a different pollinator. Similar to *P. zothecina*, which may turn out to be a variety of this taxon, but flowers more delicate, held at a greater angle from the rachis, and generally in moister streamside locations.

Platanthera zothecina (L. C. Higgins & S. L. Welsh) Kartesz & Gandhi. Alcove bog orchid. Plants similar to *P. sparsiflora*. **LEAVES** also similar in size, but initially broadly rounded, later acute; sometimes with a thick, succulent-like texture. **INFL** similar to *P. sparsiflora*, but generally more robust, with occasionally a greater number of flowers, often more than 10. **FLOWERS** similar, generally larger, light to deep green, with a distinctive cucumber-musty scent, held at a 45° angle or less to rachis; lip 1.5 times longer than that of *P. sparsiflora*, the basal median ridge lacking, spur greater than 1.5 times lip length. **FRUIT** as in *P. sparsiflora*. [*Habenaria zothecina* L. C. Higgins & S. L. Welsh; *Limnorchis zothecina* (L. C. Higgins & S. L. Welsh) W. A. Weber]. The most distinctive feature about this orchid is its distribution primarily in hanging gardens, alcoves, and under protective ledges of the Colorado and San Juan river drainages. It is locally abundant, but inconspicuous in discontinuous, cool, very shady, mesic pockets of vegetation protected by high sandstone cliffs often with seasonal seeps, springs, or drip pools, and occasionally in quite dry, but cool locations screened by dense vegetation surrounded by hot, dry, arid or semiarid habitat, such as near Lake Powell. ARIZ: Apa, Nav; UTAH. 1100–2195 m (3600–7200′). Flowering: Jun–Aug. Fruit: Jun–Oct. Known only from the Colorado River drainage of

Arizona, Colorado, and Utah. Our pollinator exclusion studies of this orchid in Natural Bridges National Monument (Reeves and Reeves 1993) have suggested that a pollinator is necessary for seed set, but many hours of observation have failed to discover the culprit, probably a midge or mosquito. Studies of *Malaxis paludosa* (L.) Sm. (Reeves and Reeves 1984) have revealed a similar scent and fungus gnats as pollinators, but these flowers are far too large for insects of that size to carry the large, multigrained pollinaria.

Spiranthes Rich. Ladies' Tresses, Rein Orchid

(Greek *speira*, coil, + *anthos*, flower) Short and delicate to tall and robust herbs; roots fleshy, tuberous. **LEAVES** linear to broadly ovate, basal or cauline, becoming sheathing bracts below and on the inflorescence. **INFL** a congested, spirally twisted spike, the longitudinal rows in several ranks. **FLOWERS** mostly white, cream, greenish white, or yellow, occasionally other colors and/or or faintly striped, often highly fragrant; sepals and lateral petals forming a tube with apices often spreading to recurved; lip concave and grooved near the base, simple or lobed, spreading to strongly recurved; column short; pollinia 2 with a common viscidium in a pollinarium. **FRUIT** ellipsoid to ovoid. A genus with about 300 species, mainly in the New World and Europe, with a single species in the western Pacific. Hybridization is common in this genus. Pollination in many species is accomplished by small bees that visit the lower flowers, first obtaining nectar from fully mature flowers. They then work upward, encountering fresh flowers without nectar and unavailable stigmas, but are quickly burdened with pollinaria. The bees continue to search for nectar on a new spike, again beginning at the lowest flower, where they deposit the pollinia on the elevated stigma and collect abundant nectar.

Spiranthes romanzoffiana Cham. (named in honor of Nicholas Romanzoff, 1754–1826, a Russian minister of state) Hooded ladies' tresses. Plant erect, robust to somewhat slender, glabrous below, pubescent above, to 50 cm tall, roots enlarged, fascicled and fleshy. **LEAVES** 3–7, 5–25 cm long, 5–15 mm wide, linear to obovate, present during and after flowering, bright green, mostly basal, forming bracts upward. **INFL** a very congested 3-ranked spike with up to 70 flowers. **FLOWERS** creamy white; sepals and lateral petals lanceolate, with apices not to slightly reflexed, forming a hood over the column and lip; lip base with small tuberosities, constricted above the middle, apex finely dentate, dilated and recurved; column small; pollinia 2, soft, in one pollinarium. **FRUIT** erect, ellipsoid, 1 cm long × 0.4 cm wide, often many. $2n = 60$. [*Triorchis stricta* (Rydb.) Nieuwl.; *Spiranthes stricta* (Rydb.) A. Nelson; *Neottia gemmipara* Sm.]. Locally confined to higher-elevation open, boggy, grassy meadows near mountain streams. ARIZ: Apa; COLO: Hin, LPl, Min, RGr, SJn. 2440–3475 m (8000–11400′). Flowering: Jul–Oct. Fruit: Aug–Oct. Alaska and southern Canada east to the Hebrides and Ireland, south to California, Nevada, Utah, Colorado, New Mexico, Arizona, Iowa, Illinois, Ohio east to New England. Halictine bees (*Halictus* spp., *Chlorhalictus* spp.) and bumblebees (*Bombus* spp.) reported to be pollinators.

OROBANCHACEAE Vent. BROOMRAPE FAMILY

Kenneth D. Heil and J. Mark Porter

Herbaceous annuals or perennials, mostly caulescent or subacaulesent; parasitic on the roots of other plants. **STEMS** simple or branched at the base, erect to decumbent. **LEAVES** alternate or mostly basal, entire, toothed to bipinnatifid or reduced to scalelike. **INFL** a spike, raceme, corymbose panicle, or of solitary flowers. **FLOWERS** tubular or bilabiate; calyx 4-cleft, 2–5-lobed, lobes equal or unequal; corolla yellow, reddish, purple, or white; stamens 4, didynamous; ovary superior, 1–2-loculed, style 1, stigma capitate, peltate, or lobed. **FRUIT** a capsule. **SEEDS** few to many. Approximately 65 genera and about 1990 species in the Northern Hemisphere, especially temperate and subtropical regions. *Orobanche* is used as a drug by the Navajo for wounds, open sores, and food. *Castilleja* is used as an ornamental, drug, and food. *Pedicularis* is used as an ornamental and by the Navajo in certain ceremonies.

1. Upper corolla lip 2-lobed ....(2)
1′ Upper corolla lip entire, occasionally upper lip toothed ....(3)

2. Plants with chlorophyll (green); stems neither fleshy nor succulent; corolla dehiscing when senescent..... ***Agalinis***
2′ Plants lacking chlorophyll (not green); stems fleshy or succulent; corolla persistent when withered.... ***Orobanche***

3. Anther cells equal; galea rounded adaxially, often extending into a narrow beak or truncate, sometimes with 2 teeth just below the apex ....(4)
3′ Anther cells unequal, the larger cell attached by its middle and the smaller by its apex or absent; galea straight, sometimes rounded at the very tip in *Orthocarpus* and *Cordylanthus*, neither toothed nor extending into a beak; leaves entire to pinnately 3- to 7-lobed ....(5)

4. Leaves opposite, toothed; calyx 4-toothed, becoming inflated with prominent veins, enclosing the fruit ..***Rhinanthus***
4′ Leaves alternate or basal, crenate, serrate, pinnatifid, or pinnately compound; calyx incised on one or both sides, becoming distended but not bladdery-inflated .... ***Pedicularis***

5. Plants perennial, or if annual (*C. exilis*), the galea much exceeding the lower lip and the bracts entire ....***Castilleja***
5′ Plants annual, the galea subequal to or only slightly longer than the lower lip; bracts 3- or more-parted with linear lobes .... (6)

6. Calyx tubular-campanulate, cleft less than halfway on the abaxial side and to about halfway on the adaxial, forming 2 lateral primary lobes, each divided into 2 segments; bracts leaflike, only 1 subtending each flower ..... ***Orthocarpus***
6′ Calyx bractlike, cleft to or nearly to the base on abaxial side, forming a single ligulate, entire or bifid structure in back; bracts of 2 types, the outer leaflike and the inner resembling and opposing the ligular calyx ....***Cordylanthus***

Agalinis Raf. Gerardia

(wonder-flax, referring to a superficial similarity to *Linum*) Annual (ours) or perennial herbs, often hemiparasitic. **STEMS** erect, usually 4-angled due to decurrent ridges from the leaf bases, branching, glabrous or with ascending-scabrous pubescence. **LEAVES** opposite or subopposite, sessile, entire, linear to filiform (in ours), sometimes lanceolate, scabridulous or scabrous on the upper surface and glabrous below, often revolute, the leaves of the fascicles and branchlets smaller and narrower. **INFL** a raceme; bracts leaflike below and progressively reduced upward, sometimes becoming alternate on the branches; pedicels often expanded at the receptacular attachment. **FLOWERS** weakly bilabiate; calyx nearly regular, somewhat accrescent, campanulate, sometimes shortly so, 5-lobed or 5-toothed, lobes usually shorter than the tube, triangular to lanceolate or subulate, acute or acuminate, the adaxial one often slightly larger than the others; corolla 5-lobed, pink to magenta or pale purple, with 2 yellow lines and purplish or reddish spots within the throat abaxially, the tube campanulate, often somewhat distended abaxially, the throat open, the lobes shorter than the tube, slightly irregular, ciliate, the upper 2 lobes usually somewhat smaller, arched, spreading to somewhat reflexed or sometimes projecting, the lower 3 lobes spreading, external in bud; stamens 4, in 2 pairs, the lower pair longer, usually included, the filaments pubescent, at least toward the base, anthers 2-celled, the cells parallel, obtuse to caudate at the base, villous; stigma solitary, flattened, somewhat elongate. **FRUIT** a globose or subglobose capsule, sometimes ellipsoid; loculicidal dehiscence. **SEEDS** numerous, triangular, reticulate. [*Gerardia* Benth.] *Agalinis* includes about 45 species formerly included in Scrophulariaceae. In our area, only one species occurs as somewhat of an outlier to its range (Canne 1979, 1981; Pennell 1928, 1929).

Agalinis tenuifolia (Vahl) Raf. (thin or narrow leaf) Threadleaved gerardia, small-flowered gerardia. Glabrous or subglabrous, often turning black upon drying. **STEMS** to about 50 cm tall, branched, axillary fascicles present (in ours) or absent, dark green. **LEAVES** spreading to arched-ascending, entire, linear, acuminate, dark green, 3–7 cm long, 1–2 mm wide, finely scabrous to glabrous. **INFL** a raceme, pedicels widely divaricate, filiform, 7–20 mm long. **FLOWER** calyx hemispherical, 3.5–5.5 mm long, the lobes broadly triangular, apiculate, sometimes subulate, 1–2 mm long, the sinuses broadly rounded; corolla 10–15 mm long, glabrate or glabrous, lobes 3–5 mm long, the margins ciliate, the upper lip concave-arched, projecting forward over the anthers and stigma; anthers 1.5–2.2 mm long, densely to sparingly (in ours) villous. **FRUIT** a globose capsule, 4–6 mm long. **SEEDS** dark brown to blackish, 0.7–0.9 mm long, bluntly trapezoid or triangular, the reticulations irregularly elongate. $2n = 28$. [*Gerardia tenuifolia* Vahl var. *parviflora* Nutt.] Damp and sandy soils. NMEX: RAr. 1370–1680 m (4500–5500′). Flowering: Aug–Oct. Fruit: late Aug–Nov. Southern Quebec to southern Manitoba, Canada; south to Florida, eastern Texas, Oklahoma, eastern Colorado, eastern New Mexico, and a disjunct area in northwestern New Mexico. In our area, this species is represented by **var. *parviflora*** (Nutt.) Pennell (small-flowered).

Castilleja Mutis ex L. f. Indian Paintbrush, Painted-cup

Mark Egger

(for Spanish botanist Domingo Castillejo, 1744–1793) Root-hemiparasitic herbs to occasionally small shrubs, annual or more commonly perennial, often with more or less woody base. **STEMS** single to clumped, branched or unbranched, strongly decumbent to erect. **LEAVES** mostly alternate along stems, sessile to rarely short-stipitate or auriculate, entire to pinnately or rarely somewhat palmately divided. **INFL** a compact to elongate spike or spicate, rarely secund, raceme; often compact when young, strongly elongating with age. **BRACTS** usually differing gradually upward from leaves, often becoming shorter, broader, more deeply divided, and differently colored, at least on upper portions; colors predominantly shades of red-orange, yellow, white, pink, violet, magenta, or purple, usually more conspicuous than or contrasting with the colors of the corolla; in a few species, green with brighter colors borne on calyx and corolla. **FLOWER** calyx sessile to long-pedicellate, tubular, divided above in several strongly differing and usually diagnostic patterns, either cleft into 4 subequal segments or with the segments variously fused and unequal in length, forming 2 lateral primary lobes, these entire, emarginate, or cleft into shorter secondary segments; entirely or more often upper portions conspicuously colored, usually as in bracts, but contrastingly in a few species; corolla long and narrow, usually greenish to less commonly yellowish to pale orange-red, tubular below, strongly bilabiate above and divided into an upper beak, its lobes joined to the tip and often differently and more brightly colored along the thin margins, and a highly variable lower lip, in ours either of greatly reduced and somewhat incurved greenish teeth or of unpouched to moderately pouched, subpetaloid, more brightly colored lobes with small apical teeth. **FRUIT** a somewhat irregular, loculicidal capsule, ovate to oblong, acuminate. **SEEDS** small, numerous, with a loose, reticulate, pitted coat. $n = 12$ (rarely $n = 10, 11$), about half of all species exhibiting various levels of polyploidy, including $n =$ 24, 36, 48, 60, 72. About 200 species. Widespread in the Western Hemisphere, with most species in mountainous regions of western North America, including Mexico. A smaller number of species found in eastern North America, Central America, Andean South America, and boreal-Arctic regions of Eurasia.

While most of our species are fairly well defined and easily identified, many species of this genus, especially outside our area, show substantial phenotypic variation both within local populations and across edaphic and geographic discontinuities. Some species are taxonomically difficult, probably due at least in part to allopolyploidy. Synonymy lists for many *Castilleja* species are extensive, and only those relevant to our region are included here. Species with red to red-orange inflorescences are often visited and pollinated by nectar-seeking hummingbirds, but an even larger proportion, including some red to red-orange species, are pollinated by pollen-collecting insects, mostly hymenopterans, especially *Bombus*. A few species outside our area are self-pollinating. Some species are important larval food sources for various Lepidoptera, especially of the genera *Euphydryas* and *Platyptilia*, and as a nectar source for hawkmoths (*Hyles*). Several species are used (with some difficulty) as ornamentals, as natural dyes, for nectar, and by Native Americans for medicinal and ceremonial purposes. A number of species are known to contain alkaloids, including some assimilated from parasitized hosts via the haustorial bridge. Some species are known to serve as intermediate hosts for certain rust fungi (*Cronartium*, *Endocronartium*) and for hummingbird flower mites (Mesostigmata: Ascidae). The inflorescence and herbage of *Castilleja* are used as an adult food source by a variety of insects, including Coleoptera, Heteroptera, and Homoptera. *Castilleja* species are also grazed by a range of mammals, including voles (*Lemmiscus*), packrats (*Neotoma*), porcupines, deer, cattle, and goats. Many species known to decrease in numbers when subjected to overgrazing by domestic livestock, including the extinction of one species (*C. guadalupensis*) and the near extinction of several others. None of our species is of grave conservation concern, though several, such as *C. minor* and *C. lineata*, are range- and habitat-restricted and should be monitored. The ranges of two species, *C. lineata* and *C. haydenii*, occur primarily within our area, though no species is strictly endemic.

1. Plants annual; bracts always entire, narrowly lanceolate, apically more or less attenuate, and brightly colored only at the tips ....(2)

1′ Plants perennial; bracts various but never with the combination of characters above ....(3)

2. Lobes of lower lip of corolla always dull white to pale yellow, spreading, with short, obtuse teeth; plants tending to be larger and more robust, with coarse, often broadly lanceolate leaves; pubescence of stem spreading, densely hirsute to pilose; fairly common in our area but absent from the extreme southern portion .... ***C. exilis***

2′ Lower lip of corolla always red to red-violet, spreading-ascending, with small, acute, ascending teeth; plants tending to be smaller and more slender, with softer textured, narrowly lanceolate to linear leaves; pubescence of stem less dense, lax, short-pilose; rare in our area, limited to extreme southern portion and absent from Utah and Colorado .... ***C. minor***

3. Inflorescence predominantly pale white to pale yellow or pale yellow-green, fairly often suffused with pale pink or dull red, often from base upward, usually then with white to yellow apices ....(4)

3′ Inflorescence predominantly various shades of red, orange, pink, magenta, red-violet, or purple, occasional individuals or localized swarms colored otherwise ....(6)

4. Bracts and at least upper leaves deeply cleft from below middle upward into 3–7 long, spreading, subpinnate, linear-lanceolate lobes; pubescence of herbage mostly woolly to somewhat pilose .... ***C. lineata***

4′ Bracts and leaves entire or bracts relatively shallowly cleft from above the middle into 3–5 (7) shorter, mostly apical and palmate, linear-lanceolate to triangular lobes; pubescence of herbage lacking to pilose, never woolly ....(5)

5. Plants 1–2 dm tall; stems usually decumbent; limited to moderately dry alpine habitats .... ***C. occidentalis***
5′ Plants 2–8 dm tall; stems ascending to erect; mostly in montane and subalpine meadows, occasionally higher .... ***C. septentrionalis***

6. Calyx cleft far more deeply in front (about 2/3) than in back (2/3 or less), with segments often upcurved; at maturity corolla often curved pendulously outward through front calyx cleft; calyx and corolla greatly exceeding bracts in length when fully developed; calyx and corolla usually much more conspicuously colored than bracts .. ***C. linariifolia***
6′ Calyx divided more or less equally in front and in back or only slightly (1–3 mm) more deeply on one side or the other, with segments straight to occasionally slightly upcurved; corolla more or less parallel to calyx and usually not at all pendulous; calyx and sometimes corolla not greatly exceeding bracts in length when fully developed; bracts usually much more conspicuously colored than calyx and corolla .... (7)

7. Inflorescence predominantly various shades of crimson, pink, magenta, red-violet, or purple, occasional individuals or localized swarms colored otherwise .... (8)
7′ Inflorescence predominantly various shades of scarlet, red, or orange, occasional individuals or localized swarms colored otherwise .... (9)

8. Upper leaves and bracts almost always deeply cleft into 3–7 spreading lobes; lower lip of corolla about 1/3 length of corolla beak, greenish at the base, becoming white to rose-pink on apical teeth .... ***C. haydenii***
8′ Upper leaves usually entire, sometimes with a pair of ascending lobes, bracts entire or with 3–5 short, ascending lobes; lower lip of corolla 1/4 length of corolla beak or less, uniformly green .... ***C. rhexiifolia***

9. Pubescence of stems and herbage usually either lacking or short-woolly and appressed to occasionally pilose, not at all hispid; leaves usually entire to occasionally divided into 3 lanceolate, acute lobes .... (10)
9′ Pubescence of stems and herbage strongly hirsute; at least upper leaves usually divided into 3–5 narrowly lanceolate to oblanceolate, acute to obtuse lobes .... (11)

10. Stems and herbage usually short-woolly with somewhat appressed hairs; leaves linear to linear-lanceolate, usually with strongly involute margins; bracts usually entire or occasionally divided into 3 acute lobes; calyx segments usually triangular to lanceolate and acute .... ***C. integra***
10′ Pubescence of stems and herbage lacking or occasionally pilose on upper stems, not at all woolly-appressed; leaves usually lanceolate, margins not strongly involute; bracts usually divided into 3–5 sharply acute lobes; calyx segments usually linear-lanceolate and more or less acuminate .... ***C. miniata***

11. Stems weakly decumbent, usually 1–2 dm tall; reduced scalelike leaves of lower stem extending upward well beyond the base; bract lobes usually narrowly lanceolate and acute-tipped; pubescence of herbage densely cinereous-hirsute; plants of sandstone or clay substrates .... ***C. scabrida***
11′ Stems ascending to erect, usually 1.5–4 dm tall; reduced scalelike leaves of lower stem lacking or confined to extreme basal portion; bract lobes usually narrowly oblanceolate and more or less obtuse; pubescence of herbage densely hirsute but not cinereous; plants not limited to sandstone or clay substrates .... ***C. chromosa***

Castilleja chromosa A. Nelson (colorful, referring to the brightly colored bracts) Desert paintbrush. Perennial herb, 1.5–4.5 dm tall. **STEMS** clustered, ascending to erect, usually several, usually unbranched above, hirsute. **LEAVES** 2.5–7 cm long, linear to narrowly lanceolate, hirsute; lobes 1–5, spreading. **INFL** 4–15 cm long, a fairly densely flowered spike, elongating with age, hirsute to occasionally pilose. **BRACTS** lanceolate, red or red-orange (rarely yellowish); lobes (1) 3–7, usually narrowly oblanceolate and obtuse. **FLOWER** calyx 15–25 mm long, colored as in bracts, divided 1/4–1/3 in front and back, about 1/7 on sides; segments obtuse to rounded; corolla 20–32 mm long; beak about 1/2 of corolla length, yellowish green with red margins; lower lip 2–3 mm, dark green, incurved; stigma bilobed. n = 12, 24. [*C. angustifolia* (Nutt.) G. Don var. *dubia* A. Nelson; *C. eremophila* Wooton & Standl.]. Sagebrush deserts and plains to arid, rocky woodlands. ARIZ: Apa, Coc, Nav; COLO: Arc, Dol, LPl, Mon, SJn, SMg; NMEX: McK, RAr, San, SJn; UTAH. 1125–2700 m (3700–8800′). Flowering: Mar–Jun (–Sep). Southeastern Oregon to central Wyoming, south to southeastern California, northern Arizona, and northern New Mexico. Widespread and variable species closely associated with sagebrush. This species is sometimes either confused with or subsumed under the more northerly *C. angustifolia* (Nutt.) G. Don, from which it can be distinguished most easily by color and floral measurements. True *C. angustifolia* does not occur in our area. The name *C. applegatei* Fernald subsp. *martinii* (Abrams) T. I. Chuang & Heckard has been misapplied to *C. chromosa* in some recent southwestern species lists, but the former name applies to a taxon endemic to California, western Nevada, and northern Baja California, and it is unknown from our region. Known to take up selenium from soil when available. Utilized by a variety of Native American groups as a dermatological aid for the treatment of sores and spider bites, as an aid for gastrointestinal and menstrual problems, and for sore eyes. Also used as a dye.

Castilleja exilis A. Nelson **CP** (thin, referring to its slender aspect) Alkaline paintbrush. Annual herb, 3–15 dm tall. **STEMS** ascending to usually erect, usually unbranched, single to few, densely pilose to hirsute and strongly stipitate-glandular. **LEAVES** 4–10 cm long, lanceolate and attenuate at apex, 4–15 mm wide near base, coarse-textured, entire. **INFL** 10–40 cm long, a narrow, elongating raceme with long internodes below, pilose to hirsute and stipitate-glandular. **BRACTS** 20–50 mm long, leaflike, lanceolate, green below with upper 1/3–1/4 red to red-orange (rarely yellow), entire, ascending. **FLOWER** calyx 14–20 mm long, subequally cleft about 2/3 in front and back, segments 1–3 mm long, acute; corolla 15–22 mm long; beak about 1/3 of corolla length, yellow, scarcely to not at all exserted above calyx; lower lip 1–2 mm long, whitish to pale yellow. n =12. [*C. stricta* Rydb.]. Alkaline marshes and stream banks. ARIZ: Apa, Nav; COLO: LPl; NMEX: RAr, SJn; UTAH. 1600–2100 m (5300–6900′). Flowering: May–Sep. Eastern Washington east to southern Montana and western Wyoming, south on eastern side of Cascade-Sierra crest to south-central California, extreme northern Arizona, and extreme northwestern New Mexico. This species is very closely related to *C. minor*, but the ranges of these two forms are essentially allopatric, and the morphological differences are geographically consistent. They are probably best regarded as varieties of a single species, with *C. minor* being the name of priority. However, some recent treatments subsume *C. exilis* entirely; this treatment separates the entities pending further study. This is the more common and widespread of the two taxa in our area.

Castilleja haydenii (A. Gray) Cockerell **CP** (for geologist and archaeologist F. V. Hayden, 1829–1887) Hayden's paintbrush. Perennial herb, 10–20 cm tall. **STEMS** clustered, spreading to ascending, few to many, unbranched, glabrous below, usually becoming short-pubescent above, often purplish. **LEAVES** 2–8 cm long, linear and entire, often becoming broader and divided above into 3–5 linear lobes, glabrous to finely hairy. **INFL** at first a densely flowered spike, elongating somewhat with age, more or less pilose below. **BRACTS** broad and deeply cleft into 3–7 linear to linear-lanceolate, acute lobes, greenish below, becoming crimson to rose-pink or lilac-purplish above, glabrous to pilose, especially along the veins. **FLOWER** calyx 1.8–2.2 cm long, divided about 1/2 in front and back, segments divided about 1/4 on each side, acute, colored as in bracts; corolla 2–2.5 cm long, with only the beak and sometimes the lower lip exserted from the calyx, if at all; beak 7–8 mm long, greenish with rose-colored margins, lower lip about 1/3 the length of the beak, with incurved teeth, greenish at the base, becoming white to rose-pink on the tips; stigma exserted, slightly bilobed. n =12. [*C. pallida* (L.) Spreng. var. *haydenii* A. Gray]. Rocky alpine slopes and ridges, usually with bunchgrasses. COLO: Con, Hin, LPl, Min, Mon, SJn; UTAH. 3400–3900 m (11000–12800′). Flowering: Jul–Sep. Fairly uncommon, endemic to alpine regions of southwestern Colorado and northwestern New Mexico, extending southward from our region only in the high elevations of the Sangre de Cristo Mountains. In Utah, known only from an old collection in the upper La Sal Mountains. Plants in the western portions of its range tend to have less-divided leaves.

Castilleja integra A. Gray (entire, referring to the often entire bracts) Entire-leaved paintbrush, southwestern paintbrush. Perennial herb, 1–5 dm tall. **STEMS** erect to ascending, 1 to several, sometimes branched above, finely and somewhat appressed-woolly. **LEAVES** 2–6 (rarely 8) cm long, linear to linear-lanceolate, spreading to somewhat ascending, margins usually strongly involute, usually entire. **INFL** 2–10 cm long, pilose. **BRACTS** 20–40 mm long, obovate to oblanceolate, pilose below, minutely glandular-pubescent above, orange-red to scarlet (rarely yellow); lobes 1–3, central lobe broad, obtuse, lateral lobes short, narrower, acute. **FLOWER** calyx 21–34 (rarely 38) mm long, short-pilose, colored as in bracts, divided about 1/3 in back, slightly less in front, divided on sides almost as deeply as in front, segments obtuse to rounded or acute; corolla 26–45 mm long, ascending; beak about 1/3 of corolla length, greenish with margins colored as in bracts, minutely pubescent, partially to strongly exserted from calyx; lower lip 1.3–2.8 mm long, reduced, dark green, incurved; stigma headlike. n = 12, 24. [*C. gloriosa* Britton; *C. integra* var. *gloriosa* (Britton) Cockerell]. Dry, gravelly soils on plains and slopes from open grasslands through juniper and pine-oak communities into subalpine meadows. ARIZ: Apa, Nav; COLO: Arc, LPl, Mon; NMEX: McK, RAr, San, SJn. 1525–2200 m (5000–7800′). Flowering: (Apr–) May–Sep. Northern Arizona, northern New Mexico, southern Colorado, and western Texas south through northern Mexico to Durango. Widespread and one of our more common paintbrushes. Important to Native Americans as a source of dyes, a preservative for storage of foods, for ceremonial purposes, and for medicinal treatments for burns, stomach complaints, blood purification, and easing of labor in women.

Castilleja linariifolia Benth. (linear-leaved) Wyoming paintbrush. Perennial herb to subshrub, 3–10 dm tall. **STEMS** clustered, ascending to erect, few to many, often branched above, glabrous to short-pubescent, usually gray-green, becoming purplish with age. **LEAVES** 2–6 (rarely 8) cm long, linear to filiform (narrowly lanceolate), often with involute edges; lobes 1–3, spreading. **INFL** 5–20 cm, a loosely flowered spike, becoming racemose with age, hirsute to pilose. **BRACTS** 15–30 mm long, lanceolate, short-pubescent, lowermost green, becoming red to red-orange (rarely yellowish, white); lobes 3, linear to linear-lanceolate. **FLOWER** calyx 18–35 mm long, curved forward near base, then upward near tips, short-pubescent, colored as in bracts, divided 1/3 in back, 2/3 in front, about 1/8 on sides, segments acute; corolla 25–45 mm long; beak about 1/2 of corolla length, yellow-green with red margins; lower lip 2–3 mm, green, incurved; stigma slightly bilobed. $n = 12, 24$. [f. *omnipubescens* Pennell; var. *omnipubescens* (Pennell) Clokey]. Dry, sandy to rocky plains and hillsides, usually with sagebrush. ARIZ: Apa, Nav; COLO: Arc, Dol, Hin, LPl, Min, Mon, SJn, SMg; NMEX: McK, RAr, San, SJn; UTAH. 1100–3150 m (3600–10400′). Flowering: (Apr) May–Oct. Central Oregon to southern Montana south to southern California, central Arizona, and central New Mexico. Widespread and a common species often associated with sagebrush. An important plant to several southwestern Native American groups as a source of a variety of dyes, paints, decorations, and ceremonial materials, as well as for medicinal purposes, including contraception, menstrual relief, several types of pain relief, as a blood purifier, physic, and emetic, and in the treatment of venereal diseases. The Hopi also use the flowers as a food.

Castilleja lineata Greene (marked with parallel veins, referring to the prominent veins of the bracts) Lineated paintbrush. Perennial herb, 1–4 dm tall. **STEMS** clustered, ascending to erect, few to several, usually unbranched above, pilose to densely woolly. **LEAVES** 2–5 cm long, linear, ascending, strongly veined; lobes 1–5 (rarely 7). **INFL** 5–15 cm, a loosely flowered spike, densely woolly. **BRACTS** 20–35 mm, broadly lanceolate, densely woolly, pale yellow to pale yellow-green; lobes 3–7, linear, acute. **FLOWER** calyx 15–20 mm long, densely woolly, yellow, divided about 1/2 in back and front, about 1/3 on sides, segments linear, acute; corolla 18–22 mm long, ascending; beak 1/5–1/4 of corolla length, pale greenish with straw-colored margins, minutely woolly and slightly glandular, tip barely exserted from calyx; lower lip 1–3 mm long, straw-colored, ovate-acuminate, incurved; stigma distinctly bilobed. $n = 12$. Dry to moderately moist slopes and meadows. ARIZ: Apa; COLO: Arc, Hin; NMEX: RAr. 2150–2900 m (7100–9500′). Flowering: Jun–Aug. Uncommon and endemic to mountainous areas of northeastern Arizona, northwestern New Mexico, and southwestern Colorado; not recorded from Utah. Used by the Navajo for stomach complaints and for its sweet nectar.

Castilleja miniata Douglas ex Hook. (scarlet, referring to its most common color form) Scarlet paintbrush. Perennial herb, 2.5–7 (rarely 10) dm tall. **STEMS** erect to ascending, usually branched above, few to several, glabrous to (in many of ours) variously pilose, green to dark purple. **LEAVES** 3–8 cm long, linear to lanceolate, weakly to strongly ascending; entire to sometimes 3-lobed above. **INFL** 3–15 cm long, pilose with short glandular pubescence beneath, often with a whitish, powdery coating. **BRACTS** 15–35 mm long, lanceolate, green below, upper 2/3–3/4 scarlet to orange-red (rarely yellow); lobes 1–5, central lobe usually broad, rounded to acute, short lateral lobes narrow, acute. **FLOWER** calyx 10–30 mm long, short-pilose, colored as in bracts at least on upper 2/3, divided about 1/2–2/3 in front, slightly less in back, about 1/4 on sides, segments linear to lanceolate, acute to acuminate; corolla 20–44 mm long, ascending; beak 2/5–1/2 of corolla length, yellow-green with margins colored as in bracts, minutely pubescent, partially to fully exserted from calyx; lower lip 1–2 mm long, dark green, incurved; stigma indistinctly bilobed. $n = 12, 24, 36, 48, 60$. [*C. confusa* Greene]. Moist stream banks to dry woods and meadows from pine-fir communities to subalpine. ARIZ: Apa; COLO: Arc, Hin, LPl, Min, Mon, SJn; NMEX: RAr, SJn; UTAH. 2550–3400 m (8400–11200′). Flowering: Jun–Oct (–Nov). Much of western North America, from southern Alaska and central Yukon east to western Ontario, then south to Baja California Norte, northern Arizona, and northern New Mexico. A widespread, highly variable, and complex species with numerous regional forms. Ours referable to **var. *miniata***. Widely used by many Native American groups for its putative medicinal value in treating hemorrhaging, coughs, sore eyes, kidney, lung, and orthopedic ailments and as a purgative and diuretic. Additional native uses include dyeing, especially for animal skins, consumption of its nectar, as a decoration, and in a variety of ceremonial and magical applications.

Castilleja minor (A. Gray) A. Gray (lesser, referring to its slender aspect) Seep paintbrush. Annual herb, 3–15 dm tall. **STEMS** erect, usually unbranched, single to few, moderately to sparsely pilose with lax hairs, moderately stipitate-glandular. **LEAVES** 4–10 cm long, linear-attenuate, 2–5 mm wide near base, moderately soft-textured, entire. **INFL** 10–40 cm long, narrow, with long internodes below, short-pilose and stipitate-glandular. **BRACTS** 20–50 mm long, leaflike, linear to linear-lanceolate, green below with upper 1/3–1/4 red to red-orange; entire, ascending. **FLOWER** calyx 13–18 mm long, short-pilose, greenish, divided about subequally 2/3–3/4 in front and back, segments about 1–3 mm, linear to triangular, acute; corolla 14–19 mm long, ascending; beak about 1/4–1/3 of corolla length, yellow, scarcely to not at all exserted above calyx; lower lip 1–2 mm long, with strongly back-curved red-violet teeth; stigma indistinctly bilobed. $n = 12$. Grassy seeps, bogs, and stream banks in foothills and mountains. ARIZ: Apa; NMEX: McK. 1500–2450 m (4800–8000′). Flowering: Jun–early Sep. North-central Arizona and north-central New Mexico south to northeastern Sonora and northwestern Chihuahua, Mexico. Uncommon and local throughout its fairly limited range. See comments under *C. exilis*. Root bark used by White Mountain Apaches as dye for animal skins.

Castilleja occidentalis Torr. (western, referring to early impressions of its distribution) Western paintbrush, alpine paintbrush. Perennial herb, 0.7–2 dm tall. **STEMS** decumbent to ascending, unbranched, few to many, pilose and often viscid, especially above. **LEAVES** 2–5 cm long, linear to lanceolate, reduced below to sometimes with 1 pair of lobes above, usually short-pilose. **INFL** a relatively short, dense spike, strongly viscid-villous. **BRACTS** 15–20 mm long, ovate to broadly lanceolate, usually entire but fairly often with 1–3 pairs of short, triangular to lanceolate, ascending lobes on upper portion, pale yellow to sometimes suffused with dull reddish brown to dull purplish, especially below, viscid-pilose. **FLOWER** calyx 12–20 mm long, cleft about 1/2 in front and slightly less deeply behind, segments 1–4 mm long, more or less triangular, acute to obtuse or rounded, colored as in bracts, viscid-pilose; corolla 16–25 mm long; beak about 1/4 to 1/3 of corolla length, green with pale yellow margins; lower lip reduced, 1.3 mm long, triangular, incurved, greenish; stigma reduced, headlike. $n = 12, 24$. Alpine and upper subalpine, consolidated talus slopes and ridges to well above timberline. COLO: Con, Hin, LPl, Min, Mon, RGr, SJn. 3050–3900 m (10100–12800′). Flowering: Jun–Sep. Southern half of Canadian Rocky Mountains of Alberta and British Columbia south into northern Montana, then recurring again in the Rocky Mountains of Colorado southward into the Sangre de Cristo Mountains of northern New Mexico and the La Sal Mountains of southwestern Utah. It is often thought that this is an alpine form of the widespread *C. septentrionalis* (see below), but in several regions in the distribution of the latter, *C. occidentalis* is missing, even when extensive areas of suitable habitat are available.

Castilleja rhexiifolia Rydb. (*Rhexia*-leaved) Rhexia-leaved paintbrush. Perennial herb, 2–7 dm tall. **STEMS** ascending to erect, usually unbranched, few to several, glabrous to more often short-pilose, especially above. **LEAVES** 3–7 cm long, spreading to ascending, linear to linear-lanceolate, entire. **INFL** a fairly short, compact spike, elongating somewhat with age, viscid-pilose. **BRACTS** broadly lanceolate to ovate, sparsely to densely viscid-pilose, crimson to rose-pink or purplish, entire or with 1–2 pairs of short, triangular to lanceolate, ascending lobes. **FLOWER** calyx 15–25 mm long, cleft about 1/2, subequally to slightly more deeply in front, colored as in bracts, segments 2–7 mm long, triangular to lanceolate, obtuse to rounded or sometimes acute; corolla 20–34 mm long; beak about 1/3 of corolla length, green with margins colored as in bracts; lower lip reduced, 1–3 mm, long, with triangular, incurved, greenish teeth; stigma reduced, headlike. $n = 12, 24, 48$. Montane to alpine meadows, open woods, slopes and ridges. COLO: Arc, Con, Hin, LPl, Min, Mon, RGr, SJn. 3050–4000 m (10000–13000′). Flowering: Jun–Sep. Rocky Mountains and adjacent ranges from southern Canada southward to southern Utah and northern New Mexico, with outlying populations in the mountains of northeastern Oregon. A common and well-known paintbrush very closely related to and sometimes intergrading with *C. septentrionalis* and *C. occidentalis* but usually readily separable from them by color, minor habitat differences, and their only partially overlapping ranges.

Castilleja scabrida Eastw. (roughened, referring to the harsh pubescence) Rough paintbrush, sandstone paintbrush. Perennial herbs, 0.7–2.2 dm tall. **STEMS** decumbent and often somewhat sprawling, few to many, cinereous-hirsute. **LEAVES** 1.4–5 cm long, often slightly auriculate, mostly lanceolate, entire below, often deeply divided above into 1 to occasionally 2 pairs of spreading, linear-lanceolate lobes, cinereous-hirsute, often tinged with dull reddish purple on leaf surfaces; leaves on about lower 1/4 of stem strongly reduced and scalelike. **INFL** a densely flowered spike, hirsute. **BRACTS** deeply divided into 1–2 pairs of spreading, linear-lanceolate to lanceolate, acute lobes, hirsute, bright red to occasionally red-orange or yellow. **FLOWER** calyx 18–33 mm long, hirsute to pilose, colored as in bracts, cleft slightly less than 1/2 in front, slightly more than 1/2 in back, segments 5–9 mm long, lanceolate, acute; corolla 28–44 mm long; beak about 1/2 length of corolla, bright green with bright red margins; lower lip reduced to 3 incurved, greenish teeth; stigma headlike. $n = 12$. Largely limited to xeric habitats with sandstone or occasionally clay substrates. [*C. zionis* Eastw.]. ARIZ: Apa; COLO: Mon; NMEX: SJn; UTAH. 1550–2250 m (5100–7300′). Flowering: Mar–Jun. Rare in extreme northeastern Arizona and extreme northwestern New Mexico, fairly common in southern and eastern Utah and southwestern Colorado. Ours referable to **var. *scabrida***.

Castilleja septentrionalis Lindl. (northern, referring to a major portion of its distribution) Northern paintbrush, sulphur paintbrush. Perennial herbs, 2.5–6 dm tall. **STEMS** mostly ascending to erect, often branched above, few to many, glabrous or short-hirsute below, becoming pilose above. **LEAVES** 2–7 cm long, linear-lanceolate to broadly lanceolate, spreading to ascending, glabrous to short-hirsute. **INFL** a densely flowered spike, elongating with age, pilose and short glandular-hairy. **BRACTS** broadly lanceolate to ovate, entire or with 1–2 pairs of short-lanceolate to triangular lobes on upper margin, usually green below and yellow, pale yellow, or white above, occasionally suffused with various shades of pink to pale red, especially above. **FLOWER** calyx 13–25 mm long, colored as in bracts, cleft about 3/5 in front, slightly less deeply behind, segments 1–4 mm long, acute to obtuse; corolla 16–30 mm long; beak about 1/4–1/3 of corolla length, green with pale yellow margins; lower lip reduced, 1–2.5 mm long, greenish; stigma weakly bilobed. *n* =12, 24, 48. Montane to lower alpine meadows, open woods, slopes and ridges, often in fairly mesic habitats. [*C. luteovirens* Rydb.; *C. sulphurea* Rydb.; *C. rhexiifolia* Rydb. var. *sulphurea* (Rydb.) N. D. Atwood]. COLO: Arc, Hin, LPl, Min, Mon, RGr, SJn; NMEX: McK, RAr; UTAH. 2300–3700 m (7700–12200′). Flowering: Jun–Sep. North America from Hudson Bay west to the southern Canadian Rockies and south to northern New England, then westward through northern Michigan, Minnesota, and the Dakotas to the Rocky Mountain states and southward through mountainous portions of Montana, Wyoming, Idaho, Utah, Colorado, and northern New Mexico. Widespread and variable species, occasionally forming hybrid swarms with the closely related *C. rhexiifolia* and *C. miniata* but usually easily recognizable. Long known in western states as *C. sulphurea*, a later name.

Cordylanthus Nutt. ex Benth. Bird's Beak

(club flower, referring to the clublike appearance of the distally expanding tubular corolla) Hemiparasitic annuals. **STEMS** single, usually branched, erect. **LEAVES** alternate, entire to more often pinnately or palmately divided with narrow segments, cauline. **INFL** a spike or sometimes reduced to a solitary flower, the spikes often congested into small, few-flowered heads, usually with 2 kinds of bracts, the outer bract(s) usually leaflike and seldom colored, the inner floral bracts usually resembling the calyx. **FLOWER** calyx cleft to or nearly to the base abaxially, forming a single adaxial lanceolate segment, entire or bifid apically, opposite the floral bracts, giving the appearance of a bilobed calyx; corolla dull yellow or purple, tubular, bilabiate, the upper lip galeate (hooded) and enclosing the anthers, often adaxially hooked at the apex, the lower lip about equaling or shorter than the galea, somewhat inflated, the 3 lobes short or obsolete, external to the upper in bud, fertile stamens 4 or 2, attached above the middle of the corolla tube, the anther cells unequally placed, the outer one attached by its middle and the inner suspended by its apex and smaller, sometimes obsolete; stigma solitary and minute, style swollen near the apex and bent forward at a right angle. **FRUIT** a loculicidal capsule, somewhat asymmetric, turgid, glabrous. **SEEDS** numerous to rather few, the coat deeply to shallowly reticulate (in ours). 2*n* = 28, 30, 42. A genus of 20 species, occurring in western North America, formerly included in Scrophulariaceae (Chuang and Heckard 1986).

1. Inner bracts pinnately 3–7-divided; herbage and inflorescence noticeably pilose-pubescent; calyx cleft 1.2–3 mm deep; upper anther cell 1–1.3 mm long; barely entering our area in San Juan County, Utah .... ***C. kingii***

1′ Inner bracts entire or rarely pinnately divided; herbage finely puberulent to glabrous; calyx cleft 0.5–1 mm deep; upper anther cell 1.3–1.6 mm long; southeast Utah, southward and eastward .... ***C. wrightii***

Cordylanthus kingii S. Watson (honoring Clarence R. King, geologist, explorer, 1842–1901) King's bird's beak. Annual, 5–30 cm tall. **STEMS** branched; densely to sparsely glandular-pubescent. **LEAVES** 1–4.5 cm long, pinnately 3–5-parted or entire, the lower segments sometimes again dichotomously divided, the divisions filiform. **INFL** a 2- to 4- or more flowered capitate spike, several terminating the branches; bracts of 2 kinds, the outer ones, when present, usually subtending the spike, not the individual flowers, usually smaller than the leaves, palmately 3–5-parted, with the lateral segments sometimes dichotomously divided from near the base, the inner bract subtending each flower, 7–22 (36) mm long, lanceolate, pinnately 3–7-parted, sometimes the bracts of the upper flowers entire, purplish violet. **FLOWER** calyx 14–26 mm long, narrowly lanceolate, bidentate, the cleft 1.2–3 mm deep; corolla 15–25 (34) mm long, the lips subequal, the galea violet with a greenish yellow tip, the lower lip purple, densely retrorsely villous; stamens 4, the anthers unequally 2-celled, the upper cell 1–1.3 mm long and the lower 0.6–1 mm long, the filaments pubescent to nearly glabrous. **FRUIT** a narrow capsule, 7–11 mm long. **SEEDS** 1.5–2.3 mm long. Sagebrush, desert scrub, piñon-juniper woodlands, mountain mahogany, and ponderosa pine communities. ARIZ: Coc; UTAH (Navajo Mountain). 1500–3000 m (4900–9900′). Flowering: Jun–Sep. Fruit: Jul–Oct. Western Nevada and adjacent California, to eastern Utah.

Cordylanthus wrightii A. Gray (honoring Charles Wright, botanist, 1811–1885) Wright's bird's beak, ke'lhokya (Hopi). Annual, 8–35 (50) cm tall. **STEMS** branched, the upper branches best developed; glandular-pubescent to glabrous. **LEAVES** 1.2–4 cm long, 3-parted, or if 5- or more parted, complexly palmate with the lower lateral divisions dichotomously divided near the base, the divisions filiform. **INFL** a 2- to 10-flowered capitate spike, several terminating the branches; bracts of 2 kinds, outer 1 or more subtending each spike, 10–25 mm long, complexly palmately divided as

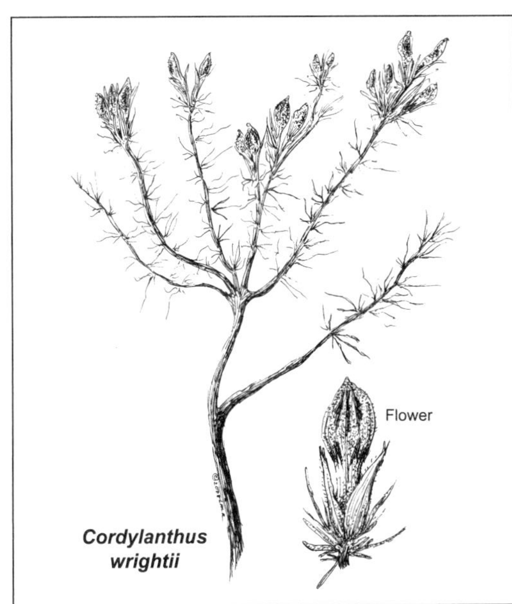

Cordylanthus wrightii

the leaves but usually more so, the segments linear, the inner bract subtending each flower 15–25 mm long, narrowly lanceolate, entire or sometimes pinnately divided into 3–5 or more lobes apically, the divisions short and narrow. **FLOWER** calyx 15–20 (30) mm long, narrowly lanceolate to linear, bifid, the cleft 0.5–1 mm deep; corolla 16–24 (35) mm long, the lips subequal, yellow to lavender or lavender-marked, sparsely pubescent; stamens 4, the anthers villous-ciliate, unequally 2-celled, the upper cell 1.3–1.6 mm long and the lower 0.7–1 mm, the filaments villous. **FRUIT** a slender capsule, 10–12 (16) mm long. **SEEDS** 1.7–2 mm long. Dry sagebrush, desert scrub, resinbush, juniper, piñon-juniper woodlands, oak, and ponderosa pine communities; usually in sandy soil. ARIZ: Apa, Coc, Nav; COLO: Arc, Dol, LPl, Mon, SMg; NMEX: McK, RAr, San, SJn; UTAH. 1300–2500 m (4200–8200′). Flowering: mid-Jul–Sep (Oct). Fruit: Aug–Oct. Southeastern Utah and southwestern Colorado, south to Trans-Pecos Texas and Chihuahua, Mexico. The Hopi boiled the plant and used it as a bath to bleach the skin. Kayenta Navajo used *C. wrightii* to treat a prolapsed uterus.

Orobanche L. Broomrape
Turner Collins

(Greek *orobos*, vetch, + *anchein*, to strangle, referring to the parastic habit) Fleshy, succulent, root-parasitic herbs, lacking chlorophyll, yellowish white, brownish, or purple, glandular-pubescent aboveground. **STEMS** with scales, mostly underground. **LEAVES** reduced to appressed scales. **INFL** solitary or clustered in a terminal spike, raceme, or panicle. **FLOWERS** zygomorphic, perfect, subtended by a broad bract, calyx 5-lobed; corolla tubular, bilabiate with palatal folds in throat, tube curved; stamens 4, mostly included, anther cells distinct; pistil 1, the ovary 1-loculed, with 2–4 parietal placentae. **FRUIT** a 2-valved capsule. **SEEDS** many, small, less than 1 mm. $x = 12$. About 100 species (about 15 species in North America), mostly Eurasian in mainly subtropical and warm-temperate regions. Plants root-parasitic mostly on Asteraceae, frequently on *Artemisia* species.

1\. Flowers borne on long slender pedicels without bractlets, 3–25 cm long; flowers solitary or several on a stem (sect. *Gymnocaulis*) ....(2)
1′ Flowers numerous, sessile or on short pedicels < 3 cm long; inflorescence a spikelike raceme or corymb; calyx with 2 subtending bractlets, lower corolla lobes usually with purple veins (sect. *Myzorrhiza*) ....(3)

2\. Stems slender, bearing 1–3 flowers; pedicel many times longer than the stem; calyx lobes longer than the calyx tube .... ***O. uniflora***
2′ Stems stout, bearing 3–12 flowers; pedicel equal to the stem or shorter; calyx lobes as long as or shorter than the calyx tube .... ***O. fasciculata***

3\. Inflorescence compactly corymbose; anthers woolly; pedicel 4–40 mm long; corolla 18–30 mm; plant 5–16 cm tall; placentae 2 .... ***O. corymbosa***
3′ Inflorescence an elongate raceme or spike; anthers glabrous or sparingly pubescent (except woolly in *O. multiflora*); plant 10–40 cm tall; placentae 4 ....(4)

4\. Corolla lobes pointed, triangular-acute or with apical tooth, 3–5 mm long; plant mostly dark purple ... ***O. cooperi***
4′ Corolla lobes rounded (sometimes pointed in *O. ludoviciana*), 3–9 mm long; plant purple, rose-purple, lavender, or yellowish .... (5)

5\. Anthers woolly; plant grayish pubescent; flower 22–35 mm long, the lobes broadly rounded, calyx 14–20 mm long .... ***O. multiflora***
5′ Anthers glabrous or slightly pubescent; plant not grayish pubescent; flower 15–20 mm long, the lobes narrower than above, calyx 8–14 mm long .... ***O. ludoviciana***

Orobanche cooperi (A. Gray) A. Heller (for James Graham Cooper, 1830–1902, ornithologist, surgeon-naturalist, and collector of plants) Desert broomrape. Fleshy, dark purplish aboveground. **STEMS** 10–40 cm tall, simple or branched, often forming several stems from 1 base, yellowish belowground. **INFL** a spike or raceme, viscid glandular-pubescent. **FLOWERS** sessile or the lower pedicels less than 3 cm long, subtended by 2 narrow bractlets; calyx 8–12 mm long, lobes triangular, acuminate; corolla 18–32 mm, strongly constricted, purple, glandular, upper lip 5–10 mm long, the lobes triangular-acute or with an apiculate tooth, erect or reflexed; anthers variously pubescent or glabrous; stigma lobes 2, thin, recurved. $2n = 24, 48, 72$. [*O. ludoviciana* Nutt. var. *cooperi* (A. Gray) Beck]. Sand dune, desert scrub, desert grassland, and piñon-juniper woodland communities. ARIZ: Apa, Coc, Nav; NMEX: San, SJn; UTAH. 1500–1980 m (4910–6500′). Flowering: May–Sep. Northern Arizona, southern Utah and Nevada, and northwest New Mexico. Our material belongs to **var. *cooperi***. Hosts are species of *Gutierrezia* in our area.

Orobanche corymbosa (Rydb.) R. S. Ferris (growing in clusters) Flat-topped broomrape, corymbose cankerroot. Fleshy, purplish or yellowish herbs. **STEMS** clustered or single, mostly below ground level, yellowish, 5–16 cm tall, erect, glandular-pubescent. **INFL** corymbose, sometimes narrow. **FLOWER** pedicel 4–40 mm long, the lower ones much longer than the upper, subtended by 2 narrow bractlets, perfect; calyx 12–20 mm long, the tube 2–7 mm long, the lobes 9–16 mm long; corolla 18–30 mm long, slightly constricted, light purple with pink guidelines on the lower lobes, upper lip 3–8 mm long, shallowly cleft; anthers woolly-pubescent. **FRUIT** a capsule, 8–14 mm long. $2n = 48, 96$. [*Myzorrhiza corymbosa* Rydb.]. Shadscale, sagebrush, rabbitbrush, and piñon-juniper communities. COLO: Mon; NMEX: SJn. 1500–1970 m (5000–6500′). Flowering: May–Jul. Fruit: Jun–Aug. Washington and Oregon to Wyoming, south to Utah, Nevada, and California. Hosts are various species of *Artemisia.*

Orobanche fasciculata Nutt. **CP** (growing in bundles) Cluster broomrape, cluster cankerroot. Fleshy, purplish or yellowish herbs. **STEMS** solitary or clustered, mostly below ground level, 5–17 cm long, glandular-puberulent. **INFL** a raceme, pedicels 3–12 per stem, 3–25 cm long, equal to or shorter than the stem, bractlets absent. **FLOWERS** perfect; calyx 7–11 mm long, the lobes triangular, 3–5 mm long, shorter than the tube; corolla 15–30 mm long, purple to pinkish or yellow, the tube curved, slightly constricted, the lobes rounded, 2–5 mm long; anthers glabrous to woolly. **FRUIT** a capsule, 10–13 mm long. $2n = 24, 48$. [*Phelipaea lutea* Desf.]. Rocky, sandy, or clay soils in saltbush, greasewood, sagebrush, rabbitbrush, cottonwood, piñon-juniper, and mountain brush communities. ARIZ: Apa, Coc, Nav; COLO: Arc, LPl, Mon, SMg; NMEX: McK, RAr, San, SJn; UTAH. 1600–2320 m (5300–7650′). Flowering: May–early Jul. Fruit: Jun–Aug. Yukon Territory, British Columbia, south to California and Mexico, east to Oklahoma, Illinois, Indiana, and Michigan. Hosts are various species of *Artemisia*.

Orobanche ludoviciana Nutt. (of Louisiana) Louisiana broomrape, Louisiana cankerroot. Fleshy, yellowish or purplish herbs. **STEMS** solitary or clustered, 7–40 cm long, yellowish and glabrous belowground. **INFL** a raceme or spike, viscid glandular-pubescent, 4–20 cm long, purplish to rose. **FLOWERS** sessile or the lower ones on pedicels to 20 mm long; calyx 7–14 mm long, tube 2–5 mm long, the lobes 5–12 mm long; corolla 14–20 mm long, strongly constricted, purple, the throat pale, the lobes rounded or sometimes broadly triangular-acute, 3–7 mm long; anthers glabrous to sparingly pubescent. **FRUIT** a capsule, 7–10 mm long. $2n = 24, 38, 48, 72, 96$. [*O. ludoviciana* var. *arenosa* (Suksd.) Cronquist]. Sandy soils in desert scrub, salt desert scrub, blackbrush, sagebrush, piñon-juniper woodland, and mountain brush communities. ARIZ: Apa; COLO: Mon; NMEX: RAr, San, SJn; UTAH. 1500–2270 m (5000–7500′). Flowering: May–Sep. Throughout the Great Plains in the United States and Canada, isolated populations in Washington, Oregon, Idaho, and western Colorado and New Mexico. Hosts are various species of *Artemisia* and other Asteraceae.

Orobanche multiflora Nutt. (many flowers) Manyflower broomrape. Fleshy, yellowish or purplish herbs. **STEMS** solitary or clustered, 10–40 cm long, glabrous and yellow belowground. **INFL** a dense raceme or thyrsoid, grayish viscid glandular-pubescent. **FLOWERS** sessile or on a short pedicel, densely pubescent; calyx 10–17 mm long; lobes unequal; corolla dark to light purple, the throat pale, 20–35 mm long, strongly constricted, upper lip 5–9 mm long, deeply cleft, lobes broadly rounded at the apex, often with purple veins in lower lobes; anthers woolly. **FRUIT** a capsule. Sandy soils in desert grassland, juniper, and piñon-juniper woodland communities. COLO: Mon; NMEX: McK, RAr, San, SJn. 1400–2800 m (4600–8500′). Flowering: Aug–Sep. Central and western New Mexico to southern Colorado.

Orobanche uniflora L. (single flower) Naked broomrape, one-flower cankerroot. Small, pale yellowish herbs. **STEMS** usually solitary, mostly below ground level, scaly, 0.5–5 cm long. **INFL** a raceme, glandular-pubescent. **FLOWER** pedicels bractless, 1–3 per stem, mostly 3–15 cm long, much longer than the stem; perfect; calyx 6–12 mm long, the lobes narrowly triangular-lanceolate, 4–9 mm long, longer than the tube; corolla 15–35 mm long, ochroleucous to purple, the tube curved, the lobes rounded, 2–7 mm long; anthers glabrous or woolly-pubescent. **FRUIT** a capsule, 8–10 mm long, ovoid. $2n = 36, 48, 72$. [*Aphyllon uniflorum* (L.) A. Gray var. *occidentale* Greene]. Mostly rocky, moist places in sagebrush, piñon-juniper, mountain brush, and ponderosa pine communities. COLO: Mon; NMEX: SJn.

1600–2300 m (5300–7600′). Flowering: late Apr–Jul. Fruit: May–Aug. British Columbia south to California, northern Nevada, Utah, northern Arizona, western Colorado, and northern New Mexico. Our material belongs to **var**. ***occidentalis*** (Greene) Roy L. Taylor & MacBryde (western).

Orthocarpus Nutt. Owl-clover

(straight fruit, referring to the symmetrical capsules) Herbaceous annuals, mostly hemiparasitic. **STEMS** slender, erect. **LEAVES** alternate, sessile, entire to pinnately divided, wholly cauline. **INFL** a short to elongate spike or spicate raceme; bracts usually gradually differing from the leaves, some abruptly differing (wider), green to colored. **FLOWER** calyx tubular-campanulate, equally or unequally cleft into 4 lobes, the lobes often partly connate in lateral segments; corolla elongate and narrow, bilabiate, the upper lip galeate (hooded), beaklike, lobes united to the tip and enclosing the anthers, the lower lip somewhat saccate-inflated, less than or equal to the galea, usually 3-toothed, external to the upper in bud; stamens 4, in 2 sets of 2 stamens, attached near the summit of the corolla tube; anthers 2-celled, cells boat-shaped, unequal in size and placement, the outer one attached by its middle, the other suspended by its apex and smaller, sometimes obsolete; stigmas united and capitate. **FRUIT** a loculicidal capsule, ovoid to obovoid, more or less symmetrical, glabrous or rarely puberulent. **SEEDS** few to many, the seed coat reticulate or alveolate, often loose. $2n = 24$. A genus of about 25 species, of the western U.S., one species in Andean South America, formerly included in Scrophulariaceae (Keck 1927).

1. Corolla 9–14 mm long, yellow; galea more or less equaling the lower lip, the apex not hooked; stems strict or seldom branched; seeds 1.2–1.5 mm long .... ***O. luteus***
1′ Corolla 15–20 mm long, violet and white; galea noticeably longer than the lower lip, the apex projected into a small, forward-projecting beak; stems mostly with corymbose branching above; seeds 1.5–2.5 mm long ***O. purpureo-albus***

Orthocarpus luteus Nutt. (yellow) Yellow owl-clover. Herbaceous annuals, slender, erect. **STEMS** simple, rarely branched above, 8–30 (50) cm tall, glandular-pubescent and also with eglandular villous hairs. **LEAVES** 1.5–3.5 cm long, linear to linear-lanceolate, entire or the uppermost 3-lobed, glandular-puberulent. **INFL** a narrow spike, becoming elongate, the flowers usually short-pedicelled, glandular-puberulent; bracts leaflike at first, becoming shorter and wider, 3-parted, green. **FLOWER** calyx 6–8 mm long, the lateral clefts 1–2.5 mm deep, the median clefts unequal, 2–3.5 mm deep abaxially and 3–5 mm adaxially; corolla 9–12 (14) mm long, puberulent, yellow, the galea 2.5–4 mm long, short and broad, the lower lip and galea equal, simple-saccate, minutely 3-toothed, the tube gradually expanded into the pouch; anthers 2-celled, cells subequal, the upper 0.6–1.2 mm long, the lower 0.4–0.8 mm long, pubescent. **FRUIT** an elliptic capsule, 4.5–7 mm long. **SEEDS** several, 1.2–1.5 mm long, with a reticulate, wrinkled seed coat. $2n = 28$. Meadows, grasslands, stream edges, wetlands; sagebrush, oak, ponderosa pine, and aspen communities. ARIZ: Apa, Nav; COLO: Arc, Con, Hin, LPl, Min, Mon, RGr, SJn; NMEX: McK, RAr, SJn; UTAH. 1500–3000 m (4900–9900′). Flowering: Jul–Sep. Fruit: late Jul–Sep. This, the most widespread species of *Orthocarpus*, occurs from British Columbia, Canada, to central California, east to Michigan, Minnesota, Nebraska, Colorado, and northern New Mexico.

Orthocarpus purpureo-albus A. Gray ex S. Watson (purple-white) Purple owl-clover. Herbaceous annual, erect. **STEMS** simple or usually branched above, 8–35 (55) cm tall; glandular-pubescent, purplish. **LEAVES** 1.5–3.5 cm long, linear to filiform, 3-parted, or the lowermost entire, the segments similar in shape, glandular-puberulent. **INFL** an elongate spike, many-flowered; bracts 3-parted from near the base, glandular-pubescent, green. **FLOWER** calyx 5–8 (9) mm long, glandular-pubescent, the lateral and abaxial clefts about 0.82 mm deep, the adaxial clefts 1.8–3 mm deep, vasculature dark green, becoming thick; corolla 15–20 mm long, the galea 3.5–5 mm long, puberulent, hooked at the tip, the lower lip 2.5–4.5 mm long from the base of the cleft, simple-saccate, without teeth; lower corolla violet, upper corolla violet with a white pouch; anthers 2-celled, cells subequal, the upper 0.9–1.2 mm long, the lower 0.7–1.1 mm long, puberulent. **FRUIT** a narrowly elliptic capsule, 6–7.5 mm long. **SEEDS** few, 1.8–2.5 mm long, with a deeply reticulate, angular seed coat. $2n = 28$. Sagebrush meadows and openings in piñon-juniper woodland, Gambel's oak, and ponderosa pine communities. ARIZ: Apa, Nav; COLO: Arc, Dol, LPl, Mon, SMg; NMEX: McK, RAr, SJn; UTAH. 1830–2700 m (6000–8900′). Flowering: Jun–Sep. Fruit: Aug–Sep. Endemic to the Colorado Plateau of southeastern Utah, southwestern Colorado, northern and eastern Arizona, and western New Mexico.

Pedicularis L. Lousewort

(Latin *pediculys*, a louse) Presumed to become lice on sheep that contact them. Herbaceous perennials, hemiparasitic from fibrous or tuberous roots. **STEMS** erect, simple or branched. **LEAVES** alternate or basal, toothed to bipinnatifid. **INFL** a bracteate spike or spicate raceme. **FLOWER** calyx 2–5-lobed, the lobes unequally cleft; corolla bilabiate, red, purple, yellow, or white, the galea hooded, the lower lip 3-lobed; stamens 4, didynamous, glabrous; stigmas united, capitate. **FRUIT** a loculicidal capsule, glabrous. **SEEDS** several, slightly winged. $x = 8$. About 400 species. Northern Hemisphere, mostly in the mountains of central and eastern Asia, one in the Andes of South America.

1. Flowers over 2 cm long; plants in loose clusters at ground level, often under piñon or juniper trees; blooming in early spring .... ***P. centranthera***
1′ Flowers less than 2 cm long; plants with erect stems; plants with basal or alternate leaves, not forming loose mats; wet mountain valleys, spruce-fir and alpine communities; blooming in summer .... (2)

2. Calyx lobes 2; leaves simple, serrate to crenate .... ***P. racemosa***
2′ Calyx lobes 5; leaves pinnatifid or bipinnatifid .... (3)

3. Corolla pink or purple .... (4)
3′ Corolla yellowish or white .... (5)

4. Galea prolonged into a slender curved beak, the flower resembling an elephant's head; inflorescence glabrous or nearly so; flowers in an elongated spike .... ***P. groenlandica***
4′ Galea not beaked, with lateral broadly triangular teeth; inflorescence usually woolly-pubescent, sometimes glabrous; flowers crowded in a short spike .... ***P. scopulorum***

5. Leaves shallowly or deeply pinnatifid, leaflets joined together by a common winged rachis; upper lip of corolla terminating in a prominent sickle-shaped beak; plants 0.5–3 dm tall .... ***P. parryi***
5′ Leaves compound, fernlike; corolla beakless; plants 3–11 dm tall .... (6)

6. Plants much less than 1 m tall; corolla 20–26 mm long; calyx 7–10 mm long; lower lip not reaching the tip of the galea .... ***P. bracteosa***
6′ Plants up to 1 m tall; corolla 25–36 mm long; calyx 10–16 mm long; lower lip almost reaching the tip of the galea .... ***P. procera***

Pedicularis bracteosa Benth. (having bracts) Bracteate lousewort. **STEMS** erect, glabrous, leafy, 3–9 dm tall. **LEAVES** pinnatifid, 7–16 cm long, linear to oblong-lanceolate, mostly cauline, the cauline sessile to short-petiolate, the basal long-petiolate. **INFL** a densely flowered spike, 5–45 cm long, the bracts 8–16 mm long, simple, lanceolate. **FLOWER** calyx 7–10 mm long, lobes 5, 3–5 mm long, the posterior one short, the anterior cleft the deepest; corolla 20–26 mm long, yellow or yellowish, the galea 9–12 mm long, extended above the lower lip, abruptly truncate, not beaked, the lower lip 4–6 mm long; anthers 1.8–2.5 mm long. **FRUIT** a capsule, 10–12 mm long, asymmetrical. $n = 16$. Engelmann spruce and corkbark fir, willow-grass, and alpine communities. COLO: Con, Hin, LPl, Min, Mon, RGr, SJn. 3300–3950 m (11000–13000′). Flowering: Jul–Aug. Fruit: Aug–Sep. Four Corners material belongs to **var. *paysoniana*** (Pennell) Cronquist [*P. paysoniana* Pennell]. Wyoming, Idaho, Montana, and Utah.

Pedicularis centranthera A. Gray (*centrum*, a prickle or sharp point, referring to the long-spurred anthers) Dwarf lousewort, piñon-juniper lousewort. **STEMS** from a tuberous root, shorter than the leaves, glabrous, 4–7 cm tall. **LEAVES** petiolate, pinnatifid, 6–15 cm long, the segments double-crenate to dentate. **INFL** a compressed spicate raceme, bracts leaflike, becoming reduced upward. **FLOWER** calyx 15–22 mm long, lobes 5, 6–9 mm long, ciliate; corolla 30–35 mm long, purple or yellowish, glabrous, anthers 3.8–4.7 mm long, aristate at the base. **FRUIT** a capsule, 10–13 mm long, ovoid, nearly symmetrical. **SEEDS** 3–4.5 mm long, curved or twisted. $n = 16$. Piñon-juniper, ponderosa pine, Douglas-fir, white pine, aspen, and Engelmann spruce communities. ARIZ: Apa, Nav; COLO: Dol, LPl, Mon, SMg; NMEX: McK, SJn; UTAH. 1950–3140 m (6400–10380′). Flowering: Mar–May. Fruit: Apr–Jun. The dwarf lousewort is found from near the base of Navajo Mountain to the top at 10380′. Oregon, Nevada, and Wyoming. The Shoshoni give this drug to children for stomachaches.

Pedicularis groenlandica Retz. **CP** (for Greenland) Elephant-head. **STEMS** often clustered, glabrous, 1.5–7 dm tall. **LEAVES** mostly basal, petiolate, pinnatifid, 5–25 cm long, the segments narrow, dentate to crenate. **INFL** densely spicate, bracts mostly shorter than the flowers. **FLOWER** calyx 3.5–5.5 mm long, lobes 5, about 1 mm long, entire, with prominent veins; corolla 10–15 mm long, violet to purple, strongly hooded, prolonged into an upturned beak, the lobes deflexed, resembling an elephant's head; anthers 1.5–2 mm long. **FRUIT** a capsule, 7–9 mm long, asymmetrical. **SEEDS** 3–3.5 mm, winged, reticulate. $2n = 16$. Mostly in wet areas and along rivulets in sedge-grass, spruce-fir, and alpine tundra communities. COLO: Arc, Con, Hin, LPl, Min, Mon, RGr, SJn. 3330–3940 m (11000–13000′). Flowering: Jun–Sep. Fruit: Aug–Oct. Across North America south to California and New Mexico. Powdered leaves and stems are taken for coughs by the Cheyenne.

Pedicularis parryi A. Gray (for Charles Christopher Parry, physician, botanist, and explorer) Parry's lousewort. **STEMS** clustered on a thick caudex, glabrous to villous, 0.5–3 dm tall. **LEAVES** basal and cauline, glabrous, the basal ones long-petiolate, pinnatifid, mostly 15–30 cm long, with irregularly serrate segments. **INFL** spicate, 15–35 cm long, villous, the bracts linear, entire to toothed, the lower ones longer than the flowers. **FLOWER** calyx 10–16 mm long, lobes 5, 3–5 mm long, the posterior one shorter; corolla 25–36 mm long, pale yellow, sometimes streaked with red, the galea 9–16 mm long, beakless but with 2 lateral teeth below the apex, the lower lip 7–12 mm long. **FRUIT** a capsule, 10–16 mm long, ovoid, subsymmetrical. $2n = 32$. Spruce-fir, sedge-forb, and alpine meadow communities. COLO: Con, Hin, LPl, Min, Mon, SJn. 3180–4090 m (10500–13500′). Flowering: Jul–Aug. Fruit: Aug–Sep. Wyoming and Utah, to Colorado, Arizona, and New Mexico. Our plants belong to **var. *parryi***.

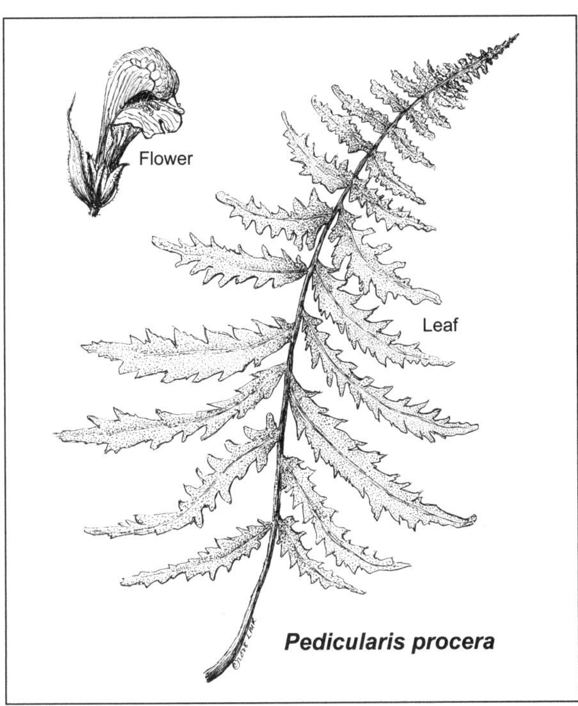

Pedicularis procera

Pedicularis procera A. Gray (tall) Gray's lousewort. **STEMS** erect, sometimes branched above, glabrous, 5–11 dm tall. **LEAVES** mostly basal, petiolate, pinnatifid, mostly 3–12 cm long, the segments toothed or incised. **INFL** spicate-racemose, 5–25 cm long, the bracts pinnately 3–7-lobed, glabrous to villous. **FLOWER** calyx 7–12 mm long, lobes 5, more deeply cleft in front; corolla 16–22 mm long, pink to pale yellow, the galea curved downward at the tip, 6–10 mm long, the beak straight; anthers 2–2.8 mm long. **FRUIT** a capsule, 10–14 mm long, asymmetrical. **SEEDS** 1.8–2.2 mm long. $2n = 16$. [*Pedicularis grayi* A. Nelson] Moderately moist aspen-fir, Gambel's oak–aspen, Douglas-fir, aspen–white fir, and spruce-fir communities. ARIZ: Apa; COLO: Arc, Hin, LPl, Min, Mon, SJn; NMEX: RAr, SJn; UTAH. 2455–3090 m (8100–10200′). Flowering: Jul–Aug. Fruit: Aug–Sep. Wyoming and the Black Hills to Nebraska, south to Arizona and New Mexico. Sometimes used in Navajo ceremonies.

Pedicularis racemosa Douglas ex Benth. (raceme) Leafy lousewort. **STEMS** usually clustered, from a woody caudex, glabrous, 1.5–5 dm tall. **LEAVES** mostly cauline, short-petiolate, simple, serrate to crenate, 4–7 cm long. **INFL** a relatively loose raceme, some flowers in the upper leaf axils, bracts foliose below, reduced upward. **FLOWER** calyx 5–7.5 mm long, the lobes 2, obliquely ovate, 1.5–3 mm long, more deeply cleft in front; corolla 9–15 mm long, white or pale yellow, the galea 5–8 mm long, strongly arching and tapering into a slender decurved beak, the beak 5–7 mm long, the lower lip prominent, deflexed-spreading; anthers 1.8–2.4 mm long. **FRUIT** a capsule, 10–16 mm long, asymmetrical. **SEEDS** 1.6–2.1 mm long. $2n = 16$. Mostly in subalpine spruce-fir, aspen communities, and meadows. COLO: Arc, Con, Hin, LPl, Min, Mon, RGr, SJn; UTAH. 2700–3640 m (8900–12000′). Flowering: Jun–Sep. Fruit: Jul–Oct. Western Canada south through Washington, Oregon, Idaho, Montana, and Wyoming to Arizona and New Mexico. Our plants belong to **var. *alba*** (Pennell) Cronquist.

Pedicularis scopulorum A. Gray (of the Rockies) Sudetic lousewort. **STEMS** erect, glabrous, 1–2 dm tall. **LEAVES** basal and cauline, petiolate, pinnately parted not quite to the midrib, the divisions lanceolate-toothed to incised, 3–8 cm long. **INFL** spicate, 3–20 cm long, bracts narrow, usually with narrow lateral divisions, lower and often longer than the flowers, the upper shorter. **FLOWER** calyx 8–10 mm long, arachnoid-villous, the lobes triangular-subulate; corolla 15–20 mm long, rose to purple, the galea about 10 mm long, curved near the end, with 2 broadly triangular teeth near the apex, not beaked. **FRUIT** a capsule, 10–15 mm long, asymmetrical. **SEEDS** mostly 3 mm long. [*Pedicularis sudetica* Willd. subsp. *scopulorum* (A. Gray) Hultén]. Wet areas in alpine communities. COLO: Con, Hin, Min, SJn. 3370–3940 m (11130–13000′). Flowering: Jul–Aug. Fruit: Aug–Sep. Wyoming, Colorado, and New Mexico.

Rhinanthus L. Rattleweed, Yellowrattle

(nose or snout flower, referring to the beaked upper lip of a flower; not relevant as the species that this refers to has been removed from the genus) Caulescent, herbaceous annuals; scabrous. **STEMS** erect, 4-angled, branching. **LEAVES** opposite, sessile, rather thick and rigid, scabrous, linear to lanceolate, sharply serrate or dentate. **INFL** a dense, subspicate raceme, often secund upward, bracts leafy. **FLOWER** calyx 4-toothed, becoming bladderlike but compressed, with prominent veins in fruit; corolla yellow, at least the tube, bilabiate, the upper lip arched (galeate) and minutely 2-toothed below apex, lower lip shorter, with 3 spreading lobes; stamens 2 differing sets, ascending under the upper lip, anthers pilose, anther sacs equal, parallel; stigma united and capitate, at the summit of the downcurved (abaxially) style. **FRUIT** an orbicular-compressed capsule, dehiscence loculicidal. **SEEDS** several, winged. A genus of 45–50 species, formerly included in Scrophulariaceae.

Rhinanthus minor L. (smaller, inferior, or lesser) Lesser rattleweed. Herbaceous annual, hemiparasitic. **STEMS** (14) 20–80 (180) cm tall, often pilose in longitudinal lines, simple or branching, frequently black-spotted. **LEAVES** linear to lanceolate, 2–3.5 cm long, acute, serrate to dentate, strigose, sessile. **INFL** a secund, leafy spike, puberulent to glandular-villous. **FLOWER** calyx 9–12 mm long, becoming bladdery-inflated, with prominent veins; corolla 7–10 mm long, light yellow, upper lip (galea) with 2 purple or whitish teeth. **FRUIT** a capsule, 5–12 mm long, orbicular, flat. **SEEDS** orbicular, flattened, 2.5–3.5 mm wide. [*Rhinanthus borealis* Chabert; *R. rigidus* Chabert] Meadows, clearings, and wooded slopes. COLO: Arc, Con, Hin, LPl, Min, Mon; NMEX: RAr. 2130–2750 m (7000–9000′). Flowering: Jun–Sep. Fruit: Jul–Oct. Alberta south to Colorado, westward to Alaska, and south to Arizona and Washington. Our material belongs to **subsp**. ***borealis*** (Sterneck) Á. Löve (northern).

OXALIDACEAE R. Br. WOODSORREL FAMILY

Eve Emshwiller, Kenneth D. Heil, and J. M. Porter

Ours annual or perennial herbs with sour juice; some with creeping rhizomes or bulbs. Worldwide includes subshrubs, shrubs, and even trees. **LEAVES** palmately or pinnately compound or monofoliolate, usually with petioles. **INFL** cymes, sometimes umbellate, or solitary on axillary peduncles. **FLOWERS** perfect, regular; most heterostylous; 5 distinct sepals; 5 distinct petals; stamens 10, united at the base; pistil 1, superior, ours with 5 carpels, the locules equaling carpels in number; styles 5, distinct. **FRUIT** a capsule in ours. **SEEDS** appear arillate, but aril is actually outer integument, which explodes to disperse the seeds. $x = 5$–12. Five genera, over 600 species; worldwide, tropical to temperate. This family is much more diverse elsewhere than in the Four Corners. Includes *Averrhoa*, *Sarcotheca*, and *Lepidobotrys*.

Oxalis L. Woodsorrel

(Greek *oxys*, sour) Ours are perennial or annual herbs, some with bulbs or underground rhizomes, caulescent or acaulescent. **LEAVES** petiolate, palmately compound, 3–11 obcordate leaflets. **FLOWERS** perfect, tristylous, distylous, or homostylous, the petals clawed and coherent near the base; stamens 10, obdiplostemonous; pistil 5-carpelled; styles 5, free, the ovary superior. About 575 species. The juice contains oxalic acid but may be used in salads in moderation. Many species are used as ornamentals; however, some have become weeds. During collection, note flower color and be sure to collect underground parts to ensure identification (Ornduff and Denton 1998; Denton 1973; Eiten 1963). Worldwide the genus includes subshrubs and small shrubs, but not trees. Found mostly in southern Africa and South America.

1. Plants with basal leaves originating from a bulb; orange calli (or black when dry) at apices of sepals and often on leaves; flowers purple, blue, pink, or rarely white, not yellow .... (2)

1′ Plants with aerial stems, without bulbs; calli absent; flowers yellow .... (3)

2. Bulb scales usually with 5 or more (up to 30) nerves, rarely 3; leaves on mature plants with 5–13 leaflets (3–4 in juvenile plants), lacking calli on leaflets .... ***O. decaphylla***

2′ Bulb scales 3-nerved; 3–5 leaflets, with or without calli .... (4)

3. Septate patent hairs (at 40×) present on stems, petioles, or pedicels; erect stems arising from a slender fleshy underground rhizome, lacking stipules; lacking taproot; pedicels may be horizontal or erect, but not deflexed in fruit .... ***O. stricta***

3′ Septate hairs (at 40×) absent from stems, petioles, or pedicels; erect, prostrate, or decumbent stems arising from a thickened taproot; stipules present, pedicels in fruit usually deflexed .... (5)

4. Flanges extending along petiole bases 0.5–1.5 cm above the bulb; leaflets 3–5, calli absent, or present only at distal end of lobes .... ***O. caerulea***

4′ Flanges absent above bulb; leaflets 3, calli absent or randomly distributed .... ***O. alpina***

5. Seeds brown, rarely with whitish lines on the transverse ridges; stems creeping, often rooting at nodes; hairs on stems variously patent, ascending, retrorse, or absent; in gardens, lawns, and greenhouse soil ..... ***O. corniculata***

5′ Seeds with conspicuous whitish lines on the transverse ridges; stems mostly erect, or decumbent at the base, not rooting at nodes; white hairs on stems appressed upward .... ***O. dillenii***

Oxalis alpina (Rose) R. Knuth (growing in the alpine) Alpine woodsorrel. Bulbs 0.8–2 cm long, up to 20 bulblets. **LEAVES** basal, 4–30 cm tall; leaflets 3, 1–3 cm wide, usually wider than long. **INFL** 1–7-flowered; pedicels less than 4 cm long. **FLOWERS** usually tristylous, some distylous or homostylous, purple, pink, or white, petals up to 27 mm long. **FRUIT** an ellipsoid capsule, 5–12 mm long. **SEEDS** 1–1.5 mm long. [*O. metcalfei* Small]. $2n = 14, 28, 42, 56, 84$. Moist, rocky places in pine-oak and coniferous forests. COLO: Arc, Hin, LPl, Min; NMEX: McK. 1800–2700 m (6000–9000′). Flowering: Jul–Sep. Fruit: Aug–Oct. New Mexico, Arizona, Texas south to Guatemala. Sometimes mistaken for *O. violacea* L., a species that is more common in the eastern half of the United States, but which is not confirmed in the study area. The capsule differs in *O. alpina* in being longer, 5–12 mm, and cylindrical, whereas in *O. violacea* it is 4–6 mm and globose. The seeds of *O. alpina* have more prominent longitudinal ridges and those of *O. violacea* have more prominent transverse ridges. The calli on the sepal apices of *O. alpina* do not occupy the entire sepal apex whereas on *O. violacea* they do.

Oxalis caerulea (Small) R. Knuth (blue) Blue woodsorrel. Bulbs 0.8–1.5 cm long, bulblets rarely formed, bulb scales 3-nerved. Flanges at petiole bases extend 0.5–1.5 cm above the bulb. **STEMS** absent. **LEAVES** basal, leaflets 3–5, 5–12 mm long and equally wide, notched 1/5 to 2/5 of their length, calli absent or only on distal end of each lobe. **INFL** 1–5-flowered, pedicels 6–14 mm long, glabrous. **FLOWERS** distylous, corolla 9–14 mm long, rose-red to pinkish lavender, yellow-green at base; sepals with 2 calli 0.4–0.8 mm long, confluent or separate. **FRUIT** an ellipsoid capsule, 3.5–6 mm long, glabrous. **SEEDS** 1 mm long, with both transverse and longitudinal ridges. Rocky hillsides, pine forests. ARIZ: Apa; NMEX: McK. 2000–2300 m (6560–7010′). Flowering: Apr–Aug. Fruit: Jul–Aug. Arizona, New Mexico, and Mexico.

Oxalis corniculata L. (with small hornlike appendages) Bulbs absent. **STEMS** creeping or decumbent, to 40 cm long, often rooting at the nodes, hairs pointed in all directions. **LEAVES** cauline; leaflets 3, to 2 cm long, often purplish. **INFL** 2–5-flowered; pedicels less than 1 cm long. **FLOWERS** homostylous, yellow, petals 4–8 mm long. **FRUIT** a cylindric capsule, angled, 10–20 mm long, hairs spreading, mostly short with few long hairs intermingled. **SEEDS** 1.2–1.5 mm long. $2n = 24, 28, 44, 48$. Common in lawns, gardens, disturbed sites, greenhouses, and occasionally escaping to woodlands and grasslands. COLO: Arc, Hin, Mon; NMEX: McK, SJn. 1525–2440 m (5000–8000′). Flowers all year; pantropical, introduced weed in temperate zones worldwide. Used as a drug, food, and a yellow dye.

Oxalis decaphylla Kunth (with ten leaflets) Tenleaf woodsorrel. Bulbs 1–2 cm or more long; bulblets seldom formed, fewer than 10 when present; bulb scales (3) 5–20 (–30)-nerved. **STEMS** absent. Septate hairs on the scapes and the petioles. **LEAVES** cauline, leaflets 5–13 (3–4 on juvenile plants), usually 1.5–13 times longer than wide, rarely as wide as long, leaflets lobed at least 1/6 of length, lacking calli. **INFL** 6–11-flowered, pedicels 11–27 mm long. **FLOWERS** distylous; calli on sepals 2, 0.2–1 mm long, orange, usually confluent; corolla 9–17 mm long, petals pinkish purple, pink, or lavender, tubes yellow or yellow-green. **FRUIT** an ellipsoid capsule, 3–11 mm long, **SEEDS** 0.8–1.2 mm long, with transverse and faint longitudinal ridges. Disturbed habitats near oak, pine, and fir forests, grasslands, and thorn-scrub communities. 1830–2440 m (6000–8000′). ARIZ: Apa, Nav. Flowering: May–Oct. Fruit: Jun–Oct. Southwestern United States to central Mexico.

Oxalis dillenii Jacq. (for J. J. Dillenius) Bulbs absent. **STEMS** erect or decumbent, to 40 cm long, hairs antrorse, pointed, nonseptate. **LEAVES** cauline; leaflets 3, 4–18 mm long. **INFL** umbelliform, 1–5-flowered; pedicels 10–30 mm long. **FLOWERS** yellow, petals 5–12 mm long. **FRUIT** a cylindric capsule, 8–25 mm long, strigose. **SEEDS** 1–1.5 mm long, brown with white spots or lines on transverse ridges. $2n = 16, 18, 20, 22$. Grassy openings in coniferous forest. COLO: Arc, Hin. 2270–2600 m (7500–8500′). Flowering: Jun–Sep. Fruit: Jul–Sep.

Oxalis stricta L. (very upright) Bulbs absent. Main root not thicker than stem. Mature plants with herbaceous underground rhizomes. **STEMS** usually erect, unbranched, to 75 cm tall, with spreading septate trichomes. **LEAVES** cauline; leaflets 3, 5–20 mm long. **INFL** 2–7-flowered; regularly cymose (dichasial), peduncles 8–10 cm long. **FLOWERS** may be homostylous or heterostylous; petals yellow, 3.5–11 mm long. **FRUIT** cylindrical, 8–15 mm long, with spreading septate hairs. **SEEDS** 0.8–1.3 mm long. $2n = 18, 24$. Montane grasslands. COLO: Hin; NMEX: McK, SJn. 1980–2575 m (6500–8500′). Flowering: Jun–Aug. Fruit: Jul–Sep. Used as a drug, food, and a yellow dye.

PAPAVERACEAE Juss. POPPY FAMILY

Lynn M. Moore

Perennial herbs with milky or colored juice. **LEAVES** basal or alternate, dissected, toothed, or lobed. **INFL** solitary or few to several in cymes or racemes. **FLOWERS** regular; perfect; sepals 2 or 3, separate, completely enclosing bud before anthesis, caducous; petals 4–6, separate; ovary superior, carpels 2 to many; stigma large, discoid, lobed, and dry; stamens numerous, distinct in several whorls, anthers 2-loculed. **FRUIT** a poricidal or septicidal capsule. x = 5, 6, 7, 8–11, or 19. Worldwide distribution mainly in the Northern Hemisphere. Seventeen genera and 63 species in North America (26 genera and 250 species worldwide). All members of the Papaveraceae produce latex. The family is famously known for the opium poppy, *Papaver somniferum*, an important plant in medicine and problematic as an illegal drug (Kiger 1997; Heywood 1993; Zomlefer 1989).

1. Petals white, sometimes lavender; leaf blades and capsules stoutly armed with prickles .... ***Argemone***
1′ Petals yellow; leaf blades and capsules not prickly, although pubescent with hispid hairs .... ***Papaver***

Argemone L. Prickly Poppy

(Greek *argemos*, "a white spot (cataract) on the eye" which this plant was once supposed to cure) Plants annual or perennial herbs or subshrubs from taproots; sap yellow or orange; caulescent, leafy, and branching. **LEAVES** sessile, basal rosulate, cauline alternate; blade unlobed or lobed; dentate, each tooth spine-tipped; glaucous; unarmed or prickly; glabrous or hispid. **INFL** terminal cyme; bracteate. **FLOWER** sepals 2–3, prickly; petals 6 in 2 whorls of 3; white, sometimes pale lavender; carpels 3–4 (5), locule 1, style short, stigma 3–4 (5)-lobed; stamens 20–250 or more. **FRUIT** capsule erect, 3–5 (7)-valved, grooved over sutures, prickly. **SEEDS** numerous; subglobose; minutely pitted. Tropical and temperate North America, South America, and Hawaii; introduced elsewhere. Thirty-two species (three in our flora). A revision is currently under production by Andrea Schwarzbach of Kent State University. The revision is working through significant problems with species boundaries and hybridization. Molecular data have been helpful in elucidating some of these problems and will likely play a significant role in working through the revision. All prickle characters that deal with density or presence or absence of prickles do not work well in *Argemone* because exceptions exist for all of these characters and species. Gerald Ownbey spent many years investigating this genus and wrote a 1958 revision of *Argemone* in North America and the West Indies. Unfortunately, Ownbey's keys do not adequately separate the species that occur in the San Juan River drainage. Upon completion of the revision, new insight will hopefully be available to further separate this genus in a satisfactory way; until then it is recommended that fresh material and observations of shape in the field be noted to aid in identification (Schwarzbach, personal communication 2006; Ownbey 1958).

1. Prickles along primary veins on both upper and lower leaf surfaces; leaf blades thick and leathery; latex bright orange when fresh; sepal horns sometimes flattened .... ***A. corymbosa***
1′ Prickles on upper surface absent or sparingly armed along veins; leaf blades succulent; latex yellow when fresh; sepal horns terete .... (2)

2. Sepal horns wider at the base, gradually narrowing from the main body of sepals, sepal horns often devoid of prickles; during early stages of flowering, shape of plant rectangular and leaf length uniform throughout plant .... ***A. albiflora***
2′ Sepal horns distinctly narrowing from the main body of sepals; sepal horns with prickles; during early stages of flowering, shape of plant triangular and lower leaves longer than upper leaves .... ***A. pleiacantha***

Argemone albiflora Hornem. (white-flowered) Bluestem prickly poppy. Plants annual or biennial herbs with bright yellow latex (when fresh). **STEMS** 1–5; 4–8 (12) dm tall; cymosely branched; shape during early stages of flowering rectangular. **LEAVES** glaucous, succulent; oblanceolate; lobed to 2/3 the distance to the midrib; prickles along primary veins on lower surface only; leaf length during early stages of flowering uniform throughout the plant. **FLOWERS** 7–10 cm in diameter; sepal horns 6–10 (15) mm long; terete, wider at the base, gradually narrowing from the main body of sepals; sepal horns often devoid of prickles. **FRUIT** capsule 35–50 mm long; elliptic; spines stout, widely spaced, spreading to recurved, the largest 8–10 mm long. **SEEDS** about 2 mm across. x = 14. [*A. alba* F. Lestib.]. Prairies, foothills, and mesas, on roadsides, sandy areas, and often disturbed sites. ARIZ: Apa; COLO: Dol, LPl, Mon; NMEX: SJn; UTAH. 1125–1900 m (3700–6000′). Flowering: Apr–Jun. Native to North America from the Midwest to the Great Basin. Work by Andrea E. Schwarzbach shows very little if any difference between *A. albiflora* and *A. polyanthemos*. She

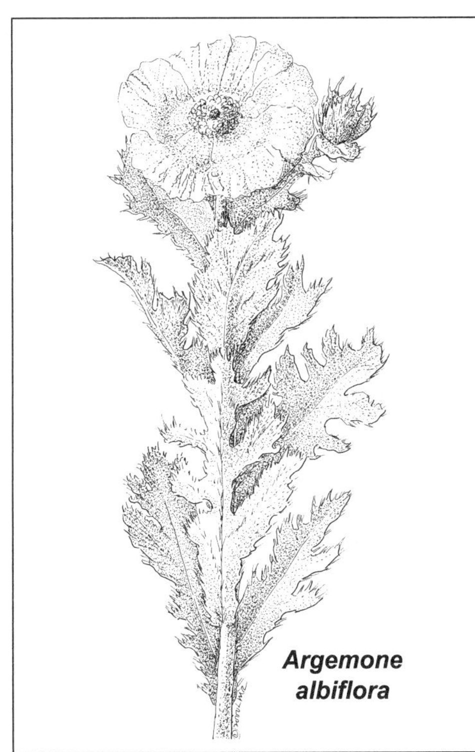

proposes that *A. polyanthemos* is a synonym for *A. albiflora*. Molecular data support this as well. Ownbey's keys in the *Flora of North America* provide for a wide range of overlap in characters between the two taxa and application of the keys to our material is not satisfactory. This taxon has also been called *A. intermedia*, but this name has been misapplied and is a nomen confusum. This name cannot be applied to any species of *Argemone* with assurance. Accurate identification of this species and *A. pleiacantha* is difficult. Of the two taxa, *A. albiflora* is more common in the Four Corners flora. Fresh material and field observations of overall shape in the early stages of flowering (i.e., some flowers, some buds, but no ripe fruits) and differences in leaf length throughout the plant will certainly aid identification. Dried specimens are problematic, as important characters such as the shape and the relative thickness of the sepal horns from the base of the horn to the tip of the horn can be lost in the pressing process. Four Corners material belongs to **subsp. *albiflora***.

Argemone corymbosa Greene (from the stem; meaning cluster of fruit or flowers) Mojave prickly poppy. Plants perennial herbs with orange latex (when fresh). **STEMS** 1–5; 2–6 dm tall; moderately branched. **LEAVES** glaucous or green; blades thick and leathery; elliptic, lanceolate to oblanceolate; lobes 1/2 to 4/5 the distance to the midrib; prickles along primary veins on both upper and lower surfaces. **FLOWERS** 4–8 cm in diameter; sepal horns 5–7 mm long, sometimes flattened in a tangential plane. **FRUIT** capsule 25–35 mm long; ovate to lanceolate; spines stout, equal-sized, speading, the largest 5–6 mm long. **SEEDS** bluish black, about 2 mm across. $x = 14$. [*Argemone corymbosa* var. *arenicola* (G. B. Ownbey) Shinners]. Dry desert valleys and plains in stabilized or actively moving sand dunes. UTAH. 1125–1500 m (3700–5000′). Flowering: May–Jun, also blooming in fall under favorable conditions. Endemic to Arizona and Utah. Apparently restricted to actively moving or stabilized rust-colored sand. This taxon has been called *A. intermedia* but this name has been misapplied and is a nomen confusum. This name cannot be applied to any species of *Argemone* with assurance. This taxon is more readily identifiable than the other two. The leaves consistently have stout, even-sized spines along the veins on both sides of the leaf surfaces, although more so beneath than above. Fresh material will aid in identification as this species has bright orange latex whereas the other two taxa have yellow latex. Our material belongs to **subsp. *arenicola*** G. B. Ownbey (living in sand).

Argemone pleiacantha Greene (with radiating spines) Southwestern prickly poppy, cowboy fried eggs. Plants annual or perennial herbs with bright yellow latex (when fresh). **STEMS** 1–several from the base; 5–15 dm tall; sparingly branched in the upper 1/2; shape during early stages of flowering triangular. **LEAVES** glaucous, succulent; oblanceolate to elliptic-lanceolate; lobes 4/5 the distance to the midrib; prickles on upper surface absent or sparingly armed along veins; leaf length during early stages of flowering longer in the lower leaves than the upper ones. **FLOWERS** 8–12 (16) cm in diameter; sepal horns 4–12 mm long; terete or flattened; more or less equal in diameter at the base and at the tip, distinctly narrowing from the main body of sepals; sepal horns with prickles. **FRUIT** capsule 25–45 mm long; ovate to elliptic-lanceolate; mostly closely prickly, surface partially obscured, spines straight or incurved; the largest 4–6 mm long. **SEEDS** 2–2.5 mm across. $2n = 28$. Prairies, foothills, and mesas, on roadsides, sandy areas, and often disturbed sites. ARIZ: Apa. 1525–2200 m (5000–7300′). Flowering: May–July. Native to the southwestern United States and northern Mexico. Our material belongs to **subsp. *pleiacantha***. *Argemone pleiacantha* subsp. *pleiacantha* can be confused with *A. albiflora* because of the smooth upper surface of the leaves. They can sometimes be distinguished by the sepal horns; *A. albiflora* has horns devoid of prickles whereas *A. pleiacantha* subsp. *pleiacantha* has sepal horns with prickles. But this character does not hold true for all specimens. The relative thickness of the sepal horns from the base of the horn to the tip of the horn is apparently a more consistent character and is best discerned when fresh. This species, like *A. albifora*, is difficult to identify. Fresh material and field observations of overall shape in the early stages of flowering (i.e., some flowers, some buds, but no ripe fruits) and differences in leaf length throughout the plant will certainly aid identification. Dried specimens are problematic, as important characters such as the shape and relative diameter of the sepal horns are lost in the pressing process.

Papaver L. Poppy, Pavot

(classical Latin for the poppy flower) Plants annual, biennial, or perennial herbs with taproots; sap white; scapose. **LEAVES** cauline, basal rosettes; subsessile, petiolate; blades lobed, entire, dentate, crenate, or incised; hispid with light brown trichomes. **INFL** cymose on long scapes or peduncles; bracteate, buds nodding. **FLOWER** sepals 2–3, distinct; petals 4–6; yellow; carpels 3–18 (22); locule 1, sometimes multilocular by placental intrusion; style absent; stigma 3–18 (22); stamens numerous. **FRUIT** capsule erect; 3–18 (22) valves; pubescent. **SEEDS** many; minutely pitted. Temperate and Arctic North America. Seventy to 160 species (one in our flora). The poppy is known for its alkaloid compounds, primarily opiates. Native North American poppies are scapose and occur primarily in Arctic and alpine regions.

Papaver radicatum Rottb. **CP** (from stem, meaning to take root) Alpine poppy, Arctic poppy. Plants perennial herbs with milky sap. **STEMS** loosely caespitose; 12–15 cm tall; persistent leaf bases. **LEAVES** gray to blue-green on both sides; 2–8 cm long; pinnately lobed into 5 unequal elliptic segments. **INFL** scapose; peduncles with short, light brown hairs. **FLOWER** petals 4 (6); yellow (pink-tinged or brick-red); 2 cm broad or less; stamens approximately 25; filiform; anthers yellow. **FRUIT** capsule ellipsoid-subglobose to oblong-obconic; strigose; hairs light brown. $2n = 42, 56$. [*P. kluanensis* D. Löve]. Dry, rocky, alpine or Arctic ridges. Known from a single collection on Mount Windom, San Juan Mountains, by C. W. Pendland in 1933. COLO: LPl. 4055 m (13300′). Flowering: Jun–Aug. Native to North America with disjunct occurrences from Alaska, across the Yukon, Alberta, British Columbia, Idaho, Montana, Wyoming, Utah, Colorado, and New Mexico (Murray 1995; Löve 1969). The San Juan Mountain material belongs to **subsp. *kluanense*** (D. Löve) D. F. Murray (for the Kluane Range, Yukon).

PARNASSIACEAE Martinov GRASS-OF-PARNASSUS FAMILY

Roy Taylor

(for Mt. Parnassus, Greece) Perennial herbs, glabrous, somewhat succulent. **LEAVES** mostly basal, petiolate, the blade simple, palmately veined; the cauline leaves (if any) reduced, sessile, often clasping, alternate; stipules lacking. **INFL** of solitary flowers. **FLOWERS** showy, regular, perfect; sepals 5, connate at the base; petals 5, sessile or clawed, the margin entire to fimbriate; androecium of 5 staminodia alternating with 5 functional stamens, alternate with the petals; gynoecium of (3) 4 carpels, the ovary 1-locular, superior or partly inferior; ovules numerous on each placenta. **FRUIT** a loculicidal capsule. **SEEDS** numerous, small, with a winglike testa (Cronquist, Holmgren, and Holmgren 1997a).

Parnassia L. Grass-of-Parnassus

Perennial, glabrous herb from short rootstock. **LEAVES**: basal leaves entire. **STEMS** with or without solitary, entire bract. **FLOWERS** solitary, terminal, perfect, regular; hypanthium calyxlike, sometimes nearly obsolete; sepals 5; petals 5, white; stamens 5, alternating with clusters of gland-tipped staminodia; ovary superior or slightly inferior, 1 locule, styles short or lacking; stigmas 3 or 4, sessile or subsessile. **FRUIT** a capsule, loculicidal. **SEEDS** numerous, cellular-reticulate.

1. Petals fringed or comblike on lower 1/2 .... ***P. fimbriata***
1′ Petals not fringed or comblike on lower 1/2, entire .... (2)

2. Flowering stems bractless or with near basal bract; petals 1–3 (5)-veined .... ***P. kotzebuei***
2′ Flowering stems usually with bract above level of basal leaves, petals usually 7–13-veined .... ***P. palustris***

Parnassia fimbriata K. D. Koenig (fringed) Fringed grass-of-Parnassus. Perennial, glabrous herb from short rootstocks with fibrous roots. **FLOWERING STEMS** 1–several, 15–30 cm tall; single bract cordate, clasping stem, 5–17 mm long, inserted at or considerably above middle of scape. **LEAVES** with petioles 2–16 cm long, blades reniform, orbicular, or broadly ovate, cordate or truncate at base; blade 12–45 mm wide and long. **FLOWERS** solitary, terminal; hypanthium nearly obsolete; sepals 4–7 mm long, 2–4 mm wide, sometimes fimbriate; petals 8–15 mm long, including clawlike base, strongly fimbriate below middle; staminodia thickened, scalelike, flared above middle. **FRUIT** a capsule to 1 cm long. [*P. rivularis* Osterh. ex Rydb.]. Montane to alpine, tundra; mixed forest, often along moist seeps, creeks. COLO: Arc, Hin, LPl, Min, Mon, RGr, SJn; UTAH. 2745–3500 m (9000–11485′). Flowering: Jul–Sep. Alaska, Yukon, and western Northwest Territories, east to Alberta, south to California and New Mexico.

Parnassia kotzebuei Cham. ex. Spreng. (for Otto von Kotzebue, 1787–1846, Russian navigator and commander of northern Pacific exploring expeditions) Perennial herb. **FLOWERING STEMS** mostly single, 2 or 3, rarely as many as 7; usually bractless, if present near base, bract ovate to lanceolate, nonclasping, to 15 mm long. **LEAVES** with petiole of basal leaves equal to or considerably longer than blade; ovate or deltoid-ovate to nearly elliptic, 5–20 mm long;

base of leaf from truncate to tapered. **FLOWERS** 1 per scape; calyx adnate to ovary for 1–4 mm; sepals to 7 mm long, usually 3-nerved; petals elliptic-lanceolate or elliptic-ovate, mostly equal to calyx lobes, 1–3-veined; filaments slender, equaling the sepals, much longer than the staminodia. **FRUIT** a capsule up to 1 cm long. COLO: SJn. 3655 m (12000′). Flowering: Jul–Aug. Alaska, Yukon, and Northwest Territories to Greenland and Labrador, south in the western mountains to Washington, Montana, Idaho, Wyoming, Colorado, and Nevada.

Parnassia palustris L. (marsh) Marsh parnassia. Perennial, glabrous herb. **FLOWERING STEMS** 1–several, erect, 8–44 cm tall, 1 sessile leaf 3–22 mm long, linear to ovate, sometimes clasping, borne mostly below middle of stem. **LEAVES** with petioles 0.7–4 cm long; blades 7–27 mm long, 5–20 mm wide, ovate to nearly orbicular; mostly cuneate or obtuse at base, rarely truncate. **FLOWERS** solitary, terminal; hypanthium about 2 mm long; sepals 3–10 mm long; petals 6–14 mm long, including narrow claw, entire, white; staminodia with thickened scalelike base, flared upward, divided into 5–11 slender filamentous segments, terminating in capitate knobs. **FRUIT** an ovoid capsule 8–12 mm long. [*P. parviflora* DC.; *P. montanensis* Fernald & Rydb.]. Montane to subalpine, tundra; mixed conifer/deciduous forest; along water, with forbs, grasses. ARIZ: Apa; COLO: Arc, Hin, LPl, SJn; NMEX: SJn. 2250–3070 m (7390–10075′). Flowering: Jul–Aug. Quebec south to Colorado, Nevada, and California.

PEDALIACEAE R. Br. SESAME FAMILY (includes MARTYNIACEAE)

Linda Mary Reeves

Herbs and a few shrubs, with stalked, broad-headed, glandular pubescence, often giving the plant surface a sticky appearance. **LEAVES** opposite, occasionally alternate, simple, entire to toothed or shallowly lobed. **INFL** a terminal or axillary raceme, cyme, or solitary. **FLOWERS** perfect; bracts often present beneath the calyx; sepals 4–5, often somewhat united; corolla zygomorphic, occasionally weakly spurred at base, lobed and often distinctly marked; stamens attached to the corolla, 1 or more of the stamens represented by a staminode; stigma bifid; ovary superior, bilocular or unilocular and secondarily quadrilocular. **FRUIT** a capsule, drupe, or nut, often with various hooks or prickles that aid in animal distribution. About 20 genera and 80 species, primarily dry and shore areas of the Old World tropics or subtropics. *Sesamum indicum*, native to tropical Asia, produces sesame seeds and oil. *Harpagophytum procumbens*, the grapple plant of South Africa, is also a member of this family.

Proboscidea Schmidel Unicorn Plant, Devil's Claw, Cinco Llagas

(Greek for proboscis-like fruit) Herbs with a pungent odor and sticky, mucilaginous secretions on all green parts. **STEMS** generally upright, but sometimes clambering and viny. **LEAVES** mostly opposite, glandular-pubescent; bract beneath petiole small, often deciduous. **INFL** in terminal racemes; pedicels lengthening and thickening as fruit develops on inflorescence. **FLOWERS** with calyx membranous, 5-lobed; corolla 5-lobed, weakly bilabiate; functional stamens 4, paired; ovary initially unilocular, becoming secondarily quadrilocular. **FRUIT** a capsule with initially fleshy, deciduous exocarp and a single stout, incurved beak, the endocarp becoming woody and dehiscent, ventrally crested, the beak splitting into 2 sharp, curved horns that become tangled in animal feet and hooves, allowing seeds to be dispersed from the opening between the horns. $x = 8, 13, 14, 15, 18$. Young fruits sometimes picked and eaten by many tribes. About 10 species, most in warm-temperate or subtropical North America.

1. Inflorescence with 8–20 flowers per raceme; flowers very inflated and open, 3.5–5.5 cm long and strongly colored .... ***P. louisiana***

1′ Inflorescence with up to 10 flowers per raceme; flowers 2.5–3.5 cm long, mostly dull or pale colored .... ***P. parviflora***

Proboscidea louisiana (Mill.) Thell. (from Louisiana, alluding to southern United States distribution) Unicorn plant, ram's horn. Weedy, clambering annual 20–25 cm tall. **INFL** a raceme with up to 20 or more flowers. **FLOWERS** conspicuous; calyx deciduous; corolla zygomorphic, inflated, open, 3.5–5.5 mm long, pale pink, yellow, and white, often with darker pink, bluish or purplish, and/or yellow markings. **FRUIT** often with "horns" 2 to 3 times longer than mature fruit body. [*Martynia louisiana* Mill.]. Fields, cattle feedlots or grazing areas, probably introduced with livestock feed. COLO: Mon; UTAH. 1700–2500 m (5500–8300′). Flowering: Jun–Sep. Fruit: Jul–Sep. Southern United States to California, south to Mexico and Central America.

Proboscidea parviflora (Wooton) Wooton & Standl. (small-flowered) Devil's claw, unicorn plant, doubleclaw. Taprooted annual, blooming when a young, upright plant, but continuing to bloom while growth continues as the plant becomes clambering and sprawling in habit, depending upon moisture available, up to 60 cm tall and 2 m wide. All green surfaces covered with viscid pubescence. **LEAVES** cordate to triangulate; entire to weakly toothed or lobed; to 20 cm long, about as wide as long; petioles as long as blade. **INFL** a raceme with up to 10 flowers, often much fewer. **FLOWERS** zygomorphic; calyx subtended by deciduous bracts; corolla 2.5–3.5 cm long, 1–1.5 cm wide, pale to dull yellowish, pinkish, or purplish, often with darker purplish splotch on upper lobe. **FRUIT** dark gray or black, 15–35 cm long, 1.5–2.5 cm wide, fimbriate-crested; "horns" quite variable in length, 2.5–25 cm. [*Martynia parviflora* Wooton]. Blackbrush, desert scrub, roadsides, and fields. ARIZ: Apa; NMEX: SJn. 1500–1900 m (5000–6300'). Flowering: Jun–Sep. Fruit: Jul–Oct. Utah to New Mexico, south to Sonora, Sinaloa, and Chihuahua. The "horns" of the dry, mature fruit easily catch in the feet of hoofed animals, which scatter the seeds most effectively by walking. Often grown as an ornamental or used in basketry, especially by Tohono O'odham (Papago) people of southern Arizona. The cultivars have exceptionally long, thin "claws" that produce a long, fine black fiber for basket weaving by Papagos. It has been suggested that the cultivar represents a different species or variety. Baskets made of this fiber are often waterproof and have been used to hold milk.

PHRYMACEAE Schauer MONKEY-FLOWER FAMILY (**SCROPHULARIACEAE** IN PART)

J. Mark Porter

Annual or perennial, mostly herbs (ours), some shrubs. **LEAVES** opposite, petiolate to sessile, simple, margins entire or toothed. **INFL** racemose, sometimes a spike, terminal or axillary. **FLOWER** calyx bilabiate, tubular, 5-toothed, 5-nerved or subplicate-ribbed; corolla more or less funnelform, zygomorphic, bilabiate, the upper lip 2-lobed, external in bud, the lower lip 3-lobed, the throat with 2 elevated ridges forming a palate, this sometimes partially or completely closing the orifice, usually bearded, rarely actinomorphic; stamens 4, anthers in 2 sets, subreniform, 2-celled, the cells confluent; ovary (1) 2-carpellate, locules 1 (pseudomonomerous) or 2; placentation basal or axial, style solitary, apical or lateral; stigma broadly 1- or 2-lobed, the lobes ligulate, receptive only on the adaxial surface, sensitive (thigmotropic), the lobes closing together upon contact. **FRUIT** a capsule, indehiscent and nutlike (achene), or (in ours) dehiscing loculicidally only at the apex, calyx persistent in fruit. **SEEDS** solitary or numerous, small, oblong to oval or fusiform, reticulate to almost smooth; endosperm present, cotyledons folded. $n = 7+$. Phrymaceae includes about 16 genera, with 234 species. The largest genus is *Mimulus*, and with nearly 170 species, it is the only representative of this family in our area. The family is nearly worldwide, with two centers of diversity, in North America and Australia; however, there are also representatives in the New World tropics, in Mexico and Central America. The family was traditionally included within a broad Scrophulariaceae (Olmstead et al. 2001; Oxelman et al. 2005; Beardsley and Olmstead 2002).

Mimulus L. Monkey-flower

(Latin, little clown or comic, referring to the facelike form of the corolla) Annual or perennial herbs, some rhizomatous or stoloniferous, the annuals often flowering from near the base; often glandular-pubescent to slightly viscid or pilose, less often glabrous. **LEAVES** opposite, petiolate or sessile, entire or toothed, sometimes laciniate. **INFL** consisting of flowers borne in pairs from the axils of opposite leaves, or less often a bracteate raceme; pedicels nonbracteolate. **FLOWER** calyx tubular to campanulate, often inflated, the tube strongly 5-angled (plicate), the lobes decidedly shorter than the tube, equal to unequal; corolla yellow, orange, red to purple, or bluish, campanulate, slightly to strongly bilabiate, the upper lip 2-lobed, erect to reflexed, external in bud, the lower lip 3-lobed, spreading or deflexed, the throat with 2 elevated, abaxial ridges forming a palate, this sometimes partially or completely closing the orifice, often bearded; stamens 4, in 2 sets, all with well-developed anther cells, the anthers 2-celled, divergent; stigmas 2, lamelliform, sensitive to touch, the spreading receptive stigmas closing to become appressed. **FRUIT** a loculicidal capsule, sometimes also splitting down the septum, membranous to coriaceous, glabrous (ours). **SEEDS** numerous, small, oblong to oval or fusiform, reticulate to almost smooth, the areolas arranged in longitudinal rows, mostly yellowish. $n = 7, 8, 14, 15, 16$. A large, complex genus of about 150–170 species, occurring mostly in western North America, particularly California; some species native to Chile, parts of eastern Asia, southern Africa, New Zealand, and Australia (Campbell 1950; Grant 1924; Pennell 1951).

1. Corolla 30–57 mm long, red or orange-red to pink, violet, or sometimes yellowish; calyx 16–30 mm long; the stems scandent to pendulous from ceiling of overhanging ledges .... ***M. eastwoodiae***

1' Corolla less than 30 mm long, or if up to 30 mm, then yellow or blue; calyx 3–16 (20) mm long; stems creeping-decumbent, ascending to erect, but not scandent to pendulous .... (2)

2. Fruiting calyx strongly inflated, the upper lobe much longer than the others ....(3)
2′ Fruiting calyx cylindrical to campanulate, not or only moderately inflated, the lobes more or less equal ....(5)

3. Corolla throat more or less open; lateral calyx teeth blunt and relatively short, sometimes nearly absent, not folded inward in fruit; corolla 7.5–20 mm long; rare in the region .... ***M. glabratus***
3′ Corolla throat closed by the well-developed palate; lateral calyx teeth more or less acute and tending to fold inward in fruit; corolla 9–30 mm long ....(4)

4. Stems usually several-flowered; corolla usually less than 20 mm long when few; plants often over 20 cm tall, annuals or perennials; low and middle elevations in the mountains and higher valleys; the most common species in the region .... ***M. guttatus***
4′ Stems only 1–3 (5)-flowered; corolla large, 17–30 mm long; plants low, 20 cm or less tall, rhizomatous or stoloniferous perennials; high elevations in the taller mountains of the region .... ***M. tilingii***

5. Leaves petiolate, petioles 1–12 mm long; calyx lobes triangular, acute to acuminate; leaves ovate to lanceolate, rounded to cordate at base .... ***M. floribundus***
5′ Leaves sessile, except sometimes the lowermost; calyx lobes rounded, often mucronate; leaves linear to lanceolate ....(6)

6. Calyx lobes usually ciliate; plants open, loose, simple or few-branched, the internodes usually longer than the leaves; pedicels 7–22 mm long, arched in fruit; corolla yellow or often pinkish to violet .... ***M. rubellus***
6′ Calyx lobes not ciliate; plants compressed, much-branched, the internodes usually shorter than the leaves; pedicels 2–7 (10) mm long, becoming sigmoid-curved in fruit; corolla usually yellow .... ***M. suksdorfii***

Mimulus eastwoodiae Rydb. **CP** (honoring Alice Eastwood, botanist, 1859–1953) Eastwood's monkey-flower. Herbaceous perennial, with stolons bearing small leaves less than 1 cm long, producing new fertile plants wherever stolons root. **STEMS** 5–30 cm long, scandent to pendent from ceilings of overhanging cliffs; glandular-puberulent, sometimes viscid-villous. **LEAVES** sessile, the lower 1.7–2.3 cm long, 10–20 mm wide, flabellate, the upper 2.5–4 (5) cm long, (5) 10–20 mm wide, obovate to oblanceolate, acute, upper and lower .coarsely toothed and palmately 3- to 5-veined. **FLOWER** calyx 16–23 (27) mm long, the lobes subequal, 4–7 mm long, lanceolate, acuminate, ciliate; pedicels 1–3 (4) cm long; corolla dropping soon after anthesis, 30–40 mm long, scarlet to orange-red, strongly bilabiate, the upper lip erect or arched, with nearly completely fused lobes, the lower lip of 3 subequal, spreading or reflexed, emarginate lobes, the tube 23–27 mm long; anthers exserted, villous on back, the cells 1–1.5 mm long. **FRUIT** a capsule, included in the calyx, 6–9 mm long, elliptic, acuminate. $2n = 16$. Moist areas, seeps, shaded places in crevices of overhanging sandstone cliffs. ARIZ: Apa, Nav; NMEX: SJn; UTAH. 1220–1980 m (4000–6500′). Flowering: May–Sep. Fruit: Jun–Sep. Endemic to the Canyonlands region of southeastern Utah and adjacent Colorado, Arizona, and New Mexico.

Mimulus floribundus Douglas ex Lindl. (profusely flowering) Floriferous monkey-flower, seep monkey-flower. Small annual, 3–22 (40) cm tall. **STEMS** erect to decumbent, simple or branched; glandular-pubescent, sometimes viscid. **LEAVES** petiolate, the blade (0.3) 0.8–2 (3) cm long, (1) 5–13 mm wide, ovate to lanceolate, rounded to cordate at the base, sparingly dentate, pinnately to subpalmately veined, thin, the petiole 1–12 mm long. **FLOWER** pedicels 5–14 (18) mm long; calyx cylindrical, 3.5–7 mm long at anthesis, to 9 mm in fruit, glandular-pubescent, at least on the ribs, the lobes subequal, 0.8–1.6 (2) mm long, triangular, acute or obtuse, ciliate; corolla dropping soon after anthesis, 7–14 mm long, yellow, tubular, slightly bilabiate, the lobes of the upper lip slightly longer, the palate pubescent, often with reddish spots; anthers included in the tube, glabrous, the cells about 0.3 mm long. **FRUIT** a capsule, included in the calyx, 3.5–5 mm long, obovoid to elliptic, acuminate. **SEEDS** 0.3–0.4 mm long. $2n = 32$. Vernally moist places, along streams, and hanging gardens in piñon-juniper woodlands, spruce-fir, and aspen communities. COLO: Hin, RGr. 1220–2600 m (4000–8500′). Flowering: late May–Jul. Fruit: Jun–Aug. Southern British Columbia to Montana and South Dakota, south to California, Arizona, western New Mexico, and Chihuahua.

Mimulus glabratus Kunth (made nearly glabrous) Smooth monkey-flower. Herbaceous perennial. **STEMS** 10–50 cm long, creeping, decumbent or more or less ascending, rooting at the lower nodes; glabrous to sparsely viscid-hispid or glandular-pubescent. **LEAVES** short-petiolate below, sessile above, the blade usually broader than long, 1.3–2.4 (2.8) cm long, 15–30 mm wide, broadly ovate to orbicular, rounded above, rounded to cordate below, dentate, undulate, or entire, palmately 3–5-veined. **FLOWER** pedicels (1) 1.5–3 (6) cm long; calyx campanulate, 5–11 (16) mm long, glabrous, sometimes splotched with red, the lobes unequal, the adaxial lobe more than twice as long as the others, blunt, the lateral lobes very short, sometimes obsolete, not folding inward in fruit; corolla dropping soon after anthesis, 7.5–20 mm long, yellow, distinctly bilabiate, the throat only partially closed by the palate, the palate bearded, often red-brown-spotted; anthers with cells about 0.6 mm long. **FRUIT** a capsule, included in the calyx, 5–9 mm long, broadly ovate, rounded. **SEEDS** 0.4–0.6 mm long. $2n$ = 28, 30, 60 (subsp. *fremontii*); $2n$ = 28, 30 (subsp. *utahensis*). Aquatic habitats, wet, often submerged soils of streams and springs; piñon-juniper, oak. Manitoba and Ontario south through North Dakota, Minnesota, Wisconsin, and Michigan to Mexico, extending west in Montana, Colorado, Utah, Nevada, and Arizona; also in Central America and South America. Pennell (1935) subdivided *M. glabratus* into four North American subspecies, two of which are in our region.

1. Pedicels 1–2 (3) cm long; calyx 5–10 mm long; leaves suborbicular .... **subsp. *fremontii***
1' Pedicels 2–6 cm long; calyx 7–16 mm long; leaves oval .... **subsp. *utahensis***

subsp. *fremontii* (Benth.) A. L. Grant (honoring John Charles Frémont, explorer, 1813–1890) Mud of streams, ditches, springs, and seeps; piñon-juniper, mountain brush, and aspen. ARIZ: Apa; COLO: Arc, Hin, LPl, Min, Mon, RGr; NMEX: McK, RAr, SJn; UTAH. 1220–2600 m (4000–8500'). Flowering: Jun–Aug. Fruit: Jul–Sep. Extending west from the Great Plains to Arizona and southwestern Utah.

subsp. *utahensis* Pennell (geographic reference to Utah) Aquatic habitats of streams, springs, or seeps; piñon-juniper, mountain brush. UTAH. 1675–1830 m (5500–6000'). Flowering: Jun–Aug. Fruit: Jul–Sep. *Mimulus glabratus* subsp. *utahensis* is easily distinguished from the other subspecies of *M. glabratus*, but it can easily be confused with *M. guttatus* DC., from which it can be distinguished by its short calyx lobes and its presumably open corolla throat.

Mimulus guttatus DC. (spotted) Common monkey-flower, yellow monkey-flower. Annual or perennial herb, rooting at the lower nodes, rarely rhizomatous or stoloniferous, 5–55 (90) cm tall. **STEMS** stout and fistulous or slender, erect or decumbent, simple or branched; glabrous to pubescent. **LEAVES** variable, petiolate below and often sessile above, the blade (0.5) 1.5–5.5 (10) cm long, (5) 10–40 (85) mm wide, obovate or broadly ovate to orbicular, rounded above, rounded or cordate below, coarsely and irregularly toothed, palmately or subpinnately 5–7 (9)-veined, the petiole 2–40 (55) mm long, sometimes with lobed margins (pinnatifid). **FLOWER** pedicels 1–3.5 (5.5) cm long; calyx campanulate, 6–16 mm long at anthesis, to 20 mm and much inflated in fruit, glabrous or puberulent, tinged or splotched with red, the lobes broadly triangular, the adaxial lobe 2–3 times longer than the others, lanceolate, obtuse, the lateral and abaxial pair of lobes usually folded over the orifice in fruit; corolla dropping soon after anthesis, 9–23 (30) mm long, yellow, distinctly bilabiate, the margins of the upper lip reflexed, the lobes of the lower lip longer and spreading, the palate raised and closing the throat, densely hairy and spotted with red; anther with cells 0.5–1 mm long. **FRUIT** a capsule, included in the calyx, (7) 9–12 mm long, oblong-obovate, rounded distally, narrowed to a stipitate base. **SEEDS** 0.4–0.5 mm long. $2n$ = 16, 26, 28, 30, 32, 48, 56. Springs, seeps, and streams, wet or marshlike areas; piñon-juniper, ponderosa pine, Douglas-fir, spruce-fir communities. ARIZ: Apa, Nav; COLO: Arc, Con, Dol, Hin, LPl, Min, Mon, RGr, SJn; NMEX: McK, RAr, SJn; UTAH. 1220–2400 (3000) m (4000–7900' [9000']). Flowering: Jun–Aug. Fruit: Jul–Sep. Alaska and Yukon, south throughout the western mountains to northern Mexico; adventive in New England and Europe. *Mimulus guttatus* is the most common species of monkey-flower in our region; it is an extremely complex species morphologically, with several recognized varieties. A careful study of this species is needed.

Mimulus rubellus A. Gray (reddish) Little red monkey-flower, reddish monkey-flower. Slender herbaceous annual, 1–22 cm tall. **STEMS** simple or loosely branched, the internodes relatively elongate; glandular-puberulent. **LEAVES** sessile, 0.3–1.5 (2.2) cm long, 2–4 (7) mm wide, narrowly lanceolate to linear, often connate at the base, entire to shallowly sinuate-denticulate, 1–3-veined. **FLOWER** pedicels 7–18 (22) mm long, becoming somewhat arched in fruit; calyx tubular, 4–7 mm long at anthesis, to 9 mm in fruit, the ribs reddish, the lobes subequal, 0.5–1 mm long, broadly ovate, rounded, ciliate; corolla dropping soon after anthesis, 6–8 (10) mm long, yellow with maroon dots, or the limb sometimes pinkish or purplish, the throat narrow, weakly bilabiate, the subequal lobes rounded to emarginate, the palate puberulent; anthers slightly exserted or in the throat, glabrous, the cells about 0.2 mm long. **FRUIT** a capsule, included in the calyx, 4–6.5 mm long, ovoid, acuminate, sometimes stipitate. **SEEDS** 0.4–0.5 mm long. Moist areas, often along intermittent streams. ARIZ: Apa, Coc, Nav; COLO: LPl, Mon; NMEX: McK, SJn; UTAH. 1500–2750 m (4900–9000'). Flowering: Apr–Jun. Fruit: May–Jun (Jul). From southeastern California across Nevada and southern Utah to southwestern Colorado, Arizona, New Mexico, western Texas, and northern Mexico.

Mimulus suksdorfii A. Gray (honoring Wilhelm N. Suksdorf, botanist, 1850–1932) Suksdorf's monkey-flower. Small, compressed, herbaceous annual, 1–7 (11) cm tall. **STEMS** much-branched; glandular-puberulent. **LEAVES** sessile,

0.5–1.6 (2) cm long, 1–3 (4) mm wide, lanceolate to linear, entire or nearly so, the lower leaves sometimes oblanceolate, subpetiolate, 1- to 3-veined. **FLOWER** pedicels 2–7 (10) mm long, becoming sigmoid-curved in fruit; calyx 3–5 mm long at anthesis, hardly expanding in fruit, glandular-pubescent, reddish, at least on the ribs, the lobes subequal, 0.2–0.7 mm long, broadly ovate, rounded, mucronate, the margins not ciliate; corolla dropping soon after anthesis, 4–6.5 mm long, yellow, tubular-funnelform, weakly bilabiate, the lobes subequal, emarginate, the palate puberulent, red-spotted; anthers slightly exserted or in the throat, glabrous, the cells 0.15–0.2 mm long. **FRUIT** a capsule, included to slightly exserted, (3.5) 4.5–5 mm long, ovoid, acuminate. **SEEDS** about 0.4 mm long. Moist to moderately dry, exposed places or streamsides in sagebrush, piñon-juniper woodland, and ponderosa pine communities. NMEX: SJn; UTAH. 1200–2440 m (4000–8000'). Flowering: May–Jun (Aug). Fruit: Jun–Aug. Southern Washington, eastern Oregon, and eastern California, east across Idaho, Nevada, and Utah to western Wyoming, northern Colorado, and northern Arizona.

Mimulus tilingii Regel (honoring Heinrich S. T. Tiling, botanist, 1818–1871) Subalpine monkey-flower, Tiling's monkey-flower. Low, herbaceous perennial, bearing stolons or rhizomes, 7–20 cm tall. **STEMS** often decumbent at the base, simple or branched below; glabrous to puberulent. **LEAVES** petiolate or the upper sometimes sessile, the blade 1–2.6 cm long, 5–15 (25) mm wide, ovate or broadly ovate to suborbicular, dentate, sometimes obscurely so, palmately 3- to 5-veined, the petiole 1–12 mm long. **FLOWER** pedicels 1–4 (6) cm long; calyx campanulate, 7–15 mm long at anthesis, to 20 mm and much inflated in fruit, pale yellow-green with red-brown spots, the lobes unequal, the adaxial lobe longer than the others, 3–7 mm long, triangular to lanceolate, obtuse, rounded, the lateral and abaxial pairs of lobes usually folded over the orifice in fruit; corolla dropping soon after anthesis, 17–30 mm long, yellow with pale red spots on the palate, broadly funnelform, distinctly bilabiate, the palate densely yellow-bearded, raised and nearly closing the throat; anther with cells 1–1.5 mm long. **FRUIT** a capsule, included in the calyx, 7–10 mm long, oblong or oval, rounded distally, narrowing to a stipitate base. **SEEDS** 0.5–0.6 mm long. $2n = 28, 30, 36, 48, 50, 56$. Wet places around springs, shores of ponds, stream banks, and mountain meadows; ponderosa pine, aspen, Douglas-fir, spruce-fir, and alpine communities. COLO: Arc, Con, Hin, LPl, Min, Mon, RGr, SJn; UTAH. 2300–4260 m (7600–14000'). Flowering: Jul–Aug. Fruit: late Jul–Sep. Southern British Columbia and southwestern Alberta south to California, across Nevada, northern Arizona, and northern New Mexico.

PLANTAGINACEAE Juss. SNAPDRAGON FAMILY or VERONICA FAMILY

(much of traditional Scrophulariaceae)

J. Mark Porter and Kenneth D. Heil

Annuals and perennials, herbs, vines (*Maurandya*), rooted aquatics, or terrestrial plants, some shrubs (not ours); taprooted, short-rhizomatous, or fibrous-rooted; caulescent or acaulescent. **LEAVES** opposite or alternate, sometimes more or less basal, exstipulate, simple to dissected, petiolate or sessile, sometimes with sheathing petioles. **INFL** a raceme, spike, head, panicle (*Veronica*), or in some the flower solitary in leaf axil; bracts foliaceous to reduced. **FLOWERS** perfect, rarely imperfect (*Callitriche*); calyx of 4 or 5 (as few as 1 or 2 in *Synthyris*, absent in *Callitriche*) distinct or united sepals, regular or irregular; corolla sympetalous (absent in *Callitriche*), slightly or usually evidently zygomorphic, some spurred or saccate at the base, 4- or 5-lobed, often 2-lipped, tubular, campanulate or sometimes rotate, rarely wanting; stamens epipetalous, 4 and in 2 sets, 4 fertile and 1 sterile (referred to as a staminode), or 2; anthers 2-celled or 1-celled (*Limosella*); ovary superior, 2-carpellate, 2-locular, rarely 1-locular in upper part (*Limosella*), the anatropous ovules on axile placentae; style solitary and entire or sometimes forked into 2 stigmatic lobes. **FRUIT** a capsule (ours), dehiscence septicidal, loculicidal, or both (4-valved), circumscissile or via pores or ruptures. **SEEDS** usually many and small, rarely few, wingless or winged, the embryo small in a well-developed endosperm. $n = 6$ or more. Plantaginaceae is composed of about 90 genera and 1700 species; all but *Plantago* and *Callitriche* were formerly included in Scrophulariaceae (Olmstead et al. 2001; Oxelman et al. 2005; Albach, Meudt, and Oxelman 2005).

1. Leaves arranged in whorls, linear; aquatic, stems wholly or partly submerged .... ***Hippuris***
1' Leaves not whorled, variously shaped; aquatic or terrestrial .... (2)

2. Flowers unisexual; staminate flowers with 1 stamen; small aquatic herbs .... ***Callitriche***
2' Flowers bisexual (perfect); flowers with 2 or more stamens, terrestrial or aquatic plants .... (3)

3. Corolla 5-lobed, upper corolla lip 2-lobed, sometimes only shallowly so, overlapping the lower lip in bud; capsule dehiscence usually septicidal, often secondarily loculicidal as well ....(4)
3' Corolla 4-lobed (occasionally fewer), upper corolla lip entire, overlapped by lower lip in bud or buds imbricate; capsule dehiscence loculicidal or circumscissile ....(11)

4. Style forked with 2 distinct stigmatic lobes, these often lamellate; leaves opposite ....(5)
4' Style entire with a single stigma, this sometimes slightly 2-lobed; leaves opposite or alternate ....(6)

5. Fertile stamens 4; leaves broadly obovate to rotund, entire, sessile; plants aquatic or subaquatic, floating or prostrate; corolla nearly regular, the lobes subequal and about as long as the tube ....***Bacopa***
5' Fertile stamens 2; leaves lanceolate to ovate; plants terrestrial, ascending to erect; corolla bilabiate, the lobes irregular and shorter than the tube ....***Gratiola***

6. Plants acaulescent, the leaves in basal tufts; corolla nearly regular; anthers 1-celled ....***Limosella***
6' Plants usually caulescent and leafy throughout; corolla bilabiate; anthers 2-celled ....(7)

7. Corolla tube neither gibbous nor spurred; staminode (a sterile stamen) present ....(8)
7' Corolla tube gibbous or spurred at the base; staminode absent or glandlike ....(9)

8. Leaf bases of opposing basal leaves connate, forming a scarious sheath; calyx funnelform to tubular, 5-lobed, lobes short and unequal ....***Chionophila***
8' Leaf bases sessile or petioled, rarely connate, but never producing a scarious sheath; calyx with 5 relatively long, subequal lobes (divided nearly to the base) ....***Penstemon***

9. Leaves all opposite; corolla pouched at the base on the adaxial side; anthers not coherent and the cells not divergent; capsule primarily septicidal ....***Collinsia***
9' Leaves, at least the upper, alternate; corolla gibbous to spurred at the base on the abaxial side; anthers coherent in pairs and the cells divergent; capsule dehiscing by loculicidal pores or irregular rupture ....(10)

10. Corolla gibbous at the base, blue, purple, violet, or red; stems twining ....***Maurandya***
10' Corolla spurred at the base, yellow; stems erect ....***Linaria***

11. Corolla imbricate in bud; stamens 4; stigma 2-lobed; fruit a circumscissile capsule ....***Plantago***
11' Upper corolla lobe overlapped by lower in bud; stamens 2; stigma capitate; fruit a loculicidal capsule ....(12)

12. Leaves wholly cauline, opposite (not to be confused with the alternate, leaflike floral bracts in some); corolla subrotate; inflorescence a terminal raceme or composed of lateral racemes ....***Veronica***
12' Leaves both basal and cauline, the basal well developed and the cauline reduced and bractlike, alternate; corolla more or less tubular; inflorescence a dense, cylindrical spike or spikelike raceme ....***Synthyris***

Bacopa Aubl. Water Hyssop

(aboriginal name, used by the natives of French Guiana) Disc water hyssop. Aquatic or subaquatic, caulescent perennial (sometimes annual) herbs. **STEMS** creeping or floating (in ours). **LEAVES** opposite, mostly sessile, ours broadly obovate to rotund, entire and palmately veined. **INFL** reduced, flowers borne singly or in pairs in the axils of the leaves; pedicels not bracteolate. **FLOWER** calyx of 5 dissimilar, nearly distinct lobes, accrescent, the adaxial lobe broadest and the adaxial lateral pair (innermost) narrowest; corolla tubular-campanulate, 5-lobed, weakly bilabiate, the upper lip shallowly 2-cleft, external in bud, the lower lip 3-lobed; stamens 4, 2 sets of different lengths, all fertile; anthers distinctly 2-celled, the cells more or less parallel; style forked, each fork terminating in a dilated stigmatic lobe. **FRUIT** a capsule, septicidal and loculicidal dehiscence (4-valved), membranous. **SEEDS** numerous, reticulate, the areolas usually arranged in longitudinal rows. A genus of 56–60 species, occurring mostly in tropical and warm-temperate regions of the world, centering in South America. Formerly included in Scrophulariaceae. One species occurring in our area (Barrett and Strother 1978; Pennell 1946).

Bacopa rotundifolia (Michx.) Wettst. (almost round-leaved) Aquatic or subaquatic herbs, rooting at the lower nodes, prostrate or floating. **STEMS** pilose to hispid when young. **LEAVES** 12–27 (40) mm long, (8) 12–23 (30) mm wide, obovate to suborbicular, sessile, entire, clasping, palmately veined. **INFL** reduced, flowers solitary or in pairs in leaf axils, the pedicels 6–17 (23) mm long, usually shorter than the subtending leaves, pilose to hispid when young. **FLOWER** calyx 3–4.5 mm long at anthesis, to 6 mm in fruit, the 3 outer segments ovate to subrotund and the 2 inner lanceolate; corolla 4.5–7 mm long, campanulate, the lobes subequal, weakly bilabiate, white with a yellow throat; stamens 4, the anther cells 0.6–0.7 mm long, versatile, dark, the filaments white. **FRUIT** a subglobose capsule, 3.5–5.5 mm long. **SEEDS** about 0.5 mm long, ellipsoid to cylindroid, reticulate. $2n = 56$. Rooted in bottom of shallow water of ponds and slow-moving streams and on muddy banks. COLO: LPl. 1370–2440 m (4500–8000'). Flowering: July–Sep. Fruit: Aug–Oct. Common from North Dakota, Minnesota, Illinois, and Indiana, south to Texas, Louisiana, and Mississippi and at scattered localities westward in central Montana, southwest Idaho, Colorado, and California.

Callitriche L. Water-starwort

C. Barre Hellquist

(*calli*, beautiful, + *trich*, hair) Annual or perennial, aquatic and terrestrial herbs. **LEAVES** opposite, lacking stipules, margin entire. **INFL** with staminate flowers 1–3 in foliar leaf axils. **FLOWERS** unisexual, perianth lacking, each flower subtended by a pair of bracteoles or lacking in some; styles 2, often longer then the ovary. **FRUIT** a schizocarp splitting into 4 achenelike mericarps; these 1-seeded, flattened, winged, margined, or smooth (Fassett 1951; Philbrick 1989). Around 40 species. *Callitriche hermaphroditica* L. is to be looked for in our flora.

1. Fruit as long as broad, rounded at base and on edges .... ***C. heterophylla***

1' Fruit longer than broad, narrowed to base, sharply keeled or narrowly winged above .... ***C. verna***

Callitriche heterophylla Pursh (diverse-leaved) Perennial, aquatic herbs with compressed filiform stems. **LEAVES** submersed and floating; submersed leaves linear, notched at apex, 1-veined, and a rosette of floating obovate leaves, 3–5-veined; plants with linear to obovate to oblong leaves. **FRUIT** sessile; 0.6–1.2 mm long, the height approximately equaling the width; carpel more broadly rounded at the apex than at the base, therefore the fruit is slightly heart-shaped, thickest just above the base; margins of carpel wingless or with a narrow wing to apex; styles 1–6 mm long, persistent or falling off early. COLO: SJn; NMEX: McK. 2425–2750 m (8800–9000'). Flowering: Jul–Aug. Fruit: Jul–Aug. British Columbia east across southern Canada to Newfoundland, south to California, New Mexico, Texas, and Florida; Mexico.

Callitriche verna L. (spring) Perennial, aquatic herbs. **LEAVES** submersed and floating; lower submersed leaves linear, clasping at base, 1-veined with a shallow notch at tip; upper submersed leaves spatulate and petioled, 3-veined, 1–3 (–5) cm long; floating leaves forming a rosette, petioles broad, blades ovate to orbicular, 3–5-veined, rounded or notched at apex; bracteoles conspicuous, linear, oblong, or obliquely oval. **FLOWERS**: staminate with filaments 1–3 mm long. **FRUIT** sessile, suborbicular, 0.6–1.4 mm wide, with height always exceeding width (drying may conceal this trait); face of mericarp reticulate, the reticulations appearing in vertical rows; margin of carpel with a scarious wing, widest at the summit. [*C. palustris* L.; *C. palustris* var. *bolanderi* Jeps.; *C. palustris* var. *stenocarpa* Jeps.]. Still waters of lakes, ponds, streams, and ditches, often terrestrial along pond margins. ARIZ: Apa; COLO: Arc, Con, Hin, Mon, SJn; NMEX: RAr, SJn. 2425–3580 m (8000–11825'). Flowering: Jul–Aug. Fruit: Jul–Aug. Alaska east to Newfoundland and Greenland, south to California, New Mexico, Nebraska, Minnesota, Illinois, West Virginia, and New England.

Chionophila Benth. Snow-lover

(Greek, snow-loving) Perennial herbs, from shortly rhizomatous and slightly tuberous roots. **STEMS** decumbent to erect, unbranched. **LEAVES** opposite, mostly basal with a few reduced cauline. **INFL** a raceme, sometimes dense and spikelike, few- to many-flowered; bracts reduced and imbricate. **FLOWER** calyx funnelform to tubular, membranous in texture, lobes 5, triangular, scarious to chartaceous in age (in ours) or lobes divided nearly to the base and glandular; corolla cream-colored or dull greenish white, tubular, throat slightly dilated, bilabiate, upper lip obscurely bilobed, lower 3-lobed, the upper lobes external to the lower in bud; fertile stamens 4, in 2 sets of 2; anthers 2-celled, cells divaricate and confluent; staminode filamentous, glabrous; stigmas united and minutely capitate. **FRUIT** a loculicidal capsule, splitting septicidally at the apex as well, asymmetrical, oblongoid, glabrous. **SEEDS** numerous, seed coat loose, areolate-reticulate. A small genus of two species, restricted to the mountains of the western United States, formerly included in Scrophulariaceae.

Chionophila jamesii Benth. **CP** (honoring Edwin James, naturalist, 1797–1861) Rocky Mountain snow-lover. Perennial herbs, from short rhizomes, 3.5–15 cm tall, clothed in marcescent leaf bases (sheaths). **STEMS** decumbent to erect; minutely puberulent. **LEAVES** opposite, basal with a few reduced cauline leaves, the basal entire, blade 1–7.3 cm long, 1.5–11.5 mm wide, spatulate, oblanceolate to lanceolate, thick, leaf bases of opposite leaves connate, forming a scarious sheath, 0.7–2.8 cm long; the cauline reduced, lanceolate to linear, 1.2–3.5 cm long, 1.4–4 mm wide, 1–2 pairs, rarely alternate; leaves often turning dark upon drying. **INFL** a spikelike raceme, dense, mostly secund; bracts opposite and connate but imbricate, reduced, lanceolate, 7–10 mm long, 2–4 mm wide, glabrescent, somewhat ciliate. **FLOWER** calyx funnelform to tubular, somewhat inflated, 6–12 mm long (to 14 mm in fruit), more or less oblique, 5-lobed, lobes 1.5–4.5 mm long, triangular-acute, becoming scarious or chartaceous; corolla cream-colored or dull greenish white, sometimes suffused with brown or purple (often drying dark), tubular, throat slightly dilated, bilabiate, upper lip erect and slightly concave, 2.5–4.5 mm long, bilobed, the notch about 1 mm deep, lower lip convex, 5–6 mm long, densely bearded, 3-lobed, the lobes recurved, 1–2.5 mm long, the upper lobes external to the lower in bud; fertile stamens 4, in 2 sets, the adaxial pair diverging from the corolla near the base, the abaxial pair diverging 4.5–6 mm above the base; anthers 2-celled, 1–2 mm long, black, cells divaricate and confluent, glabrous; staminode short, 4.8–6.5 mm long, affixed 2–3 mm above the base of the corolla tube, glabrous; stigmas united and minute, style 8–11 mm long, ovary 2–2.5 mm long, 0.5–1 mm wide, glabrous. **FRUIT** a loculicidal capsule, splitting septicidally at the apex as well, 6–7.5 mm long, 3–4.5 mm wide, asymmetrical, oblongoid, acute, glabrous. **SEEDS** numerous, 1–2.2 mm long, 0.8–1.5 mm wide, oblongoid, angular, sometimes flattened with a marginal wing, seed coat loose, areolate-reticulate, dark brown with a metallic sheen. Slopes in talus, scree, or clay; alpine tundra. COLO: Con, Hin, LPl, Min, Mon, RGr, SJn. 3450–3810 m (11000–13100′). Flowering: Jul–Aug. Fruit: late Jul–Sep. Restricted to Colorado and southern Wyoming.

Collinsia Nutt. Blue-eyed Mary

(honoring Zacchaeus Collins, botanist, 1764–1831) Annual herbs. **STEMS** erect, simple or branching. **LEAVES** opposite (or whorled), entire, crenulate, or toothed, the upper sessile or clasping. **INFL** a raceme, the flowers in pairs or fascicles in the axils of opposite foliaceous bracts. **FLOWER** calyx campanulate, irregularly 5-lobed, united below; corolla somewhat pealike, strongly bilabiate, the upper lip 2-lobed, erect, external in bud, the lower lip 3-lobed, extended horizontally, the middle lobe forming a pouch that encloses the style and stamens, shorter than the lateral lobes, the throat often occluded by a fold or palate; stamens 4, of 2 lengths, the anther cells divaricate, confluent at the tip, the lower pair inserted higher in the corolla tube than the upper pair; staminode present, but much reduced, glandlike; stigmas united and capitate or slightly 2-lobed. **FRUIT** a capsule, septicidal and loculicidal dehiscence, 4-valved, thin-walled. **SEEDS** 1–many in each cell, flattened to concavo-convex, smooth to reticulate, more or less winged. $2n = 14$. A genus of 18–20 species of North America, largely Californian. Formerly included in Scrophulariaceae (Newsom 1929).

Collinsia parviflora Douglas ex Lindl. (little flower) Small-flowered blue-eyed Mary. Annual from a taproot, 3–20 (50) cm tall. **STEMS** decumbent to ascending or subscandent, simple or branched; puberulent, often glandular above and glabrous below. **LEAVES** connected by a line around the stem, sessile or narrowed to a petiolate base, the lower leaves small, orbicular, petiolate, the upper sometimes whorled; blade entire, 1.5–3 (5) cm long, 3–5 (12) mm wide, oblanceolate, obtuse to rounded. **INFL** a raceme with foliaceous bracts, appearing as solitary flowers in axils of leaves, sometimes in whorls in the upper axils; pedicels 3–30 mm long, becoming reflexed from the base. **FLOWER** calyx 3.5–5.5 mm long, to 7 mm in fruit, glabrous, the teeth 2–3 mm long, lanceolate, acuminate, the sinuses rounded; corolla blue with a white upper lip, 4–5.5 (7) mm long, the lobes of the upper lip erect, the lobes of the lower lip projecting, the tube gibbous on upper side, near the base; anthers 0.3–0.4 mm long. **FRUIT** a capsule, 4–5.5 mm long, oblong, ovoid to ellipsoid. **SEEDS** 1.5–2.2 mm long, ovoid or elliptic, grooved on one side, smooth, brown, 2 per locule. $2n = 14$, 28. Moderately moist slopes of woodlands, forests, and meadows; sagebrush, piñon-juniper, oak, ponderosa pine, aspen, and spruce communities. ARIZ: Apa; COLO: Arc, Dol, Hin, LPl, Mon, SJn, SMg; NMEX: McK, RAr, San, SJn; UTAH. 1980–2740 m (6500–9300′). Flowering: Apr–Jun. Fruit: May–Jul. Southern Alaska to eastern California, eastern Washington, eastern Oregon, Nevada, and northern Arizona to Ontario, Canada; Michigan, South Dakota, Colorado, and New Mexico. Ground and combined with *Sisymbrium*, the smoke was given to a horse to increase running speed (Kayenta Navajo).

Gratiola L. Hedge Hyssop

(Latin: grace or favor, referring to its supposed medicinal properties) Annual, biennial, or perennial caulescent herbs. **STEMS** low, erect or diffuse. **LEAVES** opposite, lanceolate to ovate, sessile, entire or toothed, palmately veined. **INFL** reduced, flowers borne singly or in pairs in the axils of the leaves; pedicels usually bearing (0–1) 2 bractlets just below the calyx. **FLOWER** calyx of 5 distinct or somewhat united lobes, subequal to unequal; corolla tubular or narrowly campanulate, the tube quadrangular, bilabiate, the upper lip shallowly 2-lobed, flat, external in bud, the lower lip 3-lobed; fertile stamens 2, included, the anther cells parallel on a membranous expansion of the connective, the abaxial pair of stamens wanting or represented by vestigial filaments near the base of the corolla tube; style divided into 2 lamelliform stigmas. **FRUIT** a capsule, septicidal and generally also loculicidal dehiscence (4-valved), membranous.

SEEDS numerous, reticulate, the areolas arranged in longitudinal rows. $2n = 16$. A genus of about 20 species, occurring in the temperate regions and in the mountains of tropical regions in both hemispheres, formerly included in Scrophulariaceae.

Gratiola neglecta Torr. (overlooked, neglected) Common American hedge hyssop. Small annual, 5–22 (30) cm tall. **STEMS** ascending to decumbent, simple or branched, fistulose below; glandular-puberulent. **LEAVES** 1–3.5 (5) cm long, oblanceolate to obovate, acute, denticulate in upper 1/2 or sometimes entire, slightly clasping at base. **INFL** reduced, solitary flowers in axils of one or both opposite leaves, glandular-pubescent, the pedicels 8–22 mm long, 2 bracteoles, the bractlets similar to the calyx segments or longer. **FLOWER** calyx 3.2–5.5 mm long at anthesis, to 7 mm in fruit, the lobes lanceolate, acute to obtuse, rarely rounded, united at the very base, subequal; corolla 7–10 (12) mm long, the 2 lobes of the upper lip nearly completely united, the 3 lobes of the lower lip broad and emarginate, pilose within the tube, especially on the adaxial side, the tube yellow, the limb white to pale lavender; stamens included, the anthers 0.7–1 mm long. **FRUIT** a capsule, 3.5–5.5 (6.5) mm long, globose-ovoid, acuminate. **SEEDS** 0.5–0.6 mm long, oblong-ovoid, reticulate, light brown. $2n = 16$. Moist places, often in mud at the edge of lake shores, ponds, streams, and drying springs. COLO: Arc. 2130–2440 m (7000–8000′). Flowering: Jun–Aug. Fruit: Jul–Sep. Southern British Columbia to Quebec, Canada; south to east-central California, northern Arizona, northern New Mexico, eastern Texas, Louisiana, Mississippi, northern Alabama, and northern Georgia.

Hippuris L. Mare's-tail

C. Barre Hellquist

(Greek: of a horse) Perennial aquatic or amphibious herbs. **LEAVES** whorled, sessile, blade margin entire. **INFL** forming a long terminal spike along middle to upper stem. **FLOWERS** minute, pistillate or perfect, epigynous; hypanthium enclosing the ovary and bearing perfect flowers at the tip; perianth lacking; ovary 1-celled and 1-ovuled. **FRUIT** nutlike, hard, indehiscent, 1-seeded (McCully and Dale 1961a, 1961b). Three species worldwide.

Hippuris vulgaris L. (common) Mare's-tail. Rhizomatous aquatic herbs. **STEMS** submersed and flaccid, or erect, hollow, simple. **LEAVES** 6–12 in a whorl, linear-attenuate, firm and thick, or flaccid. **INFL** in axils of middle to upper portions of stem. **FLOWERS** with petals absent; stamen 1, inserted on anterior edge of calyx; ovary inferior. **FRUIT** 2–3 mm long, 1-celled and 1-seeded. Ponds, lakes, and streams. ARIZ: Apa; COLO: Arc, Hin, LPl; NMEX: McK, RAr, SJn. 2200–2750 m (7300–9100′). Flowering: Jun–Jul. Fruit: Jul–Aug. Alaska east to Newfoundland, south to Arizona, Nebraska, Minnesota, Illinois, New York, New Hampshire, and Maine. This is a fairly common plant of high-elevation ponds.

Limosella L. Mudwort

(Latin for mud) Small, caespitose, scapose annual or perennial herbs, usually stoloniferous, often submerged; glabrous. **LEAVES** in a basal tuft, entire, mostly long-petiolate, palmately veined. **INFL** long, slender pedicels arising from the tuft of leaves; pedicels nonbracteolate. **FLOWER** calyx with 5 equal lobes, the tube as long as or longer than the lobes; corolla rotate-campanulate, white or pinkish, essentially regular, the 5 spreading lobes shorter than the tube, the upper 2 lobes external in bud; stamens 4, subequal, the anthers 1-celled; stigmas united and capitate. **FRUIT** a septicidal capsule, membranous, distally 1-celled (the septum not extending to the top). **SEEDS** numerous, reticulate, the areolas arranged in longitudinal rows. $2n = 20$. A genus of about 12 species, occurring throughout the world. Formerly included in Scrophulariaceae (Glück 1934).

Limosella aquatica L. (of water) Small, caespitose, stoloniferous perennial, often submersed with floating leaves. **STEMS** very short. **LEAVES** slender, long-petiolate, some petioles as long as the water is deep and the blades floating, the blade 1–3 cm long, 3–12 mm wide, narrow-spatulate to broadly oblong-elliptic, rounded, 3–5-veined, the petiole 3–10 (20) cm long, wider at the base. **INFL** of many elongate pedicels arising from the axils of the tufted leaves; pedicels usually shorter than the leaves. **FLOWER** calyx campanulate, 2–2.5 mm long, pale green with purplish spots beneath the sinuses, the lobes equal, 0.5–0.8 mm long, broadly triangular, apiculate; corolla slightly longer than the calyx, more or less regular, white to pink, lobes equal, 0.8–1.2 mm long, oblong, acute, scarcely pilose within; stamens 4, the filaments flattened, the anthers circular, 0.3 mm across, often purple; stigma capitate. **FRUIT** a capsule, 3–3.5 mm long, ovoid-spherical, membranous. **SEEDS** 0.6–0.7 mm long, dark brown. $2n = 36, 40$. Mud flats and shallow water; drying borders of ponds and springs. ARIZ: Apa; COLO: Arc, Hin, Min, LPl, Mon; NMEX: McK, SJn. 2130–2740 m (7000–9000′). Flowering: late Jun–Aug. Fruit: Jul–Sep. Widespread over North America, Andean South America, and mountainous regions of the Old World.

Linaria Mill. Toadflax

(flax, referring to similarity to flax) Annual or perennial herbs, glabrous. **STEMS** erect, simple or branched at the base. **LEAVES** opposite or whorled below, alternate above, sessile, entire to dentate or lobed, pinnately veined. **INFL** a raceme or spike. **FLOWER** calyx 5-lobed, lobes mostly distinct, subequal; corolla 5-lobed, 2-lipped, yellow to blue, violet, or white, the tube spurred at the base, strongly bilabiate, the upper lip external in bud, the lower with reflexed lobes, the throat nearly closed; functional stamens 4, in 2 pairs, the anthers distinct, glabrous, 2-celled; stigmas united, small, capitate. **FRUIT** a symmetrical capsule, dehiscent by transverse loculicidal slits, ovoid to cylindrical or subglobose, thin-walled. **SEEDS** many, flat and winged, or angled, sometimes tuberculate. $2n = 12$. A genus of more than 100 species, mostly in Eurasia, only one species native to North America, though many are cultivated. Formerly included in Scrophulariaceae (Alex 1962).

1. Leaves lanceolate to ovate, 10–35 mm wide; calyx 5–9 mm long .... ***L. dalmatica***
1' Leaves linear, 2–6 mm wide; calyx 2.5–3.2 mm long.. ***L. vulgaris***

Linaria dalmatica (L.) Mill. (Latinized geographic reference to Dalmatia) Broom-leaved toadflax, Dalmatian toadflax. Perennial herb, 40–70 (100) cm tall. **STEMS** erect, branched above, spreading by horizontal rootstocks; glabrous and glaucous. **LEAVES** 2–5 cm long, 10–16 (35) mm wide, ovate to lance-ovate, acute, sessile, clasping, palmately veined, rigid. **INFL** an elongate raceme, the pedicels 2–4 mm long at anthesis, to 6 mm in fruit. **FLOWER** calyx lobes subequal, 5–7.5 (9) mm long, broadly lanceolate to ovate, acute, rigid; corolla 14–24 mm long, excluding the 9–17 mm spur, strongly bilabiate, the upper lip (7) 10–15 mm long, the lower lip 5–11 mm long, palate occluding the throat, bright yellow, glandular-pubescent on the sides, the palate densely white- to orange-bearded; anthers 1.2–1.3 mm across after dehiscence. **FRUIT** a capsule 6–7 (8) mm long, subglobose, dehiscing by irregular splitting at the apex of each carpel. **SEEDS** 1.2–2 mm long, tetragonal to discoid-compressed, rugulose, the angles winged. $2n = 12$. [*L. genistifolia* (L.) Mill. subsp. *dalmatica* (L.) Maire & Petitm.]. Waste places and roadsides; sagebrush, oak, ponderosa pine, aspen, and spruce-fir communities. ARIZ: Apa; COLO: Arc, Dol, LPl, Mon; NMEX: McK, SJn; UTAH. 2000–2800 m (6600–9200′). Flowering: May–Sep. Fruit: late Jul–Sep. A native of the Mediterranean region.

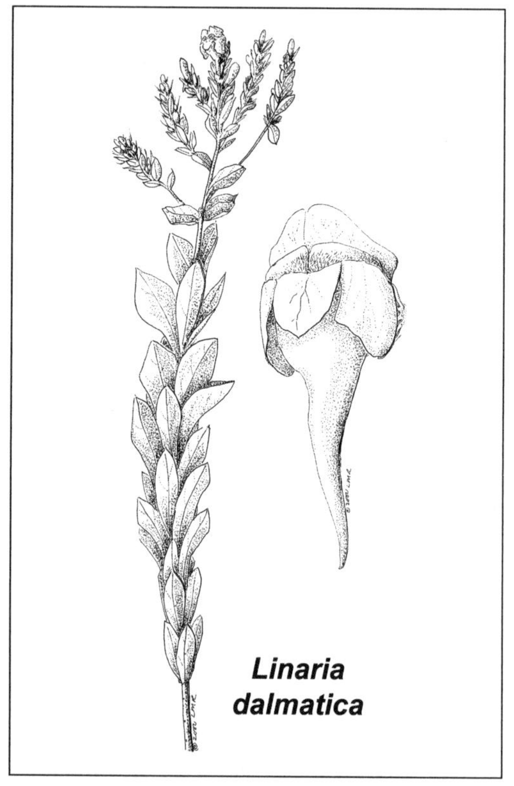
Linaria dalmatica

Linaria vulgaris Mill. (common, ordinary) Butter-and-eggs. Perennial herb, 30–60 (80) cm tall; taprooted. **STEMS** erect to ascending, 1–several, simple to branched, glabrous to sparsely pilose. **LEAVES** 2.5–4 (5) cm long, 2–6 mm wide, linear. **INFL** a congested raceme, glabrous to glandular-puberulent, pedicels 1–4 mm long. **FLOWER** calyx lobes subequal, 2.5–3.2 mm long, lanceolate to broadly lanceolate; corolla 10–14 (18) mm long excluding the nearly straight 8–14 mm long spur, strongly bilabiate, the upper lip 8–12 mm long, the lower lip 6–9 mm long, palate occluding the throat, bright yellow, the crest of the palate orange-bearded; anthers 0.8–1.2 mm across after anthesis. **FRUIT** a capsule, 5–9 mm long, subglobose, dehiscing by irregular splitting at the apex of each carpel. **SEEDS** 1.8–2.3 mm long, discoid with broad wings, the central portion tuberculate. $2n = 12$. Waste places, roadsides, stream edges, and meadows; sagebrush, oak, ponderosa pine, aspen, and spruce-fir communities. COLO: Arc, Dol, Hin, LPl, Min, Mon, SJn; NMEX: RAr, SJn. 1700–3200 m (5600–10500′). Flowering: May–Sep. Fruit: Jul–Sep. A native of Eurasia and widely naturalized in temperate North America.

Maurandya Ortega Twining Snapdragon

(honoring Catherine P. Maurandy, botanist) Climbing or prostrate perennial herbs. **STEMS** twining. **LEAVES** alternate to subopposite, petioles twining, coarsely toothed or hastate, palmately veined and lobed. **INFL** a raceme or of solitary flowers (in ours) in the axils of leaves. **FLOWER** calyx of 5 distinct, subequal segments; corolla bilabiate, lower lip forming a gibbous or saccate corona, with 2 prominent abaxial ridges, the upper lip external in bud, the throat closed (in ours) or open; fertile stamens 4, filaments with 2 rows of tack-shaped glands; stigmas united, small, subcapitate. **FRUIT** a capsule, globose, symmetrical to oblique, 1–2 locules, dehiscence by irregular rupturing near the apex. **SEEDS** oblong, pitted, winged or more often (in ours) tuberculate. $2n = 24$. A genus of about 20 species, occurring in the southwestern United States and Mexico. Formerly included in Scrophulariaceae (Pennell 1935; Munz 1926; Elisens 1985).

Maurandya wislizeni Engelm. ex A. Gray (for Friedrich Adolph Wislizenus, 1810–1889, a St. Louis physician who collected plants in the West) Viny snapdragon. Perennial vine. **STEMS** twining; glabrous; woody-based; 30–80 dm long, slender, usually much-branched, especially at the base. **LEAVES** 2–6 cm, usually longer than wide, hastate or triangular-hastate, the petioles flexuous, about as long as the leaves. **INFL** reduced to solitary flowers in the axils of the leaves; the pedicels 1–4 cm long. **FLOWER** calyx 5-lobed, strongly expanded in fruit, becoming strongly carinate and reticulate, closely enclosing the capsule; corolla 20–30 mm long, 2-lipped, usually blue, purple, or nearly white, slightly gibbous or saccate at the base; stamens 4, included, the filaments glandular or puberulent. **FRUIT** a capsule, 6–8 mm long, globose, subsymmetrical, dehiscing by 2 slits. **SEEDS** 0.8–1.5 mm long, dark brown, narrowly winged, corky-tuberculate. $2n = 24$. Sandy soils; desert grassland. NMEX: SJn. 1520–1800 m (5000–5900′). Flowering: late May–Jul. Fruit: Jun–Sep.

Penstemon Mitch. Beardtongue, Penstemon
David L. Bleakly

(Greek *pente*, five, + *stemon*, thread; probably referring to the distinctive fifth "stamen" or staminode; in the 19th and early 20th centuries, the genus was usually spelled *Pentstemon*, or rarely *Pentastemon*) Glabrous (sometimes glaucous), pubescent, or glandular-pubescent perennial herbs, or rarely shrubs with a taproot. **STEMS** decumbent or erect, solitary or several, from a herbaceous or woody caudex. **LEAVES** opposite, simple, entire or rarely serrate, linear to much broader, the lower usually petiolate, the upper typically sessile, reduced and bractlike or leaflike in inflorescence, rarely connate-perfoliate, often with a basal rosette. **INFL** axillary; in verticillate, cymose, or thyrsoid panicles (sometimes called simply cymes or thyrses), grouped at each node in verticillasters; sometimes secund. **FLOWERS** complete, generally unscented (scented in *P. palmeri*); sepals 5, fused at base, lobes equal, sometimes growing longer in fruit, entire to erose and/or scarious; corolla sympetalous, zygomorphic (only slightly so in *P. ambiguus*), bilabiate, the upper lip 2-lobed and the lower 3-lobed, less often nearly regular; tubular to ampliate (inflated); usually some shade of blue, less commonly purple, red, pink, or white, proximal tubular part, throat, and lobes often different colors; throat often with nectar guides (lines in contrasting colors), sometimes plicate; palate bearded or not; fertile stamens 4; distinct; included or exserted; epipetalous, inserted near base of corolla, filaments typically curving toward upper side of the flower; anthers divided into 2 sacs, these usually dehiscing the entire length of cells and often opening completely (explanate), but sometimes opening only at bases (and across connective) or tips, usually with minute teeth on sutures, sacs mostly opposite or divaricate (in line), but rarely the pair curved into a horseshoe shape; surface glabrous or pubescent; pistil stigma fused, capitate; style slender, elongate, curved toward top of flower. **STAMINODE** (sterile stamen) 1; conspicuous; epipetalous, inserted on upper surface of corolla near level of top of ovary and typically curving so that apex lies at base of lower lobes; usually as long as the fertile stamens, included or exserted, often dilated and bearded at apex (hence the origin of the common name). **FRUIT** a capsule with 2 locules, septicidal, often loculicidal at apex, fruiting immediately after flowering. **SEEDS** numerous, irregularly angled. A genus of about 250–270 species, primarily in western North America, centered in Utah, but ranging from Alaska to the eastern United States and south to Central America (one species in northeastern Asia). It is one of the largest genera in the Four Corners region and the largest genus of the family in North America. Penstemons are very popular garden plants because of their beautiful but predominantly odorless flowers. This is a complex genus with many closely related and interfertile species, although natural hybrids are apparently uncommon. Species with sucrose-rich nectars and usually tubular flowers tend to be hummingbird-pollinated, while insects normally pollinate flowers that are more open and hexose-rich. Plants may usually be identified by careful attention to small but mostly well-marked differences visible on mature plants. Vestiture or lack of it, glaucescence, leaf size and shape, architecture of the stems and inflorescence (arrangement of parts), and characteristics of the calyx, corolla, stamens, and staminodes are all of critical importance for correct identification of specimens. Several features that are necessary for keying are often difficult to determine on pressed plants and should be noted in the field, including glaucescence, fresh flower color, presence of folds or ridges in the throat of the corolla, and whether the inflorescence is secund or not. Pistils, fruits, and seeds are of little value for identification. Many species are narrowly endemic, often to specific substrates, which should also be noted in the field. Measurements in the key and descriptions are from pressed specimens with mature flowers and from naturally open, mature anther sacs (Nisbet and Jackson 1960).

1. Plants shrubby, stem woody well above base, glabrous; leaves linear and greater than 35 mm long; corolla pink externally, 15–24 mm long, white on the face of the lobes; upper lobes reflexed, lower lobes projecting, tube narrow and curved; sandy soils .... ***P. ambiguus***

1′ Plants herbaceous (stem bases of *P. crandallii* and *P. linarioides* often somewhat woody); leaves linear and longer or broader; corollas red, pink, bluish, or purplish (occasionally white) .... (2)

2. Corollas red or pink .... (3)

2′ Corollas blue, bluish, purple, purplish, rarely pink or pinkish or white .... (8)

3. Corollas definitely red; leaves never connate-perfoliate or serrate; anthers explanate or not .... (4)

3′ Corollas pink to rose, glandular; upper stem leaves connate-perfoliate, serrate; anthers explanate .... (7)

4. Both anther sacs curved as a pair into horseshoe shape .... (5)

4′ Anther sacs not curved into horseshoe shape, typically opposite or divaricate (in a line) .... (6)

5. Anther sacs dehiscent by a short slit across the connective, the tips remaining closed; corolla 22–33 mm long, slightly glandular outside, glabrous within, strongly bilabiate, upper lip projecting, lower reflexed; staminode included, glabrous .... ***P. rostriflorus***

5′ Anther sacs dehiscent at tips only, connective intact, minutely puberulent; corolla barely bilabiate, palate glabrous; staminode glabrous to slightly bearded at tip .... ***P. eatonii***

6. Anthers becoming explanate; corolla tubular-salverform, more or less bilabiate, lower lip projecting, upper reflexed, glandular outside and in; herbage glabrous and glaucous .... ***P. utahensis***

6′ Anthers dehiscent distal 2/3–3/4, but not across connective and not explanate; corolla more or less tubular, strongly bilabiate, upper lobes projecting, lower lobes long, narrow, reflexed, glabrous outside, palate sometimes sparsely bearded .... ***P. barbatus***

7. Corolla throat inflated, palate sparsely hairy; staminode exserted, densely bearded with yellow hairs .... ***P. palmeri***

7′ Corolla throat gradually expanding, palate glabrous; staminode included, glabrous .... ***P. pseudospectabilis* subsp. *connatifolius***

8. Leaves linear to linear-oblanceolate, or linear-lanceolate and short (less than 35 mm long); calyx and exterior of corolla (sparsely) glandular; staminode bearded; inflorescence secund .... (9)

8′ Leaves not linear, or if so, then much longer than 35 mm, entire; inflorescence secund or not .... (10)

9. Pubescence of leaves of minute, retrorsely appressed, flat, scalelike hairs, stem hairs slender, terete, erect or retrorsely curved; stems erect or ascending, older stems sometimes subwoody, prostrate, rooting, and forming mats; corolla ampliate, throat not plicate, palate lightly bearded; staminode sparsely bearded with short hairs, longer golden hairs in apical tuft .... ***P. linarioides***

9′ Pubescence of slender, terete, erect or retrorsely curved hairs throughout; stems prostrate, ascending, or erect, often forming mats from spreading rootstocks; corolla not ampliate, throat plicate .... ***P. crandallii***

10. Foliage slightly to heavily glaucous and glabrous; leaves usually thickened or fleshy; staminode tip bearded with golden hairs, usually expanded, recurved .... (11)

10′ Foliage glabrous, puberulent, and/or glandular, but not glaucous; leaves not thickened .... (12)

11. Lowest 2–4 inflorescence bracts prominent; inflorescence not secund, either open or congested, usually rather elongate; corolla 15–20 mm long; anthers completely dehiscent (including connective) but not explanate .... ***P. angustifolius***

11′ Only the lowest 1 or 2 inflorescence bracts relatively large; inflorescences usually distinctly to more or less secund, often open, peduncles and pedicels usually elongate .... ***P. lentus***

12. Inflorescence and calyx (and often corolla) glandular-pubescent externally; staminode usually densely bearded most of length or only at apex .... (13)

12′ Inflorescence, calyx, and corolla not glandular-pubescent; glabrous or puberulous; staminode various .... (18)

13. Anther sacs explanate, glabrous ....(14)
13′ Anther sacs not explanate, but widely dehiscent, glabrous ....(17)

14. Corolla dull purple (rarely white), lower lobes projecting 3–5 mm longer than upper lobes, palate plicate; staminode exserted, apex with tuft of yellow hairs .... ***P. whippleanus***
14′ Corolla violet-blue, blue-purple, lower lobes not projecting, not noticeably longer than upper lobes ....(15)

15. Stems prostrate, 5–15 cm long; inflorescence secund; leaves almost all cauline, rather broad; corolla lilac-purple, 15–20 mm long; alpine, above 3200 m (10500′) .... ***P. harbourii***
15′ Stems erect, longer; not alpine ....(16)

16. Corolla 5–6 mm wide, 12–20 mm long, orifice as high or higher than wide; throat not to moderately inflated; lower lip not glandular within; anther sacs 0.6–1 mm long; staminode not or barely exserted .... ***P. breviculus***
16′ Corolla 8–19 mm wide, 14–22 mm long, orifice much wider than high; throat much inflated; lower lip glandular within; anther sacs 0.8–1.2 mm long; staminode usually prominently exserted .... ***P. ophianthus***

17. Inflorescence secund; corolla (14) 18–24 mm long, bluish purple to violet-purple; throat glabrous; staminode often exserted; habitat rocky alpine slopes near and above timberline, above 3200 m (10000′) .... ***P. hallii***
17′ Inflorescence not secund; corolla 8–14 mm long; throat sparsely white-bearded; anther sacs 1–1.2 mm long; staminode included or barely exserted .... ***P. parviflorus***

18. Inflorescence not at all secund, in dense fascicles usually separated by long internodes; corolla 10–14 mm long; anthers glabrous, explanate; plants usually puberulent .... ***P. rydbergii***
18′ Inflorescence secund (slightly or strongly), verticillasters not in dense fascicles ....(19)

19. Plants entirely glabrous; corolla 19–23 mm long, abruptly expanded, palate sparsely white-bearded; anthers villous or glabrous; endemic to upper Navajo Mountain, Utah .... ***P. navajoa***
19′ Plants puberulent, at least below; anthers always at least slightly villous; grows elsewhere ....(20)

20. Anthers densely villous; calyx 8–10 mm long, segments usually lanceolate, acuminate, or caudate, lower margins scarious, erose; staminode short-bearded on distal 1/2; corolla pale blue to lavender, 21–30 (35) mm long .... ***P. strictiformis***
20′ Anthers sparsely to densely villous; calyx 3–6 (8) mm long, segments usually ovate; staminode glabrous or with a few hairs at the tip ....(21)

21. Inflorescence narrow, cymes 1–2-flowered on short, usually appressed peduncles and pedicels; corolla often deep blue, 18–32 mm long, anthers less densely pubescent than *P. comarrhenus*, occasionally sparse .... ***P. strictus***
21′ Inflorescence usually broader, cymes often much-branched, peduncles and pedicels elongate and divaricate; corolla pale blue to lavender, 25–38 mm long .... ***P. comarrhenus***

Penstemon ambiguus Torr. **CP** (uncertain, perhaps referring to its relationship to other more typical, herbaceous penstemons or to its nearly actinomorphic corolla) Sand, bush, moth, phlox, or gilia beardtongue. **STEMS** 2–6 dm tall, profusely branched candelabra-like, woody well above the base; plants glabrous. **LEAVES** entire, 6–30 mm long, linear, mucronate, usually inrolled, margins not or remotely and very minutely scabrescent. **INFL** narrow, not secund, not glandular; 1–2 flowers per peduncle. **FLOWER** calyx 2–3 mm long; lobes ovate, acute; margins scarious; corolla 15–24 mm long; glabrous outside, throat pubescent within; tubular, expanding slightly, curved, pale to deep pink, nectar guidelines prominent; faces of lobes glistening white, in single plane, upper lobes reflexed, lower projecting; stamens included; anthers ultimately explanate, glabrous. **STAMINODE** included, glabrous, tip not expanded. $2n = 16$. [*Leiostemon ambiguus* (Torr.) Greene]. Sandy soil, mostly low to mid-elevation shrublands and woodlands. ARIZ: Apa, Nav; UTAH. 1460–1600 m (4800–5500′). Flowering: May–Jun. Northern Mexico, southern New Mexico, and Trans-Pecos Texas to southern Utah, northern Arizona, and southern Nevada. This is the only woody beardtongue in our area. Our material belongs to **var.** ***laevissimus*** (D. D. Keck) N. H. Holmgren (smoothest).

Penstemon angustifolius Nutt. ex Pursh (narrow leaf) Broadbeard, narrowleaf, or taperleaf beardtongue. **STEMS** 1–several, 2–5 dm tall, stout; plants glabrous and glaucous. **LEAVES** entire, 2–9 cm long, 2–25 (35) mm wide, fleshy; basal usually shorter and narrower than cauline, lanceolate to oblanceolate, short-petiolate; cauline broadly lanceolate, acuminate. **INFL** glabrous, not secund; usually elongate; verticillaster usually dense and many-flowered but widely separated; lowest 2–4 inflorescence bracts prominent, lower large, lanceolate or broader, tip acuminate or caudate (tip often folded over when pressed). **FLOWER** calyx 4–7 mm long; lobes lanceolate to narrowly ovate, acute to acuminate; margins broadly scarious at base; corolla 17–23 mm long; usually glabrous but occasionally with a few short hairs on palate; tube gradually expanding, blue or bluish purple to lavender or pinkish; lobes nearly equal, spreading; stamens included; anthers completely dehiscent but not explanate, glabrous. **STAMINODE** bearded with short, deep yellow hairs, more numerous near dilated tip. $2n = 16$. Sandy soils in shrublands and piñon-juniper woodlands. 1500–1860 m (4900–6100′). Flowering: late Apr–Jun. Southern Great Plains, northern New Mexico, southwest Colorado, northeast Arizona, and southeast Utah. Similar to *P. lentus*, but in *P. angustifolius* the cauline leaves are typically wider, the inflorescence is not at all secund, the lowest two to four inflorescence bracts are conspicuous and relatively broad, if smaller than the cauline leaves and increasingly smaller upward, and the peduncles and pedicels are short, making the flowers clustered or clumped. Except for flower color, the following varieties are not different.

1. Corollas pale blue to bluish purple .... **var. *caudatus***
1′ Corollas lavender to pinkish .... **var. *venosus***

var. *caudatus* (A. Heller) Rydb. (with a tail-like appendage, referring to the caudate inflorescence bracts) Corolla blue to bluish purple. COLO: LPl; NMEX: McK, RAr, San, SJn. 1585–1860 m (5200–6100′). Flowering: late Apr–Jun. Southern Great Plains, northern New Mexico, and southwest Colorado.

var. *venosus* (D. D. Keck) N. H. Holmgren (conspicuously veined) Corolla lavender to pink. ARIZ: Apa, Coc, Nav; COLO: Dol, Mon; NMEX: SJn; UTAH. 1500–1800 m (4900–5900′). Flowering: late Apr–Jun. Central Colorado Plateau, northwest New Mexico, southwest Colorado, northeast Arizona, and southeast Utah.

Penstemon barbatus (Cav.) Roth (bearded, referring to the hairy throat of var. *barbatus*) Torrey or scarlet beardtongue. **STEMS** 1–few, 3–11 dm tall; plants glabrous or puberulous below, glabrous above. **LEAVES** entire, 5–10+ cm long, basal and lower cauline wider than upper leaves (10–25 mm wide, mostly oblanceolate, and petiolate versus 1–10 mm wide and linear or narrowly lanceolate). **INFL** glabrous; more or less secund; verticillasters usually widely separated, peduncles and pedicels typically elongate and often branched. **FLOWER** calyx 3–7 mm long; lobes lanceolate to ovate, acute to short-acuminate, usually glabrous; margins entire, often scarious; corolla scarlet red, 25–35 mm long; externally glabrous, throat glabrous or sparsely white-bearded, essentially tubular, bent at base, markedly bilabiate, the upper lobes projecting, lower lobes reflexed; stamens exserted; anther sacs opening partially from the distal end, not across the connective, glabrous (lightly villous in var. *trichander*). **STAMINODE** included, glabrous, tip slightly expanded. $2n = 16$. [*Chelone barbata* Cav.]. Dry hillsides, from piñon-juniper and oak woodlands into ponderosa pine forests. 1740–2620 m (5700–8600′). Flowering: Jun–Aug. Mexico north to Trans-Pecos Texas, New Mexico, eastern Arizona, southwest and south-central Colorado, and southern Utah. Inflorescence architecture and corolla color and shape are similar to *P. rostriflorus*, but the anthers are not horseshoe-shaped and open in the distal end only (not across the connective).

1. Anthers glabrous .... **var. *torreyi***
1′ Anthers villous .... **var. *trichander***

var. *torreyi* (Benth. ex DC.) A. Gray (for John Torrey, 1796–1873, famous American botanist) Anthers glabrous; palate glabrous to sparsely villous. COLO: Mon, SJn; NMEX: RAr. 2290–2620 m (7500–8600′). Flowering: late Jun–Aug.

var. *trichander* A. Gray (hairy anther) Scarlet or beardlip beardtongue. Anthers lightly villous; palate glabrous. ARIZ: Apa; COLO: Arc, Dol, LPl, Mon; NMEX: McK, RAr, San, SJn; UTAH. 1740–2315 m (5700–7600′). Flowering: Jun–Aug.

Penstemon breviculus (D. D. Keck) G. T. Nisbet & R. C. Jacks. (diminutive, referring to the short stature and/or small flowers) Shortstem, little, or narrowmouth beardtongue. **STEMS** 1–few, 1–3.5 dm tall; plants puberulent or glabrous below, usually retrorsely puberulent and glandular above. **LEAVES** entire (rarely irregularly toothed), 2–8 cm long, 4–16 mm wide; mostly narrowly oblanceolate, lower wider and petiolate, upper sessile. **INFL** glandular, subsecund, narrow due to short peduncles and pedicels. **FLOWER** calyx glandular; 5–7 mm long; lobes lanceolate, acute, scarious at base; corolla 12–20 mm long; glandular outside; dark blue to bluish purple with conspicuous dark guidelines, funnel-shaped; palate with long, pale yellow hairs, not glandular; stamens included; anthers glabrous, explanate. **STAMINODE** mostly included, densely bearded with threadlike, yellow hairs. [*Penstemon jamesii* Benth. subsp. *breviculus* D. D. Keck] Sandy, gravelly, or clay soils in sagebrush and piñon-juniper woodland. ARIZ: Apa; COLO: LPl, Mon; NMEX: RAr, SJn; UTAH. 1525–1980 m (5000–6500′). Flowering: late Apr–early Jun. Northwest New Mexico, northeast Arizona, southwest

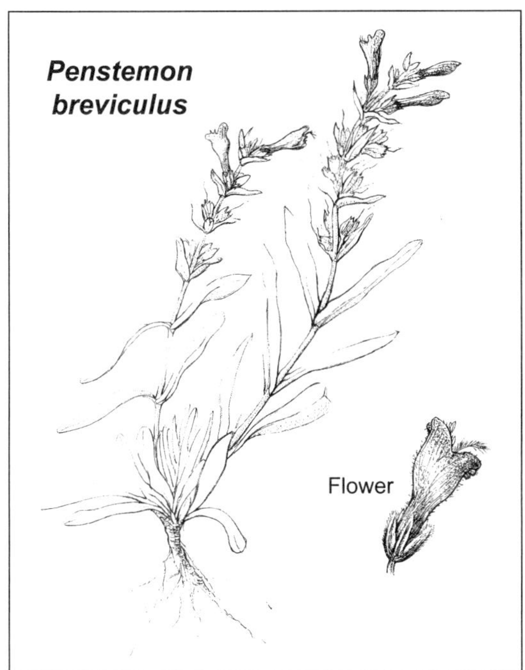

Colorado, and southeast Utah. This species is very similar to and sometimes difficult to distinguish from *P. ophianthus* Pennell. The range of *P. breviculus* is relatively small (near the Four Corners monument) and is essentially entirely included within but at the eastern edge of the distribution of *P. ophianthus*. *Penstemon breviculus* is overall smaller than *P. ophianthus* and it is usually relatively distinct in its core area, but many plants in the Four Corners region exhibit characters that overlap with *P. ophianthus*. Hartman and Nelson (2001) subsume it within the latter, perhaps with justification.

Penstemon comarrhenus A. Gray (hair + male, referring to the densely hairy anthers) Dusty beardtongue. **STEMS** 1–few, 3–8 (12) dm tall; plants glabrous or rarely puberulent below. **LEAVES** entire; 3–12 cm long, 2–20+ mm wide; mostly oblanceolate, lower wider and petiolate (and more or less winged), upper narrower to linear and sessile. **INFL** glabrous, usually somewhat secund, open, with lower peduncles and pedicels typically elongate and divaricate. **FLOWER** calyx glabrous; 3–6 (10) mm long; ovate, obtuse to apiculate, margins scarious, erose; corolla 25–35 mm long; glabrous outside and inside; tube pale bluish white; throat markedly expanded; pale pinkish or bluish lavender with dark guidelines; stamens exserted; anthers usually densely villous, mostly dehiscent (opening at distal 3/4 or more). **STAMINODE** essentially included, usually glabrous (rarely sparsely bearded), tip moderately dilated. In sagebrush, oak and piñon-juniper woodlands, ponderosa pine forests. ARIZ: Apa, Nav; COLO: Arc, Dol, LPl, Mon, SMg; NMEX: RAr, SJn; UTAH. 1890–2530 m (6200–8300′). Flowering: Jun–Jul. Northwest New Mexico, northeast Arizona, southwest Colorado, and southern Utah. Very similar to *P. strictus* and *P. strictiformis*, and the ranges of all three species overlap in the Four Corners area. The inflorescences of all three are at least somewhat secund and the classic open inflorescence architecture in *P. comarrhenus* is not always apparent. The anther sacs are usually densely villous, but not always so, and the corolla tends to be pale. Correct identification of any of these three species may be difficult in depauperate, immature, or atypical specimens.

Penstemon crandallii A. Nelson (for Charles Spencer Crandall, 1852–1929, Colorado collector and professor at the Agricultural College and Experiment Station, now Colorado State University, Fort Collins, 1889–1899) Crandall's beardtongue. **STEMS** usually several, 5–25 cm long, ascending to decumbent or sometimes prostrate; older, woodier stems often rooting; plants usually puberulent throughout with fine, terete, erect or retrorse hairs (rarely nearly glabrous), glandular-puberulent in inflorescence. **LEAVES** entire; essentially linear, sometimes slightly wider, 10–35 mm long, 1–3 mm wide, mucronate. **INFL** secund; glandular-puberulent. **FLOWER** calyx 5–8 mm long; lobes ovate to lanceolate, acuminate to caudate; margins scarious only in lowest; corolla 14–23 mm long; glandular-puberulent outside, palate sparsely bearded; throat plicate, only slightly expanded; blue to bluish purple or reddish purple; stamens reaching to near orifice; sacs dehiscing completely but not explanate; glabrous. **STAMINODE** included; densely bearded with golden hairs most of length. $2n = 16$. Oak and piñon-juniper woodlands, ponderosa pine to spruce/fir forests. 1920–2710 m (6300–8900′). Flowering: May–early Jul. Northern New Mexico, southwest, south-central, and central Colorado, and southeast Utah. In our area, leaf vestiture will consistently differentiate this species from *P. linarioides*, which otherwise is very similar. The hairs on *P. crandallii* are slender, terete, and erect or retrorse throughout, while those on the leaves of *P. linarioides* subsp. *coloradoensis* are flat and appressed and those on the stems are usually slender and erect or retrorse. Additionally, the leaves of *P. crandallii* are usually essentially linear while those on *P. linarioides* are typically broader, but this is not entirely consistent. The ranges of the following subspecies overlap somewhat in southwestern Colorado, with subsp. *glabrescens* occurring slightly eastward of the other.

1\. Leaves sparsely scabrid on upper surface, usually denser on lower .... **subsp. *crandallii***
1′ Leaves glabrous on upper surface .... **subsp. *glabrescens***

subsp. *crandallii* Leaves sparsely to moderately puberulent on upper surface and typically denser on lower surface, linear-oblanceolate, moderately crowded; corolla 14–20 mm long. COLO: Arc, Dol, Mon, SMg; UTAH. 2255–2710 m (7400–8900′). Flowering: late May–early Jul. Colorado and Utah.

subsp. *glabrescens* (Pennell) D. D. Keck (nearly or becoming smooth) Leaves glabrous on upper surface but usually puberulent on lower surface, linear or nearly so, more densely crowded; plants rarely nearly glabrous; corolla 15–23 mm long. COLO: Arc, LPl, SMg; NMEX: RAr, SJn. 1920–2380 m (6300–7800′). Flowering: May–Jun. Colorado and New Mexico.

Penstemon eatonii A. Gray (for Daniel Cady Eaton, 1834–1895, botanist at Yale University) Eaton's or firecracker beardtongue, scarlet bugler. **STEMS** few–several, 3–10 dm tall; plants robust, glabrous or puberulent, not glaucous. **LEAVES** entire, sometimes wavy; 3–10+ cm long, 1–5 cm wide; lower wider and petiolate; upper somewhat narrower, sessile, sometimes subcordate and/or clasping. **INFL** glabrous or puberulent; secund; verticillasters usually separated; peduncles and pedicels short. **FLOWER** calyx 3–6 mm long; lobes ovate, obtuse or acute, occasionally short-acuminate; glabrous; margins scarious, entire to slightly erose; corolla cardinal red; 20–30 mm long; glabrous inside and out, including palate; essentially tubular, often decurved near base, obscurely bilabiate, the lobes short, nearly equal, and mostly projecting; stamens included to exserted; anthers horseshoe-shaped, dehiscing at tips only; minutely puberulent. **STAMINODE** included; glabrous (rarely sparsely bearded at tip). $2n = 16$. Dry, rocky or sandy areas in shrublands, piñon-juniper woodlands, ponderosa pine forests. 1460–2500 m (4800–8200′). Flowering: May–Jun (Aug). Northwest New Mexico, southwest Colorado, northern Arizona, Utah, southern Nevada, and southern California. Presence or absence of pubescence has been used to define two varieties. However, both glabrous and puberulent plants occur in proximity in our area and the differences seem trivial. One of four red-flowered beardtongues in our area, each of which has a unique set of characters that make it readily identifiable. This species is the only one in the region with horseshoe-shaped anther sacs opening at the tips.

1. Herbage glabrous throughout .... **var. *eatonii***
1′ Herbage puberulent .... **var. *undosus***

var. *eatonii* CP Plants glabrous. ARIZ: Apa, Coc, Nav; COLO: Mon; NMEX: SJn; UTAH. 1585–2500 m (5200–8200′). Flowering: May–Jun. Northwestern New Mexico, southwest Colorado, northern Arizona, Utah west to southern California.

var. *undosus* M. E. Jones (wavy) Plants puberulent. ARIZ: Apa, Coc, Nav; COLO: Dol, Mon; NMEX: RAr, SJn; UTAH. 1460–2500 m (4800–2165′). Flowering: May–Jun (Aug). Northwestern New Mexico, southwest Colorado, Arizona, southern Utah to southern California.

Penstemon hallii A. Gray (for Elihu Hall, 1820–1882, an early plant collector in Colorado and associate of J. P. Harbour) Hall's beardtongue. **STEMS** short (10–20 cm), several to many, sometimes forming mats; plants glabrous below, sparsely glandular-puberulent above. **LEAVES** entire, glabrous, 2–6 cm long, 3–8 mm wide; basal leaves conspicuous, mostly narrowly oblanceolate, lower larger than upper. **INFL** secund, glandular-pubescent, condensed or elongate. **FLOWER** calyx 5–9 mm, lobes broadly ovate, usually acute, with wide scarious margins; corolla 18–24 mm long, minutely glandular-pubescent or glabrous outside, glabrous within, including palate, abruptly ampliate, blue, pink, or violet-purple; stamens dehiscing completely, including across connective, but not explanate; cells glabrous (minutely puberulent or rarely with a few longer hairs). **STAMINODE** usually exserted; short-bearded entire length, denser at widened tip. Rocky or gravelly soils near and above timberline. COLO: Arc, Con, Hin, Min, Mon, RGr, SJn. 3530–3840 m (11600–12600′). Flowering: Jul–Aug. This plant is endemic to most of the high mountains in Colorado.

Penstemon harbourii A. Gray **CP** (for J. P. Harbour, an early plant collector in Colorado and associate of Elihu Hall) Scree or Harbour's beardtongue. **STEMS** short (5–15 cm), slender, ascending, from long, creeping rootstocks, sometimes forming matlike mounds; plants puberulent below, glandular-puberulent above. **LEAVES** entire, 1–2 cm long, about 5–10 mm wide; no basal rosette; lower mostly petiolate, broader (spatulate to oblanceolate), and longer; upper sessile, narrower, shorter, and extending into inflorescence. **INFL** short (2–4 cm), secund, leafy, glandular-pubescent, mostly few-flowered and rather crowded. **FLOWER** calyx 6–9 mm, lobes linear-lanceolate, margins entire, not scarious; corolla 15–20 mm long, glandular-pubescent outside, palate densely bearded, throat slightly expanded, plicate, reddish purple to lilac and pale blue; stamens dehsicing completely, becoming explanate, glabrous. **STAMINODE** densely bearded most of length, tip dilated. Rocky, open places in alpine habitat. COLO: Hin, LPl, Min, Mon, RGr, SJn. 3475–3810 m (11400–12500′). Flowering: late Jun–Aug. This readily recognizable plant is endemic to most of the high mountains in Colorado.

Penstemon lentus Pennell (pliant) Pliant, Abajo, or handsome beardtongue. **STEMS** 1–few, 3–5 dm tall, plants glabrous and glaucous throughout. **LEAVES** entire, fleshy, 2–10 cm long, 1–2 cm wide; mostly obovate, petiolate,

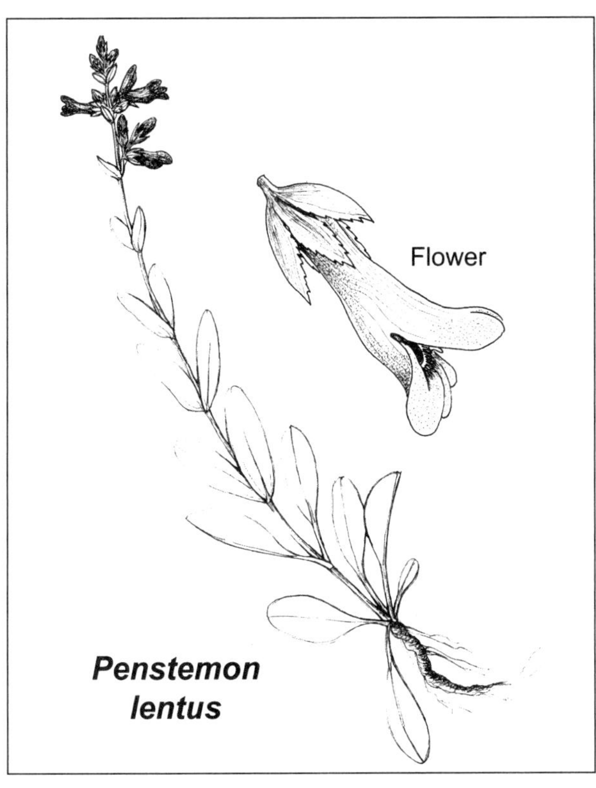

longer, and wider below, usually lanceolate, short, and narrower above. **INFL** glabrous, more or less secund, verticillasters usually separated, lower peduncles and pedicels often elongate; only lowest 1 or 2 inflorescence bracts prominent. **FLOWER** calyx 4–7 mm long, lobes usually ovate to lanceolate, acute or obtuse, margins scarious (about 0.5 mm wide), entire or erose; corolla 17–22 mm long; glabrous outside, palate sometimes sparsely white-bearded, moderately ampliate, blue, bluish violet, or white (var. *albiflorus*); stamens included, sacs dehiscing completely including across connective and usually divergent, but not explanate, glabrous. **STAMINODE** essentially included; bearded with short hairs in distal 1/2; tip dilated. Sandy or gravelly soils in sagebrush, piñon-juniper, oak, and ponderosa pine habitats. 1735–2410 m (5700–7900′). Flowering: May–Jun. Four Corners region. Similar to *P. angustifolius*, but in *P. lentus* the cauline leaves are usually narrower, the inflorescence is more or less secund, only the lowest one or two inflorescence bracts are somewhat large, and the peduncles and pedicels are usually long, causing the inflorescence to appear relatively open. Weber and Wittmann (2001) subsume *P. lentus* under *P. osterhoutii*, a species endemic to north-central Colorado, despite the apparently discontinuous ranges and somewhat different features of the two.

1. Corolla white; limited to the west side of the Abajo Mountains and the Natural Bridges National Monument area, Utah .... **var. *albiflorus***
1′ Corolla bluish violet; widespread in Four Corners area .... **var. *lentus***

var. *albiflorus* (D. D. Keck) Reveal (white-flowered) White-flowered beardtongue. Corolla white. ARIZ: Apa; UTAH. 2000–2350 m (6560–7700′). Flowering: May–Jun. Endemic to the west side of the Abajo Mountains, Utah, and the vicinity immediately to the south (including Natural Bridges National Monument).

var. *lentus* Corolla blue to bluish violet. ARIZ: Apa, Nav; COLO: Arc, Dol, LPl, Mon, SMg; NMEX: McK, SJn; UTAH. 1735–2410 m (5700–7900′). Flowering: May–Jun. Northwest New Mexico, southwest Colorado, northeast Arizona, and southeast Utah.

Penstemon linarioides A. Gray [from the genus *Linaria* (toadflax), referring to the similar leaf shape] Colorado toadflax beardtongue. **STEMS** few to many, 15–30 cm tall; sometimes decumbent and rooting at nodes, thus occasionally forming small mats; plants with tiny, retrorsely appressed, scalelike hairs on leaves, stems usually with terete, slender, erect or retrorse hairs; inflorescence glandular-puberulent. **LEAVES** entire, nearly linear or more often narrowly oblanceolate, mucronate; mostly 1–2 (3.5) cm long, 1–2.5 mm wide, similar in size and shape throughout; stems leafy but leaves also usually crowded near stem bases. **INFL** secund, glandular-puberulent. **FLOWER** calyx 5–8 mm, lobes ovate-lanceolate, acute or more commonly acuminate, margins scarious usually more than 1/2 of length; corolla 15–20 mm long, glandular-puberulent outside, palate lightly bearded, throat moderately ampliate, plicate, with prominent dark nectar guidelines, tube and throat pale-colored, elsewhere reddish lavender to bluish purple; stamens reaching orifice, cells dehiscing completely including across connective but not explanate, glabrous. **STAMINODE** included, bearded most of length, longer hairs at slightly widened tip. $2n = 16$. Dry, sandy or rocky slopes in sagebrush, shrublands, and woodlands. ARIZ: Apa, Nav; COLO: Arc, Dol, LPl, Mon, SMg; NMEX: McK, RAr, SJn; UTAH. 2040–2650 m (6700–8700′). Flowering: May–Jun. Northwest New Mexico, northeast Arizona, southwest Colorado, and southeast Utah. Our material belongs to **subsp. *coloradoensis*** (A. Nelson) D. D. Keck (of Colorado). The shape of the leaf hairs will consistently distinguish this species from *P. crandallii*. The ranges of the two species overlap in southwest Colorado. Other varieties of *P. linarioides* occur south and west of this taxon.

Penstemon navajoa N. H. Holmgren (for Navajo Mountain, Utah, the type locality) Navajo Mountain beardtongue. **STEMS** 1–few, 2–4 dm tall, plants glabrous. **LEAVES** entire, 2–9 cm long, 1–15 mm wide; narrowly (ob)lanceolate, broader and petiolate toward base of plant, narrower to linear and sessile near top of stems. **INFL** glabrous, more or less secund. **FLOWER** calyx glabrous, 3.5–5 mm long, lobes ovate, acute or obtuse, scarious margins less than 0.5 mm wide; corolla 17–25 mm long, glabrous outside, palate sparsely white-bearded, tube light blue to white, throat blue, ampliate; stamens exserted, sacs usually villous, sometimes glabrous, dehiscent except for connective. **STAMINODE** glabrous, tip expanded slightly. Rocky, open places in ponderosa pine, Douglas-fir, and subalpine habitats. UTAH. 2250–3170 m (7400–10400′). Flowering: Jul–Aug. Grows only on the upper part of Navajo Mountain, San Juan County, Utah; a Four Corners regional endemic.

Penstemon ophianthus Pennell (snake + flower) Coiled anther or Loa beardtongue. **STEMS** 1–several, 1–4 dm tall, plants glabrous or puberulent below, glandular-pubescent above. **LEAVES** entire to sometimes sinuate-dentate, 3–12 cm long, 6–22 mm wide; wider and petiolate below, narrower above. **INFL** glandular, more or less secund. **FLOWER** calyx 6–10 mm long, lobes (narrowly) lanceolate, acute to acuminate, herbaceous, sometimes narrowly scarious-margined at base; corolla 15–22 mm long, 7–10 mm wide when pressed, moderately ampliate, opening wider than tall, glandular-puberulent outside, palate with numerous long, whitish, villous hairs, also sometimes with short, glandular hairs, bluish to pale lavender or blue-violet; stamens included or slightly exserted, sacs glabrous, becoming explanate. **STAMINODE** clearly exserted, densely bearded with long yellow hairs. $2n = 16$ [*P. jamesii* Benth. subsp. *ophianthus* (Pennell) D. D. Keck]. Sandy or gravelly soils in sagebrush, piñon-juniper, oak, and ponderosa pine habitats. ARIZ: Apa; COLO: Mon; NMEX: McK, SJn; UTAH. 1400–2410 m (4600–7900′). Flowering: late May–early Jul. Northwest and western New Mexico, northeast Arizona, southwest Colorado, southeast and south-central Utah. Closely related to *P. jamesii* and *P. breviculus*. *Penstemon jamesii* has larger flowers and occurs in central and eastern New Mexico. *Penstemon breviculus* is partly sympatric with and not always distinguishable from *P. ophianthus*. Hartman and Nelson (2001) consider *P. breviculus* and *P. parviflorus* to be synonyms of *P. ophianthus*, which may be correct.

Penstemon palmeri A. Gray (for Edward Palmer, 1821–1911, Western plant collector) Palmer's beardtongue, pink wild snapdragon, balloon flower. **STEMS** few to several, robust, 5–14 dm tall, plants glaucous, glabrous below, glandular-pubescent in inflorescence. **LEAVES** usually serrate, fleshy, 5–12 cm long, 2–5 cm wide; lower petiolate, becoming connate-perfoliate above. **INFL** glandular-pubescent, more or less secund, typically elongate. **FLOWER** calyx 4–6 mm long, lobes ovate, acute; corolla sweetly fragrant, 25–35 mm long, abruptly expanded, throat inflated, white to pink, dark guidelines conspicuous, palate with sparse white hairs; stamens included, cells dehiscing completely including across connective and becoming explanate, glabrous. **STAMINODE** exserted, densely bearded with yellow hairs. $2n = 32$. Washes and roadsides in shrublands, piñon-juniper woodlands, and ponderosa pine forests. ARIZ: Apa, Coc, Nav; COLO: LPl, Mon; NMEX: San, SJn; UTAH. 1370–1830 m (4500–6000′). Flowering: May–Jul. A distinctive, easily recognizable plant. Probably only native in the extreme western portion of the San Juan drainage. Widely planted in the Southwest for revegetation on roadsides. This species is a popular garden plant because of its robust growth habit, drought resistance, and its beautiful, large, fragrant flowers.

Penstemon parviflorus Pennell (small flower) Small-flowered penstemon. **STEMS** short (15–20 cm tall); plants glandular-puberulent. **LEAVES** linear-lanceolate, 3–5 cm long, and 3–5 mm wide. **INFL** narrow, more or less secund. **FLOWER** calyx lobes about 8 mm long and linear-lanceolate; corolla about 12 mm long and glandular-puberulent outside, palate pubescent, throat ampliate, apparently purplish blue; stamen sacs minutely puberulent and dehiscing completely but not explanate. **STAMINODE** more or less included and densely bearded most of its length. COLO: Mon. Flowering: May–Jul. Known from one collection by Alice Eastwood in 1890 near Mancos, Colorado. Although the plants in the Mancos, Colorado, area have been rather thoroughly collected since 1890, *P. parviflorus* has never been relocated. Hartman and Nelson (2001) consider it to be a synonym of *P. ophianthus*, perhaps with some justification.

Penstemon pseudospectabilis M. E. Jones **subsp. *connatifolius*** (A. Nelson) D. D. Keck (Greek *pseudo*, false, and Latin *spectabilis*, showy, referring to this plant's similarity, in some respects, to *P. spectabilis* Thurb. ex A. Gray, a species from southern California and Baja California, Mexico). Desert, spectacular, pink showy, or canyon beardtongue. **STEMS** several to many, 4–10 dm tall, from a woody base; plants glabrous (except corolla), glaucous. **LEAVES** serrate; petiolate below, becoming connate-perfoliate above; upper perfoliate leaves to 18 cm long and 7.5 cm wide. **INFL** often 1/2 the height of the plant; more or less secund; glabrous; of several verticillasters, each with 1–5 flowers; the lower peduncles often elongate. **FLOWER** calyx 5–7 mm long, lobes typically ovate, apices acute to acuminate, margins narrowly scarious; corolla pale to deep rose or rose-magenta, with darker guidelines, 22–33 mm long, to 12 mm wide when pressed, glandular-puberulent externally and internally, throat gradually expanding, palate glabrous; stamens included, anther sacs explanate, glabrous. **STAMINODE** included, glabrous, narrow. [*Penstemon connatifolius* A. Nelson; *Penstemon* *spectabilis* Wooton & Standl., not Thurb.]. Piñon-juniper woodlands (to ponderosa pine forests elsewhere in its range). UTAH. 1060 m (4800′). Flowering about Jun. Often common in Arizona from the southeast to the northwest (it apparently has not been collected in the northeastern part of the state) and southwestern New Mexico, but very uncommon in southern Utah (only a handful of collections). This subspecies is readily recognizable by its pinkish flowers; connate-perfoliate upper leaves; glabrous palate; explanate, hairless anthers; and glabrous staminode. Other subspecies grow elsewhere in Arizona and southern California; subsp. *connatifolius* is the most widespread.

Penstemon rostriflorus Kellogg (beak flower) Beak-flowered, beaked, or Bridges' beardtongue. **STEMS** few to several, 3–10 dm tall, plants glabrous or obscurely puberulent and sometimes glaucous below, glandular-pubescent above. **LEAVES** entire, mostly cauline, 2–9 cm long, 2–8 mm wide; lower oblanceolate and petiolate, upper linear or narrow and sessile. **INFL** glandular-pubescent, secund, typically open, lower peduncles sometimes elongate and branched. **FLOWER** calyx glandular-pubescent, 4–6 mm long, lobes mostly lanceolate, margins not or slightly scarious; corolla red to orange-red (rarely yellow), 23–33 mm long, lightly glandular-pubescent outside, palate pale yellow, glabrous, narrowly funnel-shaped, often decurved near base, markedly bilabiate, the lower lobes reflexed, the upper lobes nearly united and projecting, slightly glandular outside, stamens exserted, anthers horseshoe-shaped, dehiscing near connective only, glabrous. **STAMINODE** more or less included, glabrous, tip not expanded. $2n = 16$; 42. [*Penstemon bridgesii* A. Gray]. Many habitats including shrublands, oak and piñon-juniper woodlands, and ponderosa pine and montane conifer forests. ARIZ: Apa, Coc, Nav; COLO: Dol, LPl, Mon; NMEX: SJn; UTAH. 1890–2770 m (6200–9100′). Flowering: Jun–Sep. Northwest and western New Mexico, northern Arizona, southwest Colorado, southern Utah, southern Nevada, California. Inflorescence architecture and corolla color and shape are similar to *P. barbatus*, but the anthers are horseshoe-shaped and open only at the connectve in *P. rostriflorus*.

Penstemon rydbergii A. Nelson (for Per Axel Rydberg, 1860–1931, well-known American botanist) Rydberg's beardtongue. **STEMS** few to several, 2–5 dm tall, plants usually puberulent. **LEAVES** entire; 3–12 cm long, 3–20 mm wide, basal and lower petiolate, longer, and wider, upper sessile, shorter, and narrower. **INFL** typically puberulent but not glandular, not secund, usually in distinctly separated, densely flowered verticillasters. **FLOWER** calyx 4–6 mm long, lobes ovate, acuminate, margins broadly scarious and erose; corolla 12–20 mm long, glabrous outside, palate bearded, throat moderately ampliate, blue to purple; stamens slightly exserted, sacs dehiscing completely including across connective but not explanate, glabrous; blackish. **STAMINODE** reaching orifice, usually densely yellow-bearded. $2n = 16$. Relatively moist to dry places in meadows, shrublands, and forests in mountains. ARIZ: Apa; COLO: Arc, Dol, Hin, Mon; NMEX: SJn. 2740–3230 m (9000–10600′). Flowering: Jul–Aug. Southeastern Arizona, northern New Mexico, Colorado, central Utah, Wyoming, southwestern Montana, Idaho, northeast Oregon, southeast Washington. One of the more widespread species of *Penstemon*, with two or three varieties. Our material belongs to ***var. rydbergii***.

Penstemon strictiformis Rydb. (tight shape) Mancos beardtongue. **STEMS** 1–few, mostly 2–6 dm tall, plants glabrous. **LEAVES** entire; 2–15 cm long, 5–15 mm wide, mostly narrowly (ob)lanceolate and similar width throughout; basal and lower cauline petiolate and longer, upper sessile, shorter, and often folded. **INFL** glabrous; more or less secund, rather narrow or contracted, peduncles and pedicels short. **FLOWER** calyx 5–8 mm, lobes lanceolate, acuminate or caudate, lower margins scarious; corolla (15) 20–30 mm long, glabrous outside and inside, throat ampliate, pale bluish lavender or bluish purple; stamens exserted, anthers densely to sparsely villous, but cells not obscured, cells mostly dehiscent (opening at distal 3/4 or more) but not across connective. **STAMINODE** included, glabrous to sparsely villous, tip slightly expanded. [*P. strictus* Benth. subsp. *strictiformis* (Rydb.) D. D. Keck]. In sagebrush, oak and piñon-juniper woodlands, ponderosa pine forests. ARIZ: Apa, Nav; COLO: Arc, Mon; NMEX: RAr, SJn; UTAH. 2100–2560 m (6900–8400′). Flowering: late May–Jul (Sep). Northwest New Mexico, northeast Arizona, southwest Colorado, and southeast Utah. Very similar to *P. comarrhenus* and *P. strictus*, and the ranges of all three species overlap in the Four Corners area. Many plants are difficult to place and exhibit intermediate characters. Anther sacs are sometimes densely villous but just as often are sparsely so. Corollas are generally smaller than the other two species and the calyx is usually longer. This seems to be the least well-defined taxon of the three. Weber and Wittmann (2001) subsume this species within *P. strictus* with some justification.

Penstemon strictus Benth. (close, tight, narrow) Rocky Mountain, porch, or stiff beardtongue. **STEMS** 1–few, 3–8 dm tall, plants glabrous or slightly puberulent in lower parts. **LEAVES** entire, 5–15 cm long, 4–20 mm wide, mostly narrowly oblanceolate; basal and lower petiolate; upper sessile, somewhat shorter and narrower, often folded. **INFL** strongly secund, narrow, peduncles 3–15, appressed. **FLOWER** calyx 2.5–5 mm long, lobes ovate, usually obtuse (sometimes apiculate), margins scarious; corolla 23–32 mm long, glabrous outside and in, throat expanded, blue to dark blue or blue-purple; stamens exserted, sacs villous, mostly dehiscent (opening at distal 3/4 or more), but not across connective. **STAMINODE** essentially included; sparsely villous or glabrous. $2n = 16$. In sagebrush, oak and piñon-juniper woodlands, ponderosa pine forests. ARIZ: Apa, Nav; COLO: Arc, Dol, Hin, LPl, Min, Mon, SJn, SMg; NMEX: McK, RAr, SJn; UTAH. 2100–2865 m (6900–9400′). Flowering: late May–early Aug. Northern New Mexico, northeast Arizona, western Colorado, Utah, and southwest Wyoming. Closely related and very similar to *P. strictiformis* and *P. comarrhenus*, and the ranges of all three species overlap in the Four Corners area. Although this species is usually readily identifiable, some specimens cannot be satisfactorily placed due to overlap in characters among the three species.

Penstemon utahensis Eastw. **CP** (of Utah) Utah beardtongue. **STEMS** 1–few, 1.5–6 dm tall, plants glabrous (except for glandular-pubescent corollas) and glaucous. **LEAVES** entire, thick, leathery or fleshy, 1–10 cm long, 5–20 mm wide; lower petiolate, mostly oblanceolate, and wider and longer than upper, which are elliptic or lanceolate and sessile. **INFL** subsecund, elongate. **FLOWER** calyx 2.5–5 mm long, lobes broadly ovate, acute or mucronate, margins broadly scarious; corolla 16–25 mm long; typically red-rose, mostly tubular-salverform, bilabiate, limbs conspicuous, more or less in same plane, often at oblique angle with lower lip projecting, glandular outside and in but palate not bearded, stamens included; sacs glabrous, becoming explanate. **STAMINODE** included; glabrous or papillate at tip. Sandy soils in shrublands and piñon-juniper habitats. ARIZ: Apa, Coc, Nav; COLO: Mon; UTAH. 1200–2000 m (3900–6600′). Flowering: Apr–May. Southwestern Colorado, northeast Arizona, southern Utah, southern Nevada, and southeast California. One of four red-flowered beardtongues in our area each of which has a unique set of characters that make it readily identifiable. This is the only one with spreading corolla lobes and explanate anther sacs.

Penstemon whippleanus A. Gray (for Amiel Weeks Whipple, 1816–1863, explorer with the U.S. Army Corps of Topographical Engineers) Whipple's or dusky beardtongue. **STEMS** 1–several, 2–10 dm tall, plants glabrous below, glandular-pubescent above. **LEAVES** entire or dentate, 2–10 cm long, 10–30+ mm wide; cluster of well-developed basal leaves usually present, lower wider than upper, with long petioles, upper narrower (but not linear) and sessile or clasping; mostly lanceolate throughout. **INFL** glandular-pubescent, secund, cymes usually in dense clusters. **FLOWER** calyx 6–11 mm long, lobes (narrowly) lanceolate, acuminate, margins scarious at base; corolla 20–30 mm long, glandular outside, variable in color but often very dark (blackish purple) or violet, (dark) blue, lavender to whitish, lower lip projecting well beyond the short upper, throat ampliate, plicate, palate with long white hairs; stamens included or slightly exserted, sacs glabrous, becoming explanate. **STAMINODE** usually exserted beyond the upper lobe, white, tip with tuft of whitish hairs. $2n = 16$. Open and forested areas in mountains, from subalpine into alpine habitats. COLO: Arc, Con, Hin, LPl, Min, Mon, RGr, SJn; NMEX: RAr, SJn; UTAH. 3350–3810 m (11000–12500′). Flowering: Jul. Widespread in the western United States; central and north-central New Mexico, central Colorado, central Utah, Wyoming, southeast Idaho, and southwest Montana. This species is quite distinctive and usually readily recognizable.

Plantago L. Plantain

Tina J. Ayers

(Latin *planta*, flat and spread out, + *ago*, kind of) Annual or perennial, acaulescent herbs from taproots. **STEMS** extremely short, disclike, simple or branched. **LEAVES** simple, spirally arranged, usually all basal, estipulate, the blades entire or variously toothed, venation parallel. **INFL** a pedunculate, bracteate spike. **FLOWERS** regular, perfect or imperfect, several to many, inconspicuous, each subtended by a bract; sepals 4 (2 fused in *P. lanceolata*), distinct, usually with overlapping scarious margins; corolla 4-lobed, whitish, scarious or membranous, persistent, the lobes with a thickened or colored basal spot; stamens 2–4, exserted; ovary superior, of 2 fused carpels, with 2 locules, the placentation axile; style 1, exserted; stigma 2-lobed. **FRUIT** a circumscissile capsule with 2–many seeds, included in the sepals or exserted, often purplish brown. **SEEDS** mucilaginous when wet, flat, concave or irregularly angled, the outer surface often patterned. About 255 species, worldwide. *Plantago afra* and *P. ovata* are used as laxatives (Morris 1900, 1901).

1. Plants annual; leaves threadlike to linear-lanceolate, rarely more than 1 cm wide .... (2)
1′ Plants perennial; leaves lanceolate to ovate, rarely less than 1 cm wide .... (4)

2. Leaves threadlike; bracts keeled, bases extended and saclike; seeds usually 4 per capsule .... ***P. elongata***
2′ Leaves linear to linear-lanceolate; bracts flat to slightly keeled, bases not extended and saclike; seeds 2 per capsule .... (3)

3. Bracts linear-triangular, 2–16 mm long, longer than sepals; seeds light brown to reddish brown; spring-flowering .... ***P. patagonica***
3′ Bracts triangular, 2–3.5 mm long, about as long as sepals; seeds dark brown to black; summer- to fall-flowering .... ***P. argyrea***

4. Leaves broadly ovate, abruptly narrowing into a petiole; seeds 6–many per capsule .... ***P. major***
4′ Leaves lanceolate, oblanceolate, or elliptic, gradually tapering into the petiole; seeds 2–4 per capsule .... (5)

5. Outer pair of sepals (those adjacent to the bract) connate, appearing as a solitary, 2-veined, apically notched or entire sepal .... ***P. lanceolata***
5′ Outer pair of sepals (those adjacent to the bract) distinct .... ***P. eriopoda***

Plantago argyrea E. Morris (silvery) Saltmeadow plantain. Annual herbs to 30 cm tall. **LEAVES** without a distinct petiole; blades linear to linear-lanceolate; 2.4–17.4 cm long, 0.2–0.5 cm wide, acute to acuminate at apex, densely villous (rarely woolly), distinctly 3-veined below, margins entire. **INFL** peduncle 5–26 cm long, hirsute, with most hairs appressed-ascending; spike 1–9.5 cm long, interrupted near base at maturity; bracts broadly triangular, 2–3 mm long, about as long as sepals, scarious-margined at base or to near middle, sparsely to densely villous, with tufts of white hairs in axils, midvein ciliate. **FLOWERS** perfect; sepals obovate, 2.3–3 mm long, scarious-margined, villous; corolla lobes spreading or reflexed, broadly ovate, 1.6–2.3 mm long; stamens 4. **FRUIT** breaking at or slightly below middle. **SEEDS** 2, narrowly elliptic, concave, 2.5 mm long, 1.2 mm wide, dark brown to black. [*Plantago purshii* Roem. & Schult. var. *argyrea* Poe]. Locally common in dry piñon-juniper woodlands and ponderosa pine forest communities. ARIZ: Apa, Nav; NMEX: McK, SJn. 2250–2685 m (7400–8800′). Flowering: Jun–Sep. Arizona and New Mexico above the Mogollon Rim. The spelling presented here agrees with that found in the original publication of the species (Morris 1900). Some treatments use the correct spelling (e.g., Wooton and Standley 1915; McDougall 1973). In *Arizona Flora* (Kearney and Peebles 1969), *A Flora of New Mexico* (Martin and Hutchins 1981), and many Web resources the species name is incorrectly spelled *P. argyraea*.

Plantago elongata Pursh (elongated) Prairie plantain. Annual herbs to 10 cm tall. **LEAVES** without a distinct petiole; blades threadlike to linear, 1.5–7 cm long, less than 1 mm wide, attenuate at base, acute at apex, sparsely pubescent, with no evidence of venation, margins entire. **INFL** peduncle 2.6–16 cm long, sparsely to densely villous, usually becoming densely villous just below the inflorescence; spike 0.3–5.5 cm long, sometimes interrupted near base; bracts ovate, 1.5–2.2 mm long, broadly scarious-margined, bases extended and saclike, glabrous. **FLOWERS** perfect; sepals broadly ovate, 1.2–2 mm long, 1/2 as long as capsule, broadly scarious-margined; midvein thick, glabrous; corolla lobes spreading or reflexed, triangular to narrowly ovate, 0.5–0.7 mm long; stamens 2. **FRUIT** breaking at middle. **SEEDS** generally 4, narrowly elliptic, nearly flat, acute at both ends, 1.2 mm long, 0.3 mm wide, olive-green to dark brown. Desert grasslands and washes. COLO: Arc, Mon; NMEX: McK, RAr, SJn. 1860–2075 m (6100–6800′). Flowering: May–Jul. Widespread in western North America. Our plants are **var. *elongata***.

Plantago eriopoda Torr. (woolly-foot, referring to the stem) Redwool plantain. Perennial herbs to 40 cm tall. **LEAVES** petiolate; blades lanceolate, 5–12 cm long, 1–2.5 cm wide, attenuate at base, acute at apex, sparsely villous to glabrate, distinctly 5-veined below, the margins with a few remote, shallow teeth. **INFL** peduncle 5.5–16 cm long, white-villous; spike 2–10 cm long, interrupted near base; bracts ovate, (1.5) 2 mm long, ciliate, with tufts of tan hairs in axils. **FLOWERS** perfect; sepals broadly elliptic, about 2.5 mm long, broadly scarious-margined; corolla lobes spreading or reflexed, subulate to narrowly ovate, 1.2–1.8 mm long; stamens 4. **FRUIT** breaking well below middle. **SEEDS** 2, ellipsoid, concave, 2 mm long, 0.8–1 mm wide, light brown to reddish brown with clear rim at base. Alkaline marshes. ARIZ: Apa; COLO: Arc, LPl; NMEX: SJn. 1700–2240 m (5600–7350′). Flowering: Jun–Aug. Widespread in western North America.

Plantago lanceolata L. (lance-shaped) Narrowleaf plantain. Perennials to 65 (85) cm tall. **LEAVES** petiolate; blades lanceolate, 4–24.5 cm long, 0.5–4.2 cm wide, long-attenuate at base, acute at apex, sparsely pubescent to glabrate, distinctly 5 (7)-veined, with margins occasionally having a few remote teeth. **INFL** peduncle 13–86 cm, strigose; spikes

1.5–7 cm long, continuously flowered without interruption; bracts broadly triangular, (1.5) 2 mm long. **FLOWERS** perfect; sepals obovate, 1.5–2.5 mm long, broadly scarious-margined, the 2 outermost characteristically fused, appearing as a solitary, 2-veined, apically notched or entire sepal; midvein thin, ciliate at apex; corolla lobes spreading or reflexed, subulate to narrowly ovate, 2–2.5 mm long; stamens 4. **FRUIT** breaking well below middle. **SEEDS** 2, ellipsoid, concave, 1.5–2.5 mm long, 1 mm wide, brown. Moist soils in disturbed sites. ARIZ: Apa, Nav; COLO: Arc, Dol, Hin, LPl, Mon, SJn; NMEX: McK, RAr, San, SJn; UTAH. 1600–2856 m (5470–9370′). Flowering: May–Aug. Native of Eurasia, now a cosmopolitan weed.

Plantago major L. (larger) Common plantain. Perennials to 42 cm tall. **LEAVES** petiolate; blades broadly ovate, 3.5–15 cm long, 2–9 cm wide, attenuate at base, acute at apex, sparsely pubescent, becoming glabrate, distinctly 3 (5)-veined, the margins with a few shallow lobes near base. **INFL** peduncle 4–20 cm long, sparsely pubescent to glabrous, sometimes appearing ridged to 4-sided; spikes 3–24 cm long, interrupted near base; bracts broadly ovate, 1.5–4.5 mm long, with broad scarious margins; midvein glabrous. **FLOWERS** perfect; sepals broadly ovate to elliptic, 1.2–2.3 mm long, broadly scarious-margined, midvein glabrous; corolla lobes spreading or reflexed, ovate, 0.7–1.5 mm long; stamens 4. **FRUIT** breaking well below middle. **SEEDS** 6–many, irregular in shape, 1 mm long, 0.5 mm wide, olive-green to dark brown. Weed of wet areas throughout the *Flora* area. ARIZ: Apa, Nav; COLO: Arc, Dol, Hin, LPl, Min, Mon, SJn, SMg; NMEX: McK, RAr, SJn; UTAH. 1730–3820 m (5690–12530′). Flowering: May–Oct. Cosmopolitan. A cold infusion of the plant was taken for "lightning infection" and as a "life medicine" by the Ramah Navajo. The Laguna and Acoma used the young leaves for food.

Plantago patagonica Jacq. (from Patagonia) Woolly plantain. Annual herbs to 21 cm tall. **LEAVES** without distinct petiole; blades linear to linear-lanceolate, 1–15 cm long, 0.1–0.7 cm wide, acute to acuminate at apex, sparsely to densely villous, distinctly 3-veined below, the margins entire. **INFL** peduncle 2–22 cm long, hirsute, with most hairs appressed-ascending; spike 1–13 cm long, continuously flowered without interruption; bracts linear-triangular to subulate, 2–16 mm long, longer than sepals, green throughout or scarious-margined near base, sparsely to densely villous, with tufts of white hairs in axils, with midvein long-ciliate. **FLOWERS** perfect; sepals obovate, 2–5 mm long, broadly scarious-margined, villous; corolla lobes spreading or reflexed, broadly ovate, 1.2–2.5 mm long; stamens 4. **FRUIT** breaking at or slightly below middle. **SEEDS** 2, ellipsoid, concave, 2.2 mm long, 1.2 mm wide, reddish brown to light brown (rarely dark brown). [*P. purshii* Roem. & Schult.]. Deserts and desert grasslands. ARIZ: Apa, Coc, Nav; COLO: Arc, Dol, LPl, Mon; NMEX: McK, RAr, San, SJn; UTAH. 1690–2225 m (5550–7300′). Flowering: Feb–Jun. British Columbia to Saskatchewan, Canada; south to California, east to Texas; Argentina and Chile. Used by the Hopi to make a person more agreeable. Used by the Navajo as a gastrointestinal aid, laxative, and seeds given to babies with colic or constipation. A cold infusion of plant parts taken to reduce appetite and prevent obesity by the Ramah Navajo. The Ramah Navajo would also use an infusion of root, leaves, and seeds blown over a rattle at the beginning of the "hoof rattle song." Porridge seeds made into mush and used for food by the Kayenta Navajo. The Hopi would use the plant as a ceremonial item for participants of the "clowning crew."

Synthyris Benth. Kittentails, Besseya

(together or joined little door, after the manner in which the capsule dehisces) Subscapose perennial herbs with fibrous roots. **LEAVES**: basal well developed, usually maturing after anthesis, petiolate, the blades ovate to rotund, cordate, cordate-ovate to oblong, crenate, crenate-serrate, serrate to pinnatifid; cauline leaves reduced, bractlike, sessile, alternate. **INFL** a dense raceme, cylindrical spike, or spikelike raceme. **FLOWER** calyx with (1) 2–4 lobes, distinct or variously united; corolla blue, pink, lavender, violet-purple, yellow, or white, campanulate to subrotate, subequally 4-lobed, the upper lobe entire, slightly larger, the lower 3 lobes external to the upper in bud, or corolla wanting, vestigial, or present and strongly bilabiate, the upper lip entire, concave, much larger than the lower, the lower lip more or less 3-lobed, external to the upper in bud; stamens 2, exserted, exserted filaments conspicuously colored in flowers lacking a corolla, the anther cells parallel; stigmas united and capitate. **FRUIT** a loculicidal capsule, plump to somewhat compressed, notched apically or entire. **SEEDS** 2–many per locule, flattened or with incurved margins; $2n = 24$. [*Besseya* Rydb.] A genus of 16 species, occurring in the mountains of the western United States. Formerly included in Scrophulariaceae (Pennell 1933; Hufford and McMahon 2004).

1. Corolla dark violet-purple; the upper lip of the corolla with the middle lobe about 1/2 as long as the entire lip; plants generally less than 15 cm tall; leafy bracts 4–6 .... ***S. alpina***

1′ Corolla white or yellow, sometimes pink-purple tinged; the upper lip of corolla with the middle lobe less than 1/2 as long as the entire lip; plants usually more than 15 cm tall; leafy bracts more than 6.......... (2)

2. Corolla lemon-yellow; the upper lip ciliate, the lower lip of corolla with lobes no more than 1/5 as long as the entire lip; lateral calyx lobes united for 1/2 their total length .... ***S. ritteriana***
2' Corolla white or pinkish purple; the upper lip entire, the lower lip of corolla with lobes more than 1/2 the length of the entire lip; lateral calyx lobes united for less than 1/3 their total length .... ***S. plantaginea***

Synthyris alpina A. Gray (growing in the alpine) Alpine kittentails. Subscapose perennial herb, 5–15 (20) cm tall. **STEMS** simple; woolly to glabrate. **LEAVES**: basal petiolate, the blades (2.5) 4–6 cm long, (1) 2.5–4 cm wide, elliptic, ovate to broadly ovate, rounded to cordate basally, crenate, incisions 0.2–1 mm deep, the primary veins pinnate, petioles 3–6 (10) cm long; cauline leaves alternate, few, 4–6, bractlike, sessile. **INFL** subspicate, dense, villous, the bracts spatulate to obovate. **FLOWER** calyx 4–6 mm long, the 4 lobes elliptic-lanceolate, white-villous, especially on the margins; corolla 5–8 mm long, violet, purple; stamens exserted, with inconspicuously colored filaments. **FRUIT** a capsule 3–5 mm long and about as wide, orbicular, entire or scarcely notched, the style about 3.5–4 mm long. **SEEDS** 1–1.5 mm long, orbicular, flat, 4 per locule. $2n = 24$. [*Besseya alpina* (A. Gray) Rydb.]. Moist, rocky alpine tundra, meadows, and talus. COLO: Hin, LPl, Min, Mon, SJn. 3000–4300 m (9900–14200'). Flowering: Jul–Aug. Fruit: late Jul–Sep. High Rocky Mountains of south-central Wyoming, Colorado, and northern New Mexico, and in the La Sal Mountains of southeastern Utah.

Synthyris plantaginea (E. James) Benth. (referring to the similarity in form with *Plantago*) Plantain kittentails. Subscapose perennial herb, (10) 18–48 cm tall, tomentose but becoming glabrate. **LEAVES**: basal tomentulose, petiole usually 1.5–6.5 cm long; the blades 2–20 cm long, 1.5–4 cm wide, acute apically, ovate to obovate or oblanceolate, obtuse at base, broadly to narrowly cuneate basally, rarely subcordate, crenate, the incisions 0.2–0.8 mm deep, the primary veins pinnate, cauline leaves alternate, bractlike, sessile to subpetiolate below, broadly lanceolate to ovate, entire to serrate, pilose. **INFL** a subspicate raceme, 7–13 cm long, tomentulose, flowers dense. **FLOWER** calyx 4–7 mm long, 4- to 5-lobed, with lateral lobes connate for less than 1/3 their length, villous-ciliate; corolla about 4.5–8 mm long, white to pinkish purple, the upper (adaxial) lip slightly to well exserted from the calyx, the lower (abaxial) lip 1/2 as long as the upper, 2- to 3-lobed, the lobes over 1/3 the length of the lip; stamens exserted, with white or pale filaments, not conspicuously pigmented, 7–11 mm long. **FRUIT** a capsule, 4.5–6 mm long, 3.5–5 mm wide, pubescent to glabrous, notched at the apex, the style 3–5 mm long. **SEEDS** 1.2–3 mm long, orbicular, flat, mostly 4 per locule. [*Besseya plantaginea* (E. James) Rydb.]. Hills and mountains, often on moist wooded slopes and meadows. 1830–3050 m (6000–10000'). Wyoming, south to New Mexico and Arizona, but apparently absent from Utah. Two subspecies occur within our study area:

1. Largest leaf blades not more than 8 cm long, rounded or subcordate basally; corolla mostly less than 5 mm long; capsule acute to rounded at the apex, 4.5–5.2 mm long .... **subsp. *arizonica***
1' Largest leaf blades up to 20 cm long, cuneate basally; corolla 5–8 mm long; capsule notched at the apex, 5–6 mm long .... **subsp. *plantaginea***

subsp. *arizonica* (Pennell) J. M. Porter (of Arizona) Arizona kittentails. Subscapose perennial herb. **STEMS** 1 to several, 10–35 cm tall, tomentose but becoming strigose or glabrate. **LEAVES**: basal tomentulose abaxially, strigose to glabrate adaxially, petiole usually 1.5–6.5 cm long; the blades 2–8 cm long, 1.5–4 cm wide, acute apically, ovate to obovate, obtuse at base, broadly cuneate to subcordate basally, crenate, the incisions 0.2–0.6 mm deep, rarely entire, the primary veins pinnate, cauline leaves alternate, bractlike, sessile to subpetiolate below, broadly lanceolate to ovate, entire to serrate, pilose. **INFL** a subspicate raceme, 7–10 cm long, tomentulose, flowers dense. **FLOWER** calyx 4–7 mm long, 5-lobed, rarely 4-lobed, with lateral lobes connate for less than 1/5 their length, villous-ciliate; corolla about 4.5–5.3 mm long, white to pinkish purple, the upper (adaxial) lip slightly exserted from the calyx, the lower (abaxial) lip 1/2 as long as the upper, 2- to 3-lobed, the lobes over 1/3 the length of the lip; stamens exserted, with white or pale filaments, not conspicuously pigmented, 7–10 mm long. **FRUIT** a capsule, 3–4 mm long, 3.5–4.5 mm wide, glabrous, acute to rounded at the apex, the style 3–4 mm long. **SEEDS** 1.2–2.4 mm long, orbicular, flat, 4–6 per locule. Moist meadows; ponderosa pine and Douglas-fir communities. ARIZ: Apa; NMEX: McK, SJn. 2130–2900 m (7000–9500'). Flowering May–Jun. Fruit: Jun–Jul. Restricted to Arizona and New Mexico. Recent systematic work suggests that *S. plantaginea* and *Besseya arizonica* are conspecific. Because of the morphological and ecological differences between the two taxa, we choose to recognize *B. arizonica* at the subspecific rank, rather than reducing it to synonymy.

subsp. *plantaginea* Subscapose perennial herb. **STEMS** 18–48 cm tall, tomentose but becoming glabrate. **LEAVES**: basal tomentulose, petiolate, the blades 5–20 cm long, 2–4 cm wide, acute apically, ovate to obovate or oblanceolate, obtuse at base, broadly to narrowly cuneate basally, crenate, the incisions 0.3–0.8 mm deep, rarely simple, the primary veins pinnate, petiole usually 2–5 cm long; cauline leaves alternate, bractlike, sessile to subpetiolate below, broadly lanceolate to ovate, entire to serrate, pilose. **INFL** a subspicate raceme, 7–13 cm long, tomentulose, flowers dense. **FLOWER** calyx 4–7 mm long, 5-lobed, rarely 4-lobed, with lateral lobes connate for less than 1/3 their length, villous-ciliate; corolla about 5–8 mm long, white to pinkish purple, the upper (adaxial) lip slightly exserted from the calyx, the lower (abaxial) lip 1/2 as long as the upper, 2- to 3-lobed, the lobes over 1/3 the length of the lip; stamens exserted, with white or pale filaments, not conspicuously pigmented, 7–11 mm long. **FRUIT** a capsule, 3–6 mm long, 3.5–5 mm wide, pubescent to glabrous, notched at the apex, the style 3–4.5 mm long. **SEEDS** 2–3 mm long, orbicular, flat, mostly 4 per locule. Hills and mountains, often on moist wooded slopes; sagebrush, piñon, oak, ponderosa pine, and mixed conifer communities. COLO: Arc, Hin, LPl, Mon, SMg; NMEX: RAr. 1830–3050 m (6000–10000′). Flowering: Jun–Jul. Fruit: Jul–Aug. Wyoming, south to New Mexico and Arizona, but apparently absent from Utah.

Synthyris ritteriana Eastw. **CP** (honoring B. W. Ritter, rancher and plant collector of Durango, Colorado) Ritter's kittentails. Subscapose perennial herb, (15) 20–35 cm tall. **STEMS** simple; pubescent to glabrate. **LEAVES**: basal often sparsely pubescent on the underside (abaxial), glabrate or ciliate on the upper (adaxial), the petioles 2.5–12 cm long; the blades 5–15 cm long, 1.5–7 cm wide, elliptic to oblong, acute and cuneate basally, acute apically, crenate to serrate, the incisions 0.3–1.4 mm deep, the primary veins pinnate, cauline leaves alternate, bractlike, sessile to subpetiolate below, broadly lanceolate to ovate, entire to serrate, sparsely villous. **INFL** a subspicate raceme, villous, flowers dense. **FLOWER** calyx 5–7.5 mm long, 3- or 4-lobed (5-lobed), orbicular to ovate, lateral lobes connate more than 1/2 the length of the calyx, villous-ciliate; corolla 5.5–6.5 mm long, lower lip of corolla with lobes short, only about 1/5 as long as the lip, lemon-yellow; stamens exserted, with white or pale filaments, not conspicuously pigmented, 6–8 mm long. **FRUIT** a capsule, 3.5–5 mm long, 3–4.5 mm wide, glabrous, rounded to notched at the apex, the style 5–6 mm long. **SEEDS** 2–3.2 mm long, orbicular, flat, mostly 4 per locule. [*Besseya ritteriana* (Eastw.) Rydb.; *S. reflexa* Eastw.; *S. ritteriana* subsp. *obtusa* A. Nelson]. Cliffsides and moist banks or meadows near timberline. COLO: Hin, LPl, Mon, SJn. 1230–3750 m (7000–12300′). Flowering: May–Aug. Fruit: late Jun–Sep. Endemic to west-central and southwestern Colorado.

Veronica L. Speedwell

(origin uncertain, possibly to honor St. Veronica) Annual, biennial, or perennial herbs. **STEMS** erect, decumbent, or prostrate. **LEAVES** opposite, cauline (not to be confused with the alternate, foliose floral bracts), entire, crenate, or serrate. **INFL** a terminal raceme or composed of lateral racemes from axils of upper leaves; bracts foliose or reduced, alternate or sometimes subopposite. **FLOWER** calyx of 4 essentially distinct lobes, sometimes united forming a short tube; corolla blue, violet to pink, or white, subrotate, the tube very short, irregularly 4-lobed, the adaxial (upper) lobe largest, the lower 3 lobes external to the upper in bud, the lowermost lobe the smallest; stamens 2, anther cells parallel; stigmas united and capitate. **FRUIT** a loculicidal capsule, strongly compressed to plump, ovoid, orbicular or often emarginate to deeply notched into a heart shape. **SEEDS** few to many, flattened, plano-convex or cup-shaped, smooth or rarely roughened. $2n = 14, 16, 18$. A large genus of between 180 and 250 species, in north-temperate Eurasia and North America, a few species in the Southern Hemisphere. Formerly included in Scrophulariaceae. Two of the following seven species are introduced from the Old World (Pennell 1921; Thieret 1955; Walters and Webb 1972).

1. Racemes axillary and opposite, never terminating the main stem; glabrous; perennial .... (2)
1′ Racemes terminal, rarely with additional axillary racemes (then often alternate); at least the stems pubescent; annual or perennial .... (4)

2. Leaves all petiolate; pedicels 5–10 mm long; leaves lanceolate to ovate, acute to obtuse; capsule entire or scarcely notched; racemes 10–25-flowered; widespread .... ***V. americana***
2′ Leaves sessile, lower leaves sometimes subpetiolate .... (3)

3. Leaves broadly lanceolate to ovate, 1.5–3 times as long as wide, somewhat succulent; racemes often more than 30-flowered, the pedicels upcurved to ascending; capsule slightly longer than wide, scarcely notched; corolla 5–10 mm across, blue or pale violet with purplish guidelines; widespread .... ***V. anagallis-aquatica***
3′ Leaves lanceolate, (2.5) 3–5 times as long as wide, thin; racemes less than 25-flowered, the pedicels more or less straight and spreading; capsule slightly wider than long with a more evident notch 0.1–0.3 mm deep; corolla 3–5 mm across, whitish to pink or pale blue .... ***V. catenata***

4. Perennial, rhizomatous; raceme somewhat congested; montane .... (5)
4′ Annual, fibrous roots or taproot; raceme mostly elongate; valleys and foothills .... (6)

5. Capsule as long as or longer than wide, 4.5–6.5 mm long, exceeding the calyx; style 0.8–1.3 mm long; corolla 6–10 mm across .... *V. nutans*
5′ Capsule wider than long, 2.8–3.7 mm long, not exceeding the calyx; style 2.2–3 mm long; corolla 4–8 mm across .... *V. serpyllifolia*

6. Pedicels short, 0.5–1.5 (2) mm long; calyx 3–6 mm long, the segments lanceolate; capsule 3.5–5 mm wide; seeds many, 0.5–1 mm long, more or less strongly flattened and smooth .... *V. peregrina*
6′ Pedicels 15–30 mm long; calyx (3) 4–12 mm long, the segments broadly lanceolate to ovate, often with pronounced veins and leaflike; capsule 4–8 mm wide, more or less plump; seeds few, 5–11 per locule, 1.2–2 mm long, cup-shaped, roughened .... *V. persica*

Veronica americana Schwein. ex Benth. (of America) American brooklime. Aquatic, rhizomatous perennial, 5–35 (60) cm tall. **STEMS** erect to ascending, decumbent at base, rooting at lower nodes, often branched; glabrous. **LEAVES** petiolate, the blade (0.5) 1.5–3 (5) cm long, (3) 7–20 (30) mm wide, lanceolate to ovate, crenate-serrate or subentire below, acute to obtuse apically. **INFL** an axillary raceme, 10- to 25-flowered, glabrous; bracts reduced, linear; pedicels 5–10 mm long, divaricate. **FLOWER** calyx 2.5–4.5 (5.5) mm long, the lobes oblanceolate to ovate; corolla 5–10 mm across, blue. **FRUIT** a septicidally and loculicidally dehiscing capsule, 2.5–3.8 mm long, 3–4 mm wide, entire or scarcely notched, the style 1.7–3 (4) mm long. **SEEDS** numerous, 0.5–0.7 mm long, brownish, plano-convex. $2n = 36$. Moist soils of springs, seeps, bogs, hanging gardens, slow-flowing streams, lake shores, or meadows. ARIZ: Apa; COLO: Arc, Con, Hin, LPl, Min, Mon, RGr; NMEX: McK, RAr, SJn; UTAH. 1400–3300 m (4600–10800′). Flowering: Jul–Aug; Fruit: late Jul–Sep. Common throughout most of North America, but absent from the southeastern U.S. and only spotty in the higher mountains of Mexico.

Veronica anagallis-aquatica L. (water pimpernel) Water speedwell, brook-pimpernel. Aquatic, rhizomatous perennial, 9–65 (100) cm tall. **STEMS** erect, ascending, simple to much-branched at the base; glabrous. **LEAVES** somewhat clasping, 2–8 cm long, 0.5–4 cm wide, elliptic, upper leaves sessile, clasping, crenate to serrate, acute or acuminate, elliptic-lanceolate to ovate; lower leaves subpetiolate or sessile, oblanceolate to obovate, obtuse to acute, cordate, rounded, or truncate basally. **INFL** an axillary raceme, glabrous to glandular-puberulent, often more than 30-flowered, the bracts reduced, lanceolate to linear; pedicels 4–8 mm long, upcurved or ascending. **FLOWER** calyx 3–5.5 mm long, the lobes broadly lanceolate; corolla 5–10 mm across, blue, pale violet with purplish guidelines. **FRUIT** a septicidally and loculicidally dehiscing capsule, slightly longer than wide, 2.8–4 mm long, 2.5–3.8 mm wide, plump, scarcely notched, the style 1.5–3 mm long. **SEEDS** numerous, 0.3–0.5 mm long, more or less plano-convex. $2n = 18, 36$. *Veronica micromera* Wooton & Standl., described from ditches in the vicinity of Shiprock, New Mexico, and known only from the type collection, is assigned here. It was purported to differ in its oval to obovate, entire leaves that are less than 30 mm long and narrowed at the base, and the sepals being conspicuously longer than the capsule. No recent collections conform to this description. Wet meadows, stream banks, or in water of seeps, springs, and slow-moving streams. ARIZ: Apa, Nav; COLO: Arc, Hin, LPl, Mon; NMEX: McK, RAr, SJn; UTAH. 1200–2500 m (4000–8200′). Flowering: Jun–Aug. Fruit: Jul–Sep. Native to Europe, Africa, and Asia, but widely naturalized in the New World.

Veronica catenata Pennell (chained, chainlike) Aquatic, rhizomatous perennial, 8–28 (60) cm tall. **STEMS** erect to ascending, branched; glabrous (ours) to pubescent. **LEAVES** sessile, clasping, blade 2.5–5 (9) cm long, 0.5–1.5 (2.5) cm wide, lanceolate, acute, entire or subcrenate, appearing thin in pressed specimens. **INFL** an axillary raceme, glabrous, 15–25-flowered, the bracts much reduced, narrowly lanceolate; pedicels more or less straight and spreading, 3–7 mm long. **FLOWER** calyx 2.5–3.5 mm long, the lobes broadly lanceolate to ovate, obtuse to acute; corolla 3–5 mm across, white to pink or pale bluish. **FRUIT** a septicidally and loculicidally dehiscing capsule, 2.5–3 mm long, 3–4 mm wide, plump, the notch evident, 0.1–0.3 mm deep, the style about 1.3–2 mm long. **SEEDS** numerous, 0.3–0.7 mm long, yellow-brown, plano-convex. $2n = 36$. Wet meadows, gravel bars, irrigation channels, banks of lakes and ponds, and slow-flowing streams. ARIZ: Apa; COLO: Arc, LPl; NMEX: RAr, SJn. 1370–1830 m (4500–6000′). Flowering: Jun–Aug. Fruit: Jul–Sep. United States and adjacent Canada.

Veronica nutans Bong. (nodding) American alpine speedwell, western alpine speedwell. Rhizomatous perennial, 8–25 (40) cm tall. **STEMS** ascending, erect, occasionally decumbent at the base, usually simple; villous-hispid with loosely spreading hairs, leaves sometimes glabrous. **LEAVES** sessile, 2–3 (4) cm long, 0.8–1.8 cm wide, elliptic to broadly lanceolate, obtuse to acute, crenate to entire. **INFL** a compact to interrupted, terminal raceme, 2–4 (9) cm long, viscid-villous or glandular, the bracts linear to lanceolate; pedicels 2–6 mm long. **FLOWER** calyx 3.5–5.5 mm long, the lobes lanceolate to oblanceolate; corolla 6–10 mm across, deep blue. **FRUIT** a glandular-pubescent capsule, longer than the calyx, longer than wide, 4.5–6.5 mm long, 2.8–4 mm wide, the notch 0.1–0.5 mm deep, the style 0.8–1.3 mm long. **SEEDS** numerous, small, 0.7–0.9 mm long, pale brown. 2*n* = 18. [*Veronica wormskjoldii* Roem. & Schult., an eastern, tetraploid species, misapplied to our diploid species]. Moist subalpine and alpine meadows, tundra, stream banks, and shores of ponds and lakes. COLO: Arc, Con, Hin, LPl, Min, Mon, RGr, SJn; NMEX: RAr, SJn; UTAH. 2600–4260 m (8500–14000′). Flowering: Jul–Aug. Fruit: Aug–Sep. North American, from Alaska to Greenland, south to the mountains of California, northern Arizona, northern Mexico, and in the east to New Hampshire.

Veronica peregrina L. (foreign, wandering) Purslane speedwell, neckweed, necklace-weed. Annual, 5–22 (30) cm tall, with short taproot. **STEMS** erect, ascending, simple or branched at the base, occasionally branched inflorescence; glandular-pubescent; leaves and bracts often glabrous. **LEAVES** sessile or the lower ones narrowed to a petiolar base, the blades 5–22 mm long, 0.5–5 mm wide, narrowly oblong to oblanceolate, entire or irregular, crenate-serrate. **INFL** an elongate, glandular-puberulent, terminal raceme, the bracts foliaceous at the base, gradually reduced upward; pedicels 0.5–1.5 mm long. **FLOWER** calyx 3–6 mm long, the lobes subequal, narrowly elliptic to lanceolate; corolla inconspicuous, 2–3 mm across, whitish. **FRUIT** a capsule, 3–4.5 mm long, 3.5–5 mm wide, more or less obcordate with a broad notch 0.2–0.5 mm deep, the style very short, 0.1–0.4 mm long. **SEEDS** numerous, 0.5–1 mm long, plano-convex. 2*n* = 52. Moderately moist to muddy meadows, stream banks, and lake or pond edges. ARIZ: Apa, Coc, Nav; COLO: Arc, Con, Dol, Hin, LPl, Min, Mon, SJn, SMg; NMEX: McK, RAr, SJn; UTAH. 1500–2900 m (4900–9500′). Flowering: Jun–Aug. Fruit: Jul–Sep. Throughout temperate North and South America, eastern Asia, and Europe. Our plants belong to **subsp. *xalapensis*** (Kunth) Pennell (of Xalapa, Mexico), with pubescent stems, whereas subsp. *peregrina* is glabrous. Subspecies *xalapensis* ranges from Alaska and Yukon to Mexico and east to the northern midwestern U.S. and the St. Lawrence River Valley.

Veronica persica Poir. (of Persia) Persian speedwell, bird's-eye speedwell. Annual, 9–30 cm tall, with a short taproot. **STEMS** loosely ascending, usually decumbent at the base and sometimes rooting at the lower nodes, simple or branched at the base; pilose. **LEAVES** short-petiolate, the blades 0.7–2.5 cm long, 0.5–2 cm wide, ovate, broadly ovate, or suborbicular, deeply crenate-dentate, obtuse to rounded apically, broadly cordate to truncate basally, sometimes the lowermost entire. **INFL** an elongate, lax, terminal raceme, the bracts foliose at the lower nodes, gradually reduced upward; pedicels 15–27 (30) mm long. **FLOWER** calyx 4.5–8 mm long, the lobes prominent, broadly lanceolate, prominent veins, ciliate; corolla relatively large, 8–12 mm across, blue. **FRUIT** a capsule, 3.5–4.5 mm long, 5–7 mm wide, the lobes divergent, the notch prominent, 0.7–1.2 mm deep, the style 2–3 mm long. **SEEDS** 5–11 per locule, 1.2–2 mm long, cup-shaped, transversely rugose, brown. 2*n* = 28. [*Pocilla persica* Fourr.]. Weed in lawns and cultivated fields. COLO: Arc, LPl, Mon; NMEX: RAr. 1220–2440 m (4000–8000′). Flowering: May–Jun. Fruit: Jun–Aug. A native of western Asia and southern Europe. Naturalized throughout Europe and most of North America.

Veronica serpyllifolia L. (wild thyme–leaved) Thyme-leaf speedwell. Rhizomatous perennial, 8–24 (30) cm tall. **STEMS** ascending, sometimes decumbent to procumbent at the base, simple or short-branched below; finely puberulent or glabrous. **LEAVES** short-petiolate below, subsessile above, blade 1–2.5 cm long, 0.8–1.5 cm wide, elliptic to broadly ovate, rounded to obtuse, obscurely crenate to entire. **INFL** a somewhat congested terminal raceme, 2.5–12 cm long, glandular-pubescent, the bracts lanceolate, foliaceous below; pedicels 2.5–5 (7.5) mm long. **FLOWER** calyx 2.5–4 mm long, the lobes ovate or oblong, slightly unequal; corolla 4–8 mm across, blue (in ours) or white, the tube pubescent within. **FRUIT** a capsule, slightly glandular-pubescent, shorter than the calyx, 2.8–3.7 mm long, 3.5–5 mm wide, the notch evident, 0.3–0.8 mm deep, the style 2.2–3 mm long. **SEEDS** numerous, 0.6–0.8 mm long, pale brown. 2*n* = 14, 28. [*Veronicastrum serpyllifolium* (L.) Fourr. subsp. *humifusum* (Dickson) W. A. Weber]. Moist stream banks and gravel bars, meadows, and shores of lakes and ponds, muddy areas; Douglas-fir to subalpine communities. ARIZ: Apa; COLO: Arc, Hin, LPl, Min, Mon, SJn; NMEX; SJn; UTAH. 1700–3600 m (5550–11800′). Flowering: Jun–Aug. Fruit: Jul–Sep. Widespread, particularly in the mountains of the Northern Hemisphere, native of North America from Alaska and Newfoundland, south to California, Nevada, northern Arizona, and northern New Mexico in the western United States, to Vermont in the east, and the mountains of Mexico. Our material belongs to **subsp. *humifusa*** (Dicks.) Syme.

PLUMBAGINACEAE Juss. SEA LAVENDER or SEA THRIFT FAMILY

Linda Mary Reeves

(Latin for leadlike, purplish blue coloring) Annual or perennial herbs, shrubs, and vines. **LEAVES** in a basal rosette or alternate on aerial branched stems; simple; without stipules. **INFL** a cyme, panicle, raceme, spike, or dense head, often with bracts that sometimes form an involucre. **FLOWERS** perfect; actinomorphic or tubular, generally 5-parted; sepals

5, forming a lobed persistent tube, often papery, sometimes colored; petals free, often connate at the base or fused; stamens 5, opposite petals, mostly free; ovary superior, carpels 5, locule 1; ovule 1; styles and/or stigmas 5. **FRUIT** indehiscent, enclosed by the calyx. About 10 genera and about 560 species, worldwide distribution, often in dry and saline or seaside environments. *Limonium*, *Armeria*, *Ceratostigma*, and *Plumbago* have many important horticultural species and cultivars.

Limonium Mill. Sea Lavender, Marsh Rosemary

(Latin name for sea lavender) Suffrutescent perennial. **LEAVES** petioled, fleshy. **INFL** a panicle or spike, with flowers often arranged in smaller 1-sided spikelets. **FLOWERS** often appearing tubular; calyx funnelform, membranous, persistent; corolla of 5 somewhat distinct petals with long claws; stamens attached to petal bases; styles 5, rarely 3. **FRUIT** membranous, enclosed by calyx. About 300 widely distributed species.

Limonium latifolium Moench (one-sided, referring to the inflorescence) Sea lavender. Perennial herb. **LEAVES** basal, fleshy, simple; glabrous to glaucous; spatulate to obovate; margin entire; apex rounded; base tapered. **INFL** cymose, with flowers arranged in groups of 2–4 spikelets; often many flowering stalks present, giving the flowering plant a cloud-like appearance. **FLOWERS** with calyx persistent, white, 5-lobed; corolla tubular, papery, pink, purple, blue, or white. Introduced from the Mediterranean. Along roadsides. COLO: Mon. 1800 m (6000′). Flowering: May–Sep. Fruit: Jun–Oct. One specimen known from our area near Cortez, but may prove to be more widely established.

POACEAE Barnhart (Gramineae Juss.) GRASS FAMILY

Kelly W. Allred

Annual or perennial herbs, tufted, rhizomatous, or stoloniferous. **CULMS** mostly terete, simple or branched, with conspicuous nodes, the internodes hollow or pithy. **LEAVES** entire, alternate, 2-ranked, differentiated into sheath, ligule, and blade; sheaths tubular, enclosing the culm, the margins usually free from one another and overlapping (open sheath), or sometimes connate edge to edge (closed sheath), the upper margins sometimes prolonged into sheath auricles; ligules (appendages produced on the adaxial surface of the leaf at the junction of sheath and blade) membranous, composed of hairs, or a combination of a basal membrane with a ciliate fringe, absent only in *Echinochloa* in our plants; blades elongate, narrow, with parallel venation, flat to variously rolled, the lower margins at junction with sheath sometimes expanded into auricles. **INFL** a spike, raceme, or panicle, the flowers arranged in spikelets. **SPIKELETS** consisting of a central axis (rachilla) and associated bracts, glumes, lemmas, and paleas, the flowers borne within the lemma and palea and this unit termed a floret; florets 1–many per spikelet, the basal portion often hardened into a callus; glumes 2 (rarely 1 or absent), 1- to several-nerved, awned or awnless; lemmas 1- to several-nerved, awned or awnless; paleas 2-nerved and keeled, rarely awned; disarticulation above the glumes and the florets falling separately, above the glumes but the florets falling together, below the glumes and the spikelet falling entire, or within the inflorescence and the spikelets falling with accessory structures. **FLOWERS** highly reduced, hidden within the spikelets and florets; perianth represented by 2 (sometimes 1 or absent) delicate lodicules; stamens usually 3, sometimes fewer or none, the anthers

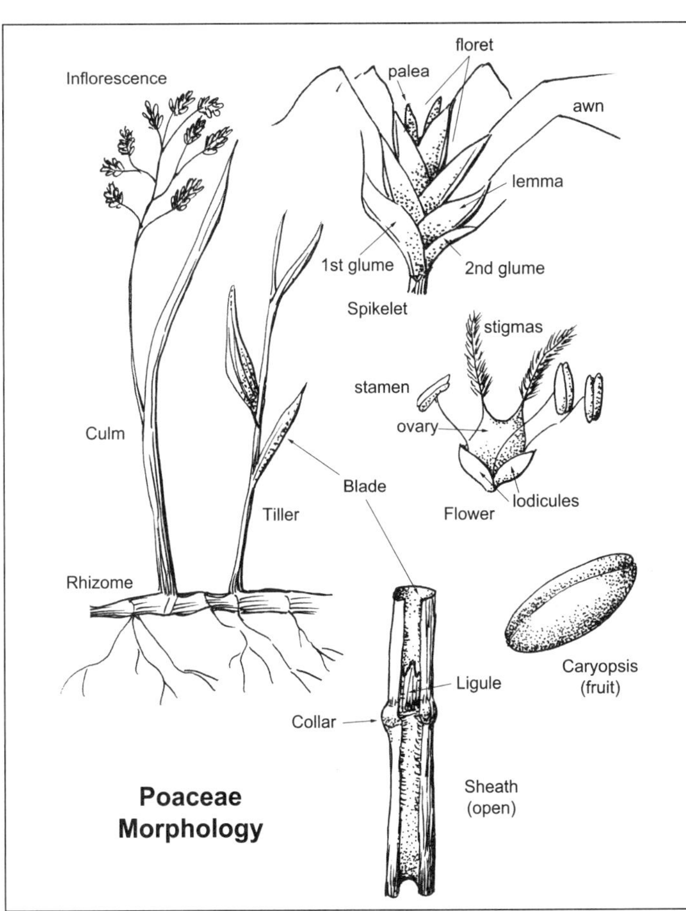

Poaceae Morphology

with 2 chambers; pistil 1 per flower, with a superior ovary, 1 ovule, and usually 2 plumose stigmas. **FRUIT** a 1-seeded, dry, indehiscent grain (caryopsis), the ovary and ovule walls fused (rarely separate). About 730 genera and 8000 species worldwide, in nearly all habitats. The grass family is the fifth largest plant family in the world in terms of number of species (behind Asteraceae, Orchidaceae, Fabaceae, and Rubiaceae). It is easily the most essential and important plant family to humans, in terms of food products and economic value. Important species include corn, wheat, oats, rice, sugarcane, barley, and millet, among many others (Allred 2005a; Clayton and Renvoize 1986; Gould and Shaw 1983; Grass Phylogeny Working Group 2001).

1. All or some of the spikelets concealed and hidden from view within modified structures, such as spiny burs, involucres, and detachable clusters of hard bracts .... **GROUP A**
1′ Spikelets not concealed and not hidden within modified structures, but evident and easily seen, sometimes closely subtended by foliage leaves or covered by hairs .... (2)

2. One or more bristles (sterile branchlets) borne immediately below the spikelets, the spikelets falling and leaving the bristles on the plant .... ***Setaria***
2′ Bristles not borne immediately below the spikelets .... (3)

3. Lemmas each with 9 plumose awns .... ***Enneapogon***
3′ Lemmas with 1–3 awns or awnless .... (4)

4. Flowering shoots 2 m or more tall .... **GROUP B**
4′ Flowering shoots less than 2 m tall .... (5)

5. All or many of the spikelets sessile and borne on the main axis; inflorescence branches absent, the inflorescence a spike, spicate raceme, or dense headlike cluster of spikelets .... **GROUP C**
5′ All or most of the spikelets borne on branches, the inflorescence a panicle, or if branches absent then all the spikelets with evident pedicels and few (if any) sessile .... (6)

6. Glumes mostly hardened and completely enclosing the florets, dorsally compressed; disarticulation below the glumes and nearly always in units consisting of a sessile spikelet with attached rachis joint and pedicel (the pedicelled spikelet present or absent); spikelets borne in pairs, one spikelet sessile or subsessile and one spikelet pedicelled (sometimes the pedicelled spikelet absent, but the pedicel always present); lemmas very thin and translucent, delicate, awned or awnless; Andropogoneae Tribe .... **GROUP D**
6′ Combination of features other than above .... (7)

7. Spikelets with a single floret only .... **GROUP E**
7′ Spikelets with at least 2 florets, some may be small and poorly developed (look carefully) .... (8)

8. Spikelets with 2 florets only, the upper bisexual and usually with a hardened lemma at maturity, the lower male or neuter; lemma of the lower floret similar to the second glume in size and texture; disarticultion below the glumes; spikelets dorsally compressed; Paniceae Tribe. .... **GROUP F**
8′ Combination of features other than above .... (9)

9. Lemmas with 3 nerves, the nerves usually prominent .... **GROUP G**
9′ Lemmas with 5–many nerves, at least at the base, or the nerves not discernible .... **GROUP H**

GROUP A Spikelets variously concealed

1. Spikelets enclosed in a bur of sharp, stiff spines, the bur falling entire .... ***Cenchrus***
1′ Spikelets not enclosed in a bur of sharp spines, but in bony clusters hidden among the foliage .... ***Buchloë***

GROUP B Flowering shoots 2 m or more tall

1. Grasses cultivated for ornament, landscaping, or as a harvested crop, occasionally escaping around fields or dwellings .... (2)
1′ Grasses wild or weedy, or seeded for range or pasture improvement, but not crop or ornamental plants .... (4)

2. Plants not in large tussocks, the shoots single, or if clustered, then with strong, vigorous rhizomes .... ***Sorghum***
2′ Plants growing in large, thick tussocks with numerous flowering shoots; rhizomes lacking .... (3)

3. Blades scabrous to smooth on the margins; spikelets borne in pairs on spicate branches, with no florets extending beyond the stiff glumes .... ***Saccharum***
3′ Blades sharply saw-toothed on the margins; spikelets borne singly on rebranching branches of the inflorescence, with several florets extending beyond thin glumes .... ***Cortaderia***

4. Plants tufted, not developing rhizomes .... (5)
4′ Plants developing rhizomes .... (9)

5. Inflorescence a spike, no branches developed .... ***Elymus***
5′ Inflorescence a panicle with branches .... (6)

6. Disarticulation above the glumes; spikelets awned .... (7)
6′ Disarticulation below the glumes; spikelets awned or awnless; sheaths mostly rounded .... (8)

7. Basal sheaths compressed-keeled; spikelets purplish; awns less than 1.5 cm long .... ***Muhlenbergia***
7′ Basal sheaths round; spikelets greenish or tawny; awns 2–3 cm long .... ***Achnatherum robustum***

8. Inflorescence branches 2–5 in number and mostly not rebranched, clustered toward the tip of the shoot .... ***Andropogon gerardii***
8′ Inflorescence branches numerous and rebranched, not clustered toward the tip of the shoot .... ***Panicum***

9. Disarticulation below the glumes, the spikelets falling entire .... (10)
9′ Disarticulation above the glumes, the glumes remaining on the plant and the florets falling .... (13)

10. Inflorescence a panicle of 2–5 spicate, unbranched primary branches clustered at the tip of the shoot, sometimes a few of the branches rebranching .... ***Andropogon gerardii***
10′ Inflorescence a rebranched panicle, the numerous primary branches always rebranching .... (11)

11. Outer bracts of the spikelet (glumes) membranous, thin and flexible, not hardened; upper floret hardened at maturity; spikelets awnless .... ***Panicum***
11′ Outer bracts of the spikelet (glumes) stiff, hardened; inner floret very thin and delicate, not at all hardened; spikelets awned, at least when young .... (12)

12. Spikelets dull, fuzzy-hairy, the hairs standing out from the spikelet; awns persistent through maturity .... ***Sorghastrum***
12′ Spikelets somewhat shiny, glabrous or slightly pubescent, the hairs pressed against the spikelet; awns early-deciduous or absent .... ***Sorghum***

13. Spikelets with a single floret .... ***Calamovilfa***
13′ Spikelets with several florets .... ***Phragmites***

GROUP C

Inflorescence a spike, spicate raceme, or dense headlike cluster, all or many of the spikelets sessile on the main axis, branches absent from the inflorescence

1. Disarticulation below the glumes, the spikelets falling entire or in clusters, no spikelet parts left on the axis (tardily so in the annual *Crypsis*) .... (2)
1′ Disarticulation above the glumes, the glumes often remaining on the inflorescence .... (15)

2. Main axis of the inflorescence breaking apart at maturity .... (3)
2′ Main axis of the inflorescence remaining intact .... (10)

3. Spikelets borne in pairs of 1 sessile and 1 pedicelled (sometimes only the pedicel present); glumes mostly enclosing the spikelet, the florets mostly not visible .... ***Schizachyrium***
3′ Spikelets borne other than above; glumes may be longer than, but not enclosing the spikelet, the florets usually visible .... (4)

4. Spikelets 3 at each node of the main axis, the lateral pair pedicelled, the central spikelet sessile; spikelets with 1 floret .... ***Hordeum***
4′ Spikelets mostly 1 or 2 at each node of the main axis, if 3 then not otherwise as above; spikelets with 2 to many florets .... (5)

5. Spikelets mostly 1 at each node of the main axis ....(6)
5′ Spikelets mostly 2 at each node of the main axis ....(9)

6. Plants annual ....(7)
6′ Plants perennial ....(8)

7. Spikes 0.6–2 cm long .... ***Eremopyrum***
7′ Spikes 5–10 cm long .... ***Aegilops***

8. Spikes 3–7 (8) cm long, very dense; glumes 6–8 mm long, 1–3 (5)-nerved .... ***Elymus scribneri***
8′ Spikes 8–16 cm long, not especially dense; glumes 7–10 mm long, (3) 5–7-nerved ... these are hybrids and hybrid derivatives between *Elymus trachycaulus* and *E. scribneri*, *E. longifolius*, or *E. canadensis*. The high-elevation populations generally above 2750 m (9000′) have been called *E. bakeri* (E. E. Nelson) Á. Löve.

9. Glumes 3–7 mm long; anthers 4–5 mm long .... ***Psathyrostachys***
9′ Glumes 25–100 mm long; anthers, when present, about 2 mm long .... ***Elymus***

10. Plants strongly rhizomatous perennials .... ***Pleuraphis***
10′ Plants tufted annuals or perennials, not rhizomatous ....(11)

11. First glume with 2 or 3 awns; lower stems angled or flattened somewhat .... ***Lycurus***
11′ First glume with a single awn or awnless; lower stems rounded ....(12)

12. Ligules hairy; sheaths prominently inflated; blades widely spreading to reflexed; inflorescence dense and headlike, the base included in the sheath; much-branched annuals .... ***Crypsis***
12′ Plants not as above in all respects ....(13)

13. Glumes awnless; lemmas awned (use a lens) .... ***Alopecurus***
13′ Glumes awned ....(14)

14. Glumes strongly flattened laterally, ciliate on the keeled midnerve .... ***Phleum***
14′ Glumes rounded on the back, not keeled, not ciliate on the midnerve but may be pubescent elsewhere .... ***Polypogon***

15. Lemmas with 3 awns .... ***Aristida***
15′ Lemmas with 1 awn or awnless ....(16)

16. Spikelets with 1 floret only ....(17)
16′ Spikelets with more than 1 floret, some may be poorly developed, rudimentary, or vestigial ....(19)

17. Plants annual; leaves with prominent, clawlike auricles 2–6 mm long; awns 50–160 mm long .... ***Hordeum***
17′ Plants perennial; leaves without auricles, or occasionally with small rounded auricles about 1 mm long; awns 1–4 mm ....(18)

18. Spikelets strongly compressed; glumes flattened, keeled on the midnerve, completely enclosing the floret .... ***Phleum***
18′ Spikelets not strongly compressed; glumes rounded on the back, only slightly keeled, not completely enclosing the floret .... ***Muhlenbergia***

19. Spikelets in dense, sessile, headlike clusters that are mostly surpassed by and nestled within the foliage ....(20)
19′ Spikelets not in dense, headlike clusters, or if so then elevated well above the foliage ....(21)

20. Plants annual; blades flat; glumes shorter than the lower lemma .... ***Munroa***
20′ Plants perennial; blades rolled; glumes longer than the lower lemma .... ***Dasyochloa***

21. Lemmas with 3 conspicuous nerves ....(22)
21′ Lemmas with 1 or 5–several nerves ....(23)

22. Lemmas conspicuously pubescent; spikelets with several well-developed florets; blades white-margined. ***Erioneuron***
22′ Lemmas glabrous or scabrous; spikelets with 1 well-developed floret and 1–3 rudiments above it; blades not white-margined .... ***Bouteloua***

23. Spikelets 2 or more per node of the rachis ....(24)
23′ Spikelets mostly 1 per node of the rachis ....(28)

24. Rhizomes present, evident, creeping .... ***Leymus***
24′ Rhizomes absent, occasionally short rhizomes developed but the plants still forming dense clumps .... (25)

25. Glumes absent or reduced to 1 or 2 minute bristles; spikelets horizontally spreading or ascending at maturity .... ***Elymus***
25′ Glumes present; spikelets rarely horizontally spreading .... (26)

26. Glumes 2–10 cm long .... ***Elymus***
26′ Glumes shorter than 1.5 cm .... (27)

27. Glumes 2- to 5-nerved; anthers 1.5–3 mm long .... ***Elymus***
27′ Glumes 1-nerved; anthers 3–5 mm long .... ***Psathyrostachys***

28. Spikelets placed edgewise to the rachis, the first glume absent on all but the terminal spikelets .... ***Lolium***
28′ Spikelets placed flatwise to the rachis; both glumes present on all spikelets .... (29)

29. Plants annual .... (30)
29′ Plants perennial .... (32)

30. Spikes very short, 0.6–2 cm long; plants usually less than 30 cm tall .... ***Eremopyrum***
30′ Spikes longer, mostly 5–15 cm long; plants usually much more than 30 cm tall .... (31)

31. Glumes narrow, linear, 1-nerved; spikelets with 2 florets .... ***Secale***
31′ Glumes broad, oblong to ovate, 3- to several-nerved; spikelets mostly with 3–5 florets .... ***Triticum***

32. Glumes linear, needlelike, 1-nerved (occasionally broader at the base and 3-nerved) .... ***Leymus***
32′ Glumes lanceolate or broader, usually 3- to 7-nerved .... (33)

33. Spikelets spreading away from the rachis, placed very close together on the main axis; rachis internodes between the spikelets 0.3–3 mm long in the middle of the spike .... ***Agropyron***
33′ Spikelets mostly pressed against the rachis, or curving outward toward the tip of the spikelet; rachis internodes between the spikelets 4–25 mm long .... ***Elymus***

GROUP D Andropogoneae Tribe

1. Each inflorescence composed of a single unbranched spicate raceme without branches, subtended by a somewhat inflated bladeless sheath (spathe), the flowering shoot usually bearing numerous such inflorescences .... ***Schizachyrium***
1′ Each inflorescence composed of numerous branches, with or without inflated sheaths subtending the inflorescence (spathes) .... (2)

2. Spikelets all similar in appearance and size .... (3)
2′ Spikelets not all similar, the pedicelled ones often smaller in size or different in appearance when compared to the sessile ones .... (8)

3. Pedicels without a spikelet borne at the tip .... (4)
3′ Pedicels with a spikelet borne at the tip .... (5)

4. Flowering shoots mostly with 1 or a few large, terminal panicles 10 cm or more long .... ***Sorghastrum***
4′ Flowering shoots with numerous small panicles clustered together, each less than 3 cm long and each with a subtending spathe .... ***Andropogon glomeratus***

5. Panicles 30–70 cm long; plants escaped from cultivation and growing in large, tall tussocks .... ***Saccharum***
5′ Panicles 4–16 cm long; plants otherwise, usually in natural habitats .... (6)

6. Pedicels and rame segments (rachis joints) with a central longitudinal groove or membrane, flattened in cross section; plants tufted .... ***Bothriochloa***
6′ Pedicels and rame segments without a central groove or membrane, nearly round in cross section; plants tufted or rhizomatous .... (7)

7. Panicles narrow and spikelike, with soft silky hairs, 1–3 cm wide and 8–18 cm long, the branches scarcely noticeable at arm's length .... ***Imperata***
7′ Panicles not as above, usually wider and/or shorter or the branches obvious at arm's length .... ***Andropogon***

8. Pedicels and rame segments (rachis joints) with a central groove or membrane running lengthwise, flattened in cross section .... ***Bothriochloa***
8′ Pedicels and rame segments without a central groove or membrane, nearly round in cross section, at least at the apex .... (9)

9. Inflorescence an open panicle with numerous (more than 5) rebranched branches; spikelets ovoid to nearly globose .... ***Sorghum***
9′ Inflorescence a panicle with 2–5 nearly digitate and mostly unbranched branches; spikelets lanceolate .... ***Andropogon gerardii***

GROUP E Spikelets with a single floret

1. Glumes and lemmas awnless .... (2)
1′ Glumes and/or lemmas awned .... (18)

2. Inflorescence a panicle of evident, unbranched, spicate primary branches .... (3)
2′ Inflorescence a panicle of rebranched branches, or dense and spikelike .... (6)

3. Panicle branches all attached at the tip of the main axis .... ***Cynodon***
3′ Panicle branches attached along the length of the main axis, not only at the tip .... (4)

4. Glumes equal in length or nearly so; spikelets nearly round in outline .... ***Beckmannia***
4′ Glumes unequal, the first glume shorter than the second; spikelets lanceolate in outline .... (5)

5. Spikelets widely spaced, rarely overlapping, appearing embedded in the branches; blades spirally twisted .... ***Schedonnardus***
5′ Spikelets very closely spaced, overlapping, not at all appearing embedded in the branches; blades not spirally twisted .... ***Spartina***

6. Disarticulation below the glumes .... (7)
6′ Disarticulation above the glumes .... (10)

7. Ligules hairy; sheaths prominently inflated; blades widely spreading to reflexed; inflorescence dense and headlike, the base included in the sheath; much-branched annuals .... ***Crypsis***
7′ Plants not as above in all respects .... (8)

8. Spikelets nearly round in outline, the glumes somewhat inflated or puffy-looking .... ***Beckmannia***
8′ Spikelets mostly lanceolate in outline, the glumes not at all inflated or puffy-looking .... (9)

9. Glumes softly pubescent on the midnerves; inflorescence dense and spikelike, rarely lobed .... ***Alopecurus***
9′ Glumes glabrous to scabrous, not softly pubescent; inflorescence usually lobed at least below .... ***Polypogon***

10. Lemma hardened at maturity, enclosing the palea and flower .... (11)
10′ Lemma remaining thin and flexible, not hardened, not enclosing the palea .... (13)

11. Lemma with 1 or 2 slender bracts, bristles, or scales at the base of the floret, these sometimes pubescent and often difficult to see without dissecting carefully .... ***Phalaris***
11′ Lemma without any bracts, bristles, or scales at the base of the floret .... (12)

12. Florets dorsally compressed; lemma margins not overlapping, the palea exposed, at least in part .... ***Piptatherum***
12′ Florets terete; lemma margins slightly overlapping, the palea hidden .... ***Oryzopsis***

13. Lemma with a single nerve; ligule a ring of hairs .... (14)
13′ Lemma with 3 or more nerves; ligule a membrane .... (15)

14. Lemma with a tuft of hairs at the base .... ***Calamovilfa***
14′ Lemma without a tuft of hairs at the base .... ***Sporobolus***

15. Lemma and palea nerves densely pubescent .... ***Blepharoneuron***
15′ Palea nerves glabrous or scabrous; lemma nerves not densely pubescent but may be short-pubescent .... (16)

16. Sheath margins fused together for 1/2 their length or more .... ***Catabrosa***
16′ Sheath margins overlapping most of their length .... (17)

17. Palea about as long as the lemma; body of the glumes (not including awn tips) shorter than the lemma; lemma mostly 3-nerved .... ***Muhlenbergia***
17′ Palea 1/2 or less as long as the lemma; body of the glumes longer than the lemma; lemma obscurely nerved .... ***Agrostis***

18. Inflorescence a panicle of several evident, unbranched, spicate, primary branches .... (19)
18′ Inflorescence a panicle of rebranched branches, or a raceme, or in some the pedicels and branches poorly developed and the inflorescence spikelike .... (21)

19. Spikelets nearly round in outline, the glumes somewhat inflated .... ***Beckmannia***
19′ Spikelets lanceolate in outline, the glumes not at all inflated .... (20)

20. Panicle branches all less than 2 cm long .... ***Bouteloua***
20′ Panicle branches mostly longer than 2 cm .... ***Spartina***

21. Lemma hard at maturity, usually enclosing the palea and flower, mostly with a well-developed and pointed callus (22)
21′ Lemma not hard (somewhat so in *Apera* but then the rachilla prolonged beyond the palea), not enclosing the flower and palea; mostly without a well-developed callus .... (27)

22. Ligule a ring of hairs; lemma terminating in 3 awns, the 2 lateral awns occasionally shortened and inconspicuous .... ***Aristida***
22′ Ligule a membrane; lemma terminating in a single awn, this may be deciduous .... (23)

23. Lemma margins strongly overlapping; palea less than 1/3 the length of the lemma, glabrous, lacking veins .... ***Nassella***
23′ Lemma margins not or only slightly overlapping; palea 1/3 to equaling the length of the lemma, always pubescent when short, sometimes glabrous when longer, 2-veined .... (24)

24. Awns 6–20 cm long or more; glumes longer than 1.8 cm .... ***Hesperostipa***
24′ Awns 0.5–7.5 cm long, if longer than 6 cm then the glumes 1–1.5 cm long .... (25)

25. Palea pubescent, the apex flat, the veins terminating below the apex; lemma coriaceous at maturity but not strongly indurate .... ***Achnatherum***
25′ Palea glabrous or pubescent, the apex appearing prow-tipped or pinched, the veins extending to the apex; lemma indurate at maturity .... (26)

26. Florets dorsally compressed; lemma margins not overlapping, the palea exposed, at least in part .... ***Piptatherum***
26′ Florets terete; lemma margins slightly overlapping, the palea hidden .... ***Oryzopsis***

27. Inflorescence spikelike or headlike, the branches absent or highly shortened .... (28)
27′ Inflorescence a panicle with evident branches .... (31)

28. First glume 2-nerved with 2 or 3 awns; lower stems angled or flattened somewhat .... ***Lycurus***
28′ First glume 1-nerved with a single awn or awnless; lower stems rounded .... (29)

29. Glumes awnless; lemma awned .... ***Alopecurus***
29′ Glumes awned .... (30)

30. Glumes strongly flattened laterally, ciliate on the keeled midnerve .... ***Phleum***
30′ Glumes rounded, not keeled, not ciliate on the midnerve, but may be pubescent on the body .... ***Polypogon***

31. Disarticulation below the glumes .... (32)
31′ Disarticulation above the glumes .... (37)

32. First glume with 2 or 3 awns; spikelets falling in pairs .... (33)
32′ First glume with a single awn or awnless .... (34)

33. Plants perennial; awn of second glume longer than the body .... ***Lycurus***
33′ Plants annual; awn of second glume shorter than the body .... ***Muhlenbergia***

34. Spikelets nearly circular in outline; glumes and lemma awnless (glumes with a tiny point, but not awned) .... ***Beckmannia***
34′ Spikelets elongate, not circular in outline; glumes and/or lemmas awned .... (35)

35. Glumes awnless; lemma awned .... ***Alopecurus***
35′ Glumes awned .... (36)

36. Glumes strongly flattened laterally, ciliate on the keeled midnerve .... ***Phleum***
36′ Glumes rounded, not keeled, not ciliate on the midnerve, but may be pubescent on the body .... ***Polypogon***

37. Glumes strongly flattened laterally, ciliate on the keeled midnerve .... ***Phleum***
37′ Glumes rounded, not keeled, not ciliate on the midnerve .... (38)

38. Lemma awned from the back, at about the middle or below .... (39)
38′ Lemma awned from the apex or just below .... (40)

39. Floret with a tuft of hairs at the base; rachilla prolonged beyond the palea as a slender bristle .... ***Calamagrostis***
39′ Floret without a tuft of hairs at the base; rachilla not prolonged beyond the palea .... ***Agrostis***

40. Rachilla prolonged beyond the palea as a slender bristle; plants annual .... ***Apera***
40′ Rachilla not prolonged beyond the palea; plants annual or perennial .... ***Muhlenbergia***

GROUP F Paniceae Tribe

1. Spikelets subtended by 1 or more bristles or enclosed in a bur of spines .... (2)
1′ Spikelets not subtended by bristles or enclosed in spines .... (3)

2. Spikelets subtended by 1 to several bristles, these remaining on the plant when the spikelets fall .... ***Setaria***
2′ Spikelets enclosed in a bur of flattened spines, the spikelets falling within the bur and not remaining on the plant .... ***Cenchrus***

3. First glume usually less than 0.5 mm long, absent or vestigial .... (4)
3′ First glume usually more than 0.5 mm long, well developed, evident .... (5)

4. Spikelets rounded on one side and flattened on the other, orbicular to ovate in outline; margins of the lemma of the upper floret firm and hard when mature, the apex rounded .... ***Paspalum***
4′ Spikelets not rounded and flattened as above, lanceolate in outline; margins of the lemma of the upper floret thin and translucent when mature, the apex acute to acuminate .... ***Digitaria***

5. Ligule absent, the ligular region glabrous (but sometimes discolored); plants annual .... ***Echinochloa***
5′ Ligule present, the ligular region often pubescent; plants annual or perennial .... (6)

6. Inflorescence a panicle of simple or nearly simple spicate branches; spikelets nearly sessile; back of upper lemma wrinkled or bumpy, not smooth; plants annual .... ***Urochloa***
6′ Inflorescence an open rebranched panicle, or if with simple branches (*Panicum obtusum*), then the plants stoloniferous perennials; spikelets often pedicelled; back of upper lemma smooth and shiny; plants annual or perennial ... (7)

7. Plants perennial, with 2 distinct growth phases: during the cool season producing a basal rosette of short broad blades and terminal panicles; during the warm season producing much-branched lateral shoots with small axillary panicles .... ***Dichanthelium***
7′ Plants annual or perennial, with a single growth phase; basal rosettes not produced; flowering during the warm season only .... ***Panicum***

GROUP G Lemmas 3-nerved; florets more than one

1. Spikelets in dense, sessile, headlike clusters closely subtended and mostly surpassed by the leaves .... (2)
1′ Spikelets not in dense, sessile, headlike clusters, and/or elevated well above the leaves .... (4)

2. Disarticulation below the glumes, the spikelets in bony clusters and falling together; plants strongly stoloniferous perennials .... ***Buchloë***
2′ Disarticulation above the glumes, the spikelets not falling in bony clusters; plants annual or perennial, stoloniferous or tufted .... (3)

3. Plants annual; blades mostly flat .... ***Munroa***
3′ Plants perennial; blades mostly rolled and needlelike .... ***Dasyochloa***

4. Inflorescence a panicle of definite and obvious spicate or racemose unbranched primary branches ....(5)
4′ Inflorescence a raceme, or a panicle of rebranched primary branches ....(10)

5. Spikelets all male, 2-flowered, with orange-red anthers; lemmas awnless .... ***Buchloë***
5′ Combination of features otherwise ....(6)

6. Panicle branches all digitate or in whorls near the apex of the main axis ....(7)
6′ Panicle branches distributed all along the main axis and most not in whorls, or with a single branch only ....(8)

7. Spikelets awnless, glabrous or at least scarcely hairy to the naked eye .... ***Cynodon***
7′ Spikelets awned, prominently and obviously hairy .... ***Chloris***

8. Spikelets with a single fertile, well-developed floret and with 1–3 smaller, rudimentary florets above..... ***Bouteloua***
8′ Spikelets with usually 3–many fertile, well-developed florets ....(9)

9. Axils of primary panicle branches with tufts of long hairs; spikelets mostly few and widely spaced on each branch .... ***Eragrostis***
9′ Axils of primary panicle branches glabrous; spikelets mostly numerous and usually crowded on each branch .... ***Leptochloa***

10. Sheath margins fused together for 1/2 or more their length ....(11)
10′ Sheath margins overlapping for most of their length, fused only at the base if at all ....(12)

11. Spikelets less than 5 mm long .... ***Catabrosa***
11′ Spikelets usually more than 10 mm long .... ***Bromus***

12. Lemmas pubescent on the nerves or at the base, the midnerve usually exserted as an awn or short point (except *Poa*) ....(13)
12′ Lemmas glabrous on the nerves and at the base, awnless ....(15)

13. Ligules membranous; lemma midnerves not exserted as a small point .... ***Poa***
13′ Ligules a ring of hairs, or if membranous then the lemma midnerve exserted as a small point ....(14)

14. Blades with white margins .... ***Erioneuron***
14′ Blades not white-margined .... ***Tridens***

15. Ligule a membrane ....(16)
15′ Ligule a ring of hairs ....(17)

16. Spikelets sessile or nearly so, the pedicels much shorter than the spikelets; plants tufted .... ***Koeleria***
16′ Spikelets on long pedicels mostly much longer than the spikelets; plants spreading from stolons or rhizomes .... ***Muhlenbergia***

17. Panicles dense, congested, spikelike, usually light greenish or whitish; lemmas notched at the apex with a minute point; plants perennial .... ***Tridens***
17′ Panicles usually open, loose, often olive- or dark-colored; lemmas lacking a minute notch and point; plants annual or perennial .... ***Eragrostis***

GROUP H Lemmas with 5–many nerves; florets more than one

1. Glumes and lemmas stiff-ciliate on the midnerves and keels; spikelets arranged in dense, 1-sided clusters at the branch tips; sheath margins fused together .... ***Dactylis***
1′ Glumes and lemmas glabrous or variously pubescent but not ciliate on the midnerves and keels; spikelets not so arranged; sheath margins fused or overlapping ....(2)

2. Sheath margins fused together 3/4 or more their length ....(3)
2′ Sheath margins free from each other, overlapping, or fused only at the lower 1/3 or less ....(7)

3. Callus of the floret with a prominent tuft of stiff hairs (otherwise glabrous) and lemmas prominently awned .... ***Schizachne***
3′ Callus of the floret lacking a tuft of hairs and/or lemmas awnless ....(4)

4. Nerves of the lemma 7 in number, nearly parallel, not converging at the truncate or rounded apex .... ***Glyceria***
4' Nerves of the lemma 3–11 in number, converging at the obtuse to acute apex, if parallel then less than 7 in number .... (5)

5. Spikelets awned, or if awnless then longer than 15 mm; palea and grain strongly adherent to each other when mature .... ***Bromus***
5' Spikelets awnless and shorter than 15 mm; palea and grain free from each other when mature .... (6)

6. Spikelets on mostly racemose unbranched primary branches, hanging like flags away from the axis; upper florets empty, inrolled and represented by a club-shaped rudiment .... ***Melica***
6' Spikelets variously arranged, but mostly on rebranched primary branches; upper florets usually not empty nor as above .... ***Poa***

7. Disarticulation below the glumes, which are noticeably dissimilar in shape, one narrowly lanceolate and the other obovate or spatulate .... ***Sphenopholis***
7' Disarticulation above the glumes, which may be similar or dissimilar .... (8)

8. Spikelets (glumes and/or lemmas) awned .... (9)
8' Spikelets (glumes and lemmas) awnless or at most with an awn tip no more than 1 mm long .... (22)

9. Inflorescence a panicle of unbranched, spicate primary branches all clustered toward the apex of the stalk; plants annual .... ***Chloris***
9' Inflorescence a panicle, but the main branches rebranched or the spikelets on obvious pedicels; plants annual or perennial .... (10)

10. Florets dissimilar, some awned, some awnless .... (11)
10' All florets alike and awned .... (12)

11. Glumes large, more than 15 mm long .... ***Avena***
11' Glumes small, less than 12 mm long .... ***Arrhenatherum***

12. Glumes not extending beyond the lowermost floret .... (13)
12' Glumes, at least the second, equal to or surpassing the lowermost floret .... (15)

13. Spikelets 2 (4)-flowered; awn arising from the back of the lemma or from a deeply cleft apex .... ***Trisetum***
13' Spikelets mostly 3- to many-flowered; awn arising from an entire apex .... (14)

14. Plants perennial; flowers with 3 stamens .... ***Festuca***
14' Plants annual; flowers with 1 stamen .... ***Vulpia***

15. Lemmas awned from the back or base .... (16)
15' Lemmas awned from an entire or cleft apex, if cleft, the awn arising from the sinus at the tip of the midnerve, or lemmas awnless .... (20)

16. Spikelets not large, the glumes 2–8 mm long .... (17)
16' Spikelets large, the glumes 10–30 mm long .... (19)

17. Glumes dissimilar in size and shape, the first shorter and narrower than the second .... ***Trisetum***
17' Glumes similar in size and shape, about the same length and breadth .... (18)

18. Glumes exceeding the florets and often the awns; lemmas awned from near the middle; blades flat, some 4–8 mm wide or more .... ***Vahlodea***
18' Glumes mostly slightly shorter than 1 floret and the awns; lemmas awned from below the middle; blades mostly folded, rarely more than 3 mm wide .... ***Deschampsia***

19. Plants annual; glumes 18–30 mm long .... ***Avena***
19' Plants perennial; glumes 10–15 mm long .... ***Helictotrichon***

20. Awns of the lemma minute and nearly obsolete, scarcely visible even with a hand lens .... ***Schismus***
20' Awns of the lemma well developed, easily visible with the naked eye .... (21)

21. Spikelets mostly 2-flowered, 3.5–6.5 mm long; rachilla extending beyond the uppermost floret .... ***Trisetum***
21' Spikelets 3- to 7-flowered, 6–15 mm long; rachilla not extending beyond the uppermost floret .... ***Danthonia***

22. Glumes mostly longer than 2 cm and longer than the florets .... ***Avena***
22′ Glumes shorter than 2 cm and/or shorter than the florets .... (23)

23. Spikelets appearing 1-flowered, but the large fertile floret subtended by 1 or 2 smaller scales or bristles representing rudimentary florets, these often appressed to the fertile floret and not immediately apparent .... ***Phalaris***
23′ Spikelets not as above .... (24)

24. Glumes and lemmas at maturity stiff, firm, greenish to straw-colored; leaves distichous, the lower ones bladeless as the stems grade into rhizomes; lemmas 7- to 11-nerved, the nerves obscure; plants strongly rhizomatous, dioecious perennials of alkaline areas and floodplains .... ***Distichlis***
24′ Glumes and lemmas pliable, thin, often greenish to purplish; leaves not distichous, the lower ones usually with well-developed blades; lemmas generally 5- to 7-nerved (9-nerved in the annual *Schismus*); plants annual or perennial, of various habitats .... (25)

25. First glume 5- to 7-nerved; blades threadlike; small tufted annuals of sandy desert areas .... ***Schismus***
25′ First glume 1- to 3-nerved; blades threadlike to much broader; annuals and perennials of various habitats ..... (26)

26. Glumes, at least the second, equaling or surpassing the lowermost floret .... (27)
26′ Glumes, at least 1 but usually both, not extending beyond the lowermost floret .... (29)

27. Florets 3 in number, the lower (outer) 2 as large as the upper (middle) one but male, their margins prominently ciliate, the upper (middle) floret fertile, somewhat hardened, and pubescent at the tip .... ***Hierochloë***
27′ Florets not as above .... (28)

28. Second glume broadened above the middle; palea colorless, scarious, white .... ***Koeleria***
28′ Second glume broadened below the middle; palea colored, at least on the nerves .... ***Trisetum wolfii***

29. Lemmas awned or narrowing at the apex to an awn tip .... ***Festuca***
29′ Lemmas completely awnless, often blunt .... (30)

30. Second glume broadened above the middle; palea colorless, scarious, white; pedicels puberulent .... ***Koeleria***
30′ Second glume, palea, and pedicels not all as above .... (31)

31. Plants rhizomatous and dioecious; blades strongly striate-nerved, usually flat .... ***Leucopoa***
31′ Plants not rhizomatous and dioecious and with strongly striate blades .... (32)

32. Sheath margins fused at least at the base; nerves of the lemma converging toward the acute apex; base of lemma with or without a tuft of cobwebby hair .... ***Poa***
32′ Sheath margins overlapping at the base; nerves of the lemma more or less parallel, not converging toward the truncate apex; base of lemma never with a tuft of cobwebby hairs .... (33)

33. Nerves of the lemma conspicuous; plants with creeping rhizomes; blades mostly flat, 4–15 mm wide; plants of freshwater habitats .... ***Torreyochoa***
33′ Nerves of the lemma obscure; plants tufted, lacking rhizomes; blades rolled, or if flat then 1–3 (4) mm wide; plants of usually alkaline or saline habitats .... ***Puccinellia***

Achnatherum P. Beauv. Needlegrass

(Greek *achne*, scale, and *ather*, awn, referring to the awned lemmas) Tufted perennials with hollow culms. **LEAVES** with open sheaths, membranous ligules, flat to rolled blades, and lacking auricles. **INFL** a narrow to open panicle, the branches rebranched. **SPIKELETS** 1-flowered, the florets terete, mostly indurate, disarticulating above the glumes; glumes longer than the florets, membranous, 1- to 5-nerved; lemmas convolute, but the margins scarcely overlapping, 5-nerved, the midnverve extended into an awn, this sometimes early-deciduous, with a strongly developed and hairy callus; palea 2-nerved, hyaline to membranous, glabrous to hairy. **STAMENS** 3. $x = 11$. Formerly included in the genus *Stipa*, which is now strictly Eurasian (Barkworth 1993).

1. Awns short, 3–5 mm long, quickly deciduous; panicle widely spreading at maturity, with dichotomous branches .... ***A. hymenoides***
1′ Awns longer than 6 mm, persistent or deciduous; panicle narrow, with ascending branches .... (2)

2. Basal segment of the once-bent awns plumose with long hairs 4–8 mm long .... ***A. speciosum***
2′ Basal segment of the awns glabrous or with hairs less than 2 mm long .... (3)

3. Awns 3–7.5 cm long, obscurely bent, the terminal segment flexuous or curving ..... ***A. aridum***
3′ Awns 1–3 cm long, usually plainly bent, the terminal segment more or less straight ..... (4)

4. Leaf sheaths evidently ciliate-villous at the summit ..... (5)
4′ Leaf sheaths glabrous at the summit, or only obscurely ciliolate on some of the lowermost sheaths ..... (7)

5. Lemma apex with hairs less than 2 mm long, about the same length as on the body of the lemma ..... ***A. robustum***
5′ Lemma apex with hairs 2–3 mm long, noticeably longer than those on the body of the lemma ..... (6)

6. Palea less than 1/2 the length of the lemma ..... ***A. scribneri***
6′ Palea about 2/3 the length of the lemma ..... ***A. coronatum***

7. Awns mostly 2–3 cm long; blades flat, 3–7 mm wide ..... ***A. nelsonii***
7′ Awns mostly 1–2 cm long; blades flat to tightly rolled, 0.5–3 mm wide ..... (8)

8. Glumes about equal, 6–9 mm long; blades mostly involute-filiform, 0.2–2 mm wide; palea nearly as long as the lemma ..... ***A. lettermanii***
8′ Glumes unequal, the first 10–15 mm long, the second 1–3 mm shorter; blades mostly flat, most at least 2–3 mm wide; palea 1/2–2/3 the length of the lemma ..... ***A. perplexum***

Achnatherum aridum (M. E. Jones) Barkworth (arid) Mormon needlegrass. **CULMS** 30–80 cm tall, slender. **LIGULES** less than 1 mm long. **BLADES** involute-filiform, less than 2.5 mm wide even when flat. **PANICLES** 10–30 cm long, spicate. **GLUMES** narrowly lanceolate with an acuminate-awned tip, shiny, mostly 3-nerved, the first 9–15 mm long, the second 7–10 mm long. **LEMMAS** 4–6 mm long, sparsely short-pilose nearly throughout, occasionally glabrous above, the hairs less than 0.7 mm long; awn 3–7.5 cm long, mostly nearly straight or obscurely geniculate. **PALEAS** about 1/2 as long as the lemmas, hairy. **ANTHERS** 2–3 mm long. [*Stipa arida* M. E. Jones; *S. mormonum* Mez]. Rocky ground in the lower desert regions; black sage, shadscale, sagebrush, and Utah juniper communities. ARIZ: Nav; NMEX: SJn; UTAH. 1200–1400 m (4000–4600′). Flowering: May–Jul. California east to Colorado and south through Arizona and northwestern New Mexico.

Achnatherum coronatum (Thurb.) Barkworth (crowned) Crested needlegrass. **CULMS** 20–80 cm tall, erect. **LIGULES** 0.5–1 mm long. **BLADES** flat, 20–60 cm long, 6 mm or less wide. **PANICLES** 10–26 cm long, spicate. **GLUMES** narrowly lanceolate, tapering to a slender point, mostly 3-nerved, the first 12–18 mm long, the second 11–15 mm long. **LEMMAS** 5–8 mm long, the body densely appressed-villous with hairs about 1 mm long, the apex with hairs 2–3 mm long; awn 1.2–2.5 cm long, once- or weakly twice-geniculate. **PALEAS** shorter than the lemmas, glabrous or sparsely pubescent below. **ANTHERS** 3.5–5 mm long. [*Stipa parishii* Vasey]. Dry, rocky slopes and ridges in juniper woodland regions; not common. UTAH. 1500–1850 m (5000–6000′). Flowering: Jun–Jul. Our material belongs to **var.** ***parishii*** (Vasey) S. L. Welsh (for S. B. Parish, California botanist). Southern California, Nevada, Utah, and Arizona.

Achnatherum hymenoides (Roem. & Schult.) Barkworth (membranelike) Indian ricegrass. **CULMS** 30–70 cm tall, erect, the internodes hollow. **LIGULES** 2.5–7.5 mm long. **BLADES** firm and strongly involute, about 1 mm long, elongate and often forming a fountainlike clump. **PANICLES** open, dichotomously branched, mostly 5–20 cm long and about 1/2 as wide. **GLUMES** ovate-acuminate, sometimes awn-tipped to 2 mm, mostly 3–5-nerved, at least at the base, the first 5–8 mm long, the second slightly shorter. **LEMMAS** fusiform to nearly globose, shiny and dark brown to black in age, 3–5 mm long, covered with long white pilose hairs 2–4 mm long, sometimes becoming glabrous; awn 3–3.5 mm long, straight, quickly deciduous and mostly absent in maturity. **PALEAS** similar to the lemmas in color, texture, and length. **ANTHERS** 0.8–1.5 mm long, with an apical tuft of hairs. $2n = 28, 48, 65, 130$. [*Oryzopsis hymenoides* (Roem. & Schult.) Ricker; *Stipa hymenoides* Roem. & Schult.]. Common in dry, open, mostly sandy sites at lower elevations, but extending upward exceptionally to nearly subalpine zones; desert scrub, piñon-juniper woodland, mountain brush, and pine communities. ARIZ: Apa, Coc, Nav; COLO: Arc, Dol, LPl, Mon, SJn, SMg; NMEX: McK, RAr, San, SJn; UTAH. 1200–2500 m (4000–8200′). Flowering: May–Aug. Western Canada south through the western United States; Chile. Originally named in the genus *Stipa*, then transferred to *Oryzopsis* because of its plump florets, Indian ricegrass now finds a suitable home in *Achnatherum*, with whose other species it shows close relationship, as evidenced by its frequent hybridization with several of the species. Seeds are high in protein and the plants provide important forage on rangelands.

Achnatherum lettermanii (Vasey) Barkworth (for George Letterman, 1840–1913, Missouri botanist) Letterman's needlegrass. **CULMS** erect, 25–80 cm tall, much-branched below. **LIGULES** 0.2–2 mm long, truncate-rounded. **BLADES** strongly involute, filiform, only rarely flat and then less than 2 mm wide, short-hispid on the inner surface, mostly glabrous on the outer. **PANICLES** narrow, 8–20 cm long, with erect branches and relatively few spikelets. **GLUMES** acuminate to an awn tip, often purplish, 3-nerved, subequal, 5–10 mm long. **LEMMAS** only slightly indurate, sericeous, the hairs at the summit (1–1.5 mm) longer than those on the body (0.2–0.5 mm), 4.5–6.5 mm long; awn 12–22 mm long, twice-geniculate. **PALEAS** about 2/3 as long as the lemmas, pubescent, the hairs at the tip longer than those below, sometimes exposed. **ANTHERS** 1.5–2.5 mm long. $2n$ = 32, 66, 68. [*Stipa lettermanii* Vasey]. Sagebrush, mountain brush and woodlands, conifer and aspen forests at middle to high elevations in the mountains. ARIZ: Apa, Nav; COLO: Arc, Hin, LPl, Min, Mon, SMg; NMEX: McK, RAr, SJn; UTAH. 2250–3300 m (7500–10800′). Flowering: Jul–Sep. Most of the western United States. Frequently confused with *A. nelsonii* of similar habitats, but that species has wider blades, longer awns, and shorter paleas.

Achnatherum nelsonii (Scribn.) Barkworth (for Aven Nelson, 1859–1952, Wyoming botanist) Nelson's needlegrass. **CULMS** stout, erect, 30–110 cm tall. **LIGULES** 0.2–1.5 mm long, truncate, often longer on the sides. **BLADES** mostly flat when fresh and involute upon drying, 2–7 mm wide, often strongly striate. **PANICLES** narrow, 10–36 cm long, the branches erect, sometimes interrupted below. **GLUMES** long-acuminate to awn tips, 3-nerved, subequal, 7–14 mm long. **LEMMAS** appressed-pilose, the hairs longer at the tip (to 1.5 mm) than on the body (to 0.5 mm), 5.5–8 mm long, the apical lobes sometimes prolonged as short lobes to 0.5 mm long; awn twice-geniculate, 2–3 (4) cm long. **PALEAS** less than 1/2 the length of the lemmas, pubescent. **ANTHERS** 2–3 mm long. $2n$ = 36, 44. [*Stipa nelsonii* Scribn.]. Sagebrush, mountain brush and woodlands, conifer and aspen forests at middle to high elevations in the mountains. ARIZ: Apa; COLO: Arc, Hin, LPl, Mon, SJn; NMEX: RAr. 1700–2900 m (5800–9500′). Flowering: Jun–Sep. Alaska and western Canada, south through most of the western United States, east to western Texas. Frequently confused with *A. lettermanii* of similar habitats, but that species has narrower blades, shorter awns, and longer paleas. This has been confused with *A. pinetorum* by some, but that species has long white spreading hairs 2–4 mm long throughout the lemma body and occurs at middle to high elevations in the mountains of the Gunnison Basin; it is not known from our region.

Achnatherum perplexum Hoge & Barkworth (perplexing) Perplexing needlegrass. **CULMS** erect, 35–90 cm tall. **LIGULES** 1–4 mm long. **BLADES** 1–3 mm wide, flat to involute upon drying. **PANICLES** narrow, the branches erect, 10–25 cm long. **GLUMES** long-acuminate, 3-nerved, unequal, the first 10–15 mm long, the second 1–4 mm shorter and somewhat outward-curving. **LEMMAS** appressed-pubescent, the hairs at the tip 1–2 mm long, the hairs on the body about 1 mm long; awns 10–20 mm long, once- to twice-geniculate. **PALEAS** 1/2–2/3 as long as the lemmas, pubescent. **ANTHERS** 2.5–4 mm long. $2n$ = unknown. Mountain grasslands, clearings, and dry slopes. NMEX: RAr. 2100–2750 m (6900–9000′). Flowering: Jul–Sep. Southwestern United States and northern Mexico. Most similar to *A. nelsonii*, but that species has broader blades, equal glumes, longer awns, shorter paleas, and flowers earlier.

Achnatherum robustum (Vasey) Barkworth (robust) Sleepygrass. **CULMS** erect, stout, 2–4.5 mm thick, usually over 1 m tall. **LIGULES** truncate to rounded, 1–2 mm long, or up to 4 mm long on the upper leaves. **BLADES** mostly flat, 6–10 mm wide. **PANICLES** narrow, 15–30 cm long, with appressed branches, densely flowered. **GLUMES** acuminate, subequal, 9–12 mm long. **LEMMAS** fusiform-cylindrical, appressed-pilose, the hairs at the tip somewhat longer than those on the body, 6–10 mm long; awn twice-geniculate, 2–3.4 cm long. **PALEAS** 2/3 to 3/4 the length of the lemmas, pubescent. **ANTHERS** 4–5 mm long. $2n$ = 64. [*Stipa robusta* (Vasey) Scribn.]. Dry plains, hillsides, forest clearings, and roadsides. COLO: Arc, Mon; NMEX: McK, RAr, SJn. 1650–2600 m (5400–8500′). Flowering: Jul–Sep. Wyoming to Colorado and New Mexico; northern Mexico. Some plants, perhaps only those from the mountains of south-central New Mexico, contain a narcotic that induces torpor in grazing horses but is not lethal. Recent studies suggest the toxin is produced by endophytic fungi infesting the plants. *Achnatherum inebrians* (Hance) Keng of Mongolia, Tibet, and China produces the same effect and contains similar fungi.

Achnatherum scribneri (Vasey) Barkworth (for Frank Lamson Scribner, 1851–1938, renowned USDA agrostologist) Scribner's needlegrass. **CULMS** erect, 35–80 cm tall. **LIGULES** irregularly truncate, finely ciliolate, 0.2–2 mm long. **BLADES** involute to flat, 1.5–5 mm wide. **PANICLES** narrow, 7–20 cm long, with appressed branches. **GLUMES** linear-lanceolate, acuminate to awn tips, 3-nerved, unequal, the first 12–18 mm long and somewhat curving or sickle-shaped, the second 1.5–2 mm shorter. **LEMMAS** pale, 5–8 mm long, appressed-pilose, the hairs at the tip 2–3 mm long and noticeably longer than those on the body; awn mostly once-geniculate, 1.4–2.2 cm long. **PALEAS** 1/3 to 1/2 the length of the lemmas, pubescent. **ANTHERS** about 4 mm long. $2n$ = 40. [*Stipa scribneri* Vasey]. Mesas, rocky plains, and slopes in desert regions. ARIZ: Apa; NMEX: McK, SJn. 1400–1650 m (4550–5400′). Flowering: Jun–Sep. Colorado, northern Arizona, and New Mexico. Grains often have a patchy sticky secretion, causing them to adhere to objects (such as teasing needles or birds' beaks).

Achnatherum speciosum (Trin. & Rupr.) Barkworth (showy) Desert needlegrass. **CULMS** stout and densely tufted, erect, 35–65 cm tall. **LIGULES** short, less than 0.5 mm long, finely ciliolate, with a conspicuous tuft of hairs at the shoulders. **BLADES** tightly involute, firm, 1 mm or less wide, elongate, glaucous. **PANICLES** narrow and compact, 8–20 cm long, with numerous spikelets, scarcely exceeding the leaves. **GLUMES** long-acuminate to an awn tip, 11–25 mm long, the second equal or somewhat shorter. **LEMMAS** fusiform, 7–10 mm long, short appressed-pilose to nearly glabrous upward; awn once-geniculate, 3–4.5 cm long, the terminal portion straight and scaberulous, the lower portion densely pilose with hairs 3–9 mm long. **PALEAS** glabrous, about 1/2 the length of the lemmas. **ANTHERS** 1.5–4 mm long. $2n = 60, 64, 66$. [*Jarava speciosa* (Trin. & Rupr.) Peñailillo, *Stipa speciosa* Trin. & Rupr.]. Sagebrush deserts and piñon-juniper woodlands in rocky or sandy ground. ARIZ: Apa, Coc, Nav; COLO: Min, Mon; NMEX: SJn; UTAH. 1500–1800 m (5000–5900′). Flowering: May–Jun. Southwestern United States, northern Mexico, South America. This species has affinities to some in South America, where it is disjunct, and is placed by some in the genus *Jarava*.

Aegilops L. Goatgrass

(Greek *aegiles*, preferred by goats, and *ops*, appearing like; a name used by Theophrastus for a grass that was similar to *Aegiles*, which is an unknown plant apparently palatable to goats) Annual, tufted herbs. **LEAVES** with open sheaths, ciliate auricles, and membranous ligules. **INFL** a thick spike, with a single spikelet per node. **SPIKELETS** with 2–7 florets, the upper often sterile; disarticulation below the glumes, the spikelets falling attached to the adjacent internodes.

Aegilops cylindrica Host (cylindric, alluding to the spike) Jointed goatgrass. **CULMS** erect to decumbent at the base, usually branched, 14–50 cm tall. **LIGULES** short, to 0.5 mm long. **BLADES** flat, 4–15 cm long, 2–5 mm wide. **SPIKES** 3–14 cm long, 2–4 mm wide, narrowly cylindrical, with numerous spikelets. **GLUMES** hardened, 9- to 13-nerved, with an awn on one side and a lobe on the other, the awns of the lower spikelets 3–10 mm long, the awns of the terminal spikelet 30–80 mm long. **LEMMAS** asymmetrical, 5- to 7-nerved, 8–11 mm long, the awns of the lower spikelets to about 20 mm long, the awns of the upper spikelets 30–70 mm long. **PALEAS** nearly as large as the lemmas. **ANTHERS** about 2.8 mm long. $2n = 28$. [*Triticum cylindricum* (Host) Ces., Pass. & Gibelli]. Disturbed sites, crop fields, roadsides, waste places. ARIZ: Apa, Nav; COLO: Arc, Dol, LPl, Mon, SMg; NMEX: McK, RAr, San, SJn; UTAH. 1500–2350 m (4800–7800′). Flowering: May–Jun. Widely established in North America; native to southern Europe and central Asia.

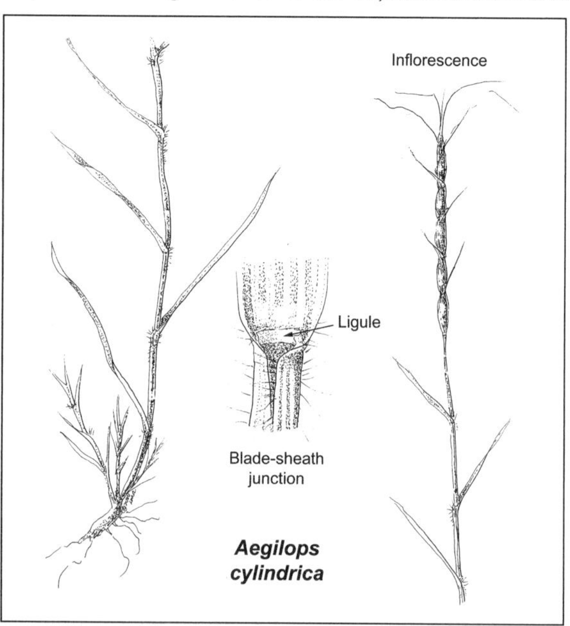

Agropyron Gaertn. Wheatgrass

(Greek *agrios*, wild, + *puros*, wheat, the original species being weeds in wheat fields) Perennial, tufted herbs. **LEAVES** with or without auricles, the basal ones with closed sheaths, the upper ones open, and with membranous ligules. **INFL** a spike, the spikelets placed closely together and spreading from the rachis. **SPIKELETS** with several fertile florets, mostly awned; disarticulation above the glumes and between the florets. **GLUMES** unequal, shorter than the florets, 1- to 4-nerved, often keeled, acute to awn-tipped. **LEMMAS** 5-nerved, awned or awnless. **PALEAS** subequal to the lemmas, 2-nerved. **STAMENS** 3. $x = 7$ (Dewey 1983). A small genus of one to a few species, depending upon interpretation. We recognize a single species in our flora, with three subspecies.

Agropyron cristatum (L.) Gaertn. (crested) Crested wheatgrass. **CULMS** erect, 30–70 cm tall. **LIGULES** mostly less than 0.5 mm long, finely ciliolate. **BLADES** flat, 2–6 mm wide, sometimes more. **SPIKES** 3–10 cm long, 5–20 mm wide, the spikelets ascending to widely spreading. **GLUMES** 3-nerved, narrowly ovate, 3–5 mm long, the midnerve often extended into an awn 1–3 mm long or awnless. **LEMMAS** firm, narrow, 5-nerved, 5–8 mm long; awn 1–3 mm long or awnless. **PALEAS** about as large as the lemmas. **ANTHERS** 2–5 mm long. $2n = 14, 28, 42$. [*Agropyron pectiniforme* Roem. & Schult.] Disturbed and revegetated sites, roadsides, in desert scrub, sagebrush, mountain brush, piñon-juniper woodlands, ponderosa pine communities. 1350–2350 m (4500–7700′). Flowering: Jun–Aug. Widely distributed throughout much of the U.S. except for the southeast region; introduced from Russia for range restoration and forage. We have three subspecies, distinguished by the following:

1. Spikelets diverging from the rachis at an angle of more than 40°; glumes widespread, forming an angle of more than 120°, giving the spike a bristly appearance; spikes 8–20 mm broad .... **subsp. *cristatum***
1' Spikelets diverging from the rachis at an angle of less than 35°; glumes spreading at a narrow angle; spikes 5–10 mm broad .... (2)

2. Lemmas with an awn 1–2 (4) mm long; glumes forming an angle of approximately 60° .... **subsp. *desertorum***
2' Lemmas awnless, sometimes mucronate; glumes forming an angle of approximately 45° .... **subsp. *fragile***

subsp. *cristatum* Fairway crested wheatgrass. ARIZ: Apa, Nav; COLO: Arc, Dol, LPl, Mon, SJn, SMg; NMEX: McK, RAr, San, SJn; UTAH.

subsp. *desertorum* (Fisch. ex Link) Á. Löve (of the desert) Desert crested wheatgrass. ARIZ: Apa; COLO: Dol, LPl, Mon, SMg; NMEX: McK, RAr, SJn.

subsp. *fragile* (Roth) Á. Löve (brittle or easily broken) Siberian crested wheatgrass. NMEX: McK.

Agrostis L. Bentgrass

(Greek *agros*, a field) Tufted, rhizomatous, or stoloniferous annuals and perennials (ours) with hollow culms. **LEAVES** with open sheaths, membranous ligules, and lacking auricles. **INFL** a panicle with often capillary branches, diffuse to contracted and spikelike. **SPIKELETS** small, 1-flowered; disarticulation above the glumes, the rachilla mostly not prolonged beyond the palea. **GLUMES** 1-nerved, usually exceeding the floret. **LEMMAS** obscurely 3- to 5-nerved, awned or awnless. **PALEAS** hyaline, well developed or obsolete. **STAMENS** 3. $x = 7$ (Björkman 1960). A genus of about 100 species in temperate and Arctic regions of the world.

1. Palea well developed, 0.5–2 mm long, 1/2 to 3/4 the length of the lemma .... (2)
1' Palea obsolete or a small scale less than 0.4 mm long, never as much as 1/2 the length of the lemma .... (4)

2. Panicle dense, compact, interrupted; spikelets usually disarticulating below the glumes ......see ***Polypogon viridis***
2' Panicle open or closed but not dense or compact; spikelets disarticulating above the glumes .... (3)

3. Anthers (0.8) 1–1.6 mm long; rachilla not extended as a minute bristle; plants of low- to high-elevation wet places .... ***A. stolonifera***
3' Anthers 0.4–0.8 mm long; rachilla extended as a minute bristle (20×), lying against the base of the palea; plants of subalpine to alpine bogs, meadows, and tundra .... ***A. humilis***

4. Panicle open to diffuse, often less than 3 times longer than broad, nearly always more than 4 cm wide .... ***A. scabra***
4' Panicle narrow, contracted, several times longer than broad, rarely more than 2 cm wide .... (5)

5. Spikelets and panicles greenish, maturing to straw-colored; plants mostly much more than 20 cm tall; panicles 4–20 cm long; blades mostly 2–10 mm wide .... ***A. exarata***
5' Spikelets and panicles purplish to brownish; plants usually less than 20 cm tall; panicles 2–6 (8) cm long; blades 1–2 (3) mm wide .... ***A. variabilis***

Agrostis exarata Trin. (plowed up, alluding to furrows between the nerves) Spike bentgrass. **CULMS** erect or sometimes decumbent and rooting at the nodes, rarely developing short rhizomes in wet ground, 20–100 cm tall or more. **LIGULES** 2–7 mm long, truncate to obtuse, lacerate. **BLADES** green, flat, 2–10 mm wide. **PANICLES** compressed and at least somewhat spikelike, 4–20 cm long or more, rarely more than 2 cm wide, greenish, maturing to straw-colored. **GLUMES** 1.8–2.7 mm long, acute. **LEMMAS** 1.2–2.5 mm long, awnless. **PALEAS** absent or minute. **ANTHERS** 0.3–0.6 mm long. $2n = 28, 42, 56$. Woodland, forest, and alpine communities in the mountains and upper foothills; moist to sometimes wet ground, often along streams and in shade. ARIZ: Apa, Nav; COLO: Arc, Con, Dol, Hin, LPl, Min, Mon, SJn; NMEX: McK, SJn; UTAH. 1850–3100 m (6150–10100'). Flowering: Jul–Aug. Alaska south through the western U.S. into Mexico, east to South Dakota and Texas; Siberia. Some choose to recognize the widespread variation of this species at the varietal level; our material would correspond to **var. *minor*** Hook., having smaller spikelets than the typical form.

Agrostis humilis Vasey (low-growing) Mountain bentgrass. **CULMS** tufted, erect, 5–50 cm tall. **LIGULES** 0.5–2 mm long, truncate to obtuse. **BLADES** flat to folded, 0.5–4 mm wide. **PANICLES** loosely contracted, short, 1–14 cm long,

1–2.5 cm wide, the branches typically terete and nearly smooth. **GLUMES** subequal, 1.5–2.2 mm long, purplish. **LEMMAS** lanceolate, 1.5–2.2 mm long; awnless. **PALEAS** 1–1.6 mm long; rachilla vestige very short to sometimes absent. **ANTHERS** 0.4–0.8 mm long. $2n = 14$. [*Agrostis thurberiana* Hitchc.]. Subalpine to alpine meadows and bogs, not common. COLO: Min, SJn. 2900–3800 m (9800–12500′). Flowering: May–Aug. Most of western North America. The dwarf form of this species is easily overlooked and can be confused in the key with *A. variabilis* if the palea is not noticed. This and *A. thurberiana* (included herein in synonymy) are sometimes treated in a separate genus, *Podagrostis*, based mostly on the prolongation of the rachilla. Here a more traditional view is taken awaiting more conclusive studies.

Agrostis scabra Willd. (rough) Ticklegrass, rough bentgrass. **CULMS** tufted, erect, mostly 30–80 cm tall, sometimes shorter. **LIGULES** 1–3 mm long, truncate to obtuse, erose. **BLADES** flat, sometimes folded, usually narrow, 0.8–1.5 (2) mm wide. **PANICLES** diffuse, 10–25 cm long, mostly 5–18 cm wide, narrower when young, the branches subcapillary, flexuous, scaberulous, mostly branched above the middle. **GLUMES** purplish, unequal, the first 2.5–3.2 mm long and usually slightly but noticeably longer than the second, sometimes awn-tipped. **LEMMAS** lanceolate, 1–2 mm long; awnless in ours. **PALEAS** absent or to 0.3 mm long. **ANTHERS** 0.3–0.6 mm long. $2n = 28, 42, 43$. Meadows, stream banks, open slopes, mountain grasslands, above the juniper zone. ARIZ: Apa; COLO: Arc, Hin, LPl, Min, Mon, SJn; NMEX: McK, SJn. 1950–3100 m (6500–10200′). Flowering: Jul–Aug. Most of North America; Asia; naturalized in Europe. The key to *A. scabra* is deliberately broad, to include plants intermediate between *A. scabra* and *A. variabilis*, which have been referred to *A. ×idahoensis* Nash, a catchall name without biological significance applied to apparent hybrid derivatives between the two. These intermediates are scattered within populations of *A. scabra* and *A. variabilis* and typically are recognized by culms 10–25 cm tall and panicles 5–12 cm long with ascending branches forking below their midlength, though they may have mixed combinations of various features.

Agrostis stolonifera L. (bearing stolons) Redtop, creeping bentgrass. Plants rhizomatous or stoloniferous. **CULMS** erect to geniculate to decumbent-based, 40–100 cm tall or more. **LIGULES** truncate to obtuse, erose, 2–6 (8) mm long. **BLADES** mostly flat or folded, those of the innovations involute when young, 2–6 mm wide. **PANICLES** open and pyramidal to contracted and spikelike, green to purplish, the branches naked or spikelet-bearing near the base, 8–22 cm long or more. **GLUMES** lanceolate, acute, scabrous on the keel, the first 2–2.5 mm long, the second slightly shorter, often open and gaping after the floret falls. **LEMMAS** 3- to 5-nerved, 1.5–2 mm long; awnless. **PALEAS** about 1/2 the length of the lemmas. **ANTHERS** 0.5–1.4 mm long. $2n = 28–46, 56$. [*A. alba* of numerous authors]. Stream banks, meadows, marshes, wet places, from low to high elevations, from desert to subalpine communities. Two intergrading forms may be found; these are often regarded as separate species but are here treated as subspecies:

1. Panicles open both during and after anthesis, more than 1.5 cm broad, the branches ascending to widely spreading; plants with well-developed rhizomes bearing more than 3 scale leaves, not stoloniferous, erect at the base .... **subsp. *gigantea***

1′ Panicles open during anthesis but contracted thereafter and when mature, mostly 1–1.5 cm broad, the branches erect-appressed; plants often stoloniferous and decumbent at the base, if short rhizomes developed then these bearing no more than 3 scale leaves .... **subsp. *stolonifera***

subsp. *gigantea* (Roth) Beldie [*Agrostis stolonifera* L.]. ARIZ: Apa; COLO: Arc, Hin, LPl, Min, Mon, SJn; NMEX: SJn. 1450–3550 m (4800–11600′). Flowering: Jun–Oct. Widespread throughout North America.

subsp. *stolonifera* [*Agrostis palustris* Huds.]. ARIZ: Apa, Nav; COLO: Arc, Dol, Hin, LPl, Min, Mon, RGr, SMg; NMEX: McK, RAr, SJn; UTAH. 1600–2700 m (5300–8900′). Flowering: Jun–Oct. Native to Eurasia and north Africa, introduced throughout the cool-temperate regions of the world as a pasture and forage grass, meadows, and lawns; widespread throughout North America.

Agrostis variabilis Rydb. (Variable) mountain bentgrass. Plants densely tufted. **CULMS** erect, 8–25 cm tall, rarely more. **LIGULES** truncate to obtuse, ciliolate, 0.5–4 mm long. **BLADES** mostly flat or folded, those of the innovations involute, 0.5–2 mm wide. **PANICLES** contracted and loosely spikelike, purplish, 2–7 cm long. **GLUMES** lanceolate, acute, scaberulous on the keel, 1-nerved, the first 2–2.5 mm long, the second slightly shorter. **LEMMAS** 5-nerved, 1.8–2.3 mm long, mostly awnless or rarely with an awn from midlength. **PALEAS** minute or absent. **ANTHERS** 0.4–0.7 mm long. $2n = 28$. [*Agrostis rossiae* of numerous authors, not Vasey]. Moist meadows and slopes, subalpine to alpine communities. COLO: LPl, SJn. Flowering: Jul–Aug. Widespread throughout the mountain regions of western North America, and previously confused with *Agrostis rossiae* Vasey, an endemic of hot springs in Yellowstone National Park.

Alopecurus L. Foxtail

(Greek *alopex*, fox, + *oura*, tail, referring to the narrow, sometimes bristly, panicle) Tufted annuals or perennials with hollow culms. **LEAVES** with open sheaths, membranous ligules, flat blades, lacking auricles. **INFL** a very compact, cylindrical, spikelike panicle, the branches rudimentary, shortened and obscure. **SPIKELETS** 1-flowered, strongly flattened; glumes subequal, exceeding the floret and pilose-ciliate, connate below; lemmas connate below, awned from the back from below the middle; disarticulation below the glumes, the spikelets falling entire. Temperate regions of the Northern Hemisphere. About 25 species.

1. Awn only slightly exserted beyond the glumes, scarcely visible without magnification .... ***A. aequalis***
1′ Awn well exserted beyond the glumes, easily visible without magnification ....(2)

2. Spikelets large, 5–6 mm long .... ***A. pratensis***
2′ Spikelets smaller, 2–4 mm long ....(3)

3. Glumes pubescent throughout with prominent villous hairs; panicles densely woolly-looking, 1–4 (5) cm long...... .... ***A. alpinus***
3′ Glumes pubescent mostly on the keels; panicles not woolly-looking, 2–10 cm long .... ***A. geniculatus***

Alopecurus aequalis Sobol. (equal, perhaps alluding to the glumes) Short-awn foxtail. Plants perennial. **CULMS** tufted, sometimes decumbent and rooting at the nodes, 10–50 cm tall. **LIGULES** attenuate, lacerate or entire, 3.5–7 mm long. **BLADES** flat, 1.5–4 mm wide. **PANICLES** narrowly cylindrical, 2–8 cm long, 4–5 mm wide. **GLUMES** 1.7–2.5 mm long, pubescent on the keels and nerves. **LEMMAS** lance-ovate, 1.8–2.2 mm long, glabrous; awn 0.7–2.5 mm long, included or only slightly exserted from the glumes. **PALEAS** absent. **ANTHERS** 0.5–1 mm long. $2n = 14$. Edges of ponds, along streams, wet meadows, in the mountains up to subalpine communities. ARIZ: Apa; COLO: Arc, Dol, Hin, LPl, Min, Mon, SJn; NMEX: McK, RAr, SJn. 2050–3150 m (6800–10300′). Flowering: Jun–Aug. Circumboreal, south to California and New Mexico, east across Kansas to Pennsylvania; Argentina, Eurasia. This has been confused with *A. carolinianus* Walter, which is not found in our area, and which can be distinguished by its annual habit, ligules 2–3.5 mm long, awns 3–5 mm long and easily visible without magnification, and anthers 0.3–0.5 mm long.

Alopecurus alpinus Sm. (alpine) Alpine foxtail. Plants rhizomatous or stoloniferous. **CULMS** erect, 25–80 (100) cm tall. **LIGULES** truncate, finely erose, 2–4 mm long. **BLADES** flat, 3–6 mm wide. **PANICLES** woolly, ovoid, 1–4 cm long, 6–10 mm wide. **GLUMES** densely villous throughout, 2.8–4 mm long. **LEMMAS** ovate, 2.5–4 mm long; awn 3.5–7 mm long. **PALEAS** absent. **ANTHERS** 1.8–2.2 mm long. $2n = 98, 100, 105, 110, 112–130$. Stream banks, wet meadows, riparian areas, subalpine to alpine vegetation. COLO: Arc, Hin, SJn. 2950–3800 m (9800–12400′). Flowering: Jun–Aug. Alaska across northern Canada, south along the Rocky Mountains to Colorado; Eurasia.

Alopecurus geniculatus L. (bent sharply like a knee) Water foxtail. Plants perennial. **CULMS** spreading to decumbent and rooting at the nodes, 20–60 cm long. **LIGULES** acute, lacerate-tipped, 2–8 mm long. **BLADES** flat, 2–6 mm wide. **PANICLES** compact and spikelike, 2–10 cm long. **GLUMES** pubescent mostly on the keels only, 2–3.5 mm long. **LEMMAS** 2.5–3 mm long; awn 3.5–6 mm long. **PALEAS** absent. **ANTHERS** 1–2.2 mm long. $2n = 14, 28$. Wet meadows, edges of ponds, in aspen, ponderosa, and spruce-fir communities. COLO: LPl, Mon; NMEX: SJn. 1800–2700 m (6000–8900′). Flowering: Jun–Aug. Native to Eurasia, scattered locales in North America, through most of the western and eastern states.

Alopecurus pratensis L. (of meadows) Meadow foxtail. Plants perennial. **CULMS** erect to decumbent and sometimes rooting at the nodes, 30–80 cm long. **LIGULES** truncate to obtuse, ciliolate, 2–4 mm long. **BLADES** flat, 3–6 mm wide. **PANICLES** 3.5–10 cm long, 5–8 mm wide. **GLUMES** ovate-lanceolate, (4.5) 5–6 mm long, villous-ciliate on the keels and lateral nerves, scarcely pubescent between. **LEMMAS** 3–5 mm long, glabrous; awn 5–8 mm long, exserted and easily visible. **PALEAS** absent. **ANTHERS** 2–3.5 mm long. $2n = 14, 28, 42$. Meadows, ditch and stream banks, roadsides, often in drier sites than the other species. COLO: LPl, SJn; UTAH. 2050–2600 m (6800–8600′). Flowering: Jun–Jul. Native to Eurasia, introduced to North America as a pasture grass, escaped in much of the West and northeastern states.

Andropogon L. Bluestem

(Greek *andros*, man, + *pogon*, beard, referring to the hairy pedicelled spikelet that is usually staminate) Robust perennials with or without rhizomes; culms solid. **LEAVES** with open sheaths, membranous ligules, flat or folded blades, and lacking auricles. **INFL** a panicle of simple or slightly rebranching branches, with few panicles present on the flowering shoot and terminating the culms, or with numerous small panicles produced in subtending spathes and clustered together on the flowering shoots. **SPIKELETS** 2-flowered, the lower floret highly reduced, the upper perfect (in the sessile spikelet), borne in repeating pairs of spikelets: one pedicelled, staminate or neuter, and awnless, the other sessile, perfect, and awned; disarticulation below the glumes, in a unit consisting of the sessile spikelet, rachis internode, and pedicelled spikelet; glumes subequal, hardened, enclosing the florets; lemmas highly reduced, entire or bifid, extended into the awn (in the sessile spikelet); palea tiny or absent; stamens 3. $x = 10$ (Campbell 1986; Gould 1967).

1. Flowering shoots little-branched above, the few inflorescences terminating the culms; pedicelled spikelets present, nearly as large as the sessile one; sessile spikelets at least 6 mm long .... ***A. gerardii***
1' Flowering shoots much-branched above, with numerous inflorescences clustered together, broomlike; pedicelled spikelets vestigial or absent; sessile spikelets less than 4 mm long .... ***A. glomeratus***

Andropogon gerardii Vitman (for John Gerard, 1546–1612, gardener-surgeon-barber, and author of an early important herbal) Big bluestem. Plants tufted or rhizomatous, robust. **CULMS** 50–130 cm or more tall, sparingly branched above. **LIGULES** 0.5–2.5 mm long. **BLADES** flat or loosely involute, 4–8 mm wide. **PANICLES** brownish to purplish, of 2–6 spicate, subdigitate, erect branches 4–10 cm long, the spikelets subtended by copious white to yellow hairs 2–6 mm long. **SESSILE SPIKELETS** 6–12 mm long, the second glume thinner and shorter than the first; awns absent or to 20 mm long. **PEDICELLED SPIKELETS** 3–8 mm long. **ANTHERS** 3–5 mm long. $2n = 20, 40, 60, 80–86$. Dry, open hills and slopes, sandy to rocky ground, ponderosa grasslands and clearings, roadsides. ARIZ: Apa, Nav; COLO: Arc, Dol, LPl, Mon; NMEX: McK, San, SJn; UTAH. 1600–2300 m (5400–7600′). Flowering: Jul–Oct. Plains and grasslands of North America, south to Arizona and Texas; Mexico. We have three intergrading subspecies:

1. Awn of sessile spikelet 0–5 mm long; rhizomes well developed; foliage glaucous .... **subsp. *hallii***
1' Awn of sessile spikelet 8–20 mm long; rhizomes absent or well developed; foliage generally green .... (2)

2. Hairs of panicle branch internodes (rachis joints) copious, 3–4 mm long and usually yellow or golden; rhizomes well developed .... **nothosubsp. *chrysocomus***
2' Hairs of panicle branch internodes sparse to copious, 1–2 mm long; rhizomes absent or short ... **subsp. *gerardii***

nothosubsp. *chrysocomus* (Nash) Wipff (golden-haired) Sand bluestem. COLO: Mon; NMEX: McK, San, SJn; UTAH.

subsp. *gerardii* ARIZ: Apa, Nav; COLO: Arc, Dol, LPl, Mon; NMEX: McK, RAr, San, SJn; UTAH.

subsp. *hallii* (Hack.) Wipff (for Elihu Hall, botanical collector from Illinois) Hall's bluestem, sand bluestem. [*Andropogon gerardii* Vitman var. *paucipilus* (Nash) Fernald; *A. hallii* Hack.]. ARIZ: Apa, Nav; NMEX: McK.

Andropogon glomeratus (Walter) Britton, Sterns & Poggenb. (wound up, glomerate) Bushy bluestem. Plants tufted, perennial. **CULMS** erect, much-branched above, 50–100 cm tall or more. **LIGULES** a firm membrane 0.4–1 mm long, accompanied by long hairs attached at the base of the blade. **BLADES** flat or folded, 2–6 mm wide, involute toward the long-acuminate tips. **PANICLES** of 2 delicate branches subtended by a reddish spathe, these tightly clustered and terminating numerous branches of the flowering shoots, which appear bushy and broomlike. **SESSILE SPIKELETS** 3–4.5 mm long; awns 1–2 cm long. **PEDICELLED SPIKELETS** typically absent, the pedicel remaining. **ANTHERS** 0.5–1.5 mm long. $2n = 20$. Hanging gardens and seeps of the Glen Canyon Recreation Area, with *Populus* and *Salix*. UTAH. 1100–1160 m (3600–3800′). Flowering: Aug–Oct. Throughout the southern states, west to Utah and Arizona; Mexico. Our material is referred to **var. *scabriglumis*** C. S. Campb.

Apera Adans. Silky-bent

(Greek *a*, not, + *peros*, maimed, apparently alluding to presence of the long awn) Plants annual, tufted, with hollow culms. **LEAVES** with open sheaths, membranous ligules, and lacking auricles. **INFL** an open to contracted panicle with numerous spikelets. **SPIKELETS** 1-flowered, the rachilla prolonged beyond the palea; disarticulation above the glumes; glumes longer than the floret, acuminate (ours) to aristate, 1- to 3-nerved; lemma becoming hardened at maturity, rounded on the back, faintly 5-nerved, awned from just below the tip; palea well developed, 2-nerved; stamens 3. $x = 7$ (Björkman 1960). A small genus of three species, native to cool-temperate regions of Europe and Asia.

Apera interrupta (L.) P. Beauv. (severed, interrupted, alluding to the panicle) Dense silky-bent. **CULMS** erect to decumbent at the base, 10–60 cm tall. **LIGULES** acute, erose, 1.5–4 mm long. **BLADES** flat to involute, 1–3 mm wide. **PANICLES** narrow and densely flowered, interrupted below, the branches ascending-erect, 6–18 cm long. **GLUMES** lanceolate, the first 1.7–2.2 mm long, the second 2–2.8 mm long. **LEMMAS** lanceolate, acute, 1.6–2.2 mm long; awn 5–15 mm long. **PALEAS** about 1/2 the length of the lemmas; rachilla prolonged as a bristle to 0.6 mm long. **ANTHERS**

0.3–0.4 mm long. $2n = 14, 28$. [*Agrostis interrupta* L.]. Disturbed sites in at least somewhat moist ground, lower elevations. ARIZ: Apa. 1400–1700 m (4700–5600′). Flowering: May–Jul. Native to Eurasia; adventive in much of the United States.

Aristida L. Threeawn

(Latin *arista*, awn) Tufted annuals or perennials with hollow culms. **LEAVES** with open sheaths, hairy or membranous-ciliate ligules, usually involute blades, and lacking auricles. **INFL** an open or contracted panicle. **SPIKELETS** 1-flowered; glumes equal or unequal, relatively large, thin, 1- to 3-nerved; lemmas fusiform-terete, convolute (ours) or involute, 3-nerved, enveloping the palea and flower, indurate at maturity, sometimes narrowed at the apex into a beak, the nerves extended into 3 awns, these sometimes wound together into an awn column, the lateral 2 awns sometimes reduced to absent; palea about 1/2 the length of the lemma; disarticulation above the glumes. **STAMENS** 3 or 1. $x = 11$ (Allred 1984; Allred and Valdes-Reyna 1997). A genus of some 300 species in the arid and semiarid regions of the world.

1. Plants annual, often branching from the lower to middle nodes ..... (2)
1′ Plants perennial, usually branching only from the very basal nodes ..... (3)

2. Awns mostly 1–2 cm long; glumes mostly 5–12 mm long ..... ***A. adscensionis***
2′ Awns 2–7 cm long; glumes mostly 20 mm or more long ..... ***A. oligantha***

3. Glumes equal or nearly so; blades usually flat and curling like wood shavings in age ..... ***A. arizonica***
3′ Glumes noticeably unequal; blades usually rolled and not curling like wood shavings, but sometimes arcuate ..... ***A. purpurea***

Aristida adscensionis L. (from Ascension Island in the South Atlantic) Six-weeks threeawn. Plants annual, tufted. **CULMS** 10–75 cm tall or more, commonly geniculate and branched in the lower 1/2. **LIGULES** a short ciliate membrane about 0.5 mm long or less. **BLADES** involute on smaller plants to flat on larger plants, 0.5–2 mm wide. **PANICLES** narrow, congested to loose, 3–25 cm long, the branches usually ascending-erect, rarely spreading and with flexuous pedicels. **GLUMES** lanceolate to narrowly ovate, acute to acuminate, 1-nerved, often straw-colored, keeled, unequal, the first 4–7.5 mm long, the second 6–11 mm long, awnless or rarely short-mucronate. **LEMMAS** 5–9 mm long from base to awns; awns subequal with lateral slightly shorter, 9–25 mm long, rarely the lateral awns 1/2 or less the length of the central. **PALEAS** 1/2 or less the length of the lemmas. **ANTHERS** 0.6–1 mm long. $2n = 22$. [*Aristida bromoides* Kunth]. Desert scrub to piñon-juniper woodland, dry hills, slopes, valleys, and mesas. ARIZ: Apa; NMEX: SJn. 1300–1700 m (4300–5600′). Flowering: Mar–Oct. Southwestern United States, east to Missouri, south into Mexico; South America; Africa; Eurasia. Plant size varies in response to moisture, from tiny plants a few cm high, to robust, much-branched ones 3/4 m tall.

Aristida arizonica Vasey (from Arizona) Arizona threeawn. Plants perennial, tufted. **CULMS** 25–70 cm tall, branching only at the very base to form the tillers. **LIGULES** tiny, 0.2–0.3 mm long. **BLADES** mostly flat, curling like wood shavings in age, 1–4 mm wide. **PANICLES** narrow, congested, the branches and pedicels mostly erect-appressed, 8–18 cm long, less than 3 cm wide (including the awns). **GLUMES** subequal, 10–16 mm long. **LEMMAS** 12–18 mm long from base to awns, the terminal beak 3–6 mm long; awns 20–35 mm long, the lateral slightly shorter than the central. **PALEAS** about 1/2 the length of the lemmas. **ANTHERS** 1.3–2 mm long. $2n = 22$. Upper piñon woodlands to ponderosa forest understory, not common in the *Flora* region. NMEX: McK, SJn. 1850–2200 m (6100–7300′). Flowering: Aug–Oct. Arizona to Texas, south through the highland mountains of Mexico. This species has been confused with *A. purpurea* var. *nealleyi*, and nearly all of the numerous specimens from our flora region identified as *A. arizonica* belong to that other species. *Aristida arizonica* is not known to the author from Utah or Colorado and would be especially out of place in desert scrub communities as reported in *A Utah Flora*.

Aristida oligantha Michx. (few-flowered) Old-field threeawn. Plants annual, tufted. **CULMS** 25–55 cm tall, highly branched, geniculate at the base. **LIGULES** less than 0.5 mm long. **BLADES** flat or loosely involute, 0.5–1.5 mm wide. **PANICLES** spicate or racemose, scarcely branched if at all, the spikelets on very short pedicels directly on the main axis and divergent from axillary pulvini, 7–20 cm long, 2–4 cm wide. **GLUMES** unequal, the first 3- to 7-nerved, 10–22 mm long, with a delicate awn 1–10 mm long, the second 1-nerved, 8–20 mm long. **LEMMAS** 10–22 mm long. **PALEAS** short, less than 1/3 the length of the lemmas. **ANTHERS** usually 1 and less than 0.5 mm long, rarely 3 and 2–4 mm long. $2n = 22$. Waste places, disturbed sites, roadsides, fields, usually in sandy soil, below 2000 m (6500′). ARIZ: Nav; NMEX: SJn. 1600–1800 m (5300–5800′). Found throughout much of the central and eastern states, and on the west coast; our plants seem to be adventive here, the result of infrequent introductions, perhaps with hay or livestock. The branched culms and nearly sessile spikelets render this species unmistakable.

Aristida purpurea Nutt. (purple) Purple threeawn. Plants perennial, tufted. **CULMS** 20–100 cm tall, branching only at the very base to form tillers. **LIGULES** less than 0.5 mm long. **BLADES** tightly involute to flat, 1–1.5 mm wide, not curling like wood shavings in age, but sometimes arcuate (while involute). **PANICLES** sparingly branched, often racemose, 5–30 cm long, 2–12 cm wide, the branches erect-appressed to spreading, drooping, or flexuous; axillary pulvini absent (in our species). **GLUMES** 1-nerved, typically strongly unequal, at least the majority, the first 4–12 mm long, the second 7–25 mm long. **LEMMAS** glabrous, scaberulous, or tuberculate, 6–16 mm long from base to awns, the apex 0.1–0.8 mm wide, the beak absent or less than 3 mm long; awns 1–12 cm long (rarely longer), subequal to rarely markedly unequal, the central not or only slightly thicker than the lateral ones. **PALEAS** 1/3–1/2 the length of the lemmas. **ANTHERS** 0.7–2 mm long. $2n = 22, 44, 66, 88$. Widely distributed in a variety of habitats, both disturbed and natural, from the low deserts to ponderosa forests. Flowering: Mar–Oct. Western Canada south throughout most of the western and central states into Mexico. We have four intergrading varieties, often intermingled in the same locale:

1. Awns 4–10 cm long ..... (2)
1' Awns 1–4 cm long ..... (3)

2. Second glume mostly less than 16 mm long; awns delicate, mostly 0.2 mm wide or less at their bases, 4–5 cm long ..... **var. *purpurea***
2' Second glume 14–25 mm long; awns usually stout, more than 0.2 mm wide at the base, 4–10 cm long ..... **var. *longiseta***

3. Panicle branches drooping at the tips; awns purplish ..... **var. *purpurea***
3' Panicle branches stiffly erect; awns mostly brownish or straw-colored ..... (4)

4. Panicle straw-colored; summit of lemma mostly less than 0.2 mm broad; awns delicate, mostly less than 0.2 mm wide at their bases ..... **var. *nealleyi***
4' Panicle dark brown or olive-colored; summit of lemma mostly broader than 0.2 mm; awns stout, mostly 0.2 mm wide or more at their bases ..... **var. *fendleriana***

var. *fendleriana* (Steud.) Vasey (for Augustus Fendler, German-born botanical collector for Asa Gray) Fendler's threeawn [*Aristida fendleriana* Steud.]. Open slopes, hills, and sandy flats. ARIZ: Apa, Coc, Nav; COLO: Arc, LPl, Mon; NMEX: McK, RAr, San, SJn; UTAH. 1600–2350 m (5300–7700'). Western United States into Mexico.

var. *longiseta* (Steud.) Vasey (long-awned) Red threeawn. [*Aristida longiseta* Steud.]. Sandy or rocky slopes and plains, and in barren soils of disturbed ground. ARIZ: Apa, Coc, Nav; COLO: Arc, Dol, LPl, Mon; NMEX: McK, San, SJn; UTAH. 1100–2350 m (3700–7600'). Canada to northern Mexico.

var. *nealleyi* (Vasey) Allred (for Greenleaf Cilley Nealley, USDA botanical collector) Nealley's threeawn [*Aristida glauca* (Nees) Walp.]. Dry slopes and plains, frequently in desert grassland vegetation. ARIZ: Apa, Coc, Nav; COLO: Arc, Mon; NMEX: RAr, SJn; UTAH. 1100–1900 m (3700–6200'). Southwestern United States into Mexico.

var. *purpurea* Purple threeawn. Sandy to clay soils, along rights of way, or on dry slopes and mesas. ARIZ: Coc; UTAH. 1100–1500 m (3700–4800'). Canada to Mexico and Cuba.

Arrhenatherum P. Beauv. Oatgrass

(Greek *arrhen*, masculine, + *ather*, awn, referring to the awned staminate floret) Mostly tufted perennials with hollow culms. **LEAVES** with open sheaths, membranous ligules, flat blades, and lacking auricles. **INFL** a narrow panicle, the branches short or slightly developed and the panicle racemose. **SPIKELETS** 2-flowered, the upper floret perfect, the lower staminate; rachilla prolonged beyond the upper floret as a bristle; disarticulation above the glumes, the florets falling together; glumes thin, unequal, the first shorter than the lower floret, the second equaling the upper floret; lemmas 5- to 7-nerved, awned from the back or not; paleas slightly shorter than the lemmas; stamens 3. $x = 7$. About six species native to Eurasia.

Arrhenatherum elatius (L.) J. Presl & C. Presl (taller) Tall oatgrass. Plants loosely tufted or occasionally with short rhizomes. **CULMS** 50–140 cm tall, sometimes bulbous (with corms) at the base. **LIGULES** obtuse to acute, ciliate, 1–3 mm long. **BLADES** 3–10 mm wide. **PANICLES** 8–28 cm long, 2–6 cm wide, the branches erect, whorled, usually spikelet-bearing to the base. **GLUMES** lanceolate-elliptic, unequal, the first 4–7 mm long, the second 7–10 mm long. **LEMMAS** 5–10 mm long; awn of lower lemma 10–20 mm long, bent, borne from the back at about midlength; awn of upper lemma absent or to 5 mm long, borne just below the apex. **PALEAS** slightly shorter than the lemmas. **ANTHERS** 3.5–6 mm long. $2n = 14, 28, 42$. Meadows, stream banks, moist roadsides in the mountains; sometimes planted for erosion control. UTAH. 2150–2900 (7100–9500′). Flowering: Jul–Aug. Native to Eurasia, widespread in eastern and western North America.

Avena L. Oats

(Latin *avena*, the ancient name for oats, possibly from *aveo*, desire, because it was sought by cattle) Tufted, robust annuals with hollow culms. **LEAVES** with open sheaths, flat blades, membranous ligules, lacking auricles. **INFL** an open panicle, the branches drooping, racemose in less developed plants. **SPIKELETS** with 2–3 florets, large, drooping, the rachilla prolonged beyond the terminal floret and often bearing a tiny rudiment; glumes large, several-nerved, papery, acuminate; lemmas hardened when mature, rounded, glabrous to hirsute, awned from the back (to awnless in cultivated forms); palea shorter than the lemma; stamens 3. $x = 7$ (Baum 1977). About 10–15 species native to Europe and Asia, widely cultivated throughout the world.

1. Awns usually well developed and bent abruptly; florets separating and falling separately, leaving a circular scar or "sucker-mouth" at the callus .... ***A. fatua***
1′ Awns usually absent or short and straight; florets falling together, when broken apart mechanically a small portion of the rachilla remaining attached to the callus .... ***A. sativa***

Avena fatua L. (foolish or tasteless) Wild oats. **CULMS** 30–100 cm tall or more. **LIGULES** acute, 4–6 mm long. **BLADES** flat, 3–12 mm wide. **PANICLES** 10–30 cm long, 6–20 cm wide. **GLUMES** subequal, 18–32 mm long, 9- to 11-nerved. **LEMMAS** 14–22 mm long, typically densely strigose below the middle; awns 23–42 mm long. **ANTHERS** about 3 mm long. $2n = 42$. Roadsides, fields, among crops, disturbed sites. ARIZ: Apa; COLO: Arc, Mon; NMEX: SJn. 1250–2100 m (4200–6800′). Flowering: May–Sep. Native to Eurasia.

Avena sativa L. (planted or sown) Common oats. **CULMS** 40–140 cm tall or more. **LIGULES** acute, 3–8 mm long. **BLADES** 4–20 mm wide. **PANICLES** 15–40 cm long, 5–15 cm wide. **GLUMES** subequal, 20–32 mm long, 9- to 11-nerved. **LEMMAS** typically glabrous, 14–18 mm long; awns usually absent, 15–30 mm when present and mostly straight. **ANTHERS** 2–4 mm long. $2n = 42$. [*A. fatua* L. var. *sativa* (L.) Hausskn.]. Widely cultivated and then found along adjacent roadsides and fields, rarely persisting for long. ARIZ: Nav; COLO: LPl; NMEX: McK, SJn; UTAH. 1120–2150 m (3700–7000′). Flowering: May–Sep. Native to Eurasia. Sometimes considered merely a subspecies or variety of *A. fatua*.

Beckmannia Host Sloughgrass

(for Johann Beckmann, 1739–1811, German botanist) Plants annual and tufted (ours), or perennial and rhizomatous. **LEAVES** with open sheaths, membranous ligules, flat blades, and lacking auricles. **INFL** a panicle of spicate, unilateral branches, secondary branches infrequent and poorly developed; disarticulation below the glumes, the spikelets falling entire. **SPIKELETS** laterally compressed, circular-ovate in outline, subsessile, with 1–2 florets, the rachilla not prolonged; glumes subequal, keeled, inflated, D-shaped in outline; lemmas lanceolate, awn-tipped; paleas similar to the lemmas in size and shape; stamens 3. $x = 7$.

Beckmannia syzigachne (Steud.) Fernald (scissorlike, referring to the glumes) American sloughgrass. Plants annual. **CULMS** solitary or loosely tufted, (20) 30–100 cm or more tall. **LIGULES** obtuse to acuminate, entire to lacerate, 4–11 mm long. **BLADES** 4–12 mm wide, scabrous. **PANICLES** narrow, congested, 8–30 cm long, the branches erect to ascending. **GLUMES** 3-nerved, apiculate, strongly keeled, 2–3 mm long. **LEMMAS** 2.5–3.5 mm long, slightly exceeding the glumes. **ANTHERS** 0.5–1.5 mm long. $2n = 14$. [*Beckmannia eruciformis* (L.) Host]. Marshes, floodplains, edges of ponds, streams, and ditches, low to high elevations. ARIZ: Apa; COLO: Arc, Hin, LPl, Min, Mon; NMEX: RAr, SJn; UTAH. 2000–2650 m (6500–8700′). Flowering: Jul–Sep. Alaska and Canada, throughout the western United States, east to the Great Lakes region; Eurasia. The panicles and spikelets are instantly recognizable. Spikelets occasionally have two florets. Our material is referred to **subsp. *baicalensis*** (Kusn.) T. Koyama & Kawano.

Blepharoneuron Nash Pine Dropseed

(Greek *blepharis*, eyelash, + *neuron*, nerve, alluding to the ciliate nerves of the lemmas) Plants annual or perennial (ours), with hollow culms. **LEAVES** with open sheaths, membranous ligules, flat to involute blades, lacking auricles. **INFL** a panicle of capillary, flexuous branches, the pedicels minutely glandular just below the spikelets. **SPIKELETS** 1-flowered, slightly laterally compressed, olive-colored; disarticulation above the glumes; glumes subequal, 1-nerved, glabrous, ovate to ovate-lanceolate; lemmas 3-nerved, slightly longer than the glumes, densely ciliate-sericeous on the 3 nerves; paleas villous between the 2 nerves; stamens 3. $x = 8$ (Peterson and Annable 1990). A genus of two species, an annual from northern Mexico, and a perennial (ours) from North America.

Blepharoneuron tricholepis (Torr.) Nash (hairy scale) Pine dropseed. Densely tufted perennials. **CULMS** erect, 15–70 cm tall. **LIGULES** truncate to rounded, 0.5–2 mm long, entire. **BLADES** 0.5–2.5 mm wide, usually folded to involute. **PANICLES** mostly 5–25 cm long, 2–10 cm wide, the branches erect-appressed to widely spreading; pedicels 2–9 mm long, capillary. **GLUMES** 1.8–2.8 mm long. **LEMMAS** 2–4 mm long. **ANTHERS** 1.2–2 mm long. $2n = 16$. Dry soils in ponderosa pine–Gambel's oak forests. ARIZ: Apa; COLO: Arc, Hin, LPl, Min, Mon, SJn; NMEX: McK, RAr, SJn; UTAH. 2100–3650 m (7000–12000′). Flowering: Jul–Oct. Utah and Colorado south to southern Mexico.

Bothriochloa Kuntze Bluestem

(Greek *bothros*, trench or pit, + *chloë*, grass, alluding either to the groove in the pedicels or to the pit in the lower glumes of some species) Plants perennial, tufted (ours) or stoloniferous, with pithy culms. **LEAVES** with open sheaths, membranous ligules, flat blades, and lacking auricles. **INFL** a terminal panicle of subdigitate to racemosely arranged branches (rames); rames with pairs of 1 sessile and 1 pedicelled spikelet; internodes with a translucent or opaque longitudinal groove, often villous on the margins; disarticulation in the rame, of the sessile and pedicelled spikelets with pedicel and rachis internode. **SPIKELETS** dorsally compressed; sessile spikelets with 2 florets, the lower reduced, hyaline and awnless, the upper fertile and awned (ours); glumes hardened, enclosing the florets except for the awn. **STAMENS** 3; pedicelled spikelets sterile or staminate, as large as or much smaller than the sessile spikelets, awnless. $x = 10$ (Allred 1983; Gould 1967). About 35 species of tropical to warm-temperate regions of the world, nearly all of which were formerly included in an expanded *Andropogon*.

1. Panicle branches subdigitate, most borne at the tip of the stem; spikelets typically purplish or reddish, with relatively fewer hairs, the spikelets scarcely obscured; pedicelled spikelets about as long and broad as the sessile ones .... ***B. ischaemum***
1′ Panicle branches mostly scattered along the main axis; spikelets typically greenish, silvery, or straw-colored, with copious hairs, the spikelets obscured; pedicelled spikelets definitely smaller than the sessile ones .... (2)

2. Awns less than 18 mm long; sessile spikelets less than 4.5 mm long; foliage often glaucous .... ***B. laguroides***
2′ Awns longer than 18 mm; sessile spikelets 4.5–7 mm long; foliage rarely glaucous .... ***B. barbinodis***

Bothriochloa barbinodis (Lag.) Herter (hairy-noded) Cane bluestem. **CULMS** 60–120 cm tall, erect to geniculate basally, often branched above in age; nodes hirsute with stiff, tan to whitish hairs 3–4 mm long. **LIGULES** 1–2 mm long, often erose. **BLADES** 2–7 mm wide, not glaucous. **PANICLES** 5–16 cm long on the larger shoots, oblong to somewhat fan-shaped, silvery; rachis 5–10 cm long, with numerous branches 4–9 cm long. **SESSILE SPIKELETS** 4.5–7 mm long, the lower glume with or without a dorsal pit; awns 20–35 mm long. **PEDICELLED SPIKELETS** 3–4 mm long, sterile, awnless. **ANTHERS** 0.5–1 mm long. $2n = 180, 220$. [*Andropogon barbinodis* Lag.]. Desert scrub and piñon-juniper woodlands, typically in dry sites below the ponderosa zone. ARIZ: Nav; UTAH. 1650–1800 m (5400–5800′). Flowering: Apr–Oct. Southwestern United States, south through Mexico and Central America, to Bolivia and Argentina.

Bothriochloa ischaemum (L.) Keng (resembling the genus *Ischaemum*) Yellow bluestem. Plants usually tufted, occasionally stoloniferous. **CULMS** 30–80 cm tall; nodes glabrous or short-hirsute. **LIGULES** 0.5–1.5 mm long. **BLADES** 2–4.5 mm wide, flat to folded, glabrous or with long scattered hairs at the base of the blade. **PANICLES** reddish purple, 5–10 cm long, with 2–8 branches clustered toward the tip of the flowering shoot. **SESSILE SPIKELETS** 3–4.5 mm long, narrowly ovate, the lower glume lacking a dorsal pit; awns 9–17 mm long. **PEDICELLED SPIKELETS** about as long as and nearly as broad as the sessile, sterile or staminate. **ANTHERS** 1–2 mm long. $2n = 40, 50, 60$. [*Andropogon ischaemum* L.]. Disturbed ground, roadsides, in desert scrub and grassland communities. ARIZ: Apa; NMEX: McK, RAr, SJn; UTAH. 1700–2300 m (5600–7400′). Flowering: Aug–Sep. Native to southern Europe and Asia, introduced for erosion control and forage in the Southwest, and spreading into other central and southern states. The groove in the pedicels and rachis internodes is not translucent and does not extend to the other side of the joint, which is rounded on the back. Two minor variants have been named: var. *songarica* (Rupr. ex Fisch. & C. A. Mey.) Celarier & J. R. Harlan (King Ranch bluestem) with pubescent nodes, and var. *ischaemum* with glabrous nodes.

Bothriochloa laguroides (DC.) Herter (resembling the genus *Lagurus*) Silver bluestem. **CULMS** erect or geniculate at the base, branched in age, 35–110 cm tall; nodes sparsely shortly hirsute or pilose with erect hairs, or glabrous. **LIGULES** 1–3 mm long. **BLADES** and foliage often glaucous, 2–7 mm wide, flat to folded, mostly glabrous. **PANICLES** narrowly oblong or lanceolate, silvery white or light tan, 4–12 cm long, with more than 10 branches, rarely with axillary pulvini, the lower branches shorter than the main axis. **SESSILE SPIKELETS** ovate, somewhat glaucous, with blunt apices, 2.5–4 mm long (rarely longer), the lower glume only rarely with a dorsal pit; awns 8–16 mm long. **PEDICELLED**

SPIKELETS 1.5–2.5 (3) mm long, sterile, distinctly smaller than the sessile. **ANTHERS** 0.6–1.4 mm long. 2*n* = 60. [*Andropogon laguroides* DC.; *Bothriochloa saccharoides* of earlier works]. Roadsides and other disturbed moist sites in desert scrub and grassland communities. ARIZ: Nav; NMEX: McK, San, SJn; UTAH. 1650–3000 m (5400–6800′). Flowering: Aug–Oct. Great Plains and adjacent regions, south into Mexico; South America. Our plants belong to **subsp**. ***torreyana*** (Steud.) Allred & Gould.

Bouteloua Lag. Grama

(for the brothers Claudio, 1774–1842, and Estéban, 1776–1813, Boutelou Agraz, Spanish botanists) Plants annual or perennial, tufted, stoloniferous, or rhizomatous, with pithy or hollow culms. **LEAVES** with open sheaths, ciliate ligules, flat, folded, or involute blades, and lacking auricles. **INFL** a panicle of spicate primary branches; disarticulation above the glumes or below with the branch. **SPIKELETS** sessile, 2- to 4-flowered, the lower floret perfect, the others staminate or vestigial; glumes unequal, 1-nerved; lower (fertile) lemmas membranous, 3-nerved, awned or awnless; upper lemmas reduced in size, often awned; stamens 3. *x* = 10 (Gould 1979). About 40 species of the Western Hemisphere.

1. Stem internodes (not the sheaths) woolly-pubescent ..... ***B. eriopoda***
1′ Stem internodes glabrous ..... (2)

2. Inflorescence branches deciduous at maturity; spikelets 1–16 per branch ..... (3)
2′ Inflorescence branches and glumes persistent on the plant; spikelets usually 20–60 per branch ..... (4)

3. Plants annual; panicles with 1–15 branches ..... ***B. aristidoides***
3′ Plants perennial; panicles with (12) 20–80 branches ..... ***B. curtipendula***

4. Plants annual ..... (5)
4′ Plants perennial ..... (6)

5. Inflorescence reduced to a single branch ..... ***B. simplex***
5′ Inflorescence with 4–several branches, only rare culms with a single branch ..... ***B. barbata***

6. Panicle branch bristlelike, extending 4–10 mm beyond the attachment of the terminal spikelet ..... ***B. hirsuta***
6′ Panicle branch not bristlelike and not extending past the last spikelet, but terminating in a spikelet ..... ***B. gracilis***

Bouteloua aristidoides (Kunth) Griseb. (resembling the genus *Aristida*) Needle grama. Plants annual, tufted. **CULMS** 5–45 cm tall, often geniculate at the bases. **LIGULES** 0.2–0.5 mm long. **BLADES** flat or folded, 1–2 mm wide, margins usually with papillose-based hairs near the ligule. **PANICLES** 2–10 cm long, with (1) 4–15 branches, these deciduous and densely pubescent at the base, the branch extending beyond the terminal spikelet. **SPIKELETS** with 1–2 florets, the upper sterile and awned, 1–6 per branch. **GLUMES** unequal. **FERTILE FLORET** 5–7 mm long, acute or awn-tipped. **STERILE FLORETS** reduced to a short column with 3 awns 2–7 mm long. **PALEAS** subequal to the lemmas. **ANTHERS** about 2.5 mm long, yellow or yellow and red. 2*n* = 40. Dry, sandy ground in desert scrub vegetation, roadsides, not common. ARIZ: Apa. 1450–1700 m (4850–5500′). Flowering: Aug–Oct. California to Texas; Mexico; South America.

Bouteloua barbata Lag. (bearded) Six-weeks grama. Low, tufted annuals. **CULMS** numerous, sprawling to erect, 5–25 cm long, often geniculate, branching. **LIGULES** 0.5–1.2 mm long. **BLADES** mostly flat, 1–2 mm wide, often with whitish margins. **PANICLES** 3–10 cm long, with 2–8 branches, these often arcuate; disarticulation above the glumes. **SPIKELETS** with 1 fertile floret and 1–2 sterile rudimentary florets, crowded on the branches. **GLUMES** 1–2.5 mm long, unequal, awn-tipped to 0.7 mm long. **FERTILE FLORET** 1.5–2.5 mm long, appressed-pubescent; awns to 2.5 mm long. **PALEAS** subequal to the lemmas. **STERILE FLORETS** 0.5–1 mm long, globose, empty, awns to 3 mm long. **ANTHERS** about 0.5 mm long, yellow. 2*n* = 20, 40. Dry, open places of deserts and lower foothills, washes and arroyos, roadsides and disturbed ground. ARIZ: Apa, Coc, Nav; COLO: LPl, Mon; NMEX: McK, RAr, SJn; UTAH. 1100–1850 m (3700–6100′). Flowering: Aug–Oct. Southern California through southern Utah and Colorado, south to Mexico.

Bouteloua curtipendula (Michx.) Torr. (short-hanging, referring to the branches) Sideoats grama. Tufted or short-rhizomatous perennials. **CULMS** erect, slender, 25–100 cm tall. **LIGULES** 0.2–1 mm long. **BLADES** typically flat, 2–4 mm wide. **PANICLES** long and narrow, 12–30 cm long, with (12) 20–80 branches, these 1–3 cm long and falling as a unit. **SPIKELETS** with 1–3 florets, the lower fertile and the others vestigial, 2–5 per branch. **GLUMES** unequal, the first 2.5–5 mm long, the second 4.5–7.5 mm long. **FERTILE FLORET** 3–6.5 mm long, short-awned to 6 mm. **PALEAS** subequal to the lemmas, unawned. **STERILE FLORETS** 0.4–3.5 mm long, with 3 unequal awns to 7 mm long. **ANTHERS** 1.5–3.5 mm long, yellow, orange, red, or purple. 2*n* = 20, 40, 41–103. Widespread in grassy, desert scrub, woodland, and ponderosa pine habitats, mostly at lower elevations. ARIZ: Apa, Coc, Nav; COLO: Arc, LPl, Mon; NMEX: McK, RAr, San, SJn; UTAH. 1100–2150 m (3700–7100′). Flowering: Jul–Oct. Southern Canada, most of the United States; Mexico; South America.

Bouteloua eriopoda (Torr.) Torr. (woolly-footed) Black grama. Tufted perennials from knotty bases, often stoloniferous with long internodes. **CULMS** sprawling to ascending, the internodes with short, dense, woolly hairs, 25–70 cm long. **LIGULES** 0.1–0.4 mm long. **BLADES** flat when fresh, 1–2.2 mm wide. **PANICLES** 2–16 cm long, with 1–8 widely spaced branches, these 1.5–4 cm long and with a prominent tuft of hairs at the base; disarticulation above the glumes. **SPIKELETS** with 1 fertile floret and 1 rudimentary floret, 8–14 per branch. **GLUMES** unequal, the lower 2–4.5 mm long, the upper 4.5–8 mm long. **FERTILE FLORET** 4–7 mm long, the awns 0.5–4 mm long. **PALEAS** subequal to the lemmas. **STERILE FLORETS** rudimentary, terminating in 3 awns 4–9 mm long. **ANTHERS** 1.5–3 mm long, yellow to orange. $2n$ = 20, 21, 28. Desert grasslands, scrublands, and into the lower foothills, sandy to rocky ground. ARIZ: Apa, Nav; COLO: Mon; NMEX: McK, SJn; UTAH. 1450–2050 m (4800–6800′). Flowering: Jul–Sep. Western United States into northern Mexico. This is the only grass in our flora with woolly (not just pubescent) internodes.

Bouteloua gracilis (Kunth) Lag. ex Griffiths (slender) Blue grama. Plants densely tufted, sometimes with short rhizomes. **CULMS** erect to decumbent-based, 20–70 cm long, unbranched above the base. **LIGULES** 0.1–0.4 mm long, often with a tuft of hair at the edges. **BLADES** flat to involute, 0.5–2.5 mm long, with scattered hairs at the base. **PANICLES** 2–12 cm long, with 1–6 widely spaced branches, these 1.5–6 cm long; disarticulation above the glumes. **SPIKELETS** with 1 fertile floret and 1 rudimentary floret, 40–100 or more per branch. **GLUMES** sometimes with papillose-based hairs, unequal, the lower 1.5–3.5 mm long, the upper 3.5–6 mm long. **FERTILE FLORET** 3.5–6 mm long, with awns 1–3 mm long. **PALEAS** about 5 mm long, awn-tipped. **STERILE FLORETS** vestigial, 1–3 mm long, with 3 awns 1–3 mm long. **ANTHERS** 1.7–3 mm long, yellow or purple. $2n$ = 20, 28, 35, 40, 42, 60, 61, 77, 84. Very widely distributed and common from desert plains through woodlands and forests to middle or rarely high elevations in the mountains, usually in open, dry sites. ARIZ: Apa, Coc, Nav; COLO: Arc, Dol, Hin, LPl, Min, Mon, SJn, SMg; NMEX: McK, RAr, San, SJn; UTAH. 1550–2450 m (5150–8000′). Flowering: Jul–Oct. Southwestern Canada through the southwestern and central United States to Mexico, rare eastward. Similar to *B. hirsuta*, but lacking the continuation of the rachis beyond the terminal spikelet. Sometimes the terminal spikelet will stick outward, mimicking a bristle; close inspection will reveal this. This is the state grass of New Mexico, and perhaps the most common grass in the Four Corners region.

Bouteloua hirsuta Lag. (shaggy or bristly) Hairy grama. Plants perennial, densely or loosely tufted, sometimes somewhat stoloniferous. **CULMS** erect to decumbent-based, rarely branched above, 15–70 cm long. **LIGULES** 0.2–0.5 mm long. **BLADES** flat to involute, 1–2.5 mm wide, often with papillose-based hairs. **PANICLES** 1–15 cm long, with 1–6 widely spaced branches, these 1–4 cm long; disarticulation above the glumes. **SPIKELETS** with 1 fertile floret and 1–2 rudimentary florets, 20–50 per branch. **GLUMES** unequal, the midnerve with papillose-based hairs, the first 1.5–3.5 mm long, the second 3–6 mm long. **FERTILE FLORET** 2–4.5 mm long, with awn(s) 0.2–2.5 mm long. **PALEAS** ovate, not awned. **STERILE FLORETS** 0.5–2 mm long, the 3 awns 2–6 mm long. **ANTHERS** 2–3.5 mm long, cream or yellow. $2n$ = 20, 40, 50, 60. Relatively infrequent on bajadas and foothills, sandy and rocky ground. ARIZ: Apa; NMEX: McK, RAr, SJn; UTAH. 1500–1800 m (5000–5900′). Flowering: Jul–Sep. Great Plains south to Arizona and Texas; Mexico.

Bouteloua simplex Lag. (simple or single, referring to the branching) Mat grama. Plants annual. **CULMS** mostly geniculate-decumbent, 3–35 cm long. **LIGULES** tiny, 0.1–0.2 mm long. **BLADES** flat to involute, 0.5–1.5 mm wide, often pilose at the base near the sheath. **PANICLES** with a single branch terminating the culm, rarely some culms of a clump with 2–3 branches, these 1–4 cm long, straight to arcuate; disarticulation above the glumes. **SPIKELETS** crowded, with 1 fertile floret and 1–2 rudimentary florets. **GLUMES** glabrous, unequal, the first 1.5–2.5 mm long, the second 3.5–5 mm long. **FERTILE FLORET** ciliate-pilose on the nerves, 2.5–3.5 mm long; awns 3, 1–3 mm long. **PALEAS** obovate, awnless. **STERILE FLORETS** rudimentary, a short column with 3 awns 5–6 mm long. **ANTHERS** 0.4–0.6 mm long. $2n$ = 20. Grassy and brushy plains and hills, roadsides, disturbed ground. ARIZ: Apa, Nav; COLO: Arc, LPl, Min, Mon, SMg; NMEX: McK, RAr, San, SJn; UTAH. 1600–2250 m (5300–7400′). Flowering: Aug–Oct. Wyoming and Colorado south to Arizona and Texas; Mexico.

Bromus L. Bromegrass

(Greek *bromo*, food, the ancient name for oats) Plants annual to perennial, tufted or rhizomatous, with hollow culms. **LEAVES** with closed sheaths, membranous ligules, mostly flat blades, and with or without auricles. **INFL** a panicle, often lax and drooping, usually rebranching but sometimes poorly developed and racemelike with few spikelets. **SPIKELETS** large, several-flowered, strongly compressed to terete; glumes unequal, shorter than the lower lemma, typically awnless; lemmas (3) 5- to 9-nerved, awned or awnless from the slightly bifid apex; palea shorter than to as long as the lemma, strongly ciliate on the 2 nerves, adhering to the caryopsis at maturity; stamens usually 3. $x = 7$ (Allred 1993; Peterson et al. 2001; Wagnon 1952). About 100 species in the temperate regions of the world. Some recognize several segregate genera (*Anisantha*, *Bromopsis*, *Ceratochloa*), but a more traditional, practical circumscription is maintained here.

1. Plants perennial .... (2)
1′ Plants annual .... (9)

2. Rhizomes present .... ***B. inermis***
2′ Rhizomes absent .... (3)

3. Spikelets strongly flattened, the lemmas V-shaped in cross section; second (upper) glume 5- to 9-nerved .... (4)
3′ Spikelets not strongly flattened, but more or less terete, the lemmas rounded on the back in cross section; second (upper) glume 3-nerved .... (5)

4. Lemma awns 0–2.5 mm long .... ***B. catharticus***
4′ Lemma awns 3–8 mm long (rarely as short as 2 mm) .... ***B. carinatus***

5. First glume 3-nerved .... (6)
5′ First glume 1 (2)-nerved .... (7)

6. Glumes and pedicels puberulent; blades erect, the midrib not narrowed below the collar .... ***B. porteri***
6′ Glumes and pedicels glabrous; blades mostly lax or spreading, the midrib often narrowed below the collar .... ***B. frondosus***

7. Sheaths densely lanate, the hairs spreading from the sheath but becoming matted at the tips .... ***B. lanatipes***
7′ Sheaths glabrous to lightly pilose or hirtellous, if pubescent then not becoming matted .... (8)

8. Sheaths with crinkled hairs at the corner of the collars; lemmas densely hairy on the margins but glabrous or nearly so on the median portion across the back; anthers mostly shorter than 2.2 mm; lemmas mostly longer than 8.6 mm .... ***B. ciliatus***
8′ Sheaths glabrous at the collars; lemmas hairy across the back, not glabrous on the median portion; anthers mostly longer than 1.8 mm; lemmas mostly shorter than 9.2 mm .... ***B. anomalus***

9. Lemma awns 0–2.5 mm long .... ***B. catharticus***
9′ Lemma awns longer than 3 mm .... (10)

10. Lemma body 6–9 (10) mm long at maturity .... (11)
10′ Lemma body (9) 10–30 mm long at maturity .... (13)

11. Awns mostly less than 5 mm long; lemmas rounded, the margins usually rolled around the grain; plants mostly glabrous .... ***B. secalinus***
11′ Awns mostly more than 5 mm long; lemmas somewhat flattened, the margins not rolled around the grain; plants pubescent .... (12)

12. Panicles dense, compact, 3–8 (10) cm long, the branches stiffly erect and shorter than the spikelets .... ***B. hordeaceus***
12′ Panicles open, 6–20 cm long, the branches spreading, at least some longer than the spikelets .... ***B. japonicus***

13. First glume 3- to 5-nerved; awns 4–8 mm long .... ***B. carinatus***
13′ First glume mostly 1-nerved (occasionally 3-nerved in *B. diandrus*); awns (7) 10–60 mm long .... (14)

14. Panicle dense, compact, ovoid; panicle branches stout, erect, and mostly much shorter than 2 cm....... ***B. rubens***
14′ Panicle loose, open, elongate; panicle branches often spreading or drooping, and mostly much longer than 2 cm .(15)

15. Awns mostly 3–6 cm long; lemmas 20–35 mm long .... ***B. diandrus***
15′ Awns mostly 1–3 cm long; lemmas 9–20 mm long .... ***B. tectorum***

Bromus anomalus Rupr. ex E. Fourn. (irregular) Nodding brome. Plants perennial, tufted. **CULMS** erect, unbranched above the base, 45–110 cm long. **LIGULES** 1 mm or less long. **BLADES** flat, 3–7 mm wide, glabrous at the collar, the midrib narrowed just below the collar. **PANICLES** 8–24 mm long, open, the branches ascending to divaricate, lax. **SPIKELETS** not much compressed. **GLUMES** unequal, the first 1-nerved and 5–6 mm long, the second 3-nerved and

6–8 mm long. **LEMMAS** 6–11 mm long, shortly pilose on the margins and across the back between the nerves, occasionally on the margins only; awns 1–4 mm long. **ANTHERS** 1.7–4.5 mm long. $2n$ = 14. [*Bromopsis anomala* (Rupr. ex E. Fourn.) Holub]. Mountain brush, ponderosa parklands, mountain meadows, and edges of coniferous forests, often in shaded understory; not common in the *Flora* area. NMEX: SJn; UTAH. 2600–3100 m (8500–10200′). Flowering: Jul–Sep. Southern Rocky Mountains south into Mexico.

Bromus carinatus Hook. & Arn. (keeled) Mountain brome. Plants perennial, tufted, sometimes flowering the first season. **CULMS** erect, 40–120 cm tall. **LIGULES** 1–3 mm long, erose. **BLADES** flat, 3–10 mm wide. **PANICLES** 10–25 cm long, open, the branches typically spreading to stiffly divaricate. **SPIKELETS** strongly compressed. **GLUMES** unequal, strongly keeled, the first 7–11 mm long and 3- to 5-nerved, the second 9–13 mm long and 5- to 7-nerved. **LEMMAS** keeled on the midnerve, glabrous to short-pilose, 11–15 mm long; awns 4–8 mm long. **ANTHERS** 1.5–3.5 mm long. $2n$ = 28, 42, 56. [*B. marginatus* Nees ex Steud.; *B. polyanthus* Scribn.; *Ceratochloa carinata* (Hook. & Arn.) Tutin; *Ceratochloa marginata* (Nees ex Steud.) W. A. Weber; *Ceratochloa polyantha* (Scribn.) Tzvelev]. Mountain brush, coniferous forest, aspen glades, often in disturbed habitats. ARIZ: Apa; COLO: Arc, LPl, Mon, RGr, SJn; NMEX: McK; UTAH. 2250–2700 m (7400–8800′). Flowering: Jun–Sep. Widespread in western North America, from Alaska to Mexico. This is a highly variable, polyploid, agamic complex, the intergrading forms more or less corresponding to ploidy levels: *B. carinatus* = 28, 56; *B. marginatus* = 42; *B. polyanthus* = 56. The following key may allow recognition of some of these populations or plants:

1. Plants annual or biennial, 30–100 cm tall; awns usually more than 7 mm long .... **the *carinatus* phase**
1′ Plants perennial, 60–120 cm tall; awns usually less than 7 mm long .... (2)

2. Sheaths and/or lemmas pubescent .... **the *marginatus* phase**
2′ Sheaths and lemmas glabrous or merely scabrous .... **the *polyanthus* phase**

Bromus catharticus Vahl (cleansing, cathartic) Rescuegrass. Plants tufted annuals to short-lived perennials. **CULMS** erect to ascending, 20–100 cm long. **LIGULES** 2–5 mm long. **BLADES** flat, 2–10 mm wide. **PANICLES** 10–26 cm long, loosely contracted to open and pyramidal, the branches erect-ascending to divaricately drooping. **SPIKELETS** strongly compressed, often greenish yellow. **GLUMES** lanceolate, unequal, the first 5- to 7-nerved and 6–11 mm long, the second 7- to 9-nerved and 7–13 mm long. **LEMMAS** 9–15 mm long, scaberulous to typically glabrous; awns 0–2.5 mm long. **ANTHERS** about 3 mm long. $2n$ = 28, 42, 56. [*B. unioloides* Kunth; *B. willdenowii* Kunth; *Ceratochloa cathartica* (Vahl) Herter]. Disturbed ground, roadsides, lawns, ditch banks, irrigated pastures. ARIZ: Nav; NMEX: SJn. 1500–1800 m (5000–6000′). Flowering: Mar–Jun. Native to South America, introduced for forage and escaping. Matua grass is a popular cultivar of this species planted for improved pastures.

Bromus ciliatus L. (fringed) Fringed brome. Plants perennial, tufted. **CULMS** erect, unbranched above the base, 50–140 cm tall. **LIGULES** 0.5–3.5 mm long. **BLADES** flat, 3–9 mm wide, glabrous or occasionally sparsely hirtellous on the upper surface, the midrib narrowed below the collar, the collar with a tuft of hairs at the corners. **PANICLES** 6–18 cm long, open or loosely contracted, the branches spreading to divaricate, occasionally ascending to erect. **SPIKELETS** not strongly compressed. **GLUMES** unequal, the first 1 (2)-nerved and 5–8 mm long, the second 3-nerved and 10–13 mm long. **LEMMAS** 7–15 mm long, pubescent on the margins and glabrous across the back, only rarely pubescent across the back; awns 3–5 mm long. **ANTHERS** 1.2–2.8 mm long. $2n$ = 14, 28. [*Bromopsis ciliata* (L.) Holub; *Bromus canadensis* Michx.; *B. richardsonii* Link]. Sparingly in oak or piñon-juniper woodlands; more common at higher elevations in piñon pine and ponderosa pine forests, mountain meadows, and grassy slopes in the spruce-fir zone. ARIZ: Apa, Nav; COLO: Arc, Hin, LPl, Min, Mon, RGr, SJn; NMEX: McK, RAr, SJn. UTAH. 2100–3350 m (6900–11000′). Flowering: Jul–Oct. Northern North America south through the western states through Mexico. Populations with longer anthers and glumes have been segregated as *B. richardsonii* Link, but the distinctions fail to hold in our material.

Bromus diandrus Roth (with two stamens) Ripgut brome. Plants stiffly pilose to hirsute, annual. **CULMS** erect to decumbent-based, 20–80 cm long. **LIGULES** obtuse, 2–3 mm long. **BLADES** flat, 1–8 mm wide. **PANICLES** 12–25 cm long, the branches stiffly erect to spreading. **SPIKELETS** somewhat laterally compressed, but not especially strongly so. **GLUMES** narrowly lanceolate, unequal, the first 13–20 mm long and 1-nerved, the second 20–26 mm long and 3-nerved. **LEMMAS** 20–25 mm long, 7-nerved, tapering to a bifid apex with teeth 3–5 mm long; awns 30–65 mm long. **ANTHERS** 0.5–1 mm long. $2n$ = 42, 56. [*Anisantha diandra* (Roth) Tutin; *B. rigidus* Roth]. Dry, disturbed sites, waste areas, roadsides. ARIZ: Apa; NMEX: SJn; UTAH. 1450–1700 m (4800–5600′). Flowering: Apr–Jun. Native to Eurasia, adventive to established in North America from British Columbia through much of the western United States to Mexico, rare eastward; South America. The stiff awns may cause severe injury to the nose, eyes, or underbelly of a grazing animal.

Bromus frondosus (Shear) Wooton & Standl. (leafy) Weeping brome. Plants tufted perennials. **CULMS** erect, unbranched above the base, 26–90 cm tall. **LIGULES** 0.5–1 mm long. **BLADES** flat or loosely folded, 2.5–5 mm wide, the midrib usually narrowed above the collar on the back. **PANICLES** well exserted from the sheath, open to loosely contracted, 8–28 cm long, the branches ascending to spreading. **SPIKELETS** terete to moderately laterally compressed, the pedicels glabrous. **GLUMES** glabrous, somewhat unequal, the first 5–8 mm long and the second 6–9 mm long. **LEMMAS** 7–10 mm long, pubescent across the back or on the margins only; awns 2–5 mm long. **ANTHERS** 1.5–3.5 mm long. $2n = 14$. [*Bromopsis frondosa* (Shear) Holub]. Mountain brush, oak-juniper woodlands, up into the ponderosa zone, mostly below 2500 m (8000′). ARIZ: Apa; COLO: Min, Mon; NMEX: McK, RAr, San, SJn. 1900–2600 m (6200–8500′). Flowering: Jun–Sep. Colorado, Arizona, and New Mexico, south into Mexico. Weeping brome intergrades with *B. anomalus* and *B. ciliatus* and is often confused with those species.

Bromus hordeaceus L. (resembling the genus *Hordeum*) Soft brome. Plants annual or biennial, softly pilose. **CULMS** erect to ascending, 5–60 cm tall. **LIGULES** obtuse, 1–1.5 mm long. **BLADES** flat, 1–4 mm wide. **PANICLES** ovoid, rather dense, 3–13 cm long, the branches stiffly erect. **SPIKELETS** terete to slightly compressed. **GLUMES** glabrous to pilose, unequal, the first 5–7 mm long and 3- to 5-nerved, the second 6.5–8 mm long and 5- to 7-nerved. **LEMMAS** 6.5–11 mm long, 7- to 9-nerved; awns 6–8 mm long, arising from just below the apex. **ANTHERS** 0.6–1.5 mm long. $2n = 28$. [*B. molliformis* Lloyd; *B. mollis* L.]. Disturbed ground and waste places at lower elevations. COLO: Dol, LPl; NMEX: SJn. 1500–1700 m (5000–5600′). Flowering: May–Jul. Native to Eurasia, now found throughout most of North America.

Bromus inermis Leyss. (unarmed, awnless) Smooth brome. Plants perennial, rhizomatous, mostly glabrous. **CULMS** erect to spreading, 50–130 cm long. **LIGULES** truncate, erose, 1–3 mm long. **BLADES** flat, 5–15 mm wide, rarely pubescent. **PANICLES** loosely open, 10–20 cm long, the branches ascending. **SPIKELETS** terete, sometimes purplish. **GLUMES** glabrous, awnless, unequal, the first 5–9 mm long and 1 (3)-nerved, the second 6–10 mm long and 3-nerved. **LEMMAS** elliptic to lanceolate, glabrous to scaberulous, 9–13 mm long, sometimes with a purplish brownish band at the apex; awns 0–3 mm long. **ANTHERS** 3.5–6 mm long. $2n = 28, 56$. [*Bromopsis inermis* (Leyss.) Holub]. Widespread in improved pastures, moist mountain slopes, roadsides, swales, and ditches, from lower valleys to upper mountains. ARIZ: Apa, Nav; COLO: Arc, Con, Dol, Hin, LPl, Min, Mon, SJn, SMg; NMEX: McK, RAr, San, SJn; UTAH. 1280–2860 m (4200–9400′). Flowering: Jun–Sep. Native to Eurasia, adventive nearly throughout the cool-temperate regions. Smooth brome is the exotic counterpart to our native rhizomatous species, *B. pumpellianus* Scribn., which should be looked for in natural communities in the mountains, and which can be distinguished by having usually pilose nodes, leaves, and lemmas, and awns 1–6 mm long.

Bromus japonicus Thunb. ex Murray (of Japan) Japanese brome. Plants annual, tufted, usually densely pilose. **CULMS** erect to ascending, 25–70 cm long. **LIGULES** obtuse, lacerate, 1–2.2 mm long. **BLADES** flat, 2–4 mm wide. **PANICLES** open, 10–22 cm long, the branches ascending to spreading, often drooping. **SPIKELETS** terete to moderately compressed. **GLUMES** glabrous to scabrous, subequal, 5–8 mm long, 5- to 7-nerved. **LEMMAS** 7–9 mm long, mostly 9-nerved; awns 8–13 mm long, straight to divergent, arising just below the apex of the lemma. **ANTHERS** 1–2 mm long. $2n = 14, 28, 56$. [*B. commutatus* Schrad.]. Dry, weedy sites in the valleys and foothills, very common. ARIZ: Apa, Nav; COLO: Arc, Dol, Hin, LPl, Min, Mon, RGr, SMg; NMEX: McK, RAr, SJn; UTAH. 1350–2700 m (4500–8900′). Flowering: Jun–Sep. Native to Eurasia, now adventive nearly throughout North America. European agrostologists, desperate for diversity, recognize several closely related species in this complex, including *B. arvensis* L. and *B. racemosus* L., which have been reported from our flora. All of the material that I have seen from our flora corresponds to *B. japonicus*, but those desiring a taxonomic challenge may look for the following:

1. Awns arising 1.5 mm or more below the lemma apices, straight to strongly divergent; palea clearly shorter than the lemma .... ***B. japonicus***
1′ Awns arising less than 1.5 mm below the lemma apices, straight or weakly divaricate; palea subequal to the lemma .... (2)

2. Anthers 2.5–5 mm long; panicles up to 30 cm long; lower leaf sheaths softly appressed-hairy .... ***B. arvensis***
2′ Anthers 1–3 mm long; panicles less than 16 cm long; lower leaf sheaths with stiff hairs .... (3)

3. Anthers 0.7–1.7 mm long; florets attached 1.5–2 mm apart on the rachilla; lemmas 8–12 mm long, the margins bluntly angled .... ***B. commutatus***
3′ Anthers 1.5–3 mm long; florets attached 1–1.5 mm apart on the rachilla; lemmas 6.5–8 mm long, the margins rounded .... ***B. racemosus***

Bromus lanatipes (Shear) Rydb. (woolly-footed, referring to the sheaths) Shaggy brome. Plants perennial, tufted. **CULMS** erect, unbranched above the base. **SHEATHS** lanate or sometimes only densely pilose, the hairs spreading away from the sheath but matted at the tips. **LIGULES** 1–2 mm long. **BLADES** flat, glabrous, 2–7 mm wide, the midrib narrowed below the collar. **PANICLES** loosely contracted to somewhat open, 5–15 cm long, the branches erect to ascending. **SPIKELETS** terete to somewhat compressed. **GLUMES** glabrous to sparsely short-pilose, unequal, the first 5–7 mm long and 1-nerved, the second 6–9 mm long and 3-nerved. **LEMMAS** pubescent across the back or sometimes on the margins only, 6–9 mm long; awns 2–4 mm long. **ANTHERS** 1.2–3.7 mm long. $2n = 28$. Dry habitats in piñon-juniper-oak woodlands, ponderosa pine, Douglas-fir. COLO: Arc, LPl, Mon; NMEX: McK. 1800–2150 m (6000–7000′). Flowering: Jul–Sep. Southwestern states south into northern Mexico. Shaggy brome has been confused with various other species that have pilose (rather than lanate) sheaths. The sheath pubescence of *B. lanatipes* is quite distinctive, the hairs extending at nearly right angles away from the sheath and becoming matted and tangled at the ends.

Bromus porteri (J. M. Coult.) Nash (for Thomas Conrad Porter, 1822–1901, Pennsylvanian botanist, poet, classicist) Porter's brome. Plants perennial, tufted, glabrous to sometimes puberulent. **CULMS** erect, unbranched above the base, 35–120 cm long. **LIGULES** 1–2 mm long. **BLADES** flat, mostly stiffly erect, 3–12 mm wide, the midrib not narrowed below the collar. **PANICLES** open or sometimes loosely congested, 7–15 cm long, the branches ascending or loosely spreading. **SPIKELETS** terete to slightly compressed, the pedicels puberulent, rarely only slightly so. **GLUMES** puberulent, unequal, the first 5–7 mm long and 1-nerved, the second 6–10 mm long and 3-nerved. **LEMMAS** pubescent across the back, 8–11 mm long; awns 1.5–4 mm long. **ANTHERS** 2–3.3 mm long. $2n = 14$. [*Bromopsis porteri* (J. M. Coult.) Holub]. Ponderosa parklands, aspen groves, mixed conifer forests, high mountain meadows, and openings in spruce-fir forests, generally above 2100 m (7000′). ARIZ: Apa; COLO: Arc, Dol, LPl, Mon, SJn; NMEX: McK, RAr, SJn. 1850–2900 m (6200–9500′). Flowering: Jun–Sep. Mostly Rocky Mountains and associated states, west into California, south into northern Mexico. Stiffly erect blades and puberulent pedicels and glumes are distinctive.

Bromus rubens L. (reddish) Foxtail brome, red brome. Plants annual, tufted, with dense retrorse pubescence. **CULMS** erect, 10–40 cm long. **LIGULES** obtuse, lacerate, 1–3 mm long. **BLADES** flat, 1–5 mm wide. **PANICLES** dense, compact, ovoid, 2–10 cm long, 2–5 cm wide, the branches stiffly erect and 1 cm or less long. **SPIKELETS** laterally compressed. **GLUMES** pilose with hyaline margins, unequal, the first 5–8 mm long, 1 (3)-nerved, the second 8–12 mm long and 3- to 5-nerved. **LEMMAS** linear-lanceolate, pubescent, 10–15 mm long, 7-nerved; awns 8–20 mm long, straight, reddish. **ANTHERS** 0.5–1 mm long. $2n = 14, 28$. [*Anisantha rubens* (L.) Nevski; *Bromus madritensis* L. subsp. *rubens* (L.) Husn.]. Dry, disturbed ground in the desert. ARIZ: Apa, Coc; COLO: Mon; NMEX: McK, SJn; UTAH. 1125–1900 m (3700–6300′). Flowering: Mar–Jun. Native to Eurasia, adventive mostly west of the Rocky Mountains.

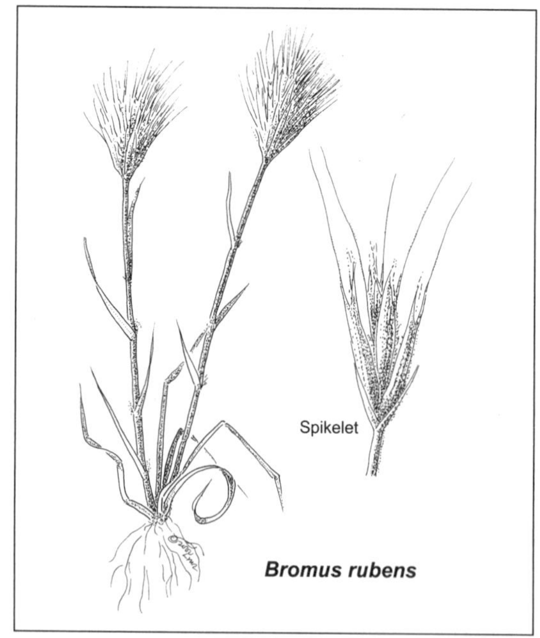

Bromus rubens

Bromus secalinus L. (resembling the genus *Secale*) Rye chess. Plants annual, tufted, glabrous to loosely pilose. **CULMS** 20–80 cm long. **LIGULES** obtuse, 2–3 mm long. **BLADES** flat, 2–4 mm wide. **PANICLES** open, nodding, 5–20 cm long, the branches ascending to spreading and drooping. **SPIKELETS** ovoid, somewhat turgid, moderately laterally compressed. **GLUMES** glabrous, unequal, the first 4–6 mm long and 3- to 5-nerved, the second 6–7 mm long and 7-nerved. **LEMMAS** 6–9 (10) mm long, rounded across the back, glabrous or sometimes pubescent on the margins, the margins inrolled around the grain; awns 0–6 mm long. **ANTHERS** 1–2 mm long. $2n = 28$. Dry, disturbed ground, waste places, not common. COLO: Mon. 1550–1900 m (5200–6200′). Flowering: Jun–Jul. Native to Eurasia, now adventive throughout much of the United States, including Utah, though many are old collections and the species is becoming less common. Specimens of *B. japonicus* have been confused with rye chess, but that species is consistently pilose, has longer awns, and lemma margins that are not rolled around the grain.

Bromus tectorum L. (of roofs) Cheatgrass. Plants annual, tufted, mostly densely pilose. **CULMS** erect to geniculate, 5–90 cm long. **LIGULES** obtuse, lacerate, 2–3 mm long. **BLADES** flat, 1–6 mm wide. **PANICLES** loose, open, often

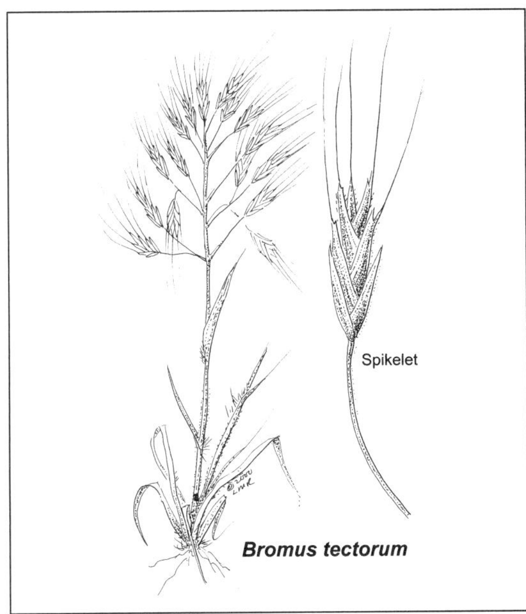

Bromus tectorum

drooping, the branches and pedicels capillary and drooping. **SPIKELETS** laterally compressed, expanding toward the tip. **GLUMES** glabrous to pubescent, unequal, the first 4–9 mm long and 1-nerved, the second 7–13 mm long, 3- to 5-nerved. **LEMMAS** lanceolate, glabrous to pubescent, 9–12 mm long, 5- to 7-nerved, the margins hyaline, the apex bifid with slender teeth 1–3 mm long; awns 10–18 mm long, straight. **ANTHERS** 0.5–1 mm long. $2n = 14$. [*Anisantha tectorum* (L.) Nevski]. Widely distributed in dry, disturbed sites, plains, valleys, foothills, open ground, and roadsides in the mountains. ARIZ: Apa, Coc, Nav; COLO: Arc, Dol, Hin, LPl, Min, Mon, SJn, SMg; NMEX: McK, RAr, San, SJn; UTAH. 1280–2450 m (4200–8100′). Flowering: Apr–Aug. Native to Eurasia, now widely adventive throughout North America, replacing many valuable rangeland grasses.

Buchloë **Engelm. Buffalograss**

(Greek *bukalos*, buffalo, and *chloë*, grass, a Greek rendering of the common name) A genus of one species (Columbus 1999).

Buchloë dactyloides (Nutt.) Engelm. (resembling the genus *Dactylis*) Buffalograss. Plants perennial, strongly stoloniferous, dioecious, often mat-forming. **CULMS** 1–30 cm tall, mostly unbranched above the base, the internodes solid or pithy, the nodes glabrous. **LEAVES** with open sheaths and ciliate-membranous ligules. **BLADES** flat, often curling when dry, sparsely pilose, 1–2.5 mm wide. **LIGULES** a ciliate membrane, 0.5–1 mm long. **STAMINATE INFL** a terminal panicle raised above the leaves, of 1–4 spicate, unilateral, pectinate, primary branches 5–15 mm long; glumes unequal, glabrous, 1- to 2-nerved; lemmas glabrous, 3-nerved, awnless; anthers mostly salmon-colored, also brownish to reddish; disarticulation above the glumes. **PISTILLATE INFL** partially hidden with bracteate leaf sheaths below the foliage leaves, burlike, with 3–5 spikelets; second glume hardened, whitish, 3-toothed; lemmas glabrous, membranous, 3-nerved, sometimes 3-awned. $x = 10$. $2n = 20, 40, 56, 60$. [*Bouteloua dactyloides* (Nutt.) Columbus]. Not common in the Four Corners region; dry plains and grasslands just above the desert floor. NMEX: McK, SJn. 1650–1950 m (5500–6400′). Flowering: Jul–Sep. Most of the central states, Montana and North Dakota south to Arizona and Texas. Recent molecular and morphologic studies indicate a close relationship to *Bouteloua*, and some have proposed that buffalograss be merged with that genus, a proposal not without considerable merit.

Calamagrostis **Adans. Reedgrass**

(Greek *calamos*, a reed, + *agrostis*, a grass) Plants usually tufted and with at least short rhizomes, the culms hollow. **LEAVES** with open sheaths, membranous ligules, flat to involute blades, and lacking auricles. **INFL** a rebranching panicle, open or contracted. **SPIKELETS** with a single floret, the rachilla prolonged as a tiny bristle; disarticulation above the glumes; glumes subequal, longer than the floret; lemmas 5-nerved, the callus with a dense tuft of pilose hairs, awned from the back below the middle; palea slightly shorter than the lemma; stamens 3. $x = 7$ (Greene 1984). A genus of more than 100 species of cool-temperate regions in both hemispheres. Hybridization, vivipary, polyploidy, and apomixis contribute to the taxonomic difficulty in the genus.

1. Awns exserted well beyond the glumes, easily visible, 4.5–8 mm long .... ***C. purpurascens***
1′ Awns scarcely if at all exserted beyond the glumes, less than 4.5 mm long .... (2)

2. Pedicels glabrous or nearly so .... ***C. scopulorum***
2′ Pedicels evidently and distinctly scabrous .... (3)

3. Glumes oblong, the apex abruptly acute and not drawn out to an awn tip; blades 1–4 mm wide, usually rolled and stiffly ascending; lemmas not translucent on the upper 1/3; callus hairs 1/2 to 2/3 as long as the lemma ...... ***C. stricta***
3′ Glumes lance-ovate, the apex of especially the first drawn out to an awn tip; blades 3–10 mm wide, mostly flat and lax; lemmas translucent on the upper 1/3; callus hairs 2/3 to as long as the lemma .... ***C. canadensis***

Calamagrostis canadensis (Michx.) P. Beauv. (of Canada) Canada reedgrass, bluejoint reedgrass. Plants perennial, in clumps and with rhizomes to 15 cm long. **CULMS** 50–150 cm tall or more, often branching above the base. **LIGULES** 3–10 mm long, lacerate. **BLADES** flat, lax, 3–10 mm wide, the upper surface strongly scabrous. **PANICLES** contracted when young and open and nodding when mature, 8–20 cm long, the branches ascending to spreading-drooping, scabrous; pedicels scabrous. **GLUMES** 2.5–5 mm long. **LEMMAS** 2–4.5 mm long, slightly shorter than the glumes, translucent on the upper 1/3; callus hairs 2–4.5 mm long; awn 1–3 mm long, often hidden among the hairs, straight. **ANTHERS** 1–2 mm long. $2n$ = 42–66. [*Calamagrostis scribneri* Beal]. Wet meadows, seeps, marshy ground, and other wet sites in the mountains. COLO: Arc, Con, Hin, LPl, Min, Mon, RGr, SJn; NMEX: RAr. 2500–3700 m (8200–12000′). Flowering: Jul–Sep. Circumboreal, throughout North America except in the southern states; Eurasia. Plants with a pubescent line across the collar and shorter callus hairs have been called *C. scribneri*, but this is a tenuous distinction and not recognized herein.

Calamagrostis purpurascens R. Br. (purplish) Purple reedgrass. Plants perennial, strongly tufted, with short rhizomes 1–4 cm long. **CULMS** 30–80 cm long, usually unbranched above the base. **LIGULES** 2–5 mm long. **BLADES** flat to involute, 2–5 mm wide. **PANICLES** narrow, contracted, 5–15 cm long, 1–2.5 cm wide, often purplish, the branches scabrous. **GLUMES** 5–7 mm long. **LEMMAS** 3.5–4.5 mm long, slightly shorter than the glumes; callus hairs 1–2 mm long; awn 4–8 mm long, exserted and easily visible. **ANTHERS** 1.5–2.5 mm long. $2n$ = 42–58, 84. Subalpine to alpine slopes, meadows, forest openings, often in rocky areas. COLO: Hin, Min, SJn. 3000–4100 m (9800–13500′). Flowering: Jul–Aug. Alaska south through the Rocky Mountains to Nevada and New Mexico.

Calamagrostis scopulorum M. E. Jones (of the Rocky Mountains) Jones' reedgrass. Plants perennial, tufted, with short rhizomes to 2 cm long. **CULMS** 50–90 cm long. **LIGULES** obtuse to lacerate, 3–7 mm long. **BLADES** flat, 3–5 mm wide. **PANICLES** narrow, contracted, 6–16 cm long, 1–2.5 cm wide, pale green to purplish. **GLUMES** 4–6 mm long. **LEMMAS** 3.5–5 mm long, shorter than the glumes; callus hairs 2–3 mm long; awn 1–2 mm long, not exserted, easily overlooked when short. **ANTHERS** 2–3 mm long. $2n$ = 28. Hanging gardens and similar rocky sites at lower elevations, and rocky slopes and cliffs in subalpine to alpine vegetation in the mountains. COLO: Arc, Min; NMEX: SJn; UTAH. 1050–3350 m (3500–11000′). Flowering: Jul–Aug. Montana south to Arizona and northern New Mexico.

Calamagrostis stricta (Timm) Koeler (constricted) Slender reedgrass. Plants perennial, tufted, with slender rhizomes 1–5 cm long. **CULMS** 35–100 cm long, usually unbranched above the base. **LIGULES** truncate to obtuse, 1–5 mm long. **BLADES** usually flat, 1–4 mm wide. **PANICLES** contracted, pale green to purplish, 8–18 cm long, 1–3 cm wide. **GLUMES** 3–4 (5) mm long. **LEMMAS** 2.5–3.5 (4) mm long, slightly shorter than the glumes; callus hairs 2.5–4 mm long; awn 2–2.5 mm long, straight, usually stout, easily distinguished from the callus hairs. **ANTHERS** 1–2 mm long, often sterile. $2n$ = 28, 56, 58, 70, 84–ca. 120. Moist meadows, stream banks, ponderosa to spruce-fir forests. ARIZ: Apa; COLO: Arc, Min, RGr, SJn. 2350–2750 m (7800–9000′). Flowering: Jul–Aug. Circumboreal, south throughout the north-eastern and western states. Our western material, with more robust growth and larger spikelet parts, is referred to **subsp. *inexpansa*** (A. Gray) C. W. Greene.

Calamovilfa Hack. Sandreed

(Greek *calamos*, reed, + *Vilfa*, a similar-looking genus) Plants perennial with rhizomes, the culms pithy. **LEAVES** with open sheaths, hairy ligules, and lacking auricles. **INFL** a rebranching panicle, open or contracted. **SPIKELETS** with a single floret, the rachilla not prolonged beyond the floret; disarticulation above the glumes; glumes unequal, the first shorter, 1-nerved, awnless; lemmas longer than the glumes, stiff-papery, 1-nerved, awnless, the callus bearded with long white hairs; palea equaling the lemma; stamens 3. $x = 10$ (Thieret 1966). A genus of five species, endemic to North America.

Calamovilfa gigantea (Nutt.) Scribn. & Merr. (gigantic) Big sandreed. Plants tough, robust perennials from strong rhizomes. **CULMS** erect, thick, 1–1.3 m tall. **LIGULES** about 1 mm long. **BLADES** flat to mostly involute, 5–10 mm wide at the base, glabrous. **PANICLES** pyramid-shaped, open, the branches stiffly spreading, 25–50 cm long. **GLUMES** glabrous, 1-nerved, unequal, the first 5–6.5 mm long, the second 6–8 mm long. **LEMMAS** lanceolate, 1-nerved, 6.5–8 mm long, awnless, villous on the midnerve; callus copiously villous, the hairs 3–4 mm long. **PALEAS** equaling the lemmas, villous on the 2 nerves. **ANTHERS** 3–5.5 mm long. $2n$ = 60. Sand dunes and deep sandy soil in the desert and foothill regions. ARIZ: Apa, Nav; UTAH. 1500–1700 m (4900–5600′). Flowering: Jul–Sep. Arizona and Colorado, east to the southern Great Plains.

Catabrosa P. Beauv. Brookgrass

(Greek *catabrosis*, devoured, alluding to the chewed appearance of the glume and lemma apices) Aquatic perennials from rhizomes and stolons, the culms decumbent and rooting freely at the nodes, hollow. **LEAVES** with closed sheaths, flat blades, membranous ligules, and lacking auricles. **INFL** a rebranching open panicle, the branches whorled.

SPIKELETS 1- to 5-flowered (ours mostly 2-flowered); disarticulation above the glumes and between the florets; glumes unequal, membranous, flat, erose at the tip, nerveless or faintly 1-nerved, shorter than the lowermost floret; lemmas 3-nerved, the nerves parallel and not converging at the apex, truncate and erose at the tip; palea slightly shorter than the lemma; stamens 3. $x = 10$. A genus of seven species, mostly Eurasian.

Catabrosa aquatica (L.) P. Beauv. (aquatic) Brookgrass. **CULMS** decumbent at the base and erect above, rooting at the nodes and forming stolons/rhizomes, 20–50 cm long. **LIGULES** 2–6 mm long. **BLADES** 3–8 mm wide, succulent. **PANICLES** 8–20 cm long, open, the branches stiffly spreading, whorled, and spikelet-bearing to near the base. **SPIKELETS** 1- to 2-flowered, but some 3-flowered spikelets occasionally encountered. **GLUMES** flat, ovate-lanceolate, 1-nerved, the first 0.6–1.5 mm long, the second 1.2–2.5 mm long. **LEMMAS** 2–3 mm long, scarious. **ANTHERS** 1.3–1.8 mm long. $2n = 20, 30$. Seeps, springs, slow-moving water of streams and ponds, ponderosa, aspen, and spruce zones; not common. COLO: LPl. 2650–2900 m (8800–9600′). Flowering: Jun–Sep. Throughout much of cool-temperate North America, also Argentina, Europe, and Asia.

Cenchrus L. Sandbur

(Greek *kenchros*, a kind of millet) Pesky annual and perennial plants, the culms geniculate to erect, pithy. **LEAVES** with open sheaths, flat or folded blades, hairy ligules, and lacking auricles. **INFL** a highly modified panicle, the spikelets borne in burs or involucres formed by fused bristles or spines (sterile branchlets), these often retrorsely barbed; disarticulation below the bur, which falls entire. **SPIKELETS** 2-flowered, dorsally compressed, awnless, the lower reduced, the upper perfect; glumes unequal, the first much smaller than the second; lower floret staminate or sterile, the lemma membranous and similar in size and texture to the second glume; upper floret perfect, the lemma and palea forming a hard seed case around the flower and grain; stamens 3. $x = 9$ (DeLisle 1963). A genus of about 16 species, mostly in the New World, but some in Africa, Asia, and Australia, and some worldwide.

1. Burs mostly with 45–75 spines, the inner bristles 0.5–1 mm wide; upper floret of the spikelets 5–7.6 mm long; both margins of the blade of uppermost leaf conspicuously crinkled at the base .... ***C. longispinus***

1′ Burs mostly with 8–40 spines, the inner bristles 1–2 mm wide; upper floret of the spikelets 3.5–5 mm long; only one margin of the blade of uppermost leaf crinkled at the base .... ***C. spinifex***

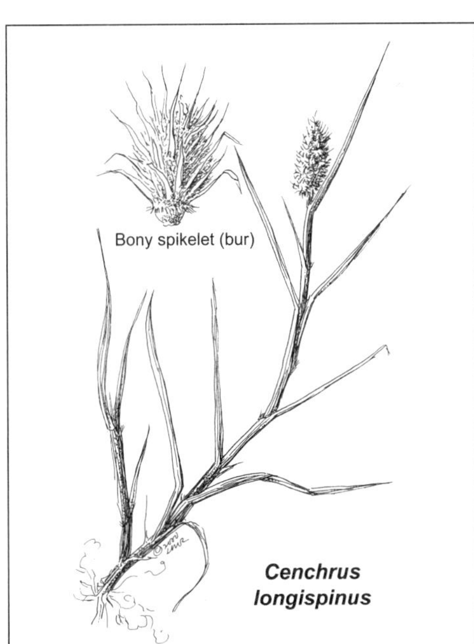

Cenchrus longispinus (Hack.) Fernald (long-spined) Innocent-weed, long-spine sandbur. Plants annual, tufted. **CULMS** ascending to decumbent, 20–80 cm tall. **LIGULES** 0.6–1.8 mm long. **BLADES** 1.5–6 mm wide. **PANICLES** 2–10 cm long. **SPIKELETS** 2–3 (4) per bur, 5–7.8 mm long, the first glume 0.8–3 mm long, the second glume 4–6 mm long. **LOWER FLORETS** 4–6.5 mm long, the anthers 1.5–2 mm long. **UPPER FLORETS** 4–7.5 mm long, the anthers 0.7–1 mm long. $2n = 34$. [*Cenchrus pauciflorus* Benth. in part]. Sandy ground and waste areas, roadsides, gardens, disturbed sites. ARIZ: Apa, Nav; COLO: Mon; NMEX: McK, SJn; UTAH. 1350–2000 m (4500–6500′). Flowering: Jul–Sep. Throughout most of the United States, southward to Venezuela.

Cenchrus spinifex Cav. (alluding to the genus *Spinifex*, which has spiny leaf blades) Common sandbur. Plants annual (rarely short-lived perennial), tufted. **CULMS** geniculate-based, 25–80 cm tall. **LIGULES** 0.5–1.5 mm long. **BLADES** 2–7 mm wide. **PANICLES** 3–9 cm long. **SPIKELETS** 2–4 per bur, 3.5–5.9 mm long, the first glume 1–3.3 mm long, the second glume 3–5 mm long. **LOWER FLORETS** 3–5.8 mm long, the anthers 1.3–1.6 mm long. **UPPER FLORETS** 3.5–5 (5.8) mm long, the anthers 0.5–1.2 mm long. $2n = 34$. [*C. incertus* M. A. Curtis; *C. pauciflorus* Benth. in part]. Sandy ground and waste areas, roadsides, gardens, disturbed sites. NMEX: SJn. 1350–2000 m (4500–6500′). Flowering: Jul–Sep. Throughout the southern United States and south into South America.

Chloris Sw. Windmillgrass

(Greek *Chloris*, the goddess of flowers) Plants annual or perennial, of various habits, but ours tufted, with pithy culms. **LEAVES** with open and keeled sheaths, flat or folded blades, and ciliolate membranous ligules. **INFL** a panicle of few to many spikelike primary branches, these digitate or whorled; disarticulation above the glumes and the florets falling together. **SPIKELETS** sessile, in 2 rows on one side of the primary branches, 2- to 3-flowered, usually only the lowermost floret perfect, the others staminate or sterile; glumes unequal, shorter than the florets, usually awnless; lemmas of fertile florets 3-nerved, the marginal nerves pubescent, awned (ours) or awnless; paleas shorter than the lemmas, 2-nerved; stamens 3. $x = 10$ (Anderson 1974). A genus of about 55 species in the warm tropical to subtropical or arid regions of the world.

1. Panicle branches typically in several whorls along an axis 2 cm or more long; apex of lower lemma glabrous or with minute hairs less than 0.2 mm long; plants perennial .... ***C. verticillata***
1′ Panicle branches in a single terminal whorl, or if in several whorls then the axis less than 2 cm long; apex of lower lemma with conspicuous pilose hairs 1–3 mm long; plants annual .... ***C. virgata***

Chloris verticillata Nutt. (whorled) Tumble windmillgrass. Plants tufted perennials. **CULMS** mostly erect, sometimes geniculate or even decumbent and rooting at the nodes, 15–40 cm tall. **LIGULES** 0.7–1.3 mm long, ciliolate. **BLADES** 2–3 mm wide. **PANICLES** with 10–16 primary branches in at least 2 whorls, the branches 5–15 cm long, spreading to divaricate at maturity, spikelet-bearing to the base. **SPIKELETS** with 1 fertile floret and 1 sterile floret. **GLUMES** slightly unequal, the first 2–3 mm long, the second 2.8–3.5 mm long. **FERTILE FLORET** 2–3.5 mm long, glabrous or minutely puberulent just below the apex, with an awn 4.5–10 mm long. **STERILE FLORET** 1.1–2.3 mm long, somewhat inflated, with an awn 3.2–7 mm long. $2n = 40$. Roadsides, disturbed ground, waste areas, valleys, and foothills. ARIZ: Apa, Nav; NMEX: McK, SJn. 1550–1850 m (5200–6000′). Flowering: May–Sep. Central and southwestern United States, south into Mexico.

Chloris virgata Sw. (broomlike) Showy windmillgrass. Plants tufted annuals. **CULMS** 10–100 cm or more tall, erect. **LIGULES** 1–4 mm long, erose to ciliolate. **BLADES** 3–15 mm wide. **PANICLES** with 4–20 digitate primary branches in a single whorl at the apex of the stem, the branches 5–10 cm long, erect to ascending, spikelet-bearing to the base. **SPIKELETS** with 1 fertile floret and 1–2 sterile florets. **GLUMES** unequal, the first 1.5–2.5 mm long, the second 2.5–4.3 mm long. **FERTILE FLORET** 2.5–4.2 mm long, with a prominent tuft of pilose hairs just below the apex, with an awn 2.5–25 mm long. **STERILE FLORETS** 1.5–3 mm long, with an awn 3–10 mm long. $2n = 20, 30, 40$. Roadsides, disturbed ground, waste areas, valleys, and foothills. ARIZ: Apa, Nav; COLO: LPl, Mon; NMEX: McK, RAr, San, SJn; UTAH. 1300–2150 m (4300–7000′). Flowering: Aug–Oct. Throughout much of the southern and southwestern United States, South America, Asia, Africa.

Cortaderia Stapf Pampasgrass

(name from the Spanish *cortada*, cutting, referring to the sharply serrate blades) Plants dioecious or monoecious perennials, forming often huge tussocks, with hollow culms. **LEAVES** primarily basal, with open sheaths, flat to folded blades, hairy ligules, and lacking auricles. **INFL** terminal, a large plumose panicle, subtended by a long, ciliate bract. **SPIKELETS** somewhat laterally compressed, usually unisexual, with 2–9 florets; disarticulation above the glumes and below the florets; glumes unequal, nearly as long as the spikelets, hyaline, 1-veined; lemmas 3–5 (7)-veined, long-acuminate, bifid and awned or entire and mucronate, usually long-sericeous, the callus pilose; paleas about 1/2 as long as the lemmas, 2-veined; stamens 3. $x = 9$. A genus of about 25 species, the majority from South America, some also in New Zealand (these might represent a different lineage and would be taken out of *Cortaderia*).

Cortaderia selloana (Schult. & Schult. f.) Asch. & Graebn. (for Friedrich Sellow, 1789–1831, German botanist who collected in South America in 1814) Pampasgrass. Plants usually dioecious. **CULMS** 2–4 m tall. **LIGULES** 1–2 mm long. **BLADES** to 2 m long, 3–8 cm wide, mostly flat, ascending to arching, bluish green, with a dense tuft of hairs at the base at the collar. **PANICLES** 30–130 cm long, only slightly elevated above the foliage, whitish or pinkish when young. **SPIKELETS** 15–17 mm long. **GLUMES** subequal, 1-nerved, nearly as long as the spikelet. **LEMMAS** shorter than the glumes, 3-nerved, translucent, narrowly lanceolate and tapering to an awn 2.5–8 mm long, the hairs on the back to 10 mm long. **PALEAS** no more than 1/2 as long as the lemmas. $2n = 36$. Grown for ornament and sometimes persisting around old dwellings, but not escaping to the wild. NMEX: SJn. 1550–1700 m (5100–5600′). Flowering: Aug–Sep. Introduced from South America and widely grown as an ornamental in the warmer parts of North America.

Crypsis Aiton Pricklegrass

(Greek *kryptos*, hidden, alluding to the partially concealed inflorescences) Plants annual, the culms prostrate to erect, pithy. **LEAVES** with open sheaths, flat to rolled leaves, hairy ligules, and lacking auricles. **INFL** a short, dense, spikelike or headlike panicle, usually partially hidden in the sheath. **SPIKELETS** small, 1-flowered, disarticulating above or below the glumes (ours, and tardily so); glumes shorter than the spikelet, keeled, 1-nerved, awnless; lemma thin, 1-nerved, awnless (ours) or awned, the callus glabrous; palea broad and slightly shorter than the lemma, 2-nerved;

stamens 3. $x = 9$ (Hammel and Reeder 1979). A genus of eight species native to the Mediterranean region and China, three of which are exotic in the United States.

Crypsis schoenoides (L.) Lam. (resembling the genus *Schoenus*) Swampgrass, swamp pricklegrass. Mat-forming annuals, the shoots much-branched, the sheaths prominently inflated. **CULMS** 3–50 cm long, prostrate to erect. **LIGULES** 0.5–1 mm long. **BLADES** 2–10 cm long, 2–6 mm wide, widely spreading to reflexed. **PANICLES** 0.5–7 cm long, 5–10 mm wide, at least the bases usually enclosed in the subtending sheath even when mature. **GLUMES** slightly unequal, the first 1.8–2.3 mm long, the second 2.2–2.7 mm long. **LEMMAS** 2.4–3 mm long, glabrous. **PALEAS** subequal to the lemmas. **ANTHERS** 0.7–1 mm long. $2n = 36$. Sandy ground of lake shores and edges of ponds. ARIZ: Apa; NMEX: McK. 1550–1950 m (5200–6400′). Flowering: Jul–Sep. Exotic in California and several other states across the continent, native to Eurasia and northern Africa. The inflated sheaths, widely spreading blades, and dense, ovoid panicles are distinctive.

Cynodon Rich. Bermudagrass

(Greek *cynos*, dog, + *odos*, tooth, taken from *chien-dent*, the French common name meaning dog's tooth) Plants perennial, often either stoloniferous or rhizomatous or both and forming dense turf, with hollow or pithy culms. **LEAVES** with open sheaths, flat to involute blades, hairy ligules, and lacking auricles, produced in a pattern of 2–3 short internodes followed by a long internode, giving the appearance of several leaves clustered together on the shoot. **INFL** a panicle of digitate primary branches at the tip of the shoot; disarticulation above the glumes. **SPIKELETS** with 1 fertile floret, the rachilla prolonged and usually bearing a minute rudiment, borne in 2 rows on 2 sides of a 3-sided rachis; glumes 1-nerved, subequal, slightly shorter than the floret; lemmas somewhat hardened, minutely pubescent on the 3 nerves, awnless; palea about as long as the lemma, 2-nerved; stamens 3. $x = 9$ (Mitich 1989). A genus of nine species native to the tropical regions of the Eastern Hemisphere, several species introduced for lawn and forage throughout the warm regions of the world.

Cynodon dactylon (L.) Pers. (a finger) Bermudagrass. Perennials, with well-developed rhizomes and stolons. **CULMS** 20–80 cm or more long, vegetative ones decumbent and rooting at the nodes, the reproductive ones often erect and leafy. **LIGULES** about 0.5 mm long. **BLADES** flat, 1.5–3 mm wide, long hairs at the collars. **PANICLES** with (2) 4–6 (9) branches, these 2–6 cm long in a single terminal whorl. **SPIKELETS** 2–3.2 mm long. **GLUMES** subequal, the first 1.5–2 mm long, the second 1.4–2.3 mm long with a mucro. **LEMMAS** ovate, blunt, 1.8–2.2 mm long, the nerves glabrous or pubescent. **ANTHERS** 0.6–1 mm long. $2n = 18, 36$. Lawns, roadsides, pastures, fields, disturbed ground. ARIZ: Apa; NMEX: McK, SJn; UTAH. 1100–2050 m (3600–6600′). Flowering: Apr–Oct. Widely introduced in North America and South America; native to Eurasia.

Dactylis L. Orchardgrass

(Greek *daktylos*, a finger). A genus of 1–5 species, depending on interpretation.

Dactylis glomerata L. (wound up, as in a ball of yarn, alluding to the densely clustered spikelets) Orchardgrass. Tufted perennials with hollow culms. **CULMS** erect, 30–100 cm tall or more. **LEAVES** with open sheaths. **BLADES** flat to folded, 4–8 mm wide, lax. **LIGULES** membranous, prominent, 4–8 mm long, rounded to lacerate. **INFL** a panicle of several primary branches with dense clusters of spikelets at the tips, pyramidal, 4–20 cm long. **SPIKELETS** (2) 3 (5)-flowered, subsessile or with short pedicels, laterally compressed, midnerves and keels of the bracts ciliate; disarticulation above the glumes and between the florets; glumes acute, keeled, 1-nerved, subequal, shorter than the lowermost lemma, 3–6 mm long, awn-tipped; lemmas lanceolate, 5-nerved, 4–8 mm long, awn-tipped; paleas subequal to the lemmas, 2-nerved. **STAMENS** 3, the anthers 2–3.5 mm long. $2n = 14, 21, 27–31, 42$. Common in pastures, meadows, roadsides in moist places, canyons, and similar sites. ARIZ: Apa, Nav; COLO; Arc, Dol, Hin, LPl, Min, Mon, RGr, SJn, SMg; NMEX: McK, RAr, SJn; UTAH. 1600–3050 m (5300–10000′). Flowering: Jun–Sep. Native to Eurasia and Africa, now introduced throughout most of the cool-temperate regions of the world for forage and pasture.

Danthonia DC. Danthonia, Oatgrass

(for Étienne Danthoine, French botanist) Plants perennial, tufted, the culms hollow. **LEAVES** with open sheaths; flat to rolled blades, hairy ligules, and lacking auricles. **INFL** a much reduced panicle or raceme, sometimes with a single spikelet. **SPIKELETS** 2- to 12-flowered, the terminal floret reduced; disarticulation above the glumes and between the florets; glumes subequal, mostly longer than the florets, papery, 1- to 7-nerved; lemmas 7- to 11-nerved, the margins short-pilose, awned between 2 lobes or teeth; awns geniculate and twisted; paleas about as long as the lemmas, 2-nerved; stamens 3. $x = 12$. Cleistogamous spikelets are usually produced in the axils of the lower sheaths. A genus of about 20 species native to Europe, North Africa, and the Americas.

1. Pedicels and panicle branches erect-appressed, the inflorescence less than 1.5 cm wide....................................(2)
1′ Pedicels and panicle branches spreading outward at maturity, the inflorescence mostly 3–10 cm wide ....(3)

2. Pedicels markedly puberulent-scabrous; lemmas 6–10 mm long, with awns 8–15 mm long; spikelets mostly straw-colored; clump of old basal sheaths persistent and prominent, 3–8 cm long..***D. parryi***
2′ Pedicels glabrous or scarcely scaberulous; lemmas 3–6 mm long, with awns 6–8 mm long; spikelets often greenish purplish; clump of old sheaths inconspicuous, 1–3 cm long ....***D. intermedia***

3. Upper blades usually strongly divergent or reflexed downward at the collar region; pedicels mostly much longer than the spikelets; lemmas glabrous or sparsely hairy across the back; mature culms breaking at the nodes ....***D. californica***
3′ Upper blades erect or ascending at the collar region; pedicels shorter than to as long as the spikelets; lemmas pilose over the back, at least basally; mature culms not breaking apart ....***D. parryi***

Danthonia californica Bol. (of California) California danthonia. Plants tufted perennials. **CULMS** 30–100 cm tall or more, breaking easily at the nodes when mature. **LIGULES** about 0.5 mm long. **BLADES** 2–5 mm wide, flat to rolled. **PANICLES** with 3–6 spikelets, 2–5 cm long; pulvini in the axils of the pedicels. **SPIKELETS** 12–26 mm long, 5- to 8-flowered. **LEMMAS** 5–10 mm long, glabrous or sparsely hairy across the back, the margins usually pubescent; teeth 2.5–5 mm long; awns 8–12 mm long. **ANTHERS** 2.5–4 mm long. $2n = 36$. Somewhat dry meadows in forests. COLO: LPl. 2350–2600 m (7800–8600′). Flowering: Jun–Aug. Mountains of western Canada and the United States. *Danthonia californica* has been reported for the *Flora* area, but I have not located specimens to verify this and its occurrence is doubtful. *Danthonia parryi* is easily mistaken for this species by general appearance, but its cauline blades are erect-ascending at the collar.

Danthonia intermedia Vasey (intermediate) Timber danthonia. Plants tufted perennials. **CULMS** 20–50 cm tall, not breaking at the nodes when mature. **LIGULES** 0.2–0.5 mm long. **BLADES** 1–3.5 mm wide, flat to rolled. **PANICLES** with 4–10 spikelets, 3–8 cm long; pulvini absent from the axils, the branches and spikelets erect-appressed. **SPIKELETS** 11–16 mm long, 3- to 6-flowered. **LEMMAS** 3–6 mm long, glabrous across the back, densely pilose on the margins; teeth 1.5–2.5 mm long; awns 6–8 mm long. **ANTHERS** 1–3 mm long. $2n = 36, 98$. Dry to moist meadows in the mountains, aspen glens, rocky slopes, up to timberline. COLO: Arc, Con, Hin, LPl, Min, Mon, SJn. 2700–3500 m (8800–11500′). Flowering: Jul–Aug. Alaska and Canada, south through the mountains of the western United States.

Danthonia parryi

Danthonia parryi Scribn. (for Charles Christopher Parry, 1823–1890, botanical explorer of the West) Parry's danthonia. Plants tufted perennials. **CULMS** 30–80 cm or more tall. **LIGULES** 0.2–0.3 mm long. **BLADES** 2–4 mm wide, flat to rolled. **PANICLES** with 3–11 spikelets, 3–10 cm long. **SPIKELETS** 16–24 mm long, 3- to 6-flowered. **LEMMAS** 5.5–10 mm long, usually pilose across the back and densely so on the margin; teeth 3–8 mm long; awns 12–15 mm long. **ANTHERS** 4–6.5 mm long. $2n = 36$. Dry mountain meadows and grasslands. COLO: Con, Hin, LPl, Min, Mon. 2500–2850 m (8300–9400′). Flowering: Jul–Aug. Endemic to the Rocky Mountains of southern Canada and the western United States.

Dasyochloa Willd. ex Rydb. Fluffgrass

(Greek *dasys*, shaggy, + *chloa*, grass, referring to the densely ciliate spikelets) A genus of a single species endemic to North America (Valdés and Hatch 1997).

Dasyochloa pulchella (Kunth) Willd. ex Rydb. (beautiful) Fluffgrass. Low, tufted, stoloniferous perennials. **CULMS** 5–15 cm tall, pithy, consisting of an elongate internode topped by a cluster of leaves, which droops over and roots. **LEAVES** with open sheaths, hairy ligules, involute blades, and lacking auricles. **LIGULES** 3–5 mm long. **BLADES** 0.5–1 mm wide (rolled), arcuate. **INFL** terminal, a dense, ovoid panicle of short, spikelike branches, each subtended by leafy bracts and nestled among the leaves, 1–2.5 cm long and 1–1.5 cm wide, densely pilose, light green or purplish. **SPIKELETS** 4- to 10-flowered, 6–10 mm long; disarticulation above the glumes; glumes shorter than the spikelet, subequal to the adjacent lemma, glabrous, 1-nerved, mucronate to awn-tipped, 6–9 mm long; lemmas 3–5.5 mm long, 3-nerved, densely pilose on the nerves, 2-lobed, the lobes about 1/2 the length of the lemma body, the midnerve extended into an awn 2–4 mm long; palea about as long as the lemma. **STAMENS** 3, the anthers 0.2–0.5 mm long. $x = 8$, $2n = 16$. Rocky soils and open ground in the desert. ARIZ: Nav; COLO: Mon; NMEX: RAr, SJn; UTAH. 1500–1650 m (4900–5400′). Flowering: May–Oct. Southwestern United States to central Mexico.

Deschampsia P. Beauv. Hairgrass

(for Jean Louise Auguste Deschamps, 1766–1842, French surgeon-naturalist on the sailing vessel *La Recherche*) Tufted annuals and perennials with hollow culms. **LEAVES** with open sheaths, flat to involute blades, membranous ligules, and lacking auricles. **INFL** a panicle of rebranching branches, sometimes becoming racemose. **SPIKELETS** mostly 2-flowered (sometimes with 1 or 3 florets), the rachilla prolonged beyond the palea; disarticulation above the glumes and between the florets; glumes subequal, 1- to 3-nerved, exceeding the lowermost floret; lemmas 5- to 7-nerved, truncate and toothed at the apex, bearded on the callus; awn attached below the middle of the lemma, slender; palea nearly as long as the lemma. **STAMENS** 3. $x = 7$. A genus of about 30 species in temperate and cold regions of the world.

1. Glumes exceeding the florets and often the awns; lemmas awned from near the middle; blades flat, some 4–8 mm wide or more .... see ***Vahlodea***

1′ Glumes mostly slightly shorter than 1 floret and the awns; lemmas awned from below the middle; blades mostly folded, rarely more than 3 mm wide .... ***D. cespitosa***

Deschampsia cespitosa (L.) P. Beauv. (tufted) Tufted hairgrass. Plants perennial. **CULMS** 20–80 cm tall, erect. **LIGULES** acuminate, lacerate, 3–8 mm long. **BLADES** firm, flat or folded, 1–3 (5) mm wide. **PANICLES** loose and open, often nodding, 5–25 cm long, the branches capillary. **SPIKELETS** 2 (3)-flowered, usually shiny, purplish. **GLUMES** lanceolate, acute, glabrous or scaberulous on the nerves, subequal, 2.5–5.2 mm long. **LEMMAS** 2.4–4 mm long, purplish basally and whitish apically, 5-nerved, with 4 apical teeth; awns 2–6 mm long, attached below midlength. **ANTHERS** 1.3–2.2 mm long. $2n = 24–28, 52$. Moist to wet meadows and slopes, banks of slow-moving streams and ponds, subalpine to alpine areas in the mountains. COLO: Arc, Con, Hin, LPl, Min, Mon, RGr, SJn; NMEX: McK, SJn. 2700–4000 m (8800–13100′). Flowering: Jul–Sep. Circumboreal, south in the western United States into Mexico; Eurasia, Africa, introduced in Argentina. There have been reports of *D. elongata* (Hook.) Munro in the *Flora* region, but I have seen no specimens of this. It can be distinguished from *D. cespitosa* by having narrow, contracted panicles with erect-appressed branches, filiform basal blades, and tiny anthers 0.3–0.5 mm long.

Dichanthelium (Hitchc. & Chase) Gould Rosettegrass

(Greek *dicha*, paired, + *anthele*, inflorescence, referring to the two kinds of panicles) Plants perennial, producing a winter rosette of short, ovate-lanceolate blades; in the spring producing a simple flowering shoot with a terminal panicle; in the summer and fall producing highly branched shoots with numerous axillary panicles; culms hollow. **LEAVES** with open sheaths, mostly hairy ligules, flat blades, and lacking auricles. **SPIKELETS** plano-convex, dorsally compressed, awnless, 2-flowered, the lower floret staminate or sterile, the upper floret perfect; disarticulating below the glumes; glumes unequal, the first small and often clasping the base of the spikelet, the second as large as the spikelet and similar to the lowermost lemma; first floret staminate or sterile, the lemma several-nerved, membranous, longer than the upper floret, often with a hyaline palea; second floret fertile, the lemma clasping the palea and the two forming an indurate, smooth, shiny seedcase; stamens 3. $x = 9$. A genus of about 72 species, formerly included as a subgenus of *Panicum*. These species are not common in the flora, but unmistakable because of the rosettes.

1. Spikelets 1–2 mm long, usually puberulent .... ***D. acuminatum***

1′ Spikelets 2.7–3.5 mm long, glabrous to sparsely puberulent .... ***D. oligosanthes***

Dichanthelium acuminatum (Sw.) Gould & C. A. Clark (long-tapered, acuminate) Woolly rosettegrass. Plants tufted, with well-developed basal rosettes. **CULMS** 15–80 cm tall, erect to spreading, sometimes decumbent. **LIGULES** 1–5 mm long. **BLADES** 3–12 mm wide, broadly lanceolate to narrowly ovate, with subcordate bases, ascending to spreading from the culm, densely to sparsely pubescent. **PANICLES** of spring phase 4–12 cm long, usually open with spreading branches and branchlets, well exserted. **SPIKELETS** 1.1–2.1 mm long, ± obovoid, puberulent. **GLUMES** unequal, the first 1/4–1/2 the length of the second. **LOWER FLORET** sterile, the lemma as long as the second glume. **UPPER FLORET** 1.1–1.7 mm long, minutely apiculate. $2n = 18$. Moist to wet ground along lakes, ponds, and streams and around springs. ARIZ: Apa; UTAH. 1100–1700 m (3600–5600′). Flowering: May–Sep. Throughout much of North and Central America.

Dichanthelium oligosanthes (Schult.) Gould (few-flowered) Rosettegrass. Plants tufted, with basal rosettes. **CULMS** 20–50 cm tall, geniculate-based, stiffly erect above. **LIGULES** 1–1.5 mm long. **BLADES** 6–15 mm wide, narrowly ovate, with cordate bases, ascending to spreading. **PANICLES** of spring phase 5–9 cm long, 3–6 cm wide, partly enclosed to long-exserted. **SPIKELETS** 2.7–3.5 mm long, obovoid, glabrous to sparsely puberulent. **GLUMES** unequal, the first about 1/2 as long as the second, the second 1.4–3 mm long and with a prominent orange to purplish spot at the base. **LOWER FLORET** sterile, the lemma as long as the second glume, with a hyaline palea about 1/2 as long as the lemma. **UPPER FLORET** 2.5–3 mm long. $2n = 18$. Sandy woodlands, stream and creek banks, wet ground around springs. NMEX: SJn. 1900–2100 m (6200–6900′). Flowering: Jun–Sep. Our material belongs to **var. *scribnerianum*** (Nash) Gould (for Frank Lamson Scribner, renowned USDA agrostologist) Scribner's rosettegrass. Throughout much of southern Canada and the United States; not yet reported from Mexico.

Digitaria Haller Crabgrass

(Latin *digitus*, finger, alluding to the fingerlike arrangement of the panicle branches) Plants annual and perennial, sometimes stoloniferous, with hollow culms. **LEAVES** with open sheaths, flat blades, membranous ligules, and lacking auricles. **INFL** a panicle of few to several unbranched, spikelike branches, these digitate or spaced along the rachis. **SPIKELETS** slightly plano-convex, lanceolate to elliptic, dorsally compressed, borne in pairs of one subsessile and the other short-pedicelled, or borne singly or in groups of 3–5; awnless, 2-flowered, the lower floret staminate or sterile, the upper floret perfect; disarticulation below the glumes; glumes unequal, the first minute or absent, the second nearly as large as the spikelet and similar in texture to the lowermost lemma, 3-nerved; first floret staminate or sterile, awnless, the lemma membranous, 5-nerved, the nerves prominent, the palea absent or vestigial; second floret fertile, awnless, cartilaginous when mature but not strongly indurate, the lemma clasping the palea by hyaline margins and the two forming a smooth, shiny seedcase; stamens 3. $x = 9$ (Henrard 1950; Webster 1987). A genus of about 200 species of tropical to warm-temperate regions.

1. Sheaths essentially glabrous, sometimes with sparse hairs; spikelets in groups of 3 at midbranch; second glume about as long as the spikelet and nearly covering completely the upper (fertile) floret .... ***D. ischaemum***

1′ Sheaths prominently hairy, often with stout, papilla-based hairs; spikelets in pairs at midbranch; second glume 1/3–1/2 as long as the spikelet and easily exposing the upper floret .... ***D. sanguinalis***

Digitaria ischaemum (Schreb.) Muhl. (resembling the genus *Ischaemum*) Smooth crabgrass. Plants annual. **CULMS** 20–60 cm tall, decumbent, often rooting at the lower nodes. **SHEATHS** glabrous or sparsely pubescent. **LIGULES** 0.5–2.5 mm long. **BLADES** 3–5 mm wide, sparsely papillose-pilose toward the base. **PANICLES** terminal and axillary; terminal panicles with 2–7 branches, these 4–10 cm long, the axes wing-margined; axillary panicles nearly always present in the lower sheaths, mostly concealed. **SPIKELETS** in groups of 3, 1.7–2.3 mm long. **GLUMES** strongly unequal, the first absent or a tiny nerveless rim or membrane, the second nearly equaling or equaling the lower lemma, pubescent. **LOWER LEMMA** 1.7–2.3 mm long, 7-nerved. **UPPER LEMMA** dark brown when mature, obscured by the second glume. **ANTHERS** 0.4–0.6 mm long. $2n = 36$. Lawns, gardens, and fields, urban areas. NMEX: SJn. 1550–1700 m (5100–5500′). Flowering: Aug–Sep. Native to Eurasia, now found nearly throughout the warm-temperate regions of the world.

Digitaria sanguinalis (L.) Scop. (pertaining to blood, alluding to the sometimes reddish foliage) Hairy crabgrass. Plants annual. **CULMS** 20–75 cm long, often decumbent and rooting at the lower nodes. **SHEATHS** keeled and sparsely papillose-pilose. **LIGULES** 0.5–2.5 mm long. **BLADES** 3–8 mm wide, usually papillose-pilose on both surfaces, sometimes glabrous. **PANICLES** terminal, with 4–12 primary branches 3–18 cm long, these digitate or on a short rachis up to 6 cm long, the axes wing-margined. **SPIKELETS** in pairs at midbranch, one subsessile and one pedicellate, 1.7–3.4 mm long. **GLUMES** strongly unequal, the first 0.2–0.4 mm long, nerveless, the second 3-nerved and 1/3–1/2 as long as the adjacent lemma. **LOWER LEMMA** usually slightly shorter than the spikelet, 7-nerved, the lateral nerves scaberulous at least on the upper 1/2. **UPPER LEMMA** 1.7–3 mm long, yellow, gray, or brown, easily visible. **ANTHERS** 0.5–0.9 mm long. $2n = 28, 34, 36, 54$. Lawns, sidewalks, gardens, and fields, mostly urban areas. ARIZ: Nav; NMEX: McK, SJn. 1600–1750 m (5300–5700′). Flowering: Jul–Sep. Native to Eurasia, now found nearly throughout the world.

Distichlis Raf. Saltgrass

(Greek *distichos*, two-ranked, referring to the arrangement of the florets or leaves) Plants perennial, mostly monoecious, strongly rhizomatous (ours) and/or stoloniferous, with solid culms. **LEAVES** distichous, the sheaths closely overlapping, with open sheaths, ciliolate-membranous ligules, stiff pungent blades, and lacking auricles. **INFL** a terminal panicle or raceme, sometimes exceeding the leaves. **SPIKELETS** large, few- to many-flowered, laterally compressed, awnless, the pistillate and staminate spikelets similar; disarticulation of the pistillate spikelets above the glumes and between the florets, staminate spikelets not disarticulating; glumes 3- to 7-nerved, shorter than the florets; lemmas coriaceous, 7- to 11-nerved, the staminate thinner than the pistillate; paleas 2-nerved, about as long as the lemmas. **STAMENS** 3. $x = 10$. A genus of about five species of alkaline areas in the Western Hemisphere and Australia.

Distichlis spicata (L.) Greene (spikelike) Saltgrass. Plants rhizomatous and sometimes stoloniferous. **CULMS** usually erect, 10–60 cm tall. **LIGULES** 0.2–0.6 mm long. **BLADES** flat to folded or rolled, stiff and rigid to more lax, 2–4 mm wide. **PANICLES** 2–8 cm long, congested. **SPIKELETS** 3- to 12-flowered, 8–16 mm long, glabrous. **GLUMES** unequal, stramineous, the first 2–3 mm long, the second 3–4 mm long. **LEMMAS** 7- to 9-nerved, 4–8 mm long. **PALEAS** slightly shorter than (pistillate florets) or longer than (staminate florets) the lemmas. **ANTHERS** 2–5 mm long. $2n = 40$. [*Distichlis spicata* (L.) Greene var. *stricta* (Torr.) Beetle]. Moist, alkaline sites in playas, salt flats, plains, ditch banks. ARIZ: Apa, Nav; COLO: Arc, Dol, LPl, Mon; NMEX: McK, RAr, SJn; UTAH. 1350–2000 m (4500–6500′). Flowering: May–Sep. Alkaline soils through much of the Western Hemisphere, Australia. Two or more varieties have been proposed, with var. *stricta* being the inland variety and var. *spicata* being coastal, but the differences seem to be inconsistent, difficult to ascertain, overlapping in expression, and unnecessary for our purposes.

Echinochloa P. Beauv. Barnyardgrass

(Greek *echinos*, hedgehog, + *chloa*, grass) Plants coarse annuals and perennials, tufted or rhizomatous, with pithy or hollow culms. **LEAVES** with compressed open sheaths, hairy or lacking (ours) ligules, flat blades, and lacking auricles. **INFL** a terminal exserted panicle of mostly unbranched spicate primary branches. **SPIKELETS** plano-convex, mostly 2-flowered, the lower floret staminate or sterile, the upper floret perfect; disarticulation below the glumes; glumes membranous, 3 (5)-nerved, unequal, the first acute and 1/4–1/2 the length of the second, the second as long as the spikelet, short- to long-awned; lower lemma similar to the second glume in size and texture, awnless to long-awned; upper (fertile) floret coriaceous to indurate, rounded on the back, smooth, awnless. **STAMENS** 3. $x = 9$ (Gould, Ali and Fairbrothers 1972). A genus of 40–50 species of tropical to warm-temperate regions of the world, usually in moist places.

1. Hairs of the spikelets and panicle branches not bulbous-based, not very stiff if at all; panicle branches 1–2 (3) cm long; spikelets awnless, 2.5–3 mm long .... ***E. colonum***

1′ Hairs of the spikelets and panicle branches bulbous-based, stiff; panicle branches usually longer than 2 cm; spikelets awned or awnless, 2.8–4 mm long .... (2)

2. Shiny apical portion of the fertile lemma with a line of minute hairs (use a lens); hairs of the panicle branches, at least some, longer than 3 mm .... ***E. crus-galli***

2′ Shiny apical portion of the fertile lemma without a line of minute hairs; hairs of the panicle branches absent to rarely longer than 3 mm .... ***E. muricata***

Echinochloa colonum (L.) Link. (of the farmers, a contraction of *colonorum*, often rendered *colona*) Jungle-rice. Plants annual. **CULMS** erect to decumbent and rooting at the lower nodes, 10–70 cm tall. **BLADES** 3–8 mm wide, mostly glabrous. **PANICLES** 2–12 cm long, with 5–10 erect to ascending primary branches 1–2 (3) cm long. **SPIKELETS** 2–3 mm long, awnless, puberulent to softly hispidulous, the hairs not very stiff. **GLUMES** unequal, the first about 1/2 the length of the spikelet, the second as long as the spikelet. **LOWER LEMMA** sterile, similar to the second glume in size and texture. **UPPER LEMMA** 2.6–2.9 mm long, the apex withering. **ANTHERS** 0.7–0.8 mm long. $2n = 54$. Uncommon in moist disturbed ground. NMEX: SJn. 1450–1600 m (4900–5200′). Flowering: May–Sep. Native to Old World tropical regions and widespread elsewhere in the world.

Echinochloa crus-galli (L.) P. Beauv. (resembling a cock's foot) Barnyardgrass. Plants annual. **CULMS** mostly erect, 30–200 cm tall. **BLADES** 5–30 mm wide, mostly glabrous. **PANICLES** 5–25 cm long, with several ascending to spreading primary branches 2–10 cm long, the axis usually beset with bulbous-based hairs, these often longer than the spikelets. **SPIKELETS** 2.5–4 mm long, awnless to long-awned, hispid with bulbous-based stiff hairs. **GLUMES** unequal, the first about 1/4–1/2 as long as the spikelet, the second as long as the spikelet. **LOWER LEMMA** sterile, similar to the second glume, awnless to awned, the awn to 5 cm long. **UPPER LEMMA** 2–3 mm long, the apex early-withering and with a line of minute hairs. **ANTHERS** 0.5–1 mm long. $2n = 54$. Common in moist to wet waste places and disturbed ground. ARIZ: Apa, Nav; COLO: Arc, LPl, Mon; NMEX: McK, RAr, SJn; UTAH. 1550–1950 m (5100–6400′). Flowering: Jul–Sep. Native to Eurasia, now found throughout most tropical and warm-temperate regions of the world.

Echinochloa muricata (P. Beauv.) Fernald (beset with sharp points) Cockspur. Plants annual. **CULMS** 50–150 cm tall, erect to sprawling. **BLADES** (2) 3–30 cm wide, glabrous. **PANICLES** 7–35 cm long, with 2–8 spreading branches, the axes glabrous to hispid, the hairs shorter than the spikelets. **SPIKELETS** 3–5 mm long, hispid with bulbous-based hairs. **GLUMES** unequal, the first 1/4–1/2 as long as the spikelet, the second as long as the spikelet. **LOWER LEMMA** sterile, similar to the second glume, awnless to awned, the awn to 16 mm long. **UPPER LEMMA** 2–3 mm long, the apex glabrous. **ANTHERS** 0.4–1.1 mm long. $2n = 36$. Common in moist to wet waste places and disturbed ground. ARIZ: Apa, Nav; COLO: Arc, Dol, LPl, Mon, SJn; NMEX: RAr, SJn; UTAH. 1550–2600 m (5200–8500′). Flowering: Jul–Sep. Widespread in North America from southern Canada to Mexico. Our plants belong to **var.** ***microstachya*** Wiegand, common west of the Mississippi River.

Elymus L. Wildrye, Wheatgrass

(Greek *elymos*, a name for a kind of millet or grain) Tufted to rhizomatous perennials, with hollow culms. **LEAVES** with open sheaths, flat to rolled blades, membranous ligules, and with or without auricles. **INFL** a spike, the rachis persistent or disarticulating, bearing 1–many spikelets per node. **SPIKELETS** 2- to many-flowered, sessile, awned or awnless, disarticulating above or below the glumes; glumes and lemmas variable in size, shape, and nervation; paleas well developed. **STAMENS** 3. $x = 7$. A diverse genus of approximately 150 species or so, including many species formerly placed in the genera *Agropyron* and *Sitanion*. *Elymus* is treated herein in a broad sense and includes species sometimes placed in the segregate genera *Elytrigia*, *Lophopyrum*, *Pascopyrum*, *Pseudoroegneria*, *Thinopyrum*, and *Trichopyrum*. Conversely, some species formerly placed in *Elymus* are found herein in *Leymus* and *Psathyrostachys*. Hybridization occurs readily among many of the species, obscuring the boundaries and making identification difficult. This treatment must be considered tentative, awaiting more intensive study of the variation of our plants by some valiant botanist.

1. Spikelets mostly solitary at each node of the rachis .... (2)
1′ Spikelets 2 or more at each node of the rachis .... (15)

2. Spikelets (glumes and/or lemmas) long-awned, the awns prominent and mostly greater than 10 mm long .... (3)
2′ Spikelets (glumes and lemmas) awnless or nearly so, any awns usually less than 5 mm long .... (7)

3. Awns erect-appressed or nearly so; glumes 3/4 to equaling the length of the spikelet .... ***E. trachycaulus***
3′ Awns spreading moderately outward to reflexed; glumes 1/2 to 3/4 the length of the spikelet .... (4)

4. Anthers 4–6 mm long; spikelets widely spaced and hardly overlapping .... (5)
4′ Anthers 1–2 mm long; spikelets at least moderately congested and overlapping .... (6)

5. Spikes 15–30 cm long, often nodding; blades 4–6 mm wide .... ***E. arizonicus***
5′ Spikes 8–15 cm long, usually erect; blades 1–2 mm wide .... ***E. spicatus***

6. Spikes 3–7 (8) cm long, very dense; glumes (4) 6–8 mm long, 1- to 3 (5)-nerved .... ***E. scribneri***
6′ Spikes 9–16 cm long, not especially dense; glumes (9) 10–12 mm long, (3) 5- to 7-nerved .... ***E. bakeri***

7. Glumes blunt, nearly truncate, thick and very firm; sheaths typically ciliate on at least one margin .... (8)
7′ Glumes acute to acuminate, thin and membranous to stiff, but not thick; sheaths rarely ciliate .... (9)

8. Plants with evident, long-creeping rhizomes; glumes with a tiny awn tip or mucro .... ***E. hispidus***
8′ Plants densely tufted, lacking evident rhizomes; glumes exactly truncate .... ***E. elongatus***

9. Anthers 1–2 mm long .... (10)
9′ Anthers (3) 4–16 mm long .... (12)

10. Glumes 1- to 2 (3)-nerved; rachis tending to break apart at maturity; sterile hybrid plants; these are *E. trachycaulus* × *E. longifolius* hybrids, occurring where the two parents grow together
10′ Glumes (3) 5-nerved; rachis remaining intact; fertile to sterile plants .... (11)

11. Plants mostly with rhizomes; glumes 1/2 to 2/3 the length of the spikelet .... ***E. ×pseudorepens***
11′ Plants tufted; glumes mostly 3/4 or greater the length of the spikelet .... ***E. trachycaulus***

12. Plants lacking evident rhizomes, occasionally rhizomes weakly developed and short .... ***E. spicatus***
12′ Plants with evident, long-creeping rhizomes .... (13)

13. Glumes acuminate, asymmetrical or somewhat sickle-shaped, gradually tapering to an awn tip; blades somewhat rigid and prominently ridged above .... ***E. smithii***
13′ Glumes acute to acuminate, symmetrical, not gradually tapering; blades often lax, not prominently ridged above. (14)

14. Blades flat, mostly 5–15 mm wide, dark green, often with a circular constriction toward the tip; anthers (3) 4–7 mm long .... ***E. repens***
14′ Blades rolled or less than 4 mm wide when flat, usually glaucous, lacking a circular constriction toward the tip; anthers 3–5 mm long .... ***E. lanceolatus***

15. Rachis fragile and breaking apart at maturity .... (16)
15′ Rachis persistent, not breaking apart at maturity .... (17)

16. Lemma awns 4–17 mm long; rachis internodes 2.5–7 mm long; these are *Elymus longifolius* hybrids [*Elymus* ×*saundersii* Vasey, *Agropyron* ×*saundersii* (Vasey) Hitchc.]
16′ Lemma awns 20–80 mm long; rachis internodes mostly 5–12 mm long .... ***E. longifolius***

17. Glumes nearly subulate, 1- to 2-nerved .... ***E. longifolius***
17′ Glumes narrowly lanceolate, mostly conspicuously 3- to 7-nerved .... (18)

18. At maturity, the spikes usually nodding or curved and the awns spreading outward; glumes 20 mm or more long; lemmas scabrous to short-hairy (rarely glabrous) .... ***E. canadensis***
18′ At maturity, the spikes erect and the awns erect-appressed; glumes mostly less than 20 mm long; lemmas glabrous to scaberulous .... ***E. glaucus***

Elymus arizonicus (Scribn. & J. G. Sm.) Gould (of Arizona) Arizona wheatgrass. Plants perennial, tufted. **CULMS** erect, 40–80 cm tall. **LIGULES** to 1 mm long. **BLADES** flat, 4–6 mm wide. **SPIKES** often nodding, 15–30 cm long, the rachis somewhat zigzag and slender. **SPIKELETS** 1 per node, distant, scarcely overlapping; disarticulation above the glumes and between the florets. **GLUMES** subequal, elliptic, 3- to 7-nerved, 5–10 mm long, 1/2 to nearly as long as the lemma, short-awned. **LEMMAS** elliptic, 5-nerved, 8–12 mm long; awns 20–30 mm long, ultimately diverging (rarely nearly awnless). **ANTHERS** 3–5 mm long. $2n = 28$. [*Agropyron arizonicum* Scribn. & J. G. Sm.; *Elytrigia arizonica* (Scribn. & J. G. Sm.) D. R. Dewey; *Pseudoroegneria arizonica* (Scribn. & J. G. Sm.) Á. Löve]. Rocky slopes in the lower mountains. Not common in our area. NMEX: McK. 2050–2250 m (6800–7400′). Flowering: Jul–Aug. West Texas to California, northern Mexico. Similar to *E. spicatus*, but more robust, with longer, flexuous spikes and slightly longer and stiffer awns.

Elymus bakeri (E. E. Nelson) Á. Löve (for Charles Fuller Baker, 1872–1927, Colorado botanist, entomologist, and teacher) Baker's wheatgrass. Plants tufted, lacking rhizomes. **CULMS** ascending to erect, 30–50 cm tall. **LIGULES** 0.5–1 mm long. **BLADES** flat to rolled, 2–4 mm wide, stiff. **SPIKES** 9–16 cm long, dense but the rachis usually easily visible. **SPIKELETS** 1 per node, overlapping; disarticulation above the glumes and between the florets. **GLUMES** (9) 10–12 mm long, (3) 5- to 7-nerved, 2–2.5 mm wide; awns 2–8 mm long. **LEMMAS** 5-nerved, 8–12 mm long; awns 10–30 mm long, arcuate to recurved. **ANTHERS** 0.8–1.5 mm long. $2n = 28$. [*Agropyron bakeri* E. E. Nelson]. Rocky, often grassy, slopes and ridges in high subalpine areas, close to timberline. COLO: Arc, Con, LPl, Min, SJn. 2750–3750 m (9000–12300′). Flowering: Jul–Aug. Rocky Mountains of the United States. This is easily confused with *E. scribneri*, which has similar-appearing spikes with divergent awns; that latter species has shorter and more densely flowered spikes and shorter glumes with fewer nerves, as in the key. *Elymus bakeri* may represent hybrids and hybrid derivatives between *E. trachycaulus* and *E. scribneri*, *E. longifolius*, or *E. canadensis*.

Elymus canadensis L. (of Canada) Canada wildrye. Plants perennial, tufted. **CULMS** 80–150 cm tall. **LIGULES** 0.3–1.4 mm long. **BLADES** flat, 3–10 mm wide or more; auricles well developed, 1–2 mm long. **SPIKES** curved or nodding to drooping, 8–20 cm long, sometimes longer, bristly, sometimes interrupted below. **SPIKELETS** mostly 2 (3) per node, overlapping; disarticulation above the glumes and between the florets. **GLUMES** equal, narrowly lanceolate, 3- to 5-nerved, 10–25 mm long, scabrous to ciliolate on the nerves, narrowing to an awn. **LEMMAS** 5- to 7-nerved, 8–15 mm long, long-scabrous, narrowing to a spreading awn 15–30 mm long. **ANTHERS** 2.5–3 mm long. $2n = 28, 42$. Stream banks, moist ground in meadows and along roads and ditches. ARIZ: Apa, Nav; COLO: Arc, Hin, LPl, Min, Mon; NMEX: RAr, San, SJn; UTAH. 1600–2400 m (5300–7800′). Flowering: Jul–Aug. Nearly throughout Canada and the United States except for the extreme southeast. Hybrids are known with *E. trachycaulus* and *E. longifolius*.

Elymus elongatus (Host) Runemark (elongated) Tall wheatgrass. Plants perennial, tufted, often glaucous. **CULMS** erect, robust, densely tufted, but sometimes producing short, thick rhizomes, 70–200 cm tall. **SHEATHS** typically ciliate on at least one margin, with erect auricles 1–2 mm long. **LIGULES** less than 1 mm long. **BLADES** flat to loosely rolled, 3–5 mm wide, the margins thick-veined and indurate. **SPIKES** 15–30 cm long, stiff, erect. **SPIKELETS** awnless, long, 15–20 mm long, widely spaced, often not overlapping at all, becoming arcuate in age and curving away from the rachis; disarticulation above the glumes and between the florets. **GLUMES** oblong, thick and indurate, the apex truncate and lacking a mucro, 5- to 7-nerved, the first 6–9 mm long, the second 7–10 mm long, shorter than the lowermost lemma. **LEMMAS** broadly lanceolate, 5-nerved, 8–11 mm long. **ANTHERS** 4–5.5 mm long. $2n$ = 14, 28, 42, 56, 70. [*Agropyron elongatum* (Host) P. Beauv.; *Elytrigia elongata* (Host) Nevski; *Thinopyrum elongatum* (Host) D. R. Dewey]. Disturbed ground along roads, ditches, and agricultural fields. ARIZ: Apa; COLO: Dol, LPl, Mon, SMg; NMEX: SJn. UTAH. 1500–2200 m (5000–7200′). Flowering: Jul–Aug. Native to Eurasia, introduced in much of the western and central United States. Plants are sometimes mistaken for *E. hispidus*, which has creeping rhizomes and mucronate glumes.

Elymus glaucus Buckley (glaucous) Blue wildrye. Plants perennial, tufted, rarely with short stolons, green to glaucous, glabrous or sparsely short-pilose. **CULMS** erect to geniculate-based, 70–100 cm tall or more. **LIGULES** 0.3–1 mm long. **BLADES** flat, 4–12 mm wide, glabrous to lightly pilose above, with well-developed auricles about 2 mm long. **SPIKES** 5–15 cm long, erect. **SPIKELETS** mostly 2 per node, closely overlapping or somewhat distant below, 2- to 4-flowered; disarticulation above the glumes and between the florets. **GLUMES** subequal, 8–15 mm long, 3- to 5-nerved, almost parallel and concealing the base of the lowermost floret, tapering to a short awn. **LEMMAS** 8–12 mm long, 5-nerved, glabrous to scabrous, tapering to an awn 10–20 mm long. **ANTHERS** 1.5–3 mm long. $2n$ = 28. Open woods, shady meadows, streamsides in the mountains. COLO: Arc, Hin, LPl, Min, Mon, SJn; NMEX: SJn; UTAH. 2250–2500 m (7400–8200′). Flowering: Jul–Aug. Wide, flat blades, parallel glumes, and erect awns help to distinguish this species of shaded mountain sites. Western Canada south into Mexico.

Elymus hispidus (Opiz) Melderis (spiny, hispid) Intermediate wheatgrass. Plants perennial, rhizomatous, usually glaucous. **CULMS** erect, 50–100 cm tall. **LIGULES** about 0.5 mm long. **BLADES** flat to loosely folded or rolled with drying, 2–6 mm wide, the auricles well developed. **SPIKES** erect, slender, stiff, 5–18 cm long. **SPIKELETS** 1 per node, rather distant but overlapping somewhat, 3- to 8-flowered, 8–15 mm long, glabrous or hirsute, awnless. **GLUMES** elliptic-lanceolate, thick and rigid, obtuse, mucronate, the first 4–7 mm long, the second 6–9 mm long. **LEMMAS** broadly lanceolate, faintly 3- to 5-nerved, 7–10 mm long, obtuse. **ANTHERS** 3–5 mm long. $2n$ = 42. [*Agropyron hispidum* Opiz; *Agropyron intermedium* (Host) P. Beauv.; *Elytrigia intermedia* (Host) Nevski]. Roadsides, pastures, fields, reseeded sites, waste ground, in woodland to subalpine communities. COLO: Arc, Dol, Hin, LPl, Mon, SJn, SMg; NMEX: RAr; UTAH. 1900–3000 (6500–9800′). Flowering: Jun–Aug. Native to Eurasia, introduced in much of the western United States. Large plants are sometimes mistaken for *E. elongatus*, which is tufted and has exactly truncate glumes. Plants with hispid spikelets have been referred to subsp. *barbulatus* (Schur) Melderis.

Elymus lanceolatus (Scribn. & J. G. Sm.) Gould (lance-shaped) Thickspike wheatgrass. Plants perennial, with creeping rhizomes, the herbage usually glaucous, glabrous or pilose. **CULMS** 40–80 cm or more tall. **LIGULES** less than 0.5 mm long. **BLADES** mostly rolled, 1–3.5 mm wide, stiff, the auricles well developed. **SPIKES** slender, erect, 6–20 cm long. **SPIKELETS** mostly 1 per node (sometimes 2 per node), awnless, 4- to 8-flowered, 10–18 mm long. **GLUMES** broadly lanceolate, acute to acuminate, symmetrical, glabrous to pubescent, faintly 3- to 5-nerved, about 1/2–2/3 as long as the lowermost floret. **LEMMAS** 7–10 mm long, acute to awn-tipped, villous to glabrous. **ANTHERS** 4–5 mm long. $2n$ = 28. [*Agropyron dasystachyum* (Hook.) Scribn. (not *Elymus dasystachys* Trin. ex Ledeb.); *Agropyron lanceolatum* Scribn. & J. G. Smith; *Elytrigia dasystachya* (Hook.) Á. Löve & D. Löve]. Sagebrush plains, juniper woodlands, mountain brush. ARIZ: Apa; COLO: Dol, Mon; NMEX: McK; UTAH. 1750–2400 m (5800–7800′). Flowering: Jun–Aug. Across Canada and south through the western and central United States into Mexico. This species is characterized by having creeping rhizomes, flat blades, short glumes, and short anthers. It is not well expressed in our region, being more distinct and more common northward. In *Agropyron* and *Elytrigia* the basionym *dasystachyum* has priority, but this epithet is preempted in *Elymus* by *E. dasystachys* Trin. ex Ledeb. Across southern Canada, south through the western United States.

Elymus longifolius (J. G. Sm.) Gould (long-leaved) Longleaf squirreltail. Plants perennial, tufted, glabrous to puberulent. **CULMS** erect to spreading, 15–50 cm tall. **LIGULES** less than 0.5 mm long. **BLADES** flat to folded or involute, 1–5 mm wide, mostly glabrous. **SPIKES** 5–15 cm long (excluding the awns), stiffly erect; disarticulation in the rachis, the spikelets falling with the subtending rachis internode, sometimes remaining intact and the disarticulation above the glumes. **SPIKELETS** 2 per node (sometimes 1 or 3), with 2–6 florets, the lowermost floret fertile and well developed.

GLUMES subulate and awnlike, stiff, not bifid, 3–8 cm long, 2-nerved but sometimes appearing as a single channel or nerve down the middle of the glume, spreading to reflexed when mature. **LEMMAS** 7–10 mm long, 3- to 5-nerved, tapering to an awn 2–8 cm long, spreading to reflexed when mature. **ANTHERS** 1–2 mm long. $2n = 28$. [*Sitanion brevifolium* J. G. Sm.; *S. caespitosum* J. G. Sm.; *S. hystrix* (Nutt.) J. G. Sm. in part; *S. longifolium* J. G. Sm.]. Widespread and common throughout the Four Corners region in plains, grasslands, woodlands, mountain slopes, and forest clearings, roadsides. ARIZ: Apa, Coc, Nav; COLO: Arc, Dol, Hin, LPl, Min, Mon, SJn, SMg; NMEX: McK, RAr, San, SJn; UTAH. 1120–3400 m (3700–11200′). Flowering: May–Aug. Western and central Canada and United States, south into Mexico. Longleaf squirreltail commonly forms hybrids with *E. trachycaulus* and *E. canadensis*, among other species. A similar species, *E. elymoides* (Raf.) Swezey s.s., is not known from our area; it can be distinguished by having the glumes bifid and the lowermost lemma reduced and subulate so it resembles an extra glume.

Elymus* ×*pseudorepens (Scribn. & J. G. Sm.) Barkworth & D. R. Dewey (false *Elymus repens*) False quackgrass. Plants perennial, with creeping rhizomes, sometimes these short. **CULMS** 40–80 cm tall. **LIGULES** less than 1 mm long. **BLADES** flat or folded, 2–4 mm wide; auricles short or lacking. **SPIKES** slender, 5–20 cm long. **SPIKELETS** 1 per node, overlapping about 1/2, 3- to 5-flowered; disarticulation above the glumes and between the florets. **GLUMES** lanceolate, acute to short-awned, 5-nerved, 1/2–2/3 the length of the spikelet. **LEMMAS** 3- to 5-nerved, awnless or short-awned. **ANTHERS** 1–2 mm long. $2n = 28$. Washes, foothills, mountains slopes and meadows, moist roadsides. ARIZ: Apa, Nav; COLO: Arc, Dol, LPl, Min; NMEX: McK, RAr; UTAH. 1550–2400 m (5200–7800′). Flowering: Jul–Sep. Widespread in the West. These plants are at least partially fertile hybrids and derivatives of *E. trachycaulus* × *E. lanceolatus*, *E. repens*, or *E. smithii* crosses that have become well established in various habitats. This causes a bewildering array of variation among these species, rendering their circumscription tenuous and difficult. Because the plants are so common, I apply the hybrid name herein.

Elymus repens (L.) Gould (creeping) Quackgrass. Plants perennial, strongly rhizomatous, green. **CULMS** 45–85 cm tall or more, erect to geniculate-based. **LIGULES** about 0.5 mm long. **BLADES** flat, mostly 5–15 mm wide, some might be narrower, often with a constriction toward the apex. **SPIKES** erect, 10–18 cm long. **SPIKELETS** 1 per node, 3- to 8-flowered, stiffly spreading somewhat, overlapping; disarticulation above the glumes and between the florets. **GLUMES** lanceolate, stiff, 5- to 7-nerved, acute and usually short-awned, the awn mostly less than 8 mm long. **LEMMAS** 5-nerved, 7–10 mm long, acute to awn-tipped, the awn mostly less than 5 mm long. **ANTHERS** (3) 4–7 mm long. $2n = 42$. [*Agropyron repens* (L.) P. Beauv.; *Elytrigia repens* (L.) Nevski]. Mesic ground along ditches and roadsides, pastures, valley bottoms, not as common in our area as northward. ARIZ: Nav; COLO: Arc, LPl, Mon, RGr; NMEX: McK, RAr, SJn. 1700–2350 m (5600–7700′). Flowering: Jun–Aug. Native to Eurasia, now nearly throughout the United States. It seems that many plants identified as *E. repens* belong more accurately to hybrids of *E. trachycaulus* and other species, perhaps including *E. repens*.

Elymus scribneri (Vasey) M. E. Jones (for Frank Lamson Scribner, 1851–1938, prominent North American agrostologist) Scribner's wheatgrass. Plants perennial, tufted. **CULMS** decumbent, geniculate, or ascending, the shoots rarely taller than 30 cm. **LIGULES** about 0.5 mm long. **BLADES** narrow, involute to sometimes flat, 1–3 mm wide, short auricles present. **SPIKES** short, 3–7 (8) cm long, erect to curving abruptly downward. **SPIKELETS** 1 per node, crowded, 3- to 6-flowered, often purplish; disarticulation above the glumes and between the florets but also in the rachis in age. **GLUMES** narrowly lanceolate, 2- to 3 (5)-nerved, the body 6–8 mm long, tapering to a slender awn 12–20 mm long and curving outward. **LEMMAS** 7–10 mm long, faintly 5-nerved, tapering to a scabrous divergent awn 15–25 mm long. **ANTHERS** 1–1.8 mm long. $2n = 28$. [*Agropyron scribneri* Vasey]. Rocky, often grassy, slopes and ridges in alpine and high subalpine areas, nearly always above timberline. COLO: Arc, Hin, LPl, Min, RGr, SJn. 3700–3900 m (12200–12700′). Flowering: Jul–Aug. High mountains of western Canada and the United States. This is a distinctive grass of open rocky areas above timberline. Similar-looking plants with longer, looser spikes and longer glumes with several nerves are referred to *E. bakeri*.

Elymus smithii (Rydb.) Gould (for Jared Gage Smith, 1866–1925, prominent USDA botanist) Western wheatgrass. Plants perennial, rhizomatous, usually prominently bluish glaucous. **CULMS** 30–80 cm tall, in clumps scattered along the rhizome. **LIGULES** about 0.5 mm long. **BLADES** flat, 2–5 mm wide, firm, prominently ridged above; auricles well developed, to 2 mm long. **SPIKES** erect, stiff, 5–18 cm long. **SPIKELETS** mostly 1 per node, overlapping, sometimes 2 per node and crowded, 4- to 8-flowered, erect to curving outward; disarticulation above the glumes and between the florets. **GLUMES** linear-lanceolate, rigid, 3- to 5-nerved, asymmetric and tapering to an awn tip, about 1/2 the length of the spikelet, subequal, the first 7–9 mm long, the second slightly longer. **LEMMAS** 5-nerved, 8–12 mm long, glabrous to hairy, awn-tipped to 4 mm. **ANTHERS** 2–4 mm long. $2n$ = 56. [*Agropyron molle* (Scribn. & J. G. Sm.) Rydb.; *A. palmeri* (Scribn. & J. G. Sm.) Rydb.; *A. smithii* Rydb.; *Elytrigia smithii* (Rydb.) Nevski; *Pascopyrum smithii* (Rydb.) Á. Löve]. Plains, swales, grassy hills and slopes, bottomlands. ARIZ: Apa, Nav; COLO: Arc, Dol, Hin, LPl, Min, Mon, SJn, SMg; NMEX: McK, RAr, San, SJn; UTAH. 1550–2800 m (5100–9100′). Flowering: Jun–Sep. Throughout western Canada and United States. Western wheatgrass often forms thick bluish stands along roads and swales.

Elymus spicatus (Pursh) Gould (spiked) Bluebunch wheatgrass. Plants perennial, tufted, rarely with short rhizomes, green to glaucous. **CULMS** erect, slender, 40–75 cm tall. **LIGULES** less than 1 mm long. **BLADES** flat to loosely rolled, 1–2 mm wide; auricles developed. **SPIKES** slender, 6–15 cm long. **SPIKELETS** 1 per node, scattered along the axis, scarcely overlapping, 4- to 6-flowered, awned or sometimes awnless; disarticulation above the glumes and between the florets. **GLUMES** narrowly oblong, blunt to acute, 4- to 5-nerved, 1/2 the length of the spikelet, the first 4–8 mm long, the second 5–9 mm long. **LEMMAS** 5-nerved, 8–10 mm long, usually with divergent awns 10–18 mm long. **ANTHERS** 4–6 mm long. $2n$ = 14, 28. [*Agropyron inerme* (Scribn. & J. G. Sm.) Rydb.; *A. spicatum* (Pursh) Scribn. & J. G. Sm.; *Elytrigia spicata* (Pursh) D. R. Dewey; *Pseudoroegneria spicata* (Pursh) Á. Löve]. Dry mountain slopes and foothills, but not common in our area. ARIZ: Nav; COLO: LPl, Mon, SJn; NMEX: McK, SJn. 2100–2500 m (6950–8200′). Flowering: Jun–Aug. Western Canada and the United States.

Elymus trachycaulus (Link) Gould ex Shinners (rough stem) Slender wheatgrass. Plants perennial, mostly glabrous, tufted, rarely with rhizomes. **CULMS** erect, 30–100 cm tall or more. **LIGULES** scarcely 0.5 mm long. **BLADES** flat, green, 2–8 mm wide; auricles absent. **SPIKES** 5–20 cm long, compact. **SPIKELETS** 1 per node, overlapping, 3- to 7-flowered, awned or awnless. **GLUMES** broadly elliptic, 3/4 or more the length of the spikelet, prominently 5- to 7-nerved, the first 7–10 mm long, the second 7–12 mm long. **LEMMAS** 7–12 mm long, mostly 5-nerved, awnless or with an awn to 25 mm long. **ANTHERS** 1–2 mm long. $2n$ = 28. [*Agropyron latiglume* (Scribn. & J. G. Sm.) Rydb.; *A. novae-angliae* Scribn.; *A. pauciflorum* (Schwein.) Hitchc.; *A. tenerum* Vasey; *Agropyron trachycaulum* (Link) Malte; *A. violaceum* (Hornem.) Lange; *A. unilaterale* Cassidy; *E. trachycaulus* subsp. *novae-angliae* (Scribn.) Tzvelev; *E. trachycaulus* subsp. *subsecundus* (Link) Á. Löve & D. Löve; *E. trachycaulus* subsp. *violaceus* (Hornem.) Á. Löve & D. Löve]. Moist areas in the foothills to alpine regions. ARIZ: Apa, Nav; COLO: Arc, Con, Dol, Hin, LPl, Min, Mon, RGr, SJn; NMEX: McK, RAr, San, SJn; UTAH. 1600–3850 m (5300–12600′). Flowering: Jun–Sep. Nearly throughout North America to northern Mexico, scarce or absent in the southeastern states; Siberia, exotic in Eurasia. This is an extremely variable species, difficult to interpret or circumscribe. My concept includes populations of tufted plants with flat blades and large, prominently nerved glumes. Numerous names have been proposed, and some infraspecific taxa may be warranted. Awned forms have been called subsp. *subsecundus*.

Enneapogon Desv. ex P. Beauv. Pappusgrass

(Greek *ennea*, nine, + *pogon*, beard, referring to the awns of the floret) Plants annual or perennial (ours), short-hairy throughout, with hollow culms and hairy nodes. **LEAVES** with open sheaths, flat to rolled blades, hairy ligules, and lacking auricles. **INFL** a weakly developed panicle, often spikelike; disarticulation above the glumes, the florets falling together. **SPIKELETS** 3- to 6-flowered, only the lowermost fertile and well developed; glumes nearly as long as the entire spikelet, few- to several-nerved; lowermost floret fertile, villous on the lower 1/2, the lemma 9-nerved and 9-awned, the awns plumose and pappuslike; upper florets progressively reduced, sterile; paleas slightly longer than the lemmas, 2-nerved, awnless. **STAMENS** 3. x = 10. About 28 species in the warm regions of the world, especially Africa and Australia.

Enneapogon desvauxii P. Beauv. (for Nicaise Auguste Desvaux, 1784–1856, French botanist) Spike pappusgrass. Plants tufted perennials, hairy throughout, with a knotty base. **CULMS** erect, 20–45 cm tall. **LIGULES** about 0.5 mm long. **BLADES** flat, rolling in age, 1–2 mm wide. **PANICLES** spikelike, 2–10 cm long, grayish, the short branches poorly developed and appressed. **SPIKELETS** 3- to 4-flowered, only the lower fertile, the uppermost scarcely developed; awnless cleistogamous spikelets often present in the lower sheaths. **GLUMES** nearly equal, 3–5 mm long, thin. **LEMMAS** 1.5–2 mm long; awns 3–4 mm long. **ANTHERS** 0.3–0.5 mm long. $2n$ = 20. [*Pappophorum wrightii* S. Watson]. Desert hills and plains, disturbed ground. ARIZ: Apa, Nav; COLO: Mon; NMEX: McK, SJn; UTAH. 1550–1800 m (5200–5850′). Flowering: Aug–Sep. Southwestern United States and Mexico; Peru, Bolivia, Argentina; Africa and Asia (Chase 1946).

Eragrostis Wolf Lovegrass

(Greek *eros*, God of love, and *agrostis*, a grass; a concoction prepared from *E. cilianensis* was thought to act as a love potion) Plants annual or perennial, mostly tufted, some rhizomatous, with pithy or hollow culms (ours). **LEAVES** with open sheaths, flat to rolled blades, hairy ligules, and lacking auricles. **INFL** a panicle with usually rebranched branches. **SPIKELETS** 3- to many-flowered, awnless; disarticulation above the glumes, the rachilla and palea persistent, the glumes and lemmas falling; glumes shorter than the lowermost floret, mostly 1-nerved; lemmas 3-nerved; paleas 2-nerved. **STAMENS** 2–3. $x = 10$. A large genus of 300 or more species from the tropical, subtropical, and warm regions of the world, some species adapted to cool-temperate areas.

1. Plants perennial, 50–130 cm tall or more; panicles 20–30 cm long .... ***E. curvula***
1′ Plants annual, mostly much less than 50 cm tall; panicles 5–25 cm long .... (2)

2. Grains with a prominent groove on the side opposite the embryo .... ***E. mexicana***
2′ Grains lacking a groove .... (3)

3. Culms prostrate and rooting at the nodes, forming mats .... ***E. hypnoides***
3′ Culms erect or only geniculate-based, not forming mats .... (4)

4. Plants lacking any glandular depressions or rings; panicle branchlets usually appressed along the primary branch .... ***E. pectinacea***
4′ Plants with glandular depressions or rings below the nodes or on the blade margins, or on the keels of the glumes or lemmas; panicles various, but often the branchlets divergent from the primary branch .... (5)

5. Keels of glumes and lemmas with small craterlike glands (use a lens); spikelets 2–4 mm wide; anthers yellow .... ***E. cilianensis***
5′ Keels of glumes and lemmas lacking glands; spikelets 1–2 mm wide; anthers reddish brown or purplish .... (6)

6. Panicles narrow, 0.5–2 cm wide, the branchlets and spikelets erect-appressed; culms with glandular pits below the nodes, but not a solid band or ring; bases of blades with glandular pits or craterlike glands on the margins .... ***E. lutescens***
6′ Panicles 2–18 cm wide, open, the branchlets diverging; culms with bands or rings of glandular tissue below the nodes; bases of blades lacking glands .... ***E. barrelieri***

Eragrostis barrelieri Daveau (for Jacques Barrelier, 1606–1673, French botanist) Mediterranean lovegrass. Plants tufted annuals. **CULMS** 10–40 cm tall, with a band of glandular tissue below the nodes, the band usually shiny and yellowish. **LIGULES** 0.2–0.5 mm long. **BLADES** flat, 1–4 mm wide, mostly glabrous, sometimes with scattered long pilose hairs on the upper surface (also tufts of hair at the collar), lacking craterlike glands on the margins. **PANICLES** 5–15 cm long, 3–8 cm wide, ovate, usually open, with yellowish glandular bands below the nodes; branches and pedicels usually with axillary pulvini. **SPIKELETS** 7- to 12-flowered, 1–2.2 mm wide, the pedicels lacking glandular bands. **GLUMES** ovate, 1-nerved, subequal, 1–1.5 mm long, shorter than the lowermost florets, early-deciduous. **LEMMAS** ovate, 1.4–1.8 mm long. **PALEAS** slightly shorter than the lemmas, hyaline. **ANTHERS** 3, tiny, 0.1–0.2 mm long, reddish brown. $2n = 40$. Roadsides, sandy ground, disturbed sites. ARIZ: Apa, Coc, Nav; NMEX: McK, SJn; UTAH. 1550–1950 m (5100–6350′). Flowering: Jul–Sep. Native to Europe; scattered locales throughout the southwestern United States and northern Mexico. *Eragrostis minor* Host may occur in the region. It can be distinguished from *E. barrelieri* by having spotty glandular areas below the nodes (not rings or bands), two anthers, and blade margins and pedicels often with glands.

Eragrostis cilianensis (All.) Vignolo ex Janch. (from the Ciliani Estate, Italy) Stinkgrass. Plants tufted annuals. **CULMS** erect or decumbent-based, 15–45 cm tall, sometimes with craterlike glands below the nodes. **LIGULES** 0.4–0.8 mm long. **BLADES** flat to rolled in age or upon drying, mostly 2–6 mm wide, sometimes narrower or wider, mostly glabrous or scaberulous, sometime lightly pilose, usually lacking glands. **PANICLES** oblong to ovate, 5–18 cm long, usually open when mature, the pedicels lacking glandular bands. **SPIKELETS** 10- to 40-flowered, 2–4 mm wide, lead-green in color. **GLUMES** ovate to lanceolate, somewhat unequal, the first 1.2–2 mm long and 1-nerved, the second 1.2–2.6 mm long and 3-nerved. **LEMMAS** ovate, 2–2.8 mm long, the keeled midvein with 1–3 craterlike glands. **PALEAS** shorter than the lemmas. **ANTHERS** 3, 0.2–0.5 mm long, yellow (drying whitish). $2n = 20$. [*E. megastachya* (Koeler) Link]. Disturbed and weedy ground, roadsides, fields, pastures. ARIZ: Apa, Nav; COLO: Dol, LPl; NMEX: McK, RAr, San, SJn; UTAH. 1300–2150 m (4300–7000′). Flowering: Jul–Sep. Native to Eurasia; widespread from southern Canada to northern Mexico. When actively growing, the glands give plants a distinctive odor, hence the common name.

Eragrostis curvula (Schrad.) Nees (curved, weeping) Weeping lovegrass. Plants tufted perennials. **CULMS** erect, robust, 50–150 cm tall. **LIGULES** 0.5–1.5 mm long. **BLADES** flat to rolled, 1–3 mm wide, elongate-attenuate to fine tips. **PANICLES** oblong, lead-greenish, 15–35 cm long, the branches ascending to spreading, pilose in the axils of the primary branches. **SPIKELETS** 4- to 10-flowered, 1–2 mm wide, the rachilla persistent. **GLUMES** lanceolate, 1-nerved, 1.5–3 mm long, the first slightly shorter than the second. **LEMMAS** ovate, 2–3 mm long, the lateral nerves prominent. **PALEAS** about the same size as the lemmas and falling with them. **ANTHERS** 3, 0.6–1.2, reddish brown. $2n$ = 40, 50. Foothills and woodlands, roadsides and reseeded ground. NMEX: SJn. 1750–1950 m (5800–6300′). Flowering: Jul–Aug. Native to South Africa, introduced for forage and rangeland restoration throughout much of the southern and southwestern states.

Eragrostis hypnoides (Lam.) Britton, Sterns & Poggenb. (resembling the moss *Hypnum*) Teal lovegrass. Plants annual, stoloniferous, often forming low mats, lacking glands. **CULMS** decumbent, rooting at the nodes, the erect portion 5–15 cm tall. **LIGULES** 0.3–0.6 mm long. **BLADES** flat to rolled, 1–2 mm wide. **PANICLES** both terminal and axillary, ovate, 1–4 cm long and about as wide. **SPIKELETS** 12- to 35-flowered, 1–1.5 mm wide. **GLUMES** lanceolate, hyaline, 1-nerved, the first 0.4–0.7 mm long, the second about twice as long. **LEMMAS** ovate, 1.4–2 mm long, the nerves greenish and prominent. **PALEAS** shorter than the lemmas. **ANTHERS** 2, tiny, 0.2–0.3 mm long, brownish. $2n$ = 20. Sandy shores of ponds and lakes, not common in our area. ARIZ: Coc; NMEX: McK. 1800–2000 m (5900–6600′). Flowering: Jul–Aug. Eastern Canada and United States; scattered locales in the West.

Eragrostis lutescens Scribn. (yellowish) Six-weeks lovegrass. Plants tufted annuals. **CULMS** erect, 8–25 cm tall, with yellowish glandular pits below the nodes, but lacking bands of glandular tissue. **LIGULES** 0.2–0.5 mm long. **BLADES** flat to rolled, 1–3 mm wide, the bases with glandular pits. **PANICLES** narrowly elliptic, 4–12 cm long, 1–2 cm wide, the branches erect-appressed, the rachises and branchlets with glandular pits. **SPIKELETS** 6- to 12-flowered, 1–2 mm wide, narrowly ovate. **GLUMES** ovate-lanceolate, 1-nerved, subequal, 1–1.8 mm long. **LEMMAS** ovate, 1.5–2.2 mm long, the nerves prominent. **PALEAS** slightly shorter than the lemmas, persistent. **ANTHERS** 3, 0.2–0.3 mm long, purplish. $2n$ = not known. Moist alkaline flats, sandy shores and banks, not common in our area. ARIZ: Nav. 1525–1650 m (5000–5400′). Flowering: Jul–Sep. Scattered locales in the western United States.

Eragrostis mexicana (Hornem.) Link (from Mexico) Mexican lovegrass. Plants tufted annuals, lacking glands. **CULMS** erect to geniculate-based, 12–80 cm tall or more. **LIGULES** 0.2–0.5 mm long. **BLADES** flat, 2–7 mm wide, glabrous to scaberulous or lightly pilose near the base. **PANICLES** ovate, 10–40 cm long, 4–18 cm wide, the branches usually spreading to divaricate. **SPIKELETS** 5- to 12-flowered, 1–2.4 mm wide, ovate-lanceolate, grayish green to purplish green. **GLUMES** ovate-lanceolate, nearly equal, 0.7–2.2 mm long, 1-nerved. **LEMMAS** ovate, 1.2–2.4 mm long, the nerves prominent. **PALEAS** shorter than the lemmas. **ANTHERS** 3, 0.2–0.5 mm long, purplish. **GRAINS** with a prominent groove on the side opposite the embryo. $2n$ = 60. [*Eragrostis neomexicana* Vasey]. Roadsides, moist disturbed sites and fields. ARIZ: Apa, Nav; NMEX: McK, SJn. 1800–2300 m (6000–7500′). Flowering: Jul–Sep. Western United States, south to Argentina. Our material seems to belong to **subsp. *mexicana***, with subsp. *virescens* found in Nevada and California.

Eragrostis pectinacea (Michx.) Nees (comblike) Carolina lovegrass. Plants tufted annuals, lacking glands. **CULMS** erect to geniculate- or decumbent-based, 10–80 cm long. **LIGULES** 0.2–0.5 mm long. **BLADES** flat to rolled, 1–5 mm wide, glabrous to scaberulous. **PANICLES** pyramidal when mature, usually open with spreading branches, 5–25 cm long, 3–12 cm wide, the branchlets and spikelets usually appressed along the primary branches. **SPIKELETS** 6- to 22-flowered, lanceolate, 1–2.5 mm wide. **GLUMES** linear to lanceolate, 1-nerved, shorter than the adjacent lemma, unequal, the first 0.5–1.5 mm long, the second 1–1.7 mm long. **LEMMAS** ovate-lanceolate, 1–2.2 mm long. **PALEAS** slightly shorter than the lemmas. **ANTHERS** 3, 0.2–0.7, purplish. $2n$ = 60. [*Eragrostis diffusa* Buckley]. Widespread in moist weedy sites, roadsides, along ditches. ARIZ: Apa, Coc, Nav; COLO: Arc, LPl, Min, Mon; NMEX: McK, San, SJn; UTAH. 1130–2550 m (3700–8400′). Flowering: Jul–Sep. From southern Canada to Argentina. This is our most common annual lovegrass, easily distinguished from the others by the absence of glands, appressed branchlets, and a nongrooved grain.

Eremopyrum (Ledeb.) Jaub. & Spach Annual Wheatgrass

(Greek *eremos*, desert, + *pyros*, wheat, alluding to the arid environment) Plants tufted annuals, with hollow, geniculate culms. **LEAVES** with open sheaths, membranous ligules, flat blades, and auricles. **INFL** a short spike with 1 spikelet per node; disarticulation in the rachis, but tardily so. **SPIKELETS** very crowded on the rachis, spreading to divaricate, 3- to 6-flowered; glumes linear-lanceolate, awn-tipped or awnless, firm; lemmas lanceolate, awn-tipped; paleas shorter than the lemmas, 2-keeled; anthers 3, yellow. x = 7. A genus of five to eight species native to central Asia and the Mediterranean.

Eremopyrum triticeum (Gaertn.) Nevski (resembling the genus *Triticum*) Annual wheatgrass. **CULMS** erect, 5–30 cm tall. **LIGULES** 0.2–1 mm long, lacerate. **BLADES** 1–4 mm wide. **SPIKES** bristly, 1–2 cm long, ovate, the peduncle extending well beyond the sheath. **SPIKELETS** 2- to 4-flowered. **GLUMES** firm, the margins hyaline, inflated at the base, 1-nerved, 5–7 mm long. **LEMMAS** lanceolate, 5–7 mm long, faintly 3- to 5-nerved, acuminate to a 1–2 mm long awn. **PALEAS** shorter than the lemmas, bifid at the apex. **ANTHERS** 0.8–1 mm long. $2n = 14$. [*Agropyron triticeum* Gaertn.]. Disturbed ground in desert areas. ARIZ: Apa, Nav; COLO: Dol, Mon; NMEX: McK, RAr, SJn; UTAH. 1450–2100 m (4800–6900′). Flowering: Apr–Jul. Native to central Asia; scattered locales in southern Canada and the western United States. Plants look like miniature clumps of crested wheatgrass.

Erioneuron Nash Tridens

(Greek *erio*, woolly, + *neuron*, nerve, alluding to the hairy nerves) Plants perennial, tufted (ours) or occasionally stoloniferous, with pithy, erect culms. **LEAVES** with open sheaths, hairy ligules, folded and white-margined blades, and lacking auricles. **INFL** a terminal, dense, few-flowered panicle raised above the foliage, often racemose. **SPIKELETS** laterally compressed, with 4–20 florets, the terminal florets staminate or sterile; disarticulation above the glumes and between the florets; glumes thin, acute to acuminate, 1-nerved; lemmas 3-nerved and conspicuously pilose on the nerves at least basally, the apex bifid and awned or mucronate from the midnerve; paleas 2-keeled, shorter than the lemmas. **STAMENS** 1 or 3. $x = 8$. An American genus of only three species, formerly including *Dasyochloa* and included in *Tridens*.

Erioneuron pilosum (Buckley) Nash (hairy, pilose) Hairy tridens. Plants tufted perennials. **CULMS** 4–25 cm tall. **LIGULES** 2–3.5 mm long. **BLADES** 1–2 mm wide, glabrous or sparsely pilose. **PANICLES** 1–5 cm long. **SPIKELETS** 6- to 12-flowered. **GLUMES** shorter than the lowermost lemma, equal, 4–7 mm long. **LEMMAS** lanceolate, 3–6 mm long; awns 0.5–2.5 mm long. **PALEAS** 3–4 mm long. **ANTHERS** usually 3, 0.3–1 mm long. $2n = 16$. [*Tridens pilosus* (Buckley) Hitchc.]. Dry, rocky hills and plains, desert and juniper woodland communities. ARIZ: Apa, Nav; COLO: Mon; NMEX: SJn; UTAH. 1450–1800 m (4800–5800′). Flowering: May–Jun. Southwestern United States, California to Kansas, south to central Mexico.

Festuca L. Fescue

(ancient Latin name for grass or straw, from the Celtic *fest*, pasture or food) Plants perennial, mostly densely tufted but sometimes with short rhizomes within the clump, with hollow culms. **LEAVES** with open or closed sheaths, membranous ligules, flat to involute blades, and with or without auricles. **INFL** a terminal panicle. **SPIKELETS** laterally compressed, 3- to 15-flowered, the upper few florets sometimes staminate or sterile; disarticulation above the glumes and between the florets; glumes lance-linear, unequal, 1- or 3-nerved, acute to short-awned; lemmas lanceolate to ovate, 3- to 7-nerved, tapered to an awn or awnless; paleas 2-keeled, about equal to the lemmas; stamens 3. $x = 7$. About 100 species in temperate and cool-temperate regions of the world. Annual species with a single stamen are treated in *Vulpia* (Aiken and Darbyshire 1990; Allred 2005b; Frederiksen 1982).

1. Blades mostly wider than 3 mm, usually at least somewhat lax and flat when fresh ....(2)
1′ Blades mostly less than 3 mm wide, usually rolled or folded and somewhat stiff ....(4)

2. Spikelets 2- to 4-flowered, 8–11 mm long; auricles absent; panicle branches spreading, at least below ***F. sororia***
2′ Spikelets (4) 5- to 9-flowered, 10–17 mm long; small auricles usually developed; panicle branches usually ascending ....(3)

3. Auricles lacking cilia (use a lens); 2 panicle branches borne at the lowermost node, together rarely bearing more than 6 spikelets; old sheaths brown, decaying to fibers; blades 3–6 (7) mm wide .... ***F. pratensis***
3′ Auricles with minute cilia (use a lens); 2 or 3 panicle branches borne at the lowermost node, together usually bearing 5–15 (30) spikelets; old sheaths pale straw-colored, generally remaining intact; blades 3–12 mm wide .... ***F. arundinacea***

4. Ligules 2.5–5 (9) mm long; lemma awns 0–0.3 mm long; nodes usually visible and conspicuous; plants generally more than 50 cm tall .... ***F. thurberi***
4′ Ligules less than 2 mm long; lemma awns usually more than 0.5 mm long, occasionally shorter; nodes often not visible or conspicuous; plant height various ....(5)

5. Anthers 2–4 mm long ....(6)
5′ Anthers 0.4–1.7 mm long, rarely longer ....(8)

6. Basal sheaths reddish and separating into threadlike fibers (the whitish veins) in age; shoots loosely clustered, usually with short rhizomes that break through the old leaf sheaths; sheath margins fused together to near the summit (most easily seen on young tiller shoots) .... ***F. rubra***
6′ Basal sheaths usually not reddish or separating into threadlike fibers; shoots densely clustered, lacking rhizomes; sheath margins fused only in the lowermost portion .... (7)

7. Blades, especially the older ones, strongly laterally compressed, thickened and stiff, 0.5–1 mm wide; awn of lower lemma 0.5–2.5 mm long; ligules 0.1–0.3 mm long .... ***F. trachyphylla***
7′ Blades, even the older ones, at least somewhat terete, not thickened, but threadlike, 0.4–0.6 mm wide; awn of lower lemma 2–7 mm long; ligules 0.3–0.6 mm long .... ***F. idahoensis***

8. Plants 3–10 cm tall .... (9)
8′ Plants over 10 cm tall, usually 15–50 cm tall .... (10)

9. Lemma body 2–3 mm long, with an awn 0.5–1.5 mm long; spikelets with 2, occasionally 3, florets; panicle branches at lowest node usually 2–3; ovary and grain apex pubescent .... ***F. minutiflora***
9′ Lemma body 3–5.5 mm long, with an awn 2–3.6 mm long; spikelets with 3–4 florets, occasionally only 2; panicle branches at lowest node 1; ovary and grain apex glabrous .... ***F. brachyphylla***

10. Basal sheaths reddish and splitting into threadlike fibers (the whitish veins) in age; ovary and grain apex pubescent .... ***F. earlei***
10′ Basal sheaths mostly straw-colored to brownish, not splitting into threadlike fibers in age (occasionally so in *F. brachyphylla*); ovary and grain apex glabrous .... (11)

11. Blades soft, striate from the veins showing, somewhat wrinkled in drying, with little or no sclerenchyma tissue; spikelets and foliage greenish; culms usually less than twice the height of the leaves; anthers 0.5–1.3 mm long; rachilla internodes of middle florets 0.6–0.8 mm long .... ***F. brachyphylla***
11′ Blades stiff, terete or sulcate, not striate or wrinkled, the veins generally not visible because of a buildup of sclerenchyma tissue; spikelets and foliage often glaucous; culms usually twice the height of the leaves or more; anthers 1–1.7 mm long (rarely longer); rachilla internodes of middle florets 0.9–1.1 mm long .... ***F. saximontana***

Festuca arizonica Vasey (from Arizona) Arizona fescue. **CULMS** erect, densely clustered, 35–100 cm tall or more, the bases not purplish. **LIGULES** 0.5–0.2 mm long. **BLADES** yellowish green, 15–40 cm long, 0.3–0.5 mm wide. **PANICLES** 4–20 cm long, the peduncle markedly scabrous, the primary branches usually spreading but sometimes erect. **SPIKELETS** 4- to 8-flowered. **GLUMES** subequal, the first 3–5 mm long and 1-nerved, the second 4.5–7 mm long and 3-nerved. **LEMMAS** 5.5–9 mm long (lowermost lemma); awn of lower lemma 0.4–2.7 mm long. **ANTHERS** 3–3.8 mm long. **GRAINS** 3–3.5 mm long, the apex pubescent. $2n = 42$. [*F. ovina* L. var. *arizonica* (Vasey) Hack. ex Beal]. Pine forests to subalpine slopes and meadows in the mountains, often in open parklands with ponderosa pine. ARIZ: Apa; COLO: Arc, Hin, LPl, Min, Mon; NMEX: McK, SJn. 2350–3450 m (7700–11400′). Flowering: Jul–Aug. Mountains of the southwestern United States.

Festuca arundinacea Schreb. (reedlike) Tall fescue. **CULMS** 30–15 cm tall, the bases not or only slightly purplish. **LIGULES** 0.4–1.2 mm long. **BLADES** flat or folded, 4–12 mm wide; auricles present, with tiny cilia. **PANICLES** 10–35 cm long, with 2–3 panicle branches at the lowermost node, these together bearing 5–15 (30) spikelets. **SPIKELETS** 4- to 8-flowered. **GLUMES** subequal, the first 3–6 mm long, the second 4–7 mm long. **LEMMAS** 6–10 mm long (lowermost lemma); awn 0–2 (4) mm long. **ANTHERS** 2.7–4 mm long. **GRAINS** 2–3 mm long, glabrous at the apex. $2n = 28$, 42, 56, 63, 70. [*Festuca elatior* L.; *Lolium arundinaceum* (Schreb.) Darbyshire; *Schedonorus phoenix* (Scop.) Holub]. Widely planted for improved pasture, erosion control, turf, and habitat restoration and found in a variety of habitats. ARIZ: Apa, Nav; COLO: Arc, Hin, LPl, Mon, SMg; NMEX: McK, SJn; UTAH. 1550–2600 m (5200–8550′). Flowering: May–Aug. Native to Europe, now widespread throughout much of North America. Potentially toxic to grazing animals, sometimes causing summer fescue toxicosis and fescue foot. This and *F. pratensis* are closely related to *Lolium perenne* L., being interfertile, and they are sometimes merged with that genus or placed in the genus *Schedonorus*.

Festuca brachyphylla Schult. & Schult. f. (short-leaved) Shortleaf fescue. **CULMS** densely tufted, 3–12 cm tall, the bases purplish or not. **LIGULES** shorter than 0.5 mm long. **BLADES** 2–6 cm long, 0.3–1 mm wide, soft and somewhat wrinkled upon drying, striate because of the veins showing. **PANICLES** 1–6 cm long. **SPIKELETS** 2- to 4-flowered. **GLUMES** unequal, the first 1.8–3 mm long and 1-nerved, the second 2.3–4.3 mm long and 3-nerved. **LEMMAS** 3.3–4.5 mm long (lowermost lemma); awn 0.7–3.6 mm long. **ANTHERS** 0.4–1.4 mm long. **GRAINS** 1.8–2.5 mm long, the apex glabrous. $2n$ = 28. [*Festuca ovina* L. var. *brachyphylla* of NM reports; *Festuca ovina* L. var. *brevifolia* of NM reports]. Subalpine to alpine slopes, ledges, grassy screes, and cliffs in the mountains. ARIZ: Apa; COLO: Arc, Con, Hin, LPl, Min, Mon, SJn; UTAH. 3050–3850 m (10000–12600′). Flowering: Jun–Aug. Western mountains of the United States, California to New Mexico and Colorado. *Festuca brachyphylla* subsp. *coloradensis* can be confused with *F. saximontana*, but the latter is usually at least somewhat glaucous rather than green, with comparatively stiffer blades and longer anthers, and may grow at lower elevations. Shortleaf fescue also grows with *F. minutiflora*, which is distinguished by having fewer florets, shorter spikelets, shorter awns, and a pubescent rather than glabrous ovary/grain apex. Our material belongs to **subsp.** ***coloradensis*** Fred. (of Colorado).

Festuca earlei Rydb. (for Franklin Sumner Earle, 1856–1929, USDA agrostologist) Earle's fescue. **CULMS** 15–45 cm tall, densely tufted, the bases usually purplish in age as the sheaths split into threadlike segments. **LIGULES** 0.1–1 mm long. **BLADES** usually rolled or folded, 6–12 cm long, 0.5–1 mm wide (1–3 mm wide when flat). **PANICLES** 3–8 cm long, narrow, the branches ascending-erect. **SPIKELETS** 2- to 4-flowered. **GLUMES** slightly unequal, the first 2.2–3 mm long and with 1–3 nerves, the second 2.7–3.7 mm long and 3-nerved. **LEMMAS** 3.2–4.5 mm long (lowermost lemma); awn 0.4–1.6 mm long. **ANTHERS** 0.6–1.4 mm long. **GRAINS** 2.5–3 mm long, the apex pubescent. $2n$ = unknown. Subalpine to alpine outcrops and rocky slopes, talus, and grasslands in spruce forests. 3400–3800 m (11100–12500′). COLO: Con, Hin. Flowering: Jul–Aug. Southwestern mountains. Like *F. rubra*, its sheaths split into threadlike segments, but that species produces rhizomes and has longer anthers and a glabrous ovary apex.

Festuca idahoensis Elmer (from Idaho) Idaho fescue. **CULMS** 30–70 cm tall, densely tufted, the bases usually not very purplish. **LIGULES** 0.3–0.6 mm long. **BLADES** flat or folded, to 35 cm long, 0.4–0.6 mm wide. **PANICLES** 5–20 cm long, the peduncle glabrous or nearly so, not densely scaberulous. **SPIKELETS** 3- to 9-flowered. **GLUMES** unequal, the first 3.5–4.5 mm long and 1–3-nerved, the second 5–8 mm long and 3-nerved. **LEMMAS** 4.5–5.5 mm long (lowermost lemma); awn 2–7 mm long. **ANTHERS** 2.5–4.5 mm long. **GRAINS** 4–5 mm long, the apex glabrous. $2n$ = 28. [*Festuca ovina* L. var. *ingrata* Hack. ex Beal]. Open subalpine grasslands in the mountains, not a common species. COLO: Hin, Min, RGr; NMEX: McK. 2700–3350 m (8850–11000′). Flowering: Jun–Aug. Throughout much of mountainous North America.

Festuca minutiflora Rydb. (tiny-flowered) Small-flowered fescue. **CULMS** 3–10 cm tall, densely tufted, the bases generally not purplish. **LIGULES** 0.1–0.3 mm long. **BLADES** 1–5 cm long, 0.3–0.4 mm wide. **PANICLES** 1–4 cm long, less than 1 cm wide. **SPIKELETS** 2 (3)-flowered. **GLUMES** subequal, 1.5–3.4 mm long. **LEMMAS** 2–3.3 mm long (lowermost lemma); awn 0.5–1.5 mm long. **ANTHERS** 0.8–1.1 mm long. **GRAINS** 2–2.5 mm long, the apex pubescent with a few hairs. $2n$ = 28. Alpine ledges, outcrops, grassy slopes, often on mountain peaks. COLO: Arc, Con, Hin, LPl, Min, Mon, RGr. 3450–3650 m (11400–12000′). Flowering: Jun–Aug. Alpine areas of much of western North America. This species is always very short, with short panicles, few spikelets, and few florets. It can be confused only with *F. brachyphylla*, but that species has larger spikelet parts and a glabrous apex on the ovary and grain.

Festuca pratensis Huds. (of meadows) Meadow fescue. **CULMS** 30–125 cm tall, tufted but not densely so, the bases purplish or not, the sheaths decaying into fibers. **LIGULES** 0.2–0.5 mm long. **BLADES** flat or folded, 10–20 cm long, 3–6 (7) mm wide; auricles present, clawlike, glabrous. **PANICLES** 8–22 cm long, with 1–2 branches at the lowermost node, together rarely bearing more than 6 spikelets. **SPIKELETS** 4- to 8-flowered. **GLUMES** subequal, 2.5–5 mm long. **LEMMAS** 5–8 mm long (lowermost lemma); awn 0–2 mm long. **ANTHERS** 2.5–4 mm long. **GRAINS** 2–3 mm long, the apex glabrous. $2n$ = 28, 42, 56, 63, 70. [*Lolium pratense* (Huds.) Darbysh.; *Schedonorus pratensis* (Huds.) P. Beauv.]. Occasionally planted for improved pasture, erosion control, turf, and habitat restoration and found in a variety of habitats; not common in our region. COLO: LPl, Mon; NMEX: SJn. 1550–2600 m (5200–8550′). Flowering: Jun–Jul. Native to Europe; throughout most of cool-temperate North America. This and *F. arundinacea* are closely related to *Lolium perenne* and form fertile hybrids. The hybrid of *F. pratensis* × *Lolium perenne* has been named × *Festulolium loliaceum* (Huds.) P. Fourn.—it produces a panicle with the upper spikelets sessile and arranged in a spike and the lower spikelets stalked and on branches.

Festuca rubra L. (red) Red fescue. **CULMS** 20–80 cm tall, curved at the base, loosely clustered from short rhizomes, which break through the basal sheaths. **SHEATHS** closed to near the summit, splitting between the veins into threadlike segments in age. **LIGULES** 0.2–0.5 mm long. **BLADES** mostly folded, 1–4 mm wide; auricles absent or poorly developed. **PANICLES** 5–20 cm long, with 1–4 branches at the lowermost node. **SPIKELETS** 4- to 9-flowered. **GLUMES** unequal, the first 1.5–4 mm long, the second 3–5 mm long. **LEMMAS** 5–8 mm long (lowermost lemma); awn 1–4.5 mm long. **ANTHERS** 2.3–4.2 mm long. **GRAINS** 2.3–4.5 mm long, the apex glabrous. $2n$ = 14, 21, 28, 42, 49,

53, 56, 64, 70. Mountain meadows, high-elevation slopes, and outcrops. COLO: Arc, Dol, Hin, LPl, Min, Mon, SJn. 2300–3500 m (7600–11500′). Flowering: Jul–Aug. Widespread in North America, also Eurasia. Young tillers of *F. rubra* have reddish sheaths fused to near the top. The old sheaths of red fescue split between the veins into threadlike segments, a feature also found in *F. earlei* (nearly always), and *F. pratensis* (indistinctly).

Festuca saximontana Rydb. (of rocky mountains) Mountain fescue. **CULMS** 20–50 cm tall (sometimes shorter), often glaucous but sometimes green, densely tufted, the basal sheaths persistent and decaying to fibers. **LIGULES** 0.2–0.5 mm long. **BLADES** 0.3–1 mm wide; auricles poorly developed. **PANICLES** 3–15 cm long. **SPIKELETS** 3- to 5-flowered. **GLUMES** unequal, the first 1.5–3.5 mm long, the second 2.5–4.5 mm long. **LEMMAS** 3–5.5 mm long (lowermost lemma); awn 0.5–2.5 mm long. **ANTHERS** 1–1.7 mm long. **GRAINS** 2.2–2.8 mm long, the apex glabrous. $2n = 42$. [*Festuca ovina* L. var. *rydbergii* St.-Yves; *Festuca ovina* L. subsp. *saximontana* (Rydb.) St.-Yves]. Mountain grasslands and forest clearings. ARIZ: Apa; COLO: Arc, Con, Hin, LPl, Min, Mon, SJn; NMEX: RAr, San; UTAH. 2650–3800 m (8700–12500′). Flowering: Jul–Aug. Widespread in the foothills and mountains of central and western North America.

Festuca sororia Piper (a sister, referring to a related species) Ravine fescue. **CULMS** 40–120 cm long, weak and loosely clustered, the foliage deep green; basal sheaths open, weakly splitting between the veins into threadlike segments. **LIGULES** 0.5–1.5 mm long. **BLADES** flat, 3–8 mm wide; auricles absent. **PANICLES** 15–30 cm long, the branches weak and widely spreading. **SPIKELETS** 2- to 4-flowered. **GLUMES** unequal, the first 3–4 mm long, 1-nerved and narrow, the second 5–6 mm long, 3-nerved and broader. **LEMMAS** 5.5–8.5 mm long (lowermost lemma); awn 0.2–2 mm long. **ANTHERS** 1.8–2.2 mm long. **GRAINS** 4–6 mm long, the apex hairy. $2n = 28$. Damp, shaded ravines and creek bottoms in the mountains. COLO: Arc, Min; NMEX: SJn; UTAH. 2600–3200 m (8500–10500′). Flowering: Jul–Aug. Mountains of the southwestern United States.

Festuca thurberi Vasey (for George Thurber, 1821–1890, botanist with the United States-Mexico boundary survey in the mid-1800s) Thurber's fescue. **CULMS** 45–150 cm tall, in thick dense clumps, the foliage green; basal sheaths not splitting between the veins. **LIGULES** 2.5–8 mm long. **BLADES** flat or more commonly folded, 25–45 cm long, 0.6–1.2 mm wide; auricles absent. **PANICLES** 8–17 cm long, the branches ascending-spreading. **SPIKELETS** 3- to 5-flowered. **GLUMES** unequal, the first 4–5 mm long and 1-nerved, the second 4.5–7 mm long and 1- or 3-nerved. **LEMMAS** 6.5–10 mm long (lowermost lemma); awn 0–0.3 mm long. **ANTHERS** 3–4.5 mm long. **GRAINS** 4–4.5 mm long, the apex hairy. $2n = 28, 42$. Grasslands, meadows, grassy slopes, and openings in coniferous forests in the mountains, to subalpine areas. COLO: Arc, Hin, LPl, Min, Mon, RGr, SJn; NMEX: RAr; UTAH. 2550–3700 m (8400–12200′). Flowering: Jul–Aug. Mountains of the western states. *Festuca thurberi* commonly grows with *F. arizonica*, sometimes completely intermingled within a community, but plants of the latter are easily recognized by their rolled glaucous leaves, short ligules, and short-awned (but the awns evident) florets.

Festuca trachyphylla (Hack.) Krajina (rough-leaved) Hard fescue. **CULMS** 25–65 cm tall, the foliage yellowish green, gray-green, or glaucous, the bases somewhat purplish, the basal sheaths persistent and not splitting between the veins. **LIGULES** 0.1–0.3 mm long. **BLADES** folded, stiff when mature, 0.5–1.2 mm wide; auricles scarcely developed. **PANICLES** 3–12 cm long, with a single branch at the lowermost node, narrow, the branches erect-ascending. **SPIKELETS** 3- to 5-flowered. **GLUMES** unequal, the first 2–4 mm long, the second 3–5.5 mm long. **LEMMAS** 4–5.5 mm long (lowermost lemma); awn 0.5–2.5 mm long. **ANTHERS** 2–3.4 mm long. **GRAIN** 2.5–3.5 mm long, the apex glabrous. $2n = 42$. Introduced for range restoration, erosion control, and mine reclamation in scattered locales. COLO: Arc, Con, SJn; NMEX: RAr, SJn. 1700–2100 m (5500–7000′). Flowering: Jun–Aug. Native to Eurasia, widespread throughout much of cool-temperate North America. When mature, the wide, stiff, folded blades are quite distinctive, but other diagnostic features include long anthers, short awns, and glabrous ovary apices. Old sheaths remain at the base of the tuft in a characteristic grayish clump.

Glyceria R. Br. Mannagrass

(Greek *glykys*, sweet, referring to the seed of some species) Plants marsh or aquatic perennials, usually rhizomatous, with hollow culms. **LEAVES** with closed sheaths, membranous ligules, and lacking auricles. **INFL** a panicle, the branches often weak and drooping. **SPIKELETS** several-flowered, linear to ovate, awnless; disarticulating above the glumes and between the florets; glumes unequal, shorter than the lowermost floret, 1-nerved; lemmas prominently 7- to 9-nerved, the nerves parallel and not converging toward the apex, the apex often erose; palea sometimes exceeding the lemma, 2-nerved; stamens 2 or 3. $x = 10$. About 30 species in the Northern Hemisphere.

1. Spikelets linear, nearly round in cross section, 9–15 (18) mm long, 8- to 12-flowered; lemmas 3.3–4 mm long .... *G. borealis*
1′ Spikelets ovate or oblong, somewhat compressed, 2.5–7 mm long, 3- to 6 (7)-flowered; lemmas 1.5–3 mm long ....(2)

2. Blades narrow, 2–5 mm wide; first glume less than 1 mm long; plants usually much less than 1 m tall....*G. striata*
2′ Blades broad, 5–15 mm wide; first glume more than 1 mm long; plants often 1 m or more tall..........................(3)

3. First glume 1.5 mm or more long; spikelets 4–8 mm long; lower valleys .... *G. grandis*
3′ First glume 0.8–1.2 mm long; spikelets 2.5–5 mm long; subalpine............ *G. elata*

Glyceria borealis (Nash) Batch. (northern) Northern mannagrass. **CULMS** 75–150 cm tall, erect or decumbent-based, loosely clumped together along the rhizome. **LIGULES** 4–12 mm long. **BLADES** flat or folded, 2–5 (7) mm wide. **PANICLES** 16–40 cm long, the branches mostly ascending to appressed, the base enclosed in the sheath. **SPIKELETS** linear, nearly round in cross section, 9–18 mm long, 8- to 12-flowered. **GLUMES** unequal, the first 1.2–2.2 mm long, nerveless, the second 2.2–3.5 mm long, 1-nerved. **LEMMAS** prominently (5) 7- to 9-nerved, erose at the apex. **PALEAS** nearly as long as the lemmas. **ANTHERS** 3, 0.5–0.8 mm long. $2n = 20$. Shallow water at the borders of lakes and ponds and wetlands, midelevations to subalpine. ARIZ: Apa, Nav; COLO: Arc, Hin, LPl, SJn; NMEX: McK, SJn. 2400–2800 m (8000–9200′). Flowering: Jul–Aug. Canada south across the northern United States and in the southwestern states.

Glyceria elata (Nash) M. E. Jones (elevated) Tall mannagrass. **CULMS** erect or decumbent-based, 75–150 cm tall. **LIGULES** truncate, erose, 2–6 mm long. **BLADES** flat, 5–12 mm wide. **PANICLES** open, lax, 15–25 cm long, the branches spreading. **SPIKELETS** 4- to 6-flowered. **GLUMES** 1–1.5 mm long, the first slightly shorter than the second. **LEMMAS** ovate, 1.7–2.2 mm long, prominently (5) 7-nerved. **PALEAS** equal to or slightly longer than the lemmas. **ANTHERS** usually 2, 0.5–0.8 mm long. $2n = 20, 28$. In water of ponds and wet meadows at high elevations, not common. COLO: Arc, Hin. 2850–3500 m (9500–11500′). Flowering: Jul–Aug. Western North America.

Glyceria grandis S. Watson (large) American mannagrass. **CULMS** erect to decumbent-based, 75–150 cm tall or more. **LIGULES** truncate to slightly obtuse, 4–6 mm long. **BLADES** flat or folded, 5–15 mm wide. **PANICLES** open, lax, 15–35 cm long, the branches spreading. **SPIKELETS** 4- to 6-flowered. **GLUMES** 1.5–2.5 mm long, the first slightly shorter than the second. **LEMMAS** ovate, 2–3 mm long, prominently (5) 7-nerved. **PALEAS** equal to or slightly longer than the lemmas. **ANTHERS** (2) 3, 0.5–0.9 mm long. $2n = 20$. Stream, pond, and ditch banks in valleys at lower elevations. COLO: Hin, LPl, Min, Mon; NMEX: SJn. 1650–1900 m (5400–6200′). Flowering: Jun–Aug. Canada and the northern United States.

Glyceria striata (Lam.) Hitchc. (furrowed, striped) Fowl mannagrass. **CULMS** erect to decumbent-based, 30–100 cm tall. **LIGULES** truncate, 1–3 mm long. **BLADES** flat or folded, relatively narrow, 2–5 mm wide. **PANICLES** open, lax, 6–20 cm long. **SPIKELETS** 3- to 6-flowered. **GLUMES** subequal, 0.5–1 mm long. **LEMMAS** ovate, 1.5–2 mm long, prominently 5- to 7-nerved. **PALEAS** equal to the lemmas. **ANTHERS** 2, 0.3–0.6 mm long. $2n = 20, 28$. Marshes and springs in the mountains, middle to high elevations. ARIZ: Apa, Nav; COLO: Arc, Hin, LPl, Min, Mon, RGr, SJn; NMEX: McK, SJn; UTAH. 1650–3050 m (5500–10000′). Flowering: Jul–Aug. Throughout much of temperate North America. This is the most common of our mannagrasses.

Helictotrichon Besser Alpine Oat

(Greek *helicos*, twisted, and *trichos*, hair, referring to the awns) Tufted perennials, with hollow culms. **LEAVES** with open sheaths, membranous ligules, flat to rolled blades, and lacking auricles. **INFL** a narrow, spikelike panicle. **SPIKELETS** 3- to 7-flowered; disarticulation above the glumes and between the florets; glumes large, thick, equal, the first 1-nerved, the second 3-nerved, exceeding the lowermost floret; lemmas 3- to several-nerved, firm below and membranous above, often toothed; awn from the midnerve, diverging at about midlength, stout, twisted; palea shorter than the lemma, 2-keeled. **STAMENS** 3, the anthers large. $x = 7$. A genus of 30 or so species, mostly Eurasian and African, with two species native to North America; sometimes split into two genera (*Avenula* and *Helictotrichon*).

Helictotrichon mortonianum (Scribn.) Henrard (for J. Sterling Morton, 1832–1903, Secretary of Agriculture in late 1880s) Alpine oat. **CULMS** 5–20 tall. **LIGULES** 0.5–1 mm long. **BLADES** filiform, rolled, 1–2 mm wide. **PANICLES** 2–8 cm long, few-flowered. **SPIKELETS** pale or purplish, mostly 2-flowered, the rachilla densely villous and extending beyond the upper floret. **GLUMES** hyaline, at least on the margins, acuminate, 8–11 mm long, the first 1-nerved, the second 3-nerved. **LEMMAS** firm, 3-nerved, 7–10 mm long, cleft into 4 bristlelike teeth at the apex; awn 10–16 mm long, twisted and geniculate, dark brown. **PALEAS** shorter than the lemmas. **ANTHERS** 1.5–2 mm long. $2n$ = 14, 28. Alpine and subalpine meadows. COLO: Arc. 3100–3800 m (10200–12500′). Flowering: Jul–Aug. Rocky Mountains.

Hesperostipa (M. K. Elias) Barkworth Needle-and-thread

(Greek *hesperos*, western, + *Stipa*, alluding to its distribution in western North America) Plants tufted perennials, with hollow culms. **LEAVES** with open sheaths, membranous ligules, usually tightly rolled leaves, and lacking auricles. **INFL** a terminal panicle. **SPIKELETS** with 1 floret, the rachilla not prolonged; disarticulation above the glumes; glumes large, papery or membranous, exceeding the floret; lemmas indurate when mature, fusiform, the margins overlapping, the apex fused into a ciliate crown below the terminal awn; awn twice geniculate, the lower portion twisted; callus sharp, densely hairy; palea equal to the lemma, leathery, pubescent, 2-nerved, the apex indurate and prow-shaped. **STAMENS** 3. x = 11. A North American endemic genus of five species, formerly included in *Stipa*, which is now strictly Eurasian (Barkworth 1993).

1. Terminal segment of awn plumose, with feathery hairs 2–3 mm long .... ***H. neomexicana***
1′ Terminal segment of awn not plumose, any hairs present shorter than 1 mm .... (2)

2. Lemmas 10–18 mm long above the callus; lower ligules rounded to truncate, thick, not cut or torn; margins of lower sheaths often ciliate .... ***H. spartea***
2′ Lemmas 5–11 mm long above the callus; lower ligules usually acute, thin, often cut or torn; margins of lower sheaths mostly glabrous .... ***H. comata***

Hesperostipa comata (Trin. & Rupr.) Barkworth (long-haired, referring to the awns) Needle-and-thread. **CULMS** 20–80 cm tall or more. **LIGULES** 2–7 mm long, acute. **BLADES** 0.5–2 mm wide. **PANICLES** 10–28 cm long, often still in the sheath below. **GLUMES** 16–35 mm long, acuminate, 3- to 5-nerved, the second glume slightly shorter than the first. **LEMMAS** 7–13 mm long (including callus); awns 7–22 cm long, the terminal portion straight or curling. **ANTHERS** 5–6 mm long. $2n$ = 38, 44, 46. [*Stipa comata* Trin. & Rupr.]. Plains, mesas, foothills, mountain slopes, clearings in coniferous forests. ARIZ: Apa, Coc, Nav; COLO: Arc, Dol, Hin, LPl, Mon, SMg; NMEX: McK, RAr, San, SJn; UTAH. 1525–2500 m (5000–8200′). Flowering: May–Jul. Throughout western North America. We have two subspecies, with overlapping distributions, though subspecies *comata* seems to be more common:

1. Terminal portion of the awns 4–12 cm long, sinuous to curled at maturity; lower cauline nodes often concealed by the sheaths; panicles often partially enclosed in the uppermost sheath at maturity .... **subsp. *comata***
1′ Terminal portion of the awns 3–8 cm long, straight; lower cauline nodes usually exposed; panicles usually completely exserted at maturity .... **subsp. *intermedia*** (Scribn. & Tweedy) Barkworth

Hesperostipa neomexicana (Thurb.) Barkworth (of New Mexico) New Mexico feathergrass. **CULMS** 40–100 cm tall. **LIGULES** ciliolate, 1–3 mm long, those of the lower leaves shorter than the upper. **BLADES** 0.5–1 mm wide. **PANICLES** 10–30 cm long, narrow, the lower portion included in the sheath. **GLUMES** 30–60 mm long, acuminate and awn-tipped, 4- to 7-nerved. **LEMMAS** 15–18 mm long (including callus); callus 4–5 mm long; awns 12–22 cm long, pilose, the terminal portion with hairs 1–3 mm long. **ANTHERS** 6–7 mm long. $2n$ = 44. Plains, grassy slopes, mesas, canyon slopes, often on limestone and in well-drained soils, middle elevations. ARIZ: Apa; COLO: LPl, Mon; NMEX: McK, SJn; UTAH. 1750–1950 m (5800–6400′). Flowering: May–Jun. Southwestern United States and northern Mexico.

Hesperostipa spartea (Trin.) Barkworth (a rope or cord, the grass being a source of fiber) Porcupinegrass. **CULMS** 15–45 cm tall, the lower nodes usually crossed by lines of pubescence. **LIGULES** 1–7 mm long, those of the lower blades about 1/2 as long as those of the upper. **BLADES** 1.5–4 mm wide. **PANICLES** 10–25 cm long. **GLUMES** 22–45 mm long, acuminate and awn-tipped. **LEMMAS** 15–25 mm long; callus 3–6 mm long; awn 9–19 cm long, the terminal portion straight. **ANTHERS** 3–4 mm long. $2n$ = 44, 46. Grassy plains and clearings in ponderosa pine forests, not common. NMEX: McK, SJn. 1800–2100 m (5900–6800′). Flowering: Aug. Plains and grasslands of North America.

Hierochloë R. Br. Holygrass, Sweetgrass

(Greek *hieros*, sacred, + *chloë*, grass, alluding to its use in religious festivals) Rhizomatous perennials with hollow culms and fragrant herbage. **LEAVES** with open sheaths, membranous ligules, flat to rolled blades, and lacking auricles. **INFL** an open or narrow panicle. **SPIKELETS** with 3 florets, the lower 2 staminate but large, the upper (middle) floret perfect; disarticulation above the glumes, all florets falling together; glumes equal, about as long as the florets, broad and thin, shiny, 1- to 3-nerved, awnless; lemmas all about equal size, entire to bifid, 3- to 5-nerved, awned or awnless; paleas 2-nerved (staminate florets) or 1- or 3-nerved (perfect floret). **STAMENS** 3 (staminate florets) or 2 (perfect floret). $x = 7$. A small genus of 15 or so species in the cool-temperate regions of the world; recently merged into the genus *Anthoxanthum* by some authors (Weimarck 1971).

Hierochloë odorata (L.) P. Beauv. (fragrant) Northern sweetgrass. **CULMS** erect, 15–60 cm tall. **LIGULES** 2–4 mm long, acute, lacerate. **BLADES** flat, 3–6 mm wide. **PANICLES** pyramidal, 4–10 cm long, about 2/3 as wide. **SPIKELETS** awnless. **GLUMES** ovate, membranous and shiny, 3–6 mm long. **STAMINATE LEMMAS** subequal, 3–5 mm long, ovate, appressed-pubescent, 5- to 7-nerved. **PERFECT LEMMA** 2–4 mm long, lanceolate, pubescent above and glabrous below. **PALEAS** about as long as the lemmas. **ANTHERS** 1.5–2.5 mm long. $2n = 28, 42, 56$. [*Anthoxanthum nitens* (Weber) Y. Schouten & Veldkamp; *Hierochloë hirta* (Schrank) Borbás subsp. *arctica* (J. Presl) G. Weim.]. Wet meadows and stream banks, mostly at high elevations. ARIZ: Apa, Nav; COLO: Arc, LPl, SJn; NMEX: SJn. 2300–2900 m (7600–9600′). Flowering: Jun–Aug. Western United States, circumboreal, Eurasia.

Hordeum L. Barley, Foxtail

(the classical Latin name for barley) Tufted annuals and perennials lacking rhizomes, with hollow culms. **LEAVES** with open sheaths, membranous ligules, flat blades, and with or without auricles. **INFL** a dense spike or spicate raceme, with both pedicelled and sessile spikelets borne on the main axis. **SPIKELETS** 1-flowered, borne in threes: the central spikelet sessile and perfect, and 2 lateral spikelets usually pedicelled and staminate or sterile (fertile in *H. vulgare*); disarticulation in the rachis and all 3 spikelets falling together; glumes usually subulate, awnlike; lemmas faintly 5-nerved, tapering to a long awn; palea shorter than the lemma. **STAMENS** 3. $x = 7$. About 35 species of Eurasia and the New World (Baum and Bailey 1990; von Bothmer et al. 1995).

1. Rachis persistent, not breaking apart when mature; plants annual .... ***H. vulgare***
1′ Rachis breaking apart when mature; plants annual or perennial .... (2)

2. Glumes of the central spikelet with conspicuous ciliate margins; auricles usually well developed, mostly longer than 1 mm .... ***H. murinum***
2′ Glumes of the central spikelet without ciliate margins, at most scabrous; auricles usually lacking or weakly developed and less than 1 mm long .... (3)

3. Glumes of the central spikelet flattened at the base; plants annual or biennial, sometimes short-lived perennial under very favorable circumstances .... ***H. pusillum***
3′ Glumes of the central spikelet setaceous, not flattened; plants perennial .... (4)

4. Glumes 7–20 mm long; awns of the lemmas 5–10 (20) mm long .... ***H. brachyantherum***
4′ Glumes 20–150 mm long; awns of the lemmas 10–70 mm long .... ***H. jubatum***

Hordeum brachyantherum Nevski (short-spiked) Meadow barley. Plants perennial. **CULMS** erect or sometimes ascending-spreading, 30–65 cm tall. **LIGULES** truncate, ciliolate, less than 1 mm long. **BLADES** flat, 2–7 mm wide, auricles lacking. **SPIKES** 2–8 cm long, erect, the rachis disarticulating. **CENTRAL SPIKELETS** sessile; glumes 7–18 mm long, setaceous; lemma awns 5–10 (20) mm long; anthers 0.8–2.5 mm long, yellow to violet. **LATERAL SPIKELETS** pedicelled, rudimentary or staminate; lemma awns to 2–6 mm long. $2n = 28$. [*Critesion brachyantherum* (Nevski) Barkworth & D. R. Dewey]. Moist, grassy slopes and meadows, middle to high elevations. ARIZ: Apa; COLO: Arc, Hin, LPl, Min, RGr, SJn, SMg; NMEX: McK, RAr, San, SJn; UTAH. 1850–3250 m (6100–10600′). Flowering: Jun–Aug. Western North America north of Mexico. Meadow barley hybridizes with *H. jubatum*, producing stabilized hybrid plants referred to *H. jubatum* subsp. *intermedium*; see below.

Hordeum jubatum L. (having a mane) Foxtail barley. Plants perennial, but sometimes flowering the first year, glabrous to velvety pubescent. **CULMS** erect to geniculate-based, (20) 30–60 cm tall. **LIGULES** truncate, 0.5–1 mm long. **BLADES** flat to rolled, 1–4 mm wide, auricles usually lacking or very weakly developed. **SPIKES** 4–10 cm long, nodding, the rachis disarticulating. **CENTRAL SPIKELETS** sessile; glumes 20–80 (150) mm long or more, setaceous, spreading when mature; lemma awns 10–70 mm long; anthers 0.6–1 mm long, yellow. **LATERAL SPIKELETS** pedicelled, staminate or sterile; lemma awns 2–7 mm long. $2n = 14, 28, 42$. [*Critesion jubatum* (L.) Nevski]. Moist meadows, roadsides, weedy ground, low to high elevations. 1520–2700 (5000–9000′). Flowering: Jun–Aug. Across cool-temperate North America. Stabilized hybrids with *H. brachyantherum* are referred to subsp. *intermedium*, distinguished as follows:

1. Glumes of the central spikelets (including awns) 20–30 mm long; lemma of central spikelet (including awn) 20–40 mm long .... **subsp. *intermedium***
1′ Glumes of the central spikelets (including awns) 30 mm or more long; lemma of central spikelet (including awn) 30 mm or more long .... **subsp. *jubatum***

subsp. ***intermedium*** Bowden **CP** (intermediate) [*Hordeum caespitosum* Scribn.]. ARIZ: Apa; COLO: Arc, Hin, LPl, Min, RGr, SMg; NMEX: McK, RAr, San, SJn; UTAH.

subsp. ***jubatum*** ARIZ: Apa, Nav; COLO: Arc, Dol, Hin, LPl, Min, Mon, SJn, SMg; NMEX: McK, RAr, San, SJn; UTAH.

Hordeum murinum L. (mouselike) Mouse barley, wall barley. Plants annual. **CULMS** erect to widely spreading and sometimes prostrate, 10–40 cm tall. **LIGULES** truncate, less than 1 mm long. **BLADES** flat to folded, 2–5 mm wide; auricles well developed, 1–4 mm long. **SPIKES** 3–8 cm long, erect, somewhat flattened, the rachis disarticulating. **CENTRAL SPIKELETS** sessile; glumes compressed, ciliate; lemma awns 2–4 cm long; anthers 0.2–2.8 mm long. **LATERAL SPIKELETS** well developed, pedicelled, staminate; glumes flattened, ciliate; lemma awns 2–5 cm long; anthers 0.4–3.2 mm long. $2n$ = 14, 28, 42. [*Critesion murinum* (L.) Á. Löve]. Disturbed ground, roadsides, fields. Flowering: Apr–Jul. Native to Europe and the Mediterranean region; scattered across the western states, South America. We have two subspecies, of which subsp. *glaucum* is the more common:

1. Anthers of the central spikelet blackish, 0.2–0.6 mm long, those of the lateral spikelets 1.2–1.8 mm long; prolongation of the rachilla of the lateral spikelets stout, orange-brown when mature .... **subsp. *glaucum***

1′ Anthers of the central and lateral spikelet yellowish, 0.9–3.2 mm long; prolongation of the rachilla of the lateral spikelets slender, greenish when mature .... **subsp. *leporinum***

subsp. ***glaucum*** (Steud.) Tzvelev (gray-green appearance) [*Hordeum glaucum* Steud.; *H. stebbinsii* Covas]. ARIZ: Apa, Nav; COLO: Dol, Mon; NMEX: McK, SJn; UTAH.

subsp. ***leporinum*** (Link) Arcang. (hare) Hare barley [*Hordeum leporinum* Link]. UTAH.

Hordeum pusillum Nutt. (little) Little barley. Plants annual or biennial (perennial). **CULMS** erect, 10–40 cm tall, rarely taller, in loose tufts. **LIGULES** truncate, less than 0.6 mm long. **BLADES** flat, 2–4 mm wide, glabrous to hairy, auricles lacking. **SPIKES** 2–7 cm long, erect, somewhat squarish, the rachis disarticulating. **CENTRAL SPIKELETS** sessile; glumes flattened at the base, 9–15 mm long, up to 1 mm wide; lemma awns 5–9 mm long; anthers 0.7–1.5 mm long, yellow. **LATERAL SPIKELETS** pedicelled, usually sterile; glumes different in shape, one setaceous and the other broadly flattened at the base. $2n$ = 14. [*Critesion pusillum* (Nutt.) Á. Löve]. Weedy sites, pastures, roadsides, disturbed grasslands. ARIZ: Apa; COLO: Arc, LPl, Mon; NMEX: McK, RAr, SJn; UTAH. 1550–2700 m (5200–8900′). Flowering: May–Aug. Most of the United States except some western regions, extending into Canada and Mexico in scattered locales.

Hordeum vulgare L. (common) Barley. Plants robust annuals. **CULMS** erect, 60–120 cm tall or more. **LIGULES** 0.5–1.2 mm long. **BLADES** flat, 5–12 mm wide, the auricles well developed and up to 6 mm long. **SPIKES** stout, 6–10 cm long, erect, the rachis persistent and not breaking apart. **CENTRAL** and **LATERAL SPIKELETS** sessile and fertile (lateral spikelets pedicelled and staminate in *distichon* phase); glumes 6–20 mm long, widened below and 3-nerved; lemma awns 6–16 cm long (awnless and 3-lobed at the tip in *trifurcatum* phase); anthers 2–2.5 mm long. $2n$ = 14, 28. [*Hordeum distichon* L.; *Hordeum trifurcatum* (Schltdl.) Wender]. Introduced crop also used for erosion control along highways, adventive along fields and roadsides. NMEX: McK, SJn. 1550–2070 m (5200–6800′). Flowering: May–Jun. Native to Eurasia, now cultivated throughout temperate areas of the world.

Imperata Cirillo Satintail

(for Ferrante Imperato, 1550–1625, Italian apothecary) Rhizomatous perennials with pithy culms, sometimes clumped. **LEAVES** with open sheaths, membranous ligules, flat blades, and lacking auricles. **INFL** a spikelike, plumose panicle with appressed branches. **SPIKELETS** short-pedicelled in groups of 2 or 3, 2-flowered, the lower floret neuter, the upper floret perfect; glumes about equal to the spikelet and upper floret, 3- to 9-nerved, beset with long white hairs; lemmas and paleas hyaline. **STAMENS** 1 or 2. x = 10. About nine species in the warm regions of the world.

Imperata brevifolia Vasey (short-leaved) Satintail. Plants with short scaly rhizomes between clumps of flowering shoots. **CULMS** tufted, erect, 50–150 cm tall. **LIGULES** 0.7–3 mm long. **BLADES** flat, 5–18 mm wide, sometimes densely pilose on the upper surface near the ligule, otherwise glabrous or scaberulous. **PANICLES** 10–30 cm long, 2–3 cm wide, silvery white. **SPIKELETS** awnless, obscured by the long hairs. **GLUMES** 2.5–4 mm long. **LEMMAS** 2.5–3.5 mm long. **ANTHERS** 1.3–2.3 mm long, yellow to orange. $2n$ = 20. Wet sites along streams in sandstone canyons along the San Juan arm of Lake Powell. UTAH. 1130–1150 m (3700–3770′). Flowering: Aug–Sep. Once known in scattered locales across the Desert Southwest, but now nearly absent from former sites.

Koeleria Pers. Junegrass

(for Georg Ludwig Koeler, 1765–1807, German botanist) Tufted annuals and perennials, some rhizomatous, with hollow culms. **LEAVES** with open sheaths, membranous ligules, flat to rolled blades, and lacking auricles. **INFL** a panicle, often spikelike, the rachis and branchlets puberulent. **SPIKELETS** 2- to 4-flowered, laterally compressed, the florets perfect or bearing a terminal vestigial floret; disarticulation above the glumes and between the florets; glumes unequal in shape, the first 1-nerved, the second broader and faintly 3-nerved, nearly reaching to the tip of the lemmas; lemmas 5-nerved, awnless or short-awned from the tip; palea 2-nerved, nearly as long as the lemma, hyaline and colorless, visible. **STAMENS** 3. $x = 7$. About 35 species of temperate and Arctic regions of the Northern Hemisphere. Annual members of the genus are sometimes removed to *Lophochloa* or *Rostraria* (Arnow 1994).

Koeleria macrantha (Ledeb.) Schult. (large-flowered) Junegrass. Tufted perennials with mostly basal foliage. **CULMS** 20–75 cm tall. **LIGULES** 0.5–2 mm long. **BLADES** 0.5–4 mm wide, flat or rolled upon drying, with prow-shaped tips. **PANICLES** 5–25 cm long, normally 1–2 cm wide, but up to 8 cm wide during anthesis (exceptional). **SPIKELETS** 4–5 mm long, mostly 2-flowered. **GLUMES** 3–5 mm long, the second broader toward the tip. **LEMMAS** 3–6 mm long, awn-tipped to an awn 1 mm long. **PALEAS** slightly shorter than the lemmas. **ANTHERS** 1–3 mm long. $2n = 14, 28$. [*Koeleria cristata* of many authors; *Koeleria nitida* Nutt.]. Mountain slopes, foothills, plains, clearings in ponderosa forests, sparingly in subalpine communities. ARIZ: Apa; COLO: Arc, Dol, Hin, LPl, Min, Mon, SJn, SMg; NMEX: McK, RAr, San, SJn; UTAH. 2150–2850 m (7150–9350′). Flowering: Jun–Aug. Widely distributed throughout central and western North America, Eurasia.

Leptochloa P. Beauv. Sprangletop

(Greek *leptos*, slender, + *chloa*, grass) Tufted annuals and perennials, with hollow culms. **LEAVES** distributed up the culms, with open sheaths, membranous ligules, flat to rolled blades, lacking auricles. **INFL** a panicle of unbranched racemose primary branches. **SPIKELETS** 2- to several-flowered, short-pedicelled in 2 rows; disarticulation above the glumes and between the florets; glumes acute, awnless to mucronate, 1- to 3-nerved; lemmas prominently 3-nerved, usually hairy on the nerves, awnless to awned; paleas slightly shorter than the lemmas. **STAMENS** 1–3. $x = 10$. About 32 species in the tropical and temperate regions of the world (Snow 1998).

Leptochloa fusca (L.) Kunth (dark brown) Bearded sprangletop. Loosely tufted annuals. **CULMS** 10–100 cm tall, prostrate to erect. **LIGULES** 2–8 mm long. **BLADES** 2–7 mm wide, flat or rolled. **PANICLES** 5–45 cm long, the base often included in the sheath, the branches ascending to spreading. **SPIKELETS** 5- to 12-flowered. **GLUMES** 1-nerved, the first 2–4 mm long, the second 3.5–5.5 mm long and awn-tipped. **LEMMAS** 3.5–5 mm long, the nerves appressed-hairy, awned from a bifid apex, the awn to 3.5 mm long. **PALEAS** equal to the lemmas, hairy on the nerves. **ANTHERS** 1–3, 0.2–0.5 mm long. $2n = 20$. [*L. fascicularis* (Lam.) A. Gray]. Wet or drying margins of ponds and lakes, irrigated ground, wet disturbed sites. ARIZ: Apa, Nav; COLO: Mon; NMEX: McK, SJn; UTAH. 1450–1800 m (4800–5800′). Flowering: Jul–Sep. Throughout much of the United States, south through Mexico; South America. Our material belongs to **subsp.** ***fascicularis*** (Lam.) N. W. Snow (in a bundle).

Leucopoa Griseb. Spike Fescue

(Greek *leucos*, white, and *poa*, bluegrass) Dioecious, rhizomatous perennials with hollow culms. **LEAVES** with open sheaths, membranous ligules, flat blades, and lacking auricles. **INFL** a narrow or open panicle. **SPIKELETS** mostly unisexual, with rudiments of either stamens or pistils, 3- to 5-flowered, laterally compressed; disarticulation above the glumes and between the florets; glumes lance-ovate, semitransparent, whitish, shorter than the lowermost lemma, the first 1-nerved, the second 3-nerved; lemmas 5-nerved, acute to short-awned; palea subequal to the lemma, 2-keeled. **STAMENS** 3 in the staminate florets. $x = 7$. A small genus of about 10 species, nearly all of which are Asian; sometimes treated in *Festuca* or *Hesperochloa* (Weber 1966).

Leucopoa kingii (S. Watson) W. A. Weber (for Clarence King, 1842–1901, American geologist and mountaineer) Strongly rhizomatous and coarse perennials, forming large rings to 2 m in diameter. **CULMS** stout, 30–100 cm tall, arising from dense clumps of the old sheaths. **LIGULES** 1–4 mm long, erose-ciliate. **BLADES** firm, flat, 2–5 mm wide, striate. **PANICLES** narrow, congested, 5–20 cm long, the branches spikelet-bearing to the base. **SPIKELETS** with mostly 3–4 florets, sometimes fewer or more. **GLUMES** mostly unequal, the first 3–6 mm long, the second 4–7 mm long. **LEMMAS** 4.5–8 mm long, awnless or mucronate, scabrous or short-hirsute. **PALEAS** about equaling the lemmas. **ANTHERS** 3–6 mm long. $2n = 56$. [*Festuca kingii* (S. Watson) Cassidy; *Hesperochloa kingii* (S. Watson) Rydb.]. Rocky slopes and mesas, infrequent. COLO: LPl; NMEX: SJn. 1900–2150 m (6200–7000′). Flowering: Jun–Aug. Western United States.

Leymus Hochst. Wildrye

(anagram of *Elymus*, without meaning) Rhizomatous perennials, with hollow culms. **LEAVES** with open sheaths, membranous ligules, flat to convolute blades, and with or without auricles. **INFL** a spike, with 1–several spikelets per node; disarticulation above the glumes and between the florets. **SPIKELETS** 3- to 7-flowered, laterally compressed to only slightly compressed; glumes mostly subulate and 1- to 3-nerved; lemmas 5- to 7-nerved, awned or awnless; palea nearly equaling the lemma, 2-keeled. **STAMENS** 3. x = 7. About 30 species of temperate regions of North and South America and Eurasia (Barkworth and Atkins 1984).

1. Plants in giant clumps to 2 m or more tall, usually much taller than 100 cm; blades flat, 5–15 mm wide; spikelets usually 3–6 per node .... ***L. cinereus***
1′ Plants mostly shorter than 100 cm (*L. triticoides* to 120 cm tall); blades flat to involute, 2.5–5 mm wide; spikelets 1–2 per node .... (2)

2. Plants strongly rhizomatous, not bunch-forming .... ***L. triticoides***
2′ Plants tufted or with short rhizomes but still bunch-forming .... ***L. salina***

Leymus cinereus (Scribn. & Merr.) Á. Löve (ash-colored) Great Basin wildrye. Robust tussocky perennials, often forming clumps to 1 m across, typically without rhizomes. **CULMS** mostly 1–2 m tall or even more, sometimes as short as 70 cm. **LIGULES** 2–7 mm long. **BLADES** mostly flat, 5–15 mm wide, the auricles strongly developed to nearly absent. **SPIKES** 10–20 cm long, 1–3 cm wide, stiffly erect, with 3–6 spikelets per node. **SPIKELETS** 3- to 5-flowered. **GLUMES** nearly subulate, 7–14 mm long, nearly as long as the spikelet. **LEMMAS** lanceolate, 5- to 7-nerved, 8–10 mm long, awnless or short-awned to 5 mm. **ANTHERS** 4–6 mm long. $2n$ = 28, 56. [*Elymus cinereus* Scribn. & Merr.]. Canyon bottoms, open grassy slopes and plains, openings in woodlands, roadsides. ARIZ: Nav; COLO: LPl, Mon, SJn; NMEX: SJn; UTAH. 1650–2050 m (5400–6700′). Flowering: Jun–Jul. Western and central North America.

Leymus salina (M. E. Jones) Á. Löve (from Salina Pass, Utah) Salina wildrye. Densely tufted perennials with short rhizomes. **CULMS** 35–70 cm tall, sometimes taller. **LIGULES** 0.2–0.7 mm long. **BLADES** flat to rolled, 2–4 mm wide, the auricles well developed and clasping the culm. **SPIKES** 5–12 cm long, mostly less than 1 cm wide, erect, slender, with 1–2 spikelets per node. **SPIKELETS** 2- to 4-flowered. **GLUMES** subulate, 4–8 mm long, about 1/2 to nearly as long as the spikelet. **LEMMAS** lanceolate, 5-nerved, 8–10 mm long, awnless or short-awned to 2 mm. **ANTHERS** 4–8 mm long. $2n$ = 28, 56. [*Elymus salina* M. E. Jones]. Dry, sandy ground in desert scrub and woodland communities, clay hills. ARIZ: Nav; COLO: Dol, LPl, Mon; NMEX: SJn; UTAH. 1500–2300 m (4900–7600′). Flowering: Jun–Jul. Intermountain regions in the western United States.

Leymus triticoides (Buckley) Pilg. (resembling the genus *Triticum*) Creeping wildrye. Glaucous to green perennials from strongly creeping rhizomes. **CULMS** 30–70 cm tall, rarely up to 1 m or more tall, nearly single or few-clumped. **LIGULES** 0.2–0.7 mm long. **BLADES** flat to rolled, 2–5 mm wide, the auricles well developed and clasping the culm. **SPIKES** 3–8 cm long, 1–2 cm wide, erect, with 1–2 spikelets per node. **SPIKELETS** 3- to 6-flowered. **GLUMES** narrowly lanceolate to subulate, 6–10 mm long, with 1–3 nerves, 1/2 to 3/4 as long as the spikelet. **LEMMAS** lanceolate, 5- to 7-nerved, 6–9 mm long, often shiny and smooth, awnless or short-awned to 6 mm. **ANTHERS** 0.5–2.5 mm long. $2n$ = 28. [*Elymus simplex* Scribn. & T. A. Williams; *Elymus triticoides* Buckley]. Benches, clay flats, desert scrub communities. ARIZ: Apa, Nav; COLO: Dol, SMg. 1300–1650 m (4400–5400′). Flowering: May–Aug. Western United States and northern Mexico.

Lolium L. Ryegrass

(the ancient Latin name, referred to by Virgil as a troublesome weed) Annuals or perennials with hollow culms. **LEAVES** with open sheaths, membranous ligules, flat or folded blades, and well-developed auricles. **INFL** a spike, without branches. **SPIKELETS** solitary at the nodes, attached edgewise to the rachis, several-flowered; disarticulation above the glumes and between the florets; glumes solitary, the first glume absent except on the terminal spikelet, present only on the side away from the rachis, equaling or exceeding the lowermost floret, 5- to 9-nerved, awnless; lemmas 5-nerved, awned or awnless; palea as long as the lemma, 2-keeled. **STAMENS** 3. x = 7. A small genus of about five species, native to Europe, Asia, and North Africa. Very closely related to the flat-leaved species of *Festuca* (*Schedonorus*), with which they hybridize (Soreng and Terrell 1997).

Lolium perenne L. (perennial) Perennial ryegrass. Plants tufted perennials, sometimes annual, glabrous. **CULMS** erect, 25–80 cm or more tall. **LIGULES** 0.5–1.5 mm long. **BLADES** 2–8 mm wide, the auricles to 2 mm long. **SPIKES** 5–25 cm long, the rachis sometimes flexuous. **SPIKELETS** 4- to 16-flowered. **GLUMES** 4–15 mm long, nearly as long as the lowermost floret. **LEMMAS** 4–9 mm long, awnless or awned to about 12 mm. **ANTHERS** 2–4 mm long. $2n$ = 14.

[*Lolium multiflorum* Lam.; *Lolium perenne* L. subsp. *italicum* (A. Braun) Syme]. Moist, disturbed areas, roadsides, flower beds, escaped from lawn and pasture plantings. ARIZ: Apa, Nav; COLO: Arc, Dol, Hin, LPl, Mon; NMEX: RAr, SJn; UTAH. 1350–2750 m (4500–9000′). Flowering: Jun–Aug. Native to Europe, now widely introduced throughout North America. We have two intergrading races: long-awned plants with 8–16 florets and rolled leaves in young shoots may be assigned to var. *aristatum* Willd.; awnless plants with 3–10 florets and folded leaves in young shoots may be assigned to var. *perenne*.

Lycurus Kunth Wolftail

(Greek *lycos*, wolf, and *oura*, tail, referring to the bristly inflorescence) Tufted perennials with puberulent, pithy culms. **LEAVES** with open sheaths, membranous ligules, flat or folded blades, and lacking auricles. **INFL** bristly, spikelike panicles with highly shortened branches. **SPIKELETS** 1-flowered, borne in pairs which fall as a unit, usually 1 of the pair staminate or sterile; glumes subequal, the lower 2-awned, the upper 1-awned; lemmas 3-nerved, hairy on the nerves, tapering to an awn; palea about as long as the lemma, 2-keeled. **STAMENS** 3. $x = 10$. A small genus of three species native to amphitropical North America. Our species can only be distinguished by vegetative means.

1. Upper leaves acute or with a mucro or bristle 1–3 mm long; ligules 1.5–3 mm long, with evident narrow triangular lobes 1.5–3 mm long on the sides; culms erect to ascending, often geniculate ....***L. phleoides***

1′ Upper leaves terminating in a fragile, awnlike tip 4–12 mm long; ligules 3–12 mm long, elongate, acute or acuminate, sometimes with a small cleft on either side; culms erect ....***L. setosus***

Lycurus phleoides Kunth (resembling the genus *Phleum*) Wolftail. **CULMS** 20–50 cm long, often geniculate-based. **LIGULES** 1.5–3 mm, commonly acute to acuminate, with narrow triangular lobes 1.5–3 (4) mm long extending from the sides of the sheaths. **BLADES** 4–8 cm long, 1–1.5 mm wide, acute or mucronate, the apex sometimes extending up to 3 mm as a short bristle. **PANICLES** 4–10 cm long, less than 1 cm wide. **GLUMES** 1–2 mm long, the first 2-nerved, the awns 1–3 mm long. **LEMMAS** 3–4 mm long, the awn 1–3 mm long. **ANTHERS** 1.5–2 mm long, yellow. $2n = 40$. Gravelly or rocky hills and slopes. ARIZ: Apa, Coc, Nav; NMEX: McK. 1650–2250 m (5400–7300′). Flowering: Jun–Aug. Southwestern United States and Mexico, northern South America.

Lycurus setosus (Nutt.) C. Reeder (bristly) Bristly wolftail. **CULMS** 25–50 cm long, erect. **LIGULES** 3–12 mm, hyaline, acuminate, sometimes shortly cleft on the sides. **BLADES** 4–10 cm long, 1–2 mm wide, glabrous, the apex fragile, easily broken, awnlike, 3–12 mm long. **PANICLES** 4–10 cm long, less than 1 cm wide. **GLUMES** 1–2 mm long, the first 2-nerved, the awns 1–3 mm long. **LEMMAS** 3–4 mm long, the awn 1–3 mm long. **ANTHERS** 1.5–2 mm long, yellow. $2n = 40$. [*Lycurus phleoides* Kunth var. *glaucifolius* Beal]. Slickrock community with stunted ponderosa pine. NMEX: McK. 2125–2300 m (7000–7500′). Flowering: Aug–Sep. Southwestern United States and northern Mexico, Bolivia, Argentina. This may be more aptly treated as a variety of *Lycurus phleoides*.

Melica L. Melica

(Greek *meli*, honey) Tufted or rhizomatous perennials, with hollow culms, often from bulbous bases. **LEAVES** with closed sheaths, membranous ligules, flat to rolled blades, and lacking auricles. **INFL** a terminal panicle, sometimes weakly developed and racemose; disarticulation below or above the glumes. **SPIKELETS** with 1–6 fertile florets, terminated by 1–4 sterile florets clustered together (the rudiment); glumes papery, 1- to 7-nerved; lemmas membranous, 7- to 15-nerved, awnless or 1-awned; palea 1/2 to as long as the lemma, 2-keeled. **STAMENS** 3. $x = 9$. About 80 species in the temperate regions of the world (except Australia).

Melica porteri Scribn. (for Thomas Conrad Porter, 1822–1901, American botanist, poet, and classicist) Porter's melica. Plants with short rhizomes, sometimes loosely tufted. **CULMS** 50–85 cm tall (sometimes taller), lacking corms basally. **LIGULES** 1–7 mm long, lacerate. **BLADES** flat, lax, 2–5 mm wide. **PANICLES** 12–25 cm long, the branches erect and with 1–2 spikelets; disarticulation below the glumes, sometimes the florets falling out also. **SPIKELETS** on sharply reflexed pedicels, 2- to 5-flowered, 8–16 mm long. **GLUMES** unequal, the first 3–6 mm long, the second 5–8 mm long. **LEMMAS** 6–10 mm long, glabrous, 5- to 11-nerved, awnless. **PALEAS** about 2/3 the length of the lemmas. **ANTHERS** 1–2.5 mm long. **TERMINAL RUDIMENT** 2–5 mm long. $2n = 18$. Moist, rocky slopes and talus. COLO: Arc, Hin, LPl, Min, Mon, SJn. 2300–2900 m (7500–9500′). Flowering: Aug–Sep. Southwestern United States and northern Mexico.

Muhlenbergia Schreb. Muhly

(for Gotthilf Heinrich Ernst Muhlenberg, 1753–1815, Lutheran pastor and pioneering botanist of Pennsylvania) Delicate annuals to robust perennials, rhizomatous or tufted, with hollow or pithy culms. **LEAVES** with open sheaths, membranous ligules, flat to rolled blades, and lacking auricles. **INFL** a terminal and sometimes axillary panicle, diverse; disarticulation mostly above the glumes. **SPIKELETS** 1-flowered (rarely with 2–3 florets); glumes mostly 1- or 2-nerved, much shorter to longer than the floret; lemmas 3-nerved, awnless or awned from the apex; paleas shorter than or equaling the lemma. **STAMENS** 3 (sometimes fewer). $x = 10$. About 155 species, mostly in the New World. (Peterson and Annable 1991.)

1. Plants annual ....(2)
1′ Plants perennial ....(8)

2. First glume prominently 2-nerved, usually cleft; panicle branches falling as a unit, bearing 2–3 (4) spikelets ......(3)
2′ First glume 1-nerved; panicle branches persistent ....(4)

3. Glumes about 1/2 the length of the floret; spikelets 3.5–6 mm long; lemma awns 10–20 mm long.......... ***M. brevis***
3′ Glumes and floret about equal in length; spikelets 2.5–3.5 mm long; lemma awns 0.5–5 (10) mm long .... ***M. depauperata***

4. Lemma awns 10–30 mm long .... ***M. tenuifolia***
4′ Lemma awns 0–5 mm long ....(5)

5. Panicle narrow, contracted, the branches appressed to the main axis; culms decumbent-based and often rooting at the nodes .... ***M. filiformis***
5′ Panicle open, the branches spreading; culms erect to somewhat decumbent, rarely rooting ....(6)

6. Pedicels 0.3–1 mm long, stout, of equal thickness throughout; blades lacking thickened white margins .... ***M. ramulosa***
6′ Pedicels 2–8 mm long, capillary, narrowed downward; blades with or without thickened white margins ....(7)

7. Glumes minutely pubescent to long-pubescent, at least at the apex (use a lens); blade margins somewhat white-thickened, but this often obscure .... ***M. minutissima***
7′ Glumes glabrous; blade margins prominently white-thickened .... ***M. fragilis***

8. Second glume evidently 3-nerved, often 3-toothed; lower sheaths flattened, ribbonlike ....(9)
8′ Second glume 1-nerved, entire or fringed; lower sheaths usually not ribbonlike ....(10)

9. Ligules 2–5 mm long; lemma awns 1–5 mm long; stems and blades very slender and narrow; plants usually 15–30 cm tall .... ***M. filiculmis***
9′ Ligules 10–20 mm long and the tip often shredded; lemma awns 6–25 mm long; stems and blades more robust; plants 25–80 cm tall .... ***M. montana***

10. Stems stiff, wiry, much-branched, the plants twiggy .... ***M. porteri***
10′ Stems not as above, the plants not twiggy ....(11)

11. Plants with evident, slender, creeping rhizomes ....(12)
11′ Plants tufted, or sometimes the bases decumbent and spreading, but lacking creeping rhizomes ....(20)

12. Callus hairs very copious, as long as the body of the lemma .... ***M. andina***
12′ Callus hairs long-pubescent to glabrous, but the hairs much shorter than the body of the lemma ....(13)

13. Panicles open, loosely flowered with usually spreading to divergent branches at maturity ....(14)
13′ Panicles contracted, narrow and usually densely flowered, the branches mostly erect to appressed....................(15)

14. Awns 1–1.5 (2) mm long; panicle branches attached in clusters; blades markedly stiff and pungent ... ***M. pungens***
14′ Awns 0–0.3 mm long; panicle branches not clustered; blades lax or somewhat stiff .... ***M. asperifolia***

15. Blades 3–6 mm wide, flat, dark green .... ***M. racemosa***
15′ Blades 0.5–2 mm wide, flat to rolled, often grayish ....(16)

16. Lemmas short-pilose on the lower 1/3 to 3/4, the hairs 0.5–1.5 mm long ....(17)
16′ Lemmas glabrous or scaberulous only ....(19)

17. Lemma awns 0–1 mm long; blades narrowed at the base where they join the sheath .... ***M. thurberi***
17′ Lemma awns 1–20 mm long; blades not narrowed at the base ....(18)

18. Ligules 1–2 mm long, with lateral lobes about 1 mm or so long; lemma awns 4–20 mm long; blades commonly rolled and mostly ascending........ ***M. arsenei***
18′ Ligules 0.2–0.6 mm long, lacking lateral lobes; lemma awns 1–6 mm long; many blades flat and stiffly diverging at right angles .... ***M. curtifolia***

19. Inflorescence usually included in the sheath at least below, with 9 nodes or fewer; ligules 0.5–1.5 mm long; lemmas mostly 3–4 mm long........ ***M. repens***
19′ Inflorescence usually well exserted from the sheath, with 11–12 nodes; ligules 1–3 mm long; lemmas 2–2.8 mm long........ ***M. richardsonis***

20. Sheaths (at least the lower) compressed-keeled; blades flat or folded .... ***M. wrightii***
20′ Sheaths rounded on the back; blades usually becoming rolled........ (21)

21. Lemma awns 0–4 (5) mm long........ (22)
21′ Lemma awns 6–40 mm long........ (25)

22. Glumes, excluding the awn, 3/4 or more the length of the floret .... ***M. rigens***
22′ Glumes, excluding the awn, 2/3 or less the length of the floret........ (23)

23. Mature panicles narrow, densely flowered, 0.5–2 cm wide, the primary branches erect to appressed .. ***M. wrightii***
23′ Mature panicles open, loosely flowered, 4–15 cm wide, at least the primary branches widely spreading .... (24)

24. Blades strongly arcuate, curving, less than 1 mm wide, 1–3 (4) cm long; leafy portion 1/16 to 1/8 the length of the plant; lateral pedicels commonly longer than the spikelets........ ***M. torreyi***
24′ Blades rather straight, 1–2 mm wide, 3–15 cm long; leafy portion 1/3 to 1/2 the length of the plant; lateral pedicels commonly shorter than the spikelets .... ***M. arenicola***

25. Awns 7–10 mm long........ (26)
25′ Awns 10–40 mm long........ (27)

26. Blades mostly 1–4 (5) cm long; glumes acute; lemmas and paleas sparsely but noticeably short-pilose on the lower 1/2; lateral lobes of ligules less than 1.5 mm long .... ***M. arsenei***
26′ Blades mostly 4–14 cm long; glumes acuminate to aristate; lemmas and paleas glabrous or minutely scaberulous; lateral lobes of ligules 1.5–3 mm long........ ***M. pauciflora***

27. Lemmas essentially glabrous, with only a few closely appressed callus hairs; ligules with lateral lobes 1.5–3 mm long .... ***M. pauciflora***
27′ Lemmas pubescent on the lower 1/2; ligules without lateral lobes .... ***M. tenuifolia***

Muhlenbergia andina (Nutt.) Hitchc. (of the Andes or other high mountains) Foxtail muhly. Rhizomatous perennials. **CULMS** 25–70 cm tall, erect. **LIGULES** 0.5–1.5 mm long, truncate, lacerate. **BLADES** flat, 2–4 mm wide, pubescent above. **PANICLES** contracted, densely flowered, plumose, 2–15 cm long, less than 3 cm wide. **SPIKELETS** 2–4 mm long. **GLUMES** as long as or longer than the florets, 1-nerved, narrow. **LEMMAS** 2–3.5 mm long, lanceolate, copiously long-hairy, the hairs as long as the floret; awns 1–10 mm long. **PALEAS** equaling the lemmas. **ANTHERS** 0.4–1.5 mm long, yellow. $2n = 20$. Wet meadows, marshes, gravel bars in the mountains. ARIZ: Apa, Nav; COLO: Arc, Hin, LPl, Min, Mon, SJn; NMEX: SJn; UTAH. 1500–2500 m (5000–8000′). Flowering: Jul–Aug. Western United States.

Muhlenbergia arenicola Buckley (sand-loving) Sand muhly. Tufted perennials. **CULMS** 20–60 cm tall, slightly decumbent-based, otherwise erect. **LIGULES** 2–9 mm long, lacerate. **BLADES** 1–2 mm wide, not arcuate, flat to rolled or folded, not white-margined. **PANICLES** 12–25 cm long, open, to 18 cm wide, the lateral pedicels commonly shorter than the spikelet. **SPIKELETS** 2.5–4 mm long. **GLUMES** 1.5–2.5 mm long, 1-nerved, acute to acuminate, awn-tipped. **LEMMAS** 2.5–4 mm long, often purplish, appressed-puberulent on the lower 1/2; awn 0.4–4 mm long. **PALEAS** slightly shorter than the lemmas, with 2 tiny awns. **ANTHERS** 1.5–2 mm long, greenish. $2n = 80, 82$. Sandy slopes and mesas in grassland regions. NMEX: San. 2000–2150 m (6800–7100′). Flowering: Jul–Sep. Southwestern United States to central Mexico, Argentina.

Muhlenbergia arsenei Hitchc. (for Brother Gustave Arsène, 1867–1938, French clergyman-botanist who lived in Santa Fe, New Mexico) Arsène's muhly. Rhizomatous perennials, the rhizomes sometimes short and the plants tufted. **CULMS** 15–50 cm tall, decumbent-based. **LIGULES** 1–2 mm long, with lateral lobes, the lobes less than 1.5 mm longer than the central portion. **BLADES** 1–2 mm wide, flat to rolled. **PANICLES** 4–12 cm long, 1–3 cm wide, narrow. **SPIKELETS** 3.5–5 mm long (excluding awn). **GLUMES** subequal, 2–4 mm long, 1-nerved, acuminate to awn-tipped. **LEMMAS** 3.5–5 mm long, purplish with green nerves, hairy on the lower 1/2 or less, the hairs to 1.5 mm long; awn 4–16 mm long. **PALEAS** equaling the lemmas. **ANTHERS** 1.3–3 mm long, purplish. $2n$ = unknown. Rocky outcrops and washes of limestone and travertine parent material; piñon-juniper woodland and ponderosa pine forest communities. ARIZ: Apa, Nav; NMEX: McK. 1900–2200 m (6300–7300′). Flowering: Aug–Sep. Southwestern United States, northern Mexico.

Muhlenbergia asperifolia (Nees & Meyen ex Trin.) Parodi (rough-leaved) Scratchgrass. Rhizomatous, bushy perennials, sometimes stoloniferous. **CULMS** 20–60 cm long, erect to decumbent-spreading. **LIGULES** 0.2–1 mm long. **BLADES** 1–3 mm wide, flat. **PANICLES** ovoid, 6–20 cm long, 4–15 cm wide. **SPIKELETS** 1–2 mm long. **GLUMES** equal, 0.6–1.7 mm long, 1-nerved, purplish. **LEMMAS** lanceolate, 1.2–2.1 mm long, acute to mucronate, the mucro to 0.3 mm long. **PALEAS** equaling the lemmas. **ANTHERS** 1–1.3 mm long. $2n$ = 20, 22, 28. Wet meadows, around seeps and springs, stream banks. ARIZ: Apa, Nav; COLO: Arc, Dol, LPl, Mon; NMEX: McK, RAr, SJn; UTAH. 1600–2400 m (5200–7900′). Flowering: Jul–Aug. Throughout western North America.

Muhlenbergia brevis C.O. Goodd. (short) Short muhly. Tufted annuals. **CULMS** 3–20 cm tall, erect, congested. **LIGULES** 1–3 mm long. **BLADES** 0.8–2 mm wide, the margins whitish-thickened. **PANICLES** 3–10 cm long, narrow, less than 2 cm wide. **SPIKELETS** 3–6 mm long, in pairs; disarticulation below the spikelet pair. **GLUMES** unequal, about 1/2 as long as the lemma, the first 2–3.5 mm long and 2-nerved, bifid with teeth or awns to 2 mm long, the second 2.4–4 mm long, 1-nerved, awned to 2 mm. **LEMMAS** 3.5–6 mm long, appressed-pubescent; awn stiff, 10–20 mm long. **PALEAS** equaling the lemmas. **ANTHERS** 0.5–1 mm long. $2n$ = 20. Rocky outcrops and barrens amid grasslands and woodlands. ARIZ: Apa. 1650–1900 m (5400–6300′). Flowering: Jul–Aug. Southwestern United States to central Mexico.

Muhlenbergia curtifolia Scribn. (short-leaved) Utah muhly. Rhizomatous perennials, the foliage glabrous to hirtellous. **CULMS** 5–45 cm tall. **LIGULES** 0.2–0.6 mm long, lacking lateral lobes. **BLADES** mostly 1–3 cm long, 1–2 mm wide, flat at the base, rolled or folded above, conspicuously diverging at right angles or reflexed. **PANICLES** 2–8 cm long, to 1 cm wide. **SPIKELETS** 3–4 mm long. **GLUMES** equal, 1-nerved, 2.5–4 mm long. **LEMMAS** lanceolate, 2.8–4 mm long, short-pilose on the lower 3/4, the hairs 1–1.5 mm long; awn 1–6 (12) mm long. **PALEAS** equaling and pubescent like the lemmas. **ANTHERS** 1.2–1.6 mm long, yellow-purple. $2n$ = unknown. Along moist vertical wall joints and fractures, and talus slopes among sandstone boulders and cliffs. ARIZ: Apa, Nav; UTAH. 1120–2450 m (3700–8000′). Flowering: Jul–Sep. Southwestern United States. This has been incorrectly merged with the sometimes sympatric *M. thurberi* by some authors, but that species can be distinguished by its ascending, rolled blades and awns less than 1 mm long.

Muhlenbergia depauperata Scribn. (impoverished) Six-weeks muhly. Tufted, short-lived annuals. **CULMS** 3–15 cm tall. **LIGULES** 1.5–2.5 mm long, with lateral lobes. **BLADES** 1–3 cm long, 0.5–1.5 mm wide, flat or rolled, the margins whitish-thickened. **PANICLES** 2–8 cm long, narrow, less than 1 cm wide. **SPIKELETS** 2.5–5 mm long, in pairs; disarticulation below the spikelet pair. **GLUMES** unequal, as long as or longer than the floret, the first 2.3–4 mm long and 2-nerved, bifid with teeth to 1.3 mm long, the second 3–5 mm long and 1-nerved, acuminate. **LEMMAS** 2.5–4.5 mm long, appressed-pubescent; awn stiff, 6–15 mm long. **PALEAS** slightly shorter than the lemmas. **ANTHERS** 0.4–0.8 mm long, purplish to yellowish. $2n$ = 20. Sandy banks, gravelly rock outcrops and flats, usually on calcareous or volcanic soils. ARIZ: Apa, Coc, Nav; COLO: Mon; NMEX: McK, SJn; UTAH. 1500–1900 m (5000–6200′). Flowering: Aug–Sep. Southwestern United States to southern Mexico.

Muhlenbergia filiculmis Vasey (threadlike stems) Slimstem muhly. Tufted perennials with erect foliage and flattened sheaths. **CULMS** 5–30 cm tall. **LIGULES** 2–5 mm long. **BLADES** 0.4–1.6 mm wide, tightly rolled, stiff, with pungent tips. **PANICLES** 2–7 cm long, narrow, less than 1 cm wide. **SPIKELETS** 2.2–3.5 mm long. **GLUMES** equal, 1–2.5 mm long, the first 1-nerved and awned to 1.6 mm long, the second 3-nerved and 3-toothed. **LEMMAS** 2.2–3.5 mm long, yellowish and mottled with green, sparsely appressed-pubescent on the lower portion, the hairs less than 0.5 mm long; awn 1–5 mm long. **PALEAS** as long as the lemmas. **ANTHERS** 1.5–2 mm long, purplish. Forest clearings and grasslands in the mountains. NMEX: SJn; UTAH. 2500–3500 m (8200–11500′). Flowering: Aug–Sep. Southern Rocky Mountains.

Muhlenbergia filiformis (Thurb. ex S. Watson) Rydb. (threadlike) Pull-up muhly. Tufted annuals, the bases sometimes clumped and appearing perennial. **CULMS** 5–25 cm long, erect to commonly geniculate-based and rooting at the nodes. **LIGULES** 1–3.5 mm long. **BLADES** flat to rolled, 0.6–1.6 mm wide. **PANICLES** spikelike, 2–6 cm long, less than

1 cm wide, exserted from the sheath. **SPIKELETS** 1.5–3.2 mm long. **GLUMES** 1-nerved, rounded to obtuse, subequal, 0.6–1.7 mm long. **LEMMAS** 1.5–3 mm long, appressed-hairy, the hairs less than 0.3 mm long, awnless to mucronate. **PALEAS** equaling the lemmas. **ANTHERS** 0.5–1.2 mm long, purplish. 2*n* = 18. Wet meadows, shores of ponds and springs, in the mountains. ARIZ: Apa; COLO: Min, SJn; NMEX: McK, SJn. 2200–2600 m (7200–8600′). Flowering: Jun–Aug. Western United States to northern Mexico.

Muhlenbergia fragilis Swallen (brittle) Delicate muhly. Tufted annuals. **CULMS** 10–30 cm tall, mostly erect. **LIGULES** 1–3 mm long, with lateral pointed lobes. **BLADES** flat, 0.5–2 mm wide, the margins whitish-thickened. **PANICLES** open, diffuse, 10–22 cm long, 4–10 cm wide. **SPIKELETS** about 1 mm long, on delicate straight pedicels. **GLUMES** 0.5–1 mm long, 1-nerved, glabrous or minutely and obscurely puberulent, obtuse. **LEMMAS** 1–1.2 mm long, glabrous to puberulent, awnless. **PALEAS** equaling the lemmas. **ANTHERS** 0.3–0.5 mm long, purplish. 2*n* = 20. Moist, sandy ground and rocky clearings, from desert to pine forest communities. ARIZ: Nav; NMEX: McK; UTAH. 1950–2200 m (6400–7200′). Flowering: Jul–Sep. Southwestern United States to southern Mexico.

Muhlenbergia minutissima (Steud.) Swallen (very tiny) Least muhly. Tufted annuals. **CULMS** erect, 5–40 cm tall. **LIGULES** 1–2.5 mm long, with or without lateral lobes. **BLADES** flat to rolled, 1–2 mm wide, the margins not or obscurely whitish-thickened. **PANICLES** 5–18 cm long, 2–6 cm wide, open. **SPIKELETS** 0.8–1.5 mm long, on capillary pedicels. **GLUMES** 1-nerved, sparsely hairy at least near the apex, subequal, 0.5–0.9 mm long, obtuse. **LEMMAS** 0.8–1.5 mm long, glabrous or minutely hairy, awnless. **PALEAS** as long as the lemmas. **ANTHERS** 0.2–0.7 mm long, purplish. 2*n* = 60, 80. Moist, sandy or gravelly sites, roadsides, open ground. ARIZ: Apa, Coc, Nav; COLO: Arc, Hin, LPl, Min, Mon, SJn; NMEX: McK, RAr, SJn. 1900–2700 m (6200–8800′). Flowering: Jul–Sep. Western United States to southern Mexico.

Muhlenbergia montana (Nutt.) Hitchc. (of mountains) Mountain muhly. Tufted perennials. **CULMS** 20–80 cm tall, with curling foliage and flattened sheaths. **LIGULES** (4) 10–20 mm long, acuminate, the tip often shredded. **BLADES** flat to rolled, 1–2.5 mm wide. **PANICLES** 5–22 cm long, narrow but loose, 2–5 cm wide. **SPIKELETS** 3–7 mm long. **GLUMES** 2–4 mm long, subequal, the first 1-nerved and sometimes awn-tipped, the second 3-nerved and 3-toothed to 1.5 mm long. **LEMMAS** 3–5 mm long, loosely short-pilose, the hairs less than 1 mm long; awn 5–25 mm long, flexuous. **PALEAS** as long as the lemmas. **ANTHERS** 1.5–2.3 mm long, purplish. 2*n* = 20, 40. Grassy parklands and slopes in the mountains, often in pine forests, to subalpine communities. ARIZ: Apa; COLO: Arc, Hin, Min, Mon, SJn, SMg; NMEX: McK, SJn; UTAH. 2200–2800 m (7350–9100′). Flowering: Jul–Sep. Western United States south to Guatemala.

Muhlenbergia pauciflora Buckley (few-flowered) New Mexico muhly. Tufted perennials, sometimes shortly rhizomatous, the foliage sometimes with brownish necrotic spots. **CULMS** 30–70 cm tall, geniculate-based and sometimes rooting at the lower nodes. **LIGULES** 1–3 mm long, with lateral lobes 1.5–3 mm long. **BLADES** 0.5–1.5 mm wide, flat or rolled. **PANICLES** 4–15 cm long, 1–3 cm wide, narrow but loosely flowered. **SPIKELETS** 3.5–5.5 mm long. **GLUMES** equal, 1.5–3.5 mm long, 1-nerved, acuminate to awn-tipped to 2 mm. **LEMMAS** 3–5.5 mm long, glabrous or with sparse short hairs only at the base; awn 6–25 mm long, flexuous. **PALEAS** as long as the lemmas. **ANTHERS** 1.5–2.1 mm long, yellowish or purplish. 2*n* = unknown. [*M. neomexicana* Vasey]. Rocky slopes and hills, outcrops, and canyons in the forests. ARIZ: Apa; COLO: LPl; NMEX: McK, SJn; UTAH. 1850–2400 m (6000–7800′). Flowering: Aug–Sep. Southwestern United States to central Mexico.

Muhlenbergia porteri Scribn. ex Beal (for Thomas Conrad Porter, 1822–1901, American botanist, poet, and classicist) Loosely tufted perennials from knotty, twiggy bases, not rhizomatous, forming rounded bushy growths. **CULMS** 25–100 cm tall, twiggy, branching at nearly all the nodes, the branches stiff and divergent. **LIGULES** 1–3 mm long, lacerate. **BLADES** 2–8 cm long, 0.5–2 mm wide, flat or folded. **PANICLES** ovoid, purplish, 4–15 cm long and about as wide. **SPIKELETS** 3–4.5 mm long, on pedicels 3–15 mm long. **GLUMES** 2–3 mm long, equal, shorter than the lemma, acute to mucronate. **LEMMAS** 3–4.5 mm long, appressed-pubescent on the lower 1/2 or so; awn 2–12 mm long, straight. **PALEAS** as long as the lemmas. **ANTHERS** 1.5–2.3 mm long, yellowish to purplish. 2*n* = 20, 23, 24, 40. Desert plains and mesas, often among shrubs and boulders. ARIZ: Apa, Coc, Nav; NMEX: McK, SJn; UTAH. 1650–1900 m (4800–6200′). Flowering: Jul–Sep. Southwestern United States to northern Mexico. Bush muhly is highly prized by livestock and will disappear from heavily grazed rangelands.

Muhlenbergia pungens Thurb. ex A. Gray (sharp, referring to the blades) Sandhill muhly. Rhizomatous perennials, forming circular clumps. **CULMS** 15–70 cm tall, decumbent-based, hairy below the nodes. **LIGULES** 0.2–1 mm long, densely ciliolate. **BLADES** 2–8 cm long, 1–2 mm wide, flat to tightly rolled, stiff and pungent-tipped. **PANICLES** 8–16 cm long, 4–12 cm wide, open when mature, appearing fascicled when young. **SPIKELETS** 2.5–4.4 mm long. **GLUMES** 1.2–3 mm long, equal, 1-nerved, awn-tipped to 1 mm. **LEMMAS** 2.5–4.5 mm long, purplish, scaberulous only, not noticeably hairy; awn 1–2 mm long, straight. **PALEAS** as long as the lemmas, 2-awned to 2 mm. **ANTHERS** 1.8–2.6 mm long, purplish. $2n$ = 26, 42, 60. Sandy soils and dunes in desert to piñon-juniper woodland communities. ARIZ: Apa, Coc, Nav; COLO: LPl; NMEX: McK, RAr, San, SJn; UTAH. 1350–2250 m (4500–7400′). Flowering: Jul–Sep. Central and western United States. Blades are often stiff and sharp enough to puncture the skin as one reaches into a clump.

Muhlenbergia racemosa (Michx.) Britton, Sterns & Poggenb. (racemelike) Green muhly. Rhizomatous perennials. **CULMS** stiffly erect, 30–110 cm tall, branched above the middle, internodes strongly keeled. **LIGULES** 0.6–1.5 mm long, truncate, ciliolate. **BLADES** flat, 3–6 mm wide, some a bit narrower, dark green. **PANICLES** 2–16 cm long, 0.5–2 cm wide, narrow and densely flowered, slightly bristly. **SPIKELETS** 3–8 mm long, nearly sessile. **GLUMES** 1-nerved, equal, 3–8 mm long (including the awn), 1–2 times as long as the lemma, acuminate to an awn to 5 mm long. **LEMMAS** 2.2–3.8 mm long, pilose on the lower 1/2, the hairs to 1 mm long, unawned or awned to 1 mm. **PALEAS** about as long as the lemmas. **ANTHERS** 0.4–0.8 mm long, yellowish. $2n$ = 40. Meadows, along ditches and roadsides. ARIZ: Apa; COLO: Arc; NMEX: McK, SJn; UTAH. 1800–2250 m (5900–7300′). Flowering: Jul–Sep. Central Canada and United States, to western states and northern Mexico. The flat, green blades and profuse branching are distinctive.

Muhlenbergia ramulosa (Kunth) Swallen (branched) Red muhly. Tufted annuals. **CULMS** 3–20 cm tall, with erect branches from the base. **LIGULES** 0.2–0.5 mm long. **BLADES** flat to rolled, 0.8–1.2 mm wide, the margins not whitish-thickened. **PANICLES** open, ovoid, 2–9 cm long, 1–3 cm wide. **SPIKELETS** 0.8–1.3 mm long, on stout thick pedicels. **GLUMES** equal, 0.4–0.7 mm long, shorter than the floret, glabrous, awnless. **LEMMAS** 0.8–1.3 mm long, glabrous to sparsely puberulent, awnless. **PALEAS** slightly shorter than the lemmas. **ANTHERS** 0.2–0.3 mm long, purplish. $2n$ = 20. [*Muhlenbergia wolfii* (Vasey) Rydb.]. Open, sandy meadows, edges of ponds and lakes, washes, and roadcuts, pine forests. COLO: LPl, Mon, SJn; NMEX: SJn. 2250–2400 m (7400–7900′). Flowering: Aug–Sep. Southwestern United States, Mexico, Guatemala, Costa Rica, Argentina.

Muhlenbergia repens (J. Presl) Hitchc. (creeping) Creeping muhly. Rhizomatous perennials, forming low, nearly turflike, mats. **CULMS** 5–30 cm tall, decumbent-based. **LIGULES** 0.1–1.5 mm long, truncate. **BLADES** 0.5–5 cm long, 0.5–1.5 mm wide, rolled. **PANICLES** few-flowered, narrow, 1–9 cm long, less than 0.6 cm wide, the base often included in the subtending sheath. **SPIKELETS** 2.6–4.2 mm long, occasionally 2-flowered. **GLUMES** subequal, 1–3.5 mm long, 1/2 to as long as the lemma, mostly 1-nerved, acute, unawned. **LEMMAS** (2.8) 3–4 mm long, glabrous to appressed-pubescent with tiny hairs, acuminate to mucronate. **PALEAS** shorter than the lemmas. **ANTHERS** 0.7–1.4 mm long, yellowish to purplish. $2n$ = 60, 70, 71, 72. Moist plains, flats, swales, and roadsides. ARIZ: Apa; NMEX: McK, SJn; UTAH. 1750–2500 m (5800–8200′). Flowering: Jul–Sep. Southwestern United States to southern Mexico.

Muhlenbergia richardsonis

Muhlenbergia richardsonis (Trin.) Rydb. (for Sir John Richardson, 1787–1865, Scottish naturalist) Mat muhly. Rhizomatous perennials, mat-forming. **CULMS** 10–30 cm tall, decumbent-based. **LIGULES** 0.8–3 mm long, erose. **BLADES** 1–6 cm long, 0.5–3 mm wide, flat to rolled. **PANICLES** few-flowered, narrow, 1–15 cm long, less than 2 cm wide, usually exserted from the sheath. **SPIKELETS** 1.7–3.1 mm long, occasionally 2-flowered. **GLUMES**

subequal, 0.6–2 mm long, 1/3 to 1/2 as long as the lemma, mostly 1-nerved, acute to mucronate. **LEMMAS** 1.7–2.8 mm long, glabrous, acuminate to mucronate to 0.5 mm. **PALEAS** about as long as the lemmas. **ANTHERS** 0.9–1.6 mm long, yellowish to purplish. 2*n* = 40. Mountain meadows and cienegas, shores of ponds, alkaline flats, foothills to the mountains. ARIZ: Apa; COLO: LPl, Mon, SJn; NMEX: McK, SJn. 2300–2700 m (7600–8800′). Flowering: Jul–Sep. Throughout western Canada and United States.

Muhlenbergia rigens (Benth.) Hitchc. (stiff, rigid) Deergrass. Tufted, tussocky perennials, lacking rhizomes. **CULMS** erect, 40–150 cm tall. **LIGULES** 0.5–3 mm long, truncate. **BLADES** stiff, flat or rolled, 15–50 cm long, 1.5–5 mm wide. **PANICLES** spikelike, 15–50 cm long, 0.5–1 cm wide. **SPIKELETS** 2.5–4 mm long. **GLUMES** 1.8–3.2 mm long, nearly as long as the florets, 1-nerved, acute to mucronate to 0.5 mm. **LEMMAS** 2.4–4 mm long, short-hairy only at the base, awnless. **PALEAS** only slightly shorter than the lemmas. **ANTHERS** 1.3–1.8 mm long, yellowish to purplish. 2*n* = 40. [*Muhlenbergia mundula* I. M. Johnst.]. Rocky ravines and washes, along woodland streams. UTAH. 1800–1950 m (5900–6400′). Flowering: Jul–Sep. Southwestern United States to southern Mexico.

Muhlenbergia tenuifolia (Kunth) Trin. (thin-leaved) Mesa muhly. Tufted annuals or short-lived perennials, lacking rhizomes. **CULMS** 20–60 cm tall, erect or decumbent-based. **LIGULES** 1–4 mm long, acute and lacerate. **BLADES** flat or loosely rolled, 1.2–2.5 mm wide. **PANICLES** terminal and axillary, several per shoot, mostly 5–10 cm long, narrow, 0.5–2 cm wide, often nodding. **SPIKELETS** 2–4 mm long. **GLUMES** 1.2–2.8 mm long, shorter than the florets, 1-nerved, awnless to awn-tipped. **LEMMAS** 2–4 mm long, hairy on the lower 1/2; awns 10–30 mm long, sinuous. **PALEAS** slightly shorter than the lemmas. **ANTHERS** 0.9–1.5 mm long, yellowish. 2*n* = 20, 40. [*M. monticola* Buckley]. Rocky outcrops and ledges, canyon walls. ARIZ: Apa; COLO: Mon; NMEX: McK, SJn. Flowering: Jul–Sep. Southwestern United States, south through Mexico to northern South America.

Muhlenbergia thurberi (Scribn.) Rydb. (for George Thurber, 1821–1890, botanist with the United States-Mexico boundary survey in the mid-1800s) Thurber's muhly. Green to glaucous perennials, with slender creeping rhizomes. **CULMS** wiry, 10–40 cm tall, clumped together, decumbent-based. **LIGULES** 1–1.2 mm long. **BLADES** tightly involute, 1–4 cm long, 0.2–1 mm wide, the margins not whitish-thickened. **PANICLES** narrow, 1–6 cm long, less than 1 cm wide. **SPIKELETS** 2.6–4 mm long. **GLUMES** subequal, 1.6–3 mm long, awnless. **LEMMAS** 2.6–4 mm long, pubescent on the lower 3/4, awnless or awn-tipped to 1 mm. **PALEAS** as long as the lemmas. **ANTHERS** 2.1–2.3 mm long, yellowish purple. 2*n* = unknown. Seeps and springs, often at the base of cliffs and ledges, juniper to ponderosa pine zones, imperfectly known from the *Flora* region. NMEX: San. 2050–2200 m (6800–7200′). Flowering: Aug–Sep. Southwestern United States. This has been incorrectly expanded by some authors to include the sometimes sympatric *M. curtifolia*, but that species can be distinguished by its widely spreading flat blades and awns 1–12 mm long.

Muhlenbergia torreyi (Kunth) Hitchc. ex Bush (for John Torrey, 1796–1873, American physician-botanist) Ring muhly. Tufted, ring-forming perennials, lacking creeping rhizomes. **CULMS** 10–35 cm tall (occasionally taller), with decumbent bases. **LIGULES** 2–5 mm long, acuminate. **BLADES** 1–4 cm long, less than 1 mm wide, tightly rolled, arcuate, the margins not whitish-thickened. **PANICLES** diffuse, 7–20 cm long, 3–15 cm wide. **SPIKELETS** 2–3.5 mm long. **GLUMES** equal, 1-nerved, 1.3–2.5 mm long, acuminate, awn-tipped to 1 mm. **LEMMAS** 2.3–3.5 mm long, hairy on the lower 1/2–3/4; awn 0.5–4 mm long. **PALEAS** as long as the lemmas. **ANTHERS** 1.2–2.1 mm long, greenish. 2*n* = 20. Desert and woodland grasslands, mesas, and plains. ARIZ: Apa; NMEX: McK, San, SJn. 1350–1700 m (4500–5600′). Flowering: Jul–Sep. Southwestern United States, Argentina.

Muhlenbergia wrightii Vasey ex J. M. Coult. (for Charles Wright, 1811–1885, American botanical collector) Spike muhly. Tufted perennials, lacking rhizomes. **CULMS** erect, 15–50 cm tall, compressed, as are the sheaths. **LIGULES** 1–4 mm long, truncate. **BLADES** flat to folded, 1–3 mm wide. **PANICLES** spikelike, interrupted below, 5–15 cm long, 0.3–1.2 cm wide. **SPIKELETS** 2–3 mm long, dark. **GLUMES** equal, 1-nerved, 0.5–1.6 mm long, less than 3/4 the length of the lemma, acuminate and awn-tipped. **LEMMAS** 2–3 mm long, hairy on the lower 1/2–3/4; awn 0.3–1 mm long. **PALEAS** as long as the lemmas. **ANTHERS** 1.3–1.8 mm long, greenish. 2*n* = unknown. Grassy, rocky hills and slopes in woodlands and forests. ARIZ: Apa; COLO: Arc, Dol, Hin, LPl, Min, Mon; NMEX: McK, RAr, SJn; UTAH. 2000–2750 m (6600–9000′). Flowering: Aug–Sep. Southwestern United States to northern Mexico.

Munroa Torr. False Buffalograss

(for William Munro, 1818–1880, British botanist and general in the Indian colonial army) Stoloniferous, mat-forming annuals with hollow culms. **LEAVES** clustered at the nodes, with open sheaths, hairy ligules, stiff pungent blades, and lacking auricles. **INFL** a dense cluster of spikelets almost hidden in the fascicle of leaves at the branch tips.

SPIKELETS 3- to 5-flowered, disarticulating above the glumes and between the florets; glumes 1-nerved, narrow, shorter than the lowermost floret; lemmas 3-nerved, lanceolate, mucronate to short-awned, hairy on the nerves; palea shorter than the lemma, 2-keeled. **STAMENS** 3. $x = 7, 8$. A small genus of five species in the New World. The genus name was originally spelled "Monroa," but this is an obvious orthographic error that need not be perpetuated.

Munroa squarrosa (Nutt.) Torr. (curled back) False buffalograss. **CULMS** 3–15 cm long. **LIGULES** 0.5–1 mm long. **BLADES** flat to folded, white-margined, 1–5 cm long, 1–2.5 mm wide. **SPIKELETS** 6–8 mm long, 3- to 5-flowered. **GLUMES** of lower 1–2 spikelets subequal, 2.5–4 mm long, those of upper spikelets unequal, the first reduced or absent. **LEMMAS** 3.5–5 mm long, with a tuft of hairs at the base; awn 0.5–2 mm long. **ANTHERS** 1–1.5 mm long. $2n = 16$. [*Munroa squarrosa* (Nutt.) Torr. var. *floccuosa* Vasey ex Beal]. Sandy plains and flats, disturbed ground, desert plains, and foothills. ARIZ: Apa, Coc, Nav; COLO: Arc, LPl, Mon; NMEX: McK, RAr, San, SJn; UTAH. 1250–2200 m (4200–7200′). Flowering: Jun–Sep. Western Canada and United States, to northern Mexico. Some plants are beset with a white-woolly covering of unsubstantiated origin: leavings of a species of woolly aphid, or water-soluble crystals that mimic hairs.

Nassella É. Desv. Needlegrass

(diminutive of the Latin *nassa*, a wicker basket with a narrow neck, referring to the shape of the lemma) Tufted perennials with hollow culms. **LEAVES** with open sheaths, membranous ligules, flat or folded blades, and lacking auricles. **INFL** a panicle or rebranching branches. **SPIKELETS** 1-flowered, the rachilla not extended; glumes usually longer than the floret, (1) 3- to 5-nerved, acuminate, often purplish-based; lemmas leathery, convolute, the margins overlapping, the apex often solid and necklike; awn twisted or not, persistent to deciduous; palea to 1/3 the length of the lemma, glabrous, without nerves. **STAMENS** 1 or 3, when 3 then of 2 different lengths. $x = 11$. A New World genus of about 70 species, most of which are found in Argentina, Bolivia, and northern Chile; formerly included in the genus *Stipa*, which is now strictly Eurasian (Barkworth 1993).

Nassella viridula (Trin.) Barkworth (somewhat green) Green needlegrass. Tufted perennials with green foliage. **CULMS** 45–100 cm or more tall. **SHEATHS** with a line of hairs on the margins. **LIGULES** 1–3 mm long. **BLADES** flat to rolled, 2–6 mm wide. **PANICLES** narrow, contracted, 10–25 cm long, mostly less than 2 cm wide. **GLUMES** subequal, translucent-membranous, 8–13 mm long, with 3 bright green nerves. **LEMMAS** 5–6.5 mm long, brownish, appressed-pubescent below the neck, with a tuft of hairs at the callus; awn 20–35 mm long, twice-geniculate, twisted. **PALEAS** membranous, rounded at the apex. **ANTHERS** 2–3 mm long or vestigial. $2n = 82, 88$. Roadsides, edges of grassy clearings in woodlands and forests, foothills and mountains. ARIZ: Apa; COLO: Arc, Dol, Min, Mon, SJn. 2000–2250 m (6600–7300′). Flowering: Jul–Sep. Central and Rocky Mountain United States and Canada. Confused with *Achnatherum robustum*, but that species lacks the bright green nerves on the glumes, and the lemma margins are scarcely overlapping if at all.

Oryzopsis Michx. Ricegrass

(Greek *oryza*, rice, and *opsis*, similar to, alluding to a supposed similarity to rice) Following Barkworth (1993), *Oryzopsis* is treated here as a monotypic genus endemic to North America. Species formerly found in *Oryzopsis* are herein treated in *Achnatherum* and *Piptatherum* (Barkworth 1993).

Oryzopsis asperifolia Michx. (rough-leaved) Mountain ricegrass. Tufted, glabrous perennials. **CULMS** hollow, erect to spreading, 25–65 cm tall. **SHEATHS** open. **LIGULES** membranous, 0.2–1 mm long, truncate or rounded. **BASAL BLADES** flat, 25–60 cm long, 3–10 mm wide, the bases twisted so the lower (abaxial) surface is uppermost. **UPPER BLADES** rudimentary. **INFL** a contracted panicle, 4–13 cm long. **SPIKELETS** 1-flowered, disarticulating above the glumes and the floret falling out. **GLUMES** subequal, 6- to 10-nerved, 5–7.5 mm long. **LEMMAS** terete to laterally compressed, 3- to 5-nerved or more, the margins strongly overlapping, leathery or indurate when mature, appressed-pubescent, 5–7 mm long; awn 7–15 mm long, deciduous. **PALEAS** similar to the lemmas, concealed, 2-nerved. **ANTHERS** 3, 2–4 mm long. $x = 11, 12$; $n = 46$. Forests, under aspen and pines. COLO: Arc, Hin, LPl, Min; NMEX: RAr. 1775–2700 m (5800–8800′). Flowering: Jul–Aug. Across Canada and the northern United States.

Panicum L. Panicum, Panic-grass

(Latin *panus*, an ear of millet, or *panis*, bread) Annuals and perennials with hollow or pithy culms. **LEAVES** with open sheaths, membranous and usually ciliate ligules, flat to rolled blades, and lacking auricles; basal leaves not forming a winter rosette. **INFL** terminal and axillary panicles, sterile branchlets and bristles absent, disarticulating mostly below the glumes. **SPIKELETS** 2-flowered, dorsally compressed, awnless, the lower floret reduced, the upper floret perfect;

glumes unequal, the first much smaller than the second to almost obsolete; lower floret staminate or sterile, the lemma membranous and similar in size and texture to the second glume; upper floret perfect, the lemma and palea forming a hard seedcase around the flower and grain. **STAMENS** 3. $x = 9, 10$. A large genus of 300 to 400 species of tropical to warm-temperate regions of the world. Members of subgenus *Dichanthelium* are herein treated as a separate genus.

1. Plants annual ....(2)
1' Plants perennial ....(6)

2. Lemma of the upper floret wrinkled; spikelets nearly sessile on simple or nearly simple primary branches .... see ***Urochloa***
2' Lemma of the upper floret smooth, not wrinkled; spikelets pedicelled in a usually open, freely rebranched panicle.(3)

3. First glume about 1/4 as long as the spikelet, obtuse or rounded at the tip; stems as much as 1 m long, coarse and often somewhat trailing .... ***P. dichotomiflorum***
3' First glume more than 1/4 as long as the spikelet, acute to acuminate at the tip; stems various ....(4)

4. Spikelets 4.5–5 mm long; panicle nodding at maturity .... ***P. miliaceum***
4' Spikelets less than 4 mm long; panicle usually not nodding....................(5)

5. Mature panicles less than 1/2 the length of the entire plant; panicle axils glabrous .... ***P. hirticaule***
5' Mature panicles more than 1/2 the length of the entire plant; panicle axils pubescent .... ***P. capillare***

6. First glume about as long as the second; primary panicle branches mostly unbranched; long stolons developed .... ***P. obtusum***
6' First glume shorter than the second; primary panicle branches often rebranched; stolons not developed ....(7)

7. Culms with bulbous, cormlike bases; sheaths keeled; fertile florets dull, transversely rugose.................... ***P. bulbosum***
7' Culms lacking bulbous, cormlike bases; sheaths not keeled; fertile florets shiny, not rugose ....(8)

8. Plants rhizomatous; blades not curling; lower florets staminate.................... ***P. virgatum***
8' Plants tufted, lacking rhizomes; blades usually curling in age; lower florets sterile .... ***P. hallii***

Panicum bulbosum Kunth (bulbous) Bulb panicum. Tufted perennials with short rhizomes. **CULMS** erect or geniculate-based, 50–100 cm tall or more, with bulbous corms at the base. **LIGULES** 0.5–2 mm long, ciliate. **BLADES** flat, 2–12 mm wide. **PANICLES** open, pyramidal, 10–35 cm long, 2–12 cm wide. **SPIKELETS** 2.8–4.5 mm long. **FIRST GLUME** 3- to 5-nerved, 1/2 to nearly as long as the spikelet. **SECOND GLUME** 5- to 7-nerved, often longer than the fertile floret. **LOWER FLORET** sterile or staminate, glabrous. **UPPER FLORET** 3–4 mm long, dull, finely transversely rugose. $2n = 36, 54, 70, 72$. Moist sites in canyons and rocky hills, often with ponderosa pine. ARIZ: Apa; COLO: LPl; NMEX: McK. 1900–2300 m (6300–7600′). Flowering: Jul–Sep. Southwestern United States to central Mexico.

Panicum capillare L. (hairlike) Common witchgrass. Tufted, stiff-hairy annuals, the hairs bulbous-based. **CULMS** 15–70 cm tall, erect, often much-branched. **LIGULES** 1–2 mm long, ciliate. **BLADES** flat, 3–15 mm wide. **PANICLES** diffuse, oblong, 12–40 cm long, about 1/2 as wide, the base often included in the sheath, breaking at the base of the peduncle and rolling as a tumbleweed. **SPIKELETS** 2–4 mm long. **FIRST GLUME** 1- to 3-nerved, 1/3–1/2 as long as the spikelet. **SECOND GLUME** 7- to 9-nerved, 1.8–3.1 mm long. **LOWER FLORET** sterile, exceeding the fertile floret. **UPPER FLORET** shiny, 1.5–2 mm long. $2n = 18$. Roadsides, moist disturbed ground, deserts and valleys. ARIZ: Apa, Nav; COLO: Arc, Dol, LPl, Min, Mon, SJn; NMEX: McK, RAr, San, SJn; UTAH. 1300–2550 m (4300–8300′). Flowering: Jun–Sep. Throughout temperate North America, South America, naturalized in much of Eurasia.

Panicum dichotomiflorum Michx. (flower divided in two) Fall panicum. Tufted, sprawling annual (perennial in tropical regions). **CULMS** 10–100 cm or more tall, erect to decumbent- or geniculate-based, rooting at the lower nodes in wet soil, often zigzag and compressed, branched below. **LIGULES** 0.5–2 mm long. **BLADES** flat, 3–20 mm wide. **PANICLES** terminal and axillary, open, 5–30 cm long or more. **SPIKELETS** 1.8–3.8 mm long, narrowly elliptic. **FIRST GLUME** mostly 3-nerved, 0.6–1.2 mm long, less than 1/3 as long as the spikelet. **SECOND GLUME** 7- to 9-nerved, as long as the spikelet. **LOWER FLORET** sterile, 7- to 9-nerved, with or without a palea. **UPPER FLORET** 1.4–2.5 mm long, narrowly elliptic, smooth and shiny. $2n = 36, 54$. Moist to wet disturbed ground, roadsides, ditches. ARIZ: Apa; NMEX: SJn; UTAH. 1300–1700 m (4400–5500′). Flowering: Aug–Sep. Native to the eastern United States and Canada, introduced westward; Eurasia.

Panicum hallii Vasey (for Elihu Hall, 1822–1882, botanical collector from Illinois) Hall's panicum. Tufted perennials. **CULMS** 12–50 cm tall, erect, scarcely branched. **LIGULES** 0.6–2 mm long. **BLADES** flat, 2–10 mm wide, curling in age (conspicuous in winter), glaucous, with thickened margins. **PANICLES** open, 10–25 cm long, 4–12 cm wide. **SPIKELETS** 2.1–4.2 mm long. **FIRST GLUME** 1/2–3/4 as long as the spikelet. **SECOND GLUME** 7- to 11-nerved, as long as the spikelet. **LOWER FLORET** sterile, similar to the second glume, with a palea. **UPPER FLORET** 1.5–2.4 mm long, smooth, shiny, elliptic. $2n = 18$. Roadsides, grassy hills, deserts, and woodlands. ARIZ: Apa. 1350–1550 m (4400–5100′). Flowering: Aug–Sep. Southwestern United States to Guatemala.

Panicum hirticaule J. Presl (hairy-stemmed) Mexican witchgrass. Tufted annuals, glabrous or hispid with bulbous-based hairs. **CULMS** 6–55 cm tall, mostly erect. **LIGULES** 1.5–3.5 mm long. **BLADES** flat, 3–6 mm wide, with bulbous-based hairs. **PANICLES** open, 8–25 cm long, 4–8 cm wide. **SPIKELETS** 2–4 mm long. **FIRST GLUME** 3- to 5-nerved, 1/2–3/4 as long as the spikelet. **SECOND GLUME** 7- to 11-nerved, 1.8–3.3 mm long. **LOWER FLORET** sterile, the lemma similar to the second glume. **UPPER FLORET** 1.5–2.4 mm long, smooth or papillose, shiny, often with a lunate scar at the base. $2n = 18, 36$. Gravelly washes, roadsides, rocky hills and slopes, commonly in desert regions. ARIZ: Nav; NMEX: SJn. 1550–1700 m (5200–5600′). Flowering: Jul–Sep. Southwestern United States south to Central America, western South America, and Argentina.

Panicum miliaceum L. (milletlike) Broomcorn millet. Tufted annuals with bulbous-based hairs. **CULMS** 20–75 cm tall or more, sometimes branching below. **LIGULES** 1–3 mm long. **BLADES** 5–20 mm wide. **PANICLES** 6–20 cm long, 4–11 cm wide, the base often included in the sheath, often nodding, the spikelets toward the ends of the branches. **SPIKELETS** 4–6 mm long, ovoid, usually glabrous. **FIRST GLUME** 5- to 7-nerved, 1/2–3/4 the length of the spikelet. **SECOND GLUME** 11- to 13-nerved, exceeding the upper floret. **LOWER FLORET** sterile, 9- to 13-nerved, similar to the second glume. **UPPER FLORET** 3–3.8 mm long, smooth or striate, shiny, remaining in the spikelet or falling before the spikelet disarticulates. $2n = 36, 40, 42, 49, 54, 72$. Weedy, disturbed ground at lower elevations. COLO: Arc; NMEX: SJn; UTAH. 1550–1950 m (5200–6400′). Flowering: Jul–Aug. Native to Asia, escaped to scattered locales throughout North America. Widely cultivated throughout the world as a crop; also used in birdseed mixes.

Panicum obtusum Kunth (blunt) Vine mesquite. Stoloniferous perennials, occasionally with short rhizomes. **CULMS** 20–70 cm tall; stolon internodes to 35 cm or more long, the nodes swollen and densely pilose. **LIGULES** 0.2–2 mm long, truncate. **BLADES** flat, 2–7 mm wide, glaucous, stiff and ascending. **PANICLES** narrow, 5–15 cm long, 1–2 cm wide, the branches ascending to appressed, usually simple. **SPIKELETS** 2.8–4.4 mm long, blunt, glabrous, nearly terete. **FIRST GLUME** 5- to 7-nerved, 3/4 to nearly as long as the entire spikelet. **SECOND GLUME** 5- to 9-nerved, as long as the spikelet. **LOWER FLORET** staminate, similar to the second glume. **UPPER FLORET** 2.5–3.2 mm long, broadly elliptic, smooth and shiny. $2n = 20, 36, 40$. Heavy soils of swales, flats, playas, and low spots. NMEX: SJn; UTAH. 1350–1650 m (4500–5400′). Flowering: Aug–Sep. Southwestern United States to central Mexico.

Panicum virgatum L. (broomlike) Switchgrass. Robust, rhizomatous perennials. **CULMS** 40–150 cm tall or more, erect, thick, not much-branched. **LIGULES** 2–6 mm long. **BLADES** flat, 3–15 mm wide. **PANICLES** stiffly open, 10–45 cm long, 5–20 cm wide. **SPIKELETS** 2.5–8 mm long, pointed. **FIRST GLUME** 5- to 9-nerved, 1/2 to nearly as long as the spikelet. **SECOND GLUME** 7- to 11-nerved, exceeding the upper floret. **LOWER FLORET** staminate, with a well-developed palea. **UPPER FLORET** 2.3–3 mm long, narrowly ovoid, smooth and shiny. $2n = 18, 21, 25, 30, 32, 35, 36, 67–72, 74, 77, 90, 108$. Moist sites along roads, drainages, hanging gardens, ditch banks, at lower elevations. COLO: Dol, LPl, Mon; NMEX: San; UTAH. 1125–2100 m (3700–6800′). Flowering: Jun–Sep. Southern Canada throughout much of the United States, especially east of the Rocky Mountains; Mexico and Central America.

Paspalum L. Paspalum

(Greek *paspalos*, a kind of millet) Annuals and perennials, tufted, rhizomatous, or stoloniferous, with pithy culms. **LEAVES** with open sheaths, membranous ligules, flat blades, and lacking auricles. **INFL** a panicle of 1 to many unbranched unilateral primary branches. **SPIKELETS** 2-flowered, dorsally compressed, awnless, the lower floret reduced, the upper floret perfect; disarticulation below the glumes; glumes unequal, the first mostly absent; lower floret sterile, the lemma membranous and similar in size and texture to the second glume, the palea absent or rudimentary; upper floret perfect, the lemma and palea forming a hard seedcase around the flower and grain. **STAMENS** 3. $x = 10, 12$. A large genus of 300–400 species throughout the world, most well developed in tropical to warm-temperate regions.

Paspalum distichum L. (two-rowed, referring to the spikelets) Knotgrass. Stoloniferous or rhizomatous perennials, glabrous to densely pilose. **CULMS** 20–100 cm long, erect or decumbent, rooting at the lower nodes in wet soil. **LIGULES** 1–2 mm long. **BLADES** flat or folded, 2–10 mm wide. **PANICLES** terminal, composed of 2 branches paired at the tip of the peduncle, sometimes a third branch present below, the branches 2–7 cm long, winged. **SPIKELETS** 2.4–3.2 mm long, solitary. **FIRST GLUME** absent or rarely to 1 mm long. **SECOND GLUME** 3-nerved, as long as the spikelet, puberulent. **LOWER FLORET** 3-nerved, glabrous. **UPPER FLORET** 2.2–2.8 mm long, ovate, smooth and shiny. $2n$ = 20, 30, 40, 48, 60. [*Paspalum distichum* var. *indutum* Shinners]. Along streams, canals, and ditch banks. NMEX: SJn. 1550–1700 m (5100–5600′). Flowering: Aug–Sep. Native to warm regions of the world. Forms with densely pilose sheaths have been referred to var. *indutum* Shinners, but this seems to be an arbitrary expression without taxonomic strength.

Phalaris L. Canarygrass

(Greek *phalaris*, a coot, which has a white spot on the head, referring to a grain enclosed in white scales) Tufted or rhizomatous annuals and perennials with hollow culms. **LEAVES** with open sheaths, membranous ligules, flat blades, and lacking auricles. **INFL** a contracted panicle, often spikelike. **SPIKELETS** with 1 perfect well-developed floret and (0) 1–2 sterile rudimentary florets below, laterally compressed; disarticulation above the glume; glumes large, exceeding the floret, often keeled on the midnerve, 3- to 7-nerved, awnless; sterile florets much reduced, scalelike or bristlelike; fertile floret hard and shiny, usually hairy, the lemma permanently enclosing the palea and grain, obscurely 5-nerved. **STAMENS** 3. x = 6, 7. About 22 species nearly throughout the temperate regions of the world (Anderson 1961). The common name, canarygrass, originates from the Canary Islands. The islands were named not for the birds but for their aboriginal dogs, the name deriving from the Latin *Insulae Canariae*, dog islands, incorrectly anglicized to Canary Islands.

1. Plants perennial, rhizomatous; panicles 5–40 cm long .... ***P. arundinacea***
1′ Plants annual, tufted; panicles 1–5 cm long .... ***P. canariensis***

Phalaris arundinacea L. (reedlike) Reed canarygrass. Robust, rhizomatous perennials. **CULMS** erect, 40–200 cm tall or more. **LIGULES** 4–10 mm long. **BLADES** 10–30 cm long, 5–20 mm wide, serrate. **PANICLES** elongate, 5–40 cm long, 1–2 cm wide, dense and contracted, the branches erect-appressed. **SPIKELETS** 4–8 mm long. **GLUMES** as long as the spikelet, the keels scarcely winged. **STERILE FLORETS** 2, subequal, 1.5–2 mm long, less than 1/2 the length of the fertile floret, hairy. **FERTILE FLORET** 2.5–4 mm long, the lemma hairy above and on the margins, shiny pale brown when mature. **ANTHERS** 2.5–3 mm long. $2n$ = 27, 28, 29, 30, 31, 35. [*Phalaroides arundinacea* (L.) Rauschert]. Marshy ground, sloughs, wet meadows, forming dense thickets. COLO: Arc, Dol, LPl, Min, Mon, SJn; NMEX: SJn. 1550–2750 m (5150–8950′). Flowering: Jun–Aug. Throughout cool-temperate North America, Mexico, South America.

Phalaris canariensis L. (of the Canary Islands) Common canarygrass. Tufted annuals. **CULMS** erect, 25–100 cm tall. **LIGULES** 3–6 mm long. **BLADES** 3–25 cm long, 2–10 mm wide. **PANICLES** ovoid, 1.5–5 cm long, 1–2 cm wide, dense and contracted. **SPIKELETS** 7–10 mm long. **GLUMES** as long as the spikelet, the keels winged. **STERILE FLORETS** 2, subequal, 2–4.5 mm long, chaffy and sparsely hairy. **FERTILE FLORET** 4.5–7 mm long, ovate, densely hairy, shiny. **ANTHERS** 2–4 mm long. $2n$ = 12. Moist to wet, weedy ground, often grown for birdseed. COLO: LPl. 1900–2200 m (6250–7200′). Flowering: Jul–Aug. Native to southern Europe, now widespread throughout the world.

Phleum L. Timothy

(Greek *phleos*, the name for some marsh reed; the common name of timothy remembers Timothy Hansen, who promoted its use in Virginia and the Carolinas about 1720) Tufted annuals and perennials with hollow culms, sometimes with short rhizomes. **LEAVES** with open sheaths, membranous ligules, flat blades, and with absent or inconspicuous auricles. **INFL** a dense, spikelike panicle, the branches much reduced and shortened. **SPIKELETS** 1-flowered; disarticulation above (and sometimes below) the glumes; glumes 3-nerved, subequal, strongly laterally compressed, ciliate on the keels, abruptly short-awned from the truncate apex; lemma much shorter than the glumes, 3- to 5-nerved, not keeled, awnless; palea subequal to the lemma, 2-nerved. **STAMENS** 3. x = 7. About 15 species of the temperate regions of the world, most native to Eurasia.

1. Panicles subglobose or broadly cylindrical, only 2 or 3 times longer than wide, 1–5 cm long and 8–12 mm wide; awns of glumes 1.5–2.5 mm long .... ***P. alpinum***
1′ Panicles cylindrical, several times longer than wide, 4–16 cm long and 5–8 mm wide; awns of glumes 1–1.5 mm long .... ***P. pratense***

Phleum alpinum L. (alpine) Alpine timothy. Tufted perennials, sometimes with short rhizomes. **CULMS** erect to decumbent-based, 15–50 cm tall, the bases not bulbous. **LIGULES** 1–4 mm long, truncate. **SHEATH** of flag leaf inflated. **BLADES** 4–7 mm wide, auricles not developed. **PANICLES** subglobose or broadly cylindrical, 1–5 (6) cm long, (5) 8–12 mm wide. **GLUMES** 2.5–3.5 mm long, the keels ciliate-hispid; awn of the midnerve 1.5–2.5 mm long. **LEMMAS** 1.7–2.5 mm long, puberulent. **PALEAS** about as long as the lemmas. **ANTHERS** 1–2 mm long. $2n = 14, 28$. [*Phleum commutatum* Gaudin]. Mountain meadows, wet slopes, and moist roadsides, subalpine to alpine vegetation. COLO: Arc, Con, Hin, LPl, Min, Mon, RGr, SJn; NMEX: McK, SJn. 2750–3950 m (9000–13000′). Flowering: Jun–Aug. Circumboreal in the Northern Hemisphere, South America, Eurasia.

Phleum pratense L. (of meadows) Timothy. Tufted perennials. **CULMS** erect, 30–150 cm tall, the bases often bulbous. **LIGULES** 2–4 mm long. **SHEATH** of flag leaf not inflated. **BLADES** 4–8 mm wide. **PANICLES** cylindrical, 4–16 cm long, 5–8 mm wide. **GLUMES** 3–4 mm long, the keels ciliate-hispid. **LEMMAS** 1.5–2 mm long, usually puberulent. **PALEAS** about as long as the lemmas. **ANTHERS** 1.6–2.3 mm long. $2n = 42$ (21, 35, 36, 49, 56, 63, 70, 84). Meadows, roadsides, pastures, moist fields, from juniper woodlands to subalpine zones. ARIZ: Apa; COLO: Arc, Dol, Hin, LPl, Min, Mon, RGr, SJn, SMg; NMEX: McK, RAr, SJn; UTAH. 2000–3200 m (6500–10500′). Flowering: Jun–Sep. Native to Eurasia, introduced widely in cool-temperate regions of the world.

Phragmites Adans. Reed

(Greek *phragma*, a hedge, referring to its growth habit) Tall perennials with long-reaching rhizomes or stolons, forming dense thickets or hedges, the culms hollow. **LEAVES** mostly glabrous, with open sheaths, membranous-ciliate ligules, flat to folded blades, and lacking auricles. **INFL** terminal, a plumose panicle. **SPIKELETS** 3- to 8-flowered, weakly laterally compressed, the lower 1–2 florets staminate, the terminal 1–2 florets rudimentary, the remaining florets perfect; rachilla internode sericeous; disarticulation above the glumes and between the florets; glumes unequal, shorter than the florets, glabrous, 1- to 3-nerved; lemmas 3-nerved, linear, glabrous, acuminate, awnless; palea shorter than the lemma. **STAMENS** 1–3. $x = 12$. A small genus of one to three species. Recent investigations indicate the presence of at least three races in North America (Saltonstall, Peterson, and Soreng 2004).

Phragmites australis (Cav.) Trin. ex Steud. (southern) Common reed. **CULMS** 1–4 m tall, 0.5–1.5 cm thick. **LIGULES** 0.4–2 mm long. **BLADES** to 40 cm long, 2–4 cm wide. **PANICLES** ovoid to lanceolate, 15–35 cm long, 8–20 cm wide, often purplish when young, straw-colored in age. **SPIKELETS** long-hairy on the rachilla, the hairs 6–10 mm long. **GLUMES** unequal, the first 3–7 mm long, the second 5–10 mm long. **LEMMAS** 8–15 mm long, the margins inrolled. **PALEAS** 3–4 mm long. **ANTHERS** 1.5–2 mm long, purplish. $2n = 36, 42, 44, 46, 48, 49–54, 72, 84, 96$. [*P. communis* Trin.; *P. berlandieri* E. Fourn.; *P. phragmites* H. Karst.]. Along streams, rivers, canals, and ditches, wet ground of springs and seeps. ARIZ: Apa, Nav; COLO: LPl, Mon; NMEX: RAr, SJn; UTAH. 1200–2300 m (4000–7500′). Flowering: Jul–Sep. Essentially throughout the world. Roots, shoots, and seeds have been used for food throughout the world. Known as *carrizo* in the Southwest. We have both native and exotic races in the Four Corners region, which are told apart with difficulty.

1. Ligules 1–1.7 mm long; lower glumes 3–6.5 mm long; upper glumes 5.5–11 mm long; lemmas 8–13.5 mm long; leaf sheaths deciduous in age, the culms exposed in the winter, smooth and shiny; native .... **subsp. *americanus*** Saltonstall, P. M. Peterson & Soreng
1′ Ligules 0.4–0.9 mm long; lower glumes 2.5–5 mm long; upper glumes 4.5–7.5 mm long; lemmas 7.5–12 mm long; leaf sheaths not deciduous in age, the culms not exposed, smooth and shiny or ridged and not shiny .... (2)

2. Culm internodes smooth and shiny; native .... **var. *berlandieri*** (E. Fourn.) C. F. Reed
2′ Culm internodes ridged and not shiny; exotic .... **subsp. *australis***

Piptatherum P. Beauv. Ricegrass

(Greek *pipto*, to fall, and *ather*, awn, alluding to deciduous awns) Tufted perennials with hollow culms. **LEAVES** with open sheaths, membranous ligules, flat to involute blades, and lacking auricles. **INFL** terminal, a branching panicle. **SPIKELETS** 1-flowered; glumes subequal, obtuse to acute, with evident nerves; disarticulation above the glumes; floret hardened, fusiform; lemma convolute, but the margins separated at maturity and not overlapping, the awn deciduous; palea similar to the lemma. **STAMENS** 3. $x = 11$. About 30 species in arid, temperate, and subtropical regions of the world. Formerly included in the genus *Oryzopsis* (Barkworth 1993).

Piptatherum micranthum (Trin. & Rupr.) Barkworth (small-flowered) Littleseed ricegrass. **CULMS** densely tufted, 25–70 cm tall. **LIGULES** 0.4–2 mm long, truncate. **BLADES** flat to involute, less than 2 mm wide. **PANICLES** 5–18 cm long, narrow, the branches ascending to appressed. **GLUMES** 3- to 5-nerved, 2.7–4 mm long, ovate-acute, hyaline above. **LEMMAS** 1.8–2.8 mm long, glabrous, shiny, becoming indurate and brownish in age; awn 2–9 mm long, straight to sinuous, early-deciduous. **PALEAS** similar to the lemmas. **ANTHERS** about 1 mm long. $2n = 22$. [*Oryzopsis micrantha* (Trin. & Rupr.) Thurb.]. Moist, shaded, often rocky ground in the foothills and mountains. ARIZ: Apa, Nav; COLO: Arc, LPl, Mon, SJn, SMg; NMEX: McK, RAr, SJn; UTAH. 1700–2600 m (5500–8500′). Flowering: Jun–Aug. Western Canada and the United States.

Pleuraphis Torr. Galleta

(Greek *pleuron*, rib, and *raphis*, needle, referring to the short-awned nerve of glumes and lemmas) Strongly rhizomatous perennials forming large clumps, tufts, or bushes, with pithy culms. **LEAVES** with open sheaths, membranous-ciliate ligules, flat to rolled blades, and lacking auricles. **INFL** a spikelike panicle with extremely reduced branches; rachis zigzag or wavy. **SPIKELETS** in clusters of 3, with a tuft of hairs at the base, the central fertile, the lateral staminate. **CENTRAL SPIKELET** subsessile; glumes equal, 3- to 9-nerved, shorter than the florets, ciliate, the tip deeply 2-lobed, awned from the nerves; florets 1–2, the lower perfect, the upper (if present) perfect or staminate; lemma lanceolate, 3-nerved, ciliate, the tip 2-lobed and awned from the sinus. **LATERAL SPIKELETS** sessile; glumes shorter than to equaling the florets, ciliate, usually lobed, the first glume asymmetric and awned from about the middle near the margin; florets 1–4, usually staminate; lemma 3-nerved, ciliate at the tip. A small genus of three species of the western United States. Formerly included in *Hilaria*, which differs (among numerous features) in being stoloniferous and not rhizomatous.

Pleuraphis jamesii Torr. (for Edwin James, 1797–1861, surgeon-naturalist with the 1820 Long Expedition) Galleta. **CULMS** 15–50 cm tall, forming large clumps from the thick rhizomes, appearing tufted. **SHEATHS** glabrous to scaberulous, often with a few long hairs at the summit. **LIGULES** 1–3 mm long, with lateral erect lobes. **BLADES** 2–3 mm wide, mostly glabrous. **SPIKES** 3–7 cm long. $2n = 36, 38$. [*Hilaria jamesii* (Torr.) Benth.]. Desert flats and gravelly slopes of foothills. ARIZ: Apa, Coc, Nav; COLO: Arc, Dol, LPl, Mon, SMg; NMEX: McK, RAr, San, SJn; UTAH. 1550–2200 m (5100–7200′). Flowering: May–Aug. Southwestern United States.

Poa L. Bluegrass

(ancient Greek name for grass or fodder) Annuals and perennials, tufted, rhizomatous, or stoloniferous, with hollow culms. **LEAVES** with open to closed sheaths, membranous ligules, flat to folded (sometimes rolled) blades, and lacking ligules. **INFL** a terminal panicle, open to contracted, generally elevated well above the leaves. **SPIKELETS** (1) 2- to 8-flowered, relatively small, the florets usually perfect, some species dioecious; disarticulation above the glumes and between the florets; glumes membranous, shorter than the lowermost floret, the first mostly 1-nerved, the second often 3-nerved; lemmas membranous, (3) 5-nerved, glabrous to variously pubescent, some with a conspicuous tuft of cobwebby hairs on the callus; palea slightly shorter than the lemma, 2-keeled. **STAMENS** 3. $x = 7$. A difficult genus of 150–200 species, mostly of temperate to Arctic regions, the identification complicated by hybridization and apomixis. Care should be taken to collect mature, complete specimens, paying particular attention to the presence of rhizomes, basal growth, and mature panicles (Kellogg 1985a; Soreng 1985, 1991).

1. Florets modified and forming small leafy plantlets; stems slightly to strongly bulblike at the base .... ***P. bulbosa***
1′ Florets not modified into small leafy plantlets; stems rarely somewhat bulblike .... (2)

2. Stems and nodes strongly flattened; plants with creeping rhizomes; lemma short-villous on the nerves toward the base, longer cobwebby hairs absent or scant .... ***P. compressa***
2′ Stems and nodes terete or nearly so; combination of features other than above .... (3)

3. Panicles closed and contracted at maturity, the branches erect, appressed to ascending .... (4)
3′ Panicles open at maturity, the branches spreading to reflexed .... (12)

4. Florets with a conspicuous tuft of cobwebby hairs at the base, these short-kinky to long-sinuous and longer than any other hairs on the lemma .... (5)
4′ Florets glabrous or hairy, but not with a conspicuous tuft of cobwebby hairs at the base as above, any hairs present all about the same length .... (6)

5. Plants tufted, annual or short-lived perennial; anthers less than 1 mm long .... ***P. bigelovii***
5′ Plants with creeping rhizomes, perennial; anthers mostly longer than 1 mm .... ***P. pratensis***

6. Plants with short-creeping rhizomes; ligules 1.5–3.5 mm long, conspicuous .... ***P. arida***
6' Plants tufted, lacking creeping rhizomes; ligules various .... (7)

7. Spikelets rounded and little compressed; glumes and lemmas not keeled or only obscurely so at the tip .... ***P. secunda***
7' Spikelets compressed; glumes and lemmas mostly keeled to the base .... (8)

8. Lemmas glabrous to minutely scabrous, the hairs scarcely visible with magnification; plants of subalpine to alpine rocky ridges and talus .... (9)
8' Lemmas noticeably pubescent or puberulent on the keels, nerves, or body, though the hairs may be short .... (10)

9. Dwarf plants, 3–10 cm tall; lemmas 2–3 mm long; glumes nearly as long as the spikelet; sheaths open their full length .... ***P. lettermanii***
9' Taller plants, 15–45 cm tall; lemmas 4–4.5 mm long; glumes shorter than the spikelet; sheaths open only about 1/2 their length, closed below .... ***P. cusickii***

10. Plants unisexual, all the spikelets of a plant either male or female; uppermost stem blade very reduced, often absent; lemmas 4–7 mm long .... ***P. fendleriana***
10' Plants perfect-flowered, the spikelets with both anthers and pistil in a single floret; uppermost stem blade well developed; lemmas 2–5 mm long .... (11)

11. Blades stiffly erect, 1–5 cm long; sheaths not elongate or papery; panicle stiff; plants forming tight clumps .... ***P. glauca***
11' Blades gracefully curved, 5–10 cm long; sheaths elongate and papery, persistent; panicle soft and lax; plants forming loose spreading clumps .... ***P. abbreviata***

12. Plants annual, the remains of old shoots not present .... ***P. annua***
12' Plants perennial, with some remains of old shoots .... (13)

13. Florets with a conspicuous tuft of cobwebby hairs at the base, these short-kinky to long-sinuous and longer than other hairs on the lemma .... (14)
13' Florets glabrous or hairy, but not with a conspicuous tuft of cobwebby hairs at the base as above, any hairs present all about the same length .... (24)

14. Plants with creeping rhizomes; panicle branches glabrous to moderately scabrous, round .... (15)
14' Plants tufted, lacking rhizomes (in wet habitats occasionally producing decumbent stems that root at the nodes); panicle branches distinctly scabrous, mostly angled .... (16)

15. Glumes distinctly keeled, scabrous on the nerves, the second glume plainly shorter than the first lemma; panicles often with 4 or more branches at the lowermost node (some occasionally vestigial); ligules mostly 1–2 mm long .... ***P. pratensis***
15' Glumes weakly keeled, nearly glabrous, the second glume subequal to or longer than the first lemma; panicles usually with fewer than 4 branches at the lowermost node; ligules 2–4 mm long .... ***P. arctica***

16. Sheaths densely and conspicuously scabrous with downward-pointing hairs .... (17)
16' Sheaths glabrous to sparsely scaberulous .... (18)

17. Anthers mostly 2–3 mm long; sheaths rarely scabrous .... ***P. tracyi***
17' Anthers mostly 0.5–1 mm long; sheaths nearly always scabrous .... ***P. occidentalis***

18. Anthers mostly less than 1 mm long; blades flat .... (19)
18' Anthers mostly longer than 1 mm; blades various, flat to rolled or folded .... (21)

19. First glume mostly 3-nerved; sheath margins fused together only at the very base .... ***P. palustris***
19' First glume mostly 1-nerved; sheath margins fused together 1/4 to 2/3 their length .... (20)

20. First glume linear-lanceolate, much narrower than the second; paleas glabrous to scabrous on the keels .... ***P. leptocoma***
20' First glume about the same shape and width as the second, both broadly lanceolate; paleas short-pubescent on the keels .... ***P. reflexa***

21. Panicles mostly 10–30 cm long, abundantly rebranched; lowermost panicle branches mostly 3–5 per node....(22)
21′ Panicles mostly less than 12 cm long, sparingly rebranched; lowermost panicle branches mostly 1–2 per node.....(23)

22. Anthers mostly 2–3 mm long; first glume 1-nerved .... ***P. tracyi***
22′ Anthers less than 1.5 mm long; first glume mostly 3-nerved .... ***P. palustris***

23. Lemmas glabrous between the nerves, with a copious web on the callus; leaves green .... ***P. interior***
23′ Lemmas mostly pubescent between the nerves, with a scant web on the callus; leaves glaucous .... ***P. glauca***

24. Basal blades short and broad, 2–5 mm wide; spikelets rounded or almost cordate at the base; stem bases enclosed in persistent, thick, closely overlapping sheaths .... ***P. alpina***
24′ Basal blades elongate and narrow, 1–3 mm wide; spikelets not broadly rounded at the base; stem bases not enclosed in persistent sheaths as above ....(25)

25. Plants with creeping rhizomes ....(26)
25′ Plants tufted, lacking rhizomes ....(27)

26. Lower sheaths minutely retrorse-hairy and purplish; lemmas nearly glabrous on the nerves .... ***P. wheeleri***
26′ Lower sheaths smooth, greenish; lemmas puberulent on the nerves .... ***P. arctica***

27. Plants annual or short-lived perennial, with lax succulent blades; spikelets strongly compressed, the glumes and lemmas mostly keeled to the base; lemmas prominently short-pilose on the nerves, but not between .... ***P. annua***
27′ Plants perennial, with relatively stiff, firm blades; spikelets rounded and little compressed, the glumes and lemmas not keeled or only obscurely so at the tip; lemmas not pilose on the nerves, but rather uniformly puberulent across the back and between the nerves .... ***P. secunda***

Poa abbreviata R. Br. (shortened) Short bluegrass. Densely tufted alpine perennials, with numerous basal leaves and persistent elongate sheaths. **CULMS** 5–30 cm tall. **LIGULES** 1–3 mm long. **BLADES** folded, lax and curving, mostly less than 10 cm long, 1–2 mm wide. **PANICLES** narrow and loosely contracted, 3–9 cm long. **SPIKELETS** 2- to 4-flowered, nearly sessile, strongly compressed. **GLUMES** somewhat unequal, the first 2–4 mm long and 1-nerved, the second 3–5 mm long and 1- to 3-nerved and often equaling the lowermost lemma. **LEMMAS** 2.5–5 mm long, short-pubescent on the keel and lateral nerves, also puberulent between the nerves, usually purplish banded at the tip. **ANTHERS** 0.8–1.8 mm long. $2n = 42$. Subalpine to alpine slopes, ridges, and meadows, mostly near timberline, uncommon. COLO: Hin, LPl. 3400–3800 m (11100–12500′). Flowering: Jul–Aug. Western Rocky Mountains, Canada south to Colorado. Our material belongs to **subsp. *pattersonii*** (Vasey) Á. Löve, D. Löve & B. M. Kapoor (for Henry Norton Patterson, 1853–1919, Illinois botanist of the late 19th century). Its most conspicuous feature is the persistent, papery sheaths.

Poa alpina L. (alpine) Alpine bluegrass. Densely tufted perennials with thick mats of basal leaves, the sheaths closely overlapping and persistent. **CULMS** erect, 10–30 cm tall. **LIGULES** 1–4 mm long, truncate to obtuse. **BLADES** usually flat, the basal ones 3–5 cm long and 2–5 mm wide. **PANICLES** 3–5 cm long and wide, broadly pyramidal, the branches spreading to horizontal. **SPIKELETS** mostly 4- to 5-flowered, purplish, the bases rounded or subcordate. **GLUMES** ovate, 3-nerved, the first 2.3–3.2 mm long, the second slightly longer. **LEMMAS** faintly nerved, 3–4 mm long, strongly scarious-margined, short-villous on the keel and marginal nerves, lacking a web on the callus. **ANTHERS** 1.2–2 mm long. $2n = 14, 21–74$. Subalpine to alpine meadows and rocky or gravelly slopes. COLO: Arc, Con, Hin, LPl, Min, Mon, SJn; UTAH. 2400–3800 m (8000–12500′). Flowering: Jul–Aug. Circumboreal, south into the Rocky Mountains; Eurasia. This is one of the easier *Poa* species to identify, with its broad, almost rosettelike, basal leaves and cordate-based spikelets, always at high elevations.

Poa annua L. (annual) Annual bluegrass. Tufted, bright green, often winter-annuals or short-lived perennials. **CULMS** erect to prostrate and then sometimes rooting at the nodes and mat-forming, 3–25 cm tall. **LIGULES** 1–3 mm long. **BLADES** lax, folded to flat, 1–3 mm wide. **PANICLES** 2–5 (8) cm long, open. **SPIKELETS** 3- to 6-flowered, the rachilla often visible. **GLUMES** subequal, oblanceolate, broadest above the middle, the first 1-nerved and 1.5–2.2 mm long, the second 1- to 3-nerved and slightly longer. **LEMMAS** 2.5–3.5 mm long, (3) 5-nerved, short-pilose on the nerves, lacking a web on the callus. **ANTHERS** 0.7–1.1 mm long. $2n = 14, 24–26, 28$. Moist, weedy areas, stream banks, ditches, lake margins, gardens, and lawns. ARIZ: Apa, Coc, Nav; COLO: Hin, LPl, Mon, SJn; NMEX: RAr, SJn; UTAH. 1450–2800 m (4800–9100′). Flowering: Apr–Sep. Native to Europe, throughout much of North America.

Poa arctica R. Br. (arctic) Arctic bluegrass. Tufted and rhizomatous perennials. **CULMS** in small tufts and also usually producing short rhizomes, 10–45 cm tall. **LIGULES** 2–4 mm long, acuminate. **BLADES** flat or folded, 1.5–3.5 mm wide. **PANICLES** open, 4–12 cm long, the lower branches spreading. **SPIKELETS** 2- to 5-flowered, often purplish. **GLUMES** subequal, the first 2.5–4.5 mm long, the second slightly longer. **LEMMAS** 3.5–5 mm long, short-pilose on the nerves, usually with a scant web on the callus. **ANTHERS** 1.4–2.5 mm long. $2n$ = 38–92. [*Poa aperta* Scribn. & Merr.; *Poa grayana* Vasey]. Upper subalpine and alpine ridges, open slopes, and meadows. COLO: Arc, Con, Hin, LPl, Min, SJn; UTAH. 2700–4000 m (9000–13000′). Flowering: Jul–Sep. Circumboreal, south in the Rocky Mountains to New Mexico.

Poa arida Vasey (arid) Plains bluegrass. Tufted and rhizomatous perennials. **CULMS** in small tufts along rhizomes, 12–50 cm tall. **LIGULES** 1.5–3.5 mm long, conspicuous. **BLADES** flat to rolled or folded, 1–3 mm wide. **PANICLES** narrow, contracted, 3–12 cm long, the branches erect. **SPIKELETS** 2- to 6-flowered. **GLUMES** subequal, 1- to 3-nerved, the first 2.5–4.2 mm long, the second 3–5 mm long. **LEMMAS** 3.2–5 mm long, densely short-villous on the nerves, lacking a web on the callus. **ANTHERS** 1.2–2.5 mm long. $2n$ = 63–103. [*Poa glaucifolia* Scribn. & T. A. Williams]. Low flats in the valleys, poorly understood in our area. ARIZ: Apa; COLO: Dol, LPl. 1550–1900 m (5100–6200′). Flowering: Jun–Aug. Central plains of southern Canada and the United States, west into the Rocky Mountain region.

Poa bigelovii Vasey & Scribn. (for John Milton Bigelow, 1804–1878, surgeon-botanist on early boundary surveys) Bigelow's bluegrass. Tufted winter annuals. **CULMS** erect, leafy, 10–55 cm tall. **LIGULES** 1–3 mm long. **BLADES** lax, flat, 1–4 mm wide. **PANICLES** elongate, narrow, 4–18 cm long, less than 2 cm wide, the branches erect. **SPIKELETS** 4- to 6-flowered. **GLUMES** acuminate, mostly 3-nerved, subequal, the first 2–3.5 mm long, the second slightly longer. **LEMMAS** 2.7–4 mm long, short-villous on the nerves, with a web on the callus. **ANTHERS** 0.4–0.8 mm long. $2n$ = 28, 29. Seasonally moist washes, woodlands, desert plains, and foothills. NMEX: McK, San; UTAH. 1150–2130 m (3800–7000′). Flowering: Feb–May. Southwestern United States from Oklahoma to California.

Poa bulbosa L. (having a bulb) Bulbous bluegrass. Densely tufted perennials. **CULMS** usually bulbous and purplish at the base, 25–50 cm tall. **LIGULES** 1.5–5 mm long, acuminate. **BLADES** flat to folded, 1.5–4 mm wide. **PANICLES** 4–10 cm long, usually narrow and contracted, the branches ascending to erect. **SPIKELETS** 4- to 6-flowered, these nearly always transformed into conspicuous vegetative bulblets. **GLUMES** 2–3.5 mm long. **LEMMAS** 3–6 mm long, normal lemmas rarely present and with a web on the callus. **ANTHERS** 1–2.5 mm long when present. $2n$ = 14, 21, 28, 39, 40–58. Moist, weedy ground in the foothills and mountains, pastures, meadows, stream banks. ARIZ: Apa, Nav; COLO: Arc, Dol, Mon, SMg; NMEX: McK, SJn; UTAH. 1500–2950 m (4900–9700′). Flowering: Apr–Aug. Native to Eurasia and North Africa, common in western North America. The viviparous bulblets in the inflorescence are unmistakable, and perhaps confusing at first sight.

Poa compressa L. (flattened) Canada bluegrass. Rhizomatous, bluish green perennials. **CULMS** 20–70 cm tall, nearly always prominently flattened, including the nodes, decumbent- or geniculate-based, remaining green after the sheaths have faded. **LIGULES** 0.5–2 mm long, rounded. **BLADES** flat to folded, 1.5–3.5 mm long. **PANICLES** narrow and contracted to somewhat open, but never diffuse, the branches erect-appressed to spreading. **SPIKELETS** 3- to 7-flowered. **GLUMES** subequal, 1.8–3 mm long, the first 1- to 3-nerved, the second broader and 3-nerved. **LEMMAS** 2–3 mm long, firm and almost leathery, usually bronze-tipped, short-villous on the nerves, without a web on the callus, or the web quite scant. **ANTHERS** 0.9–1.7 mm long. $2n$ = 14, 35, 39, 42, 45, 49, 50, 56. Quite common in forest clearings, disturbed meadows, roadsides in the mountains, often with *P. pratensis*. ARIZ: Apa, Nav; COLO: Arc, Dol, Hin, LPl, Min, Mon, SJn, SMg; NMEX: McK, RAr, San, SJn; UTAH. 2050–2750 m (6700–9000′). Flowering: Jun–Aug. Native to Eurasia, now found nearly throughout cool-temperate regions of the world. The florets of *P. compressa* are distinctive, being tannish purplish with a rough, almost leathery surface and obscure nerves, and quite similar to those of *P. palustris*. The flattened culms and sheaths are diagnostic.

Poa cusickii Vasey (for W. C. Cusick, 1842–1922, Oregon botanist) Skyline bluegrass. Loosely tufted perennials, lacking rhizomes. **CULMS** erect, generally not decumbent- or geniculate-based, 15–45 cm tall. **LIGULES** 2–5 mm long. **BLADES** flat or folded, bright green, 1–3 mm wide. **PANICLES** narrow and contracted, the branches appressed, 3–8 cm long. **SPIKELETS** 2- to 5-flowered, strongly compressed, usually pistillate. **GLUMES** subequal, 1-nerved, 2.4–5 mm long, the second slightly longer and sometimes 3-nerved. **LEMMAS** 4–6 mm long, glabrous to minutely scabrous, 5-nerved, lacking a web on the callus. **ANTHERS** about 2.5 mm long when present. $2n$ = 56, 84. [*Poa epilis* Scribn.]. Alpine or upper subalpine rocky ridges and talus, not commonly collected. COLO: Hin, Min, Mon, LPl, SJn; UTAH. 3000–3750 m (9800–12400′). Flowering: Jun–Aug. Western Canada and United States, south to Colorado

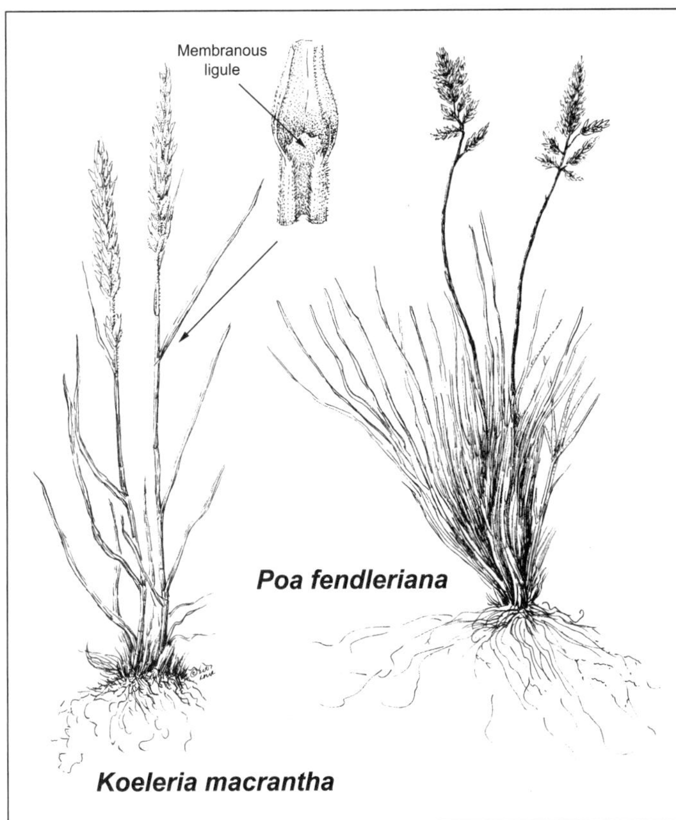

and California. Four Corners material belongs to **subsp**. ***epilis*** (Scribn.) W. A. Weber (hairless). Related to *P. fendleriana*, and merged by some into that species, but our populations are distinguished by the glabrous florets (Soreng 1991).

Poa fendleriana (Steud.) Vasey (for Augustus Fendler, 1813–1883, German-born botanical collector) Muttongrass. Densely tufted perennials, sometimes with short rhizomes, often glaucous. **CULMS** erect, commonly with a curving or geniculate base, 25–65 cm tall. **LIGULES** 1–10 mm long, highly variable. **BLADES** firm, flat to commonly folded or rolled, 1.5–4 mm wide, the uppermost blade often reduced to nothing. **PANICLES** narrow, oblong, 4–15 cm long, 1–2 cm wide. **SPIKELETS** mostly pistillate, 3- to 6-flowered, large and strongly compressed, appearing papery. **GLUMES** subequal, 2.8–5 mm long, the second broader than the first. **LEMMAS** 4–7 mm long, short-pilose on the nerves, lacking a web on the callus. **ANTHERS** usually abortive, less than 1 mm long. $2n$ = 29, 56. [*Poa longiligula* Scribn. & T. A. Williams]. Common throughout our region in foothill woodlands to subalpine forests and grassy slopes. ARIZ: Apa, Coc, Nav; COLO: Arc, Con, Dol, Hin, LPl, Min, Mon, SJn, SMg; NMEX: McK, RAr, San, SJn; UTAH. 1370–3700 m (4500–12200′). Flowering: May–Sep. Western half of Canada and the United States, northern Mexico. Plants with long ligules (2–10 mm) have been referred to subspecies *longiligula* (Scribn. & T. A. Williams) Soreng.

Poa glauca Vahl (bluish) Timberline bluegrass. Densely tufted perennials. **CULMS** erect, rigid, 5–25 cm tall. **LIGULES** 0.5–2 mm long, rounded. **BLADES** rolled, short and stiff, 15 cm long, 1–2 mm wide. **PANICLES** narrow, compact, often purplish, 1–7 cm long, the branches appressed. **SPIKELETS** 2- to 4-flowered, strongly compressed. **GLUMES** subequal, 3-nerved, 2–3.5 mm long. **LEMMAS** 2.5–4.5 mm long, short-villous on the nerves and also between the nerves, lacking a web on the callus. **ANTHERS** 1–1.5 mm long. $2n$ = 44, 56–58, 62, 70–72, 75, 78. Alpine and subalpine ridges, rocky grassy slopes, and mossy ledges often above timberline. COLO: Con, Hin, LPl, Min, Mon, SJn. 3600–3900 m (11800–12800′). Flowering: Jul–Aug. Western Canada and United States, south to New Mexico and California. Our material is referred to as **subsp**. ***rupicola*** (Nash) W. A. Weber (growing on rocks). **Subspecies *glauca*** may also be present, which is sparsely webbed on the callus.

Poa interior Rydb. (interior) Inland bluegrass. Tufted perennials. **CULMS** stiffly erect, 20–50 cm tall. **LIGULES** 0.5–2 mm long, truncate. **BLADES** crowded at the base, lax, flat to rolled, 1–2 mm wide. **PANICLES** 4–14 cm long, pyramidal, the branches spreading, the lower branches 2–5 per node. **SPIKELETS** mostly 2- to 3-flowered. **GLUMES** subequal, 1- to 3-nerved, 1.8–3 mm long. **LEMMAS** 2.5–3.5 mm long, strongly keeled, short-villous on the nerves, with a copious web on the callus. **ANTHERS** 1–1.5 mm long. $2n$ = 28, 42, 43, 56. [*Poa nemoralis* L. subsp. *interior* (Rydb.) W. A. Weber]. Subalpine to alpine meadows, ledges, and forest clearings in the mountains. COLO: Arc, Dol, LPl, Mon, SJn, SMg; NMEX: SJn. 2900–4000 m (9500–13000′). Flowering: Jun–Aug. Canada, northern and western United States.

Poa leptocoma Trin. (weakly hairy) Bog bluegrass. Loosely tufted perennials, lacking rhizomes. **CULMS** erect to decumbent-based and rooting at the nodes, solitary or a few together, 15–55 cm tall. **LIGULES** 1–4 mm long, truncate. **BLADES** lax, flat, green, 1–4 mm wide. **PANICLES** delicate, lax, the tip often nodding, 4–14 cm long, the branches capillary, spreading to drooping. **SPIKELETS** 2- to 5-flowered, strongly compressed, purplish. **GLUMES** unequal, the first 2–3.5 mm long, mostly 1-nerved and narrower than the second, the second 2.5–4 mm long and 3-nerved. **LEMMAS** 3–5 mm long, short-sericeous on the nerves, with a web on the callus. **ANTHERS** 0.5–1 mm long (rarely longer). $2n$ = 42. Subalpine to alpine bogs, springs, and wet meadows. COLO: Arc, Hin, LPl, Min, Mon, SJn. 3000–3900 m (9850–12800′). Flowering: Jul–Aug. Western Canada and United States.

Poa lettermanii Vasey (for George Washington Letterman, 1840–1913, Missouri botanist) Letterman's bluegrass. Small, densely tufted perennials. **CULMS** erect, 3–12 cm tall, the leaves often exceeding the panicle. **LIGULES** 1–2 mm long, truncate. **BLADES** flat to folded, 0.5–2 mm wide. **PANICLES** narrow, few-flowered, 1–2 cm long, to 1 cm wide, the branches erect. **SPIKELETS** (1) 2 (3)-flowered, purplish, compressed. **GLUMES** subequal, nearly as long as the spikelet, 2–3 mm long, 1- to 3-nerved. **LEMMAS** 2–3 mm long, glabrous, lacking a web on the callus. **ANTHERS** about 0.3–0.8 mm long. $2n$ = not known. Rocky ridges and ledges above timberline. COLO: SJn. 3900–4050 m (12700–13200′). Flowering: Jul–Aug. Western Canada and United States, south to California and Colorado. Dwarf and often overlooked, *P. lettermanii* is easily distinguished by its diminutive stature, few glabrous florets, and relatively large glumes.

Poa occidentalis Vasey (western) New Mexico bluegrass. Tufted perennials, lacking rhizomes. **CULMS** erect, 25–75 cm tall or more, terete. **LIGULES** 1–6 mm long, acute. **BLADES** flat, lax, blue-green, 2–5 mm wide. **PANICLES** broadly pyramidal, open, 8–30 cm long or more, nodding, the branches widely spreading. **SPIKELETS** 2- to 7-flowered. **GLUMES** nearly equal, 2–3.5 mm long, the first 1-nerved, the second 3-nerved. **LEMMAS** 2.6–4.2 mm long, strongly keeled, short-villous on the nerves, with a web on the callus. **ANTHERS** 0.4–1 mm long. $2n$ = 14, about 28. Forest clearings and moist woods, montane to subalpine forests. COLO: LPl, Min, Mon. 2300–3000 m (7500–9800′). Flowering: Jul–Aug. Colorado and New Mexico.

Poa palustris L. (of marshes) Fowl bluegrass. Loosely tufted perennials. **CULMS** decumbent-based and often rooting at the lower nodes in wet soil, 25–90 cm tall or more, often purplish based. **LIGULES** 1–5 mm long, acute. **BLADES** lax, flat or folded, 1–3.5 mm wide. **PANICLES** open and loosely flowered, pyramidal, 7–25 dm long, yellowish green or purplish, the lower branches spreading. **SPIKELETS** 2- to 4-flowered, strongly compressed. **GLUMES** subequal, 1.8–3 mm long, 3-nerved. **LEMMAS** 2.5–3.5 mm long, bronze-tipped, obscurely nerved, short-villous on the nerves, with a web on the callus. **ANTHERS** 0.8–1.2 mm long. $2n$ = 28, 42. Moist meadows, marshy ground, sloughs, and ditch banks. COLO: Arc, Con, Dol, Hin, LPl, Min, Mon, SJn; NMEX: McK, RAr, SJn; UTAH. 2250–3500 m (7450–11500′). Flowering: Jun–Sep. Circumboreal throughout much of North America, Europe, and Eurasia, introduced in South America. The spikelets are quite similar to those of *P. compressa*.

Poa pratensis L. (of meadows) Kentucky bluegrass. Densely tufted perennials with creeping rhizomes, often forming a thick sod. **CULMS** erect to geniculate- or decumbent-based, 20–80 cm tall or more, freely branching below. **LIGULES** 0.5–1.7 mm long, truncate. **BLADES** soft or firm, flat to folded or rolled, 1–4 mm wide, mostly glabrous above (in our subspecies). **PANICLES** open, often pyramidal, sometimes contracted and dense, 4–12 cm long, the lower branches in whorls of 4 or 5. **SPIKELETS** 2- to 5-flowered, strongly compressed. **GLUMES** keeled, scabrous on the nerves, unequal, the first 1.8–2.5 mm long and 1-nerved, the second 2.2–3.3 mm long and 3-nerved. **LEMMAS** 3- to 5-nerved, the lateral nerves faint, mostly glabrous on the marginal nerves and short-villous on the midnerve, with a web on the callus. **ANTHERS** 1–2 mm long. $2n$ = 21–147. In a variety of habitats throughout the *Flora* region; roadsides, meadows, grassy slopes, clearings in forests, ditch banks, seeps, cultivated as lawn grass. Expected for all counties in the Four Corners region. 1150–2900 m (3800–9500′). We have two subspecies, one native and the other exotic:

1. Stems erect at the base; basal leaves pale grayish blue-green, 0.8–2 mm broad, folded and somewhat revolute, strongly ribbed on the back with the ribs almost touching, the sheaths remaining intact through several seasons; upper stem blade 1–3 cm long; lowermost panicle branches usually 2–3 at the node (up to 5), ascending; spikelets mostly 2-flowered; upper glume 2–2.7 mm long; lowermost lemma only slightly cobwebby ...**subsp. *agassizensis***

1′ Stems geniculate at the base; basal leaves bright green, 2–3 mm broad, flat or channeled, ribbed on the back with well-separated ridges, the sheaths withering and disintegrating after a season or two; upper stem blade 3–8 cm long; lowermost panicle branches usually 5 at the node, spreading; spikelets mostly 3- to 5-flowered; upper glume 3–3.5 mm long; lowermost lemma very cobwebby at the base ....**subsp. *pratensis***

subsp. *agassizensis* (B. Boivin & D. Löve) Roy L. Taylor & MacBryde (of ancient glacial Lake Agassiz, Manitoba, Canada) [*Poa agassizensis* B. Boivin & D. Löve]. This is the native race, common in drier, upland meadows and grassy understories of open forests. ARIZ: Apa; COLO: Dol, Hin, LPl, Mon, SMg; NMEX: McK, RAr, SJn.

subsp. *pratensis* Introduced from Europe (and called smooth meadow grass there) for improved pastures, meadow reseeding, and lawns, escaping to similar moist sites in natural habitats. ARIZ: Apa, Nav; COLO: Dol, Hin, LPl, Mon, SMg; NMEX: McK, RAr, SJn.

Poa reflexa Vasey & Scribn. (bent back) Nodding bluegrass. Loosely tufted perennials, lacking rhizomes. **CULMS** erect to decumbent-based and rooting at the nodes, in small tufts, 15–50 cm tall. **LIGULES** 1–3 mm long, truncate.

BLADES flat, 1–3 mm wide. **PANICLES** open, 4–14 cm long, the branches capillary and spreading to reflexed. **SPIKELETS** 2- to 4-flowered, strongly compressed. **GLUMES** subequal, 1.8–3 mm long, the first 1-nerved, the second 3-nerved. **LEMMAS** 2.2–3.5 mm long, purplish below a scarious tip, short-villous on the nerves, with a web on the callus. **ANTHERS** 0.5–1 mm long. $2n = 28$. Subalpine to alpine bogs, springs and seeps, rocky ridges and ledges. COLO: Arc, Con, Hin, LPl, Min, Mon, RGr, SJn. 3100–3600 m (10200–11800′). Flowering: Jul–Aug. Rocky Mountains in southern Canada and the United States, south to New Mexico.

Poa secunda J. Presl (to one side) Sandberg's bluegrass. Tufted perennials, sometimes with short rhizomes but these not well developed, yellowish green to glaucous. **CULMS** erect, 20–100 cm tall or more. **LIGULES** 1–6 mm long, truncate to acute. **BLADES** mostly folded, rolled upon drying, 0.5–3.5 mm wide. **PANICLES** narrow, contracted, only rarely open in infrequent individuals, 3–25 cm long or more, the branches ascending to erect-appressed (rarely spreading). **SPIKELETS** 3- to 6-flowered, rounded on the back and only slightly compressed if at all. **GLUMES** subequal, the first 2–5 mm long, the second slightly longer, shorter to longer than the lowermost floret. **LEMMAS** 3–6 mm long, rounded on the back, often bronze-tipped, densely even-pubescent across the back to glabrous, lacking a web on the callus. **ANTHERS** 1.2–3.5 mm long. $2n = 44, 56, 61–72, 74, 78, 81–106$. Forest clearings, sagebrush plains, dry meadows, disturbed ground. ARIZ: Apa; COLO: Dol, Hin, Mon; NMEX: McK, RAr, SJn; UTAH. 1400–3800 m (4600–12500′). Flowering: Apr–Aug. Most of Canada, south through the western United States, into Mexico, South America (Kellogg 1985a, 1985b). A widely varying and difficult group, with numerous races that have been named as species. One may attempt to recognize two subspecies:

1. Lemmas glabrous to minutely scabrous on the back .... **subsp. *juncifolia*** (Scribn.) Soreng [*Poa ampla* Merr.; *P. nevadensis* Vasey].
1′ Lemmas prominently crisp-puberulent across the back toward the base .... **subsp. *secunda*** [*Poa canbyi* (Scribn.) Howell; *P. sandbergii* Vasey; *P. scabrella* (Thurb.) Vasey].

Poa tracyi Vasey (for Samuel Mills Tracy, 1847–1920, USDA agronomist and forage botanist) Tracy's bluegrass. Tufted perennials, sometimes with short rhizomes. **CULMS** erect, 30–100 cm tall or more. **LIGULES** 1–4 mm long, acute. **BLADES** cauline, flat, 1.5–5 mm wide. **PANICLES** open, pyramidal, 8–25 cm long, the branches capillary, spreading to divergent. **SPIKELETS** compressed, 2- to 5-flowered. **GLUMES** unequal, the first 1.6–3.5 mm long and 1-nerved, the second 2.2–5 mm long and 1- to faintly 3-nerved. **LEMMAS** strongly keeled, 3- to faintly 5-nerved, 2.6–5 mm long, short-sericeous on the nerves, with a web on the callus (rarely scant). **ANTHERS** 1–3 mm long. $2n = 28$. Rich humus and moist loam of forests and woodlands in the mountains. COLO: Arc, SJn. Flowering: May–Jul. Colorado, Utah (Grand County), and New Mexico (Soreng and Hatch 1983).

Poa wheeleri Vasey (for George Montague Wheeler, 1842–1905, of the U.S. Army Corps of Engineers, director of the western surveys of 1869–1879, mapping almost 1/3 of the land west of the 100th meridian; namesake of Wheeler Peak, highest point in New Mexico) Wheeler's bluegrass. Shortly rhizomatous to sometimes loosely tufted perennials. **CULMS** erect or with weakly decumbent bases, 35–80 cm tall, terete or weakly compressed. **LIGULES** 0.5–2 mm long, minutely fringed, truncate. **BLADES** flat or folded, lax, 2–3.5 mm wide. **PANICLES** pyramidal to loosely contracted, 5–15 cm long, the branches lax and ascending to spreading. **SPIKELETS** 2- to 7-flowered, compressed. **GLUMES** 1/4 to 2/3 as long as the adjacent lemmas, subequal, 1- to 3-nerved. **LEMMAS** 3–6 mm long, distinctly keeled, mostly glabrous, sometimes puberulent between the nerves, glabrous on the callus. **ANTHERS** mostly vestigial and 0.1–0.2 mm long, or sometimes large and up to 2 mm long. [*Poa nervosa* (Hook.) Vasey var. *wheeleri* (Vasey) C. L. Hitchc.]. Moist, coniferous woods, montane to subalpine habitats. COLO: Hin, LPl. Flowering: May–Aug. Western United States.

Polypogon Desf. Polypogon

(Greek *poly*, many, and *pogon*, beard, alluding to the numerous awns) Tufted to stoloniferous annuals and perennials, often of wet places, with hollow culms. **LEAVES** with open sheaths, membranous ligules, flat blades, and lacking auricles. **INFL** a terminal, dense panicle; disarticulation below the glumes, the spikelet falling entire and often with a portion of the pedicel. **SPIKELETS** 1-flowered, perfect, the rachilla not prolonged; glumes exceeding the floret, awned or awnless; lemmas 1- to 3-nerved, usually awned, the awn terminal or subterminal, sometimes near midlength; palea much shorter than to equaling the lemma. **STAMENS** 3. $x = 7$. A genus of about 18 species of the tropical and warm-temperate regions of the world (Björkman 1960).

1. Glumes awnless .... ***P. viridis***
1′ Glumes awned .... (2)

2. Awns 1–3 (5) mm long; perennial .... *P. interruptus*
2' Awns 4–12 mm long; annual .... *P. monspeliensis*

Polypogon interruptus Kunth (severed, interrupted) Ditch polypogon. Perennials, sometimes flowering the first year. **CULMS** decumbent-based and often rooting at the nodes, 20–80 cm long or more. **LIGULES** 2–6 mm long. **BLADES** 3–6 mm wide. **PANICLES** compact but often lobed or interrupted, 3–15 cm long, 1–3 cm wide, greenish. **GLUMES** 2–3 mm long, scabrous, acute and entire to minutely cleft at the apex; awns 1–3 (5) mm long. **LEMMAS** 0.8–1.5 mm long, smooth and shiny, the awn 1–2 mm long. **PALEAS** about 3/4 the length of the lemmas. **ANTHERS** 0.5–0.7 mm long. $2n = 28, 42$. [×*Agropogon littoralis* (Sm.) C. E. Hubb.; *Polypogon littoralis* Small]. Wet ground, ditches, seeps, and springs in disturbed sites at lower elevations. ARIZ: Apa, Nav; COLO: Mon; NMEX: SJn; UTAH. 1500–1900 m (4900–6250′). Flowering: May–Aug. Native to South America, exotic in western United States and southward.

Polypogon monspeliensis (L.) Desf. (of Montpellier, France) Rabbitfootgrass. Tufted annuals. **CULMS** erect to ascending, often geniculate-based, 5–70 cm tall or more. **LIGULES** 2.5–16 mm long. **BLADES** 1–7 mm wide. **PANICLES** dense, elliptic, 2–17 cm long. **GLUMES** 1–3 mm long, hispidulous, the apex rounded and minutely lobed; awns 4–12 mm long. **LEMMAS** 0.5–1.5 mm long, smooth and shiny, the awn 0.5–4 mm long. **PALEAS** subequal to the lemmas. **ANTHERS** 0.2–1 mm long. $2n = 14, 28, 35, 42$. Ditch banks, seeps, wet disturbed ground, widespread. ARIZ: Apa, Nav; COLO: Dol, LPl, Mon, SMg; NMEX: McK, RAr, SJn; UTAH. 1550–2250 m (5050–7400′). Flowering: Jun–Sep. Native to Eurasia, widespread in North America, Mexico, South America. The furry-feeling panicles can scarcely be confused with any other species in the *Flora* region.

Polypogon viridis (Gouan) Breistr. (green) Water polypogon, water bentgrass. Tufted perennials. **CULMS** erect to decumbent-based and rooting at the nodes, 10–90 cm long. **LIGULES** 1–5 mm long. **BLADES** 1–6 mm wide. **PANICLES** oblong to narrowly pyramidal, interrupted, 2–10 cm long, 1–3 cm wide, greenish to purplish. **GLUMES** 1.5–2 mm long, scabrous, awnless, the apex obtuse to truncate. **LEMMAS** 0.7–1.3 mm long, erose-tipped, awnless. **PALEAS** subequal to the lemmas. **ANTHERS** 0.3–0.5 mm long. $2n = 28, 42$. [*Agrostis semiverticillata* (Forssk.) C. Chr.; *Polypogon semiverticillatus* (Forssk.) Hyl.]. Wet ground of springs, seeps, ponds, and ditch banks. ARIZ: Apa, Nav; COLO: Mon; NMEX: SJn; UTAH. 1120–1750 m (3700–5800′). Flowering: May–Aug. Native to Eurasia, now found in scattered locales throughout the western United States, Mexico, South America. Because of the absence of awns, this species was formerly classed in *Agrostis*, but its disarticulation below the glumes reveals a relationship to *Polypogon*, where most botanists now treat it.

Psathyrostachys Nevski Wildrye

(Greek *psathyros*, shattering, and *stachys*, spike) Tufted perennials, sometimes rhizomatous or stoloniferous, with hollow culms. **LEAVES** with closed sheaths, membranous ligules, and with or without auricles. **INFL** a spike; disarticulation in the rachis of the spike. **SPIKELETS** 2–3 per node of the rachis, 1- to 3-flowered and often with additional vestigial florets; glumes subulate, ± equal, 1-nerved; lemmas 5- to 7-nerved, narrowly elliptic, awnless to awned; palea equaling to slightly longer than the lemma, 2-keeled. **STAMENS** 3, yellow or violet. $x = 7$. A small genus of eight species native to central Asia (Baden 1991).

Psathyrostachys juncea (Fisch.) Nevski (rushlike) Russian wildrye. Densely tufted perennials. **CULMS** erect to decumbent-based, 30–80 cm tall. **LIGULES** tiny, 0.2–0.3 mm long. **BLADES** flat or rolled, 1–5 mm wide; auricles 0.5–1.5 mm long. **SPIKES** stiffly erect, 5–15 cm long, 5–15 mm wide, the rachis puberulent. **SPIKELETS** strongly overlapping. **GLUMES** 4–10 mm long, scabrous to puberulent-ciliate. **LEMMAS** lanceolate, 5.5–7.5 mm long, glabrous to stiff-puberulent; awn 1–4 mm long. **ANTHERS** 2.5–5 mm long. $2n = 14, 28$. Introduced for range restoration and reseeding, woodlands and ponderosa pine forests. COLO: LPl, Mon; NMEX: SJn. 1850–2150 m (6100–7000′). Flowering: Jun–Jul. Native to central Asia, introduced in scattered locales in western North America.

Puccinellia Parl. Alkaligrass

(for Benedetto Puccinelli, 1808–1850, Italian botanist) Tufted annuals and perennials of wetlands, with hollow culms. **LEAVES** with open sheaths, membranous ligules, flat to rolled blades, and lacking auricles. **INFL** a panicle, the branches usually rebranched. **SPIKELETS** 2- to 9-flowered, disarticulating above the glumes and between the florets; glumes rounded on the back, the first 1-nerved, the second 3- to 5-nerved, shorter than the lowermost lemma, awnless; lemmas rounded on the back or rarely keeled, obscurely (3) 5- to 7-nerved, the nerves parallel or nearly so, the tip erose, awnless; palea nearly as long as the lemma. **STAMENS** 3. $x = 7$. About 30 species. *Puccinellia pauciflora* has been removed to the genus *Torreyochloa* and is distinguished from *Puccinellia* by its conspicuous nerves, creeping rhizomes, very broad flat blades, and freshwater habitats. Plants are often more easily identified in the field than from herbarium specimens. Distinctions among the perennial species are difficult and this treatment must be considered tentative (Church 1949).

1. Plants annual; lemma nerves puberulent on the lower 1/2, glabrous between the nerves .... ***P. parishii***
1′ Plants perennial; lemma nerves glabrous, the body of the lemma may be minutely hairy between the nerves ...(2)

2. Florets coriaceous, firmer than the glumes; panicle branches spikelet-bearing to the base or nearly so; known only from Apache County, Arizona .... ***P. fasciculata***
2′ Florets membranous, similar to the glumes in texture; panicle branches naked below; widespread ....(3)

3. Plants with yellow-green herbage and erect culms; panicles 10–28 cm long, the branches as much as 15 cm long; lemmas 2–3.5 mm long; anthers 0.7–2 mm long........................ ***P. nuttalliana***
3′ Plants with blue-green herbage and geniculate-based culms; panicles 5–14 cm long, the branches to about 8 cm long; lemmas 1.5–2.5 mm long; anthers 0.3–1 mm long........................ ***P. distans***

Puccinellia distans (L.) Parl. (separated, apart) Weeping alkaligrass. Weakly tufted perennials with blue-green herbage, the culms sometimes solitary. **CULMS** usually geniculate-based, 10–70 cm tall. **LIGULES** 1–2 mm long. **BLADES** flat to involute, 1–5 mm wide. **PANICLES** open at maturity, 6–20 cm long, the branches divergent to reflexed and loosely spaced, to 8 cm long. **SPIKELETS** 5- to 9-flowered. **GLUMES** ovate-lanceolate, unequal, the first 0.5–1.5 mm long, the second 1–2.5 mm long. **LEMMAS** ovate to oblong, similar to the glumes in texture, 1.5–3 mm long, subtruncate and minutely erose at the apex. **ANTHERS** 0.3–1 mm long. $2n$ = 14, 28, 42. Alkaline and saline lowlands in desert shrub communities. ARIZ: Apa, Nav; COLO: Dol, LPl, Mon; NMEX: RAr, SJn; UTAH. 1350–2250 m (4400–7400′). Flowering: Jun–Aug. Alaska across southern Canada, south through the western United States.

Puccinellia fasciculata (Torr.) E. P. Bicknell (clustered in bundles) Torrey's alkaligrass. Tufted perennials, the herbage usually glaucous. **CULMS** erect to decumbent-based, 6–60 cm tall. **LIGULES** 0.5–2.5 mm long. **BLADES** flat to commonly involute, 1–4 mm wide. **PANICLES** mostly narrow and contracted, the branches spikelet-bearing to near the base, sometimes diverging at maturity, to 7 cm long. **SPIKELETS** 2- to 6-flowered. **GLUMES** ovate, unequal, the first 0.5–1.5 mm long and 1-nerved, the second 1–2 mm long and 3-nerved. **LEMMAS** firmer than the glumes, somewhat coriaceous, ovate to oblong, 1.5–2.5 mm long, faintly nerved, glabrous throughout. **ANTHERS** 0.5–1 mm long. $2n$ = 28. Alkaline and saline lowlands in desert shrub communities. ARIZ: Apa. 1200–1700 m (4200–5600′). Flowering: May–Jul. Intermountain region of the United States.

Puccinellia nuttalliana (Schult.) Hitchc. (for Thomas Nuttall, 1786–1859, early botanist in North America) Nuttall's alkaligrass. Tufted perennials with blue-green herbage, the culms clustered or solitary. **CULMS** strictly erect, 30–100 cm tall. **LIGULES** 1–3.1 mm long. **BLADES** flat to involute, 0.5–4 mm wide. **PANICLES** open when mature, 6–30 cm long, the branches divergent, to 15 cm long. **SPIKELETS** 3- to 7-flowered. **GLUMES** ovate-lanceolate, unequal, the first 0.5–1.8 mm long, the second 1–2.5 mm long. **LEMMAS** ovate to oblong, similar to the glumes in texture, 1.8–3.5 mm long. **ANTHERS** 0.7–2 mm long. $2n$ = 42, 56. Alkaline and saline lowlands in desert shrub communities. ARIZ: Apa, Nav; COLO: Arc, Dol, LPl, Mon, SMg; NMEX: RAr, SJn; UTAH. 1200–2050 m (4200–6700′). Flowering: May–Aug. Alaska across southern Canada, south through the western United States.

Puccinellia parishii Hitchc. (for Samuel Bonsall Parish, 1838–1928, southern California botanist) Parish's alkaligrass. Dwarf tufted annuals. **CULMS** erect, solitary or few together, 3–15 cm tall. **LIGULES** 0.5–1 mm long. **BLADES** flat to involute, less than 1 mm wide. **PANICLES** narrow and contracted, the branches erect or sometimes spreading in fruit, 1–8 cm long. **SPIKELETS** 3- to 6-flowered. **GLUMES** unequal, 1–2.3 mm long, the first about 2/3 the length of the second. **LEMMAS** broadly oblong, 1.5–2 mm long, puberulent on the lower 1/2 of the nerves. **ANTHERS** 0.5–0.8 mm long. $2n$ = unknown. Alkaline and saline lowlands in desert shrub communities, frequently with *Distichlis spicata*. ARIZ: Apa, Nav; NMEX: SJn. 1200–1700 m (4200–5600′). Flowering: May–Jun. New Mexico to California, Colorado.

Saccharum L. Sugar Cane

(Latin *saccharum*, sugar) Large, coarse, tufted perennials, sometimes with short rhizomes, with pithy to hollow culms. **LEAVES** cauline, with open sheaths, membranous or hairy ligules, flat blades, and lacking auricles. **INFL** a terminal plumose panicle, often very large. **SPIKELETS** all alike and perfect, in pairs of 1 sessile and 1 pedicelled, with a tuft of long silky hair at the base, 2-flowered, the lower floret sterile, the upper floret perfect; disarticulation below the glumes, the sessile spikelet falling with attached rachis joint and pedicel, the pedicelled spikelet falling separately; glumes large and firm, exceeding the florets, subequal, glabrous or villous; lemmas hyaline. **STAMENS** 2 or 3. $x = 10$. A tropical to subtropical genus of 25–40 species. Awned species have been placed in the genus *Erianthus*, which is herein included in *Saccharum*.

Saccharum ravennae (L.) L. (from the valley of Ravenna, Italy) Ravenna-grass. **CULMS** 2–4 m tall, glabrous. **LIGULES** hairy, 0.3–1.2 mm long. **BLADES** to 1 m long, 5–15 mm wide, strongly scabrous on the margins and usually long-hairy at the base near the ligule. **PANICLES** 30–70 cm long, the branches erect to spreading and 6–20 cm long. **SPIKELETS** 4–7 mm long, straw-colored, the basal hairs to about 7 mm long and obscuring the spikelet. **GLUMES** 4- to 5-nerved (lower) or 3-nerved (upper), acute to acuminate, awn-tipped. **LOWER LEMMA** 3–5 mm long, 1-nerved. **UPPER LEMMA** without nerves, the awn 2–8 mm long, flat. **ANTHERS** about 2 mm long. $2n = 20$. [*Erianthus ravennae* (L.) P. Beauv.]. Grown as an ornamental and escaping and persisting, often becoming noxious in wetlands and along riparian areas. NMEX: SJn. 1600–1800 m (5200–5800′). Flowering: Jul–Aug. Native to southern Europe and western Asia.

Schedonnardus Steud. Tumblegrass

(Greek *schedon*, near, and *Nardus*, a genus of grass that Steudel took to be a close relative) A monotypic genus.

Schedonnardus paniculatus (Nutt.) Branner & Coville (having a panicle) Tumblegrass. Tufted perennials with open sheaths and mostly basal foliage. **CULMS** hollow to pithy, strongly angled, 10–45 cm tall, ascending to geniculate-based. **LIGULES** membranous, 1–3.5 mm long. **BLADES** flat to folded, 1–10 cm long, 1–2 mm wide, white-margined, twisting upon drying; auricles lacking. **PANICLES** with a few unbranched spicate branches spread along the rachis, 10–25 cm long, the branches stiff and widely spreading, 4–10 cm long, often breaking at the base and rolling as a tumbleweed. **SPIKELETS** 1-flowered, disarticulating above the glumes. **GLUMES** acuminate-lanceolate, 1-nerved, stiff, unequal, the first 1.5–3 mm long, the second 2.5–5 mm long and as long as the lemma. **LEMMAS** 3-nerved, awnless or awn-tipped, 3–5 mm long. **PALEAS** similar to the lemmas and about as long. **ANTHERS** 3, 0.5–1.2 mm long. $x = 10$; $2n = 20, 30$. Dry plains and grasslands in the valleys and foothills, often with disturbance. ARIZ: Apa; COLO: LPl, Mon; NMEX: McK, RAr, San, SJn. 1700–2000 m (5500–6500′). Flowering: Jun–Sep. Prairies and plains of Canada, the United States, and northwestern Mexico. The twisted, white-margined blades are unmistakable.

Schismus P. Beauv. Mediterranean Grass

(Greek *schizein*, to split, alluding to the cleft lemma) Tufted annuals or short-lived perennials, with hollow culms. **LEAVES** with open sheaths, membranous or hairy ligules, flat or folded blades, and lacking auricles. **INFL** a terminal dense panicle. **SPIKELETS** 4- to 10-flowered, awnless; glumes subequal, about as long as the spikelet, 3- to 7-nerved; lemmas 7- to 9-nerved, the apex cleft to shallowly notched; palea 2-keeled, shorter than the lemma. **STAMENS** 3. $x = 6$. A small genus of five species native to Africa and Asia (Conert and Turpe 1974).

Schismus arabicus Nees (of Arabia) Arabian Mediterranean grass. Tightly tufted winter-annuals. **CULMS** 2–30 cm tall. **LIGULES** hairy, 0.5–1.5 mm long. **BLADES** 0.5–2 mm wide. **PANICLES** 1–4 cm long, usually less than 2 cm wide, congested. **SPIKELETS** 5–7 mm long. **GLUMES** 4–6.5 mm long. **LEMMAS** 1.8–2.6 mm long, densely short-sericeous between the nerves, the apex narrowly notched, the teeth narrowly triangular and longer than wide. **PALEAS** 1.5–2.2 mm long, reaching at most to the middle of the lemma teeth and usually to only the base of the cleft. **ANTHERS** 0.2–0.5 mm long. **CARYOPSIS** 0.5–0.8 mm long, remarkably glossy-translucent. $2n = 12$. Disturbed sites in desert communities. ARIZ: Coc; NMEX: SJn; UTAH. 1125–1675 m (3700–5500′). Flowering: Feb–May. Native to Eurasia.

Schizachne Hack. False Melic

(Greek *schizein*, to split, and *achne*, chaff, referring to the cleft lemmas) A monotypic genus of North America and eastern Asia.

Schizachne purpurascens (Torr.) Swallen (purplish) Tufted perennials. **CULMS** hollow, erect or slightly geniculate-based, loosely tufted, 35–80 cm tall or more. **LEAVES** with closed sheaths and lacking auricles. **LIGULES** membranous, 0.5–1.5 mm long, sometimes split and appearing as auricles. **BLADES** flat to involute, 2–5 mm wide, largely glabrous.

INFL a few-flowered panicle or raceme, open or closed, 7–15 cm long, the spikelets drooping. **SPIKELETS** 3- to 7-flowered, disarticulating above the glumes and between the florets. **GLUMES** membranous, 2- to 5-nerved, awnless, shorter than the lowermost floret, the first 4–6 mm long, the second 6–9 mm long. **LEMMAS** 7- to 13-nerved with the nerves converging, 8–12 mm long, awned from just below the bifid apex, the awns 8–15 mm long and twisted or geniculate; callus strongly bearded. **PALEAS** much shorter than the lemmas. **STAMENS** 3, the anthers 1.4–2 mm long. $x = 10$; $2n = 20$. Moist woods, pine forests, streamsides, and meadows. COLO: Arc, LPl, SJn. 2350–2600 m (7800–8500′). Flowering: Jun–Aug. Across Canada and the eastern United States and south through the mountainous states.

Schizachyrium Nees Bluestem

(Greek *schizein*, to split, + *achyron*, chaff, alluding to the cleft apex of the fertile lemma) Tufted or rhizomatous annuals and perennials, with hollow culms. **LEAVES** with open sheaths, membranous ligules, flat to rolled blades, and lacking auricles. **INFL** a cluster of spicate racemes (rame), each subtended by a narrow spathe and composed of repeating pairs of sessile and pedicelled spikelets; disarticulation below the glumes, a unit composed of the sessile spikelet, rachis joint, pedicel, and pedicelled spikelet. **SPIKELETS** in pairs of 1 sessile and 1 pedicelled. **SESSILE SPIKELETS** perfect and fertile, with 2 florets; glumes exceeding the florets, stiff-membranous; lower floret reduced to a sterile hyaline lemma; upper floret fertile, hyaline, 2-cleft, awned from the sinus. **STAMENS** 3. **PEDICELLED SPIKELETS** usually shorter and smaller than the sessile spikelet, with 1 floret, sterile or staminate, often disarticulating at maturity; lemma hyaline, awnless or short-awned. $x = 10$. About 60 species in the tropical, subtropical, and warm-temperate regions of the world (Gould 1967).

Schizachyrium scoparium (Michx.) Nash (broomlike) Little bluestem. Tufted perennials (ours), sometimes producing short rhizomes. **CULMS** erect, often grooved above the nodes, 45–100 cm tall or more. **LIGULES** 0.5–3 mm long. **BLADES** flat or folded, 1.5–5 mm wide. **FLOWERING SHOOTS** about 1/2–2/3 as long as the culms; racemes 3–5 cm long, with 7–12 spikelets. **SESSILE SPIKELETS** 6–11 mm long, the callus hairs to 3 mm long; awn 5–17 mm long. **PEDICELLED SPIKELETS** 1–6 mm long, sterile, mostly lacking a lemma and awnless. $2n = 40$. [*Andropogon scoparius* Michx.; *Andropogon scoparius* Michx. var. *frequens* F. T. Hubb.]. Canyons and rocky slopes. ARIZ: Apa, Nav; COLO: Arc, Dol, Hin; NMEX: McK, RAr, San, SJn; UTAH. 1700–2350 m (5600–7700′). Flowering: Jul–Sep. Canada to central Mexico, absent from the far western states. Our plants belong to **var.** ***scoparium***.

Secale L. Rye

(classical Latin name for rye) Tufted annuals (ours) and perennials, with hollow culms. **LEAVES** with open sheaths, membranous ligules, flat blades, and prominent auricles. **INFL** a spike, the rachis persistent or disarticulating. **SPIKELETS** 1 per node, 2-flowered, the rachilla prolonged beyond the upper floret and sometimes bearing a rudiment; disarticulation in the rachis or above the glumes and between the florets (ours mostly) or not at all; glumes narrow, rigid, shorter than the lemmas; lemmas 5-nerved, asymmetrically keeled, strongly ciliate, tapering to a long scabrous awn; palea 2-nerved. **STAMENS** 3. $x = 7$. A small genus of three species native to Eurasia (Frederiksen and Petersen 1998).

Secale cereale L. (a grain) Rye. Robust tufted annuals or occasionally biennials. **CULMS** 40–120 cm or more tall. **LIGULES** 0.5–2 mm long. **BLADES** 3–10 mm wide, the auricles about 1 mm long. **SPIKES** 4–18 cm long, often nodding when mature. **GLUMES** 1-nerved, ciliolate on the keel, equal, 8–20 mm long, nearly linear. **LEMMAS** 12–18 mm long, strongly ciliate on the keel; awns 10–50 mm long. **PALEAS** about as long as the lemmas. **ANTHERS** 7–9 mm long. $2n = 14, 16, 27–29$. Introduced as a cultivated crop, and also widely used for erosion control along roadsides, occasionally escaping around fields, but not persisting long. ARIZ: Nav; COLO: LPl, Mon; NMEX: SJn; UTAH. 1400–2400 m (4600–7800′). Flowering: May–Jul. Native to Eurasia, now found widely throughout Canada and the United States, and somewhat in Mexico, also South America. This is one of the world's most important cereal grasses but is also susceptible to ergot infestation by the fungus *Claviceps purpurea*. Hybrids between this and wheat are known as ×*Triticosecale* (triticale).

Setaria P. Beauv. Bristlegrass

(Latin *seta*, bristle, + *aria*, possessing) Tufted annuals and perennials (rarely rhizomatous), with hollow or pithy culms. **LEAVES** with open sheaths, membranous or hairy ligules, flat to rolled blades, and lacking auricles. **INFL** a terminal panicle, the branches usually highly reduced and the panicle spikelike; disarticulation below the glumes, the bristles persistent on the plant. **SPIKELETS** subtended by 1–several bristles (sterile branchlets), 2-flowered, dorsally compressed, awnless, the lower floret reduced, the upper floret perfect; glumes unequal, the first much smaller than the second; lower floret staminate or sterile, the lemma membranous and similar in size and texture to the second glume;

upper floret perfect, the lemma and palea forming a hard seedcase around the flower and grain. **STAMENS** 3. $x = 9$. About 140 species of tropical to warm-temperate regions of the world. *Setaria vulpiseta* Roem. & Schult. has been reported from the *Flora* area, but that name is restricted to South America and should not be confused with our North American *Setaria* (Rominger 1962; Toolin and Reeder 2000).

1. Margins of the sheaths glabrous; second glume about 1/2 the length of the adjacent upper lemma, which is easily visible; upper lemma strongly wrinkled .... ***S. pumila***

1' Margins of the sheaths pubescent; second glume about the same length as the adjacent upper lemma, which is scarcely visible; upper lemma only faintly wrinkled .... ***S. viridis***

Setaria pumila (Poir.) Roem. & Schult. (dwarfish, pygmy) Yellow bristlegrass. Tufted annuals. **CULMS** erect to geniculate-based or decumbent, sometimes rooting at the lower nodes, 25–100 cm tall or more. **LIGULES** a ciliate membrane, 0.3–1.2 mm long. **BLADES** flat, 3–8 mm wide, usually papillose-pilose near the throat. **PANICLES** yellowish, erect, densely spicate, 3–15 cm long; bristles 4–12 below each spikelet, 3–8 mm long. **SPIKELETS** 2–3.4 mm long, turgid. **FIRST GLUME** about 1/3 the length of the spikelet, 3-nerved. **SECOND GLUME** about 1/2 the length of the spikelet, 5-nerved. **LOWER FLORET** sterile or often staminate, the lemma equaling the upper lemma. **UPPER FLORET** conspicuously exposed, strongly rugose, 2.3–3.2 mm long; anthers 0.5–1.5 mm long. $2n = 36, 72$. [*Setaria glauca* of numerous authors; *Setaria lutescens* (Weigel) F. T. Hubbard]. Weedy ground along roads, fields, and in lawns. ARIZ: Nav; NMEX: SJn; UTAH. 1300–1700 m (4300–5600′). Flowering: Jul–Sep. Native to Europe and common throughout North America and warm-temperate regions of the world.

Setaria viridis (L.) P. Beauv. (green) Green bristlegrass. Tufted annuals. **CULMS** decumbent-based and ascending above, 10–100 cm tall or more. **LIGULES** a ciliate membrane, 1–2 mm long. **BLADES** flat, 4–20 mm wide, glabrous near the throat. **PANICLES** green, nodding only at the tip, densely spicate, 3–20 cm long; bristles 1–3 below each spikelet, 5–10 mm long. **SPIKELETS** 1.8–2.2 mm long. **FIRST GLUME** about 1/3 the length of the spikelet, 3-nerved. **SECOND GLUME** about as long as the spikelet, 5- to 6-nerved. **LOWER FLORET** sterile, the lemma slightly exceeding the upper lemma. **UPPER FLORET** finely rugose, pale green, 1.8–2.5 mm long; anthers 0.5–0.7 mm long. $2n = 18$, 35, 36. A common weed of roadsides, lawns, flower beds, and other disturbed areas, widespread. ARIZ: Apa, Nav; COLO: Arc, Dol, LPl, Min, Mon, SJn; NMEX: McK, RAr, San, SJn; UTAH. 1500–2400 m (4900–7900′). Flowering: Jun–Sep. Native to Europe and common throughout North America and warm-temperate regions of the world.

Sorghastrum Nash

(*Sorghum*, a grass genus, and Latin *astrum*, a poor imitation of, alluding to the resemblance) Tufted perennials, with or without rhizomes, the culms hollow or pithy. **LEAVES** with open sheaths, the upper margins projecting into erect sheath auricles, and flat blades. **INFL** a rebranching panicle, narrow but loose. **SPIKELETS** borne in pairs of 1 sessile and 1 pedicelled, the pedicelled spikelet reduced and completely absent, all the spikelets present thus sessile, 2 pedicels present at the ends of the branches; disarticulation below the glumes and consisting of the sessile spikelet, rachis joint, and pedicel. **SESSILE SPIKELETS** perfect, 2-flowered; glumes coriaceous, exceeding the florets, the first glume pubescent on the back, the second glume glabrous; lower floret sterile and reduced to a hyaline lemma; upper floret perfect, the lemma hyaline, bifid, and awned from the sinus. **STAMENS** 3. $x = 10$. About 18 species, all but 2 native to the Americas.

Sorghastrum nutans (L.) Nash (nodding) Indiangrass. Tufted perennials from short, stout rhizomes. **CULMS** erect, 50–150 cm tall or more, the nodes hirsute with erect hairs. **LIGULES** 2–6 mm long. **BLADES** 1–4 mm wide. **PANICLES** 20–75 cm long, yellowish to brownish. **SPIKELETS** 5–9 mm long. **FIRST GLUME** 7- to 9-nerved, 5–8 mm long, pubescent across the back. **SECOND GLUME** 5-nerved, 5–8 mm long, glabrous. **LOWER FLORET** sterile, 4–5 mm long. **UPPER FLORET** 4.5–6.5 mm long; awn 10–30 mm long, once-geniculate. $2n = 20, 40, 80$. Open woods, grassy plains, moist rocky hillsides. ARIZ: Apa; COLO: Arc, LPl; NMEX: McK, RAr, SJn; UTAH. 1700–2200 m (5600–7200′). Flowering: Aug–Sep. From Canada to Mexico in the Rocky Mountain states and eastward, one of the main grasses of the tallgrass prairie.

Sorghum Moench Sorghum, Johnson Grass

(Latinized from *sorgo*, the Italian name of the plant) Tufted annuals to rhizomatous perennials with hollow to pithy culms. **LEAVES** with open sheaths, membranous ligules, flat blades, and lacking auricles. **INFL** a large rebranching panicle, open to compact. **SPIKELETS** borne in pairs of 1 sessile and 1 pedicelled, 2 pedicels present at the ends of the branches; disarticulation below the glumes and consisting of the sessile spikelet, rachis joint, and pedicelled spikelet. **SESSILE SPIKELETS** perfect, 2-flowered, dorsally compressed; glumes coriaceous, exceeding the florets, the first glume 5- to 15-nerved and awnless, the second glume 2-keeled and sometimes awned; lower floret sterile and reduced to a hyaline lemma; upper floret pistillate or perfect, the lemma hyaline, and sometimes awned. **STAMENS** 3. **PEDICELLED SPIKELETS** staminate or sterile, nearly as large as the sessile spikelet. $x = 10$. About 25 species of tropical and subtropical regions, most native to the Old World (Doggett 1970; Spangler 2000).

1. Plants robust annuals, lacking rhizomes; panicle branch segments persistent and not breaking apart easily, or tardily disarticulating .... ***S. bicolor***

1′ Plants perennial, with strong rhizomes; panicle branch segments breaking apart easily .... ***S. halepense***

Sorghum bicolor (L.) Moench (two-colored) Sorghum, milo. Robust tufted annuals. **CULMS** 75–200 cm tall or more. **LIGULES** 1.5–5.5 mm long. **BLADES** flat, 15–50 mm wide. **PANICLES** 10–60 cm long, open or densely contracted, 5–25 cm wide. **SESSILE SPIKELETS** perfect, lanceolate to ovate, 3–9 mm long; glumes glabrous to pubescent; lower floret sterile, 3.8–5 mm long; upper floret fertile; awn 5–20 mm long; anthers 2–2.8 mm long. **PEDICELLED SPIKELETS** staminate or sterile, usually shorter and thinner than the sessile spikelets. $2n = 20, 40$. A cultivated crop and sometimes escaping but not persisting long. NMEX: SJn. 1400–1700 m (4600–5500′). Flowering: Jul–Sep. Native to Eurasia and cultivated in many parts of the world. There are numerous races and cultivated forms, all interfertile; these include grain sorghum, sweet sorghum, sorgo, pop sorghum, broomcorn sorghum, and sudangrass.

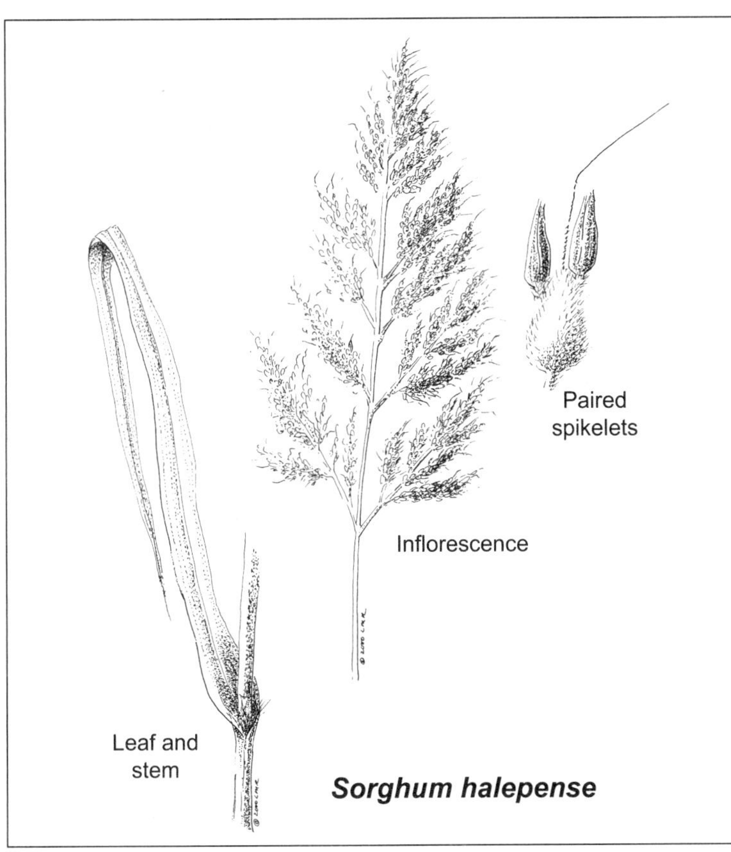

Sorghum halepense

Sorghum halepense (L.) Pers. (from Aleppo, Turkey) Johnson grass. Plants rhizomatous perennials. **CULMS** 50–150 cm tall or more. **LIGULES** 2–6 mm long, conspicuously ciliate. **BLADES** flat, 8–40 mm wide. **PANICLES** 10–50 cm long, open, 5–25 cm wide. **SESSILE SPIKELETS** lanceolate to ovate, 3.5–6.5 mm long; glumes appressed-pubescent; lower floret sterile, 4–4.5 mm long; upper floret fertile, awnless or with an awn to 13 mm long; anthers 1.9–2.7 mm long. **PEDICELLED SPIKELETS** staminate, nearly as long as the sessile spikelets, but narrower, 3.6–5.6 mm long; awnless. $2n = 20, 40$. An aggressive weed of fields, pastures, ditches, and moist waste places. ARIZ: Apa; COLO: LPl; NMEX: McK, San, SJn; UTAH. 1200–2050 m (3900–6700′). Flowering: Jul–Sep. Native to Eurasia and now found throughout the warm regions of the world. Johnson grass may accumulate cyanide under stress conditions and become toxic to livestock, but it can also provide valuable forage if cured properly. The name remembers William Johnson of Alabama, who was active in promoting this grass for forage in the mid-1800s.

Spartina Schreb. Cordgrass

[Greek *spartine*, a cord made from *Spartium junceum* (a legume), alluding to the cordlike blades of some species] Strongly rhizomatous perennials with hollow culms. **LEAVES** with open sheaths, hairy ligules, flat to rolled blades, and lacking auricles. **INFL** a panicle of few to many spicate unbranched branches. **SPIKELETS** 1-flowered, strongly compressed, disarticulating above the glumes; glumes unequal, 1- to 3-nerved, awnless or awned, the first shorter than the floret, the second longer than the floret; lemma 1- to 3-nerved, awnless; palea 2-neved. **STAMENS** 3. $x = 10$. About 15 or so species, all from wetlands, marshes, and estuaries, both coastal and interior. Some species can be invasive.

Spartina gracilis Trin. (slender) Alkali cordgrass. Strongly rhizomatous perennials. **CULMS** not clustered but usually solitary, 30–75 cm tall. **LIGULES** 0.5–1.5 mm long. **BLADES** flat, becoming involute upon drying, 3–5 mm wide, scabrous on the margins. **PANICLES** 8–20 cm long; branches appressed to the main axis, 2–8 in number, 2–5 cm long. **SPIKELETS** densely crowded and appressed on the branch, 18–25 per branch. **GLUMES** noticeably ciliate on the keels, the first 3.5–6 mm long, the second 7–10 mm long. **LEMMAS** 1-nerved, 6–8 mm long, blunt, the keel ciliate. **PALEAS** subequal to the lemmas. **ANTHERS** 2.5–5 mm long. $2n$ = 40, 42. Wet meadows and alkaline lakeshores. UTAH. 1350–1550 m (4400–5100′). Flowering: Jun–Sep. Central Canada westward and southward to California and New Mexico.

Sphenopholis Scribn. Wedgescale

(Greek *sphen*, a wedge, + *pholis*, scale, alluding to the shape of the second glume) Tufted annuals or short-lived perennials with hollow culms. **LEAVES** with open sheaths, membranous ligules, flat blades, and lacking auricles. **INFL** a panicle, usually contracted. **SPIKELETS** (1) 2- to 3-flowered, the florets perfect, the rachilla prolonged beyond the upper floret, awnless; disarticulation below the glumes, the spikelet falling as a unit; glumes dissimilar in size and shape, the first glume narrow and 1-nerved, the second glume broader, oblanceolate, and 3-nerved, shorter than the lowermost lemma; lemmas faintly 5-nerved, rounded on the back, awnless; palea shorter than the lemma, 2-nerved. **STAMENS** 3. x = 7. A small genus of four species in North America (Erdman 1965).

Sphenopholis obtusata (Michx.) Scribn. (blunt) Prairie wedgescale. Tufted annuals or short-lived perennials. **CULMS** erect, 30–70 cm tall. **LIGULES** erose-ciliate, 1.5–3.5 mm long. **BLADES** flat, 1.5–6 mm wide. **PANICLES** narrow, spicate, interrupted below, green, 4–18 cm long, about 2 cm wide, the rachis and pedicels glabrous. **SPIKELETS** 1.5–5 mm long. **GLUMES** unequal, the first 1.7–3 mm long and linear, the second 2–3 mm long and obovate, somewhat hood-shaped, about 2/3 the length of the spikelet. **LEMMAS** lanceolate, 2–3 mm long. **PALEAS** shorter than the lemmas. **ANTHERS** 0.3–0.7 mm long. $2n$ = 14. Moist or wet ground along springs, seeps, streams, ditches, and canals. ARIZ: Apa, Nav; COLO: Arc, LPl, Mon; NMEX: SJn; UTAH. 1370–2100 m (4500–6800′). Flowering: Jul–Sep. Throughout North America. This is often confused with *Koeleria macrantha*, but that species has puberulent panicle rachises and pedicels, much larger glumes, and usually tan-colored panicles.

Sporobolus R. Br. Dropseed

(Greek *spora*, seed, + *ballein*, to throw, alluding to the fact that the grain is not fused to the ovary, allowing it to drop from the spikelet) Tufted annuals and perennials, lacking rhizomes, with pithy or hollow culms. **LEAVES** with open sheaths, hairy or membranous ligules fringed with hairs, flat to rolled blades, and lacking auricles. **INFL** a rebranched panicle, narrow and spikelike to open and diffuse. **SPIKELETS** 1-flowered, small, awnless, the rachilla not prolonged beyond the floret, mostly disarticulating above the glumes and also the seed separating from the pericarp; glumes 1-nerved, scarious, often unequal, at least one shorter than the floret; lemma 1-nerved, scarious; palea 2-nerved, nearly as long as the lemma. **STAMENS** 3. x = 6, 9. A large genus of 160 or more species growing throughout the world in tropical to temperate habitats.

1. Plants small, slender annuals or short-lived perennials, mostly less than 35 cm tall; culms geniculate ....(2)
1′ Plants usually stout perennials, mostly more than 35 cm tall; culms erect to decumbent-based.........................(3)

2. Lower nodes of the panicle with 7–12 whorled branches; glumes very unequal; panicles narrow when in flower and open at maturity..................... ***S. pyramidatus***
2′ Lower nodes of the panicle with 1–3 branches; glumes equal or nearly so; panicles narrow, the lower branches often included in the subtending sheath..................... ***S. neglectus***

3. Exserted portion of the panicle dense and spikelike, 1–3 (4) cm wide, the branches erect-appressed, often scarcely discernible at arm's length ....(4)
3′ Exserted portion of the panicle loosely contracted to diffuse, 3–30 cm wide, the branches ascending to widely diverging, easily discernible at arm's length ....(6)

4. Flowering shoots with both terminal and axillary panicles, the latter often hidden in the sheaths; spikelets large, 4–6 mm long; anthers 1.5–2.5 mm long..................... ***S. compositus***
4′ Flowering shoots with only terminal panicles; spikelets smaller, 1–3 mm long; anthers 0.4–1.5 mm long............(5)

5. Culms robust, 1–2 m tall, 3–8 mm thick at the base; anthers 0.6–1 mm long .... ***S. giganteus***
5′ Culms more slender, mostly less than 1 m tall, 1.5–3.5 (4) mm thick at the base; anthers 0.3–0.5 mm long..................... ***S. contractus***

6. Sheaths with many long hairs at the summit; plants more slender, the shoots easily pulled from the ground, the basal sheaths not shiny, often darkened, the roots thin ....(7)
6' Sheaths glabrous or with only a few long hairs at the summit; plants robust, the shoots difficult to pull from the ground, the basal sheaths shiny and cream-colored, the roots thick ....(8)

7. Mature panicle branches and pedicels divaricate and flexuous, usually tangled with other branches or other panicles; branch pulvini pubescent; spikelets loosely arranged on the branches .... ***S. flexuosus***
7' Mature panicle branches erect to spreading but not flexuous or tangled; branch pulvini glabrous; spikelets crowded on the branches .... ***S. cryptandrus***

8. Panicles 10–45 cm long; branchlets naked below, the pedicels 0.5–2 mm long, often spreading .... ***S. airoides***
8' Panicles 20–60 cm long; branchlets densely flowered to the base, the pedicels less than 0.5 mm long, appressed to the branchlets .... ***S. wrightii***

Sporobolus airoides (Torr.) Torr. (resembling the grass genus *Aira*) Alkali sacaton. Densely tufted, robust perennials without rhizomes, the basal sheaths thick and shiny, cream-colored. **CULMS** 35–150 cm tall, usually hollow. **LIGULES** 0.1–0.3 mm long. **BLADES** flat to involute, 2–5 mm wide, the collar glabrous or sparsely hairy. **PANICLES** diffuse, 10–45 cm long, 10–25 cm wide, the branches widely spreading, the branchlets naked below, the pulvini glabrous. **SPIKELETS** 1.3–2.8 mm long, greenish or purple-tinged. **GLUMES** unequal, the first 0.5–1.8 mm long, the second 1.1–2.8 mm long and at least 2/3 as long as the floret. **LEMMAS** 1.2–2.5 mm long, glabrous. **PALEAS** subequal to the lemmas. **ANTHERS** 1.1–1.8 mm long, yellowish to purplish. $2n = 80, 90, 108, 126$. Low alkaline plains, flats, playas, and washes. ARIZ: Apa, Nav; COLO: Arc, Dol, LPl, Mon; NMEX: McK, RAr, San, SJn; UTAH. 1450–2080 m (4800–6800′). Flowering: Jun–Aug. Central and western plains and mountain states, into Mexico.

Sporobolus compositus (Poir.) Merr. (put together, joined) Tall dropseed. Stout tufted perennials, frequently with sharp-pointed basal buds, without short rhizomes (ours). **CULMS** erect, stout, 30–130 cm tall, 2–5 mm thick. **LIGULES** 0.1–0.5 mm long. **BLADES** flat to involute, 2–5 mm wide, the collars sparsely hairy. **PANICLES** terminal and axillary, usually spikelike, 8–30 cm long, less than 2 cm wide, the branches erect-appressed, the pulvini glabrous. **SPIKELETS** 4–6 mm long. **GLUMES** subequal, the midnerves usually greenish, 2–5 mm long, about as long as the lemma. **LEMMAS** 2.5–6 mm long, glabrous, mostly 1-nerved, but sometimes with 2 or 3 nerves. **PALEAS** about as long as the lemmas. **ANTHERS** 0.5–3.2 mm long, yellow to orange. $2n = 54, 88, 108$. [*S. asper* (P. Beauv.) Kunth]. Sandy plains and roadsides, deserts and foothills. NMEX: SJn. 1450–1900 m (4800–6200′). Flowering: Aug–Sep. Throughout most of the United States from the eastern coast to Utah and Arizona.

Sporobolus contractus Hitchc. (contracted) Spike dropseed. Tufted perennials. **CULMS** 40–100 (120) cm tall, 2–4 mm thick near the base. **LIGULES** 0.4–1 mm long. **BLADES** flat to rolled, 3–8 mm wide, the collars with conspicuous tufts of long hairs. **PANICLES** all terminal, spikelike, 15–45 cm long, less than 1 cm wide, the base usually included in the sheath, the pulvini glabrous. **SPIKELETS** 1.7–3.2 mm long. **GLUMES** unequal, the first 0.7–1.7 mm long, the second 2–3.2 mm long, at least 2/3 the length of the lemma. **LEMMAS** 2–3.2 mm long, glabrous. **PALEAS** slightly shorter than the lemmas. **ANTHERS** 0.3–0.5 mm long. $2n = 36$. Sandy ground in desert and foothill communities. ARIZ: Apa, Coc, Nav; NMEX: McK, RAr, San, SJn; UTAH. 1400–2025 m (4600–6600′). Flowering: Jun–Sep. Southwestern United States to northern Mexico.

Sporobolus cryptandrus (Torr.) A. Gray (hidden man or stamen) Sand dropseed. Tufted perennials, the bases not hard and knotty. **CULMS** 30–100 cm or more tall, erect to decumbent-based. **LIGULES** 0.5–1 mm long. **BLADES** flat to rolled, 2–6 mm wide, the collars with conspicuous tufts of long hair. **PANICLES** terminal, spikelike when emerging from the sheath, ultimately loosely open when mature, 15–40 cm long, 3–12 cm wide, the base often included in the sheath or the entire panicle sometimes enclosed in the sheath, the pulvini glabrous. **SPIKELETS** 1.5–2.7 mm long. **GLUMES** unequal, the first 0.6–1.1 mm long, the second 1.5–2.7 mm long and at least 2/3 the length of the lemma. **LEMMAS** 1.4–2.5 mm long, glabrous. **PALEAS** slightly shorter than the lemmas. **ANTHERS** 0.5–1 mm long, yellowish to purplish. $2n = 36, 38, 72$. Sandy plains and mesas, roadsides, waste places, deserts to the mountains. ARIZ: Apa, Coc, Nav; COLO: Arc, Dol, LPl, Min, Mon, SMg; NMEX: McK, RAr, San, SJn; UTAH. 1200–2500 m (4000–8200′). Flowering: Jun–Sep. Nearly throughout North America.

Sporobolus flexuosus (Thurb. ex Vasey) Rydb. (bent, curved, flexuous) Mesa dropseed. Tufted perennials, the bases not hard and knotty. **CULMS** 30–100 cm or more tall, erect to decumbent-based. **LIGULES** 0.5–1 mm long. **BLADES** flat to rolled, 2–5 mm long, the collars with conspicuous tufts of long hair. **PANICLES** terminal, open, 10–30 cm long,

4–12 cm wide, the base often included in the sheath, the pulvini prominent and pubescent. **SPIKELETS** 1.8–2.5 mm long. **GLUMES** unequal, the first 1–1.5 mm long, the second 1.4–2.5 mm long and about as long as the lemma. **LEMMAS** 1.4–2.5 mm long. **PALEAS** as long as the lemmas. **ANTHERS** 0.4–0.7 mm long, yellowish. 2*n* = 36, 38. Sandy to gravelly plains and mesas, roadsides, desert scrub and woodland communities. ARIZ: Apa, Coc, Nav; COLO: Mon; NMEX: McK, SJn; UTAH. 1350–1750 m (4400–5800′). Flowering: Jun–Sep. Southwestern United States and northern Mexico.

Sporobolus giganteus Nash (gigantic, mighty) Giant dropseed. Tufted robust perennials. **CULMS** 1–2 m tall, 4–10 mm thick near the base. **LIGULES** 0.5–1.5 mm long. **BLADES** flat, 4–10 mm wide, the collars with conspicuous tufts of long hair. **PANICLES** terminal, spikelike, dense, 25–75 cm long, 1–4 cm wide, the base often included in the sheath, the pulvini glabrous. **SPIKELETS** 2.5–3.5 mm long. **GLUMES** unequal, the first 0.6–2 mm long, the second 2–3.5 mm long, as long as the lemma. **LEMMAS** 2.5–3.4 mm long, glabrous. **PALEAS** about as long as the lemmas. **ANTHERS** 0.6–1 mm long, yellowish. 2*n* = 36. Sandy hills and plains, dunes, desert communities. ARIZ: Apa, Nav; COLO: Mon; NMEX: McK, San, SJn; UTAH. 1200–1900 m (4000–6200′). Flowering: Jun–Sep. Southwestern United States and northern Mexico.

Sporobolus neglectus Nash (disregarded, not chosen) Puff-sheath dropseed. Tufted annuals, often much-branched, with inflated sheaths. **CULMS** geniculate, wiry, 10–40 cm tall. **LIGULES** 0.1–0.3 mm long. **BLADES** flat to loosely folded, 0.6–2 mm wide, the collars glabrous to sparsely hairy. **PANICLES** terminal and axillary, narrow and spikelike, 2–5 cm long, less than 1 cm wide, the base often included in the sheath, the lower nodes with 1–3 branches spikelet-bearing to the base. **SPIKELETS** 1.6–3 mm long. **GLUMES** subequal, shorter than the floret, 1.5–2.5 mm long. **LEMMAS** 1.6–3 mm long, glabrous. **PALEAS** as long as the lemmas. **ANTHERS** 1–1.6 mm long, purplish. 2*n* = 36. Sandy fields, floodplains, disturbed ground. ARIZ: Apa; COLO: Arc, LPl; NMEX: SJn. 1550–2555 m (5100–8390′). Flowering: Aug–Sep. Scattered locales throughout southern Canada and the United States, most common in the central states. Plants are typically much-branched, with exposed nodes and numerous terminal and axillary panicles.

Sporobolus pyramidatus (Lam.) Hitchc. (pyramid-shaped) Whorled dropseed. Low, tufted annuals (ours) or short-lived perennials. **CULMS** in tight clumps, geniculate to ascending, 7–35 cm tall. **LIGULES** 0.3–1 mm long. **BLADES** flat, 2–5 mm wide, the collars glabrous to sparsely hairy. **PANICLES** terminal, closed during anthesis, open at maturity, 4–15 cm long, 0.5–6 cm wide, the primary branches with elongated glands, branchlets and spikelets appressed on the primary branch, the pulvini glabrous. **SPIKELETS** 1.2–1.8 mm long. **GLUMES** unequal, the first 0.3–0.7 mm long, the second 1.2–1.8 mm long and as long as the lemma. **LEMMAS** 1.2–1.7 mm long, translucent. **PALEAS** as long as the lemmas. **ANTHERS** 0.2–0.4 mm long, yellowish to purplish. **GRAINS** noticeably squarish, brownish. 2*n* = 24, 36, 54. [*Sporobolus pulvinatus* Swallen]. Sandy or clay soils, disturbed ground, low flats and plains in desert communities. ARIZ: Apa, Nav; COLO: LPl; NMEX: McK, San, SJn; UTAH. 1500–1850 m (4900–6200′). Flowering: Jul–Sep. South-central and southwestern United States to Argentina. Our plants correspond to what was called *S. pulvinatus*, a smaller annual plant of the Southwest, as opposed to the more eastern and southern *S. pyramidatus*, which can be a short-lived perennial.

Sporobolus wrightii Munro ex Scribn. (for Charles Wright, 1811–1885, American botanical collector) Giant sacaton. Densely tufted, robust perennials without rhizomes, the basal sheaths thick and shiny, cream-colored. **CULMS** 1–2.5 m tall, usually hollow. **LIGULES** 0.5–1 mm long. **BLADES** mostly flat, 3–10 mm wide, the collar glabrous or sparsely hairy. **PANICLES** open and exserted, 20–60 cm long, 12–26 cm wide, the branches spreading, the branchlets appressed, the pulvini glabrous. **SPIKELETS** 1.5–2.5 mm long, greenish or purple-tinged. **GLUMES** unequal, the first 0.5–1 mm long, the second 0.8–2 mm long and at least 2/3 as long as the floret. **LEMMAS** 1.2–2.5 mm long, glabrous. **PALEAS** subequal to the lemmas. **ANTHERS** 1.1–1.3 mm long, yellowish to purplish. 2*n* = 36. [*S. airoides* (Torr.) Torr. var. *wrightii* (Munro ex Scribner) Gould]. Alkaline flats and washes. NMEX: McK, SJn. 1500–1750 m (5000–5800′). Flowering: Jun–Aug. Southwestern United States to central Mexico. Small plants can be confused with tall *S. airoides*, but the panicle features usually hold.

Stipa L. Needlegrass

Recent studies in the *Stipeae* have resulted in the relegation of all North American members of *Stipa* to six segregate genera, keyed below, of which five occur in our flora. The genus *Stipa* remains solely a Eurasian taxon.

1. Palea hardened, longitudinally grooved and slightly longer than the lemma, protruding from between the lemma margins as a small point; lemma margins involute, fitting into the grooves of the palea .... ***Piptochaetium*** (not in our flora)

1′ Palea usually membranous, not grooved, shorter than or equaling the lemma, not protruding as a small point; lemma margins flat .... (2)

2. Lemma margins strongly overlapping; palea less than 1/3 the length of the lemma, glabrous, lacking veins .... ***Nassella***
2′ Lemma margins not or only slightly overlapping; palea 1/3 to equaling the length of the lemma, always pubescent when short, sometimes glabrous when longer, 2-veined .... (3)

3. Awns 6–20 cm long or more; glumes longer than 1.8 cm .... ***Hesperostipa***
3′ Awns 0.5–7.5 cm long, if longer than 6 cm then the glumes 1–1.5 cm long .... (4)

4. Palea pubescent, the apex flat, the veins terminating below the apex; lemma coriaceous at maturity but not strongly indurate .... ***Achnatherum***
4′ Palea glabrous or pubescent, the apex appearing prow-tipped or pinched, the veins extending to the apex; lemma indurate at maturity .... (5)

5. Florets dorsally compressed; lemma margins not overlapping, the palea exposed, at least in part .... ***Piptatherum***
5′ Florets terete; lemma margins slightly overlapping, the palea hidden .... ***Oryzopsis***

Torreyochloa G. L. Church Mannagrass

(for John Torrey, 1796–1873, celebrated New England botanist) Rhizomatous perennials of nonsaline wet places, with hollow culms and cross-septate sheaths, decumbent-based and often rooting at the nodes. **LEAVES** with open sheaths, membranous ligules, flat blades, and lacking auricles. **INFL** a terminal panicle. **SPIKELETS** with 2–8 florets, laterally compressed to nearly terete, pedicelled, awnless; disarticulation above the glumes and between the florets; glumes nearly equal to unequal, shorter than the lowermost lemma, 1- to 3-nerved; lemmas 7- to 9-nerved, the nerves ± parallel and not converging at the apex and scaberulous; palea about as long as the lemma, 2-nerved. **STAMENS** usually 3. $x = 7$. A small genus of four species, two from North America and two from Asia; formerly included in *Glyceria* and *Puccinellia* (Church 1952).

Torreyochloa pallida (Torr.) G. L. Church (pale) Weak mannagrass. **CULMS** 25–120 cm or more long. **LIGULES** 3–9 mm long. **BLADES** flat and broad, 4–10 mm or more wide. **PANICLES** loose, 8–25 cm long, the lax branches 2–3 per node and ascending to drooping. **SPIKELETS** 4- to 8-flowered. **GLUMES** ovate, 1–2 mm long, the first only slightly smaller than the second. **LEMMAS** strongly 5- to 7-nerved, 1.5–4 mm long, truncate to acute, purplish-banded. **ANTHERS** 0.5–0.7 mm long. $2n = 14$. Freshwater marshes, ponds, seeps, streams, and bogs, coniferous forests, montane to subalpine. ARIZ: Apa, Nav; COLO: Hin, Min, SJn. 2550–3000 m (8400–9800′). Flowering: Jun–Sep. Western North America, from Alaska to New Mexico. Our material belongs to **var. *pauciflora*** (J. Presl) J. I. Davis (few-flowered). The other varieties grow east of the Mississippi.

Tridens Roem. & Schult. Tridens

(Latin *tres*, three, + *dens*, tooth, referring to the tip of the lemma) Tufted perennials, often with knotty bases and short rhizomes, with pithy culms. **LEAVES** with open sheaths, hairy ligules (a long-fringed minute rim), flat to rolled blades, and lacking auricles. **INFL** an open or contracted panicle, few- to many-flowered. **SPIKELETS** several-flowered, awnless, disarticulating above the glumes and between the florets; glumes shorter than to equaling the lowermost floret, 1- to 3-nerved; lemmas 3-nerved, the nerves usually hairy, the apex emarginate to bilobed and with a tiny mucro; palea slightly shorter than the lemma. **STAMENS** 3, reddish purple. $x = 10$. A genus of 14 species native to North America; *Dasyochloa* and *Erioneuron* have been segregated from *Tridens* (Tateoka 1961).

Tridens muticus (Torr.) Nash (awnless) Slim tridens. Tufted perennials with hard bases. **CULMS** 20–50 cm tall (ours). **LIGULES** 0.5–1 mm long. **BLADES** 1–2 mm wide. **PANICLES** narrow and contracted, but interrupted and not spike-like, 6–20 cm long, less than 1 cm wide. **SPIKELETS** 5- to 11-flowered, purplish. **GLUMES** glabrous, the first 3–8 mm long, the second 4–10 mm long and 1-nerved (ours). **LEMMAS** 3.5–7 mm long, short-pilose on the nerves. **PALEAS** 1–2 mm shorter than the lemmas. **ANTHERS** 1–1.5 mm long. $2n = 40$. Desert scrub communities. UTAH. 1300–1550 m (4300–5100′). Flowering: May–Sep. Arid southwestern United States and adjacent Mexico.

Trisetum Pers. Trisetum

(Latin *tres*, three, + *seta*, bristle, alluding to the awn and two lateral teeth at the lemma apex) Tufted annuals and perennials with hollow culms. **LEAVES** with open sheaths, membranous ligules, flat to rolled blades, and lacking auricles. **INFL** a terminal panicle, usually contracted. **SPIKELETS** 2- to 4-flowered, perfect, the rachilla usually prolonged beyond the uppermost floret; disarticulation above the glumes and between the florets (ours), or below the glumes in some; glumes awnless, usually as long as the lowermost floret or longer; lemmas 3- to 7-nerved, awnless or awned from below the bifid apex; palea 2-nerved, the nerves extended as bristles. **STAMENS** 3. $x = 7$. Seventy-five species occurring in the cool-temperate to Arctic regions of the world (Finot et al. 2005; Hultén 1959).

1. Lemmas awnless or more commonly with short awns less than 2 mm long, scarcely visible .... *T. wolfii*
1' Lemmas with awns longer than 3 mm, easily visible .... (2)

2. Panicles dense and spikelike, the branches mostly less than 1 cm long and erect-appressed; stems 5–50 cm tall; leaves mostly basal .... *T. spicatum*
2' Panicles loose and more or less open, the branches mostly 2–6 cm long and spreading; stems usually (30) 40–80 cm tall; leaves mostly cauline .... *T. montanum*

Trisetum montanum Vasey (of the mountains) Rocky Mountain trisetum. Loosely tufted, slender perennials. **CULMS** 30–80 cm tall, erect to ascending or even almost sprawling. **LIGULES** 1–4 mm long, erose-ciliate. **BLADES** flat, lax, 1.5–5 mm wide. **PANICLES** loose and open, 8–18 cm long, often drooping, the branches mostly erect-ascending and 2–6 cm long. **SPIKELETS** 2-flowered, usually greenish. **GLUMES** lanceolate, glabrous except for the scabrous keel, the first 3–5.5 mm long and 1-nerved, the second 4–6 mm long and 3-nerved. **LEMMAS** 4.5–5.5 mm long, 5-nerved, the lateral nerves prolonged into teeth, bifid at the apex with narrow teeth; awn 4–6 mm long, geniculate, arising below the apex of the lemma. **PALEAS** slightly shorter than the lemmas. **ANTHERS** 0.8–1.2 mm long. $2n$ = unknown. [*Trisetum spicatum* (L.) K. Richt. subsp. *montanum* (Vasey) W. A. Weber]. Mountain woodlands and forests, clearings, grassy slopes, often in shady understory, mostly at lower elevations than *T. spicatum*. ARIZ: Apa; COLO: Hin, LPl, Min, Mon, SJn; NMEX: RAr. 2270–2700 m (7450–8850′). Flowering: Jul–Sep. Mountains of the southwestern United States. Some authors have merged this within *T. spicatum*, but the most recent treatment of the group (Finot et al. 2005) maintains its distinction at the specific level. Our populations are easily identified and not confused with *T. spicatum*.

Trisetum spicatum (L.) K. Richt. (spikelike) Spike trisetum. Tightly tufted perennials, variously pubescent to glabrous. **CULMS** erect, 5–50 cm tall (our populations). **LIGULES** 1–3 mm long, erose-ciliate. **BLADES** flat to folded, 1–4 mm wide. **PANICLES** dense and spikelike, 2–14 cm long, the branches erect-appressed and less than 1 cm long. **SPIKELETS** 2- to 3-flowered, often purplish. **GLUMES** lanceolate, the first 3–5 mm long and 1-nerved, the second 4.5–6 mm long and 3-nerved and about as long as the upper floret. **LEMMAS** 4–5.5 mm long, 5-nerved, the lateral nerves prolonged into teeth, bifid at the apex with narrow teeth; awn 4–6 mm long, geniculate, arising below the apex of the lemma. **PALEAS** slightly shorter than the lemmas. **ANTHERS** 0.7–1.7 mm long. $2n$ = 14, 28, 42. Subalpine to alpine ridges, slopes, and forest clearings. COLO: Arc, Con, Hin, LPl, Min, Mon, RGr, SJn; UTAH. 2900–3650 m (9500–12000′). Flowering: Jul–Sep. Eastern and western Canada, south through the Rocky Mountains to New Mexico; Asia, Central and South America, Europe, Australia, and New Zealand. Throughout its distribution, *T. spicatum* exhibits enormous variability, and numerous infraspecific taxa have been proposed, for which there exists current disagreement as to their validity.

Trisetum wolfii Vasey (for John Wolf, 1820–1897, Illinois botanist and naturalist) Wolf's trisetum. Tufted perennials, with short rhizomes. **CULMS** stout, erect, 50–100 cm tall. **LIGULES** 1.5–2 mm long, erose. **BLADES** flat, lax, 3–7 mm wide, the lower blades pilose on the outer surface. **PANICLES** contracted, erect, 8–15 cm long, 1–2 cm wide, purplish tinged. **SPIKELETS** 2- to 3-flowered. **GLUMES** nearly as long as the spikelet, the first 4–6 mm long and 1-nerved, the second 5–6.5 mm long and 3-nerved. **LEMMAS** 5–5.5 mm long, 5-nerved, the lateral nerves not prolonged into teeth, the apex acute, awnless or with a short mucro to 2 mm long borne below the apex. **PALEAS** shorter than the lemmas. **ANTHERS** 0.7–0.8 mm long. $2n$ = unknown. [*Graphephorum wolfii* (Vasey) Vasey ex Coult.]. Marshy ground around seeps and springs at high elevations. COLO: LPl, Min, RGr, SJn. 2650–4000 m (8800–13000′). Flowering: Jul–Sep. Endemic to the western United States. Recent authors have segregated this into the small genus *Graphephorum*, with an entire lemma apex and the dorsal awn reduced to a subapical mucro. This may have some merit, but we maintain the traditional alliance out of practicality.

Triticum L. Wheat

(classical Latin name for wheat, derived from *tero*, to grind) Tufted annuals or biennials with hollow culms. **LEAVES** with open sheaths, membranous ligules, flat blades, and well-developed auricles. **INFL** a terminal spike, the rachis persistent or disarticulating. **SPIKELETS** sessile and placed flatwise on the rachis, 2- to 5-flowered, slightly laterally flattened but thick, disarticulation above the glumes and between the florets; glumes thick and firm, 3- to several-nerved; lemmas 5- to 7-nerved, the nerves not converging at the apex, keeled asymmetrically, awned or awnless; palea shorter than the lemma. **STAMENS** 3. $x = 7$. A small genus of 15–20 species native to the cool-temperate regions of Eurasia.

Triticum aestivum L. (of the summer) Wheat. Tufted annuals. **CULMS** erect, 30–150 cm or more tall. **LIGULES** 0.5–1.5 mm long. **BLADES** flat, 2–10 mm or more wide, the auricles prominent. **SPIKES** 5–12 cm long, the rachis continuous. **SPIKELETS** 3- to 5-flowered. **GLUMES** subequal, hardened, asymmetrical, broad, 5- to 7-nerved but the lateral nerves often faint, awn-tipped to 4 mm. **LEMMAS** 7–12 mm long, similar to the glumes, hardened but with scarious margins, essentially awnless to long-awned to 70 mm long. **ANTHERS** 2–3 mm long. $2n = 42$. Cultivated crop throughout the region, sometimes escaping along fields and roadsides but not persisting for long. ARIZ: Nav; COLO: LPl, Mon, SMg; NMEX: McK, RAr, SJn; UTAH. 1400–2200 m (4600–7200′). Flowering: Apr–Jul. Native to Eurasia. Both awned (bearded) and awnless (beardless) forms exist in cultivation.

Urochloa P. Beauv. Signalgrass

(Greek *oura*, tail, + *chloa*, grass, in reference to the tail-like bristle terminating the upper lemma of *U. panicoides*) Tufted annuals and perennials, sometimes mat-forming, with pithy culms. **LEAVES** with open sheaths, membranous-ciliate ligules (the basal membrane minute), flat blades, and lacking auricles. **INFL** terminal and axillary, a panicle of spikelike or rebranching branches. **SPIKELETS** 2-flowered, dorsally compressed, awnless, the lower floret reduced, the upper floret perfect; glumes unequal, the first much smaller than the second; lower floret staminate or sterile, the lemma membranous and similar in size and texture to the second glume; upper floret perfect, the lemma cross-wrinkled or bumpy, the apex round to mucronate to awned, the lemma and palea forming a hard seedcase around the flower and grain. **STAMENS** 3. $x = 7, 8, 9, 10$. About 100 species of tropical to subtropical or warm-temperate regions of the world. Many of the species were formerly included in *Brachiaria* or *Panicum*.

Urochloa arizonica (Scribn. & Merr.) Morrone & Zuloaga (of Arizona) Arizona signalgrass. Tufted annuals. **CULMS** 15–60 cm tall, often branched below. **LIGULES** 1–1.6 mm long. **BLADES** flat, 3–10 mm wide, glabrous. **PANICLES** ovoid, 6–20 cm long, 2–5 cm wide, spikelets and branchlets mostly appressed along the primary branches. **SPIKELETS** 3–4 mm long. **FIRST GLUME** 1.5–2 mm long and 1/2 or less the length of the spikelet. **SECOND GLUME** 2.5–3.2 mm long and as long as the spikelet, usually not cross-veined. **LOWER FLORET** staminate or sterile, 2.5–3.2 mm long, the palea present. **UPPER FLORET** 2.8–3 mm long. $2n = 36$. [*Brachiaria arizonica* (Scribn. & Merr.) S. T. Blake; *Panicum arizonicum* Scribn. & Merr.]. Disturbed ground and rocky slopes in desert communities. ARIZ: Apa. 1550–1650 m (5200–5400′). Flowering: Aug–Sep. Southwestern United States and northern Mexico, introduced in the southeastern United States.

Vahlodea Fr. Mountain Hair-grass

(for Jens Laurentius Moestue Vahl, 1796–1854, the son of botanist Martin Vahl) A monotypic genus, discontinuously circumboreal, also South America.

Vahlodea atropurpurea (Wahlenb.) Fr. ex Hartm. (dark purple) Mountain hair-grass. Loosely tufted perennials with open sheaths. **CULMS** hollow, 15–80 cm tall. **LIGULES** membranous, 1–4 mm long, often lacerate. **BLADES** flat, 2–8 mm wide, those of the culm often conspicuously erect. **PANICLES** open to contracted, 5–20 cm long, the branches capillary and naked below. **SPIKELETS** 2-flowered, the rachilla prolonged beyond the upper floret, but this hidden among the callus hairs and easily missed; disarticulation above the glumes and between the florets. **GLUMES** exceeding the florets, 4–6 mm long, acuminate, dull purplish (rarely pale), the first 4–6 mm long and 1-nerved, the second about the same length and 3-nerved. **LEMMAS** included within the glumes, 5-nerved, 1.8–3 mm long, with callus hairs about 1/2 the length of the floret; awn arising at about midlength, 2–4 mm long, exserted to scarcely visible. **PALEAS** slightly shorter than the lemmas. **STAMENS** 3, the anthers 0.5–1.2 mm long. $x = 7$; $2n = 14$. [*Deschampsia atropurpurea* (Wahlenb.) Scheele]. Woods and wet meadows, subalpine to alpine habitats. COLO: Min, SJn. 3000–4100 m (9800–13500′). Flowering: Aug–Sep. Easily confused with large-flowered *Deschampsia cespitosa* until one recognizes the large, dull glumes enclosing the florets and the awn attached at midlength. The erect culm blades are also quite distinctive, but this feature may not be consistent.

Vulpia C. C. Gmel. Six-weeks Fescue

(for J. S. Vulpius, 1760–1840, a pharmacist-botanist of Baden, Germany) Tufted annuals (rarely long-lived) with hollow culms. **LEAVES** with open sheaths, membranous-ciliate ligules, flat to rolled blades, and lacking auricles. **INFL** a terminal or axillary panicle, sometimes weakly developed and racemose. **SPIKELETS** 3- to 17-flowered, on short pedicels, laterally compressed, the terminal florets reduced; disarticulation above the glumes and below the florets; glumes shorter than the florets, the first 1-nerved, the second 3-nerved; lemmas 3- to 5-nerved, awned or awnless; palea slightly shorter than the lemma. **STAMENS** 1 (rarely 3 in cleistogamous spikelets). $x = 7$. About 30 species, many native to Europe and the Mediterranean region. Formerly included in *Festuca*.

1. First glume less than 1/2 the length of the second glume, often nearly absent .... ***V. myuros***
1' First glume more than 1/2 the length of the second glume .... ***V. octoflora***

Vulpia myuros (L.) C. C. Gmel. (mouse-tail) Rat-tail fescue. Loosely tufted annuals. **CULMS** 10–65 cm tall. **LIGULES** 0.3–0.5 mm long. **BLADES** usually rolled, 0.4–2 mm wide. **PANICLES** 3–20 cm long, narrow and contracted, less than 2 cm wide, with a single branch per node. **SPIKELETS** 3- to 7-flowered, the rachilla internodes 0.7–2 mm long. **GLUMES** unequal, the first 0.5–2 mm long and less than 1/2 the length of the second. **LEMMAS** 5-nerved, 4.5–7 mm long; awns 5–15 mm long. **ANTHERS** 0.5–2 mm long. $2n = 14, 42$. [*Festuca myuros* L.]. Dry, disturbed ground in the desert. ARIZ: Apa. 1500–1600 m (4900–5300'). Flowering: Apr–Jul. Native to Europe and North Africa, throughout western North America.

Vulpia octoflora (Walter) Rydb. (eight-flowered) Six-weeks fescue. Loosely tufted annuals. **CULMS** 5–50 cm tall. **LIGULES** 0.3–1 mm long. **BLADES** flat to rolled, 0.5–1 mm wide. **PANICLES** 1–10 (15) cm long, narrow and contracted, less than 2 cm wide, with 1–2 branches per node. **SPIKELETS** 5- to 15-flowered, the rachilla internodes 0.5–0.7 mm long. **GLUMES** unequal, the first 1.7–4.5 mm long, 1/3 to 1/2 the length of the second. **LEMMAS** 5-nerved, 2.7–6.5 mm long; awns 0.3–9 mm long. **ANTHERS** 0.3–1.5 mm long. $2n = 14$. [*Festuca octoflora* Walter]. Grassland and desert plains, open woodlands, rocky slopes, disturbed roadsides. 1450–2100 m (4800–7000'). Flowering: Jul–Sep. Nearly throughout North America. We have three weakly defined varieties, not satisfactorily distinguished:

1. Spikelets typically 4–6.5 mm long; awn of the lowermost lemma 0.3–3 mm long .... **var. *glauca***
1' Spikelets typically 5.5–13 mm long; awn of the lowermost lemma 2.5–9 mm long .... (2)

2. Lemmas scabrous to pubescent .... **var. *hirtella***
2' Lemmas glabrous or slightly scabrous, often scabrous on the margins .... **var. *octoflora***

var. *glauca* (Nutt.) Fernald (bluish green) [*Festuca tenella* Willd.; *F. octoflora* Walter var. *glauca* (Nutt.) Fernald]. NMEX: McK, RAr; UTAH.

var. *hirtella* (Piper) Henrard (a little hairy) [*Festuca octoflora* Walter subsp. *hirtella* Piper]. ARIZ: Apa.

var. *octoflora* ARIZ: Apa, Coc, Nav; COLO: Arc, Dol, LPl, Mon; NMEX: McK, RAr, San, SJn; UTAH.

POLEMONIACEAE Juss. PHLOX FAMILY

J. Mark Porter

Annuals, perennial herbs, or shrubs, pubescent, rarely glabrous. **LEAVES** alternate to opposite, simple to compound; stipules absent. **INFL** cymose, open to congested, the cymes usually arranged in panicles; flowers rarely solitary. **FLOWERS** perfect, actinomorphic to zygomorphic, hypogynous; sepals usually 5, united, with equal to unequal lobes, the tube usually with herbaceous ribs separated by hyaline membranes that are distended or ruptured in fruit; corolla tubular, with 5 lobes; stamens usually 5, epipetalous, alternate with the corolla lobes, equally to unequally inserted; filaments equal to unequal in length; ovary usually with 3 locules; style simple, included to exserted, with (2) 3 stigmatic lobes. **FRUIT** a capsule, rarely indehiscent or circumscissile, with 1–many ovules and seeds in each locule, the mature seeds in most taxa becoming gelatinous when wet, rarely largely unchanged when wet. Approximately 26 genera, about 380 species, chiefly of western North America, but also in South America and temperate Eurasia (Grant 1959; Porter and Johnson 2000).

1. Calyx tube apparently herbaceous throughout, becoming membranous and not ruptured in fruit .... (2)
1' Calyx tube with herbaceous ribs (costae) separated by thin, translucent intercostal membranes, these usually distended or ruptured in fruit .... (3)

2. Annual; leaves entire to toothed .... ***Collomia***
2' Perennial; leaves pinnately compound .... ***Polemonium***

3. Cauline leaves mostly opposite, upper ones sometimes alternate, either simple or palmately lobed to compound .... (4)
3' Cauline leaves mostly alternate, simple to pinnately lobed or compound .... (7)

4. Stamens inserted at different levels on the tube; corolla salverform; leaves simple, linear to linear-oblong or narrowly lanceolate ....(5)
4′ Stamens inserted at the same level on the tube; corolla rotate, campanulate, or funnelform, if salverform then the leaves somewhat rigid and spinulose; leaves simple and filiform to palmately compound, the blades or lobes filiform to linear ....(6)

5. Annual; corolla tube 4–8 mm long .... ***Microsteris***
5′ Perennial; corolla tube (5) 10–30 mm long .... ***Phlox***

6. Hyaline membranes separating the calyx lobes narrow and somewhat obscure; uppermost leaves opposite; corolla diurnal, and often open at night, corolla tube glandular-pilose externally .... ***Leptosiphon***
6′ Hyaline membranes separating the calyx lobes prominent; uppermost leaves often alternate; corolla nocturnal, closed during the day, corolla tube glabrous externally .... ***Linanthus***

7. Leaves deeply palmately or subpalmately lobed, the lobes rigid ....(8)
7′ Leaves entire or pinnately lobed to toothed, the lobes usually soft and herbaceous (rigid in *Navarretia*) ....(9)

8. Corolla white to cream-colored or yellowish, opening nocturnally, usually closed during the day .... ***Linanthus***
8′ Corolla bright blue, with yellow anthers, opening from noon to sundown, closed during the night .... ***Giliastrum***

9. Inflorescence capitate to densely corymbose, terminal at tips of ascending to erect branches; flowers subsessile to sessile ....(10)
9′ Inflorescence open and diffuse, with 2–5 flowers at the branch tips, or composed of several compact to dense, lateral clusters; flowers short- to long-pedicellate ....(13)

10. Calyx lobes unequal; floral bracts pinnately lobed, the apices firm or rigid ....(11)
10′ Calyx lobes equal; floral bracts usually entire to toothed, the apices acute to short-aristate but not sharply rigid..(12)

11. Upper leaves and floral bracts woolly or with entangled hairs; anthers sagittate .... ***Eriastrum***
11′ Upper leaves and floral bracts glabrous or short glandular-pubescent; anthers elliptic .... ***Navarretia***

12. Corolla campanulate to funnelform, blue or with white to bluish lobes and a yellow tube or throat .... ***Gilia***
12′ Corolla salverform, uniformly white to lavender .... ***Ipomopsis***

13. Corolla salverform, uniformly white to bluish or red; flowers congested at tips of lateral branches .... ***Ipomopsis***
13′ Corolla funnelform to salverform, if salverform then usually with white to bluish lobes and a yellow tube or throat (crimson in *Aliciella subnuda*); flowers 2–5 (7) in loose, terminal clusters ....(14)

14. Corolla broadly funnelform, rotate, lobes deep blue, anthers bright yellow; leaves gradually reduced upward, pinnatifid below, subpalmate above, the lobes needlelike .... ***Giliastrum***
14′ Corolla narrowly funnelform to salverform, the lobes white to red or bluish, the throat or tube often yellow, anthers white to blue; leaves abruptly reduced above a basal rosette (except *Aliciella latifolia*), the lobes usually acute or mucronate ....(15)

15. Glandular hairs on calyx, pedicels, and upper leaves colorless to yellowish; basal and lower leaves glabrous or mostly glandular; seeds not conspicuously gelatinous when wet .... ***Aliciella***
15′ Glandular hairs on calyx, pedicels, and leaves, if present, dark or reddish; basal and lower leaves with short, curled hairs or cobwebby hairs; seeds gelatinous when wet .... ***Gilia***

Aliciella Brand. Aliciella

(honoring Alice Eastwood, botanist, 1859–1953) Annual or perennial herbs, simple to branched, leafy to scapose, mostly glandular, the hairs uniseriate with terminal, multicellular glands, the glands colorless. **LEAVES** basal to alternate, entire to deeply pinnately lobed, the lobes completely confluent with the rachis, flat to terete. **INFL** a terminal panicle, open to congested, the basic unit composed of 2–7 pedicelled flowers subtended by a single bract. **FLOWERS** actinomorphic; calyx tube membranes usually ruptured in fruit, rarely remaining intact; corolla funnelform to salverform, the tube, throat, and lobes similar in color or with different hues; stamens equally or rarely unequally inserted on the corolla tube or throat; filaments equal or unequal in length; anthers included to exserted; style included to exserted. **FRUIT** a capsule, ovoid to spheroid; seeds 2–many per locule, unchanged or only weakly gelatinous when wet. (Porter 1998.) Twenty-two species restricted to western North America. Formerly included in *Gilia*; however, *Aliciella* has been shown to be more closely related to *Ipomopsis* and *Loeselia* than to *Gilia*.

1. Stamens more or less unequally inserted in the upper tube; flowers subsessile to short-pedicelled, in terminal, corymbose or subcapitate clusters; corolla watermelon-red to crimson .... ***A. subnuda***
1' Stamens equally inserted at the sinuses or midtube (*A. latifolia*); most flowers pedicelled, the clusters spreading to ascending in open inflorescences; corolla white, bluish, lavender, or magenta .... (2)

2. Leaves oblanceolate to broadly ovate, dentate, the teeth broad, acuminate-aristate; filaments unequal in length and papillose below the anthers .... ***A. latifolia***
2' Leaves linear to oblanceolate, pinnately lobed, the teeth acute to at most mucronate; filaments smooth below the anthers .... (3)

3. Corolla lobes 3-toothed, the central one longest; corolla somewhat salverform, narrower at the orifice than at midtube .... ***A. triodon***
3' Corolla lobes obtuse to acuminate; corolla somewhat funnelform, orifice broader than midtube .... (4)

4. Anthers long-exserted from the corolla tube, nearly as long as the corolla lobes .... (5)
4' Anthers not or only shortly exserted from the corolla tube, less than 1/2 as long as the corolla lobes .... (6)

5. Leaves all simple, entire and terete; calyx usually suffused with blue, corolla tube 1.4–4.3 mm long .. ***A. sedifolia***
5' Leaves pinnatifid, the lobes much longer than the width of the rachis; calyx sometimes suffused with purple, but not blue; corolla tube 4–9 mm long .... ***A. pinnatifida***

6. Basal and lower leaves usually twice pinnately lobed; corolla tube conspicuously glandular externally on the lower tube .... ***A. hutchinsifolia***
6' Basal and lower leaves entire, dentate or once pinnately lobed; corolla tube glabrous externally, sometimes with a few glandular hairs near the orifice, external to the insertion of the stamens .... (7)

7. Corolla tube 3–6 mm long, the lobes 1–3 mm long; annual; corolla lobes obtuse to nearly truncate, with a cuspidate apex .... ***A. leptomeria***
7' Corolla tube 8–15 mm long, lobes 3.5–6 mm long; short-lived perennials, with 1–3 rosettes .... (8)

8. Leaves linear, entire .... ***A. formosa***
8' Leaves spatulate to narrowly oblong, dentate or once pinnately lobed .... (9)

9. Corolla blue when fresh, drying pale blue; corolla lobes narrowly oblanceolate, 1.9–3.5 mm wide ..... ***A. cliffordii***
9' Corolla magenta when fresh, drying blue or magenta; corolla lobes oblanceolate, 2–4.2 mm wide..... ***A. haydenii***

Aliciella cliffordii J. M. Porter (honoring Arnold Clifford, botanist) Clifford's Diné-star. Short-lived perennial, 12–50 cm tall, usually branched throughout; stems glabrous to glandular-pubescent. **LEAVES** forming a rosette, gradually to abruptly reduced above the rosette; basal and lower sparsely short-pubescent above, glabrous beneath, lobed to toothed, the rachis broader than the lobes; cauline usually glabrous, toothed to entire. **INFL** open, with 1–5 pedicelled flowers on tips of spreading to ascending branches. **FLOWER** calyx 4–5 mm long, glandular-pubescent to nearly glabrous in Arizona, the lobes acute; corolla narrowly funnelform, 6–10 mm long, blue, sometimes pale; stamens inserted at the sinuses of the corolla lobes; anthers located at the orifice or slightly exserted; stigma located above to below the anthers. **FRUIT** a capsule 3–5 mm long, ovoid. Sandy clay soils, badlands, piñon-juniper and ponderosa pine communities. ARIZ: Apa; NMEX: McK, SJn. 1525–1980 m (5000–6500′). Flowering: May–Jun. Fruit: May–Jul. Endemic to the eastern slopes of the Chuska Mountains. Formerly but incorrectly identified as *Gilia haydenii*. Pollinated primarily by bee flies.

Aliciella formosa (Greene ex Brand) J. M. Porter (beautiful) Beautiful gilia. Long-lived perennial, 5–15 cm tall, from branched, woody caudex, sparsely and coarsely glandular. **LEAVES** forming a rosette, entire, linear, 1–4.5 cm long, 1–1.5 mm wide, glandular and crisp-puberulent with white hairs; cauline leaves linear, entire, gradually to abruptly reduced in size. **INFL** few-flowered, open, cymose panicle, the flowers mostly crowded at the tips of the branches. **FLOWER** calyx 3.5–6.1 mm long, tube slightly longer than the lobes, glandular; corolla narrowly funnelform-salverform, 14.7–27 mm long, rose-purple, magenta, or pink-lavender, rarely white, but drying to a lead-blue, stamens inserted at the sinuses of the lobes, anthers slightly exserted, corolla sparsely glandular externally, at filament insertion. **FRUIT** a capsule 3.5–7 mm long, ovoid. [*Gilia formosa* Greene ex Brand]. Sandy clay soils, badlands, piñon-juniper woodland and desert scrub communities. NMEX: SJn. 1640–1980 m (5000–6500′). Flowering: May–Jun. Fruit: late May–Jul. Endemic to the Four Corners region. Pollinated primarily by hawkmoths.

Aliciella haydenii (A. Gray) J. M. Porter (honoring Ferdinand Hayden, geologist and explorer) Hayden's gilia. Short-lived perennial, 12–50 cm tall, usually branched throughout. **STEMS** glabrous to glandular-pubescent. **LEAVES** forming a rosette, gradually to abruptly reduced above basal rosette; basal and lower sparsely short-pubescent above, glabrous beneath, lobed to toothed, the rachis broader than the lobes; cauline usually glabrous, toothed to entire. **INFL** open, with 1–5 pedicelled flowers on tips of spreading to ascending branches. **FLOWER** calyx 4–5 mm long, glandular-pubescent to nearly glabrous in Arizona, the lobes acute; corolla funnelform, 6–10 mm long, magenta; stamens inserted at the sinuses of the corolla lobes; anthers located at the orifice or slightly exserted; stigma located above to below the anthers. **FRUIT** a capsule 3–5 mm long, ovoid. $2n = 16, 18$. [*Gilia haydenii* A. Gray]. Two subspecies. Southeastern Utah to south-central Colorado, south to northeastern Arizona and north-central to northwestern New Mexico.

1. Corolla 17–26 mm, the lobes 6–9 mm long, corolla tube glandular externally; corolla drying pink; plants primarily of higher-elevation piñon-juniper, oak woodlands, and ponderosa pine communities .... **subsp. *crandallii***

1' Corolla 11–20 mm, the lobes 3.5–5.5 mm long, corolla tube glabrous or only a few glands externally at the point where the filaments are attached; corolla drying dull blue; plants primarily of lower-elevation piñon-juniper, saltbush, and desert scrub communities .... **subsp. *haydenii***

subsp. *crandallii* (Rydb.) J. M. Porter (for Cletus Crandall, forester) Crandall's gilia. Corolla (16) 17–26 mm long, rose-purple to magenta, corolla tube uniformly glandular externally. Rocky clay soils, clay badlands, piñon-juniper and ponderosa pine woodland communities. COLO: LPl, Mon; NMEX: RAr, SJn. 1700–2260 m (5600–7500'). Flowering: May–Jun. Fruit: late May–early Jul. Southwestern and south-central Colorado, and northwestern and north-central New Mexico.

subsp. *haydenii* Corolla 10–20 mm long, magenta, with a few glands externally at the point of filament attachment. Rocky clay soils, clay badlands, saltbush scrub, and piñon-juniper woodland communities. ARIZ: Apa; COLO: LPl, Mon; NMEX: McK, RAr, SJn; UTAH. 1220–1980 m (4000–6500'). Flowering: May–Jun. Fruit: late May–early Jul. Southeastern Utah to south-central Colorado, south to northeastern Arizona and northwestern New Mexico.

Aliciella hutchinsifolia (Rydb.) J. M. Porter (*Hutchinsia*, genus of Brassicaceae; *folius*, leaves) Broadlobe gilia, Hutchinsia-leaved gilia. Annual, 5–30 cm tall, usually branched throughout. **STEMS** glandular. **LEAVES** glandular and short-pilose, reduced above the basal rosette; basal and lower lobed once or twice, the lobes entire to toothed; cauline basally lobed to entire. **INFL** open, with 1–2 pedicelled flowers at branch tips. **FLOWER** calyx 2–3 mm long, glandular-pubescent, the lobes acuminate to attenuate; corolla narrowly funnelform to salverform, 7–15 mm long, the tube and lobes white to lavender, the throat yellow; stamens inserted on upper throat; anthers located slightly above the throat; stigma located slightly below or among the anthers. **FRUIT** a capsule 3–6 mm long, ovoid. $2n = 18$. [*Gilia hutchinsifolia* Rydb.]. Washes, bajadas, desert shrublands, and woodlands. ARIZ: Apa, Nav; UTAH. 1150–1700 m (3800–5600'). Flowering: Apr–Jun. Fruit: May–Jun. Southern California, Nevada to southern Utah, south to Arizona.

Aliciella latifolia (S. Watson) J. M. Porter (broad leaf) Broadleaf gilia. Annual, (4) 8–30 cm tall, simple to branched throughout. **STEMS** glandular. **LEAVES** glandular and pilose, gradually reduced upward, simple, oblanceolate to rotund, dentate to shallowly lobed, the teeth or lobes acuminate. **INFL** open, with 1–2 pedicelled to subsessile flowers at branch tips or in axils. **FLOWER** calyx 2.8–4.8 mm long, sparsely to moderately glandular-pubescent, the lobes equal to or longer than the tube, spinulose; corolla funnelform, bright pink to pink-purple, the tube and throat 6–11 mm long, the lobes 3–8 mm long; stamens inserted on the lower throat; anthers located in the upper throat; stigmas slightly exceeding the stamens. **FRUIT** a capsule 3–4.5 mm long, ovoid to oblong. $2n = 36$. [*Gilia latifolia* S. Watson]. Our material is **subsp. *imperialis*** (S. L. Welsh) J. M. Porter (Imperial Valley). Cataract gilia. Calyx 2.8–4.8 mm long vs. 4–6 mm in subsp. *latifolia*; fruits 3–4.5 mm long vs. 4–7 mm in subsp. *latifolia*; plants frequently over 25 cm tall. Sandy washes, restricted to the Colorado River drainage, UTAH. 1160–1590 m (3800–5200'). Flowering: Mar–May. The species as a whole ranges from southern California to southern Utah, and south to northwestern Mexico.

Aliciella leptomeria (A. Gray) J. M. Porter (thin parts) Common gilia, lobeleaf gilia, Great Basin gilia. Annual, 6–30 cm tall, simple to branched. **STEMS** usually glandular. **LEAVES** glandular, rarely glabrous, abruptly reduced above the basal rosette; basal and lower dentate to shallowly lobed, the teeth or lobe length equal to the rachis width; cauline dentate to entire. **INFL** open, with 1–3 pedicelled flowers at branch tips, the terminal pedicel subsessile or often shorter than those below. **FLOWER** calyx 1–3 mm long, usually glabrous, the lobes obtuse to cuspidate; corolla salverform to narrowly campanulate, 3–7 mm long, the tube equal to or slightly longer than the calyx, the lobes acuminate;

the tube and lobes white, pink, or lavender, the throat yellow; stamens inserted on the upper throat; anthers slightly exserted; stigmas slightly exceeding the anthers. **FRUIT** a capsule 3–5 mm long, narrowly ovoid. 2*n* = 34, 36. [*Gilia leptomeria* A. Gray]. Washes, rocky slopes, blackbrush, desert scrub, sagebrush, mountain brush, piñon-juniper, and ponderosa pine communities. ARIZ: Apa, Coc, Nav; COLO: LPl, Mon; NMEX: McK, SJn; UTAH. 1150–2300 m (3800–7500′). Flowering: Apr–Jun. Washington, Idaho, and Wyoming, south to California, Arizona, and New Mexico. Populations in the study area differ significantly (though cryptically) from populations near the type locality (the Great Salt Lake, Utah). It is likely that our material represents a different but unnamed taxon. Used as a poultice and applied to scorpion stings or worm bites by the Kayenta Navajo.

Aliciella pinnatifida (Nutt. ex A. Gray) J. M. Porter (*pinnate*, divided leaf, + *fidus*, divided in the distal third) Pinnate gilia. Biennial or short-lived perennial (annual), 15–70 cm tall, densely glandular or largely glabrous, with only a few scattered glands. **LEAVES** forming a basal rosette, persistent or sometimes deciduous, deeply pinnatifid with narrow to sometimes elliptic lobes, coarsely few-toothed or entire, 1.5–8 cm long, 7–30 mm wide; cauline leaves deeply pinnatifid, often as large as basal, progressively reduced above, often undivided, ultimately linear bracts. **INFL** loose and open, but the ultimate pedicels short, (0) 1–5 mm long, often sympodial. **FLOWER** calyx 2.5–4.5 mm long, the lobes slightly shorter than the tube; corolla tube pale, short and broad, 4–9 mm long, lobes blue, loosely to widely spreading, 3–5 mm long, from slightly longer to barely more than 1/2 as long as tube; anthers well-exserted from the tube; style exserted, equal to or longer than anthers. **FRUIT** a capsule 2.5–4 mm long, narrowly ovoid. 2*n* = 16. [*Gilia pinnatifida* Nutt. ex A. Gray]. Dry, open places, sagebrush, piñon-juniper, mixed conifer, and subalpine meadow communities. COLO: Arc, Hin, Min, RGr, SJn; NMEX: RAr. 1550–2900 m (5100–9500′). Flowering: May–Sep. Fruit: Jun–Sep. Colorado and southern Wyoming, to western Nebraska and northern New Mexico.

Aliciella sedifolia (Brandegee) J. M. Porter (*Sedum*-like leaves) Sedum-leaved aliciella. Monocarpic, short-lived perennial, 4–12 cm tall. **STEMS** glandular-pubescent, simple and erect, becoming more or less thyrsoid in flower. **LEAVES** forming a rosette, linear, entire, 0.6–1.7 cm long, 1–2.6 mm wide, terete and succulent, cauline leaves gradually reduced in size, becoming bractlike. **INFL** strict, thyrsoid, cymose panicle. **FLOWER** calyx 2.4–4.5 mm long, bearing dense glandular trichomes, the lobes shorter than the tube; corolla 4–8.5 mm long, mostly blue to lavender, glabrous externally, salverform to narrowly campanulate; the tube pale, shorter than the calyx, 1.4–4.3 mm long, lobes oval to orbicular, 2.3–4.6 mm long, 1.4–2 mm wide; stamens affixed in the sinus of the corolla lobes, the free portion as long as the fused portion, 1.8–4.3 mm long, glabrous, anthers 0.7–1.8 mm long, shortly exserted; nectary an undulate disc at the base of the ovary; ovary ovoid to oblongoid, glabrous, about 1.6 mm long and 1.1 mm wide at the base, the style 3–4.2 mm long, stigmatic lobes approximately 0.5 mm long. **FRUIT** a capsule 3–6.5 mm long. **SEEDS** 1–5 per locule, about 1.5 mm long, lenticular to angular, narrowly winged. Alpine flats covered with tuff or rhyolitic gravel. COLO: Hin. 3650–4120 m (12000–13500′). Flowering: Jul–Aug. Fruit: Aug. Collections are known only from Sheep Mountain and Half Peak. Though not collected in the study area, it is known to occur within 3 km and there is ample unexplored habitat. One of the rarest species in the family.

Aliciella subnuda (Torr. ex A. Gray) J. M. Porter **CP** (somewhat naked) Carmine gilia. Biennial or short-lived perennial, 10–70 cm tall. **STEMS** branching only within the inflorescence, glandular. **LEAVES** forming a basal rosette, glandular, abruptly reduced above the basal rosette; basal oblanceolate, dentate to shallowly lobed distally; cauline narrowly oblong to linear, entire. **INFL** congested, usually corymbose, with short-pedicelled flowers at top of stem. **FLOWER** calyx 4–6 mm long, glandular short-pubescent; corolla salverform, 10–20 (25) mm long, deep scarlet to crimson, the tube 2–3 times the calyx; stamens inserted on the upper tube; anthers included in the tube or slightly exserted; stigma located below to above the anthers. **FRUIT** a capsule 3–5 mm long, narrowly ovoid. 2*n* = 16. [*Gilia subnuda* Torr. ex A. Gray]. Sandy soils, sandstone outcrops, desert scrub, piñon-juniper, and ponderosa pine communities. ARIZ: Apa; Coc, Nav; UTAH. 1500–2400 m (5000–7800′). Flowering: Apr–Jun (Aug). Fruit: May–Aug. Northern Arizona to southern Utah (previously reported from Colorado and Nevada in error). The flowers were ground and eaten to ensure a healthy pregnancy and ease labor by the Kayenta Navajo.

Aliciella triodon (Eastw.) Brand (three teeth, referring to corolla lobes) Three-toothed aliciella. Annual, 5–15 (20) mm tall, simple to branched. **STEMS** glandular. **LEAVES** glandular, abruptly reduced above the basal rosette; basal dentate to shallowly lobed, the rachis broader than the teeth or lobes; cauline usually entire, linear. **INFL** open, with 1–3 short-pedicelled flowers on distal branches. **FLOWER** calyx 1–3 mm long, usually glabrous, the lobes acuminate; corolla narrowly funnelform, 3–6 mm long, the tube exserted, white to pink, the throat light yellow, the lobes 3-toothed, white to pink; stamens inserted on the upper throat; anthers slightly exserted; stigma slightly exceeding the anthers. **FRUIT** a capsule, 3–4 mm long, narrowly ovoid. $2n = 18$. [*Gilia triodon* Eastw.]. Sandy, adobe, or gravelly soils, sagebrush scrubland and piñon-juniper woodland communities. ARIZ: Apa, Coc, Nav; COLO: LPl, Mon; NMEX: SJn; UTAH. 1150–1740 m (3800–5700'). Flowering: May–Jun. Fruit: late May–Jun. Southeastern California, northern Arizona, and northwestern New Mexico to Nevada, Utah, western Colorado, and western Wyoming.

Collomia Nutt. Collomia

(glue, referring to the wetted seeds) Annual or perennial herbs. **LEAVES** mostly alternate, simple to lobed. **INFL** terminal and axillary, compact, subcapitate or rarely solitary, bracteate. **FLOWERS** perfect, actinomorphic, subsessile; calyx herbaceous, becoming membranous, not ruptured in fruit, the membranes distended or carinate in fruit, the ribs and lobes with prominent veins in fruit, acute to acuminate; corolla funnelform to salverform, blue, red, yellow, or white; stamens equally or unequally inserted in the corolla tube; filaments mostly unequal in length; anthers included to exserted; style included to exserted. **FRUIT** a capsule, ovoid to ellipsoid, explosively dehiscent; seeds (in our area) 1 per locule, gelatinous when wet. $2n = 16$ (Wherry 1944; Chuang, Hsieh, and Wilken 1978; Wilken et al. 1982). Subtropical and temperate forests or shrublands, 15 species of North America, one species in South America.

1. Corolla 15–30 mm long, yellow to salmon-orange; pollen blue; calyx lobes acute .... ***C. grandiflora***
1' Corolla 8–15 mm long, white to bluish violet; pollen cream to blue; calyx lobes acuminate .... ***C. linearis***

Collomia grandiflora Douglas ex Lindl. (large-flowered) Large-flowered collomia, large collomia. Annual. **STEMS** 1–5 dm. **LEAVES** subsessile, entire to toothed, 1–6 cm long, the lower cauline leaves linear to narrowly elliptic, the upper leaves lanceolate to ovate. **INFL** terminal or axillary, compact, the bracts ovate to lanceolate, 10–22 mm long, 4–10 mm wide, the lower margins short-pubescent. **FLOWER** calyx lobes acute; corollas of normal flowers open, 15–25 (30) mm long, orange to salmon or (in cleistogamous flowers) unopened, restricted to the lower axillary cymes, less than 5 mm long, white to light yellow; pollen blue; style equal to the corolla tube. **FRUIT** an explosively dehiscing capsule 5–6 mm long. Dry streambeds, shaded slopes, pine forest, oak woodland, mountain brush, piñon-juniper woodland, sagebrush, aspen, and spruce-fir communities. COLO: Arc, LPl, Mon. 1400–2570 m (4500–8200'). Flowering: Apr–Jun (Aug). Fruit: May–Sep. British Columbia, Washington to Idaho, south to California and east to Colorado and Arizona.

Collomia linearis Nutt. (narrow, referring to leaves) Linear-leaf collomia, small collomia. Annual. **STEMS** 1–5 dm tall. **LEAVES** subsessile, entire, 1–6 cm long, the lower cauline leaves linear to lanceolate, the upper leaves narrowly lanceolate. **INFL** terminal, rarely axillary, compact, the bracts ovate to lanceolate, 8–20 mm long, 3–6 mm wide, the lower margins glandular. **FLOWER** calyx lobes acuminate; corollas salverform, 8–15 mm long, bluish violet to nearly white; pollen cream to blue; style equal to the corolla tube. **FRUIT** a capsule 3–5 mm long. Open sites, meadows, pine forest, sagebrush, mountain brush, aspen, and spruce-fir communities. COLO: Arc, Dol, Hin, LPl, Min, Mon, SJn, SMg; NMEX: McK, RAr, SJn; UTAH. 2000–2500 m (6500–8000'). Flowering: Jun–Jul, rarely into Sep. Fruit: late Jun–Sep. Western Canada to California and east to Colorado, New Mexico, and Wyoming; introduced to central and eastern U.S.

Eriastrum Wooton & Standl. Star-gilia, Woolly-star, Eriastrum

(wool star) Annual (in our area) or perennial. **STEMS** solitary to much-branched, erect to prostrate, leafy. **LEAVES** basal to alternate, simple to deeply pinnately lobed, the lobes confluent with the rachis, flat to terete, spinulose. **INFL** terminal, compact, bracteate, short-woolly, the outer bracts deeply lobed. **FLOWERS** sessile, actinomorphic to slightly irregular; calyx tube membranes usually ruptured in fruit, the lobes subequal to strongly unequal, acute to attenuate, often spinulose; corolla actinomorphic, funnelform to salverform, or zygomorphic with at least 2 lobes unequal, white to deep bluish lavender; stamens equally to unequally inserted on the throat; anthers usually exserted; filaments equal to subequal in length; style included to exserted. **FRUIT** a capsule, ellipsoid to ovoid; seeds 1–several per locule. $n = 7$ (Mason 1945; Harrison 1972; Patterson 1993). Fourteen to 17 species of the arid western U.S. and northern Mexico.

Eriastrum diffusum (A. Gray) H. Mason (widely spreading) Spreading eriastrum, spreading starflower, miniature woolly-star. Annual, 3–35 cm tall, erect and simple to diffusely branching. **LEAVES** subglabrous to sparsely woolly, entire or with 1–2 pairs of lobes near the base of the rachis, 1–3 cm long. **FLOWER** calyx 6–7 mm long; corolla actinomorphic, narrowly funnelform to slightly zygomorphic, the throat white to yellow, the lobes white to pale blue or bluish lavender, the tube and throat 4–7 mm long, slightly longer than the calyx tube, the lobes 3–5 mm long; stamens inserted on the throat near the sinuses, less than the length of the corolla lobes; filaments slightly unequal in length; pistil 5–7 mm long; style included in the tube or throat. **FRUIT** a capsule, 2–4 mm long. $2n = 14$. Open sites; desert shrublands, blackbrush, and piñon-juniper woodland communities. COLO: Mon, SMg; UTAH. 1220–1710 m (4000–5600'). Flowering: May–Jun. Fruit: late May–Jul. California south to northern Mexico, east to Texas. Our material is closest to **subsp.** ***jonesii*** H. Mason, with corollas 10–12 mm long and anthers 0.7–1 mm long. The typical subsp. *diffusum* is of the Chihuahuan Desert and has shorter corollas and anthers less than 0.8 mm long.

Gilia Ruiz & Pav. Gilia

(for Philippo Gili, Italian naturalist) Annual herbs, simple to branched, leafy to scapose, glabrous, pubescent to glandular, the hairs sometimes loosely tangled like cobwebs. **LEAVES** basal to alternate, entire to deeply pinnately lobed, the lobes completely confluent with the rachis, flat to terete. **INFL** terminal, paniculate, open to congested, the basic unit composed of 2–7 pedicelled flowers subtended by a single bract, rarely solitary. **FLOWERS** actinomorphic; calyx tube membranes usually ruptured in fruit, rarely remaining intact; corolla funnelform to salverform, the tube, throat, and lobes often with different hues; stamens equally to subequally inserted on the corolla tube or throat; filaments equal or unequal in length; anthers included to exserted; style included to exserted. **CAPSULES** ovoid to spheroid; seeds 2–many per locule (Day 1993). About 40 species; western North America and southern South America.

1. Flowers subsessile to short-pedicelled, in capitate terminal clusters .... ***G. capitata***
1′ Most flowers pedicelled, spreading to ascending in open inflorescences .... (2)

2. Stems near base glabrous and glaucous; basal leaves usually glabrous; cauline leaves clasping; rachis of basal leaves strap-shaped .... ***G. sinuata***
2′ Stems near base and basal leaf axils cobwebby-pubescent; cauline leaves not clasping; basal leaf rachis ± linear .. (3)

3. Corolla 4–8 mm long, throat white to pale blue, lobes equal to or longer than the throat; calyx lobes acute .... ***G. clokeyi***
3′ Corolla 7–12 mm long, throat yellow, lobes shorter than the throat; calyx lobes acuminate ...... ***G. ophthalmoides***

Gilia capitata Sims (headlike) Ball gilia, capitate gilia, ornamental gilia. Annual, 10–50 cm tall, branched throughout. **STEMS** glabrous to glandular or sparsely tomentose. **LEAVES** pubescent, gradually reduced upward, deeply lobed once or twice, the lobes linear. **INFL** usually globose, with many subsessile flowers at the branch tips. **FLOWER** calyx 4–5 mm long, pubescent, the lobes acuminate; corolla campanulate, 7–12 mm long, blue to bluish violet, the tube equal to the calyx, the throat exserted; stamens inserted on the throat; anthers exserted; stigmas located among or above the anthers. **CAPSULES** 5–6 mm long, ovoid to globose. $2n = 18$. [*G. achilleifolia* Benth. subsp. *staminea* (Greene) H. Mason & A. D. Grant]. Canyons, washes; piñon-juniper woodland communities. NMEX: SJn. 1600–1900 m (5000–5600′). Flowering: Mar–May. Fruit: Apr–May. Our plants belong to **subsp.** ***staminea*** (Greene) V. E. Grant. Native to the San Joaquin Valley and adjacent mountains of California. Escape from cultivation.

Gilia clokeyi H. Mason (for Ira Clokey, botanist) Clokey's gilia. Annual, 8–20 (35) cm tall, branched throughout. **STEMS** cobwebby-pubescent below, glandular above. **LEAVES** cobwebby-pubescent, reduced above the basal rosette; basal and lower deeply lobed, the lobes linear to oblong; cauline basally lobed or entire. **INFL** open, with 1–2 pedicelled flowers at branch tips. **FLOWER** calyx 2–5 mm long, the lobes acute; corolla funnelform, 4–7 mm long, the tube equal to the calyx, the throat pale blue to white, the lobes light to deep violet, yellow-spotted at base; stamens inserted on the throat; anthers located above the throat; stigma situated among the anthers. **CAPSULES** 3–6.5 mm long, globose to ovoid. $2n = 18$. Open, sandy soils; blackbrush, sagebrush, saltbush, piñon-juniper woodland, and mountain brush communities. ARIZ: Apa, Nav; COLO: Arc, Dol, LPl, Mon, SMg; NMEX: McK, RAr, SJn; UTAH. 1136–1900 m (3725–6200′). Flowering: Mar–Apr. California to Wyoming, south to Arizona and New Mexico. In most previous local floras, incorrectly referred to *G. inconspicua* (Sm.) Sweet, a species of the Columbia River valley, Oregon and Washington.

Gilia ophthalmoides Brand (eyelike, referring to the yellow throat) Eye-gilia. Annual, 8–30 cm tall, branched throughout. **STEMS** cobwebby-pubescent below, glandular above. **LEAVES** cobwebby-pubescent, reduced above the basal rosette; basal and lower leaves deeply lobed once or twice; cauline basally lobed or entire. **INFL** open, with 1–2 pedicelled flowers at the branch tips. **FLOWER** calyx 3–5 mm long, glabrous, the lobes acuminate; corolla funnelform, 7–12 mm long, the tube usually exserted, purple, the throat yellow, the lobes pink; stamens inserted on the throat; anthers exserted; stigma located among the anthers. **CAPSULES** 4–6 mm long, ovoid to subglobose. $2n = 36$. Adobe hills and flats; desert scrub and piñon-juniper woodland communities. ARIZ: Apa, Coc, Nav; COLO: Arc, Dol, LPl, Mon, SMg; NMEX: McK, RAr, San, SJn; UTAH. 1136–2150 m (3725–7000′). Flowering: Apr–Jun. Southeast California to Utah, south to Arizona and New Mexico. Incorrectly referred to *G. inconspicua* (Sm.) Sweet in most previous local floras.

Gilia sinuata Douglas ex Benth. (strongly waved, referring to the leaf margins) Sinuate-leaf gilia, sinuate gilia. Annual, 9–30 (35) cm tall, simple or branched above rosette, glabrous and glaucous below, glandular above. **LEAVES** cobwebby-pubescent on upper surface, abruptly reduced above the basal rosette; basal deeply lobed once, the lobes oblong; cauline clasping, dentate to entire. **INFL** open, with 1–3 short-pedicelled flowers at branch tips. **FLOWER** calyx 3–5 mm long, glandular, the lobes short-acuminate; corolla funnelform, 7–12 mm long, the tube exserted, purple and white striate, the throat yellow or purple-tinged below, the lobes white to lavender; stamens inserted on the throat; anthers exserted; stigmas located among or slightly above the anthers. **CAPSULES** 4–7 mm long, ovoid. $2n = 36$. Sandy soils, mesas; desert scrub and piñon-juniper woodland communities. ARIZ: Apa, Coc, Nav; COLO: Mon; NMEX: McK, SJn; UTAH. 1136–1900 m (3725–6200′). Flowering: Mar–May. Washington and Idaho, south to southern California, Arizona, and New Mexico. Incorrectly referred to *G. inconspicua* (Sm.) Sweet in most previous local floras. Our material differs significantly in floral morphology from the material around the type locality in Okanogan County, Washington. Grant (1956) has further demonstrated that it is reproductively isolated from the typical *G. sinuata*. Ours is likely an undescribed, cryptic species.

Giliastrum (Brand.) Rydb. Giliastrum

(*Gilia*, genus of Polemoniaceae, + *astrum*, referring to similarity with *Gilia*) Perennial herbs, sometimes flowering the first year, simple to branched, leafy to subscapose, glabrous or sparsely pubescent. **LEAVES** basal to alternate, gradually reduced upward, entire to pinnately lobed, the lobes completely confluent with the rachis, usually flat. **INFL** terminal, paniculate, open to congested, the basic unit composed of 2–5 pedicelled flowers subtended by a single bract, rarely solitary. **FLOWERS** actinomorphic; calyx tube membranes ruptured in fruit; corolla rotate to broadly funnelform, the tube, throat, and lobes often with the same hues; stamens equally inserted on the lower corolla tube; filaments equal or unequal in length; anthers included to exserted; style included to exserted. **FRUIT** a capsule, ovoid to spheroid; seeds 2–many per locule. Arid grasslands, shrublands, and woodlands. Nine species of western North America and the Andes of Argentina, South America (Porter 1998b).

Giliastrum acerosum (A. Gray) Rydb. (needle-shaped, referring to the leaves) Bluebowls. Perennial, 6–15 cm tall, branched at the base. **STEMS** 5–12, spreading to erect, woody below, glandular throughout. **LEAVES** slightly reduced upward; lower deeply pinnately lobed, the lobes needlelike; upper subpalmately lobed, the lobes 3–5. **INFL** open, with 1–3 pedicelled flowers at branch tips. **FLOWER** calyx 7–8 mm long, the lobes attenuate, needlelike; corolla broadly funnelform, 7–10 mm long, deep blue, the tube usually shorter than the calyx; stamens inserted on the lower tube; anthers located above the throat, bright yellow; stigma located among the anthers or slightly above. **CAPSULES** 4–5 mm long, broadly ovoid. $2n = 18$. [*Gilia rigidula* Benth. subsp. *acerosa* (A. Gray) Wherry]. Gravelly soils, rocky slopes, canyons; desert scrub and piñon-juniper woodland communities. ARIZ: Apa; NMEX: McK, SJn. 1300–1950 m (4200–6500′). Flowering: May–Sep. Fruit: Jun–Sep. Arizona to Texas and southward to northern Mexico.

Ipomopsis Michx. Skyrocket, Gilia

(*Ipomoea*, a genus of Convolvulaceae, + *opsis*, like) Annuals, perennial herbs, or subshrubs, simple to branched. **LEAVES** basal to alternate, entire to deeply pinnately lobed, gradually reduced upward, the lobes completely confluent with the rachis, flat to terete, usually linear to oblong. **INFL** terminal, paniculate, open to congested, the basic unit composed of 2–7 pedicelled flowers subtended by a single bract, these sometimes arranged along one side of the rachis. **FLOWERS** actinomorphic to slightly zygomorphic; calyx tube membranes usually ruptured in fruit; corolla rotate to salverform, white to purplish or red; stamens unequally inserted on the corolla tube or throat; filaments equal or unequal in length; anthers included to exserted; style included to exserted. **CAPSULES** ovoid. **SEEDS** 1–many per locule (Day 1980; Grant and Wilken 1986, 1988). Approximately 30 species of western North America, one species in the southeastern U.S., and one species in Andean South America.

1. Flowers in terminal, capitate clusters; corolla tube 3–8 mm long; stamens inserted on the upper throat between the lobes ....(2)
1′ Flowers usually in clusters at tips of lateral branches, the inflorescence a diffuse to congested panicle (open in *I. longiflora*); corolla tube 5–50 mm long (5–15 mm long in *I. multiflora*); stamens inserted on the tube or base of throat ....(6)

2. Perennials, often branched and woody at base; stems 5–60 cm tall; seeds 1 per locule, sometimes locules empty..(3)
2′ Annuals, usually branching from the primary axis, 5–30 cm tall; seeds several per locule (except sometimes in *I. gunnisonii*) ....(4)

3. Filaments shorter than the anthers, inserted at the sinuses of the corolla lobes .... ***I. roseata***
3′ Filaments longer than the anthers, inserted at the sinuses of the corolla lobes .... ***I. congesta***

4. Lower cauline leaves linear and entire; seeds 1–2 per locule .... ***I. gunnisonii***
4′ Lower cauline leaves with 3–5 teeth or ovate to oblong lobes; seeds 2–3 (4) per locule .... (5)

5. Outer inflorescence bracts leaflike, toothed; stems with short curly hairs; corolla tube 3–5 mm long, lobes 1–1.5 (1.8) mm long .... ***I. polycladon***
5′ Outer inflorescence bracts reduced, entire; stems with woolly hairs; corolla tube 4–8 mm long, lobes 2–3.5 (4) mm long .... ***I. pumila***

6. Inflorescence diffusely branched, the flowers pedicelled, solitary or in pairs .... ***I. longiflora***
6′ Inflorescence open to narrow, often 1-sided, the flowers short-pedicelled to subsessile, in lateral, pedunculate clusters .... (7)

7. Corollas scarlet to magenta .... ***I. aggregata***
7′ Corollas white, peach, lavender, or pale violet to purplish .... (8)

8. Corolla tube 5–15 mm long; stamens inserted at the same level on the upper tube or throat .... ***I. multiflora***
8′ Corolla tube 15–50 mm long; stamens inserted unequally on the tube .... (9)

9. Corollas pale lavender to white, the throat 4–6 mm wide, the lobes obovate, the apices rounded to apiculate .... ***I. polyantha***
9′ Corollas pale peach to lavender, the throat 2–4 mm wide, the lobes lanceolate, the apices acute to attenuate .... ***I. tenuituba***

Ipomopsis aggregata (Pursh) V. E. Grant (clustered, referring to flowers) Skyrocket, scarlet gilia. Short-lived perennial, 20–100 cm tall, simple to branched at base. **STEMS** with short, glandular hairs, often with short, curly, eglandular hairs below. **LEAVES** subglabrous to short-pilose, deeply lobed. **INFL** diffuse to 1-sided, with subsessile to short-pedicelled flowers on lateral branches. **FLOWER** calyx 3–8 mm long, shortly glandular-pubescent, the lobes shorter than or equal to the tube and acuminate in Arizona; corolla usually scarlet, the tube 15–30 mm long, the throat 3–6 mm wide, the lobes lanceolate, acuminate, often with dark red flecks; stamens inserted unequally above the midtube; filaments unequal; anthers located in the throat or exserted; stigma slightly exceeding the anthers. **CAPSULES** 8–12 mm long. **SEEDS** 5–10 per locule. [*Gilia aggregata* (Pursh) Spreng.]. Approximately seven subspecies; western North America. The Kayenta Navajo used the plant as a cathartic, for spider bites, as an emetic, and for stomach disease. The Ramah applied an infusion to the body of a hunter and weapons for good luck. The Hopi used it for a beverage, dye, and for decoration. The Ute boiled the plant for glue.

1. Corolla pale, pink (rarely red), anthers included, from the orifice to about 4 mm below the orifice ..... **subsp. *collina***
1′ Corolla various shades of red, usually crimson, anthers exserted or close to the orifice .... (2)

2. Anthers all well exserted from the orifice of the corolla tube, 3–5 mm beyond the orifice .... **subsp. *formosissima***
2′ Anthers all close to the orifice of the corolla tube, from 2 mm beyond to 2 mm below the orifice ... **subsp. *aggregata***

subsp. *aggregata* Corolla scarlet with yellow spotting at the orifice, tube 18–20 mm long, the throat 3–4 mm wide; anthers close to the orifice, from 2 mm beyond to 2 mm below the orifice, stigma usually at the orifice. $2n = 14$. Open sites; ponderosa pine and other coniferous forest communities. ARIZ: Apa, Nav; COLO: Arc, Dol, Hin, LPl, Min, Mon, RGr, SJn, SMg; NMEX: McK, RAr, SJn; UTAH. 1680–2840 m (5000–10500′). Flowering: May–Sep. Fruit: June–Sep.

subsp. *collina* (Greene) Wilken & Allard (pertaining to hills) Corolla pink or red with pink or white spotting at the orifice, tube 20–25 mm long, the throat 2–4 mm wide; anthers at the orifice to 4 mm below the orifice, stigma usually at the orifice. $2n = 14$. Open sites; ponderosa pine, Douglas-fir, and spruce-fir communities. ARIZ: Apa; COLO: Arc, LPl; NMEX: RAr. 1680–2840 m (6000–9000′). Flowering: May–Sep. Fruit: Jun–Sep.

subsp. *formosissima* Wherry (greatly beautiful) Corolla scarlet with yellow spotting at the orifice, tube 18–25 mm long, the throat 4–8 mm wide; anthers and stigma usually exserted, longest anthers 3–5 mm beyond the orifice. $2n = 14$. [*Gilia aggregata* (Pursh) Spreng. var. *maculata* M. E. Jones; *G. aggregata* var. *texana* (Greene) I. M. Johnst.]. Open sites; piñon-juniper woodland, ponderosa pine, Douglas-fir, and spruce-fir communities. ARIZ: Apa, Nav; COLO: Arc, Dol, LPl, Mon, SMg; NMEX: McK, RAr, SJn; UTAH. 1680–2840 m (5500–9300′). Flowering: May–Sep. Fruit: Jun–Sep.

Ipomopsis congesta (Hook.) V. E. Grant (crowded together, referring to flowers) Ballhead gilia. Perennial, herbaceous to woody at base, 20–60 cm tall, simple to branched at base. **STEMS** with short, white, curly or woolly hairs. **LEAVES**

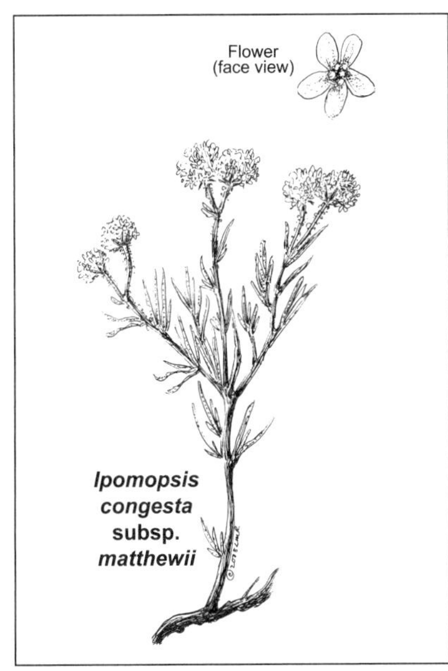

subglabrous to short-woolly, entire to pinnately or palmately lobed. **INFL** a dense terminal head, bracteate; flowers subsessile to sessile. **FLOWER** calyx 3–4 mm long, the lobes lanceolate, acuminate; corolla white, the tube 3–5 mm long, the throat 0.5–1 mm wide, the lobes rounded, sometimes with pinkish flecks; stamens inserted on the throat between the lobes; filaments subequal; anthers and stigma slightly exserted. **CAPSULES** 1.5–2 mm long. **SEEDS** 1–2 per locule. $2n$ = 14. Eight to 10 subspecies. Western North America.

1. Stems mostly simple, tall and wandlike; some cauline leaves entire but others pinnatifid; flowers white; 2 ovules per locule .... **subsp.** ***frutescens***
1′ Stems short and branched; all cauline leaves simple and entire; flowers cream-colored; 1 ovule per locule .... **subsp.** ***matthewii***

subsp. ***frutescens*** (Rydb.) A. G. Day **STEMS** tall and wandlike, several, woody at base, with short, curly hairs. **LEAVES** entire. Rocky outcrops, gravelly soils; piñon-juniper woodland communities. ARIZ: Nav. 1525–1620 m (5000–5300′) Flowering: Jul–Aug. Fruit: late Jul–Sep. Southwestern Utah, south to northwestern Arizona.

subsp. ***matthewii*** J. M. Porter **CP** (honoring Matthew Heil, botanist) **STEMS** short, several, woody at base, with short, curly hairs. **LEAVES** entire to pinnatifid. Rocky outcrops, gravelly soils; piñon-juniper woodland communities. ARIZ: Apa; COLO: Arc, LPl, Mon; NMEX: SJn; UTAH. 1525–1700 m (5000–6000′). Flowering: Jul–Aug. Fruit: late Jul–Sep. Western Colorado to eastern Utah, south to northwestern New Mexico and northeastern Arizona.

Ipomopsis gunnisonii (Torr. & A. Gray) V. E. Grant (for John W. Gunnison, surveyor) Gunnison's gilia. Annual, 3–30 cm tall, simple to branched. **STEMS** glabrous or with short glandular hairs. **LEAVES** glabrous to sparsely short-pilose, entire to remotely few-toothed. **INFL** with flowers congested in terminal heads, bracteate; flowers subsessile. **FLOWER** calyx 3–4.5 mm long, glabrous to sparsely glandular, the lobes acuminate; corolla white to pale lavender, the tube 4–7 mm long, throat 0.5–1 mm wide, the lobes rounded; stamens inserted on the upper tube or throat; filaments subequal; anthers and stigma slightly exserted. **CAPSULES** 3–4 mm long. **SEEDS** 1–2 per locule. $2n$ = 14. [*Gilia gunnisonii* Torr. & A. Gray]. Sandy soils; desert scrub, shadscale, sagebrush, and piñon-juniper woodland communities. ARIZ: Apa, Nav; COLO: Mon; NMEX: SJn; UTAH. 1280–2200 m (4200–7200′). Flowering: May–Jul (Sep). Fruit: Jun–Sep. Southern Utah to southwest Colorado, south to Arizona and New Mexico. Used as a blood purifier and poultice for sores by the Kayenta Navajo.

Ipomopsis longiflora (Torr.) V. E. Grant (long flower) Annual in Arizona or biennial, 25–100 cm tall, simple to branched. **STEMS** glabrous to sparsely short-pubescent. **LEAVES** glabrous to sparsely short-pilose, deeply lobed. **INFL** diffuse, with 1–3 subsessile to long-pedicelled flowers at tips of branches. **FLOWER** calyx 5–10 mm long, shortly glandular-pubescent, the lobes lanceolate to ovate, acuminate, glabrous to sparsely short-pubescent, the calyx 6–11 mm long in fruit; corolla white to bluish, tube 30–50 mm long, the throat 2–3 mm wide, the lobes ovate, rounded to acuminate; stamens inserted on the tube; filaments unequal; anthers included to exserted; stigma slightly exceeding the anthers. **CAPSULES** 10–15 mm long. **SEEDS** 8–15 per locule. $2n$ = 14. [*Gilia longiflora* (Torr.) G. Don]. Three subspecies, Utah to South Dakota, south to Texas and northern Mexico. Open sites, washes, sandy soils; desert scrub, sagebrush, piñon-juniper. ARIZ: Apa, Coc, Nav; COLO: Arc, LPl, Mon; NMEX: McK, RAr, San, SJn; UTAH. 1136–1980 m (3725–6500′). Flowering: May–Sep. Fruit: Jun–Sep. Utah to Colorado, south to northern Sonora and northern Chihuahua, Mexico. In our area we find only **subsp.** ***neomexicana*** Wilken (of New Mexico). The Hopi used a decoction of leaves for stomachache. For the Navajo it is used as a ceremonial medicine plant in the Wind and Female Shooting Chants; a decoction of pounded plant taken to vomit; infusion of flowers mixed with feed and given to sheep for stomach problems.

Ipomopsis multiflora (Nutt.) V. E. Grant (many flowers) Many-flowered gilia. Short-lived perennial, 15–50 cm tall, simple to branched at base; stems with short to long, glandular to nonglandular hairs. **LEAVES** glabrous to sparsely short-pilose or glandular, the lower deeply lobed, the upper entire to few-lobed. **INFL** somewhat diffuse to 1-sided, with subsessile flowers crowded on short, lateral branches. **FLOWER** calyx 4–8 mm long, shortly glandular-pubescent, the lobes short-aristate; corolla pale violet to purplish, the tube 5–15 mm long, the throat 1–2.5 mm wide, the lobes subequal, the lower 3 partly united, often with purple flecks; stamens inserted on the upper tube or throat; filaments unequal; anthers exserted; stigma slightly exceeding the anthers. **CAPSULES** 4.5–7 mm long. **SEEDS** 2–8 per locule. [*Gilia multiflora* Nutt.]. Open sites; desert grasslands, piñon-juniper woodland, and ponderosa pine communities. ARIZ: Apa; NMEX: McK, RAr, San, SJn. 1525–2590 m (5000–8500′). Flowering: Jul–Oct. Fruit: Aug–Oct. Southern Colorado, New Mexico, and Arizona (perhaps also in northern Chihuahua, Mexico). The Ramah Navajo used a decoction of the plant as a ceremonial medicine. The Zuni powdered the plant and applied it to the face for headache.

Ipomopsis polyantha (Rydb.) V. E. Grant **CP** (many flowers) Pagosa skyrocket. Short-lived monocarpic perennial, 25–60 cm tall, simple, branched toward the apex. **STEMS** with short to long, glandular to nonglandular hairs. **LEAVES** 3–4 cm long, pinnatifid, glabrous to sparsely short-pilose or glandular. **INFL** narrow, somewhat diffuse, with short-pedicelled flowers crowded on short, lateral branches. **FLOWER** calyx 4–5 mm long, shortly glandular-pubescent, the lobes cuspidate-pungent; corolla 0.9–1.4 cm long, white to pale violet, the tube 4.5–6.5 mm long, the throat 1.5–2.5 mm wide, the lobes subequal, widely spreading, often with purple flecks; stamens inserted on the upper tube or throat; filaments unequal; anthers well exserted; stigma slightly exceeding the anthers. **CAPSULES** 4.5–6 mm long. **SEEDS** 2–8 per locule. $2n = 14$. [*Gilia polyantha* Rydb.]. Clay soils of roadsides and open forests; ponderosa pine and mountain brush communities. COLO: Arc. 1980–2300 m (6500–7500′). Flowering: May–Aug. Fruit: Jun–Aug. This species is endemic to the Pagosa Springs area of Archuleta County.

Ipomopsis polycladon (Torr.) V. E. Grant (many branches) Spreading gilia, internode gilia. Annual, 4–12 cm tall, usually with ascending to spreading branches. **STEMS** with short glandular hairs and some eglandular curly hairs. **LEAVES** subglabrous to shortly glandular-pubescent, coarsely toothed to lobed, the lobes ovate. **INFL** a congested terminal head, bracteate, the outer bracts leaflike; flowers sessile to subsessile. **FLOWER** calyx 3–6 mm long, the lobes lanceolate, acuminate; corolla white, the tube 3–5 mm long, the throat 0.5–1 mm wide, the lobes rounded; stamens inserted on the throat between the lobes; filaments subequal; anthers and stigma slightly exserted. **CAPSULES** 4–5 mm long. **SEEDS** 1–3 per locule. $2n = 14$. [*Gilia polycladon* Torr.]. Sandy or gravelly soils, washes; desert scrub, blackbrush, and piñon-juniper woodland communities. ARIZ: Apa, Nav; COLO: Mon; NMEX: McK, RAr, SJn; UTAH. 1136–1920 m (3725–6300′). Flowering: Apr–Jun. Fruit: May–Jul. Eastern California to southern Idaho, south to Texas and northern Mexico. Used as a sedative and tonic by the Kayenta Navajo.

Ipomopsis pumila (Nutt.) V. E. Grant (dwarf, short) Dwarf gilia, eriastrum gilia. Annual, 3–18 cm tall, simple to branched. **STEMS** short-glandular to short-woolly. **LEAVES** subglabrous to short-woolly, often glandular, lower deeply lobed, upper lobed to entire. **INFL** a congested terminal head, bracteate, the bracts entire to toothed; flowers subsessile. **FLOWER** calyx 3–6 mm long, the lobes lanceolate, acuminate; corolla lavender to purplish, the tube 4–8 mm long, the throat 0.5–1 mm wide, the lobes acute to rounded; stamens inserted on the throat between the lobes; filaments subequal; anthers and stigma slightly exserted. **CAPSULES** 3–5.5 mm long. **SEEDS** 2–5 per locule. $2n = 14$. [*Gilia pumila* Nutt.]. Sandy soils; desert scrub, sagebrush, and piñon-juniper woodland communities. ARIZ: Apa, Nav; COLO: LPl, Mon; NMEX: McK, RAr, San, SJn; UTAH. 1490–1920 m (4900–6300′). Flowering: Mar–Jun (Oct). Fruit: Apr–Oct. Utah to Wyoming, south to Texas and New Mexico.

Ipomopsis roseata (Rydb.) V. E. Grant (provided with a rose color) Roseate gilia. Perennial, 10–30 cm tall, simple to branched at base; stems often woody at base, with white curled hairs, or short glandular hairs, or glabrous. **LEAVES** with white curled hairs to sparsely glandular, or subglabrous, deeply pinnatifid to entire. **INFL** sparse to dense cymose heads, with subsessile flowers, terminating the branches. **FLOWER** calyx 5.5–8.5 mm long, the lobes spinulose-tipped; corolla cream to white, the tube 7–9 mm long, the throat 1.5–3 mm wide, the lobes 2.5–4 mm long, ovate to obovate, rounded; stamens inserted at the sinuses of the lobes; filaments equal to or shorter than the anthers; anthers partially exserted; stigma included. **CAPSULES** 3–6 mm long; seeds 1 per locule. $2n = 28$. [*Gilia roseata* Rydb.]. Sandstone outcrops and rocky slopes; blackbrush, desert scrub, sagebrush, and piñon-juniper communities. UTAH. 1190–2290 m (4000–7300′). Flowering: May–Jun. Fruit: late May–Jul. Colorado and Utah.

Ipomopsis tenuituba (Rydb.) V. E. Grant (thin, fine tube) Short-lived perennial, 35–100 cm tall, simple to branched at base. **STEMS** often woody at base, with short glandular hairs, eglandular below. **LEAVES** subglabrous to sparsely short-pilose, deeply lobed. **INFL** narrow, 1-sided, with subsessile to short-pedicelled flowers on lateral branches. **FLOWER** calyx 4–6 mm long, shortly glandular-pubescent, the lobes acute to acuminate; corolla white to lavender, the tube 25–35 mm long, the throat 2–4 mm wide, the lobes lanceolate to ovate, acuminate to apiculate, often with pink to lavender flecks; stamens inserted unequally on the tube; filaments unequal; anthers included to slightly exserted; stigma slightly exserted. **CAPSULES** 5–8 mm long. **SEEDS** 6–12 per locule. $2n = 14$. [*Gilia tenuituba* Rydb.]. Open sites, meadows, ponderosa pine, and spruce-fir communities. COLO: LPl, Mon; UTAH. 2140–2760 m (7000–9050′). Flowering:

Jul–Sep. Fruit: Aug–Sep. Oregon to Idaho, south to California, Colorado, and Arizona. In our area this species is represented by **subsp**. ***tenuituba***. Hybridization occurs with *I. aggregata* subsp. *formosissima*, the hybrids being intermediate in corolla size and color.

Leptosiphon Benth. Leptosiphon

(narrow tube) Annuals or perennials. **STEMS** simple to much-branched, erect to decumbent, leafy. **LEAVES** opposite, rarely alternate above, palmately lobed, the lobes confluent with the rachis, linear, flat to terete, acute to weakly spinulose. **INFL** terminal or axillary, compact and capitate, glabrous, pubescent, or glandular. **FLOWERS** pedicelled, actinomorphic; calyx tube membranes very narrow, usually ruptured or distended in fruit, the lobes equal, linear to attenuate, often mucronate; corolla white to yellowish or lavender; stamens equally inserted on the corolla throat or tube; anthers slightly exserted or included; filaments equal in length; style included to exserted. **CAPSULES** ovoid to oblong. **SEEDS** 1–several per locule (Patterson 1977). About 30 species in southwestern North America and one species in South America; one species in our area.

Leptosiphon nuttallii (A. Gray) J. M. Porter & L. A. Johnson (honoring Thomas Nuttall, naturalist) Suffrutescent perennial to 3 dm tall, branching at the base; stems erect, glabrous to short-pubescent. **LEAVES**: lobes (4) 5–9, linear to oblong, spinulose, 10–20 mm long, short-pubescent. **INFL** compact, the flowers 2–5 in terminal, bracteate clusters. **FLOWERS** subsessile; calyx glabrous to pubescent, narrowly campanulate, 7–10 mm long, the lobes longer than the tube, the membranes mostly herbaceous, the hyaline part narrow, often obscure; corolla diurnal, salverform, 8–15 mm long, the tube and lobes white, the throat yellow; stamens inserted on the throat; style slightly exserted. $2n = 18$. [*Linanthus nuttallii* (A. Gray) Greene ex Milliken; *Linanthastrum nuttallii* (A. Gray) Ewan]. Sandy to rocky soils; meadows, sagebrush, piñon-juniper woodland, ponderosa pine, aspen, spruce-fir forests, and oak woodland communities. ARIZ: Apa; COLO: Arc, Con, Hin, Min, RGr; NMEX: McK, RAr; UTAH. 1650–2600 m (5400–8500′). Flowering: May–Sep. Fruit: June–Sep. Washington to Colorado, south to California, New Mexico, and northwestern Mexico. Our material is referred to **subsp**. ***nuttallii***, with leaf lobes linear to linear-lanceolate.

Linanthus Benth. Linanthus

(flax-flower) Annuals or perennials. **STEMS** simple to much-branched, erect to decumbent, leafy. **LEAVES** mostly opposite, sometimes alternate above, simple to deeply palmately lobed, the lobes confluent with the rachis, linear, flat to terete, acute to weakly spinulose. **INFL** terminal and compact or axillary, glabrous to glandular. **FLOWERS** sessile to short-pedicelled, actinomorphic; calyx tube membranes ruptured or distended in fruit, the lobes equal, linear to attenuate, often mucronate; corolla white to yellow or lavender, sometimes bluish; stamens equally inserted on the corolla throat or tube; anthers exserted or included; filaments equal in length; style included to exserted. **CAPSULES** ovoid to oblong. **SEEDS** 1–several per locule (Bell and Patterson 2000). About 24 species in western North America.

1. Plants with erect stems, 10–80 cm tall; lower leaves opposite, upper leaves usually alternate; sepals, petals, and stamens 5 .... ***L. pungens***

1′ Plants loosely matted or sprawling, stems 5–11 cm tall; leaves usually all opposite; sepals, petals, and stamens often 6 .... ***L. watsonii***

Linanthus pungens (Torr.) J. M. Porter & L. A. Johnson (sharp-pointed) Prickly phlox. Subshrub, 1–4 dm tall, branching mostly below the middle; stems ascending to erect, glandular and pilose. **LEAVES** opposite below, alternate above, palmately to pinnately lobed, 5–12 mm long, the 3–9 lobes linear to narrowly oblong, glabrous to pubescent, spinulose, the upper leaves subtending clusters of short leaves. **FLOWER** calyx 6–10 mm long, 5-merous, glabrous to sparsely pubescent, the lobes slightly unequal, usually shorter than the tube; corolla nocturnal, closed during the day, 5-merous, salverform, 14–20 (25) mm long, cream to salmon, the throat often tinged lavender or purple; stamens 5, inserted on the upper tube; stigmas 3; ovary with 3 locules. [*Leptodactylon pungens* (Torr.) Rydb.]. Sandy to rocky soils; sagebrush, piñon-juniper woodland, mountain brush, and ponderosa pine communities. ARIZ: Apa, Nav; COLO: Arc, Dol, LPl, Mon; NMEX: McK, RAr, San, SJn; UTAH. 1700–2100 m (5600–7000′). Flowering: May–Sep. Fruit: late May–Sep. British Columbia to Montana, south to California and New Mexico. Used for scorpion stings and kidney disease by the Kayenta Navajo. A decoction of the plant was used for "snake infection" by the Ramah Navajo.

Linanthus watsonii (Torr.) J. M. Porter & L. A. Johnson (for Sereno Watson, geologist, explorer, botanist) Watson's leptodactylon. Sprawling to loosely matted subshrub, 5–12 cm tall, branching mostly at the base. **STEMS** ascending to erect, glandular and puberulent. **LEAVES** opposite, 6–20 mm long, palmately lobed, the 3–9 lobes linear to narrowly oblong, glandular-pubescent, spinulose, often somewhat rigid. **FLOWER** calyx 6–10 mm long, glabrous to sparsely pubescent, the lobes distinctly unequal, shorter than the tube, usually 6-merous; corolla nocturnal, closed during the day, usually 6-merous, salverform, 14–20 (25) mm long, cream to salmon, the throat often tinged lavender or purple; stamens usually 6, inserted on the upper tube; stigmas 3 or 4; ovary with 3–4 locules. [*Leptodactylon watsonii* (Torr.) Rydb.]. Crevices in granite and sandstone, sandy to rocky soils; blackbrush, sagebrush, pinon-juniper, mountain brush, and ponderosa pine communities. ARIZ: Apa; COLO: Arc, Hin, LPl, Min, Mon, RGr, SJn; UTAH. 1700–2100 m (5600–7000′). Flowering: May–Aug. Fruit: Jun–Sep. Idaho south to Nevada and northern Arizona, east to western Wyoming, Utah, and Colorado.

Microsteris Greene Microsteris

(tiny star, referring to the small, starlike corolla) Annuals. **STEMS** erect to ascending, often much-branched. **LEAVES** mostly opposite, often alternate above and in the inflorescence, simple, entire, linear to elliptic or lanceolate. **INFL** terminal, the flowers 1–3. **FLOWERS** pedicelled to subsessile; calyx tube membranes ruptured in fruit, the lobes equal; corolla salverform, white to bluish; stamens unequally inserted on the tube; filaments short, usually equal in length, glabrous; anthers and style usually included. **CAPSULES** ovoid to ellipsoid. **SEEDS** 1 (2–3) per locule, conspicuously gelatinous when wet. $2n = 14$. One or two species. North America and South America.

Microsteris gracilis (Douglas ex Hook.) Greene (slender) Little polecat. Annual with 1–3 erect stems, 3–15 cm tall, often branched below, the lower stems ascending or spreading. **LEAVES** linear to elliptic or narrowly oblanceolate, pubescent and glandular, 1–2 (4) cm long, 2–5 mm wide. **INFL** flowers 1–2, subsessile to pedicelled at tips of terminal branches, sometimes axillary; pedicels glandular. **FLOWER** calyx 3–8 mm long; corolla white to bluish lavender, the throat sometimes yellow-tinged, the tube 4–8 mm long, the lobes 1–2 mm long, obtuse to retuse; stamens inserted on the upper tube and throat; stigmas located among the lower stamens. **SEEDS** gelatinous when wet. [*Phlox gracilis* (Hook.) Greene]. Open, sandy to gravelly sites; sagebrush, piñon-juniper woodlands, mountain brush, and ponderosa pine communities. ARIZ: Apa, Coc, Nav; COLO: Arc, Dol, Hin, LPl, Min, Mon, SMg; NMEX: McK, RAr, San, SJn; UTAH. 1220–2450 m (4000–8000′). Flowering: Mar–Aug. Fruit: Apr–Aug. Southern British Columbia to Montana, south to Arizona and New Mexico; Baja California, northern Sonora, Mexico.

Navarretia Ruiz & Pav. Navarretia

(for Francisco Fernandez Navarrete, Spanish botanist, physician) Annuals. **STEMS** erect to prostrate. **LEAVES** mostly alternate, sometimes opposite below, twice pinnately lobed in Arizona, the lobes narrow, rigid, and spinulose. **INFL** terminal and axillary, compact, capitate, bracteate, the bracts pinnately to palmately lobed, villous to glandular, spinulose. **FLOWERS** actinomorphic, sessile; calyx membranes usually remaining intact in fruit, the lobes usually unequal, linear to acuminate, spinulose; corolla (4) 5-merous, salverform to funnelform, white, yellow, or blue; stamens inserted at the same level on the upper tube or throat; filaments mostly equal in length; anthers included in the throat or exserted; style included or slightly exserted, the stigmatic lobes 2–3, short or obscure. **CAPSULES** ovoid to ellipsoid, dehiscent to indehiscent; seeds 1–7 per locule, gelatinous when wet (Day 1993b). About 30 species of western North America; one in Chile.

1. Corolla yellow, vascular traces in the upper tube 3 per lobe; stigma lobes 3; calyx lobe sinus U-shaped, intercostal region truncate .... ***N. breweri***

1′ Corolla white to pale lavender or blue, vascular traces in the upper tube 1 per lobe; stigma lobes 2; calyx lobe sinus V-shaped .... ***N. intertexta***

Navarretia breweri (A. Gray) Greene (for William H. Brewer, botanist, geologist) Brewer's pincushion. Plants 3–7 cm tall. **STEMS** mostly erect, simple or branching above, reddish brown, densely pubescent. **LEAVES** 8–18 mm long, glabrous to short-pubescent, the rachis flattened, the lobes linear, the lateral lobes 2–6, mostly simple. **INFL** terminal, compact, subcapitate, the bracts similar to the cauline leaves. **FLOWERS** subsessile; calyx 7–10 mm long, the tube glandular short-pubescent, the lobes simple; corolla salverform, yellow, 6–8 mm long; stamens inserted on the upper tube below the sinuses; filaments 1–2 mm long; anthers exserted; style exserted; stigmatic lobes mostly 3, situated below the anthers. **CAPSULES** broadly ovoid, mostly 3-locular, dehiscent, 3–5 (8) mm long; seeds 1–2 (3) per locule. $2n = 18$. Infrequent on rocky or shallow clay soils of dry meadows and flats; sagebrush, piñon-juniper woodland, and ponderosa pine communities. ARIZ: Coc; COLO: Arc, Dol, LPl, Mon, SMg; UTAH. 2290–2560 m (7500–8400′). Flowering: Jun–Aug. Fruit: Jul–Sep. Washington and Idaho, east to Alberta and Colorado, south to Baja California and northern Arizona.

Navarretia intertexta (Benth.) Hook. (between woven) Plants simple or branched, 3–11 cm tall. **STEMS** puberulent to more or less villous in the inflorescence. **LEAVES** up to 3 cm long, pinnatifid, rachis narrow, terminal segment generally elongate. **INFL** terminal, compact, subcapitate, the bracts with broader, firmer rachis than leaves. **FLOWERS** subsessile; calyx white-hairy at the internal orifice, the intercostal membranes V-shaped, the lobes more or less unequal, the larger ones often trifid; corolla white to pale lavender or pale blue, 4–7 mm long, usually surpassed by the calyx; filaments mostly 0.5–2 mm long; stigma lobes 2 (very rarely 3); ovary 2-locular, with (2) 3–5 ovules and seeds per locule. **CAPSULES** ovoid, indehiscent, or tardily dehiscent. $2n = 18$. Ephemerally moist sites, edges of drying ponds, lakes, or reservoirs; meadows, sagebrush, and ponderosa pine communities. COLO: Arc, Hin, LPl, Mon. 1830–2560 m (6000–8400′). Flowering: Jun–Sep (Oct). Fruit: Jul–Oct. Southern Canada to Colorado and Arizona, south to Baja California, Mexico. Our material belongs to **subsp**. ***propinqua*** (Suksd.) A. G. Day (near, referring to similarity to *N. intertexta*).

Phlox L. Phlox

(flame, referring to the brightly colored corollas) Perennial, often caespitose. **STEMS** erect to decumbent, often much-branched. **LEAVES** mostly opposite, often alternate only in the inflorescence, simple, entire, linear to elliptic or lanceolate. **INFL** terminal, the flowers 1–3. **FLOWERS** pedicelled to sessile; calyx tube membranes ruptured in fruit, the lobes equal; corolla salverform, white to red, blue, or purple; stamens unequally inserted on the tube; filaments short, usually equal in length, glabrous; anthers and style usually included. **CAPSULES** ovoid to ellipsoid. **SEEDS** 1 (2–3) per locule, usually not gelatinous when wet. $2n = 14, 28, 42$. Seventy species in North America, one in Siberia (Wherry 1955). The genus *Phlox* is both complex and very poorly understood. Wherry, the most recent monographer of *Phlox*, recognized as many as 22 taxa within the boundaries of the *Flora*. Based upon recent floristic treatments in adjacent states, most current authors would recognize only 10 taxa. Actual diversity likely lies somewhere between these two estimates. Following is a relatively conservative treatment recognizing 13 taxa in our area.

1. Plant habit relatively loose and erect, the stems scattered or few together, often more than 10 cm tall, most of the leaves well spaced and not forming a basal cluster or mat; flowers born on pedicels, in loosely branched, terminal cymes ..... (2)
1′ Plant habit more compact, the stems numerous and low, generally less than 10–15 cm tall, forming loose to dense mats or mounds; flowers sessile or inconspicuously short-pedicelled, 1–3 (6) together at the stem tips ..... (8)

2. Intercostal membranes of the calyx plicate or strongly bulged-carinate toward the base ..... (3)
2′ Intercostal membranes of the calyx flat or inconspicuously wrinkled, not keeled or bulged ..... (5)

3. Calyx intercostal membrane plicate; corolla tube 19–25 mm long; style 15–21 mm long ..... ***P. standsburyi***
3′ Calyx intercostal membrane bulged-carinate; corolla tube 11–18 mm long; style 8–15 (16) mm long ..... (4)

4. Stems generally with 4 leaf-bearing nodes below the inflorescence; leaves thick-textured, 12–30 (35) mm long, (2) 3–6 (8) mm wide ..... ***P. grayi***
4′ Stems generally with 5–6 leaf-bearing nodes below the inflorescence; leaves moderate to thin-textured, 25–90 mm long, 1–5 mm wide ..... ***P. longifolia***

5. Corolla lobes retuse, evidently notched at the tip, the notch 1–2 mm deep, so that collectively the lobes appear 10-pointed; styles 1–2.5 mm long; stems several from a woody base ..... ***P. woodhousii***
5′ Corolla lobes entire, wavy, erose, or slightly emarginate, but not retuse; styles 2–18 mm long; stems not from a woody base ..... (6)

6. Stamens inserted in the middle to lower tube; style short, 2–3 mm long; corolla tube glandular-pilose .... ***P. nana***
6′ Stamens inserted in the upper tube; style long, 8–18 mm long; corolla tube glabrous ..... (7)

7. Stems scattered on a creeping rhizome, forming widely spaced colonies of vegetative and reproductive stems; floral pedicels glandular-pubescent, 8–18 mm long ..... ***P. cluteana***
7′ Stems loosely clustered on a taproot and branching caudex, or the plant with a branching, woody base; floral pedicels eglandular, up to 25 mm long ..... ***P. caryophylla***

8. Intercostal membranes of the calyx carinate, though sometimes weakly so ..... (9)
8′ Intercostal membranes of the calyx flat or inconspicuously wrinkled, not carinate (sometimes obscured by the pubescence) ..... (10)

9. Keel of the intercostal membrane conspicuous; inflorescence densely, often appressed pubescent to glabrescent ..... ***P. austromontana***
9′ Keel of the intercostal membrane weak and inconspicuous, low, linear (but distinct); inflorescence finely pubescent to glabrescent ..... ***P. diffusa***

10. Pubescence of the calyx tangled-woolly, usually copious and obscuring the intercostal membranes, never glandular .... (11)
10' Pubescence of the calyx various or absent, when present, not tangled-woolly or obscuring the intercostal membranes, often glandular .... (12)

11. Leaves firm and pungent, narrowly linear, mostly (2.5) 4–10 (13) mm long and 0.5–1 mm wide near the middle, usually somewhat arachnoid-woolly, but varying to essentially glabrous; common in the foothills and lowlands, seldom ascending to middle elevations in the mountains .... ***P. canescens***
11' Leaves softer and larger, mostly 5–20 mm long and (0.5) 1–2 mm wide, green and glabrous except for the often basally arachnoid-ciliate margins; more montane .... ***P. diffusa***

12. Style elongate, 5.5–8 mm long; calyx glabrous or seldom inconspicuously arachnoid-puberulent; leaves relatively soft and lax, rather sparsely granular-scaberulous, otherwise essentially glabrous, the better-developed ones mostly (10) 12–30 mm long; montane .... ***P. multiflora***
12' Style short, 2–5 mm long; calyx usually glandular-hairy, seldom eglandular or even glabrous; leaves linear-subulate, firm, erect and appressed or arcuate, glandular, pilose or glabrous, mostly 5–12 (15) mm long .... (13)

13. Stems tending to be interlaced; leaves arcuate and spreading; styles 2.5–5.5 mm long .... ***P. pulvinata***
13' Stems tending to be discrete and subparallel; leaves erect and appressed; styles 1.5–3 mm long .. ***P. condensata***

Phlox austromontana Cov. (southern mountains) Desert phlox. Compact, matted, sometimes more open and spreading, taprooted perennial, with 8–many decumbent to ascending stems. **STEMS** 6–10 (15) cm long, the lower internodes obscured by the leaves. **LEAVES** linear, firm, thick, grayish green, mucronate, mostly glabrous on the lower surface, glabrous to more often pubescent on the upper surface, 8–15 (18) mm long, 1–2 mm wide, midrib and margins thickened, giving the lower surface a 2-grooved appearance. **INFL** flowers solitary, rarely to 5, subsessile. **FLOWER** calyx 6–10 (12) mm long, the membranes weakly ridged to distended, glabrous to densely white-pubescent; corolla white to light lavender, rarely blue, the tube 8–18 mm long, the lobes 5–8 mm long, obovate, obtuse; stamens inserted on the upper tube; stigmas located below most of the stamens, subequal to the calyx. **FRUIT** obovate, apex acute, 3.5–6 mm long, 1.5–2.5 mm wide. **SEEDS** elliptic, acute on one end, 2.8–3.5 mm long, 1.2–1.5 mm wide. Rocky soils, slopes; coniferous forest and woodland communities. 1250–2800 m (4100–9000′). Flowering: Apr–Jul. Nevada and Idaho south to Baja California, Mexico. The plant was crushed and placed in cavity for toothaches by the Kayenta Navajo.

1. Stems open and more or less spreading, internodes apparent .... **subsp. *prostrata***
1' Stems compact, internodes obscured by leaves .... (2)

2. Flowers white or less often pink to blue; calyx turbinate to subcylindric .... **subsp. *austromontana***
2' Flowers cream, fading lemon-yellow; calyx campanulate .... **subsp. *lutescens***

subsp. *austromontana* Desert scrub, piñon-juniper woodland, sagebrush, and ponderosa pine communities. ARIZ: Apa, Coc, Nav; COLO: Arc, Hin, LPl, Mon; NMEX: RAr, SJn; UTAH. 1500–3000 m (4920–9840′). Flowering: May–Jun. Fruit: Jun–Jul.

subsp. *lutescens* (Welsh) J. M. Porter (becoming yellow) Yellow desert phlox. Blackbrush, desert scrub, and piñon-juniper woodland communities. ARIZ: Apa; UTAH. 1300–2050 m (4265–6725′). Flowering: May–Jun. Fruit: (late May) Jun–Jul.

subsp. *prostrata* (E. E. Nelson) Wherry (reclined) Silver reef phlox. Mountain brush and piñon-juniper woodland communities. UTAH. 1250–1300 m (4100–4265′). Flowering: May–Jun. Fruit: Jun–Jul.

Phlox canescens Torr. & A. Gray (becoming gray) Compact, matted perennial with many decumbent stems. **STEMS** 2–5 cm tall, the internodes obscured by the leaves, densely pubescent to woolly. **LEAVES** linear, firm, thick, green, subulate, 6–10 (12) mm long, 1–2 mm wide, the margins and midrib often thick, glabrous to densely cobwebby-woolly below. **INFL** flowers solitary, sessile to subsessile. **FLOWER** calyx 5–9 mm long, the membranes flat or wrinkled, the lobe margins glabrous, glandular, or woolly; corolla white to light lavender, the tube 8–12 mm long, the lobes 4–7 mm long, obtuse; stamens inserted from the middle to the upper tube; stigma located among the stamens, subequal to much longer than the calyx. **FRUIT** obovoid, apex acute, 3.5–4.5 mm long, 1.8–2.2 mm wide. **SEEDS** 1 or fewer per locule, 2.5–3 mm long, 0.8–1.2 mm wide, oblong. [*P. hoodii* Richardson subsp. *canescens* (Torr. and A. Gray) Wherry].

Gravelly soils, rocky slopes and ledges; desert grassland, desert scrub, piñon-juniper woodland, and ponderosa pine communities. ARIZ: Apa, Coc, Nav; COLO: Arc, LPl, Mon; NMEX: McK, RAr, SJn; UTAH. 1950–2700 m (5500–8800′). Flowering: Apr–May (Aug). Fruit: May–Sep. Washington south to California and east to Colorado and northwestern New Mexico.

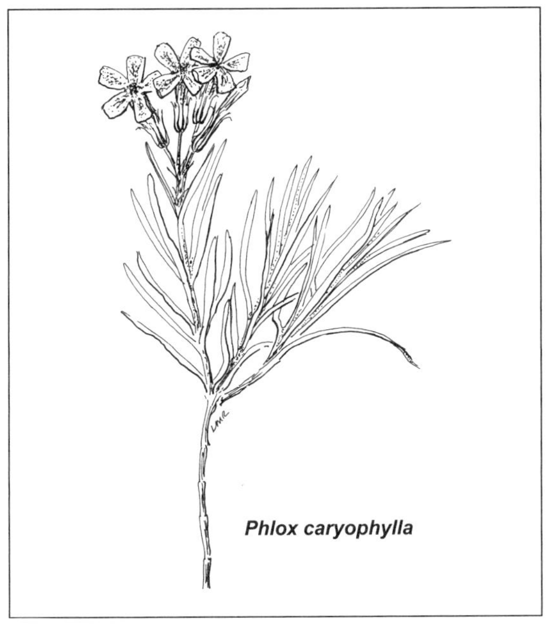
Phlox caryophylla

Phlox caryophylla Wherry (nut leaf, referring to the clove or carnation-like fragrance of the flowers) Clove phlox. Open-tufted, woody-based perennials. **STEMS** 10–25 cm long, pilose-pubescent, eglandular, internodes elongate. **LEAVES** linear, oblong, or narrowly elliptic, acute to acuminate, 3–5.5 cm long and 2–4 mm wide. **INFL** composed of 3–12 flowers, the pedicels up to 25 mm long. **FLOWER** calyx 11–15 mm long, lobes subequal to slightly shorter than the tube, membrane broad, flat to slightly wrinkled, but not carinate, pilose-puberulent; corolla with bright purple to pink lobes, the tube pale, 13–17 mm long, the lobes 7–12 mm long, obtuse, entire to emarginate or slightly retuse; stamens inserted on the upper tube and near the orifice; uppermost anthers slightly exserted; style 8.5–13 mm long, stigmas located in the upper tube or slightly exserted, among the upper stamens. **FRUIT** not observed. **SEEDS** not observed. Sparsely wooded slopes and roadsides; sagebrush, mountain brush, ponderosa pine, and spruce-fir communities. COLO: Arc, Hin, LPl; NMEX: RAr. 1980–2130 m (6500–7800′). Flowering: Apr–Jun. Fruit: May–Jun. Endemic to the Four Corners region.

Phlox cluteana A. Nelson **CP** (for William N. Clute, botanist) Perennial with 1–2 stems from long, slender rhizomes, forming colonies of scattered flowering and vegetative shoots. **STEMS** 8–15 (20) cm tall, the internodes elongate, the upper internodes often longer than the leaves, glabrous to glandular and short-villous. **LEAVES** linear, elliptic, or narrowly lanceolate, flat, acute to obtuse, glabrous to ciliate, sometimes glandular, 1–5 cm long, 2–5 mm wide, sparsely ciliate and pilose. **INFL** flowers 6–12, composed of 2- to 3-flowered cymes; pedicels glandular-pubescent, 8–18 mm long. **FLOWER** calyx 6–10 mm long, glandular-pubescent, the membranes flat; corolla pink to deep red-purple, the tube 15–17 (20) mm long, the lobes 8–10 mm long, rounded, emarginate or slightly retuse; stamens inserted on the upper tube and near the orifice; uppermost anthers slightly exserted; stigmas located in the upper tube among the stamens, sometimes slightly exserted. **FRUIT** subglobose to ovoid, apex obtuse, 3.2–4.2 mm long, 2.5–3.5 mm wide. **SEEDS** not observed. Open sites in sagebrush, ponderosa pine, and aspen communities. ARIZ: Apa, Coc; NMEX: SJn; UTAH. 1950–3140 m (6400–10300′). Flowering: (Apr) Jun–Jul. Fruit: Jun–Aug. South-central and southeast Utah and adjacent north-central and northeastern Arizona and northwestern New Mexico. A Four Corners regional endemic.

Phlox condensata (A. Gray) E. Nelson **CP** (condensed) Compact phlox. Densely pulvinate-caespitose, taprooted plants, forming cushion-mats or low mounds. **STEMS** numerous, short, erect, 1–4.5 (5) cm long, subparallel and discrete, internodes obscured by leaves. **LEAVES** linear to oblong, 4–9.5 mm long, 0.5–1.5 mm wide, erect and appressed, sharply bristle-pointed, thick and firm, 2-grooved along the back, hirsute-ciliate along the margins, at least below the middle, glandular-hairy or minutely granular-glandular to essentially glabrous on the surfaces. **INFL** flowers solitary and subsessile. **FLOWER** calyx 4–7.5 mm long, slightly to strongly glandular or glandular-puberulent to glandular-villous, intercostal membranes flat; corolla white to pink or blue-lavender, the tube 6–10 mm long, the lobes 3–5 mm; stamens inserted on the upper tube; style 1.5–3.2 mm long, shorter than or subequal to the lower stamen. Open, rocky places at high elevation, mostly in alpine tundra. COLO: Arc, Con, Hin, LPl, Min, Mon, RGr, SJn. 2700–3700 m (9000–12200′). Flowering: Jul–Aug. Fruit: Aug–Sep.

Phlox diffusa Bentham (spreading) Spreading phlox, Sierra phlox. Caespitose and commonly mat-forming, taprooted perennial, occasionally looser. **STEMS** 5–10 cm tall. **LEAVES** green and pilose to glabrescent, with basally arachnoid-ciliate margins, 8–15 (20) mm long, (0.5) 1–2 mm wide. **INFL** flowers solitary to 3, subsessile, the pedicels 0.5–3 mm long, pilose to glabrescent, but eglandular. **FLOWER** calyx 7–11 mm long, commonly slightly arachnoid-villous; calyx lobes linear-subulate and often thickened; corolla white to pinkish or light bluish, the tube 9–17 mm long, less than twice as long as the calyx, the lobes 4.5–9 mm long, obovate; stamens inserted in the upper tube, style 3.5–7 mm long, subequal to the lower anthers. $2n = 28$. Forests and open, rocky slopes; ponderosa pine and Douglas-fir communities.

ARIZ: Apa. 1980–2590 m (6500–8500′). Flowering: May–Aug. Fruit: Jun–Aug (Sep). Cascade-Sierra Nevada ranges from southern British Columbia to southern California, extending east, less commonly, to Montana (west of the Continental Divide) and central Idaho, south to northwestern New Mexico and adjacent Colorado, and northern Arizona. Our material belongs to **subsp.** ***subcarinata*** Wherry (loosely spreading). *Phlox diffusa* subsp. *subcarinata* has recently been treated as conspecific with *P. austromontana* in regional floras. However, this taxon appears intermediate between *P. austromontana* and *P. diffusa*, showing some similarity to *P. canescens* in our area. Whether this represents a unique taxon or a widespread hybrid is not known, but without recognition, it is highly unlikely that this question will be investigated and resolved.

Phlox grayi Wooton & Standl. (for Asa Gray, botanist, 1810–1888) Erect to ascending, often suffrutescent, taprooted perennial. **STEMS** 3–8 (10), 8–18 (25) cm long, sometimes loosely clumped, the internodes evident, often 4, glandular and short-villous. **LEAVES** linear to oblong-ovate, 12–35 (40) mm long, 2–8 mm wide, pilose and ciliate. **INFL** flowers 3–12, composed of 2- to 3-flowered cymes; pedicels 8–50 mm, glandular to short-pilose. **FLOWER** calyx 7–12 mm long, the membranes carinate to ridged at the base; corolla purple to deep pink, less often white, the tube 12–18 mm long, the lobes 5–10 (12) mm long, obovate, obtuse; stamens inserted on the upper tube; stigmas located among the stamens. Gravelly slopes; piñon-juniper woodland and ponderosa pine communities. ARIZ: Nav; COLO: LPl, Mon; NMEX: McK, RAr, SJn. 1500–2280 m (5000–7500′). Flowering: Apr–Jun. Fruit: May–Jul. Historically and generally treated as conspecific with *P. longifolia* [*P. longifolia* var. *brevifolia*], differing in its short, broad, and stiff leaves. Expected throughout the study area; however, few populations correspond to this taxon.

Phlox longifolia Nutt. (long leaves) Erect to ascending, often suffrutescent, taprooted perennial. **STEMS** 3–8 (10), 9–30 (40) cm tall, sometimes loosely clumped, the internodes evident, glabrous to glandular and short-villous. **INFL** flowers 2–3, pedicels glandular to short-pilose. **FLOWER** calyx 7–12 (14) mm long, the membranes carinate to ridged at the base; corolla white to deep pink, the tube 12–30 mm long, the lobes 5–12 (15) mm long, obovate, obtuse; stamens inserted on the upper tube; stigmas located among the stamens. **FRUIT** globose, ovoid to obovoid, apex acute to obtuse, 3–6 mm long, 2–4.2 mm wide. **SEEDS** ovoid, somewhat irregular, dark brown, 2.2–3.6 mm long, 1–1.7 mm wide. [subspp. *calva* Wherry, *compacta* (A. Brand) Wherry, *cortezana* (A. Nelson) Wherry, *longipes* Wherry]. Sandy to rocky soils, open sites; sagebrush, desert scrub, piñon-juniper woodlands, ponderosa pine, aspen, and spruce-fir communities. ARIZ: Apa, Coc, Nav; COLO: Arc, Dol, Hin, LPl, Min, Mon, SMg; NMEX: McK, RAr, San, SJn; UTAH. 1125–2070 m (3700–6800′). Flowering: Apr–Jun. Fruit: May–Jul. British Columbia to Montana, south to eastern California and New Mexico. Geographic and ecological variation with respect to habit, leaf size, vesture, and floral morphology is complex throughout the range of the species and in much need of thorough investigation. In Wherry's monographic study, two species (one glandular, the other eglandular) and several subspecies were recognized within what is here referred to as *P. longifolia*.

Phlox multiflora A. Nelson (many flowers) Rocky Mountain phlox. Loosely matted, taprooted perennials. **STEMS** numerous, short, suberect, rarely rising as much as 5–10 (15) cm tall. **LEAVES** linear, thin, (10) 12–30 mm long and 1–2 mm wide, slightly ciliate, inconspicuously granular-scaberulous to virtually glabrous. **INFL** flowers 1–3 at the ends of the stems, sessile to short-pedicelled. **FLOWER** calyx glabrous or seldom inconspicuously tomentose, the intercostal membranes flat, the lobes rather narrow and often thickened, but not firm; corolla white to sometimes pink or bluish, the tube 10–14 mm long, the limb relatively large, with lobes 6–11 mm long; filaments inserted in the upper tube, style 5.5–8 (10) mm long, the same height as the anthers. [*P. multiflora* subsp. *depressa* (E. Nelson) Wherry; *P. multiflora* subsp. *patula* (A. Nelson) Wherry]. Open or wooded, often rocky places, from the higher foothills to above timberline; sagebrush, ponderosa pine, and Douglas-fir communities. COLO: Arc, Hin, LPl, Min, Mon, SJn. 2200–3265 m (7215–10710′). Flowering: Jun–Aug. Fruit: Jul–Aug (Sep). Idaho and western Montana, south to northeastern Nevada and northern Utah, Wyoming, Colorado, and north-central New Mexico

Phlox nana Nutt. (dwarf) Canyon phlox. Erect to ascending, often clumped, taprooted perennial. **STEMS** 10–25 (30), 10–30 (35) cm tall, the internodes evident but shorter than the leaves. **LEAVES** 25–45 mm long, 2–5 mm wide, linear, narrowly elliptic to lanceolate, sparsely to moderately glandular-pubescent. **INFL** 2–6 (15) flowers, composed of 2- to 3-flowered cymes, pedicels 3–30 mm long, densely glandular-pubescent. **FLOWER** calyx 8–17 mm long, the intercostal membranes narrow and flat; corolla purple, lilac to deep pink, or white, rarely yellow, the tube 12–18 mm long, the lobes 9–15 mm long, 8–10 mm wide, orbicular to obovate; stamens inserted in the middle to lower tube; style short, stigmatic lobes subequal to the stamens. Mountain meadows and slopes and piñon-juniper woodland communities. NMEX: RAr. 1525–2500 m (5000–8200′). Flowering: May–Aug. Fruit: Jun–Sep. New Mexico, southeastern Arizona, Trans-Pecos Texas, and Mexico.

Phlox pulvinata (Wherry) Cronq. (cushion-shaped) Cushion phlox. Caespitose, loosely to densely matted or low-mounded, taprooted perennials. **STEMS** 2.5–5 (12) cm long, tending to spread out along the substrate, sometimes also spreading belowground, the internodes short, obscured by the leaves. **LEAVES** crowded, rather firm, mostly 5–12 (15) mm long and 1–2.5 mm wide, erect but arcuate and spreading, the surfaces glabrous to often glandular or hairy, the margins slightly thickened, usually ciliate toward the base. **INFL** flowers solitary (to 3), sessile or short-pedicelled, the pedicels glandular-pubescent. **FLOWER** calyx 6–11 mm long, usually glandular-hairy, occasionally eglandular or even glabrous, the intercostal membranes flat, the lobes flattened, with inconspicuous midvein; corolla white or light bluish, the tube 8–13 mm long, up to twice as long as the calyx, the lobes 4–7 mm long, obovate, obtuse; stamens inserted from midtube to the upper tube; style 2–5 mm long. Open, rocky places, clay slopes, meadows, mountain slopes; spruce-fir, krummholz, subalpine, and alpine meadows. COLO: Arc, Con, Hin, LPl, Min, Mon, RGr. 3500–3700 m (11500–12200′). Flowering: Jul–Aug. Fruit: Aug–Sep. Southwestern Montana to north-central New Mexico, west to northeastern Oregon and the Sierra Nevada of California.

Phlox standsburyi (Torr.) Heller. (for Howard Standsbury, 1806–1863) Erect to ascending, often suffrutescent, taprooted perennial. **STEMS** 9–14 (20) cm tall, sometimes loosely clumped, the internodes evident, glandular and puberulent to short-villous. **LEAVES** linear to oblong-lanceolate, 35–82 mm long, 3–6 mm wide, puberulent and glandular. **INFL** 3- to 21-flowered, composed of 2- to 3-flowered cymes, pedicels 5–25 mm, glandular-pubescent. **FLOWER** calyx 7–14 mm long, the membranes carinate to plicate; corolla white to deep pink, the tube 19–25 mm long, the lobes 7–12 (15) mm long, narrowly obovate, obtuse; stamens inserted on the upper tube; stigmas located among the stamens. Sandy to rocky soils, open sites; sagebrush, desert scrub, piñon-juniper woodland, ponderosa pine, aspen, and spruce-fir communities. ARIZ: Nav; UTAH. 1200–1900 m (4000–6000′). Flowering: Apr–Jun. Fruit: May–July. British Columbia to Montana, south to eastern California and New Mexico. The Ramah Navajo made a decoction of leaves that was taken during menstruation as a contraceptive. The leaves were used as a lotion for sores. The Hopi used an infusion to keep grasshoppers, rabbits, and pack rats from eating corn.

Phlox woodhousii (Torr. ex A. Gray) E. E. Nelson (for S. W. Woodhouse, horticulturist, physician) Perennial. **STEMS** 1–4 from a deep-seated, creeping rhizome, sometimes suffrutescent; 6–15 cm tall, the lower internodes evident. **LEAVES** linear-oblong to narrowly lanceolate, flat, acute to obtuse, glandular and short-villous, 1.5–4 cm long, 3–5 mm wide. **INFL** flowers 2–3, pedicels glandular, short-villous. **FLOWER** calyx 7–9 mm long, the membrane flat; corolla usually bright pink, the tube 10–15 mm long, the lobes 6–10 mm long, retuse to emarginate; stamens subequally inserted on the middle tube; style short, 1–2.5 mm long, stigmas located well below the stamens. [var. *oculata* A. Nelson]. Open sites, rocky slopes; ponderosa pine communities. ARIZ: Apa. 1200–2450 m (4000–8000′). Flowering: May–Jul (Aug). Fruit: Jun–Aug. Central Arizona to northwest New Mexico.

Polemonium L. Jacob's Ladder, Sky Pilot

(origin ambiguous: possibly honoring Polemon, Greek philosopher, perhaps Greek *polemos*, strife) Perennial herbs (in our area), one annual. **STEMS** erect to decumbent. **LEAVES** alternate, pinnately lobed to compound; leaflets sessile, entire or divided into 2–5 lobes. **INFL** terminal, open to compact and capitate or rarely axillary and then solitary. **FLOWERS** actinomorphic; calyx herbaceous, becoming membranous, not ruptured in fruit, the lobes rounded to attenuate; corolla rotate to funnelform or salverform, white, yellow, blue, or bluish violet; stamens equally inserted on the corolla tube; filaments equal in length, basally pilose; anthers included to exserted; style included to exserted, with 3 stigmatic branches. **CAPSULES** globose to ovoid, dehiscent; seeds 1–20 per locule, gelatinous or not when wet. $2n = 18, 36$. About 28 species. Western North America, one species in southern South America, and several in Eurasia (Davidson 1950; Grant 1989). Recent population genetics studies indicate that species boundaries of the tufted, alpine polemoniums are far more complex than previously imagined. However, no taxonomic changes have been proposed to date.

1. Corolla campanulate or broader, about as wide as long, or wider, the lobes from a little shorter than to about twice as long as the tube; leaflets opposite or offset, ordinarily undivided and not appearing verticillate; plants of various habit and habitat, glandular and mephitic, but not strongly so ....(2)

1′ Corolla funnelform or tubular-funnelform to subsalverform, distinctly longer than wide, the lobes distinctly shorter than the tube; leaflets, or many of them, so deeply 2- to 5-cleft as to appear verticillate; plants small, up to 30 cm tall, mostly alpine and subalpine, very strongly glandular and mephitic, with numerous stems from a taproot and much-branched, sometimes elongate caudex ....(4)

2. Rhizomatous, the rhizomes simple, upturned distally, stem solitary, erect, (15) 38–100 cm tall; styles exserted, twice as long as anthers; plants of very wet places ..... ***P. occidentale***
2′ Caudex present, branched, mostly clustered, rarely solitary, sometimes elongate and somewhat rhizomelike; styles included, plants of moist to dry, often rocky places ..... (3)

3. Plants relatively large and coarse, (30) 40–120 cm tall, leafy throughout, lacking a tuft of basal leaves ..... ***P. foliosissimum***
3′ Plants relatively small and compact, 5–20 (25) cm tall, leafy chiefly toward the base, often with tufted basal leaves ..... ***P. pulcherrimum***

4. Flowers ochroleucous to pale yellow; leaflets tending to be rather remote, often many of them undivided; inflorescence loose and somewhat elongate ..... ***P. brandegei***
4′ Flowers blue-violet (very rarely white); leaflets crowded, most or all of them deeply 2- to 5-cleft; inflorescence dense and compact, fan-shaped to globose ..... (5)

5. Corolla broadly funnelform, mostly 17–26 mm long, 10–15 mm wide, pale blue; inflorescence globose ..... ***P. confertum***
5′ Corolla narrowly funnelform, mostly 15–30 mm long, 5–10 mm wide, deep blue-violet; inflorescence fan-shaped ..... ***P. viscosum***

Polemonium brandegei (A. Gray) Greene (for Townshend S. Brandegee, botanist, 1843–1925) Pale sky pilot. Low herbaceous perennial with a stout taproot and branched caudex, 8–24 (32) cm tall. **LEAVES** generally basal, 1.7–20 cm long, densely stipitate-glandular or glandular-villous and strongly mephitic; leaflets numerous, crowded, mostly or all 2- to 3-cleft to the base, appearing verticillate, the individual segments 4–10 (12) mm long and 1–4 mm wide, fairly well spaced. **INFL** loosely cymose-capitate or spicate, usually elongating a little in fruit. **FLOWER** calyx 7–12 mm long, lobes acuminate, shorter than the tube, densely glandular-villous; corolla ochroleucous, cream-colored to pale yellow, narrowly funnelform or tubular-funnelform, mostly 21–31 mm long, longer than wide, throat 5–8.5 mm wide; stamens distinctly equaling the corolla lobes; style exserted, longer than the stamens. **FRUIT** a capsule, 3.8–5.5 mm long, 2.8–3.2 mm wide, ovoidal to subglobose. **SEEDS** 1.8–3.1 mm long, 0.7–1.2 mm wide, obovoid, angular, winged at one end, dark reddish brown, not becoming mucilaginous when wetted, (1) 3–5 per locule. $2n = 18$. [*P. viscosum* Nutt. subsp. *mellitum* (A. Gray) Davidson]. Alpine scree, rock slides, and crevices. COLO: Arc, Con, Hin, LPl, Min, Mon, RGr, SJn. 3050–4115 m (10000–13500′). Flowering: Jun–Aug. Fruit: Aug–Sep. Wyoming and western South Dakota, south to northern Utah; the Tushar Mountains, near Marysvale, Utah; Colorado; and northern New Mexico. Similar to *P. viscosum*, but leaflets longer, less crowded, frequently offset, and many of them undivided. The inflorescence is also longer, looser, more thyrsoid or almost racemose or spicate, and the flowers are ochroleucous to yellow.

Polemonium confertum A. Gray (densely crowded) Gray's sky pilot. Low herbaceous perennial with a taproot and branched caudex, 7–26 cm tall. **LEAVES** mostly basal, up to 20–40 cm long; densely stipitate-glandular or glandular-villous, and strongly mephitic, leaflets numerous, crowded, mostly 2- to 3-cleft, appearing verticillate, the individual segments 4–10 mm long and 2–4 mm wide. **INFL** densely cymose-capitate, usually elongating a little in fruit. **FLOWER** calyx 7–12 mm long at anthesis, the narrow, acute lobes shorter than the tube; corolla blue-violet, more or less funnelform or tubular-funnelform, mostly 17–26 mm long, longer than wide, often very conspicuously so, 10–15 mm wide at the orifice, the lobes shorter than the tube; stamens distinctly shorter than to sometimes nearly equaling the corolla lobes, anthers (pollen) orange-yellow; style exserted beyond the anthers. **FRUIT** a capsule, 3.8–5.5 mm long, 2.8–3.2 mm wide, ovoidal to subglobose. **SEEDS** 1.8–3.1 mm long, 0.7–1.2 mm wide, obovoid, angular, winged at one end, dark reddish brown, not becoming mucilaginous when wetted. Alpine scree slopes. COLO: Arc, Con, Hin, Min, SJn. 3350–3720 m (11000–12200′). Flowering: Jun–Aug. Fruit: Aug–Sep. Rocky Mountains of central Colorado, barely extending into our study area.

Polemonium foliosissimum A. Gray (many leaves) Leafy Jacob's ladder. Perennial herbs with 1–5 leafy stems, 20–90 (110) cm tall, glabrous, pilose, or villous. **LEAVES** 3–10 cm long, 0.8–2 cm wide, glabrous to sparsely glandular, little reduced; leaflets 11–27, lanceolate to elliptic, ovate to narrowly lanceolate, 5–25 (35) mm long, 5–7 (10) mm wide, the terminal (3) 5 leaflets confluent. **INFL** cymose, open to congested, often branched, short, wide, flat-topped corymb. **FLOWER** pedicels 2–8 mm long; calyx 5–8 (10) mm long, the tube and lobes equal in length, glandular-pubescent; corollas bluish violet to lavender, campanulate, 10–15 mm long, the lobes rounded to elliptic, acute, nearly twice as long as the tube; anthers exserted, but shorter than the corolla lobes; style exserted, usually exceeding the stamens.

FRUIT a capsule, (2.5) 4–6 mm long, (2) 3.7–4.3 mm wide, globose to ovoid. **SEEDS** 2–2.5 mm long, 0.8–1 mm wide, (1) 3–5 per locule, oblong-ovoid, angular, dark brown, often with a partial wing, becoming mucilaginous when wetted. [*P. foliosissimum* subsp. *molle* (Greene) Wherry]. Our material belongs to **subsp.** ***foliosissimum***. Wooded areas, brushy slopes, streamsides, and mountain meadows in aspen, spruce-fir forests, Douglas-fir, and oak woodland communities. COLO: Arc, Hin, LPl, Min, Mon, RGr, SJn; NMEX: RAr; UTAH. (1520) 2130–3350 m [(5000) 7000–11000′]. Flowering: (May) Jun–Aug. Fruit: Jul–Aug. Nevada and Idaho, east to Colorado, and south to Arizona and western New Mexico.

Polemonium occidentale Greene (western) Western Jacob's ladder. Herbaceous perennial. **STEMS** solitary from a fairly short and simple, horizontal rhizome, (15) 40–100 cm tall, glandular-villous. **LEAVES** both basal and lower cauline, well developed and rather long-petiolate, 3–25 cm long, with 9–27 lanceolate to elliptic and acute leaflets, (5) 10–40 mm long and (0.5) 2–13 mm wide, the 3 terminal ones often confluent; middle and upper leaves more or less reduced. **INFL** strongly glandular-villous, an elongate and narrow panicle or thyrse, flowers typically crowded. **FLOWER** calyx 5–8 mm long, the lobes equaling or commonly a little shorter than the tube, glandular-puberulent; corolla mostly sky-blue or nearly so (rarely white), 10–16 mm long, the lobes rounded to elliptic, twice as long as the tube; stamens usually about equal to or shorter than the corolla lobes; style generally exserted and conspicuously surpassing the stamens. **FRUIT** a capsule, 3–6 mm long, 2–4 mm wide, globose to ovoid. **SEEDS** 2–3 mm long, 0.7–1 mm wide, obovoid, angular, winged at one end, dark reddish brown, not becoming mucilaginous when wetted. $2n$ = 18. [*P. caeruleum* L. subsp. *amygdalinum* (Wherry) Munz]. Our material belongs to **subsp.** ***amygdalinum*** Wherry. At middle elevations in the mountains, occasionally in the foothills or near timberline, swamps, mossy subalpine birch-willow bogs, stream edges. COLO: Arc, Hin, LPl, Min, Mon, RGr, SJn. 1830–3100 m (6000–10200′). Flowering: Jun–Aug. Fruit: Jul–Aug. From Yukon and eastern Alaska to California, Nevada, Utah, and Colorado, also rarely in northern Minnesota. Most frequently confused with *P. foliosissimum*; however, the two species are distinguished as follows: *P. foliosissimum* has a short, wide, flat-topped inflorescence; purple flowers; style subequal to or slightly longer than the corolla lobes; plant from rhizomes. *Polemonium occidentale* has a long and narrow inflorescence; blue flowers; style much longer than the corolla lobes; and the plants are from a woody caudex.

Polemonium pulcherrimum Hook. (beautiful) Short Jacob's ladder, pretty Jacob's ladder. Perennial herbs with several stems, 10–25 cm tall, glabrous to sparsely pilose-glandular, from a subrhizomatous caudex. **LEAVES** few per stem, mostly 5–14 cm long, villous to glandular-pilose; leaflets 17–25, ovate to elliptic, 5–15 mm long, 3–7 mm wide, the 3 terminal confluent. **INFL** congested to open and elongate, a corymb. **FLOWER** pedicels 1–8 mm long; calyx tube 4–7 mm long, the lobes lanceolate to ovate, 2–3 mm long, mostly longer than the tube; corollas blue-violet, with a yellow or white throat, rotate, 8–14 mm long, the tube 2–5 mm long, the lobes obovate; anthers slightly exserted, subequal to the corolla lobes; style included to slightly exserted, usually equaling the stamens. **FRUIT** a capsule, 2.5–3.2 mm long, 1.3–1.8 mm wide. **SEEDS** 1–3 per locule, 1.3–2.2 mm long, 0.6–1.2 mm wide, pyriform, ovoid, obovoid, or lenticular, often acute at one end, reddish brown, not becoming mucilaginous when wetted. $2n$ = 18. [*P. delicatum* Rydb.]. Three subspecies, Alaska south to Arizona and New Mexico. In our area, only **subsp.** ***delicatum*** (Rydb.) A. Brand (addicted to pleasure, dainty) is represented. Shaded forest floors, aspen, spruce-fir communities. ARIZ: Apa, Nav; COLO: Arc, Con, Hin, LPl, Min, Mon, RGr, SJn; NMEX: RAr; UTAH. 2315–3510 m (7600–12500′). Flowering: Jun–Aug. Fruit: Jul–Sep. Eastern Nevada to Wyoming, south to Arizona and New Mexico.

Polemonium viscosum Nutt. (viscous, glutinous) Sky pilot. Low herbaceous perennial with a stout taproot and branched caudex, 6–24 (30) cm tall. **LEAVES** generally basal, 1.7–15 (25) cm long, densely stipitate-glandular or glandular-villous and strongly mephitic; leaflets numerous, crowded, mostly or all 2- to 5-cleft to the base, appearing verticillate, the individual segments 1.5–6 (9) mm long and 1–3 mm wide. **INFL** cymose-capitate and fan-shaped, usually elongating a little in fruit. **FLOWER** calyx 7–12 mm long, lobes acuminate, shorter than the tube; corolla blue-violet, narrowly funnelform or tubular-funnelform, mostly 11–28 (30) mm long, longer than wide, throat 6–10 mm wide; stamens distinctly shorter than to sometimes nearly equaling the corolla. **FRUIT** a capsule, 3.8–5.5 mm long, 2.8–3.2 mm wide, ovoid to subglobose. **SEEDS** 1.8–3.1 mm long, 0.7–1.2 mm wide, obovoid, angular, winged at one end, dark reddish brown, not becoming mucilaginous when wetted, (1) 3–5 per locule. $2n$ = 18. Stable alpine tundra, rocky meadows. COLO: Arc, Con, Hin, LPl, Min, Mon, RGr, SJn. 3050–4010 m (11200–13200′). Flowering: Jun–Aug. Fruit: Jul–Sep. Rocky Mountain region, from southwestern Alberta to northern New Mexico, west to Okanogan County, Washington, eastern Oregon, and the mountains of central Nevada and northern Arizona.

POLYGALACEAE R. Br. MILKWORT FAMILY

Tom Wendt

Annual herbs, trees, lianas, achlorophyllous herbs or (in ours) shrubs, subshrubs, and perennial herbs. **LEAVES** simple, alternate (in ours), verticillate, or opposite; margin entire (in ours) or lightly denticulate; stipules absent. **INFL** a bracteate raceme (in ours), spike, or panicle. **FLOWERS** perfect, zygomorphic, often (in ours) papilionoid (appearing like a pea flower) in aspect but not in structural detail; sepals 5, free in ours, the 2 lateral (internal) pairs frequently much larger and petaloid ("wings"); petals 3 (in ours) or 5, distinct from each other but adnate to the staminal tube, often (in ours) differentiated into a lower keel petal and 2 straplike upper petals; stamens 3–10 (most frequently 8), usually united at least at the base, forming a tube open on the upper side, the anthers usually dehiscing by a longitudinal slit that is most often (in ours) reduced to a short apical porelike slit; intrastaminal annual disc often present, often reduced to an adaxial gland; ovary superior, 1–8-locular (2 in ours), each locule with 1 ovule (apically attached in ours), style 1. **FRUIT** a loculicidal capsule (in ours), drupe, samara, or nut. **SEEDS** 1 per locule. $x = 5$–11+ (Blake 1924). About 18 genera and 800 species. Mostly tropical and subtropical, with *Polygala* reasonably well represented also in temperate areas. The family can be diverse locally, and many species have attractive flowers, but the family is rarely of economic or ecological importance. It is uncommon in our area and represented only by the largest genus, *Polygala.*

Polygala L. Milkwort

(Greek *polys*, much, + *gala*, milk, for purported increase in lactation caused by a European species) **FLOWERS** papilionoid in aspect; lateral (inner) sepals enlarged as wings; lower (abaxial) petal differentiated as a keel which typically has a straplike base adnate to the staminal column and a distal pouchlike part, this latter frequently apically appendaged or ornamented, the 2 upper (adaxial) petals straplike and distally more or less reflexed; stamens 6–8 (8 in ours); ovary 2-locular, style with 2 dissimilar lobes distally. **FRUIT** a bilocular, loculicidally dehiscent, usually flattened capsule. **SEEDS** fusiform to ovoid, usually hairy and with a 2-lobed white aril, the testa black. A genus of about 500 species, cosmopolitan except for New Zealand and Polynesia (where introduced), and high latitudes. Represented in our area by three more or less thorny shrubs and subshrubs. A fourth species, *Polygala alba* Nutt. (white milkwort), a perennial herb, is known from just outside our area in McKinley County, New Mexico, and is included in the key.

1. Perennial herb, unarmed; wings 2–3.5 mm long, white; keel distally ornamented with a finely many-lobed fimbriate crest; inflorescence a dense conical raceme of many more than 15 flowers and buds. (Suspected but not documented for the study area)........*P. alba*

1′ Shrub, subshrub, or suffrutescent perennial, thorny (sometimes weakly so); wings (outer sepals) 2.5–12 mm long, cream-colored or rose; keel distally unornamented or with an undivided cylindrical fingerlike beak; inflorescence few-flowered, rarely with more than 15 flowers and buds, lax or dense, thorn-tipped, never with an elongate regular conical aspect........(2)

2. Wings 7 mm or more in length, rose or pink; young twigs minutely puberulent with erect hairs distinctly less than 0.1 mm long, to subglabrous........***P. subspinosa***

2′ Wings 5 mm or less in length, cream-colored; young twigs densely pilose to tomentose with hairs distinctly longer than 0.1 mm........(3)

3. Leaves and branchlets densely beset with short spreading hairs; pedicel and abaxial surface of outer sepals with spreading hairs........***P. acanthoclada***

3′ Leaves with incurved hairs, branchlets with dense matted or shaggy tomentum of appressed, incurved, or occasionally inclined or divergent hairs; pedicel glabrous; outer sepals glabrous or only ciliate, occasionally with a few incurved hairs distally........***P. intermontana***

Polygala acanthoclada A. Gray (thorn-branched) Desert polygala. Subshrub or shrub to 1 m, usually 15–70 cm tall, sometimes erect but usually low, unkempt, spreading and often semirhizomatous; twigs with dense, short, spreading pubescence, tip eventually becoming a thorn-tipped raceme. **LEAVES** oblanceolate or narrowly elliptic to narrowly obovate, 5–25 mm long, 1–5 mm wide, with spreading hairs on both surfaces, sessile or nearly so. **INFL** a terminal raceme on a branchlet, loosely to densely (1) 2–15-flowered, up to 2.5 cm long, or flowers fasciculate; pedicels mostly 1.5–5 mm. **FLOWERS** 3–5.3 mm long; outer sepals ovate to elliptic, 1.6–3.5 mm long; wings obovate, 3–5 mm long, cream-colored; free portion of upper petals 1.8–3 mm long, cream-colored, the bifid tips usually with rose markings;

keel 2.7–3.8 mm long (including the adnate base), beakless or with a minute beak up to 0.7 mm. **CAPSULES** oval or elliptic or slightly obovate, emarginate, 4–6 mm long, when ripe the walls slightly succulent, more or less translucent, prominently veined, glabrous. **SEEDS** (including aril and hairs) 3.2–4.2 mm long; testa evenly to unevenly hairy; aril 1–1.7 mm long. *n* = 9. Mostly on sandy soils near washes in desert shrubland, locally common but very restricted in range in our area to the Clay Hills/Paiute Farms area of the lower San Juan River drainage. UTAH. 1100–1500 m (3600–4900′). Flowering: May–Oct. Southern California and west-central Arizona northeastward along the Colorado River drainage into southeastern Utah. See comments under *P. intermontana.*

Polygala intermontana T. Wendt (referring to the Intermountain region) Great Basin polygala. Intricately branched shrub to 1 m, spreading to 1.4 m across; twigs with white tomentum of densely matted or shaggy, appressed, incurved, or occasionally irregularly inclined or divergent hairs, tips eventually becoming thorn-tipped racemes. **LEAVES** oblanceolate or linear to obovate, 3–25 mm long, 0.8–3.5 mm wide, with incurved hairs on both surfaces, sessile or nearly so. **INFL** a terminal raceme on branchlets, loosely to densely 1–7-flowered, to 1.5 cm long, or flowers fasciculate; pedicels mostly 3–7 mm. **FLOWERS** 2.5–5.2 mm long; outer sepals ovate to elliptic or narrowly so, 1.3–3.3 mm long; wings obovate, 2.5–4.9 mm long, cream-colored; free portion of upper petals 1.5–2.7 mm long, cream-colored, the bifid tips usually with rose markings; keel 2–3.4 mm long (including the adnate base), beakless or with a minute beak up to 0.5 mm. **CAPSULES** oval or ovate to nearly orbicular, emarginate, 4–6 mm long, when ripe the walls slightly succulent, more or less translucent, prominently veined, glabrous. **SEEDS** (including aril and hairs) 2.8–4.2 mm long; testa sparsely hairy to nearly glabrous; aril 1.2–2.3 mm long. *n* = 9. [*P. acanthoclada* A. Gray var. *intricata* Eastw.]. Sandy soils to rimrock, desert scrub to piñon-juniper, restricted in our area to the lower San Juan River drainage and the Monument Valley region. ARIZ: Nav; UTAH. 1400–1900 m (4600–6200′). Flowering: May–Sept. Mountains of west-central Nevada and adjacent California, eastward through the Great Basin to southeastern Utah and northern Arizona. Our area is essentially the only region in which the ranges of this species and the closely related, more southern *P. acanthoclada* overlap. A few populations in the Clay Hills area display stages of morphological intermediacy between these two species. While both species are diploid (*n* = 9) with regular meiosis, the intermediate populations are tetraploid (*n* = 18) and display irregular meiosis and are thus best considered to be hybrids rather than intergrading populations, supporting the specific status of the two species (Wendt 1978).

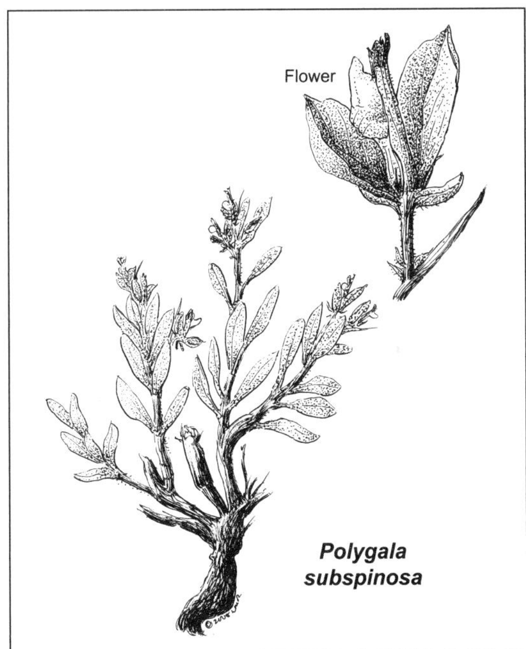

Polygala subspinosa S. Watson **CP** (somewhat spiny) Spiny polygala. Suffrutescent perennial or subshrub to 15 (–25) cm tall or infrequently a shrub to 60 cm, occasionally subrhizomatous; twigs glabrous or puberulent with spreading or slightly incurved hairs, tips eventually becoming a thorn-tipped raceme. **LEAVES** oblanceolate to obovate or elliptic, 4–31 mm long, 2–11 mm wide, puberulent with spreading hairs to subglabrous. **INFL** a terminal raceme on branchlets, loosely 1–8 (–30)-flowered, to 4 (12.5) cm long; pedicels mostly 3–10 mm. **FLOWERS** (6–) 8–13 mm long; outer sepals ovate or elliptic-lanceolate, 2–7.2 mm long; wings obovate, (5–) 7–12 mm long, rose or rose-pink; free portion of upper petals 3–6.6 mm long, silvery rose to mottled rose and white, the usually bifid tips dark rose; keel 5.5–10.5 mm long (including the adnate base), with a rather prominent flattened-cylindrical apical yellow beak 1–3 mm long. **CAPSULES** with short stipe 0.1–1.6 mm long, above the stipe elliptic or obovate, distally rounded to emarginate, 5.5–10 mm long (including stipe), wall opaque and not succulent, prominently veined, glabrous or puberulent. **SEEDS** (including aril and hairs) 3.3–4.9 mm long; testa more or less evenly hairy; aril 1.2–3.1 mm long. *n* = 9, 18. Usually in loose, rocky to sandy soils, breaks, and outcrops in shrubland and open piñon-juniper woodland. COLO: LPl, Mon; NMEX: SJn; UTAH. 1500–2400 m (4900–7800′). Flowering: Apr–Jul. Northwestern California, Nevada, Utah, western Colorado, northwestern New Mexico, and northern Arizona. This is a smaller, less shrubby species than the other two in our area, often having short herbaceous stems arising from a stout rootstock, and is thus frequently less obviously thorny.

POLYGONACEAE Juss. KNOTWEED FAMILY

James L. Reveal

Shrubs, subshrubs, or herbs, occasionally trees and vines, annual, biennial, or polycarpic (rarely monocarpic) perennial. **STEMS** prostrate to erect, sometimes scandent or scapose, solid or hollow, glabrous or variously pubescent, sometimes glandular; nodes smooth (subfam. Eriogonoideae) or swollen (subfam. Polygonoideae). **LEAVES** deciduous or persistent, estipulate (subfam. Eriogonoideae) or stipulate (having fused stipules called ocreae; subfam. Polygonoideae), these persistent or deciduous, cylindric or funnelform, entire or 2-lobed, chartaceous, membranous, coriaceous, or foliaceous; petioles present or absent; blades basal or basal and cauline, occasionally merely cauline, rosulate, alternate, opposite, or whorled, simple, usually entire. **INFL** cymose, paniculate, racemose, spicate, umbellate, or capitate; bracts scalelike to foliaceous (subfam. Eriogonoideae) or absent (subfam. Polygonoideae); peduncles present or absent; cluster of flowers subtended by 1–several involucral lobes or enclosed in a nonmembranous tube of fused lobes (involucre; subfam. Eriogonoideae), or subtended by fused bracteoles forming a membranous tube (ocreolae; subfam. Polygonoideae). **FLOWERS** actinomophic, hypogynous, imperfect, 1–many, pedicellate, sometimes stipitate, usually bisexual, occasionally bisexual and unisexual on the same or different plant, rarely unisexual only; tepals petaloid or sepaloid, rarely coriaceous, 2–6, distinct or connate proximally and forming a tube, usually in 2 whorls, dimorphic or monomorphic; nectary a disc at base of ovary or glands at base of filaments; stamens (1) 6–9; filaments distinct or connate proximally; anthers dehiscing by longitudinal slits; pistil (1) 2–3 (4)-carpellate; ovary superior, 1-locular; ovules 1; styles 1–3, distinct or proximally connate; stigmas 1 per style, usually capitate, occasionally fimbriate, peltate, or penicillate. **FRUIT** an achene, trigonous or less often lenticular, rarely quadrangular, winged or unwinged; seed 1; endosperm abundant, mealy; embryo straight or curved (Brandbyge 1993; Freeman and Reveal 2005; Graham and Wood 1965; Lamb Frye and Kron 2003; Reveal 1978; Roberty and Vautier 1964; Ronse Decraene and Akeroyd 1988). Forty-eight genera, about 1200 species. Worldwide. Many important weed species belong to this family. *Fagopyrum* is the source of buckwheat; seeds of *Coccoloba* (seaside grape) are eaten in the tropics. *Rheum* (rhubarb) is an important medicinal in eastern Asia; its petioles are edible. The cultivated ornamental *Antigonon* (coral vine) may be seen in warm, protected sites. The traditional, broad definition of *Polygonum* is abandoned here in favor of recognizing smaller, monophyletic genera.

1. Ocreae absent; nodes not swollen; flowers enclosed in nonmembranous involucres (subfam. Eriogonoideae Arn.) .... (2)
1′ Ocreae present, persistent or deciduous; nodes usually swollen; flowers enclosed in membranous ocreolae (subfam. Polygonoideae Arn.) .... (5)

2. Involucral structure not awn-tipped .... (3)
2′ Involucral structure awn-tipped .... (4)

3. Involucral structure tubular .... ***Eriogonum***
3′ Involucral structure of 2 whorls of 3 involucral lobes .... ***Stenogonum***

4. Involucre with awns at both distal and proximal ends of involucral tube .... ***Centrostegia***
4′ Involucre with awns only at distal end of involucral tube .... ***Chorizanthe***

5. Tepals 6; stamens 6 .... ***Rumex***
5′ Tepals 4 or 5; stamens (2) 3–8 .... (6)

6. Tepals 4; achenes winged, lenticular; leaves basal .... ***Oxyria***
6′ Tepals 4 or 5; achenes not winged, trigonous, lenticular, or sometimes biconvex; leaves cauline or basal and cauline .... (7)

7. Leaves mostly basal, some cauline; inflorescence spicate; flowering stems unbranched .... ***Bistorta***
7′ Leaves cauline; inflorescences terminal or axillary; stems usually branched .... (8)

8. Stems scandent or sprawling vines; leaf blades cordate-ovate, cordate-hastate, or sagittate; flowers stipitate proximally .... ***Fallopia***
8′ Stems prostrate to erect, not scandent or sprawling vines; leaf blades linear to nearly round, not hastate or sagittate; flowers not stipitate proximally .... (9)

9. Ocreae chartaceous, usually tan, brown, or red, glabrous or scabrous to variously pubescent, not 2-lobed distally, tearing but not fibrous with age .... ***Persicaria***
9′ Ocreae often hyaline, silvery, glabrous, 2-lobed distally, often becoming fibrous with age .... ***Polygonum***

Bistorta (L.) Scop. Bistort

Rodney Myatt

(Latin *bis*, twice, + *tortus*, a twist, alluding to the knotty rhizomes of some species) Plants perennial herbs; rhizomes often contorted. **STEMS** erect, glabrous; flowering stems unbranched. **LEAVES** mostly basal, some cauline, alternate, sometimes petiolate; ocrea persistent or disintegrating with age, chartaceous entirely or distally; blades linear-lanceolate or narrowly elliptic to ovate. **INFL** terminal, spikelike; peduncle slender, ascending or spreading. **FLOWERS** bisexual, 1–2 per ocreate fascicle, not stipitate proximally, pedicellate; perianth nonaccrescent, white, greenish, pink, red, or purple, glabrous; tepals 5, petaloid, connate 1/5 their length, monomorphic or slightly dimorphic; stamens 5–8; filaments distinct or connate proximally; styles 3, erect or spreading, distinct or connate proximally; stigmas capitate. **ACHENES** trigonous, brown to dark brown, not winged, glabrous; embryo curved. $x = 11, 12$ (Freeman and Hinds 2005). About 50 species of Arctic and temperate North America (four species) and Eurasia.

1. Racemes short-cylindric to ovoid, (8) 12–25 mm wide, without bulblets; flowers white or pinkish .... ***B. bistortoides***
1′ Racemes narrowly elongate-cylindric, 4–8 (10) mm wide, bearing pyriform, pink to brown or purple bulblets in place of some of the lower flowers; flowers greenish, rarely red .... ***B. vivipara***

Bistorta bistortoides (Pursh) Small (similar to *Polygonum bistorta* of the Old World) American bistort. Plants (10) 20–60 (75) cm tall; rhizomes contorted. **LEAVES** basal and cauline; ocrea brown, 9–25 (32) mm long, glabrous; petiole attached to sheath, 1–3.5 (5) cm long, wingless, absent on distal cauline blades; basal blade linear to linear-lanceolate to elliptic, (3) 5–20 (22) cm long, (0.5) 0.8–3.5 (4.8) cm wide, not revolute, glabrous or pubescent and glaucous abaxially, glabrous adaxially; cauline blade linear-lanceolate to lanceolate or elliptic. **INFL** short-cylindric to ovoid, (1) 2–4 (5) cm long, (0.8) 1.2–2.5 cm wide; peduncle 1–10 cm long. **FLOWERS** 1–2 per ocreate fascicle; pedicel 2–8 (11) mm long; perianth white or pale pink, 4–5 mm long; tepals oblong; bulblets absent; stamens exserted; anthers yellow. **ACHENES** yellowish brown to olive-brown, 3–4 mm, shiny, smooth. $n = 12$. [*Polygonum bistortoides* Pursh; *Persicaria bistortoides* (Pursh) H. R. Hinds]. Moist flats and slopes in subalpine and alpine meadows or along stream and lakesides. COLO: Arc, Con, Hin, LPl, Min, Mon, RGr, SJn; UTAH. 2745–3810 m (9000–12500′). Flowering: Jun–Aug. British Columbia and Alberta to Montana, south to New Mexico and California.

Bistorta vivipara (L.) Delarbre (bearing live young, as to reproduction by bulblets) Serpent-grass. Plants (2) 5–30 (45) cm tall; rhizomes sometimes contorted. **LEAVES** basal and cauline; ocrea brown, 4–22 (27) mm long, glabrous; petiole attached to sheath, 0.6–4.5 (6) cm long, wingless, absent on distal cauline blades; basal blade linear to narrowly elliptic or narrowly ovate, (1) 2.5–6 (10) cm long, 0.5–2 (2.3) cm wide, often revolute, pubescent and glaucous abaxially, glabrous adaxially; cauline blade linear to linear-lanceolate. **INFL** narrowly elongate-cylindric, (1.5) 2–9 cm long, 0.4–1 cm wide; peduncle 1–5 cm long. **FLOWERS** 1–2 per ocreate fascicle; pedicel (1) 2–5 mm long; perianth greenish proximally, white or pink distally, rarely red, 2–4 mm long; tepals obovate; bulblets present at lower nodes, pyriform, pink to brown or purple; stamens included or exserted; anthers reddish or purplish. **ACHENES** dark brown, 2–3 mm long, dull, granular. $n =$ mainly 48, 60. [*Polygonum viviparum* L.; *Persicaria vivipara* (L.) Ronse Decr.]. Moist flats and slopes in subalpine forest communities, subalpine and alpine meadows, or along stream and lakesides. COLO: Arc, Con, Hin, LPl, Min, Mon, RGr, SJn. 2590–3960 m (8500–13000′). Flowering: Jun–Aug. Greenland to Alaska, south (in the West) to Arizona and New Mexico; Eurasia. Reproduction is mainly by bulblets, as achenes are rarely produced. Exceedingly variable, with numerous chromosome numbers ranging from $2n = 66$ to 132. In the Old World, the species is sometimes subdivided into varieties, but none seems worthy of taxonomic recognition.

Centrostegia A. Gray ex Benth. Red Triangles

James L. Reveal

(Greek *kentron*, spur, + *stegion*, roof, for the arched saccate spurs at base of the involucre) Plants annual herbs from a slender taproot. **STEMS** 1 to many, erect, sparsely glandular; caudex stems absent; aerial flowering stems usually erect, slender, solid, not brittle, arising directly from a root. **LEAVES** basal, rosulate, estipulate, usually petiolate; blade oblong to broadly spatulate, sparsely glandular. **INFL** cymose; branches spreading, open to diffuse, dichotomously branched, sparsely glandular; bracts semifoliaceous, 3, connate 1/4 their length, positioned to side of node, awn-tipped. **PEDUNCLES** absent. **INVOLUCRES** tubular, prismatic, 3-angled, saccate and awn-tipped proximally; teeth 5, erect, awn-tipped. **FLOWERS** bisexual, 2 per involucre, not stipitate proximally, pedicellate; perianth nonaccresent, white to pink, pubescent; tepals 6, petaloid, oblanceolate, monomorphic, bilobed apically; stamens 9; filaments distinct, glabrous; anthers pink to red, oblong; styles 3, erect to spreading, distinct; stigmas capitate. **ACHENES** trigonous, brown, not winged, glabrous; embryo curved. x = 19 (Goodman 1957; Reveal 1989, 2005a). Monospecific, southwestern North America.

Centrostegia thurberi A. Gray ex Benth. (George Thurber, 1821–1890, naturalist, author, and botanist with United States-Mexico Boundary Survey, 1850–1853) Plants 0.3–2 (3) dm tall, (0.6) 1–4 (5) dm across, sparsely glandular. **LEAVES** with blade (0.5) 1–3.5 (4) cm long, 0.3–0.8 (1) cm wide. **INFL** bracts (1) 2–6 (10) mm long, linear to linear-lanceolate, mostly spreading, commonly acerose; awns 1–2 mm long. **INVOLUCRES** (2) 3–6 (8) mm long; basal spurs with awns 0.2–2 mm long; teeth 3, flattened, keeled, each terminated by a short, erect awn 0.3–1 mm long. **FLOWERS** 2–3 (3.5) mm long; tepals oblanceolate, bilobed apically; filaments 1–3 mm long. **ACHENES** 2–2.5 mm long. n = 19. Sandy to gravelly flats and slopes in mixed grassland, saltbush, blackbrush, and chaparral communities, and in pine-oak and montane conifer woodlands. UTAH. 1125–2400 m (3700–7875′). Flowering: Mar–Jun. Mojave and Sonoran deserts of southern California, southern Nevada, and southern Utah, south through Arizona to Baja California and Sonora, Mexico.

Chorizanthe R. Br. ex Benth. Spineflower
James L. Reveal

(Greek *chorizo*, to divide, + *anthos*, referring to the tepals of the flower) Plants annual (ours) or perennial herbs or subshrubs; root a slender to woody taproot. **STEMS** 1 to many, prostrate or decumbent to erect, glabrous or pubescent, sometimes glandular; caudex stems absent (ours) or woody, compact to spreading; aerial flowering stems usually erect or nearly so (ours), sometimes prostrate to decumbent, slender, solid, sometimes brittle and breaking into segments, arising from nodes of caudex branches, at distal nodes of aerial caudex branches, or directly from a root (ours). **LEAVES** basal or basal and cauline, alternate, estipulate, usually petiolate; blade linear to oblanceolate or spatulate, variously pubescent. **INFL** cymose (ours), sometimes capitate; branches spreading to erect, open to diffuse, tomentose to floccose or glabrous, occasionally glandular; bracts foliaceous to subulate or linear, mostly 2 and opposite, connate proximally, proximal to node, occasionally awn-tipped. **PEDUNCLES** absent. **INVOLUCRES** tubular, cylindric to urceolate or turbinate to campanulate, 3–6-ribbed, sometimes with a membranaceous or scarious margin; teeth 3, 5, or 6, erect, awn-tipped. **FLOWERS** bisexual, 1 (2) per involucre, not stipitate proximally, pedicellate; perianth nonaccresent, white to yellow, pink to rose-pink, or red to maroon or purple, glabrous or pubescent; tepals (5) 6, petaloid, linear to ovate, connate 2/3 their lengths, monomorphic or dimorphic, entire to lobed or variously fringed; stamens 3, 6, or 9, or variously 3–9; filaments distinct or connate into staminal tube, sometimes adnate to floral tube; anthers maroon to red or cream to white or yellow, oblong to oval; styles 3, erect to spreading, distinct; stigmas capitate. **ACHENES** lenticular or rarely trigonous, light to dark brown or black, not winged, glabrous; embryo straight or curved. x = 10 (Goodman 1934; Hardham 1989; Reveal and Hardham 1989; Reveal 2005b). About 50 species of temperate western North America (33 species) and temperate South America (mainly Chile).

Chorizanthe brevicornu Torr. (alluding to the short, divergent involucral teeth) Plants annual, spreading to erect, 0.5–3 (5) cm tall and 0.5–3 dm across. **STEMS** brittle and breaking into segments in mature, dry plants. **LEAVES** basal; blade oblanceolate to narrowly elliptic (ours) or spatulate, (1) 1.5–3 (4) cm long, 0.1–1 cm wide, pubescent; petiole 0.5–2 cm long, pubescent. **INFL** cymose; branches thinly pubescent, often with appressed hairs, infrequently somewhat strigose or glabrate, brittle and often breaking into segments, green to reddish; bracts 2, opposite, similar to the proximal blades only more reduced, 0.3–1 (1.5) cm long, 0.1–0.25 cm wide, becoming scalelike at distal nodes, linear and acicular; awns 0.2–0.5 mm long. **INVOLUCRES** cylindric, 3-angled but 6-ribbed, 3–5 mm long, green, not corrugate, thinly strigose; teeth 6, spreading, divergent, 0.4–1.2 mm long; awns uncinate, 0.2–0.5 mm long. **FLOWERS** greenish white to white or pale yellowish white, 2–4 mm long, glabrous; tepals monomorphic, connate 3/4 their length, linear to narrowly oblanceolate; stamens 3; filaments distinct, adnate at top of floral tube, 2–3.5 mm long, glabrous. **ACHENES** dark brown, lenticular, 3–4 mm long. Sandy to gravelly flats and slopes in mixed grassland, saltbush, creosote bush, blackbrush, and sagebrush communities, and in piñon and/or juniper woodlands. UTAH. 1125–1220 m (3700–4000′). Flowering: Feb–Jul. Great Basin of southeastern Oregon and southern Idaho south through Nevada and Utah to northern Arizona, and in the Mojave and Sonoran deserts of southern California, Arizona, and Mexico (Baja California and Sonora). Our plants are the southern **var.** ***brevicornu***, primarily of the Mojave and Sonoran deserts, with leaves 0.1–0.3 (0.5) cm wide. Dried specimens tend to fragment due to the brittle nature of the stems and branches.

Eriogonum Michx. Wild Buckwheat

James L. Reveal

(Greek *erion*, wool, + *gonu*, knee, referring to the tomentose nodes of the type species, *Eriogonum tomentosum*) Plants annual or perennial herbs, subshrubs, or shrubs; root a slender to woody taproot, rarely somewhat tuberous and chambered. **STEMS** 1 to many, prostrate or decumbent to erect, herbaceous or woody, glabrous or pubescent, sometimes glandular; caudex stems woody, compact to spreading; aerial flowering stems herbaceous, usually erect or nearly so, slender to stout, solid or hollow and sometimes fistulose, arising from nodes of caudex branches, at distal nodes of aerial caudex branches, or directly from a root. **LEAVES** basal or basal and cauline, sometimes sheathing, alternate, opposite, or whorled, estipulate, usually petiolate; blades linear to orbicular, variously pubescent, glandular or glabrous. **INFL** cymose, racemose, simple or compound umbellate, or capitate; branches spreading to erect, open to diffuse, variously tomentose to floccose or glabrous, occasionally glandular, rarely scabrellous; bracts foliaceous to scalelike, usually 3, connate proximally, proximal to node, not awn-tipped. **PEDUNCLES** absent or capillary or filiform to slender or stout, straight or curved, erect, horizontal, or deflexed, glabrous or variously pubescent, sometimes glandular. **INVOLUCRES** tubular, cylindric or narrowly turbinate to broadly campanulate or hemispheric; teeth 3–10 and erect, or 5–10-lobed and recurved, not awn-tipped. **FLOWERS** bisexual or polygamodioecious, (2) 6–100 per involucre, occasionally stipitate proximally, pedicellate; perianth nonaccresent, mostly white to red or yellow, glabrous to variously pubescent or glandular; tepals 6, petaloid, oblanceolate to oval, connate 1/5–1/2, monomorphic or dimorphic, entire; stamens 9; filaments distinct; anthers usually red to cream or yellow, mostly oblong; styles 3, usually erect, distinct; stigmas capitate. **ACHENES** lenticular or trigonous, light brown to brown, maroon, or black, glabrous or pubescent at least on beak, smooth, 3-angled, -ridged, or -winged; embryo curved or straight. $x = 10$ (Reveal 2004, 2005c). About 250 species of temperate North America (225 north of Mexico). The largest endemic North American genus in terms of numbers of species. A paraphyletic group ultimately to be enlarged to include all genera of the tribe *Eriogoneae* (some 325 species) or reduced to two species. Several species are worthy of cultivation; an important source of honey, especially in the chaparral of California.

1. Plants perennial ....(2)
1′ Plants annual ....(21)

2. Flowers stipitate ....(3)
2′ Flowers not stipitate ....(5)

3. Flowers glabrous .... ***E. umbellatum***
3′ Flowers pubescent ....(4)

4. Flowers white to cream; inflorescences compound umbellate .... ***E. jamesii***
4′ Flowers yellow; inflorescences capitate or umbellate .... ***E. arcuatum***

5. Achenes winged; plants monocarpic .... ***E. alatum***
5′ Achenes not winged; plants polycarpic (flowering and fruiting annually) ....(6)

6. Leaves short-hirsute on both surfaces .... ***E. inflatum***
6′ Leaves variously tomentose to floccose at least abaxially ....(7)

7. Plants caespitose or matted ....(8)
7′ Plants not caespitose or matted ....(10)

8. Flowers and achenes glabrous .... ***E. ovalifolium***
8′ Flowers and achenes pubescent ....(9)

9. Involucres (2) 4–6 mm long, deeply 5–10-toothed; perianth (ours) mostly white or rose to yellow; northern Arizona and northern New Mexico .... ***E. shockleyi***
9′ Involucres (2) 3–4 mm long, 5-toothed; perianth yellow; northeastern Arizona and northern New Mexico .... ***E. lachnogynum***

10. Plants erect herbs with racemose inflorescences ....(11)
10′ Plants shrubs or subshrubs with nonracemose inflorescences or involucres racemose only at tips of distal inflorescence branches ....(12)

11. Stems and inflorescence branches pubescent, not fistulose .... ***E. racemosum***
11′ Stems and inflorescence branches glabrous, fistulose .... ***E. zionis***

12. Leaf blades tightly revolute, narrow .... (13)
12′ Leaves not tightly revolute, plane or merely slightly rolled, narrow to elliptic or nearly orbicular .... (16)

13. Leaves (1.5) 2–6 cm long, glabrous and green adaxially; inflorescences compact, bright green; inflorescence branches glabrous; plants rounded shrubs .... ***E. leptophyllum***
13′ Leaves 0.5–1.5 (1.8) cm long, tomentose to glabrous and green or grayish adaxially; inflorescences open, gray; inflorescence branches tomentose to floccose or glabrous; plants various .... (14)

14. Involucres 2–3 mm long; leaves floccose adaxially; open subshrubs to shrubs .... ***E. microthecum***
14′ Involucres (3.5) 4–6 mm long; leaves mostly glabrous adaxially; compact to open subshrubs .... (15)

15. Flowers 3–3.5 mm long; leaves in fascicles; outer tepals obovate to nearly fan-shaped; stems and inflorescence branches hairy .... ***E. clavellatum***
15′ Flowers 4–5 mm long; leaves not in fascicles; outer tepals narrowly oblong; stems and inflorescence branches glabrous .... ***E. lonchophyllum***

16. Leaf apices rounded or nearly so, the blades 1–4 cm long; flowers white or yellow; plants large, rounded shrubs .... ***E. corymbosum***
16′ Leaf apices acute, the blades (1) 1.5–4 (6) cm long; flowers white; plants various .... (17)

17. Leaves 1–2 mm wide, 0.5–1.8 (2.5) cm long; inflorescences compact and often flat-topped; plants subshrubs to shrubs, (2) 4–15 tall .... ***E. microthecum***
17′ Leaves 2–20 mm wide; inflorescences mostly open; plants herbaceous perennials to shrubs .... (18)

18. Plants grayish to reddish brown tomentose to floccose, with numerous stems and branches, the tomentum often drying blackish; leaves cauline, (1) 1.5–3 cm long, 0.2–0.7 cm wide .... ***E. effusum***
18′ Plants greenish or whitish floccose or glabrous, with few to many stems and branches, the tomentum not drying blackish; leaves basal or cauline, 0.5–2 cm long, 0.2–2 cm wide .... (19)

19. Inflorescences without involucres racemosely arranged distally; herbs or subshrubs; peduncles 1–8 mm, glabrous; aerial flowering stems glabrous, grayish .... ***E. lonchophyllum***
19′ Inflorescences with involucres racemosely arranged distally; subshrubs or shrubs; peduncles absent; aerial stems tomentose to floccose or, if glabrous, usually greenish .... (20)

20. Aerial flowering stems and inflorescence branches floccose or glabrous and greenish; plants mainly of moving sand habitats .... ***E. leptocladon***
20′ Aerial flowering stems and inflorescence branches tomentose to floccose and grayish; plants mainly of gravelly to rocky places .... ***E. wrightii***

21. Leaves hirsute or pilose and greenish on both surfaces .... (22)
21′ Leaves tomentose to floccose at least abaxially .... (24)

22. Leaves basal and cauline; plants spreading .... ***E. divaricatum***
22′ Leaves basal; plants erect .... (23)

23. Stems fistulose; flowers hirsute, yellow .... ***E. fusiforme***
23′ Stems not fistulose; flowers glabrous, white .... ***E. gordonii***

24. Stems and inflorescence branches tomentose to floccose, whitish or grayish .... (25)
24′ Stems and inflorescence branches glabrous or scabrellous, green, reddish, or grayish .... (27)

25. Flowers densely pubescent adaxially; peduncles erect, 1–5 mm long; tepals not fan-shaped; plants biennial or late-season annual .... ***E. annuum***
25′ Flowers glabrous adaxially; peduncles absent; tepals fan-shaped; plants annual .... (26)

26. Leaves basal; leaf blades suborbicular to cordate .... ***E. palmerianum***
26′ Leaves cauline; leaf blades narrowly oblanceolate to broadly elliptic .... ***E. polycladon***

27. Outer tepals monomorphic .... (28)
27′ Outer tepals dimorphic .... (29)

28. Involucres 4-lobed; flowers yellowish to red; plants spreading .... ***E. wetherillii***
28′ Involucres 5-lobed; flowers white to rose; plants erect .... ***E. subreniforme***

29. Outer tepals pandurate; tepal margins crisped .... ***E. cernuum***
29′ Outer tepals obovate to ovate; tepal margins not crisped .... (30)

30. Stems and inflorescence branches floccose and scabrellous .... ***E. scabrellum***
30′ Stems and inflorescence branches glabrous .... (31)

31. Flowers white .... ***E. deflexum***
31′ Flowers yellow .... ***E. hookeri***

Eriogonum alatum Torr. (winged, as to the winged achenes) Winged wild buckwheat. Plants tall, erect, monocarpic herbaceous perennials, 5–13 (17) dm high. **STEMS** mostly 1, erect, 2–13 dm long, strigose. **LEAVES** basal, infrequently cauline; petiole 2–5 cm, strigose; basal blade lanceolate to oblanceolate, (3) 5–15 cm long, 0.3–1.5 cm wide, strigose or glabrous except for margins and midvein; margin entire, plane; cauline blade 1–6 cm long. **INFL** cymose, 20–100 cm long; branches strigose; bracts 3, foliaceous and linear to linear-lanceolate, 2–9 mm long, 1–3 mm wide proximally, scalelike and triangular, 0.8–5 mm long, 0.5–2 mm wide distally. **PEDUNCLES** erect, straight or curving upward, 5–35 mm long, strigose. **INVOLUCRES** turbinate to campanulate, 2–4 (4.5) mm long, 2–4 (4.5) mm wide; teeth 5, 1–1.8 mm long. **FLOWERS** bisexual, nonstipitate; perianth yellow to yellowish green, 1.5–2.5 mm in anthesis, 3–6 mm and often reddish in fruit, glabrous; tepals monomorphic, lanceolate, connate 1/4 their length; stamens exserted, 1.5–3 mm long; filaments glabrous. **ACHENES** broadly trigonous, yellowish green to reddish brown, 5–8 mm long, 3–6 mm wide, glabrous, 3-winged. $n = 20$. [*E. alatum* var. *mogollense* S. Stokes ex. M. E. Jones; *E. alatum* subsp. *mogollense* (S. Stokes ex. M. E. Jones) S. Stokes; *E. alatum* subsp. *triste* (S. Watson) S. Stokes; *E. triste* S. Watson; *Pterogonum alatum* (Torr.) H. Gross]. Sandy to gravelly flats and slopes in mixed grassland, saltbush, and sagebrush communities, and in oaks, piñon and/or juniper, and montane conifer woodlands. ARIZ: Apa, Nav; COLO: Arc, Dol, Hin, LPl, Mon, SMg; NMEX: McK, RAr, San, SJn; UTAH. 1350–3100 m (4430–10170′). Flowering: Jun–Oct. Northern Arizona and eastern Utah east through Colorado and southeastern Wyoming to southwest Nebraska, western Kansas, New Mexico, western Oklahoma, and northern and western Texas, to Chihuahua, Mexico. Our plants as described here are **var. *alatum***. Used for pain, food, and ceremonial items by the Navajo. Kayenta Navajo used it as a lotion for rashes and as a "life medicine." Ramah Navajo used it as a ceremonial medicine, cough medicine, and dermatological aid.

Eriogonum annuum Nutt. (annual, the first named nonperennial of the genus) Plants tall, leafy, erect, herbaceous biennials and late-season annuals, 5–20 dm tall, 5–10 dm across. **STEMS** usually 1, erect, 4–10 (15) dm, densely tomentose to floccose. **LEAVES** basal in first-year rosettes, otherwise cauline and solitary or in fascicles; rosette petiole 0.3–1.2 cm long; cauline petiole 0.2–0.5 cm long, mostly tomentose; blade oblanceolate to oblong, 1–7 cm long, 0.3–1.5 cm wide, densely tomentose abaxially, floccose and grayish adaxially; margin mostly entire, infrequently slightly revolute. **INFL** cymose, 30–100 cm long, 20–70 cm wide; branches densely tomentose to floccose; bracts 3, scalelike, 1–4 mm long, 1–3 mm wide. **PEDUNCLES** erect, straight, stoutish, 1–5 mm long, usually absent distally, tomentose to floccose. **INVOLUCRES** turbinate to campanulate, 2.5–4 mm long, 2–3 mm wide, tomentose; teeth 5–6, erect, 0.4–1 mm long. **FLOWERS** polygamodioecious, nonstipitate; perianth white to rose, 1–2.5 mm long, glabrous abaxially, densely pubescent adaxially; tepals dimorphic, those of outer whorl obovate, those of inner whorl narrowly ovate to oblong, connate about 1/4–1/3 their length; stamens mostly included, 1–2 mm long; filaments pilose proximally. **ACHENES** trigonous, brown, 1.5–2 mm long, glabrous. $n = 20$. Sandy flats, slopes, and banks in mixed grassland, and in desert scrub, oak, and conifer woodlands. NMEX: McK, RAr, San. 1900–2300 m (6235–7545′). Flowering: Apr–Nov. Primarily of the Great Plains from North Dakota and eastern Montana south through eastern New Mexico and western Texas to Mexico. A potentially aggressive weed species.

Eriogonum arcuatum Greene (bow-shaped) Plants low, spreading, herbaceous perennials, 0.2–2.5 dm tall, (1) 2–4 (6) dm across. **STEMS** 1–several, erect or nearly so, (0.2) 0.5–2 dm long, floccose. **LEAVES** basal; petiole 0.5–2 cm long; blade oblanceolate to elliptic, (0.5) 1–3 cm long, 0.5–1.5 cm wide, densely tomentose abaxially, mostly floccose adaxially; margin entire, plane. **INFL** umbellate or compound umbellate, occasionally capitate, 3–20 cm long and wide; branches floccose; bracts 3–8, semifoliaceous, narrowly elliptic at first node, (0.3) 0.5–2 cm long, (0.2) 0.4–1 cm wide. **PEDUNCLES** absent. **INVOLUCRES** campanulate, 3–7 mm long, (3) 4–8 mm wide, floccose; teeth 5–8, erect, 0.1–0.5

mm long. **FLOWERS** bisexual, stipitate; perianth yellow, (4) 5–8 mm including 0.7–2 mm long stipe, densely pubescent; tepals dimorphic, those of outer whorl lanceolate to elliptic and 1.5–5 mm long, 1–3 mm wide, those of inner whorl lanceolate to fan-shaped, 1.5–7 mm long, 2–4 mm wide, connate 1/4–1/3 their length; stamens exserted, 2–4 mm long; filaments pilose proximally. **ACHENES** trigonous, light brown to brown, 4–5 mm long, glabrous except for sparsely pubescent beak. Sandy, clayey, gravelly, or rocky flats, slopes, and outcrops, ledges and cliffs in mixed grassland, saltbush, blackbrush, and sagebrush communities, and in piñon, juniper, or conifer woodlands. 1220–4200 m (3700–13780′). Flowering: Jun–Oct. Southeastern Wyoming, eastern Utah, and north-central Arizona eastward to central Colorado and northwestern New Mexico.

1. Involucres (3) 4–8 mm wide; inflorescences umbellate to compound umbellate, rarely capitate; flowers (4) 5–8 mm long .... **var. *arcuatum***
1′ Involucres 3–5 mm wide; inflorescences capitate; flowers 6–7 mm long .... **var. *xanthum***

var. *arcuatum* (bent, as to curved inflorescence branches) Baker's wild buckwheat. Plants (1) 2–3 (6) dm across. **FLOWERING STEMS** mostly erect, (0.3) 0.5–2 dm. **LEAF BLADES** oblanceolate to elliptic, (0.5) 1–3 cm long, 0.5–1.5 cm wide. **INFL** umbellate and branched 1–3 times, rarely capitate; bracts (3) 5–20 mm long, (3) 4–10 mm wide. **INVOLUCRES** 3–7 mm long, (3) 4–8 mm wide. **FLOWERS** (4) 5–8 mm long including stipe. [*E. bakeri* Greene; *E. jamesii* Benth. var. *arcuatum* (Greene) S. Stokes; *E. jamesii* subsp. *bakeri* (Greene) S. Stokes; *E. jamesii* var. *flavescens* S. Watson; *E. jamesii* subsp. *flavescens* (S. Watson) S. Stokes; *E. jamesii* var. *higginsii* S. L. Welsh]. Sandy, clayey, gravelly, or rocky flats, slopes, and outcrops, ledges and cliffs in mixed grassland, saltbush, blackbrush, and sagebrush communities, and in piñon, juniper, or conifer woodlands. ARIZ: Apa, Nav; COLO: Arc, Hin, Min; NMEX: McK, San, SJn; UTAH. 1220–3000 m (3700–9840′). Flowering: Jun–Oct. Southeastern Wyoming, eastern Utah, and north-central Arizona eastward to southwestern Wyoming, central Colorado, and northwestern New Mexico.

var. *xanthum* (Small) Reveal (yellow, as to color of the flowers) Ivy League wild buckwheat. Plants 3–10 dm across. **FLOWERING STEMS** somewhat erect, 0.2–0.8 dm. **LEAF BLADES** oblanceolate to oblong, 1–2.5 cm long, 0.5–1 cm wide. **INFL** capitate, often reduced to a single involucre; bracts 3–10 mm long, 2–7 mm wide. **INVOLUCRES** 4–6 mm long, 3–5 mm wide. **FLOWERS** 6–7 mm long including stipe. [*E. chloranthum* Greene; *E. flavum* Nutt. subsp. *chloranthum* (Greene) S. Stokes; *E. flavum* var. *xanthum* (Small) S. Stokes; *E. jamesii* Benth. var. *xanthum* (Small) Reveal; *E. xanthum* Small]. Sandy, gravelly to rocky or talus slopes in high-elevation grassland, sagebrush, conifer woodlands, and tundra communities. COLO: Hin, SJn. 3000–4200 m (9840–13780′). Flowering: Jul–Sep. Rocky Mountains of central Colorado.

Eriogonum cernuum Nutt. (nodding, as to downwardly curved peduncle) Nodding wild buckwheat. Plants spreading to erect annuals, 0.5–6 dm high, 0.5–6 (10) dm across. **STEMS** 1–several, mostly erect, 0.3–2 dm long, glabrous, tomentose to floccose among sheathing leaves. **LEAVES** basal, sometimes sheathing up stems; petiole 1–4 cm long, tomentose; blade round-ovate to orbicular, (0.5) 1–2 (2.5) cm long, (0.5) 1–2 (2.5) cm wide, white to grayish tomentose abaxially, tomentose to floccose or glabrate and greenish adaxially; margin entire and plane. **INFL** cymose, open to sometimes diffuse, 0.5–5 dm high, 0.5–6 dm wide; bracts 3, scalelike, 1–2 mm long, 1–2.5 mm wide. **PEDUNCLES** cernuous, spreading to ascending or deflexed, straight or curved, slender, (1) 2–25 mm, glabrous, sometime lacking at distal nodes. **INVOLUCRES** turbinate, (1) 1.5–2 mm long, 1–1.5 mm wide, glabrous; teeth 5, erect, 0.4–0.7 mm long. **FLOWERS** bisexual, nonstipitate; perianth white to pinkish, becoming rose to red in fruit, 1–2 mm long, glabrous; tepals dimorphic, those of outer whorl panduriform with crisped margins and truncate bases, those of inner whorl obovate, connate proximally; stamens mostly excluded, 1–2 mm long; filaments pilose proximally. **ACHENES** trigonous, light brown to brown, 1.5–2 mm long, glabrous. [*E. cernuum* var. *psammophilum* S. L. Welsh; *E. cernuum* var. *tenue* Torr. & A. Gray; *E. cernuum* subsp. *tenue* (Torr. & A. Gray) S. Stokes]. Sandy to gravelly flats and slopes in mixed grassland, saltbush, sagebrush, and mountain mahogany communities, and in oak and conifer woodlands. ARIZ: Apa, Coc, Nav; COLO: LPl, Mon; NMEX: McK, RAr, San, SJn; UTAH. 1220–2285 m (4000–7500′). Flowering: Apr–Oct. Eastern Washington south to central California, east to southern Saskatchewan, Montana, South Dakota, western Nebraska, Colorado, and northern Mexico. Kayenta Navajo used the plant for rashes and kidney disease, and the achenes were made into a mush for food. Ramah Navajo chewed the leaves and applied the poultice to red ant bites.

Eriogonum clavellatum Small **CP** (having nature of a small club, as to shape of the leaf blades) Comb Wash wild buckwheat. Plants low, heavily branched subshrubs, 1–2.5 dm tall, 3–8 dm across. **STEMS** numerous; caudex stems rather compact; aerial stems spreading to erect, 0.06–0.2 dm long, thinly floccose to glabrous, dark green. **LEAVES** cauline, in fascicles; petiole 0.05–0.1 cm long, tomentose to floccose, rarely glabrous; blade oblanceolate, 3–15 mm long,

0.5–2 mm wide, densely white-tomentose abaxially, thinly tomentose and grayish to (rarely) glabrous and green adaxially; margin entire, tightly revolute. **INFL** umbellate to cymose, compact, 0.5–1.5 cm long, 1–2 dm wide; branches thinly floccose or glabrous, dark green; bracts 3, scalelike, 1.5–2.5 (3) mm long, 1–2.5 mm wide. **PEDUNCLES** erect, straight, stout, 1.5–8 mm long, glabrous. **INVOLUCRES** turbinate-campanulate, (3) 3.5–4.5 (5) mm long, 2.5–4.5 mm wide, glabrous; teeth 5, erect, 0.6–0.9 mm long. **FLOWERS** bisexual, nonstipitate; perianth white, (2.5) 3–3.5 mm long, glabrous; tepals dimorphic, those of the outer whorl obovate to nearly fan-shaped, 2–2.5 mm wide, those of the inner whorl slightly shorter and oblanceolate to spatulate, connate 1/4–1/3 their length; stamens long-exserted, 3–6 mm long; filaments sparsely pilose proximally. **ACHENES** trigonous, light brown, 3–3.5 mm long, glabrous. Sandy to heavy clay washes and slopes; saltbush communities. ARIZ: Apa; COLO: Mon; NMEX: SJn; UTAH. 1350–1750 m (4430–5740′). Flowering: May–Jul. A Four Corners regional endemic.

Eriogonum corymbosum Benth. (bunch of flowers, as to the corymbose inflorescence) Plants low subshrubs to large, rounded, diffusely branched and spreading shrubs, 2–8 (1.2) dm tall, 3–15 (25) dm across, grayish to reddish brown tomentose to floccose or glabrous and grayish or greenish. **STEMS** numerous; caudex stems spreading to erect (ours), rarely matted; aerial flowering stems spreading to erect (ours), rarely scapose, (0.5) 1–2 dm, grayish to reddish brown tomentose to floccose or glabrous and grayish or greenish. **LEAVES** cauline, not in fascicles; petioles 0.2–1.5 cm long, tomentose to floccose; blade lanceolate to oblanceolate or elliptic to nearly orbicular, (0.5) 1–3 (4.5) cm long, (0.3) 0.5–3 (3.5) cm wide, densely white-, tannish-, or brownish-tomentose on both surfaces or less so to nearly glabrous and green abaxially; margin often crisped, occasionally crenulate. **INFL** cymose, diffuse to rather open (ours), rarely capitate, (1) 3–20 cm long, 2–25 (30) cm wide; branches tomentose to glabrous; bracts 3, scalelike, 1–3 (6) mm long, mostly triangular, or foliaceous, 1–2.5 cm long and similar to the blades. **PEDUNCLES** lacking. **INVOLUCRES** turbinate, 1.5–3.5 mm long, 1–2 (2.5) mm wide, tomentose to floccose, occasionally glabrate or glabrous; teeth 5, erect, 0.3–1 mm long. **FLOWERS** bisexual, nonstipitate; perianth white or pale yellow to yellow, 2–3.5 (4) mm long, glabrous (ours), rarely sparsely pilose; tepals essentially monomorphic, oblanceolate to spatulate, connate 1/4–1/3 their length; stamens mostly included to slightly exserted, 1–4 (5) mm long; filaments typically pilose proximally. **ACHENES** trigonous, brown, 2–2.5 (3) mm long, glabrous except for the occasionally papillate beak. Sandy to gravelly or clayey flats, washes, slopes, outcrops, and cliffs; saltbush, blackbrush, and sagebrush communities, piñon-juniper and montane conifer woodlands. 1200–2800 m (3935–9185′). Southern Nevada and northwestern Arizona eastward onto the Colorado Plateau from southwestern Wyoming south to northwestern New Mexico, disjunct in northwestern Texas. A species of eight varieties; we have three. The Hopi boiled the plant (probably var. *glutinosum*), mixed it with water and cornmeal, and then baked it into bread.

1. Flowers in a population mainly pale yellow to yellow; common .... **var. *glutinosum***
1′ Flowers in a population mainly white or brownish white .... (2)

2. Leaves mostly broadly elliptic to oblong or ovate, densely white-tomentose abaxially, floccose and brownish or yellowish white adaxially; plants brownish white tomentose; flowers mostly 2–2.5 mm; common .... **var. *velutinum***
2′ Leaves mostly broadly ovate to nearly orbicular, densely grayish tomentose abaxially, often subglabrous to glabrous and green adaxially; plants greenish tomentose; flowers mostly 2.5–3 mm; San Juan County, Utah .... **var. *orbiculatum***

var. *glutinosum* (M. E. Jones) M. E. Jones (sticky, as to a mistaken observation that inflorescences are viscid) Sticky wild buckwheat. Plants subshrubs or shrubs, 2–10 dm tall, 3–10 dm across. **LEAF BLADES** lanceolate to oblanceolate or elliptic, 1–4 cm long, 0.5–1.5 cm wide, usually densely tomentose on both surfaces, sometimes less so and greenish adaxially. **INFL** 3–10 cm long, tomentose to floccose. **INVOLUCRES** 1–2 mm long, 1–1.5 (2) mm wide. **PERIANTH** pale yellow to yellow in a population, sometimes with white-flowered individuals, 1.5–2.5 mm long. [*E. aureum* M. E. Jones var. *glutinosum* M. E. Jones; *E. microthecum* Nutt. var. *crispum* (L. O. Williams) S. Stokes]. Sandy to gravelly flats, washes, and slopes in mixed grassland, saltbush, blackbrush, and infrequently in sagebrush communities, and occasionally in piñon-juniper woodlands. ARIZ: Apa, Coc, Nav; UTAH. 1600–2300 m (5250–7545′). Flowering: Jul–Oct. Northern Arizona and southern Utah.

var. *orbiculatum* (S. Stokes) Reveal & Brotherson (having the nature of a circle, as to leaf shape) Orbicular-leaf wild buckwheat. Plants mostly large shrubs, (3) 5–15 dm tall, 5–15 dm across. **LEAF BLADES** broadly ovate to nearly orbicular, 1–3 cm long and wide, floccose to tomentose on both surfaces, or floccose adaxially. **INFL** 3–10 cm long, compact and rather flat-topped, densely tomentose. **INVOLUCRES** 2–3.5 mm long, 1.5–2.5 mm wide. **PERIANTHS** white, 2.5–3 mm long. [*E. effusum* Nutt. subsp. *orbiculatum* S. Stokes]. Sandy to gravelly flats, washes, and slopes in mixed grassland, saltbush, blackbrush, and infrequently in sagebrush communities, and occasionally in piñon-juniper woodlands. COLO: Dol, Mon; UTAH. 1600–2300 m (5250–7545′). Flowering: Aug–Oct. Southeastern Utah and northeastern Arizona east into western Colorado.

var. velutinum Reveal (velvety, as to the hairs on the abaxial surface of leaves) Velvety wild buckwheat. Plants large shrubs, 5–10 dm tall, 8–15 (20) dm wide. **LEAF BLADES** broadly elliptic to oblong, 1.5–3 (3.5) cm long, (1) 1.5–2.5 (3.5) cm wide, densely white-tomentose abaxially, less so to brownish floccose adaxially. **INFL** 3–10 cm long, rather open, densely tomentose. **INVOLUCRES** 2–3.5 mm long, 1.5–2.5 mm wide. **PERIANTHS** white, 2–2.5 (3) mm long. Sandy to gravelly or clayey flats, washes, and slopes; mixed grassland, saltbush, and sagebrush communities, and in piñon-juniper woodlands. ARIZ: Apa, Nav; COLO: Dol, LPl, Mon; NMEX: McK, RAr, San, SJn; UTAH. 1250–2300 m (4100–7545′). Flowering: Aug–Oct. Eastern Utah and western Colorado south to northeastern Arizona and northwestern New Mexico, disjunct in northern Texas. This is our common morphology, but it can be difficult to distinguish from var. *orbiculatum*.

Eriogonum deflexum Torr. (pointing down, as to position of involucre on branch) Skeleton weed. Plants erect to spreading annuals, (0.5) 1–5 (20) dm high. **STEMS** mostly 1, erect, 0.3–3 (4) dm long, glabrous, not fistulose. **LEAVES** basal; petiole 1–7 cm, mostly floccose; blade cordate to reniform or nearly orbicular, 1–2.5 (4) cm long, 2–4 (5) cm wide, densely white-tomentose abaxially, less so to floccose or subglabrous and grayish to greenish adaxially; margin entire, plane. **INFL** cymose, open to diffuse, often flat-topped, 1–4.5 (16) dm high; branches glabrous; bracts 3, scalelike, 1–3 mm long, 0.5–1.5 mm. **PEDUNCLES** deflexed, rarely some slightly erect, slender to stout, 0–5 mm, glabrous. **INVOLUCRES** turbinate, 1.5–2.5 mm long, 1–2 mm wide, glabrous; teeth 5, erect, (0.2) 0.5–1 mm. **FLOWERS** bisexual, nonstipitate; perianth white to pink, becoming pinkish to reddish in fruit, 1–2 mm, glabrous; tepals dimorphic, those of the outer whorl ovate with cordate bases, those of the inner whorl lanceolate to narrowly ovate; stamens included, 1–1.5 mm long; filaments glabrous or sparsely pilose proximally. **ACHENES** trigonous, brown to dark brown, (1.5) 2–3 mm, glabrous. $n = 20$. [*E. deflexum* f. *stenopetala* H. Gross; *E. deflexum* subsp. *insigne* (S. Watson) S. Stokes; *E. deflexum* var. *insigne* (S. Watson) M. E. Jones; *E. deflexum* var. *turbinatum* (Small) Reveal; *E. insigne* S. Watson; *E. turbinatum* Small]. Sandy to gravelly washes, flats, and slopes in saltbush, greasewood, blackbrush, and sagebrush communities, and in piñon-juniper woodlands. UTAH. 1850–2000 m (6070–6560′). Flowering: (Jan) May–Oct (Dec). Southeastern California across southern Nevada to southern Utah, Arizona, and southwestern New Mexico, south into northern Baja California and northwestern Sonora, Mexico. Our plants, as described here, are **var. *deflexum***. Some individuals in the Lake Powell region have involucres with more or less erect peduncles.

Eriogonum divaricatum Hook. (spread apart, as to spreading branching pattern) Divergent wild buckwheat. Plants low, spreading annuals, 1–2 (3) dm high. **STEMS** decumbent to spreading, 0.3–0.5 dm long, puberulent to short-pilose. **LEAVES** basal and cauline; basal petiole 2–4 cm, mostly short-pilose; cauline petiole (0) 0.1–2 cm long; basal blade elliptic-oblong to orbicular, 1–3 cm long and wide, puberulent to short-pilose and green on both surfaces; cauline blade opposite, oblanceolate to oblong or elliptic, 0.3–1 (1.5) cm long, 0.2–0.8 (1.2) cm wide, similar to the basal blades. **INFL** cymose, distally uniparous due to suppression of secondary branches, diffuse, 5–25 cm long, 10–45 cm wide; branches puberulent; bracts 3, scalelike, 1–3 (5) mm long, 1–2 mm wide. **PEDUNCLES** absent. **INVOLUCRES** turbinate, 1–2 mm long, 1–1.5 (2) mm wide, pilose; teeth 5, spreading to somewhat reflexed, 0.7–1.5 mm long. **FLOWERS** bisexual, nonstipitate; perianth yellow, rarely pale yellow, (1) 1.5–2 mm, hispidulous and glandular with yellowish white hairs; tepals monomorphic, oblong-lanceolate to oblong-ovate, connate proximally; stamens included, 0.7–1.5 mm long; filaments pilose proximally. **ACHENES** trigonous, light brown, 1.5–2 mm long, glabrous. Heavy clay flats and slopes in saltbush, greasewood, and sagebrush communities, and in piñon-juniper woodlands. ARIZ: Apa, Coc, Nav; COLO: Mon; NMEX: McK, San, SJn; UTAH. 1150–2300 m (3770–7545′). Flowering: May–Oct. Colorado Plateau of southwestern Wyoming south through eastern Utah and western Colorado to northern Arizona and northwestern New Mexico; disjunct in west-central Utah. *Eriogonum divaricatum* was found in Argentina twice in 1899–1900. These plants were named *E. ameghinoi* Speg. and ultimately became the only species of the genus *Sanmartinia* Buchinger. The plants, probably introduced by migrating birds, did not persist in South America (Reveal 1981). Kayenta Navajo used the plant as a ceremonial medicine, orthopedic aid, and snakebite remedy.

Eriogonum effusum Nutt. (pour out, as to many spreading branches) Spreading wild buckwheat. Plants large, spreading, diffusely branched shrubs, (1.5) 2–5 (7) dm tall, 5–15 dm across. **STEMS** numerous; caudex compact to spreading; aerial flowering stems 0.3–0.8 dm long, floccose to glabrous, grayish. **LEAVES** cauline, not in fascicles; petiole 0.2–0.7 cm, tomentose to floccose; blade oblanceolate to oblong or obovate, (1) 1.5–3 cm long, (0.2) 0.3–0.7 cm wide, densely white-tomentose abaxially, white-floccose and grayish to glabrate or glabrous and greenish adaxially;

margin entire, plane. **INFL** cymose, diffuse and often congested, 10–30 (40) cm long, 10–40 cm wide; branches usually slender, white-floccose to glabrate and greenish; bracts 3, scalelike, 0.5–2 (5) mm long, 0.5–2 mm wide. **PEDUNCLES** erect, straight, slender, 3–25 mm long, floccose, often absent distally. **INVOLUCRES** turbinate, 1.5–2.5 (3) mm long, 1–2 mm wide, tomentose to floccose; teeth 5, erect, 0.3–0.6 mm long. **FLOWERS** bisexual, nonstipitate; perianth white, becoming reddish, 2–4 mm long, glabrous; tepals essentially monomorphic, elliptic to obovate, connate 1/4 their length; stamens mostly exserted, 2–4.5 mm long; filaments sparsely pilose proximally. **ACHENES** trigonous, brown, 2–2.5 mm long, glabrous. [*E. microthecum* Nutt. var. *effusum* (Nutt.) Torr. & A. Gray]. Sandy to rocky slopes and flats mainly in grassland and sagebrush communities, and in juniper or infrequently montane conifer woodlands. NMEX: McK, San, SJn. 1640–2500 m (5000–8200′). Flowering: Jun–Sep (Oct). Mainly of the Great Plains and eastern slope of the Rocky Mountains from southwestern South Dakota, southwestern Nebraska, and southeastern Wyoming south through central and eastern Colorado to northeastern New Mexico, then scattered westward to our area. Difficult to distinguish from the more westerly *E. nummulare* M. E. Jones of the Intermountain region just to our west.

Eriogonum fusiforme Small (shape of a spindle, as to fistulose stems and branches) Grand Valley desert trumpet. Plants spreading annuals, (0.3) 0.5–4 dm high. **STEMS** usually 1, erect, 0.5–1.5 dm long, conspicuously fistulose, glabrous, villous proximally. **LEAVES** basal; petiole 1–3 cm, hirsute; blade ovate to round, 0.5–3 cm long, 0.5–2.5 cm wide, short-hirsute and greenish on both surfaces; margin entire, plane. **INFL** cymose, open, 5–30 cm long, 5–30 cm wide; branches glabrous, fistulose; bracts 3, scalelike, 1–2 mm long, 1–1.5 mm wide. **PEDUNCLES** erect, straight, filiform to capillary, 10–20 mm long, glabrous. **INVOLUCRES** turbinate, 1–1.2 mm long, 0.7–1 mm wide, glabrous; teeth (4) 5, erect, 0.4–0.6 mm long. **FLOWERS** bisexual, nonstipitate; perianth yellow, 1.3–1.6 mm, densely hirsute with coarse curved hair; tepals monomorphic, ovate; stamens exserted, 1.3–1.8 mm long; filaments sparsely pubescent proximally. **ACHENES** lenticular to trigonous, light brown, 1.3–1.8 (2) mm long, glabrous. n = 16. [*E. inflatum* Torr. & Frém. var. *fusiforme* (Small) Reveal]. Heavy clay, sometimes gravelly, flats and slopes in saltbush and greasewood communities. UTAH. 1150–2000 m (3770–6560′). Flowering: Apr–Jul. Colorado Plateau of southwestern Wyoming, eastern Utah, and western Colorado.

Eriogonum gordonii Benth. (Alexander Gordon, 1813–1873, British collector with Sir William Drummond Stewart in the Rocky Mountains in 1843, later collected seeds in Colorado and northern New Mexico for George Charlwood, a nurseryman) Gordon's wild buckwheat. Plants erect to spreading annuals, (0.5) 1–5 (7) dm high. **STEMS** usually 1, erect, 0.5–1.5 (3) dm long, glabrous, sometimes sparsely hispid proximally, grayish. **LEAVES** basal; petiole 1–5 cm long, glabrous to sparsely hirsute; blade obovate to round or reniform, 1–5 cm long and wide, sparsely villous to hirsute and green on both surfaces, becoming glabrous with age; margin entire, plane. **INFL** cymose, open to diffuse, 5–40 (50) cm long, 5–50 cm wide; branches glabrous, rarely sparsely hispid, not fistulose; bracts 3, scalelike, 0.5–2 (3) mm long, 0.5–2.5 mm wide. **PEDUNCLES** erect, straight, slender, 5–20 mm long, glabrous. **INVOLUCRES** campanulate, 0.6–1.5 mm long, 0.8–1.5 mm wide, glabrous; teeth 5, erect, 0.2–0.4 mm long. **FLOWERS** bisexual, nonstipitate; perianth white, rarely yellowish, becoming pink to rose in fruit, 1–2.5 mm, glabrous; tepals monomorphic, oblong to narrowly ovate; stamens included to exserted, 1–1.8 mm long; filaments glabrous. **ACHENES** trigonous, shiny light brown to brown, 2–2.5 mm long, glabrous. n = 20. Sandy to clayey flats and slopes in saltbush, greasewood, and sagebrush communities, and in piñon and/or juniper or montane conifer woodlands. ARIZ: Apa; COLO: Arc, LPl, Mon; NMEX: SJn; UTAH. 1220–2200 m (4000–7220′). Flowering: May–Oct. Southwestern South Dakota to Wyoming south through eastern Utah and western Colorado to northern Arizona and northwestern New Mexico.

Eriogonum hookeri S. Watson (Joseph Dalton Hooker, 1817–1911, British botanist at Kew, visited western United States in 1877 with Asa Gray) Hooker's wild buckwheat. Plants erect annuals, 1–6 dm high. **STEMS** mostly 1, erect, 0.1–0.4 dm long, glabrous, not fistulose. **LEAVES** basal; petiole 1–5 cm long, tomentose; blade cordate to subreniform, (1) 2–5 cm long, 2–6 cm wide, densely white felty-tomentose abaxially, tomentose and grayish adaxially; margin occasionally wavy. **INFL** cymose, open to diffuse, spreading to subglobose or flat-topped to umbrella-shaped, 5–35 cm long, 5–50 cm wide; branches glabrous; bracts 3, scalelike, 1–3 mm long, 0.5–1.5 mm wide. **PEDUNCLES** absent. **INVOLUCRES** deflexed, broadly campanulate to hemispheric, 1–2 mm long, 1.5–3 (3.5) mm wide, glabrous; teeth 5, erect, 0.5–0.7 mm long. **FLOWERS** bisexual, nonstipitate; perianth yellow to reddish yellow, 1.5–2 mm long, glabrous; tepals dimorphic, those of the outer whorl orbicular with cordate bases, those of the inner whorl narrowly ovate; stamens mostly included, 1.3–1.5 mm long; filaments glabrous. **ACHENES** trigonous, light brown, 2–2.5 mm long, glabrous. n = 20. [*E. deflexum* Torr. var. *gilvum* S. Stokes; *E. deflexum* subsp. *hookeri* (S. Watson) S. Stokes]. Sandy washes, flats, and slopes in saltbush, greasewood, sagebrush, and mountain mahogany communities, and in piñon-juniper woodlands.

ARIZ: Apa; COLO: Mon; NMEX: SJn; UTAH. 1300–2135 m (4265–7000′). Flowering: Jun–Oct. East-central California and south-central Oregon east across Nevada to southwestern Idaho, southwestern Wyoming, western Colorado, south to northern Arizona and northwestern New Mexico. Hopi used the plant for flavoring mush.

Eriogonum inflatum Torr. & Frém. (referring to the fistulose stems and branches) Desert trumpet, Indian pipeweed. Plants erect, first-year flowering annuals or, more commonly, long-lived herbaceous perennials, 1–10 (15) dm high. **STEMS** 1–several, erect, (0.2) 2–5 dm long, usually fistulose, glabrous, occasionally hirsute proximally. **LEAVES** basal; petioles 2–6 cm long, hirsute; blades oblong-ovate to oblong or rounded to reniform, (0.5) 1–2.5 (3) cm long, (0.5) 1–2 (2.5) cm wide, short-hirsute and greenish on both surfaces, sometimes less so or glabrous and green adaxially; margin occasionally undulate. **INFL** cymose, open, 5–70 cm long; branches glabrous, occasionally fistulose; bracts 3, scalelike, 1–2.5 (5) mm long, 1–2.5 mm wide. **PEDUNCLES** erect, straight, filiform to capillary, 5–20 mm long, glabrous. **INVOLUCRES** turbinate, 1–1.5 mm long, 1–1.8 mm wide, glabrous; teeth 5, erect, 0.4–0.6 mm. **FLOWERS** bisexual, nonstipitate; perianth yellow, (1) 2–2.5 (3) mm long, densely hirsute with coarse curved hairs; tepals monomorphic, narrowly ovoid to ovate; stamens exserted, 1.3–2.5 mm long; filaments glabrous or sparsely pubescent proximally. **ACHENES** lenticular to trigonous, light brown to brown, 2–2.5 mm long, glabrous. $n = 16$. [*E. inflatum* var. *deflatum* I. M. Johnst.]. Sandy to gravelly washes, flats, and slopes in mixed grassland, saltbush, and sagebrush communities, and occasionally in piñon and/or juniper woodlands. ARIZ: Apa, Nav; COLO: Mon; NMEX: SJn; UTAH. 1125–1675 m (3700–5500′). Flowering: Mar–Oct (Dec). Eastern California east across central and southern Nevada to central and southern Utah and southwestern Colorado, south through Arizona and northwestern New Mexico to northern Sonora and southern Baja California, Mexico. Kayenta Navajo used the plant as a lotion for bear or dog bites.

Eriogonum jamesii Benth. (Edwin James, 1797–1861, surgeon-naturalist with the Long Expedition to the Rocky Mountains, 1819–1820) Antelope sage. Plants low, spreading, matted, herbaceous perennials, 1–4 dm tall, 3–8 dm across. **STEMS** several; caudex stems absent or more often matted, spreading; aerial flowering stems erect or nearly so, 0.5–1.5 dm long, tomentose to floccose. **LEAVES** basal; petiole 0.5–6 cm long; blade mostly narrowly elliptic, 1–3 (3.5) cm long, (0.3) 0.5–1 (1.2) cm wide, densely tomentose abaxially, less so to thinly tomentose and grayish to greenish adaxially; margin plane (ours) or undulate and crisped. **INFL** compound umbellate, 10–30 cm long, 10–25 cm wide; branches tomentose to floccose; bracts 3–9, semifoliaceous at proximal node, 5–20 mm long, 3–10 mm wide, becoming scalelike distally. **PEDUNCLES** lacking. **INVOLUCRES** turbinate, 1.5–7 mm long, 2–5 mm wide, tomentose to floccose; teeth 5–8, erect, 0.1–0.5 mm long. **FLOWERS** bisexual, stipitate; perianth white to cream, 3–8 mm including 0.7–2 mm long stipe, densely pubescent; tepals dimorphic, those of the outer whorl lanceolate to elliptic, 2–5 mm long, 1–3 mm wide, those of the inner whorl lanceolate to fan-shaped, 1.5–6 mm long, 2–4 mm, connate 1/4–1/3 their length; stamens exserted, 2–4 mm long; filaments pilose proximally. **ACHENES** trigonous, light brown to brown, 4–5 mm long, glabrous except for sparsely pubescent beak. $n = 20$. Sandy to gravelly or infrequently rocky flats and slopes in mixed grassland, saltbush, blackbrush, and sagebrush communities, and in oak, piñon and/or juniper, and montane conifer woodlands. ARIZ: Apa, Coc, Nav; COLO: Arc, LPl, Mon; NMEX: McK, RAr, San, SJn. 1465–2285 m (4800–7500′). Flowering: Jun–Oct. Northeastern Arizona, southern Colorado, southwestern Kansas, and western Oklahoma south through Arizona, New Mexico, and western Texas to central Mexico. Our plants, as described here, are **var. *jamesii***. Kayenta Navajo smoked the plant when disturbed by dreams of tobacco worms. Ramah Navajo used a decoction to ease labor pains. Also used as a contraceptive, gastrointestinal aid, and as a "life medicine."

Eriogonum lachnogynum Torr. ex Benth. (woolly fruit) Woolly-cup buckwheat. Plants hummock-forming and matlike, 0.5–1.5 dm tall, 0.5–3 dm wide. **STEMS** numerous, compact, aerial flowering stem 1–5 (6.5) cm long, silky-tomentose. **LEAVES** basal, 1 per node or fasciculate in terminal tufts; petiole 0.1–2.5 (3) cm long, tomentose to floccose; blade narrowly elliptic, 0.4–1.2 cm long, 1.5–3.5 cm wide, densely white or silvery tomentose on both sufaces; margins entire. **INFL** capitate (ours) or open and divided 1–3 times; bracts 3, scalelike, triangular, 0.5–1.5 mm long, 0.5–2 mm wide; branches absent. **INVOLUCRES** broadly campanulate, 2–5 per cluster, 2–3.5 mm long, 3–6 mm wide, tomentose; teeth 5, erect to spreading, 1–1.5 mm long. **FLOWERS** bisexual, nonstipitate, perianth yellow, 2.5–5 mm long, densely white-pubescent; tepals monomorphic, lanceolate to broadly lanceolate; stamens exserted, 3.5–5 mm long; filaments glabrous. **ACHENES** trigonous, brown to dark brown, 3–4 mm long, villous to tomentose. Widespread from northeastern Arizona and eastern Colorado to southwestern Kansas, western Oklahoma, and northern Texas. Our material belongs to **var. *sarahiae*** (N. D. Atwood & A. Clifford) Reveal (for Sarah Charley of Beclabito, New Mexico, a master weaver of Navajo rugs, as well as a sheepherder and herbalist). Limestone flats and mesa tops in piñon-juniper woodlands. ARIZ: Apa; NMEX: McK. 2285 m (7500′). Flowering: late June–July. A Four Corners endemic.

Eriogonum leptocladon Torr. & A. Gray. (thin or delicate stems) Sand wild buckwheat. Plants erect to spreading, diffusely branched shrubs, (2) 3–10 (12) dm tall, 5–15 (20) dm across. **STEMS** numerous; caudex stems spreading, often extensively so in moving sand; aerial flowering stems 0.3–1 dm long, mostly floccose to glabrate or glabrous, greenish to green. **LEAVES** cauline, not in fascicles; petiole 0.2–0.5 cm long; blade linear-lanceolate or linear-oblanceolate to narrowly oblong, 1.5–4 (6) cm long, 0.2–0.8 (1) cm wide, densely white-tomentose abaxially, less so and greenish adaxially; margin entire or infrequently somewhat revolute. **INFL** cymose, open, 5–40 cm long, 10–50 cm wide; branches white-tomentose to (more commonly) floccose and greenish, or occasionally glabrous and green, often with involucres racemosely arranged distally on branch tips; bracts 3, scalelike, 1–3 (6) mm long, 1–3.5 mm wide. **PEDUNCLES** absent. **INVOLUCRES** erect, turbinate to turbinate-campanulate, 1.5–3 mm long, 1–2 mm wide, tomentose to floccose or glabrous; teeth 5, erect, 0.4–0.7 mm long. **FLOWERS** bisexual, nonstipitate; perianth white (ours) or pale yellow to yellow, (2) 2.5–3.5 mm long, glabrous; tepals essentially monomorphic, oblong to broadly obovate, connate 1/4–1/3 their length; stamens slightly exserted, 2–4 mm long; filaments sparsely pilose proximally. **ACHENES** trigonous, light brown, 2.5–3.5 mm, glabrous except for minutely papillate beak. [*E. effusum* Nutt. subsp. *leptocladon* (Torr. & A. Gray) S. Stokes]. Sandy or infrequently gravelly or clayey flats, washes, and slopes in mixed grassland, saltbush, and sagebrush communities, and in piñon-juniper woodlands. 1280–2070 m (4200–6790′). Eastern Utah and northeastern Arizona east to southwestern Colorado and northwestern New Mexico. The yellow-flowered var. *leptocladon* occurs to our north in the San Rafael Desert of central eastern Utah.

1. Flowering stems and inflorescences essentially glabrous .... **var. *papiliunculi***
1′ Flowering stems and inflorescences tomentose to floccose .... **var. *ramosissimum***

var. *papiliunculi* Reveal (of small butterflies, honoring Oakley Shields, b. 1941, who found a new butterfly on this plant) Butterfly wild buckwheat. Plants (3) 4–10 dm tall, 5–15 dm across. **FLOWERING STEMS** glabrous or nearly so. **LEAF BLADES** lanceolate to narrowly elliptic, 2.5–3.5 cm long, (0.4) 0.5–0.8 cm wide; margin entire, plane. **INFL** glabrous. **PERIANTH** white. White or red blow sand on flats, washes, and slopes in mixed grassland, blackbrush, and sagebrush communities, and in piñon-juniper woodlands. ARIZ: Apa, Nav; NMEX: SJn; UTAH. 1200–2140 m (3940–7020′). Central Utah south to northern Arizona.

var. ramosissimum (Eastw.) Reveal (full of small branches) San Juan wild buckwheat. Plants 2–8 dm tall, 5–15 dm across. **FLOWERING STEMS** white-tomentose. **LEAF BLADES** linear-lanceolate to narrowly oblong, 1.5–3 cm long, (0.2) 0.3–1 cm wide; margin plane to slightly revolute. **INFL** tomentose, rarely floccose. **PERIANTH** white. [*E. effusum* Nutt. subsp. *pallidum* (Small) S. Stokes; *E. pallidum* Small; *E. ramosissimum* Eastw.]. White or infrequently red blow sand on flats, washes, and slopes in mixed grassland, saltbush, and sagebrush communities, and in piñon-juniper woodlands. ARIZ: Apa, Nav; COLO: Mon; NMEX: San, McK, RAr, SJn; UTAH. 1280–2070 m (4200–6310′). Flowering: Jun–Oct. Eastern Utah and southwestern Colorado south to northeastern Arizona and northwestern New Mexico.

Eriogonum leptophyllum (Torr.) Wooton & Standl. (thin leaf, as to leaf blades) Slender-leaf wild buckwheat. Plants spreading to rounded subshrubs or shrubs, 2–8 (13) dm tall, (1) 3–15 (18) dm across. **STEMS** numerous; caudex stems mostly spreading; aerial flowering stems (0.05) 0.1–0.8 dm long, thinly pubescent or glabrous, green, yellowish green, or infrequently grayish. **LEAVES** cauline, occasionally in fascicles; petiole 0.05–0.1 cm long, thinly tomentose to floccose or glabrous; blade linear to linear-oblanceolate, (0.5) 2–6 cm long, (0.03) 0.1–0.3 cm wide, densely to thinly white-tomentose abaxially, thinly so to glabrous and green adaxially; margin entire, tightly revolute. **INFL** cymose, densely compact, (1) 2–12 (15) cm long, (1) 4–15 (30) cm wide; branches thinly pubescent or glabrous and green; bracts 3, scalelike, 1–4 mm long, 1–3 mm wide. **PEDUNCLES** lacking. **INVOLUCRES** narrowly turbinate, 2–4 (4.5) mm long, 1–2 mm wide, glabrous; teeth 5, erect, 0.3–0.7 mm long. **FLOWERS** bisexual, nonstipitate; perianth white to pinkish, 2.5–4 mm, glabrous; tepals essentially monomorphic, oblong to narrowly obovate, connate less than 1/4 their length; stamens long-exserted, 3–6 mm long; filaments subglabrous to sparsely puberulent proximally. **ACHENES** trigonous, brown, (2.5) 3.5–4 mm, glabrous. n = 20. Clay flats, slopes, and outcrops, mainly in mixed grassland and sagebrush communities, and in piñon-juniper or conifer woodlands. ARIZ: Apa, Nav; COLO: LPl, Mon; NMEX: McK, RAr, San, SJn; UTAH. 1530–2300 m (5020–7545′). Flowering: Jul–Oct. Northeastern and southeastern Utah east to southwestern Colorado and northwestern New Mexico. Used as a "life medicine" by Ramah Navajo.

Eriogonum lonchophyllum Torr. & A. Gray (spear leaf, as to leaf shape) Spear-leaf wild buckwheat. Plants low, spreading to erect subshrubs or spreading herbaceous perennials, (1) 1.5–5 dm tall, 2–5 (8) dm across. **STEMS** few to many; caudex stems mostly spreading; aerial flowering stems spreading to erect, 0.3–3 dm long, glabrous (ours) or rarely floccose to tomentose, remaining tomentose among leaves, grayish. **LEAVES** basal or cauline on the proximal 1/2 of stem, not in fascicles; petiole 0.5–2 cm, tomentose to floccose or glabrous; blade narrowly lanceolate or oblanceolate to elliptic, 1.5–7 (9) cm long, 0.2–2 cm wide, densely white-tomentose abaxially, sparsely tomentose to thinly floccose and grayish or glabrous and green adaxially; margin plane or occasionally crenulate. **INFL** cymose, dense to more commonly open, 2–25 × 2–20 cm; branches glabrous (ours), rarely floccose, grayish; bracts 3, scalelike, 1–3 mm long, 1–3 mm wide, occasionally foliaceous and 0.8–3 cm long. **PEDUNCLES** erect, straight, stout, 1–8 mm, glabrous, proximal, usually absent. **INVOLUCRES** 1 or 2–5 in a cluster, turbinate to turbinate-campanulate, 2.5–4 mm long, (1.3) 1.5–3.5 (4) mm long, glabrous; teeth 5, erect, 0.4–0.9 mm long. **FLOWERS** bisexual, nonstipitate; perianth white, 2–3.5 (4) mm, glabrous; tepals monomorphic, oblanceolate, elliptic to oblong, or obovate, connate 1/4–1/3 their length; stamens exserted, 2–4 mm long; filaments pilose proximally. **ACHENES** trigonous, light brown to brown, 2–3 mm, glabrous except (typically) for slightly papillate beaks. [*E. corymbosum* Benth. var. *humivagans* (Reveal) S. L. Welsh; *E. effusum* Nutt. subsp. *salicinum* (Greene) S. Stokes; *E. humivagans* Reveal; *E. lonchophyllum* var. *saurinum* (Reveal) S. L. Welsh; *E. nudicaule* (Torr.) Small subsp. *scoparium* (Small) S. Stokes; *E. nudicaule* subsp. *tristichum* (Small) S. Stokes; *E. salicinum* Greene; *E. saurinum* Reveal; *E. scoparium* Small; *E. tristichum* Small]. Heavy gumbo clay soil or (at higher elevations) sandy loam to gravelly or rocky soil on flats, slopes, and outcrops in mixed grassland, saltbush, blackbrush, and sagebrush communities, piñon-juniper and montane conifer woodlands. COLO: Arc, Dol, LPl, Mon; NMEX: RAr, San, SJn; UTAH. 1400–2900 m (4595–9515'). Flowering: Jun–Oct. Eastern Utah, western and central Colorado south to northern New Mexico.

Eriogonum microthecum Nutt. (small cup, as to the small involucre) Plants erect to spreading subshrubs or shrubs, rarely matted, 0.5–1.5 dm tall, 6–13 (16) dm wide. **STEMS** many, spreading to erect; caudex stems absent or spreading, occasionally matted; aerial flowering stems erect to spreading, slender, solid, not fistulose, 0.1–1 dm long, tomentose, floccose, or glabrous. **LEAVES** cauline, usually in fascicles; petiole 0.1–0.5 cm, tomentose to floccose or glabrous; blade mostly elliptic, sometimes linear to obovate, 0.3–3.5 (4) cm long, 0.1–2 cm wide, tomentose abaxially, less so or glabrous and greenish adaxially; margin entire, occasionally revolute. **INFL** cymose, compact, often flat-topped, 0.5–6 (12) cm long, 1–10 (13) cm wide; branches whitish- to brownish- or reddish-tomentose to floccose, infrequently green or gray and subglabrous or glabrous; bracts 3, scalelike, 1–5 mm long, 1–3 mm wide. **PEDUNCLES** mostly erect, straight, slender, 3–15 mm long, tomentose to floccose, usually absent distally. **INVOLUCRES** turbinate, (1.5) 2–3.5 (4) mm long, 1.3–2.5 (3) mm wide, tomentose or glabrous; teeth 5, erect, (0.3) 0.5–1 (1.7) mm long. **FLOWERS** bisexual, nonstipitate; perianth yellow or white, becoming orange or rose to red, 1.5–3 (4) mm long, glabrous; tepals connate 1/5–2/5 their length, essentially monomorphic, oblong to obovate; stamens mostly exserted, 2.5–4 mm; filaments sparsely to densely puberulent proximally. **ACHENES** trigonous, brown, 1.5–3 mm, glabrous. Widespread throughout most of the western United States from Washington to Montana south to southern California, northern Arizona, and northern New Mexico. A species of 13 varieties; the two most common ones are in our area.

1. Leaf margins plane, flowering stems and inflorescences floccose or glabrous; just entering our area in San Juan County, Utah .... **var. *laxiflorum***

1' Leaf margins revolute or nearly so; flowering stems and inflorescences tomentose to floccose; common and widespread in our area .... **var. *simpsonii***

var. *laxiflorum* Hook. (loose leaf, as to the well-spaced cauline leaves) Great Basin wild buckwheat. Plants shrubs, (1) 2–4 (5) dm tall, 3–8 dm wide. **FLOWERING STEMS** 0.2–0.6 (0.8) dm long, floccose or tomentose, rarely glabrous. **LEAF BLADES** mostly elliptic, (0.5) 1–2 (2.5) cm long, (0.1) 0.2–0.6 (0.8) cm, densely to sparsely whitish tomentose abaxially, less so to sparsely whitish floccose adaxially; margin not revolute. **INFL** (1) 2–4 (8) cm long and wide; branches floccose to glabrate, rarely glabrous. **INVOLUCRES** 2–3 (3.5) mm long, subglabrous, glabrous or floccose between angled ridges. **PERIANTH** white to pink or rose, 2–3 mm long. **ACHENES** 2–3 mm long. [*E. confertiflorum* Benth.; *E. microthecum* subsp. *confertiflorum* (Benth.) S. Stokes; *E. microthecum* subsp. *laxiflorum* (Hook.) S. Stokes]. Sandy to gravelly flats and slopes in mixed grassland, saltbush, blackbrush, and sagebrush communities, piñon-juniper and montane conifer woodlands. ARIZ: Apa, Nav; UTAH. 1750–2315 m (5735–7600'). Flowering: Jun–Oct. Washington to Montana south to east-central California, north-central Arizona, southern Utah, and northwestern Colorado.

var. *simpsonii* (Benth.) Reveal (Lieutenant James Harvey Simpson, 1813–1883, U.S. Topographical Engineers, co-discoverer of Chaco Canyon during Navajo Expedition to New Mexico in 1849) Simpson's wild buckwheat. Plants subshrubs to shrubs, (1) 2–15 dm tall, 4–16 dm across. **FLOWERING STEMS** erect or nearly so, 0.2–0.7 dm long, densely lanate to tomentose, sometimes floccose or rarely glabrous, whitish to grayish. **LEAF BLADES** mostly narrowly elliptic, 0.5–1.8 (2.5) cm long, 0.1–0.2 cm wide, densely white-tomentose abaxially, mostly white-floccose or rarely glabrous and green adaxially; margins usually revolute; petioles 1–5 mm long. **INFL** (1.5) 2–4 (6) cm tall and wide; branches tomentose to floccose or rarely glabrate, grayish to greenish. **INVOLUCRES** 2–3 mm long, tomentose to floccose or glabrate. **PERIANTH** white, becoming pink to rose, 2–3 mm. **ACHENES** 2–3 mm long. [*E. effusum* Nutt. var. *foliosum* Torr. & A. Gray; *E. effusum* subsp. *nelsonii* (L. O. Williams) S. Stokes; *E. microthecum* var. *foliosum* (Torr. & A. Gray) Reveal; *E. microthecum* var. *friscanum* (M. E. Jones) S. Stokes; *E. microthecum* var. *macdougalii* (Gand.) S. Stokes; *E. simpsonii* Benth.] Clayey to gravelly or occasionally sandy washes, flats, and slopes in mixed grassland, saltbush, blackbrush, and sagebrush communities, and in piñon-juniper woodlands or infrequently in montane conifer woodlands. ARIZ: Apa, Coc, Nav; COLO: Arc, LPl, Mon, SMg; NMEX: McK, RAr, San, SJn; UTAH. 1400–2300 m (4595–7545′). Flowering: Jun–Oct. Southeastern California east across central and southern Nevada through Utah and northern Arizona to southwestern Wyoming, western and south-central Colorado, and northern New Mexico.

Eriogonum ovalifolium Nutt. **CP** (oval-leaved) Cushion wild buckwheat. Plants low, pulvinate, herbaceous perennials, 0.5–3 dm tall, 2.5–4 dm across. **STEMS** numerous; caudex stems matted, compact to somewhat spreading; aerial flowering stems scapose, mostly erect, 0.4–2 dm long, tomentose, whitish, grayish, greenish, or tawny. **LEAVES** basal; petiole 1–10 cm long, mostly tomentose; blade spatulate, oblong or obovate to oval or round, 0.5–2 cm long, 0.4–1.5 (2) cm wide, tomentose, sometimes less so to floccose and whitish to grayish or tawny adaxially. **INFL** capitate, 1.5–3.5 cm across; bracts 3, scalelike, 0.8–4 mm long, 0.5–3 mm wide. **PEDUNCLES** absent. **INVOLUCRES** turbinate, 3.5–5 mm long, 2–4 mm wide, tomentose to floccose; teeth 5, erect, 0.4–1 mm. **FLOWERS** bisexual, nonstipitate; perianth white to cream or pinkish, rose, red, or purplish, 4–5 mm, glabrous; tepals dimorphic, those of the outer whorl mostly oval to orbicular, 2–4 mm long and wide, those of the inner whorl oblanceolate to elliptic, 3–5 mm long, 0.8–1.5 mm wide; stamens mostly included, 1–3 mm; filaments pilose proximally. **ACHENES** trigonous, light brown to brown, 2–3 mm long, glabrous. $n = 20$. Sandy to gravelly flats, washes, slopes, and ridges in mixed grassland, saltbush, and sagebrush communities, and in piñon and/or juniper and montane conifer woodlands. ARIZ: Apa, Coc, Nav; COLO: LPl, Mon; NMEX: RAr, SJn; UTAH. 1525–2285 m (5000–7500′). Flowering: Apr–Aug. Widespread in western North America from southwestern Canada south to northern Arizona and New Mexico. *Eriogonum ovalifolium* is a variable species of some 11 varieties found mainly to the north and west of us. Our material, as described here, is ***var. purpureum*** (Nutt.) Durand (purple, as to color of some flowers in fruit). [*Eucycla purpurea* Nutt.; *Eriogonum purpureum* (Nutt.) Benth.; *E. orthocaulon* Small; *E. ovalifolium* subsp. *purpureum* (Nutt.) S. Stokes]. The var. *ovalifolium* (with yellow flowers) occurs just to our north. Used by the Ute for an unspecified purpose.

Eriogonum palmerianum Reveal (Edward Palmer, 1831–1911, archaeological, zoological, and botanical collector in the southwestern United States and Mexico) Palmer's wild buckwheat. Plants low, spreading to erect annuals, (0.5) 1–3 dm high. **STEMS** few, mostly erect, mostly floccose to tomentose, grayish to tawny. **LEAVES** basal; petioles 1–4 cm, floccose; blade suborbicular to cordate, 0.5–1.5 (1.8) cm long, 0.5–2 cm wide, densely white to grayish tomentose abaxially, less so to glabrate and often greenish adaxially; margin entire, plane. **INFL** cymose, distally uniparous due to suppression of secondary branches, forming open crowns, 10–30 cm; branches mostly few, floccose to tomentose, spreading to outwardly curved; bracts 3, scalelike, 0.5–3 mm long, 1–3 mm wide. **PEDUNCLES** absent. **INVOLUCRES** appressed to branches, campanulate, 1.5–2 mm long and wide; teeth 5, erect, 0.2–0.3 mm long. **FLOWERS** bisexual, nonstipitate; perianth white to pink or rarely pale yellowish, becoming red, 1.5–2 mm long, glabrous; tepals dimorphic, those of the outer whorl narrowly obovate and narrowly fan-shaped, those of the inner whorl oblanceolate; stamens included, 1–1.5 mm long; filaments pilose proximally. **ACHENES** trigonous, brown, 1.5–1.8 mm long, glabrous. $n = 20$. Sandy to gravelly washes, flats, and slopes in saltbush, greasewood, blackbrush, and sagebrush communities, and in piñon and/or juniper woodlands. ARIZ: Coc, Nav; COLO: Mon; UTAH. 1125–1525 m (3700–5000′). Flowering: Mar–Oct. Eastern California east across Nevada, Arizona, and western and southern Utah to western New Mexico.

Eriogonum polycladon Benth. (many branches, as to the highly branched inflorescence) Plants erect annuals, (0.5) 1–6 dm tall. **STEMS** mostly 1, erect, (0.3) 1–3 dm long, tomentose. **LEAVES** cauline; petioles 0.3–1.5 cm long, tomentose; blade narrowly oblanceolate to broadly elliptic, (0.7) 1–3 cm long, 0.5–1.5 cm wide, densely tomentose and whitish to grayish on both surfaces; margin entire, plane or slightly crisped. **INFL** virgate, narrow and strict, (5) 10–50 cm long, 5–25 cm wide; branches tomentose; bracts 3, scalelike, 1.5–3 mm long, 1–2.5 mm wide. **PEDUNCLES** absent. **INVOLUCRES** turbinate, 1.5–2.5 mm long, 1–2 mm wide, tomentose; teeth 5, 0.4–1 mm long. **FLOWERS** bisexual, nonstipitate; perianth white to pink or red, (1) 1.5–2 mm long, glabrous; tepals dimorphic, those of outer whorl obovate and broadly fan-shaped, those of inner whorl oblanceolate; stamens included, 1–1.5 mm long; filaments pilose proximally. **ACHENES** trigonous, dark brown, 1–1.3 mm long, glabrous. $n = 13$. [*E. densum* Greene; *E. vimineum* Douglas ex Benth. subsp. *polycladon* (Benth.) S. Stokes; *E. vimineum* var. *densum* (Greene) S. Stokes]. Sandy to gravelly washes, flats, and slopes in saltbush, greasewood, blackbrush, and sagebrush communities, and in oak and piñon and/or juniper or occasionally montane conifer woodlands. ARIZ: Apa, Nav; NMEX: McK, San, SJn. 1125–2150 m (3700–7055′). Flowering: Sep–Oct. Arizona, New Mexico, and western Texas south to northern Mexico, disjunct and probably introduced but now extirpated in southeastern California. Locally weedy in places.

Eriogonum racemosum Nutt. (full of clusters, as to racemose inflorescence) Red-root wild buckwheat. Plants erect herbaceous perennials, 3–8 (10) dm tall, 0.5–1.5 dm across. **STEMS** usually 1; caudex stems compact; aerial flowering stems erect to slightly spreading, slender to stout, not fistulose, (1) 1.5–2.5 (3) dm long, tomentose to floccose (ours) or rarely glabrous, mostly grayish. **LEAVES** basal, not in fascicles; petiole (2) 3–10 (15) cm, tomentose to floccose; blade elliptic to ovate or oval to nearly rotund, (1.5) 2–6 (10) cm long, 1–4 (5) cm wide, lanate to thinly tomentose and grayish abaxially, floccose to glabrous and green adaxially; margin entire, plane. **INFL** cymose, virgate, or racemose, 15–50 cm long, 5–20 cm wide; branches tomentose (ours), rarely glabrous, with 5–20 (30) racemosely arranged involucres distally; bracts 3, scalelike, triangular and (1) 2.5–7 mm long, 1–4 mm wide, or foliaceous and linear-oblanceolate or oblanceolate to elliptic, 10–40 mm long, 5–20 (25) mm wide. **PEDUNCLES** erect, straight, stout, 3–40 mm, tomentose, often absent distally or occasionally entirely. **INVOLUCRES** turbinate-campanulate, (2) 3–5 mm long, (2) 2.5–4 mm wide, tomentose to floccose; teeth 5, erect, (0.1) 0.2–0.5 mm long. **FLOWERS** bisexual, nonstipitate; perianth white to pinkish, (2) 2.5–5 mm long, glabrous; tepals monomorphic, oblong; stamens exserted, 2–5 mm long; filaments pilose proximally. **ACHENES** trigonous, light brown, 3–4 mm long, glabrous. n = 18. [*E. racemosum* var. *obtusum* (Benth.) S. Stokes; *E. racemosum* var. *orthocladon* (Torr.) S. Stokes]. Sandy to gravelly flats and slopes in mixed grass, sagebrush, and mountain mahogany communities, and in scrub oak, piñon, juniper, and conifer woodlands. ARIZ: Apa, Coc, Nav; COLO: Arc, Dol, Hin, LPl, Min, Mon, SMg; NMEX: McK, RAr, San, SJn; UTAH. 1830–2900 m (6000–9515′). Flowering: Jun–Oct. Western Colorado, Utah, and southern Nevada south to northern Arizona and northwestern New Mexico. Kayenta Navajo used the plant for backaches and side aches. Ramah Navajo used an infusion of the whole plant for blood poisoning or internal injuries.

Eriogonum scabrellum Reveal (somewhat roughened, as to scabrellous stems and branches) Westwater wild buckwheat. Plants low to somewhat rounded, spreading annuals, 1–3 (5) dm high. **STEMS** mostly 1, erect, 0.5–1.5 dm long, floccose and scabrellous. **LEAVES** basal and sheathing up the stems 1–3 cm; petiole 1–4 (5) cm, floccose; blade cordate, 1–3 (4) cm long and wide, densely white-tomentose abaxially, sparsely floccose and greenish to green or reddish brown adaxially; margin crisped and wavy. **INFL** cymose, open, spreading, often flat-topped, 5–40 cm high, 20–100 dm wide; branches scabrellous; bracts 3, scalelike, 1–1.5 mm long, 0.4–0.9 mm wide. **PEDUNCLES** lacking. **INVOLUCRES** horizontal, turbinate, 1.5–2.5 mm long, 1.5–2 mm wide, scabrellous; teeth 5, erect, 0.4–0.6 mm. **FLOWERS** bisexual, nonstipitate; perianth white, becoming pink to rose or deep red in fruit, 1–1.5 mm, minutely pustulose; tepals dimorphic, those of the outer whorl obovate with truncate bases, those of the inner whorl ovate; stamens excluded, 1–1.5 mm long; filaments glabrous. **ACHENES** trigonous, light brown to brown, 2 mm long, glabrous. n = 20. Clayey to gravelly washes, flats, and slopes in saltbush, blackbrush, and sagebrush communities, and in piñon-juniper woodlands. ARIZ: Apa; COLO: Mon; NMEX: SJn: UTAH. 1400–2300 m (4595–7545′). Flowering: Aug–Nov. Colorado Plateau of northwestern and southwestern Colorado and eastern Utah south into northeastern Arizona and northwestern New Mexico.

Eriogonum shockleyi S. Watson (William Hillman Shockley, 1855–1925, mining engineer and botanical collector in Nevada and eastern California) Shockley's wild buckwheat. Plants low, caespitose to pulvinate-matted, herbaceous perennials, 0.3–0.5 (0.7) dm tall, (0.5) 1–4 (10) dm across. **STEMS** numerous; caudex stems matted, compact; aerial stems scapose, erect or nearly so, (0.5) 1–3 cm long, floccose to tomentose, greenish or grayish, infrequently absent. **LEAVES** basal, not in fascicles; petiole 0.2–0.5 cm long, tomentose to floccose; blade oblanceolate to elliptic or spatulate, (0.2) 0.3–0.8 (1.2) cm long, 0.2–0.4 (0.6) cm wide, tomentose to floccose and whitish to grayish or tawny on both surfaces, infrequently greenish; margin entire, plane. **INFL** capitate, 0.8–2 cm across; branches absent; bracts 3–5, ± scalelike, linear to linear-lanceolate, 1.5–4 mm long, 0.6–1 mm wide. **PEDUNCLES** absent. **INVOLUCRES** campanulate, (2) 2.5–5 (6) mm long, 3–6 (7) mm wide, tomentose; teeth 5–10, sometimes ± lobelike, erect to spreading, (0.5) 1–3 mm long. **FLOWERS** bisexual, nonstipitate; perianth white to rose (ours) or yellow, 2.5–4 mm long, densely pilose; tepals monomorphic, oblong to obovate; stamens exserted, 2.5–5 mm long; filaments subglabrous to sparsely pilose proximally. **ACHENES** trigonous, light brown to brown, 2.5–3 mm long, tomentose. [*E. pulvinatum* Small; *E. shockleyi* subsp. *candidum* (M. E. Jones) S. Stokes; *E. shockleyi* subsp. *longilobum* (M. E. Jones) S. Stokes; *E. shockleyi* var. *longilobum* (M. E. Jones) Reveal]. Gravelly or clayey (rarely sandy) flats, washes, and slopes in saltbush, blackbrush, and sagebrush communities, and in piñon-juniper woodlands. ARIZ: Apa, Coc, Nav; COLO: Mon; NMEX: McK, SJn; UTAH. 1370–2135 m (4500–7000′). Flowering: May–Aug. Great Basin and Colorado Plateau, eastern California, southern

Idaho, Nevada, Utah, and western Colorado south to northern Arizona and northwestern New Mexico. Our plants form small, compact mats not more than 4 dm across and generally have spreading, lobelike involucral teeth.

Eriogonum subreniforme S. Watson (somewhat kidneylike, as to leaf shape) Kidney-shaped wild buckwheat. Plants tall, erect to slightly spreading annuals, 1–5 (7) dm high. **STEMS** mostly 1, (0.2) 0.5–1.5 dm long, not fistulose, glabrous except floccose to glabrescent near leaves, usually greenish. **LEAVES** basal; petiole (1) 2–6 cm, mostly floccose; blade reniform to orbicular, (0.5) 1–3.5 (4) cm long, (0.5) 1–4 (5) cm wide, densely white-tomentose abaxially, hirsute to floccose or glabrous and greenish adaxially; margin entire, plane. **INFL** cymose, open to slightly diffuse, 5–40 (50) cm long, 5–35 cm wide; branches glabrous, greenish; bracts 3, scalelike, 1–3 mm long, 1.5–2.5 mm wide. **PEDUNCLES** erect or nearly so, straight, filiform, 5–25 mm long, glabrous or rarely floccose at or before anthesis. **INVOLUCRES** turbinate, 0.5–1 mm long, 0.6–0.9 mm wide, glabrous; teeth 5, erect, 0.2–0.4 mm. **FLOWERS** bisexual, nonstipitate; perianth white to rose or rarely yellowish, 0.6–1.6 (2) mm long, glabrous or sparsely hirsute; tepals monomorphic, lanceolate to spatulate or elliptic to ovate; stamens included to slightly excluded, 0.7–1.4 mm long; filaments glabrous. **ACHENES** trigonous, light to dark brown, 1.7–2 mm long, glabrous. [*E. filicaule* S. Stokes]. Sandy to gypsophilous or clayey flats and slopes in saltbush, greasewood, and blackbrush communities, and in oak and piñon-juniper woodlands. ARIZ: Apa, Nav; NMEX: McK, SJn. 1370–2100 m (4500–6890′). Flowering: Apr–Oct. Southwestern Utah, northern Arizona, and northwestern New Mexico.

Eriogonum umbellatum Torr. (having an umbel, as to inflorescence) Sulphur flower. Plants low, caespitose, matted, herbaceous perennials to large spreading shrubs, 1–12 (20) dm tall, 3–12 (20) dm across. **STEMS** numerous; caudex stems matted to spreading; aerial flowering stems erect or nearly so, arising at nodes of caudex and at distal nodes of short, nonflowering aerial stems, (0.3) 0.5–3 (4) dm long, tomentose to floccose or glabrous, greenish to green. **LEAVES** in loose to compact basal rosettes; petioles 0.1–3 (4) cm long, tomentose to floccose, rarely glabrous; blades elliptic to round, 0.3–3 (4) cm long, 0.2–3 cm wide, tomentose to thinly floccose or glabrous on one or both surfaces; margins entire, plane. **INFL** umbellate to compound umbellate, 3–25 cm long, 2–18 cm wide; branches tomentose to floccose or glabrous; bracts 3–several, semifoliaceous at first node, 3–15 mm long, 2–10 mm wide, mostly scalelike and 1–5 mm long, 0.5–3 mm wide distally. **PEDUNCLES** mostly absent. **INVOLUCRES** turbinate to campanulate, 1–6 mm long, (1) 1.5–10 mm wide, tomentose to floccose or glabrous; teeth 6–12, lobelike, reflexed, 1–4 (6) mm long. **FLOWERS** bisexual or unisexual, often polygamodioecious, stipitate; perianth yellow, cream to red, or purple, 2.5–12 mm long including the (0.7) 1.3–2 mm long stipe, glabrous; tepals monomorphic, mostly spatulate to obovate, connate 1/4–1/3 their length; stamens exserted, 2–8 mm; filaments pilose proximally. **ACHENES** trigonous, light brown to brown, 2–7 mm, glabrous except for sparsely pubescent beak. Southern Canada south to southern California, northern Arizona, and northern New Mexico. A large and complex species with more than 40 varieties. Kayenta Navajo used the plant (probably var. *subaridum*) as a fumigant.

1. Inflorescences compound umbellate .... **var. *subaridum***
1′ Inflorescences umbellate or capitate .... (2)

2. Leaf blades tomentose abaxially; plants mainly below 2500 m (8200′) .... **var. *umbellatum***
2′ Leaf blades glabrous on both surfaces; plants mainly above 2500 m (8200′) .... **var. *porteri***

var. *porteri* (Small) S. Stokes (Thomas Conrad Porter, 1822–1901, professor of botany and zoology at Lafayette College, Pennsylvania, and occasional collector in Colorado) Plants caespitose to a low subshrub, 0.2–0.6 dm tall, 1–5 dm across. **FLOWERING STEMS** erect, 0.1–0.5 dm long, floccose to nearly glabrous. **LEAVES** in tight rosettes; blade elliptic to spatulate, 0.4–1.1 cm long, 0.3–0.8 cm wide, glabrous on both surfaces. **INFL** umbellate and subcapitate to capitate; branches less than 0.5 cm long, glabrous. **INVOLUCRAL TUBES** 2–3 mm long; lobes 2–3 mm long, reflexed. **PERIANTH** bright yellow, 3–6 mm long. [*E. porteri* Small]. Rocky slopes and ridges, high-elevation sagebrush and meadow communities, subalpine and alpine communities. COLO: Hin, LPl. 2400–3700 m (7875–12140′). Flowering: Jul–Sep. Mainly in the Rocky Mountains of Colorado, with disjunct populations in the Uinta and Wasatch mountains of Utah and in scattered ranges in central and northeastern Nevada.

var. *subaridum* S. Stokes [somewhat like subsp. *aridum* (var. *dicrocephalum*), a related variety] Ferris' sulphur flower. Plants mostly low to rounded subshrubs or shrubs, 2–7 dm tall, 3–9 (12) dm across. **FLOWERING STEMS** erect, (0.5) 1–2 dm long, floccose to glabrous. **LEAVES** in loose rosettes; blade elliptic, 1–3 cm long, 0.5–2 cm wide, thinly floccose on both surfaces or glabrous adaxially, rarely glabrous on both surfaces or tomentose abaxially. **INFL** compound umbellate, divided 2–5 times, floccose or glabrous. **INVOLUCRAL TUBES** 2–3 (3.5) mm long; lobes 1–3 mm long, spreading to slightly reflexed. **PERIANTH** bright yellow, 3–7 mm long. [*E. biumbellatum* Rydb.; *E. umbellatum* subsp. *ferrissii* (A. Nelson) S. Stokes; *E. umbellatum* subsp. *subaridum* (S. Stokes) Munz]. Sandy to gravelly flats and slopes in mixed grassland, saltbush, and sagebrush communities, and in oak, piñon-juniper, and montane conifer woodlands. ARIZ: Apa, Coc, Nav; COLO: LPl, Mon; UTAH. 1525–2590 m (5000–8500′). Flowering: Jun–Oct. Southern California east across southern Nevada and northern Arizona to Utah and southwestern Colorado.

var. *umbellatum* Common sulphur flower. Plants compact mats 1–3.5 dm tall, 2–6 dm across. **FLOWERING STEMS** erect, mostly 1–3 dm long, tomentose to floccose. **LEAVES** in loose rosettes; blades mostly elliptic to ovate, 1–2.5 (3) cm long, 0.5–1.5 (2) cm wide, white- to gray-lanate or -tomentose abaxially, less so to floccose or more commonly glabrous and green adaxially. **INFL** umbellate; branches 0.3–2.5 (10) cm long, floccose. **INVOLUCRAL TUBES** 2–3 mm long; lobes 1.5–3 mm long, strongly reflexed. **PERIANTH** bright yellow, 4–7 (8) mm long. Sandy to gravelly flats and slopes in mixed grassland and sagebrush communities, and in scrub oak and montane conifer woodlands. COLO: Arc, Dol, LPl, Mon, SMg; NMEX: RAr; UTAH. 1525–2590 m (5000–8500′). Flowering: Jun–Sep. Eastern Idaho and western Montana south through western and southern Wyoming into southeastern Utah and southern Colorado.

Eriogonum wetherillii Eastw. (Benjamin Alfred Wetherill, 1861–1950, rancher, who, along with brother Richard, 1858–1910, located Mesa Verde in 1888 and guided Alice Eastwood in 1895) Wetherill's wild buckwheat. Plants low, spreading annuals, 0.5–2.5 dm high. **STEMS** mostly numerous, 0.1–0.5 dm long, not fistulose, glabrous except villous near leaves, usually reddish. **LEAVES** basal; petioles 1–5 cm, tomentose to floccose; blade oblong to orbicular, (0.5) 1–4 cm long, (0.5) 1–3 cm wide, densely white-tomentose abaxially, floccose to glabrate and reddish or greenish adaxially; margin entire, plane. **INFL** cymose, flat-topped to spreading or rounded, diffuse, 5–20 cm long, 10–35 cm wide; branches glabrous, greenish or reddish; bracts 3, scalelike, 0.5–2 mm long, 1–2 mm wide. **PEDUNCLES** erect, straight, filiform, (3) 5–10 mm long, rarely absent distally. **INVOLUCRES** turbinate, (0.3) 0.5–1 mm long, (0.2) 0.5–1 mm wide, glabrous; teeth 4, erect, 0.1–0.3 mm long. **FLOWERS** bisexual, nonstipitate; perianth yellowish to red and 0.5–1.5 mm, glabrous; tepals monomorphic, elliptic to obovate; stamens included to slightly excluded, 0.7–1 mm long; filaments glabrous. **ACHENES** lenticular, dark brown to black, 0.6–1 mm long, glabrous. [*E. filiforme* L. O. Williams; *E. sessile* S. Stokes ex M. E. Jones]. Sandy to clayey flats, washes, and slopes in saltbush, greasewood, and blackbrush communities, and in oak and piñon and/or juniper woodlands. ARIZ: Apa, Coc, Nav; COLO: Mon; NMEX: SJn; UTAH. 1125–1980 m (3700–6500′). Flowering: Apr–Oct. Colorado Plateau: southeastern Utah and southwestern Colorado south to northeastern Arizona and northwestern New Mexico.

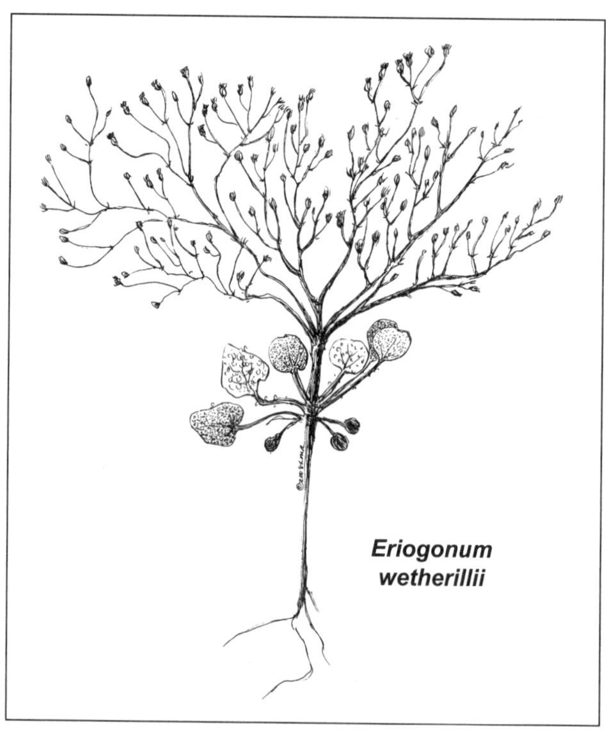
Eriogonum wetherillii

Eriogonum wrightii Torr. ex Benth. (Charles Wright, 1811–1885, western collector on numerous boundary and railroad surveys) Bastard-sage. Plants spreading to rounded subshrubs or shrubs, (1) 1.5–5 (7.5) dm tall, 1–12 (18) dm across. **STEMS** numerous; caudex stems absent; aerial flowering stems erect to spreading, (0.1) 0.5–4 (6) dm long, tomentose, whitish. **LEAVES** cauline, usually in fascicles; petiole 0.2–0.5 (1) cm long, tomentose to floccose; blade oblanceolate to elliptic, 0.5–1.5 cm long, 0.2–0.5 (0.7) cm wide, tomentose to floccose on both surfaces; margin entire, plane. **INFL** virgate or nearly cymose with racemosely arranged involucres distally, usually diffuse, 5–15 cm long, 10–20 cm wide; branches tomentose to floccose; bracts 3, mostly scalelike, 0.5–3.5 mm long and wide. **PEDUNCLES** absent. **INVOLUCRES** turbinate, 2–2.5 mm long, 1–2 mm wide; teeth 5, erect, 0.3–1 mm long. **FLOWERS** bisexual, nonstipitate; perianth white to pink or rose, 2.5–3.5 mm long, glabrous; tepals monomorphic, obovate; stamens exserted, 2.5–4 mm long; filaments usually sparsely pilose proximally. **ACHENES** trigonous, light brown to brown, 2.5–3 mm long, glabrous. [*E. trachygonum* Torr. ex Benth. subsp. *wrightii* (Torr. ex Benth.) S. Stokes]. Gravelly to rocky (often calcareous) slopes in mixed grassland, saltbush, and blackbrush communities, and in oak or piñon and/or juniper woodlands. ARIZ: Apa, Nav; NMEX: SJn. 1525–2200 m (5000–7215′). Flowering: Jul–Oct. Southeastern California across southernmost Nevada to western Texas, south to central Mexico. Our plants, as described here, are **var. *wrightii***. Kayenta Navajo used the plant as an emetic.

Eriogonum zionis J. T. Howell (for Zion Canyon, Utah) Zion wild buckwheat. Plants erect to slightly spreading, herbaceous perennials, 3–8 (10) dm high, 0.5–1.5 dm across. **STEMS** usually 1; caudex stems compact; aerial flowering stems erect to slightly spreading, usually fistulose, 1–2.5 (3) dm long, glabrous (ours), rarely tomentose, greenish. **LEAVES** basal, not in fascicles; petiole 3–6 (10) cm, tomentose to floccose; blade elliptic or oblong to ovate, 2–4.5 (6) cm long, 1.5–2.5 (5) cm wide, lanate to densely tomentose abaxially, thinly floccose to glabrous and green adaxially; margin plane, entire. **INFL** narrowly cymose, virgate, or racemose, 20–50 (60) cm long, 5–25 cm wide; branches glabrous (ours), rarely tomentose, with (8) 10–15 (20) racemosely arranged involucres distally; bracts 3, scalelike, triangular and 2–7 mm long, or foliaceous and elliptic, 10–25 mm long, 5–12 mm wide. **PEDUNCLES** absent. **INVOLUCRES** erect, turbinate to turbinate-campanulate, 1.5–3 mm long, 1.5–2.5 mm wide, tomentose; teeth 5, erect, 0.2–0.4 mm long. **FLOWERS** bisexual, nonstipitate; perianth white to yellow, 2–3.5 (4) mm long, glabrous; tepals monomorphic, oblong; stamens exserted, 2–5 mm; filaments pilose proximally. **ACHENES** trigonous, light brown, 3–4 mm, glabrous. $n = 20$. [*E. racemosum* Nutt. var. *nobilis* S. L. Welsh & N. D. Atwood; *E. racemosum* var. *zionis* (J. T. Howell) S. L. Welsh]. Deep sandy soil in sagebrush communities. UTAH. 1310 m (4300′). Flowering: Jul–Oct. Southern Utah south to north-central Arizona. Our plants, as described here, are **var. *zionis*** and are known from a single location near Bluff, Utah.

Fallopia Adans. False-buckwheat
Rodney Myatt

(for Gabriello Fallopio, 1532–1562, an Italian anatomist) Plants annual or perennial vines or herbs; roots fibrous or woody; rhizomes occasionally present. **STEMS** erect to scandent, rarely procumbent, glabrous or pubescent; flowering stems branched. **LEAVES** cauline, alternate, petiolate; ocrea persistent or deciduous, chartaceous; blade broadly ovate to triangular, sometimes hastate or sagittate. **INFL** terminal or terminal and axillary, paniculate, racemose, or spicate; peduncles present or absent. **FLOWERS** bisexual or bisexual and unisexual, occasionally some plants pistillate, 1–5 per ocreate fascicle, stipitate proximally, pedicellate; perianth usually accrescent, pale green or white to pink, glabrous or (ours) with hyaline hairs; tepals 5, petaloid, connate proximally to nearly throughout, dimorphic; stamens 6–8; filaments distinct, free; styles 3, spreading, connate; stigma capitate, fimbriate, or peltate. **ACHENES** trigonous, brown to black, not winged, glabrous; embryo straight. $x = 10, 11$ (Freeman and Hinds 2005b; Kim, Park, and Park 2000). About 12 species of North America (8 species), South America, Eurasia, Africa, and scattered Pacific islands.

Fallopia convolvulus (L.) Á. Löve (bindweedlike) Black-bindweed. Plants annual vines, 5–10 dm long; rhizomes absent. **STEMS** scandent or sprawling, puberulent. **LEAVES** cauline; ocrea 2–4 mm long, glabrous or scabrid on margin; petiole 0.5–5 cm long, puberulent; blade broadly ovate with a cordate base, hastate or sagittate, 2–6 (15) cm long, 2–5 (10) cm wide, glabrous abaxially, scabrous adaxially with epidermal projections along veins. **INFL** axillary, erect or spreading, spicate, 2–10 (15) cm long; peduncle (0) 0.1–10 cm long, glabrous or scabrid. **FLOWERS** bisexual, 3–6 per ocreate fascicle; pedicel ascending to spreading, articulated distally, 1–3 mm long, glabrous or scabrid; perianth nonaccrescent, greenish white with a pinkish or purplish base, 3–5 mm long including stipe, glabrous or outer 3 with blunt, hyaline hairs; tepals elliptic to ovate, connate 1/3 their length, outer 3 obscurely keeled; stamens 8; anthers cream; styles connate distally; stigma capitate. **ACHENES** black, 4–6 mm long, dull, minutely granular. $n = 20$. [*Polygonum convolvulus* L.; *Bilderdykia convolvulus* (L.) Dumort.; *Reynoutria convolvulus* (L.) Shinners]. Weedy in dry fields and along roadsides, often climbing on fences and other structures. COLO: Arc; NMEX: SJn. 1615–1890 m (5300–6195′). Flowering: Jun–Aug. Introduced and widely naturalized throughout North America, South America, Africa, Australia, and Pacific Islands; native of Eurasia.

Oxyria Hill Mountain-sorrel
Rodney Myatt

(Greek *oxys*, sour, + *aria*, possession, alluding to acidic-tasting leaves) Plants synoecious, perennial herbs; roots thick, fleshy, fibrous; rhizomes present or absent. **STEMS** erect, glabrous; flowering stems unbranched. **LEAVES** basal, rarely cauline, alternate, petiolate; ocrea persistent or sometimes deciduous, chartaceous; blade reniform to cordate-orbicular. **INFL** terminal, paniculate or racemose; peduncles slender, erect or nearly so. **FLOWERS** bisexual, (1) 3–7 per ocreate fascicle, stipitate proximally, pedicellate; perianth nonaccrescent, greenish to reddish, glabrous; tepals 4, petaloid, distinct, dimorphic; margin entire or slightly wavy; tubercle absent; stamens 6; filaments distinct, free; styles 2, spreading, distinct; stigmas 2, penicillate. **ACHENES** lenticular, yellowish to tan, 2-winged, glabrous; embryo straight or curved. $x = 7$ (Chrtek and Šourková 1992; Freeman and Packer 2005). About four species of Arctic and alpine areas in western North America (one species) and Eurasia. The "pedicel" length given here includes both the pedicel itself and the elongated stipe.

Oxyria digyna (L.) Hill. (two ovaries) Mountain-sorrel. Plants (3) 5–30 (50) cm tall. **LEAVES** basal; ocrea hyaline or brownish, 2.5–10 mm long, glabrous; petiole 1–15 cm long; blade reniform, (0.5) 1–6.5 cm long, 0.5–6 (8) cm wide. **INFL** (1) 2–20 cm long; peduncle 1–17 cm long. **FLOWERS** 2–6 per ocreate fascicle; pedicel (1) 3–5 mm long; perianth 1–2.5 mm long; tepals dimorphic, those of outer whorl lanceolate, 1.2–1.7 mm long, 0.5–1 mm wide, spreading in fruit, those of inner tepals broadly oblanceolate to ovate or orbicular, 1.4–2.5 mm long, 0.7–1.6 mm wide, appressed to fruit; stamens exserted, 1.5–2 mm long; anthers white to cream. **ACHENES** brown, 3–4.5 mm long, 2.5–5 mm wide including reddish or pinkish wings, dull, granular. $n = 7$. [*Rumex digynus* L.]. Rocky flats and slopes mainly in alpine communities, especially near melting snow. COLO: Arc, Con, Hin, LPl, Min, Mon, RGr, SJn. 2895–3810 m (9500–12500′). Flowering: Jun–Aug. Greenland to Alaska, south to New Hampshire, south in the Rocky Mountains and the Sierra Nevada to California, Arizona, and New Mexico; Eurasia.

Persicaria Mill. Smartweed

Rodney Myatt

(Latin *persica*, peach, + *aria*, pertaining to, referring to leaf shape of some species) Plants annual or perennial herbs (ours), rarely suffrutescent; roots fibrous or woody; rhizomes or stolons sometimes present. **STEMS** erect or sometimes prostrate to scandent, glabrous or pubescent, rarely with recurved prickles; flowering stems usually branched. **LEAVES** mainly cauline, alternate, petiolate or sessile; ocrea persistent or disintegrating with age, usually deciduous entirely or distally, entire, tannish to brownish or reddish, charataceous or foliaceous, rarely coriaceous, glabrous or scabrous to variously pubescent; blade narrowly lanceolate to elliptic, ovate, hastate, or sagittate. **INFL** terminal or terminal and axillary, spicate, paniculate, or capitate; peduncle present. **FLOWERS** bisexual but functionally unisexual in some, 1–14 per ocreate fascicle, not stipitate proximally, pedicellate or not; perianth accrescent or nonaccrescent, white, greenish, pink, red, or purple, glabrous or glandular-punctate; tepals 4–5, petaloid, sometimes fleshy in fruit, connate 1/4–2/3 their length, dimorphic; stamens 5–8; filaments distinct or connate proximally, some occasionally adnate to perianth; styles 2–3, erect to spreading or reflexed; stigmas capitate. **ACHENES** lenticular or trigonous, brown to dark brown or black, not winged, glabrous; embryo curved. $x = 10, 11, 12$ (Hinds and Freeman 2005). About 100 species of both aquatic and terrestrial places, occasionally weedy, throughout North America north of Mexico (26 species) and nearly worldwide. All of our species belong to sect. *Persicaria*.

1. Plants perennial, of aquatic or subaquatic habitat; inflorescences terminal or nearly so, usually short and rather broad; flowers pink or red .... ***P. amphibia***
1′ Plants annual or rarely perennial, of terrestrial habitat, never floating; inflorescences terminal and axillary, usually elongate and narrow; flowers pink, white, or green .... (2)

2. Ocrea with marginal ciliate hairs 1–12 mm long .... (3)
2′ Ocrea without marginal cilia or with hairs 1 mm long or less .... (5)

3. Flowers not glandular-punctate .... ***P. maculosa***
3′ Flowers glandular-punctate .... (4)

4. Achenes minutely granular, dull; cleistogamous flowers of sessile, axillary inflorescences sometimes enclosed by ocreae .... ***P. hydropiper***
4′ Achenes smooth, shiny; inflorescences not enclosed by ocreae .... ***P. punctata***

5. Peduncle with yellowish glands; perianth mostly greenish white, with midvein of outer tepal anchor-shaped; inflorescence 3–8 cm long, arching or nodding .... ***P. lapathifolia***
5′ Peduncle with reddish purple glands; perianth mostly pink, with midvein of outer tepal not anchor-shaped; inflorescence 0.5–5 cm long, usually erect .... ***P. pensylvanica***

Persicaria amphibia (L.) Gray (growing in water) Water smartweed. Plants perennial, floating to emergent, aquatic to terrestrial herbs, up to 30 dm tall in aquatic settings; rhizomes horizontal. **STEMS** prostrate to ascending or erect, often floating, glabrous or strigose to hirsute. **LEAVES** cauline; ocrea 0.5–5 cm long, with a smooth or ciliate margin bearing hairs 0.5–4.5 mm long, glabrous to pubescent or hirsute, not enclosing sessile, axillary inflorescences; petiole (0) 0.1–3 (7) cm long, glabrous or pubescent to hirsute; blade without a triangular or lunate blotch adaxially, broadly lanceolate to elliptic or narrowly oblong, (2) 4–15 (20) cm long, 1–6 (8) cm wide, the floating blades broader and longer than nonfloating ones, glabrous or slightly strigose, often coriaceous, minutely punctate; margin entire. **INFL** terminal, rarely axillary, ascending to erect, (1) 2–5 cm long, 0.8–1.5 (2) cm wide; peduncle 1–5 cm long, glabrous or variously pubescent, often stipitate-glandular. **FLOWERS** bisexual or functionally unisexual, 1–3 (4) per ocreate fascicle, heterostylous; pedicel ascending, 0.5–1.5 mm long; perianth dark pink to red, rarely white, 4–6 mm, glabrous, slightly accrescent; tepals 5, connate 1/3 their length, obovate to elliptic; stamens 5, included or exserted; anthers pink or red; styles 2, connate 1/2–2/3 their length. **ACHENES** brown, biconvex, 2–3 mm long, shiny or dull, smooth or minutely granular. *n* = 33, 48, 66. [*Polygonum amphibium* L.; *P. hartwrightii* A. Gray; *Persicaria natans* (Michx.) Eaton; *Pe. hartwrightii* (A. Gray) Greene]. Mostly aquatic in lakes, ponds, and streams, occasionally emergent in shallow waters or terrestrial in wet sites. ARIZ: Apa; COLO: Arc, Hin, LPl, Mon, SJn; NMEX: RAr, SJn. 1370–2975 m (4500–9760′). Flowering: Jun–Aug. Labrador to Alaska, south to Mexico; South America, Eurasia, Africa. Particularly robust, emergent plants with lance-ovate leaves are sometimes recognized as the taxonomically dubious var. *emersa* (Michx.) J. C. Hickman [*P. coccinea* (Muhl. ex Willd.) Greene].

Persicaria hydropiper (L.) Spach (water pepper, a classical name of unknown meaning) Mild water-pepper smartweed. Plants annual, terrestrial herbs, (2) 4–8 (10) dm tall; rhizomes and stolons absent. **STEMS** decumbent, ascending to erect, glabrous, glandular-punctate. **LEAVES** cauline; ocrea (8) 10–15 mm long, with a ciliate margin bearing hairs 1–4 mm long, glabrous, or hispidulous, sometimes strigose, sometimes enclosing axillary inflorescences of mostly cleistogamous flowers; petiole (0) 0.1–0.8 cm long, glandular-punctate; blade without a triangular or lunate blotch adaxially, lanceolate to narrowly rhombic, (1.5) 4–10 (15) cm long, 0.4–2.5 cm wide, reduced distally, usually densely brownish and often with ciliate margins, glabrous or slightly scabrous on midvein. **INFL** terminal and axillary, erect or nodding, 3–18 cm long, 0.5–1 cm wide; peduncle (0) 1–5 cm long, glabrous, glandular-punctate. **FLOWERS** bisexual, 1–3 (5) per ocreate fascicle, homostylous; pedicel ascending, 1–3 mm long; perianth greenish proximally, white or pink distally, 2–3.5 mm long, glandular-punctate, scarcely accrescent; tepals (4) 5, connate 1/3 their length, obovate; stamens 6 (8), included; anthers pink to red; styles 2–3, connate proximally. **ACHENES** dark brown to brownish black, biconvex or trigonous, 2–3 mm long, dull, minutely granular. *n* = 10. [*Polygonum hydropiper* L.]. Weedy along moist ditches, streams, and lakesides. COLO: Mon, SJn; NMEX: SJn. 1525–1675 m (5000–5500′). Flowering: Jun–Aug. Introduced and widely naturalized from Newfoundland to Alaska, south to Mexico and Central America; native of Europe but widely introduced elsewhere in Asia, Africa, Australia, and on many Pacific islands.

Persicaria lapathifolia (L.) Gray (leaves like a sorrel) Dock-leaf smartweed. Plants annual, terrestrial herbs, (0.5) 1–10 dm tall; rhizomes and stolons absent. **STEMS** ascending to erect, glabrous, rarely appressed-pubescent distally, occasionally glandular-punctate or stipitate-glandular. **LEAVES** cauline; ocrea 4–25 (35) mm long, with a smooth or ciliate margin bearing hairs 1 mm long, glabrous, rarely strigose, not enclosing sessile, axillary inflorescences; petiole 0.1–1.6 cm long, glabrous or more often hispidulous to strigose; blade sometimes with a triangular or lunate blotch adaxially, narrowly to broadly lanceolate, (2) 4–12 (22) cm long, (0.3) 0.5–4 (6) cm wide, glabrous or tomentose abaxially, glandular-punctate adaxially, occasionally tomentose on both surfaces, strigose on main veins. **INFL** terminal and occasionally axillary, mostly arching or nodding at maturity, 3–8 cm long, 0.5–1.2 cm wide; peduncle 2–2.5 mm long, stipitate-glandular with yellowish glands. **FLOWERS** bisexual, 4–14 per ocreate fascicle, homostylous; pedicel ascending, 0.5–2.5 mm long; perianth greenish white to pink, 2.5–3 mm long, glabrous, glandular-punctate only on tube and inner tepals, scarcely accrescent; tepals 4 (5), connate 1/4–1/3 their length, obovate to elliptic; stamens (5) 6, included; anthers pink or red; styles 2 (3), connate proximally. **ACHENES** brown to black, discoid, rarely trigonous, 1.5–3 mm long, shiny or dull, smooth. *n* = 11. [*Polygonum lapathifolium* L.; *P. lapathifolium* var. *salicifolium* Sibth.; *P. nodosum* Pers.]. Wet habitats along streams, lakesides, and in wet meadows where occasionally weedy. ARIZ: Apa; COLO: Arc, LPl, Mon; NMEX: McK, RAr, SJn; UTAH. 1525–2285 m (5000–7500′). Flowering: Jun–Aug. Throughout most of North America from Alaska to Greenland, to South America, Eurasia, Africa, and on scattered Pacific islands where introduced. The primary vein of each outer tepal is prominently bifurcate distally resulting in an anchor-shaped midrib. Plants with varying degrees of tomentum on the leaves have been segregated as varieties; none is accepted here. The characteristic nodding inflorescence is seen only in late anthesis and in fruit. Ramah Navajo prepared a cold infusion as a ceremonial chant lotion. Zuni used the plant as an emetic and cathartic drug.

Persicaria maculosa Gray, nom. cons. (spotted, as to leaf blade) Spotted lady's-thumb. Plants annual, terrestrial herbs, (0.5) 1–7 (13) dm tall; rhizomes and stolons absent. **STEMS** procumbent to ascending or erect, glabrous or appressed-pubescent. **LEAVES** cauline; ocrea 4–10 (15) mm long, with a ciliate margin bearing hairs 1–3.5 (5) mm long, glabrous or strigose, not enclosing sessile, axillary inflorescences; petiole 0.2–2 cm long, glabrous or appressed-pubescent, occasionally strigose; blade often with a triangular or lunate blotch adaxially, lanceolate to narrowly ovate, (1) 5–10 (18) cm long, (0.2) 0.5–2.5 (4) cm wide, glabrous or appressed-pubescent, occasionally strigose, sometimes glandular-punctate abaxially. **INFL** terminal and axillary, erect, 1–4.5 (6) cm long, 0.7–1.2 cm wide; peduncle 1–5 cm long, glabrous or glandular-pubescent. **FLOWERS** bisexual, 4–14 per ocreate fascicle, homostylous; pedicel ascending, 1–2.5 (3) mm long; perianth greenish white proximally, dark pink distally, sometimes entirely pink to reddish or even purplish, 2–3.5 mm long, glabrous, scarcely accrescent; tepals (4) 5, connate 1/3 their length, oblong to obovate; stamens 4–8, included; anthers yellow or pink; styles 2 (3), connate proximally. **ACHENES** dark brown to black, discoid, biconvex or rarely trigonous, 2–2.7 mm long, shiny, smooth. n = 11, 22. [*Polygonum persicaria* L.]. Weedy in moist, open, disturbed places. COLO: Arc, LPl, Min; NMEX: McK, SJn. 1525–2135 m (5000–7000′). Flowering: Jun–Aug. Introduced and naturalized in Greenland to Yukon south to Mexico; Eurasia where both native and naturalized, also introduced in Africa and on scattered Pacific islands. Our plants, as described here, are **var. *maculosa***.

Persicaria pensylvanica (L.) M. Gómez (of Pennsylvania) Pinkweed. Plants annual, terrestrial herbs, 1–20 dm tall; rhizomes and stolons absent. **STEMS** ascending to erect, glabrous or appressed-pubescent distally, often stipitate-glandular distally. **LEAVES** cauline; ocrea 5–20 mm long, with a smooth or ciliate margin bearing hairs 0.5 mm long, glabrous or appressed-pubescent, not enclosing sessile, axillary inflorescences; petiole 0.1–2 (3) cm long, glabrous or appressed-pubescent; blade sometimes with a triangular or lunate blotch adaxially, narrow to broadly lanceolate, 4–20 (23) cm long, (0.5) 1–5 cm wide, glabrous or appressed-pubescent, glandular-punctate at least abaxially and sometimes adaxially. **INFL** terminal and axillary, erect or rarely nodding, 0.5–5 cm long, 0.5–1.5 cm wide; peduncle 1–5.5 (7) cm long, glabrous or pubescent, often stipitate-glandular with reddish purple glands. **FLOWERS** bisexual, 2–14 per ocreate fascicle, homostylous; pedicel ascending, 1.5–4.5 mm long; perianth greenish white to dark pink, 2.5–5 mm long, glabrous, accrescent; tepals 5, connate 1/4–1/3 their length, obovate to elliptic; stamens 6–8, included; anthers pink to red; styles 2 (3), connate proximally. **ACHENES** dark brown to black, discoid or rarely trigonous, slightly indented on one side, 2–3.5 mm long, shiny, smooth. n = 44. [*Polygonum pensylvanicum* L.; *P. omissum* Greene]. Open, moist, disturbed places where occasionally weedy. COLO: LPl, Mon; NMEX: McK. 1525–2135 m (5000–7000′). Nova Scotia to Minnesota, south to Florida and Mexico; disjunct and introduced in Alaska, Ecuador, England, Spain, and probably elsewhere. Our plants consistently have stipitate-glandular peduncles.

Persicaria punctata (Elliott) Small (glandularly spotted) Dotted smartweed. Plants annual or perennial, terrestrial herbs, 1.5–12 dm tall; rhizomes slender; stolons absent. **STEMS** ascending to erect, glabrous, glandular-punctate; flowering stems branched. **LEAVES** mostly cauline; ocrea (4) 9–18 mm long, chartaceous, swollen proximally, with a ciliate margin bearing hairs 2–11 mm long, glabrous or strigose, glandular-punctate, not enclosing sessile, axillary inflorescences; petiole 0.1–1 cm long, glandular-punctate; blade without a triangular or lunate blotch adaxially, lanceolate to broadly lanceolate or nearly rhombic, 4–10 (15) cm long, 0.6–2.5 cm wide, glabrous or scabrous, glandular-punctate. **INFL** terminal, sometimes also axillary, erect, 5–20 cm long, 0.4–0.8 cm wide; peduncle 3–6 cm long, glabrous, glandular-punctate. **FLOWERS** bisexual, 2–6 per ocreate fascicle, homostylous; pedicel ascending, 1–4 mm long; perianth greenish proximally, white distally, rarely tinged with pink, 3–3.5 mm long, glandular-punctate, scarcely accrescent; tepals 5, connate 1/3 their length, obovate; stamens 6–8, included; anthers pink or red; styles 2–3, connate proximally. **ACHENES** black, usually trigonous, (1.8) 2–3 mm long, shiny, smooth. n = 22. [*Polygonum punctatum* Elliott]. Along lakes and streams, in marshes, and on floodplains. ARIZ: Nav. 1370–1675 m (4500–5500′). Flowering: Jun–Aug. British Columbia to Nova Scotia south throughout the United States to South America; disjunct in Hawaii. None of the numerous varieties proposed for this species can be justified taxonomically.

Polygonum L., nom. cons. Knotweed

Rodney Myatt

(Greek *poly*, many, + probably *gone*, seeds, rather than *gony*, knee) Plants annual, biennial, or perennial herbs (ours) or subshrubs; roots fibrous or woody; rhizomes and stolons absent. **STEMS** prostrate to erect, glabrous, smooth or sometimes minutely scabrous; flowering stems branched. **LEAVES** cauline, alternate, rarely opposite, petiolate or sessile; ocrea distally persistent but otherwise disintegrating into fibers or completely deciduous, usually hyaline, 2-lobed, white or silvery, chartaceous, glabrous; blade linear to nearly round, entire. **INFL** axillary or axillary and terminal, spicate or as solitary flowers; peduncle absent. **FLOWERS** bisexual, 1–7 (10) per ocreate fascicle, not stipitate proximally; pedicels present or absent; perianth nonaccrescent, white or greenish white to pink, glabrous; tepals 5, petaloid or sepaloid, connate proximally or up to 3/4 their length, monomorphic or rarely dimorphic; stamens 3–8, occasionally some reduced to staminodes; filament distinct, free or adnate to perianth; styles (2) 3, spreading; stigmas capitate. **ACHENES** lenticular or trigonous, light brown to black, not winged, glabrous; embryo curved. $x = 10$ (Costea, Tardif, and Hinds 2005). About 65 species found nearly worldwide including North America north of Mexico (33 species), of which many are introduced and weedy. As noted in the key, the introduced and aggressive weed *P. argyrocoleon* may be found in our area.

1. Stems generally sharply 4-angled especially near nodes, ribs obscure or absent; petiole articulated to ocrea; leaf venation parallel, secondary veins obscure; anthers pink to purple; achenes of one size and forming during summer; natives in natural habitats (sect. *Duravia* S. Watson) ....(2)
1′ Stems more or less rounded, 8–16 ribs on stem somewhat elevated and distinct; petiole articulated to proximal part of ocrea; leaf venation pinnate, secondary veins obvious; anthers whitish yellow; achenes of two sizes, summer achenes brown, smooth or tuberculate, fall achenes olivaceous, 2–5 times longer, and smooth; native or adventive weeds of disturbed sites (sect. *Polygonum*) ....(5)

2. Achenes 1.3–1.7 mm long, yellow to greenish brown; stamens 3 .... ***P. kelloggii***
2′ Achenes 1.8–4.5 mm long, black; stamens (3) 8 ....(3)

3. Leaves ovate to elliptic or nearly round, not reduced distally .... ***P. minimum***
3′ Leaves linear to oblanceolate, reduced distally ....(4)

4. Pedicels reflexed; achenes 3–4.5 mm long; inflorescences spicate .... ***P. douglasii***
4′ Pedicels erect; achenes 2.5–3 mm long; inflorescences racemose .... ***P. sawatchense***

5. Flowers axillary throughout stem; stems generally prostrate to ascending, forming loose to rather compact mats; leaves flat, with lateral veins more or less distinct .... ***P. aviculare***
5′ Flowers mostly in upper 1/2 of stems; inflorescence appearing somewhat terminal; leaves reduced upward, bractlike, often absent within the inflorescence; stems mostly erect ....(6)

6. Tepal margins pink; achenes 1.3–2.3 mm long; pedicels 1–2 mm long, hidden by ocrea; stamens 8; introduced and often weedy, to be expected in our area .... *P. argyrocoleon* Steud. ex Kunze
6′ Tepal margins greenish yellow or yellow, occasionally white or pink; achenes (2.3) 2.5–3.5 mm long; pedicels 2.5–6 mm long, exserted beyond ocrea; stamens 3–6; native, not weedy .... ***P. ramosissimum***

Polygonum aviculare L. (small bird) Yard knotweed. Plants annual, homophyllous. **STEMS** prostrate to ascending, generally forming mats, (0.5) 1–5 (10) dm long, glabrous, several branches from base, round with evident ribs. **LEAVES** articulated to proximal part of ocrea; ocrea 3–12 mm long, glabrous, distally disintegrating into straight fibers or completely deciduous; petiole 0.3–3 mm long; blades pinnately veined with obvious secondary veins, green, oblanceolate to narrowly elliptic or elliptic, (0.6) 0.8–2.5 (3.5) cm long, (0.15) 0.2–0.7 (1) cm wide, distally reduced and bractlike, often overtopping flowers. **INFL** axillary, uniformly distributed or aggregated at tips of branches. **FLOWERS** closed, 2–7 per ocreate fascicle; pedicel 1–2.5 mm long, hidden by ocrea; perianth green or reddish brown with white margins, (1.8) 2–3.5 (4) mm long; tepals not keeled, petaloid, oblong, connate about 1/2 their length, midvein unbranched; stamens 5–7; anthers whitish yellow. **ACHENES** slightly exserted beyond perianth, light to dark brown, lenticular or trigonous, 1.5–4 (summer) or 2–5 (fall) mm long, dull, usually coarsely striate or obscurely tuberculate. $n = 20, 30$. [*P. arenastrum* Boreau]. Fields, roadsides, waste places, and disturbed, open habits. ARIZ: Apa, Coc, Nav; COLO: Arc, Dol, Hin, LPl, Mon, SJn; NMEX: McK, RAr, San, SJn; UTAH. 1220–3050 m (4000–10000′). Flowering: Jul–Sep. Introduced and naturalized throughout North America and nearly worldwide. Our plants, as described here, are the naturalized weed **var. *depressum*** Meisn. (low, as to stature); native to Europe. Additional varieties are to be expected in our area. Ramah Navajo took a warm infusion of the entire plant for stomachache.

Polygonum douglasii Greene (David Douglas, 1798–1834, Scottish botanical explorer in North America) Douglas' knotweed. Plants annual, homophyllous. **STEMS** erect, 5–8 dm long, glabrous or sparsely minutely scabrous, simple or branched, 4-angled, not ribbed. **LEAVES** articulated directly to ocrea; ocrea 6–12 mm long, glabrous or minutely scabrous, distally lacerate; petiole 0.1–2 mm long; blade 1-veined, green to dark green, linear to oblanceolate or narrowly oblong, 1.5–5.5 cm long, 0.2–0.8 (–1.2) cm wide; margin revolute, smooth or minutely denticulate; apex acute to mucronate. **INFL** axillary and terminal, widely spaced. **FLOWERS** closed, 2–4 per ocreate fascicle; pedicel 2–6 mm long, mostly exserted beyond ocreae; perianth green or tannish with white or pink margins, 3–4.5 mm long; tepals not keeled, petaloid, oblong, connate 1/4 their length, midvein branched; stamens 8; anthers pink to purple. **ACHENES** enclosed by perianth, black, trigonous, 3–4 (–4.5) mm long, shiny or dull, smooth or minutely tuberculate. [*P. douglasii* var. *montanum* Small; *P. montanum* (Small) Greene]. Open sites along trails, meadows, and in forest clearings. ARIZ: Apa; COLO: Arc, Dol, Hin, LPl, Min, Mon, RGr, SJn; NMEX: McK, RAr, SJn; UTAH. 2135–2925 m (7000–9600′). Flowering: Jun–Sep. Yukon to Labrador south to California, Arizona, New Mexico, and Virginia.

Polygonum douglasii

Polygonum kelloggii Greene (Albert Kellogg, 1813–1887, founder of the California Academy of Sciences) Kellogg's knotweed. Plants annual, homophyllous. **STEMS** erect, 0.2–1.5 dm long, glabrous, branched, 4-angled, not ribbed. **LEAVES** articulated directly to ocrea; ocrea 4–8 mm long, glabrous, distally lacerate; petiole absent; blade inconspicuously 3-veined, green, linear, 0.7–4 cm long, 0.1–0.2 cm wide; margins flat or revolute, smooth; apex mucronate. **INFL** mostly terminal, crowded. **FLOWERS** closed, 1–3 per ocreate fascicle; pedicel (0) 0.1–2 mm long, hidden by ocreae; perianth white to pink or red, 1.5–2.5 mm long; tepals not keeled, petaloid, narrowly lanceolate, connate 1/3 their length, midvein unbranched; stamens 8; anthers pink to purple. **ACHENES** enclosed by perianth, light yellow to greenish brown, trigonous, 1.3–1.7 mm long, shiny, smooth or slightly reticulate. [*P. polygaloides* Meisn. subsp. *kelloggii* (Greene) J. C. Hickman]. Mountain meadows and dry montane slopes. COLO: Arc, Mon; NMEX: SJn; UTAH. 1830–2720 m (6005–8920′). Flowering: Jul–Sep. British Columbia to Saskatchewan south to California and New Mexico. Our plants, as described here, are **var. *kelloggii***, which occurs from British Columbia to Montana, south to California and New Mexico.

Polygonum minimum S. Watson (small, as to size of plant) Zigzag knotweed. Plants annual, homophyllous. **STEMS** prostrate to erect, 0.2–3 dm long, scabrous, simple or branched, 4-angled, not ribbed. **LEAVES** articulated directly to ocrea; ocrea 1–4 mm long, minutely scabrous, distally slightly lacerate; petiole 0.1–3 mm long; blade 1-veined, green, narrowly elliptic to ovate or nearly round, 0.6–2.5 cm long, 0.3–0.8 cm wide; margins flat, smooth, irregularly thickened or minutely denticulate; apex apiculate. **INFL** axillary, widely spaced, sometimes crowded near tips of branches. **FLOWERS** closed or somewhat opened, 1–3 per ocreate fascicle; pedicel 2–3 mm long, hidden by ocreae; perianth green with white or pink margins, 1.8–2.5 mm long; tepals not keeled, sepaloid, oblong, connate 1/4 their length, midvein unbranched; stamens 8; anthers pink to purple. **ACHENES** enclosed by perianth or tip just exserted, black, trigonous, 1.8–2.3 mm long, shiny, smooth. Subalpine to alpine flats and slopes, often on nearly barren soil. COLO: RGr, SJn. 3100–3250 m (10150–10600′). Flowering: Jun–Sep. Alaska, British Columbia, and western Alberta south through Idaho, Montana, and Wyoming to Nevada, Utah, and Colorado; disjunct in the Black Hills of South Dakota.

Polygonum ramosissimum Michx. (with many branches) Yellow-flowered knotweed. Plants annual, heterophyllous. **STEMS** ascending to erect, 3–20 dm long, glabrous, several branches from base, round with evident ribs. **LEAVES** articulated to proximal part of ocrea; ocrea 6–12 (15) mm long, glabrous, distally disintegrating into straight fibers; petiole 2–4 mm long; blade pinnately veined with obvious secondary veins, yellowish green, lanceolate to narrowly elliptic, rarely ovate, 3.5–7 cm long, 0.7–2 (3.5) cm wide, distally reduced and bractlike, occasionally overtopping flowers. **INFL** axillary and terminal, aggregated at tips of branches. **FLOWERS** closed, 2–5 per ocreate fascicle; pedicel 2.5–6 mm long, exserted beyond ocrea; perianth green with yellow or rarely white or pink margins, (2.5) 3–4 mm long; tepals not keeled, petaloid or sepaloid, elliptic to oblong, connate about 1/4 their length, midvein unbranched; stamens 3–6; anthers whitish yellow. **ACHENES** enclosed or exserted beyond perianth, dark brown, trigonous, 2.5–3.5 (summer) or 5–12 (fall) mm long, shiny or dull, smooth or obscurely tuberculate. $n = 30$. Disturbed places, occasionally on saline soils. COLO: Hin, LPl, Mon; NMEX: San, SJn; UTAH. 1530–2285 m (5025–7500′). Flowering: Jul–Sep. Widespread throughout much of North America north of Mexico. Our plants, as described here, are **var. *ramosissimum***. Ramah Navajo took a warm infusion of the entire plant for stomachache and as a "life medicine."

Polygonum sawatchense Small (Sawatch Mountains, Colorado) Sawatch knotweed. Plants annual, homophyllous. **STEMS** erect, 0.4–5 dm long, glabrous, simple or branched, 4-angled, not ribbed. **LEAVES** articulated directly to ocrea; ocrea 4–10 mm long, glabrous, distally fibrous; petiole 0.1–2 mm long; blade 1-veined, linear-lanceolate to oblanceolate or narrowly oblong, 1.5–4.5 cm long, 0.2–0.8 (1.2) cm wide; margin usually revolute, smooth; apex mucronate. **INFL** axillary or axillary and terminal, widely spaced. **FLOWERS** closed, (1) 2–4 per ocreate fascicle; perianth

1–4 mm long, enclosed in or sometimes exserted beyond ocreae; perianth greenish or infrequently reddish, with white or pink margins, (2.5) 3–3.5 mm long; tepals not keeled, petaloid or sepaloid, narrowly oblong to oblong, connate 1/4–1/3 their length, midvein mostly unbranched; stamens 3–8; anthers pink to purple. **ACHENES** enclosed by perianth, black, trigonous, 2.5–3.5 mm long, shiny, smooth. [*P. douglasii* Greene subsp. *johnstonii* (Munz) J. C. Hickman]. Open sites along trails and lakesides in montane forests. ARIZ: Apa, Nav; COLO: Arc, Dol, Min; NMEX: McK, RAr, SJn; UTAH. 1720–2745 m (5640–9000′). Flowering: Jun–Sep. British Columbia to Manitoba, south to Nebraska, California, Arizona, and New Mexico. Our plants, as described here, are **var. *sawatchense***, with a slightly constricted range from what is given above; they are closely related to and often confused with *P. douglasii*. Ramah Navajo used the plant as snuff for nose troubles.

Rumex L. Dock

Rodney Myatt

(classical Latin name for *Rumex*, probably from *rumo*, to suck, as in sucking leaves to relieve thirst) Plants synoecious or dioecious, occasionally polygamomonoecious, annual, biennial, or perennial herbs; roots a taproot or rootstock; caudex short or absent; rhizomes or stolons occasionally present. **STEMS** erect to ascending or decumbent to prostrate, glabrous or minutely papillose; flowering stems usually branched. **LEAVES** basal in a rosette and cauline or just cauline, alternate, petiolate at least on basal and lower cauline blades; ocrea persistent or partially deciduous, membranous; blade linear to elliptic or ovate to nearly orbicular, reduced distally. **INFL** terminal, occasionally terminal and axillary, paniculate; peduncle absent. **FLOWERS** bisexual or unisexual, (1) 4–30 per ocreate fascicle, stipitate proximally, pedicellate; perianth accrescent, usually winglike and distinctly veined in fruit, green, pinkish, or red, glabrous; tepals (5) 6, sepaloid, distinct, dimorphic; margin entire or erose to distinctly toothed; tubercle absent or 1–3, lanceolate to broadly ovate or rounded, about 1/3 width of tepals; stamens 6; filaments distinct, free; styles 3, spreading or reflexed, distinct; stigmas 3, fimbriate or plumose. **ACHENES** trigonous, tan or brown to black, unwinged or only weakly so, glabrous, dull or shiny; embryo straight. $x = 7, 8, 9, 10$ (Mosyakin 2005; Rechinger 1937; Sarkar 1958). Perhaps 200 species throughout North America north of Mexico (63 species) and most of the world. Many species are adventive weeds, and some are escaped garden plants. Correct identification depends on collections which include some portion of root or underground rhizome system, basal leaves if present, and mature fruits. The "pedicel" length given here includes both the pedicel itself and the elongated stipe.

1. Plants dioecious; leaf blade hastate, narrow; stem slender; pedicel articulated near base of tepals (subg. *Acetosella* (Meisn.) Rech. f.) .... ***R. acetosella***

1′ Plants synoecious, youngest flowers rarely seemingly staminate; leaf blade not hastate; stem generally stout; pedicel with or without articulation (subg. *Rumex*) .... (2)

2. Leaves cauline, not developing a basal rosette (sect. *Axillares*) .... ***R. triangulivalvis***

2′ Leaves basal and cauline, developing a basal rosette (sect. *Rumex*) .... (3)

3. Tubercles 1–3, usually enlarged and evident in fruit .... (4)

3′ Tubercles absent or not evident .... (6)

4. Inner tepal margin with few and obscure teeth .... ***R. crispus***

4′ Inner tepal margin with evident teeth, 1 mm or longer .... (5)

5. Inflorescence elongated, lax, and interrupted, glabrous; leaf blades narrowly elliptic to oblong or narrowly ovate .... ***R. dentatus***

5′ Inflorescence dense distally, interrupted only proximally, papillose; leaf blades lanceolate .... ***R. fueginus***

6. Inner tepals (8) 11–25 (30) mm wide; root tuberous .... ***R. hymenosepalus***

6′ Inner tepals 4–8 (11) mm wide; root not tuberous .... (7)

7. Plants with creeping rhizomes; inner tepals entire or minutely dentate near base; plants of subalpine habitats .... ***R. densiflorus***

7′ Plants with vertical rootstock; inner tepals entire or weakly erose; plants of often moist, shady, or disturbed habitats .... ***R. occidentalis***

Rumex acetosella L. (a classic generic name meaning little sorrel) Common sheep sorrel. Plants dioecious, perennial, 1–4 dm tall, glabrous; rootstock slender, vertical; rhizomes creeping. **STEMS** erect or ascending, several from base, often reddish. **LEAVES** with ocrea 0.3–0.6 cm long, brownish proximally, silvery proximally, lacerate; petiole (0.3) 1–4 cm long; blade hastate and usually lanceolate to narrowly obovate, infrequently linear to narrowly lanceolate, 2–6 cm long, 0.3–2 cm wide, glabrous; margin entire, flat or nearly so. **INFL** terminal, upper 1/2–2/3 of stem, usually lax, narrowly to broadly paniculate. **PEDICELS** articulated near base of tepals, filiform, 1–3 mm long. **FLOWERS** unisexual, staminate and pistillate on different plants, (3) 5–8 (10) in whorls, 1.2–2 mm long; inner tepals not or only slightly enlarged, equaling or slightly wider than achene, margins entire; tubercles absent. **ACHENES** brown to dark brown, 0.9–1.5 mm long. *n* = 7, 14, 21. [*Acetosella vulgaris* (W. D. J. Koch) Fourr.]. Weedy in moist habitats, often in disturbed places. ARIZ: Apa; COLO: Arc, Hin, LPl, Min, Mon, SJn; NMEX: RAr, SJn. 2395–3310 m (7855–10860′). Flowering: Apr–Sep. Introduced and now widespread throughout North America; both native and naturalized in Eurasia, introduced almost worldwide. Our plants are the hexaploids (*n* = 21) native to the Mediterranean region, which are sometimes distinguished as subsp. *pyrenaicus* (Pourr. ex Lapeyr.) Akeroyd [*R. acetosella* subsp. *angiocarpus* (Murb.) Murb.]. This phase may be distinguished by accrescent inner tepals which hold the achene, a feature of dubious taxonomic significance.

Rumex crispus L. (curly, as to leaf margins) Curly dock. Plants synoecious, biennial or more often perennial, 4–10 (15) dm tall, glabrous or slightly papillose; rootstock fusiform, vertical; rhizomes absent. **STEMS** erect, stout, branched distal to middle, greenish to reddish. **LEAVES** basal and cauline, developing a basal rosette; ocrea 1–7 cm long, usually deciduous, tan, membranous; petiole (3) 5–10 (15) cm long; blade linear-lanceolate to lanceolate, (6) 10–30 (35) cm long, (1) 2–6 cm wide, glabrous, sometimes slightly papillose on veins; base cuneate to truncate or weakly cordate; margin entire or nearly so, strongly undulate and crisped. **INFL** terminal, distal 1/2 of stem, dense to open with interrupted clusters of flowers proximally, paniculate. **PEDICELS** articulated proximal 1/3, filiform, enlarged distally, (3) 4–8 mm long. **FLOWERS** bisexual, 10–25 in whorls, 3.5–6 mm long, 3–6 mm wide at maturity; inner tepals broadly ovate to nearly orbicular; margins entire or nearly so to weakly erose; tubercles (1–2) 3, 1 larger than the others, 1/3 or less the width of tepal, verrucose. **ACHENES** reddish brown, 2–3 mm long. *n* = 30. Mainly in disturbed, low-lying, wet meadows, pastures, or riparian borders. ARIZ: Apa; COLO: Arc, Dol, Hin, LPl, Min, Mon, SJn; NMEX: McK, RAr, SJn; UTAH. 1615–3050 m (5300–10000′). Introduced and widespread in North America; Eurasia where both native and naturalized, now established throughout most of the world. A stimulant used for fainting by the Navajo. Used as a ceremonial emetic by the Ramah Navajo.

Rumex densiflorus Osterh. (densely flowered) Dense-flower dock. Plants synoecious, perennial, 5–10 dm tall, glabrous or faintly papillose; rootstock absent; rhizomes creeping, horizontal. **STEMS** erect, branched distally above middle, greenish or reddish. **LEAVES** basal and cauline, developing a basal rosette; ocrea 0.5–4 cm long, deciduous or partially persistent, tannish, membranous; petiole (1) 5–20 cm long; blade narrowly oblong to oblong, (20) 30–40 (50) cm long, 8–12 (18) cm wide, glabrous, papillose on veins; base cuneate to truncate; margins entire, flat. **INFL** terminal, distal 1/2 of stem, dense, narrowly paniculate. **PEDICELS** indistinctly articulated in proximal 1/3–1/2, filiform, not enlarged, (3) 6–16 (17) mm long. **FLOWERS** bisexual, 10–20 in whorls, 4–7 mm long, 4.5–6 mm wide at maturity; inner tepals broadly triangular to subcordate; margins entire, erose, or proximally denticulate; tubercles absent. **ACHENES** brown to reddish brown, 2.5–4 (4.5) mm long. *n* = 60. [*R. orthoneurus* Rech. f.; *R. pycnanthus* Rech. f.]. Along stream banks in subalpine and alpine communities. COLO: Arc, Hin, LPl, Mon. 2745–3200 m (9000–10500′). Nevada, Utah, and Colorado south to Sonora, Mexico.

Rumex dentatus L. (with teeth) Toothed dock. Plants synoecious, perennial, 2–7 (8) dm tall, glabrous or faintly papillose on leaf veins; rootstock fusiform, vertical; rhizomes absent. **STEMS** erect, often flexuous in inflorescence, branched, greenish or reddish. **LEAVES** basal and cauline, developing a basal rosette; ocrea 0.2–1 cm long, deciduous or partially persistent, tannish to reddish brown, membranous; petiole 5–20 cm long; blade usually narrowly elliptic to oblong or narrowly ovate, 3–8 (12) cm long, 2–5 cm wide, glabrous or faintly papillose at least on veins in some; base truncate to slightly cordate; margins entire, flat or weakly undulate. **INFL** terminal, distal 1/2 of stem, lax and interrupted, occasionally more dense distally, broadly paniculate. **PEDICELS** articulated proximal 1/3, filiform, enlarged distally, 2–5 mm long. **FLOWERS** bisexual, 10–20 in whorls, 3–6 mm long, 2–5 mm wide at maturity; inner tepals broadly triangular to deltoid; margins dentate with 2–4 (5), sharp teeth 1.5–4 (5) mm long on each side of the tepal; tubercles (1) 3, thick, about 1/3 width of tepal, distinctly verrucose. **ACHENES** reddish brown, 2–3 mm long. *n* = 20. Waste places, roadsides, and shores. ARIZ: Nav; NMEX: SJn; UTAH. 1125–1775 m (3700–5820′). Flowering: Jul–Aug. Introduced and widely naturalized in scattered locations mainly in the West; native of Europe, Africa, and tropical Asia; introduced elsewhere. Exactly which European variety we have has not been determined as the taxa there are unsettled.

Rumex fueginus Phil. (of the Fuegina region near Tierra del Fuego, Chile) American golden dock. Plants annual, rarely biennial, 1.5–6 dm tall, usually papillose especially in inflorescence; rootstock fusiform, vertical; rhizomes absent.

STEMS erect, rarely decumbent, branched distal 2/3, occasionally to near base. **LEAVES** basal and cauline, developing a basal rosette; ocrea 0.3–2 cm long, deciduous, occasionally partially persistent proximally, tannish, membranous; petiole 0–5 cm long, papillose; blade narrowly lanceolate to lanceolate, 0.5–25 (30) cm long, (1) 1.5–3 (4) cm wide, papillose at least abaxially; base nearly truncate; margins entire or nearly so, undulate and minutely crisped. **INFL** terminal, distal 1/2 of stem, dense distally, interrupted proximally, broadly paniculate. **PEDICELS** articulated at base or proximal 1/3, filiform, not enlarged, 3–7 (9) mm long. **FLOWERS** bisexual, 15–30 (40) in whorls, 1.5–2 mm long, 2–4.5 mm wide at maturity; inner tepals narrowly to broadly triangular; margins entire or subentire, more often dentate with 2–3 teeth 1–3 mm long on each side of tepal; tubercles 3, thin, about 1/4 width of tepal or less. **ACHENES** light brown, 1–1.5 mm long. *n* = 20. [*R. maritimus* of United States authors, not L.; *R. maritimus* var. *fueginus* (Phil.) Dusén]. Riparian borders, shorelines, and marginally wet habitats. NMEX: McK, RAr, SJn; UTAH. 1165– 2730 m (3830–8960′). Widespread in North and South America; introduced and uncommon in Europe. Some of our plants have entire or subentire inner tepals; these have been separated as **var. *athrix*** (H. St. John) Rech. f. Infusion of plant taken for blood, and seeds made into a mush and used for food by the Kayenta Navajo.

Rumex hymenosepalus Torr. (membranous tepals) Sand dock. Plants perennial, 2.5–10 dm tall, glabrous or faintly papillose; rootstock tuberous, vertical; rhizomes thick, horizontal. **STEMS** usually erect, branched above middle, greenish or reddish. **LEAVES** basal and cauline, developing a basal rosette; ocrea 0.5–4 (5) cm long, persistent, white or silvery white to tannish, membranous; petiole (1) 3–7 cm long; blade broadly lanceolate to elliptic or oblong, (5) 8–30 cm long, 2–8 (12) cm wide, glabrous or faintly papillose; base cuneate; margin entire, flat or faintly crisped. **INFL** terminal, distal 1/2 of stem, dense, narrowly paniculate. **PEDICELS** articulated proximal 1/3–1/2, filiform, not enlarged, 5–15 (20) mm long. **FLOWERS** bisexual, 5–20 in whorls, 10–25 mm long, 8–30 mm wide at maturity; inner tepals broadly cordate to orbiculate-cordate; margins entire; tubercles absent. **ACHENES** brown to reddish brown, 4–8 mm long. *n* = 20. [*Rumex hymenosepalus* var. *salinus* (A. Nelson) Rech. f.]. Sandy to rocky flats and low hills, occasionally in alkaline soil. ARIZ: Apa, Coc, Nav; COLO: LPl, Mon; NMEX: McK, RAr, San, SJn; UTAH. 1500–2135 m (4920–7000′). Flowering: Apr–Aug. California, Montana, and Wyoming south to Texas and Mexico. Used by the Hopi as a cold remedy and as a dye. Used by the Navajo for medicine and the brown roots were boiled to make dye for yarn. Used as a "life medicine" by the Kayenta Navajo and a ceremonial medicine by the Ramah Navajo.

Rumex occidentalis S. Watson (from the West) Western dock. Plants synoecious, perennial, 5–10 (14) dm tall, glabrous or slightly papillose; rootstock fusiform, vertical or oblique; rhizomes absent. **STEMS** erect, stout, unbranched below inflorescence, greenish to reddish. **LEAVES** basal and cauline, developing a basal rosette; ocrea 0.5–4 (5) cm long, persistent or partially deciduous, white or dark tan, membranous; petiole (1) 3–10 cm long; blade narrowly lanceolate to broadly lanceolate or ovate-triangular, 10–35 cm long, 5–12 cm wide, glabrous, sometimes slightly papillose on veins; base cordate to truncate or rounded; margin entire, undulate or somewhat crisped. **INFL** terminal, distal 2/3 of stem, dense or open with interrupted cluster of flowers, paniculate. **PEDICELS** obscurely articulated proximal 1/3, filiform, not enlarged distally, 5–13 (17) mm long. **FLOWERS** bisexual, 12–25 in a cluster, 5–10 (12) mm long, 5–8 (11) mm wide at maturity; inner tepals much wider and longer than achene, ovate to triangular or orbiculate; margins entire or nearly so to slightly erose; tubercle absent. **ACHENES** reddish brown, 3–4.5 mm long. *n* = 60. [*R. aquaticus* L. var. *fenestratus* of southern Rocky Mountains references, not (Greene) Dorn; *R. aquaticus* subsp. *occidentalis* (S. Watson) Hultén]. Wet meadows and bogs or shallow water in montane forests. COLO: Arc, Hin, Min, Mon, SJn; NMEX: McK, RAr, McK; UTAH. 1825–3660 m (6000–12000′). Flowering: Apr–Sep. Alaska to Labrador south to California, Arizona, New Mexico, South Dakota, and Iowa.

Rumex triangulivalvis (Danser) Rech. f. (triangle-shaped tepals) White willow dock. Plants synoecious, perennial, (3) 4–10 dm tall, glabrous; rootstock stout, vertical; rhizomes creeping. **STEMS** ascending to erect, 1 to several axillary shoots from near base, greenish. **LEAVES** cauline, not developing a basal rosette; ocrea 0.3–2 cm long, persistent or partially deciduous, tan, membranous; petiole 0.5–3 cm, often indistinct; blade narrowly lanceolate, 6–17 cm long, 1–4 (5) cm wide, glabrous; base cuneate; margins entire, flat or undulate. **INFL** terminal and axillary, terminal 1/5–1/3 of stem, rather dense or somewhat interrupted, narrowly to broadly paniculate. **PEDICELS** articulated proximal 2/3, filiform, slightly thickened distally, 4–8 mm long. **FLOWERS** bisexual, 10–25 in whorls, (2) 2.5–3.5 (4) mm long, (2) 2.5–3 (3.5) mm wide at maturity; inner tepals distinctly enlarged, wider and longer than achene, broadly triangular; margins entire or faintly erose near base; tubercles (1) 3, thick, evident, lance-ovate, about 1/3 width of tepal, glabrous or minutely verrucose. **ACHENES** brown to reddish brown, 1.7–2.2 mm long. *n* = 10. [*R. salicifolius* Weinm. var. *triangulivalvis* (Danser) J. C. Hickman; *R. mexicanus* misapplied, not Meisn.]. Occasional along valleys and streams from desert scrub to montane conifer and aspen communities, sometimes weedy in waste places. COLO: Mon, SJn;

NMEX: SJn; UTAH. 1220–2440 m (4000–8000′). Thoughout much of North America (except the southeastern states) and Europe.

Stenogonum Nutt. Two-whorl Buckwheat

James L. Reveal

(Greek *stenos*, narrow, + *gonos*, seed, as to the slender achenes) Plants annual herbs; root a slender taproot. **STEMS** numerous, slender, glabrous, minutely strigose or glandular; caudex absent; aerial flowering stems spreading to somewhat erect, slender, solid, not fistulose, arising direct from a root. **LEAVES** basal or basal and cauline, alternate, estipulate, petiolate; blade narrowly lanceolate to oblanceolate or spatulate to orbicular, green and glabrous to sparsely and minutely strigose or glandular. **INFL** cymose; branches spreading, open to diffuse, mostly dichotomously branched, glabrous, minutely strigose or glandular; bracts scalelike, 3, connate proximally, proximal to node, not awned. **PEDUNCLES** ascending, straight or flexed, slender to capillary, glabrous. **INVOLUCRES** reduced to a series of 2 whorls, each of 3 lanceolate, foliaceous, bractlike lobes; lobes not awn-tipped. **FLOWERS** bisexual, (6) 9–15 per involucral cluster, not stipitate proximally, pedicellate; perianth nonaccresent, yellow, pilose; tepals 6, petaloid, lanceolate, monomorphic, entire apically; stamens 9; filaments adnate proximally, glabrous; anthers yellow, oval; styles 3, erect to spreading, distinct; stigmas capitate. **ACHENES** trigonous, light brown, not winged, glabrous; embryo curved. x = 20 (Reveal and Ertter 1977; Reveal 2005d). Two species, mainly on the Colorado Plateau.

1. Leaves basal; peduncle flexed; plants erect, sparsely glandular .... ***S. flexum***
1′ Leaves basal and cauline; peduncle straight; plants spreading, glabrous .... ***S. salsuginosum***

Stenogonum flexum (M. E. Jones) Reveal & J. T. Howell (bent, as to the peduncle) Bent two-whorl buckwheat. Plants mostly erect, (0.5) 1–3 dm tall. **LEAVES** basal; petioles 1–4 cm long; blade mostly orbicular, 0.5–2 cm long and wide, minutely strigose when young, glabrous at maturity. **INFL** erect or slightly spreading, 0.5–2.5 dm long; branches glandular at nodes and lower internode, green to yellow-green; bracts 0.5–2 mm long, glabrous or sparsely glandular. **PEDUNCLES** erect to spreading, flexed, filiform, 1–3 cm long, glandular to middle. **INVOLUCRAL LOBES** 2–3 mm long, 2–4 mm wide, glabrous or sparsely glandular. **FLOWERS** yellow to reddish yellow, 1.5–3.5 mm long, pilose; tepals lanceolate; stamens 1.5–2 mm long. **ACHENES** 2–2.5 mm long. n = 20. [*Eriogonum flexum* M. E. Jones]. Clay hills and flats primarily in saltbush communities. ARIZ: Apa; COLO: Mon; NMEX: SJn; UTAH. 1350–2000 m (4430–6560′). Flowering: Apr–Sep. Primarily Colorado Plateau of eastern Utah, western Colorado, north-central Arizona, and northwestern New Mexico.

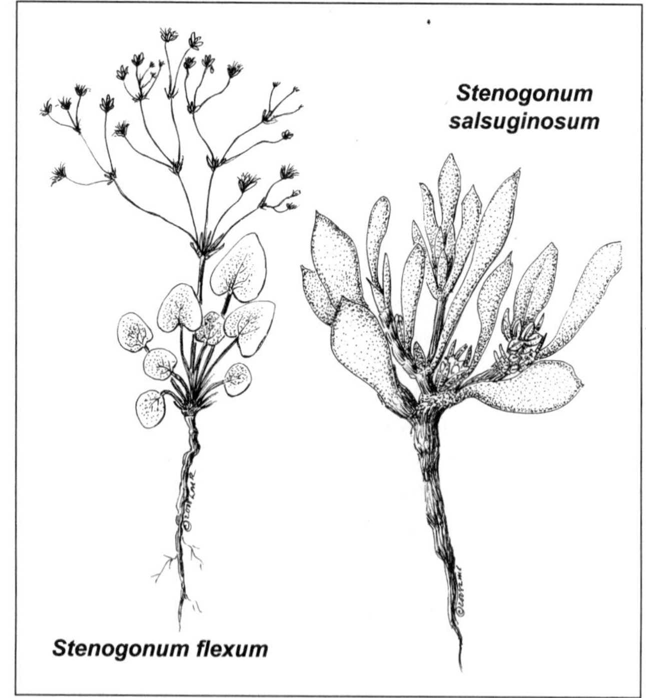
Stenogonum salsuginosum
Stenogonum flexum

Stenogonum salsuginosum Nutt. (of salty places, as to habitat) Smooth two-whorl buckwheat. Plants mostly spreading, 0.5–2 dm tall. **LEAVES** basal and cauline, glabrous; petiole 0.5–2 cm long on basal blade, otherwise absent; basal blade spatulate, (1) 2–4 cm long, (0.5) 1–2.5 cm wide; cauline blade narrowly lanceolate to oblanceolate, 0.5–4.5 cm long, 0.2–1 cm wide. **INFL** spreading, 0.5–2.5 (3) dm long; branches glabrous, green; bracts 0.5–4 mm long, glabrous. **PEDUNCLES** erect to spreading, straight, slender to filiform, 0–4 cm long, glabrous, often absent distally. **INVOLUCRAL LOBES** 2–8 mm long, 2–3 mm wide, glabrous. **FLOWERS** yellow, 1.5–3 mm, pilose; tepals lanceolate; stamens 1.5–2 mm long. **ACHENES** 2–2.5 mm long. n = 20. [*Eriogonum salsuginosum* (Nutt.) Hook.]. Clay hills and flats primarily in saltbush communities. ARIZ: Apa, Nav; COLO: Mon; NMEX: McK, San, SJn; UTAH. 1300–2150 m (4265–7055′). Flowering: Apr–Sep. Primarily Colorado Plateau of southwestern Wyoming south through eastern Utah and western Colorado to northern Arizona and northwestern New Mexico.

PORTULACACEAE Juss. PURSLANE FAMILY

David J. Ferguson

Ours annual or perennial herbs; roots fibrous to tuberous. **STEMS** reduced to well developed, succulent, base sometimes suffrutescent and persistent. **LEAVES** opposite, subopposite, or alternate and sometimes secund, sometimes rosulate or subrosulate; fleshy to succulent, mostly entire, mostly glabrous; stipules none (*Portulaca* with nodal or axillary hairs sometimes regarded as stipular). **INFL** axillary or terminal, racemose, cymose, or with flowers solitary. **FLOWERS** mostly actinomorphic or slightly irregular; sepals 2 (–9); petals usually free, sometimes fused at extreme base, 5 (2–20); stamens 1–many, often opposite to petals, sometimes basally adnate to petals or each other; gynoecium 3 (2–9)-carpelled; ovary 1, superior (perigynous in *Portulaca*), 1-locular at maturity, placentation basal or free-central, ovules 1–many; style present, simple; stigma lobes 1–9. **FRUIT** a capsule, valvate or circumscissal. **SEEDS** smooth or sculptured, with or without investing testicular pellicle, strophioles, or elaiosomes. The family Portulacaceae as traditionally defined (excluding closely related satellite groups Basellaceae, Cactaceae, and Didiereaceae) consists of two distinct groups based on morphological and molecular evidence (Applequist and Wallace 2001) and might best be considered as two distinct families. Within the study area, *Portulaca* belongs to one group and the remaining genera to the other. *Montia chamissoi* (Ledeb. ex Spreng.) Greene, "miners-lettuce," has been found near the study area and is to be expected.

1. Leaf axils usually with hairs or bristles; ovary 1/2 inferior to inferior .... ***Portulaca***
1′ Lacking axillary hairs or bristles; ovary superior .... (2)

2. Capsule circumscissile, valves longitudinally dehiscent from the base .... ***Lewisia***
2′ Capsule 2- or 3-valved, dehiscing from the apex .... (3)

3. Sepals mostly deciduous; inflorescence not secund; leaves articulate at the base .... ***Phemeranthus***
3′ Sepals persistent; inflorescence somewhat to markedly secund; leaves not articulate at base .... (4)

4. Cauline leaves 2 (–3), distinct or partially or completely connate; ovules 3 or 6 .... ***Claytonia***
4′ Cauline leaves more than 2, distinct; ovules 3 .... *Montia chamissoi* (Ledeb. ex Spreng.) Greene

Claytonia L. Spring-beauty

(for John Clayton, British botanist, 1686–1773) Annual or perennial, succulent or semisucculent, glabrous herbs; roots various, mostly with thick taproot or bearing rounded tubers. **STEMS** reduced, thick, not well differentiated from rootstock. **LEAVES** basal, few to several in rosettes or clusters, petiolate, blade flat, mostly somewhat succulent. **INFL** scapose, racemose, or cymose; lowermost bracts paired, leaflike; uppermost bracts paired or more often single, small, mostly scarious. **FLOWERS** relatively showy; sepals persistent, foliaceous, unequal; petals 5; stamens 5, adnate to petal bases; ovary oblong to globose, ovules 3 or 6; style 1, often 3-lobed; stigma lobes 3. **FRUIT** a 3-valved capsule, longitudinally dehiscent from apex, sometimes explosively, valves not deciduous, margins involute. **SEEDS** 3–6, black, mostly orbicular, shiny and smooth to tuberculate, with white elaiosome; seeds dispersed ballistically and by ants. Approximately 26 species in North America and Asia; adventive elsewhere.

1. Numerous leaves crowded in a basal rosette from a thick, woody taproot, leaves thick, midrib often obscure, petiole short and ill-defined; inflorescences several, spreading laterally; alpine .... ***C. megarhiza***
1′ Few leaves, or 2 pairs on a scape, leaves relatively thin with ribs obvious, petiole prominent; 1 to few inflorescences; below alpine .... (2)

2. Foliaceous bracts lanceolate to broadly lanceolate, prominently 3-ribbed; petals retuse to emarginate .... ***C. lanceolata***
2′ Foliaceous bracts linear to narrowly lanceolate, 1-ribbed or indistinctly 3-ribbed; petals rounded to acute at apex .... ***C. rosea***

Claytonia lanceolata Pursh (lance-shaped, referring to the leaf) Western spring-beauty, lance-leaf spring-beauty. Herbaceous, succulent perennial, with globose, cormlike tubers 5–20 mm diameter. **LEAVES** in a basal cluster, 1–6, often absent on flowering tubers, 5–10 cm long; blade mostly shorter than petiole, linear to lanceolate, 2–6 × 0.6–2.5 cm. **INFL** 10–23 cm long, mostly 1–3 per tuber, racemose, 5–15-flowered, multibracteate; lower bracts leaflike, paired, sessile, ovate to narrowly lanceolate, 3–5-ribbed, 3–6 × 1–2 cm long; upper bract(s) unpaired, reduced to membranous ovate scales; pedicels 1–3 cm, recurved in fruit. **FLOWER** sepals elliptic to ovate, obtuse, 4–5 mm; petals white to pink or magenta, often striped, 10–12 mm long, ovate, retuse to emarginate; ovules 6. **FRUIT** broadly ovoid, ca. 4 mm long. **SEEDS** orbicular, ca. 2 mm long, shiny dark brown to black; elaiosome 1–2 mm. 2*n* = 12, 16, 24, 32, 44, 48, 52, 54, 74, about 90. Piñon-juniper, mountain brush, aspen, ponderosa pine, and Douglas-fir communities. COLO: Arc, Dol, Hin, LPl, Min, Mon, SMg; NMEX: RAr; UTAH. 1980–2590 m (6500–8500'). Flowering: May–Jun. Widespread in mountainous areas, primarily west of the Continental Divide and south from southern Canada to Colorado, Utah, Nevada, and California. Pressed herbarium material is often difficult to determine between *C. lanceolata* and *C. rosea*, especially if key features are not represented.

Claytonia megarhiza (A. Gray) Parry ex S. Watson **CP** (large root) Alpine spring-beauty, big-root spring-beauty, fell-fields claytonia. Herbaceous, succulent, rosulate perennial; rootstock stout, thick, roughly napiform, usually branching, semiwoody taproot. **LEAVES** in crowded basal rosette, succulent; petiole short, broad, winged; blade oblanceolate, obovate, rhomboid to spatulate, obtuse, 4–10 × 0.5–3 cm. **INFL** usually several, 5–15 (25) cm, spreading laterally to an ascending, corymbiform cyme; lower paired (rarely alternate) bracts sessile, unequal, oblanceolate to narrowly spatulate, 1–3 cm; upper bracts single, obtuse, subtending pedicels. **FLOWER** sepals 4–8 mm, unequal, ovate, broadly acute; petals white to rose-pink, 5–20 × 3–8 mm, obovate, slightly retuse, clawed; ovules 6. **FRUIT** ovoid, 4–6 mm. **SEEDS** 3–6, 2–2.5 mm long, rounded-oval, shiny, smooth, black; elaiosome 1 mm or less. 2*n* = 12, 16, 24, 32, 34, 36. Talus slopes in the alpine tundra. COLO: Arc, Con, Hin, LPl, Min, Mon, SJn. 3350–4115 m (11000–13500'). Flowering: Jun–Aug. Washington to Alberta, south to Nevada, Colorado, and New Mexico.

Claytonia rosea Rydb. (pink, for the flower color) Rose spring-beauty, Rocky Mountain spring-beauty. Herbaceous, succulent perennial, with globose, cormlike tubers 5–15 mm diameter. **LEAVES** (0) 1–6, in a basal cluster, petiole 5–10 cm long; blade about equal to petiole, mostly oblong, ovate, or spatulate, 1–7 × 0.4–2 cm, apex mostly obtuse. **INFL** 12–15 cm, racemose, mostly 5–10-flowered, multibracteate; lower bracts leaflike, paired, sessile, linear to linear-lanceolate, 1- or weakly 3-ribbed, 2–5 cm × 5–12 mm, apex acute to obtuse; upper bracts unpaired, reduced to membranous scales; pedicels 2–3 cm long, recurved in fruit. **FLOWERS** 8–14 mm diameter; sepals round-ovate, 3–5 mm long; petals pink to magenta, sometimes white or striped, obovate, rounded to acute, 8–10 mm; ovules 6. **FRUIT** ovoid, ca. 4 mm long. **SEEDS** orbicular, 2–3 mm long, shiny black; elaiosome 1–2 mm. 2*n* = 16. [*C. lanceolata* Pursh var. *rosea* (Rydb.) R. J. Davis]. Well-drained, shallow soils. Piñon-juniper, mountain brush, aspen, fir, lodepole pine, and spruce-fir communities. COLO: Arc, LPl, Mon; NMEX: McK, RAr, SJn; UTAH. 1830–2895 m (6000–9500'). Flowering: Apr–Jun. Scattered populations in Arizona, New Mexico, Utah, and Colorado; Mexico. *Claytonia rosea*, along with several other putative species, is sometimes considered a variant of *C. lanceolata*; however, based on morphological distinctness and lack of intermediacy among wild populations, it is maintained as distinct here.

Lewisia Pursh Bitterroot, Lewisia

(for Captain Meriwether Lewis, of the Lewis and Clark Expedition, 1774–1809) Herbaceous, succulent perennials; rootstock usually a thick and fleshy taproot, gradually ramified distally, napiform or globose. **STEMS** occasionally well developed, succulent, slender (not our species); or thick, short, inconspicuous, not well differentiated as separate from rootstock, often subterranean, and usually obscured by densely crowded leaves. **LEAVES** succulent, deciduous (ours) to evergreen, mostly basal or basal and cauline; basal leaves tightly crowded in rosettes or appearing caespitose, variable from nearly terete to flat and broad, from entire to toothed or lobed, from epetiolate to long-petiolate. **INFL** scapose, cymose, or occasionally racemose or single-flowered; bracts persistent, usually small, mostly in subequal pairs, sometimes single, margins entire, toothed, or glandular-toothed, herbaceous or scarious. **FLOWERS** pedicellate or sessile; sepals persistent, 2 (–9), equal or subequal when paired, margins entire, toothed, or glandular-toothed, herbaceous or scarious; petals 5–20, mostly persistent after drying; stamens 1–50, distinct to scarcely connate and/or adnate to petals basally; ovules 1–50; style branched; stigma lobes 2–8. **FRUIT** caplike, dehiscent by circumscissal split near base, mostly splitting acropetally by partial longitudinal slits. **SEEDS** 1–50, brown or black, smooth or minutely sculpted; no pellicle, elaiosome, or strophiole. Approximately 16 species in mountainous regions of the western United States and Canada. The number of species in *Lewisia* is debated due to morphological variability and differing species concepts. Natural hybridization may occur between some species. In cultivation related species hybridize easily, and several species and hybrids are cultivated by alpine/rockery plant enthusiasts.

1. Leaves nearly terete; uppermost bracts 4–8 in a whorl, closely subtending sepals; flowers usually over 3 cm in diameter, mostly 1 per inflorescence; sepals 4–9; petals 10–20 .... ***L. rediviva***
1' Leaves roughly plano-convex; bracts paired or single; flowers under 3 cm in diameter; sepals 2; petals 5–10 .... (2)

2. Flowers usually solitary, mostly 2 cm or more diameter, slightly zygomorphic; sepal margins mostly entire, sometimes obscurely or irregularly toothed, not glandular, apex acute to subacute; petals white, rarely pinkish .... ***L. nevadensis***

2′ Flowers usually multiple per inflorescence, mostly under 2 cm diameter, actinomorphic, sepal margins regularly toothed, usually glandular-toothed (rarely ± entire), apex truncate or sometimes rounded, obtuse, subacute, or apiculate; petals mostly pink, rarely white .... ***L. pygmaea***

Lewisia nevadensis (A. Gray) B. L. Rob. **CP** (of Nevada) Nevada bitterroot. Herbaceous, succulent perennial; rootstock napiform to short-fusiform. **STEMS** crowning and not obviously distinct from root, short, often subterranean, with leaves tightly crowded. **LEAVES** succulent, narrowly linear to linear-oblanceolate, dorsally flattened to slightly concave, gradually narrowed to broad petiole, 4–15 cm long, 2–6 mm wide, margins entire, apex obtuse to subacute; withering at or soon after anthesis. **INFL** with flowers solitary, rarely 2–3-flowered; bracts 2, opposite, linear-lanceolate, 6–18 mm long, margins entire, apex acute. **FLOWERS** slightly zygomorphic, usually appearing slightly laterally "pinched," 0.5–2 cm diameter; sepals 2, broadly ovate, 5–13 mm long, herbaceous at anthesis, margins entire or with few shallow, nonglandular teeth, apex acute to subacute; petals 5–10, white or rarely pinkish, elliptic to oblanceolate, 10–15 (-20) mm long; stamens 6–15; stigma lobes 3–6; pedicel 10–40 mm long. **FRUIT** 5–10 mm long, base persistent after dehiscence. **SEEDS** 20–50, 1.3 mm long, shiny, muricate. $2n = 56$. [*Calandrinia nevadensis* A. Gray; *Claytonia grayana* Kuntze; *Lewisia bernardina* Davidson; *L. pygmaea* (A. Gray) B. L. Rob. var. *nevadensis* (A. Gray) Fosberg; *Oreobroma nevadense* (A. Gray) Howell]. Near springs and wet, grassy slopes; upper piñon-juniper woodlands, ponderosa pine, and Douglas-fir communities. ARIZ: Apa; COLO: Arc, LPl, Mon; NMEX: McK, RAr, SJn; UTAH. 2135–2590 m (7000–8500′). Flowering: Apr–Aug. Washington to Idaho, south to California, Nevada, Utah, Colorado, Arizona, and New Mexico.

Lewisia pygmaea (A. Gray) B. L. Rob. (very small) Pygmy bitterroot, alpine bitterroot, alpine lewisia. Succulent, herbaceous perennial; rootstock fleshy, fusiform to narrowly napiform, usually branching apically. **STEMS** ± prostrate or suberect, becoming reflexed in fruit, 1–6 cm long. **LEAVES** basal, crowded, succulent, linear to linear-oblanceolate, plano-convex to slightly concavo-convex, occasionally nearly terete, no obvious petiole, blade linear to linear-lanceolate, 3–9 cm, margins entire, apex acute to obtuse. **INFL** 2–7-flowered cyme; lower bracts 2, opposite; upper bracts 1 subtending each flower, linear-oblong, linear-lanceolate, or lanceolate, 2–10 mm long, margins glandular-toothed, sometimes eglandular-toothed or nearly entire, apex acute; pedicel 2–10 mm. **FLOWERS** rotate, 1.5–2 cm diameter; sepals 2, suborbiculate, broadly ovate, or obovate, 2–6 mm, herbaceous at anthesis, margins usually glandular-toothed, sometimes eglandular-toothed or rarely ± entire, apex usually truncate, sometimes rounded, obtuse, subacute, or apiculate; petals 5–9, white, pink, or magenta, sometimes green at base, narrowly oblong, elliptic, or oblanceolate, 4–10 mm; stamens (4–) 5–8; stigma lobes 3–6. **FRUIT** 4–5 mm long, base persistent after dehiscence. **SEEDS** 15–24, 1–2 mm long, shiny, smooth. $2n =$ about 66. [*Talinum pygmaeum* A. Gray; *Calandrinia grayi* Britton; *C. pygmaea* (A. Gray) A. Gray; *Lewisia exarticulata* H. St. John; *L. glandulosa* (Rydb.) S. Clay; *L. minima* (A. Nelson) A. Nelson; *L. pygmaea* var. *aridorum* Bartlett; *L. pygmaea* subsp. *glandulosa* (Rydb.) R. S. Ferris; *L. sierrae* R. S. Ferris; *Oreobroma aridorum* (Bartlett) A. Heller; *O. exarticulatum* (H. St. John) Rydb.; *O. glandulosum* Rydb.; *O. grayi* (Britton) Rydb.; *O. minimum* A. Nelson; *O. pygmaeum* (A. Gray) Howell]. Mostly exposed bare places or in short turf, on gravelly or rocky substrates; aspen, spruce-fir, and alpine tundra. COLO: Arc, Con, Dol, Hin, LPl, Min, Mon, SJn; UTAH. 2300–4200 m (7500–13800′). Flowering: Apr–Aug. Southern Alaska to southern California, west to Montana, South Dakota, Colorado, Arizona, and New Mexico.

Lewisia rediviva Pursh (coming back to life, referring to the ability of a plant to grow after being dried) Bitterroot. Herbaceous, succulent perennial; rootstock thick, fleshy, roughly napiform, most often branching. **STEMS** procumbent to erect, 1–3 cm. **LEAVES** basal, withering at or soon after anthesis, sessile, blade linear to narrowly clavate, subterete to slightly concave above, 0.5–5 cm, margins entire, apex obtuse to subacute. **INFL** 1-flowered; bracts 3–8, whorled, subulate to linear-lanceolate, 4–10 mm, margins entire, apex acuminate; pedicel mostly 5–20 mm, disarticulating with flower remains and fruit at maturity. **FLOWERS** showy, sepals 4–9, imbricate, broadly elliptic to ovate, 10–25 mm long, scarious after anthesis, margins entire to somewhat erose, apex obtuse to rounded; petals 10–19, white or pink to magenta, sometimes pale in center or striped, elliptic, oblong, or narrowly oblanceolate, 15–35 mm; stamens 12–50, united at base; stigma lobes 4–9. **FRUIT** 5–6 mm long, ovoid. **SEEDS** 6–25, 2–2.5 mm long, orbicular-reniform, shiny dark brown to black, minutely tuberculate. $2n = 26, 28$. [*L. alba* Kellogg]. Exposed, often unconsolidated gravelly to rocky substrates; sagebrush, juniper, black sagebrush, and piñon-juniper woodland communities. UTAH. 1900–2450 m (6000–8000′). Flowering: Mar–May. Our material belongs to **var. *rediviva***. Alberta, British Columbia south to California, east to Idaho, Montana, Nevada, and south to Utah and Arizona. Native Americans commonly harvested and ate the boiled or dried roots. Colonies are spectacular when in full flower.

Phemeranthus Raf. Fameflower, Rockpink, Sunbright

(Greek *ephemeros*, for living for one day, + *anthos*, flower) Herbaceous annuals or perennials, caulescent, glabrous; rootstock thickened, tuberous, often suffrutescent, horizontal to vertical, globose to fusiform, or branching. **STEMS** prostrate to erect, simple or branching, sometimes basally suffrutescent. **LEAVES** alternate, sessile, semiterete to terete, 1–3 mm wide, 1–10 cm long, succulent. **INFL** cymose; single- to many-flowered; with peduncles suberect, long, slender, and scapelike; or, 1- to few-flowered with peduncles lateral, short, and succulent, often not exceeding leaves. **FLOWERS** often showy, pedicellate, open 1–4 hours at various times of day or night depending upon species; sepals 2, opposite, scarious to foliaceous, deciduous after anthesis, or persistent into fruiting; petals 5 (10); stamens 5–many, distinct or with filaments basally shortly coherent in several clusters; anthers 2-locular, oblong; ovary normally 3-carpellate, placentation free-central; style 1; stigma lobes 3 (–5). **FRUIT** a capsule, mostly erect, usually longitudinally basipetally dehiscent, sometimes acropetally; valves rigid, deciduous or briefly persistent. **SEEDS** many, black or dark brown, compressed, nearly smooth or with raised concentric ridges, estrophiolate, circular-reniform, small, seed coat shiny, covered with pale white or gray, thin, translucent, chartaceous pellicle. $x = 12$. About 25 species. Temperate North America and western South America. *Phemeranthus* has been traditionally subsumed into the distantly related and very different, primarily subtropical to tropical genus *Talinum*. That genus is separable by numerous characters from *Phemeranthus*, including its flattened leaves with visible midribs, acropetally dehiscent pendent capsules with immediately deciduous valves, and in having the endocarp and exocarp differentiated and often physically separating at maturity.

1. Stems prostrate to procumbent; leaves short (mostly under 2.5 cm), crowded along stem and upturned; inflorescences short (mostly under 5 cm), mostly appearing axillary and spreading laterally, 1- to few-flowered, with at least lowest flowers among leaves .... ***P. brevifolius***

1' Stems erect; leaves longer (mostly over 2.5 cm), spreading equally around stem; inflorescence long (mostly over 5 cm), usually appearing terminal, erect, many-flowered, with lowest flowers usually above leaves ..... ***P. confertiflorus***

Phemeranthus brevifolius (Torr.) Hershkovitz (short-leaved) Canyonland fameflower, canyonland rockpink, small-leaf fameflower, small-leaf rockpink, pygmy fameflower. Herbaceous, succulent perennial, resembling *Sedum*; rootstock thickened, often nearly horizontal, elongate, fleshy-suffrutescent. **STEMS** up to 10 cm long, usually much less, herbaceous-succulent, spreading, procumbent to ascending, sometimes branching basally. **LEAVES** sessile, subterete, obtuse, up to 1.5 (2) cm long, usually light glaucous, sometimes dark green or purplish. **INFL** 1-flowered (rarely cymulose and few-flowered), not or barely exceeding leaves; peduncle short, not scapelike, up to 5 mm long. **FLOWERS** variably polygamomonoecious (occurence inconsistent even within one plant, and apparently affected by growth conditions); sepals scarious and early-deciduous, oval to orbiculate, up to 5 mm long, apically rounded to obtuse, not exceeding capsule; petals white to magenta, usually pink to lavender, obovate, rounded to obtuse, up to 12 mm long; stamens 10–25, color mostly matching or lighter than petals; pistil greatly exceeding stamens, with 3 stigma lobes, sublinear. **FRUIT** a capsule, subglobose to ellipsoid, apically rounded to obtuse, 3.5–5 mm long. **SEEDS** without arcuate ridges, 1–1.4 mm long. Most often in small local populations in slickrock country; primarily shallow, fine sand pockets overlying noncalcareous, often reddish sandstones; mostly within piñon-juniper woodland communities. ARIZ: Apa, Coc, Nav; NMEX: McK, SJn; UTAH. 1555–2130 m (5100–6990′). Flowering: Apr–Sep. Arizona to Texas.

Phemeranthus confertiflorus (Greene) Hershkovitz (densely flowered) Rocky Mountain fameflower, Rocky Mountain rockpink, Rocky Mountain sunbright, New Mexico fameflower. Herbaceous, succulent perennial; rootstock thick, fleshy-suffrutescent, vertical, irregularly and few-branched. **STEMS** 2–10 (15) cm, ± erect, simple or branching, base often perennial, suffrutescent, shiny orange-brown to dark reddish or brown, remainder of stem annual, succulent-herbaceous, green to slightly purplish, usually glaucous. **LEAVES** succulent, sessile, terete, sometimes slightly broadened at base, up to 2.5 (4) cm long; mostly glaucous gray-green, often purplish apically, especially under stress; dried bases often persistent on stems. **INFL** cymose, exceeding leaves; peduncle scapelike, to 10 (15) cm long. **FLOWERS** subsessile to short-pedicellate; sepals foliaceous becoming scarious, persistent, ovate, up to 4 mm long, apex often purplish, acuminate to acute; corolla white to light pink, rarely purplish, elliptic to obovate, acuminate to acute, petals up to 6 mm long; stamens 5 (–10), color mostly matching or lighter than petals; pistil not or little exceeding stamens; stigma subcapitate, 3-lobed. **FRUIT** a capsule, broadly ellipsoid, obtusely trigonous, usually broadly acute apically, 3–6 mm long. **SEEDS** nearly smooth, black, but pellicle imparting gray or blue-gray color, 0.8–1.2 mm long. $2n = 48$.

[*Talinum confertiflorum* Greene; *T. gracile* Rose & Standl.; *T. rosei* P. Wilson; *T. gooddingii* P. Wilson; *T. fallax* Poelln.]. Sandy, gravelly, rocky, or shallow soil pockets overlying rock; black sagebrush, piñon-juniper, and ponderosa pine communities. ARIZ: Apa, Coc, Nav; COLO: Arc, LPl, Mon; NMEX: McK, RAr, SJn; UTAH. 1450–2685 m (4755–8185′). Flowering: Apr–Oct. Arizona and Idaho to North Dakota, south to Texas; northern Mexico. Persistently confused with *P. parviflorus* (Nutt.) Kiger, but that species distinguished by obtuse, early-deciduous sepals; capsules usually more broadly rounded, obtuse; seeds deep brown to black, even with pellicle intact; more easterly distribution; $2n = 24$. Used by the Ramah Navajo as a poultice of root bark, applied to sores and used as a lotion.

Portulaca L. Purslane, Mossrose

(Latin *portula*, little door, in reference to the lid of the capsule) Annual or perennial succulent herbs; root fleshy-fibrous to tuberous, mostly with a single, often tuberous taproot and laterals. **STEMS** prostrate to erect, usually branched, sometimes suffrutescent; hairlike trichomes in inflorescence or stem, nodes absent or present, otherwise glabrous. **LEAVES** succulent, alternate or subopposite, congested and involucre-like immediately proximal to inflorescence; blade terete, subterete, or flattened. **INFL** terminal or axillary on short branches, reduced, subtended by several densely crowded leaves, multiflowered and tightly crowded. **FLOWERS** sessile or subsessile, diurnal but open only a short time each day; sepals broadly clasping at base, herbaceous to scarious; petals 5–7, usually distinct, margins usually entire; stamens 4–many; ovary 1/2 inferior to inferior, plurilocular proximally to 1-locular distally, placentation free-central; style 1, stigma lobes 3–18. **FRUIT** a capsule, circumscissally dehiscent before dry, remaining base persistent or deciduous after drying. **SEEDS** many, dark brown to black, reniform to orbicular or globose, smooth to tuberculate; with tightly investing pellicle apparently of epidermal origin, often giving metallic blue or gray appearance; testa with raised, rounded-stellate pattern cells. Species about 100 nearly worldwide, primarily tropical and subtropical; also temperate.

1. Leaves flat, spatulate; axils not or inconspicuously pilose .... ***P. oleracea***
1′ Leaves linear, terete or nearly so; axils pilose .... (2)

2. Flowers usually over 5 mm wide, usually magenta; taproot thickened-tuberous or napiform .... ***P. mundula***
2′ Flowers usually less than 5 mm wide, yellow to yellow-orange; roots not thickened .... ***P. halimoides***

Portulaca halimoides L. (having the form of or a resemblance to the genus *Halimium*) Sand mossrose, dwarf purslane, silkcotton purslane. Succulent annual; root fibrous, usually 1 taproot with laterals. **STEMS** prostrate to suberect, usually many-branched, often reddish in stress, 1–15 (–25) cm long, leaf axils and inflorescence moderately villous, this more conspicuous as the plant ages. **LEAVES** glabrous, linear, terete to broadly subulate, 2–15 × 0.4–1.5 mm, apex obtuse to acute; involucre-like leaves 4–10. **FLOWERS** 1–8 mm diameter; corolla yellow to orange, rarely red, obovate, 1–4 × 1–2.5 mm; stamens 5–20; stigma lobes 3–6. **FRUIT** a capsule, ovoid, usually short-stipitate, 1–3 mm diameter. **SEEDS** gray to nearly black, 0.3–0.6 mm wide, stellate-tubercular. $2n = 18$. [*P. parvula* A. Gray]. Sandy soil in mostly open saltbush, ephedra, cliffrose, and piñon-juniper communities. ARIZ: Apa, Coc, Nav; COLO: LPl, Mon; NMEX: McK, RAr, San, SJn; UTAH. 1370–1895 m (4500–8000′). Flowering: May–Sep, depending upon temperature and precipitation. California, Arizona to Oklahoma and Texas, Arkansas, east to Virginia and Delaware; Mexico; West Indies; Central America; South America. Used as a potherb before introduction of *P. oleracea* L.

Portulaca mundula I. M. Johnst. (trim or neat) Rock mossrose, purple mossrose, kiss-me-quick. Succulent annual to short-lived perennial; root a taproot, tuberous, ovoid to napiform, usually unbranched. **STEMS** suberect to erect, ascending in age, usually few-branched, often reddish in stress; leaf axils and inflorescence conspicuously white-villous; branches 2–10 (15) cm. **LEAVES** often glaucous, linear, terete to broadly subulate, 5–15 (20) × 1–2 mm, apex obtuse to acute; involucre-like leaves 4–10. **FLOWERS** 4–15 mm diameter; corolla pink, magenta, purple, or rarely white, obovate, 2–8 × 1.5–3.5 mm; stamens 5–15 (30); stigma lobes 3–6. **FRUIT** a capsule, ovoid, usually short-stipitate, 1.5–6 mm diameter. **SEEDS** usually gray to metallic, due to slightly separated investing pellicle, blackish if rubbed, 0.5–1.2 mm long. $2n = 8, 16$. [*P. pilosa* L. var. *mundula* (I. M. Johnst.) D. Legrand]. Shallow soils over rock, or exposed gravelly to rocky substrates, mostly ledges, slopes, hilltops. COLO: Mon (Mesa Verde National Park); NMEX: SJn. 2135 m (7000′). Flowering: May–Oct, depending upon temperature and precipitation. Southern Colorado, southern Kansas west to eastern Arizona, east to Missouri, North Carolina, south to Florida; northern Mexico.

Portulaca oleracea L. (herbaceous) Common purslane, pursley, little hogweed. Glabrous; annual (ours); root fibrous, usually from slender main taproot; with globular to ovoid, tuberous taproot. **STEMS** prostrate to ascending, succulent; trichomes in axils and inflorescence absent or inconspicuous; branches up to 60 cm wide. **LEAVES** 1–6, obovate or spatulate, usually truncate to rounded, occasionally retuse, flattened, 4–30 × 2–15 mm, apex round to retuse or nearly truncate; involucre-like. **FLOWERS** 3–10 mm diameter; petals yellow to purple (yellow to orange in ours), oblong, 3–4.6 × 1.8–3 mm; stamens 3–20; stigma lobes 3–6. **FRUIT** a capsule, ovoid to globose or rounded-turbinate, 4–9 mm long. **SEEDS** orbiculate, somewhat flattened, 0.4–1.5 mm long, gray to black, minutely punctate. $2n$ = 18, 36, 54. Open, sunny areas, a weed of cultivated areas, roadsides, and indigenous plant communities. ARIZ: Apa, Coc, Nav; COLO: Arc, Dol, Hin, LPl, Mon, SMg; NMEX: McK, RAr, San, SJn; UTAH. 1125–2440 m (3700–8000′). Flowering: late spring–early fall, depending primarily upon availability of moisture. Widely distributed in North America; cosmopolitan. Plant used for pain by the Navajo. The Ramah Navajo used the leaves as a potherb. The Kayenta Navajo used the plant as a lotion for scarlet fever.

POTAMOGETONACEAE Bercht. & J. Presl PONDWEED FAMILY

C. Barre Hellquist

Perennial or annual herbs; rhizomatous or nonrhizomatous; caulescent; turions absent or present. **LEAVES** stipulate, simple, alternate, submersed, or submersed and floating, sessile or petiolate. **INFL** a spike, capitate spike, or panicle of spikes borne axillary or terminal, submersed or emersed. **FLOWERS** with 4 tepals in 1 series, stamens (2–) 4, epipetalous in 1 series; anthers dehiscing vertically; pistils 1–4, mostly not stipitate, ovules marginal. **FRUIT** a drupe. **SEEDS** 1 with curved embryo. x = 13, 14 (Fernald 1932; Hagström 1916). Three genera and about 100 species worldwide.

1. Submersed leaves lacking adnate stipular sheaths; peduncles stiff, inflorescences submersed or erect if on the surface .... ***Potamogeton***
1′ Submersed leaves with stipular sheaths adnate for 2/3 or more their length; peduncles flexible, inflorescences floating on the surface .... ***Stuckenia***

Potamogeton L. Pondweed

(Greek *potamos*, river neighbor) Perennial or annual herbs, rhizomes present or absent, no tubers formed, turions formed on some species. **STEMS** terete to compressed, nodes with oil glands present or absent. **LEAVES** submersed or submersed and floating, alternate, some nearly opposite, stipule connate or convolute, tubular, sheathing the stem; submersed leaves sessile, petiolate, or perfoliate, apex subulate to obtuse, 1–35 veins; margins entire or serrate, stipules either free or adnate to the leaf base for less than 1/2 the length of the stipule; floating leaves petiolate, rarely sessile, stipule free from leaf base; blades elliptic to ovate, leathery, base cuneate, rounded to cordate, 1–51 veins. **INFL** a spike or panicle of spikes, submersed and/or emersed, capitate, or cylindric; peduncles stiff, often projecting above the water surface. **FLOWERS** with pistils 1 or 4. **FRUIT** abaxially rounded or keeled, beaked, embryo coiled 1 or more times. x = 13, 14 (Haynes 1975; Ogden 1943). About 90 species worldwide. *Potamogeton* plants are extremely important food sources for waterfowl and habitat for numerous aquatic animals. Whenever possible, plants collected should be in fruit. These are important for species identification.

1. Plants producing floating leaves .... (2)
1′ Plants lacking floating leaves .... (5)

2. Submersed leaves linear, 0.8–2 mm wide, petiole of floating leaves pale at junction with blade .... ***P. natans*** (in part)
2′ Submersed leaves broadly linear-oblong, lanceolate to elliptic, petiole of floating leaves not pale at junction with blade .... (3)

3. Submersed leaves petiolate; fruits 3.5–4.3 mm long .... ***P. nodosus*** (in part)
3′ Submersed leaves sessile; fruits 1.7–3.5 mm long .... (4)

4. Submersed leaves with distinct reticulate portion along midrib, leaf tips obtuse or somewhat acute, fruit stalked .... ***P. alpinus*** (in part)
4′ Submersed leaves lacking reticulate portion along midrib, leaf tips sharply acute or awl-shaped .... ***P. gramineus*** (in part)

5. Leaf margins distinctly serrate, fruit beaks 2–3 mm long .... ***P. crispus***
5′ Leaf margins entire, fruit beaks lacking or less than 2 mm long .... (6)

6. Submersed leaves linear, 2 mm or less wide ....(7)
6′ Submersed leaves broadly linear-oblong to lanceolate to elliptic, 10–75 mm wide; fruits 1.7–3.5 mm long ....(9)

7. Submersed leaves phyllodial, 0.8–2 mm wide (young plants), lacking distinct veins .... ***P. natans*** (in part)
7′ Submersed leaves flattened, less than 2 mm wide, with 3–5 distinct veins ....(8)

8. Nodal glands present at the base of leaves; peduncles terminal or terminal and axillary, straight .... ***P. pusillus***
8′ Nodal glands lacking at the base of leaves; peduncles axillary, recurved .... ***P. foliosus***

9. Submersed leaves clasping the stem ....(10)
9′ Submersed leaves sessile (not clasping) or petiolate ....(11)

10. Rhizomes spotted rusty red; stems with a zigzag appearance, the stem changing direction at each node; leaf tips boat-shaped, splitting when pressed .... ***P. praelongus***
10′ Rhizomes white, unspotted; stems straight, lacking the zigzag appearance; leaf tips flat, not splitting when pressed .... ***P. richardsonii***

11. Submersed leaves petiolate; fruits 3.5–4.3 mm long .... ***P. nodosus*** (in part)
11′ Submersed leaves sessile; fruits 1.9–3.5 mm long ....(12)

12. Leaf apex obtuse to somewhat acute, leaves with reticulate portion along midrib, fruit stalked ...... ***P. alpinus*** (in part)
12′ Leaf apex sharply acute or awl-shaped, leaves lacking reticulate portion along midrib, fruit lacking a stalk .... ***P. gramineus*** (in part)

Potamogeton alpinus Balb. (alpine) Alpine pondweed. Perennial from rhizomes. **STEMS** terete. **LEAVES** submersed and floating or only submersed; submersed leaves sessile, blades reddish green, oblong-linear to linear-lanceolate, 4.5–18 (–25) × 0.5–2 cm, base rounded, apex obtuse to acute, lacunae 0–6 rows on each side of the midvein, often transitioning into floating leaves, stipules persistent, convolute, apex blunt, 1.2–4 cm; veins 7 (–9); floating leaves reddish green, elliptic, oblanceolate, obovate to oblong-linear, 4–7 (–10) × 1–2.5 (–4) cm, base gradually tapering into petiole; apex obtuse or acute, veins (7–) 9–13 (–15). **INFL** unbranched, emersed, peduncles terminal or axillary, erect, 3–10 (–16) cm, spikes cylindric, 10–35 mm. **FRUIT** olive-green, with or without keels, 2.5–3.5 × 1.7–2.4 mm, beak curved, 0.5–0.9 mm. $2n = 52$. [*P. alpinus* subsp. *tenuifolius* (Raf.) Hultén; *P. alpinus* var. *subellipticus* (Fernald) Ogden; *P. alpinus* var. *tenuifolius* (Raf.) Ogden; *P. tenuifolius* Raf.; *P. tenuifolius* var. *subellipticus* Fernald]. Cold, neutral to alkaline streams, lakes, and ponds, often at inlets. ARIZ: Apa; COLO: Con, Hin, Mon, RGr, SJn. 2330–3425 m (7700–11300′). Flowering: Jun–Aug. Fruit: Jul–Aug. Alaska and Nunavut east to Newfoundland, south in the mountains of California and Colorado, Minnesota, Michigan, and Maine; Eurasia. It is known to hybridize with *P. gramineus*, *P. nodosus*, and *P. praelongus* from our area.

Potamogeton crispus L. (crimped) Curly-leaved pondweed, curled pondweed. Turions common, axillary and terminal, hard. Rhizomes absent. **STEMS** flattened. **LEAVES** submersed, linear, 4–10 mm wide, base obtuse to rounded, margins serrate, apex round to round-acute, lacunae in 2–5 rows on each side of midvein. **INFL** emersed; peduncles terminal, rarely axillary, erect to ascending, cylindric, 2.5–4 cm; spikes cylindric, 10–15 mm. **FRUIT** red to reddish brown, keeled, beak 2–3 mm. $2n = 52$. Alkaline or polluted waters of streams, lakes, and ponds. NMEX: McK, SJn. 1550–1900 m (5115–6300′). Flowering: Jun–Jul. Fruit: July. British Columbia east to Saskatchewan, Ontario, and Quebec, south to California, Arizona, Texas, Louisiana, and Florida; South America; Eurasia; Australia. This is the only pondweed introduced into North America. It was introduced in the mid-1880s in the eastern United States. This species reaches sexual maturity early in the season, usually May and June, but observed later in the Southwest. It occasionally produces fruit and regularly produces hard, almost woody, turions. The plant deteriorates in about a month and the turions are released. These root during the summer and the plants remain small until spring, when they grow rapidly.

Potamogeton foliosus Raf. (leafy) Leafy pondweed. Turions uncommon, lateral and terminal, soft, inner leaves rolled into a hardened, fusiform structure; rhizomes absent or present (in shallow streams). **STEMS** slightly compressed, glands usually absent, if present 0.3 mm. **LEAVES** submersed, linear, 0.3–2.3 mm wide (wider in flowing streams), base slightly tapering, apex usually acute, rarely apiculate, lacunae 0–2, usually none, on each side of the midvein, veins 1–3 (–5); stipules disintegrating with age, convolute, delicate to fibrous, apex obtuse, 0.2–2.2 cm. **INFL** unbranched, emersed; peduncles in leaf axils, recurved, clavate, 0.3–1.1 (–3.7) cm, spikes capitate to cylindric, 1.5–7 mm. **FRUIT** olive-green or brown, obovate to orbicular, concave, 1.5–2.7 × 1.2–2.2 mm, keel 0.2 mm high, undulate, beak 0.2–0.6 mm. $2n = 28$. [*P. curtissii* Morong; *P. foliosus* var. *macellus* Fernald]. Neutral to alkaline streams, ponds, and lakes. ARIZ:

Apa; COLO: LPl; NMEX: McK, SJn. 1400–2750 m (4620–9075′). Flowering: Jun–Aug. Fruit: Jul–Aug. Alaska east to Nova Scotia, south to California, Texas, and Florida; Central America. Two subspecies occur in the United States; ours is **subsp.** ***foliosus***. Plants are often more robust in faster-flowing waters.

Potamogeton gramineus L. (grasslike) Variable pondweed. Perennial from rhizomes. **STEMS** terete. **LEAVES** submersed and floating, or only submersed, sessile, blade elliptic, apex acuminate, lacunae in 1–2 rows on either side of the midvein, 3–9 veins; apex acute to obtuse; floating leaves with petioles 3–4.5 cm long, blades elliptic to ovate, 3.5–4 × 1.6–2 cm, base rounded, apex acuminate, veins 11–13; stipules persistent, convolute, 1.3–1.6 cm. **INFL** unbranched, emersed; peduncles both axillary and terminal, erect to ascending, cylindric, 3.2–7.7 cm, spikes cylindric, 1.5–3.5 cm. **FRUIT** greenish brown, ovoid, laterally compressed, keeled, 1.9–2.3 × 1.8–2 mm, beak 0.3–0.5 mm. $2n = 52$. [*P. gramineus* var. *maximus* Morong; *P. gramineus* var. *myriophyllus* J. W. Robbins]. Acid to alkaline waters in streams, lakes, ponds, and bogs. ARIZ: Apa; COLO: Arc, Hin, LPl, Mon, SJn; NMEX: McK, SJn. 2325–2850 m (7675–9400′). Flowering and fruiting: Jul–Aug. Alaska east to Newfoundland, south to California, Colorado, Kentucky, and Pennsylvania; Eurasia. *Potamogeton gramineus* is a highly variable species that takes on many different forms. The narrowest forms occur in acid waters, particularly bogs. The broader-leaved plants often approach *P. illinoensis* Morong in stature. From the Four Corners region it has been documented hybridizing with *P. alpinus*, *P. natans*, *P. nodosus*, *P. praelongus*, and *P. richardsonii*.

Potamogeton natans L. (swimming) Floating-leaf pondweed, floating brownleaf. Perennial from rhizomes. **STEMS** terete. **LEAVES** submersed and floating, submersed leaves sessile, phyllodial, 0.7–2.2 mm wide, apex obtuse, 3–5 veins; stipules persistent, convolute, free from blade 4.5–10 cm; floating leaves with petiole a lighter color at the junction with the leaf blades, blades elliptic to ovate, 3.5–11 cm, veins 17–37. **INFL** unbranched, emersed, peduncles terminal, cylindric, 25–50 mm. **FRUIT** sessile, green to greenish brown, obovoid, unkeeled, 3.5–5 × 2–3 mm, beak 0.4–0.8 mm. $2n = 52$. Acid to alkaline waters of lakes, ponds, and slow-flowing streams. ARIZ: Apa; COLO: LPl, RGr, SJn; NMEX: SJn. 2200–2875 m (7290–9450′). Flowering: Jul–Aug. Alaska, Yukon, and Manitoba, east to Newfoundland, south to California, New Mexico, Indiana, and West Virginia; Eurasia. The Navajo have used this species as a ceremonial emetic.

Potamogeton nodosus Poir. (knotty) Long-leaf pondweed. Perennial from rhizomes. **STEMS** terete. **LEAVES** submersed and floating, or only submersed, submersed leaves with petioles 2–13 cm, blade linear-lanceolate to lance-elliptic, 9–20 × 1–3.5 cm, base and apex acute, 2–5 rows of lacunae on each side of midvein, stipules mostly persistent, convolute, free from blade, 3–9 cm; floating leaf petioles 3.5–26 cm, blades lenticular to elliptic, 3–11 × 1.5–4.5 cm, bases mostly rounded, apex acute to rounded, veins 9–21. **INFL** unbranched, emersed, peduncles terminal, erect to ascending, cylindric, 3–15 cm, spikes cylindric, 20–70 mm. **FRUIT** red to reddish brown, keeled, laterally ridged, 2.7–4.3 × 2.5–3 mm, beak erect. $2n = 52$. [*P. americanus* Cham. & Schltdl.]. Fast-flowing streams of low alkalinity or ponds, lakes, and slow-flowing streams. COLO: Arc, LPl, SJn; NMEX: SJn; UTAH. 1625–2080 m (5350–6875′). Flowering and fruiting: Jul–Aug. British Columbia east to New Brunswick, south to California, Texas, and Florida; Mexico; West Indies; Central America; South America; Eurasia. This is one of the most common species in North America and the world. It has been documented in North America hybridizing with *P. alpinus*, *P. gramineus*, *P. illinoensis*, *P. natans*, and *P. richardsonii*.

Potamogeton praelongus Wulfen (greatly prolonged) White-stemmed pondweed, muskie-weed. Rhizomes present. **STEMS** white, terete. **LEAVES** submersed, linear-lanceolate, base clasping, apex hooded or boat-shaped, obtuse, splitting when pressed, lacunae absent, 1.1–4.6 cm wide; stipules persistent, conspicuous, convolute, 3–8.1 cm, fibrous, shredding at apex. **INFL** emersed, unbranched, terminal or axillary, erect, cylindric, 9.5–53 cm, spikes cylindric, 3.4–7.5 cm. **FRUIT** greenish brown, obovoid, keeled, 4–5.7 × 3.2–4 mm, beak 0.5–1 mm. $2n = 52$. Usually in deep, neutral to moderately alkaline waters of ponds and lakes. COLO: SJn. 3275 m (10744′). Flowering: Jun–Jul. Fruit: Jul–Aug. Alaska east to Newfoundland, south to California, Colorado, Nebraska, Indiana, and New Jersey. This species flowers and fruits early in the season. It often grows in deep water. It is easily distinguished by the zigzag pattern of the stem reversing direction at each node and the large, persistent stipules. Hybrids have been documented in North America with *P. alpinus*, *P. crispus, and P. gramineus*.

Potamogeton pusillus L. (very small) Rhizomes absent or present. Turions common, lateral and terminal, soft, 2-ranked, inner leaves rolled into hard fusiform structure. **STEMS** terete to slightly compressed, nodal glands present on most nodes, 0.05 mm diameter. **LEAVES** submersed, sessile, blade linear, 0.5–1.9 mm wide, base slightly tapered, lacunae 0–2 on each side of the midvein, veins (1–) 3, apex acute, rarely apiculate, stipules persistent, connate, apex obtuse, 3–9 mm. **INFL** unbranched; peduncles 1–3 per plant, filiform to cylindric, erect to slightly clavate; spikes capitate to cylindric, interrupted, 1.5–10.1 cm. **FRUIT** obovoid, sides centrally concave, 1.5–2.2 × 1.2–1.6 mm, beak 0.1–0.6 mm. 2*n* = 26. [*P. panormitanus* Biv.; *P. pusillus* var. *minor* (Biv.) Fernald & B. G. Schub.]. Neutral to alkaline waters of lakes, ponds, and streams. ARIZ: Apa; COLO: Dol, SJn; NMEX: McK, SJn; UTAH. 1600–2725 m (5280–9000′). Flowering and fruiting: Jul–Sep. Yukon east to Nova Scotia, south to California, New Mexico, Texas, and Florida; South America; Eurasia; Africa. Three subspecies occur in the United States, of which only **subsp.** ***pusillus*** occurs in our area.

Potamogeton richardsonii (A. Benn.) Rydb. (for Sir John Richardson, 1787–1865) Clasping-leaved pondweed, bassweed, Richardson's pondweed. Rhizomes present. **STEMS** terete. **LEAVES** submersed, clasping, ovate-lanceolate to narrowly lanceolate, margins crisped; apex acute to obtuse, lacunae absent, 0.5–2.8 cm wide, veins 3–35; stipules persistent, conspicuous, convolute, disintegrating to persistent fibers, obtuse until splitting. **INFL** emersed, unbranched, peduncles terminal or axillary, erect to recurved, clavate, 1.5–14.8 cm, spikes cylindric, 13–47 mm. **FRUIT** greenish brown, obovoid, turgid to concave, keeled, 2.2–4.2 × 1.7–2.9 mm; beak 0.4–0.7 mm. 2*n* = 52. [*P. perfoliatus* L. var. *richardsonii* A. Benn., *P. perfoliatus* subsp. *richardsonii* (A. Benn.) Hultén]. Alkaline waters of lakes, ponds, and rivers. ARIZ: Apa; COLO: Arc. 2080–2750 m (6850–9050′). Flowering and fruiting: Jul–Aug. Alaska east to Newfoundland, south to California, Colorado, Michigan, Pennsylvania, and Connecticut. Two hybrids have been documented in North America, hybridizing with *P. gramineus* and *P. nodosus*.

Stuckenia Börner Pondweed

(derivation unknown) Perennial herbs from rhizomes, often with tubers. **LEAVES** submersed, alternate, channeled, turgid, margins entire, apex acute, obtuse, veins 1–5; stipules adnate to leaf base for 2/3 or more of the length of the stipule, terminating in a free ligule. **INFL** a capitate or cylindric spike, submersed, peduncles flexible, floating on the water surface. **FLOWER** pistils 4. **FRUIT** abaxially rounded, beaked or lacking beak, embryo with less than 1 full coil. $x = 13$ (Börner 1912; Holub 1997; St. John 1916). About six species worldwide.

1. Leaf apex acute, apiculate, cuspidate, or rarely rounded; fruit with distinct beak ....(2)
1′ Leaf apex retuse, blunt, or apiculate; fruit with minute beak or beakless ....(3)

2. Leaves 0.2–1 mm wide; leaf apex acute (rarely apiculate in young plants) .... ***S. pectinata***
2′ Leaves 0.8–1.7 mm wide; leaf apex apiculate, cuspidate, or rounded .... ***S. striata***

3. Free portion of stipule forming a distinct ligule up to 20 mm long; tips of midstem stipular sheaths distinctly inflated; fruits 3–3.5 mm long .... ***S. vaginata***
3′ Free portion of stipule forming a small ligule up to 1 mm long; tips of midstem stipular sheath slightly inflated; fruits 2–3 mm long .... ***S. filiformis***

Stuckenia filiformis (Pers.) Börner (threadlike) Threadleaf pondweed. **STEMS** branching, subterete. **LEAVES** on main stem slightly larger than those on branches, lower stipular sheaths often inflated, 1–4.5 (–9.5) cm, blade filiform to linear, 0.2–2 (–3.7) mm, apex notched, blunt, or short-apiculate, veins 1–3. **INFL** peduncles terminal, erect, filiform to slender, 2–10 (–15) cm; spikes cylindric to moniliform, 5–55 mm, 2–6 (–9) whorls, often adjacent. **FRUIT** dark brown, 2–3 × 1.5–2.4 mm, beak obscure. 2*n* = 78. [*Potamogeton filiformis* Pers.]. Calcareous, alkaline waters of rivers, streams, and shallow waters of shorelines of lakes and ponds. COLO: Arc, LPl, SJn; NMEX: SJn; UTAH. 1970–3700 m (6500–12180′). Flowering: Jun–Aug. Fruit: Jul–Aug. Alaska, Northwest Territories, and Nunavut east to Quebec and Newfoundland, south to California, New Mexico, Nebraska, Michigan, New York, and Pennsylvania; Eurasia. Subspecies five, three in North America, one (possibly two) in our range. Most of our material is **subsp.** ***alpina*** (Blytt) R. R. Haynes, Les & M. Král (alpine), which are short plants, typically with narrow leaves. Some material in our area is larger and more taxonomic study is needed on this taxon.

Stuckenia pectinata (L.) Börner (comblike) Sago pondweed. **STEMS** branched, usually distally, terete to slightly compressed. **LEAVES** of main stem only slightly larger than those of the branches; stipular sheaths not inflated, 0.8–1.1 cm long, ligule to 0.8 mm, blade linear, 0.2–1.1 mm wide, apex acute, mucronate or blunt on young stems, veins 1–3. **INFL** peduncles terminal or axillary, cylindric, 4.5–11.4 cm; spikes moniliform to cylindric, 14–22 mm, 3–5 whorls. **FRUIT** yellow-brown to brown, 3.8–4.5 × 2.5–3.1 mm, beak erect, 0.5–1.1 mm. 2*n* = 78. [*Coleogeton pectinatus* (L.) Les & R. R. Haynes; *Potamogeton interruptus* Kit.; *P. pectinatus* L.]. Calcareous, saline, and alkaline waters of ponds, lakes, springs, streams, and rivers. ARIZ: Apa; COLO: Hin, LPl, Mon, SJn, SMg; NMEX: SJn; UTAH. 1600–2750 m (5365–7075′). Flowering: Jul–Aug. Alaska east to Newfoundland, south throughout the United States; Mexico; almost worldwide. The tubers produced are a valuable waterfowl food.

Stuckenia striata (Ruiz & Pav.) Holub (furrowed, channeled) Nevada pondweed. **STEMS** branched distally, 5-ridged to terete. **LEAVES** on stems 2 times or more those on branches, stipular sheaths not inflated, 1.2–3.4 cm long, ligule 0.2–1.1 cm; blade linear, 0.4–5.1 (–8.5) mm wide (0.8–1.7 mm on plants in our range); apex apiculate, cuspidate, rarely rounded, veins 3–5. **INFL** peduncles axillary, rarely terminal, erect to ascending, cylindric, 1.2–5.2 cm; spikes cylindric, occasionally moniliform, 13–45 mm; 4–9 whorls. **FRUIT** brown to reddish brown, 3–3.9 × 2.8–3 mm, beak erect, rarely curved, 0.2–0.3 mm. [*Potamogeton striatus* Ruiz & Pav.; *P. latifolius* (J. W. Robbins) Morong]. Calcareous and alkaline waters of rivers, canals, irrigation ditches, and ponds. NMEX: SJn. 1600 m (5275′). Flowering and fruiting unknown in our range. Oregon east to Colorado, south to California, Texas; Central America; South America. This species is uncommon throughout its range in the western United States. This was previously known as *Potamogeton latifolius* but was found to be the same taxon as *S. striata.*

Stuckenia vaginata (Turcz.) Holub. (sheathed) Bigsheath pondweed. **STEMS** readily branched, terete. **LEAVES** on stem only slightly larger than those on branches, stipulular sheaths inflated 3–5 times the stem thickness, 2–9 cm long, ligule absent or to 0.2 mm. Blade filiform to linear, 0.2–2.9 mm wide, apex rounded, obtuse, or slightly notched, veins 1(–3). **INFL** peduncles terminal, erect, 3–15; spikes moniliform, 10–80 mm, 3–12 whorls. **FRUIT** brown, 3–3.8 × 2–2.9 mm, beak small. $2n = 78$. [*Potamogeton vaginatus* Turcz.]. Calcareous, alkaline waters of lakes and ponds. NMEX: SJn. 2740 m (9042′). Flowering: Jul–Aug. Fruit: Jul–Aug. Alaska and Northwest Territories east to Nunavut, and Ontario south to California, New Mexico, Colorado, and Minnesota; Eurasia. One record for this taxon is known in our range from the Chuska Mountains. This is the southernmost record for this species in North America.

PRIMULACEAE Batsch ex Borkh. PRIMROSE FAMILY

Sylvia Kelso

Herbs, perennial or annual. **LEAVES** alternate, opposite, or whorled, often basal, entire to lobed. **INFL** a panicle, raceme, umbel, or with flowers solitary, usually subtended by bracts. **FLOWERS** perfect; sometimes heterostylous (only in *Primula* in our region); sepals usually 5, more or less united; corolla regular; petals 5, united or rarely lacking (*Glaux*); stamens 5, opposite the corolla lobes; pistil 1, ovary superior to partly inferior, 1-loculed, 5-carpelled; style 1; stigma capitate. **FRUIT** a capsule. In its traditional sense, the primrose family contains about 22 genera and over 1000 species worldwide, mostly occurring in temperate and mountainous regions of the Northern Hemisphere. The center of diversity for the family lies in Asia, notably in western China and the Himalayas. Recent molecular analyses have suggested that some genera traditionally placed in the Primulaceae may be more closely related to the primarily tropical families Myrsinaceae and Theophrastaceae. In this view, *Glaux* and *Lysimachia* would be transferred to a more inclusive Myrsinaceae. Other molecular data show a close relationship between *Dodecatheon* and *Primula*; based on this evidence and the floral similarity of the two genera, some botanists suggest that *Dodecatheon* might be placed into a broadly defined genus *Primula*. The nomenclature provided here is the traditional circumscription of the Primulaceae to parallel the most generally accepted treatment for North American floras. Many species of *Androsace* and *Primula* are cultivated in gardens or as potted plants (Källersjö, Bergqvist, and Anderberg 2000; Mast et al. 2004).

1. Plants with leafy stems, diminutive, usually less than 10 cm tall ..... ***Glaux***
1′ Plants with basal leaves only, or if with leafy stems, then taller than 10 cm ..... (2)

2. Flowers axillary, yellow, on slender pedicels ..... ***Lysimachia***
2′ Flowers not as above ..... (3)

3. Corolla lobes reflexed ..... ***Dodecatheon***
3′ Corolla lobes erect or spreading ..... (4)

4. Flowers white; the corolla tube shorter than or barely equal to the calyx ..... ***Androsace***
4′ Flowers shades of pink, rose, purple, rarely white except in mutants; the corolla tube equal to or noticeably longer than the calyx ..... ***Primula***

Androsace L. Rock Jasmine

(Greek *androsac*, man, male, or husband, the reference obscure) **LEAVES** in a basal rosette, single or forming mats or compact cushions, vegetative parts completely efarinose (lacking a mealy powder on the surface). **INFL** an umbel subtended by 1–several bracts, with several to many flowers per umbel. **FLOWERS** pedicellate, calyces campanulate to subglobose, shallowly to deeply lobed; corollas white (in North American species), often fading to pinkish, eye yellow, tube shorter than or barely equal to the calyx, limb 5-lobed, lobes entire to emarginate; homostylous. **FRUIT** subglobose, shorter than the calyx. **SEEDS** few to numerous, somewhat quadrate, usually less than 1 × 1 mm. $x = 10$. About 100 species, widely distributed across the temperate zone of the Northern Hemisphere.

1. Plants perennial, with multiple rosettes, mat-forming .... ***A. chamaejasme***
1' Plants annual or biennial, with a single rosette, not mat-forming .... (2)

2. Involucral bracts broad and leaflike .... ***A. occidentalis***
2' Involucral bracts narrow, lanceolate to subulate .... ***A. septentrionalis***

Androsace chamaejasme Wulfen ex Host (creeping or mat-forming) Boreal rock jasmine. **LEAVES** in multiple rosettes forming a loose or dense mat, margins with prominent ciliate hairs, blade ligulate to oblong, 0.3–1.5 cm long, lacking petioles. **INFL** scape 2–7 (10) cm long, bracts broadly leaflike and often overlapping. **FLOWERS** 3–6 per umbel, pedicel less than 1 cm; calyx green, broadly campanulate to subglobose, about 3 mm; corolla limb 6–9 mm. **FRUIT** subglobose. $2n = 20$. Alpine tundra, in open gravelly areas. COLO: SJn; UTAH. 3000–3500 m (9850–10500′). Flowering: Jun–Jul. Fruit: Jul–Aug. Western North America, Eurasia. The distribution of this species in the region is unclear. It is extremely common throughout northern and central Colorado and documented throughout the high elevations of the mountain front of the southern Rockies into Taos County, New Mexico, on the border of Rio Arriba County. Similarly, it occurs in Utah in the La Sal Mountains on the northwestern edge of the Four Corners region. It is very likely that *A. chamaejasme* occurs in the higher peaks of the San Juan Mountains, where abundant appropriate habitat exists, but no specimens have yet been collected to verify its occurrence in the Four Corners region of Colorado. The Rocky Mountain representatives of this species have previously been classified as subsp. *carinata* (Torr.) Hultén on the basis of a single character: an apparently prominent keel on the leaf blade. Although this character does appear often on plants growing in exposed sites on the tundra, it disappears when these plants are grown in protected sites and transplant gardens. Material collected at the type locality on Pikes Peak in Colorado shows a carinate leaf at high elevation, but in the same material transplanted to lower elevations, the character disappears as the leaves flatten and expand in less environmentally stressful habitats, becoming identical to those described as subsp. *lehmanniana* from northwestern North America and Asia. Since this character is environmentally induced and there are no other reliable characters to distinguish these putative infraspecific taxa, all of our North American material should be treated as **subsp. *lehmanniana*** (Spreng.) Hultén.

Androsace occidentalis Pursh (western) Western rock jasmine. **LEAVES** in a single rosette, lacking petioles, sparsely pubescent, broadly lanceolate or oblanceolate, entire or somewhat toothed, 0.5–3 cm long. **INFL** scape 2–10 cm long, sometimes multiple, bracts lanceolate to elliptic or almost ovate, with 3–10 flowers per umbel, pedicels 0.3–3 cm, slender, erect. **FLOWER** calyx green, broadly campanulate, ridged, about 0.5 cm; corolla limb 3–5 mm. **FRUIT** globose. $2n = 20$. Open areas with sandy or gravelly soil, in grassland, shrub, and open forest communities. ARIZ: Apa; COLO: Arc, Dol, Hin, LPl, Mon, SMg; NMEX: McK, RAr, SJn. 1525–2225 m (5000–7300′). Flowering: May–Jun. Fruit: Jul. Western North America. Used by the Ramah Navajo as a decoction for birth injury.

Androsace septentrionalis L. (northern) Northern rock jasmine, pygmy flower rock jasmine. **LEAVES** in a single rosette, lacking petioles, sparsely or densely pubescent, oblanceolate to spatulate, usually toothed, 0.5–6 cm long. **INFL** scape 0–20 cm long, sometimes multiple, bracts narrowly lanceolate to linear-lanceolate, with 5–over 20 flowers per umbel; pedicels 0.3–6 cm, slender, erect. **FLOWER** calyx green, narrowly campanulate, about 0.5 cm; corolla limb about 0.3 cm. **FRUIT** globose or subglobose. $2n = 20$. Open areas with sandy or gravelly soil in grassland, shrub, forest, and tundra communities. ARIZ: Apa, Nav; COLO: Arc, Con, Dol, Hin, LPl, Min, Mon, RGr, SJn; NMEX: McK, RAr, SJn; UTAH. 1980–3505 m (6500–11500′). Flowering: May–Sep. Fruit: Jun–Oct. Western North America and throughout the Northern Hemisphere. This is our most common species of *Androsace*, extremely variable in its morphology, depending on elevation, exposure, and amount of light. The variation, which has resulted in many infraspecific names being applied over the years, appears to be entirely plastic, depending on the age of individuals and environmental factors. High-elevation individuals may be only 1 cm in height, while lowland individuals can be well over 10 cm; shaded areas produce long pedicels, exposed areas produce very short ones. Location and degree of pubescence and glands also vary greatly between and within populations. The morphological plasticity coupled with frequent selfing produces abundant local variants, none of which appear to be coherent enough to justify taxonomic recognition. Accordingly, *A. septentrionalis* is best treated as a single, highly variable species. Used by the Ramah Navajo for the following: cold infusion for internal pain, "life medicine," taken before sweat bath for venereal disease, and as a lotion to give protection from witches.

Dodecatheon L. Shooting Star

(ancient Greek, "flower of the twelve gods," the classical pantheon of deities in Greek mythology) **LEAVES** usually in a single basal rosette, simple, broadly lanceolate to elliptic, glandular-pubescent or glabrous, efarinose. **INFL** an umbel subtended by 1–several bracts, with 2 to more than 15 flowers per umbel. **FLOWER** calyces broadly campanulate, more or less 5-angled, lobes 4–5, divided about 1/4 the length of the calyx, corolla magenta (rarely white, in mutants), campanulate, limb (4) 5-lobed, usually entire, prominently reflexed at anthesis, homostylous. **FRUIT** cylindric, or somewhat ovoid-elliptic in our species, equal to or longer than the calyx. **SEEDS** numerous, somewhat quadrate, ovate or oblong, usually about 1 × 1 mm. $x = 11$. About 14 species, mostly occurring in North America north of Mexico, with one species in the Russian Far East.

Dodecatheon pulchellum (Raf.) Merr. **CP** (beautiful, referring to the flower) Darkthroat shooting star. **LEAVES** thick, somewhat succulent, tapering to obscure petioles, oblanceolate, 3–20 cm. **INFL** scape 15–50 cm with 1–many flowers per umbel. **FLOWER** corolla tube yellow, 0.8–1.2 cm long, lobes 5, pink to purple, 1–2 cm long; stigma lobe not conspicuously enlarged, little wider than the style. **FRUIT** 0.8–1.8 cm long. $2n = 44$. [*Dodecatheon radicatum* Greene]. Moist montane meadows. COLO: Arc, Mon. 2042–3414 m (6700–11200′). Flowering: Jun–Jul. Fruit: Jul–Aug. Widely distributed across western North America; however, quite rare in the Four Corners region.

Glaux L. Sea Milkwort

(Greek *glaucus*, blue-green, for the color of the leaves and stem) **LEAVES** occurring throughout the stem, usually opposite below, becoming alternate above, simple, oval to narrowly oblong, efarinose. **INFL** with flowers solitary in the axils of the leaves. **FLOWER** calyces campanulate, with 4–5 petal-like lobes; corolla lacking. **FRUIT** globose. **SEEDS** few, somewhat flattened, about 1 × 1 mm. $x = 15$. A single species, widely distributed in the temperate zone of the Northern Hemisphere.

Glaux maritima L. (seaside, referring to the saline habitat) Sea milkwort. Plants diminutive, growing from shallow, slender rhizomes, vegetative parts efarinose. **LEAVES** appearing throughout the stem, sessile, 0.3–2 × 0.1–0.5 cm, elliptic to oblanceolate. **INFL** scape 3–25 cm, with flowers appearing in the leaf axils, sessile or subsessile. **FLOWER** calyx white or pinkish, campanulate, 3–4 mm, 4–5-lobed, petaloid, corolla lacking, homostylous. **FRUIT** globose, 2–3 mm long, more or less equal to the calyx. $2n = 30$. Moist, alkaline or saline soils, seeps, springs, stream banks, and meadows. Known from one location near Navajo Dam. NMEX: SJn. 1830 m (6000′). Flowering: Jun. Fruit: Jul–Sep. Widely distributed in North America and throughout the Northern Hemisphere.

Lysimachia L. Loosestrife

(from the name of the ancient Greek king Lysimachos, believed to have tamed a wild bull with this herb) **LEAVES** occurring throughout the stem, opposite, sessile to long-petiolate, ovate to ovate-lanceolate. **INFL** a terminal raceme or solitary and pedicellate in axils of the leaves. **FLOWER** calyces campanulate, deeply 5 (–9)-parted nearly to the base; corollas rotate, deeply 5 (rarely 3–9)-parted nearly to the base; homostylous. **FRUIT** globose to ovoid. **SEEDS** few to many, oblong to trigonous. About 180 species, mostly distributed in the temperate zone of the Northern Hemisphere, but a few species also occurring in Africa, South America, and Australia.

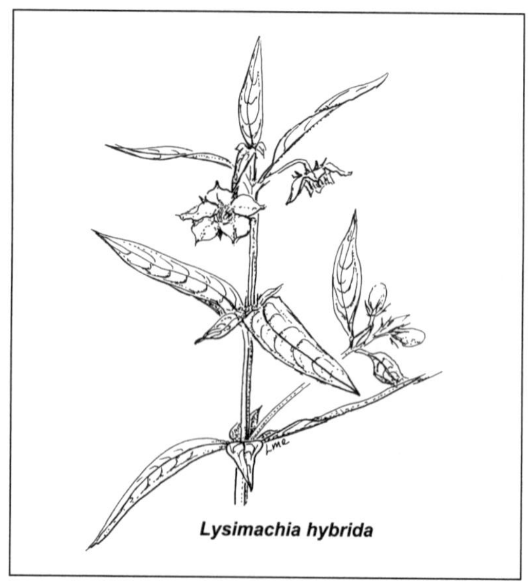

Lysimachia hybrida

Lysimachia hybrida Michx. (referring to an intermediate morphology between *L. ciliata* and *L. lanceolata*) Plants erect to somewhat reclining, stems simple or sparingly branched, to about 1 m. **LEAVES** opposite, sessile above but distinctly petiolate on lower portion of stem, petioles only obscurely ciliate, ovate-lanceolate to lanceolate, 4–7 × 0.5–2.4 cm; base broadly rounded to subcordate. **INFL** pedicellate, with flowers solitary in the axils of the leaves. **FLOWER** calyces green, 0.5–1 cm; corollas yellow, broad. **FRUIT** globose, 4–6 mm long. $2n = 34$. [*L. ciliata* L. var. *validula* (Greene) Kearney & Peebles; *Steironema validulum* Greene]. Moist meadows, stream banks, and shorelines. NMEX: McK, SJn. 1800–2500 m (5900–8100′). Flowering: Jun. Fruit: Jul–Sep. Widely distributed across the eastern half of North America but uncommon in the Southwest. *Lysimachia hybrida* shows a clear relationship to the widespread species *L. ciliata*, from which it differs in having narrower, less broadly ovate leaves, and less prominently ciliate petioles. Although both are currently recognized at the species level, future genetic analysis may suggest that *L. hybrida* be better treated as a subspecies or variety of *L. ciliata*.

Primula L. Primrose

(Latin *primus*, the first, referring to the early spring blooming time) **LEAVES** usually in a single basal rosette, simple, linear, broadly lanceolate to cuneate, pubescent or glabrous, farinose in some species. **INFL** an umbel subtended by 1–several bracts, solitary to more than 15 flowers per umbel, some species heterostylous with pin (long-styled) and thrum (short-styled) flowers, other species homostylous; pedicellate. **FLOWER** calyces broadly campanulate to cylindric, more or less 5-angled, lobed from 1/4 the length to near the middle; corollas pink, rose, violet, or rarely white in mutants, campanulate, limb 5-lobed, lobes 2-cleft to entire, margin usually entire in North American species. **FRUIT** cylindric or ellliptic in our species, equal to or longer than the calyx. **SEEDS** numerous, somewhat quadrate, ovate or oblong, usually less than 1 × 1 mm. $x = 9, 11$ (in North American species). About 500 species, mostly indigenous to the north-temperate zone, with only a few outliers in the mountains of Africa (Ethiopia), tropical Asia, Central and South America.

1. Leaves somewhat fleshy, broadly or narrowly lanceolate to spatulate; involucral bracts unequal in size; plants usually over 20 cm tall and rankly aromatic; higher mountains ..... ***P. parryi***

1′ Leaves thin, spatulate to elliptic; involucral bracts more or less equal in size; plants less than 20 cm tall; hanging garden communities ..... ***P. specuicola***

Primula parryi A. Gray (for Charles Parry, early Colorado botanist) Parry's primrose. Plants robust, rosettes single but often clumped; vegetative parts lacking farina. **LEAVES** somewhat succulent, with a skunky odor, oblanceolate to oblong-obovate, almost entire to remotely denticulate, petioles broadly winged. **INFL** scape 15–50 cm long, 5–25-flowered, pedicels often unequal, 0.3–1.5 cm. **FLOWERS** heterostylous, calyx green, often purple-tinged, cylindrical to campanulate, 0.8–1.5 cm; corolla magenta, tube 0.5–2 cm, limb 1–2.5 cm. **FRUIT** broadly cylindrical, equal to the calyx. $2n = 44$. Subalpine and alpine bogs, streamsides, and wet meadows. COLO: Arc, Con, Hin, LPl, Min, Mon, RGr, SJn; UTAH. 2392–4115 m (7900–13500′). Flowering: Jun–Jul. Fruit: Aug–Sep. Western United States from Idaho to Arizona, east to Colorado and New Mexico. This is the most widespread species of primrose in the western United States; it is particularly abundant in Colorado. *Primula parryi* is relatively uncommon in Arizona and New Mexico, where it occurs only on the highest peaks.

Primula specuicola Rydb. (cave, referring to its preferred habitat of alcove hanging gardens) Cave primrose, Easterflower. Plants of medium stature, rosettes single, vegetative parts heavily farinose. **LEAVES** with petioles broadly winged, thin, spatulate, sinuate-dentate on the margins. **INFL** scape to 25 cm long, 6–25-flowered, pedicels erect, 1–3 cm. **FLOWERS** heterostylous; calyx green, campanulate, 3–5 mm; corolla lavender, tube 0.8–1 cm, 2× length of calyx, limb 1–1.6 cm. **FRUIT** elliptic, more or less equal to calyx, occasionally exserted. $2n = 18$. [*P. hunnewellii* Fernald]. Moist seepage areas on carbonate-rich sandstones in canyons along the Colorado River and its tributaries. ARIZ: Apa; UTAH. 1006–2057 m (3300–6800′). Flowering: Apr–Jun. Fruit: Aug–Sep. Arizona and Utah.

RANUNCULACEAE Juss. CROWFOOT or BUTTERCUP FAMILY

Bruce D. Parfitt and Alan T. Whittemore

Perennial or annual herbs, sometimes woody or herbaceous climbers, often rhizomatous. **LEAVES** with blades entire, toothed, or lobed, simple or variously compound, basal and/or cauline, alternate or opposite (rarely whorled); petioles usually present, often proximally sheathing; stipules present or absent. **INFL** terminal or axillary, racemes, cymes, umbels, panicles, or spikes, or flowers solitary, pedicels present or absent. **FLOWERS** actinomorphic or zygomorphic, bisexual (sometimes unisexual in *Thalictrum*, *Clematis*, and some Asian genera), inconspicuous or showy; sepals 3–6 (–20), distinct, often petaloid and colored, occasionally spurred; petals 0–26, distinct, greenish, white, blue, purple, yellow, pink, or red, usually bearing nectaries, planar, cup-shaped, funnel-shaped, or spurred, showy or reduced to minute stalked nectaries; stamens 5–many, distinct, free; staminodes present only in *Aquilegia* and *Clematis*; pistils 3–many per flower (1 in *Actaea* and genera not in the flora), ovary superior, styles and stigmas 0–1 per pistil, often persistent as a beak in fruit. **FRUIT** an aggregate of achenes or follicles (rarely utricles, capsules, or solitary berry), achenes often in globose to cylindric heads. **SEEDS** 1–many per ovary. About 60 genera with 1700 species worldwide. Species cultivated in *Aquilegia*, *Delphinium*, *Clematis*, *Trollius*, and other genera; weedy species in *Ceratocephalus* and

Ranunculus. Flowers may begin to open before the floral organs have achieved mature size and morphology; use only mature flowers with dehisced anthers to determine diagnostic characteristics (especially measurements) (Flora of North America Editorial Committee 1997).

1. Flowers zygomorphic, blue to white; pistils 3 (–5) per flower; fruits several-seeded follicles ....(2)
1' Flowers actinomorphic, blue, white, yellow, green, or red; pistils 1 or 4–many per flower (2–6 in *Thalictrum alpinum*); fruit an aggregate of 1-seeded achenes or utricles, or several-seeded follicles or a berry ....(3)

2. Upper (adaxial) sepal hood- or helmet-shaped; petals completely hidden by colored calyx ....***Aconitum***
2' Upper (adaxial) sepal spurred; petals at least partly exserted from colored calyx ....***Delphinium***

3. Fruit a berry, pistil 1 per flower ....***Actaea***
3' Fruit an aggregate of achenes, utricles, or follicles, pistils 2–many per flower ....(4)

4. Plants annual, scapose, from a slender taproot; flowers solitary, relatively inconspicuous; fruit an aggregate of achenes, many (more than 20) per flower...(5)
4' Plants usually perennial, aerial (or aquatic) stems elongated, leafy (or if scapose or nearly so and perennial, then inflorescence a raceme or flowers showy with yellow petals); flowers few to many per stem, or if solitary then showy; fruits achenes, utricles, follicles, or a berry, 1–many per flower..(6)

5. Plants gray-tomentose; leaves ternate to biternate with linear lobes; achenes in globose to thick-cylindric aggregate ....***Ceratocephalus***
5' Plants glabrous; leaves entire, linear or very narrowly oblanceolate; achenes in long, tapered, slender aggregate.. ....***Myosurus***

6. Petals prominent, spurred; fruit aggregate of follicles, 5 per flower..***Aquilegia***
6' Petals (if present) planar, sometimes with a cupped nectary near the base; fruit an aggregate of achenes, utricles, or follicles...(7)

7. Fruit a several-seeded aggregate of follicles, 4–15 per flower...(8)
7' Fruit an aggregate of 1-seeded achenes or utricles, usually many per flower (2–6 in *Thalictrum alpinum*; 15–16 in *Trautvetteria*) ....(9)

8. Leaves simple, blade often lobed 1/2–3/4 its length, margins entire, crenate, or toothed; petals absent.....***Caltha***
8' Leaves palmately compound or divided to base; petals usually inconspicuous...***Trollius***

9. Sepals valvate, 4; leaves all cauline and opposite; stems ± woody vines, or if not woody then sepals petaloid, violet-blue or pink (rarely white), more than 25 mm long ....***Clematis***
9' Sepals imbricate, 3–9, if sepals 4 then less than 15 mm long and/or not violet-blue or pink; leaves basal, or basal and cauline, alternate (leaflike involucral bracts alternate, opposite, or whorled); stems herbaceous ....(10)

10. Plants with whorled or paired (opposite) involucral bracts, these leaflike (sepals petaloid, white, violet-blue, or rarely white) ....***Anemone***
10' Plants without whorled or paired involucral bracts (rarely a pair of opposite, unlobed leaves in *Ranunculus alismifolius* and *R. flammula* with green sepals and showy yellow petals); cauline leaves (if present) alternate ....(11)

11. Petals present, white or yellow; inflorescences simple or compound cymes or flowers solitary............***Ranunculus***
11' Petals absent; inflorescences panicles, racemes, or corymbs...(12)

12. Leaves pinnately or ternately compound, leaflets less than 3 cm wide; flowers commonly unisexual (bisexual in scapose, racemose *T. alpinum*) ....***Thalictrum***
12' Leaves simple, 8–30 (–40) cm wide; flowers bisexual ....***Trautvetteria***

Aconitum L. Monkshood, Aconite, Wolfsbane

(for the ancient Black Sea port Aconis) Perennial herbs with tubers or elongate, fascicled roots, glabrous to pubescent. **STEMS** erect or reclining. **LEAVES** basal and cauline, alternate; blade pentagonal in outline, deeply palmately divided into 3–7 segments; ultimate segments narrowly elliptic or lanceolate to linear, margins incised and toothed. **INFL** terminal, sometimes also axillary, 2–32+-flowered racemes or panicles, 4–45 cm long, or flowers solitary; bracts leaflike, not forming involucre. **FLOWERS** zygomorphic, bisexual; sepals 5, usually blue (sometimes white to yellowish); sepals 5, lower sepals (pendents) 2, planar, 6–20 mm long; lateral sepals 2, round-reniform; upper sepal (hood) 1, saccate, arched, crescent-shaped or hemispheric to rounded-conic or tall and cylindric, usually beaked, 10–50 mm long; petals 2–5, blue to whitish, bearing near apex a capitate to coiled spur with a nectary concealed inside sepal-hood; stamens 25–50; filaments with base expanded; staminodes absent between stamens and pistils; pistils 3 (–5), each with 10–20 ovules; styles filiform. **FRUIT** aggregate, sessile, oblong follicles, fruit walls prominently transversely veined; beak straight to curved, 2–3 mm. $x = 8$. About 100 species in North America south into Mexico, Eurasia, and North Africa. The distinctive "monk's hood" upper sepal distinguishes *Aconitum* from all other genera. *Aconitum* contains poisonous diterpene alkaloids historically used as medicines and as arrow or spear poisons. *Aconitum* extracts appear to have been the most widely used poison to hunt game.

Aconitum columbianum Nutt. **CP** (of the Columbia River region) Columbine monkshood. Roots tuberous, tuber to 60 × 15 mm, parent tuber connected with 1 (rarely 2) daughter tubers by very short rhizome. **STEMS** erect and stout to weak and reclining, 2–30 dm long, glabrous below inflorescence; bulbils absent from leaf axils and inflorescences. **LEAVES** (cauline) with blade 5–20 cm wide, deeply 3–5 (–7)-divided, segment margins variously incised and toothed. **INFL** pilose. **FLOWERS** commonly blue, sometimes white, cream-colored, or blue-tinged at sepal margins, 18–50 mm long from tips of pendent sepals to top of hood; pendent sepals 6–16 mm long; hood conic-hemispheric, hemispheric, or crescent-shaped, 11–34 mm from receptacle to top of hood, 6–26 mm wide from receptacle to beak apex; pistils 3 per flower. **FRUIT** 12–20 mm long. $2n = 16, 18$. [*A. columbianum* var. *ochroleucum* A. Nelson; *A. columbianum* subsp. *pallidum* Pipe; *A. geranioides* Greene; *A. infectum* Greene; *A. leibergii* Greene; *A. mogollonicum* Greene; *A. noveboracense* A. Gray]. Moist places in riparian, meadow, or forest of alder-willow, pine–Douglas-fir, oak, aspen, spruce-fir. ARIZ: Apa; COLO: Arc, Hin, LPl, Min, Mon, RGr, SJn; NMEX: McK, RAr, SJn; UTAH. 2300–3550 m (7600–11600′). Flowering: Jun–Sep. British Columbia to South Dakota, south to California, Arizona, New Mexico, and Mexico; disjunct in Iowa, Wisconsin, Ohio, and New York. Ours are **subsp. *columbianum***, extremely variable within populations and groups of populations.

Actaea L. Baneberry, Necklaceweed, Cohosh, Bugbane

(*aktea*, ancient Greek name for elder, a plant with compound leaves) Perennial herbs from caudices about 1 cm thick, glabrous to puberulent. **STEMS** erect to ascending, 0–few-branched. **LEAVES**: basal leaves reduced to clasping scales 0.8–4 cm long; cauline leaves alternate, compound, blades broadly ovate to reniform in outline, 1–3 times ternate or 1–3 times pinnate, leaflets ovate to narrowly elliptic, 0–3-lobed, margins sharply cleft, irregularly coarsely dentate. **INFL** usually terminal, 20–50-flowered racemes, 2–17 cm long; bracts leaflike or absent, not forming an involucre. **FLOWERS** actinomorphic, bisexual; sepals 3–5, whitish green, planar, orbiculate, 2–4.5 mm long; petals 4–10, cream-colored, planar, spatulate to obovate, 2–4.5 mm long, glabrous, nectary absent; stamens 15–50, filaments filiform; staminodes absent between stamens and pistils; pistil 1, with several ovules; style absent or nearly so. **FRUIT** a sessile, broadly ellipsoid to nearly globose berry, fruit wall smooth; beak none, stigma appearing essentially sessile on fruit (short beak apparent on immature fruit). $x = 8$. About eight species; temperate to cool forests throughout the Northern Hemisphere.

Actaea rubra (Aiton) Willd. (red) Red baneberry. **LEAVES** with leaflets abaxially glabrous or pubescent. **INFL** at anthesis ± densely flowered, as long as wide or longer; elongating in flower. **FLOWER** petals acute to obtuse at apex; stigma nearly sessile, 0.7–1.2 mm diameter during anthesis, much narrower than ovary. **FRUIT** red or white, widely ellipsoid to spherical, 5–11 mm long; pedicel dull green or brown, filiform, much thinner than axis of raceme, 0.3–0.7 mm diameter. **SEEDS** 2.9–3.6 mm long. $2n = 16$. [*A. spicata* L. var. *rubra* Aiton; *A. arguta* Nutt.; *A. eburnea* Rydb.; *A. rubra* subsp. *arguta* (Nutt.) Hultén; *A. neglecta* Gillman; *A. rubra* var. *dissecta* Britton; *A. viridiflora* Greene]. Moist, shaded areas in forests of aspen, ponderosa pine, Douglas-fir, spruce, spruce-fir, riparian areas. ARIZ: Apa; COLO: Arc, Hin, LPl, Min, Mon, SJn; NMEX: McK, RAr, SJn; UTAH. 2300–3400 m (7500–11200′). Flowering: May–Jun. Alaska to Maine, south to California, Arizona, New Mexico, and New Jersey.

Anemone L. Windflower, Anemone, Thimbleweed

(Greek *anemos*, wind) Perennial herbs with rhizomes, caudices, or tubers, glabrous to hairy. **STEMS** erect, usually unbranched below inflorescence. **LEAVES** basal (involucre also leafy), simple or compound; blades reniform to obtriangular or lanceolate in outline, lobed or parted or undivided, segments or lobes filiform to broadly ovate, oblanceolate, or rhombic, margins entire, divided, or variously toothed. **INFL** a terminal 2–9-flowered cyme to 60 cm long, or an umbel, or flower solitary; involucres leaflike, opposite or whorled in 1–3 tiers, the primary tier subtending inflorescences, the secondary and tertiary tiers (when present) subtending branches or single flowers, involucral leaves (bracts) 2–7 (–9), closely subtending or distant from flowers. **FLOWERS** actinomorphic, bisexual; sepals 4–20 (–27), white, purple, blue, green, yellow, pink, or red, planar, linear to oblong or ovate to obovate, 3.5–40 mm long, glabrous to puberulent, sericeous or pilose; petals usually absent (reduced petals in *A. patens*, planar, obovate to elliptic, 1.5–2 mm long); nectary present; stamens 10–200; filaments filiform or somewhat broadened at base; staminodes absent; pistils many, each with 1 ovule; style linear-subulate to filiform. **FRUIT** aggregate, sessile or stalked head of ovoid to obovoid achenes, fruit wall not veined; beak straight or curved, 0.5–40 (–50) mm long (sometimes rudimentary), often plumose. x = 7, 8. [*Anemonastrum* Holub; *Pulsatilla* Mill.]. About 150 species nearly worldwide, especially in cooler temperate and Arctic regions. An acrid oil in many species of *Anemone* causes severe topical and gastrointestinal irritation but becomes harmless upon drying.

1. Sepals (18–) 20–40 mm long; involucral leaves connate, sessile; achene beak plumose, 20–40 mm long (on mature fruit) .... ***A. patens***
1′ Sepals 5–12 (–15) mm long; involucral leaves distinct, petioled or sessile; achene beak glabrous, 0.3–6 mm (on mature fruit) .... (2)

2. Involucral leaves clearly petioled, petioles terete, often with an adaxial groove, uninterrupted portion of blades (4–) 6 mm wide or more .... ***A. cylindrica***
2′ Involucral leaves sessile, or if petioled then petioles flattened or winged at least distally, uninterrupted portion of blades 1.5–3 (–4.3) mm wide or less .... ***A. multifida***

Anemone cylindrica A. Gray (cylindric, for the shape of the heads of achenes) Long-headed anemone, thimbleweed, candle anemone. **STEMS** (14–) 28–70 (–80) cm tall. **LEAVES** (basal) (2–) 5–10 (–13), ternate; petiole 9–21 cm long; terminal leaflet broadly rhombic to oblanceolate, crenate, or serrate and deeply incised on distal 1/2, strigose, more so abaxially; lateral leaflets 1–2 times parted and -lobed; uninterrupted portion of blades 4–10 (–13) mm wide. **INFL** a 2–8-flowered cyme or flower solitary, sometimes appearing umbel-like; involucral leaves 3–7 (–9), distinct, ± similar to basal leaves, 2 (–3)-tiered (can appear 1-tiered), ternately compound or divided, petioled; petioles terete, often with an adaxial groove; leaflets or major segments oblanceolate, cuneate, 2–3-lobed, lacerate, serrate, puberulous, more so abaxially, the various lobes and teeth very unequal in length, uninterrupted portion of blades (4–) 6–10 (–15) mm wide. **FLOWER** sepals 4–5 (–6), green to whitish, 5–12 (–15) × 3–6 mm; petals absent; stamens 50–75. **FRUIT** a head of cylindric achenes; achene body ovoid, (1.8–) 2–3 × 1.5–2 mm, woolly; beak usually recurved, (0.3–) 0.5–1 mm, hidden by achene indument, glabrous. $2n$ =16. Montane forest, spruce-fir. COLO: Arc, LPl, Hin. 2350–2600 m (7700–8600′). Flowering: Jun–Jul. British Columbia to Quebec and Maine, south to Pennsylvania, Illinois, Kansas, New Mexico, and Arizona. Used by Native Americans for headaches, sore eyes, and bad burns; also as a psychological aid and relief for tuberculosis.

Anemone multifida Poir. (much cleft, referring to the leaves) Cut-leaved anemone. **STEMS** 3–70 cm tall. **LEAVES** (basal) 3–8 (–10), 1–2-ternate; petiole (1.5–) 4–14 cm long; terminal leaflet broadly and irregularly rhombic to obovate in outline, deeply divided into narrow lobes, abaxially hairy, adaxially glabrous to hairy; lateral leaflets (2–) 3 times parted; uninterrupted portion of blades (1.5–) 2–3.5 (–5) mm wide. **INFL** a 2–7-flowered cyme or flower solitary; involucral leaves usually 3–5, distinct, ± similar to basal leaves, 1–2-tiered, ± ternately divided, sessile, or if petioled then petioles flattened or winged at least distally; blade segments linear to narrowly elliptic, usually again deeply twice-lobed, adaxially glabrous to hairy; the lobes usually unequal in length, uninterrupted portion of blades 1.5–3 (–4.3) mm wide. **FLOWER** sepals 5–9, usually (60%) purple, red, or yellow and red, sometimes (30%) green to yellow, white, or possibly blue, 5–12 × (3.5–) 5–7 (–9) mm; petals absent; stamens 50–80. **FRUIT** a spheric head of achenes; achene body irregularly elliptic, flat, 3–4 × 1.5–2 mm, woolly; beak ± straight and distally recurved or strongly hooked, 1–6 mm, glabrous. $2n$ =32. [*A. globosa* Nutt. ex A. Nelson; *A. multifida* var. *hirsuta* C. L. Hitchc.; *A. multifida* var. *hudsoniana* DC.] Rock outcrops, gravelly hills and ridges, montane meadows, aspen, spruce-fir, and alpine. COLO: Arc, Hin, LPl, RGr, SJn; UTAH. 1850–3650 m (6100–12000′). Flowering: May–Jul. Alaska to Newfoundland, south to New York, Wisconsin, Nebraska, New Mexico, Utah, Nevada, California; South America. Although var. *multifida* and var. *saxicola* B. Boivin might be expected in our area, we are unable to consistently distinguish varieties among the available specimens. Used by Native Americans as an antirheumatic, cold remedy, nosebleed cure, general panacea, and to relieve headaches and kill lice and fleas.

Anemone patens L. (spreading) Pasqueflower, prairie-crocus. **STEMS**, including inflorescence, 5–40 (–60) cm tall. **LEAVES** (basal) (1–) 3–8 (–10), primarily 3-foliolate; petioles 5–13 (–16) cm long; terminal leaflet obovate in outline, deeply divided into very narrow lobes, hairy (rarely glabrous); lateral leaflets 3–4 times parted (± dichotomously); uninterrupted portion of blades 2–4 mm wide. **INFL** 1-flowered; involucral leaves usually 3, connate-clasping, dissimilar to basal leaves, 1-tiered, simple, divided almost to the base, sessile; blade segments linear-filiform, entire except sometimes with 1–2 lateral lobes, sericeous or often shaggy-villous (rarely glabrous or nearly so), the lobes ± equal in length, uninterrupted portion of blades 1–2 (–3) mm wide. **FLOWER** sepals 5–8, blue or purple to rarely nearly white,

(18–) 20–40 × (8–) 10–16 mm; petals 1.5–2 mm long; stamens 150–200. **FRUIT** a spheric to ovoid head of achenes; achene body ellipsoid to obovoid, 3–4 (–6) × about 1 mm, hairy; beak curved, 20–40 mm long, plumose. $2n = 16$. [*A. patens* var. *nuttalliana* (DC.) A. Gray; *A. patens* var. *wolfgangiana* (Besser) Koch; *Pulsatilla patens* (L.) Mill. subsp. *multifida* (Pritz.) Zämelis]. Meadows and clearings; sagebrush, Gambel's oak, ponderosa pine. COLO: Arc, LPl; NMEX: RAr. 2000–2400 m (6600–7800'). Flowering: May–Jun. Alaska to Ontario, south to Illinois, Nebraska, and New Mexico; Eurasia. Ours is **var. *multifida*** Pritz., not to be confused with the name of a very different taxon, *A. multifida*, which has mature achenes and sepals less than half the length of those in *A. patens*. Used by Native Americans to treat rheumatism, neuralgia, headaches, and lung problems; also as poultices.

Aquilegia L. Columbine

(origin uncertain; possibly Latin *aqua*, water, and *legere*, to draw or collect, from the wet habitats or abundant nectar in the spurs; or Latin *aquila*, eagle, for the talon-shaped, curved spurs of some European species) Perennial herbs with slender woody rhizomes. **STEMS** erect, usually unbranched below inflorescence, glabrous or proximally sparsely pilose. **LEAVES** basal and cauline, alternate; blades broadly ovate to reniform in outline, 1–3 times ternately compound, leaflets crenate, glabrous or sparsely pilose. **INFL** an open, terminal, 1–10-flowered cyme to 30 cm long, often viscid-pubescent, involucres absent. **FLOWERS** actinomorphic, bisexual; sepals 5, white to blue, yellow, or red, planar, narrowly ovate to oblong-lanceolate, 7–51 mm long, glabrous; petals 5, white to blue, yellow, or red, each basally forming a backward-pointing tubular spur, apex forming a planar, oblong to rounded or spatulate blade, 0–30 mm long, glabrous, nectary in the enlarged tip of spur, without a scale; stamens numerous, filaments filiform; inner stamens often modified as scalelike staminodia; pistils 5–10, each with many ovules, styles filiform. **FRUIT** aggregate, sessile cylindrical follicles, fruit wall prominently veined, styles persistent but scarcely modified in fruit. About 70 species across North America and Eurasia. The species of *Aquilegia* are polymorphic and difficult to define adequately. Some of this variability is due to introgressive hybridization. Even distantly related species of columbine are often freely interfertile, and many cases of natural hybridization and introgression are known from North America. Only the most important are mentioned below. In arid areas, *Aquilegia* species tend to form small populations that are often completely isolated from one another. This leads to local fixation of recessive genes and therefore increased variability in species such as *A. micrantha* and *A. desertorum*. In addition, mutations giving rise to spurless petals are occasionally found in many species. The floral organs (especially the spurs and stamens) continue to elongate up to anthesis, so measurements should only be taken from flowers with open anthers (Grant 1952; Miller 1978; Munz 1946; Pelton 1958).

1. Sepals 28–43 × 8–23 mm; flowers erect, blue or white .... ***A. coerulea***
1' Sepals 7–20 × 3–6 mm; flowers erect to pendent, white, cream, blue, pink, or red .... (2)

2. Sepals erect, scarcely longer than petal blades, green distally; spurs and distal part of sepals red; leaflets not viscid .... ***A. elegantula***
2' Sepals spreading perpendicularly, much longer than petal blades, not green distally; spurs and sepals white, cream, blue, or pink; leaflets viscid .... ***A. micrantha***

Aquilegia coerulea E. James **CP** (blue) Colorado blue columbine. **STEMS** 15–80 cm tall. **LEAVES** (basal) 2–3-ternate, 9–37 cm long, much shorter than stems; leaflets green adaxially, to 13–42 (–61) mm long, not viscid; primary petiolules (10–) 20–70 mm long (therefore leaflets not crowded), glabrous or occasionally pilose (very rarely very minutely viscid). **FLOWERS** erect; sepals spreading perpendicularly, white to medium or deep blue or pink, elliptic-ovate to lance-ovate, 28–51 mm long, 8–26 mm wide, apex obtuse to acute or acuminate; petals with spurs white, blue, or sometimes pink, straight to divergent, (25–) 34–70 mm long, slender, evenly tapered from base; terminal blade white, oblong or spatulate, (17–) 20–24 mm long, 5–14 mm wide; stamens 13–19 mm long. **FRUIT** 20–30 mm long, styles 8–12 mm long. $2n = 14$. [*A. coerulea* var. *daileyae* Eastw.]. Across its broad range, *A. coerulea* consists of four varieties that are mostly allopatric, with limited overlap in the Four Corners region and a few other places. Populations with the small flowers of the Rocky Mountain var. *coerulea* and the pale color of the Great Basin var. *pinetorum* are probably the result of local hybridization between these two taxa, which are otherwise distinct over their broad ranges. *Aquilegia coerulea* has been reported to hybridize with *A. elegantula* (Pelton 1957; Miller 1978) and *A. flavescens* S. Watson (Grant 1952). Most authors have spelled the epithet *caerulea*. However, *coerulea* is the original spelling.

1. Sepals medium to deep blue; petal spurs 34–48 mm (population mean 39–45 mm); stamens 13–19 mm .... **var. *coerulea***
1' Sepals white, pale blue, or pink; petal spurs 45–70 mm (population mean 50–58 mm); stamens 18–24 mm .... **var. *pinetorum***

var. *coerulea* **LEAVES** 2-ternate. **FLOWER** sepals medium or deep blue, 28–43 mm long; petal spurs 34–48 mm long, petal blades (17–) 20–24 mm long; stamens 13–19 mm long. Rocky slopes or near streams, in open woodland or herbland, Gambel's oak–ponderosa pine, aspen, Douglas-fir, spruce-fir, subalpine to alpine. COLO: Arc, Con, Hin, LPl, Min, Mon, RGr, SJn; NMEX: SJn; UTAH. 1850–3950 m (6000–13000′). Flowering: Jun–Aug. New Mexico to southern Wyoming.

var. *pinetorum* (Tidestr.) Payson ex Kearney & Peebles (of the pines) **LEAVES** 2–3-ternate. **FLOWER** sepals white, pale blue, or pink, (22–) 29–51 mm long; petal spurs 43–72 mm long, petal blades 20–28 mm long; stamens 17–24 mm long. Riparian; aspen-fir forest. ARIZ: Apa; COLO: LPl; UTAH. 2560–2680 m (8400–8800′). Flowering: Jun–Jul. Colorado to Utah and Arizona.

Aquilegia elegantula Greene (little elegant one) Western red columbine. **STEMS** 10–60 cm tall. **LEAVES** (basal) 2-ternate, 7–30 cm long, usually shorter than stems; leaflets green adaxially, to 11–33 mm long, not viscid; primary petiolules 17–58 mm long (therefore leaflets not crowded), glabrous or pilose. **FLOWERS** pendent; sepals erect, red proximally, yellow-green distally, elliptic-ovate, 7–11 mm long, 4–5 mm wide, apex rounded to acute; petals with spurs red, straight, 16–23 mm long, stout (at least proximally), abruptly narrowed near the middle; terminal blade yellow-green, oblong or rounded, 6–8 mm long, 3–4 mm wide; stamens 8–14 mm long. **FRUIT** 13–20 mm long, styles 13–15 mm long. Moist coniferous forests, especially along streams. Ponderosa pine and subalpine spruce-fir forests. ARIZ: Apa; COLO: Arc, Con, LPl, Hin, Min, Mon, SJn; NMEX: RAr, SJn. 2250–3950 m (7400–13000′). Flowering: May–Jul. Northern Utah and Colorado to northern Mexico. Reported to hybridize with *A. coerulea* (Pelton 1957; Miller 1978) and *A. chrysantha* A. Gray (Munz 1946).

Aquilegia micrantha Eastw. **CP** (small-flowered) Alcove columbine. **STEMS** 30–60 cm tall. **LEAVES** (basal) 2–3-ternate, 10–35 cm long, much shorter than stems; leaflets green adaxially, to 13–32 mm long, viscid; primary petiolules 21–64 mm long (therefore leaflets not crowded), glandular-pubescent or glandular. **FLOWERS** erect or nodding; sepals spreading perpendicularly, white, cream, blue, or pink, oblong-lanceolate, 8–20 mm long, 3–6 mm wide, apex acuminate to obtuse; petals with spurs white or colored like sepals, straight or divergent, 15–30 mm long, slender, evenly tapered from base or occasionally abruptly narrowed near the middle; terminal blade white or cream, oblong, 6–10 mm long, 3–7 mm wide; stamens 9–14 mm long. **FRUIT** 10–20 mm long, styles 8–10 mm long. [*A. flavescens* S. Watson var. *rubicunda* (Tidestr.) S. L. Welsh; *A. micrantha* var. *mancosana* Eastw.; *A. navajonis* A. Nelson]. Sandy soils in hanging garden communities, seepy rock walls of canyons. ARIZ: Apa, Nav; UTAH. 1150–2050 m (3700–6700′). Flowering: Apr–Jun. Arizona to northern Utah and Colorado.

Caltha L. Marsh-marigold

(Greek name for some yellow-flowered plants) Perennial herbs with thick caudices 0.5–2 cm diameter or slender stolons, glabrous. **STEMS** erect or ascending, 0–few-branched. **LEAVES** basal and often cauline, alternate, simple; blade oblong-ovate to orbiculate-reniform or cordate, unlobed, margins entire, dentate, or crenate. **INFL** a terminal or axillary, 2–6-flowered cyme to 30 cm long, or flower solitary; bracts leaflike, not forming involucre. **FLOWERS** actinomorphic, bisexual; sepals 5–12, white, pinkish, yellow, or orange, planar, oval-orbiculate to narrowly obovate, 4–23 mm long; petals and nectary absent; stamens 10–40, filaments filiform; staminodes absent between stamens and pistils; pistils 5–many, each with 15–35 ovules; style stout, recurved at tip. **FRUIT** sessile or stipitate, linear-oblong to ellipsoid follicles, fruit walls prominently veined or not; beak straight or weakly curved, 0.2–2 mm long. $x = 8$. Ten species, mainly temperate wetlands, worldwide.

Caltha leptosepala DC. **CP** (having narrow sepals) White marsh-marigold. Plants 2–30 cm tall. **STEMS** with aboveground portion leafless or with 1 leaf, erect. **LEAVES** usually all basal; blade oblong-ovate to orbiculate-reniform, largest 1.5–11.5 (–15) × 1–13 cm, margins entire or crenate to dentate. **INFL** 1–2 (–4)-flowered. **FLOWERS** 15–40 mm diameter; sepals adaxially white, abaxially bluish, 8.5–23 mm long. **FRUIT** 4–15, ± spreading, short-stipitate or sessile, linear-oblong; bodies 10–20 × 3–4.5 mm. **SEEDS** elliptic, 1.9–2.5 mm long. $2n = 48, 96$. [*Caltha biflora* DC.; *C. uniflora* Rydb.; *Psychrophila leptosepala* (DC.) W. A. Weber]. Open riparian, wet meadows, and marshy edges of lakes, montane to alpine. COLO: Arc, Con, Hin, LPl, Min, Mon, RGr, SJn. 3100–3900 m (10200–12800′). Flowering: May–Sep. Alaska to Arizona, California, and New Mexico.

Ceratocephalus Pers. Curveseed, Butterwort

Alan T. Whittemore and Bruce D. Parfitt

(spiky head) Annual herbs, scapose, with a slender taproot, tomentose. **STEMS** absent. **LEAVES** all basal; blade broadly spatulate in outline, 1–2 times dissected; ultimate segments linear, margins entire. **INFL** a solitary flower at the end of a scape without involucre. **FLOWERS** actinomorphic, bisexual; sepals 5, green (sometimes almost hidden in white tomentum), planar, elliptic, 3–6 mm long, villous; petals 5, yellow, planar, elliptic to oblanceolate or obovate, 3–5 mm long, glabrous, nectary covered by a basally attached scale; stamens 10–20, filaments filiform; staminodia absent between stamens and pistils; pistils 20–50, each with 1 ovule, styles linear-subulate. **FRUIT** sessile, ellipsoid, 3-chambered achenes in long-cylindrical heads, fruit wall thick, not veined, beak straight, lanceolate, spine-tipped, 2–2.5 mm long. x = 7. Ten species, Eurasia and North Africa; introduced in Australia and North America. Often treated as a subgenus of *Ranunculus*, but may be more closely related to *Myosurus*. The structure of the achene in *Ceratocephalus* is unique, with two empty lateral sacs adjacent to the seed.

Ceratocephalus testiculatus (Crantz) Roth (having testicles, actually a pair of empty, round sacs at base of achene) Curveseed butterwort, bur butterwort, little bur, testiculate buttercup. **STEMS** erect or ascending, not rooting nodally, 0.5–8 cm tall, villous, not bulbous-based; roots never tuberous. **LEAVES** broadly spatulate in outline, 0.9–3.8 cm long, 0.5–1.5 cm wide, 1–2 times dissected, segments linear, margins entire, apices obtuse to acuminate. **FLOWER** sepals spreading, 3–6 mm long, 1–4 mm wide, villous; petals 5, yellow, 3–5 mm long, 1–3 mm wide. **FRUIT** with achenes in cylindrical heads 9–16 (–20) mm long, 8–10 (–14) mm wide; the body of each achene 1.6–2 mm long, 1.8–2 mm wide, tomentose, the beak persistent, lanceolate, 3.5–4.5 (–6) mm long; receptacles glabrous. [*C. orthoceras* DC.; *Ranunculus testiculatus* Crantz]. Weedy in disturbed areas in desert scrub, sagebrush, open piñon-juniper, or oak-aspen-ponderosa pine. ARIZ: Apa, Coc, Nav; COLO: Arc, Dol, LPl, Mon, SMg; NMEX: McK, RAr, San, SJn; UTAH. 1700–2600 m (5600–8500′). Flowering: Mar–Jun. California east to Ohio and north to southern Canada; native to Eurasia. *Ceratocephalus testiculatus* is expanding its range rapidly in arid and semiarid areas of North America.

Clematis L. Clematis, Leather Flower

(Greek *clema*, plant shoot, ancient name of a vine) Perennial herbs with elongate rhizomes, or ± woody vines, sometimes woody only at base, glabrous or variously pubescent. **STEMS** erect or prostrate to climbing by means of tendril-like petioles, leaf rachises, and petiolules. **LEAVES** cauline, opposite, simple or compound; blade ovate in outline, undivided or 1–3-pinnately or -ternately compound; leaflets linear to lanceolate, oblanceolate, ovate, oblong, or orbiculate, lobed or unlobed, margins entire or toothed. **INFL** an axillary and/or terminal 2–many-flowered cyme or panicle to 15 cm long or flower solitary or in a fascicle; bracts leaflike or ± scalelike or absent, not forming involucre. **FLOWERS** actinomorphic, bisexual or unisexual; sepals 4 (–5), white, blue, violet, red, yellow, or greenish, planar, ovate to obovate or linear, 6–60 mm long; petals absent; stamens many; filaments filiform to flattened; sometimes staminodes with sterile anthers in pistillate flowers or between sepals and stamens of bisexual flowers; pistils 5–150, each with 1 ovule; style filiform. **FRUITS** aggregate, sessile, lenticular, nearly terete, or flattened-ellipsoid achenes, fruit wall not prominently veined; beak straight or curved, 12–110 mm long, commonly plumose. x = 8. About 300 species worldwide: temperate, subarctic, subalpine, or tropical. Often split into genera based on its diverse floral and vegetative morphology. Many ornamental species; *Clematis orientalis* could escape from cultivation in our area (Johnson 2001).

1. Stems herbaceous, ± erect, 0.15–0.65 m long; petioles and petiolules not tendril-like; perianth narrowly bell- to urn-shaped, ± thick, leathery .... ***C. hirsutissima***
1′ Stems ± woody, climbing or clambering vines 0.5–8 (–20) m long; petioles and petiolules often tendril-like, i.e., curling around slender objects; perianth widely bell-shaped to rotate, thin .... (2)

2. Flowers solitary, terminal on short shoots or rarely long shoots; sepals 2.5–6 cm long; perianth violet-blue, rarely white, widely bell-shaped or tardily rotate .... ***C. columbiana***
2′ Flowers in 7–30 (–65)-flowered compound cymes; sepals 0.6–1.2 cm long; perianth white to cream, rotate, wide-spreading .... ***C. ligusticifolia***

Clematis columbiana (Nutt.) Torr. & A. Gray (of the Columbia River region) Rock clematis. **STEMS** ± woody, viny, climbing or trailing, 0.5–1.5 (–3.5) m long. **LEAVES** with blades 2–3-ternate; leaflets usually very deeply 2–3-lobed, mostly lanceolate to ovate, 15–48 × 8–35 mm; lateral lobes often nearly distinct, 3–12 mm wide, glabrous or with a few long hairs near the midvein and base of blade. **INFL** terminal on short shoots or rarely terminal on long shoots, 1-flowered, with long peduncles subtended by 1–2 pairs of leaves. **FLOWERS** bisexual, ± nodding; perianth widely bell-shaped to rotate; sepals thin, ascending or tardily spreading, not connivent or connivent only in proximal 1/4, violet-blue (rarely white), lance-ovate to ovate, 25–60 mm long, abaxially nearly glabrous to sparsely pilose at least near base, adaxially glabrous, margins sparsely pilose; petaloid staminodes present between stamens and sepals. **FRUIT** flattened; beak 3.5–5 cm long. [*Atragene columbiana* Nutt.; *C. pseudoalpina* (Kuntze) A. Nelson] Rocky, open woods and thickets; riparian, montane grassland, montane shrubland, piñon-juniper, ponderosa pine, oak, Douglas-fir, aspen, spruce-fir. ARIZ: Apa, Nav; COLO: Arc, Hin, LPl, Mon, SMg; NMEX: McK, RAr, SJn; UTAH. 1350–2600 m (4400–8500′). Flowering: May–Jul. Montana south to Arizona, New Mexico, and Texas. The name *C. pseudoalpina* often has been used improperly in place of *C. columbiana.* Two intergrading varieties in western North America; if accepted, ours would be **var**. ***columbiana***. Used by the Thompson Indians for scabs and eczema.

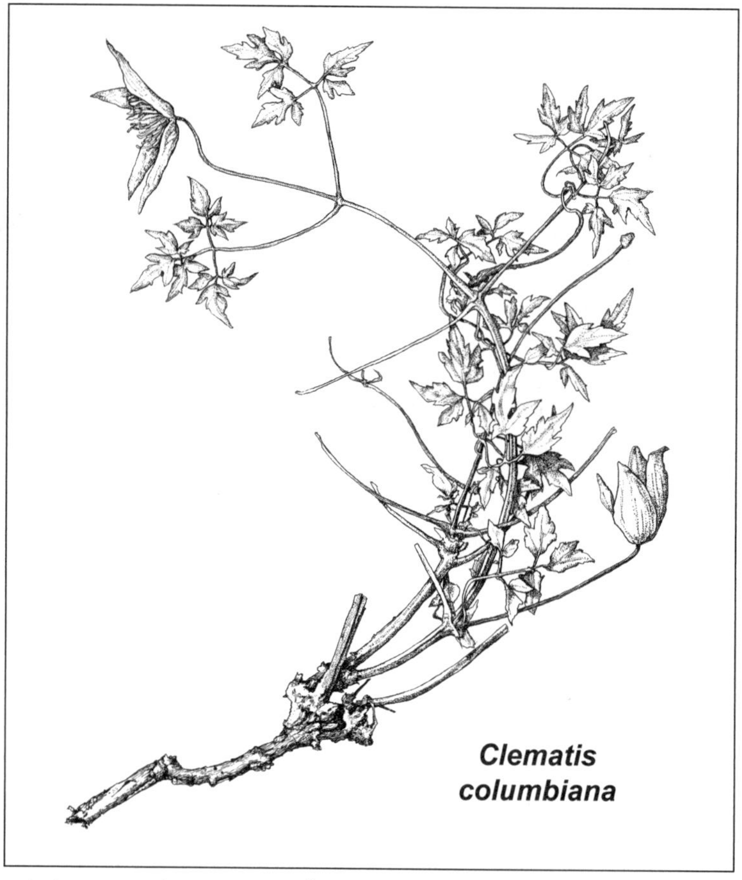

Clematis columbiana

Clematis hirsutissima Pursh (very hairy) Hairy clematis. ± herbaceous perennials. **STEMS** erect, not branched or viny, 1.5–6.5 m long. **LEAVES** with blades 2–3-pinnate; leaflets often deeply 2–several-lobed; lateral lobes narrowly linear to narrowly elliptic-lanceolate, 10–60 × 0.5–8 (–10) mm, nearly glabrous to densely silky-hirsute. **INFL** terminal, 1-flowered. **FLOWERS** bisexual, nodding, perianth broadly cylindric to urn-shaped; sepals thick, usually leathery, connivent at least proximally and usually much of length, very dark violet-blue or rarely pink or white, oblong-lanceolate, 25–45 mm long, abaxially hirsute, usually densely so, adaxially glabrous, margins distally ± crisped, tomentose. **FRUIT** flattened; beak 4–9 cm long. $2n = 16$. [*C. hirsutissima* var. *arizonica* (A. Heller) R. O. Erickson; *Coriflora hirsutissima* (Pursh) W. A. Weber; *Viorna arizonica* (A. Heller) A. Heller; *V. bakeri* (Greene) Rydb.; *V. eriophora* Rydb.; *V. jonesii* (Kuntze) Rydb.; *V. wyethii* (Nutt.) Rydb.]. Sagebrush, mountain brush, piñon-juniper, ponderosa pine, Douglas-fir, aspen-willow, spruce-fir, alpine. ARIZ: Apa; COLO: Arc, Dol, LPl, Mon, SMg; NMEX: McK, RAr, SJn; UTAH. 1750–3300 m (5700–10900′). Flowering: May–Jun. Washington to Montana south to Arizona and New Mexico. Density of leaf pubescence is highly variable. Ours have been treated as **var. *hirsutissima***, which intergrades extensively with var. *scottii* (Porter) R. O. Erickson, occurring to the east of our area and northward. Leaflets and larger lobes of var. *scottii* tend to be more widely spaced and broader (5–15 mm wide), but plants of *hirsutissima* from the opposite (western) side of the Four Corners region sometimes have leaf lobes as wide as 10 mm. Some Native Americans have used finely crushed leaves of this species as snuff or have used the roots to stimulate exhausted horses.

Clematis ligusticifolia Nutt. (leaves of privet, *Ligustrum*) Virgin's-bower, western white clematis, old man's beard, hierba de chivo. **STEMS** woody, viny, clambering or climbing, to 6 (–20) m long. **LEAVES** with blades pinnately 5-foliolate, 2-ternately 9-foliolate, or bipinnately 8–15-foliolate; leaflets lobed or unlobed, lanceolate to broadly ovate, (10–) 30–90 × 9–72 mm; abaxially glabrous or sparsely pilose or silky, especially on veins. **INFL** axillary on current year's stems, usually 7–30 (–65)-flowered compound cymes, often distinctly corymbiform. **FLOWERS** unisexual, staminate and pistillate on different plants (plants dioecious), not nodding; perianth rotate; sepals thin, wide-spreading, not recurved, not connivent, white to cream, obovate to oblanceolate, 6–12 mm long, abaxially and adaxially pilose, margins densely pilose. **FRUIT** flattened or nearly terete; beak 3–4.5 cm long. $2n = 16$. [*C. ligusticifolia* var. *brevifolia* Nutt.; *C. ligusticifolia* var. *californica* S. Watson; *C. neomexicana* Wooton & Standl.; *C. suksdorfii* B. L. Rob.]. Piñon-juniper, montane or riparian woodland and forest edges, canyon bottoms, hanging gardens, or scrub, clearings and pastures usually on moist slopes. ARIZ: Apa, Coc; COLO: Arc, Dol, LPl, Mon; NMEX: McK, RAr, SJn; UTAH. 1550–2650 m (5100–8700′). Flowering: Jun–Sep. British Columbia to Manitoba south to northwestern Mexico and east to the Dakotas. Common and widespread. Often cultivated. Used by Native Americans for boils and skin eruptions, backaches or swollen limbs, colds and sore throats, stomachaches and cramps, and protection against witches.

Delphinium L. Larkspur, Delphinium

(Greek *delphin*, for the resemblance of the spurred flower to classical Greek sculptures of dolphins) Perennial herbs, from dry and fibrous or fasciculate and ± tuberous roots. **LEAVES** basal and/or cauline, alternate; blades round to pentagonal or reniform in outline, deeply palmately divided into 3 (–7) primary segments; ultimate segments elongate, linear to oblanceolate, margins apically entire or crenate to lacerate, lobes of basal blades wider and fewer than those of cauline blades. **INFL** terminal, 2–100+-flowered racemes or few-branched panicles, 5–40 cm long or more, bracteate, not forming involucre. **FLOWERS** zygomorphic, bisexual; sepals 5, usually blue to purplish (rarely white to pink; red or yellow in species outside our area); lateral and lower sepals 4, planar, lanceolate to ovate or elliptic, 8–18 mm long; upper sepal 1, spurred, 8–24 mm long; petals 5, upper petals 2, often white, spurred, bearing a nectary concealed inside sepal-spur; lower petals 2, often colored as the sepals, planar, blade ± ovate, ± 2-lobed, 2–12 mm long, conspicuously hairy, lacking nectaries; stamens 25–40, filaments proximally dilated; staminodes absent between stamens and pistils; pistils 3 (–5), each with 8–20 ovules; style filiform. **FRUIT** sessile cylindric follicles, fruit walls prominently veined or not; beak not developed, style scarcely modified in fruit. $x = 8$. About 300 species in northern temperate and Arctic habitats, also subtropical and tropical mountains in Africa. At least three Eurasian species have been cultivated in North America.

1. Proximal internodes similar in length to those of midstem; leaves largest near midstem, gradually reduced upward, midstem blades 7–10 cm broad; pedicels densely velutinous, with spreading yellow hairs, some hairs minutely glandular (use 30× lens) at apex .... ***D. barbeyi***

1′ Proximal internodes much shorter than those of midstem; leaves largest near base of stem (but sometimes absent at anthesis), often abruptly reduced upward, midstem blades 0.5–5.5 cm broad; pedicels glabrous or pubescent with recurved or arching, white hairs, not glandular (rarely some spreading, yellow, minutely glandular hairs among the recurved, white hairs in *D. nuttallianum*) .... (2)

2. Plants pubescent (except sometimes the leaves and stem are glabrous); base of stem covered by few membranous leaf bases or scale leaves, not fibrous; roots fascicled, short, tuberous, not fibrous but often with numerous roots tomentose with root hairs .... ***D. nuttallianum***

2′ Plants glabrous or nearly so (except the flowers, bracts, and leaf blades may be pubescent); base of stem covered by coarse vein fibers persisting from previous years' leaf bases; roots (at least the main one) elongate, fibrous, dry .... ***D. scaposum***

Delphinium barbeyi (Huth) Huth (for William Barbey, 1842–1914, Swiss botanist) Subalpine larkspur, tall larkspur, Barbey's larkspur. Roots twisted, fibrous, dry. **STEMS** 50–150 cm tall; base green, glabrous or sparsely pubescent, lower stem portion often not collected. **LEAVES** cauline, usually absent from lower 1/5 of stem at anthesis; midstem blades lacerate, 7–15 cm broad, segments usually cuneate. **INFL** a few-branched panicle or raceme; pedicels densely golden-velutinous with finely glandular hairs (use 30× lens); bracteoles 1–4 (–8) mm from flowers, 5–14 mm long. **FLOWER** sepals dark bluish purple, sparsely puberulent; lateral sepals lanceolate, sharply acuminate; lower petal blades 4–7 mm long, hairs centered at or below proximal end of notch, sparse elsewhere, white. **FRUIT** 17–22 mm long. $2n = 16$. [*D. occidentale* (S. Watson) S. Watson var. *barbeyi* (Huth) S. L. Welsh]. Subalpine and alpine, wet soils, riparian, ponderosa pine, Douglas-fir, spruce-fir, tundra, scree slopes. COLO: Arc, Con, Hin, LPl, Min, Mon, RGr, SJn. UTAH. 2550–3900 m (8300–12700′). Flowering: Jun–Aug. Utah and Wyoming, south to Arizona and New Mexico. Known to hybridize with closely related species including *D. glaucum* S. Watson, which closely approaches our area from the north. Similar species could occur in our area, but they cannot be confidently segregated.

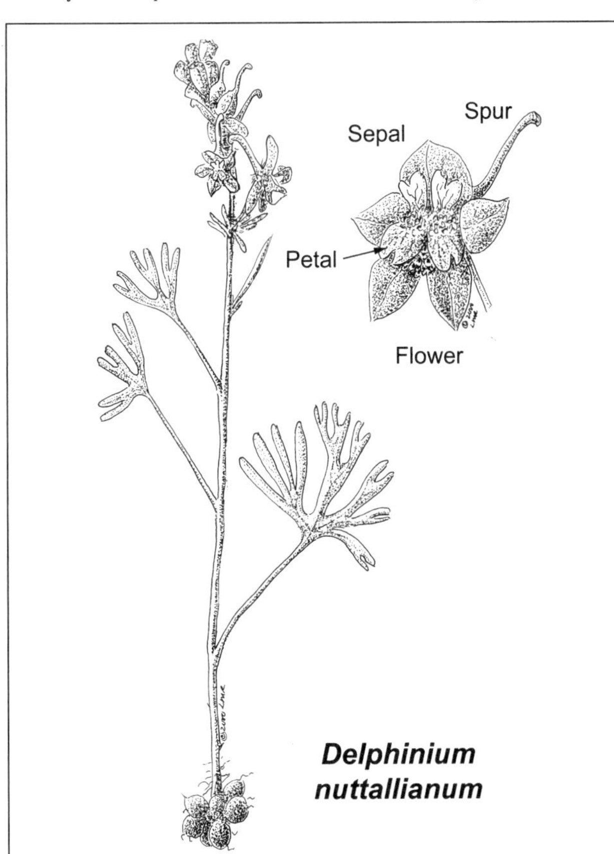

Delphinium nuttallianum

Delphinium nuttallianum Pritz. (for Thomas Nuttall, early 19th-century botanical explorer) Two-lobed larkspur. Roots fascicled, tuberous, not fibrous but often with numerous roots covered by root hairs. **STEMS** 11–45

(–70) cm tall, pubescence variable, white-recurved hairs throughout or in part; lower stem portion reddish (often whitish underground), usually long-tapered to the deep root crown, base surrounded by few membranous, brownish scale-leaves or leaf bases, not fibrous. **LEAVES** basal and cauline, present in lower 1/5 of stem at anthesis; midstem blades dissected, 1–5.5 cm broad, segments linear. **INFL** a raceme; pedicels puberulent with hairs white, short, recurved (rarely mixed with or replaced by spreading, yellow, minutely glandular hairs); bracteoles 2–8 mm from flowers, 3–7 mm long. **FLOWER** sepals usually bluish purple (rarely white) to pink, puberulent; lateral sepals oblanceolate, obtuse to acute; lower petal blades 4–11 mm long; hairs white, usually curly, and centered at proximal end of notch. **FRUIT** 7–17 mm long, 3.5–5 times longer than wide. $2n = 16$. Open coniferous woods, grassy sage scrub, meadow edges, and well-drained streamsides (generally not in very wet sites), desert scrub, sage, piñon-juniper, ponderosa pine, aspen, Gambel's oak, spruce-fir. ARIZ: Apa, Nav; COLO: Arc, Dol, Hin, LPl, Min, Mon, SJn; NMEX: McK, RAr, San, SJn; UTAH. 1450–2950 m (4800–9700′). Flowering: (Apr–) May–Jun (–Aug). British Columbia to Montana, south to California, Arizona, New Mexico. Highly variable and often difficult to identify.

Delphinium scaposum Greene **CP** (scapose, stem naked of leaves) Tall mountain larkspur. Roots (at least the main one) elongate, fibrous, dry. **STEMS** 17–61 cm tall, glabrous; lower stem portion glaucous, usually reddish; base surrounded by ascending coarse vein fibers persisting from previous years' leaf bases. **LEAVES** all basal or nearly so; leaves present on lower 1/5 of stem (or less) at anthesis; midstem blades absent, uppermost cauline blades lobed, 0.3–1.2 cm broad, segments short-spatulate. **INFL** a raceme (panicled in one individual); pedicel glabrous (rarely with a few short, curly hairs between sepals and bracteoles); bracteoles 2–5 mm from flowers, 2–4 mm long. **FLOWER** sepals bright dark blue (rarely ± pale), glabrous to very sparsely hairy; lateral sepals oblanceolate, rounded to acute; lower petal blades 5–8 mm long, hairs nearly as dense on lobes as on the rest of blade, curly. **FRUIT** 12–16 mm long, barrel-shaped. $2n = 16$. [*D. andersonii* A. Gray var. *scaposum* (Greene) S. L. Welsh]. Desert scrub, sagebrush, grassland, piñon-juniper, or ponderosa-piñon woodland. ARIZ: Apa, Coc, Nav; COLO: Arc, Dol, LPl, Mon; NMEX: McK, RAr, San, SJn; UTAH. 1500–2350 m (4900–7700′). Flowering: May–Jun. Arizona, Colorado, New Mexico, Utah. Used by Navajo and Hopi for a wash after childbirth.

Myosurus L. Mousetail

(Greek *mus*, mouse, + *oura*, tail) Annual herbs, scapose, with a slender taproot. **STEMS** absent. **LEAVES** all basal, simple, blades linear or very narrowly oblanceolate, tapering to a filiform base, entire, glabrous. **INFL** a solitary flower at the end of a scape without involucre. **FLOWERS** actinomorphic, bisexual; sepals (3–) 5 (–8), green or with scarious margins, spurred, oblong to elliptic, lanceolate or oblanceolate, 1.5–4 mm long, glabrous; petals 0–5, white, planar, linear to very narrowly spatulate, 1–2.5 mm long, glabrous, nectary present, scale absent; stamens 5–25, filaments filiform; staminodia absent between stamens and pistils; pistils 10–400, each with 1 ovule, styles filiform. **FRUIT** aggregate, sessile prismatic achenes, exposed face of achene forming a planar outer surface, sides faceted or curved by compression against adjacent achenes, fruit wall not veined, beak straight, 0.05–1.8 mm long, glabrous. $x = 8$. Fifteen species, worldwide in temperate regions.

1. Outer face of achene circular to square or broadly rhombic, 1–1.3 times higher than wide; petal claw 3 times longer than blade .... ***M. nitidus***
1′ Outer face of achene elliptic or oblong to linear, 1.5–5 times higher than wide; petal claw 1–2 times longer than blade .... (2)

2. Beak of achene 0.05–0.4 mm, parallel to outer face of achene (strongly appressed), head of achenes thus appearing smooth .... ***M. minimus***
2′ Beak of achene 0.6–1.8 mm, diverging at a narrow angle, head of achenes thus conspicuously roughened with projecting achene beaks .... ***M. apetalus***

Myosurus apetalus Gay (without petals) Bristly mousetail. Plants 1.5–12.5 cm tall. **LEAVES** linear or narrowly oblanceolate, 0.9–4 cm long. **INFL** scape 0.9–10.5 cm long. **FLOWER** sepals faintly 3-nerved, scarious margins narrow; petal claw 1–2 times longer than blade. **FRUIT** a head of achenes 11–33 mm long, 1.5–2.5 mm wide; achene outer face narrowly rhombic, 1–2.2 mm high, 0.4–1 mm wide, 2–5 times higher than wide; beak 0.6–1.4 mm long, 0.4–1 times longer than achene body, diverging from outer face of achene, head of achenes thus strongly roughened by projecting achene beaks. $2n = 16$. [*M. minimus* L. subsp. *montanus* G. R. Campb.]. Wet meadows, margins of lakes and intermittent ponds, mudflats and moist, sandy soils. ARIZ: Apa; COLO: Dol, Hin, LPl, Mon, SMg; NMEX: McK, RAr, SJn. 2000–2700 m (6600–8900′). Flowering: May–Jun. Fruit: May–Jul. California and Oregon, east to New Mexico, north to southern Saskatchewan. Ours are **var. *montanus*** (G. R. Campb.) Whittem. (of Montana), for which the illegitimate name *M. aristatus* Benth. subsp. *montanus* (G. R. Campb.) D. E. Stone ex H. Mason has been used. Our most common mousetail.

Myosurus minimus L. (very small, minimal) Tiny mousetail. Plants 4–16.5 cm tall. **LEAVES** narrowly oblanceolate or linear, 2.2–11.5 cm long. **INFL** scape 1.8–12.8 cm long. **FLOWER** sepals faintly or distinctly 3–5-nerved, scarious margins narrow or none; petal claw 1–2 times longer than blade. **FRUIT** a head of achenes 16–50 mm long, 1–3 mm wide; achene outer face narrowly rhombic to elliptic or oblong, 0.8–1.4 mm high, 0.2–0.6 mm wide, 1.5–5 times higher than wide; beak 0.05–0.4 mm long, 0.05–0.3 times longer than achene body, parallel to outer face of achene (appressed), head of achenes thus appearing smooth. $2n = 16$. Muddy margins of lakes, stock ponds, and dry washes; greasewood, piñon-juniper. ARIZ: Coc; COLO: Arc, LPl, Mon; NMEX: McK, RAr, San, SJn. 1850–2300 m (6000–7500′). Flowering: Apr–Jun. Fruit: May–Jun. Southern Canada to northern Mexico and across the continent, except the Great Lakes states, New England, the Appalachians, and Florida. Common. May be found growing with *M. apetalus*.

Myosurus nitidus Eastw. (shiny) Western mousetail. Plants 1–4 cm tall. **LEAVES** linear or very narrowly oblanceolate, 0.6–2.4 cm long. **INFL** scape 0.5–2.4 cm long. **FLOWER** sepals sometimes strongly nerved, scarious margin narrow or broad; petal claw 3 times longer than blade. **FRUIT** with heads of achenes 8–17 mm long, 2–3 mm wide; achene outer face circular to rhombic, 0.8–1 mm high, 0.6–0.9 mm wide, 1–1.3 times higher than wide; beak 0.7–1 mm long, 0.8–1.2 times longer than outer face of achene, diverging (sometimes weakly) from outer face of achene, head of achenes thus ± roughened by projecting achene beaks. [*M. egglestonii* Wooton & Standl.]. Ephemeral wet places under sagebrush or in shallow depressions in bedrock; woodlands with ponderosa pine and Gambel's oak. ARIZ: Apa; COLO: LPl, Mon. 2100–2500 m (7400–8100′). Flowering: May or Jun. Northern Arizona and New Mexico, southwestern edge of Colorado. Very infrequently collected.

Ranunculus L. Buttercup, Crowsfoot

(Latin *rana*, frog, + *unculus*, little, alluding to the wet places where many species occur) Annual or perennial herbs with fibrous or tuberous roots, caudices, rhizomes, stolons, or bulbous stem bases. **STEMS** erect to prostrate. **LEAVES** basal, cauline, or both, alternate (or a distal pair opposite in *R. alismifolius* and *R. flammula*), simple and entire or variously cleft or divided, or once ternately or pinnately compound. **INFL** terminal or sometimes axillary in creeping plants, the flowers solitary or in an open few-flowered cyme. **FLOWERS** actinomorphic, bisexual; sepals 3–5 (–6), green, yellow, or purple, planar, narrowly to broadly elliptic, ovate, or nearly round; petals 0–26 (usually 5), commonly yellow, sometimes white or red, planar, obovate to oblanceolate or elliptic, with a basal adaxial nectary covered by a scale, or the scale sometimes poorly developed; stamens (5 or) 10–many, filaments filiform; without staminodia between stamens and pistils; pistils 5–many, each with 1 ovule, styles linear, sometimes lacking. **FRUIT** aggregate, sessile, globose or discoid achenes, fruit wall not prominently veined (transversely wrinkled in one var. of *R. sceleratus*), styles scarcely modified in fruit. About 300 species, worldwide except the lowland tropics. Most *Ranunculus* species are poisonous to stock; when abundant, they may be troublesome to ranchers. A few species have acrid juice and were formerly used as vesicatories in medicine (Cook 1966; Johansson 1998).

1. Sepals covered with long, dense, brown hairs; distal leaves and bracts 3-crenate or shallowly 3-lobed apically, otherwise entire ..... ***R. macauleyi***
1′ Sepals glabrous or with colorless hairs; leaves not as above ..... (2)

2. All leaves simple and unlobed ..... (3)
2′ Some or all leaves deeply lobed or compound ..... (6)

3. Achene walls papery, longitudinally ribbed, glabrous; leaf apices broadly rounded to truncate, margins crenate ..... ***R. cymbalaria***
3′ Achene walls thick, smooth, glabrous or sometimes pubescent; leaf apices acuminate to rounded-obtuse, margins entire or finely toothed ..... (4)

4. Stems rooting nodally, erect to prostrate ..... ***R. flammula***
4′ Stems never rooting nodally, erect or ascending ..... (5)

5. Petals 2–3.8 mm broad; nectary scales glabrous; head of achenes 3–7 mm long ..... ***R. alismifolius***
5′ Petals 5–12 mm broad; nectary scales usually ciliate; head of achenes 7–12 (–20) mm long ..... ***R. glaberrimus***

6. Stems creeping and rooting at nodes, or floating in water (then rootless) ..... (7)
6′ Stems upright, or if decumbent then rooting only at the base ..... (11)

7. Basal and lower cauline leaves 3-foliolate; upper cauline leaves deeply parted or compound; achene discoid, much thinner than broad, 2.4–3.2 mm long ..... (8)

7′ All leaves 3-parted or filiform-dissected, never clearly 3-foliolate; without differentiated basal leaves; achenes compressed-ellipsoidal to discoid, plump, 1–1.8 mm long .... (9)

8. Petals 6–18 mm long, 5–12 mm wide; achene beak curved .... ***R. repens***
8′ Petals 4–6 mm long, 3.5–5 mm wide; achene beak straight .... ***R. macounii***

9. Petals white or white with yellow claws; achene with strong transverse ridges; stems floating in water; lower cauline leaves (or all leaves) divided into capillary segments .... ***R. aquatilis***
9′ Petals yellow; achene smooth; stem reclining on drying mud or sometimes floating in water; basal leaves absent, cauline leaf blades reniform to circular or broadly flabellate, deeply 3-lobed or 5-parted .... (10)

10. Leaf blades 0.3–1.2 cm long, deeply 3-lobed or -parted, terminal segment entire or distally crenulate; style 0.1–0.2 mm .... ***R. hyperboreus***
10′ Leaf blades 0.6–7.3 cm long, 3-parted, terminal segment again lobed or dissected; style 0.2–1.2 mm .... ***R. gmelinii***

11. Style absent, stigma sessile; achene margins thick and corky; emergent aquatic, sometimes also found on very wet soil .... ***R. sceleratus***
11′ Style present; achene margins ridges or narrow wings, or else differentiated margins not evident; wet or dry soil (*R. macounii* sometimes emergent from shallow water) .... (12)

12. Achene strongly flattened, discoid, width 3–15 times thickness; nectary scale free from the petal for at least 1/2 of its length, forming a free flap; basal leaves lobed, divided, or compound .... (13)
12′ Achene weakly compressed, thick-lenticular, width 1.2–2 times thickness; nectary scale joined to the petal on 3 sides, forming a pocket; basal leaves simple, lobed or divided, never compound .... (15)

13. Achene beak 1–2.5 mm long, strongly curved; basal leaves 3-parted (rarely 3-foliolate) .... ***R. uncinatus***
13′ Achene beak 0.6–1.2 mm long, straight; basal leaves 3-foliolate .... (14)

14. Petals 2–4 mm long, 1–2.5 mm wide; heads of achenes short-cylindrical, 5–7 mm wide .... ***R. pensylvanicus***
14′ Petals 4–6 mm long, 3.5–5 mm wide; heads of achenes globose to ovoid, 7–10 mm wide .... ***R. macounii***

15. Some or all basal leaves simple and unlobed .... (16)
15′ All basal leaves deeply 3-lobed or -parted .... (19)

16. Head of achenes globose; stems prostrate or ascending; basal leaves entire (rarely with 3 broad, shallow crenations) .... ***R. glaberrimus***
16′ Head of achenes ovoid or cylindrical; stems erect or nearly so; basal leaves crenate (with more than 3 crenations), the innermost basal leaves sometimes lobed or parted .... (17)

17. Sepals glabrous; petals 1.5–3.5 mm long, 1–2 mm wide .... ***R. abortivus***
17′ Sepals pilose; petals 4–13 mm long, 2–13 mm wide .... (18)

18. Sepals 3–5 mm long, 2–3 mm wide; leaf base acute to rounded; nectary scale glabrous .... ***R. inamoenus***
18′ Sepals 5–8 mm long, 3–7 mm wide; leaf base cordate to broadly obtuse; nectary scale usually ciliate .... ***R. cardiophyllus***

19. Flowering stems 0.6–3.5 cm long (sometimes longer in fruit); petals 1.2–3.5 mm long .... ***R. pygmaeus***
19′ Flowering stems 4–27 cm long; petals 6–16 mm long .... ***R. eschscholtzii***

Ranunculus abortivus L. (abortive or incomplete) Littleleaf buttercup. Roots never tuberous, sometimes enlarged basally. **STEMS** erect or suberect, 10–60 cm long, glabrous, never rooting nodally. **LEAVES** with basal blades reniform or circular, 1.4–4.2 cm long, 2–5.2 cm wide, crenulate to crenate-lobulate or sometimes the innermost 3-parted or -foliate, bases shallowly to deeply cordate, apices rounded to rounded-obtuse; upper cauline leaves lobed. **INFL** with pedicels glabrous or nearly so. **FLOWER** sepals spreading or reflexed from base, 2.5–4 mm long, 1–2 mm wide, glabrous; petals 5, yellow, 1.5–3.5 mm long, 1–2 mm wide, nectary scale attached on 3 sides, forming a pocket enclosing the nectary (sometimes with apex free, forming a flap shorter than the pocket), glabrous; style present. **FRUIT** an ovoid head of achenes, 3–6 mm long, 2.5–5 mm wide; achene body thick-lenticular to compressed-globose, 1.4–1.6 mm long, 1–1.5 mm wide, not or weakly flattened, achene wall thick, firm, smooth, glabrous, margin a low, narrow ridge, often inconspicuous, beak subulate, curved, 0.1–0.2 mm long; receptacle sparsely to very sparsely pilose. $2n = 16$. [*R. abortivus* subsp. *acrolasius* (Fernald) B. M. Kapoor & Á. Löve; *R. abortivus* var. *acrolasius* Fernald]. Woods, meadows, and clearings. COLO: Min. 2550 m (8400′). Flowering: Jun. Across temperate and subarctic North America south to Colorado.

Ranunculus alismifolius Geyer ex Benth. (with leaves like *Alisma*) Caltha-flowered buttercup, plantainleaf buttercup. Roots ± fusiform-thickened proximally. **STEMS** erect or ascending, 15–40 cm long, glabrous, never rooting nodally. **LEAVES** with lower cauline blades lanceolate to narrowly ovate or elliptic, 2.4–9.7 cm long, 0.9–2.2 (–3) cm wide, undivided, bases acute to rounded-obtuse, margins entire (rarely serrulate), apices obtuse to acuminate; bracts lanceolate; upper cauline leaves undivided. **INFL** with pedicels glabrous to appressed-hairy. **FLOWER** sepals spreading or reflexed from base, 2–6 mm long, 1–4 mm wide, glabrous or hirsute; petals 5–12, yellow, 5–9 mm long, (1.5–) 2–3.8 mm wide, nectary scale attached on 3 sides, forming a pocket enclosing the nectary (sometimes with apex free, forming a flap shorter than the pocket), glabrous; style present. **FRUIT** a hemispherical to globose head of achenes, heads 3–7 mm long, 4–8 mm wide; achene body globose-lenticular to globose, 1.6–2.8 mm long, 1.2–2 mm wide, not or weakly flattened, achene wall thick, firm, smooth, glabrous, margin a low, narrow ridge, often inconspicuous, beak lance-subulate, straight or weakly curved (stigma curled up at the tip), 0.4–1.2 mm long; receptacle glabrous. $2n = 16$. Wet meadows, stream banks, and pond margins; ponderosa pine or aspen to spruce-fir and alpine tundra. ARIZ: Apa; COLO: Arc, Con, LPl, Hin, Min, Mon, RGr, SJn; NMEX: SJn; UTAH. 2500–3950 m (8000–13000′). Flowering: May–Sep. New Mexico and Utah, north to Idaho. Ours are **var. *montanus*** S. Watson.

Ranunculus aquatilis L. (living in water) Roots never tuberous. **STEMS** creeping or floating, length indeterminate, glabrous, rooting nodally. **LEAVES**: basal leaves none; cauline leaf blades simple and laminate (with blade distinctly flattened) or completely dissected into filiform segments, or both present with floating laminate leaves and submerged filiform-dissected leaves; laminate leaves reniform, 3-parted, segments obovate or fan-shaped, shallowly cleft, margins crenate. **INFL** with pedicels glabrous. **FLOWER** sepals spreading or reflexed, 2–4 mm long, 1–2 mm wide, glabrous; petals 5, white or white with yellow claws, 4–7 mm long, 1–5 mm wide, nectary scale attached on 3 sides, forming a ridge or shallow pocket not covering the nectary, glabrous; style present. **FRUIT** a hemispherical to ovoid head of achenes, 2–4 mm long, 2–5 mm wide; achene body ellipsoidal or flattened-ellipsoidal, 1–2 mm long, 0.8–1.4 mm wide, not or weakly flattened, achene wall thick, firm, with coarse, transverse ridges, glabrous or hispid, margin a low, narrow ridge, beak filiform, 0.1–1.2 mm long; receptacle hispid. $2n = 16, 32, 48$. Aquatic or on drying mud; streams, irrigation ditches, lakes, and ponds in meadows, riparian woodland, juniper, oak savanna, ponderosa pine, spruce. ARIZ: Apa, Nav; COLO: Arc, Dol, LPl, Hin, Min, Mon, SMg, SJn; NMEX: McK, RAr, San, SJn; UTAH. 1750–2950 m (5700–9600′). Flowering: May–Sep. Throughout nearly all of North America; almost worldwide. We have two varieties that probably should be distinct species (Cook 1966). Unfortunately, deep-water plants may flower without producing any floating leaves. Before such plants can be identified, they must be cultured in shallower water until floating leaves develop. Herbarium specimens with all the leaves filiform-dissected are not identifiable to variety because there is no way to know if the upper leaves were floating or submerged when the plant was alive. Plants with floating leaves laminate and submerged leaves filiform-dissected are **var. *aquatilis*** [*R. aquatilis* var. *hispidulus* Drew; *R. trichophyllus* (Chaix) Bosch var. *hispidulus* (Drew) W. B. Drew]. Those with all leaves (floating and submerged) filiform-dissected are **var. *diffusus*** With. [*Batrachium circinatum* (Sibth.) Spach subsp. *subrigidum* (W. B. Drew) Á. Löve & D. Löve; *B. longirostris* (Godr.) F. W. Schultz; *B. trichophyllum* (Chaix) Bosch; *R. aquatilis* var. *capillaceus* (Thuill.) DC.; *R. aquatilis* var. *longirostris* (Godr.) Lawson; *R. aquatilis* var. *subrigidus* (W. B. Drew) Breitung; *R. circinatus* Sibth. var. *subrigidus* (W. B. Drew) L. D. Benson; *R. subrigidus* W. B. Drew; *R. trichophyllus* (Chaix) Bosch].

Ranunculus cardiophyllus Hook. (heart-shaped leaves) Heartleaf buttercup. Roots never tuberous. **STEMS** erect, 11–53 cm long, pilose or glabrous, never rooting nodally. **LEAVES** with basal blades ovate or elliptic, 2.2–6.9 cm long, 1.8–4.5 cm wide, crenate but undivided or the innermost 3–5-parted, bases cordate to broadly obtuse, apices rounded to broadly acute; upper cauline leaves lobed. **INFL** with pilose pedicels. **FLOWER** sepals spreading or reflexed from base, 5–8 mm long, 3–7 mm wide, pilose; petals 5–10, yellow, 6–13 mm long, 4–13 mm wide, nectary scale attached on 3 sides, forming a pocket enclosing the nectary (sometimes with apex free, forming a flap shorter than the pocket), ciliate or sometimes glabrous; style present. **FRUIT** an ovoid or cylindrical head of achenes, 5–16 mm long, 5–9 mm wide; achene body thick-lenticular to compressed-globose, 1.8–2.2 mm long, 1.5–2 mm wide, not or weakly flattened, achene wall thick, firm, smooth, margin a low, narrow ridge, often inconspicuous, beak subulate, curved or straight, 0.6–1.2 mm long; receptacle canescent. $2n = 32$. [*R. cardiophyllus* var. *coloradensis* L. D. Benson; *R. cardiophyllus* var. *subsagittatus* (A. Gray) L. D. Benson; *R. pedatifidus* Sm. var. *cardiophyllus* (Hook.) Britton]. Wet meadows and moist slopes of open parks, and subalpine. ARIZ: Apa; COLO: Hin, Min, SJn; NMEX: RAr, SJn. 2350–3500 m (8000–11500′). Flowering: May–Jul. Arizona and New Mexico, north to the Northwest Territories. *Ranunculus cardiophyllus* is very variable.

Throughout most of its range, the leaves always have rounded marginal crenae and cordate or truncate bases, the stems are often densely pilose (but may sometimes be sparsely pilose or glabrous), and the achene beaks are curved. However, in plants from Arizona and New Mexico the leaves sometimes have obtuse marginal crenae or broadly obtuse bases, the stems are never densely pilose, and the achene beaks are sometimes straight. Forms showing some or all of these characters are often separated as *R. cardiophyllus* var. *subsagittatus*, but the characters are poorly correlated and taxonomic recognition is not warranted.

Ranunculus cymbalaria Pursh (cymbal-like) Marsh buttercup, alkali buttercup. Roots never tuberous. **STEMS** dimorphic, flowering stems erect or ascending, stolons prostrate, 9–41 cm long, glabrous or sparsely hirsute, rooting nodally. **LEAVES** with basal blades oblong to cordate or circular, 0.5–3.8 cm long, 0.6–3.8 cm wide, simple and undivided, bases rounded to cordate, margins crenate or crenate-serrate (sometimes weakly so, rarely nearly entire), apices rounded; upper cauline leaves small and undivided. **INFL** with pedicels glabrous to sparsely pilose. **FLOWER** sepals spreading, 2.5–6 mm long, 1.5–3 mm wide, glabrous (rarely sparsely hirsute); petals 5–7 (–8), yellow, 2–7 mm long, 1–3 mm wide, nectary scale attached basally, forming a flap over the nectary, glabrous; style present. **FRUIT** a long-ovoid or cylindrical head of achenes, 6–12 mm long, 4–5 (–9) mm wide; achene body discoid to flattened-lenticular, oblong to obovate in outline, 1.4 (–2.2) mm long, 0.8–1.2 mm wide, moderately flattened, achene wall papery, longitudinally nerved, glabrous, margin a prominent narrow ridge, beak conical, straight, 0.1–0.2 mm long; receptacle hispid or glabrous. $2n = 16$. [*Halerpestes cymbalaria* (Pursh) Greene; *Ranunculus cymbalaria* var. *alpina* Hook.; *R. cymbalaria* var. *saximontanus* Fernald]. Marshes, ditches, muddy areas, often saline; hanging gardens, seeps, marshes, streams, and ponds. ARIZ: Apa, Coc, Nav; COLO: Arc, Dol, LPl, Mon, SMg; NMEX: McK, RAr, SJn; UTAH. 1350–2700 m (4400–8900′). Flowering: May–Sep. Circumboreal, south to South America. *Ranunculus cymbalaria* is very different from our other *Ranunculus* species, especially in the thin, papery walls of its achenes, which split open along the adaxial margin like a follicle when squeezed end-to-end. It has often been treated as a separate small genus, *Halerpestes*, but recent research suggests that it is actually very closely related to the genus *Trautvetteria* (Johansson 1998), which also has a papery, strongly veined fruit wall (fruit considered a utricle). Further research is needed before the genera can be redefined in a more natural way.

Ranunculus eschscholtzii Schltdl. (for Johann F. Gustav von Eschscholtz, 1793–1831, an Estonian surgeon and botanist) Eschscholtz's buttercup. Roots never tuberous. **STEMS** erect or decumbent from prominent caudices, 4–27 cm long, glabrous, never rooting nodally. **LEAVES** with basal blades reniform or cordate, 1–2.3 cm long, 1.5–3.7 cm wide, 3-parted with at least the lateral segments again lobed, bases truncate or cordate, apices rounded in outline; upper cauline leaves lobed. **INFL** with glabrous pedicels. **FLOWER** sepals spreading or reflexed from base, 4–8 mm long, 2–6 mm wide, glabrous or pilose; petals 5–8, yellow, 6–12 mm long, 4–16 mm wide, nectary scale attached on 3 sides, forming a pocket enclosing the nectary (sometimes with apex free, forming a flap shorter than the pocket), glabrous; style present. **FRUIT** a cylindrical or ovoid head of achenes, 5–10 mm long, 4–7 mm wide; achene body thick-lenticular to compressed-globose, 1.4–2 mm long, 1–1.6 mm wide, not or weakly flattened, achene wall thick, firm, smooth, glabrous, margin a low, narrow ridge, often inconspicuous, beak lanceolate or subulate, straight (sometimes curved when immature), 0.6–1.8 mm long; receptacle glabrous or sparsely pilose. $2n = 32, 48$. Riparian meadows, open rocky slopes, spruce, timberline, snowmelt area; high subalpine. COLO: Hin, LPl, Mon, SJn. 3300–3600 m (10800–11800′). Flowering: Jul. New Mexico west to California, north to Alaska. Our plants are **var. *eschscholtzii.***

Ranunculus flammula L. (a little flame) Greater creeping spearwort. Roots not thickened. **STEMS** erect to prostrate or sometimes ascending, 5–30 cm long, glabrous or sparsely strigose, usually rooting nodally. **LEAVES** with lower cauline blades lance-elliptic to lanceolate or linear, 0.8–3.3 cm long, 0.1–0.9 cm wide, undivided, bases acute to filiform, margins entire or serrulate, apices acute to filiform; bracts lanceolate to oblanceolate; upper cauline leaves undivided. **FLOWER** sepals spreading or weakly reflexed, 2–3 mm long, 1–2 mm wide, glabrous or appressed-hispid; petals 5–6, yellow, 3–5 mm long, 2–3 mm wide, nectary scale attached on 3 sides, forming a pocket enclosing the nectary (sometimes with apex free, forming a flap shorter than the pocket), glabrous; style present. **FRUIT** a globose or hemispherical head of achenes, 2–4 mm long, 3–4 mm wide; achene body globose-lenticular to globose, 1.2–1.6 mm long, 1–1.4 mm wide, not or weakly flattened, achene wall thick, firm, smooth, glabrous, margin a low, narrow ridge, often inconspicuous, beak lanceolate to linear, straight or curved, 0.1–0.6 mm long; receptacle glabrous. $2n = 32$. [*R. reptans* L. var. *ovalis* (J. M. Bigelow) Torr. & A. Gray]. Muddy ground or shallow water at margins of ponds, lakes, and streams in meadows, ponderosa pine, Douglas-fir, and spruce. ARIZ: Apa; COLO: Hin; NMEX: McK, SJn. 2500–3350 m (8100–11000′). Flowering: Jun–Sep. Across cool-temperate and subarctic North America, south to California, Arizona, New Mexico, and Pennsylvania. Our plants are **var. *ovalis*** (J. M. Bigelow) L. D. Benson (oval), which, as currently understood, is heterogeneous across its range, but material from the study area is relatively uniform. Biosystematic study of *R. flammula* as a whole, concentrating on more northern areas, will be needed before the nomenclature of our plants is really stable.

Ranunculus glaberrimus Hook. (most glabrous) Sagebrush buttercup. Roots thick but not tuberous. **STEMS** prostrate or ascending, 4–15 cm long, glabrous, never rooting nodally. **LEAVES** with basal blades elliptic to oblong or reniform, 0.7–1.9 cm long, 1–1.7 cm wide, entire to deeply 3-crenate, bases obtuse to truncate, apices rounded; upper cauline leaves deeply 3-crenate to unlobed. **INFL** with pedicels glabrous or nearly so. **FLOWER** sepals spreading or reflexed from base, 5–8 mm long, 3–7 mm wide, glabrous or sparsely pilose; petals 5–10, yellow, 8–13 mm long, 5–12 mm wide, nectary scale attached on 3 sides, forming a pocket enclosing the nectary (sometimes with apex free, forming a flap shorter than the pocket), glabrous or ciliate; style present. **FRUIT** a globose head of achenes, 7–12 (–20) mm long, 6–11 (–20) mm wide; achene body thick-lenticular to compressed-globose, 1.4–2.2 mm long, 1.1–1.8 mm wide, not or weakly flattened, achene wall thick, firm, smooth, usually finely pubescent, margin a low, narrow ridge, often inconspicuous, beak subulate or lance-subulate, straight or curved, 0.4–1 mm long; receptacle glabrous. $2n$ = 80. [*R. ellipticus* Greene; *R. oreogenes* Greene]. Moist, seepy slopes, dry meadows, Gambel's oak–ponderosa pine. ARIZ: Apa; COLO: Arc, Dol, Mon, SMg. 2300–2700 m (7500–8800′). Flowering: Apr–May. Arizona and New Mexico north to Alberta. Ours are **var. *ellipticus*** (Greene) Greene (elliptic).

Ranunculus gmelinii DC. (for Johann Georg Gmelin, 1709–1755, a German naturalist) Gmelin's buttercup. Roots never tuberous. **STEMS** reclining or sometimes floating, 2–60 cm long, glabrous or hirsute, rooting nodally. **LEAVES**: basal leaves none; cauline leaf blades reniform to circular, 0.6–6.5 cm long, 1.1–9 cm wide, blade 3-parted, segments again 1–3 times lobed to dissected, entire or crenate, bases cordate, apices rounded to filiform. **INFL** with glabrous pedicels. **FLOWER** sepals spreading or reflexed from base, 2–5 mm long, 2–4 mm wide, glabrous or sparsely pilose; petals 4–14, yellow, 3–7 mm long, 2–5 mm wide, nectary scale well developed and attached basally, forming a flap with nectary displaced on its outer surface, glabrous; style present. **FRUIT** a globose or ovoid head of achenes, 3–7 mm long, 2–5 mm wide; achene body thick-lenticular or compressed-ellipsoidal to discoid, 1–1.6 mm long, 1–1.2 mm wide, not or weakly flattened, achene wall thick, firm, smooth, glabrous, margin a broad, low or high corky band or ridge, beak narrowly lanceolate or filiform, 0.4–0.8 mm long; receptacle sparsely hispid. $2n$ = 16, 32, 64. [*R. gmelinii* subsp. *purshii* (Richardson) Hultén; *R. gmelinii* var. *hookeri* (D. Don) L. D. Benson; *R. gmelinii* var. *limosus* (Nutt.) H. Hara; *R. gmelinii* var. *prolificus* (Fernald) H. Hara; *R. purshii* Richardson]. Shallow water or drying mud, meadows and margins of ponds and lakes in oak–ponderosa pine to spruce or spruce-fir. ARIZ: Apa; COLO: Arc; NMEX: SJn. 2300–2750 m (7500–9000′). Flowering: Jun–Sep. Circumboreal, south to Nevada, Utah, New Mexico, Wisconsin.

Ranunculus hyperboreus Rottb. (extreme northern) High northern buttercup. Roots never tuberous. **STEMS** prostrate or floating, length indeterminate, glabrous, rooting nodally. **LEAVES**: basal leaves none; cauline leaf blades wider than long, 0.3–1.2 cm long, 0.4–2.1 cm wide, deeply 3-lobed or 3-parted, sometimes with a shallow tooth on each lateral lobe, bases cordate to broadly rounded, apices rounded. **INFL** with glabrous pedicels. **FLOWER** sepals spreading or slightly reflexed from base, 2–4 mm long, 1–3 mm wide, glabrous; petals 5, yellow, 2–4 mm long, 1–3 mm wide, nectary on petal surface, scale poorly developed and forming a crescent-shaped ridge surrounding the nectary; style 0.1–0.2 mm. **FRUIT** a ± spheric head of achenes, 3–5 mm long, 2–5 mm wide; achene body compressed-ellipsoid, 1–1.5 mm long, 0.8–1.2 mm wide, not or weakly flattened, achene wall thick, firm, smooth, glabrous, margin a low, poorly developed ridge, beak minute, hooked, 0.1–0.4 mm long; receptacle glabrous. [*R. natans* C. A. Mey. var. *intertextus* (Greene) L. D. Benson]. Subalpine ponds, peat bogs, and spruce-fir communities. COLO: SJn. 2950–3050 m (9600–10100′). Flowering: Jun–Jul. Circumboreal, south to Nevada, Utah, Colorado, Newfoundland.

Ranunculus inamoenus Greene (unlovely) Graceful buttercup, unlovely buttercup. Roots never tuberous. **STEMS** erect, 5–33 cm long, pilose or glabrous, never rooting nodally. **LEAVES** with basal blades ovate, obovate, or orbicular (rarely reniform), 1–3.7 cm long, 1.1–3.5 cm wide, crenate and undivided or the innermost with 2 clefts or partings near apex, bases acute to rounded, apices rounded; upper cauline leaves lobed. **INFL** with appressed-pubescent pedicels. **FLOWER** sepals spreading or reflexed from base, 3–5 mm long, 2–3 mm wide, pilose; petals 5, yellow, 4–9 mm long, 2–5 mm wide, nectary scale attached on 3 sides, forming a pocket enclosing the nectary (sometimes with apex free, forming a flap shorter than the pocket), glabrous; style present. **FRUIT** a cylindrical head of achenes, 7–17 mm long, 5–8 mm wide; achene body thick-lenticular to compressed-globose, 1.5–2 mm long, 1.3–1.8 mm wide, not or weakly flattened, achene wall thick, firm, smooth, glabrous, margin a low, narrow ridge, often inconspicuous, beak subulate, straight or hooked, 0.4–0.9 mm long; receptacle pilose or glabrous. $2n$ = 32, 48. [*R. inamoenus* var. *alpeophilus* (A. Nelson)

L. D. Benson]. Meadows, margins of streams, ponds, and lakes; slopes, occasionally aspen, Gambel's oak, ponderosa pine, Douglas-fir, and spruce-fir communities. ARIZ: Apa; COLO: Arc, Hin, LPl, Min, Mon, SJn; NMEX: RAr, SJn; UTAH. 2250–3750 m (7400–12300′). Flowering: Apr–Sep. North to Alberta, east to Nebraska. Our plants are **var.** ***inamoenus.***

Ranunculus macauleyi A. Gray (for Charles Adam Hoke McCauley, 1843–1913, soldier and naturalist) Rocky Mountain buttercup. Roots never tuberous. **STEMS** erect from short caudices, 4–15 cm long, glabrous or sometimes pilose, never rooting nodally. **LEAVES** with basal blades narrowly elliptic to lanceolate or oblanceolate, 1.5–4.5 cm long, 0.5–1.1 (–2.8) cm wide, undivided and entire except for apex, bases acute or long-attenuate, apices truncate or rounded and 3 (–5)-toothed; upper cauline leaves simple or apically lobed. **INFL** with pedicels glabrous or brown (or white)-pilose. **FLOWER** sepals spreading or reflexed from base, 6–12 mm long, 2.5–8 mm wide, densely brown-pilose; petals 5 (–8), yellow, 10–19 mm long, 6–21 mm wide, nectary scale attached on 3 sides, forming a pocket enclosing the nectary (sometimes with apex free, forming a flap shorter than the pocket), glabrous; style present. **FRUIT** an ovoid or cylindrical head of achenes, 5–10 mm long, 4–5.5 mm wide; achene body thick-lenticular to compressed-globose, 1.5–1.7 mm long, 1.2–1.3 mm wide, not or weakly flattened, achene wall thick, firm, smooth, glabrous, margin a low, narrow ridge, often inconspicuous, beak slender, straight or recurved, 0.5–1.5 (–2.2) mm long; receptacle glabrous. Sunny, open soil of alpine meadows and slopes. COLO: Arc, Con, Hin, LPl, Min, Mon, SJn. 3600–3950 m (11800–13000′). Flowering: Jun–Aug. Colorado, New Mexico; east to the west slope of the Rockies. Two aberrant specimens (*M. Heil 22, L. Herring 88*) have 7–11 teeth 1–3 mm long in the distal half of the basal and cauline leaves.

Ranunculus macounii Britton (for James M. Macoun, 1862–1920, Canadian botanist and ornithologist) Macoun's buttercup. Roots never tuberous. **STEMS** prostrate to erect, 13–38 cm long, hirsute or glabrous, often rooting nodally. **LEAVES** with basal blades cordate to reniform in outline, 3.7–7.5 cm long, 4.5–9.5 cm wide, 3-foliolate, leaflets 3-lobed or -parted, ultimate segments elliptic or lance-elliptic, toothed or lobulate, apices acute to broadly acute; upper cauline leaves deeply parted or compound. **INFL** with hirsute pedicels. **FLOWER** sepals spreading or reflexed about 1 mm above base, 4–6 mm long, 1.5–3 mm wide, glabrous or hirsute; petals 5, yellow, 4–6 (–7.5) mm long, 3.5–5 mm wide, nectary scale attached basally, forming a flap covering the nectary, glabrous; style present. **FRUIT** a globose or ovoid head of achenes, 7–11 mm long, 7–10 mm wide; achene body discoid, 2.4–3 mm long, 2–2.4 mm wide, much thinner than broad, achene wall thick, firm, smooth, glabrous, margin a narrow rib, beak lanceolate to broadly lanceolate, straight or nearly so, 1–1.2 mm long; receptacle hirsute. $2n = 32, 48$. Wet soil or emergent from shallow water, meadows, ditches, seeps, edges of ponds and streams in cottonwood, willow, Gambel's oak, ponderosa pine, Douglas-fir, and spruce-fir communities. ARIZ: Apa; COLO: Arc, Hin, LPl, Mon; NMEX: McK, RAr, SJn. 1700–2700 m (5600–8900′). Flowering: May–Jul (–Sep). North to Alaska, east to Newfoundland.

Ranunculus pensylvanicus L. f. (of Pennsylvania) Pennsylvania buttercup. Roots never tuberous. **STEMS** erect, 45 cm long, hispid, never rooting nodally. **LEAVES** with basal blades broadly cordate in outline, 1.6–7 cm long, 3–9 cm wide, 3-foliolate, leaflets cleft (usually deeply so), ultimate segments narrowly elliptic, toothed, apices acute; upper cauline leaves deeply parted or compound. **INFL** hirsute pedicels. **FLOWER** sepals reflexed about 1 mm above base, 3–5 mm long, 1.5–2 mm wide, ± hispid; petals 5, yellow, 2–3 mm long, 1–2.5 mm wide, nectary scale attached basally, forming a flap covering the nectary, glabrous; style present. **FRUIT** a cylindrical head of achenes, 9–12 mm long, 5–7 mm wide; achene body discoid, 1.8–2.8 mm long, 1.6–2 mm wide, much thinner than broad, achene wall thick, firm, smooth, glabrous, margin a narrow rib, beak broadly lanceolate or subdeltoid, straight or nearly so, 0.6–0.8 mm long; receptacle hirsute. $2n = 16$. Riparian. COLO: Arc, LPl, SJn. 1850–2700 m (6100–8900′). Flowering: Jul. North to Alaska, east to Newfoundland.

Ranunculus pygmaeus Wahlenb. (dwarf) Pygmy buttercup. Roots never tuberous. **STEMS** erect or ascending from short caudices, 0.6–3.5 cm long, glabrous or sparsely pilose, never rooting nodally. **LEAVES** with basal blades reniform to transversely elliptic or semicircular, 0.45–0.9 cm long, 0.6–1.3 cm wide, 3-parted or -divided, at least lateral segments again lobed, bases truncate or subcordate, apices rounded to obtuse; upper cauline leaves lobed. **INFL** with glabrous or pubescent pedicels. **FLOWER** sepals spreading or reflexed from base, 2–4 mm long, 1.2–1.6 mm wide, sparsely hairy; petals 5, yellow, 1.2–3.5 mm long, 1.1–2.8 mm wide, nectary scale attached on 3 sides, forming a pocket enclosing the nectary (sometimes with apex free, forming a flap shorter than the pocket), glabrous; style present. **FRUIT** a subglobose to cylindrical head of achenes, 2.5–7 mm long, 2.5–5 mm wide; achene body thick-lenticular to compressed-globose, 1–1.2 mm long, 0.8–1.1 mm wide, not or weakly flattened, achene wall thick, firm, smooth, glabrous, margin a low, narrow ridge, often inconspicuous, beak subulate, straight or curved, 0.3–0.7 mm long; receptacle glabrous. $2n = 16$. [*R. pygmaeus* var. *langianus* Nathorst]. Snowmelt areas; alpine meadows and slopes. COLO: SJn.

4250 m (13900′). Flowering: Aug. North to Alaska, east to Greenland and Spitzbergen. Inconspicuous; the San Juan County site is one of two known from the West Slope of the Colorado Rockies; the other is outside our area.

Ranunculus repens L. (creeping) Creeping buttercup. Roots never tuberous. **STEMS** decumbent or creeping, 25 cm long, hispid to strigose or almost glabrous, rooting nodally. **LEAVES** with basal blades ovate to reniform in outline, 1–8.5 cm long, 1.5–10 cm wide, 3-foliolate, leaflets lobed, parted, or parted and again lobed, ultimate segments obovate to elliptic or sometimes narrowly oblong, toothed, apices obtuse to acuminate; upper cauline leaves deeply parted or compound. **INFL** with appressed-hairy pedicels. **FLOWER** sepals spreading or reflexed from base, 4–7 (–10) mm long, 1.5–3 (–4) mm wide, hispid or sometimes glabrous; petals 5 (–150), yellow, 6–18 mm long, 5–12 mm wide, nectary scale attached basally, forming a flap covering the nectary, glabrous; style present. **FRUIT** a globose or ovoid head of achenes, 5–10 mm long, 5–8 mm wide; achene body discoid, 2.6–3.2 mm long, 2–2.8 mm wide, much thinner than broad, achene wall thick, firm, smooth, glabrous, margin a narrow rib, beak lanceolate to lance-filiform, curved, 0.8–1.2 mm long; receptacle hispid or rarely glabrous. $2n = 14, 32$. [*R. repens* var. *erectus* DC.; *R. repens* var. *glabratus* DC.; *R. repens* var. *linearilobus* DC.; *R. repens* var. *pleniflorus* Fernald; *R. repens* var. *villosus* Lamotte]. Meadows, borders of marshes, pastures, streams in shaded ravines. COLO: Arc, Mon. 1850–2300 m (6100–7600′). Flowering: Jun–Jul. Native to Eurasia, naturalized in temperate areas worldwide. Plants with sparse pubescence have been called *R. repens* var. *glabratus*. Horticultural forms with the outer stamens transformed into numerous extra petals occasionally become established and have been called *R. repens* var. *pleniflorus*. These variants have no taxonomic significance.

Ranunculus sceleratus L. (hard) Blister buttercup. Roots never tuberous. **STEMS** erect, 8–32 cm long, glabrous, only very rarely rooting at lower nodes. **LEAVES** with basal and lower cauline blades reniform to semicircular in outline, 1–5 cm long, 1.6–6.8 cm wide, 3-lobed or -parted (sometimes deeply so), segments again lobed or parted, sometimes undivided, crenate or lobulate, bases truncate to cordate, apices rounded or occasionally obtuse; upper cauline leaves deeply 3-parted or compound. **INFL** with glabrous pedicels. **FLOWER** sepals reflexed from at or near the base, 2–5 mm long, 1–3 mm wide, glabrous or sparsely hirsute; petals 3–5, yellow, 2–5 mm long, 1–3 mm wide, nectary scale poorly developed and forming a crescent-shaped or circular ridge surrounding but not covering the nectary, glabrous; style none. **FRUIT** an ellipsoidal or cylindrical head of achenes, 3–9 mm long, 3–5 mm wide, achene body thick-lenticular or compressed-ellipsoidal to discoid, 1–1.2 mm long, 0.8–1 mm wide, not or weakly flattened, achene wall thick, firm, smooth or finely transversely wrinkled, glabrous, margin a broad, low or high corky band or ridge, beak triangular, usually straight, 0.1 mm long; receptacle pubescent or glabrous. [*Hecatonia scelerata* (L.) Fourr.]. Semiaquatic; wet ground or shallow water, in bogs, ponds, shores of lakes and rivers. ARIZ: Apa; COLO: Arc, LPl; NMEX: McK, RAr, SJn; UTAH. 1100–2350 m (3600–8000′). Flowering: May–Sep. North to Alaska, east to Quebec. Two weak varieties in the study area: **var.** ***sceleratus*** with fruits smooth (longitudinal striations visible only at high magnification), and **var.** ***multifidus*** Nutt. (multiply divided) with achenes finely transversely wrinkled (visible at 10×).

Ranunculus uncinatus D. Don (hooked) Woodland buttercup. Roots never tuberous. **STEMS** erect, hispid or glabrous, never rooting nodally. **LEAVES** with basal blades cordate to reniform in outline, 1.8–5.6 cm long, 2.8–8.3 cm wide, 3-parted or sometimes 3-foliolate, segments again lobed, ultimate segments elliptic to lanceolate, toothed or crenate-toothed, apices acute to rounded-obtuse; upper cauline leaves deeply parted or compound. **INFL** with appressed-pubescent pedicels. **FLOWER** sepals reflexed or sometimes spreading, 2–4.5 mm long, 1–2 mm wide, pubescent; petals 5, yellow, 2–5 mm long, 1–2 (–3) mm wide, nectary scale attached basally or on 3 sides, forming a flap or pocket covering the nectary, glabrous; style present. **FRUIT** a globose or hemispherical head of achenes, 4–7 mm long, 4–8 mm wide; achene body discoid, 2–3 mm long, 1.6–2 mm wide, much thinner than broad, achene wall thick, firm, smooth, glabrous, margin a narrow rib, beak lanceolate, curved and hooked, 1–2.5 mm long; receptacle glabrous. $2n = 28$. [*R. bongardii* Greene; *R. bongardii* var. *tenellus* (A. Gray) Greene; *R. uncinatus* var. *earlei* (Greene) L. D. Benson; *R. uncinatus* var. *parviflorus* (Torr.) L. D. Benson]. Moist riparian, meadows, seeps, and wet depressions in woodlands of aspen-fir or spruce-fir communities. COLO: Mon, SJn; UTAH. 2150–2900 m (7000–9500′). Flowering: Jun–Jul. Arizona, New Mexico, north to Alaska, west to southern California. Plants with hispid stems and achenes are often separated as *R. uncinatus* var. *parviflorus*, but these two characters are poorly correlated and sometimes vary between plants in a single collection.

Thalictrum L. Meadow-rue

(*Thaliktron*, an ancient name used by Dioscorides) Perennial herbs with woody rhizomes, caudices, or tuberous roots. **STEMS** erect or ascending, usually branched in large species, unbranched in *T. alpinum*. **LEAVES** basal and cauline, alternate; blade ovate to reniform in outline, 1–4 times ternately or pinnately compound; leaflets cordate-reniform, obovate, lanceolate, or linear, margins often 3–5+-lobed, crenate or entire. **INFL** a terminal, sometimes also axillary,

(1–) 2–200-flowered panicle, raceme, corymb, or umbel, to 41 cm long, or flower solitary; involucres absent or involucral bracts 2–3 (these compound, when sessile resembling a whorl of 6–9 simple bracts), leaflike, not closely subtending flowers. **FLOWERS** actinomorphic, all bisexual, bisexual and unisexual on same plant, or all unisexual on monoecious or dioecious plants; sepals 4–10, whitish to greenish yellow or purplish, planar, lanceolate to reniform or spatulate, 1–18 mm long; petals absent; stamens 7–30, filaments filiform to clavate or distally dilated; staminodes absent between stamens and pistils; pistils 1–16, each with 1 ovule; style elongate with decurrent style, or absent. **FRUIT** sessile or stipitate, ovoid to obovoid, falcate, or discoid aggregate achenes, fruit walls prominently veined or ribbed; beak straight to coiled or absent, 0–4 mm long. $x = 7$. One hundred twenty to 200 species, mostly temperate, nearly worldwide. Several species are cultivated as ornamentals. Approaching our area, or reported from it, are *T. dasycarpum* Fisch. & Avé-Lall. (leaflet with the length up to 3 times the width, acutely 0–3 (–5)-lobed, lobes entire, margins revolute; hairs multicellular, nonglandular, rare glands minute and sessile), *T. occidentale* A. Gray (fruits all deflexed against pedicel), *T. revolutum* DC. (leaflet shape and margins as in *T. dasycarpum*, but all hairs gland-tipped and unicellular), *T. sparsiflorum* Turcz. ex Fisch. & C. A. Mey. (flowers bisexual, stigmatic beak of achene 1–1.5 mm long, anther blunt-tipped, shorter than filament), and *T. venulosum* Trel. (achenes terete or subterete, also smaller—3–4 (–6) mm—and narrower than in *T. fendleri*). Although *T. alpinum* is distinctive, most other western North American species are more similar, variable, and taxonomically confusing. From this group a single species, *T. fendleri*, from San Juan drainage material, has been identified. Future collections should include both male and female plants (with mature fruits) taken on the same day from the same population, and whole plants (not merely a piece of the panicle).

1. Plants scapose or with a single cauline leaf; inflorescence racemose; flowers bisexual .... ***T. alpinum***
1′ Plants with leafy stems; inflorescence paniculate; flowers unisexual (very rarely also bisexual) .... ***T. fendleri***

Thalictrum alpinum L. (alpine) Arctic meadow-rue, dwarf meadow-rue. **STEMS** erect, scapose or nearly so, 6–20 (–30) cm tall, with very slender rhizomes. **LEAVES** all basal or a single cauline leaf near the base, leaf blade twice ternately or twice pinnately compound with proximal primary divisions ternate; leaflets 2–10 mm long, obovate to orbiculate, apically 3–5-lobed, the lobes often again lobed. **INFL** an elongate raceme, few-flowered. **FLOWERS** bisexual, sepals 5, ovate or elliptic; anthers acute or apiculate; filaments filiform. **FRUITS** 2–6 per flower, reflexed or appearing so because of a sharp bend near pedicel apex; body of achene lance-obovoid, 2–3.5 mm, thick-veined, 2–3-ribbed, stigmatic beak about 0.7 mm. $2n = 14$ (all from Asia or Greenland), 21. Wet meadows, damp, rocky ledges and slopes, and cold (often calcareous) bogs in willow-sedge, lodgepole pine, and spruce-fir, outcrops, subalpine and alpine. COLO: Hin, LPl, SJn. 2350–4200 m (7700–13700′). Flowering: Jul. Circumpolar, isolated in alpine zones south to California, Utah, New Mexico.

Thalictrum fendleri Engelm. ex A. Gray (for Augustus Fendler) Fendler's meadow-rue. **STEMS** mostly erect, sometimes reclining, leafy, (20–) 30–60 (–160) cm tall, from rhizomes or branched caudices. **LEAVES** mainly cauline, leaf blade (2–) 3–4-ternately compound, glandular or glabrous; leaflets (5–) 10–20 mm long, obliquely orbiculate, apically 3+-lobed, crenate, margins sometimes slightly revolute but inconsistently so on any specimen. **INFL** a leafy panicle, many-flowered. **FLOWERS** unisexual (plants dioecious), very rarely intermixed with some bisexual flowers, sepals 4 (–6), in pistillate flowers ovate to rhombic or broadly lanceolate (ovate to elliptic in staminate flowers); anthers apiculate, filaments filiform. **FRUITS** 7–11 (–14) per flower, spreading; body of achene oblique (lopsided), oblanceolate to obovate-elliptic, strongly angled and bilaterally compressed, length less than twice the width, 4–6 mm long, 2.5–3 mm wide, conspicuous veins 3–4 (–5) on each side, veins prominently raised forming the angles, converging toward ends (rarely branched or sinuous), not anastomosing-reticulate; stigmatic beak (1.5–) 2.5–4 mm. [*T. fendleri* var. *platycarpum* Trel.; *T. fendleri* var. *wrightii* Trel.] Gambel's oak, chokecherry, ponderosa pine, Douglas-fir, aspen, and spruce-fir communities. ARIZ: Apa, Nav; COLO: Arc, Hin, LPl, Min, Mon, RGr, SJn; NMEX: McK, RAr, SJn; UTAH. 1750–3350 m (5700–11000′). Flowering: (May–) Jun–Sep. Fruit: Jul–Sep. Oregon to South Dakota, south to California, Arizona, New Mexico, Texas, and Mexico. Common. An extremely variable species, commonly with minute sessile or stipitate glands especially on leaves and fruits. This species has petiolate, leaflike bracts subtending panicle branches. The panicles of other species often confused with *T. fendleri* have only sessile leaflike bracts, appearing to be three leaves at a node because of their ternate architecture. Native tribes prepared decoctions from the roots to cure colds and gonorrhea.

Trautvetteria Fisch. & C. A. Mey. False Bugbane, Tassel-rue

(for Ernst Rudolph von Trautvetter, Russian botanist) Perennial herbs with short, slender rhizomes. **STEMS** erect, branched. **LEAVES** basal and cauline, alternate, simple; blades palmately 5–11-lobed; segments broadly cuneate, margins lacerate to serrate or sharply toothed. **INFL** terminal, ± many-flowered corymb, 2.5–43 cm long; bracts inconspicuous, linear-lanceolate, not forming involucre. **FLOWERS** actinomorphic, bisexual; sepals 3–5 (–7), greenish white, concave-cupped, broadly ovate to obovate, 3–6 mm long; petals and nectaries absent; stamens about 50–100, outer filaments spatulate, often distally wider than anthers, inner filaments not dilated; staminodes absent between stamens and pistils; pistils 15–16, simple; ovule 1 per pistil; style short, hooked. **FRUIT** sessile, ellipsoid to obovoid aggregate utricles, 4-angled in cross section, fruit walls prominently veined; beak terminal, curved to hooked, 0.4–0.8 mm long. $x = 8$. One species, North America, eastern Asia.

Trautvetteria caroliniensis (Walter) Vail (of Carolina) False bugbane, tassel-rue. Plants 0.5–1.5 m tall. Rhizomes with fascicles of fibrous roots. **STEMS** 1–several, erect, usually unbranched below distalmost bract, glabrous or glabrate. **LEAVES** (basal) with petiole 7–45 cm long, blade 8–30 (–40) cm wide, lobes acute to acuminate; cauline leaves reduced toward apex of stem. **INFL** peduncle 10–80 cm long; pedicels sparsely pubescent with minute, curled trichomes. **FLOWER** stamens white, 5–10 mm long. **FRUIT** papery, veins prominent along angles and on 2 adaxial faces. $2n = 16$. [*T. caroliniensis* var. *borealis* (H. Hara) T. Shimizu; *T. caroliniensis* var. *occidentalis* (A. Gray) C. L. Hitchc.; *T. grandis* Nutt.; *T. palmata* Fisch. & C. A. Mey.]. Riparian and other wet areas in montane spruce-fir forests to subalpine and alpine meadows. COLO: Arc, Hin, LPl, SJn; UTAH. 2350–3900 m (8000–12800′). Flowering: Jun–Aug. Distribution disjunct: Pacific Northwest and northern Rocky Mountains (British Columbia and Montana to northwestern Wyoming and California); southern Rocky Mountains (Arizona, New Mexico, and Mexico); southeastern United States (Illinois and Pennsylvania south to Arkansas and Florida); eastern Asia.

Trollius L. Globe-flower

(German *trollblume*, globe-flower) Perennial herbs with short caudices. **STEMS** erect or ascending, unbranched. **LEAVES** basal and cauline, alternate; blade pentagonal to orbicular in outline, deeply palmately divided into (3–) 5–7 segments; segments obovate, ± 3-lobed, margins coarsely toothed, often incised. **INFL** terminal, 1–3-flowered (to 7-flowered in Eurasian species), open cyme, 2–30 cm long; bracts leaflike, not forming involucre. **FLOWERS** actinomorphic, bisexual; sepals (4–) 5–9 (–30 in species outside our area), white (to orange-yellow, orange-red, or purplish outside our area), ± planar (strongly concave and incurved outside our area), elliptic, orbiculate, or obovate, 10–30 mm long; petals 5–25, yellow or orange, planar, linear-oblong (ovate outside our area), 2–10 (–40 in other species) mm long; nectary within pocketlike base of blade; stamens 20–75, filaments filiform; staminodes absent between stamens and pistils; pistils 5–28 (–50 outside our area), each with 4–5 (–9) ovules; style attenuate. **FRUIT** sessile, aggregate, oblong follicles, fruit walls transversely veined; beak terminal, straight, 2–4 mm long. $x = 8$. About 30 species, North America, Europe, Asia. Several ornamental species cultivated in North America.

Trollius albiflorus (A. Gray) Rydb. **CP** (white-flowered) White globe-flower. **STEMS** 4–55 cm tall (to 80 cm in fruit), base with few petioles persistent from previous year. **LEAVES**: basal leaves with petioles 4–25 cm long, some leaves reduced to ovate, sessile, membranous scales; cauline leaves 1–3 (–5), with broad, clasping, membranous sheaths. **FLOWERS** 2.5–5 cm diameter; sepals 5–9, spreading, white when fresh (pale yellow to greenish white before anthesis), nearly orbiculate to obovate or ovate, 10–20 mm long; petals 15–25, yellow, 1/2–2/3 stamen length when pollen is shed, 3–6 mm long. **FRUIT** usually 11–14, 8–16 mm long including beak; beak often somewhat recurved, sometimes straight. $2n = 16$. [*T. laxus* Salisb. var. *albiflorus* A. Gray]. Wetlands, along streams, subalpine to alpine meadows. COLO: Arc, Hin, LPl, Min, Mon, SJn. 3000–3850 m (9800–12700′). Flowering: May–Aug. Southwestern Canada south to Washington, Idaho, Utah, and Colorado. Sometimes confused with *Anemone narcissiflora* subsp. *zephyra*, which has sepals yellow (not white), achenes (not follicles), cauline leaves whorled (not alternate), and the stems and leaves pilose to villous (not glabrous).

RHAMNACEAE Juss. BUCKTHORN FAMILY

John R. Spence

Shrubs or small trees. **STEMS** armed or unarmed, typically strongly branched, bark smooth to somewhat shreddy, sometimes furrowed, bud scales present or absent. **LEAVES** deciduous to evergreen, alternate to opposite, solitary or in fascicles, petiolate, simple or rarely compound, pinnately veined or with several distinct veins rising from leaf base, stipules present, small. **INFL** a terminal or axillary cyme or cymose panicle, raceme, or umbel, or sometimes flowers solitary. **FLOWERS** actinomorphic, perigynous or epigynous, perfect or unisexual, 4–5-merous; sepals usually deciduous, triangular, free; petals (0) 4 or 5, free, concave or hoodlike, often clawed at base, showy or inconspicuous; stamens 4–5, opposite the petals, filaments adnate to petals, anthers often enclosed in petal hood, dehiscing longitudinally;

nectary disc well developed, alternate with filaments, adnate to hypanthium, free from ovary to sometimes enveloping entire ovary; ovary of 2–5 carpels, unilocular or plurilocular, ovules solitary, placentation basal, style lobed or deeply cleft. **FRUIT** a capsule or drupe with 1–3 stones enclosing seeds. A family of 60 genera and about 900 species, distributed worldwide, but most common in the tropics. The buckthorn family includes a few species with edible fruits, such as *Ziziphus jujuba*, jujube, from Europe, as well as valuable honey plants. Many species are used medicinally in various parts of the world, and some are commonly cultivated as ornamentals, especially in *Ceanothus*.

1. Plants spiny; low-growing shrubs .... ***Ceanothus*** (in part)
1' Plants unarmed; low to more often tall shrubs .... (2)

2. Leaves opposite or nearly so; plants to 5 dm; leaves small, 8–17 mm long, thick; in dry sites .... ***Ceanothus*** (in part)
2' Leaves alternate or fascicled; plants of various heights; leaves 20–120 mm long, thin; in shaded, mesic to wet sites .... (3)

3. Flowers 4-merous, unisexual, styles deeply cleft; bud scales present .... ***Rhamnus***
3' Flowers 5-merous, bisexual, styles lobed but not deeply cleft; bud scales absent .... ***Frangula***

Ceanothus L. Buckbrush, Mountain Lilac, P–ih pit–a', Tinéctc'il

(Greek *keanothus*, for a kind of thorny plant) Erect to somewhat sprawling low shrubs. **STEMS** spinose-armed or unarmed, typically strongly branched, spreading to erect, bark gray-green, gray to brown, glabrous or pubescent, bud scales present. **LEAVES** deciduous to evergreen, alternate or opposite, petiolate, stipules persistent or deciduous, pubescent, pinnately veined or with 3 distinct veins rising from leaf base or venation inconspicuous, margins smooth or finely serrate above the middle. **INFL** a terminal or axillary cymose panicle, on a short to long peduncle. **FLOWERS** perfect, pedicellate, 5-merous; sepals short, petaloid; petals longer than sepals, showy, blue, purple, or white, concave or hoodlike, clawed at base, lobes strongly hooded; anthers enclosed in petal hood; ovary 3-carpellate, unilocular, style typically 3-cleft about 1/2 its length. **FRUIT** appearing drupelike when immature, becoming a septicidal 3-valved capsule at maturity; each valve with or without a horn. **SEEDS** single per locule. A genus of 50 species endemic to North America, reaching south to Guatemala, but best represented in California, where species are conspicuous components of coastal chaparral. Many species are cultivated as ornamentals and for their honey. Indigenous tribes used the leaves as a tea, a poultice, and to soothe lung ailments.

1. Leaves alternate, mostly deciduous, with 3 distinct veins arising from the base; plants spiny; flowers typically white .... ***C. fendleri***
1' Leaves opposite, evergreen, venation inconspicuous, pinnate; plants unarmed; flowers typically blue .... ***C. vestitus***

Ceanothus fendleri A. Gray **CP** (for botanist A. Fendler) Fendler's ceanothus. Sprawling, low shrub, 0.4–1.5 m. **STEMS** strongly branched, spreading to erect, with lateral branches ending in sharp spinose tips, young twigs green or gray-green, pubescent, becoming brown with age. **LEAVES** alternate, deciduous at higher elevations to evergreen at lower elevations; petiole short, 2–4 mm, stipules small, deciduous, blades 10–18 mm, with 3 distinct veins rising from leaf base or sometimes venation inconspicuous, elliptic to elliptic-obovate, dull green above, pale green or whitish below, pubescent, margins smooth or finely serrate, flat, not conspicuously thickened. **INFL** terminal or axillary in axils of 1–3-year-old branches, umbellate to racemose. **FLOWER** sepals white to green-white, 0.3–0.6 mm; petals white or sometimes pink-tinged, 1.2–2 mm. **FRUIT** viscid when young, 3-lobed, 3–5 mm wide, brown to black, lacking horns. Common in open, montane conifer woodlands and forests and in shrub communities. ARIZ: Apa; COLO: Arc, Dol, Hin, LPl, Min, Mon, SMg; NMEX: McK, RAr, SJn; UTAH. 2285–2900 m (7500–9500'). Flowering: May–Jul. Fruit: Jun–Sep. Colorado and Utah south to Texas, Arizona, New Mexico, and Mexico. A very common shrub of the ponderosa pine community.

Ceanothus vestitus Greene (hairs on the leaves) Desert ceanothus. Erect, intricately branched, rounded shrub, 0.5–1 m. **STEMS** unarmed, strongly branched, erect, young twigs gray, gray-brown, or sometimes nearly white, pubescent, becoming gray-brown with age. **LEAVES** opposite, evergreen, petiole 1–2 mm or nearly absent, stipules evergreen, conspicuous, 1–2 mm, with a swollen base, blades 5–15 mm, obscurely pinnately veined, elliptic to elliptic-obovate, green or gray-green above, pale green or gray-green below, pubescent when young, margins smooth or with a few serrations, curled, concave above to rolled, thick. **INFL** terminal or axillary in axils of 1–3-year-old branches, umbellate, cymose panicle. **FLOWERS** with short to long pedicels, sepals white, 1–1.5 mm; petals blue to lavender or sometimes white, 2–3 mm. **FRUIT** not especially viscid when young, globose to 3-lobed, 4–5 mm wide, brown, each valve with a short, erect horn, 0.5–2 mm long. [*C. greggii* A. Gray]. Rare and local on sandstone ridges and rims in semiarid shrub communities and piñon-juniper woodlands above the San Juan and Colorado rivers. UTAH. 1525–1830 m (5000–6000').

Flowering: May–Jun. Fruit: Jul–Aug. True *C. greggii* occurs in Mexico only (Fross and Wilken 2006). Our material is **var.** ***franklinii*** (S. L. Welsh) J. R. Spence & K. D. Heil (for Ben Franklin, Utah state botanist), which is a distinctive variety with predominantly blue flowers endemic to the Colorado Plateau.

Frangula Mill. Alder Buckthorn

(medieval name for the genus) Erect to sprawling shrubs or small trees. **STEMS** unarmed, strongly branched, spreading to erect, bark gray-brown, red-brown, or gray, glabrous or pubescent, bud scales absent. **LEAVES** deciduous to evergreen, alternate, petiolate, stipules persistent, pubescent, pinnately veined, margins smooth or finely serrate; **INFL** in an axillary cyme, on a short to long, sometimes branched peduncle. **FLOWERS** perfect, pedicellate, 5-merous; sepals petaloid, often deciduous and falling away with upper hypanthium; petals shorter than sepals, lacking hood, often folded, clawed or not at base, anthers not enclosed in a petal hood; ovary of 3 carpels, each of 1 locule, style 2–3-lobed but not deeply cleft. **FRUIT** a purple-black spherical drupe with 2–3 stones. A genus of about 20 species, not always separated from *Rhamnus*, distributed primarily in the north-temperate zone.

Frangula betulifolia (Greene) Grubov (birchlike leaves) Birch-leaved buckthorn. Erect to sprawling shrub or small tree, 1–4 m. **STEMS** unarmed, strongly branched, erect or ascendent, occasionally prostrate, young twigs brown or gray-brown, sparsely pubescent, becoming gray with age; naked buds densely pubescent. **LEAVES** large, broadly elliptic to nearly orbicular, 4–12 cm long, 3–8 cm wide, tip abruptly rounded to indented, strongly pinnately veined, sparsely pubescent when young, becoming glabrous with age, dark green above, paler below, margins crenulate. **INFL** an axillary cyme of 4–12 flowers, on a long peduncle that is sometimes forked. **FLOWERS** small, inconspicuous, campanulate; sepals 1–2.5 mm, eventually deciduous; petals dull brown or green-brown, 0.8–1.2 mm; style short, to 1 mm. **FRUIT** a glabrous, purple or purple-black, spherical to obovate drupe, 6–10 mm long. [*Rhamnus betulifolia* Greene]. Common around springs and in shaded dense riparian vegetation and other damp protected sites. ARIZ: Apa, Nav; COLO: LPl; NMEX: McK; UTAH. 1160–2130 m (3800–7000′). Flowering: Apr–May. Fruit: Jun–Jul. The large leaves of this species are popular with leaf-cutter bees (Megachilidae), which cut out pieces of leaf around the margins for their nests.

Rhamnus L. Buckthorn

(ancient Greek name for the genus) Erect to somewhat sprawling shrubs or small trees. **STEMS** unarmed, strongly branched, ascending to erect, bark gray-green, gray to brown, glabrous or pubescent, bud scales present. **LEAVES** deciduous to evergreen, alternate or opposite, petiolate, stipules persistent, simple, pinnately veined or with 3 distinct veins rising from leaf base or venation inconspicuous, margins smooth or finely serrate above the middle. **INFL** a terminal or axillary cymose panicle, on short to long pedicels. **FLOWERS** unisexual, short- to long-pedicelled, 4-merous; sepals triangular; petals 0 or 4, not showy, often shorter than sepals, lacking hood, often folded, clawed or not at base; anthers not enclosed in petal hood; ovary of 2–4 carpels, each of 1 locule, style typically deeply cleft into 2–4 segments about 1/2 its length. **FRUIT** a red, purple, or black spherical drupe with 2 stones. A genus of about 100 species, widespread in temperate and subtropical regions of the world. *Rhamnus cathartica* has been used as a purgative for centuries in Eurasia, and many other species have purgative chemicals in their bark. The drupes of some species are sources of yellow and green dyes.

Rhamnus smithii Greene (named for the collector Benjamin H. Smith) Smith's buckthorn. Erect, strongly branched shrub, 1–3 m. **STEMS** unarmed, erect, young twigs gray-yellow, pubescent, becoming glabrous with age; bud scales tan-brown, ciliate-margined. **LEAVES** alternate, in fascicles or rarely nearly opposite, evergreen, elliptic to ovate or oblong-lanceolate, blades 2–7 cm long, 1–3 cm wide, apex acute to obtuse, distinctly pinnately veined, green above, pale green to yellow-green below, pubescent or glabrous, margins serrulate to crenulate. **INFL** an axillary cyme with 2–4 flowers on short peduncles, sometimes branched. **FLOWERS** small, inconspicuous; sepals pale green-yellow, 1.5–2.5 mm; petals green-yellow to white, 0.7–1.5 mm, sometimes absent; styles deeply cleft into 2–4 segments. **FRUIT** a black spherical drupe, 5–9 mm wide, typically with 2 stones. Uncommon on slopes in woodlands and forests, along riparian zones, in shaded moist sites, and sometimes on midelevation clay barrens. COLO: Arc, LPl, Mon; NMEX: RAr. 2000–2500 m (6500–8200′). Flowering: Jun. Fruit: Jul–Aug. Colorado and New Mexico.

ROSACEAE Juss. ROSE FAMILY

Genus descriptions by Kenneth Robertson
Species descriptions by Kenneth D. Heil and Noel H. Holmgren
Genus key by Kenneth D. Heil

(ancient Latin name for roses) Trees, shrubs, perennial or annual herbs, sometimes armed with thorns or prickles. **LEAVES** mostly stipulate, simple to palmately or pinnately compound, variously toothed or lobed, often with gland-tipped teeth, alternate. **INFL** bracteate, corymbs, racemes, spikes, panicles, solitary. **FLOWERS** actinomorphic, mostly perfect, hypanthium variously flat, cup-shaped, cylindrical, campanulate, turbinate, or urceolate, often enlarging in fruit, nectar ring inside; calyx lobes 5, sometimes with alternating epicalyx lobes, often persistent; petals 5, usually subcircular with irregularly erose margins and a short claw; stamens mostly 15 or more, rarely 5, perigynous on hypanthium, usually persistent; pistil 1–many-carpellate, distinct or infrequently connate, free or adnate to hypanthium; ovary superior or inferior, ovules 1 or 2 or rarely several per carpel, styles same number as carpels. **FRUIT** follicles, achenes, drupes, drupelets, or pomes. **SEEDS** usually lacking endosperm. x = 7, 8, 9, 17. About 100 genera, 3000 species, most abundant in north-temperate zone, especially western North America and eastern Asia. Likely the third most economically important plant family in North America, many species with edible fruit and also numerous ornamental plants (Cronquist, Holmgren, and Holmgren 1997a).

1. Plants herbs, sometimes woody at the very base ....(2)
1′ Plants shrubs or small trees ....(11)

2. Petals lacking; inflorescence a dense spike, flowers numerous; leaves pinnately compound .... ***Sanguisorba***
2′ Petals present (minute in *Sibbaldia*); inflorescence either not a dense spike or flowers not numerous; leaves various ....(3)

3. Leaves simple, crenate; petals 8–10, white; fruit a plumose achene; flowers solitary on scapose peduncles .... ***Dryas***
3′ Leaves compound or variously divided; petals mostly 5; fruit not a plumose achene; flowers usually more than 1 ...(4)

4. Flowers yellow (or dark red in *Potentilla thurberi*) ....(5)
4′ Flowers white or pink ....(6)

5. Petals minute, narrow; leaflets toothed at apex; plants prostrate or mat-forming; of high elevations .... ***Sibbaldia***
5′ Petals showy, not narrow; leaflets entire, lobed, or serrate; plants not prostrate or mat-forming; plants not limited to high elevations .... ***Potentilla***

6. Leaves trifoliolate; plants with well-developed stolons; receptacle ripening into an accessory fruit .... ***Fragaria***
6′ Leaves usually with more than 3 leaflets; plants lacking stolons; receptacle not ripening ....(7)

7. Leaflets very numerous, mostly less than 6 mm long; petals usually clawed .... ***Ivesia***
7′ Leaflets mostly 3–15, usually more than 6 mm long; petals mostly sessile ....(8)

8. Flowers in a narrow raceme; upper 1/2 of mature hypanthium covered with hooked bristles; rare in our area .... ***Agrimonia***
8′ Flowers in a branched inflorescence; hypanthium without hooks ....(9)

9. Leaves pinnately lobed or compound or more usually lyrate-pinnatifid; styles at maturity elongate and conspicuous .... ***Geum***
9′ Leaves palmately or pinnately lobed or compound, not lyrate-pinnatifid; styles at maturity not elongate and conspicuous ....(10)

10. Style attached to near top of the ovary; leaves either palmately compound or pinnately compound with narrow leaflets .... ***Potentilla***
10′ Style attached near the base of the ovary; leaves pinnately compound with broadly oval leaflets .... ***Drymocallis***

11. Leaves compound ....(12)
11′ Leaves simple ....(15)

12. Flowers yellow; leaves pinnately compound, crowded and appearing to be palmate .... ***Dasiphora***
12′ Flowers rose, white, or pink; leaves distinctly pinnate ....(13)

13. Leaflets 11–15; small trees with orange berries (pomes); stems and/or leaves lacking prickles .... ***Sorbus***
13' Leaflets 5–7; shrubs without orange berries; stems and/or leaves with prickles .... (14)

14. Leaves glaucous beneath; flowers white; fruit an aggregate of drupelets (a raspberry) .... ***Rubus***
14' Leaves green, often pale; flowers rose or rarely white; fruit a hip .... ***Rosa***

15. Leaves opposite; petals 0; low desert .... ***Coleogyne***
15' Leaves alternate; petals present; plants of various habitats .... (16)

16. Low mat-forming shrubs; flowers solitary or in dense spikes on leafless or merely bracteate scapes .... (17)
16' Shrubs or small trees, never mat-forming; flowers various but not scapose or subscapose .... (18)

17. Flowers solitary, mostly 8 petals; leaves simple, crenate, white beneath; alpine tundra on limestone .... ***Dryas***
17' Flowers in dense spikes, mostly 5 petals; leaves entire, in tight rosettes forming extensive hard mats on cliffs .... ***Petrophytum***

18. Ovary or ovaries superior, free from the hypanthium .... (19)
18' Ovary inferior, 2–5 carpels fused and the fruit a pome .... (24)

19. Petals lacking, flowers inconspicuous; leaves entire, evergreen (except *C. montanus*) .... ***Cercocarpus***
19' Petals present; flowers showy or small; leaves mostly toothed or lobed, not evergreen .... (20)

20. Pistil 1; fruit a drupe; leaves usually with glands at the base of the blade or on the petiole .... ***Prunus***
20' Pistils 1 to many; fruit not a drupe; leaves without glands .... (21)

21. Flowers large, 2–5 cm across, in few-flowered cymes; leaves 2–15 cm wide; fruit an aggregate .... ***Rubus***
21' Flowers less than 2 cm across, solitary or in corymbs; leaves up to 2 cm wide; fruit an achene, plumose-tailed or not .... (22)

22. Flower petals white, in terminal, many-flowered pyramidal clusters; usually on talus slopes and cliff faces .... ***Holodiscus***
22' Flower petals white, cream, or pale yellowish; borne singly or in few-flowered cymes; various habitats .... (23)

23. Pistils numerous; petals white; leaf lobes tightly revolute; often in wash bottoms and roadsides .... ***Fallugia***
23' Pistils 1–5; petals white to cream or pale yellow; leaf lobes not tightly revolute; various habitats .... ***Purshia***

24. Stems armed with thorns .... ***Crataegus***
24' Stems unarmed (rarely so in *Malus*) .... (25)

25. Leaves narrowly elliptic, entire or indistinctly toothed .... ***Peraphyllum***
25' Leaves serrate to doubly serrate .... (26)

26. Flowers white, in racemes; plants indigenous, rarely cultivated; leaves toothed at the apex .... ***Amelanchier***
26' Flowers white or otherwise, in corymbs or short panicles; plants cultivated, sometimes escaping; leaves toothed or lobed .... ***Malus***

Agrimonia L. Agrimony

(corruption of *Argemone*, a name used by Pliny) Perennial herbs. **STEMS** unbranched below inflorescences. **LEAVES** imparipinnately compound, basal ones small, not persistent, cauline leaves interrupted, with small leaflets interspersed with larger leaflets, leaflet margins serrate, stipules persistent, resembling smallest leaflets, sometimes coarsely toothed. **INFL** of racemes terminating stems, pedicels short. **FLOWERS** perfect, yellow, small; hypanthium turbinate to hemispheric, throat constricted, outer rim with numerous hooked bristles that elongate and become indurate in fruit; sepals 5, spreading at anthesis, becoming incurved and forming a beak on the fruit, epicalyx absent; petals small, orbicular, deciduous; stamens 5–15; carpels 2, free from each other and from hypanthium, the ovary superior, inserted at base of hypanthium, styles terminal, stigmas terminal, ovaries superior. **FRUIT** an accessory, 2 achenes enclosed within enlarged, indurate, 10-ribbed hypanthium, ring of hooked bristles around top of hypanthium. $x = 7$. About 18 species, 10 in North America.

Agrimonia striata Michx. (furrowed or channeled) Herbaceous perennial with rhizomes. **STEMS** erect and branched above, 30–100 cm tall, hirsute and somewhat glandular. **LEAVES** with 7–13 leaflets, 3–12 cm long, crenate-serrate, resinous-glandular beneath and sparingly hirsute only on the veins. **INFL** a raceme, 20–40 cm long. **FLOWER** calyx tube in fruit mostly 5 mm long; petals obovate, about 5 mm long, yellow; stamens 5–15; pistils 1 or 2, stigmas 2-lobed. **FRUIT** 1 or 2 achenes. [*A. brittoniana* E. P. Bicknell; *A. brittoniana* var. *occidentalis* Bickn.]. Ponderosa pine communities. ARIZ: Apa; COLO: Arc, Hin, LPl; NMEX: SJn. 1675–2590 m (5500–8500′). Flowering: Jul–Sep. Nova Scotia to British Columbia, south to West Virginia, Colorado, Arizona, and New Mexico.

Amelanchier Medik. Serviceberry

(derived from *amelanche*, the French name of the European *A. ovalis*) Deciduous, unarmed shrubs or small trees. **STEMS** woody, bark smooth, pale, with shallow longitudinal fissures. **LEAVES** deciduous, alternate, simple, petiolate, toothed, pinnately veined; stipules linear, caducous. **INFL** racemose, leafy, sometimes reduced to a few-flowered cluster, appearing with the leaves, bracts linear, caducous. **FLOWERS** perfect, white, showy, large nectar ring present; hypanthium campanulate to urceolate, lower part adnate to the carpels at least at the base; sepals 5, persistent, epicalyx absent; petals 5, narrow, not clawed, flaccidly spreading, deciduous; stamens usually 20; carpels 2–5, fully connate, partially adnate to hypanthium, ovary about 1/2 inferior, styles connate below. **FRUIT** a fleshy pome, purple or orange to yellow or nearly white, the core thin or cartilaginous, false septa present, topped by persistent, recurved sepals. $x = 17$. About 33 species, mostly North American with a few species in Europe and eastern Asia.

1. Larger leaf blades (2) 2.5–6 cm long, sparsely pubescent to glabrous with age; styles (4) 5, united into a column below; petals 7–13 mm long; pome becoming fleshy and purplish at maturity .... ***A. alnifolia***
1′ Leaf blades 1–3 cm long; permanently pubescent; styles 2 or 3 (4 or 5), usually distinct to the base; petals 5.5–9 (10) mm long; pome more or less dry or mealy, orangish, yellowish, or whitish, but sometimes with a purplish tinge .... ***A. utahensis***

Amelanchier alnifolia (Nutt.) Nutt. ex M. Roem. (leaves like the genus *Alnus*, alder) Saskatoon serviceberry, western serviceberry, shadbush. Shrub 1–4.5 m tall, bark of the young stems smooth, reddish brown, eventually becoming gray; herbage glabrous to sparsely puberulent when young. **LEAVES** with petioles 0.7–1.8 (2.8) cm long, stipules filiform, caducous, the blade (ob)ovate to suborbicular, (2) 2.5–6 cm long, 1.5–4.5 cm wide, obtuse, rounded, truncate, or subtruncate basally, obtuse, rounded, or sometimes retuse apically, serrate or coarsely toothed, usually from or beyond the middle, the veins often prominent beneath, the surface often green above and glaucous beneath. **INFL** a short, erect, 5–15-flowered raceme, variously pubescent to glabrous. **FLOWERS** fragrant; hypanthium brownish within; sepals becoming reflexed at anthesis, deltate-lanceolate to narrowly lanceolate, 1.8–4 (5.5) mm long, persistent; petals oblanceolate, 7–18 mm long, 2.5–6.7 mm wide, the claw pilosulous at the base, white; stamens 20, the filaments white, the anthers 0.5–0.8 mm long, brownish; styles (4) 5, united into a column below. **FRUIT** a globose pome, 7–10 mm thick, densely tomentose at the apex or sometimes completely glabrous, becoming fleshy and purplish at maturity, usually 10-seeded, edible. **SEEDS** 3.5–5 mm long, dark brown. $2n = 68$.

1. Herbage and inflorescence pubescent, glabrescent with age, the top of the ovary remaining densely hairy through development into a pome .... **var. *alnifolia***
1′ Herbage and inflorescence glabrous, including the top of the ovary (pome) .... **var. *pumila***

var. *alnifolia* [*Pyrus alnifolia* (Nutt.) Lindl ex Ser.; *A. bakeri* Greene]. Gambel's oak, mountain mahogany, ponderosa pine, Douglas-fir, aspen, and spruce-fir communities. ARIZ: Apa, Nav; COLO: LPl, Min, Mon, SJn; UTAH. 2135–2745 m (7000–9000′). Flowering: May–Jun. Alaska, Yukon, British Columbia, Alberta east to Manitoba, south through Washington, Oregon, northern California, Idaho, Montana, and the Dakotas, Utah, northern Arizona, Colorado, and western Nebraska. Fruit eaten for food by the Navajo.

var. *pumila* (Torr. & A. Gray) C. K. Schneid. (dwarf) Dwarf serviceberry. [*A. pumila* Nutt ex Torr. & A. Gray; *A. polycarpa* Greene]. Gambel's oak, mountain mahogany, ponderosa pine, Douglas-fir, aspen, and spruce-fir communities. ARIZ: Apa; COLO: Arc, Dol, Hin, LPl, Min, SJn; NMEX: McK, RAr, SJn. 2285–3350 m (7500–11000′). Flowering: May–Jun. Southern Canada south to Washington, Oregon, and northern California, east to Idaho, Montana, Wyoming, and south to northern Arizona and northern New Mexico.

Amelanchier utahensis Koehne (of Utah) Utah serviceberry. Shrub up to 3 dm tall, often much-branched, bark reddish when young, soon becoming gray. **LEAVES** with petioles 0.2–0.9 (1.4) cm long, stipules filiform, caducous, the blade (ob)ovate to suborbicular, (1) 1.5–3 dm long, 0.8–2 (2.3) cm wide, rounded basally, obtuse, truncate, or sometimes retuse apically, serrate or crenate-serrate, usually prominently tomentose on both surfaces, the veins prominent beneath. **INFL** a 3–6 (10)-flowered raceme, the pedicels, hypanthium, and sepals usually tomentose. **FLOWER** sepals becoming reflexed at anthesis, lanceolate to narrowly lanceolate, 1.5–4.2 mm long, acute or acuminate, persistent; petals oblanceolate, sometimes narrowly so, 5.5–9 (10) mm long, 1.7–3.7 (4.2) mm wide, narrowly cuneate below, white; stamens 15–20, the filaments white, the anthers 0.6–0.9 mm long, pale yellow, turning brown; styles 2 or 3 (4 or 5) distinct to the base or nearly so. **FRUIT** a pome, globose to more or less pyriform, 5–10 mm thick, finely pubescent or glabrous, dry or at least not juicy at maturity, the flesh sometimes mealy, orangish, yellowish, or nearly white, often with a purplish tinge, 3–6-seeded. **SEEDS** 3.8–5.6 mm long, dark brown. [*A. prunifolia* Greene; *A. crenata* Greene; *A. rubescens* Greene; *A. oreophila* A. Nelson]. Streamsides, foothills, and mountain slopes; grassland, mountain mahogany, Gambel's oak, mountain brush, piñon-juniper woodland, aspen, and ponderosa pine communities. ARIZ: Apa, Coc, Nav; COLO: Arc, Dol, LPl, Mon, SMg; NMEX: McK, RAr, SJn; UTAH. 1370–2460 m (4500–8065′). Flowering: mid-Apr–May. Southern Washington, central Idaho, southwestern Montana, south through eastern Oregon, southern Idaho, western and southern Wyoming to eastern California, Nevada, northern Arizona, Utah, western Colorado, and northwestern New Mexico; Baja California. Our material belongs to **var.** ***utahensis***. However, *A. utahensis* var. *covillei* (Standl.) N. H. Holmgren may be the same as what is called *A. prunifolia* Greene in Colorado and New Mexico. More collections and more study are needed to determine if var. *covillei* is found in the study area. The plant was used during labor and delivery and berries were dried for winter use by the Navajo. Hopi used the plant to make pahos (prayer sticks), and to make bows and arrows. The Ramah Navajo used the leaves as emetics in various ceremonies as a "life medicine," ate the berries raw or cooked, and used the stem as a ceremonial item (to make "Evil-way hoop").

Cercocarpus Kunth Mountain Mahogany

(Greek *kerkos*, tail, + *karpos*, fruit) Evergreen shrubs or small trees, unarmed. **STEMS** woody, bark smooth, gray to reddish brown. **LEAVES** simple, clustered at tips of short shoots, entire or dentate, usually thick or coriaceous; lower part of stipules adnate to bases of petioles, the free parts caducous. **INFL** of solitary flowers or few-flowered clusters terminating short shoots; bracts small, membranaceous, caducous. **FLOWERS** perfect, sessile, nectar disc absent, not showy; hypanthium of 2 parts, lower part narrowly tubular, persistent, upper part cup-shaped, deciduous; petals absent; sepals 5, small, deciduous with the upper hypanthium; stamens 10–45; carpel 1, free from hypanthium, style terminal, persistent. **FRUIT** an achene, included in hypanthium, style greatly elongate, plumose. $x = 9$. About 13 species of western North America and Mexico.

1. Leaves linear; a highly and intricately branched low shrub; slickrock sites and cliffs .... ***C. intricatus***
1′ Leaves broader; a tall, open-branched shrub; various sites .... (2)

2. Leaves entire, strongly revolute; often a small tree; rare in our area .... ***C. ledifolius***
2′ Leaves crenate-serrate; not strongly revolute; tall shrubs; widespread .... ***C. montanus***

Cercocarpus intricatus S. Watson (entangled) Littleleaf mountain mahogany, dwarf mountain mahogany. Evergreen, small shrub, intricately branched, spinescent; up to 2.5 m high, young branches reddish brown, finely villous with crinkly white hairs or glabrous, older stems ashy gray. **LEAVES** with petioles 0.8–1.5 mm long, stipules lanceolate, brown, acute; blades linear, 5–15 mm long, 0.5–2 mm wide, mucronulate-tipped, entire, revolute-margined, dark green above. **INFL** of 1–3 flowers terminating short, lateral branches. **FLOWERS** subsessile; hypanthium tube persistent, 4–6 mm long, green, becoming pale brown with brown nerves with age, hypanthium cup rim 1–2 mm high, campanulate, reddish; sepals recurved, broadly deltate, 0.7–1.2 mm long, obtuse, pubescent within; stamens 10–15, anthers 0.8–0.9 mm long, glabrous. **ACHENES** about 6 mm long, ascending-pubescent, mature style 3–4.5 cm long, spirally coiled, plumose. [*C. ledifolius* Nutt. var. *intricatus* (S. Watson) M. E. Jones; *C. arizonicus* M. E. Jones; *C. intricatus* var. *villosus* C. K. Schneid.]. Cracks and crevices of sandstone outcrops, or in shallow rocky soils; desert scrub and piñon-juniper woodland communities. ARIZ: Apa, Coc, Nav; COLO: Mon; NMEX: SJn; UTAH. 1495–2725 m (4900–8940′). Flowering: Apr–Jun. Southeastern California, Nevada, Utah, southwestern Colorado, northern Arizona, and northwestern New Mexico. Used for ceremonial items by the Hopi.

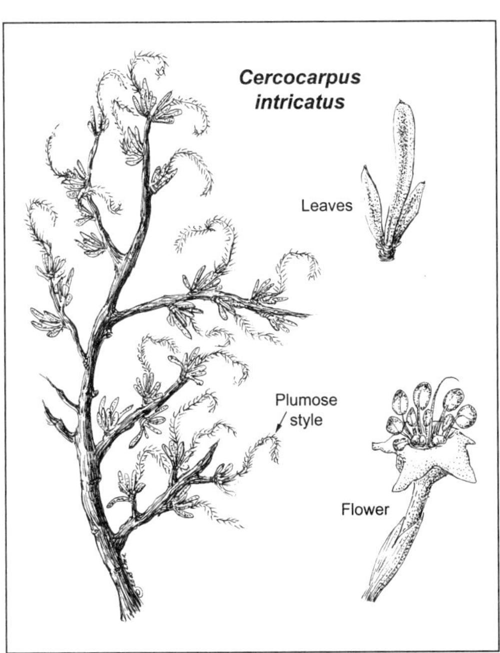

Cercocarpus ledifolius Nutt. (curl leaf) Curl-leaf mountain mahogany. Small tree or evergreen shrub, erect, up to 8 m tall, sharp, stiff branches, elongating branches reddish, villous or tomentose with crinkly white hairs, older branches ashy gray. **LEAVES** nonfascicled on elongating branches, with petioles 1–6 mm long, stipules brown, with a thick midvein, blade lance-elliptic, narrowly lanceolate, or linear, 1–4 cm long, mostly 2–8 mm wide, cuneate basally, acute apically, entire and revolute-margined, dark resinous green, sparsely lanate above when young, moderately to densely lanate beneath, midrib prominent. **FLOWERS** subsessile; hypanthium tube persistent, up to 13 mm long, lanate, hypanthium cup 1.5–3 mm high, reddish; sepals broadly deltate, mostly 1.2–2.4 mm long, rounded, recurved, woolly-pubescent; stamens 15–25, anthers 0.6–1.3 mm long, glabrous. **ACHENES** 7–9 mm long, mature style 4–10 cm long, bent or spirally coiled, plumose. $2n = 18$. [*C. ledifolia* Nutt. ex Hook. & Arn.; *C. ledifolius* var. *intercedens* C. K. Schneid.; *C. hypoleucus* Rydb.]. Mountain slopes; sagebrush, piñon pine, oak, and ponderosa pine communities. ARIZ: Apa; UTAH. 1920–2775 m (6300–9110′). Flowering: May–Jun. Southeastern Washington, southern Oregon, south to California, southern Idaho, Nevada, Utah, northern Arizona, and east to southwestern Montana, Wyoming, and western Colorado. Our material belongs to **var. *intermontanus*** N. H. Holmgren (between mountains).

Cercocarpus montanus Raf. (of mountains) Alderleaf mountain mahogany, birchleaf mountain mahogany. Deciduous shrub, up to 4 m tall, young branches reddish brown, older bark gray to brown, smooth. **LEAVES** with petioles 2.5–6 mm, stipules short, lanceolate, acute, light brown; blade ovate, rhombic, 12–50 mm long, mostly 8–20 mm wide, cuneate basally, rounded apically, margins crenate-serrate, plane or slightly revolute, dark green with impressed veins, pale beneath. **FLOWER** hypanthium cup 3–8 mm long; hypanthium tube persistent, 6.5–11 mm long, sericeous, becoming reddish brown; sepals deltate, 1–1.8 mm long, rounded, recurved; stamens 22–44, anthers 0.6–1 mm long, hirsute. **ACHENES** 9–12 mm long, mature style 5–9 cm long, spirally coiled, densely plumose. $2n = 18$. [*C. parvifolius* Nutt. ex Hook. & Arn.; *C. macrourus* Rydb.; *C. douglasii* Rydb.; *C. rotundifolius* Rydb.; *C. flabellifolius* Rydb.]. Sagebrush, piñon-juniper woodland, Gambel's oak, ponderosa pine, and Douglas-fir communities. ARIZ: Apa, Nav; COLO: Arc, Dol, LPl, Mon, SMg; NMEX: McK, RAr, San, SJn; UTAH. 1525–2285 m (5000–7500′). Flowering: late Apr–May. Oregon east to Nebraska and South Dakota, and south through California, Nevada, Utah, Colorado, Arizona, New Mexico, western Oklahoma, and western Texas; Mexico. Four Corners material belongs to **var. *montanus***.

Coleogyne Torr. Blackbrush

(Greek *koleos*, sheath, + *gyne*, ovary, alluding to the hypanthium enclosing the pistil) Deciduous shrub with spine-tipped branches. **STEMS** woody, bark gray, with thin longitudinal fissures. **LEAVES** simple, crowded in fascicles terminating opposite short shoots, entire, coriaceous; stipules small, partially adnate to leaf bases, persistent. **INFL** of solitary flowers terminating short shoots; bracts persistent. **FLOWERS** perfect, pale to bright yellow, showy, nectar disc absent; hypanthium short, urceolate; sepals 4, petaloid; petals generally absent; stamens 30–40, inserted near the outer base of a peculiar membranaceous sheath that separates the stamens from the carpel; carpel 1, free from hypanthium. **FRUIT** a relatively large, leathery achene. $x = 8$. Monotypic, southwestern North America.

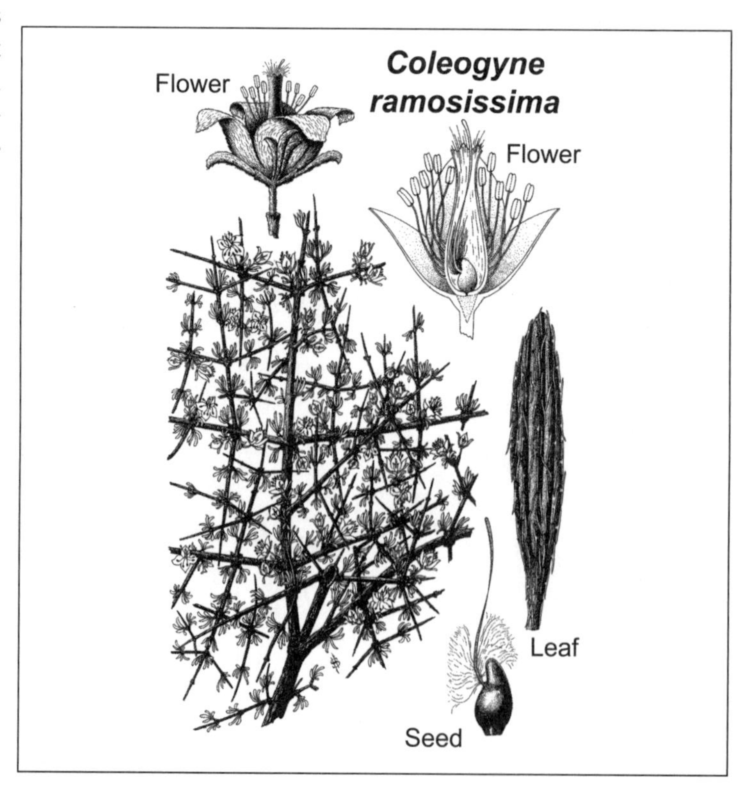

Coleogyne ramosissima Torr. (much-branched) Blackbrush. Shrub up to 2 m tall, divergent and often with spinescent branches, bark exfoliating, gray. **LEAVES** in fasicles on short lateral branches or in the axil of opposite leafless stipules, main leaves narrowly oblanceolate, 5–10 mm long, obtuse or mucronate apically, entire, strigose with appressed malpighian hairs. **INFL** of solitary flowers terminating the lateral branchlets; pedicel, hypanthium, and outside of the sepals strigose. **FLOWER** hypanthium short, turbinate; sepals arranged in 2 pairs, outer pair ovate, 4.5–7 mm long, acute, inner pair broadly ovate, 5–7.5 mm long, all yellow and glabrous on the adaxial surface, green or purplish, strigose abaxially; petals rare, opposite the sepals, up to 12 mm long, yellow; torus narrowly pyriform, 4–6 mm long, with 5 subulate teeth at the apex, glabrous on the outside, white-villous inside; stamens 20–40, anthers 1–1.3 mm long; pistil 1, style persistent, exserted from the torus. **ACHENES** 3–4 mm long, glabrous. $2n = 16$. Often the dominant species in our area, commonly found on the Morrison Formation. ARIZ: Apa, Coc, Nav; UTAH. 1125–1905 m (3700–6255′). Flowering: May–Jun. Southern California, southern Nevada, southern Utah, northern Arizona, and southwestern Colorado.

Crataegus L. Hawthorn

(Greek *krataigos*, ancient name for a flowering thorn) Small trees and shrubs, usually armed with stout thorns. **STEMS** woody, young bark smooth and often reddish, becoming gray, rough, and scaly or checked. **LEAVES** deciduous, alternate, petiolate, stipules often glandular-toothed, small and caducous, sometimes large and persistent on long shoots; blade simple, entire to variously toothed or lobed. **INFL** a corymb, cyme, or panicle terminating short, leafy shoots, bracts linear, glandular-toothed, caducous. **FLOWERS** perfect, white, showy, large nectar ring present; hypanthium cup-shaped to urceolate, lower part fully adnate to the carpels; sepals 5, reflexed at anthesis, entire or glandular-serrate, persistent or deciduous, epicalyx absent; petals more or less circular, deciduous; stamens 5–25; carpels 2–5, fully connate, fully adnate to the hypanthium and the ovary inferior, styles mostly free. **FRUIT** a fleshy pome containing pyrenes, 1 pyrene per carpel, red, purple, black, orange, or yellow, usually topped by persistent sepals. $x = 17$. Estimating the number of species is made difficult by apomictic complexes, but a current estimate is 186 species of North and Central America and Eurasia.

1. Thorns 1–3 cm long; mature pome purplish black .... ***C. douglasii***
1′ Thorns 2.5–6 cm long; mature pome dark red or reddish orange to dark brown .... (2)

2. Thorns averaging more than 3.5 cm in length; petioles and lower surface of leaves pubescent at maturity, especially along the veins, teeth not strongly glandular-tipped; rare in the study area .... ***C. macracantha***
2′ Thorns averaging between 2.5–3.5 cm in length; petioles and lower surface of leaves glabrous at maturity, teeth with black glandular tips; common in the study area .... ***C. erythropoda***

Crataegus douglasii Lindl. (for David Douglas, 1799–1834, Scottish botanical explorer) River hawthorn. Large shrub or small tree, 2–7 m tall, bark reddish or orange, smooth at first, becoming gray, armed with shiny, dark red, relatively short thorns 1–3 cm long; herbage pubescent to glabrate. **LEAVES** with petioles mostly 0.5–2 cm long, stipules linear, glandular-toothed, caducous, the leaf blade lanceolate to rhombic-oblanceolate, 3–8.5 cm long, 2–6 cm wide, mostly less than 2 times as long as wide, cuneate basally, acuminate, acute, obtuse, or rounded apically, once- or double-serrate. **INFL** a few-flowered, flat-topped, corymbose panicle or raceme. **FLOWER** sepals recurved at anthesis, deltate-caudate, 2.7–3.6 mm long, entire or glandular-toothed, slightly villous on the upper surface, glabrous beneath; petals obovate to orbicular, 4.5–7.5 mm long, 5–7.8 mm wide, white; stamens 10 or more, filaments 3.5–6 mm long, white, anthers 0.8–1.1 mm long, pink; styles 3–5, 3.5–4.5 mm long, pale green. **FRUIT** a pome, broadly ovoid or globose, 8–10 mm in diameter when dried, bony, purplish black, glaucous; nutlets with a cavity on the ventral side. $2n = 34, 48, 68$. [*C. rivularis* Nutt.; *C. wheeleri* A. Nelson]. Mostly found in canyons along streams in piñon-juniper woodland and ponderosa pine communities. COLO: Arc, Hin, LPl, Mon; NMEX: SJn. 1830–2530 m (6000–8300′). Flowering: May–early Jun. Our material belongs to **var. *rivularis*** (Nutt.) Sarg. (riparian). Northeastern Nevada, southeastern Idaho, southwestern Wyoming, Utah, central Arizona, western Colorado, and western and northern New Mexico.

Crataegus erythropoda Ashe (red-based) Rocky Mountain hawthorn. Shrub or small tree, 2–5 m tall, the bark of the young branches reddish, armed with relatively slender, recurved, dark brown thorns, mostly 2.5–3.5 cm long. **LEAVES** with petioles 1–2.2 cm long, usually winged distally, the blade rhombic-ovate, 4–8 cm long, 3–5.5 cm wide, broadly cuneate at the base, acute apically, coarsely double-serrate or often shallowly lobed in upper 1/2 or 2/3, teeth with black glandular tips, sparsely appressed-pubescent and shiny green above, glabrous and pale beneath, the veins impressed above and prominent beneath. **INFL** a glabrous (2) 5–10-flowered corymb. **FLOWER** sepals narrowly lanceolate or caudate, 3–4.3 mm long, often glandular-toothed; petals suborbicular, 5–8.5 mm long, clawed, white; stamens 5–8 (10), anthers 0.9–1 mm long, pink, rose, or purple; styles usually 5. **FRUIT** a subglobose pome, 7–10 mm in diameter, reddish orange or dark brown, glabrous; nutlets 5, plane on ventral surface. [*C. cerronis* A. Nelson]. Along streams and canyons. COLO: Arc, Hin, LPl, Min, Mon; NMEX: RAr, SJn. 1675–2895 m (5500–9500′). Flowering: May–Jun. South-central Wyoming, western 2/3 of Colorado, northwestern and north-central New Mexico, east-central Arizona, and San Juan County, Utah. This is our most common hawthorn.

Crataegus macracantha Lodd. (big-thorned) Big-spine hawthorn. Shrub or small tree up to 7 m tall, with reddish bark, armed with thorns 3–7 cm long. **LEAVES** with petioles up to 2.5 cm long, pubescent at maturity, especially on the veins, blade broadly (ob)ovate, 4–7 cm long, 2–5 cm wide, weakly lobed above the widest point, once- or twice-serrate teeth, not strongly glandular-tipped, villous-pubescent on the lower surface, at least near the veins. **INFL** a flat-topped corymb. **FLOWER** sepals narrowly lanceolate, 4–6 mm long; petals obovate to orbicular, 6–8 mm long, white; stamens 20 or less, ours not over 10, filaments 4–5 mm long, anthers 1–1.2 mm long, white to rose or purple. **FRUIT** a globose to subglobose pome, about 7–11 mm in diameter, bright red to purplish, pubescent to glabrous, with 2–4 nutlets. [*C. florifera* Sarg.; *C. succulenta* Schrad. ex Link var. *occidentalis* (Britton) E. J. Palmer; *C. coloradensis* A. Nelson; *C. columbiana* Howell var. *occidentalis* (Britton) Dorn]. Hillsides and canyons. COLO: Arc; NMEX: RAr. 2000–2210 m (6560–7250′). Flowering: May. Eastern Canada to Saskatchewan, south to Pennsylvania, Nebraska, Colorado, and northwestern New Mexico. Our material belongs to **var.** ***occidentalis*** (Britton) Eggl. (western). Apparently rare in the Four Corners region.

Dasiphora Raf. Shrubby Cinquefoil

(Greek *dasy*, woolly, + *phor*, carry, for the hairy achenes) Deciduous shrubs, unarmed. **STEMS** woody, bark shiny brown, exfoliated when older. **LEAVES** with stipules adnate to base of petiole, extending as almost foliaceous free lobes, persistent; blade pinnately compound, 3–7-foliolate, leaflets serrate. **INFL** of solitary axillary flowers or cymose, bracts reduced leaves, persistent. **FLOWERS** perfect, yellow (in ours), showy; hypanthium saucer-shaped, free from carpels, nectar disc present; sepals 5 alternating with 5 epicalyx lobes, persistent; petals 5, obovate to orbicular, short claw at base, deciduous; stamens 20–25; carpels numerous, free from each other and from hypanthium, the ovaries superior, inserted on base of receptacle and a short hemispheric projection, style inserted laterally near base of ovary. **FRUIT** numerous achenes. Five to six species in the Northern Hemisphere, mostly in temperate regions. $x = 7$. A segregate of *Potentilla*, s.l.

Dasiphora fruticosa (L.) Rydb. (shrubby) Shrubby cinquefoil. Shrub, mostly 1–15 dm tall; current year's growth villous, the stem becoming glabrous with shreddy, reddish brown bark; leaves sericeous beneath. **LEAVES** with stipules sheathing, broadly lanceolate to ovate, acuminate; petioles up to 15 mm long; blade pinnate, mostly 5-foliolate; leaflets narrowly elliptic, 0.8–1.8 cm long, entire. **INFL** cymose or of solitary axillary flowers. **FLOWER** bractlets narrowly lanceolate, 4–6 mm long; sepals deltate-ovate, 4.5–6.5 mm long; petals cuneate-obovate to orbicular, 7–12 mm long, yellow; stamens 20–25, anthers 1–1.2 mm long; pistils numerous; style 0.8–1.5 mm long, laterally attached; stigma capitate. **FRUIT** hirsute. $2n = 14, 28$. [*Potentilla fruticosa* L.; *Sasiphora riparia* Raf.; *Pentaphylloides fruticosa* (L.) O. Schwarz; *Potentilla floribunda* Pursh; *Pentaphylloides floribunda* (Pursh) Á. Löve]. Moist meadows and along mountain streams. ARIZ: Apa; COLO: Arc, Dol, Hin, LPl, Min, Mon, SJn; NMEX: McK, RAr, SJn. 2285–3810 m (7500–12500′). Flowering: Jun–Sep. Alaska to Labrador, south to California, Nevada, northern and eastern Arizona, and New Mexico, and east to New Jersey; Eurasia. Popular as an ornamental.

Dryas L. Mountain Avens

(for the dryads of Greek mythology, wood nymphs) Low, evergreen, mat-forming undershrubs. **STEMS** much-branched, prostrate, rooting at nodes. **LEAVES** with stipules adnate to base of petiole, persistent; blade simple, coriaceous, entire, crenate, or subpinnatifid, white below. **INFL** scapose, flowers solitary; bracts absent or small. **FLOWERS** perfect, white to cream, showy; hypanthium saucer-shaped to campanulate, free from carpels, with a nectar ring; sepals 8–10, persistent, epicalyx absent; petals 8–10, elliptic to obovate, persistent; stamens many; carpels several to many, free from each other and the hypanthium, the ovaries superior, inserted on a flat receptacle, styles terminal, persistent. **FRUIT** an achene with greatly elongated plumose style. $x = 9$. Perhaps two to six species, depending on species delimitation; circumpolar, mostly in Arctic and alpine regions.

Dryas octopetala L. **CP** (eight-petaled) Mountain avens. Subshrub, mat-forming, scapose. **LEAVES** with stipules lanceolate, acute; petioles with black stipitate glands, 1–2.5 cm long; blades up to 2 cm long, mostly 4–8 mm wide, lanceolate or elliptic, rounded basally, obtuse apically, veins conspicuously impressed above, margins revolute, upper surface green, glabrous or with a few hairs along the midnerve, lower surface white-tomentose. **INFL** a solitary flower, scape 3–6 cm long, stipitate-glandular, tomentose. **FLOWER** hypanthium cupulate to subdiscoid; sepals 8–10, lanceolate, 4.5–8 mm long, glabrous on the inner surface; petals 8–10, elliptic or oblanceolate, 10–13 mm long, clawed, rounded apically, cream to white; anthers 0.5–0.6 mm long. **ACHENES** 2.5–4 mm long, style 2–3 mm long, silky plumose. $2n = 18, 36$. [*Ptilotum octopetalum* (L.) Dulac; *Dryadaea octopetala* (L.) Kuntze; *Geum octopetalum* (L.) E. H. L. Krause; *D. hookeriana* Juz.]. Limestone (ours of alpine tundra communities). COLO: LPl, SJn. 3625–3870 m (11900–12700′). Flowering: Jul–Aug. Alaska and Yukon south to the Cascades of Washington, northeast Oregon, and the Rocky Mountains of Idaho, northeast Utah, and Colorado. Very rare in our area.

Drymocallis Fourr. ex Rydb. Sticky Cinquefoil

(Greek *drymos*, woodland, + *calli*, trail) Ours perennial herbs up to 10 dm tall; herbage brownish viscid, some hairs glandular. **LEAVES** pinnate with a well-defined terminal leaflet. **FLOWER** petals white to cream; style fusiform, basally attached. Four species in North America.

Drymocallis arguta (Pursh) Rydb. (sharply toothed) Tall cinquefoil. Perennial herb, 5–10 dm tall; herbage brownish viscid-villous, the pubescence on the stems and rachis of the leaves shaggy. **STEMS** mostly unbranched below the inflorescence. **LEAVES** pinnate, 7–9-foliolate, 1–2.5 dm long; upper cauline leaves few and reduced upward. **INFL** a narrow, mostly flat-topped cyme, viscid-villous; pedicels mostly 2–20 mm long, erect. **FLOWER** bractlets lanceolate, 3–5 mm long; sepals ovate, 5–10 mm long; petals obovate to orbicular, 4–8 mm long, white, cream, or pale yellow; stamens 25–30; anthers 0.6–1.1 mm long; pistils numerous; style 0.7–1.1 mm long, laterally attached below the middle of the ovary. **FRUIT** an achene 0.8–1.5 mm long, glabrous. $2n = 14$. [*Potentilla arguta* Pursh; *P. convallaria* Rydb.]. Ponderosa pine, Gambel's oak, aspen, and spruce-fir communities. COLO: Arc, Hin, Mon; NMEX: RAr; UTAH. 2465–2925 m (8085–9600′). Flowering: May–Jul. Southern Canada south to northeastern Nevada, Utah, northeastern Arizona, northern New Mexico, east to New Jersey. Our material comes closest to **subsp. *convallaria*** (Rydb.) Soják; however, more research and collections are needed.

Fallugia Endl. Apache Plume

(named after Italian botanist Virgilio Fallugi) Deciduous or semievergreen shrubs, unarmed. **STEMS** woody, much-branched, erect; bark exfoliating, light gray with reddish fissures. **LEAVES** with small stipules, caducous; blades deeply pinnatifid, often in fascicles. **INFL** of solitary flowers or few-flowered loose cymes terminating stems; bracts few, reduced, leaf-like. **FLOWERS** mostly imperfect and the plants functionally dioecious; white, showy; hypanthium hemispheric, nectar disc absent; sepals 5, alternating with 5 linear epicalyx lobes, persistent; petals 5, subcircular, deciduous; stamens many; carpels many, free from each other and the hypanthium, the ovaries superior, style terminal. **FRUIT** an achene with a greatly elongated plumose style. $x = 7$. Monotypic, southwestern North America.

Fallugia paradoxa (D. Don) Endl. ex Torr. **CP** (not of the expected) Apache plume. Shrub, up to 2 m tall, with slender branches, sometimes evergreen. **STEMS** of the current season pale brownish tomentose, bark of the older branches light gray, exfoliating. **LEAVES** with small stipules, triangular, persistent; in fascicles, on short lateral branches, larger ones 7–20 mm long, pinnately 3–5 (9)-lobed, the larger lobes sometimes divided again, blades thick, strongly revolute-margined, midvein prominent, yellow-brown tomentose beneath. **INFL** of 1–3 flowers borne at the tip of slender, elongating branches. **FLOWERS** mostly unisexual, sometimes perfect; female flowers with sterile anthers; male flowers with sterile ovaries; hypanthium hemispheric, 2–4 mm deep, yellow-brown tomentose; bractlets lanceolate, 4–7 mm long, sometimes bifid, revolute margins; sepals ovate, 4–7.5 mm long, rounded, apiculate, glabrate and whitish on the inside, reflexed in fruit; petals spreading, suborbicular, 9–17 mm long, rounded apically, white; stamens about 100, in 3 series, filaments broadened at the base, cream-colored, anthers 0.9–1 mm long, yellow; receptacle densely short-hirsute. **ACHENES** 2.5–2.8 mm long, fusiform, densely pubescent; style elongating to 3–4 mm long, pink to purplish, sinuously curved, densely plumose. $2n = 14, 28$. [*Sieversia paradoxa* D. Don; *Geum paradoxum* (D. Don) Steud.; *F. paradoxa* var. *acuminata* Wooton; *F. micrantha* Cockerell]. Canyon bottoms and wash banks; blackbrush and desert grassland communities. ARIZ: Coc; NMEX: McK; UTAH. 1435–2170 m (4710–7120′). Flowering: late Apr–Sep. Southeastern California, southern Nevada, southern Utah, Colorado, through Arizona, New Mexico, and western Texas; northern Mexico. Extremely rare in the Four Corners region. Used as witchcraft to cause insanity by the Kayenta Navajo. An infusion of leaves used as a ceremonial lotion and ceremonial emetic by the Ramah Navajo.

Fragaria L. Strawberry

(Latin *fraga*, strawberry, the adjectival form used by Linnaeus referring to the fruit's fragrance) Perennial herbs. **STEMS** scaly crowns terminating underground rhizomes, producing stolons that root at nodes. **LEAVES** in rosettes, trifoliolate or simple in inflorescence, long-petiolate, stipules adnate to bases of petiole, persistent, forming scales on crowns; leaflets coarsely serrate. **INFL** few- to several-flowered cymes or short racemes, long-pedunculate, bracts often foliaceous, persistent. **FLOWERS** perfect or partially to wholly imperfect, white, showy; hypanthium saucer-shaped, free of carpels, lined with nectar disc; sepals 5 alternating with 5 lobes of an epicalyx, persistent; petals 5, circular or ovate, short-clawed at base, deciduous; stamens 20–35; carpels many, free from each other and from hypanthium, the ovaries superior, inserted on conelike receptacle, styles inserted adaxially near base of carpels. **FRUIT** accessory with many achenes on surface of enlarged, fleshy, red, fragrant receptacle, subtended by calyx and epicalyx. $x = 7$. About 12 species primarily of the north-temperate zone, extending southward to Patagonia, also Hawaii. Fruit eaten by many tribes.

1. Terminal tooth of the leaflets well developed, projecting beyond the adjacent lateral pair; inflorescence equal to or exserted beyond the leaves; leaves green or yellowish green on the upper surface, veiny .... ***F. vesca***
1' Terminal tooth of the leaflets relatively small, usually shorter than the adjacent lateral pair; inflorescence usually surpassed by the leaves; leaves glaucous on the upper surface, not so veiny .... ***F. virginiana***

Fragaria vesca L. (little) Woodland strawberry. Stoloniferous, scapose herb from a branched caudex; pubescence of the scape, petiole, and pedicels spreading to ascending-pilose, leaves spreading-villous on the veins. **LEAVES** 3-foliolate, petiole 3–18 cm long, stipules lanceolate to broadly lanceolate, acuminate; leaflets ovate, mostly 3–6 cm long, 2–4.5 cm wide, terminal leaflet the largest, coarsely crenate-serrate in the upper 2/3, terminal tooth projecting beyond the adjacent lateral teeth, all leaflets glaucous beneath and green to yellowish green above, veins well developed. **INFL** a cyme, exceeding the leaves, 0.5–2 dm long, 3–15-flowered. **FLOWER** bractlets lanceolate to broadly lanceolate, mostly 4.5–7 mm long, acuminate, spreading to ascending in fruit; sepals lanceolate to broadly lanceolate, 4.5–8 mm long, acuminate, spreading to ascending in fruit; petals broadly obovate, mostly 5–8 mm long, rounded, white with some pinkish tinge; stamens 20, anthers 0.7–0.9 mm long. **FRUIT** with receptacle becoming fleshy, up to 1 cm broad. **ACHENES** 1.3–1.4 mm long. $2n = 14$. [*F. helleri* Holz.; *F. bracteata* A. Heller; *F. americana* (Porter) Britton]. Shady and moist coniferous woodland and aspen communities. ARIZ: Apa; COLO: Arc, Hin, LPl, Min, Mon, RGr, SJn; NMEX: McK, RAr, SJn. 2285–3900 m (7500–12800′). Flowering: May–Jul. Southern British Columbia, southern Alberta south to California, Arizona, and New Mexico.

Fragaria virginiana Mill. **CP** (of Virginia) Mountain strawberry. Stoloniferous, scapose herb from a branched caudex; scapes 0.2–2.5 dm long, usually surpassed by the leaves; stolons, scape, petiole, and pedicels sparsely to densely appressed- or spreading-pilose; sparsely to abundantly silky-villous beneath. **LEAVES** 3-foliolate, petiole 2–14 cm long, stipules lanceolate to ovate, acuminate; leaflets narrowly to broadly obovate, 2.2–7 cm long, 1.3–4.5 cm wide, cuneate below, coarsely crenate-serrate in the upper 1/3–2/3, the terminal tooth reduced and often surpassed by the adjacent lateral teeth. **INFL** a cyme, 2–15-flowered. **FLOWER** bractlets lanceolate, 3–6.5 mm long; sepals lanceolate, 4–8 mm long, acuminate, spreading to ascending in fruit; petals obovate, 5–12 mm long, rounded, white or sometimes pinkish; stamens 20, anthers 0.7–1 mm long. **FRUIT** with receptacle fleshy, mostly 1 cm broad. **ACHENES** 1.2–1.4 mm long. $2n = 56$. [*Potentilla ovalis* Lehm.; *F. glauca* (S. Watson) Rydb.; *F. sibbaldifolia* Rydb.; *F. truncata* Rydb.; *F. platypetala* Rydb.; *F. prolifica* C. F. Baker & Rydb.; *F. pumila* Rydb.; *F. pauciflora* Rydb.; *F. firma* Rydb.]. Moist soils; meadows and coniferous forests. ARIZ: Apa; COLO: Arc, Hin, LPl, Min, Mon, RGr, SJn; NMEX: RAr, SJn. 2255–3655 m (7400–12000′). Flowering: Jun–Jul. Alaska, Canada, south to California, west to Oregon, Nevada, Utah, northern Arizona, Colorado, and New Mexico. Our material belongs to **var.** ***glauca*** S. Watson (bluish gray).

Geum L. Avens

(a name used by Pliny, perhaps for *G. urbanum*) Perennial herbs with rosettes terminating short vertical rootstocks or elongated horizontal rhizomes, scaly from persistent leaf bases. **STEMS** leafy or scapose. **LEAVES** basal and sometimes also cauline, pinnately compound, often lyrate in outline, leaflets toothed, of different sizes, small ones interspersed with larger ones along rachis, the terminal one larger than lower ones; stipules persistent, those of basal leaves fully adnate to the bases of the petioles, forming membranaceous wings, those of cauline leaves free. **INFL** several- to many-flowered open cymes or corymbs; bracts reduced leaves. **FLOWERS** perfect, white to pink or yellow, showy; hypanthium saucer-shaped, discoid, or turbinate, free from carpels, with a small to prominent nectar ring; sepals 5, alternating with 5 epicalyx lobes, persistent; petals 5, broadly obovate to elliptic or circular, longer than or shorter than sepals, deciduous; stamens 20–many; carpels several to many, free from each other and the hypanthium, the ovaries superior, inserted on a hemispheric to cylindrical receptacle, styles terminal, of 2 types, entire and wholly persistent and becoming plumose on the fruit or jointed and geniculate near the middle, the apical part deciduous and leaving a hooked beak on the fruit. **FRUIT** an aggregate of achenes, the receptacle not enlarging but sometimes elevated above calyx on a stipe. $x = 7$, most species hexaploids with $n = 21$. About 60 species mostly of the Northern Hemisphere, many in Arctic, alpine, or moist boreal forest habitats.

1. Style jointed, the lower part persistent, with a terminal hook; leaves pinnately compound, with few unequal leaflets .... (2)
1' Style continuous, without a hook; leaves with many narrow segments .... (4)

2. Sepals purple; petals violet .... ***G. rivale***
2' Sepals green; petals yellow .... (3)

3. Portion of style below the hook glabrous or minutely pubescent, not glandular; terminal leaf segment not enlarged .... ***G. aleppicum***
3′ Portion of style below the hook with stalked glands; terminal leaf segment greatly enlarged ...... ***G. macrophyllum***

4. Flowers nodding; petals pinkish .... ***G. triflorum***
4′ Flowers erect; petals yellow .... ***G. rossii***

Geum aleppicum Jacq. (of Aleppos, Syria) Yellow avens, Aleppo avens. Perennial herb from a thick, fleshy crown; herbage and inflorescence hirsute, the hairs bulbose-based; stems and petioles finely puberulent. **STEMS** few, erect, 4–12 dm tall, leafy. **LEAVES**: basal well developed, 15–28 cm long, petiolate, pinnately divided, blade obovate, terminal leaflet cuneate basally and often decurrent on the rachis, 3–5-lobed, double crenate-toothed, cauline leaves 3–5-foliolate with stipules up to 2 cm long, ovate. **INFL** a few- to several-flowered cyme on long pedicels. **FLOWERS** erect; hypanthium more or less discoid; bractlets linear to linear-lanceolate, 1.2–3.5 mm long, obtuse; sepals reflexed at anthesis, deltate-ovate, 3–7.5 mm long, acute, finely tomentulose on the inside; petals spreading, suborbicular or obovate, 4–7 mm long, obtuse to rounded, short-clawed, yellow; stamens numerous, anthers 0.4–0.6 mm long; receptacle clavate, up to 2 mm long; pistils 200–250, hairy. **ACHENES** elliptic, flattened, 3.5–4 mm long, mature style strongly geniculate and jointed, the lower segment 4–6 mm long, glabrous or sparsely eglandular-hirsute, hooked at the apex. $2n = 42$. [*G. strictum* Aiton; *G. decurrens* Rydb.]. Wet meadows and along streams. ARIZ: Apa; COLO: Arc, LPl, Min, Mon, RGr, SJn. 1965–3505 m (6450–11500′). Flowering: Jun–Aug. British Columbia to eastern Canada and northeastern and midwestern United States, and Washington, Oregon, and northern California, south to Arizona and New Mexico.

Geum macrophyllum Willd. (large-leaved) Large-leaved avens, bigleaf avens. Perennial herb from a thick, scaly crown; herbage and inflorescence hirsute, stems partly glandular-puberulent. **STEMS** few, erect, up to 12 dm tall. **LEAVES**: basal well developed, mostly 10–25 cm long, petiolate, pinnately divided, the terminal leaflet cordate-reniform, 3–5-lobed, the lobes coarsely toothed, many times larger than the main lateral leaflets; cauline leaves alternate, smaller, 3 (7)-foliolate, leaflets oblanceolate, serrate, stipules leaflike, 6–18 mm long. **INFL** a divergently branched 4–9-flowered cyme with long pedicels. **FLOWERS** erect; hypanthium discoid; bractlets linear to lanceolate, 0.5–2.5 m long, obtuse; sepals reflexed at anthesis, deltate, mostly 3–6 mm long, finely tomentulose adaxially; petals spreading, obovate, 3.5–6.5 mm long, yellow; stamens numerous, anthers 0.4–0.6 mm long; receptacle clavate, 4–6 mm long; pistils about 250. **ACHENES** ovoid-elliptic, flattened, 2.2–3.2 mm long, mature style geniculate and jointed, glandular-puberulent, hooked at the apex. $2n = 42$. [*Geum urbanum* L. subsp. *oregonense* Scheutz; *G. perincisum* Rydb.]. Wet meadows and stream banks; ponderosa pine, Douglas-fir, aspen, white fir, spruce-fir communities. ARIZ: Apa; COLO: RGr, SJn; UTAH. 2335–3505 m (7670–11500′). Flowering: late May–Aug. Alaska to Greenland, south to California, Nevada, northern Arizona, northern New Mexico, and Michigan.

Geum rivale L. (growing by streams) Purple avens. Perennial. **STEMS** few, 25–60 cm tall; hirsute and glandular-pilose. **LEAVES** basal, lyrate-pinnate; leaflets obovate, 2–10 cm long, doubly serrate to deeply incised, uneven in size; cauline leaves ternate. **INFL** 1–4-flowered cyme. **FLOWER** bractlets less than 1/2 as long as the sepals, narrowly linear; sepals 8–12 mm long, lanceolate to ovate-lanceolate, not reflexed, densely pilose, purplish; petals 6–10 mm long, flesh-colored or yellow-tinged, purple-veined, clawed, flabelliform; receptacle short-hirsute, more or less stalked in fruit; mature style geniculate, the lower segment about 7–9 mm long, hirsute below and more or less glandular-pubescent, upper segment 4–5 mm long, hirsute, articulated to lower by a curved and hooked joint. **ACHENES** mostly 4 mm long, hirsute. Meadows in the subalpine. COLO: SJn. 3200 m (10500′). Flowering: Jun–Aug. Known from one location along Clear Creek Trail (*M. Douglas 54-411*). Northern United States from Washington to Maine, Wyoming, and Colorado.

Geum rossii (R. Br.) Ser. (for Captain J. C. Ross, Arctic explorer) Alpine avens, Ross' avens. Perennial herb from a thick caudex forming dense clumps up to 3 dm broad or greater; herbage glabrous to sericeous or villous. **STEMS** erect, subscapose, 0.5–2.5 dm tall, simple or branched. **LEAVES**: basal well developed, 3–14 cm long, pinnately divided, the lateral leaflets mostly 7–15 mm long, cleft 1/2 their length into narrow segments; cauline leaves reduced, sessile, alternate, pinnately toothed, with stipules. **INFL** a 1–4-flowered cyme. **FLOWER** hypanthium turbinate, 2.5–6 mm long, purple-tinged; bractlets lanceolate, 2.5 mm long, acute to obtuse; sepals erect, deltate-lanceolate, 3.5–5.5 mm long, acute to acuminate; petals spreading, broadly obovate to suborbicular, 5–9 mm long, short-clawed, yellow; stamens numerous, anthers 0.5–0.8 mm long; receptacle short, rounded; pistils few to many. **ACHENES** 2.5–4 mm long, fusiform-lanceolate, pubescent, mature style not jointed or hooked, persistent, glabrous, about as long as the achene. $2n = 56, 70, 112$. [*Sieversia rossii* R. Br.; *Potentilla nivalis* Torr.; *G. sericeum* Greene; *P. gracilipes* Piper; *Sieversia scapoidea* A. Nelson; *Acomastylis rossii* (R. Br.) Greene]. Alpine and subalpine meadows. COLO: Arc, Con, Hin, LPl, Min, Mon, RGr,

SJn. 3220–3960 m (10560–13000′). Flowering: Jul–Aug. Alaska south to Washington and Oregon, northeastern Nevada, Utah, north-central Arizona, east to southwest Montana, Wyoming, Colorado, and New Mexico; Asia. Four Corners regional material belongs to **var**. ***turbinatum*** (Rydb.) C. L. Hitchc.

Geum triflorum Pursh (three-flowered) Old man's whiskers. Perennial herb from a thick caudex, branching to form clumps up to 3 dm broad or greater; herbage hirsute or pilose, often densely so. **STEMS** erect, 1.5–4 dm tall, subscapose with a pair of opposite, much-reduced leaves at about the middle. **LEAVES**: basal well developed, mostly 6–15 cm long, oblanceolate in outline, petiolate, pinnately lobed, becoming somewhat pinnatifid above, the lateral leaflets 13–22 mm long. **INFL** a 1–3 (9)-flowered cyme, the flowers nodding on long pedicels. **FLOWER** hypanthium hemispheric; bractlets linear to narrowly lanceolate, 6–15 mm long, acute; sepals erect, lanceolate, 7–13 mm long, acute; petals erect, narrowly to broadly elliptic, 8–15 mm long, white to pinkish; stamens numerous, from a ring within the hypanthium, anthers 0.8–1 mm long; receptacle short-clavate, to 2 mm long, short-pubescent; pistils numerous. **ACHENES** mostly 3 mm long, pyriform, the mature style slightly geniculate and jointed, the persistent lower segment 2–3.4 mm long, purplish, not hooked at the apex. $2n = 42$. [*G. ciliatum* Pursh; *Erythrocoma brevifolia* Greene; *E. arizonica* Greene; *E. grisea* Greene; *E canescens* Greene; *E. tridentata* Greene]. Montane meadows to alpine tundra. ARIZ: Apa; COLO: Arc, Hin, LPl, Min, Mon, SJn; NMEX: McK, RAr, SJn. 2180–3780 m (7150–12400′). Flowering: late May–Jul. British Columbia to Newfoundland, south through the mountain states to eastern California, Nevada, northern Arizona, and northern New Mexico, and in the Midwest, Great Lakes region, and New York. Our material belongs to **var**. ***ciliatum*** (Pursh) Fassett. A decoction of roots was given to stimulate tired horses by the Paiute.

Holodiscus (K. Koch) Maxim. Oceanspray

(Greek *holos*, entire, + *discos*, referring to the hypanthium) Deciduous, unarmed shrub. **STEMS** woody, bark reddish when young, becoming gray. **LEAVES** simple, entire, toothed, or shallowly lobed; stipules absent. **INFL** large, showy, diffuse, many-flowered panicles or racemes; bracts small. **FLOWERS** perfect, white to cream, small; hypanthium saucer-shaped, free from carpels, with a prominent nectar ring; sepals 5, persistent; petals 5, ovate, short claw at base; stamens 15–20; carpels 5, free from each other and the hypanthium, the ovaries superior, styles terminal. **FRUIT** an achene. $x = 9$. About five to eight species of western North America southward to northern South America.

Holodiscus dumosus (S. Watson) A. Heller (bushy) Bush oceanspray. Shrub, up to 2 m tall, the terminal branches often becoming weakly spinescent. **STEMS** of the current year reddish tan. **LEAVES** with oblanceolate to obovate blades, 2.5–3.5 cm long, mostly 0.5–1.2 cm wide, toothed above the middle, pilose above, lower surface less pubescent and with sessile, glandular hairs; short petioles. **INFL** a diffuse panicle, flowers subtended by bractlets. **FLOWER** sepals ovate, 1.2–2 mm long, petals spatulate, 1.7–2.5 mm long, white to cream; hypanthium discoid to shallowly cupulate, 1.8–2.3 mm across; stamens 20, anthers about 0.3 mm long; stigma discoid. **ACHENES** 1.5–2 mm long, densely pilose with spreading hairs. $2n = 36$. [*Spiraea dumosa* Nutt. ex Hook.; *Spiraea discolor* Pursh var. *glabrescens* Greenm.; *H. microphyllus* Rydb.]. Rocky sites, often steep canyon cliff faces; piñon-juniper woodland and ponderosa pine communities. ARIZ: Apa, Coc, Nav; COLO: Arc, Dol, Mon; NMEX: McK, SJn; UTAH. 1870–3010 m (6145–9875′). Flowering: Jun–Aug. Eastern Oregon, southern Idaho, and Wyoming, south through California, Nevada, Utah, Arizona, New Mexico, Trans-Pecos Texas, and Mexico. Our material belongs to **var. *dumosus***. A decoction of the leaves is taken for influenza by the Ramah Navajo.

Ivesia Torr. & A. Gray Mousetail, Ivesia

(for Eli Ives, pharmacologist at Yale University) Perennial herbs with thick, simple to much-branched caudex, persistent leaf bases crowded at top. **STEMS** and other aboveground parts mostly covered with resinous, fragrant glands. **LEAVES** pinnately compound, mostly in basal rosette, cauline reduced upward, leaflets deeply lobed, tightly overlapping, the terminal leaflet usually reduced and confluent with rachis, sometimes individual leaflets barely distinguishable; stipules adnate to base of petiole, persistent, sometimes sheathing. **INFL** scapose, lax to condensed cymes; bracts reduced leaves. **FLOWERS** perfect, white to pale or greenish yellow; hypanthium saucer- to cup-shaped, campanulate, or turbinate, free from carpels, lined with nectar disc; sepals 5, usually alternating with 5 epicalyx lobes, persistent; petals mostly oblanceolate or linear to broadly obovate, often clawed at base, larger than to much smaller than sepals, deciduous; stamens 5 or 15–35; carpels few in most species, 1–20, free from each other and the hypanthium, the ovaries superior, the style subapical, more or less basally rough-thickened or swollen, the thin upper part deciduous. **FRUIT** an achene. $x = 7$. About 30 taxa of the western United States and northern Baja California.

Ivesia gordonii (Hook.) Torr. & A. Gray (for Alexander Gordon, an English horticulturist and nurseryman who traveled in the western United States) Gordon's ivesia, alpine ivesia. Subscapose herb from a thick caudex. **STEMS** erect, spreading, or prostrate, 0.6–4 dm long, with 1 or 2 reduced leaves, glandular-puberulent. **LEAVES** basal, odd-pinnate, 3–24 cm long, pilose to glabrate; leaflets sometimes glandular-punctate, lateral leaflets 10–25 pairs, 2–17 mm long. **INFL** a capitate cyme, bracts palmately 5–7-parted, pedicels 1–2 mm long. **FLOWERS** campanulate or turbinate, 2–4.5 mm deep, yellowish green; bractlets narrowly elliptic to linear, mostly 1.3–3.3 mm long, obtuse or rounded, pale green; sepals deltate-lanceolate, 1–5 mm long, acute or acuminate, yellowish green; petals spatulate, 1.3–2.7 mm long, yellow, sometimes drying white; stamens 5, yellow, anthers 0.4–0.8 mm long; pistils mostly 2, sometimes 1–6, style yellow. **ACHENES** mostly 1.7–2 mm long, smooth. $2n = 28$. [*Horkelia gordonii* Hook.; *Potentilla gordonii* (Hook.) Greene]. Shallow, rocky, limestone soils; subalpine to alpine communities. COLO: LPl, SJn. 3585 m (11760′). Flowering: Jun–Aug. Montana and Washington, south to Colorado and California. Known from one collection on the Lime Ridge Trail northeast of Durango, Colorado (*Heil 24634*) and a 1935 collection near Lime Creek.

Malus Mill. Apple

(classical name of the apple) Deciduous trees, unarmed or with thorns. **STEMS** woody, bark smooth with horizontal lenticels or with scaly plates. **LEAVES** deciduous, simple, toothed or lobed; stipules small, deciduous. **INFL** of corymbs, panicles, or umbels; bracts small, caducous. **FLOWERS** perfect, white to pink, showy, large nectar ring present; hypanthium saucer- to cup-shaped, the lower part fully adnate to the carpels; sepals 5, persistent or deciduous, epicalyx absent; petals circular or broadly elliptic, slightly to distinctly clawed at base, deciduous; stamens 15–30; carpels 3–5, fully connate and fully adnate to the hypanthium, the ovary inferior, styles fused at base. **FRUIT** a large, fleshy pome, red to yellow or green, the core cartilaginous. $x = 17$. About 50 species of Eurasia and North America. *Malus ioensis* and related species are sometimes placed in the segregate genus *Chloromeles*.

1. Leaves on elongated shoots lobed or notched .... ***M. ioensis***
1′ Leaves on elongated shoots neither lobed nor notched .... ***M. pumila***

Malus ioensis (Alph. Wood) Britton (refers to Iowa) Prairie crab apple. Small tree up to 9 m tall, branchlets tomentose, sometimes spinescent. **LEAVES** ovate to oblong, 2.5–10 cm long, tomentose on both sides. **FLOWER** petals white to pink, 12–25 mm long. **FRUIT** 2–3 cm in diameter. [*Pyrus ioensis* (Alph. Wood) L. H. Bailey]. Cultivated tree that persists and escapes. COLO: LPl, Min, Mon; NMEX: SJn; UTAH. 1350–2315 m (4425–7600′). Flowering: Apr–May. Widespread throughout North America.

Malus pumila Mill. (dwarf) Common apple. Tree up to 10 m tall, branchlets tomentose when young, becoming glabrous in age. **LEAVES** ovate to oblong or elliptic, 1.5–10 cm long, tomentose on 1 or both sides. **FLOWER** petals mostly white, often pink dorsally, 12–25 mm long. **FRUIT** mostly 2.5–12 cm in diameter, red, reddish purple, or yellow. [*M. domestica* Borkh.]. Widely cultivated throughout the Four Corners region. ARIZ: Apa; COLO: Arc, Mon, SJn; NMEX: SJn; UTAH. 1615–2850 m (5300–9350′). Flowering: Apr–May. Widespread throughout North America.

Peraphyllum Nutt. Peraphyllum

(from the Greek *pera*, excessive, + *phyllon*, leaf, for the abundant leaves) Deciduous, unarmed shrub. **STEMS** woody, bark reddish when young, becoming gray, smooth. **LEAVES** simple, faintly toothed; stipules small, caducous. **INFL** reduced, 1- to few-flowered fascicle, bracts small, caducous. **FLOWERS** perfect, white to pinkish, showy, nectar disc present; hypanthium subglobose to turbinate, lower part fully adnate to the carpels; sepals 5, persistent, epicalyx absent; petals 5, obovate to circular, deciduous; stamens 15–20; carpels 2–3, fully connate and fully adnate to the hypanthium, the ovary inferior, styles mostly free. **FRUIT** a small fleshy pome, globose, yellow to orange or red, the core cartilaginous, false septa present. $x = 17$. Monotypic, western North America, a distinct relative of *Amelanchier*.

Peraphyllum ramosissimum Nutt. (many branches) Squaw apple. Shrub 0.5–2 m tall, bark reddish on young growth, becoming gray, herbage appressed-pubescent. **LEAVES** clustered on short shoots or alternate on elongating branches, petiole 1.5–4.5 mm long, stipules reddish, deciduous, the blade narrowly elliptic or narrowly oblanceolate, 2–5 cm long, 0.4–1.2 cm wide, narrowly cuneate basally, gradually tapering to the petiolate base, acuminate or acute or sometimes rounded or obtuse apically, entire or shallowly serrulate. **INFL** a fascicle of solitary or 2 or 3 flowers, terminating on short shoots appearing with the mature leaves, pedicels 0.4–1.3 cm long. **FLOWERS** fragrant; hypanthium campanulate or cylindrical, the free part 2–4 mm long; sepals spreading to reflexed, lanceolate to deltate, 2–5.5 mm long, white-tomentose on the inside, sometimes also on the outside; petals broadly obovate, 5–10 mm long, clawed, white to pinkish; stamens 15–20, the filaments 4–6.5 mm long, the anthers 0.9–1.2 (1.4) mm long; styles 2 (3). **FRUIT** a globose pome, 8–10 (12) mm in diameter, yellowish to reddish, with a bitter taste. **SEEDS** about 5 mm long, bluish purple, ovoid. $2n = 34$. Sagebrush, serviceberry, piñon-juniper woodland, and Gambel's oak communities. COLO: Arc,

Dol, LPl, Mon, SMg; NMEX: RAr, SJn; UTAH. 1870–2345 m (6130–7700′). Flowering: May–Jun. Eastern California, eastern Oregon, and adjacent Idaho, Nevada, southern and central Utah, western Colorado, and northwestern New Mexico.

Petrophytum Rydb. Rock-spiraea

(Greek *petra*, rock, + *phyton*, plant, in reference to the rocky habitat) Prostrate, mat-forming shrubs. **STEMS** woody, branched, condensed, covered with persistent leaf bases. **LEAVES** simple, entire, densely crowded on stems; stipules absent. **INFL** a compact spikelike raceme on an elongated scape bearing reduced leaves. **FLOWERS** perfect, white, small, nectar disc present; hypanthium cup-shaped to turbinate, free from carpels; sepals 5, persistent, epicalyx absent; petals 5, oblanceolate, persistent; stamens 20–40; carpels usually 5, free from each other and the hypanthium, the ovaries superior, styles terminal. **FRUIT** a follicle, dehiscent on both sutures. $x = 9$. Four species of western North America, a distinct segregate of *Spiraea*.

Petrophytum caespitosum (Nutt.) Rydb. (tufted) Rock-spiraea, tufted rockmat. A mat-forming shrub; herbage sericeous. **STEMS** much-branched, up to 25 cm long. **LEAVES** sessile, linear-spatulate to oblanceolate, 3–30 mm long, mostly 2–4 mm wide, apically obtuse. **INFL** a dense cylindrical to globular raceme, 2–4 cm wide, compound below with short lateral branches; pedicels 0.5–1.5 mm long. **FLOWERS** turbinate or cupulate, 0.5–1 mm deep, sericeous; sepals ovate to deltate, 1–1.6 mm long, acute; petals oblanceolate, 1.5–2 mm long, rounded to truncate, white; stamens mostly 20, filaments 2–3 mm long, anthers 0.3–0.5 mm long; style 2–3 mm long. **FOLLICLES** 1.2–1.5 mm long, pilose. $2n = 18$. [*Spiraea caespitosa* Nutt.; *S. caespitosa* var. *elatior* S. Watson; *P. elatum* (S. Watson) A. Heller; *P. acuminatum* Rydb.]. In our region, on steep, exposed, andesite porphyry and sandstone rock surfaces; rooting in the cracks of the rock. ARIZ: Apa, Nav; UTAH. 1215–2075 m (3700–6810′). Flowering: Aug–Sep. South Dakota, Montana, and Idaho south to Trans-Pecos Texas, Arizona, New Mexico, and northern Mexico. The Kayenta Navajo used the plant as a charm or prayer in the "Pleiades rite," and as a narcotic.

Potentilla L. Cinquefoil

(diminutive of Latin *potens*, powerful, for the supposed medicinal properties of *P. anserina*) Perennial or rarely annual herbs. **STEMS** ascending to erect or spreading, rarely decumbent, often with long rhizomes with persistent leaf bases, plants in rosettes, caespitose, or with runners rooting at nodes. **LEAVES** cauline and basal, palmately or pinnately compound with 3–15 leaflets, lower leaves long-petiolate; leaflets entire, lobed, or serrate; stipules of basal leaves adnate to the petioles, those of cauline leaves often free. **INFL** a narrow or open cyme, 1- to several-flowered, or solitary at the nodes; pedicels up to 2.5 cm long, bracts often foliaceous, persistent. **FLOWERS** perfect, mostly pale to bright yellow, rarely dark red, showy; hypanthium saucer-shaped, free of carpels, lined with nectar disc; sepals 5 alternating with 5 lobes of an epicalyx, persistent; petals 5, circular, obovate, or cuneate, short-clawed at base, deciduous; stamens 10–30, often 20; carpels many, free from each other and from hypanthium, the ovaries superior, inserted on conelike receptacle, styles terminal, inserted laterally to nearly basal on adaxial side of carpels. **FRUIT** accessory with many achenes on the dry receptacle, surrounded or enclosed by persistent, sometimes accrescent calyx. $x = 7$. A taxonomically diverse and difficult genus with several hundred species mostly of the north-temperate zone. In this treatment, *Potentilla* is treated largely sensu lato; segregate genera recognized here are *Dasiphora*, *Drymocallis*, and *Ivesia*.

1\. Annual, biennial, or short-lived perennial; mostly weedy plants .... (2)
1′ Perennial; native plants .... (3)

2\. Lowermost leaves 5–9-foliolate; achene with a protuberance on the ventral suture .... ***P. paradoxa***
2′ Lowermost leaves 3-foliolate; achene lacking a protuberance .... ***P. norvegica***

3\. Style shorter than 1.2 mm, mostly thick and conical, thick just below the stigma, shorter than the achene .... (4)
3′ Style longer than 1.2 mm, usually much longer, usually thin just below the stigma and longer than the achene .... (6)

4\. Principal basal leaves 5–15-foliolate, with revolute margins; montane to alpine .... ***P. pensylvanica***
4′ Basal leaves 3–5 (7)-foliolate, margins not revolute; alpine .... (5)

5\. Basal leaves 5–7-foliolate; petioles spreading-pilose .... ***P. rubricaulis***
5′ Basal leaves 3 (5)-foliolate; petioles densely tomentose .... ***P. nivea***

6\. Leaves pinnate with (5) 7 or more leaflets .... (7)
6′ Leaves digitate or subdigitate, or occasionally 5-digitate above with 1 or 2 extra leaflets separated by a long rachis internode .... (14)

7. Plants stoloniferous; flowers solitary at the nodes .... *P. anserina*
7′ Plants not stoloniferous; flowers few to several in cymes .... (8)

8. Leaves subdigitate, with 7 leaflets .... (9)
8′ Leaves definitely pinnate, with 9 or more leaflets .... (10)

9. Leaves with 3 terminal leaflets and 2 lower pairs, leaflets narrowly toothed; mostly alpine .... *P. subjuga*
9′ Leaves highly variable, some pinnate and some digitate; upper leaflets mostly 5; lower montane to higher montane .... *P. pulcherrima × hippiana*

10. Leaves sericeous-strigose to glabrate, not tomentose, often appearing green .... *P. ovina*
10′ Leaves densely hirsute to tomentose, appearing silvery or grayish green on the lower surface .... (11)

11. Leaves folded, 3-toothed in the upper 1/3, narrow, silvery hirsute .... *P. crinita*
11′ Leaves not folded, several-toothed throughout their length, broader, densely tomentose below; often bicolored .... (12)

12. Leaves densely tomentose on lower surface .... *P. hippiana*
12′ Leaves green on upper and lower surface, merely strigose .... (13)

13. Stems 6–7 dm tall; leaflets coarsely serrate, not folded .... *P. ambigens*
13′ Stems 1–4 dm tall; leaflets toothed at apex, folded .... *P. crinita*

14. Leaves strictly digitate, no rachis between the leaflets; bicolored or green on both surfaces .... (15)
14′ Leaves subdigitate, with at least a short rachis visible between the leaflets on some leaves; the surfaces various .... (18)

15. Leaflets green on both surfaces .... *P. diversifolia*
15′ Leaflets green above and white-tomentose beneath .... (16)

16. Stems 2–8 dm tall; leaflets mostly 7 .... *P. pulcherrima*
16′ Stems 0.2–2.8 dm tall; leaflets mostly 5 .... (17)

17. Leaves legumelike; leaflets mostly 2- or 3-toothed at the apex .... *P. bicrenata*
17′ Leaves not legumelike; leaflets toothed to shallowly lobed .... *P. concinna*

18. Petals rose-red or dark red .... *P. thurberi*
18′ Petals yellow .... (19)

19. Leaves with 3 terminal leaflets and 2 lower pairs, leaflets narrowly toothed; mostly alpine .... *P. subjuga*
19′ Leaves highly variable, some pinnate and some digitate; upper leaflets mostly 5; lower montane to higher montane .... *P. pulcherrima × hippiana*

Potentilla ambigens Greene (wandering) Silkyleaf cinquefoil. Perennial. **STEMS** stout, silky-villous, up to 7 dm tall. **LEAVES**: basal pinnate, up to 20 cm long; leaflets 9–15-foliolate, larger leaflets 3–6 cm long, coarsely serrate, silky-villous below, more or less glabrous above, color contrast not striking. **INFL** a narrow cyme. **FLOWER** bractlets as long or longer than the sepals; sepals mostly 6–7 mm long, lanceolate, strigose; petals about 8 mm long, yellow; style filiform near the apex. **ACHENES** glabrous. Meadows. COLO: Arc. 2285 m (7500′). Flowering: Jun–Aug. Wyoming, Colorado, and New Mexico. Known from one location south of Pagosa Springs along Rito Blanco Creek (*Heil 10095*).

Potentilla anserina L. (goose) Stoloniferous perennial, the stolons up to 50 cm long and producing a rosette of leaves at each node. **LEAVES** pinnate, 8–20 cm long, 9–25-foliolate, larger leaflets interspersed with smaller ones, the petioles hairy; larger leaflets oblanceolate to cuneate-obovate, 1–4 cm long, toothed or lobed, sparsely appressed-villous to glabrate above, white-tomentose beneath. **FLOWERS** solitary at the stolon nodes; pedicels 3–15 cm long, silky-tomentose; hypanthium saucer-shaped; bractlets lanceolate or ovate-elliptic; 4–7 mm long; sepals deltate or ovate, 3–5 mm long, acuminate to acute, erect with fruit development; petals obovate, 5.5–11 mm long, yellow, rounded; stamens 20–25, filaments flattened, broad at the base, anthers 0.7–1 mm long; pistils numerous on a flat receptacle, style 1.5–2.5 mm long. **ACHENES** ovoid, about 2 mm long, light brown. $2n = 28, 42$. [*Fragaria anserina* (L.) Crantz; *Argentina anserina* (L.) Rydb.]. Floodplains, lakeshores, meadows, and other wet areas. ARIZ: Apa; COLO: Arc, Hin, SJn; NMEX: McK, RAr, SJn. 1700–3125 m (5575–10225′). Flowering: May–Sep. Alaska south to California, northern two-thirds of Nevada, Arizona, and New Mexico; Eurasia.

Potentilla bicrenata Rydb. (two notches, pertaining to the notched leaflets) Elegant cinquefoil. Perennial herb from a branched caudex. **STEMS** ascending at anthesis, prostrate in fruit; 0.2–0.8 dm long. **LEAVES** mostly basal, digitate, 5–7 (9)-foliolate; petiole 2–10 cm long, leaflets oblanceolate, 10–40 mm long, terminal the longest, entire or 2- or 3-toothed at the apex, green above and white-tomentose beneath. **INFL** mostly 2- or 3-flowered cyme; pedicels 1.5–2 cm long. **FLOWER** bracts lanceolate, 2–4 mm long, acute; sepals dentate-lanceolate, 2.5–5.5 mm long, acute; petals obovate, 4–6 mm long, yellow; stamens 20, anthers about 0.5–1 mm long; pistils numerous, style 1.5–2.5 mm long. **ACHENES** 1.7–2 mm long. [*P. concinna* var. *bicrenata* (Rydb.) S. L. Welsh & B. C. Johnston]. Ponderosa pine and sagebrush communities. ARIZ: Apa; NMEX: SJn; UTAH. 2065–2805 m (6775–9200′). Flowering: May–Jun. Eastern Utah, northeastern Arizona, and northwestern New Mexico.

Potentilla concinna Richardson (elegant) Elegant cinquefoil. Perennial herb from a branched caudex. **STEMS** numerous, ascending in flower, prostrate in fruit, 0.5–3 dm long. **LEAVES** mostly basal, digitate, 5–7 (9)-foliolate, sometimes subpinnate, petiole mostly 1.5–3.5 cm long; leaflets 7–20 mm long, the terminal one longest, toothed to shallowly lobed, oblanceolate to obovate, often folded. **INFL** a 1–7-flowered cyme, divaricately branched, pedicels 0.5–2.5 cm long, often becoming recurved in fruit. **FLOWER** branchlets oblong-lanceolate, mostly 2.5–4 mm long, acute, obtuse, or rounded; sepals 3.5–5 mm long, acute; petals obovate to oblanceolate, 4–6.5 mm long, yellow; stamens 20, anthers 0.6–1 mm long; pistils numerous, style 1.3–2.5 mm long. **ACHENES** 1.7–2 mm long. $2n = 70$. Spruce-fir communities to alpine. COLO: Arc, Hin, LPl, Min, Mon, RGr, SJn. 3210–3720 m (10540–12200′). Flowering: Jun–Aug. Alberta and Saskatchewan south through Montana and the Dakotas to eastern California, Nevada, Utah, Arizona, and Colorado. Within the Four Corners region this cinquefoil is one of the most common in the high mountains of Colorado.

Potentilla crinita A. Gray (long-haired) Bearded cinquefoil. Perennial herb; 2–5 dm tall from a branched caudex. **STEMS** ascending to erect. **LEAVES** strigose-sericeous, mostly basal, pinnate, mostly 11–13-foliolate, 4–20 cm long; leaflets oblanceolate to narrowly oblanceolate, 10–30 mm long, shallowly toothed beyond the middle or at the apex only, entire, the upper lateral pair decurrent on the rachis; cauline leaves few, reduced. **INFL** a divaricately branched cyme, strigose-sericeous. **FLOWER** bractlets lanceolate, 1.7–4.5 mm long, acute to obtuse; sepals lanceolate, mostly 4–6 mm long, acute or acuminate; petals oblanceolate, mostly 4.5–7.5 mm long, yellow; stamens 20, anthers about 0.5–1 mm long; pistils numerous, styles about 1.8–2.5 mm long. **ACHENES** 1.5–1.7 mm long, smooth to weakly rugulose. $2n = 84$. [*P. lemmoni* (S. Watson) Greene; *P. vallicola* Greene]. Meadows; piñon-juniper woodland, ponderosa pine, aspen, and Gambel's oak communities. ARIZ: Apa; COLO: Arc; NMEX: RAr, SJn. 2040–2680 m (6700–8800′). Flowering: Jun–Aug. Southern Utah and Nevada, southwestern Colorado, northeastern Arizona, and northwestern New Mexico. Used by the Ramah Navajo as a "life medicine."

Potentilla diversifolia Lehm. (many-shaped leaves) Mountain meadow cinquefoil, diverse-leaved cinquefoil. Perennial herb from a much-branched caudex. **STEMS** decumbent to ascending, 0.5–3 dm long. **LEAVES** mostly basal, digitate (rarely subpinnate), 5–7-foliolate, the petiole mostly 2–6 cm long; leaflets oblanceolate to obovate, 0.8–3 cm long, the terminal one usually the longest, cuneate, toothed in the distal 1/2, sparsely to densely sericeous, ciliate, or glabrate, greenish. **INFL** a many-flowered, open cyme, tomentose to villous. **FLOWER** bractlets lanceolate to ovate, mostly 2.5–4 mm long, obtuse; sepals ovate to deltate-lanceolate, 3.5–5.5 mm long, acute; petals oblanceolate to broadly obovate, 4.8–9 mm long, rounded, yellow; stamens 20, anthers 0.35–0.7 mm long; pistils numerous, style 1.4–2.3 mm long. **ACHENES** 1.2–1.6 mm long. $2n = 42–101$. Wet meadows, streamsides, lake margins; from subalpine to alpine communities. COLO: Arc, Con, Hin, LPl, Min, Mon, RGr, SJn; NMEX: SJn. 3110–3840 m (10200–12600′). Flowering: Jun–Sep. British Columbia and Alberta south through eastern Oregon, Idaho, western Montana, Wyoming, to eastern California, Nevada, Utah, Arizona, Colorado, and New Mexico. Often hybridizes with *P. pulcherrima*.

Potentilla hippiana Lehm. (for Carl Frederick Hippio, a colleague of Lehmann) Woolly cinquefoil, Hippio's cinquefoil. Perennial herb from a branched caudex. **STEMS** 0.6–5 dm tall, ascending to erect, sericeous. **LEAVES** mostly basal, crowded, pinnate, 5–13-foliolate, 3–14 cm long; leaflets densely white-tomentose beneath, 12–40 mm long, the upper 3 the largest, oblanceolate, cuneate basally, rounded or acute apically, deeply cleft to nearly halfway to the midrib or less, the upper lateral pair decurrent on the rachis; cauline leaves few, ovate to lanceolate. **INFL** an open cyme, branches straight and ascending. **FLOWER** bractlets lanceolate, 2.5–5 mm long, acute; sepals lanceolate, 4–6.5 mm long, acuminate; petals obovate, 4–6.2 mm long, slightly longer than the sepals, yellow; stamens about 20, anthers 0.5–1 mm long; pistils numerous, style 1.7–2.3 mm long. **ACHENES** about 1.5 mm long, smooth. $2n = 42, 70, 77, 84, 98$. Meadows; ponderosa pine, Douglas-fir, aspen, and spruce-fir woodland communities. ARIZ: Apa; COLO: Arc, Hin, LPl, Mon, SJn; NMEX: McK, SJn; UTAH. 2485–3330 m (8000–10930′). Flowering: Jun–Aug. British Columbia, Alberta,

Saskatchewan, south through Montana, Wyoming, the Dakotas, and Nebraska, south to Utah, northeast Arizona, Colorado, and New Mexico. Often hybridizes with *P. pulcherrima*. Used as a lotion for burns, applied to sores caused by a bear, and to expedite childbirth by the Kayenta Navajo. Used by the Ramah Navajo as a poultice for injuries, and as a "life medicine."

Potentilla nivea L. (of the snow) Snow cinquefoil. Mat-forming perennial herb from a much-branched, dark caudex. **STEMS** arachnoid-lanate or spreading-hirsute, decumbent to ascending, 0.2–2.5 dm long. **LEAVES** green above, white-tomentose below, mostly basal, 3-foliolate; petiole 0.7–6.5 cm long; leaflets obovate to flabellate, 5–15 mm long, terminal leaflet larger than the lateral pair, coarsely toothed; cauline leaves reduced. **INFL** a 1- to few-flowered cyme, divaricately branched. **FLOWER** bractlets lanceolate, 4–6 mm long; petals obovate, 4–5.5 mm long, yellow; stamens about 20, anthers 0.5–0.7 mm long; pistils numerous, style 0.8–1.1 mm long, thickened at the base. **ACHENES** about 1.5 mm long, smooth. $2n$ = 14, 28, 54, 56, 63, 70. [*P. nipharga* Rydb.]. Rocky slopes and ridges; alpine communities. COLO: Arc, Hin, Min, SJn. 3535–3900 m (11600–12800′). Flowering: late June–Aug. Alaska to Quebec, south through the Rocky Mountains to Utah, Colorado, and northern Arizona.

Potentilla norvegica L. (of Norway) Norwegian cinquefoil. Erect annual, biennial, or short-lived perennial herb; 1–7 dm tall, from a slender taproot; herbage hirsute with stiff spreading hairs. **STEMS** leafy, few to several, usually branched in the upper part. **LEAVES** mostly cauline, reduced upward, 3-foliolate; petiole up to 6 cm long; leaflets oblanceolate to obovate, 1.5–4 cm long, shallowly lobed to toothed or coarsely serrate, the terminal leaflet longer than the lateral pair. **INFL** a leafy cyme; flowers inconspicuous. **FLOWER** bractlets broadly lanceolate, 3–7 mm long, acute to obtuse; sepals 3.5–8.5 mm long; petals mostly shorter than the sepals, broadly obovoid, 2–4 mm long, yellow; stamens 15–20, anthers 0.3–0.5 mm long; pistils numerous, style 0.7–0.8 mm long, thickened below the middle, subterminally attached. **ACHENES** 0.8–1.1 mm long, often becoming rugulose at maturity, brown. $2n$ = 42, 56, 63, 70. Lakeshores, irrigation ditches, meadows, and other moist sites. ARIZ: Apa; COLO: Arc, Hin, LPl, Min, Mon, SJn; NMEX: McK, RAr, SJn; UTAH. 1920–2715 m (6300–8905′). Throughout most of North America; Europe and Asia. The Ramah Navajo made a cold infusion of the whole plant for pain and fumes from the plant were used for sexual infection.

Potentilla ovina Macoun ex J. M. Macoun (of sheep) Sheep cinquefoil. Mat-forming perennial herb from a much-branched elongate caudex; herbage pubescent. **STEMS** decumbent, 0.5–2.5 dm long, usually several clustered, sericeous-hirsute. **LEAVES** mostly basal, 2–14 cm long, crowded, pinnate with linear segments, the upper leaflet sometimes decurrent, sericeous-strigose; cauline leaves few. **INFL** a 3–7 divaricately branched cyme; pedicels 0.5–4 cm long, more or less straight, spreading to ascending in fruit. **FLOWER** bractlets lanceolate, mostly 2.5–3.5 mm long, obtuse to rounded; sepals lanceolate, 3.5–5.2 mm long, acute to acuminate; petals obovate, 4–6 mm long, yellow; stamens 15–20, anthers 0.5–0.8 mm long; pistils numerous on a conical receptacle, style 1.8–2.5 mm long. **ACHENES** 1.5–2 mm long, smooth. [*P. diversifolia* Lehm. var. *pinnatisecta* S. Watson]. Rocky slopes and ridges; alpine communities. COLO: LPl, Mon. 3505–3655 m (11500–12000′). Flowering: late Jun–Aug. Our material belongs to **var. *decurrens*** (S. Watson) S. L. Welsh & B. C. Johnst. (running down). In the Four Corners region, sheep cinquefoil is known only from the La Plata Mountains.

Potentilla paradoxa Nutt. (not of the expected) Contrary cinquefoil. Annual or short-lived perennial herb from a slender taproot; herbage villous-hirsute with spreading hairs. **STEMS** leafy, spreading to ascending, 1–5 dm long, often branched in the upper part, divaricate. **LEAVES** mostly cauline, reduced upward, pinnate, 5–9-foliolate, 4–10 cm long including the petiole; leaflets lanceolate to ovate, 10–27 mm long, the terminal leaflet usually longer. **INFL** a divaricately branched cyme. **FLOWER** bractlets lance-elliptic, mostly 2.5–4.5 mm long; sepals deltate-ovate, 2.8–6 mm long, acute; petals obovate, 2.5–3 mm long, yellow; stamens 20 or greater, anthers about 0.35–0.4 mm long; pistils numerous, style 0.5–0.6 mm long, thickened at the base. **ACHENES** 0.7–1 mm long, brown, rugose, with a protuberance on the ventral suture. $2n$ = 28. [*P. supina* L. subsp. *paradoxa* (Nutt.) Soják]. Shorelines along lakes and reservoirs. ARIZ: Apa, Nav; COLO: Arc; NMEX: SJn; UTAH. 1640–1890 m (5390–6195′). Flowering: Jun–Jul. Widespread in North America and Asia.

Potentilla pensylvanica L. (of Pennsylvania) Pennsylvania cinquefoil. Perennial herb from a branched caudex. **STEMS** few to several, erect to decumbent, 0.5–3 dm long, spreading-villous and substrigose, grayish green. **LEAVES** mostly basal, 5–15 cm long, with 2–6 (7) lateral pairs of leaflets, green, strigose above and grayish tomentulose beneath, upper 3 largest, oblanceolate, 0.8–3 cm long, the upper lateral pair often confluent, pinnatifid, cleft halfway to the midrib, the lobes 2–5 mm long, rounded, margins revolute; cauline leaves few and small to well developed. **INFL** a several-flowered glomerate or open cyme; pedicels 5–20 mm long. **FLOWER** bractlets elliptic-lanceolate, mostly 3.5–4.5 mm long, rounded; sepals ovate, 3–6 mm long, obtuse; petals suberect, obovate, 3–6 mm long, pale yellow; stamens mostly 20, filaments yellow, anthers 0.5–0.8 mm long; pistils numerous, style 0.8–1.2 mm long, shorter than the achene, thickened at base, yellow. **ACHENES** 1.1–1.4 mm long. $2n$ = 28. [*P. strigosa* (Pursh) Pall. ex Tratt.]. Rocky sites; ponderosa pine, black sage, aspen, and spruce-fir communities. ARIZ: Apa; NMEX: RAr, SJn; UTAH. 2255–3200 m (7400–10500′). Flowering: Jun–Sep. Alaska east to Hudson Bay, south to Arizona, New Mexico, and Mexico.

Potentilla pulcherrima Lehm. (beautiful) Beautiful cinquefoil. Perennial herb from a branched caudex. **STEMS** erect or ascending, 2–8 dm tall, sparsely to densely pubescent with spreading or appressed hairs. **LEAVES** mostly basal, digitate, 5–9-foliolate; petiole mostly 7–20 cm long; leaflets oblanceolate to obovate, mostly 2.5–6 cm long, the terminal one the largest; shallowly toothed to pinnatifid; leaflets green above and whitish lanate beneath. **INFL** a 2–3 (8)-flowered cyme; the pedicels 1.3–1.8 cm long. **FLOWER** bractlets 1.8–4 mm, lanceolate, acute; sepals deltate-lanceolate, mostly 2.6–5 mm long, acute; petals obovate to suborbicular, 4–6.5 mm long, yellow; stamens 20, anthers 0.6–0.9 mm long; pistils numerous, style 1.5–2.5 mm long. **ACHENES** 1.7–2.1 mm long. [*P. hippiana* Lehm. var. *pulcherrima* (Lehm.) S. Watson; *P. pulcherrimum* (Lehm.) Nieuwl.; *P. gracilis* Douglas ex Hook. var. *pulcherrima* (Lehm.) Fernald]. Moist meadows and open woods to alpine communities. COLO: Arc, Con, Dol, Hin, LPl, Min, Mon, RGr, SJn, SMg; NMEX: RAr; UTAH. 2135–3718 m (7000–12200′). Flowering: late May–Aug. British Columbia to Manitoba south to South Dakota, New Mexico, and Arizona. Hybridizes with *P. diversifolia*, *P. hippiana*, and *P. concinna.*

Potentilla rubricaulis Lehm. (red-stemmed) Red stem cinquefoil. Perennial herb from a dark brown caudex; vestiture often with a few small glandular hairs. **STEMS** decumbent to erect, 0.5–1.5 dm long, villous. **LEAVES** mostly basal, digitate, 5 (7)-foliolate, sometimes subpinnate; petiole 1–4 cm long, villous; leaflets 9–20 mm long, the terminal one the longest, obovate-cuneate, deeply pinnately divided, segments lanceolate, revolute-margined, green-sericeous above and densely white-tomentose beneath; upper cauline leaves 3-foliolate. **INFL** a 3–6-flowered, congested to open cyme; pedicels 2–10 mm long. **FLOWER** bractlets narrowly elliptic to narrowly lanceolate, mostly 2–3 mm long, obtuse; sepals lanceolate to ovate, 2.2–4 mm long, acute; petals 3–5 mm long, yellow; stamens 20, anthers 0.3–0.5 mm long; pistils numerous, style 0.7–1 mm long. **ACHENES** 1–1.2 mm long. $2n = 56$. [*P. dissecta* var. *rubricaulis* (Lehm.) Rydb.; *P. saximontana* Rydb.; *P. modesta* Rydb.; *P. paucijuga* Rydb.]. Rocky soils; alpine communities. COLO: SJn. 3840 m (12600′). Flowering: July–Aug. Known from one collection (*K. Heil 9142*) at Stony Pass summit. Alaska and Greenland, south to Nevada, Wyoming, Utah, and Colorado.

Potentilla subjuga Rydb. (somewhat paired) Perennial from a woody caudex. **STEMS** 1–3 dm tall, tufted, silky-villous. **LEAVES** digitate, (3) 5-foliolate with an additional smaller pair of leaflets on what appears to be the petiole; leaflets 1–4 cm long, oblong or oblanceolate to obovate, deeply incised, silky, denser below; cauline leaves reduced, trifoliate. **FLOWER** bractlets 1/4–1/3 shorter than the sepals; sepals 5–6 mm long, strigose; petals larger than the sepals, yellow; styles filiform, greater than 1.2 mm long. **ACHENES** glabrous. Alpine communities. COLO: Arc, Hin, SJn. 3570–3915 m (11715–12850′). Flowering: Jul–Aug. Wyoming, Colorado, and New Mexico.

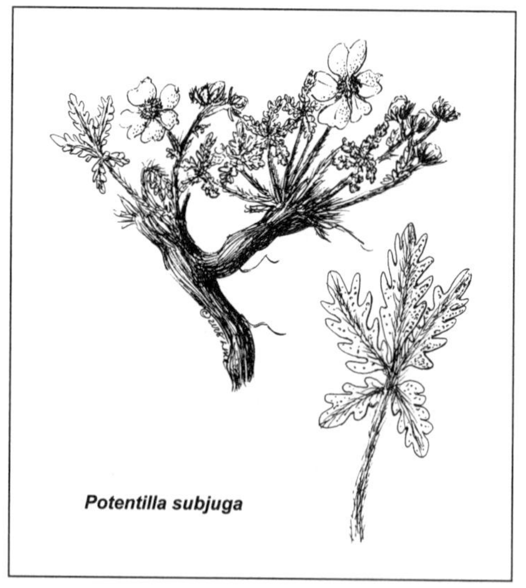
Potentilla subjuga

Potentilla thurberi A. Gray (for George Thurber, botanist, naturalist, author, and editor) Thurber's cinquefoil. Perennial herb with horizontal rhizomes. **STEMS** 2–6 dm tall, erect or nearly so, sparsely villous. **LEAVES** palmate, with 5–7 leaflets, petioles up to 12 cm long; leaflets 1–5 cm long, oblong to oblanceolate, toothed. **FLOWER** bractlets 2–2.5 mm long; sepals villous, 4–5 mm long, acute; petals broad, longer than the sepals, dark red. Ponderosa pine communities. COLO: SJn. 2635–2680 m (8640–8800′). Flowering: Jul–Aug. Southern New Mexico and Arizona. Known from a single collection by Cyndie Holmes (*1683*) along the railroad between Durango and Silverton, Colorado. Probably an escape.

Prunus L. Cherry, Plum

(ancient Latin name of the plum) Deciduous (in ours) trees or shrubs, unarmed or with thorns. **STEMS** woody, bark often reddish, thin, smooth, with elongate horizontal lenticels. **LEAVES** simple, usually serrate, with prominent gland on petioles or base of blades; stipules small, caducous. **INFL** umbel, corymb, raceme, or flowers solitary; bracts small, caducous. **FLOWERS** perfect, white to pink, showy, large nectar disc present; hypanthium mostly cup-shaped, free from carpels, the ovary superior, usually circumscissile, deciduous; sepals 5, deciduous with the hypanthium; petals spreading, circular to elliptic, quickly falling; stamens usually 15–20; carpel 1, free from hypanthium, the ovary superior, style terminal. **FRUIT** a drupe, red to purple or black, rarely yellow; pits very hard, indehiscent, smooth or variously textured. $x = 8$. Perhaps 200 species of the north-temperate zone, but also the subtropics.

1. Inflorescence consisting of 15 or more flowers in an elongate raceme; a common native species ..... *P. virginiana*
1' Inflorescence consisting of 15 or fewer flowers in a corymbose raceme, or umbel, or solitary; a rare native or cultivated species ..... (2)

2. Inflorescence consisting of 3–15 flowers in a corymbose raceme; leaves finely crenate with a glandular hair in each sinus; native ..... *P. emarginata*
2' Inflorescence of a solitary flower or a few in an umbellate cluster; escaped, cultivated species ..... (3)

3. Drupe glabrous, neither furrowed nor glaucous, the stone (endocarp) globose or ovoid; cherries ..... *P. avium*
3' Drupe usually glaucous or pubescent, longitudinally furrowed, the stone compressed ..... (4)

4. Drupe glabrous, distinctly pedicellate, the pedicel usually remaining with the fruit at maturity; stone not prominently furrowed; plums ..... *P. americana*
4' Drupe pubescent; pedicel short to nearly absent, not falling with the fruit; stone usually prominently furrowed on the margin; apricots and peaches ..... (5)

5. Leaves (ob)ovate; sepals mostly glabrous; apricots ..... *P. armeniaca*
5' Leaves (ob)lanceolate to narrowly oblanceolate; sepals pubescent, at least on the margins ..... *P. persica*

Prunus americana Marshall (of America) American plum. Mostly shrubs, rarely treelike, to 5 m tall, glabrous. **LEAVES** up to 7 cm long, mostly 0.5–3 cm wide, elliptic to ovate or lanceolate, serrate, acuminate to a long-attenuate apex, acute to obtuse basally. **INFL** of 1–4 sessile or subsessile umbels, appearing before the leaves; pedicels up to 12 mm long. **FLOWER** hypanthium puberulent; sepals puberulent, spreading; petals white, 5–7 mm long, 2.5–3 mm wide. **FRUIT** a yellow to red plum, glabrous, distinctly pedicellate. Cultivated and escaped. ARIZ: Apa; COLO: Mon; NMEX: RAr, SJn. 1770–2080 m (5810–6820′). Flowering: late Apr–Jun. Widespread throughout North America. Cultivated by many southwestern tribes.

Prunus armeniaca L. (Armenian, apricot-colored) Apricot. Small trees up to 8 m tall; branchlets green to brown. **LEAVES** with blades 1.5–7 cm long, 1–6 cm wide, cordate-ovate to (ob)ovate, serrate, attenuate, obtuse to cordate basally, glabrous. **INFL** a solitary flower, appearing before the leaves, pedicel short. **FLOWER** hypanthium glabrous except basally; sepals glandular; petals white to pink, 8–12 mm long. **FRUIT** pubescent, fleshy, the stone longitudinally furrowed, compressed. Cultivated and escaped. ARIZ: Nav; NMEX: SJn. 1675–1830 m (5500–6000′). Flowering: late Apr–Jun. Widespread throughout North America.

Prunus avium (L.) L. (bird-attracting) Sweet cherry. Trees up to 8 m tall; branchlets brown. **LEAVES** with blades up to 15 cm long, 2.5–8 cm wide, oblanceolate to obovate, serrate or doubly serrate, attenuate, obtuse to rounded basally, glabrous to long-hairy. **INFL** of 2–4 flowers per bud, appearing with early leaves. **FLOWER** hypanthium glabrous; sepals glabrous; petals white to pink, 8–14 mm long. **FRUIT** glabrous, red to black, the stone globose or ovoid. Cultivated and escaped. ARIZ: Apa; NMEX: SJn. 1495–1700 m (4900–5575′). Flowering: Apr–May. Widespread throughout North America.

Prunus emarginata (Douglas ex Hook.) D. Dietr. (with a notched margin) Bitter cherry. Shrub up to 6 m tall, the branches erect to spreading, the young stems with smooth reddish bark; herbage glabrous. **LEAVES** in fascicles on short lateral branches or spread along elongating stems, petioles 1–10 mm long, occasionally with 1 or 2 glands near the summit, the stipules linear to lanceolate, deeply and finely lobed, the blade broadly oblanceolate to obovate, 1.7–4.5 cm long, 0.5–2.5 cm wide, cuneate basally, obtuse or usually rounded apically, finely crenate with a glandular hair in each sinus. **INFL** a 6–10 (12)-flowered corymbose raceme, on short lateral shoots, the pedicels 3–14 mm long. **FLOWER** hypanthium campanulate, 2–3.2 mm deep; sepals reflexed, ovate-rounded, 1.5–2.5 mm long; petals broadly (ob)ovate, 3–4.5 mm long, clawed, white; stamens about 20, the filaments 2.3–4.2 mm long, anthers 0.4–0.6 mm long; style deciduous, about as long as the filaments. **FRUIT** a drupe, 7–8 mm thick, bright red with pulpy exocarp, glabrous. [*Cerasus emarginata* Douglas]. Rocky slopes, often with aspen. ARIZ: Apa. 2745 m (9000′). Flowering: Jun–early Jul. Our one record is from the Carrizo Mountains, about two miles south-southeast of Pastora Peak in a thick aspen grove (*Heil & Clifford 17663*). British Columbia to Idaho, south to California and Arizona.

Prunus persica (L.) Batsch (from Persia) Peach. Trees up to 4 m tall, branchlets green, aging to gray, glabrous. **LEAVES** with blades up to 15 cm long, 0.5–5.5 cm wide, oblong-lanceolate to narrowly (ob)lanceolate, serrate, attenuate, obtuse to acute basally, glabrous. **INFL** solitary, appearing before the leaves, the pedicel short. **FLOWER** sepals villous at the margin; petals white to pink, or red. **FRUIT** pubescent, fleshy at maturity, the stone usually furrowed on the margin, somewhat compressed. Cultivated and escaped. ARIZ: Apa; COLO: LPl. 1770–1995 m (5810–6540′). Flowering: late Apr–May. Widespread throughout North America.

Prunus virginiana L. **CP** (of Virginia) Western chokecherry, black chokecherry. Shrub or small tree up to 8 m tall, young stems puberulent, greenish at first, soon becoming glabrous, reddish, older stems ashy gray with a reddish brown undertone. **LEAVES** with petioles 1–2.5 cm long, with a pair of distal reddish glands, these glands sometimes on the blade near the base, stipules small, blade (ob)ovate or elliptic, 4–12 cm long, 1.5–4.5 (7.5) cm wide or usually smaller on the raceme-bearing branches, obtuse, rounded or subcordate basally, acute or acuminate apically, finely serrulate with ascending teeth, dark green above and pale green beneath, glabrous or with a few tufts of hairs in the axils of the lateral veins beneath, the midrib impressed above, prominent beneath. **INFL** an elongate raceme, 5–14 (20) cm long, with numerous flowers, the pedicels 2.5–7 mm long, each subtended by a bract, the lower flowers often in the axils of the upper leaves. **FLOWER** hypanthium 2–3 mm deep; sepals reflexed, broadly rounded or deltate, 0.6–1.3 (2) mm long, erose or usually fimbriate with reddish, glandular, clavate hairs; petals broadly ovate or suborbicular, 3–5 (6.5) mm long, white; stamens about 25, the filaments 2–5 mm long, the anthers 0.5–0.8 mm long. **FRUIT** 6–8 mm diameter, becoming dark red or bluish purple to nearly black. $2n = 16, 26, 32$. [*Cerasus demissa* Nutt. var. *melanocarpa* A. Nelson; *P. melanocarpa* (A. Nelson) Rydb.; *P. demissa* (Nutt.) D. Dietr. var. *melanocarpa* (A. Nelson) A. Nelson]. Along streams, canyon bottoms; sagebrush, piñon-juniper woodland, oak-serviceberry, ponderosa pine, and aspen communities. ARIZ: Apa, Nav; COLO: Arc, Dol, Hin, LPl, Min, Mon, SJn, SMg; NMEX: McK, RAr, SJn; UTAH. 1700–2895 m (5575–9500′). Flowering: May–Jun. Four Corners material belongs to **var. *melanocarpa*** (A. Nelson) Sarg. (black-fruited). British Columbia and Alberta south to eastern California, Nevada, Utah, Colorado, the Dakotas, Arizona, and New Mexico. The Ramah Navajo used a cold infusion of dried fruit for stomachache, an emetic in various ceremonies, and as a "life medicine"; the fruit was ground and made into small cakes or eaten fresh. The Navajo cooked the fruits into a gruel with cornmeal.

Purshia DC. ex Poiret Bitterbrush, Cliffrose

(for Frederick T. Pursh, Saxon explorer and plant collector) Fragrant shrubs or small trees, mostly deciduous. **STEMS** much-branched, woody, bark thin and gray to brown or shredding into long strips revealing orange underbark. **LEAVES** simple, pinnatifid or apically 3-lobed, coriaceous, crowded into fascicles terminating short shoots; stipules adnate to the leaf bases, forming a joint from which leaf blades dehisce, part below joint persistent. **INFL** of solitary flowers terminating short shoots; bracts small, persistent. **FLOWERS** mostly perfect, white to cream or yellow, fragrant, showy; hypanthium turbinate to funnelform, free from carpels, lined with nectar disc; sepals 5, epicalyx absent; petals 5, orbicular or spatulate, basally clawed, deciduous; stamens numerous; carpels 1–2 or 4–10, free from each other and the hypanthium, the ovaries superior, styles terminal, persistent. **FRUIT** an achene, either large with a short, nonplumose style or smaller with a long, plumose style. $x = 9$. About seven species and one hybrid taxon of western North America. The often-segregated genus *Cowania* is included here.

1. Carpels several; style long, plumose .... ***P. stansburyana***
1′ Carpel 1 (2–3); style short, not plumose .... ***P. tridentata***

Purshia stansburyana (Torr.) Henrickson **CP** (for Howard Stansbury, 1806–1863, explorer and naturalist) Cliffrose, cowania. Shrubs up to 7.5 m tall, with ascending branches, reddish brown bark becoming gray and exfoliating in age. **LEAVES** fasciculate on short lateral branches, revolute-margined, the larger ones 15 mm long, pinnately (3) 5–7-lobed, veins prominent beneath, green, finely pubescent along the midrib above, white-tomentose beneath, stipules persistent. **INFL** of solitary flowers terminating on short lateral branches; pedicels and hypanthium stipitate-glandular. **FLOWERS** fragrant, perfect or rarely staminate or pistillate; hypanthium narrowly obconic to narrowly campanulate, mostly 5–6 mm long, tapering into the pedicel; sepals ovate, 3–5 mm long, rounded, glabrate on the surface, tomentose on the margins; petals obovate to broadly oblanceolate or narrowly rhombic, 7–14 mm long, cuneate basally, rounded to obtuse apically, white, cream, or pale yellow; stamens numerous in 2 series, anthers 0.8–1 mm long; pistils 4–10, style elongating. **ACHENES** 4–12, narrowly turbinate, 5–7 mm long, glabrate at the base, densely pubescent toward the apex, mature style 2–4.5 cm long, densely plumose. $2n = 18$. [*Cowania stansburyana* Torr.; *C. mexicana* D. Don var. *stansburyana* (Torr.) Jeps.; *C. mexicana* var. *dubia* Brandegee; *C. alba* Goodd.; *C. davidsonii* Rydb.]. Rocky foothills, along washes and canyon bottoms; salt desert scrub, blackbrush, piñon-juniper woodland, and ponderosa pine communities. ARIZ: Apa, Coc, Nav; COLO: Dol, Mon, SMg; NMEX: McK, San, SJn; UTAH. 1340–2135 m (4400–7000′). Flowering: May–Jun. Nevada and southeastern California, Utah, southwestern Colorado, northern Arizona, northwestern New Mexico; Mexico. *Purshia stansburyana* often hybridizes with *P. tridentata*. The Navajo used the plant for deer and livestock forage; pounded the leaves and stems and mixed with pounded juniper to make a brown or yellow-brown dye, and used inner bark for baby diapers. Used for arrows and dermatological aid by the Hopi. The Ramah Navajo used the leaves as an emetic in various ceremonies.

Purshia tridentata (Pursh) DC. (three-toothed) Bitterbrush, antelope bitterbrush. Shrub, upright or low and prostrate with trailing branches, gray bark. **LEAVES** fasciculate on short lateral branches, 10–23 mm long, cuneate, 3-lobed apically, margins revolute, veins impressed above and prominent beneath, blade thinly pubescent and green above and greenish white-tomentose beneath; stipules small, triangular, persistent. **INFL** of solitary flowers terminating on short, leafy branchlets; pedicels and hypanthium with stipitate-glandular hairs, tomentose. **FLOWERS** fragrant, perfect; hypanthium funnelform-turbinate, mostly 3–5.5 mm long, tapering into the pedicel; sepals ovate, 1.8–3 mm long, rounded or obtuse, lightly tomentose on the back; petals oblanceolate or obovate, 4–10 mm long, clawed, rounded apically, yellow; stamens about 25, anthers 0.9–1.5 mm long; pistils 1 (2). **ACHENES** 7–11 mm long, narrowly obovoid-fusiform, densely pubescent, leathery, style nonplumose. **SEEDS** 6–8 mm long, black. $2n = 18$. [*Tigarea tridentata* Pursh; *Kunzia tridentata* (Pursh) Spreng.]. Often in sandy soils; sagebrush, juniper, piñon-juniper woodland, and ponderosa pine communities. ARIZ: Apa, Coc, Nav; COLO: Arc, Dol, LPl, Mon; NMEX: McK, RAr, San, SJn; UTAH. 1495–2300 m (4900–7550′). Flowering: late Apr–Jun. Southern British Columbia south to Washington, Oregon, and northeast California, east to western Montana, Wyoming, western Colorado, northern Arizona, and New Mexico; Black Hills of South Dakota. Navajo uses include gynecological aid (plant taken during confinement), considered as an important browse plant, bark used for diapers, and used to make arrows. Ramah Navajo used the leaves in various ceremonies, hunters chewed the leaves for "deer infection" and for good luck in hunting, and twig and leaf ash used for "Evilway" blackening.

Rosa L. Rose, Briar

Walter H. Lewis and Barbara Ertter

(ancient Latin *rosa* for the rose) Deciduous shrubs armed with prickles and/or aciculi; roots woody, "mostly rhizomatous turions," prickly. **STEMS** erect to spreading, climbing, or trailing, strict or branched; prickles straight, curved, or hooked, small and round to large, broad-based, and flat, often paired and infrastipular, aciculi mixed with prickles or alone on stolons and basally, upper stems and branches sometimes lacking armature. **LEAVES** alternate, petiolate, stipules conspicuous, paired, entire to pinnatifid, adnate along base of petiole and forming flared auricles apically, blade imparipinnate or rarely trifoliate, indument when present abaxially of simple hairs, sessile or stipitate glands, pricklets common to rachis and petiole; leaflets uniserrate or multiserrate, or a combination of both, often gland-tipped, short-petiolulate, longer for terminal leaflet. **INFL** a 1- to many-flowered determinate corymb or panicle, mostly bracteate. **FLOWERS** perfect, 5-merous, strongly perigynous, showy, often aromatic, borne at ends of lateral branches or occasionally main stems; hypanthium globose, oval, or urceolate, often constricting toward apex, constricted orifice, disc forming ring around opening; sepals (calyx lobes) 5, persistent or sometimes (in introduced *R. multiflora* in this *Flora*) deciduous, outer lobes sometimes toothed or pinnatifid, apices acute to long-attenuate or prolonged into foliaceous tips, outer surface eglandular or glandular, inner tomentose; petals 5, or by transformation of stamens numerous, light pink to deep rose, often fading with age, or white (*R. multiflora*), obovate or obcordate, apices usually emarginate, inserted on outer edge of disc; stamens numerous in several whorls inserted on disc, outer ones longer than inner, filaments slender, persistent, anthers small, yellow; pistils few or usually with many carpels inserted at base and/or inner walls of hypanthium, styles connate (*R. multiflora*) or free, included or long-exserted (*R. multiflora*), stigmas mostly capitate, grouped around the hypanthial orifice; ovaries hairy, few to mostly numerous, free, each with a single ovule. **FRUIT** accessory (hips), fleshy or pulpy, enlarged at maturity, usually red or orange, containing few or numerous bony achenes, seed coat thin, embryo filling the seed, radicle superior (Lewis 1959). *Rosa* consists of 100–150 species, primarily in north-temperate America and Eurasia, infrequently extending to subtropical areas. Widespread in Canada and the United States and occasionally northern Mexico; many species of Eurasian origin occur throughout the continent. In the *Flora* area we recognize four species, three indigenous and one introduced. Because of hybridization, polyploidy, and inherent diversity throughout widespread and multiple habitats, species boundaries are often obscured and their identities difficult to determine even using a well-constructed key. Subspecies are recognized for two indigenous species, *R. nutkana* and *R. woodsii*, the two most widely distributed and diverse species in western North America. Their subspecies are characterized by well-defined core features within ecogeographic patterns. Intermediate plants may occur where such zones coalesce within the overall distributional continuum of each species. Such plants can be treated as part of the subspecies they most closely approximate or identified only to the binomial level.

Fleshy or dried rose hips are a good source of digestible energy food for birds and other wildlife in the fall and winter when high crude protein food is scarce. Upper stems and leaves are also browsed by wildlife and livestock from spring to fall, and their protein content is sufficient to meet the maintenance requirements of sheep and cattle during the growing season. Native Americans have made extensive use of all parts of roses for food and therapeutic materials. In addition to *R. multiflora*, other self-perpetuating naturalized roses could be anticipated in our area, particularly *R. canina* L. (rare in Utah), *R. eglanteria* L. (Utah, rare in New Mexico), and *R. rugosa* Thunb. (Utah). *Rosa foetida* Herrm. (Austrian briar rose) or hybrids (probably Harison's yellowrose, *R.* ×*harisonii* Rivers) with semidouble yellow, sometimes red-tinged, petals occur around current or abandoned dwellings as individual clumps or hedges, with totally abortive hips in the few maturing collections seen. No example of the species or hybrid(s) naturalizing and reproducing has been found in our area. Tradition has it that such plants were carried across the plains by Mormon pioneers (Cronquist et al. 1997b), and another tradition takes the rose to Texas, where it gained the appellation Yellow Rose of Texas. The hybrid ×*harisonii* (possibly *R. foetida* Herrm. × *R. spinosissima* L.) originated in 1830 in the New York City garden of attorney and amateur rose grower George Harison.

1. Stems arching and sprawling to 5 m long, forming dense, impenetrable clumps; petals white, very small (7–10 mm long); sepals deciduous, styles connate and exserted beyond the hypanthium (hip); introduced species, rare .... ***R. multiflora***
1′ Stems upright, short to infrequently 3 m long; petals pink to rose, petals greater than 10 mm; sepals persistent; styles free and ± inserted in the hypanthium (hip); native species .... (2)

2. Prickles common, various, usually straight and/or curved (plants occasionally unarmed but not with the combination of characters below); leaflets obovate to elliptic; flowers small, 30–35 mm across; sepals at base ± 2 mm wide; hips commonly small, less than 10 mm wide .... ***R. woodsii***
2′ Prickles either straight, dense, needlelike aciculi to stem apex or straight to curved, broad-based, paired infrastipular prickles, rarely unarmed; leaflets elliptic and often large; flowers large, (35–) 40–60 mm across; sepals at base 2.5–4 mm wide; hips large, greater than 10 mm wide .... (3)

3. Stems ± slender and flexible, armed with dense, straight, fine aciculi to apex; outer surface of sepals typically stipitate-glandular; rare in our area .... ***R. acicularis***
3′ Stems armed with stout, straight or curved/hooked, mostly paired infrastipular prickles, occasionally unarmed; outer surface of sepals eglandular, rarely glandular; frequent at high elevations .... ***R. nutkana***

Rosa acicularis Lindl. (needlelike prickles or aciculi) Prickly rose. **STEMS** 1 m tall or commonly shorter, armed to apices of floral branches with aciculi or rarely with aciculi and small round prickles, occasionally with few bristles. **LEAVES** with stipules edged with few or numerous glands, flared auricle averaging 4.6 mm, blades (3–) 5–7-foliolate, leaflets elliptic to oval, large, 25–40 (–50) × 12–21 (–30) mm, abaxially mostly puberulent to pubescent, frequently slightly glandular, uniserrate or biserrate; petioles with or without pricklets, few or many stipitate glands, glabrous to pubescent. **INFL** solitary or rarely 2–3-flowered; pedicel 20–30 mm long, often curving as hip matures, eglandular or rarely with stalked glands. **FLOWERS** 35–50 mm across; hypanthium elliptic, globose to pyriform, often with a distinct neck, 2.5–4 mm across, eglandular; sepals persistent in fruit, 19 × 3 mm basally, often with flared tips, outer surfaces typically stipitate-glandular; petals pink to rose, obovate or obcordate, ± 20 × 20 mm; pistils with many carpels, styles free, included in hypanthium. **FRUIT** orange-red at maturity, elliptic, globose, or elongated with extended neck, 13–18+ × 9–12 mm, eglandular, sepals upright or lateral, many achenes. $2n = 42$, hexaploid. [*R. acicularis* var. *bourgeauiana* Crép.; *R. engelmannii* S. Watson; *R. sayi* Schwein.]. Edge of spruce forest communities. COLO: Min. 2900 m (9400′). Flowering unknown. Our material belongs to **subsp.** ***sayi*** (Schwein.) W. Lewis (for Thomas Say, 1787–1834, distinguished American zoologist of early 19th century) Say's or American prickly rose. Subspecies *sayi* occurs from eastern Siberia, Alaska, and northern Canada east to Quebec and south to West Virginia, in the prairies south to South Dakota, and westerly in the Rocky Mountains south to southeastern and tentatively south-central Colorado.

Rosa multiflora Thunb. (many-flowered) Japanese rose, multiflora rose. **STEMS** few to many, forming dense, impenetrable clumps reaching 10 m across, erect and arching to trailing or sprawling to 5 m long, 1.5–3 (–5) m tall, armed with stout, curved prickles. **LEAVES** with petioles finely puberulent to pubescent; stipules pinnatifid, stipitate-glandular or eglandular, blade (5–) 7–9 (–11)-foliolate, 10–35 × 8–20 mm, abaxially glabrous to usually pubescent. **INFL** few to usually many flowers in a conical corymb; pedicel short, 1–1.5 cm, with stipitate glands or eglandular; bracts narrow and long (5–9 mm), fimbriate, early-deciduous, resembling small, free stipules. **FLOWERS** small, 15–20 mm across; hypanthium small, 1–1.5 mm wide, orifice minute; sepals ± triangular, about 6 × 1.5–2 mm, shorter than petals, early-deciduous, pinnatifid, outer surfaces glabrous or stipitate-glandular; petals white, rarely pink, obovate, about 7–10 × 5–7 mm; styles connate above, exserted from hypanthial orifice. **FRUIT** small, ovoid to globose, less than 7 mm diameter, red indument, with few yellowish achenes ± 4 mm long, flattened. $2n = 14$. Infrequent in disturbed areas. NMEX: SJn. 1650 m (5400′). Flowering: May. Widely naturalized in North America from East Asia, particularly in the East,

Midwest, and Pacific Coast areas of Canada and the United States. Introduced into North America to establish dense hedges and fencerows and to serve as understock for budding roses in the large commercial rose industry, the species has spread rapidly by its stolons and achenes (up to 500,000 seeds per plant) to become a major weedy pest species. Plants invade pastures, reducing grazing areas, and can cause severe eye and skin irritations in cattle.

Rosa nutkana C. Presl (for the Nootka tribe of west-central Vancouver Island, British Columbia) Nootka rose. **STEMS** stout, few to many from base, much-branched, 0.5–3 m tall, prickly throughout or unarmed above, stout, broad-based, infrastipular prickles with scattered prickles internodally, straight and/or curved to hooked. **LEAVES** with petioles 20–30 mm long, glabrous or pubescent, eglandular or glandular, pricklets common; stipules broad, 10–18 mm long, usually glandular-dentate, flared auricles 4–6 mm, blades 5–7 (–9)-foliolate, leaflets elliptic to oval, 25–35 × (6–) 15–20 mm, base ± rounded, apex ± obtuse, uniserrate blade or commonly biserrate or multiserrate with glandular teeth, abaxially glabrous or pubescent, eglandular or glandular. **INFL** solitary or few-flowered; pedicel 10–20 mm long, glabrous or glandular. **FLOWERS** large, to 50–60 mm across, flowering at ends of lateral branches; hypanthium globose or subglobose, 5–7 mm in diameter, glabrous or infrequently glandular-hispid to bristly; sepals persistent in fruit, 15–40 mm × 3–4 mm basally, often constricted in the middle and expanded distally, outer surfaces glabrous, eglandular or occasionally glandular-hispid to bristly; petals broadly obcordate, 20–30 × 20–35 mm, pink to deep rose; styles free, included in hypanthium. **FRUIT** large, 12–20 (–24) × 10–18 mm, subglobose, globose to pyriform, sometimes with neck, red to purplish, numerous large 6 × 4 mm achenes. $2n = 42$. Wooded regions or open areas from sea level to moderate or high mountainous elevations. Higher montane in aspen, fir, spruce, and/or pine forests. COLO: Arc, LPl, Min, Mon, RGr, SJn; UTAH. 2600–3200 m (8450–10400′). Flowering: late Jun–Jul. Fruit: (Jul–) Aug–Sep. Coastal Alaska south to northern California inland from British Columbia to Montana, south and west to Colorado, northern New Mexico, and Utah. Three subspecies are recognized: subsp. *nutkana* from coastal Alaska to northern California [*R. aleutiensis* Crép.; *R. muriculata* Greene]; subsp. *macdougalii* (Holz.) Piper of the intermontane region from British Columbia to Nevada and Utah [*R. macdougalii* Holz.; *R. spaldingii* Crép. ex Rydb.]; and our material belongs to **subsp. *melina*** (Greene) W. H. Lewis & Ertter [*R. melina* Greene] of north-central New Mexico, southwest and central Colorado, northern and southeast Utah, and south-central Idaho. It is distinguished from other subspecies by its high-elevation habitat, populations of compact shrubs with short stems, 0.3–1.3 m tall, having curved and/or hooked prickles, sometimes combined with straight prickles, or infrequently unarmed in upper branches, sepals with outer surfaces typically stipitate-glandular. An additional "nutkanoid" type with large, greatly elongated hips and lacking prickles in upper branches is tentatively placed here [*O'Kane Jr. et al.* (SJNM); 3200 m (9100′)].

Rosa woodsii Lindl. (for Joseph Woods, 1776–1864, British botanist widely recognized for his *Synopsis of the British Species of Rosa*) Woods' rose. **STEMS** much-branched, from 2–3 dm to 3 (–4) m tall, usually forming thickets, commonly with infrastipular prickles, differing from those of internodes where prickles are often dense, variable, and with aciculi, mostly straight, less frequently curved or hooked (subsp. *arizonica* and subsp. *manca*), occasionally unarmed, reddish brown to grayish bark with age. **LEAVES** with petioles glabrous or pubescent, eglandular or glandular, and often with pricklets; stipules 6–14 × 0.5–2 mm, often glandular-margined, auricles flared, 2–5 mm wide, blades 5–7 (–9)-foliolate, leaflets elliptic to oval or obovate, variable in size, 10–30 × 6–20 mm, sharply uniserrate, occasionally biserrate or multiserrate with gland tips, abaxial surface glabrous to pubescent, eglandular to glandular. **INFL** 1- to few-flowered or a corymb with 5–10+ flowers; pedicels 6–14 mm, glabrous, eglandular. **FLOWERS** mostly borne at ends of lateral branches, 30–35 mm across; hypanthium globose to oval, often with short neck, glabrous, eglandular; sepals persistent in fruit, short and ovate or long with foliaceous tip, sometimes pinnatifid, 10–16 × about 2 mm basally, outer surfaces eglandular or less commonly glandular, glabrous to pubescent, margins usually pubescent, inner tomentose; petals obovate or obcordate, 20–25 × 9–17 mm, pink to deep rose; pistils with many carpels, styles free and included in hypanthium. **FRUIT** fleshy, subglobose, globose to urceolate, with or without neck, 7–12 × 7–10 mm, numerous small achenes about 4 × 2 mm. Flowering: May–early Aug. Fruit: Jul–Oct. Central Alaska and northern Canada to western Ontario, south to Iowa, Oklahoma, western Texas, and Arizona, northeast of the Cascade and Sierra Nevada mountains to eastern California and central British Columbia; the commonest species in our area represented by three subspecies. $2n = 14$. A fourth subspecies, *R. woodsii* subsp. *ultramontana* (S. Watson) Roy L. Taylor & MacBryde occurs in the intermontane region north and west of our area. Used as a pink or light orange dye for wool by the Navajo.

1. Stems frequently tall, 1–2 (–3) m, not compact or highly branched, internodes long, prickles mostly curved/hooked; terminal leaflets mostly elliptic; low riparian and wetland areas, Southwest .... **subsp. *arizonica***

1′ Stems short to infrequently tall, 0.2–1 (–2) m, compact and markedly branched shrubs, internodes short, prickles straight or curved/hooked; terminal leaflets (some to many) obovate; various habitats ....(2)

2. Prickles straight; prairies, plains, to limited high montane regions, mostly central Rocky Mountains.... **subsp. *woodsii***
2' Prickles mostly curved/hooked; high montane regions, mostly southern Rocky Mountains.............. **subsp. *manca***

subsp. *arizonica* (Rydb.) W. H. Lewis & Ertter (for the state of Arizona) Found along creeks and other riparian habitats. [*R. arizonica* Rydb.; R. *granulifera* Rydb.; *R. neomexicana* Cockerell]. ARIZ: Apa, Coc, Nav; COLO: Dol; NMEX: SJn; UTAH. 1560–2200 m (5400–7150′). Flowering: Jun–mid-Jul. Fruit: Jul–Oct. Southern Nevada, northern and central Arizona, western New Mexico, southern, central, and northeast Utah, southern Idaho, and southwest Colorado.

subsp. *manca* (Greene) W. H. Lewis & Ertter (for Mancos Canyon area of southwest Colorado) In montane meadows, hillsides, and other openings and edges of aspen groves and pine/spruce/fir forests [*R. manca* Greene]. ARIZ: Apa, Coc, Nav; COLO: Arc, Dol, Hin, LPl, Min, Mon, SMg, SJn; NMEX: McK, RAr, SJn; UTAH. 2350–3050+ m (7700–10000+′). Flowering: Jun–Jul. Fruit: late Jul–Sep. Western Wyoming, Colorado, eastern Utah, north-central and northeast Arizona, and northern New Mexico.

subsp. *woodsii* In prairies, high plains, to montane regions, often in wetland and riparian habitats. [*R. macounii* Greene; *R. maximiliani* Nees]. ARIZ: Apa; COLO: Arc, Dol, Hin, LPl, Min, Mon, SMg, SJn; NMEX: McK, RAr, San, SJn. 1600–3150 m (5300–10350′). Flowering: late May–early Aug. Fruit: Jul–Sep. Central North America from Alaska to Iowa and Oklahoma, extending through the Rocky Mountains from Montana to New Mexico and eastern Arizona.

Rubus L. Blackberry, Raspberry

(Latin name for some species of the genus; allied with *ruber*, red) Mostly erect to trailing shrubs or scramblers from rootstocks or creeping stems, usually abundantly armed with prickles, sometimes unarmed. **STEMS** mostly biennial, sterile first year's growth, primocanes, and fertile second-year stems, floricanes. **LEAVES** often different on primocanes and floricanes; simple and then palmately lobed or compound with 3–7 leaflets, the leaflets arranged palmately or pinnately; stipules linear, small to conspicuous, free or adnate to the base of the petiole, persistent or caducous. **INFL** simple or compound cymes, racemes, or panicles, or flowers solitary; bracts foliaceous to linear, caducous or persistent. **FLOWERS** perfect, white to pink or rose-purple, showy, nectar disc present; hypanthium short, saucer-shaped; sepals 5, spreading to ascending, persistent, epicalyx absent; petals 5, elliptic to spatulate or circular, deciduous; stamens many; carpels many, free from each other and from hypanthium, the ovaries superior, inserted on a conelike to nearly flat receptacle, styles terminal, often deciduous. **FRUIT** red to purple or black, an aggregate of many drupelets, drupelets fleshy or rarely nearly dry, falling individually or coalescent and falling from the receptacle or with it; pits small, hard, indehiscent, variously textured. $x = 7$. Perhaps 700 species found in most parts of the world, a notoriously difficult genus with apomictic complexes.

1. Leaves simple; stems unarmed .... (2)
1' Leaves compound, 3–5-foliolate; stems armed with prickles .... (3)

2. Leaves 10–25 cm long; style glabrous; fruit fleshy, widespread throughout the Four Corners region....... ***R. parviflorus***
2' Leaves 2.7–8.5 cm long; style hairy; fruit more or less dry; entering our range in the southern Chuska Mountains .... ***R. neomexicanus***

3. Prickles slender, mostly straight, narrow-based, terete to subterete, very numerous, present on the hypanthium and sepals; anthers 0.7–0.8 mm long; pedicels glandular-pubescent; fruit red; widespread .... ***R. idaeus***
3' Prickles stout, often hooked, broad-based, laterally flattened, not numerous, not on the flowers; anthers 0.4–0.6 mm long; pedicels finely villous, rarely with glandular hairs; fruit dark red to purple; Navajo Mountain region .... ***R. leucodermis***

Rubus idaeus L. (from Mount Ida) American red raspberry, wild raspberry. Prickly shrub. **STEMS** up to 2 m long, bark yellow to yellowish brown, exfoliating; prickles numerous, slender, narrow-based, terete to subterete, straight or nearly so; present on stems, petioles, petiolules, and main leaf veins beneath, also in the inflorescence and the pedicels, hypanthium, and sepals; primocanes with 3- or 5-foliolate leaves, leaflets often lobed. **LEAVES** of the floricanes 3-foliolate, petiole 1.7–4.5 cm long, the terminal leaflet largest, lanceolate to ovate, 3–5 cm long, 1.5–3.2 cm wide, truncate to subcordate basally, rounded, acute, or usually acuminate apically, coarsely double-serrate, gray-tomentose beneath, green and sparsely pubescent to glabrate above. **INFL** a raceme, few-flowered thyrse, or 1 or 2 flowers in the leaf axils; pedicel, hypanthium, and sepals glandular-pubescent and with prickles. **FLOWER** sepals lanceolate-acuminate or caudate, 6–11 mm long, becoming reflexed; petals oblanceolate or spatulate, 3–7 mm long, white; stamens numerous, filaments broad, flat, anthers 0.7–0.8 mm long; pistils numerous, style 2–4 mm long. **FRUIT** an aggregate (of weakly coherent drupelets), tomentulose, red, reticulate-pitted. $2n = 14$. [*R. strigosus* Michx.; *R. melanolasius* Dieck; *Batidaea strigosa* (Michx.) Greene subsp. *acalyphacea* Greene]. Moist slopes and canyon bottoms; ponderosa pine, Douglas-fir, white fir, and spruce-fir communities. ARIZ: Apa; COLO: Arc, Hin, LPl, Mon, SJn; NMEX: RAr, SJn; UTAH. 1985–3305 m (6520–10850′). Flowering: Jun–early Aug. Circumboreal; from Alaska to Newfoundland south to Oregon, eastern Nevada, Arizona, New Mexico, east to North Carolina. North American plants belong to **var. *strigosus*** (Michx.) Maxim. (bristled).

Rubus leucodermis Douglas ex Torr. & A. Gray (white bark) Western black raspberry, whitebark raspberry. Prickly shrub. **STEMS** arched, 1–3 m long, bark yellowish at first, becoming reddish brown, often with a whitish bloom; primocanes green to yellow; floricanes purplish, prickles stout, on the stems, petioles, and some leaf veins; stems and petioles with a whitish bloom. **LEAVES** of primocanes 3- or 5-foliolate, the 5-foliolate ones with 3 petiolulate leaflets and 2 sessile; of floricanes 3-foliolate, petioles 2–5 cm long, terminal leaflet largest, lanceolate to ovate, 3–5 cm long, 1.5–4.5 cm wide, obtuse, truncate, or subcordate basally, acute or acuminate apically, sharp-serrate, double-serrate, sometimes lobed, whitish tomentose below, dark green and sparsely pubescent above. **INFL** a 2–7 (10)-flowered corymb, sometimes reduced to solitary flowers in the upper leaf axils; the pedicel, hypanthium, and sepals finely villous; scarcely if at all glandular; peduncle and pedicel with prickles. **FLOWER** sepals reflexed at anthesis, lanceolate-caudate, 5–9 mm long, to 12 mm in fruit; petals oblanceolate or spatulate, 5–6 mm long, white; stamens numerous, filaments flat, anthers 0.4–0.6 mm long; pistils numerous, styles glabrous. **FRUIT** an aggregate (of coherent drupelets), subglobose, to 12 mm diameter, drupelets densely tomentose basally, dark red to purple; reticulate-pitted. $2n = 14$. [*R. idaeus* L. var. *gracilipes* M. E. Jones]. Slopes, stream banks, and moist canyons. ARIZ: Coc; UTAH. 1855–2950 m (6090–9685′). Flowering Jun–Jul. British Columbia and Alberta south to Washington, western Montana, Arizona, and Utah (Navajo Mountain).

Rubus neomexicanus A. Gray (of New Mexico) New Mexico raspberry. Unarmed shrub, erect. **STEMS** up to 2 m tall, becoming woody, bark reddish or orange-brown. **LEAVES** simple, petiolate, stipules lanceolate, petiole 1.5–3.8 cm long, crisped-pubescent, the blade 2.7–8.5 cm long and about as wide, palmately 3-lobed, sometimes 5–7 lobes, cordate, double-serrate, lobes obtuse to rounded, green, sparsely pilose above, pale, soft pubescence below. **INFL** of 1 or 2 flowers in the leaf axils or terminating branches. **FLOWER** sepals attenuate or caudate, 5–22 mm long, sometimes glandular-pubescent; petals (8) 11–28 mm long, white; stamens numerous, anthers 0.5–1 mm long; pistils numerous, style short, hairy. **FRUIT** an aggregate, red, about 15 mm diameter, more or less dry. [*Oreobatus neomexicanus* (A. Gray) Rydb.; *O. deliciosus* (Torr.) Rydb. subsp. *neomexicanus* (A. Gray) W. A. Weber]. Ours on rocky slopes in a ponderosa pine community. NMEX: McK. 2675 m (8775′). Flowering: Jun. Known from one collection at the southern end of the Chuska Mountains (*Heil 25419*). Southwestern and northeastern Nerw Mexico, southeastern Arizona, and southeastern Utah (Lake Powell).

Rubus parviflorus Nutt. (small-flowered) Thimbleberry. Unarmed shrub, erect, up to 3 m tall, gray, flaky bark; herbage glandular-pubescent. **LEAVES** simple, petiolate, stipules lanceolate; petiole 2–11 cm long, blade mostly 10–20 cm long and about as wide, palmately (3) 5 (7)-lobed, cordate, irregularly serrate, lobes obtuse or acute. **INFL** a (1) 2–4 (7)-flowered raceme. **FLOWERS** with 5 (6–7) sepals and petals; sepals ovate-caudate, 10–18 mm long, glandular-pubescent, villous; petals ovate to obovate, 8–20 mm long, white, sometimes pinkish-tinged; stamens numerous, anthers 0.8–1.1 mm long; pistils numerous, ovary pubescent, style glabrous, 1–1.5 mm long. **FRUIT** an aggregate (of coherent drupelets), thimble-shaped. $2n = 14$. [*R. nutkanus* Moc. ex Ser.; *R. parviflorus* var. *hypomalacus* Fernald]. Moist places, often near streams; ponderosa pine, Douglas-fir, white fir, and spruce-fir communities. ARIZ: Apa; COLO: Arc, Hin, LPl, Min, Mon, RGr, SJn; NMEX: RAr, SJn; UTAH. 2150–3230 m (7050–10600′). Flowering: Jun–Aug. Alaska south throughout the mountain states to northern Mexico, east to Ontario and Michigan.

Sanguisorba L. Burnet

(Latin *sanguis*, blood, + *sorbere*, to absorb, referring to the use of *S. officinalis* as an astringent) Perennial or annual herbs. **STEMS** from taproots. **LEAVES** pinnately compound, basal and cauline, leaflets deeply dissected or toothed; stipules persistent, mostly adnate to bases of petioles, the apices expanded and dissected, those of basal leaves adnate to and partially sheathing stems. **INFL** compact to dense indeterminate heads or spikes, long-pedunculate; bracts small, usually 3 subtending each flower. **FLOWERS** perfect or imperfect and the plants then monoecious (or dioecious), green, reddish, or white, small; hypanthium urceolate, nectar ring at top; sepals 4, somewhat petaloid, persistent, epicalyx absent; petals absent; stamens 2–48; carpels 1 or 2, enclosed by but free from hypanthium, the ovaries superior, stigmas much-branched with filiform segments, reddish, showy. **FRUIT** accessory, of 1 or 2 achenes enclosed within hardened, dry, winged hypanthium. $x = 7$. About 18 species of the Northern Hemisphere.

Sanguisorba minor Scop. (smaller) Burnet. Garden burnet, small burnet. Perennial from a branched caudex. **STEMS** simple or branched above, up to 7 dm tall; herbage glabrous or sparsely pilose with moniliform hairs. **LEAVES** mostly basal, pinnate, basal and lower cauline leaves mostly 10–15 cm long, 11–21-foliolate, blade ovate to ovate-flabellate; leaflets 10–25 mm long, crenate-toothed, pinnately veined. **INFL** a globose to ovoid spike, 10–25 mm long, 7–12 mm thick. **FLOWERS** perfect or lower of staminate, upper of pistillate; bracts ovate, ciliate; hypanthium urceolate; sepals 4, 4–5 mm long, cuneate, mucronate, greenish brown, often red-tinged; petals none; stamens mostly 12, filaments long-filiform, anthers 0.5–0.9 mm long; pistils 2; fruiting hypanthium pyriform, 3–5 mm long. $2n = 28, 56$. [*Poterium sanguisorba* L.]. Disturbed sites, often in piñon-juniper woodland and sagebrush communities. COLO: Dol, LPl, Mon; NMEX: RAr, San, SJn; UTAH. 1890–2365 m (6200–7760′). Flowering: May–Jun. Scattered throughout the United States; native to Eurasia.

Sibbaldia L. Sibbaldia

(for Sir Robert Sibbald, a Scottish physician and a founder of the Royal Botanic Garden, Edinburgh) Low, perennial herbs, forming mats or cushions. **STEMS** arising from a branched caudex covered in persistent leaf bases. **LEAVES** trifoliolate (in ours), with 3–5 apical teeth; stipules adnate to the bases of the petioles, forming a narrow wing. **INFL** few-flowered cyme, scapose, bracts small, persistent. **FLOWERS** perfect, yellow, small; hypanthium saucer-shaped, small, nectar disc present; sepals 5, alternating with 5 epicalyx lobes, persistent; petals 5, oblanceolate, much shorter than sepals, deciduous; stamens 5; carpels 5–10, free from each other and from the hypanthium, the ovaries superior, styles lateral. **FRUIT** an achene. $x = 7$. About 20 species, except for *S. procumbens* all Eurasian, mostly China to India.

Sibbaldia procumbens L. (prostrate) Creeping sibbaldia. Mat- or cushion-forming, rhizomatous, perennial herb; herbage appressed-pilose. **LEAVES** basal, 3-foliolate; leaflets 1–2.5 cm long, the terminal one slightly larger than the lateral pair, cuneate-obovate, (2) 3 (5)-toothed apically, teeth rounded, petiole 1–5.5 cm long; stem leaves smaller. **INFL** a few-flowered cyme, usually glandular-pubescent, strigose; pedicels 1–5 mm long. **FLOWER** hypanthium shallowly cupulate, mostly 2.2–3 mm across; bractlets linear to narrowly lanceolate, mostly 1.3–2.3 mm long; sepals broadly lanceolate to ovate, mostly 2.5–4 mm long; petals oblanceolate, 1.3–1.6 mm long, rounded to obtuse, pale yellow; stamens 5, anthers 0.3 mm long; pistils 5–15 (20). **ACHENES** 1.2–1.6 mm long, ovoid, smooth. $2n = 14$. [*Potentilla procumbens* (L.) Clairv.; *P. sibbaldia* Kurtz]. Rocky slopes and ridges; subalpine and alpine communities. COLO: Arc, Con, Hin, LPl, Min, Mon, RGr, SJn; UTAH. 3105–3890 m (10200–12760′). Flowering: mid-June–Aug. Circumpolar; south throughout the western United States to northern New Mexico.

Sorbus L. Mountain-ash

(ancient Latin name, perhaps for *S. aucuparia*) Deciduous, unarmed trees or shrubs. **STEMS** woody, bark smooth, lustrous, with elongated horizontal lenticels. **LEAVES** pinnately compound, leaflets usually serrate; stipules small, caducous. **INFL** large, convex, showy panicles; bracts small, caducous. **FLOWERS** perfect, white to cream, medium-sized, nectar ring present; hypanthium saucer-shaped, lower part fully adnate to hypanthium; sepals 5, reflexed at anthesis, persistent, epicalyx absent; petals ovate, elliptic, or obovate, clawed at base, spreading, deciduous; stamens mostly 20; carpels 3–5, incompletely connate, fully adnate to the hypanthium, styles fused at base. **FRUIT** a small, fleshy pome, red to orange (in ours), topped by persistent, incurved sepals. $x = 17$. In the strict sense, about 90 species of the north-temperate zone.

1. Rachis of leaf densely to sparsely white-hairy; leaflets 9–14, 18–45 mm long; Chuska Mountains ....***S. dumosa***
1′ Rachis of leaf glabrous or sparsely pilose; leaflets 11–15, 40–75 mm long; widespread.......................***S. scopulina***

Sorbus dumosa Greene (bushy) Arizona mountain-ash. Shrub, 1–5 m tall; slender branches, the young twigs more or less villous-pubescent, the stems becoming sparsely so; winter buds densely white-hairy. **LEAVES** 9–14-pinnate, 7–15 cm long, the rachis densely to sparsely white-hairy; leaflets 18–45 mm long, lanceolate, sharply serrate, rounded or cuneate at the base, acute or acuminate at the apex, green above, dull green below. **INFL** a corymb, 3–15 cm across, mostly 10–40-flowered, branches hairy. **FLOWER** sepals 1–2 mm long; petals oval, 5–6 mm long. **FRUIT** a globose pome, red at maturity. Moist sites in ponderosa pine and Douglas-fir communities. ARIZ: Apa; NMEX: SJn; UTAH. 2100–2800 m (6890–9185′). Flowering: late Jun–early Jul. Arizona and New Mexico.

Sorbus scopulina Greene (of rocks) Rocky Mountain ash. Shrub or small tree, 1–4 m tall, bark yellowish to reddish purple, becoming grayish red; herbage strigose-pilose, the stems becoming sparsely so, leaves becoming glabrate; winter buds glutinous brown, glossy, glabrous to sparsely whitish pubescent. **LEAVES** 11–15-pinnate, 14–21 cm long, the rachis grooved above and glandular-pubescent near the nodes, petiole 2–3.5 (4.5) cm long; leaflets broadly lanceolate or elliptic, 4–7.5 (8.5) cm long, 1.5–2.5 cm wide, rounded to abruptly cuneate basally, acute or usually acuminate apically, finely serrate almost to the base, dark, glossy green above, dull, pale green and sparsely pubescent beneath. **INFL** a corymb, more or less flat-topped, 80–200-flowered, white appressed-pilose on the pedicels, hypanthium, and sepals. **FLOWERS** fragrant; sepals deltate, 0.7–1.8 mm long, acute or obtuse; petals (ob)ovate, 3–5 (6) mm long, white or cream; stamens 20, the filaments 2.5–4 mm long, anthers 0.5–0.7 mm long; styles 3 or 4. **FRUIT** a globose pome, 8–10 mm in diameter, bright glossy orange-red, drying purplish, of bitter taste. **SEEDS** oblong, 3.5–4 mm long, flattened, light brown. Mountain slopes; ponderosa pine, Douglas-fir, white fir, aspen, and spruce-fir communities. COLO: Arc, Hin, LPl, Min, Mon, SJn. 2265–3220 m (7440–10565′). Flowering: late Jun–early Jul. Alaska, Yukon, south to British Columbia, western Alberta, Washington, Oregon, northern California, Idaho, western Montana, Wyoming, northeastern Nevada, Utah, northeastern Arizona, and New Mexico; Black Hills of South Dakota. Our material belongs to ***var. scopulina***.

RUBIACEAE Juss. MADDER (COFFEE) FAMILY

Linda Mary Reeves

(Latin *rub*, red or reddish, referring to the flowers or fruit) Annual or perennial herbs, shrubs, vines, or trees. **LEAVES** entire, mostly opposite or appearing whorled due to presence of leaflike stipules, sometimes forming a sheath. **INFL** paniculate, cymose, headlike, clustered or solitary. **FLOWERS** actinomorphic, perfect, occasionally functionally imperfect; sepals generally 4–5, united, the lobes sometimes reduced or absent; petals generally 4–5, united, white, greenish, blue, purple, yellowish, pink, or red, often with striped markings; stamens epipetalous, alternate with corolla lobes; ovary inferior, 2–4-loculed, style often lobed. **FRUIT** 2–4 nutlets, berries, drupes, or schizocarps. About 450 genera with about 6500 species; worldwide, mostly tropical. Many important cultivated genera and species, including *Coffea arabica* (coffee), *Gardenia jasminoides* (gardenia), *Cinchona* spp. (quinine), *Ixora*, *Bouvardia*, *Pentas*, and other genera of ornamental note. The Asian genera *Myrmecodia* and *Hydnophytum* contain noted myrmecophyte species (Dempster 1995).

1. Ovules several in each carpel; fruit a capsule; flowers conspicuous, showy .... ***Houstonia***
1′ Ovule 1 per carpel; fruit a nutlet, mericarp, or berry; flowers small or inconspicuous .... (2)

2. Leaves opposite .... ***Kelloggia***
2′ Leaves appearing whorled .... ***Galium***

Galium L. Cleavers, Bedstraw

(Greek *gala*, milk, from use of *Galium verum* for curdling; the common name bedstraw from old use of vinelike plants with uncinate hairs to grip and bind up straw used for bedding) Annual or perennial herbs, small vines, subshrubs, or shrubs. **STEMS** squared in cross section. **LEAVES** generally opposite, but may appear whorled with presence of leaflike stipules. **INFL** a panicle or axillary cyme, often a subcyme of 3 flowers or solitary in leaf axils. **FLOWERS** perfect or imperfect; calyx absent; corolla usually rotate, sometimes campanulate or weakly tubular, usually white, cream, greenish to pink, or dark red, the petal lobes usually 4; style 1, deeply bifid. **FRUIT** a schizocarp of 2 nutlets, mericarps, or 2 berries; variously hirsute to glabrous. The fruits are critical in identifying the species. About 400 species worldwide, mostly from temperate regions. Two or more species often grow together in the same habitat and are often collected and mounted on the same herbarium sheet.

1. Fruit of 2 spherical, hard mericarps, black when mature, weakly joined; leaf apices rounded .... ***G. trifidum***
1′ Fruit various, but not as above; leaves not round at apex .... (2)

2. Leaves generally, on a single plant, more than 4 per node .... (3)
2′ Leaves 4 or fewer per node .... (4)

3. Perennial, generally prostrate; leaves narrowly to broadly ovoid, vanilla-scented; flowers white to cream .... ***G. triflorum***
3′ Annual, erect or climbing; leaves linear-spatulate or spatulate .... (7)

4. All or some flowers imperfect; inflorescences narrow and diffuse; fruits with straight, straw-colored hairs .... (5)
4′ All flowers perfect; inflorescence/fruit variable .... (6)

5. Plants dioecious; flowers white, cream, or pinkish .... ***G. coloradoense***
5′ Plants polygamous; inflorescence diffuse; flowers generally dark red, rarely pinkish or striped .... ***G. wrightii***

6. Inflorescence large, with numerous somewhat congested, whitish flowers .... ***G. boreale***
6' Inflorescence with solitary flower in leaf axil; pedicel long, reflexed .... ***G. bifolium***

7. Plants generally climbing; leaves linear-spatulate .... ***G. aparine***
7' Plants erect or ascending; lower leaves spatulate .... ***G. proliferum***

Galium aparine L. (old generic name meaning "to cling") Bedstraw, catchweed bedstraw, cleavers, cleaverwort, goose-grass, stickywilly. Clambering or prostrate annual, scabrous with prickly hairs. **STEMS** 10–90 cm or more long; nodes often tomentose. **LEAVES** 6–8 per node, ovate-spatulate to linear-oblanceolate, 13–31 mm long, bases narrow, apices mucronate, margins ciliate. **INFL** few-flowered on indeterminate lateral branchlets, 1–2 in a small cyme or solitary and axillary. **FLOWERS** perfect; corolla rotate, lobes obtuse or acute, white or yellowish, especially with age. **FRUIT** with short, uncinate, upturned hairs. $2n = 20$ (typical in our area), 22, 42, 44, 63, 64, 66, 86, 88. [*G. spurium* L.; *G. vaillantii* DC.]. Moist places in blackbrush, sagebrush, willow, piñon-juniper, aspen in shade or edges of forests; talus. Relatively uncommon in our area. COLO: Arc, Hin, Min, Mon; NMEX: SJn; UTAH. 1130–2600 m (3700–8400'). Flowering: Apr–Jun. Fruit: Jun–Jul. Western North America from southern British Columbia through the United States and Alaska. Possibly native to Europe.

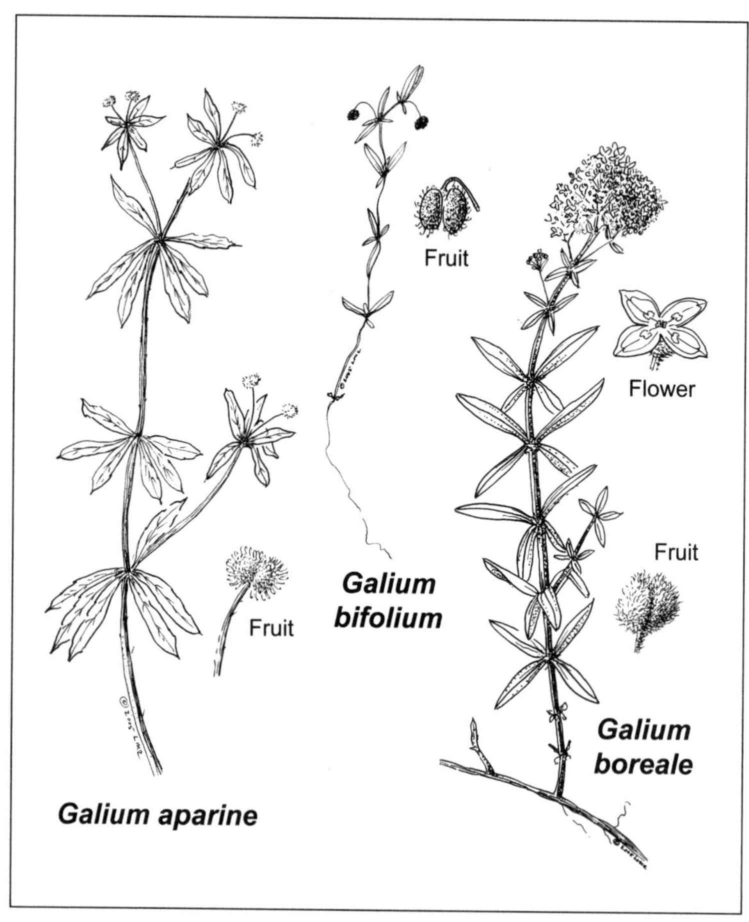

Galium bifolium S. Watson (two-leaved) Low mountain bedstraw, twinleaf bedstraw. Erect, slender, glabrous annuals. **STEMS** 5–18 cm high. **LEAVES** lanceolate or narrowly elliptic, acute, mainly 4 per node, in unequal pairs, the upper leaves occasionally in pairs; the larger leaves 0.4–2 cm long. **FLOWERS** perfect; solitary in leaf axil; pedicel long, reflexed below the fruit; corolla 3-lobed, lobes glabrous, cleft about 2/3 of length, white, ascending, obtuse, shorter than ovaries. **FRUIT** with short uncinate hairs; mericarps nearly separate at maturity. Shaded slopes and meadows. Rare, but may be mixed and/or misidentified with other *Galium* collections. ARIZ: Apa; COLO: Mon; NMEX: SJn. 1700–2800 m (5500–9200'). Flowering: May–Jun. Fruit: Jun–Aug. Western North America from British Columbia and Montana to southern California, Arizona, and New Mexico.

Galium boreale L. (northern) Northern bedstraw. Erect perennial, 3–6 dm high, nearly glabrous. **LEAVES** 4 per node, 13–31 mm long, linear to broadly lanceolate, 3-veined, minutely scabrous, apex obtuse. **INFL** a congested, terminal, nearly leafless, more or less pyramidal panicle of cymules. **FLOWERS** perfect, numerous; corolla cream or white, sometimes with faint purplish markings which fade in dried specimens, rotate or a little cupped at base, the 4 lobes ovate, the apices blunt; ovaries generally densely hairy with short, upwardly curved hairs. **FRUIT** dry, with short bristles, the mericarps becoming reniform. $2n = 44, 66$. [*G. septentrionale* Roem. & Schult.; *G. utahense* Eastw.]. Sagebrush, meadows, aspen, spruce-fir, streamsides in moist shade. ARIZ: Apa; COLO: Arc, Hin, LPl, Min, Mon, RGr; NMEX: RAr, SJn. 1770–3050 m (5800–10000'). Flowering: May–Sep. Fruit: Jul–Oct. Montane North America, east to Virginia, north to Maine; not in the southeastern United States; east Asia, Europe. One of our most common species of *Galium*. Plants from North America and east Asia are hexaploid ($2n = 66$), differing from those tetraploids of Europe and west Asia ($2n = 44$). Ours are **subsp. *septentrionale*** (Roem. & Schult.) H. Hara. Roots were used as a dye by Great Basin tribes.

Galium coloradoense W. F. Wright (of Colorado) Shrubby bedstraw, Colorado bedstraw. Plants perennial, mostly erect, sometimes shrubby, to 30 cm tall. **STEMS** slender, several from woody base, arising from creeping rhizomes, to about

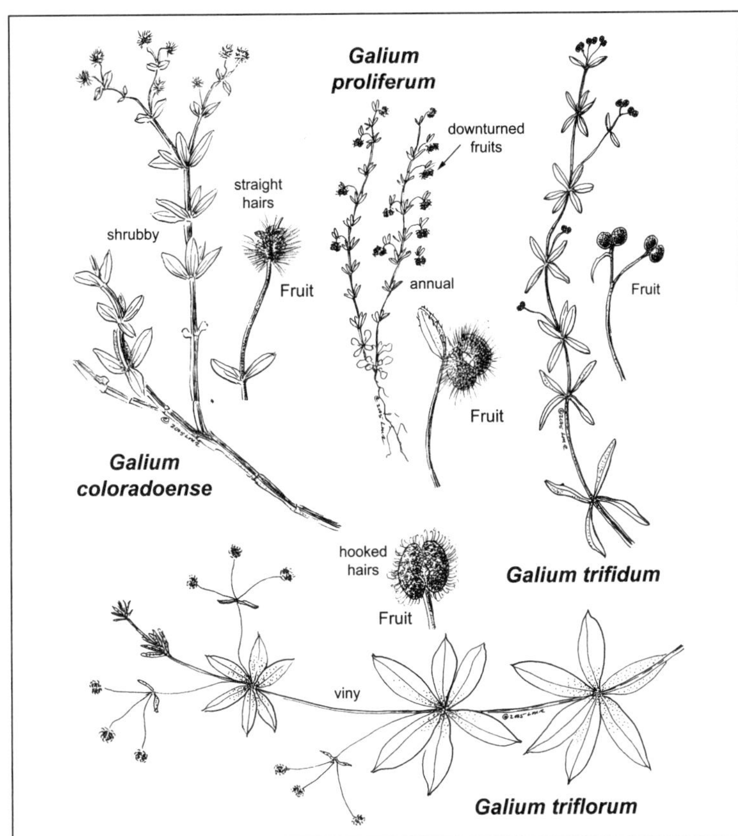

40 cm long, often woody and darker toward the base. **LEAVES** mostly 4 in whorl-like clusters, sessile, mostly linear to lanceolate, apex acute, often reflexed, 5–25 mm long, 2–3 mm wide, with prominent, often lighter midrib below, glabrous to scabrous. **INFL** small and few-flowered, borne on short axillary branches, occasionally larger and more congested. **FLOWERS** dioecious, pedicellate; corolla greenish yellow, inconspicuous, sometimes with purplish markings which fade in dried specimens, 2–4 mm wide, often scabrous-hispid externally. **FRUIT** with long, spreading, flattened, white or straw-colored hairs 1–3 mm long. $2n = 22$. [*G. multiflorum* Kellogg var. *coloradoense* (W. F. Wright) Cronquist]. Shaded, rocky or sandstone crevices and cliff locations in desert scrub, sagebrush, mountain brush, and piñon-juniper. Relatively common. ARIZ: Apa, Nav; COLO: Dol, LPl, Mon; NMEX: SJn; UTAH. 1460–3100 m (4800–10100′). Flowering: May–Aug. Fruit: May–Oct. Southern Wyoming, eastern Utah, western Colorado, northeastern Arizona, northwestern New Mexico. A highly variable species.

Galium proliferum A. Gray (bearing many offspring) Limestone bedstraw. Annual to 30 cm high, erect to stiffly ascending, glabrous to hairy. **LEAVES** mostly 4 per node; 3–9 mm long; the lower petiolate and spatulate; the upper more or less sessile, linear to obovate, more or less remote. **INFL** a small cyme with short branches at nodes or flowers solitary. **FLOWERS** perfect, minute; short-pedicelled; corolla tiny; lobes erect, apices rounded; white, often tipped with pink. **FRUIT** with 2 dark, kidney-shaped mericarps, downturned from 2 erect leaves; hairs long, straight or weakly uncinate. Damp locations near streambeds, washes, hanging gardens, or under shrubs. Uncommon. UTAH. 1620 m (5300′). Flowering: Mar–Jun. Fruit: Jun–Sep. Southern California east to Utah, Arizona, New Mexico, the Edwards Plateau and Trans-Pecos region of Texas, south through Baja California and Sonora, Mexico.

Galium trifidum L. (three-parted) Threepetal bedstraw. **STEMS** usually perennial, thin, lax; internodes widely spaced, 6–12 cm long; glabrous. **LEAVES** 2–6 per node, mostly 4, sessile, apex rounded, 4–12 mm long; linear to obovate; glabrous. **INFL** with 1–4 flowers, sometimes subtended with a minute bract or scale. **FLOWERS** perfect; pedicel slender, long; corolla rotate, white to greenish white, rarely striped with pink, generally 3-lobed, lobes ovate, ascending. **FRUIT** with 2 mericarps, spherical, glabrous or with a few short, flat hairs or scales, black or dark brown. $2n = 24$. [*G. brandegeei* A. Gray; *G. columbianum* Rydb.]. Margins of lakes, ponds, streams, mountain brush, ponderosa pine, spruce-fir. May be locally common at higher elevations. Often grows with or near *G. boreale*. ARIZ: Apa; COLO: Arc, Hin, LPl, Min, Mon, SJn; NMEX: SJn. 2470–3400 m (8100–11100′). Flowering: Jun–Jul. Fruit: Jun–Oct. Circumpolar, extending southward in North America along the Sierra Nevada and Rocky Mountains; Texas and Louisiana, east to Virginia. Our plants belong to **var. *pusillum*** A. Gray (very small).

Galium triflorum Michx. (three-flowered) Fragrant bedstraw, sweet-scented bedstraw. **STEMS** perennial, herbaceous, 10–50 cm long, decumbent or viny, growth radiating from a central rhizome or root. **LEAVES** generally 5–6 per node, 5–40 mm long, elliptic to ovate-obovate, base gradually narrowing, apex apiculate or mucronate, glabrous to scabrous. **INFL** 1–5-flowered, usually 3, pedicellate in axillary, open, often divaricate cymules with 2 subtending bracts. **FLOWERS** perfect; corolla rotate, cream-colored or greenish, ovary about as long as or longer than corolla. **FRUIT** covered with soft white, cream, or golden brown hooked hairs; mericarps dark-papillate or furrowed, curved. $2n = 22, 44, 66$. [*G. brachiatum* Pursh]. Shady forest floors or moist streamsides under brush, ponderosa pine, Douglas-fir, and aspen. A common species in our area. Often grows with or near *G. boreale*. ARIZ: Apa, Nav; COLO: Arc, Hin, LPl, Min, Mon, SJn; NMEX: RAr, SJn; UTAH. 1650–3290 m (5400–10800′). Flowering: May–Aug. Fruit: Jun–Sep. Circumboreal, all states including Alaska, and mountainous Mexico.

Galium wrightii A. Gray (for Charles Wright, noted botanist) Wright's bedstraw. Polygamous, erect, subshrubby perennials. **STEMS** suffrutescent, several to many, 5–50 cm high, glabrous to scabrous. **LEAVES** 4 per node, 2 often below flowers, 7–20 mm long, linear to oblanceolate with broad insertions, apex acute; glabrous, scabrous, or hispid, with hookless hairs. **INFL** diffuse, branchlets spreading. **FLOWERS** sometimes subtended by a single bract; corolla dark red to pink, rarely pale or striped, rotate, lobe tips slender, acute. Helpful to see flowers in this species. **FRUIT** often with reddish papillae, hairs few to many, long, straight. 2*n* = 22. [*G. rothrockii* A. Gray]. Moist, semishady slopes under ponderosa pine, Douglas-fir, and aspen, hanging gardens, cliffs. Uncommon, but may be locally abundant. ARIZ: Apa; NMEX: SJn. 1980–3820 m (6500–9900′). Flowering: Jun–Sep. Fruit: Jul–Oct. Texas to California, Nevada, and Utah, south to Baja California, Coahuila, and Sonora, Mexico.

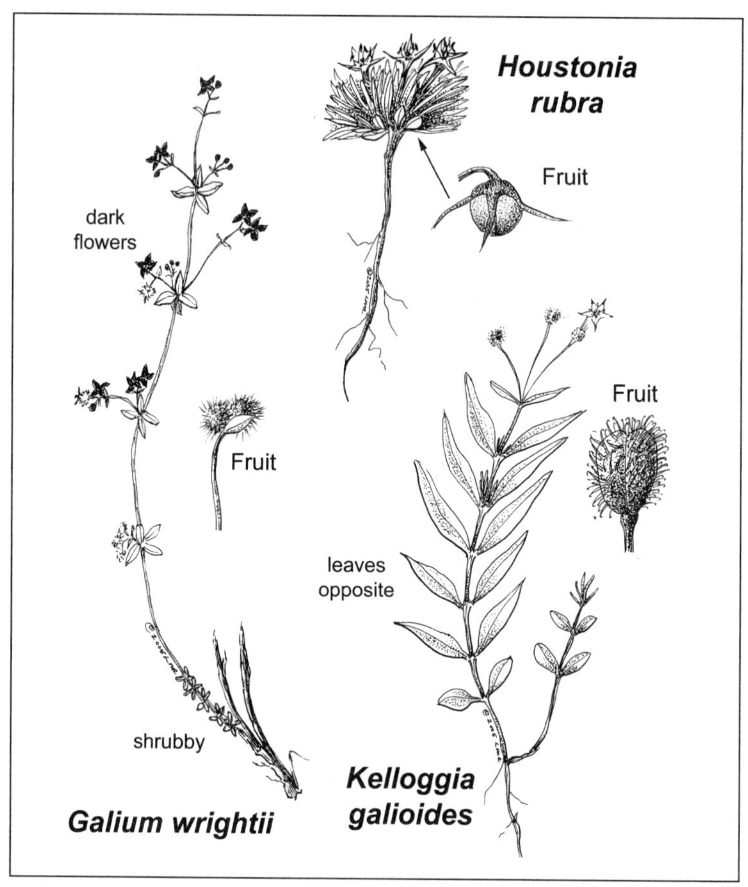

Houstonia L. Bluet

(for William Houston, who collected in tropical America) Small annual or perennial herbs. **STEMS** upright. **LEAVES** opposite, with interpetiolar stipules. **INFL** cymose or flowers solitary. **FLOWERS** perfect, calyx 4-lobed; corolla 4-lobed, funnelform or salverform; stamens 4; stigmas bifid. **FRUIT** a capsule, bilocular, inferior. **SEEDS** crateriform. Fifty species, United States and Canada to Mexico. Closely allied to the more tropical genus *Hedyotis*.

Houstonia rubra Cav. **CP** (red) Red bluet. Caespitose perennial with small woody taproots. **STEMS** 1–10 cm tall, slender. **LEAVES** linear to oblanceolate, 5–30 mm long, 0.5–4 mm wide, erect or ascending. **FLOWERS** usually 1 per node, sessile or with a short pedicel to 4 mm long, corolla 8–40 mm long, salverform, red or pink to purple or rarely, white; heterostylous. **FRUIT** 2–3.5 mm long, 2–5 mm wide, walls thick, on recurved pedicel. 2*n* = 22. [*Hedyotis rubra* (Cav.) A. Gray; *Houstonia saxicola* Eastw.]. Blackbrush, rocky grassland, piñon-juniper, washes. ARIZ: Apa, Nav; UTAH. 1680–2650 m (5500–8700′). Flowering: Apr–Oct. Fruit: May–Oct. Texas to Utah, south to southern Mexico. Puebloan tribes used an infusion of the plant for sore eyes and stomach upsets. Navajos used a decoction of the plant for menstrual problems.

Kelloggia Torr. ex Benth. Kelloggia

(for Albert Kellogg, 19th-century California botanist) Perennial herbs. **LEAVES** opposite, with small interpetiolar stipules. **INFL** a loose terminal cyme. **FLOWERS** 4–5-merous; corolla funnelform; ovary 2-loculed, each with 1 ovule. **FRUIT** a schizocarp splitting into 2 mericarps. Two species, one from western North America, the other from China.

Kelloggia galioides Torr. (resembling *Galium*) Milk kelloggia. **STEMS** erect, several, 15–40 cm long. **LEAVES** opposite, fascicled, 19–38 mm long, lanceolate to narrowly ovate, minutely serrulate, apex acute; stipules hyaline, 0.5–2 mm long, lanceolate, bifid or erose. **INFL** few-flowered. **FLOWERS** with pedicel filiform, 6–25 mm long, swelling below the junction with flower, becoming divergent; calyx curved inward toward the corolla; corolla 4–8 mm long, pink or white, outer surface pubescent, inner surface with tufts of hairs, tube slender, lobes spreading. **FRUIT** obovoid, 3 mm long, covered with flat, transparent, hooked hairs. Moist slopes, under piñon-juniper, sagebrush, mountain brush, Douglas-fir, ponderosa pine, or in meadows. ARIZ: Apa; NMEX: McK, SJn (Falling Iron Cliffs); UTAH. 1770–2670 m (5800–8700′). Flowering: Jun–Sep. Fruit: July–Oct. Washington south to Baja California along the coast; Rocky Mountains from Montana south to Arizona and New Mexico.

RUPPIACEAE Horan. DITCH-GRASS FAMILY

C. Barre Hellquist

Annual, rarely perennial, submerged aquatic herbs usually of brackish or saline water. Rhizomes lacking. **LEAVES** alternate or subopposite, submersed, sessile, vein 1, blade linear. **INFL** terminal, capitate, with subtending spathe; pedunculate, peduncle following fertilization elongates and coils. **FLOWERS** bisexual, perianth absent, stamens 2, anthers distinct; pistils 4–16, stipitate. **FRUIT** drupaceous. **SEEDS** 1 (Fernald and Wiegand 1914; Haynes 1978). One genus with about 10 species worldwide.

Ruppia L. Ditch-grass, Widgeon-grass

(for German botanist Heinrich Bernhard Ruppius, 1688–1719) Herbs rooting at proximal nodes. **LEAVES** with blades minutely serrulate distally, apex obtuse to acute; vein 1. **FLOWERS** bisexual, anthers 2-loculed, stipe elongating after anthesis. **FRUIT** with long, stipitate beak. Two species in North America. Seeds have been made into a meal.

Ruppia cirrhosa (Petagna) Grande (with many tendrils, curls) **STEMS** 0.1–0.3 mm wide. **LEAVES** 3.2–45.1 cm, blade 0.2–0.5 mm wide. **INFL** peduncle with 5–30 coils, 30–300 × 0.5 mm. **FLOWERS** with pistils 4–6. **FRUIT** 1.5–2 × 1.1–1.5 mm; on a gynophore 2–5 cm; beak 0.5–1 mm. $2n = 40$ (in Europe). [*Buccaferrea cirrhosa* Petagna; *Ruppia cirrhosa* subsp. *occidentalis* (S. Watson) Á. Löve & D. Löve; *R. occidentalis* S. Watson]. Brackish waters, or waters of high sulfur and/or calcium content, in ditches, ponds, and lakes. COLO: Arc, Dol, SMg; NMEX: RAr, SJn. 1700–2270 m (5600–7500′). Alaska east to Yukon, Michigan, and Ohio south to California, Arizona, New Mexico, and Texas; West Indies; Central America; South America; and Europe. Flowering: Apr–Jul. The inland *R. cirrhosa* has recently been recognized as different from the coastal *R. maritima* L. and conspecific with the European plants.

SALICACEAE Mirb. WILLOW FAMILY

Robert D. Dorn

Trees or shrubs. **LEAVES** simple, alternate (rarely opposite), entire or toothed, rarely lobed. **INFL** a catkin. **FLOWERS** unisexual, the plants dioecious or very rarely monoecious; perianth none or vestigial, each flower subtended by a bract and/or cupular disc or 1 or 2 nectar glands; stamens (1) 2–80; pistil 1, this 2–4-carpellate; ovary superior, sessile or stipitate, locule 1, style 1 (rarely none), stigmas 2–4. **FRUIT** a 2–4-valved capsule. **SEEDS** usually many, comose (long-hairy for wind dispersal). $x = 19$. The only family of order Salicales, with two genera and about 480 species. Worldwide except for Antarctica, Australasia, and Malaysia, mostly in Arctic and temperate regions of the Northern Hemisphere. Most species are pioneers on newly created, open, moist habitats where there is little competition. Once established, they can persist for a long time. Proliferation of the introduced saltcedar (*Tamarix* spp.) and Russian olive (*Elaeagnus angustifolia*) has greatly reduced the possibilities for willow and cottonwood reproduction along most lower-elevation streams in the Four Corners region. Heavy grazing in some areas has also made seedling establishment difficult. Water diversions and pond construction since settlement have changed habitats, eliminating some but creating others.

1. Bud scales more than 1; bracts subtending flowers usually fringed (often deciduous); catkins mostly hanging downward; stamens 6–80; trees .... ***Populus***

1′ Bud scale 1; bracts subtending flowers usually entire, rarely toothed or slightly lobed (sometimes deciduous); catkins mostly erect to spreading or drooping; stamens 2–8 (12); trees or shrubs .... ***Salix***

Populus L. Cottonwood, Aspen, Poplar, T'iis (Navajo), Söhövi (Hopi), Sho-ap (Ute), Álamo (Spanish)

(the classical Latin name for poplar) Trees. Bud scales several, often resinous. **LEAVES** mostly toothed, rarely lobed. **INFL** a catkin, mostly pendulous. **FLOWERS** usually subtended by a fringed bract as well as a cup-shaped disc; stamens 6–80. **FRUIT** a 2–4-valved capsule (Eckenwalder 1977, 1992). About 30 species, mostly in the Northern Hemisphere worldwide, about a dozen in North America. Two species are commonly planted at lower elevations in the Four Corners region but are not known to escape cultivation. Carolina poplar (*P.* ×*canadensis* Moench) is similar to *P. deltoides*, but the leaves are not nearly so deltoid and have finer teeth and a more acuminate tip. Lombardy poplar (*P. nigra* L. var. *italica* Du Roi) has suberect branches forming a narrowly conical growth form. The leaves are similar to those of *P. deltoides* but not nearly so deltoid and with finer teeth. Additional hybrids are sold in the horticultural trade. Cottonwoods were put to many uses by native peoples including fire drills, tinder boxes, firewood, saddle frames, cradleboards, snowshoes, shovels, hoes, lances, prayer sticks, sweathouse frames, carved ceremonial figurines, dice, scrapers, awls, bows, roof beams, loom frames, drums, and kachina dolls. Buds were used as chewing gum and food. Aspen was also used for some of these items. The inner bark, sap, and catkins were sometimes eaten.

1. Leaf blades white-tomentose on underside, margins coarsely toothed or lobed .... *P. alba*
1' Leaf blades usually glabrous or glabrate on underside, margins mostly toothed .... (2)

2. Leaf blades suborbicular or cordate to deltoid; petioles strongly laterally flattened just below blade .... (3)
2' Leaf blades mostly lanceolate or ovate; petioles usually not flattened .... (5)

3. Leaf blades mostly suborbicular or cordate; bark smooth, whitish green to nearly white .... *P. tremuloides*
3' Leaf blades mostly deltoid or nearly so; bark rough, usually dark .... (4)

4. Capsules ovoid, with saucer-shaped discs mostly 1.5–3.5 mm wide; young branchlets glabrous or sparsely hairy; blades of later-developing leaves often wider than long .... *P. deltoides*
4' Capsules mostly globose or ellipsoid, with cup-shaped discs mostly 3.5–9 mm wide; young branchlets often densely hairy; blades of later-developing leaves about as wide as long .... *P. fremontii*

5. Petioles mostly less than 1/3 the blade length; leaf blades mostly lanceolate or lance-ovate .... *P. angustifolia*
5' Petioles mostly over 1/3 the blade length; leaf blades mostly ovate .... *P. acuminata*

Populus acuminata Rydb. (narrowly pointed, referring to the leaf tip) Lanceleaf cottonwood. Tree to 25 m high, bark furrowed. **LEAVES** mostly ovate, crenate-serrate, the blades 5–9 (13) cm long, 3–7 cm wide. **CATKINS** 6.5–9 (16) cm long; stamens 25–40. **CAPSULES** 5–7 mm long, ovoid, 2- or 3-valved, stipes 1–2 (4) mm long. $2n = 38$. Stream banks, shores, and washes. ARIZ: Apa; COLO: Arc, Dol, LPl, Mon, SMg; NMEX: McK, RAr, SJn; UTAH. 1650–2600 m (5400–8500′). Flowering: May–Jun. Fruit: Jun. Alberta to western North Dakota south to northeast Arizona and Texas. Usually treated as a hybrid between *P. angustifolia* and *P. deltoides* but often occurring in the absence of one or both reputed parents. Probably better treated as a hybrid derived species.

Populus alba L. (white, referring to the tomentum of the leaves) Silver poplar. Tree to 30 m high, bark furrowed below. **LEAVES** ovate, undulate-toothed to deeply 3–5-lobed, usually white-tomentose on underside, the blades 3–5 (7) cm long, 2–4 (6) cm wide. **CATKINS** 2–9 cm long; stamens 6–10. **CAPSULES** 2–5 mm long, ovoid, 2- or 3-valved, stipes 0.5–1 (2) mm long. $2n = 38, 57$. Introduced from Europe and occasionally becoming naturalized, mostly along roadsides. ARIZ: Apa; COLO: Arc, Dol, LPl, Mon; NMEX: McK, SJn; UTAH. 1550–2000 m (5100–6600′). Flowering: May–Jun. Fruit: Jun.

Populus angustifolia E. James (narrow leaf) Narrowleaf cottonwood. Tree to 20 m high, bark furrowed. **LEAVES** lanceolate to ovate, finely crenate-serrate, the blades 2–7 (10) cm long, 0.8–2.5 (4) cm wide. **CATKINS** 3–8 cm long; stamens 10–20. **CAPSULES** 3–8 mm long, ovoid or orbiculoid, 2-valved, stipes 0.5–1.5 (3) mm long. $2n = 38$. Stream banks, seeps, and mountain slopes. ARIZ: Apa, Nav; COLO: Arc, Dol, Hin, LPl, Min, Mon, RGr, SJn, SMg; NMEX: McK, RAr, SJn; UTAH. 1550–2950 m (5100–9700′). Flowering: May–Jun. Fruit: Jun–Jul. Southern Alberta south to central Chihuahua and northwest Coahuila, mostly in or along the western cordillera.

Populus deltoides W. Bartram ex Marshall (Greek *delta*, triangle, + *oides*, resembling) Cottonwood. Tree to 20 m high, bark deeply furrowed; branchlets glabrous or nearly so. **LEAVES** deltoid to deltoid-ovate, coarsely toothed to almost lobed, the blades 3–8 cm long, usually slightly wider, acute to short-acuminate. **CATKINS** 5–13 (21) cm long. Stamens 50–80. **CAPSULES** (4) 8–16 mm long, ovoid, 3- or 4-valved, floral disc 1–3.5 (4) mm wide, saucer-shaped (shallowly cup-shaped), stipes (1) 5–15 mm long. $2n = 38$. Stream banks, shores, and washes. ARIZ: Apa, Nav; COLO: Arc, Dol, LPl, Mon, SMg; NMEX: McK, RAr, San, SJn; UTAH. 1150–2150 m (3800–7100′). Flowering: Apr–Jun. Fruit: May–Jun. Southwest Alberta to southern Quebec south to northern Chihuahua and Florida. Our plants are **var.** ***wislizeni*** (S. Watson) Dorn (for Friedrich Adolph Wislizenus, 1810–1889, a St. Louis physician who collected the variety in New Mexico). This is the variety from northeast Utah to south-central Wyoming south to northern Chihuahua and southwest Texas. Basin cottonwood.

Populus fremontii S. Watson (for John Charles Frémont, 1813–1890, army officer and western explorer who collected the species in California) Fremont cottonwood. Tree to 30 m high, bark deeply furrowed; year-old branchlets mostly pubescent. **LEAVES** deltoid or deltoid-cordate (deltoid-ovate), mostly coarsely round-toothed to almost lobed, the blades (4) 5–11 (14) cm long and about as wide, acute to short-acuminate (long-acuminate). **CATKINS** 4–13 cm long; stamens 30–70; floral disc 3.5–9 mm wide, deeply cup-shaped, stipes about (1) 3–4 mm long. **CAPSULES** 6–10 mm long, globose or ellipsoid (ovoid), usually (3) 4-valved, $2n = 38$. Stream banks, shores, and washes. UTAH. 1150–1250 m (3800–4100′). Flowering: Apr–May. Fruit: May–Jun. Our plants are **var.** ***fremontii***, the variety from northern California to northern Utah south to Baja California Sur and central Sonora. Another variety extends south to the Valley of Mexico. Presence of this species in the study area needs to be confirmed with a pistillate collection. There is some apparent intergrading with *P. deltoides*.

Populus tremuloides Michx. (referring to the "trembling" or "quaking" of the leaves in the slightest breeze) Quaking aspen. Tree to 30 m high but usually to 1/2 that, forming clones; bark mostly smooth and whitish green to nearly white. **LEAVES** cordate (ovate) to suborbicular, finely crenate-serrate, the blades 2–7 cm long and about as wide. **CATKINS** 2–10 cm long; stamens 6–12. **CAPSULES** 4–7 mm long, lanceoloid, 2-valved, stipes 1–2 mm long. $2n = 38, 57, 76$. Moist areas in the mountains and in steep shaded canyons at lower elevations. ARIZ: Apa, Nav; COLO: Arc, Hin, LPl, Min, Mon, RGr, SJn; NMEX: McK, RAr, San, SJn; UTAH. 1900–3250 m (6200–10700′). Flowering: May–Jun. Fruit: Jun–Jul. Throughout much of temperate North America from within the Arctic Circle to the mountains of central Mexico.

Salix L. Willow, K'ai' (Navajo), Qahavi (Hopi), Kan-nab (Ute), Sauce (Spanish)

(the classical Latin name for willow) Trees (lower elevations) to shrubs to tiny, creeping, Arctic-alpine subshrubs. Bud scales solitary, nonresinous. **LEAVES** entire or toothed, usually stipulate but the stipules often deciduous. **INFL** a mostly spreading or drooping to erect catkin, the catkins sessile or on floriferous branchlets (peduncles) that are usually leafy. **FLOWERS** subtended by an entire or rarely toothed to erose bract and 1 or 2 nectaries; stamens (1) 2–8 (12). **FRUIT** a 2-valved capsule (Argus 1995, 1997; Dorn 1995, 1997, 1998, 2000). About 450 species worldwide but lacking from Antarctica, Australasia, and Malaysia, only one species from South America, nearly 100 in North America. Two species are often planted at lower elevations in the Four Corners region but are not known to escape cultivation: *S. babylonica* L., the weeping willow, is a tree with long, pendulous branches and lance-linear or narrowly lanceolate, glaucous leaves; *S. matsudana* Koidz., the globe willow (cultivar 'Umbraculifera'), is a small tree with a graceful, spherical crown and narrowly lanceolate, glaucous leaves. Except for growth form, the two are very similar and some authors consider them the same species. *Salix alba* L. var. *vitellina* (L.) Stokes has been sparingly planted, as along Chinle Wash at Many Farms, Apache County, Arizona. It differs from *S. babylonica* in having shorter, somewhat drooping branchlets rather than long, pendent branchlets, broader leaves, and the pistillate floral bracts are deciduous in fruit rather than persistent. It is separated from *S. fragilis* by its persistently pubescent leaves and shorter styles to 0.2 mm long. The specimens of *S. exilifolia* Dorn (*S. taxifolia* Humb., Bonpl. & Kunth) reported from San Juan County, New Mexico, were actually collected in Cochise County, Arizona. The use of willows by native peoples is not well documented by species, probably due to confusion of identity. The coyote willow, *S. exigua*, is the species most mentioned, likely because of its abundance and wide distribution at lower elevations. It is also one of the more easily identifiable willows. Navajos have used it in the ceremonials of the Lightning Way and Big Star Way as a medicine and tobacco. In addition, the leaves were used for medicinal purposes by soaking in water and the liquid then used as an emetic. The wood was used for weaving sticks and arrow shafts. The wood was also used for firewood when foods were cooked in ashes. Ancestral Puebloans used the wood for textile loom anchors, rods to control the weaving rhythm, for finishing needles, bows, arrow points, pot rests, scrapers, basket weaving, and cradle parts. The Utes used the branches for snowshoe frames and the basketry covering on the cradleboard. Mats were also made from willow branches. Leaves and roots have been used to produce dyes. Medicinal teas brewed from willow bark have been used to treat upset stomachs, headaches, fevers, and inflammations. The active ingredient is likely salicylic acid, a derivative of which is the main ingredient of aspirin. Other willow uses included prayer sticks, lances, and roof beams. Willow sticks were used for the Night Chant and Mountain Chant by the Diné (Navajo).

For best results in the following keys, avoid sucker shoots with vigorous growth, severely browsed shoots, young catkins, and leaves in early stages of development. *Salix arizonica* Dorn and *S. boothii* Dorn have both been collected within about 16 km of the study area in Conejos County, Colorado. These two species will key to *S. wolfii*, which has entire, persistently pubescent leaves rather than glandular-serrate (occasionally entire in *S. boothii*), glabrous or glabrate leaves when mature. The catkins of *S. wolfii* are generally shorter.

1. Vegetative material to be used for identification .... **Vegetative Key**
1′ Reproductive material to be used for identification .... (2)

2. Reproductive material with female catkins .... **Pistillate Key**
2′ Reproductive material with male catkins .... **Staminate Key**

Pistillate Key

1. Capsules hairy .... (2)
1′ Capsules glabrous .... (12)

2. Plants creeping shrubs 1–10 cm high with branchlets usually rooting, near or above timberline .... (3)
2′ Plants erect shrubs or trees mostly well over 10 cm high, only occasionally above timberline .... (4)

3. Peduncles prominent, usually without leaves; leaf tip usually rounded, the blade somewhat leathery and prominently reticulate-veined on underside; styles less than 0.5 mm long .... ***S. reticulata***
3′ Peduncles usually leafy or very short; leaf tip usually pointed, the blade not leathery, venation reticulate but not especially prominent; styles 0.5–2 mm long .... ***S. arctica***

4. Flower bracts yellowish, greenish, whitish, or tawny, deciduous in fruit; leaves linear to narrowly lanceolate or elliptic, nonglaucous .... (5)
4′ Flower bracts only occasionally yellowish, greenish, whitish, or tawny, usually brown or black, persistent in fruit; leaves often broader, generally glaucous or glaucescent on underside .... (6)

5. Petioles prominent, slender; bud scales split down the side toward branch with the free margins overlapping; leaves mostly finely serrulate .... ***S. gooddingii***
5′ Petioles short and thick (rarely to 6 mm long) or lacking; bud scales caplike, not split down the side; leaves remotely denticulate or serrulate to entire .... ***S. exigua***

6. Branchlets of previous year, and sometimes those of current season, glaucous, sometimes only apparent at nodes, especially behind buds .... (7)
6′ Branchlets not glaucous .... (8)

7. Catkins 15–60 mm long, sessile or nearly so, densely flowered; leaves often densely silver-hairy on underside, glabrous or glabrate on upper side; flower bracts usually dark .... ***S. drummondiana***
7′ Catkins 6–20 (25) mm long, on leafy branchlets 1–18 mm long, loosely flowered; leaves sparsely to moderately sericeous on 1 or both sides; flower bracts tawny or light brown in fruit .... ***S. geyeriana***

8. Stipes mostly 2–5 mm long; styles 0.4 mm or less long; flower bracts light brown or tawny; buds often with depressed margins .... ***S. bebbiana***
8′ Stipes 2 mm or less long, or if as long as 3 mm, the styles often over 0.4 mm long; flower bracts tawny or dark; buds without depressed margins .... (9)

9. Plants mostly to 1.5 m high; catkins appearing with the leaves on leafy branchlets 2–35 mm long; leaves and branchlets often conspicuously hairy .... (10)
9′ Plants sometimes over 1.5 m high; catkins appearing before or with the leaves, sessile or nearly so or sometimes on branchlets to 13 mm long; leaves and branchlets often sparsely hairy to glabrous .... (11)

10. Catkins 0.5–2 (3) cm long; stipes less than 0.5 mm long; expanded leaf blades 2–4 cm long, petioles mostly 1–3 mm long .... ***S. brachycarpa***
10′ Catkins (2) 3–5 cm long; stipes 0.2–1.8 mm long; expanded leaf blades 3–8 cm long, petioles mostly over 3 mm long .... ***S. glauca***

11. Stipes 0–1 mm long; leaves elliptic, usually very shiny on upper surface; year-old branchlets often reddish and shiny; stigmas usually less than 0.5 mm long; plants of wet places .... ***S. planifolia***
11′ Stipes 0.8–2.8 mm long; leaves predominantly oblanceolate to obovate, not especially shiny; year-old branchlets mostly yellowish to reddish brown, dull; stigmas usually over 0.5 mm long; plants of drier upland areas .... ***S. scouleriana***

12. Flower bracts yellowish, greenish, whitish, or tawny, deciduous in fruit; catkins on leafy floriferous branchlets; styles mostly less than 1 mm long .... (13)
12′ Flower bracts often blackish or brown, persistent in fruit; catkins sessile or on leafy floriferous branchlets; styles 0.3–2 mm long .... (19)

13. Bud scales split down the side toward branch, with the free margins overlapping; native trees at lower elevations .... (14)
13′ Bud scales caplike, not split down the side; mostly shrubs or an introduced tree with very brittle branchlets .... (16)

14. Leaves not glaucous on underside .... ***S. gooddingii***
14' Leaves glaucous on underside .... (15)

15. Year-old branchlets mostly yellowish to grayish; branchlets of year mostly glabrous; later-expanded leaves mostly lanceolate to ovate with acuminate tips .... ***S. amygdaloides***
15' Year-old branchlets mostly reddish, purplish, or brownish; branchlets of year sometimes hairy; later-expanded leaves lanceolate to narrowly elliptic with mostly acute (acuminate) tips .... ***S. bonplandiana***

16. Leaf blades predominantly linear to narrowly elliptic, remotely denticulate or serrulate to entire, usually hairy; petioles rather short and thick (rarely to 6 mm long) or lacking; catkins sometimes 2 to several in a terminal cluster in addition to lateral solitary catkins .... ***S. exigua***
16' Leaf blades predominantly lanceolate to ovate or broadly elliptic, mostly closely serrate or serrulate, often glabrous; petioles generally well developed and slender; catkins all lateral and solitary .... (17)

17. Plants naturalized trees with very brittle branchlets that are easily broken off; catkins mostly slender and elongate, mostly 1 cm or less wide .... ***S. fragilis***
17' Plants native shrubs, the branchlets not especially brittle; catkins often thick and short, mostly 1 cm or more wide .... (18)

18. Capsules mostly (6) 7–12 mm long when mature, somewhat shiny, maturing in phenological summer; catkins (1) 2–4 (5) cm long; leaves glabrous even when young (except first one emerging from the bud); stipules lacking or merely glands .... ***S. serissima***
18' Capsules mostly 7 mm or less long, usually dull, maturing in phenological spring; catkins 1.7–10 cm long; young leaves often hairy; stipules usually developed, occasionally deciduous, rarely reduced to glands .... ***S. lasiandra***

19. Leaves not glaucous on underside .... ***S. wolfii***
19' Leaves glaucous on underside (catkins sometimes appearing before the leaves) .... (20)

20. Year-old branchlets glaucous (sometimes obscurely); flower bracts mostly obovate to suborbicular and densely fringed with long, relatively straight hairs; catkins mostly appearing before the leaves, sessile or on branchlets to 5 mm long; expanded leaves glabrous or essentially so .... ***S. irrorata***
20' Year-old branchlets not glaucous; flower bracts mostly narrower with often sparser, shorter, or curly hairs or glabrate; catkins appearing before or with the leaves, sessile or on branchlets to 17 mm long; expanded leaves glabrous or hairy .... (21)

21. Flower bracts densely fringed with relatively straight, untangled hairs; catkins mostly appearing before the leaves, sessile or on branchlets to 7 mm long; typical buds short and plump, barely longer than wide; expanded leaf blades oblong, narrowly elliptic, or oblanceolate (obovate), usually hairy at least on underside .... ***S. lasiolepis***
21' Flower bracts with often sparse hairs or with curly or tangled hairs; catkins appearing before or with the leaves, usually on prominent branchlets; buds often elongate; expanded leaf blades mostly lanceolate or elliptic to ovate (obovate or oblong), sometimes glabrous or glabrate .... (22)

22. Styles averaging 0.7 mm or more long; leaf blades tending to be less than 3 times as long as wide; hairs of floral bracts long and straight or curly .... ***S. monticola***
22' Styles averaging 0.7 mm or less long; leaf blades tending to be over 3 times as long as wide; hairs of floral bracts crinkly and tangled mostly toward base of bract or sometimes bracts glabrate .... ***S. eriocephala***

Staminate Key

1. Stamens 3–8 per flower .... (2)
1' Stamens 2 per flower .... (6)

2. Bud scales split down the side toward branch, with the free margins overlapping; catkins generally slender and lax; trees at lower elevations .... (3)
2' Bud scales caplike, not split down the side; catkins generally thick and stiff; shrubs .... (5)

3. Leaves not glaucous .... ***S. gooddingii***
3' Leaves glaucous on underside .... (4)

4. Year-old branchlets mostly yellowish to grayish; branchlets of current year mostly glabrous .... ***S. amygdaloides***
4′ Year-old branchlets mostly reddish, purplish, or brownish; branchlets of current year sometimes hairy .... ***S. bonplandiana***

5. Leaves glabrous even when young (except first one emerging from the bud); stipules lacking or merely glands; catkins (1) 2–4 (5) cm long .... ***S. serissima***
5′ Leaves often hairy when young; stipules usually developed, occasionally deciduous, rarely reduced to glands; catkins 1.7–10 cm long .... ***S. lasiandra***

6. Plants creeping shrubs to 10 cm high with branchlets usually rooting, near or above timberline .... (7)
6′ Plants upright trees or shrubs mostly well over 10 cm high, only occasionally above timberline .... (8)

7. Catkins pseudoterminal opposite the terminal leaf, the prominent peduncle usually without leaves; leaves mostly oval to suborbicular .... ***S. reticulata***
7′ Catkins often borne laterally, the peduncle usually leafy or very short; leaves mostly elliptic .... ***S. arctica***

8. Leaves glaucous on underside (catkins sometimes appearing before the leaves) .... (9)
8′ Leaves not glaucous .... (20)

9. Year-old branchlets glaucous, sometimes apparent only behind buds, branchlets of current year sometimes also glaucous .... (10)
9′ Year-old branchlets and branchlets of current year not glaucous .... (12)

10. Catkins mostly appearing with the leaves, on leafy branchlets; flower bracts mostly tawny or light brown .... ***S. geyeriana***
10′ Catkins appearing before the leaves, sessile or subsessile; flower bracts usually dark .... (11)

11. Flower bracts mostly obovate (oblong), dark brown with rather dense, long, relatively straight hairs conspicuously exceeding bract tip; leaves soon becoming glabrous or nearly so .... ***S. irrorata***
11′ Flower bracts often narrower or darker or less hairy; leaves remaining mostly densely hairy on underside .... ***S. drummondiana***

12. Plants naturalized trees with very brittle branchlets that are easily broken off; flower bracts yellowish, greenish, whitish, or tawny; expanded leaf blades mostly lance-elliptic or lanceolate and glabrous .... ***S. fragilis***
12′ Plants usually shrubs, the branchlets usually not very brittle; flower bracts often blackish or brownish, occasionally lighter; expanded leaf blades variously shaped .... (13)

13. Flower bracts mostly brown to black, obovate to oval (oblong), densely fringed with straight, untangled hairs; leaves hairy at least on underside; typical buds mostly short and plump, barely longer than wide; anthers mostly 0.8 mm or less long; lowlands .... ***S. lasiolepis***
13′ Flower bracts sometimes lighter, mostly narrower or pointed at tip, with often sparse hairs or with curly or tangled hairs; leaves sometimes glabrous or glabrate; buds variable; anthers sometimes over 0.8 mm long .... (14)

14. Nectaries generally 2 or more, both dorsal and ventral; expanded leaves mostly hairy; catkins coetaneous, terminating leafy branchlets; mostly high mountains .... (15)
14′ Nectaries generally 1, ventral; leaves sometimes glabrous or glabrate; catkins sometimes precocious or subprecocious and mostly sessile or subsessile; sometimes lower in mountains .... (16)

15. Petioles mostly 1–3 mm long; catkins mostly less than twice as long as wide, 0.5–2 (3) cm long; expanded leaf blades 2–3 (4) cm long .... ***S. brachycarpa***
15′ Petioles mostly 3–16 mm long; catkins mostly over twice as long as wide, (2) 3–4 (5) cm long; expanded leaf blades generally 3–6 (8) cm long .... ***S. glauca***

16. Young emerging leaves with some reddish hairs, especially on margins or at tip; flower bracts often with tawny hairs; mostly wet sites .... ***S. planifolia***
16′ Young emerging leaves with white hairs, or if with some reddish, then flower bracts with white hairs; sometimes in drier upland sites .... (17)

17. Bracts at base of catkin densely fringed with long hairs, often tawny; filaments often 8 mm or more long; plants of drier uplands .... ***S. scouleriana***

17′ Bracts at base of catkin becoming greenish and glabrous or glabrate, at least at tip, rarely lacking; filaments rarely as much as 8 mm long; mostly wet sites ....(18)

18. Flower bracts pale; buds generally with depressed margins; 2-year-old branchlets usually with cracked bark giving a white-streaking appearance .... ***S. bebbiana***
18′ Flower bracts usually dark but occasionally pale; buds without depressed margins; 2-year-old branchlets variable ....(19)

19. Leaf blades tending to be less than 3 times as long as wide; flower bracts generally dark and prominent with long, straight or curly hairs .... ***S. monticola***
19′ Leaf blades tending to be over 3 times as long as wide; flower bracts often brownish and inconspicuous with short, tangled, crinkly hairs mostly toward base of bract or sometimes bracts glabrate (hairs on catkin rachis may be long and tangled) .... ***S. eriocephala***

20. Flower bracts yellowish, whitish, greenish, or tawny; expanded leaf blades predominantly linear or linear-elliptic, remotely denticulate or serrulate to entire; catkins occasionally 2 to several in a terminal cluster in addition to lateral solitary catkins; petioles short and thick (rarely to 6 mm long) or none .... ***S. exigua***
20′ Flower bracts blackish or brownish; expanded leaf blades predominantly lanceolate, oblanceolate, or elliptic, entire; catkins all lateral and solitary; petioles mostly well developed and slender .... ***S. wolfii***

Vegetative Key

1. Leaves glaucous on underside, or rarely underside much lighter from dense hairs that obscure leaf surface ....(2)
1′ Leaves not glaucous ....(22)

2. Plants creeping shrubs to 10 cm high with usually rooting branchlets, near or above timberline ....(3)
2′ Plants upright trees or shrubs usually well over 10 cm high, only occasionally above timberline ....(4)

3. Leaf blades predominantly oval to suborbicular with prominent reticulate venation on underside .... ***S. reticulata***
3′ Leaf blades predominantly elliptic with reticulate venation not especially prominent .... ***S. arctica***

4. Bud scales split down the side toward branch, with the free margins overlapping; native lowland trees ....(5)
4′ Bud scales caplike, not split down the side; mostly shrubs or naturalized trees ....(6)

5. Year-old branchlets mostly yellowish to grayish; branchlets of current year mostly glabrous; expanded leaf blades mostly lanceolate to ovate and acuminate .... ***S. amygdaloides***
5′ Year-old branchlets mostly reddish, purplish, or brownish; branchlets of current year sometimes hairy; expanded leaf blades mostly lanceolate to narrowly elliptic and acute (acuminate) .... ***S. bonplandiana***

6. Plants naturalized trees with very brittle branchlets that are easily broken off; expanded leaf blades predominantly lanceolate or lance-elliptic .... ***S. fragilis***
6′ Plants usually shrubs, the branchlets not especially brittle; expanded leaf blades variable ....(7)

7. Year-old branchlets, and sometimes branchlets of current year, glaucous, sometimes only apparent behind buds ....(8)
7′ Year-old branchlets and branchlets of current year not glaucous ....(10)

8. Expanded leaf blades glabrous or essentially so .... ***S. irrorata***
8′ Expanded leaf blades hairy at least on underside ....(9)

9. Expanded leaf blades with silver hairs on underside that are often so dense as to obscure the leaf surface; leaf margins slightly revolute, the blades 1–2.6 cm wide .... ***S. drummondiana***
9′ Expanded leaf blades sparsely to moderately sericeous at least on underside; leaf margins not revolute, the blades 0.6–1.5 cm wide .... ***S. geyeriana***

10. Branchlets of current year usually red-purple and appressed-hairy; bark of 2-year-old branchlets cracked, giving a white-streaking appearance; buds often with depressed margins .... ***S. bebbiana***
10′ Branchlets and buds not as above ....(11)

11. Plants with mostly oblanceolate to obovate leaf blades; freshly stripped bark of living branchlets of previous year usually with a "skunky" odor; large shrubs of drier upland sites .... ***S. scouleriana***
11′ Plants not as above ....(12)

12. Most leaves entire or nearly so ....(13)
12′ Most leaves toothed ....(17)

13. Expanded leaves glabrous or nearly so ....(14)
13′ Expanded leaves usually obviously hairy, rarely glabrate ....(15)

14. Year-old branchlets usually reddish and shiny; upper side of leaf surface shiny .... ***S. planifolia***
14′ Year-old branchlets mostly reddish brown, yellowish, or greenish and dull; upper side of leaf surface dull .... ***S. eriocephala***

15. Leaf blades mostly oblong or narrowly elliptic to oblanceolate (obovate), mostly 3.5–12.5 cm long; lowland shrub .... ***S. lasiolepis***
15′ Leaf blades mostly elliptic to oblanceolate or elliptic-obovate, mostly 2–6 (8) cm long; mountain shrubs ....(16)

16. Expanded leaf blades mostly 3–8 cm long, usually glabrous to moderately hairy; petioles mostly over 3 mm long .... ***S. glauca***
16′ Expanded leaf blades mostly 2–4 cm long, usually densely hairy; petioles mostly less than 3 mm long .... ***S. brachycarpa***

17. Leaf blades mostly elliptic, dark green and shiny on upper side; year-old branchlets usually reddish and shiny .... ***S. planifolia***
17′ Leaf blades mostly lanceolate to ovate or obovate, if elliptic, the leaves not shiny on upper side and the year-old branchlets not reddish and shiny ....(18)

18. Petioles usually with glands near base of leaf blade; leaf tips mostly acuminate ....(19)
18′ Petioles usually lacking glands; leaf tips mostly acute to rounded ....(20)

19. Leaves glabrous even when young (except first one emerging from the bud); stipules lacking or merely glands .... ***S. serissima***
19′ Leaves often hairy at least when young; stipules usually developed, occasionally deciduous, rarely reduced to glands .... ***S. lasiandra***

20. Petioles often reddish; leaf blades usually ovate, obovate, or broadly elliptic; branchlets of current year hairy .... ***S. monticola***
20′ Petioles usually green; leaf blades often lanceolate, elliptic, or oblanceolate; branchlets of current year often glabrous ....(21)

21. Leaf blades oblong or narrowly elliptic to oblanceolate (obovate), usually hairy at least on underside; lowland shrubs .... ***S. lasiolepis***
21′ Leaf blades predominantly lanceolate or lance-ovate to elliptic, occasionally oblong, glabrous or glabrate when expanded; shrubs mostly in foothills and mountains .... ***S. eriocephala***

22. Bud scales split down the side toward branch, with the free margins overlapping; lower-elevation tree... ***S. gooddingii***
22′ Bud scales caplike, not split down the side; mostly shrubs ....(23)

23. Leaf blades mostly linear or linear-elliptic, remotely denticulate or serrulate to entire; petioles short and thick (rarely to 6 mm long) or none .... ***S. exigua***
23′ Leaf blades mostly broader, closely toothed to entire; petioles usually well developed and slender ....(24)

24. Petioles usually lacking glands; leaves mostly persistently hairy, the tips mostly acute to obtuse .... ***S. wolfii***
24′ Petioles usually with glands near base of leaf blade; leaves usually glabrous when expanded, the tips mostly acuminate ....(25)

25. Leaves glabrous even when young (except first one emerging from the bud); stipules lacking or reduced to glands .... ***S. serissima***
25′ Leaves often hairy at least when young; stipules usually developed, occasionally deciduous, rarely reduced to glands .... ***S. lasiandra***

Salix amygdaloides Andersson (resembling a peach) Peachleaf willow. Tree to 15 (30) m high; year-old branchlets mostly yellowish to grayish, glabrous. **LEAVES** mostly lanceolate to ovate; petioles 5–21 mm long; the blades 5.5–13 cm long, (0.7) 1–3.7 cm wide, base rounded to acute, tip acuminate, serrulate, glaucous on underside, glabrous or becoming so. **CATKINS** coetaneous, 2.5–11 cm long, rather lax, on leafy branchlets 0.4–3.5 (6) cm long; flower bracts greenish, whitish, or yellowish to tawny, deciduous in fruit, with long, crinkly hairs denser toward base; stamens 5–8. **CAPSULES** glabrous, 3–5.5 mm long, styles 0.3–0.6 mm long, stipes 1.2–3.2 mm long. $2n = 38$. Stream banks, floodplains, shores, marshes, and seeps. ARIZ: Apa; COLO: Arc, Dol, LPl, Mon; NMEX: McK, SJn; UTAH. 1150–2350 m (3800–7700′). Flowering: Apr–Jun. Fruit: May–Jun. British Columbia to southern Quebec, south to Arizona, northern Chihuahua, and western Kentucky. Hybridizes with *S. gooddingii* (*S.* ×*wrightii* Andersson).

Salix arctica Pall. (northern, referring to its Arctic location) Arctic willow. Creeping shrub less than 10 cm high; year-old branchlets brown or reddish brown (yellow-brown), glabrous or glabrate. **LEAVES** mostly elliptic (oval); petioles 2–6 (15) mm long; the blades (1) 1.5–4 cm long, 0.4–1.5 (2) cm wide, base acute to obtuse, tip acute to obtuse, entire, glaucous on underside, either glabrous or pubescent mostly on underside. **CATKINS** coetaneous, (0.7) 1–5.5 cm long, on leafy branchlets (0.5) 1–3 (5.5) cm long; flower bracts brown (black), with long, relatively straight hairs; stamens 2. **CAPSULES** pubescent, 3–6 (7) mm long, styles 0.5–2 mm long, stipes 0–0.8 mm long. $2n = 76, 114$. [*S. petrophila* Rydb.]. Alpine and subalpine. COLO: Arc, Con, Hin, LPl, Min, Mon, RGr, SJn. 3350–3950 m (11000–13000′). Flowering: Jun–Aug. Fruit: Jul–Sep. Alaska to Greenland, south to California, northern New Mexico, and Quebec; Eurasia. Our plants are **var.** ***petraea*** (Andersson) Bebb (rock, referring to its rocky habitat). Alpine willow. $2n = 76$. The variety from British Columbia and Alberta south to California and northern New Mexico.

Salix bebbiana Sarg. (for Michael S. Bebb, 1833–1895, an American willow specialist) Bebb willow. Shrub or small tree to 10 m high; year-old branchlets reddish brown or reddish purple, pubescent to glabrate. **LEAVES** elliptic or sometimes ovate to obovate; petioles 2–15 mm long; the blades (2) 4–8 cm long, 1–3.3 cm wide, base acute to obtuse or rounded, tip acute or rarely obtuse, crenate or irregularly serrate to entire, glaucous on underside, pubescent to glabrate. **CATKINS** coetaneous or subprecocious, 0.6–6 cm long, on leafy branchlets 0.1–6 cm long; flower bracts light brown or tawny, with straight or slightly wavy hairs; stamens 2. **CAPSULES** pubescent, strongly beaked, 5–9 mm long, styles 0.1–0.4 mm long, stipes 2–5 mm long. $2n = 38$. Swamp edges, moist woods, stream banks, and meadows. ARIZ: Apa; COLO: Arc, Hin, LPl, Min, Mon, SJn; NMEX: McK, RAr, SJn; UTAH. 2150–3300 m (7000–10800′). Flowering: May–Jun. Fruit: Jun–Aug. Alaska to Newfoundland, south to northern California, New Mexico, and Maryland; Eurasia.

Salix bonplandiana Kunth (for A. J. Alexandre Bonpland, 1773–1858, a French naturalist who collected the species in Mexico) Bonpland willow. Tree to 15 (20) m high. Year-old branchlets reddish, purplish, or brownish (greenish), glabrous or pubescent. **LEAVES** narrowly elliptic to lanceolate; petioles 3–14 (18) mm long; the blades 5–17 (19) cm long, (0.7) 1–3 (4) cm wide, base acute to subcordate, tip acute or acuminate, serrulate to entire, glaucous on underside, glabrous or becoming so. **CATKINS** coetaneous, (1.5) 3–7.5 (9) cm long, often not very lax, on leafy branchlets (0.4) 0.5–4 cm long; flower bracts greenish, whitish, or yellowish to tawny, deciduous in fruit, with crinkly hairs especially toward base; stamens 5–8. **CAPSULES** glabrous, 3–6 mm long, styles 0–0.4 mm long, stipes 0.8–2.8 (3.5) mm long. $2n = 38$. [*S. laevigata* Bebb]. Stream banks, washes, shores, seeps, and ditch banks. NMEX: McK, SJn. 2010–2070 m (6600–6800′). Flowering: Apr–May. Fruit: Jun–Jul. Southwest Oregon to Utah, south to Guatemala. Our plants are **var.** ***laevigata*** (Bebb) Dorn (smooth and polished, probably referring to the glossy upper leaf surface). Red willow. The variety in southwest Oregon, California, Nevada, Utah, Arizona, northwest New Mexico, and Baja California Norte.

Salix brachycarpa Nutt. (Greek *brachy*, short, + *carpa*, fruit, in this case referring to the catkin) Shortfruit willow. Shrub to 1.5 (3) m high; year-old branchlets reddish brown, pubescent. **LEAVES** elliptic to elliptic-obovate, oblong, or oval; petioles 1–3 (4) mm long; the blades 2–3 (4) cm long, 0.6–1.6 cm wide, base rounded to acute, tip acute to obtuse, entire or with a few glands near base, glaucous on underside, pubescent on both sides **CATKINS** coetaneous, 0.5–2 (3) cm long, on leafy branchlets 0.2–2 cm long; flower bracts brown (black), with relatively straight hairs; stamens 2. **CAPSULES** pubescent, (4) 5–7 mm long, styles 0.1–0.8 (1.5) mm long, stipes 0–0.5 mm long. $2n = 38$. Alaska to Quebec, south to eastern California and northern New Mexico. Meadows, slopes, bogs, wet alkaline barrens, and stream banks. COLO: Con, Hin, LPl, Min, Mon, RGr, SJn. 2500–4050 m (8200–13300′). Flowering: Jun–Jul. Fruit: Jul–Aug. Our plants are **var.** ***brachycarpa***. The variety from Yukon Territory to Quebec, south to eastern California and northern New Mexico.

Salix drummondiana Barratt ex Hook. (for Thomas Drummond, 1780–1835, Scottish naturalist with the second Franklin Expedition who collected the species in the Canadian Rockies) Drummond willow. Shrub to 6 m high; year-old branchlets reddish brown, glabrous or glabrate, usually glaucous at least behind buds. **LEAVES** elliptic or oblong (lanceolate) to elliptic-obovate; petioles 2–12 mm long; the blades 4–11 cm long, 1–2.6 cm wide, base and tip acute, entire to rarely crenate, glaucous on underside, glabrous or sparsely pubescent on upper side, silver-sericeous to rarely glabrate on underside. **CATKINS** precocious (subprecocious), 1.5–6 (11) cm long, sessile or subsessile; flower bracts brown (black), with relatively straight hairs; stamens 2. **CAPSULES** pubescent, 3–5.6 mm long, styles 0.4–1.8 mm long, stipes 0.1–2 mm long. $2n$ = 38, 76. Stream banks, swamps, and thickets. COLO: Arc, Hin, LPl, Min, Mon, RGr, SJn; UTAH. 2250–3350 m (7400–11000′). Flowering: May–Jul. Fruit: Jun–Jul. Yukon Territory to Newfoundland, south to California, New Mexico, Michigan, and New Hampshire.

Salix eriocephala Michx. (wool head, referring to the hairy catkins). Heartleaf willow. Shrub, or rarely treelike, to 8 m high; year-old branchlets reddish brown to greenish yellow (ashy white), glabrous (pubescent). **LEAVES** elliptic, oblong, or lanceolate to lance-ovate, rarely oblanceolate or elliptic-obovate; petioles 3–15 (25) mm long; the blades (3.5) 4–10 (12) cm long, (0.8) 1–3 (4.5) cm wide, base mostly rounded (acute or subcordate), tip mostly acute, serrulate (serrate) to entire, glaucous on underside, glabrous or glabrate. **CATKINS** subprecocious to coetaneous, 1–6 cm long, sessile or on leafy branchlets to 0.9 (1.7) cm long; flower bracts brown or black, with tangled, crinkly hairs especially toward base, sometimes glabrate; stamens 2. **CAPSULES** glabrous, 3–6 mm long, styles 0.1–0.7 mm long, stipes 0.5–4 (4.5) mm long. $2n$ = 38. [*S. rigida* Muhl.]. Southern Yukon to Newfoundland, south to California and northwest Florida. Six varieties, two in the Four Corners region.

1. Year-old branchlets predominantly reddish or reddish brown (avoid severely browsed shoots); stipes 0.5–2 (2.5) mm long .... **var. *ligulifolia***

1′ Year-old branchlets predominantly yellowish or greenish; stipes (1) 1.5–4 (4.5) mm long .... **var. *watsonii***

var. *ligulifolia* (C. R. Ball) Dorn (strap leaf, referring to the leaf shape) Strapleaf willow. Year-old branchlets predominantly reddish brown. **LEAVES** lanceolate to elliptic, occasionally oblong, rarely oblanceolate; petioles 3–12 (15) mm long; the blades 5–10 (12) cm long, 1–2.5 (3.5) cm wide. **CATKINS** 2–6 cm long, sessile or on leafy branchlets to 0.9 cm long. **CAPSULES** 3.5–6 mm long, styles 0.1–0.7 mm long, stipes 0.5–2 (2.5) mm long. $2n$ = 38. [*S. ligulifolia* C. R. Ball ex C. K. Schneid.]. Stream banks, shores, swamps, and other moderately wet areas. ARIZ: Apa; COLO: Arc, Hin, LPl, Min, Mon; NMEX: RAr, SJn; UTAH. 1700–2750 m (5600–9000′). Flowering: Apr–May. Fruit: May–Jun. The variety from southern Oregon to southeast Wyoming, south to California and New Mexico but skipping Nevada.

var. *watsonii* (Bebb) Dorn (for Sereno Watson, 1826–1892, American botanist who collected the variety in Nevada) Yellow willow. Year-old branchlets predominantly greenish yellow (ashy white). **LEAVES** lanceolate to lance-ovate or broadly elliptic, rarely elliptic-obovate; petioles 4–15 (25) mm long; the blades (3.5) 4–8 (12) cm long, (0.8) 1–3 (4.5) cm wide. **CATKINS** 1–6 cm long, sessile or on leafy branchlets to 0.7 (1.7) cm long. **CAPSULES** 3–5.5 mm long, styles 0.2–0.7 mm long, stipes (1) 1.5–4 (4.5) mm long. $2n$ = 38. [*S. lutea* Nutt.]. Stream banks, shores, swamps, and other moderately wet areas. UTAH. 1350–2450 m (4500–8000′). Flowering: May–Jun. Fruit: Jun. The variety from southeast Washington to Montana, south to California, northern Arizona, and northwest Colorado. There is some intergrading with var. *ligulifolia*.

Salix exigua Nutt. (Latin *exiguus*, weak or small) Sandbar willow. Shrub to 5 m high. Year-old branchlets grayish, brownish, or reddish brown (yellowish, purplish), glabrous or sometimes pubescent. **LEAVES** mostly linear (narrowly elliptic); petioles 0.5–6 mm long; the blades (3) 4–16 cm long, 0.3–1.1 cm wide, base and tip mostly acute, entire or serrulate with the teeth widely spaced, nonglaucous, pubescent (glabrous). **CATKINS** coetaneous or serotinous, (1) 1.5–10 cm long, on leafy branchlets 0.5–18 cm long; flower bracts greenish, whitish, or yellowish to tawny, deciduous in fruit, with straight or wavy hairs at least toward base (glabrate); stamens 2. **CAPSULES** 3–5 (7) mm long, glabrous or pubescent, styles 0–0.2 mm long, stipes 0–1.5 (2) mm long. $2n$ = 38. Stream banks, floodplains, washes, shores, and ditch banks. ARIZ: Apa, Nav; COLO: Arc, Dol, Hin, LPl, Min, Mon, SJn, SMg; NMEX: McK, RAr, San, SJn; UTAH. 1150–3000 m (3800–9900′). Flowering: Apr–Aug. Fruit: May–Aug. Alaska to New Brunswick, south to Baja California Norte, Louisiana, and Maryland. Our plants are **var. *exigua***, coyote willow. The variety from southern British Columbia and Alberta south to Baja California Norte and western Texas.

Salix fragilis L. (fragile, referring to the brittle branchlets) Crack willow. Introduced tree to 15 (25) m high. Year-old branchlets mostly yellow-brown, glabrous or sparsely pilose. **LEAVES** lanceolate or lance-elliptic; petioles 7–20 mm long; the blades (7) 10–17 cm long, 1.7–3.5 cm wide, base acute to obtuse, tip acute or acuminate, serrate, glaucous on underside, glabrous. **CATKINS** coetaneous, (2) 4–8 cm long, on leafy branchlets 1–5 cm long; flower bracts greenish, whitish, or yellowish to tawny, deciduous in fruit, with straight or wavy hairs; stamens 2. **CAPSULES** glabrous, 4–5.5 mm long, styles 0.3–0.8 mm long, stipes 0.5–1 mm long. $2n$ = 38–114. Introduced from Eurasia and occasionally escaping on stream banks, ditch banks, and shores. COLO: Arc, LPl, Mon; NMEX: SJn. 1550–2350 m (5100–7700′).

Flowering: Apr–May. Fruit: May–Jun. Frequently confused with *S. alba* L., another introduced species sparingly planted in the Four Corners region, but which does not escape cultivation.

Salix geyeriana Andersson (for Carl Andreas Geyer, 1809–1853, German botanist who collected the species in Idaho) Geyer willow. Shrub to 7 m high; year-old branchlets reddish brown or purplish brown (yellow-brown), usually glaucous at least behind buds, glabrous. **LEAVES** mostly lance-elliptic or elliptic; petioles 2–9 mm long; the blades 2–8 cm long, 0.6–1.5 cm wide, base and tip mostly acute, entire or nearly so, mostly glaucous or glaucescent on underside, usually pubescent on both sides. **CATKINS** coetaneous, 0.6–2 (2.5) cm long, on leafy branchlets 0.1–1.2 (1.8) cm long; flower bracts tawny or light brown, with short, straight or slightly wavy hairs; stamens 2. **CAPSULES** pubescent, 3–6 mm long, styles 0.1–0.8 mm long, stipes 1–3 mm long. $2n = 38$. Edges of swamps, moist meadows, and stream banks, mostly montane. COLO: Hin, SJn. Common around Silverton, Colorado. 2150–3250 m (7100–10700′). Flowering: May–Jun. Fruit: Jul. Southern British Columbia to Montana, south to California and New Mexico.

Salix glauca L. (gray, referring to the leaf color) Gray willow. Shrub to 1.5 (4) m high; year-old branchlets reddish brown to grayish, pubescent to glabrate. **LEAVES** mostly elliptic to oblanceolate (obovate); stipules inconspicuous; the blades 3–6 (8) cm long, 0.7–3.5 cm wide, base acute to rounded, tip acute to rounded, entire or sometimes serrulate toward base, glaucous on underside, pubescent to glabrous, petioles 2–10 (16) mm long. **CATKINS** coetaneous, (2) 3–4 (5) cm long, on leafy branchlets 0.5–3.5 cm long; flower bracts brown (greenish), with relatively straight (slightly wavy) hairs; stamens 2. **CAPSULES** pubescent, 4–9 mm long, styles 0.3–1 mm long, stipes 0.2–1.8 mm long. $2n =$ 76–190. Stream banks and subalpine slopes. COLO: Con, Hin, LPl, Min, Mon, RGr, SJn. 3080–3800 m (10100–12500′). Flowering: Jul–Aug. Fruit: Jul–Sep. Alaska to Greenland, south to northern New Mexico, Manitoba, and Nova Scotia; Eurasia. Our plants are **var**. ***villosa*** Andersson (Latin *villosus*, long-hairy, referring to the leaves). $2n = 114$. This is the variety from Yukon to Nunavut and Manitoba south to Utah and northern New Mexico; sometimes appearing to intergrade with *S. brachycarpa*.

Salix gooddingii C. R. Ball (for Leslie Newton Goodding, 1880–1967, western American botanist who collected the species in Nevada) Goodding willow. Tree to 15 (30) m high; year-old branchlets gray, yellowish, or pale yellowish brown, glabrous, sometimes tardily so. **LEAVES** linear or oblong to lance-linear or narrowly elliptic; petioles 3–10 mm long; the blades 6–13 cm long, 0.8–1.6 cm wide, base acute, tip acuminate, serrulate or serrate, nonglaucous, glabrous or becoming so. **CATKINS** coetaneous, 2.2–8 cm long, rather lax, on leafy branchlets 0.4–3 cm long; flower bracts greenish, whitish, or yellowish to tawny, deciduous in fruit, with short, wavy hairs; stamens 5–8. **CAPSULES** glabrous or pubescent, 3–7 mm long, styles 0–0.4 mm long, stipes 1–3.2 mm long. $2n = 38$. Stream banks, shores, floodplains, washes, and seeps. ARIZ: Apa, Nav; NMEX: McK, SJn; UTAH. 1150–2010 m (3800–6600′). Flowering: Apr–May. Fruit: May–Jun. California to western Colorado, south to western Texas and Mexico (Baja California Norte and Sinaloa).

Salix irrorata Andersson (moistened with dew, probably referring to the glaucous branchlets) Bluestem willow. Shrub to 7 m high; year-old branchlets reddish brown or purplish, usually glaucous, glabrous. **LEAVES** oblong, narrowly elliptic, or oblanceolate; petioles 3–15 mm long; the blades 4.7–12 cm long, (0.5) 0.8–2.2 cm wide, base acute, tip acute to obtuse, entire to serrate or crenate, glaucous on underside, glabrous or rarely sparsely pubescent. **CATKINS** precocious or subprecocious, 1.5–4.2 cm long, sessile or on leafy branchlets to 0.5 cm long; flower bracts brown, with long, relatively straight hairs; stamens 2. **CAPSULES** glabrous, 3–5 mm long, styles 0.2–0.9 mm long, stipes 0.3–1.2 mm long. $2n = 38$. Rocky stream banks and washes, mostly low montane. ARIZ: Apa; COLO: Arc, Hin, LPl, Min; NMEX: McK. 1980–2400 m (6500–7900′). Flowering: Apr–May. Fruit: May. Southeast Wyoming, Colorado, New Mexico, and eastern Arizona.

Salix lasiandra Benth. (hairy male, referring to the hairy stamens) Longleaf willow. Shrub or tree to 15 m high; year-old branchlets yellow-brown to reddish brown or gray-brown, pubescent or glabrous. **LEAVES** lanceolate or sometimes elliptic; petioles 5–30 mm long; the blades 2.4–17 (20) cm long, 0.9–4.3 cm wide, base acute to rounded, tip acute or acuminate, serrate or serrulate, glaucous or not on underside, glabrous to occasionally pilose. **CATKINS** coetaneous, 1.7–10 cm long, on leafy branchlets 0.8–6.5 cm long; flower bracts greenish, whitish, or yellowish to tawny, often toothed or lobed, deciduous in fruit, with short, wavy hairs near base or glabrate; stamens 5–8. **CAPSULES** glabrous, 4–7 mm long, styles 0.2–1 mm long, stipes 0.5–2 (4) mm long. $2n = 76$. Alaska to western Manitoba, south to California and New Mexico. Two varieties, both in the Four Corners region, are sometimes treated as subspecies of *S. lucida* Muhl. Definite intergrading with that eastern species has not been demonstrated.

1. Leaves not glaucous on underside .... **var. *caudata***
1′ Leaves glaucous on underside .... **var. *lasiandra***

var. *caudata* (Nutt.) Sudw. (tailed, referring to the acuminate, tail-like leaf tip) Whiplash willow. **LEAVES** not glaucous. 2*n* = 76. Stream banks, shores, wet meadows, and seeps. COLO: Arc, Dol, LPl, SJn. 1950–2900 m (6400–9500′). Flowering: May–Jun. Fruit: Jun–Jul. The variety from southern British Columbia to southwest Saskatchewan, south to eastern California and northern New Mexico. Intermediates to var. *lasiandra* can be expected on the northern edge of the Four Corners region.

var. *lasiandra* Longleaf willow. **LEAVES** glaucous on underside. 2*n* = 76. Stream banks, shores, wet meadows, and seeps. ARIZ: Apa, Nav; COLO: Arc, Hin, LPl, Min, Mon, SJn; NMEX: McK, RAr, SJn; UTAH. 1650–3150 m (5400–10300′). Flowering: Apr–Jul. Fruit: Jun–Jul. The variety from Alaska to western Manitoba, south to California and New Mexico but replaced by var. *caudata* in much of the Rocky Mountain and Great Basin regions.

Salix lasiolepis Benth. (hairy scale, referring to the floral bracts) Arroyo willow. Shrub (ours) or tree to 12 m high; year-old branchlets yellow, green, or reddish brown, pubescent to glabrate (glabrous). **LEAVES** oblong, narrowly elliptic, or oblanceolate to obovate; petioles 2–16 (20) mm long; the blades (2.5) 3.5–12.5 (22) cm long, 0.6–3.2 (4.5) cm wide, base acute, tip acute to obtuse, entire to irregularly serrate or crenate, glaucous on underside, densely pubescent to glabrate on both sides, or becoming glabrous on upper side (and lower side). **CATKINS** precocious or subprecocious, 1.5–7 (9) cm long, sessile or on leafy branchlets to 0.7 cm long; flower bracts brown (black), with dense, straight to slightly wavy hairs; stamens 2. **CAPSULES** glabrous, 2.5–5.5 mm long, styles 0.2–0.8 (1) mm long, stipes 0.5–1.8 (2.2) mm long. 2*n* = 76. Stream banks, washes, seeps, and shores. ARIZ: Apa, Nav; NMEX: McK, SJn; UTAH. 1550–2470 m (5100–8100′). Some specimens suggest possible gene exchange with *S. irrorata*. Flowering: Apr–May. Fruit: Apr–Jun. Washington and western Idaho south to Mexico (Baja California Sur, Jalisco, and San Luis Potosí).

Salix monticola Bebb (mountain-dwelling, referring to the mountain habitat) Mountain willow. Shrub to 6 m high; year-old branchlets reddish brown or yellowish, glabrous or rarely pubescent. **LEAVES** ovate, lanceolate, or elliptic to obovate; petioles 5–14 mm long; the blades 3–8 (9.5) cm long, (1) 1.5–3.5 cm wide, base acute to rounded, tip acute, crenate or serrate or serrulate (subentire), glaucous on underside, glabrous. **CATKINS** precocious to coetaneous, 1–5 (6) cm long, subsessile or on leafy branchlets to 0.8 (1.7) cm long; flower bracts brown (black), with long, straight or curly hairs; stamens 2. **CAPSULES** glabrous, 3–6 mm long, styles 0.6–1.5 (1.8) mm long, stipes 0.3–1.5 (2) mm long. 2*n* = 114. Stream banks and wet meadows in or near mountains. ARIZ: Apa, Nav; COLO: Arc, Hin, LPl, Min, Mon, RGr, SJn; NMEX: McK, RAr, SJn; UTAH. 1800–3600 m (5900–11800′). Flowering: May–Jun. Fruit: May–Jul. Southern Wyoming, eastern Utah, Colorado, eastern Arizona, and northern New Mexico.

Salix monticola

Salix planifolia Pursh (flat leaf) Planeleaf willow. Shrub to 5 m high; year-old branchlets dark brown or reddish brown to red, glabrous or glabrate, occasionally lightly glaucous. **LEAVES** elliptic (lance-elliptic to obovate); petioles 3–13 mm long; the blades (2) 3.5–5 (8) cm long, 0.9–1.5 (2.2) cm wide, base and tip acute, entire to sometimes crenate or serrate, shiny and glabrous or glabrate on upper side, sparsely pubescent often with some reddish hairs (glabrous) and glaucous on underside. **CATKINS** precocious or subprecocious, (1) 1.5–6 cm long, sessile or subsessile; flower bracts brown (tawny), with long, relatively straight, often tawny hairs; stamens 2. **CAPSULES** pubescent, (3.5) 5–6 mm long, styles 0.4–1.8 mm long, stipes 0–1 mm long. 2*n* = 76. Wet meadows, subalpine slopes, fens, and willow swamps. COLO: Con, Hin, LPl, Min, Mon, RGr, SJn. 3050–4000 m (10000–13100′). Flowering: Jun–Jul. Fruit: Jul–Aug. Our plants are **var. *planifolia***, the variety from Yukon to Newfoundland, south to eastern California, northern New Mexico, northern Minnesota, and New Hampshire. Another variety is endemic on the Lake Athabasca sand dunes, Saskatchewan.

Salix reticulata L. (netted, referring to the network of veins in the leaves) Net willow. Creeping shrub less than 10 cm high; year-old branchlets light brown to reddish brown (yellow-brown), glabrous. **LEAVES** mostly oval to suborbicular (elliptic); petioles 1–15 (28) mm long; the blades 0.4–2.5 (3.6) cm long, 0.3–1.5 (2.3) cm wide, base obtuse to rounded (subcordate), tip obtuse to rounded (retuse), entire, slightly revolute, glaucous on underside, glabrous or with sparse, long hairs mostly on underside. **CATKINS** serotinous, 0.5–2 (3) cm long, on naked branchlets 2–20 mm long; flower bracts greenish (tawny or brown), glabrous externally, with wavy hairs on inner surface; stamens 2. **CAPSULES** pubescent, 1.5–4 mm long, styles 0.1–0.4 mm long, stipes 0–0.5 mm long. 2*n* = 38. [*S. nivalis* Hook.]. Alpine and subalpine. COLO:

Arc, Con, Hin, LPl, Min, Mon, SJn. 3450–4050 m (11300–13300′). Flowering: Jun–Aug. Fruit: Jul–Sep. Alaska to Newfoundland, south to eastern California, northern New Mexico, and northern Quebec; Eurasia. Our plants are **var.** ***nana*** Andersson (dwarf, referring to the small size of the plants). Snow willow. This is the variety from British Columbia and Alberta south to eastern California and northern New Mexico.

Salix scouleriana Barratt ex Hook. (for John Scouler, 1804–1871, Scottish surgeon-naturalist on a Hudson's Bay Company voyage who collected the species around the Columbia River) Scouler willow; shrub or small tree to 15 (20) m high; year-old branchlets reddish brown to yellow-brown, usually pubescent. **LEAVES** elliptic to obovate; petioles 3–10 (18) mm long; the blades 3–8 (10) cm long, (1.3) 2–3 cm wide, base acute, tip acute to rounded, entire to irregularly serrulate or somewhat crenate, glaucous on underside, pubescent at least along midrib on underside. **CATKINS** precocious, 1.5–5 (7) cm long, sessile or on leafy branchlets to 1.3 cm long. flower bracts brown (black), with relatively straight hairs; stamens 2. **CAPSULES** pubescent, strongly beaked, 4.5–11 mm long, styles 0.2–1.1 mm long, stipes 0.8–2.8 mm long. $2n$ = 76, 114. Woods, slopes, and meadows, rarely on shores. ARIZ: Apa; COLO: Arc, Hin, LPl, Min, Mon, RGr, SJn; NMEX: McK, SJn; UTAH. 2350–3250 m (7700–10700′). Flowering: Apr–Jun. Fruit: May–Jul. Alaska to Manitoba, south to California, New Mexico, and Coahuila.

Salix serissima (L. H. Bailey) Fernald **CP** (very late, referring to the late-maturing fruits) Autumn willow. Shrub to 5 m high; year-old branchlets reddish brown to tan, glabrous. **LEAVES** lanceolate or elliptic; petioles 4–11 mm long; the blades 4–10 cm long, 1.5–2.7 (3.5) cm wide, base acute to rounded, tip acute or acuminate, serrulate, glaucescent or pale on underside, glabrous. **CATKINS** serotinous, (1) 2–5 cm long, on leafy branchlets 1–5 cm long; flower bracts greenish, whitish, or yellowish to tawny, deciduous in fruit, with straight to slightly wavy hairs; stamens 5–8. **CAPSULES** glabrous, (6) 7–12 mm long, maturing in summer, styles 0.1–0.8 mm long, stipes 0.8–2 mm long. $2n$ = 76. Swamps and bogs. COLO: LPl. West side of Haviland Lake at 2500 m (8200′). Flowering: May–Jul. Fruit: Jul–Sep. Northwest Territories to Newfoundland, south to Colorado and New Jersey.

Salix wolfii Bebb (for John Wolf, 1820–1897, botanist with the Wheeler Expedition who collected the species in Colorado) Wolf willow. Shrub to 1 (2) m high; year-old branchlets reddish brown (yellow-brown), pubescent or glabrous. **LEAVES** mostly elliptic, lanceolate, or oblanceolate; petioles 2–10 mm long; the blades 2–6 cm long, 0.5–1.5 (2) cm wide, base rounded or obtuse (acute), tip acute (obtuse), entire, nonglaucous, pubescent on both sides. **CATKINS** coetaneous, 0.8–2 (3) cm long, subsessile or on leafy branchlets to 1.2 cm long; flower bracts brown or black, with long (short), straight to wavy hairs; stamens 2. **CAPSULES** glabrous, 3.5–5 mm long, styles 0.2–1.3 mm long, stipes 0–0.8 mm long. $2n$ = 38. Wet meadows and stream banks, subalpine to upper montane. COLO: Hin, Min, RGr, SJn. 3110–3550 m (10200–11700′). Flowering: Jun–Jul. Fruit: Jul–Aug. Northeast Oregon to Montana, south to northeast Nevada, Utah, and northern New Mexico. Our plants are **var.** ***wolfii***. This is the variety from southwest Montana to north-central Wyoming, south to Utah and northern New Mexico.

SANTALACEAE R. Br. SANDALWOOD FAMILY

Linda Mary Reeves

(Latin *santalum*, sandalwood) Trees, shrubs, or herbs, photosynthetic, but most semi- or root-parasitic, primarily absorbing water and minerals through haustoria, terrestrial, occasionally epiphytic branch parasites. **LEAVES** opposite, occasionally alternate or spirally arranged, simple, entire. **INFL** a spike, raceme, corymb, or head. **FLOWERS** actinomorphic; perfect or unisexual; greenish, pinkish, or whitish; perianth of 1 whorl, tubular with 3–8, usually 4–5, lobes, sometimes fleshy; stamens fused to perianth tube and opposite the lobes; nectary disc often surrounding the ovary or fused to perianth tube; ovary inferior or semi-inferior, generally unilocular, of 2–5 carpels; stigma lobed or capitate; ovule pendulous. **FRUIT** a nut or drupe. **SEEDS** solitary, with copious fleshy or oily endosperm. A family of about 35 genera, about 400 species, primarily in dry tropical or subtropical habitats. *Santalum album*, the sandalwood tree, yields a fragrant wood useful for carving. The oil is used in soap, incense, and perfumes.

Comandra Nutt. Bastard Toadflax

(Greek *come*, hair, + *aner*, man, alluding to hairs of perianth lobes attached to the anthers) Perennial, erect, glabrous, rhizomatous herbs or shrubs 7–50 cm tall, partially root-parasitic on other angiosperms. **STEMS** branching freely at base, 5–45 cm tall; rhizome cortex blue when fresh, turning black when dried. **LEAVES** alternate, subsessile to short-petiolate, linear-lanceolate to ovate-oblong, entire, acute, somewhat fleshy, 5–60 mm long, 1–16 mm wide; base mostly acute; apex acute to obtuse. **INFL** a cluster of 3–6 flowered cymes, each subtended by a bract. **FLOWERS** perfect, 3–7 mm wide; pedicel 0–4 mm long, subtended by 1–3 bracts; perianth rotate to campanulate, with 3–7 (usually 5) lobes, white to pinkish to purplish, spreading, with long hairs on the inner surface adjacent to the anthers; style filiform, 2–3 mm long; stigma capitate. **FRUIT** an ovoid or globose drupe, glabrous to slightly rough, 4–8 mm long, yellow to brown when mature; mesocarp somewhat fleshy. $2n = 28$. Monospecific genus with several subspecies.

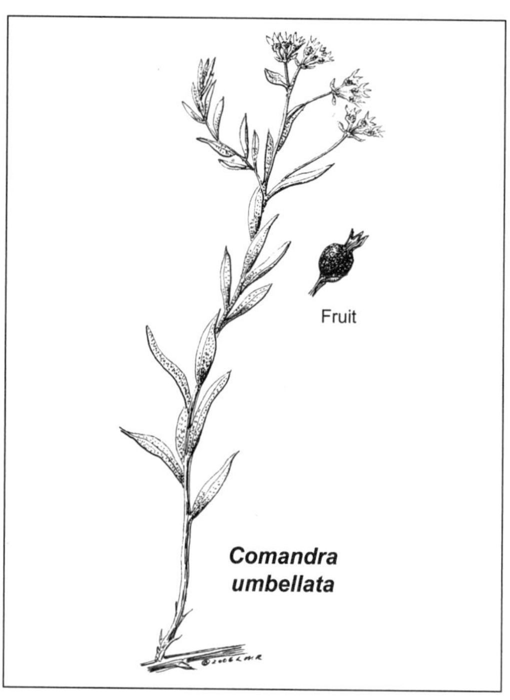

Comandra umbellata

Comandra umbellata (L.) Nutt. (umbellate, referring to an inflorescence similar to the Apiaceae or Umbelliferae) Pale bastard toadflax. Herbs 5–34 cm tall, herbage dying down in winter. **LEAVES** linear or lanceolate, grayish green, fleshy, glaucous, 0.5–4.5 cm long. **FLOWER** perianth lobes lanceolate, 2.5–4 mm long. **FRUIT** 6–9 mm diameter, slightly roughened or ridged. Open slopes, blackbrush, desert scrub, piñon-juniper, ponderosa pine, and spruce-fir, often in sandy soils, sometimes along creeks, commonly associated with sandstones. ARIZ: Apa, Nav; COLO: Arc, Dol, Hin, LPl, Min, Mon, SMg; NMEX: McK, RAr, San, SJn; UTAH. 1400–2700 m (4600–8900′). Flowering: Apr–Aug. Fruit: May–Sep. Most of the United States (except Louisiana and Florida), southern Canada, northern Mexico, and the Balkans. Our specimens belong to **subsp. *pallida*** (A. DC.) Piehl, which is restricted to the western United States.

SAURURACEAE Rich. ex T. Lestib. LIZARD-TAIL FAMILY

Kenneth D. Heil

(Greek *saur*, lizard) Perennial herbs, rhizomatous or stoloniferous, more or less aromatic. **STEMS** simple or branched, often scapelike. **LEAVES** mostly basal, alternate, simple, petiolate, stipules adnate to the petiole. **INFL** a dense spike or lax raceme, sometimes subtended by involucral bracts. **FLOWERS** small, perfect, perianth lacking; stamens (3–) 6–8, hypogynous or epigynous; anthers bilocular; pistil 1, of 3–5 (–7) carpels, styles and stigmas distinct. **FRUIT** a schizocarp or in ours, a capsule. **SEEDS** 1–40. $x = 11$. Five genera, seven species. North and Central America, eastern Asia.

Anemopsis Hook. & Arn. Yerba Mansa

(Greek *anemo*, wind) Plants 1–5 dm tall, glabrous or pubescent. **LEAVES** basal and cauline; basal leaves petioled, the blades 3.5–10 cm long, elliptic-oblong with a truncate or cordate base, entire, petioles as long or longer than the blades; cauline leaves sessile and clasping; 1–3 small, petioled leaves in the axils. **INFL** a terminal, congested, conic spike of 75–100 (–150) flowers, subtended by 4–9 white to reddish petaloid bracts, 0.5–3.5 cm long. **FLOWERS** each subtended by an adnate, white, ovate, clawed bract; stamens 6 (–8), epigynous; pistil 1, of 3 (–4) carpels, ovary inferior. **FRUIT** a capsule, the carpels adnate but easily separated. **SEEDS** 6–10, brown, reticulate. One species in the western United States.

Anemopsis californica Hook. & Arn. (of California) $2n = 22$. [*A. californica* var. *subglabra* Kelso]. In alkaline soils of wet meadows, and along streams. ARIZ: Apa. 1650 m (5450′). Flowering: May–Jun. Fruit: Aug–Sep. Oregon and California to Kansas and Texas, south to Mexico. In the San Juan River drainage, yerba mansa is known only from Canyon de Chelly. The major uses are a drug for colds, disinfectant, blood medicine, and dermatological and gastrointestinal aid. The pulverized seeds have been used in making bread.

SAXIFRAGACEAE Juss. SAXIFRAGE FAMILY

Roy L. Taylor

(Latin for stone breaking, referring to purported ability of medicinals to treat kidney stones) Perennial herbs. **LEAVES** alternate, sometimes all basal, or less often opposite, simple or pinnately or palmately compound, stipules absent. **INFL** a cyme, raceme, or flowers solitary. **FLOWERS** regular or somewhat irregular, perfect or rarely unisexual, perigynous to often partly or wholly epigynous; sepals 3–10, mostly 5; petals 3–10, mostly 5, often clawed, sometimes cleft or dissected, well developed or relatively small and inconspicuous; stamens as many as or twice as many as the petals;

nectary sometimes present as an intrastaminal disc or annulus; gynoecium of 2–5 (mostly 2) carpels; ovules several to usually numerous on each placenta. **FRUIT** usually a capsule. **SEEDS** more or less numerous, small. x = 7–17, mostly 7. About 35 genera and 650 species, cosmopolitan in distribution. (Cronquist, Holmgren, and Holmgren 1997a; Welsh et al. 2003.)

1. Leaves all basal, distinctly and often abruptly petioled, not lobed more than 1/2 the distance to the midrib; flowers on naked scapes or these with a solitary bract; stamens 5 or 10 ....(2)
1' Leaves not all basal, or sometimes so in depauperate plants, the blades not distinctly petioled or else lobed more than 1/2 the distance to the midrib; flowers often not scapose; stamens 10 ....(4)

2. Flowers in very narrow, spikelike, ebracteate racemes; petals parted or divided into filiform segments; leaves toothed and lobed .... ***Mitella***
2' Flowers not in spikelike racemes or, if so, bracteate; petals entire; leaves various ....(3)

3. Stamens 5; leaves crenate-toothed and lobed; plants of cliffs and rocky outcrops .... ***Heuchera***
3' Stamens 10; leaves subentire, crenate, or very coarsely dentate but not lobed; stamens 10; plants mostly of dry meadows or wet places .... ***Saxifraga***

4. Leaves parted or divided to the midrib, the basal ones abruptly constricted into slender petioles, 0.5–10 cm long; petals deeply lobed or cleft .... ***Lithophragma***
4' Leaves entire, toothed, or lobed, but not divided more than 1/2 the distance to the midrib or, if so, sessile or nearly so; petals entire .... ***Saxifraga***

Heuchera L. Alum-root

Perennial herb from scaly, woody, branched caudex or rootstock. **FLOWERING STEMS** 1–several; simple or several branches, with or without bracts. **LEAVES**: basal stipulate. **INFL** a raceme or scalelike bracteate cyme, few flowers. **FLOWERS** perfect, regular; hypanthium calyxlike, partly fused to ovary; petals lobed or toothed; sepals 5; petals 5, stamens 10; pistil 1; ovary superior to 1/2 inferior, placentas 3; styles 3. **FRUIT** a capsule. **SEEDS** several. Plants indigenous to North America. About 55 species throughout North America.

1. Stamens shorter than sepals; petals about 1–2 mm long; sepals not reddish; pedicels 1–2 mm long or obsolete .... ***H. parvifolia***
1' Stamens exserted 1–4 mm beyond sepals; petals 3–4 mm long; sepals often reddish; pedicels 2–7 mm long .... ***H. rubescens***

Heuchera parvifolia Nutt. ex Torr. & A. Gray (little leaf) Little leaf alum-root. Slender, perennial, caulescent herb. **FLOWERING STEMS** 2–5 dm tall, unbranched, arising from basal rosette; without or with 1–3 inconspicuous bracts, finely glandular, puberulent; hairs, if present, often unequal. **LEAVES**: basal leaves glabrous to densely pubescent, deeply 3-lobed to palmately compound, lobes lobed, teeth acute; hairs, when present, often unequal. **INFL** 4–14-flowered, a nodding raceme. **FLOWERS** pedicellate, hypanthium long-obconic-elongate at anthesis, becoming elongate in fruit; corolla widely spreading; sepals triangular; petals white, occasionally pink, usually with prominent vein markings, 7–16 mm long, always 3-cleft; ovary inferior. **FRUIT** an elongate capsule. **SEEDS** smooth or wrinkled. [*Heuchera flavescens* Rydb.; *H. utahensis* Rydb.; *H. flabellifolia* Rydb.; *H. duranii* Bacig.]. Often in rocky places; piñon-juniper, sagebrush, mountain brush, ponderosa pine, aspen, Douglas-fir, and spruce-fir communities. ARIZ: Apa, Nav; COLO: Arc, Con, Dol, Hin, LPl, Min, Mon, RGr, SJn; NMEX: McK, RAr, SJn; UTAH. 1775–3960 m (5825–13000'). Flowering: May–Sep. Fruit: Jul. Alberta to New Mexico and west to Idaho and Nevada. Plant used for rat bites by the Kayenta Navajo. The Ramah Navajo used a decoction of root for stomachache and as a "life medicine."

Heuchera rubescens

Heuchera rubescens Torr. (reddish) Red alum-root. Slender, perennial, caulescent herb. **FLOWERING STEMS** 1.5–3 dm tall, 1–several, scapose, unbranched, arising from basal rosette herbage. **LEAVES**: basal leaves light green, sparsely pubescent, orbicular, short-petiolate to 8 cm long, blade simple to irregularly 3–5-lobed to almost pinnatifid; cauline leaves 2, palmately compound, much reduced, more highly dissected than basal leaves, always appearing pinnatifid; stipules broad, fimbriate, decurrent on petiolar base. **INFL** a 3–12-flowered, compact raceme. **FLOWERS** slightly pendulous; hypanthium campanulate or hemispheric, becoming elongate-campanulate in fruit; sepals triangular, valvate in bud, becoming widespread at anthesis; corolla widely spreading, open at throat, petals pink, occasionally white, ovate, 3–7 mm long, palmately 5-parted; ovary less than 1/2 inferior. **FRUIT** an elongate capsule. **SEEDS** smooth or wrinkled. [*Heuchera versicolor* Greene; *H. rubescens* var. *versicolor* (Greene) M. G. Stewart; *H. clutei* A. Nelson]. Cliff faces and other rocky sites; piñon-juniper, sagebrush, mountain brush, and ponderosa pine communities. ARIZ: Apa; NMEX: McK, SJn; UTAH. 2030–3050 m (6665–10000'). Flowering: May–Aug. Fruit: Sep. Uncommon in study area. California, Oregon, Idaho, south through Nevada, Utah, Arizona, New Mexico, southern Colorado, and Texas to northern Mexico.

Lithophragma (Nutt.) Torr. & A. Gray Woodland Star

Slender, perennial, small herbs from underground rhizomes that may bear bulblets. **FLOWERING STEMS** 1–several; simple or several-branched; 1–10 stem leaves, usually 2; some with numerous bulbils in axils. **LEAVES**: basal leaves petioled, glandular-pubescent, lobed; cauline leaves alternate, petiolate. **INFL** a compact, few-flowered, bracteate cyme or raceme, or solitary flowers. **FLOWERS** perfect; regular; hypanthium calyxlike, partly fused to ovary; sepals 5, triangular; petals 5, narrowly clawed, lobed or toothed, white to pink or purple-tinged; pistil 1; ovary superior to 1/2 inferior, placentas 3; styles 3; valves 3; stamens 10. **FRUIT** a 3-beaked capsule. **SEEDS** several, usually dark brown. About nine species endemic to western North America.

1. Plants with few to several purple bulblets in the inflorescence and usually in axils of the upper leaves; lower pedicels 1.5–3 times longer than the hypanthium .... ***L. glabrum***
1' Plants without bulblets in the inflorescence or in leaf axils; lower pedicels about 0.5–1.5 times longer than the hypanthium .... ***L. tenellum***

Lithophragma glabrum Nutt. (glabrous) Slender woodland star, rock star. Slender, perennial, caulescent herb; herbage sparingly pubescent to nearly glabrous. **FLOWERING STEMS** several, usually reddish; 0.8–3.5 dm tall, arising from basal rosette; hirsute, stipitate-glandular, 2–7 leaves. **LEAVES**: basal leaves orbicular, short-petiolate; 1–4 cm; blade usually digitately trifoliate; cauline with petioles 1.5–5.5 cm long, glandular; blades 1.2–6 cm wide, about as long; reniform or orbicular, truncate or cordate at the base, crenate, more or less shallowly lobed; bracts similar to upper leaves but smaller, less toothed, upper ones usually entire; often red-purple bulbils in leaf axils. **FLOWER** sepals 3–4.5 mm long, glandular; petals subequal to sepals or shorter, pinkish or reddish. **HYPANTHIUM** 4–7 mm long, campanulate, glandular to pilose-glandular, often reddish or purplish. **FRUIT** a capsule. **SEEDS** 0.5–0.7 mm long, tuberculate, brown. [*L. bulbiferum* Rydb.]. Sandy to rocky soils; aspen, Gambel's oak, sagebrush, piñon-juniper, mountain brush, and ponderosa pine communities. COLO: Arc, Dol, Mon, SJn, SMg; UTAH. 2260–2475 m (7410–8115'). Flowering: May–Jun. British Columbia to California, east to Alberta, the Dakotas, and Colorado.

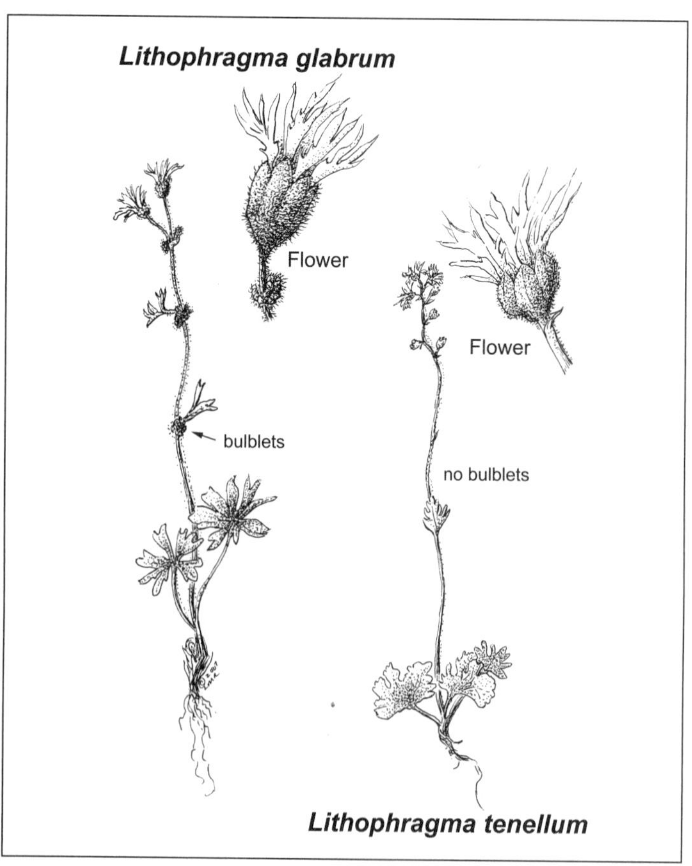

Lithophragma tenellum Nutt. (tender) Slender fringecup. Slender, perennial, caulescent herb, herbage light green, sparsely pubescent. **FLOWERING STEMS** 1.5–3 dm tall, 1–several, unbranched, arising from a basal rosette; 2 (1–3)

leaves. **LEAVES**: basal orbicular, short-petiolate, to 8 cm long, blade simple to irregularly 3–5-lobed to almost pinnatifid; cauline leaves 2, palmately compound, much reduced, more highly dissected than basal leaves, always appearing pinnatifid; stipules broad, fimbriate, decurrent on petiolar base. **INFL** a 3–12-flowered, compact raceme. **FLOWERS** slightly pendulous; sepals triangular, valvate in bud, becoming widespread at anthesis; corolla wide; spreading, open at throat, petals pink, occasionally white, ovate, 3–7 mm long, palmately 5-parted; ovary less than 1/2 inferior. **HYPANTHIUM** campanulate or hemispheric, becoming elongate-campanulate in fruit. **FRUIT** a campanulate capsule. **SEEDS** smooth or wrinkled. [*L. australis* Rydb.]. Piñon-juniper woodlands, sagebrush, mountain brush, ponderosa pine, aspen, and riparian communities. ARIZ: Apa, Nav; COLO: Arc, Dol, LPl, Min, Mon, SMg; NMEX: McK, RAr, SJn; UTAH. 2225–3165 m (7300–10385′). Flowering: Apr–Jun. Washington to Arizona, east to Montana and south to New Mexico.

Mitella L. Bishop's Cap, Mitrewort

(Latin for headband) Perennial herb from scaly, woody, branched caudex or rootstock. **FLOWERING STEMS** 1–several; scapose, simple or several-branched. **LEAVES** all basal. **INFL** borne in compact, few-flowered, bracteate cymes; scale-like bracts. **FLOWERS** with a calyxlike hypanthium, partly fused to ovary; petiolate, regular, sepals 5; petals 5, lobed or toothed; stamens 10; pistil 1; ovary superior to 1/2 inferior; chamber 1, placentas 3; styles 3, free or connate below. **FRUIT** a 3-valved capsule. **SEEDS** several. About 20 species in western North America.

1. Racemes not secund; pedicels 2–8 mm long; petals pinnately divided; stamens opposite petals; sepals greenish .... ***M. pentandra***

1′ Racemes secund; pedicels absent or to 3 mm long; petals with 2–3 lobes; stamens opposite sepals; sepals whitish or purplish .... ***M. stauropetala***

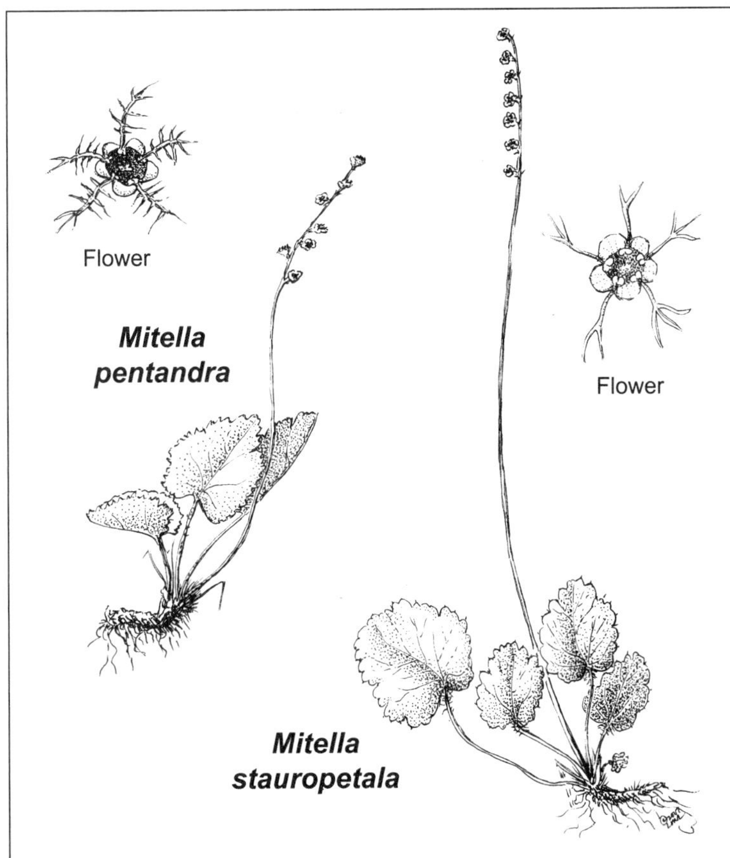

Mitella pentandra Hook. **CP** (five stamens) Five-star mitrewort. Slender, perennial, caulescent herb; herbage from glabrous to densely pubescent. **FLOWERING STEMS** 2–5 dm tall, unbranched, arising from basal rosette. **LEAVES** deeply 3-lobed to palmately compound, lobes lobed, teeth acute. **INFL** 4–14-flowered, in nodding, open raceme. **FLOWER** hypanthium long-obconic-elongate at anthesis, becoming elongate in fruit; pedicellate; sepals triangular; corolla widely spreading, petals white, occasionally pink, usually with prominent vein markings, 7–16 mm long, always 3-cleft; ovary inferior. **FRUIT** an elongate capsule. **SEEDS** smooth or wrinkled. Often where shaded; aspen, spruce-fir, and willow communities. COLO: Arc, Con, Hin, LPl, Min, Mon, RGr, SJn; UTAH. 2650–3640 m (8700–11940′). Flowering: Jun–Aug. Alaska south to northern California, east to Alberta and Colorado.

Mitella stauropetala Piper (cross-shaped petals) Side-flowered mitrewort. Slender, perennial, caulescent herb; herbage light green, sparsely pubescent. **FLOWERING STEMS** 1.5–3 dm tall; 1–several, unbranched; arising from basal rosette base. **LEAVES** short-petiolate, to 8 cm long, blade orbicular, simple to irregularly 3–5-lobed, almost pinnatifid; stipules acute to rounded. **INFL** a 3–12-flowered, compact, secund raceme. **FLOWER** hypanthium campanulate or hemispheric, becoming elongate-campanulate in fruit; slightly pendulous; sepals triangular, valvate in bud, becoming widespread at anthesis; corolla widely spreading, open at throat, petals pink, occasionally white, ovate, 3–7 mm long, palmately 5-parted; ovary less than 1/2 inferior. **FRUIT** a campanulate capsule. **SEEDS** smooth or wrinkled. Mountain brush, aspen, ponderosa pine, and spruce-fir communities. COLO: Arc, Hin, LPl, Min, Mon, SJn; UTAH. 2620–3640 m (8600–11940′). Flowering: Jun–Aug. Washington and Oregon to Montana, and south to Colorado.

Saxifraga L. Saxifrage

(Latin for stone breaking) Mostly small perennial herbs; some short-lived, a few annual; some reproducing by stolons. **FLOWERING STEMS** may be scapose; many with evergreen leaves; alternate; commonly caespitose or mat-forming; bulbils in leaf axils or inflorescence. **LEAVES** alternate or basal. **INFL** of solitary flowers or modified cymes, often appearing racemose or open, variously branched. **FLOWERS** perfect, with or without hypanthium, sepals 5, erect to deflexed; petals 5, clawless or clawed, usually white, pink to purple, yellow, or with red spotting, deciduous or persistent; stamens 10, filaments petaloid, linear or flattened; ovary superior or partly to wholly inferior, carpels 2–5. **FRUIT** a capsule dehiscent across top, or follicle. **SEEDS** small, numerous.

1. Flower petals yellow; plants with spreading stolons from slender rhizomes ....(2)
1' Flower petals white to cream, or pink to purple, sometimes yellow ....(3)

2. Plants with naked, spreading stolons; flowering stems to 15 cm, without leaves .... ***S. flagellaris***
2' Plants mat-forming, stolons absent; with numerous tufted-leafy, erect flowering stems to 20 cm .... ***S. serpyllifolia***

3. Leaves all basal; flowering stems leafless, usually single, inflorescence a cyme ....(4)
3' Flowering stems with leaves, inflorescence open, of 1 to many flowers ....(6)

4. Leaves circular in outline, cordate at the base; cymose panicle loose and open .... ***S. odontoloma***
4' Leaves lanceolate or broader, toothed or entire, tapering to a petiole; inflorescence a dense head or spike .....(5)

5. Spikes usually simple, headlike in anthesis; leaves rhomboid, short; foothills to lower tundra............... ***S. rhomboidea***
5' Spikes usually compound; leaves elongate, narrowly oblanceolate; wet meadows and streamsides...... ***S. oregana***

6. Low, mat-forming perennials; produced from branched woody caudex ....(7)
6' Plants from delicate caudex, often short-lived perennials ....(8)

7. Plants covered with persistent sessile, rigid, entire, closely appressed leaves, imbricate at base, with setose tips, coarsely ciliate margins; flowering stems to 20 cm; petals white, 3.5–4.5 mm long, purplish red, with yellow or orange spots .... ***S. bronchialis***
7' Leaves mostly in crowded basal rosette, distinctly 3-lobed apically, densely pubescent, glandular-ciliate; flowering stems erect, to 15 cm; petals white or cream, not spotted .... ***S. cespitosa***

8. Plants from slender caudex; flowering stems erect, 10–15 cm tall; 4–14 flowers; petals 7–16 mm long, white, occasionally pink, without colored spots.. ***S. adscendens***
8' Plants delicate, with lobed or toothed, petiolate leaves, blades to 15 mm long; flowering stems branched or not; 1–5 terminal flowers; petals white..(9)

9. Reddish bulblets present in the upper leaf axils; often along snow-runoff rivulets.................................. ***S. cernua***
9' Bulblets absent; alpine ....(10)

10. Inflorescence with erect pedicels; pedicels with short, straight, glandular hairs; calyx lobes often shorter than the hypanthium; hypanthium narrowly campanulate .... ***S. hyperborea***
10' Inflorescence with spreading pedicels; pedicels with long, crinkly, glandular hairs; calyx lobes equaling or exceeding the hypanthium; hypanthium broadly campanulate .... ***S. rivularis***

Saxifraga adscendens L. (ascending) Wedgeleaf saxifrage. Short-lived perennial, caulescent herb; herbage glabrous to densely pubescent. **FLOWERING STEMS** usually 1–few, 2–8 cm tall, unbranched; arising from basal rosette; strongly glandular-pubescent; leaves gradually tapering to broadly winged petiole-like base. **LEAVES**: basal leaves 5–15 mm long, sessile, wedgelike, deeply 3-lobed to palmately compound lobes, apically 2–5 acute teeth. **INFL** a 4–14-flowered, nodding raceme. **FLOWER** hypanthium long-obconic-elongate at anthesis, becoming elongate in fruit; pedicellate, sepals triangular; corolla widely spreading, petals white, occasionally pink, usually with prominent vein markings, 7–16 mm long, always 3-cleft; ovary inferior. **FRUIT** a capsule, 3–5 mm long. **SEEDS** smooth or wrinkled. [*Muscaria adscendens* (L.) Small]. Along streams; subalpine to alpine, tundra. COLO: Arc, Con, Hin, LPl, SJn. 3505–4145 m (11500–13600'). Flowering: Jul–Aug. British Columbia east to Alberta, south to Colorado and Utah.

Saxifraga bronchialis L. **CP** (windpipe; reference is unknown) Yellow dot saxifrage. Slender, perennial, caulescent herb. **FLOWERING STEMS** 1–several, 1.5–3 dm tall, unbranched, arising from basal rosette; leaves 2. **LEAVES**: basal leaves light green, sparsely pubescent, orbicular, short, petiolate, to 8 cm long, blade simple to irregularly 3–5-lobed to almost pinnatifid; cauline leaves palmately compound, much reduced, more highly dissected than basal leaves, always appearing pinnatifid; stipules broad, fimbriate, decurrent on petiolar base. **INFL** a 3–12-flowered, compact raceme. **FLOWER** hypanthium campanulate or hemispheric, becoming elongate-campanulate in fruit; slightly pendulous; sepals triangular, valvate in bud, becoming widespread at anthesis; corolla widely spreading, open at throat; petals white to cream, with dark red spots in upper 1/2, yellow or orange spots in lower 1/2, ovate, 3–7 mm long, ovary less than 1/2 inferior. **FRUIT** a capsule 4–6 mm long. **SEEDS** smooth or wrinkled. [*Ciliaria austromontana* (Wiegand) W. A. Weber]. Open rocky slopes and meadows; subalpine to alpine. COLO: Arc, Con, Hin, LPl, Min, Mon, RGr, SJn. 2530–3720 m (8300–12200′). Flowering: Jun–Sep. Alaska and Yukon south to Oregon, Idaho, and New Mexico. Our plants belong to **var.** ***austromontana*** (Wiegand) M. Peck (of southern mountains).

Saxifraga cespitosa L. (tufted) Tufted alpine saxifrage. Small, usually densely caespitose perennial from woody-branched caudex with taproot. **FLOWERING STEMS** caudex branches ascending, covered with densely imbricate leaves; scape erect, up to 25 cm tall, often densely glandular-pubescent. **LEAVES** mostly in a crowded basal rosette, narrowly cuneate with a broad petiole, almost always distinctly 3-lobed; cauline leaves 1–10 mm long, upper leaves progressively smaller, usually 3-lobed, bractlike. **INFL** a 1–5-flowered cyme, pedicels up to 10 mm long, glandular-pubescent. **FLOWER** sepals 1–3 mm long, erect to spreading, triangular, obtuse, glandular-pubescent, often purple; petals 1–5 mm long, obovate, gradually narrowed to broad base, only slightly clawed, with 3 veins, white or tawny; ovary inferior at anthesis. **FRUIT** a capsule to 7 mm long, more or less obovate, dark brown. [*Muscaria monticola* Small; *S. caespitosa* subsp. *exaratoides* var. *purpusii* Engl. & Irmsch.]. Fell-fields and rocky slopes; subalpine and alpine tundra communities. COLO: Hin, LPl, Min, Mon, SJn. 3535–4235 m (11600–13900′). Flowering: Jul–Aug. Circumboreal, south to Nevada, Arizona, and New Mexico.

Saxifraga cernua L. (nodding) Nodding saxifrage. Delicate perennial herbs from fibrous root system. **FLOWERING STEMS** 1–3, erect, 10–20 cm tall, glandular-pubescent and viscid-villous, often rusty villous at base; leaves distributed along stem, more numerous, larger at base; upper leaves smaller, fewer lobes, entire, sessile. **LEAVES**: basal leaves petiolate, arranged in loose rosette, blades wider than long, 4–20 mm wide, 3–15 mm long, petioles 5–9 cm long; leaf blades subrotund, 3–9-lobed, upper stem leaves reduced, entire at top of stem, bearing small clusters of red-purple to black bulbils in axils; basal leaves bearing cream-colored, ricelike bulbils at petiole base. **INFL** with 1–3 flowers produced at top, often only bulbils on the stem. **FLOWER** hypanthium up to 2 mm long, turbinate; petals 3–12 mm long, obovate-cuneate, rounded, not clawed, white with 3–5 pink or purple veins; ovary superior, only inferior at the base. **FRUIT** a capsule that rarely develops. Rocky talus. COLO: Hin, LPl, SJn. 3320–4265 m (10900–14000′). Flowering: Jul–Aug. Circumboreal, south to New Mexico.

Saxifraga flagellaris Willd. ex Sternb. (with flagella, referring to the stolons) Whiplash saxifrage. Stoloniferous perennial or biennial herbs from slender rhizomes; stolons leafless, up to 25 cm long, terminating in bud that develops roots and tufts of glandular-ciliate leaves. **FLOWERING STEMS** with solitary scapes, erect, to 15 cm long. **LEAVES** with a basal rosette; glandular-pubescent with purplish gland-tipped hairs; 5–20 mm long, 1–5 mm wide, oblanceolate to narrowly spatulate, acuminate, setose-tipped, entire, coarsely glandular-ciliate; cauline leaves numerous, similar to basal leaves, but reduced as produced up flowering stem. **INFL** of 1–3 flowers, occasionally more. **FLOWERS** with a showy bright yellow corolla; pedicels 5–9 mm long, stipitate-glandular; sepals green to red-purple, oblong to lanceolate or ovate, 2–5 mm long, stipitate-glandular, glands purplish; petals 6–12 mm long, yellow, 7–9-veined; stamens longer than sepals, filaments subulate, ovary 1/4 inferior; hypanthium 1–4 mm long, turbinate-campanulate, completely adnate to ovary, acute to rounded, stipitate-glandular. **FRUIT** a capsule 4–8 mm long. **SEEDS** about 0.6 mm long, ovoid-lanceolate, nearly smooth, brown. [*Hirculus platysepalus* (Trautv.) W. A. Weber subsp. *crandallii* (Gand.) W. A. Weber]. Alpine; tundra, krummholz, rocky; forbs, grasses. COLO: Con, Hin, LPl, Min, Mon, RGr, SJn. 3505–4205 m (11500–13800′). Flowering: Jun–Sep. Circumboreal, south to Arizona and New Mexico.

Saxifraga hyberborea R. Br. (extreme northern) Pygmy saxifrage. A loosely tufted and weak perennial. **FLOWERING STEMS** 4–14 cm tall, glabrous or nearly so. **LEAVES** with blades 6–17 mm wide, suborbicular or reniform, usually 5-palmately lobed; stem leaves lobed or entire, basal and lower cauline long-petioled. **INFL** of 1–3 flowers. **FLOWER** calyx tube 2.5–4 mm long; calyx lobes 2–3 mm long, shorter than the tube, ovate or oblong-ovate; petals 3–7 mm long, white, oblong, somewhat clawed at the base; ovary partly inferior. **FRUIT** a 5–7 mm long capsule. [*Saxifraga debilis* Engelm. ex A. Gray]. Subalpine to alpine; usually in dry, shaded alcoves and under boulders. COLO: Con, LPl, SJn. 2895–4265 m (9500–14000′). Flowering: Jul–Aug. Montana south to Colorado and Utah. Our material belongs to **var.** ***debilis*** (Engelm.) Á. Löve, D. Löve & B. M. Kapoor (debilitated or weak).

Saxifraga odontoloma Piper, (toothed margin) Brook saxifrage. Scapose perennial herb from stout, horizontal rootstock. **FLOWERING STEMS** with usually a single scape, 15–80 cm tall. **LEAVES** all basal, somewhat fleshy, glabrous, blades 1–10 cm wide and long, orbicular to reniform, cordate or truncate at base, margins coarsely dentate or crenate; petioles 2–3 cm long; leaves on scapes glabrous below, glandular above. **INFL** open-spreading, a several-flowered cyme, stipitate-glandular. **FLOWER** sepals 2–3 mm long, strongly deflexed in anthesis, purplish; petals 2–4 mm long, white, spreading; stamens equalling or exceeding petals. **FRUIT** a capsule 5–9 mm long. **SEEDS** 0.7–1.2 mm long, fusiform, weakly longitudinally ribbed, pale brown. [*S. arguta* D. Don, misapplied; *Micranthes odontoloma* (Piper) W. A. Weber]. Often along stream banks, meadows, and seeps; subalpine and alpine communities. COLO: Arc, Con, Hin, LPl, Min, Mon, RGr, SJn; UTAH. 2775–3870 m (9100–12700′). Flowering: Jul–Sep. Alaska south to California, east to Alberta, and south to New Mexico.

Saxifraga oregana Howell (of Oregon) Oregon saxifrage. Scapose perennial herbs from a stout, fleshy, simple caudex, often rhizomatous. **FLOWERING STEMS** scapose, single, 20–120 cm tall, pubescent with white hairs tipped with yellow or pink to red viscid glands. **LEAVES** all basal, thick, somewhat succulent; petiole tapering gradually, broad-oblanceolate, sinuate-denticulate margin, up to 20 cm long, up to 7 cm wide. **INFL** a many-branched, glandular-pubescent cyme, 20–200 flowers in loose clusters or heads. **FLOWER** hypanthium turbinate to cupulate, completely adnate to ovary; pedicels 0.5–4 mm long; sepals 1.2–3 mm long, usually deflexed, deltate-ovate, obtuse, glabrous, green; petals 1.5–4 mm long, 0.5–2.5 mm wide, deciduous, usually clawless, white; filaments 1–2 mm long, inserted at edge of disc, white, greenish white, or purplish; anthers 0.5–0.7 mm long, pale orange. **FRUIT** a follicular capsule, usually up to 7 mm long, green to purplish. **SEEDS** 1 mm long, ellipsoid. [*Micranthes oregana* (Howell) Small]. Talus, meadows, and streamsides; subalpine to alpine. COLO: SJn. 3655–3960 m (12000–13000′). Flowering: Jul–Aug. Washington and Oregon, south to California, east to Montana, and south to Colorado.

Saxifraga rhomboidea Greene (diamond-shaped) Diamond leaf saxifrage. Scapose perennial herbs to 30 cm tall from short rootstock, fibrous roots. **FLOWERING STEMS** usually single, sometimes several, with scapes 3–30 cm tall, upper definitely glandular-pubescent. **LEAVES** all basal, petioles 0.3–2.5 cm long, usually pilose-ciliate fringed, broad, flattened, gradually transposed into blade, blades 9–35 mm long, 3–20 mm wide, rhombic, obovate, or ovate, dentate or crenate to entire, not lobed, glabrous or ciliate. **INFL** cymose-paniculate, globose, densely congested; glandular; 10–40 flowers. **FLOWER** hypanthium adnate to ovary; sepals 1–2 mm long, triangular, not reflexed; petals 2–4 mm long, white; stamens longer than sepals, subequal to petals. **FRUIT** a follicular capsule 4–6 mm long. **SEEDS** less than 1 mm long, ellipsoidal, wrinkled longitudinally. [*Micranthes rhomboidea* (Greene) Small]. Meadows; ponderosa pine, Douglas-fir, spruce-fir, and alpine communities. ARIZ: Apa; COLO: Arc, Con, Hin, LPl, Min, Mon, RGr, SJn; NNEX: McK, SJn; UTAH. 2590–3750 m (8500–12300′). Flowering: May–Sep. British Columbia south to New Mexico.

Saxifraga rivularis L. (from a streamside) Bract saxifrage. Small, delicate, perennial herb from a fibrous root system; herbage often somewhat reddish. **FLOWERING STEMS** 1–several, erect, up to 15 cm tall, simple to sparingly branched, glabrous or sparsely to densely pubescent. **LEAVES** flabelliform to reniform, blades 3–8 mm long, 5–15 mm wide, 3–7-lobed, base cuneate to subcordate; petioles slender, 5–45 mm long, cauline leaves often reduced, entire; basal leaves sometimes bearing cream to brownish, ricelike bulbils in axils; cauline leaves more numerous at base, reduced, broadly ovate to lanceolate, 3–5 (–7) lobes to entire, bractlike. **INFL** a raceme; 1–5 flowers. **FLOWER** hypanthium 1–1.4 mm long, 2–3 mm wide at anthesis, glandular-puberulent; sepals oblong-ovate, 2–2.5 mm long, acute or rounded at apex; petals white to pale rose, sometimes purplish, 3–5 mm long. **FRUIT** an ovoid capsule 4–5 mm long, glabrous, strongly nerved. **SEEDS** dark brown to dark gray, 0.3–0.5 mm long, testa wrinkled, minutely reticular-foveolate. Often near melting snowbanks; talus slopes and meadows; Engelmann spruce and alpine communities. COLO: Arc, Con, Hin, LPl, Min, Mon, RGr, SJn. 2960–4220 m (9715–13840′). Flowering: Jul–Oct. British Columbia south to California, east to Montana and Arizona.

Saxifraga serpyllifolia Pursh (thyme-leaved) Gold bloom saxifrage. Low, tufted, perennial plant from slender rhizomes. **FLOWERING STEMS** 1–7 cm tall, glandular-pubescent, hairs often purple-tipped. glabrous or glandular when present. **LEAVES**: basal leaves mostly fleshy, pubescent above, glabrous below; 4–12 mm wide, linear-oblanceolate to narrowly spatulate, rounded, entire, glabrous; 1–3 cauline leaves may be lacking, but occasionally 1–5, alternate. **INFL** 1–3-flowered; peduncle 5–25 mm long, glandular-pubescent. **FLOWER** hypanthium short, adnate to ovary; sepals 2–4 mm long, becoming sharply deflexed, ovate, rounded, externally pubescent; petals 4–8 mm long, 2–5 mm wide, orbicular to broadly obovate, narrowed to a short claw; 5–9-nerved, golden yellow, often darker at base when aged; stamens inserted around edge of ovary, longer than sepals but shorter than petals; ovaries slightly inferior. **FRUIT** a capsule 6–8 mm long, ovoid. **SEEDS** 1 mm long, reticulate-wrinkled [*S. chrysantha* A. Gray; *Hirculus serpyllifolius* (Pursh) W. A. Weber subsp. *chrysanthus* (A. Gray) W. A. Weber]. Fell-fields, rock stripes; alpine tundra communities. COLO: SJn. 3200–4115 m (10500–13500′). Our material belongs to **var. *chrysantha*** (A. Gray) Dorn (yellow-flowered). Wyoming south to New Mexico.

SCROPHULARIACEAE Juss. FIGWORT FAMILY

(ALSO SEE OROBANCHACEAE, PHRYMACEAE, and PLANTAGINACEAE)

J. Mark Porter

Annuals and perennials, herbs (ours) or shrubs. **LEAVES** opposite, the petioles sometimes connate, or alternate, sometimes whorled or largely basal, without stipules or with stipulelike structures, simple to dissected. **INFL** raceme or spike, bracts foliaceous to reduced. **FLOWERS** perfect; calyx of 4 or 5 distinct or united sepals, regular or irregular; corolla sympetalous, actinomorphic to evidently zygomorphic, 4- or 5-lobed, often 2-lipped, tubular, campanulate or sometimes rotate, rarely wanting; stamens epipetalous, often 4, or 4 fertile and 1 sterile (referred to as a staminode, e.g., *Scrophularia*), rarely 5 fertile (*Verbascum*), alternate with the corolla lobes; anthers 2-celled with equal cells; ovary superior, 2-carpelled, 2-locular, the anatropous ovules on axile placentae; style solitary and entire, capitate or flattened. **FRUIT** a capsule (ours), dehiscence septicidal, sometimes loculicidal at the apex, or both (4-valved), berry, drupe, or schizocarp. **SEEDS** usually many and small, rarely few, wingless or winged, the embryo small, endosperm abundant or absent. n = 6 or more. Scrophulariaceae sensu stricto includes 65 genera and about 1700 species; however, only *Scrophularia* and *Verbascum* occur in our area. The traditional circumscription of Scrophulariaceae sensu lato (see for example the floras of Harrington, McDougall, Martin and Hutchins, and Weber) includes many genera that are not particularly close in their relationship to one another. In fact, some are far more closely related to other well-recognized families than they are to the genus *Scrophularia*, where the family name is attached. Within the context of our flora, members of Scrophulariaceae sensu lato are placed into four families: Orobanchaceae (7 genera), Phrymaceae (1 genus), Plantaginaceae (13 genera), and Scrophulariaceae (sensu stricto, 2 genera) (Olmstead et al. 2001; Oxelman et al. 2005).

1. Corolla globular to urceolate; fertile stamens 4, staminode flattened, scalelike; leaves opposite, rarely subopposite; inflorescence an elongate thyrsoid panicle .... ***Scrophularia***

1′ Corolla rotate; fertile stamens 5, staminode absent; leaves alternate on the stem and clustered in a basal rosette; inflorescence a congested spikelike panicle .... ***Verbascum***

Scrophularia L. Figwort

(Latin *scrofula*, a rare disease characterized by swelling of the neck glands, reputedly cured by a European species of *Scrophularia*) Perennial herbs or rarely shrubs. **STEMS** quadrangular. **LEAVES** opposite, decussate, petiolate, the blades irregularly serrate or divided, lanceolate to triangular-ovate. **INFL** a thyrsoid panicle; bracts lanceolate, reduced, alternate or subopposite, shorter than pedicels. **FLOWERS** small; calyx deeply and subequally 5-parted, the lobes broad, obtuse to rounded; corolla greenish yellow or greenish purple to dark reddish brown, the tube globular or urceolate, bilabiate, the upper lip 2-lobed, flat, erect, external in bud, the lower lip shorter, with erect lateral lobes and a deflexed middle lobe; staminode (sterile filament) flattened, narrowly clavate to broadly fan-shaped (in ours), attached to the corolla on the distal, upper tube, appressed; fertile stamens 4, the anthers 1-celled, usually included at the throat, filaments attached at the tube base (proximally); stigmas united, capitate. **FRUIT** a septicidal capsule, the walls firm. **SEEDS** many, oblong-ovoid, turgid, furrowed. $2n$ = 18, 20, 24, 26, 36, 40. A genus of 150–200 species, mostly Eurasian, but also in New World temperate and tropical regions. Ten species are native to North America; one occurs in our area (Shaw 1962).

Scrophularia lanceolata Pursh (armed like a spear, pointed, referring to the leaf shape) Lanceleaf figwort. Perennial herb, glandular-puberulent. **STEMS** 80–150 cm tall, clustered. **LEAVES** with petioles 2–4 cm long, blades 8–14 cm long, 3.5–7 cm wide, lanceolate to lance-ovate, apex acuminate, base rounded, truncate, or cordate, serrate. **INFL** a narrow, elongate panicle. **FLOWER** calyx 2–3.8 mm long, the lobes ovate to broadly rounded, the margins scarious and slightly erose; corolla 8–14 mm long, pale reddish brown or greenish brown, the tube urceolate, throat slightly constricted; staminode fan-shaped, 1–2 mm wide, wider than long, yellow-green, sometimes purplish; anthers 0.8–1.1 mm wide. **FRUIT** a capsule, 5.5–10 mm long, ovoid, acuminate. **SEEDS** 0.8–0.9 mm long. $2n$ = 92–96. Semimoist stream banks, thickets, woods, roadsides, and fencerows; sagebrush, aspen, and spruce-fir. ARIZ: Apa; COLO: Arc, Hin,

LPl, Min, Mon; NMEX: RAr; UTAH. 1400–2800 (3300) m (4600–10800′). Flowering: Jun–Jul. Fruit: Jul–Aug. Southern British Columbia, Washington, Oregon, and northern California (mostly east of the Cascade Range), east across northeast Nevada, and through Idaho, Utah, Montana, Wyoming, Colorado, and northern New Mexico to the northeastern United States.

Verbascum L. Mullein

(*barbascum*, ancient name used for some species) Biennial or perennial, taprooted herbs, or rarely small shrubs. **STEMS** tall, coarse, simple or branched. **LEAVES** basal and/or cauline, basal ones forming a rosette, the cauline alternate, simple, sessile, clasping or somewhat decurrent, pinnately veined. **INFL** a raceme, spike, or congested panicle; bracts reduced. **FLOWER** calyx deeply 5-parted, the lobes equal; corolla, ours usually yellow, sometimes white or occasionally purple, rotate, slightly irregularly 5-lobed, the upper 2 lobes external in bud and slightly shorter than the lower 3, the tube very short; stamens 5 and usually all anther-bearing, the filaments usually villous and more or less alike, or the lower pair glabrous to sparsely hairy; anthers 1-celled; staminode absent; style flattened distally, the stigmas united and capitate. **FRUIT** a septicidal capsule, walls firm, ovoid to globose. **SEEDS** numerous, densely packed around the placentae, ours obconic, with thick, sinuous, longitudinal ribs which sometimes anastomose, forming areolas. $2n$ = 30, 32, 36. A large genus of about 360 species, occurring chiefly in southeastern Europe and southwestern Asia. Several species have been introduced and naturalized in North America, including our only representative.

Verbascum thapsus L. (long) Common mullein, woolly mullein, wupaviva (Hopi). Stout biennial, producing a rosette of leaves the first year and a tall, stout, flowering stem the second, 30–200 cm tall. **STEMS** erect, winged by the decurrent leaf bases; densely yellow-tomentose with stellate or dendritic hairs. **LEAVES**: basal 8–50 cm long, 2.5–14 cm wide, oblanceolate to obovate, obtuse, tapering to a petiolate base, entire to shallowly crenate, sometimes persisting for 2 years; cauline leaves progressively reduced upward, sessile, decurrent, oblanceolate. **INFL** a compact spikelike panicle; pedicels very short. **FLOWER** calyx (5) 8–12 mm long, the segments lanceolate; corolla 12–30 (35) mm in diameter, yellow or rarely white, the lobes scurfy-pubescent on back, sometimes ciliate with stellate hairs; stamens of 2 distinctly dissimilar types, the lower 2 with glabrous to sparsely villous filaments and decurrent anthers, the upper 3 with yellow-villous filaments. **FRUIT** capsule 7–10 mm long, broadly ovoid, densely tomentose with stellate or dendritic hairs, becoming scurfy at maturity. **SEEDS** 0.7–0.8 mm long. $2n$ = 32, 34, 36. Moist, gravelly roadsides and waste places; sagebrush, piñon-juniper, ponderosa pine, and spruce-fir. ARIZ: Apa, Nav; COLO: Arc, Dol, Hin, LPl, Min, Mon, SJn, SMg; NMEX: McK, RAr, SJn; UTAH. 1220–2810 m (4000–9200′). Flowering: Jun–Jul. Fruit: late June–Aug. A native of Eurasia, but naturalized throughout most of temperate North America. Leaves were gathered by the Hopi, then dried and smoked in combination with *Macromeria viridiflora* A. DC. to treat mental disturbances.

SIMAROUBACEAE DC. QUASSIA FAMILY

Linda Mary Reeves

Primarily trees or shrubs, often dioecious, most herbage with bitter quassinoids. **LEAVES** alternate, rarely opposite; pinnately compound, rarely simple. **INFL** a raceme, panicle, or solitary. **FLOWERS** perfect or imperfect, actinomorphic; sepals 3–8, often connate; petals 3–8, often connate; disc present; stamens numerous, usually twice as many as petals; styles 2–5; carpels 2–5, often partially connate, ovary superior. **FRUIT** a capsule, often schizocarpic, drupe or samara-like. x = 8–13+. About 25 genera, 150 species, primarily tropical and subtropical. Several species important medicinally, especially *Quassia* and *Picramnia. Simarouba glauca*, tree-of-paradise, is a native of the U.S. (Florida).

Ailanthus Desf. Tree of Heaven

(Moluccan name meaning tree of heaven) Trees or large shrubs; polygamodioecious; bark gray. **LEAVES** deciduous, mostly large, odd-pinnately compound, pungent when crushed. **INFL** a large, terminal panicle. **FLOWERS** often pungent; calyx lobes 5–6; petals 5–6; stamens 10–12 in staminate flowers, 2–3 in perfect flowers; ovary 2–5-lobed. **FRUIT** samara-like, 1–5 per flower, 1 per carpel, pendulous. About 15 species, primarily eastern Asia and northeastern Australia.

Ailanthus altissima (Mill.) Swingle (very tall) Tree of heaven, homestead tree, copal tree. Slender polygamodioecious tree or large shrub rapidly gaining height to 20 m tall, often in thickets of clones from rhizomes. **STEMS** green, glandular-puberulent, bark smooth, wood soft, with large pith. **LEAVES** glandular-puberulent; dark green with light, prominent venation; 10–100 cm long; leaflets 9–31, 5–15 cm long, elliptic-oblong to lanceolate, acuminate, often falcate, with 1–6 large glands on basal lower margin. **INFL** 10–40 cm long. **FLOWERS** small, greenish to whitish or yellowish; staminate flowers malodorous; calyx less than 1 mm long; petals 2–3 mm long, enfolded with woolly pubescence on

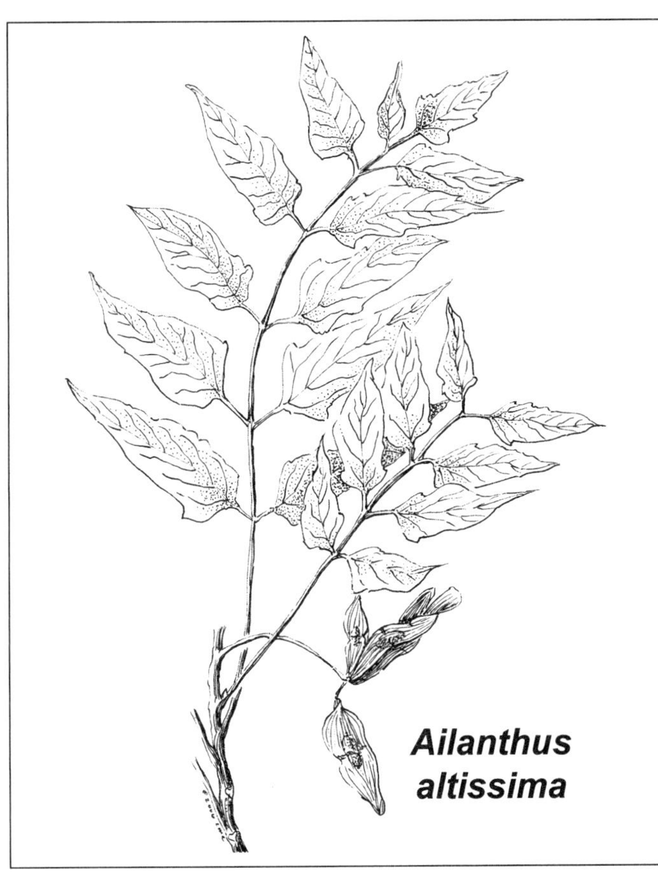
Ailanthus altissima

or near margin; disc lobed; filaments basally pubescent; ovary 2–5 flat, unicarpellate sections. **FRUIT** samara-like, 3–5 cm long, 3–7 mm wide, linear, curved or oblique at base, seed near the center. $2n = 64$. [*A. glandulosa* Desf.] Weedy cultivated tree, escaped near buildings, in alleyways, vacant lots, and persisting at old homesteads and mine sites. ARIZ: Nav; COLO: LPl, Mon; NMEX: McK, SJn; UTAH. 1300–2000 m (4400–6500′). Flowering: May–Jul. Fruit: Jun–Jul. Extensively planted in the United States and Canada from Ontario and Maine south and west to Florida, Washington, and California, commonly escaping, weedy and persisting. Native to China. Resistant to pollution and extremely hardy even in dry conditions. The starring plant in the play *A Tree Grows in Brooklyn*.

SOLANACEAE Juss. POTATO FAMILY

William J. Litzinger

Annual, biennial, perennial herbs or shrubs; glabrous to strongly pubescent. **LEAVES** simple or sometimes pinnately divided, alternate though may become subopposed in the inflorescence, estipulate, usually petioled. **INFL** determinate, cymose, axillary to extra-axillary, often reduced to a few or solitary flowers in the axils. **FLOWERS** usually actinomorphic to slightly zygomorphic, perfect, hypogynous, often showy, (4) 5 (7) to numerous; calyx sympetalous with usually 5 lobes or teeth, persistent and often enlarging in fruit; corolla sympetalous with usually 5 lobes, rotate, funnelform or salverform, variously colored, plicate or convolute; androecium of 5 stamens, alternate with the corolla lobes, epipetalous; filaments distinct; anthers basifixed, sometimes connivent, sometimes with enlarged connective, equal or unequal, dehiscing longitudinally or by apical pores or both; gynoecium of 1 pistil, 2-carpellate; ovary superior, 2-locular (or sometimes falsely 3–5), often surrounded at base by a fleshy disc; ovules numerous, anatropous to campylotropous, placentation axile or basal; style single; stigma 2-lobed. **FRUIT** a fleshy to dry berry, sometimes enclosed in a persistent inflated calyx, or capsule. **SEEDS** numerous, flattened and discoid, smooth to pitted or alveolate, endosperm copious, fleshy; embryo straight or bent. $x = 7–12$. Cosmopolitan, with approximately 75 genera and over 3000 species; most abundant in Central and South America. The family is important worldwide for humans as food, medicines, ornamentals, and narcotic and toxic plants. Many are invasive, weedy plants that have greatly extended their range by human activity. Use of plants in this family is extensively documented in the contemporary and historic ethnobotanical literature of Southwestern indigenous peoples. Remains, especially charred seeds, of several species in this family are frequently reported from archaeological excavations.

1. Woody shrubs; fruit fleshy or dry and bony; corolla funnelform .... ***Lycium***
1′ Herbaceous annuals, biennials, or perennials; fruit various; corolla various .... (2)

2. Flowers large, over 10 cm long, white .... ***Datura***
2′ Flowers various, but smaller than above; colors various .... (3)

3. Fruit a circumscissile capsule; calyx enlarged at maturity .... ***Hyoscyamus***
3′ Fruit a berry or capusule, not circumscissile; calyx not enlarged at maturity .... (4)

4. Corolla rotate, or nearly so, anthers connivent, mostly exserted .... ***Solanum***
4′ Corolla rotate or tubular, anthers not connivent, mostly not exserted .... (5)

5. Corolla rotate, with tomentose pads alternating with the stamens; calyx herbaceous, not inflated, closely vesting the berry but open above .... ***Chamaesaracha***
5′ Corolla rotate or tubular, but without tomentose pad; calyx not as above .... (6)

6. Calyx inflated, papery in fruit, nearly closed at top .... ***Physalis***
6' Calyx not as above .... (7)

7. Flowers in racemes or panicles .... ***Nicotiana***
7' Flowers solitary or axillary .... ***Petunia***

Chamaesaracha (A. Gray) Benth. False Nightshade

(*chamae*, on the ground, + *saracha*, for Isidore Saracha, a Spanish monk) Rhizotamous to taprooted herbaceous perennial. **STEMS** decumbent, branched from base. **LEAVES** entire to deeply pinnately lobed, scarcely to densely pubescent, hairs simple, dendritic, or stellate, sometimes gland-tipped, with margined petioles. **INFL** in axillary clusters, 1–2 (4)-flowered. **FLOWERS** small, 5-lobed; calyx sometimes accrescent and closely investing and partially to largely concealing the fruit; corolla rotate, white to yellowish to greenish; androecium sometimes purple-tinged, glandular to velvety tomentose in fornices between stamens; anthers free, yellow, generally smaller than the slender, terete filaments, dehiscing by lengthwise slits; gynoecium with style 1. **FRUIT** a small, usually whitish, rather dry berry, spheric, seed-bearing only on the basal part of the placentae. **SEEDS** more or less flat, reniform, strongly alveolate to rugose-favose, the embryo curved. About nine species, especially in the southwest United States and Mexico.

Chamaesaracha coronopus (Dunal) A. Gray (horny base) **STEMS** many from base, 10–50 cm long; decumbent or prostrate, especially when young or grazed, when protected becoming more erect at maturity, hairs becoming sparse with age, eglandular, branched and mostly stalked; sessile to subsessile. **LEAVES** (1) 2–6 (15) mm wide, oblong-lanceolate to linear, entire to deeply lobed, margin sinuate-dentate. **INFL** with pedicels ± 1 cm, in fruit 2 cm, reflexed. **FLOWER** calyx 3–5 mm, in fruit 5–10 mm, lobes deltoid, about equal to the corolla tube, in fruit 2 mm, strongly pubescent, the hairs varying from all stellate and either sessile or stalked, to all elongate and simple, or often mixed; corolla greenish white, sometimes dull purplish speckled, the fornices white-tomentose, anthers (1.5) 2 mm long. **FRUIT** 5–8 mm wide, with calyx closely enclosing except at the top. **SEEDS** about 2.5 to 3 mm long, very prominently rugose-favose. $2n = 48, 72$. Blackbrush and desert scrub communities; weedy. ARIZ: Apa, Coc, Nav; COLO: LPl, Mon; NMEX: McK, SJn; UTAH. 1125–1830 m (3700–6000′). Flowering: Apr–Jul. California to Texas, Colorado, Kansas, and New Mexico.

Datura L. Jimsonweed, Thorn-apple

(ancient Hindu name) Coarse, rank-smelling herbs. **STEMS** erect or spreading. **LEAVES** alternate, short-petioled, entire to sinuate-dentate or lobed. **FLOWERS** solitary, large, 8–20 cm long, erect on short peduncles in the forks of branching stems, green-tinged when plicate, opening whitish purple to violet, opening in the evening at least in some species; calyx tubular, 5-toothed, circumscissile near the base in ours, with the lower part persistent as a collar below the capsule; corolla funnelform, convolute-plicate in bud; stamens included in ours, filaments filiform, the anthers dehiscent by lateral slits, stigma 2-lobed; ovary 2- or falsely 4-celled. **FRUIT** in ours a spiny capsule, dehiscing irregularly. **SEEDS** numerous, large. About 25 species, in warmer parts of all continents. The plants are poisonous and have narcotic properties. Extensively used by indigenous peoples of the Southwest and worldwide.

1. Fruit erect; flowers 5–10 cm long; annual .... ***D. quercifolia***
1' Fruit pendent; flowers mostly longer than 10 cm; perennial .... ***D. wrightii***

Datura quercifolia Kunth (oak leaf) Oak-leaved thorn-apple. Annual, glabrate to pubescent. **LEAVES** ovate, 6–16 cm long, 3–6 cm wide, the upper surface glabrescent, the lower surface pubescent, mostly along veins, acute, pinnately lobed, lobes often toothed; base cuneate to subcordate; petioles sometimes as long as or longer than the blade. **FLOWERS** with calyx 1.5–4 cm long, teeth acute to acuminate, 2–6 mm long; corolla white, light violet to purple, tube 3.8–7.7 cm long, the limb 1–2 cm wide. **FRUIT** erect, dehiscing along 4 sutures to base, green, ovoid to ellipsoid, 3–4.5 cm long, terminal spines 2–3.5 mm long; pericarp glabrate to puberulent. **SEEDS** reniform to discoid, 3–4.5 mm long, 2.5–3.5 mm wide, rugose. Roadsides and waste ground. NMEX: SJn. 1675–1830 m (5500–6000′). Flowering: May–Oct. Southeast California, southern Arizona, New Mexico, and Texas south to central Mexico.

Datura wrightii Regel **CP** (for Charles Wright, botanical explorer of the Southwest) Perennial erect herb. **STEMS** widely branched, 0.5–1.5 m tall and broad, minutely grayish pubescent throughout with fine, eglandular, more or less crisp hairs less than 0.2 mm long, the underside of leaves somewhat velvety to touch. **LEAVES** unequally ovate, (4–) 12–20 (25) cm long, irregularly repand to subentire, on somewhat short petioles. **INFL** 1-flowered, in the forks of branching stems, erect, becoming downturned in fruit. **FLOWERS** open in the early evening and usually persist for 1 day; calyx 7–10 cm long, initially enclosing the corolla, teeth initially connivent and somewhat plicate to convolute in bud, the

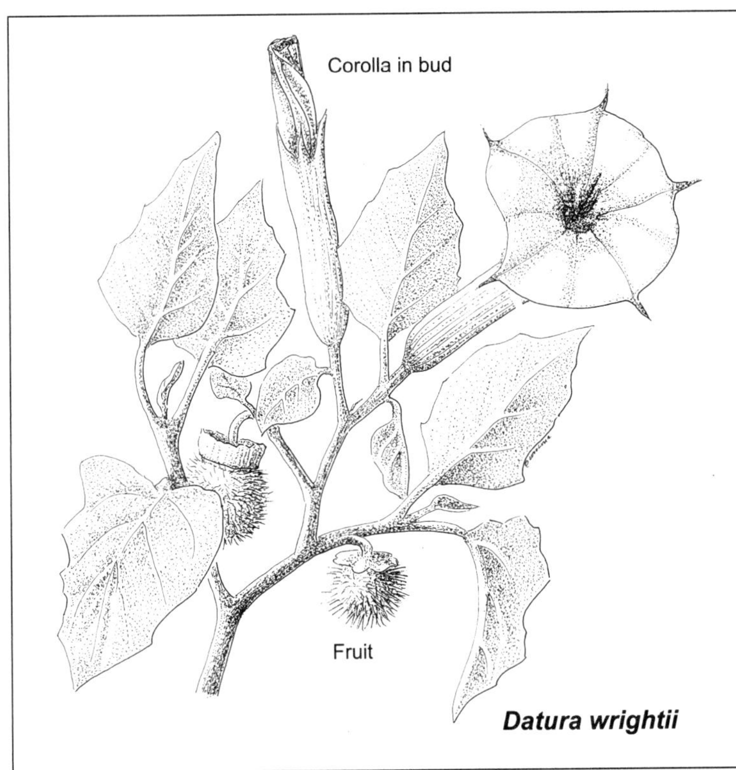

Datura wrightii

persistent base usually rotate, sometimes reflexed; corolla green in bud, opening to white, suffused with violet, (8) 12–20 cm long, corolla teeth becoming tendril-like. **FRUIT** a capsule, nodding, 2.5–3 cm, densely spiny and usually puberulent; with irregular dehiscence; spines 5–12 mm long. **SEEDS** subreniform, buff to light brown, about 5 mm long, flattened, smooth, with a cordlike margin. 2*n* = 24. [*Datura metel* L. var. *quinquecuspida* Torr.; *Datura meteloides* Dunal]. Blackbrush, desert scrub, sagebrush, and piñon-juniper woodland communities. ARIZ: Apa, Nav; COLO: LPl, Mon; NMEX: McK, SJn; UTAH. 1125–2070 m (3700–6800′). Flowering: May–Oct. Utah, Colorado, California to Texas, and northern Mexico.

Hyoscyamus L. Henbane

(*hys*, sow, + *kyamos*, bean) Annual or biennial herbaceous plants. **STEMS** leafy. **LEAVES** alternate, lobed or pinnatifid. **INFL** solitary in the upper axils and in terminal racemes, these more or less 1-sided to helicoid. **FLOWER** calyx urn-shaped to campanulate, 5-cleft, enlarging and becoming more or less reticulate in fruit, enclosing the capsule; corolla greenish yellow to whitish with dark rose or purple veins, funnelform, with slightly oblique 5-lobed limb; stamens exserted, anthers opening by longitudinal slits; ovary 2-celled, stigma capitate. **FRUIT** a capsule, circumscissile above the middle, included in the calyx. About 15 species in Europe, North Africa, to southwest and central Asia.

Mature calyx

Hyoscyamus niger

Hyoscyamus niger L. (black) Black henbane. **STEMS** 30–100 cm tall, stout, viscid short-villous. **LEAVES** 6–20 cm long, oblong-ovate to lanceolate, irregularly lobed, cleft, or pinnatifid, sessile to upper clasping, viscid and short-villous. **FLOWER** calyx 2–2.5 cm long in fruit; corolla mostly 1.5–2 cm long. **FRUIT** a capsule 10–14 mm long. A weed of roadsides and waste places in piñon-juniper woodland communities. NMEX: SJn. 1970 m (6000′). Flowering: Jun–Aug. Naturalized from Europe and widespread in North America. All parts of the plant are poisonous.

Lycium L. Wolfberry, Box-thorn

(named for Lycia, in Asia Minor) Armed shrubs, erect or spreading, sometimes scrambling over supports. **LEAVES** often fascicled, entire to minutely dentate, glabrous and glaucous to glandular or pubescent, fleshy in some. **FLOWERS** solitary or 2–4 in the axils; calyx campanulate to tubular, commonly ruptured by the fruit; corolla whitish to purplish or greenish purple, regular, tubular to funnelform, 4–6-lobed; anthers affixed near their middle. **FRUIT** 2-celled, from dry and bony to fleshy and juicy. **SEEDS** 2–many; embryo coiled. About 100 species of temperate to tropical zones; dry to humid regions.

1. Corolla tube less than 2.5 mm .... ***L. andersonii***
1′ Corolla tube greater than 2.5 mm .... (2)

2. Corolla tube 4–7 mm; uncommon; spreading or viny, of waste places .... ***L. barbatum***
2′ Corolla tube greater than 12 mm; widespread; low shrub-forming colonies .... ***L. pallidum***

Lycium andersonii A. Gray (for Frederick W. Anderson, botanist) Water jacket. Plants glabrous or subglabrous, branches armed, stiffly spreading to erect, spreading. **LEAVES** (1) 2–15 (20) mm, more or less linear, more or less ternate, fleshy. **FLOWER** calyx 1.5–3 mm, cup-shaped, lobes (2) 4–5, from 1/4 to nearly as long as the tube; corolla narrowly funnelform, white to yellowish, with lobes often bluish purple to violet-tinged, tube slender, 5–10 mm, the lobes (4) 5, 1–1.5 mm long, ciliate to glabrous; androecium with stamens more or less included to slightly exserted, unequal, attached at 1/3 from base, hairy on the free base; gynoecium with style equaling to slightly exceeding the stamens. **FRUIT** 3–8 mm, red or orange, juicy. **SEEDS** several to many, about 1.5 mm long. Blackbrush, desert scrub, sagebrush, and piñon-juniper communities; sandy washes and rocky slopes. ARIZ: Nav; UTAH. 1125–1525 m (3700–5000′). Flowering: Mar–May. California to southern Nevada and Utah, western New Mexico south to Mexico.

Lycium barbarum L. (foreign or strange) Matrimony vine. Plants glabrous, usually armed but may be unarmed, the thorns short if present, branches curved or arched, recumbent or climbing shrubs. **LEAVES** 2–6 cm, lanceolate to oblanceolate. **FLOWER** calyx 3–4 mm, bell-shaped, lobes (1) 2–3 long; corolla rotate-campanulate to short-funnelform, purple to lavender, lilac, rose-tinged, fading tan, tube 4–6 mm, abruptly expanded to the throat, lobes generally spreading; stamens unequal, attached at the middle of the tube, hair-tufted at base. **FRUIT** about 1 cm, fleshy, bright salmon-red. **SEEDS** 10–20. $2n = 24$. Gardens, escaping along roadsides, and old homesites. [*L. halimifolium* Mill.]. COLO: LPl, Mon. 1525–2135 m (5000–7000′). Flowering: Mar–Oct. An escaped cultivar found widely in North America; native to Eurasia. *L. chinense* Mill. is another cultivated species that may be expected in our area. It has corolla tubes with shorter lobes that are more or less hidden by the calyx.

Lycium pallidum Miers (pale) Pale wolfberry. Intricately branched shrub, 1–2 m tall, upright, but with spreading branches, rhizomatous, forming clonal clumps or "stands" of several to many square meters. **STEMS**: young stems glabrous to sparsely pubescent, maturing to reddish or silver-gray, the older bark becoming fissured. **LEAVES** 1–4 cm long, strongly glaucous when young, green, glabrous, in fascicles on older branches, alternate on new growth. ovate-elliptic, oblong-oblanceolate to oblong-spatulate to lanceolate, upper leaves often lanceolate. **FLOWER** calyx 5–8 mm long, the lobes as long as or longer than the tube; corolla 15–25 mm long, plicate in bud, narrowly funnelform, sometimes slightly lipped, greenish or tinged with purple, sometimes slightly reticulately purple-veined, fading tan; stamens usually slightly exserted, 1–2 mm. **FRUIT** about 1 cm wide, glaucescent, red to reddish blue, rarely yellow, ovoid to nearly pyriform. **SEEDS** 2–many. $2n = 24$. Blackbrush, desert scrub, piñon-juniper, and ponderosa pine communities. ARIZ: Apa, Coc, Nav; COLO: LPl, Mon; NMEX: McK, RAr, San, SJn; UTAH. 1220–1875 m (4000–6150′). Flowering: Mar–May. California through southern Nevada, southern Utah, Colorado, to southwest Texas and south to Mexico. Our material belongs to **subsp. *pallidum***. Pale wolfberry shows considerable variability, including forms with gray or reddish stems at maturity; forms that have fleshy leaves rather than the more common thin leaves; forms with distinctly zygomorphic flowers and calyces that are open on one side, cupping the flower and fruit; and forms with reduced numbers of seeds, typically 1–8, and occasionally seedless. Pale wolfberry flowers infrequently or may flower but not produce viable seed; it exhibits masting behavior, producing abundant fruit crops in cycles of (1) 3–7 or more years. It is frequently found on prehistoric Pueblo archaeological sites, near old sheep camps, animal pens, and along roadsides where it appears to reproduce by spreading rhizomes, often aggressively invading graded or heavily disturbed areas from existing stands. Occasionally it is found in large stands covering several hectares.

Nicotiana L. Tobacco

(for Jean Nicot, French ambassador to Portugal who introduced tobacco into France) Heavy-scented annual or perennial herb, usually viscid-pubescent. **LEAVES** large, alternate, entire or repand. **INFL** in terminal panicles or racemes. **FLOWER** calyx tubular-campanulate or ovoid, 5-cleft, persistent; corolla funnelform, salverform, or nearly tubular, the limb usually shallowly 5-lobed and spreading; filaments filiform; ovary 2 (4)-celled. **FRUIT** a capsule, 2- or 4-valved at summit. **SEEDS** many, small, ovoid to reniform, minutely reticulate-punctate. Approximately 60 species, mostly in North and South America, some important in commerce and for ornamentals.

1. Annual; may have glandular hairs, but not viscid-villous; all leaves petioled, never hastate .... ***N. attenuata***
1′ Perennial; viscid-villous throughout; lower leaves petioled, upper hastate .... ***N. obtusifolia***

Nicotiana attenuata Torr. (weak or meager) Mountain tobacco. Annual herb, erect, simple or branched, glandular-pubescent to glabrate, 30–200 cm tall. **LEAVES** 5–15 (25) cm long, ovate to lanceolate-ovate, mostly petioled. **INFL** in a raceme or panicle. **FLOWERS** about 2.5–3 cm long; calyx ovoid-campanulate, 6–8 mm long, the teeth deltoid; corolla white to cream to reddish, often infused with purple, sometimes purple-banded or with a green tube and white lobes. **FRUIT** a capsule, about 8–12 mm long, rarely to 20 mm. Solitary or in dense populations in disturbed sites, roadsides, abandoned fields, and for 1–3 years after fires. ARIZ: Apa, Coc, Nav; COLO: LPl, Mon; NMEX: McK, RAr, San, SJn; UTAH. 1220–2285 m (4000–7500′). Flowering: Jun–Sep. Used by aboriginal peoples for smoking. British Columbia to northern Baja California, Mexico; Idaho to Texas and northern Mexico.

Nicotiana obtusifolia M. Martens & Galeotti (blunt-lobed) Desert tobacco. Perennial herb, viscid-pubescent, 20–50 cm tall, erect, simple or few-branched. **LEAVES**: lower petioled, upper ovate to lanceolate, sessile and auricled, 2–8 cm long. **INFL** loosely paniculate or racemose, pedicels 5–10 mm long. **FLOWER** calyx 6–12 mm long; campanulate, with lanceolate-subulate lobes approximately as long as the corolla tube; corolla greenish white, 18–22 mm long, constricted at orifice, the limb 8–10 mm, broad, sometimes slightly zygomorphic. **FRUIT** a capsule, 2-valved, 8–10 mm long. **SEEDS** dark brown, about 0.6 mm long. [*N. trigonophylla* Dunal]. Cliff faces, in rocks, along arroyos in blackbrush and desert scrub communities. ARIZ: Nav; UTAH. 1125–1220 m (3700–4000′). Flowering: Mar–Nov. California, Arizona, Utah, and New Mexico to Texas and northern Mexico.

Petunia Juss. Petunia

(*petun*, name for tobacco) Viscid annual or perennial herb. **LEAVES** entire, the upper often subopposed. **INFL** terminal, solitary, or axillary. **FLOWER** calyx 5-parted, 10-nerved; corolla funnelform, minute or showy; stamens 5, unequal, 4 being didynamous, the fifth shortest, with a hypogynous fleshy disc. **FRUIT** an ovoid capsule, with 2 undivided valves. **SEEDS** minute, spherical or angled. About 40 species, largely South American; some of horticultural value.

Petunia parviflora Juss. (small-flowered) Prostrate, diffusely branched annual, glandular-puberulent. **STEMS** 10–40 cm long. **LEAVES** oblong-linear to oblong-spatulate, 0.4–1.2 cm long, rather fleshy. **FLOWER** calyx subsessile, 3–4 mm long, purplish with a whitish tube, the short lobes somewhat unequal. **FRUIT** a capsule 2–3 mm long. **SEEDS** about 0.6 mm long, pale brown, favose-reticulate. Streamsides and lake margins, often among cottonwood, *Baccharis*, and saltcedar stands. ARIZ: Apa; UTAH. 1220–1525 m (4000–5000′). Flowering: Apr–Sep. Southern California to Arizona, Texas, and Florida.

Physalis L. Ground-cherry, Husk Tomato

(bladder, referring to the inflated calyx) Annual to somewhat suffrutescent perennial herbs, subglabrous to pubescent with simple to dendritic or stellate hairs. **LEAVES** simple, alternate or somewhat unequally geminate, petiolate to subsessile, entire to sinuate-pinnatifid. **INFL** regular, usually solitary or sometimes in 2s or 3s, on mostly drooping pedicels. **FLOWER** calyx campanulate to cup-shaped, shallowly 5-lobed, shorter that the corolla at anthesis, but becoming greatly inflated, dry, and membranous and completely enclosing the fruit; corolla usually yellowish, but sometimes white or bluish, often darker spotted at the base of the limb, 5-lobed; stamens 5, the filaments terete to flattened and broadly clavate in outline; anthers yellow or sometimes blue, dehiscing by full-length slits. **FRUIT** a fleshy, many-seeded berry. **SEEDS** flattened, the embryo curved. About 80 species.

1. Corolla blue to purple (rarely white), with a white, woolly star at the center; herbage scurfy with crystalline vesicles .... ***P. lobata***
1′ Corolla yellow, yellowish green, or white, with or without darker spots; plant surfaces generally not scurfy, without crystalline vesicles .... (2)

2. Fruiting calyx with 5 sharp folds; corolla 6–7 mm long; anthers blue .... ***P. subulata***
2′ Fruiting calyx not sharply folded; corolla more than 7 mm long; anthers yellow .... (3)

3. Pubescence of forked to stellate or dendritic hairs .... ***P. hederifolia***
3′ Pubescence of simple hairs only .... (4)

4. Plants annual, never rhizomatous .... ***P. philadelphica***
4′ Plants perennial, often rhizomatous .... (5)

5. Stems freely branched and mostly strongly zigzag, rather brittle; restricted in our area to desert scrub communities; simple hairs throughout .... ***P. crassifolia***
5′ Stems not zigzag-branched, not brittle; not found in desert scrub communities; pubescence of the leaves minute or none, almost wholly restricted to the main veins .... ***P. longifolia***

Physalis crassifolia Benth. (thick-leaved) Perennial or subshrub less than 80 cm tall; hairs simple, dense, short, generally glandular. **STEMS** mostly zigzag, rigid. **LEAVES** 1–3 cm long, generally ovate, fleshy, entire or more or less wavy; petiole more or less equal to the blade. **INFL** with pedicels 15–30 mm long in fruit, smaller in flower. **FLOWER** calyx 20–25 mm long, weakly angled; corolla 15–20 mm in diameter, widely bell-shaped, yellow; anthers 2–3 mm long, yellow. [*Physalis genucaulis* A. Nelson]. Open desert and rocky ravines and slopes in desert scrub communities. UTAH. 1125–1220 m (3700–4000′). Flowering: Mar–May. California, southern Nevada, northwest Arizona, and southern Utah south to Mexico.

Physalis hederifolia A. Gray (ivy-leaved) Ivy-leaved ground-cherry. Herbaceous perennial to subshrub, 10–80 cm tall; from fleshy rhizomes; hairs sometimes branched or glandular, sometimes found only on the calyx. **LEAVES** ovate, entire to coarsely toothed, 2–4 cm long, generally appearing green-gray, base tapered or subcordate. **INFL** pedicels 3–5 (10) mm in fruit. **FLOWER** calyx 7–10 mm long at anthesis, up to 30 mm in fruit; lobes triangular to lanceolate, with 10 green veins; corolla 10–15 mm long, widely bell-shaped, yellow, often with 5 purple-brown spots inside the base. $2n = 24$.

1. Hairs short, not multicellular, forked to dendritic, nonglandular .... **var. *fendleri***
1′ Hairs partially or wholly long and multicellular, sometimes branched and sometimes glandular .... **var. *hederifolia***

var *fendleri* (A. Gray) Cronquist (for Augustus Fendler, German-American botanical collector) With hairs generally branched, generally nonglandular. [*Physalis fendleri* A. Gray; *P. fendleri* var. *cordifolia* A. Gray; *P. hederifolia* var. *cordifolia* (A. Gray) Waterf.]. Sagebrush, piñon-juniper, ponderosa pine, and Gambel's oak communities. ARIZ: Apa, Nav; COLO: Arc, Dol, LPl, Mon; NMEX: McK, San, SJn; UTAH. 1370–2285 m (4500–7500′). Flowering: May–Oct. California, southern Nevada, southern Utah, Arizona, and New Mexico, to Oklahoma and Texas, and south to Mexico.

var. hederifolia With hairs sometimes branched or glandular. Sagebrush, piñon-juniper, Gambel's oak, and mountain brush communities. ARIZ: Apa; COLO: Mon; NMEX: McK, SJn. 1600–2000 m (5250–6560′). Flowering: May–Oct. Southwest Utah, eastern Arizona, New Mexico, and south to Mexico.

Physalis lobata Torr. (lobed) Purple ground-cherry. Perennial herb, rhizomatous, often forming patches. **STEMS** spreading or prostrate, 10–50 cm long, usually diffusely branched, sparsely hairy with rounded, white, unicellular, pustulate or papillate hairs. **LEAVES** lanceolate to ovate, 1–7 cm long, entire or lobed, gradually tapering into a petiole that may be shorter or longer than the blade. **INFL** with flowers commonly in pairs in the leaf axils, sometimes solitary or in small clusters, with pedicels 3–4.5 mm, not generally longer in fruit. **FLOWER** calyx 3–6 mm long at anthesis, in fruit 15–20 mm, the lobes about 1/2 as long as the tube; corolla 15–20 mm in diameter, blue to violet or purple with a white, woolly star at the center; in fruit about 2 cm long and about as wide, sharply 5-angled, sunken at base. **FRUIT** greenish. Uncommon to rare, found mostly in desert grassland and piñon-juniper communities. NMEX: SJn; UTAH. 1525–1830 m (5000–6000′). Flowering: May–Oct. Kansas to Texas, Colorado, Nevada, Arizona, and northern Mexico.

Physalis longifolia Nutt. (long-leaved) Longleaf ground-cherry. Perennial herb with rather thick rhizomes, 30–60 (–100) cm tall, glabrous to somewhat pubescent with short, opaque, mostly appressed, obscurely stellate hairs on the upper part of the stem and on the pedicels and veins of the lower side of the leaves, and especially along the 10 principal nerves of the young calyx, often glabrate at maturity. **LEAVES** firm-textured, lanceolate or elliptic-lanceolate, 3–9 cm long, or narrowly rhombic and generally tapering to the petiole, entire or coarsely and irregularly blunt-toothed, the blade 4–10 cm long. **INFL** mostly solitary, with pedicels 1–4 cm long. **FLOWER** calyx campanulate, 7–10 mm long at anthesis, 25–35 (40) mm long in fruit, the lobes triangular or ovate, 3–4 mm long at anthesis, tube minutely strigose, mostly in 10 narrow longitudinal strips, with minute, appressed, simple or obscurely septate hairs, rarely as much as 0.5 mm long; corolla yellow, blotched with purple or reddish purple at the base of the limb, 11–20 mm long, 10–15 mm wide; anthers yellowish or bluish, 3–4 mm long. **SEEDS** about 2 mm long, minutely pitted. [*P. virginiana* Mill. var. *sonorae* (Torr.) Waterf.; *P. virginiana* var. *subglabrata* (Mack. & Bush) Waterf.]. A weed in fields and waste places; sagebrush, piñon-juniper woodland, and mountain brush communities. ARIZ: Apa; COLO: Mon; NMEX: McK, RAr, San, SJn; UTAH. 1370–2285 m (4500–7500′). Flowering: Apr–Aug. The Colorado Plateau east to the Atlantic. Four Corners material belongs to **var. *longifolia***.

Physalis philadelphica Lam. (of Philadelphia) Annual, much-branched, erect or spreading, 30–90 cm tall, glabrous, or the younger parts puberulent. **LEAVES** with blades ovate, cuneate, or cordate at base, entire to sinuate-dentate, 1.5–6 cm long, on equally long petioles. **INFL** pedicels 4–8 mm long. **FLOWER** calyx in flower 3.5–4.5 mm long, the deltoid lobes shorter than the tube, in fruit calyx ovoid, 1.5–2 cm long; corolla yellow with a dark center, 8–15 mm broad. **FRUIT** purple at maturity. $2n = 24$. [*P. ixocarpa* Brot. ex Hornem. var. *philadelphica*]. An orchard weed; escapes from gardens, and mostly found in disturbed sites. NMEX: McK. 1220–1980 m (4000–6500′). Flowering: Jun–Oct. Northern California to the Atlantic coast and Mexico. This species is sometimes cultivated for the fruits.

Physalis subulata Rydb. (awl-shaped) Chihuahuan ground-cherry. Annual, 10–60 cm tall, foliage with usually dense, more or less yellowish or brownish, capitate-glandular hairs. **STEMS** usually branched. **LEAVES** with blades 3–6 cm long, ovate to oblong-ovate or lanceolate-ovate, margins toothed, or sometimes sinuate-dentate; petioles 1/2 to 3/4 as long as the blades. **FLOWER** calyx in flower 3–3.5 mm long on peduncles usually 1.5–3 mm; in fruit 2–3 cm long; with 5 sharp folds, the teeth acuminate and more or less ovate in outline; peduncles mostly 4–7 mm long; corolla 6–7 mm long, bluish spotted; anthers (0.3) 1–1.5 (2) mm long, bluish, filaments filiform. [*Physalis foetens* Poir. var. *neomexicana* (Rydb.) Waterf.; *Physalis neomexicana* Rydb.]. Often in cultivated fields; piñon-juniper woodland, and ponderosa pine communities. NMEX: McK. 1830–2135 m (6000–7000′). Flowering: Aug–Sep. New Mexico and adjacent Arizona and Colorado. Our material belongs to **var. *neomexicana*** (Rydb.) Waterf. ex Kartesz & Gandhi.

Solanum L. Nightshade

(classical Latin name used by Pliny) Annual to shrub or vine, glabrous to pubescent or tomentose, often glandular, sometimes prickly. **LEAVES** alternate to subopposite, often unequal, entire to deeply pinnately lobed. **INFL** commonly umbels or cymes, often 1-sided. **FLOWER** calyx 5-cleft or -toothed, more or less bell-shaped; corolla more or less rotate, 5-angled or -lobed, plaited in bud, white to purple, yellow; stamens 5, inserted on the corolla tube; filaments short, with anthers free, larger than the filaments, oblong or tapered, opening by 2 pores of short slits near the tip; ovaries 2-chambered, style 1, stigma small, capitate or bilobed. **FRUIT** a spheric berry or dry capsule. **SEEDS** many, compressed, generally reniform. More than 1500 species worldwide, especially in the tropics. Many species cultivated for food, including the potato (*S. tuberosum*) and the tomato (*S. lycopersicum*). Some species are grown as ornamentals and many are toxic.

1. Stems and leaves with prickles .... (2)
1′ Stems and leaves without prickles .... (4)

2. Anthers all alike; leaves linear .... ***S. elaeagnifolium***
2′ Anthers dissimilar; leaves once- or twice-pinnatifid .... (3)

3. Flowers yellow .... ***S. rostratum***
3′ Flowers violet .... ***S. heterodoxum***

4. Leaves with blades deeply lobed or pinnatifid .... (5)
4′ Leaves with blades entire to shallowly lobed .... (7)

5. Leaf blades deeply pinnatifid, with acute, triangular lobes .... ***S. triflorum***
5′ Leaf blades less deeply pinnatifid, with broad, rounded lobes .... (6)

6. Flowers purple; a somewhat woody vine .... ***S. dulcamara***
6′ Flowers white; low herb with small round tubers .... ***S. jamesii***

7. Corolla shallowly lobed, lobes shorter than the tube .... ***S. xanti***
7′ Corolla deeply lobed, lobes generally longer than the tube .... (8)

8. Leaves thick; plants hirsute or glandular-villous; calyx enlarged at maturity and enclosing the fruit base .... ***S. physalifolium***
8′ Leaves thin, translucent; plants glabrous or nearly so; calyx remaining small, not covering part of the fruit at maturity .... ***S. americanum***

Solanum americanum Mill. (of America) Black nightshade. Annual herb or subshrub, more or less glabrous, or with hairs short, curved and more or less appressed. **STEMS** 30–80 (100) cm tall, slender, usually divergently branched, glabrous or nearly so. **LEAVES** with blades ovate, ovate-deltoid to lanceolate, entire or sinuately dentate, pale green, translucent to transmitted light, 2–10 cm long. **INFL** an umbel or more or less racemelike with slender peduncles 1–3 cm long, the pedicels usually shorter, becoming reflexed in fruit. **FLOWER** calyx 1–2 mm long, with lobes spreading to reflexed in fruit; corolla 3–6 mm wide, deeply lobed, white, sometimes faintly tinged with purple, 4–6 mm in diameter; anthers yellow, 1.4–2.2 mm long, style 2.5–4 mm long. **FRUIT** a berry, 5–8 mm in diameter, greenish or black, shiny. **SEEDS** pale cream, many, reticulate-pitted, 1–1.5 mm in diameter. $x = 12$. [*S. nodiflorum* Jacq.; *S. nigrum* L.]. A weed of cultivated ground; low valleys. ARIZ: Apa, Nav; COLO: Mon. 1525–1980 m (5000–6500′). Flowering: Jul–Oct. Canada to the eastern United States and south to Mexico.

Solanum dulcamara L. (bittersweet) Bittersweet, climbing nightshade. Perennial herb to subshrub. **STEMS** generally less than 1 m tall, climbing; more or less glabrous or with soft simple hairs, sometimes dendritic. **LEAVES** more or less cordate to deeply 1–2-lobed near the base, 5–12 cm long. **INFL** generally 8–30-flowered, in umbel-like clusters in the forked axils, with peduncles 1.5–4 mm long and pedicels 5–10 mm long. **FLOWER** calyx 3–4 mm long, with lobes broadly triangular and about 1 mm long; corolla 8–12 mm wide, deeply lobed, the lobes lanceolate, 6–8 mm long, spreading to reflexed, about 3 times as long as the tube, purple or violet, lobed base with 5 pairs of green spots; anthers about 5 mm long. **FRUIT** a berry, 8–12 mm in diameter, more or less ovoid, red. **SEEDS** 2–3 mm long. $2n$ = 24. A nonnative weed often found along fencerows. COLO: LPl, Mon; NMEX: SJn. 1585–2135 m (5200–7000′). Flowering: May–Sep. Canada to the eastern United States; native of Eurasia.

Solanum elaeagnifolium Cav. (marsh plant) Horse-nettle. Herbaceous perennial with tough, deep-seated rhizomes; stellate-silvery canescent throughout. **STEMS** up to 1 m tall; the stems, petioles, and midribs of the leaves usually sparsely to densely prickly with slender yellow spines 1–5 mm long. **LEAVES** with blades linear to oblong or lanceolate, 3–10 cm long, repand-dentate margin, or the smaller leaves entire, veins prominent below. **INFL** cymose. **FLOWER** calyx ovate to linear, about 1 cm long, calyces usually armed with straight yellow spines, little if at all accrescent in fruit, the lobes linear, shorter to slightly longer than the tube; corolla violet or blue (rarely white), 2–3 cm in diameter, with lobes ovate, but abruptly acute, longer than the tube; anthers 7–9 mm long, yellow, all alike. **FRUIT** a berry, globose, 9–14 mm in diameter, yellow or brown, smooth, glabrous. **SEEDS** flattened, about 3 mm in diameter, brown. $2n$ = 24. A native weed found in disturbed sites, especially roadsides. Blackbrush, sagebrush, and piñon-juniper woodland communities. ARIZ: Apa, Nav; COLO: Mon; NMEX: McK, SJn; UTAH. 1525–2135 m (5000–7000′). Flowering: Apr–Aug. Washington, Idaho, south to California, Nevada, Arizona, Colorado, Utah, New Mexico, east to North Carolina, south to Mexico; South America.

Solanum heterodoxum Dunal (*hetero*, other, + *doxum*, glory) Melon-leaf nightshade. Herbaceous annual with pubescence of glandular-tipped hairs mixed with a few 5-rayed bristles, and copiously armed with yellow spines, especially on the petiole, stems, and calyces. **LEAVES** resemble those of the watermelon; blades pinnatifid with rounded lobes, segments ovate to obovate and obtuse. **STEMS** branched, 15–80 cm tall. **INFL** cymose. **FLOWER** calyx enlarging and closely investing the fruit, armed with long, straight, straw-colored, sharp spines; corolla about 35 mm in diameter; violet; anthers yellow, the lowest one larger than the others, spreading, and tinged with purple. **FRUIT** a berry closely invested by the accrescent, spiny calyx. Waste places and roadsides. NMEX: SJn. 1830 m (6000′). Flowering: Jul–Sep. Arizona, New Mexico, Colorado (El Paso County), New Mexico, and west Texas to Mexico. Our material belongs to **var.** ***novomexicanum*** Bartlett (of New Mexico).

Solanum jamesii Torr. (for Edwin James, surgeon-naturalist) Wild-potato. Perennial herb with rhizomes bearing tubers about 10–15 mm in diameter. **STEMS** 10–45 cm long, erect or spreading, glabrous or nearly so to sparsely pilose. **LEAVES** pinnate with 5–9 leaflets, the leaflets lanceolate-oblong, entire or nearly so, glabrous or nearly so, 5–60 mm long, the terminal leaflet the largest, lateral leaflets smaller. **INFL** in a few- to several-flowered cyme. **FLOWER** calyx enlarging somewhat but not investing the fruit; corolla white, 12–18 mm in diameter; anthers yellow, about 5–6 mm long. $2n$ = 24. Piñon-juniper woodlands and mountain brush communities. ARIZ: Apa; COLO: Arc, LPl, Mon; NMEX: McK, San, SJn; UTAH. 1675–2285 m (5500–7500′). Flowering: Jul–Sep. Southern Utah, Arizona to northern Mexico, east to Colorado and New Mexico. The tubers are edible.

Solanum physalifolium Rusby (leaf like a *Physalis*) Hairy nightshade. Annual with herbage strongly viscid-villous with somewhat flattened hairs throughout. **STEMS** prostrate to decumbent or ascendingly branched, 10–50 (–100) cm tall. **LEAVES** with blades narrowly to broadly deltoid to oblong-lanceolate, (2) 3–6 (8) cm long, subentire to deeply sinuate-dentate, obtuse or acute at the apex, gradually or abruptly narrowed at the base, the petioles mostly 10–15 mm long. **INFL** corymbose, sometimes branched, with flowers on stout peduncles, 5–15 mm long or more, the pedicels shorter and swollen immediately below the calyx. **FLOWER** calyx with lobes deltoid-lanceolate, 1–1.5 mm long, shorter than the tube, becoming enlarged in fruit, 4–6 (9) mm long, the connate portion expanded and cupping the lower 1/2 of the berry; corolla white to greenish white, 3–5 mm in diameter, the lobes spreading, narrowly deltoid-lanceolate, (2) 2.5–3 mm long, white-villous near the tip outside; anthers erect, about 2 mm long, yellow; style barely exceeding the stamens. **FRUIT** a berry, globose, 6–7 mm in diameter, greenish or yellowish at maturity, sometimes brownish or reddish, many-seeded. **SEEDS** yellowish, finely reticulate-pitted, about 2 mm long. $2n$ = 24. [*S. sarrachoides* Sendtn.;

Bosleria nevadensis A. Nelson]. A weedy species found in fields, roadsides, and gardens. ARIZ: Apa, Nav; COLO: Mon; NMEX: McK. 1525–1980 m (5000–6500′). Flowering: May–Oct. A native of South America, it is now widespread from British Columbia south to California and Arizona.

Solanum rostratum Dunal (curved at the end) Buffalo-bur. Stout annual, densely stellate-pubescent throughout, armed with yellow spines 3–12 mm long. **STEMS** 20–80 cm tall, branched. **LEAVES** with blades oval to ovate, irregularly pinnately lobed or twice-pinnatifid, up to 5 cm wide, 12 cm long. **INFL** in few-flowered racemes with stout peduncles, 1–5 cm long. **FLOWER** calyx 5–7 mm long at anthesis, the lobes about as long as the tube; corolla yellow, 18–28 mm in diameter; anthers yellow, with 4 of them 5–6 mm long, and the fifth about twice as long, curved in the distal 1/3 and darker. **FRUIT** a berry, globose, 9–10 mm in diameter, smooth, glabrous, brown. **SEEDS** about 3 mm. $2n = 24$. A weed of disturbed sites. ARIZ: Apa; COLO: Arc, LPl, Mon; NMEX: McK, RAr, San, SJn; UTAH. 1525–2135 m (5000–7000′). Flowering: Jun–Sep. North Dakota to Wyoming, south to Mexico. Native east of the Rocky Mountains.

Solanum triflorum Nutt. (three-flowered) Cut-leaved nightshade. Annual, sparsely to copiously pubescent and often somewhat scurfy, with appressed hairs. **STEMS** prostrate to decumbent, 10–40 cm long, spreading up to 1 m broad. **LEAVES** numerous, with blades ovate, 2–4 (5) cm long, 1 cm broad, oblong-ovate to elliptic, pinnatifid, deeply lobed, the lobes lanceolate to linear, extending more than halfway to the midrib, the sinuses between the lobes rounded, on slender petioles, 10–25 cm long. **INFL** a cyme with 2–3 umbellate flowers, peduncles short, 1–1.5 (2) cm, and with stout pedicels, often about the same length as the peduncles. **FLOWERS** 1–3 together, 5–15 mm long, calyx 2–3 mm long at anthesis, in fruit enlarging up to 6 mm; strongly accrescent, not investing the fruit but persisting at its base, not at all spiny; corolla white, 7–9 mm in diameter; anthers all alike, about 3 mm long, the filaments glabrous. **FRUIT** a berry, globose, 6–12 mm in diameter, greenish, smooth. **SEEDS** about 2 mm long, yellowish to pale brown, finely reticulate-pitted. $2n = 24$. Blackbrush, sagebrush, piñon-juniper woodland, and mountain brush communities. ARIZ: Apa, Coc, Nav; COLO: Arc, Dol, LPl, Mon; NMEX: McK, RAr, San, SJn; UTAH. 1525–2285 m (5000–7500′). Flowering: May–Sep. Considered to be native to the Great Plains of North America and now widespread as a weed in southwestern Canada and throughout the western states, where it may or may not be native.

Solanum xanti A. Gray (for John Xantus de Vesey, a Hungarian lawyer) Chaparral nightshade. Suffrutescent perennial, usually with many slender stems. **STEMS** spreading, 20–100 cm long, the herbage strongly to sparsely short-villous throughout with a mixture of simple, 1-celled hairs, dendritic (forked or branched), and glandular hairs. **LEAVES** with blades ovate to lanceolate-ovate, 2–4 cm long or more, entire to subentire, rounded or acute to hastately lobed at the base. **INFL** mostly in 4–10 subumbellate cymes; peduncles 5–12 mm long, pedicels slender, basally cupulate-bracteolate, 5–15 (20) mm long. **FLOWER** calyx 3.5–5 (6) mm long at anthesis, the lobes deltoid, usually shorter than the tube; corolla deep blue to violet or dark lavender, 15–25 mm in diameter, lobed about 1/3 the length of the lobes, spreading, more or less abruptly acute; anthers about 4 mm long, yellow, filaments about 2 mm long. **FRUIT** a berry, globose, 6–12 mm in diameter, greenish, smooth. **SEEDS** about 2 mm long, yellowish to pale brown. $2n = 24$. Desert scrub and piñon-juniper woodland communities. ARIZ: Apa, Nav. 1525–1830 m (5000–6000′). Flowering: May–Jul. Oregon, California, eastward to eastern Nevada and south to Baja California.

TAMARICACEAE Link TAMARISK FAMILY

Kelly W. Allred

Much-branched shrubs or small trees of desert and saline habitats. **LEAVES** simple, stipules absent, blades entire, alternate, scalelike, often completely covering the twigs, with salt-excreting glands. **INFL** solitary or commonly in dense scaly spikes, racemes, or panicles. **FLOWERS** actinomorphic, small, 4- or 5-merous, perfect, with a hypogynous nectar disc; sepals 4–5, distinct; petals 4–5, distinct, alternating with the sepals; stamens 4–10 or many, distinct or connate at the base, the anthers opening by longitudinal slits; pistil of 2–5 united carpels; locule 1; ovules few to many and borne on parietal or basal placentae; ovary superior; style usually free; stigmas 2–5. **FRUIT** a loculicidal capsule. **SEEDS** with an apical tuft of hair; embryo straight, typically without endosperm. The family consists of four genera and 78 species native to Eurasia. The largest genus is *Tamarix*, with 54 species, several of which have been introduced widely in western North America.

Tamarix L. Saltcedar, Tamarisk

(ancient Latin name, possibly named for the Tamaris River in Spain) Deciduous or evergreen (ours) shrubs. **LEAVES** as in the family. **INFL** a bracteate raceme or spike, these sometimes aggregated into a terminal panicle. **FLOWERS** small, numerous; sepals 4–5, persistent; petals white to pink or reddish, sometimes persistent; stamens 4–5 (ours) or numerous; stigmas 3–4. **FRUIT** a capsule separating into 3–4 valves, many-seeded. **SEEDS** with a tuft of hair at the terminal end. $x = 12$. About 54 species native to Eurasia and Africa. The twigs of *T. mannifera* produce a sweet, gummy, white substance considered by some to be the source of the biblical manna (Allred 2002; Baum 1967, 1978; Di Tomaso 1998).

Tamarix chinensis Lour. (of China) Saltcedar. Multistemmed shrubs or small trees, to 6 m tall, forming dense thickets; bark of larger trunks reddish brown to dark brown or blackish. **LEAVES** scalelike, 1.5–3.5 mm long, lanceolate-triangular, deciduous with the branchlets. **INFL** of numerous racemes, each 2–7 cm long, 3–5 mm wide in flower (wider in fruit), often aggregated into terminal panicles. **FLOWERS** 5-merous, pedicel about as long as the sepals; sepals 0.5–1.3 mm long, entire or minutely toothed; petals 1–2.3 mm long, persistent in fruit; stamens 5, the filaments emerging from between the emarginate or rounded lobes of the nectary disc, which may be variable in shape, symmetry, and filament insertion. **FRUIT** a capsule, 3–4 mm long. $2n = 24$. [*T. pentandra* Pall., nom. illeg.; *T. ramosissima* Ledeb.]. Along watercourses, riverbeds, washes, and roadsides in desert and foothill communities. ARIZ: Apa, Nav; COLO: Arc, Dol, LPl, Mon, SMg; NMEX: McK, RAr, San, SJn; UTAH. Flowering: Apr–Sep (later farther south). Native to Eurasia, now introduced throughout western North America, southern Canada to northern Mexico. This is the common saltcedar that has overrun vast tracts of wetlands, riparian areas, and disturbed ground throughout the Southwest. All recent studies in the genus support the sinking of *T. ramosissima* into synonymy; populations in North America have hybridized extensively and cannot be differentiated, even though they remain distinct in their native ranges.

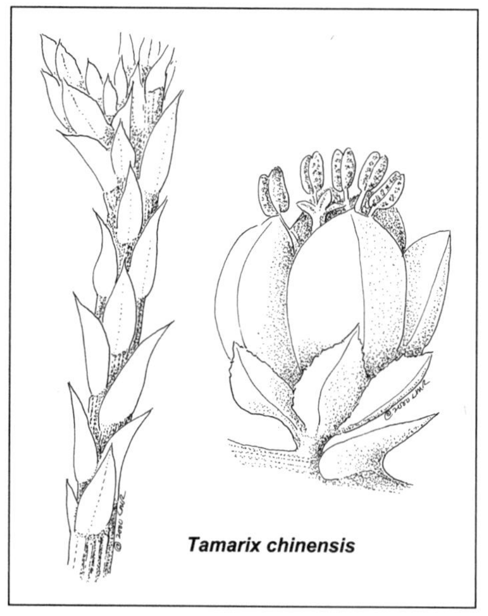
Tamarix chinensis

TYPHACEAE Juss. CATTAIL FAMILY

(including SPARGANIACEAE Hanin BUR-REED FAMILY)

C. Barre Hellquist

Perennial, monoecious, rhizomatous, caulescent, emergent or floating herbs. **LEAVES** basally disposed, cauline mostly 2-ranked; sheaths open, tapering into blade, auriculate or nonauriculate; blades flat or keeled, some twisted, apex acute, obtuse, or retuse, aerenchyma present. **INFL** 1, terminal, erect, emergent or floating, below or exceeding leaves, spikelike or globose; staminate flowers above pistillate flowers, deciduous, with a persistent aril; pistillate heads or spikes often subtended by a persistent or deciduous bract, axillary or supra-axillary. **FLOWERS** unisexual, staminate and pistillate on the same plant, wind-pollinated; staminate flowers stipitate or sessile, stamens 1–several; pistillate flowers hypogynous, sessile to stipitate, pistil 1, ovaries 1–2 (–3)-locular. **FRUIT** an achenelike drupe, or follicle. **SEEDS** with starchy or mealy embryo, cylindrical or straight (Hotchkiss and Dozier 1949; Smith 1987; Stuckey and Salamon 1987). Two genera, 22–27 species. *Typha* has been used by Native Americans of the Four Corners area for food and beverages, for smoking by the Hopi, Navajo, and Southern Paiute, for medicine by the Navajo, and for textiles by the Navajo and Ute.

1. Fruits in globose, burlike heads, green-brown; stems keeled or flattened, floating or erect .... ***Sparganium***
1′ Fruits in cylindric, spikelike heads, brown-cinnamon; stems flattened, twisted, erect.................................... ***Typha***

Sparganium L. Bur-reed

(Greek, probably a name used by Dioscorides, derived from *sparganon*, swaddling band, for strap-shaped leaves) Herbaceous, monoecious; freshwater, emergent or floating; rhizomatous. **LEAVES** flat, plano-convex, or keeled, spongy. **INFL** single, terminal, erect or floating; heads globose, sessile or peduncled; pistillate heads subtended by axillary or supra-axillary bracts. **FLOWERS**: staminate sessile, whitish; stamens 2–8, exceeding tepals; pistillate flowers sessile to stipitate, pistil 1, exceeding tepals; stigmas 1–2, white to greenish, linear, ovate, or subcapitate. **FRUIT** an achenelike drupe, often constricted at or near center, sessile or stipitate; tepals persistent, attached at base in most species. **SEEDS** 1–2 (–3), slender-ovoid. $x = 15$ (Cook and Nicholls 1986, 1987; Kaul 2002). Fourteen species. There appears to be no record of southwestern tribes utilizing bur-reeds. Unspecified species have been used as human food and cattle feed.

1. Staminate heads 1; pistillate heads 8–12 mm in diameter .... ***S. natans***
1′ Staminate heads 2 or more; pistillate heads 10–35 mm in diameter .... (2)

2. Stems submersed and floating; leaves flat, not keeled; fruit beak 1.5–2.2 mm long .... ***S. angustifolium***
2′ Stems erect; leaves keeled; fruit beak 2–4.5 mm long .... ***S. emersum***

Sparganium angustifolium Michx. (narrow-leaved) **STEMS** slender, floating, up to 2 m long. **LEAVES** flat or plano-convex, 2–5 (–10) mm wide, middle and upper leaves dilated with subinflated base. **INFL** unbranched, with 1 or more of the 1–3 pistillate heads supra-axillary, 1–3 cm in diameter; staminate heads 1–4 (–6). **FLOWERS** with tepals, stigma 1, lance-ovate. **FRUIT** usually reddish at base, 3–7 mm long, 1.2–1.7 mm thick, constricted at center, beak 1.5–2.2 mm long. **SEEDS** 1. $2n = 30$. [*S. angustifolium* var. *multipedunculatum* (Morong) Brayshaw; *S. emersum* Rehmann var. *multipedunculatum* (Morong) Reveal; *S. multipedunculatum* (Morong) Rydb.; *S. simplex* Huds. var. *multipedunculatum* Morong]. Acid waters of ditches, lakes, ponds, and rivers. COLO: Arc, LPl, Min, Mon, SJn; NMEX: McK, RAr, SJn; UTAH. 2630 to 2750 m (8675–9075′). Flowering: Jun–Jul. Fruit: Jul–Aug. Alaska east to Quebec and Newfoundland, south to California, New Mexico, Minnesota, Connecticut, New Hampshire, and Maine.

Sparganium emersum Rehmann (emergent) **STEMS** robust to slender, erect, to 0.8 (–2) m, some floating. **LEAVES** flat or slightly keeled, slightly dilated at base, 2–12 mm wide. **INFL** usually simple; pistillate heads 1–6, supra-axillary, sessile, 16–35 mm in diameter, the lowest borne 16–65 cm above the base; staminate heads 3–7 (–10), contiguous or not. **FLOWERS** with tepals linear-lanceolate, stigma 1. **FRUIT** green to reddish brown, lustrous, 3–4 mm long, 1.5–2 mm thick, slightly constricted below center, beak straight or curved, 2–4.5 mm long. **SEEDS** 1. $2n = 30$. [*S. angustifolium* Michx. subsp. *emersum* (Rehmann) Brayshaw; *S. chlorocarpum* Rydb.; *S. chlorocarpum* var. *acaule* (Beeby ex Macoun) Fernald; *S. emersum* subsp. *acaule* (Beeby ex Macoun) C. D. K. Cook & M. S. Nicholls]. Neutral to alkaline waters of ditches, ponds, lakes, and river shores. COLO: Hin, LPl, SJn; NMEX: McK, RAr, SJn. 2470–2850 m (8100–9350′). Flowering: Jun–Jul. Fruit: Jul–Aug. Alaska east to Quebec and Newfoundland, south to California, New Mexico, Nebraska, Ohio, and North Carolina.

Sparganium natans L. (swimming) Northern bur-reed. **STEMS** slender, floating or suberect when stranded, to 60 cm. **LEAVES** flat, unkeeled, 1.5–8 mm wide. **INFL** unranked, with 1–3 (–4) axillary pistillate heads, 8–12 mm in diameter; staminate head 1, terminal. **FLOWERS** with tepals, stigma 1, lance-ovate. **FRUIT** dark green or brown, subsessile, body ellipsoid to obovoid, slightly constricted at center, 3–3.5 mm long, beak 0.5–1.5 mm long. **SEEDS** 1. $2n = 30$. [*S. minimum* (L.) Fr.]. Acid, neutral, and slightly alkaline waters of ponds, lakes, and ditches. COLO: Arc, Hin, LPl, SJn; NMEX: SJn. 2630–2675 m (8675–9080′). Flowering: Jun–Jul. Fruit: Jul–Aug. Alaska east to Quebec and Newfoundland, south to California, Colorado, New Mexico, Illinois, Pennsylvania, and Rhode Island.

Typha L. Cattail

(Greek, possibly from *typhein*, to smoke or emit smoke, possibly used in maintaining smoky fires, or from the smoky color of the mature spikes) Herbs of fresh, brackish, or slightly saline water, emergent from horizontal rhizomes. **LEAVES** linear, persistent, twisted, slightly oblanceolate, plano-convex or concavo-convex. **INFL** a spike with staminate scales shorter than or exceeding flowers; pistillate spikes brown, orange-brown to cinnamon-colored at spike surface. **FLOWERS** with pistillate hairs exceeded by the stigmas. $x = 15$. About 8–13 species worldwide.

1. Staminate heads usually contiguous with pistillate heads; pistillate heads 24–36 mm thick, dark brown .... ***T. latifolia***
1′ Staminate heads separated from pistillate heads; pistillate heads 6–20 mm thick, medium brown to cinnamon brown, 13–25 mm thick .... (2)

2. Pistillate spikes medium brown; leaf sheath with terminal membranous auricles; flowering stems 2–3 mm thick at inflorescence .... ***T. angustifolia***
2′ Pistillate spikes cinnamon to orange-brown; leaves tapered to blades, occasionally with membranous auricles; flowering stems 3–4 mm thick at inflorescence .... ***T. domingensis***

Typha angustifolia L. (narrow-leaved) Narrow-leaved cattail. **STEMS** 0.75–1.5 (–3) m tall, flowering shoots 2–3 mm thick near the spike. **LEAVES** 3–8 mm broad, slightly convex on back, leaf sheath with terminal membranous auricles disintegrating late in season. **INFL** with staminate spike 3–15 cm long, pistillate spike 6–20 cm long, 5–6 mm thick, medium brown, with the pistillate spike separated by 1–8 (–12) cm, usually above the height of the leaves. **FLOWERS**: staminate, 4–6 mm, anthers 1.5–2 mm, pollen single or in clumps; pistillate flowers 2 mm, pistillate hairs green when young, straw-colored with orange-brown spots when dried. $2n = 30$. Fresh, brackish, or slightly saline wetlands and shores of ponds, lakes, and rivers. ARIZ: Apa; COLO: Arc, Mon, SJn; NMEX: SJn; UTAH. 1530–2015 m (5050–6650′). Flowering: Jun–Jul. Fruit: Jul–Aug. British Columbia, east to Quebec and Newfoundland, south to California, New Mexico, Louisiana, and South Carolina; Eurasia. Stuckey and Salamon (1987) believed that *T. angustifolia* was introduced from Europe into the coastal waters of the eastern United States. In more recent years it expanded its range in roadside ditches and disturbed habitats. This species has been documented hybridizing with *T. latifolia* as *T.* ×*glauca* Godr. and with *T. domingensis.* This species has been used as cordage, fiber, and fine matting by prehistoric native tribes. Modern tribes have used it as a kidney medicine, the mature heads chewed with tallow as gum, and the pollen used for baking. The tender young shoots are eaten raw. The fiber has been used in weaving baskets and for baby diapers, and the pollen as ceremonial skin decorations.

Typha domingensis Pers. (of Santo Domingo) Southern cattail. **STEMS** 1.5–4 m tall, flowering shoots 3–4 cm thick near the spike. **LEAVES** 6–18 mm broad, flat, leaf sheath with persistent, membranous auricles. **INFL** with staminate spike 6–35 cm long, 5–6 mm thick, cinnamon-brown; pistillate spike separated by (0–) 1–8 cm, usually above the height of the leaves. **FLOWERS**: staminate 5 mm, anthers 2–2.5 mm, pollen grains single; pistillate flowers 2 mm, pistil hair tips straw-colored to orange-brown in mass. $2n = 30$. Fresh or brackish wetlands, and shores of lakes, ponds, and rivers. ARIZ: Apa, Nav; COLO: Mon; NMEX: RAr, SJn; UTAH. 1350–1975 m (4775–6475′). Flowering: Jun–Aug. California, Wyoming, Nebraska, Illinois, and Delaware, south to Arizona, Texas, Louisiana, and Florida; Mexico; Central America, South America, West Indies; Eurasia, Africa; Pacific Islands; Australia. Flowering: Jul–Aug. Fruit: Jul–Oct. This is a highly aggressive species found throughout the United States and worldwide north and south of 40° latitude. Hybrids with *T. angustifolia* and *T. latifolia* occur throughout its range. Fruiting heads eaten like corn by various tribes nationwide. Fiber was widely used in roofing, mats, and bedding.

Typha latifolia L. (broad-leaved) Common cattail. **STEMS** stout, 1–3 m tall; flowering shoots 3–7 mm thick near the spike. **LEAVES** 6–23 mm broad, flat; leaf sheath with terminal membranous auricles. **INFL** with staminate spike 7–13 cm long, pistillate spike 2.5–20 cm long, 12–35 mm thick, dark brown, 24–36 mm thick in fruit, with the pistillate spike usually contiguous, rarely with 4 (–8) cm separation, usually at the same height as the leaves. **FLOWERS**: staminate 5–12 mm, anthers 1–3 mm, pollen in tetrads; pistillate flowers 2–3 mm, pistil hairs straw-colored. $2n = 30$. Fresh to slightly brackish wetlands, and shores of ponds, lakes, and rivers. ARIZ: Apa, Nav; COLO: Arc, Dol, Hin, LPl, Min, Mon, SJn, SMg; NMEX: McK, RAr, SJn; UTAH. 1950–2150 m (5400–7050′). Alaska east to Quebec and Newfoundland, south throughout the United States; worldwide. This is one of the most common aquatic plants worldwide. Flowering: Jun–Aug. Fruit: Jul–Oct. It has been introduced into parts of Australia. This species has been documented hybridizing with *T. angustifolia* L. as *T.* ×*glauca* Godr. and also with *T. domingensis*. *Typha latifolia* has been used as food and beverage by most of the Pueblos and for construction by the southern Tiwa. The Mescalero Apache have used the pollen as medicine. The Navajo have used the plant as a ceremonial medicine and emetic, and the leaves for baskets and water jugs. The Apache cooked the rootstalks with meat in the early growing season and used the fibers for matting and roofing material.

ULMACEAE Mirb. ELM FAMILY

Jane Mygatt

Trees or shrubs, deciduous. Bark grayish to brown, warty to deeply furrowed; sap watery. **LEAVES** alternate, simple; stipules deciduous, petiole present; blade often oblique at the base, margin entire or serrate; venation pinnate to palmate-pinnate. **INFL** few-flowered and axillary, or in cymose fascicles. **FLOWERS** hypogynous, bisexual or unisexual, small and inconspicuous, actinomorphic or nearly so; sepals persistent, usually 5; petals absent; stamens usually as many as calyx lobes and opposite them; pistil 1, of 2 carpels; styles 2, distinct. **FRUIT** a drupe or samara. About 18 genera, about 150 species in tropical and north-temperate regions. Plants of this family are wind-pollinated.

1. Leaf blade palmately veined at base, pinnately veined over remainder of blade; flowers usually unisexual; fruit a drupe .... ***Celtis***
1′ Leaf blade pinnately veined; flowers bisexual; fruit a samara .... ***Ulmus***

Celtis L. Hackberry, Sugarberry

Trees or shrubs to 8 m; crown spreading. Bark usually gray, with corky ridges. **LEAVES** pinnately veined but somewhat 3-veined from the base, upper surface usually conspicuously roughened to the touch; leaf blade deltate to ovate, thick; base cordate or oblique; margin entire or serrate-dentate. **FLOWERS** usually unisexual, staminate and pistillate on the same plants, along with a few bisexual flowers; pedicellate on branches of current year, appearing in midspring or late spring; staminate flowers with filaments incurved in bud, exserted after anthesis; gynoecium minute, rudimentary; pistillate flowers with calyx 4 (–5)-lobed; ovary sessile, ovoid; styles short, divided into 2 divergent, elongate, reflexed lobes. **FRUIT** a globose drupe; ripening in autumn, persisting after leaves fall. $x = 10$. About 60 species of tropical and temperate regions, worldwide.

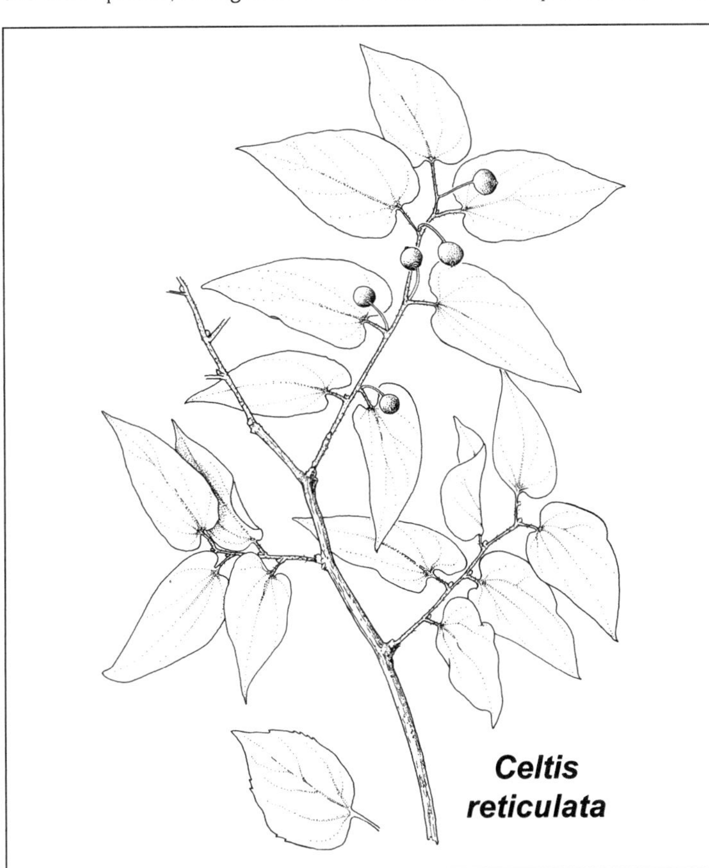
Celtis reticulata

Celtis reticulata Torr. (netted) Netleaf hackberry, palo blanco. Branches villous when young. **LEAVES** with petiole 3–8 mm long; leaf blade ovate, 3–6.5 × 2–4.5 cm, thick, rigid, base cordate or inequilateral, margins entire or serrate above the middle, apex obtuse, acute, or acuminate; surfaces pubescent, abaxially yellow-green, adaxially gray-green, grooved, ± scabrous; prominently reticulate-veined on lower surface. **INFL** of 1–4 flowers in axils of young leaves. **FRUIT** a drupe, orange-reddish when ripe, orbicular, averaging 8–10 mm diameter, beaked; pedicel 10–14 mm. [*C. brevipes* S. Watson; *C. laevigata* Willd. var. *reticulata* (Torr.) L. D. Benson; *C. douglasii* Planch.; *C. occidentalis* L. var. *reticulata* (Torr.) Sarg.; *C. reticulata* Torr. var. *vestita* Sarg.]. Along streams and rocky canyons; widespread throughout the Southwest. ARIZ: Apa, Nav; COLO: LPl, Mon; NMEX: RAr, SJn; UTAH. 1060–1830 m (3500–6000′). Flowering: Apr–May. Fruit: Jun. Navajos use *C. reticulata* medicinally in the treatment of indigestion.

Ulmus L. Elm, Orme, Olmo

(Latin for elm) Trees to 30 m; crowns open, with slender unarmed branches; twigs pubescent. Bark gray to brown. **LEAVES** late-deciduous; stipules early-deciduous; leaf blade ovate to elliptic, base usually oblique, margins serrate, venation pinnate. **INFL** in fascicles, subtended by 2 bracts, on branches of previous season, appearing in spring before leaves. **FLOWERS** bisexual, pedicellate or sessile; calyx 3–9-lobed; stamens 3–9; styles persistent, deeply 2-lobed. **FRUIT** a samara, broadly winged. Twenty to 40 species of northern temperate regions, most in Eurasia. Siberian elm (*U. pumila*) sometimes has been mistakenly called Chinese elm. Chinese elm (*U. parvifolia*) is a cultivated tree which flowers in late summer–early fall. *Ulmus parvifolia* is not included in this treatment, as herbarium records do not indicate this species has naturalized in the region.

Ulmus pumila L. (dwarf) Siberian elm. Introduced small to medium-sized tree. Bark rough and furrowed. **LEAVES** with petiole 2–7 mm long; leaf blade ovate to narrowly elliptic, 2–6.5 × 2–3.5 cm, base ± oblique; margins serrate, with many straight side veins, apex acute. **INFL** in clusters; 3 mm wide; greenish. **FLOWERS** with calyx shallowly lobed, lobes 4–5, glabrous; stamens 4–8, anthers brownish red; stigmas green, lobes exserted. **FRUIT** a samara, 10–15 mm diameter, flat, orbiculate, with notched wing 1/3 to 1/2 length; several clustered. $2n = 28$. [*U. campestris* L. var. *pumila* Maxim.; *U. manshurica* Nakai; *U. turkestanica* Req.]. Disturbed areas and roadsides, canyons, forested and riparian areas. ARIZ: Apa, Nav; COLO: Arc, Dol, LPl, Mon, SMg; NMEX: McK, RAr, San, SJn; UTAH. 1600–2500 m (5300–8200′). Flowering: Jan–Mar. Fruit: Apr–May. Siberian elm has been widely introduced in the United States and commonly escapes cultivation. It is on state noxious weed lists for 42 states.

URTICACEAE Juss. NETTLE FAMILY

Jane Mygatt

Herbs, ours annuals or rhizomatous perennials. Stems and leaves usually pubescent, often with specialized stinging hairs, and with calcium carbonate cystoliths commonly present, often independently of the hairs. **LEAVES** opposite or alternate, simple, petioles present; stipules present or absent; lamina margins entire or serrate to dentate. **INFL** usually an axillary cymose panicle or raceme. **FLOWERS** small, inconspicuous; plants monoecious, dioecious, or polygamous, bisexual or unisexual; anemophilous; petals absent; perfect flowers (when present) with 4 sepals; stamens 4; staminate flowers often pedicellate, white or green; sepals 4–5; stamens 4–5 (4 in ours), filaments inflexed in bud, reflexed when pollen is shed; pistillate flowers often sessile, green or reddish; sepals 2–4 (4 in ours), hypogynous; pistil 1, ovule 1; style present or stigma sessile. **FRUIT** an achene, surrounded by persistent perianth (Woodland 1982). Genera about 48, species about 700–800; widely distributed primarily in tropical and subtropical regions.

1. Stinging hairs absent; leaves alternate, margin entire; stipules absent .... ***Parietaria***
1′ Stinging hairs present; leaves opposite, margin dentate to serrate; stipules present .... ***Urtica***

Parietaria L. Pellitory

(Latin *parietarius*, of walls, where the plant likes to grow) Herbs, annual (in our flora) or perennial, sparsely to densely pubescent; lacking stinging hairs. **LEAVES** alternate, simple, stipules absent, margins entire, lanceolate to round. **INFL** a few-flowered panicle or spike, subtended by an involucre of 1–3 bracts. **FLOWERS** perfect, staminate, or pistillate; lower flowers usually perfect and staminate, upper pistillate; sepals fused basally. **FRUIT** shiny, hard, ovate, and stipitate, enclosed by sepals. Twenty to 30 species worldwide.

Parietaria pensylvanica Muhl. ex Willd. (of Pennsylvania) Pennsylvania pellitory. **STEMS** 4–60 cm, decumbent to erect, simple or branched. **LEAVES** linear-lanceolate to ovate, 2–9 cm × 0.4–3 cm, base tapered to obtuse, apex acuminate to long-tapered, lowest lateral veins arising from midrib. **FLOWERS** subtended by involucral bracts 1.8–5 mm long and exceeding the sepals; sepals 1.5–2 mm long. **FRUIT** brownish, 0.9–1.3 mm long × 0.6–0.9 mm wide, apex obtuse. $2n = 14, 16$. [*P. obtusa* Rydb.; *P. occidentalis* Rydb.; *P. pensylvanica* var. *obtusa* (Rydb. ex Small) Shinners]. Moist banks along margins of rivers to ledges in canyons, and shady areas. ARIZ: Apa; UTAH. 1730–2200 m (5700–7200′). Flowering: Apr–Jul. British Columbia to Quebec, and most of the United States into Mexico. This native species is reported in the USDA Plants database as a noxious weed of Nebraska and the Great Plains.

Urtica L. Nettle

(Latin *uro*, to burn, referring to the stinging hairs) Monoecious or dioecious herbs (in our flora) to shrubs with stinging and nonstinging hairs. Stems simple or branched, erect to sprawling. **LEAVES** opposite, stipules present; lanceolate to broadly ovate, margins dentate to serrate; cystoliths round to elongate. **INFL** an axillary, cymose raceme or panicle. **FLOWERS**: staminate with 4 equal sepals; stamens 4; pistillate with 4 sepals, outer 2 smaller; style absent; stigma tufted. **FRUIT** sessile, lenticular, loosely enclosed by inner sepals. Forty-five species, nearly worldwide.

Urtica dioica L. (two houses, each plant containing either male or female flowers) Stinging nettle, ortie. Perennial herbs, 0.5–3 m, rhizomatous, erect or sprawling. **LEAVES** lanceolate to widely ovate, margin serrate, 4–18 cm long; base tapered to cordate, apex acute. **INFL** an elongate panicle or raceme, staminate ascending, pistillate recurved. **FLOWERS**: pistillate lax, with outer sepals 0.8–1.2 mm long, inner sepals 1.4–1.8 mm long; staminate ascending. **FRUIT** 1–1.3 mm long. $2n = 26$. Three species, North America, Mexico, Eurasia. Our two subspecies are indigenous in North America. Nettles have mild diuretic properties. While the medicinal uses have not been verified by modern scientific methods, traditionally the leaves were used in herbal remedies as a cleansing tonic and blood purifier, astringent, antiallergenic, and for insufficient lactation. The root is used in a prostate tonic. Externally, the plant is used to treat skin complaints, arthritic pain, gout, sciatica, neuralgia, hemorrhoids, and hair problems. The fresh leaves of nettles have been rubbed or beaten onto the skin in the treatment of rheumatism. Termed "urtification," this procedure causes intense irritation to the skin as it is stung by the nettles.

1. Stems and lower leaf blades nearly glabrous, except for a few stinging hairs .... **subsp. *gracilis***
1' Stems and lower leaf blades densely pubescent .... **subsp. *holosericea***

subsp. *gracilis* (Aiton) Selander (slender) California nettle. [*Urtica dioica* var. *gracilis* (Aiton) Roy L. Taylor & MacBryde; *U. dioica* var. *procera* (Muhl. ex Willd.) Wedd.; *U. gracilis* Aiton]. Riparian, deciduous forests along canyon bottoms to mixed conifer forests. ARIZ: Apa; COLO: Arc, Hin, LPl, Min, Mon, SJn; NMEX: McK, RAr, SJn; UTAH. 1550–3300 m (5100–10800′). Flowering: May–Sep. Widely distributed in North America.

subsp. *holosericea* (Nutt.) Thorne (silky) Stinging nettle. [*Urtica holosericea* Nutt.; *U. dioica* var. *occidentalis* S. Watson]. Oak and mixed conifer woodlands. ARIZ: Apa. 2100–3100 m (6900–10000′). Flowering: May–Sep. Western United States to northern Mexico.

VALERIANACEAE Batsch VALERIAN FAMILY

Fred Barrie

Perennial, hermaphroditic, dioecious or gynodioecious herbs in our flora, elsewhere occasionally shrubs or annual or biennial herbs, commonly with a distinctive fetid odor, particularly in dried specimens. **LEAVES** opposite, decussate; petiolate or spatulate; simple to pinnatifid or pinnately compound; the bases sometimes sheathing; stipules absent. **INFL** a compound cyme. **FLOWERS** irregular; perfect or unisexual; calyx variously modified or obsolete; corolla sympetalous, often gibbous or spurred, with 5 imbricate lobes; anthers 1 to 4, epipetalous, alternate with the corolla lobes, thecae 2, dehiscing longitudinally; ovary inferior, carpels 3, the 2 abaxial locules sterile, the adaxial locule fertile, bearing a single pendent, anatropous ovule; style 1, stigmas 3. **FRUIT** an achene. A small family with seven genera and approximately 325 species, divided almost equally between the north-temperate regions of North America, Europe, and Asia and the montane and temperate regions of Central and South America. The family includes several cultivated genera and species, including *Valerianella* (lamb's lettuce; corn salad), *Centranthus ruber* (red valerian), and *Valeriana officinalis* (garden heliotrope; garden valerian). The latter is the commercial source of valerian root (actually the rhizome), used medicinally and in teas as a calmative. Throughout its range, many *Valeriana* species are used medicinally by indigenous peoples. The rhizome of *Nardostachys jatamansi* (spikenard) is the source of an essential oil used medicinally and in perfumes and incense.

Valeriana L. Valerian

(Latin *valeo*, to be strong, a reference to the medicinal properties of the genus) Perennial herbs (ours), rhizomatous or taprooted, rarely shrubs, occasionally scandent. **INFL** a many-flowered, compound or aggregate dichasium. **FLOWERS** with a calyx of 8–30 plumose setae, inrolled at anthesis, unfurling when the fruit matures to form a pappuslike structure persistent in fruit; corolla rotate to funnel-shaped or tubular, gibbous near the base, the corolla of female flowers commonly 1/3 to 1/2 the size of that of perfect flowers; anthers 3, exserted in perfect flowers, vestigial or absent in female flowers; ovary with the 2 sterile locules reduced or vestigial. **FRUIT** an achene with 3 veins on the abaxial side, 1 on the adaxial side, and 2 along the margin; the pappose calyx persistent (ours) or rarely absent. (Meyer 1951; Cronquist et al. 1997b.) *Valeriana arizonica* A. Gray, a species known from montane ponderosa pine parklands from central Arizona to the Guadalupe Mountains in New Mexico and Texas, has been reported as occurring in San Juan County, New Mexico, but the specimens seen that had been identified as such are actually collections of *V. acutiloba*. *Valeriana arizonica* differs most obviously in having truncate leaf bases and perfect flowers with tubular corollas 10–20 mm long.

1. Plants with a taproot .... ***V. edulis***
1' Plants rhizomatous, the roots fibrous .... (2)

2. Basal leaves 3–14 cm long; corollas of perfect flowers funnelform, 5–9 mm long .... ***V. acutiloba***
2' Basal leaves 10–25 cm long; corollas of perfect flowers rotate, 2–5 mm long .... ***V. occidentalis***

Valeriana acutiloba Rydb. (acutely lobed, referring to the leaves) Plants rhizomatous, gynodioecious, the stems, flowers, and/or fruits lightly pubescent or less commonly glabrous. **STEMS** 20–80 cm, arising from the ends of the rhizome branches. **LEAVES** mostly basal, with 1–3 cauline pairs, the basal leaves 3–14 cm long, simple or less commonly pinnatifid with 1–3 pairs of reduced lateral lobes, elliptic to narrowly elliptic or oblanceolate, the base cuneate to attenuate, the apex acute, the tip often mucronulate; the cauline leaves reduced, pinnatifid. **INFL** terminal, capitate in flower or with 1–2 pairs of reduced, lateral branches, expanding in fruit. **FLOWER** corolla white or pink, perfect flowers 5–9 mm long, funnel-shaped, female flowers 3–5 mm long, the lobes about 1/3 the length of the tube. **FRUIT** lance-ovate to narrowly oblong, 3–5 mm long; calyx limbs 12–16, about 5 mm long. [*V. capitata* Pall. ex Link subsp. *acutiloba* (Rydb.) F. G. Mey.]. Pine forests and wet meadows. ARIZ: Apa, Nav; COLO: Arc, Con, Dol, Hin, LPl, Min, Mon, RGr, SJn; NMEX: McK, RAr, SJn; UTAH. 1900–3600 m (6500–12000′). Flowering: Apr–Aug. Fruit: May–Sep. Oregon and California east to Montana and New Mexico. Cronquist et al. (1977b) recognized three varieties in *V. acutiloba*, not recognized here, which he described as "geographically distinct but morphologically ill-defined and confluent." Most local populations would fall under **var. *acutiloba*** in his system.

Valeriana edulis Nutt. ex Torr. & A. Gray (edible) Plant taprooted, gynodioecious, glabrous or the leaf margins and/or fruits pubescent. **STEMS** 1–4 from a branching caudex, 0.5–1.5 m tall. **LEAVES** basal or with 1–3 cauline pairs, 10–50 cm long, 0.5–5 cm wide, narrowly elliptic to oblanceolate or obovate-spatulate, occasionally with a weakly defined petiole, margins entire or with 2–10 narrow lobes; basal leaves with the base free or connate-perfoliate, the apex acute; the lobes, when present, linear to oblanceolate, the margins spreading or antrorse-ciliate, rarely glabrous; surfaces glabrous or with scattered simple hairs, restricted to the veins. **INFL** paniculiform, 7–60 cm long. **FLOWERS** perfect or female, white to greenish white or ivory, rotate, corolla 1–5 mm long; stamens 3, weakly to strongly exserted. **FRUIT** tan to purple-maculate, ovate to elliptic, 2–4.5 mm long, calyx limbs 8–14, 5–8 mm long. $2n = 64$. [*V. furfurescens* A. Nelson; *V. trachycarpa* Rydb.]. Wet meadows and alpine valleys. ARIZ: Apa; COLO: Arc, Con, Dol, Hin, LPl, Min, Mon, RGr, SJn, SMg; NMEX: McK, RAr, SJn; UTAH. 1700–3500 m (5900–11600′). Flowering: May–Aug. Fruit: May–Aug. Southern British Colombia east to South Dakota, Colorado, and New Mexico, south to Chihuahua and Nuevo León, Mexico, and in the Great Lakes region from Minnesota to Ohio and Ontario, Canada. The eastern populations, sometimes segregated as var. or subsp. *ciliata*, are morphologically indistinguishable from the western populations. Individuals with perfect flowers rarely set seed and are functionally male. The root of *V. edulis* was eaten by a number of western tribes.

Valeriana edulis

Valeriana occidentalis A. Heller (western) Plant rhizomatous, gynodioecious, glabrous or with a few hairs at the leaf and inflorescence nodes and on the fruits. **STEMS** 40–100 cm tall, arising from the ends of the rhizome branches. **LEAVES** basal and cauline, the basal leaves 10–25 cm long, simple or pinnatifid with 1–4 pairs of lateral leaflets, the blade or terminal lobe elliptic, 4–8 cm long, the base cuneate or rounded, the apex acute, the lateral lobes 1/4–1/2 the size of the terminal lobe and similar in shape; cauline leaves in 1–4 pairs, the lowermost 1–2 pairs often well developed, 5–15 cm long, pinnatifid, the uppermost leaf pairs usually reduced. **INFL** capitate to somewhat conical in flower, expanding in fruit. **FLOWERS** perfect or female, the corolla white, rotate, 2–5 mm long, the lobes approximately equaling the length of the tube. **FRUIT** lanceolate, 3–4 mm long, often purple-maculate, the calyx limbs 10–14, 4–5 mm long. $2n = 64$. Wet meadows; sagebrush, mountain brush, aspen, spruce-fir, sedge, and alpine communities. COLO: Dol, LPl, Min, Mon, SJn; NMEX: RAr; UTAH. 2200–3500 m (7500–11600′). Flowering: May–Aug. Fruit: May–Aug. Northern Idaho and Montana to northwest California, south to Arizona, New Mexico, and Colorado.

VERBENACEAE J. St.-Hil. VERVAIN FAMILY

Susan C. Barber

Annual or perennial, often strongly aromatic. **LEAVES** opposite, simple. **INFL** a spike or spicate raceme. **FLOWERS** zygomorphic, perfect; sepals 5, fused; petals 4 or 5, fused; stamens 4, didynamous; pistil 1, 2-carpellate; ovary superior, 4-lobed, locules 2, but appearing 4-locular due to false septa, placentae 4, axile; style 1, apical, stigma 1, often lobed. **FRUIT** a schizocarp forming 2 or 4 nutlets, enclosed by persistent calyx. $x = 5$ (*Glandularia*) and $x = 7$ (*Verbena*) (Lewis and Oliver 1961). Tropical, subtropical, and temperate distribution. Many tropical species are woody, but temperate species tend to be herbaceous. There are many herbal remedies found in this family. Teakwood (*Tectona grandis*) is one of the most economically important species in the family, which also includes horticulturally important *Vitex*, *Clerodendrum*, and *Lantana*. Circumscription of this family is in question as scientists continue to research the traditional relationships within the Lamiales. Based on a paper by Umber (1979), *Glandularia* is recognized as a segregate genus from *Verbena*. The key is written to reflect the segregation.

1. Fruits of 2 nutlets; corolla 4-lobed, distinctly 2-lipped .... ***Phyla***
1′ Fruits of 4 nutlets; corolla 5-lobed, slightly bilateral .... (2)

2. Spikes including bracts 10–20 mm wide; calyces 2 times longer than length of nutlets; styles 6–20 mm long; nutlets completely enclosed within closed calyces .... ***Glandularia***
2′ Spikes including bracts 3–8 mm wide; calyces equal to length of nutlets; styles 2–3 mm long; nutlets partially enclosed within open calyces .... ***Verbena***

Glandularia J. F. Gmel. Mock Vervain

(Latin *gland*, probably referring to the glandular pubescence) Plants annual or perennial herbs; stems often semiwoody; taproots present. **LEAVES** opposite, simple, entire, serrate, lobed, or pinnatifid; pubescence stiff, often glandular. **INFL** a spike, corymbose at anthesis, terminal. **FLOWERS** perfect, zygomorphic, calyces tubular, 5-toothed by vein extensions; corolla salverform, 5-lobed, tube exceeding calyces, blue or purple or lavender or pink or white, fused, minutely barbed hairs inside the tubes; stamens 4, didynamous; style inserted at the top of the ovary. **FRUIT** dry, schizocarpic, splitting into 4 1-seeded nutlets, usually black at maturity, commissural faces typically covered with papillae, outer surface reticulate. A New World genus of mesic to desert habitats. Origins are thought to be South American, where many species are thought to be undescribed. Hybridization is common. $n = 5, 10, 15, 20$.

Glandularia bipinnatifida (Nutt.) Nutt. (twice split) Plants annual or perennial, prostrate, decumbent to ascending, often rooting at the lower nodes, branched from base, pubescent, not glandular. **STEMS** 5–60 cm tall, hispid-hirsute. **LEAVES** petiolate, deeply incised to bipinnatifid, lobes linear to oblong, hirsute, appressed on both abaxial and adaxial surfaces, 2–6 cm long, 1–6 cm wide. **INFL** a pedunculate spike, short at anthesis, elongating at maturity, bractlets longer than the calyces. **FLOWER** corolla pink to lavender or purple, limb 7–10 mm wide, lobes emarginate, producing nectar, calyces 7–10 mm long, lobes unequal, hispid-hirsute. **FRUIT** 4 subcylindric, 1-seeded nutlets, 3 mm long. $n = 10, 15$. [*Verbena bipinnatifida* Nutt.]. Widespread along roadsides, prairies, meadows, dry plains, pastures, and in ponderosa pine communities. COLO: Arc, LPl. 1980–2285 m (6500–7500′). Flowering: Mar–Oct. Widespread from California east to Georgia, Texas north to South Dakota.

Phyla Lour. Frog-fruit

(Greek *phyle*, a clan or tribe, referring to the many flowers in a head) Perennial, erect or prostrate herbs with trailing or ascending stems, often woody at the base, glabrate to canescent-strigulose. **LEAVES** opposite, dentate apically. **INFL** axillary, pedunculate, a dense subglobose to cylindric spike. **FLOWERS** sessile, small, borne singly in axils of cuneate-obovate bractlets.

Phyla nodiflora (L.) Greene (flowers from nodes) Common frog-fruit. Plants perennial herbs. **STEMS** rooting at the nodes, prostrate or ascending. Stem hairs malpighian and canescent. **LEAVES** opposite, 17–20 mm long, 6–25 mm wide, spatulate or obovate or oblanceolate to elliptic, broadest above the middle, tapering into a petiole, sharply serrate above the middle, glabrous or strigillose-puberulent. **INFL** a densely flowered, pedunculate, axillary spike, globose when young, becoming cylindrical and greatly elongated in fruit, peduncles longer than leaves in fruit, up to 12 cm long. **FLOWERS** sessile, single in axils of bractlets, bractlets imbricate; corolla rose or purple or white, often with a yellow center, 2–2.5 mm long, corolla tube slightly exserted from calyces; stamens 4, paired; ovaries with 2 1-celled chambers, stigmas thickened. **FRUIT** 2 nutlets, included in and sometimes adnate to calyces. Moist soils, typically disturbed. NMEX: McK. 1675–1980 m (5500–6500′). Flowering: May–Oct. California to Texas; South America.

Verbena L. Vervain

(Latin name for any of certain sacred boughs) Plants annual or perennial herbs, prostrate, procumbent, ascending, or erect; glabrous or variously pubescent. **LEAVES** opposite, dentate or variously lobed, incised, or pinnatifid. **INFL** a spike, terminal, usually densely many-flowered, but sometimes greatly elongate and sparsely flowered. **FLOWERS** perfect, zygomorphic, 5-lobed, in the axil of a bract, corolla salverform or funnelform, blue or purple or lavender or pink or white, fused; calyces tubular, 5-angled, 5-ribbed, 5-toothed; stamens 4, didynamous, inserted in the upper 1/2 of the corolla tube, usually included; anthers ovate. **FRUIT** a dry schizocarp splitting into 4 1-seeded nutlets. A New World genus mostly of temperate and tropical America. Origins are thought to be North American. Hybridization is common. $n = 7, 14, 21$.

1. Bracts longer than the calyx tubes, reflexed at maturity; plants prostrate; leaves deeply incised to pinnatifid or cleft .... ***V. bracteata***

1′ Bracts equal to or shorter than the calyx tubes, erect at maturity; plants erect; leaves serrate-dentate .... ***V. macdougalii***

Verbena bracteata Lag. & Rodr. (with bracts) Plants prostrate or decumbent perennial (can flower as an annual). **STEMS** 9–55 cm tall; coarsely hirsute. **LEAVES** usually 3-lobed, the central lobe cuneate-obovate and larger than the other 2, rarely pinnately incised, narrowed into a petiole 2–20 mm long, hirsute. **INFL** a terminal spike, sessile, 5–15 cm long, 2–10 mm wide, bracts 8–15 mm long, conspicuous, much longer than the calyx, recurved at maturity, coarsely hirsute. **FLOWER** corolla bluish to lavender or purple, inconspicuous, limb to 3 mm wide, tubes slightly exserted, calyces 3–4 mm long. **FRUIT** 4 subcylindric, 1-seeded nutlets, 2–2.5 mm long. $n = 7, 14$. [*V. imbricata* Wooton & Standl.]. Disturbed areas throughout North America; widespread. ARIZ: Apa, Nav; COLO: Arc, Dol, LPl, Min, Mon, SJn, SMg; NMEX: McK, RAr, San, SJn; UTAH. 1125–2285 m (3700–7500′). Flowering: Apr–Oct. Widely distributed in North America and Mexico. Hybridizes readily with several species.

Verbena macdougalii A. Heller (for Daniel T. Macdougal, plant physiologist) Plants perennial. **STEMS** erect, to 1 m tall; villous-hirsute with spreading hairs. **LEAVES** short-petiolate or narrowed into a subpetiolar base, blade oblong-elliptic to ovate, 6–10 cm long, coarsely and prominently rugose, irregularly serrate-dentate. **INFL** a spike, typically solitary or in 3s, 7–10 mm wide, compact, hirsute; bracts usually longer than the calyx. **FLOWER** calyx 4–5 mm long, densely pubescent, obtuse lobes with subulate teeth; corolla deep purple, barely exserted, limbs 6 mm wide; **FRUIT** 4 1-seeded, trigonous nutlets, 2.5 mm long. $n = 7$. Mountain meadows and ponderosa pine forests. ARIZ: Apa; COLO: Arc, LPl; NMEX: RAr, SJn. 1850–2300 m (6070–7545′). Flowering: Jun–Oct. Colorado and Wyoming west to Utah and south to Arizona and New Mexico.

VIOLACEAE Batsch VIOLET FAMILY

R. John Little

Treelike, small shrubs, vines, or herbs (ours), perennial or annual. **LEAVES** simple, entire to deeply lobed or compound. **INFL** axillary from upper leaf axils or scapose from rhizome; flowers solitary in ours or in racemes. **FLOWERS** perfect, zygomorphic in ours; sepals 5, the base of sepals auriculate (with earlike lobes) or not; petals 5, the base of lower petal elongated into a spur or gibbous (swollen); stamens 5, alternate with petals, surrounding the ovary, the filaments short, wide, the 2 lowest anthers often with nectaries at their base extending into the spur; pistil 1, the ovary superior, unilocular, carpels 3–5, the placentae parietal, the ovules 1–2 or numerous on each placenta, anatropous, the style 1. **FRUIT** a capsule in ours, loculicidal, sometimes explosively dehiscent. **SEEDS** usually arillate. Twenty-three genera, 830 species. Chiefly temperate, subtropical, and tropical regions worldwide. Cultivated species of horticultural importance include *Viola tricolor* L. (Johnny jump-up) and *V.* ×*wittrockiana* Gams (pansy) (Baker 1957; Ballard 1992; Clausen 1964; Davidse 1976; Fabijan, Packer, and Denford 1987; Little 2001; McKinney 1992).

Viola L. Violet

(Latin *viola*, classical name) Caulescent or acaulescent annual or perennial herbs. **LEAVES** simple, dissected or compound; basal, cauline, or both, stipulate, petiolate, alternate or lower sometimes opposite. **INFL** axillary or scapose, flowers solitary. **FLOWERS** zygomorphic; sepals subequal, auriculate at base; petals unequal; lower petal shorter or longer than others, the base elongated into a spur or gibbous; lateral 2 petals equal, usually spreading, usually bearded (with hairs) at base; upper 2 equal, erect or reflexed; lower 2 stamens with nectaries projecting into the spur. **FRUIT** a 3-valved, 1-celled capsule, ovoid to oblong, glabrous or hairy; apetalous, cleistogamous flowers present in many species. **SEEDS** about 15–20, usually ovoid, many species arillate with a prominent caruncle by which they are ant-dispersed (myrmecochory). About 400 species worldwide in temperate areas. Flower color should be recorded when collected because petal colors fade in specimen preparation and storage.

1. Plant with branched stems (caulescent); petioles arise from lower and upper stem nodes ....(2)
1′ Plant with unbranched stems (acaulescent); petioles arise only from rootstalk or rhizome ....(5)

2. Flower violet or blue .... *V. adunca*
2′ Flower white or yellow ....(3)

3. Flower white .... *V. canadensis*
3′ Flower yellow ....(4)

4. Lower petal (including spur) 6–13 mm; glabrous to minutely puberulent; capsule 4–9 mm; seed caruncle usually a narrow row of tissue beginning at the micropyle (narrow end of seed), flattened perpendicular to the seed and forming a sharp edge, covering 1/3 to entire length of seed .... *V. nuttallii*
4′ Lower petal (including spur) 12–20 mm; glabrous to very hairy; capsule 6–13 mm; seed caruncle a globular to variously shaped mass of tissue near the micropyle, covering about 1/3 length of seed .... *V. praemorsa*

5. Plant with divided leaves; dry habitats .... *V. pedatifida*
5′ Plant with entire leaves; wet habitats ....(6)

6. Plant mat-forming, with stolons; petals white .... *V. macloskeyi*
6′ Plant erect, not mat-forming, without stolons; petals blue .... *V. sororia*

Viola adunca Sm. (hooked) Western dog violet, hooked violet, early blue violet, sand violet. Perennial herb 2–30 cm tall. **STEMS** branched, at first short, usually elongating later in season, glabrous to variously pubescent, clustered on thin, usually vertical, woody rootstalks. **LEAVES** simple, basal and cauline, glabrous to pubescent; basal leaf petiole 5–70 mm long, blade ovate to ovate-triangular, 5–47 mm long to 4.5 cm wide, base cordate, truncate, or attenuate, margins usually crenate, often entire in *V. adunca* var. *bellidifolia*, apex acute to obtuse; cauline petioles winged, leaves similar to basal but usually smaller, blade 6–42 mm long to 3.3 cm wide. **INFL** axillary from upper leaf axils, pedicels to 10.3 cm long. **FLOWER** sepals to 10 mm long, auricles usually small; petals pale to deep violet, the lower 3 purple-veined, white-bearded at the base, the lateral pair white-bearded at the base, the lowest petal (including spur) 7–20 mm long, the spur usually elongated, conspicuous, straight, curved, pointed, or hooked at tip, 1–15 mm long to 3 mm wide. **FRUIT** 6–11 mm long, glabrous. $2n = 20$. [*V. adunca* var. *ashtoniae* M. S. Baker; *V. adunca* var. *oxyceras* (S. Watson) Jeps.; *V. adunca* var. *uncinulata* (Greene) C. L. Hitchc.]. Alaska south to Canada, west to the eastern United States. A highly polymorphic species with many named variants. Two subspecies recognized in our area; they are often confused with *V. sororia* subsp. *affinis*, an acaulescent species that lacks branching stems, and *V. labradorica* Schrank, a mostly glabrous species that lacks winged petioles and does not occur in our area.

1. Plant 3.5–30 cm tall; basal leaf blades 1.3–4.8 cm long, 1.2–3.3 cm wide; cauline leaf blades 1.2–4.2 cm long, 1.1–3.3 cm wide; lower petal (including spur) 9–20 mm long; pedicels 2.9–10.3 cm long .... **var. *adunca***
1′ Dwarf plant 1.8–6.5 cm tall, appearing acaulescent; basal leaf blades 0.5–1.7 cm long, 4–14 mm wide; cauline leaf blades 6–14.5 mm long, 4–14 mm wide; lower petal (including spur) 7–14 mm long; pedicels to 5 cm long; known only from Colorado in our area .... **var. *bellidifolia***

var. *adunca* Damp banks, meadow edges in coniferous forests, usually in shade. ARIZ: Apa; COLO: Arc, Hin, LPl, Min, Mon, SJn; NMEX: RAr, SJn; UTAH. 3050–3750 m (7600–10000′). Flowering: May–Aug. New Brunswick to Alaska, south to Vermont, New Mexico, and California.

var. *bellidifolia* (Greene) H. D. Harr. (with beautiful leaves) Subalpine meadows to alpine tundra, above 3050 m (10000′). COLO: Arc, Hin, LPl, Min, Mon, RGr, SJn. Flowering: late Jun–early Aug. British Columbia south to Colorado.

Viola canadensis L. (of Canada) Canada violet, tall white violet. Perennial herb 3–38 cm tall. **STEMS** erect, branched above, glabrous to puberulent, from vertical or horizontal rhizomes, or stolons. **LEAVES** simple, basal and cauline, with sparse hairs along major veins or not, the margins ciliate or not; basal leaf petioles 1–14.8 cm long, puberulent, blades ovate-reniform, 0.7–7.5 cm long to 9.2 cm wide, base cordate, margins serrate, apex acuminate to acute; cauline petioles 0.4–9.6 cm long, blades ovate to triangular, 1–6.5 cm long to 5.8 cm wide, base cordate, truncate, or attenuate, margins serrate, apex acute. **INFL** axillary from upper leaf axils, pedicels to 6 cm long, glabrous to puberulent. **FLOWER** sepals to 9 mm long; petals white, the lower 3 with yellow at base or not, purple-veined, the lateral pair white-bearded, the outer side of petals often purple-tinged, the lowest petal (including spur) 5.5–15 mm long, the spur short, gibbous. **FRUIT** 3–10 mm long, glabrous. $2n = 24$. A complex group in need of study, consisting of about five subspecies, two in our area.

1. Plant 11–38 cm tall, not tufted; basal leaf blades 2.4–7.5 cm long, 2.2–9.2 cm wide; cauline blades 2.5–6.5 cm long, 9–58 mm wide; lowest petal (including spur) 11–15 mm long; pedicels 1.4–6 cm long.......... **var. *canadensis***

1′ Plant 3–18 cm tall, tufted; basal leaf blades 0.7–3.1 cm long, 0.9–3.8 cm wide; cauline blades 1–2.8 cm long, 8–70 mm wide; lowest petal (including spur) 5.5–9 mm long; pedicels 0.9–3.1 cm long.......... **subsp. *scopulorum***

var. *canadensis* [*V. muriculata* Greene; *V. canadensis* var. *neomexicana* (Greene) House]. Riparian habitats in coniferous or aspen forests, or moist, shaded slopes, in sandy, rich, or rocky soil, roadcuts. ARIZ: Apa, Coc, Nav; COLO: Arc, Hin, LPl, Min, Mon, SJn; NMEX: RAr, SJn; UTAH. 1310–3540 m (4300–11600′). Flowering: Apr–Oct. Alaska south to British Columbia, Washington, Oregon, east to Labrador, south through the Rocky Mountains, Idaho to Mexico, to the eastern United States except Texas, Louisiana, Alabama, and Florida.

subsp. *scopulorum* (A. Gray) House (of the rocks) [*V. scopulorum* (A. Gray) Greene]. Damp, wooded sites. ARIZ: Apa; COLO: Arc, Hin, Min, Mon, SJn. 2285–3505 m (7500–11500′). Flowering: Apr–Oct. Arizona, Colorado, and New Mexico.

Viola macloskeyi F. E. Lloyd (for G. Macloskey) Macloskey's violet, wild white violet, smooth white violet. Perennial herb 2.5–22 cm tall, forming a dense patch. **STEMS** none; plants arise from thin, creeping rhizomes that form late-season stolons in a dense patch. **LEAVES** simple, basal, erect; stipules reddish; petioles 2–17 cm long, blades round to ovate, base cordate to truncate, blades 1–6.5 cm long, about as wide, margins entire, crenate, or crenate-serrate, apex obtuse, rarely acute, glabrous to hairy, thin. **INFL** scapose from rhizomes, pedicels to 21 cm long, sometimes reddish when mature. **FLOWER** petals white, the tips acute, the lower 3 petals with purple veins, the lateral pair usually white-bearded; the lowest petal (including spur) 6–17 mm long, lateral petals twisted at base, upper often reflexed backward; sepal lobes 1–2 mm long, earlike. **FRUIT** 5–9 mm long, glabrous. $2n = 24$. [*V. blanda* Willd. var. *macloskeyi* (F. E. Lloyd) Jeps.]. Bogs, wet meadows, seeps, lake margins, streamsides, floodplains, and mesic roadside depressions, often growing among mosses. COLO: Hin, SJn. 2190–3600 m (7200–11850′). Flowering: May–Jul. Circumboreal, northern Europe and Asia, Alaska to Labrador, Washington to southern California, east to Arizona, Colorado, New Mexico, New England. A variable species needing study. Plants with inconspicuously crenate leaves, usually less than 2.5 cm wide, have been called subsp. *macloskeyi*; plants with prominently crenate leaves often over 2.5 cm wide have been called subsp. *pallens* (Banks ex DC.) M. S. Baker. Although occurring in all adjacent states, there are no Utah records. Often confused with *V. palustris* L., which has pale blue to almost white flowers, but does not form dense patches.

Viola nuttallii Pursh (for Thomas Nuttall) Nuttall's violet. Perennial herb 2–27 cm tall. **STEMS** branched, clustered on short, vertical rootstalks, at first short, usually elongating later in season, erect to spreading or compressed, suberect in shade, depressed in sun; usually partly buried; upper stems usually puberulent, the lower glabrate. **LEAVES** simple, basal and cauline, glabrous to puberulent; basal leaf petioles 1.9–17 cm long, basal blades lanceolate, ovate, or elliptic, 1–9 cm long, 6–25 mm wide, base attenuate, margins entire, sinuate, or subserrate, ciliate, apex obtuse to acute; cauline leaf petioles 7–70 mm long, blades ovate, lanceolate to elliptic, 1.4–7.2 cm long to 18 mm wide, base attenuate, margins entire, sinuate, or subserrate, apex acute. **INFL** axillary from upper leaf axils, pedicels to 13 cm long. **FLOWER** sepals linear, acute, to 8 mm long, auricles with or without pubescence; petals yellow on the face, the lower 3 brown- to purple-veined, the upper 2 usually brownish purple on back, the lateral pair bearded; the lowest petal (including spur) 6–13 mm long, the spur gibbous. **FRUIT** 4–9 mm long, glabrous. **SEED** caruncle usually a narrow row of tissue beginning at the micropyle (narrow end of seed), flattened perpendicular to the seed and forming a sharp edge, covering 1/3 to entire length of seed. $2n = 24$. Sagebrush flats and piñon-juniper woodlands. ARIZ: Apa; COLO: Arc, LPl, Mon; NMEX: SJn. 1870–2550 m (6150–8400′). Flowering: Apr–Jun. Southern Alberta, Saskatchewan, and Manitoba, Canada; Idaho, Utah, Montana, Minnesota, North Dakota, South Dakota, Wyoming, Nebraska, and Kansas.

Viola pedatifida G. Don (leaves cleft like a bird's foot) Prairie violet, larkspur violet, crowfoot violet. Perennial herb 4–30 cm tall. **STEMS** none; plants arising from a short, vertical, fleshy rhizome, glabrous to sparsely strigose. **LEAVES** basal, erect, glabrous to short-hirsute, margins ciliate; petioles 3–16 cm long; blades palmately divided into 5–9 linear, spatulate, obovate, or falcate segments, each cleft or parted toward apex; 2.3–4.5 cm long, 3.5–7.3 cm wide, bases truncate, reniform, or cordate, apices obtuse, acute to mucro-tipped. **INFL** scapose from rhizome; pedicels to 18 cm long; chasmogamous and cleistogamous peduncles ascending to erect. **FLOWER** sepals to 7 mm long, auricles up to 1/2 as long as sepals; petals light to dark blue-violet on face, the lower 3 bearded, the lowest petal (including spur) 13–25 mm long. **FRUIT** 8–13 mm long, elliptic, glabrous. $2n = 54$. Normally of prairie or grassy habitats, rarely in open woods (as in Arizona). COLO: Arc, Hin. 1770–2380 m (5800–7800′). Flowering: May–Jun. Southern Alberta east to southern Ontario, south to Ohio and west to Colorado and Arizona. Two subspecies; ours are **subsp**. ***pedatifida***.

Viola praemorsa Douglas ex Lindl. (bitten on the end) Astoria violet. Perennial herb 6–30 cm tall. **STEMS** usually several, spreading from a mostly vertical, woody rootstock; appearing stemless early in season, but later elongating. **LEAVES** simple, basal and cauline, glabrous to very hairy with 1 mm wavy hairs; basal petioles to 19.2 cm long, usually longer than blades, blades narrow to broadly ovate, oblong-lanceolate, or narrow to broadly elliptic, 2.4–12.5 cm long, to 6.7 cm wide, base attenuate or truncate, sometimes oblique, margins usually irregularly crenate/serrate, sometimes entire or wavy, apex acute or obtuse; cauline petioles to 13.5 cm long, blades similar to basal, 1.9–14.3 cm long, to 5.3 cm wide, much longer than wide. **INFL** axillary from upper leaf axils, pedicels to 19.5 cm long, shorter or longer than leaves. **FLOWER** petals deep yellow, lower and lateral veined brown-purple, the lateral pair bearded, the backs of the upper 2 often colored maroon or brownish due to numerous colored veins, the lowest petal (including spur) 12–20 mm, the spur short, blunt. **FRUIT** 6–13 mm long, glabrous to hairy. **SEED** caruncle a globular to variously shaped mass of tissue near the micropyle (narrow end of seed), covering about 1/3 length of seed. $2n = 36, 48$. [*V. linguifolia* Nutt.; *V. nuttallii* Pursh var. *praemorsa* (Douglas) S. Watson; *V. nuttallii* subsp. *praemorsa* (Douglas) Piper; *V. nuttallii* var. *linguifolia* (Nutt.) Piper; *V. nuttallii* subsp. *linguaefolia* Piper; *V. praemorsa* subsp. *major* (Hook.) M. S. Baker]. Various habitats, moist to dry soil, grassy slopes, meadows, yellow pine forests. COLO: SJn. 3810 m (12500′). Flowering: Jun. Saanich Peninsula of Vancouver Island south to California, southwest Alberta, south to Idaho, western Montana, Utah, Nevada, Wyoming. A complex group needing study; three subspecies; ours are **subsp**. ***linguifolia*** (Nutt.) M. S. Baker. One record from summit of Cinnamon Pass Road, Colorado.

Viola sororia Willd. (sisterly, as in resembling other species) Common blue violet, woolly blue violet, downy blue violet. Perennial herb 2–50 cm tall. **STEMS** none; plants arising from short, thick, fleshy rhizomes with long, fibrous roots. **LEAVES** simple, all basal; basal petioles 1.8–23 cm long, midseason blades cordiform, reniform, to very widely ovate, 2–5.5 cm long to 10 cm wide, base cordate to reniform, margins uniformly serrate/crenate, ciliate or not, apex acute to slightly caudate, glabrous to strigose throughout. **INFL** scapose from rhizome, pedicels to 5–25 cm long. **FLOWER** sepals to 8 mm long, petals light to dark blue-violet, the lower 3 purple-veined, the lateral pair white-bearded; the lowest petal spurred or not, 10–20 mm long (including spur), the spur blunt, straight, 1–5 mm long, 3 mm wide; sepal auricles of cleistogamous flowers elongated or not. **FRUIT** 5–12 mm long, elliptic, glabrous. $2n = 54$. [*V. nephrophylla* Greene; *V. nephrophylla* var. *arizonica* (Greene) Kearney & Peebles; *V. nephrophylla* var. *cognata* (Greene) C. L. Hitchc.; *V. papilionacea* Pursh]. In wet or damp soil along creeks, streams, and shady hillsides of coniferous forests, prairies, pastures, marshes, mossy bogs, gravelly shores. ARIZ: Apa, Nav; COLO: Arc, Hin, LPl, Min, Mon, SJn; NMEX: McK, SJn; UTAH. 1370–3000 m (4500–9850′). Flowering: Mar–Sep. Alaska, throughout Canada, south throughout the western United States, east through Iowa and Wisconsin to the New England states. Four subspecies; ours are **subsp**. ***affinis*** (Leconte) R. J. Little (similar to). Often confused with *V. adunca*, which has branching stems, flowers in upper axils, and woody roots. Plant (cited as *V. nephrophylla*) used as a ceremonial emetic by the Ramah Navajo.

VISCACEAE Batsch MISTLETOE FAMILY

John R. Spence

Aerial parasites on branches or trunks of host, producing haustoria that penetrate the host, dioecious or monoecious "subshrubs," with or without chlorophyll. **STEMS** brittle, disarticulating at nodes, branching verticillate to flabellate-dichotomous, variously colored red-brown, orange, olive, green, or yellow, glabrous or sometimes pubescent when young, nodes swollen. **LEAVES** opposite, reduced (ours), evergreen, scalelike, entire, glabrous, free or connate at base; stipules absent. **INFL** an axillary bracteate spike or cyme or flower solitary. **FLOWERS** actinomorphic, epigynous, unisexual, 3-merous (carpellate flowers) or 2–4-merous (staminate flowers), sessile or short-pedicellate, subtended by a scalelike bract; perianth of 1 whorl, tepals yellow or green, often reduced to teeth; stamens of staminate flowers 4, adnate to and opposite tepals, anthers with 1–2 thecae, filaments short or absent, carpellate flowers lacking stamens; ovary of 3–4 carpels, with 1 locule, style short or absent, stigma short; staminate flowers sometimes nectariferous. **FRUIT** a fleshy drupe or berry, sessile or on a straight to curved pedicel, often brightly colored, sometimes dehiscing explosively, seed 1 (2), lacking a testa, surrounded or capped at one end by sticky fleshy material. A small family of eight genera and about 450 species, distributed worldwide but most common in the tropics and subtropics. The mistletoes are related to the Santalaceae (sandalwood family), also parasites. Recent research has tended to place Viscaceae within the Santalaceae, although the interrelationships of these and other families such as the Loranthaceae are complex and only partially resolved at present. The Eurasian genus *Viscum* is the traditional Christmas mistletoe and is a sacred plant of Druids. Mistletoes can seriously affect host plant growth and survival and can eventually kill the host. Dwarf mistletoes (*Arceuthobium*) are widespread and can cause significant damage to western coniferous forests.

1. Parasitic on Pinaceae (*Pinus* or *Pseudotsuga* in ours); perianth of pistillate flowers 2-merous; anthers unilocular; fruit blue-green or blue-purple, compressed, on a recurved pedicel .... ***Arceuthobium***
1′ Parasitic on Cupressaceae (*Juniperus* in ours) and on angiosperms; perianth of pistillate flowers 3-merous; anthers bilocular; fruit white or pink, globose, sessile, pedicel straight or recurved .... ***Phoradendron***

Arceuthobium M. Bieb. Dwarf Mistletoe

(Greek *arceuthos*, juniper, + *bios*, life) Aerial, parasitic, dioecious subshrub, with chlorophyll, on branches or trunks of host. **STEMS** mostly erect or rarely pendent, brittle, disarticulating at nodes, round to sometimes square or angled, branching dichotomous and flabellate, variously colored red-brown, olive, green, orange, or yellow, glabrous, nodes swollen. **LEAVES** reduced to opposite connate scales. **INFL** an axillary bracteate spike of numerous flowers. **FLOWERS** 3- or 4-merous (both carpellate and staminate flowers), sessile or pedicellate; perianth of staminate flowers (2) 3–4 (7) inconspicuous persistent tepals, often reduced to teeth; anthers sessile, adnate to tepals, with 1 theca; ovary of carpellate flowers fused with perianth, forming a bilobed cap. **FRUIT** a pedicellate, fleshy, bicolored berry, pale tan to brown above, glaucous, blue or blue-purple below, capped by persistent tepals, dehiscing explosively, pedicel erect to recurved when mature, seeds 1, mucilaginous. A small genus of either 26 or 42 species, distributed on members of the Pinaceae and Cupressaceae in North America, Eurasia, and Africa (Hawksworth and Wiens 1996; Nickrent et al. 2004). Hawksworth and Wiens (1996) recognized 42 species of *Arceuthobium*, with 8 species endemic to the Old World and the remainder endemic to the New World. Species are morphologically highly variable and difficult to distinguish; hence there has been an over-reliance on hosts to define them. Nickrent et al. (2004), based on a phylogenetic analysis of DNA sequences, recognized 26 species, all 8 of the Old World species and 18 in the New World. The *A. campylopodum* complex was reduced in their study to two species, *A. cyanocarpum*, with multiple hosts, and *A. divaricatum*, primarily infesting piñon pines. Morphological diversity in the genus does not appear to be strongly correlated with DNA sequence divergence, suggesting potential recent dispersal and minor ecotypic differentiation on different hosts. The fruits of *Arceuthobium* species split open explosively, throwing seeds as far as 50 feet. The seeds are coated with a viscous substance that acts as a cement, attaching the seeds to leaves and branches of the host. After a rain, the cement of seeds attached to needles hydrates and lubricates, causing the seeds to slide toward the needle bases, where they germinate. They can also germinate directly on young twigs if the bark is not too thick. Seeds are consumed by wildlife, and many stick to the feathers and feet of birds, allowing them to be dispersed large distances. The Navajo make a tea of the stems that is used in various ceremonies. All species, however, are probably at least slightly toxic.

1. Carpellate plants small, less than 4 cm tall, stems yellow-green to olive-green; parasitic on *Pseudotsuga menziesii* (Douglas-fir) .... ***A. douglasii***
1′ Carpellate plants typically larger, often >5 cm tall; stems yellow, brown, orange, or green; parasitic on *Pinus* (pines) .... (2)

2. Stems yellow-orange to orange-brown; plants flowering in spring; parasitic on *Pinus ponderosa* or rarely *P. contorta* (lodgepole pine) .... ***A.vaginatum***
2′ Stems dull brown or green to olive-green, lacking orange or yellow tints; plants flowering in late summer; parasitic on *Pinus edulis* (piñon pine) .... ***A. divaricatum***

Arceuthobium divaricatum Engelm. **CP** (strongly spreading-branching) Piñon dwarf mistletoe. Aerial parasitic subshrub on branches and rarely the trunks of piñon pine, not or rarely producing systemic witches'-brooms. **STEMS** 5–10 cm long, central one 1.5–4 mm thick at base, strongly branching, opposite branches often strongly divergent, divaricate and flabellate, dark green to olive-green or brown, lacking yellow tints, third branch internode 5–7 times as long as thick. **FLOWERS** the same color as shoots; staminate flowers 2–3 mm across, tepals 3, unequal, 2 somewhat asymmetrically keeled and hooded, the third innermost flat and deflexed at anthesis. **FRUIT** brown above, glaucous blue below, 3–4 mm long, 2 mm wide, elliptic, on a short, erect to recurved pedicel. Common on *Pinus edulis* (piñon pine). ARIZ: Apa, Nav; COLO: Dol, LPl, Mon, SMg; NMEX: McK, RAr, San, SJn; UTAH. 1525–2130 m (5000–7000′). Flowering: Jul–Sep. Fruit: Sep–Oct of following year. This species is most closely related to *A. vaginatum*, which differs in its reddish to orange-yellow shoots, and spring flowering.

Arceuthobium douglasii Engelm. (for David Douglas, 1798–1834, botanical explorer) Douglas-fir dwarf mistletoe. Aerial parasitic subshrub on branches and trunks of *Pseudotsuga menziesii*, commonly forming systemic witches'-brooms. **STEMS** 0.3–5 cm, central ones 1–1.5 mm thick at base, branching divaricate and flabellate, yellow-green to olive-green, third branch internode 3–4 times as long as thick. **FLOWERS** same color as shoots on outside, staminate flowers 2–3 mm across, tepals 3, purple to red on inside, equal to subequal, somewhat asymmetrically keeled, spreading at anthesis. **FRUIT** olive-green throughout, 3–5 mm long, 1.5–2.5 mm wide, obovate, on short, erect to recurved pedicel. Common on *Pseudotsuga menziesii* (Douglas-fir). ARIZ: Apa; COLO: Arc, Hin, Min, Mon; NMEX: SJn; UTAH. 2320–2750 m (7000–9000′). Flowering: Apr–May. Fruit: Aug–Sep of following year.

Arceuthobium vaginatum (Willd.) J. Presl (*vagina*, sheath, for the persistent tepals sheathing the ovary) Southwestern dwarf mistletoe. Aerial parasitic subshrub on branches and trunks of pines, primarily *Pinus ponderosa* but also *P. contorta*, sometimes forming systemic witches'-brooms. **STEMS** 5–25 cm long, central ones 3–8 mm thick at base, branching divaricate and flabellate, yellow-orange, dull reddish brown or sometimes nearly black, third branch internode 2–3 times as long as thick. **FLOWERS** same color as shoots, staminate flowers 2–3 mm across, tepals 3, unequal, 2 somewhat asymmetrically keeled and hooded, the third innermost flat and deflexed at anthesis. **FRUIT** brown or tan above, glaucous blue below, 4–6 mm long, 2–3 mm wide, elliptic, on a long, strongly recurved pedicel. [*A. cryptopodum* Engelm.]. Our material belongs to **var. *cryptopodum*** (Engelm.) Cronquist (hidden foot). Common on *Pinus ponderosa*, rare on *P. contorta* (lodgepole pine). ARIZ: Apa; COLO: Arc, Dol, Hin, LPl, Min, Mon, SMg; NMEX: McK, RAr, San, SJn; UTAH. 2320–2750 m (7000–9000′). Flowering: May–June. Fruit: Aug–Sep of following year. Used as a ceremonial medicine by the Ramah Navajo.

Phoradendron Nutt. Mistletoe

(Greek *phor*, thief, + *dendron*, tree) Aerial, parasitic, dioecious subshrub, with chlorophyll, on branches of host. **STEMS** erect to pendent, brittle, disarticulating at nodes, round to sometimes square or angled, branching dichotomously, variously colored red-brown, green, or yellow, glabrous or pubescent when young, nodes swollen. **LEAVES** well developed to scalelike (ours), free or connate at base. **INFL** an axillary bracteate spike, carpellate spike reduced to 2 flowers, staminate spike with 2–many flowers. **FLOWERS** 3- or 4-merous (both carpellate and staminate flowers), sessile; perianth of 3–4 inconspicuous persistent tepals, often reduced to teeth; anthers sessile, adnate to tepal, with 2 thecae; ovary fused with perianth. **FRUIT** a brightly colored, sessile, fleshy, drupelike berry, white, white-pink to red, capped by persistent tepals, on straight or recurved pedicel, not dehiscing explosively, seeds 1. A large genus of 235 species, endemic to the New World, with a large concentration in the tropics and subtropics. Mistletoes are distributed by birds, who consume the fruits and either wipe the seed off with their bills or void the seed on perches. Although only a single species occurs in the flora, several others approach our area from the south, including *P. bolleanum* (parasitic on Cupressaceae), *P. californicum* (parasitic on Fabaceae and *Larrea*), and *P. flavescens* (parasitic on various riparian angiosperm trees), and may eventually be found in the Four Corners region on planted hosts.

Phoradendron juniperinum A. Gray (on *Juniperus*) Juniper mistletoe, taa'tshaa'. Aerial parasitic subshrub on branches of host. **STEMS** brittle, 5–20 cm long, somewhat angular, branching dichotomous to irregular, often branching flattened in 1 plane, green, olive, or yellow-green, glabrous, somewhat shiny. **LEAVES** 0.5–1.5 mm long, triangular, acute, erect to somewhat recurved. **INFL** an axillary spike or cyme, carpellate inflorescence reduced to 1 segment with 2 flowers, staminate inflorescence of 4–10 flowers. **FLOWERS** inconspicuous, sessile, tepals 3 (4), 0.5–2 mm, pale yellow to green. **FRUIT** a shiny, somewhat translucent, white, pink, or pale red berry, 3–5 mm, elliptic to globose, on recurved pedicel, capped by pale, persistent, erect tepals. Common on *Juniperus osteosperma* (Utah juniper), *J. monosperma* (single-seed juniper), and *J. scopulorum* (Rocky Mountain juniper). ARIZ: Apa, Coc, Nav; COLO: Arc, Dol, LPl, Mon, SMg; NMEX: McK, RAr, San, SJn; UTAH. 1675–2460 m (5500–8200′). Flowering: Jul–Aug. Fruit: Aug–Oct. This species is reported to rarely parasitize *Pinus edulis* (piñon pine). The Navajos ate the berries, produced a medicinal tea from the branches, and used the plant in ceremonies, such as hanging the plant at the entrance to the hogan to ward off lightning. The Hopi used the plant to make a tea as a gastrointestinal aid and used the branches as a textile dye.

VITACEAE Juss. GRAPE FAMILY

John R. Spence

Woody vines, sprawling or climbing over other vegetation, unarmed. **STEMS** often with lenticels; tendrils opposite the leaves, weakly or strongly branched, with or without adhesive discs. **LEAVES** alternate, blade petiolate, simple to palmately lobed or compound, glabrous to pubescent, deciduous or sometimes persistent; stipules deciduous. **INFL** terminal or lateral opposite the leaves, cymose, umbellate, or paniculate. **FLOWERS** inconspicuous, perfect or unisexual, regular, hypogynous, 4- or 5-merous; calyx small, toothed or lobed or occasionally reduced, free or connate; corolla valvate, petals distinct below, sometimes coherent above; stamens distinct, as many as petals and opposite; nectar disc present between stamens, annular to distinctly lobed; ovary superior, carpels and locules 2, united, style simple or sometimes absent, stigma discoid to capitate, ovules 2, basally attached. **FRUIT** a fleshy to dry berry. **SEEDS** 2–4, deeply grooved adaxially, with an abaxial knot. A family of about 15 genera and 850 species distributed worldwide in the tropics, subtropics, and north-temperate zone. Many species are cultivated for edible fruit, particularly the wine grapes (cultivars of *Vitis labrusca*, Concord grape, especially *Vitis vinifera*, the wine grape), and some as landscape ornamentals. The family's systematic position is controversial at present. In the past, relationships have been suggested with the buckthorn family (Rhamnaceae) or various families in the Rosalean complex. Its origin and relationships remain enigmatic despite recent DNA studies.

1. Leaves palmately compound .... ***Parthenocissus***
1' Leaves simple, sometimes weakly palmately lobed .... ***Vitis***

Parthenocissus Planch. Thicket Creeper, Creeper, Woodbine

(Greek *partheno*, virgin, + *cissus*, ivy) Climbing, sprawling, or creeping woody vines. **STEMS** freely branching, with smooth bark, resinous lenticels prominent; tendrils mostly bifurcate, with or without adhesive discs. **LEAVES** palmately compound, deciduous, leaflets 3–7, margins serrate. **INFL** of lateral or terminal cymose panicles, opposite the leaves. **FLOWERS** either unisexual or perfect but often functionally unisexual, 5-merous; calyx small, 5-tipped; petals separate above; stamens 5, reduced in female flowers; pistil vestigial in male flowers, style absent or short; nectary disc obsolete. **FRUIT** a somewhat dry to fleshy berry. **SEEDS** 4 or fewer by abortion. About 15 species, distributed in the temperate zone of North America and Asia.

1. Leaflets dull green above; inflorescence with a strongly zigzagging main axis; tendrils ending in adhesive discs .... ***P. quinquefolia***
1' Leaflets dark shiny green above; inflorescence dichotomously branching; tendrils lacking distinct adhesive discs .... ***P. vitacea***

Parthenocissus quinquefolia (L.) Planch. (five-parted leaves) Virginia creeper. Clambering or sprawling to strongly climbing vine. **STEMS** pale tan to brown, mostly round, young bark glabrous, older bark becoming flaky, resinous lenticels common on older stems; tendrils strongly branched, with distinct adhesive discs. **LEAVES** dull green above, glabrous below, the central leaflet 4–14 cm long and 3–7 cm wide, lateral leaflets somewhat smaller, leaves turning red in autumn. **INFL** of terminal or axillary cymose clusters opposite leaves, with a strongly developed main axis that zigzags, divergent branches with small umbellate clusters of flowers. **FLOWERS** unisexual, small, inconspicuous, green. **FRUIT** blackish, somewhat fleshy, 4–8 mm wide [*P. inserta* (J. Kern.) Fritsch]. Introduced cultivated vine, often persisting or sometimes escaping in damp riparian areas and around buildings, native to eastern North America. ARIZ: Apa, Nav; COLO: Mon; NMEX: McK, SJn; UTAH. 1500–2200 m (5000–7200'). Flowering: May–Jun. Fruit: Jun–Jul. The name *P. inserta* has been applied to western plants, but it is a synonym of the eastern United States species *P. quinquefolia* (L.) Planch. This species is commonly grown as an ornamental in the West and can persist around old structures and abandoned farms.

Parthenocissus vitacea (Knerr) Hitchc. (affinity to *Vitis*) Thicket creeper, tc'ilna'at–'ó'íh, píla'astla'íh. Weakly climbing, clambering or sprawling vine. **STEMS** pale tan to brown, mostly round, young bark glabrous or rarely minutely whitish scabrous, older bark becoming flaky, resinous lenticels common on older stems; tendrils weakly branched, lacking adhesive discs although slightly swollen at tip. **LEAVES** shiny dark green above, glabrous to pubescent below, the central leaflet 4–15 cm long and 3–8 cm wide, lateral leaflets somewhat smaller, leaves turning red in autumn. **INFL** of terminal cymose clusters opposite leaves, dichotomously forked at the top of the peduncle with subsequent dichotomous forks with clusters of numerous flowers. **FLOWERS** unisexual, small, inconspicuous, green. **FRUIT** purple to

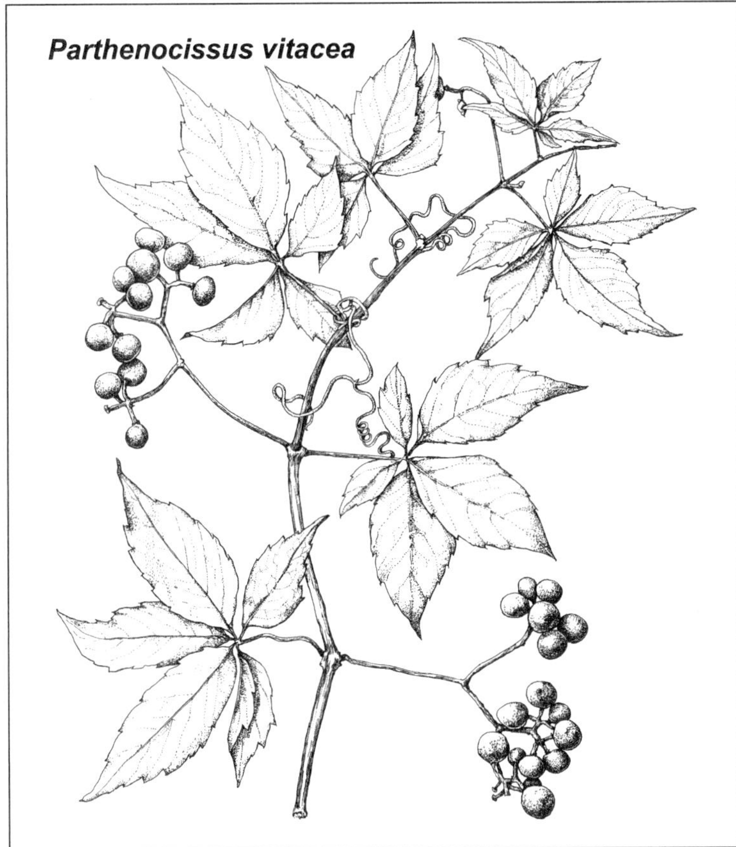

black, somewhat fleshy, 5–10 mm wide [*P. inserta* (J. Kern.) Fritsch, misapplied in some literature.] Native vine of springs, seeps, and wet, shaded riparian zones. ARIZ: Apa, Nav; COLO: Arc, LPl, Mon, SMg; NMEX: McK, RAr, SJn; UTAH. 1150–2400 m (3800–8000′). Flowering: May–Jun. Fruit: Jul. The native thicket creeper of the west, this species has been used by the Navajo as a ceremonial and medicinal plant, for food (the berries), and for thatching ramadas.

Vitis L. Grape

(classical Greek name for wine grape) Strongly climbing to sprawling, freely branching woody vines. **STEMS** with shredding or flaking bark, lenticels obscured or not prominent; tendrils mostly bifurcate, lacking adhesive discs. **LEAVES** simple, palmately veined, margins smooth to serrate, blades often palmately lobed. **INFL** of lateral cymose panicles opposite the leaves. **FLOWERS** functionally or structurally unisexual, 5-merous; calyx mostly obsolete, sometimes a connate ring present; petals separate below but coherent above; stamens 5, reduced in female flowers; pistil vestigial in male flowers; style absent or short; nectary disc usually well developed, at base of ovary, typically 5-lobed. **FRUIT** a juicy berry. **SEEDS** 4 or fewer by abortion. A genus of 60–70 species, primarily distributed in the temperate zone of the Northern Hemisphere. Most species of *Vitis* have been used for food and beverages, and the wine grape, *V. vinifera*, native to the Mediterranean and Middle East and one of the staples of western civilization, has been widely cultivated for wine production in virtually all suitable habitats worldwide. The species first appeared in cultivation sometime prior to 5000 B.C.E.

Vitis arizonica Engelm. **CP** (of Arizona) Arizona grape, tc'ilna'at–'ó'íh. Climbing or clambering woody vines. **STEMS** with shredding or flaking bark, freely branching, young twigs tomentose, becoming mostly glabrous with age; tendrils bifurcate, lacking adhesive discs. **LEAVES** long-petiolate, blades cordate-ovate, simple, palmately veined, cottony pubescent on both sides when young, hairs persisting on lower leaf veins although sometimes becoming glabrous with age, margins with sharp teeth, blades sometimes palmately lobed, large, 5–12 cm long. **INFL** a loose, open, strongly branched panicle, opposite the leaves. **FLOWERS** functionally or structurally unisexual, 5-merous; calyx obsolete; petals white, separate below but coherent above; stamens 5, reduced in female flowers, long-erect in male flowers; pistil vestigial in male flowers; style short, stigma discoid; nectary disc present, 5-lobed. **FRUIT** a juicy purple-blue to black berry, skin readily separated from pulp. Rare in southern parts of the region, around springs and seeps, and in dense, shaded, damp, riparian woodlands. ARIZ: Apa; NMEX: SJn. 1500–2100 m (5000–7000′). Flowering: May. Fruit: Jun. *Vitis arizonica* is highly variable in pubescence and leaf shape, with forms occurring that are nearly glabrous. These have been named var. *glabra* Munson. The berries are edible and are relished by both wildlife and humans. Apparently the combination *V. arizonica* was never validly published by Engelmann, according to Eric Wada (pers. comm. 2006), who is currently revising the western grapes. The wine grape, *V. vinifera*, is widely cultivated in the region and can persist after abandonment. This species differs from *V. arizonica* in that the bark is not very shreddy or flaky on mature stems, the flowers are bisexual, and the skin of the fruit cannot be separated easily from the pulp.

ZANNICHELLIACEAE Chevall. HORNED-PONDWEED FAMILY

C. Barre Hellquist

Annual, rarely perennial herbs; rhizomes present or absent; caulescent aquatic, submerged. **LEAVES** alternate, opposite, and pseudowhorled, often on the same plant, submersed, sessile, sheath not persisting; blade linear. **INFL** an axillary cyme, without subtending spathe, sessile. **FLOWERS** unisexual, staminate and pistillate flowers on same plant; subtending bracts and perianth absent; staminate flowers with 1 stamen; anthers dehiscing longitudinally; pistillate flowers with (1–) 4–5 (–8) pistils, distinct, short-stipitate; ovules pendulous. **FRUIT** drupaceous. **SEED** 1; embryo curved (Haynes and Holm-Nielsen 1987). Four genera. About 10–12 species worldwide.

Zannichellia L. Horned-pondweed

(for Gian Girolamo Zannichelli, 1662–1729, Venetian apothecary and botanist) Root single or paired at nodes. **LEAVES** 2–10 cm long, up to 2 mm wide with an entire blade, veins 1 (–3), nearly terete. **INFL** usually of 2 flowers, 1 staminate, 1 pistillate. **FLOWERS** short-pedicellate. **FRUIT** minutely dentate along convex margin; endocarp coarsely papillose; stipe developing into a stalk supporting the pistil (podogyne). Species about five worldwide.

Zannichellia palustris L. (of marshes) Horned-pondweed. Submersed herbs. **LEAVES** 3.5–4.2 cm × 0.2–1 mm, apex acute. **FLOWERS**: staminate with filaments 1.5–2 mm; pistillate with 4 or 5 pistils; style 0.4–0.7 mm. **FRUIT** 1.7–2.8 × 0.6–0.9 mm; rostrum 0.7–2 mm; podogyne 0.1–1.5 mm; pedicel 0.3–1.2 mm. Brackish and freshwater ponds, lakes, streams, and ditches. ARIZ: Apa, Nav; COLO: LPl; NMEX: SJn. 1475–2750 m (4925–9100′). Flowering: Jul–Aug. Fruit: Jul–Sep. Alaska east to Quebec and Newfoundland, south throughout the United States; Mexico; West Indies; Central America; South America; Eurasia; Africa; Australia.

ZYGOPHYLLACEAE R. Br. LIGNUM VITAE (CALTROP) FAMILY

Linda Mary Reeves

Primarily shrubs, herbs, and a few trees. **STEMS** sometimes jointed. **LEAVES** mostly fleshy or leathery; often pinnately compound; opposite, occasionally alternate; with stipules often becoming spiny. **INFL** solitary-flowered or paired. **FLOWERS** perfect; actinomorphic; sepals imbricate, 4–5; petals 4–5, distinct, imbricate; stamens in 1–3 whorls of 5, the outer series opposite the petals, filaments often scaled at the base; disc present; ovary superior, carpels 2–12, usually 5, fused, winged, locules 5; placentation axile. **FRUIT** a capsule, often splitting into 5 sections, occasionally a berry or drupelet. About 25 genera with about 240 species. Worldwide dry or seaside tropical, subtropical, and temperate zones. *Guaiacum officinale* is the source of the decongestant guaifenesin and, with *G. sanctum* and *Balansaea* sp., the source of valuable lignum vitae timber, the heaviest known wood. Some species provide a valuable oil used in perfumes. *Peganum harmala*, a noxious weed of Mediterranean origin, produces the dye Turkey red.

1. Leaves alternate; fruit a glabrous loculicidal capsule .... ***Peganum***
1′ Leaves opposite; fruit at maturity separating into 5 or 10 indehiscent mericarps .... (2)

2. Mericarps 10, tubercled, 1-seeded, with persistent beak .... ***Kallstroemia***
2′ Mericarps 5, spiny, 3–5-seeded; fruit beak deciduous .... ***Tribulus***

Kallstroemia Scop. Desert Poppy, Caltrop

Prostrate to ascending, diffusely branching, annual or perennial herbs. **STEMS** pubescent to glaucous or glabrous. **LEAVES** opposite; even-pinnate; leaflets obovate to elliptic; 3–8 pairs, 1 of each pair sometimes alternately smaller or abortive or often obtuse. **FLOWERS** solitary, axillary from smaller leaves; sepals 5, pubescent, imbricate, usually persistent; petals 5, orange to yellow, cream, or white, apex rounded or notched; stamens 10, in 2 whorls of 5 each, outer whorl slightly longer than inner whorl; style persistent, ovary globose to ovoid, 10-loculed, pubescent to glabrous; ovules 1 per locule. **FRUIT** at maturity separating into 10 indehiscent, 1-seeded mericarps. About 17 species of New World subtropical and tropical areas, in disturbed or alluvial soils.

Kallstroemia parviflora Norton (small-flowered) Caltrop. Prostrate to decumbent annual. **STEMS** hirsute, sericeous to glabrous, to 1 m long. **LEAVES** 1–6 cm long; leaflets 3–5 pairs; elliptic to oval, hirsute, 8–19 mm long, 3.5–9 mm wide. **FLOWER** pedicels longer than subtending leaves; sepals lanceolate, 4–7 mm long, 1–2 mm wide; petals orange, narrowly obovate to oval, 5–11 mm long, 3.5–6 mm wide. **FRUIT** ovoid, strigose, 3–4 mm long, 4–6 mm wide; beak persistent, longer than fruit body; mericarps rugose to tubercled, 3–4 mm high, about 1 mm wide. [*K. intermedia* Rydb.].

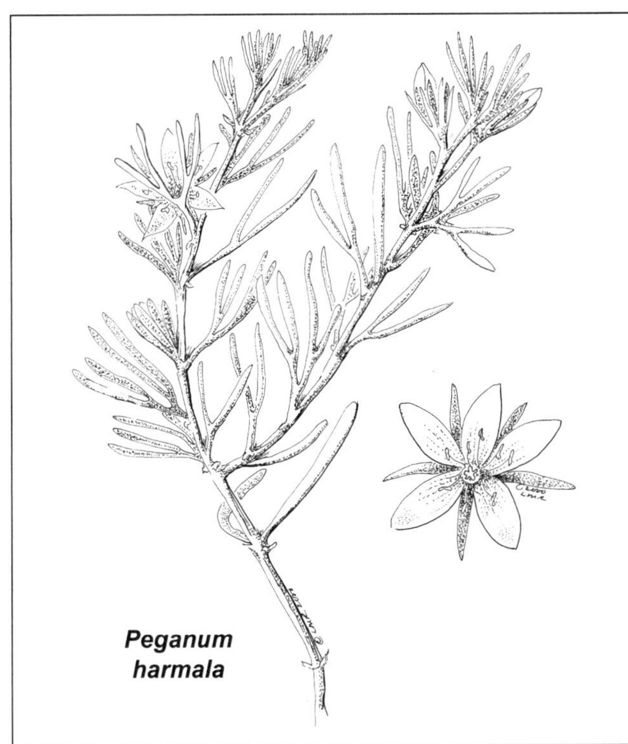
Peganum harmala

Primarily alluvial soils and disturbed sites. ARIZ: Nav; NMEX: SJn. 1400–2000 m (4600–6500′). Flowering: Jul–Oct. Fruit: Aug–Nov. Central and northern to western Texas through New Mexico, Colorado, Arizona, southern Nevada to southern California, north through Oklahoma to Illinois, and south from Sonora, Chihuahua, Coahuila, Nuevo León to Durango, Aguascalientes, and San Luis Potosí.

Peganum L. African Rue, Garbancillo

Dichotomously branched, prostrate to ascending, often globose herbs, height less than 1.5 m. **STEMS** herbaceous, pubescent to glabrous, arising from a perennial rootstock. **LEAVES** congested, alternate, fleshy, irregularly pinnatifid; lobes linear, acute to apiculate; pubescent to glabrous; stipules hairlike. **FLOWERS** solitary; sepals 4–5, irregularly pinnatifid to entire, segments linear, acute, valvate in bud, persistent; petals 4–5, white or cream to yellow, subequal, imbricate in bud; stamens 12–15; filaments with bases dilated; ovary globose, 2–4-loculed. **FRUIT** an irregularly dehiscent capsule with beak as long as or longer than fruit body. **SEEDS** triangular, dark, curved, numerous. A genus of three species, one from the Chihuahuan Desert of North America, the other two from the arid regions of the Mediterranean and Asia.

Peganum harmala L. (possibly named after Harmala, a city in Syria; also an old plant name in Arabia) African rue. **STEMS** prostrate to ascending-globose, pubescent to glabrous, to 90 cm long and 30 cm high. **LEAVES** glabrous to slightly pubescent. **FLOWERS** with pedicels twice as long as petals; sepals longer than petals; petals white to yellow, oblong-elliptic, to 15 mm long; filaments dilated basally; disc cupulate. **FRUIT** a capsule; peduncles longer than fruit; to 15 mm in diameter. Known from one disturbed site near Aztec, New Mexico. NMEX: SJn. 2000 m (6000′). Flowering: Apr–Nov. Fruit: May–Nov. An introduced weed from Asian deserts. Originally introduced in Luna County, New Mexico, in the 1930s and extending its range. Known from Nevada, Arizona, southern New Mexico, and western Texas.

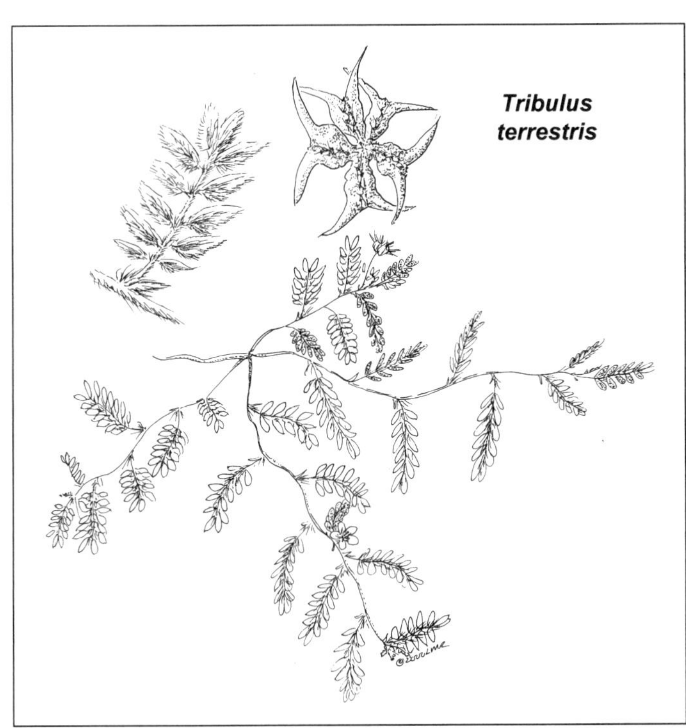
Tribulus terrestris

Tribulus L. Goathead, Caltrop

(three-pointed, referring to the mericarp) Prostrate or ascending-spreading annual or perennial herbs from a central taproot, to 3 m long. **STEMS** diffusely branched, appressed, sericeous-hirsute. **LEAVES** opposite; even-pinnate; 1 of each pair alternately smaller or abortive; leaflets 3–7 pairs, oblong to ovate, tending toward falcate. **FLOWERS** solitary, pedicels emerging from smaller leaf axils; sepals 5, pubescent, deciduous; petals 5, bright to light yellow, occasionally white; obovate, apex generally rounded; stamens 10, in 2 whorls of 5 each, outer whorl adnate to petals, with glands; ovary with 5 locules, hirsute; ovules 15–25. **FRUIT** of 5 indehiscent mericarps, each with 2–4 dorsal spines and smaller spines and/or tubercles; beak deciduous. A desert genus of the Old World with about 70 species, 3 of which are introduced in North America.

Tribulus terrestris L. (of the land) Goathead, caltrop, puncture weed, puncture vine, arrojo de flor amarilla, cadillo. Prostrate annual. **LEAVES** 10–45 mm long, leaflets 3–6 pairs, oblong to ovate, 4–10 mm long, 1–5 mm wide. **FLOWERS** 5–10 mm diameter; sepals 2–3 mm long, ovate; petals 3–5 mm long; intrastaminal glands free. **FRUIT** to about 2.5 cm wide, including spines, largest spines about 3–7 mm long; separating into 5 crested segments. Typical inhabitant of disturbed soils, especially sandy and gravelly substrates near agricultural areas, homes, and roadsides. $2n = 12, 24, 36, 48$. ARIZ: Apa, Coc, Nav; COLO: Arc, Dol, LPl, Mon; NMEX: McK, RAr, San, SJn; UTAH. 1400–2100 m (4600–7000′). Flowering: Apr–Sep. Fruit: May–Nov. Introduced from Asian deserts. A scourge of dry, sandy habitats, backyards, and gardens that destroys bicycle tires and scars animal paws as well as the feet of those foolish enough to walk without shoes. Unfortunately, seeds have been known to be viable for up to eight years.

Aconitum columbianum

Aliciella subnuda

Anaphalis margaritacea

Apocynum cannabinum

Aquilegia coerulea

Aquilegia micrantha

Arceuthobium divaricatum

Astragalus chuskanus

Astragalus heilii

Astragalus missouriensis var. *humistratus*

Astragalus preussii var. *latus*

Calochortus aureus

Caltha leptosepala

Campanula parryi

Castilleja exilis

Castilleja haydenii

Ceanothus fendleri

Chionophila jamesii

Cirsium scopulorum

Claytonia megarhiza

Commelina dianthifolia

Corydalis caseana subsp. *brandegei*

Cryptantha fulvocanescens var. *fulvocanescens*

Cryptantha paradoxa

Cymopterus constancei

Cymopterus longilobus

Datura wrightii

Delphinium scaposum

Dodecatheon pulchellum

Draba graminea

Dryas octopetala

Echinocereus engelmannii var. *variegatus*

Echinocereus triglochidiatus var. *triglochidiatus*

Echinocereus viridiflorus var. *viridiflorus*

Elodea bifoliata

Erigeron abajoensis

Erigeron canaani

Erigeron compositus

Erigeron utahensis

Eriogonum clavellatum

Eriogonum ovalifolium var. *purpureum*

Eriophorum scheuchzeri

Eritrichium nanum

Escobaria missouriensis

Escobaria vivipara var. arizonica

Evolvulus nuttallianus

Fallugia paradoxa

Fendlera rupicola

Fragaria virginiana var. *glauca*

Heliotropium convolvulaceum

Hordeum jubatum subsp. *intermedium*

Houstonia rubra

Hymenoxys grandiflora

Ipomopsis congesta subsp. *matthewii*

Ipomopsis polyantha

Lewisia nevadensis

Lithospermum multiflorum

Maianthemum racemosum

Mentzelia filifolia

Mentzelia sivinskii

Menyanthes trifoliata

Mertensia lanceolata var. *nivalis*

Mimulus eastwoodiae

Minuartia obtusiloba

Mirabilis multiflora

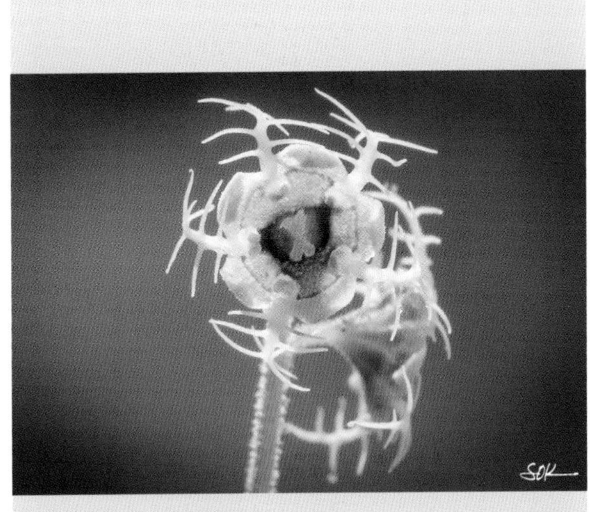

Mitella pentandra

Monarda fistulosa

Moneses uniflora

Noccaea fendleri

Oenothera cespitosa

Opuntia macrorhiza

Opuntia phaeacantha

Opuntia polyacantha var. *hystricina*

Orobanche fasciculata

Oxytropis deflexa var sericea

Papaver radicatum

Pedicularis groenlandica

Pediocactus knowltonii

Pediocactus simpsonii

Penstemon ambiguus var. *laevissimus*

Penstemon eatonii var. *eatonii*

Penstemon harbourii

Penstemon utahensis

Phacelia ivesiana

Phacelia sericea

Phlox cluteana

Phlox condensata

Physaria acutifolia

Physaria scrotiformis

Polygala subspinosa

Prunus virginiana var. *melanocarpa*

Pterospora andromedea

Purshia stansburiana

Pyrola picta

Pyrrocoma crocea

Salix serissima

Saxifraga bronchialis

Scabrethia scabra subsp. *canescens*

Sclerocactus cloveriae subsp. *cloveriae*

Sclerocactus mesae-verdae

Sclerocactus parviflorus subsp. *parviflorus*

Sedum lanceolatum

Sisyrinchium montanum var. *montanum*

Sophora stenophylla

Sphaeralcea leptophylla

Stanleya pinnata var. *pinnata*

Stellaria longipes

Symphoricarpos oreophilus

Synthyris ritteriana

Townsendia glabella

Townsendia incana

Trifolium brandegeei

Trifolium nanum

Trollius albiflorus

Vitis arizonica

Yucca baileyi

Zeltnera calycosa

Zigadenus elegans

FERNS AND FERN ALLIES

Aspleniaceae
Asplenium resiliens LMR*
Asplenium septentrionale LMR
Asplenium trichomanes LMR
Asplenium viride LMR

Dryopteridaceae
Athyrium alpestre LMR
Athyrium filix-femina LMR
Cystopteris fragilis LMR
Cystopteris montana LMR
Cystopteris reevesiana LMR
Cystopteris tenuis LMR
Cystopteris utahensis LMR
Dryopteris expansa LMR
Dryopteris filix-mas LMR
Gymnocarpium dryopteris LMR
Polystichum lonchitis LMR
Woodsia neomexicana LMR
Woodsia oregana LMR
Woodsia plummerae LMR
Woodsia scopulina LMR

Ophioglossaceae
Botrychium campestre LMR
Botrychium echo LMR
Botrychium hesperium LMR
Botrychium lanceolatum LMR
Botrychium lunaria LMR
Botrychium minganense LMR
Botrychium multifidum LMR
Botrychium pallidum LMR
Botrychium pinnatum LMR
Botrychium simplex LMR

GYMNOSPERMS

Cupressaceae
Juniperus scopulorum United States Department of Agriculture Forest Service Collection, Hunt Institute for Botanical Documentation, Carnegie Mellon University, Pittsburgh, PA.

Ephedraceae
Ephedra torreyana Beth Dennis, used with permission.

Pinaceae
Abies concolor Frederick Andrews Walpole Collection, courtesy of the Hunt Institute for Botanical Documentation, Carnegie Mellon University, Pittsburgh, PA, on indefinite loan from the Smithsonian Institution.
Picea engelmannii Frederick Andrews Walpole Collection, courtesy of the Hunt Institute for Botanical Documentation, Carnegie Mellon University, Pittsburgh, PA, on indefinite loan from the Smithsonian Institution.
Pinus ponderosa Dwight DeWitt Ivey. 2003. Flowering Plants of New Mexico, 4th ed. p. 533. Used with permission.
Pinus strobiformis Beth Dennis, used with permission.
Pseudotsuga menziesii Frederick Andrews Walpole Collection, courtesy of the Hunt Institute for Botanical Documentation, Carnegie Mellon University, Pittsburgh, PA, on indefinite loan from the Smithsonian Institution (as *Pseudotsuga taxifolia*).

ANGIOSPERMS

Aceraceae
Acer glabrum LMR
Acer grandidentatum LMR
Acer negundo LMR

Adoxaceae
Adoxa moschatellina LMR
Sambucus racemosa LMR

Anacardiaceae
Toxicodendron rydbergii LMR

Apiaceae
Angelica grayi LMR
Cicuta bulbifera LMR
Conium maculatum LMR
Cymopterus alpinus and *Cymopterus bakeri* (1 plate) LMR
Ligusticum porteri LMR
Lomatium triternatum LMR
Podistera eastwoodiae LMR

Araliaceae
Aralia racemosa subsp. *bicrenata* LMR

Asclepiadaceae
Asclepias subverticillata LMR

Asteraceae
Asteraceae morphology LMR
Acroptilon repens LMR
Artemisia spinescens Beth Dennis, used with permission.
Artemisia tridentata Frederick Andrews Walpole Collection, courtesy of the Hunt Institute for Botanical Documentation, Carnegie Mellon University, Pittsburgh, PA, on indefinite loan from the Smithsonian Institution.
Bahia dissecta LMR
Balsamorhiza sagittata LMR
Brickellia leaves LMR (2 plates)
Carduus nutans LMR
Centaurea diffusa LMR
Centaurea solstitialis LMR

* LMR=Linda Mary Reeves

Frankeniaceae
Frankenia jamesii LMR
Gentianaceae
Gentianodes algida LMR
Gentianopsis barbellata LMR
Geraniaceae
Geranium richardsonii United States Department of Agriculture Forest Service Collection, Hunt Institute for Botanical Documentation, Carnegie Mellon University, Pittsburgh, PA.
Grossulariaceae
Ribes inerme Marjorie C. Leggit, used with permission.
Hydrophyllaceae
Hydrophyllum fendleri LMR
Phacelia howelliana LMR
Phacelia indecora LMR
Juglandaceae
Carya illinoinensis LMR
Juglans major LMR
Juncaginaceae
Triglochin maritima Frederick Andrews Walpole Collection, courtesy of the Hunt Institute for Botanical Documentation, Carnegie Mellon University, Pittsburgh, PA, on indefinite loan from the Smithsonian Institution.
Lamiaceae
Poliomintha incana Beth Dennis, used with permission.
Liliaceae
Eremocrinum albomarginatum LMR
Leucocrinum montanum LMR
Lloydia serotina LMR
Zigadenus paniculatus LMR
Zigadenus venenosus LMR
Linaceae
Linum australe LMR
Moraceae
Maclura pomifera LMR
Morus alba LMR
Oleaceae
Fraxinus anomala Marjorie C. Leggit, used with permission.
Onagraceae
Gayophytum diffusum subsp. *parviflorum* LMR
Orchidaceae
Calypso bulbosa LMR
Coeloglossum viride LMR
Corallorhiza maculata LMR
Corallorhiza striata LMR
Corallorhiza trifida LMR
Corallorhiza wisteriana LMR
Cypripedium parviflorum LMR
Epipactis gigantea LMR
Goodyera oblongifolia LMR
Goodyera repens LMR
Platanthera dilatata LMR
Platanthera huronensis LMR
Orobanchaceae
Cordylanthus wrightii LMR
Pedicularis procera LMR
Papaveracee
Argemone albiflora LMR
Plantaginaceae
Linaria dalmatica LMR
Penstemon breviculus LMR
Penstemon lentus LMR
Plantago argyrea LMR
Poaceae
Poaceae morphology LMR
Aegilops cylindrica LMR
Bromus rubens LMR
Bromus tectorum LMR
Cenchrus longispinus LMR
Danthonia parryi Hitchcock-Chase Collection of Grass Drawings, courtesy of the Hunt Institute for Botanical Documentation, Carnegie Mellon University, Pittsburgh, PA, on indefinite loan from the Smithsonian Institution.
Koeleria macrantha and *Poa fendleriana* (1 plate) LMR
Muhlenbergia richardsonis Hitchcock-Chase Collection of Grass Drawings, courtesy of the Hunt Institute for Botanical Documentation, Carnegie Mellon University, Pittsburgh, PA, on indefinite loan from the Smithsonian Institution.
Puccinellia parishii Hitchcock-Chase Collection of Grass Drawings, courtesy of the Hunt Institute for Botanical Documentation, Carnegie Mellon University, Pittsburgh, PA, on indefinite loan from the Smithsonian Institution.
Sorghum halepense LMR
Polemoniaceae
Ipomopsis congesta subsp. *matthewii* LMR
Phlox caryophylla LMR
Polygalaceae
Polygala subspinosa LMR
Polygonaceae
Eriogonum wetherillii LMR
Polygonum douglasii LMR
Stenogonum flexum and *Stenogonum salsuginosum* (1 plate) LMR
Primulaceae
Lysimachia hybrida LMR
Ranunculaceae
Clematis columbiana Marjorie C. Leggit, used with permission.
Delphinium nuttallianum LMR
Rosaceae
Cercocarpus intricatus LMR

Coleogyne ramosissima USDA Forest Service Collection, Hunt Institute for Botanical Documentation, Carnegie Mellon University, Pittsburgh, PA.
Potentilla subjuga LMR

Rubiaceae
Galium aparine LMR
Galium bifolium LMR
Galium boreale LMR
Galium coloradoense LMR
Galium proliferum LMR
Galium trifidum LMR
Galium triflorum LMR
Galium wrightii LMR
Houstonia rubra LMR
Kelloggia galioides LMR

Salicaceae
Salix monticola LMR

Santalaceae
Comandra umbellata subsp. *pallida* LMR

Saxifragaceae
Heuchera rubescens LMR
Lithophragma glabrum and *Lithophragma tenellum* (1 plate) LMR
Mitella pentandra and *Mitella stauropetala* (1 plate) LMR

Simaroubaceae
Ailanthus altissima LMR

Solanaceae
Datura wrightii LMR
Hyoscyamus niger LMR
Solanum heterodoxum LMR
Solanum rostratum LMR

Tamaricaceae
Tamarix chinensis LMR

Ulmaceae
Celtis reticulata Marjorie C. Leggit, used with permission.

Valerianaceae
Valeriana edulis Frederick Andrews Walpole Collection, courtesy of the Hunt Institute for Botanical Documentation, Carnegie Mellon University, Pittsburgh, PA, on indefinite loan from the Smithsonian Institution.

Vitaceae
Parthenocissus vitacea Marjorie C. Leggit, used with permission.

Zygophyllaceae
Peganum harmala LMR
Tribulus terrestris LMR

GLOSSARY

Abaxial The side away from the axis, usually the underside.

Abcission zone The layer of cells, often in a leaf petiole, that becomes corky and dry under certain circumstances, such as a change in hormones, and allows the leaf or other structure to become deciduous.

Abortive Failing to mature, undeveloped.

Acaulescent Without an obvious stem; plants with basal leaves only are considered acaulescent.

Accessory Fruit derived from parts other than the ovary, often the receptacle, as in *Fragaria*. Most of the fruit, especially the exocarp, may include some accessory-derived structures.

Accrescent A structure that becomes larger with age, such as a calyx in some flowers.

Acequia An irrigation canal ("ditch") originally constructed or planned by Spanish settlers in New Mexico, but later applying to all managed farm irrigation ditches.

Acerose Needle-shaped, as in the leaves of conifers.

Achene (akene) Small, dry, one-seeded, indehiscent fruit developing from a superior ovary, as in *Carex*. The ovary wall is usually thin (pericarp) and not fused to the seed layers (the seed is loose).

Achenial Of the achene.

Achlorophyllous Without chlorophyll, as in nongreen parasitic plants.

Acicula (aciculus) A small needle-shaped structure or the end of a sharp prickle, thorn, or spine.

Acicular Needle-shaped.

Acropetal Closer to the tip than the base.

Acroscopic Facing or on the side toward the apex.

Actinomorphic (regular) Flowers with whorls of parts radially symmetrical, all parts in each whorl alike; typically based on petals unless otherwise stated.

Aculeate Sharply pointed; beset with prickles or spines.

Acuminate Gradually tapering to a point with slightly concave margins.

Acute Forming an angle of less than 90°.

Adaxial The side toward the axis, usually the upper side.

Adhesive disc Disc of plant tissue growing from the stem of some vines and lianas that attaches to the substrate.

Adnate Union or fusion of two or more unlike parts, as stamens adnate to petals.

Adventive Not native; introduced and populations increasing.

Aerenchyma A type of parenchyma, usually chlorenchyma, which contains many air spaces, such as in *Nymphaea*.

Aerial Borne above ground level, usually stems.

Agamospermous Able to form a seed without fertilization.

Aggregate Fruit formed from a cluster of pistils that were distinct in a single flower, as in *Fragaria*.

Alcove Shallow cave-like depression in a cliff wall, usually in sandstone, and which often has seeps in the rear.

Allopatric Occupying different geographic locations.

Allopolyploidy The condition of having complete chromosome sets from two or more different species.

Allotetraploid An allopolyploid with four chromosome sets, or $4n$.

Alluvial Composed of alluvium substrate, such as mud, sand, gravel, or cobble deposited by water.

Alluvium Soil substrate that has been moved into place by moving water.

Alpine Above tree line, generally 3500 m (11500′) in our area, depending upon exposure and other factors.

Alpine tundra The treeless plant community of matted, tufted, and prostrate plants found above timberline.

Alternate Having one leaf at a node, with successive leaves on opposite sides of the stem.

Alveolate As a honeycomb, with depressions separated by thin partitions or ridges.

Ament (see **catkin**) A dense inflorescence of unisexual flowers.

Amphibious Plants that live both in water and on land, often shore dwellers.

Amphimictic Able to interbreed freely, producing fertile offspring.

Amphitropic Usually an ovule that is half inverted and straight, with a lateral hilum.

Amplexicaul Referring to leaf bases that clasp the stem.

Ampliate Enlarged or expanded.

Anastomose Netlike, as in many leaf veins.

Anatropous An ovule that is inverted, with the micropyle opening axis parallel to the funiculus.

Androgynous Staminate flowers borne above pistillate flowers.

Andromonoecious Having staminate and perfect flowers on the same plant.

Angled With a sharp corner or corners.

Angustiseptate A fruit that is flattened at right angles to the septum, as in some Brassicaceae.

Annual Plants that grow from seed, flower, produce seed, and die in the same year, sometimes within a few weeks.

Annulus A ring-shaped structure; a row of specialized, thick-walled cells along one side of a fern sporangium that aids in spore dispersal by contraction.

Anther The portion of the stamen that produces and bears pollen.

Anthesis The time when a flower is fully mature.

Anthocyanic With a reddish or purplish tinge due to pigments called anthocyanins.

Antisepalous Located opposite the sepals.

Ant lion The larva of a neuropteran insect that builds a conical trap in sand to catch ants as prey.

Antrorse Forward or upward in direction.

Aperturate With an opening, typically referring to the openings in pollen grains.

Apetalous Without petals.

Apex Tip of the leaf blade or other structure.

Apical At the apex or tip.

Apiculate With a short, flexible tip.

Apiculum (see **apiculus**).

Apiculus A small, sharp point.

Apogamous Relating to an embryo produced without fertilization.

Apomictic (see **apomixis**).

Apomixis Seed production without fertilization (agamospermy).

Apophysis (**apophyses**) A projection on a structure; the portion of a cone scale exposed when the cone is still closed.

Appendage A petal-like extension of the gland in many species, as in *Chamaesyce* and *Euphorbia*; may be obsolete to conspicuously petaloid; if present, usually white or less often pink or even reddish, usually entire, sometimes crenulate or divided, rarely fimbriate or pubescent.

Appressed Lying flat and close against the ground or another part.

Aquatic Growing in water, either emergent, submerged, or floating.

Arachnoid With cobwebby, tangled hairs.

Arachnose (see **arachnoid**).

Archegonium The structure in plants that produces and contains the female gamete; a vase-shaped structure in bryophytes, ferns, and fern allies that produces and contains the egg on the gametophyte.

Arcto-Tertiary Geoflora The ancient temperate flora in place in much of North America at the beginning of the Tertiary. This flora included many gymnosperm and angiosperm trees and shrubs.

Arcuate Curved into an arch; bowed.

Arenophilous Sand-loving.

Areole The node of a cactus plant that bears spines, flowers, or both (rarely leaves) and may produce new stems.

Aril A structure (usually fleshy) growing at or in the vicinity of the hilum of a seed; may be substantial and provide food for a distributing organism.

Arillate With an aril.

Arista A bristle.

Aristate With an elongate, bristlelike tip.

Armature Thorns, spines, prickles, or sharp trichomes, barbs.

Ascendant (see **ascending**).

Ascending Growing upright, often in a curving fashion.

Assurgent (see **ascending**).

Atomiferous Bearing sessile or subsessile glands.

Attenuate Gradually narrowed and tapering to a point.

Auct. nonn. Of some authors.

Auricle A small, ear-shaped appendage.

Auriculate Having an auricle.

Autogamy Self-fertilization.

Autonym When an author names a new subspecies or variety, it is given the new rank based on the original species type and duplicating the specific epithet; i.e., *Platanthera dilatata* var. *dilatata*.

Autopollination Self-pollination.

Autopolyploidy The condition where an organism has two or more complete chromosome sets derived from the same species.

Awn A bristlelike appendage.

Axial (see **axile**).

Axil The upper angle between a stem and a leaf or another stem.

Axile Located in the axis, such as axile placentation.

Axis The central line or stem along which plant parts are organized or arranged; the imaginary central line in an organ or structure.

Baccate Berrylike; soft.

Backcross A genetic cross in which offspring are crossed with parents or other ancestors.

Ballistic dispersal Dispersal of seeds or other propagules, usually under pressure, which scatters the materials a distance from their origin.

Barbed Bearing sharp, rigid, reflexed points like those on a fishhook.

Barbellate Bearing short, stiff hairs.

Barbellulate With very minute, stiff hairs or barbs.

Basal Related to or located at the base, usually of the leaf or petal.

Basifixed Hairs or scales fixed at the base.

Basionym The first name published for a taxon.

Basipetal Closer to the base rather than the tip; produced sequentially from the apex toward the base.

Basiscopic Facing or on the side toward the base.

Beak A prolonged, usually narrowed tip of a thicker organ, as in some fruits and petals.

Bermuda High A high-pressure subtropical atmospheric air mass.

Berry Fleshy, indehiscent, many-seeded fruit with no true stone (pit) or core, embedded in pulp (includes pepo and hesperidium).

Bicarinate Two-keeled.

Bicolored Of two distinct colors.

Bicornute Two-horned.

Biennial Plants that live two years, usually flowering and fruiting the second year. The first year is typically represented by a rosette of leaves.

Bifid Split into two parts or lobes, as in some styles.

Bifurcate Forked; divided into two branches.

Bilabiate With two lips, as in many zygomorphic flowers.

Bipinnate Pinnately compound with each segment also divided pinnately; twice compound in a featherlike fashion.

Bipinnatifid Cleft or deeply lobed bipinnately.

Biserrate Doubly serrate; the individual teeth are also serrate.

Bisexual (see **perfect**).

Bladder A thin-walled, inflated structure.

Blade The expanded portion of a leaf or petal.

Bolt To produce flowers quickly or prematurely, often due to excessive heat or low water availability.

Bony Hard, stonelike.

Bract A leaflike structure usually associated with inflorescences.

Bracteole An especially small bract, especially subtending a flower.

Bractlet A small bract.

Branched Divide into two or more branches from a common stalk.

Breccia A sedimentary rock composed of angular fragments that have undergone lithification.

Bristle A hairlike structure, often representing an extremely reduced leaf, sepal, or petal.

Bud An undeveloped leaf or flower shoot typically enclosed by bud scales.

Bud scale Specialized leaves (bracts) that enclose a bud.

Bulb An underground bud with flat discoid stem, roots, and thickened storage leaves (as in an onion).

Bulbil An aerial bulb or reproductive stem.

Bulblets Small to tiny bulbs.

Bulbous Bulblike or pertaining to bulbs; rounded.

Bunchgrass Grasses that grow in caespitose tufts, usually without stolons or long rhizomes.

Bur-like (**burlike**) Spiny or prickly and often rounded.

Caducous Early-deciduous compared to similar structures in other plants.

Caespitose Growth in tufts or dense clumps from a common point.

Calcareous Containing deposits of calcium salts, often calcium carbonate.

Caldera A large depression in the summit or summit region of a volcano caused by collapse or ejection of volcanic materials.

Callous Hardened or thickened.

Callus (plural **calli**) A hard thickening or protuberance; the basal extension of the lemma in grasses.

Calyculus (**calycule** or **calycle**) Small bracts surrounding the calyx, appearing as a second calyx whorl.

Calyx (plural **calyces**) Outermost whorl of flower parts; collectively, all the sepals.

Campanulate Bell-shaped.

Canescent With an overall grayish aspect due to hairs.

Capillary Hairlike, very fine, slender.

Capitate Head-shaped or in a head inflorescence.

Capitulescence A grouping of heads into an inflorescence.

Capitulum A head, a dense inflorescence of small, sessile flowers, as in the Asteraceae.

Capsule Dry, dehiscent fruit of more than one carpel and opening on more than one line of dehiscence.

Carcinogenic A chemical, material, or organism capable of causing cell transformation and/or cancer.

Carinal Pertaining to a ridge or series of ridges.

Carinate (see **keel**).

Carpel A simple pistil, consisting of a single ovary, style, and stigma and having marginal placentation. The basic unit of the gynoecium; compound pistils are formed from the fusion of two or more carpels.

Carpophore A thin extension of the receptacle forming a central axis between the carpels, as in the Apiaceae and Geraniaceae.

Carpopodium (see **gynophore**) A stipe supporting an ovary.

Carr A meadowlike wetland, often with willows.

Cartilaginous Tough and firm like cartilage.

Caruncle A protuberance or appendage near the hilum of a seed; often conspicuous in subgenus *Esula* (Euphorbiaceae) and in *Croton texensis* (also many other *Croton* species). The caruncle often provides food for other organisms and aids dispersal of seeds, often by ants.

Caryopsis (grain) A dry fruit typical of the Poaceae, usually with the ovary wall fused to the seed layers.

Castaneous Dark reddish brown; chestnutlike in color.

Cataphyll Bract present in catkins and subtending the flowers.

Cathartic A substance or chemical causing vomiting and/or diarrhea.

Catkin Spikelike, commonly pendulous inflorescence of unisexual flowers.

Caudate Bearing a tail or tail-like appendage.

Caudex (plural **caudices**) The woody and often enlarged base of a perennial.

Caudicle Small stalk connecting the pollinium to the stipe in the pollinarium or directly to the viscidium in hemipollinaria of orchids.

Caulescent With an obvious stem.

Cauline On the stem.

Central spine One of the innermost spines of an areole.

Cernuous Nodding or drooping, as in some flowers.

Chaff Thin, dry scales or bracts, as on the receptacles of flowers in the Asteraceae.

Chaffy Having or possessing chaff.

Chalaza (chalazal) In the ovary, where integuments are connected to the nucellus at the opposite end of the seed's (or ovule's) micropyle.

Chalazal collar In the Onagraceae, a ridge or flap surrounding the chalaza (as opposed to a micropylar collar).

Chalazal knot A bump or protuberance on one side of the seed, as in some Vitaceae.

Channeled Possessing one or more deep longitudinal grooves over the length of the structure, such as a seed.

Chaparral From the Spanish word "chaparro," which means scrub oak; a semiarid vegetation type characterized by thickets of often spiny shrubs.

Chartaceous With a stiff, papery texture.

Chasmogamous Flowers that open before fertilization and are generally cross-pollinated.

Chinle Formation A reddish, yellowish, or grayish clay-containing formation of sandstones or siltstones of Triassic age. This formation contains numerous endemics.

Ciliate With marginal hairs or scales; thinly and finely fringed.

Ciliolate With short marginal hairs or scales; thinly and finely fringed.

Cinereous Ash-colored; grayish due to a covering of short hairs.

Circinate vernation When new fern fronds emerge rolled or coiled.

Circular (orbicular, round) Flat and round in outline.

Circumboreal Distribution circling around the earth's northern pole.

Circumpolar Distribution circling around one of the earth's poles, usually in boreal and/or north-temperate regions.

Circumscissile A fruit, sporangium, or anther that opens by dehiscence around the top, forming a lidlike structure.

Cirque An amphitheater-shaped structure at the head of a glaciated valley.

Clade A monophyletic lineage.

Cladode (cladophyll) A stem with the morphology and photosynthesis function of a leaf.

Clambering Growing in a vining manner on the substrate.

Clasping Grasping or partially surrounding, but attached on one side; usually a leaf grasping a stem, often with basal lobes.

Clavate Club-shaped; widening toward the apex.

Claw A narrow petal base, often petiole-like.

Clay barren Alkaline and clay soils, usually of Morrison, Mancos, Chinle, and Fruitland formations, often with bentonite clays that shrink and swell according to the amount of water absorbed or lost. Swelling may be seven times the original volume due to water absorption. This is an area with low vegetation diversity, but those plants that are able to survive may be specifically adapted and endemic to clay soils.

Clearcut Places where all trees or shrubs have been cut or logged.

Cleft Incompletely divided or split nearly to the middle.

Cleistogamous Flowers that do not open and generally self-fertilize.

Club-cholla Cholla cacti with jointed, club-shaped stems; the genus *Corynopuntia* in our area.

Coalescent Grown together to form a unit.

Cobwebby Arachnoid; referring to hairs, such as in some Polemoniaceae or Asteraceae; a type of phylogeny in which species are interrelated or sometimes indistinct.

Coetaneous Flowers developing at approximately the same time as leaves.

Coiled cyme A determinate, coiled inflorescence.

Colluvium (hillwash) A heterogeneous mixture of rock and soil material that has moved down a slope or cliff by gravitational action and collected at the base of a cliff; such as avalanches, mudslides, rockslides, etc.

Columella An axis to which a carpel of a compound pistil may be attached, which may remain in the open dry fruit.

Column The united filament(s) and style of the Orchidaceae; the union of staminal filaments, as in the Malvaceae.

Columnar More or less vertical; pertaining to the column.

Coma A tuft of hairs, usually on the tip of a fruit or seed, as in the Asteraceae; the head of a tree.

Commissure The face by which two carpels join one another, as in the Apiaceae.

Comose With a coma or comalike.

Compound Leaves composed of separate leaflike parts called leaflets.

Concave Forming an inward-curving space on a structure, such as a seed or floral part.

Concavo-convex Concave on one side and convex on the other.

Concolored All parts of uniform color; grayish brown in color.

Conduplicate Folded together lengthwise, as in many grasses.

Cone A usually woody strobilus in nonflowering plants; a dense cluster of sporophylls on an axis; any structure with a conelike form.

Cone scale The woody, leaflike structure on a conifer cone that holds seeds or pollen.

Confluent The blending of one part into another.

Congested With the units grouped closely together, as in a congested inflorescence.

Conifer savanna A parklike vegetation association where, in our area, conifers, often junipers or piñon pines, are interspersed among grassland species.

Connate Union of like adjacent structures.

Connivent Converging, stuck together but not fused or united.

Conniving Becoming connivent.

Consimilar Similar to one another.

Convergent Meeting together.

Convex Bulging out, as in seeds or floral structures.

Convolute Parts overlapping like roof shingles; rolled up longitudinally.

Coralloid Coral-shaped, such as the underground roots of *Corallorhiza* or cycads.

Cordate Heart-shaped, with base rounded and prominently notched; sides tapering to a narrowed apex.

Coriaceous Leathery in texture.

Cork Dead-celled outer layer of bark tissue produced by cork cambium in woody plants; a layer of bark infused with waxy suberin.

Corky Possessing cork tissue; unusually thick layer of tissue.

Corm A short, sometimes disclike, vertical underground stem covered with small, thin, papery leaves.

Cormose Cormlike.

Corolla Second from outside whorl of flower parts; the collective name for the petals.

Corona Petal-like crown structure such as in some *Asclepias* or *Nasturtium* flowers.

Corona scale Papillose scales forming the corona in the Boraginaceae; scales forming a circle around the eye in the Onagraceae.

Coroniform Crown-shaped.

Corpusculum A gland associated with the pollinarium in the Asclepiadaceae.

Cortex A layer of storage parenchyma tissue interior to the epidermis and exterior to the phloem in roots and stems; part of the bark in woody plants.

Corymb Nearly flat-topped, indeterminate inflorescence; lower and outer pedicels longest; simple or compound.

Corymbiform An inflorescence with the general appearance of a corymb, but not necessarily the form.

Costa (plural **costae**) A rib or prominent midvein.

Costellate With many small ribs.

Cotyledon A seed leaf of the plant embryo, often containing nutrients for the growth of the young plant.

Cowl A gently folded-over structure, often forming a rim.

Crassulacean Acid Metabolism (**CAM**) A type of carbon fixation in many succulent plants which fixes carbon dioxide at night into organic acids, allowing stomata to remain closed during the hot day.

Crateriform Shallowly cup-shaped.

Creeping With prostrate growth, often producing roots at nodes.

Crenate With broad, rounded teeth; scalloped.

Crested Cristate or with a tuft or projection.

Cretaceous Geological period from 135 to 65 million years ago; several geologic formations from this time period are found throughout the Four Corners region.

Crisped Curled, wavy, or crinkled.

Cristate (see **crested**).

Cross section A section obtained by cutting across a long axis.

Crown The perennial stem base of a herbaceous plant. Also, the leafy top of a tree.

Crown scale A type of scale found on the corona of gentians; a small scale opposite the nectaries, often attached at the base of the filament, in *Frasera* at the base of the ovary.

Crozier A young, curled fern frond.

Cruciform In the form of a cross, as in species of *Marsilea* or flowers of the Brassicaceae (Cruciferae).

Cucullate Hooded or hood-shaped.

Cuesta An asymmetrical ridge with a steep slope on one side and a gentle slope on the other. The gentle slope is at a lower angle than that of hogbacks.

Culm A stem, often hollow or pithy, especially in the Poaceae or Cyperaceae.

Cultivar A variety that has been artificially selected and grown horticulturally.

Cuneate Wedge-shaped or triangular, with the point of attachment at the narrowest part.

Cupular With a small cup or cuplike structure.

Cupulate Cup-shaped; with a small cup-shaped structure.

Cuspidate Tipped with a short, sharp, abrupt point.

Cyanobacteria Photosynthetic prokaryotes responsible for much of the nitrogen fixation in an ecosystem and which often affect calcium carbonate deposition in an aquatic environment and can be symbionts with certain plants, as in the Myristicaceae or Orchidaceae; "blue-green algae."

Cyathial lobes Five (or appearing more) short, variably shaped and/or variably margined extensions of the cyathial rim that are the tips of the fused floral bracts; the glands and lobes typically alternate along the rim in the Euphorbiaceae.

Cyathium A greatly reduced, highly modified inflorescence of the Euphorbiaceae that is composed primarily of an involucre of fused bracts (that is more or less cup-shaped) containing separate male and female flowers and on its margins glands and sometimes petaloid appendages, all of which in aggregate resemble a single small flower; each cyathium encloses few to many staminate flowers (each consisting of a single stamen) and one pistillate flower; one to five glands, often bearing appendages, typically alternate with the cyathial lobes (the distal vestiges of the floral bracts); the cyathium may have hairs externally and/or internally.

Cylindric (**cylindrical**) Elongate and round in cross section.

Cylindroid Tending toward cylindrical.

Cyme Determinate inflorescence with central flowers opening first.

Cypsela (plural **cypselae**) Small, dry, one-seeded, indehiscent fruit developing from an inferior ovary, as in the Asteraceae. The ovary wall is usually thin (pericarp) and not fused to the seed layers (the seed is loose).

Dakota Sandstone A type of sandstone, sometimes arkosic, composed of sand grains and a matrix.

Deciduous Plant parts that are shed or fall off either seasonally or after a certain stage of development.

Declined Curved downward.

Decoction An extract, usually of plants, made by gently boiling the materials in water. The strained liquor is usually the decoction.

Decompound More than once compound; the leaflets subdivided.

Decumbent Flat or resting on the ground, but with growth tips ascending.

Decurrent Extending downward to the point of insertion.

Decussate Opposite pairs, usually leaves, that alternate perpendicular to the previous and following pair.

Deflexed Turned abruptly downward.

Dehiscing Opening at maturity, of such structures as dry fruits and anthers.

Deltate (see **deltoid**).

Deltoid Equilaterally triangular.

Dentate With pointed teeth facing outward.

Denticle A small tooth or toothlike projection.

Denticulate Minutely dentate or toothed.

Depauperate A specimen that is poorly developed or a plant community with poor expression of characters or diversity.

Depressed Low and/or flattened from above perspective.

Descending Pointing downward at an angle.

Determinate An inflorescence where the terminal (most distal) flower opens first; generally the inflorescence produces a finite number of buds.

Detrital Consisting of decayed and/or partially decomposed materials.

Diad A group of two structures, free from others.

Diatreme A volcanic feature formed by gas or volatile magma, usually in the lower crust or upper mantle.

Dichasial (see **dichasium**).

Dichasium A cymose inflorescence in which each axis produces two opposite axes; equally dividing in structure.

Dichotomous Forked in equal branching pairs.

Dicot (**dicotyledon**) A paraphyletic group of angiosperms that possess two cotyledons (food storage leaves) in the seed and young seedling.

Didymous Found in pairs.

Didynamous With two pairs of stamens of unequal length.

Diffuse Finely and widely spread out, as in a diffuse inflorescence.

Dilated Broadened, but remaining flat.

Dimorphic Existing in two forms or sizes.

Diné Literally, "the people," referring to people of the Navajo tribe.

Dioecious (see **imperfect**) With staminate and pistillate flowers on different plants.

Dioecy Being dioecious; with male and female flowers or cones on separate plants, common in tropical trees and some gymnosperms.

Diorite A type of volcanic rock that cools rapidly and has large crystals.

Disarticulating Coming apart, as some parts of Poaceae inflorescences.

Disc A fleshy structure surrounding the base of the ovary; the tubular flowers of the Asteraceae.

Disc floret A regular, usually tubular flower in a head of the Asteraceae.

Disciform In the form of a disc.

Discoid Resembling a disc; with disc flowers, as in the inflorescence of a member of the Asteraceae that lacks ray flowers.

Disjunct species A species that occurs as a population, often in a novel or atypical habitat, strongly isolated from the remainder of the species range. Disjuncts result either from persistence (i.e., they are relictual) or from recent long-distance dispersal.

Dissected Leaf surface deeply cut into numerous fine divisions.

Distal Toward the end of the organ opposite attachment.

Distinct Not united with adjacent similar parts; separate.

Dithecal Usually referring to anthers, where there are only two anther cells.

Diuretic Promoting the excretion of urine.

Diurnal Active or open during daylight hours.

Divaricate Widely diverging or spreading, as in some stigmas.

Dolabriform (see **malpighian**).

Drainage The area of land that a river or other body of water drains; a watershed.

Drupaceous Drupelike; bearing drupes.

Drupe Stone fruit; usually fleshy and indehiscent, with a stony endocarp and one seed, as in *Prunus*.

Drupelet A small drupelike fruit, often in a head or aggregate fruit.

Echinate With stout, sometimes blunt, prickles or spines.

Eciliate Without cilia.

Ecotype Populations adapted to a particular set of environmental conditions.

Ectomycorrhizae Mutualistic fungi that surround a plant's root and/or cells (but do not enter them) and provide nutrition from other plants or decaying soil materials. Often forms a covering of hyphae on the root called a Hartig net.

Edaphic Pertaining to the soil or substrate.

Efarinose Not covered with mealy particles, or losing these particles if they were present at one time.

Eglandular Without glands.

Elaiosome (elaisome) A gland that secretes an oil, volatile material, fragrance, or nutrient reward that rewards or attracts ants.

Elaminate Without a leaf blade; without the expanded portion of a sepal or petal.

Elator Structures attached to spores that aid in their dispersal; such as elongated flaps attached to *Equisetum* spores.

Eligulate Without a ligule, as in some Asteraceae.

Ellipsoid Three-dimensionally elliptic.

Elliptic Longer than broad, tapering equally at both ends; widest in the middle.

Emarginate Deeply and broadly notched.

Embedded A structure that is located entirely within another.

Emergent (see **emersed**).

Emersed Rising from, or standing out of water.

Emetic A substance or chemical that causes vomiting.

Encinal Small, shrubby oaks, from the Spanish "encino," referring to a chaparral scrub oak.

Endemic Found only in a specific geographic area or ecological situation.

Endocarp Inner layer of a fruit pericarp, sometimes stony, as in some *Prunus* species.

Endomycorrhizae Mycorrhizal fungi that enter spaces between cells in plant roots.

Endosperm A typically $3n$ or $5n$ nutritious food storage tissue surrounding an embryo in a seed; the result of the fertilization of polar nuclei by a sperm cell in the embryo sac.

Entire Smooth, not toothed, indented, or lobed.

Epaleate Without chaffy scales or bracts, as in some genera of the Asteraceae; without paleae, as in some grasses.

Epetiolate Sessile; without a petiole.

Ephemeral Not long-lasting, usually referring to pools or streams; annual plants that germinate and flower whenever sufficient moisture is present in any season.

Epidermis The layer of cells that forms the outermost protective tissue on a plant.

Epigynous Attached at the top of the ovary; when the ovary is completely inferior.

Epiparasitic Plants that obtain all or most of their carbon from other plants via mycorrhizal fungi.

Epipetalous On top of or attached to the petals.

Epiphyte A plant that grows on another plant for support but is not a parasite.

Eradiate Not radiating.

Erect Upright growth or position.

Erose (see **lacerate**) Irregularly indented; appearing gnawed or torn.

Erosulate More or less erose.

Estipulate Without stipules.

Estrophiolate Without a strophiole or rim-aril (seed appendage developing from the distal end of the funiculus), as in some seeds of the Fabaceae or Portulacaceae. Often contains oil droplets.

Ethologically Investigating an organism or process in a behavioral fashion.

Evanescent Remaining only a very short time.

Evaporite A sedimentary rock formed of material deposited from solution by evaporation of water.

Evapotranspiration The combination of water loss from both evaporation and plant transpiration.

Excluded Outside a structure or surface.

Excurrent Sticking out from the main structure.

Exfoliate Peeling off in flakes or plates, as in the bark of some trees.

Exine The outer hard layer, often with various sculpting, on a pollen grain or spore.

Exocarp The outer layer or skin of a fruit pericarp.

Explanate Flat; spreading.

Explosively dehiscent Opening all at once, usually expelling the contents quickly and scattering them a good distance, as in some fruits; the outer wall is usually obviously valvate.

Exserted Protruding, not included within the main structure.

Exstipulate Without stipules.

Extirpated Eliminated from an environment or extinct.

Extrusive A rock type that cooled at or near the surface; basalt.

Eye Generally a ring of differing color around the entrance to the corolla tube, as in some Primulaceae.

f. Author associates, such as children or siblings of a well-known authority.

Facultatively Able to live under varying conditions.

Falcate Sickle-shaped.

False indusium A structure covering the sori or a sorus in ferns that is derived from the inrolling of the frond margin.

False raceme Appearing as a raceme, but not a true raceme.

False septa Partially divided, the septa not completely meeting.

Farina A mealy yellowish, orange, or whitish material often on leaf or frond undersides.

Farinose Covered with meal-like particles.

Fascicle A tight bundle of leaves, stems, flowers, fruits, or roots, sometimes surrounded by a containing structure, as in some *Pinus* species.

Fell-field An alpine or subalpine soil of loose rock and gravel with less than 50% vegetation cover.

Female cone Typically a woody cone of gymnosperms, which produces seeds.

Fen A calcareous, marshy wetland.

Ferruginous Rust-colored.

Filament A threadlike structure; the stalk of an anther.

Filiform Threadlike.

Fimbria (fimbriae) A fringelike structure or projection, such as from a leaf or petal.

Fimbriate Fringed; thicker and less delicate than ciliate.

Fistulose Hollow and cylindrical.

Flabellate Fan-shaped.

Flaccid Limp; not turgid.

Flag With a short, slender extension, usually at the apex or top, as in some stamens; an ecological growth form in response to high winds, usually at high elevations, where the crown of a tree or shrub is in the form of a flag in wind.

Flange A rim or edge.

Fleshy Thickened and at least somewhat succulent.

Flexuous With curves or bends; zigzag.

Floccose Bearing tufts of long, soft, tangled hairs.

Floral tube An elongate tube primarily of the perianth, but may include other floral parts.

Floret A small flower, usually in a group or head of many, such as in the Asteraceae; a reduced flower in the Poaceae consisting of the flower proper and the lemma and palea.

Floricane A flowering and fruiting cane (shoot) of *Rubus*, usually second year or later.

Floricaulous Flowers borne on main stem or trunk; typical of many tropical genera, such as *Ficus*.

Floristics The study of plant occurrences and distribution.

Fluvial Pertaining to river sites.

Foliaceous Leaflike in form, color, and texture; pertaining to leaves; bearing leaves.

Foliose Leaflike; bearing leaves.

Follicle Dry, one-carpelled fruit opening along one longitudinal suture, as in *Asclepias*.

Forb A nongrass, herbaceous plant.

Fornix (plural **fornices**) One of a set of small scales or crests in the throat of a corolla, as in the Boraginaceae.

Fovea (**foveae**) A pit on a surface.

Foveate Pitted, with foveae.

Foveola (plural **foveolae**) A very small pit.

Foveolate With foveolae; minutely pitted.

Free Not joined to other similar structures or organs.

Free-central placentation In a unilocular ovary, ovules attached to a free-standing central structure.

Frugivorous Fruit-eating.

Fulvous Dull yellowish brown or gray; sometimes reddish brown.

Funicle (see **funiculus**).

Funiculus A stalk connecting an ovule to a placenta or seed to a fruit.

Funnelform In the shape of a funnel.

Fusiform Spindle-shaped, thicker in the middle, tapering at either end.

Galea The hoodlike upper lip of a two-lipped corolla that often encloses the style and stigma.

Galeate Structured with a galea.

Gelatinous Jellylike.

Gametophyte The 1*n* generation of a plant (or algal) life cycle that produces gametes; independent in mosses, liverworts, ferns, and fern allies, not free-living in gymnosperms and angiosperms.

Gemma (plural **gemmae**) A cluster of cells or a budlike structure that asexually propagates offspring plants from the parent plant.

Gemmate Having gemmae and/or reproducing by gemmae.

Gemmiferous Producing gemmae or buds.

Geniculate Possessing angular bends and joints.

Gibbous (see **ventricose**) Enlarged on one side.

Girdle A band of tissue around the middle of a fruit, flower, etc.

Glabrate Becoming glabrous in age.

Glabrescent (see **glabrate**).

Glabrous Completely hairless.

Gland A depression, cavity, protuberance, hair, or structure that secretes and accumulates an oil, wax, or sticky fluid.

Glaucescent Slightly glaucous or becoming glaucous.

Glaucous With a waxy covering, often appearing waxy.

Globose More or less spherical.

Globular (see **globose**).

Glochid Small, fine spine that often easily detaches from the areole, such as in *Opuntia*.

Glochidiate Barbed at the tip or bearing glochids.

Glomerule A headlike, often axillary, cyme as in some Asteraceae or Lamiaceae; a dense cluster.

Glume One of the pair of basal bracts on a spikelet in the Poaceae; a chaffy bract in the Cyperaceae.

Glutinous Sticky or covered with a gummy exudate.

Grain (caryopsis) The fruit of the grass family, the Poaceae (Gramineae), where the pericarp is usually fused to the seed coat.

Graminoid Grasslike.

Granite An igneous rock composed of coarse-grained, light-colored silicates.

Granitic Composed of or mostly composed of granite.

Granular With a grainlike texture, knobs, or swelling on a surface.

Granulate (see **granular**).

Great Basin Desert The intermountain cool, high-elevation desert that covers southwestern Oregon, California, Idaho, Nevada, Utah, western Colorado, northern Arizona, and northeastern New Mexico.

Groove A small furrow, usually in cacti, on a tubercle.

Guild An ecological group of organisms using environmental resources in the same or a similar way.

Gymnocarpy Bare seeds protruding from the cone or fruit.

Gynaecandrous Pistillate flowers borne above staminate flowers.

Gynobase An elongation of the receptacle, such as in the Boraginaceae.

Gynoecious Having only female flowers or cones, as in some cultivars.

Gynoecium The aggregate female parts of a flower; all the pistils of a flower.

Gynoecy The tendency to have female or pistillate flowers or cones.

Gynophore A stalk between the receptacle and the pistil, of receptacular origin.

Gynostegium A structure formed from the fusion of the androecium and gynoecium, as in the Asclepiadaceae/ Apocynaceae.

Gypsophile A plant that grows on gypsum deposits (magnesium sulfate).

Habitat Where an organism lives or grows.

Half inferior A hypanthium that is fused to the lower half of the ovary; the flower parts appear to arise from the equator of the ovary.

Halophyte A plant that grows in a salty or hypertonic environment.

Hanging garden A local area in cliffs, or adjacent to a cliff, with water and more mesic, aquatic, and/or semiaquatic flora surrounded by dry vegetation associations.

Hastate With basal lobes spreading.

Haustoria Specialized roots that absorb water from the roots or stems of other plants, such as those possessed by mistletoes.

Hawkmoth A sphingid moth with a plump, hairy body and tapered, hairy wings, sometimes rather large, with a long proboscis. It is a major pollinator of large, light-colored, sweet-scented vespertine flowers and bright, low-growing tubular flowers, such as many species of *Mirabilis*.

Head (capitulum) Dense cluster of sessile or subsessile flowers on an expanded peduncle or receptacle.

Helicoid Coiled in a helix or spiral; such as in one-sided cyme inflorescences in the Hydrophyllaceae or Boraginaceae.

Hemiparasite Partial parasite in which the parasitic plant has chlorophyll and thus can photosynthesize, producing carbohydrates, such as in mistletoes.

Hemipollinaria Half a pollinarium, as in some species of *Platanthera*.

Herb A plant with soft, green growth that grows without producing wood or woodlike structures; a plant used for food, medical, or other ethnobotanical purposes.

Herbaceous (see **herb**).

Heterogeneity Possessing differences, as in an organism with differing alleles at a gene locus.

Heterophyllous A plant possessing different kinds of leaves.

Heterosporous An individual with two or more morphologically different spores, such as in some selaginellas and other ferns.

Heterostylous Possessing differing style forms.

Heterotrophic Drawing nutrition from digesting other organisms.

Heterozygote An organism (or cell) that contains two genetic forms of an allele.

Hilum A scar on a seed that indicates previous attachment to a placenta within the ovary.

Hippocrepiform Horseshoe-shaped.

Hirsute Stiff hairs that are ascending or spreading.

Hirtellous Minutely hirsute.

Hispid Rough, with bristly hairs.

Histology The study of plant tissues or plant anatomy.

Holocene A geological epoch from 10,000 years ago to the present.

Holoparasitic Completely parasitic; getting all nutrition and water from the host plant.

Homophyllous A plant possessing similar leaves.

Homosporous Having the same size spores; not differentiated into microspores and megaspores, resulting in one type of gametophyte with both male and female sex organs, as in some ferns.

Homostylous Possessing the same style forms.

Hook A curved tip, often sharp in cactus spines.

Horn Projection from flower parts or fruit as in *Proboscidea*; part of the corolla in *Asclepias*; long or short hornlike extensions from the gland ends of some species of *Euphorbia* subgenus *Esula*.

Hyaline Translucent; letting light through, but not transparent.

Hybridy Exhibiting the production of hybrids.

Hydroxy fatty acid A fatty acid (has a long chain of $-CH_2$s) with an extra hydroxyl group (-OH).

Hypanthium A cuplike structure on which floral parts are inserted, formed from the fusion of the calyx, corolla, and stamens.

Hypersupinate Twisted 360°, such as in some orchid flowers, with the lip uppermost.

Hypertonic An environment where the concentration of water is lower (and solute concentration higher) than inside a cell or plant.

Hypocotyl The portion of the stem in plant embryos just below the cotyledon(s).

Hypogaean (see **hypogeous**).

Hypogeous Beneath the ground.

Hypogynous With stamens, petals, and sepals attached below the ovary, with the ovary superior.

Imbricate With overlapping units, as imbricate petals.

Imparipinnate Odd-pinnate.

Imperfect (see **unisexual**) Having flowers that are either male or female.

Incised Irregularly cut with sharp teeth.

Included Not projecting beyond surrounding structures, such as stamens within a corolla tube.

Indehiscent Not opening along suture lines or pores when mature, as in some dry fruits.

Indeterminate An inflorescence that may continue to flower at the distal end of a stem or axis.

Indolizidine Alkaloids with one or more indole group(s).

Indument (**indumentum**) Collectively, the hairs, scales, waxes, etc., covering the epidermis of a plant.

Indurate Hardened.

Indusium (plural **indusia**) A thin flap that covers the sorus in ferns, derived from leaf epidermis; in the Onagraceae, a cuplike flap of tissue at the base of the stigma lobes.

Inferior ovary An ovary that is beneath other parts of the flower.

Infrastaminal Located below the stamens.

Infrastipular A structure located below the stipules.

Infusion A liquid produced from a plant part that is steeped or soaked without boiling.

Inrolled Curled inward; involute.

Insertion Where one structure is attached to another.

Integument The layer or tissue that covers the ovule and that will eventually become the seed coat.

Intercostal Between the ribs or nerves.

Intermountain Region That region in the western United States that encompasses the Great Basin Desert as well as included and adjacent mountain ranges and plateaus.

Internode The part of the stem between nodes.

Interpetiolar Between adjacent petioles; usually of whorled leaves.

Interspecific Subdivisions of a species above the level of varieties; between species.

Intracalycine Within the calyx, as in the membrane found in the Gentianaceae.

Intraspecific Within a species.

Intrastaminal Within the stamen.

Intrataxon Within the taxon.

Introgression The movement of genes between species.

Introrse Turning or opening inward; toward the axis.

Intrusive Magma that cools slowly below the earth's surface; forming a coarse-grained rock such as granite.

Investing aril Outgrowths of a seed from the chalazal region.

Involucel A small or secondary involucre, as in secondary umbels in the Apiaceae.

Involucre A subtending whorl of bracts (or phyllaries) subtending an inflorescence.

Involute (see **inrolled**).

Irregular Flower parts in series that are somewhat varied, but not to the extent of zygomorphy, e.g., *Gladiolus*.

Joint The node on a stem; a stem unit in certain Cactaceae, especially chollas.

Jointed With the obvious appearance of real or apparent articulation.

Kayenta Navajo Navajo (Diné) people living or originating in the vicinity of Kayenta, Arizona, or on the western portion of the Navajo Nation.

Keel A prominent dorsal ridge; the two lower united petals of a papilionaceous (pealike) corolla.

Knot (see **chalazal knot**).

Kranz anatomy A specialized leaf anatomy that promotes a more efficient photosynthesis in C_4 metabolizing plants.

Krummholz (see **flag**) Woody alpine or subalpine vegetation that is stunted and contorted due to extreme weather conditions.

Labellum (lip) A liplike petal, often modified as a landing platform for pollinators or other functions.

Laccolith A massive igneous body intruded between preexisting strata and subsequently remaining as a mountain or plateau when weaker strata are eroded. Laccoliths found within the San Juan drainage include the Abajo Mountains, La Plata Mountains, Carrizo Mountains, Navajo Mountain, and Ute Mountain.

Lacerate Appearing torn; irregularly cut.

Laciniate Slashed into numerous narrow lobes.

Lacuna (plural **lacunae**) A gap within a tissue; an air space.

Lacunose Possessing lacunae.

Lacustrine Pertaining to lakes; of a lake; a sedimentary deposit typical of a lake.

Lamelliform Having the form of a thin plate or lamella.

Lamina (plural **laminae**) The blade or expanded portion of a structure, often referring to a petal.

Lanate Woolly; densely covered with long tangled hairs.

Lanceolate At least three times longer than wide, tapering to apex, widest below the middle.

Lanceoloid Somewhat lanceolate.

Lateral Located at the side of a structure.

Lateral spines Spines on the side or margin of the areole, usually thick, without a bulbous base but often similar in color (dark) to central spines.

Laticifer A channel through which flows latex or an opaque latexlike substance.

Latiseptate Possessing a broad septum.

Lax With the components, usually flowers or stems, separate and loose.

Leaflet A segment or leaflike portion of a compound leaf.

Leaf scar The scar left on a stem from a deciduous leaf.

Leathery With a leatherlike, thick, but flexible texture, usually referring to leaves.

Legume (see **pod**) A fruit, the product of a simple pistil, that dehisces along two sutures, typical of the Fabaceae (Leguminosae).

Lemma The lower of the two bracts that subtend a grass flower.

Lens A structure that is convex on both sides.

Lenticel A somewhat corky, sometimes lens-shaped area on the surface of a stem which serves as a gas exchange location.

Lenticular Lens-shaped.

Lepidote Covered with small, scurfy scales.

Liana A woody vine.

Ligneous Woody.

Ligulate Strap-shaped.

Ligule A strap-shaped organ; the flattened part of a ray floret corolla in the Asteraceae; a membranous structure arising from the inner surface of the leaf at the junction with the leaf sheath in the Poaceae (Gramineae) and many sedges; a projection at the base of the leaves above the sporangia in *Isoetes*. The expanded flat part of an organ, especially the flat part of a sympetalous corolla.

Linear Long and narrow, with parallel sides, as in many grass leaves; a straight, narrow line.

Lip (see **labellum**).

Lithophyte A plant that grows on rock; rupicolous.

Loam A friable type of soil composed of a more or less equal mixture of sand, silt, and clay, which drains well, yet retains moisture.

Lobed Indented, parted, divided; leaf may be shallowly, deeply, pinnately, or palmately lobed.

Locoism The tendency of certain ungulates to exhibit unusual behavior after eating certain plants, such as locoweeds (*Oxytropis*) and milkvetches (*Astragalus*).

Locule The cavity of an organ, primarily pistils and anthers.

Loculicidal Longitudinally dehiscent through the middle back, into the cavity, usually of a capsule or other hollow, dry fruit.

Lodicule Paired vestigial or rudimentary scales at the base of an ovary in grass florets.

Lunate Moon- or crescent-shaped.

Lyrate Lyre-shaped; pinnatifid with the terminal lobe large and the lower lobes much smaller or shallower.

Maculate Spotted or dotted.

Madro-Tertiary Geoflora An arid-adapted flora that first appeared in the Tertiary and expanded rapidly in the Miocene and Pliocene in southwestern North America.

Male cone (see **microsporangia**) A pollen-producing cone in conifers and other gymnosperms.

Malpighian (**dolabriform**) Straight hairs attached near the middle and tapering at both ends; shaped like a rock pick.

Mammiform Breast-shaped.

Mammillate With nipplelike structures on a surface.

Marcescent Withering but persistent, as in some basal leaves.

Margin The edge, such as the edge of a leaf blade or petal.

Marginal band A band of tissue at or near the margin of a leaf, petal, etc.

Marginal cell Cell located on the margin of a leaf blade, petal, or sepal; a cell of an outer wood structure.

Matutinal Flowers that open in the morning.

Mealy (see **farinose**) Covered with meal-like particles or secretions, as in goosefoot (*Chenopodium*).

Membranous Thin, soft, and pliable or papery.

Mephitic Skunklike, usually referring to smell; strongly foul-smelling or irritating.

Mericarp Unicarpellate segment of a schizocarp.

-merous Number of parts, e.g., a three-merous corolla.

Mesic A moist environment, as opposed to xeric (dry) or hydric (saturated).

Metaconglomerate Sedimentary rock that has experienced some metamorphic changes.

Metafixed (**dolabriform**, **malpighian**) Hairs that are attached to a surface at or near their midpoint.

Metamorphic Rock formed by the alteration of preexisting solid rock deep within the earth by pressure, heat, and/or chemicals. The rock does not become fully liquid in the process.

Microphyll A small, narrow, single-veined leaf of lower vascular plants, such as *Lycopodium*.

Micropylar (see **micropyle**).

Micropyle The opening in the integuments of the ovule through which the pollen tube enters.

Microsporangia (singular **microsporangium**) A microspore-producing structure.

Microspore Precursor to a male gametophyte and sperm.

Microsporophyll A stamen; a modified leaf or leaflike structure that bears one or more microsporangia.

Midpetaline band A band located midway along the petal, such as hairs or coloration.

Misappl. Designation of a name that has been or is incorrect in that it is not applied correctly to the taxon in question.

Mojave Desert The desert located in east-central California, southern Nevada, southwestern Utah, and northwestern Arizona. The indicator plant is *Yucca brevifolia* (Joshua tree).

Monad A single, free, individual structure; usually referring to pollen grains that are shed singly.

Moniliform Cylindrical, but constricted at regular intervals like beads on a string.

Monocarpic Flowering and fruiting only once in the life of the plant or shoot, which then dies, such as in some species of *Agave*.

Monocephalous A plant or inflorescence with one head or capitulum.

Monochasial With the form of a monochasium.

Monochasium A type of cymose inflorescence possessing a single main axis.

Monochazial (see **monochasial**).

Monoclinous Perfect, the male and female parts found within the same flower.

Monocot (**monocotyledon**) Plants that have only one cotyledon or seed leaf when germinating. Generally with parallel veins in leaves and flower parts in threes or sixes; includes grasses, orchids, palms, sedges, etc.

Monoecious Having staminate and pistillate imperfect flowers on the same plant; meaning "of one house" in Greek.

Monofoliolate With only one leaf.

Monomorphic With only one form; especially referring to flowers or leaves.

Monophyletic A clade (lineage) that includes its common ancestor and all of the ancestor's descendants.

Monophyly (see **monophyletic**).

Monotypic Having only one representative.

Monsoon A Southwest season in summer characterized by frequent thunderstorms with often heavy downpours and flash floods.

Moraine A formation composed of boulders, rocks, and/or other debris deposited by a glacier.

Morphology The overall form of a plant or its parts, such as leaves, flowers, sepals, indument, etc.

Mucilaginous Having a mucuslike substance.

Mucro A sharp tip.

Mucronate (**cuspidate**) With a short, abrupt, firm tip.

Mucronulate Tipped with a very small point or mucro.

Mudstone A fine-grained sedimentary rock formed from silt and clay and without laminations.

Multigeneric A hybrid formed of many genera, such as human-made hybrids in the Orchidaceae.

Multiple Fruit formed from closely clustered ovaries of many separate flowers, as in *Morus* or *Ficus*.

Muricate Possessing a small, sharp projection.

Murication A small, sharp projection or point.

Muriculate Very finely muricate.

Mutagenic Causing genetic changes in cells, which can lead to cancer or deformities.

Mycoheterotrophic (see **mycotrophic**).

Mycorrhizal fungi Fungi that form a mutualistic relationship with plant roots either as endomycorrhizae or ectomycorrhizae.

Mycotrophic (**saprophytic**) Obtaining nutrition with the aid of fungi that break down organic soil materials.

Myrmecochory The seed dispersal of plants by ants.

Myrmecophyte A plant that has a mutualistic relationship with ants, often providing specialized structures to maintain them.

Myxogenic Papillose hairs or apical achenial or cypselal pericarp hairs in the Asteraceae; mucilaginous cells.

n The haploid chromosome number; counted in developing pollen cells undergoing meiosis.

Napiform Turnip-shaped.

Naturalized Introduced plants, but now established and reproducing without human intervention.

Navajoan Desert A southern and eastern portion of the Great Basin Desert in Arizona, Colorado, New Mexico, and Utah that has differing geology and climate from the main unit. *Coleogyne ramosissima* (blackbrush) is the indicator plant.

Nebulose Indistinct, as in some fine inflorescences.

Necrotic Decaying or turning black.

Nectar A sugary solution secreted by plants, which may contain other materials such as amino acids. Often contained in nectaries, spurs, or bowl-like depressions in the perianth.

Nectar bowl A depression in a lip or other perianth segment that secretes and accumulates nectar for pollinators, such as in some species of *Malaxis*.

Nectar disc A disc or ring often at the base of the ovary or stamens that secretes nectar.

Nectar guide A line or lines of splotches on a petal or other flower structure that guides a pollinator to a nectary. The guides may be in ultraviolet pigmentation, detectable mainly by insects, but also by some bats; minimally by some hummingbirds.

Nectariferous Producing nectar.

Nectary A tissue or group of cells that secrete nectar. Nectaries may be found inside or outside flowers that attract ants to aid in a plant's defense.

Nectary disc A round structure, often below the ovary, that secretes nectar.

Needlelike (see **acerose**, **acicular**).

Neotropical Geoflora The ancient tropical forests that existed in tropical regions of North, Central, and South America at the beginning of the Tertiary.

Neotropics Tropical regions of the Western Hemisphere.

Nerve Major large vein or midrib.

Neuter Lacking both functional stamens and pistils.

Nivation Related to late-lying snowbanks.

Nocturnal Blooming or active at night, such as *Datura wrightii*.

Nodal Occurring at the node or nodes; pertaining to a node or nodes.

Node The stem joint; where buds, leaves, inflorescences, and new stems arise.

Nodulose Finely knobby.

Nom. cons. Name conserved, sometimes over a name with priority.

Nomen confusum Confused name.

Nonaccrescent Not accrescent; not becoming larger with age.

Noncircinate Not coiled from the tip down, as in some young fern fronds.

Nut A one-seeded, indehiscent fruit with a hard or woody pericarp, as in acorns of *Quercus*.

Nutlet A small nut; a section of a mature ovary in the Boraginaceae, Verbenaceae, and Lamiaceae (Labiatae).

Obconic Cone-shaped, with the attachment at the narrow end.

Obcordate As cordate, but with attachment at narrowed end and notch at apex.

Obdeltoid Deltoid, with the attachment at the narrow end.

Obdiplostemonous Having two whorls of stamens, the outer whorl opposite the petals, the inner whorl opposite the sepals.

Oblanceolate As lanceolate, but tapering to the base and widest above the middle.

Oblate Generally spheroid, but flattened at the poles.

Oblique Asymmetrical by having unequal sides.

Oblong Longer than broad, with margins parallel and ends rounded.

Oblongoid Tending toward oblong.

Obovate As ovate, but tapering to the base and widest above the middle.

Obovoid Obovate, but three-dimensional.

Obpyrimidal Appearing as an upside-down pyramid, three-dimensional triangle.

Obscure Minute or difficult to observe; not immediately obvious.

Obsolete Rudimentary or vestigial.

Obtrullate Trowel-shaped, with attachment at the narrow end.

Obtuse Forming an angle of greater than 90°.

Ochroleucous Off-white, buff.

Ocrea (plural **ocreae**) A sheath around the stem formed by stipules, as in some Polygonaceae.

Ocreate With ocreae.

Ocreola (plural **ocreolae**) A minute, stipular sheath around the secondary inflorescence divisions in some Polygonaceae.

Oil tube Narrow ducts in the fruit of many Apiaceae which contain volatile oils.

Olivaceous Olivelike or olive-green in color; like olivine or its geological relatives.

Opiate Alkaloids commonly found in sap, resin, or latex of the Papaveraceae and other families.

Opposite Arranged with two opposing structures, as in opposite leaves where there are two at a node on opposing sides; one part in front of another, as when sepals and petals align with one another.

Opuntioid A cactus in the subfamily Opuntioideae including the genera *Cylindropuntia*, *Grusonia*, and *Opuntia*; prickly pears, club-chollas, and chollas.

Orifice A mouth or opening, as in the opening of a tubular corolla.

Orographic Precipitation that occurs due to a movement of an air mass to higher elevations and cooler temperatures, resulting in lower moisture-holding capability of the air mass (higher relative humidity) and deposition of precipitation, such as on mountaintops.

Ovate Egg-shaped in outline, less than three times as long as wide, tapering to the apex and widest below the middle.

Ovoid Ovate, but three-dimensional, egg-shaped.

Oxalic acid An organic acid produced by certain succulents and members of the Oxalidaceae or Chenopodiaceae; an intermediate metabolite in photosynthesis and respiration, especially important in CAM and C_4 metabolism.

Palatal Possessing a palate in the floral structure.

Palate A raised structure on the lower lip of a corolla that almost completely obscures the throat, as in some flowers of *Penstemon*.

Palea (plural **paleae**) A chaffy scale or bract; the uppermost of two bracts that subtend a floret in the Poaceae; a chaffy scale or bract on the receptacles of some members of the Asteraceae.

Palmate Arranged in a palmlike fashion, spreading.

Palmatifid Palmately cleft or lobed.

Palustrine Growing in wet meadows or marshes.

Pandurate Fiddle-shaped.

Panduriform (see **pandurate**).

Panicle Compound or branched raceme.

Paniculiform With the basic appearance of a panicle, but not necessarily the morphology or developmental sequence of a panicle.

Pannose With a short, dense, felt-like, tomentose covering.

Papilionaceous Pealike floral morphology, with an upper petal (the banner or standard), two lateral petals (the wings), and a forward-projecting keel of two apically fused petals.

Papilionoid Having a pealike zygomorphic flower in conformation, such as in papilionoid plants of the Fabaceae.

Papillae Pimplelike or blisterlike structures, usually on the epidermis.

Papillate Having papillae.

Papilliform (see **papillae**).

Papillose With minute papillae.

Pappus (plural **pappi**) Modified calyx on florets of the Asteraceae. These may be awns, scales, or bristles at the achene or cypsela apex.

Paraphyletic A clade (lineage) that has a single common ancestor but which does not include all of the ancestor's descendants.

Parasite (see **hemiparasite**) An organism that obtains its food and water from another organism.

Parenchyma A type of tissue that is generalized in cell type, has a thin cell wall, and is often used for food storage.

Parietal Attached to the outer wall of the ovary, as in parietal placentation, where the ovary has only one locule and the placentae are arranged around the outside of the locule.

Park A large montane meadow.

Parkland Patchy forested land with clumps or individual trees, usually separated by grassy meadows or expanses of much shorter vegetation.

Pectinate Comblike; with closely spaced appendages or hairs often in a single row.

Pedate Palmately divided, with lateral lobes two-cleft.

Pedatisect Palmately lobed, with lateral lobes two-cleft.

Pedicel The stalk of a single flower in an inflorescence or of a spikelet in grasses.

Peduncle The stalk subtending an inflorescence.

Pellicle A thin, membranous, or skinlike covering.

Pellucid Transparent or translucent.

Peltate With stalk attached to lower surface of leaf blade or other structure well inside the margin.

Pendulous Hanging downward, usually somewhat loosely.

Penicil A tuft of short hairs, usually in brushlike form.

Penicillate With a brushlike tuft of short hairs; with a penicil or penicils.

Penultimate The next to the last.

Pepo A type of berry that has a hard or leathery rind exocarp, many seeds, and a usually fleshy (sometimes fibrous) mesocarp; usually one-loculed; typical of the Cucurbitaceae.

Perennate Renewing or regrowing, as in the regrowth of shoots.

Perennial A plant that lives for three or more years.

Perfect (bisexual, hermaphrodite) Having both pistil(s) and stamen(s) on the same flower.

Perfoliate The margins (usually of a leaf) entirely surrounding the stem, giving the impression of the stem passing through the leaf.

Perforate With holes or other small openings.

Perianth Collectively, the calyx and corolla.

Pericarp The wall of a fruit, usually derived from ovary tissue, but sometimes derived from an accessory structure, often with exocarp, mesocarp, and endocarp.

Pericarpel A structure consisting of the upper portion of the flower-bearing stem, the receptacle, and the lower portion of the pistil.

Periderm A layer of outer tissue, often corky, in plant stems.

Perigynium A sac often enclosing an achene in the Cyperaceae.

Perigynous A flower possessing a tubular hypanthium on which occur stamens, petals, and sepals. The calyx tube surrounds, but is not attached to, the superior ovary.

Perisperm Food storage arising from the nucellus, or maternal tissue, in some seeds; around or near the seed.

Perispore The outer covering of a spore.

Persistent Remaining attached, as a style on a fruit.

Petal The individual segment of the corolla, the second whorl (moving inward) of the flower perianth.

Petal hood The structure formed when a petal arches or curves to form a hood; when a sepal arches over a smaller petal to form a hood.

Petaloid Petal-like; may be a stamen, sepal (as in many Cactaceae), style (as in *Iris*), other flower part, an entire flower (as a ray floret in the Asteraceae), or a bract (as in *Poinsettia*).

Petiole The stem or stalk attached to the leaf blade in a leaf.

Petiolule The stem or stalk attached to a leaflet.

Phenological spring The time period when leaves emerge from deciduous plants, later at higher and cooler elevations, sooner at lower and warmer ones.

Phenological summer The time period when photosynthesis in plants is at its greatest rate.

Phenology The sequence of events in a plant's life; usually referring to flowering and fruiting events.

Phloem The sugar-containing conductive tissue in plants; part of the inner bark.

Phyllary A bract of the involucre, such as in the Asteraceae.

Phyllode A leaflike petiole lacking a true leaf blade.

Phyllodial Pertaining to a phyllode.

Piki A type of very thin, membranous flatbread (not leavened) often made from ground corn and other ingredients by Puebloan peoples. Originally, the batter was spread and toasted on a hot, flat rock.

Pilose (includes villous, velutinous) Soft, distinct, straight hairs that are ascending or spreading.

Pilosulous With minute, soft, straight hairs.

Pin A flower that has an elongated style.

Pinna (plural **pinnae**) A lobe of a pinnatifid leaf, petal, etc.

Pinnate Arranged in a fashion similar to a feather; fernlike.

Pinnatifid Pinnately lobed almost to the point of being pinnately compound.

Pinnatisect Pinnately cleft to the midrib of a leaf.

Pinnule The ultimate division of the blade in a bipinnately or more compound leaf.

Pistil The gynoecium, female parts of the flower: ovary, stigma, and style.

Pistillate With pistils; imperfect female flowers or the inflorescences that contain only this flower type.

Pit A small depression; the endocarp of a drupe, such as in *Prunus*; small opening in the walls of tracheids and vessel elements that allows water to move from one cell to another in xylem.

Pithy With spongy tissue in the central area of roots and stems, as in some Poaceae.

Placenta (plural **placentae**) The attachment place of the ovule.

Placentation The arrangement or configuration of the placentae in the ovary.

Plane Three-dimensionally flat, such as a leaf without an undulate margin.

Plano Flat.

Plano-convex Flat on one side and convex, or extruded, on the other.

Pleistocene A geological epoch which extended from 1.8 million to 10 thousand years ago.

Plica (plural **plicae**) A fold or braidlike structure.

Plicate Plaited, pleated, or folded similar to a fan, such as the leaf blades of *Calypso bulbosa* or the corolla in many Gentianaceae and Solanaceae.

Plumose Plumelike; feathery with hairs or fine bristles on both sides of the main axis.

Plurilocular Containing many locules or compartments.

Pod (see **legume**).

Podogyne (see **carpophore**) A stipe supporting an ovary.

Pollen cone A cone that produces pollen; a male cone, usually deciduous.

Pollen sac The portion of an anther or cone scale that contains the pollen.

Pollinarium (plural **pollinaria**) A structure in the Orchidaceae and Asclepiadaceae (Apocynaceae) that contains pollinia and other structures that facilitate pollen transport by pollinating insects, birds, etc. Usually one anther in orchids (two in *Cypripedium*); may be two half-anthers in milkweeds.

Pollinium (plural **pollinia**) A mass of pollen grains forming a transportable unit with other parts of a pollinarium, as in the Orchidaceae or Asclepiadaceae (Apocynaceae).

Polyad A group of pollen grains; chromosomes that group more than two at meiosis.

Polycarpic Flowering more than once.

Polygamodioecious Primarily dioecious, but with a few perfect flowers.

Polygamomonoecious Primarily monoecious, but with a few perfect flowers.

Polygamous Bearing imperfect and perfect flowers on the same plant.

Polymorphic Having many different forms.

Polypetalous A corolla of completely separate petals.

Polyphyletic An artificial taxon having more than one evolutionary origin, despite the similarity of characters.

Polyploidy Having more than three haploid sets of chromosomes.

Polysporangiate Giving rise to many sporangia.

Poricidal Opening by pores, as in capsules of the Papaveraceae.

Porrect Resembling a parrot beak and extending forward; usually a flower part.

Pouch A petal or other structure that has a compartment with an opening on top, such as the labellum of *Cypripedium*.

Precocious Flowering before the appearance of leaves.

Prickle Epidermal outgrowth, easily removable, without vasculature and more substantial than most trichomes, as in *Rosa*.

Pricklet Small prickle.

Primocane The first-year cane or shoot, often nonflowering, in many species of *Rubus*.

Priority The taxonomic rule that the earliest valid name has precedence in use over later names.

Proboscis Nose or noselike projection.

Procumbent Lying on the ground, but not rooting in it.

Prokaryote Single-celled organism that lacks a nucleus or other cell organelles.

Pro sp. As a species.

Prostrate Growing or lying flat on the ground or substrate.

Protandrous A type of flower in which the anthers release pollen before the stigma is receptive.

Prothallium A gametophyte, often heart-shaped or carrot-shaped, in ferns and fern allies, such as selaginellas and equisetums.

Protogynous A condition in which the stigma(s) are receptive before the anthers release pollen.

Prow-tipped Possessing an upward-curving spine or other structure at the tip of an organ.

Proximal Toward the base or axis.

Pruinose Conspicuously glaucous with a waxy, whitish, grayish, or bluish bloom (coating) on the surface.

Psammophyte A plant that grows in sand.

Pseudobulb An enlarged stem that is able to store water and food, especially in orchids.

Pseudocereal A grainlike fruit that is not in the Poaceae but another family, such as Amaranthaceae.

Pseudohermaphrodite False perfect flower, where one sex is nonfunctional chemically or anatomically.

Pseudomonomerous A structure that appears to be simple, though derived from the fusion of two or more separate structures, as a cypsela derived from ovary and calyx tissue.

Pseudoscape Where not all leaves are truly basal.

Pseudostamen A false stamen, usually on the petal, often bright yellow; used to attract insects, as in *Calypso*.

Pseudostaminode (plural **pseudostaminodia**) A false staminode, or false stamenlike structure.

Pseudoterminal Falsely terminal.

Pseudowhorled Falsely appearing whorled, as the leaves of some species of *Galium* where some of the "leaves" are actually stipules.

Puberulent Minutely pubescent.

Puberulous (see **puberulent**).

Pubescent Bearing hairs.

Pulvinate Cushionlike or matlike.

Pulvinus (plural **pulvini**) A swelling, usually at the node of a stem or the base of a petiole or petiolule, especially in the Poaceae or Caryophyllaceae.

Punctate Dotted with holes, depressions, spots, or translucent glands.

Pustulate (see **pustulose**).

Pustulose With small blisters or pustules, often at the base of a hair, as in many species of *Cryptantha*.

Putative Generally regarded as true, although facts may be inconclusive.

Pyramidal Three-dimensional equivalent of triangular.

Pyriform Pear-shaped.

Pyxis A capsule with the top coming off lidlike; a circumscissile capsule.

Quadrate Squared.

Quadrilocular With four cavities or compartments.

Quassinoid A bitter alkaloid typically found in the Simaroubaceae.

Quinolizidine alkaloid An alkaloid containing quinolizidine ring(s). Often colored.

Raceme Indeterminate inflorescence bearing several pedicelled solitary flowers along a central axis.

Racemiform Racemelike in overall appearance, but not actually in morphology or development.

Rachilla A small rachis, usually the axis of a grass spikelet or sedge inflorescence.

Rachis The axis or stem of a spike, raceme, or a compound leaf.

Radial spine One of the outermost spines of an areole, often radiating or appressed, usually fine and lighter colored than central spines.

Radiate With parts spreading from a single point; when some of the flowers in a head are ligulate, especially on the outer whorl as in the Asteraceae.

Ramada (shade) A shaded outdoor working area usually built of pole logs and leafy branches (palm leaves in warmer climates) or branches with bark.

Ramah Navajo Navajo people who live or originate on the Ramah Navajo Reservation near Ramah, New Mexico, southwest of Gallup; apart from the "big reservation" to the north and west.

Rame A branch, usually herbaceous; a cluster of spicate racemes.

Rank Row, as in three-ranked; also refers to category, i.e., genus vs. family, etc.

Raphe A ridge located on a seed formed by part of the funiculus fused to seed coat tissue.

Raphial cavity A cavity within the raphe.

Ray The straplike part of a ray or ligulate floret or the flower itself in the Asteraceae; a branch of an umbel, especially in the Apiaceae.

Ray floret A flower in a head of an Asteraceae inflorescence that has petals fused into a straplike structure, often in the outer whorl of florets.

rDNA Ribosomal DNA, the genes that code for ribosomal RNA (ribonucleic acid).

Receptacle Terminal portion of the flower stalk, often inflated, from which arise the flower parts.

Reclining More or less horizontal in form.

Reflexed Bent downward.

Relict taxon A taxon that appears "out of place," with no known close relatives. Typically these taxa and their relatives have a long evolutionary history in the region and have persisted for long periods of time, such as *Berberis* and *Paxistima*.

Relictual An "out of place" population or taxon of plants. The population may be a "leftover" from a retreating ice age or other geoclimatological event or may be an evolutionary relict, unrelated or not obviously related to other known plants in the area.

Remote Spaced distantly.

Reniform Kidney-shaped.

Repand With a slightly wavy, sinuate, or undulate margin.

Replum Partition between two locules of siliques or silicles.

Resin A liquid defensive material secreted by plants, often translucent, common in gymnosperms and the Burseraceae.

Resupinate The flower twisted with the lip uppermost; the flower with the lip lowermost, twisted 180° (see **hypersupinate**), as in some orchids.

Reticulate (reticulation) Net-veined; with a network, or netlike markings or structure.

Retting A process of partial bacterial and fungal decomposition of, primarily, flax (Linaceae) stems to a usable and strong fiber.

Retuse Shallowly notched.

Revolute Rolled backward from both margins toward the underside.

Rhizomatous Possessing rhizomes; rhizomelike.

Rhizome An underground stem possessing nodes.

Rhizophore In selaginellas, a downward-growing, leafless shoot that may produce roots at its apex.

Rhombic Rhomboid-shaped in cross section or outline; a polygon with four equal-length sides, two equal angles greater than 90°, and two equal angles less than 90°.

Rhyolite An igneous volcanic rock of felsic composition; fine-grained.

Rib Ridge of stem tissue parallel to the stem axis in cacti with areoles; the woody underlying skeleton of fleshy ribs in cacti; a raised line of tissue, such as veins or a line in dry fruits.

Rim Edge of a rock outcrop or bench, often of sandstone.

Rimrock A sandstone ledge.

Ring chromosome A group of chromosomes, as in some Onagraceae, that form a ring at meiosis.

Riparian Vegetation found in a river valley or other flowing water community; sometimes referring to seasonally dry wash vegetation.

Robust Vigorous in growth and anatomy; sometimes stocky, not thin.

Root The portion of a plant usually belowground, lacking nodes, leaves, and buds at nodes.

Rootstock The part of a root from which the main stem arises.

Rosette A cluster of radiating leaves arising from a short or underground stem close to ground level.

Rostellum A small beak; an often clawlike extension of the stigmatic upper edge on an orchid column.

Rostrum A beaklike structure such as that on the column of Orchidaceae flowers.

Rosulate Leaves arranged in basal rosettes; the stem short or absent.

Rotate Wheel-shaped; usually referring to the limb of a corolla.

Rotund Rounded.

Rounded Forming a semicircular shape.

Ruderal Weedy or growing in regularly disturbed habitats.

Rudiment Vestigial or imperfect in development; a part of a structure.

Ruga (plural **rugae**) A fold or wrinkle.

Rugose Wrinkled or furrowed.

Rugulose Minutely wrinkled or furrowed.

Runcinate Sharply pinnatifid or cleft, with the segments pointing downward.

Saccate With a sac; bag-shaped.

Sagittate Shaped like an arrow, with two backward-directed lobes.

Salverform A narrow tube, usually a corolla, abruptly expanding to a flattened portion.

Samara Dry, indehiscent, winged fruit, as in *Acer* or *Ulmus*. May be schizocarpic, as in *Acer*.

Sandstone A sedimentary rock formed from sand granules (usually quartz) and a cementing agent.

Scaberulous Intermediate between scabrous and minutely pilose.

Scabrescent Minutely scabrous.

Scabrid Roughened.

Scabridous Possessing a scabrid surface.

Scabrous Rough to the touch.

Scale A thin, small bract, vestigial leaf, sepal, or petal; a flattened, wide trichome.

Scale leaf A small, scalelike true leaf, as in junipers.

Scandent Climbing or slightly viny.

Scape A leafless peduncle arising from the ground in acaulescent plants.

Scapose Bearing a scape.

Scarious Thin, dry, membranous; not green.

Schizocarp Dry fruit splitting into one- or few-seeded segments, as in many Apiaceae.

Sclereid A type of sclerenchyma cell with a very thick, ligneous cell wall and a roundish or many-sided shape.

Sclerenchyma A type of plant tissue with a very thick secondary cell wall, usually containing lignin. Typical cell types include fibers and sclereids.

Scorpioid A coiled cyme, shaped like a scorpion's tail, as in some plants of the Boraginaceae and Hydrangeaceae; a zigzag rachis of a determinate inflorescence.

Scree Loose rock debris covering a slope; the accumulation of sloping loose rock debris at the base of a cliff or steep incline.

Scrotiform Scrotumlike in appearance; a drooping pouch or sac.

Scurfy (lepidote) Covered with small scales.

Secondary partition Not a major partition; usually thinner-walled or partially walled.

Secund Arranged on one side of a rachis or axis, such as in some species of *Penstemon*.

Seed cone The female cone, usually woody and larger than the pollen cone, such as in conifers.

Seleniferous Soil or structure containing high amounts of selenium or its salts.

Selenophyte A plant that grows in an area of high soil selenium concentration, as some species of *Astragalus*.

Semifoliaceous Somewhat leaflike, such as some bracts or petals.

Seminiferous Seed-bearing; semenlike scent in some fly-pollinated orchids.

Semisucculent Somewhat fleshy.

Sensu In the sense of.

Sensu lato In the broad sense.

Sensu stricto In the narrow sense; narrowly.

Sepal A segment of the calyx; the outer whorl of the perianth.

Sepaloid Sepal-like in color and/or texture, but not actually a sepal, or part of the calyx.

Septate Divided by one or more partitions or septa.

Septicidal Dehiscence of a capsule between locules.

Septum (plural **septa**) Partition separating cavities, as in an ovary.

Sericeous (see **silky**).

Serotinous Remaining unopened for a long time, especially as referring to the cones of some pines that require the heat of a fire to open.

Serrate With pointed teeth facing toward apex.

Serrulate Minutely serrate.

Sessile Attached directly by the base, not stalked.

Seta (plural **setae**) A bristle or bristlelike.

Setiform Possessing a bristlelike morphology.

Setose Possessing setae or bristles.

Sheath Tubular basal portion of a leaf that encloses a stem in the Poaceae and Cyperaceae; in *Opuntia* and *Cylindropuntia*, a papery, tubular, epidermis-derived layer that encloses a spine, which sometimes separates from the spine at the tip; the papery bract or scale that encloses bundles of needles in the Pinaceae.

Short-hirsute (**hirsutulous**, **hirtellous**, **hirtellate**, **hispidulous**) With short, stiff hairs.

Short-pilose (pilosulous, villosulous) With short, soft, ascending hairs.

Short-strigose (strigulose, strigillose) With short, stiff hairs that are closely appressed.

Short-woolly With short, soft, dense hairs.

Showy Obvious; large and often colorful.

Shrub A woody perennial plant with several or more stems.

Sigmoid S-shaped.

Silicle (silicula) A dry fruit less than three times longer than broad, dehiscing by two valves, with seeds attached to a central, persistent rim (**replum**) usually spanned by a membrane, as in the Brassicaceae.

Silique (siliqua) A dry fruit more than three times longer than broad, dehiscing by two valves, with seeds attached to a central, persistent rim (**replum**) usually spanned by a membrane, as in the Brassicaceae.

Silky Soft, distinct, straight hairs that are closely appressed.

Siltstone A fine-grained sedimentary rock of lignified silt.

Simple Unbranched; without leaflets or not compound.

Sinuate Wavy on margins only, not three-dimensional.

Sinus A cleft between two lobes of an organ such as a leaf (or petal).

Sister relationship Two taxa that are each other's closest relative.

Slickrock Smooth sandstone polished by wind and water.

Sobole Elongated caudex branch or a shoot arising from the stem base or an underground stem.

Sodgrass A grass that spreads vegetatively by rhizomes or stolons.

Solifluction Mass soil movement resulting from freeze-thaw action.

Solitary Borne singly.

Sordid Muddy, brownish gray, or any dull color.

Spathe A large, often colorful or showy bract subtending and often surrounding a usually spikelike inflorescence, such as in the Araceae.

Spatulate Oblanceolate, but apex broadly rounded, tapering at the base, like a spatula.

Spherical Round in outline; three-dimensional.

Spicate Spikelike.

Spiciform (spiceriform) An inflorescence that has the appearance of a spike, but not necessarily the true morphology or development of a spike.

Spike Indeterminate inflorescence bearing sessile or subsessile solitary flowers along a central axis.

Spikelet A section of a larger inflorescence similar in conformation to a spike; the ultimate flower cluster in grasses which consists of two glumes and one or more florets; flower cluster in sedges.

Spine Modified leaf or leaf parts, including stipules, as in the Cactaceae.

Spinescent Bearing a spine or spines.

Spine sheath A layer of epidermis found in some cacti which may become detached from the spine.

Spinose (spinous) Bearing spines.

Spinulose Bearing a small spine or spines.

Sporangium (plural **sporangia**) A structure that produces spores in plants, usually by meiosis.

Sporeling A young plant just arising from a spore.

Sporocarp A structure on some ferns, such as *Marsilea*, that contains sori and sporangia.

Sporophyll A spore- or sporangium-bearing leaf; a fertile leaf; a structure containing sporangia.

Sporophyte The generation in plants that produces spores, usually $2n$.

Spreading Branching in close to horizontal fashion.

Spur An enclosed, often nectar-containing extension from the corolla or calyx.

Spurred Bearing a spur or spurs.

Squamella (plural **squamellae**) A small scale or squama.

Squarrose Rough or scurfy, with spreading processes, recurved at the tips, as in some involucres of bracts.

Stamen The male pollen-bearing portion of the flower; the microsporophyll; composed of anther and filament.

Staminate Containing stamens; an imperfect male flower or the inflorescence that contains only this flower type.

Staminode (plural **staminodia**) A sterile stamen, sometimes petaloid, resembling another structure, or otherwise modified.

Stereome A border of hyaline cells; the central area of the phyllary in the Asteraceae.

Sterile Not capable of producing viable pollen, ovules, gametes, or spores.

Stigma The receptive sticky or feathery part of the pistil on which the pollen is deposited and germinates.

Stigmatic cap The tissue that seals the top of the inrolled developing pistil, often with projections or papillae; especially in plants with "dry" stigmas.

Stipe (see **gynophore**).

Stipiform Stalklike; stipelike.

Stipitate With a stipe or gynophore.

Stipulate With small stipulelike or substipulelike projections; bearing stipules.

Stipule One of the pair of leaflike appendages at the base of some leaves.

Stolon Thin, usually long, often trailing stems that produce new plants at the tip or nodes.

Stoloniferous Bearing a stolon or stolons.

Stoma (plural **stomata**) A gas exchange pore, usually on the underside of the leaf, generally flanked by two guard cells.

Stomatal band A row of stomata, usually on leaves or needles in conifers.

Stramineous Strawlike in texture or color.

Striate Striped with longitudinal lines or furrows.

Strict Straight and upright, not spreading; in a somewhat narrow sense.

Strigillose Minutely strigose.

Strigose Stiff hairs that are closely appressed.

Strobilus A cone that produces spores and/or seeds and/or pollen and/or gametophytes.

Strophiole In some seeds, such as in the Portulacaceae, an appendage at the hilum.

Stylar Pertaining to the style.

Style The portion of the pistil between the ovary and the stigma.

Stylopodium A disclike structure at the base of the styles in the Apiaceae.

Sub- Slightly less or almost.

Submersed Submerged.

Subprecocious Flowering slightly before the appearance of leaves.

Subshrub (see **suffrutescent**) A perennial plant with a woody lower portion.

Subspecies A unit of classification inferior in rank to a species.

Subspecific Referring to a rank below species, such as variety and subspecies.

Subsume Decrease the importance of.

Subtended Immediately below and close to another structure. Typically, a bract.

Subulate Long, narrow, tapering gradually to a rigid apex; awl-shaped.

Succulent Juicy or fleshy, as in the Cactaceae or families such as the Crassulaceae or Agavaceae.

Succurent Extending upward and adnate to an axis.

Suffrutescent (see **subshrub**) Shrublike; slightly woody.

Suffruticose Somewhat shrubby; having a stem that is woody only at the base.

Suffrutose (see **suffrutescent**).

Sulcate With grooves or furrows, usually longitudinal.

Superior ovary An ovary that is free from or above other organs in the flower.

Suture On a fruit or other structure, a line that may result in dehiscence or is the result of union between or among various parts.

Sympatric (**sympatry**) (see **allopatric**). Occupying the same geographical region, as in sympatric species.

Sympodial Consisting of a number of short axillary branches, often underground.

Synapomorphy A derived trait that is shared by two or more taxa and which reflects their shared ancestry.

Synconium A fleshy multiple accessory fruit typical of *Ficus*, with the receptacle enlarged and enclosing a cavity with many small, partially embedded, achenelike structures with seeds.

Synoecious When a head or other inflorescence contains both staminate and pistillate flowers.

Synonym Name of a plant family, genus, species, etc., which, for various reasons, is no longer valid but generally represents the equivalent taxon.

Synsepal Two sepals fused together in cypripedioid orchids.

Syrphid fly A fly pollinator that behaves ecologically like a bee or wasp.

Talus A layer of broken rock, individually boulders to small in size, usually on a slope or below a cliff.

Taproot A thick, fleshy, main or primary root that functions in food and water storage.

Tardily Produced later than usual.

Tarn A small mountain lake usually formed from a glacier; a cirque filled with water.

Taxon (plural **taxa**) A representative of a unit of taxonomic classification, such as a species.

Tendril Part or all of a stem, leaf, or petiole modified to form a threadlike structure, which may coil around objects for support.

Tepal Sterile sepaloid or petaloid structure of the flower when the perianth parts are not differentiated into calyx and corolla, as in many monocots and many cactus flowers.

Terete Round in cross section.

Terminal At the end of a stem or structure.

Ternate A leaf divided into three leaflets, usually where it is difficult to determine if it is pinnately compound or palmately compound.

Terpene An organic compound, often a component of plant resins.

Terrestrial Growing in or on the ground; not aquatic, epiphytic, or lithophytic.

Tertiary A geological period extending from 65 to 1.6 million years ago.

Tessellate With a checkered pattern.

Testa The seed coat, which develops from maternal sporophytic tissue; the integuments.

Testicular Resembling testicles; relating to the testa.

Tetrad A group of four, usually referring to pollen that is shed in groups of four grains.

Tetradynamous Flowers with four long and two short stamens, as in the Brassicaceae.

Tetraploid Possessing an extra complete diploid set of chromosomes.

Theca (plural **thecae**) A pollen sac or anther cell.

Thecum One half of the anther cell.

Thigmotropic Responding to touch, as in *Mimosa pudica*; a sensitive plant.

Thorn Modified, sharp stem, such as in *Condalia* or *Elaeagnus*.

Throat The portion of the corolla between the limb and the tube; the opening or orifice in a zygomorphic corolla; the upper margin of the leaf sheath in grasses.

Thrum (see **pin**) A flower with a short style.

Thyrse A panicle that is more compact and oval in shape than usual and with an indeterminate main axis.

Thyrsiform Thyrselike, but may not be an actual thyrse.

Thyrsoid Thyrselike.

Tiller A generally erect shoot of underground or basal origin, usually referring to grasses.

Timberline The line marking the upper limit of normal tree growth.

Tomentose With rather short, densely matted, soft, whitish, woolly hairs.

Tomentum Woolly, short, soft covering of hairs.

Torulose Slightly constricted at intervals, with a generally cylindric form.

Toxicosis A diseased condition resulting from poisoning.

Tracheid A thin, tapered, end-perforated cell wall (the cell itself is dead) in xylem which functions to conduct water; considered to be less evolved than vessel elements.

Translator Part of the pollinarium of the Asclepiadaceae (Apocynaceae, in part) which connects the pollinia of adjacent anthers.

Travertine A type of limestone deposited in watery environments by bacteria, typically cyanobacteria.

Tree A woody perennial with one or a few main trunks.

Triangular (see **deltoid**, **deltate**) Having the form of an equilateral triangle.

Trichome A hairlike, scaly, or prickly outgrowth of the epidermis.

Trichotomous Three-forked.

Tridentate Three-toothed.

Trifid Three-cleft or three-parted.

Trifoliolate (**trifoliate**) With three leaves or three leaflets.

Trigonate (**trigonous**) Three-angled.

Trilete A trifurcate mark on spores that indicates where the three other spores in a tetrad attached during development.

Tripartite Three-parted.

Tripinnate A leaf pinnately compound three times, with each pinnule pinnately compound.

Truncate Cut off squarely.

Tube A hollow, cylindrical structure, as in the corolla of some flowers.

Tuber A short, usually thickened, underground storage stem with buds at the nodes, such as in potatoes.

Tubercle Short, usually rounded, nodule or projection, as in the Cactaceae; a persistent style base in the Cyperaceae.

Tubercule (see **tubercle**) Small rounded propagules or other rounded projecting structures; a small tuber.

Tuberoid A thickened root that resembles a very small tuber.

Tuberosity A structure or protuberance that resembles a tuber in shape.

Tuberous Resembling a tuber in form.

Tubular Forming a tubelike hollow structure, usually fused parts of a calyx or corolla.

Tuff A volcanic rock composed of compacted volcanic ash.

Turbinate Top-shaped.

Turf Densely packed grasses or grasslike plants.

Turgid Swollen or inflated.

Turion A small shoot that often overwinters, as in some Cyperaceae and *Epilobium*.

Tussock A large, tufted growth of sedges or grasses.

Twining Coiling around a support and climbing.

$2n$ The diploid chromosome number; counted in root tip cells undergoing mitosis.

Ultimate The final or top section of a structure.

Umbel Indeterminate inflorescence with pedicels usually of similar length arising from one point, sometimes flat-topped, simple or compound.

Umbellate Umbel-like.

Umbellet An umbellate cluster of an umbel.

Umbilicus A navel-like structure, such as a hilum.

Umbo A blunt protuberance, as on the ends of some pine seed cones.

Uncinate Hooked at the tip; clawlike.

Undulate Wavy, including leaf surface; three-dimensional.

Ungulate A hoofed mammal.

Unicarpellate With a single, free carpel.

Unilocular With a single locule or compartment, as in some ovaries.

Uniparous With only a single branch produced at each branching, such as in some cymes.

Uniseriate In a single row or series.

Unisexual (see **imperfect**).

Urceolate Pitcherlike, hollow, but contracted near the mouth.

Urticating Itch-producing, irritating, and usually rash-producing, as in the hairs in *Cypripedium*.

Utricle A small, thin-walled, one-seeded, inflated dry fruit.

Vallecula (plural **valleculae**) A furrow, groove, or depression.

Vallecular Pertaining to the vallecula or valleys between ridges, as in *Equisetum* stems.

Valvate Opening by sutures that separate the sections of the fruit at maturity, as in loculicidal capsules; a flower in which the petals and/or sepals lie edge to edge along their length, but do not overlap.

Valve One of the segments into which a dehiscent capsule or legume separates.

Variety A unit of classification below the species and subspecies level (q.v.).

Velamen The often thick, spongy, white, water-absorbing layer of epidermis that is dead at maturity and that covers some orchid roots.

Venation The arrangement of the veins of a leaf or petal.

Ventral Lower surface or underside.

Ventricose Inflated or swollen on one side, as in some salvias or penstemons.

Vernation Leaf arrangement in a bud; the configuration of an unopened fern frond.

Vernicose Having a brilliantly polished surface, as some leaves.

Verrucate (see **verrucose**) Slightly reticulate, as in a spore or pollen exine.

Verrucose Covered with wartlike structures.

Versatile Attached near the middle instead of at one end; mobile anthers of this construction, as in many *Lilium*.

Verticil A whorl; an arrangement of similar parts around a central axis.

Verticillaster A pair of axillary cymes arising from opposite leaves or bracts, forming a false whorl.

Vesicle A small inflated structure.

Vespertine Flowering or pollinated during dawn and dusk hours, such as some Onagraceae.

Vessel element A lignified, stout cell wall often with spiral or other inner wall sculpting which conducts water in xylem in higher vascular plants. Many vessel elements have completely open ends.

Vestigial Reduced to a trace of an organ or part, which formerly was much larger or more usefully formed.

Vestiture Collectively, all of the epidermal covering of a plant, including hairs, scales, secretions, and prickles.

Virgate Erect, straight, and slender; usually long and wandlike.

Viscid Sticky, such as in the leaves of *Proboscidea* species.

Viscidium A sticky pad or gluelike substance of a pollinarium which attaches it to a pollinator in the Orchidaceae.

Viscin Gluelike or elasticlike substance.

Vivipary Offspring that develop from germinating seeds or as asexually forming plantlets on the mother plant before dispersal, as in *Rhizophora* or in some grasses and agaves.

Wanting Lacking.

Wetlands Communities that have water as a major component most of the growing year; including swamps and marshes.

Whip leaf A leaf intermediate between a juvenile needle leaf and an adult scale leaf, usually found on "whip" stems in the Cupressaceae.

Whorl A ring of organs or structures.

Whorled Leaves or other structures arranged in a whorl or ring at the same node.

Wing A thin extension bordering an organ; a lateral petal of a papilionaceous (pea) flower.

Woolly Soft, dense, commonly matted hairs.

x The base chromosome number for a plant family or genus. Individual species' chromosome numbers may be identical to or be multiples of this number.

Xeromorphic Exhibiting phenotypic adaptations for a dry environment, such as succulence, dry season deciduousness, etc.

Xylem The system of lignified tubes (dead cell walls) or a formerly active system which conducts water in a vascular plant, usually consisting of tracheids and vessel elements.

Zygomorphic Flowers with a series of parts definitely bilaterally symmetrical, with not all parts in each series alike; usually based on petals unless otherwise stated.

Adams, R. P. 1975. Numerical-chemosystematic studies of infraspecific variation in *Juniperus pinchotii*. Biochem. Syst. Ecol. 3:71–74.

Adams, R. P. 1993. *Juniperus*. Pp. 412–420 *in* Flora of North America Editorial Committee (editors), Flora of North America, Vol. 2. Oxford University Press, New York.

Adams, R. P., and R. N. Pandey. 2003. Analysis of *Juniperus communis* and its varieties based on DNA fingerprinting. Biochem. Syst. Ecol. 31:1271–1278.

Aedo, C. 2001. The genus *Geranium* L. (Geraniaceae) in North America II. Perennial species. Anales Jard. Bot. Madrid 59:3–65.

Aiken, S. G. 1981. A conspectus of *Myriophyllum* (Haloragaceae) in North America. Brittonia 33:57–69.

Aiken, S. G. and S. J. Darbyshire. 1990. Fescue grasses of Canada. Agriculture Canada Publication No. 1844/E. Canadian Government Publishing Centre, Ottawa. 113 pp.

Albach, D. C., H. M. Meudt, and B. Oxelman. 2005. Piecing together the "new" Plantaginaceae. Amer. J. Bot. 92: 297–315.

Alex, J. F. 1962. The taxonomy, history, and distribution of *Linaria dalmatica*. Canad. J. Bot. 40:295–307.

Allred, K. W. 1983. Systematics of the *Bothriochloa saccharoides* complex (Poaceae: Andropogoneae). Syst. Bot. 8:168–184.

Allred, K. W. 1984. Morphologic variation and classification of the North American *Aristida purpurea* complex (Gramineae). Brittonia 36:382–395.

Allred, K. W. 1993. *Bromus*, section *Pnigma*, in New Mexico, with a key to the bromegrasses of the state. Phytologia 74:319–345.

Allred, K. W. 1997. A Field Guide to the Grasses of New Mexico, 2nd ed. Agricultural Experiment Station, New Mexico State University, Las Cruces.

Allred, K. W. 2002. Identification and taxonomy of *Tamarix* (Tamaricaceae) in New Mexico. Desert Pl. 18:26–32.

Allred, K. W. 2005a. A Field Guide to the Grasses of New Mexico, 3rd ed. Agricultural Experiment Station, New Mexico State University, Las Cruces.

Allred, K. W. 2005b. Perennial *Festuca* (Gramineae) of New Mexico. Desert Pl. 21:3–12.

Allred, K. W., and J. Valdés-Reyna. 1997. The *Aristida pansa* complex and a key to the Divaricatae group of North America (Gramineae: Aristideae). Brittonia 49:54–66.

Al-Shehbaz, I. A., K. Mummenhoff, and O. Appel. 2002. *Cardaria*, *Coronopus*, and *Stroganowia* are united with *Lepidium* (Brassicaceae). Novon 12:5–11.

Al-Shehbaz, I. A., and S. L. O'Kane, Jr. 2002. *Lesquerella* is united with *Physaria* (Brassicaceae). Novon 12:319–329.

Anderson, D. E. 1961. Taxonomy and distribution of the genus *Phalaris*. Iowa State Coll. J. Sci. 36:1–96.

Anderson, D. E. 1974. Taxonomy of the genus *Chloris* (Gramineae). Brigham Young Univ. Sci. Bull., Biol. Ser. 19: 1–133.

Anderson, E. F. 2001. The Cactus Family. Timber Press, Portland.

Anderson, L. C. 1963. Studies on *Petradoria* (Compositae): Anatomy, cytology, taxonomy. Trans. Kansas Acad. Sci. 66:632–684.

Appel, O., and I. A. Al-Shehbaz. 2003. Cruciferae. Pp. 75–174 *in* K. Kubitzki (editor), The Families and Genera of Vascular Plants, Vol. 5. Springer-Verlag, Berlin.

Applequist, W. L., and R. S. Wallace. 2001. Phylogeny of the portulacaceous cohort based on *ndh*F sequence data. Syst. Bot. 26:406–419.

Arber, A. 1931. Studies in floral morphology III. On the Fumarioideae with special reference to the androecium. New Phytol. 30:317–354.

Argus, G. W. 1995. Salicaceae: Willow family. Part 2: *Salix* L. J. Arizona-Nevada Acad. Sci. 29:39–62.

Argus, G. W. 1997. Infrageneric classification of *Salix* (Salicaceae) in the New World. Syst. Bot. Monogr. 52.

Arnow, L. 1994. *Koeleria macrantha* and *K. pyramidata* (Poaceae): Nomenclatural problems and biological distinctions. Syst. Bot. 19:6–20.

Atwood, N. D., and S. L. Welsh. 2002. Overview of *Sphaeralcea* (Malvaceae) in southern Utah and northern Arizona, U.S.A., and description of a new species. Novon 12:159–166.

Austin, D. F. 1986. Convolvulaceae (morning glory). Pp. 652–661 *in* T. M. Barkley (editor), Flora of the Great Plains. University Press of Kansas, Lawrence.

Austin, D. F. 1990. Annotated checklist of New Mexican Convolvulaceae. Sida 14:273–286.

Austin, D. F. 1991. Annotated checklist of Arizona Convolvulaceae. Sida 14:443–457.

Austin, D. F. 1998. Convolvulaceae: Morning glory family. J. Arizona-Nevada Acad. Sci. 30:61–83.

Austin, D. F. 2000. Bindweed (*Convolvulus arvensis*, Convolvulaceae) in North America, from medicine to menace. J. Torrey Bot. Soc. 127:172–177.

Austin, D. F., and R. S. Bianchini. 1998. Additions and corrections in American *Ipomoea* (Convolvulaceae). Taxon 47:833–838.

Austin, D. F., G. M. Diggs, and B. L. Lipscomb. 1997. *Calystegia* (Convolvulaceae) in Texas. Sida 17:837–840.

Austin, D. F., and Z. Huáman. 1996. A synopsis of *Ipomoea* (Convolvulaceae) in the Americas. Taxon 45:3–38.

Baars, D. L. 1995. Navajo Country: A Geology and Natural History of the Four Corners Region. University of New Mexico Press, Albuquerque.

Baars, D. L. 2000. Geology of Canyonlands National Park. Pp. 61–83 *in* D. A. Sprinkel, T. C. Chidsey, Jr., and P. B. Anderson (editors), Geology of Utah's Parks and Monuments. Utah Geol. Assoc. Publ. 28.

Babcock, E. B. 1947a. The genus *Crepis*, Pt. I: The taxonomy, phylogeny, distribution, and evolution of *Crepis*. Univ. Calif. Publ. Bot. 21.

Babcock, E. B. 1947b. The genus *Crepis*, Pt. II: Systematic treatment. Univ. Calif. Publ. Bot. 22.

Bacigalupi, R. 1931. A monograph of the genus *Perezia*, section *Acourtia*, with a provisional key to the section *Euperezia*. Contr. Gray Herb. 97:3–81.

Baden, C. 1991. A taxonomic revision of *Psathyrostachys* (Poaceae). Nordic J. Bot. 11:3–26.

Bailey, R. G. 1995. Description of the ecoregions of the United States, 2nd ed. Misc. Publ. U.S.D.A. Forest Serv. No. 1391 (revised).

Baird, R. J. 1996. *Agoseris*. Ph.D. Dissertation. University of Texas at Austin, Austin.

Baker, M. S. 1957. Studies in western violets, VIII. The Nuttallianae continued. Brittonia 9:217–230.

Baker, W. L. 1983. Alpine vegetation of Wheeler Peak, New Mexico, U.S.A.: Gradient analysis, classification, and biogeography. Arctic Alpine Res. 15:223–240.

Baldwin, B. G., B. L. Wessa, and J. L. Panero. 2002. Nuclear rDNA evidence for major lineages of helenioid Heliantheae (Compositae). Syst. Bot. 27:161–198.

Ballard, H. E. 1992. Systematics of *Viola* section *Viola* in North America North of Mexico. M.S. Thesis, Central Michigan University, Mount Pleasant.

Barkworth, M. E. 1993. North American Stipeae (Gramineae): Taxonomic changes and other comments. Phytologia 74:1–25.

Barkworth, M. E., and R. J. Atkins. 1984. *Leymus* Hochst. (Gramineae: Triticeae) in North America: Taxonomy and distribution. Amer. J. Bot. 71:609–625.

Barneby, R. C. 1952. A revision of the North American species of *Oxytropis* DC. Proc. Calif. Acad. Sci. 27:177–312.

Barneby, R. C. 1964a. Atlas of North American *Astragalus*. Part I. The Phacoid and Homaloboid Astragali. Mem. New York Bot. Gard. 13.

Barneby, R. C. 1964b. Atlas of North American *Astragalus*. Part II. The Cercidothrix, Hypoglottis, Piptoloboid, Trimeniaeus and Orophaca Astragali. Mem. New York Bot. Gard. 13.

Barrett, S. C. H., and J. L. Strother. 1978. Taxonomy and natural history of *Bacopa* (Scrophulariaceae) in California. Syst. Bot. 3:408–419.

Bassett, I. J., and C. W. Crompton. 1982. The genus *Chenopodium* in Canada. Canad. J. Bot. 60:586–610.

Baum, B. R. 1967. Introduced and naturalized tamarisks in the United States and Canada. (Tamaricaceae). Baileya 15:19–25.

Baum, B. R. 1977. Oats: Wild and Cultivated. A Monograph of the Genus *Avena* L. (Poaceae). Biosystematics Research Institute, Canada Department of Agriculture and Agri-Food, Ottawa.

Baum, B. R. 1978. The Genus *Tamarix*. Israel Academy of Sciences and Humanities, Jerusalem. 209 pp.

Baum, B. R., and L. G. Bailey. 1990. Key and synopsis of North American *Hordeum* species. Canad. J. Bot. 68: 2433–2442.

Baum, D. A., K. J. Sytsma, and P. C. Hoch. 1994. A phylogenetic analysis of *Epilobium* (Onagraceae) based on nuclear ribosomal DNA sequences. Syst. Bot. 19:363–388.

Bayer, R. J. 1990. A phylogenetic reconstruction of *Antennaria* (Asteraceae: Inuleae). Canad. J. Bot. 68:1389–1397.

Bayer, R. J., D. E. Soltis, and P. S. Soltis. 1996. Phylogenetic inferences in *Antennaria* (Asteraceae: Gnaphalieae: Cassiniinae) based on sequences from nuclear ribosomal DNA internal transcribed spacers (ITS). Amer. J. Bot. 83:516–527.

Bayer, R. J., and G. L. Stebbins. 1987. Chromosome numbers, patterns of distribution, and apomixis in *Antennaria* (Asteraceae: Inuleae). Syst. Bot. 12:305–319.

Bayer, R. J., and G. L. Stebbins. 1993. A synopsis with keys for the genus *Antennaria* (Asteraceae: Inuleae: Gnaphaliinae) of North America. Canad. J. Bot. 71:1589–1604.

Beal, E. O. 1956. Taxonomic revision of the genus *Nuphar* Sm. of North America and Europe. J. Elisha Mitchell Sci. Soc. 72:317–346.

Beardsley, P. M., and R. G. Olmstead. 2002. Redefining Phrymaceae: The placement of *Mimulus*, tribe Mimuleae, and *Phryma*. Amer. J. Bot. 89:1093–1102.

Beatley, J. C. 1973. Russian-thistle (*Salsola*) species in western United States. J. Range Managem. 26:225–226.

Bell, C. D., and R. W. Patterson. 2000. Molecular phylogeny and biogeography of *Linanthus* (Polemoniaceae). Amer. J. Bot. 87:1857–1870.

Benson, L. 1943. Revisions of status of southwestern desert trees and shrubs. Amer. J. Bot. 30:630–632.

Benson, L. 1982. The Cacti of the United States and Canada. Stanford University Press, Palo Alto.

Biddulph, S. F. 1944. A revision of the genus *Gaillardia*. Res. Stud. State Coll. Wash. 12:195–256.

Bierner, M. W. 1972. Taxonomy of *Helenium* sect. *Tetrodus* and a conspectus of North American *Helenium* (Compositae). Brittonia 24:331–355.

Bierner, M. W. 1994. Submersion of *Dugaldia* and *Plummera* in *Hymenoxys* (Asteraceae: Heliantheae: Gaillardiinae). Sida 16:1–8.

Bierner, M. W. 2001. Taxonomy of *Hymenoxys* subgenus *Picradenia* and a conspectus of the subgenera of *Hymenoxys* (Asteraceae: Helenieae: Tetraneurinae). Lundellia 4:37–63.

Bierner, M. W. 2004. Taxonomy of *Hymenoxys* subgenus *Macdougalia* (Asteraceae: Helenieae: Tetraneurinae). Sida 21:657–663.

Bierner, M. W. 2005. Taxonomy of *Hymenoxys* subgenus *Rydbergia* (Asteraceae: Helenieae: Tetraneurinae). Lundellia 8:28–37.

Bierner, M. W., and R. K. Jansen. 1998. Systematic implications of DNA restriction site variation in *Hymenoxys* and *Tetraneuris* (Asteraceae: Helenieae: Gaillardiinae). Lundellia 1:17–26.

Bierner, M. W., and B. L. Turner. 2003. Taxonomy of *Tetraneuris* (Asteraceae: Helenieae: Tetraneurinae). Lundellia 6: 44–96.

Björkman, S. O. 1960. Studies in *Agrostis* and related genera. Symb. Bot. Upsal. 17:1–112.

Blake, S. F. 1917. The varieties of *Chimaphila umbellata*. Rhodora 19:237–244.

Blake, S. F. 1924. Polygalaceae. N. Amer. Fl. 25:305–379.

Bleakly, D. 1996. A Checklist of the Vascular Flora of El Morro National Monument. M.S. Thesis, University of New Mexico, Albuquerque.

Blum, W., M. Lange, W. Rischer, and J. Rutow. 1998. *Echinocereus* Fa. Proost N. V., Turnhout, Belgium.

Bogin, C. 1955. Revision of the genus *Sagittaria* (Alismataceae). Mem. New York Bot. Gard. 9:179–233.

Boivin, B. 1951. The genus *Chenopodium* in Canada. Canad. Field-Naturalist 65:17.

Boivin, B. 1962. Etudes sur les *Oxytropis* DC. *Oxytropis deflexa* (Pallas) DC., Vol. I. Svensk Bot. Tidskr. 56:496–500.

Börner, C. 1912. Botanisch-systematische Notizen. Abh. Naturwiss. Vereins Bremen 21:245–282.

Boufford, D. E. 1982. The systematics and evolution of *Circaea* (Onagraceae). Ann. Missouri Bot. Gard. 69:804–994.

Boufford, D. E., J. V. Crisci, H. Tobe, and P. C. Hoch. 1990. A cladistic analysis of *Circaea* (Onagraceae). Cladistics 6:171–182.

Boulos, L. 1973. Révision systématique du genre *Sonchus* L. s.l.: IV. Sous-genre: 1. *Sonchus*. Bot. Not. 126:155–196.

Bowman, J. L., H. Brüggemann, J.-Y. Lee, and K. Mummenhoff. 1999. Evolutionary changes in floral structure within *Lepidium* L. (Brassicaceae). Int. J. Pl. Sci. 160:917–929.

Brandbyge, J. 1993. Polygonaceae. Pp. 531–544 *in* K. Kubitzki, J. G. Rohwer, and V. Bittrich (editors), The Families and Genera of Vascular Plants, Vol. 2. Springer-Verlag, Berlin.

Brasher, J. W. 2001. Betulaceae: Birch family. J. Arizona-Nevada Acad. Sci. 33:1–7.

Breshears, D. D., N. S. Cobb, P. M. Rich, K. P. Price, C. D. Allen, R. G. Balice, W. H. Romme, J. H. Kastens, M. Lisa-Floyd, J. Belnap, J. J. Anderson, O. B. Myers, and C. W. Meyer. 2005. Regional vegetation die-off in response to global-change-type drought. Proc. Natl. Acad. Sci. U.S.A. 102:15144–15148.

Bricker, J. S., G. K. Brown, and T. L. Patts Lewis. 2000. Status of *Descurainia torulosa* (Brassicaceae). W. N. Amer. Naturalist 60:426–432.

Brooks, R. E., and S. E. Clemants. 2000. *Juncus*. Pp. 211–255 *in* Flora of North America Editorial Committee (editors), Flora of North America, Vol. 22. Oxford University Press, New York.

Brouillet, L., L. Urbatsch, and R. P. Roberts. 2004. *Tonestus kingii* and *T. aberrans* are related to *Eurybia* and the Machaerantherinae (Asteraceae: Astereae) based on *nr*DNA (ITS and ETS) data: Reinstatement of *Herrickia* and a new genus *Triniteurybia*. Sida 21:889–900.

Brown, D. E. (editor). 1982. Biotic communities of the American Southwest—United States and Mexico. Desert Pl., Vol. 4.

Brown, D. E., C. H. Lowe, and C. P. Pase. 1980. A digitized systematic classification for ecosystems with an illustrated summary of the natural vegetation of North America. U.S. Department of Agriculture, Forest Service General Technical Report RM-73.

Brown, G. K. 1983. Chromosome numbers in *Platyschkuhria* Rydberg (Compositae) and their systematic significance. Amer. J. Bot. 70:591–601.

Brown, R. C. 1978. Biosystematics of *Psilostrophe* (Compositae: Helenieae). II. Artificial hybridization and systematic treatment. Madroño 25:187–201.

Brummitt, R. K. 1965. New contributions in North American *Calystegia*. Ann. Missouri Bot. Gard. 52:214–216.

Brummitt, R. K. 1992. Vascular Plant Families and Genera. Royal Botanic Gardens, Kew.

Brummitt, R. K., and C. E. Powell. 1992. Authors of Plant Names. Royal Botanic Gardens, Kew.

Burns, R. M., and B. H. Honkala. 1990. Silvics of North America 1. Conifers. U.S. Department of Agriculture, Forest Service Agricultural Handbook 654. Washington, D.C.

Cabrera, R. L. 1992. Systematics of *Rzedowskiela* gen. nov. (Asteraceae: Nassauviinae). Ph.D. Dissertation, University of Texas at Austin, Austin.

Cabrera, R. L. 2001. Six new species of *Acourtia* (Asteraceae) and a historical account of *Acourtia mexicana*. Brittonia 53:416–429.

Campbell, C. S. 1986. Phylogenetic reconstructions and two new varieties in the *Andropogon virginicus* complex (Poaceae: Andropogoneae). Syst. Bot. 11:280–292.

Campbell, G. R. 1950. *Mimulus guttatus* and related species. El Aliso 2:319–335.

Canne, J. M. 1979. A light and scanning electron microscope study of seed morphology in *Agalinis* (Scrophulariaceae) and its taxonomic significance. Syst. Bot. 4:281–296.

Canne, J. M. 1981. Chromosome counts in *Agalinis* and related taxa (Scrophulariaceae). Canad. J. Bot. 59:1111–1116.

Cantino, P. D. 1992. Evidence for a polyphyletic origin of the Labiatae. Ann. Missouri Bot. Gard. 79:361–379.

Carlquist, S. 1955. Tribal interrelationships and phylogeny of the Asteraceae. Aliso 8:465–492.

Carter, J. L. 1997. Trees and Shrubs of New Mexico. Mimbres Publishing, Silver City, New Mexico.

Chambers, K. L. 1955. A biosystematic study of the annual species of *Microseris*. Contr. Dudley Herb. 4: 207–312.

Chase, A. 1946. *Enneapogon desvauxii* and *Pappophorum wrightii*, an agrostological detective story. Madroño 8: 187–189.

Cholewa, A. F., and D. M. Henderson. 1984. Biosystematics of *Sisyrinchium* section *Bermudiana* (Iridaceae) of the Rocky Mountains. Brittonia 36:342–363.

Chrtek, J., and M. Šourková. 1992. Variation in *Oxyria digyna*. Preslia 64:207–210.

Chuang, T. I., and L. R. Heckard. 1986. Systematics and evolution of *Cordylanthus*. Syst. Bot. Monogr. 10:1–105.

Chuang, T. I., W. C. Hsieh, and D. H. Wilken. 1978. Contribution of pollen morphology to systematics in *Collomia* (Polemoniaceae). Amer. J. Bot. 65:450–458.

Church, G. L. 1949. A cytotaxonomic study of *Glyceria* and *Puccinellia*. Amer. J. Bot. 36:155–165.

Church, G. L. 1952. The genus *Torreyochloa*. Rhodora 54:197–200.

Clausen, J. 1964. Cytotaxonomy and distributional ecology of western North American violets. Madroño 17:173–197.

Clayton, W. D., and S. A. Renvoize. 1986. Genera *Graminum*: Grasses of the world. Kew Bull., Addit. Ser. XIII:1–389.

Clemants, S. E. 2003a. *Chenopodium berlandieri*. P. 294 *in* Flora of North America Editorial Committee (editors), Flora of North America, Vol. 4. Oxford University Press, New York.

Clemants, S. E. 2003b. *Guilleminea*. Pp. 437–438 *in* Flora of North America Editorial Committee (editors), Flora of North America, Vol. 4. Oxford University Press, New York.

Clemants, S. E., and S. L. Mosyakin. 2003. Chenopodiaceae. P. 296 *in* Flora of North America Editorial Committee (editors), Flora of North America, Vol. 4. Oxford University Press, New York.

Cobb, B., E. Farnsworth, and C. Lowe. 2005. A Field Guide to Ferns and Their Related Families of Northeastern and Central North America. 2nd ed. Houghton Mifflin, New York.

Coleman, R. A. 2002. The Wild Orchids of Arizona and New Mexico. Cornell University Press, Ithaca.

Columbus, J. T. 1999. An expanded circumscription of *Bouteloua* (Gramineae: Chlorideae): New combinations and names. Aliso 18:61–65.

Comer, C. W., R. P. Adams, and D. F. van Haverbeke. 1982. Intra- and interspecific variation of *Juniperus virginiana* and *Juniperus scopulorum* seedlings based on volatile oil composition. Biochem. Syst. Ecol. 10:297–306.

Conert, H. J., and A. M. Turpe. 1974. Revision der gattung *Schismus* (Poaceae: Arundinoideae: Danthonieae). Abh. Senckenberg. Naturf. Ges. 532:1–81.

Constance, L. 1937. A systematic study of the genus *Eriophyllum* Lag. Univ. Calif. Publ. Bot. 18:69–136.

Cook, C. D. K. 1966. A monographic study of *Ranunculus* subgenus *Batrachium* (DC.) A. Gray. Mitt. Bot. Staatssamml. München 6:47–237.

Cook, C. D. K., and M. S. Nicholls. 1986. A monographic study of the genus *Sparganium* (Sparganiaceae). Part 1. Subgenus *Xanthosparganium* Holmberg. Bot. Helv. 96: 213–267.

Cook, C. D. K., and M. S. Nicholls. 1987. A monographic study of the genus *Sparganium* (Sparganiaceae). Part 2. Subgenus *Sparganium*. Bot. Helv. 97:1–44.

Cook, C. D. K., and K. Urmi-König. 1985. A revision of the genus *Elodea* (Hydrocharitaceae). Aquatic Bot. 21:111–156.

Copeland, H. F. 1947. Observations on the structure and classification of the Pyroleae. Madroño 9:65–102.

Correll, D. S., and M. C. Johnston. 1979a. *Chenopodium chenopodioides*. P. 533 *in* Manual of the Vascular Plants of Texas. University of Texas, Dallas.

Correll, D. S., and M. C. Johnston. 1979b. Juglandaceae. Pp. 457–462 *in* Manual of the Vascular Plants of Texas. University of Texas, Dallas.

Costea, M., G. L. Nesom, and S. Stefanovi. 2006. Taxonomy of the *Cuscuta salina-californica* complex (Convolvulaceae). Sida 22:177–195.

Costea, M., G. L. Nesom, and F. J. Tardif. 2005. Taxonomic status of *Cuscuta nevadensis* and *C. veatchii* (Convolvulaceae) in North America. Brittonia 57:264–272.

Costea, M., F. J. Tardif, and H. R. Hinds. 2005. *Polygonum*. Pp. 547–571 *in* Flora of North America Editorial Committee (editors), Flora of North America, Vol. 5. Oxford University Press, New York.

Cox, B. J. 1970. A Monograph of the *Lupinus ornatus* Complex. M.S. Thesis, University of Missouri, Columbia.

Cribb, P. J. 1997. The Genus *Cypripedium*. Timber Press, Portland.

Critchfield, W. B., and E. L. Little. 1966. Geographic distribution of the pines of the world. U.S. Department of Agriculture Forest Service Misc. Publ. 991, Washington, D.C.

Cronquist, A., A. H. Holmgren, N. H. Holmgren, J. L. Reveal, and P. K. Holmgren. 1977. Intermountain Flora: Vascular Plants of the Intermountain West, U.S.A., Vol. 6. The Monocotyledons. The New York Botanical Garden, Bronx.

Cronquist, A., A. H. Holmgren, N. H. Holmgren, J. L. Reveal, and P. K. Holmgren. 1989. Intermountain Flora: Vascular Plants of the Intermountain West, U.S.A., Vol. 3, Part B. Fabales. The New York Botanical Garden, Bronx.

Cronquist, A., A. H. Holmgren, and P. K. Holmgren. 1997a. Intermountain Flora: Vascular Plants of the Intermountain West, U.S.A., Vol. 3, Part A. Subclass Rosidae (Except Fabales). The New York Botanical Garden, Bronx.

Cronquist, A., A. H. Holmgren, N. H. Holmgren, J. L. Reveal, and P. K. Holmgren. 1997b. Intermountain Flora: Vascular Plants of the Intermountain West, U.S.A., Vol. 4. Subclass Asteridae (Except Asteraceae). The New York Botanical Garden, Bronx.

Crow, G. E., and C. B. Hellquist. 1983. Aquatic vascular plants of New England. Part 6. Trapaceae, Haloragaceae and Hippuridaceae. New Hampshire Agric. Exp. Sta. Bull. 524.

Crow, G. E., and C. B. Hellquist. 1985. Aquatic vascular plants of New England. Part 8. Lentibulariaceae. New Hampshire Agric. Exp. Sta. Bull. 528.

Crow, G. E., and C. B. Hellquist. 2000. Aquatic and Wetland Plants of Northeastern North America. Angiosperms: Monocotyledons. University of Wisconsin Press, Madison.

Cruden, R. W. 1981. New *Echeandia* (Liliaceae) from Mexico. Sida 9:139–146.

Cutler, H. C. 1939. Monograph of the North American species of the genus *Ephedra*. Ann. Missouri Bot. Gard. 26:373–428.

Daubs, E. H. 1965. A monograph of Lemnaceae. Illinois Biol. Monogr. 34:1–118.

Davidse, G. 1976. A study of some intermountain violets (*Viola* sect. *Chamaemelanium*). Madroño 23: 274–283.

Davidson, J. F. 1950. The genus *Polemonium* (Tourn.) L. Univ. Calif. Publ. Bot. 23:209–282.

Day, A. G. 1980. Nomenclatural changes in *Ipomopsis congesta* (Polemoniaceae). Madroño 27:111–112.

Day, A. G. 1993a. *Gilia*. Pp. 828–836 in J. C. Hickman (editor), The Jepson Manual: Higher Plants of California. University of California Press, Berkeley.

Day, A. G. 1993b. *Navarretia*. Pp. 844–849 in J. C. Hickman (editor), The Jepson Manual: Higher Plants of California. University of California Press, Berkeley.

DeLisle, D. G. 1963. Taxonomy and distribution of the genus *Cenchrus*. Iowa State Coll. J. Sci. 37:259–351.

Dempster, L. M. 1995. Rubiaceae. J. Arizona-Nevada Acad. Sci. 29:29–38.

Denton, M. F. 1973. A monograph of *Oxalis* section *Ionoxalis* (Oxalidaceae) in North America. Publ. Mus. Michigan State Univ., Biol. Ser. 4, No. 10:455–615.

Detling, L. E. 1939. A revision of the North American species of *Descurainia*. Amer. Midl. Naturalist 22: 481–520.

DeVelice, R. L., J. A. Ludwig, W. H. Moir, and F. Ronco, Jr. 1986. A classification of forest habitat types of northern New Mexico and southern Colorado. U.S. Department of Agriculture Forest Service General Technical Report RM-131.

Dewey, D. R. 1983. Historical and current taxonomic perspectives of *Agropyron*, *Elymus* and related genera. Crop Sci. 23:637–642.

Dick-Peddie, W. A. 1993. New Mexico Vegetation: Past, Present, and Future. University of New Mexico Press, Albuquerque.

Dietrich, W., and W. L. Wagner. 1988. Systematics of *Oenothera* section *Oenothera* subsection *Raimannia* and subsection *Nutantigemma* (Onagraceae). Syst. Bot. Monogr. 24:1–91.

Dietrich, W., W. L. Wagner, and P. H. Raven. 1997. Systematics of *Oenothera* sect. *Oenothera* subsect. *Oenothera* (Onagraceae). Syst. Bot. Monogr. 50:1–234.

Dishman, L. 1982. Ranching and farming in the lower Dolores River Valley. Pp. 23–41 *in* G. D. Kendrick (editor), The River of Sorrows: The History of the Lower Dolores River Valley. U.S. Department of the Interior, National Park Service and Bureau of Reclamation. Printed by U.S. Government Printing Office, Denver. 74 pp.

Di Tomaso, J. M. 1998. Impact, biology and ecology of salt cedar (*Tamarix* spp.) in the southwestern United States. Weed Technol. 12:326–336.

Doggett, H. 1970. *Sorghum*. Longmans, London. 403 pp.

Donoghue, M. J., T. Erikkson, P. A. Reeves, and R. G. Olmstead. 2001. Phylogeny and phylogenetic taxonomy of Dipsacales, with special reference to *Sinadoxa* and *Tetradoxa* (Adoxaceae). Harvard Pap. Bot. 6:459–480.

Dorn, R. D. 1995. A taxonomic study of *Salix* section *Cordatae* subsection *Luteae* (Salicaceae). Brittonia 47:160–174.

Dorn, R. D. 1997. Rocky Mountain region willow identification field guide. U.S. Department of Agriculture, Forest Service Rocky Mtn. Region Renewable Resources R2-RR-97-01.

Dorn, R. D. 1998. A taxonomic study of *Salix* section *Longifoliae* (Salicaceae). Brittonia 50:193–210.

Dorn, R. D. 2000. A taxonomic study of *Salix* sections *Mexicanae* and *Viminella* subsection *Sitchenses* (Salicaceae) in North America. Brittonia 52:1–19.

Downie, S. R., and K. E. Denford. 1988. Taxonomy of *Arnica* (Asteraceae) subgenus *Arctica*. Rhodora 90:245–275.

Dressler, R. L. 1981. The Orchids: Natural History and Classification. Harvard University Press, Cambridge.

Dressler, R. L. 1993. Phylogeny and Classification of the Orchid Family. Dioscorides Press, Portland.

DuBois, C. 1903. Report on the proposed San Juan Forest Reserve, Colorado. Unpublished report, on file at the supervisor's office, San Juan National Forest, Durango, Colorado.

Dunn, D. B. 1956. Leguminosae of Nevada, II—*Lupinus*. Contr. Fl. Nevada 39:1–64.

Dunn, D. B. 1957. A revision of the *Lupinus arbustus* complex of the Laxiflori. Madroño 14:54–73.

Dunn, D. B. 1959. *Lupinus pusillus* and its relationship. Amer. Midl. Naturalist 62:500–510.

Dunn, D. B. 1964. Pp. 140–143 *in* S. L. Welsh, M. Treshow, and G. Moore, Common Utah Plants. Brigham Young University Press, Provo, Utah.

Eckenwalder, J. E. 1977. North American cottonwoods (*Populus*, Salicaceae) of sections *Abaso* and *Aigeiros*. J. Arnold Arbor. 58:194–208.

Eckenwalder, J. E. 1992. Salicaceae: Willow family. J. Arizona-Nevada Acad. Sci. 26:29–33.

Eiten, G. 1963. Taxonomy and regional variation of *Oxalis* section *Corniculatae* 1. Introduction, keys and synopsis of the species. Amer. Midl. Naturalist 69:257–309.

Elisens, W. J. 1985. Monograph of Maurandyinae (Scrophulariaceae-Antirrhineae). Syst. Bot. Monogr. 5:1–97.

Ellison, W. L. 1964. A systematic study of the genus *Bahia* (Compositae). Rhodora 66:67–86, 177–215, 281–311.

Ellison, W. L. 1971. Taxonomy of *Platyschkuhria* (Compositae). Brittonia 23:269–279.

Epling, C. 1934. Preliminary revision of American *Stachys*. Repert. Spec. Nov. Regni Veg. Beih. 80:1–75.

Erdman, K. S. 1965. Taxonomy of the genus *Sphenopholis* (Gramineae). Iowa State Coll. J. Sci. 39:289–336.

Ernst, W. R. 1963. The genera of Capparaceae and Moringaceae in the southeastern United States. J. Arnold Arbor. 44:81–95.

Fabijan, D. M., J. G. Packer, and K. E. Denford. 1987. The taxonomy of the *Viola nuttallii* complex. Canad. J. Bot. 65:2562–2580.

Farjon, A., and B. T. Styles. 1997. *Pinus* (Pinaceae). Fl. Neotrop. Monogr. 75. New York Botanical Garden, Bronx.

Farrar, D. R. 2005. Systematics of western moonworts-*Botrychium* subgenus *Botrychium in* S. J. Popovich (editor), unpublished report, U.S. Department of Agriculture, U.S. Forest Service Moonwort Workshop. Arapaho-Roosevelt National Forests and Pawnee National Grassland, Fort Collins, Colorado, July 13–15.

Fassett, N. C. 1951. *Callitriche* in the New World. Rhodora 53:137–155, 161–182, 185–194, 210–222.

Ferguson, A. M. 1901. Crotons of the United States. Report (Annual) Missouri Bot. Gard. 12:33–73.

Fernald, M. L. 1932. The linear-leaved North American species of *Potamogeton* section *Axillares*. Mem. Amer. Acad. Arts 17:1–183.

Fernald, M. L., and K. M. Wiegand. 1914. The genus *Ruppia* in eastern North America. Rhodora 16:119–127.

Finot, V. L., P. M. Peterson, R. J. Soreng, and F. O. Zuloaga. 2005. A revision of *Trisetum* and *Graphephorum*. Sida 21:1419–1453.

Fleak, L. S., and D. B. Dunn. 1971. Nomenclature of the *Lupinus sericeus* L. complex (Papilionaceae). Trans. Missouri Acad. Sci. 5:85–88.

Fleischner, T. L. 1994. Ecological costs of livestock grazing in western North America. Conservation Biol. 8:629–664.

Flora of North America Editorial Committee. 1993. Flora of North America North of Mexico, Vol. 2. Pteridophytes and Gymnosperms. Oxford University Press, New York.

Flora of North America Editorial Committee. 1997. Flora of North America North of Mexico, Vol. 3. Magnoliophyta: Magnoliidae and Hamamelidae. Oxford University Press, New York.

Flora of North America Editorial Committee. 2000. Flora of North America North of Mexico, Vol. 22. Magnoliophyta: Alismatidae, Arecidae, Commelinidae (in part), and Zingiberidae. Oxford University Press, New York.

Flora of North America Editorial Committee. 2002. Flora of North America North of Mexico, Vol. 23. Magnoliophyta: Commelinidae (in part): Cyperaceae. Oxford University Press, New York.

Flora of North America Editorial Committee. 2003. Flora of North America North of Mexico, Vol. 4. Magnoliophyta: Caryophyllidae, part 1. Oxford University Press, New York.

Flora of North America Editorial Committee. 2006. Flora of North America North of Mexico, Volume 21. Magnoliophyta: Asteridae (in part): Asteraceae, part 3. Oxford University Press, New York.

Flores, H., and J. I. Davis. 2001. A cladistic analysis of Atripliceae (Chenopodiaceae) based on morphological data. J. Torrey Bot. Soc. 128:297–319.

Floyd, M. L. 2003. Ancient Piñon-Juniper Woodlands: A Natural History of Mesa Verde Country. University Press of Colorado, Boulder.

Floyd, M. L., D. D. Hanna, and W. H. Romme. 2004. Historical and recent fire regimes in piñon-juniper woodlands on Mesa Verde, Colorado, USA. Forest Ecol. Managem. 198:269–289.

Floyd, M. L., W. H. Romme, and D. Hanna. 2000. Fire history and vegetation pattern in Mesa Verde National Park. Ecol. Applic. 10:1666–1680.

Floyd-Hanna, L., A. W. Spencer, and W. H. Romme. 1996. Biotic communities of the semiarid foothills and valleys. Pp. 143–158 *in* R. Blair (editor), The Western San Juan Mountains: Their Geology, Ecology, and Human History. University Press of Colorado, Boulder.

Ford, B. A., and P. W. Ball. 1988. A reevaluation of the *Triglochin maritimum* complex (Juncaginaceae) in eastern and central North America and Europe. Rhodora 90:313–337.

Frederiksen, S. 1983. *Festuca brachyphylla*, *F. saximontana* and related species in North America. Nordic J. Bot. 2: 525–536.

Frederiksen, S., and G. Petersen. 1998. A taxonomic revision of *Secale* L. (Triticeae, Poaceae). Nordic J. Bot. 18: 399–420.

Freeman, C. C., and H. R. Hinds. 2005a. *Bistorta*. Pp. 594–597 *in* Flora of North America Editorial Committee (editors), Flora of North America, Vol. 5. Oxford University Press, New York.

Freeman, C. C., and H. R. Hinds. 2005b. *Fallopia*. Pp. 541–546 *in* Flora of North America Editorial Committee (editors), Flora of North America, Vol. 5. Oxford University Press, New York.

Freeman, C. C., and J. G. Packer. 2005. *Oxyria*. Pp. 533–534 *in* Flora of North America Editorial Committee (editors), Flora of North America, Vol. 5. Oxford University Press, New York.

Freeman, C. C., and J. L. Reveal. 2005. Polygonaceae. Pp. 216–601 *in* Flora of North America Editorial Committee (editors), Flora of North America, Vol. 5. Oxford University Press, New York.

Fross, D., and D. Wilken. 2006. *Ceanothus.* Timber Press, Portland.

Freudenstein, J. 1999. Relationships and character transformation in Pyroloideae (Ericaceae) based on ITS sequences, morphology, and development. Syst. Bot. 24:398–408.

Gaiser, L. O. 1946. The genus *Liatris*. Rhodora 48:165–183, 216–263, 273–326, 331–382, 393–412.

Gentry, J. L., and R. L. Carr. 1976. A revision of the genus *Hackelia* (Boraginaceae) in North America north of Mexico. Mem. New York Bot. Gard. 26:121–227.

Gillett, J. M. 1965. Taxonomy of *Trifolium*: Five American species of section *Lupinaster* (Leguminosae). Brittonia 17:121–136.

Gillett, J. M. 1969. Taxonomy of *Trifolium* (Leguminosae) II. The *T. longipes* complex in North America. Canad. J. Bot. 47:93–113.

Gillett, J. M. 1971. Taxonomy of *Trifolium* (Leguminosae) III. *T. eriocephalum*. Canad. J. Bot. 49:395–405.

Gillett, J. M. 1972. Taxonomy of *Trifolium* (Leguminosae) IV. The American species of section *Lupinaster* (Adanson) Seringe. Canad. J. Bot. 50:1975–2007.

Glück, H. 1934. *Limosella*-Studien. Bot. Jahrb. Syst. 66:488–566.

Gontcharova, S. B., E. V. Artyukova, and A. A. Gontcharov. 2006. Phylogenetic relationships among members of the subfamily Sedoideae (Crassulaceae) inferred from the ITS region sequences of nuclear rDNA. Russ. J. Genet. 42:654–661.

Goodman, G. J. 1934. A revision of the North American species of the genus *Chorizanthe*. Ann. Missouri Bot. Gard. 21:1–102.

Goodman, G. J. 1957. The genus *Centrostegia*, tribe Eriogoneae. Leafl. W. Bot. 8:125–128.

Gould, F. W. 1967. The grass genus *Andropogon* in the United States. Brittonia 19:70–76.

Gould, F. W. 1979. The genus *Bouteloua* (Poaceae). Ann. Missouri Bot. Gard. 66:348–416.

Gould, F. W., M. A. Ali, and D. E. Fairbrothers. 1972. A revision of *Echinochloa* in the United States. Amer. Midl. Naturalist 87:36–59.

Gould, F. W., and R. B. Shaw. 1983. Grass Systematics, 2nd ed. Texas A&M University Press, College Station.

Graham, S. A., and C. E. Wood, Jr. 1965. The genera of Polygonaceae in the southeastern United States. J. Arnold Arbor. 46:91–121.

Grant, A. L. 1924. A monograph of the genus *Mimulus*. Ann. Missouri Bot. Gard. 11:99–388.

Grant, V. 1952. Isolation and hybridization between *Aquilegia formosa* and *A. pubescens*. Aliso 2:341–360.

Grant, V. 1956. The genetic structure of races and species in *Gilia*. Advances Genet. 8:55–87.

Grant, V. 1959. Natural History of the Phlox Family. Martinus Nijhoff, The Hague.

Grant, V. 1989. Taxonomy of the tufted alpine and subalpine Polemoniums (Polemoniaceae). Bot. Gaz. 150:158–169.

Grant, V., and D. H. Wilken. 1986. Taxonomy of the *Ipomopsis aggregata* group (Polemoniaceae). Bot. Gaz. 147: 359–371.

Grant, V., and D. H. Wilken. 1988. Racial variation in *Ipomopsis tenuituba* (Poemoniaceae). Bot. Gaz. 149:443–449.

Grass Phylogeny Working Group. 2001. Phylogeny and subfamilial classification of the grasses (Poaceae). Ann. Missouri Bot. Gard. 88:373–457.

Greene, C. W. 1984. Sexual and apomictic reproduction in *Calamagrostis* (Gramineae) from eastern North America. Amer. J. Bot. 71:285–293.

Greene, E. L. 1900. Type collections of *Trifolium*. Pittonia 4:137.

Greer, L. F. 1997. *Thelesperma curvicarpum* (Asteraceae), an achene form in populations of *T. simplicifolium* var. *simplicifolium* and *T. filifolium* var. *filifolium*. Southw. Naturalist 42:242–244.

Griffith, M. P. 2002. *Grusonia pulchella* classification and its impacts on the genus *Grusonia*: Morphological and molecular evidence. Haseltonia 9:86–93.

Grissino-Mayer, H. D., W. H. Romme, M. L. Floyd, and D. D. Hanna. 2004. Climatic and human influences on fire regimes of the southern San Juan Mountains, Colorado, USA. Ecology 85:1708–1724.

Grossman, D. H., D. Faber-Langendoen, A. S. Weakley, M. Anderson, P. Bourgeron, R. Crawford, K. Goodin, S. Landaal, K. Metzler, K. Patterson, M. Pyne, M. Reid, and L. Sneddon. 1998. Terrestrial Vegetation of the

United States, Vol. 1, The National Vegetation Classification System: Development, Status, and Applications. The Nature Conservancy, Arlington, Virginia.

Haber, E. 1974. Pyrolaceae: Wintergreen family. Canad. J. Bot. 52:877–883.

Haber, E. 1987. Variability distribution and systematics of *Pyrola picta* s.l. Ericaceae in western North America. Syst. Bot. 12:324–335.

Haber, E. 1992. Monotropaceae: Indian pipe family. J. Arizona-Nevada Acad. Sci. 26:15–16.

Haddow, D., R. Musselman, T. Blett, and R. Fisher. 1998. Guidelines for evaluating air pollution impacts on wilderness within the Rocky Mountain region: Report of a workshop, 1990. U.S. Department of Agriculture, Forest Service Gen. Tech. Rep. RMRS-GTR-4, Fort Collins, Colorado.

Hagström, J. O. 1916. Critical researches on the potamogetons. Kungl. Svenska Vetenskapsakad. Handl. 55:1–281.

Hall, H. M., and F. E. Clements. 1923. The phylogenetic method in taxonomy: The North American species of *Artemisia*, *Chrysothamnus*, and *Atriplex*. Publ. Carnegie Inst. Wash., Publication no. 326.

Hall, J. C., K. J. Sytsma, and H. H. Iltis. 2002. Phylogeny of Capparaceae and Brassicaceae based on chloroplast sequence data. Amer. J. Bot. 89:1826–1842.

Hammel, B. E., and J. R. Reeder. 1979. The genus *Crypsis* (Gramineae) in the United States. Syst. Bot. 4:267–280.

Hansen, C. J., L. Allphin, and M. D. Windham. 2002. Biosystematic analysis of the *Thelesperma subnudum* complex (Asteraceae). Sida 20:71–96.

Hanser, C., and M. M. Iljin. 1936. *Chenopodium*. Pp. 41–73 *in* V. L. Komarov and B. K. Shishkin (editors), Flora U.S.S.R., Vol. 6, 5th ed. Publishing House, Academy of Sciences, Moscow.

Hanson, C. A. 1962. Perennial *Atriplex* of Utah and the Northern Deserts. M.S. Thesis, Brigham Young University, Provo, Utah.

Hardham, C. B. 1989. Chromosome numbers in some annual species of *Chorizanthe* and related genera (Polygonaceae: Eriogonoideae). Phytologia 66:89–94.

Harrison, H. K. 1972. Contributions to the study of the genus *Eriastrum*. Brigham Young Sci. Bull., Biol. Ser. 16:1–26.

Hartman, R. L. 2005. *Drymaria*. Pp. 9–14 *in* Flora of North Americal Editorial Committee (editors), Flora of North America, Vol. 5. Oxford University Press, New York.

Hartman, R. L. 2006. New combinations in the genus *Cymopterus* (Apiaceae) of the southwestern United States. Sida 22:955–957.

Hartman, R. L., and B. E. Nelson. 2001. A Checklist of the Vascular Plants of Colorado. http://www.rmh.uwyo.edu/data/co_checklist.php, accessed 24 April 2012.

Hartog, C. den, and F. van der Plas. 1970. A synopsis of the Lemnaceae. Blumea 18:355–368.

Hauke, R. L. 1963. A taxonomic monograph of the genus *Equisetum* subgenus *Hippochaete*. Beih. Nova Hedwigia, Vol. 8. 123 pp.

Hauke, R. L. 1978. A taxonomic monograph of *Equisetum* subgenus *Equisetum*. Beih. Nova Hedwigia 30:385–455.

Hawksworth, F. G., and D. Wiens. 1996. Dwarf mistletoes: Biology, pathology and systematics. Agricultural Handbook 709, U.S. Department of Agriculture, Forest Service. Washington, D.C.

Haynes, R. R. 1975. A revision of North American *Potamogeton* subsection *Pusilli* (Potamogetonaceae). Rhodora 76:564–649.

Haynes, R. R. 1978. The Potamogetonaceae in the southeastern United States. J. Arnold Arbor. 59:170–191.

Haynes, R. R. 1979. Revision of North and Central American *Najas* (Najadaceae). Sida 8:34–56.

Haynes, R. R., and C. B. Hellquist. 2000. Juncaginaceae. Pp. 43–46 *in* Flora of North America Editorial Committee (editors), Flora of North America, Vol. 22. Oxford University Press, New York.

Haynes, R. R., and L. B. Holm-Nielsen. 1987. The Zannichelliaceae in the southeastern United States. J. Arnold Arbor. 68:259–268.

Heil, K. D., B. Armstrong, and D. Schleser. 1981. A review of the genus *Pediocactus*. Cact. Succ. J. (Los Angeles) 53:17–39.

Heil, K. D., and S. L. O'Kane, Jr. 2005. Catalog of the Four Corners Flora: Vascular Plants of the San Juan River Drainage, Arizona, Colorado, New Mexico and Utah, 9th ed. San Juan College, Farmington, New Mexico.

Heil, K. D., and J. M. Porter. 1994. *Sclerocactus* (Cactaceae): A revision. Haseltonia 2:20–46.

Heiser, C. B. 1944. Monograph of *Psilostrophe*. Ann. Missouri Bot. Gard. 31:279–300.

Henderson, D. M. 1976. A biosystematic study of Pacific Northwestern blue-eyed grasses (*Sisyrinchium*, Iridaceae). Brittonia 28:149–176.

Hendricks, A. J. 1957. A review of the genus *Alisma* (Dill) L. Amer. Midl. Naturalist 58:470–493.

Henrard, J. T. 1950. Monograph of the genus *Digitaria*. Universitaire Pers Leiden, Leiden, The Netherlands.

Henrickson, J. 1987. A taxonomic reevaluation of *Gossypianthus* and *Guilleminea* (Amaranthaceae). Sida 12:307–337.

Hereford, R., R. H. Webb, and S. Graham. 2002. Precipitation history of the Colorado Plateau region, 1900–2000. U.S. Department of the Interior, U.S. Geological Survey Fact Sheet 119–02.

Hermann, F. J. 1953. A botanical synopsis of the cultivated clovers (*Trifolium*). U.S. Department of Agriculture, Agricultural Monograph 22. Washington, D.C.

Hermann, F. J. 1960. Vetches of the United States: Native, naturalized, and cultivated. U.S. Department of Agriculture, Agricultural Handbook No. 168, pp. 1–84, Washington, D.C.

Hermann, F. J. 1970. Manual of the carices of the Rocky Mountains and Colorado Basin. U.S. Department of Agriculture, Agricultural Handbook No. 374. Washington, D.C.

Hermann, F. J. 1975. Manual of the rushes (*Juncus* spp.) of the Rocky Mountains and Colorado Basin. Gen. Techn. Rep. RM-18, U.S. Department of Agriculture, Forest Service, Fort Collins, Colorado.

Hess, L. W. 1969. The Biosystematics of the *Lupinus argenteus* Complex and Allies. Ph.D. Dissertation, University of Missouri, Columbia.

Heywood, V. H. 1993. Flowering Plants of the World, Updated Edition. Oxford University Press, New York.

Higgins, L. C. 1971. A revision of *Cryptantha* subgenus *Oreocarya*. Brigham Young Univ. Sci. Bull., Biol. Ser. 13:1–63.

Higgins, L. C. 1979. Boraginaceae of the southwestern United States. Great Basin Naturalist 39:293–350.

Hinds, H. R., and C. C. Freeman. 2005. *Persicaria*. Pp. 574–594 *in* Flora of North America Editorial Committee (editors), Flora of North America, Vol. 5. Oxford University Press, New York.

Hitchcock, C. L. 1936. The genus *Lepidium* in the United States. Madroño 3:265–300.

Hitchcock, C. L. 1945. The South American species of *Lepidium*. Lilloa 11:75–134.

Hitchcock, C. L. 1950. On the subspecies of *Lepidium montanum*. Madroño 10:155–158.

Hitchcock, C. L. 1952. A revision of the North American species of *Lathyrus*. Univ. Wash. Publ. Biol. 15:1–104.

Hitchcock, C. L. 1957. A study of the perennial species of *Sidalcea*. Part I. Taxonomy. Univ. Wash. Publ. Biol. 18:1–79.

Hitchcock, C. L., A. Cronquist, M. Ownbey, and J. Thompson. 1969. Vascular Plants of the Pacific Northwest, Part 1. University of Washington Press, Seattle.

Hitchcock, C. L., and B. Maguire. 1947. A revision of the North American species of *Silene*. Univ. Wash. Publ. Biol. 13.

Ho, T. N., and S. W. Liu. 1990. The infrageneric classification of Gentiana (Gentianaceae). Bull. Brit. Mus. (Nat. Hist.) Bot. 20:169–192.

Hoggard, G. D., P. J. Kores, M. Molvray, and R. K. Hoggard. 2004. The phylogeny of *Gaura* (Onagraceae) based on ITS, ETS, and *trnL-F* sequence data. Amer. J. Bot. 91:139–148.

Holm, R. W. 1950. The American species of *Sarcostemma* R. Br. (Asclepiadaceae). Ann. Missouri Bot. Gard. 37: 477–560.

Holmgren, N. H. 1988. *Glossopetalon* (Crossosomataceae) and a new variety of *G. spinescens* from the Great Basin, U.S.A. Brittonia 40:269–274.

Holmgren, N. H., P. K. Holmgren, and A. Cronquist. 2005. Intermountain Flora: Vascular Plants of the Intermountain West, U.S.A., Vol. 2, part B. Subclass Dilleniidae. The New York Botanical Garden, Bronx.

Holub, J. 1997. *Stuckenia* Börner. The correct name for *Coleogeton* (Potamogetonaceae). Preslia 69:361–366.

Hotchkiss, N., and H. L. Dozier. 1949. Taxonomy and distribution of North American cattails. Amer. Midl. Naturalist 41:237–254.

Howell, J. T. 1971. A new name for "winter fat." Wasmann J. Biol. 29:105.

Hufford, L. 1989. The structure and potential loasaceous affinities of *Schismocarpus*. Nordic J. Bot. 9:217–227.

Hufford, L., and M. McMahon. 2004. Morphological evolution and systematics of *Synthyris* and *Besseya* (Veronicaceae): A phylogenetic analysis. Syst. Bot. 29:716–736.

Hultén, E. 1959. The *Trisetum spicatum* complex. Svensk. Bot. Tidskr. 53:203–228.

Hurd, E. G., S. Goodrich, and N. L. Shaw. 1994. Field guide to intermountain rushes. Gen. Techn. Rep. INT-306, U.S. Department of Agriculture, Forest Service, Ogden, Utah.

Hurd, E. G., N. L. Shaw, J. Mastrogiuseppe, L. C. Smithman, and S. Goodrich. 1998. Field guide to intermountain sedges. Gen. Techn. Rep. RMRS-GTR-10, U.S. Department of Agriculture, Forest Service, Ogden, Utah.

Iltis, H. H. 1954. Studies in the Capparidaceae. I. *Polanisia dodecandra* (L.) DC., the correct name for *Polanisia graveolens* Rafinesque. Rhodora 56:65–70.

Iltis, H. H. 1957. Studies in the Capparidaceae. III. Evolution and phylogeny of the western North American Cleomoideae. Ann. Missouri Bot. Gard. 44:77–119.

Iltis, H. H. 1958. Studies in the Capparidaceae. IV. *Polanisia* Raf. Brittonia 10:33–58.

Iltis, H. H. 1966. Studies in the Capparidaceae. VIII. *Polanisia dodecandra* (L.) DC. Rhodora 68:41–47.

IPCC (Intergovernmental Panel on Climate Change). 1998. The Regional Impacts of Climate Change. Cambridge University Press, Cambridge.

Isely, D. 1978. New varieties and combinations in *Lotus*, *Baptista*, *Thermopsis*, and *Sophora* (Leguminosae). Brittonia 30:466–472.

Isely, D. 1998. Native and Naturalized Leguminosae (Fabaceae) of the United States (exclusive of Alaska and Hawaii). Monte L. Bean Life Science Museum Press, Brigham Young University, Provo, Utah.

Isely, D., and F. J. Peabody. 1984. *Robinia* (Leguminosae: Papilionoidea). Castanea 49:187–202.

Jacobs, D. L. 1947. An ecological life-history of *Spirodela polyrhiza* (greater duckweed) with emphasis on the turion phase. Ecol. Monogr. 17:437–469.

Jacobs, S. W. L. 2001. Review of leaf anatomy and ultrastructure in the Chenopodiaceae (Caryophyllales). J. Torrey Bot. Soc. 128:236–253.

Jefferies, J. A. M. 1972. A Revision of the Genus *Sphaeralcea* (Malvaceae) for the State of Utah. Unpublished Thesis. Brigham Young University, Provo, Utah.

Johansson, J. T. 1998. Chloroplast DNA restriction site mapping and the phylogeny of *Ranunculus* (Ranunculaceae). Pl. Syst. Evol. 213:1–19.

Johnson, M. 2001. The Genus *Clematis*. Magnus Johnsons Plantskola AB. Södertälje, Sweden. 896 pp.

Johnston, B. C. 2001. Field guide to sedge species of the Rocky Mountain Region: The genus *Carex* in Colorado, Wyoming, western South Dakota, western Nebraska, and western Kansas. USDA Forest Service, Rocky Mountain Region, Denver.

Johnston, I. M. 1923. Studies in the Boraginaceae, 1. Restoration of the genus *Hackelia*. 2. The genus *Antiphytum*. 3. Novelties and new combinations in the genus *Cryptantha*. 4. A synopsis and redefinition of *Plagiobothrys*. Contr. Gray Herb., New Ser. No. 68:57–80.

Johnston, I. M. 1925. Studies in the Boraginaceae, Part IV. The North American species of *Cryptantha*. Contr. Gray Herb., New Ser. No. 74:1–114.

Johnston, I. M. 1927. Studies in the Boraginaceae, Part VI. A revision of the South American Boraginoideae. Contr. Gray Herb. No. 78:2–118.

Johnston, I. M. 1952. Studies in the Boraginaceae, Part XXIII. A survey of the genus *Lithospermum*. J. Arnold Arbor. 33:299–366.

Jones, G. N. 1940. A monograph of the genus *Symphoricarpos*. J. Arnold Arbor. 21:201–252.

Judd, W. S., C. S. Campbell, E. A. Kellogg, P. F. Stevens, and M. J. Donoghue. 2008. Plant Systematics: A Phylogenetic Approach. Sinauer Associates, Sunderland, Massachusetts.

Judd, W. S., R. W. Sanders, and M. J. Donoghue. 1994. Angiosperm family pairs: Preliminary phylogenetic analyses. Harvard Pap. Bot. 5:1–51.

Källersjö, M., G. Bergqvist, and A. A. Anderberg. 2000. Generic realignment in primuloid families of the Ericales s.l.: A phylogenetic analysis based on DNA sequences from three chloroplast genes and morphology. Amer. J. Bot. 87:1325–1341.

Kaul, R. B. 2002. *Sparganium*. Pp. 270–277 *in* Flora of North America Editorial Committee (editors), Flora of North America, Vol. 22. Oxford University Press, New York.

Kay, M. A. 1996. Healing with Plants in the American and Mexican West. University of Arizona Press, Tucson.

Kearney, T. H. 1935. The North American species of *Sphaeralcea*, subgenus *Eusphaeralcea*. Univ. Calif. Publ. Bot. 19:1–128.

Kearney, T. H., R. H. Peebles, and collaborators. 1969. Arizona Flora, 2nd ed. (with supplement). University of California Press, Berkeley.

Keck, D. D. 1927. A revision of the genus *Orthocarpus*. Proc. Calif. Acad. Sci. IV, 16:517–571.

Keeler, K. H. 1980. The extrafloral nectaries of *Ipomoea leptophylla* (Convolvulaceae). Amer. J. Bot. 67:216–222.

Keeley, J. E., and S. C. Keeley. 1988. Chaparral. Pp. 165–207 *in* M. G. Barbour and W. D. Billings (editors), North American Terrestrial Vegetation. Cambridge University Press, Cambridge.

Keen, R. A. 1996. Weather and climate. Pp. 113–126 *in* R. Blair (editor), The Western San Juan Mountains: Their Geology, Ecology, and Human History. University Press of Colorado, Boulder.

Keil, D. J. 1977. A revision of *Pectis* section *Pectothrix* (Compositae: Tageteae). Rhodora 79:32–78.

Keller, S. 1979. A revision of the genus *Wislizenia* (Capparidaceae) based on population studies. Brittonia 31: 333–351.

Kellogg, E. A. 1985a. A biosystematic study of the *Poa secunda* complex. J. Arnold Arbor. 66:201–242.

Kellogg, E. A. 1985b. Variation and names in the *Poa secunda* complex. J. Range Managem. 38(6):516–521.

Kiger, R. W. 1997. Papaveraceae. Pp. 300–339 *in* Flora of North America Editorial Committee (editors), Flora of North America, Vol. 3. Oxford University Press, New York.

Kiger, R. W. 2006. *Cosmos*. Pp. 203–205 *in* Flora of North America Editorial Committee (editors), Flora of North America, Vol. 21. Oxford University Press, New York.

Kim, H., D. J. Loockerman, and R. K. Jansen. 2002. Systematic implications of *ndhF* sequence variation in the Mutisieae (Asteraceae). Syst. Bot. 27:598–609.

Kim, M. H., J. H. Park, and C. W. Park. 2000. Flavonoid chemistry of *Fallopia* section *Fallopia* (Polygonaceae). Biochem. Syst. Ecol. 28:433–441.

Kingsbury, J. M. 1964. Poisonous Plants of the United States and Canada. Prentice Hall, New Jersey.

Knudsen, J. T., and L. Tollsten. 1991. Floral scent and intrafloral scent differentiation in *Moneses* and *Pyrola* (Pyrolaceae). Pl. Syst. Evol. 177:81–91.

Koch, M. B., B. Haubold, and T. Mitchell-Olds. 2001. Molecular systematics of the Brassicaceae. Amer. J. Bot. 88:534–544.

Kron, K. A., W. S. Judd, P. F. Stevens, D. M. Crayn, A. A. Anderberg, P. A. Gadek, C. J. Quinn, and J. L. Luteyn. 2002. Phylogenetic classification of Ericaceae: Molecular and morphological evidence. Bot. Rev. (Lancaster) 68:335–423.

Kuchler, A. W. 1985. Potential natural vegetation (map) of the Conterminous United States, revised. National Atlas of the United States, U.S. Geological Survey, Reston, Virginia.

Laferrière, J. E. 1994. Juglandaceae: Walnut family. J. Arizona-Nevada Acad. Sci. 27: 219.

LaFrankie, Jr., J. V. 1986. Transfer of the species of *Smilacina* to *Maianthemum* (Liliaceae). Taxon 35:584–589.

Lamb Frye, A. S., and K. A. Kron. 2003. *rbcL* phylogeny and character evolution in Polygonaceae. Syst. Bot. 28: 326–332.

Landolt, E. 1986. The family of Lemnaceae: A monographic study, Vol. 1. Veröff. Geobot. Inst. ETH Stiftung Rübel Zürich 71.

Landolt, E. 2000. Lemnaceae. Pp. 143–153 *in* Flora of North America Editorial Committee (editors), Flora of North America, Vol. 22. Oxford University Press, New York.

Landolt, E., and R. Kandeler. 1987. The family of Lemnaceae: A monographic study, Vol. 2. Veröff. Geobot. Inst. ETH Stiftung Rübel Zürich 95.

Landrum, L. R. 1995. Aceraceae: Maple family. J. Arizona-Nevada Acad. Sci. 29:2–3.

Lane, M. A. 1985. Taxonomy of *Gutierrezia* (Compositae: Astereae) in North America. Syst. Bot. 10:7–28.

Les, D. H. 1986. The phytogeography of *Ceratophyllum demersum* and *C. echinatum* (Ceratophyllaceae) in glaciated North America. Canad. J. Bot. 64:498–509.

Les, D. H. 1988. The origin and affinities of the Ceratophyllaceae. Taxon 37:326–345.

Les, D. H., and D. J. Crawford. 1999. *Landoltia* (Lemnaceae), a new genus of duckweeds. Novon 9:530–533.

Levin, R. A. 2000. Phylogenetic relationships within Nyctaginaceae tribe Nyctagineae: Evidence from nuclear and chloroplast genomes. Syst. Bot. 25:738–750.

Levin, R. A., W. L. Wagner, P. C. Hoch, M. Nepokroeff, J. C. Pires, and E. A. Zimmer. 2004. Paraphyly in tribe Onagreae: Insights into phylogenetic relationships of Onagraceae based on nuclear and chloroplast sequence data. Syst. Bot. 29:147–164.

Levin, R. A., W. L. Wagner, P. C. Hoch, M. Nepokroeff, J. C. Pires, E. A. Zimmer, and K. J. Sytsma. 2003. Family-level relationships of Onagraceae based on chloroplast *rbcL* and *ndhL* data. Amer. J. Bot. 90:107–115.

Lewis, H., and J. Szweykowski. 1964. The genus *Gayophytum* (Onagraceae). Brittonia 16:343–391.

Lewis, P. P., and G. P. Oliver. 1961. Cytogeography and phylogeny of North American species of Verbena. Amer. J. Bot. 48:638–643.

Lewis, W. H. 1957. A monograph of the genus *Rosa* in North America east of the Rocky Mountains. Ph.D. Dissertation, University of Virginia, Charlottesville.

Lewis, W. H. 1959. A monograph of the genus *Rosa* in North America. I. *R. acicularis*. Brittonia 11:1–24.

Liede, S. 1996. *Sarcostemma* (Asclepiadaceae). A controversial generic circumscription reconsidered: Morphological evidence. Syst. Bot. 21:31–44.

Ling, Y. R. 1982. On the system of the genus *Artemisia* L. and the relationship with its allies. Bull. Bot. Lab. N. E. Forest. Inst., Harbin 2:1–60.

Ling, Y. R. 1995. The New World of *Artemisia* L. Pp. 225–281 *in* D. J. N. Hind, C. Jeffrey, and G. V. Pope. Advances in Compositae Systematics, Royal Botanic Gardens, Kew.

Little, E. L. 1979. Checklist of United States trees (native and naturalized). U.S. Department of Agriculture, Forest Service Agriculture Handbook No. 541. 375 pp.

Little, R. J. 2001. Violaceae: Violet family. J. Arizona-Nevada Sci. 33:73–82.

Lloyd, F. E. 1935. *Utricularia*. Biol. Rev. 10:72–100.

Löve, Á., and D. Löve. 1958. Biosystematics of *Triglochin maritimum* Agg. Naturaliste Canad. 85:156–165.

Löve, D. 1969. *Papaver* at high altitudes in the Rocky Mountains. Brittonia 21:1–10.

Lowden, R. M. 1978. Studies on the submerged genus *Ceratophyllum* L. in the neotropics. Aquatic Bot. 4: 127–142.

Lowden, R. M. 1986. Taxonomy of the genus *Najas* L. (Najadaceae) in the neotropics. Aquatic Bot. 24:147–184.

Luer, C. A. 1975. The Native Orchids of the United States and Canada Excluding Florida. The New York Botanical Garden, Bronx.

Lynch, D. L., W. H. Romme, and M. L. Floyd. 2000. Forest restoration in southwestern ponderosa pine. J. Forest. 98:17–24.

Maguire, B. 1943. A monograph of the genus *Arnica* (Senecioneae, Compositae). Brittonia 4:386–510.

Mahler, W. F., and U. T. Waterfall. 1964. *Baccharis* (Compositae) in Oklahoma, Texas, and New Mexico. Southw. Naturalist 9:189–202.

Mansion, G., and L. Struwe. 2004. Generic delimitation and phylogenetic relationships within the subtribe Chironiinae (Chironieae: Gentianaceae) with special reference to *Centaurium*: Evidence from nrDNA and cpDNA sequences. Molec. Phylogen. Evol. 32:951–977.

Martin, J. S. 1946. Notes on *Trifolium eriocephalum* Nuttall. Madroño 8:152–157.

Martin, W. C., and C. R. Hutchins. 1980. A Flora of New Mexico, Vol. 1. J. Cramer, Vaduz, Germany.

Martin, W. C., and C. R. Hutchins. 1981. A Flora of New Mexico, Vol. 2. J. Cramer, Vaduz, Germany.

Mason, Jr., C. T. 1999. Cannabaceae: Hemp family. J. Arizona-Nevada Acad. Sci. 32:53–54.

Mason, H. L. 1945. The genus *Eriastrum* and the influence of Bentham and Gray upon the problem of generic confusion in Polemoniaceae. Madroño 8:33–59.

Mason, H. L. 1957. A Flora of the Marshes of California. University of California Press, Berkeley.

Mast, A. R., D. M. S. Feller, S. Kelso, and E. Conti. 2004. Buzz-pollinated *Dodecatheon* originated from within the heterostylous *Primula* subgenus *Auriculastrum* (Primulaceae): A seven-region cpDNA phylogeny and its implications for floral evolution. Amer. J. Bot. 91:926–942.

Mathias, M. E., and L. Constance. 1944–45. Umbelliferae. N. Amer. Fl. 28B:143–148.

Mayes, R. A. 1976. A cytotaxonomic and chemosystematic study of the genus *Pyrrocoma* (Asteraceae: Astereae). Ph.D. Dissertation, University of Texas, Austin.

Mayuzumi, S., and H. Ohba. 2004. The phylogenetic position of eastern Asian Sedoideae (Crassulaceae) inferred from chloroplast and nuclear DNA sequences. Syst. Bot. 29:587–598.

McArthur, E. D., B. L. Welch, and S. C. Sanderson. 1988. Natural and artificial hybridization between big sagebrush (*Artemisia tridentata*) subspecies. J. Heredity 79:268–276.

McCabe, G. J., M. A. Palecki, and J. L. Betancourt. 2004. Pacific and Atlantic Ocean influences on multidecadal drought frequency in the United States. Proc. Natl. Acad. Sci. U.S.A. 101:4136–4141.

McCully, M. E., and H. M. Dale. 1961a. Heterophylly in *Hippuris*, a problem in identification. Canad. J. Bot. 39: 1099–1116.

McCully, M. E., and H. M. Dale. 1961b. Variations in leaf numbers in *Hippuris*. Canad. J. Bot. 39:611–625.

McDermott, L. F. 1910. An illustrated key to the North American species of *Trifolium*. Cunningham, Curtis and Welch, San Francisco. 325 pp.

McDougall, W. B. 1973. Seed Plants of Northern Arizona with Keys and Detailed Descriptions for the Identification of Families, Genera and Species. The Museum of Northern Arizona, Flagstaff.

McIntosh, R. P. 1993. The continuum continued: John T. Curtis's influence on ecology. Pp. 95–122 *in* J. S. Fralish, R. P. McIntosh, and O. L. Loucks (editors), John T. Curtis: Fifty Years of Wisconsin Plant Ecology. Wisconsin Academy of Sciences, Arts, and Letters, Madison.

McKelvey, S. D. 1938. Yuccas of the Southwestern United States, Part 1. The Arnold Arboretum of Harvard University, Jamaica Plain, Massachusetts.

McKelvey, S. D. 1947. Yuccas of the Southwestern United States, Part 2. The Arnold Arboretum of Harvard University, Jamaica Plain, Massachusetts.

McKinney, L. E. 1992. A taxonomic revision of the acaulescent blue violets (*Viola*) of North America. Sida, Bot. Misc. 7.

McNeill, J., F. R. Barrie, H. M. Burdet, V. Demoulin, D. L. Hawksworth, K. Marhold, D. H. Nicolson, J. Prado, P. C. Silva, J. E. Skog, J. H. Wiersema, and N. J. Turland (editors). 2006. International Code of Botanical Nomenclature (Vienna Code). Regnum Veg. 146.

Mears, J. A. 1967. Revision of *Guilleminea* (*Brayulinea*) including *Gossypianthus* (Amaranthaceae). Sida 3:137–152.

Merriam, C. H. 1890. Results of a biological survey of the San Francisco Mountain region and desert of the Little Colorado, Arizona. N. Amer. Fauna 3:1–136.

Merriam, C. H. 1898. Life zones and crop zones of the United States. U.S. Department of Agriculture, Division of Biological Survey, Bulletin No. 10. Washington, D.C.

Meyer, F. G. 1951. *Valeriana* in North America and the West Indies (Valerianaceae). Ann. Missouri Bot. Gard. 38:377–503.

Michener, J. 1960. The high altitude vegetation of the Needle Mountains of southwestern Colorado. M.S. Thesis, University of Colorado, Boulder.

Michener-Foote, J., and T. Hogan. 1999. The flora and vegetation of the Needle Mountains, San Juan Range, southwestern Colorado. University of Colorado Museum Herbarium, Boulder. 39 pp.

Miller, R. B. 1978. The pollination ecology of *Aquilegia elegantula* and *A. caerulea* (Ranunculaceae) in Colorado. Amer. J. Bot. 65:406–414.

Mitich, L. W. 1989. History and taxonomy of bermudagrass. Proc. Calif. Weed Conf. 41:181–188.

Moench, C. 1794. *Morocarpus foliosus* Moench. Methodus (Moench) 342.

Mooring, J. S. 1965. Chromosome studies in *Chaenactis* and *Chamaechaenactis* (Compositae, Helenieae). Brittonia 17:17–25.

Mooring, J. S. 1980. A cytogeographic study of *Chaenactis douglasii* (Compositae, Helenieae). Amer. J. Bot. 67: 1304–1319.

Morgan, D. R., and R. L. Hartman. 2003. A synopsis of *Machaeranthera* (Asteraceae), with recognition of segregate genera. Sida 20:1387–1416.

Morris, E. L. 1900. A revision of the species of *Plantago* commonly referred to as *Plantago patagonica* (Jacquin). Bull. Torrey Bot. Club 27:105–119.

Morris, E. L. 1901. North American Plantaginaceae II. Bull. Torrey Bot. Club 28:112–122.

Morton, J. K. 2005a. *Cerastium*. Pp. 74–93 *in* Flora of North America Editorial Committee (editors), Flora of North America, Vol 5. Oxford University Press, New York.

Morton, J. K. 2005b. *Silene*. Pp. 166–214 *in* Flora of North America Editorial Committee (editors), Flora of North America, Vol. 5. Oxford University Press, New York.

Morton, J. K. 2005c. *Stellaria*. Pp. 96–214 in Flora of North America Editorial Committee (editors), Flora of North America, Vol. 5. Oxford University Press, New York.

Mosquin, T. 1966. A new taxonomy for *Epilobium angustifoliuim* L. (Onagraceae). Brittonia 18:167–188.

Mosyakin, S. L. 1995. New taxa of *Corispermum* L. (Chenopodiaceae), with preliminary comments on the taxonomy of the genus in North America. Novon 5:340–353.

Mosyakin, S. L. 2005. *Rumex*. Pp. 489–533 *in* Flora of North America Editorial Committee (editors), Flora of North America, Vol. 5. Oxford University Press, New York.

Muller, C. H. 1954. A new species of *Quercus* in Arizona. Madroño 12:140–145.

Mulligan, G. A. 1980. The genus *Cicuta* in North America. Canad. J. Bot. 58:1755–1767.

Mulligan, G. A., and C. Frankton. 1962. Taxonomy of the genus *Cardaria* with particular reference to the species introduced into North America. Canad. J. Bot. 40:1411–1425.

Mummenhoff, K., H. Brüggemann, and J. L. Bowman. 2001. Chloroplast DNA phylogeny and biogeography of *Lepidium* (Brassicaceae). Amer. J. Bot. 88:2051–2063.

Munz, P. A. 1926. The Antirrhinoideae-Antirrhineae of the New World. Proc. Calif. Acad. Sci. Ser. IV, 15:323–397.

Munz, P. A. 1946. *Aquilegia*: The cultivated and wild columbines. Gentes Herb. 7:1–150.

Munz, P. A. 1965. Onagraceae. N. Amer. Fl. II, 5:1–278.

Munz, P. A., and D. D. Keck. 1973. A California Flora, with Supplement. University of California Press, Berkeley.

Murray, D. F. 1995. New names in *Papaver* section *Meconella* (Papaveraceae). Novon 5:294–295.

Navaro, A. M., and W. H. Blackwell. 1990. A revision of *Paxistima* (Celastraceae). Sida 14:231–249.

Nesom, G. L. 1983. Taxonomy of *Erigeron concinnus* (Asteraceae) and its separation from *E. pumilus*. Sida 10: 159–166.

Nesom, G. L. 1988. Synopsis of *Chaetopappa* (Compositae: Asteraceae) with a new species and the inclusion of *Leucelene*. Phytologia 64:448–456.

Nesom, G. L. 1989. New species, new sections, and a taxonomic overview of American *Pluchea* (Compositae: Inuleae). Phytologia 67:158–167.

Nesom, G. L. 1990a. Studies in the systematics of Mexican and Texan *Grindelia* (Asteraceae: Astereae). Phytologia 68:303–332.

Nesom, G. L. 1990b. Taxonomy of the genus *Laënnecia* (Asteraceae: Astereae). Phytologia 68:205–228.

Nesom, G. L. 1991. Taxonomy of *Isocoma* (Asteraceae: Astereae). Phytologia 70:69–114.

Nesom, G. L. 1994. Review of the taxonomy of *Aster* sensu lato (Asteraceae: Astereae) emphasizing the New World species. Phytologia 77:141–297.

Nesom, G. L. 1997. Review. A revision of *Heterotheca* sect. *Phyllotheca* (Nutt.) Harms (Compositae: Astereae) by J. C. Semple. Phytologia 83:7–21.

Nesom, G. L. 2002. New combination in *Xylorhiza* (Asteraceae: Astereae). Sida 20:145–147.

Nesom, G. L. 2004. *Pseudognaphalium canescens* (Asteraceae: Gnaphalieae) and putative relatives in western North America. Sida 21:781–789.

Nesom, G. L. 2005. Infrageneric classification of *Liatris* (Asteraceae: Eupatorieae). Sida 21:1305–1321.

Nesom, G. L. 2006. Taxonomic overview of the *Heterotheca villosa* complex (Asteraceae: Astereae). Sida 22:367–380.

Nesom, G. L., Y. Suh, D. R. Morgan, S. D. Sundberg, and B. B. Simpson. 1991. *Chloracantha*, a new genus of North American Astereae (Asteraceae). Phytologia 70:371–380.

Newsom, V. M. 1929. A revision of the genus *Collinsia* (Scrophulariaceae). Bot. Gaz. 87:260–301.

Nickrent, D. L., M. A. Garcia, M. P. Martin, and R. L. Mathiasen. 2004. A phylogeny of all species of *Arceuthobium* (Viscaceae) using nuclear and chloroplast DNA sequences. Amer. J. Bot. 91:125–138.

Niles, W. E. 1970. Taxonomic investigations in the genera *Perityle* and *Laphamia* (Compositae). Mem. New York Bot. Gard. 21:1–82.

Nisbet, G. T., and R. C. Jackson. 1960. The genus *Penstemon* in New Mexico. Univ. Kansas Sci. Bull. 41:691–759.

Northstrom, T. E. 1974. The Genus *Hedysarum* in North America. Ph.D. Dissertation, Brigham Young University, Provo, Utah.

Northstrom, T. E., and S. L. Welsh. 1970. Revision of the *Hedysarum boreale* complex. Great Basin Naturalist 30: 109–130.

Noyes, R. D. 2000. Biogeographical and evolutionary insights on *Erigeron* and allies (Asteraceae) from ITS sequence data. Pl. Syst. Evol. 220:93–114.

Ockendon, D. J. 1965. A taxonomic study of *Psoralea* subgenus *Pediomelum* (Leguminosae). Southw. Naturalist 10:81–124.

Ogden, E. C. 1943. The broad-leaved species of *Potamogeton* of North America north of Mexico. Rhodora 45:57–105, 119–163, 171–214.

O'Kane, Jr., S. L., K. D. Heil, and G. L. Nesom. 2010. Variation in *Erigeron grandiflorus* (Asteraceae: Astereae) in southwestern Colorado. Phytoneuron 44:1–6.

Olmstead, R. G., C. W. dePamphilis, A. D. Wolfe, N. D. Young, W. J. Elisens, and P. A. Reeves. 2001. Disintegration of the Scrophulariaceae. Amer. J. Bot. 88:348–361.

Ooststroom, S. J. van. 1934. A monograph of the genus *Evolvulus*. Meded. Bot. Mus. Herb. Rijks Univ. Utrecht 14: 1–267.

Orchard, A. E. 1981. A revision of South American *Myriophyllum* (Haloragaceae) and its repercussions on some Australian and North American species. Brunonia 4:27–65.

Ornduff, R., and M. Denton. 1998. Oxalidaceae: Oxalis family. J. Arizona-Nevada Acad. Sci. 30:115–119.

Ottley, A. M. 1944. The American Loti with special consideration of a proposed new section, *Simpeteria*. Brittonia 5:81–123.

Ownbey, G. B. 1958. Monograph of the genus *Argemone* for North America and the West Indies. Mem. Torrey Bot. Club, Vol. 21.

Ownbey, M. 1940. A monograph of the genus *Calochortus*. Ann. Missouri Bot. Gard. 27:371–560.

Ownbey, M. 1947. The genus *Allium* in Arizona. Res. Stud. State Coll. Wash. 15:211–232.

Ownbey, M., and H. C. Aase. 1955. Cytotaxonomic studies in *Allium*. I. The *Allium canadense* alliance. Res. Stud. State Coll. Wash., Monogr. Suppl. 1:1–106.

Oxelman, B., P. Kornhall, R. G. Olmstead, and B. Bremer. 2005. Further disintegration of the Scrophulariaceae. Taxon 54:411–425.

Padgett, D. 1997. *Nuphar*. Ph.D. Dissertation, University of New Hampshire, Durham.

Parfitt, B. D. 1980. Origin of *Opuntia curvospina* (Cactaceae). Syst. Bot. 5:408–418.

Parfitt, B. D. 1998. New nomenclatural combinations in the *Opuntia polyacantha* complex. Cact. Succ. J. 70:188.

Patterson, R. 1977. A revision of *Linanthus* sect. *Siphonella* (Polemoniaceae). Madroño 24:36–48.

Patterson, R. 1993. *Eriastrum*. Pp. 826–828 *in* J. C. Hickman (editor), The Jepson Manual: Higher Plants of California. University of California Press, Berkeley.

Paulsen, Jr., H. A. 1975. Range management in the central and southern Rocky Mountains: A summary of the status of our knowledge by range ecosystems. U.S. Department of Agriculture, Forest Service Research Paper RM–154, Fort Collins, Colorado.

Payson, E. B. 1921. A monograph of the genus *Lesquerella*. Ann. Missouri Bot. Gard. 8:103–236.

Payson, E. B. 1922. A synoptical revision of the genus *Cleomella*. Univ. Wyoming Publ. Sci., Bot. 1:29–46.

Pedersen, K. L. 1994. Modern and Pleistocene climatic patterns in the West. Pp. 27–53 *in* K. T. Harper, L. L. St. Clair, K. H. Thorne, and W. M. Hess (editors), Natural History of the Colorado Plateau and Great Basin. University Press of Colorado, Niwot.

Pelton, J. S. 1958. Evidence of introgressive hybridization and mutation in certain Colorado populations of *Aquilegia*. Proc. Indiana Acad. Sci. 67:292–296.

Pennell, F. W. 1921. *Veronica* in North and South America. Rhodora 23:1–22, 29–41.

Pennell, F. W. 1928. *Agalinis* and allies in North America. I. Proc. Acad. Nat. Sci. Philadelphia 80:339–449.

Pennell, F. W. 1929. *Agalinis* and allies in North America. II. Proc. Acad. Nat. Sci. Philadelphia 81:111–249.

Pennell, F. W. 1933. A revision of *Synthyris* and *Besseya*. Proc. Acad. Nat. Sci. Philadelphia 85:77–106.

Pennell, F. W. 1935. The Scrophulariaceae of eastern temperate North America. Monogr. Acad. Nat. Sci. Philadelphia 1:314.

Pennell, F. W. 1946. Reconsideration of the *Bacopa-Herpestis* problem of the Scrophulariaceae. Proc. Acad. Nat. Sci. Philadelphia 98:83–98.

Pennell, F. W. 1951. *Mimulus*. Pp. 688–731 *in* Illustrated Flora of the Pacific States, Vol. 3, Stanford University Press, Palo Alto.

Perry, L. M. 1937. Notes on *Silphium*. Rhodora 39:281–297.

Peterson, P. M., and C. R. Annable. 1990. A revision of *Blepharoneuron* (Poaceae: Eragrostideae). Syst. Bot. 15: 515–525.

Peterson, P. M., and C. R. Annable. 1991. Systematics of the annual species of *Muhlenbergia* (Poaceae: Eragrostideae). Syst. Bot. Monogr. 31:1–109.

Peterson, P. M., J. Cayouette, Y. S. N. Ferdinandez, B. Coulman, and R. E. Chapman. 2001. Recognition of *Bromus richardsonii* and *B. ciliatus*: Evidence from morphology, cytology, and DNA fingerprinting (Poaceae: Bromeae). Aliso 20:21–36.

Philbrick, C. T. 1989. Systematic Studies of *Callitriche* (Callitrichaceae). Ph.D. Dissertation, University of Connecticut, Storrs.

Pinkava, D. J. 1999a. Cactaceae: Cactus family, part 3. *Cylindropuntia*. J. Arizona-Nevada Acad. Sci. 32: 32–47.

Pinkava, D. J. 1999b. Cactaceae: Cactus family, part 4. *Grusonia*. J. Arizona-Nevada Acad. Sci. 32:48–52.

Porter, J. M. 1998a. *Aliciella*, a recircumscribed genus of Polemoniaceae. Aliso 17:23–46.

Porter, J. M. 1998b. Nomenclatural changes in Polemoniaceae. Aliso 17:83–85.

Porter, J. M., and L. A. Johnson. 2000. A phylogenetic classification of Polemoniaceae. Aliso 19:55–91.

Potter, L. D., R. C. Reynolds, and E. T. Louderbough. 1985a. Mancos shale and plant community relationships—Field observations. J. Arid Environm. 9:137–145.

Potter, L. D., R. C. Reynolds, and E. T. Louderbough. 1985b. Mancos shale and plant community relationships—Analysis of shale, soil, and vegetation transects. J. Arid Environm. 9:147–165.

Powell, A. M. 1973. Taxonomy of *Perityle* section *Laphamia* (Compositae-Helenieae-Peritylinae). Sida 5:61–128.

Preece, S. J., and B. L. Turner. 1953. A taxonomic study of the genus *Chamaechaenactis* Rydberg (Compositae). Madroño 12:97–103.

Pridgeon, A. M., R. M. Bateman, A. V. Cox, J. R. Hapeman, and M. W. Chase. 1997. Phylogenetics of subtribe Orchidinae (Orchidoideae, Orchidaceae) based on nuclear ITS sequences. 1. Intergeneric relationships and polyphyly of *Orchis* sensu lato. Lindleyana 12:89–109.

Rabeler, R. K., and R. L. Hartman. 2005. Caryophyllaceae. Pp. 3–215 *in* Flora of North America Editorial Committee (editors), Flora of North America, Vol. 5. Oxford University Press, New York.

Ralphs, M. H., S. L. Welsh, and D. R. Gardner. 2002. Distribution of locoweed toxin swainsonine in populations of *Oxytropis lambertii*. J. Chem. Ecol. 28:701–707.

Rataj, K. 1972. Revision of the genus *Sagittaria*, part 2 (The species of the West Indies, Central and South America). Annot. Zool. Bot. 78. 61 pp.

Raven, P. H. 1964. The generic subdivision of Onagraceae, tribe Onagreae. Brittonia 16:276–288.

Raven, P. H. 1969. A revision of the genus *Camissonia* (Onagraceae). Contr. U.S. Natl. Herb. 37:161–396.

Raven, P. H. 1979. A survey of reproductive biology in Onagraceae. New Zealand J. Bot. 17:575–593.

Raven, P. H., and D. P. Gregory. 1972. A revision of the genus *Gaura* (Onagraceae). Mem. Torrey Bot. Club 23:1–96.

Rechinger, K. H. 1937. The North American species of *Rumex* (Vorarbeiten zu einer Monographie der Gattung *Rumex* 5). Publ. Field Mus. Nat. Hist., Bot. Ser. 17:1–150.

Redders, J. S. 2003. Vegetation of the San Juan National Forest: Existing and potential natural community types. Working white paper, updated 11/1/03.

Reeves, L., and T. Reeves. 1984. Life history and reproduction of *Malaxis paludosa* in Minnesota. Amer. Orchid Soc. Bull. 53:1280–1291.

Reeves, L., and T. Reeves. 1993. On composition, distribution, abundance and habitat requirements of endangered, threatened and rare plant species in the southeast Utah group of national parks. National Park Service, Flagstaff.

Reeves, T. 1979. A Monograph of the Fern Genus *Cheilanthes* subgenus *Physapteris* (Adiantaceae). Ph.D. Dissertation, Arizona State University, Tempe.

Reveal, J. L. 1972. Botanical explorations in the Intermountain Region. Pp. 40–72 *in* A. Cronquist, A. H. Holmgren, N. H. Holmgren, and J. L. Reveal (editors), Intermountain Flora. Vascular Plants of the Intermountain West, U.S.A., Vol. 1, Geological and Botanical History of the Region, Its Plant Geography and a Glossary. The New York Botanical Garden, Bronx.

Reveal, J. L. 1978. Distribution and phylogeny of the Eriogonoideae (Polygonaceae). Great Basin Naturalist Mem. 2:169–190.

Reveal, J. L. 1981. *Eriogonum divaricatum* Hook. (Polygonaceae), an intermountain species in Argentina. Great Basin Naturalist 41:143–146.

Reveal, J. L. 1989. Notes on selected genera related to *Chorizanthe* (Polygonaceae: Eriogonoideae). Phytologia 66:199–220.

Reveal, J. L. 2004. Nomenclatural summary of Polygonaceae subfamily Eriogonoideae. Harvard Pap. Bot. 9:143–230.

Reveal, J. L. 2005a. *Centrostegia*. P. 473 *in* Flora of North America Editorial Committee (editors), Flora of North America, Vol. 5. Oxford University Press, New York.

Reveal, J. L. 2005b. *Chorizanthe*. Pp. 445–470 *in* Flora of North America Editorial Committee (editors), Flora of North America, Vol. 5. Oxford University Press, New York.

Reveal, J. L. 2005c. *Eriogonum*. Pp. 221–430 *in* Flora of North America Editorial Committee (editors), Flora of North America, Vol. 5. Oxford University Press, New York.

Reveal, J. L. 2005d. *Stenogonum*. Pp. 431–432 *in* Flora of North America Editorial Committee (editors), Flora of North America, Vol. 5. Oxford University Press, New York.

Reveal, J. L., and B. J. Ertter. 1977. Re-establishment of *Stenogonum* (Polygonaceae). Great Basin Naturalist 36: 272–280.

Reveal, J. L., and C. B. Hardham. 1989. A revision of the annual species of *Chorizanthe* (Polygonaceae: Eriogonoideae). Phytologia 66:98–198.

Reveal, J. L., and R. M. King. 1973. Re-establishment of *Acourtia* D. Don (Asteraceae). Phytologia 27:228–232.

Richardson, A. 1977. Monograph of the genus *Tiquilia* (*Coldenia* sensu lato), Boraginaceae: Ehretioideae. Rhodora 79:467–572.

Rink, G. 2005. A checklist of the vascular flora of Canyon de Chelly National Monument, Apache County, Arizona. J. Torrey Bot. Soc. 132:510–532.

Robertson, K. R., and S. E. Clemants. 2003. Amaranthaceae. Pp. 405–456 *in* Flora of North America Editorial Committee (editors), Flora of North America, Vol. 4. Oxford University Press, New York.

Roberty, G. E., and S. Vautier. 1964. Les genres de Polygonacées. Boissiera 10:7–128.

Robinson, H., and J. Cuatrecasas. 1973. The generic limits of *Pluchea* and *Tessaria*. Phytologia 27:277–285.

Rollins, R. C. 1939. The cruciferous genus *Physaria*. Rhodora 41:392–415.

Rollins, R. C. 1940. Studies in the genus *Hedysarum* in North America. Rhodora 42:217–239.

Rollins, R. C. 1993. The Cruciferae of Continental North America: Systematics of the Mustard Family from the Arctic to Panama. Stanford University Press, Palo Alto.

Rollins, R. C., and E. A. Shaw. 1973. The Genus *Lesquerella* (Cruciferae) in North America. Harvard University Press, Cambridge.

Rominger, J. M. 1962. Taxonomy of *Setaria* (Gramineae) in North America. Illinois Biol. Monogr. 29:1–132.

Romme, W. H., M. L. Floyd, and D. Hanna. 2003a. Landscape condition analysis for the South Central Highlands Section, southwestern Colorado and northwestern New Mexico. Draft final report to the U.S. Department of Agriculture, Forest Service, San Juan National Forest, Durango, Colorado.

Romme, W. H., M. L. Floyd-Hanna, and D. D. Hanna. 2003b. Ancient piñon-juniper forests of Mesa Verde and the West: A cautionary note for forest restoration programs. Pp. 335–352 *in* P. N. Omi and L. Joyce (editors), Fire, Fuel Treatments, and Ecological Restoration, Conference Proceedings, April 2002, Fort Collins, Colorado. U.S. Department of Agriculture, Forest Service Proceedings RMRS-P-29.

Ronse Decraene, L. P., and J. Akeroyd. 1988. Generic limits in *Polygonum* and related genera (Polygonaceae) on the basis of floral characters. Bot. J. Linn. Soc. 98:321–371.

Rosatti, T. J. 1987. Field and garden studies of *Arctostaphylos uva-ursi* (Ericaceae) in North America. Syst. Bot. 12: 61–77.

Rossbach, R. P. 1940. *Spergularia* in North and South America. Rhodora 42:57–83, 105–143, 158–193, 203–213.

Roush, E. M. F. 1931. A monograph of the genus *Sidalcea*. Ann. Missouri Bot. Gard. 18:117–244.

Rowlands, P. G. 1994. Colorado Plateau vegetation assessment and classification manual. Technical Report NPS/NAUCPRS/NRTR-94/06, National Park Service, U.S. Department of the Interior, Colorado Plateau Research Station at Northern Arizona University, Flagstaff.

Rydberg, P. A. 1906. Flora of Colorado. Agricultural Experiment Station, Colorado Agricultural College, Fort Collins.

Rydberg, P. A. 1919a. *Parryella*. N. Amer. Fl. 24:25–26.

Rydberg, P. A. 1919b. *Pediomelum*. N. Amer. Fl. 24:17–24.

Rydberg, P. A. 1919c. *Psoralidium*. N. Amer. Fl. 24:12–17.

Rydberg, P. A. 1919d. *Psorothamnus*. N. Amer. Fl. 24:41–48.

Sa'ad, F. 1967. The *Convolvulus* species of the Canary Islands, the Mediterranean Region and the Near and Middle East. Meded. Bot. Mus. Herb. Rijks Univ. Utrecht 281:1–288.

Saltonstall, K., P. M. Peterson, and R. J. Soreng. 2004. Recognition of *Phragmites australis* subsp. *americanus* (Poaceae: Arundinoidae) in North America: Evidence from morphological and genetic analysis. Sida 21:683–692.

Sánchez del Pino, I., and S. E. Clemants. 2003. *Tidestromia*. Pp. 439–443 *in* Flora of North America Editorial Committee (editors), Flora of North America, Vol. 4. Oxford University Press, New York.

Sarkar, N. M. 1958. Cytotaxonomic studies on *Rumex* sect. *Axillares*. Canad. J. Bot. 36:947–996.

Sauer, J. D. 1955. Revision of the dioecious amaranths. Madroño 13:5–46.

Sauer, J. D. 1967. The grain amaranths and their relatives: A revised taxonomic and geographic survey. Ann. Missouri Bot. Gard. 54:103–137.

Sauer, J. D. 1972. The dioecious amaranths: A new species name and major range extensions. Madroño 21:426–434.

Savage, M., and T. W. Swetnam. 1990. Early 19th-century fire decline following sheep pasturing in a Navajo ponderosa pine forest. Ecology 71:2374–2378.

Schenk, J. J. 2009. A systematic monograph of *Mentzelia* section *Bartonia* (Loasaceae): Phylogeny, diversity, and divergence times. Ph.D. Dissertation, Washington State University, Pullman.

Schenk, J. J., and L. Hufford. 2010. Taxonomic novelties from western North America in *Mentzelia* section *Bartonia* (Loasaceae). Madroño 57:246–260.

Scott, A. J. 1978. A revision of the Camphorosmioideae (Chenopodiaceae). Feddes Repert. 89:101–119.

Semple, J. C. 1996. A revision of *Heterotheca* sect. *Phyllotheca* (Nutt.) Harms (Compositae: Astereae): The prairie and montane goldenasters of North America. Univ. Waterloo Biol. Ser. 37.

Semple, J. C., V. Blok, and P. Heiman. 1980. Morphological, anatomical, habit, and habitat differences among the goldenaster genera *Chrysopsis*, *Heterotheca* and *Pityopsis* (Compositae: Astereae). Canad. J. Bot. 58:147–163.

Semple, J. C., and J. G. Chmielewski. 1987. Revision of the *Aster lanceolatus* complex, including *A. simplex* and *A. hesperius* (Compositae: Asteraceae): A multivariate morphometric study. Canad. J. Bot. 65:1047–1062.

Shaw, G. R. 1914. The Genus *Pinus*. Publ. Arnold Arbor. Riverside Press, Cambridge.

Shaw, R. J. 1962. The biosystematics of *Scrophularia* in western North America. Aliso 5:147–178.

Sherff, E. E. 1937. The genus *Bidens*. Publ. Field Mus. Nat. Hist., Bot. 16:1–709.

Shinners, L. H. 1946. Revision of the genus *Chaetopappa* DC. Wrightia 1:63–81.

Shinwari, Z. K., R. Terauchi, F. H. Utech, and S. Kawano. 1994. Recognition of the New World *Disporum* section *Prosartes* as *Prosartes* (Liliaceae) based on the sequence data of the rbcL gene. Taxon 43:353–366.

Shishkin, B. K. (editor). 1970. Flora of the U.S.S.R. (Flora SSSR). Vol. 6, Centrospermae. Israel Program for Scientific Translations.

Sieren, D. 1981. The taxonomy of the genus *Euthamia*. Rhodora 83:551–579.

Silba, J. 1986. Encyclopedia coniferae. Phytologia Mem. 8:1–127.

Simpson, M. G. 2010. Plant Systematics, 2nd ed. Elsevier, Amsterdam.

Small, E. 1968. Systematics of autopolyploidy in *Epilobium latifolium* (Onagraceae). Brittonia 20:169–181.

Smith, J. P. 1977. Vascular Plant Families. Mad River Press, Eureka, California.

Smith, S. G. 1987. *Typha*: Its taxonomy and the ecological significance of hybrids. Arch. Hydrobiol. 27:129–138.

Snow, N. W. 1998. Nomenclatural changes in *Leptochloa* P. Beauvois sensu lato (Poaceae, Chloridoideae). Novon 8:77–80.

Soltis, D. E., and P. S. Soltis. 1989. Allopolyploid speciation in *Tragopogon*: Insights from chloroplast DNA. Amer. J. Bot. 76:1119–1124.

Soreng, R. J. 1985. *Poa* L. in New Mexico, with a key to middle and southern Rocky Mountain species (Poaceae). Great Basin Naturalist 45:395–422.

Soreng, R. J. 1991. Systematics of the "Epiles" group of *Poa* (Poaceae). Syst. Bot. 16:507–528.

Soreng, R. J., and S. L. Hatch. 1983. A comparison of *Poa tracyi* and *Poa occidentalis* (Poaceae: Poeae). Sida 10: 123–141.

Soreng, R. J., and E. E. Terrell. 1997. Taxonomic notes on *Schedonorus*. Phytologia 83:85–88.

Sosa, V., and M. W. Chase. 2003. Phylogenetics of Crossosomataceae based on rbcL sequence data. Syst. Bot. 28:96–105.

Spangler, R. E. 2000. Andropogoneae systematics and generic limits in *Sorghum*. Pp. 167–170 *in* S. W. L. Jacobs and J. Everett (editors), Grasses: Systematics and Evolution. International Symposium on Grass Systematics and Evolution. CSIRO Publishing, Collingwood, Victoria, Australia. 408 pp.

Spellenberg, R. W. 2003. *Mirabilis*. Pp. 40–56 *in* Flora of North America Editorial Committee (editors), Flora of North America, Vol. 4. Oxford University Press, New York.

Spence, J. R. 2001. Climate of the central Colorado Plateau, Utah and Arizona: Characterization and recent trends. Pp. 189–203 *in* C. van Riper III, K. A. Thomas, and M. A. Stuart (editors), Proceedings of the Fifth Biennial Conference on Research on the Colorado Plateau. U.S. Department of the Interior, Geological Survey Report Series USGSFRESC/COPL/2001/24.

Spence, J. R. 2002. The Spence/Romme/Floyd-Hanna/Rowlands (SRFR) classification, Version 4.0. Unpublished report, Glen Canyon National Recreation Area, P.O. Box 1507, Page, Arizona.

Spence, J. R., and N. R. Henderson. 1993. Tinaja and hanging garden vegetation of Capitol Reef National Park, south-central Utah. J. Arid Environm. 24:21–36.

Spence, J. R., W. H. Romme, L. Floyd-Hanna, and P. G. Rowlands. 1995. A preliminary vegetation classification for the Colorado Plateau. Pp. 193–213 *in* C. van Riper III (editor), Proceedings of the Second Biennial Conference on Research in Colorado Plateau National Parks. National Park Service Transactions and Proceedings Series NPS/NRNAU/NRTP-95/11.

Spencer, A. W., and W. H. Romme. 1996. Ecological patterns. Pp. 129–142 in R. Blair (editor), The Western San Juan Mountains: Their Geology, Ecology, and Human History. University Press of Colorado, Boulder.

Stahelin, R. 1943. Factors influencing the natural re-stocking of high altitude burns by coniferous trees in the central Rocky Mountains. Ecology 24:19–30.

Standley, P. C. 1916. *Chenopodium salinum* Standley. N. Amer. Fl. 21:29.

Steyermark, J. A. 1934. Studies in *Grindelia* II. A monograph of the North American species of the genus *Grindelia*. Ann. Missouri Bot. Gard. 21:433–608.

Steyermark, J. A. 1937. Studies in *Grindelia* III. Ann. Missouri Bot. Gard. 24:225–262.

St. John, H. 1916. A revision of the North American species of *Potamogeton* of the section *Coleophylli*. Rhodora 18:121–138.

St. John, H. 1929. Notes on northwestern ferns. Amer. Fern J. 19:11–16.

St. John, H. 1962. Monograph of the genus *Elodea* (Hydrocharitaceae): Part I. Res. Stud. State Coll. Wash. 30:19–44.

Stockwell, P. 1940. A revision of the genus *Chaenactis*. Contr. Dudley Herb. 3:89–167.

Strother, J. L. 1969. Systematics of *Dyssodia* Cavanilles (Compositae: Tageteae). Univ. Calif. Publ. Bot. 48:1–88.

Strother, J. L. 1974. Taxonomy of *Tetradymia* (Compositae: Senecioneae). Brittonia 26:177–202.

Strother, J. L. 1986. Renovation of *Dyssoidea* (Compositae: Tageteae). Sida 11:371–378.

Strother, J. L. 2006. *Thelesperma*. Pp. 199–203 *in* Flora of North America Editorial Committee (editors), Flora of North America, Vol. 21. Oxford University Press, New York.

Strother, J. L., and G. E. Pilz. 1975. Taxonomy of *Psathyrotes* (Compositae: Senecioneae). Madroño 23:24–40.

Strother, J. L., and M. A. Wetter. 2006. *Grindelia*. Pp. 424–436 *in* Flora of North America Editorial Committee (editors), Flora of North America, Vol. 20. Oxford University Press, New York.

Struwe, L., and V. A. Albert (editors). 2002. Gentianaceae: Systematics and Natural History. Cambridge University Press, Cambridge.

Stuckey, R. L. 1972. Taxonomy and distribution of the genus *Rorippa* (Cruciferae) in North America. Sida 4:279–430.

Stuckey, R. L., and D. P. Salamon. 1987. *Typha angustifola* in North America: A foreigner masquerading as a native. Ohio J. Sci. Abstr. 87:2.

Stutz, H. C., and S. C. Sanderson. 1983. Evolutionary studies of *Atriplex*: Chromosome races of *A. confertifolia* (shadscale). Amer. J. Bot. 70:1536–1547.

Sun, F.-J., and S. R. Downie. 2004. A molecular systematic investigation of *Cymopterus* and its allies (Apiaceae) based on phylogenetic analyses of nuclear (ITS) and plastid (rps16 intron) DNA sequences. S. African J. Bot. 70:407–416.

Sun, F.-J., S. R. Downie, and R. L. Hartman. 2004. An ITS-based phylogenetic analysis of the perennial, endemic Apiaceae subfamily Apioideae of western North America. Syst. Bot. 29:419–431.

Sun, F.-J., G. A. Levin, and S. R. Downie. 2005. A multivariate analysis of *Cymopterus glomeratus*, formerly known as *C. acaulis* (Apiaceae). Rhodora 107:359–385.

Sundberg, S. D. 1991. Infraspecific classification of *Chloracantha spinosa* (Benth.) Nesom (Asteraceae) Astereae. Phytologia 70:382–391.

Sundell, E. 1990. Notes on Arizona *Asclepias* (Asclepiadaceae) with a new combination. Phytologia 69:265–270.

Taschereau, P. M. 1972. Taxonomy and distribution of *Atriplex* species in Nova Scotia. Canad. J. Bot. 50:1571–1594.

Tateoka, T. 1961. A biosystematic study of *Tridens* (Gramineae). Amer. J. Bot. 48:565–573.

Taylor, N. P. 1983. Die Arten der Gattung *Escobaria* Britton et Rose: [1]-5. Kakt. And. Sukk. 34:76–79, 120–123, 136–140, 154–158, 184–188.

Taylor, N. P. 1985. The Genus *Echinocereus*. A Kew Magazine Monograph. Timber Press, Portland.

Taylor, P. 1989. The genus *Utricularia*: A taxonomic monograph. Kew Bull. Addit. Ser. 14:1–724.

Taylor, W. C., N. T. Luebke, D. M. Britton, R. J. Hickey, and D. F. Brunton. 1993. Isoëtaceae. Pp. 64–75 *in* Flora of North America Editorial Committee (editors), Flora of North America, Vol. 2, Oxford University Press, New York.

Terry, R. G., R. S. Nowak, and R. J. Tausch. 2000. Genetic variation in chloroplast and nuclear ribosomal DNA in Utah juniper (*Juniperus osteosperma*, Cupressaceae): Evidence for interspecific gene flow. Amer. J. Bot. 87:250–258.

Theobald, W. L., C. C. Tseng, and M. E. Mathias. 1964. A revision of *Aletes* and *Neoparrya* (Umbelliferae). Brittonia 16:296–315.

Thieret, J. W. 1955. The seeds of *Veronica* and allied genera. Lloydia 18:37–45.

Thieret, J. W. 1966. Synopsis of the genus *Calamovilfa* (Gramineae). Castanea 31:145–152.

Thieret, J. W. 1993. Pinaceae. Pp. 352–398 *in* Flora of North America Editorial Committee (editors), Flora of North America, Vol. 2. Oxford University Press, New York.

Thilenius, J. F. 1975. Alpine range management in the western United States—Principles, practices, and problems: The status of our knowledge. USDA Forest Service Research Paper RM-157.

Thorne, K. H. 1977. A Revision of the Herbaceous Members of the Genus *Atriplex* (Chenopodiaceae) for the State of Utah. Unpublished M.S. Thesis, Brigham Young University, Provo, Utah.

Thorne, R. F. 1974. A phytogenetic classification of the Annoniflorae. Aliso 8:147–209.

Thornthwaite, C. W. 1948. An approach toward a rational classification of climate. Geogr. Rev. 38:55–94.

Tidestrom, I., and T. Kittell. 1941. A Flora of Arizona and New Mexico. Catholic University of America Press, Washington, D.C.

Tomb, A. S. 1980. Taxonomy of *Lygodesmia* (Asteraceae). Syst. Bot. Monogr. 1:1–51.

Toolin, L., and J. R. Reeder. 2000. The status of *Setaria macrostachya* and its relationship to *S. vulpiseta* (Gramineae). Syst. Bot. 25:26–32.

Torrell, M., N. Garcia-Jacas, A. Susanna, and J. Vallés. 1999. Phylogeny in *Artemisia* (Asteraceae, Anthemideae) inferred from nuclear ribosomal DNA (ITS) sequences. Taxon 48:721–736.

Towner, H. F. 1977. The biosystematics of *Calylophus* (Onagraceae). Ann. Missouri Bot. Gard. 64:48–120.

Tryon, Jr., R. M. 1941. A revision of the genus *Pteridium*. Rhodora 43:1–31, 37–67.

Tryon, Jr., R. M. 1955. *Selaginella rupestris* and its allies. Ann. Missouri Bot. Gard. 42:1–99.

Tryon, Jr., R. M., and A. F. Tryon. 1982. Ferns and Allied Plants with Special Reference to Tropical America. Springer-Verlag, New York.

Tucker, G. C. 1986. The genera of Elatinaceae in the southeastern United States. J. Arnold Arbor. 67:471–483.

Tucker, J. M. 1961. Studies in the *Quercus undulata* complex. I. A preliminary statement. Amer. J. Bot. 48:202–208.

Tucker, J. M. 1970. Studies in the *Quercus undulata* complex. IV. The contribution of *Q. havardii*. Amer. J. Bot. 57:71–84.

Tucker, J. M. 1971. Studies in the *Quercus undulata* complex. V. The type of *Quercus undulata*. Amer. J. Bot. 58: 329–341.

Turner, B. L. 1956. A cytotaxonomic study of the genus *Hymenopappus* (Compositae). Rhodora 58:163–186, 208–242, 259–269, 295–308.

Turner, B. L. 1978. Taxonomic study of the scapiform species of *Acourtia* (Asteraceae–Mutisiieae). Phytologia 38: 456–468.

Turner, B. L. 1987. Taxonomic study of *Machaeranthera*, sections *Machaeranthera* and *Hesperastrum* (Asteraceae). Phytologia 62:207–266.

Turner, B, L. 1993. New taxa, new combinations, and nomenclatural comments on the genus *Acourtia* (Asteraceae, Mutisiieae). Phytologia 74:385–412.

Turner, M. W. 1993. Systematic study of the genus *Baileya* (Asteraceae: Helenieae). Sida 15:491–508.

Umber, R. E. 1979. The genus *Glandularia* (Verbenaceae) in North America. Syst. Bot. 4:72–102.

Urbatsch, L. E., R. P. Roberts, and K. M. Neubig. 2004. *Cuniculotinus* and *Lorandersonia*, two new genera of Asteraceae: Astereae and new combinations in *Chrysothamnus*. Sida 21:1615–1632.

Valdés-Reyna, J., and S. L. Hatch. 1997. A revision of *Erioneuron* and *Dasyochloa* (Poaceae: Eragrostideae). Sida 17:645–666.

Vallès, J., and E. D. McArthur. 2001. *Artemisia* systematics and phylogeny: Cytogenetic and molecular insights. Pp. 67–74 *in* E. D. McArthur and D. J. Fairbanks (compilers), Shrubland Ecosystem Genetics and Biodiversity: Proceedings. 13–15 June 2000, Provo, Utah. U.S. Department of Agriculture, Forest Service Proceedings RMRS-P-21.

Vallès, J., M. N. Torrell, T. Garnatje, N. Garcia-Jacas, R. Vilatersana, and A. Susanna. 2003. The genus *Artemisia* and its allies: Phylogeny of the subtribe Artemisiinae (Asteraceae, Anthemideae) based on nucleotide sequences of nuclear ribosomal DNA internal transcribed spacers (ITS). Plant Biol. 5:274–284.

Vander Kloet, S. P. 1992. On the etymology of *Vaccinium* L. Rhodora 94:371–373.

Vander Kloet, S. P., and T. A. Dickinson. 1999. The taxonomy of *Vaccinium* sect. *Myrtillus* (Ericaceae). Brittonia 51: 231–254.

Vasek, F. C. 1966. The distribution and taxonomy of three western junipers. Brittonia 18:350–372.

von Bothmer, R., N. Jacobsen, C. Baden, R. B. Jørgensen, and I. Linde-Laursen. 1995. An ecogeographical study of the genus *Hordeum*, 2nd ed. Systematic and Ecogeographic Studies on Crop Genepools 7. International Plant Genetic Resources Institute, Rome.

von Hagen, K. B., and J. W. Kadereit. 2002. Phylogeny and flower evolution of the Swertiinae (Gentianaceae-Gentianeae): Homoplasy and the principle of variable proportions. Syst. Bot. 27:548–572.

von Schreber, J. C. D. 1804. *Trifolium pratense* var. *sativum* Deutschl. Fl. Abt. 1, Band 4, Heft 15.

Wagner, Jr., W. H., and J. M. Beitel. 1992. Generic classification of modern North American Lycopodiaceae. Ann. Missouri Bot. Gard. 79:676–686.

Wagner, Jr., W. H., R. C. Moran, and C. R. Werth. 1993. Aspleniaceae. Pp. 228–245 *in* Flora of North America Editorial Committee (editors), Flora of North America, Vol. 2. Oxford University Press, New York.

Wagner, Jr., W. H., and F. S. Wagner. 1983. Two moonworts of the Rocky Mountains, *Botrychium hesperium* and a new species formerly confused with it. Amer. Fern. J. 73:53–62.

Wagner, Jr., W. H., and F. S. Wagner. 1986. Three new species of moonworts (*Botrychium* subg. *Botrychium*) endemic in western North America. Amer. Fern. J. 76:33–47.

Wagner, Jr., W. H., and F. S. Wagner. 1990. Notes on the fan-leaflet group of moonworts in North America with descriptions of two new members. Amer. Fern J. 80:73–81.

Wagner, Jr., W. H., and F. S. Wagner. 1993. Ophioglossaceae. Pp. 85–106 *in* Flora of North America Editorial Committee (editors), Flora of North America, Vol. 2. Oxford University Press, New York.

Wagner, W. L. 1986. New taxa in *Oenothera* (Onagraceae). Ann. Missouri Bot. Gard. 73:475–480.

Wagner, W. L., P. C. Hoch, and P. H. Raven. 2007. Revised classification of the Onagraceae. Syst. Bot. Monogr. 83: 1–240.

Wagner, W. L., R. Stockhouse, and W. M. Klein. 1985. The Systematics and Evolution of the *Oenothera caespitosa* Species Complex (Onagraceae). Monogr. Syst. Bot. Missouri Bot. Gard. 12.

Wagnon, H. K. 1952. A revision of the genus *Bromus*, section *Bromopsis*, of North America. Brittonia 7:415–480.

Wahl, H. A. 1954. A preliminary study of the genus *Chenopodium* in North America. Bartonia 27:1–46.

Wallace, G. D. 1975. Studies on the Monotropoideae (Ericaceae) taxonomy and distribution. Wasmann J. Biol. 33: 1–88.

Walters, S. M., and D. A. Webb. 1972. *Veronica*. Pp. 242–251 *in* T. G. Tutin, V. H. Heywood, N. A. Burges, and D. H. Valentine (editors), Flora Europaea 3. Cambridge University Press, Cambridge.

Warwick, S. I., and I. A. Al-Shehbaz. 2006. Brassicaceae: Chromosome number index and database on CD-ROM. Pl. Syst. Evol. 259:237–248.

Warwick, S. I., I. A. Al-Shehbaz, R. A. Price, and C. A. Sauder. 2002. Phylogeny of *Sisymbrium* (Brassicaceae) based on ITS sequences of nuclear ribosomal DNA. Canad. J. Bot. 80:1002–1017.

Warwick, S. I., I. A. Al-Shehbaz, C. A. Sauder, D. F. Murray, and K. Mummenhoff. 2004. Phylogeny of *Smelowskia* and related genera (Brassicaceae) based on nuclear ITS DNA and chloroplast *trn*L intron DNA sequences. Ann. Missouri Bot. Gard. 91:99–123.

Watson, F. D., and J. E. Eckenwalder. 1993. Cupressaceae Bartlett: Redwood or cypress family. Pp. 399–422 *in* Flora of North America Editorial Committee (editors), Flora of North America, Vol. 2. Oxford University Press, New York.

Watson, L., P. Bates, T. Evans, M. Unwin, and J. Estes. 2002. Molecular phylogeny of subtribe Artemisiinae (Asteraceae), including *Artemisia* and its allied and segregate genera. BMC Evol. Biol. 2:17.

Watson, Jr., T. J. 1977. The taxonomy of *Xylorhiza* (Asteraceae–Astereae). Brittonia 29:199–216.

Webber, J. M. 1953. Yuccas of the Southwest. Agric. Monogr. U.S.D.A., No. 17, Washington, D.C., 1097 pp.

Weber, W. A. 1966. *Leucopoa kingii* (S. Watson). Univ. Colorado Stud. Ser. Biol. 23:2.

Weber, W. A. 1987. *Ligularia holmii* (Greene) Weber. P. 139 *in* Colorado Flora: Western Slope. University Press of Colorado, Boulder.

Weber, W. A., and R. C. Wittmann. 1990. *Ligularia holmii* (Greene) Weber. P. 96 *in* Colorado Flora: Eastern Slope. University Press of Colorado, Boulder.

Weber, W. A., and R. C. Wittmann. 1992. Catalog of the Colorado Flora. University Press of Colorado, Boulder.

Weber, W. A., and R. C. Wittmann. 1996. Colorado Flora: Western Slope, revised ed. University Press of Colorado, Boulder.

Weber, W. A., and R. C. Wittmann. 2001. Colorado Flora: Western Slope, 3rd ed. University Press of Colorado, Boulder.

Webster, G. L. 1994a. Classification of the Euphorbiaceae. Ann. Missouri Bot. Gard. 81:3–32.

Webster, G. L. 1994b. Synopsis of the genera and suprageneric taxa of Euphorbiaceae. Ann. Missouri Bot. Gard. 81:33–144.

Webster, G. L., and K. I. Miller. 1963. The genus *Reverchonia* (Euphorbiaceae). Rhodora 65:193–207.

Webster, R. D. 1987. Taxonomy of *Digitaria* section *Digitaria* in North America (Poaceae: Paniceae). Sida 12:209–222.

Weimarck, G. 1971. Variation and taxonomy of *Hierochloë* (Gramineae) in the Northern Hemisphere. Bot. Not. 124: 129–175.

Wells, P. V. 2000. The Manzanitas of California: Also of Mexico and the World. University of Kansas Department of Ecology and Evolutionary Biology, Lawrence.

Welsh, S. L. 1965. Legumes of Utah. III. *Lathyrus* L. Proc. Utah Acad. Sci. 42:214–221.

Welsh, S. L. 1974. Anderson's Flora of Alaska and Adjacent Parts of Canada. Brigham Young University Press, Provo, Utah.

Welsh, S. L. 1978. Utah flora: Fabaceae (Leguminosae). Great Basin Naturalist 38:225–367.

Welsh, S. L. 1980. Utah flora: Malvaceae. Great Basin Naturalist 40:27–37.

Welsh, S. L. 1983. Utah flora: Compositae (Asteraceae). Great Basin Naturalist 43:179–357.

Welsh, S. L. 1984. Utah flora: Chenopodiaceae. Great Basin Naturalist 44:183–209.

Welsh, S. L. 1989. On the distribution of Utah's hanging gardens. Great Basin Naturalist 49:1–30.

Welsh, S. L. 1991. *Oxytropis* DC.: Names, basionyms, types, and synonyms—Flora North America Project. Great Basin Naturalist 51:377–396.

Welsh, S. L. 1995a. Names and types of *Hedysarum* L. (Fabaceae) in North America. Great Basin Naturalist 55:66–73.

Welsh, S. L. 1995b. North American types of *Oxytropis* DC. (Leguminosae) at the Natural History Museum and Royal Botanic Garden, England, with nomenclatural comments and a new variety. Great Basin Naturalist 55:271–281.

Welsh, S. L. 2001. Revision of the North American species of *Oxytropis* de Candolle (Leguminosae). EPS, Orem, Utah.

Welsh, S. L., N. D. Atwood, S. Goodrich, and L. C. Higgins (editors). 1993. A Utah Flora, 2nd ed., rev. Brigham Young University Press, Provo, Utah.

Welsh, S. L., N. D. Atwood, S. Goodrich, and L. C. Higgins (editors). 2003. A Utah Flora, 3rd ed. Brigham Young University Press. Provo, Utah.

Welsh, S. L., and J. A. Erdman. 1964. Annotated checklist of the plants of Mesa Verde, Colorado. Brigham Young Univ. Sci. Bull., Biol. Ser. 4:1–32.

Wemple, D. K. 1970. Revision of the genus *Petalostemon* (Leguminosae). Iowa State J. Sci. 45:1–102.

Wendt, T. 1978. A Systematic Study of *Polygala* section *Rhinotropis* (Polygalaceae). Ph.D. Dissertation, University of Texas, Austin.

Werth, C. R., and M. D. Windham. 1991. A model for divergent, allopatric speciation of polyploid pteridophytes resulting from silencing of duplicate-gene expression. Amer. Naturalist 137:515–526.

West, N. E. 1988. Intermountain deserts, shrub steppes, and woodlands. Pp. 209–230 *in* M. G. Barbour and W. D. Billings (editors), North American Terrestrial Vegetation. Cambridge University Press, New York.

Wheeler, L. C. 1941. *Euphorbia* subgenus *Chamaesyce* in Canada and the United States exclusive of southern Florida. Rhodora 43:97–154, 168–205, 223–286.

Wherry, E .T. 1944. Review of the genera *Collomia* and *Gymnosteris*. Amer. Midl. Nat. 31:216–231.

Wherry, E. T. 1955. The genus *Phlox*. Morris Arbor. Monogr. 3.

Whittaker, R. H. 1975. Communities and Ecosystems, 2nd ed. MacMillan, New York.

Whittemore, A. T., and B. D. Parfitt. 1997. Ranunculaceae. Pp. 85–271 *in* Flora of North America Editorial Committee (editors), Flora of North America, Vol. 3. Oxford University Press, New York.

Wiggins, I. L. 1936. A resurrection and revision of the genus *Iliamna* Greene. Contr. Dudley Herb. 1:213–227.

Wilken, D. H. 1971. Seasonal dimorphism in *Baccharis glutinosa* (Compositae). Madroño 21:113–119.

Wilken, D. H., D. M. Smith, J. B. Harborne, and W. Glennie. 1982. Flavonoid and anthocyanin patterns and the systematic relationships in *Collomia*. Biochem. Syst. Ecol. 10:239–243.

Wilkowske, C. D., D. V. Allen, and J. V. Phillips. 2003. Drought conditions in Utah during 1999–2002: A historical perspective. USDI U.S. Geological Survey Fact Sheet 037–03.

Windham, M. D., and I. A. Al-Shehbaz. 2006. New and noteworthy species of *Boechera* (Brassicaceae) I: Sexual diploids. Harvard Pap. Bot. 11:61–88.

Wolf, S. J., and K. E. Denford. 1984. Taxonomy of *Arnica* (Compositae) subgenus *Austromontana*. Rhodora 86: 239–309.

Wood, Jr., C. E. 1959. The genera of the Nymphaeaceae and Ceratophyllaceae in the southeastern United States. J. Arnold Arbor. 40:94–112.

Woodland, D. W. 1982. Biosystematics of the perennial North American taxa of *Urtica*. II. Taxonomy. Syst. Bot. 7:282–290.

Woodson, Jr., R. E. 1928. Studies in the Apocynaceae. III. A monograph of the genus *Amsonia*. Ann. Missouri Bot. Gard. 15:379–434.

Woodson, Jr., R. E. 1954. The North American species of *Asclepias*. Ann. Missouri Bot. Gard. 41:1–211.

Wooton, E. O., and P. C. Standley. 1915. Flora of New Mexico. Contr. U.S. Natl. Herb. 19.

Wu, R. 1999. Fire history and forest structure in the mixed conifer forests of southwest Colorado. M.S. Thesis, Department of Forest Sciences, Colorado State University, Fort Collins.

Zanoni, T. A. 1978. The American junipers of the section *Sabina* (*Juniperus*, Cupressaceae): A century later. Phytologia 38:433–454.

Zomlefer, W. B. 1989. Papaveraceae. Pp. 45–49 *in* Guide to Flowering Plant Families. University of North Carolina Press, Chapel Hill.

Zomlefer, W. B., and W. S. Judd. 2002. Resurrection of segregates of the polyphyletic genus *Zigadenus* s.l. (Liliales: Melanthiaceae) and resulting new combinations. Novon 12:299–308.

Allred, K. 2008. Flora Neomexicana 1: The Vascular Plants of New Mexico. An Annotated Checklist to the Names of Vascular Plants with Synonymy and Bibliography. New Mexico State University, Las Cruces.

Benson, L., and R. A. Darrow. 1981. Trees and Shrubs of the Southwest Deserts, 3rd ed. University of Arizona Press, Tucson.

Bridson, G. D. R., and E. R. Smith. 1991. B-P-H/S. Botanico-Periodicum-Huntianum/Supplementum. Hunt Institute for Botanical Documentation, Carnegie Mellon University, Pittsburgh.

Cooke, S. (editor). 1997. A Field Guide to the Common Wetland Plants of Western Washington and Northwestern Oregon. Seattle Audubon Society, Seattle.

Correll, D. S., and H. B. Correll. 1975. Aquatic and Wetland Plants of Southwestern United States, Vol. I and II. Stanford University Press, Palo Alto.

Correll, D. S., and M. C. Johnston. 1979. Manual of the Vascular Plants of Texas. The University of Texas at Dallas, Dallas.

Cronquist, A., A. H. Holmgren, N. H. Holmgren, and J. L. Reveal. 1972. Intermountain Flora: Vascular Plants of the Intermountain West, U.S.A., Vol. 1. Geological and Botanical History of the Region, Its Plant Geography and a Glossary. The Vascular Cryptogams and the Gymnosperms. The New York Botanical Garden, Bronx.

Cronquist, A., A. H. Holmgren, N. H. Holmgren, J. L. Reveal, and P. K. Holmgren. 1994. Intermountain Flora: Vascular Plants of the Intermountain West, U.S.A., Vol. 5. Asterales. The New York Botanical Garden, Bronx.

Crow, G. E., and C. B. Hellquist. 2000. Aquatic and Wetland Plants of Northeastern North America: Angiosperms, Monocotyledons. University of Wisconsin Press, Madison.

Dorn, R. D. 2001. Vascular Plants of Wyoming, 3rd ed. Mountain West Publishing, Santa Fe.

Dunmire, W. W., and G. D. Tierney. 1995. Wild Plants of the Pueblo Province: Exploring Ancient and Enduring Uses. Museum of New Mexico Press, Santa Fe.

Dunmire, W. W., and G. D. Tierney. 1997. Wild Plants and Native Peoples of the Four Corners. Museum of New Mexico Press, Santa Fe.

Elliott, B. A. 2009. Handbook of Edible and Poisonous Plants of Western North America. Elliott Environmental Consulting, Laramie, Wyoming.

Ewan, J. 1950. Rocky Mountain Naturalists. University of Denver Press, Denver.

Ewan, J., and N. D. Ewan. 1981. Biographical Dictionary of Rocky Mountain Naturalists. Dr. W. Junk Publishers, The Hague.

Flora of North America Editorial Committee. 1993. Flora of North America North of Mexico, Vol. 1. Introduction. Oxford University Press, New York.

Flora of North America Editorial Committee. 2002. Flora of North America North of Mexico, Vol. 26. Magnoliophyta: Liliidae: Liliales and Orchidales. Oxford University Press, New York.

Flora of North America Editorial Committee. 2003. Flora of North America North of Mexico, Vol. 25. Magnoliophyta: Commelinidae (in part): Poaceae, part 2. Oxford University Press, New York.

Flora of North America Editorial Committee. 2006. Flora of North America North of Mexico, Vol. 19. Magnoliophyta: Asteridae (in part): Asteraceae, part 1. Oxford University Press, New York.

Flora of North America Editorial Committee. 2006. Flora of North America North of Mexico, Vol. 20. Magnoliophyta: Asteridae (in part): Asteraceae, part 2. Oxford University Press, New York.

Flora of North America Editorial Committee. 2006. Flora of North America North of Mexico, Vol. 21. Magnoliophyta: Asteridae (in part): Asteraceae, part 3. Oxford University Press, New York.

Harrington, H. D. 1954. Manual of the Plants of Colorado. For the Identification of the Ferns and Flowering Plants of the State. Sage Publications, Thousand Oaks, California.

Harris, J. G., and M. Woolf Harris. 2004. Plant Identification Terminology, An Illustrated Glossary, 2nd ed. Spring Lake Publishing, Spring Lake, Utah.

Heil, K. D. 1995. Endangered, Threatened and Sensitive Plant Field Guide. Bureau of Land Management, Farmington, New Mexico.

Heil, K. D. 2000. Four Corners Invasive and Poisonous Plant Field Guide. Bureau of Land Management, Farmington, New Mexico.

Heil, K. D., and S. L. O'Kane Jr. 2003. Catalog of the Four Corners Flora: Vascular Plants of the San Juan River Drainage: Arizona, Colorado, New Mexico and Utah. Harvard Pap. Bot. 7:321–379.

Hermann, F. J. 1970. Manual of the Carices of the Rocky Mountains and Colorado Basin. USDA Forest Service, Washington, D.C.

Hickman, J. C. (editor). 1993. The Jepson Manual: Higher Plants of California. University of California Press, Berkeley.

Humphrey, H. B. 1961. Makers of North American Botany. Ronald Press, New York.

Ivey, R. D. 2003. Flowering Plants of New Mexico, 4th ed. R. D. & I. Ivey, Albuquerque.

Judd, W. S., C. S. Campbell, E. A. Kellogg, P. F. Stevens, and M. J. Donoghue. 2008. Plant Systematics, 3rd ed. Sinauer Assoc., Sunderland, Massachusetts. 611 pp.

Kartesz, J. 2008. The Biota of North America Program (BONAP). University of North Carolina, Chapel Hill.

Komarek, S. 1994. Flora of the San Juans. A Field Guide to the Mountain Plants of Southwestern Colorado. Kivaki Press, Asheville, North Carolina.

Mabberley, D. J. 1989. The Plant-Book. Cambridge University Press, Cambridge.

Mayes, V. O., and B. Bayless Lacy 1989. Nanise': A Navajo Herbal: One Hundred Plants from the Navajo Reservation. Navajo Community College Press, Tsaile, Arizona.

Moerman, D. E. 1998. Native American Ethnobotany. Timber Press, Portland.

Nabhan, G. 1985 Gathering the Desert. University of Arizona Press, Tucson.

Peattie, N. 1991. A Natural History of Western Trees. Houghton Mifflin, Boston.

Ricketts, T. H., E. Dinerstein, D. M. Olson, C. J. Loucks, W. Eichbaum, D. DellaSala, K. Kavanaugh, P. Hedao, P. T. Hurley, K. M. Carney, R. Abell, and S. Walters. 1999. Terrestrial Ecoregions of North America: A Conservation Assessment. Island Press, Washington, D.C.

Simpson, M. G. 2010. Plant Systematics, 2nd ed. Elsevier (Academic Press), Amsterdam.

Stearn, W. T. 1992. Botanical Latin. Timber Press, Portland.

Tai-yien, C., L. Lian-li, Y. Guang, and I. Al-Shehbaz. 2001. Brassicaceae (Cruciferae). Pp. 1–193 *in* Z. Y. Wu and P. H. Raven (editors), Flora of China, Vol. 8, Science Press, Beijing, and Missouri Botanical Garden Press, St. Louis.

USDA (United States Department of Agriculture) Natural Resources Conservation Service. 2008. Plants Database. http://plants.usda.gov/topics.html.

von Bothmer, R., N. Jacobsen, C. Baden, R. B. Jørgensen, and I. Linde-Laursen. 1995. Systematic and Ecogeographic Studies on Crop Genepools 7. International Plant Genetic Resources Institute, Rome.

Wagner, W. L., P. C. Hoch, and P. H. Raven. 2007. Systematic Botany Monographs: Revised Classification of the Onagraceae, Vol. 83. The American Society of Plant Taxonomists, Ann Arbor.

Williams, R. L. 2003. A Region of Astonishing Beauty: The Botanical Exploration of the Rocky Mountains. Roberts Rinehart, Lanham, Maryland.

INDEX

Page numbers in italics refer to illustrations.

N